ALGEBRA

Lines

Slope of the line through $P_1 = (x_1, y_1)$ and $P_2 = (x_2, y_2)$:

$$m = \frac{y_2 - y_1}{x_2 - x_1}$$

Slope-intercept equation of line with slope m and y-intercept b:

$$y = mx + b$$

Point-slope equation of line through $P_1 = (x_1, y_1)$ with slope m:

$$y - y_1 = m(x - x_1)$$

Point-point equation of line through $P_1 = (x_1, y_1)$ and $P_2 = (x_2, y_2)$:

$$y - y_1 = m(x - x_1) \quad \text{where } m = \frac{y_2 - y_1}{x_2 - x_1}$$

Lines of slope m_1 and m_2 are parallel if and only if $m_1 = m_2$.
Lines of slope m_1 and m_2 are perpendicular if and only if $m_1 = -\frac{1}{m_2}$.

Circles

Equation of the circle with center (a, b) and radius r:

$$(x - a)^2 + (y - b)^2 = r^2$$

Distance and Midpoint Formulas

Distance between $P_1 = (x_1, y_1)$ and $P_2 = (x_2, y_2)$:

$$d = \sqrt{(x_2 - x_1)^2 + (y_2 - y_1)^2}$$

Midpoint of $\overline{P_1 P_2}$: $\left(\dfrac{x_1 + x_2}{2}, \dfrac{y_1 + y_2}{2} \right)$

Laws of Exponents

$$x^m x^n = x^{m+n} \qquad \frac{x^m}{x^n} = x^{m-n} \qquad (x^m)^n = x^{mn}$$

$$x^{-n} = \frac{1}{x^n} \qquad (xy)^n = x^n y^n \qquad \left(\frac{x}{y} \right)^n = \frac{x^n}{y^n}$$

$$x^{1/n} = \sqrt[n]{x} \qquad \sqrt[n]{xy} = \sqrt[n]{x}\,\sqrt[n]{y} \qquad \sqrt[n]{\frac{x}{y}} = \frac{\sqrt[n]{x}}{\sqrt[n]{y}}$$

$$x^{m/n} = \sqrt[n]{x^m} = \left(\sqrt[n]{x} \right)^m$$

Special Factorizations

x^2
x^3
x^3

Binomial Theorem

$$(x + y)^2 = x^2 + 2xy + y^2$$
$$(x - y)^2 = x^2 - 2xy + y^2$$
$$(x + y)^3 = x^3 + 3x^2 y + 3xy^2 + y^3$$
$$(x - y)^3 = x^3 - 3x^2 y + 3xy^2 - y^3$$

$$(x + y)^n = x^n + nx^{n-1}y + \frac{n(n-1)}{2}x^{n-2}y^2$$
$$+ \cdots + \binom{n}{k}x^{n-k}y^k + \cdots + nxy^{n-1} + y^n$$

where $\displaystyle \binom{n}{k} = \frac{n(n-1)\cdots(n-k+1)}{1 \cdot 2 \cdot 3 \cdots \cdot k}$

Quadratic Formula

If $ax^2 + bx + c = 0$, then $x = \dfrac{-b \pm \sqrt{b^2 - 4ac}}{2a}$.

Inequalities and Absolute Value

If $a < b$ and $b < c$, then $a < c$.

If $a < b$, then $a + c < b + c$.

If $a < b$ and $c > 0$, then $ca < cb$.

If $a < b$ and $c < 0$, then $ca > cb$.

$|x| = x \quad$ if $x \geq 0$

$|x| = -x \quad$ if $x \leq 0$

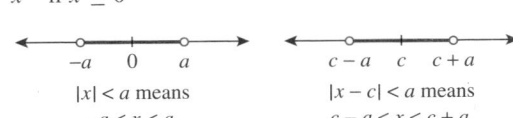

$|x| < a$ means
$-a < x < a$.

$|x - c| < a$ means
$c - a < x < c + a$.

GEOMETRY

Formulas for area A, circumference C, and volume V

Triangle	Circle	Sector of Circle	Sphere	Cylinder	Cone	Cone with arbitrary base
$A = \frac{1}{2}bh$	$A = \pi r^2$	$A = \frac{1}{2}r^2\theta$	$V = \frac{4}{3}\pi r^3$	$V = \pi r^2 h$	$V = \frac{1}{3}\pi r^2 h$	$V = \frac{1}{3}Ah$
$= \frac{1}{2}ab\sin\theta$	$C = 2\pi r$	$s = r\theta$	$A = 4\pi r^2$		$A = \pi r\sqrt{r^2 + h^2}$	where A is the area of the base
		(θ in radians)				

Pythagorean Theorem: For a right triangle with hypotenuse of length c and legs of lengths a and b, $c^2 = a^2 + b^2$.

TRIGONOMETRY

Angle Measurement

π radians $= 180°$

$1° = \dfrac{\pi}{180}$ rad 1 rad $= \dfrac{180°}{\pi}$

$s = r\theta$ (θ in radians)

Right Triangle Definitions

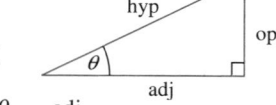

$\sin \theta = \dfrac{\text{opp}}{\text{hyp}}$ $\cos \theta = \dfrac{\text{adj}}{\text{hyp}}$

$\tan \theta = \dfrac{\sin \theta}{\cos \theta} = \dfrac{\text{opp}}{\text{adj}}$ $\cot \theta = \dfrac{\cos \theta}{\sin \theta} = \dfrac{\text{adj}}{\text{opp}}$

$\sec \theta = \dfrac{1}{\cos \theta} = \dfrac{\text{hyp}}{\text{adj}}$ $\csc \theta = \dfrac{1}{\sin \theta} = \dfrac{\text{hyp}}{\text{opp}}$

Trigonometric Functions

$\sin \theta = \dfrac{y}{r}$ $\csc \theta = \dfrac{r}{y}$

$\cos \theta = \dfrac{x}{r}$ $\sec \theta = \dfrac{r}{x}$

$\tan \theta = \dfrac{y}{x}$ $\cot \theta = \dfrac{x}{y}$

$\displaystyle\lim_{\theta \to 0} \dfrac{\sin \theta}{\theta} = 1$ $\displaystyle\lim_{\theta \to 0} \dfrac{1 - \cos \theta}{\theta} = 0$

$P = (r\cos \theta, r\sin \theta)$

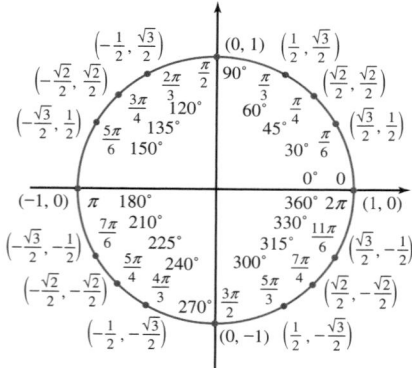

Fundamental Identities

$\sin^2 \theta + \cos^2 \theta = 1$ $\sin(-\theta) = -\sin \theta$

$1 + \tan^2 \theta = \sec^2 \theta$ $\cos(-\theta) = \cos \theta$

$1 + \cot^2 \theta = \csc^2 \theta$ $\tan(-\theta) = -\tan \theta$

$\sin\left(\dfrac{\pi}{2} - \theta\right) = \cos \theta$ $\sin(\theta + 2\pi) = \sin \theta$

$\cos\left(\dfrac{\pi}{2} - \theta\right) = \sin \theta$ $\cos(\theta + 2\pi) = \cos \theta$

$\tan\left(\dfrac{\pi}{2} - \theta\right) = \cot \theta$ $\tan(\theta + \pi) = \tan \theta$

The Law of Sines

$\dfrac{\sin A}{a} = \dfrac{\sin B}{b} = \dfrac{\sin C}{c}$

The Law of Cosines

$a^2 = b^2 + c^2 - 2bc \cos A$

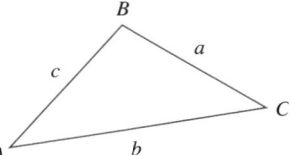

Addition and Subtraction Formulas

$\sin(x + y) = \sin x \cos y + \cos x \sin y$

$\sin(x - y) = \sin x \cos y - \cos x \sin y$

$\cos(x + y) = \cos x \cos y - \sin x \sin y$

$\cos(x - y) = \cos x \cos y + \sin x \sin y$

$\tan(x + y) = \dfrac{\tan x + \tan y}{1 - \tan x \tan y}$

$\tan(x - y) = \dfrac{\tan x - \tan y}{1 + \tan x \tan y}$

Double-Angle Formulas

$\sin 2x = 2 \sin x \cos x$

$\cos 2x = \cos^2 x - \sin^2 x = 2\cos^2 x - 1 = 1 - 2\sin^2 x$

$\tan 2x = \dfrac{2 \tan x}{1 - \tan^2 x}$

$\sin^2 x = \dfrac{1 - \cos 2x}{2}$ $\cos^2 x = \dfrac{1 + \cos 2x}{2}$

Graphs of Trigonometric Functions

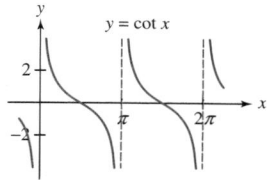

CALCULUS
Early Transcendentals

CALCULUS
Early Transcendentals

JON ROGAWSKI
University of California, Los Angeles

W. H. Freeman and Company
New York

Publisher: *Craig Bleyer*
Executive Editor: *Ruth Baruth*
Senior Acquisitions Editor: *Terri Ward*
Development Editor: *Tony Palermino*
Development Editor: *Bruce Kaplan*
Associate Editor: *Brendan Cady*
Market Development: *Steve Rigolosi*
Executive Marketing Manager: *Robin O'Brien*
Senior Media Editor: *Roland Cheyney*
Assistant Editor: *Laura Capuano*
Photo Editor: *Ted Szczepanski*
Photo Researcher: *Julie Tesser*
Design Manager: *Blake Logan*
Project Editor: *Vivien Weiss*
Illustrations: *Network Graphics*
Illustration Coordinator: *Susan Timmins*
Production Coordinator: *Paul W. Rohloff*
Composition: *Integre Technical Publishing Co.*
Printing and Binding: *RR Donnelley*

Library of Congress Control Number 2006936454
ISBN-13: 978-1-4292-1073-7
ISBN-10: 1-4292-1073-7

ISBN-13: 978-1-4292-0841-3
ISBN-10: 1-4292-0841-4

Printed in the United States of America

Tenth printing

W. H. Freeman and Company, 41 Madison Avenue, New York, NY 10010
Houndmills, Basingstoke RG21 6XS, England
www.whfreeman.com

To Julie

CONTENTS | CALCULUS

Early Transcendentals

As a successful teacher for more than 25 years, Jon Rogawski has listened to and learned much from his own students. These valuable lessons have made an impact on his thinking, his writing, and his shaping of a calculus text.

Jon Rogawski received his undergraduate degree and simultaneously a master's degree in mathematics from Yale University and a Ph.D. in mathematics from Princeton University, where he studied under Robert Langlands. Before joining the Department of Mathematics at UCLA in 1986, where he is currently a full professor, he held teaching positions at Yale University, University of Chicago, the Hebrew University in Jerusalem, and visiting positions at the Institute for Advanced Study, the University of Bonn, and the University of Paris at Jussieu and at Orsay.

Jon's areas of interest are number theory, automorphic forms, and harmonic analysis on semisimple groups. He has published numerous research articles in leading mathematical journals, including a research monograph titled *Automorphic Representations of Unitary Groups in Three Variables* (Princeton University Press). He is the recipient of a Sloan Fellowship and an editor of the *Pacific Journal of Mathematics*.

Jon and his wife, Julie, a physician in family practice, have four children. They run a busy household and, whenever possible, enjoy family vacations in the mountains of California. Jon is a passionate classical music lover and plays the violin and classical guitar.

PREFACE

◼ ABOUT *CALCULUS* by Jon Rogawski

On Teaching Mathematics

As a young instructor, I enjoyed teaching but didn't appreciate how difficult it is to communicate mathematics effectively. Early in my teaching career, I was confronted with a student rebellion when my efforts to explain epsilon-delta proofs were not greeted with the enthusiasm I anticipated. Experiences of this type taught me two basic principles:

1. We should try to teach students as much as possible, but not more so.
2. As math teachers, how we say it is as important as what we say.

When a concept is wrapped in dry mathematical formalism, the majority of students cannot assimilate it. The formal language of mathematics is intimidating to the uninitiated. By presenting the same concept in everyday language, which may be more casual but no less precise, we open the way for students to understand the underlying idea and integrate it into their way of thinking. Students are then in a better position to appreciate the need for formal definitions and proofs and to grasp their logic.

My most valuable resource in writing this text has been my classroom experience of the past 25 years. I have learned how to teach from my students, and I hope my text reflects a true sensitivity to student needs and capabilities. I also hope it helps students experience the joy of understanding and mastering the beautiful ideas of the subject.

On Writing a New Calculus Text

Calculus has a deservedly central role in higher education. It is not only the key to the full range of scientific and engineering disciplines; it is also a crucial component in a student's intellectual development. I hope my text will help to open up this multifaceted world of ideas for the student. Like a great symphony, the ideas in calculus never grow stale. There is always something new to appreciate and delight in, and I enjoy the challenge of finding the best way to communicate these ideas to students.

My text builds on the tradition of several generations of calculus authors. There is no perfect text, given the choices and inevitable compromises that must be made, but many of the existing textbooks reflect years of careful thought, meticulous craft, and accumulated wisdom, all of which cannot and should not be ignored in the writing of a new text.

I have made a sustained effort to communicate the underlying concepts, ideas, and "reasons why things work" in language that is accessible to students. Throughout the text, student difficulties are anticipated and addressed. Problem-solving skills are systematically developed in the examples and problem sets. I have also included a wide range of applications, both innovative and traditional, which provide additional insight into the mathematics and communicate the important message that calculus plays a vital role in the modern world.

My textbook follows a largely traditional organization, with a few exceptions. Two such exceptions are the structure of Chapter 2 on limits and the placement of Taylor polynomials in Chapter 8.

The Limit Concept and the Structure of Chapter 2

Chapter 2 introduces the fundamental concepts of limit and convergence. Sections 2.1 and 2.2 give students a good intuitive grasp of limits. The next two sections discuss limit laws (Section 2.3) and continuity (Section 2.4). The notion of continuity is then used to justify the algebraic approach to limits, developed in Section 2.5. I chose to delay the epsilon-delta definition until Section 2.8, when students have a strong understanding of limits. This placement allows you to treat this topic, if you wish, whenever you would like.

Placement of Taylor Polynomials

Taylor polynomials appear in Chapter 8, before infinite series in Chapter 10. My goal is to present Taylor polynomials as a natural extension of the linear approximation. When I teach infinite series, the primary focus is on convergence, a topic that many students find challenging. By the time we have covered the basic convergence tests and studied the convergence of power series, students are ready to tackle the issues involved in representing a function by its Taylor series. They can then rely on their previous work with Taylor polynomials and the error bound from Chapter 8. However, the section on Taylor polynomials is written so that you can cover this topic together with the materials on infinite series if this order is preferred.

Careful, Precise Development

W. H. Freeman is committed to high quality and precise textbooks and supplements. From this project's inception and throughout its development and production, quality and precision have been given significant priority. We have in place unparalleled procedures to ensure the accuracy of the text.

These are the steps we took to ensure an accurate first edition for you:

- **Exercises and Examples** Rather than waiting until the book was finished before checking it for accuracy (which is often the practice), we have painstakingly checked all the examples, exercises, and their solutions for accuracy in every draft of the manuscript and each phase of production.
- **Exposition** A team of 12 calculus instructors acted as accuracy reviewers, and made four passes through all exposition, confirming the accuracy and precision of the final manuscript.
- **Figures** Tom Banchoff of Brown University verified the appropriateness and accuracy of all figures throughout the production process.
- **Editing** The author worked with an editor with an advanced degree in mathematics to review each line of text, exercise, and figure.
- **Composition** The compositor used the author's original LaTeX files to prevent the introduction of new errors in the production process.
- **Math Clubs** We engaged math clubs at twenty universities to accuracy check all the exercises and solutions.

Together, these procedures far exceed prior industry standards to safeguard the quality and precision of a calculus textbook.

■ SUPPLEMENTS

For Instructors

- Instructor Solutions Manual
  Brian Bradie, Christopher Newport University, and Greg Dresden, Washington and Lee University
  Single Variable: 0-7167-9591-4
  Multivariable: 0-7167-9592-2
  Contains worked-out solutions to all exercises in the text.

- Printed Test Bank
  Calculus: 0-7167-9598-1
  Includes multiple-choice and short-answer test items.

- Test Bank CD-ROM
  Calculus: 0-7167-9895-6
  Available online or on a CD-ROM.

- Instructor's Resource Manual
  Vivien Miller, Mississippi State University; Len Miller, Mississippi State University; and Ted Dobson, Mississippi State University
  Calculus: 0-7167-9589-2
  Offers instructors support material for each section of every chapter. Each section includes suggested class time and content emphasis, selected key points, lecture material, discussion topics and class activities, suggested problems, worksheets, and group projects.

- Instructor's Resource CD-ROM
  Calculus: 1-429-20043-X
  Your one-stop resource. Search and export all resources by key term or chapter. Includes text images, Instructor's Solutions Manual, Instructor's Resource Manual, Lecture PowerPoint slides, Test Bank.

- Instructor's Resource Manual for AP Calculus
  Calculus: 0-7167-9590-6
  In conjunction with the text, this manual provides the opportunity for instructors to prepare their students for the AP Exam.

For Students

- Student Solutions Manual
  Brian Bradie, Christopher Newport University, and Greg Dresden, Washington and Lee University
  Single Variable: 0-7167-9594-9
  Multivariable: 0-7167-9880-8
  Offers worked-out solutions to all odd-numbered exercises in the text.

- Companion website at www.whfreeman.com/rogawski

CalcPortal

One click. One place. For all the tools you need.
CalcPortal is the digital gateway to Rogawski's *Calculus*, designed to enrich your course and improve your students' understanding of calculus. For students, CalcPortal integrates review, diagnostic, and tutorial resources right where they are needed in the eBook. For instructors, CalcPortal provides a powerful but easy-to-use course management system complete with a state-of-the-art algorithmic homework and assessment system.

NEW! Next-Generation eBook

CalcPortal is organized around three main teaching and learning tools: (1) The online eBook is a complete version of Rogawski's *Calculus* and includes a Personalized Study Plan that connects students directly to what they need to learn and to the resources on CalcPortal, including important prerequisite content; (2) the homework and assessment center, with an algorithm problem generator; and (3) the CalcResource center with Just-in-Time algebra and precalculus tutorials.

FEATURES

Pedagogical Features in Rogawski's *CALCULUS*

Conceptual Insights
encourage students to develop a conceptual understanding of calculus by explaining important ideas clearly but informally.

> **CONCEPTUAL INSIGHT Are Limits Really Necessary?** It is natural to ask whether limits are really necessary. Since the tangent line is so easy to visualize, is there perhaps a better or simpler way to find its equation? History gives one answer. The methods of calculus based on limits have stood the test of time and are used more widely today than ever before.
>
> History aside, there is a more intuitive argument for why limits are necessary. The slope of a line can be computed if the coordinates of *two* points $P = (x_1, y_1)$ and $Q = (x_2, y_2)$ on the line are known:
>
> $$\text{Slope of line} = \frac{y_2 - y_1}{x_2 - x_1}$$
>
> We cannot use this formula to find the slope of the tangent line because we have only one point on the tangent line to work with, namely $P = (a, f(a))$. Limits provide an ingenious way around this obstacle. We choose a point $Q = (a + h, f(a + h))$ on the graph near P and form the secant line. The slope of this secant line is only an approximation to the slope of the tangent line:
>
> $$\text{Slope of secant line} = \frac{f(a + h) - f(a)}{h} \approx \text{slope of tangent line}$$
>
> But this approximation improves as $h \to 0$, and by taking the limit, we convert our approximations into an exact result.

Ch. 3, p. 123

Graphical Insights
enhance students' visual understanding by making the crucial connection between graphical properties and the underlying concept.

> **GRAPHICAL INSIGHT** Keep the graphical interpretation of limits in mind. In Figure 4(A), $f(x)$ approaches L as $x \to c$ because for any $\epsilon > 0$, we can make the gap less than ϵ by taking δ sufficiently small. The function in Figure 4(B) has a jump discontinuity at $x = c$ and the gap cannot be made small, no matter how small δ is taken. Therefore, the limit does not exist.

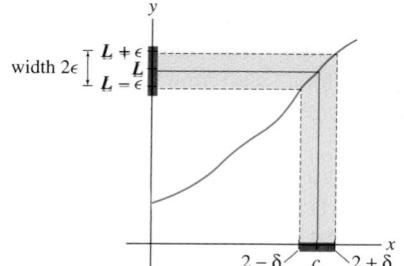

(A) The function is continuous at $x = c$. By taking δ sufficiently small, we can make the gap smaller than ϵ.

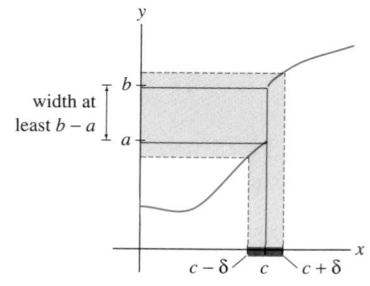

(B) The function is not continuous at $x = c$. The gap is always larger than $(b - a)/2$, no matter how small δ is.

FIGURE 4

Ch. 2, p. 112

■ **EXAMPLE 4** Rotating About the x-Axis   Use the Shell Method to compute the volume V of the solid obtained by rotating the area under $y = 9 - x^2$ over $[0, 3]$ about the x-axis.

Solution   Since we are rotating about the x-axis rather than the y-axis, the Shell Method gives us an integral with respect to y. Therefore, we solve $y = 9 - x^2$ to obtain $x^2 = 9 - y$ or $x = \sqrt{9 - y}$.

The cylindrical shells are generated by *horizontal* segments. The segment $\overline{AB}$ in Figure 8 generates a cylindrical shell of radius y and height $\sqrt{9 - y}$ (we still use the term "height" although the cylinder is horizontal). Using the substitution $u = 9 - y$, $du = -dy$ in the resulting integral, we obtain

◀·· **REMINDER** *After making the substitution $u = 9 - y$, the limits of integration must be changed. Since $u(0) = 9$ and $u(9) = 0$, we change $\int_0^9$ to $\int_9^0$.*

$$V = 2\pi \int_0^9 (\text{radius})(\text{height})\, dy = 2\pi \int_0^9 y\sqrt{9 - y}\, dy = -2\pi \int_9^0 (9 - u)\sqrt{u}\, du$$

$$= 2\pi \int_0^9 (9u^{1/2} - u^{3/2})\, du = 2\pi \left(6u^{3/2} - \frac{2}{5} u^{5/2} \right) \Big|_0^9 = \frac{648}{5}\pi \qquad ■$$

Ch. 6, p. 404

Reminders are margin notes that link back to important concepts discussed earlier in the text to give students a quick review and make connections with earlier concepts.

CAUTION Do not *use the Quotient Rule to compute the partial derivative in Eq. (1). The denominator does not depend on y, so we treat it as a constant when differentiating with respect to y.*

$$g_y(x, y) = \frac{\partial}{\partial y} \frac{y^2}{(1 + x^2)^3} = \frac{1}{(1 + x^2)^3} \frac{\partial}{\partial y} y^2 = \frac{2y}{(1 + x^2)^3} \qquad \boxed{1}$$

$$g_y(1, 3) = \frac{2(3)}{(1 + 1^2)^3} = \frac{3}{4}$$

Ch. 14, p. 811

Caution Notes
warn students of common pitfalls they can encounter in understanding the material.

Historical Perspectives
are brief vignettes that place key discoveries and conceptual advances in their historical context. They give students a glimpse into past accomplishments of great mathematicians and an appreciation for their significance.

HISTORICAL PERSPECTIVE

We take it for granted that the basic laws of physics are best expressed as mathematical relationships. Think of $F = ma$ or the universal law of gravitation. However, the fundamental insight that mathematics could be used to formulate laws of nature (and not just for counting or measuring) developed gradually, beginning with the philosophers of ancient Greece and culminating some 2,000 years later in the discoveries of Galileo and Newton. Archimedes (287–212 BCE) was one of the first scientists (perhaps the first) to formulate a precise physical law. He considered the following question: If weights of mass m_1 and m_2 are placed at the ends of a weightless lever, where should the fulcrum P be located so that the lever does not tip to one side? Suppose that the distance from P to m_1 and m_2 is L_1 and L_2, respectively (Figure 16). Archimedes's Law states that the lever will balance if

$$\boxed{L_1 m_1 = L_2 m_2}$$

In our terminology, what Archimedes had discovered was the center of mass P of the system of weights (see Exercises 40 and 41).

Ch. 8, p. 498

Assumptions Matter uses short explanations and well-chosen counterexamples to help students appreciate why hypotheses are needed in theorems.

■ **EXAMPLE 10** Assumptions Matter: Choosing the Right Comparison   Determine whether $\int_1^\infty \dfrac{dx}{\sqrt{x} + e^{3x}}$ converges or diverges.

Solution Let's try using the inequality

$$\frac{1}{\sqrt{x} + e^{3x}} \le \frac{1}{e^{3x}} \quad (x \ge 0)$$

$$\int_1^\infty \frac{dx}{e^{3x}} = \lim_{R \to \infty} -\frac{1}{3} e^{-3x} \bigg|_1^R = \lim_{R \to \infty} \frac{1}{3}(e^{-3} - e^{-3R}) = e^{-3} \quad \text{(converges)}$$

The Comparison Test tells us that our integral converges:

$$\int_1^\infty \frac{dx}{e^{3x}} \quad \text{converges} \quad \Rightarrow \quad \int_1^\infty \frac{dx}{\sqrt{x} + e^{3x}} \quad \text{also converges}$$

Had we not been thinking, we might have tried to use the inequality

$$\frac{1}{\sqrt{x} + e^{3x}} \le \frac{1}{\sqrt{x}}$$

4

The integral $\int_1^\infty \dfrac{dx}{\sqrt{x}}$ diverges by Theorem 1 (with $p = \frac{1}{2}$), but this tells us nothing about our integral which, by (4), is smaller (Figure 10). ■

FIGURE 10 The divergence of $\int_1^\infty \dfrac{dx}{\sqrt{x}}$ says nothing about the divergence or convergence of the smaller integral $\int_1^\infty \dfrac{dx}{\sqrt{x} + e^{3x}}$.

Ch. 7, p. 472

Section Summaries summarize a section's key points in a concise and useful way and emphasize for students what is most important in the section.

Section Exercise Sets offer a comprehensive set of exercises closely coordinated with the text. These exercises vary in levels of difficulty from routine, to moderate, to more challenging. Also included are problems marked with icons that require the student to give a written response 📖 or require the use of technology GU CAS:

> **Preliminary Exercises** begin each exercise set and need little or no computation. They can be used to check understanding of key concepts of a section before problems from the exercise set are attempted.
> **Exercises** offer numerous problems from routine drill problems to moderately challenging problems. These are carefully graded and include many innovative and interesting geometric and real-world applications.
> **Further Insights and Challenges** are more challenging problems that require a deeper level of conceptual understanding and sometimes extend a section's material. Many are excellent for use as small-group projects.

Chapter Review Exercises offer a comprehensive set of exercises closely coordinated with the chapter material to provide additional problems for self-study or assignments.

■ ACKNOWLEDGMENTS

Jon Rogawski and W. H. Freeman and Company are grateful to the many instructors from across the United States and Canada who have offered comments that assisted in the development and refinement of this book. These contributions included class testing, manuscript reviewing, problems reviewing, and participating in surveys about the book and general course needs.

ALABAMA Tammy Potter, *Gadsden State Community College*; David Dempsey, *Jacksonville State University*; Douglas Bailer, *Northeast Alabama Community College*; Michael Hicks, *Shelton State Community College*; Patricia C. Eiland, *Troy University, Montgomery Campus*; James L. Wang, *The University of Alabama*; Stephen Brick, *University of South Alabama*; Joerg Feldvoss, *University of South Alabama* **ALASKA** Mark A. Fitch, *University of Alaska Anchorage*; Kamal Narang, *University of Alaska Anchorage*; Alexei Rybkin, *University of Alaska Fairbanks*; Martin Getz, *University of Alaska Fairbanks* **ARIZONA** Stefania Tracogna, *Arizona State University*; Bruno Welfert, *Arizona State University*; Daniel Russow, *Arizona Western College*; Garry Carpenter, *Pima Community College, Northwest Campus*; Katie Louchart, *Northern Arizona University*; Donna M. Krawczyk, *The University of Arizona* **ARKANSAS** Deborah Parker, *Arkansas Northeastern College*; J. Michael Hall, *Arkansas State University*; Kevin Cornelius, *Ouachita Baptist University*; Hyungkoo Mark Park, *Southern Arkansas University*; Katherine Pinzon, *University of Arkansas at Fort Smith*; Denise LeGrand, *University of Arkansas at Little Rock*; John Annulis, *University of Arkansas at Monticello*; Erin Haller, *University of Arkansas Fayetteville*; Daniel J. Arrigo, *University of Central Arkansas* **CALIFORNIA** Harvey Greenwald, *California Polytechnic State University, San Luis Obispo*; John M. Alongi, *California Polytechnic State University, San Luis Obispo*; John Hagen, *California Polytechnic State University, San Luis Obispo*; Colleen Margarita Kirk, *California Polytechnic State University, San Luis Obispo*; Lawrence Sze, *California Polytechnic State University, San Luis Obispo*; Raymond Terry, *California Polytechnic State University, San Luis Obispo*; James R. McKinney, *California State Polytechnic University, Pomona*; Charles Lam, *California State University, Bakersfield*; David McKay, *California State University, Long Beach*; Melvin Lax, *California State University, Long Beach*; Wallace A. Etterbeek, *California State University, Sacramento*; Mohamed Allali, *Chapman University*; George Rhys, *College of the Canyons*; Janice Hector, *DeAnza College*; Isabelle Saber, *Glendale Community College*; Peter Stathis, *Glendale Community College*; Kristin Hartford, *Long Beach City College*; Eduardo Arismendi-Pardi, *Orange Coast College*; Mitchell Alves, *Orange Coast College*; Yenkanh Vu, *Orange Coast College*; Yan Tian, *Palomar College*; Donna E. Nordstrom, *Pasadena City College*; Don L. Hancock, *Pepperdine University*; Kevin Iga, *Pepperdine University*; Adolfo J. Rumbos, *Pomona College*; Carlos de la Lama, *San Diego City College*; Matthias Beck, *San Francisco State University*; Arek Goetz, *San Francisco State University*; Nick Bykov, *San Joaquin Delta College*; Eleanor Lang Kendrick, *San Jose City College*; Elizabeth Hodes, *Santa Barbara City College*; William Konya, *Santa Monica College*; John Kennedy, *Santa Monica College*; Peter Lee, *Santa Monica College*; Richard Salome, *Scotts Valley High School*; Norman Feldman, *Sonoma State University*; Elaine McDonald, *Sonoma State University*; Bruno Nachtergaele, *University of California, Davis*; Boumediene Hamzi, *University of California, Davis*; Peter Stevenhagen, *University of California, San Diego*; Jeffrey Stopple, *University of California, Santa Barbara*; Guofang Wei, *University of California, Santa Barbara*; Rick A. Simon, *University of La Verne*; Mohamad A. Alwash, *West Los Angeles College* **COLORADO** Tony Weathers, *Adams State College*; Erica Hastert, *Arapahoe Community College*; Karen Walters, *Arapahoe Community College*; Joshua D. Laison, *Colorado College*; Gerrald G. Greivel, *Colorado School of Mines*; Jim Thomas, *Colorado State University*; Eleanor Storey, *Front Range Community College*; Larry Johnson, *Metropolitan State College of Denver*; Carol Kuper, *Morgan Community*

College; Larry A. Pontaski, *Pueblo Community College*; Terry Reeves, *Red Rocks Community College*; Debra S. Carney, *University of Denver* CONNECTICUT Jeffrey McGowan, *Central Connecticut State University*; Ivan Gotchev, *Central Connecticut State University*; Charles Waiveris, *Central Connecticut State University*; Christopher Hammond, *Connecticut College*; Anthony Y. Aidoo, *Eastern Connecticut State University*; Kim Ward, *Eastern Connecticut State University*; Joan W. Weiss, *Fairfield University*; Theresa M. Sandifer, *Southern Connecticut State University*; Cristian Rios, *Trinity College*; Melanie Stein, *Trinity College* DELAWARE Patrick F. Mwerinde, *University of Delaware* DISTRICT OF COLUMBIA Jeffrey Hakim, *American University*; Joshua M. Lansky, *American University*; James A. Nickerson, *Gallaudet University* FLORIDA Abbas Zadegan, *Florida International University*; Gerardo Aladro, *Florida International University*; Gregory Henderson, *Hillsborough Community College*; Pam Crawford, *Jacksonville University*; Penny Morris, *Polk Community College*; George Schultz, *St. Petersburg College*; Jimmy Chang, *St. Petersburg College*; Carolyn Kistner, *St. Petersburg College*; Aida Kadic-Galeb, *The University of Tampa*; Heath M. Martin, *University of Central Florida*; Constance Schober, *University of Central Florida*; S. Roy Choudhury, *University of Central Florida*; Kurt Overhiser, *Valencia Community College* GEORGIA Thomas T. Morley, *Georgia Institute of Technology*; Ralph Wildy, *Georgia Military College*; Shahram Nazari, *Georgia Perimeter College*; Alice Eiko Pierce, *Georgia Perimeter College Clarkson Campus*; Susan Nelson, *Georgia Perimeter College Clarkson Campus*; Shahram Nazari, *Georgia Perimeter College Dunwoody Campus*; Laurene Fausett, *Georgia Southern University*; Scott N. Kersey, *Georgia Southern University*; Jimmy L. Solomon, *Georgia Southern University*; Allen G. Fuller, *Gordon College*; Marwan Zabdawi, *Gordon College*; Carolyn A. Yackel, *Mercer University*; Shahryar Heydari, *Piedmont College*; Dan Kannan, *The University of Georgia* HAWAII Shuguang Li, *University of Hawaii at Hilo*; Raina B. Ivanova, *University of Hawaii at Hilo* IDAHO Charles Kerr, *Boise State University*; Otis Kenny, *Boise State University*; Alex Feldman, *Boise State University*; Doug Bullock, *Boise State University*; Ed Korntved, *Northwest Nazarene University* ILLINOIS Chris Morin, *Blackburn College*; Alberto L. Delgado, *Bradley University*; John Haverhals, *Bradley University*; Herbert E. Kasube, *Bradley University*; Brenda H. Alberico, *College of DuPage*; Ayse Sahin, *DePaul University*; Marvin Doubet, *Lake Forest College*; Marvin A. Gordon, *Lake Forest Graduate School of Management*; Richard J. Maher, *Loyola University Chicago*; Joseph H. Mayne, *Loyola University Chicago*; Marian Gidea, *Northeastern Illinois University*; Miguel Angel Lerma, *Northwestern University*; Mehmet Dik, *Rockford College*; Tammy Voepel, *Southern Illinois University Edwardsville*; Rahim G. Karimpour, *Southern Illinois University*; Thomas Smith, *University of Chicago* INDIANA Julie A. Killingbeck, *Ball State University*; John P. Boardman, *Franklin College*; Robert N. Talbert, *Franklin College*; Robin Symonds, *Indiana University Kokomo*; Henry L. Wyzinski, *Indiana University Northwest*; Melvin Royer, *Indiana Wesleyan University*; Gail P. Greene, *Indiana Wesleyan University*; David L. Finn, *Rose-Hulman Institute of Technology* IOWA Nasser Dastrange, *Buena Vista University*; Mark A. Mills, *Central College*; Karen Ernst, *Hawkeye Community College*; Richard Mason, *Indian Hills Community College*; Robert S. Keller, *Loras College*; Eric Robert Westlund, *Luther College* KANSAS Timothy W. Flood, *Pittsburg State University*; Sarah Cook, *Washburn University*; Kevin E. Charlwood, *Washburn University* KENTUCKY Alex M. McAllister, *Center College*; Sandy Spears, *Jefferson Community & Technical College*; Leanne Faulkner, *Kentucky Wesleyan College*; Donald O. Clayton, *Madisonville Community College*; Thomas Riedel, *University of Louisville*; Manabendra Das, *University of Louisville*; Lee Larson,

University of Louisville; Jens E. Harlander, *Western Kentucky University*
LOUISIANA William Forrest, *Baton Rouge Community College*; Paul Wayne Britt, *Louisiana State University*; Galen Turner, *Louisiana Tech University*; Randall Wills, *Southeastern Louisiana University*; Kent Neuerburg, *Southeastern Louisiana University* **MAINE** Andrew Knightly, *The University of Maine*; Sergey Lvin, *The University of Maine*; Joel W. Irish, *University of Southern Maine*; Laurie Woodman, *University of Southern Maine* **MARYLAND** Leonid Stern, *Towson University*; Mark E. Williams, *University of Maryland Eastern Shore*; Austin A. Lobo, *Washington College* **MASSACHUSETTS** Sean McGrath, *Algonquin Regional High School*; Norton Starr, *Amherst College*; Renato Mirollo, *Boston College*; Emma Previato, *Boston University*; Richard H. Stout, *Gordon College*; Matthew P. Leingang, *Harvard University*; Suellen Robinson, *North Shore Community College*; Walter Stone, *North Shore Community College*; Barbara Loud, *Regis College*; Andrew B. Perry, *Springfield College*; Tawanda Gwena, *Tufts University*; Gary Simundza, *Wentworth Institute of Technology*; Mikhail Chkhenkeli, *Western New England College*; David Daniels, *Western New England College*; Alan Gorfin, *Western New England College*; Saeed Ghahramani, *Western New England College*; Julian Fleron, *Westfield State College*; Brigitte Servatius, *Worcester Polytechnic Institute* **MICHIGAN** Mark E. Bollman, *Albion College*; Jim Chesla, *Grand Rapids Community College*; Jeanne Wald, *Michigan State University*; Allan A. Struthers, *Michigan Technological University*; Debra Pharo, *Northwestern Michigan College*; Anna Maria Spagnuolo, *Oakland University*; Diana Faoro, *Romeo Senior High School*; Andrew Strowe, *University of Michigan–Dearborn*; Daniel Stephen Drucker, *Wayne State University* **MINNESOTA** Bruce Bordwell, *Anoka-Ramsey Community College*; Robert Dobrow, *Carleton College*; Jessie K. Lenarz, *Concordia College–Moorhead Minnesota*; Bill Tomhave, *Concordia College–Moorhead Minnesota*; David L. Frank, *University of Minnesota*; Steven I. Sperber, *University of Minnesota*; Jeffrey T. McLean, *University of St. Thomas*; Chehrzad Shakiban, *University of St. Thomas*; Melissa Loe, *University of St. Thomas* **MISSISSIPPI** Vivien G. Miller, *Mississippi State University*; Ted Dobson, *Mississippi State University*; Len Miller, *Mississippi State University*; Tristan Denley, *The University of Mississippi* **MISSOURI** Robert Robertson, *Drury University*; Gregory A. Mitchell, *Metropolitan Community College-Penn Valley*; Charles N. Curtis, *Missouri Southern State University*; Vivek Narayanan, *Moberly Area Community College*; Russell Blyth, *Saint Louis University*; Blake Thornton, *Saint Louis University*; Kevin W. Hopkins, *Southwest Baptist University* **MONTANA** Kelly Cline, *Carroll College*; Richard C. Swanson, *Montana State University*; Nikolaus Vonessen, *The University of Montana* **NEBRASKA** Edward G. Reinke Jr., *Concordia University, Nebraska*; Judith Downey, *University of Nebraska at Omaha* **NEVADA** Rohan Dalpatadu, *University of Nevada, Las Vegas*; Paul Aizley, *University of Nevada, Las Vegas* **NEW HAMPSHIRE** Richard Jardine, *Keene State College*; Michael Cullinane, *Keene State College*; Roberta Kieronski, *University of New Hampshire at Manchester* **NEW JERSEY** Paul S. Rossi, *College of Saint Elizabeth*; Mark Galit, *Essex County College*; Katarzyna Potocka, *Ramapo College of New Jersey*; Nora S. Thornber, *Raritan Valley Community College*; Avraham Soffer, *Rutgers The State University of New Jersey*; Chengwen Wang, *Rutgers The State University of New Jersey*; Stephen J. Greenfield, *Rutgers The State University of New Jersey*; John T. Saccoman, *Seton Hall University*; Lawrence E. Levine, *Stevens Institute of Technology* **NEW MEXICO** Kevin Leith, *Central New Mexico Community College*; David Blankenbaker, *Central New Mexico Community College*; Joseph Lakey, *New Mexico State University*; Jurg Bolli, *University of New Mexico*; Kees Onneweer, *University of New Mexico* **NEW YORK** Robert C. Williams, *Alfred University*; Timmy G. Bremer, *Broome Community College State*

University of New York; Joaquin O. Carbonara, *Buffalo State College*; Robin Sue Sanders, *Buffalo State College*; Daniel Cunningham, *Buffalo State College*; Rose Marie Castner, *Canisius College*; Sharon L. Sullivan, *Catawba College*; Camil Muscalu, *Cornell University*; Maria S. Terrell, *Cornell University*; Margaret Mulligan, *Dominican College of Blauvelt*; Robert Andersen, *Farmingdale State University of New York*; Leonard Nissim, *Fordham University*; Jennifer Roche, *Hobart and William Smith Colleges*; James E. Carpenter, *Iona College*; Peter Shenkin, *John Jay College of Criminal Justice/CUNY*; Gordon Crandall, *LaGuardia Community College/CUNY*; Gilbert Traub, *Maritime College, State University of New York*; Paul E. Seeburger, *Monroe Community College Brighton Campus*; Abraham S. Mantell, *Nassau Community College*; Daniel D. Birmajer, *Nazareth College*; Sybil G. Shaver, *Pace University*; Margaret Kiehl, *Rensselaer Polytechnic Institute*; Carl V. Lutzer, *Rochester Institute of Technology*; Michael A. Radin, *Rochester Institute of Technology*; Hossein Shahmohamad, *Rochester Institute of Technology*; Thomas Rousseau, *Siena College*; Jason Hofstein, *Siena College*; Leon E. Gerber, *St. Johns University*; Christopher Bishop, *Stony Brook University*; James Fulton, *Suffolk County Community College*; John G. Michaels, *SUNY Brockport*; Howard J. Skogman, *SUNY Brockport*; Cristina Bacuta, *SUNY Cortland*; Jean Harper, *SUNY Fredonia*; Kelly Black, *Union College*; Thomas W. Cusick, *University at Buffalo/The State University of New York*; Gino Biondini, *University at Buffalo/The State University of New York*; Robert Koehler, *University at Buffalo/The State University of New York* **NORTH CAROLINA** Jeffrey Clark, *Elon University*; William L. Burgin, *Gaston College*; Manouchehr H. Misaghian, *Johnson C. Smith University*; Legunchim L. Emmanwori, *North Carolina A&T State University*; Drew Pasteur, *North Carolina State University*; Demetrio Labate, *North Carolina State University*; Mohammad Kazemi, *The University of North Carolina at Charlotte*; Richard Carmichael, *Wake Forest University*; Gretchen Wilke Whipple, *Warren Wilson College* **NORTH DAKOTA** Anthony J. Bevelacqua, *The University of North Dakota*; Richard P. Millspaugh, *The University of North Dakota* **OHIO** Christopher Butler, *Case Western Reserve University*; Pamela Pierce, *The College of Wooster*; Tzu-Yi Alan Yang, *Columbus State Community College*; Greg S. Goodhart, *Columbus State Community College*; Kelly C. Stady, *Cuyahoga Community College*; Brian T. Van Pelt, *Cuyahoga Community College*; David Robert Ericson, *Miami University, Middletown Campus*; Frederick S. Gass, *Miami University*; Thomas Stacklin, *Ohio Dominican University*; Vitaly Bergelson, *The Ohio State University*; Darry Andrews, *The Ohio State University*; Robert Knight, *Ohio University*; John R. Pather, *Ohio University, Eastern Campus*; Teresa Contenza, *Otterbein College*; Ali Hajjafar, *The University of Akron*; Jianping Zhu, *The University of Akron*; Ian Clough, *University of Cincinnati Clermont College*; Atif Abueida, *University of Dayton*; Judith McCrory, *The University at Findlay*; Thomas Smotzer, *Youngstown State University*; Angela Spalsbury, *Youngstown State University* **OKLAHOMA** Michael McClendon, *University of Central Oklahoma*; Teri Jo Murphy, *The University of Oklahoma* **OREGON** Lorna TenEyck, *Chemeketa Community College*; Angela Martinek, *Linn-Benton Community College*; Tevian Dray, *Oregon State University* **PENNSYLVANIA** John B. Polhill, *Bloomsburg University of Pennsylvania*; Russell C. Walker, *Carnegie Mellon University*; Jon A. Beal, *Clarion University of Pennsylvania*; Kathleen Kane, *Community College of Allegheny County*; David A. Santos, *Community College of Philadelphia*; David S. Richeson, *Dickinson College*; Christine Marie Cedzo, *Gannon University*; Monica Pierri-Galvao, *Gannon University*; John H. Ellison, *Grove City College*; Gary L. Thompson, *Grove City College*; Dale McIntyre, *Grove City College*; Dennis Benchoff, *Harrisburg Area Community College*; William A. Drumin, *King's College*; Denise Reboli, *King's College*; Chawne Kimber, *Lafeyette College*; David L. Johnson, *Lehigh University*; Zia

Uddin, *Lock Haven University of Pennsylvania*; Donna A. Dietz, *Mansfield University of Pennsylvania*; Samuel Wilcock, *Messiah College*; Neena T. Chopra, *The Pennsylvania State University*; Boris A. Datskovsky, *Temple University*; Dennis M. DeTurck, *University of Pennsylvania*; Jacob Burbea, *University of Pittsburgh*; Mohammed Yahdi, *Ursinus College*; Timothy Feeman, *Villanova University*; Douglas Norton, *Villanova University*; Robert Styer, *Villanova University*   RHODE ISLAND   Thomas F. Banchoff, *Brown University*; Yajni Warnapala-Yehiya, *Roger Williams University*; Carol Gibbons, *Salve Regina University*   SOUTH CAROLINA   Stanley O. Perrine, *Charleston Southern University*; Joan Hoffacker, *Clemson University*; Constance C. Edwards, *Coastal Carolina University*; Thomas L. Fitzkee, *Francis Marion University*; Richard West, *Francis Marion University*; Douglas B. Meade, *University of South Carolina*; George Androulakis, *University of South Carolina*; Art Mark, *University of South Carolina Aiken*   SOUTH DAKOTA   Dan Kemp, *South Dakota State University*   TENNESSEE   Andrew Miller, *Belmont University*; Arthur A. Yanushka, *Christian Brothers University*; Laurie Plunk Dishman, *Cumberland University*; Beth Long, *Pellissippi State Technical Community College*; Judith Fethe, *Pellissippi State Technical Community College*; Andrzej Gutek, *Tennessee Technological University*; Sabine Le Borne, *Tennessee Technological University*; Richard Le Borne, *Tennessee Technological University*; Jim Conant, *The University of Tennessee*; Pavlos Tzermias, *The University of Tennessee*; Jo Ann W. Staples, *Vanderbilt University*   TEXAS   Sally Haas, *Angelina College*; Michael Huff, *Austin Community College*; Scott Wilde, *Baylor University and The University of Texas at Arlington*; Rob Eby, *Blinn College*; Tim Sever, *Houston Community College–Central*; Ernest Lowery, *Houston Community College–Northwest*; Shirley Davis, *South Plains College*; Todd M. Steckler, *South Texas College*; Mary E. Wagner-Krankel, *St. Mary's University*; Elise Z. Price, *Tarrant County College, Southeast Campus*; David Price, *Tarrant County College, Southeast Campus*; Michael Stecher, *Texas A&M University*; Philip B. Yasskin, *Texas A&M University*; Brock Williams, *Texas Tech University*; I. Wayne Lewis, *Texas Tech University*; Robert E. Byerly, *Texas Tech University*; Ellina Grigorieva, *Texas Woman's University*; Abraham Haje, *Tomball College*; Scott Chapman, *Trinity University*; Elias Y. Deeba, *University of Houston Downtown*; Jianping Zhu, *The University of Texas at Arlington*; Tuncay Aktosun, *The University of Texas at Arlington*; John E. Gilbert, *The University of Texas at Austin*; Jorge R. Viramontes-Olivias, *The University of Texas at El Paso*; Melanie Ledwig, *The Victoria College*; Gary L. Walls, *West Texas A&M University*; William Heierman, *Wharton County Junior College*   UTAH   Jason Isaac Preszler, *The University of Utah*   VIRGINIA   Verne E. Leininger, *Bridgewater College*; Brian Bradie, *Christopher Newport University*; Hongwei Chen, *Christopher Newport University*; John J. Avioli, *Christopher Newport University*; James H. Martin, *Christopher Newport University*; Mike Shirazi, *Germanna Community College*; Ramon A. Mata-Toledo, *James Madison University*; Adrian Riskin, *Mary Baldwin College*; Josephine Letts, *Ocean Lakes High School*; Przemyslaw Bogacki, *Old Dominion University*; Deborah Denvir, *Randolph-Macon Woman's College*; Linda Powers, *Virginia Tech*; Gregory Dresden, *Washington and Lee University*; Jacob A. Siehler, *Washington and Lee University*   VERMONT   David Dorman, *Middlebury College*; Rachel Repstad, *Vermont Technical College*   WASHINGTON   Jennifer Laveglia, *Bellevue Community College*; David Whittaker, *Cascadia Community College*; Sharon Saxton, *Cascadia Community College*; Aaron Montgomery, *Central Washington University*; Patrick Averbeck, *Edmonds Community College*; Jeff Eldridge, *Edmonds Community College*; Tana Knudson, *Heritage University*; Kelly Brooks, *Pierce College*; Shana P. Calaway, *Shoreline Community College*; Abel Gage, *Skagit Valley College*; Scott MacDonald, *Tacoma Community College*; Martha A. Gady, *Whitworth College*

WEST VIRGINIA Ralph Oberste-Vorth, *Marshall University*; Suda Kunyosying, *Shepard University*; Nicholas Martin, *Shepherd University*; Rajeev Rajaram, *Shepherd University*; Xiaohong Zhang, *West Virginia State University*; Sam B. Nadler, *West Virginia University* **WYOMING** Claudia Stewart, *Casper College*; Charles Newberg, *Western Wyoming Community College* **WISCONSIN** Paul Bankston, *Marquette University*; Jane Nichols, *Milwaukee School of Engineering*; Yvonne Yaz, *Milwaukee School of Engineering*; Terry Nyman, *University of Wisconsin–Fox Valley*; Robert L. Wilson, *University of Wisconsin–Madison*; Dietrich A. Uhlenbrock, *University of Wisconsin–Madison*; Paul Milewski, *University of Wisconsin–Madison*; Donald Solomon, *University of Wisconsin–Milwaukee*; Kandasamy Muthuvel, *University of Wisconsin–Oshkosh*; Sheryl Wills, *University of Wisconsin–Platteville*; Kathy A. Tomlinson, *University of Wisconsin–River Falls* **CANADA** Don St. Jean, *George Brown College*; Len Bos, *University of Calgary*

We would also like to acknowledge the following people for their help in reviewing the Math Marvels and Calculus Chronicles cards:

Gino Biondini, *University at Buffalo/The State University of New York*; Russell Blyth, *Saint Louis University*; Christopher Butler, *Case Western Reserve University*; Timmy G. Bremer, *Broome Community College, The State University of New York*; Robert E. Byerly, *Texas Tech University*; Timothy W. Flood, *Pittsburg State University*; Kevin W. Hopkins, *Southwest Baptist University;* Herbert E. Kasube, *Bradley University*; Scott N. Kersey, *Georgia Southern University*; Jessie K. Lenarz, *Concordia College–Moorehead Minnesota*; Ralph Oberste-Vorth, *Marshall University*; Rachel Repstad, *Vermont Technical College*; Chehrzad Shakiban, *University of St. Thomas*; James W. Thomas, *Colorado State University*; Chengwen Wang, *Rutgers, The State University of New Jersey*; Robert C. Williams, *Alfred University*; Jianping Zhu, *University of Akron*

Finally, we would like to thank the Math Clubs at the following schools for their help in checking the accuracy of the exercises and their solutions:

Arizona State University; California Polytechnic State University, San Luis Obispo; University of California, Berkeley; University of California, Santa Barbara; Florida Atlantic University; University of South Florida; Indiana University; Kansas State University; Louisiana State University; Worcester Polytechnic Institute; University of Missouri–Columbia; Mississippi State University; University of Montana; North Carolina State University; University of South Carolina; Vanderbilt University; Texas State University–San Marcos; University of Vermont; University of Wyoming

It is a pleasant task to thank the many people whose guidance and support were crucial in bringing this book to fruition. First, I thank Tony Palermino, my development editor, for his wisdom and dedication. Tony is responsible for improvements too numerous to detail. Special thanks also go to Professor Thomas Banchoff, who gave generously of his expertise in the development of the graphics throughout the text. I am grateful to the many mathematicians who provided valuable insights, constructive criticism, and innovative exercises. In particular, I thank Professors John Alongi, Chris Bishop, Brian Bradie, Dennis DeTurck, Ted Dobson, Greg Dresden, Mark Fitch, James McKinney, Len Miller, Vivien Miller, Aaron Montgomery, Vivek Narayanan, Emma Previato, Michael Radin, Todd Ruskell, David Wells, and my students Shraddha Chaplot and Sharon Hori. I also thank Rami Zelingher and Dina Lavi of JustAsk!

The production of a first-edition textbook is a major undertaking for author and publisher alike. I thank the staff of W. H. Freeman and Company for unwavering confidence in the project and outstanding editorial support. It is my pleasure to thank Craig Bleyer for signing the project and standing behind it, Terri Ward for skillfully organizing the complex production process, Ruth Baruth and Bruce Kaplan for invaluable knowledge and publishing experience, Steve Rigolosi for expert market development, and Brendan Cady and Laura Capuano for editorial assistance. My thanks are also due to the superb production team: Blake Logan, Bill Page, Paul Rohloff, Ted Szczepanski, Susan Timmins, and Vivien Weiss, as well as Don DeLand and Leslie Galen of Integre Technical Publishing, for their expert composition.

To my wife, Julie, I owe more than I can say. Thank you for everything. To our wonderful children Rivkah, Dvora, Hannah, and Akiva, thank you for putting up with the calculus book, a demanding extra sibling that dwelled in our home for so many years. And to my mother Elise and late father Alexander Rogawski, MD z"l, thank you for love and support from the beginning.

TO THE STUDENT

Although I have taught calculus for more than 25 years, I still feel excitement when I enter the classroom on the first day of a new semester, as if a great drama is about to unfold. Does the word "drama" seem out of place in a discussion of mathematics?

There is no doubt that calculus is useful—it is used throughout the sciences and engineering, from weather prediction and space flight to nanotechnology and financial modeling. But what is dramatic about it?

For me, one part of the drama lies in the conceptual and logical development of the subject. Starting with just a few basic concepts such as limits and tangent lines, we gradually build the tools for solving innumerable problems of great practical importance. Along the way, there are high points and moments of suspense—computing a derivative using limits for the first time or learning how the Fundamental Theorem of Calculus unifies differential and integral calculus. We also discover that calculus provides the right language for expressing the universal laws of nature, not just Newton's laws of motion to which it was first applied, but also the laws of electromagnetism and even the strange laws of quantum mechanics.

Another part of the drama is the learning process itself—the personal voyage of discovery. Certainly, one important part of learning calculus is the acquisition of technical skills. You will learn how to compute derivatives and integrals, solve optimization problems, and so on. These are the skills you need to apply calculus in practical situations. But when you study calculus, you also learn the language of science. You gain access to the thoughts of Newton, Euler, Gauss, and Maxwell, the greatest scientific thinkers, all of whom expressed their insights using calculus. The distinguished mathematician I. M. Gelfand put it this way: "The most important thing a student can get from the study of mathematics is the attainment of a higher intellectual level."

This text is designed to develop both skills and conceptual understanding. In fact, the two go hand-in-hand. As you become proficient in problem solving, you will come to appreciate the underlying ideas. And it is equally true that a solid understanding of the concepts will make you a more effective problem solver. You are likely to devote much of your time to studying examples and working exercises. However, the text also contains numerous down-to-earth explanations, sometimes under the heading "Conceptual Insight" or "Graphical Insight." They are designed to show you how and why calculus works. I urge you to take the time to read these explanations and think about them.

A major challenge for me in writing this textbook was to present calculus in a style that students would find comprehensible and interesting. As I wrote, I continually asked myself: Can it be made simpler? Have I assumed something the student may not be aware of? Can I explain the deeper significance of a concept without confusing the student who is learning the subject for the first time?

I'm afraid that no textbook can make learning calculus entirely painless. According to legend, Alexander the Great once asked the mathematician Menaechmus to show him an easy way to learn geometry. Menaechmus replied, "There is no royal road to geometry." Even kings must work hard to learn geometry, and the same is true of calculus—achieving mastery requires time and effort.

However, I hope my efforts have resulted in a textbook that is "student friendly" and that also encourages you to appreciate the big picture—the beautiful and elegant ideas that hold the entire structure of calculus together. Please let me know if you have comments or suggestions for improving the text. I look forward to hearing from you.

Best wishes and good luck!

Jon Rogawski

CALCULUS

Early Transcendentals

1 | PRECALCULUS REVIEW

Functions are one of our most important tools for analyzing phenomena. Biologists have studied the antler weight of male red deer as a function of age (see p. 6).

Additional properties of real numbers are discussed in Appendix B.

FIGURE 1 The set of real numbers represented as a line.

Calculus builds on the foundation of algebra, analytic geometry, and trigonometry. In this chapter we review some of the basic concepts, facts, and formulas that are used throughout the text. In the last section, we discuss some ways technology can be used to enhance your visual understanding of functions and their properties.

1.1 Real Numbers, Functions, and Graphs

We begin our review with a short discussion of real numbers. This gives us the opportunity to recall some basic properties and standard notation.

A **real number** is a number represented by a decimal or "decimal expansion." There are three types of decimal expansions: finite, repeating, and infinite but nonrepeating. For example,

$$\frac{3}{8} = 0.375, \qquad \frac{1}{7} = 0.142857142857\ldots = 0.\overline{142857}$$

$$\pi = 3.141592653589793\ldots$$

The number $\frac{3}{8}$ is represented by a finite decimal, whereas $\frac{1}{7}$ is represented by a *repeating* or *periodic* decimal. The bar over 142857 indicates that this sequence repeats indefinitely. The decimal expansion of π is infinite but nonrepeating.

The set of all real numbers is denoted by the boldface **R**. When there is no risk of confusion, we refer to a real number simply as a *number*. We also use the standard symbol $\in$ for the phrase "belongs to." Thus,

$$a \in \mathbf{R} \qquad \text{reads} \qquad \text{``}a \text{ belongs to } \mathbf{R}\text{''}$$

The set of integers is commonly denoted by the letter **Z** (this choice comes from the German word *zahl*, meaning "number"). Thus, $\mathbf{Z} = \{\ldots, -2, -1, 0, 1, 2, \ldots\}$. A **whole number** is a nonnegative integer, i.e., one of the numbers $0, 1, 2, \ldots$.

A real number is called **rational** if it can be represented by a fraction p/q, where p and q are integers with $q \neq 0$. The set of rational numbers is denoted **Q** (for "quotient"). Numbers such as π and $\sqrt{2}$ that are not rational are called **irrational**. We can tell whether or not a number is rational from its decimal expansion: Rational numbers have finite or repeating decimal expansions and irrational numbers have infinite, nonrepeating decimal expansions. Furthermore, the decimal expansion of a number is unique, apart from the following exception: Every finite decimal is equal to an infinite decimal in which the digit 9 repeats. For example,

$$1 = 0.999\ldots, \qquad \frac{3}{8} = 0.375 = 0.374999\ldots, \qquad \frac{47}{20} = 2.35 = 2.34999\ldots$$

We visualize real numbers as points on a line (Figure 1). For this reason, real numbers are often referred to as **points**. The number 0 is called the **origin** of the real line.

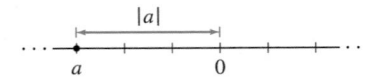

FIGURE 2 The absolute value $|a|$.

FIGURE 3 The distance between a and b is $|b - a|$.

The **absolute value** of a real number a, denoted $|a|$, is defined by (Figure 2)

$$|a| = \text{distance from the origin} = \begin{cases} a & \text{if } a \geq 0 \\ -a & \text{if } a < 0 \end{cases}$$

For example, $|1.2| = 1.2$ and $|-8.35| = 8.35$. The absolute value satisfies

$$|a| = |-a|, \qquad |ab| = |a|\,|b|$$

The **distance** between two real numbers a and b is $|b - a|$, which is the length of the line segment joining a and b (Figure 3).

Two real numbers a and b are close to each other if $|b - a|$ is small, and this is the case if their decimal expansions agree to many places. More precisely, *if the decimal expansions of a and b agree to k places (to the right of the decimal point), then the distance $|b - a|$ is at most 10^{-k}.* Thus, the distance between $a = 3.1415$ and $b = 3.1478$ is at most 10^{-2} because a and b agree to two places. In fact, the distance is exactly $|3.1415 - 3.1478| = 0.0063$.

Beware that $|a + b|$ is not equal to $|a| + |b|$ unless a and b have the same sign or at least one of a and b is zero. If they have opposite signs, cancellation occurs in the sum $a + b$ and $|a + b| < |a| + |b|$. For example, $|2 + 5| = |2| + |5|$ but $|-2 + 5| = 3$, which is less than $|-2| + |5| = 7$. In any case, $|a + b|$ is never larger than $|a| + |b|$ and this gives us the simple but important **triangle inequality**:

$$|a + b| \leq |a| + |b| \qquad \boxed{1}$$

We use standard notation for intervals. Given real numbers $a < b$, there are four intervals with endpoints a and b (Figure 4). Each of these intervals has length $b - a$, but they differ according to whether or not one or both endpoints are included.

The **closed interval** $[a, b]$ is the set of all real numbers x such that $a \leq x \leq b$:

$$[a, b] = \{x \in \mathbf{R} : a \leq x \leq b\}$$

We usually write this more simply as $\{x : a \leq x \leq b\}$, it being understood that x belongs to $\mathbf{R}$. The **open** and **half-open intervals** are the sets

$$\underbrace{(a, b) = \{x : a < x < b\}}_{\text{Open interval (endpoints excluded)}}, \qquad \underbrace{[a, b) = \{x : a \leq x < b\}}_{\text{Half-open}}, \qquad \underbrace{(a, b] = \{x : a < x \leq b\}}_{\text{Half-open}}$$

FIGURE 4 The four intervals with endpoints a and b.

Closed interval $[a, b]$ (endpoints included)

Open interval (a, b) (endpoints excluded)

Half-open interval $[a, b)$

Half-open interval $(a, b]$

The infinite interval $(-\infty, \infty)$ is the entire real line $\mathbf{R}$. A half-infinite interval may be open or closed. A closed half-infinite interval contains its finite endpoint:

$$[a, \infty) = \{x : a \leq x < \infty\}, \qquad (-\infty, b] = \{x : -\infty < x \leq b\}$$

FIGURE 5 Closed half-infinite intervals.

$[a, \infty)$

$(-\infty, b]$

FIGURE 6 The interval
$(-r, r) = \{x : |x| < r\}$.

FIGURE 7 $(a, b) = (c - r, c + r)$, where

$$c = \frac{a + b}{2}, \qquad r = \frac{b - a}{2}$$

FIGURE 8 The interval $[7, 13]$ is described by $|x - 10| \leq 3$.

FIGURE 9 The set
$S = \left\{x : \left|\frac{1}{2}x - 3\right| > 4\right\}$.

In Example 2 we use the notation $\cup$ to denote "union": The union $A \cup B$ of sets A and B consists of all elements that belong to either A or B (or both).

Open and closed intervals may be described by inequalities. For example (Figure 6),

$$\boxed{|x| < r \quad \Leftrightarrow \quad -r < x < r \quad \Leftrightarrow \quad x \in (-r, r)} \qquad \boxed{2}$$

More generally, for any c (Figure 7),

$$\boxed{|x - c| < r \quad \Leftrightarrow \quad c - r < x < c + r \quad \Leftrightarrow \quad x \in (c - r, c + r)} \qquad \boxed{3}$$

A similar statement holds for closed intervals with $<$ replaced by $\leq$. We refer to r as the **radius** and c as the **midpoint** or **center**. The intervals (a, b) and $[a, b]$ have midpoint $c = \frac{1}{2}(a + b)$ and radius $r = \frac{1}{2}(b - a)$ (Figure 7).

■ **EXAMPLE 1** Describing Intervals via Inequalities Describe the intervals $(-4, 4)$ and $[7, 13]$ using inequalities.

Solution We have $(-4, 4) = \{x : |x| < 4\}$. The midpoint of the interval $[7, 13]$ is $c = \frac{1}{2}(7 + 13) = 10$ and its radius is $r = \frac{1}{2}(13 - 7) = 3$ (Figure 8). Therefore

$$[7, 13] = \left\{x \in \mathbf{R} : |x - 10| \leq 3\right\} \qquad ■$$

■ **EXAMPLE 2** Describe the set $S = \left\{x : \left|\frac{1}{2}x - 3\right| > 4\right\}$ in terms of intervals.

Solution It is easier to consider the opposite inequality $\left|\frac{1}{2}x - 3\right| \leq 4$ first. By (2),

$$\left|\frac{1}{2}x - 3\right| \leq 4 \quad \Leftrightarrow \quad -4 \leq \frac{1}{2}x - 3 \leq 4$$

$$-1 \leq \frac{1}{2}x \leq 7 \qquad \text{(add 3)}$$

$$-2 \leq x \leq 14 \qquad \text{(multiply by 2)}$$

Thus, $\left|\frac{1}{2}x - 3\right| \leq 4$ is satisfied when x belongs to $[-2, 14]$. The set S is the *complement*, consisting of all numbers x *not in* $[-2, 14]$. We can describe S as the union of two intervals: $S = (-\infty, -2) \cup (14, \infty)$ (Figure 9). ■

Graphing

Graphing is a basic tool in calculus, as it is in algebra and trigonometry. Recall that rectangular (or Cartesian) coordinates in the plane are defined by choosing two perpendicular axes, the x-axis and the y-axis. To a pair of numbers (a, b) we associate the point P located at the intersection of the line perpendicular to the x-axis at a and the line perpendicular to the y-axis at b [Figure 10(A)]. The numbers a and b are the x- and y-**coordinates** of P. The **origin** is the point with coordinates $(0, 0)$.

The axes divide the plane into four quadrants labeled I–IV, determined by the signs of the coordinates [Figure 10(B)]. For example, quadrant III consists of points (x, y) such that $x < 0$ and $y < 0$.

The distance d between two points $P_1 = (x_1, y_1)$ and $P_2 = (x_2, y_2)$ is computed using the Pythagorean Theorem. In Figure 11, we see that $\overline{P_1 P_2}$ is the hypotenuse of a right triangle with sides $a = |x_2 - x_1|$ and $b = |y_2 - y_1|$. Therefore,

$$d^2 = a^2 + b^2 = (x_2 - x_1)^2 + (y_2 - y_1)^2$$

We obtain the distance formula by taking square roots.

The x-coordinate is sometimes called the "abscissa" and the y-coordinate the "ordinate."

The term "Cartesian" refers to the French philosopher and mathematician René Descartes (1596–1650), whose Latin name was Cartesius. He is credited (along with Pierre de Fermat) with the invention of analytic geometry. In his great work La Géométrie, Descartes used the letters x, y, z for unknowns and a, b, c for constants, a convention that has been followed ever since.

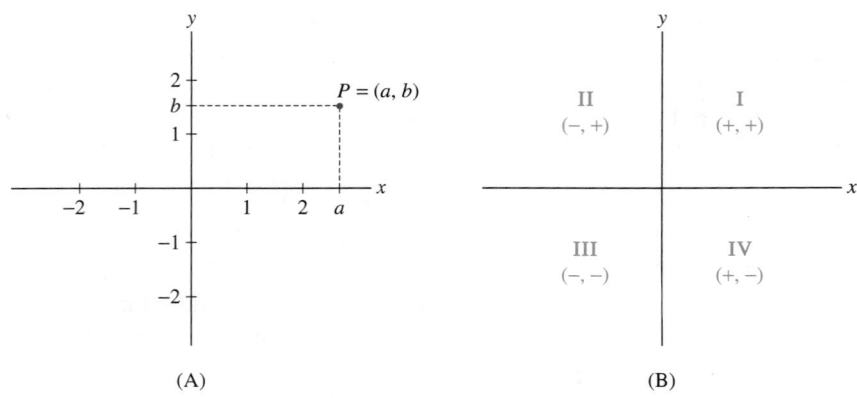

(A)

(B)

FIGURE 10 Rectangular coordinate system.

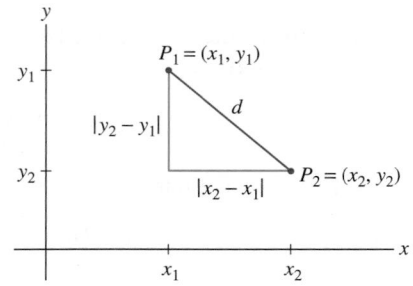

FIGURE 11 Distance d is given by the distance formula.

Distance Formula   The distance d between two points $P_1 = (x_1, y_1)$ and $P_2 = (x_2, y_2)$ is equal to

$$d = \sqrt{(x_2 - x_1)^2 + (y_2 - y_1)^2}$$

Once we have the distance formula, we can easily derive the equation of a circle of radius r and center (a, b) (Figure 12). A point (x, y) lies on this circle if the distance from (x, y) to (a, b) is r:

$$\sqrt{(x - a)^2 + (y - b)^2} = r$$

Squaring both sides, we obtain the standard equation of the circle:

$$(x - a)^2 + (y - b)^2 = r^2$$

We now review some definitions and notation concerning functions.

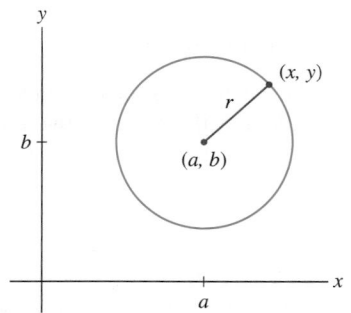

FIGURE 12 Circle with equation $(x - a)^2 + (y - b)^2 = r^2$.

A function $f : D \to Y$ is also called a "map." The sets D and Y can be arbitrary. For example, we can define a map from the set of living people to the set of whole numbers by mapping each person to his or her year of birth. The range of this map is the set of numbers that occur as years of birth of a living person.

DEFINITION Function   A **function** f between two sets D and Y is a rule that assigns, to each element x in D, a unique element $y = f(x)$ in Y. Symbolically, we write

$$f : D \to Y$$

For $x \in D$, $f(x)$ is called the **value** of f at x (Figure 13). We call the set D the **domain** of $f(x)$. The **range** R of f is the subset of Y consisting of all values $f(x)$:

$$R = \{y \in Y : f(x) = y \text{ for some } x \in D\}$$

Informally, we think of f as a "machine" that produces an output y for every input x in the domain D (Figure 14).

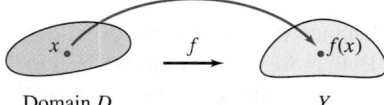

Domain D Y

FIGURE 13 A function is a rule that assigns an element $y = f(x)$ in Y to each $x \in D$.

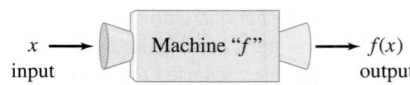

FIGURE 14 Think of f as a "machine" that takes the input x and produces the output $f(x)$.

In multivariable calculus, we deal with functions having a variety of domains and ranges. The domain might be a set of points in three-dimensional space and the range a set of numbers, points, or vectors.

The first part of this text deals with *numerical* functions f, where both the domain and the range are sets of real numbers. We refer to such a function interchangeably as f or $f(x)$. The letter x denotes the **independent variable** that can take on any value in the domain D. We write $y = f(x)$ and refer to y as the **dependent variable** (because its value depends on the choice of x).

When f is defined by a formula, its natural domain is the set of real numbers x for which the formula is meaningful. For example, the function $f(x) = \sqrt{9 - x}$ has domain $D = \{x : x \le 9\}$ because $\sqrt{9 - x}$ is defined if $9 - x \ge 0$. Here are some other examples of domains and ranges.

$f(x)$	Domain D	Range R
x^2	$\mathbf{R}$	$\{y : y \ge 0\}$
$\cos x$	$\mathbf{R}$	$\{y : -1 \le y \le 1\}$
$\dfrac{1}{x + 1}$	$\{x : x \ne -1\}$	$\{y : y \ne 0\}$

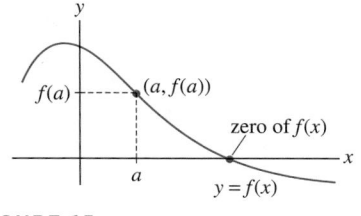

FIGURE 15

The **graph** of a function $y = f(x)$ is obtained by plotting the points $(a, f(a))$ for a in the domain D (Figure 15). If you start at $x = a$ on the x-axis, move up to the graph and then over to the y-axis, you arrive at the value $f(a)$. The value $|f(a)|$ is thus the distance of the graph above or below the x-axis [depending on whether $f(a) \ge 0$ or $f(a) < 0$].

A **zero** or **root** of a function $f(x)$ is a number c such that $f(c) = 0$. The zeros are the values of x where the graph intersects the x-axis.

In Chapter 4, we will use calculus to sketch and analyze graphs. At this stage, to sketch a graph by hand, we make a table of function values, plot the corresponding points (including any zeros), and connect them by a smooth curve.

■ **EXAMPLE 3** Find the roots and sketch the graph of $f(x) = x^3 - 2x$.

Solution First, we solve $x^3 - 2x = x(x^2 - 2) = 0$. The roots of $f(x)$ are $x = 0$ and $x = \pm\sqrt{2}$. To sketch the graph, we plot the roots and a few values listed in Table 1 and join them by a curve (Figure 16). ■

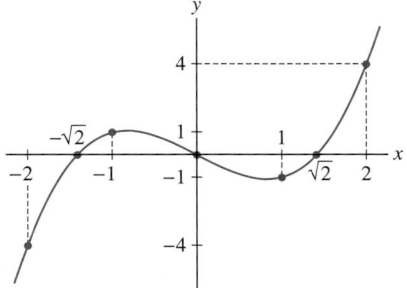

FIGURE 16 Graph of $f(x) = x^3 - 2x$.

	TABLE 1
x	$x^3 - 2x$
-2	-4
-1	1
0	0
1	-1
2	4

Functions arising in applications are not always given by formulas. Data collected from observation or experiment define functions that can be displayed either graphically or by a table of values. Figure 17 and Table 2 display data collected by biologist Julian Huxley (1887–1975) in a study of the antler weight W of male red deer as a function of age t. We will see that many of the tools from calculus can be applied to functions constructed from data in this way.

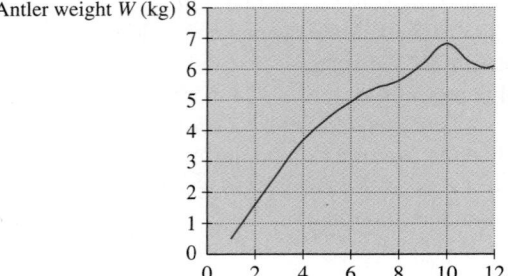

t (year)	W (kg)	t (year)	W (kg)
1	0.48	7	5.34
2	1.59	8	5.62
3	2.66	9	6.18
4	3.68	10	6.81
5	4.35	11	6.21
6	4.92	12	6.1

TABLE 2

FIGURE 17 Average antler weight of a male red deer as a function of age.

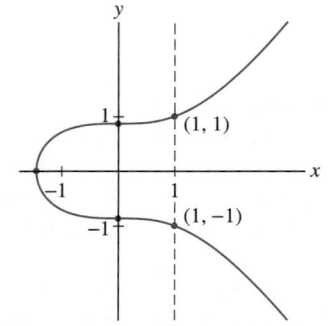

FIGURE 18 Graph of $4y^2 - x^3 = 3$. This graph fails the Vertical Line Test, so it is not the graph of a function.

We can graph any equation relating y and x. For example, to graph the equation $4y^2 - x^3 = 3$, we plot the pairs (x, y) satisfying the equation (Figure 18). This curve is not the graph of a function because there are two values of y for a given x-value such as $x = 1$. In general, a curve is the graph of a function exactly when it passes the **Vertical Line Test:** Every vertical line $x = a$ intersects the curve in at most one point.

We are often interested in whether a function is increasing or decreasing. Roughly speaking, a function $f(x)$ is increasing if its graph goes up as we move to the right and decreasing if its graph goes down [Figures 19(A) and (B)]. More precisely, we define the notion of increase/decrease on an interval:

- **Increasing** on (a, b) if $f(x_1) \leq f(x_2)$ for all $x_1, x_2 \in (a, b)$ such that $x_1 \leq x_2$
- **Decreasing** on (a, b) if $f(x_1) \geq f(x_2)$ for all $x_1, x_2 \in (a, b)$ such that $x_1 \leq x_2$

The function in Figure 19(D) is neither increasing nor decreasing (for all x), but it is decreasing on the interval (a, b). We say that $f(x)$ is **strictly increasing** if $f(x_1) < f(x_2)$ for $x_1 < x_2$. Strictly decreasing functions are defined similarly. The function in Figure 19(C) is increasing but not strictly increasing.

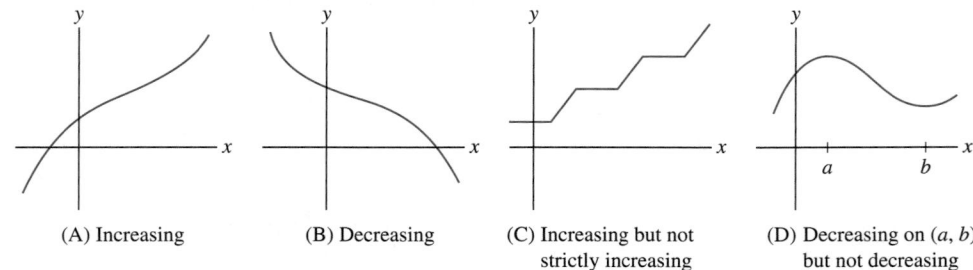

| (A) Increasing | (B) Decreasing | (C) Increasing but not strictly increasing | (D) Decreasing on (a, b) but not decreasing everywhere |

FIGURE 19

Another important property is **parity**, which refers to whether a function is even or odd:

- $f(x)$ is **even** if $f(-x) = f(x)$
- $f(x)$ is **odd** if $f(-x) = -f(x)$

The graphs of functions with even or odd parity have a special symmetry. The graph of an even function is symmetric with respect to the y-axis (we also say "symmetric *about* the y-axis"). This means that if $P = (a, b)$ lies on the graph, then so does $Q = (-a, b)$

[Figure 20(A)]. The graph of an odd function is symmetric with respect to the origin, which means that if $P = (a, b)$ lies on the graph, then so does $Q = (-a, -b)$ [Figure 20(B)]. A function may be neither even nor odd [Figure 20(C)].

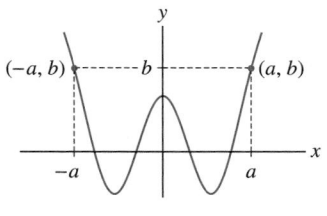

(A) Even function: $f(-x) = f(x)$
Graph is symmetric
about the y-axis.

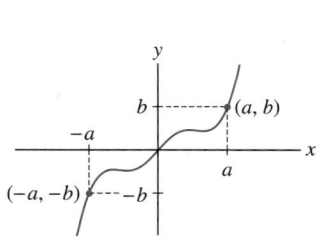

(B) Odd function: $f(-x) = -f(x)$
Graph is symmetric
about the origin.

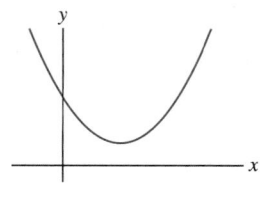

(C) Neither even nor odd

FIGURE 20

■ **EXAMPLE 4** Determine whether the function is even, odd, or neither.

(a) $f(x) = x^4$ **(b)** $g(x) = x^{-1}$ **(c)** $h(x) = x^2 + x$

Solution

(a) $f(-x) = (-x)^4 = x^4$. Thus, $f(x) = f(-x)$ and $f(x)$ is even.

(b) $g(-x) = (-x)^{-1} = -x^{-1}$. Thus, $g(-x) = -g(x)$, and $g(x)$ is odd.

(c) $h(-x) = (-x)^2 + (-x) = x^2 - x$. We see that $h(-x)$ is not equal to $h(x) = x^2 + x$ or $-h(x) = -x^2 - x$. Therefore, $h(x)$ is neither even nor odd. ■

■ **EXAMPLE 5** **Graph Sketching Using Symmetry** Sketch the graph of $f(x) = \dfrac{1}{x^2 + 1}$.

Solution The function $f(x) = \dfrac{1}{x^2 + 1}$ is positive [$f(x) > 0$ for all x] and it is even since $f(x) = f(-x)$. Therefore, the graph of $f(x)$ lies above the x-axis and is symmetric with respect to the y-axis. Furthermore, $f(x)$ is decreasing for $x \geq 0$ (since a larger value of x makes for a larger denominator). We use this information and a short table of values (Table 3) to sketch the graph (Figure 21). Note that the graph approaches the x-axis as we move to the right or left since $f(x)$ is small when $|x|$ is large. ■

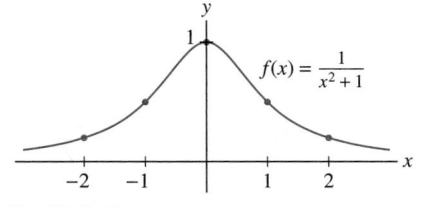

FIGURE 21

TABLE 3

x	$\dfrac{1}{x^2 + 1}$
0	1
±1	$\frac{1}{2}$
±2	$\frac{1}{5}$

Two important ways of modifying a graph are **translation** (or **shifting**) and **scaling**. Translation consists of moving the graph horizontally or vertically:

DEFINITION Translation (Shifting)

- **Vertical translation** $y = f(x) + c$: This shifts the graph of $f(x)$ *vertically* c units. If $c < 0$, the result is a downward shift.
- **Horizontal translation** $y = f(x + c)$: This shifts the graph of $f(x)$ *horizontally* c units *to the right* if $c < 0$ and c units to the left if $c > 0$.

Figure 22 shows the effect of translating the graph of $f(x) = \dfrac{1}{x^2 + 1}$ vertically and horizontally.

Remember that $f(x) + c$ and $f(x + c)$ are different. The graph of $y = f(x) + c$ is a vertical translation and $y = f(x + c)$ a horizontal translation of the graph of $y = f(x)$.

(A) $y = f(x) = \dfrac{1}{x^2 + 1}$ (B) $y = f(x) + 1 = \dfrac{1}{x^2 + 1} + 1$ (C) $y = f(x + 1) = \dfrac{1}{(x + 1)^2 + 1}$

FIGURE 22

■ **EXAMPLE 6** Figure 23(A) is the graph of $f(x) = x^2$, and Figure 23(B) is a horizontal and vertical shift of (A). What is the equation of graph (B)?

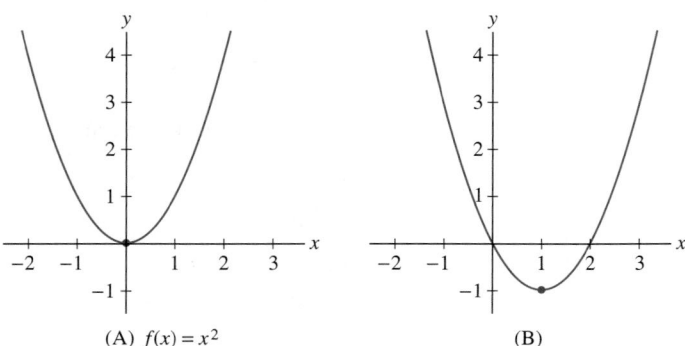

(A) $f(x) = x^2$ (B)

FIGURE 23

Solution Graph (B) is obtained by shifting graph (A) one unit to the right and one unit down. We can see this by observing that the point $(0, 0)$ on the graph of $f(x)$ is shifted to $(1, -1)$. Therefore, (B) is the graph of $g(x) = (x - 1)^2 - 1$.   ■

Scaling (also called dilation) consists of compressing or expanding the graph in the vertical or horizontal directions:

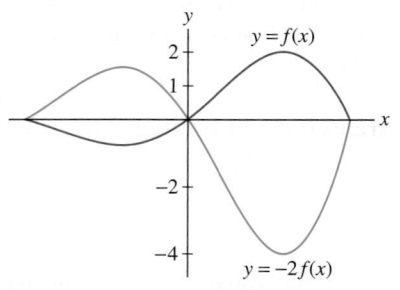

FIGURE 24 Negative vertical scale factor $k = -2$.

DEFINITION Scaling

- **Vertical scaling** $y = kf(x)$: If $k > 1$, the graph is expanded vertically by the factor k. If $0 < k < 1$, the graph is compressed vertically. When the scale factor k is negative $(k < 0)$, the graph is also reflected across the x-axis (Figure 24).
- **Horizontal scaling** $y = f(kx)$: If $k > 1$, the graph is compressed in the horizontal direction. If $0 < k < 1$, the graph is expanded. If $k < 0$, then the graph is also reflected across the y-axis.

We refer to the vertical size of a graph as its **amplitude**. Thus, vertical scaling changes the amplitude by the factor $|k|$.

Remember that $kf(x)$ and $f(kx)$ are different. The graph of $y = kf(x)$ is a vertical scaling and $y = f(kx)$ a horizontal scaling of the graph of $y = f(x)$.

■ **EXAMPLE 7** Sketch the graphs of $f(x) = \sin(\pi x)$ and its two dilates $f(3x)$ and $3f(x)$.

Solution The graph of $f(x) = \sin(\pi x)$ is a sine curve with period 2 [it completes one cycle over every interval of length 2—see Figure 25(A)].

- The graph of $y = f(3x) = \sin(3\pi x)$ is a compressed version of $y = f(x)$ [Figure 25(B)]. It completes three cycles over intervals of length 2 instead of just one cycle.
- The graph of $y = 3f(x) = 3\sin(\pi x)$ differs from $y = f(x)$ only in amplitude: It is expanded in the vertical direction by a factor of 3 [Figure 25(C)]. ∎

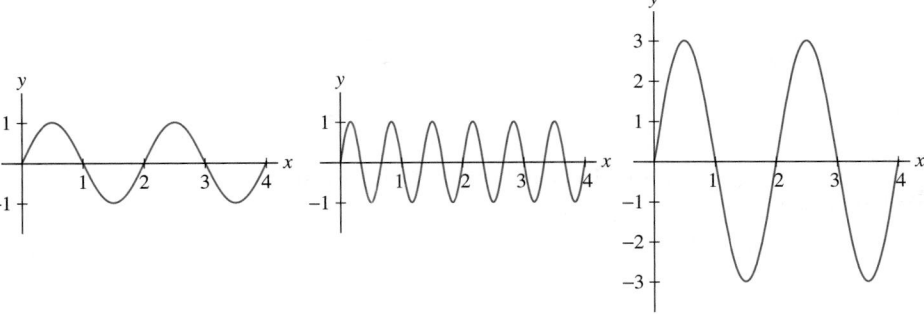

FIGURE 25 Horizontal and vertical scaling of $f(x) = \sin(\pi x)$.

(A) $y = f(x) = \sin(\pi x)$

(B) Horizontal compression: $y = f(3x) = \sin(3\pi x)$

(C) Vertical expansion: $y = 3f(x) = 3\sin(\pi x)$

1.1 SUMMARY

- Absolute value: $|a| = \begin{cases} a & \text{if } a \geq 0 \\ -a & \text{if } a < 0 \end{cases}$

- Triangle inequality: $|a + b| \leq |a| + |b|$
- There are four types of intervals with endpoints a and b:

$$(a, b), \qquad [a, b], \qquad [a, b), \qquad (a, b]$$

- We can express open and closed intervals using inequalities:

$$(a, b) = \{x : |x - c| < r\}, \qquad [a, b] = \{x : |x - c| \leq r\}$$

where $c = \frac{1}{2}(a + b)$ is the midpoint and $r = \frac{1}{2}(b - a)$ is the radius.

- Distance d between (x_1, y_1) and (x_2, y_2): $d = \sqrt{(x_2 - x_1)^2 + (y_2 - y_1)^2}$.
- Equation of circle of radius r with center (a, b): $(x - a)^2 + (y - b)^2 = r^2$.
- A zero or root of a function $f(x)$ is a number c such that $f(c) = 0$.
- Vertical Line Test: A curve in the plane is the graph of a function if and only if each vertical line $x = a$ intersects the curve in at most one point.
- Even function: $f(-x) = f(x)$ (graph is symmetric about the y-axis).
- Odd function: $f(-x) = -f(x)$ (graph is symmetric about the origin).
- Four ways to transform the graph of $f(x)$:

$f(x) + c$	Shifts graph vertically c units
$f(x + c)$	Shifts graph horizontally c units (to the left if $c > 0$)
$kf(x)$	Scales graph vertically by factor k; if $k < 0$, graph is reflected across x-axis
$f(kx)$	Scales graph horizontally by factor k (compresses if $k > 1$); if $k < 0$, graph is reflected across y-axis

1.1 EXERCISES

Preliminary Questions

1. Give an example of numbers a and b such that $a < b$ and $|a| > |b|$.

2. Which numbers satisfy $|a| = a$? Which satisfy $|a| = -a$? What about $|-a| = a$?

3. Give an example of numbers a and b such that $|a + b| < |a| + |b|$.

4. What are the coordinates of the point lying at the intersection of the lines $x = 9$ and $y = -4$?

5. In which quadrant do the following points lie?

(a) $(1, 4)$ **(b)** $(-3, 2)$

(c) $(4, -3)$ **(d)** $(-4, -1)$

6. What is the radius of the circle with equation $(x - 9)^2 + (y - 9)^2 = 9$?

7. The equation $f(x) = 5$ has a solution if (choose one):

(a) 5 belongs to the domain of f.

(b) 5 belongs to the range of f.

8. What kind of symmetry does the graph have if $f(-x) = -f(x)$?

Exercises

1. Use a calculator to find a rational number r such that $|r - \pi^2| < 10^{-4}$.

2. Let $a = -3$ and $b = 2$. Which of the following inequalities are true?

(a) $a < b$ **(b)** $|a| < |b|$ **(c)** $ab > 0$

(d) $3a < 3b$ **(e)** $-4a < -4b$ **(f)** $\dfrac{1}{a} < \dfrac{1}{b}$

In Exercises 3–8, express the interval in terms of an inequality involving absolute value.

3. $[-2, 2]$ **4.** $(-4, 4)$ **5.** $(0, 4)$

6. $[-4, 0]$ **7.** $[1, 5]$ **8.** $(-2, 8)$

In Exercises 9–12, write the inequality in the form $a < x < b$ for some numbers a, b.

9. $|x| < 8$ **10.** $|x - 12| < 8$

11. $|2x + 1| < 5$ **12.** $|3x - 4| < 2$

In Exercises 13–18, express the set of numbers x satisfying the given condition as an interval.

13. $|x| < 4$ **14.** $|x| \leq 9$

15. $|x - 4| < 2$ **16.** $|x + 7| < 2$

17. $|4x - 1| \leq 8$ **18.** $|3x + 5| < 1$

In Exercises 19–22, describe the set as a union of finite or infinite intervals.

19. $\{x : |x - 4| > 2\}$ **20.** $\{x : |2x + 4| > 3\}$

21. $\{x : |x^2 - 1| > 2\}$ **22.** $\{x : |x^2 + 2x| > 2\}$

23. Match the inequalities (a)–(f) with the corresponding statements (i)–(vi).

(a) $a > 3$ **(b)** $|a - 5| < \dfrac{1}{3}$

(c) $\left|a - \dfrac{1}{3}\right| < 5$ **(d)** $|a| > 5$

(e) $|a - 4| < 3$ **(f)** $1 < a < 5$

(i) a lies to the right of 3.

(ii) a lies between 1 and 7.

(iii) The distance from a to 5 is less than $\frac{1}{3}$.

(iv) The distance from a to 3 is at most 2.

(v) a is less than 5 units from $\frac{1}{3}$.

(vi) a lies either to the left of -5 or to the right of 5.

24. Describe the set $\left\{x : \dfrac{x}{x + 1} < 0\right\}$ as an interval.

25. Show that if $a > b$, then $b^{-1} > a^{-1}$, provided that a and b have the same sign. What happens if $a > 0$ and $b < 0$?

26. Which x satisfy $|x - 3| < 2$ and $|x - 5| < 1$?

27. Show that if $|a - 5| < \frac{1}{2}$ and $|b - 8| < \frac{1}{2}$, then $|(a + b) - 13| < 1$. *Hint:* Use the triangle inequality.

28. Suppose that $|a| \leq 2$ and $|b| \leq 3$.

(a) What is the largest possible value of $|a + b|$?

(b) What is the largest possible value of $|a + b|$ if a and b have opposite signs?

29. Suppose that $|x - 4| \leq 1$.

(a) What is the maximum possible value of $|x + 4|$?

(b) Show that $|x^2 - 16| \leq 9$.

30. Prove that $|x| - |y| \leq |x - y|$. *Hint:* Apply the triangle inequality to y and $x - y$.

31. Express $r_1 = 0.\overline{27}$ as a fraction. *Hint:* $100r_1 - r_1$ is an integer. Then express $r_2 = 0.2666\ldots$ as a fraction.

32. Represent 1/7 and 4/27 as repeating decimals.

33. The text states the following: *If the decimal expansions of two real numbers a and b agree to k places, then the distance $|a - b| \leq 10^{-k}$.* Show that the converse is not true, that is, for any k we can find real numbers a and b whose decimal expansions *do not agree at all* but $|a - b| \leq 10^{-k}$.

34. Plot each pair of points and compute the distance between them:
(a) $(1, 4)$ and $(3, 2)$ **(b)** $(2, 1)$ and $(2, 4)$
(c) $(0, 0)$ and $(-2, 3)$ **(d)** $(-3, -3)$ and $(-2, 3)$

35. Determine the equation of the circle with center $(2, 4)$ and radius 3.

36. Determine the equation of the circle with center $(2, 4)$ passing through $(1, -1)$.

37. Find all points with integer coordinates located at a distance 5 from the origin. Then find all points with integer coordinates located at a distance 5 from $(2, 3)$.

38. Determine the domain and range of the function

$$f : \{r, s, t, u\} \rightarrow \{A, B, C, D, E\}$$

defined by $f(r) = A$, $f(s) = B$, $f(t) = B$, $f(u) = E$.

39. Give an example of a function whose domain D has three elements and range R has two elements. Does a function exist whose domain D has two elements and range has three elements?

In Exercises 40–48, find the domain and range of the function.

40. $f(x) = -x$ **41.** $g(t) = t^4$

42. $f(x) = x^3$ **43.** $g(t) = \sqrt{2 - t}$

44. $f(x) = |x|$ **45.** $h(s) = \dfrac{1}{s}$

46. $f(x) = \dfrac{1}{x^2}$ **47.** $g(t) = \sqrt{2 + t^2}$

48. $g(t) = \cos\dfrac{1}{t}$

In Exercises 49–52, find the interval on which the function is increasing.

49. $f(x) = |x + 1|$ **50.** $f(x) = x^3$

51. $f(x) = x^4$ **52.** $f(x) = \dfrac{1}{x^2 + 1}$

In Exercises 53–58, find the zeros of the function and sketch its graph by plotting points. Use symmetry and increase/decrease information where appropriate.

53. $f(x) = x^2 - 4$ **54.** $f(x) = 2x^2 - 4$

55. $f(x) = x^3 - 4x$ **56.** $f(x) = x^3$

57. $f(x) = 2 - x^3$ **58.** $f(x) = \dfrac{1}{(x - 1)^2 + 1}$

59. Which of the curves in Figure 26 is the graph of a function?

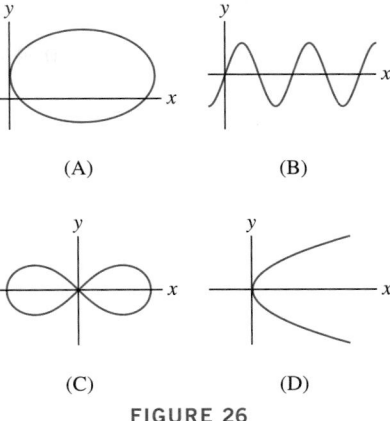

(A)   (B)

(C)   (D)

FIGURE 26

60. State whether the function is increasing, decreasing, or neither.
(a) Surface area of a sphere as a function of its radius
(b) Temperature at a point on the equator as a function of time
(c) Price of an airline ticket as a function of the price of oil
(d) Pressure of the gas in a piston as a function of volume

In Exercises 61–66, let $f(x)$ be the function whose graph is shown in Figure 27.

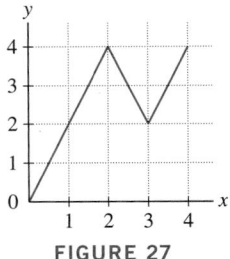

FIGURE 27

61. What are the domain and range of $f(x)$?

62. Sketch the graphs of $f(x + 2)$ and $f(x) + 2$.

63. Sketch the graphs of $f(2x)$, $f\left(\frac{1}{2}x\right)$, and $2f(x)$.

64. Sketch the graphs of $f(-x)$ and $-f(-x)$.

65. Extend the graph of $f(x)$ to $[-4, 4]$ so that it is an even function.

66. Extend the graph of $f(x)$ to $[-4, 4]$ so that it is an odd function.

67. Suppose that $f(x)$ has domain $[4, 8]$ and range $[2, 6]$. What are the domain and range of:
(a) $f(x) + 3$ **(b)** $f(x + 3)$
(c) $f(3x)$ **(d)** $3f(x)$

68. Let $f(x) = x^2$. Sketch the graphs of the following functions over $[-2, 2]$:

(a) $f(x + 1)$ (b) $f(x) + 1$

(c) $f(5x)$ (d) $5f(x)$

69. Suppose that the graph of $f(x) = \sin x$ is compressed horizontally by a factor of 2 and then shifted 5 units to the right.

(a) What is the equation for the new graph?

(b) What is the equation if you first shift by 5 and then compress by 2?

(c) [GU]   Verify your answers by plotting your equations.

70. Figure 28 shows the graph of $f(x) = |x| + 1$. Match the functions (a)–(e) with their graphs (i)–(v).

(a) $f(x - 1)$ (b) $-f(x)$ (c) $-f(x) + 2$

(d) $f(x - 1) - 2$ (e) $f(x + 1)$

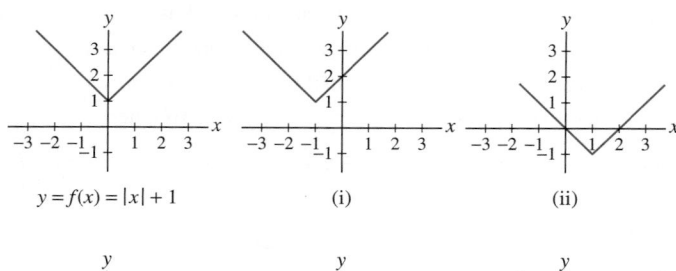

$y = f(x) = |x| + 1$ (i)   (ii)

(iii)   (iv)   (v)

FIGURE 28

71. Sketch the graph of $f(2x)$ and $f\left(\frac{1}{2}x\right)$, where $f(x) = |x| + 1$ (Figure 28).

72. Find the function $f(x)$ whose graph is obtained by shifting the parabola $y = x^2$ three units to the right and four units down as in Figure 29.

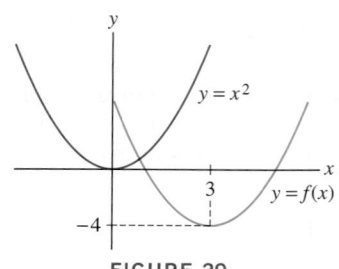

FIGURE 29

73. Define $f(x)$ to be the larger of x and $2 - x$. Sketch the graph of $f(x)$. What are its domain and range? Express $f(x)$ in terms of the absolute value function.

74. For each curve in Figure 30, state whether it is symmetrical with respect to the y-axis, the origin, both, or neither.

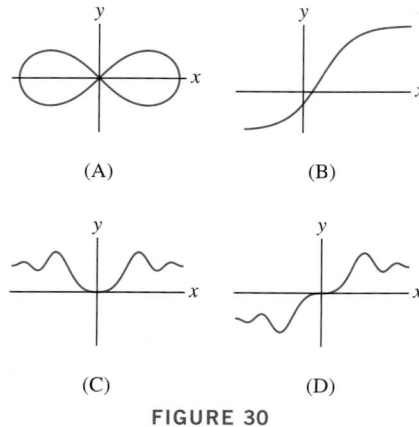

(A)   (B)

(C)   (D)

FIGURE 30

75. Show that the sum of two even functions is even and the sum of two odd functions is odd.

76. Suppose that $f(x)$ and $g(x)$ are both odd. Which of the following functions are even? Which are odd?

(a) $f(x)g(x)$ (b) $f(x)^3$

(c) $f(x) - g(x)$ (d) $\dfrac{f(x)}{g(x)}$

77. Give an example of a curve that is symmetrical with respect to both the y-axis and the origin. Can the graph of a function have both symmetries? *Hint:* Prove algebraically that $f(x) = 0$ is the only such function.

Further Insights and Challenges

78. Prove the triangle inequality by adding the two inequalities

$$-|a| \le a \le |a|, \qquad -|b| \le b \le |b|$$

79. Show that if $r = a/b$ is a fraction in lowest terms, then r has a *finite* decimal expansion if and only if $b = 2^n 5^m$ for some $n, m \ge 0$. *Hint:* Observe that r has a finite decimal expansion when $10^N r$ is an integer for some $N \ge 0$ (and hence b divides 10^N).

80. Let $p = p_1 \ldots p_s$ be an integer with digits $p_1, \ldots, p_s$. Show that

$$\frac{p}{10^s - 1} = 0.\overline{p_1 \ldots p_s}$$

Use this to find the decimal expansion of $r = \dfrac{2}{11}$. Note that

$$r = \frac{2}{11} = \frac{18}{10^2 - 1}$$

81. A function $f(x)$ is symmetrical with respect to the vertical line $x = a$ if $f(a - x) = f(a + x)$.

(a) Draw the graph of a function that is symmetrical with respect to $x = 2$.

(b) Show that if $f(x)$ is symmetrical with respect to $x = a$, then $g(x) = f(x + a)$ is even.

82. Formulate a condition for $f(x)$ to be symmetrical with respect to the point $(a, 0)$ on the x-axis.

1.2 Linear and Quadratic Functions

Linear functions are the simplest of all functions and their graphs (lines) are the simplest of all curves. However, linear functions and lines play an enormously important role in calculus. For this reason, it is important to be thoroughly familiar with the basic properties of linear functions and the different ways of writing the equation of a line.

Let's recall that a **linear function** is a function of the form

$$f(x) = mx + b \quad (m \text{ and } b \text{ constants})$$

The graph of $f(x)$ is a line of slope m, and since $f(0) = b$, the graph intersects the y-axis at the point $(0, b)$ (Figure 1). The number b is called the y-intercept, and the equation $y = mx + b$ for the line is said to be in **slope-intercept form**.

We use the symbols Δx and Δy to denote the *change* (or *increment*) in x and $y = f(x)$ over an interval $[x_1, x_2]$ (Figure 1):

$$\Delta x = x_2 - x_1, \qquad \Delta y = y_2 - y_1 = f(x_2) - f(x_1)$$

The slope m of a line is the ratio

$$m = \frac{\Delta y}{\Delta x} = \frac{\text{vertical change}}{\text{horizontal change}} = \frac{\text{rise}}{\text{run}}$$

This follows from the formula $y = mx + b$:

$$\frac{\Delta y}{\Delta x} = \frac{y_2 - y_1}{x_2 - x_1} = \frac{(mx_2 + b) - (mx_1 + b)}{x_2 - x_1} = \frac{m(x_2 - x_1)}{x_2 - x_1} = m$$

The slope m measures the *rate of change* of y with respect to x. In fact, by writing

$$\Delta y = m\Delta x$$

we see that a one-unit increase in x (i.e., $\Delta x = 1$) produces an m-unit change Δy in y. For example, if $m = 5$, then y increases by five units per unit increase in x. The rate-of-change interpretation of the slope is fundamental in calculus. We discuss it in greater detail in Section 2.1.

Graphically, the slope m measures the steepness of the line $y = mx + b$. Figure 2(A) shows lines through a point of varying slope m. Note the following properties:

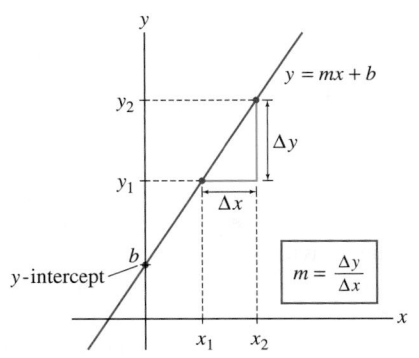

FIGURE 1 The slope m is the ratio "rise over run."

- **Steepness**: The larger the absolute value $|m|$, the steeper the line.
- **Negative slope**: If $m < 0$, the line slants downward from left to right.
- $f(x) = mx + b$ is strictly increasing if $m > 0$ and strictly decreasing if $m < 0$.
- The **horizontal line** $y = b$ has slope $m = 0$ [Figure 2(B)].
- A **vertical line** has equation $x = c$, where c is a constant. Informally speaking, the slope of a vertical line is "infinite," so it is not possible to write the equation of a vertical line in slope-intercept form $y = mx + b$.

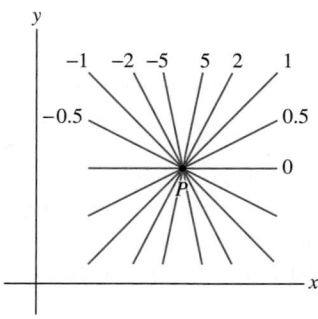

(A) Lines of varying slopes through P

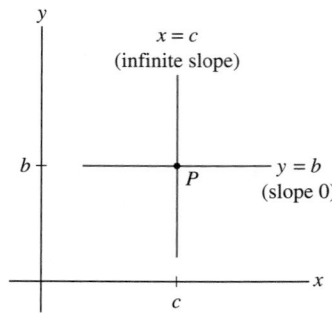

(B) Horizontal and vertical lines through P

FIGURE 2

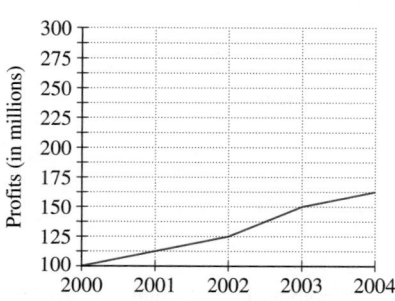

FIGURE 3 Growth of company profits.

CAUTION: Graphs are often plotted using different scales for the x- and y-axes. This is necessary to keep the sizes of graphs within reasonable bounds. However, when the scales are different, lines do not appear with their true slopes.

Scale is especially important in applications because the steepness of a graph depends on the choice of units for the x- and y-axes. We can create very different *subjective* impressions by changing the scale. Figure 3 shows the growth of company profits over a four-year period. The two plots convey the same information, but the upper plot makes the growth look more dramatic.

Next, we recall the relation between the slopes of parallel and perpendicular lines (Figure 4):

- Lines of slopes m_1 and m_2 are **parallel** if and only if $m_1 = m_2$.
- Lines of slopes m_1 and m_2 are **perpendicular** if and only if $m_1 = -1/m_2$ (or $m_1 m_2 = -1$).

(A) Parallel lines

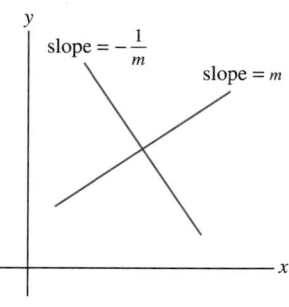

(B) Perpendicular lines

FIGURE 4 Parallel and perpendicular lines.

As mentioned above, it is important to be familiar with the standard ways of writing the equation of a line. The general **linear equation** is

$$ax + by = c \qquad \boxed{1}$$

where a and b are not *both* zero. For $b = 0$, this gives the vertical line $ax = c$. When $b \neq 0$, we can rewrite (1) in slope-intercept form. For example, $-6x + 2y = 3$ can be rewritten as $y = 3x + \frac{3}{2}$.

Two other forms we will use frequently are **point-slope** and **point-point** form. Given a point $P = (a, b)$ and a slope m, the equation of the line through P with slope m is $y - b = m(x - a)$. Similarly, the line through two distinct points $P = (a_1, b_1)$ and $Q = (a_2, b_2)$ has slope (Figure 5)

$$m = \frac{b_2 - b_1}{a_2 - a_1}$$

Therefore, we can write its equation as $y - b_1 = m(x - a_1)$.

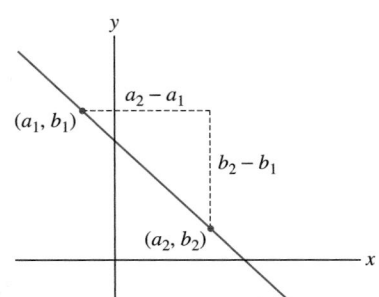

FIGURE 5 Slope of the line between $P = (a_1, b_1)$ and $Q = (a_2, b_2)$ is $m = \dfrac{b_2 - b_1}{a_2 - a_1}$.

Equations for Lines

1. **Point-slope form:** The line through $P = (a, b)$ with slope m has equation

$$y - b = m(x - a)$$

2. **Point-point form:** The line through $P = (a_1, b_1)$ and $Q = (a_2, b_2)$ has equation

$$y - b_1 = m(x - a_1) \qquad \text{where } m = \frac{b_2 - b_1}{a_2 - a_1}$$

■ **EXAMPLE 1** Line of Given Slope Through a Given Point   Find the equation of the line through $(9, 2)$ with slope $-\frac{2}{3}$.

Solution   We can write the equation directly in point-slope form:

$$y - 2 = -\frac{2}{3}(x - 9)$$

In slope-intercept form: $y = -\frac{2}{3}(x - 9) + 2$ or $y = -\frac{2}{3}x + 8$. See Figure 6.   ■

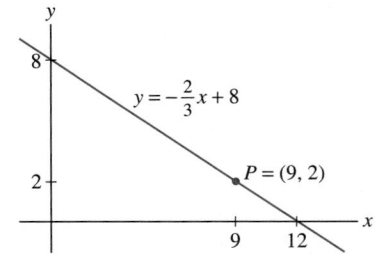

FIGURE 6 Line through $P = (9, 2)$ with slope $m = -\frac{2}{3}$.

■ **EXAMPLE 2** Line Through Two Points   Find the equation of the line $\mathcal{L}$ through $(2, 1)$ and $(9, 5)$.

Solution   The line $\mathcal{L}$ has slope

$$m = \frac{5 - 1}{9 - 2} = \frac{4}{7}$$

Since $(9, 5)$ lies on $\mathcal{L}$, its equation in point-slope form is $y - 5 = \frac{4}{7}(x - 9)$.   ■

CONCEPTUAL INSIGHT   We can define the increments Δx and Δy over an interval $[x_1, x_2]$ for any function $f(x)$ (linear or not), but in general, the ratio $\Delta y / \Delta x$ depends on the interval. The characteristic property of a linear function $f(x) = mx + b$ is that

$\Delta y/\Delta x$ has the same value m for every interval (Figure 7). In other words, y has a constant rate of change with respect to x. We can use this property to test if two quantities are related by a linear equation (see Example 3).

   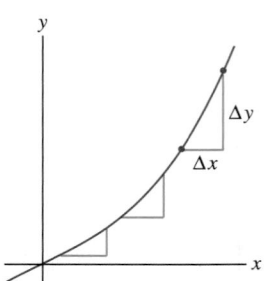

Graph of a linear function.
The ratio $\Delta y/\Delta x$ is the same
over all intervals.

Graph of a nonlinear function.
The ratio $\Delta y/\Delta x$ changes,
depending on the interval.

FIGURE 7

TABLE 1

Temperature (°F)	Pressure (lb/in.^{2})
70	187.42
75	189
85	192.16
100	196.9
110	200.06

Real experimental data are unlikely to reveal perfect linearity, even if the data points do essentially lie on a line. The method of "linear regression" is used to find the linear function that best fits the data.

■ **EXAMPLE 3** Testing for a Linear Relationship   Table 1 gives the pressure reading P of a gas at different temperatures T. Do the data suggest a linear relation between P and T?

Solution We calculate $\dfrac{\Delta P}{\Delta T}$ at successive data points and check whether this ratio is constant:

(T_1, P_1)	(T_2, P_2)	$\dfrac{\Delta P}{\Delta T}$
$(70, 187.42)$	$(75, 189)$	$\dfrac{189 - 187.42}{75 - 70} = 0.316$
$(75, 189)$	$(85, 192.16)$	$\dfrac{192.16 - 189}{85 - 75} = 0.316$
$(85, 192.16)$	$(100, 196.9)$	$\dfrac{196.9 - 192.16}{100 - 85} = 0.316$
$(100, 196.9)$	$(110, 200.06)$	$\dfrac{200.06 - 196.9}{110 - 100} = 0.316$

Because $\Delta P/\Delta T$ has the constant value 0.316, the data points lie on a line with slope $m = 0.316$ (this is confirmed in the plot in Figure 8). The line passes through the first data point $(70, 187.42)$, so its equation in point-slope form is

$$P - 187.42 = 0.316(T - 70)$$ ■

A **quadratic function** is a function defined by a quadratic polynomial

$$f(x) = ax^2 + bx + c \quad (a, b, c, \text{ constants with } a \neq 0)$$

The graph of $f(x)$ is a **parabola** (Figure 9). The parabola opens upward if the leading coefficient a is positive and downward if a is negative. The **discriminant** of $f(x)$ is the

FIGURE 8 Line through
pressure-temperature data points.

quantity

$$D = b^2 - 4ac$$

The roots of $f(x)$ are given by the **quadratic formula** (see Exercise 56):

$$\text{Roots of } f(x) = \frac{-b \pm \sqrt{b^2 - 4ac}}{2a} = \frac{-b \pm \sqrt{D}}{2a}$$

The sign of D determines whether or not $f(x)$ has real roots (Figure 9). If $D > 0$, then $f(x)$ has two real roots, and if $D = 0$, it has one real root (a "double root"). If $D < 0$, then $\sqrt{D}$ is imaginary and $f(x)$ has no real roots.

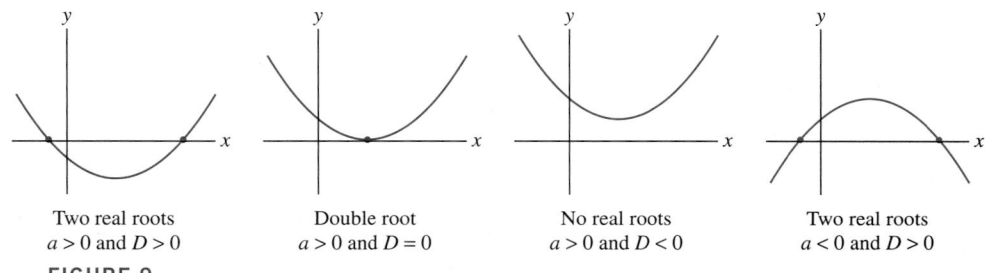

| Two real roots | Double root | No real roots | Two real roots |
| $a > 0$ and $D > 0$ | $a > 0$ and $D = 0$ | $a > 0$ and $D < 0$ | $a < 0$ and $D > 0$ |

FIGURE 9

When $f(x)$ has two real roots r_1 and r_2, then $f(x)$ factors as

$$f(x) = a(x - r_1)(x - r_2)$$

For example, $f(x) = 2x^2 - 3x + 1$ has discriminant $D = b^2 - 4ac = 9 - 8 = 1 > 0$, and by the quadratic formula, its roots are $(3 \pm 1)/4$ or 1 and $\frac{1}{2}$. Therefore,

$$f(x) = 2x^2 - 3x + 1 = 2(x - 1)\left(x - \frac{1}{2}\right)$$

The technique of **completing the square** consists of writing a quadratic polynomial as a multiple of a square plus a constant:

$$ax^2 + bx + c = a\underbrace{\left(x + \frac{b}{2a}\right)^2}_{\text{Square term}} + \underbrace{\frac{4ac - b^2}{4a}}_{\text{Constant}}$$ 2

It is not necessary to memorize this formula, but you should know how to carry out the process of completing the square.

Cuneiform texts written on clay tablets show that the method of completing the square was known to ancient Babylonian mathematicians who lived some 4,000 years ago.

■ **EXAMPLE 4** Completing the Square   Complete the square for the quadratic polynomial $4x^2 - 12x + 3$.

Solution First factor out the leading coefficient:

$$4x^2 - 12x + 3 = 4\left(x^2 - 3x + \frac{3}{4}\right)$$

Then complete the square for the term $x^2 - 3x$:

$$x^2 + bx = \left(x + \frac{b}{2}\right)^2 - \frac{b^2}{4}, \qquad x^2 - 3x = \left(x - \frac{3}{2}\right)^2 - \frac{9}{4}$$

Therefore,

$$4x^2 - 12x + 3 = 4\left(\left(x - \frac{3}{2}\right)^2 - \frac{9}{4} + \frac{3}{4}\right) = 4\left(x - \frac{3}{2}\right)^2 - 6 \qquad \blacksquare$$

The method of completing the square can be used to find the minimum or maximum value of a quadratic function.

■ **EXAMPLE 5** Finding the Minimum of a Quadratic Function Complete the square and find the minimum value of the quadratic function $f(x) = x^2 - 4x + 9$.

Solution We have

$$f(x) = x^2 - 4x + 9 = (x - 2)^2 - 4 + 9 = \overbrace{(x-2)^2}^{\text{This term is} \geq 0} + 5$$

Thus, $f(x) \geq 5$ for all x and the minimum value of $f(x)$ is $f(2) = 5$ (Figure 10). ■

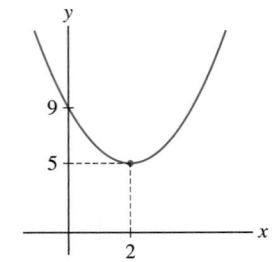

FIGURE 10 Graph of $f(x) = x^2 - 4x + 9$.

1.2 SUMMARY

- A function of the form $f(x) = mx + b$ is called a linear function.
- The general equation of a line is $ax + by = c$. The line $y = c$ is horizontal and $x = c$ is vertical.
- There are three convenient ways of writing the equation of a nonvertical line:

  - Slope-intercept form: $y = mx + b$ (slope m and y-intercept b)
  - Point-slope form: $y - b = m(x - a)$ [slope m, passes through (a, b)]
  - Point-point form: The line through two points $P = (a_1, b_1)$ and $Q = (a_2, b_2)$ has slope $m = \dfrac{b_2 - b_1}{a_2 - a_1}$ and equation $y - b_1 = m(x - a_1)$.

- Two lines of slopes m_1 and m_2 are parallel if and only if $m_1 = m_2$, and they are perpendicular if and only if $m_1 = -1/m_2$.
- The roots of a quadratic function $f(x) = ax^2 + bx + c$ are given by the quadratic formula $x = (-b \pm \sqrt{D})/2a$, where $D = b^2 - 4ac$ is the discriminant. Furthermore, $f(x)$ has distinct real roots if $D > 0$, a double root if $D = 0$, and no real roots if $D < 0$.
- The technique of completing the square consists of writing a quadratic function as a multiple of a square plus a constant.

1.2 EXERCISES

Preliminary Questions

1. What is the slope of the line $y = -4x - 9$?

2. Are the lines $y = 2x + 1$ and $y = -2x - 4$ perpendicular?

3. When is the line $ax + by = c$ parallel to the y-axis? To the x-axis?

4. Suppose $y = 3x + 2$. What is Δy if x increases by 3?

5. What is the minimum of $f(x) = (x + 3)^2 - 4$?

6. What is the result of completing the square for $f(x) = x^2 + 1$?

Exercises

In Exercises 1–4, find the slope, the y-intercept, and the x-intercept of the line with the given equation.

1. $y = 3x + 12$

2. $y = 4 - x$

3. $4x + 9y = 3$

4. $y - 3 = \frac{1}{2}(x - 6)$

In Exercises 5–8, find the slope of the line.

5. $y = 3x + 2$

6. $y = 3(x - 9) + 2$

7. $3x + 4y = 12$

8. $3x + 4y = -8$

In Exercises 9–20, find the equation of the line with the given description.

9. Slope 3, y-intercept 8

10. Slope -2, y-intercept 3

11. Slope 3, passes through $(7, 9)$

12. Slope -5, passes through $(0, 0)$

13. Horizontal, passes through $(0, -2)$

14. Passes through $(-1, 4)$ and $(2, 7)$

15. Parallel to $y = 3x - 4$, passes through $(1, 1)$

16. Passes through $(1, 4)$ and $(12, -3)$

17. Perpendicular to $3x + 5y = 9$, passes through $(2, 3)$

18. Vertical, passes through $(-4, 9)$

19. Horizontal, passes through $(8, 4)$

20. Slope 3, x-intercept 6

21. Find the equation of the perpendicular bisector of the segment joining $(1, 2)$ and $(5, 4)$ (Figure 11). *Hint:* The midpoint Q of the segment joining (a, b) and (c, d) is $\left(\dfrac{a + c}{2}, \dfrac{b + d}{2} \right)$.

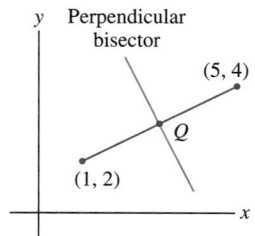

FIGURE 11

22. Intercept-intercept Form   Show that if $a, b \neq 0$, then the line with x-intercept $x = a$ and y-intercept $y = b$ has equation (Figure 12)

$$\frac{x}{a} + \frac{y}{b} = 1$$

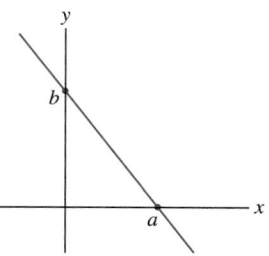

FIGURE 12 Line with equation $\dfrac{x}{a} + \dfrac{y}{b} = 1$.

23. Find the equation of the line with x-intercept $x = 4$ and y-intercept $y = 3$.

24. A line of slope $m = 2$ passes through $(1, 4)$. Find y such that $(3, y)$ lies on the line.

25. Determine whether there exists a constant c such that the line $x + cy = 1$

(a) Has slope 4

(b) Passes through $(3, 1)$

(c) Is horizontal

(d) Is vertical

26. Assume that the number N of concert tickets which can be sold at a price of P dollars per ticket is a linear function $N(P)$ for $10 \leq P \leq 40$. Determine $N(P)$ (called the demand function) if $N(10) = 500$ and $N(40) = 0$. What is the decrease ΔN in the number of tickets sold if the price is increased by $\Delta P = 5$ dollars?

27. Materials expand when heated. Consider a metal rod of length L_0 at temperature T_0. If the temperature is changed by an amount ΔT, then the rod's length changes by $\Delta L = \alpha L_0 \Delta T$, where α is the thermal expansion coefficient. For steel, $\alpha = 1.24 \times 10^{-5}\,^\circ\text{C}^{-1}$.

(a) A steel rod has length $L_0 = 40$ cm at $T_0 = 40^\circ$C. What is its length at $T = 90^\circ$C?

(b) Find its length at $T = 50^\circ$C if its length at $T_0 = 100^\circ$C is 65 in.

(c) Express length L as a function of T if $L_0 = 65$ in. at $T_0 = 100^\circ$C.

28. Do the points $(0.5, 1)$, $(1, 1.2)$, $(2, 2)$ lie on a line?

29. Find b such that $(2, -1)$, $(3, 2)$, and $(b, 5)$ lie on a line.

30. Find an expression for the velocity v as a linear function of t that matches the following data.

t (s)	0	2	4	6
v (m/s)	39.2	58.6	78	97.4

31. The period T of a pendulum is measured for pendulums of several different lengths L. Based on the following data, does T appear to be a linear function of L?

L (ft)	2	3	4	5
T (s)	1.57	1.92	2.22	2.48

32. Show that $f(x)$ is linear of slope m if and only if

$$f(x+h) - f(x) = mh \quad \text{(for all } x \text{ and } h\text{)}$$

33. Find the roots of the quadratic polynomials:

(a) $4x^2 - 3x - 1$ (b) $x^2 - 2x - 1$

In Exercises 34–41, complete the square and find the minimum or maximum value of the quadratic function.

34. $y = x^2 + 2x + 5$ **35.** $y = x^2 - 6x + 9$

36. $y = -9x^2 + x$ **37.** $y = x^2 + 6x + 2$

38. $y = 2x^2 - 4x - 7$ **39.** $y = -4x^2 + 3x + 8$

40. $y = 3x^2 + 12x - 5$ **41.** $y = 4x - 12x^2$

42. Sketch the graph of $y = x^2 - 6x + 8$ by plotting the roots and the minimum point.

43. Sketch the graph of $y = x^2 + 4x + 6$ by plotting the minimum point, the y-intercept, and one other point.

44. If the alleles A and B of the cystic fibrosis gene occur in a population with frequencies p and $1 - p$ (where p is a fraction between 0 and 1), then the frequency of heterozygous carriers (carriers with both alleles) is $2p(1 - p)$. Which value of p gives the largest frequency of heterozygous carriers?

45. For which values of c does $f(x) = x^2 + cx + 1$ have a double root? No real roots?

46. ✍ Let $f(x)$ be a quadratic function and c a constant. Which is correct? Explain graphically.

(a) There is a unique value of c such that $y = f(x) - c$ has a double root.

(b) There is a unique value of c such that $y = f(x - c)$ has a double root.

47. Prove that $x + \dfrac{1}{x} \geq 2$ for all $x > 0$. *Hint:* Consider $(x^{1/2} - x^{-1/2})^2$.

48. Let $a, b > 0$. Show that the *geometric mean* $\sqrt{ab}$ is not larger than the *arithmetic mean* $\dfrac{a+b}{2}$. *Hint:* Use a variation of the hint given in Exercise 47.

49. If objects of weights x and w_1 are suspended from the balance in Figure 13(A), the cross-beam is horizontal if $bx = aw_1$. If the lengths a and b are known, we may use this equation to determine an unknown weight x by selecting w_1 so that the cross-beam is horizontal. If a and b are not known precisely, we might proceed as follows. First balance x by w_1 on the left as in (A). Then switch places and balance x by w_2 on the right as in (B). The average $\bar{x} = \frac{1}{2}(w_1 + w_2)$ gives an estimate for x. Show that $\bar{x}$ is greater than or equal to the true weight x.

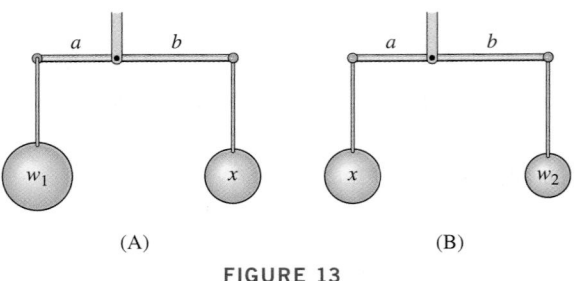

(A)   (B)

FIGURE 13

50. Find numbers x and y with sum 10 and product 24. *Hint:* Find a quadratic polynomial satisfied by x.

51. Find a pair of numbers whose sum and product are both equal to 8.

52. Show that the graph of the parabola $y = x^2$ consists of all points P such that $d_1 = d_2$, where d_1 is the distance from P to $(0, \frac{1}{4})$ and d_2 is the distance from P to the horizontal line $y = -\frac{1}{4}$ (Figure 14).

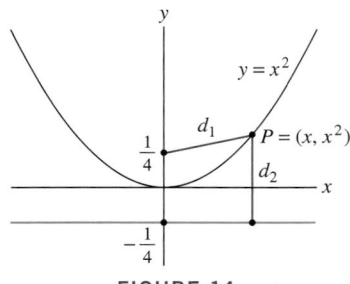

FIGURE 14

Further Insights and Challenges

53. Show that if $f(x)$ and $g(x)$ are linear, then so is $f(x) + g(x)$. Is the same true of $f(x)g(x)$?

54. Show that if $f(x)$ and $g(x)$ are linear functions such that $f(0) = g(0)$ and $f(1) = g(1)$, then $f(x) = g(x)$.

55. Show that the ratio $\Delta y / \Delta x$ for the function $f(x) = x^2$ over the interval $[x_1, x_2]$ is not a constant, but depends on the interval. Determine the exact dependence of $\Delta y / \Delta x$ on x_1 and x_2.

56. Use Eq. (2) to derive the quadratic formula for the roots of $ax^2 + bx + c = 0$.

57. Let $a, c \neq 0$. Show that the roots of $ax^2 + bx + c = 0$ and $cx^2 + bx + a = 0$ are reciprocals of each other.

58. Complete the square to show that the parabolas $y = ax^2 + bx + c$ and $y = ax^2$ have the same shape (show that the first parabola is congruent to the second by a vertical and horizontal translation).

59. Prove **Viète's Formulas**, which state that the quadratic polynomial with given numbers α and β as roots is $x^2 + bx + c$, where $b = -\alpha - \beta$ and $c = \alpha\beta$.

1.3 The Basic Classes of Functions

It would be impossible (and useless) to describe all possible functions $f(x)$. Since the values of a function can be assigned arbitrarily, a function chosen at random would likely be so complicated that we could neither graph it nor describe it in any reasonable way. However, calculus makes no attempt to deal with all possible functions. The techniques of calculus, powerful and general as they are, apply only to functions that are sufficiently "well behaved" (we will see what well-behaved means when we study the derivative in Chapter 3). Fortunately, such functions are adequate for a vast range of applications. In this text, we deal mostly with the following important and familiar classes of well-behaved functions:

polynomials   rational functions   algebraic functions

exponential functions   trigonometric functions

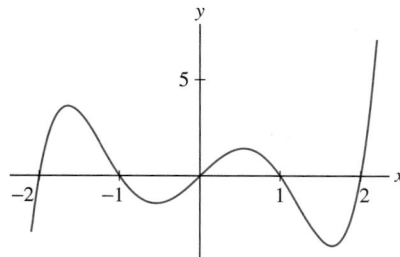

FIGURE 1 The polynomial $y = x^5 - 5x^3 + 4x$.

- **Polynomials:** For any real number m, the function $f(x) = x^m$ is called the **power function** with exponent m. A polynomial is a sum of multiples of power functions with whole number exponents (Figure 1):

$$f(x) = x^5 - 5x^3 + 4x, \qquad g(t) = 7t^6 + t^3 - 3t - 1$$

Thus, the function $f(x) = x + x^{-1}$ is not a polynomial because it includes a power function x^{-1} with a negative exponent. The general polynomial in the variable x may be written

$$P(x) = a_n x^n + a_{n-1} x^{n-1} + \cdots + a_1 x + a_0$$

- The numbers $a_0, a_1, \ldots, a_n$ are called **coefficients**.
- The **degree** of $P(x)$ is n (assuming that $a_n \neq 0$).
- The coefficient a_n is called the **leading coefficient**.
- The domain of $P(x)$ is **R**.

- **Rational functions:** A rational function is a *quotient* of two polynomials (Figure 2):

$$f(x) = \frac{P(x)}{Q(x)} \quad [P(x) \text{ and } Q(x) \text{ polynomials}]$$

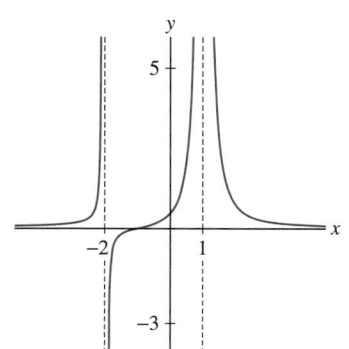

FIGURE 2 The rational function $f(x) = \dfrac{x+1}{x^3 - 3x + 2}$.

Every polynomial is also a rational function [with $Q(x) = 1$]. The domain of a rational function $\dfrac{P(x)}{Q(x)}$ is the set of numbers x such that $Q(x) \neq 0$. For example,

$$f(x) = \frac{1}{x^2} \qquad \qquad \text{domain } \{x : x \neq 0\}$$

$$h(t) = \frac{7t^6 + t^3 - 3t - 1}{t^2 - 1} \qquad \text{domain } \{t : t \neq \pm 1\}$$

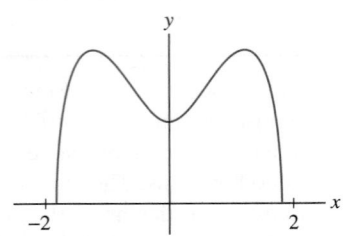

FIGURE 3 The algebraic function $f(x) = \sqrt{1 + 3x^2 - x^4}$.

Any function that is not algebraic is called "transcendental." Exponential and trigonometric functions are examples. Other transcendental functions, such as the gamma and Bessel functions, appear in advanced applications to physics, engineering, and statistics. The word "transcendental" was used to describe functions of this type in the 1670s by Gottfried Wilhelm Leibniz (1646–1716).

- **Algebraic functions:** An algebraic function is produced by taking sums, products, and quotients of *roots* of polynomials and rational functions (Figure 3):

$$f(x) = \sqrt{1 + 3x^2 - x^4}, \qquad g(t) = (\sqrt{t} - 2)^{-2}, \qquad h(z) = \frac{z + z^{-5/3}}{5z^3 - \sqrt{z}}$$

  A number x belongs to the domain of f if each term in the formula for f is defined and the result does not involve division by zero. For example, $g(t)$ is defined if $t \geq 0$ and $\sqrt{t} \neq 2$, so the domain of $g(t)$ is $D = \{t : t \geq 0 \text{ and } t \neq 4\}$. More generally, algebraic functions are defined by polynomial equations between x and y. In this case, we say that y is **implicitly defined** as a function of x. For example, the equation $y^4 + 2x^2y + x^4 = 1$ defines y implicitly as a function of x.
- **Exponential functions:** The function $f(x) = b^x$, where $b > 0$, is called the exponential function with base b. Some examples are

$$y = 2^x, \qquad y = 10^x, \qquad y = \left(\frac{1}{3}\right)^x, \qquad y = (\sqrt{5})^x$$

  Exponential functions and their *inverses*, the **logarithm functions**, are treated in greater detail in Section 1.6.
- **Trigonometric functions:** Functions built from $\sin x$ and $\cos x$ are called trigonometric functions. These functions are discussed in the next section.

Constructing New Functions

If f and g are functions, we may construct new functions by forming the sum, difference, product, and quotient functions:

$$(f + g)(x) = f(x) + g(x), \qquad (f - g)(x) = f(x) - g(x)$$

$$(fg)(x) = f(x)\,g(x), \qquad \left(\frac{f}{g}\right)(x) = \frac{f(x)}{g(x)} \quad \text{(where } g(x) \neq 0\text{)}$$

For example, if $f(x) = x^2$ and $g(x) = \sin x$, then

$$(f + g)(x) = x^2 + \sin x, \qquad (f - g)(x) = x^2 - \sin x$$

$$(fg)(x) = x^2 \sin x, \qquad \left(\frac{f}{g}\right)(x) = \frac{x^2}{\sin x}$$

We can also multiply functions by constants. A function of the form

$$c_1 f(x) + c_2 g(x) \quad (c_1, c_2 \text{ constants})$$

is called a **linear combination** of $f(x)$ and $g(x)$.

Composition is another important way of constructing new functions. The composition of f and g is the function $f \circ g$ defined by $(f \circ g)(x) = f(g(x))$, defined for values of x in the domain of g such that $g(x)$ lies in the domain of f.

■ **EXAMPLE 1** Compute the composite functions $f \circ g$ and $g \circ f$ and discuss their domains, where

$$f(x) = \sqrt{x}, \qquad g(x) = 1 - x$$

Example 1 shows that the composition of functions is not commutative: The functions $f \circ g$ and $g \circ f$ may be different.

Solution We have

$$(f \circ g)(x) = f(g(x)) = f(1 - x) = \sqrt{1 - x}$$

The square root $\sqrt{1-x}$ is defined if $1-x \geq 0$ or $x \leq 1$, so the domain of $f \circ g$ is $\{x : x \leq 1\}$. On the other hand,

$$(g \circ f)(x) = g(f(x)) = g(\sqrt{x}) = 1 - \sqrt{x}$$

The domain of $g \circ f$ is $\{x : x \geq 0\}$.   ∎

Elementary Functions

Inverse functions are discussed in Section 1.5.

We have reviewed some of the most basic and familiar functions of mathematics. All of these can be found on any scientific calculator. New functions may be produced using the operations of addition, multiplication, division, as well as composition, extraction of roots, and taking inverses. It is convenient to refer to a function constructed in this way from the basic functions listed above as an **elementary function**. The following functions are elementary:

$$f(x) = \sqrt{2x + \sin x}, \qquad f(x) = 10^{\sqrt{x}}, \qquad f(x) = \frac{1 + x^{-1}}{1 + \cos x}$$

1.3 SUMMARY

- For any real number m, the function $f(x) = x^m$ is the power function with exponent m. A polynomial $P(x)$ is a sum of multiples of power functions x^m, where m is a whole number:

$$P(x) = a_n x^n + a_{n-1} x^{n-1} + \cdots + a_1 x + a_0$$

This polynomial has degree n (if we assume that $a_n \neq 0$) and a_n is called the leading coefficient.
- A rational function is a quotient $P(x)/Q(x)$ of two polynomials.
- An algebraic function is produced by taking sums, products, and nth roots of polynomials and rational functions.
- Exponential function: $f(x) = b^x$, where $b > 0$ (b is called the base).
- The composite function $f \circ g$ is defined by $(f \circ g)(x) = f(g(x))$. The domain of $f \circ g$ is the set of x in the domain of g such that $g(x)$ belongs to the domain of f.

1.3 EXERCISES

Preliminary Questions

1. Give an example of a rational function.

2. Is $|x|$ a polynomial function? What about $|x^2 + 1|$?

3. What is unusual about the domain of $f \circ g$ for $f(x) = x^{1/2}$ and $g(x) = -1 - |x|$?

4. Is $f(x) = \left(\frac{1}{2}\right)^x$ increasing or decreasing?

5. Give an example of a transcendental function.

Exercises

In Exercises 1–12, determine the domain of the function.

1. $f(x) = x^{1/4}$

2. $g(t) = t^{2/3}$

3. $f(x) = x^3 + 3x - 4$

4. $h(z) = z^3 + z^{-3}$

5. $g(t) = \dfrac{1}{t + 2}$

6. $f(x) = \dfrac{1}{x^2 + 4}$

7. $G(u) = \dfrac{1}{u^2 - 4}$

8. $f(x) = \dfrac{\sqrt{x}}{x^2 - 9}$

9. $f(x) = x^{-4} + (x-1)^{-3}$

10. $F(s) = \sin\left(\dfrac{s}{s+1}\right)$

11. $g(y) = 10^{\sqrt{y}+y^{-1}}$

12. $f(x) = \dfrac{x + x^{-1}}{(x-3)(x+4)}$

In Exercises 13–24, identify each of the following functions as polynomial, rational, algebraic, or transcendental.

13. $f(x) = 4x^3 + 9x^2 - 8$

14. $f(x) = x^{-4}$

15. $f(x) = \sqrt{x}$

16. $f(x) = \sqrt{1-x^2}$

17. $f(x) = \dfrac{x^2}{x + \sin x}$

18. $f(x) = 2^x$

19. $f(x) = \dfrac{2x^3 + 3x}{9 - 7x^2}$

20. $f(x) = \dfrac{3x - 9x^{-1/2}}{9 - 7x^2}$

21. $f(x) = \sin(x^2)$

22. $f(x) = \dfrac{x}{\sqrt{x} + 1}$

23. $f(x) = x^2 + 3x^{-1}$

24. $f(x) = \sin(3^x)$

25. Is $f(x) = 2^{x^2}$ a transcendental function?

26. Show that $f(x) = x^2 + 3x^{-1}$ and $g(x) = 3x^3 - 9x + x^{-2}$ are rational functions (show that each is a quotient of polynomials).

In Exercises 27–34, calculate the composite functions $f \circ g$ and $g \circ f$, and determine their domains.

27. $f(x) = \sqrt{x}, \quad g(x) = x + 1$

28. $f(x) = \dfrac{1}{x}, \quad g(x) = x^{-4}$

29. $f(x) = 2^x, \quad g(x) = x^2$

30. $f(x) = |x|, \quad g(\theta) = \sin\theta$

31. $f(\theta) = \cos\theta, \quad g(x) = x^3 + x^2$

32. $f(x) = \dfrac{1}{x^2 + 1}, \quad g(x) = x^{-2}$

33. $f(t) = \dfrac{1}{\sqrt{t}}, \quad g(t) = -t^2$

34. $f(t) = \sqrt{t}, \quad g(t) = 1 - t^3$

35. The population (in millions) of a country as a function of time t (years) is $P(t) = 30 \cdot 2^{kt}$, with $k = 0.1$. Show that the population doubles every 10 years. Show more generally that for any nonzero constants a and k, the function $g(t) = a2^{kt}$ doubles after $1/k$ years.

36. Find all values of c such that the domain of

$$f(x) = \dfrac{x+1}{x^2 + 2cx + 4}$$

is **R**.

Further Insights and Challenges

In Exercises 37–43, we define the first difference δf of a function $f(x)$ by $\delta f(x) = f(x+1) - f(x)$.

37. Show that if $f(x) = x^2$, then $\delta f(x) = 2x + 1$. Calculate δf for $f(x) = x$ and $f(x) = x^3$.

38. Show that $\delta(10^x) = 9 \cdot 10^x$, and more generally, $\delta(b^x) = c \cdot b^x$ for some constant c.

39. Show that for any two functions f and g, $\delta(f + g) = \delta f + \delta g$ and $\delta(cf) = c\delta(f)$, where c is any constant.

40. First differences can be used to derive formulas for the sum of the kth powers. Suppose we can find a function $P(x)$ such that $\delta P = (x+1)^k$ and $P(0) = 0$. Prove that $P(1) = 1^k$, $P(2) = 1^k + 2^k$, and more generally, for every whole number n,

$$P(n) = 1^k + 2^k + \cdots + n^k \qquad \boxed{1}$$

41. First show that

$$P(x) = \dfrac{x(x+1)}{2}$$

satisfies $\delta P = (x+1)$. Then apply Exercise 40 to conclude that

$$1 + 2 + 3 + \cdots + n = \dfrac{n(n+1)}{2}$$

42. Calculate $\delta(x^3)$, $\delta(x^2)$, and $\delta(x)$. Then find a polynomial $P(x)$ of degree 3 such that $\delta P = (x+1)^2$ and $P(0) = 0$. Conclude that

$$P(n) = 1^2 + 2^2 + \cdots + n^2$$

43. This exercise combined with Exercise 40 shows that for all k, there exists a polynomial $P(x)$ satisfying Eq. (1). The solution requires proof by induction and the Binomial Theorem (see Appendix C).

(a) Show that

$$\delta(x^{k+1}) = (k+1)x^k + \cdots$$

where the dots indicate terms involving smaller powers of x.

(b) Show by induction that for all whole numbers k, there exists a polynomial of degree $k+1$ with leading coefficient $1/(k+1)$:

$$P(x) = \dfrac{1}{k+1}x^{k+1} + \cdots$$

such that $\delta P = (x+1)^k$ and $P(0) = 0$.

1.4 Trigonometric Functions

We begin our trigonometric review by recalling the two systems of angle measurement: **radians** and **degrees**. They are best described using the relationship between angles and rotation. As is customary, we often use the lowercase Greek letter θ ("theta") to denote angles and rotations.

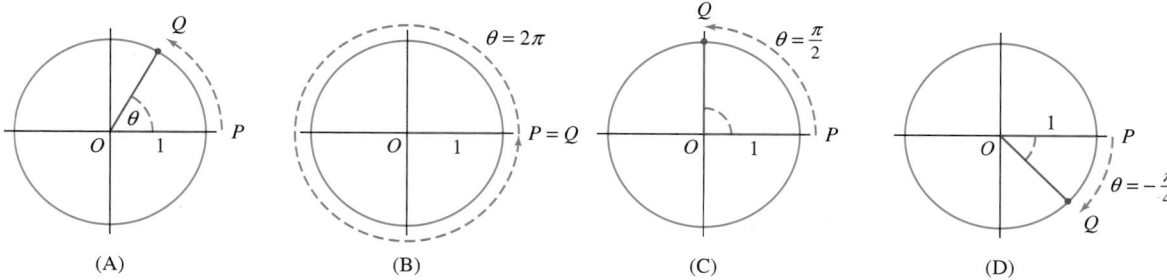

(A) (B) (C) (D)

FIGURE 1 The radian measure θ of a counterclockwise rotation is the length of the arc traversed by P as it rotates into Q.

TABLE 1

Rotation Through	Radian Measure
Two full circles	4π
Full circle	2π
Half circle	π
Quarter circle	$2\pi/4 = \pi/2$
One-sixth circle	$2\pi/6 = \pi/3$

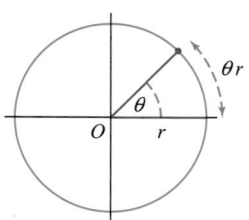

FIGURE 2 On a circle of radius r, the arc traversed by a counterclockwise rotation of θ radians has length θr.

Radians	Degrees
0	$0°$
$\dfrac{\pi}{6}$	$30°$
$\dfrac{\pi}{4}$	$45°$
$\dfrac{\pi}{3}$	$60°$
$\dfrac{\pi}{2}$	$90°$

Figure 1(A) shows a unit circle with radius $\overline{OP}$ rotating counterclockwise into radius $\overline{OQ}$. *The radian measure θ of this rotation is the length of the circular arc traversed by P as it rotates into Q.*

The unit circle has circumference 2π. Therefore, a rotation through a full circle has radian measure $\theta = 2\pi$ [Figure 1(B)]. The radian measure of a rotation through one-quarter of a circle is $\theta = 2\pi/4 = \pi/2$ [Figure 1(C)] and, in general, the rotation through one-nth of a circle has radian measure $2\pi/n$ (Table 1). A negative rotation (with $\theta < 0$) is a rotation in the *clockwise* direction [Figure 1(D)]. On a circle of radius r, *the arc traversed by a counterclockwise rotation of θ radians has length θr* (Figure 2).

Now consider angle $\angle POQ$ in Figure 1(A). The radian measure of angle $\angle POQ$ is defined as the radian measure of a rotation that carries $\overline{OP}$ to $\overline{OQ}$. Notice that every rotation has a unique radian measure but the radian measure of an angle is not unique. For example, the rotations through θ and $\theta + 2\pi$ both carry $\overline{OP}$ to $\overline{OQ}$. Although the rotation $\theta + 2\pi$ takes an extra trip around the circle, θ and $\theta + 2\pi$ represent the same angle. In general, *two angles coincide if the corresponding rotations differ by an integer multiple of 2π.* For example, the angle $\pi/4$ can be represented as $9\pi/4$ or $-15\pi/4$ because

$$\frac{\pi}{4} = \frac{9\pi}{4} - 2\pi = -\frac{15\pi}{4} + 4\pi$$

Every angle has a unique radian measure satisfying $0 \le \theta < 2\pi$. With this choice, the angle subtends an arc of length θr on a circle of radius r.

Degrees are defined by dividing the circle (not necessarily the unit circle) into 360 parts. A degree is $(1/360)$th of a circle. A rotation through θ degrees (denoted $\theta°$) is a rotation through the fraction $\theta/360$ of the complete circle. For example, a rotation through $90°$ is a rotation through the fraction $90/360$, or $\frac{1}{4}$ of a circle.

As with radians, every rotation has a unique degree measure but the degree measure of an angle is not unique. Two angles coincide if their degree measures differ by an integer multiple of 360. Every angle has a unique degree measure θ with $0 \le \theta < 360$. For example, the angles $-45°$ and $675°$ coincide since $675 = -45 + 720$.

To convert between radians and degrees, remember that 2π rad is equal to $360°$. Therefore, 1 rad equals $360/2\pi$ or $180/\pi$ degrees.

Radian measurement is usually the better choice for mathematical purposes, but there are good practical reasons for using degrees. The number 360 has many divisors (360 = 8 · 9 · 5) and, consequently, many fractional parts of the circle can be expressed as an integer number of degrees. For example, one-fifth of the circle is 72°, two-ninths is 80°, three-eighths is 135°, etc.

- To convert from radians to degrees, multiply by $180/\pi$.
- To convert from degrees to radians, multiply by $\pi/180$.

■ **EXAMPLE 1** Convert (a) 55° to radians and (b) 0.5 rad to degrees.

Solution

(a) $55° \times \dfrac{\pi}{180} \approx 0.9599$ rad

(b) 0.5 rad $\times \dfrac{180}{\pi} \approx 28.648°$ ■

Convention: *Unless otherwise stated, we always measure angles in radians.*

The trigonometric functions $\sin\theta$ and $\cos\theta$ are defined in terms of right triangles. Let θ be an acute angle in a right triangle and let us label the sides as in Figure 3. Then

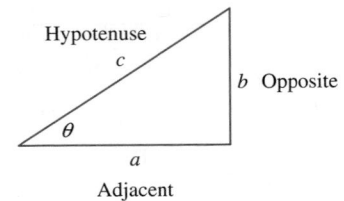

FIGURE 3

$$\sin\theta = \frac{b}{c} = \frac{\text{opposite}}{\text{hypotenuse}}, \qquad \cos\theta = \frac{a}{c} = \frac{\text{adjacent}}{\text{hypotenuse}}$$

A disadvantage of this definition is that it only makes sense if θ lies between 0 and $\pi/2$ (because an angle in a right triangle cannot exceed $\pi/2$). However, sine and cosine can also be defined in terms of the unit circle and this definition is valid for all angles. Let $P = (x, y)$ be the point on the unit circle corresponding to the angle θ as in Figure 4(A). Then we define

$$\cos\theta = x\text{-coordinate of } P, \qquad \sin\theta = y\text{-coordinate of } P$$

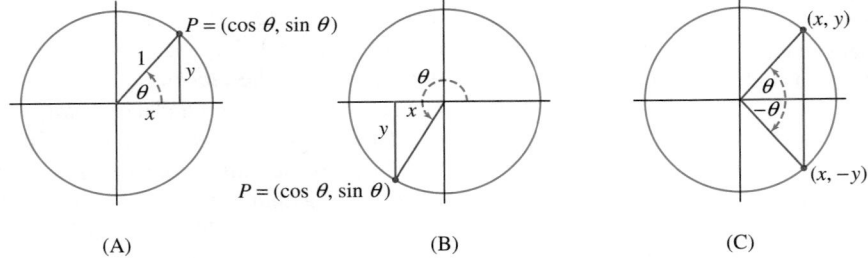

FIGURE 4 The unit circle definition of sine and cosine is valid for all angles θ.

This agrees with the right-triangle definition when $0 < \theta < \pi/2$. Furthermore, we see from Figure 4(C) that $\sin\theta$ is an odd function and $\cos\theta$ is an even function:

$$\sin(-\theta) = -\sin\theta, \qquad \cos(-\theta) = \cos\theta$$

Although we use a calculator to evaluate sine and cosine for general angles, the standard values listed in Figure 5 and Table 2 appear often and should be memorized.

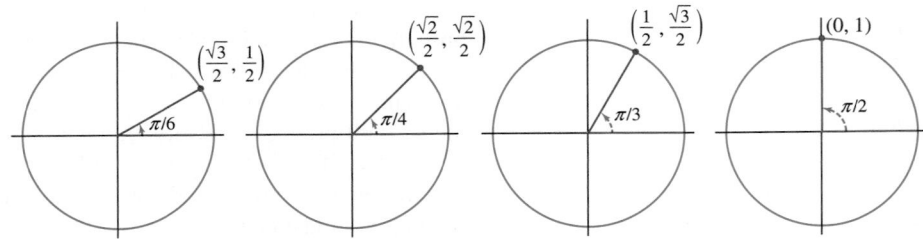

FIGURE 5 Four standard angles: The x- and y-coordinates of the points are $\cos\theta$ and $\sin\theta$.

TABLE 2

θ	0	$\dfrac{\pi}{6}$	$\dfrac{\pi}{4}$	$\dfrac{\pi}{3}$	$\dfrac{\pi}{2}$	$\dfrac{2\pi}{3}$	$\dfrac{3\pi}{4}$	$\dfrac{5\pi}{6}$	π
$\sin\theta$	0	$\dfrac{1}{2}$	$\dfrac{\sqrt{2}}{2}$	$\dfrac{\sqrt{3}}{2}$	1	$\dfrac{\sqrt{3}}{2}$	$\dfrac{\sqrt{2}}{2}$	$\dfrac{1}{2}$	0
$\cos\theta$	1	$\dfrac{\sqrt{3}}{2}$	$\dfrac{\sqrt{2}}{2}$	$\dfrac{1}{2}$	0	$-\dfrac{1}{2}$	$-\dfrac{\sqrt{2}}{2}$	$-\dfrac{\sqrt{3}}{2}$	-1

The functions $\sin\theta$ and $\cos\theta$ are defined for all real numbers θ, and it is not necessary to think of θ as an angle. We often write $\sin x$ and $\cos x$, using x instead of θ. Depending on the application, the angle interpretation may or may not be appropriate.

The graph of $y = \sin\theta$ is the familiar "sine wave" shown in Figure 6. Observe how the graph is generated by the y-coordinate of a point moving around the unit circle. The graph of $y = \cos\theta$ has the same shape, but is shifted to the left $\frac{\pi}{2}$ units (Figure 7). The signs of $\sin\theta$ and $\cos\theta$ vary as the point $P = (\cos\theta, \sin\theta)$ on the unit circle changes quadrant (Figure 7).

FIGURE 6 A graph of $y = \sin\theta$ is generated as a point moves around the unit circle.

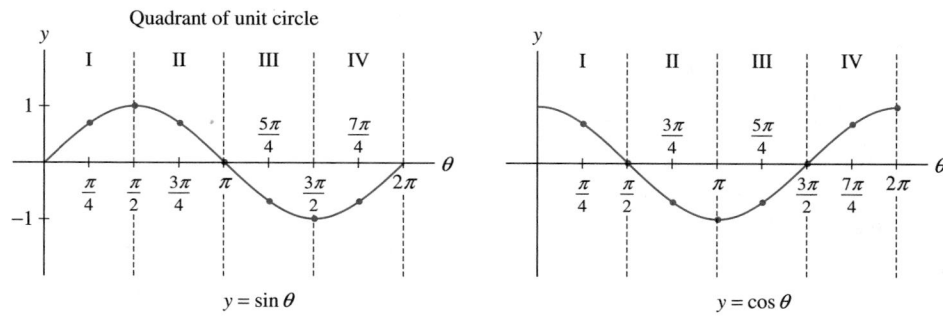

FIGURE 7 Graphs of $y = \sin\theta$ and $y = \cos\theta$ over one *period* of length 2π.

A function $f(x)$ is called **periodic** with period T if $f(x + T) = f(x)$ (for all x) and T is the smallest positive number with this property. The sine and cosine functions are periodic with period $T = 2\pi$ since angles that differ by an integer multiple $2\pi k$ correspond to the same point on the unit circle (Figure 8):

$$\sin x = \sin(x + 2\pi k), \qquad \cos x = \cos(x + 2\pi k)$$

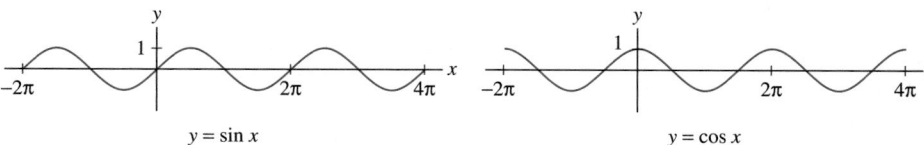

FIGURE 8 Sine and cosine have period 2π.

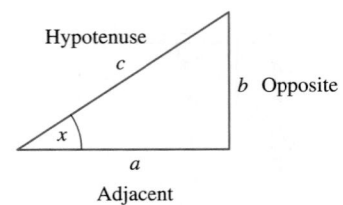

FIGURE 9

Recall that there are four other standard trigonometric functions, each defined in terms of $\sin x$ and $\cos x$ or as ratios of sides in a right triangle (Figure 9):

$$\text{Tangent:} \quad \tan x = \frac{\sin x}{\cos x} = \frac{b}{a}, \qquad \text{Cotangent:} \quad \cot x = \frac{\cos x}{\sin x} = \frac{a}{b}$$

$$\text{Secant:} \quad \sec x = \frac{1}{\cos x} = \frac{c}{a}, \qquad \text{Cosecant:} \quad \csc x = \frac{1}{\sin x} = \frac{c}{b}$$

These functions are periodic (Figure 10): $y = \tan x$ and $y = \cot x$ have period π, $y = \sec x$ and $y = \csc x$ have period 2π (see Exercise 51).

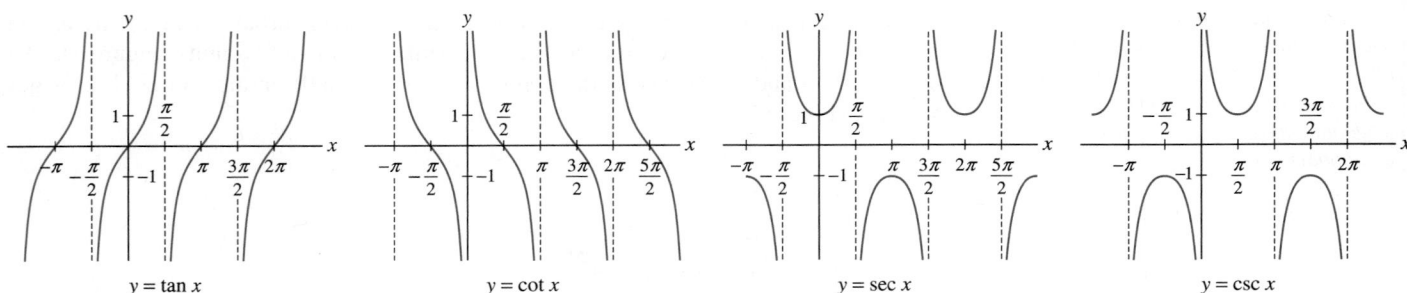

FIGURE 10 Graphs of the standard trigonometric functions.

■ **EXAMPLE 2** Computing Values of Trigonometric Functions Find the values of the six trigonometric functions at $x = 4\pi/3$.

Solution The point P on the unit circle corresponding to the angle $x = 4\pi/3$ lies opposite the point with angle $\pi/3$ (Figure 11). Thus, we see that (refer to Table 2)

$$\sin \frac{4\pi}{3} = -\sin \frac{\pi}{3} = -\frac{\sqrt{3}}{2}, \qquad \cos \frac{4\pi}{3} = -\cos \frac{\pi}{3} = -\frac{1}{2}$$

The remaining values are

$$\tan \frac{4\pi}{3} = \frac{\sin 4\pi/3}{\cos 4\pi/3} = \frac{-\sqrt{3}/2}{-1/2} = \sqrt{3}, \qquad \cot \frac{4\pi}{3} = \frac{\cos 4\pi/3}{\sin 4\pi/3} = \frac{\sqrt{3}}{3}$$

$$\sec \frac{4\pi}{3} = \frac{1}{\cos 4\pi/3} = \frac{1}{-1/2} = -2, \qquad \csc \frac{4\pi}{3} = \frac{1}{\sin 4\pi/3} = \frac{-2\sqrt{3}}{3} \qquad ■$$

$P = \left(-\frac{1}{2}, -\frac{\sqrt{3}}{2}\right)$

FIGURE 11

■ **EXAMPLE 3** Find the angles x such that $\sec x = 2$.

Solution Since $\sec x = 1/\cos x$, we must solve $\cos x = \frac{1}{2}$. From Figure 12 we see that $x = \frac{\pi}{3}$ and $x = -\frac{\pi}{3}$ are solutions. We may add any integer multiple of 2π, so the general solution is $x = \pm\frac{\pi}{3} + 2\pi k$ for any integer k. ■

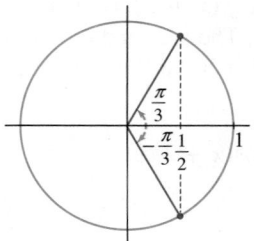

FIGURE 12 $\cos x = 1/2$ for $x = \pm\pi/3$

■ **EXAMPLE 4** Sketch the graph of $f(x) = 3 \cos\left(2\left(x + \frac{\pi}{2}\right)\right)$ over $[0, 2\pi]$.

Solution The graph is obtained by scaling and shifting the graph of $y = \cos x$ in three steps (Figure 13):

• Compress horizontally by a factor of 2: $y = \cos 2x$

Note: To shift the graph of $y = \cos 2x$ to the left $\frac{\pi}{2}$ units, we must replace x by $x + \frac{\pi}{2}$ to obtain $\cos\left(2(x + \frac{\pi}{2})\right)$. It is incorrect to take $\cos\left(2x + \frac{\pi}{2}\right)$

- Shift to left $\dfrac{\pi}{2}$ units: $\qquad\qquad y = \cos\left(2\left(x + \dfrac{\pi}{2}\right)\right)$

- Expand vertically by factor of 3: $\qquad y = 3\cos\left(2\left(x + \dfrac{\pi}{2}\right)\right)$ ∎

   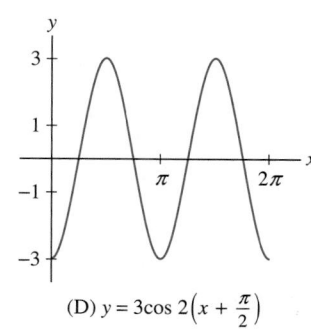

(A) $y = \cos x$ | (B) $y = \cos 2x$ (periodic with period π) | (C) $y = \cos 2\left(x + \frac{\pi}{2}\right)$ | (D) $y = 3\cos 2\left(x + \frac{\pi}{2}\right)$

FIGURE 13

Trigonometric Identities

A key feature of trigonometric functions is that they satisfy a large number of identities. First and foremost, sine and cosine satisfy the following fundamental identity, which is equivalent to the Pythagorean Theorem:

The expression $(\sin\theta)^k$ is usually denoted $\sin^k\theta$. For example, $\sin^2\theta$ is the square of $\sin\theta$. We use similar notation for the other trigonometric functions.

$$\boxed{\sin^2\theta + \cos^2\theta = 1}$$ $\boxed{1}$

Equivalent versions are obtained by dividing Eq. (1) by $\cos^2\theta$ or $\sin^2\theta$:

$$\boxed{\tan^2\theta + 1 = \sec^2\theta, \qquad 1 + \cot^2\theta = \csc^2\theta}$$ $\boxed{2}$

Here is a list of some other commonly used identities.

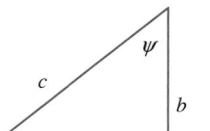

FIGURE 14 For complementary angles θ and $\psi = \dfrac{\pi}{2} - \theta$, we have $\sin\theta = \cos\psi$ and $\sin\psi = \cos\theta$.

Basic Trigonometric Identities

Complementary angles: $\quad \sin\left(\dfrac{\pi}{2} - x\right) = \cos x, \quad \cos\left(\dfrac{\pi}{2} - x\right) = \sin x$

Addition formulas: $\quad \sin(x + y) = \sin x \cos y + \cos x \sin y$

$\qquad\qquad\qquad\quad \cos(x + y) = \cos x \cos y - \sin x \sin y$

Double-angle formulas: $\quad \sin^2 x = \dfrac{1}{2}(1 - \cos 2x), \quad \cos^2 x = \dfrac{1}{2}(1 + \cos 2x)$

$\qquad\qquad\qquad\qquad \cos 2x = \cos^2 x - \sin^2 x, \quad \sin 2x = 2\sin x \cos x$

Shift formulas: $\quad \sin\left(x + \dfrac{\pi}{2}\right) = \cos x, \quad \cos\left(x + \dfrac{\pi}{2}\right) = -\sin x$

■ **EXAMPLE 5** Suppose that $\cos\theta = \dfrac{2}{5}$. Calculate $\tan\theta$ if (a) $0 \le \theta < \dfrac{\pi}{2}$ and (b) $\pi \le \theta < 2\pi$.

Solution We use the identity $\cos^2 \theta + \sin^2 \theta = 1$ to determine $\sin \theta$:

$$\sin \theta = \pm\sqrt{1 - \cos^2 \theta} = \pm\sqrt{1 - \frac{4}{25}} = \pm\frac{\sqrt{21}}{5}$$

(a) If $0 \le \theta < \pi/2$, then $\sin \theta$ is positive and we take the positive square root:

$$\tan \theta = \frac{\sin \theta}{\cos \theta} = \frac{\sqrt{21}/5}{2/5} = \frac{\sqrt{21}}{2}$$

A good way to visualize this computation is to draw a right triangle with an angle θ such that $\cos \theta = \frac{2}{5}$ as in Figure 15. The opposite side then has length $\sqrt{21} = \sqrt{5^2 - 2^2}$ by the Pythagorean Theorem.

(b) If we assume that $\pi \le \theta < 2\pi$, then $\sin \theta$ is negative and $\tan \theta = -\dfrac{\sqrt{21}}{2}$. ■

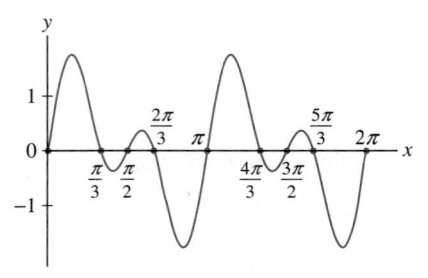

Hypotenuse 5

Opposite $\sqrt{21}$

θ

Adjacent 2

FIGURE 15

■ **EXAMPLE 6** **A Trigonometric Equation**   Solve the equation $\sin 4x + \sin 2x = 0$ for $0 \le x < 2\pi$.

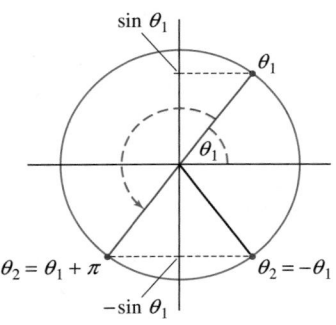

$\sin \theta_1$

θ_1

θ_1

$\theta_2 = \theta_1 + \pi$

$\theta_2 = -\theta_1$

$-\sin \theta_1$

FIGURE 16 $\sin \theta_2 = -\sin \theta_1$ when $\theta_2 = -\theta_1$ or $\theta_2 = \theta_1 + \pi$.

Solution We must find the angles x such that $\sin 4x = -\sin 2x$. First, let's determine when angles θ_1 and θ_2 satisfy $\sin \theta_2 = -\sin \theta_1$. Figure 16 shows that this occurs if $\theta_2 = -\theta_1$ or $\theta_2 = \theta_1 + \pi$. Since the sine function is periodic with period 2π,

$$\sin \theta_2 = -\sin \theta_1 \quad \Leftrightarrow \quad \theta_2 = -\theta_1 + 2\pi k \quad \text{or} \quad \theta_2 = \theta_1 + \pi + 2\pi k$$

where k is an integer. Taking $\theta_2 = 4x$ and $\theta_1 = 2x$, we see that

$$\sin 4x = -\sin 2x \quad \Leftrightarrow \quad 4x = -2x + 2\pi k \quad \text{or} \quad 4x = 2x + \pi + 2\pi k$$

The first equation gives $6x = 2\pi k$ or $x = \frac{\pi}{3}k$ and the second equation gives $2x = \pi + 2\pi k$ or $x = \frac{\pi}{2} + \pi k$. We obtain eight solutions in $[0, 2\pi)$ (Figure 17)

$$x = 0, \quad \frac{\pi}{3}, \quad \frac{2\pi}{3}, \quad \pi, \quad \frac{4\pi}{3}, \quad \frac{5\pi}{3} \quad \text{and} \quad x = \frac{\pi}{2}, \quad \frac{3\pi}{2} \quad ■$$

We conclude this section by quoting the **Law of Cosines**, which is a generalization of the Pythagorean Theorem (Figure 18).

y

1

$\frac{2\pi}{3}$ π

$\frac{5\pi}{3}$

2π

x

0

$\frac{\pi}{3}$ $\frac{\pi}{2}$

$\frac{4\pi}{3}$ $\frac{3\pi}{2}$

-1

$y = \sin 4x + \sin 2x$

FIGURE 17 Solutions of $\sin 4x + \sin 2x = 0$.

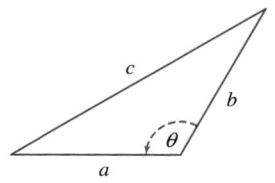

FIGURE 18

> **THEOREM 1 Law of Cosines**   If a triangle has sides a, b, and c, and θ is the angle opposite side c, then
>
> $$c^2 = a^2 + b^2 - 2ab \cos \theta$$

If $\theta = 90°$, then $\cos \theta = 0$ and the Law of Cosines reduces to the Pythagorean Theorem.

1.4 SUMMARY

- An angle of θ radians subtends an arc of length θr on a circle of radius r.

- To convert from radians to degrees, multiply by $\dfrac{180}{\pi}$.

- To convert from degrees to radians, multiply by $\dfrac{\pi}{180}$.

- Unless otherwise stated, all angles in this text are given in radians.

- The functions $\cos \theta$ and $\sin \theta$ are defined in terms of right triangles for acute angles and as coordinates of a point on the unit circle for general angles (Figure 19):

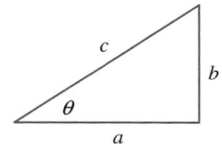

$$\sin \theta = \frac{b}{c} = \frac{\text{opposite}}{\text{hypotenuse}}, \qquad \cos \theta = \frac{a}{c} = \frac{\text{adjacent}}{\text{hypotenuse}}$$

- Basic properties of sine and cosine:

  - Periodicity: $\sin(\theta + 2\pi) = \sin \theta, \quad \cos(\theta + 2\pi) = \cos \theta$

  - Parity: $\sin(-\theta) = -\sin \theta, \quad \cos(-\theta) = \cos \theta$

  - Basic identity: $\sin^2 \theta + \cos^2 \theta = 1$

FIGURE 19

- The four additional trigonometric functions:

$$\tan \theta = \frac{\sin \theta}{\cos \theta}, \qquad \cot \theta = \frac{\cos \theta}{\sin \theta}, \qquad \sec \theta = \frac{1}{\cos \theta}, \qquad \csc \theta = \frac{1}{\sin \theta}$$

1.4 EXERCISES

Preliminary Questions

1. How is it possible for two different rotations to define the same angle?

2. Give two different positive rotations that define the angle $\frac{\pi}{4}$.

3. Give a negative rotation that defines the angle $\frac{\pi}{3}$.

4. The definition of $\cos \theta$ using right triangles applies when (choose the correct answer):

(a) $0 < \theta < \dfrac{\pi}{2}$ (b) $0 < \theta < \pi$ (c) $0 < \theta < 2\pi$

5. What is the unit circle definition of $\sin \theta$?

6. How does the periodicity of $\sin \theta$ and $\cos \theta$ follow from the unit circle definition?

Exercises

1. Find the angle between 0 and 2π that is equivalent to $13\pi/4$.

2. Describe the angle $\theta = \frac{\pi}{6}$ by an angle of negative radian measure.

3. Convert from radians to degrees:

(a) 1
(b) $\frac{\pi}{3}$
(c) $\frac{5}{12}$
(d) $-\frac{3\pi}{4}$

4. Convert from degrees to radians:

(a) $1°$
(b) $30°$
(c) $25°$
(d) $120°$

5. Find the lengths of the arcs subtended by the angles θ and ϕ radians in Figure 20.

6. Calculate the values of the six standard trigonometric functions for the angle θ in Figure 21.

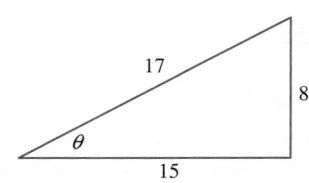

FIGURE 20 Circle of radius 4.   **FIGURE 21**

7. Fill in the remaining values of $(\cos\theta, \sin\theta)$ for the points in Figure 22.

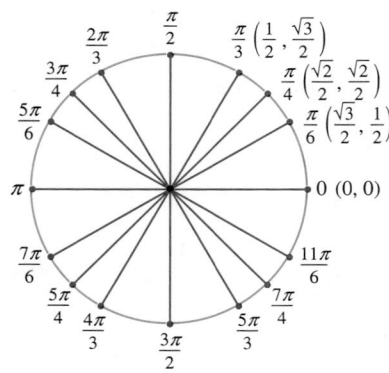

FIGURE 22

8. Find the values of the six standard trigonometric functions at $\theta = 11\pi/6$.

In Exercises 9–14, use Figure 22 to find all angles between 0 and 2π satisfying the given condition.

9. $\cos\theta = \frac{1}{2}$
10. $\tan\theta = 1$
11. $\tan\theta = -1$

12. $\csc\theta = 2$
13. $\sin x = \frac{\sqrt{3}}{2}$
14. $\sec t = 2$

15. Fill in the following table of values:

θ	$\frac{\pi}{6}$	$\frac{\pi}{4}$	$\frac{\pi}{3}$	$\frac{\pi}{2}$	$\frac{2\pi}{3}$	$\frac{3\pi}{4}$	$\frac{5\pi}{6}$
$\tan\theta$							
$\sec\theta$							

16. Complete the following table of signs for the trigonometric functions:

θ	sin	cos	tan	cot	sec	csc
$0 < \theta < \frac{\pi}{2}$	+	+				
$\frac{\pi}{2} < \theta < \pi$						
$\pi < \theta < \frac{3\pi}{2}$						
$\frac{3\pi}{2} < \theta < 2\pi$						

17. Show that if $\tan\theta = c$ and $0 \le \theta < \pi/2$, then $\cos\theta = 1/\sqrt{1+c^2}$. *Hint:* Draw a right triangle whose opposite and adjacent sides have lengths c and 1.

18. Suppose that $\cos\theta = \frac{1}{3}$.

(a) Show that if $0 \le \theta < \pi/2$, then $\sin\theta = 2\sqrt{2}/3$ and $\tan\theta = 2\sqrt{2}$.

(b) Find $\sin\theta$ and $\tan\theta$ if $3\pi/2 \le \theta < 2\pi$.

In Exercises 19–24, assume that $0 \le \theta < \pi/2$.

19. Find $\sin\theta$ and $\tan\theta$ if $\cos\theta = \frac{5}{13}$.

20. Find $\cos\theta$ and $\tan\theta$ if $\sin\theta = \frac{3}{5}$.

21. Find $\sin\theta$, $\sec\theta$, and $\cot\theta$ if $\tan\theta = \frac{2}{7}$.

22. Find $\sin\theta$, $\cos\theta$, and $\sec\theta$ if $\cot\theta = 4$.

23. Find $\cos 2\theta$ if $\sin\theta = \frac{1}{5}$.

24. Find $\sin 2\theta$ and $\cos 2\theta$ if $\tan\theta = \sqrt{2}$.

25. Find $\cos\theta$ and $\tan\theta$ if $\sin\theta = 0.4$ and $\pi/2 \le \theta < \pi$.

26. Find $\cos\theta$ and $\sin\theta$ if $\tan\theta = 4$ and $\pi \le \theta < 3\pi/2$.

27. Find the values of $\sin\theta$, $\cos\theta$, and $\tan\theta$ at the eight points in Figure 23.

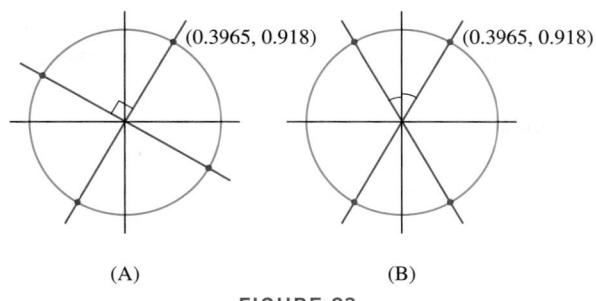

(A)   (B)

FIGURE 23

28. Refer to Figure 24(A). Express the functions $\sin \theta$, $\tan \theta$, and $\csc \theta$ in terms of c.

29. Refer to Figure 24(B). Compute $\cos \psi$, $\sin \psi$, $\cot \psi$, and $\csc \psi$.

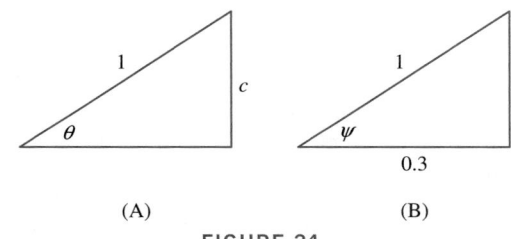

(A)   (B)

FIGURE 24

30. Express $\cos\left(\theta + \dfrac{\pi}{2}\right)$ and $\sin\left(\theta + \dfrac{\pi}{2}\right)$ in terms of $\cos \theta$ and $\sin \theta$. *Hint:* Find the relation between the coordinates (a, b) and (c, d) in Figure 25.

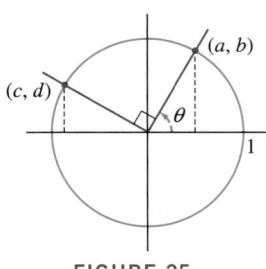

FIGURE 25

In Exercises 31–34, sketch the graph over $[0, 2\pi]$.

31. $2 \sin 4\theta$

32. $\cos\left(2\left(\theta - \dfrac{\pi}{2}\right)\right)$

33. $\cos\left(2\theta - \dfrac{\pi}{2}\right)$

34. $\sin\left(2\left(\theta - \dfrac{\pi}{2}\right) + \pi\right) + 2$

35. How many points lie on the intersection of the horizontal line $y = c$ and the graph of $y = \sin x$ for $0 \leq x < 2\pi$? *Hint:* The answer depends on c.

36. How many points lie on the intersection of the horizontal line $y = c$ and the graph of $y = \tan x$ for $0 \leq x < 2\pi$?

In Exercises 37–40, solve for $0 \leq \theta < 2\pi$ *(see Example 6).*

37. $\sin 2\theta + \sin 3\theta = 0$

38. $\sin \theta = \sin 2\theta$

39. $\cos 4\theta + \cos 2\theta = 0$

40. $\sin \theta = \cos 2\theta$

In Exercises 41–50, derive the identities using the identities listed in this section.

41. $\cos 2\theta = 2 \cos^2 \theta - 1$

42. $\cos^2 \dfrac{\theta}{2} = \dfrac{1 + \cos \theta}{2}$

43. $\left|\sin \dfrac{\theta}{2}\right| = \sqrt{\dfrac{1 - \cos \theta}{2}}$

44. $\sin(\theta + \pi) = -\sin \theta$

45. $\cos(\theta + \pi) = -\cos \theta$

46. $\tan x = \cot\left(\dfrac{\pi}{2} - x\right)$

47. $\tan(\pi - \theta) = -\tan \theta$

48. $\tan 2x = \dfrac{2 \tan x}{1 - \tan^2 x}$

49. $\tan x = \dfrac{\sin 2x}{1 + \cos 2x}$

50. $\sin^2 x \cos^2 x = \dfrac{1 - \cos 4x}{8}$

51. Use Exercises 44 and 45 to show that $\tan \theta$ and $\cot \theta$ are periodic with period π.

52. Use the addition formula to compute $\cos \dfrac{\pi}{12}$, noting that $\dfrac{\pi}{12} = \dfrac{\pi}{3} - \dfrac{\pi}{4}$.

53. Use the Law of Cosines to find the distance from P to Q in Figure 26.

FIGURE 26

Further Insights and Challenges

54. Use the addition formula to prove

$$\cos 3\theta = 4 \cos^3 \theta - 3 \cos \theta$$

55. Use the addition formulas for sine and cosine to prove

$$\tan(a + b) = \dfrac{\tan a + \tan b}{1 - \tan a \tan b}$$

$$\cot(a - b) = \dfrac{\cot a \cot b + 1}{\cot b - \cot a}$$

56. Let θ be the angle between the line $y = mx + b$ and the x-axis [Figure 27(A)]. Prove that $m = \tan \theta$.

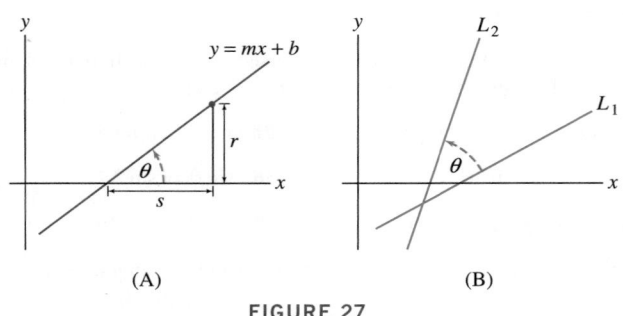

(A)   (B)

FIGURE 27

57. Let L_1 and L_2 be the lines of slope m_1 and m_2 [Figure 27(B)]. Show that the angle θ between L_1 and L_2 satisfies $\cot \theta = \dfrac{m_2 m_1 + 1}{m_2 - m_1}$.

58. Perpendicular Lines   Use Exercise 57 to prove that two lines with nonzero slopes m_1 and m_2 are perpendicular if and only if $m_2 = -1/m_1$.

59. Apply the double-angle formula to prove:

(a) $\cos \dfrac{\pi}{8} = \dfrac{1}{2}\sqrt{2 + \sqrt{2}}$

(b) $\cos \dfrac{\pi}{16} = \dfrac{1}{2}\sqrt{2 + \sqrt{2 + \sqrt{2}}}$

Guess the values of $\cos \dfrac{\pi}{32}$ and of $\cos \dfrac{\pi}{2^n}$ for all n.

1.5 Inverse Functions

Many important functions, such as logarithms or the arcsine, may be defined as **inverses** of other functions. In this section, we review inverse functions and their graphs, and we discuss the inverse trigonometric functions.

The inverse of $f(x)$, denoted $f^{-1}(x)$, is the function that *reverses* the effect of $f(x)$ (Figure 1). For example, the inverse of $f(x) = x^3$ is the cube root function $f^{-1}(x) = x^{1/3}$. Given a table of function values for $f(x)$, we obtain a table for $f^{-1}(x)$ by interchanging the x and y columns:

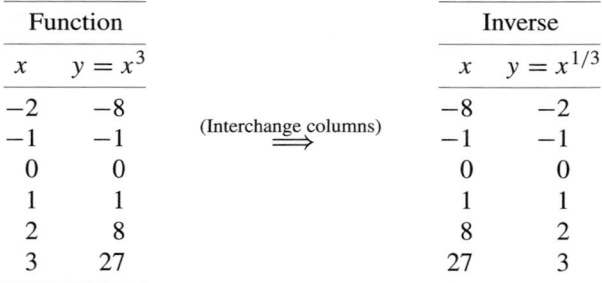

FIGURE 1 A function and its inverse.

Function			Inverse	
x	$y = x^3$		x	$y = x^{1/3}$
-2	-8		-8	-2
-1	-1	(Interchange columns) $\Longrightarrow$	-1	-1
0	0		0	0
1	1		1	1
2	8		8	2
3	27		27	3

Thus, if we apply both f and f^{-1} to a number x in either order, we get back x. For instance,

$$\text{Apply } f \text{ and then } f^{-1}: \quad 2 \xrightarrow{\text{(Apply } x^3)} 8 \xrightarrow{\text{(Apply } x^{1/3})} 2$$

$$\text{Apply } f^{-1} \text{ and then } f: \quad 8 \xrightarrow{\text{(Apply } x^{1/3})} 2 \xrightarrow{\text{(Apply } x^3)} 8$$

These two properties are used in the formal definition of the inverse function.

←·· **REMINDER** The "domain" is the set of numbers x such that $f(x)$ is defined (the set of allowable inputs), and the "range" is the set of all values $f(x)$ (the set of outputs).

DEFINITION Inverse   Let $f(x)$ have domain D and range R. The inverse function $f^{-1}(x)$ (if it exists) is the function with domain R such that

$$f^{-1}(f(x)) = x \quad \text{for } x \in D \qquad \text{and} \qquad f(f^{-1}(x)) = x \quad \text{for } x \in R$$

If f^{-1} exists, then f is called **invertible**.

■ **EXAMPLE 1** Finding the Inverse   Show that $f(x) = 2x - 18$ is invertible. What are the domain and range of $f^{-1}(x)$?

Solution We show that $f(x)$ is invertible by computing f^{-1} in two steps.

Step 1. **Solve the equation $y = f(x)$ for x in terms of y.**

$$y = 2x - 18$$

$$y + 18 = 2x$$

$$x = \frac{1}{2}y + 9$$

This gives us the inverse as a function of the variable y: $f^{-1}(y) = \frac{1}{2}y + 9$.

Step 2. **Interchange variables.**

We usually prefer to write the inverse as a function of x, so we interchange the roles of x and y (Figure 2):

$$f^{-1}(x) = \frac{1}{2}x + 9$$

To check our calculation, let's verify that $f^{-1}(f(x)) = x$ and $f(f^{-1}(x)) = x$:

$$f^{-1}\big(f(x)\big) = f^{-1}(2x - 18) = \frac{1}{2}(2x - 18) + 9 = (x - 9) + 9 = x$$

$$f\big(f^{-1}(x)\big) = f\left(\frac{1}{2}x + 9\right) = 2\left(\frac{1}{2}x + 9\right) - 18 = (x + 18) - 18 = x$$

Since f^{-1} is a linear function, its domain and range are **R**.   ■

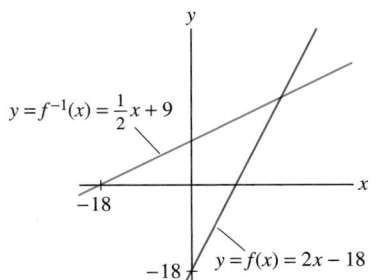

$y = f^{-1}(x) = \frac{1}{2}x + 9$

$y = f(x) = 2x - 18$

FIGURE 2 Graph of $f(x) = 2x - 18$ and its inverse $f^{-1}(x) = \frac{1}{2}x + 9$.

Some functions do not have inverses. For example, when we interchange the columns in a table of values for $f(x) = x^2$ (which should give us a table of values for f^{-1}), the resulting table does not define a function:

Function			Inverse (?)		
x	$y = x^2$		x	y	
−2	4	(Interchange columns) ⟹	4	−2	$f^{-1}(1)$ has
−1	1		1	−1	two values:
0	0		0	0	1 and −1.
1	1		1	1	
2	4		4	2	

Since 1 occurs twice as an *output* of x^2 in the first table, it also occurs twice as an *input* in the second table, and thus we obtain $f^{-1}(1) = 1$ or $f^{-1}(1) = -1$. However, neither value satisfies the inverse property. For instance, if we set $f^{-1}(1) = 1$, then $f^{-1}(f(-1)) = f^{-1}(1) = 1$, but an inverse would have to satisfy $f^{-1}(f(-1)) = -1$. So when does a function $f(x)$ have an inverse? The answer is: If $f(x)$ is **one-to-one**, which means that $f(x)$ takes on each value at most once (Figure 3).

An equivalent definition of one-to-one: if $a, b \in D$ and $f(a) = f(b)$, then $a = b$.

DEFINITION One-to-One Functions   A function $f(x)$ is one-to-one (on its domain D) if for every value c, the equation $f(x) = c$ has at most one solution for $x \in D$.

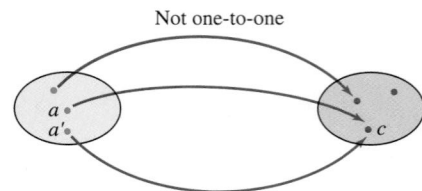

FIGURE 3 A one-to-one function takes on each value at most once.

$f(x) = c$ has at most one solution for all c

$f(x) = c$ has two solutions: $x = a$ and $x = a'$

Think of a function as a device for "labeling" members of the range by members of the domain. When f is one-to-one, this labeling is unique and f^{-1} maps each number in the range back to its label.

When $f(x)$ is one-to-one on its domain D, the inverse f^{-1} exists and its domain is equal to the range R of f (Figure 4). Indeed, for all $c \in R$, there is a unique $a \in D$ such that $f(a) = c$ and we may define the inverse by setting $f^{-1}(c) = a$. With this definition, $f(f^{-1}(c)) = f(a) = c$. Now consider $a \in D$ and set $c = f(a)$. By definition, $f^{-1}(c) = a$ and thus $f^{-1}(f(a)) = f^{-1}(c) = a$ as required. This proves the following theorem.

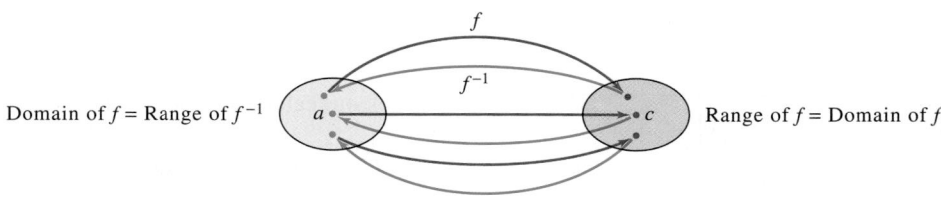

Domain of f = Range of f^{-1}

Range of f = Domain of f^{-1}

FIGURE 4 In passing from f to f^{-1}, the domain and range are interchanged.

> **THEOREM 1 Existence of Inverses**   If $f(x)$ is one-to-one on its domain D, then f is invertible. Furthermore,
>
> - Domain of f = range of f^{-1}.
> - Range of f = domain of f^{-1}.

■ **EXAMPLE 2** Finding the Inverse   Show that $f(x) = \dfrac{3x + 2}{5x - 1}$ is invertible. Determine the domain and range of f and f^{-1}.

Solution The domain of $f(x)$ is $D = \{x : x \neq \frac{1}{5}\}$ (Figure 5). Assume that $x \in D$ and let's solve $y = f(x)$ for x in terms of y:

$$y = \frac{3x + 2}{5x - 1}$$

$$y(5x - 1) = 3x + 2$$

$$5xy - y = 3x + 2$$

$$5xy - 3x = y + 2 \qquad \text{(gather terms involving } x\text{)}$$

$$x(5y - 3) = y + 2 \qquad \text{(factor out } x \text{ in order to solve for } x\text{)} \qquad \boxed{1}$$

$$x = \frac{y + 2}{5y - 3} \qquad\qquad\qquad\qquad\qquad\qquad \boxed{2}$$

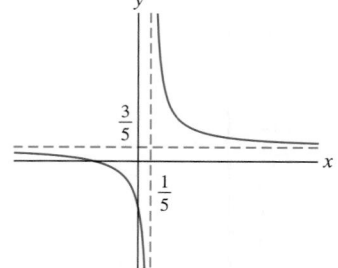

FIGURE 5 Graph of $f(x) = \dfrac{3x + 2}{5x - 1}$.

It is often impossible to solve for x in the equation $y = f(x)$, for example, if $f(x) = xe^x$. In such cases, the inverse may exist but we must make do without an explicit formula for it.

In the last step, we divided by $5y - 3$. This is valid if $y \neq \frac{3}{5}$. However, y cannot equal $\frac{3}{5}$ because Eq. (1) would then give $0 = \frac{3}{5} + 2$, which is false. We conclude that $f(x)$ is invertible on its domain with inverse function $f^{-1}(x) = \dfrac{x + 2}{5x - 3}$. The domain of f^{-1} is

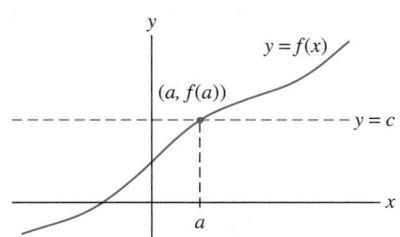

FIGURE 6 The line $y = c$ intersects the graph of $y = f(x)$ at points where $f(a) = c$.

the set $R = \{x : x \neq \frac{3}{5}\}$. Because the domain and range are interchanged under inversion, we conclude that $f(x)$ has range R and f^{-1} has range D. ∎

We can tell if a function $f(x)$ is one-to-one by looking at its graph. The horizontal line $y = c$ intersects the graph at the points $(a, f(a))$ such that $f(a) = c$ (Figure 6). If every horizontal line intersects the graph in at most one point, then f is one-to-one.

Horizontal Line Test   A function $f(x)$ is one-to-one if and only if every horizontal line intersects the graph of f in at most one point.

Observe that $f(x) = x^3$ is one-to-one and passes the Horizontal Line Test [Figure 7(A)], whereas $f(x) = x^2$ fails the test [Figure 7(B)].

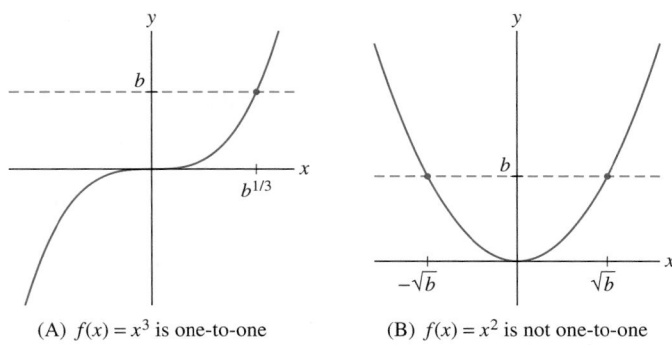

(A) $f(x) = x^3$ is one-to-one   (B) $f(x) = x^2$ is not one-to-one

FIGURE 7 A function is one-to-one if every horizontal line intersects its graph in at most one point.

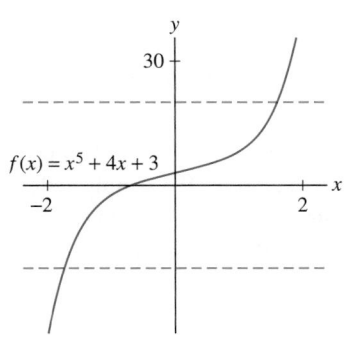

FIGURE 8 The strictly increasing function $f(x) = x^5 + 4x + 3$ satisfies the Horizontal Line Test.

■ **EXAMPLE 3** Strictly Increasing Functions Are One-to-One   Show that strictly increasing functions are one-to-one. Then show that $f(x) = x^5 + 4x + 3$ is one-to-one.

Solution If $f(x)$ is not one-to-one, then $f(a) = f(b)$ for some $a < b$. But a strictly increasing function satisfies $f(a) < f(b)$, so this cannot occur.

Now observe that

• The functions cx^n for n odd and $c > 0$ are strictly increasing.
• A sum of strictly increasing functions is strictly increasing.

Thus, x^5, $4x$, and hence $x^5 + 4x$ are strictly increasing. It follows that the function $f(x) = x^5 + 4x + 3$ is strictly increasing and therefore one-to-one. This is also clear from the graph of $f(x)$ (Figure 8), which, as we observe, satisfies the Horizontal Line Test. ∎

Often, it is possible to make a function one-to-one by restricting the domain.

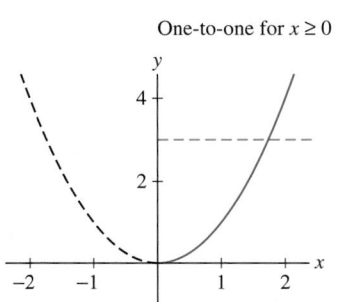

FIGURE 9 $f(x) = x^2$ satisfies the Horizontal Line Test on the domain $\{x : x \geq 0\}$.

■ **EXAMPLE 4** Restricting the Domain   Find a domain on which $f(x) = x^2$ is one-to-one and determine its inverse on this domain.

Solution The function $f(x) = x^2$ is one-to-one on the domain $D = \{x : x \geq 0\}$, for if $a^2 = b^2$ and a and b are both nonnegative, then $a = b$ (Figure 9). The inverse of $f(x)$ on D is the positive square root $f^{-1}(x) = \sqrt{x}$. Alternatively, we may restrict $f(x)$ to the domain $\{x : x \leq 0\}$, on which the inverse is $f^{-1}(x) = -\sqrt{x}$. ∎

Next we describe the graph of the inverse function. We define the **reflection** of a point (a, b) through the line $y = x$ to be the point (b, a) (Figure 10). If we draw the x- and y-axes to the same scale, then (a, b) and (b, a) are equidistant from the line $y = x$ and the segment joining them is perpendicular to $y = x$.

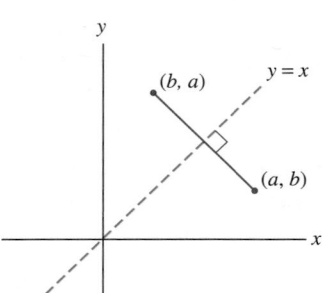

FIGURE 10 Reflection of the point (a, b) through the line $y = x$.

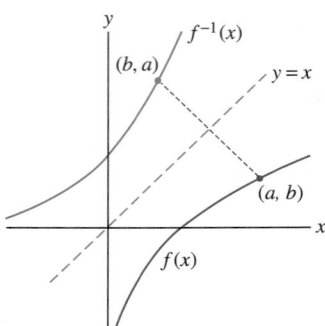

FIGURE 11 Graph of $f^{-1}(x)$ is the reflection of the graph of $f(x)$ through the line $y = x$.

The graph of f^{-1} is obtained by reflecting the graph of f through $y = x$. To check this, note that (a, b) lies on the graph of f if $f(a) = b$. In this case, $f^{-1}(b) = a$ and the reflection $(b, f^{-1}(b)) = (b, a)$ lies on the graph of f^{-1} (Figure 11).

■ **EXAMPLE 5** **Sketching the Graph of the Inverse** Sketch the graph of the inverse of $f(x) = \sqrt{4 - x}$.

Solution Let $g(x) = f^{-1}(x)$. Observe that $f(x)$ has domain $\{x : x \le 4\}$ and range $\{x : x \ge 0\}$. We do not need a formula for $g(x)$ to draw its graph. We simply reflect the graph of f through the line $y = x$ as in Figure 12. If desired, however, we can easily solve $y = \sqrt{4 - x}$ to obtain $x = 4 - y^2$ and thus $g(x) = 4 - x^2$ with domain $\{x : x \ge 0\}$. ■

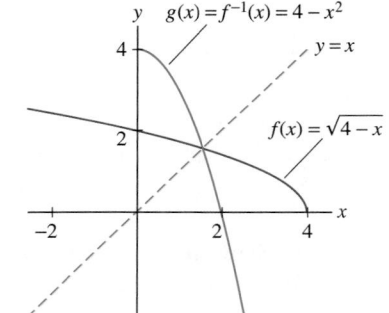

FIGURE 12 Graph of the inverse $g(x)$ of $f(x) = \sqrt{4 - x}$.

Inverse Trigonometric Functions

We have seen that the inverse $f^{-1}(x)$ exists if $f(x)$ is one-to-one on its domain. The trigonometric functions are not one-to-one. Therefore, we define their inverses by restricting them to domains on which they are one-to-one.

First consider the sine function. Figure 13 shows that $f(\theta) = \sin \theta$ is one-to-one on $[-\pi/2, \pi/2]$. When $\sin \theta$ is restricted to the interval $[-\pi/2, \pi/2]$, its inverse is called the

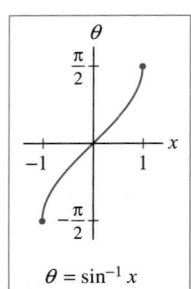

sin θ with restricted domain

$\theta = \sin^{-1} x$

FIGURE 13 $f(\theta) = \sin \theta$ is one-to-one on $[-\pi/2, \pi/2]$ and $\theta = \sin^{-1} x$ is its inverse on $[-\pi/2, \pi/2]$.

Do not confuse the inverse $\sin^{-1} x$ *with the reciprocal*

$$(\sin x)^{-1} = \frac{1}{\sin x}$$

The inverse functions $\sin^{-1} x$, $\cos^{-1} x$, ... *are often denoted* arcsin x, arccos x, *etc.*

Summary of inverse relation between the sine and the arcsine:

$\sin(\sin^{-1} x) = x$ for $-1 \le x \le 1$

$\sin^{-1}(\sin \theta) = \theta$ for $-\dfrac{\pi}{2} \le \theta \le \dfrac{\pi}{2}$

arcsine function and is denoted $\theta = \sin^{-1} x$. By definition,

$$\theta = \sin^{-1} x \text{ is the unique angle in } \left[-\frac{\pi}{2}, \frac{\pi}{2}\right] \text{ such that } \sin \theta = x$$

A table of values for the arcsine (Table 1) is obtained by reversing the columns in a table of values for $\sin x$.

TABLE 1

x	$\theta = \sin^{-1} x$
-1	$-\frac{\pi}{2}$
$-\frac{\sqrt{3}}{2}$	$-\frac{\pi}{3}$
$-\frac{\sqrt{2}}{2}$	$-\frac{\pi}{4}$
$-\frac{1}{2}$	$-\frac{\pi}{6}$
0	0
$\frac{1}{2}$	$\frac{\pi}{6}$
$\frac{\sqrt{2}}{2}$	$\frac{\pi}{4}$
$\frac{\sqrt{3}}{2}$	$\frac{\pi}{3}$
1	$\frac{\pi}{2}$

EXAMPLE 6 (a) Show that $\sin^{-1}\left(\sin\left(\frac{\pi}{4}\right)\right) = \frac{\pi}{4}$.

(b) Explain why $\sin^{-1}\left(\sin\left(\frac{5\pi}{4}\right)\right) \ne \frac{5\pi}{4}$.

Solution The equation $\sin^{-1}(\sin \theta) = \theta$ is valid only if θ lies in $[-\pi/2, \pi/2]$.

(a) Since $\frac{\pi}{4}$ lies in the required interval, $\sin^{-1}\left(\sin\left(\frac{\pi}{4}\right)\right) = \frac{\pi}{4}$.

(b) Let $\theta = \sin^{-1}\left(\sin\left(\frac{5\pi}{4}\right)\right)$. By definition, θ is the angle in $[-\pi/2, \pi/2]$ such that $\sin \theta = \sin\left(\frac{5\pi}{4}\right)$. By the identity $\sin \theta = \sin(\pi - \theta)$ (Figure 14),

$$\sin\left(\frac{5\pi}{4}\right) = \sin\left(\pi - \frac{5\pi}{4}\right) = \sin\left(-\frac{\pi}{4}\right)$$

The angle $-\frac{\pi}{4}$ lies in the required interval, so $\theta = \sin^{-1}\left(\sin\left(\frac{5\pi}{4}\right)\right) = -\frac{\pi}{4}$. ∎

FIGURE 14 $\sin\left(\frac{5\pi}{4}\right) = \sin\left(-\frac{\pi}{4}\right)$.

The cosine function is one-to-one on $[0, \pi]$ rather than $[-\pi/2, \pi/2]$ (Figure 15). When $f(\theta) = \cos \theta$ is restricted to the domain $[0, \pi]$, the inverse is called the **arccosine function** and is denoted $\theta = \cos^{-1} x$. By definition,

$$\theta = \cos^{-1} x \text{ is the unique angle in } [0, \pi] \text{ such that } \cos \theta = x$$

Summary of inverse relation between the cosine and arccosine:

$\cos(\cos^{-1} x) = x$ for $-1 \le x \le 1$

$\cos^{-1}(\cos \theta) = \theta$ for $0 \le \theta \le \pi$

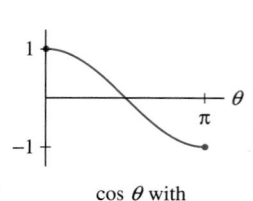

cos θ with
restricted domain

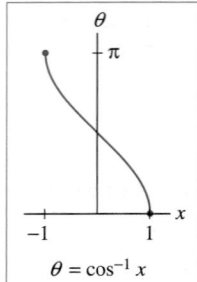

$\theta = \cos^{-1} x$

FIGURE 15 $f(\theta) = \cos \theta$ is one-to-one on $[0, \pi]$ and $\theta = \cos^{-1} x$ is its inverse on $[0, \pi]$.

When we study the calculus of inverse trigonometric functions in Section 3.9, we will need to simplify composite expressions such as $\cos(\sin^{-1} x)$ and $\tan(\sec^{-1} x)$. This can be done in two ways: by referring to the appropriate right triangle or by using trigonometric identities.

■ **EXAMPLE 7**   Simplify $\cos(\sin^{-1} x)$ and $\tan(\sin^{-1} x)$.

Solution   This problem asks for the values of $\cos \theta$ and $\tan \theta$ at the angle $\theta = \sin^{-1} x$. Consider a right triangle with hypotenuse of length 1 and angle θ such that $\sin \theta = x$ as in Figure 16. By the Pythagorean Theorem, the adjacent side has length $\sqrt{1 - x^2}$. Now we can read off the values from Figure 16:

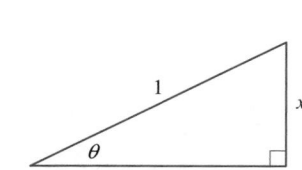

FIGURE 16 Right triangle constructed so that $\sin \theta = x$.

$$\cos(\sin^{-1} x) = \cos \theta = \frac{\text{adjacent}}{\text{hypotenuse}} = \sqrt{1 - x^2}$$

$$\tan(\sin^{-1} x) = \tan \theta = \frac{\text{opposite}}{\text{adjacent}} = \frac{x}{\sqrt{1 - x^2}}$$

| 3 |

Alternatively, we may argue using trigonometric identities. Because $\sin \theta = x$,

$$\cos(\sin^{-1} x) = \cos \theta = \sqrt{1 - \sin^2 \theta} = \sqrt{1 - x^2}$$

We take the positive square root because $\theta = \sin^{-1} x$ lies in $[-\pi/2, \pi/2]$ and $\cos \theta$ is positive in this interval.   ■

We now address the remaining trigonometric functions. The function $f(\theta) = \tan \theta$ is one-to-one on $(-\pi/2, \pi/2)$, and $f(\theta) = \cot \theta$ is one-to-one on $(0, \pi)$. We define their inverses by restricting to these domains:

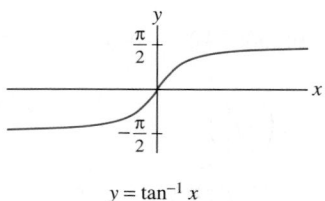

$y = \tan^{-1} x$

$\theta = \tan^{-1} x$ is the unique angle in $\left(-\dfrac{\pi}{2}, \dfrac{\pi}{2}\right)$ such that $\tan \theta = x$

$\theta = \cot^{-1} x$ is the unique angle in $(0, \pi)$ such that $\cot \theta = x$

The range of both $\tan \theta$ and $\cot \theta$ is the set of all real numbers **R**. Therefore, $\theta = \tan^{-1} x$ and $\theta = \cot^{-1} x$ have domain **R** (Figure 17).

The function $f(\theta) = \sec \theta$ is not defined at $\theta = \pi/2$, but we see in Figure 18 that it is one-to-one on both $[0, \pi/2)$ and $(\pi/2, \pi]$. Similarly, $f(\theta) = \csc \theta$ is not defined at $\theta = 0$, but it is one-to-one on $[-\pi/2, 0)$ and $(0, \pi/2]$. We define the inverse functions:

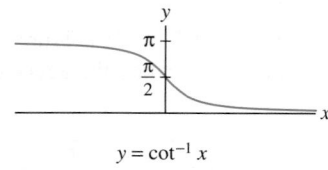

$y = \cot^{-1} x$

FIGURE 17

$$\theta = \sec^{-1} x \text{ is the unique angle in } \left[0, \frac{\pi}{2}\right) \cup \left(\frac{\pi}{2}, \pi\right] \text{ such that } \sec \theta = x$$

$$\theta = \csc^{-1} x \text{ is the unique angle in } \left[-\frac{\pi}{2}, 0\right) \cup \left(0, \frac{\pi}{2}\right] \text{ such that } \csc \theta = x$$

Figure 18 shows that the range of $f(\theta) = \sec \theta$ is the set of real numbers x such that $|x| \geq 1$. The same is true of $f(\theta) = \csc \theta$. It follows that both $\theta = \sec^{-1} x$ and $\theta = \csc^{-1} x$ have domain $\{x : |x| \geq 1\}$.

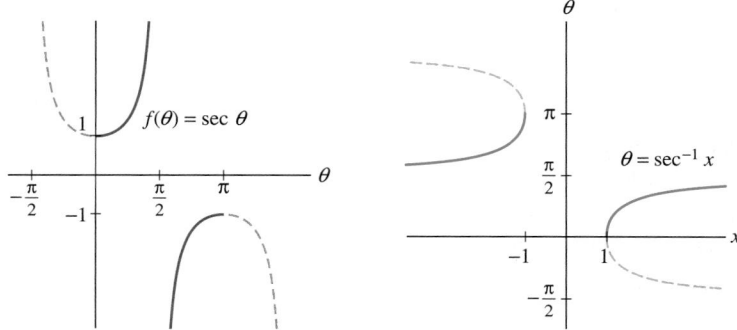

FIGURE 18 $f(\theta) = \sec \theta$ is one-to-one on the interval $[0, \pi]$ (with $\pi/2$ removed).

1.5 SUMMARY

- A function $f(x)$ is *one-to-one* on a domain D if, for all values c, $f(x) = c$ has at most one solution for $x \in D$.
- Let $f(x)$ have domain D and range R. The *inverse* $f^{-1}(x)$ (if it exists) is the function with domain R and range D satisfying $f(f^{-1}(x)) = x$ and $f^{-1}(f(x)) = x$.
- The inverse of $f(x)$ exists if and only if $f(x)$ is one-to-one on its domain.
- To find the inverse of $f(x)$, solve $y = f(x)$ for x in terms of y to obtain $x = g(y)$. The inverse is the function $g(x)$.
- *Horizontal Line Test:* $f(x)$ is one-to-one if and only if every horizontal line intersects the graph of $f(x)$ in at most one point.
- The graph of $f^{-1}(x)$ is obtained by reflecting the graph of $f(x)$ through the line $y = x$.
- The *arcsine* and *arccosine* are defined for $-1 \leq x \leq 1$:

$$\theta = \sin^{-1} x \text{ is the unique angle in } \left[-\frac{\pi}{2}, \frac{\pi}{2}\right] \text{ such that } \sin \theta = x$$

$$\theta = \cos^{-1} x \text{ is the unique angle in } [0, \pi] \text{ such that } \cos \theta = x$$

- $\tan^{-1} x$ and $\cot^{-1} x$ are defined for all x:

$$\theta = \tan^{-1} x \text{ is the unique angle in } \left(-\frac{\pi}{2}, \frac{\pi}{2}\right) \text{ such that } \tan \theta = x$$

$$\theta = \cot^{-1} x \text{ is the unique angle in } (0, \pi) \text{ such that } \cot \theta = x$$

- $\sec^{-1} x$ and $\csc^{-1} x$ are defined for $|x| \geq 1$:

$$\theta = \sec^{-1} x \text{ is the unique angle in } \left[0, \frac{\pi}{2}\right) \cup \left(\frac{\pi}{2}, \pi\right] \text{ such that } \sec \theta = x$$

$$\theta = \csc^{-1} x \text{ is the unique angle in } \left[-\frac{\pi}{2}, 0\right) \cup \left(0, \frac{\pi}{2}\right] \text{ such that } \csc \theta = x$$

1.5 EXERCISES

Preliminary Questions

1. Which of the following satisfy $f^{-1}(x) = f(x)$?

(a) $f(x) = x$ **(b)** $f(x) = 1 - x$

(c) $f(x) = 1$ **(d)** $f(x) = \sqrt{x}$

(e) $f(x) = |x|$ **(f)** $f(x) = x^{-1}$

2. The graph of a function looks like the track of a roller coaster. Is the function one-to-one?

3. Consider the function f that maps teenagers in the United States to their last names. Explain why the inverse of f does not exist.

4. View the following fragment of a train schedule for the New Jersey Transit System as defining a function f from towns to times. Is f one-to-one? What is $f^{-1}(6:27)$?

Trenton	6:21
Hamilton Township	6:27
Princeton Junction	6:34
New Brunswick	6:38

5. A homework problem asks for a sketch of the graph of the *inverse* of $f(x) = x + \cos x$. Frank, after trying but failing to find a formula for $f^{-1}(x)$, says it's impossible to graph the inverse. Bianca hands in an accurate sketch without solving for f^{-1}. How did Bianca complete the problem?

6. Which of the following quantities is undefined?

(a) $\sin^{-1}\left(-\frac{1}{2}\right)$ **(b)** $\cos^{-1}(2)$

(c) $\csc^{-1}\left(\frac{1}{2}\right)$ **(d)** $\csc^{-1}(2)$

7. Give an example of an angle θ such that $\cos^{-1}(\cos \theta) \neq \theta$. Does this contradict the definition of inverse function?

Exercises

1. Show that $f(x) = 7x - 4$ is invertible by finding its inverse.

2. Is $f(x) = x^2 + 2$ one-to-one? If not, describe a domain on which it is one-to-one.

3. What is the largest interval containing zero on which $f(x) = \sin x$ is one-to-one?

4. Show that $f(x) = \dfrac{x-2}{x+3}$ is invertible by finding its inverse.

(a) What is the domain of $f(x)$? The range of $f^{-1}(x)$?

(b) What is the domain of $f^{-1}(x)$? The range of $f(x)$?

5. Verify that $f(x) = x^3 + 3$ and $g(x) = (x - 3)^{1/3}$ are inverses by showing that $f(g(x)) = x$ and $g(f(x)) = x$.

6. Repeat Exercise 5 for $f(t) = \dfrac{t+1}{t-1}$ and $g(t) = \dfrac{t+1}{t-1}$.

7. The escape velocity from a planet of mass M and radius R is $v(R) = \sqrt{\dfrac{2GM}{R}}$, where G is the universal gravitational constant. Find the inverse of $v(R)$ as a function of R.

In Exercises 8–15, find a domain on which f is one-to-one and a formula for the inverse of f restricted to this domain. Sketch the graphs of f and f^{-1}.

8. $f(x) = 3x - 2$ **9.** $f(x) = 4 - x$

10. $f(x) = \dfrac{1}{x+1}$ **11.** $f(x) = \dfrac{1}{7x - 3}$

12. $f(s) = \dfrac{1}{s^2}$ **13.** $f(x) = \dfrac{1}{\sqrt{x^2 + 1}}$

14. $f(z) = z^3$ **15.** $f(x) = \sqrt{x^3 + 9}$

16. For each function shown in Figure 19, sketch the graph of the inverse (restrict the function's domain if necessary).

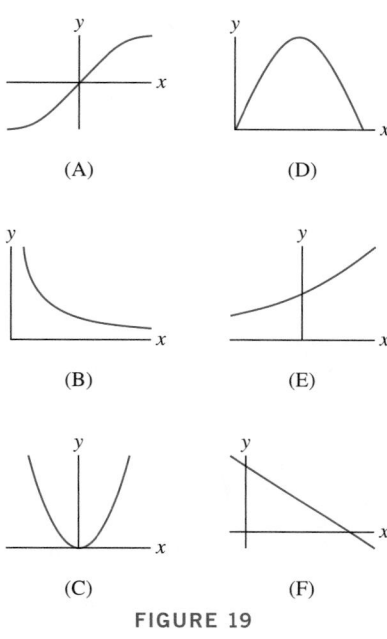

(A)

(D)

(B)

(E)

(C)

(F)

FIGURE 19

17. Which of the graphs in Figure 20 is the graph of a function satisfying $f^{-1} = f$?

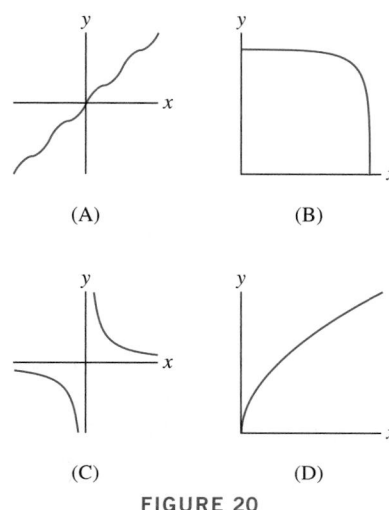

(A)

(B)

(C)

(D)

FIGURE 20

18. Let n be a non-zero integer. Find a domain on which $f(x) = (1 - x^n)^{1/n}$ coincides with its inverse. *Hint:* The answer depends on whether n is even or odd.

19. Let $f(x) = x^7 + x + 1$.

(a) Show that f^{-1} exists (but do not attempt to find it). *Hint:* Show that f is strictly increasing.

(b) What is the domain of f^{-1}?

(c) Find $f^{-1}(3)$.

20. Show that $f(x) = (x^2 + 1)^{-1}$ is one-to-one on $(-\infty, 0]$ and find a formula for f^{-1} for this domain.

21. Let $f(x) = x^2 - 2x$. Determine a domain on which f^{-1} exists and find a formula for f^{-1} on this domain.

22. Show that $f(x) = x + x^{-1}$ is one-to-one on $[1, \infty)$ and find a formula for f^{-1} on this domain. What is the domain of f^{-1}?

In Exercises 23–28, evaluate without using a calculator.

23. $\cos^{-1} 1$ **24.** $\sin^{-1} \frac{1}{2}$

25. $\cot^{-1} 1$ **26.** $\sec^{-1} \frac{2}{\sqrt{3}}$

27. $\tan^{-1} \sqrt{3}$ **28.** $\sin^{-1}(-1)$

In Exercises 29–38, compute without using a calculator.

29. $\sin^{-1}\left(\sin \frac{\pi}{3}\right)$ **30.** $\sin^{-1}\left(\sin \frac{4\pi}{3}\right)$

31. $\cos^{-1}\left(\cos \frac{3\pi}{2}\right)$ **32.** $\sin^{-1}\left(\sin \left(-\frac{5\pi}{6}\right)\right)$

33. $\tan^{-1}\left(\tan \frac{3\pi}{4}\right)$ **34.** $\tan^{-1}(\tan \pi)$

35. $\sec^{-1}(\sec 3\pi)$ **36.** $\sec^{-1}\left(\sec \frac{3\pi}{2}\right)$

37. $\csc^{-1}(\csc(-\pi))$ **38.** $\cot^{-1}\left(\cot \left(-\frac{\pi}{4}\right)\right)$

In Exercises 39–42, simplify by referring to the appropriate triangle or trigonometric identity.

39. $\tan(\cos^{-1} x)$ **40.** $\cos(\tan^{-1} x)$

41. $\cot(\sec^{-1} x)$ **42.** $\cot(\sin^{-1} x)$

In Exercises 43–50, refer to the appropriate triangle or trigonometric identity to compute the given value.

43. $\cos\left(\sin^{-1} \frac{2}{3}\right)$ **44.** $\tan\left(\cos^{-1} \frac{2}{3}\right)$

45. $\tan\left(\sin^{-1} 0.8\right)$ **46.** $\cos\left(\cot^{-1} 1\right)$

47. $\cot\left(\csc^{-1} 2\right)$ **48.** $\tan\left(\sec^{-1}(-2)\right)$

49. $\cot\left(\tan^{-1} 20\right)$ **50.** $\sin\left(\csc^{-1} 20\right)$

Further Insights and Challenges

51. Show that if $f(x)$ is odd and $f^{-1}(x)$ exists, then $f^{-1}(x)$ is odd. Show, on the other hand, that an even function does not have an inverse.

52. A cylindrical tank of radius R and length L lying horizontally as in Figure 21 is filled with oil to height h. Show that the volume $V(h)$ of oil in the tank as a function of height h is

$$V(h) = L \left(R^2 \cos^{-1}\left(1 - \frac{h}{R}\right) - (R - h)\sqrt{2hR - h^2} \right)$$

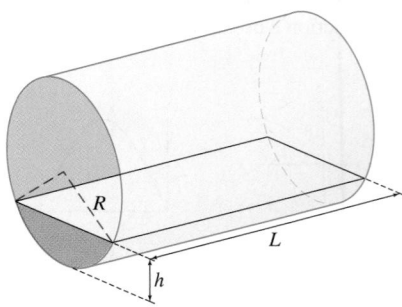

FIGURE 21 Oil in the tank has level h.

1.6 Exponential and Logarithmic Functions

An **exponential function** is a function of the form $f(x) = b^x$, where $b > 0$ and $b \neq 1$. The number b is called the **base**. Some examples are 2^x, $(1.4)^x$, and 10^x. We exclude the case $b = 1$ because $f(x) = 1^x$ is a constant function. Calculators give good decimal approximations to values of exponential functions:

$$2^4 = 16, \quad 2^{-3} = 0.125, \quad (1.4)^3 = 2.744, \quad 10^{4.6} \approx 39{,}810.717$$

Three properties of exponential functions should be singled out from the start (see Figure 2 for the case $b = 2$):

- *Exponential functions are positive*: $b^x > 0$ for all x.
- $f(x) = b^x$ is *increasing* if $b > 1$ and *decreasing* if $0 < b < 1$.
- The *range* of $f(x) = b^x$ is the set of all positive real numbers.

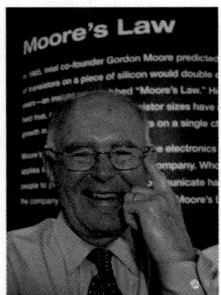

FIGURE 1 Gordon Moore (1929–).

Moore (who later became chairman of Intel Corporation) predicted that in the decades following 1965, the number of transistors per integrated circuit would grow "exponentially." This prediction has held up for nearly four decades and may well continue for several more years. Moore has said, "Moore's Law is a term that got applied to a curve I plotted in the mid-sixties showing the increase in complexity of integrated circuits versus time. It's been expanded to include a lot more than that, and I'm happy to take credit for all of it."

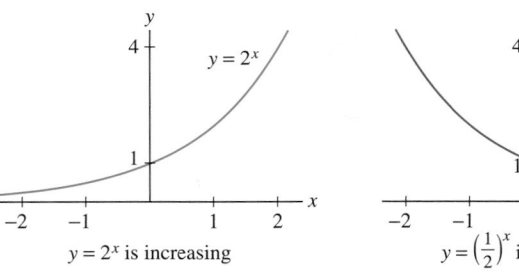

FIGURE 2

Another important feature of $f(x) = b^x$ (for $b > 1$) is that it increases rapidly. Although "rapid increase" is a subjective term, the following precise statement is true: $f(x) = b^x$ increases more rapidly than every polynomial function (see Section 4.7). For example, Figure 3 shows that $f(x) = 3^x$ eventually overtakes and increases faster than the power functions x^3, x^4, and x^5. Table 1 compares 3^x and x^5.

We now review the laws of exponents. The most important law is

$$\boxed{b^x b^y = b^{x+y}}$$

In other words, *under multiplication, the exponents add*, provided that the bases are the same. This law does not apply to a product such as $3^2 \cdot 5^4$.

TABLE 1		
x	x^5	3^x
1	1	3
5	3,125	243
10	100,000	59,049
15	759,375	14,348,907
25	9,765,625	847,288,609,443

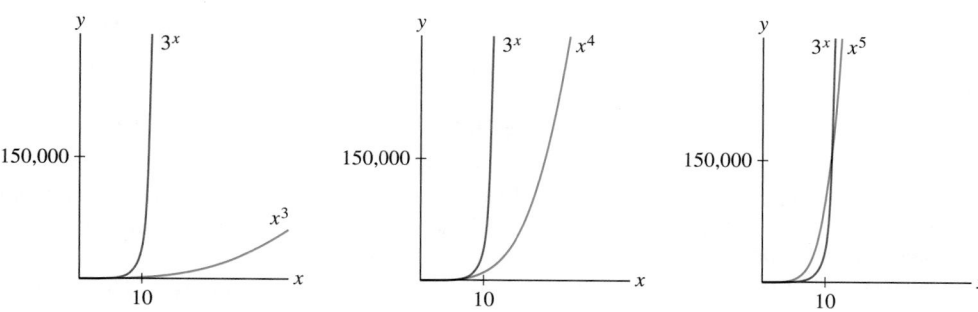

FIGURE 3 Comparison of 3^x and power functions.

THEOREM 1 Laws of Exponents ($b > 0$)

	Rule	Example
Exponent zero	$b^0 = 1$	
Products	$b^x b^y = b^{x+y}$	$2^5 \cdot 2^3 = 2^{5+3} = 2^8$
Quotients	$\dfrac{b^x}{b^y} = b^{x-y}$	$\dfrac{4^7}{4^2} = 4^{7-2} = 4^5$
Negative exponents	$b^{-x} = \dfrac{1}{b^x}$	$3^{-4} = \dfrac{1}{3^4} = \dfrac{1}{81}$
Power to a power	$(b^x)^y = b^{xy}$	$(3^2)^4 = 3^{2(4)} = 3^8$
Roots	$b^{1/n} = \sqrt[n]{b}$	$5^{1/2} = \sqrt{5}$

Be sure you are familiar with the laws of exponents. They are used throughout this text.

■ **EXAMPLE 1** Rewrite as a whole number or fraction (without using a calculator):

(a) $16^{-1/2}$ **(b)** $27^{2/3}$ **(c)** $\dfrac{9^3}{3^7}$ **(d)** $(4^3)^6 \, 4^{-16}$

Solution

(a) $16^{-1/2} = \dfrac{1}{16^{1/2}} = \dfrac{1}{\sqrt{16}} = \dfrac{1}{4}$ **(b)** $27^{2/3} = (27^{1/3})^2 = 3^2 = 9$

(c) $\dfrac{9^3}{3^7} = \dfrac{(3^2)^3}{3^7} = \dfrac{3^6}{3^7} = 3^{-1} = \dfrac{1}{3}$ **(d)** $(4^3)^6 \cdot 4^{-16} = 4^{18} \cdot 4^{-16} = 4^2 = 16$ ■

In the next example, we use the fact that $f(x) = b^x$ is one-to-one. In other words, if $b^x = b^y$, then $x = y$.

■ **EXAMPLE 2** Solve for the unknown:

(a) $2^{3x+1} = 2^5$ **(b)** $b^3 = 5^6$ **(c)** $7^{t+1} = \left(\frac{1}{7}\right)^{2t}$

Solution

(a) If $2^{3x+1} = 2^5$, then $3x + 1 = 5$ and thus $x = \frac{4}{3}$.

(b) Raise both sides of $b^3 = 5^6$ to the $\frac{1}{3}$ power. By the "power to a power" rule, we get

$$b = \left(b^3\right)^{1/3} = \left(5^6\right)^{1/3} = 5^{6/3} = 5^2 = 25$$

(c) Since $\frac{1}{7} = 7^{-1}$, the right-hand side of the equation is $\left(\frac{1}{7}\right)^{2t} = \left(7^{-1}\right)^{2t} = 7^{-2t}$. The equation becomes $7^{t+1} = 7^{-2t}$. Therefore, $t + 1 = -2t$, or $t = -\frac{1}{3}$. ∎

The Number e

Although written references to the number π go back more than 4,000 years, mathematicians first became aware of the special role played by e in the seventeenth century. The notation e was introduced by Leonhard Euler, who discovered many fundamental properties of this important number.

In Chapter 3, we will use calculus to study exponential functions. One of the surprising insights of calculus is that the most convenient or "natural" base for an exponential function is not $b = 10$ or $b = 2$, as one might think at first, but rather a certain irrational number that is denoted by e. The value of e is approximately 2.7. More accurate approximations may be obtained using a calculator or computer algebra system:

$$e \approx 2.718281828459\ldots$$

Similarly, we may use a calculator to evaluate specific values $f(x) = e^x$. For example,

$$e^3 \approx 20.0855, \qquad e^{-1/4} \approx 0.7788$$

How is e defined? There are many different definitions, but they all require the calculus concept of a limit. We shall discuss the definition of e in detail in Section 3.2. For now, we mention two graphical ways of describing e based on Figure 4.

- Using Figure 4(A): Among all exponential functions $y = b^x$, the base $b = e$ is the unique base for which the slope of the tangent line to the graph of $y = e^x$ at the point $(0, 1)$ is equal to 1.
- Using Figure 4(B): The number e is the unique number such that the area of the region under the hyperbola $y = \frac{1}{x}$ for $1 \le x \le e$ is equal to 1.

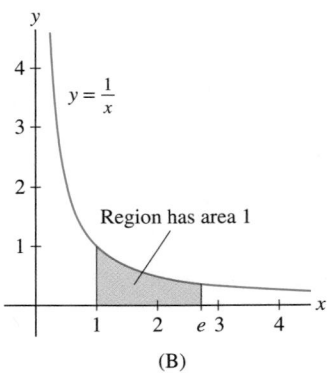

FIGURE 4

We will see how these graphical properties of e arise when we analyze exponential and logarithm functions using calculus. Also, note that the tangent line in Figure 4(A) is the line that touches the graph in one point without crossing the graph. The precise definition of tangent lines also requires calculus and is discussed in Sections 2.1 and 3.1.

Logarithms

Logarithm functions are inverses of exponential functions. More precisely, if $b > 0$ and $b \ne 1$, then the *logarithm to the base b*, denoted $\log_b x$, is the inverse of $f(x) = b^x$. By definition, $y = \log_b x$ if $b^y = x$.

$$b^{\log_b x} = x \qquad \text{and} \qquad \log_b(b^x) = x$$

Thus, $\log_b x$ is the number to which b must be raised in order to get x. For example,

$$\log_2(8) = 3 \quad \text{because} \quad 2^3 = 8$$

$$\log_{10}(1) = 0 \quad \text{because} \quad 10^0 = 1$$

$$\log_3\left(\frac{1}{9}\right) = -2 \quad \text{because} \quad 3^{-2} = \frac{1}{3^2} = \frac{1}{9}$$

The logarithm to the base e, denoted $\ln x$, plays a special role and is called the **natural logarithm**. We use a calculator to evaluate logarithms numerically. For example,

$$\ln 17 \approx 2.83321 \quad \text{because} \quad e^{2.83321} \approx 17$$

Recall that the domain of b^x is **R** and its range is the set of positive real numbers $\{x : x > 0\}$. Since the domain and range are reversed in the inverse function,

- The *domain* of $\log_b x$ is $\{x : x > 0\}$.
- The *range* of $\log_b x$ is the set of all real numbers **R**.

If $b > 1$, then $\log_b x$ is positive for $x > 1$ and negative for $0 < x < 1$. Figure 6 illustrates these facts for the base $b = e$. Note that the logarithm of a negative number does not exist. For example, $\log_{10}(-2)$ does not exist because $10^y > 0$ for all y and thus $10^y = -2$ has no solution.

For each law of exponents, there is a corresponding law for logarithms. The rule $b^{x+y} = b^x b^y$ corresponds to the rule

$$\log_b(xy) = \log_b x + \log_b y$$

In words: *The log of a product is the sum of the logs.* To verify this rule, we write xy as an exponential with base b in two ways:

$$b^{\log_b(xy)} = xy = b^{\log_b x} \cdot b^{\log_b y} = b^{\log_b x + \log_b y}$$

Therefore, the exponents $\log_b(xy)$ and $\log_b x + \log_b y$ are equal as claimed. The remaining logarithm laws are collected in the following table.

FIGURE 5 Renato Solidum, director of the Philippine Institute of Volcanology and Seismology, checks the intensity of the October 8, 2004, Manila earthquake, which registered 6.2 on the Richter scale. The Richter scale is based on the logarithm (to base 10) of the amplitude of seismic waves. Each whole-number increase in Richter magnitude corresponds to a 10-fold increase in amplitude and around 31 times more energy.

In this text, the natural logarithm is denoted $\ln x$. Other common notations are $\log x$ or $\mathrm{Log}\,x$.

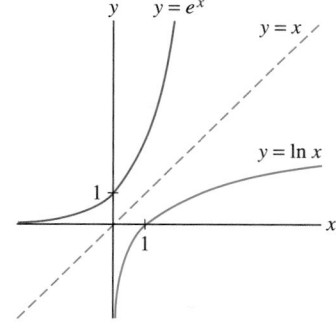

FIGURE 6 $y = \ln x$ is the inverse of $y = e^x$.

Laws of Logarithms

	Law	Example
Log of 1	$\log_b(1) = 0$	
Log of b	$\log_b(b) = 1$	
Products	$\log_b(xy) = \log_b x + \log_b y$	$\log_5(2 \cdot 3) = \log_5 2 + \log_5 3$
Quotients	$\log_b\left(\dfrac{x}{y}\right) = \log_b x - \log_b y$	$\log_2\left(\dfrac{3}{7}\right) = \log_2 3 - \log_2 7$
Reciprocals	$\log_b\left(\dfrac{1}{x}\right) = -\log_b x$	$\log_2 \dfrac{1}{7} = -\log_2 7$
Powers (any n)	$\log_b(x^n) = n \log_b x$	$\log_{10}(8^2) = 2 \cdot \log_{10} 8$

We also note that all logarithm functions are proportional. More precisely, the following **change of base** formula holds (see Exercise 44):

$$\log_b x = \frac{\log_a x}{\log_a b}, \qquad \log_b x = \frac{\ln x}{\ln b}$$

$\boxed{1}$

■ **EXAMPLE 3** Using the Logarithm Laws   Evaluate without using a calculator:

(a) $\log_6 9 + \log_6 4$ (b) $\ln\left(\dfrac{1}{\sqrt{e}}\right)$ (c) $10\log_b(b^3) - 4\log_b(\sqrt{b})$

Solution

(a) $\log_6 9 + \log_6 4 = \log_6(9 \cdot 4) = \log_6(36) = \log_6(6^2) = 2$

(b) $\ln\left(\dfrac{1}{\sqrt{e}}\right) = \ln(e^{-1/2}) = -\dfrac{1}{2}\ln(e) = -\dfrac{1}{2}$

(c) $10\log_b(b^3) - 4\log_b(\sqrt{b}) = 10(3) - 4\log_b(b^{1/2}) = 30 - 4\left(\dfrac{1}{2}\right) = 28$ ■

■ **EXAMPLE 4** Solving an Exponential Equation   The bacteria population in a bottle at time t (in hours) has size $P(t) = 1{,}000e^{0.35t}$. After how many hours will there be 5,000 bacteria?

Solution We must solve $P(t) = 1{,}000e^{0.35t} = 5{,}000$ for t (Figure 7):

$$e^{0.35t} = \frac{5{,}000}{1{,}000} = 5$$

$$\ln(e^{0.35t}) = \ln 5 \qquad \text{(take logarithms of both sides)}$$

$$0.35t = \ln 5 \qquad [\text{because } \ln(e^a) = a]$$

$$t = \frac{\ln 5}{0.35} \approx 4.6 \text{ hours} \qquad ■$$

Bacteria population P

FIGURE 7 Bacteria population as a function of time.

Hyperbolic Functions

The hyperbolic functions are certain special combinations of e^x and e^{-x} that play a role in engineering and physics (see Figure 8 for a real-life example). The hyperbolic sine and cosine, often called "cinch" and "cosh," are defined as follows:

$$\sinh x = \frac{e^x - e^{-x}}{2}, \qquad \cosh x = \frac{e^x + e^{-x}}{2}$$

As the terminology suggests, there are similarities between the hyperbolic and trigonometric functions. Here are some examples:

• **Parity:** The trigonometric functions and their hyperbolic analogs have the same parity. Thus, $\sin x$ and $\sinh x$ are both odd, and $\cos x$ and $\cosh x$ are both even (Figure 9):

$$\sinh(-x) = -\sinh x, \qquad \cosh(-x) = \cosh x$$

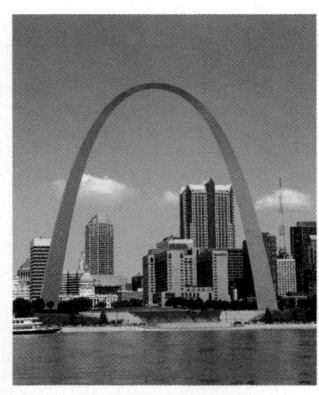

FIGURE 8 The St. Louis Arch has the shape of an inverted hyperbolic cosine.

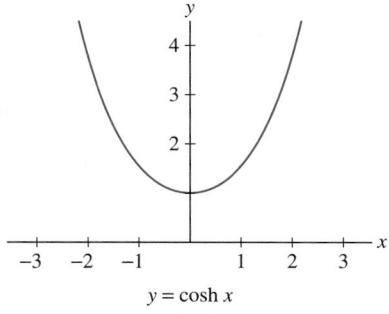

FIGURE 9 $y = \sinh x$ is an odd function, $y = \cosh x$ is an even function.

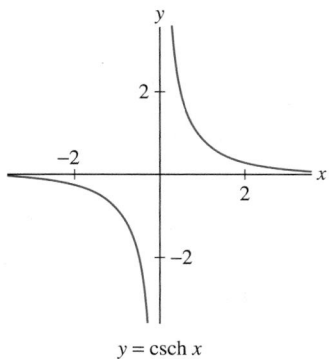

FIGURE 11 The hyperbolic secant and cosecant.

- **Identities:** The basic trigonometric identity $\sin^2 x + \cos^2 x = 1$ has a hyperbolic analog:

$$\boxed{\cosh^2 x - \sinh^2 x = 1}$$ $\boxed{2}$

The addition formulas satisfied by $\sin\theta$ and $\cos\theta$ also have hyperbolic analogs:

$$\sinh(x + y) = \sinh x \cosh y + \cosh x \sinh y$$
$$\cosh(x + y) = \cosh x \cosh y + \sinh x \sinh y$$

■ **EXAMPLE 5** Verifying the Basic Identity   Verify Eq. (2).

Solution It follows from the definitions that

$$\cosh x + \sinh x = e^x, \qquad \cosh x - \sinh x = e^{-x}$$

We obtain Eq. (2) by multiplying these two equations together:

$$\cosh^2 x - \sinh^2 x = (\cosh x + \sinh x)(\cosh x - \sinh x) = e^x \cdot e^{-x} = 1 \qquad ■$$

- **Hyperbola instead of the circle:** The identity $\sin^2 t + \cos^2 t = 1$ tells us that the point $(\cos t, \sin t)$ lies on the unit circle $x^2 + y^2 = 1$. Similarly, the identity $\cosh^2 t - \sinh^2 t = 1$ says that the point $(\cosh t, \sinh t)$ lies on the *hyperbola* $x^2 - y^2 = 1$ (Figure 10).

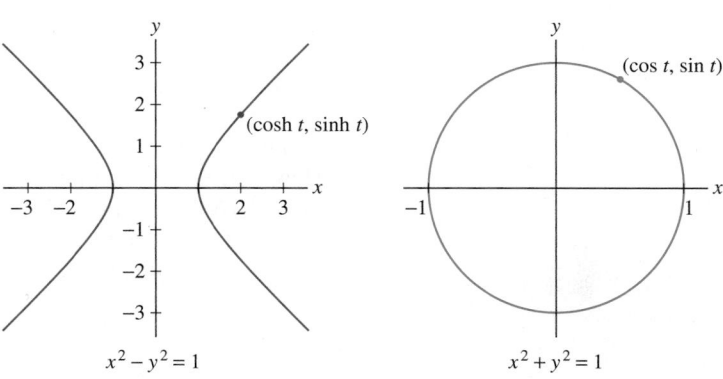

FIGURE 10

- **Other hyperbolic functions:** The hyperbolic tangent, cotangent, secant, and cosecant functions (see Figures 11 and 12) are defined like their trigonometric counterparts:

$$\tanh x = \frac{\sinh x}{\cosh x} = \frac{e^x - e^{-x}}{e^x + e^{-x}}, \qquad \operatorname{sech} x = \frac{1}{\cosh x} = \frac{2}{e^x + e^{-x}}$$

$$\coth x = \frac{\cosh x}{\sinh x} = \frac{e^x + e^{-x}}{e^x - e^{-x}}, \qquad \operatorname{csch} x = \frac{1}{\sinh x} = \frac{2}{e^x - e^{-x}}$$

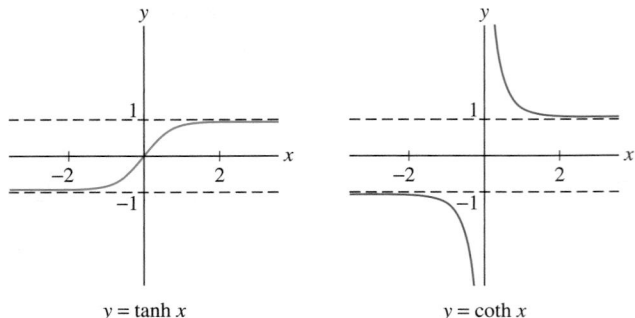

$y = \tanh x$ $y = \coth x$

FIGURE 12 The hyperbolic tangent and cotangent.

Inverse Hyperbolic Functions

Function	**Domain**
$y = \sinh^{-1} x$	all x
$y = \cosh^{-1} x$	$x \geq 1$
$y = \tanh^{-1} x$	$|x| < 1$
$y = \coth^{-1} x$	$|x| > 1$
$y = \operatorname{sech}^{-1} x$	$0 < x \leq 1$
$y = \operatorname{csch}^{-1} x$	$x \neq 0$

Einstein's Law (3) reduces to Galileo's Law, $w = u + v$, when u and v are small relative to the velocity of light c. See Exercise 45 for another way of expressing (3).

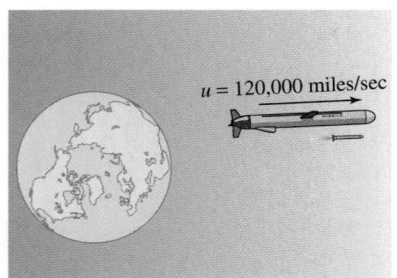

$u = 120{,}000$ miles/sec

FIGURE 13 What is the missile's velocity relative to the earth?

Inverse Hyperbolic Functions

Each of the hyperbolic functions, except $y = \cosh x$ and $y = \operatorname{sech} x$, is one-to-one on its domain and therefore has a well-defined inverse. The functions $y = \cosh x$ and $y = \operatorname{sech} x$ are one-to-one on the restricted domain $\{x : x \geq 0\}$. We let $\cosh^{-1} x$ and $\operatorname{sech}^{-1} x$ denote the corresponding inverses.

Einstein's Law of Velocity Addition

The hyperbolic tangent plays a role in the Special Theory of Relativity, developed by Albert Einstein in 1905. One consequence of this theory is that no object can travel faster than the speed of light, $c \approx 186{,}000$ miles/sec. Einstein realized that this contradicts a law stated by Galileo more than 250 years earlier, namely that *velocities add*. Imagine a train traveling at $u = 100$ mph and a man walking down the aisle in the train at $v = 3$ mph. According to Galileo, the man's velocity relative to the ground is $u + v = 100 + 3$ or 103 mph. This agrees with our everyday experience. But now imagine an (unrealistic) rocket traveling away from the earth at $u = 120{,}000$ miles/sec and suppose that the rocket fires a missile with velocity $v = 80{,}000$ miles/sec (relative to the rocket). If Galileo's Law were correct, the velocity of the missile relative to the earth would be $u + v = 200{,}000$ miles/sec, which exceeds Einstein's maximum speed limit of $c \approx 186{,}000$ miles/sec.

However, Einstein's theory replaces Galileo's Law with a new law stating that the *inverse hyperbolic tangents of the velocities add*. More precisely, if u is the rocket's velocity relative to the earth and v is the missile's velocity relative to the rocket, then the velocity of the missile relative to the earth (Figure 13) is w, where

$$\tanh^{-1}\left(\frac{w}{c}\right) = \tanh^{-1}\left(\frac{u}{c}\right) + \tanh^{-1}\left(\frac{v}{c}\right) \qquad \boxed{3}$$

■ **EXAMPLE 6** A space ship travels away from the earth at a velocity of 100,000 mph. An astronaut in the ship fires a gun that emits an electron which travels at 140,000 miles per second in the same direction as the space ship. Use Einstein's Law to find the velocity w of the electron relative to the earth.

Solution The space ship's velocity is $100{,}000/3600 \approx 28$ miles/sec. According to Einstein's Law,

$$\tanh^{-1}\left(\frac{w}{c}\right) = \tanh^{-1}\left(\frac{28}{186{,}000}\right) + \tanh^{-1}\left(\frac{140{,}000}{186{,}000}\right)$$

$$\approx 0.00015 + 0.97913 = 0.97928$$

Therefore, $\dfrac{w}{c} \approx \tanh(0.97928) \approx 0.75275$, and

$$w \approx 0.75275c \approx 0.75275(186,000) \approx 140,012 \text{ miles/sec}$$

If we had used Galileo's Law, we would have obtained $w = 140,028$ miles/sec. ■

■ **EXAMPLE 7** Low Velocities A plane traveling at 600 mph fires a missile at a velocity of 400 mph. Calculate the missile's velocity w relative to the earth (in mph) using both Einstein's Law and Galileo's Law.

Solution To apply Einstein's law, we convert all velocities to miles/sec:

$$\tanh^{-1}\left(\frac{w/3600}{186,000}\right) = \tanh^{-1}\left(\frac{600/3600}{186,000}\right) + \tanh^{-1}\left(\frac{400/3600}{186,000}\right)$$

$$\approx 1.493428912784 \cdot 10^{-6}$$

We obtain $w \approx (3600)(186,000)\tanh(1.493428912784 \cdot 10^{-6}) \approx 1000 - 6 \cdot 10^{-10}$ mph. This is practically indistinguishable from $w = 600 + 400 = 1000$ mph, obtained using Galileo's Law. ■

1.6 SUMMARY

- $f(x) = b^x$ is the *exponential function* with base b (where $b > 0$ and $b \neq 1$).
- $f(x) = b^x$ is increasing if $b > 1$ and decreasing if $b < 1$.
- The number $e \approx 2.718$.
- For $b > 0$ with $b \neq 1$, the *logarithm function* $\log_b x$ is the inverse of b^x;

$$x = b^y \quad \Leftrightarrow \quad y = \log_b x$$

- The *natural logarithm* is the logarithm to the base e and is denoted $\ln x$.
- Important logarithm laws:

    (i) $\log_b(xy) = \log_b x + \log_b y$ **(ii)** $\log_b\left(\dfrac{x}{y}\right) = \log_b x - \log_b y$

(iii) $\log_b(x^n) = n \log_b x$ **(iv)** $\log_b 1 = 0$ and $\log_b b = 1$

- The *hyperbolic sine and cosine:*

$$\sinh x = \frac{e^x - e^{-x}}{2} \quad \text{(odd function)}, \qquad \cosh x = \frac{e^x + e^{-x}}{2} \quad \text{(even function)}$$

The remaining hyperbolic functions:

$$\tanh x = \frac{\sinh x}{\cosh x}, \qquad \coth x = \frac{\cosh x}{\sinh x}, \qquad \operatorname{sech} x = \frac{1}{\cosh x}, \qquad \operatorname{csch} x = \frac{1}{\sinh x}$$

- Basic identity: $\cosh^2 x - \sinh^2 x = 1$.
- The inverse hyperbolic functions:

$$\sinh^{-1} x, \text{ for all } x \qquad\qquad \coth^{-1} x, \text{ for } |x| > 1$$
$$\cosh^{-1} x, \text{ for } x \geq 1 \qquad\qquad \operatorname{sech}^{-1} x, \text{ for } 0 < x \leq 1$$
$$\tanh^{-1} x, \text{ for } |x| < 1 \qquad\qquad \operatorname{csch}^{-1} x, \text{ for } x \neq 0$$

1.6 EXERCISES

Preliminary Questions

1. Which of the following equations is incorrect?

(a) $3^2 \cdot 3^5 = 3^7$

(b) $(\sqrt{5})^{4/3} = 5^{2/3}$

(c) $3^2 \cdot 2^3 = 1$

(d) $(2^{-2})^{-2} = 16$

2. Compute $\log_{b^2}(b^4)$.

3. When is $\ln x$ negative?

4. What is $\ln(-3)$? Explain.

5. Explain the phrase "the logarithm converts multiplication into addition."

6. What are the domain and range of $\ln x$?

7. Which hyperbolic functions take on only positive values?

8. Which hyperbolic functions are increasing on their domains?

9. Describe three properties of hyperbolic functions that have trigonometric analogs.

Exercises

1. Rewrite as a whole number (without using a calculator):

(a) 7^0

(b) $10^2(2^{-2} + 5^{-2})$

(c) $\dfrac{(4^3)^5}{(4^5)^3}$

(d) $27^{4/3}$

(e) $8^{-1/3} \cdot 8^{5/3}$

(f) $3 \cdot 4^{1/4} - 12 \cdot 2^{-3/2}$

In Exercises 2–10, solve for the unknown variable.

2. $9^{2x} = 9^8$

3. $e^{2x} = e^{x+1}$

4. $e^{t^2} = e^{4t-3}$

5. $3^x = \left(\frac{1}{3}\right)^{x+1}$

6. $(\sqrt{5})^x = 125$

7. $4^{-x} = 2^{x+1}$

8. $b^4 = 10^{12}$

9. $k^{3/2} = 27$

10. $(b^2)^{x+1} = b^{-6}$

In Exercises 11–22, calculate directly (without using a calculator).

11. $\log_3 27$

12. $\log_5 \frac{1}{25}$

13. $\log_2(2^{5/3})$

14. $\log_2(8^{5/3})$

15. $\log_{64} 4$

16. $\log_7(49^2)$

17. $\log_8 2 + \log_4 2$

18. $\log_{25} 30 + \log_{25} \frac{5}{6}$

19. $\log_4 48 - \log_4 12$

20. $\ln(\sqrt{e} \cdot e^{7/5})$

21. $\ln(e^3) + \ln(e^4)$

22. $\log_2 \frac{4}{3} + \log_2 24$

23. Write as the natural log of a single expression:

(a) $2\ln 5 + 3\ln 4$

(b) $5\ln(x^{1/2}) + \ln(9x)$

24. Solve for x: $\ln(x^2 + 1) - 3\ln x = \ln(2)$.

In Exercises 25–30, solve for the unknown.

25. $7e^{5t} = 100$

26. $6e^{-4t} = 2$

27. $2^{x^2 - 2x} = 8$

28. $e^{2t+1} = 9e^{1-t}$

29. $\ln(x^4) - \ln(x^2) = 2$

30. $\log_3 y + 3\log_3(y^2) = 14$

31. Use a calculator to compute $\sinh x$ and $\cosh x$ for $x = -3, 0, 5$.

32. Compute $\sinh(\ln 5)$ and $\tanh(3\ln 5)$ without using a calculator.

33. For which values of x are $y = \sinh x$ and $y = \cosh x$ increasing and decreasing?

34. Show that $y = \tanh x$ is an odd function.

35. The population of a city (in millions) at time t (years) is $P(t) = 2.4e^{0.06t}$, where $t = 0$ is the year 2000. When will the population double from its size at $t = 0$?

36. According to the Gutenberg–Richter law, the number N of earthquakes worldwide of Richter magnitude M approximately satisfies a relation $\ln N = 16.17 - bM$ for some constant b. Find b, assuming that there are 800 earthquakes of magnitude $M = 5$ per year. How many earthquakes of magnitude $M = 7$ occur per year?

37. The energy E (in joules) radiated as seismic waves from an earthquake of Richter magnitude M is given by the formula $\log_{10} E = 4.8 + 1.5M$.

(a) Express E as a function of M.

(b) Show that when M increases by 1, the energy increases by a factor of approximately 31.

38. ✏️🖊 Refer to the graphs to explain why the equation $\sinh x = t$ has a unique solution for every t and $\cosh x = t$ has two solutions for every $t > 1$.

39. Compute $\cosh x$ and $\tanh x$, assuming that $\sinh x = 0.8$.

40. Prove the addition formula for $\cosh x$.

41. Use the addition formulas to prove

$$\sinh(2x) = 2\cosh x \sinh x$$

$$\cosh(2x) = \cosh^2 x + \sinh^2 x$$

42. An (imaginary) train moves along a track at velocity v, and a woman walks down the aisle of the train with velocity u in the direction of the train's motion. Compute the velocity w of the woman relative to the ground using the laws of both Galileo and Einstein in the following cases.

(a) $v = 100$ mph and $u = 5$ mph. Is your calculator accurate enough to detect the difference between the two laws?

(b) $v = 140,000$ mph and $u = 100$ mph.

Further Insights and Challenges

43. Show that $\log_a b \log_b a = 1$.

44. Verify the formula $\log_b x = \dfrac{\log_a x}{\log_a b}$ for $a, b > 0$.

45. (a) Use the addition formulas for $\sinh x$ and $\cosh x$ to prove

$$\tanh(u + v) = \frac{\tanh u + \tanh v}{1 + \tanh u \tanh v}$$

(b) Use (a) to show that Einstein's Law of Velocity Addition [Eq. (3)] is equivalent to

$$w = \frac{u + v}{1 + \dfrac{uv}{c^2}}$$

46. Prove that every function $f(x)$ is the sum of an even function $f_+(x)$ and an odd function $f_-(x)$. [*Hint:* $f_\pm(x) = \frac{1}{2}(f(x) \pm f(-x))$.] Express $f(x) = 5e^x + 8e^{-x}$ in terms of $\cosh x$ and $\sinh x$.

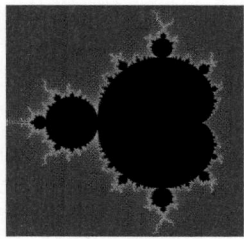

FIGURE 1 Computer-generated image of the Mandelbrot Set, which occurs in the mathematical theory of chaos and fractals.

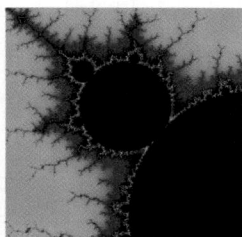

FIGURE 2 Even greater complexity is revealed when we zoom in on a portion of the Mandelbrot Set.

1.7 Technology: Calculators and Computers

Computer technology has vastly extended our ability to calculate and visualize mathematical relationships. In applied settings, computers are indispensable for solving complex systems of equations and analyzing data, as in weather prediction and medical imaging. Mathematicians use computers to study complex structures such as the Mandelbrot set (Figures 1 and 2). We take advantage of this technology to explore the ideas of calculus visually and numerically.

When we plot a function with a graphing calculator, the graph is contained within a **viewing rectangle**, the area determined by the range of x and y values in the plot. We write $[a, b] \times [c, d]$ to denote the rectangle where $a \le x \le b$ and $c \le y \le d$.

The appearance of the graph depends heavily on the choice of viewing rectangle. Different choices may convey very different impressions, which are sometimes misleading. Compare the three viewing rectangles for the graph of $f(x) = 12 - x - x^2$ in Figure 3. Only (A) successfully displays the shape of the graph as a parabola. In (B), the graph is cut off, and no graph at all appears in (C). Keep in mind that the scales along the axes may change with the viewing rectangle. For example, the unit increment along the y-axis is larger in (B) than (A), and therefore, the graph in (B) is steeper.

There is no single "correct" viewing rectangle. The goal is to select the viewing rectangle that displays the properties you wish to investigate. This usually requires experimentation.

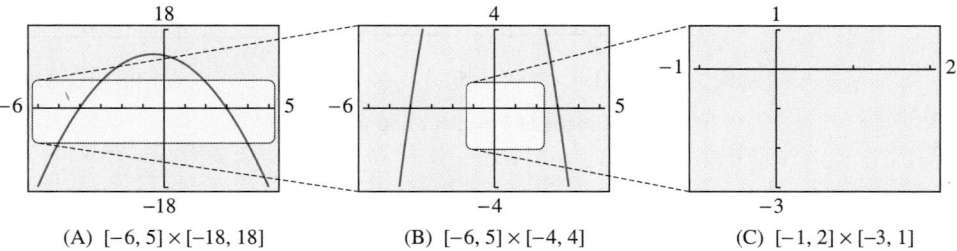

FIGURE 3 Viewing rectangles for the graph of $f(x) = 12 - x - x^2$.

(A) $[-6, 5] \times [-18, 18]$ (B) $[-6, 5] \times [-4, 4]$ (C) $[-1, 2] \times [-3, 1]$

■ **EXAMPLE 1 How Many Roots and Where?** How many real roots does the function $f(x) = x^9 - 20x + 1$ have? Find their approximate locations.

Solution We experiment with several viewing rectangles (Figure 4). Our first attempt (A) displays a cut off graph, so we try a viewing rectangle that includes a larger range of y values. Plot (B) shows that the roots of $f(x)$ lie somewhere in the interval $[-3, 3]$, but it does not reveal how many real roots there are. Therefore, we try the viewing rectangle in (C). Now we can see clearly that $f(x)$ has three roots. A further zoom in (D) shows

(A) [−12, 12] × [−10, 10]

(B) [−12, 12] × [−10,000, 10,000]

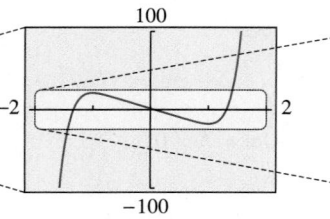
(C) [−2, 2] × [−100, 100]

(D) [−2, 2] × [−20, 25]

FIGURE 4 Graphs of $f(x) = x^9 - 20x + 1$.

that these roots are located near $-1.5, 0.1$, and 1.5. Further zooming would provide their locations with greater accuracy. ■

■ **EXAMPLE 2** **Does a Solution Exist?** Does $\cos x = \tan x$ have a solution? Describe the set of all solutions.

Solution The solutions of $\cos x = \tan x$ are the x-coordinates of the points where the graphs of $y = \cos x$ and $y = \tan x$ intersect. Figure 5(A) shows that there are two solutions in the interval $[0, 2\pi]$. By zooming in on the graph as in (B), we see that the first positive root lies between 0.6 and 0.7 and the second positive root lies between 2.4 and 2.5. Further zooming shows that the first root is approximately 0.67 [Figure 5(C)]. Continuing this process, we find that the first two roots are $x \approx 0.666$ and $x \approx 2.475$.

Since $\cos x$ and $\tan x$ are periodic, the picture repeats itself with period 2π. All solutions are obtained by adding multiples of 2π to the two solutions in $[0, 2\pi]$:

$$x \approx 0.666 + 2\pi k \qquad \text{and} \qquad x \approx 2.475 + 2\pi k \quad \text{(for any integer } k)$$ ■

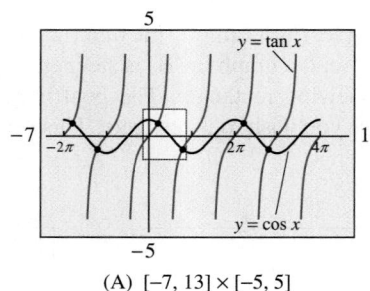
(A) [−7, 13] × [−5, 5]

(B) [0, 3] × [−5, 5]

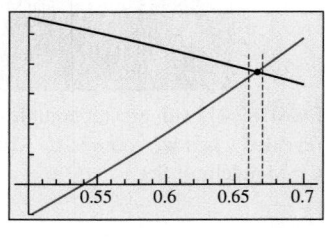
(C) [0.5, 0.7] × [0.55, 0.85]

FIGURE 5 Graphs of $y = \cos x$ and $y = \tan x$.

■ **EXAMPLE 3** **Functions with Asymptotes** Plot the function $f(x) = \dfrac{1 - 3x}{x - 2}$ and describe its asymptotic behavior.

Solution First, we plot $f(x)$ in the viewing rectangle $[-10, 20] \times [-5, 5]$ as in Figure 6(A). The vertical line $x = 2$ is called a **vertical asymptote**. Many graphing calculators display this line, but it is *not* part of the graph (and it can usually be eliminated by choosing a smaller range of y-values). We see that $f(x)$ tends to ∞ as x approaches 2 from the left, and to $-\infty$ as x approaches 2 from the right. To display the horizontal asymptotic behavior of $f(x)$, we use the viewing rectangle $[-10, 20] \times [-10, 5]$ [Figure 6(B)]. Here we see that the graph approaches the horizontal line $y = -3$, called a **horizontal asymptote** (which we have added as a dashed horizontal line in the figure). ■

(A) [−10, 20] × [−5, 5]   (B) [−10, 20] × [−10, 5]

FIGURE 6 Graphs of $f(x) = \dfrac{1 - 3x}{x - 2}$.

Calculators give us the freedom to experiment numerically. For instance, we can explore the behavior of a function by constructing a table of values. In the next example, we investigate a function related to exponential functions and compound interest (see Section 7.5).

We will prove in Section 5.8 that $f(n)$ approaches the number e as n tends to infinity. This provides yet another definition of e as the limit *of the function $f(n)$.*

■ **EXAMPLE 4** Investigating the Behavior of a Function How does $f(n) = \left(1 + \dfrac{1}{n}\right)^n$ behave for large whole number values of n? Does $f(n)$ tend to infinity as n gets larger?

Solution First, we make a table of values of $f(n)$ for larger and larger values of n. Table 1 suggests that $f(n)$ does not tend to infinity. Rather, $f(n)$ appears to get closer to some value near 2.718 as n grows larger. This is an example of limiting behavior that we will discuss in Chapter 2. Next, replace n by the variable x and plot the function $f(x) = \left(1 + \dfrac{1}{x}\right)^x$. The graphs in Figure 7 confirm that $f(x)$ approaches a limit of approximately 2.7. ■

TABLE 1

n	$\left(1 + \dfrac{1}{n}\right)^n$
10	2.59374
10^2	2.70481
10^3	2.71692
10^4	2.71815
10^5	2.71827
10^6	2.71828

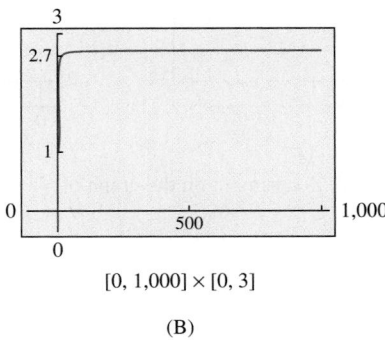

[0, 10] × [0, 3]   [0, 1,000] × [0, 3]

(A)   (B)

FIGURE 7 Graphs of $f(x) = \left(1 + \dfrac{1}{x}\right)^x$.

■ **EXAMPLE 5** Bird Flight: Finding a Minimum Graphically According to one model of bird flight, the power consumed by a pigeon flying at velocity v (in meters per second) is $P(v) = 17v^{-1} + 10^{-3}v^3$ (in joules per second). Use a graph of $P(v)$ to find the velocity that minimizes power consumption.

Solution The velocity that minimizes power consumption corresponds to the lowest point on the graph of $P(v)$. We first plot $P(v)$ in a large viewing rectangle (Figure 8). This figure reveals the general shape of the graph, and we see that $P(v)$ takes on a minimum value for v somewhere between $v = 8$ and $v = 9$. In the viewing rectangle $[8, 9.2] \times [2.6, 2.65]$, we see that the minimum occurs at approximately $v = 8.65$ m/s. ■

FIGURE 8 Power consumption $P(v)$ as a function of velocity v.

(A) $[0, 20] \times [0, 12]$

(B) $[8, 9.2] \times [2.6, 2.65]$

An important concept in calculus is **local linearity**, the idea that many functions are *nearly linear* over small intervals. Local linearity can be effectively illustrated with a graphing calculator.

▮ **EXAMPLE 6** Illustrating Local Linearity Illustrate local linearity for the function $f(x) = x^{\sin x}$ at $x = 1$.

Solution First, we plot $f(x) = x^{\sin x}$ in the viewing window of Figure 9(A). The graph moves up and down and appears very wavy. However, as we zoom in, the graph straightens out. Figures (B)–(D) show the result of zooming in on the point $(1, f(1))$. When viewed up close, the graph looks like a straight line. This illustrates local linearity for the graph of $f(x)$ at $x = 1$.   ▮

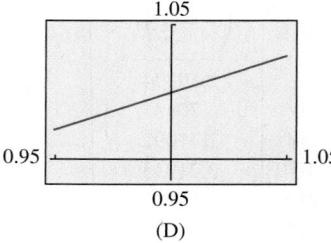

(A)

(B)

(C)

(D)

FIGURE 9 Zooming in on the graph of $f(x) = x^{\sin x}$ near $x = 1$.

1.7 SUMMARY

• The appearance of a graph on a graphing calculator depends on the choice of viewing rectangle. Experiment with different viewing rectangles until you find one that displays the information you want. Keep in mind that the scales along the axes may change as you vary the viewing rectangle.
• The following are some ways in which graphing calculators and computer algebra systems can be used in calculus:

 – Visualizing the behavior of a function
 – Finding solutions graphically or numerically
 – Conducting numerical or graphical experiments
 – Illustrating theoretical ideas (such as local linearity)

1.7 EXERCISES

Preliminary Questions

1. Is there a definite way of choosing the optimal viewing rectangle, or is it best to experiment until you find a viewing rectangle appropriate to the problem at hand?

2. Describe the calculator screen produced when the function $y = 3 + x^2$ is plotted with viewing rectangle:
(a) $[-1, 1] \times [0, 2]$ **(b)** $[0, 1] \times [0, 4]$

3. According to the evidence in Example 4, it appears that $f(n) = (1 + 1/n)^n$ never takes on a value greater than 3 for $n \geq 0$. Does this evidence *prove* that $f(n) \leq 3$ for $n \geq 0$?

4. How can a graphing calculator be used to find the minimum value of a function?

Exercises

The exercises in this section should be done using a graphing calculator or computer algebra system.

1. Plot $f(x) = 2x^4 + 3x^3 - 14x^2 - 9x + 18$ in the appropriate viewing rectangles and determine its roots.

2. How many solutions does $x^3 - 4x + 8 = 0$ have?

3. How many *positive* solutions does $x^3 - 12x + 8 = 0$ have?

4. Does $\cos x + x = 0$ have a solution? Does it have a positive solution?

5. Find all the solutions of $\sin x = \sqrt{x}$ for $x > 0$.

6. How many solutions does $\cos x = x^2$ have?

7. Let $f(x) = (x - 100)^2 + 1{,}000$. What will the display show if you graph $f(x)$ in the viewing rectangle $[-10, 10]$ by $[-10, -10]$? What would be an appropriate viewing rectangle?

8. Plot the graph of $f(x) = \dfrac{8x + 1}{8x - 4}$ in an appropriate viewing rectangle. What are the vertical and horizontal asymptotes (see Example 3)?

9. Plot the graph of $f(x) = \dfrac{x}{4 - x}$ in a viewing rectangle that clearly displays the vertical and horizontal asymptotes.

10. Illustrate local linearity for $f(x) = x^2$ by zooming in on the graph at $x = 0.5$ (see Example 6).

11. Plot $f(x) = \cos(x^2) \sin x$ for $0 \leq x \leq 2\pi$. Then illustrate local linearity at $x = 3.8$ by choosing appropriate viewing rectangles.

12. A bank pays $r = 5\%$ interest compounded monthly. If you deposit P_0 dollars at time $t = 0$, then the value of your account after N months is $P_0 \left(1 + \dfrac{0.05}{12}\right)^N$. Find to the nearest integer N the number of months after which the account value doubles.

In Exercises 13–18, investigate the behavior of the function as n or x grows large by making a table of function values and plotting a graph (see Example 4). Describe the behavior in words.

13. $f(n) = n^{1/n}$

14. $f(n) = \dfrac{4n + 1}{6n - 5}$

15. $f(n) = \left(1 + \dfrac{1}{n}\right)^{n^2}$

16. $f(x) = \left(\dfrac{x + 6}{x - 4}\right)^x$

17. $f(x) = \left(x \tan \dfrac{1}{x}\right)^x$

18. $f(x) = \left(x \tan \dfrac{1}{x}\right)^{x^2}$

19. The graph of $f(\theta) = A \cos \theta + B \sin \theta$ is a sinusoidal wave for any constants A and B. Confirm this for $(A, B) = (1, 1)$, $(1, 2)$, and $(3, 4)$ by plotting $f(\theta)$.

20. Find the maximum value of $f(\theta)$ for the graphs produced in Exercise 19. Can you guess the formula for the maximum value in terms of A and B?

21. Find the intervals on which $f(x) = x(x + 2)(x - 3)$ is positive by plotting a graph.

22. Find the set of solutions to the inequality $(x^2 - 4)(x^2 - 1) < 0$ by plotting a graph.

Further Insights and Challenges

23. CAS Let $f_1(x) = x$ and define a sequence of functions by $f_{n+1}(x) = \frac{1}{2}(f_n(x) + x/f_n(x))$. For example, $f_2(x) = \frac{1}{2}(x + 1)$. Use a computer algebra system to compute $f_n(x)$ for $n = 3, 4, 5$ and plot $f_n(x)$ together with $\sqrt{x}$ for $x \geq 0$. What do you notice?

24. Set $P_0(x) = 1$ and $P_1(x) = x$. The **Chebyshev polynomials** (useful in approximation theory) are defined successively by the formula $P_{n+1}(x) = 2x P_n(x) - P_{n-1}(x)$.

(a) Show that $P_2(x) = 2x^2 - 1$.

(b) Compute $P_n(x)$ for $3 \leq n \leq 6$ using a computer algebra system or by hand, and plot each $P_n(x)$ over the interval $[-1, 1]$.

(c) Check that your plots confirm two interesting properties of Chebyshev polynomials: (a) $P_n(x)$ has n real roots in $[-1, 1]$ and (b) for $x \in [-1, 1]$, $P_n(x)$ lies between -1 and 1.

CHAPTER REVIEW EXERCISES

1. Express $(4, 10)$ as a set $\{x : |x - a| < c\}$ for suitable a and c.

2. Express as an interval:
(a) $\{x : |x - 5| < 4\}$ (b) $\{x : |5x + 3| \leq 2\}$

3. Express $\{x : 2 \leq |x - 1| \leq 6\}$ as a union of two intervals.

4. Give an example of numbers x, y such that $|x| + |y| = x - y$.

5. Describe the pairs of numbers x, y such that $|x + y| = x - y$.

6. Sketch the graph of $y = f(x + 2) - 1$, where $f(x) = x^2$ for $-2 \leq x \leq 2$.

In Exercises 7–10, let $f(x)$ be the function whose graph is shown in Figure 1.

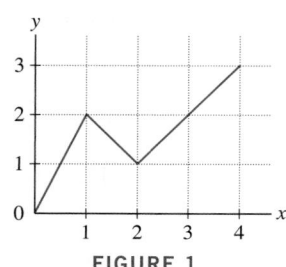

FIGURE 1

7. Sketch the graphs of $y = f(x) + 2$ and $y = f(x + 2)$.

8. Sketch the graphs of $y = \frac{1}{2}f(x)$ and $y = f(\frac{1}{2}x)$.

9. Continue the graph of $f(x)$ to the interval $[-4, 4]$ as an even function.

10. Continue the graph of $f(x)$ to the interval $[-4, 4]$ as an odd function.

In Exercises 11–14, find the domain and range of the function.

11. $f(x) = \sqrt{x + 1}$ **12.** $f(x) = \dfrac{4}{x^4 + 1}$

13. $f(x) = \dfrac{2}{3 - x}$ **14.** $f(x) = \sqrt{x^2 - x + 5}$

15. Determine whether the function is increasing, decreasing, or neither:
(a) $f(x) = 3^{-x}$ (b) $f(x) = \dfrac{1}{x^2 + 1}$
(c) $g(t) = t^2 + t$ (d) $g(t) = t^3 + t$

16. Determine whether the function is even, odd, or neither:
(a) $f(x) = x^4 - 3x^2$
(b) $g(x) = \sin(x + 1)$
(c) $f(x) = 2^{-x^2}$

In Exercises 17–22, find the equation of the line.

17. Line passing through $(-1, 4)$ and $(2, 6)$

18. Line passing through $(-1, 4)$ and $(-1, 6)$

19. Line of slope 6 through $(9, 1)$

20. Line of slope $-\dfrac{3}{2}$ through $(4, -12)$

21. Line through $(2, 3)$ parallel to $y = 4 - x$

22. Horizontal line through $(-3, 5)$

23. Does the following table of market data suggest a linear relationship between price and number of homes sold during a one-year period? Explain.

Price (thousands of $)	180	195	220	240
No. of homes sold	127	118	103	91

24. Does the following table of yearly revenue data for a computer manufacturer suggest a linear relation between revenue and time? Explain.

Year	1995	1999	2001	2004
Revenue (billions of $)	13	18	15	11

25. Find the roots of $f(x) = x^4 - 4x^2$ and sketch its graph. On which intervals is $f(x)$ decreasing?

26. Let $h(z) = 2z^2 + 12z + 3$. Complete the square and find the minimum value of $h(z)$.

27. Let $f(x)$ be the square of the distance from the point $(2, 1)$ to a point $(x, 3x + 2)$ on the line $y = 3x + 2$. Show that $f(x)$ is a quadratic function and find its minimum value by completing the square.

28. Prove that $x^2 + 3x + 3 \geq 0$ for all x.

In Exercises 29–34, sketch the graph by hand.

29. $y = t^4$ **30.** $y = t^5$

31. $y = \sin \dfrac{\theta}{2}$ **32.** $y = 10^{-x}$

33. $y = x^{1/3}$ **34.** $y = \dfrac{1}{x^2}$

35. Show that the graph of $y = f(\frac{1}{3}x - b)$ is obtained by shifting the graph of $y = f(\frac{1}{3}x)$ to the right $3b$ units. Use this observation to sketch the graph of $y = |\frac{1}{3}x - 4|$.

36. Let $h(x) = \cos x$ and $g(x) = x^{-1}$. Compute the composite functions $h(g(x))$ and $g(h(x))$, and find their domains.

37. Find functions f and g such that the function

$$f(g(t)) = (12t + 9)^4$$

38. Sketch the points on the unit circle corresponding to the following three angles and find the values of the six standard trigonometric functions at each angle:

(a) $\dfrac{2\pi}{3}$ **(b)** $\dfrac{7\pi}{4}$ **(c)** $\dfrac{7\pi}{6}$

39. What is the period of the function $g(\theta) = \sin 2\theta + \sin \dfrac{\theta}{2}$?

40. Assume that $\sin \theta = \dfrac{4}{5}$, where $\dfrac{\pi}{2} < \theta < \pi$. Find:

(a) $\tan \theta$ **(b)** $\sin 2\theta$ **(c)** $\csc \dfrac{\theta}{2}$

41. Give an example of values a, b such that

(a) $\cos(a + b) \neq \cos a + \cos b$ **(b)** $\cos \dfrac{a}{2} \neq \dfrac{\cos a}{2}$

42. Let $f(x) = \cos x$. Sketch the graph of $y = 2f\left(\dfrac{1}{3}x - \dfrac{\pi}{4}\right)$ for $0 \leq x \leq 6\pi$.

43. Solve $\sin 2x + \cos x = 0$ for $0 \leq x < 2\pi$.

44. How does $h(n) = \dfrac{n^2}{2^n}$ behave for large whole number values of n? Does $h(n)$ tend to infinity?

45. $\boxed{\text{GU}}$ Use a graphing calculator to determine whether the equation $\cos x = 5x^2 - 8x^4$ has any solutions.

46. $\boxed{\text{GU}}$ Using a graphing calculator, find the number of real roots and estimate the largest root to two decimal places:
(a) $f(x) = 1.8x^4 - x^5 - x$
(b) $g(x) = 1.7x^4 - x^5 - x$

47. Match each quantity (a)–(d) with (i), (ii), or (iii) if possible, or state that no match exists.

(a) $2^a 3^b$ **(b)** $\dfrac{2^a}{3^b}$

(c) $(2^a)^b$ **(d)** $2^{a-b} 3^{b-a}$

(i) 2^{ab} **(ii)** 6^{a+b} **(iii)** $\left(\dfrac{2}{3}\right)^{a-b}$

48. Match each quantity (a)–(d) with (i), (ii), or (iii) if possible, or state that no match exists.

(a) $\ln\left(\dfrac{a}{b}\right)$ **(b)** $\dfrac{\ln a}{\ln b}$

(c) $e^{\ln a - \ln b}$ **(d)** $(\ln a)(\ln b)$

(i) $\ln a + \ln b$ **(ii)** $\ln a - \ln b$ **(iii)** $\dfrac{a}{b}$

49. Find the inverse of $f(x) = \sqrt{x^3 - 8}$ and determine its domain and range.

50. Find the inverse of $f(x) = \dfrac{x - 2}{x - 1}$ and determine its domain and range.

51. Find a domain on which $h(t) = (t - 3)^2$ is one-to-one and determine the inverse on this domain.

52. Show that $g(x) = \dfrac{x}{x + 1}$ is equal to its inverse on the domain $\{x : x \neq -1\}$.

53. Suppose that $g(x)$ is the inverse of $f(x)$. Match the functions (a)–(d) with their inverses (i)–(iv).

(a) $f(x) + 1$ **(b)** $f(x + 1)$
(c) $4f(x)$ **(d)** $f(4x)$

(i) $g(x)/4$ **(ii)** $g(x/4)$
(iii) $g(x - 1)$ **(iv)** $g(x) - 1$

54. $\boxed{\text{GU}}$ Plot $f(x) = xe^{-x}$ and use the zoom feature to find two solutions of $f(x) = 0.5$.

2 | LIMITS

Calculus is usually divided into two branches, differential and integral, partly for historical reasons. The subject was developed in the seventeenth century to solve two important geometric problems: finding tangent lines to curves (differential calculus) and computing areas under curves (integral calculus). However, calculus is a very broad subject with no clear boundaries. It includes other topics, such as the theory of infinite series, and it has an extraordinarily wide range of applications, especially in the sciences and in engineering. What makes these methods and applications part of calculus is that they all rely ultimately on the concept of a limit. We will see throughout the text how limits allow us to make computations and solve problems that cannot be solved using algebra alone.

This chapter introduces the limit concept and focuses on those aspects of limits that will be used in Chapter 3 to develop differential calculus. The first section, intended as motivation, discusses how limits arise in the study of rates of change and tangent lines.

Computer simulation of limiting behavior in a fluid flow with fractal properties.

2.1 Limits, Rates of Change, and Tangent Lines

Rates of change play a role whenever we study the relationship between two changing quantities. Velocity is a familiar example (the rate of change of position with respect to time), but there are many others, such as:

- The infection rate of an epidemic (*newly infected individuals per month*)
- Mileage of an automobile (*miles per gallon*)
- Rate of change of atmospheric temperature with respect to altitude

Roughly speaking, if y and x are related quantities, the rate of change should tell us how much y changes in response to a unit change in x. However, this has a clear meaning only if the rate of change is constant. If a car travels at a constant velocity of 40 mph, then its position changes by 40 miles for each unit change in time (the unit being 1 hour). If the trip lasts only half an hour, its position changes by 20 miles, and in general, the change in position is $40t$ miles, where t is the length of time in hours. In other words,

$$\boxed{\text{Change in position} = \text{velocity} \times \text{time}}$$

This simple formula is no longer valid if the velocity is changing and, in fact, it is not even meaningful. Indeed, if the automobile is accelerating or decelerating, which velocity would we use in the formula? It is possible to compute the change in position when velocity is changing, but this involves both differential and integral calculus. Differential calculus uses the limit concept to define the *instantaneous velocity* of an object, and integral calculus tells us how to compute the change in position in terms of instantaneous velocity. This is just one example of how calculus provides the concepts and tools for modeling real-world phenomena.

In this section, we discuss velocity and other rates of change, emphasizing the graphical interpretation in terms of *tangent lines*. At this stage, we cannot define tangent lines precisely. This will have to wait until Chapter 3. In the meantime, you can think of a tangent line as a line that *skims* a curve at a point as in Figures 1(A) and (B) (but not (C)).

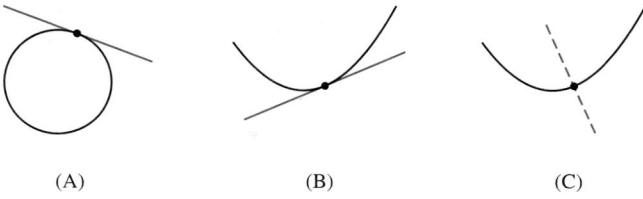

(A)   (B)   (C)

FIGURE 1 The line is tangent in (A) and (B) but not in (C).

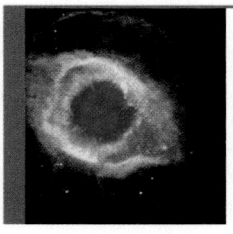
The scientific revolution of the sixteenth and seventeenth centuries reached its high point in the work of Isaac Newton (1643–1727), who was the first scientist to show that the physical world, despite its complexity and diversity, is governed by a small number of universal laws, such as the laws of motion and gravitation. One of Newton's great insights was that the universal laws do not describe the world as it is, now or at any particular time, but rather how the world *changes over time* in response to forces.

To express the laws in mathematical form, Newton invented calculus, which he understood to be the mathematics of change.

More than fifty years before the work of Newton, the astronomer Johannes Kepler (1571–1630) discovered his three laws of planetary motion, including the law stating that the path of a planet around the sun is an ellipse. Kepler found these laws through a painstaking analysis of astronomical data, but he could not explain why they were true. Newton's laws explain the motion of any object—planet or pebble—in terms of the forces acting on it. According to Newton, the planets, if left undisturbed, will travel in straight lines. Since their paths are elliptical, some force—in this case, the gravitational force of the sun—must be acting to make them change direction continuously. In a show of unmatched mathematical skill, Newton proved that Kepler's laws follow from Newton's own inverse square law of gravity. Ironically, the proof Newton chose to publish in 1687 in his magnum opus *Principia Mathematica* uses geometrical arguments rather than the calculus Newton himself had invented.

Velocity

Consider an object traveling in a straight line (linear motion). When we speak of its velocity, we usually mean the *instantaneous* velocity, which tells us the speed and direction of the object at a particular moment. As suggested above, however, instantaneous velocity does not have a straightforward definition. It is simpler to define the **average velocity** over a given time interval as the ratio

$$\text{Average velocity} = \frac{\text{change in position}}{\text{length of time interval}}$$

For example, if a car travels 200 miles in 4 hours, then its average velocity during this 4-hour period is $\frac{200}{4} = 50$ mph. At any given moment the car may be going faster or slower than the average.

Unlike average velocity, we cannot define instantaneous velocity as a ratio because we would have to divide by the length of the time interval (which is zero). However,

The terms "speed" and "velocity," though sometimes used interchangeably, do not have the same meaning. For linear motion, velocity may be positive or negative (indicating the direction of motion). Speed is the absolute value of velocity and is always positive.

we should be able to estimate instantaneous velocity by computing average velocity over successively smaller time intervals. The guiding principle is the following: *Average velocity over a very small time interval is very close to instantaneous velocity.* To explore this idea further, we introduce some notation.

The Greek letter Δ (delta) is commonly used to denote the *change* in a function or variable. If $s(t)$ is the position of an object (distance from the origin) at time t and $[t_0, t_1]$ is a time interval, we set

$$\Delta s = s(t_1) - s(t_0) = \text{change in position}$$

$$\Delta t = \quad t_1 \quad - \quad t_0 \quad = \text{change in time (length of interval)}$$

For $t_1 \neq t_0$,

$$\boxed{\text{Average velocity over } [t_0, t_1] = \frac{\Delta s}{\Delta t} = \frac{s(t_1) - s(t_0)}{t_1 - t_0}}$$

The change in position Δs is also called the **displacement,** or net change in position.

One motion we will study is the motion of an object falling to earth under the influence of gravity (assuming no air resistance). Galileo discovered that if the object is released at time $t = 0$ from a state of rest (see Figure 2), then the distance traveled after t seconds is given by the formula

$$s(t) = 16t^2 \text{ ft}$$

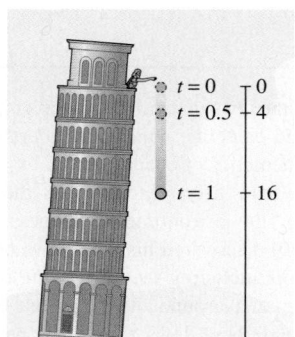

FIGURE 2 Distance traveled by a falling object after t seconds is $s(t) = 16t^2$ feet.

■ **EXAMPLE 1** A stone is released from a state of rest and falls to earth. Estimate the instantaneous velocity at $t = 0.5$ s by calculating average velocity over several small time intervals.

Solution The distance traveled by the stone after t seconds is $s(t) = 16t^2$ by Galileo's formula. We use this to compute the average velocity over the five time intervals listed in Table 1. To find the stone's average velocity over the first interval $[t_0, t_1] = [0.5, 0.6]$, we compute the changes:

$$\Delta s = s(0.6) - s(0.5) = 16(0.6)^2 - 16(0.5)^2 = 5.76 - 4 = 1.76 \text{ ft}$$

$$\Delta t = 0.6 - 0.5 = 0.1 \text{ s}$$

The average velocity over $[0.5, 0.6]$ is the ratio

$$\frac{\Delta s}{\Delta t} = \frac{s(0.6) - s(0.5)}{0.6 - 0.5} = \frac{1.76}{0.1} = 17.6 \text{ ft/s}$$

Table 1 shows the results of similar calculations for several intervals of successively shorter lengths. It is clear that the average velocities get closer to 16 ft/s as the time interval shrinks:

$$17.6, \quad 16.8, \quad 16.16, \quad 16.08, \quad 16.0016$$

We express this by saying that the average velocity *converges* to 16 or that 16 is the *limit* of average velocity as the length of the time interval shrinks to zero. This suggests that the instantaneous velocity at $t = 0.5$ is 16 ft/s.   ■

The conclusion in this example is correct, but we need to base it on more formal and precise reasoning. This will be done in Chapter 3, after we have developed the limit concept.

TABLE 1

Time Interval	Average Velocity
[0.5, 0.6]	17.6
[0.5, 0.55]	16.8
[0.5, 0.51]	16.16
[0.5, 0.505]	16.08
[0.5, 0.5001]	16.0016

Note that there is nothing special about the particular time intervals shown in Table 1. We are looking for a trend and we could have chosen any intervals $[0.5, t]$ for values of t approaching 0.5. We could also have chosen intervals $[t, 0.5]$ for $t < 0.5$.

Graphical Interpretation of Velocity

The idea that average velocity converges to instantaneous velocity as we shorten the time interval has a vivid interpretation in terms of secant lines where, by definition, a **secant line** is a line through two points on a curve.

Consider the graph of position $s(t)$ for an object traveling in a straight line as in Figure 3. The ratio we use to define average velocity over $[t_0, t_1]$ is nothing more than the slope of the secant line through the points $(t_0, s(t_0))$ and $(t_1, s(t_1))$. For $t_1 \neq t_0$,

$$\text{Average velocity} = \text{slope of secant line} = \frac{\Delta s}{\Delta t} = \frac{s(t_1) - s(t_0)}{t_1 - t_0}$$

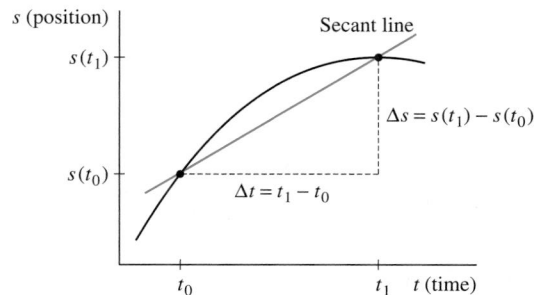

FIGURE 3 Average velocity over $[t_0, t_1]$ is equal to the slope of the secant line.

By interpreting average velocity as a slope, we can visualize what happens as the time interval gets smaller. Figure 4 shows the graph of $s(t) = 16t^2$ together with the secant lines for the time intervals $[0.5, 0.6]$, $[0.5, 0.55]$, and $[0.5, 0.51]$. We see that as the time interval shrinks, the secant lines get closer to—*and seem to rotate into*—the tangent line at $t = 0.5$.

Time Interval	Average Velocity
$[0.5, 0.6]$	17.6
$[0.5, 0.55]$	16.8
$[0.5, 0.51]$	16.16

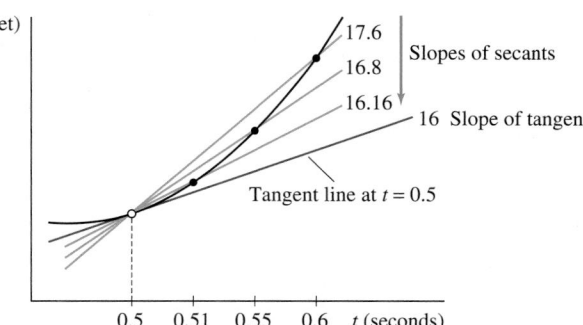

FIGURE 4 The secant line "rotates into" the tangent line as the time interval shrinks. *Note:* The figures are not drawn to scale.

Since the secant lines approach the tangent line, the slopes of the secant lines get closer and closer to the slope of the tangent line. In other words, the statement

> As the time interval shrinks to zero, the average velocity approaches the instantaneous velocity.

has the graphical interpretation

> As the time interval shrinks to zero, the slope of the secant line approaches the slope of the tangent line.

We are thus led to the following interpretation: *Instantaneous velocity is equal to the slope of the tangent line*. This conclusion and its generalization to other rates of change are fundamental to almost all aspects of differential calculus and its applications.

Other Rates of Change

Velocity is only one of many examples of a rate of change (ROC). Our reasoning applies to any quantity y that is a function of a variable x, say, $y = f(x)$. For any interval $[x_0, x_1]$, we set

The word "instantaneous" is often dropped. When we use the term "rate of change," it is understood that the instantaneous rate is intended.

$$\Delta f = f(x_1) - f(x_0), \qquad \Delta x = x_1 - x_0$$

For $x_1 \neq x_0$, the **average rate of change** of y with respect to x over $[x_0, x_1]$ is the ratio

Sometimes, we write Δy and $\frac{\Delta y}{\Delta x}$ instead of Δf and $\frac{\Delta f}{\Delta x}$.

$$\text{Average ROC} = \frac{\Delta f}{\Delta x} = \underbrace{\frac{f(x_1) - f(x_0)}{x_1 - x_0}}_{\text{Slope of secant line}}$$

The **instantaneous rate of change** is the limit of the average rates of change. We estimate the instantaneous rate of change at $x = x_0$ by computing the average rate of change over smaller and smaller intervals. In the previous example, we considered only right-hand intervals $[x_0, x_1]$. In the next example, we compute the average ROC for intervals lying to both the left and right of x_0.

An airplane traveling faster than the speed of sound produces a sonic boom. Because temperature decreases linearly with altitudes up to approximately 37,000 ft, pilots of supersonic jets can estimate the speed of sound in terms of altitude.

■ **EXAMPLE 2 Speed of Sound in Air** The formula $v = 20\sqrt{T}$ provides a good approximation to the speed of sound v in dry air (in m/s) as a function of air temperature T (in kelvins).

(a) Compute the average ROC of v with respect to T over the interval $[273, 300]$. What are the units of this ROC? Draw the corresponding secant line.

(b) Estimate the instantaneous ROC when $T = 273$ K by computing average rates over small intervals lying to the left and right of $T = 273$.

Solution

(a) The average ROC over the temperature interval $[273, 300]$ is

$$\frac{\Delta v}{\Delta T} = \frac{20\sqrt{300} - 20\sqrt{273}}{300 - 273} \approx \frac{15.95}{27} \approx 0.59 \text{ m/s per K}$$

Since this is the average rate of increase in speed per degree increase in temperature, the appropriate units are *meters per second per kelvin*. This ROC is equal to the slope of the line through the points where $T = 273$ and $T = 300$ on the graph in Figure 5.

FIGURE 5 The graph of $v = 20\sqrt{T}$.

FIGURE 6 Secant lines for intervals lying to the left of $T = 273$.

FIGURE 7 Secant lines for intervals lying to the right of $T = 273$.

TABLE 2 Left-Hand Intervals		TABLE 3 Right-Hand Intervals	
Temperature Interval	Average ROC	Temperature Interval	Average ROC
[272.5, 273]	0.60550	[273, 273.5]	0.60495
[272.8, 273]	0.60534	[273, 273.2]	0.60512
[272.9, 273]	0.60528	[273, 273.1]	0.60517
[272.99, 273]	0.60523	[273, 273.01]	0.60522

(b) To estimate the instantaneous ROC at $T = 273$, we compute the average ROC for several intervals lying to the left (Figure 6) and right (Figure 7) of $T = 273$. The results are shown in Tables 2 and 3. Both tables suggest that the instantaneous rate is approximately 0.605 m/s per K. ∎

To conclude this section, we recall an important point discussed in Section 1.2: For any linear function $f(x) = mx + b$, *the average rate of change over every interval is equal to the slope m* (Figure 8). Indeed, for the interval $[x_0, x_1]$ with $x_1 \neq x_0$,

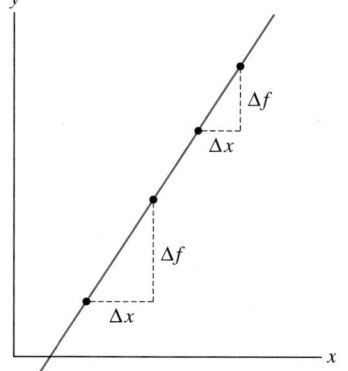

FIGURE 8 For a linear function, the ratio $\Delta f / \Delta x$ is equal to the slope m for every interval.

$$\frac{\Delta f}{\Delta x} = \frac{f(x_1) - f(x_0)}{x_1 - x_0} = \frac{(mx_1 + b) - (mx_0 + b)}{x_1 - x_0} = \frac{m(x_1 - x_0)}{x_1 - x_0} = m$$

The instantaneous rate of change at $x = x_0$ is also equal to m because it is the limit of average rates of change. This result makes sense graphically because all secant lines and all tangent lines to the graph of $f(x)$ coincide with the graph itself.

2.1 SUMMARY

- The average rate of change (ROC) of $y = f(x)$ over an interval $[x_0, x_1]$ is

$$\text{Average ROC} = \frac{\Delta f}{\Delta x} = \frac{f(x_1) - f(x_0)}{x_1 - x_0} \qquad (x_1 \neq x_0)$$

Average ROC is interpreted graphically as the slope of the secant line through the points $(x_0, f(x_0))$ and $(x_1, f(x_1))$ on the graph of $f(x)$.
- We estimate the instantaneous rate of change at $x = x_0$ by computing the average ROC over intervals $[x_0, x_1]$ (or $[x_1, x_0]$) for x_1 close to x_0. The instantaneous ROC is the limit of the average ROCs.
- Instantaneous ROC is interpreted graphically as the slope of the tangent line at x_0.
- The velocity of an object traveling in a straight line is the ROC of $s(t)$, where $s(t)$ is the position of the object at time t.
- If $f(x) = mx + b$ is a linear function, then the average rate of change over every interval and the instantaneous rate of change at every point are equal to the slope m.

2.1 EXERCISES

Preliminary Questions

1. Average velocity is defined as a ratio of which two quantities?

2. Average velocity is equal to the slope of a secant line through two points on a graph. Which graph?

3. Can instantaneous velocity be defined as a ratio? If not, how is instantaneous velocity computed?

4. What is the graphical interpretation of instantaneous velocity at a moment $t = t_0$?

5. What is the graphical interpretation of the following statement: The average ROC approaches the instantaneous ROC as the interval $[x_0, x_1]$ shrinks to x_0?

6. The ROC of atmospheric temperature with respect to altitude is equal to the slope of the tangent line to a graph. Which graph? What are possible units for this rate?

Exercises

1. A ball is dropped from a state of rest at time $t = 0$. The distance traveled after t seconds is $s(t) = 16t^2$ ft.

(a) How far does the ball travel during the time interval $[2, 2.5]$?

(b) Compute the average velocity over $[2, 2.5]$.

(c) Compute the average velocity over time intervals $[2, 2.01]$, $[2, 2.005]$, $[2, 2.001]$, $[2, 2.00001]$. Use this to estimate the object's instantaneous velocity at $t = 2$.

2. A wrench is released from a state of rest at time $t = 0$. Estimate the wrench's instantaneous velocity at $t = 1$, assuming that the distance traveled after t seconds is $s(t) = 16t^2$.

3. Let $v = 20\sqrt{T}$ as in Example 2. Estimate the instantaneous ROC of v with respect to T when $T = 300$ K.

4. Compute $\Delta y / \Delta x$ for the interval $[2, 5]$, where $y = 4x - 9$. What is the instantaneous ROC of y with respect to x at $x = 2$?

In Exercises 5–6, a stone is tossed in the air from ground level with an initial velocity of 15 m/s. Its height at time t is $h(t) = 15t - 4.9t^2$ m.

5. Compute the stone's average velocity over the time interval $[0.5, 2.5]$ and indicate the corresponding secant line on a sketch of the graph of $h(t)$.

6. Compute the stone's average velocity over the time intervals $[1, 1.01]$, $[1, 1.001]$, $[1, 1.0001]$ and $[0.99, 1]$, $[0.999, 1]$, $[0.9999, 1]$. Use this to estimate the instantaneous velocity at $t = 1$.

7. With an initial deposit of \$100, the balance in a bank account after t years is $f(t) = 100(1.08)^t$ dollars.

(a) What are the units of the ROC of $f(t)$?

(b) Find the average ROC over $[0, 0.5]$ and $[0, 1]$.

(c) Estimate the instantaneous rate of change at $t = 0.5$ by computing the average ROC over intervals to the left and right of $t = 0.5$.

8. The distance traveled by a particle at time t is $s(t) = t^3 + t$. Compute the average velocity over the time interval $[1, 4]$ and estimate the instantaneous velocity at $t = 1$.

In Exercises 9–16, estimate the instantaneous rate of change at the point indicated.

9. $P(x) = 4x^2 - 3$; $x = 2$

10. $f(t) = 3t - 5$; $t = -9$

11. $y(x) = \dfrac{1}{x+2}$; $x = 2$

12. $y(t) = \sqrt{3t + 1}$; $t = 1$

13. $f(x) = e^x$; $x = 0$

14. $f(x) = e^x$; $x = e$

15. $f(x) = \ln x$; $x = 3$

16. $f(x) = \tan^{-1} x$; $x = \frac{\pi}{4}$

17. The atmospheric temperature T (in °F) above a certain point on earth is $T = 59 - 0.00356h$, where h is the altitude in feet (valid for $h \leq 37{,}000$). What are the average and instantaneous rates of change of T with respect to h? Why are they the same? Sketch the graph of T for $h \leq 37{,}000$.

18. The height (in feet) at time t (in seconds) of a small weight oscillating at the end of a spring is $h(t) = 0.5\cos(8t)$.

(a) Calculate the weight's average velocity over the time intervals [0, 1] and [3, 5].

(b) Estimate its instantaneous velocity at $t = 3$.

19. The number $P(t)$ of E. coli cells at time t (hours) in a petri dish is plotted in Figure 9.

(a) Calculate the average ROC of $P(t)$ over the time interval [1, 3] and draw the corresponding secant line.

(b) Estimate the slope m of the line in Figure 9. What does m represent?

FIGURE 9 Number of E. coli cells at time t.

20. ✏️   Calculate $\sqrt{x+1} - \sqrt{x}$ for $x = 1, 10, 100, 1,000$. Does $\sqrt{x}$ change more rapidly when x is large or small? Interpret your answer in terms of tangent lines to the graph of $y = \sqrt{x}$.

21. ✏️   Assume that the period T (in seconds) of a pendulum (the time required for a complete back-and-forth cycle) is $T = \frac{3}{2}\sqrt{L}$, where L is the pendulum's length (in meters).

(a) What are the units for the ROC of T with respect to L? Explain what this rate measures.

(b) Which quantities are represented by the slopes of lines A and B in Figure 10?

(c) Estimate the instantaneous ROC of T with respect to L when $L = 3$ m.

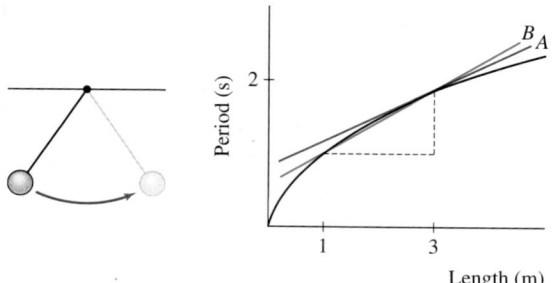

FIGURE 10 The period T is the time required for a pendulum to swing back and forth.

22. The graphs in Figure 11 represent the positions of moving particles as functions of time.

(a) Do the instantaneous velocities at times t_1, t_2, t_3 in (A) form an increasing or decreasing sequence?

(b) Is the particle speeding up or slowing down in (A)?

(c) Is the particle speeding up or slowing down in (B)?

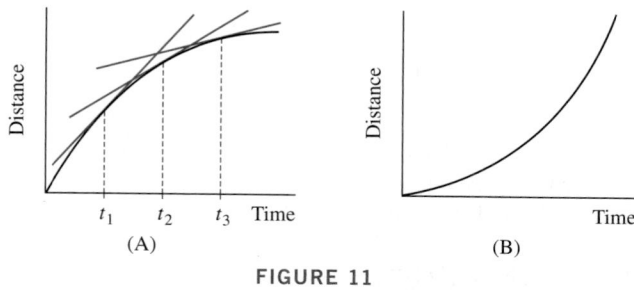

FIGURE 11

23. The graphs in Figure 12 represent the positions s of moving particles as functions of time t. Match each graph with one of the following statements:

(a) Speeding up

(b) Speeding up and then slowing down

(c) Slowing down

(d) Slowing down and then speeding up

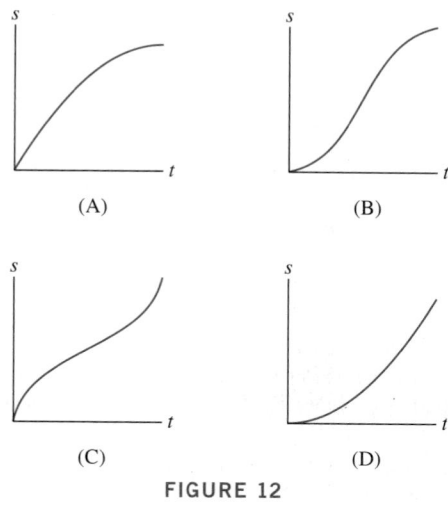

FIGURE 12

24. An epidemiologist finds that the percentage $N(t)$ of susceptible children who were infected on day t during the first three weeks of a measles outbreak is given, to a reasonable approximation, by the formula

$$N(t) = \frac{100t^2}{t^3 + 5t^2 - 100t + 380}$$

A graph of $N(t)$ appears in Figure 13.

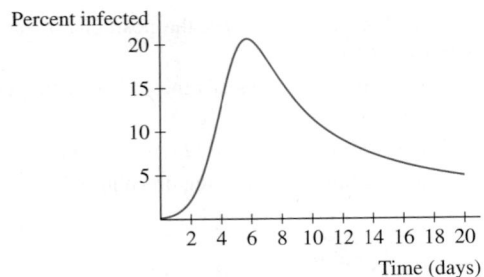

FIGURE 13 Graph of $N(t)$.

(a) Draw the secant line whose slope is the average rate of increase in infected children over the intervals between days 4 and 6 and between days 12 and 14. Then compute these average rates (in units of percent per day).

(b) Estimate the ROC of $N(t)$ on day 12.

25. The fraction of a city's population infected by a flu virus is plotted as a function of time (in weeks) in Figure 14.

(a) Which quantities are represented by the slopes of lines A and B? Estimate these slopes.

(b) Is the flu spreading more rapidly at $t = 1, 2,$ or 3?

(c) Is the flu spreading more rapidly at $t = 4, 5,$ or 6?

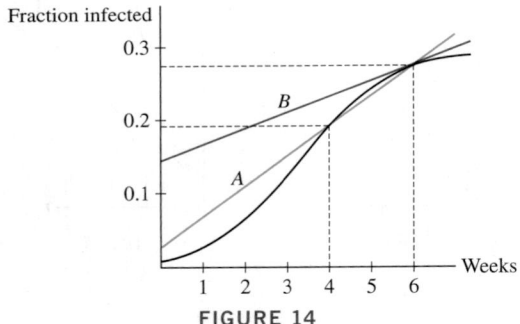

FIGURE 14

26. The fungus *fusarium exosporium* infects a field of flax plants through the roots and causes the plants to wilt. Eventually, the entire field is infected. The percentage $f(t)$ of infected plants as a function of time t (in days) since planting is shown in Figure 15.

(a) What are the units of the rate of change of $f(t)$ with respect to t? What does this rate measure?

(b) Use the graph to rank (from smallest to largest) the average infection rates over the intervals $[0, 12]$, $[20, 32]$, and $[40, 52]$.

(c) Use the following table to compute the average rates of infection over the intervals $[30, 40]$, $[40, 50]$, $[30, 50]$:

Days	0	10	20	30	40	50	60
Percent infected	0	18	56	82	91	96	98

(d) Draw the tangent line at $t = 40$ and estimate its slope. Choose any two points on the tangent line for the computation.

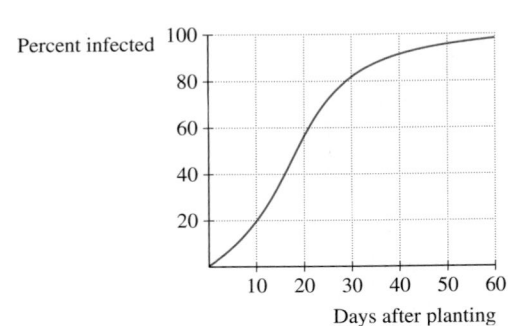

FIGURE 15

27. Let $v = 20\sqrt{T}$ as in Example 2. Is the ROC of v with respect to T greater at low temperatures or high temperatures? Explain in terms of the graph.

28. If an object moving in a straight line (but with changing velocity) covers Δs feet in Δt seconds, then its average velocity is $v_0 = \Delta s / \Delta t$ ft/s. Show that it would cover the same distance if it traveled at constant velocity v_0 over the same time interval of Δt seconds. This is a justification for calling $\Delta s / \Delta t$ the *average velocity*.

29. Sketch the graph of $f(x) = x(1 - x)$ over $[0, 1]$. Refer to the graph and, without making any computations, find:

(a) The average ROC over $[0, 1]$

(b) The (instantaneous) ROC at $x = \frac{1}{2}$

(c) The values of x at which the ROC is positive

30. Which graph in Figure 16 has the following property: For all x, the average ROC over $[0, x]$ is greater than the instantaneous ROC at x? Explain.

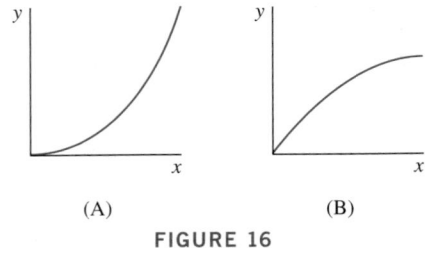

FIGURE 16

Further Insights and Challenges

31. The height of a projectile fired in the air vertically with initial velocity 64 ft/s is $h(t) = 64t - 16t^2$ ft.

(a) Compute $h(1)$. Show that $h(t) - h(1)$ can be factored with $(t - 1)$ as a factor.

(b) Using part (a), show that the average velocity over the interval $[1, t]$ is $-16(t - 3)$.

(c) Use this formula to find the average velocity over several intervals $[1, t]$ with t close to 1. Then estimate the instantaneous velocity at time $t = 1$.

32. Let $Q(t) = t^2$. As in the previous exercise, find a formula for the average ROC of Q over the interval $[1, t]$ and use it to estimate the instantaneous ROC at $t = 1$. Repeat for the interval $[2, t]$ and estimate the ROC at $t = 2$.

33. Show that the average ROC of $f(x) = x^3$ over $[1, x]$ is equal to $x^2 + x + 1$. Use this to estimate the instantaneous ROC of $f(x)$ at $x = 1$.

34. Find a formula for the average ROC of $f(x) = x^3$ over $[2, x]$ and use it to estimate the instantaneous ROC at $x = 2$.

35. Let $T = \frac{3}{2}\sqrt{L}$ as in Exercise 21. The numbers in the second column of Table 4 are increasing and those in the last column are decreasing. Explain why in terms of the graph of T as a function of L. Also, explain graphically why the instantaneous ROC at $L = 3$ lies between 0.4329 and 0.4331.

TABLE 4   Average Rates of Change of T with Respect to L

Interval	Average ROC	Interval	Average ROC
[3, 3.2]	0.42603	[2.8, 3]	0.44048
[3, 3.1]	0.42946	[2.9, 3]	0.43668
[3, 3.001]	0.43298	[2.999, 3]	0.43305
[3, 3.0005]	0.43299	[2.9995, 3]	0.43303

2.2 Limits: A Numerical and Graphical Approach

As we noted in the introduction to this chapter, the limit concept plays a fundamental role in all aspects of calculus. In this section, we define limits and study them using numerical and graphical techniques. We begin with the following question: *How do the values of a function $f(x)$ behave when x approaches a number c, whether or not $f(c)$ is defined?*

To explore this question, we'll experiment with the function

$$f(x) = \frac{\sin x}{x} \quad (x \text{ in radians})$$

The value $f(0)$ is not defined because

The undefined expression $\frac{0}{0}$ is also referred to as an "indeterminate form."

$$f(0) = \frac{\sin 0}{0} = \frac{0}{0} \quad \text{(undefined)}$$

Nevertheless, $f(x)$ can be computed for values of x close to 0. When we do this, a clear trend emerges (see the table in the margin and Table 1 on the next page).

To describe the trend, we use the phrases "x approaches 0" or "x tends to 0" to indicate that x takes on values (both positive and negative) that get closer and closer to 0. The notation for this is $x \to 0$. We write:

x	$\dfrac{\sin x}{x}$
1	0.841470985
0.5	0.958851077
0.1	0.998334166
0.05	0.999583385
0.01	0.999983333
0.005	0.999995833
0.001	0.999999833
$x \to 0+$	$f(x) \to 1$

- $x \to 0+$ if x approaches 0 from the right (through positive values).
- $x \to 0-$ if x approaches 0 from the left (through negative values).

Now consider the values listed in Table 1. The table gives the unmistakable impression that $f(x)$ gets closer and closer to 1 as x approaches 0 through positive and negative values.

This conclusion is supported by the graph of $f(x)$ in Figure 1. The point $(0, 1)$ is missing from the graph because $f(x)$ is not defined at $x = 0$, but the graph approaches this missing point as x approaches 0 from the left and right. We say that the *limit* of $f(x)$ as $x \to 0$ is equal to 1, and we write

$$\lim_{x \to 0} f(x) = 1$$

We also say that $f(x)$ *approaches* or *converges to* 1 as $x \to 0$.

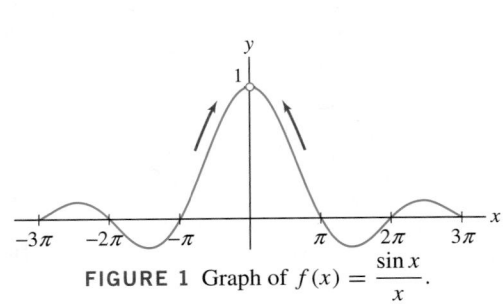

FIGURE 1 Graph of $f(x) = \dfrac{\sin x}{x}$.

	TABLE 1		
x	$\dfrac{\sin x}{x}$	x	$\dfrac{\sin x}{x}$
1	0.841470985	−1	0.841470985
0.5	0.958851077	−0.5	0.958851077
0.1	0.998334166	−0.1	0.998334166
0.05	0.999583385	−0.05	0.999583385
0.01	0.999983333	−0.01	0.999983333
0.005	0.999995833	−0.005	0.999995833
0.001	0.999999833	−0.001	0.999999833
$x \to 0+$	$f(x) \to 1$	$x \to 0-$	$f(x) \to 1$

CONCEPTUAL INSIGHT The numerical and graphical evidence may have convinced us that $f(x) = \dfrac{\sin x}{x}$ converges to 1 as $x \to 0$. But since $f(0) = 0/0$, could we not have arrived at this conclusion more simply by saying that $0/0 = 1$? The answer is no. It is not correct to say that $0/0$ equals 1 or any other number. In fact, if $0/0$ were equal to 1, we could prove the absurd equality $2 = 1$:

$$2\left(\frac{0}{0}\right) = \frac{2(0)}{0} = \frac{0}{0}$$

$$2 = 1$$

The expression $0/0$ is undefined, and in our example, the limit of $f(x)$ happens to be 1. We will encounter other examples where the formula for a function $f(x)$ produces $0/0$ and the limit is not equal to 1.

Definition of a Limit

We now define limits more formally. Recall that the distance between two numbers a and b is the absolute value $|a - b|$, so we can express the idea that $f(x)$ is close to L by saying that $|f(x) - L|$ is small.

Although fundamental in calculus, the limit concept was not fully clarified until the nineteenth century. The French mathematician Augustin-Louis Cauchy (1789–1857, pronounced Koh-shee) gave the following verbal definition: "When the values successively attributed to the same variable approach a fixed value indefinitely, in such a way as to end up differing from it by as little as one could wish, this last value is called the limit of all the others. So, for example, an irrational number is the limit of the various fractions which provide values that approximate it more and more closely." (Translated by J. Grabiner.)

DEFINITION Limit Assume that $f(x)$ is defined for all x near c (i.e., in some open interval containing c), but not necessarily at c itself. We say that

the limit of $f(x)$ as x approaches c is equal to L

if $|f(x) - L|$ becomes arbitrarily small when x is any number sufficiently close (but not equal) to c. In this case, we write

$$\lim_{x \to c} f(x) = L$$

We also say that $f(x)$ *approaches* or *converges to L as $x \to c$* (and we write $f(x) \to L$).

If the values of $f(x)$ do not converge to any limit as $x \to c$, we say that $\lim_{x \to c} f(x)$ *does not exist*. It is important to note that the value $f(c)$ itself, which may or may not be defined, plays no role in the limit. All that matters are the values $f(x)$ for x close to c. Furthermore, if $f(x)$ approaches a limit as $x \to c$, then the limiting value L is unique.

■ **EXAMPLE 1** Use the definition above to verify the following limits:

(a) $\lim_{x \to 7} 5 = 5$

(b) $\lim_{x \to 4} (3x + 1) = 13$

Solution

(a) Let $f(x) = 5$. To show that $\lim_{x \to 7} f(x) = 5$, we must show that $|f(x) - 5|$ becomes arbitrarily small when x is sufficiently close (but not equal) to 7. But $|f(x) - 5| = |5 - 5| = 0$ *for all* x, so what we are required to show is automatic (and it is not necessary to take x close to 7).

(b) Let $f(x) = 3x + 1$. To show that $\lim_{x \to 4} (3x + 1) = 13$, we must show that $|f(x) - 13|$ becomes arbitrarily small when x is sufficiently close (but not equal) to 4. We have

$$|f(x) - 13| = |(3x + 1) - 13| = |3x - 12| = 3|x - 4|$$

Since $|f(x) - 13|$ is a multiple of $|x - 4|$, we can make $|f(x) - 13|$ arbitrarily small by taking x sufficiently close to 4. ■

Reasoning as in Example 1 but with arbitrary constants, we obtain the following simple but important results:

THEOREM 1   For any constants k and c,   (a) $\lim_{x \to c} k = k$ and   (b) $\lim_{x \to c} x = c$.

Here is a way of stating the limit definition precisely: $\lim_{x \to c} f(x) = L$ if, for any n, we have $|f(x) - L| < 10^{-n}$, provided that $0 < |x - c| < 10^{-m}$ (where the choice of m depends on n).

To deal with more complicated limits in a rigorous manner, it is necessary to make the phrases "arbitrarily small" and "sufficiently close" more precise using inequalities. We do this in Section 2.8.

Graphical and Numerical Investigation

Our goal in the rest of this section is to develop a better intuitive understanding of limits by investigating them graphically and numerically.

Graphical Investigation Use a graphing utility to produce a graph of $f(x)$. The graph should give a visual impression of whether or not a limit exists. It can often be used to estimate the value of the limit.

Numerical Investigation To investigate $\lim_{x \to c} f(x)$,

(i) Make a table of values of $f(x)$ for x close to but less than c.

(ii) Make a second table of values of $f(x)$ for x close to but greater than c.

(iii) If both tables indicate convergence to the same number L, we take L to be an estimate for the limit.

Keep in mind that graphical and numerical investigations provide evidence for a limit but they do not prove that the limit exists or has a given value. This is done using the Limit Laws established in the following sections.

The tables should contain enough values to reveal a clear trend of convergence to a value L. If $f(x)$ approaches a limit, the successive values of $f(x)$ will generally agree to more and more decimal places as x is taken closer to c. If no pattern emerges, then the limit may not exist.

■ **EXAMPLE 2** Investigate graphically and numerically $\lim_{x \to 9} \dfrac{x - 9}{\sqrt{x} - 3}$.

Solution The graph of $f(x) = \dfrac{x-9}{\sqrt{x}-3}$ in Figure 2 has a gap at $x = 9$ because $f(9)$ is not defined:

$$f(9) = \frac{9-9}{\sqrt{9}-3} = \frac{0}{0} \quad \text{(undefined)}$$

However, the graph indicates that $f(x)$ approaches 6 as $x \to 9$. For numerical evidence, we consider a table of values of $f(x)$ for x approaching 9 from both the left and right. The values listed in Table 2 confirm our impression that

$$\lim_{x \to 9} \frac{x-9}{\sqrt{x}-3} = 6 \qquad \blacksquare$$

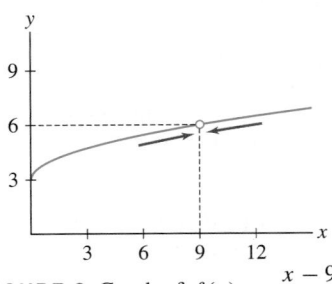

FIGURE 2 Graph of $f(x) = \dfrac{x-9}{\sqrt{x}-3}$.

	TABLE 2		
x	$\dfrac{x-9}{\sqrt{x}-3}$	x	$\dfrac{x-9}{\sqrt{x}-3}$
8.9	5.98329	9.1	6.01662
8.99	5.99833	9.01	6.001666
8.999	5.99983	9.001	6.000167
8.9999	5.9999833	9.0001	6.0000167

■ **EXAMPLE 3** Limit Equals Function Value   Investigate graphically and numerically $\lim\limits_{x \to 4} x^2$.

Solution Let $f(x) = x^2$. Figure 3 and Table 3 both suggest that $\lim\limits_{x \to 4} x^2 = 16$. Of course, in this case $f(4)$ is defined and $f(4) = 16$, so *the limit is equal to the function value*. This pleasant conclusion is valid whenever $f(x)$ is a *continuous* function, a concept treated in Section 2.4. ■

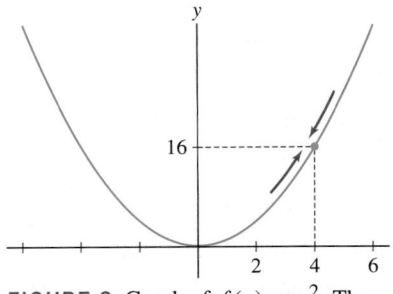

FIGURE 3 Graph of $f(x) = x^2$. The limit is equal to the function value $f(4) = 16$.

	TABLE 3		
x	x^2	x	x^2
3.9	15.21	4.1	16.81
3.99	15.9201	4.01	16.0801
3.999	15.992001	4.001	16.008001
3.9999	15.99920001	4.0001	16.00080001

The limit in Example 4 may be taken as the definition of the number e (see Section 3.2). In other words,

$$\lim_{h \to 0} \frac{b^h - 1}{h} = 1 \quad \text{if and only if } b = e$$

■ **EXAMPLE 4** Defining Property of e Verify numerically that $\lim\limits_{h \to 0} \dfrac{e^h - 1}{h} = 1$.

Solution The function $f(h) = \dfrac{e^h - 1}{h}$ is undefined at $h = 0$, but both Figure 4 and Table 4 suggest that $\lim\limits_{h \to 0} \dfrac{e^h - 1}{h} = 1$. ■

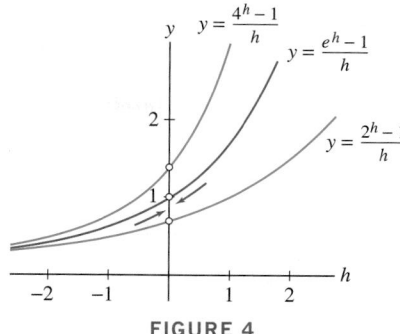

FIGURE 4

$h \to 0+$	$\dfrac{e^h - 1}{h}$	$h \to 0-$	$\dfrac{e^h - 1}{h}$
	TABLE 4		
0.02	1.0101	−0.02	0.990
0.005	1.00250	−0.005	0.99750
0.001	1.000500	−0.001	0.999500
0.0001	1.00005000	−0.0001	0.99995000

■ **EXAMPLE 5** A Limit That Does Not Exist Investigate graphically and numerically $\lim\limits_{x \to 0} \sin \dfrac{1}{x}$.

Solution The function $f(x) = \sin \dfrac{1}{x}$ is not defined at $x = 0$, but Figure 5 shows that it oscillates between $+1$ and -1 infinitely often as $x \to 0$. It appears, therefore, that $\lim\limits_{x \to 0} \sin \dfrac{1}{x}$ does not exist. This impression is confirmed by Table 5, which suggests that the values of $f(x)$ bounce around and do not tend toward any limit L as $x \to 0$. ■

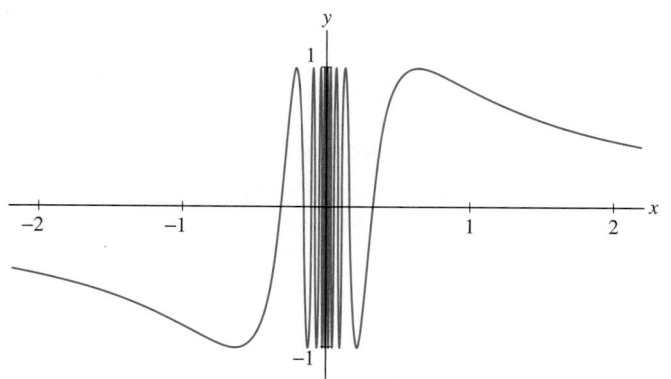

FIGURE 5 Graph of $f(x) = \sin \dfrac{1}{x}$.

	TABLE 5 The Function $f(x) = \sin \frac{1}{x}$ Does Not Approach a Limit as $x \to 0$		
x	$\sin \dfrac{1}{x}$	x	$\sin \dfrac{1}{x}$
0.1	0.5440	−0.1	−0.5440
0.05	0.9129	−0.05	−0.9129
0.01	−0.5064	−0.01	0.5064
0.001	0.8269	−0.001	−0.8269
0.0005	0.9300	−0.0005	−0.9300
0.00002	−0.9998	−0.00002	0.9998
0.00001	0.0357	−0.00001	−0.0357

One-Sided Limits

The limits we have discussed so far are *two-sided*. To show that $\lim\limits_{x \to c} f(x) = L$, it is necessary to check that $f(x)$ converges to L as x approaches c through values both larger and smaller than c. In some instances, $f(x)$ may approach L from one side of c without necessarily approaching it from the other side, or $f(x)$ may be defined on only one side of c. For this reason, we define the one-sided limits

$$\lim_{x \to c-} f(x) \quad \text{(left-hand limit)}, \qquad \lim_{x \to c+} f(x) \quad \text{(right-hand limit)}$$

The limit itself exists if both one-sided limits exist and are equal.

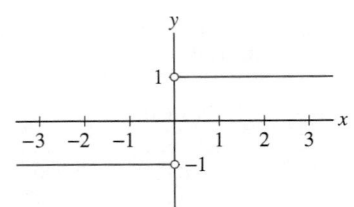

FIGURE 6 Graph of $y = \dfrac{x}{|x|}$.

■ **EXAMPLE 6**   Left- and Right-Hand Limits Not Equal   Investigate the one-sided limits of $f(x) = \dfrac{x}{|x|}$ as $x \to 0$. Does $\lim\limits_{x \to 0} f(x)$ exist?

Solution   The graph in Figure 6 shows what is going on. For $x > 0$, we have

$$f(x) = \frac{x}{|x|} = \frac{x}{x} = 1$$

Therefore, $f(x) \to 1$ as $x \to 0+$. But for $x < 0$,

$$f(x) = \frac{x}{|x|} = \frac{x}{-x} = -1$$

and thus $f(x) \to -1$ as $x \to 0-$. We see that the one-sided limits exist:

$$\lim_{x \to 0+} f(x) = 1, \qquad \lim_{x \to 0-} f(x) = -1$$

However, these one-sided limits are not equal, so $\lim\limits_{x \to 0} f(x)$ does not exist.   ■

■ **EXAMPLE 7**   The function $f(x)$ shown in Figure 7 is not defined at $c = 0, 2, 4$. Do the one- or two-sided limits exist at these points? If so, find their values.

Solution

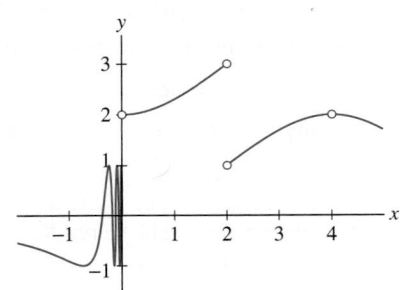

FIGURE 7

- $c = 0$: The left-hand limit $\lim\limits_{x \to 0-} f(x)$ does not seem to exist because $f(x)$ appears to oscillate infinitely often to the left of 0. On the other hand, $\lim\limits_{x \to 0+} f(x) = 2$.
- $c = 2$: The one-sided limits exist but are not equal:

$$\lim_{x \to 2-} f(x) = 3 \qquad \text{and} \qquad \lim_{x \to 2+} f(x) = 1$$

Therefore, $\lim\limits_{x \to 2} f(x)$ does not exist.

- $c = 4$: The one-sided limits exist and both have the value 2. Therefore, the two-sided limit exists and $\lim\limits_{x \to 4} f(x) = 2$.   ■

Infinite Limits

Some functions tend to ∞ or $-\infty$ as x approaches a value c. In this case, the limit $\lim\limits_{x \to c} f(x)$ does not exist, but we say that f has an *infinite limit*. More precisely, we write

$$\lim_{x \to c} f(x) = \infty$$

if $f(x)$ increases without bound as $x \to c$. If $f(x)$ tends to $-\infty$ (i.e., $f(x)$ becomes negative and $|f(x)| \to \infty$), then we write

$$\lim_{x \to c} f(x) = -\infty$$

When using this notation, keep in mind that ∞ and $-\infty$ are not numbers. One-sided infinite limits are defined similarly. In the next example, we use the notation $x \to c\pm$ to indicate that the left- and right-hand limits are to be considered separately.

■ **EXAMPLE 8**   One- and Two-Sided Infinite Limits   GU   Investigate the one-sided limits graphically:

(a) $\lim\limits_{x \to 2\pm} \dfrac{1}{x-2}$ **(b)** $\lim\limits_{x \to 0\pm} \dfrac{1}{x^2}$ **(c)** $\lim\limits_{x \to 0+} \ln x$

Solution

When $f(x)$ approaches ∞ or $-\infty$ as x approaches c from one or both sides, we say that the line $x = c$ is a vertical asymptote. For example, in Figure 8(A), $x = 2$ is a vertical asymptote. Asymptotic behavior is discussed further in Section 4.5.

(a) The function $f(x) = \dfrac{1}{x-2}$ is negative for $x < 2$ and positive for $x > 2$. As we see in Figure 8(A), $f(x)$ approaches $-\infty$ as x approaches 2 from the left and ∞ as x approaches 2 from the right:

$$\lim_{x \to 2-} \frac{1}{x-2} = -\infty, \qquad \lim_{x \to 2+} \frac{1}{x-2} = \infty$$

(b) The function $f(x) = \dfrac{1}{x^2}$ is positive for all $x \neq 0$ and becomes arbitrarily large as $x \to 0$ from either side [Figure 8(B)]. Therefore, $\lim\limits_{x \to 0} \dfrac{1}{x^2} = \infty$.

(c) The function $f(x) = \ln x$ is negative for $0 < x < 1$ and it tends to $-\infty$ as $x \to 0+$ [Figure 8(C)]. Therefore, $\lim\limits_{x \to 0+} \ln x = -\infty$. ■

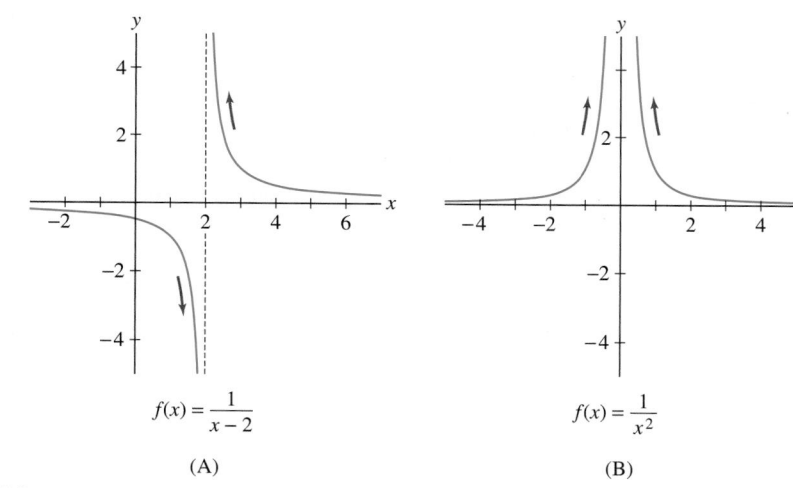

$f(x) = \dfrac{1}{x-2}$

(A)

$f(x) = \dfrac{1}{x^2}$

(B)

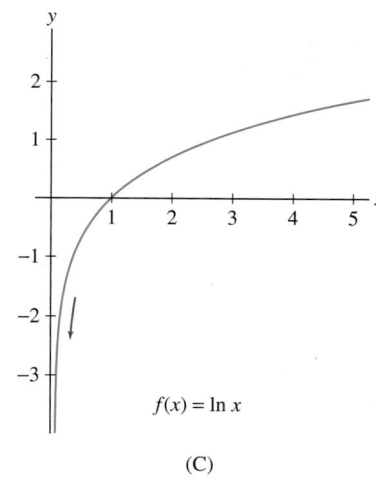

$f(x) = \ln x$

(C)

FIGURE 8

CONCEPTUAL INSIGHT You should not think of an infinite limit as a true limit. The notation $\lim\limits_{x \to c} f(x) = \infty$ is a shorthand way of saying that no finite limit exists but $f(x)$ increases beyond all bounds as x approaches c. We must be careful about this because ∞ and $-\infty$ are not numbers, and contradictions may arise if we manipulate them as numbers. For example, if ∞ were a number larger than any finite number, then presumably $\infty + 1 = \infty$. This would give

$$\infty + 1 = \infty$$

$$(\infty + 1) - \infty = \infty - \infty$$

$$1 = 0 \qquad \text{(contradiction!)}$$

For this reason, the Limit Laws discussed in the next section are valid only for finite limits and cannot be applied to infinite limits.

2.2 SUMMARY

- By definition, $\lim_{x \to c} f(x) = L$ if $|f(x) - L|$ becomes arbitrarily small when x is sufficiently close (but not equal) to c. We then say that *the limit of $f(x)$ as x approaches c is L*. Alternate terminology: $f(x)$ *approaches* or *converges to L* as x approaches c.
- If $f(x)$ approaches a limit as $x \to c$, then the limiting value L is unique.
- If $f(x)$ does not approach any limit as $x \to c$, we say that $\lim_{x \to c} f(x)$ does not exist.
- The limit may exist even if $f(c)$ is not defined, as in the example $\lim_{x \to 0} \dfrac{\sin x}{x}$.
- One-sided limits: $\lim_{x \to c-} f(x) = L$ if $f(x)$ converges to L as x approaches c through values less than c and $\lim_{x \to c+} f(x) = L$ if $f(x)$ converges to L as x approaches c through values greater than c.
- The limit exists if and only if both one-sided limits exist and are equal.
- We say that $\lim_{x \to c} f(x) = \infty$ if $f(x)$ increases beyond bound as x approaches c, and $\lim_{x \to c} f(x) = -\infty$ if $f(x)$ becomes arbitrarily large (in absolute value) but negative as x approaches c.

2.2 EXERCISES

Preliminary Questions

1. What is the limit of $f(x) = 1$ as $x \to \pi$?

2. What is the limit of $g(t) = t$ as $t \to \pi$?

3. Can $f(x)$ approach a limit as $x \to c$ if $f(c)$ is undefined? If so, give an example.

4. Is $\lim_{x \to 10} 20$ equal to 10 or 20?

5. What does the following table suggest about $\lim_{x \to 1-} f(x)$ and $\lim_{x \to 1+} f(x)$?

x	0.9	0.99	0.999	1.1	1.01	1.001
$f(x)$	7	25	4317	3.0126	3.0047	3.00011

6. Is it possible to tell if $\lim_{x \to 5} f(x)$ exists by only examining values $f(x)$ for x close to but *greater* than 5? Explain.

7. If you know in advance that $\lim_{x \to 5} f(x)$ exists, can you determine its value just knowing the values of $f(x)$ for all $x > 5$?

8. Which of the following pieces of information is sufficient to determine whether $\lim_{x \to 5} f(x)$ exists? Explain.

(a) The values of $f(x)$ for all x

(b) The values of $f(x)$ for x in [4.5, 5.5]

(c) The values of $f(x)$ for all x in [4.5, 5.5] other than $x = 5$

(d) The values of $f(x)$ for all $x \geq 5$

(e) $f(5)$

Exercises

In Exercises 1–6, fill in the tables and guess the value of the limit.

1. $\lim_{x \to 1} f(x)$, where $f(x) = \dfrac{x^3 - 1}{x^2 - 1}$.

x	$f(x)$	x	$f(x)$
1.002		0.998	
1.001		0.999	
1.0005		0.9995	
1.00001		0.99999	

2. $\lim_{t \to 0} h(t)$, where $h(t) = \dfrac{\cos t - 1}{t^2}$. Note that $h(t)$ is even, that is, $h(t) = h(-t)$.

t	± 0.002	± 0.0001	± 0.00005	± 0.00001
$h(t)$				

3. $\lim_{y \to 2} f(y)$, where $f(y) = \dfrac{y^2 - y - 2}{y^2 + y - 6}$.

y	$f(y)$	y	$f(y)$
2.002		1.998	
2.001		1.999	
2.0001		1.9999	

4. $\displaystyle\lim_{\theta\to 0} f(\theta)$, where $f(\theta) = \dfrac{\sin\theta - \theta}{\theta^3}$.

θ	± 0.002	± 0.0001	± 0.00005	± 0.00001
$f(\theta)$				

5. $\displaystyle\lim_{x\to 0} f(x)$, where $f(x) = \dfrac{e^x - x - 1}{x^2}$.

x	± 0.5	± 0.1	± 0.05	± 0.01
$f(x)$				

6. $\displaystyle\lim_{x\to 0+} f(x)$, where $f(x) = x\ln x$.

x	1	0.5	0.1	0.05	0.01	0.005	0.001
$f(x)$							

7. Determine $\displaystyle\lim_{x\to 0.5} f(x)$ for the function $f(x)$ shown in Figure 9.

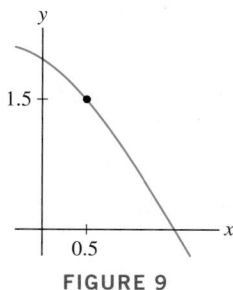

FIGURE 9

8. Do either of the two oscillating functions in Figure 10 appear to approach a limit as $x \to 0$?

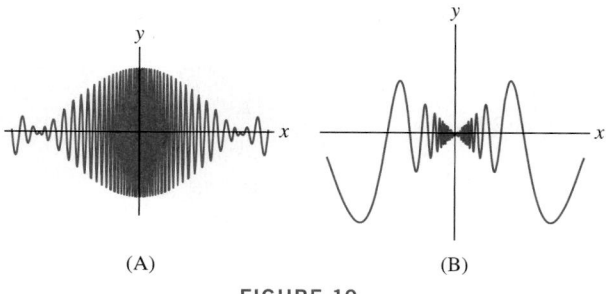

(A)   (B)

FIGURE 10

In Exercises 9–10, evaluate the limit.

9. $\displaystyle\lim_{x\to 21} x$

10. $\displaystyle\lim_{x\to 4.2} \sqrt{3}$

In Exercises 11–20, verify each limit using the limit definition. For example, in Exercise 11, show that $|2x - 6|$ can be made as small as desired by taking x close to 3.

11. $\displaystyle\lim_{x\to 3} 2x = 6$

12. $\displaystyle\lim_{x\to 3} 4 = 4$

13. $\displaystyle\lim_{x\to 2} (4x + 3) = 11$

14. $\displaystyle\lim_{x\to 3} (5x - 7) = 8$

15. $\displaystyle\lim_{x\to 9} (-2x) = -18$

16. $\displaystyle\lim_{x\to -5} (1 - 2x) = 11$

17. $\displaystyle\lim_{x\to 0} x^2 = 0$

18. $\displaystyle\lim_{x\to 0} (2x^2 + 4) = 4$

19. $\displaystyle\lim_{x\to 0} (x^2 + 2x + 3) = 3$

20. $\displaystyle\lim_{x\to 0} (x^3 + 9) = 9$

In Exercises 21–36, estimate the limit numerically or state that the limit does not exist.

21. $\displaystyle\lim_{x\to 1} \dfrac{\sqrt{x} - 1}{x - 1}$

22. $\displaystyle\lim_{x\to -3} \dfrac{2x^2 - 18}{x + 3}$

23. $\displaystyle\lim_{x\to 2} \dfrac{x^2 + x - 6}{x^2 - x - 2}$

24. $\displaystyle\lim_{x\to 3} \dfrac{x^3 - 2x^2 - 9}{x^2 - 2x - 3}$

25. $\displaystyle\lim_{x\to 0} \dfrac{\sin 2x}{x}$

26. $\displaystyle\lim_{x\to 0} \dfrac{\sin 5x}{x}$

27. $\displaystyle\lim_{x\to 0} \dfrac{\sin x}{x^2}$

28. $\displaystyle\lim_{\theta\to 0} \dfrac{\cos\theta - 1}{\theta}$

29. $\displaystyle\lim_{h\to 0} \cos \dfrac{1}{h}$

30. $\displaystyle\lim_{h\to 0} \sin h \cos \dfrac{1}{h}$

31. $\displaystyle\lim_{h\to 0} \dfrac{2^h - 1}{h}$

32. $\displaystyle\lim_{x\to 0} \dfrac{e^x - 2^x}{x}$

33. $\displaystyle\lim_{x\to 1+} \dfrac{\sec^{-1} x}{\sqrt{x - 1}}$

34. $\displaystyle\lim_{x\to 0} |x|^x$

35. $\displaystyle\lim_{x\to 0} \dfrac{\tan^{-1} x}{\sin^{-1} x - x}$

36. $\displaystyle\lim_{x\to 0} \dfrac{\tan^{-1} x - x}{\sin^{-1} x - x}$

37. Determine $\displaystyle\lim_{x\to 2+} f(x)$ and $\displaystyle\lim_{x\to 2-} f(x)$ for the function shown in Figure 11.

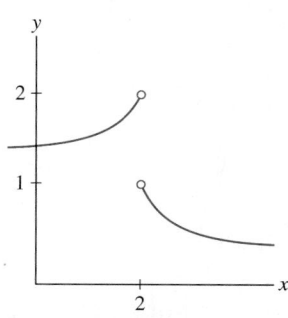

FIGURE 11

38. Determine the one-sided limits at $c = 1, 2, 4, 5$ of the function $g(t)$ shown in Figure 12 and state whether the limit exists at these points.

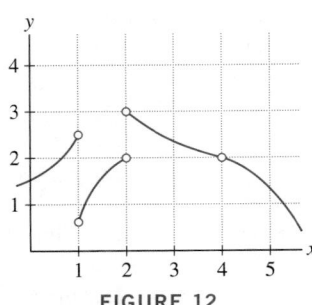

FIGURE 12

39. The greatest integer function is defined by $[x] = n$, where n is the unique integer such that $n \leq x < n + 1$. See Figure 13.

(a) For which values of c does $\lim_{x \to c-} [x]$ exist? What about $\lim_{x \to c+} [x]$?

(b) For which values of c does $\lim_{x \to c} [x]$ exist?

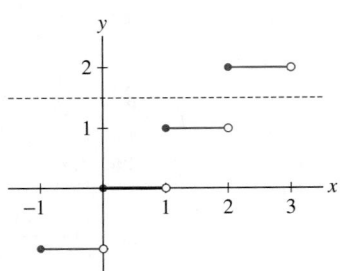

FIGURE 13 Graph of $y = [x]$.

40. Draw a graph of $f(x) = \dfrac{x - 1}{|x - 1|}$ and use it to determine the one-sided limits $\lim_{x \to 1+} f(x)$ and $\lim_{x \to 1-} f(x)$.

In Exercises 41–43, determine the one-sided limits numerically.

41. $\lim_{x \to 0\pm} \dfrac{\sin x}{|x|}$

42. $\lim_{x \to 0\pm} |x|^{1/x}$

43. $\lim_{x \to 0\pm} \dfrac{x - \sin(|x|)}{x^3}$

44. Determine the one- or two-sided infinite limits in Figure 14.

(A)   (B)

FIGURE 14

45. Determine the one-sided limits of $f(x)$ at $c = 2$ and $c = 4$, for the function shown in Figure 15.

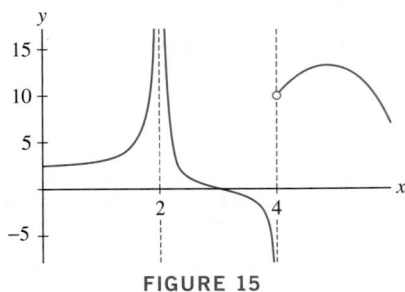

FIGURE 15

46. Determine the infinite one- and two-sided limits in Figure 16.

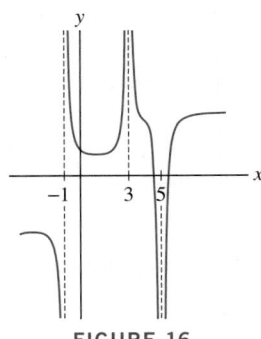

FIGURE 16

In Exercises 47–50, draw the graph of a function with the given limits.

47. $\lim_{x \to 1} f(x) = 2, \quad \lim_{x \to 3-} f(x) = 0, \quad \lim_{x \to 3+} f(x) = 4$

48. $\lim_{x \to 1} f(x) = \infty, \quad \lim_{x \to 3-} f(x) = 0, \quad \lim_{x \to 3+} f(x) = -\infty$

49. $\lim_{x \to 2+} f(x) = f(2) = 3, \quad \lim_{x \to 2-} f(x) = -1,$
$\lim_{x \to 4} f(x) = 2 \neq f(4)$

50. $\lim_{x \to 1+} f(x) = \infty, \quad \lim_{x \to 1-} f(x) = 3, \quad \lim_{x \to 4} f(x) = -\infty$

[GU]   *In Exercises 51–56, graph the function and use the graph to estimate the value of the limit.*

51. $\lim_{\theta \to 0} \dfrac{\sin 3\theta}{\sin 2\theta}$

52. $\lim_{x \to 0} \dfrac{2^x - 1}{4^x - 1}$

53. $\lim_{x \to 0} \dfrac{2^x - \cos x}{x}$

54. $\lim_{\theta \to 0} \dfrac{\sin^2 4\theta}{\cos \theta - 1}$

55. $\lim_{\theta \to 0} \dfrac{\cos 3\theta - \cos 4\theta}{\theta^2}$

56. $\lim_{\theta \to 0} \dfrac{\cos 3\theta - \cos 5\theta}{\theta^2}$

Further Insights and Challenges

57. Light waves of frequency λ passing through a slit of width a produce a **Fraunhofer diffraction pattern** of light and dark fringes (Figure 17). The intensity as a function of the angle θ is given by

$$I(\theta) = I_m \left(\frac{\sin(R \sin \theta)}{R \sin \theta} \right)^2$$

where $R = \pi a / \lambda$ and I_m is a constant. Show that the intensity function is not defined at $\theta = 0$. Then check numerically that $I(\theta)$ approaches I_m as $\theta \to 0$ for any two values of R (e.g., choose two integer values).

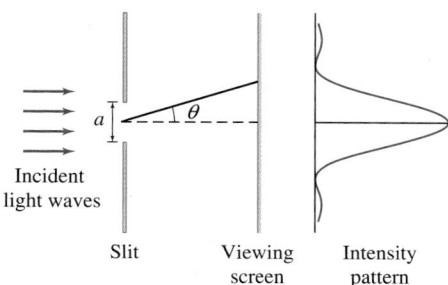

FIGURE 17 Fraunhofer diffraction pattern.

58. Investigate $\displaystyle\lim_{\theta \to 0} \frac{\sin n\theta}{\theta}$ numerically for several values of n and then guess the value in general.

59. Show numerically that $\displaystyle\lim_{x \to 0} \frac{b^x - 1}{x}$ for $b = 3, 5$ appears to equal $\ln 3, \ln 5$, where $\ln x$ is the natural logarithm. Then make a conjecture (guess) for the value in general and test your conjecture for two additional values of b.

60. Investigate $\displaystyle\lim_{x \to 1} \frac{x^n - 1}{x^m - 1}$ for (m, n) equal to $(2, 1)$, $(1, 2)$, $(2, 3)$, and $(3, 2)$. Then guess the value of the limit in general and check your guess for at least three additional pairs.

61. Find by experimentation the positive integers k such that $\displaystyle\lim_{x \to 0} \frac{\sin(\sin^2 x)}{x^k}$ exists.

62. [GU] Sketch a graph of $f(x) = \dfrac{2^x - 8}{x - 3}$ with a graphing calculator. Observe that $f(3)$ is not defined.

(a) Zoom in on the graph to estimate $L = \displaystyle\lim_{x \to 3} f(x)$.

(b) Observe that the graph of $f(x)$ is increasing. Explain how this implies that

$$f(2.99999) \le L \le f(3.00001)$$

Use this to determine L to three decimal places.

63. [GU] The function $f(x) = \dfrac{2^{1/x} - 2^{-1/x}}{2^{1/x} + 2^{-1/x}}$ is defined for $x \ne 0$.

(a) Investigate $\displaystyle\lim_{x \to 0+} f(x)$ and $\displaystyle\lim_{x \to 0-} f(x)$ numerically.

(b) Produce a graph of f on a graphing utility and describe its behavior near $x = 0$.

64. Show that $f(x) = \dfrac{\sin x}{x}$ is equal to the slope of a secant line through the origin and the point $(x, \sin x)$ on the graph of $y = \sin x$ (Figure 18). Use this to give a geometric interpretation of $\displaystyle\lim_{x \to 0} f(x)$.

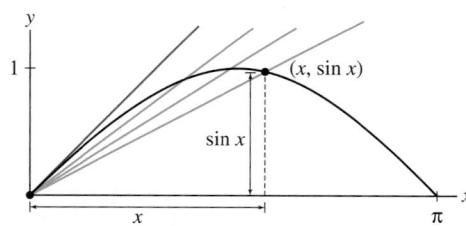

FIGURE 18 Graph of $y = \sin x$.

2.3 Basic Limit Laws

In Section 2.2 we relied on graphical and numerical approaches to investigate limits and estimate their values. In the next four sections we go beyond this intuitive approach and develop tools for computing limits in a precise way. This section discusses the **Basic Limit Laws** used for computing limits of functions constructed as sums, multiples, products, or quotients of other functions.

The proofs of these laws are discussed in Section 2.8 and Appendix D. However, we can understand the underlying idea by arguing informally. For example, to prove the Sum Law, we observe that if $f(x)$ is close to L and $g(x)$ is close to M when $|x - c|$ is sufficiently small, then $f(x) + g(x)$ must be close to $L + M$ when $|x - c|$ is sufficiently small. Similarly, $f(x)g(x)$ must be close to LM, etc.

THEOREM 1 Basic Limit Laws Assume that $\lim_{x \to c} f(x)$ and $\lim_{x \to c} g(x)$ exist. Then:

(i) Sum Law:

$$\lim_{x \to c} \left(f(x) + g(x) \right) = \lim_{x \to c} f(x) + \lim_{x \to c} g(x)$$

(ii) Constant Multiple Law: For any number k,

$$\lim_{x \to c} kf(x) = k \lim_{x \to c} f(x)$$

(iii) Product Law:

$$\lim_{x \to c} f(x)g(x) = \left(\lim_{x \to c} f(x) \right) \left(\lim_{x \to c} g(x) \right)$$

(iv) Quotient Law: If $\lim_{x \to c} g(x) \neq 0$, then

$$\lim_{x \to c} \frac{f(x)}{g(x)} = \frac{\lim_{x \to c} f(x)}{\lim_{x \to c} g(x)}$$

Before proceeding to the examples, we make two useful remarks. First, the Sum and Product Laws are valid for any number of functions. For example,

$$\lim_{x \to c} \left(f_1(x) + f_2(x) + f_3(x) \right) = \lim_{x \to c} f_1(x) + \lim_{x \to c} f_2(x) + \lim_{x \to c} f_3(x)$$

Second, the Sum Law has a counterpart for differences:

$$\lim_{x \to c} \left(f(x) - g(x) \right) = \lim_{x \to c} f(x) - \lim_{x \to c} g(x).$$

This is not listed as a separate law because it follows by combining the Sum and Constant Multiple Laws:

$$\lim_{x \to c} \left(f(x) - g(x) \right) = \lim_{x \to c} f(x) + \lim_{x \to c} \left(- g(x) \right) = \lim_{x \to c} f(x) - \lim_{x \to c} g(x)$$

The Limit Laws are applied step by step in the solution to Example 1 to illustrate how they are used. In practice, we take the Limit Laws for granted and skip the intermediate steps.

■ EXAMPLE 1 Use the Limit Laws to evaluate the following limits:

(a) $\lim_{x \to 3} x^2$ **(b)** $\lim_{t \to -1} (4t^3 - t - 5)$ **(c)** $\lim_{t \to 2} \frac{t + 5}{3t}$

Solution

(a) By Theorem 1 of Section 2.2, $\lim_{x \to c} x = c$ for all c. Since x^2 is equal to the product $x^2 = x \cdot x$, we may apply the Product Law

$$\lim_{x \to 3} x^2 = \left(\lim_{x \to 3} x \right) \left(\lim_{x \to 3} x \right) = 3(3) = 9$$

(b) First use the Sum Law (for three functions):

$$\lim_{t \to -1} (4t^3 - t - 5) = \lim_{t \to -1} 4t^3 - \lim_{t \to -1} t - \lim_{t \to -1} 5$$

Then evaluate each limit using the Constant Multiple and Product Laws:

$$\lim_{t \to -1} 4t^3 = 4 \lim_{t \to -1} t^3 = 4 \left(\lim_{t \to -1} t \right) \left(\lim_{t \to -1} t \right) \left(\lim_{t \to -1} t \right) = 4(-1)^3 = -4$$

$$\lim_{t \to -1} t = -1$$

$$\lim_{t \to -1} 5 = 5$$

We obtain

$$\lim_{t \to -1} (4t^3 - t - 5) = -4 - (-1) - 5 = -8$$

(c) Use the Quotient, Sum, and Constant Multiple Laws:

$$\lim_{t \to 2} \frac{t + 5}{3t} = \frac{\lim_{t \to 2}(t + 5)}{\lim_{t \to 2} 3t} = \frac{\lim_{t \to 2} t + \lim_{t \to 2} 5}{3 \lim_{t \to 2} t} = \frac{2 + 5}{3 \cdot 2} = \frac{7}{6} \qquad \blacksquare$$

The next example reminds us that the Limit Laws apply only when the limits of $f(x)$ and $g(x)$ exist.

■ **EXAMPLE 2** Assumptions Matter Show that the Product Law cannot be applied to $\lim_{x \to 0} f(x)g(x)$ if $f(x) = x$ and $g(x) = x^{-1}$.

Solution The product function is $f(x)g(x) = x \cdot x^{-1} = 1$ for $x \neq 0$, so the limit of the product exists:

$$\lim_{x \to 0} f(x)g(x) = \lim_{x \to 0} 1 = 1$$

However, $\lim_{x \to 0} x^{-1}$ does not exist because $g(x) = x^{-1}$ approaches ∞ as $c \to 0+$ and $-\infty$ as $c \to 0-$, so the Product Law cannot be applied and its conclusion does not hold:

$$\left(\lim_{x \to 0} f(x) \right) \left(\lim_{x \to 0} g(x) \right) = \left(\lim_{x \to 0} x \right) \underbrace{\left(\lim_{x \to 0} x^{-1} \right)}_{\text{Does not exist}} \qquad \blacksquare$$

2.3 SUMMARY

- The Limit Laws state that if $\lim_{x \to c} f(x)$ and $\lim_{x \to c} g(x)$ both exist, then:

(i) $\lim_{x \to c} \left(f(x) + g(x) \right) = \lim_{x \to c} f(x) + \lim_{x \to c} g(x)$

(ii) $\lim_{x \to c} kf(x) = k \lim_{x \to c} f(x)$

(iii) $\lim_{x \to c} f(x)\, g(x) = \left(\lim_{x \to c} f(x) \right) \left(\lim_{x \to c} g(x) \right)$

(iv) If $\lim_{x \to c} g(x) \neq 0$, then

$$\lim_{x \to c} \frac{f(x)}{g(x)} = \frac{\lim_{x \to c} f(x)}{\lim_{x \to c} g(x)}$$

- If $\lim_{x \to c} f(x)$ or $\lim_{x \to c} g(x)$ does not exist, then the Limit Laws cannot be applied.

2.3 EXERCISES

Preliminary Questions

1. State the Sum Law and Quotient Law.

2. Which of the following is a verbal version of the Product Law?
(a) The product of two functions has a limit.
(b) The limit of the product is the product of the limits.
(c) The product of a limit is a product of functions.
(d) A limit produces a product of functions.

3. Which of the following statements are incorrect (k and c are constants)?
(a) $\lim_{x \to c} k = c$
(b) $\lim_{x \to c} k = k$

(c) $\lim_{x \to c^2} x = c^2$
(d) $\lim_{x \to c} x = x$

4. Which of the following statements are incorrect?
(a) The Product Law does not hold if the limit of one of the functions is zero.

(b) The Quotient Law does not hold if the limit of the denominator is zero.

(c) The Quotient Law does not hold if the limit of the numerator is zero.

Exercises

In Exercises 1–22, evaluate the limits using the Limit Laws and the following two facts, where c and k are constants:

$$\lim_{x \to c} x = c, \qquad \lim_{x \to c} k = k$$

1. $\lim_{x \to 9} x$

2. $\lim_{x \to -3} x$

3. $\lim_{x \to 9} 14$

4. $\lim_{x \to -3} 14$

5. $\lim_{x \to -3} (3x + 4)$

6. $\lim_{y \to -3} 14y$

7. $\lim_{y \to -3} (y + 14)$

8. $\lim_{y \to -3} y(y + 14)$

9. $\lim_{t \to 4} (3t - 14)$

10. $\lim_{x \to -5} (x^3 + 2x)$

11. $\lim_{x \to \frac{1}{2}} (4x + 1)(2x - 1)$

12. $\lim_{x \to -1} (3x^4 - 2x^3 + 4x)$

13. $\lim_{x \to 2} x(x + 1)(x + 2)$

14. $\lim_{x \to 2} (x + 1)(3x^2 - 9)$

15. $\lim_{t \to 9} \dfrac{t}{t + 1}$

16. $\lim_{t \to 4} \dfrac{3t - 14}{t + 1}$

17. $\lim_{x \to 3} \dfrac{1 - x}{1 + x}$

18. $\lim_{x \to -1} \dfrac{x}{x^3 + 4x}$

19. $\lim_{t \to 2} t^{-1}$

20. $\lim_{x \to 5} x^{-2}$

21. $\lim_{x \to 3} (x^2 + 9x^{-3})$

22. $\lim_{z \to 1} \dfrac{z^{-1} + z}{z + 1}$

23. Use the Quotient Law to prove that if $\lim_{x \to c} f(x)$ exists and is nonzero, then

$$\lim_{x \to c} \frac{1}{f(x)} = \frac{1}{\lim_{x \to c} f(x)}$$

24. Assume that $\lim_{x \to 6} f(x) = 4$ and compute:

(a) $\lim_{x \to 6} f(x)^2$
(b) $\lim_{x \to 6} \dfrac{1}{f(x)}$
(c) $\lim_{x \to 6} xf(x)$

In Exercises 25–28, evaluate the limit assuming that $\lim_{x \to -4} f(x) = 3$ and $\lim_{x \to -4} g(x) = 1$.

25. $\lim_{x \to -4} f(x)g(x)$

26. $\lim_{x \to -4} (2f(x) + 3g(x))$

27. $\lim_{x \to -4} \dfrac{g(x)}{x^2}$

28. $\lim_{x \to -4} \dfrac{f(x) + 1}{3g(x) - 9}$

29. Can the Quotient Law be applied to evaluate $\lim_{x \to 0} \dfrac{\sin x}{x}$? Explain.

30. Show that the Product Law cannot be used to evaluate the $\lim_{x \to \pi/2} (x - \pi/2) \tan x$.

31. Give an example where $\lim_{x \to 0} (f(x) + g(x))$ exists but neither $\lim_{x \to 0} f(x)$ nor $\lim_{x \to 0} g(x)$ exists.

32. Assume that the limit $L_a = \lim_{x \to 0} \dfrac{a^x - 1}{x}$ exists and that $\lim_{x \to 0} a^x = 1$ for all $a > 0$. Prove that $L_{ab} = L_a + L_b$ for $a, b > 0$. *Hint:* $(ab)^x - 1 = a^x(b^x - 1) + (a^x - 1)$. Verify numerically that $L_{12} = L_3 + L_4$.

33. ✏ Use the Limit Laws and the result $\lim_{x \to c} x = c$ to show that $\lim_{x \to c} x^n = c^n$ for all whole numbers n. If you are familiar with induction, give a formal proof by induction.

34. ✏ Extend Exercise 33 to negative integers.

Further Insights and Challenges

35. Show that if both $\lim\limits_{x\to c} f(x)\,g(x)$ and $\lim\limits_{x\to c} g(x)$ exist and $\lim\limits_{x\to c} g(x)$ is nonzero, then $\lim\limits_{x\to c} f(x)$ exists. *Hint:* Write $f(x) = (f(x)\,g(x))/g(x)$ and apply the Quotient Law.

36. Show that if $\lim\limits_{t\to 3} tg(t) = 12$, then $\lim\limits_{t\to 3} g(t)$ exists and equals 4.

37. Prove that if $\lim\limits_{t\to 3} \dfrac{h(t)}{t} = 5$, then $\lim\limits_{t\to 3} h(t) = 15$.

38. Assuming that $\lim\limits_{x\to 0} \dfrac{f(x)}{x} = 1$, which of the following statements is necessarily true? Why?

(a) $f(0) = 0$ **(b)** $\lim\limits_{x\to 0} f(x) = 0$

39. Prove that if $\lim\limits_{x\to c} f(x) = L \ne 0$ and $\lim\limits_{x\to c} g(x) = 0$, then $\lim\limits_{x\to c} \dfrac{f(x)}{g(x)}$ does not exist.

40. Suppose that $\lim\limits_{h\to 0} g(h) = L$.

(a) Explain why $\lim\limits_{h\to 0} g(ah) = L$ for any constant $a \ne 0$.

(b) If we assume instead that $\lim\limits_{h\to 1} g(h) = L$, is it still necessarily true that $\lim\limits_{h\to 1} g(ah) = L$?

(c) Illustrate the conclusions you reach in (a) and (b) with the function $f(x) = x^2$.

41. Show that if $\lim\limits_{h\to 0} \dfrac{f(h)}{h} = L$, then $\lim\limits_{h\to 0} \dfrac{f(ah)}{h} = aL$ for all $a \ne 0$. *Hint:* Apply the result of Exercise 40 to $g(h) = \dfrac{f(h)}{h}$.

42. In Section 2.2, we mentioned that the number e is characterized by the property $\lim\limits_{h\to 0} \dfrac{e^h - 1}{h} = 1$. Assuming this, show that for $b > 0$,

$$\lim_{h\to 0} \frac{b^h - 1}{h} = \ln b$$

Hint: Apply Exercise 41 to $f(h) = \dfrac{e^h - 1}{h}$ and $a = \ln b$.

43. Assume that the limits

$$L = \lim_{x\to c} g(x) \quad \text{and} \quad M = \lim_{x\to L} f(x)$$

exist and that $M = f(L)$. Which of the following statements is correct? Give an intuitive explanation.

(a) $\lim\limits_{x\to c} g(f(x)) = g(c)$

(b) $\lim\limits_{x\to c} f(g(x)) = f(L)$

(c) $\lim\limits_{x\to L} f(g(x)) = f(c)$

Let $f(x) = x^{-1}$. Use the correct version to evaluate

$$\lim_{x\to 2} f(g(x)), \quad \text{where} \quad \lim_{x\to 2} g(x) = 4$$

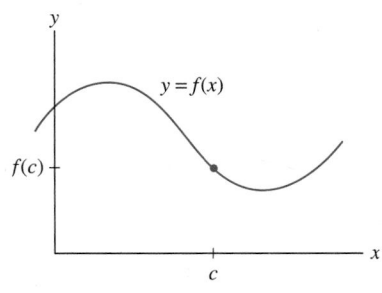

FIGURE 1 $f(x)$ is continuous at $x = c$.

FIGURE 2 $g(x)$ has a discontinuity at $x = c$.

2.4 Limits and Continuity

In everyday speech, the word "continuous" means having no breaks or interruptions. In calculus, continuity is used to describe functions whose graphs have no breaks. If we imagine the graph of a function f as a wavy metal wire, then f is continuous if its graph consists of a single piece of wire as in Figure 1. A break in the wire as in Figure 2 is called a **discontinuity**.

Now observe that the function $g(x)$ in Figure 2 has a discontinuity at $x = c$ and that $\lim\limits_{x\to c} g(x)$ does not exist (the left- and right-hand limits are not equal). By contrast, in Figure 1, $\lim\limits_{x\to c} f(x)$ exists and is equal to the function value $f(c)$. This suggests the following definition of continuity in terms of limits.

> **DEFINITION Continuity at a Point** Assume that $f(x)$ is defined on an open interval containing $x = c$. Then f is **continuous** at $x = c$ if
>
> $$\lim_{x\to c} f(x) = f(c)$$
>
> If the limit does not exist, or if it exists but is not equal to $f(c)$, we say that f has a **discontinuity** (or is **discontinuous**) at $x = c$.

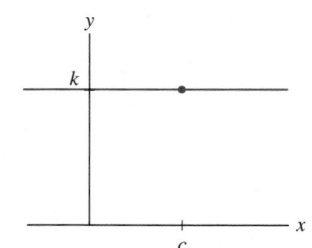

FIGURE 3 Graph of $f(x) = k$.

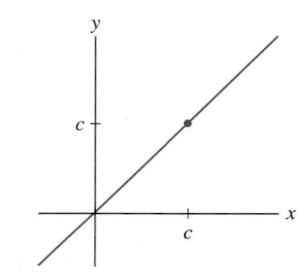

FIGURE 4 Graph of $g(x) = x$.

A function $f(x)$ may be continuous at some points and discontinuous at others. If $f(x)$ is continuous at all points in an interval I, then $f(x)$ is said to be continuous on I. Here, if I is an interval $[a, b]$ or $[a, b)$ that includes a as a left-endpoint, we require that $\lim_{x \to a+} f(x) = f(a)$. Similarly, we require that $\lim_{x \to b-} f(x) = f(b)$ if I includes b as a right-endpoint b. If $f(x)$ is continuous at all points in its domain, then $f(x)$ is simply called continuous.

■ **EXAMPLE 1** Show that the following functions are continuous:

(a) $f(x) = k$ (k any constant)   **(b)** $g(x) = x$

Solution

(a) Since $f(x) = k$ for all x,

$$\lim_{x \to c} f(x) = \lim_{x \to c} k = k = f(c)$$

The limit exists and is equal to the function value, so $f(x)$ is continuous at $x = c$ for all c (Figure 3).

(b) Since $g(x) = x$ for all x,

$$\lim_{x \to c} g(x) = \lim_{x \to c} x = c = g(c)$$

Again, the limit exists and is equal to the function value, so $g(x)$ is continuous at c for all c (Figure 4).   ■

Examples of Discontinuities

To understand continuity better, let's consider some ways that a function can fail to be continuous. Keep in mind that continuity at a point requires more than just the existence of a limit. For $f(x)$ to be continuous at $x = c$, three conditions must hold:

1. $\lim_{x \to c} f(x)$ exists.   **2.** $f(c)$ exists.   **3.** They are equal.

Otherwise, $f(x)$ is discontinuous at $x = c$.

   If the first two conditions hold but the third fails, we say that f has a **removable discontinuity** at $x = c$. The function in Figure 5(A) has a removable discontinuity as $c = 2$ because

$$\underbrace{\lim_{x \to 2} f(x) = 5 \qquad \text{but} \qquad f(2) = 10}_{\text{Limit exists but is not equal to function value}}$$

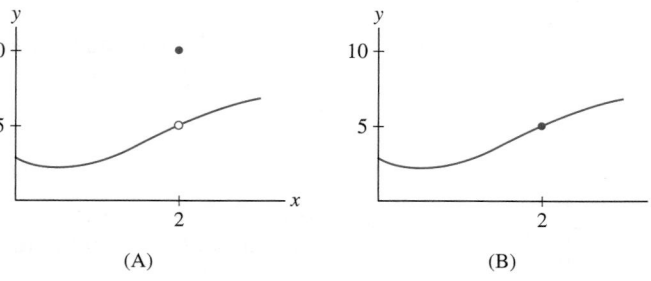

FIGURE 5 A removable discontinuity: The discontinuity can be removed by redefining $f(2)$.

(A)   (B)

Removable discontinuities are "mild" in the following sense: We can make f continuous at $x = c$ by redefining $f(c)$. In Figure 5(B), the value $f(2)$ has been redefined as $f(2) = 5$ and this makes f continuous at $x = 2$.

A "worse" type of discontinuity is a **jump discontinuity**, which occurs if the one-sided limits $\lim\limits_{x \to c-} f(x)$ and $\lim\limits_{x \to c+} f(x)$ exist but are not equal. Figure 6 shows two functions with jump discontinuities at $c = 2$. Unlike the removable case, we cannot make $f(x)$ continuous by redefining $f(c)$.

In connection with jump discontinuities, it is convenient to define *one-sided continuity*.

DEFINITION One-Sided Continuity   A function $f(x)$ is called:

• **Left-continuous** at $x = c$ if $\lim\limits_{x \to c-} f(x) = f(c)$

• **Right-continuous** at $x = c$ if $\lim\limits_{x \to c+} f(x) = f(c)$

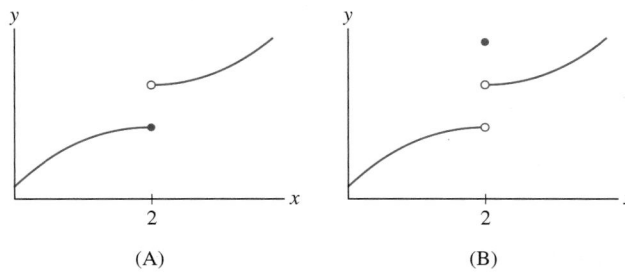

FIGURE 6 Jump discontinuities.

(A)   (B)

In Figure 6 above, the function in (A) is left-continuous but the function in (B) is neither left- nor right-continuous. The next example explores one-sided continuity using a piecewise-defined function, that is, a function defined by different formulas on different intervals.

■ **EXAMPLE 2** Piecewise-Defined Function   Discuss the continuity of the function $F(x)$ defined by

$$F(x) = \begin{cases} x & \text{for } x < 1 \\ 3 & \text{for } 1 \leq x \leq 3 \\ x & \text{for } x > 3 \end{cases}$$

Solution The functions $f(x) = x$ and $g(x) = 3$ are continuous, so $F(x)$ is also continuous, except possibly at the transition points $x = 1$ and $x = 3$ where the formula for $F(x)$ changes (Figure 7). We observe that $F(x)$ has a jump discontinuity at $x = 1$ because the one-sided limits exist but are not equal:

$$\lim_{x \to 1-} F(x) = \lim_{x \to 1-} x = 1, \qquad \lim_{x \to 1+} F(x) = \lim_{x \to 1+} 3 = 3$$

Furthermore, the right-hand limit equals the function value $F(1) = 3$, so $F(x)$ is *right-continuous* at $x = 1$. At $x = 3$,

$$\lim_{x \to 3-} F(x) = \lim_{x \to 3-} 3 = 3, \qquad \lim_{x \to 3+} F(x) = \lim_{x \to 3+} x = 3$$

Both the left- and right-hand limits exist and are equal to $F(3)$, so $F(x)$ is *continuous* at $x = 3$.   ■

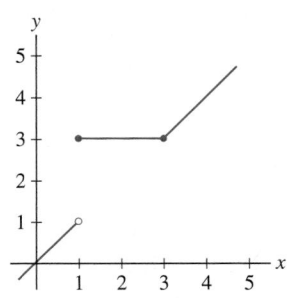

FIGURE 7 Piecewise-defined function $F(x)$ in Example 2.

We say that $f(x)$ has an **infinite discontinuity** at $x = c$ if one or both of the one-sided limits is infinite (even if $f(x)$ itself is not defined at $x = c$). Figure 8 illustrates three types of infinite discontinuities occurring at $x = 2$. Notice that $x = 2$ does not belong to the domain of the function in cases (A) and (B).

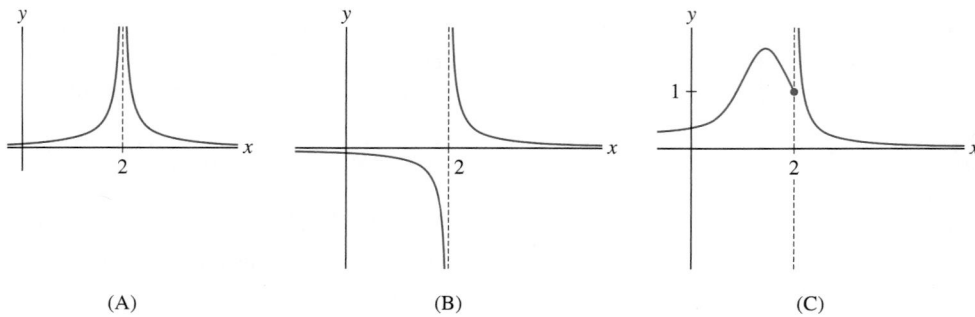

(A)   (B)   (C)

FIGURE 8 Functions with an infinite discontinuity at $x = 2$.

We should mention that some functions have more "severe" types of discontinuity than those discussed above. For example, $f(x) = \sin\dfrac{1}{x}$ oscillates infinitely often between $+1$ and -1 as $x \to 0$ (Figure 9). Neither the left- nor the right-hand limits exist at $x = 0$, so this discontinuity is not a jump discontinuity. See Exercises 90 and 91 for even stranger examples. Although of interest from a theoretical point of view, these discontinuities rarely arise in practice.

Building Continuous Functions

Having studied some examples of discontinuities, we focus again on continuous functions. How can we show that a function is continuous? One way is to use the **Laws of Continuity**, which state, roughly speaking, that a function is continuous if it is built out of functions that are known to be continuous.

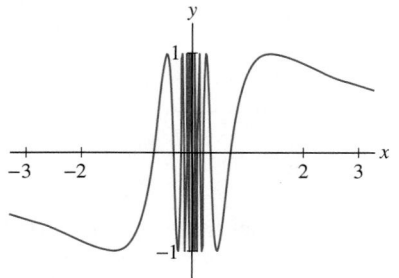

FIGURE 9 Graph of $y = \sin\dfrac{1}{x}$. The discontinuity at $x = 0$ is more "severe." It is neither a jump discontinuity nor a removable or infinite discontinuity.

THEOREM 1 Laws of Continuity   Assume that $f(x)$ and $g(x)$ are continuous at a point $x = c$. Then the following functions are also continuous at $x = c$:

(i) $f(x) + g(x)$ and $f(x) - g(x)$ (iii) $f(x)\,g(x)$

(ii) $kf(x)$ for any constant k (iv) $f(x)/g(x)$ if $g(c) \neq 0$

Proof   These laws follow directly from the corresponding Limit Laws (Theorem 1, Section 2.3). We illustrate by proving the first part of (i) in detail. The remaining laws are proved similarly. By definition, we must show that $\lim\limits_{x \to c}(f(x) + g(x)) = f(c) + g(c)$. Because $f(x)$ and $g(x)$ are both continuous at $x = c$, we have

$$\lim_{x \to c} f(x) = f(c), \qquad \lim_{x \to c} g(x) = g(c)$$

The Sum Law for limits yields the desired result:

$$\lim_{x \to c}(f(x) + g(x)) = \lim_{x \to c} f(x) + \lim_{x \to c} g(x) = f(c) + g(c) \qquad \blacksquare$$

In Section 2.3, we remarked that the Limit Laws for Sums and Products are valid for the sums and products of an arbitrary number of functions. The same is true for continuity, namely, if $f_1(x), \dots, f_n(x)$ are continuous, then so are the functions

$$f_1(x) + f_2(x) + \cdots + f_n(x), \qquad f_1(x) \cdot f_2(x) \cdots f_n(x)$$

We now use the Laws of Continuity to study the continuity of some basic classes of functions.

> **THEOREM 2 Continuity of Polynomial and Rational Functions**   Let $P(x)$ and $Q(x)$ be polynomials. Then:
>
> - $P(x)$ is continuous on the real line.
> - $\dfrac{P(x)}{Q(x)}$ is continuous at all values c such that $Q(c) \neq 0$.

When a function $f(x)$ is defined and continuous for all values of x, we say that $f(x)$ is continuous on the real line.

◀┈ *REMINDER A "rational function" is a quotient of two polynomials $\dfrac{P(x)}{Q(x)}$.*

Proof We use the Laws of Continuity together with the result of Example 1, according to which constant functions and the function $f(x) = x$ are continuous. For any whole number $m \geq 1$, the power function x^m can be written as a product $x^m = x \cdot x \cdots x$ (m factors). Since each factor is continuous, x^m itself is continuous. Similarly, ax^m is continuous for any constant a. Now consider a polynomial

$$P(x) = a_n x^n + a_{n-1} x^{n-1} + \cdots + a_1 x + a_0$$

where $a_0, a_1, \ldots, a_n$ are constants. Each term $a_j x^j$ is continuous, so $P(x)$ is a sum of continuous functions and hence is continuous. Finally, if $Q(x)$ is a polynomial, then by the Continuity Law (iv), $\dfrac{P(x)}{Q(x)}$ is continuous at $x = c$ provided that $Q(c) \neq 0$. ■

This result shows, for example, that $f(x) = 3x^4 - 2x^3 + 8x$ is continuous for all x and that

$$g(x) = \frac{x + 3}{x^2 - 1}$$

is continuous for $x \neq \pm 1$. Furthermore, if n is a whole number, then $f(x) = x^{-n}$ is continuous for $x \neq 0$ since $f(x) = x^{-n} = 1/x^n$ is a rational function.

Most of the basic functions are continuous on their domains. For example, the sine and cosine, exponential and logarithm, and nth root functions are all continuous on their domains, as stated formally in the next theorem. This should not be too surprising because the graphs of these functions have no visible breaks (see Figure 10). However, the formal proofs of continuity are somewhat technical and are thus omitted.

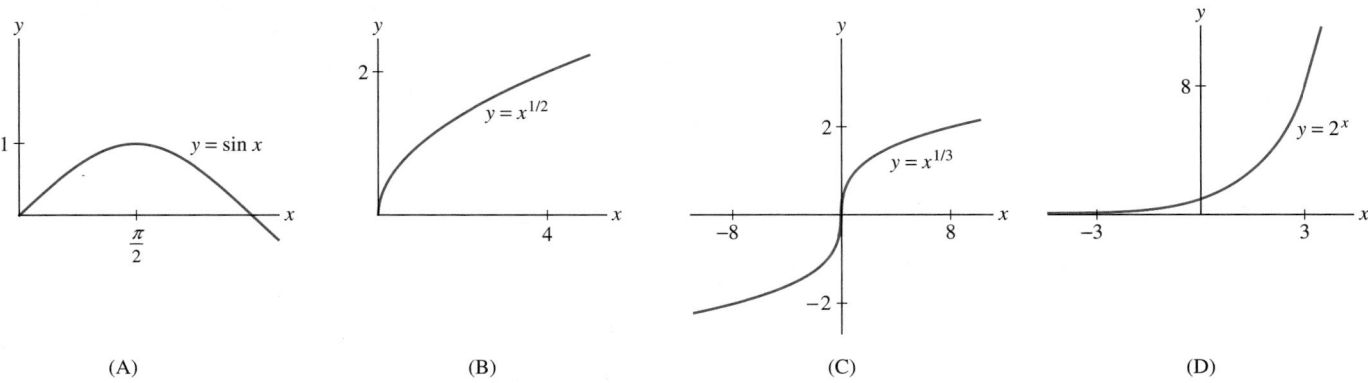

(A)   (B)   (C)   (D)

FIGURE 10 As the graphs suggest, these functions are continuous on their domains.

← ·· **REMINDER** *The domain of* $y = x^{1/n}$ *is the real line if n is odd and the half-line* $[0, \infty)$ *if n is even.*

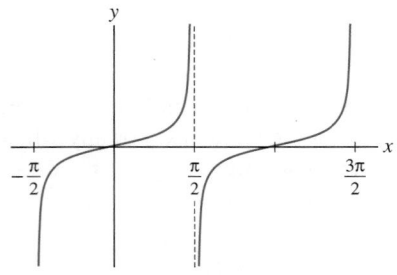

FIGURE 11 Graph of $y = \tan x$.

> **THEOREM 3 Continuity of Some Basic Functions**
> - $y = \sin x$ and $y = \cos x$ are continuous on the real line.
> - For $b > 0$, $y = b^x$ is continuous on the real line.
> - For $b > 0$ and $b \neq 1$, $\log_b x$ is continuous for $x > 0$.
> - If n is a natural number, then $y = x^{1/n}$ is continuous on its domain.

Because $\sin x$ and $\cos x$ are continuous, Continuity Law (iv) for Quotients implies that the other standard trigonometric functions are continuous on their domains, that is, at all values of x where their denominators are nonzero:

$$\tan x = \frac{\sin x}{\cos x}, \qquad \cot x = \frac{\cos x}{\sin x}, \qquad \sec x = \frac{1}{\cos x}, \qquad \csc x = \frac{1}{\sin x}$$

They have infinite discontinuities at points where their denominators are zero. For example, $\tan x$ has infinite discontinuities at the points (Figure 11)

$$x = \pm\frac{\pi}{2}, \quad \pm\frac{3\pi}{2}, \quad \pm\frac{5\pi}{2}, \ldots.$$

According to the next theorem, the inverse of a continuous function $f(x)$ is continuous. This is to be expected because the graph of the inverse $f^{-1}(x)$ is obtained by reflecting the graph of $f(x)$ through the line $y = x$. If the graph of $f(x)$ "has no breaks," the same ought to be true of the graph of $f^{-1}(x)$. We omit the formal proof here.

> **THEOREM 4 Continuity of the Inverse Function**   If $f(x)$ is a continuous function on an interval I with range R and if the inverse $f^{-1}(x)$ exists, then $f^{-1}(x)$ is continuous on the domain R.

We may use this theorem to conclude that the inverse trigonometric functions $\sin^{-1} x$, $\cos^{-1} x$, and $\tan^{-1} x$, etc. are all continuous on their domains.

Many functions of interest are composite functions, so it is important to know that a composition of continuous functions is again continuous. The following theorem is proved in Appendix D.

> **THEOREM 5 Continuity of Composite Functions**   Let $F(x) = f(g(x))$ be a composite function. If g is continuous at $x = c$ and f is continuous at $x = g(c)$, then $F(x)$ is continuous at $x = c$.

For example, $F(x) = (x^2 + 9)^{1/3}$ is continuous because it is the composite of the continuous functions $f(x) = x^{1/3}$ and $g(x) = x^2 + 9$. The function $F(x) = \cos(x^{-1})$ is the composite of $f(x) = \cos x$ and $g(x) = x^{-1}$, so it is continuous for all $x \neq 0$. Similarly, $e^{\sin x}$ is continuous for all x because it is the composite of two functions that are continuous for all x, namely e^x and $\sin x$.

Substitution: Evaluating Limits Using Continuity

It is easy to evaluate a limit when the function in question is known to be continuous. In this case, by definition, the limit is equal to the function value:

$$\lim_{x \to c} f(x) = f(c)$$

We call this the **Substitution Method** because the limit is evaluated by "plugging in" $x = c$.

■ **EXAMPLE 3** Evaluate (a) $\lim\limits_{y \to \frac{\pi}{3}} \sin y$ and (b) $\lim\limits_{x \to -1} \dfrac{3^x}{\sqrt{x+5}}$.

Solution

(a) Since $f(y) = \sin y$ is continuous, we may evaluate the limit using substitution:

$$\lim_{y \to \frac{\pi}{3}} \sin y = \sin \frac{\pi}{3} = \frac{\sqrt{3}}{2}$$

(b) The function $f(x) = \dfrac{3^x}{\sqrt{x+5}}$ is continuous at $x = -1$ because the numerator and denominator are both continuous at $x = -1$ and the denominator $\sqrt{x+5}$ is nonzero at $x = -1$. Therefore, we may evaluate the limit using substitution:

$$\lim_{x \to -1} \frac{3^x}{\sqrt{x+5}} = \frac{3^{-1}}{\sqrt{-1+5}} = \frac{1}{6}$$ ■

■ **EXAMPLE 4** **Assumptions Matter** Can substitution be used to evaluate $\lim\limits_{x \to 2} [x]$, where $[x]$ is the greatest integer function?

Solution Let $f(x) = [x]$ (Figure 12). Although $f(2) = 2$, we cannot conclude that $\lim\limits_{x \to 2} [x]$ is equal to 2. In fact, $f(x)$ is not continuous at $x = 2$, since the one-sided limits are not equal:

$$\lim_{x \to 2+} [x] = 2 \quad \text{and} \quad \lim_{x \to 2-} [x] = 1$$

Therefore, $\lim\limits_{x \to 2} [x]$ does not exist and substitution does not apply. ■

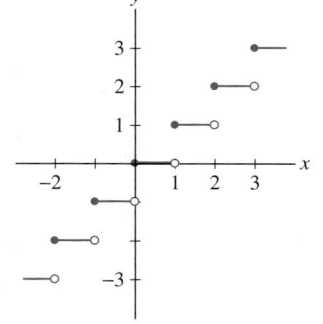

FIGURE 12 Graph of $f(x) = [x]$.

CONCEPTUAL INSIGHT **Real-World Modeling by Continuous Functions** Continuous functions are often used to represent physical quantities such as velocity, temperature, or voltage. This reflects our everyday experience that change in the physical world tends to occur continuously rather than through abrupt transitions. However, mathematical models are at best approximations to reality and it is important to be aware of their limitations.

For example, in Figure 13, atmospheric temperature is represented as a continuous function of altitude. This is justified for large-scale objects such as the earth's atmosphere because the reading on a thermometer appears to vary continuously as altitude changes. However, temperature is a measure of the average kinetic energy of molecules and thus, at the microsopic level, it would not be meaningful to treat temperature as a quantity that varies continuously from point to point.

Population is another quantity that is often treated as a continuous function of time. The size $P(t)$ of a population at time t is a whole number that changes by ± 1 when an individual is born or dies, so strictly speaking, $P(t)$ is not continuous. If the population is large, the effect of an individual birth or death is small and it is reasonable to treat $P(t)$ as a continuous function of time.

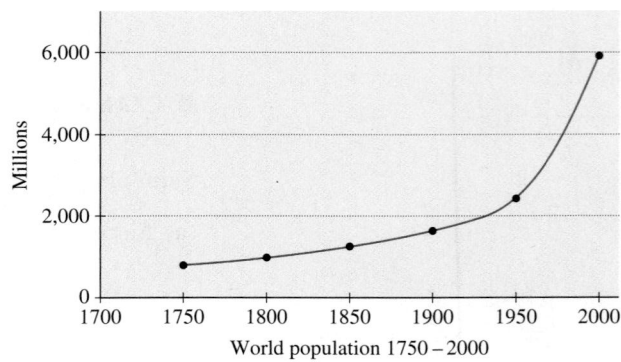

FIGURE 13 Atmospheric temperature and world population are represented by continuous graphs.

2.4 SUMMARY

- By definition, $f(x)$ is *continuous* at $x = c$ if $\lim_{x \to c} f(x) = f(c)$.
- If this limit does not exist, or if it exists but does not equal $f(c)$, then f is *discontinuous* at $x = c$.
- If $f(x)$ is continuous at all points in its domain, f is simply called *continuous*.
- $f(x)$ is *right-continuous* at $x = c$ if $\lim_{x \to c+} f(x) = f(c)$ and is *left-continuous* at $x = c$ if $\lim_{x \to c-} f(x) = f(c)$.
- There are three common types of discontinuities: *removable discontinuity* [$\lim_{x \to c} f(x)$ exists but does not equal $f(c)$], *jump discontinuity* (the one-sided limits both exist but are not equal), and *infinite discontinuity* (the limit is infinite as x approaches c from one or both sides).
- The Laws of Continuity state that sums, products, multiples, and composites of continuous functions are again continuous. The same holds for a quotient $\dfrac{f(x)}{g(x)}$ at points where $g(x) \neq 0$.
- Polynomials, rational functions, trigonometric functions, inverse trigonometric functions, exponential functions, and logarithmic functions are continuous on their domains.
- Substitution Method: If $f(x)$ is known to be continuous at $x = c$, then the value of the limit $\lim_{x \to c} f(x)$ is $f(c)$.

2.4 EXERCISES

Preliminary Questions

1. Which property of $f(x) = x^3$ allows us to conclude that $\lim_{x \to 2} x^3 = 8$?

2. What can be said about $f(3)$ if f is continuous and $\lim_{x \to 3} f(x) = \frac{1}{2}$?

3. Suppose that $f(x) < 0$ if x is positive and $f(x) > 1$ if x is negative. Can f be continuous at $x = 0$?

4. Is it possible to determine $f(7)$ if $f(x) = 3$ for all $x < 7$ and f is right-continuous at $x = 7$?

5. Are the following true or false? If false, state a correct version.

(a) $f(x)$ is continuous at $x = a$ if the left- and right-hand limits of $f(x)$ as $x \to a$ exist and are equal.

(b) $f(x)$ is continuous at $x = a$ if the left- and right-hand limits of $f(x)$ as $x \to a$ exist and equal $f(a)$.

(c) If the left- and right-hand limits of $f(x)$ as $x \to a$ exist, then f has a removable discontinuity at $x = a$.

(d) If $f(x)$ and $g(x)$ are continuous at $x = a$, then $f(x) + g(x)$ is continuous at $x = a$.

(e) If $f(x)$ and $g(x)$ are continuous at $x = a$, then $f(x)/g(x)$ is continuous at $x = a$.

Exercises

1. Find the points of discontinuity of the function shown in Figure 14 and state whether it is left- or right-continuous (or neither) at these points.

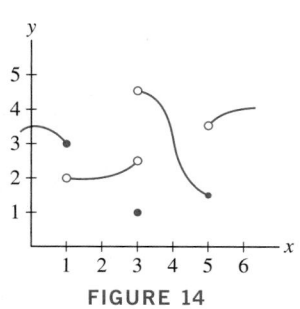

FIGURE 14

In Exercises 2–4, refer to the function $f(x)$ in Figure 15.

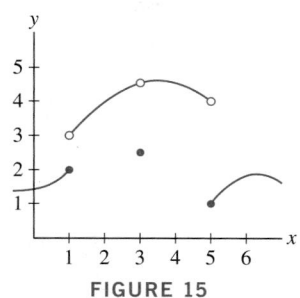

FIGURE 15

2. Find the points of discontinuity of $f(x)$ and state whether $f(x)$ is left- or right-continuous (or neither) at these points.

3. At which point c does $f(x)$ have a removable discontinuity? What value should be assigned to $f(c)$ to make f continuous at $x = c$?

4. Find the point c_1 at which $f(x)$ has a jump discontinuity but is left-continuous. What value should be assigned to $f(c_1)$ to make f right-continuous at $x = c_1$?

5. (a) For the function shown in Figure 16, determine the one-sided limits at the points of discontinuity.

(b) Which of these discontinuities is removable and how should f be redefined to make it continuous at this point?

FIGURE 16

6. Let $f(x)$ be the function (Figure 17):

$$f(x) = \begin{cases} x^2 + 3 & \text{for } x < 1 \\ 10 - x & \text{for } 1 \le x \le 2 \\ 6x - x^2 & \text{for } x > 2 \end{cases}$$

Show that $f(x)$ is continuous for $x \ne 1, 2$. Then compute the right- and left-hand limits at $x = 1$ and 2, and determine whether $f(x)$ is left-continuous, right-continuous, or continuous at these points.

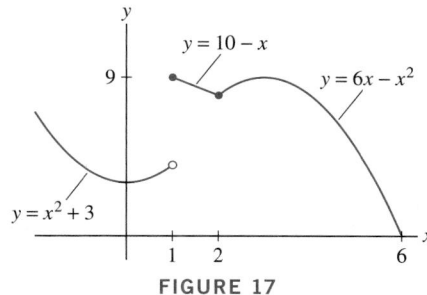

FIGURE 17

In Exercises 7–16, use the Laws of Continuity and Theorems 2–3 to show that the function is continuous.

7. $f(x) = x + \sin x$

8. $f(x) = x \sin x$

9. $f(x) = 3x + 4 \sin x$

10. $f(x) = 3x^3 + 8x^2 - 20x$

11. $f(x) = \dfrac{1}{x^2 + 1}$

12. $f(x) = \dfrac{x^2 - \cos x}{3 + \cos x}$

13. $f(x) = \dfrac{3^x}{1 + 4^x}$

14. $f(x) = 10^x \cos x$

15. $f(x) = e^x \cos 3x$

16. $f(x) = \ln(x^4 + 1)$

In Exercises 17–36, determine the points at which the function is discontinuous and state the type of discontinuity: removable, jump, infinite, or none of these.

17. $f(x) = \dfrac{1}{x}$

18. $f(x) = |x|$

19. $f(x) = \dfrac{x - 2}{|x - 1|}$

20. $f(x) = \dfrac{x - 2}{|x - 2|}$

21. $f(x) = [x]$

22. $f(x) = \left[\dfrac{1}{2}x\right]$

23. $g(t) = \dfrac{1}{t^2 - 1}$

24. $f(x) = \dfrac{x + 1}{4x - 2}$

25. $f(x) = 3x^{3/2} - 9x^3$

26. $g(t) = 3t^{-3/2} - 9t^3$

27. $h(z) = \dfrac{1 - 2z}{z^2 - z - 6}$

28. $h(z) = \dfrac{1 - 2z}{z^2 + 9}$

29. $f(x) = \dfrac{x^2 - 3x + 2}{|x - 2|}$

30. $g(t) = \tan 2t$

31. $f(x) = \csc x^2$

32. $f(x) = \cos\dfrac{1}{x}$

33. $f(x) = \tan(\sin x)$

34. $f(x) = x - |x|$

35. $f(x) = \dfrac{1}{e^x - e^{-x}}$

36. $f(x) = \ln|x - 4|$

In Exercises 37–50, determine the domain of the function and prove that it is continuous on its domain using the Laws of Continuity and the facts quoted in this section.

37. $f(x) = \sqrt{9 - x^2}$

38. $f(x) = \sqrt{x^2 + 9}$

39. $f(x) = \sqrt{x}\,\sin x$

40. $f(x) = \dfrac{x^2}{x + x^{1/4}}$

41. $f(x) = x^{2/3}2^x$

42. $f(x) = x^{1/3} + x^{3/4}$

43. $f(x) = x^{-4/3}$

44. $f(x) = \ln(9 - x^2)$

45. $f(x) = \tan^2 x$

46. $f(x) = \cos(2^x)$

47. $f(x) = (x^4 + 1)^{3/2}$

48. $f(x) = e^{-x^2}$

49. $f(x) = \dfrac{\cos(x^2)}{x^2 - 1}$

50. $f(x) = 9^{\tan x}$

51. Suppose that $f(x) = 2$ for $x > 0$ and $f(x) = -4$ for $x < 0$. What is $f(0)$ if f is left-continuous at $x = 0$? What is $f(0)$ if f is right-continuous at $x = 0$?

52. Sawtooth Function Draw the graph of $f(x) = x - [x]$. At which points is f discontinuous? Is it left- or right-continuous at those points?

In Exercises 53–56, draw the graph of a function on $[0, 5]$ with the given properties.

53. $f(x)$ is not continuous at $x = 1$, but $\lim\limits_{x\to 1+} f(x)$ and $\lim\limits_{x\to 1-} f(x)$ exist and are equal.

54. $f(x)$ is left-continuous but not continuous at $x = 2$ and right-continuous but not continuous at $x = 3$.

55. $f(x)$ has a removable discontinuity at $x = 1$, a jump discontinuity at $x = 2$, and

$$\lim_{x\to 3-} f(x) = -\infty, \qquad \lim_{x\to 3+} f(x) = 2$$

56. $f(x)$ is right- but not left-continuous at $x = 1$, left- but not right-continuous at $x = 2$, and neither left- nor right-continuous at $x = 3$.

57. Each of the following statements is *false*. For each statement, sketch the graph of a function that provides a counterexample.

(a) If $\lim\limits_{x\to a} f(x)$ exists, then $f(x)$ is continuous at $x = a$.

(b) If $f(x)$ has a jump discontinuity at $x = a$, then $f(a)$ is equal to either $\lim\limits_{x\to a-} f(x)$ or $\lim\limits_{x\to a+} f(x)$.

(c) If $f(x)$ has a discontinuity at $x = a$, then $\lim\limits_{x\to a-} f(x)$ and $\lim\limits_{x\to a+} f(x)$ exist but are not equal.

(d) The one-sided limits $\lim\limits_{x\to a-} f(x)$ and $\lim\limits_{x\to a+} f(x)$ always exist, even if $\lim\limits_{x\to a} f(x)$ does not exist.

58. According to the Laws of Continuity, if $f(x)$ and $g(x)$ are continuous at $x = c$, then $f(x) + g(x)$ is continuous at $x = c$. Suppose that $f(x)$ and $g(x)$ are discontinuous at $x = c$. Is it true that $f(x) + g(x)$ is discontinuous at $x = c$? If not, give a counterexample.

In Exercises 59–78, evaluate the limit using the substitution method.

59. $\lim\limits_{x\to 5} x^2$

60. $\lim\limits_{x\to 8} x^3$

61. $\lim\limits_{x\to -1} (2x^3 - 4)$

62. $\lim\limits_{x\to 2} (5x - 12x^{-2})$

63. $\lim\limits_{x\to 0} \dfrac{x + 9}{x - 9}$

64. $\lim\limits_{x\to 3} \dfrac{x + 2}{x^2 + 2x}$

65. $\lim\limits_{x\to \pi} \sin\left(\dfrac{x}{2} - \pi\right)$

66. $\lim\limits_{x\to \pi} \sin(x^2)$

67. $\lim\limits_{x\to \frac{\pi}{4}} \tan(3x)$

68. $\lim\limits_{x\to \pi} \dfrac{1}{\cos x}$

69. $\lim\limits_{x\to 4} x^{-5/2}$

70. $\lim\limits_{x\to 2} \sqrt{x^3 + 4x}$

71. $\lim\limits_{x\to -1} (1 - 8x^3)^{3/2}$

72. $\lim\limits_{x\to 2} \left(\dfrac{7x + 2}{4 - x}\right)^{2/3}$

73. $\lim\limits_{x\to 3} 10^{x^2 - 2x}$

74. $\lim\limits_{x\to \frac{\pi}{2}} 3^{\sin x}$

75. $\lim\limits_{x\to 1} e^{x^2 - x}$

76. $\lim\limits_{x\to 5} \ln(ex + e)$

77. $\lim\limits_{x\to 4} \sin^{-1}\left(\dfrac{x}{4}\right)$

78. $\lim\limits_{x\to 0} \tan^{-1}(e^x)$

In Exercises 79–82, sketch the graph of the given function. At each point of discontinuity, state whether f is left- or right-continuous.

79. $f(x) = \begin{cases} x^2 & \text{for } x \le 1 \\ 2 - x & \text{for } x > 1 \end{cases}$

80. $f(x) = \begin{cases} x + 1 & \text{for } x < 1 \\ \dfrac{1}{x} & \text{for } x \ge 1 \end{cases}$

81. $f(x) = \begin{cases} |x - 3| & \text{for } x \le 3 \\ x - 3 & \text{for } x > 3 \end{cases}$

82. $f(x) = \begin{cases} x^3 + 1 & \text{for } -\infty < x \le 0 \\ -x + 1 & \text{for } 0 < x < 2 \\ -x^2 + 10x - 15 & \text{for } x \ge 2 \end{cases}$

In Exercises 83–84, find the value of c that makes the function continuous.

83. $f(x) = \begin{cases} x^2 - c & \text{for } x < 5 \\ 4x + 2c & \text{for } x \geq 5 \end{cases}$

84. $f(x) = \begin{cases} 2x + 9 & \text{for } x \leq 3 \\ -4x + c & \text{for } x > 3 \end{cases}$

85. Find all constants a, b such that the following function has no discontinuities:

$$f(x) = \begin{cases} ax + \cos x & \text{for } x \leq \dfrac{\pi}{4} \\ bx + 2 & \text{for } x > \dfrac{\pi}{4} \end{cases}$$

86. Which of the following quantities would be represented by continuous functions of time and which would have one or more discontinuities?

(a) Velocity of an airplane during a flight

(b) Temperature in a room under ordinary conditions

(c) Value of a bank account with interest paid yearly

(d) The salary of a teacher

(e) The population of the world

87. ✎ In 1993, the amount $T(x)$ of federal income tax owed on an income of x dollars was determined by the formula

$$T(x) = \begin{cases} 0.15x & \text{for } 0 \leq x < 21{,}450 \\ 3{,}217.50 + 0.28(x - 21{,}450) & \text{for } 21{,}450 \leq x < 51{,}900 \\ 11{,}743.50 + 0.31(x - 51{,}900) & \text{for } x \geq 51{,}900 \end{cases}$$

Sketch the graph of $T(x)$ and determine if it has any discontinuities. Explain why, if $T(x)$ had a jump discontinuity, it might be advantageous in some situations to earn *less* money.

Further Insights and Challenges

88. ✎ If $f(x)$ has a removable discontinuity at $x = c$, then it is possible to redefine $f(c)$ so that $f(x)$ is continuous at $x = c$. Can $f(c)$ be redefined in more than one way?

89. Give an example of functions $f(x)$ and $g(x)$ such that $f(g(x))$ is continuous but $g(x)$ has at least one discontinuity.

90. Function Continuous at Only One Point Let $f(x) = x$ for x rational, $f(x) = -x$ for x irrational. Show that f is continuous at $x = 0$ and discontinuous at all points $x \neq 0$.

91. Let $f(x) = 1$ if x is rational and $f(x) = -1$ if x is irrational. Show that $f(x)$ is discontinuous at all points, whereas $f(x)^2$ is continuous at all points.

2.5 Evaluating Limits Algebraically

In the previous section, we used substitution to calculate limits. This method applies when the function in question is known to be continuous. For example, $f(x) = x^{-2}$ is continuous at $x = 3$ and therefore

$$\lim_{x \to 3} x^{-2} = 3^{-2} = \frac{1}{9}$$

When we study derivatives in Chapter 3, we will be faced with limits $\lim_{x \to c} f(x)$, where $f(c)$ is not defined. In such cases, substitution cannot be used directly. However, some of these limits can be evaluated using substitution, provided that we first use algebra to rewrite the formula for $f(x)$.

To illustrate, consider the limit

$$\lim_{x \to 4} \frac{x^2 - 16}{x - 4}$$

The function $f(x) = \dfrac{x^2 - 16}{x - 4}$ (Figure 1) is not defined at $x = 4$ because

$$f(4) = \frac{4^2 - 16}{4 - 4} = \frac{0}{0} \quad \text{(undefined)}$$

However, the numerator of $f(x)$ factors and

$$\frac{x^2 - 16}{x - 4} = \frac{(x + 4)(x - 4)}{x - 4} = x + 4 \quad \text{(valid for } x \neq 4\text{)}$$

FIGURE 1 Graph of $f(x) = \dfrac{x^2 - 16}{x - 4}$.

In other words, $f(x)$ coincides with the *continuous* function $x + 4$ for all $x \neq 4$. Since the limit depends only on the values of $f(x)$ for $x \neq 4$, we have

$$\lim_{x \to 4} \frac{x^2 - 16}{x - 4} = \underbrace{\lim_{x \to 4} (x + 4) = 8}_{\text{Evaluate by substitution}}$$

Other indeterminate forms are 1^∞, ∞^0, and 0^0. These are treated in Section 4.7.

In general, we say that $f(x)$ has an **indeterminate form** at $x = c$ if, when $f(x)$ is evaluated at $x = c$, we obtain an undefined expression of the type

$$\frac{0}{0}, \quad \frac{\infty}{\infty}, \quad \infty \cdot 0, \quad \infty - \infty$$

We also say that f is **indeterminate** at $x = c$. Our strategy is to *transform $f(x)$ algebraically if possible into a new expression that is defined and continuous at $x = c$, and then evaluate by substitution* ("plugging in"). As you study the following examples, notice that the critical step in each case is to cancel a common factor from the numerator and denominator at the appropriate moment, thereby removing the indeterminacy.

■ **EXAMPLE 1** Calculate $\lim\limits_{x \to 3} \dfrac{x^2 - 4x + 3}{x^2 + x - 12}$.

Solution The function is indeterminate at $x = 3$ because

$$\frac{3^2 - 4(3) + 3}{3^2 + 3 - 12} = \frac{0}{0} \quad \text{(indeterminate)}$$

Step 1. **Transform algebraically and cancel.**
Factor the numerator and denominator, and cancel the common factors:

$$\frac{x^2 - 4x + 3}{x^2 + x - 12} = \underbrace{\frac{(x - 3)(x - 1)}{(x - 3)(x + 4)}}_{\text{Cancel common factor}} = \underbrace{\frac{x - 1}{x + 4}}_{\text{Continuous at } x = 3} \quad \text{(if } x \neq 3\text{)} \qquad \boxed{1}$$

Step 2. **Substitute (evaluate using continuity).**
Since the expression on the right in Eq. (1) is *continuous* at $x = 3$,

$$\lim_{x \to 3} \frac{x^2 - 4x + 3}{x^2 + x - 12} = \underbrace{\lim_{x \to 3} \frac{(x - 1)}{(x + 4)} = \frac{2}{7}}_{\text{Evaluate by substitution}}$$
■

■ **EXAMPLE 2** The Form $\frac{\infty}{\infty}$ Calculate $\lim\limits_{x \to \frac{\pi}{2}} \dfrac{\tan x}{\sec x}$.

Solution Both the numerator and denominator of $f(x) = \dfrac{\tan x}{\sec x}$ have infinite discontinuities at $x = \frac{\pi}{2}$:

$$\frac{\tan \dfrac{\pi}{2}}{\sec \dfrac{\pi}{2}} = \frac{1/0}{1/0} \quad \text{(indeterminate)}$$

We say that $f(x)$ has the indeterminate form $\frac{\infty}{\infty}$ at $x = \frac{\pi}{2}$.

Step 1. **Transform algebraically and cancel.**

$$\frac{\tan x}{\sec x} = \frac{(\sin x)\left(\dfrac{1}{\cos x}\right)}{\dfrac{1}{\cos x}} = \sin x \quad (\text{if } \cos x \neq 0)$$

Step 2. **Substitution (evaluate using continuity).**

Since $\sin x$ is continuous,

$$\lim_{x \to \frac{\pi}{2}} \frac{\tan x}{\sec x} = \lim_{x \to \frac{\pi}{2}} \sin x = \sin \frac{\pi}{2} = 1 \qquad \blacksquare$$

The next example illustrates the algebraic technique of "multiplying by the conjugate," which we use to treat some indeterminate forms involving square roots.

■ **EXAMPLE 3** **Multiplying by the Conjugate** Evaluate:

(a) $\displaystyle\lim_{x \to 4} \frac{\sqrt{x} - 2}{x - 4}$

(b) $\displaystyle\lim_{h \to 5} \frac{h - 5}{\sqrt{h + 4} - 3}$.

Solution

(a) The function $f(x) = \dfrac{\sqrt{x} - 2}{x - 4}$ is indeterminate at $x = 4$ since

$$f(4) = \frac{\sqrt{4} - 2}{4 - 4} = \frac{0}{0} \quad (\text{indeterminate})$$

Step 1. **Transform algebraically and cancel.**

The conjugate of $\sqrt{x} - 2$ is $\sqrt{x} + 2$, so we have

In Step 1, observe that
$(\sqrt{x} - 2)(\sqrt{x} + 2) = x - 4.$

$$\left(\frac{\sqrt{x} - 2}{x - 4}\right)\left(\frac{\sqrt{x} + 2}{\sqrt{x} + 2}\right) = \frac{x - 4}{(x - 4)(\sqrt{x} + 2)} = \frac{1}{\sqrt{x} + 2} \quad (\text{if } x \neq 4)$$

Step 2. **Substitute (evaluate using continuity).**

Since $\dfrac{1}{\sqrt{x} + 2}$ is continuous at $x = 4$,

$$\lim_{x \to 4} \frac{\sqrt{x} - 2}{x - 4} = \lim_{x \to 4} \frac{1}{\sqrt{x} + 2} = \frac{1}{4}$$

(b) The function $f(h) = \dfrac{h - 5}{\sqrt{h + 4} - 3}$ is indeterminate at $h = 5$ since

$$f(5) = \frac{5 - 5}{\sqrt{5 + 4} - 3} = \frac{5 - 5}{3 - 3} = \frac{0}{0} \quad (\text{indeterminate})$$

Step 1. **Transform algebraically.**

The conjugate of $\sqrt{h + 4} - 3$ is $\sqrt{h + 4} + 3$, and

$$\frac{h - 5}{\sqrt{h + 4} - 3} = \left(\frac{h - 5}{\sqrt{h + 4} - 3}\right)\left(\frac{\sqrt{h + 4} + 3}{\sqrt{h + 4} + 3}\right) = \frac{(h - 5)(\sqrt{h + 4} + 3)}{(\sqrt{h + 4} - 3)(\sqrt{h + 4} + 3)}$$

The denominator is equal to

$$(\sqrt{h + 4} - 3)(\sqrt{h + 4} + 3) = (\sqrt{h + 4})^2 - 9 = h - 5$$

Thus, for $h \neq 5$,

$$f(h) = \frac{h-5}{\sqrt{h+4}-3} = \frac{(h-5)\left(\sqrt{h+4}+3\right)}{h-5} = \sqrt{h+4}+3$$

Step 2. Substitute (evaluate using continuity).

$$\lim_{h \to 5} \frac{h-5}{\sqrt{h+4}-3} = \lim_{h \to 5} \left(\sqrt{h+4}+3\right) = \sqrt{9}+3 = 6 \qquad \blacksquare$$

■ **EXAMPLE 4** The Form $\infty - \infty$ Calculate $\lim\limits_{x \to 1} \left(\dfrac{1}{x-1} - \dfrac{2}{x^2-1} \right)$.

Solution This example gives rise to the indeterminate form $\infty - \infty$ because

$$\frac{1}{1-1} - \frac{2}{1^2-1} = \frac{1}{0} - \frac{2}{0} \quad \text{(indeterminate)}$$

Step 1. Transform algebraically and cancel.
To rewrite the expression, combine the fractions and simplify (for $x \neq 1$):

$$\frac{1}{x-1} - \frac{2}{x^2-1} = \frac{x+1}{x^2-1} - \frac{2}{x^2-1} = \frac{x-1}{x^2-1} = \frac{x-1}{(x-1)(x+1)} = \frac{1}{x+1}$$

Step 2. Substitute (evaluate using continuity).

$$\lim_{x \to 1} \left(\frac{1}{x-1} - \frac{2}{x^2-1} \right) = \lim_{x \to 1} \frac{1}{x+1} = \frac{1}{1+1} = \frac{1}{2} \qquad \blacksquare$$

■ **EXAMPLE 5** Function Is Infinite but Not Indeterminate Calculate $\lim\limits_{x \to 2} \dfrac{x^2 - x + 5}{x-2}$.

Solution Let $f(x) = \dfrac{x^2 - x + 5}{x-2}$. At $x = 2$ we have

$$f(2) = \frac{2^2 - 2 + 5}{2-2} = \frac{7}{0} \quad \text{(undefined, but not an indeterminate form)}$$

This is *not* an indeterminate form. In fact, Figure 2 shows that the one-sided limits are infinite:

$$\lim_{x \to 2-} \frac{x^2 - x + 5}{x-2} = -\infty, \qquad \lim_{x \to 2+} \frac{x^2 - x + 5}{x-2} = \infty$$

The limit itself does not exist. $\blacksquare$

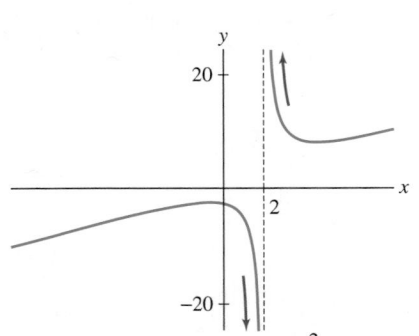

FIGURE 2 Graph of $f(x) = \dfrac{x^2 - x + 5}{x-2}$.

As preparation for the derivative in Chapter 3, we illustrate how to evaluate a limit that involves a symbolic constant.

■ **EXAMPLE 6** Limit Involving a Symbolic Constant Calculate $\lim\limits_{h \to 0} \dfrac{(h+a)^2 - a^2}{h}$, where a is a constant.

Solution The limit is indeterminate at $h = 0$ because

$$\frac{(0+a)^2 - a^2}{0} = \frac{a^2 - a^2}{0} = \frac{0}{0} \quad \text{(indeterminate)}$$

Step 1. Transform algebraically and cancel.
We expand the numerator and simplify (for $h \neq 0$):

$$\frac{(h+a)^2 - a^2}{h} = \frac{(h^2 + 2ah + a^2) - a^2}{h} = \frac{h^2 + 2ah}{h}$$

$$= \frac{h(h + 2a)}{h} \qquad \text{(factor out } h\text{)}$$

$$= h + 2a \qquad \text{(cancel } h\text{)}$$

Step 2. Substitute (evaluate using continuity).
The function $h + 2a$ is continuous (since h is continuous and a is a constant), so

$$\lim_{h \to 0} \frac{(h+a)^2 - a^2}{h} = \lim_{h \to 0} (h + 2a) = 2a \qquad \blacksquare$$

2.5 SUMMARY

- When $f(x)$ is known to be continuous at $x = c$, the limit can be evaluated by substitution: $\lim_{x \to c} f(x) = f(c)$.
- We say that $f(x)$ is *indeterminate* (or has an *indeterminate form*) at $x = c$ if, when we evaluate $f(c)$, we obtain an undefined expression of the type

$$\frac{0}{0}, \quad \frac{\infty}{\infty}, \quad \infty \cdot 0, \quad \infty - \infty$$

- Substitution Method: If $f(x)$ is indeterminate at $x = c$, try using algebra to transform $f(x)$ into a new function that coincides with $f(x)$ near c and is continuous at $x = c$. Then evaluate the limit by applying substitution to the continuous function.

2.5 EXERCISES

Preliminary Questions

1. Which of the following is indeterminate at $x = 1$?

$$\frac{x^2 + 1}{x - 1}, \qquad \frac{x^2 - 1}{x + 2}, \qquad \frac{x^2 - 1}{\sqrt{x + 3} - 2}, \qquad \frac{x^2 + 1}{\sqrt{x + 3} - 2}$$

2. Give counterexamples to show that each of the following statements is false:

(a) If $f(c)$ is indeterminate, then the right- and left-hand limits as $x \to c$ are not equal.

(b) If $\lim_{x \to c} f(x)$ exists, then $f(c)$ is not indeterminate.

(c) If $f(x)$ is undefined at $x = c$, then $f(x)$ has an indeterminate form at $x = c$.

3. Although the method for evaluating limits discussed in this section is sometimes called "simplify and plug in," explain how it actually relies on the property of continuity.

Exercises

In Exercises 1–4, show that the limit leads to an indeterminate form. Then carry out the two-step procedure: Transform the function algebraically and evaluate using continuity.

1. $\lim_{x \to 5} \dfrac{x^2 - 25}{x - 5}$

2. $\lim_{x \to -1} \dfrac{x + 1}{x^2 + 2x + 1}$

3. $\lim_{t \to 7} \dfrac{2t - 14}{5t - 35}$

4. $\lim_{h \to -3} \dfrac{h + 3}{h^2 - 9}$

In Exercises 5–34, evaluate the limit or state that it does not exist.

5. $\lim_{x \to 8} \dfrac{x^2 - 64}{x - 8}$

6. $\lim_{x \to 8} \dfrac{x^2 - 64}{x - 9}$

7. $\lim\limits_{x\to2}\dfrac{x^2-3x+2}{x-2}$

8. $\lim\limits_{x\to2}\dfrac{x^3-4x}{x-2}$

9. $\lim\limits_{x\to2}\dfrac{x-2}{x^3-4x}$

10. $\lim\limits_{x\to2}\dfrac{x^3-2x}{x-2}$

11. $\lim\limits_{h\to0}\dfrac{(1+h)^3-1}{h}$

12. $\lim\limits_{h\to4}\dfrac{(h+2)^2-9h}{h-4}$

13. $\lim\limits_{x\to2}\dfrac{3x^2-4x-4}{2x^2-8}$

14. $\lim\limits_{t\to-2}\dfrac{2t+4}{12-3t^2}$

15. $\lim\limits_{y\to2}\dfrac{(y-2)^3}{y^3-5y+2}$

16. $\lim\limits_{x\to16}\dfrac{\sqrt{x}-4}{x-16}$

17. $\lim\limits_{h\to0}\dfrac{\frac{1}{3+h}-\frac{1}{3}}{h}$

18. $\lim\limits_{h\to0}\dfrac{\frac{1}{(h+2)^2}-\frac{1}{4}}{h}$

19. $\lim\limits_{h\to0}\dfrac{\sqrt{2+h}-2}{h}$

20. $\lim\limits_{x\to10}\dfrac{\sqrt{x-6}-2}{x-10}$

21. $\lim\limits_{x\to2}\dfrac{x-2}{\sqrt{x}-\sqrt{4-x}}$

22. $\lim\limits_{x\to0}\dfrac{\sqrt{1+x}-\sqrt{1-x}}{x}$

23. $\lim\limits_{x\to2}\dfrac{\sqrt{x^2-1}-\sqrt{x+1}}{x-3}$

24. $\lim\limits_{x\to4}\dfrac{\sqrt{5-x}-1}{2-\sqrt{x}}$

25. $\lim\limits_{x\to4}\left(\dfrac{1}{\sqrt{x}-2}-\dfrac{4}{x-4}\right)$

26. $\lim\limits_{x\to0+}\left(\dfrac{1}{\sqrt{x}}-\dfrac{1}{\sqrt{x^2+x}}\right)$

27. $\lim\limits_{x\to0}\dfrac{\cot x}{\csc x}$

28. $\lim\limits_{\theta\to\frac{\pi}{2}}\dfrac{\cot\theta}{\csc\theta}$

29. $\lim\limits_{x\to\frac{\pi}{4}}\dfrac{\sin x-\cos x}{\tan x-1}$

30. $\lim\limits_{\theta\to\frac{\pi}{2}}\left(\sec\theta-\tan\theta\right)$

31. $\lim\limits_{x\to0}\dfrac{e^x-e^{2x}}{1-e^x}$

32. $\lim\limits_{x\to0}\dfrac{3^{2x}-1}{3^x-1}$

33. $\lim\limits_{x\to\frac{\pi}{3}}\dfrac{2\cos^2 x+3\cos x-2}{2\cos x-1}$

34. $\lim\limits_{\theta\to0}\dfrac{\cos\theta-1}{\sin\theta}$

35. [GU] Use a plot of $f(x)=\dfrac{x-2}{\sqrt{x}-\sqrt{4-x}}$ to estimate $\lim\limits_{x\to2}f(x)$ to two decimal places. Compare with the answer obtained algebraically in Exercise 21.

36. [GU] Use a plot of $f(x)=\dfrac{1}{\sqrt{x}-2}-\dfrac{4}{x-4}$ to evaluate $\lim\limits_{x\to4}f(x)$ numerically. Compare with the answer obtained algebraically in Exercise 25.

In Exercises 37–42, use the identity $a^3-b^3=(a-b)(a^2+ab+b^2)$.

37. $\lim\limits_{x\to1}\dfrac{x^3-1}{x-1}$

38. $\lim\limits_{x\to2}\dfrac{x^3-8}{x^2-4}$

39. $\lim\limits_{x\to1}\dfrac{x^2-3x+2}{x^3-1}$

40. $\lim\limits_{x\to2}\dfrac{x^3-8}{x^2-6x+8}$

41. $\lim\limits_{x\to1}\dfrac{x^4-1}{x^3-1}$

42. $\lim\limits_{x\to27}\dfrac{x-27}{x^{1/3}-3}$

In Exercises 43–52, evaluate the limits in terms of the constants involved.

43. $\lim\limits_{x\to0}(2a+x)$

44. $\lim\limits_{h\to-2}(4ah+7a)$

45. $\lim\limits_{t\to-1}(4t-2at+3a)$

46. $\lim\limits_{h\to0}\dfrac{(3a+h)^2-9a^2}{h}$

47. $\lim\limits_{x\to0}\dfrac{2(x+h)^2-2x^2}{h}$

48. $\lim\limits_{x\to a}\dfrac{(x+a)^2-4x^2}{x-a}$

49. $\lim\limits_{x\to a}\dfrac{\sqrt{x}-\sqrt{a}}{x-a}$

50. $\lim\limits_{h\to0}\dfrac{\sqrt{a+2h}-\sqrt{a}}{h}$

51. $\lim\limits_{x\to0}\dfrac{(x+a)^3-a^3}{x}$

52. $\lim\limits_{h\to a}\dfrac{\frac{1}{h}-\frac{1}{a}}{h-a}$

Further Insights and Challenges

In Exercises 53–54, find all values of c such that the limit exists.

53. $\lim\limits_{x\to c}\dfrac{x^2-5x-6}{x-c}$

54. $\lim\limits_{x\to1}\dfrac{x^2+3x+c}{x-1}$

55. For which sign $\pm$ does the following limit exist?
$$\lim\limits_{x\to0}\left(\dfrac{1}{x}\pm\dfrac{1}{x(x-1)}\right)$$

56. For which sign $\pm$ does the following limit exist?
$$\lim\limits_{x\to0+}\left(\dfrac{1}{x}\pm\dfrac{1}{|x|}\right)$$

2.6 Trigonometric Limits

In our study of the derivative, we will need to evaluate certain limits involving transcendental functions such as sine and cosine. The algebraic techniques of the previous section are often ineffective for such functions and other tools are required. One such tool is the

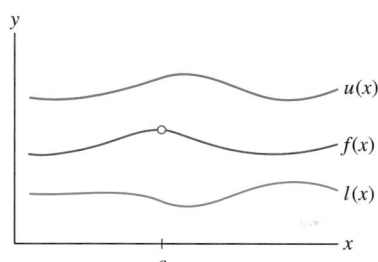

FIGURE 1 $f(x)$ is trapped between $l(x)$ and $u(x)$ (but not squeezed at $x = c$).

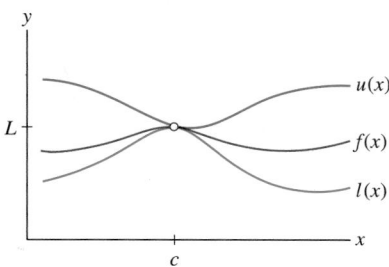

FIGURE 2 $f(x)$ is squeezed by $l(x)$ and $u(x)$ at $x = c$.

Squeeze Theorem, which we discuss in this section and use to evaluate the trigonometric limits needed in Section 3.6.

The Squeeze Theorem

Consider a function $f(x)$ that is trapped between two functions $l(x)$ and $u(x)$ on an interval I. In other words,

$$l(x) \le f(x) \le u(x) \quad \text{for all } x \in I$$

In this case, the graph of $f(x)$ lies between the graphs of $l(x)$ and $u(x)$ (Figure 1), with $l(x)$ as the lower and $u(x)$ as the upper function.

The Squeeze Theorem applies when $f(x)$ is not just trapped, but actually **squeezed** at a point $x = c$ by $l(x)$ and $u(x)$ (Figure 2). By this we mean that for all $x \ne c$ in some open interval containing c,

$$l(x) \le f(x) \le u(x) \quad \text{and} \quad \lim_{x \to c} l(x) = \lim_{x \to c} u(x) = L$$

We do not require that $f(x)$ be defined at $x = c$, but it is clear graphically that $f(x)$ must approach the limit L (Figure 2). We state this formally in the next theorem. A proof is given in Appendix D.

THEOREM 1 Squeeze Theorem Assume that for $x \ne c$ (in some open interval containing c),

$$l(x) \le f(x) \le u(x) \quad \text{and} \quad \lim_{x \to c} l(x) = \lim_{x \to c} u(x) = L$$

Then $\lim_{x \to c} f(x)$ exists and

$$\lim_{x \to c} f(x) = L$$

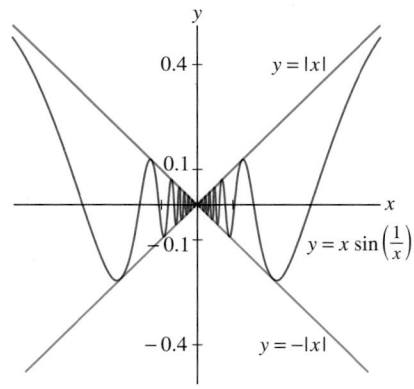

FIGURE 3

■ **EXAMPLE 1** Show that $\lim_{x \to 0} x \sin \dfrac{1}{x} = 0$.

Solution Although $f(x) = x \sin \dfrac{1}{x}$ is a product of two functions, we cannot use the Product Law because $\lim_{x \to 0} \sin \dfrac{1}{x}$ does not exist. Instead, we apply the Squeeze Theorem. The sine function takes on values between 1 and -1 and therefore $\left| \sin \dfrac{1}{x} \right| \le 1$ for all $x \ne 0$. Multiplying by $|x|$, we obtain $\left| x \sin \dfrac{1}{x} \right| \le |x|$ and conclude that (see Figure 3)

$$-|x| \le x \sin \dfrac{1}{x} \le |x|$$

Since

$$\lim_{x \to 0} |x| = 0 \quad \text{and} \quad \lim_{x \to 0} (-|x|) = -\lim_{x \to 0} |x| = 0$$

we can apply the Squeeze Theorem to conclude that $\lim_{x \to 0} x \sin \dfrac{1}{x} = 0$.   ■

The next theorem states two important limits that will be used to develop the calculus of trigonometric functions in Section 3.6.

> **THEOREM 2 Important Trigonometric Limits**
>
> $$\lim_{\theta \to 0} \frac{\sin \theta}{\theta} = 1, \qquad \lim_{\theta \to 0} \frac{1 - \cos \theta}{\theta} = 0$$

Note that both $\dfrac{\sin \theta}{\theta}$ and $\dfrac{\cos \theta - 1}{\theta}$ are indeterminate at $\theta = 0$, so Theorem 2 cannot be proved by substitution.

We use the Squeeze Theorem to evaluate the first limit in Theorem 2. To do so, we must find functions that squeeze $(\sin \theta)/\theta$ at $\theta = 0$. These are provided by the next theorem (see Figure 4).

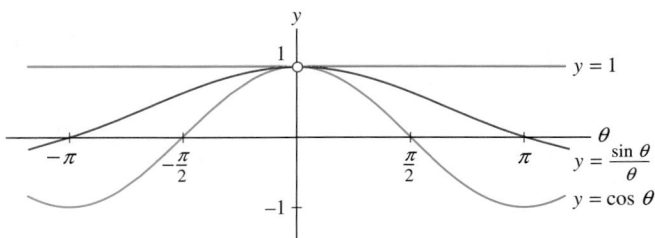

FIGURE 4 Graph illustrating the inequalities of Theorem 3.

> **THEOREM 3**
>
> $$\cos \theta \le \frac{\sin \theta}{\theta} \le 1 \quad \text{for} \quad -\frac{\pi}{2} < \theta < \frac{\pi}{2}, \ \theta \ne 0$$
>
> $\boxed{1}$

Proof Assume first that $0 < \theta < \frac{\pi}{2}$. Our proof is based on the following relation between the areas in Figure 5:

$$\text{Area of } \triangle OAB < \text{area of sector } BOA < \text{area of } \triangle OAC \qquad \boxed{2}$$

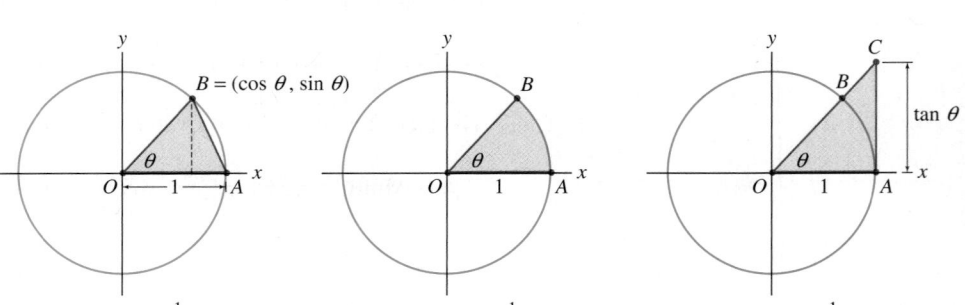

Area of triangle = $\frac{1}{2} \sin \theta$ Area of sector = $\frac{1}{2} \theta$ Area of triangle = $\frac{1}{2} \tan \theta$

FIGURE 5

Let's compute these three areas. First, $\triangle OAB$ has base 1 and height $\sin \theta$, so its area is $\frac{1}{2} \sin \theta$. Next, recall that a sector of angle θ has area $\frac{1}{2}\theta$. Finally, to compute the area of $\triangle OAC$, we observe that

$$\tan \theta = \frac{\text{opposite side}}{\text{adjacent side}} = \frac{AC}{OA} = \frac{AC}{1} = AC$$

◄·· **REMINDER** Let's recall why a sector of angle θ in a circle of radius r has area $\frac{1}{2}r^2\theta$. Since the entire circle is a sector of angle 2π, a sector of angle θ represents a fraction $\frac{\theta}{2\pi}$ of the entire circle. The circle has area πr^2, and hence the sector has area $\left(\frac{\theta}{2\pi}\right)\pi r^2 = \frac{1}{2}r^2\theta$. In the unit circle $(r = 1)$, the sector has area $\frac{1}{2}\theta$.

Thus, $\triangle OAC$ has base 1 and height $\tan \theta$ and its area is $\frac{1}{2} \tan \theta$. We have therefore shown that

$$\underbrace{\frac{1}{2} \sin \theta}_{\text{Area } \triangle OAB} \leq \underbrace{\frac{1}{2}\theta}_{\text{Area of sector}} \leq \underbrace{\frac{1}{2}\frac{\sin \theta}{\cos \theta}}_{\text{Area } \triangle OAC} \qquad \boxed{3}$$

The first inequality gives $\sin \theta \leq \theta$, and since $\theta > 0$, we obtain

$$\frac{\sin \theta}{\theta} \leq 1 \qquad \boxed{4}$$

Next, multiply the second inequality in (3) by $(2 \cos \theta)/\theta$ to obtain

$$\cos \theta \leq \frac{\sin \theta}{\theta} \qquad \boxed{5}$$

The combination of (4) and (5) gives us (1) when $0 < \theta < \frac{\pi}{2}$. However, the functions in (1) do not change when θ is replaced by $-\theta$ because both $\cos \theta$ and $\frac{\sin \theta}{\theta}$ are even functions. Indeed, $\cos(-\theta) = \cos \theta$ and

$$\frac{\sin(-\theta)}{-\theta} = \frac{-\sin \theta}{-\theta} = \frac{\sin \theta}{\theta}$$

Therefore, (1) automatically holds for $-\frac{\pi}{2} < \theta < 0$ as well. This completes the proof of Theorem 3. ∎

Proof of Theorem 2 According to Theorem 3,

$$\cos \theta \leq \frac{\sin \theta}{\theta} \leq 1$$

Furthermore,

$$\lim_{\theta \to 0} \cos \theta = \cos 0 = 1 \qquad \text{and} \qquad \lim_{\theta \to 0} 1 = 1$$

Therefore, the Squeeze Theorem gives $\displaystyle\lim_{\theta \to 0} \frac{\sin \theta}{\theta} = 1$ as required. For a proof that $\displaystyle\lim_{\theta \to 0} \frac{1 - \cos \theta}{\theta} = 0$, see Exercises 42 and 51. ∎

In the next example, we evaluate another trigonometric limit. The key idea is to rewrite the function of h in terms of the new variable $\theta = 4h$.

h	$\dfrac{\sin 4h}{h}$
± 14	-0.75680
± 0.5	1.81859
± 0.2	3.58678
± 0.1	$3.89\,418$
± 0.05	$3.97\,339$
± 0.01	$3.99\,893$
± 0.005	$3.999\,73$

■ **EXAMPLE 2** Evaluating a Limit by Changing Variables Investigate $\displaystyle\lim_{h \to 0} \frac{\sin 4h}{h}$ numerically and then evaluate it exactly.

Solution The table of values at the left suggests that the limit is equal to 4. To evaluate the limit exactly, we rewrite it in terms of the limit of $(\sin \theta)/\theta$ so that Theorem 2 can be applied. Thus, we set $\theta = 4h$ and write

$$\frac{\sin 4h}{h} = 4\left(\frac{\sin 4h}{4h}\right) = 4\frac{\sin \theta}{\theta}$$

The new variable θ tends to zero as $h \to 0$ since θ is a multiple of h. Therefore, we may change the limit as $h \to 0$ into a limit as $\theta \to 0$ to obtain

$$\lim_{h \to 0} \frac{\sin 4h}{h} = \lim_{\theta \to 0} 4 \frac{\sin \theta}{\theta} = 4 \left(\lim_{\theta \to 0} \frac{\sin \theta}{\theta} \right) = 4(1) = 4 \qquad \boxed{6}$$

∎

2.6 SUMMARY

- We say that $f(x)$ is *squeezed* at $x = c$ if there exist functions $l(x)$ and $u(x)$ such that $l(x) \le f(x) \le u(x)$ for all $x \ne c$ in an open interval I containing c, and

$$\lim_{x \to c} l(x) = \lim_{x \to c} u(x) = L$$

The Squeeze Theorem states that in this case, $\lim_{x \to c} f(x) = L$.

- Two important trigonometric limits:

$$\lim_{\theta \to 0} \frac{\sin \theta}{\theta} = 1, \qquad \lim_{\theta \to 0} \frac{1 - \cos \theta}{\theta} = 0$$

2.6 EXERCISES

Preliminary Questions

1. Assume that $-x^4 \le f(x) \le x^2$. What is $\lim_{x \to 0} f(x)$? Is there enough information to evaluate $\lim_{x \to \frac{1}{2}} f(x)$? Explain.

2. State the Squeeze Theorem carefully.

3. Suppose that $f(x)$ is squeezed at $x = c$ on an open interval I by two *constant* functions $u(x)$ and $l(x)$. Is $f(x)$ necessarily constant on I?

4. If you want to evaluate $\lim_{h \to 0} \frac{\sin 5h}{3h}$, it is a good idea to rewrite the limit in terms of the variable (choose one):

(a) $\theta = 5h$ **(b)** $\theta = 3h$ **(c)** $\theta = \frac{5h}{3}$

Exercises

1. In Figure 6, is $f(x)$ squeezed by $u(x)$ and $l(x)$ at $x = 3$? At $x = 2$?

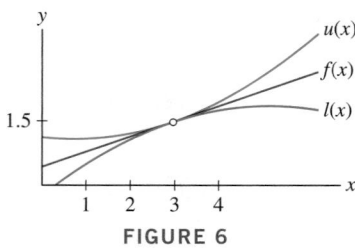

FIGURE 6

2. What is $\lim_{x \to 0} f(x)$ if $f(x)$ satisfies $\cos x \le f(x) \le 1$ for x in $(-1, 1)$?

3. What information about $f(x)$ does the Squeeze Theorem provide if we assume that the graphs of $f(x)$, $u(x)$, and $l(x)$ are related as in Figure 7 and that $\lim_{x \to 7} u(x) = \lim_{x \to 7} l(x) = 6$? Note that the inequality $f(x) \le u(x)$ is not satisfied for all x. Does this affect the validity of your conclusion?

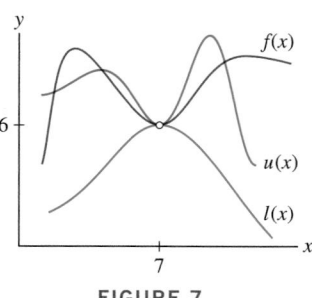

FIGURE 7

4. State whether the inequality provides sufficient information to determine $\lim_{x \to 1} f(x)$ and, if so, find the limit.

(a) $4x - 5 \leq f(x) \leq x^2$

(b) $2x - 1 \leq f(x) \leq x^2$

(c) $2x + 1 \leq f(x) \leq x^2 + 2$

In Exercises 5–8, use the Squeeze Theorem to evaluate the limit.

5. $\lim\limits_{x \to 0} x \cos \dfrac{1}{x}$

6. $\lim\limits_{x \to 0} x^2 \sin \dfrac{1}{x}$

7. $\lim\limits_{x \to 0+} \sqrt{x}\, e^{\cos(\pi/x)}$

8. $\lim\limits_{x \to 1} (x - 1) \sin \dfrac{\pi}{x - 1}$

In Exercises 9–16, evaluate the limit using Theorem 2 as necessary.

9. $\lim\limits_{x \to 0} \dfrac{\sin x \cos x}{x}$

10. $\lim\limits_{t \to 0} \dfrac{\sqrt{t + 2}\, \sin t}{t}$

11. $\lim\limits_{t \to 0} \dfrac{\sin^2 t}{t}$

12. $\lim\limits_{x \to 0} \dfrac{\tan x}{x}$

13. $\lim\limits_{x \to 0} \dfrac{x^2}{\sin^2 x}$

14. $\lim\limits_{t \to \frac{\pi}{2}} \dfrac{1 - \cos t}{t}$

15. $\lim\limits_{t \to \frac{\pi}{4}} \dfrac{\sin t}{t}$

16. $\lim\limits_{t \to 0} \dfrac{\cos t - \cos^2 t}{t}$

17. Let $L = \lim\limits_{x \to 0} \dfrac{\sin 10x}{x}$.

(a) Show, by letting $\theta = 10x$, that $L = \lim\limits_{\theta \to 0} 10 \dfrac{\sin \theta}{\theta}$.

(b) Compute L.

18. Evaluate $\lim\limits_{h \to 0} \dfrac{\sin 2h}{\sin 5h}$. *Hint:* $\dfrac{\sin 2h}{\sin 5h} = \left(\dfrac{2}{5}\right)\left(\dfrac{\sin 2h}{2h}\right)\left(\dfrac{5h}{\sin 5h}\right)$.

In Exercises 19–40, evaluate the limit.

19. $\lim\limits_{h \to 0} \dfrac{\sin 6h}{h}$

20. $\lim\limits_{h \to 0} \dfrac{6 \sin h}{6h}$

21. $\lim\limits_{h \to 0} \dfrac{\sin 6h}{6h}$

22. $\lim\limits_{h \to 0} \dfrac{\sin h}{6h}$

23. $\lim\limits_{x \to 0} \dfrac{\sin 7x}{3x}$

24. $\lim\limits_{x \to \frac{\pi}{4}} \dfrac{x}{\sin 11x}$

25. $\lim\limits_{x \to 0} \dfrac{\tan 4x}{9x}$

26. $\lim\limits_{x \to 0} \dfrac{x}{\csc 25x}$

27. $\lim\limits_{t \to 0} \dfrac{\tan 4t}{t \sec t}$

28. $\lim\limits_{h \to 0} \dfrac{\sin 2h \sin 3h}{h^2}$

29. $\lim\limits_{z \to 0} \dfrac{\sin(z/3)}{\sin z}$

30. $\lim\limits_{\theta \to 0} \dfrac{\sin(-3\theta)}{\sin(4\theta)}$

31. $\lim\limits_{x \to 0} \dfrac{\tan 4x}{\tan 9x}$

32. $\lim\limits_{t \to 0} \dfrac{\csc 8t}{\csc 4t}$

33. $\lim\limits_{x \to 0} \dfrac{\sin 5x \sin 2x}{\sin 3x \sin 5x}$

34. $\lim\limits_{x \to 0} \dfrac{\sin 3x \sin 2x}{x \sin 5x}$

35. $\lim\limits_{h \to 0} \dfrac{1 - \cos 2h}{h}$

36. $\lim\limits_{h \to 0} \dfrac{\sin(2h)(1 - \cos h)}{h}$

37. $\lim\limits_{t \to 0} \dfrac{1 - \cos t}{\sin t}$

38. $\lim\limits_{\theta \to 0} \dfrac{\cos 2\theta - \cos \theta}{\theta}$

39. $\lim\limits_{h \to \frac{\pi}{2}} \dfrac{1 - \cos 3h}{h}$

40. $\lim\limits_{\theta \to 0} \dfrac{1 - \cos 4\theta}{\sin 3\theta}$

41. Calculate (a) $\lim\limits_{x \to 0+} \dfrac{\sin x}{|x|}$ and (b) $\lim\limits_{x \to 0-} \dfrac{\sin x}{|x|}$.

42. Prove the following result stated in Theorem 2:

$$\lim\limits_{\theta \to 0} \dfrac{1 - \cos \theta}{\theta} = 0 \qquad \boxed{7}$$

Hint: $\dfrac{1 - \cos \theta}{\theta} = \dfrac{1}{1 + \cos \theta} \cdot \dfrac{1 - \cos^2 \theta}{\theta}$.

43. Show that $-|\tan x| \leq \tan x \cos \dfrac{1}{x} \leq |\tan x|$. Then evaluate $\lim\limits_{x \to 0} \tan x \cos \dfrac{1}{x}$ using the Squeeze Theorem.

44. Evaluate $\lim\limits_{x \to 0} \tan x \cos\left(\sin \dfrac{1}{x}\right)$ using the Squeeze Theorem.

45. [GU] Plot the graphs of $u(x) = 1 + |x - \frac{\pi}{2}|$ and $l(x) = \sin x$ on the same set of axes. What can you say about $\lim\limits_{x \to \frac{\pi}{2}} f(x)$ if $f(x)$ is squeezed by $l(x)$ and $u(x)$ at $x = \frac{\pi}{2}$?

46. Use the Squeeze Theorem to prove that if $\lim\limits_{x \to c} |f(x)| = 0$, then $\lim\limits_{x \to c} f(x) = 0$.

Further Insights and Challenges

47. [GU] Investigate $\lim\limits_{h \to 0} \dfrac{1 - \cos h}{h^2}$ numerically (and graphically if you have a graphing utility). Then prove that the limit is equal to $\frac{1}{2}$. *Hint:* See the hint for Exercise 42.

48. Evaluate $\lim\limits_{h \to 0} \dfrac{\cos 3h - 1}{h^2}$. *Hint:* Use the result of Exercise 47.

49. Evaluate $\lim\limits_{h \to 0} \dfrac{\cos 3h - 1}{\cos 2h - 1}$.

50. Use the result of Exercise 47 to prove that for $m \neq 0$,

$$\lim\limits_{x \to 0} \dfrac{\cos mx - 1}{x^2} = -\dfrac{m^2}{2}$$

51. Using a diagram of the unit circle and the Pythagorean Theorem, show that

$$\sin^2 \theta \leq (1 - \cos \theta)^2 + \sin^2 \theta \leq \theta^2$$

Conclude that $\sin^2 \theta \leq 2(1 - \cos \theta) \leq \theta^2$ and use this to give an alternate proof of Eq. (7) in Exercise 42. Then give an alternate proof of the result in Exercise 47.

52. (a) Investigate $\lim\limits_{x \to c} \dfrac{\sin x - \sin c}{x - c}$ numerically for the five values $c = 0, \frac{\pi}{6}, \frac{\pi}{4}, \frac{\pi}{3}, \frac{\pi}{2}$.

(b) Can you guess the answer for general c? (We'll see why in Chapter 3.)

(c) Check if your answer to part (b) works for two other values of c.

53. 📖   Let $A(n)$ be the area of a regular n-gon inscribed in a unit circle (Figure 8).

(a) Prove that $A(n) = \dfrac{1}{2} n \sin\left(\dfrac{2\pi}{n}\right)$.

(b) Intuitively, why might we expect $A(n)$ to converge to the area of the unit circle as $n \to \infty$?

(c) Use Theorem 2 to evaluate $\lim\limits_{n \to \infty} A(n)$.

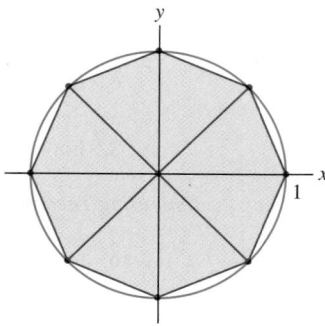

FIGURE 8 Regular n-gon inscribed in a unit circle.

2.7 Intermediate Value Theorem

The **Intermediate Value Theorem (IVT)** is a basic result which states that *a continuous function on an interval cannot skip values.* Consider a plane that takes off and climbs from 0 to 20,000 ft in 20 min. The plane must reach every altitude between 0 and 20,000 during this 20-min interval. Thus, at some moment, the plane's altitude must have been exactly 12,371 ft. Of course, we assume that the plane's motion is continuous, so its altitude cannot jump abruptly, say, from 12,000 to 13,000 ft.

To state this conclusion formally, let $A(t)$ be the plane's altitude at time t. The IVT then asserts that for every altitude M between 0 and 20,000, there is a time t_0 between 0 and 20 such that $A(t_0) = M$. In other words, the graph of $A(t)$ must intersect the horizontal line $y = M$ [Figure 1(A)].

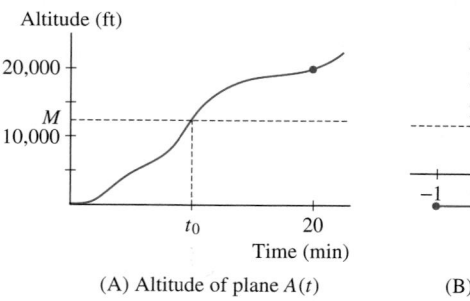

(A) Altitude of plane $A(t)$ (B) Graph of $f(x) = [x]$

FIGURE 1

By contrast, a discontinuous function can skip values. The greatest integer function $f(x) = [x]$ satisfies $[1] = 1$ and $[2] = 2$ [Figure 1(B)] but does not take on the value 1.5 (or any other value between 1 and 2).

| A proof of the IVT is given in Appendix B.

THEOREM 1 Intermediate Value Theorem   If $f(x)$ is continuous on a closed interval $[a, b]$ and $f(a) \neq f(b)$, then for every value M between $f(a)$ and $f(b)$, there exists at least one value $c \in (a, b)$ such that $f(c) = M$.

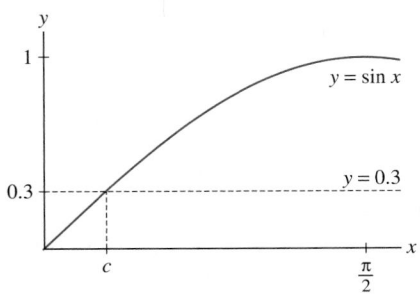

FIGURE 2

A zero or root of a function is a value c such that $f(c) = 0$. Sometimes the word "root" is reserved to refer specifically to the zero of a polynomial.

■ EXAMPLE 1 Prove that the equation $\sin x = 0.3$ has at least one solution.

Solution We may apply the IVT since $\sin x$ is continuous. We choose an interval where we suspect that a solution exists. The desired value 0.3 lies between the two function values

$$\sin 0 = 0 \qquad \text{and} \qquad \sin \frac{\pi}{2} = 1$$

so the interval $[0, \frac{\pi}{2}]$ will work (Figure 2). The IVT tells us that $\sin x = 0.3$ has at least one solution in $(0, \frac{\pi}{2})$. Since $\sin x$ is periodic, $\sin x = 0.3$ actually has infinitely many solutions. ■

The IVT can be used to show the existence of zeros of functions. If $f(x)$ is continuous and takes on both positive and negative values, say, $f(a) < 0$ and $f(b) > 0$, then the IVT guarantees that $f(c) = 0$ for some c between a and b.

COROLLARY 2 Existence of Zeros If $f(x)$ is continuous on $[a, b]$ and if $f(a)$ and $f(b)$ are nonzero and have opposite signs, then $f(x)$ has a zero in (a, b).

We can locate zeros of functions to arbitrary accuracy using the **Bisection Method**, as illustrated in the next example.

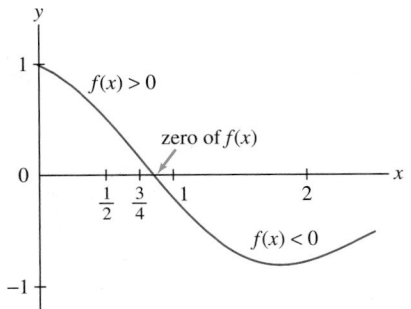

FIGURE 3 Graph of $f(x) = \cos^2 x - 2 \sin \dfrac{x}{4}$.

Computer algebra systems have built-in commands for finding roots of a function or solving an equation numerically. These systems use a variety of methods, including more sophisticated versions of the Bisection Method. Notice that to use the Bisection Method, we must first find an interval containing a root.

■ EXAMPLE 2 The Bisection Method   Show that $f(x) = \cos^2 x - 2 \sin \dfrac{x}{4}$ has a zero in $(0, 2)$. Then locate the zero more accurately using the Bisection Method.

Solution Using a calculator, we find that $f(0)$ and $f(2)$ have opposite signs:

$$f(0) = 1 > 0, \qquad f(2) \approx -0.786 < 0$$

Corollary 2 guarantees that $f(x) = 0$ has a solution in $(0, 2)$ (Figure 3).

We can locate a zero more accurately by dividing $[0, 2]$ into two intervals $[0, 1]$ and $[1, 2]$. One of these must contain a zero of $f(x)$. To determine which, we evaluate $f(x)$ at the midpoint $m = 1$. A calculator gives $f(1) \approx -0.203 < 0$, and since $f(0) = 1$, we see that

$$f(x) \text{ takes on opposite signs at the endpoints of } [0, 1]$$

Therefore, $(0, 1)$ must contain a zero. We discard $[1, 2]$ because both $f(1)$ and $f(2)$ are negative. The Bisection Method consists of continuing this process until we narrow down the location of the zero to the desired accuracy. In the following table, the process is carried out three times:

Interval	Midpoint of Interval	Function Values	Conclusion
$[0, 1]$	$\frac{1}{2}$	$f\left(\frac{1}{2}\right) \approx 0.521$ $f(1) \approx -0.203$	Zero lies in $\left(\frac{1}{2}, 1\right)$
$\left[\frac{1}{2}, 1\right]$	$\frac{3}{4}$	$f\left(\frac{3}{4}\right) \approx 0.163$ $f(1) \approx -0.203$	Zero lies in $\left(\frac{3}{4}, 1\right)$
$\left[\frac{3}{4}, 1\right]$	$\frac{7}{8}$	$f\left(\frac{7}{8}\right) \approx -0.0231$ $f\left(\frac{3}{4}\right) \approx 0.163$	Zero lies in $\left(\frac{3}{4}, \frac{7}{8}\right)$

We conclude that $f(x)$ has a zero c satisfying $0.75 < c < 0.875$. The Bisection Method can be continued to locate the root with any degree of accuracy. ∎

> **CONCEPTUAL INSIGHT**   The IVT seems to state the obvious, namely that a continuous function cannot skip values. Yet its proof (given in Appendix B) is quite subtle because it depends on the *completeness property* of real numbers. To highlight the subtlety, observe that the IVT is *false* for functions defined only on the *rational numbers*. For example, $f(x) = x^2$ does not have the intermediate value property if we restrict its domain to the rational numbers. Indeed, $f(0) = 0$ and $f(2) = 4$ but $f(c) = 2$ *has no solution* for c rational. The solution $c = \sqrt{2}$ is "missing" from the set of rational numbers because it is irrational. From the beginnings of calculus, the IVT was surely regarded as obvious. However, it was not possible to give a genuinely rigorous proof until the completeness property was clarified in the second half of the nineteenth century.

2.7 SUMMARY

- The Intermediate Value Theorem (IVT) says that a continuous function does not *skip* values. More precisely, if $f(x)$ is continuous on $[a, b]$ with $f(a) \neq f(b)$, and if M is a number between $f(a)$ and $f(b)$, then $f(c) = M$ for some $c \in (a, b)$.
- Existence of zeros: If $f(x)$ is continuous on $[a, b]$ and if $f(a)$ and $f(b)$ take opposite signs (one is positive and the other negative), then $f(c) = 0$ for some $c \in (a, b)$.
- Bisection Method: Assume f is continuous. Suppose that $f(a)$ and $f(b)$ have opposite signs, so that f has a zero in (a, b). Then f has a zero in $[a, m]$ or $[m, b]$, where $m = (a + b)/2$ is the midpoint of $[a, b]$. To determine which, compute $f(m)$. A zero lies in (a, m) if $f(a)$ and $f(m)$ have opposite signs and in (m, b) if $f(m)$ and $f(b)$ have opposite signs. Continuing the process, we can locate a zero with arbitrary accuracy.

2.7 EXERCISES

Preliminary Questions

1. Explain why $f(x) = x^2$ takes on the value 0.5 in the interval $[0, 1]$.

2. The temperature in Vancouver was 46° at 6 AM and rose to 68° at noon. What must we assume about temperature to conclude that the temperature was 60° at some moment of time between 6 AM and noon?

3. What is the graphical interpretation of the IVT?

4. Show that the following statement is false by drawing a graph that provides a counterexample: *If $f(x)$ is continuous and has a root in $[a, b]$, then $f(a)$ and $f(b)$ have opposite signs.*

Exercises

1. Use the IVT to show that $f(x) = x^3 + x$ takes on the value 9 for some x in $[1, 2]$.

2. Show that $g(t) = \dfrac{t}{t+1}$ takes on the value 0.499 for some t in $[0, 1]$.

3. Show that $g(t) = t^2 \tan t$ takes on the value $\frac{1}{2}$ for some t in $[0, \frac{\pi}{4}]$.

4. Show that $f(x) = \dfrac{x^2}{x^7 + 1}$ takes on the value 0.4.

5. Show that $\cos x = x$ has a solution in the interval $[0, 1]$. *Hint:* Show that $f(x) = x - \cos x$ has a zero in $[0, 1]$.

6. Use the IVT to find an interval of length $\frac{1}{2}$ containing a root of $f(x) = x^3 + 2x + 1$.

In Exercises 7–16, use the IVT to prove each of the following statements.

7. $\sqrt{c} + \sqrt{c+1} = 2$ for some number c.

8. For all integers n, $\sin nx = \cos x$ for some $x \in [0, \pi]$.

9. $\sqrt{2}$ exists. *Hint: Consider $f(x) = x^2$.*

10. A positive number c has an nth root for all positive integers n. (This fact is usually taken for granted, but it requires proof.)

11. For all positive integers k, there exists x such that $\cos x = x^k$.

12. $2^x = bx$ has a solution if $b > 2$.

13. $2^x = b$ has a solution for all $b > 0$ (treat $b \geq 1$ first).

14. $\tan x = x$ has infinitely many solutions.

15. The equation $e^x + \ln x = 0$ has a solution in $(0, 1)$.

16. $\tan^{-1} x = \cos^{-1} x$ has a solution.

17. Carry out three steps of the Bisection Method for $f(x) = 2^x - x^3$ as follows:

(a) Show that $f(x)$ has a zero in $[1, 1.5]$.

(b) Show that $f(x)$ has a zero in $[1.25, 1.5]$.

(c) Determine whether $[1.25, 1.375]$ or $[1.375, 1.5]$ contains a zero.

18. Figure 4 shows that $f(x) = x^3 - 8x - 1$ has a root in the interval $[2.75, 3]$. Apply the Bisection Method twice to find an interval of length $\frac{1}{16}$ containing this root.

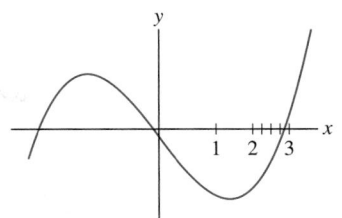

FIGURE 4 Graph of $y = x^3 - 8x - 1$.

19. Find an interval of length $\frac{1}{4}$ in $[0, 1]$ containing a root of $x^5 - 5x + 1 = 0$.

20. Show that $\tan^3 \theta - 8 \tan^2 \theta + 17 \tan \theta - 8 = 0$ has a root in $[0.5, 0.6]$. Apply the Bisection Method twice to find an interval of length 0.025 containing this root.

In Exercises 21–24, draw the graph of a function $f(x)$ on $[0, 4]$ with the given property.

21. Jump discontinuity at $x = 2$ and does not satisfy the conclusion of the IVT.

22. Jump discontinuity at $x = 2$, yet does satisfy the conclusion of the IVT on $[0, 4]$.

23. Infinite one-sided limits at $x = 2$ and does not satisfy the conclusion of the IVT.

24. Infinite one-sided limits at $x = 2$, yet does satisfy the conclusion of the IVT on $[0, 4]$.

25. Corollary 2 is not foolproof. Let $f(x) = x^2 - 1$ and explain why the corollary fails to detect the roots $x = \pm 1$ if $[a, b]$ contains $[-1, 1]$.

Further Insights and Challenges

26. Take any map (e.g., of the United States) and draw a circle on it anywhere. Prove that at any moment in time there exists a pair of diametrically opposite points on that circle corresponding to locations where the temperatures at that moment are equal. *Hint: Let θ be an angular coordinate along the circle and let $f(\theta)$ be the difference in temperatures at the locations corresponding to θ and $\theta + \pi$.*

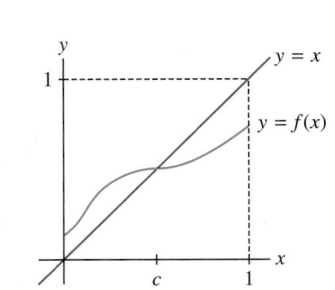

FIGURE 5 A function satisfying $0 \leq f(x) \leq 1$ for $0 \leq x \leq 1$.

27. Assume that $f(x)$ is continuous and that $0 \leq f(x) \leq 1$ for $0 \leq x \leq 1$ (see Figure 5). Show that $f(c) = c$ for some c in $[0, 1]$.

28. Use the IVT to show that if $f(x)$ is continuous and one-to-one on an interval $[a, b]$, then $f(x)$ is either an increasing or a decreasing function.

29. **Ham Sandwich Theorem** Figure 6(A) shows a slice of ham. Prove that for any angle θ ($0 \leq \theta \leq \pi$), it is possible to cut the slice of ham in half with a cut of incline θ. *Hint:* The lines of inclination θ are given by the equations $y = (\tan \theta)x + b$, where b varies from $-\infty$ to ∞. Each such line divides the ham slice into two pieces (one of which may be empty). Let $A(b)$ be the amount of ham to the left of the line minus the amount to the right and let A be the total area of the ham. Show that $A(b) = -A$ if b is sufficiently large and $A(b) = A$ if b is sufficiently negative. Then use the IVT. This works if $\theta \neq 0$ or $\frac{\pi}{2}$. If $\theta = 0$, define $A(b)$ as the amount of ham above the line $y = b$ minus the amount below. How can you modify the argument to work when $\theta = \frac{\pi}{2}$ (in which case $\tan \theta = \infty$)?

30. Figure 6(B) shows a slice of ham on a piece of bread. Prove that it is possible to slice this open-faced sandwich so that each part has equal amounts of ham and bread. *Hint:* By Exercise 29, for all $0 \le \theta \le \pi$ there is a line $L(\theta)$ of incline θ (which we assume is unique) that divides the ham into two equal pieces. For $\theta \ne 0$ or π, let $B(\theta)$ denote the amount of bread to the left of $L(\theta)$ minus the amount to the right. Notice that $L(\pi) = L(0)$ since $\theta = 0$ and $\theta = \pi$ give the same line, but $B(\pi) = -B(0)$ since left and right get interchanged as the angle moves from 0 to π. Assume that $B(\theta)$ is continuous and apply the IVT. (By a further extension of this argument, one can prove the full "Ham Sandwich Theorem," which states that if you allow the knife to cut at a slant, then it is possible to cut a sandwich consisting of a slice of ham and two slices of bread so that all three layers are divided in half.)

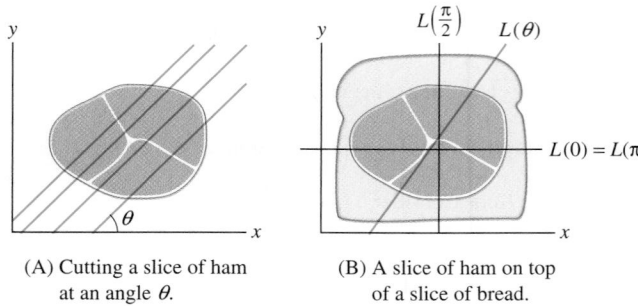

(A) Cutting a slice of ham at an angle θ.

(B) A slice of ham on top of a slice of bread.

FIGURE 6

2.8 The Formal Definition of a Limit

In this section, we reexamine the definition of a limit in order to state it in a rigorous and precise fashion. Why is this necessary? In Section 2.3, we defined limits by saying that $\lim_{x \to c} f(x) = L$ if $|f(x) - L|$ becomes arbitrarily small when x is sufficiently close (but not equal) to c. The problem with this definition lies in the phrases "arbitrarily small" and "sufficiently close." We must find a way to specify just how close is sufficiently close.

The formal limit definition eliminates the vagueness of the informal approach. It is used to develop calculus in a precise and rigorous fashion. Some proofs using the formal limit definition are included in Appendix D, but for a more complete development, we refer to textbooks on the branch of mathematics called "analysis."

The Size of the Gap

Recall that the distance from $f(x)$ to L is $|f(x) - L|$. It is convenient to refer to the quantity $|f(x) - L|$ as the *gap* between the value $f(x)$ and the limit L.

Let's reexamine the basic trigonometric limit

$$\lim_{x \to 0} \frac{\sin x}{x} = 1 \qquad \boxed{1}$$

In this example, $f(x) = \dfrac{\sin x}{x}$ and $L = 1$, so (1) tells us that the gap $|f(x) - 1|$ gets arbitrarily small when x is sufficiently close but not equal to 0.

Suppose we want the gap $|f(x) - 1|$ to be less than 0.2. How close to 0 must x be? Figure 1(B) shows that $f(x)$ lies within 0.2 of $L = 1$ for all values of x in the interval $[-1, 1]$. In other words, the following statement is true:

$$\left| \frac{\sin x}{x} - 1 \right| < 0.2 \qquad \text{if} \qquad 0 < |x| < 1$$

If we insist instead that the gap be smaller than 0.004, we can check by zooming in on the graph as in Figure 1(C) that

$$\left| \frac{\sin x}{x} - 1 \right| < 0.004 \qquad \text{if} \qquad 0 < |x| < 0.15$$

It would seem that this process can be continued: By zooming in on the graph, we can find a small interval around $c = 0$ where the gap $|f(x) - 1|$ is smaller than any prescribed positive number.

To express this in a precise fashion, we follow time-honored tradition and use the Greek letters ϵ (epsilon) and δ (delta) to denote small numbers specifying the size of the

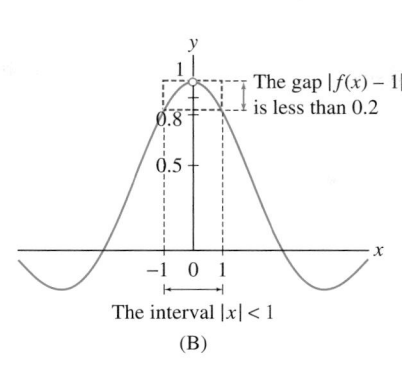

Enlarged view of
graph near $x = 0$

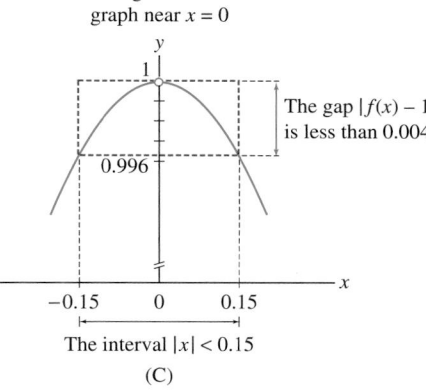

(A)

(B)

(C)

FIGURE 1 Graphs of $y = \frac{\sin x}{x}$. To shrink the gap from 0.2 to 0.004, we require that x lie within 0.15 of 0.

gap and the quantity $|x - c|$, respectively. In our case, $c = 0$ and $|x - c| = |x - 0| = |x|$. The precise meaning of Eq. (1) is that for every choice of $\epsilon > 0$, there exists some δ (depending on ϵ) such that

$$\left| \frac{\sin x}{x} - 1 \right| < \epsilon \quad \text{if} \quad 0 < |x| < \delta$$

The number δ tells us how close is sufficiently close for a given ϵ. With this motivation, we are ready to state the formal definition of the limit.

The formal definition of a limit is often called the ϵ-δ definition. The tradition of using the symbols ϵ and δ originates in the writings of Augustin-Louis Cauchy on calculus and analysis in the 1820s.

FORMAL DEFINITION OF A LIMIT Suppose that $f(x)$ is defined for all x in an open interval containing c (but not necessarily at $x = c$). Then

$$\boxed{\lim_{x \to c} f(x) = L}$$

if for all $\epsilon > 0$, there exists $\delta > 0$ such that

$$\boxed{|f(x) - L| < \epsilon \quad \text{if} \quad 0 < |x - c| < \delta}$$

If the symbols ϵ and δ seem to make this definition too abstract, keep in mind that we may take $\epsilon = 10^{-n}$ and $\delta = 10^{-m}$. Thus, $\lim_{x \to c} f(x) = L$ if, for any n, we have $|f(x) - L| < 10^{-n}$, provided that $0 < |x - c| < 10^{-m}$ (where the choice of m depends on n).

The condition $0 < |x - c| < \delta$ in this definition excludes $x = c$. As in our previous informal definition, we formulate it this way so that the limit depends only on values of $f(x)$ near c but not on $f(c)$ itself. As we have seen, in many cases the limit exists even when $f(c)$ is not defined.

■ **EXAMPLE 1** Let $f(x) = 8x + 3$.

(a) Show that $\lim_{x \to 3} f(x) = 27$ using the formal definition of the limit.

(b) Find values of δ that work for $\epsilon = 0.2$ and 0.001.

Solution

(a) We break the proof into two steps.

Step 1. Relate the gap to $|x - c|$.
  We must find a relation between two absolute values: $|f(x) - L|$ for $L = 27$ and $|x - c|$ for $c = 3$. Observe that

$$\underbrace{|f(x) - 27|}_{\text{Size of gap}} = |(8x + 3) - 27| = |8x - 24| = 8|x - 3|$$

Thus, the gap is 8 times as large as $|x - 3|$.

Step 2. Choose δ (in terms of ϵ).

We can now see how to make the gap small: If $|x - 3| < \frac{\epsilon}{8}$, then the gap is smaller than $8|x - 3| < 8\left(\dfrac{\epsilon}{8}\right) = \epsilon$. Therefore, for any $\epsilon > 0$, we choose $\delta = \frac{\epsilon}{8}$. The following statement will then hold:

$$|f(x) - 27| < \epsilon \quad \text{if} \quad 0 < |x - 3| < \delta, \quad \text{where } \delta = \frac{\epsilon}{8}$$

Since we have specified δ for all $\epsilon > 0$, we have fulfilled the requirements of the formal definition of the limit, thus proving rigorously that $\lim_{x \to 3} (8x + 3) = 27$.

(b) For the particular choice $\epsilon = 0.2$, we may take $\delta = \dfrac{\epsilon}{8} = \dfrac{0.2}{8} = 0.025$:

$$|f(x) - 27| < 0.2 \quad \text{if} \quad 0 < |x - 3| < 0.025$$

This statement is illustrated in Figure 2. We note that *any positive δ smaller than* 0.025 *will also work*. For example, the following statement is also true, although it places an unnecessary restriction on x:

$$|f(x) - 27| < 0.2 \quad \text{if} \quad 0 < |x - 3| < 0.019$$

Similarly, to make the gap less than $\epsilon = 0.001$, we may take

$$\delta = \frac{\epsilon}{8} = \frac{0.001}{8} = 0.000125 \qquad \blacksquare$$

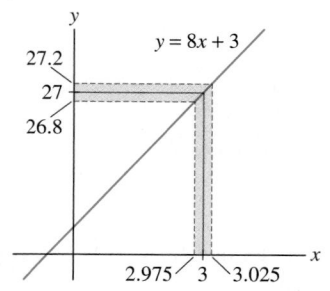

FIGURE 2 To make the gap less than 0.2, we may take $\delta = 0.025$ (not drawn to scale).

The difficulty in applying the limit definition frequently lies in trying to relate $|f(x) - L|$ to $|x - c|$. The next two examples illustrate how this can be done in special cases.

■ **EXAMPLE 2** A Limit of x^2 Use the limit definition to prove that $\lim_{x \to 2} x^2 = 4$.

Solution Let $f(x) = x^2$.

Step 1. Relate the gap to $|x - c|$.

In this case, we must relate the gap $|f(x) - 4| = |x^2 - 4|$ to the quantity $|x - 2|$ (see Figure 3). This is more difficult than in the previous example because the gap is not a

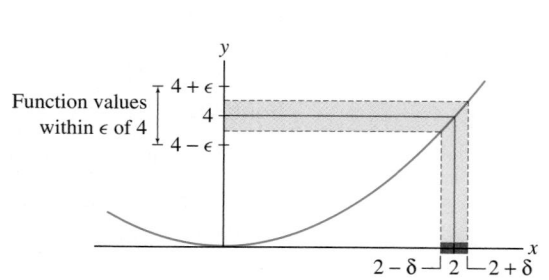

FIGURE 3 Graph of $f(x) = x^2$. We may choose δ so that $f(x)$ lies within ϵ of 4 for all x in $[2 - \delta, 2 + \delta]$.

constant multiple of $|x - 2|$. To proceed, consider the factorization

$$|x^2 - 4| = |x + 2| \, |x - 2|$$

Since we are going to require that $|x - 2|$ be small, we may as well assume from the outset that $|x - 2| < 1$, which means that $1 < x < 3$. In this case, $|x + 2|$ is less than 5 and the gap satisfies

$$|x^2 - 4| = |x + 2| \, |x - 2| < 5 \, |x - 2| \qquad \text{if} \quad |x - 2| < 1 \qquad \boxed{2}$$

Step 2. **Choose δ (in terms of ϵ).**

We see from (2) that if $|x - 2|$ is smaller than both $\frac{\epsilon}{5}$ and 1, then the gap satisfies

$$|x^2 - 4| < 5|x - 2| < 5 \left(\frac{\epsilon}{5} \right) = \epsilon$$

Therefore, the following statement holds for all $\epsilon > 0$:

$$\boxed{\; |x^2 - 4| < \epsilon \qquad \text{if} \qquad 0 < |x - 2| < \delta, \quad \text{where } \delta \text{ is the smaller of } \tfrac{\epsilon}{5} \text{ and } 1 \;}$$

We have specified δ for all $\epsilon > 0$, so we have fulfilled the requirements of the formal limit definition, thus proving rigorously that $\lim\limits_{x \to 2} x^2 = 4$. ∎

■ **EXAMPLE 3** Use the limit definition to prove that $\lim\limits_{x \to 3} \dfrac{1}{x} = \dfrac{1}{3}$.

Solution

Step 1. **Relate the gap to $|x - c|$.**

The gap is equal to

$$\left| \frac{1}{x} - \frac{1}{3} \right| = \left| \frac{3 - x}{3x} \right| = |x - 3| \left| \frac{1}{3x} \right|$$

Since we are going to require that $|x - 3|$ be small, we may as well assume from the outset that $|x - 3| < 1$, that is, $2 < x < 4$. Now observe that if $x > 2$, then $3x > 6$ and $\frac{1}{3x} < \frac{1}{6}$, so the following inequality is valid if $|x - 3| < 1$:

◄·· **REMINDER** If $a > b > 0$, then $\frac{1}{a} < \frac{1}{b}$.
Thus, if $3x > 6$, then $\dfrac{1}{3x} < \dfrac{1}{6}$.

$$\left| f(x) - \frac{1}{3} \right| = \left| \frac{3 - x}{3x} \right| = \left| \frac{1}{3x} \right| \, |x - 3| < \frac{1}{6} |x - 3| \qquad \boxed{3}$$

Step 2. **Choose δ (in terms of ϵ).**

By (3), if $|x - 3| < 1$ and $|x - 3| < 6\epsilon$, then

$$\left| \frac{1}{x} - \frac{1}{3} \right| < \frac{1}{6} |x - 3| < \frac{1}{6} (6\epsilon) = \epsilon$$

Therefore, given any $\epsilon > 0$, we let δ be the smaller of the numbers 6ϵ and 1. Then

$$\boxed{\; \left| \frac{1}{x} - \frac{1}{3} \right| < \epsilon \qquad \text{if} \qquad 0 < |x - 3| < \delta, \quad \text{where } \delta \text{ is the smaller of } 6\epsilon \text{ and } 1 \;}$$

Again, we have fulfilled the requirements of the formal limit definition, thus proving rigorously that $\lim\limits_{x \to 3} \dfrac{1}{x} = \dfrac{1}{3}$. ∎

GRAPHICAL INSIGHT   Keep the graphical interpretation of limits in mind. In Figure 4(A), $f(x)$ approaches L as $x \to c$ because for any $\epsilon > 0$, we can make the gap less than ϵ by taking δ sufficiently small. The function in Figure 4(B) has a jump discontinuity at $x = c$ and the gap cannot be made small, no matter how small δ is taken. Therefore, the limit does not exist.

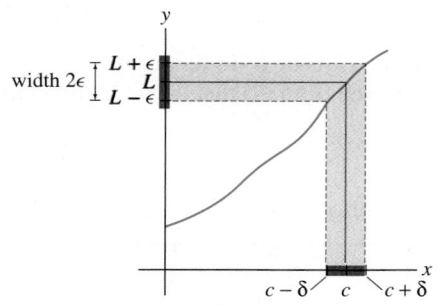

(A) The function is continuous at $x = c$.
By taking δ sufficiently small, we
can make the gap smaller than ϵ.

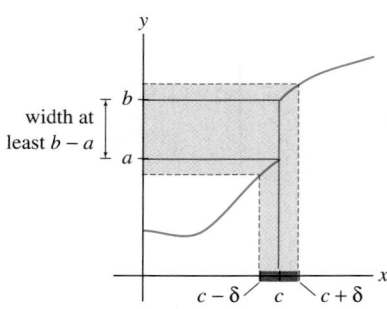

(B) The function is not continuous at $x = c$.
The gap is always larger than $(b - a)/2$,
no matter how small δ is.

FIGURE 4

Proving Limit Theorems

In practice, the formal limit definition is rarely used to evaluate limits. Most limits are evaluated using the Limit Laws or other techniques such as the Squeeze Theorem. However, the formal definition allows us to prove these laws in a rigorous fashion and thereby ensure that calculus is based on a solid foundation. We illustrate by proving the Sum Law. Other proofs are given in Appendix D.

Proof of the Sum Law   Assume that

$$\lim_{x \to c} f(x) = L \quad \text{and} \quad \lim_{x \to c} g(x) = M$$

We must prove that $\lim_{x \to c} (f(x) + g(x)) = L + M$.

By the Triangle Inequality [Eq. (1) in Section 1.1], $|a + b| \le |a| + |b|$ for all a and b. Setting $a = f(x) - L$ and $b = g(x) - M$, we obtain

$$|(f(x) + g(x)) - (L + M)| \le |f(x) - L| + |g(x) - M| \qquad \boxed{4}$$

By the limit definition, we can make each term on the right in (4) as small as desired by taking $|x - c|$ small enough. More precisely, given $\epsilon > 0$, we can choose δ such that $|f(x) - L| < \frac{\epsilon}{2}$ and $|g(x) - M| < \frac{\epsilon}{2}$ if $0 < |x - c| < \delta$ (in principle, we might have to choose different δ's for f and g, but we may then use the smaller of the two δ's). Thus, (4) gives

$$|f(x) + g(x) - (L + M)| < \frac{\epsilon}{2} + \frac{\epsilon}{2} = \epsilon \qquad \text{if} \qquad 0 < |x - c| < \delta \qquad \boxed{5}$$

This proves that

$$\lim_{x \to c} \big(f(x) + g(x)\big) = L + M = \lim_{x \to c} f(x) + \lim_{x \to c} g(x) \qquad \blacksquare$$

2.8 SUMMARY

- Informally speaking, the statement $\lim_{x \to c} f(x) = L$ means that the gap $|f(x) - L|$ tends to 0 as x approaches c. To make this precise, we introduce the *formal definition* (called the ϵ-δ definition): $\lim_{x \to c} f(x) = L$ if, for all $\epsilon > 0$, there exists a $\delta > 0$ such that

$$|f(x) - L| < \epsilon \quad \text{if} \quad 0 < |x - c| < \delta$$

2.8 EXERCISES

Preliminary Questions

1. Given that $\lim_{x \to 0} \cos x = 1$, which of the following statements is true?

(a) If $|\cos x - 1|$ is very small, then x is close to 0.

(b) There is an $\epsilon > 0$ such that $|x| < 10^{-5}$ if $0 < |\cos x - 1| < \epsilon$.

(c) There is a $\delta > 0$ such that $|\cos x - 1| < 10^{-5}$ if $0 < |x| < \delta$.

(d) There is a $\delta > 0$ such that $|\cos x| < 10^{-5}$ if $0 < |x - 1| < \delta$.

2. Suppose that for a given ϵ and δ, it is known that $|f(x) - 2| < \epsilon$ if $0 < |x - 3| < \delta$. Which of the following statements must also be true?

(a) $|f(x) - 2| < \epsilon$ if $0 < |x - 3| < 2\delta$

(b) $|f(x) - 2| < 2\epsilon$ if $0 < |x - 3| < \delta$

(c) $|f(x) - 2| < \dfrac{\epsilon}{2}$ if $0 < |x - 3| < \dfrac{\delta}{2}$

(d) $|f(x) - 2| < \epsilon$ if $0 < |x - 3| < \dfrac{\delta}{2}$

Exercises

1. Consider $\lim_{x \to 4} f(x)$, where $f(x) = 8x + 3$.

(a) Show that $|f(x) - 35| = 8|x - 4|$.

(b) Show that for any $\epsilon > 0$, $|f(x) - 35| < \epsilon$ if $|x - 4| < \delta$, where $\delta = \frac{\epsilon}{8}$. Explain how this proves rigorously that $\lim_{x \to 4} f(x) = 35$.

2. Consider $\lim_{x \to 2} f(x)$, where $f(x) = 4x - 1$.

(a) Show that $|f(x) - 7| < 4\delta$ if $|x - 2| < \delta$.

(b) Find a δ such that:

$$|f(x) - 7| < 0.01 \quad \text{if} \quad |x - 2| < \delta$$

(c) Prove rigorously that $\lim_{x \to 2} f(x) = 7$.

3. Consider $\lim_{x \to 2} x^2 = 4$ (refer to Example 2).

(a) Show that $|x^2 - 4| < 0.05$ if $0 < |x - 2| < 0.01$.

(b) Show that $|x^2 - 4| < 0.0009$ if $0 < |x - 2| < 0.0002$.

(c) Find a value of δ such that $|x^2 - 4|$ is less than 10^{-4} if $0 < |x - 2| < \delta$.

4. With regard to the limit $\lim_{x \to 5} x^2 = 25$,

(a) Show that $|x^2 - 25| < 11|x - 5|$ if $4 < x < 6$. *Hint:* Write $|x^2 - 25| = |x + 5| \cdot |x - 5|$.

(b) Find a δ such that $|x^2 - 25| < 10^{-3}$ if $|x - 5| < \delta$.

(c) Give a rigorous proof of the limit by showing that $|x^2 - 25| < \epsilon$ if $0 < |x - 5| < \delta$, where δ is the smaller of $\frac{\epsilon}{11}$ and 1.

5. Refer to Example 3 to find a value of $\delta > 0$ such that

$$\left| \frac{1}{x} - \frac{1}{3} \right| < 10^{-4} \quad \text{if} \quad 0 < |x - 3| < \delta$$

6. [GU] Plot $f(x) = \sqrt{2x - 1}$ together with the horizontal lines $y = 2.9$ and $y = 3.1$. Use this plot to find a value of $\delta > 0$ such that $|\sqrt{2x - 1} - 3| < 0.1$ if $|x - 5| < \delta$.

7. [GU] Plot $f(x) = \tan x$ together with the horizontal lines $y = 0.99$ and $y = 1.01$. Use this plot to find a value of $\delta > 0$ such that $|\tan x - 1| < 0.01$ if $|x - \frac{\pi}{4}| < \delta$.

8. [GU] The number e has the following property: $\lim_{x \to 0} \dfrac{e^x - 1}{x} = 1$. Use a plot of $f(x) = \dfrac{e^x - 1}{x}$ to find a value of $\delta > 0$ such that $|f(x) - 1| < 0.01$ if $|x - 1| < \delta$.

9. [GU] Let $f(x) = \dfrac{4}{x^2 + 1}$. Observe that the limit $L = \lim_{x \to \frac{1}{2}} f(x)$ has the value $L = f(\frac{1}{2}) = \frac{16}{5}$. Let $\epsilon = 0.5$ and, using a plot of $f(x)$, find a value of $\delta > 0$ such that $|f(x) - \frac{16}{5}| < \epsilon$ for $|x - \frac{1}{2}| < \delta$. Repeat for $\epsilon = 0.2$ and 0.1.

10. Consider $\lim_{x \to 2} \dfrac{1}{x}$.

(a) Show that if $|x - 2| < 1$, then

$$\left| \frac{1}{x} - \frac{1}{2} \right| < \frac{1}{2}|x - 2|$$

(b) Let δ be the smaller of 1 and 2ϵ. Prove the following statement:

$$\left|\frac{1}{x} - \frac{1}{2}\right| < \epsilon \quad \text{if} \quad 0 < |x - 2| < \delta$$

(c) Find a $\delta > 0$ such that $\left|\frac{1}{x} - \frac{1}{2}\right| < 0.01$ if $|x - 2| < \delta$.

(d) Prove rigorously that $\lim\limits_{x \to 2} \frac{1}{x} = \frac{1}{2}$.

11. Consider $\lim\limits_{x \to 1} \sqrt{x + 3}$.

(a) Show that $|\sqrt{x + 3} - 2| < \frac{1}{2}|x - 1|$ if $|x - 1| < 4$. *Hint:* Multiply the inequality by $|\sqrt{x + 3} + 2|$ and observe that $|\sqrt{x + 3} + 2| > 2$.

(b) Find $\delta > 0$ such that $|\sqrt{x + 3} - 2| < 10^{-4}$ for $|x - 1| < \delta$.

(c) Prove rigorously that the limit is equal to 2.

12. Based on the information conveyed in Figure 5(A), find values of L, ϵ, and $\delta > 0$ such that the following statement holds: $|f(x) - L| < \epsilon$ if $|x| < \delta$.

13. Based on the information conveyed in Figure 5(B), find values of c, L, ϵ, and $\delta > 0$ such that the following statement holds: $|f(x) - L| < \epsilon$ if $|x - c| < \delta$.

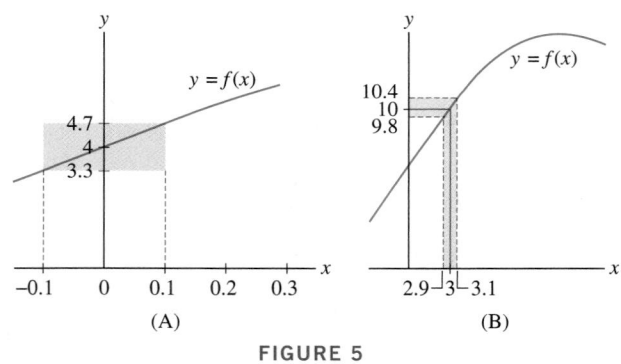

(A)　　　　　(B)

FIGURE 5

14. 📖 A calculator gives the following values for $f(x) = \sin x$:

$$f\left(\frac{\pi}{4} - 0.1\right) \approx 0.633, \quad f\left(\frac{\pi}{4}\right) \approx 0.707, \quad f\left(\frac{\pi}{4} + 0.1\right) \approx 0.774$$

Use these values and the fact that $f(x)$ is increasing on $[0, \frac{\pi}{2}]$ to justify the statement

$$\left|f(x) - f\left(\frac{\pi}{4}\right)\right| < 0.08 \quad \text{if} \quad \left|x - \frac{\pi}{4}\right| < 0.1$$

Then draw a figure like Figure 3 to illustrate this statement.

15. Adapt the argument in Example 1 to prove rigorously that $\lim\limits_{x \to c} (ax + b) = ac + b$, where a, b, c are arbitrary.

16. Adapt the argument in Example 2 to prove rigorously that $\lim\limits_{x \to c} x^2 = c^2$ for all c.

17. Adapt the argument in Example 3 to prove rigorously that $\lim\limits_{x \to c} x^{-1} = \frac{1}{c}$ for all c.

In Exercises 18–23, use the formal definition of the limit to prove the statement rigorously.

18. $\lim\limits_{x \to 4} \sqrt{x} = 2$

19. $\lim\limits_{x \to 1} (3x^2 + x) = 4$

20. $\lim\limits_{x \to 1} x^3 = 1$

21. $\lim\limits_{x \to 2} x^{-2} = \frac{1}{4}$

22. $\lim\limits_{x \to c} |x| = |c|$

23. $\lim\limits_{x \to 0} x \sin \frac{1}{x} = 0$

24. Let $f(x) = \dfrac{x}{|x|}$. Prove rigorously that $\lim\limits_{x \to 0} f(x)$ does not exist. *Hint:* Show that no number L qualifies as the limit because there always exists some x such that $|x| < \delta$ but $|f(x) - L| \geq \frac{1}{2}$, no matter how small δ is taken.

25. Prove rigorously that $\lim\limits_{x \to 0} \sin \frac{1}{x}$ does not exist.

26. Use the identity

$$\sin x + \sin y = 2 \sin\left(\frac{x + y}{2}\right) \cos\left(\frac{x - y}{2}\right)$$

to verify the relation

$$\sin(a + h) - \sin a = h \frac{\sin(h/2)}{h/2} \cos\left(a + \frac{h}{2}\right) \qquad \boxed{6}$$

Conclude that $|\sin(a + h) - \sin a| < |h|$ for all a and prove rigorously that $\lim\limits_{x \to a} \sin x = \sin a$. You may use the inequality $\left|\dfrac{\sin x}{x}\right| \leq 1$ for $x \neq 0$.

27. Use Eq. (6) to prove rigorously that

$$\lim\limits_{h \to 0} \frac{\sin(a + h) - \sin a}{h} = \cos a$$

Further Insights and Challenges

28. Uniqueness of the Limit Show that a function converges to at most one limiting value. In other words, use the limit definition to show that if $\lim\limits_{x \to c} f(x) = L_1$ and $\lim\limits_{x \to c} f(x) = L_2$, then $L_1 = L_2$.

In Exercises 29–31, prove the statement using the formal limit definition.

29. The Constant Multiple Law [Theorem 1, part (ii) in Section 2.3, p. 80]

30. The Squeeze Theorem. (Theorem 1 in Section 2.6, p. 99)

31. The Product Law [Theorem 1, part (iii) in Section 2.3, p. 80]. *Hint:* Use the identity

$$f(x)g(x) - LM = (f(x) - L)g(x) + L(g(x) - M)$$

32. Let $f(x) = 1$ if x is rational and $f(x) = 0$ if x is irrational. Prove that $\lim_{x \to c} f(x)$ does not exist for any c.

33. Here is a function with strange continuity properties:

$$f(x) = \begin{cases} \dfrac{1}{q} & \text{if } x \text{ is the rational number } \dfrac{p}{q} \text{ in lowest terms} \\ 0 & \text{if } x \text{ is an irrational number} \end{cases}$$

(a) Show that $f(x)$ is discontinuous at c if c is rational. *Hint:* There exist irrational numbers arbitrarily close to c.

(b) Show that $f(x)$ is continuous at c if c is irrational. *Hint:* Let I be the interval $\{x : |x - c| < 1\}$. Show that for any $Q > 0$, I contains at most finitely many fractions $\frac{p}{q}$ with $q < Q$. Conclude that there is a δ such that all fractions in $\{x : |x - c| < \delta\}$ have a denominator larger than Q.

CHAPTER REVIEW EXERCISES

1. A particle's position at time t (s) is $s(t) = \sqrt{t^2 + 1}$ m. Compute its average velocity over $[2, 5]$ and estimate its instantaneous velocity at $t = 2$.

2. The price p of natural gas in the United States (in dollars per 1,000 ft^3) on the first day of each month in 2004 is listed in the table below.

J	F	M	A	M	J
9.7	9.84	10	10.52	11.61	13.05

J	A	S	O	N	D
13.45	13.79	13.29	11.68	11.44	11.11

Compute the average rate of change of p (in dollars per 1,000 ft^3 per month) over the quarterly periods January–March, April–June, and July–September.

3. For a whole number n, let $P(n)$ be the number of *partitions* of n, that is, the number of ways of writing n as a sum of one or more whole numbers. For example, $P(4) = 5$ since the number 4 can be partitioned in five different ways: $4, 3 + 1, 2 + 2, 2 + 1 + 1, 1 + 1 + 1 + 1$. Treating $P(n)$ as a continuous function, use Figure 1 to estimate the rate of change of $P(n)$ at $n = 12$.

4. The average velocity v (m/s) of an oxygen molecule in the air at temperature T (°C) is $v = 25.7\sqrt{273.15 + T}$. What is the average

speed at $T = 25°$ (room temperature)? Estimate the rate of change of average velocity with respect to temperature at $T = 25°$. What are the units of this rate?

In Exercises 5–8, estimate the limit numerically to two decimal places or state that the limit does not exist.

5. $\lim_{x \to 0} \dfrac{1 - \cos^3(x)}{x^2}$

6. $\lim_{x \to 1} x^{1/(x-1)}$

7. $\lim_{x \to 2} \dfrac{x^x - 4}{x^2 - 4}$

8. $\lim_{x \to 2} \dfrac{3^x - 9}{5^x - 25}$

In Exercises 9–43, evaluate the limit if possible or state that it does not exist.

9. $\lim_{x \to 4} (3 + x^{1/2})$

10. $\lim_{x \to 1} \dfrac{5 - x^2}{4x + 7}$

11. $\lim_{x \to -2} \dfrac{4}{x^3}$

12. $\lim_{x \to -1} \dfrac{3x^2 + 4x + 1}{x + 1}$

13. $\lim_{x \to 1} \dfrac{x^3 - x}{x - 1}$

14. $\lim_{x \to 1} \dfrac{x^3 - 2x}{x - 1}$

15. $\lim_{t \to 9} \dfrac{\sqrt{t} - 3}{t - 9}$

16. $\lim_{t \to 9} \dfrac{t - 6}{\sqrt{t} - 3}$

17. $\lim_{x \to 3} \dfrac{\sqrt{x + 1} - 2}{x - 3}$

18. $\lim_{s \to 0} \dfrac{1 - \sqrt{s^2 + 1}}{s^2}$

19. $\lim_{h \to 0} \dfrac{2(a + h)^2 - 2a^2}{h}$

20. $\lim_{t \to -1+} \dfrac{1}{x + 1}$

21. $\lim_{t \to 3} \dfrac{1}{t^2 - 9}$

22. $\lim_{x \to b} \dfrac{x^3 - b^3}{x - b}$

23. $\lim_{a \to b} \dfrac{a^2 - 3ab + 2b^2}{a - b}$

24. $\lim_{x \to 0} \dfrac{e^{3x} - e^x}{e^x - 1}$

25. $\lim_{x \to 1} \dfrac{x^3 - ax^2 + ax - 1}{x - 1}$

FIGURE 1 Graph of $P(n)$.

26. $\lim\limits_{x\to 0}\left(\dfrac{1}{3x}-\dfrac{1}{x(x+3)}\right)$

27. $\lim\limits_{x\to 1+}\left(\dfrac{1}{\sqrt{x-1}}-\dfrac{1}{\sqrt{x^2-1}}\right)$

28. $\lim\limits_{x\to 1.5}\dfrac{[x]}{x}$

29. $\lim\limits_{x\to 4.3}\dfrac{1}{x-[x]}$

30. $\lim\limits_{t\to 0-}\dfrac{[x]}{x}$

31. $\lim\limits_{t\to 0+}\dfrac{[x]}{x}$

32. $\lim\limits_{\theta\to\frac{\pi}{4}}\sec\theta$

33. $\lim\limits_{\theta\to\frac{\pi}{2}}\theta\sec\theta$

34. $\lim\limits_{\theta\to 0}\dfrac{\cos\theta-2}{\theta}$

35. $\lim\limits_{\theta\to 0}\dfrac{\sin 5\theta}{\theta}$

36. $\lim\limits_{\theta\to 0}\dfrac{\sin 4x}{\sin 3x}$

37. $\lim\limits_{t\to 0}\dfrac{\sin^2 t}{t^3}$

38. $\lim\limits_{x\to\frac{\pi}{2}}\tan x$

39. $\lim\limits_{t\to 0}\cos\dfrac{1}{t}$

40. $\lim\limits_{t\to 0+}\sqrt{t}\cos\dfrac{1}{t}$

41. $\lim\limits_{x\to 0}\dfrac{\sin 7x}{\sin 3x}$

42. $\lim\limits_{x\to 0}\dfrac{\cos x-1}{\sin x}$

43. $\lim\limits_{\theta\to 0}\dfrac{\tan\theta-\sin\theta}{\sin^3\theta}$

44. Find the left- and right-hand limits of the function $f(x)$ in Figure 2 at $x=0,2,4$. State whether $f(x)$ is left- or right-continuous (or both) at these points.

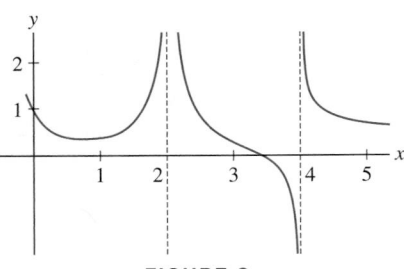

FIGURE 2

45. Sketch the graph of a function $f(x)$ such that

$$\lim_{x\to 2-}f(x)=1,\qquad \lim_{x\to 2+}f(x)=3$$

and such that $\lim\limits_{x\to 4}f(x)$ exists but does not equal $f(4)$.

46. Graph $h(x)$ and describe the type of discontinuity.

$$h(x)=\begin{cases}e^x & \text{for } x\le 0\\ \ln x & \text{for } x>0\end{cases}$$

47. Sketch the graph of a function $g(x)$ such that

$$\lim_{x\to -3-}g(x)=\infty,\qquad \lim_{x\to -3+}g(x)=-\infty,\qquad \lim_{x\to 4}g(x)=\infty$$

48. Find the points of discontinuity of $g(x)$ and determine the type of discontinuity, where

$$g(x)=\begin{cases}\cos\left(\dfrac{\pi x}{2}\right) & \text{for } |x|<1\\[2mm] |x-1| & \text{for } |x|\ge 1\end{cases}$$

49. Show that $f(x)=xe^{\sin x}$ is continuous on its domain. State the domain.

50. Find a constant b such that $h(x)$ is continuous at $x=2$, where

$$h(x)=\begin{cases}x+1 & \text{for } |x|<2\\ b-x^2 & \text{for } |x|\ge 2\end{cases}$$

With this choice of b, find all points of discontinuity.

51. Calculate the following limits, assuming that

$$\lim_{x\to 3}f(x)=6,\qquad \lim_{x\to 3}g(x)=4$$

(a) $\lim\limits_{x\to 3}(f(x)-2g(x))$

(b) $\lim\limits_{x\to 3}x^2 f(x)$

(c) $\lim\limits_{x\to 3}\dfrac{f(x)}{g(x)+x}$

(d) $\lim\limits_{x\to 3}(2g(x)^3-g(x)^2)$

52. Let $f(x)$ and $g(x)$ be functions such that $g(x)\ne 0$ for $x\ne a$, and let

$$A=\lim_{x\to a}f(x),\qquad B=\lim_{x\to a}g(x),\qquad L=\lim_{x\to a}\dfrac{f(x)}{g(x)}$$

Prove that if the limits A, B, and L exist and $L=1$, then $A=B$. *Hint:* You cannot use the Quotient Law if $B=0$, so apply the Product Law to L and B instead.

53. 📖 In the notation of Exercise 52, give an example in which L exists but neither A nor B exists.

54. Let $f(x)$ be a function defined for all real numbers. Which of the following statements *must be true*, *might be true*, or *are never true*.

(a) $\lim\limits_{x\to 3}f(x)=f(3)$.

(b) If $\lim\limits_{x\to 0}\dfrac{f(x)}{x}=1$, then $f(0)=0$.

(c) If $\lim\limits_{x\to -7}f(x)=8$, then $\lim\limits_{x\to -7}\dfrac{1}{f(x)}=\dfrac{1}{8}$.

(d) If $\lim\limits_{x\to 5+}f(x)=4$ and $\lim\limits_{x\to 5-}f(x)=8$, then $\lim\limits_{x\to 5}f(x)=6$.

(e) If $\lim\limits_{x\to 0}\dfrac{f(x)}{x}=1$, then $\lim\limits_{x\to 0}f(x)=0$

(f) If $\lim\limits_{x\to 5}f(x)=2$, then $\lim\limits_{x\to 5}f(x)^3=8$.

55. Use the Squeeze Theorem to prove that the function $f(x)=x\cos\dfrac{1}{x}$ for $x\ne 0$ and $f(0)=0$ is continuous at $x=0$.

56. Let $f(x) = x \left[\frac{1}{x} \right]$, where $[x]$ is the greatest integer function.

(a) Sketch the graph of $f(x)$ on the interval $[\frac{1}{4}, 2]$.

(b) Show that for $x \neq 0$,

$$\frac{1}{x} - 1 < \left[\frac{1}{x} \right] \le \frac{1}{x}$$

Then use the Squeeze Theorem to prove that

$$\lim_{x \to 0} x \left[\frac{1}{x} \right] = 1$$

57. Let r_1 and r_2 be the roots of the quadratic polynomial $f(x) = ax^2 - 2x + 20$. Observe that $f(x)$ "approaches" the linear function $L(x) = -2x + 20$ as $a \to 0$. Since $r = 10$ is the unique root of $L(x) = 0$, we might expect at least one of the roots of $f(x)$ to approach 10 as $a \to 0$ (Figure 3). Prove that the roots can be labeled so that $\lim_{a \to 0} r_1 = 10$ and $\lim_{a \to 0} r_2 = \infty$.

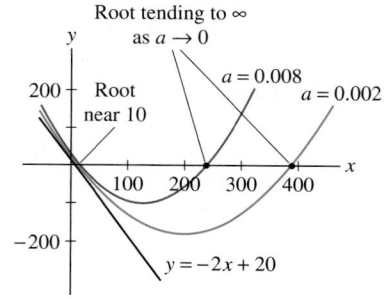

FIGURE 3 Graphs of $f(x) = ax^2 - 2x + 20$.

58. Let $f(x) = \frac{1}{x+2}$.

(a) Show that $\left| f(x) - \frac{1}{4} \right| < \frac{|x-2|}{12}$ if $|x - 2| < 1$. *Hint:* Observe that $|4(x+2)| > 12$ if $|x - 2| < 1$.

(b) Find $\delta > 0$ such that $\left| f(x) - \frac{1}{4} \right| < 0.01$ for $|x - 2| < \delta$.

(c) Prove rigorously that $\lim_{x \to 2} f(x) = \frac{1}{4}$.

59. [GU] Plot the function $f(x) = x^{1/3}$. Use the zoom feature to find a $\delta > 0$ such that $|x^{1/3} - 2| < 0.05$ for $|x - 8| < \delta$.

60. Prove rigorously that $\lim_{x \to -1} (4 + 8x) = -4$.

61. Prove rigorously that $\lim_{x \to 3} (x^2 - x) = 6$.

62. Use the IVT to prove that the curves $y = x^2$ and $y = \cos x$ intersect.

63. Use the IVT to prove that $f(x) = x^3 - \frac{x^2+2}{\cos x + 2}$ has a root in the interval $[0, 2]$.

64. Use the IVT to show that $e^{-x^2} = x$ has a solution on $(0, 1)$.

65. Use the Bisection Method to locate a root of $x^2 - 7 = 0$ to two decimal places.

66. Give an example of a (discontinuous) function that does not satisfy the conclusion of the IVT on $[-1, 1]$. Then show that the function

$$f(x) = \begin{cases} \sin \dfrac{1}{x} & \text{for } x \neq 0 \\ 0 & \text{for } x = 0 \end{cases}$$

satisfies the conclusion of the IVT on every interval $[-a, a]$, even though f is discontinuous at $x = 0$.

3 | DIFFERENTIATION

Calculus and physics are used to design rides like this one, called Superman the Escape, at Six Flags Magic Mountain in Valencia, California.

The term "differential calculus" was introduced in its Latin form "calculus differentialis" by Gottfried Wilhelm Leibniz. Newton referred to calculus as the "method of fluxions" (from Latin, meaning "to flow"), but this term was never adopted universally.

Differential calculus is the study of the derivative and its many applications. What is the derivative of a function? This question has three equally important answers: A derivative is a rate of change, it is the slope of a tangent line, and more formally, it is the limit of a difference quotient, as we will explain shortly. In this chapter, we explore these three facets of the derivative and develop the basic rules of differentiation, that is, techniques for computing derivatives. When you master these techniques, you will possess one of the most useful and flexible tools that mathematics has to offer.

3.1 Definition of the Derivative

Let's begin with two questions: How can we define the tangent line to a graph and how can we compute its slope? To answer these questions, we return to the relationship between tangent lines and secant lines first mentioned in Section 2.1.

Recall the formula for the slope of the secant line between two points P and Q on the graph of $y = f(x)$. If $P = (a, f(a))$ and $Q = (x, f(x))$ with $x \neq a$, as in Figure 1(A), then

$$\text{Slope of secant line through } P \text{ and } Q = \frac{\Delta f}{\Delta x} = \frac{f(x) - f(a)}{x - a}$$

where we set

$$\Delta f = f(x) - f(a) \qquad \text{and} \qquad \Delta x = x - a$$

The expression $\dfrac{f(x) - f(a)}{x - a}$ is called the **difference quotient**.

| ←·· REMINDER A *secant line* is any line through two points on a curve or graph.

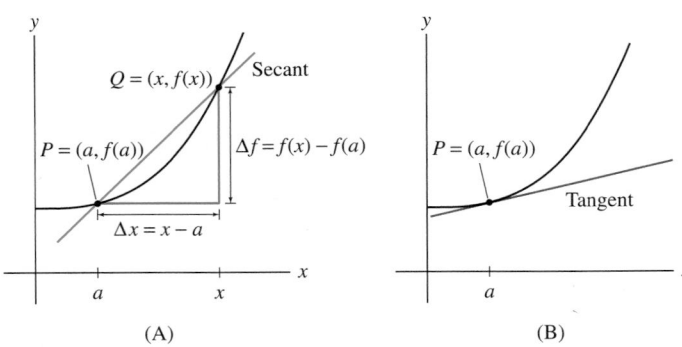

FIGURE 1 The secant line has slope $\Delta f / \Delta x$. Our goal is to compute the slope of the tangent line at $(a, f(a))$.

It is not easy to describe tangent lines accurately in words. Even the Oxford English Dictionary is less than clear when it defines a tangent as "a straight line which touches a curve, i.e. meets it at a point and being produced does not (ordinarily) intersect it at that point." Perhaps it is not surprising that mathematicians found it necessary to use limits to attain the required precision.

Now observe what happens as Q approaches P or, equivalently, as x approaches a. Figure 2 suggests that the secant lines get progressively closer to the tangent line. If we imagine Q moving toward P, then the secant line appears to rotate into the tangent line as in (D). Therefore, we may expect the slopes of the secant lines to approach the slope of the tangent line. Based on this intuition, we define the **derivative** $f'(a)$ (read "f prime of a") as the limit

$$f'(a) = \underbrace{\lim_{x \to a} \frac{f(x) - f(a)}{x - a}}_{\text{Limit of slopes of secant lines}}$$

(A)

(B)

(C)

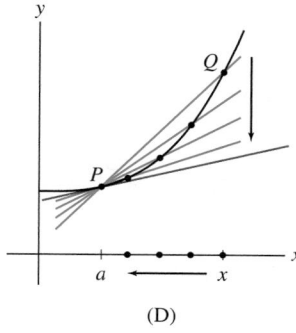

(D)

FIGURE 2 The secant lines approach the tangent line as Q approaches P.

There is another way of writing the difference quotient using the variable h:

$$h = x - a$$

We have $x = a + h$ and for $x \neq a$ (Figure 3)

$$\frac{f(x) - f(a)}{x - a} = \frac{f(a + h) - f(a)}{h}$$

The variable h approaches 0 as $x \to a$, so we can rewrite the derivative as

$$f'(a) = \lim_{h \to 0} \frac{f(a + h) - f(a)}{h}$$

Each way of writing the derivative is useful. The version using h is often more convenient in computations.

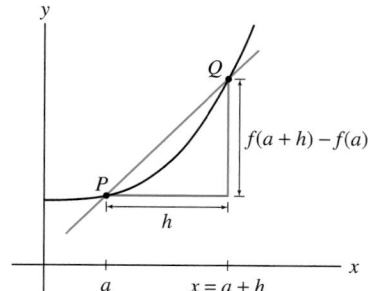

FIGURE 3 Secant line and difference quotient in terms of h.

DEFINITION The Derivative The derivative of a function f at $x = a$ is the limit of the difference quotients (if it exists):

$$f'(a) = \lim_{h \to 0} \frac{f(a + h) - f(a)}{h} \qquad \boxed{1}$$

When the limit exists, we say that f **is differentiable** at $x = a$. An equivalent definition of the derivative is

$$f'(a) = \lim_{x \to a} \frac{f(x) - f(a)}{x - a} \qquad \boxed{2}$$

We can now use the derivative to define the tangent line in a precise and unambiguous way as the line of slope $f'(a)$ through $P = (a, f(a))$.

→·· **REMINDER** *The equation in point-slope form of the line through* $P = (a, b)$ *of slope* m *is*

$$y - b = m(x - a)$$

> **DEFINITION Tangent Line**   Assume that $f(x)$ is differentiable at $x = a$. The tangent line to the graph of $y = f(x)$ at $P = (a, f(a))$ is the line through P of slope $f'(a)$. The equation of the tangent line in point-slope form is
>
> $$\boxed{y - f(a) = f'(a)(x - a)}$$ $\boxed{3}$

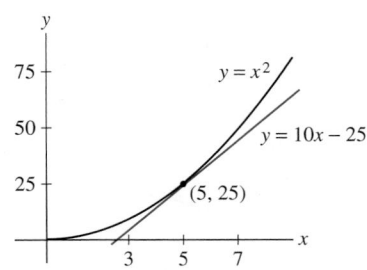

FIGURE 4 Tangent line to $y = x^2$ at $x = 5$.

■ **EXAMPLE 1** Finding an Equation of a Tangent Line   Find an equation of the tangent line to the graph of $f(x) = x^2$ at $x = 5$.

Solution We need to compute $f'(5)$ and we are free to use either Eq. (1) or (2). Using Eq. (2), we have

$$f'(5) = \lim_{x \to 5} \frac{f(x) - f(5)}{x - 5} = \lim_{x \to 5} \frac{x^2 - 25}{x - 5} = \lim_{x \to 5} \frac{(x - 5)(x + 5)}{x - 5}$$

$$= \lim_{x \to 5} (x + 5) = 10$$

Next, we apply Eq. (3) with $a = 5$. Since $f(5) = 25$, an equation of the tangent line is $y - 25 = 10(x - 5)$, or in slope-intercept form: $y = 10x - 25$ (Figure 4).   ■

The process of computing the derivative is called **differentiation**. The next two examples illustrate differentiation using Eq. (1). For clarity, we break up the computations into three steps.

■ **EXAMPLE 2** Compute $f'(3)$, where $f(x) = 4x^2 - 7x$.

Solution We use Eq. (1). The difference quotient at $a = 3$ is

$$\frac{f(a + h) - f(a)}{h} = \frac{f(3 + h) - f(3)}{h} \quad (h \neq 0)$$

Step 1. **Write out the numerator of the difference quotient.**

$$f(3 + h) - f(3) = \left(4(3 + h)^2 - 7(3 + h)\right) - \left(4(3^2) - 7(3)\right)$$

$$= \left(4(9 + 6h + h^2) - (21 + 7h)\right) - 15$$

$$= \left(15 + 17h + 4h^2\right) - 15 = 17h + 4h^2 = h(17 + 4h)$$

Step 2. **Divide by h and simplify the difference quotient.**
Simplify the difference quotient by canceling h in the numerator and denominator:

$$\frac{f(3 + h) - f(3)}{h} = \underbrace{\frac{h(17 + 4h)}{h}}_{\text{Cancel } h} = 17 + 4h$$

Step 3. **Compute the derivative by taking the limit.**

$$f'(3) = \lim_{h \to 0} \frac{f(3 + h) - f(3)}{h} = \lim_{h \to 0} (17 + 4h) = 17 \qquad ■$$

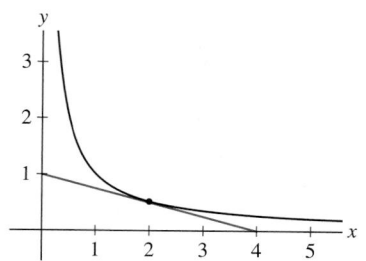

FIGURE 5 Graph of $y = \frac{1}{x}$. The equation of the tangent line at $x = 2$ is $y = -\frac{1}{4}x + 1$.

■ **EXAMPLE 3** Sketch the graph of $f(x) = \frac{1}{x}$ and draw the tangent line at $x = 2$.

(a) On the basis of the sketch, do you expect $f'(2)$ to be positive or negative?

(b) Find an equation of the tangent line at $x = 2$.

Solution The graph of $f(x)$ and the tangent line at $x = 2$ are sketched in Figure 5.

(a) The tangent line has negative slope, so we expect our calculation of $f'(2)$ to yield a negative number.

(b) We break up the calculation of $f'(2)$ into three steps as before.

Step 1. **Write out the numerator of the difference quotient.**

$$f(2+h) - f(2) = \frac{1}{2+h} - \frac{1}{2} = \frac{2}{2(2+h)} - \frac{2+h}{2(2+h)} = -\frac{h}{2(2+h)}$$

Step 2. **Divide by h and simplify the difference quotient.**

$$\frac{f(2+h) - f(2)}{h} = \frac{1}{h} \cdot \left(-\frac{h}{2(2+h)}\right) = -\frac{1}{2(2+h)}$$

Step 3. **Compute the derivative by taking the limit.**

$$f'(2) = \lim_{h \to 0} \frac{f(2+h) - f(2)}{h} = \lim_{h \to 0} \frac{-1}{2(2+h)} = -\frac{1}{4}$$

Since $f(2) = \frac{1}{2}$, the tangent line passes through $(2, \frac{1}{2})$ and has equation

$$y - \frac{1}{2} = -\frac{1}{4}(x - 2)$$

In slope-intercept form: $y = -\frac{1}{4}x + 1$. ■

If $f(x)$ is a linear function, say, $f(x) = mx + b$ (where m and b are constants), then the graph of f is a straight line of slope m. We should expect that $f'(a) = m$ for all a because the tangent line at any point coincides with the line $y = f(x)$ (Figure 6). Let's check this by computing the derivative:

$$f'(a) = \lim_{h \to 0} \frac{f(a+h) - f(a)}{h} = \lim_{h \to 0} \frac{(m(a+h) + b) - (ma + b)}{h}$$

$$= \lim_{h \to 0} \frac{mh}{h} = \lim_{h \to 0} m = m$$

If $m = 0$, then $f(x) = b$ is constant and $f'(a) = 0$ (Figure 7). In summary,

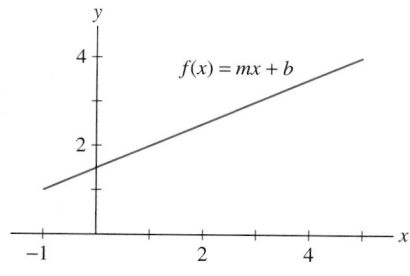

FIGURE 6 The derivative of $f(x) = mx + b$ is $f'(a) = m$.

THEOREM 1 Derivative of Linear and Constant Functions

- If $f(x) = mx + b$ is a linear function, then $f'(a) = m$ for all a.
- If $f(x) = b$ is a constant function, then $f'(a) = 0$ for all a.

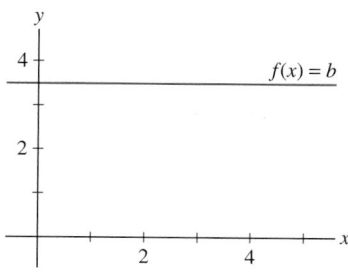

FIGURE 7 The derivative of a constant function $f(x) = b$ is $f'(a) = 0$.

■ **EXAMPLE 4** Find the derivative of $f(x) = 9x - 5$ at $x = 2$ and $x = 5$.

Solution We have $f'(a) = 9$ for all a. Hence, $f'(2) = f'(5) = 9$. ■

Estimating the Derivative

Approximations to the derivative are useful in situations where we cannot evaluate $f'(a)$ exactly. Since the derivative is the limit of difference quotients, the difference quotient should give a good numerical approximation when h is sufficiently small:

$$f'(a) \approx \frac{f(a+h) - f(a)}{h} \qquad \text{if } h \text{ is small}$$

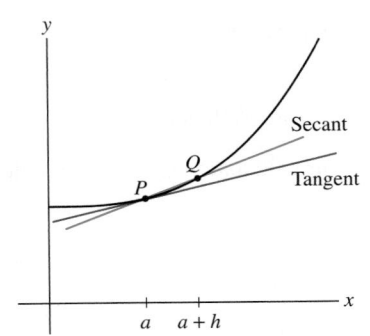

FIGURE 8 When h is small, the secant line has nearly the same slope as the tangent line.

Graphically, this says that when h is small, the slope of the secant line is nearly equal to the slope of the tangent line (Figure 8).

■ **EXAMPLE 5** Estimate the derivative of $f(x) = \sin x$ at $x = \frac{\pi}{6}$.

Solution The difference quotient is

$$\frac{\sin\left(\frac{\pi}{6} + h\right) - \sin\frac{\pi}{6}}{h} = \frac{\sin\left(\frac{\pi}{6} + h\right) - 0.5}{h}$$

We calculate the difference quotient for several small values of h. Table 1 suggests that the limit has a decimal expansion beginning 0.866. In other words, $f'\left(\frac{\pi}{6}\right) \approx 0.866$. ■

TABLE 1   Values of the Difference Quotient for Small h

$h > 0$	$\dfrac{\sin\left(\frac{\pi}{6} + h\right) - 0.5}{h}$	$h < 0$	$\dfrac{\sin\left(\frac{\pi}{6} + h\right) - 0.5}{h}$
0.01	0.863511	−0.01	0.868511
0.001	0.865775	−0.001	0.866275
0.0001	0.866000	−0.0001	0.866050
0.00001	0.8660229	−0.00001	0.8660279

In the previous example, we used a table of values to *suggest* an estimate for $f'\left(\frac{\pi}{6}\right)$. In the next example, we use graphical reasoning to determine the accuracy of this estimate.

■ **EXAMPLE 6** GU **Determining Accuracy Graphically** Let $f(x) = \sin x$. Show that the approximation $f'\left(\frac{\pi}{6}\right) \approx 0.8660$ is accurate to four decimal places by studying the relationship between the secant and tangent lines.

Solution In Figure 9 you will notice that the relationship of the secant line to the tangent line depends on whether h is positive or negative. The slope of the secant line is *smaller* than the slope of the tangent when $h > 0$, but it is *larger* when $h < 0$. This tells us that the difference quotients in the second column of Table 1 are smaller than $f'\left(\frac{\pi}{6}\right)$ and those in the fourth column are greater than $f'\left(\frac{\pi}{6}\right)$. In particular, from the last line in Table 1 we may conclude that

This technique of estimating an unknown quantity by showing that it lies between two known values occurs frequently in calculus.

$$0.866022 \leq f'\left(\frac{\pi}{6}\right) \leq 0.866028$$

The value of $f'\left(\frac{\pi}{6}\right)$ rounded off to five places is either 0.86602 or 0.86603, so the estimate 0.8660 is accurate to four decimal places. In Section 3.6, we will see that the exact value is $f'\left(\frac{\pi}{6}\right) = \cos\left(\frac{\pi}{6}\right) = \sqrt{3}/2 \approx 0.8660254$. ■

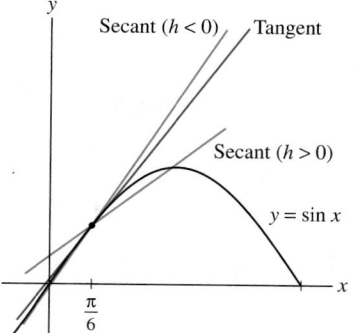

FIGURE 9 The tangent line is squeezed in *between* the secant lines with $h > 0$ and $h < 0$.

CONCEPTUAL INSIGHT **Are Limits Really Necessary?**   It is natural to ask whether limits are really necessary. Since the tangent line is so easy to visualize, is there perhaps a better or simpler way to find its equation? History gives one answer. The methods of calculus based on limits have stood the test of time and are used more widely today than ever before.

History aside, there is a more intuitive argument for why limits are necessary. The slope of a line can be computed if the coordinates of *two* points $P = (x_1, y_1)$ and $Q = (x_2, y_2)$ on the line are known:

$$\text{Slope of line} = \frac{y_2 - y_1}{x_2 - x_1}$$

We cannot use this formula to find the slope of the tangent line because we have only one point on the tangent line to work with, namely $P = (a, f(a))$. Limits provide an ingenious way around this obstacle. We choose a point $Q = (a + h, f(a + h))$ on the graph near P and form the secant line. The slope of this secant line is only an approximation to the slope of the tangent line:

$$\text{Slope of secant line} = \frac{f(a+h) - f(a)}{h} \approx \text{slope of tangent line}$$

But this approximation improves as $h \to 0$, and by taking the limit, we convert our approximations into an exact result.

3.1 SUMMARY

- The *difference quotient*

$$\frac{f(a+h) - f(a)}{h}$$

is the slope of the secant line through $P = (a, f(a))$ and $Q = (a + h, f(a + h))$ on the graph of $f(x)$.

- The *derivative* $f'(a)$ is defined by the following equivalent limits:

$$f'(a) = \lim_{h \to 0} \frac{f(a+h) - f(a)}{h} = \lim_{x \to a} \frac{f(x) - f(a)}{x - a}$$

If the limit exists, we say that f is *differentiable* at $x = a$.

- By definition, the tangent line at $P = (a, f(a))$ is the line through P with slope $f'(a)$ [assuming that $f'(a)$ exists]. The equation of the tangent line in point-slope form is

$$y - f(a) = f'(a)(x - a)$$

- To calculate $f'(a)$ using the limit definition, it is convenient to follow three steps:

Step 1. Write out the numerator of the difference quotient.

Step 2. Divide by h and simplify the difference quotient.

Step 3. Compute the derivative by taking the limit.

- For small values of h, we have the estimate $f'(a) \approx \dfrac{f(a+h) - f(a)}{h}$.

3.1 EXERCISES

Preliminary Questions

1. Which of the lines in Figure 10 are tangent to the curve?

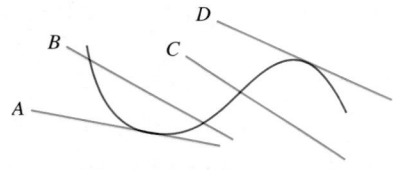

FIGURE 10

2. What are the two ways of writing the difference quotient?

3. For which value of x is

$$\frac{f(x) - f(3)}{x - 3} = \frac{f(7) - f(3)}{4}?$$

4. What do the following quantities represent in terms of the graph of $f(x) = \sin x$?

(a) $\sin 1.3 - \sin 0.9$

(b) $\dfrac{\sin 1.3 - \sin 0.9}{0.4}$

(c) $f'(1.3)$

5. For which values of a and h is $\dfrac{f(a+h) - f(a)}{h}$ equal to the slope of the secant line between the points $(3, f(3))$ and $(5, f(5))$ on the graph of $f(x)$?

6. To which derivative is the quantity

$$\frac{\tan\left(\frac{\pi}{4} + 0.00001\right) - 1}{0.00001}$$

a good approximation?

Exercises

1. Let $f(x) = 3x^2$. Show that

$$f(2 + h) = 3h^2 + 12h + 12$$

Then show that

$$\frac{f(2 + h) - f(2)}{h} = 3h + 12$$

and compute $f'(2)$ by taking the limit as $h \to 0$.

2. Let $f(x) = 2x^2 - 3x - 5$. Show that the slope of the secant line through $(2, f(2))$ and $(2 + h, f(2 + h))$ is $2h + 5$. Then use this formula to compute the slope of:

(a) The secant line through $(2, f(2))$ and $(3, f(3))$

(b) The tangent line at $x = 2$ (by taking a limit)

In Exercises 3–6, compute $f'(a)$ in two ways, using Eq. (1) and Eq. (2).

3. $f(x) = x^2 + 9x, \quad a = 0$

4. $f(x) = x^2 + 9x, \quad a = 2$

5. $f(x) = 3x^2 + 4x + 2, \quad a = -1$

6. $f(x) = 9 - 3x^2, \quad a = 0$

In Exercises 7–10, refer to the function whose graph is shown in Figure 11.

7. Calculate the slope of the secant line through the points on the graph where $x = 0$ and $x = 2.5$.

8. Compute $\dfrac{f(2.5) - f(1)}{2.5 - 1}$. What does this quantity represent?

9. Estimate $\dfrac{f(2 + h) - f(2)}{h}$ for $h = 0.5$ and $h = -0.5$. Are these numbers larger or smaller than $f'(2)$? Explain.

10. Estimate $f'(2)$.

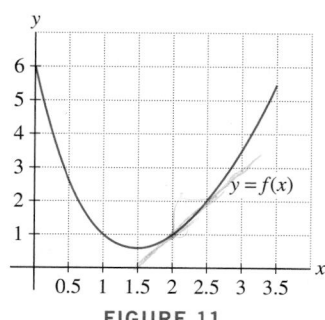

FIGURE 11

In Exercises 11–14, let $f(x)$ be the function whose graph is shown in Figure 12.

FIGURE 12 Graph of $f(x)$.

11. Determine $f'(a)$ for $a = 1, 2, 4, 7$.

12. Estimate $f'(6)$.

13. Which is larger: $f'(5.5)$ or $f'(6.5)$?

14. Show that $f'(3)$ does not exist.

In Exercises 15–18, use the limit definition to find the derivative of the linear function. (Note: The derivative does not depend on a.)

15. $f(x) = 3x - 2$ **16.** $f(x) = 2$

17. $g(t) = 9 - t$ **18.** $k(z) = 16z + 9$

19. Let $f(x) = \dfrac{1}{x}$. Does $f(-2 + h)$ equal $\dfrac{1}{-2 + h}$ or $\dfrac{1}{-2} + \dfrac{1}{h}$?
Compute the difference quotient for $f(x)$ at $a = -2$ with $h = 0.5$.

20. Let $f(x) = \sqrt{x}$. Does $f(5 + h)$ equal $\sqrt{5 + h}$ or $\sqrt{5} + \sqrt{h}$?
Compute the difference quotient for $f(x)$ at $a = 5$ with $h = 1$.

21. Let $f(x) = \sqrt{x}$. Show that

$$\frac{f(9 + h) - f(9)}{h} = \frac{1}{\sqrt{9 + h} + 3}$$

Then use this formula to compute $f'(9)$ (by taking the limit).

22. First find the slope and then an equation of the tangent line to the graph of $f(x) = \sqrt{x}$ at $x = 4$.

In Exercises 23–40, compute the derivative at $x = a$ using the limit definition and find an equation of the tangent line.

23. $f(x) = 3x^2 + 2x, \quad a = 2$

24. $f(x) = 3x^2 + 2x, \quad a = -1$

25. $f(x) = x^3, \quad a = 2$

26. $f(x) = x^3, \quad a = 3$

27. $f(x) = x^3 + x, \quad a = 0$

28. $f(t) = 3t^3 + 2t, \quad a = 4$

29. $f(x) = x^{-1}, \quad a = 3$ **30.** $f(x) = x^{-2}, \quad a = 1$

31. $f(x) = 9x - 4, \quad a = -7$ **32.** $f(t) = \sqrt{t + 1}, \quad a = 0$

33. $f(x) = \dfrac{1}{x + 3}, \quad a = -2$ **34.** $f(x) = \dfrac{x + 1}{x - 1}, \quad a = 3$

35. $f(t) = \dfrac{2}{1 - t}, \quad a = -1$ **36.** $f(t) = \dfrac{1}{t + 9}, \quad a = 2$

37. $f(t) = t^{-3}, \quad a = 1$ **38.** $f(t) = t^4, \quad a = 2$

39. $f(x) = \dfrac{1}{\sqrt{x}}, \quad a = 9$

40. $f(x) = (x^2 + 1)^{-1}, \quad a = 0$

41. What is an equation of the tangent line at $x = 3$, assuming that $f(3) = 5$ and $f'(3) = 2$?

42. Suppose that $y = 5x + 2$ is an equation of the tangent line to the graph of $y = f(x)$ at $a = 3$. What is $f(3)$? What is $f'(3)$?

43. Consider the "curve" $y = 2x + 8$. What is the tangent line at the point $(1, 10)$? Describe the tangent line at an arbitrary point.

44. Suppose that $f(x)$ is a function such that $f(2 + h) - f(2) = 3h^2 + 5h$.
(a) What is $f'(2)$?
(b) What is the slope of the secant line through $(2, f(2))$ and $(6, f(6))$?

45. ⏢GU Verify that $P = \left(1, \tfrac{1}{2}\right)$ lies on the graphs of both $f(x) = \dfrac{1}{1 + x^2}$ and $L(x) = \tfrac{1}{2} + m(x - 1)$ for every slope m. Plot $f(x)$ and $L(x)$ on the same axes for several values of m. Experiment until you find a value of m for which $y = L(x)$ appears tangent to the graph of $f(x)$. What is your estimate for $f'(1)$?

46. ⏢GU Plot $f(x) = x^x$ for $0 \le x \le 1.5$.
(a) How many points x_0 are there such that $f'(x_0) = 0$?
(b) Estimate x_0 by plotting several horizontal lines $y = c$ together with the plot of $f(x)$ until you find a value of c for which the line is tangent to the graph.

47. GU   Plot $f(x) = x^x$ and the line $y = x + c$ on the same set of axes for several values of c. Experiment until you find a value c_0 such that the line is tangent to the graph. Then:

(a) Use your plot to estimate the value x_0 such that $f'(x_0) = 1$.

(b) Verify that $\dfrac{f(x_0 + h) - f(x_0)}{h}$ is close to 1 for $h = 0.01, 0.001,$ 0.0001.

48. Use the following table to calculate a few difference quotients of $f(x)$. Then estimate $f'(1)$ by taking an average of difference quotients at h and $-h$. For example, take the average for $h = 0.03$ and -0.03, etc. Recall that the average of a and b is $\dfrac{a+b}{2}$.

x	1.03	1.02	1.01	1
$f(x)$	0.5148	0.5234	0.5319	0.5403

x	0.97	0.98	0.99	
$f(x)$	0.5653	0.557	0.5486	

49. 📖   The vapor pressure of water is defined as the atmospheric pressure P at which no net evaporation takes place. The following table and Figure 13 give P (in atmospheres) as a function of temperature T in kelvins.

(a) Which is larger: $P'(300)$ or $P'(350)$? Answer by referring to the graph.

(b) Estimate $P'(T)$ for $T = 303, 313, 323, 333, 343$ using the table and the average of the difference quotients for $h = 10$ and -10:

$$P'(T) \approx \frac{P(T + 10) - P(T - 10)}{20} \qquad \boxed{4}$$

T (K)	293	303	313	323	333	343	353
P (atm)	0.0278	0.0482	0.0808	0.1311	0.2067	0.3173	0.4754

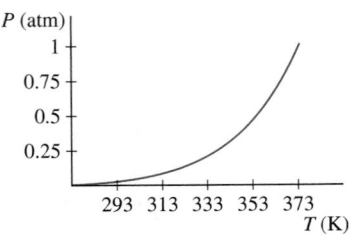

FIGURE 13 Vapor pressure of water as a function temperature T in kelvins.

In Exercises 50–51, traffic speed S along a certain road (in mph) varies as a function of traffic density q (number of cars per mile on the road). Use the following data to answer the questions:

q (density)	100	110	120	130	140
S (speed)	45	42	39.5	37	35

50. Estimate $S'(q)$ when $q = 120$ cars per mile using the average of difference quotients at h and $-h$ as in Exercise 48.

51. 📖   The quantity $V = qS$ is called *traffic volume*. Explain why V is equal to the number of cars passing a particular point per hour. Use the data to compute values of V as a function of q and estimate $V'(q)$ when $q = 120$.

52. For the graph in Figure 14, determine the intervals along the x-axis on which the derivative is positive.

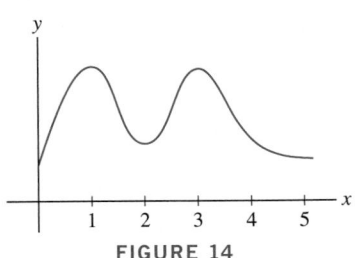

FIGURE 14

In Exercises 53–58, each of the limits represents a derivative $f'(a)$. Find $f(x)$ and a.

53. $\displaystyle \lim_{h \to 0} \frac{(5 + h)^3 - 125}{h}$

54. $\displaystyle \lim_{x \to 5} \frac{x^3 - 125}{x - 5}$

55. $\displaystyle \lim_{h \to 0} \frac{\sin(\frac{\pi}{6} + h) - .5}{h}$

56. $\displaystyle \lim_{x \to \frac{1}{4}} \frac{x^{-1} - 4}{x - \frac{1}{4}}$

57. $\displaystyle \lim_{h \to 0} \frac{5^{2+h} - 25}{h}$

58. $\displaystyle \lim_{h \to 0} \frac{5^h - 1}{h}$

59. Sketch the graph of $f(x) = \sin x$ on $[0, \pi]$ and guess the value of $f'\left(\frac{\pi}{2}\right)$. Then calculate the slope of the secant line between $x = \frac{\pi}{2}$ and $x = \frac{\pi}{2} + h$ for at least three small positive and negative values of h. Are these calculations consistent with your guess?

60. Figure 15(A) shows the graph of $f(x) = \sqrt{x}$. The close-up in (B) shows that the graph is nearly a straight line near $x = 16$. Estimate the slope of this line and take it as an estimate for $f'(16)$. Then compute $f'(16)$ and compare with your estimate.

(A) Graph of $y = \sqrt{x}$

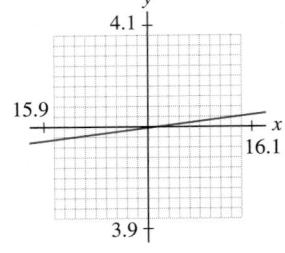

(B) Zoom view near (16, 4)

FIGURE 15

61. GU Let $f(x) = \dfrac{4}{1 + 2^x}$.

(a) Plot $f(x)$ over $[-2, 2]$. Then zoom in near $x = 0$ until the graph appears straight and estimate the slope $f'(0)$.

(b) Use your estimate to find an approximate equation to the tangent line at $x = 0$. Plot this line and the graph on the same set of axes.

62. GU Repeat Exercise 61 with $f(x) = (2 + x)^x$.

63. GU Let $f(x) = \cot x$. Estimate $f'\left(\frac{\pi}{2}\right)$ graphically by zooming in on a plot of $f(x)$ near $x = \frac{\pi}{2}$.

64. For each graph in Figure 16, determine whether $f'(1)$ is larger or smaller than the slope of the secant line between $x = 1$ and $x = 1 + h$ for $h > 0$. Explain.

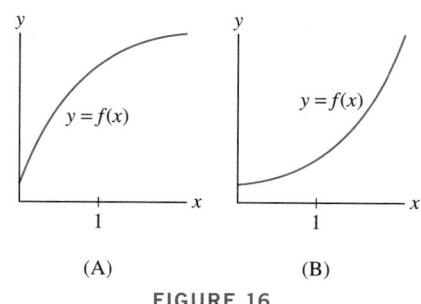

(A)   (B)

FIGURE 16

65. Apply the method of Example 6 to $f(x) = \sin x$ to determine $f'\left(\frac{\pi}{4}\right)$ accurately to four decimal places.

66. Apply the method of Example 6 to $f(x) = \cos x$ to determine $f'\left(\frac{\pi}{5}\right)$ accurately to four decimal places. Use a graph of $f(x)$ to explain how the method works in this case.

67. GU Sketch the graph of $f(x) = x^{5/2}$ on $[0, 6]$.

(a) Use the sketch to justify the inequalities for $h > 0$:

$$\frac{f(4) - f(4 - h)}{h} \leq f'(4) \leq \frac{f(4 + h) - f(4)}{h}$$

(b) Use part (a) to compute $f'(4)$ to four decimal places.

(c) Use a graphing utility to plot $f(x)$ and the tangent line at $x = 4$ using your estimate for $f'(4)$.

68. The graph of $f(x) = 2^x$ is shown in Figure 17.

(a) Referring to the graph, explain why the inequality

$$f'(0) \leq \frac{f(h) - f(0)}{h}$$

holds for $h > 0$ and the opposite inequality holds for $h < 0$.

(b) Use part (a) to show that $0.69314 \leq f'(0) \leq 0.69315$.

(c) Use the same procedure to determine $f'(x)$ to four decimal places for $x = 1, 2, 3, 4$.

(d) Now compute the ratios $f'(x)/f'(0)$ for $x = 1, 2, 3, 4$. Can you guess an (approximate) formula for $f'(x)$ for general x?

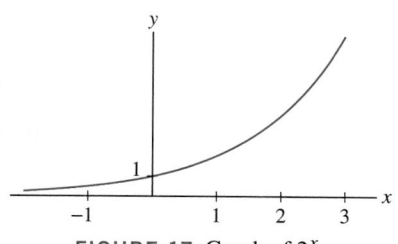

FIGURE 17 Graph of 2^x.

69. The graph in Figure 18 (based on data collected by the biologist Julian Huxley, 1887–1975) gives the average antler weight W of a red deer as a function of age t. Estimate the slope of the tangent line to the graph at $t = 4$. For which values of t is the slope of the tangent line equal to zero? For which values is it negative?

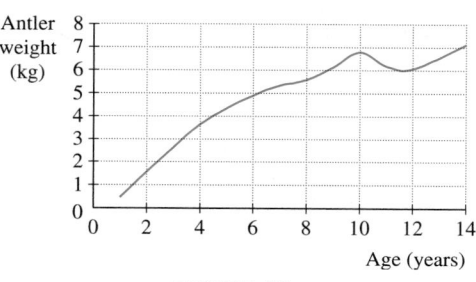

FIGURE 18

In Exercises 70–72, $i(t)$ is the current (in amperes) at time t (seconds) flowing in the circuit shown in Figure 19. According to Kirchhoff's law, $i(t) = Cv'(t) + R^{-1}v(t)$, where $v(t)$ is the voltage (in volts) at time t, C the capacitance (in farads), and R the resistance (in ohms, Ω).

FIGURE 19

70. Calculate the current at $t = 3$ if $v(t) = 0.5t + 4$ V, $C = 0.01$ F, and $R = 100\ \Omega$.

71. Use the following table to estimate $v'(10)$. For a better estimate, take the *average* of the difference quotients for h and $-h$ as described in Exercise 48. Then estimate $i(10)$, assuming $C = 0.03$ and $R = 1,000$.

t	9.8	9.9	10	10.1	10.2
$v(t)$	256.52	257.32	258.11	258.9	259.69

72. Assume that $R = 200\ \Omega$ but C is unknown. Use the following data to estimate $v'(4)$ as in Exercise 71 and deduce an approximate value for the capacitance C.

t	3.8	3.9	4	4.1	4.2
$v(t)$	388.8	404.2	420	436.2	452.8
$i(t)$	32.34	33.22	34.1	34.98	35.86

Further Insights and Challenges

In Exercises 73–76, we define the symmetric difference quotient (SDQ) at $x = a$ for $h \neq 0$ by

$$\frac{f(a+h) - f(a-h)}{2h} \qquad \boxed{5}$$

73. Explain how SDQ can be interpreted as the slope of a secant line.

74. The SDQ usually gives a better approximation to the derivative than does the ordinary difference quotient. Let $f(x) = 2^x$ and $a = 0$. Compute the SDQ with $h = 0.001$ and the ordinary difference quotients with $h = \pm 0.001$. Compare with the actual value of $f'(0)$ that, to eight decimal places, is 0.69314718.

75. Show that if $f(x)$ is a quadratic polynomial, then the SDQ at $x = a$ (for any $h \neq 0$) is *equal* to $f'(a)$. Explain the graphical meaning of this result.

76. Let $f(x) = x^{-2}$. Compute $f'(1)$ by taking the limit of the SDQs (with $a = 1$) as $h \to 0$.

77. Which of the two functions in Figure 20 satisfies the inequality

$$\frac{f(a+h) - f(a-h)}{2h} \le \frac{f(a+h) - f(a)}{h}$$

for $h > 0$? Explain in terms of secant lines.

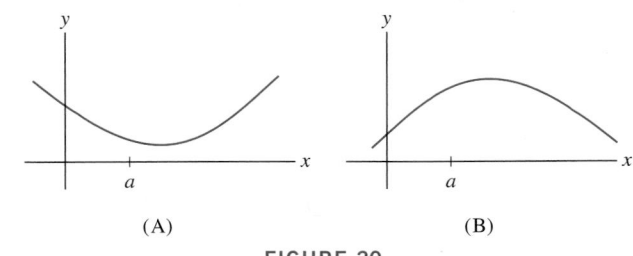

(A)   (B)

FIGURE 20

3.2 The Derivative as a Function

In the previous section, we computed the derivative $f'(a)$ for specific values of a. It is also useful to view the derivative as a function $f'(x)$ whose value at $x = a$ is $f'(a)$. The function $f'(x)$ is still defined as a limit, but the fixed number a is replaced by the variable x:

$$f'(x) = \lim_{h \to 0} \frac{f(x+h) - f(x)}{h} \qquad \boxed{1}$$

If $y = f(x)$, we also write y' or $y'(x)$ for $f'(x)$.

The domain of $f'(x)$ consists of all values of x in the domain of $f(x)$ for which the limit in Eq. (1) exists. We say that $f(x)$ is **differentiable** on (a, b) if $f'(x)$ exists for all x in (a, b). When $f'(x)$ exists for all x in the interval or intervals on which $f(x)$ is defined, we say simply that $f(x)$ is differentiable.

Often, the domain of $f'(x)$ is clear from the context. If so, we usually do not mention the domain explicitly.

■ **EXAMPLE 1** Prove that $f(x) = x^3 - 12x$ is differentiable. Compute $f'(x)$ and find an equation of the tangent line at $x = -3$.

Solution We compute $f'(x)$ in three steps as in the previous section.

Step 1. **Write out the numerator of the difference quotient.**

$$f(x+h) - f(x) = \left((x+h)^3 - 12(x+h)\right) - \left(x^3 - 12x\right)$$
$$= (x^3 + 3x^2h + 3xh^2 + h^3 - 12x - 12h) - (x^3 - 12x)$$

$$= 3x^2h + 3xh^2 + h^3 - 12h$$

$$= h(3x^2 + 3xh + h^2 - 12) \quad \text{(factor out } h\text{)}$$

Step 2. Divide by h and simplify the difference quotient.

$$\frac{f(x+h) - f(x)}{h} = \frac{h(3x^2 + 3xh + h^2 - 12)}{h} = 3x^2 + 3xh + h^2 - 12 \quad (h \neq 0)$$

Step 3. Compute the derivative by taking the limit.

$$f'(x) = \lim_{h \to 0} \frac{f(x+h) - f(x)}{h} = \lim_{h \to 0} (3x^2 + 3xh + h^2 - 12) = 3x^2 - 12$$

In this limit, x is treated as a constant since it does not change as $h \to 0$. We see that the limit exists for all x, so $f(x)$ is differentiable and $f'(x) = 3x^2 - 12$.

Now evaluate:

$$f(-3) = (-3)^3 - 12(-3) = 9, \qquad f'(-3) = 3(-3)^2 - 12 = 15$$

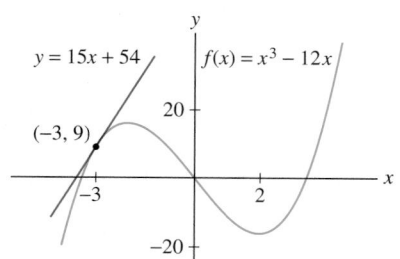

$y = 15x + 54$ $f(x) = x^3 - 12x$

$(-3, 9)$

FIGURE 1 Graph of $f(x) = x^3 - 12x$.

An equation of the tangent line at $x = -3$ is $y - 9 = 15(x + 3)$ (Figure 1). ∎

■ **EXAMPLE 2** Prove that $y = x^{-2}$ is differentiable and calculate y'.

Solution The domain of $f(x) = x^{-2}$ is $\{x : x \neq 0\}$, so assume that $x \neq 0$. We compute $f'(x)$ directly, without the separate steps of the previous example:

$$y' = \lim_{h \to 0} \frac{f(x+h) - f(x)}{h} = \lim_{h \to 0} \frac{\frac{1}{(x+h)^2} - \frac{1}{x^2}}{h}$$

$$= \lim_{h \to 0} \frac{\frac{x^2 - (x+h)^2}{x^2(x+h)^2}}{h} = \lim_{h \to 0} \frac{1}{h}\left(\frac{x^2 - (x+h)^2}{x^2(x+h)^2}\right)$$

$$= \lim_{h \to 0} \frac{1}{h}\left(\frac{-h(2x+h)}{x^2(x+h)^2}\right) = \lim_{h \to 0} -\frac{2x+h}{x^2(x+h)^2} \quad \text{(cancel } h\text{)}$$

$$= -\frac{2x+0}{x^2(x+0)^2} = -\frac{2x}{x^4} = -\frac{2}{x^3}$$

The limit exists for all $x \neq 0$, so y is differentiable and $y' = -2x^{-3}$. ∎

Leibniz Notation

The "prime" notation y' and $f'(x)$ was introduced by the French mathematician Joseph Louis Lagrange (1736–1813). There is another standard notation for the derivative due to Leibniz (Figure 2):

$$\frac{df}{dx} \quad \text{or} \quad \frac{dy}{dx}$$

In Example 2, we showed that the derivative of $y = x^{-2}$ is $y' = -2x^{-3}$. In Leibniz notation, this result is written

$$\frac{dy}{dx} = -2x^{-3} \quad \text{or} \quad \frac{d}{dx}x^{-2} = -2x^{-3}$$

FIGURE 2 Gottfried Wilhelm von Leibniz (1646–1716), German philosopher and scientist. Newton and Leibniz (pronounced "Libe-nitz") are often regarded as the inventors of calculus (working independently of each other). It is more accurate to credit them with developing calculus into a general and fundamental discipline, since many particular results of calculus had been discovered previously.

To specify the value of the derivative for a fixed value of x, say, $x = 4$, we write

$$\left.\frac{df}{dx}\right|_{x=4} \qquad \text{or} \qquad \left.\frac{dy}{dx}\right|_{x=4}$$

You should not think of $\frac{dy}{dx}$ as the fraction "dy divided by dx." The expressions dy and dx are called **differentials**. They play an important role in more advanced mathematics, but we shall treat them merely as symbols with no independent meaning.

CONCEPTUAL INSIGHT   Leibniz notation is widely used for several reasons. First, it reminds us that the derivative $\frac{df}{dx}$, although not itself a ratio, is in fact a *limit* of ratios $\frac{\Delta f}{\Delta x}$. Second, the notation specifies the independent variable, which is useful when variables other than x are used. For example, if the independent variable is t, we write $\frac{df}{dt}$. Third, we often think of $\frac{d}{dx}$ as an "operator" that performs the differentiation operation on functions. In other words, we apply the operator $\frac{d}{dx}$ to f to obtain the derivative $\frac{df}{dx}$. We will see other advantages of Leibniz notation when we discuss the Chain Rule in Section 3.7.

One of the main goals of this chapter is to develop the basic rules of differentiation. These rules are particularly useful because they allow us to find derivatives without computing limits.

THEOREM 1 The Power Rule   For all exponents n,

$$\frac{d}{dx}x^n = nx^{n-1}$$

The Power Rule is valid for all exponents. Here, we prove it for a whole number n (see Exercise 97 for a negative integer n and p. 193 for arbitrary n).

Proof Assume that n is a whole number and let $f(x) = x^n$. Then

$$f'(a) = \lim_{x \to a} \frac{x^n - a^n}{x - a}$$

To simplify the difference quotient, we need to generalize the following identities:

$$x^2 - a^2 = (x - a)(x + a)$$
$$x^3 - a^3 = (x - a)(x^2 + xa + a^2)$$
$$x^4 - a^4 = (x - a)(x^3 + x^2a + xa^2 + a^3)$$

The generalization is

$$x^n - a^n = (x - a)(x^{n-1} + x^{n-2}a + x^{n-3}a^2 + \cdots + xa^{n-2} + a^{n-1}) \qquad \boxed{2}$$

To verify Eq. (2), observe that the right-hand side is equal to

$$x\left(x^{n-1} + x^{n-2}a + x^{n-3}a^2 + \cdots + xa^{n-2} + a^{n-1}\right)$$
$$- a\left(x^{n-1} + x^{n-2}a + x^{n-3}a^2 + \cdots + xa^{n-2} + a^{n-1}\right)$$

When we carry out the multiplications, all terms cancel except the first and the last, so $x^n - a^n$ remains, as required.

The identity (2) gives us

$$\frac{x^n - a^n}{x - a} = \underbrace{x^{n-1} + x^{n-2}a + x^{n-3}a^2 + \cdots + xa^{n-2} + a^{n-1}}_{n \text{ terms}} \quad (x \neq a) \qquad \boxed{3}$$

Therefore,

$$f'(a) = \lim_{x \to a} \left(x^{n-1} + x^{n-2}a + x^{n-3}a^2 + \cdots + xa^{n-2} + a^{n-1}\right)$$
$$= a^{n-1} + a^{n-2}a + a^{n-3}a^2 + \cdots + aa^{n-2} + a^{n-1} \quad (n \text{ terms})$$
$$= na^{n-1}$$

This proves that $f'(a) = na^{n-1}$, which we may also write as $f'(x) = nx^{n-1}$.   ■

You can think of the Power Rule as "bringing down the exponent and subtracting one (from the exponent)":

$$\frac{d}{dx}x^{\text{exponent}} = (\text{exponent})\, x^{\text{exponent}-1}$$

The Power Rule is valid with any variable, not just x. Here are some examples:

$$\frac{d}{dx}x^2 = 2x, \qquad \frac{d}{dz}z^3 = 3z^2, \qquad \frac{d}{dt}t^{20} = 20t^{19}, \qquad \frac{d}{dr}r^5 = 5r^4$$

CAUTION The Power Rule applies only to the power functions $y = x^n$. It does not apply to exponential functions such as $y = 2^x$. The derivative of $y = 2^x$ **is not** $x2^{x-1}$. We will study the derivatives of exponential functions later in this section.

We also emphasize that the Power Rule holds for all exponents, whether negative, fractional, or irrational:

$$\frac{d}{dx}x^{-3/5} = -\frac{3}{5}x^{-8/5}, \qquad \frac{d}{dx}x^{\sqrt{2}} = \sqrt{2}\,x^{\sqrt{2}-1}$$

Next, we state the Linearity Rules for derivatives, which are analogous to the linearity laws for limits.

THEOREM 2 Linearity Rules   Assume that f and g are differentiable functions.

Sum Rule: The function $f + g$ is differentiable and

$$(f + g)' = f' + g'$$

Constant Multiple Rule: For any constant c, cf is differentiable and

$$(cf)' = cf'$$

Proof By definition,

$$(f+g)'(x) = \lim_{h \to 0} \frac{(f(x+h) + g(x+h)) - (f(x) + g(x))}{h}$$

This difference quotient is equal to a sum:

$$\frac{(f(x+h) + g(x+h)) - (f(x) + g(x))}{h} = \frac{f(x+h) - f(x)}{h} + \frac{g(x+h) - g(x)}{h}$$

Therefore, by the Sum Law for limits,

$$(f+g)'(x) = \lim_{h \to 0} \frac{f(x+h) - f(x)}{h} + \lim_{h \to 0} \frac{g(x+h) - g(x)}{h}$$

$$= f'(x) + g'(x)$$

as claimed. The Constant Multiple Rule is proved similarly. ■

The Difference Rule is also valid:

$$(f-g)' = f' - g'$$

However, it follows from the Sum and Constant Multiple Rules:

$$(f-g)' = \underbrace{f' + (-g)'}_{\text{Sum Rule}} = \underbrace{f' - g'}_{\text{Constant Multiple Rule}}$$

■ **EXAMPLE 3** Find the points on the graph of $f(t) = t^3 - 12t + 4$ where the tangent line is horizontal.

Solution First, we calculate the derivative:

$$\frac{df}{dt} = \frac{d}{dt}(t^3 - 12t + 4)$$

$$= \frac{d}{dt}t^3 - \frac{d}{dt}(12t) + \frac{d}{dt}4 \quad \text{(Sum and Difference Rules)}$$

$$= \frac{d}{dt}t^3 - 12\frac{d}{dt}t \quad\quad\quad \text{(Constant Multiple Rule)}$$

$$= 3t^2 - 12 \quad\quad\quad\quad\quad \text{(Power Rule)}$$

Notice that the derivative of the constant 4 is zero.

The tangent line is horizontal at points where the slope $f'(t)$ is zero, so we solve

$$f'(t) = 3t^2 - 12 = 0 \implies t = \pm 2$$

Therefore, the tangent lines are horizontal at $(2, -12)$ and $(-2, 20)$ (Figure 3). ■

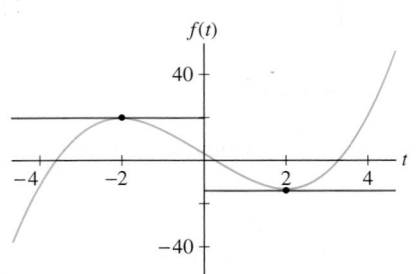

FIGURE 3 Graph of $f(t) = t^3 - 12t + 4$. Tangent lines at $t = \pm 2$ are horizontal.

In the next example, we differentiate term by term using the Power Rule without justifying the intermediate steps, as was done in Example 3.

■ **EXAMPLE 4** Calculate $\left.\dfrac{dg}{dt}\right|_{t=1}$, where $g(t) = t^{-3} + 2\sqrt{t} - t^{-4/5}$.

Solution To use the Power Rule, we write $\sqrt{t}$ as $t^{1/2}$:

$$\frac{dg}{dt} = \frac{d}{dt}\left(t^{-3} + 2t^{1/2} - t^{-4/5}\right) = -3t^{-4} + 2\left(\frac{1}{2}\right)t^{-1/2} - \left(-\frac{4}{5}\right)t^{-9/5}$$

$$= -3t^{-4} + t^{-1/2} + \frac{4}{5}t^{-9/5}$$

$$\left.\frac{dg}{dt}\right|_{t=1} = -3 + 1 + \frac{4}{5} = -\frac{6}{5}$$

∎

The derivative $f'(x)$ gives us important information about the graph of $f(x)$. For example, the sign of $f'(x)$ tells us whether the tangent line has positive or negative slope and the size of $f'(x)$ reveals how steep the slope is.

■ **EXAMPLE 5** Graphical Insight How is the graph of $f(x) = x^3 - 12x$ related to the derivative $f'(x) = 3x^2 - 12$?

Solution We observe that the derivative $f'(x) = 3x^2 - 12 = 3(x^2 - 4)$ is negative for $-2 < x < 2$ and positive for $|x| > 2$ [see Figure 4(B)]. The following table summarizes this sign information [see Figure 4(A)]:

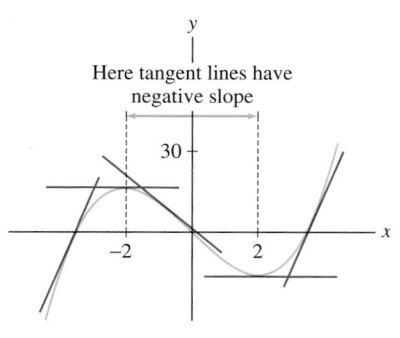

(A) Graph of $f(x) = x^3 - 12x$

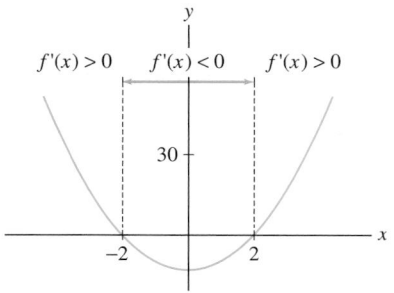

(B) Graph of the derivative $f'(x) = 3x^2 - 12$

FIGURE 4

Property of $f'(x)$	Property of the Graph of $f(x)$
$f'(x) < 0$ for $-2 < x < 2$	Tangent line has negative slope for $-2 < x < 2$
$f'(-2) = f'(2) = 0$	Tangent line is horizontal at $x = -2$ and $x = 2$
$f'(x) > 0$ for $x < -2$ and $x > 2$	Tangent line has positive slope for $x < -2$ and $x > 2$

Note also that $f'(x) \to \infty$ as $|x|$ becomes large. This corresponds to the fact that the tangent lines to the graph of $f(x)$ get steeper as $|x|$ grows large. ∎

■ **EXAMPLE 6** Identifying the Derivative The graph of $f(x)$ is shown in Figure 5. Which of (A) or (B) is the graph of $f'(x)$?

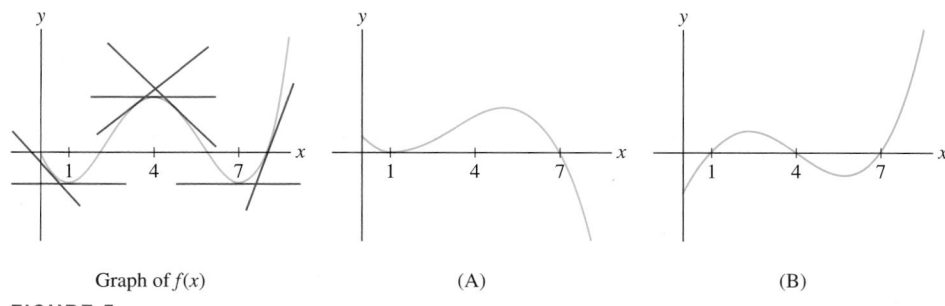

Graph of $f(x)$ (A)   (B)

FIGURE 5

Solution From Figure 5 we glean the following information about the tangent lines to the graph of $f(x)$:

Slope of Tangent Line	Where
Negative	on $(0, 1)$ and $(4, 7)$
Zero	for $x = 1, 4, 7$
Positive	on $(1, 4)$ and $(7, \infty)$

Therefore, the graph of $f'(x)$ must be negative on $(0, 1)$ and $(4, 7)$ and positive on $(1, 4)$ and $(7, \infty)$. Only (B) has these properties, so (B) is the graph of $f'(x)$. ■

The Derivative of e^x

The number e and the exponential function e^x were introduced informally in Section 1.6. We now define them more precisely using the derivative. First, let us compute the derivative of $f(x) = b^x$, where $b > 0$ and $b \neq 1$. The difference quotient (for $h \neq 0$) is

$$\frac{f(x+h) - f(x)}{h} = \frac{b^{x+h} - b^x}{h} = \frac{b^x b^h - b^x}{h} = \frac{b^x(b^h - 1)}{h}$$

and the derivative is equal to the limit:

$$f'(x) = \lim_{h \to 0} \frac{f(x+h) - f(x)}{h} = \lim_{h \to 0} \frac{b^x(b^h - 1)}{h} = b^x \lim_{h \to 0} \left(\frac{b^h - 1}{h} \right) \qquad \boxed{4}$$

Notice that we took the factor b^x outside the limit. This is legitimate because b^x does not depend on h.

The last limit on the right in Eq. (4) does not depend on x. We assume it exists and denote its value by m_b:

$$m_b = \lim_{h \to 0} \left(\frac{b^h - 1}{h} \right) \qquad \boxed{5}$$

What we have shown, then, is that *the derivative of b^x is proportional to b^x*:

$$\boxed{\frac{d}{dx} b^x = m_b\, b^x} \qquad \boxed{6}$$

What is m_b? Note, first of all, that m_b has a graphical interpretation as the slope of the tangent line to $y = b^x$ at $x = 0$ (Figure 6) since

$$\frac{d}{dx} b^x \bigg|_{x=0} = m_b \cdot b^0 = m_b$$

In Section 3.10, we will show that $m_b = \ln b$, the natural logarithm of b. Without assuming this, we may investigate m_b numerically using Eq. (5).

■ **EXAMPLE 7** Investigating m_b **Numerically** Investigate m_b numerically for $b = 2$, 2.5, 3, and 10.

Solution We create a table of values to estimate m_b numerically:

h	$\dfrac{2^h - 1}{h}$	$\dfrac{(2.5)^h - 1}{h}$	$\dfrac{3^h - 1}{h}$	$\dfrac{10^h - 1}{h}$
0.01	0.69556	0.92050	1.10467	2.32930
0.001	0.69339	0.91671	1.09921	2.30524
0.0001	0.69317	0.91633	1.09867	2.30285
0.00001	0.69315	0.916295	1.09861	2.30261
	$\boxed{m_2 \approx 0.693}$	$\boxed{m_{2.5} \approx 0.916}$	$\boxed{m_3 \approx 1.10}$	$\boxed{m_{10} \approx 2.30}$

■

These computations suggest that the numbers m_b are increasing:

$$m_2 \quad < \quad m_{2.5} \quad < \quad m_3 \quad < \quad m_{10}$$

Let us assume, as is reasonable, that m_b varies *continuously* as a function of b. Then, since $m_2 < 1$ and $m_3 > 1$, the Intermediate Value Theorem says that $m_b = 1$ for some number b between 2 and 3. This number is the number e. As mentioned in Section 1.6, e is irrational. To 40 places,

$$e = 2.7182818284590452353602874713526624977572\ldots$$

For most purposes, the approximation $e \approx 2.718$ is adequate.

Since e is defined by the property $m_e = 1$, Eq. (6) says that $(e^x)' = e^x$. In other words, $y = e^x$ *is equal to its own derivative.*

Although written reference to the number π goes back more than 4,000 years, mathematicians first became aware of the special role played by e in the seventeenth century. The notation e was introduced around 1730 by Leonhard Euler, who discovered many fundamental properties of this important number.

In many books, e^x is denoted $\exp(x)$. Because of property (7), e^x plays a greater role than the seemingly more natural exponential functions such as 2^x or 10^x. We refer to $f(x) = e^x$ as the exponential function.

> **The Number e**   There is a unique positive real number e ($e \approx 2.718$) with the property
>
> $$\frac{d}{dx}e^x = e^x$$
>
> $\boxed{7}$

GRAPHICAL INSIGHT   All exponential curves $y = b^x$ pass through $(0, 1)$. For $b > 1$, their tangent lines at $(0, 1)$, of slope m_b, become steeper as b increases (Figure 6). By definition, e is the unique base for which the slope m_e of the tangent line at $(0, 1)$ is 1.

■ **EXAMPLE 8**   Find the equation of the tangent line to the graph of $f(x) = 3e^x - 5x^2$ at $x = 2$.

Solution   We compute the derivative $f'(2)$ and the function value $f(2)$:

$$f'(x) = \frac{d}{dx}(3e^x - 5x^2) = 3\frac{d}{dx}e^x - 5\frac{d}{dx}x^2 = 3e^x - 10x$$

$$f'(2) = 3e^2 - 10(2) \approx 2.17$$

$$f(2) = 3e^2 - 5(2^2) \approx 2.17$$

The equation of the tangent line is $y = f(2) + f'(2)(x - 2)$ (Figure 7). Using the approximate values, we write the equation as

$$y = 2.17 + 2.17(x - 2) \qquad \text{or} \qquad y = 2.17(x - 1)$$

■

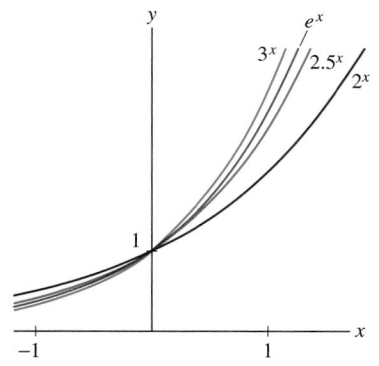

FIGURE 6 The tangent lines to $y = b^x$ at $x = 0$ grow steeper as b increases.

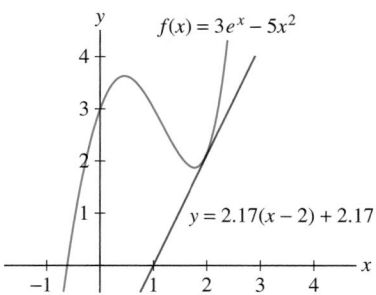

FIGURE 7

CONCEPTUAL INSIGHT   How is b^x defined when x is not a rational number? If n is a whole number, then b^n is simply the product $b \cdot b \cdots b$ (n times), and for any rational number $x = m/n$,

$$b^x = b^{m/n} = \left(b^{1/n}\right)^m = \left(\sqrt[n]{b}\right)^m$$

When x is irrational, this definition does not apply and b^x cannot be defined directly in terms of roots and powers of b. However, it makes sense to view $b^{m/n}$ as an approximation to b^x when m/n is a rational number close to x. For example, $3^{\sqrt{2}}$ should be approximately equal to $3^{1.4142} \approx 4.729$ since 1.4142 is a good rational approximation to $\sqrt{2}$. Formally, then, we may define b^x as a limit over rational numbers m/n approaching x:

$$b^x = \lim_{m/n \to x} b^{m/n}$$

It can be shown that this limit exists and the function $f(x) = b^x$ thus defined is differentiable.

Differentiability, Continuity, and Local Linearity

In the rest of this section, we examine the concept of **differentiability** more closely. We begin by proving that a differentiable function is necessarily continuous. In other words, $f'(c)$ does not exist if f has a discontinuity at $x = c$. Figure 8 shows why in the case of a jump discontinuity: Although the secant lines from the right approach the line L (which is tangent to the right half of the graph), the secant lines from the left approach the vertical (and their slopes tend to ∞).

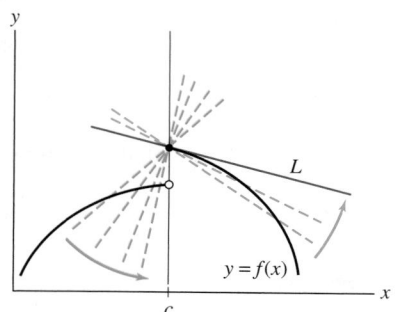

FIGURE 8 Secant lines at a jump discontinuity.

> **THEOREM 3 Differentiability Implies Continuity**   If f is differentiable at $x = c$, then f is continuous at $x = c$.

Proof By definition, if f is differentiable at $x = c$, then the following limit exists:

$$f'(c) = \lim_{x \to c} \frac{f(x) - f(c)}{x - c}$$

Our goal is to prove that f is continuous at $x = c$, which means that $\lim_{x \to c} f(x) = f(c)$. To relate the two limits, consider the equation (valid for $x \neq c$)

$$f(x) - f(c) = (x - c) \frac{f(x) - f(c)}{x - c}$$

Both factors on the right approach a limit as $x \to c$, so we may apply the Limit Laws:

$$\lim_{x \to c} \left(f(x) - f(c) \right) = \lim_{x \to c} \left((x - c) \frac{f(x) - f(c)}{x - c} \right)$$

$$= \left(\lim_{x \to c} (x - c) \right) \left(\lim_{x \to c} \frac{f(x) - f(c)}{x - c} \right)$$

$$= 0 \cdot f'(c) = 0$$

Therefore, $\lim_{x \to c} f(x) = \lim_{x \to c} (f(x) - f(c)) + \lim_{x \to c} f(c) = 0 + f(c) = f(c)$, as desired. ∎

Most of the functions encountered in this text are differentiable, but exceptions exist, as the next example shows.

Theorem 3 states that differentiability implies continuity. Example 9 shows that the converse is false: Continuity does not automatically imply differentiability.

■ **EXAMPLE 9** **Continuous But Not Differentiable** Show that $f(x) = |x|$ is continuous but not differentiable at $x = 0$.

Solution The function $f(x)$ is continuous at $x = 0$ since $\lim_{x \to 0} |x| = 0 = f(0)$. On the other hand,

$$f'(0) = \lim_{h \to 0} \frac{f(0 + h) - f(0)}{h} = \lim_{h \to 0} \frac{|0 + h| - |0|}{h} = \lim_{h \to 0} \frac{|h|}{h}$$

This limit does not exist [and hence $f(x)$ is not differentiable at $x = 0$] because

$$\frac{|h|}{h} = \begin{cases} 1 & \text{if } h > 0 \\ -1 & \text{if } h < 0 \end{cases}$$

and the one-sided limits are not equal:

$$\lim_{h \to 0+} \frac{|h|}{h} = 1 \qquad \text{and} \qquad \lim_{h \to 0-} \frac{|h|}{h} = -1 \qquad\blacksquare$$

GRAPHICAL INSIGHT Differentiability has an important graphical interpretation in terms of local linearity. We say that f is **locally linear** at $x = a$ if the graph looks more and more like a straight line as we zoom in on the point $(a, f(a))$. In this context, the adjective *linear* means "resembling a line," and *local* indicates that we are concerned only with the behavior of the graph near $(a, f(a))$. The graph of a locally linear function may be very wavy or *nonlinear*, as in Figure 9. But as soon as we zoom in on a sufficiently small piece of the graph, it begins to appear straight.

FIGURE 9 As we zoom in on a point, the graph looks more and more like the tangent line.

Not only does the graph look like a line as we zoom in on a point, but as Figure 9 suggests, the "zoom line" is the tangent line. Thus, the relation between differentiability and local linearity can be expressed as follows:

> If $f'(a)$ exists, then f is locally linear at $x = a$ in the following sense: As we zoom in on the point $(a, f(a))$, the graph becomes nearly indistinguishable from its tangent line.

Local linearity gives us a graphical way to understand why $f(x) = |x|$ is not differentiable at $x = 0$, as we proved in Example 9. Figure 10 shows that the graph of $f(x) = |x|$ has a corner at $x = 0$. Furthermore, this corner *does not disappear*, no matter how closely we zoom in on the origin. Since the graph does not straighten out under zooming, $f(x)$ is not locally linear at $x = 0$, and we cannot expect $f'(0)$ to exist.

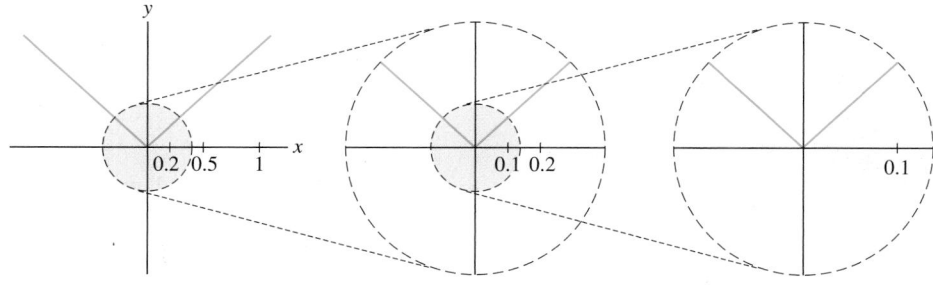

FIGURE 10 The graph of $f(x) = |x|$ is not locally linear at $x = 0$. The corner does not disappear when we zoom in on the origin.

Another way that a continuous function can fail to be differentiable is if the tangent line exists but is vertical, in which case the tangent line has infinite slope.

■ **EXAMPLE 10** Vertical Tangents Show that $f(x) = x^{1/3}$ is not differentiable at $x = 0$ (Figure 11).

Solution The limit defining $f'(0)$ is infinite:

$$\lim_{h \to 0} \frac{f(h) - f(0)}{h} = \lim_{h \to 0} \frac{h^{1/3} - 0}{h} = \lim_{h \to 0} \frac{h^{1/3}}{h} = \lim_{h \to 0} \frac{1}{h^{2/3}} = \infty$$

Note that $h^{-2/3}$ tends to infinity because $h^{2/3}$ tends to zero through positive values as $h \to 0$. Therefore $f'(0)$ does not exist. The tangent line has infinite slope. ■

As a final remark, we mention that there are more complicated ways in which a continuous function can fail to be differentiable. Figure 12 shows the graph of $f(x) = x \sin \frac{1}{x}$. If we define $f(0) = 0$, then f is continuous but not differentiable at $x = 0$. The secant lines keep oscillating and never settle down to a limiting position (see Exercise 99).

FIGURE 11 The tangent line to $f(x) = x^{1/3}$ at the origin is the (vertical) y-axis.

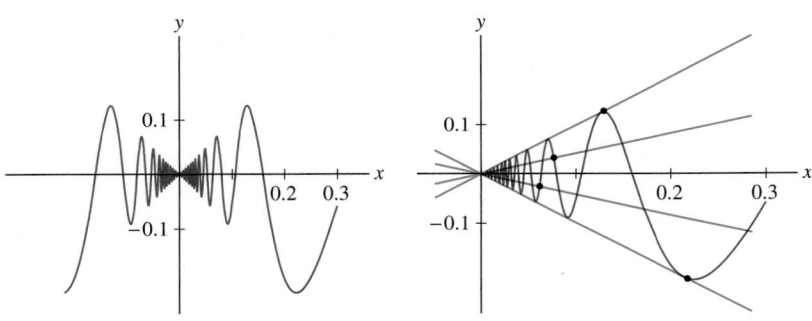

(A) Graph of $f(x) = x \sin \frac{1}{x}$

(B) Secant lines do not settle down to a limiting position.

FIGURE 12

3.2 SUMMARY

- The derivative $f'(x)$ is the function whose value at $x = a$ is the derivative $f'(a)$.
- We have several different notations for the derivative of $y = f(x)$:

$$y', \quad y'(x), \quad f'(x), \quad \frac{dy}{dx}, \quad \frac{df}{dx}$$

The value of the derivative at $x = a$ is written

$$y'(a), \quad f'(a), \quad \frac{dy}{dx}\bigg|_{x=a}, \quad \frac{df}{dx}\bigg|_{x=a}$$

- The Power Rule holds for all exponents n:

$$\frac{d}{dx}x^n = nx^{n-1} \quad \text{or} \quad \frac{d}{dx}x^{\text{exponent}} = (\text{exponent}) \, x^{\text{exponent}-1}$$

- The Linearity Rules allow us to differentiate term by term:

$$\text{Sum Rule:} \quad (f+g)' = f' + g', \qquad \text{Constant Multiple Rule:} \quad (cf)' = cf'$$

- The derivative of b^x is proportional to b^x: $\dfrac{d}{dx}b^x = m_b b^x$, where $m_b = \lim_{h \to 0} \dfrac{b^h - 1}{h}$.

- The number $e \approx 2.718$ is defined by the property $m_e = 1$, so that $\dfrac{d}{dx}e^x = e^x$.

- Differentiability implies continuity: If $f(x)$ is differentiable at $x = a$, then $f(x)$ is continuous at $x = a$. However, there exist continuous functions that are not differentiable.

- If $f'(a)$ exists, then f is locally linear in the following sense: As we zoom in on the point $(a, f(a))$, the graph becomes nearly indistinguishable from its tangent line.

3.2 EXERCISES

Preliminary Questions

1. What is the slope of the tangent line through the point $(2, f(2))$ if f is a function such that $f'(x) = x^3$?

2. Evaluate $(f - g)'(1)$ and $(3f + 2g)'(1)$ assuming that $f'(1) = 3$ and $g'(1) = 5$. Can we evaluate $(fg)'(1)$ using the information given and the rules presented in this section?

3. To which of the following does the Power Rule apply?

(a) $f(x) = x^2$
(b) $f(x) = 2^e$
(c) $f(x) = x^e$
(d) $f(x) = e^x$
(e) $f(x) = x^x$
(f) $f(x) = x^{-4/5}$

4. Which algebraic identity is used to prove the Power Rule for positive integer exponents? Explain how it is used.

5. Does the Power Rule apply to $f(x) = \sqrt[5]{x}$? Explain.

6. In which of the following two cases does the derivative not exist?

(a) Horizontal tangent
(b) Vertical tangent

7. Which property distinguishes $f(x) = e^x$ from all other exponential functions $g(x) = b^x$?

8. What is the slope of the tangent line to $y = e^x$ at $x = 0$? What about $y = 7e^x$?

Exercises

In Exercises 1–8, compute $f'(x)$ using the limit definition.

1. $f(x) = 4x - 3$
2. $f(x) = x^2 + x$

3. $f(x) = 1 - 2x^2$
4. $f(x) = x^3$

5. $f(x) = x^{-1}$
6. $f(x) = \dfrac{x}{x - 1}$

7. $f(x) = \sqrt{x}$
8. $f(x) = x^{-1/2}$

In Exercises 9–16, use the Power Rule to compute the derivative.

9. $\dfrac{d}{dx}x^4 \bigg|_{x=-2}$
10. $\dfrac{d}{dx}x^{-4} \bigg|_{x=3}$

11. $\dfrac{d}{dt}t^{2/3} \bigg|_{t=8}$
12. $\dfrac{d}{dt}t^{-2/3} \bigg|_{t=1}$

13. $\dfrac{d}{dx}x^{0.35}$
14. $\dfrac{d}{dx}x^{14/3}$

15. $\dfrac{d}{dt}t^{\sqrt{17}}$
16. $\dfrac{d}{dt}t^{-\pi^2}$

In Exercises 17–20, compute $f'(a)$ and find an equation of the tangent line to the graph at $x = a$.

17. $f(x) = x^5$, $a = 1$
18. $f(x) = x^{-2}$, $a = 3$

19. $f(x) = 3\sqrt{x} + 8x$, $a = 9$
20. $f(x) = \sqrt[3]{x}$, $a = 8$

21. Calculate:

(a) $\dfrac{d}{dx}9e^x$
(b) $\dfrac{d}{dt}(3t - 4e^t)$
(c) $\dfrac{d}{dt}e^{t+2}$

Hint for (c): Write e^{t+2} as $e^2 e^t$.

22. Find the equation of the tangent line to $y = 4e^x$ at $x = 2$.

In Exercises 23–38, calculate the derivative of the function.

23. $f(x) = x^3 + x^2 - 12$
24. $f(x) = 2x^3 - 3x^2 + 2x$

25. $f(x) = 2x^3 - 10x^{-1}$

26. $f(x) = x^5 - 7x^2 + 10x + 9$

27. $g(z) = 7z^{-3} + z^2 + 5$
28. $h(t) = 6\sqrt{t} + \dfrac{1}{\sqrt{t}}$

29. $f(s) = \sqrt[4]{s} + \sqrt[3]{s}$
30. $W(y) = 6y^4 + 7y^{2/3}$

31. $f(x) = (x + 1)^3$ (*Hint:* Expand)
32. $R(s) = (5s + 1)^2$

33. $P(z) = (3z - 1)(2z + 1)$
34. $q(t) = \sqrt{t}(t + 1)$

35. $g(x) = e^2$

36. $f(x) = 3e^x - x^3$

37. $h(t) = 5e^{t-3}$

38. $f(x) = 9 + 2e^x$

In Exercises 39–46, calculate the derivative indicated.

39. $f'(2)$, $f(x) = \dfrac{3}{x^4}$

40. $y'(16)$, $y = \dfrac{\sqrt{x}+1}{x}$

41. $\dfrac{dT}{dC}\Big|_{C=8}$, $T = 3C^{2/3}$

42. $\dfrac{dP}{dV}\Big|_{V=-2}$, $P = \dfrac{7}{V}$

43. $\dfrac{ds}{dz}\Big|_{z=2}$, $s = 4z - 16z^2$

44. $\dfrac{dR}{dW}\Big|_{W=1}$, $R = W^\pi$

45. $\dfrac{dp}{dh}\Big|_{h=4}$, $p = 7e^{h-2}$

46. $\dfrac{dy}{dt}\Big|_{t=4}$, $y = t - e^t$

47. Match the functions in graphs (A)–(D) with their derivatives (I)–(III) in Figure 13. Note that two of the functions have the same derivative. Explain why.

(A)   (B)   (C)   (D)

  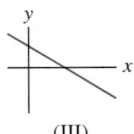

(I)   (II)   (III)

FIGURE 13

48. Assign the labels $f(x)$, $g(x)$, and $h(x)$ to the graphs in Figure 14 in such a way that $f'(x) = g(x)$ and $g'(x) = h(x)$.

  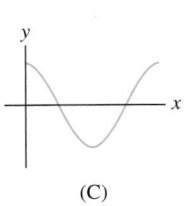

(A)   (B)   (C)

FIGURE 14

49. Sketch the graph of $f'(x)$ for $f(x)$ as in Figure 15, omitting points where $f(x)$ is not differentiable.

FIGURE 15 Graph of $f(x)$.

50. 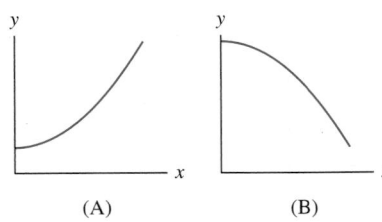 The table below lists values of a function $f(x)$. Does $f'(x)$ appear to increase or decrease as a function of x? Is (A) or (B) in Figure 16 the graph of $f'(x)$? Explain.

x	0	0.5	1	1.5	2	2.5	3	3.5	4
$f(x)$	10	55	98	139	177	210	237	257	268

(A)   (B)

FIGURE 16 Which is the graph of $f'(x)$?

51. Let R be a variable and r a constant. Compute the derivatives

(a) $\dfrac{d}{dR}R$ **(b)** $\dfrac{d}{dR}r$ **(c)** $\dfrac{d}{dR}r^2R^3$

52. Sketch the graph of $f(x) = x - 3x^2$ and find the values of x for which the tangent line is horizontal.

53. Find the points on the curve $y = x^2 + 3x - 7$ at which the slope of the tangent line is equal to 4.

54. Sketch the graphs of $f(x) = x^2 - 5x + 4$ and $g(x) = -2x + 3$. Find the value of x at which the graphs have parallel tangent lines.

55. Find all values of x where the tangent lines to $y = x^3$ and $y = x^4$ are parallel.

56. Show that there is a unique point on the graph of the function $f(x) = ax^2 + bx + c$ where the tangent line is horizontal (assume $a \neq 0$). Explain graphically.

57. Determine coefficients a and b such that $p(x) = x^2 + ax + b$ satisfies $p(1) = 0$ and $p'(1) = 4$.

58. Find all values of x such that the tangent line to the graph of $y = 4x^2 + 11x + 2$ is steeper than the tangent line to $y = x^3$.

59. Let $f(x) = x^3 - 3x + 1$. Show that $f'(x) \geq -3$ for all x, and that for every $m > -3$, there are precisely two points where $f'(x) = m$. Indicate the position of these points and the corresponding tangent lines for one value of m in a sketch of the graph of $f(x)$.

60. Show that if the tangent lines to the graph of $y = \frac{1}{3}x^3 - x^2$ at $x = a$ and at $x = b$ are parallel, then either $a = b$ or $a + b = 2$.

61. Compute the derivative of $f(x) = x^{-2}$ using the limit definition. *Hint:* Show that

$$\frac{f(x+h) - f(x)}{h} = -\frac{1}{x^2(x+h)^2}\cdot\frac{(x+h)^2 - x^2}{h}$$

62. Compute the derivative of $f(x) = x^{3/2}$ using the limit definition. *Hint:* Show that

$$\frac{f(x+h) - f(x)}{h} = \frac{(x+h)^3 - x^3}{h}\left(\frac{1}{\sqrt{(x+h)^3} + \sqrt{x^3}}\right)$$

63. Find an approximation to m_4 using the limit definition and estimate the slope of the tangent line to $y = 4^x$ at $x = 0$ and $x = 2$.

64. Let $f(x) = xe^x$. Use the limit definition to compute $f'(0)$ and find the equation of the tangent line at $x = 0$.

65. The average speed (in meters per second) of a gas molecule is $v_{avg} = \sqrt{8RT/(\pi M)}$, where T is the temperature (in kelvin), M is the molar mass (kg/mol) and $R = 8.31$. Calculate dv_{avg}/dT at $T = 300$ K for oxygen, which has a molar mass of 0.032 kg/mol.

66. Biologists have observed that the pulse rate P (in beats per minute) in animals is related to body mass (in kilograms) by the approximate formula $P = 200m^{-1/4}$. This is one of many *allometric scaling laws* prevalent in biology. Is the absolute value $|dP/dm|$ increasing or decreasing as m increases? Find an equation of the tangent line at the points on the graph in Figure 17 that represent goat ($m = 33$) and man ($m = 68$).

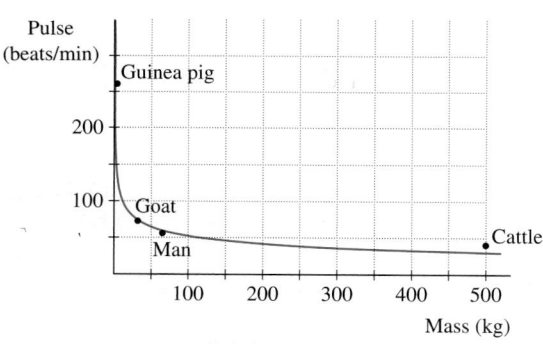

Pulse (beats/min)

FIGURE 17

67. Some studies suggest that kidney mass K in mammals (in kilograms) is related to body mass m (in kilograms) by the approximate formula $K = 0.007m^{0.85}$. Calculate dK/dm at $m = 68$. Then calculate the derivative with respect to m of the relative kidney-to-mass ratio K/m at $m = 68$.

68. The relation between the *vapor pressure* P (in atmospheres) of water and the temperature T (in kelvin) is given by the Clausius–Clapeyron law:

$$\frac{dP}{dT} = k\frac{P}{T^2}$$

where k is a constant. Use the table below and the approximation

$$\frac{dP}{dT} \approx \frac{P(T+10) - P(T-10)}{20}$$

to estimate dP/dT for $T = 303, 313, 323, 333, 343$. Do your estimates seem to confirm the Clausius–Clapeyron law? What is the approximate value of k? What are the units of k?

T	293	303	313	323	333	343	353
P	0.0278	0.0482	0.0808	0.1311	0.2067	0.3173	0.4754

69. Let L be a tangent line to the hyperbola $xy = 1$ at $x = a$, where $a > 0$. Show that the area of the triangle bounded by L and the coordinate axes does not depend on a.

70. In the notation of Exercise 69, show that the point of tangency is the midpoint of the segment of L lying in the first quadrant.

71. Match the functions (A)–(C) with their derivatives (I)–(III) in Figure 18.

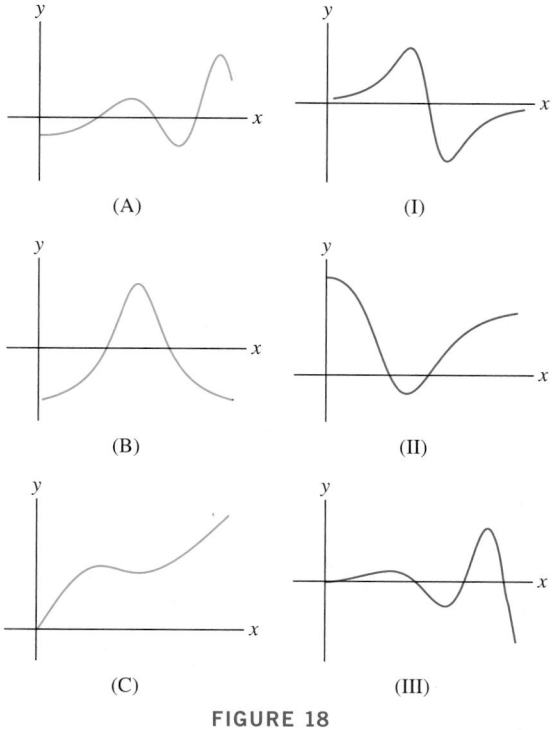

FIGURE 18

72. GU Plot the derivative $f'(x)$ of $f(x) = 2x^3 - 10x^{-1}$ for $x > 0$ (set the bounds of the viewing box appropriately) and observe that $f'(x) > 0$. What does the positivity of $f'(x)$ tell us about the graph of $f(x)$ itself? Plot $f(x)$ and confirm.

73. Make a rough sketch of the graph of the derivative of the function shown in Figure 19(A).

74. Graph the derivative of the function shown in Figure 19(B), omitting points where the derivative is not defined.

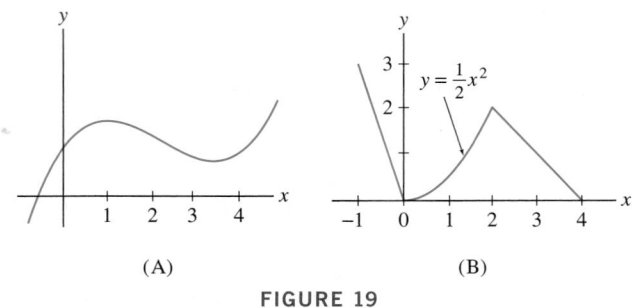

(A)   (B)

FIGURE 19

75. Of the two functions f and g in Figure 20, which is the derivative of the other? Justify your answer.

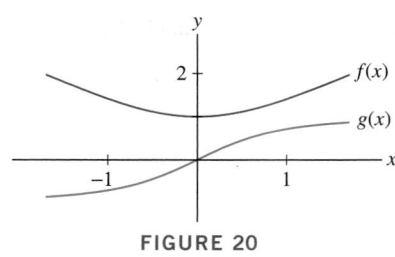

FIGURE 20

76. At which points is the function in Figure 21 discontinuous? At which points is it nondifferentiable?

In Exercises 77–82, find the points c (if any) such that $f'(c)$ does not exist.

77. $f(x) = |x - 1|$

78. $f(x) = [x]$

79. $f(x) = x^{2/3}$

80. $f(x) = x^{3/2}$

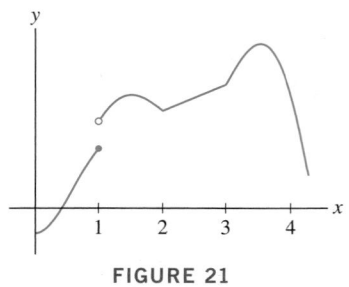

FIGURE 21

81. $f(x) = |x^2 - 1|$ **82.** $f(x) = |x - 1|^2$

GU   *In Exercises 83–88, zoom in on a plot of $f(x)$ at the point $(a, f(a))$ and state whether or not $f(x)$ appears to be differentiable at $x = a$. If nondifferentiable, state whether the tangent line appears to be vertical or does not exist.*

83. $f(x) = (x - 1)|x|$, $a = 0$

84. $f(x) = (x - 3)^{5/3}$, $a = 3$

85. $f(x) = (x - 3)^{1/3}$, $a = 3$

86. $f(x) = \sin(x^{1/3})$, $a = 0$

87. $f(x) = |\sin x|$, $a = 0$

88. $f(x) = |x - \sin x|$, $a = 0$

89. Is it true that a nondifferentiable function is not continuous? If not, give a counterexample.

90. Sketch the graph of $y = x |x|$ and show that it is differentiable for all x (check differentiability separately for $x < 0$, $x = 0$, and $x > 0$).

Further Insights and Challenges

91. Prove the following theorem of Apollonius of Perga (the Greek mathematician born in 262 BCE who gave the parabola, ellipse, and hyperbola their names): The tangent to the parabola $y = x^2$ at $x = a$ intersects the x-axis at the midpoint between the origin and $(a, 0)$. Draw a diagram.

92. Apollonius's Theorem in Exercise 91 can be generalized. Show that the tangent to $y = x^3$ at $x = a$ intersects the x-axis at the point $x = \frac{2}{3}a$. Then formulate the general statement for the graph of $y = x^n$ and prove it.

93. CAS   Plot the graph of $f(x) = (4 - x^{2/3})^{3/2}$ (the "astroid"). Let L be a tangent line to a point on the graph in the first quadrant. Use a computer algebra system to compute $f'(x)$ and show that the portion of L in the first quadrant has a constant length 8.

94. Two small arches have the shape of parabolas. The first is given by $f(x) = 1 - x^2$ for $-1 \le x \le 1$ and the second by $g(x) = 4 - (x - 4)^2$ for $2 \le x \le 6$. A board is placed on top of these arches

so it rests on both (Figure 22). What is the slope of the board? *Hint:* Find the tangent line to $y = f(x)$ that intersects $y = g(x)$ in exactly one point.

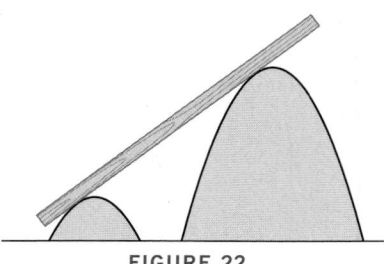

FIGURE 22

95. A vase is formed by rotating $y = x^2$ around the y-axis. If we drop in a marble, it will either touch the bottom point of the vase or be suspended above the bottom by touching the sides (Figure 23). How small must the marble be to touch the bottom?

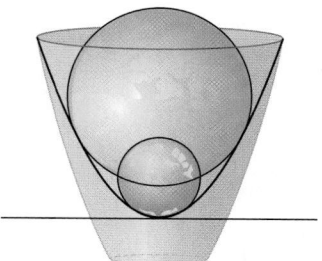

FIGURE 23

96. [icon] Let $f(x)$ be a differentiable function and set $g(x) = f(x + c)$, where c is a constant. Use the limit definition to show that $g'(x) = f'(x + c)$. Explain this result graphically, recalling that the graph of $g(x)$ is obtained by shifting the graph of $f(x)$ c units to the left (if $c > 0$) or right (if $c < 0$).

97. Negative Exponents Let n be a whole number. Use the Power Rule for x^n to calculate the derivative of $f(x) = x^{-n}$ by showing that

$$\frac{f(x + h) - f(x)}{h} = \frac{-1}{x^n(x + h)^n} \frac{(x + h)^n - x^n}{h}$$

98. Verify the Power Rule for the exponent $1/n$, where n is a positive integer, using the following trick: Rewrite the difference quotient for $y = x^{1/n}$ at $x = b$ in terms of $u = (b + h)^{1/n}$ and $a = b^{1/n}$.

99. Infinitely Rapid Oscillations Define

$$f(x) = \begin{cases} x \sin \dfrac{1}{x} & \text{if } x \neq 0 \\ 0 & \text{if } x = 0 \end{cases}$$

Show that $f(x)$ is continuous at $x = 0$ but $f'(0)$ does not exist (see Figure 12).

100. Prove that $f(x) = e^x$ is not a polynomial function. *Hint:* Differentiation lowers the degree of a polynomial by 1.

101. Consider the equation $e^x = \lambda x$, where λ is a constant.

(a) For which λ does it have a unique solution? For intuition, draw a graph of $y = e^x$ and the line $y = \lambda x$.

(b) For which λ does it have at least one solution?

3.3 Product and Quotient Rules

One of the main goals of this chapter is to develop efficient techniques for computing derivatives. This section covers the **Product Rule** and **Quotient Rule**. These two rules together with the Chain Rule and implicit differentiation (covered in later sections) make up an extremely effective "differentiation toolkit."

--- *REMINDER The product function fg is defined by $(fg)(x) = f(x) g(x)$.*

> **THEOREM 1 Product Rule** If f and g are differentiable functions, then fg is differentiable and
>
> $$(fg)'(x) = f(x) g'(x) + g(x) f'(x)$$

It may be helpful to remember the Product Rule in words: The derivative of a product is equal to *the first function times the derivative of the second function plus the second function times the derivative of the first function*:

$$\text{First} \cdot (\text{second})' + \text{second} \cdot (\text{first})'$$

We prove the Product Rule after presenting three examples.

■ **EXAMPLE 1** Find the derivative of $h(x) = 3x^2(5x + 1)$.

Solution This function is a product

$$h(x) = \overbrace{3x^2}^{\text{First}} \overbrace{(5x + 1)}^{\text{Second}}$$

144 CHAPTER 3 | DIFFERENTIATION

By the Product Rule (which we now write in Leibniz notation),

$$h'(x) = \overbrace{3x^2}^{\text{First}} \overbrace{\frac{d}{dx}(5x+1)}^{\text{Second}'} + \overbrace{(5x+1)}^{\text{Second}} \overbrace{\frac{d}{dx}(3x^2)}^{\text{First}'}$$

$$= (3x^2)5 + (5x+1)(6x)$$

$$= 15x^2 + (30x^2 + 6x) = 45x^2 + 6x$$

■ **EXAMPLE 2** Find the derivative of $y = (x^{-1}+2)(x^{3/2}+1)$.

Solution Use the Product Rule:

$$y' = \text{first} \cdot (\text{second})' + \text{second} \cdot (\text{first})'$$

$$= (x^{-1}+2)(x^{3/2}+1)' + (x^{3/2}+1)(x^{-1}+2)'$$

$$= (x^{-1}+2)(\tfrac{3}{2}x^{1/2}) + (x^{3/2}+1)(-x^{-2}) \qquad \text{(compute the derivatives)}$$

$$= \tfrac{3}{2}x^{-1/2} + 3x^{1/2} - x^{-1/2} - x^{-2} \qquad \text{(simplify)}$$

$$= \tfrac{1}{2}x^{-1/2} + 3x^{1/2} - x^{-2}$$

Note how the prime notation is used in the solution to Example 2. We write $(x^{3/2}+1)'$ to denote the derivative of $x^{3/2}+1$, etc.

In the previous two examples, we could have avoided using the Product Rule by expanding the function. For example,

$$h(x) = 3x^2(5x+1) = 15x^3 + 3x^2$$

We may use the Product Rule or differentiate directly:

$$h'(x) = \frac{d}{dx}(15x^3 + 3x^2) = 45x^2 + 6x$$

However, in the next example, the function cannot be expanded and we must use the Product Rule (or go back to the limit definition of the derivative).

■ **EXAMPLE 3** Calculate $\frac{d}{dt}t^2 e^t$.

Solution Use the Product Rule:

$$\frac{d}{dt}t^2 e^t = \text{first} \cdot (\text{second})' + \text{second} \cdot (\text{first})'$$

$$= t^2 \frac{d}{dt}e^t + e^t \frac{d}{dt}t^2$$

$$= t^2 e^t + e^t(2t) = (t^2 + 2t)e^t$$

Proof of the Product Rule We prove the Product Rule by writing the difference quotient for $f(x)g(x)$ in a clever way as a sum of two terms. The limit definition of the derivative applied to the product function gives us

$$(fg)'(x) = \lim_{h \to 0} \frac{f(x+h)g(x+h) - f(x)g(x)}{h}$$

The trick is to subtract $f(x+h)g(x)$ and add it back again in the numerator of the difference quotient:

$$f(x+h)g(x+h)\underbrace{-f(x+h)g(x)+f(x+h)g(x)}_{\text{Sum is zero}}-f(x)g(x)$$

Combining terms, we see that the numerator is equal to

$$f(x+h)\big(g(x+h)-g(x)\big)+g(x)\big(f(x+h)-f(x)\big)$$

Now write $(fg)'(x)$ as a sum of two limits:

$$(fg)'(x)=\underbrace{\lim_{h\to 0}f(x+h)\frac{g(x+h)-g(x)}{h}}_{\text{Show that this equals }f(x)g'(x)}+\underbrace{\lim_{h\to 0}g(x)\frac{f(x+h)-f(x)}{h}}_{\text{Show that this equals }g(x)f'(x)}\qquad\boxed{1}$$

The use of the Sum Law is valid, provided that each limit on the right exists. To check that the first limit exists and to evaluate it, we note that $f(x)$ is continuous (because it is differentiable) and that $g(x)$ is differentiable:

$$\lim_{h\to 0}f(x+h)\frac{g(x+h)-g(x)}{h}=\lim_{h\to 0}f(x+h)\ \lim_{h\to 0}\frac{g(x+h)-g(x)}{h}$$

$$=f(x)\,g'(x)\qquad\boxed{2}$$

The second limit is similar:

$$\lim_{h\to 0}g(x)\frac{f(x+h)-f(x)}{h}=g(x)\lim_{h\to 0}\frac{f(x+h)-f(x)}{h}=g(x)\,f'(x)\qquad\boxed{3}$$

Using (2) and (3) in (1), we conclude that fg is differentiable and that $(fg)'(x)=f(x)g'(x)+g(x)f'(x)$ as desired. ∎

CONCEPTUAL INSIGHT The Product Rule was first stated by the 29-year-old Leibniz in 1675. He had been on an extended visit to Paris, where he developed his major ideas on calculus. Perhaps Leibniz had wondered whether $(fg)'$ is equal to $f'g'$ before arriving at the correct rule. We can check that $(fg)'$ *is not equal* to $f'g'$ by taking $f(x)=g(x)=x$. Then $(fg)(x)=x^2$, but

$$(fg)'(x)=(x^2)'=2x\qquad\text{is } not \text{ equal to}\qquad f'(x)g'(x)=1\cdot 1=1$$

There is also a rule for differentiating quotients that, like the product rule, has an unexpected form.

◀‥ REMINDER *The quotient function f/g is defined by*

$$\left(\frac{f}{g}\right)(x)=\frac{f(x)}{g(x)}$$

THEOREM 2 Quotient Rule If f and g are differentiable functions, then f/g is differentiable for all x such that $g(x)\neq 0$, and

$$\left(\frac{f}{g}\right)'(x)=\frac{g(x)f'(x)-f(x)g'(x)}{g(x)^2}$$

The numerator in the Quotient Rule is equal to *the bottom times the derivative of the top minus the top times the derivative of the bottom*:

$$\frac{\text{Bottom}\cdot(\text{top})'-\text{top}\cdot(\text{bottom})'}{\text{Bottom}^2}$$

We omit the proof of the Quotient Rule because it is similar to that of the Product Rule (see Exercise 56).

■ **EXAMPLE 4** Compute the derivative of $f(x) = \dfrac{x}{x+1}$.

Solution Apply the Quotient Rule:

$$f'(x) = \frac{\overbrace{(x+1)}^{\text{Bottom}}\ \overbrace{(x)'}^{\text{Top}'} - \overbrace{(x)}^{\text{Top}}\ \overbrace{(x+1)'}^{\text{Bottom}'}}{(x+1)^2} = \frac{(x+1)1 - (x)1}{(x+1)^2} = \frac{1}{(x+1)^2} \qquad \blacksquare$$

■ **EXAMPLE 5** Calculate $\dfrac{d}{dt}\dfrac{e^t}{e^t + t}$.

Solution Use the Quotient Rule and the formula $(e^t)' = e^t$:

$$\frac{d}{dt}\frac{e^t}{e^t + t} = \frac{(e^t + t)(e^t)' - e^t(e^t + t)'}{(e^t + t)^2} = \frac{(e^t + t)e^t - e^t(e^t + 1)}{(e^t + t)^2} = \frac{(t-1)e^t}{(e^t + t)^2} \qquad \blacksquare$$

■ **EXAMPLE 6** Find an equation of the tangent line to the graph of $f(x) = \dfrac{3x^2 - 2}{4x^3 + 1}$ at $x = 1$.

Solution We first compute the derivative using the Quotient Rule:

$$f'(x) = \frac{d}{dx}\left(\frac{3x^2 - 2}{4x^3 + 1}\right) = \frac{\overbrace{(4x^3 + 1)}^{\text{Bottom}}\ \overbrace{(3x^2 - 2)'}^{\text{Top}'} - \overbrace{(3x^2 - 2)}^{\text{Top}}\ \overbrace{(4x^3 + 1)'}^{\text{Bottom}'}}{(4x^3 + 1)^2}$$

$$= \frac{(4x^3 + 1)(6x) - (3x^2 - 2)(12x^2)}{(4x^3 + 1)^2} = \frac{(24x^4 + 6x) - (36x^4 - 24x^2)}{(4x^3 + 1)^2}$$

$$= \frac{-12x^4 + 24x^2 + 6x}{(4x^3 + 1)^2} = \frac{6x(-2x^3 + 4x + 1)}{(4x^3 + 1)^2}$$

At $x = 1$, we have

$$f(1) = \frac{3-2}{4+1} = \frac{1}{5}, \qquad f'(1) = \frac{6(-2+4+1)}{5^2} = \frac{18}{25}$$

An equation of the tangent line at $(1, \frac{1}{5})$ is

$$y - \frac{1}{5} = \frac{18}{25}(x - 1) \qquad \text{or} \qquad y = \frac{18}{25}x - \frac{13}{25} \qquad \blacksquare$$

■ **EXAMPLE 7** Power Delivered by a Battery The power that a battery can supply to an apparatus such as a cell phone depends on the *internal resistance* of the battery. For a battery of voltage V and internal resistance r, the total power delivered to an apparatus

FIGURE 1 An apparatus of resistance R attached to a battery of voltage V.

of resistance R (Figure 1) is

$$P = \frac{V^2 R}{(R+r)^2}$$

(a) Calculate $\dfrac{dP}{dR}$, assuming that V and r are nonzero constants and $V > 0$.

(b) Find the value of R at which the tangent to the graph of P versus R is horizontal.

Solution

(a) The key to solving this problem is to keep in mind that P and R are variables, but V and r are constants:

$$\frac{dP}{dR} = \frac{d}{dR}\left(\frac{V^2 R}{(R+r)^2}\right) = V^2 \frac{d}{dR}\left(\frac{R}{(R+r)^2}\right) \qquad (V \text{ is a constant})$$

$$= V^2 \frac{(R+r)^2 \frac{d}{dR}R - R\frac{d}{dR}(R+r)^2}{(R+r)^4} \qquad \text{(Quotient Rule)} \qquad \boxed{4}$$

In (5), we use

$$\frac{d}{dR}R^2 = 2R$$

and, since r is a constant,

$$\frac{d}{dR}2rR = 2r\frac{d}{dR}R = 2r$$

$$\frac{d}{dR}r^2 = 0$$

We have $\dfrac{d}{dR}R = 1$ and, since r is a constant,

$$\frac{d}{dR}(R+r)^2 = \frac{d}{dR}(R^2 + 2rR + r^2)$$

$$= \frac{d}{dR}R^2 + \frac{d}{dR}2rR + \frac{d}{dR}r^2 \qquad \boxed{5}$$

$$= 2R + 2r + 0 = 2(R+r)$$

Using this in Eq. (4) and factoring out $(R+r)$, we obtain

$$\frac{dP}{dR} = V^2 \frac{(R+r)^2 - 2R(R+r)}{(R+r)^4} = V^2 \frac{(R+r)\big((R+r) - 2R\big)}{(R+r)^4}$$

$$= V^2 \frac{r - R}{(R+r)^3} \qquad \text{[cancel $(R+r)$]}$$

(b) The tangent line is horizontal when the derivative is zero. Thus, we set the derivative equal to zero and solve for R:

$$V^2 \frac{r - R}{(R+r)^3} = 0$$

This equation is satisfied only if the numerator is zero, that is, if $R = r$. ∎

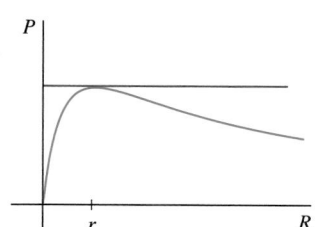

FIGURE 2 Graph of power versus resistance:

$$P = \frac{V^2 R}{(R+r)^2}$$

GRAPHICAL INSIGHT   Figure 2 shows that the point where the tangent line is horizontal is the *maximum point* on the graph. This proves an important result in circuit design: Maximum power is delivered when the resistance of the load (apparatus) is equal to the internal resistance of the battery.

3.3 SUMMARY

- This section deals with two basic rules of differentiation:

$$\text{Product Rule:} \qquad (fg)' = fg' + gf'$$

$$\text{Quotient Rule:} \qquad \left(\frac{f}{g}\right)' = \frac{gf' - fg'}{g^2}$$

- It is important to remember that the derivative of fg is *not* equal to $f'g'$. Similarly, the derivative of f/g is *not* equal to f'/g'.

3.3 EXERCISES

Preliminary Questions

1. Are the following statements true or false? If false, state the correct version.

(a) The notation fg denotes the function whose value at x is $f(g(x))$.

(b) The notation f/g denotes the function whose value at x is $f(x)/g(x)$.

(c) The derivative of the product is the product of the derivatives.

2. Are the following equations true or false? If false, state the correct version.

(a) $\dfrac{d}{dx}(fg)\Big|_{x=4} = f(4)g'(4) - g(4)f'(4)$

(b) $\dfrac{d}{dx}\dfrac{f}{g}\Big|_{x=4} = \dfrac{f(4)g'(4) + g(4)f'(4)}{(g(4))^2}$

(c) $\dfrac{d}{dx}(fg)\Big|_{x=0} = f(0)g'(0) + g(0)f'(0)$

3. What is the derivative of f/g at $x = 1$ if $f(1) = f'(1) = g(1) = 2$, and $g'(1) = 4$?

4. Suppose that $f(1) = 0$ and $f'(1) = 2$. Find $g(1)$, assuming that $(fg)'(1) = 10$.

Exercises

In Exercises 1–6, use the Product Rule to calculate the derivative.

1. $f(x) = x(x^2 + 1)$

2. $f(x) = \sqrt{x}(1 - x^4)$

3. $f(x) = e^x(x^3 - 1)$

4. $f(x) = x^2(1 + 4e^x)$

5. $\dfrac{dy}{dt}\Big|_{t=3}$, $y = (t^2 + 1)(t + 9)$

6. $\dfrac{dh}{dx}\Big|_{x=4}$, $h(x) = (x^{-1/2} + 2x)(7 - x^{-1})$

In Exercises 7–12, use the Quotient Rule to calculate the derivative.

7. $f(x) = \dfrac{x}{x - 2}$

8. $f(x) = \dfrac{x + 4}{x^2 + x + 1}$

9. $\dfrac{dg}{dt}\Big|_{t=-2}$, $g(t) = \dfrac{t^2 + 1}{t^2 - 1}$

10. $\dfrac{dw}{dz}\Big|_{z=9}$, $w = \dfrac{z^2}{\sqrt{z} + z}$

11. $g(x) = \dfrac{1}{1 + e^x}$

12. $f(x) = \dfrac{e^x}{x^2 + 1}$

In Exercises 13–16, calculate the derivative in two ways: First use the Product or Quotient Rule, then rewrite the function algebraically and apply the Power Rule directly.

13. $f(t) = (2t + 1)(t^2 - 2)$

14. $f(x) = x^2(3 + x^{-1})$

15. $g(x) = \dfrac{x^3 + 2x^2 + 3x^{-1}}{x}$

16. $h(t) = \dfrac{t^2 - 1}{t - 1}$

In Exercises 17–38, calculate the derivative using the appropriate rule or combination of rules.

17. $f(x) = (x^4 - 4)(x^2 + x + 1)$

18. $f(x) = (x^2 + 9)(2 - e^x)$

19. $\dfrac{dy}{dx}\Big|_{x=2}$, $y = \dfrac{1}{x + 4}$

20. $\dfrac{dz}{dx}\Big|_{x=-1}$, $z = \dfrac{x}{x^2 + 1}$

21. $f(x) = (\sqrt{x} + 1)(\sqrt{x} - 1)$

22. $f(x) = \dfrac{3\sqrt{x} - 2}{x}$

23. $\dfrac{dy}{dx}\Big|_{x=2}$, $y = \dfrac{x^4 - 4}{x^2 - 5}$

24. $f(x) = \dfrac{x^4 + e^x}{x + 1}$

25. $\dfrac{dz}{dx}\Big|_{x=1}$, $z = \dfrac{1}{x^3 + 1}$

26. $f(x) = \dfrac{3x^3 - x^2 + 2}{\sqrt{x}}$

27. $h(t) = \dfrac{t}{(t^4 + t^2)(t^7 + 1)}$

28. $f(x) = x^{3/2}(2x^4 - 3x + x^{-1/2})$

29. $f(t) = 3^{1/2} \cdot 5^{1/2}$

30. $h(x) = \pi^2(x - 1)$

31. $f(x) = (x + 3)(x - 1)(x - 5)$

32. $f(x) = x(x^2 + 1)(x + 4)$

33. $f(x) = \dfrac{e^x}{(e^x + 1)(x + 1)}$

34. $g(x) = \dfrac{e^{x+1} + e^x}{e + 1}$

35. $g(z) = \left(\dfrac{z^2 - 4}{z - 1}\right)\left(\dfrac{z^2 - 1}{z + 2}\right)$ *Hint: Simplify first.*

36. $\dfrac{d}{dx}\left((ax + b)(abx^2 + 1)\right)$ $(a, b$ constants$)$

37. $\dfrac{d}{dt}\left(\dfrac{xt - 4}{t^2 - x}\right)$ $(x$ constant$)$

38. $\dfrac{d}{dx}\left(\dfrac{ax + b}{cx + d}\right)$ $(a, b, c, d$ constants$)$

39. GU   Plot the derivative of $f(x) = \dfrac{x}{x^2 + 1}$ over $[-4, 4]$. Use the graph to determine the intervals on which $f'(x) > 0$ and $f'(x) < 0$. Then plot $f(x)$ and describe how the sign of $f'(x)$ is reflected in the graph of $f(x)$.

40. GU   Plot $f(x) = \dfrac{x}{x^2 - 1}$ (in a suitably bounded viewing box). Use the plot to determine whether $f'(x)$ is positive or negative on its domain $\{x : x \neq \pm 1\}$. Then compute $f'(x)$ and confirm your conclusion algebraically.

41. Let $P = \dfrac{V^2 R}{(R + r)^2}$ as in Example 7. Calculate dP/dr, assuming that r is variable and R is constant.

42. Find the value $a > 0$ such that the tangent line to $f(x) = x^2 e^{-x}$ passes through the origin (Figure 3).

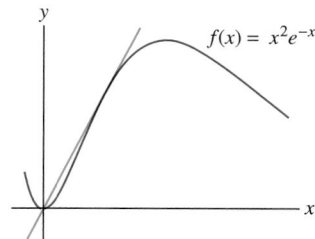

FIGURE 3

43. According to Ohm's Law, current I (in amps), voltage V (in volts), and resistance R (in ohms) in a circuit are related by $I = V/R$.

(a) Calculate $\left.\dfrac{dI}{dR}\right|_{R=6}$, assuming that V has the constant value $V = 24$.

(b) Calculate $\left.\dfrac{dV}{dR}\right|_{R=6}$, assuming that I has the constant value $I = 4$.

44. Find an equation of the tangent line to the graph $y = \dfrac{x}{x + x^{-1}}$ at $x = 2$.

45. The curve $y = \dfrac{1}{x^2 + 1}$ is called the *witch of Agnesi* (Figure 4) after the Italian mathematician Maria Agnesi (1718–1799), who wrote one of the first books on calculus. This strange name is the result of a mistranslation of the Italian word *la versiera*, meaning "that which turns." Find equations of the tangent lines at $x = \pm 1$.

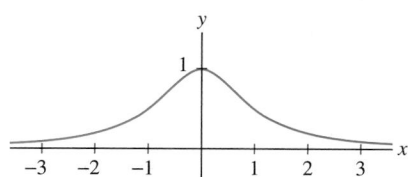

FIGURE 4 Graph of the witch of Agnesi.

46. Let $f(x) = g(x) = x$. Show that $(f/g)' \neq f'/g'$.

47. Use the Product Rule to show that $(f^2)' = 2ff'$.

48. Show that $(f^3)' = 3f^2 f'$.

In Exercises 49–52, use the following function values:

$f(4)$	$f'(4)$	$g(4)$	$g'(4)$
2	-3	5	-1

49. Find the derivative of fg and f/g at $x = 4$.

50. Calculate $F'(4)$, where $F(x) = xf(x)$.

51. Calculate $G'(4)$, where $G(x) = xg(x)f(x)$.

52. Calculate $H'(4)$, where $H(x) = \dfrac{x}{g(x)f(x)}$.

53. Calculate $F'(0)$, where

$$F(x) = \dfrac{x^9 + x^8 + 4x^5 - 7x}{x^4 - 3x^2 + 2x + 1}$$

Hint: Do not calculate $F'(x)$. Instead, write $F(x) = f(x)/g(x)$ and express $F'(0)$ directly in terms of $f(0)$, $f'(0)$, $g(0)$, $g'(0)$.

54. Proceed as in Exercise 53 to calculate $F'(0)$, where

$$F(x) = \left(1 + x + x^{4/3} + x^{5/3}\right)\dfrac{3x^5 + 5x^4 + 5x + 1}{8x^9 - 7x^4 + 1}$$

Further Insights and Challenges

55. Let f, g, h be differentiable functions. Show that $(fgh)'(x)$ is equal to

$$f(x)g(x)h'(x) + f(x)g'(x)h(x) + f'(x)g(x)h(x)$$

Hint: Write fgh as $f(gh)$.

56. Prove the Quotient Rule using the limit definition of the derivative.

57. Derivative of the Reciprocal   Use the limit definition to show that if $f(x)$ is differentiable and $f(x) \neq 0$, then $1/f(x)$ is differentiable and

$$\frac{d}{dx}\left(\frac{1}{f(x)}\right) = -\frac{f'(x)}{f^2(x)}$$ $\boxed{6}$

Hint: Show that the difference quotient for $1/f(x)$ is equal to

$$\frac{f(x) - f(x+h)}{hf(x)f(x+h)}$$

58. Derive the Quotient Rule using Eq. (6) and the Product Rule.

59. Show that Eq. (6) is a special case of the Quotient Rule.

60. Carry out the details of Agnesi's proof of the Quotient Rule from her book on calculus, published in 1748: Assume that f, g, and $h = f/g$ are differentiable. Compute the derivative of $hg = f$ using the Product Rule and solve for h'.

61. The Power Rule Revisited   If you are familiar with *proof by induction*, use induction to prove the Power Rule for all whole numbers n. Show that the Power Rule holds for $n = 1$, then write x^n as $x \cdot x^{n-1}$ and use the Product Rule.

In Exercises 62–63, let $f(x)$ be a polynomial. A basic fact of algebra states that c is a root of $f(x)$ if and only if $f(x) = (x - c)g(x)$

for some polynomial $g(x)$. We say that c is a multiple root if $f(x) = (x - c)^2 h(x)$, where $h(x)$ is a polynomial.

62. Show that c is a multiple root of $f(x)$ if and only if c is a root of both $f(x)$ and $f'(x)$.

63. Use Exercise 62 to determine whether $c = -1$ is a multiple root of the polynomials
(a) $x^5 + 2x^4 - 4x^3 - 8x^2 - x + 2$
(b) $x^4 + x^3 - 5x^2 - 3x + 2$

64. 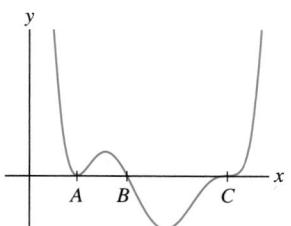 Figure 5 is the graph of a polynomial with roots at A, B, and C. Which of these is a multiple root? Explain your reasoning using the result of Exercise 62.

FIGURE 5

65. In Section 3.1, we showed that $\dfrac{d}{dx}b^x = m_b\, b^x$, where

$$m_b = \lim_{h \to 0} \frac{b^h - 1}{h}$$

Prove in two ways that the numbers m_b satisfy the relation

$$m_{ab} = m_a + m_b$$

3.4 Rates of Change

Recall the notation for the average rate of change (ROC) of a function $y = f(x)$ over an interval $[x_0, x_1]$:

$$\Delta y = \text{change in } y = f(x_1) - f(x_0)$$

$$\Delta x = \text{change in } x = x_1 - x_0$$

$$\boxed{\text{Average ROC} = \frac{\Delta y}{\Delta x} = \frac{f(x_1) - f(x_0)}{x_1 - x_0}}$$

We usually omit the word "instantaneous" and refer to the derivative simply as the rate of change (ROC). This is shorter and also more accurate when applied to general ROCs, since the term instantaneous would seem to refer to only ROCs with respect to time.

In our prior discussion in Section 2.1, limits and derivatives had not yet been introduced. Now that we have them at our disposal, we are ready to define the **instantaneous** ROC of y with respect to x at $x = x_0$:

$$\boxed{\text{Instantaneous ROC} = f'(x_0) = \lim_{\Delta x \to 0} \frac{\Delta y}{\Delta x} = \lim_{x_1 \to x_0} \frac{f(x_1) - f(x_0)}{x_1 - x_0}}$$

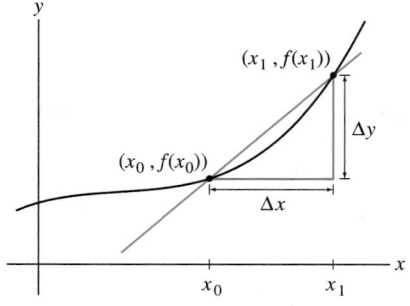

FIGURE 1 The average ROC over $[x_0, x_1]$ is the slope of the secant line.

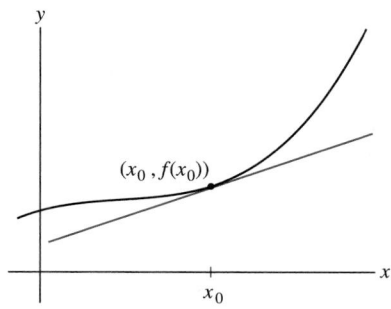

FIGURE 2 The instantaneous ROC at x_0 is the slope of the tangent line.

It is important to keep the geometric interpretation of rate of change in mind. The average ROC is the slope of the secant line (Figure 1) and the instantaneous ROC is the slope of the tangent line (Figure 2).

Whenever we measure the rate of change of a quantity y, it is always with respect to some other variable x (the independent variable). The rate dy/dx is measured in units of y per unit of x. For example, the ROC of temperature with respect to time has units such as degrees per minute and the ROC of temperature with respect to altitude has units such as degrees per kilometer.

Leibniz notation is particularly convenient when dealing with rates of change because it specifies the independent variable. The notation dy/dx indicates that we are considering the ROC of y with respect to the quantity x.

■ **EXAMPLE 1** Table 1 contains data on the temperature T (in degrees Celsius) on the surface of Mars at Martian time t, collected by the NASA Pathfinder space probe. A Martian hour is approximately 61.6 minutes.

(a) Use the data in Table 1 to calculate the average ROC of temperature over the time interval from 6:11 AM to 9:05 AM.

(b) Use Figure 3 to estimate the instantaneous ROC at $t = 12:28$ PM.

Solution

(a) The time interval from 6:11 to 9:05 has a length of 174 min or $\Delta t = 2.9$ hours. According to Table 1, the change in temperature from 6:11 to 9:05 is

$$\Delta T = -44.3 - (-71.6) = 27.3°C$$

TABLE 1   Data from Mars Pathfinder Mission, July 1997

Time	Temperature
5:42	−74.7
6:11	−71.6
6:40	−67.2
7:09	−63.7
7:38	−59.5
8:07	−53
8:36	−47.7
9:05	−44.3
9:34	−42

FIGURE 3 Temperature variation on the surface of Mars on July 6, 1997.

The average rate of temperature change is the ratio

$$\frac{\Delta T}{\Delta t} = \frac{27.3}{2.9} \approx 9.4°\text{C/h}$$

(b) The instantaneous ROC is the slope of the tangent line through the point $(12:28, -22.3)$ shown in Figure 3. To compute its slope, we must choose a second point on the tangent line. A convenient choice is $(00:00, -70)$. The time interval between these two points is 12 hours and 28 min, or approximately 12.47 hours. Therefore,

$$\text{ROC} = \text{slope of tangent line} \approx \frac{-22.3 - (-70)}{12.47} \approx 3.8°\text{C/hr} \qquad \blacksquare$$

■ **EXAMPLE 2** Let $A = \pi r^2$ be the area of a circle of radius r.

(a) Calculate the ROC of area with respect to radius.

(b) Compute dA/dr for $r = 2$ and $r = 5$, and explain why dA/dr is larger at $r = 5$.

Solution

(a) The ROC of area with respect to radius is the derivative

$$\frac{dA}{dr} = \frac{d}{dr}(\pi r^2) = 2\pi r$$

Note that the derivative of the area is equal to the circumference.

(b) At $r = 2$ and $r = 5$, we have

$$\left.\frac{dA}{dr}\right|_{r=2} = 2\pi(2) \approx 12.57 \qquad \text{and} \qquad \left.\frac{dA}{dr}\right|_{r=5} = 2\pi(5) \approx 31.42$$

Why is the derivative larger at $r = 5$? The derivative dA/dr measures how quickly the area changes when r increases. When the radius is increased from r to $r + \Delta r$, the area of the circle increases by a band of thickness Δr. We see in Figure 4 that the area of the band is greater when $r = 5$ than when $r = 2$. Thus, area increases more rapidly as a function of radius when r is larger. ■

FIGURE 4 The blue bands represent the change in area when r is increased by Δr.

The Effect of a One-Unit Change

For small values of h, the difference quotient is close to the derivative itself:

$$f'(x_0) \approx \frac{f(x_0 + h) - f(x_0)}{h} \qquad \boxed{1}$$

This approximation generally improves as h gets smaller, but in some applications, the approximation is already useful with $h = 1$. Setting $h = 1$ in Eq. (1) gives

$$f'(x_0) \approx f(x_0 + 1) - f(x_0) \qquad \boxed{2}$$

In other words, $f'(x_0)$ *is approximately equal to the change in f caused by a one-unit change in x when $x = x_0$.*

■ **EXAMPLE 3**   **Stopping Distance**   The stopping distance of an automobile after the brakes are applied (in feet) is given approximately by the function $F(s) = 1.1s + 0.05s^2$ for speeds s between 30 and 75 mph.

(a) Calculate the stopping distance and the ROC of stopping distance (with respect to speed) when $s = 30$ mph. Then estimate the change in stopping distance if the speed is increased by 1 mph.

(b) Do the same for $s = 60$ mph.

Solution The ROC of stopping distance with respect to speed is the derivative

$$F'(s) = \frac{d}{ds}(1.1s + 0.05s^2) = 1.1 + 0.1s \text{ ft/mph}$$

In this example, we use the derivative $F'(s)$ to estimate the change in stopping distance for a one-unit increase in speed. Compare the estimates with the following exact values:

Actual Change (ft)	$F'(s)$
$F(31) - F(30) = 4.15$	4.1
$F(61) - F(60) = 7.15$	7.1

(a) The stopping distance at $s = 30$ mph is $F(30) = 1.1(30) + 0.05(30)^2 = 78$ ft. Since $F'(30) = 1.1 + 0.1(30) = 4.1$ ft/mph, Eq. (2) gives us the approximation:

$$\underbrace{F(31) - F(30)}_{\text{Change in stopping distance}} \approx F'(30) = 4.1 \text{ ft}$$

Thus, when traveling at 30 mph, stopping distance increases by roughly 4 ft if the speed is increased by 1 mph.

(b) The stopping distance at $s = 60$ mph is $F(60) = 1.1(60) + 0.05(60)^2 = 246$ ft. We have $F'(60) = 1.1 + 0.1(60) = 7.1$ ft/mph. Thus, when traveling at 60 mph, stopping distance increases by roughly 7 ft if the speed is increased by 1 mph. ■

Marginal Cost in Economics

Many products can be manufactured more cheaply if produced in large quantities. The number of units manufactured is called the **production level**. To study the relation between costs and production level, economists define a rate of change called **marginal cost**. Let $C(x)$ denote the dollar cost (including labor and parts) of producing x units of a particular product. By definition, *the marginal cost at production level x_0 is the cost of producing one additional unit*:

Although $C(x)$ is meaningful only when x is whole number, we usually treat $C(x)$ as a differentiable function of x so that the techniques of calculus can be applied.

$$\text{Marginal cost} = C(x_0 + 1) - C(x_0)$$

In this setting, Eq. (2) usually gives a good approximation, so we take $C'(x_0)$ as an estimate of the marginal cost.

■ **EXAMPLE 4**   **Cost of an Air Flight**   Company data suggest that the total dollar cost of a 1,200-mile flight is approximately $C(x) = 0.0005x^3 - 0.38x^2 + 120x$, where x is the number of passengers (Figure 5).

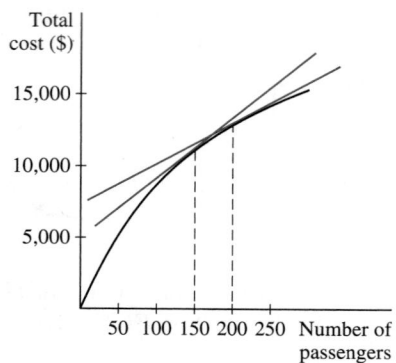

FIGURE 5 Cost of an air flight. The slopes of the tangent lines are decreasing, so marginal cost is decreasing.

(a) Estimate the marginal cost of an additional passenger if the flight already has 150 passengers.

(b) Compare your estimate with the actual cost.

(c) Is it more expensive to add a passenger when $x = 150$ or 200?

Solution   The derivative is $C'(x) = 0.0015x^2 - 0.76x + 120$.

(a) We estimate the marginal cost at $x = 150$ by the derivative

$$C'(150) = 0.0015(150)^2 - 0.76(150) + 120 = 39.75$$

Thus, it costs approximately $39.75 to add one additional passenger.

(b) The actual cost of adding one additional passenger is

$$C(151) - C(150) \approx 11{,}177.10 - 11{,}137.50 = 39.60$$

Our estimate $39.75 is very close.

(c) We estimate the marginal cost at $x = 200$ by the derivative

$$C'(200) = 0.0015(200)^2 - 0.76(200) + 120 = 28$$

Thus, it costs approximately $28 to add one additional passenger when $x = 200$. It is more expensive to add a passenger when $x = 150$.   ∎

Linear Motion

In his "Lectures on Physics," Nobel laureate Richard Feynman (1918–1988) uses the following dialogue to make a point about instantaneous velocity:

Policeman: "My friend, you were going 75 miles an hour."

Driver: "That's impossible, sir, I was traveling for only seven minutes."

Recall that *linear motion* refers to the motion of an object along a straight line. This includes the horizontal motion of a vehicle along a straight highway or the vertical motion of a falling object. Let $s(t)$ denote the position or distance from the origin at time t. Velocity is the ROC of position with respect to time:

$$v(t) = \text{velocity} = \frac{ds}{dt}$$

The *sign* of $v(t)$ indicates the direction of motion. For example, if $s(t)$ is the height above ground, then $v(t) > 0$ indicates that the object is rising. **Speed** is defined as the absolute value of velocity $|v(t)|$.

The graph of $s(t)$ conveys a great deal of information about velocity and how it changes. Consider Figure 6, showing the position of a car as a function of time. Keep in mind that the height of the graph represents the car's distance from the point of origin.

Here are some facts we can glean from the graph:

FIGURE 6 Graph of distance versus time.

- **Speeding up or slowing down?**   The tangent lines get steeper in the interval [0, 1], indicating that *the car was speeding up during the first hour*. They get flatter in the interval [1, 2], indicating that the car slowed down in the second hour.
- **Standing still**   The graph is horizontal over [2, 3] (perhaps the driver stopped at a restaurant for an hour).
- **Returning to the same spot**   The graph rises and falls in the interval [3, 4], indicating that the driver returned to the restaurant (perhaps she left her wallet there).
- **Average velocity**   The graph rises more over [0, 2] than over [3, 5], indicating that the average velocity was greater over the first two hours than the last two hours.

■ **EXAMPLE 5**   A large truck enters the off-ramp of a freeway at $t = 0$. Its position after t seconds is $s(t) = 84t - t^3$ ft for $0 \le t \le 5$.

(a) How fast is the truck going at the moment it enters the off-ramp?

(b) Is the truck speeding up or slowing down?

Solution The truck's velocity at time t is $v(t) = \dfrac{d}{dt}(84t - t^3) = 84 - 3t^2.$

(a) We have $v(0) = 84$, so the truck enters the off-ramp with a velocity of 84 ft/s.

(b) The function $v(t) = 84 - 3t^2$ is decreasing (Figure 7), so the truck is slowing down. ∎

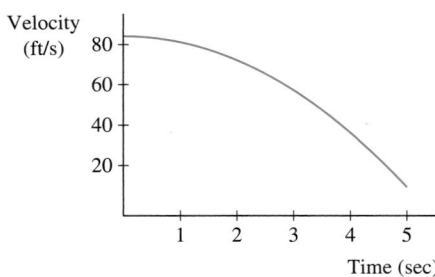

Velocity (ft/s)

Time (sec)

FIGURE 7 Graph of velocity $v(t) = 84 - 3t^2$.

Motion under the Influence of Gravity

Galileo discovered that the height $s(t)$ and velocity $v(t)$ of an object tossed vertically in the air are given as functions of time by the formulas

Galileo's formulas are valid when air resistance is negligible. We assume this to be the case in all examples here.

$$s(t) = s_0 + v_0 t - \frac{1}{2}gt^2, \qquad v(t) = \frac{ds}{dt} = v_0 - gt \qquad \boxed{3}$$

The constants s_0 and v_0 are the *initial values*:

- $s_0 = s(0)$ is the position at time $t = 0$.
- $v_0 = v(0)$ is the velocity at $t = 0$.
- $-g$ is the acceleration due to gravity on the surface of the earth (negative because the up direction is positive), where

$$g = 9.8 \text{ m/s}^2 \qquad \text{or} \qquad g \approx 32 \text{ ft/s}^2$$

A simple observation allows us to find the object's maximum height. Since velocity is positive as the object rises and negative as it falls back to earth, the object reaches its maximum height at the moment of transition, when it is no longer rising and has not yet begun to fall. At that moment, its velocity is zero. In other words, *the maximum height is attained when $v(t) = 0$.* Since $v(t) = s'(t)$, the object reaches its maximum height when the tangent line to the graph of $s(t)$ is horizontal (Figure 8).

FIGURE 8 Maximum height occurs when $s'(t) = v(t) = 0$.

■ **EXAMPLE 6** **Finding the Maximum Height** A slingshot launches a stone vertically with an initial velocity of 300 ft/s from an initial height of 6 ft.

(a) Find the stone's velocity at $t = 2$ and at $t = 12$. Explain the change in sign.

(b) What is the stone's maximum height and when does it reach that height?

Solution Apply Eq. (3) with $s_0 = 6$, $v_0 = 300$, and $g = 32$:

$$s(t) = 6 + 300t - 16t^2, \qquad v(t) = 300 - 32t$$

(a) Therefore,

$$v(2) = 300 - 32(2) = 236 \text{ ft/s}, \qquad v(12) = 300 - 32(12) = -84 \text{ ft/s}$$

At $t = 2$, the stone is rising and its velocity is positive (Figure 8). At $t = 12$, the stone is already on the way down and its velocity is negative.

(b) Since the maximum height is attained when the velocity is zero, we set the velocity equal to zero and solve for t:

$$v(t) = 300 - 32t = 0 \qquad \text{gives} \qquad t = \frac{300}{32} = 9.375$$

Therefore, the stone reaches its maximum height at $t = 9.375$ s. The maximum height is

$$s(9.375) = 6 + 300(9.375) - 16(9.375)^2 \approx 1{,}412 \text{ ft} \qquad \blacksquare$$

In the previous example, we specified the initial values of position and velocity. In the next example, the goal is to determine initial velocity.

▌ **EXAMPLE 7   Finding Initial Conditions**   A bullet is fired vertically from an initial height $s_0 = 0$. What initial velocity v_0 is required for the bullet to reach a maximum height of 2 km?

Solution   We need to derive a formula for the maximum height as a function of initial velocity v_0. Since $s_0 = 0$, the bullet's height is $s(t) = v_0 t - \frac{1}{2}gt^2$. Its maximum height is attained when the velocity is zero:

$$v(t) = v_0 - gt = 0 \qquad \text{or} \qquad t = \frac{v_0}{g}$$

We obtain the maximum height by computing $s(t)$ at $t = v_0/g$:

$$s\left(\frac{v_0}{g}\right) = v_0\left(\frac{v_0}{g}\right) - \frac{1}{2}g\left(\frac{v_0}{g}\right)^2 = \frac{v_0^2}{g} - \frac{1}{2}\frac{v_0^2}{g} = \frac{v_0^2}{2g}$$

Now we can solve for v_0 using the value $g = 9.8 \text{ m/s}^2$ (note that 2 km = 2,000 m):

$$\text{Maximum height} = \frac{v_0^2}{2g} = \frac{v_0^2}{2(9.8)} = 2{,}000 \text{ m} \qquad \boxed{4}$$

This yields $v_0 = \sqrt{(2)(9.8)2{,}000} \approx 198 \text{ m/s}$. $\qquad \blacksquare$

How important are units? In September 1999, the $125 million Mars Climate Orbiter spacecraft burned up in the Martian atmosphere before completing its scientific mission. According to Arthur Stephenson, NASA chairman of the Mars Climate Orbiter Mission Failure Investigation Board, 1999, "The 'root cause' of the loss of the spacecraft was the failed translation of English units into metric units in a segment of ground-based, navigation-related mission software."

In Eq. (4), distance must be in meters rather than kilometers since the units of g are m/s^2.

HISTORICAL PERSPECTIVE

Around 1600 Galileo Galilei (1564–1642) discovered the laws of motion for falling objects on the earth's surface. This was a major event in the history of science because it gave us, for the first time, a mathematical description of motion as a function of time. Galileo's discoveries paved the way for Newton's general laws of motion. How did Galileo arrive at his formulas? The motion of a falling object is too rapid to measure directly. Today, we can photograph an object in motion and examine the trajectory frame by frame, but Galileo had to be more resourceful. He experimented with balls rolling down an incline. For a sufficiently flat incline, the motion is slow enough to measure accurately, and Galileo found that the velocity of the rolling ball is proportional to time. He then reasoned that motion in free-fall is just a faster version of motion down an incline, and deduced the formula

FIGURE 9 Apparatus of the type used by Galileo to study the motion of falling objects.

$v(t) = -gt$ for falling objects (assuming that the initial velocity v_0 is 0). This formula has a striking consequence.

Prior to Galileo, it had been incorrectly assumed that heavy objects fall more quickly than lighter ones. Galileo realized that this was not true, as long as air resistance is negligible. The formula $v(t) = -gt$ shows clearly that the velocity depends on time but not on the weight of the object. Interestingly, 300 years later, another great physicist, Albert Einstein, was deeply puzzled by Galileo's discovery that objects fall at the same rate regardless of their weight. He called this the Principle of Equivalence and sought to understand why it was true. In 1916, after a decade of intensive work, Einstein developed the General Theory of Relativity, which finally gave a full explanation of the Principle of Equivalence in terms of the geometry of space and time.

3.4 SUMMARY

- The instantaneous rate of change (ROC) of $y = f(x)$ with respect to x at $x = x_0$ is defined as the derivative

$$f'(x_0) = \lim_{\Delta x \to 0} \frac{\Delta y}{\Delta x} = \lim_{x_1 \to x_0} \frac{f(x_1) - f(x_0)}{x_1 - x_0}$$

- The rate dy/dx is measured in *units of y per unit of unit of x.*
- For linear motion, velocity $v(t)$ is the ROC of position $s(t)$ with respect to time, that is, $v(t) = s'(t)$.
- In some applications, $f'(x_0)$ provides a good estimate of the change in f due to a one-unit increase in x when $x = x_0$:

$$f'(x_0) \approx f(x_0 + 1) - f(x_0)$$

- Marginal cost is the cost of producing one additional unit. If $C(x)$ is the cost of producing x units, then the marginal cost at production level x_0 is $C(x_0 + 1) - C(x_0)$. The derivative $C'(x_0)$ is often a good estimate for marginal cost.
- Galileo's formulas for the height and velocity at time t of an object rising or falling under the influence of gravity near the earth's surface ($s_0 =$ initial position, $v_0 =$ initial velocity) are

$$s(t) = s_0 + v_0 t - \frac{1}{2}gt^2, \qquad v(t) = v_0 - gt$$

3.4 EXERCISES

Preliminary Questions

1. What units might be used to measure the ROC of:

(a) Pressure (in atmospheres) in a water tank with respect to depth?

(b) The reaction rate of a chemical reaction (the ROC of concentration with respect to time), where concentration is measured in moles per liter?

2. Suppose that $f(2) = 4$ and the average ROC of f between 2 and 5 is 3. What is $f(5)$?

3. Two trains travel from New Orleans to Memphis in 4 hours. The first train travels at a constant velocity of 90 mph, but the velocity of the second train varies. What was the second train's average velocity during the trip?

4. Estimate $f(26)$, assuming that $f(25) = 43$ and $f'(25) = 0.75$.

5. The population $P(t)$ of Freedonia in 1933 was $P(1933) = 5$ million.

(a) What is the meaning of the derivative $P'(1933)$?

(b) Estimate $P(1934)$ if $P'(1933) = 0.2$. What if $P'(1933) = 0$?

Exercises

1. Find the ROC of the area of a square with respect to the length of its side s when $s = 3$ and $s = 5$.

2. Find the ROC of the volume of a cube with respect to the length of its side s when $s = 3$ and $s = 5$.

3. Find the ROC of $y = x^{-1}$ with respect to x for $x = 1, 10$.

4. At what rate is the cube root $\sqrt[3]{x}$ changing with respect to x when $x = 1, 8, 27$?

In Exercises 5–8, calculate the ROC.

5. $\dfrac{dV}{dr}$, where V is the volume of a cylinder whose height is equal to its radius (the volume of a cylinder of height h and radius r is $\pi r^2 h$)

6. ROC of the volume V of a cube with respect to its surface area A

7. ROC of the volume V of a sphere with respect to its radius (the volume of a sphere is $V = \frac{4}{3}\pi r^3$)

8. $\dfrac{dA}{dD}$, where A is the surface area of a sphere of diameter D (the surface area of a sphere of radius r is $4\pi r^2$)

In Exercises 9–10, refer to Figure 10, which shows the graph of distance (in kilometers) versus time (in hours) for a car trip.

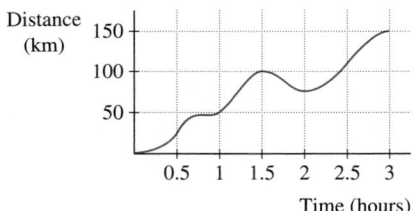

FIGURE 10 Graph of distance versus time for a car trip.

9. **(a)** Estimate the average velocity over $[0.5, 1]$.

(b) Is average velocity greater over $[1, 2]$ or $[2, 3]$?

(c) At what time is velocity at a maximum?

10. Match the description with the interval (a)–(d).

(i) Velocity increasing   **(ii)** Velocity decreasing

(iii) Velocity negative

(iv) Average velocity of 50 km/h

(a) $[0, 0.5]$ **(b)** $[0, 1]$

(c) $[1.5, 2]$ **(d)** $[2.5, 3]$

11. Figure 11 displays the voltage across a capacitor as a function of time while the capacitor is being charged. Estimate the ROC of voltage at $t = 20$ s. Indicate the values in your calculation and include proper units. Does voltage change more quickly or more slowly as time goes on? Explain in terms of tangent lines.

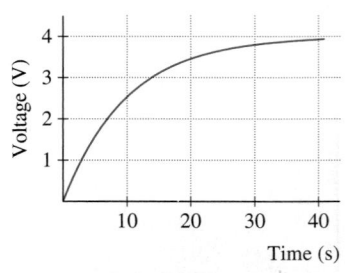

FIGURE 11

12. Use Figure 12 to estimate dT/dh at $h = 30$ and 70, where T is atmospheric temperature (in degrees Celsius) and h is altitude (in kilometers). Where is dT/dh equal to zero?

13. A stone is tossed vertically upward with an initial velocity of 25 ft/s from the top of a 30-ft building.

(a) What is the height of the stone after 0.25 s?

(b) Find the velocity of the stone after 1 s.

(c) When does the stone hit the ground?

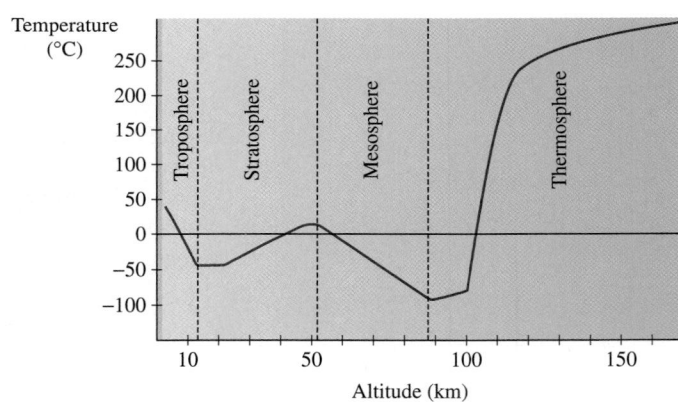

FIGURE 12 Atmospheric temperature versus altitude.

14. The height (in feet) of a skydiver at time t (in seconds) after opening his parachute is $h(t) = 2{,}000 - 15t$ ft. Find the skydiver's velocity after the parachute opens.

15. The temperature of an object (in degrees Fahrenheit) as a function of time (in minutes) is $T(t) = \frac{3}{4}t^2 - 30t + 340$ for $0 \le t \le 20$. At what rate does the object cool after 10 min (give correct units)?

16. The velocity (in centimeters per second) of a blood molecule flowing through a capillary of radius 0.008 cm is given by the formula $v = 6.4 \times 10^{-8} - 0.001r^2$, where r is the distance from the molecule to the center of the capillary. Find the ROC of velocity as a function of distance when $r = 0.004$ cm.

17. The earth exerts a gravitational force of $F(r) = \dfrac{2.99 \times 10^{16}}{r^2}$ (in Newtons) on an object with a mass of 75 kg, where r is the distance (in meters) from the center of the earth. Find the ROC of force with respect to distance at the surface of the earth, assuming the radius of the earth is 6.77×10^6 m.

18. The escape velocity at a distance r meters from the center of the earth is $v_{esc} = (2.82 \times 10^7)r^{-1/2}$ m/s. Calculate the rate at which v_{esc} changes with respect to distance at the surface of the earth.

19. The power delivered by a battery to an apparatus of resistance R (in ohms) is $P = \dfrac{2.25R}{(R+0.5)^2}$ W. Find the rate of change of power with respect to resistance for $R = 3$ and $R = 5\ \Omega$.

20. The position of a particle moving in a straight line during a 5-s trip is $s(t) = t^2 - t + 10$ cm.
(a) What is the average velocity for the entire trip?
(b) Is there a time at which the instantaneous velocity is equal to this average velocity? If so, find it.

21. By Faraday's Law, if a conducting wire of length ℓ meters moves at velocity v m/s perpendicular to a magnetic field of strength B (in teslas), a voltage of size $V = -B\ell v$ is induced in the wire. Assume that $B = 2$ and $\ell = 0.5$.
(a) Find the rate of change dV/dv.
(b) Find the rate of change of V with respect to time t if $v = 4t + 9$.

22. The height (in feet) of a helicopter at time t (in minutes) is $s(t) = -3t^3 + 400t$ for $0 \le t \le 10$.
(a) Plot the graphs of height $s(t)$ and velocity $v(t)$.
(b) Find the velocity at $t = 6$ and $t = 7$.
(c) Find the maximum height of the helicopter.

23. The population $P(t)$ of a city (in millions) is given by the formula $P(t) = 0.00005t^2 + 0.01t + 1$, where t denotes the number of years since 1990.
(a) How large is the population in 1996 and how fast is it growing?
(b) When does the population grow at a rate of 12,000 people per year?

24. According to Ohm's Law, the voltage V, current I, and resistance R in a circuit are related by the equation $V = IR$, where the units are volts, amperes, and ohms. Assume that voltage is constant with $V = 12$ V. Calculate (specifying the units):
(a) The average ROC of I with respect to R for the interval from $R = 8$ to $R = 8.1$
(b) The ROC of I with respect to R when $R = 8$
(c) The ROC of R with respect to I when $I = 1.5$

25. Ethan finds that with h hours of tutoring, he is able to answer correctly $S(h)$ percent of the problems on a math exam. What is the meaning of the derivative $S'(h)$? Which would you expect to be larger: $S'(3)$ or $S'(30)$? Explain.

26. Suppose $\theta(t)$ measures the angle between a clock's minute and hour hands. What is $\theta'(t)$ at 3 o'clock?

27. Table 2 gives the total U.S. population during each month of 1999 as determined by the U.S. Department of Commerce.
(a) Estimate $P'(t)$ for each of the months January–November.
(b) Plot these data points for $P'(t)$ and connect the points by a smooth curve.
(c) Write a newspaper headline describing the information contained in this plot.

TABLE 2   Total U.S. Population in 1999

t	$P(t)$ in Thousands
January	271,841
February	271,987
March	272,142
April	272,317
May	272,508
June	272,718
July	272,945
August	273,197
September	273,439
October	273,672
November	273,891
December	274,076

28. The tangent lines to the graph of $f(x) = x^2$ grow steeper as x increases. At what rate do the slopes of the tangent lines increase?

29. According to a formula widely used by doctors to determine drug dosages, a person's body surface area (BSA) (in meters squared) is given by the formula BSA $= \sqrt{hw}/60$, where h is the height in centimeters and w the weight in kilograms. Calculate the ROC of BSA with respect to weight for a person of constant height $h = 180$. What is this ROC for $w = 70$ and $w = 80$? Express your result in the correct units. Does BSA increase more rapidly with respect to weight at lower or higher body weights?

30. A slingshot is used to shoot a pebble in the air vertically from ground level with an initial velocity 200 m/s. Find the pebble's maximum velocity and height.

31. What is the velocity of an object dropped from a height of 300 m when it hits the ground?

32. It takes a stone 3 s to hit the ground when dropped from the top of a building. How high is the building and what is the stone's velocity upon impact?

33. A ball is tossed up vertically from ground level and returns to earth 4 s later. What was the initial velocity of the ball and how high did it go?

34. An object is tossed up vertically from ground level and hits the ground T s later. Show that its maximum height was reached after $T/2$ s.

35. A man on the tenth floor of a building sees a bucket (dropped by a window washer) pass his window and notes that it hits the ground 1.5 s later. Assuming a floor is 16 ft high (and neglecting air friction), from which floor was the bucket dropped?

36. Which of the following statements is true for an object falling under the influence of gravity near the surface of the earth? Explain.

(a) The object covers equal distance in equal time intervals.

(b) Velocity increases by equal amounts in equal time intervals.

(c) The derivative of velocity increases with time.

37. Show that for an object rising and falling according to Galileo's formula in Eq. (3), the average velocity over any time interval $[t_1, t_2]$ is equal to the average of the instantaneous velocities at t_1 and t_2.

38. A weight oscillates up and down at the end of a spring. Figure 13 shows the height y of the weight through one cycle of the oscillation. Make a rough sketch of the graph of the velocity as a function of time.

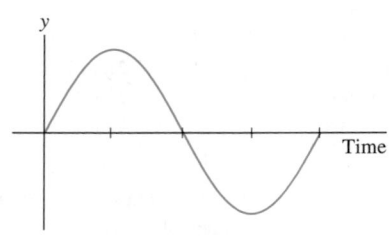

FIGURE 13

In Exercises 39–46, use Eq. (2) to estimate the unit change.

39. Estimate $\sqrt{2} - \sqrt{1}$ and $\sqrt{101} - \sqrt{100}$. Compare your estimates with the actual values.

40. Suppose that $f(x)$ is a function with $f'(x) = 2^{-x}$. Estimate $f(7) - f(6)$. Then estimate $f(5)$, assuming that $f(4) = 7$.

41. Let $F(s) = 1.1s + 0.03s^2$ be the stopping distance as in Example 3. Calculate $F(65)$ and estimate the increase in stopping distance if speed is increased from 65 to 66 mph. Compare your estimate with the actual increase.

42. According to Kleiber's Law, the metabolic rate P (in kilocalories per day) and body mass m (in kilograms) of an animal are related by a *three-quarter power law* $P = 73.3m^{3/4}$. Estimate the increase in metabolic rate when body mass increases from 60 to 61 kg.

43. The dollar cost of producing x bagels is $C(x) = 300 + 0.25x - 0.5(x/1,000)^3$. Determine the cost of producing 2,000 bagels and estimate the cost of the 2001st bagel. Compare your estimate with the actual cost of the 2001st bagel.

44. Suppose the dollar cost of producing x video cameras is $C(x) = 500x - 0.003x^2 + 10^{-8}x^3$.

(a) Estimate the marginal cost at production level $x = 5,000$ and compare it with the actual cost $C(5,001) - C(5,000)$.

(b) Compare the marginal cost at $x = 5,000$ with the average cost per camera, defined as $C(x)/x$.

45. The demand for a commodity generally decreases as the price is raised. Suppose that the demand for oil (per capita per year) is $D(p) = 900/p$ barrels, where p is the price per barrel in dollars. Find the demand when $p = \$40$. Estimate the decrease in demand if p rises to $\$41$ and the increase if p is decreased to $\$39$.

46. The reproduction rate of the fruit fly *Drosophila melanogaster*, grown in bottles in a laboratory, decreases as the bottle becomes more crowded. A researcher has found that when a bottle contains p flies, the number of offspring per female per day is

$$f(p) = (34 - 0.612p)p^{-0.658}$$

(a) Calculate $f(15)$ and $f'(15)$.

(b) Estimate the decrease in daily offspring per female when p is increased from 15 to 16. Is this estimate larger or smaller than the actual value $f(16) - f(15)$?

(c) GU   Plot $f(p)$ for $5 \leq p \leq 25$ and verify that $f(p)$ is a decreasing function of p. Do you expect $f'(p)$ to be positive or negative? Plot $f'(p)$ and confirm your expectation.

47. Let $A = s^2$. Show that the estimate of $A(s + 1) - A(s)$ provided by Eq. (2) has error exactly equal to 1. Explain this result using Figure 14.

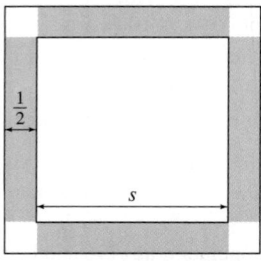

FIGURE 14

48. According to Steven's Law in psychology, the perceived magnitude of a stimulus (how strong a person feels the stimulus to be) is proportional to a power of the actual intensity I of the stimulus. Although not an exact law, experiments show that the *perceived brightness* B of a light satisfies $B = kI^{2/3}$, where I is the light intensity, whereas the *perceived heaviness* H of a weight W satisfies $H = kW^{3/2}$ (k is a constant that is different in the two cases). Compute dB/dI and dH/dW and state whether they are increasing or decreasing functions. Use this to justify the statements:

(a) A one-unit increase in light intensity is felt more strongly when I is small than when I is large.

(b) Adding another pound to a load W is felt more strongly when W is large than when W is small.

49. Let $M(t)$ be the mass (in kilograms) of a plant as a function of time (in years). Recent studies by Niklas and Enquist have suggested that for a remarkably wide range of plants (from algae and grass to palm trees), the growth rate during the life span of the organism satisfies a *three-quarter power law*, that is, $dM/dt = CM^{3/4}$ for some constant C.

(a) If a tree has a growth rate of 6 kg/year when $M = 100$ kg, what is its growth rate when $M = 125$ kg?

(b) If $M = 0.5$ kg, how much more mass must the plant acquire to double its growth rate?

Further Insights and Challenges

50. As an epidemic spreads through a population, the percentage p of infected individuals at time t (in days) satisfies the equation (called a *differential equation*)

$$\frac{dp}{dt} = 4p - 0.06p^2 \quad 0 \le p \le 100$$

(a) How fast is the epidemic spreading when $p = 10\%$ and when $p = 70\%$?

(b) For which p is the epidemic neither spreading nor diminishing?

(c) Plot dp/dt as a function of p.

(d) What is the maximum possible rate of increase and for which p does this occur?

51. The size of a certain animal population $P(t)$ at time t (in months) satisfies $\dfrac{dP}{dt} = 0.2(300 - P)$.

(a) Is P growing or shrinking when $P = 250$? when $P = 350$?

(b) Sketch the graph of dP/dt as a function of P for $0 \le P \le 300$.

(c) Which of the graphs in Figure 15 is the graph of $P(t)$ if $P(0) = 200$?

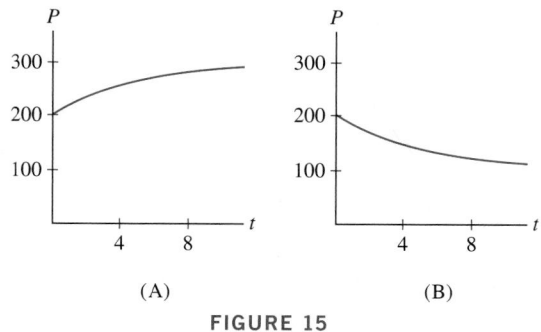

(A)   (B)

FIGURE 15

52. **CAS** Studies of internet usage show that website popularity is described quite well by Zipf's Law, according to which the nth most popular website receives roughly the fraction $1/n$ of all visits. Suppose that on a particular day, the nth most popular site had approximately $V(n) = 10^6/n$ visitors (for $n \le 15{,}000$).

(a) Verify that the top 50 websites received nearly 45% of the visits. *Hint:* Let $T(N)$ denote the sum of $V(n)$ for $1 \le n \le N$. Use a computer algebra system to compute $T(45)$ and $T(15{,}000)$.

(b) Verify, by numerical experimentation, that when Eq. (2) is used to estimate $V(n+1) - V(n)$, the error in the estimate decreases as n grows larger. Find (again, by experimentation) an N such that the error is at most 10 for $n \ge N$.

(c) Using Eq. (2), show that for $n \ge 100$, the nth website received at most 100 more visitors than the $(n+1)$st web site.

In Exercises 53–54, the average cost per unit at production level x is defined as $C_{avg}(x) = C(x)/x$, where $C(x)$ is the cost function. Average cost is a measure of the efficiency of the production process.

53. Show that $C_{avg}(x)$ is equal to the slope of the line through the origin and the point $(x, C(x))$ on the graph of $C(x)$. Using this interpretation, determine whether average cost or marginal cost is greater at points A, B, C, D in Figure 16.

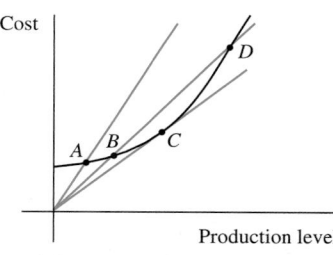

FIGURE 16 Graph of $C(x)$.

54. The cost in dollars of producing alarm clocks is

$$C(x) = 50x^3 - 750x^2 + 3{,}740x + 3{,}750$$

where x is in units of 1,000.

(a) Calculate the average cost at $x = 4, 6, 8,$ and 10.

(b) Use the graphical interpretation of average cost to find the production level x_0 at which average cost is lowest. What is the relation between average cost and marginal cost at x_0 (see Figure 17)?

FIGURE 17 Cost function $C(x) = 50x^3 - 750x^2 + 3{,}740x + 3{,}750$.

3.5 Higher Derivatives

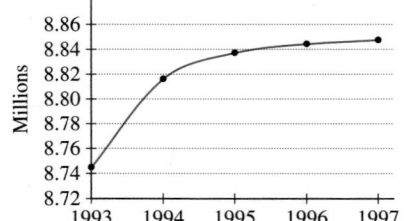

FIGURE 1 Population $P(t)$ of Sweden (in millions). The rate of increase declined in the period 1993–1997.

Higher derivatives are the functions obtained by repeatedly differentiating a function $y = f(x)$. If f' is differentiable, the **second derivative**, denoted f'' or y'', is the derivative of f':

$$f''(x) = \frac{d}{dx}\left(f'(x)\right)$$

The second derivative is the rate of change of $f'(x)$. We refer to $f(x)$ as the zeroth derivative and $f'(x)$ as the first derivative.

To clarify the distinction between the first and second derivatives, consider the population $P(t)$ of Sweden (Figure 1). Government statistics show that $P(t)$ has increased every year for more than 100 years. Therefore, the first derivative $P'(t)$ is positive. However, Table 1 shows that the rate of yearly increase declined dramatically in the years 1993–1997. So although $P'(t)$ was still positive in these years, $P'(t)$ decreased and therefore the second derivative $P''(t)$ was negative in the period 1993–1997.

TABLE 1	Population of Sweden				
Year	**1993**	**1994**	**1995**	**1996**	**1997**
Population	8,745,109	8,816,381	8,837,496	8,844,499	8,847,625
Yearly increase		71,272	21,115	7,003	3,126

Differentiation can be continued beyond the second derivative, provided that the derivatives exist. The third derivative is the derivative of $f''(x)$ and is denoted $f'''(x)$ or $f^{(3)}(x)$. More generally, the nth-order derivative is the derivative of the $(n-1)$st derivative and is denoted $f^{(n)}(x)$. In Leibniz notation, the higher derivatives are denoted

- dy/dx has units of y per unit of x.
- d^2y/dx^2 has units of dy/dx per unit of x, or units of y per unit of x-squared.

$$\frac{df}{dx}, \quad \frac{d^2 f}{dx^2}, \quad \frac{d^3 f}{dx^3}, \quad \frac{d^4 f}{dx^4}, \dots$$

■ **EXAMPLE 1** Find $f''(x)$ and $f'''(x)$ for $f(x) = x^4 + 2x - 9x^{-2}$ and evaluate $f'''(-1)$.

Solution We must calculate the derivatives, beginning with $f'(x)$:

$$f'(x) = \frac{d}{dx}\left(x^4 + 2x - 9x^{-2}\right) = 4x^3 + 2 + 18x^{-3}$$

$$f''(x) = \frac{d}{dx}\left(4x^3 + 2 + 18x^{-3}\right) = 12x^2 - 54x^{-4}$$

$$f'''(x) = \frac{d}{dx}\left(12x^2 - 54x^{-4}\right) = 24x + 216x^{-5}$$

At $x = -1$, $f'''(-1) = -24 - 216 = -240$. ■

If $f(x)$ is a polynomial of degree k, then $f^{(k)}(x)$ is a constant and, therefore, the higher derivatives $f^{(n)}(x)$ for $n > k$ are all zero. Indeed, each time we differentiate a polynomial, the degree decreases by 1. Table 2 shows the higher derivatives of $f(x) = x^5$, and we note that the nth derivative is zero for $n > 5$. By contrast, if k is not a whole number, the higher derivatives of $f(x) = x^k$ are all nonzero, as illustrated in the next example.

TABLE 2   **Derivatives of x^5**

$f(x)$	$f'(x)$	$f''(x)$	$f'''(x)$	$f^{(4)}(x)$	$f^{(5)}(x)$	$f^{(6)}(x)$
x^5	$5x^4$	$20x^3$	$60x^2$	$120x$	120	0

■ **EXAMPLE 2** Calculate the first four derivatives of $y = x^{-1}$. Then find the pattern and determine a general formula for the nth derivative $y^{(n)}$.

Solution By the Power Rule,

$$y'(x) = -x^{-2}, \quad y'' = 2x^{-3}, \quad y''' = -2(3)x^{-4}, \quad y^{(4)} = 2(3)(4)x^{-5}$$

We see that $y^{(n)}(x)$ is equal to $\pm n!\,x^{-n-1}$. Now observe that the sign alternates. Since the odd-order derivatives occur with a minus sign, the sign of $y^{(n)}(x)$ is $(-1)^n$. In general, therefore, $y^{(n)}(x) = (-1)^n n!\,x^{-n-1}$. ■

■ **EXAMPLE 3** Calculate $f'''(x)$ for $f(x) = xe^x$.

Solution We calculate the derivatives using the Product Rule, beginning with $f'(x)$:

$$f'(x) = \frac{d}{dx}(xe^x) = xe^x + e^x = (x+1)e^x$$

$$f''(x) = \frac{d}{dx}\big((x+1)e^x\big) = (x+1)e^x + e^x = (x+2)e^x$$

$$f'''(x) = \frac{d}{dx}\big((x+2)e^x\big) = (x+2)e^x + e^x = (x+3)e^x$$ ■

Acceleration, defined as the ROC of velocity, is a familiar second derivative. If an object in linear motion has position $s(t)$ at time t, then it has velocity $v(t) = s'(t)$ and acceleration $a(t) = v'(t) = s''(t)$. The unit of acceleration is the unit of velocity per unit of time or "distance per time squared." Thus, if velocity is measured in units of m/s, then acceleration has units of m/s per second or m/s².

Height (ft)

(A)

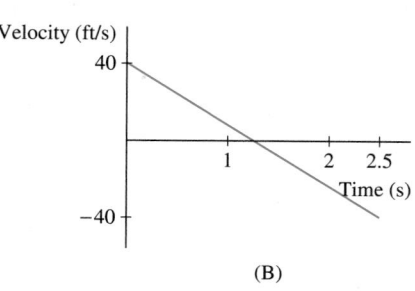

Velocity (ft/s)

(B)

FIGURE 2 Height and velocity of the ball.

■ **EXAMPLE 4** Acceleration due to Gravity   Calculate the acceleration $a(t)$ of a ball tossed vertically in the air from ground level with an initial velocity of 40 ft/s. How does $a(t)$ describe the change in the ball's velocity as it rises and falls?

Solution By Galileo's formula, the height of the ball at time t is $s(t) = s_0 + v_0 t - 16t^2$ [Figure 2(A)]. In our case, $s_0 = 0$ and $v_0 = 40$, so $s(t) = 40t - 16t^2$ ft. Therefore, $v(t) = s'(t) = 40 - 32t$ ft/s and the ball's acceleration is

$$a(t) = s''(t) = \frac{d}{dt}(40 - 32t) = -32 \text{ ft/s}^2$$

We see that the acceleration is constant with value $a = -32$ ft/s². Although the ball rises and falls, its velocity decreases from 40 to -40 ft/s at a constant rate [Figure 2(B)]. Note that a is the slope of the graph of velocity. ■

GRAPHICAL INSIGHT   Can we visualize the rate represented by a second derivative? Since the second derivative is the rate at which $f'(x)$ is changing, $f''(x)$ is large when the slopes of the tangent lines change rapidly, as in Figure 3(A). Similarly, f'' is small if the slopes of the tangent lines change slowly—in this case, the curve is relatively flat, as in Figure 3(B). If f is a linear function, the tangent line does not change at all and $f''(x) = 0$.

(A) Large second derivative: Tangent lines turn rapidly.

(B) Smaller second derivative: Tangent lines turn slowly.

(C) Second derivative is zero: Tangent line does not change.

FIGURE 3

> With this understanding, we can see how the graphs of f, f', and f'' are related. In Figure 4, the slopes of the tangent lines to the graph of f are *increasing* on the interval $[a, b]$. This is reflected in the graphs of both $f'(x)$ and $f''(x)$. Namely, $f'(x)$ is increasing on $[a, b]$ and $f''(x)$ is positive on $[a, b]$.

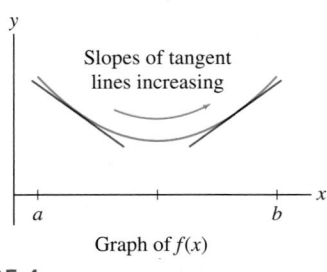

Slopes of tangent lines increasing

Graph of $f(x)$

$f'(x)$ is increasing

$f''(x)$ is positive

Graph of first two derivatives

FIGURE 4

3.5 SUMMARY

- The higher derivatives f', f'', f''', ... are defined by successive differentiation:

$$f''(x) = \frac{d}{dx} f'(x), \qquad f'''(x) = \frac{d}{dx} f''(x), \dots .$$

The nth derivative is also denoted $f^{(n)}(x)$.
- The second derivative plays an important role: It is the rate at which f' varies. Graphically, f'' measures how fast the tangent lines change direction and is thus a measure of how curved the graph is.
- If $s(t)$ is the position of an object at time t, then the first derivative $s'(t)$ is velocity, and the second derivative $s''(t)$ is acceleration.

3.5 EXERCISES

Preliminary Questions

1. An economist who announces that "America's economic growth is slowing" is making a statement about the gross national product (GNP) as a function of time. Is the second derivative of the GNP positive? What about the first derivative?

2. On September 4, 2003, the *Wall Street Journal* printed the headline "Stocks Go Higher, Though the Pace of Their Gains Slows." Rephrase as a statement about the first and second time derivatives of stock prices and sketch a possible graph.

3. Is the following statement true or false? The third derivative of position with respect to time is zero for an object falling to earth under the influence of gravity. Explain.

4. Which type of polynomial satisfies $f''(x) = 0$ for all x?

5. What is the millionth derivative of $f(x) = e^x$?

Exercises

In Exercises 1–14, calculate the second and third derivatives.

1. $y = 14x^2$

2. $y = 7 - 2x$

3. $y = x^4 - 25x^2 + 2x$

4. $y = 4t^3 - 9t^2 + 7$

5. $y = \dfrac{4}{3}\pi r^3$

6. $y = \sqrt{x}$

7. $y = 20t^{4/5} - 6t^{2/3}$

8. $y = x^{-9/5}$

9. $y = z - \dfrac{1}{z}$

10. $y = t^2(t^2 + t)$

11. $y = (x^2 + x)(x^3 + 1)$

12. $y = \dfrac{1}{1 + x}$

13. $y = 2x + e^x$

14. $y = x^4 e^x$

In Exercises 15–28, calculate the derivative indicated.

15. $f^{(4)}(1), \quad f(x) = x^4$

16. $g'''(1), \quad g(t) = 4t^{-3}$

17. $\left.\dfrac{d^2 y}{dt^2}\right|_{t=1}, \quad y = 4t^{-3} + 3t^2$

18. $\left.\dfrac{d^4 f}{dt^4}\right|_{t=1}, \quad f(t) = 6t^9 - 2t^5$

19. $h'''(9), \quad h(x) = \sqrt{x}$

20. $g'''(9), \quad g(x) = x^{-1/2}$

21. $\left.\dfrac{d^4 x}{dt^4}\right|_{t=16}, \quad x = t^{-3/4}$

22. $f'''(4), \quad f(t) = 2t^2 - t$

23. $g''(1), \quad g(x) = \dfrac{x}{x+1}$

24. $f''(1), \quad f(t) = \dfrac{1}{t^3 + 1}$

25. $h''(1), \quad h(x) = \dfrac{1}{\sqrt{x} + 1}$

26. $F''(2), \quad F(x) = \dfrac{x^2}{x - 3}$

27. $f''(0), \quad f(x) = \dfrac{x}{e^x + 1}$

28. $f'''(0), \quad f(x) = 4e^x - x^3$

29. Calculate $y^{(k)}(0)$ for $0 \le k \le 5$, where $y = x^4 + ax^3 + bx^2 + cx + d$ (with a, b, c, d the constants).

30. Which of the following functions satisfy $f^{(k)}(x) = 0$ for all $k \ge 6$?

(a) $f(x) = 7x^4 + 4 + x^{-1}$

(b) $f(x) = x^3 - 2$

(c) $f(x) = \sqrt{x}$

(d) $f(x) = 1 - x^6$

(e) $f(x) = x^{9/5}$

(f) $f(x) = 2x^2 + 3x^5$

31. Use the result in Example 2 to find $\dfrac{d^6}{dx^6} x^{-1}$.

32. Calculate the first five derivatives of $f(x) = \sqrt{x}$.
(a) Show that $f^{(n)}(x)$ is a multiple of $x^{-n+1/2}$.
(b) Show that $f^{(n)}(x)$ alternates in sign as $(-1)^{n-1}$ for $n \ge 1$.
(c) Find a formula for $f^{(n)}(x)$ for $n \ge 2$. *Hint:* Verify that the coefficient of $x^{-n+1/2}$ is $\pm 1 \cdot 3 \cdot 5 \cdots \dfrac{2n - 3}{2^n}$.

In Exercises 33–36, find a general formula for $f^{(n)}(x)$.

33. $f(x) = (x + 1)^{-1}$

34. $f(x) = x^{-2}$

35. $f(x) = x^{-1/2}$

36. $f(x) = x^{-3/2}$

37. (a) Find the acceleration at time $t = 5$ min of a helicopter whose height (in feet) is $h(t) = -3t^3 + 400t$.
(b) $\boxed{\text{GU}}$ Plot the acceleration $h''(t)$ for $0 \le t \le 6$. How does this graph show that the helicopter is slowing down during this time interval?

38. Find an equation of the tangent to the graph of $y = f'(x)$ at $x = 3$, where $f(x) = x^4$.

39. Figure 5 shows f, f', and f''. Determine which is which.

(A)   (B)   (C)

FIGURE 5

40. The second derivative f'' is shown in Figure 6. Determine which graph, (A) or (B), is f and which is f'.

$f''(x)$ (A)   (B)

FIGURE 6

41. Figure 7 shows the graph of the position of an object as a function of time. Determine the intervals on which the acceleration is positive.

42. Find the second derivative of the volume of a cube with respect to the length of a side.

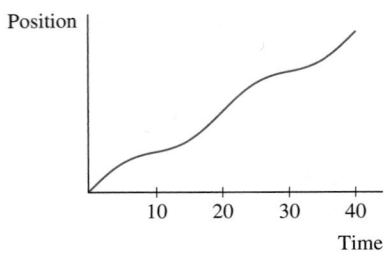

FIGURE 7

43. Find a polynomial $f(x)$ satisfying the equation $xf''(x) + f(x) = x^2$.

44. Find a value of n such that $y = x^n e^x$ satisfies the equation $xy' = (x - 3)y$.

45. 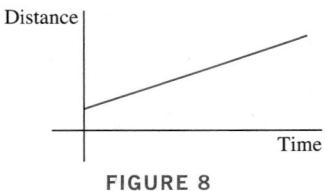 Which of the following descriptions could *not* apply to Figure 8? Explain.

(a) Graph of acceleration when velocity is constant

(b) Graph of velocity when acceleration is constant

(c) Graph of position when acceleration is zero

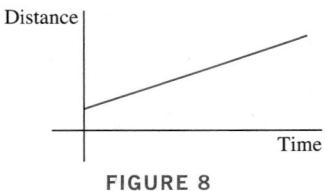

FIGURE 8

46. A servomotor controls the vertical movement of a drill bit that will drill a pattern of holes in sheet metal. The maximum vertical speed of the drill bit is 4 in./s, and while drilling the hole, it must move no more than 2.6 in./s to avoid warping the metal. During a cycle, the bit begins and ends at rest, quickly approaches the sheet metal, and quickly returns to its initial position after the hole is drilled. Sketch possible graphs of the drill bit's vertical velocity and acceleration. Label the point where the bit enters the sheet metal.

In Exercises 47–48, refer to the following. In their 1997 study, Boardman and Lave related the traffic speed S on a two-lane road to traffic density Q (number of cars per mile of road) by the formula $S = 2{,}882Q^{-1} - 0.052Q + 31.73$ for $60 \le Q \le 400$ (Figure 9).

47. Calculate dS/dQ and d^2S/dQ^2.

48. (a) 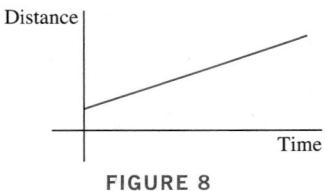 Explain intuitively why we should expect that $dS/dQ < 0$.

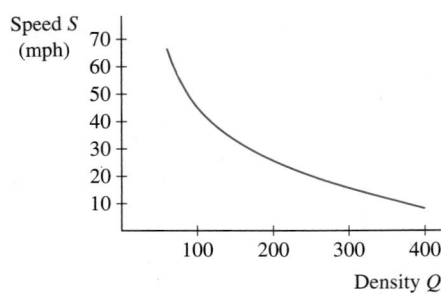

FIGURE 9 Speed as a function of traffic density.

(b) Show that $d^2S/dQ^2 > 0$. Then use the fact that $dS/dQ < 0$ and $d^2S/dQ^2 > 0$ to justify the following statement: *A one-unit increase in traffic density slows down traffic more when Q is small than when Q is large.*

(c) GU Plot dS/dQ. Which property of this graph shows that $d^2S/dQ^2 > 0$?

49. 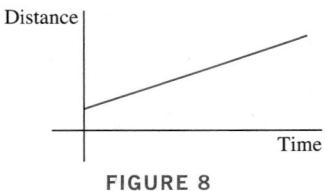 According to one model that attempts to account for air resistance, the distance $s(t)$ (in feet) traveled by a falling raindrop satisfies

$$\frac{d^2s}{dt^2} = g - \frac{0.0005}{D}\left(\frac{ds}{dt}\right)^2$$

where D is the raindrop diameter and $g = 32$ ft/s^2. Terminal velocity v_{term} is defined as the velocity at which the drop has zero acceleration (one can show that velocity approaches v_{term} as time proceeds).

(a) Show that $v_{\text{term}} = \sqrt{2000gD}$.

(b) Find v_{term} for drops of diameter 0.003 and 0.0003 ft.

(c) In this model, do raindrops accelerate more rapidly at higher or lower velocities?

50. *CAS* Use a computer algebra system to compute $f^{(k)}(x)$ for $k = 1, 2, 3$ for the functions:

(a) $f(x) = (1 + x^3)^{5/3}$

(b) $f(x) = \dfrac{1 - x^4}{1 - 5x - 6x^2}$

51. *CAS* Let $f(x) = \dfrac{x + 2}{x - 1}$. Use a computer algebra system to compute the $f^{(k)}(x)$ for $1 \le k \le 4$. Can you find a general formula for $f^{(k)}(x)$?

Further Insights and Challenges

52. Find the 100th derivative of

$$p(x) = (x + x^5 + x^7)^{10}(1 + x^2)^{11}(x^3 + x^5 + x^7)$$

53. What is the $p^{(99)}(x)$ for $p(x)$ as in Exercise 52?

54. Use the Product Rule twice to find a formula for $(fg)''$ in terms of the first and second derivative of f and g.

55. Use the Product Rule to find a formula for $(fg)'''$ and compare your result with the expansion of $(a + b)^3$. Then try to guess the general formula for $(fg)^{(n)}$.

56. Compute

$$\Delta f(x) = \lim_{h \to 0} \frac{f(x+h) + f(x-h) - 2f(x)}{h^2}$$

for the following functions:

(a) $f(x) = x$ (b) $f(x) = x^2$ (c) $f(x) = x^3$

Based on these examples, can you formulate a conjecture about what Δf is?

57. Calculate the first four derivatives of $f(x) = x^2 e^x$. Then guess the formula for $f^{(n)}(x)$ (use induction to prove it if you are familiar with this method of proof).

3.6 Trigonometric Functions

We can use the rules developed so far to differentiate functions involving powers of x, but we cannot yet handle the trigonometric functions. What is missing are the formulas for the derivatives of $\sin x$ and $\cos x$. Fortunately, these functions have simple derivatives—each is the derivative of the other up to a sign.

Recall our convention concerning angles: *All angles are measured in radians, unless otherwise specified.*

CAUTION *Theorem 1 applies only when radians are used. The derivatives of sine and cosine with respect to degrees involves an extra, unwieldy factor of $\pi/180$ (see Example 6 in Section 3.7).*

THEOREM 1 Derivative of Sine and Cosine   The functions $y = \sin x$ and $y = \cos x$ are differentiable and

$$\frac{d}{dx} \sin x = \cos x \qquad \text{and} \qquad \frac{d}{dx} \cos x = -\sin x$$

Proof We must go back to the definition of the derivative:

$$\frac{d}{dx} \sin x = \lim_{h \to 0} \frac{\sin(x+h) - \sin x}{h} \qquad \boxed{1}$$

Unlike the derivative computations in Sections 3.1 and 3.2, it is not possible to cancel h by rewriting the difference quotient. Instead, we use the addition formula

$$\sin(x+h) = \sin x \cos h + \cos x \sin h$$

to rewrite the numerator of the difference quotient as a sum of two terms:

$$\sin(x+h) - \sin x = \sin x \cos h + \cos x \sin h - \sin x \qquad \text{(addition formula)}$$
$$= (\sin x \cos h - \sin x) + \cos x \sin h \quad \text{(rearrange)}$$
$$= \sin x (\cos h - 1) + \cos x \sin h \qquad \text{(sum of two terms)}$$

This gives us

$$\frac{\sin(x+h) - \sin x}{h} = \frac{\sin x \,(\cos h - 1)}{h} + \frac{\cos x \,\sin h}{h}$$

Now apply the Sum Law for Limits:

$$\frac{d \sin x}{dx} = \lim_{h \to 0} \frac{\sin x \,(\cos h - 1)}{h} + \lim_{h \to 0} \frac{\cos x \,\sin h}{h}$$

$$= (\sin x) \underbrace{\lim_{h \to 0} \frac{\cos h - 1}{h}}_{\text{This equals 0}} + (\cos x) \underbrace{\lim_{h \to 0} \frac{\sin h}{h}}_{\text{This equals 1}} \qquad \boxed{2}$$

Note that $\sin x$ and $\cos x$ do not depend on h and may be taken outside the limits in Eq. (2). We are justified in using the Sum Law because the final two limits exist. Indeed, by Theorem 2 in Section 2.6,

$$\lim_{h \to 0} \frac{\cos h - 1}{h} = 0 \quad \text{and} \quad \lim_{h \to 0} \frac{\sin h}{h} = 1$$

Therefore, Eq. (2) reduces to the formula $\dfrac{d}{dx} \sin x = \cos x$ as desired. The formula $\dfrac{d}{dx} \cos x = -\sin x$ is proved similarly (see Exercise 52). ∎

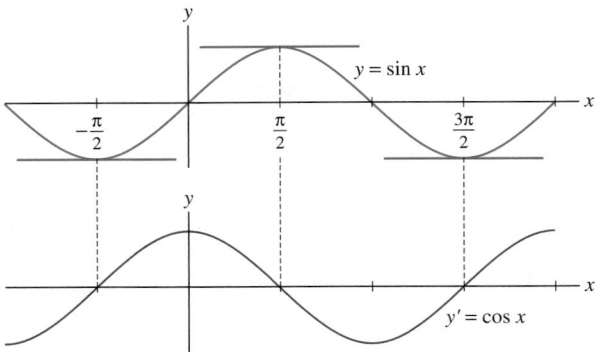

FIGURE 1 Compare the graphs of $y = \sin x$ and its derivative $y' = \cos x$.

GRAPHICAL INSIGHT   The formula $(\sin x)' = \cos x$ makes sense when we compare the graphs of sine and cosine (Figure 1). The tangent lines to the graph of $y = \sin x$ have positive slope on the interval $(-\frac{\pi}{2}, \frac{\pi}{2})$, and on this interval, the derivative $y' = \cos x$ is positive. Similarly, the tangent lines have negative slope on the interval $(\frac{\pi}{2}, \frac{3\pi}{2})$, where $y' = \cos x$ is negative. The tangent lines are horizontal at $x = -\frac{\pi}{2}, \frac{\pi}{2}, \frac{3\pi}{2}$, where $\cos x = 0$.

■ **EXAMPLE 1**   Calculate $f'(x)$ and $f''(x)$, where $f(x) = x \cos x$.

Solution   By the Product Rule,

$$f'(x) = x \frac{d}{dx} \cos x + \cos x \frac{d}{dx} x = x(-\sin x) + \cos x = \cos x - x \sin x$$

$$f''(x) = \frac{d}{dx} (\cos x - x \sin x) = \frac{d}{dx} \cos x - \frac{d}{dx} x \sin x$$

$$= -\sin x - \left(x(\sin x)' + \sin x\right) = -2 \sin x - x \cos x$$

∎

REMINDER *The standard trigonometric functions are defined in Section 1.4.*

The derivatives of the other **standard trigonometric functions** may be computed using the Quotient Rule. We derive the formula for $(\tan x)'$ in Example 2 and leave the remaining formulas for the exercises (Exercises 39–41).

THEOREM 2   Derivatives of Standard Trigonometric Functions

$$\frac{d}{dx} \tan x = \sec^2 x, \qquad \frac{d}{dx} \sec x = \sec x \tan x$$

$$\frac{d}{dx} \cot x = -\csc^2 x \qquad \frac{d}{dx} \csc x = -\csc x \cot x$$

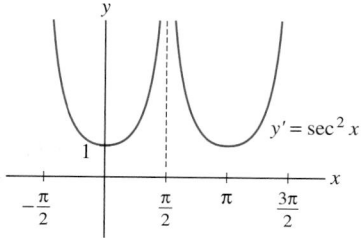

FIGURE 2 Graphs of $y = \tan x$ and its derivative $y' = \sec^2 x$.

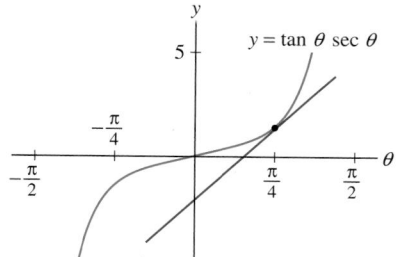

FIGURE 3 Tangent line to $y = \tan \theta \sec \theta$ at $\theta = \frac{\pi}{4}$.

■ **EXAMPLE 2** Verify the formula $\dfrac{d}{dx} \tan x = \sec^2 x$ (Figure 2).

Solution We use the Quotient Rule and the identity $\cos^2 x + \sin^2 x = 1$:

$$\frac{d}{dx} \tan x = \left(\frac{\sin x}{\cos x} \right)' = \frac{\cos x \cdot (\sin x)' - \sin x \cdot (\cos x)'}{\cos^2 x}$$

$$= \frac{\cos x \cos x - \sin x (- \sin x)}{\cos^2 x}$$

$$= \frac{\cos^2 x + \sin^2 x}{\cos^2 x} = \frac{1}{\cos^2 x} = \sec^2 x \qquad ■$$

■ **EXAMPLE 3** Find an equation of the tangent line to the graph of $y = \tan \theta \sec \theta$ at $\theta = \frac{\pi}{4}$.

Solution By the Product Rule,

$$y' = \tan \theta \, (\sec \theta)' + \sec \theta \, (\tan \theta)' = \tan \theta \, (\sec \theta \tan \theta) + \sec \theta \, \sec^2 \theta$$

$$= \tan^2 \theta \sec \theta + \sec^3 \theta$$

Now use the values $\sec \frac{\pi}{4} = \sqrt{2}$ and $\tan \frac{\pi}{4} = 1$ to compute

$$y \left(\frac{\pi}{4} \right) = \tan \left(\frac{\pi}{4} \right) \sec \left(\frac{\pi}{4} \right) = \sqrt{2}$$

$$y' \left(\frac{\pi}{4} \right) = \tan^2 \left(\frac{\pi}{4} \right) \sec \left(\frac{\pi}{4} \right) + \sec^3 \left(\frac{\pi}{4} \right) = \sqrt{2} + 2\sqrt{2} = 3\sqrt{2}$$

An equation of the tangent line (Figure 3) is $y - \sqrt{2} = 3\sqrt{2} \left(\theta - \frac{\pi}{4} \right)$. $\qquad ■$

■ **EXAMPLE 4** Calculate $f'(0)$, where $f(x) = e^x \cos x$.

Solution Use the Product Rule:

$$f'(x) = e^x \cdot (\cos x)' + \cos x \cdot (e^x)' = -e^x \sin x + \cos x \cdot e^x = e^x (\cos x - \sin x)$$

Then $f'(0) = e^0 (1 - 0) = 1$. $\qquad ■$

3.6 SUMMARY

- The basic trigonometric derivatives are

$$\frac{d}{dx} \sin x = \cos x, \qquad \frac{d}{dx} \cos x = -\sin x$$

- These two formulas are used to derive the additional formulas:

$$\frac{d}{dx} \tan x = \sec^2 x, \qquad \frac{d}{dx} \sec x = \sec x \tan x,$$

$$\frac{d}{dx} \cot x = -\csc^2 x, \qquad \frac{d}{dx} \csc x = -\csc x \cot x.$$

- Remember: *These formulas are valid only when the angle x is measured in radians.*

3.6 EXERCISES

Preliminary Questions

1. Determine the sign $\pm$ that yields the correct formula for the following:

(a) $\dfrac{d}{dx}(\sin x + \cos x) = \pm \sin x \pm \cos x$

(b) $\dfrac{d}{dx}\tan x = \pm \sec^2 x$

(c) $\dfrac{d}{dx}\sec x = \pm \sec x \tan x$

(d) $\dfrac{d}{dx}\cot x = \pm \csc^2 x$

2. Which of the following functions can be differentiated using the rules we have covered so far?

(a) $y = 3 \cos x \cot x$ **(b)** $y = \cos(x^2)$ **(c)** $y = e^x \sin x$

3. Compute $\dfrac{d}{dx}(\sin^2 x + \cos^2 x)$ without using the derivative formulas for $\sin x$ and $\cos x$.

4. How is the addition formula used in deriving the formula $(\sin x)' = \cos x$?

Exercises

In Exercises 1–4, find an equation of the tangent line at the point indicated.

1. $y = \sin x$, $x = \dfrac{\pi}{4}$

2. $y = \cos x$, $x = \dfrac{\pi}{3}$

3. $y = \tan x$, $x = \dfrac{\pi}{4}$

4. $y = \sec x$, $x = \dfrac{\pi}{6}$

In Exercises 5–26, use the Product and Quotient Rules as necessary to find the derivative of each function.

5. $f(x) = \sin x \cos x$

6. $f(x) = x^2 \cos x$

7. $f(x) = \sin^2 x$

8. $f(x) = 9 \sec x + 12 \cot x$

9. $f(x) = x^3 \sin x$

10. $f(x) = \dfrac{\sin x}{x}$

11. $f(\theta) = \tan \theta \sec \theta$

12. $g(\theta) = \dfrac{e^\theta}{\cos \theta}$

13. $h(\theta) = e^{-\theta} \cos^2 \theta$

14. $k(\theta) = \theta^2 \sin^2 \theta$

15. $f(x) = (x - x^2) \cot x$

16. $f(z) = z \tan z$

17. $f(x) = \dfrac{\sec x}{x^2}$

18. $f(x) = \dfrac{x}{\tan x}$

19. $g(t) = \sin t - \dfrac{2}{\cos t}$

20. $R(y) = \dfrac{\cos y - 1}{\sin y}$

21. $f(x) = \dfrac{x}{\sin x + 2}$

22. $f(x) = (x^2 - x - 4) \csc x$

23. $f(x) = \dfrac{1 + \sin x}{1 - \sin x}$

24. $f(x) = \dfrac{1 + \tan x}{1 - \tan x}$

25. $g(x) = \dfrac{\sec x}{x}$

26. $h(x) = \dfrac{\sin x}{4 + \cos x}$

In Exercises 27–30, calculate the second derivative.

27. $f(x) = 3 \sin x + 4 \cos x$

28. $f(x) = \tan x$

29. $g(\theta) = \theta \sin \theta$

30. $h(\theta) = \csc \theta$

In Exercises 31–38, find an equation of the tangent line at the point specified.

31. $y = x^2 + \sin x$, $x = 0$

32. $y = \theta \tan \theta$, $\theta = \dfrac{\pi}{4}$

33. $y = 2 \sin x + 3 \cos x$, $x = \dfrac{\pi}{3}$

34. $y = \theta^2 + \sin \theta$, $\theta = 0$

35. $y = \csc x - \cot x$, $x = \dfrac{\pi}{4}$

36. $y = \dfrac{\cos \theta}{1 + \sin \theta}$, $\theta = \dfrac{\pi}{3}$

37. $y = e^x \cos x$, $x = 0$

38. $y = e^\theta \sec \theta$, $\theta = \dfrac{\pi}{4}$

In Exercises 39–41, verify the formula using $(\sin x)' = \cos x$ and $(\cos x)' = -\sin x$.

39. $\dfrac{d}{dx}\cot x = -\csc^2 x$

40. $\dfrac{d}{dx}\sec x = \sec x \tan x$

41. $\dfrac{d}{dx}\csc x = -\csc x \cot x$

42. Find the values of x between 0 and 2π where the tangent line to the graph of $y = \sin x \cos x$ is horizontal.

43. Calculate the first five derivatives of $f(x) = \cos x$. Then determine $f^{(8)}$ and $f^{(37)}$.

44. Find $y^{(157)}$, where $y = \sin x$.

45. Calculate $f''(x)$ and $f'''(x)$, where $f(x) = \tan x$.

46. GU Let $g(t) = t - \sin t$.

(a) Plot the graph of g with a graphing utility for $0 \le t \le 4\pi$.

(b) Show that the slope of the tangent line is always positive and verify this on your graph.

(c) For which values of t in the given range is the tangent line horizontal?

47. CAS Let $f(x) = \dfrac{\sin x}{x}$ for $x \ne 0$ and $f(0) = 1$.

(a) Plot $f(x)$ on $[-3\pi, 3\pi]$.

(b) Show that $f'(c) = 0$ if $c = \tan c$. Use the numerical root finder on a computer algebra system to find a good approximation to the smallest *positive* value c_0 such that $f'(c_0) = 0$.

(c) Verify that the horizontal line $y = f(c_0)$ is tangent to the graph of $y = f(x)$ at $x = c_0$ by plotting them on the same set of axes.

48. ✏️ Show that no tangent line to the graph of $f(x) = \tan x$ has zero slope. What is the least slope of a tangent line? Justify your response by sketching the graph of $(\tan x)'$.

49. The height at time t (s) of a weight, oscillating up and down at the end of a spring, is $s(t) = 300 + 40 \sin t$ cm. Find the velocity and acceleration at $t = \frac{\pi}{3}$ s.

50. The horizontal range R of a projectile launched from ground level at an angle θ and initial velocity v_0 m/s is $R = (v_0^2/9.8) \sin \theta \cos \theta$. Calculate $dR/d\theta$. If $\theta = 7\pi/24$, will the range increase or decrease if the angle is increased slightly? Base your answer on the sign of the derivative.

51. ✏️ If you stand 1 m from a wall and mark off points on the wall at equal increments δ of angular elevation (Figure 4), then these points grow increasingly far apart. Explain how this illustrates the fact that the derivative of $\tan \theta$ is increasing.

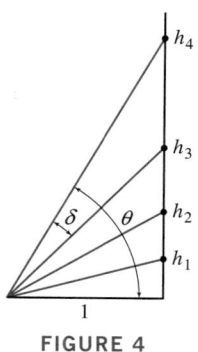

FIGURE 4

Further Insights and Challenges

52. Use the limit definition of the derivative and the addition law for the cosine to prove that $(\cos x)' = -\sin x$.

53. Show that a nonzero polynomial function $y = f(x)$ *cannot* satisfy the equation $y'' = -y$. Use this to prove that neither $\sin x$ nor $\cos x$ is a polynomial.

54. Verify the following identity and use it to give another proof of the formula $\sin' x = \cos x$:

$$\sin(x + h) - \sin x = 2 \cos \left(x + \tfrac{1}{2}h\right) \sin \left(\tfrac{1}{2}h\right)$$

Hint: Use the addition formula to prove that $\sin(a + b) - \sin(a - b) = 2 \cos a \sin b$.

55. Let $f(x) = x \sin x$ and $g(x) = x \cos x$.

(a) Show that $f'(x) = g(x) + \sin x$ and $g'(x) = -f(x) + \cos x$.

(b) Verify that $f''(x) = -f(x) + 2 \cos x$ and $g''(x) = -g(x) - 2 \sin x$.

(c) By further experimentation, try to find formulas for all higher derivatives of f and g. *Hint:* The kth derivative depends on whether $k = 4n, 4n + 1, 4n + 2,$ or $4n + 3$.

56. Show that if $\pi/2 < \theta < \pi$, then the distance along the x-axis between θ and the point where the tangent line intersects the x-axis is equal to $|\tan \theta|$ (Figure 5).

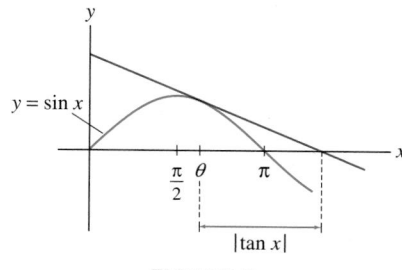

FIGURE 5

3.7 The Chain Rule

The **Chain Rule** is used to differentiate composite functions such as $y = \cos(x^2)$ and $y = \sqrt{x^3 + 1}$, which cannot be differentiated using any of the rules covered thus far. Recall that a *composite function* is obtained by "plugging" one function into another. More formally, the composite of f and g, denoted $f \circ g$, is defined by

$$(f \circ g)(x) = f\big(g(x)\big)$$

For convenience, we refer to g as the *inside* function and f as the *outside* function. It is convenient to use u rather than x as the variable for the outside function f. We then regard the composite as $f(u)$, where $u = g(x)$. For example, $y = \sqrt{x^3 + 1}$ is the function $y = \sqrt{u}$, where $u = x^3 + 1$.

A proof of the Chain Rule is given at the end of this section.

THEOREM 1 Chain Rule   If f and g are differentiable, then $(f \circ g)(x) = f(g(x))$ is differentiable and

$$\left(f(g(x))\right)' = f'\left(g(x)\right)g'(x)$$

Setting $u = g(x)$, we may also write the Chain Rule as

$$\frac{d}{dx}f(u) = f'(u)\frac{du}{dx}$$

Here is a verbal form of the Chain Rule:

$$\left(f(g(x))\right)' = \text{outside}'(\text{inside}) \cdot \text{inside}'$$

■ EXAMPLE 1   Calculate the derivative of $y = \sqrt{x^3 + 1}$.

Solution   As noted above, $y = \sqrt{x^3 + 1}$ is a composite $f(g(x))$ with

$$f(u) = \sqrt{u},$$
$$u = g(x) = x^3 + 1$$

Since $f'(u) = \frac{1}{2}u^{-1/2}$ and $g'(x) = 3x^2$, we have

$$\text{Outside}'(\text{inside}) = f'(g(x)) = \frac{1}{2}u^{-1/2} = \frac{1}{2}(x^3 + 1)^{-1/2}$$
$$\text{Inside}' = g'(x) = 3x^2$$

By the Chain Rule,

$$\frac{d}{dx}\sqrt{x^3 + 1} = \text{outside}'(\text{inside}) \cdot \text{inside}' = \underbrace{\frac{1}{2}(x^3 + 1)^{-1/2}}_{f'(g(x))} \underbrace{(3x^2)}_{g'(x)} = \frac{3x^2}{2\sqrt{x^3 + 1}}$$ ■

■ EXAMPLE 2   Calculate $\dfrac{dy}{dx}$ for   (a) $y = \cos(x^2)$ and   (b) $y = \tan\left(\dfrac{x}{x + 1}\right)$.

Solution

(a) We have $\cos(x^2) = f(g(x))$, where

	Outside:	$f(u) = \cos u,$	inside:	$u = g(x) = x^2$
	Outside':	$f'(u) = -\sin u,$	inside':	$g'(x) = 2x$

By the Chain Rule,

$$\frac{d}{dx}\cos(x^2) = f'(g(x))\,g'(x) = -\sin(x^2)\,2x = -2x\sin(x^2)$$

(b) The outside function for $y = \tan\left(\dfrac{x}{x+1}\right)$ is $f(u) = \tan u$. Since $f'(u) = \sec^2 u$, the Chain Rule gives us

$$\frac{d}{dx}\tan\left(\frac{x}{x+1}\right) = \sec^2\left(\frac{x}{x+1}\right)\underbrace{\frac{d}{dx}\left(\frac{x}{x+1}\right)}_{\substack{\text{Derivative of}\\\text{inside function}}}$$

Now, by the Quotient Rule,

$$\frac{d}{dx}\left(\frac{x}{x+1}\right) = \frac{(x+1)\dfrac{d}{dx}x - x\dfrac{d}{dx}(x+1)}{(x+1)^2} = \frac{1}{(x+1)^2}$$

We obtain

$$\frac{d}{dx}\tan\left(\frac{x}{x+1}\right) = \sec^2\left(\frac{x}{x+1}\right)\frac{1}{(x+1)^2} = \frac{\sec^2\left(\dfrac{x}{x+1}\right)}{(x+1)^2}\qquad\blacksquare$$

It is instructive to write the Chain Rule in Leibniz notation. Suppose that $u = g(x)$ and

$$y = f(u) = f(g(x))$$

Then, by the Chain Rule,

$$\frac{dy}{dx} = f'(u)\,g'(x) = \frac{df}{du}\frac{du}{dx}$$

or

$$\boxed{\frac{dy}{dx} = \frac{dy}{du}\frac{du}{dx}}$$

Christiaan Huygens (1629–1695), one of the greatest scientists of his age, was Leibniz's teacher in mathematics and physics. He greatly admired Isaac Newton but did not accept Newton's theory of gravitation, which he referred to as the "improbable principle of attraction" because it did not explain how two masses separated by a distance could influence each other.

CONCEPTUAL INSIGHT   In Leibniz notation, it appears as if we are multiplying fractions and the Chain Rule is simply a matter of "cancelling the du." Since the symbolic expressions dy/du and du/dx are not fractions, this does not make sense literally, but it does suggest that derivatives behave *as if they were fractions* (this is reasonable because a derivative is a *limit* of fractions, namely of the difference quotients). Leibniz's form also emphasizes a key aspect of the Chain Rule: *rates of change multiply*. To illustrate this point, suppose that your salary increases three times as fast as the gross domestic product (GDP) and that the GDP increases by 2% per year. Then your salary will increase at the rate of 3×2 or 6% per year. In terms of derivatives,

$$\frac{d(\text{salary})}{dt} = \frac{d(\text{salary})}{d(\text{GDP})} \times \frac{d(\text{GDP})}{dt}$$

$$6\% \text{ per year} = 3 \times 2\% \text{ per year}$$

174 CHAPTER 3 | DIFFERENTIATION

■ **EXAMPLE 3** Imagine a sphere whose radius r increases at a rate of 3 cm/s. At what rate is the volume V of the sphere increasing when $r = 10$ cm?

Solution Since we are asked to determine the rate at which V is increasing, we must find dV/dt. The rate dr/dt is given as 3 cm/s, so we use the Chain Rule to express dV/dt in terms of dr/dt:

$$\underbrace{\frac{dV}{dt}}_{\substack{\text{ROC of volume} \\ \text{with respect to time}}} = \underbrace{\frac{dV}{dr}}_{\substack{\text{ROC of volume} \\ \text{with respect to radius}}} \times \underbrace{\frac{dr}{dt}}_{\substack{\text{ROC of radius} \\ \text{with respect to time}}}$$

To compute dV/dr, we use the formula for the volume of a sphere, $V = \frac{4}{3}\pi r^3$:

$$\frac{dV}{dr} = \frac{d}{dr}\left(\frac{4}{3}\pi r^3\right) = 4\pi r^2$$

Since $dr/dt = 3$, we obtain

$$\frac{dV}{dt} = \frac{dV}{dr}\frac{dr}{dt} = 4\pi r^2(3) = 12\pi r^2$$

When $r = 10$,

$$\left.\frac{dV}{dt}\right|_{r=10} = (12\pi)10^2 = 1{,}200\pi \approx 3{,}770 \text{ cm}^3/\text{s} \qquad ■$$

We now discuss some important special cases of the Chain Rule that arise frequently.

COROLLARY 2 General Power and Exponential Rules If $u = g(x)$ is a differentiable function, then

- $\dfrac{d}{dx}g(x)^n = n(g(x))^{n-1}g'(x)$ (for any number n)

- $\dfrac{d}{dx}e^{g(x)} = g'(x)e^{g(x)}$

Proof To prove the General Power Rule, observe that $g(x)^n = f(g(x))$, where $f(u) = u^n$. Since $f'(u) = nu^{n-1}$, the Chain Rule yields

$$\frac{d}{dx}g(x)^n = f'(g(x))g'(x) = n(g(x))^{n-1}g'(x)$$

On the other hand, $e^{g(x)} = h(g(x))$, where $h(u) = e^u$. In this case, $h'(u) = h(u) = e^u$ and the Chain Rule yields

$$\frac{d}{dx}e^{g(x)} = h'(g(x))g'(x) = e^{g(x)}g'(x) = g'(x)e^{g(x)} \qquad ■$$

■ **EXAMPLE 4** Using the General Power Rule Find the derivatives:
(a) $y = (x^3 + 9x + 2)^{-1/3}$ and   **(b)** $y = \sec^4 t$.

Solution We use the General Power Rule.

(a) $$\frac{d}{dx}(x^3 + 9x + 2)^{-1/3} = -\frac{1}{3}(x^3 + 9x + 2)^{-4/3}\frac{d}{dx}(x^3 + 9x + 2)$$

$$= -\frac{1}{3}(x^3 + 9x + 2)^{-4/3}(3x^2 + 9)$$

$$= -(x^2 + 3)(x^3 + 9x + 2)^{-4/3}$$

(b) $$\frac{d}{dt}\sec^4 t = 4\sec^3 t\,\frac{d}{dt}\sec t = 4\sec^3 t\,(\sec t\,\tan t) = 4\sec^4 t\,\tan t \qquad \blacksquare$$

■ **EXAMPLE 5** Using the General Exponential Rule Differentiate **(a)** $f(x) = e^{9x}$ and **(b)** $f(x) = e^{\cos x}$.

Solution

(a) $\dfrac{d}{dx}\left(e^{9x}\right) = 9e^{9x}$
 (b) $\dfrac{d}{dx}\left(e^{\cos x}\right) = -(\sin x)e^{\cos x} \qquad \blacksquare$

The next special case of the Chain Rule deals with functions of the form $f(kx + b)$, where k and b are constants. Recall that the graph of $f(kx + b)$ is related to the graph of $f(x)$ by scaling and shifting (Section 1.1). To compute the derivative, we view $f(kx + b)$ as a composite function with inside function $u = kx + b$ and apply the Chain Rule:

$$\frac{d}{dx}f(kx + b) = f'(kx + b)\frac{d}{dx}(kx + b) = kf'(kx + b)$$

COROLLARY 3 Shifting and Scaling Rule If $f(x)$ is differentiable, then for any constants k and b,

$$\frac{d}{dx}f(kx + b) = k\,f'(kx + b)$$

In particular, $\dfrac{d}{dx}e^{kx+b} = ke^{kx+b}$.

For example,

$$\frac{d}{dx}\sin\left(2x + \frac{\pi}{4}\right) = 2\cos\left(2x + \frac{\pi}{4}\right)$$

$$\frac{d}{dx}(9x - 2)^5 = (9)(5)(9x - 2)^4 = 45(9x - 2)^4$$

$$\frac{d}{dt}e^{7-4t} = -4e^{7-4t}$$

According to our convention, $\sin x$ denotes the sine of x radians, and with this convention, the formula $(\sin x)' = \cos x$ holds. In the next example, we derive a formula for the derivative of the sine function when x is measured in degrees.

■ **EXAMPLE 6** Trigonometric Derivatives in Degrees Calculate the derivative of the sine function as a function of degrees rather than radians.

Solution It is convenient to underline a trigonometric function to indicate that the variable represents degrees rather than radians. For example,

$$\underline{\sin} x = \text{the sine of } x \text{ degrees}$$

A similar calculation shows that the factor $\pi/180$ appears in the formulas for the derivatives of the other standard trigonometric functions with respect to degrees. For example,

$$\frac{d}{dx}\underline{\tan}\,x = \left(\frac{\pi}{180}\right)\underline{\sec^2 x}$$

The functions $\sin x$ and $\underline{\sin}\,x$ are different, but they are related. In fact, for any number x, we have

$$\underline{\sin}\,x = \sin\left(\frac{\pi}{180}x\right)$$

since x degrees corresponds to $\frac{\pi}{180}x$ radians. By Corollary 3,

$$\frac{d}{dx}\underline{\sin}\,x = \frac{d}{dx}\sin\left(\frac{\pi}{180}x\right) = \left(\frac{\pi}{180}\right)\cos\left(\frac{\pi}{180}x\right) = \left(\frac{\pi}{180}\right)\underline{\cos x} \qquad \blacksquare$$

When the inside function is itself a composite function, we apply the Chain Rule more than once, as in the next example.

■ **EXAMPLE 7** Using the Chain Rule Twice Find the derivative of $y = \sqrt{x + \sqrt{x^2 + 1}}$.

Solution First apply the Chain Rule with inside function $u = x + \sqrt{x^2 + 1}$:

$$\frac{d}{dx}\sqrt{x + \sqrt{x^2 + 1}} = \frac{1}{2}(x + \sqrt{x^2 + 1})^{-1/2}\frac{d}{dx}\left(x + \sqrt{x^2 + 1}\right)$$

Then use the Chain Rule again to compute the remaining derivative:

$$\frac{d}{dx}\sqrt{x + \sqrt{x^2 + 1}} = \frac{1}{2}(x + \sqrt{x^2 + 1})^{-1/2}\left(1 + \frac{1}{2}(x^2 + 1)^{-1/2}\frac{d}{dx}(x^2 + 1)\right)$$

$$= \frac{1}{2}(x + \sqrt{x^2 + 1})^{-1/2}\left(1 + \frac{1}{2}(x^2 + 1)^{-1/2}(2x)\right)$$

$$= \frac{1}{2}(x + \sqrt{x^2 + 1})^{-1/2}\left(1 + x(x^2 + 1)^{-1/2}\right) \qquad \blacksquare$$

Proof of the Chain Rule The difference quotient for the composite $f \circ g$ is

$$\frac{f(g(x + h)) - f(g(x))}{h} \qquad (h \neq 0)$$

Our goal is to show that $(f \circ g)'$ is the product of $f'(g(x))$ and $g'(x)$, so it makes sense to write the difference quotient as a product:

$$\frac{f(g(x + h)) - f(g(x))}{h} = \frac{f(g(x + h)) - f(g(x))}{g(x + h) - g(x)} \times \frac{g(x + h) - g(x)}{h} \qquad \boxed{1}$$

This is legitimate only if the denominator $g(x + h) - g(x)$ is nonzero. Therefore, to continue our proof, we make the extra assumption that $g(x + h) - g(x) \neq 0$ for all h near but not equal to 0. This assumption is not necessary, but without it, the argument is more technical (see Exercise 103).

Under our assumption, we may use Eq. (1) to write $(f \circ g)'(x)$ as a product of two limits:

$$(f \circ g)'(x) = \underbrace{\lim_{h \to 0}\frac{f(g(x + h)) - f(g(x))}{g(x + h) - g(x)}}_{\text{Show that this equals } f'(g(x)).} \times \underbrace{\lim_{h \to 0}\frac{g(x + h) - g(x)}{h}}_{\text{This is } g'(x).} \qquad \boxed{2}$$

The second limit on the right is $g'(x)$. The Chain Rule will follow if we show that the first limit equals $f'(g(x))$. To verify this, set

$$k = g(x + h) - g(x)$$

Then $g(x + h) = g(x) + k$ and

$$\frac{f(g(x + h)) - f(g(x))}{g(x + h) - g(x)} = \frac{f(g(x) + k) - f(g(x))}{k}$$

Since $g(x)$ is differentiable, it is also continuous. Therefore, $g(x + h)$ tends to $g(x)$ and $k = g(x + h) - g(x)$ tends to zero as $h \to 0$. Thus, we may rewrite the limit in terms of k to obtain the desired result:

$$\lim_{h \to 0} \frac{f(g(x + h)) - f(g(x))}{g(x + h) - g(x)} = \lim_{k \to 0} \frac{f(g(x) + k) - f(g(x))}{k} = f'(g(x)) \quad \blacksquare$$

3.7 SUMMARY

- The Chain Rule expresses $(f \circ g)'$ in terms of f' and g':

$$(f(g(x)))' = f'(g(x))\, g'(x)$$

- In Leibniz notation: $\dfrac{dy}{dx} = \dfrac{dy}{du}\dfrac{du}{dx}$, where $y = f(u)$ and $u = g(x)$

- General Power Rule: $\dfrac{d}{dx} g(x)^n = n(g(x))^{n-1} g'(x)$.

- General Exponential Rule: $\dfrac{d}{dx} e^{g(x)} = g'(x) e^{g(x)}$.

- Shifting and Scaling Rule: $\dfrac{d}{dx} f(kx + b) = kf'(kx + b)$. In particular,

$$\frac{d}{dx} e^{kx+b} = k e^{kx+b}$$

3.7 EXERCISES

Preliminary Questions

1. Identify the outside and inside functions for each of these composite functions.

(a) $y = \sqrt{4x + 9x^2}$

(b) $y = \tan(x^2 + 1)$

(c) $y = \sec^5 x$

(d) $y = (1 + e^x)^4$

2. Which of the following can be differentiated easily *without* using the Chain Rule?

$$y = \tan(7x^2 + 2), \qquad y = \frac{x}{x + 1},$$

$$y = \sqrt{x} \cdot \sec x, \qquad y = \sqrt{x} \cos x,$$

$$y = xe^x, \qquad y = e^{\sin x}$$

3. Which is the derivative of $f(5x)$?

(a) $5f'(x)$ 　　　(b) $5f'(5x)$ 　　　(c) $f'(5x)$

4. How many times must the Chain Rule be used to differentiate each function?

(a) $y = \cos(x^2 + 1)$

(b) $y = \cos((x^2 + 1)^4)$

(c) $y = \sqrt{\cos((x^2 + 1)^4)}$

(d) $y = \tan(e^{5-6x})$

5. Suppose that $f'(4) = g(4) = g'(4) = 1$. Do we have enough information to compute $F'(4)$, where $F(x) = f(g(x))$? If not, what is missing?

Exercises

In Exercises 1–4, fill in a table of the following type:

$f(g(x))$	$f'(u)$	$f'(g(x))$	$g'(x)$	$(f \circ g)'$

1. $f(u) = u^{3/2}, \quad g(x) = x^4 + 1$

2. $f(u) = u^3, \quad g(x) = 3x + 5$

3. $f(u) = \tan u, \quad g(x) = x^4$

4. $f(u) = u^4 + u, \quad g(x) = \cos x$

In Exercises 5–6, write the function as a composite $f(g(x))$ and compute the derivative using the Chain Rule.

5. $y = (x + \sin x)^4$

6. $y = \cos(x^3)$

7. Calculate $\dfrac{d}{dx} \cos u$ for the following choices of $u(x)$:

(a) $u = 9 - x^2$ (b) $u = x^{-1}$ (c) $u = \tan x$

8. Calculate $\dfrac{d}{dx} f(x^2 + 1)$ for the following choices of $f(u)$:

(a) $f(u) = \sin u$ (b) $f(u) = 3u^{3/2}$
(c) $f(u) = u^2 - u$

In Exercises 9–14, use the General Power Rule or the Shifting and Scaling Rule to find the derivative.

9. $y = (x^2 + 9)^4$

10. $y = \sin^5 x$

11. $y = \sqrt{11x + 4}$

12. $y = (7x - 9)^5$

13. $y = e^{10 - x^2}$

14. $y = (x^4 - x + 2)^{-3/2}$

In Exercises 15–18, find the derivative of $f \circ g$.

15. $f(u) = \sin u, \quad g(x) = 2x + 1$

16. $f(u) = 2u + 1, \quad g(x) = \sin x$

17. $f(u) = e^u, \quad g(x) = x + x^{-1}$

18. $f(u) = \dfrac{u}{u - 1}, \quad g(x) = \csc x$

In Exercises 19–20, find the derivatives of $f(g(x))$ and $g(f(x))$.

19. $f(u) = \cos u, \quad g(x) = x^2 + 1$

20. $f(u) = u^3, \quad g(x) = \dfrac{1}{x + 1}$

In Exercises 21–34, use the Chain Rule to find the derivative.

21. $y = \sin(x^2)$

22. $y = \sin^2 x$

23. $y = \cot(4t^2 + 9)$

24. $y = \sqrt{1 - t^2}$

25. $y = (t^2 + 3t + 1)^{-5/2}$

26. $y = (x^4 - x^3 - 1)^{2/3}$

27. $y = \left(\dfrac{x + 1}{x - 1}\right)^4$

28. $y = \sec \dfrac{1}{x}$

29. $y = \cos^3(e^{4\theta})$

30. $y = \tan(\theta + \cos \theta)$

31. $y = \dfrac{1}{\sqrt{\cos(x^2) + 1}}$

32. $y = (\sqrt{x + 1} - 1)^{3/2}$

33. $y = e^{1/x}$

34. $y = e^{x^2}$

In Exercises 35–72, find the derivative using the appropriate rule or combination of rules.

35. $y = \tan 5x$

36. $y = \sin(x^2 + 4x)$

37. $y = x \cos(1 - 3x)$

38. $y = \sin(x^2) \cos(x^2)$

39. $y = (4t + 9)^{1/2}$

40. $y = \sin(\cos \theta)$

41. $y = (x^3 + \cos x)^{-4}$

42. $y = \sin(\cos(\sin x))$

43. $y = \sqrt{\sin x \cos x}$

44. $y = x^2 \tan 2x$

45. $y = (z + 1)^4 (2z - 1)^3$

46. $y = 3 + 2s\sqrt{s}$

47. $y = (x + x^{-1})\sqrt{x + 1}$

48. $y = \cos^2(8x)$

49. $y = (\cos 6x + \sin x^2)^{1/2}$

50. $y = \dfrac{(x + 1)^{1/2}}{x + 2}$

51. $y = \tan^3 x + \tan(x^3)$

52. $y = \sqrt{4 - 3\cos x}$

53. $y = \sqrt{\dfrac{z + 1}{z - 1}}$

54. $y = (\cos^3 x + 3\cos x + 7)^9$

55. $y = \dfrac{\cos(1 + x)}{1 + \cos x}$

56. $y = \sec(\sqrt{t^2 - 9})$

57. $y = \cot^7(x^5)$

58. $y = \dfrac{\cos(x^2)}{1 + x^2}$

59. $y = (1 + (x^2 + 2)^5)^3$

60. $y = \left(1 + \cot^5(x^4 + 1)\right)^9$

61. $y = 4e^{-x} + 7e^{-2x}$

62. $y = e^{\tan \theta}$

63. $y = (2e^{3x} + 3e^{-2x})^4$

64. $y = \dfrac{1}{1 - e^{-3t}}$

65. $y = \cos(te^{-2t})$

66. $y = \tan(e^{5 - 6x})$

67. $y = e^{(x^2 + 2x + 3)^2}$

68. $y = e^{e^x}$

69. $y = \sqrt{1 + \sqrt{1 + \sqrt{x}}}$

70. $y = \sqrt{\sqrt{x + 1} + 1}$

71. $y = \sqrt{kx + b};$ k and b any constants

72. $y = \dfrac{1}{\sqrt{kt^4 + b}};$ k, b constants, not both zero

73. Compute $\dfrac{df}{dx}$ if $\dfrac{df}{du} = 2$ and $\dfrac{du}{dx} = 6.$

74. The average molecular velocity v of a gas in a certain container is given by $v = 29\sqrt{T}$ m/s, where T is the temperature in kelvins. The temperature is related to the pressure (in atmospheres) by $T = 200P$. Find $\dfrac{dv}{dP}\bigg|_{P=1.5}$.

75. With notation as in Example 6, calculate

(a) $\dfrac{d}{d\theta} \sin \theta \bigg|_{\theta=60°}$

(b) $\dfrac{d}{d\theta} \left(\theta + \tan \theta\right) \bigg|_{\theta=45°}$

76. Assume that $f(0) = 2$ and $f'(0) = 3$. Find the derivatives of $(f(x))^3$ and $f(7x)$ at $x = 0$.

77. Compute the derivative of $h(\sin x)$ at $x = \frac{\pi}{6}$, assuming that $h'(0.5) = 10$.

78. Let $F(x) = f(g(x))$, where the graphs of f and g are shown in Figure 1. Estimate $g'(2)$ and $f'(g(2))$ from the graph and compute $F'(2)$.

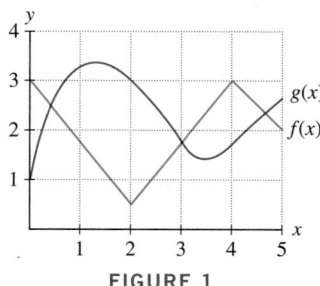

FIGURE 1

In Exercises 79–82, use the table of values to calculate the derivative of the function at the given point.

x	1	4	6
$f(x)$	4	0	6
$f'(x)$	5	7	4
$g(x)$	4	1	6
$g'(x)$	5	$\frac{1}{2}$	3

79. $f(g(x))$, $x = 6$

80. $e^{f(x)}$, $x = 4$

81. $g(\sqrt{x})$, $x = 16$

82. $f(2x + g(x))$, $x = 1$

In Exercises 83–86, compute the indicated higher derivatives.

83. $\dfrac{d^2}{dx^2} \sin(x^2)$

84. $\dfrac{d^2}{dx^2} (x^2 + 9)^5$

85. $\dfrac{d^3}{dx^3} (3x + 9)^{11}$

86. $\dfrac{d^3}{dx^3} \sin(2x)$

87. ⊂A⊆ Use a computer algebra system to compute $f^{(k)}(x)$ for $k = 1, 2, 3$ for the following functions:

(a) $f(x) = \cot(x^2)$

(b) $f(x) = \sqrt{x^3 + 1}$

88. Use the Chain Rule to express the second derivative of $f \circ g$ in terms of the first and second derivatives of f and g.

89. Compute the second derivative of $\sin(g(x))$ at $x = 2$, assuming that $g(2) = \frac{\pi}{4}$, $g'(2) = 5$, and $g''(2) = 3$.

90. An expanding sphere has radius $r = 0.4t$ cm at time t (in seconds). Let V be the sphere's volume. Find dV/dt when (a) $r = 3$ and (b) $t = 3$.

91. The power P in a circuit is $P = Ri^2$, where R is resistance and i the current. Find dP/dt at $t = 2$ if $R = 1{,}000\ \Omega$ and i varies according to $i = \sin(4\pi t)$ (time in seconds).

92. The price (in dollars) of a computer component is $P = 2C - 18C^{-1}$, where C is the manufacturer's cost to produce it. Assume that cost at time t (in years) is $C = 9 + 3t^{-1}$ and determine the ROC of price with respect to time at $t = 3$.

93. The force F (in Newtons) between two charged objects is $F = 100/r^2$, where r is the distance (in meters) between them. Find dF/dt at $t = 10$ if the distance at time t (in seconds) is $r = 1 + 0.4t^2$.

94. According to the U.S. standard atmospheric model, developed by the National Oceanic and Atmospheric Administration for use in aircraft and rocket design, atmospheric temperature T (in degrees Celsius), pressure P (kPa $= 1{,}000$ Pascals), and altitude h (meters) are related by the formulas (valid in the troposphere $h \leq 11{,}000$):

$$T = 15.04 - 0.000649h, \qquad P = 101.29 + \left(\frac{T + 273.1}{288.08}\right)^{5.256}$$

Calculate dP/dh. Then estimate the change in P (in Pascals, Pa) per additional meter of altitude when $h = 3{,}000$.

95. Conservation of Energy   The position at time t (in seconds) of a weight of mass m oscillating at the end of a spring is $x(t) = L \sin(2\pi\omega t)$. Here, L is the maximum length of the spring, and ω the frequency (number of oscillations per second). Let v and a be the velocity and acceleration of the weight.

(a) By Hooke's Law, the spring exerts a force of magnitude $F = -kx$ on the weight, where k is the *spring constant*. Use Newton's Second Law, $F = ma$, to show that $2\pi\omega = \sqrt{k/m}$.

(b) The weight has kinetic energy $K = \frac{1}{2}mv^2$ and potential energy $U = \frac{1}{2}kx^2$. Prove that the total energy $E = K + U$ is conserved, that is, $dE/dt = 0$.

Further Insights and Challenges

96. Show that if f, g, and h are differentiable, then

$$[f(g(h(x)))]' = f'(g(h(x)))g'(h(x))h'(x)$$

97.   Recall that $f(x)$ is *even* if $f(-x) = f(x)$ and *odd* if $f(-x) = -f(x)$.

(a) Sketch a graph of any even function and explain graphically why its derivative is an odd function.

(b) Show that differentiation reverses parity: If f is even, then f' is odd, and if f is odd, then f' is even. *Hint:* Differentiate $f(-x)$.

(c) Suppose that f' is even. Is f necessarily odd? *Hint:* Check if this is true for linear functions.

98. Power Rule for Fractional Exponents Let $f(u) = u^q$ and $g(x) = x^{p/q}$. Show that $f(g(x)) = x^p$ (recall the laws of exponents). Apply the Chain Rule and the Power Rule for integer exponents to show that $f'(g(x))\,g'(x) = px^{p-1}$. Then derive the Power Rule for $x^{p/q}$.

In Exercises 99–101, use the following fact (proved in Chapter 4): If a differentiable function f satisfies $f'(x) = 0$ for all x, then f is a constant function.

99. Differential Equation of Sine and Cosine Suppose that $f(x)$ satisfies the following equation (called a **differential equation**):

$$f''(x) = -f(x) \qquad \boxed{3}$$

(a) Show that $f(x)^2 + f'(x)^2 = f(0)^2 + f'(0)^2$. *Hint:* Show that the function on the left has zero derivative.

(b) Verify that $\sin x$ and $\cos x$ satisfy Eq. (3) and deduce that $\sin^2 x + \cos^2 x = 1$.

100. Suppose that both f and g satisfy Eq. (3) and have the same initial values, that is, $f(0) = g(0)$ and $f'(0) = g'(0)$. Show that $f(x) = g(x)$ for all x. *Hint:* Show that $f - g$ satisfies Eq. (3) and apply Exercise 99(a).

101. Use the result of Exercise 100 to show the following: $f(x) = \sin x$ is the unique solution of Eq. (3) such that $f(0) = 0$ and $f'(0) = 1$; and $g(x) = \cos x$ is the unique solution such that $g(0) = 1$ and $g'(0) = 0$. This result provides a means of defining and developing all the properties of the trigonometric functions without reference to triangles or the unit circle.

102. A Discontinuous Derivative Use the limit definition to show that $g'(0)$ exists but $g'(0) \neq \lim_{x \to 0} g'(x)$, where

$$g(x) = \begin{cases} x^2 \sin \dfrac{1}{x} & \text{if } x \neq 0 \\ 0 & \text{if } x = 0 \end{cases}$$

103. Chain Rule This exercise proves the Chain Rule without the special assumption made in the text. For any number b, define a new function

$$F(u) = \frac{f(u) - f(b)}{u - b} \qquad \text{for all } u \neq b$$

Observe that $F(u)$ is equal to the slope of the secant line through $(b, f(b))$ and $(u, f(u))$.

(a) Show that if we define $F(b) = f'(b)$, then $F(u)$ is continuous at $u = b$.

(b) Take $b = g(a)$. Show that if $x \neq a$, then for all u,

$$\frac{f(u) - f(g(a))}{x - a} = F(u)\frac{u - g(a)}{x - a} \qquad \boxed{4}$$

Note that both sides are zero if $u = g(a)$.

(c) Substitute $u = g(x)$ in Eq. (4) to obtain

$$\frac{f(g(x)) - f(g(a))}{x - a} = F(g(x))\frac{g(x) - g(a)}{x - a}$$

Derive the Chain Rule by computing the limit of both sides as $x \to a$.

3.8 Implicit Differentiation

To differentiate using the methods covered thus far, we must have a formula for y in terms of x, for instance, $y = x^3 + 1$. But suppose that y is determined instead by an equation such as

$$y^4 + xy = x^3 - x + 2 \qquad \boxed{1}$$

In this case, we say that y is defined *implicitly*. How can we find the slope of the tangent line at a point on the graph (Figure 1)? It may be inconvenient or even impossible to solve for y explicitly as a function of x, but we can find dy/dx using the method of **implicit differentiation**.

To illustrate, consider the equation of the unit circle:

$$x^2 + y^2 = 1$$

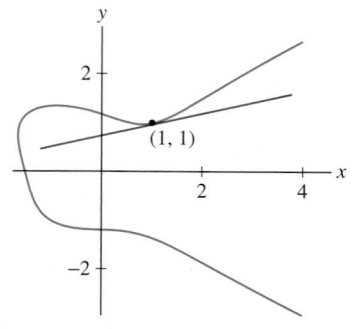

FIGURE 1 Graph of
$$y^4 + xy = x^3 - x + 2$$

To compute $\dfrac{dy}{dx}$, first take the derivative of both sides of the equation and evaluate:

$$\frac{d}{dx}\left(x^2 + y^2\right) = \frac{d}{dx}(1)$$

$$\frac{d}{dx}\left(x^2\right) + \frac{d}{dx}\left(y^2\right) = 0$$

$$2x + \frac{d}{dx}\left(y^2\right) = 0 \qquad \boxed{2}$$

What should we do with the term $\dfrac{d}{dx}(y^2)$? We use the Chain Rule. If we think of y as a function $y = f(x)$, then $y^2 = f(x)^2$, and by the Chain Rule,

$$\frac{d}{dx}y^2 = \frac{d}{dx}f(x)^2 = 2f(x)\frac{df}{dx} = 2y\frac{dy}{dx}$$

Equation (2) becomes $2x + 2y\dfrac{dy}{dx} = 0$, and we may solve for $\dfrac{dy}{dx}$ if $y \neq 0$:

$$\boxed{\frac{dy}{dx} = -\frac{x}{y}} \qquad \boxed{3}$$

■ **EXAMPLE 1** Use Eq. (3) to find the slope of the tangent line at the point $P = \left(\frac{3}{5}, \frac{4}{5}\right)$ on the unit circle (Figure 2).

Solution Set $x = \frac{3}{5}$ and $y = \frac{4}{5}$ in Eq. (3):

$$\frac{dy}{dx}\bigg|_P = -\frac{x}{y} = -\frac{\frac{3}{5}}{\frac{4}{5}} = -\frac{3}{4} \qquad \blacksquare$$

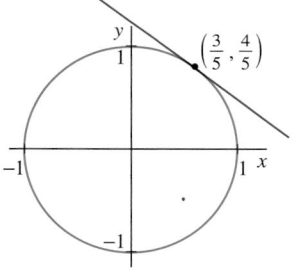

FIGURE 2 Tangent line to unit circle at $\left(\frac{3}{5}, \frac{4}{5}\right)$ has slope $-\frac{3}{4}$.

In the previous example, we could have computed dy/dx directly without using implicit differentitation. The upper semicircle is the graph of $y = \sqrt{1 - x^2}$ and

$$\frac{dy}{dx} = \frac{d}{dx}\sqrt{1 - x^2} = \frac{1}{2}\left(1 - x^2\right)^{-1/2}\frac{d}{dx}\left(1 - x^2\right)$$

$$= \frac{1}{2}\left(1 - x^2\right)^{-1/2}(-2x) = -\frac{x}{\sqrt{1 - x^2}}$$

This is a formula for the derivative in terms of x alone, whereas Eq. (3) expresses the derivative in terms of both x and y, as is typical when we use implicit differentiation. The two formulas agree because $y = \sqrt{1 - x^2}$. Although it was possible to solve for y in terms of x in this particular case, it is often difficult or impossible to do so. In such cases, implicit differentiation is the only available method for computing the derivative.

Before presenting additional examples, let's pause to examine again how the factor dy/dx arises when you differentiate an expression involving y with respect to x. It would not appear if you were differentiating with respect to y. Thus,

Notice what happens if we insist on applying the Chain Rule to $\dfrac{d}{dy}\sin y$. The extra factor appears but it is equal to 1:

$$\frac{d}{dy}\sin y = (\cos y)\frac{dy}{dy} = \cos y$$

$$\frac{d}{dy}\sin y = \cos y \qquad \text{but} \qquad \frac{d}{dx}\sin y = (\cos y)\frac{dy}{dx}$$

Similarly, we may apply the Product Rule to xy to obtain (where y' denotes dy/dx):

$$\frac{d}{dx}(xy) = x\frac{dy}{dx} + y\frac{dx}{dx} = x\frac{dy}{dx} + y = xy' + y$$

or we may apply the Quotient Rule to t/y (where now y' denotes dy/dt):

$$\frac{d}{dt}\left(\frac{t}{y}\right) = \frac{y\left(\dfrac{dt}{dt}\right) - t\left(\dfrac{dy}{dt}\right)}{y^2} = \frac{y - ty'}{y^2}$$

■ **EXAMPLE 2** Find an equation of the tangent line at the point $P = (1, 1)$ on the curve (Figure 1)

$$y^4 + xy = x^3 - x + 2$$

Solution We break up the derivative calculation into two steps.

Step 1. **Differentiate both sides of the equation with respect to x.**

$$\frac{d}{dx}y^4 + \frac{d}{dx}(xy) = \frac{d}{dx}(x^3 - x + 2)$$

$$4y^3\,y' + (xy' + y) = 3x^2 - 1 \qquad\boxed{4}$$

Step 2. **Solve for y'.**
Group the terms in Eq. (4) involving y' on the left-hand side and place the remaining terms on the right-hand side:

$$4y^3y' + xy' = 3x^2 - 1 - y$$

Then factor out y':

$$(4y^3 + x)y' = 3x^2 - 1 - y$$

and divide:

$$y' = \frac{3x^2 - 1 - y}{4y^3 + x} \qquad\boxed{5}$$

To find the derivative at $P = (1, 1)$, we apply Eq. (5) with $x = 1$ and $y = 1$:

$$\left.\frac{dy}{dx}\right|_{(1,1)} = \frac{3 \cdot 1^2 - 1 - 1}{4 \cdot 1^3 + 1} = \frac{1}{5}$$

The equation of the tangent line may be written $y - 1 = \frac{1}{5}(x - 1)$ or $y = \frac{1}{5}x + \frac{4}{5}$. ■

CONCEPTUAL INSIGHT   The graph of an equation need not define a function because there may be more than one y-value for a given value of x. However, the graph is generally made up of several pieces called **branches**, each of which does define a function. For example, the branches of the unit circle $x^2 + y^2 = 1$ are the graphs of the functions $y = \sqrt{1 - x^2}$ and $y = -\sqrt{1 - x^2}$. Similarly, the graph in Figure 3 has upper and lower branches. We have not discussed the precise conditions under which these branches are differentiable—a more advanced topic that lies beyond the scope of this text. But in the examples we present, it will be clear that y' exists either for all x or for all but a certain number of exceptional values of x that are identifiable from the graph.

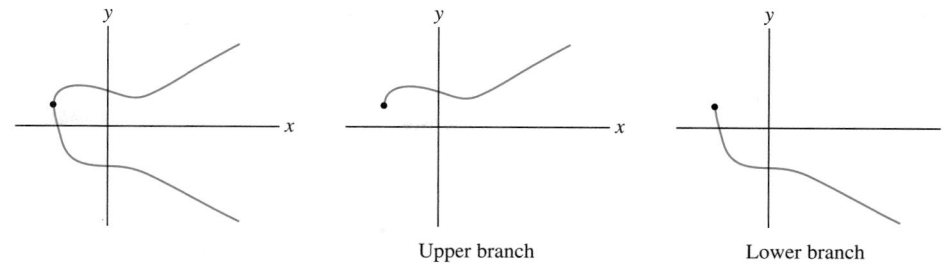

FIGURE 3 Each branch of the graph of $y^4 + xy = x^3 - x + 2$ defines a function of x.

Upper branch   Lower branch

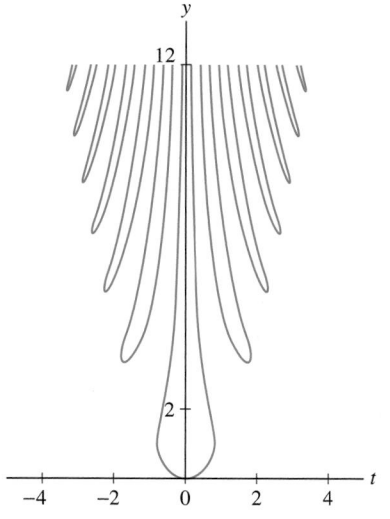

FIGURE 4 Graph of $\cos(ty) = \dfrac{t^2}{y}$.

■ **EXAMPLE 3** Find $\dfrac{dy}{dt}$, where $\cos(ty) = \dfrac{t^2}{y}$ (Figure 4).

Solution The two steps used in the previous example also apply here.

Step 1. Differentiate both sides of the equation with respect to t.
We obtain

$$\frac{d}{dt}\cos(ty) = \frac{d}{dt}\left(\frac{t^2}{y}\right) \qquad \boxed{6}$$

We calculate each derivative separately, writing y' for dy/dt:

$$\frac{d}{dt}(\cos(ty)) = -\big(\sin(ty)\big)\frac{d}{dt}(ty) = -\sin(ty)\big(ty' + y\big) \quad \text{(Chain Rule)}$$

$$\frac{d}{dt}\left(\frac{t^2}{y}\right) = \frac{y\dfrac{d}{dt}t^2 - t^2\dfrac{dy}{dt}}{y^2} = \frac{2ty - t^2 y'}{y^2} \qquad \text{(Quotient Rule)}$$

Equation (6) becomes

$$-\big(\sin(ty)\big)\big(ty' + y\big) = \frac{2ty - t^2 y'}{y^2} \qquad \boxed{7}$$

Step 2. Solve for y'.

$$-t\big(\sin(ty)\big)\,y' - y\sin(ty) = \frac{2t}{y} - \frac{t^2}{y^2}\,y' \qquad \text{[expand Eq. (7)]}$$

$$-t\sin(ty)\,y' + \frac{t^2}{y^2}\,y' = \frac{2t}{y} + y\sin(ty) \quad \text{(place all y' terms on left)}$$

$$y'\left(-t\sin(ty) + \frac{t^2}{y^2}\right) = \frac{2t}{y} + y\sin(ty) \quad \text{(factor out y')}$$

Now we can solve for y':

$$y' = \frac{\dfrac{2t}{y} + y\sin(ty)}{-t\sin(ty) + \dfrac{t^2}{y^2}} = \frac{2ty + y^3\sin(ty)}{t^2 - ty^2\sin(ty)} \qquad ■$$

EXAMPLE 4 Find the slope of the tangent line to the curve $e^{xy} = x + y$ at the point $P = (-1, 1.28)$ (the y-coordinate is approximate).

Solution We use implicit differentiation (Figure 5):

$$\frac{d}{dx}e^{xy} = \frac{d}{dx}(x + y)$$

$$(xy' + y)e^{xy} = 1 + y'$$

$$(xe^{xy} - 1)y' = 1 - ye^{xy}$$

$$y' = \frac{1 - ye^{xy}}{xe^{xy} - 1}$$

The slope of the tangent line at $P = (-1, 1.28)$ is (approximately)

$$\left.\frac{dy}{dx}\right|_{(-1,1.28)} \approx \frac{1 - (1.28)e^{-1.28}}{e^{-1.28} - 1} \approx -0.89 \qquad \blacksquare$$

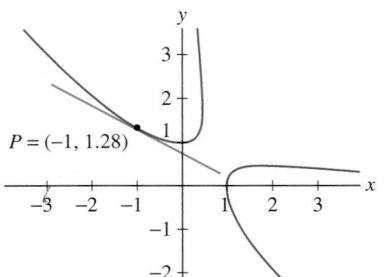

FIGURE 5 Graph of $e^{xy} = x + y$.

3.8 SUMMARY

- Implicit differentiation is used to compute dy/dx when x and y are related by an equation.

Step 1. Take the derivative of both sides of the equation with respect to x.

Step 2. Solve for y' by collecting the terms involving y' on one side and the remaining terms on the other side of the equation.

- Remember to include the factor dy/dx when differentiating expressions involving y with respect to x. For instance,

$$\frac{d}{dx}\sin y = (\cos y)\frac{dy}{dx}$$

3.8 EXERCISES

Preliminary Questions

1. Which differentiation rule is used to show $\dfrac{d}{dx}\sin y = \cos y\,\dfrac{dy}{dx}$?

2. One of (a)–(c) is incorrect. Find the mistake and correct it.

(a) $\dfrac{d}{dy}\sin(y^2) = 2y\cos(y^2)$

(b) $\dfrac{d}{dx}\sin(x^2) = 2x\cos(x^2)$

(c) $\dfrac{d}{dx}\sin(y^2) = 2y\cos(y^2)$

3. On an exam, Jason was asked to differentiate the equation

$$x^2 + 2xy + y^3 = 7$$

What are the errors in Jason's answer?

$$2x + 2xy' + 3y^2 = 0$$

4. Which of (a) or (b) is equal to $\dfrac{d}{dx}(x\sin t)$?

(a) $(x\cos t)\dfrac{dt}{dx}$

(b) $(x\cos t)\dfrac{dt}{dx} + \sin t$

Exercises

1. Show that if you differentiate both sides of $x^2 + 2y^3 = 6$, the result is $2x + 6y^2 y' = 0$. Then solve for y' and calculate dy/dx at the point $(2, 1)$.

2. Show that if you differentiate both sides of the equation $xy + 3x + y = 1$, the result is $(x + 1)y' + y + 3 = 0$. Then solve for y' and calculate dy/dx at the point $(1, -1)$.

In Exercises 3–8, differentiate the expression with respect to x.

3. $x^2 y^3$

4. $\dfrac{x^3}{y}$

5. $(x^2 + y^2)^{3/2}$

6. $\tan(xt)$

7. $z + z^2$

8. $e^{y/x}$

In Exercises 9–26, calculate the derivative of y (or other variable) with respect to x.

9. $3y^3 + x^2 = 5$

10. $y^4 - y = x^3 + x$

11. $x^2 y + 2xy^2 = x + y$

12. $\sin(xt) = t$

13. $x^2 y + y^4 - 3x = 8$

14. $x^3 R^5 = 1$

15. $x^4 + z^4 = 1$

16. $\dfrac{y}{x} + \dfrac{x}{y} = 2y$

17. $y^{-3/2} + x^{3/2} = 1$

18. $x^{1/2} + y^{2/3} = x + y$

19. $\sqrt{x+s} = \dfrac{1}{x} + \dfrac{1}{s}$

20. $\sqrt{x+y} + \sqrt{y} = 2x$

21. $y + \dfrac{x}{y} = 1$

22. $\dfrac{x}{y} + \dfrac{y^2}{x+1} = 0$

23. $\sin(x + y) = x + \cos y$

24. $\tan(x^2 y) = x + y$

25. $xe^y = 2xy + y^3$

26. $e^{xy} = x^2 + y^2$

In Exercises 27–28, find dy/dx at the given point.

27. $(x + 2)^2 - 6(2y + 3)^2 = 3,\quad (1, -1)$

28. $\sin^2(3y) = x + y,\quad \left(\dfrac{2 - \pi}{4}, \dfrac{\pi}{4}\right)$

In Exercises 29–34, find an equation of the tangent line at the given point.

29. $xy - 2y = 1,\quad (3, 1)$

30. $x^2 y^3 + 2y = 3x,\quad (2, 1)$

31. $x^{2/3} + y^{2/3} = 2,\quad (1, 1)$

32. $ye^x + xe^y = 4,\quad (4, 0)$

33. $x^{1/2} + y^{-1/2} = 2xy,\quad (1, 1)$

34. $ye^{x^2 - 4} + y^2 = 6,\quad (2, 2)$

35. Find the points on the graph of $y^2 = x^3 - 3x + 1$ (Figure 6) where the tangent line is horizontal.

(a) First show that $2yy' = 3x^2 - 3$, where $y' = dy/dx$.

(b) Do not solve for y'. Rather, set $y' = 0$ and solve for x. This gives two possible values of x where the slope may be zero.

(c) Show that the positive value of x does not correspond to a point on the graph.

(d) The negative value corresponds to the two points on the graph where the tangent line is horizontal. Find the coordinates of these two points.

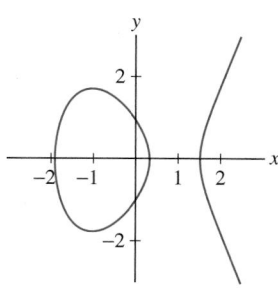

FIGURE 6 Graph of $y^2 = x^3 - 3x + 1$.

36. Find all points on the graph of $3x^2 + 4y^2 + 3xy = 24$ where the tangent line is horizontal (Figure 7).

(a) By differentiating the equation of the curve implicitly and setting $y' = 0$, show that if the tangent line is horizontal at (x, y), then $y = -2x$.

(b) Solve for x by substituting $y = -2x$ in the equation of the curve.

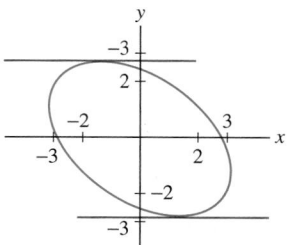

FIGURE 7 Graph of $3x^2 + 4y^2 + 3xy = 24$.

37. Show that no point on the graph of $x^2 - 3xy + y^2 = 1$ has a horizontal tangent line.

38. Figure 1 shows the graph of $y^4 + xy = x^3 - x + 2$. Find dy/dx at the two points on the graph with x-coordinate 0 and find an equation of the tangent line at $(1, 1)$.

39. If the derivative dx/dy exists at a point and $dx/dy = 0$, then the tangent line is vertical. Calculate dx/dy for the equation $y^4 + 1 = y^2 + x^2$ and find the points on the graph where the tangent line is vertical.

40. Differentiate the equation $xy = 1$ with respect to the variable t and derive the relation $\dfrac{dy}{dt} = -\dfrac{y}{x}\dfrac{dx}{dt}$.

41. Differentiate the equation $x^3 + 3xy^2 = 1$ with respect to the variable t and express dy/dt in terms of dx/dt, as in Exercise 40.

In Exercises 42–43, differentiate the equation with respect to t to calculate dy/dt in terms of dx/dt.

42. $x^2 - y^2 = 1$

43. $y^4 + 2xy + x^2 = 0$

44. The volume V and pressure P of gas in a piston (which vary in time t) satisfy $PV^{3/2} = C$, where C is a constant. Prove that

$$\frac{dP/dt}{dV/dt} = -\frac{3}{2}\frac{P}{V}$$

The ratio of the derivatives is negative. Could you have predicted this from the relation $PV^{3/2} = C$?

45. The **folium of Descartes** is the curve with equation $x^3 + y^3 = 3xy$ (Figure 8). It was first discussed in 1638 by the French philosopher-mathematician René Descartes. The name "folium" means leaf. Descartes's scientific colleague Gilles de Roberval called it the "jasmine flower." Both men believed incorrectly that the leaf shape in the first quadrant was repeated in each quadrant, giving the appearance of petals of a flower. Find an equation of the tangent line to this curve at the point $\left(\frac{2}{3}, \frac{4}{3}\right)$.

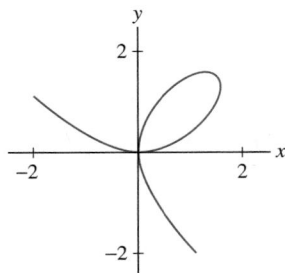

FIGURE 8 Folium of Descartes: $x^3 + y^3 = 3xy$.

46. Find all points on the folium $x^3 + y^3 = 3xy$ at which the tangent line is horizontal.

47. [GU] Plot the equation $x^3 + y^3 = 3xy + b$ for several values of b.

(a) Describe how the graph changes as $b \to 0$.

(b) Compute dy/dx (in terms of b) at the point on the graph where $y = 0$. How does this value change as $b \to \infty$? Do your plots confirm this conclusion?

48. Show that the tangent lines at $x = 1 \pm \sqrt{2}$ to the so-called *conchoid* (Figure 9) with equation $(x - 1)^2(x^2 + y^2) = 2x^2$ are vertical.

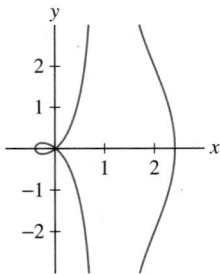

FIGURE 9 Conchoid: $(x - 1)^2(x^2 + y^2) = 2x^2$.

49. The equation $xy = x^3 - 5x^2 + 2x - 1$ defines a *trident curve* (Figure 10), named by Isaac Newton in his treatise on curves published in 1710. Find the points where the tangent to the trident is horizontal as follows.

(a) Show that $xy' + y = 3x^2 - 10x + 2$.

(b) Set $y' = 0$ in (a), replace y by $x^{-1}(x^3 - 5x^2 + 2x - 1)$, and solve the resulting equation for x.

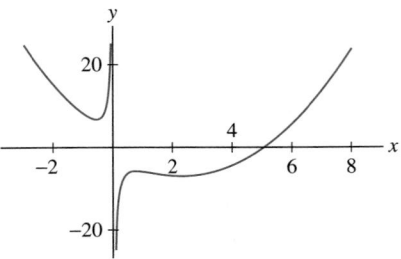

FIGURE 10 Trident curve: $xy = x^3 - 5x^2 + 2x - 1$.

50. Find an equation of the tangent line at the four points on the curve $(x^2 + y^2 - 4x)^2 = 2(x^2 + y^2)$, where $x = 1$. This curve (Figure 11) is an example of a *limaçon of Pascal*, named after the father of the French philosopher Blaise Pascal, who first described it in 1650.

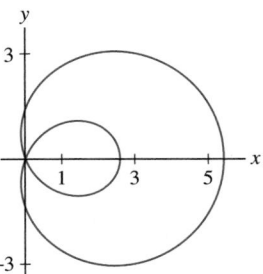

FIGURE 11 Limaçon: $(x^2 + y^2 - 4x)^2 = 2(x^2 + y^2)$.

51. Find equations of the tangent lines at the points where $x = 1$ on the so-called folium (Figure 12):

$$(x^2 + y^2)^2 = \frac{25}{4}xy^2$$

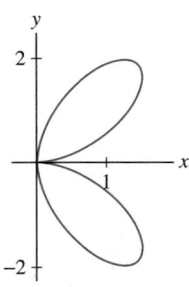

FIGURE 12

52. [CAS] Use a computer algebra system to plot $(x^2 + y^2)^2 = 12(x^2 - y^2) + 2$ for $-4 \le x, y \le 4$. How many horizontal tangent lines does the curve appear to have? Find the points where these occur.

53. *CAS*   Use a computer algebra system to plot $y^2 = x^3 - 4x$ for $-4 \le x, y \le 4$. Show that if $dx/dy = 0$, then $y = 0$. Conclude that the tangent line is vertical at the points where the curve intersects the x-axis. Does your plot confirm this conclusion?

In Exercises 54–57, use implicit differentiation to calculate higher derivatives.

54. Consider the equation $y^3 - \frac{3}{2}x^2 = 1$.

(a) Show that $y' = x/y^2$ and differentiate again to show that

$$y'' = \frac{y^2 - 2xyy'}{y^4}$$

(b) Express y'' in terms of x and y using part (a).

55. Use the method of the previous exercise to show that $y'' = -y^{-3}$ if $x^2 + y^2 = 1$.

56. Calculate y'' at the point $(1, 1)$ on the curve $xy^2 + y - 2 = 0$ by the following steps:

(a) Find y' by implicit differentiation and calculate y' at the point $(1, 1)$.

(b) Differentiate the expression for y' found in (a) and then compute y'' at $(1, 1)$ by substituting $x = 1$, $y = 1$, and the value of y' found in (a).

57. Use the method of the previous exercise to compute y'' at the point $\left(\frac{2}{3}, \frac{4}{3}\right)$ on the folium of Descartes with the equation $x^3 + y^3 = 3xy$.

Further Insights and Challenges

58. Show that if P lies on the intersection of the two curves $x^2 - y^2 = c$ and $xy = d$ (c, d constants), then the tangents to the curves at P are perpendicular.

59. The *lemniscate curve* $(x^2 + y^2)^2 = 4(x^2 - y^2)$ was discovered by Jacob Bernoulli in 1694, who noted that it is "shaped like a figure

8, or a knot, or the bow of a ribbon." Find the coordinates of the four points at which the tangent line is horizontal (Figure 13).

60. Divide the curve in Figure 14 into five branches, each of which is the graph of a function. Sketch the branches.

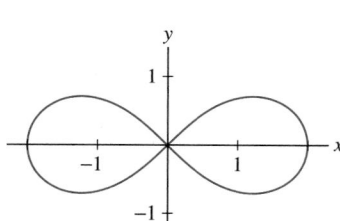

FIGURE 13 Lemniscate curve: $(x^2 + y^2)^2 = 4(x^2 - y^2)$.

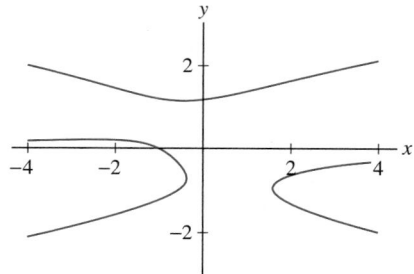

FIGURE 14 Graph of $y^5 - y = x^2y + x + 1$.

3.9 Derivatives of Inverse Functions

In this section, we begin with an important formula for the derivative of the inverse $f^{-1}(x)$ of a function $f(x)$. We then apply this formula to determine the derivatives of the inverse trigonometric functions. In the next section, we will use the formula to compute the derivatives of logarithmic functions.

THEOREM 1 Derivative of the Inverse   Assume that $f(x)$ is differentiable and one-to-one with inverse $g(x) = f^{-1}(x)$. If b belongs to the domain of $g(x)$ and $f'(g(b)) \neq 0$, then $g'(b)$ exists and

$$g'(b) = \frac{1}{f'(g(b))}$$

$\boxed{1}$

Derivative of the Inverse: *If* $g(x) = f^{-1}(x)$ *is the inverse of* $f(x)$, *then*

$$g'(x) = \frac{1}{f'(g(x))}$$

CONCEPTUAL INSIGHT   A proof of Theorem 1 is given in Appendix D, but we can explain the result graphically as follows. Recall that the graph of the inverse $g(x)$ is obtained by reflecting the graph of $f(x)$ through the line $y = x$. Now, consider a line L of slope m and let L' be its reflection through $y = x$ as in Figure 1(A). Then the slope of L' is $1/m$. Indeed, if (a, b) and (c, d) are any two points on L, then (b, a) and (d, c) lie on L' and

$$\underbrace{\text{Slope of } L = \frac{b - d}{a - c}, \qquad \text{Slope of } L' = \frac{a - c}{b - d}}_{\text{Reciprocal slopes}}$$

Figure 1(B) tells the rest of the story. Let $g(x) = f^{-1}(x)$. The reflection of the tangent line to $y = f(x)$ at $x = a$ is the tangent line to $y = g(x)$ at $x = b$ [where $b = f(a)$ and $a = g(b)$]. These tangent lines have reciprocal slopes, and thus $g'(b) = 1/f'(a) = 1/f'(g(b))$ as claimed in Theorem 1.

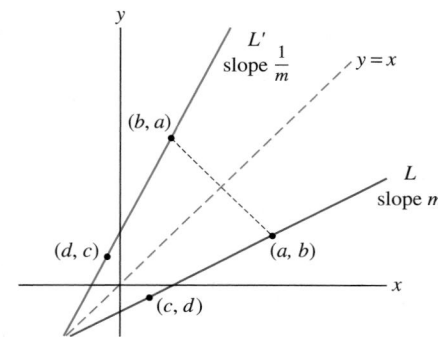

(A) If L has slope m, then its reflection L' has slope $1/m$.

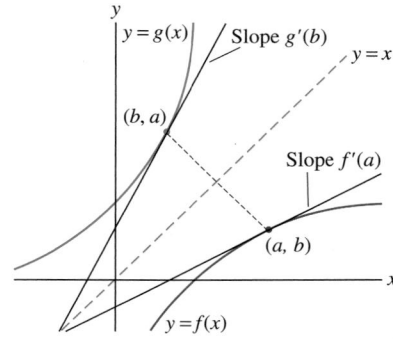

(B) The tangent line to the inverse $y = g(x)$ is the reflection of the tangent line to $y = f(x)$

FIGURE 1 Graphical illustration of the formula $g'(b) = 1/f'(g(b))$.

■ EXAMPLE 1   Calculate $g'(x)$, where $g(x)$ is the inverse of $f(x) = x^2 + 4$ on the domain $\{x : x \geq 0\}$.

Solution   We solve $y = x^2 + 4$ to obtain $x = \sqrt{y - 4}$ and thus $g(x) = \sqrt{x - 4}$. Since $f'(x) = 2x$, we have $f'(g(x)) = 2g(x)$ and by the formula for the derivative of the inverse:

$$g'(x) = \frac{1}{f'(g(x))} = \frac{1}{2g(x)} = \frac{1}{2\sqrt{x - 4}} = \frac{1}{2}(x - 4)^{-1/2}$$

We obtain the same result by differentiating $g(x) = \sqrt{x - 4}$ directly.   ■

■ EXAMPLE 2   *Calculating $g'(x)$ Without Solving for $g(x)$*   Calculate $g'(1)$, where $g(x)$ is the inverse of $f(x) = x + e^x$.

Solution   We have $f'(x) = 1 + e^x$, and thus

$$g'(1) = \frac{1}{f'(g(1))} = \frac{1}{f'(c)} = \frac{1}{1 + e^c} \qquad \text{where } c = g(1)$$

We do not have to solve for $g(x)$ (which cannot be done explicitly in this case) if we can compute $c = g(1)$ directly. By definition of the inverse, $f(c) = 1$ and we see by

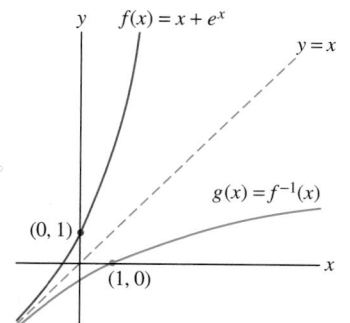

FIGURE 2 Graph of $f(x) = x + e^x$ and its inverse $g(x)$.

←·· **REMINDER** Recall from Example 7 in Section 1.5 that to simplify the composite $\cos(\sin^{-1} x)$, we refer to the right triangle in Figure 3 to obtain

$$\cos(\sin^{-1} x) = \cos\theta = \frac{\text{adjacent}}{\text{hypotenuse}}$$

$$= \sqrt{1 - x^2}$$

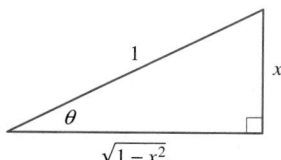

FIGURE 3 Right triangle constructed so that $\sin\theta = x$.

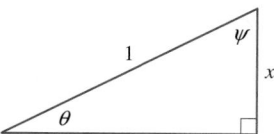

FIGURE 4 The angles $\theta = \sin^{-1} x$ and $\psi = \cos^{-1} x$ are complementary and thus sum to $\pi/2$.

inspection that $f(0) = 0 + e^0 = 1$ (Figure 2). Therefore, $c = 0$ and the formula above yields $g'(1) = 1/(1 + e^0) = \frac{1}{2}$. ■

Derivatives of Inverse Trigonometric Functions

We now apply Theorem 1 to the inverse trigonometric functions. An interesting feature of these functions is that their derivatives are not trigonometric. Rather, they involve quadratic expressions and their square roots.

THEOREM 2 Derivatives of Arcsine and Arccosine

$$\frac{d}{dx}\sin^{-1} x = \frac{1}{\sqrt{1 - x^2}}, \qquad \frac{d}{dx}\cos^{-1} x = -\frac{1}{\sqrt{1 - x^2}} \qquad \boxed{2}$$

Proof We apply Eq. (1) to $f(x) = \sin x$ and $g(x) = \sin^{-1} x$. Then $f'(x) = \cos x$, and by the equation in the margin,

$$\frac{d}{dx}\sin^{-1} x = \frac{1}{f'(g(x))} = \frac{1}{\cos(\sin^{-1} x)} = \frac{1}{\sqrt{1 - x^2}}$$

The derivative of $\cos^{-1} x$ is computed in a similar fashion (see Exercise 38 or the next example). ■

■ **EXAMPLE 3 Complementary Angles** The derivatives of $\sin^{-1} x$ and $\cos^{-1} x$ are equal up to a minus sign. Explain this by proving that

$$\sin^{-1} x + \cos^{-1} x = \frac{\pi}{2}$$

Solution In Figure 4, we have $\theta = \sin^{-1} x$ and $\psi = \cos^{-1} x$. These angles are complementary, so $\theta + \psi = \pi/2$ as claimed. Therefore,

$$\frac{d}{dx}\cos^{-1} x = \frac{d}{dx}\left(\frac{\pi}{2} - \sin^{-1} x\right) = -\frac{d}{dx}\sin^{-1} x \qquad ■$$

■ **EXAMPLE 4** Calculate $f'\left(\frac{1}{2}\right)$, where $f(x) = \arcsin(x^2)$.

Solution Recall that $\arcsin x$ is another notation for $\sin^{-1} x$. By the Chain Rule,

$$\frac{d}{dx}\arcsin(x^2) = \frac{d}{dx}\sin^{-1}(x^2) = \frac{1}{\sqrt{1 - x^4}}\frac{d}{dx}x^2 = \frac{2x}{\sqrt{1 - x^4}}$$

$$f'\left(\frac{1}{2}\right) = \frac{2\left(\frac{1}{2}\right)}{\sqrt{1 - \left(\frac{1}{2}\right)^4}} = \frac{1}{\sqrt{\frac{15}{16}}} = \frac{4}{\sqrt{15}} \qquad ■$$

The next theorem states the formulas for the derivatives of the remaining inverse trigonometric functions.

The proofs of the formulas in Theorem 3 are similar to the proof of Theorem 2. See Exercises 39 and 41.

> **THEOREM 3 Derivatives of Inverse Trigonometric Functions**
>
> $$\frac{d}{dx}\tan^{-1}x = \frac{1}{x^2+1}, \qquad \frac{d}{dx}\cot^{-1}x = -\frac{1}{x^2+1}$$
>
> $$\frac{d}{dx}\sec^{-1}x = \frac{1}{|x|\sqrt{x^2-1}}, \qquad \frac{d}{dx}\csc^{-1}x = -\frac{1}{|x|\sqrt{x^2-1}}$$

■ **EXAMPLE 5** Calculate **(a)** $\dfrac{d}{dx}\tan^{-1}(3x+1)$ and **(b)** $\dfrac{d}{dx}\csc^{-1}(e^x+1)\Big|_{x=0}$.

Solution

(a) Apply the Chain Rule using the formula $\dfrac{d}{du}\tan^{-1}u = \dfrac{1}{u^2+1}$:

$$\frac{d}{dx}\tan^{-1}(3x+1) = \frac{1}{(3x+1)^2+1}\frac{d}{dx}(3x+1) = \frac{3}{9x^2+6x+2}$$

(b) Apply the Chain Rule using the formula $\dfrac{d}{du}\csc^{-1}u = -\dfrac{1}{|u|\sqrt{u^2-1}}$:

$$\frac{d}{dx}\csc^{-1}(e^x+1) = -\frac{1}{|e^x+1|\sqrt{(e^x+1)^2-1}}\frac{d}{dx}(e^x+1)$$

$$= -\frac{e^x}{(e^x+1)\sqrt{e^{2x}+2e^x}}$$

We have replaced $|e^x+1|$ by e^x+1 since this quantity is positive. Now we have

$$\frac{d}{dx}\csc^{-1}(e^x+1)\Big|_{x=0} = -\frac{e^0}{(e^0+1)\sqrt{e^0+2e^0}} = -\frac{1}{2\sqrt{3}} \qquad ■$$

3.9 SUMMARY

• Derivative of the inverse: If $f(x)$ is differentiable and one-to-one with inverse $g(x)$, then for x such that $f'(g(x)) \neq 0$,

$$g'(x) = \frac{1}{f'(g(x))}$$

• Derivative formulas:

$$\frac{d}{dx}\sin^{-1}x = \frac{1}{\sqrt{1-x^2}},$$

$$\frac{d}{dx}\cos^{-1}x = \frac{-1}{\sqrt{1-x^2}},$$

$$\frac{d}{dx}\tan^{-1}x = \frac{1}{x^2+1}, \qquad \frac{d}{dx}\cot^{-1}x = -\frac{1}{x^2+1}$$

$$\frac{d}{dx}\sec^{-1}x = \frac{1}{|x|\sqrt{x^2-1}}, \qquad \frac{d}{dx}\csc^{-1}x = -\frac{1}{|x|\sqrt{x^2-1}}$$

3.9 EXERCISES

Preliminary Questions

1. What is the slope of the line obtained by reflecting the line $y = x/2$ through the line $y = x$?

2. Suppose that $P = (2, 4)$ lies on the graph of $f(x)$ and that the slope of the tangent line through P is $m = 3$. Assuming that $f^{-1}(x)$ exists, what is the slope of the tangent line to the graph of $f^{-1}(x)$ at the point $Q = (4, 2)$?

3. Which inverse trigonometric function $g(x)$ has the derivative
$$g'(x) = \frac{1}{x^2 + 1}?$$

4. What is the geometric interpretation of the identity
$$\sin^{-1} x + \cos^{-1} x = \frac{\pi}{2}?$$

Exercises

1. Find the inverse $g(x)$ of $f(x) = \sqrt{x^2 + 9}$ with domain $x \ge 0$ and calculate $g'(x)$ in two ways: using Theorem 1 and by direct calculation.

2. Let $g(x)$ be the inverse of $f(x) = x^3 + 1$. Find a formula for $g(x)$ and calculate $g'(x)$ in two ways: using Theorem 1 and then by direct calculation.

In Exercises 3–8, use Theorem 1 to calculate $g'(x)$, where $g(x)$ is the inverse of $f(x)$.

3. $f(x) = 7x + 6$

4. $f(x) = \sqrt{3 - x}$

5. $f(x) = x^{-5}$

6. $f(x) = 4x^3 - 1$

7. $f(x) = \dfrac{x}{x + 1}$

8. $f(x) = 2 + x^{-1}$

9. Let $g(x)$ be the inverse of $f(x) = x^3 + 2x + 4$. Calculate $g(7)$ [without finding a formula for $g(x)$] and then calculate $g'(7)$.

10. Find $g'(-\frac{1}{2})$, where $g(x)$ is the inverse of $f(x) = \dfrac{x^3}{x^2 + 1}$.

In Exercises 11–16, calculate $g(b)$ and $g'(b)$, where g is the inverse of f (in the given domain, if indicated).

11. $f(x) = x + \cos x$, $\quad b = 1$

12. $f(x) = 4x^3 - 2x$, $\quad b = -2$

13. $f(x) = \sqrt{x^2 + 6x}$ for $x \ge 0$, $\quad b = 4$

14. $f(x) = \sqrt{x^2 + 6x}$ for $x \le -6$, $\quad b = 4$

15. $f(x) = \dfrac{1}{x + 1}$, $\quad b = \dfrac{1}{4}$

16. $f(x) = e^x$, $\quad b = e$

17. Let $f(x) = x^n$ and $g(x) = x^{1/n}$. Compute $g'(x)$ using Theorem 1 and check your answer using the Power Rule.

In Exercises 18–21, compute the derivative at the point indicated without using a calculator.

18. $y = \sin^{-1} x$, $\quad x = \frac{3}{5}$

19. $y = \tan^{-1} x$, $\quad x = \frac{1}{2}$

20. $y = \sec^{-1} x$, $\quad x = 4$

21. $y = \arccos(4x)$, $\quad x = \frac{1}{5}$

In Exercises 22–37, find the derivative.

22. $y = \sin^{-1}(7x)$

23. $y = \arctan\left(\dfrac{x}{3}\right)$

24. $y = \cos^{-1}(x^2)$

25. $y = \sec^{-1}(t + 1)$

26. $y = x \tan^{-1} x$

27. $y = \dfrac{\cos^{-1} x}{\sin^{-1} x}$

28. $y = \arcsin(e^x)$

29. $y = \csc^{-1}(x^{-1})$

30. $y = \tan^{-1}\left(\dfrac{z}{1 - z^2}\right)$

31. $y = \tan^{-1}\left(\dfrac{1 + t}{1 - t}\right)$

32. $y = \cos^{-1} t^{-1} - \sec^{-1} t$

33. $y = e^{\cos^{-1} x}$

34. $y = \cos^{-1}(x + \sin^{-1} x)$

35. $y = \ln(\sin^{-1} t)$

36. $y = \sqrt{1 - t^2} + \sin^{-1} t$

37. $y = x^{\sin^{-1} x}$

38. Use Figure 5 to prove that $(\cos^{-1} x)' = -\dfrac{1}{\sqrt{1 - x^2}}$.

FIGURE 5 Right triangle with $\theta = \cos^{-1} x$.

39. Show that $(\tan^{-1} x)' = \cos^2(\tan^{-1} x)$ and then use Figure 6 to prove that $(\tan^{-1} x)' = (x^2 + 1)^{-1}$.

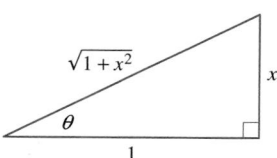

FIGURE 6 Right triangle with $\theta = \tan^{-1} x$.

40. Let $\theta = \sec^{-1} x$. Show that $\tan\theta = \sqrt{x^2-1}$ if $x \geq 1$ and $\tan\theta = -\sqrt{x^2-1}$ if $x \leq -1$. *Hint:* $\tan\theta \geq 0$ on $(0, \pi/2)$ and $\tan\theta \leq 0$ on $(\pi/2, \pi)$.

41. Use Exercise 40 to verify the formula

$$(\sec^{-1} x)' = \frac{1}{|x|\sqrt{x^2-1}}$$

Further Insights and Challenges

42. Let g be the inverse of a function f satisfying $f'(x) = f(x)$. Show that $g'(x) = x^{-1}$. We will apply this in the next section to show that the inverse of $f(x) = e^x$ (the natural logarithm) has the derivative $f'(x) = x^{-1}$.

3.10 Derivatives of General Exponential and Logarithmic Functions

In Section 3.2, we proved the formula for any base $b > 0$:

$$\frac{d}{dx}b^x = m_b\,b^x, \qquad \text{where} \quad m_b = \lim_{h \to 0}\frac{b^h - 1}{h}$$

This established the important result that the derivative of an exponential function $f(x) = b^x$ is proportional to b^x, but we were not able to identify the factor of proportionality m_b (other than to say that e is the unique number for which $m_e = 1$). We now use the Chain Rule to prove that $m_b = \ln b$. Recall that $\ln x$ is the natural logarithm, that is, $\ln x = \log_e x$.

THEOREM 1 Derivative of $f(x) = b^x$

$$\frac{d}{dx}b^x = (\ln b)b^x \qquad \text{for } b > 0 \qquad \boxed{1}$$

Proof Since $b = e^{\ln b}$, we have $b^x = e^{\ln b^x} = e^{(\ln b)x}$. Now use the Chain Rule:

$$\frac{d}{dx}b^x = \frac{d}{dx}e^{(\ln b)x} = (\ln b)e^{(\ln b)x} = (\ln b)b^x \qquad \blacksquare$$

For example, $(10^x)' = (\ln 10)10^x$.

■ **EXAMPLE 1** Find the derivative of **(a)** $f(x) = 4^{3x}$ and **(b)** $f(x) = 5^{x^2}$.

Solution

(a) The function $f(x) = 4^{3x}$ is a composite of 4^u and $u = 3x$:

$$\frac{d}{dx}4^{3x} = \frac{d}{du}4^u\,\frac{du}{dx} = (\ln 4)4^u(3x)' = (3\ln 4)4^{3x}$$

(b) The function $f(x) = 5^{x^2}$ is a composite of 5^u and $u = x^2$:

$$\frac{d}{dx}5^{x^2} = (\ln 5)5^{x^2}(x^2)' = (2\ln 5)\,x\,5^{x^2} \qquad \blacksquare$$

Next, we calculate the derivative of $y = \ln x$. Recall that if $g(x)$ is the inverse of $f(x)$, then $g'(x) = 1/f'(g(x))$. Let $f(x) = e^x$ and $g(x) = \ln x$. Then $f'(x) = f(x)$ and

$$\frac{d}{dx}\ln x = g'(x) = \frac{1}{f'(g(x))} = \frac{1}{f(g(x))} = \frac{1}{x}$$

The two most important calculus facts about exponentials and logs are

$$\frac{d}{dx}e^x = e^x, \qquad \frac{d}{dx}\ln x = \frac{1}{x}$$

THEOREM 2 Derivative of the Natural Logarithm

$$\frac{d}{dx}\ln x = \frac{1}{x} \qquad \text{for } x > 0$$

2

■ **EXAMPLE 2**   Find the derivative of   **(a)** $y = x \ln x$ and   **(b)** $y = (\ln x)^2$.

Solution

(a) Use the Product Rule:

$$\frac{d}{dx}(x \ln x) = x \cdot (\ln x)' + (x)' \cdot \ln x = x \cdot \frac{1}{x} + \ln x = 1 + \ln x$$

(b) By the General Power Rule,

$$\frac{d}{dx}(\ln x)^2 = 2 \ln x \cdot \frac{d}{dx}\ln x = \frac{2 \ln x}{x}$$ ■

In Section 3.2, we proved the Power Rule for whole number exponents. We now prove it for all real numbers n and $x > 0$ by writing x^n as an exponential:

$$x^n = (e^{\ln x})^n = e^{n \ln x}$$

$$\frac{d}{dx}x^n = \frac{d}{dx}e^{n \ln x} = \left(\frac{d}{dx}n \ln x\right)e^{n \ln x}$$

$$= \left(\frac{n}{x}\right)x^n = nx^{n-1}$$

The Chain Rule gives us a useful formula for the derivative of a composite function of the form $\ln(f(x))$. Taking $u = f(x)$, we have

$$\frac{d}{dx}\ln(f(x)) = \frac{d}{du}\ln(u)\frac{du}{dx} = \frac{1}{u} \cdot u' = \frac{1}{f(x)}f'(x)$$

$$\frac{d}{dx}\ln(f(x)) = \frac{f'(x)}{f(x)}$$

3

■ **EXAMPLE 3**   Find the derivative of   **(a)** $y = \ln(x^3 + 1)$ and   **(b)** $y = \ln(\sqrt{\sin x})$.

Solution   Use Eq. (3):

(a) $\dfrac{d}{dx}\ln(x^3 + 1) = \dfrac{(x^3 + 1)'}{x^3 + 1} = \dfrac{3x^2}{x^3 + 1}$

(b) The algebra is simpler if we write $\ln(\sqrt{\sin x}) = \ln((\sin x)^{1/2}) = \frac{1}{2}\ln(\sin x)$:

$$\frac{d}{dx}\ln(\sqrt{\sin x}) = \frac{1}{2}\frac{d}{dx}\ln(\sin x) = \frac{1}{2}\frac{(\sin x)'}{\sin x} = \frac{1}{2}\frac{\cos x}{\sin x} = \frac{1}{2}\cot x$$ ■

According to Eq. (1) in Section 1.6, we have the "change of base" formulas:

$$\log_b x = \frac{\log_a x}{\log_a b}, \qquad \log_b x = \frac{\ln x}{\ln b}$$

It follows, as in Example 4, that for any base $b > 0$, $b \neq 1$:

$$\frac{d}{dx}\log_b x = \frac{1}{(\ln b)x}$$

Logarithmic differentiation saves work when the function is a product of several factors or a quotient, as in Example 5.

■ **EXAMPLE 4   Derivative of the Logarithm to Another Base**   Use Eq. (1) from Section 1.6 to calculate $\dfrac{d}{dx}\log_{10} x$.

Solution   By Eq. (1) from Section 1.6, $\log_{10} x = \dfrac{\ln x}{\ln 10}$, so

$$\frac{d}{dx}\log_{10} x = \frac{d}{dx}\frac{\ln x}{\ln 10} = \frac{1}{\ln 10}\frac{d}{dx}\ln x = \frac{1}{(\ln 10)x}$$ ■

The next example illustrates the technique of **logarithmic differentiation**.

■ **EXAMPLE 5   Logarithmic Differentiation**   Find the derivative of

$$f(x) = \frac{(x + 1)^2(2x^2 - 3)}{\sqrt{x^2 + 1}}$$

Solution In logarithmic differentiation, we differentiate $\ln(f(x))$ rather than $f(x)$ itself. We have

$$\ln(f(x)) = \ln\left((x+1)^2\right) + \ln\left(2x^2 - 3\right) - \ln\left(\sqrt{x^2+1}\right)$$

Now use Eq. (3):

$$\frac{f'(x)}{f(x)} = \frac{d}{dx}\ln(f(x)) = \frac{d}{dx}\ln\left((x+1)^2\right) + \frac{d}{dx}\ln\left(2x^2-3\right) - \frac{d}{dx}\ln\left(\sqrt{x^2+1}\right)$$

$$= 2\frac{d}{dx}\ln\left(x+1\right) + \frac{d}{dx}\ln\left(2x^2-3\right) - \frac{1}{2}\frac{d}{dx}\ln\left(x^2+1\right)$$

$$\frac{f'(x)}{f(x)} = 2\frac{1}{x+1} + \frac{4x}{2x^2-3} - \frac{1}{2}\frac{2x}{x^2+1}$$

To obtain $f'(x)$, multiply through by $f(x)$:

$$f'(x) = \left(\frac{(x+1)^2(2x^2-3)}{\sqrt{x^2+1}}\right)\left(\frac{2}{x+1} + \frac{4x}{2x^2-3} - \frac{x}{x^2+1}\right) \qquad \blacksquare$$

■ **EXAMPLE 6** Find the derivative (for $x > 0$) of **(a)** $f(x) = x^x$ and **(b)** $g(x) = x^{\sin x}$.

Solution The two problems are similar. We illustrate two different methods (Figure 1).

(a) Method 1: Use the equation $x = e^{\ln x}$ to rewrite $f(x)$ as an exponential and apply the Chain Rule:

$$f(x) = x^x = (e^{\ln x})^x = e^{x\ln x}$$

$$f'(x) = (x\ln x)'e^{x\ln x} = (1 + \ln x)e^{x\ln x} = (1 + \ln x)x^x$$

(b) Method 2: Use logarithmic differentiation, that is, differentiate $\ln(g(x))$ rather than $g(x)$. Since $\ln(g(x)) = \ln(x^{\sin x}) = (\sin x)\ln x$, Eq. (3) yields

$$\frac{g'(x)}{g(x)} = \frac{d}{dx}\ln(g(x)) = \frac{d}{dx}(\sin x)(\ln x) = \frac{\sin x}{x} + (\cos x)\ln x$$

$$g'(x) = \left(\frac{\sin x}{x} + (\cos x)\ln x\right)g(x) = \left(\frac{\sin x}{x} + (\cos x)\ln x\right)x^{\sin x} \qquad \blacksquare$$

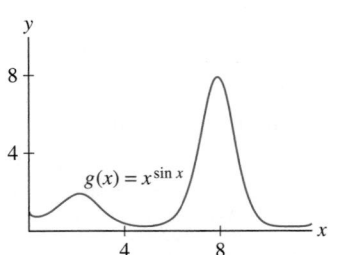

FIGURE 1 Graphs of $f(x) = x^x$ and $g(x) = x^{\sin x}$.

Derivatives of Hyperbolic Functions

Recall from Section 1.6 that the hyperbolic functions are special combinations of e^x and e^{-x} that arise in engineering and physics applications. We can now compute their derivatives easily using our rules of differentiation. First, we treat the hyperbolic sine and cosine:

$$\sinh x = \frac{e^x - e^{-x}}{2}, \qquad \cosh x = \frac{e^x + e^{-x}}{2}$$

The formulas for their derivatives are similar to those for the corresponding trigonometric functions, differing at most by a sign:

$$\frac{d}{dx}\sinh x = \cosh x, \qquad \frac{d}{dx}\cosh x = \sinh x$$

See Section 1.6 (p. 48) to review hyperbolic functions. The basic identity is

$$\cosh^2 x - \sinh^2 x = 1$$

For example,

$$\frac{d}{dx}\left(\frac{e^x - e^{-x}}{2}\right) = \left(\frac{e^x - e^{-x}}{2}\right)' = \frac{e^x + e^{-x}}{2} = \cosh x$$

The hyperbolic tangent, cotangent, secant, and cosecant functions are defined like their trigonometric counterparts:

$$\tanh x = \frac{\sinh x}{\cosh x} = \frac{e^x - e^{-x}}{e^x + e^{-x}}, \qquad \operatorname{sech} x = \frac{1}{\cosh x} = \frac{2}{e^x + e^{-x}}$$

$$\coth x = \frac{\cosh x}{\sinh x} = \frac{e^x + e^{-x}}{e^x - e^{-x}}, \qquad \operatorname{csch} x = \frac{1}{\sinh x} = \frac{2}{e^x - e^{-x}}$$

We can also calculate the derivatives of these additional hyperbolic functions directly. They too differ from their trigonometric counterparts by a sign at most.

Derivatives of Hyperbolic and Trigonometric Functions

$$\frac{d}{dx}\tanh x = \operatorname{sech}^2 x, \qquad\qquad \frac{d}{dx}\tan x = \sec^2 x$$

$$\frac{d}{dx}\coth x = -\operatorname{csch}^2 x, \qquad\qquad \frac{d}{dx}\cot x = -\csc^2 x$$

$$\frac{d}{dx}\operatorname{sech} x = -\operatorname{sech} x \tanh x, \qquad \frac{d}{dx}\sec x = \sec x \tan x$$

$$\frac{d}{dx}\operatorname{csch} x = -\operatorname{csch} x \coth x, \qquad \frac{d}{dx}\csc x = -\csc x \cot x$$

■ **EXAMPLE 7** Verify the formula $\dfrac{d}{dx}\coth x = -\operatorname{csch}^2 x$.

Solution By the Quotient Rule and the identity $\cosh^2 x - \sinh^2 x = 1$,

$$\frac{d}{dx}\coth x = \left(\frac{\cosh x}{\sinh x}\right)' = \frac{(\sinh x)(\cosh x)' - (\cosh x)(\sinh x)'}{\sinh^2 x}$$

$$= \frac{\sinh^2 x - \cosh^2 x}{\sinh^2 x} = \frac{-1}{\sinh^2 x} = -\operatorname{csch}^2 x \qquad ■$$

■ **EXAMPLE 8** Calculate **(a)** $\dfrac{d}{dx}\cosh(3x^2 + 1)$ and **(b)** $\dfrac{d}{dx}\sinh x \tanh x$.

Solution

(a) By the Chain Rule, $\dfrac{d}{dx}\cosh(3x^2 + 1) = 6x \sinh(3x^2 + 1)$

(b) By the Product Rule,

$$\frac{d}{dx}(\sinh x \tanh x) = \sinh x \operatorname{sech}^2 x + \tanh x \cosh x = \operatorname{sech} x \tanh x + \sinh x \qquad ■$$

Inverse Hyperbolic Functions

Recall that a function $f(x)$ with domain D has an inverse if it is one-to-one on D. Each of the hyperbolic functions except $\cosh x$ and $\operatorname{sech} x$ is one-to-one on its domain and therefore has a well-defined inverse. The functions $\cosh x$ and $\operatorname{sech} x$ are one-to-one on the restricted domain $\{x : x \geq 0\}$. We let $\cosh^{-1} x$ and $\operatorname{sech}^{-1} x$ denote the corresponding inverses (Figure 2). In reading the following table, keep in mind that the domain of the inverse is equal to the range of the function.

(A)

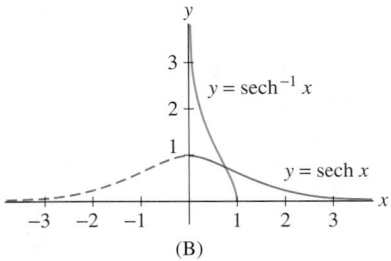

(B)

FIGURE 2

Inverse Hyperbolic Functions and Their Derivatives

Function	Domain	Derivative
$y = \sinh^{-1} x$	all x	$\dfrac{d}{dx} \sinh^{-1} x = \dfrac{1}{\sqrt{x^2 + 1}}$
$y = \cosh^{-1} x$	$x \geq 1$	$\dfrac{d}{dx} \cosh^{-1} x = \dfrac{1}{\sqrt{x^2 - 1}}$
$y = \tanh^{-1} x$	$|x| < 1$	$\dfrac{d}{dx} \tanh^{-1} x = \dfrac{1}{1 - x^2}$
$y = \coth^{-1} x$	$|x| > 1$	$\dfrac{d}{dx} \coth^{-1} x = \dfrac{1}{1 - x^2}$
$y = \operatorname{sech}^{-1} x$	$0 < x \leq 1$	$\dfrac{d}{dx} \operatorname{sech}^{-1} x = -\dfrac{1}{x\sqrt{1 - x^2}}$
$y = \operatorname{csch}^{-1} x$	$x \neq 0$	$\dfrac{d}{dx} \operatorname{csch}^{-1} x = -\dfrac{1}{|x|\sqrt{x^2 + 1}}$

■ **EXAMPLE 9** (a) Verify the formula $\dfrac{d}{dx} \tanh^{-1} x = \dfrac{1}{1 - x^2}$.

(b) The functions $y = \tanh^{-1} x$ and $y = \coth^{-1} x$ appear to have the same derivative. Does this imply that they differ by a constant?

Solution

(a) We apply the formula for the derivative of an inverse [Eq. (1) in Section 3.9]. Since $(\tanh x)' = \operatorname{sech}^2 x$,

$$\frac{d}{dx} \tanh^{-1} x = \frac{1}{\operatorname{sech}^2(\tanh^{-1} x)}$$

Now let $t = \tanh^{-1} x$. Then

$$\cosh^2 t - \sinh^2 t = 1 \qquad \text{(basic identity)}$$

$$1 - \tanh^2 t = \operatorname{sech}^2 t \qquad \text{(divide by } \cosh^2 t)$$

$$1 - x^2 = \operatorname{sech}^2(\tanh^{-1} x) \quad \text{(because } x = \tanh t)$$

This gives the desired result:

$$\frac{d}{dx} \tanh^{-1} x = \frac{1}{\operatorname{sech}^2(\tanh^{-1} x)} = \frac{1}{1 - x^2}$$

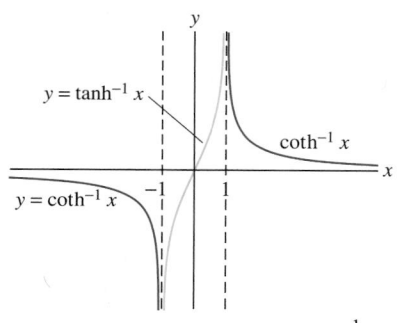

FIGURE 3 The functions $y = \tanh^{-1} x$ and $y = \coth^{-1} x$ have disjoint domains.

(b) A similar computation shows that $\dfrac{d}{dx} \coth^{-1} x = \dfrac{1}{1 - x^2}$. Thus, the derivatives of $y = \tanh^{-1} x$ and $y = \coth^{-1} x$ are given by the same formula. Functions defined on a common interval with the same derivative differ by a constant. However, the functions $y = \tanh^{-1} x$ and $y = \coth^{-1} x$ have disjoint domains and therefore do not differ by a constant (Figure 3). ∎

3.10 SUMMARY

- Derivative formulas:

$$\frac{d}{dx} e^x = e^x, \qquad \frac{d}{dx} \ln x = \frac{1}{x}, \qquad \frac{d}{dx} b^x = (\ln b) b^x, \qquad \frac{d}{dx} \log_b x = \frac{1}{(\ln b)\, x}$$

- Derivative formulas:

$$\frac{d}{dx} \sinh x = \cosh x, \qquad\qquad \frac{d}{dx} \cosh x = \sinh x$$

$$\frac{d}{dx} \tanh x = \operatorname{sech}^2 x, \qquad\qquad \frac{d}{dx} \coth x = -\operatorname{csch}^2 x$$

$$\frac{d}{dx} \operatorname{sech} x = -\operatorname{sech} x \tanh x, \qquad \frac{d}{dx} \operatorname{csch} x = -\operatorname{csch} x \coth x$$

$$\frac{d}{dx} \sinh^{-1} x = \frac{1}{\sqrt{x^2 + 1}}, \qquad\qquad \frac{d}{dx} \cosh^{-1} x = \frac{1}{\sqrt{x^2 - 1}} \quad (x > 1)$$

$$\frac{d}{dx} \tanh^{-1} x = \frac{1}{1 - x^2} \quad (|x| < 1), \qquad \frac{d}{dx} \coth^{-1} x = \frac{1}{1 - x^2} \quad (|x| > 1)$$

$$\frac{d}{dx} \operatorname{sech}^{-1} x = \frac{-1}{x\sqrt{1 - x^2}} \quad (0 < x < 1), \qquad \frac{d}{dx} \operatorname{csch}^{-1} x = -\frac{1}{|x|\sqrt{x^2 + 1}} \quad (x \neq 0)$$

3.10 EXERCISES

Preliminary Questions

1. What is the slope of the tangent line to $y = 4^x$ at $x = 0$?

2. For which $b > 0$ does the tangent line to $y = b^x$ at $x = 0$ have slope 2?

3. What is the rate of change of $y = \ln x$ at $x = 10$?

4. For which base b is $(\log_b x)' = \dfrac{1}{3x}$?

5. What are $y^{(100)}$ and $y^{(101)}$ for $y = \cosh x$?

Exercises

In Exercises 1–18, find the derivative.

1. $y = x \ln x$

2. $y = t \ln t - t$

3. $y = (\ln x)^2$

4. $y = \ln(x^2)$

5. $y = \ln(x^3 + 3x + 1)$

6. $y = \ln(2^s)$

7. $y = \ln(\sin t + 1)$

8. $y = x^2 \ln x$

9. $y = \dfrac{\ln x}{x}$

10. $y = e^{(\ln x)^2}$

11. $y = \ln(\ln x)$

12. $y = (\ln(\ln x))^3$

13. $y = \ln((\ln x)^3)$

14. $y = \ln((x + 1)(2x + 9))$

15. $y = \ln(\tan x)$

16. $y = \ln\left(\dfrac{x + 1}{x^3 + 1}\right)$

17. $y = 5^x$

18. $y = 5^{x^2 - x}$

In Exercises 19–22, compute the derivative using Eq. (1).

19. $f'(x), \quad f(x) = \log_2 x$

20. $\dfrac{d}{dx} \log_{10}(x^3 + x^2)$

21. $f'(3), \quad f(x) = \log_5 x$

22. $\dfrac{d}{dt} \log_3(\sin t)$

In Exercises 23–34, find an equation of the tangent line at the point indicated.

23. $f(x) = 4^x, \quad x = 3$

24. $f(x) = (\sqrt{2})^x, \quad x = \sqrt{2}$

25. $s(t) = 3^{7t}, \quad t = 2$

26. $f(x) = \pi^{3x+9}, \quad x = 1$

27. $f(x) = 5^{x^2 - 2x + 9}, \quad x = 1$

28. $s(t) = \ln t, \quad t = 5$

29. $s(t) = \ln(8 - 4t), \quad t = 1$

30. $f(x) = \ln(x^2), \quad x = 4$

31. $f(x) = 4\ln(9x + 2), \quad x = 2$

32. $f(x) = \ln(\sin x), \quad x = \dfrac{\pi}{4}$

33. $f(x) = \log_5 x, \quad x = 2$

34. $f(x) = \log_2(x + x^{-1}), \quad x = 1$

In Exercises 35–40, find the derivative using the methods of Example 6.

35. $f(x) = x^{2x}$

36. $f(x) = x^{\cos x}$

37. $f(x) = x^{e^x}$

38. $f(x) = x^{x^2}$

39. $f(x) = x^{2^x}$

40. $f(x) = e^{x^x}$

In Exercises 41–48, evaluate the derivative using logarithmic differentiation as in Example 5.

41. $y = (x + 2)(x + 4)$

42. $y = (x + 1)(x + 2)(x + 4)$

43. $y = \dfrac{x(x + 1)^3}{(3x - 1)^2}$

44. $y = \dfrac{x(x^2 + 1)}{\sqrt{x + 1}}$

45. $y = (2x + 1)(4x^2)\sqrt{x - 9}$

46. $y = \sqrt{\dfrac{x(x + 2)}{(2x + 1)(2x + 2)}}$

47. $y = (x^2 + 1)(x^2 + 2)(x^2 + 3)^2$

48. $y = \dfrac{x \cos x}{(x + 1)\sin x}$

49. Use the formula $(\ln f(x))' = f'(x)/f(x)$ to show that $\ln x$ and $\ln(2x)$ have the same derivative. Is there a simpler explanation of this result?

In Exercises 50–75, calculate the derivative.

50. $y = \sinh(3x)$

51. $y = \sinh(x^2)$

52. $y = \cosh(1 - 4t)$

53. $y = \tanh(t^2 + 1)$

54. $y = \sqrt{\cosh x + 1}$

55. $y = \sinh x \tanh x$

56. $y = \dfrac{\sinh t}{1 + \cosh t}$

57. $y = \ln(\cosh x)$

58. $y = \tanh(2x)$

59. $y = \sinh(\ln x)$

60. $y = e^{\tanh x}$

61. $y = \tanh(e^x)$

62. $y = \sinh(xe^x)$

63. $y = \sinh(\cosh x)$

64. $y = \text{sech}(\sqrt{x})$

65. $y = \coth(x^2)$

66. $y = \ln(\coth x)$

67. $y = \text{sech}\, x \coth x$

68. $y = \tanh(3x^2 - 9)$

69. $y = x^{\sinh x}$

70. $y = \tanh^{-1} 3x$

71. $y = \sinh^{-1}(x^2)$

72. $y = (\text{csch}^{-1} 3x)^4$

73. $y = e^{\cosh^{-1} x}$

74. $y = \sinh^{-1}(\sqrt{x^2 + 1})$

75. $y = \ln(\tanh^{-1} x)$

In Exercises 76–78, prove the formula.

76. $\dfrac{d}{dx}(\coth x) = -\text{csch}^2 x$

77. $\dfrac{d}{dt}\sinh^{-1} t = \dfrac{1}{\sqrt{t^2 + 1}}$

78. $\dfrac{d}{dt}\cosh^{-1} t = \dfrac{1}{\sqrt{t^2 - 1}} \quad \text{for } t > 1$

79. The energy E (in joules) radiated as seismic waves from an earthquake of Richter magnitude M is given by the formula $\log_{10} E = 4.8 + 1.5M$.

(a) Express E as a function of M.

(b) Show that when M increases by 1, the energy increases by a factor of approximately 31.

(c) Calculate $\dfrac{dE}{dM}$.

80. The Palermo Technical Impact Hazard Scale P is used to quantify the risk associated with the impact of an asteroid colliding with the earth:

$$P = \log_{10}\left(\dfrac{p_i E^{0.8}}{0.03T}\right)$$

where p_i is the probability of impact occurring in T years and E is the energy of impact (in megatons of TNT). The risk is greater than a random event of similar magnitude if $P > 0$ (asteroids observed in 2006 have $P \leq -2$, indicating low risk).

(a) Calculate $\dfrac{dP}{dT}$, assuming that $p_i = 2 \times 10^{-5}$ and $E = 2$ megatons.

(b) Use the derivative to estimate the approximate change in P if T increases to 26 years.

Further Insights and Challenges

81. **(a)** Show that if f and g are differentiable, then

$$\frac{d}{dx}\ln(f(x)g(x)) = \frac{f'(x)}{f(x)} + \frac{g'(x)}{g(x)}$$ 4

(b) Give a new proof of the Product Rule by observing that the left-hand side of Eq. (4) is equal to $\dfrac{(f(x)g(x))'}{f(x)g(x)}$.

82. Use the formula $\log_b x = \dfrac{\log_a x}{\log_a b}$ for $a, b > 0$ to verify the formula

$$\frac{d}{dx}\log_b x = \frac{1}{(\ln b)x}$$

3.11 Related Rates

In *related rate* problems, the goal is to calculate an unknown rate of change in terms of other rates of change that are known. One typical problem involves a ladder leaning against a wall. The question is: *How fast does the top of the ladder move if the bottom of the ladder is pulled away from the wall at constant speed?* What is interesting and perhaps surprising is that the top and bottom travel at different speeds. Figure 1 shows this clearly: the bottom travels the same distance over each time interval, but the top travels farther during the second time interval than the first. In other words, the top is speeding up while the bottom moves at a constant speed. In the next example, we use calculus to find the speed of the ladder's top.

FIGURE 1 Positions of a ladder at times $t = 0, 1, 2$.

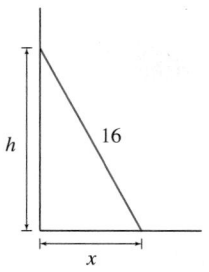

FIGURE 2 The variables x and h.

■ **EXAMPLE 1** Ladder Problem A 16-ft ladder leans against a wall. The bottom of the ladder is 5 ft from the wall at time $t = 0$ and slides away from the wall at a rate of 3 ft/s. Find the velocity of the top of the ladder at time $t = 1$.

Solution The first step in any related rate problem is to choose variables for the relevant quantities. Since we are considering how the top and bottom of the ladder change position, we use variables (Figure 2):

- $x = x(t)$ distance from the bottom of the ladder to the wall
- $h = h(t)$ height of the ladder's top

Both x and h are functions of time. The velocity of the bottom is $dx/dt = 3$ ft/s. Since the velocity of the top is dh/dt and the initial distance from the bottom to the wall is $x(0) = 5$, we can restate the problem as

> Compute $\dfrac{dh}{dt}$ at $t = 1$ given that $\dfrac{dx}{dt} = 3$ ft/s and $x(0) = 5$ ft

To solve this problem, we need an equation relating x and h (Figure 2). This is provided by the Pythagorean Theorem:

$$x^2 + h^2 = 16^2$$

To calculate dh/dt, we differentiate both sides of this equation *with respect to t*:

$$\frac{d}{dt}x^2 + \frac{d}{dt}h^2 = \frac{d}{dt}16^2$$

$$2x\frac{dx}{dt} + 2h\frac{dh}{dt} = 0$$

This yields $\dfrac{dh}{dt} = -\dfrac{x}{h}\dfrac{dx}{dt}$, and since $\dfrac{dx}{dt} = 3$ ft/s, the velocity of the top is

$$\boxed{\dfrac{dh}{dt} = -3\dfrac{x}{h} \text{ ft/s}} \qquad \boxed{1}$$

t	x	h	dh/dt
0	5	15.20	−0.99
1	8	13.86	−1.73
2	11	11.62	−2.84
3	14	7.75	−5.42

This table of values confirms that the top of the ladder is speeding up.

To apply this formula, we must find x and h at time $t = 1$. Since the bottom slides away at 3 ft/s and $x(0) = 5$, we have $x(1) = 8$ and $h(1) = \sqrt{16^2 - 8^2} \approx 13.86$. Therefore,

$$\left.\dfrac{dh}{dt}\right|_{t=1} = -3\dfrac{x(1)}{h(1)} \approx -3\dfrac{8}{13.86} \approx -1.7 \text{ ft/s} \qquad \blacksquare$$

In the next examples, we divide the solution into three steps that can be used when working the exercises.

■ **EXAMPLE 2** Filling a Rectangular Tank   Water pours into a fish tank at a rate of 3 ft³/min. How fast is the water level rising if the base of the tank is a rectangle of dimensions 2 × 3 ft?

If possible, choose variables that are either related to or traditionally associated with the quantity represented. For example, we may use V for volume, θ for an angle, h or y for height, and r for radius.

Solution   To solve a related rate problem, it is useful to draw a diagram if possible. Figure 3 illustrates our problem.

Step 1. **Assign variables and restate the problem.**
First, we must recognize that the rate at which water pours into the tank is the derivative of water volume with respect to time. Therefore, let V be the volume and h the height of the water at time t. Then

$$\dfrac{dV}{dt} = \text{rate at which water is added to the tank}$$

$$\dfrac{dh}{dt} = \text{rate at which the water level is rising}$$

Now we can restate our problem in terms of derivatives:

$$\boxed{\text{Compute } \dfrac{dh}{dt} \quad \text{given that} \quad \dfrac{dV}{dt} = 3 \text{ ft}^3/\text{min}}$$

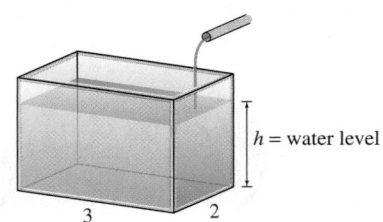

h = water level

3   2

FIGURE 3 V = water volume at time t.

Step 2. **Find an equation that relates the variables and differentiate.**
We need a relation between V and h. We have $V = 6h$ since the tank's base has area 6 ft². Therefore,

$$\dfrac{dV}{dt} = 6\dfrac{dh}{dt}$$

Step 3. **Use the data to find the unknown derivative.**
Since $dV/dt = 3$,

$$\dfrac{dh}{dt} = \dfrac{1}{6}\dfrac{dV}{dt} = \dfrac{1}{6}(3) = 0.5 \text{ ft/min}$$

Note that dh/dt has units of ft/min since h and t are in feet and minutes, respectively. ■

The set-up in the next example is similar but more complicated because the water tank has the shape of a circular cone. We use similar triangles to derive a relation between the volume and height of the water. We also need the formula $V = \frac{1}{3}\pi h r^2$ for the volume of a circular cone of height h and radius r.

■ **EXAMPLE 3** **Filling a Conical Tank** Water pours into a conical tank of height 10 ft and radius 4 ft at a rate of 10 ft³/min.

(a) How fast is the water level rising when it is 5 ft high?

(b) As time passes, what happens to the rate at which the water level rises? Explain.

Solution

(a) *Step 1.* **Assign variables and restate the problem.**

As in the previous example, let V and h be the volume and height of the water in the tank at time t. The problem in part (a) is

$$\text{Compute } \frac{dh}{dt} \text{ at } h = 5 \quad \text{given that} \quad \frac{dV}{dt} = 10 \text{ ft}^3/\text{min}$$

Step 2. **Find an equation that relates the variables and differentiate.**

The volume is $V = \frac{1}{3}\pi h r^2$, where r is the radius of the cone at height h, but *we cannot use this relation unless we eliminate the variable r*. Using similar triangles in Figure 4, we see that

$$\frac{r}{h} = \frac{4}{10} \quad \text{or} \quad r = 0.4\,h$$

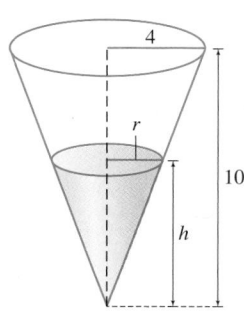

FIGURE 4 By similar triangles,

$$\frac{h}{r} = \frac{10}{4}$$

Therefore,

$$V = \frac{1}{3}\pi h(0.4\,h)^2 = \frac{1}{3}\pi(0.16)h^3 \qquad \boxed{2}$$

$$\frac{dV}{dt} = (0.16)\pi h^2 \frac{dh}{dt} \qquad \boxed{3}$$

Step 3. **Use the data to find the unknown derivative.**

Using the given data $dV/dt = 10$ in Eq. (3), we have

$$(0.16)\pi h^2 \frac{dh}{dt} = 10$$

$$\frac{dh}{dt} = \frac{10}{(0.16)\pi h^2} \approx \frac{20}{h^2} \qquad \boxed{4}$$

When $h = 5$, the level is rising at a rate of $dh/dt \approx 20/5^2 = 0.8$ ft³/min.

(b) Eq. (4) shows that dh/dt is inversely proportional to h^2. As h increases, the water level rises more slowly. This is reasonable if you consider that a thin slice of the cone of width Δh has more volume when h is large, so more water is needed to raise the level when h is large (Figure 5). ∎

> **CAUTION** *A common mistake is substituting $h = 5$ in Eq. (2). Do not set $h = 5$ until the end of the problem, after the derivatives have been computed. This applies to all related rate problems.*

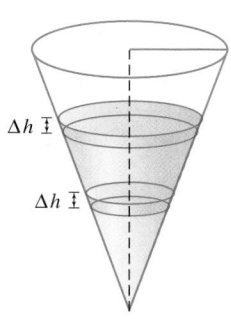

FIGURE 5 When h is larger, it takes more water to raise the level by an amount Δh.

■ **EXAMPLE 4** **Tracking a Rocket** A spy tracks a rocket through a telescope to determine its velocity. The rocket is traveling vertically from a launching pad located 10 km away, as in Figure 6. At a certain moment, the spy's instruments show that the angle between the telescope and the ground is equal to $\frac{\pi}{3}$ and is changing at a rate of 0.5 rad/min. What is the rocket's velocity at that moment?

Solution

Step 1. Assign variables and restate the problem.

Let θ be the angle between the telescope and the ground, and let y be the height of the rocket at time t. Then our goal is to compute the rocket's velocity dy/dt when $\theta = \frac{\pi}{3}$. We restate the problem as follows:

$$\text{Compute } \frac{dy}{dt}\bigg|_{\theta=\frac{\pi}{3}} \qquad \text{given that} \qquad \frac{d\theta}{dt} = 0.5 \text{ rad/min when } \theta = \frac{\pi}{3}$$

Step 2. Find an equation that relates the variables and differentiate.

We need a relation between θ and y. As we see in Figure 6,

$$\tan \theta = \frac{y}{10}$$

Now differentiate with respect to time:

$$\sec^2 \theta \frac{d\theta}{dt} = \frac{1}{10} \frac{dy}{dt}$$

$$\frac{dy}{dt} = \frac{10}{\cos^2 \theta} \frac{d\theta}{dt} \qquad \boxed{5}$$

Step 3. Use the given data to find the unknown derivative.

At the given moment, $\theta = \frac{\pi}{3}$ and $d\theta/dt = 0.5$, so Eq. (5) yields

$$\frac{dy}{dt} = \frac{10}{\cos^2(\pi/3)}(0.5) = \frac{10}{(0.5)^2}(0.5) = 20 \text{ km/min.}$$

The rocket's velocity at this moment is 20 km/min or 1,200 km/hour. ∎

For the next example, let's recall the following fact: If an object is located at $x = A$ at time $t = 0$ and moves along the x-axis with constant velocity v, then its position at time t is $x = A + vt$. For example, if $v = 3$ mph, then the object's position is $A + 3t$ miles (see Figure 7).

■ **EXAMPLE 5** One train travels west toward Denver at 120 mph, while a second train travels north away from Denver at 90 mph (Figure 8). At time $t = 0$, the first train is 10 miles east and the second train 20 miles north of the Denver station. Calculate the rate at which the distance between the trains is changing:

(a) At time $t = 0$ (b) 10 min later

Solution

Step 1. Assign variables and restate the problem.

$x = $ distance in miles from the westbound train to the station

$y = $ distance in miles from the northbound train to the station

$\ell = $ distance in miles between the two trains

The velocities of the trains are dx/dt and dy/dt, so our problem can be restated as follows:

FIGURE 6 Tracking a rocket through a telescope.

FIGURE 7 Moving along the x-axis (starting at A) with constant velocity 3.

FIGURE 8 Location and direction of trains.

Compute $\dfrac{d\ell}{dt}$ given that $\dfrac{dx}{dt} = -120$ mph and $\dfrac{dy}{dt} = 90$ mph

Note that dx/dt is negative since the train is moving west (to the left).

Step 2. Find an equation that relates the variables and differentiate.

The variables are related by the Pythagorean Theorem

$$\ell^2 = x^2 + y^2$$

Differentiate with respect to t:

$$2\ell \frac{d\ell}{dt} = 2x \frac{dx}{dt} + 2y \frac{dy}{dt}$$

$$\frac{d\ell}{dt} = \frac{x \frac{dx}{dt} + y \frac{dy}{dt}}{\ell} \qquad \boxed{6}$$

Step 3. Use the given data to find the unknown derivative.

Substituting $dx/dt = -120$ and $dy/dt = 90$ in Eq. (6), we obtain

$$\frac{d\ell}{dt} = \frac{-120x + 90y}{\ell} = \frac{-120x + 90y}{\sqrt{x^2 + y^2}} \qquad \boxed{7}$$

We can now solve the two parts of the problem.

(a) At $t = 0$, we are given $x = 10$ and $y = 20$. By (7), the distance between the trains is changing at a rate of

$$\left. \frac{d\ell}{dt} \right|_{t=0} = \frac{-120(10) + 90(20)}{\sqrt{10^2 + 20^2}} \approx \frac{600}{22.36} \approx 27 \text{ mph}$$

(b) By the remark preceding this example, the coordinates of the trains as a function of time are

$$x = 10 - 120t, \qquad y = 20 + 90t$$

After 10 min, or $t = \frac{1}{6}$ hour,

$$x = 10 - 120 \left(\frac{1}{6} \right) = -10 \text{ miles} \qquad \text{and} \qquad y = 20 + 90 \left(\frac{1}{6} \right) = 35 \text{ miles}$$

Therefore, by (7), the distance between the trains is changing at a rate of

$$\left. \frac{d\ell}{dt} \right|_{t=\frac{1}{6}} = \frac{-120(-10) + 90(35)}{\sqrt{(-10)^2 + 35^2}} \approx \frac{4,350}{36.4} \approx 120 \text{ mph} \qquad \blacksquare$$

3.11 SUMMARY

- Related rate problems present us with situations in which one or more variables are related by an equation. We are asked to compute the ROC of one of the variables in terms of the ROCs of the other variables.
- Draw a diagram if possible. It may also be useful to break the solution into three steps:

Step 1. Assign variables and restate the problem.

Step 2. Find an equation that relates the variables and differentiate.

This gives us an equation relating the known and unknown derivatives. Remember not to substitute values for the variables until after you have computed all derivatives.

Step 3. Use the given data to find the unknown derivative.

• The two facts from geometry that arise most often in related rate problems are the Pythagorean Theorem and the Theorem of Similar Triangles (ratios of corresponding sides are equal).

3.11 EXERCISES

Preliminary Questions

1. Assign variables and restate the following problem in terms of known and unknown derivatives (but do not solve it): How fast is the volume of a cube increasing if its side increases at a rate of 0.5 cm/s?

2. What is the relation between dV/dt and dr/dt if

$$V = \left(\frac{4}{3}\right)\pi r^3$$

In Questions 3–4, suppose that water pours into a cylindrical glass of radius 4 cm. The variables V and h denote the volume and water level at time t, respectively.

3. Restate in terms of the derivatives dV/dt and dh/dt: How fast is the water level rising if water pours in at a rate of 2 cm³/min?

4. Repeat the same for this problem: At what rate is water pouring in if the water level rises at a rate of 1 cm/min?

Exercises

In Exercises 1–2, consider a rectangular bathtub whose base is 18 ft².

1. How fast is the water level rising if water is filling the tub at a rate of 0.7 ft³/min?

2. At what rate is water pouring into the tub if the water level rises at a rate of 0.8 ft/min?

3. The radius of a circular oil slick expands at a rate of 2 m/min.

(a) How fast is the area of the oil slick increasing when the radius is 25 m?

(b) If the radius is 0 at time $t = 0$, how fast is the area increasing after 3 min?

4. At what rate is the diagonal of a cube increasing if its edges are increasing at a rate of 2 cm/s?

In Exercises 5–8, assume that the radius r of a sphere is expanding at a rate of 14 in./min. The volume of a sphere is $V = \frac{4}{3}\pi r^3$ and its surface area is $4\pi r^2$.

5. Determine the rate at which the volume is changing with respect to time when $r = 8$ in.

6. Determine the rate at which the volume is changing with respect to time at $t = 2$ min, assuming that $r = 0$ at $t = 0$.

7. Determine the rate at which the surface area is changing when the radius is $r = 8$ in.

8. Determine the rate at which the surface area is changing with respect to time at $t = 2$ min, assuming that $r = 3$ at $t = 0$.

9. A road perpendicular to a highway leads to a farmhouse located 1 mile away (Figure 9). An automobile travels past the farmhouse at a speed of 60 mph. How fast is the distance between the automobile and the farmhouse increasing when the automobile is 3 miles past the intersection of the highway and the road?

1

60 mph

Automobile

FIGURE 9

10. A conical tank has height 3 m and radius 2 m at the top. Water flows in at a rate of 2 m³/min. How fast is the water level rising when it is 2 m?

11. Follow the same set-up as Exercise 10, but assume that the water level is rising at a rate of 0.3 m/min when it is 2 m. At what rate is water flowing in?

12. Sonya and Isaac are in motorboats located at the center of a lake. At time $t = 0$, Sonya begins traveling south at a speed of 32 mph. At the same time, Isaac takes off, heading east at a speed of 27 mph.

(a) How far have Sonya and Isaac each traveled after 12 min?
(b) At what rate is the distance between them increasing at $t = 12$ min?

13. Answer (a) and (b) in Exercise 12 assuming that Sonya begins moving 1 minute after Isaac takes off.

14. A 6-ft man walks away from a 15-ft lamppost at a speed of 3 ft/s (Figure 10). Find the rate at which his shadow is increasing in length.

15

x l

FIGURE 10

15. At a given moment, a plane passes directly above a radar station at an altitude of 6 miles.
(a) If the plane's speed is 500 mph, how fast is the distance between the plane and the station changing half an hour later?
(b) How fast is the distance between the plane and the station changing when the plane passes directly above the station?

16. In the setting of Exercise 15, suppose that the line through the radar station and the plane makes an angle θ with the horizontal. How fast is θ changing 10 min after the plane passes over the radar station?

17. A hot air balloon rising vertically is tracked by an observer located 2 miles from the lift-off point. At a certain moment, the angle between the observer's line-of-sight and the horizontal is $\frac{\pi}{5}$, and it is changing at a rate of 0.2 rad/min. How fast is the balloon rising at this moment?

18. As a man walks away from a 12-ft lamppost, the tip of his shadow moves twice as fast as he does. What is the man's height?

In Exercises 19–23, refer to a 16-ft ladder sliding down a wall, as in Figures 1 and 2. The variable h is the height of the ladder's top at time t, and x is the distance from the wall to the ladder's bottom.

19. Assume the bottom slides away from the wall at a rate of 3 ft/s. Find the velocity of the top of the ladder at $t = 2$ if the bottom is 5 ft from the wall at $t = 0$.

20. Suppose that the top is sliding down the wall at a rate of 4 ft/s. Calculate dx/dt when $h = 12$.

21. Suppose that $h(0) = 12$ and the top slides down the wall at a rate of 4 ft/s. Calculate x and dx/dt at $t = 2$ s.

22. What is the relation between h and x at the moment when the top and bottom of the ladder move at the same speed?

23. Show that the velocity dh/dt approaches infinity as the ladder slides down to the ground (assuming dx/dt is constant). This suggests that our mathematical description is unrealistic, at least for small values of h. What would, in fact, happen as the top of the ladder approaches the ground?

24. The radius r of a right circular cone of fixed height $h = 20$ cm is increasing at a rate of 2 cm/s. How fast is the volume increasing when $r = 10$?

25. Suppose that both the radius r and height h of a circular cone change at a rate of 2 cm/s. How fast is the volume of the cone increasing when $r = 10$ and $h = 20$?

26. A particle moves counterclockwise around the ellipse $9x^2 + 16y^2 = 25$ (Figure 11).

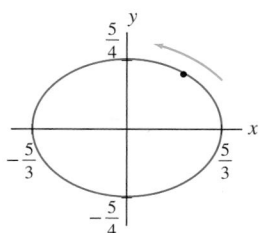

FIGURE 11

(a) In which of the four quadrants is the derivative dx/dt positive? Explain your answer.
(b) Find a relation between dx/dt and dy/dt.
(c) At what rate is the x-coordinate changing when the particle passes the point $(1, 1)$ if its y-coordinate is increasing at a rate of 6 ft/s?
(d) What is dy/dt when the particle is at the top and bottom of the ellipse?

27. A searchlight rotates at a rate of 3 revolutions per minute. The beam hits a wall located 10 miles away and produces a dot of light that moves horizontally along the wall. How fast is this dot moving when the angle θ between the beam and the line through the searchlight perpendicular to the wall is $\frac{\pi}{6}$? Note that $d\theta/dt = 3(2\pi) = 6\pi$.

28. A rocket travels vertically at a speed of 800 mph. The rocket is tracked through a telescope by an observer located 10 miles from the launching pad. Find the rate at which the angle between the telescope and the ground is increasing 3 min after lift-off.

29. A plane traveling at an altitude of 20,000 ft passes directly overhead at time $t = 0$. One minute later you observe that the angle between the vertical and your line of sight to the plane is 1.14 rad and that this angle is changing at a rate of 0.38 rad/min. Calculate the velocity of the airplane.

30. Calculate the rate (in cm^2/s) at which area is swept out by the second hand of a circular clock as a function of the clock's radius.

31. A jogger runs around a circular track of radius 60 ft. Let (x, y) be her coordinates, where the origin is at the center of the track. When the jogger's coordinates are $(36, 48)$, her x-coordinate is changing at a rate of 14 ft/s. Find dy/dt.

32. A car travels down a highway at 55 mph. An observer is standing 500 ft from the highway.
(a) How fast is the distance between the observer and the car increasing at the moment the car passes in front of the observer? Can you justify your answer without relying on any calculations?

(b) How fast is the distance between the observer and the car increasing 1 min later?

In Exercises 33–34, assume that the pressure P (in kilopascals) and volume V (in cubic centimeters) of an expanding gas are related by $PV^b = C$, where b and C are constants (this holds in adiabatic expansion, without heat gain or loss).

33. Find dP/dt if $b = 1.2$, $P = 8$ kPa, $V = 100$ cm^2, and $dV/dt = 20$ cm^3/min.

34. Find b if $P = 25$ kPa, $dP/dt = 12$ kPa/min, $V = 100$ cm^2, and $dV/dt = 20$ cm^3/min.

35. A point moves along the parabola $y = x^2 + 1$. Let $\ell(t)$ be the distance between the point and the origin. Calculate $\ell'(t)$, assuming that the x-coordinate of the point is increasing at a rate of 9 ft/s.

36. The base x of the right triangle in Figure 12 increases at a rate of 5 cm/s, while the height remains constant at $h = 20$. How fast is the angle θ changing when $x = 20$?

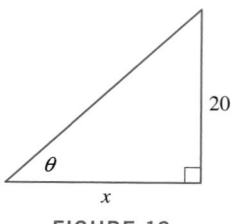

FIGURE 12

37. A water tank in the shape of a right circular cone of radius 300 cm and height 500 cm leaks water from the vertex at a rate of 10 cm^3/min. Find the rate at which the water level is decreasing when it is 200 cm.

38. Two parallel paths 50 ft apart run through the woods. Shirley jogs east on one path at 6 mph, while Jamail walks west on the other at 4 mph. If they pass each other at time $t = 0$, how far apart are they 3 s later, and how fast is the distance between them changing at that moment?

Further Insights and Challenges

39. Henry is pulling on a rope that passes through a pulley on a 10-ft pole and is attached to a wagon (Figure 13). Assume that the rope is attached to a loop on the wagon 2 ft off the ground. Let x be the distance between the loop and the pole.

(a) Find a formula for the speed of the wagon in terms of x and the rate at which Henry pulls the rope.

(b) Find the speed of the wagon when it is 12 ft from the pole, assuming that Henry pulls the rope at a rate of 1.5 ft/s.

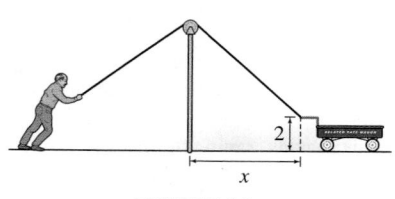

FIGURE 13

40. A roller coaster has the shape of the graph in Figure 14. Show that when the roller coaster passes the point $(x, f(x))$, the vertical velocity of the roller coaster is equal to $f'(x)$ times its horizontal velocity.

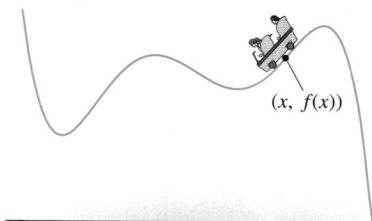

FIGURE 14 Graph of $f(x)$ as a roller coaster track.

41. Using a telescope, you track a rocket that was launched 2 miles away, recording the angle θ between the telescope and the ground at half-second intervals. Estimate the velocity of the rocket if $\theta(10) = 0.205$ and $\theta(10.5) = 0.225$.

42. Two trains leave a station at $t = 0$ and travel with constant velocity v along straight tracks that make an angle θ.

(a) Show that the trains are separating from each other at a rate $v\sqrt{2 - 2\cos\theta}$.

(b) What does this formula give for $\theta = \pi$?

43. A baseball player runs from home plate toward first base at 20 ft/s. How fast is the player's distance from second base changing when the player is halfway to first base? See Figure 15.

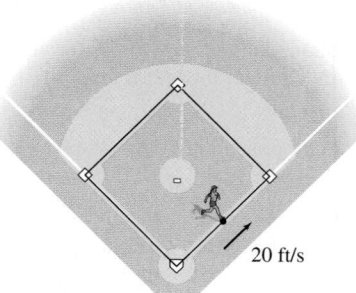

FIGURE 15 Baseball diamond.

44. As the wheel of radius r cm in Figure 16 rotates, the rod of length L attached at the point P drives a piston back and forth in a straight line. Let x be the distance from the origin to the point Q at the end of the rod as in the figure.

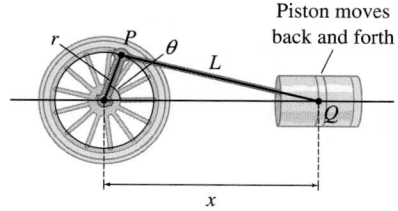

FIGURE 16

(a) Use the Pythagorean Theorem to show that

$$L^2 = (x - r\cos\theta)^2 + r^2\sin^2\theta \qquad \boxed{8}$$

(b) Differentiate Eq. (8) with respect to t to prove that

$$2(x - r\cos\theta)\left(\frac{dx}{dt} + r\sin\theta\frac{d\theta}{dt}\right) + 2r^2\sin\theta\cos\theta\frac{d\theta}{dt} = 0$$

(c) Calculate the speed of the piston when $\theta = \frac{\pi}{2}$, assuming that $r = 10$ cm, $L = 30$ cm, and the wheel rotates at 4 revolutions per minute.

45. A spectator seated 300 m away from the center of a circular track of radius 100 m watches an athlete run laps at a speed of 5 m/s. How fast is the distance between the spectator and athlete changing when the runner is approaching the spectator and the distance between them is 250 m? *Hint:* The diagram for this problem is similar to Figure 16, with $r = 100$ and $x = 300$.

46. A cylindrical tank of radius R and length L lying horizontally as in Figure 17 is filled with oil to height h.

(a) Show that the volume $V(h)$ of oil in the tank as a function of height h is

$$V(h) = L\left(R^2\cos^{-1}\left(1 - \frac{h}{R}\right) - (R - h)\sqrt{2hR - h^2}\right)$$

(b) Show that $\dfrac{dV}{dh} = 2L\sqrt{h(2R - h)}$.

(c) Suppose that $R = 4$ ft and $L = 30$ ft, and that the tank is filled at a constant rate of 10 ft^3/min. How fast is the height h increasing when $h = 5$?

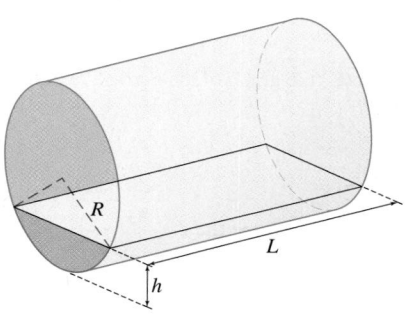

FIGURE 17 Oil in the tank has level h.

CHAPTER REVIEW EXERCISES

In Exercises 1–4, refer to the function $f(x)$ whose graph is shown in Figure 1.

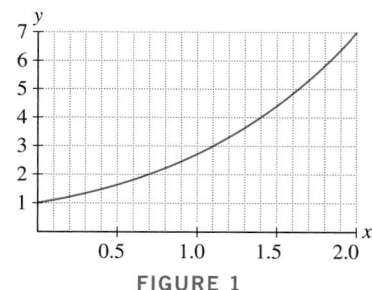

FIGURE 1

1. Compute the average ROC of $f(x)$ over $[0, 2]$. What is the graphical interpretation of this average ROC?

2. For which value of h is $\dfrac{f(0.7 + h) - f(0.7)}{h}$ equal to the slope of the secant line between the points where $x = 0.7$ and $x = 1.1$.

3. Estimate $\dfrac{f(0.7 + h) - f(0.7)}{h}$ for $h = 0.3$. Is this number larger or smaller than $f'(0.7)$?

4. Estimate $f'(0.7)$ and $f'(1.1)$.

In Exercises 5–8, compute $f'(a)$ using the limit definition and find an equation of the tangent line to the graph of $f(x)$ at $x = a$.

5. $f(x) = x^2 - x, \quad a = 1$

6. $f(x) = 5 - 3x, \quad a = 2$

7. $f(x) = x^{-1}, \quad a = 4$

8. $f(x) = x^3, \quad a = -2$

In Exercises 9–12, compute dy/dx using the limit definition.

9. $y = 4 - x^2$

10. $y = \sqrt{2x + 1}$

11. $y = \dfrac{1}{2 - x}$

12. $y = \dfrac{1}{(x - 1)^2}$

In Exercises 13–16, express the limit as a derivative.

13. $\displaystyle\lim_{h \to 0} \frac{\sqrt{1 + h} - 1}{h}$

14. $\displaystyle\lim_{x \to -1} \frac{x^3 + 1}{x + 1}$

15. $\displaystyle\lim_{t \to \pi} \frac{\sin t \cos t}{t - \pi}$

16. $\displaystyle\lim_{\theta \to \pi} \frac{\cos\theta - \sin\theta + 1}{\theta - \pi}$

17. Find $f(4)$ and $f'(4)$ if the tangent line to the graph of $f(x)$ at $x = 4$ has equation $y = 3x - 14$.

18. Each graph in Figure 2 shows the graph of a function $f(x)$ and its derivative $f'(x)$. Determine which is the function and which is the derivative.

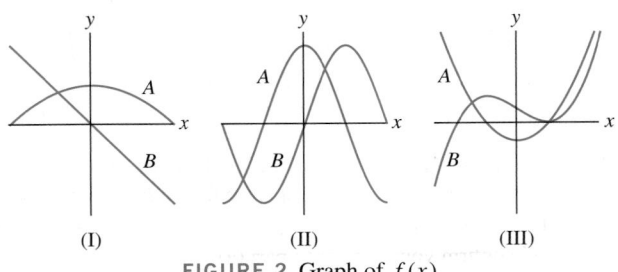

(I)   (II)   (III)

FIGURE 2 Graph of $f(x)$.

19. Is (A), (B), or (C) the graph of the derivative of the function $f(x)$ shown in Figure 3?

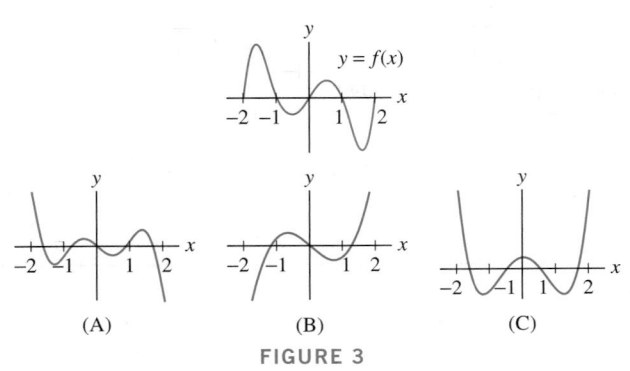

(A)   (B)   (C)

FIGURE 3

20. Let $N(t)$ be the percentage of a state population infected with a flu virus on week t of an epidemic. What percentage is likely to be infected in week 4 if $N(3) = 8$ and $N'(3) = 1.2$?

21. A girl's height $h(t)$ (in centimeters) is measured at time t (years) for $0 \le t \le 14$:

52, 75.1, 87.5, 96.7, 104.5, 111.8, 118.7, 125.2,

131.5, 137.5, 143.3, 149.2, 155.3, 160.8, 164.7

(a) What is the girl's average growth rate over the 14-year period?

(b) Is the average growth rate larger over the first half or the second half of this period?

(c) Estimate $h'(t)$ (in centimeters per year) for $t = 3, 8$.

22. A planet's period P (time in years to complete one revolution around the sun) is approximately $0.0011A^{3/2}$, where A is the average distance (in millions of miles) from the planet to the sun.

(a) Calculate P and dP/dA for Mars using the value $A = 142$.

(b) Estimate the increase in P if A is increased to 143.

In Exercises 23–24, use the following table of values for the number $A(t)$ of automobiles (in millions) manufactured in the United States in year t.

t	1970	1971	1972	1973	1974	1975	1976
$A(t)$	6.55	8.58	8.83	9.67	7.32	6.72	8.50

23. What is the interpretation of $A'(t)$? Estimate $A'(1971)$. Does $A'(1974)$ appear to be positive or negative?

24. Given the data, which of (A)–(C) in Figure 4 could be the graph of the derivative $A'(t)$? Explain.

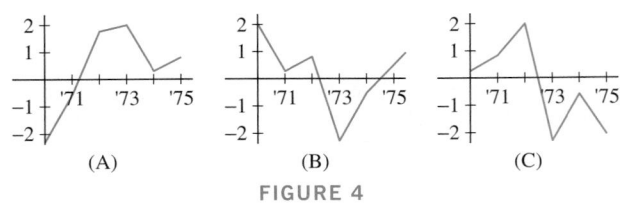

(A)   (B)   (C)

FIGURE 4

25. Which of the following is equal to $\dfrac{d}{dx}2^x$?

(a) 2^x

(b) $(\ln 2)2^x$

(c) $x2^{x-1}$

(d) $\dfrac{1}{\ln 2}2^x$

26. Describe the graphical interpretation of the relation $g'(x) = 1/f'(g(x))$, where $f(x)$ and $g(x)$ are inverses of each other.

27. Find $g'(8)$, where $g(x)$ is the inverse of a differentiable function $f(x)$ such that $f(-1) = 8$ and $f'(-1) = 12$.

In Exercises 28–53, compute the derivative.

28. $y = 3x^5 - 7x^2 + 4$

29. $y = 4x^{-3/2}$

30. $y = t^{-7.3}$

31. $y = 4x^2 - x^{-2}$

32. $y = \dfrac{x+1}{x^2+1}$

33. $y = \dfrac{3t-2}{4t-9}$

34. $y = (x^4 - 9x)^6$

35. $y = (3t^2 + 20t^{-3})^6$

36. $y = (2 + 9x^2)^{3/2}$

37. $y = (x+1)^3(x+4)^4$

38. $y = \dfrac{z}{\sqrt{1-z}}$

39. $y = \left(1 + \dfrac{1}{x}\right)^3$

40. $y = \dfrac{x^4 + \sqrt{x}}{x^2}$

41. $y = \dfrac{1}{(1-x)\sqrt{2-x}}$

42. $y = \tan(t^{-3})$

43. $y = 4\cos(2 - 3x)$

44. $y = \sin(2x)\cos^2 x$

45. $y = \sin(\sqrt{x^2 + 1})$

46. $y = \tan^3 \theta$

47. $y = \sin\left(\dfrac{4}{\theta}\right)$

48. $y = \dfrac{t}{1 + \sec t}$

49. $y = z\csc(9z + 1)$

50. $y = \dfrac{8}{1 + \cot\theta}$

51. $y = \tan(\cos x)$

52. $y = \sqrt{x + \sqrt{x + \sqrt{x}}}$

53. $y = \cos(\cos(\cos(\theta)))$

In Exercises 54–75, find the derivative.

54. $f(x) = 9e^{-4x}$

55. $f(x) = \ln(4x^2 + 1)$

56. $f(x) = \dfrac{e^{-x}}{x}$

57. $f(x) = \ln(x + e^x)$

58. $G(s) = (\ln(s))^2$

59. $G(s) = \ln(s^2)$

60. $g(t) = e^{4t - t^2}$

61. $g(t) = t^2 e^{1/t}$

62. $f(\theta) = \ln(\sin\theta)$

63. $f(\theta) = \sin(\ln\theta)$

64. $f(x) = e^{x + \ln x}$

65. $f(x) = e^{\sin^2 x}$

66. $h(y) = 2^{1 - y}$

67. $h(y) = \dfrac{1 + e^y}{1 - e^y}$

68. $G(s) = \cos^{-1}(s^{-1})$

69. $G(s) = \tan^{-1}(\sqrt{s})$

70. $f(x) = \ln(\csc^{-1} x)$

71. $f(x) = e^{\sec^{-1} x}$

72. $g(t) = \sinh(t^2)$

73. $h(y) = y\tanh(4y)$

74. $g(x) = \tanh^{-1}(e^x)$

75. $g(t) = \sqrt{t^2 - 1}\,\sinh^{-1} t$

76. Suppose that $f(g(x)) = e^{x^2}$, where $g(1) = 2$ and $g'(1) = 4$. Find $f'(2)$.

In Exercises 77–78, let $f(x) = xe^{-x}$.

77. Show that $f(x)$ has an inverse on $[1, \infty)$. Let $g(x)$ be this inverse. Find the domain and range of $g(x)$ and compute $g'(2e^{-2})$.

78. Show that $f(x) = c$ has two solutions if $0 < c < 1$.

In Exercises 79–84, use the table of values to calculate the derivative of the given function at $x = 2$.

x	$f(x)$	$g(x)$	$f'(x)$	$g'(x)$
2	5	4	-3	9
4	3	2	-2	3

79. $S(x) = 3f(x) - 2g(x)$

80. $H(x) = f(x)g(x)$

81. $R(x) = \dfrac{f(x)}{g(x)}$

82. $G(x) = f(g(x))$

83. $F(x) = f(g(2x))$

84. $K(x) = f(x^2)$

In Exercises 85–88, let $f(x) = x^3 - 3x^2 + x + 4$.

85. Find the points on the graph of $f(x)$ where the tangent line has slope 10.

86. For which values of x are the tangent lines to the graph of $f(x)$ horizontal?

87. Find all values of b such that $y = 25x + b$ is tangent to the graph of $f(x)$.

88. Find all values of k such that $f(x)$ has only one tangent line of slope k.

89. **(a)** Show that there is a unique value of a such that $f(x) = x^3 - 2x^2 + x + 1$ has the same slope at both a and $a + 1$.
(b) $\boxed{\text{GU}}$ Plot $f(x)$ together with the tangent lines at $x = a$ and $x = a + 1$ and confirm your answer to part (a).

90. The following table gives the percentage of voters supporting each of two candidates in the days before an election. Compute the average ROC of A's percentage over the intervals from day 20 to 15, day 15 to 10, and day 10 to 5. If this trend continues over the last 5 days before the election, will A win?

Days before Election	20	15	10	5
Candidate A	44.8%	46.8%	48.3%	49.3%
Candidate B	55.2%	53.2%	51.7%	50.7%

In Exercises 91–96, calculate y''.

91. $y = 12x^3 - 5x^2 + 3x$

92. $y = x^{-2/5}$

93. $y = \sqrt{2x + 3}$

94. $y = \dfrac{4x}{x + 1}$

95. $y = \tan(x^2)$

96. $y = \sin^2(x + 9)$

97. In Figure 5, label the graphs f, f', and f''.

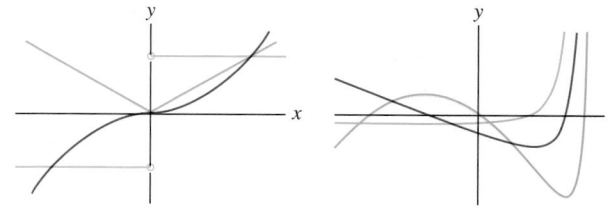

FIGURE 5

98. Let q be a differentiable function of p, say, $q = f(p)$. If p changes by an amount Δp, then the percentage change in p is $(100\Delta p)/p$, and the corresponding percentage change in q is $(100\Delta q)/q$, where $\Delta q = f(p + \Delta p) - f(p)$. The *percentage rate of change* of q with respect to p is the limit of the ratio of percentage changes:

$$\lim_{\Delta p \to 0} \frac{(100\Delta q)/q}{(100\Delta p)/p}$$

Show that this limit is equal to $\left(\dfrac{p}{q}\right)\dfrac{dq}{dp}$.

99. The number q of frozen chocolate cakes that a commercial bakery can sell per week depends on the price p. In this setting, the percentage ROC of q with respect to p is called the *price elasticity of demand* $E(p)$. Assume that $q = 50p(10 - p)$ for $5 < p < 10$.

(a) Show that $E(p) = \dfrac{2p - 10}{p - 10}$.

(b) Compute the value of $E(8)$ and explain how it justifies the following statement: If the price is set at \$8, then a 1% increase in price reduces demand by 3%.

100. The daily demand for flights between Chicago and St. Louis at the price p is $q = 350 - 0.7p$ (in hundreds of passengers). Calculate the price elasticity of demand when $p = \$250$ and estimate the number of additional passengers if the ticket price is lowered by 1%.

In Exercises 101–108, compute $\dfrac{dy}{dx}$.

101. $x^3 - y^3 = 4$

102. $4x^2 - 9y^2 = 36$

103. $y = xy^2 + 2x^2$

104. $\dfrac{y}{x} = x + y$

105. $x + y = \sqrt{3x^2 + 2y^2}$

106. $y = \sin(x + y)$

107. $\tan(x + y) = xy$

108. $(\sin x)(\cos y) = x^2 - y$

109. Find the points on the graph of $x^3 - y^3 = 3xy - 3$ where the tangent line is horizontal.

110. Find the points on the graph of $x^{2/3} + y^{2/3} = 1$ where the tangent line has slope 1.

111. For which values of α is $f(x) = |x|^\alpha$ differentiable at $x = 0$?

112. Let $f(x) = x^2 \sin(x^{-1})$ for $x \neq 0$ and $f(0) = 0$. Show that $f'(x)$ exists for all x (including $x = 0$) but that $f'(x)$ is not continuous at $x = 0$ (Figure 6).

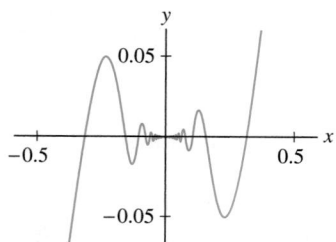

FIGURE 6 Graph of $f(x) = x^2 \sin(x^{-2})$

113. Water pours into the tank in Figure 7 at a rate of 20 m³/min. How fast is the water level rising when the water level is $h = 4$ m?

114. The minute hand of a clock is 4 in. long and the hour hand is 3 in. long. How fast is the distance between the tips of the hands changing at 3 o'clock?

FIGURE 7

115. A light moving at 3 ft/s approaches a 6-ft man standing 12 ft from a wall (Figure 8). The light is 3 ft above the ground. How fast is the tip P of the man's shadow moving when the light is 24 ft from the wall?

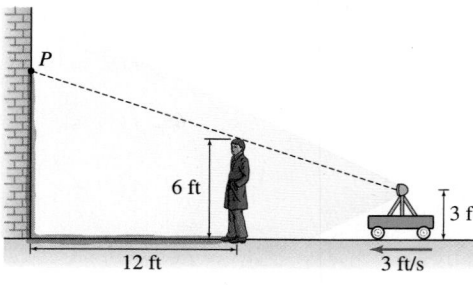

FIGURE 8

116. A bead slides down the curve $xy = 10$. Find the bead's horizontal velocity if its height at time t seconds is $y = 80 - 16t^2$ cm.

117. (a) Side x of the triangle in Figure 9 is increasing at 2 cm/s and side y is increasing at 3 cm/s. Assume that θ decreases in such a way that the area of the triangle has the constant value 4 cm². How fast is θ decreasing when $x = 4$, $y = 4$?

(b) How fast is the distance between P and Q changing when $x = 4$, $y = 4$?

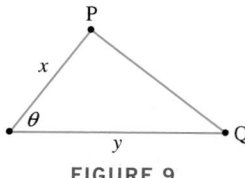

FIGURE 9

In Exercises 118–123, use logarithmic differentiation to find the derivative.

118. $y = \dfrac{(x + 1)^3}{(4x - 2)^2}$

119. $y = \dfrac{(x + 1)(x + 2)^2}{(x + 3)(x + 4)}$

120. $y = e^{(x-1)^2} e^{(x-3)^2}$

121. $y = \dfrac{e^x \sin^{-1} x}{\ln x}$

122. $y = \dfrac{e^{3x}(x - 2)^2}{(x + 1)^2}$

123. $y = x^{\sqrt{x}}(x^{\ln x})$

4 | APPLICATIONS OF THE DERIVATIVE

The previous chapter developed the basic tools of differentiation. In this chapter, we put these tools to work. We use the first and second derivatives to analyze functions and their graphs, and we develop techniques for solving optimization problems (problems that involve finding the minimum or maximum value of a function). Newton's Method in Section 4.8 employs the derivative to find numerical approximations to solutions of equations. In Section 4.9, we introduce antidifferentiation, the inverse operation to differentiation, to prepare for the study of integration in Chapter 5.

The honeycomb structure is designed to minimize the amount of wax required (see Exercise 72 in Section 4.2). As far as we know, bees found this structure without using calculus.

4.1 Linear Approximation and Applications

In some situations we are interested in determining the "effect of a small change." For example:

- A basketball player releases the ball at a certain angle θ. How does a small change in θ affect the distance of the shot? (Exercises 64–65.)
- The price of a movie ticket is $9.00. How will revenues at the box office be affected if the ticket price is increased to $9.50?
- The cube root of 27 is 3. How much larger is the cube root of 27.2? (Exercise 25.)

In each case, we have a function $f(x)$ and we're interested in the change

$$\Delta f = f(a + \Delta x) - f(a), \qquad \text{where } \Delta x \text{ is small}$$

The **Linear Approximation** uses the derivative to estimate Δf without computing it exactly. By definition, the derivative is the limit

$$f'(a) = \lim_{\Delta x \to 0} \frac{f(a + \Delta x) - f(a)}{\Delta x} = \lim_{\Delta x \to 0} \frac{\Delta f}{\Delta x}$$

So when Δx is small, we have $\Delta f / \Delta x \approx f'(a)$, and thus,

$$\Delta f \approx f'(a) \Delta x$$

←·· REMINDER The notation $\approx$ means "approximately equal to." The accuracy of the linear approximation is discussed at the end of this section.

Linear Approximation of Δf   If f is differentiable at $x = a$ and Δx is small, then

$$\boxed{\Delta f \approx f'(a) \Delta x}$$

where $\Delta f = f(a + \Delta x) - f(a)$.

The linear approximation states

$$\Delta f \approx f'(a)\,\Delta x$$

where $\Delta f = f(a + \Delta x) - f(a)$.

GRAPHICAL INSIGHT   The Linear Approximation is often called the **tangent line approximation** because of the following interpretation. The quantity Δf is the vertical change from $x = a$ to $x = a + \Delta x$ in the graph of $f(x)$ (Figure 1). Recall that for a nonvertical straight line, the vertical change is equal to the slope times the horizontal change. Since the tangent line has slope $f'(a)$, the vertical change in the tangent line is $f'(a)\Delta x$. What the Linear Approximation does, therefore, is use the vertical change in the tangent line as an approximation to the vertical change in the graph of $f(x)$. When Δx is small, the two quantities are nearly equal.

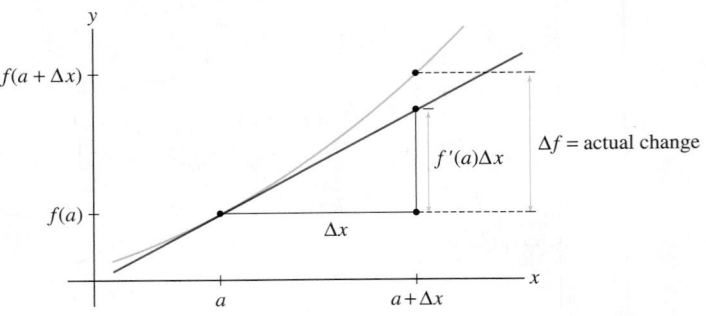

FIGURE 1 Graphical meaning of the Linear Approximation $\Delta f \approx f'(a)\Delta x$.

Keep in mind the different roles played by Δf and $f'(a)\Delta x$. The quantity of interest is Δf and we estimate it by $f'(a)\,\Delta x$. Most importantly, the Linear Approximation tells us that there is a nearly linear relationship between Δf and Δx when Δx is small.

When we use a quantity E as an estimate for another quantity A, the "error" is defined as the absolute value $|A - E|$.

■ **EXAMPLE 1**   Use the Linear Approximation to estimate $\dfrac{1}{10.2} - \dfrac{1}{10}$. How accurate is your estimate?

Solution   We apply the Linear Approximation to $f(x) = \dfrac{1}{x}$ with $a = 10$ and $\Delta x = 0.2$:

$$\Delta f = f(10.2) - f(10) = \frac{1}{10.2} - \frac{1}{10} \approx f'(10)\Delta x$$

Since $f'(x) = -x^{-2}$, $f'(10) = -0.01$ and we obtain the estimate

$$\frac{1}{10.2} - \frac{1}{10} \approx f'(10)\Delta x = -(0.01)(0.2) = -0.002$$

How accurate is this estimate? A calculator gives the value

$$\frac{1}{10.2} - \frac{1}{10} \approx -0.00196$$

The error in the linear approximation is the quantity

$$Error = \left| \Delta f - f'(a)\Delta x \right|$$

Therefore, the error is less than 10^{-4}:

$$Error \approx \left| -0.00196 - (-0.002) \right| = 0.00004 < 10^{-4}$$ ■

Recall the formula for the distance traveled when the velocity is constant:

$$\text{Distance traveled} = \text{velocity} \times \text{time}$$

The Linear Approximation tells us that this equation is *approximately* true over small time intervals even if the velocity is changing. If $s(t)$ is the position of an object at time

t and Δs is the change in position over a small time interval $[t_0, t_0 + \Delta t]$, then

$$\Delta s \approx \underbrace{s'(t_0)}_{\text{Velocity at } t_0} \times \underbrace{\Delta t}_{\text{Length of time}}$$

■ **EXAMPLE 2** **Position and Velocity** The position of an object in linear motion at time t (in seconds) is $s(t) = t^3 - 20t + 8$ m. Estimate the distance traveled over the time interval $[3, 3.025]$.

Solution We have $s'(t) = 3t^2 - 20$ and hence $s'(3) = 7$ m/s. The time interval $[3, 3.025]$ has length $\Delta t = 0.025$, so the Linear Approximation gives

$$\Delta s \approx s'(3)\Delta t = 7(0.025) = 0.175 \text{ m} \qquad ■$$

When engineers need to monitor the change in position of an object with great accuracy, they may use a cable position transducer (Figure 2). This device detects and records the movement of a metal cable attached to the object. The accuracy is affected by changes in the ambient temperature because heat causes the cable to stretch. The Linear Approximation can be used to estimate these effects.

■ **EXAMPLE 3** **Thermal Expansion** A thin metal cable has length $L = 6$ in. when the ambient temperature is $T = 70°$F. Estimate the change in length when T rises to $75°$F, assuming that

$$\frac{dL}{dT} = kL \qquad \boxed{1}$$

where $k = 9.6 \times 10^{-6°}\text{F}^{-1}$ (k is called the coefficient of thermal expansion).

Solution How does the Linear Approximation apply here? We want to estimate the change in length ΔL when T increases from $70°$ to $75°$, that is, when $\Delta T = 5°$. We use Eq. (1) to find dL/dT when $L = 6$ in.:

$$\frac{dL}{dT} = kL = (9.6 \times 10^{-6})(6) \approx 5.8 \times 10^{-5} \text{ in./}°\text{F}$$

According to the Linear Approximation, ΔL is estimated by the derivative times ΔT:

$$\Delta L \approx \frac{dL}{dT}\Delta T \approx (5.8 \times 10^{-5})(5) = 2.9 \times 10^{-4} \text{ in.} \qquad ■$$

FIGURE 2 Cable position transducers (manufactured by Space Age Control, Inc.). In one application, a transducer was used to compare the changes in throttle position on a Formula 1 race car with the shifting actions of the driver.

In the next example, we set $h = \Delta x$ and write the Linear Approximation in the equivalent form:

$$\boxed{f(a + h) - f(a) \approx f'(a)h} \qquad \boxed{2}$$

■ **EXAMPLE 4** **Weight Loss in an Airplane** Newton's Law of Gravitation can be used to show that if an object weighs w pounds on the surface of the earth, then its weight at distance x from the center of the earth is

$$W(x) = \frac{wR^2}{x^2} \qquad (\text{for } x \geq R)$$

where $R = 3{,}960$ miles is the radius of the earth. Estimate the weight lost by a 200-lb football player flying in a jet at an altitude of 7 miles.

Solution We use Eq. (2) with $a = 3{,}960$. In moving from the earth's surface ($x = 3{,}960$) to altitude h ($x = 3{,}960 + h$), an object's weight changes by the amount

$$\Delta W = W(3{,}960 + h) - W(3{,}960) \approx W'(3{,}960)h$$

We have

$$W'(x) = \frac{d}{dx}\left(\frac{wR^2}{x^2}\right) = -\frac{2wR^2}{x^3}$$

$$W'(3{,}960) = -\frac{2wR^2}{3{,}960^3} = -\frac{2w(3{,}960)^2}{3{,}960^3} \approx -0.000505w$$

The player's weight is $w = 200$, so $W'(3{,}960) \approx -(0.000505)(200) = -0.101$. By Eq. (2), $\Delta W \approx -0.101h$, or more simply, $\Delta W \approx -0.1h$. In other words, *the player loses approximately one-tenth of a pound per mile of altitude gained*. At an altitude of $h = 7$ miles, the weight loss is $\Delta W \approx -0.7$ lb. Table 1 compares ΔW with the approximation $-0.101h$. ∎

TABLE 1   Actual versus Approximate Weight Loss for a 200-lb Person

Altitude (miles)	ΔW (pounds)	Linear Approx.
0	0	0
1	−0.10097	−0.101
2	−0.20187	−0.202
3	−0.30269	−0.303
4	−0.40343	−0.404
5	−0.50410	−0.505
6	−0.60469	−0.606
7	−0.70520	−0.707

Suppose we measure the *diameter* of a circle to be D and use this result to compute the *area* of the circle. If our measurement of D is inexact, the area computation will also be inexact. What is the effect of the measurement error on the resulting area computation? This can be estimated using the Linear Approximation, as in the next example.

■ **EXAMPLE 5   Effect of an Inexact Measurement** The Bonzo Pizza Company claims that its pizzas are circular with diameter 18 in.

(a) What is the area of the pizza?

(b) Estimate the amount of pizza lost or gained if Bonzo's chefs err in the diameter by at most 0.4 in. (Figure 3).

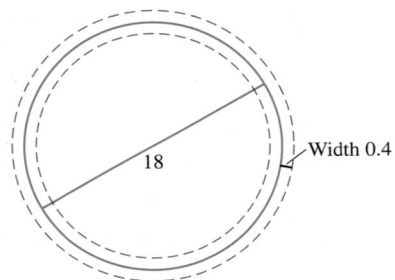

FIGURE 3 The border of the actual pizza lies between the dashed circles.

Solution First, we need a formula for the area A of a circle in terms of its diameter D. Since the radius is $r = D/2$, the area is

$$A(D) = \pi r^2 = \pi\left(\frac{D}{2}\right)^2 = \frac{\pi}{4}D^2$$

(a) The area of an 18-in. pizza is $A(18) = (\frac{\pi}{4})(18)^2 \approx 254.5$ in.^{2}.

(b) If the chefs make a mistake and the actual diameter is equal to $18 + \Delta D$, then the loss or gain in pizza area is $\Delta A = A(18 + \Delta D) - A(18)$. Observe that $A'(D) = \frac{\pi}{2}D$ and $A'(18) = 9\pi \approx 28.3$ in., so the Linear Approximation yields

$$\Delta A = A(18 + \Delta D) - A(18) \approx (28.3)\,\Delta D$$

Because ΔD is at most ± 0.4 in., the gain or loss in pizza is no more than around

$$\Delta A \approx \pm(28.3)(0.4) \approx \pm 11.3 \text{ in.}^2$$

This is a loss or gain of approximately $4\frac{1}{2}\%$. ∎

In this example, we interpret ΔA as the possible error in the computation of $A(D)$. This should not be confused with the error in the Linear Approximation. This latter error refers to the accuracy in using $A'(D)\,\Delta D$ to approximate ΔA.

Some textbooks express the Linear Approximation using the "differentials" dx and dy, where dy denotes the vertical change in the tangent line corresponding to a change dx in x. Thus, $dy = f'(x)dx$ since the tangent line has slope $f'(x)$. In this notation, the Linear Approximation is the statement $\Delta y \approx dy$ or $\Delta y \approx f'(x)dx$, where Δy is the change in $f(x)$ for a given change dx in x.

Linearization

$y = f(x)$ $y = L(x)$

P

FIGURE 4 The tangent line is a good approximation in a small neighborhood of $P = (a, f(a))$.

We can approximate $f(x)$ itself rather than the change Δf by rewriting the Linear Approximation in terms of the variable $x = a + \Delta x$. Then

$$f(a + \Delta x) - f(a) \approx f'(a)\,\Delta x$$

$$f(x) - f(a) \approx f'(a)(x - a) \qquad \text{(since } \Delta x = x - a\text{)}$$

$$f(x) \approx f(a) + f'(a)(x - a)$$

The function on the right, denoted $L(x)$, is called the **linearization** of $f(x)$ at $x = a$:

$$L(x) = f'(a)(x - a) + f(a)$$

We refer to $x = a$ as the *center* of the linearization. Notice that $y = L(x)$ is the equation of the tangent line to the graph of $f(x)$ at $x = a$ (Figure 4).

Approximating $f(x)$ by Its Linearization   Assume that f is differentiable at $x = a$. If x is close to a, then

$$f(x) \approx L(x) = f'(a)(x - a) + f(a)$$

■ **EXAMPLE 6** Compute the linearization of $f(x) = \sqrt{x}\,e^{x-1}$ at $a = 1$.

Solution We compute the derivative of $f(x)$ using the Product Rule:

$$f'(x) = x^{1/2}e^{x-1} + \frac{1}{2}x^{-1/2}e^{x-1} = \left(x^{1/2} + \frac{1}{2}x^{-1/2}\right)e^{x-1}$$

Therefore,

$$f(1) = \sqrt{1}\,e^0 = 1, \qquad f'(1) = \left(1 + \frac{1}{2}\right)e^0 = \frac{3}{2}$$

and

$$L(x) = f(1) + f'(1)(x - 1) = 1 + \frac{3}{2}(x - 1) = \frac{3}{2}x - \frac{1}{2} \qquad ■$$

The linearization can be used to approximate function values. Consider the linearization $L(x)$ of $f(x) = \sqrt{x}$ at $a = 1$. Since $f(1)$ and $f'(x) = \frac{1}{2}x^{-1/2}$, we have $f'(1) = \frac{1}{2}$ and

$$L(x) = \frac{1}{2}(x - 1) + 1 = \frac{1}{2}x + \frac{1}{2}$$

The following table compares three approximations with the value of $\sqrt{x}$ obtained from a calculator. As expected, the error is large for $x = 9$ since 9 is not close to the center $a = 1$ (Figure 5).

y

$y = L(x) = \frac{1}{2}x + \frac{1}{2}$

5

3

1

Error is large for $x = 9$

1   5   9 x

FIGURE 5 Graph of linearization of $f(x) = \sqrt{x}$ at $a = 1$.

x	$\sqrt{x}$	**Linearization at $a = 1$**	**Calculator**	**Error**
1.1	$\sqrt{1.1}$	$L(1.1) = \frac{1}{2}(1.1) + \frac{1}{2} = 1.05$	1.0488	0.0012
0.98	$\sqrt{0.98}$	$L(0.98) = \frac{1}{2}(0.98) + \frac{1}{2} = 0.99$	0.98995	$5 \cdot 10^{-5}$
9	$\sqrt{9}$	$L(9) = \frac{1}{2}(9) + \frac{1}{2} = 5$	3	2

In the next example, we compute the **percentage error**, which is often more important than the error itself. By definition,

$$\text{Percentage error} = \left| \frac{\text{error}}{\text{actual value}} \right| \times 100\%$$

■ **EXAMPLE 7** Use the linearization to estimate $\tan(\frac{\pi}{4} + 0.02)$ and compute the percentage error.

Solution We find the linearization of $f(x) = \tan x$ at $a = \frac{\pi}{4}$:

$$f\left(\frac{\pi}{4}\right) = \tan\left(\frac{\pi}{4}\right) = 1, \qquad f'\left(\frac{\pi}{4}\right) = \sec^2\left(\frac{\pi}{4}\right) = (\sqrt{2})^2 = 2$$

$$L(x) = f\left(\frac{\pi}{4}\right) + f'\left(\frac{\pi}{4}\right)\left(x - \frac{\pi}{4}\right) = 1 + 2\left(x - \frac{\pi}{4}\right)$$

The linearization yields the estimate

$$\tan\left(\frac{\pi}{4} + 0.02\right) \approx L\left(\frac{\pi}{4} + 0.02\right) = 1 + 2(0.02) = 1.04$$

A calculator gives $\tan(\frac{\pi}{4} + 0.02) \approx 1.0408$, so

$$\text{Percentage error} \approx \left| \frac{1.0408 - 1.04}{1.0408} \right| \times 100 \approx 0.08\% \qquad ■$$

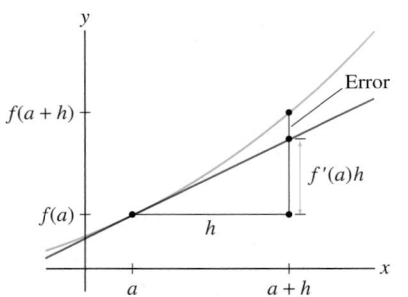

FIGURE 6 Graphical interpretation of the error in the Linear Approximation.

The Size of the Error

The examples in this section may have convinced you that the Linear Approximation gives a good approximation to $\Delta f = f(a + h) - f(a)$ when h is small, but if we want to rely on the Linear Approximation, we need to know more about the size of the error:

$$E = \text{error} = \left| \Delta f - f'(a)h \right|$$

Graphically, the error in the Linear Approximation is the vertical distance between the graph and the tangent line (Figure 6). Now compare the two graphs in Figure 7. When the graph is relatively flat as in (A), the Linear Approximation is quite accurate, especially within the dashed circle. By contrast, when the graph bends sharply as in (B), the Linear Approximation is much less accurate.

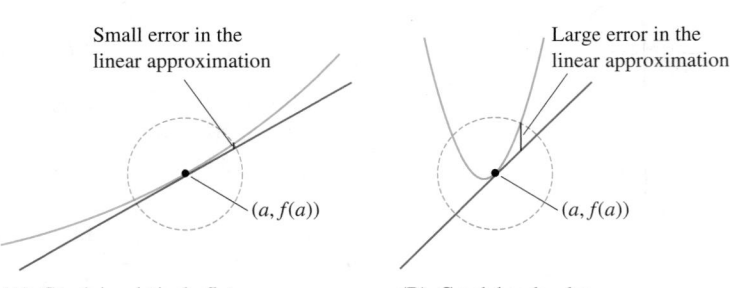

(A) Graph is relatively flat.   (B) Graph bends a lot.

FIGURE 7 The accuracy of the linear approximation depends on how much the curve bends.

The bending or curvature of a graph is determined by how quickly the tangent lines change direction, which is related to $f''(x)$. When $|f''(x)|$ is smaller, the graph is flatter and the Linear Approximation is more accurate over a larger interval around $x = a$.

In Section 8.4, we will prove that if $f''(x)$ is continuous (as is usually the case in practice), then the error satisfies the following **Error Bound**:

$$E \le \frac{1}{2} K h^2$$

3

where K is the maximum value of $|f''(x)|$ on the interval from a to $a + h$. This is a quantitative relation between the error E and the size of the second derivative. It also shows that the error E is of *order two* in h, meaning that E is no larger than a constant times h^2. Thus, if h is small, say, $h = 10^{-n}$, then E is substantially smaller because it has order of magnitude $h^2 = 10^{-2n}$.

The Error Bound also tells us that the accuracy improves as we take smaller and smaller intervals around $x = a$. This is a precise version of the "local linearity" property of differentiable functions discussed in Section 3.2.

4.1 SUMMARY

- Let $\Delta f = f(a + \Delta x) - f(a)$. The *Linear Approximation* is the estimate

$$\Delta f \approx f'(a) \Delta x$$

for small Δx. It can also be written $\Delta y \approx \dfrac{dy}{dx} \Delta x$.

- $L(x) = f'(a)(x - a) + f(a)$ is the *linearization* of f at $x = a$. The Linear Approximation can be rewritten as the estimate $f(x) \approx L(x)$ for small $|x - a|$.
- The error in the Linear Approximation is the quantity

$$\text{Error} = \left| \Delta f - f'(a) \Delta x \right|$$

In many cases, the percentage error is more relevant than the error itself:

$$\text{Percentage error} = \left| \frac{\text{error}}{\text{actual value}} \right| \times 100\%$$

4.1 EXERCISES

Preliminary Questions

1. Estimate $g(1.2) - g(1)$ if $g'(1) = 4$.

2. Estimate $f(2.1)$, assuming that $f(2) = 1$ and $f'(2) = 3$.

3. The velocity of a train at a given instant is 110 ft/s. How far does the train travel during the next half-second (use the Linear Approximation)?

4. Discuss how the Linear Approximation makes the following statement more precise: The sensitivity of the output to a small change in input depends on the derivative.

5. Suppose that the linearization of $f(x)$ at $a = 2$ is $L(x) = 2x + 4$. What are $f(2)$ and $f'(2)$?

Exercises

In Exercises 1–6, use the Linear Approximation to estimate $\Delta f = f(3.02) - f(3)$ for the given function.

1. $f(x) = x^2$

2. $f(x) = x^4$

3. $f(x) = x^{-1}$

4. $f(x) = \dfrac{1}{x+1}$

5. $f(x) = e^{2x}$

6. $f(x) = x \ln x$

7. Let $f(x) = \ln x$. Use the Linear Approximation to estimate $\Delta f = \ln(e^2 + 0.1) - \ln(e^2)$.

In Exercises 8–15, estimate Δf using the Linear Approximation and use a calculator to compute both the error and the percentage error.

8. $f(x) = x - 2x^2$, $a = 5$, $\Delta x = -0.4$

9. $f(x) = \sqrt{1 + x}$, $a = 8$, $\Delta x = 1$

10. $f(x) = \dfrac{1}{1 + x^2}$, $a = 3$, $\Delta x = 0.5$

11. $f(x) = \sin x$, $a = 0$, $\Delta x = 0.02$

12. $f(x) = \tan x$, $a = \frac{\pi}{4}$, $\Delta x = 0.013$

13. $f(x) = \cos x$, $a = \frac{\pi}{4}$, $\Delta x = 0.03$

14. $f(x) = \ln x$, $a = 1$, $\Delta x = 0.02$

15. $f(x) = x^{1/3} e^{x-1}$, $a = 1$, $\Delta x = 0.1$

In Exercises 16–24, estimate the quantity using the Linear Approximation and find the error using a calculator.

16. $\sqrt{26} - \sqrt{25}$

17. $16.5^{1/4} - 16^{1/4}$

18. $\dfrac{1}{\sqrt{101}} - \dfrac{1}{10}$

19. $\dfrac{1}{\sqrt{98}} - \dfrac{1}{10}$

20. $\tan^{-1}(1.05) - \tan^{-1} 1$

21. $\sin(0.023)$ *Hint: Estimate* $\sin(0.023) - \sin(0)$.

22. $(15.8)^{1/4}$

23. $\ln(0.97)$

24. $e^{0.02}$

25. The cube root of 27 is 3. How much larger is the cube root of 27.2? Estimate using the Linear Approximation.

26. Which is larger: $\sqrt{2.1} - \sqrt{2}$ or $\sqrt{9.1} - \sqrt{9}$? Explain using the Linear Approximation.

27. Estimate $\sin 61° - \sin 60°$ using the Linear Approximation. *Note:* You must express $\Delta \theta$ in radians.

28. A thin silver wire has length $L = 18$ cm when the temperature is $T = 30°C$. Estimate the length when $T = 25°C$ if the coefficient of thermal expansion is $k = 1.9 \times 10^{-5} °C^{-1}$ (see Example 3).

29. The atmospheric pressure P (in kilopascals) at altitudes h (in kilometers) for $11 \le h \le 25$ is approximately $P(h) = 128e^{-0.157h}$.

(a) Use the Linear Approximation to estimate the change in pressure at $h = 20$ when $\Delta h = 0.5$.

(b) Compute the actual change and compute the percentage error in the Linear Approximation.

30. The resistance R of a copper wire at temperature $T = 20°C$ is $R = 15\ \Omega$. Estimate the resistance at $T = 22°C$, assuming that $dR/dT\big|_{T=20} = 0.06\ \Omega/°C$.

31. The side s of a square carpet is measured at 6 ft. Estimate the maximum error in the area A of the carpet if s is accurate to within half an inch.

32. A spherical balloon has a radius of 6 in. Estimate the change in volume and surface area if the radius increases by 0.3 in.

33. A stone tossed vertically in the air with initial velocity v ft/s reaches a maximum height of $h = v^2/64$ ft.

(a) Estimate Δh if v is increased from 25 to 26 ft/s.

(b) Estimate Δh if v is increased from 30 to 31 ft/s.

(c) In general, does a 1 ft/s increase in initial velocity cause a greater change in maximum height at low or high initial velocities? Explain.

34. If the price of a bus pass from Albuquerque to Los Alamos is set at x dollars, a bus company takes in a monthly revenue of $R(x) = 1.5x - 0.01x^2$ (in thousands of dollars).

(a) Estimate the change in revenue if the price rises from $50 to $53.

(b) Suppose that $x = 80$. How will revenue be affected by a small increase in price? Explain using the Linear Approximation.

35. The *stopping distance* for an automobile (after applying the brakes) is approximately $F(s) = 1.1s + 0.054s^2$ ft, where s is the speed in mph. Use the Linear Approximation to estimate the change in stopping distance per additional mph when $s = 35$ and when $s = 55$.

36. Juan measures the circumference C of a spherical ball at 40 cm and computes the ball's volume V. Estimate the maximum possible error in V if the error in C is at most 2 cm. Recall that $C = 2\pi r$ and $V = \frac{4}{3}\pi r^3$, where r is the ball's radius.

37. Estimate the weight loss per mile of altitude gained for a 130-lb pilot. At which altitude would she weigh 129.5 lb? See Example 4.

38. How much would a 160-lb astronaut weigh in a satellite orbiting the earth at an altitude of 2,000 miles? Estimate the astronaut's weight loss per additional mile of altitude beyond 2,000.

39. The volume of a certain gas (in liters) is related to pressure P (in atmospheres) by the formula $PV = 24$. Suppose that $V = 5$ with a possible error of ± 0.5 L.

(a) Compute P and estimate the possible error.

(b) Estimate the maximum allowable error in V if P must have an error of at most 0.5 atm.

40. The dosage D of diphenhydramine for a dog of body mass w kg is $D = kw^{2/3}$ mg, where k is a constant. A cocker spaniel has mass $w = 10$ kg according to a veterinarian's scale. Estimate the maximum allowable error in w if the percentage error in the dosage D must be less than 5%.

In Exercises 41–50, find the linearization at $x = a$.

41. $y = \cos x \sin x$, $a = 0$

42. $y = \cos x \sin x$, $a = \dfrac{\pi}{4}$

43. $y = (1 + x)^{-1/2}$, $a = 0$

44. $y = (1 + x)^{-1/2}$, $a = 3$

45. $y = (1 + x^2)^{-1/2}$, $a = 0$

46. $y = \dfrac{\sin x}{x}$, $a = \dfrac{\pi}{2}$

47. $y = \sin^{-1} x$, $a = \frac{1}{2}$

48. $y = e^{\sqrt{x}}$, $a = 1$

49. $y = e^x \ln x, \quad a = 1$ **50.** $y = \tan^{-1} x, \quad a = 1$

51. GU Estimate $\sqrt{16.2}$ using the linearization $L(x)$ of $f(x) = \sqrt{x}$ at $a = 16$. Plot $f(x)$ and $L(x)$ on the same set of axes and determine if the estimate is too large or too small.

52. GU Estimate $1/\sqrt{15}$ using a suitable linearization of $f(x) = 1/\sqrt{x}$. Plot $f(x)$ and $L(x)$ on the same set of axes and determine if the estimate is too large or too small. Use a calculator to compute the percentage error.

In Exercises 53–61, approximate using linearization and use a calculator to compute the percentage error.

53. $\sqrt{17}$ **54.** $\dfrac{1}{\sqrt{17}}$ **55.** $(17)^{1/4}$

56. $(27.03)^{1/3}$ **57.** $(27.001)^{1/3}$ **58.** $(1.2)^{5/3}$

59. $\ln 1.07$ **60.** $e^{-0.012}$ **61.** $\cos^{-1}(0.52)$

62. GU Plot $f(x) = \tan x$ and its linearization $L(x)$ at $a = \dfrac{\pi}{4}$ on the same set of axes.

(a) Does the linearization overestimate or underestimate $f(x)$?

(b) Show, by graphing $y = f(x) - L(x)$ and $y = 0.1$ on the same set of axes, that the error $|f(x) - L(x)|$ is at most 0.1 for $0.55 \le x \le 0.95$.

(c) Find an interval of x-values on which the error is at most 0.05.

63. GU Compute the linearization $L(x)$ of $f(x) = x^2 - x^{3/2}$ at $a = 2$. Then plot $f(x) - L(x)$ and find an interval around $a = 1$ such that $|f(x) - L(x)| \le 0.1$.

In Exercises 64–65, use the following fact derived from Newton's Laws: An object released at an angle θ with initial velocity v ft/s travels a total distance $s = \frac{1}{32} v^2 \sin 2\theta$ ft (Figure 8).

64. A player located 18.1 ft from a basket launches a successful jump shot from a height of 10 ft (level with the rim of the basket), at an angle $\theta = 34°$ and initial velocity of $v = 25$ ft/s.

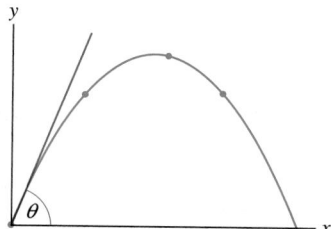

FIGURE 8 Trajectory of an object released at an angle θ.

(a) Show that the distance s of the shot changes by approximately $0.255\Delta\theta$ ft if the angle changes by an amount $\Delta\theta$. Remember to convert the angles to radians in the Linear Approximation.

(b) Is it likely that the shot would have been successful if the angle were off by 2°?

65. Estimate the change in the distance s of the shot if the angle changes from 50° to 51° for $v = 25$ ft/s and $v = 30$ ft/s. Is the shot more sensitive to the angle when the velocity is large or small? Explain.

66. Compute the linearization of $f(x) = 3x - 4$ at $a = 0$ and $a = 2$. Prove more generally that a linear function coincides with its linearization at $x = a$ for all a.

67. According to (3), the error in the Linear Approximation is of "order two" in h. Show that the Linear Approximation to $f(x) = \sqrt{x}$ at $x = 9$ yields the estimate $\sqrt{9 + h} - 3 \approx \frac{1}{6} h$. Then compute the error E for $h = 10^{-n}$, $1 \le n \le 4$, and verify numerically that $E \le 0.006h^2$.

68. Show that the Linear Approximation to $f(x) = \tan x$ at $x = \frac{\pi}{4}$ yields the estimate $\tan(\frac{\pi}{4} + h) - 1 \approx 2h$. Compute the error E for $h = 10^{-n}$, $1 \le n \le 4$, and verify that E satisfies the Error Bound (3) with $K = 6.2$.

Further Insights and Challenges

69. Show that for any real number k, $(1 + x)^k \approx 1 + kx$ for small x. Estimate $(1.02)^{0.7}$ and $(1.02)^{-0.3}$.

70. GU **(a)** Show that $f(x) = \sin x$ and $g(x) = \tan x$ have the same linearization at $a = 0$.

(b) Which function is approximated more accurately? Explain using a graph over $[0, \frac{\pi}{6}]$.

(c) Calculate the error in these linearizations at $x = \frac{\pi}{6}$. Does the answer confirm your conclusion in (b)?

71. Let $\Delta f = f(5 + h) - f(5)$, where $f(x) = x^2$, and let $E = |\Delta f - f'(5)h|$ be the error in the Linear Approximation. Verify directly that E satisfies (3) with $K = 2$ (thus E is of order two in h).

72. Let $f(x) = x^{-1}$ and let $E = |\Delta f - f'(1)h|$ be the error in the Linear Approximation at $a = 1$. Show directly that $E = h^2/(1 + h)$.

Then prove that $E \le 2h^2$ if $-\frac{1}{2} \le h \le \frac{1}{2}$. *Hint:* In this case, $\frac{1}{2} \le 1 + h \le \frac{3}{2}$.

73. Consider the curve $y^4 + x^2 + xy = 13$.

(a) Calculate dy/dx at $P = (3, 1)$ and find the linearization $y = L(x)$ at P.

(b) GU Plot the curve and $y = L(x)$ on the same set of axes.

(c) Although y is only defined implicitly as a function of x, we can use $y = L(x)$ to approximate y for x near 3. Use this idea to find an approximate solution of $y^4 + 3.1^2 + 3.1y = 13$. Based on your plot in (b), is your approximation too large or too small?

(d) CAS Solve $y^4 + 3.1^2 + 3.1y = 13$ numerically using a computer algebra system and use the result to find the percentage error in your approximation.

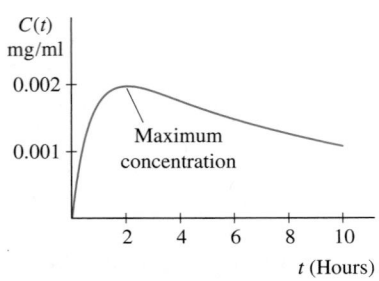

FIGURE 1 $C(t) =$ drug concentration in bloodstream.

We often drop the word "absolute" and speak simply of the min or max on an interval I. When no interval is mentioned, it is understood that we refer to the extreme values on the entire domain of the function.

4.2 Extreme Values

An important problem in many applications is to find the minimum or maximum value of a function $f(x)$. For example, a physician must determine the maximum drug concentration in a patient's bloodstream when a drug is administered. This amounts to finding the highest point on the graph of $C(t)$, the concentration at time t (Figure 1).

We refer to the maximum and minimum values (max and min for short) as **extreme values** or **extrema** (singular: extremum) and to the process of finding them as **optimization**. Sometimes, we are interested in finding the min or max for x in a particular interval I, rather than on the entire domain of $f(x)$.

> **DEFINITION Extreme Values on an Interval**   Let $f(x)$ be a function on an interval I and let $a \in I$. We say that $f(a)$ is the
>
> - **Absolute minimum** of $f(x)$ on I if $f(a) \leq f(x)$ for all $x \in I$.
> - **Absolute maximum** of $f(x)$ on I if $f(a) \geq f(x)$ for all $x \in I$.

Does every function have a minimum or maximum value? Clearly not, as we see by taking $f(x) = x$. Indeed, $f(x) = x$ increases without bound as $x \to \infty$ and decreases without bound as $x \to -\infty$. In fact, extreme values do not always exist even if we restrict ourselves to an interval I. Figure 2 illustrates what can go wrong if I is open or f has a discontinuity:

- **Discontinuity**: (A) shows a discontinuous function with no maximum value. The values of $g(x)$ get arbitrarily close to 3 from below, but 3 is not the maximum value because $g(x)$ never actually takes on the value 3.
- **Open interval:** In (B), $f(x)$ is defined on the *open* interval (a, b) and $f(x)$ has no min or max because it tends to ∞ on the right and $-\infty$ on the left.

Fortunately, the following theorem guarantees that extreme values exist when $f(x)$ is continuous and I is closed [Figure 2(C)].

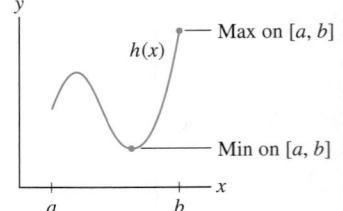

(A) $g(x)$ is discontinuous and has no max on $[a, b]$.

(B) $f(x)$ has no min or max on the open interval (a, b).

(C) $h(x)$ is continuous and $[a, b]$ is closed. Therefore, $h(x)$ has a min and max on $[a, b]$.

FIGURE 2

◀⋯ REMINDER A closed, bounded interval is an interval $I = [a, b]$ (endpoints included) where a and b are both finite. Often, we drop the word "bounded" and refer to a closed, bounded interval more simply as a closed interval. An open interval (a, b) (endpoints not included) may have one or two infinite endpoints.

> **THEOREM 1 Existence of Extrema on a Closed Interval**   If $f(x)$ is a continuous function on a closed (bounded) interval $I = [a, b]$, then $f(x)$ takes on a minimum and a maximum value on I.

CONCEPTUAL INSIGHT   The idea in Theorem 1 is that a continuous function on a closed interval must take on a min and a max because it cannot skip values or go off to $\pm\infty$, as in Figures 2(A) and (B). However, a rigorous proof relies on the *completeness property* of the real numbers (see Appendix D for a proof).

Local Extrema and Critical Points

We now focus on the problem of finding extreme values. A key concept is that of a local minimum or maximum.

DEFINITION Local Extrema   We say that $f(x)$ has a

- **Local minimum** at $x = c$ if $f(c)$ is the minimum value of f on some open interval (in the domain of f) containing c.
- **Local maximum** at $x = c$ if $f(c)$ is the maximum value of $f(x)$ on some open interval (in the domain of f) containing c.

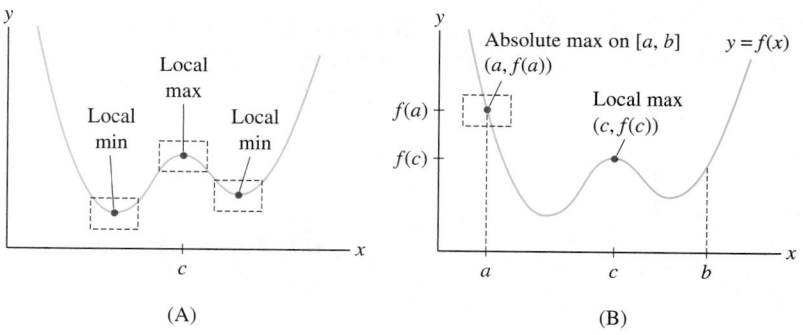

(A)   (B)

FIGURE 3

A local min or max occurs at $x = c$ if $(c, f(c))$ is the highest or lowest point on the graph within some small box [Figure 3(A)]. To be a local max, $f(c)$ must be greater than or equal to all other *nearby* values, but it does not have to be the absolute maximum value (and similarly for a local min). Figure 3(B) illustrates the difference between local and absolute extrema. Note that $f(a)$ is the absolute max on $[a, b]$ but it is not a local max because $f(x)$ takes on larger values to the left of $x = a$.

How do we find the local extrema of a function? The crucial observation is that *the tangent line at a local min or max is horizontal* [Figure 4(A)]. In other words, if $f(c)$

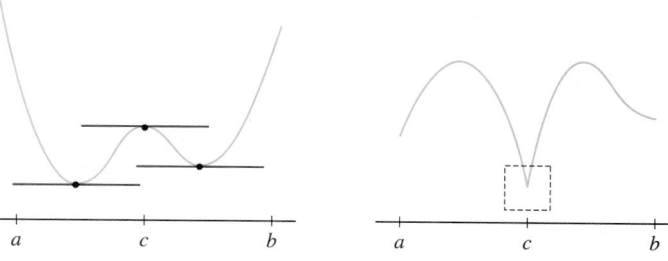

(A) Tangent line is horizontal
at the local extrema.

(B) This local minimum occurs at a point
where the function is not differentiable.

FIGURE 4

is a local min or max, then $f'(c) = 0$. However, this assumes $f(x)$ is differentiable. Otherwise, the tangent line may not exist at a local min or max, as in Figure 4(B). To take both possibilities into account, we define the notion of a critical point.

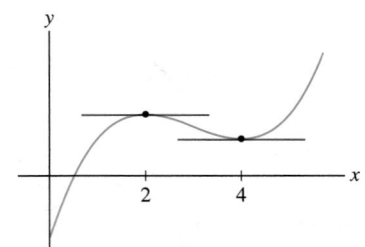

FIGURE 5 Graph of $f(x) = x^3 - 9x^2 + 24x - 10$.

> **DEFINITION Critical Points**   A number c in the domain of f is called a **critical point** if *either* $f'(c) = 0$ or $f'(c)$ does not exist.

■ **EXAMPLE 1**   Find the critical points of $f(x) = x^3 - 9x^2 + 24x - 10$ (Figure 5).

Solution The function $f(x)$ is differentiable everywhere, so the critical points are the solutions of $f'(x) = 0$:

$$f'(x) = 3x^2 - 18x + 24 = 3(x^2 - 6x + 8) = 3(x - 2)(x - 4) = 0$$

The critical points are the roots $c = 2$ and $c = 4$. ■

■ **EXAMPLE 2**   A Nondifferentiable Function   Find the critical points of $f(x) = |x|$.

Solution As we see in Figure 6, $f'(x) = -1$ for $x < 0$ and $f'(x) = 1$ for $x > 0$. Therefore, $f'(x) = 0$ has no solutions with $x \neq 0$. However, $f'(0)$ does not exist and therefore $c = 0$ is a critical point. ■

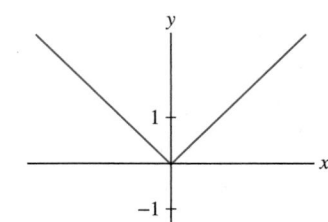

FIGURE 6 Graph of $f(x) = |x|$.

We now state a basic result that we will use repeatedly to find extreme values.

> **THEOREM 2 Fermat's Theorem on Local Extrema**   If $f(c)$ is a local min or max, then c is a critical point of f.

Proof Suppose that $f(c)$ is a local minimum (a local maximum is treated similarly). If $f'(c)$ does not exist, then c is a critical point and there is nothing more to prove. So assume that $f'(c)$ exists. We must then prove that $f'(c) = 0$.

Because $f(c)$ is a local minimum, we have $f(c + h) \geq f(c)$ for all sufficiently small $h \neq 0$. Equivalently, $f(c + h) - f(c) \geq 0$. Now divide this inequality by h:

$$\frac{f(c + h) - f(c)}{h} \geq 0 \qquad \text{if } h > 0 \qquad \boxed{1}$$

$$\frac{f(c + h) - f(c)}{h} \leq 0 \qquad \text{if } h < 0 \qquad \boxed{2}$$

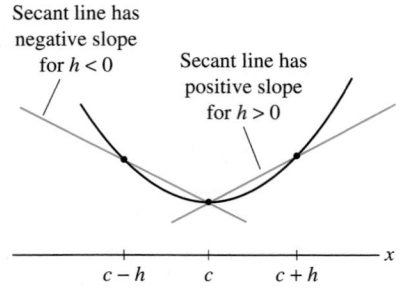

FIGURE 7

Figure 7 shows the graphical interpretation of these inequalities. Taking the one-sided limits of both sides of (1) and (2), we obtain

$$f'(c) = \lim_{h \to 0+} \frac{f(c + h) - f(c)}{h} \geq \lim_{h \to 0+} 0 = 0$$

$$f'(c) = \lim_{h \to 0-} \frac{f(c + h) - f(c)}{h} \leq \lim_{h \to 0-} 0 = 0$$

This shows that $f'(c) \geq 0$ and $f'(c) \leq 0$. The only possibility is $f'(c) = 0$ as claimed. ■

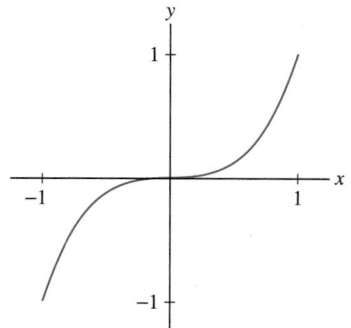

FIGURE 8 Graph of $f(x) = x^3$. The tangent line at $(0, 0)$ is horizontal.

In this section, we restrict our attention to closed intervals because in this case extreme values are guaranteed to exist (Theorem 1). Optimization on open intervals is discussed in Section 4.6.

CONCEPTUAL INSIGHT   Fermat's Theorem *does not claim* that all critical points yield local extrema. "False positives" may exist, by which we mean that $f'(c) = 0$ but $f(c)$ is not a local extremum. In this case, f may have a point of inflection at $x = c$, where the tangent line is horizontal but crosses the graph. For example, if $f(x) = x^3$, then $f'(x) = 3x^2$ and $f'(0) = 0$, but $f(0)$ is neither a local min nor a local max (Figure 8).

Optimizing on a Closed Interval

Finally, we have all the tools needed for optimizing a continuous function on a closed interval. Theorem 1 tells us that the extreme values exist, and the following theorem tells us where to find them, namely among the critical points or endpoints of the interval.

> **THEOREM 3 Extreme Values on a Closed Interval**   Assume that $f(x)$ is continuous on $[a, b]$ and let $f(c)$ be the minimum or maximum value on $[a, b]$. Then c is either a critical point or one of the endpoints a or b.

Proof   Assume first that $f(c)$ is a max on $[a, b]$. Either c is an endpoint a or b, or c belongs to the open interval (a, b). In the latter case, $f(c)$ is also a local maximum because it is the maximum value of $f(x)$ on (a, b). By Fermat's Theorem, c is a critical point. The case of a minimum is similar.   ∎

■ **EXAMPLE 3**   Find the extreme values of $f(x) = 2x^3 - 15x^2 + 24x + 7$ on $[0, 6]$.

Solution   The extreme values occur at critical points or endpoints by Theorem 3, so we can break up the problem neatly into two steps.

Step 1. **Find the critical points.**
   The function $f(x)$ is differentiable, so we find the critical points by solving

$$f'(x) = 6x^2 - 30x + 24 = 6(x - 1)(x - 4) = 0$$

The critical points are $c = 1$ and 4.

Step 2. **Calculate $f(x)$ at the critical points and endpoints and compare.**

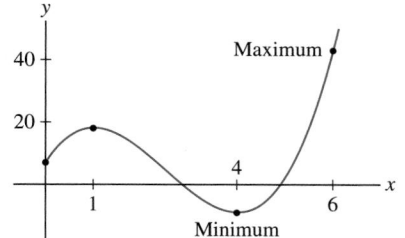

FIGURE 9 Graph of $f(x) = 2x^3 - 15x^2 + 24x + 7$.

x-value	Value of f	
1 (critical point)	$f(1) = 18$	
4 (critical point)	$f(4) = -9$	min
0 (endpoint)	$f(0) = 7$	
6 (endpoint)	$f(6) = 43$	max

The maximum of $f(x)$ on $[0, 6]$ is the largest of the values in this table, namely $f(6) = 43$ (Figure 9). Similarly, the minimum is $f(4) = -9$.   ■

■ **EXAMPLE 4**   *Function with a Cusp*   Find the maximum of $f(x) = 1 - (x - 1)^{2/3}$ on $[-1, 2]$.

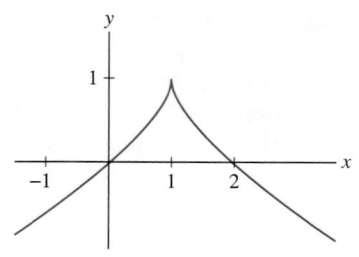

FIGURE 10 Graph of
$f(x) = 1 - (x-1)^{2/3}$.

Solution We have $f(x) = 1 - (x-1)^{2/3}$.

Step 1. Find the critical points.

The derivative is

$$f'(x) = -\frac{2}{3}(x-1)^{-1/3} = -\frac{2}{3(x-1)^{1/3}}$$

The equation $f'(x) = 0$ has no solutions since the numerator of $f'(x)$ is never zero. However, $f'(x)$ does not exist at $x = 1$, so $c = 1$ is a critical point (Figure 10).

Step 2. Calculate $f(x)$ at the critical points and endpoints and compare.

x-value	Value of f	
1 (critical point)	$f(1) = 1$	max
-1 (endpoint)	$f(-1) \approx -0.59$	min
2 (endpoint)	$f(2) = 0$	

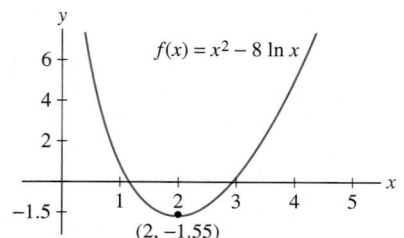

FIGURE 11

EXAMPLE 5 Find the extreme values of $f(x) = x^2 - 8\ln x$ on $[1, 4]$ (Figure 11).

Solution First, solve for the critical points:

$$f'(x) = 2x - \frac{8}{x} = 0 \quad \Rightarrow \quad 2x = \frac{8}{x} \quad \Rightarrow \quad x = \pm 2$$

There is only one critical point $c = 2$ in the interval $[1, 4]$. Next, we compare $f(x)$ at the critical points and endpoints:

x-value	Value of f	
2 (critical point)	$f(2) \approx -1.55$	
1 (endpoint)	$f(1) = 1$	min
4 (endpoint)	$f(4) \approx 4.9$	max

We see that the minimum value on $[1, 4]$ is $f(2) \approx -1.55$ and the maximum value is $f(4) \approx 4.9$.

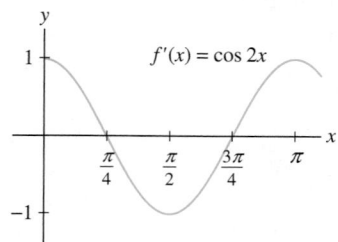

FIGURE 12 Two solutions of $\cos 2x = 0$ lie in the interval $[0, \pi]$, namely $\frac{\pi}{4}$ and $\frac{3\pi}{4}$.

EXAMPLE 6 Trigonometric Example Find the min and max of $f(x) = \sin x \cos x$ on $[0, \pi]$.

Solution First, we rewrite $f(x)$ using a trigonometric identity as $f(x) = \frac{1}{2}\sin 2x$ and solve $f'(x) = \cos 2x = 0$. This gives $2x = \frac{\pi}{2}$ or $\frac{3\pi}{2}$, so the critical points in the interval $[0, \pi]$ are $c = \frac{\pi}{4}$ and $\frac{3\pi}{4}$ (Figure 12). Next, we evaluate $f(x)$ at the critical points and endpoints to find the min and max:

x-value	Value of f	
$\frac{\pi}{4}$ (critical point)	$f\left(\frac{\pi}{4}\right) = \frac{1}{2}$	max
$\frac{3\pi}{4}$ (critical point)	$f\left(\frac{3\pi}{4}\right) = -\frac{1}{2}$	min
0 (endpoint)	$f(0) = 0$	
π (endpoint)	$f(\pi) = 0$	

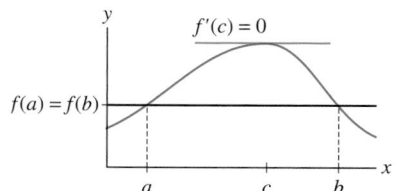

FIGURE 13 Rolle's Theorem: If $f(a) = f(b)$, then $f'(c) = 0$ for some c between a and b.

Rolle's Theorem

As an application of our optimization methods, we prove Rolle's Theorem, which states that if $f(x)$ is differentiable and $f(a) = f(b)$, then the tangent line is horizontal at some point c between a and b (Figure 13).

THEOREM 4 Rolle's Theorem Assume that $f(x)$ is continuous on $[a, b]$ and differentiable on (a, b). If $f(a) = f(b)$, then there exists a number c between a and b such that $f'(c) = 0$.

Proof Since $f(x)$ is continuous and $[a, b]$ is closed, $f(x)$ has a min and a max in $[a, b]$. Where do they occur? There are two cases. If an extreme value occurs at a point c in the *open* interval (a, b), then $f(c)$ is a local extreme value and $f'(c) = 0$ by Fermat's Theorem (Theorem 2). Otherwise, both the min and the max occur at the endpoints. However, $f(a) = f(b)$, so the maximum and minimum values coincide and $f(x)$ must be a constant function. Therefore, $f'(x) = 0$ for all $x \in (a, b)$ and Rolle's Theorem is satisfied for all $c \in (a, b)$. ∎

Pierre de Fermat
(1601–1665)

René Descartes
(1596–1650)

HISTORICAL PERSPECTIVE

We can hardly expect a more general method . . . . This method never fails and could be extended to a number of beautiful problems; with its aid we have found the centers of gravity of figures bounded by straight lines or curves, as well as those of solids, and a number of other results which we may treat elsewhere if we have the time to do so.

—From Fermat's *On Maxima and Minima and on Tangents*

Sometime in the 1630s, in the decade before Isaac Newton was born, the French mathematician Pierre de Fermat (1601–1665) invented a general method for finding extreme values. In essence, Fermat said that if you want to find extrema, you must set the derivative equal to zero and solve for the critical points, just as we have done in this section. He also described a general method for finding tangent lines that is not essentially different from our method of derivatives. For this reason, Fermat is often regarded as one of the inventors of calculus, along with Newton and Leibniz.

At around the same time, René Descartes (1596–1650) developed a different but much less effective approach to finding tangent lines. Descartes, after whom Cartesian coordinates are named, was the leading philosopher and scientist of his time in Europe and is regarded today as the father of modern philosophy. He was a profound thinker and brilliant writer, but he was also known for his sarcastic wit and disdainful attitude, believing that he had little

to learn from other people. Fermat had angered Descartes because he had once dismissed Descartes's work on optics as "groping about in the shadows." Perhaps in retaliation, Descartes attacked Fermat, claiming that his method of finding tangents was wrong. After some third-party refereeing, Descartes was forced to admit that it was correct after all. He wrote back to Fermat:

. . . Seeing the last method that you use for finding tangents to curved lines, I can reply to it in no other way than to say that it is very good and that, if you had explained it in this manner at the outset, I would have not contradicted it at all.

Unfortunately, Descartes, who was the more famous and influential of the two men, continued to try to damage Fermat's reputation. Today, Fermat is recognized as one of the greatest mathematicians of his age who made far-reaching contributions in several areas of mathematics.

■ **EXAMPLE 7**   **Illustrating Rolle's Theorem**   Verify Rolle's Theorem for

$$f(x) = x^4 - x^2 \qquad \text{on } [-2, 2]$$

Solution   First, we check that $f(2) = f(-2)$:

$$f(2) = 2^4 - 2^2 = 12, \qquad f(-2) = (-2)^4 - (-2)^2 = 12$$

Second, we note that $f(x)$ is differentiable everywhere. Therefore, the hypotheses of Rolle's Theorem are satisfied and $f'(c) = 0$ must have a solution in $(-2, 2)$. To verify this, we solve $f'(x) = 4x^3 - 2x = 2x(2x^2 - 1) = 0$. The solutions $c = 0$ and $c = \pm 1/\sqrt{2} \approx \pm 0.707$ all lie in $(-2, 2)$, so Rolle's Theorem is satisfied with three values of c. ■

■ **EXAMPLE 8**   **Using Rolle's Theorem**   Show that $f(x) = x^3 + 9x - 4$ has at most one real root.

Solution   We use proof by contradiction. If $f(x)$ had two real roots a and b, then $f(a) = f(b) = 0$ and Rolle's Theorem would imply that $f'(c) = 0$ for some $c \in (a, b)$. However, $f'(x) = 3x^2 + 9 \geq 9$, so $f'(c) = 0$ has no solutions. We conclude that $f(x)$ cannot have more than one real root (Figure 14). ■

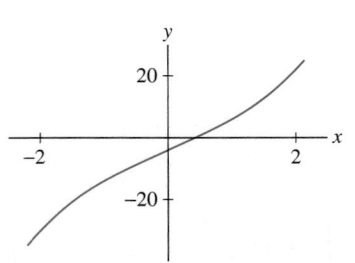

FIGURE 14 Graph of $f(x) = x^3 + 9x - 4$. This function has one real root.

4.2 SUMMARY

- The *extreme values* of $f(x)$ on an interval I are the minimum and maximum values of $f(x)$ for $x \in I$ (also called *absolute* extrema on I).
- Basic Theorem: If $f(x)$ is continuous on a closed interval $[a, b]$, then $f(x)$ has both a min and max on $[a, b]$.
- $f(c)$ is a *local minimum* if $f(x) \geq f(c)$ for all x in some open interval around c. Local maxima are defined similarly.
- $x = c$ is a *critical point* of $f(x)$ if either $f'(c) = 0$ or $f'(c)$ does not exist.
- Fermat's Theorem: If $f(c)$ is a local min or max, then c is a critical point.
- To find the extreme values of a continuous function $f(x)$ on a closed interval $[a, b]$:

Step 1. Find the critical points of $f(x)$ in $[a, b]$.

Step 2. Calculate $f(x)$ at the critical points in $[a, b]$ and at the endpoints.

The min and max on $[a, b]$ are the smallest and largest among the values computed in Step 2.

- Rolle's Theorem: If $f(x)$ is continuous on $[a, b]$ and differentiable on (a, b), and if $f(a) = f(b)$, then there exists c between a and b such that $f'(c) = 0$.

4.2 EXERCISES

Preliminary Questions

1. If $f(x)$ is continuous on $(0, 1)$, then (choose the correct statement):

(a) $f(x)$ has a minimum value on $(0, 1)$.

(b) $f(x)$ has no minimum value on $(0, 1)$.

(c) $f(x)$ might not have a minimum value on $(0, 1)$.

2. If $f(x)$ is not continuous on $[0, 1]$, then (choose the correct statement):

(a) $f(x)$ has no extreme values on $[0, 1]$.

(b) $f(x)$ might not have any extreme values on $[0, 1]$.

3. What is the definition of a critical point?

4. True or false: If $f(x)$ is differentiable and $f'(x) = 0$ has no solutions, then $f(x)$ has no local minima or maxima.

5. Fermat's Theorem *does not* claim that if $f'(c) = 0$, then $f(c)$ is a local extreme value (this is false). What *does* Fermat's Theorem assert?

6. If $f(x)$ is continuous but has no critical points in $[0, 1]$, then (choose the correct statement):

(a) $f(x)$ has no min or max on $[0, 1]$.

(b) Either $f(0)$ or $f(1)$ is the minimum value on $[0, 1]$.

Exercises

1. The following questions refer to Figure 15.

FIGURE 15

(a) How many critical points does $f(x)$ have?

(b) What is the maximum value of $f(x)$ on $[0, 8]$?

(c) What are the local maximum values of $f(x)$?

(d) Find a closed interval on which both the minimum and maximum values of $f(x)$ occur at critical points.

(e) Find an interval on which the minimum value occurs at an endpoint.

2. State whether $f(x) = x^{-1}$ (Figure 16) has a minimum or maximum value on the following intervals:

(a) $(0, 2)$ **(b)** $(1, 2)$ **(c)** $[1, 2]$

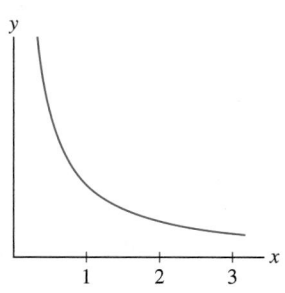

FIGURE 16 Graph of $f(x) = x^{-1}$.

In Exercises 3–14, find all critical points of the function.

3. $f(x) = x^2 - 2x + 4$

4. $f(x) = 7x - 2$

5. $f(x) = x^3 - \dfrac{9}{2}x^2 - 54x + 2$

6. $f(t) = 8t^3 - t^2$

7. $f(x) = \dfrac{x}{x^2 + 1}$

8. $f(t) = 4t - \sqrt{t^2 + 1}$

9. $f(x) = x^{1/3}$

10. $f(x) = \dfrac{x^2}{x + 1}$

11. $f(x) = x \ln x$

12. $f(x) = xe^{2x}$

13. $f(x) = \sin^{-1} x - 2x$

14. $f(x) = \sec^{-1} x - \ln x$

15. Let $f(x) = x^2 - 4x + 1$.

(a) Find the critical point c of $f(x)$ and compute $f(c)$.

(b) Compute the value of $f(x)$ at the endpoints of the interval $[0, 4]$.

(c) Determine the min and max of $f(x)$ on $[0, 4]$.

(d) Find the extreme values of $f(x)$ on $[0, 1]$.

16. Find the extreme values of $2x^3 - 9x^2 + 12x$ on $[0, 3]$ and $[0, 2]$.

17. Find the minimum value of $y = \tan^{-1}(x^2 - x)$.

18. Find the critical points of $f(x) = \sin x + \cos x$ and determine the extreme values on $[0, \frac{\pi}{2}]$.

19. Compute the critical points of $h(t) = (t^2 - 1)^{1/3}$. Check that your answer is consistent with Figure 17. Then find the extreme values of $h(t)$ on $[0, 1]$ and $[0, 2]$.

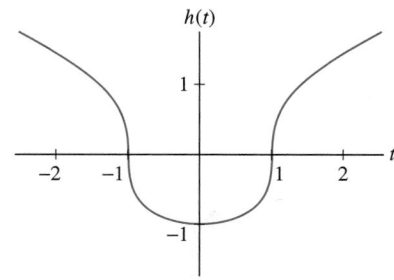

FIGURE 17 Graph of $h(t) = (t^2 - 1)^{1/3}$.

20. [GU] Plot $f(x) = 2\sqrt{x} - x$ on $[0, 4]$ and determine the maximum value graphically. Then verify your answer using calculus.

21. [GU] Plot $f(x) = \ln x - 5 \sin x$ on $[0, 2\pi]$ (choose an appropriate viewing rectangle) and approximate both the critical points and extreme values.

22. [CAS] Approximate the critical points of $g(x) = x \arccos x$ and estimate the maximum value of $g(x)$.

In Exercises 23–56, find the maximum and minimum values of the function on the given interval.

23. $y = 2x^2 - 4x + 2$, $[0, 3]$

24. $y = -x^2 + 10x + 43$, $[3, 8]$

25. $y = x^2 - 6x - 1$, $[-2, 2]$

26. $y = x^2 - 6x - 1$, $[-2, 0]$

27. $y = -4x^2 + 3x + 4$, $[-1, 1]$

28. $y = x^3 - 3x + 1$, $[0, 2]$

29. $y = x^3 - 6x + 1$, $[-1, 1]$

30. $y = x^3 - 12x^2 + 21x$, $[0, 2]$

31. $y = x^3 + 3x^2 - 9x + 2$, $[-1, 1]$

32. $y = x^3 + 3x^2 - 9x + 2$, $[0, 2]$

33. $y = x^3 + 3x^2 - 9x + 2$, $[-4, 4]$

34. $y = x^5 - x$, $[0, 2]$

35. $y = x^5 - 3x^2$, $[-1, 5]$

36. $y = 2s^3 - 3s^2 + 3$, $[-4, 4]$

37. $y = \dfrac{x^2 + 1}{x - 4}$, $[5, 6]$

38. $y = \dfrac{1 - x}{x^2 + 3x}$, $[1, 4]$

39. $y = x - \dfrac{4x}{x + 1}$, $[0, 3]$

40. $y = 2\sqrt{x^2 + 1} - x$, $[0, 2]$

41. $y = (2 + x)\sqrt{2 + (2 - x)^2}$, $[0, 2]$

42. $y = \sqrt{1 + x^2} - 2x$, $[3, 6]$

43. $y = \sqrt{x + x^2} - 2\sqrt{x}$, $[0, 4]$

44. $y = (t - t^2)^{1/3}$, $[-1, 2]$

45. $y = \sin x \cos x$, $[0, \frac{\pi}{2}]$

46. $y = x - \sin x$, $[0, 2\pi]$

47. $y = \sqrt{2}\,\theta - \sec\theta$, $[0, \frac{\pi}{3}]$

48. $y = \cos\theta + \sin\theta$, $[0, 2\pi]$

49. $y = \theta - 2\sin\theta$, $[0, 2\pi]$

50. $y = \sin^3\theta - \cos^2\theta$, $[0, 2\pi]$

51. $y = \tan x - 2x$, $[0, 1]$

52. $y = xe^{-x}$, $[0, 2]$

53. $y = \dfrac{\ln x}{x}$, $[1, 3]$

54. $y = 3e^x - e^{2x}$, $[-\frac{1}{2}, 1]$

55. $y = 5\tan^{-1} x - x$, $[1, 5]$

56. $y = x^3 - 24\ln x$, $[\frac{1}{2}, 3]$

57. Let $f(\theta) = 2\sin 2\theta + \sin 4\theta$.
(a) Show that θ is a critical point if $\cos 4\theta = -\cos 2\theta$.
(b) Show, using a unit circle, that $\cos\theta_1 = -\cos\theta_2$ if and only if $\theta_1 = \pi \pm \theta_2 + 2\pi k$ for an integer k.
(c) Show that $\cos 4\theta = -\cos 2\theta$ if and only if $\theta = \pi/2 + \pi k$ or $\theta = \pi/6 + (\pi/3)k$.
(d) Find the six critical points of $f(\theta)$ on $[0, 2\pi]$ and find the extreme values of $f(\theta)$ on this interval.
(e) $\boxed{\text{GU}}$ Check your results against a graph of $f(\theta)$.

58. $\boxed{\text{GU}}$ Find the critical points of $f(x) = 2\cos 3x + 3\cos 2x$. Check your answer against a graph of $f(x)$.

In Exercises 59–62, find the critical points and the extreme values on $[0, 3]$. In Exercises 61 and 62, refer to Figure 18.

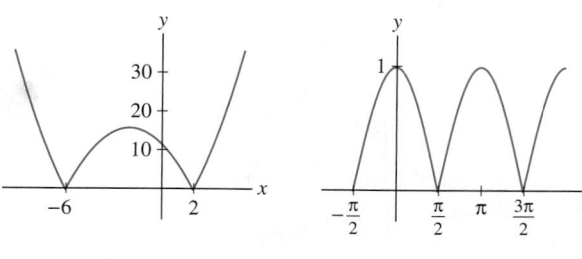

$y = |x^2 + 4x - 12|$ $y = |\cos x|$

FIGURE 18

59. $y = |x - 2|$

60. $y = |3x - 9|$

61. $y = |x^2 + 4x - 12|$

62. $y = |\cos x|$

63. Let $f(x) = 3x - x^3$. Check that $f(-2) = f(1)$. What may we conclude from Rolle's Theorem? Verify this conclusion.

In Exercises 64–67, verify Rolle's Theorem for the given interval.

64. $f(x) = x + x^{-1}$, $[\frac{1}{2}, 2]$

65. $f(x) = \sin x$, $[\frac{\pi}{4}, \frac{3\pi}{4}]$

66. $f(x) = \dfrac{x^2}{8x - 15}$, $[3, 5]$

67. $f(x) = \sin^2 x - \cos^2 x$, $[\frac{\pi}{4}, \frac{3\pi}{4}]$

68. Use Rolle's Theorem to prove that $f(x) = x^5 + 2x^3 + 4x - 12$ has at most one real root.

69. Use Rolle's Theorem to prove that $f(x) = \dfrac{x^3}{6} + \dfrac{x^2}{2} + x + 1$ has at most one real root.

70. The concentration $C(t)$ (in mg/cm^3) of a drug in a patient's bloodstream after t hours is

$$C(t) = \frac{0.016t}{t^2 + 4t + 4}$$

Find the maximum concentration and the time at which it occurs.

71. Maximum of a Resonance Curve   The size of the response of a circuit or other oscillatory system to an input signal of frequency ω ("omega") is described in suitable units by the function

$$\phi(\omega) = \frac{1}{\sqrt{(\omega_0^2 - \omega^2)^2 + 4D^2\omega^2}}$$

Both ω_0 (the natural frequency of the system) and D (the damping factor) are positive constants. The graph of ϕ is called a resonance curve, and the positive frequency $\omega_r > 0$ where ϕ takes its maximum value, if such a positive frequency exists, is the **resonant frequency**. Show that a resonant frequency exists if and only if $0 < D < \omega_0/\sqrt{2}$ (Figure 19). Then show that if this condition is satisfied, $\omega_r = \sqrt{\omega_0^2 - 2D^2}$.

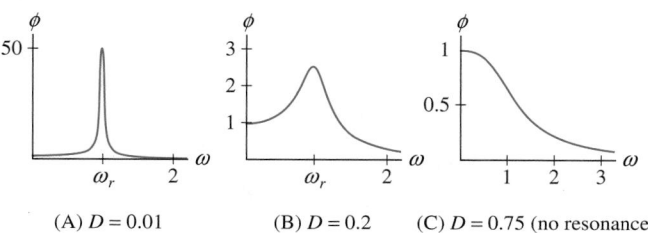

(A) $D = 0.01$ (B) $D = 0.2$ (C) $D = 0.75$ (no resonance)

FIGURE 19   Resonance curves with $\omega_0 = 1$.

72. Bees build honeycomb structures out of cells with a hexagonal base and three rhombus-shaped faces on top as in Figure 20. Using geometry, we can show that the surface area of this cell is

$$A(\theta) = 6hs + \frac{3}{2}s^2(\sqrt{3}\,\csc\theta - \cot\theta)$$

where h, s, and θ are as indicated in the figure. It is a remarkable fact that bees "know" which angle θ minimizes the surface area (and therefore requires the least amount of wax).

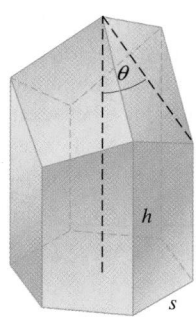

FIGURE 20   A cell in a honeycomb constructed by bees.

(a) Show that this angle is approximately 54.7° by finding the critical point of $A(\theta)$ for $0 < \theta < \pi/2$ (assume h and s are constant).

(b) GU   Confirm, by graphing $f(\theta) = \sqrt{3}\,\csc\theta - \cot\theta$, that the critical point indeed minimizes the surface area.

73. Migrating fish tend to swim at a velocity v that minimizes the total expenditure of energy E. According to one model, E is proportional to $f(v) = \dfrac{v^3}{v - v_r}$, where v_r is the velocity of the river water.

(a) Find the critical points of $f(v)$.

(b) GU   Choose a value of v_r (say, $v_r = 10$) and plot $f(v)$. Confirm that $f(v)$ has a minimum value at the critical point.

74. Find the maximum of $y = x^a - x^b$ on $[0, 1]$ where $0 < a < b$. In particular, find the maximum of $y = x^5 - x^{10}$ on $[0, 1]$.

In Exercises 75–77, plot the function using a graphing utility and find its critical points and extreme values on $[-5, 5]$.

75. GU $y = \dfrac{1}{1 + |x - 1|}$

76. GU $y = \dfrac{1}{1 + |x - 1|} + \dfrac{1}{1 + |x - 4|}$

77. GU $y = \dfrac{x}{|x^2 - 1| + |x^2 - 4|}$

78. (a) Use implicit differentiation to find the critical points on the curve $27x^2 = (x^2 + y^2)^3$.

(b) GU   Plot the curve and the horizontal tangent lines on the same set of axes.

79. Sketch the graph of a continuous function on $(0, 4)$ with a minimum value but no maximum value.

80. Sketch the graph of a continuous function on $(0, 4)$ having a local minimum but no absolute minimum.

81. Sketch the graph of a function on $[0, 4]$ having

(a) Two local maxima and one local minimum.

(b) An absolute minimum that occurs at an endpoint, and an absolute maximum that occurs at a critical point.

82. Sketch the graph of a function $f(x)$ on $[0, 4]$ with a discontinuity such that $f(x)$ has an absolute minimum but no absolute maximum.

Further Insights and Challenges

83. Show that the extreme values of $f(x) = a\sin x + b\cos x$ are $\pm\sqrt{a^2 + b^2}$.

84. Show, by considering its minimum, that $f(x) = x^2 - 2x + 3$ takes on only positive values. More generally, find the conditions on r and s under which the quadratic function $f(x) = x^2 + rx + s$ takes

on only positive values. Give examples of r and s for which f takes on both positive and negative values.

85. Show that if the quadratic polynomial $f(x) = x^2 + rx + s$ takes on both positive and negative values, then its minimum value occurs at the midpoint between the two roots.

86. Generalize Exercise 85: Show that if the horizontal line $y = c$ intersects the graph of $f(x) = x^2 + rx + s$ at two points $(x_1, f(x_1))$ and $(x_2, f(x_2))$, then $f(x)$ takes its minimum value at the midpoint

$$M = \frac{x_1 + x_2}{2}$$ (Figure 21).

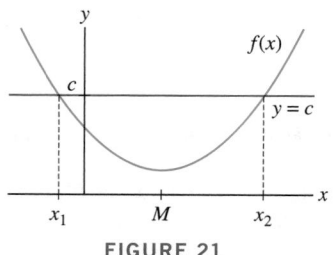

FIGURE 21

87. The graphs in Figure 22 show that a cubic polynomial may have a local min and max, or it may have neither. Find conditions on the coefficients a and b of $f(x) = \frac{1}{3}x^3 + \frac{1}{2}ax^2 + bx + c$ that ensure f has neither a local min nor max. *Hint:* Apply Exercise 84 to $f'(x)$.

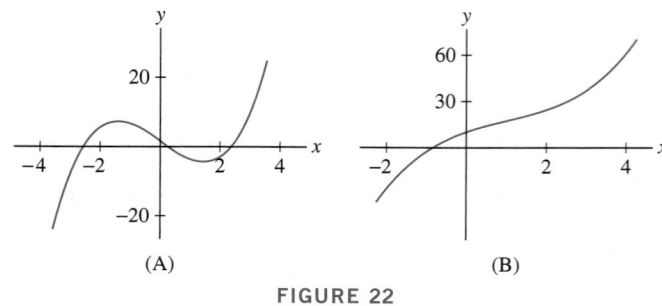

(A)   (B)

FIGURE 22

88. Find the minimum and maximum values of $f(x) = x^p(1 - x)^q$ on $[0, 1]$, where p and q are positive numbers.

89. Prove that if f is continuous and $f(a)$ and $f(b)$ are local minima where $a < b$, then there exists a value c between a and b such that $f(c)$ is a local maximum. (*Hint:* Apply Theorem 1 to the interval $[a, b]$.) Show that continuity is a necessary hypothesis by sketching the graph of a function (necessarily discontinuous) with two local minima but no local maximum.

4.3 The Mean Value Theorem and Monotonicity

We know intuitively that the derivative determines whether a function is increasing or decreasing. Since $f'(x)$ is the rate of change, $f(x)$ is increasing if the rate is positive and decreasing if the rate is negative. In this section, we prove this statement rigorously using an important theoretical result called the Mean Value Theorem (MVT). We then develop a method for "testing" critical points, that is, for determining whether they correspond to local minima or maxima.

Consider the secant line through points $(a, f(a))$ and $(b, f(b))$ on a graph as in Figure 1. According to the MVT, *there exists at least one tangent line that is parallel to the secant line in the interval (a, b)*. Because two lines are parallel if they have the same slope, what the MVT claims is that there exists a point c between a and b such that

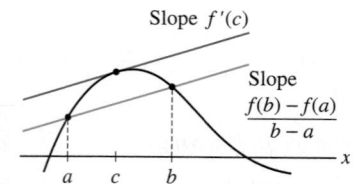

FIGURE 1 By the MVT, there exists at least one tangent line parallel to the secant line.

$$\underbrace{f'(c)}_{\text{Slope of tangent line}} = \underbrace{\frac{f(b) - f(a)}{b - a}}_{\text{Slope of secant line}}$$

THEOREM 1   The Mean Value Theorem   Assume that f is continuous on the closed interval $[a, b]$ and differentiable on (a, b). Then there exists at least one value c in (a, b) such that

$$f'(c) = \frac{f(b) - f(a)}{b - a}$$

Note that Rolle's Theorem is the special case of the MVT where $f(a) = f(b)$. In this case, $f'(c) = 0$.

FIGURE 2 Move the secant line in a parallel fashion until it becomes tangent to the curve.

GRAPHICAL INSIGHT   To see that the MVT is plausible, imagine what happens when the secant line is moved parallel to itself. Eventually, it becomes a tangent line as in Figure 2. We present a formal proof at the end of this section.

■ **EXAMPLE 1** Illustrate the MVT with $f(x) = \sqrt{x}$ and the points $a = 1$ and $b = 9$.

Solution The slope of the secant line is

$$\frac{f(b) - f(a)}{b - a} = \frac{\sqrt{9} - \sqrt{1}}{9 - 1} = \frac{3 - 1}{9 - 1} = \frac{1}{4}$$

The MVT states that $f'(c) = \frac{1}{4}$ for at least one value c in $(1, 9)$. Because $f'(x) = \frac{1}{2}x^{-1/2}$, we can find c by solving

$$f'(c) = \frac{1}{2\sqrt{c}} = \frac{1}{4}$$

This gives $2\sqrt{c} = 4$ or $c = 4$. Notice that $c = 4$ lies in the interval $(1, 9)$ as required. ■

As a first application, we prove that a function with zero derivative is constant.

COROLLARY   Assume that $f(x)$ is differentiable and $f'(x) = 0$ for all $x \in (a, b)$. Then $f(x)$ is constant on (a, b). In other words, $f(x) = C$ for some constant C.

Proof If a_1 and b_1 are any two distinct points in (a, b), then, by the MVT, there exists c between a_1 and b_1 such that

$$f'(c) = \frac{f(b_1) - f(a_1)}{b_1 - a_1} = 0$$

The last equality holds because $f'(c) = 0$. Therefore, $f(b_1) = f(a_1)$ for all a_1 and b_1 in (a, b). This shows that $f(x)$ is constant on (a, b). ■

Increasing/Decreasing Behavior of Functions

We now prove that the sign of the derivative determines whether $f(x)$ is increasing or decreasing. Recall that $f(x)$ is called:

- **Increasing on (a, b)**   if $f(x_1) < f(x_2)$ for all $x_1, x_2 \in (a, b)$ such that $x_1 < x_2$
- **Decreasing on (a, b)**   if $f(x_1) > f(x_2)$ for all $x_1, x_2 \in (a, b)$ such that $x_1 < x_2$

We say that f is "nondecreasing" if

$$f(x_1) \le f(x_2) \quad \text{for} \quad x_1 \le x_2$$

The term "nonincreasing" is defined similarly. In Theorem 2, if we assume that $f'(x) \ge 0$ (instead of > 0), then $f(x)$ is nondecreasing on (a, b). Similarly, if $f'(x) \le 0$, then $f(x)$ is nonincreasing on (a, b).

We say that $f(x)$ is **monotonic** on (a, b) if it is either increasing or decreasing on (a, b).

THEOREM 2 The Sign of the Derivative   Let f be a differentiable function on the open interval (a, b).

- If $f'(x) > 0$ for $x \in (a, b)$, then f is increasing on (a, b).
- If $f'(x) < 0$ for $x \in (a, b)$, then f is decreasing on (a, b).

Proof Suppose first that $f'(x) > 0$ for all $x \in (a, b)$. For any two points $x_1 < x_2$ in (a, b), the MVT tells us that there exists c between x_1 and x_2 such that

$$f'(c) = \frac{f(x_2) - f(x_1)}{x_2 - x_1} > 0$$

Increasing function
Tangent lines have positive slope.
Decreasing function
Tangent lines have negative slope.

FIGURE 3

Here, by our hypothesis, $f'(c) > 0$. Since $(x_2 - x_1)$ is positive, we may conclude that $f(x_2) - f(x_1) > 0$ or $f(x_2) > f(x_1)$ as required. The case $f'(x) < 0$ is similar. ∎

GRAPHICAL INSIGHT   Theorem 2 confirms our graphical intuition: $f(x)$ is increasing if the tangent lines to the graph have positive slope and decreasing if they have negative slope (Figure 3).

■ **EXAMPLE 2** Describe the graph of $f(x) = \ln x$. Is $f'(x)$ increasing or decreasing?

Solution The derivative $f'(x) = x^{-1}$ is positive on the domain $\{x : x > 0\}$, and thus $f(x) = \ln x$ is increasing. However, $f'(x) = x^{-1}$ is decreasing, so the graph of $f(x)$ grows flatter as $x \to \infty$ (Figure 4). ∎

■ **EXAMPLE 3** Find the intervals on which $f(x) = x^2 - 2x - 3$ is monotonic.

Solution The derivative $f'(x) = 2x - 2 = 2(x - 1)$ is positive for $x > 1$ and negative for $x < 1$. By Theorem 2, f is decreasing on the interval $(-\infty, 1)$ and increasing on the interval $(1, \infty)$.

Note how the graph of $f'(x)$ reflects the increasing/decreasing behavior of $f(x)$ (Figure 5). The graph of $f'(x)$ lies *below* the x-axis on the interval $(-\infty, 1)$ where f is decreasing, and it lies *above* the x-axis on the interval $(1, \infty)$ where f is increasing. ∎

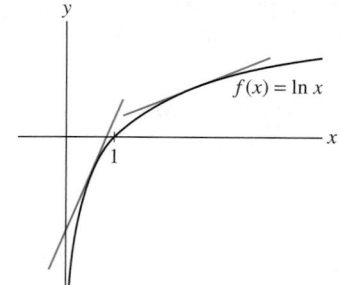

$f(x) = \ln x$

FIGURE 4 The tangent lines to $y = \ln x$ get flatter as $x \to \infty$.

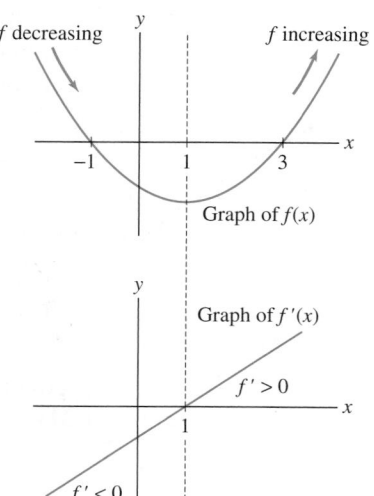

f decreasing

f increasing

Graph of $f(x)$

Graph of $f'(x)$

$f' > 0$

$f' < 0$

FIGURE 5 Graph of $f(x) = x^2 - 2x - 3$ and its derivative $f'(x) = 2x - 2$.

Testing Critical Points

There is a useful test for determining whether a critical point of a function $f(x)$ is a min or max (or neither) based on the *sign change* of $f'(x)$.

To explain the term "sign change," suppose that a function $F(x)$ satisfies $F(c) = 0$. We say that $F(x)$ *changes from positive to negative* at $x = c$ if $F(x) > 0$ to the left of c and $F(x) < 0$ to the right of c for x within a small open interval around c (Figure 6). A sign change from negative to positive is defined similarly. Another possibility is that $F(x)$ does not change sign (as in Figure 6 at $x = 5$).

Now suppose that $f'(c) = 0$ and that $f'(x)$ changes sign at $x = c$, say, from $+$ to $-$. Then $f(x)$ is increasing to the left of c and decreasing to the right; therefore, $f(c)$ is a local maximum [Figure 7(A)]. Similarly, if $f'(x)$ changes sign from $-$ to $+$, then $f(c)$ is a local minimum. If $f'(x)$ does not change sign at $x = c$ [Figure 7(B)], then $f(x)$ is either nondecreasing or nonincreasing near $x = c$, and $f(c)$ is neither a local min nor a local max [unless $f(x)$ is constant to the right or left of $x = c$].

FIGURE 6

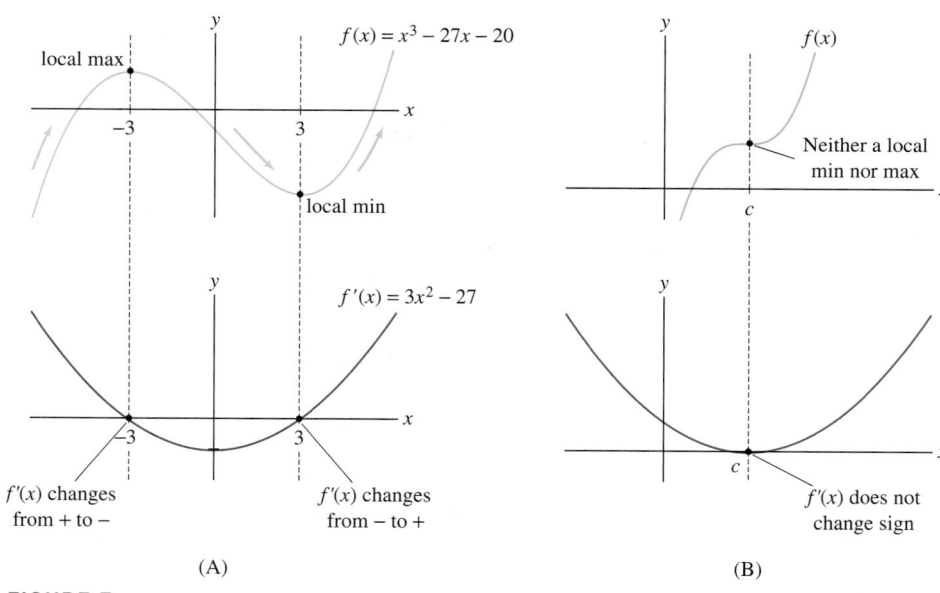

(A)   (B)

FIGURE 7

THEOREM 3 First Derivative Test for Critical Points   Assume that $f(x)$ is differentiable and let c be a critical point of $f(x)$. Then:

- $f'(x)$ changes from $+$ to $-$ at $c \Rightarrow f(c)$ is a local maximum.
- $f'(x)$ changes from $-$ to $+$ at $c \Rightarrow f(c)$ is a local minimum.

The sign of $f'(x)$ may change at a critical point, but it cannot change on the interval between two consecutive critical points [one can prove this is true even if $f'(x)$ is not assumed to be continuous]. Therefore, to determine the sign of $f'(x)$ on an interval between consecutive critical points, we need only evaluate $f'(x)$ at an arbitrarily chosen *test point* x_0 inside the interval. The sign of $f'(x_0)$ is the sign of $f'(x)$ on the entire interval.

■ **EXAMPLE 4** Analyzing Critical Points Analyze the critical points of

$$f(x) = x^3 - 27x - 20$$

Solution The derivative is $f'(x) = 3x^2 - 27 = 3(x^2 - 9)$ and the critical points are $c = \pm 3$ [Figure 7(A)]. These critical points divide the real line into three intervals

$$(-\infty, -3), \qquad (-3, 3), \qquad (3, \infty)$$

We determine the sign of $f'(x)$ on each of these intervals by evaluating $f'(x)$ at a test point inside each interval. For example, in $(-\infty, -3)$ we choose the test point -4. Because $f'(-4) = 21 > 0$, $f'(x)$ is positive on the entire interval $(-3, \infty)$. Similarly,

$$
\begin{aligned}
f'(-4) &= 21 > 0 &&\Rightarrow& f'(x) > 0 &\quad \text{for all } x \in (-\infty, -3) \\
f'(0) &= -27 < 0 &&\Rightarrow& f'(x) < 0 &\quad \text{for all } x \in (-3, 3) \\
f'(4) &= 21 > 0 &&\Rightarrow& f'(x) > 0 &\quad \text{for all } x \in (3, \infty)
\end{aligned}
$$

We chose the test points -4, 0, and 4 arbitrarily. For example, to find the sign of $f'(x)$ on $(-\infty, -3)$, we could just as well have computed $f'(-5)$ or any other value of f' in the interval in $(-\infty, -3)$.

This information is displayed in the following sign diagram:

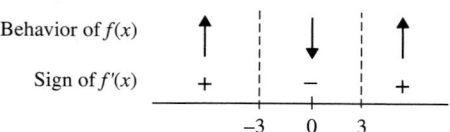

Now use the First Derivative Test:

- $c = -3$: $f'(x)$ changes from $+$ to $-$ $\Rightarrow$ $f(-3)$ is a local max.
- $c = 3$: $f'(x)$ changes from $-$ to $+$ $\Rightarrow$ $f(3)$ is a local min. ■

■ **EXAMPLE 5** Finding Intervals of Increase and Decrease Find the intervals of increase/decrease for $f(x) = \cos^2 x + \sin x$ on $[0, \pi]$.

Solution The increasing/decreasing behavior of $f(x)$ can change only at critical points, so first we find the critical points by solving

$$f'(x) = -2\cos x \sin x + \cos x = -(\cos x)(2\sin x - 1) = 0$$

This equation is satisfied if

$$\cos x = 0 \qquad \text{or} \qquad 2\sin x - 1 = 0$$

On $[0, \pi]$ we have

$$\cos x = 0 \quad \text{for } x = \frac{\pi}{2}$$

and

$$\sin x = \frac{1}{2} \quad \text{for } x = \frac{\pi}{6}, \frac{5\pi}{6}$$

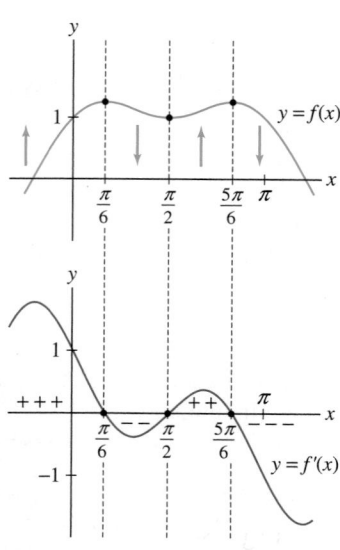

FIGURE 8 Graph of
$f(x) = \cos^2 x + \sin x$ and its derivative.

In the following table, we calculate test values to determine the sign of $f'(x)$ on the intervals between the critical points:

Interval	Test Value	Sign of $f'(x)$	Behavior of $f(x)$
$(0, \frac{\pi}{6})$	$f'(0.5) \approx 0.04$	$+$	$\uparrow$
$(\frac{\pi}{6}, \frac{\pi}{2})$	$f'(1) \approx -0.37$	$-$	$\downarrow$
$(\frac{\pi}{2}, \frac{5\pi}{6})$	$f'(2) \approx 0.34$	$+$	$\uparrow$
$(\frac{5\pi}{6}, \pi)$	$f'(3) \approx -0.71$	$-$	$\downarrow$

The behavior of $f(x)$ and $f'(x)$ is reflected in the graphs in Figure 8.   ■

The next example illustrates the case where $f'(c) = 0$ but $f'(x)$ does not change sign at $x = c$. In this case, $f(c)$ is neither a local min nor a local max.

■ **EXAMPLE 6**   A Critical Point Without a Sign Transition   Analyze the critical points of $f(x) = \frac{1}{3}x^3 - x^2 + x$.

Solution   The derivative is $f'(x) = x^2 - 2x + 1 = (x - 1)^2$, so $c = 1$ is the only critical point. However, $(x - 1)^2 \geq 0$, so $f'(x)$ does not change sign at $c = 1$ and $f(1)$ is neither a local min nor a local max (Figure 9).   ■

FIGURE 9 Graph of $f(x) = \frac{1}{3}x^3 - x^2 + x$. The function does not have an extreme value at the critical point.

Proof of the MVT

Let $m = \dfrac{f(b) - f(a)}{b - a}$ be the slope of the secant line joining $(a, f(a))$ and $(b, f(b))$. We may write the equation of the secant line in the form $y = mx + r$ for some r (Figure 10). The precise value of r is not important, but you can check that $r = f(a) - ma$. To show that $f'(c) = m$ for some c in (a, b), consider the function

$$G(x) = f(x) - (mx + r) \qquad \boxed{1}$$

As indicated in Figure 10, $G(x)$ is the vertical distance between the graph and the secant line at x (it is negative at points where the graph of f lies below the secant line). Furthermore, this distance is zero at the endpoints and therefore $G(a) = G(b) = 0$. By Rolle's Theorem (Section 4.2), there exists a point c in (a, b) such that $G'(c) = 0$. By Eq. (1), $G'(x) = f'(x) - m$, so we obtain

$$G'(c) = f'(c) - m = 0$$

Thus, $f'(c) = m$ as desired.   ■

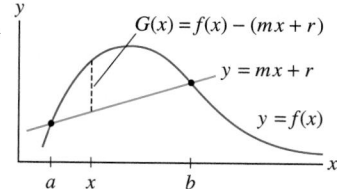

FIGURE 10 $G(x)$ is the vertical distance between the graph and the secant line.

4.3 SUMMARY

- The *sign* of $f'(x)$ determines whether $f(x)$ is increasing or decreasing:

$$f'(x) > 0 \text{ for } x \in (a, b) \quad \Rightarrow \quad f \text{ is increasing on } (a, b)$$
$$f'(x) < 0 \text{ for } x \in (a, b) \quad \Rightarrow \quad f \text{ is decreasing on } (a, b)$$

- The sign of $f'(x)$ can change only at the critical points, so $f(x)$ is *monotonic* (increasing or decreasing) on the intervals between the critical points.

- To find the sign of $f'(x)$ on the interval between two critical points, calculate the sign of $f'(x_0)$ at any test point x_0 in that interval.
- *First Derivative Test:* Assume that $f(x)$ is differentiable. If c is a critical point, then:

Sign Change of f' at c	Type of Critical Point
From $+$ to $-$	Local maximum
From $-$ to $+$	Local minimum

If f' does not change sign at $x = c$, then $f(c)$ is neither a local min nor max (unless f is constant to the left or right of c).

4.3 EXERCISES

Preliminary Questions

1. Which value of m makes the following statement correct? If $f(2) = 3$ and $f(4) = 9$, where $f(x)$ is differentiable, then the graph of f has a tangent line of slope m.

2. Which of the following conclusions does *not* follow from the MVT (assume that f is differentiable)?

(a) If f has a secant line of slope 0, then $f'(c) = 0$ for some value of c.

(b) If $f(5) < f(9)$, then $f'(c) > 0$ for some $c \in (5, 9)$.

(c) If $f'(c) = 0$ for some value of c, then there is a secant line whose slope is 0.

(d) If $f'(x) > 0$ for all x, then every secant line has positive slope.

3. Can a function that takes on only negative values have a positive derivative? Sketch an example or explain why no such functions exist.

4. (a) Use the graph of $f'(x)$ in Figure 11 to determine whether $f(c)$ is a local minimum or maximum.

(b) Can you conclude from Figure 11 that $f(x)$ is a decreasing function?

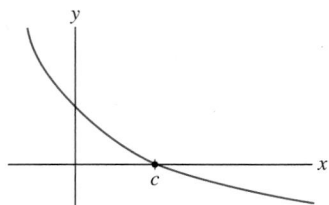

FIGURE 11 Graph of derivative $f'(x)$.

Exercises

In Exercises 1–10, find a point c satisfying the conclusion of the MVT for the given function and interval.

1. $y = x^{-1}$, $[1, 4]$

2. $y = \sqrt{x}$, $[4, 9]$

3. $y = (x - 1)(x - 3)$, $[1, 3]$

4. $y = \cos x - \sin x$, $[0, 2\pi]$

5. $y = \dfrac{x}{x + 1}$, $[3, 6]$

6. $y = x^3$, $[-3, -1]$

7. $y = x \ln x$, $[1, 2]$

8. $y = e^{-2x}$, $[0, 3]$

9. $y = \cosh x$, $[-1, 1]$

10. $y = e^x - x$, $[-1, 1]$

11. $\boxed{\text{GU}}$ Let $f(x) = x^5 + x^2$. Check that the secant line between $x = 0$ and $x = 1$ has slope 2. By the MVT, $f'(c) = 2$ for some

$c \in (0, 1)$. Estimate c graphically as follows. Plot $f(x)$ and the secant line on the same axes. Then plot the lines $y = 2x + b$ for different values of b until you find a value of b for which it is tangent to $y = f(x)$. Zoom in on the point of tangency to find its x-coordinate.

12. Determine the intervals on which $f'(x)$ is positive and negative, assuming that Figure 12 is the graph of $f(x)$.

13. Determine the intervals on which $f(x)$ is increasing or decreasing, assuming that Figure 12 is the graph of the derivative $f'(x)$.

FIGURE 12

14. $\boxed{\text{GU}}$ Plot the derivative $f'(x)$ of $f(x) = 3x^5 - 5x^3$ and describe the sign changes of $f'(x)$. Use this to determine the local extreme values of $f(x)$. Then graph $f(x)$ to confirm your conclusions.

In Exercises 15–18, sketch the graph of a function $f(x)$ whose derivative $f'(x)$ has the given description.

15. $f'(x) > 0$ for $x > 3$ and $f'(x) < 0$ for $x < 3$

16. $f'(x) > 0$ for $x < 1$ and $f'(x) < 0$ for $x > 1$

17. $f'(x)$ is negative on $(1, 3)$ and positive everywhere else.

18. $f'(x)$ makes the sign transitions $+, -, +, -$.

In Exercises 19–22, use the First Derivative Test to determine whether the function attains a local minimum or local maximum (or neither) at the given critical point.

19. $y = 7 + 4x - x^2, \quad c = 2$

20. $y = x^3 - 27x + 2, \quad c = -3$

21. $y = \dfrac{x^2}{x + 1}, \quad c = 0$

22. $y = \sin x \cos x, \quad c = \dfrac{\pi}{4}$

23. Assuming that Figure 12 is the graph of the derivative $f'(x)$, state whether $f(2)$ and $f(4)$ are local minima or maxima.

24. Figure 13 shows the graph of the derivative $f'(x)$ of a function $f(x)$. Find the critical points of $f(x)$ and determine whether they are local minima, maxima, or neither.

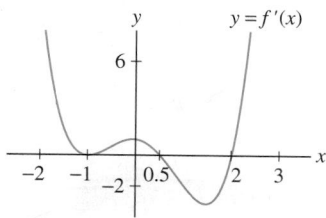

FIGURE 13 Graph of the *derivative* of $f(x)$.

In Exercises 25–52, find the critical points and the intervals on which the function is increasing or decreasing, and apply the First Derivative Test to each critical point.

25. $y = -x^2 + 7x - 17$

26. $y = 5x^2 + 6x - 4$

27. $y = x^3 - 6x^2$

28. $y = x(x + 1)^3$

29. $y = 3x^4 + 8x^3 - 6x^2 - 24x$

30. $y = x^2 + (10 - x)^2$

31. $y = \frac{1}{3}x^3 + \frac{3}{2}x^2 + 2x + 4$

32. $y = x^5 + x^3 + 1$

33. $y = x^4 + x^3$

34. $y = x^2 - x^4$

35. $y = \dfrac{1}{x^2 + 1}$

36. $y = \dfrac{2x + 1}{x^2 + 1}$

37. $y = x + x^{-1} \quad (x > 0)$

38. $y = x^4 - 4x^{3/2} \quad (x > 0)$

39. $y = x^{5/2} - x^2 \quad (x > 0)$

40. $y = x^{-2} - 4x^{-1} \quad (x > 0)$

41. $y = \sin \theta \cos \theta, \quad [0, 2\pi]$

42. $y = \cos \theta + \sin \theta, \quad [0, 2\pi]$

43. $y = \theta + \cos \theta, \quad [0, 2\pi]$

44. $y = \theta - 2 \cos \theta, \quad [0, 2\pi]$

45. $y = x + e^{-x}$

46. $y = \dfrac{e^x}{x}$ for $x > 0$

47. $y = e^{-x} \cos x, \quad [-\frac{\pi}{2}, \frac{\pi}{2}]$

48. $y = x^2 e^x$

49. $y = \tan^{-1} x - \frac{1}{2}x$

50. $y = (x^3 - 2x)e^x$

51. $y = x - \ln x$ for $x > 0$

52. $y = \dfrac{\ln x}{x}$ for $x > 0$

53. Show that $f(x) = x^2 + bx + c$ is decreasing on $(-\infty, -\frac{b}{2})$ and increasing on $(-\frac{b}{2}, \infty)$.

54. Show that $f(x) = x^3 - 2x^2 + 2x$ is an increasing function. *Hint:* Find the minimum value of $f'(x)$.

55. Find conditions on a and b that ensure that $f(x) = x^3 + ax + b$ is increasing on $(-\infty, \infty)$.

56. [GU] Suppose that $h(x) = \dfrac{x(x^2 - 1)}{x^2 + 1}$ is the derivative of a function $f(x)$. Plot $h(x)$ and use the plot to describe the local extrema and the increasing/decreasing behavior of $f(x)$. Sketch a plausible graph for $f(x)$ itself.

57. Sam made two statements that Deborah found dubious.

(a) "Although the average velocity for my trip was 70 mph, at no point in time did my speedometer read 70 mph."

(b) "Although a policeman clocked me going 70 mph, my speedometer never read 65 mph."

In each case, which theorem did Deborah apply to prove Sam's statement false: the Intermediate Value Theorem or the Mean Value Theorem? Explain.

58. Determine where $f(x) = (1,000 - x)^2 + x^2$ is increasing. Use this to decide which is larger: $1,000^2$ or $998^2 + 2^2$.

59. Show that $f(x) = 1 - |x|$ satisfies the conclusion of the MVT on $[a, b]$ if both a and b are positive or negative, but not if $a < 0$ and $b > 0$.

60. Which values of c satisfy the conclusion of the MVT on the interval $[a, b]$ if $f(x)$ is a linear function?

61. Show that if f is a quadratic polynomial, then the midpoint $c = \dfrac{a + b}{2}$ satisfies the conclusion of the MVT on $[a, b]$ for any a and b.

62. Suppose that $f(0) = 4$ and $f'(x) \le 2$ for $x > 0$. Apply the MVT to the interval $[0, 3]$ to prove that $f(3) \le 10$. Prove more generally that $f(x) \le 4 + 2x$ for all $x > 0$.

63. Suppose that $f(2) = -2$ and $f'(x) \ge 5$. Show that $f(4) \ge 8$.

64. Find the minimum value of $f(x) = x^x$ for $x > 0$.

Further Insights and Challenges

65. Show that the cubic function $f(x) = x^3 + ax^2 + bx + c$ is increasing on $(-\infty, \infty)$ if $b > a^2/3$.

66. Prove that if $f(0) = g(0)$ and $f'(x) \le g'(x)$ for $x \ge 0$, then $f(x) \le g(x)$ for all $x \ge 0$. *Hint:* Show that $f(x) - g(x)$ is nonincreasing.

67. Use Exercise 66 to prove that $\sin x \le x$ for $x \ge 0$.

68. Use Exercises 66 and 67 to establish the following assertions for all $x \ge 0$ (each assertion follows from the previous one):

(a) $\cos x \ge 1 - \frac{1}{2}x^2$

(b) $\sin x \ge x - \frac{1}{6}x^3$

(c) $\cos x \le 1 - \frac{1}{2}x^2 + \frac{1}{24}x^4$

(d) Can you guess the next inequality in the series?

69. Let $f(x) = e^{-x}$. Use the method of Exercise 68 to prove the following inequalities for $x \ge 0$:

(a) $e^{-x} \ge 1 - x$

(b) $e^{-x} \le 1 - x + \frac{1}{2}x^2$

(c) $e^{-x} \ge 1 - x + \frac{1}{2}x^2 - \frac{1}{6}x^3$

Can you guess what the next inequality in the series is?

70. Assume that f'' exists and $f''(x) = 0$ for all x. Prove that $f(x) = mx + b$, where $m = f'(0)$ and $b = f(0)$.

71. [✎] Define $f(x) = x^3 \sin(\frac{1}{x})$ for $x \ne 0$ and $f(0) = 0$.

(a) Show that $f'(x)$ is continuous at $x = 0$ and $x = 0$ is a critical point of f.

(b) [GU] Examine the graphs of $f(x)$ and $f'(x)$. Can the First Derivative Test be applied?

(c) Show that $f(0)$ is neither a local min nor max.

4.4 The Shape of a Graph

In the previous section, we studied the increasing/decreasing behavior of a function, as determined by the sign of the derivative. Another important property of a function is its concavity, which refers to the way its graph bends. Informally, a curve is called *concave up* if it bends up and *concave down* if it bends down (Figure 1).

Concave down   Concave up

FIGURE 1

To analyze concavity in a precise fashion, let's examine how concavity is related to tangent lines and derivatives. Observe in Figure 2 that when $f(x)$ is concave up, $f'(x)$ is increasing (the slopes of the tangent lines get larger as we move to the right). Similarly, when $f(x)$ is concave down, $f'(x)$ is decreasing. Thus we make the following definition.

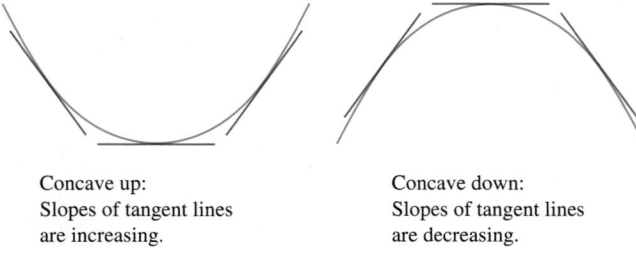

Concave up:
Slopes of tangent lines
are increasing.

Concave down:
Slopes of tangent lines
are decreasing.

FIGURE 2

> **DEFINITION Concavity** Let $f(x)$ be a differentiable function on an open interval (a, b). Then
>
> - f is **concave up** on (a, b) if $f'(x)$ is increasing on (a, b).
> - f is **concave down** on (a, b) if $f'(x)$ is decreasing on (a, b).

■ **EXAMPLE 1** Concavity and Stock Prices Two stocks, A and B, went up in value and both currently sell for $75 (Figure 3). However, one is clearly a better investment than the other. Explain in terms of concavity.

FIGURE 3

Company A Company B

Solution The growth rate (first derivative) of Stock A is declining as time goes on (since the graph is concave down) and the growth rate of Stock B is increasing (since the graph is concave up). If these trends continue, then Stock B is the better investment. ■

The concavity of a function is determined by the *sign* of its second derivative. Indeed, if $f''(x) > 0$, then $f'(x)$ is increasing and hence $f(x)$ is concave up. Similarly, if $f''(x) < 0$, then $f'(x)$ is decreasing and $f(x)$ is concave down.

> **THEOREM 1 Test for Concavity** Suppose that $f''(x)$ exists for all $x \in (a, b)$.
>
> - If $f''(x) > 0$ for all $x \in (a, b)$, then f is concave up on (a, b).
> - If $f''(x) < 0$ for all $x \in (a, b)$, then f is concave down on (a, b).

Of special interest are the points where the concavity of the graph changes. We say that $P = (c, f(c))$ is a **point of inflection** of $f(x)$ if the concavity changes from up to down or vice versa at $x = c$. In other words, $f'(x)$ is increasing on one side of $x = c$ and decreasing on the other. In Figure 4, the curve on the right is made up of two arcs—one is concave down and one is concave up (the word "arc" refers to a piece of a curve). The point P where the two arcs are joined is a point of inflection. We will denote points of inflection in graphs by a solid square ■.

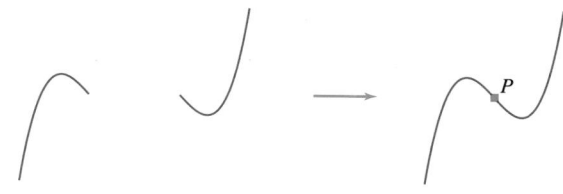

FIGURE 4 P is a point of inflection.

Concave down   Concave up

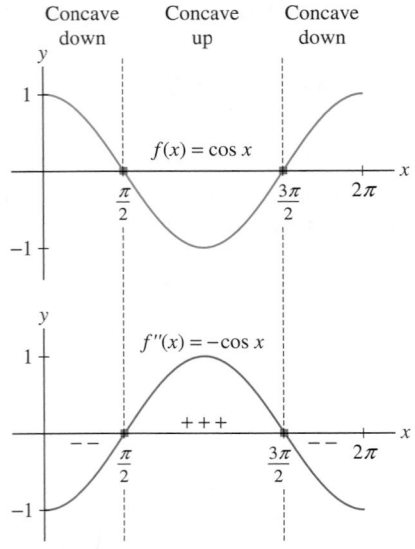

FIGURE 5

According to Theorem 1, the concavity of f is determined by the sign of f''. Thus, a point of inflection is a point where $f''(x)$ changes sign.

> **THEOREM 2 Test for Inflection Points**   Assume that $f''(x)$ exists for all $x \in (a, b)$ and let $c \in (a, b)$. If $f''(c) = 0$ and $f''(x)$ changes sign at $x = c$, then $f(x)$ has a point of inflection at $x = c$.

■ **EXAMPLE 2**   Find the points of inflection of $f(x) = \cos x$ on $[0, 2\pi]$.

Solution   We have $f'(x) = -\sin x$, $f''(x) = -\cos x$, and $f''(x) = 0$ for $x = \frac{\pi}{2}, \frac{3\pi}{2}$. Figure 5 shows that $f''(x)$ changes sign at both $x = \frac{\pi}{2}$ and $\frac{3\pi}{2}$, so $f(x)$ has a point of inflection at both points.   ■

■ **EXAMPLE 3**   **Finding Points of Inflection and Intervals of Concavity**   Find the points of inflection of $f(x) = 3x^5 - 5x^4 + 1$ and determine the intervals where $f(x)$ is concave up and down.

Solution   The first derivative is $f'(x) = 15x^4 - 20x^3$ and

$$f''(x) = 60x^3 - 60x^2 = 60x^2(x - 1)$$

The zeros $x = 0, 1$ of $f''(x)$ divide the x-axis into three intervals: $(-\infty, 0)$, $(0, 1)$, and $(1, \infty)$. We determine the sign of $f''(x)$ and the concavity of f by computing "test values" within each interval (Figure 6):

Interval	Test Value	Sign of $f''(x)$	Behavior of $f(x)$
$(-\infty, 0)$	$f''(-1) = -120$	$-$	Concave down
$(0, 1)$	$f''(\frac{1}{2}) = -\frac{15}{2}$	$-$	Concave down
$(1, \infty)$	$f''(2) = 240$	$+$	Concave up

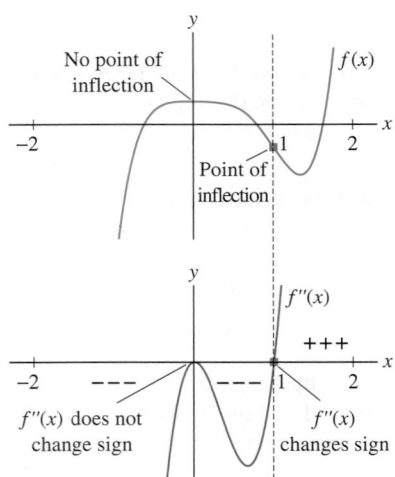

FIGURE 6 Graph of $f(x) = 3x^5 - 5x^4 + 1$ and its second derivative.

We can read off the points of inflection from this table:

- $c = 0$: no point of inflection, because $f''(x)$ does not change sign at 0.
- $c = 1$: point of inflection, because $f''(x)$ changes sign at 1.   ■

Usually, we find the inflection points by solving $f''(x) = 0$ and checking whether $f''(x)$ changes sign. However, an inflection point may also occur at a point c where $f''(c)$ does not exist.

■ **EXAMPLE 4**   **A Case Where the Second Derivative Does Not Exist**   Find the points of inflection of $f(x) = x^{5/3}$.

Solution   In this case, $f'(x) = \frac{5}{3}x^{2/3}$ and $f''(x) = \frac{10}{9}x^{-1/3}$. Although $f''(0)$ does not exist, $f''(x)$ does change sign at $x = 0$:

$$f''(x) = \frac{10}{9x^{1/3}} = \begin{cases} > 0 & \text{for } x > 0 \\ < 0 & \text{for } x < 0 \end{cases}$$

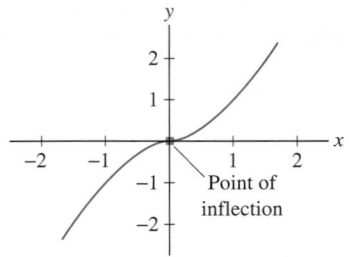

FIGURE 7 The concavity of $f(x) = x^{5/3}$ changes at $x = 0$ even though $f''(0)$ does not exist.

Therefore, the concavity of $f(x)$ changes at $x = 0$, and $(0, 0)$ is a point of inflection (Figure 7).   ■

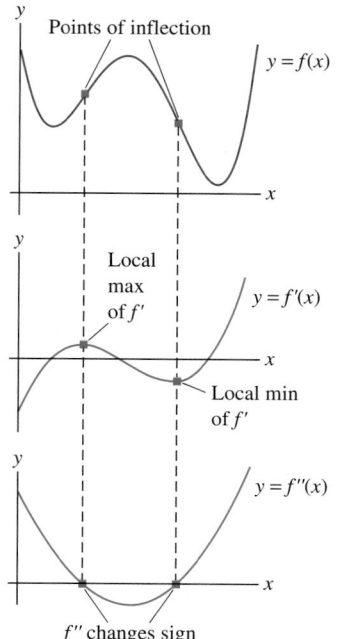

Points of inflection

$y = f(x)$

Local max of f'

$y = f'(x)$

Local min of f'

$y = f''(x)$

f'' changes sign

FIGURE 8

GRAPHICAL INSIGHT Points of inflection are easy to spot on the graph of the first derivative $f'(x)$. If $f''(c) = 0$ and $f''(x)$ changes sign at $x = c$, then the increasing/decreasing behavior of $f'(x)$ changes at $x = c$. Thus, *inflection points of f occur where $f'(x)$ has a local min or max* (Figure 8).

Second Derivative Test for Critical Points

There is a simple test for critical points based on concavity. Suppose that $f'(c) = 0$. As we see in Figure 9, if $f(x)$ is concave down, then $f(x)$ is a local max, and if $f(x)$ is concave up, then $f(x)$ is a local min. Because concavity is determined by the sign of f'', we obtain the Second Derivative Test. (See Exercise 60 for a detailed proof.)

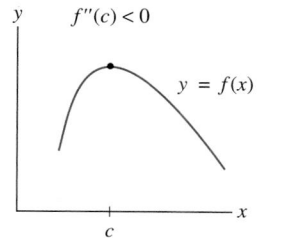
$f''(c) < 0$

$y = f(x)$

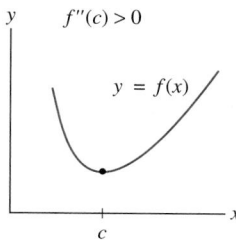
$f''(c) > 0$

$y = f(x)$

Concave down—local max   Concave up—local min

FIGURE 9 Concavity determines whether the critical point is a local minimum or maximum.

THEOREM 3 Second Derivative Test Assume that $f(x)$ is differentiable and let c be a critical point. If $f''(c)$ exists, then

- $f''(c) > 0 \quad \Rightarrow \quad f(c)$ is a local minimum
- $f''(c) < 0 \quad \Rightarrow \quad f(c)$ is a local maximum
- $f''(c) = 0 \quad \Rightarrow \quad$ inconclusive: $f(c)$ may be a local min, max, or neither

---

■ **EXAMPLE 5** Analyze the critical points of $f(x) = (2x - x^2)e^x$.

Solution First we find the critical points by solving

$$f'(x) = (2x - x^2)e^x + e^x(2 - 2x) = (2 - x^2)e^x = 0$$

The critical points are $c = \pm\sqrt{2}$ (Figure 10). Next, we determine the sign of the second derivative at the critical points:

$$f''(x) = (2 - x^2)e^x + e^x(-2x) = (2 - 2x - x^2)e^x$$

We have

$$f''(-\sqrt{2}) = (2 - 2(-\sqrt{2}) - (-\sqrt{2})^2)e^{-\sqrt{2}} = 2\sqrt{2}e^{-\sqrt{2}} \quad > 0 \quad \text{(local min)}$$

$$f''(\sqrt{2}) = (2 - 2\sqrt{2} - (\sqrt{2})^2)e^{\sqrt{2}} \quad = -2\sqrt{2}e^{\sqrt{2}} \quad < 0 \quad \text{(local max)}$$

By the Second Derivative Test, $f(x)$ has a local min at $c = -\sqrt{2}$ and a local max at $c = \sqrt{2}$ (Figure 10). ■

■ **EXAMPLE 6** **Second Derivative Test Inconclusive** Analyze the critical points of $f(x) = x^5 - 5x^4$.

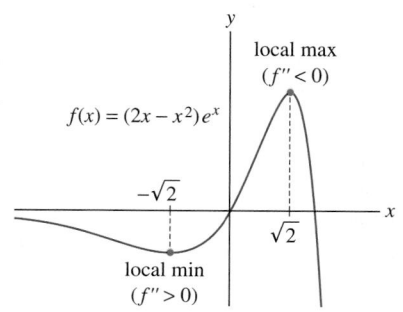

$f(x) = (2x - x^2)e^x$

local max $(f'' < 0)$

$-\sqrt{2}$

$\sqrt{2}$

local min $(f'' > 0)$

FIGURE 10 At $c = \sqrt{2}$, $f''(c) < 0$ and f has a local maximum by the Second Derivative Test.

Solution The first two derivatives of $f(x) = x^5 - 5x^4$ are

$$f'(x) = 5x^4 - 20x^3 = 5x^3(x-4)$$
$$f''(x) = 20x^3 - 60x^2$$

The critical points are $c = 0, 4$ and the Second Derivative Test yields

$$f''(0) = 0 \qquad \Rightarrow \qquad \text{Second Derivative Test fails}$$
$$f''(4) = 320 > 0 \qquad \Rightarrow \qquad f(4) \text{ is a local min}$$

The Second Derivative Test fails at $c = 0$, so we fall back on the First Derivative Test. Choosing test points to the left and right of $c = 0$, we find

$$f'(-1) = 5 + 20 = 25 > 0 \qquad \Rightarrow \qquad f'(x) \text{ is positive on } (-\infty, 0)$$
$$f'(1) = 5 - 20 = -15 < 0 \qquad \Rightarrow \qquad f'(x) \text{ is negative on } (0, 4)$$

Since $f'(x)$ changes from $+$ to $-$ at $c = 0$, $f(0)$ is a local max (Figure 11). ∎

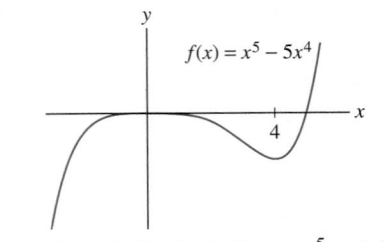

FIGURE 11 Graph of $f(x) = x^5 - 5x^4$.

4.4 SUMMARY

- A differentiable function $f(x)$ is *concave up* on (a, b) if $f'(x)$ is increasing and *concave down* if $f'(x)$ is decreasing on (a, b).
- The signs of the first two derivatives provide the following information:

First Derivative	Second Derivative
$f' > 0 \Rightarrow f$ is increasing	$f'' > 0 \Rightarrow f$ is concave up
$f' < 0 \Rightarrow f$ is decreasing	$f'' < 0 \Rightarrow f$ is concave down

- A *point of inflection* is a point where the concavity changes, from concave up to concave down or vice versa.
- If $f''(c) = 0$ and $f''(x)$ changes sign at c, then c is a point of inflection.
- Second Derivative Test: If $f'(c) = 0$ and $f''(c)$ exists, then
  - $f(c)$ is a local maximum if $f''(c) < 0$.
  - $f(c)$ is a local minimum if $f''(c) > 0$.
  - The test fails if $f''(c) = 0$.

4.4 EXERCISES

Preliminary Questions

1. Choose the correct answer: If f is concave up, then f' is:

(a) increasing   **(b)** decreasing

2. If x_0 is a critical point and f is concave down, then $f(x_0)$ is a local:

(a) min   **(b)** max   **(c)** undetermined

In Questions 3–8, state whether true or false and explain. Assume that $f''(x)$ exists for all x.

3. If $f'(c) = 0$ and $f''(c) < 0$, then $f(c)$ is a local minimum.

4. A function that is concave down on $(-\infty, \infty)$ can have no minimum value.

5. If $f''(c) = 0$, then f must have a point of inflection at $x = c$.

6. If f has a point of inflection at $x = c$, then $f''(c) = 0$.

7. If f is concave up and f' changes sign at $x = c$, then f' changes sign from negative to positive at $x = c$.

8. If $f(c)$ is a local maximum, then $f''(c)$ must be negative.

9. Suppose that $f''(c) = 0$ and $f''(x)$ changes sign from $+$ to $-$ at $x = c$. Which of the following statements are correct?

(a) $f(x)$ has a local maximum at $x = c$.

(b) $f'(x)$ has a local minimum at $x = c$.

(c) $f'(x)$ has a local maximum at $x = c$.

(d) $f(x)$ has a point of inflection at $x = c$.

Exercises

1. Match the graphs in Figure 12 with the description:

(a) $f''(x) < 0$ for all x.

(b) $f''(x)$ goes from $+$ to $-$.

(c) $f''(x) > 0$ for all x.

(d) $f''(x)$ goes from $-$ to $+$.

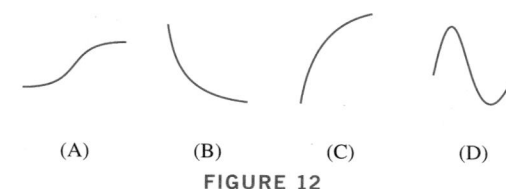

(A)   (B)   (C)   (D)

FIGURE 12

2. Match each statement with a graph in Figure 13 that represents company profits as a function of time.

(a) The outlook is great: The growth rate keeps increasing.

(b) We're losing money, but not as quickly as before.

(c) We're losing money, and it's getting worse as time goes on.

(d) We're doing well, but our growth rate is leveling off.

(e) Business had been cooling off, but now it's picking up.

(f) Business had been picking up, but now it's cooling off.

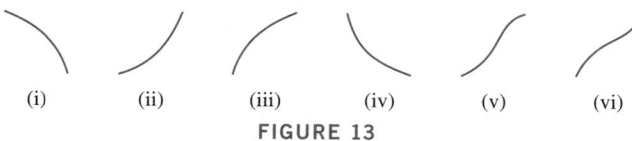

(i)   (ii)   (iii)   (iv)   (v)   (vi)

FIGURE 13

3. Sketch the graph of an increasing function such that $f''(x)$ changes from $+$ to $-$ at $x = 2$ and from $-$ to $+$ at $x = 4$. Do the same for a decreasing function.

4. If Figure 14 is the graph of a function $f(x)$, where do the points of inflection of $f(x)$ occur, and on which interval is $f(x)$ concave down?

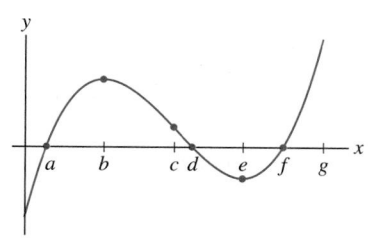

FIGURE 14

5. If Figure 14 is the graph of the *derivative* $f'(x)$, where do the points of inflection of $f(x)$ occur, and on which interval is $f(x)$ concave down?

6. If Figure 14 is the graph of the *second derivative* $f''(x)$, where do the points of inflection of $f(x)$ occur, and on which interval is $f(x)$ concave down?

In Exercises 7–18, determine the intervals on which the function is concave up or down and find the points of inflection.

7. $y = x^2 + 7x + 10$

8. $y = t^3 - 3t^2 + 1$

9. $y = x - 2\cos x$

10. $y = 4x^5 - 5x^4$

11. $y = x(x - 8\sqrt{x})$

12. $y = (x - 2)(1 - x^3)$

13. $y = \dfrac{1}{x^2 + 3}$

14. $y = x^{7/5}$

15. $y = xe^{-x}$

16. $y = \dfrac{\ln x}{x^2}$ for $x > 0$

17. $y = (x^2 - 3)e^x$

18. $y = x - \ln x$ for $x > 0$

19. Sketch the graph of $f(x) = x^4$ and state whether f has any points of inflection. Verify your conclusion by showing that $f''(x)$ does not change sign.

20. Through her website, Leticia has been selling bean bag chairs with monthly sales as recorded below. In a report to investors, she states, "Sales reached a point of inflection when I started using pay-per-click advertising." In which month did that occur? Explain.

Month	1	2	3	4	5	6	7	8
Sales	2	20	30	35	38	44	60	90

21. 📖   The growth of a sunflower during its first 100 days is modeled well by the *logistic curve* $y = h(t)$ shown in Figure 15. Estimate

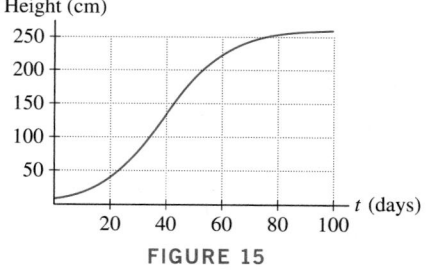

FIGURE 15

the growth rate at the point of inflection and explain its significance. Then make a rough sketch of the first and second derivatives of $h(t)$.

22. Figure 16 shows the graph of the *derivative* $f'(x)$ on $[0, 1.2]$. Locate the points of inflection of $f(x)$ and the points where the local minima and maxima occur. Determine the intervals on which $f(x)$ has the following properties:

(a) Increasing   (b) Decreasing

(c) Concave up   (d) Concave down

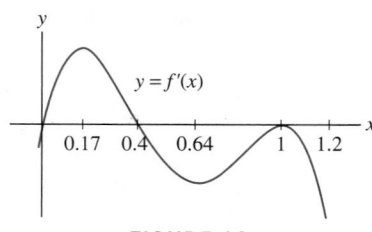

FIGURE 16

In Exercises 23–36, find the critical points of $f(x)$ and use the Second Derivative Test (if possible) to determine whether each corresponds to a local minimum or maximum.

23. $f(x) = x^3 - 12x^2 + 45x$

24. $f(x) = x^4 - 8x^2 + 1$

25. $f(x) = 3x^4 - 8x^3 + 6x^2$

26. $f(x) = x^5 - x^2$

27. $f(x) = x^5 - x^3$

28. $f(x) = \sin^2 x + \cos x$, $[0, \pi]$

29. $f(x) = \dfrac{1}{\cos x + 2}$

30. $f(x) = \dfrac{1}{x^2 - x + 2}$

31. $f(x) = 3x^{3/2} - x^{1/2}$

32. $f(x) = x^{7/4} - x$

33. $f(x) = xe^{-x^2}$

34. $f(x) = e^x + e^{-2x}$

35. $f(x) = x^3 \ln x$ for $x > 0$

36. $f(x) = \ln x + \ln(4 - x^2)$ for $0 < x < 2$

In Exercises 37–50, find the intervals on which f is concave up or down, the points of inflection, and the critical points, and determine whether each critical point corresponds to a local minimum or maximum (or neither).

37. $f(x) = x^3 - 2x^2 + x$

38. $f(x) = x^2(x - 4)$

39. $f(t) = t^2 - t^3$

40. $f(x) = 2x^4 - 3x^2 + 2$

41. $f(x) = x^2 - x^{1/2}$

42. $f(x) = \dfrac{x}{x^2 + 2}$

43. $f(t) = \dfrac{1}{t^2 + 1}$

44. $f(x) = \dfrac{1}{x^4 + 1}$

45. $f(\theta) = \theta + \sin \theta$ for $0 \le \theta \le 2\pi$

46. $f(t) = \sin^2 t$ for $0 \le t \le \pi$

47. $f(x) = x - \sin x$ for $0 \le x \le 2\pi$

48. $f(x) = \tan x$ for $-\frac{\pi}{2} < x < \frac{\pi}{2}$

49. $f(x) = e^{-x} \cos x$ for $-\frac{\pi}{2} \le x \le \frac{3\pi}{2}$

50. $f(x) = (2x^2 - 1)e^{-x^2}$

51. An infectious flu spreads slowly at the beginning of an epidemic. The infection process accelerates until a majority of the susceptible individuals are infected, at which point the process slows down.

(a) If $R(t)$ is the number of individuals infected at time t, describe the concavity of the graph of R near the beginning and end of the epidemic.

(b) Write a one-sentence news bulletin describing the status of the epidemic on the day that $R(t)$ has a point of inflection.

52. Water is pumped into a sphere at a constant rate (Figure 17). Let $h(t)$ be the water level at time t. Sketch the graph of $h(t)$ (approximately, but with the correct concavity). Where does the point of inflection occur?

53. Water is pumped into a sphere at a variable rate in such a way that the water level rises at a constant rate c (Figure 17). Let $V(t)$ be the volume of water at time t. Sketch the graph of $V(t)$ (approximately, but with the correct concavity). Where does the point of inflection occur?

FIGURE 17

54. [GU]   (Continuation of Exercise 53) When the water level in a sphere of radius R is h, the volume of water is $V = \pi(Rh^2 - \frac{1}{3}h^3)$. Assume the level rises at a constant rate $c = 1$ (i.e., $h = t$).

(a) Find the inflection point of $V(t)$. Does this agree with your conclusion in Exercise 53?

(b) [GU]   Plot $V(t)$ for $R = 1$.

55. Let $f(x) = \tan^{-1} x$. Compute $f'(x)$ and $f''(x)$, and determine the increasing/decreasing and concavity behavior of $f(x)$.

In Exercises 56–58, sketch the graph of a function $f(x)$ satisfying all of the given conditions.

56. $f'(x) > 0$ and $f''(x) < 0$ for all x.

57. (i) $f'(x) > 0$ for all x, and

(ii) $f''(x) < 0$ for $x < 0$ and $f''(x) > 0$ for $x > 0$.

58. (i) $f'(x) < 0$ for $x < 0$ and $f'(x) > 0$ for $x > 0$, and
(ii) $f''(x) < 0$ for $|x| > 2$, and $f''(x) > 0$ for $|x| < 2$.

59. [✎] Use graphical reasoning to determine if the following statements are true or false. If false, modify the statement to make it correct.

(a) If $f(x)$ is increasing, then $f^{-1}(x)$ is decreasing.

(b) If $f(x)$ is decreasing, then $f^{-1}(x)$ is decreasing.

(c) If $f(x)$ is concave up, then $f^{-1}(x)$ is concave up.

(d) If $f(x)$ is concave down, then $f^{-1}(x)$ is concave up.

Further Insights and Challenges

In Exercises 60–62, assume that $f(x)$ is differentiable.

60. Proof of the Second Derivative Test Let c be a critical point in (a, b) such that $f''(c) > 0$ (the case $f''(c) < 0$ is similar).

(a) Show that $f''(c) = \lim\limits_{h \to 0} \dfrac{f'(c + h)}{h}$.

(b) Use part (a) to show that there exists an open interval (a, b) containing c such that $f'(x) < 0$ if $a < x < c$ and $f'(x) > 0$ if $c < x < b$. Conclude that $f(c)$ is a local minimum.

61. [✎] Assume that $f''(x)$ exists and $f''(x) > 0$. Prove that the graph of $f(x)$ "sits above" its tangent lines as follows.

(a) For any c, set $G(x) = f(x) - f'(c)(x - c) - f(c)$. It is sufficient to prove that $G(x) \geq 0$ for all c. Explain why with a sketch.

(b) Show that $G(c) = G'(c) = 0$ and $G''(x) > 0$ for all x. Use this to conclude that $G'(x) < 0$ for $x < c$ and $G'(x) > 0$ for $x > c$. Then deduce, using the MVT, that $G(x) > G(c)$ for $x \neq c$.

62. [✎] Assume that $f''(x)$ exists and let c be a point of inflection of $f(x)$.

(a) Use the method of Exercise 61 to prove that the tangent line at $x = c$ *crosses the graph* (Figure 18). *Hint:* Show that $G(x)$ changes sign at $x = c$.

FIGURE 18 Tangent line crosses graph at point of inflection.

(b) [GU] Verify this conclusion for $f(x) = \dfrac{x}{3x^2 + 1}$ by graphing $f(x)$ and the tangent line at each inflection point on the same set of axes.

63. Let $C(x)$ be the cost of producing x units of a certain good. Assume that the graph of $C(x)$ is concave up.

(a) Show that the average cost $A(x) = C(x)/x$ is minimized at that production level x_0 for which average cost equals marginal cost.

(b) Show that the line through $(0, 0)$ and $(x_0, C(x_0))$ is tangent to the graph of $C(x)$.

64. Let $f(x)$ be a polynomial of degree n.

(a) Show that if n is odd and $n > 1$, then $f(x)$ has at least one point of inflection.

(b) Show by giving an example that $f(x)$ need not have a point of inflection if n is even.

65. Critical Points and Inflection Points If $f'(c) = 0$ and $f(c)$ is neither a local min or max, must $x = c$ be a point of inflection? This is true of most "reasonable" examples (including the examples in this text), but it is not true in general. Let

$$f(x) = \begin{cases} x^2 \sin \frac{1}{x} & \text{for } x \neq 0 \\ 0 & \text{for } x = 0 \end{cases}$$

(a) Use the limit definition of the derivative to show that $f'(0)$ exists and $f'(0) = 0$.

(b) Show that $f(0)$ is neither a local min nor max.

(c) Show that $f'(x)$ changes sign infinitely often near $x = 0$ and conclude that $f(x)$ does not have a point of inflection at $x = 0$.

4.5 Graph Sketching and Asymptotes

In this section, our goal is to sketch graphs using the information provided by the first two derivatives f' and f''. We will see that a useful sketch can be produced without plotting a large number of points. Although nowadays almost all graphs are produced by computer (including, of course, the graphs in this textbook), sketching graphs by hand is useful as a way of solidifying your understanding of the basic concepts in this chapter.

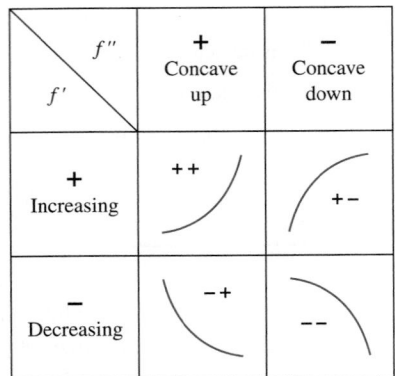

FIGURE 1 The four basic shapes.

Most graphs are made up of smaller *arcs* that have one of the four basic shapes, corresponding to the four possible sign combinations of f' and f'' (Figure 1). Since f' and f'' can each have sign $+$ or $-$, the sign combinations are

$$++ \qquad +- \qquad -+ \qquad --$$

In this notation, the first sign refers to f' and the second sign to f''. For instance, $-+$ indicates that $f'(x) < 0$ and $f''(x) > 0$.

In graph sketching, we pay particular attention to the **transition points**, where the basic shape changes due to a sign change in either f' (local min or max) or f'' (point of inflection). In this section, local extrema are indicated by solid dots and points of inflection by solid squares (Figure 2).

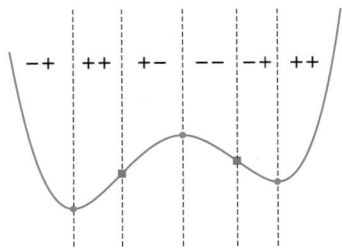

FIGURE 2 The graph of $f(x)$ with transition points and sign combinations of f' and f''.

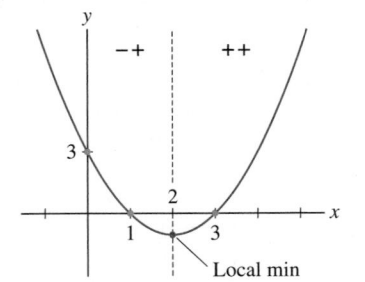

FIGURE 3 Graph of $f(x) = x^2 - 4x + 3$.

■ **EXAMPLE 1** Quadratic Polynomial Sketch the graph of $f(x) = x^2 - 4x + 3$ (Figure 3).

Solution We have $f'(x) = 2x - 4 = 2(x - 2)$, so

$$f'(x) < 0 \quad \text{for} \quad x < 2, \qquad f'(x) > 0 \quad \text{for} \quad x > 2$$

Thus, f decreases to the left and increases to the right of $x = 2$, with a local min at $x = 2$. The second derivative $f''(x) = 2$ is positive, so $f(x)$ is concave up. To sketch the graph, we plot the local minimum $(2, -1)$, the y-intercept, and the roots $x = 1, 3$ (Figure 3). ■

■ **EXAMPLE 2** A Cubic Polynomial Sketch the graph of

$$f(x) = \frac{1}{3}x^3 - \frac{1}{2}x^2 - 2x + 3$$

Solution

Step 1. **Determine the signs of f' and f''.**
First, set the derivative equal to zero to find the critical points:

$$f'(x) = x^2 - x - 2 = (x + 1)(x - 2) = 0$$

The critical points $c = -1, 2$ divide the x-axis into intervals $(-\infty, -1)$, $(-1, 2)$, and $(2, \infty)$. As in previous sections, we determine the sign of f' by computing "test values" within the intervals:

Interval	Test Value	Sign of f'
$(-\infty, -1)$	$f'(-2) = 4$	$+$
$(-1, 2)$	$f'(0) = -2$	$-$
$(2, \infty)$	$f'(3) = 4$	$+$

In this example, we sketch the graph of $f(x)$ where

$$f(x) = \frac{1}{3}x^3 - \frac{1}{2}x^2 - 2x + 3$$

$$f'(x) = (x+1)(x-2)$$

$$f''(x) = 2x - 1$$

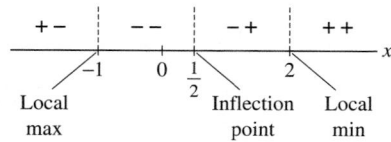

Local max   Inflection   Local
point   min

FIGURE 4 Sign combinations of f' and f''.

Next, set the second derivative equal to zero and solve: $f''(x) = 2x - 1 = 0$ has the solution $c = \frac{1}{2}$ and the sign of f'' is as follows:

Interval	Test Value	Sign of f''
$(-\infty, \frac{1}{2})$	$f''(0) = -1$	$-$
$(\frac{1}{2}, \infty)$	$f''(1) = 1$	$+$

Step 2. Note transition points and sign combinations.

This step merges the information about f' and f'' in a sign diagram (Figure 4). There are three transition points:

- $c = -1$: local max since f' changes from $+$ to $-$ at $c = -1$.
- $c = \frac{1}{2}$: point of inflection since f'' changes sign at $c = \frac{1}{2}$.
- $c = 2$: local min since f' changes from $-$ to $+$ at $c = 2$.

We plot the transition points $c = -1, \frac{1}{2}, 2$ together with the y-intercept as in Figure 5. It is necessary to compute y-values at these points:

$$f(-1) = \frac{25}{6}, \qquad f\left(\frac{1}{2}\right) = \frac{23}{12}, \qquad f(0) = 3, \qquad f(2) = -\frac{1}{3}$$

Step 3. Draw arcs of appropriate shape.

To create the sketch, it remains only to connect the transition points by arcs of the appropriate concavity as in Figure 5.

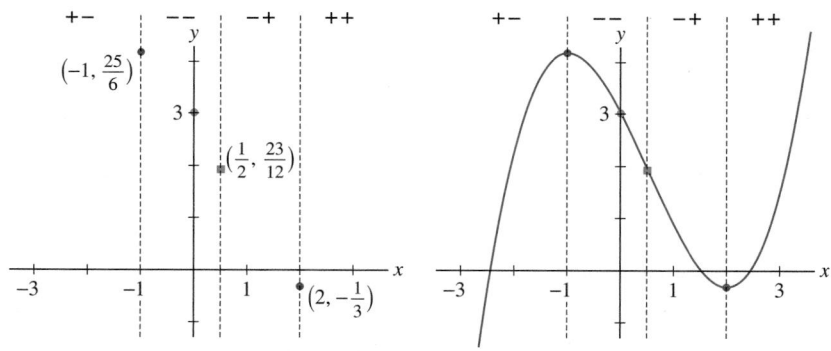

FIGURE 5 Graph of $f(x) = \frac{1}{3}x^3 - \frac{1}{2}x^2 - 2x + 3$.

Notice that the graph of our cubic is built out of four arcs, each with the appropriate increase/decrease and concavity behavior (Figure 6). ∎

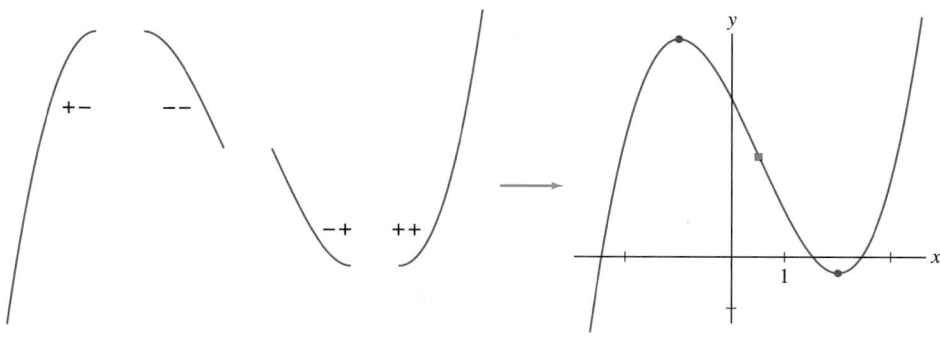

FIGURE 6 The graph of the cubic is composed of four arcs.

The graph of a polynomial $f(x)$ has the following general characteristics. It "wiggles" up and down a finite number of times and then tends to positive or negative infinity (Figure 7). To analyze the behavior of $f(x)$ as $|x|$ tends to infinity (this is called the **asymptotic behavior**), consider first the simplest polynomials x^n:

Strictly speaking, we have not yet defined limits as $x \to \pm\infty$. This is done more formally below. However,

$$\lim_{x \to \infty} f(x) = \infty$$

means that $f(x)$ gets arbitrarily large as x increases beyond all bounds. The limit $\lim_{x \to \infty} f(x) = -\infty$ has a similar meaning.

$$\lim_{x \to \infty} x^n = \infty \qquad \text{and} \qquad \lim_{x \to -\infty} x^n = \begin{cases} \infty & \text{if } n \text{ is even} \\ -\infty & \text{if } n \text{ is odd} \end{cases}$$

Now consider a general polynomial

$$f(x) = a_n x^n + a_{n-1} x^{n-1} + \cdots + a_1 x + a_0$$

where the leading coefficient a_n is nonzero. We may write $f(x)$ as a product:

$$f(x) = x^n \cdot \left(a_n + \frac{a_{n-1}}{x} + \cdots + \frac{a_1}{x^{n-1}} + \frac{a_0}{x^n} \right) \qquad \boxed{1}$$

As $|x|$ grows large, the second factor approaches a_n and thus $f(x)$ tends to infinity in the same way as its leading term $a_n x^n$ (Figure 7). The lower-order terms have no effect on the asymptotic behavior. Therefore, if $a_n > 0$,

$$\lim_{x \to \infty} f(x) = \infty \qquad \text{and} \qquad \lim_{x \to -\infty} f(x) = \begin{cases} \infty & \text{if } n \text{ is even} \\ -\infty & \text{if } n \text{ is odd} \end{cases}$$

If $a_n < 0$, these limits have opposite signs.

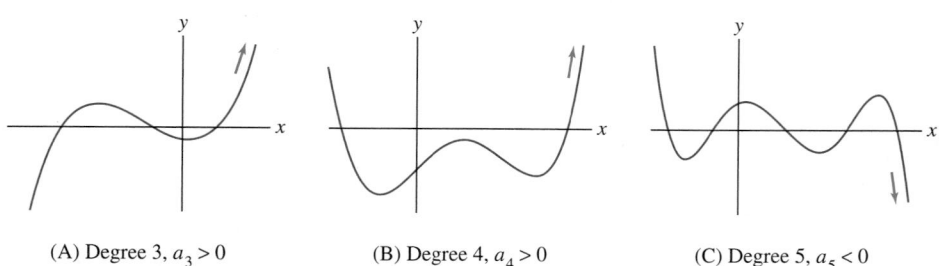

(A) Degree 3, $a_3 > 0$ (B) Degree 4, $a_4 > 0$ (C) Degree 5, $a_5 < 0$

FIGURE 7 Graphs of polynomials.

■ **EXAMPLE 3** **A Polynomial of Degree 4** Sketch the graph of

$$f(x) = 3x^4 - 8x^3 + 6x^2 + 1.$$

Solution

Step 1. **Determine the signs of f' and f''.**
First, set the derivative equal to zero:

$$f'(x) = 12x^3 - 24x^2 + 12x = 12x(x-1)^2 = 0$$

The critical points are $c = 0, 1$. Next, set the second derivative equal to zero:

$$f''(x) = 36x^2 - 48x + 12 = 12(x-1)(3x-1) = 0$$

The zeros of $f''(x)$ are $x = \frac{1}{3}$ and 1. The signs of f' and f'' are recorded in the following tables:

In this example, we sketch the graph of $f(x)$ where

$$f(x) = 3x^4 - 8x^3 + 6x^2 + 1$$

$$f'(x) = 12x(x-1)^2$$

$$f''(x) = 12(x-1)(3x-1)$$

Interval	Test Value	Sign of f'
$(-\infty, 0)$	$f'(-1) = -48$	$-$
$(0, 1)$	$f'(\frac{1}{2}) = \frac{3}{2}$	$+$
$(1, \infty)$	$f'(2) = 24$	$+$

Interval	Test Value	Sign of f''
$\left(-\infty, \frac{1}{3}\right)$	$f''(0) = 12$	$+$
$\left(\frac{1}{3}, 1\right)$	$f''(\frac{1}{2}) = -3$	$-$
$(1, \infty)$	$f''(2) = 60$	$+$

Step 2. Note transition points and sign combinations.

There are three transition points $c = 0, \frac{1}{3}, 1$, which divide the x-axis into four intervals (Figure 8). The type of sign change determines the nature of the transition point:

- $c = 0$: local min since f' changes from $-$ to $+$ at $c = 0$.
- $c = \frac{1}{3}$: point of inflection since f'' changes sign at $c = \frac{1}{3}$.
- $c = 1$: neither a local min nor max since f' does not change sign, but it is a point of inflection since $f''(x)$ changes sign at $c = 1$.

We plot the transition points $c = 0, \frac{1}{3}, 1$ in Figure 9 using function values $f(0) = 1$, $f(\frac{1}{3}) = \frac{38}{27}$, and $f(1) = 2$.

```
  - +    + +      + -       + +
 ─────┼────┼──────────┼──────────── x
      0    1          1
           3
 Local      Inflection   Inflection
  min         point        point
```

FIGURE 8

 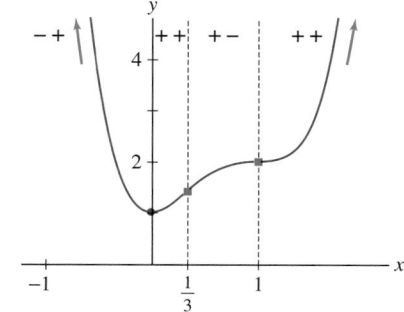

FIGURE 9 $f(x) = 3x^4 - 8x^3 + 6x^2 + 1$.

Step 3. Draw arcs of appropriate shape and asymptotic behavior.

Before connecting the transition points by arcs, we note the asymptotic behavior of $f(x)$. Since $f(x)$ is a polynomial of degree 4 whose leading term $3x^4$ has a positive coefficient, $f(x)$ tends to ∞ as $x \to \infty$ and as $x \to -\infty$. This asymptotic behavior is noted by the arrows in Figure 9. ∎

■ **EXAMPLE 4 A Trigonometric Function**   Sketch the graph of $f(x) = \cos x + \frac{1}{2}x$ over the interval $[0, \pi]$.

Solution First, we solve on $[0, \pi]$:

$$f'(x) = -\sin x + \frac{1}{2} = 0 \quad \Rightarrow \quad x = \frac{\pi}{6}, \frac{5\pi}{6}$$

$$f''(x) = -\cos x = 0 \quad \Rightarrow \quad x = \frac{\pi}{2}$$

and determine the sign combinations:

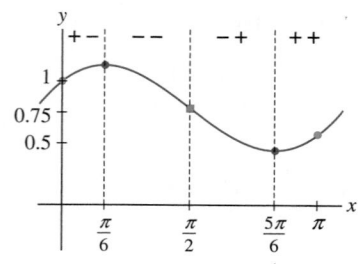

FIGURE 10 $f(x) = \cos x + \frac{1}{2}x$.

Interval	Test Value	Sign of f'
$\left(0, \frac{\pi}{6}\right)$	$f'\left(\frac{\pi}{12}\right) \approx 0.24$	$+$
$\left(\frac{\pi}{6}, \frac{5\pi}{6}\right)$	$f'\left(\frac{\pi}{2}\right) = -\frac{1}{2}$	$-$
$\left(\frac{5\pi}{6}, \pi\right)$	$f'\left(\frac{11\pi}{12}\right) \approx 0.24$	$+$

Interval	Test Value	Sign of f''
$\left(0, \frac{\pi}{2}\right)$	$f''\left(\frac{\pi}{4}\right) = -\frac{\sqrt{2}}{2}$	$-$
$\left(\frac{\pi}{2}, \pi\right)$	$f''\left(\frac{3\pi}{4}\right) = \frac{\sqrt{2}}{2}$	$+$

We record the sign changes and transition points in Figure 10, and sketch the graph using the values:

$$f(0) = 1, \quad f\left(\frac{\pi}{6}\right) \approx 1.13, \quad f\left(\frac{\pi}{2}\right) \approx 0.79, \quad f\left(\frac{5\pi}{6}\right) \approx 0.44, \quad f(\pi) \approx 0.57 \quad \blacksquare$$

■ **EXAMPLE 5** Sketching a Function Involving e^x Sketch the graph of $f(x) = xe^x$ on the interval $[-4, 2]$.

Solution As usual, the first step is to solve for the critical points:

$$f'(x) = \frac{d}{dx}\, xe^x = xe^x + e^x = (x + 1)e^x = 0$$

Since $e^x > 0$ for all x, the unique critical point is $x = -1$ and

$$f'(x) = \begin{cases} < 0 & \text{for} \quad x < -1 \\ > 0 & \text{for} \quad x > -1 \end{cases}$$

Thus, $f'(x)$ changes sign from $-$ to $+$ at $x = -1$ and $f(-1)$ is a local minimum. For the second derivative, we have

$$f''(x) = (x + 1) \cdot (e^x)' + e^x \cdot (x + 1)' = (x + 1)e^x + e^x = (x + 2)e^x$$

$$f''(x) = \begin{cases} < 0 & \text{for} \quad x < -2 \\ > 0 & \text{for} \quad x > -2 \end{cases}$$

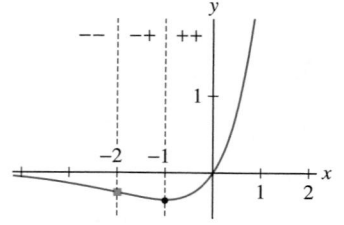

FIGURE 11 Graph of $f(x) = xe^x$. The sign combinations $--$, $-+$, $++$ indicate the signs of f' and f''.

Thus, $x = -2$ is a point of inflection and the graph changes from concave down to concave up at $x = -2$. Figure 11 shows the graph with its local minimum and point of inflection. ■

Asymptotic Behavior

Asymptotic behavior refers to the behavior of a function $f(x)$ as either x or $f(x)$ approaches $\pm\infty$. A horizontal line $y = L$ is called a **horizontal asymptote** if one or both of the following limits exist:

$$\lim_{x \to \infty} f(x) = L \qquad \text{or} \qquad \lim_{x \to -\infty} f(x) = L$$

By definition, in this first limit, $|f(x) - L|$ gets arbitrarily small as x increases without bound (that is, for any $\epsilon > 0$, there exists a number M such that $|f(x) - L| < \epsilon$ for all $x > M$). In the second limit, $|f(x) - L|$ gets arbitrarily small as x decreases without bound. In Figure 12, $\lim_{x \to \pm\infty} f(x) = 1$ and $y = 1$ is a horizontal asymptote.

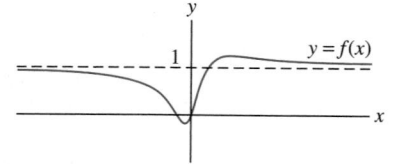

FIGURE 12 The line $y = 1$ is a horizontal asymptote.

We are also interested in vertical asymptotic behavior, where the function value $f(x)$ approaches $\pm\infty$. A vertical line $x = L$ is called a **vertical asymptote** if $f(x)$ has an infinite limit as x approaches L from the left or right (or both) (Figure 13). That is, if

$$\lim_{x \to L+} f(x) = \pm\infty \quad \text{and/or} \quad \lim_{x \to L-} f(x) = \pm\infty$$

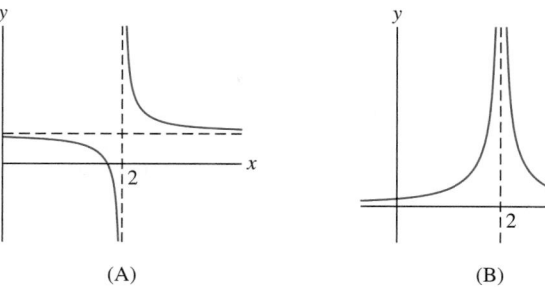

(A)   (B)

FIGURE 13 The line $x = 2$ is a vertical asymptote for both graphs.

For example, we have

$$\lim_{x \to -\infty} e^x = 0 \quad \text{and} \quad \lim_{x \to 0+} \ln x = -\infty$$

It follows that the x-axis, $y = 0$, is a horizontal asymptote of $f(x) = e^x$ and the y-axis, $x = 0$, is a vertical asymptote of $f(x) = \ln x$ (Figure 14).

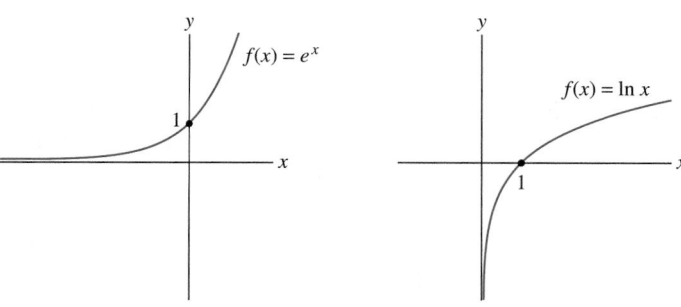

(A) The x-axis is a horizontal asymptote.   (B) The y-axis is a vertical asymptote.

FIGURE 14

■ **EXAMPLE 6** Calculate $\displaystyle\lim_{x \to \pm\infty} \frac{20x^2 - 3x}{3x^5 - 4x^2 + 5}$.

Solution It would be nice if we could apply the Quotient Law for limits, but this law is valid only if the denominator has a finite, nonzero limit. In our case, both the numerator and the denominator approach infinity:

$$\lim_{x \to \infty} (20x^2 - 3x) = \infty \quad \text{and} \quad \lim_{x \to \infty} (3x^5 - 4x^2 + 5) = \infty$$

The way around this difficulty is to divide the numerator and denominator by x^5 (the highest power of x in the denominator):

$$\frac{20x^2 - 3x}{3x^5 - 4x^2 + 5} = \frac{x^{-5}(20x^2 - 3x)}{x^{-5}(3x^5 - 4x^2 + 5)} = \frac{20x^{-3} - 3x^{-4}}{3 - 4x^{-3} + 5x^{-5}}$$

After rewriting the function in this way, we can use the Quotient Law:

$$\lim_{x\to\pm\infty} \frac{20x^2 - 3x}{3x^5 - 4x^2 + 5} = \frac{\lim_{x\to\pm\infty} \left(20x^{-3} - 3x^{-4}\right)}{\lim_{x\to\pm\infty} \left(3 - 4x^{-3} + 5x^{-5}\right)} = \frac{0}{3} = 0$$ ∎

The method of the previous example applies to any rational function:

$$f(x) = \frac{a_n x^n + a_{n-1} x^{n-1} + \cdots + a_0}{b_m x^m + b_{m-1} x^{m-1} + \cdots + b_0}$$

where $a_n \neq 0$ and $b_m \neq 0$. Divide the numerator and denominator by x^m:

$$f(x) = \frac{a_n x^{n-m} + a_{n-1} x^{n-1-m} + \cdots + a_0 x^{-m}}{b_m + b_{m-1} x^{-1} + \cdots + b_0 x^{-m}}$$

Now, in the denominator, each term after the first tends to zero, and thus

$$\lim_{x\to\infty} (b_m + b_{m-1} x^{-1} + \cdots + b_0 x^{-m}) = b_m$$

The limit of the numerator depends only on the leading term $a_n x^{n-m}$, so

$$\lim_{x\to\pm\infty} f(x) = \frac{\lim_{x\to\pm\infty} (a_n x^{n-m} + a_{n-1} x^{n-1-m} + \cdots + a_0 x^{-m})}{b_m}$$

$$= \frac{\lim_{x\to\pm\infty} a_n x^{n-m}}{b_m} = \frac{a_n}{b_m} \lim_{x\to\pm\infty} x^{n-m}$$

THEOREM 1 Asymptotic Behavior of a Rational Function The asymptotic behavior of a rational function depends only on the leading terms of its numerator and denominator. If $a_n, b_m \neq 0$, then

$$\lim_{x\to\pm\infty} \left(\frac{a_n x^n + a_{n-1} x^{n-1} + \cdots + a_0}{b_m x^m + b_{m-1} x^{m-1} + \cdots + b_0} \right) = \frac{a_n}{b_m} \lim_{x\to\pm\infty} x^{n-m}$$

The following examples illustrate this theorem:

- $n = m$:

$$\lim_{x\to\infty} \frac{3x^4 - 7x + 9}{7x^4 - 4} = \frac{3}{7} \lim_{x\to\infty} x^0 = \frac{3}{7}$$

- $n < m$:

$$\lim_{x\to\infty} \frac{3x^3 - 7x + 9}{7x^4 - 4} = \frac{3}{7} \lim_{x\to\infty} x^{-1} = 0$$

- $n > m$, $n - m$ odd:

$$\lim_{x\to-\infty} \frac{3x^8 - 7x + 9}{7x^3 - 4} = \frac{3}{7} \lim_{x\to-\infty} x^5 = -\infty$$

- $n > m$, $n - m$ even:

$$\lim_{x\to-\infty} \frac{3x^7 - 7x + 9}{7x^3 - 4} = \frac{3}{7} \lim_{x\to-\infty} x^4 = \infty$$

In the next example, we apply our method to quotients involving noninteger exponents and algebraic functions.

■ **EXAMPLE 7** Calculate the limits   **(a)** $\displaystyle\lim_{x\to\infty} \frac{3x^{5/2} + 7x^{-1/2}}{x - x^{1/2}}$ **(b)** $\displaystyle\lim_{x\to\infty} \frac{x^2}{\sqrt{x^3 + 1}}$

Solution

(a) As before, we divide the numerator and denominator by the largest power of x occurring in the denominator (multiply the top and bottom by x^{-1}):

$$\frac{3x^{5/2} + 7x^{-1/2}}{x - x^{1/2}} = \left(\frac{x^{-1}}{x^{-1}}\right)\frac{3x^{5/2} + 7x^{-1/2}}{x - x^{1/2}} = \frac{3x^{3/2} + 7x^{-3/2}}{1 - x^{-1/2}}$$

We obtain

$$\lim_{x\to\infty} \frac{3x^{3/2} + 7x^{-3/2}}{1 - x^{-1/2}} = \frac{\lim\limits_{x\to\infty} (3x^{3/2} + 7x^{-3/2})}{\lim\limits_{x\to\infty} (1 - x^{-1/2})} = \frac{\infty}{1} = \infty$$

(b) The key is to observe that the denominator of $\dfrac{x^2}{\sqrt{x^3 + 1}}$ "behaves" like $x^{3/2}$:

$$\sqrt{x^3 + 1} = \sqrt{x^3(1 + x^{-3})} = x^{3/2}\sqrt{1 + x^{-3}}$$

This suggests that we handle the quotient by dividing the numerator and denominator by $x^{3/2}$:

$$\frac{x^2}{\sqrt{x^3 + 1}} = \left(\frac{x^{-3/2}}{x^{-3/2}}\right)\frac{x^2}{x^{3/2}\sqrt{1 + x^{-3}}} = \frac{x^{1/2}}{\sqrt{1 + x^{-3}}}$$

We may now apply the Rule for Quotients:

$$\lim_{x\to\infty} \frac{x^2}{\sqrt{x^3 + 1}} = \lim_{x\to\infty} \frac{x^{1/2}}{\sqrt{1 + x^{-3}}} = \frac{\infty}{1} = \infty \qquad \blacksquare$$

The next two examples carry out graph sketching for functions with horizontal and vertical asymptotes.

■ **EXAMPLE 8** Sketch the graph of $f(x) = \dfrac{3x + 2}{2x - 4}$.

Solution The function $f(x)$ is not defined for all x. This plays a role in our analysis so we add a Step 0 to our procedure:

Step 0. **Determine the domain of f.**
   Since $f(x)$ is not defined for $x = 2$, the domain of f consists of the two intervals $(-\infty, 2)$ and $(2, \infty)$. We must analyze the behavior of f on these intervals separately.

Step 1. **Determine the signs of f' and f''.**
   A calculation shows that

$$f'(x) = -\frac{4}{(x - 2)^2}, \qquad f''(x) = \frac{8}{(x - 2)^3}$$

The derivative is not defined at $x = 2$, but we do not call $x = 2$ a critical point because 2 is not in the domain of f. The denominator of $f'(x)$ is positive, so $f'(x) < 0$ for $x \neq 2$. Therefore, $f(x)$ is decreasing for all $x \neq 2$ and $f(x)$ has no critical points.
   On the other hand, $f''(x) > 0$ for $x > 2$ and $f''(x) < 0$ for $x < 2$. Although $f''(x)$ changes sign at $x = 2$, we do not call $x = 2$ a point of inflection because 2 is not in the domain of f.

Step 2. **Note transition points and sign combinations.**
   There are no transition points in the domain of f.

$(-\infty, 2)$	$f'(x) < 0$ and $f''(x) < 0$
$(2, \infty)$	$f'(x) < 0$ and $f''(x) > 0$

Step 3. Draw arcs of appropriate shape and asymptotic behavior.
The following limits show that $y = \frac{3}{2}$ is a horizontal asymptote:

$$\lim_{x \to \pm\infty} \frac{3x + 2}{2x - 4} = \lim_{x \to \pm\infty} \frac{3 + 2x^{-1}}{2 - 4x^{-1}} = \frac{3}{2}$$

The line $x = 2$ is a vertical asymptote because $f(x)$ has infinite one-sided limits as x approaches 2:

$$\lim_{x \to 2-} \frac{3x + 2}{2x - 4} = -\infty, \qquad \lim_{x \to 2+} \frac{3x + 2}{2x - 4} = \infty$$

To verify these limits, observe that as x approaches 2, the numerator $3x + 2$ is positive, whereas the denominator $2x - 4$ is small negative if $x < 2$ and small positive if $x > 2$.

To create the sketch, first draw the asymptotes as in Figure 15(A). What does the graph look like to the right of $x = 2$? It is decreasing and concave up since $f' < 0$ and $f'' > 0$, and it approaches the asymptotes. The only possibility is the right-hand curve in Figure 15(B). To the left of $x = 2$, the graph is decreasing, concave down, and approaches the asymptotes. The x-intercept is $x = -\frac{2}{3}$ because $f(-\frac{2}{3}) = 0$ and the y-intercept is $y = f(0) = -\frac{1}{2}$. ∎

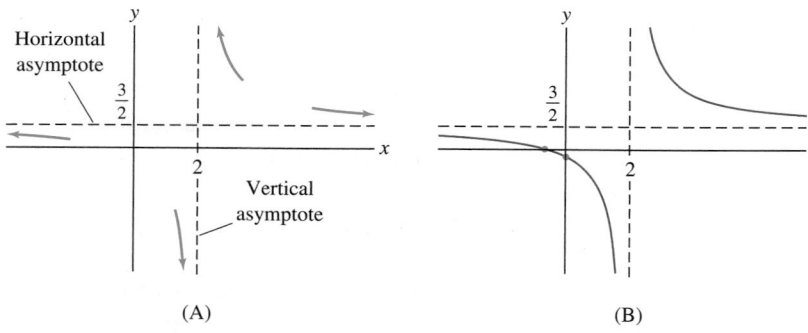

FIGURE 15 Graph of $y = \dfrac{3x + 2}{2x - 4}$.

(A)   (B)

EXAMPLE 9 Sketch the graph of $f(x) = \dfrac{1}{x^2 - 1}$.

Solution The function $f(x)$ is defined for $x \neq \pm 1$. By calculation,

$$f'(x) = -\frac{2x}{(x^2 - 1)^2}, \qquad f''(x) = \frac{6x^2 + 2}{(x^2 - 1)^3}$$

For $x \neq \pm 1$, the denominator of $f'(x)$ is positive. Therefore, $f'(x)$ and x have opposite signs:

- $f'(x) > 0$ for $x < 0$ and $f'(x) < 0$ for $x > 0$

Because $f'(x)$ changes sign from $+$ to $-$ at $x = 0$, $f(0) = -1$ is a local max. The sign of $f''(x)$ is equal to the sign of $x^2 - 1$ because $6x^2 + 2$ is positive:

- $f''(x) > 0$ for $x < -1$ or $x > 1$ and $f''(x) < 0$ for $-1 < x < 1$

Figure 16 summarizes the sign information.

In this example, we sketch the graph of $f(x)$ *where*

$$f(x) = \frac{1}{x^2 - 1}$$

$$f'(x) = -\frac{2x}{(x^2 - 1)^2}$$

$$f''(x) = \frac{6x^2 + 2}{(x^2 - 1)^3}$$

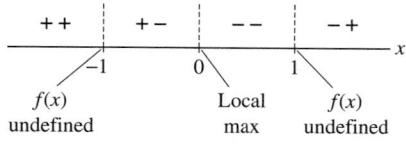

FIGURE 16

The x-axis $y = 0$ is a horizontal asymptote because

$$\lim_{x \to \infty} \frac{1}{x^2 - 1} = 0 \quad \text{and} \quad \lim_{x \to -\infty} \frac{1}{x^2 - 1} = 0$$

The function $f(x)$ has infinite one-sided limits as $x \to \pm 1$. We observe that $f(x) < 0$ if $-1 < x < 1$ and that $f(x) > 0$ if $|x| > 1$. Thus, $f(x)$ approaches $-\infty$ as $x \to \pm 1$ from within the interval $(-1, 1)$, and it approaches ∞ as $x \to \pm 1$ from outside $(-1, 1)$ (Figure 17). We obtain the sketch in Figure 18.

Vertical Asymptote	Left-Hand Limit	Right-Hand Limit
$x = -1$	$\displaystyle\lim_{x \to -1-} \frac{1}{x^2 - 1} = \infty$	$\displaystyle\lim_{x \to -1+} \frac{1}{x^2 - 1} = -\infty$
$x = 1$	$\displaystyle\lim_{x \to 1-} \frac{1}{x^2 - 1} = -\infty$	$\displaystyle\lim_{x \to 1+} \frac{1}{x^2 - 1} = \infty$

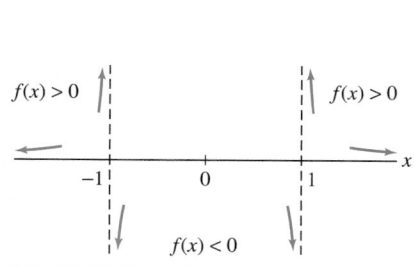

FIGURE 17 Behavior at vertical asymptotes.

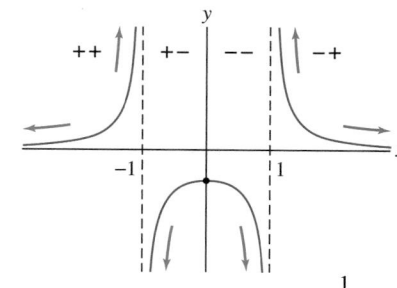

FIGURE 18 Graph of $y = \dfrac{1}{x^2 - 1}$.

4.5 SUMMARY

- Most graphs are made up of arcs that have one of the four basic shapes (Figure 19):

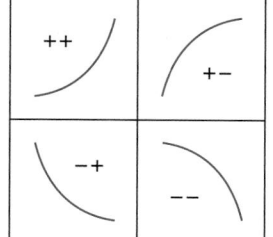

FIGURE 19 The four basic shapes.

	Sign Combination	Curve Type
++	$f' > 0, f'' > 0$	Increasing and concave up
+−	$f' > 0, f'' < 0$	Increasing and concave down
−+	$f' < 0, f'' > 0$	Decreasing and concave up
−−	$f' < 0, f'' < 0$	Decreasing and concave down

- A *transition point* is a point in the domain of f at which either f' changes sign (local min or max) or f'' changes sign (point of inflection).
- It is convenient to break up the curve sketching process into steps:

Step 0. Determine the domain of f.

Step 1. Determine the signs of f' and f''.

Step 2. Note transition points and sign combinations.

Step 3. Draw arcs of appropriate shape and asymptotic behavior.

- A horizontal line $y = L$ is a *horizontal asymptote* if

$$\lim_{x \to \infty} f(x) = L \qquad \text{and/or} \qquad \lim_{x \to -\infty} f(x) = L$$

- A vertical line $x = L$ is a *vertical asymptote* if

$$\lim_{x \to L+} f(x) = \pm\infty \qquad \text{and/or} \qquad \lim_{x \to L-} f(x) = \pm\infty$$

- For a rational function $f(x) = \dfrac{a_n x^n + a_{n-1} x^{n-1} + \cdots + a_0}{b_m x^m + b_{m-1} x^{m-1} + \cdots + b_0}$ with $a_n, b_m \neq 0$,

$$\lim_{x \to \pm\infty} f(x) = \frac{a_n}{b_m} \lim_{x \to \pm\infty} x^{n-m}$$

4.5 EXERCISES

Preliminary Questions

1. Sketch an arc where f' and f have the sign combination $++$. Do the same for $-+$.

2. If the sign combination of f' and f'' changes from $++$ to $+-$ at $x = c$, then (choose the correct answer):

(a) $f(c)$ is a local min
(b) $f(c)$ is a local max
(c) c is a point of inflection

3. What are the following limits?
(a) $\lim_{x \to \infty} x^3$
(b) $\lim_{x \to -\infty} x^3$
(c) $\lim_{x \to -\infty} x^4$

4. What is the sign of a_2 if $f(x) = a_2 x^2 + a_1 x + a_0$ satisfies $\lim_{x \to -\infty} f(x) = -\infty$?

5. What is the sign of the leading coefficient a_7 if $f(x)$ is a polynomial of degree 7 such that $\lim_{x \to -\infty} f(x) = \infty$?

6. The second derivative of the function $f(x) = (x - 4)^{-1}$ is $f''(x) = 2(x - 4)^{-3}$. Although $f''(x)$ changes sign at $x = 4$, $f(x)$ does not have a point of inflection at $x = 4$. Why not?

Exercises

1. Determine the sign combinations of f' and f'' for each interval A–G in Figure 20.

FIGURE 20

2. State the sign change at each transition point A–G in Figure 21. Example: $f'(x)$ goes from $+$ to $-$ at A.

FIGURE 21

In Exercises 3–6, draw the graph of a function for which f' and f'' take on the given sign combinations.

3. $++$, $+-$, $--$
4. $+-$, $--$, $-+$
5. $-+$, $--$, $-+$
6. $-+$, $++$, $+-$

7. Sketch the graph of $y = x^2 - 2x + 3$.

8. Sketch the graph of $y = 3 + 5x - 2x^2$.

9. Sketch the graph of the cubic $f(x) = x^3 - 3x^2 + 2$. For extra accuracy, plot the zeros of $f(x)$, which are $x = 1$ and $x = 1 \pm \sqrt{3}$ or $x \approx -0.73, 2.73$.

10. Show that the cubic $x^3 - 3x^2 + 6x$ has a point of inflection but no local extreme values. Sketch the graph.

11. Extend the sketch of the graph of $f(x) = \cos x + \frac{1}{2}x$ over $[0, \pi]$ in Example 4 to the interval $[0, 5\pi]$.

12. Sketch the graphs of $y = x^{2/3}$ and $y = x^{4/3}$.

In Exercises 13–36, sketch the graph of the function. Indicate the transition points (local extrema and points of inflection).

13. $y = x^3 + \frac{3}{2}x^2$
14. $y = x^3 - 3x + 5$

15. $y = x^2 - 4x^3$

16. $y = \frac{1}{3}x^3 + x^2 + 3x$

17. $y = 4 - 2x^2 + \frac{1}{6}x^4$

18. $y = 7x^4 - 6x^2 + 1$

19. $y = x^5 + 5x$

20. $y = x^5 - 5x^3$

21. $y = x^4 - 3x^3 + 4x$

22. $y = x^2(x - 4)^2$

23. $y = 6x^7 - 7x^6$

24. $y = x^6 - 3x^4$

25. $y = x - \sqrt{x}$

26. $y = \sqrt{x + 2}$

27. $y = \sqrt{x} + \sqrt{9 - x}$

28. $y = x(1 - x)^{1/3}$

29. $y = (x^2 - x)^{1/3}$

30. $y = (x^3 - 4x)^{1/3}$

31. $y = 4x - \ln x$

32. $y = \ln(x^2 + 1)$

33. $y = xe^{-x^2}$

34. $y = x(4 - x) - 3\ln x$

35. $y = (2x^2 - 1)e^{-x^2}$

36. $y = x - 2\ln(x^2 + 1)$

37. Sketch the graph of $f(x) = 18(x - 3)(x - 1)^{2/3}$ using the following formulas:

$$f'(x) = \frac{30(x - \frac{9}{5})}{(x - 1)^{1/3}}, \qquad f''(x) = \frac{20(x - \frac{3}{5})}{(x - 1)^{4/3}}$$

CRS *In Exercises 38-41, sketch the graph of the function. Indicate all transition points. If necessary, use a graphing utility or computer algebra system to locate the transition points numerically.*

38. $y = x^2 - 10\ln(x^2 + 1)$

39. $y = e^x - 4x - 8$

40. $y = (x^2 - 1)\sin^{-1} x \quad (-1 \le x \le 1)$

41. $y = \tan^{-1}(x^2 - x)$

In Exercises 42–47, sketch the graph over the given interval. Indicate the transition points.

42. $y = x + \sin x, \quad [0, 2\pi]$

43. $y = \sin x + \cos x, \quad [0, 2\pi]$

44. $y = 2\sin x - \cos^2 x, \quad [0, 2\pi]$

45. $y = \sin x + \frac{1}{2}x, \quad [0, 2\pi]$

46. $y = \sin x + \sqrt{3}\cos x, \quad [0, \pi]$

47. $y = \sin x - \frac{1}{2}\sin 2x, \quad [0, \pi]$
Hint for Exercise 47: We find numerically that there is one point of inflection in the interval, occurring at $x \approx 1.3182$.

48. Are all sign transitions possible? Explain with a sketch why the transitions $++ \to -+$ and $-- \to +-$ do not occur if the function is differentiable. (See Exercise 97 for a proof.)

49. Suppose that f is twice differentiable satisfying (i) $f(0) = 1$, (ii) $f'(x) > 0$ for all $x \ne 0$, and (iii) $f''(x) < 0$ for $x < 0$ and $f''(x) > 0$ for $x > 0$. Let $g(x) = f(x^2)$.
(a) Sketch a possible graph of $f(x)$.
(b) Prove that $g(x)$ has no points of inflection and a unique local extreme value at $x = 0$. Sketch a possible graph of $g(x)$.

50. Which of the graphs in Figure 22 *cannot* be the graph of a polynomial? Explain.

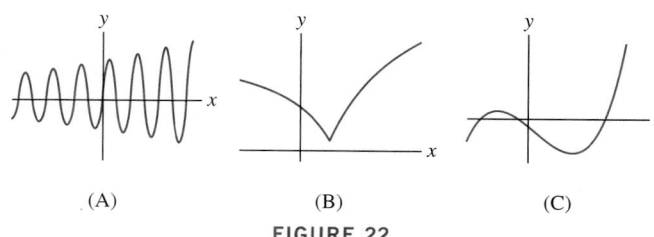

(A)   (B)   (C)

FIGURE 22

In Exercises 51–60, calculate the following limits (divide the numerator and denominator by the highest power of x appearing in the denominator).

51. $\lim\limits_{x \to \infty} \dfrac{x}{x + 9}$

52. $\lim\limits_{x \to \infty} \dfrac{3x^2 + 20x}{4x^2 + 9}$

53. $\lim\limits_{x \to \infty} \dfrac{3x^2 + 20x}{2x^4 + 3x^3 - 29}$

54. $\lim\limits_{x \to \infty} \dfrac{4}{x + 5}$

55. $\lim\limits_{x \to \infty} \dfrac{7x - 9}{4x + 3}$

56. $\lim\limits_{x \to \infty} \dfrac{7x^2 - 9}{4x + 3}$

57. $\lim\limits_{x \to -\infty} \dfrac{7x^2 - 9}{4x + 3}$

58. $\lim\limits_{x \to \infty} \dfrac{5x - 9}{4x^3 + 1}$

59. $\lim\limits_{x \to -\infty} \dfrac{x^2 - 1}{x + 4}$

60. $\lim\limits_{x \to \infty} \dfrac{2x^5 + 3x^4 - 31x}{8x^4 - 31x^2 + 12}$

In Exercises 61–70, calculate the limit.

61. $\lim\limits_{x \to \infty} \dfrac{\sqrt{x^2 + 1}}{x + 1}$

62. $\lim\limits_{x \to -\infty} \dfrac{\sqrt{x^4 + 1}}{x^3 + 1}$

63. $\lim\limits_{x \to \infty} \dfrac{x + 1}{\sqrt[3]{x^2 + 1}}$

64. $\lim\limits_{x \to \infty} \dfrac{x^{4/3} + 4x^{1/3}}{x^{5/4} - 2x}$

65. $\lim\limits_{x \to -\infty} \dfrac{x}{(x^6 + 1)^{1/3}}$

66. $\lim\limits_{x \to -\infty} \dfrac{4x - 9}{(3x^4 - 2)^{1/3}}$

67. $\lim\limits_{x \to 0+} \dfrac{\ln x}{x^2}$

68. $\lim\limits_{x \to -\infty} (x^2 + 4x - 8)e^{-x}$

69. $\lim\limits_{x \to \infty} \dfrac{e^{-x^2/2}}{x^2 - 5}$

70. $\lim\limits_{x \to \infty} (x^3 + 4x^2)e^x$

71. Which curve in Figure 23 is the graph of $f(x) = \dfrac{2x^4 - 1}{1 + x^4}$? Explain on the basis of horizontal asymptotes.

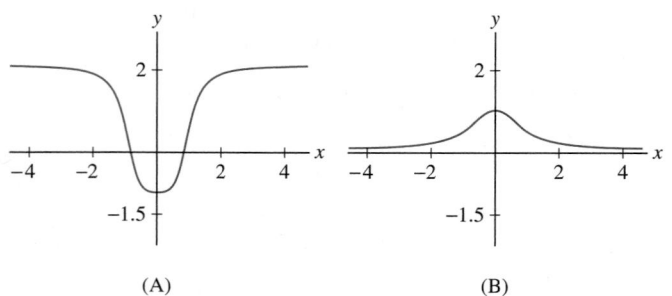

(A)   (B)

FIGURE 23

72. Match the graphs in Figure 24 with the two functions $y = \dfrac{3x}{x^2 - 1}$ and $y = \dfrac{3x^2}{x^2 - 1}$. Explain.

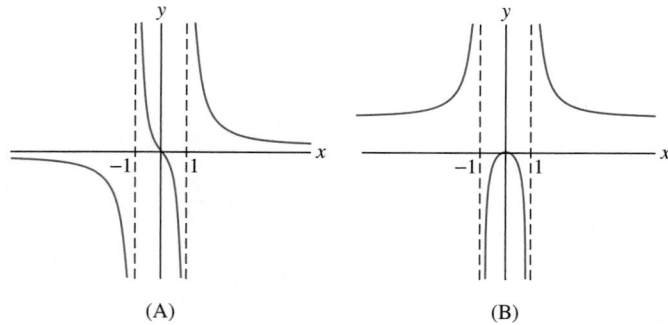

(A)   (B)

FIGURE 24

73. Match the functions with their graphs in Figure 25.

(a) $y = \dfrac{1}{x^2 - 1}$ **(b)** $y = \dfrac{x^2}{x^2 + 1}$

(c) $y = \dfrac{1}{x^2 + 1}$ **(d)** $y = \dfrac{x}{x^2 - 1}$

74. Sketch the graph of $f(x) = \dfrac{2x - 1}{x + 1}$.

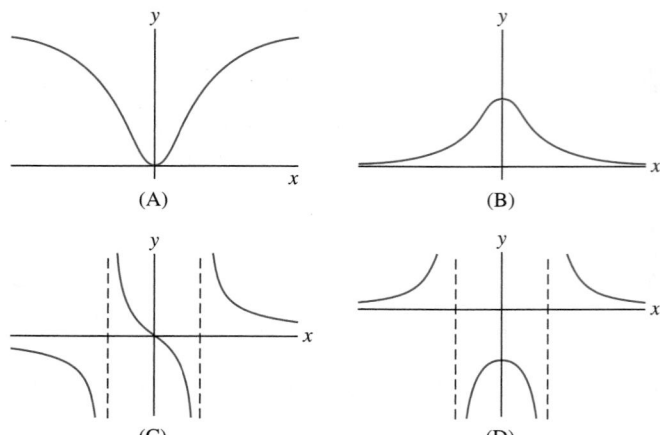

(A)   (B)

(C)   (D)

FIGURE 25

75. Sketch the graph of $f(x) = \dfrac{x}{x^2 + 1}$ using the formulas

$$f'(x) = \frac{1 - x^2}{(1 + x^2)^2}, \qquad f''(x) = \frac{2x(x^2 - 3)}{(x^2 + 1)^3}$$

In Exercises 76–91, sketch the graph of the function. Indicate the asymptotes, local extrema, and points of inflection.

76. $y = \dfrac{1}{2x - 1}$ **77.** $y = \dfrac{x}{2x - 1}$

78. $y = \dfrac{x + 1}{2x - 1}$ **79.** $y = \dfrac{x + 3}{x - 2}$

80. $y = x + \dfrac{1}{x}$ **81.** $y = \dfrac{1}{x} + \dfrac{1}{x - 1}$

82. $y = \dfrac{1}{x} - \dfrac{1}{x - 1}$ **83.** $y = 1 - \dfrac{3}{x} + \dfrac{4}{x^3}$

84. $y = \dfrac{1}{x^2 - 6x + 8}$ **85.** $y = \dfrac{1}{x^2} + \dfrac{1}{(x - 2)^2}$

86. $y = \dfrac{1}{x^2} - \dfrac{1}{(x - 2)^2}$ **87.** $y = \dfrac{4}{x^2 - 9}$

88. $y = \dfrac{1}{(x^2 + 1)^2}$ **89.** $y = \dfrac{x^2}{(x^2 - 1)(x^2 + 1)}$

90. $y = \dfrac{1}{\sqrt{x^2 + 1}}$ **91.** $y = \dfrac{x}{\sqrt{x^2 + 1}}$

Further Insights and Challenges

In Exercises 92–96, we explore functions whose graphs approach a nonhorizontal line as $x \to \infty$. A line $y = ax + b$ is called a **slant asymptote** *if*

$$\lim_{x \to \infty} (f(x) - (ax + b)) = 0$$

or

$$\lim_{x \to -\infty} (f(x) - (ax + b)) = 0$$

92. Let $f(x) = \dfrac{x^2}{x-1}$. Verify the following:

(a) f is concave down on $(-\infty, 1)$ and concave up on $(1, \infty)$ as in Figure 26.

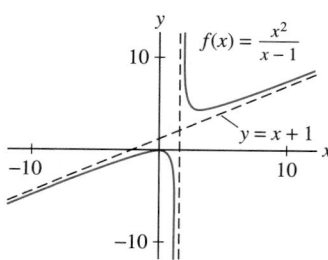

FIGURE 26

(b) $f(0)$ is a local max and $f(2)$ a local min.

(c) $\lim\limits_{x \to 1-} f(x) = -\infty$ and $\lim\limits_{x \to 1+} f(x) = \infty$.

(d) $y = x + 1$ is a slant asymptote of $f(x)$ as $x \to \pm\infty$.

(e) The slant asymptote lies above the graph of $f(x)$ for $x < 1$ and below the graph for $x > 1$.

93. If $f(x) = P(x)/Q(x)$, where P and Q are polynomials of degrees $m + 1$ and m, then by long division, we can write

$$f(x) = (ax + b) + P_1(x)/Q(x)$$

where P_1 is a polynomial of degree $< m$. Show that $y = ax + b$ is the slant asymptote of $f(x)$. Use this procedure to find the slant asymptotes of the functions:

(a) $y = \dfrac{x^2}{x+2}$

(b) $y = \dfrac{x^3 + x}{x^2 + x + 1}$

94. Sketch the graph of $f(x) = \dfrac{x^2}{x+1}$. Proceed as in the previous exercise to find the slant asymptote.

95. Show that $y = 3x$ is a slant asymptote for $f(x) = 3x + x^{-2}$. Determine whether $f(x)$ approaches the slant asymptote from above or below and make a sketch of the graph.

96. Sketch the graph of $f(x) = \dfrac{1 - x^2}{2 - x}$.

97. Assume that $f'(x)$ and $f''(x)$ exist for all x and let c be a critical point of $f(x)$. Show that $f(x)$ cannot make a transition from $++$ to $-+$ at $x = c$. *Hint:* Apply the MVT to $f'(x)$.

98. Assume that $f''(x)$ exists and $f''(x) > 0$ for all x. Show that $f(x)$ cannot be negative for all x. *Hint:* Show $f'(b) \neq 0$ for some b and use the result of Exercise 61 in Section 4.4.

4.6 Applied Optimization

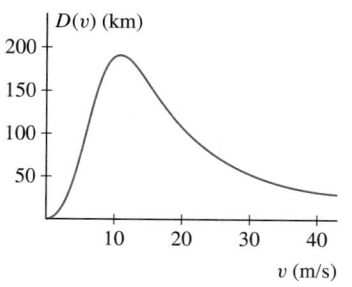

FIGURE 1 Scientists use principles of physiology and aerodynamics to obtain a plausible formula for bird migration distance $D(v)$ as a function of velocity v. The optimal velocity corresponds to the maximum point on the graph (see Exercise 64).

Optimization problems arise in a wide range of disciplines, including physics, engineering, economics, and biology. For example, ornithologists have found that migrating birds solve an optimization problem: They fly at a velocity v that maximizes the distance they can travel without stopping, given the energy that can be stored as body fat (Figure 1).

In many optimization problems, a key step is to determine the function whose minimum or maximum we need. Once we find the function, we can apply the optimization techniques developed in this chapter. We first consider problems that lead to the optimization of a continuous function $f(x)$ on a closed interval $[a, b]$. Let's recall the steps for finding extrema developed in Section 4.2:

(i) Find the critical points of $f(x)$ in $[a, b]$.

(ii) Evaluate $f(x)$ at the critical points and the endpoints a and b.

(iii) The largest and smallest of these values are the extreme values of $f(x)$ in $[a, b]$.

■ **EXAMPLE 1** A piece of wire of length L is bent into the shape of a rectangle (Figure 2). Which dimensions produce the rectangle of maximum area?

FIGURE 2

Solution If the rectangle has sides of length x and y, then its area is $A = xy$. Since A depends on two variables x and y, we cannot find the maximum until we eliminate one of the variables. This can be done because the variables are related: The perimeter of the rectangle is L and $2x + 2y = L$. This gives us $y = L/2 - x$, and we can rewrite the area as a function of x alone:

$$A(x) = x\left(\frac{L}{2} - x\right) = \left(\frac{L}{2}\right)x - x^2$$

Over which interval does the optimization take place? Since the sides of the rectangle cannot have negative length, we must require both $x \geq 0$ and $L/2 - x \geq 0$. Thus, $0 \leq x \leq L/2$ and our problem reduces to finding the maximum of $A(x)$ on the closed interval $[0, L/2]$.

Solving $A'(x) = L/2 - 2x = 0$, we find that $x = L/4$ is the critical point. It remains to calculate the values of $A(x)$ at the endpoints and critical points:

$$\text{Endpoints:} \qquad A(0) \quad = 0$$

$$A\left(\frac{L}{2}\right) = \frac{L}{2}\left(\frac{L}{2} - \frac{L}{2}\right) = 0$$

$$\text{Critical point:} \quad A\left(\frac{L}{4}\right) = \left(\frac{L}{4}\right)\left(\frac{L}{2} - \frac{L}{4}\right) = \frac{L^2}{16}$$

The largest area occurs for $x = L/4$. The corresponding value of y is

$$y = \frac{L}{2} - x = \frac{L}{2} - \frac{L}{4} = \frac{L}{4}$$

This shows that the rectangle of maximum area is the square of sides $x = y = L/4$. ∎

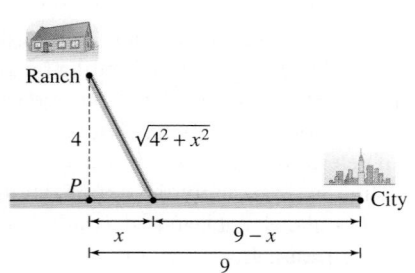

Ranch

4

$\sqrt{4^2 + x^2}$

P

City

x

$9 - x$

9

FIGURE 3

■ **EXAMPLE 2** Minimizing Travel Time Cowboy Clint wants to build a dirt road from his ranch to the highway so that he can drive to the city in the shortest amount of time (Figure 3). The perpendicular distance from the ranch to the highway is 4 miles, and the city is located 9 miles down the highway. Where should Clint join the dirt road to the highway if the speed limit is 20 mph on the dirt road and 55 mph on the highway?

Solution This problem is more complicated than the previous one, so we'll analyze it in three steps. You can follow these steps to solve other optimization problems.

Step 1. **Choose variables.**
We need to decide where the dirt road should join the highway. So let P be the point on the highway nearest the ranch and let x be the distance from P to the point where the dirt road joins the highway.

Step 2. **Find the function and the interval.**
Since we want to minimize the travel time, we need to compute the travel time $T(x)$ of the trip as a function of x. By the Pythagorean Theorem, the length of the dirt road is $\sqrt{4^2 + x^2}$. How long will it take to traverse the road at 20 mph? Remember that at constant velocity v, the distance traveled is $d = vt$. Therefore, the *time* required to travel a distance d is $t = d/v$. Applying this with $v = 20$ mph, we see that it will take

$$\frac{\sqrt{4^2 + x^2}}{20} \text{ hours to traverse the dirt road}$$

FIGURE 4 Graph of time of trip as function of x.

The strip of highway has length $9 - x$. At a speed of 55 mph, it will take

$$\frac{9 - x}{55} \text{ hours to traverse the strip of highway}$$

The total number of hours for the trip is

$$T(x) = \frac{\sqrt{16 + x^2}}{20} + \frac{9 - x}{55}$$

Over which interval does the optimization take place? Since the dirt road joins the highway somewhere between P and the city, we have $0 \le x \le 9$. So our problem is to find the minimum of $T(x)$ on $[0, 9]$ (Figure 4).

Step 3. **Optimize the function.**

We solve $T'(x) = 0$ to find the critical points:

$$T'(x) = \frac{x}{20\sqrt{16 + x^2}} - \frac{1}{55} = 0$$

$$55x = 20\sqrt{16 + x^2}$$

$$11x = 4\sqrt{16 + x^2}$$

$$121x^2 = 16(16 + x^2)$$

Thus, $105x^2 = 16^2$ or $x = 16/\sqrt{105} \approx 1.56$ miles. To find the minimum value of $T(x)$, we compute $T(x)$ at the critical point and endpoints of $[0, 9]$:

$$T(0) \approx 0.36 \text{ h}, \qquad T(1.56) \approx 0.35 \text{ h}, \qquad T(9) \approx 0.49 \text{ h}$$

We conclude that the travel time is minimized if the dirt road joins the highway at a distance $x \approx 1.56$ miles from P.   ∎

Calculus is widely used to solve optimization problems that arise in business decision making. The next example is adapted from a 1957 analysis of corn production in Kansas carried out by the American economist Earl Heady.

■ **EXAMPLE 3** **Optimization in Agriculture** Experiments show that if fertilizer made from N pounds of nitrogen and P pounds of phosphate is used on an acre of Kansas farmland, then the yield of corn is $B = 8 + 0.3\sqrt{NP}$ bushels per acre. Suppose that nitrogen costs 25 cents/lb and phosphate costs 20 cents/lb. A farmer intends to spend $30 per acre on fertilizer. Which combination of nitrogen and phosphate will produce the highest yield of corn?

Solution The formula for the yield B depends on two variables N and P, so we must eliminate one of the variables as before. The farmer will spend $30 per acre, so we have the constraint equation

$$0.25N + 0.2P = 30 \qquad \text{or} \qquad P = 150 - 1.25N$$

We substitute in the formula for B to obtain a function of N alone:

$$B(N) = 8 + 0.3\sqrt{N(150 - 1.25N)}$$

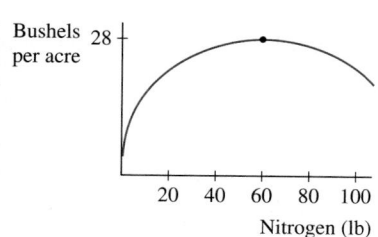

FIGURE 5 Yield of corn per acre as a function of N, with a budget of $30 per acre.

Both N and P must be nonnegative. Since $P = 150 - 1.25N \ge 0$, we require that $0 \le N \le 120$. In other words, $B(N)$ must be optimized over $[0, 120]$ (Figure 5).

To find the critical points of $B(N)$, set the derivative equal to zero:

$$\frac{dB}{dN} = \frac{0.3(150 - 2.5N)}{2\sqrt{N(150 - 1.25N)}} = 0$$

This quotient is zero if the numerator is zero:

$$150 - 2.5N = 0 \qquad \text{or} \qquad N = 60$$

The derivative is undefined at $N = 0$ and $N = 120$, so these are also critical points (which happen to be endpoints). We evaluate $B(N)$ at the critical points:

$$B(0) = 8, \qquad B(60) = 8 + 0.3\sqrt{60(150 - 1.25 \cdot 60)} \approx 28, \qquad B(120) = 8$$

The maximum occurs for $N = 60$. By the constraint equation, $P = 150 - 1.25(60) = 75$, so the combination $N = 60$, $P = 75$ is optimal.   ▮

Whenever possible, check that your answer is plausible. The values $N = 60$ and $P = 75$ are plausible for the following reason: Phosphate is cheaper so it makes sense to use more phosphate than nitrogen—but we cannot reduce the nitrogen to zero because the yield depends on the product NP.

Open Versus Closed Intervals

Some problems require us to optimize a continuous function on an open interval. In this case, there is no guarantee that a min or max exists. Remember that Theorem 1 in Section 4.2 states that a continuous function $f(x)$ has extreme values on a closed, bounded interval, but it says nothing about open intervals. Often, we can show that a min or max exists by examining $f(x)$ near the endpoints of the open interval. If $f(x)$ tends to infinity as x approaches the endpoints (see Figure 7), then $f(x)$ must take on a minimum at a critical point.

▮ **EXAMPLE 4**   Design a cylindrical can of volume 10 ft^3 so that it uses the least amount of metal (Figure 6). In other words, minimize the surface area of the can (including its top and bottom).

FIGURE 6 Cylinders with the same volume but different surface areas.

Solution

Step 1. **Choose variables.**
We must specify the can's radius and height. Therefore, let r be the radius and h the height. Let A be the surface area of the can.

Step 2. **Find the function and the interval.**
We compute A as a function of r and h:

$$A = \underbrace{\pi r^2}_{\text{Top}} + \underbrace{\pi r^2}_{\text{Bottom}} + \underbrace{2\pi r h}_{\text{Side}} = 2\pi r^2 + 2\pi r h$$

The can's volume is $V = \pi r^2 h$. Since we require that $V = 10$ ft^3, we obtain the constraint equation $\pi r^2 h = 10$. Thus $h = (10/\pi)r^{-2}$ and

$$A(r) = 2\pi r^2 + 2\pi r \left(\frac{10}{\pi r^2}\right) = 2\pi r^2 + \frac{20}{r}$$

The radius r can take on any positive value, so the interval of optimization is $(0, \infty)$.

Step 3. Optimize the function.

We observe that $A(r)$ tends to infinity as r approaches the endpoints of $(0, \infty)$ (Figure 7).

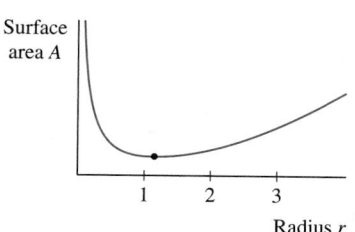

Surface area A

Radius r

FIGURE 7 Surface area increases as r tends to zero or ∞. The minimum value exists.

- $A(r) \to \infty$ as $r \to \infty$ (because of the r^2 term)
- $A(r) \to \infty$ as $r \to 0$ (because of the $1/r$ term)

We conclude that $A(r)$ takes on a minimum value at a critical point in the open interval $(0, \infty)$. Solving for the critical point, we obtain

$$\frac{dA}{dr} = 4\pi r - \frac{20}{r^2} = 0 \qquad \text{or} \qquad r^3 = \frac{5}{\pi}$$

Therefore, the value of r that minimizes the surface area is

$$r = \left(\frac{5}{\pi}\right)^{1/3} \approx 1.17 \text{ ft}$$

We also need to calculate the height:

$$h = \frac{10}{\pi} r^{-2} = 2 \left(\frac{5}{\pi}\right) r^{-2} = 2 \left(\frac{5}{\pi}\right) \left(\frac{5}{\pi}\right)^{-2/3} = 2 \left(\frac{5}{\pi}\right)^{1/3} \approx 2.34 \text{ ft}$$

Notice that the optimal dimensions satisfy $h = 2r$. In other words, the optimal can is as tall as it is wide. ∎

■ **EXAMPLE 5** Optimization Problem with No Solution Is it possible to design a cylinder of volume 10 ft^3 with the largest possible surface area?

Solution The answer is no. In the previous example, we showed that a cylinder of volume 10 and radius r has surface area

$$A(r) = 2\pi r^2 + \frac{20}{r}$$

This function has no maximum value because it tends to infinity as $r \to 0$ or $r \to \infty$ (Figure 7). In physical terms, this means that a cylinder of fixed volume has a large surface area if it is either very fat and short (r large) or very tall and skinny (r small). ∎

The **Principle of Least Distance** states that a light beam reflected in a mirror travels along the shortest path. In other words, if the beam originates at point A in Figure 8 and

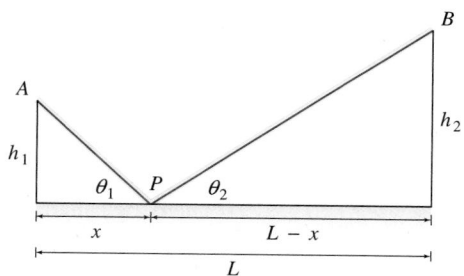

FIGURE 8 Reflection of a light beam in a mirror.

The Principle of Least Distance was known to the Greek mathematician Heron of Alexandria (c. 100 AD). See Exercise 50 for Snell's Law, a more general optical law based on the Principle of Least Time.

ends at point B, then the point of reflection P has the following minimality property:

$$AP + PB = \text{as small as possible}$$

The next example uses this principle to show that the angle of incidence θ_1 is equal to the angle of reflection θ_2.

■ **EXAMPLE 6** Angle of Incidence Equals Angle of Reflection Use the Principle of Least Distance to show that the angle of incidence θ_1 is equal to the angle of reflection θ_2.

Solution Let h_1 and h_2 be the distances from A and B to the mirror as in Figure 8, and let x be the distance from P to the left end of the mirror (Figure 8). By the Pythagorean Theorem, the length of the path from A to B is

$$f(x) = AP + PB = \sqrt{x^2 + h_1^2} + \sqrt{(L-x)^2 + h_2^2}$$

The light beam strikes the mirror at the point P for which $f(x)$ has a minimum value. The graph of $f(x)$ shows that the minimum occurs at a critical point (see Figure 9). Thus, we set the derivative equal to zero:

$$f'(x) = \frac{x}{\sqrt{x^2 + h_1^2}} - \frac{L-x}{\sqrt{(L-x)^2 + h_2^2}} = 0$$

FIGURE 9 Graph of path length for $h_1 = 10$, $h_2 = 20$, $L = 40$.

If we wanted to find the critical point (giving the actual point on the mirror where the light beam strikes), we would have to solve this equation for x. However, our goal is not to find x, but rather to show that $\theta_1 = \theta_2$. Therefore, we rewrite this equation as

$$\underbrace{\frac{x}{\sqrt{x^2 + h_1^2}}}_{\cos \theta_1} = \underbrace{\frac{L-x}{\sqrt{(L-x)^2 + h_2^2}}}_{\cos \theta_2}$$

Referring to Figure 8, we see that this equation says $\cos \theta_1 = \cos \theta_2$, and since θ_1 and θ_2 lie between 0 and $\frac{\pi}{2}$, we have $\theta_1 = \theta_2$ as required. ∎

CONCEPTUAL INSIGHT The examples in this section were selected because they lead to optimization problems where the min or max occurs at a critical point. In problems of this type, the critical point often represents the best compromise between "competing factors." In Example 3, we maximized corn yield by finding the compromise between phosphate (which is cheaper) and nitrogen (without which the yield is low). In Example 4, we minimized the surface area of a metal can by finding the best compromise between height and width. In daily life, however, we often encounter endpoint rather than critical point solutions. For example, to minimize your time in the 40-yard dash, you should run as fast as you can—the solution is not a critical point but rather an endpoint (your maximum speed).

4.6 SUMMARY

- There are usually three main steps in solving an applied optimization problem:

Step 1. Choose variables.

Determine which quantities are relevant, often by drawing a diagram, and assign appropriate variables.

Step 2. Find the function and the interval.

Restate as an optimization problem for a function f over some interval. If the function depends on more than one variable, use a *constraint equation* to write f as a function of just one variable.

Step 3. Optimize the function.

- If the interval is open, f does not necessarily take on a minimum or maximum value. But if it does, these must occur at critical points within the interval. To determine if a min or max exists, analyze the behavior of f as x approaches the endpoints of the interval.

4.6 EXERCISES

Preliminary Questions

1. The problem is to find the right triangle of perimeter 10 whose area is as large as possible. What is the constraint equation relating the base b and height h of the triangle?

2. What are the relevant variables if the problem is to find a right circular cone of surface area 20 and maximum volume?

3. Does a continuous function on an open interval always have a maximum value?

4. Describe a way of showing that a continuous function on an open interval (a, b) has a minimum value.

Exercises

1. Find the dimensions of the rectangle of maximum area that can be formed from a 50-in. piece of wire.

(a) What is the constraint equation relating the lengths x and y of the sides?

(b) Find a formula for the area in terms of x alone.

(c) Does this problem require optimization over an open interval or a closed interval?

(d) Solve the optimization problem.

2. A 100-in. piece of wire is divided into two pieces and each piece is bent into a square. How should this be done in order to minimize the sum of the areas of the two squares?

(a) Express the sum of the areas of the squares in terms of the lengths x and y of the two pieces.

(b) What is the constraint equation relating x and y?

(c) Does this problem require optimization over an open or closed interval?

(d) Solve the optimization problem.

3. Find the positive number x such that the sum of x and its reciprocal is as small as possible. Does this problem require optimization over an open interval or a closed interval?

4. The legs of a right triangle have lengths a and b satisfying $a + b = 10$. Which values of a and b maximize the area of the triangle?

5. Find positive numbers x, y such that $xy = 16$ and $x + y$ is as small as possible.

6. A 20-in. piece of wire is bent into an L-shape. Where should the bend be made to minimize the distance between the two ends?

7. Let S be the set of all rectangles with area 100.

(a) What are the dimensions of the rectangle in S with the least perimeter?

(b) Is there a rectangle in S with the greatest perimeter? Explain.

8. A box has a square base of side x and height y.

(a) Find the dimensions x, y for which the volume is 12 and the surface area is as small as possible.

(b) Find the dimensions for which the surface area is 20 and the volume is as large as possible.

9. Suppose that 600 ft of fencing are used to enclose a corral in the shape of a rectangle with a semicircle whose diameter is a side of the rectangle as in Figure 10. Find the dimensions of the corral with maximum area.

FIGURE 10   FIGURE 11

10. Find the rectangle of maximum area that can be inscribed in a right triangle with legs of length 3 and 4 if the sides of the rectangle are parallel to the legs of the triangle, as in Figure 11.

11. A landscape architect wishes to enclose a rectangular garden on one side by a brick wall costing $30/ft and on the other three sides by a metal fence costing $10/ft. If the area of the garden is 1,000 ft^2, find the dimensions of the garden that minimize the cost.

12. Find the point on the line $y = x$ closest to the point $(1, 0)$.

13. Find the point P on the parabola $y = x^2$ closest to the point $(3, 0)$ (Figure 12).

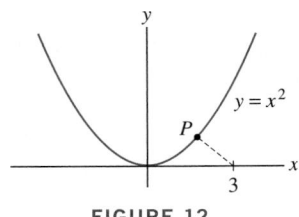

FIGURE 12

14. $\mathcal{CAS}$ Find a good numerical approximation to the coordinates of the point on the graph of $y = \ln x - x$ closest to the origin (Figure 13).

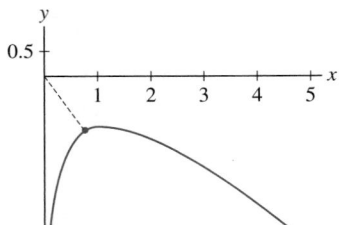

FIGURE 13 Graph of $y = \ln x - x$.

15. A box is constructed out of two different types of metal. The metal for the top and bottom, which are both square, costs $1/ft^2 and the metal for the sides costs $2/ft^2. Find the dimensions that minimize cost if the box has a volume of 20 ft^3.

16. Find the dimensions of the rectangle of maximum area that can be inscribed in a circle of radius r (Figure 14).

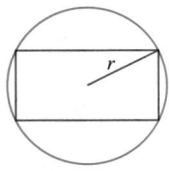

FIGURE 14

17. Problem of Tartaglia (1500–1557)   Among all positive numbers a, b whose sum is 8, find those for which the product of the two numbers and their difference is largest. *Hint:* Let $x = a - b$ and express abx in terms of x alone.

18. Find the angle θ that maximizes the area of the isosceles triangle whose legs have length ℓ (Figure 15).

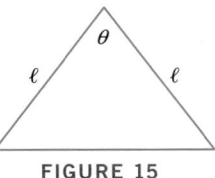

FIGURE 15

19. The volume of a right circular cone is $\frac{\pi}{3}r^2h$ and its surface area is $S = \pi r\sqrt{r^2 + h^2}$. Find the dimensions of the cone with surface area 1 and maximal volume (Figure 16).

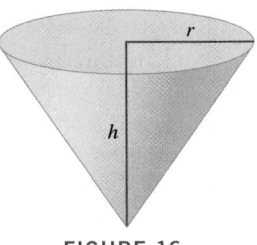

FIGURE 16

20. Rice production requires both labor and capital investment in equipment and land. Suppose that if x dollars per acre are invested in labor and y dollars per acre are invested in equipment and land, then the yield P of rice per acre is given by the formula $P = 100\sqrt{x} + 150\sqrt{y}$. If a farmer invests $40/acre, how should he divide the $40 between labor and capital investment in order to maximize the amount of rice produced?

21. Kepler's Wine Barrel Problem   The following problem was stated and solved in the work *Nova stereometria doliorum vinariorum* (New Solid Geometry of a Wine Barrel), published in 1615 by the astronomer Johannes Kepler (1571–1630). What are the dimensions of the cylinder of largest volume that can be inscribed in the sphere of radius R? *Hint:* Show that the volume of an inscribed cylinder is $2\pi x(R^2 - x^2)$, where x is one-half the height of the cylinder.

22. Find the dimensions x and y of the rectangle inscribed in a circle of radius r that maximizes the quantity xy^2.

23. Find the angle θ that maximizes the area of the trapezoid with a base of length 4 and sides of length 2, as in Figure 17.

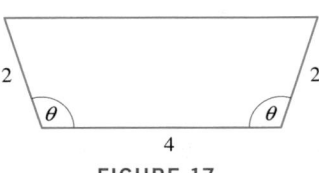

FIGURE 17

24. Consider a rectangular industrial warehouse consisting of three separate spaces of equal size as in Figure 18. Assume that the wall materials cost $200 per linear ft and the company allocates $2,400,000 for the project.

(a) Which dimensions maximize the total area of the warehouse?

(b) What is the area of each compartment in this case?

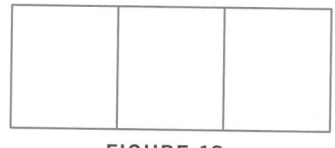

FIGURE 18

25. Suppose, in the previous exercise, that the warehouse consists of n separate spaces of equal size. Find a formula in terms of n for the maximum possible area of the warehouse.

26. The amount of light reaching a point at a distance r from a light source A of intensity I_A is I_A / r^2. Suppose that a second light source B of intensity $I_B = 4I_A$ is located 10 ft from A. Find the point on the segment joining A and B where the total amount of light is at a minimum.

27. Find the area of the largest rectangle that can be inscribed in the region bounded by the graph of $y = \dfrac{4 - x}{2 + x}$ and the coordinate axes (Figure 19).

$$y = \frac{4 - x}{2 + x}$$

FIGURE 19

28. According to postal regulations, a carton is classified as "oversized" if the sum of its height and girth (the perimeter of its base) exceeds 108 in. Find the dimensions of a carton with square base that is not oversized and has maximum volume.

29. Find the maximum area of a triangle formed in the first quadrant by the x-axis, y-axis, and a tangent line to the graph of $y = (x + 1)^{-2}$.

30. What is the area of the largest rectangle that can be circumscribed around a rectangle of sides L and H? *Hint:* Express the area of the circumscribed rectangle in terms of the angle θ (Figure 20).

31. Optimal Price Let r be the monthly rent per unit in an apartment building with 100 units. A survey reveals that all units can be rented when $r = \$900$ and that one unit becomes vacant with each

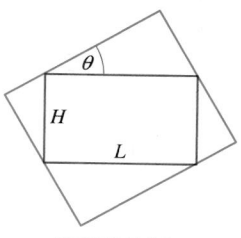

FIGURE 20

$10 increase in rent. Suppose that the average monthly maintenance per occupied unit is $100/month.

(a) Show that the number of units rented is $n = 190 - r/10$ for $900 \le r \le 1,900$.

(b) Find a formula for the net cash intake (revenue minus maintenance) and determine the rent r that maximizes intake.

32. An 8-billion-bushel corn crop brings a price of $2.40/bushel. A commodity broker uses the following rule of thumb: If the crop is reduced by x percent, then the price increases by $10x$ cents. Which crop size results in maximum revenue and what is the price per bushel? *Hint:* Revenue is equal to price times crop size.

33. Given n numbers $x_1, \dots, x_n$, find the value of x minimizing the sum of the squares:

$$(x - x_1)^2 + (x - x_2)^2 + \cdots + (x - x_n)^2$$

First solve the problem for $n = 2, 3$ and then solve it for arbitrary n.

34. Let $P = (a, b)$ be a point in the first quadrant.

(a) Find the slope of the line through P such that the triangle bounded by this line and the axes in the first quadrant has minimal area.

(b) Show that P is the midpoint of the hypotenuse of this triangle.

35. A truck gets 10 mpg (miles per gallon) traveling along an interstate highway at 50 mph, and this is reduced by 0.15 mpg for each mile per hour increase above 50 mph.

(a) If the truck driver is paid $30/hour and diesel fuel costs $P = \$3/\text{gal}$, which speed v between 50 and 70 mph will minimize the cost of a trip along the highway? Notice that the actual cost depends on the length of the trip but the optimal speed does not.

(b) [GU] Plot cost as a function of v (choose the length arbitrarily) and verify your answer to part (a).

(c) [GU] Do you expect the optimal speed v to increase or decrease if fuel costs go down to $P = \$2/\text{gal}$? Plot the graphs of cost as a function of v for $P = 2$ and $P = 3$ on the same axis and verify your conclusion.

36. Figure 21 shows a rectangular plot of size 100×200 feet. Pipe is to be laid from A to a point P on side BC and from there to C. The cost of laying pipe through the lot is $30/ft (since it must be underground) and the cost along the side of the plot is $15/ft.

(a) Let $f(x)$ be the total cost, where x is the distance from P to B. Determine $f(x)$, but note that f is discontinuous at $x = 0$ (when $x = 0$, the cost of the entire pipe is $15/ft).

FIGURE 21

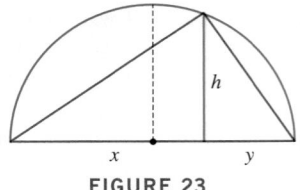

FIGURE 23

(b) What is the most economical way to lay the pipe? What if the cost along the sides is $24/ft?

37. Find the dimensions of a cylinder of volume 1 m³ of minimal cost if the top and bottom are made of material that costs twice as much as the material for the side.

38. In Example 6 in this section, find the x-coordinate of the point P where the light beam strikes the mirror if $h_1 = 10$, $h_2 = 5$, and $L = 20$.

In Exercises 39–41, a box (with no top) is to be constructed from a piece of cardboard of sides A and B by cutting out squares of length h from the corners and folding up the sides (Figure 22).

FIGURE 22

39. Find the value of h that maximizes the volume of the box if $A = 15$ and $B = 24$. What are the dimensions of the resulting box?

40. Which value of h maximizes the volume if $A = B$?

41. Suppose that a box of height $h = 3$ in. is constructed using 144 in.² of cardboard (i.e., $AB = 144$). Which values A and B maximize the volume?

42. The monthly output P of a light bulb factory is given by the formula $P = 350LK$, where L is the amount invested in labor and K the amount invested in equipment (in thousands of dollars). If the company needs to produce 10,000 units per month, how should the investment be divided among labor and equipment to minimize the cost of production? The cost of production is $L + K$.

43. 📖 Use calculus to show that among all right triangles with hypotenuse of length 1, the isosceles triangle has maximum area. Can you see more directly why this must be true by reasoning from Figure 23?

44. Janice can swim 3 mph and run 8 mph. She is standing at one bank of a river that is 300 ft wide and wants to reach a point located 200 ft downstream on the other side as quickly as possible. She will swim diagonally across the river and then jog along the river bank. Find the best route for Janice to take.

45. Optimal Delivery Schedule A gas station sells Q gallons of gasoline per year, which is delivered N times per year in equal shipments of Q/N gallons. The cost of each delivery is d dollars and the yearly storage costs are sQT, where T is the length of time (a fraction of a year) between shipments and s is a constant. Show that costs are minimized for $N = \sqrt{sQ/d}$. (*Hint:* Express T in terms of N.) Find the optimal number of deliveries if $Q = 2$ million gal, $d = \$8,000$, and $s = 30$ cents/gal-yr. Your answer should be a whole number, so compare costs for the two integer values of N nearest the optimal value.

46. (a) Find the radius and height of a cylindrical can of total surface area A whose volume is as large as possible.

(b) Can you design a cylinder with total surface area A and minimal total volume?

47. Find the area of the largest isosceles triangle that can be inscribed in a circle of radius r.

48. A billboard of height b is mounted on the side of a building with its bottom edge at a distance h from the street. At what distance x should an observer stand from the wall to maximize the angle of observation θ (Figure 24)?

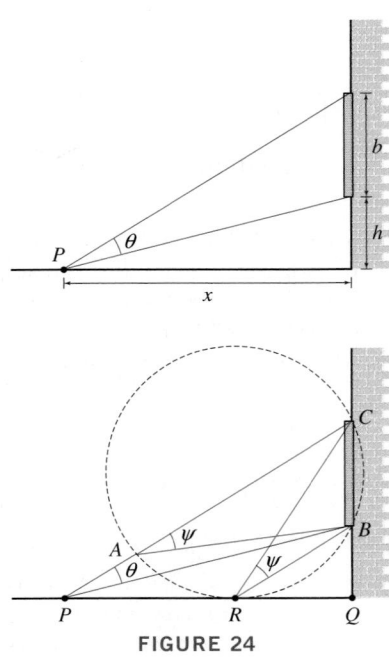

FIGURE 24

(a) Find x using calculus.

(b) Solve the problem again using geometry (without any calculation!). There is a unique circle passing through points B and C which is tangent to the street. Let R be the point of tangency. Show that θ is maximized at the point R. *Hint:* The two angles labeled ψ are, in fact, equal because they subtend equal arcs on the circle. Let A be the intersection of the circle with PC and show that $\psi = \theta + \angle PBA > \theta$.

(c) Prove that the two answers in (a) and (b) agree.

49. Use the result of Exercise 48 to show that θ is maximized at the value of x for which the angles $\angle QPB$ and $\angle QCP$ are equal.

50. Snell's Law Figure 25 represents the surface of a swimming pool. A light beam travels from point A located above the pool to point B located underneath the water. Let v_1 be the velocity of light in air and let v_2 be the velocity of light in water (it is a fact that $v_1 > v_2$). Prove Snell's Law of Refraction according to which the path from A to B that takes the *least time* satisfies the relation

$$\frac{\sin \theta_1}{v_1} = \frac{\sin \theta_2}{v_2}$$

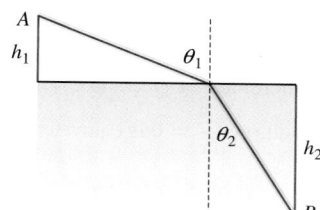

FIGURE 25

51. Snell's Law, derived in Exercise 50, explains why it is impossible to see above the water if the angle of vision θ_2 is such that $\sin \theta_2 > v_2/v_1$.

(a) Show that if this inequality holds, then there is no angle θ_1 for which Snell's Law holds.

(b) What will you see if you look through the water with an angle of vision θ_2 such that $\sin \theta_2 > v_2/v_1$?

52. A poster of area 6 ft² has blank margins of width 6 in. on the top and bottom and 4 in. on the sides. Find the dimensions that maximize the printed area.

53. Vascular Branching A small blood vessel of radius r branches off at an angle θ from a larger vessel of radius R to supply blood along a path from A to B. According to Poiseuille's Law, the total resistance to blood flow is proportional to

$$T = \left(\frac{a - b \cot \theta}{R^4} + \frac{b \csc \theta}{r^4} \right)$$

where a and b are as in Figure 26. Show that the total resistance is minimized when $\cos \theta = (r/R)^4$.

54. Find the minimum length ℓ of a beam that can clear a fence of height h and touch a wall located b ft behind the fence (Figure 27).

FIGURE 26   **FIGURE 27**

55. Let (a, b) be a fixed point in the first quadrant and let $S(d)$ be the sum of the distances from $(d, 0)$ to the points $(0, 0)$, (a, b), and $(a, -b)$.

(a) Find the value of d for which $S(d)$ is minimal. The answer depends on whether $b < \sqrt{3}a$ or $b \geq \sqrt{3}a$. *Hint:* Show that $d = 0$ when $b \geq \sqrt{3}a$.

(b) $\boxed{\text{GU}}$ Let $a = 1$. Plot $S(d)$ for $b = 0.5, \sqrt{3}, 3$ and describe the position of the minimum.

56. The force F (in Newtons) required to move a box of mass m kg in motion by pulling on an attached rope is

$$F(\theta) = \frac{fmg}{\cos \theta + f \sin \theta}$$

where θ is the angle between the rope and the horizontal, f is the coefficient of static friction, and $g = 9.8$ m/s². Find the angle θ that minimizes the required force F, assuming $f = 0.4$. *Hint:* Find the maximum value of $\cos \theta + f \sin \theta$.

FIGURE 28

57. In the setting of Exercise 56, show that for any f the minimal force required is proportional to $1/\sqrt{1 + f^2}$.

58. The problem is to put a "roof" of side s on an attic room of height h and width b. Find the smallest length s for which this is possible. See Figure 29.

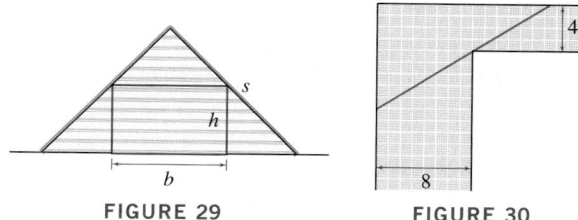

FIGURE 29   **FIGURE 30**

59. Find the maximum length of a pole that can be carried horizontally around a corner joining corridors of widths 8 ft and 4 ft (Figure 30).

60. Redo Exercise 59 for corridors of arbitrary widths a and b.

61. Find the isosceles triangle of smallest area that circumscribes a circle of radius 1 (from Thomas Simpson's *The Doctrine and Application of Fluxions*, a calculus text that appeared in 1750). See Figure 31.

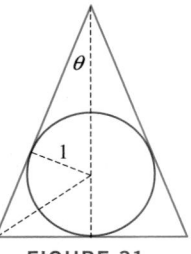

FIGURE 31

62. [icon]   A basketball player stands d feet from the basket. Let h and α be as in Figure 32. Using physics, one can show that if the player releases the ball at an angle θ, then the initial velocity required to make the ball go through the basket satisfies

$$v^2 = \frac{16d}{\cos^2 \theta (\tan \theta - \tan \alpha)}$$

(a) Explain why this formula is only meaningful for $\alpha < \theta < \frac{\pi}{2}$. Why does v approach infinity at the endpoints of this interval?

(b) [GU] Take $\alpha = \frac{\pi}{6}$ and plot v^2 as a function of θ for $\frac{\pi}{6} < \theta < \frac{\pi}{2}$. Verify that the minimum occurs at $\theta = \frac{\pi}{3}$.

(c) Set $F(\theta) = \cos^2 \theta (\tan \theta - \tan \alpha)$. Explain why v is minimized for θ such that $F(\theta)$ is maximized.

(d) Verify that $F'(\theta) = \cos(\alpha - 2\theta) \sec \alpha$ (you will need to use the addition formula for cosine) and show that the maximum value of $F(\theta)$ on $[\alpha, \frac{\pi}{2}]$ occurs at $\theta_0 = \frac{\alpha}{2} + \frac{\pi}{4}$.

(e) For a given α, the optimal angle for shooting the basket is θ_0 because it minimizes v^2 and therefore minimizes the energy required to make the shot (energy is proportional to v^2). Show that the velocity v_{opt} at the optimal angle θ_0 satisfies

$$v_{\text{opt}}^2 = \frac{32d \cos \alpha}{1 - \sin \alpha} = \frac{32 d^2}{-h + \sqrt{d^2 + h^2}}$$

(f) [GU] Show with a graph that for fixed d (say, $d = 15$ ft, the distance of a free throw), v_{opt}^2 is an increasing function of h. Use this to explain why taller players have an advantage and why it can help to jump while shooting.

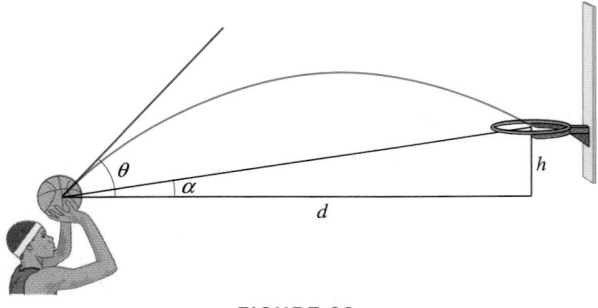

FIGURE 32

63. Tom drives with his friend Ali along a highway represented by the graph of a differentiable function $y = f(x)$ as in Figure 33. During the trip, Ali views a billboard represented by the segment $\overline{BC}$ along the y-axis. Let Q be the y-intercept of the tangent line to $y = f(x)$. Show that θ is maximized at the value of x for which the angles $\angle QPB$ and $\angle QPC$ are equal. This is a generalization of Exercise 49 [which corresponds to the case $f(x) = 0$]. *Hints:*

(a) Compute $d\theta/dx$ and check that it equals

$$(b - c) \cdot \frac{(x^2 + (xf'(x))^2) - (b - (f(x) - xf'(x)))(c - (f(x) - xf'(x)))}{(x^2 + (b - f(x))^2)(x^2 + (c - f(x))^2)}$$

(b) Show that the y-coordinate of Q is $f(x) - xf'(x)$.

(c) Show that the condition $\dfrac{d\theta}{dx} = 0$ is equivalent to

$$PQ^2 = BQ \cdot CQ$$

(d) Use (c) to conclude that triangles $\triangle QPB$ and $\triangle QCP$ are similar.

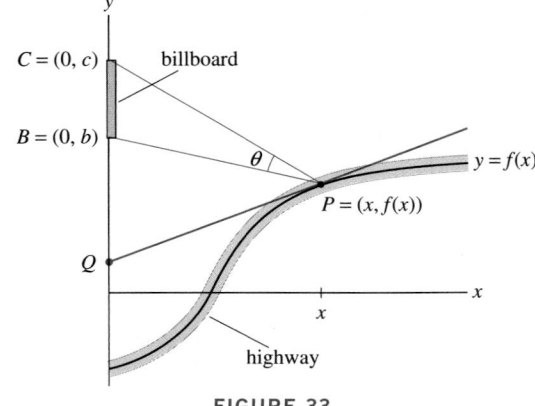

FIGURE 33

Further Insights and Challenges

64. Bird Migration   Power P is the rate at which energy E is consumed per unit time. Ornithologists have found that the power consumed by a certain pigeon flying at velocity v m/s is described well by the function $P(v) = 17v^{-1} + 10^{-3}v^3$ J/s. Assume that the pigeon can store 5×10^4 J of usable energy as body fat.

(a) Find the velocity v_{pmin} that *minimizes* power consumption.

(b) Show that a pigeon flying at velocity v and using all of its stored energy can fly a total distance of $D(v) = (5 \times 10^4)v/P(v)$.

(c) Migrating birds are smart enough to fly at the velocity that maximizes distance traveled rather than minimizes power consumption.

Show that the velocity v_{dmax} which maximizes total distance $D(v)$ satisfies $P'(v) = P(v)/v$. Show that v_{dmax} is obtained graphically as the velocity coordinate of the point where a line through the origin is tangent to the graph of $P(v)$ (Figure 34).

(d) Find v_{dmax} and the maximum total distance that the bird can fly.

FIGURE 34

Seismic Prospecting *Exercises 65–67 are concerned with determining the thickness d of a layer of soil that lies on top of a rock formation. Geologists send two sound pulses from point A to point D separated by a distance s. The first pulse travels directly from A to D along the surface of the earth. The second pulse travels down to the rock formation, then along its surface, and then back up to D (path ABCD), as in Figure 35. The pulse travels with velocity v_1 in the soil and v_2 in the rock.*

FIGURE 35

65. (a) Show that the time required for the first pulse to travel from A to D is $t_1 = s/v_1$.

(b) Show that the time required for the second pulse is

$$t_2 = \frac{2d}{v_1} \sec\theta + \frac{s - 2d\tan\theta}{v_2}$$

provided that

$$\tan\theta \le \frac{s}{2d} \qquad \boxed{1}$$

(*Note:* If this inequality is not satisfied, then point B does not lie to the left of C.)

(c) Show that t_2 is minimized when $\sin\theta = v_1/v_2$.

66. In this exercise, assume that $v_2/v_1 \ge \sqrt{1 + 4(d/s)^2}$.

(a) Show that inequality (1) holds if $\sin\theta = v_1/v_2$.

(b) Show that the minimal time for the second pulse is

$$t_2 = \frac{2d}{v_1}(1 - k^2)^{1/2} + \frac{s}{v_2}$$

where $k = v_1/v_2$.

(c) Conclude that $\dfrac{t_2}{t_1} = \dfrac{2d(1 - k^2)^{1/2}}{s} + k$.

67. Continue with the assumption of the previous exercise.

(a) Find the thickness of the soil layer, assuming that $v_1 = 0.7v_2$, $t_2/t_1 = 1.3$, and $s = 1,500$ ft.

(b) The times t_1 and t_2 are measured experimentally. The equation in Exercise 66(c) shows that t_2/t_1 is a linear function of $1/s$. What might you conclude if experiments were formed for several values of s and the points $(1/s, t_2/t_1)$ did *not* lie on a straight line?

68. Three towns A, B, and C are to be joined by an underground fiber cable as illustrated in Figure 36(A). Assume that C is located directly below the midpoint of $\overline{AB}$. Find the junction point P that minimizes the total amount of cable used.

(a) First show that P must lie directly above C. Show that if the junction is placed at a point Q as in Figure 36(B), then we can reduce the cable length by moving Q horizontally over to the point P lying above C. You may want to use the result of Example 6.

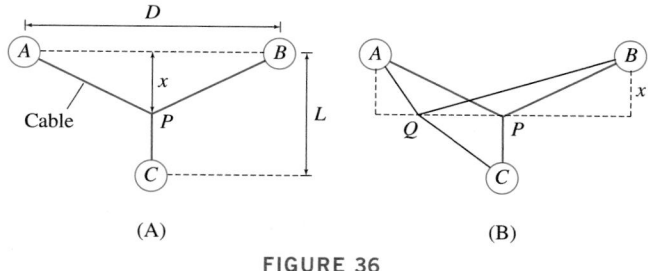

FIGURE 36

(b) With x as in Figure 36(A), let $f(x)$ be the total length of cable used. Show that $f(x)$ has a unique critical point c. Compute c and show that $0 \le c \le L$ if and only if $D \le 2\sqrt{3}\,L$.

(c) Find the minimum of $f(x)$ on $[0, L]$ in two cases: $D = 2$, $L = 4$ and $D = 8$, $L = 2$.

69. A jewelry designer plans to incorporate a component made of gold in the shape of a frustum of a cone of height 1 cm and fixed lower radius r (Figure 37). The upper radius x can take on any value between

FIGURE 37 Frustrum of height 1 cm.

0 and r. Note that $x = 0$ and $x = r$ correspond to a cone and cylinder, respectively. As a function of x, the surface area (not including the top and bottom) is $S(x) = \pi s(r + x)$, where s is the *slant height* as indicated in the figure. Which value of x yields the least expensive design [the minimum value of $S(x)$ for $0 \le x \le r$]?

(a) Show that $S(x) = \pi(r + x)\sqrt{1 + (r - x)^2}$.

(b) Show that if $r < \sqrt{2}$, then $S(x)$ is an increasing function. Conclude that the cone ($x = 0$) has minimal area in this case.

(c) Assume that $r > \sqrt{2}$. Show that $S(x)$ has two critical points $x_1 < x_2$ in $(0, r)$, and that $S(x_1)$ is a local maximum, and $S(x_2)$ is a local minimum.

(d) Conclude that the minimum occurs at $x = 0$ or x_2.

(e) Find the minimum in the cases $r = 1.5$ and $r = 2$.

(f) Challenge: Let $c = \sqrt{(5 + 3\sqrt{3})/4} \approx 1.597$. Prove that the minimum occurs at $x = 0$ (cone) if $\sqrt{2} < r < c$, but the minimum occurs at $x = x_2$ if $r > c$.

4.7 L'Hôpital's Rule

L'Hôpital's Rule is named for the French mathematician Guillaume François Antoine Marquis de L'Hôpital (1661–1704) who wrote the first textbook on calculus in 1696. The name L'Hôpital is pronounced "Lo-pee-tal."

L'Hôpital's Rule is used to evaluate limits of certain types of quotients:

$$\lim_{x \to a} \frac{f(x)}{g(x)}$$

Roughly speaking, L'Hôpital's Rule states that when $f(x)/g(x)$ has an indeterminate form at $x = a$ of type $0/0$ or ∞/∞, then *we may replace $f(x)/g(x)$ by the quotient of the derivatives $f'(x)/g'(x)$.*

> **THEOREM 1 L'Hôpital's Rule**   Assume that $f(x)$ and $g(x)$ are differentiable on an open interval containing a and that
>
> $$f(a) = g(a) = 0$$
>
> Also assume that $g'(x) \ne 0$ for x near but not equal to a. Then
>
> $$\lim_{x \to a} \frac{f(x)}{g(x)} = \lim_{x \to a} \frac{f'(x)}{g'(x)}$$
>
> provided that the limit on the right exists. This conclusion also holds if $f(x)$ and $g(x)$ are differentiable for x near (but not equal to) a and
>
> $$\lim_{x \to a} f(x) = \pm\infty \quad \text{and} \quad \lim_{x \to a} g(x) = \pm\infty$$
>
> Furthermore, these limits may be replaced by one-sided limits.

■ **EXAMPLE 1**   Use L'Hôpital's Rule to evaluate $\displaystyle\lim_{x \to 1} \frac{x^3 - 1}{x - 1}$.

Solution   Both $f(x) = x^3 - 1$ and $g(x) = x - 1$ are differentiable and $f(1) = g(1) = 0$, so the quotient is indeterminate at $x = 1$:

$$\left.\frac{x^3 - 1}{x - 1}\right|_{x=1} = \frac{1^3 - 1}{1 - 1} = \frac{0}{0} \quad \text{(indeterminate)}$$

Furthermore, $g'(x) = 1$, and thus $g'(x)$ is nonzero. Therefore, L'Hôpital's Rule applies. We may replace the numerator and denominator by their derivatives to obtain

CAUTION When using L'Hôpital's Rule, be sure to take the derivative of the numerator and denominator separately. Do not differentiate the quotient function.

$$\lim_{x \to 1} \frac{x^3 - 1}{x - 1} = \lim_{x \to 1} \frac{3x^2}{1} = 3$$ ■

■ **EXAMPLE 2**   Evaluate $\displaystyle\lim_{x \to \pi/2} \frac{\cos^2 x}{\sin x - 1}$.

Solution L'Hôpital's Rule applies because $\cos^2 x$ and $\sin x - 1$ are differentiable but

$$\frac{\cos^2 x}{\sin x - 1}\bigg|_{x=\pi/2} = \frac{\cos^2(\frac{\pi}{2})}{\sin\frac{\pi}{2} - 1} = \frac{0^2}{1 - 1} = \frac{0}{0} \quad \text{(indeterminate)}$$

We replace the numerator and denominator by their derivatives:

$$\lim_{x\to\pi/2} \frac{\cos^2 x}{\sin x - 1} = \underbrace{\lim_{x\to\pi/2} \frac{-2\cos x \sin x}{\cos x} = \lim_{x\to\pi/2}(-2\sin x)}_{\text{Simplify}} = -2 \qquad \blacksquare$$

■ **EXAMPLE 3** Indeterminate Form $\frac{\infty}{\infty}$ Evaluate $\lim_{x\to 0+} x \ln x$.

Solution This is a one-sided limit because $f(x) = x \ln x$ is not defined for $x \leq 0$. L'Hôpital's Rule is valid for one-sided limits, but $f(x)$ is not a quotient, so we rewrite it as $f(x) = (\ln x)/x^{-1}$. Since $\ln x$ approaches $-\infty$ and x^{-1} approaches ∞ as $x \to 0+$, we have the indeterminate form $-\infty/\infty$ and L'Hôpital's Rule applies:

$$\lim_{x\to 0+} x \ln x = \lim_{x\to 0+} \frac{\ln x}{x^{-1}} = \lim_{x\to 0+} \left(\frac{x^{-1}}{-x^{-2}}\right) = \lim_{x\to 0+}(-x) = 0 \qquad \blacksquare$$

■ **EXAMPLE 4** Using L'Hôpital's Rule Twice Evaluate $\lim_{x\to 0} \dfrac{e^x - x - 1}{\cos x - 1}$.

Solution L'Hôpital's Rule applies because

$$\frac{e^x - x - 1}{\cos x - 1}\bigg|_{x=0} = \frac{e^0 - 0 - 1}{\cos 0 - 1} = \frac{1 - 0 - 1}{1 - 1} = \frac{0}{0} \quad \text{(indeterminate)}$$

A first application of L'Hôpital's Rule gives

$$\lim_{x\to 0} \frac{e^x - x - 1}{\cos x - 1} = \lim_{x\to 0}\left(\frac{e^x - 1}{-\sin x}\right) = \lim_{x\to 0} \frac{1 - e^x}{\sin x}$$

This limit is again indeterminate of type $0/0$, so we apply L'Hôpital's Rule again:

$$\lim_{x\to 0} \frac{1 - e^x}{\sin x} = \lim_{x\to 0} \frac{-e^x}{\cos x} = \frac{-e^0}{\cos 0} = -1 \qquad \blacksquare$$

■ **EXAMPLE 5** Assumptions Matter Can L'Hôpital's Rule be applied to $\lim_{x\to 1} \dfrac{x^2 + 1}{2x + 1}$?

Solution The answer is no. The function does *not* have an indeterminate form because

$$\frac{x^2 + 1}{2x + 1}\bigg|_{x=1} = \frac{1^2 + 1}{2\cdot 1 + 1} = \frac{2}{3}$$

However, the limit can be evaluated directly by substitution:

$$\lim_{x\to 1} \frac{x^2 + 1}{2x + 1} = \frac{2}{3}$$

An incorrect application of L'Hôpital's Rule would lead to the following limit with a different value:

$$\lim_{x\to 1} \frac{2x}{2} = 1 \quad \text{(not equal to original limit)} \qquad \blacksquare$$

EXAMPLE 6 The Form $\infty - \infty$ Evaluate $\displaystyle\lim_{x \to 0} \left(\frac{1}{\sin x} - \frac{1}{x} \right)$.

Solution L'Hôpital's Rule does not apply directly because the indeterminate form is neither $0/0$ nor ∞/∞:

$$\left(\frac{1}{\sin x} - \frac{1}{x} \right) \bigg|_{x=0} = \frac{1}{0} - \frac{1}{0} \quad \text{(indeterminate)}$$

However,

$$\frac{1}{\sin x} - \frac{1}{x} = \frac{x - \sin x}{x \sin x}$$

This quotient has the indeterminate form $0/0$ at $x = 0$, and L'Hôpital's Rule now yields

$$\lim_{x \to 0} \left(\frac{1}{\sin x} - \frac{1}{x} \right) = \lim_{x \to 0} \frac{x - \sin x}{x \sin x} = \lim_{x \to 0} \frac{1 - \cos x}{x \cos x + \sin x}$$

The limit on the right is still indeterminate at $x = 0$, but a second application of L'Hôpital's Rule finishes the job (Figure 1):

$$\lim_{x \to 0} \frac{1 - \cos x}{x \cos x + \sin x} = \lim_{x \to 0} \frac{\sin x}{-x \sin x + 2 \cos x} = \frac{0}{2} = 0 \qquad \blacksquare$$

Limits of functions of the form $f(x)^{g(x)}$ can lead to indeterminate forms of the type 0^0, 1^∞, or ∞^0. In such cases, we may take the logarithm and then apply L'Hôpital's Rule.

EXAMPLE 7 The Form 0^0 Evaluate $\displaystyle\lim_{x \to 0+} x^x$.

Solution First, compute the limit of the logarithm. If $y = x^x$, then $\ln y = x \ln x$ and

$$\lim_{x \to 0+} \ln(x^x) = \lim_{x \to 0+} x \ln x = \lim_{x \to 0+} \frac{\ln x}{x^{-1}} \qquad \text{(rewrite as a quotient)}$$

$$= \lim_{x \to 0+} \frac{x^{-1}}{-x^{-2}} = \lim_{x \to 0+} (-x) = 0 \quad \text{(apply L'Hôpital's Rule)}$$

By the continuity of e^x, we may exponentiate to obtain the desired limit (see Figure 2):

$$\lim_{x \to 0+} x^x = \lim_{x \to 0+} e^{\ln(x^x)} = e^{\lim_{x \to 0+} \ln(x^x)} = e^0 = 1 \qquad \blacksquare$$

Comparing Growth of Functions

Sometimes, we are interested in determining which of two functions, $f(x)$ and $g(x)$, grows faster. For example, there are two standard computer algorithms for sorting data (alphabetizing, ordering according to rank, etc.): **Quick Sort** and **Bubble Sort.** The average time required to sort a list of size n has order of magnitude $n \ln n$ for Quick Sort and n^2 for Bubble Sort. Which algorithm is faster when the size n is large? Although n is a whole number, this problem amounts to comparing the growth of $f(x) = x \ln x$ and $g(x) = x^2$ as $x \to \infty$.

We say that $f(x)$ grows *faster* than $g(x)$ if

$$\lim_{x \to \infty} \frac{f(x)}{g(x)} = \infty \qquad \text{or equivalently,} \qquad \lim_{x \to \infty} \frac{g(x)}{f(x)} = 0$$

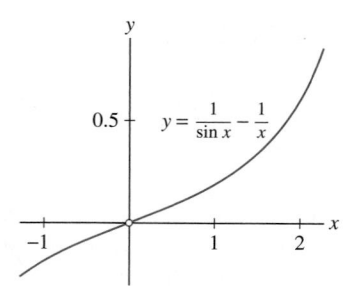

FIGURE 1 The graph confirms that $y = \dfrac{1}{\sin x} - \dfrac{1}{x}$ approaches 0 as $x \to 0$.

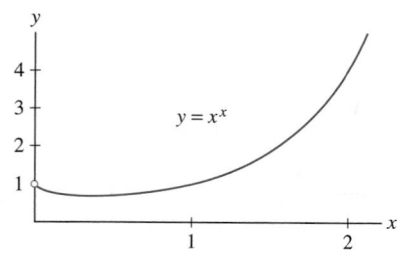

FIGURE 2 The function $y = x^x$ approaches 1 as $x \to 0$.

In this case, we write $g \ll f$. For example, $x \ll x^2$ because

$$\lim_{x \to \infty} \frac{x^2}{x} = \lim_{x \to \infty} x = \infty$$

To compare the growth of functions, we need a version of L'Hôpital's Rule that applies to limits as $x \to \infty$.

THEOREM 2 L'Hôpital's Rule for Limits as $x \to \infty$ Assume that $f(x)$ and $g(x)$ are differentiable in an interval (b, ∞) and that $g'(x) \neq 0$ for $x > b$. If $\lim_{x \to \infty} f(x)$ and $\lim_{x \to \infty} g(x)$ exist and either both are zero or both are infinite, then

$$\lim_{x \to \infty} \frac{f(x)}{g(x)} = \lim_{x \to \infty} \frac{f'(x)}{g'(x)}$$

provided that the limit on the right exists. A similar result holds for limits as $x \to -\infty$.

■ **EXAMPLE 8** Does $f(x) = x^2$ or $g(x) = x \ln x$ grow faster as $x \to \infty$?

Solution Both $f(x)$ and $g(x)$ approach infinity as $x \to \infty$, so L'Hôpital's Rule applies to the quotient:

$$\lim_{x \to \infty} \frac{f(x)}{g(x)} = \lim_{x \to \infty} \frac{x^2}{x \ln x} = \underbrace{\lim_{x \to \infty} \frac{x}{\ln x} = \lim_{x \to \infty} \frac{1}{x^{-1}}}_{\text{L'Hôpital's Rule}} = \lim_{x \to \infty} x = \infty$$

We conclude that $x \ln x \ll x^2$ (Figure 3).   ■

■ **EXAMPLE 9** Jonathan is interested in comparing two computer algorithms whose average run times are approximately $(\ln n)^2$ and $\sqrt{n}$. Which algorithm takes less time for large values of n?

Solution We replace n by the continuous variable x and apply L'Hôpital's Rule:

$$\lim_{x \to \infty} \frac{\sqrt{x}}{(\ln x)^2} = \lim_{x \to \infty} \frac{\frac{1}{2}x^{-1/2}}{2x^{-1} \ln x} \qquad \text{(L'Hôpital's Rule)}$$

$$= \lim_{x \to \infty} \frac{x^{1/2}}{4 \ln x} \qquad \text{(simplify)}$$

$$= \lim_{x \to \infty} \frac{\frac{1}{2}x^{-1/2}}{4x^{-1}} \qquad \text{(L'Hôpital's Rule again)}$$

$$= \lim_{x \to \infty} \frac{1}{8} x^{1/2} = \infty \quad \text{(simplify)}$$

This shows that $(\ln x)^2 \ll \sqrt{x}$. We conclude that the algorithm whose average time is proportional to $(\ln n)^2$ takes less time for large n.   ■

In Section 1.6, we asserted that exponential functions increase more rapidly than polynomial functions. We now prove this by showing that $x^n \ll e^x$ for every exponent n (Figure 4).

FIGURE 3

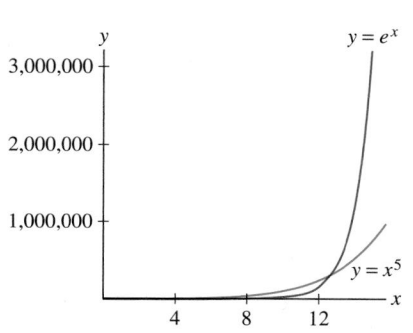

FIGURE 4 Graph illustrating $x^5 \ll e^x$.

> **THEOREM 3 Growth of e^x**
>
> $$x^n \ll e^x \qquad \text{for every exponent } n$$

Proof For $n = 1$, L'Hôpital's Rule gives

$$\lim_{x\to\infty} \frac{e^x}{x} = \lim_{x\to\infty} \frac{e^x}{1} = \infty$$

To treat higher powers, we observe that L'Hôpital's Rule reduces the limit with x^n to the limit for x^{n-1}. For example,

$$\lim_{x\to\infty} \frac{e^x}{x^2} = \lim_{x\to\infty} \frac{e^x}{2x} = \frac{1}{2} \lim_{x\to\infty} \frac{e^x}{x} = \infty$$

Similarly, having proved that $e^x/x^2 \to \infty$, we conclude

$$\lim_{x\to\infty} \frac{e^x}{x^3} = \lim_{x\to\infty} \frac{e^x}{3x^2} = \frac{1}{3} \lim_{x\to\infty} \frac{e^x}{x^2} = \infty$$

Proceeding in this way, we may prove the result for all n. A more formal proof would use the principle of induction. ∎

Proof of L'Hôpital's Rule

We prove L'Hôpital's Rule here only in the first case of Theorem 1, namely, in the case that $f(a) = g(a) = 0$. We also assume that f' and g' are continuous at $x = a$ and that $g'(a) \neq 0$. Then $g(x) \neq g(a)$ for x near but not equal to a. In this case, L'Hôpital's Rule states

$$\lim_{x\to a} \frac{f(x)}{g(x)} = \lim_{x\to a} \frac{f'(x)}{g'(x)} = \frac{f'(a)}{g'(a)}$$

To prove this, we observe that if $f(a) = g(a) = 0$, then for x near but not equal to a,

$$\frac{f(x)}{g(x)} = \frac{f(x) - f(a)}{g(x) - g(a)} = \frac{\dfrac{f(x) - f(a)}{x - a}}{\dfrac{g(x) - g(a)}{x - a}}$$

By the Quotient Law for Limits and the definition of the derivative, we obtain

$$\lim_{x\to a} \frac{f(x)}{g(x)} = \frac{\displaystyle\lim_{x\to a} \frac{f(x) - f(a)}{x - a}}{\displaystyle\lim_{x\to a} \frac{g(x) - g(a)}{x - a}} = \frac{f'(a)}{g'(a)}$$ ∎

4.7 SUMMARY

- *L'Hôpital's Rule:* Assume that f and g are differentiable near a and that

$$f(a) = g(a) = 0$$

Assume also that $g'(x) \neq 0$ for x near (but not equal to) a. Then

$$\lim_{x \to a} \frac{f(x)}{g(x)} = \lim_{x \to a} \frac{f'(x)}{g'(x)}$$

provided that the limit on the right exists.
• L'Hôpital's Rule remains valid if $\lim_{x \to a} f(x)$ and $\lim_{x \to a} g(x)$ are both infinite. It also applies to limits as $x \to \infty$.
• Limits involving the indeterminate forms 0^0, 1^∞, or ∞^0 can often be evaluated by first taking the logarithm and then applying L'Hôpital's Rule.
• In comparing the growth rates of functions, we write $g \ll f$ if

$$\lim_{x \to \infty} \frac{f(x)}{g(x)} = \infty$$

4.7 EXERCISES

Preliminary Questions

1. Which of the following two limits can be evaluated using L'Hôpital's Rule?

$$\lim_{x \to 4} \frac{3x - 12}{x^2 - 16}, \qquad \lim_{x \to 4} \frac{12x - 3}{x - 4}$$

2. What is wrong with evaluating

$$\lim_{x \to 0} \frac{x^2 - 2x}{3x - 2}$$

using L'Hôpital's Rule?

Exercises

In Exercises 1–10, show that L'Hôpital's Rule is applicable and use it to evaluate the limit.

1. $\lim_{x \to 1} \dfrac{2x^2 + x - 3}{x - 1}$

2. $\lim_{x \to 2} \dfrac{2x^2 + x - 10}{2x^5 - 40x + 16}$

3. $\lim_{x \to -1} \dfrac{6x^3 + 13x^2 + 9x + 2}{6x^3 - x^2 - 5x + 2}$

4. $\lim_{x \to 0} \dfrac{\sin 4x}{x^2 + 3x}$

5. $\lim_{x \to 9} \dfrac{\sqrt{x} - 3}{2x^2 - 17x - 9}$

6. $\lim_{x \to 0} \dfrac{x^3}{\sin x - x}$

7. $\lim_{x \to 0} \dfrac{x^2}{1 - \cos x}$

8. $\lim_{x \to 0} \dfrac{\cos 2x - 1}{\sin 5x}$

9. $\lim_{x \to 0} \dfrac{\cos x - \cos^2 x}{\sin x}$

10. $\lim_{x \to 3} \dfrac{\sqrt{x + 1} - 2}{x^3 - 7x - 6}$

In Exercises 11–16, show that L'Hôpital's Rule is applicable to the limit as $x \to \pm\infty$ and evaluate.

11. $\lim_{x \to \infty} \dfrac{3x - 1}{7 - 12x}$

12. $\lim_{x \to \infty} \dfrac{\ln x}{\sqrt{x}}$

13. $\lim_{x \to \infty} \dfrac{x}{e^x}$

14. $\lim_{x \to -\infty} \dfrac{\ln(x^4 + 1)}{x}$

15. $\lim_{x \to \infty} \dfrac{x^2}{e^x}$

16. $\lim_{x \to -\infty} x \sin \dfrac{1}{x}$

In Exercises 17–48, apply L'Hôpital's Rule to evaluate the limit. In some cases, it may be necessary to apply it more than once.

17. $\lim_{x \to 0} \dfrac{\sin 4x}{\sin 3x}$

18. $\lim_{x \to \pi/2} \dfrac{\tan 3x}{\tan 5x}$

19. $\lim_{x \to 0} \dfrac{\tan x}{x}$

20. $\lim_{x \to 1} \dfrac{\sqrt{8 + x} - 3x^{1/4}}{x^2 + 3x - 4}$

21. $\lim_{x \to 0} \left(\cot x - \dfrac{1}{x} \right)$

22. $\lim_{x \to 2} \dfrac{e^{x^2} - e^4}{x - 2}$

23. $\lim_{x \to -\infty} \dfrac{3x - 2}{1 - 5x}$

24. $\lim_{x \to \infty} \dfrac{x^2 + 4x}{9x^3 + 4}$

25. $\lim_{x \to -\infty} \dfrac{7x^2 + 4x}{9 - 3x^2}$

26. $\lim_{x \to 1} \tan \left(\dfrac{\pi x}{2} \right) \ln x$

27. $\lim_{x \to \infty} \dfrac{3x^3 + 4x^2}{4x^3 - 7}$

28. $\lim_{x \to 4} \left[\dfrac{1}{\sqrt{x} - 2} - \dfrac{4}{x - 4} \right]$

29. $\lim_{x \to 1} \dfrac{x(\ln x - 1) + 1}{(x - 1) \ln x}$

30. $\lim_{x \to \pi/2} \left(x - \dfrac{\pi}{2} \right) \tan x$

31. $\lim_{x \to 0} \dfrac{\cos(x + \frac{\pi}{2})}{\sin x}$

32. $\lim_{x \to 0} \dfrac{e^x - 1}{\sin x}$

33. $\lim_{x \to 0} \dfrac{\sin x - x \cos x}{x - \sin x}$

34. $\lim_{x \to \infty} \left(\dfrac{1}{x^2} - \csc^2 x \right)$

35. $\displaystyle\lim_{x\to 1}\frac{e^x - e}{\ln x}$

36. $\displaystyle\lim_{x\to 0}\frac{x^2}{1-\cos x}$

37. $\displaystyle\lim_{x\to \pi/2}\frac{\cos x}{\sin(2x)}$

38. $\displaystyle\lim_{x\to 0}\frac{e^{2x}-1}{x}$

39. $\displaystyle\lim_{x\to \infty} e^{-x}(x^3 - x^2 + 9)$

40. $\displaystyle\lim_{x\to 0}\frac{a^x - 1}{x}$ $(a > 0)$

41. $\displaystyle\lim_{x\to 1}\frac{x^{1/3}-1}{x^2 + 3x - 4}$

42. $\displaystyle\lim_{x\to \pi/2}(\sec x - \tan x)$

43. $\displaystyle\lim_{x\to \infty} x^{1/x}$

44. $\displaystyle\lim_{x\to 1}(1 + \ln x)^{1/(x-1)}$

45. $\displaystyle\lim_{t\to 0}(\sin t)(\ln t)$

46. $\displaystyle\lim_{x\to 0+} x^{\sin x}$

47. $\displaystyle\lim_{x\to 1}\frac{(1+3x)^{1/2}-2}{(1+7x)^{1/3}-2}$

48. $\displaystyle\lim_{x\to \infty}\left(\frac{x}{x+1}\right)^x$

In Exercises 49–52, evaluate the limit using L'Hôpital's Rule if necessary.

49. $\displaystyle\lim_{x\to 0}\frac{\sin^{-1} x}{x}$

50. $\displaystyle\lim_{x\to 0}\frac{\tan^{-1} x}{\sin^{-1} x}$

51. $\displaystyle\lim_{x\to 1}\frac{\tan^{-1} x - \frac{\pi}{4}}{\tan\frac{\pi}{4}x - 1}$

52. $\displaystyle\lim_{x\to 0}\ln x \tan^{-1} x$

53. Evaluate $\displaystyle\lim_{x\to \pi/2}\frac{\cos mx}{\cos nx}$, where m and n are nonzero whole numbers.

54. Evaluate $\displaystyle\lim_{x\to 1}\frac{x^n - 1}{x^m - 1}$ for any numbers $n, m \neq 0$.

55. GU Can L'Hôpital's Rule be applied to $\displaystyle\lim_{x\to 0+} x^{\sin(1/x)}$? Does a graphical or numerical investigation suggest that the limit exists?

56. Sketch the graph of $y = \dfrac{\ln x}{x}$ by determining the increasing/decreasing behavior, concavity, and asymptotic behavior as in Section 4.5.

57. Let $f(x) = x^{1/x}$ in the domain $\{x : x > 0\}$.

(a) Calculate $\displaystyle\lim_{x\to 0+} f(x)$ and $\displaystyle\lim_{x\to \infty} f(x)$.

(b) Find the maximum value of $f(x)$ and determine the intervals on which $f(x)$ is increasing or decreasing.

(c) Use (a) and (b) to prove that $x^{1/x} = c$ has a unique solution if $0 < c \le 1$ or $c = e^{1/e}$, two solutions if $1 < c < e^{1/e}$, and no solutions if $c > e^{1/e}$.

(d) GU Plot the graph of $f(x)$. Explain how the graph confirms the conclusions in (c).

58. Determine if $f \ll g$ or $g \ll f$ (or neither) for the functions $f(x) = \log_{10} x$ and $g(x) = \ln x$.

59. Show that $(\ln x)^2 \ll \sqrt{x}$ and $(\ln x)^4 \ll x^{1/10}$.

60. Just as exponential functions are distinguished by their rapid rate of increase, the logarithm functions grow particularly slowly. Show that $\ln x \ll x^a$ for all $a > 0$.

61. Show that $(\ln x)^N \ll x^a$ for all N and all $a > 0$.

62. Determine whether $\sqrt{x} \ll e^{\sqrt{\ln x}}$ or $e^{\sqrt{\ln x}} \ll \sqrt{x}$. *Hint:* Do not use L'Hôpital's Rule. Instead, make a substitution $u = \ln x$.

63. Show that $\displaystyle\lim_{x\to \infty} x^n e^{-x} = 0$ for all whole numbers $n > 0$.

64. Assumptions Matter Let $f(x) = x(2 + \sin x)$ and $g(x) = x^2 + 1$.

(a) Show directly that $\displaystyle\lim_{x\to \infty} f(x)/g(x) = 0$.

(b) Show that $\displaystyle\lim_{x\to \infty} f(x) = \lim_{x\to \infty} g(x) = \infty$, but $\displaystyle\lim_{x\to \infty} f'(x)/g'(x)$ does not exist.

Do (a) and (b) contradict L'Hôpital's Rule? Explain.

65. Show that $\displaystyle\lim_{t\to \infty} t^k e^{-t^2} = 0$ for all k. *Hint:* Compare with $\displaystyle\lim_{t\to \infty} t^k e^{-t} = 0$.

In Exercises 66–68, let

$$f(x) = \begin{cases} e^{-1/x^2} & \text{for } x \neq 0 \\ 0 & \text{for } x = 0 \end{cases}$$

These exercises show that $f(x)$ has an unusual property: All of its higher derivatives at $x = 0$ exist and are equal to zero.

66. Show that $\displaystyle\lim_{x\to 0}\frac{f(x)}{x^k} = 0$ for all k. *Hint:* Let $t = x^{-1}$ and apply the result of Exercise 65.

67. Show that $f'(0)$ exists and is equal to zero. Also, verify that $f''(0)$ exists and is equal to zero.

68. Show that for $k \ge 1$ and $x \neq 0$,

$$f^{(k)}(x) = \frac{P(x)e^{-1/x^2}}{x^r}$$

for some polynomial $P(x)$ and some exponent $r \ge 1$. Use the result of Exercise 66 to show that $f^{(k)}(0)$ exists and is equal to zero for all $k \ge 1$.

Further Insights and Challenges

69. Show that L'Hôpital's Rule applies to $\displaystyle\lim_{x\to \infty}\frac{x}{\sqrt{x^2 + 1}}$, but that it is of no help. Then evaluate the limit directly.

70. Use L'Hôpital's Rule to evaluate the following limit, assuming that f is differentiable and f' is continuous:

$$\lim_{x\to a}\frac{f(x) - f(a)}{x - a}$$

71. Resonance   A spring oscillates with a natural frequency $\lambda/2\pi$. If we drive the spring with a sinusoidal force $C\sin(\omega t)$, where $\omega \neq \lambda$, then the spring oscillates according to

$$y(t) = \frac{C}{\lambda^2 - \omega^2}\left(\lambda \sin(\omega t) - \omega \sin(\lambda t)\right)$$

(a) Use L'Hôpital's Rule to determine $y(t)$ in the limit as $\omega \to \lambda$.

(b) Define $y_0(t) = \lim\limits_{\omega \to \lambda} y(t)$. Show that $y_0(t)$ ceases to be periodic and that its amplitude $|y_0(t)|$ tends to infinity as $t \to \infty$ (the system is said to be in resonance; eventually, the spring is stretched beyond its limits).

(c) ⌂⌂⌂ Plot $y(t)$ for $\lambda = 1$ and $\omega = 0.5, 0.8, 0.9, 0.99,$ and 0.999. How do the graphs change? Do the graphs confirm your conclusion in (b)?

72. We expended a lot of effort to evaluate $\lim\limits_{x \to 0} \dfrac{\sin x}{x}$ in Chapter 2. Show that we could have evaluated it easily using L'Hôpital's Rule. Then explain why this method would involve *circular reasoning*.

73. Suppose that f and g are polynomials such that $f(a) = g(a) = 0$. In this case, it is a fact from algebra that $f(x) = (x - a)f_1(x)$ and $g(x) = (x - a)g_1(x)$ for some polynomials f_1 and g_1. Use this to verify L'Hôpital's Rule directly for f and g.

74. In this exercise, we verify the following limit formula for e:

$$e = \lim_{x \to 0}(1 + x)^{1/x} \qquad \boxed{1}$$

(a) Use L'Hôpital's Rule to prove that $\lim\limits_{x \to 0} \dfrac{\ln(1 + x)}{x} = 1$.

(b) Use (a) to prove that $\lim\limits_{x \to 0} \ln\left((1 + x)^{1/x}\right) = 1$.

(c) Use (b) to verify Eq. (1).

(d) Find a value of x such that $|(1 + x)^{1/x} - e| \le 0.001$.

75. Patience Required   Use L'Hôpital's Rule to evaluate and check your answers numerically:

(a) $\lim\limits_{x \to 0+} \left(\dfrac{\sin x}{x}\right)^{1/x^2}$

(b) $\lim\limits_{x \to 0} \left(\dfrac{1}{\sin^2 x} - \dfrac{1}{x^2}\right)$

4.8 Newton's Method

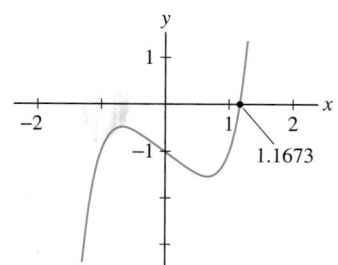

◀┈ *REMINDER A "zero" or "root" of a function $f(x)$ is a solution to the equation $f(x) = 0$.*

FIGURE 1 Graph of $y = x^5 - x - 1$. The value 1.1673 is a good numerical approximation to the root.

Newton's Method is a procedure for finding numerical approximations to zeros of functions. Numerical approximations are important because it is often impossible to find the zeros exactly. For example, the polynomial $f(x) = x^5 - x - 1$ has one real root c (see Figure 1), but we can prove, using an advanced branch of mathematics called *Galois Theory*, that there is no algebraic formula for this root. Newton's Method shows that $c \approx 1.1673$, and with enough computation, we can compute c to any desired degree of accuracy.

In Newton's Method, we begin by choosing a number x_0, which we believe is close to a root of $f(x)$. This starting value x_0 is called the **initial guess**. Newton's Method then produces a sequence $x_0, x_1, x_2, \ldots$ of successive approximations that, in favorable situations, converge to a root.

Figure 2 illustrates the procedure. Given an initial guess x_0, we draw the tangent line to the graph at $(x_0, f(x_0))$. The approximation x_1 is defined as the x-coordinate of the

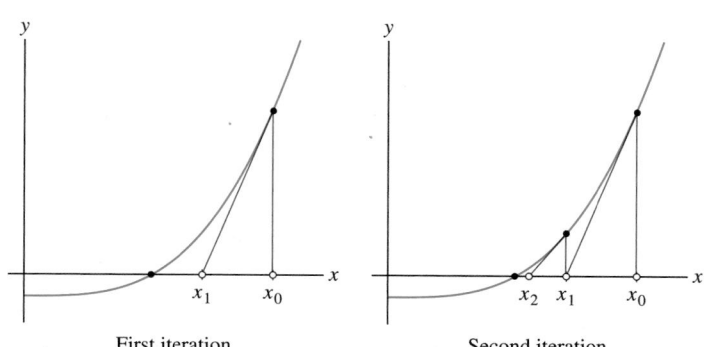

First iteration   Second iteration

FIGURE 2 The sequence produced by iteration converges to a root.

point where the tangent line intersects the x-axis. To produce the second approximation x_2 (also called the second iterate), we apply this procedure to x_1.

Let's derive a formula for x_1. The equation of the tangent line at $(x_0, f(x_0))$ is

$$y = f(x_0) + f'(x_0)(x - x_0)$$

The tangent line crosses the x-axis at x_1, where

$$y = f(x_0) + f'(x_0)(x_1 - x_0) = 0$$

If $f'(x_0) \neq 0$, we may solve for x_1 to obtain $x_1 - x_0 = -f(x_0)/f'(x_0)$, or

$$x_1 = x_0 - \frac{f(x_0)}{f'(x_0)}$$

The second iterate x_2 is obtained by applying this formula to x_1 instead of x_0:

$$x_2 = x_1 - \frac{f(x_1)}{f'(x_1)}$$

and so on. Notice in Figure 2 that x_1 is closer to the root than x_0 and x_2 is closer still. This is typical: The successive approximations usually converge to the actual root. However, there are cases where Newton's Method fails (see Figure 4).

Newton's Method is an example of an iterative procedure. To "iterate" means to repeat, and in Newton's Method we use (1) repeatedly to produce the sequence of approximations.

Newton's Method   To find a numerical approximation to a root of $f(x) = 0$:

Step 1. **Choose initial guess x_0 (close to the desired root if possible).**

Step 2. **Generate successive approximations $x_1, x_2, \dots$ where**

$$x_{n+1} = x_n - \frac{f(x_n)}{f'(x_n)} \qquad \boxed{1}$$

■ **EXAMPLE 1** Approximating $\sqrt{5}$ Calculate the first three approximations x_1, x_2, x_3 to a root of $f(x) = x^2 - 5$ using the initial guess $x_0 = 2$.

Solution We have $f'(x) = 2x$. Therefore:

$$x_1 = x_0 - \frac{f(x_0)}{f'(x_0)} = x_0 - \frac{x_0^2 - 5}{2x_0}$$

We compute the successive approximations as follows:

$$x_1 = x_0 - \frac{f(x_0)}{f'(x_0)} = 2 - \frac{2^2 - 5}{2 \cdot 2} = 2.25$$

$$x_2 = x_1 - \frac{f(x_1)}{f'(x_1)} = 2.25 - \frac{2.25^2 - 5}{2 \cdot 2.25} \approx 2.23611$$

$$x_3 = x_2 - \frac{f(x_2)}{f'(x_2)} = 2.23611 - \frac{2.23611^2 - 5}{2 \cdot 2.23611} \approx \mathbf{2.23606797789}$$

This sequence provides successive approximations to the root

$$\sqrt{5} = \mathbf{2.236067977}499789696\dots$$

Observe that x_3 is accurate to within an error of less than 10^{-9}. This is impressive accuracy for just three iterations of Newton's Method. ■

How Many Iterations Are Required?

How many iterations of Newton's Method are required to approximate a root to within a given accuracy? There is no definitive answer, but in many cases, the number of correct decimal places roughly doubles at each step. In practice, it is usually safe to assume that if x_n and x_{n+1} agree to m decimal places, then the approximation x_n is correct to these m places.

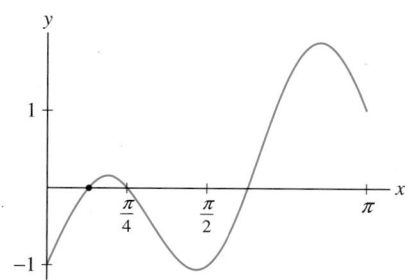

FIGURE 3 Graph of $f(x) = \sin 3x - \cos x$.

■ **EXAMPLE 2** GU   Let c be the smallest positive solution of $\sin 3x = \cos x$.

(a) Use a computer-generated graph to choose an initial guess x_0 for c.

(b) Use Newton's Method to approximate c to within an error of at most 10^{-6}.

Solution

(a) A solution of $\sin 3x = \cos x$ is a zero of the function $f(x) = \sin 3x - \cos x$. Figure 3 shows that the smallest zero is approximately halfway between 0 and $\frac{\pi}{4}$. Because $\frac{\pi}{4} \approx 0.785$, a good initial guess is $x_0 = 0.4$.

(b) Since $f'(x) = 3 \cos 3x + \sin x$, Eq. (1) yields the formula

$$x_{n+1} = x_n - \frac{\sin 3x_n - \cos x_n}{3 \cos 3x_n + \sin x_n}$$

With $x_0 = 0.4$ as the initial guess, the first four iterates are

$$x_1 \approx 0.\mathbf{3925}647447$$
$$x_2 \approx 0.\mathbf{3926990}382$$
$$x_3 \approx 0.\mathbf{3926990817}$$
$$x_4 \approx 0.\mathbf{3926990816}$$

> There is no single "correct" initial guess. In Example 2, we chose $x_0 = 0.4$ but another possible choice is $x_0 = 0$, leading to the sequence
>
> $x_1 \approx 0.3333333333$
>
> $x_2 \approx 0.3864547725$
>
> $x_3 \approx 0.3926082513$
>
> $x_4 \approx 0.3926990816$
>
> You can check, however, that $x_0 = 1$ yields a sequence converging to $\frac{\pi}{4}$, which is the second positive solution of $\sin 3x = \cos x$.

Stopping here, we can be fairly confident that 0.392699 approximates the smallest positive root c with an error of less than 10^{-6}. As a further test, we check that $\sin(3x_4) \approx \cos(x_4)$:

$$\sin(3x_4) \approx \sin(3(0.392699)) \approx 0.92387944$$
$$\cos(x_4) \approx \cos(0.392699) \approx 0.92387956$$ ■

Which Root Does Newton's Method Compute?

Sometimes, Newton's Method computes no root at all. In Figure 4, the iterates diverge to infinity. However, Newton's Method usually converges quickly and if a particular choice of x_0 does not lead to a root, the best strategy is to try a different initial guess, consulting a graph if possible. If $f(x) = 0$ has more than one root, different initial guesses x_0 may lead to different roots.

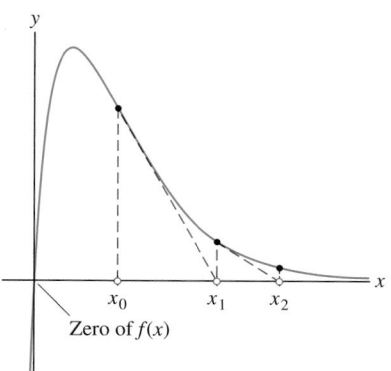

FIGURE 4 Function has only one zero but the sequence of Newton iterates goes off to infinity.

■ **EXAMPLE 3**   Figure 5 shows that $f(x) = x^4 - 6x^2 + x + 5$ has four real roots.

(a) Show that with $x_0 = 0$, Newton's Method converges to the root near -2.

(b) Show that with $x_0 = -1$, Newton's Method converges to the root near -1.

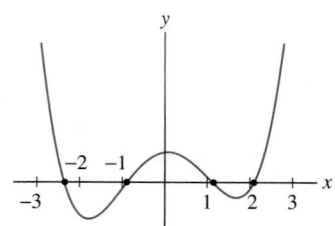

FIGURE 5 Graph of $f(x) = x^4 - 6x^2 + x + 5$.

Solution We have $f'(x) = 4x^3 - 12x + 1$ and

$$x_{n+1} = x_n - \frac{x_n^4 - 6x_n^2 + x_n + 5}{4x_n^3 - 12x_n + 1} = \frac{3x_n^4 - 6x_n^2 - 5}{4x_n^3 - 12x_n + 1}$$

(a) On the basis of Table 1, we can be confident that when $x_0 = 0$, Newton's Method converges to a root near -2.3. Notice in Figure 5 that this is not the closest root to x_0.

(b) Table 2 suggests that with $x_0 = -1$, Newton's Method converges to the root near -0.9. ∎

TABLE 1	
x_0	0
x_1	-5
x_2	-3.9179954
x_3	-3.1669480
x_4	-2.6871270
x_5	-2.4363303
x_6	-2.3572979
x_7	-2.3495000

TABLE 2	
x_0	-1
x_1	-0.8888888888
x_2	-0.8882866140
x_3	-0.88828656234358
x_4	-0.888286562343575

4.8 SUMMARY

- Newton's Method: To find a sequence of numerical approximations to a solution of $f(x) = 0$, begin with an initial guess x_0. Then construct the sequence $x_0, x_1, x_2, \ldots$ using the formula

$$x_{n+1} = x_n - \frac{f(x_n)}{f'(x_n)}$$

In favorable cases, the sequence converges to a solution. The initial guess x_0 should be chosen as close as possible to a root, possibly by referring to a graph.

- In most cases, the sequence $x_0, x_1, x_2, \ldots$ converges rapidly to a solution. If x_n and x_{n+1} agree to m decimal places, you may be reasonably confident that x_n agrees with a true solution to m decimal places.

4.8 EXERCISES

Preliminary Questions

1. How many iterations of Newton's Method are required to compute a root if $f(x)$ is a linear function?

2. What happens in Newton's Method if your initial guess happens to be a zero of f?

3. What happens in Newton's Method if your initial guess happens to be a local min or max of f?

4. Is the following a reasonable description of Newton's Method: "A root of the equation of the tangent line to $f(x)$ is used as an approximation to a root of $f(x)$ itself"? Explain.

Exercises

In Exercises 1–4, use Newton's Method with the given function and initial value x_0 to calculate x_1, x_2, x_3.

1. $f(x) = x^2 - 2$, $x_0 = 1$

2. $f(x) = x^2 - 7$, $x_0 = 2.5$

3. $f(x) = x^3 - 5$, $x_0 = 1.6$

4. $f(x) = \cos x - x$, $x_0 = 0.8$

5. Use Figure 6 to choose an initial guess x_0 to the unique real root of $x^3 + 2x + 5 = 0$. Then compute the first three iterates of Newton's Method.

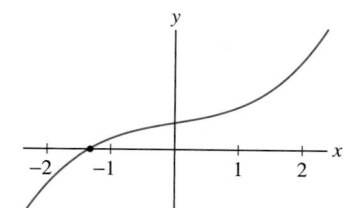

FIGURE 6 Graph of $y = x^3 + 2x + 5$.

6. Use Newton's Method to find a solution to $\sin x = \cos 2x$ in the interval $[0, \frac{\pi}{2}]$ to three decimal places. Then guess the exact solution and compare with your approximation.

7. Use Newton's Method to find the two solutions of $e^x = 5x$ to three decimal places (Figure 7).

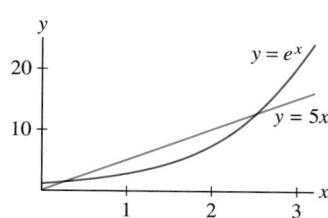

FIGURE 7 Graphs of e^x and $5x$.

8. Use Newton's Method to approximate the positive solution to the equation $\ln(x + 4) = x$ to three decimal places.

In Exercises 9–12, use Newton's Method to approximate the root to three decimal places and compare with the value obtained from a calculator.

9. $\sqrt{10}$

10. $7^{1/4}$

11. $5^{1/3}$

12. $2^{-1/2}$

13. Use Newton's Method to approximate the largest positive root of $f(x) = x^4 - 6x^2 + x + 5$ to within an error of at most 10^{-4}. Refer to Figure 5.

14. GU   Sketch the graph of $f(x) = x^3 - 4x + 1$ and use Newton's Method to approximate the largest positive root to within an error of at most 10^{-3}.

15. GU   Use a graphing calculator to choose an initial guess for the unique positive root of $x^4 + x^2 - 2x - 1 = 0$. Calculate the first three iterates of Newton's Method.

16. The first positive solution of $\sin x = 0$ is $x = \pi$. Use Newton's Method to calculate π to four decimal places.

17. GU   Estimate the smallest positive solution of $\dfrac{\sin \theta}{\theta} = 0.9$ to three decimal places. Use a graphing calculator to choose the initial guess.

18. In 1535, the mathematician Antonio Fior challenged his rival Niccolo Tartaglia to solve this problem: A tree stands 12 *braccia* high; it is broken into two parts at such a point that the height of the part left standing is the cube root of the length of the part cut away. What is the height of the part left standing? Show that this is equivalent to solving $x^3 + x = 12$ and find the height to three decimal places. Tartaglia, who had discovered the secrets of cubic equations, was able to determine the exact answer:

$$x = \left(\sqrt[3]{\sqrt{2{,}919} + 54} - \sqrt[3]{\sqrt{2{,}919} - 54} \right) \Big/ \sqrt[3]{9}$$

19. Let x_1, x_2 be the estimates to a root obtained by applying Newton's Method with $x_0 = 1$ to the function graphed in Figure 8. Estimate the numerical values of x_1 and x_2, and draw the tangent lines used to obtain them.

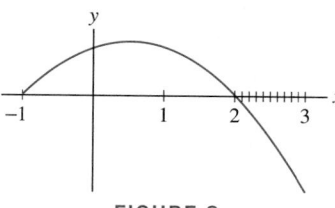

FIGURE 8

20. Find the coordinates to two decimal places of the point P in Figure 9 where the tangent line to $y = \cos x$ passes through the origin.

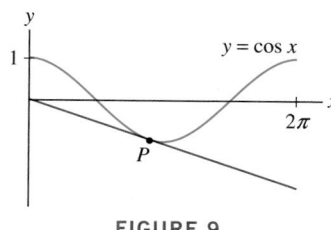

FIGURE 9

21. Find the x-coordinate to two decimal places of the first point in the region $x > 0$ where $y = x$ intersects $y = \tan x$ (draw a graph).

Newton's Method is often used to determine interest rates in financial calculations. In Exercises 22–24, r denotes a yearly interest rate expressed as a decimal (rather than as a percent).

22. If P dollars are deposited every month in an account earning interest at the yearly rate r, then the value S of the account after N years is

$$S = P \left(\frac{b^{12N+1} - b}{b - 1} \right) \qquad \text{where } b = 1 + \frac{r}{12}$$

You have decided to deposit $P = 100$ dollars per month.

(a) What is the value after 5 years if the yearly interest rate is $r = 0.07$ (i.e., 7%)?

(b) Show that to save $10,000 after 5 years, you must earn interest at a rate r determined by the equation $b^{61} - 101b + 100 = 0$. Use Newton's Method to solve for b (note that $b = 1$ is a root, but you want the root satisfying $b > 1$). Then find r.

23. If you borrow L dollars for N years at a yearly interest rate r, your monthly payment of P dollars is calculated using the equation

$$L = P\left(\frac{1 - b^{-12N}}{b - 1}\right) \qquad \text{where } b = 1 + \frac{r}{12}$$

(a) What is the monthly payment if $L = \$5,000$, $N = 3$, and $r = 0.08$ (8%)?

(b) You are offered a loan of $L = \$5,000$ to be paid back over 3 years with monthly payments of $P = \$200$. Use Newton's Method to compute b and find the implied interest rate r of this loan. *Hint:* Show that $(L/P)b^{12N+1} - (1 + L/P)b^{12N} + 1 = 0$.

24. If you deposit P dollars in a retirement fund every year for N years with the intention of then withdrawing Q dollars per year for M years, you must earn interest at a rate r satisfying $P(b^N - 1) = Q(1 - b^{-M})$, where $b = 1 + r$. Assume that $2,000 is deposited each year for 30 years and the goal is to withdraw $8,000 per year for 20 years. Use Newton's Method to compute b and then find r.

25. Kepler's Problem Although planetary motion is beautifully described by Kepler's three laws (see Section 13.6), there is no simple formula for the position of a planet P along its elliptical orbit as a function of time. Kepler developed a method for locating P at time t by drawing the auxiliary dashed circle in Figure 10 and introducing the angle θ (note that P determines θ, which is the central angle of the point B on the circle). Let $a = OA$ and $e = OS/OA$ (the eccentricity of the orbit).

(a) Show that sector BSA has area $(a^2/2)(\theta - e\sin\theta)$.

(b) It follows from Kepler's Second Law that the area of sector BSA is proportional to the time t elapsed since the planet passed point A. More precisely, since the circle has area πa^2, BSA has area $(\pi a^2)(t/T)$, where T is the period of the orbit. Deduce that

$$\frac{2\pi t}{T} = \theta - e\sin\theta$$

(c) The eccentricity of Mercury's orbit is approximately $e = 0.2$. Use Newton's Method to find θ after a quarter of Mercury's year has elapsed ($t = T/4$). Convert θ to degrees. Has Mercury covered more than a quarter of its orbit at $t = T/4$?

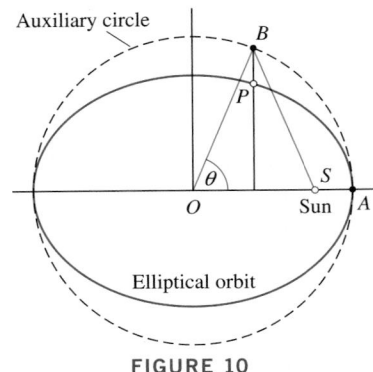

FIGURE 10

26. What happens when you apply Newton's Method to find a zero of $f(x) = x^{1/3}$? Note that $x = 0$ is the only zero.

27. What happens when you apply Newton's Method to the equation $x^3 - 20x = 0$ with the unlucky initial guess $x_0 = 2$?

Further Insights and Challenges

28. Let c be a positive number and let $f(x) = x^{-1} - c$.

(a) Show that $x - (f(x)/f'(x)) = 2x - cx^2$. Thus, Newton's Method provides a way of computing reciprocals without performing division.

(b) Calculate the first three iterates of Newton's Method with $c = 10.324$ and the two initial guesses $x_0 = 0.1$ and $x_0 = 0.5$.

(c) Explain graphically why $x_0 = 0.5$ does not yield a sequence of approximations to the reciprocal $1/10.324$.

29. The roots of $f(x) = \frac{1}{3}x^3 - 4x + 1$ to three decimal places are -3.583, 0.251, and 3.332 (Figure 11). Determine the root to which

Newton's Method converges for the initial choices $x_0 = 1.85$, 1.7, and 1.55. The answer shows that a small change in x_0 can have a significant effect on the outcome of Newton's Method.

In Exercises 30–31, consider a metal rod of length L inches that is fastened at both ends. If you cut the rod and weld on an additional m inches of rod, leaving the ends fixed, the rod will bow up into a circular arc of radius R (unknown), as indicated in Figure 12.

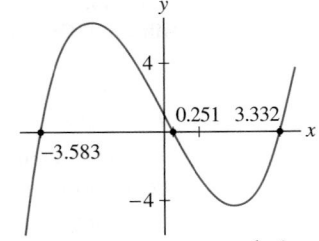

FIGURE 11 Graph of $f(x) = \frac{1}{3}x^3 - 4x + 1$.

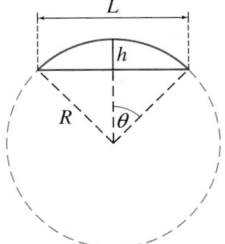

FIGURE 12 The bold circular arc has length $L + m$.

30. Let h be the maximum vertical displacement of the rod.

(a) Show that $L = 2R\sin\theta$ and conclude that

$$h = \frac{L(1 - \cos\theta)}{2\sin\theta}$$

(b) Show $L + m = 2R\theta$ and then prove

$$\frac{\sin\theta}{\theta} = \frac{L}{L + m} \qquad \boxed{2}$$

31. Let $L = 3$ and $m = 1$. Apply Newton's Method to Eq. (2) to estimate θ and use this to estimate h.

32. Quadratic Convergence to Square Roots Let $f(x) = x^2 - c$ and let $e_n = x_n - \sqrt{c}$ be the error in x_n.

(a) Show that $x_{n+1} = \frac{1}{2}(x_n + c/x_n)$ and that $e_{n+1} = e_n^2/2x_n$.

(b) Show that if $x_0 > \sqrt{c}$, then $x_n > \sqrt{c}$ for all n. Explain this graphically.

(c) Assuming that $x_0 > \sqrt{c}$, show that $e_{n+1} \le e_n^2/2\sqrt{c}$.

In Exercises 33–35, a flexible chain of length L is suspended between two poles of equal height separated by a distance $2M$ (Figure 13). By Newton's laws, the chain describes a curve (called a **catenary**) *with equation $y = a\cosh\left(\dfrac{x}{a}\right) + C$. The constant C is arbitrary and a is the number such that $L = 2a\sinh\left(\dfrac{M}{a}\right)$. The sag s is the vertical distance from the highest to the lowest point on the chain.*

33. Suppose that $L = 120$ and $M = 50$.

(a) Use Newton's Method to find a value of a (to two decimal places) satisfying $L = 2a\sinh(M/a)$.

(b) Show that the sag is given by $s = a\cosh(M/a) - a$. Compute s.

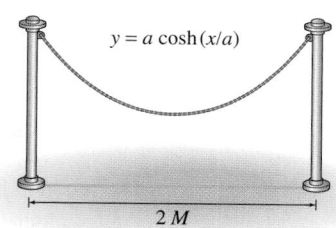

FIGURE 13 Chain hanging between two poles describes the curve $y = a\cosh(x/a)$.

34. Let M be a fixed constant. Show that the sag is given by $s = a\cosh\left(\dfrac{M}{a}\right) - a$.

(a) Calculate $\dfrac{ds}{da}$.

(b) Calculate $\dfrac{da}{dL}$ by implicit differentiation using the relation $L = 2a\sinh\left(\dfrac{M}{a}\right)$.

(c) Use (a) and (b) and the Chain Rule to show that

$$\frac{ds}{dL} = \frac{ds}{da}\frac{da}{dL} = \frac{\cosh(M/a) - (M/a)\sinh(M/a) - 1}{2\sinh(M/a) - (2M/a)\cosh(M/a)} \qquad \boxed{3}$$

35. Suppose that $L = 160$ and $M = 50$.

(a) Use Newton's Method to find a value of a (to two decimal places) satisfying $L = 2a\sinh(M/a)$.

(b) Use Eq. (3) and the Linear Approximation to estimate the increase in cable sag s if L is increased from $L = 160$ to $L = 161$ and from $L = 160$ to $L = 165$.

(c) $\boxed{CAS}$ Compute $s(161) - s(160)$ and $s(165) - s(160)$ directly and compare with your estimates in (a).

4.9 Antiderivatives

In addition to finding derivatives, there is an important "inverse" problem, namely *given the derivative, find the function itself.* In physics, for example, we may want to compute the position $s(t)$ of an object given its velocity $v(t)$. Since $s'(t) = v(t)$, this amounts to finding a function whose derivative is $v(t)$. A function $F(x)$ whose derivative is $f(x)$ is called an antiderivative of $f(x)$.

DEFINITION Antiderivatives   A function $F(x)$ is an **antiderivative** of $f(x)$ on (a, b) if $F'(x) = f(x)$ for all $x \in (a, b)$.

For example, $F(x) = \frac{1}{3}x^3$ is an antiderivative of $f(x) = x^2$ because

$$F'(x) = \frac{d}{dx}\left(\frac{1}{3}x^3\right) = x^2 = f(x)$$

Similarly, $F(x) = -\cos x$ is an antiderivative of $f(x) = \sin x$ because

$$F'(x) = \frac{d}{dx}(-\cos x) = \sin x = f(x)$$

One of the first characteristics to note about antiderivatives is that they are not unique. We are free to add a constant C: If $F'(x) = f(x)$, then $(F(x) + C)' = f(x)$ because the derivative of the constant is zero. Thus, each of the following is an antiderivative of x^2:

$$\frac{1}{3}x^3, \qquad \frac{1}{3}x^3 + 5, \qquad \frac{1}{3}x^3 - 4$$

Are there any antiderivatives of $f(x)$ other than those obtained by adding a constant to a given antiderivative $F(x)$? According to the following theorem, the answer is no if $f(x)$ is defined on an interval (a, b).

THEOREM 1 The General Antiderivative   Let $F(x)$ be an antiderivative of $f(x)$ on (a, b). Then every other antiderivative on (a, b) is of the form $F(x) + C$ for some constant C.

Proof Let $F(x)$ and $G(x)$ be antiderivatives of $f(x)$ and set $H(x) = G(x) - F(x)$. Then $H'(x) = G'(x) - F'(x) = f(x) - f(x) = 0$. By the Corollary in Section 4.3, $H(x)$ must be a constant, say, $H(x) = C$, and this gives $G(x) = F(x) + C$. ■

GRAPHICAL INSIGHT   The graph of $F(x) + C$ is obtained by shifting the graph of $F(x)$ vertically by C units. Since vertical shifting moves the tangent lines without changing their slopes, it makes sense that all of the functions $F(x) + C$ have the same derivative (Figure 1). Conversely, Theorem 1 tells us that if two graphs have parallel tangent lines, then one graph is obtained from the other by a vertical shift.

We often describe the *general* antiderivative of a function in terms of an arbitrary constant C, as in the following example.

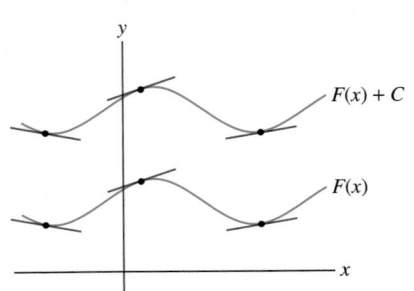

FIGURE 1 The tangent lines to the graphs of $y = F(x)$ and $y = F(x) + C$ are parallel.

■ **EXAMPLE 1**   Find two antiderivatives of $f(x) = \cos x$. Then determine the general antiderivative.

Solution The functions $F(x) = \sin x$ and $G(x) = \sin x + 2$ are both antiderivatives of $f(x)$. The general antiderivative is $F(x) = \sin x + C$, where C is any constant. ■

The process of finding an antiderivative is called **integration**. We will see why in Chapter 5, when we discuss the connection between antiderivatives and areas under curves given by the Fundamental Theorem of Calculus. Anticipating this result, we now begin using the integral sign $\int$, the standard notation for antiderivatives.

The terms "antiderivative" and "indefinite integral" are used interchangeably. In some textbooks, an antiderivative is called a "primitive function."

NOTATION Indefinite Integral   The notation

$$\int f(x)\, dx = F(x) + C \quad \text{means that} \quad F'(x) = f(x)$$

We say that $F(x) + C$ is the general antiderivative or **indefinite integral** of $f(x)$.

The function $f(x)$ appearing in the integral sign is called the **integrand**. The symbol "dx," called a *differential*, is part of the integral notation. It indicates the independent variable, but we do not assign it any other meaning. The constant C is called the *constant of integration*.

Some indefinite integrals can be evaluated by reversing the familiar derivative formulas. For example, we obtain the indefinite integral of x^n by reversing the Power Rule for derivatives.

THEOREM 2 Power Rule for Integrals

$$\int x^n \, dx = \frac{x^{n+1}}{n+1} + C \qquad \text{for } n \neq -1$$

Proof We verify the Power Rule for Integrals by taking the derivative

$$\frac{d}{dx}\left(\frac{x^{n+1}}{n+1} + C\right) = \frac{1}{n+1}((n+1)x^n) = x^n \qquad \blacksquare$$

In words, the Power Rule for Integrals says that to integrate a power of x, "add one to the exponent and then divide by the new exponent." Here are some examples:

$$\int x^5 \, dx = \frac{1}{6}x^6 + C, \qquad \int x^{-9} \, dx = -\frac{1}{8}x^{-8} + C, \qquad \int x^{3/5} \, dx = \frac{5}{8}x^{8/5} + C$$

The Power Rule is not valid for $n = -1$. In fact, for $n = -1$, we obtain the meaningless result

$$\int x^{-1} \, dx = \frac{x^{n+1}}{n+1} + C = \frac{x^0}{0} + C \qquad \text{(meaningless)}$$

Recall, however, that the derivative of the natural logarithm is $\frac{d}{dx}\ln x = x^{-1}$. This shows that $\ln x$ is an antiderivative of $y = x^{-1}$, and thus, for $n = -1$, instead of the Power Rule we have

$$\int \frac{dx}{x} = \ln x + C$$

This formula is valid for $x > 0$, where $f(x) = \ln x$ is defined. We would like to have an antiderivative of $y = x^{-1}$ on its full domain, namely on $\{x : x \neq 0\}$. To achieve this end, we extend $f(x)$ to an even function by setting $f(x) = \ln|x|$ (Figure 2). Then $f(x) = f(-x)$, and by the Chain Rule, $f'(x) = -f'(-x)$. For $x < 0$, we obtain

$$\frac{d}{dx}\ln|x| = f'(x) = -f'(-x) = -\frac{1}{-x} = \frac{1}{x}$$

This proves that $f'(x) = x^{-1}$ for all $x \neq 0$.

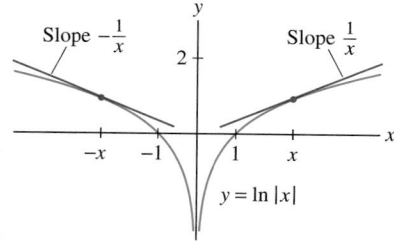

Slope $-\frac{1}{x}$ Slope $\frac{1}{x}$

$y = \ln|x|$

FIGURE 2

> **THEOREM 3** **Antiderivative of** $y = \dfrac{1}{x}$ The function $f(x) = \ln|x|$ is an antiderivative
>
> of $y = \dfrac{1}{x}$ in the domain $\{x : x \neq 0\}$, that is,
>
> $$\int \frac{dx}{x} = \ln|x| + C \qquad \boxed{1}$$

The indefinite integral obeys the usual linearity rules that allow us to integrate "term by term." This follows directly from the linearity rules for the derivative (see Exercise 81.)

> **THEOREM 4** **Linearity of the Indefinite Integral**
>
> - **Sum Rule:** $\displaystyle\int (f(x) + g(x))\,dx = \int f(x)\,dx + \int g(x)\,dx$
>
> - **Multiples Rule:** $\displaystyle\int cf(x)\,dx = c \int f(x)\,dx$

■ **EXAMPLE 2** Evaluate $\int (3x^4 - 5x^{2/3} + x^{-3})\,dx$.

Solution We integrate term by term and use the Power Rule:

$$\int (3x^4 - 5x^{2/3} + x^{-3})\,dx = \int 3x^4\,dx - \int 5x^{2/3}\,dx + \int x^{-3}\,dx \qquad \text{(Sum Rule)}$$

$$= 3\int x^4\,dx - 5\int x^{2/3}\,dx + \int x^{-3}\,dx \qquad \text{(Multiples Rule)}$$

$$= 3\frac{x^5}{5} - 5\frac{x^{5/3}}{5/3} + \frac{x^{-2}}{-2} + C \qquad \text{(Power Rule)}$$

$$= \frac{3}{5}x^5 - 3x^{5/3} - \frac{1}{2}x^{-2} + C$$

When we break up an indefinite integral into a sum of several integrals as in Example 2, it is not necessary to include separate constants of integration for each integral.

To check the answer, we verify that the derivative is equal to the integrand:

$$\frac{d}{dx}\left(\frac{3}{5}x^5 - 3x^{5/3} - \frac{1}{2}x^{-2} + C\right) = 3x^4 - 5x^{2/3} + x^{-3} \qquad ■$$

■ **EXAMPLE 3** Evaluate $\displaystyle\int \left(\frac{5}{x} - 3x^{-10}\right)\,dx$.

Solution Apply Eq. (1) and the Power Rule:

$$\int \left(\frac{5}{x} - 3x^{-10}\right)\,dx = 5\int \frac{dx}{x} - 3\int x^{-10}\,dx$$

$$= 5\ln|x| - 3\left(\frac{x^{-9}}{-9}\right) + C = 5\ln|x| + \frac{1}{3}x^{-9} + C \qquad ■$$

The differentiation formulas for the trigonometric functions give us the following integration formulas. Each formula can be checked by differentiation.

Basic Trigonometric Integrals

$$\int \sin x \, dx = -\cos x + C \qquad\qquad \int \cos x \, dx = \sin x + C$$

$$\int \sec^2 x \, dx = \tan x + C \qquad\qquad \int \csc^2 x \, dx = -\cot x + C$$

$$\int \sec x \tan x \, dx = \sec x + C \qquad\qquad \int \csc x \cot x \, dx = -\csc x + C$$

Similarly, for any constants b and k with $k \neq 0$, the formulas

$$\frac{d}{dx} \sin(kx + b) = k \cos(kx + b), \qquad \frac{d}{dx} \cos(kx + b) = -k \sin(kx + b)$$

translate to the following indefinite integral formulas:

$$\int \cos(kx + b) \, dx = \frac{1}{k} \sin(kx + b) + C$$

$$\int \sin(kx + b) \, dx = -\frac{1}{k} \cos(kx + b) + C$$

■ **EXAMPLE 4** Evaluate $\displaystyle\int \left(\sin(2t - 9) + 20 \cos 3t \right) dt$.

Solution

$$\int \left(\sin(2t - 9) + 20 \cos 3t \right) dt = \int \sin(2t - 9) \, dt + 20 \int \cos 3t \, dt$$

$$= -\frac{1}{2} \cos(2t - 9) + \frac{20}{3} \sin 3t + C \qquad\qquad ■$$

Integrals Involving e^x

The formula $(e^x)' = e^x$ says that e^x is its own derivative. But this means that e^x is also *its own antiderivative*. In other words,

$$\int e^x \, dx = e^x + C$$

More generally, for any constants b and $k \neq 0$,

$$\int e^{kx+b} \, dx = \frac{1}{k} e^{kx+b} + C$$

■ **EXAMPLE 5** Evaluate **(a)** $\displaystyle\int (3e^x - 4) \, dx$ and **(b)** $\displaystyle\int 12 e^{7-3x} \, dx$.

Solution

(a) $\int (3e^x - 4)\,dx = 3\int e^x\,dx - \int 4\,dx = 3e^x - 4x + C$

(b) $\int 12e^{7-3x}\,dx = 12\int e^{7-3x}\,dx = 12\left(\dfrac{1}{-3}e^{7-3x}\right) = -4e^{7-3x} + C$ ∎

Initial Conditions

We can think of an antiderivative as a solution to the **differential equation**

$$\frac{dy}{dx} = f(x) \qquad \boxed{2}$$

In this equation, the unknown is the function $y = y(x)$ and a solution is a function $y = F(x)$ whose derivative is $f(x)$, that is, $F(x)$ is an antiderivative of $f(x)$. The antiderivative is not unique, but we can specify a particular solution by imposing an **initial condition**, that is, by requiring that the solution satisfy a condition $y(x_0) = y_0$ for some fixed x_0 and y_0.

An initial condition is like the y-intercept of a line, which specifies that line among all lines with the same slope. Similarly, the graphs of the antiderivatives of $f(x)$ are all parallel (Figure 1), and the initial condition specifies one of them.

■ **EXAMPLE 6** Solve $\dfrac{dy}{dx} = 4x^7$ subject to the initial condition $y(0) = 4$.

Solution Since y is an antiderivative of $4x^7$, the general solution is

$$y = \int 4x^7\,dx = \frac{1}{2}x^8 + C$$

We choose C so that the initial condition $y(0) = 4$ is satisfied. Since $y(0) = 0 + C = 4$, we have $C = 4$ and the particular solution is $y = \frac{1}{2}x^8 + 4$. ∎

■ **EXAMPLE 7** Solve $\dfrac{dy}{dt} = \sin(\pi t)$ with initial condition $y(2) = 2$.

Solution The general solution is

$$y(t) = \int \sin(\pi t)\,dt = -\frac{1}{\pi}\cos(\pi t) + C$$

We solve for C by evaluating at $t = 2$:

$$y(2) = -\frac{1}{\pi}\cos(2\pi) + C = 2 \quad \Rightarrow \quad C = 2 + \frac{1}{\pi}$$

Therefore, the solution is $y(t) = -\dfrac{1}{\pi}\cos(\pi t) + 2 + \dfrac{1}{\pi}$. ∎

■ **EXAMPLE 8** At time $t = 0$, a car traveling with velocity 96 ft/s begins to slow down with constant deceleration $a = -12$ ft/s². Find the velocity $v(t)$ at time t and the distance traveled before the car comes to a halt.

Solution The derivative of velocity is acceleration, so $v'(t) = a = -12$ and (writing C_1 for the constant of integration)

$$v(t) = \int a\,dt = \int (-12)\,dt = -12t + C_1$$

The initial condition is $v(0) = C_1 = 96$, so $v(t) = -12t + 96$. Next we compute the car's position $s(t)$. Since $s'(t) = v(t)$,

$$s(t) = \int v(t)\, dt = \int (-12t + 96)\, dt = -6t^2 + 96t + C_2$$

We use the initial condition $s(0) = 0$ since we want to compute the distance traveled from time $t = 0$. This gives $C_2 = 0$ and $s(t) = -6t^2 + 96t$.

Finally, to compute how far the car traveled, we observe that the car came to a halt when its velocity was zero, so we solve:

$$v(t) = -12t + 96 = 0 \quad \Rightarrow \quad t = \frac{96}{12} = 8 \text{ s}$$

Therefore, the car came to a halt after 8 s and traveled $s(8) = -6(8^2) + 96(8) = 384$ ft.

∎

■ **EXAMPLE 9** Solve $y' = 10e^{-2x}$ with initial condition $y(0) = 12$.

Solution The general solution is

$$y = \int 10e^{-2x}\, dx = 10 \int e^{-2x}\, dx = 10\left(\frac{e^{-2x}}{-2}\right) + C = -5e^{-2x} + C$$

To determine C, we use the initial condition:

$$y(0) = -5e^0 + C = -5 + C = 12$$

Therefore, $C = 17$ and $y = -5e^{-2x} + 17$.

∎

4.9 SUMMARY

- $F(x)$ is called an *antiderivative* of $f(x)$ if $F'(x) = f(x)$.
- Any two antiderivatives of $f(x)$ on an interval (a, b) differ by a constant.
- The general antiderivative is denoted by the indefinite integral

$$\int f(x)\, dx = F(x) + C$$

- Integration formulas:

$$\int x^n\, dx = \frac{x^{n+1}}{n+1} + C \qquad (n \neq -1)$$

$$\int \sin(kx + b)\, dx = -\frac{1}{k}\cos(kx + b) + C \qquad (k \neq 0)$$

$$\int \cos(kx + b)\, dx = \frac{1}{k}\sin(kx + b) + C \qquad (k \neq 0)$$

$$\int e^{kx+b}\, dx = \frac{1}{k}e^{kx+b} + C \qquad (k \neq 0)$$

$$\int \frac{dx}{x} = \ln|x| + C$$

• To solve a differential equation $\dfrac{dy}{dx} = f(x)$ with initial condition $y(x_0) = y_0$, first find the general antiderivative $y = F(x) + C$. Then determine C using the initial condition $F(x_0) + C = y_0$.

4.9 EXERCISES

Preliminary Questions

1. Does $f(x) = x^{-1}$ have an antiderivative for $x < 0$? If so, describe one.

2. Find an antiderivative of the function $f(x) = 0$.

3. What is the difference, if any, between finding the general antiderivative of a function $f(x)$ and evaluating $\int f(x)\,dx$?

4. Jacques happens to know that $f(x)$ and $g(x)$ have the same derivative, and he would like to know if $f(x) = g(x)$. Does Jacques have sufficient information to answer his question?

5. Write any two antiderivatives of $\cos x$. Which initial conditions do they satisfy at $x = 0$?

6. Suppose that $F'(x) = f(x)$ and $G'(x) = g(x)$. Are the following statements true or false? Explain.

(a) If $f = g$, then $F = G$.

(b) If F and G differ by a constant, then $f = g$.

(c) If f and g differ by a constant, then $F = G$.

7. Determine if $y = x^2$ is a solution to the differential equation with initial condition

$$\frac{dy}{dx} = 2x, \qquad y(0) = 1$$

Exercises

In Exercises 1–8, find the general antiderivative of $f(x)$ and check your answer by differentiating.

1. $f(x) = 12x$

2. $f(x) = x^2$

3. $f(x) = x^2 + 3x + 2$

4. $f(x) = x^2 + 1$

5. $f(x) = 8x^{-4}$

6. $f(x) = \cos x + 3\sin x$

7. $f(x) = 5e^x + x^2$

8. $f(x) = 3x^3 + \frac{4}{x}$

In Exercises 9–12, match the function with its antiderivative (a)–(d).

(a) $F(x) = \cos(1 - x)$

(b) $F(x) = -\cos x$

(c) $F(x) = -\frac{1}{2}\cos(x^2)$

(d) $F(x) = \sin x - x\cos x$

9. $f(x) = \sin x$

10. $f(x) = x\sin(x^2)$

11. $f(x) = \sin(1 - x)$

12. $f(x) = x\sin x$

In Exercises 13–42, evaluate the indefinite integral.

13. $\displaystyle \int (x + 1)\,dx$

14. $\displaystyle \int (9 - 5x)\,dx$

15. $\displaystyle \int (t^5 + 3t + 2)\,dt$

16. $\displaystyle \int 8s^{-4}\,ds$

17. $\displaystyle \int t^{-9/5}\,dt$

18. $\displaystyle \int (5x^3 - x^{-2} - x^{3/5})\,dx$

19. $\displaystyle \int 2\,dx$

20. $\displaystyle \int \frac{1}{\sqrt{x}}\,dx$

21. $\displaystyle \int (5t - 9)\,dt$

22. $\displaystyle \int (x^3 + 4x^{-2})\,dx$

23. $\displaystyle \int x^{-2}\,dx$

24. $\displaystyle \int \sqrt{x}\,dx$

25. $\displaystyle \int (x + 3)^{-2}\,dx$

26. $\displaystyle \int (4t - 9)^{-3}\,dt$

27. $\displaystyle \int \frac{3}{z^5}\,dz$

28. $\displaystyle \int \frac{3}{x^{3/2}}\,dx$

29. $\displaystyle \int \sqrt{x}(x - 1)\,dx$

30. $\displaystyle \int (x + x^{-1})(3x^2 - 5x)\,dx$

31. $\displaystyle \int \frac{t - 7}{\sqrt{t}}\,dt$

32. $\displaystyle \int \frac{x^2 + 2x - 3}{x^4}\,dx$

33. $\displaystyle \int (4\sin x - 3\cos x)\,dx$

34. $\displaystyle \int \sin 9x\,dx$

35. $\displaystyle \int \cos(6t + 4)\,dt$

36. $\displaystyle \int (4\theta + \cos 8\theta)\,d\theta$

37. $\displaystyle \int \cos(3 - 4t)\,dt$

38. $\displaystyle \int 18\sin(3z + 8)\,dz$

39. $\displaystyle \int (\cos x - e^x)\,dx$

40. $\displaystyle \int \left(\frac{8}{x} + 3e^x\right)dx$

41. $\displaystyle \int 25e^{5x}\,dx$

42. $\displaystyle \int (2x + e^{14 - 2x})\,dx$

43. In Figure 3, which of (A) or (B) is the graph of an antiderivative of $f(x)$?

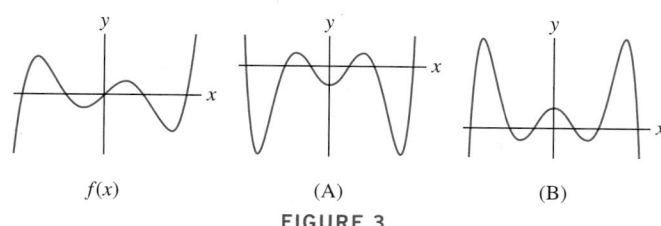

$f(x)$ (A)   (B)

FIGURE 3

44. In Figure 4, which of (A), (B), (C) is not the graph of an antiderivative of $f(x)$? Explain.

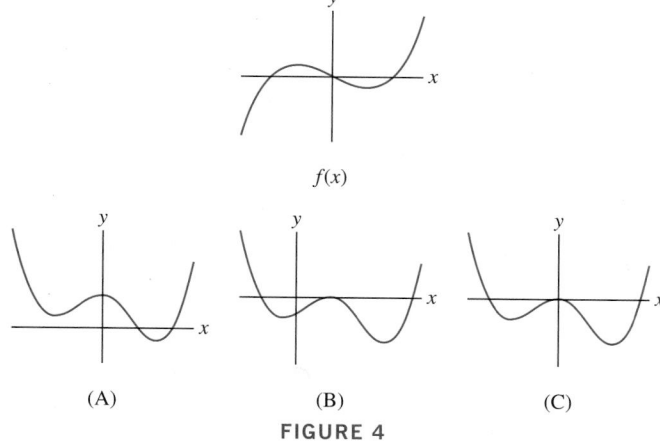

$f(x)$

(A)   (B)   (C)

FIGURE 4

45. Use the formulas for the derivatives of $f(x) = \tan x$ and $f(x) = \sec x$ to evaluate the integrals.

(a) $\displaystyle\int \sec^2(3x)\,dx$

(b) $\displaystyle\int \sec(x+3)\tan(x+3)\,dx$

46. Use the formulas for the derivatives of $f(x) = \cot x$ and $f(x) = \csc x$ to evaluate the integrals.

(a) $\displaystyle\int \csc^2 x\,dx$

(b) $\displaystyle\int \csc x \cot x\,dx$

In Exercises 47–62, solve the differential equation with initial condition.

47. $\dfrac{dy}{dx} = \cos 2x, \quad y(0) = 3$

48. $\dfrac{dy}{dx} = x^3, \quad y(0) = 2$

49. $\dfrac{dy}{dx} = x, \quad y(0) = 5$

50. $\dfrac{dy}{dt} = 0, \quad y(3) = 5$

51. $\dfrac{dy}{dt} = 5 - 2t^2, \quad y(1) = 2$

52. $\dfrac{dy}{dx} = 8x^3 + 3x^2 - 3, \quad y(1) = 1$

53. $\dfrac{dy}{dt} = 4t + 9, \quad y(0) = 1$

54. $\dfrac{dy}{dt} = \sqrt{t}, \quad y(1) = 1$

55. $\dfrac{dy}{dx} = \sin x, \quad y\left(\dfrac{\pi}{2}\right) = 1$

56. $\dfrac{dy}{dz} = \sin 2z, \quad y\left(\dfrac{\pi}{4}\right) = 4$

57. $\dfrac{dy}{dx} = \cos 5x, \quad y(\pi) = 3$

58. $\dfrac{dy}{dx} = \sec^2 3x, \quad y\left(\dfrac{\pi}{4}\right) = 2$

59. $\dfrac{dy}{dx} = e^x, \quad y(0) = 4$

60. $\dfrac{dy}{dx} = e^x, \quad y(2) = 4$

61. $\dfrac{dy}{dx} = e^{5x}, \quad y(0) = -3$

62. $\dfrac{dy}{dx} = e^{-x}, \quad y(1) = 1$

In Exercises 63–68, first find f' and then find f.

63. $f''(x) = x, \quad f'(0) = 1, \quad f(0) = 0$

64. $f''(x) = x^3 - 2x + 1, \quad f'(0) = 1, \quad f(0) = 0$

65. $f''(x) = x^3 - 2x + 1, \quad f'(1) = 0, \quad f(1) = 4$

66. $f''(t) = t^{-3/2}, \quad f'(4) = 1, \quad f(4) = 4$

67. $f''(\theta) = \cos\theta, \quad f'\left(\dfrac{\pi}{2}\right) = 1, \quad f\left(\dfrac{\pi}{2}\right) = 6$

68. $f''(t) = t - \cos t, \quad f'(0) = 2, \quad f(0) = -2$

69. Show that $f(x) = \tan^2 x$ and $g(x) = \sec^2 x$ have the same derivative. What can you conclude about the relation between f and g? Verify this conclusion directly.

70. Show, by computing derivatives, that $\sin^2 x = -\tfrac{1}{2}\cos 2x + C$ for some constant C. Find C by setting $x = 0$.

71. A particle located at the origin at $t = 0$ begins moving along the x-axis with velocity $v(t) = \tfrac{1}{2}t^2 - t$ ft/s. Let $s(t)$ be its position at time t. State the differential equation with initial condition satisfied by $s(t)$ and find $s(t)$.

72. Repeat Exercise 71, but replace the initial condition $s(0) = 0$ with $s(2) = 3$.

73. A particle moves along the x-axis with velocity $v(t) = 25t - t^2$ ft/s. Let $s(t)$ be the position at time t.

(a) Find $s(t)$, assuming that the particle is located at $x = 5$ at time $t = 0$.

(b) Find $s(t)$, assuming that the particle is located at $x = 5$ at time $t = 2$.

74. A particle located at the origin at $t = 0$ moves in a straight line with *acceleration* $a(t) = 4 - \tfrac{1}{2}t$ ft/s^2. Let $v(t)$ be the velocity and $s(t)$ the position at time t.

(a) State and solve the differential equation for $v(t)$ assuming that the particle is at rest at $t = 0$.

(b) Find $s(t)$.

75. A car traveling 84 ft/s begins to decelerate at a constant rate of 14 ft/s^2. After how many seconds does the car come to a stop and how far will the car have traveled before stopping?

---

76. Beginning at rest, an object moves in a straight line with constant acceleration a, covering 100 ft in 5 s. Find a.

77. A 900-kg rocket is released from a spacecraft. As the rocket burns fuel, its mass decreases and its velocity increases. Let $v(m)$ be the velocity (in meters per second) as a function of mass m. Find the velocity when $m = 500$ if $dv/dm = -50m^{-1/2}$. Assume that $v(900) = 0$.

78. As water flows through a tube of radius $R = 10$ cm, the velocity of an individual water particle depends on its distance r from the center

of the tube according to the formula $\dfrac{dv}{dr} = -0.06r$. Determine $v(r)$, assuming that particles at the walls of the tube have zero velocity.

79. Find constants c_1 and c_2 such that $F(x) = c_1 x \sin x + c_2 \cos x$ is an antiderivative of $f(x) = x \cos x$.

80. Find the general antiderivative of $(2x + 9)^{10}$.

81. Verify the linearity properties of the indefinite integral stated in Theorem 4.

Further Insights and Challenges

82. Suppose that $F'(x) = f(x)$ and $G'(x) = g(x)$. Is it true that $F(x)G(x)$ is an antiderivative of $f(x)g(x)$? Confirm or provide a counterexample.

83. Suppose that $F'(x) = f(x)$.
(a) Show that $\frac{1}{2}F(2x)$ is an antiderivative of $f(2x)$.
(b) Find the general antiderivative of $f(kx)$ for any constant k.

84. Find an antiderivative for $f(x) = |x|$.

85. Let $F(x) = \dfrac{x^{n+1} - 1}{n + 1}$.

(a) Show that if $n \neq -1$, then $F(x)$ is an antiderivative of $y = x^n$.
(b) Use L'Hôpital's Rule to prove that

$$\lim_{n \to -1} F(x) = \ln x$$

In this limit, x is fixed and we treat n as a variable tending to -1. This shows that although the Power Rule breaks down for $n = -1$, we may view the antiderivative of $y = x^{-1}$ as a limit of antiderivatives of x^n as $n \to -1$.

CHAPTER REVIEW EXERCISES

In Exercises 1–6, estimate using the Linear Approximation or linearization and use a calculator to compute the error.

1. $8.1^{1/3} - 2$

2. $\dfrac{1}{\sqrt{4.1}} - \dfrac{1}{2}$

3. $625^{1/4} - 624^{1/4}$

4. $\sqrt{101}$

5. $\dfrac{1}{1.02}$

6. $\sqrt[5]{33}$

In Exercises 7–12, find the linearization at the point indicated.

7. $y = \sqrt{x}, \quad a = 25$

8. $v(t) = 32t - 4t^2, \quad a = 2$

9. $A(r) = \frac{4}{3}\pi r^3, \quad a = 3$

10. $V(h) = 4h(2 - h)(4 - 2h), \quad a = 1$

11. $P(x) = e^{-x^2/2}, \quad a = 1$

12. $f(x) = \ln(x + e), \quad a = e$

In Exercises 13–17, use the Linear Approximation.

13. The position of an object in linear motion at time t is $s(t) = 0.4t^2 + (t + 1)^{-1}$. Estimate the distance traveled over the time interval $[4, 4.2]$.

14. A bond that pays \$10,000 in 6 years is offered for sale at a price P. The percentage yield Y of the bond is

$$Y = 100\left(\left(\frac{10{,}000}{P}\right)^{1/6} - 1\right)$$

Verify that if $P = \$7{,}500$, then $Y = 4.91\%$. Estimate the drop in yield if the price rises to \$7,700.

15. A store sells 80 MP3 players per week when the players are priced at $P = \$75$. Estimate the number N sold if P is raised to \$80, assuming that $dN/dP = -4$. Estimate N if the price is lowered to \$69.

16. The circumference of a sphere is measured at $C = 100$ cm. Estimate the maximum percentage error in V if the error in C is at most 3 cm.

17. Show that $\sqrt{a^2 + b} \approx a + \frac{b}{2a}$ if b is small. Use this to estimate $\sqrt{26}$ and find the error using a calculator.

18. Use the Intermediate Value Theorem to prove that $\sin x - \cos x = 3x$ has a solution and use Rolle's Theorem to show that this solution is unique.

19. Show that $f(x) = 2x^3 + 2x + \sin x + 1$ has precisely one real root.

20. Verify the MVT for $f(x) = \ln x$ on $[1, 4]$.

21. Suppose that $f(1) = 5$ and $f'(x) \geq 2$ for $x \geq 1$. Use the MVT to show that $f(8) - f(1) \geq 19$.

22. Use the MVT to prove that if $f'(x) \le 2$ for $x > 0$ and $f(0) = 4$, then $f(x) \le 2x + 4$ for all $x \ge 0$.

In Exercises 23–26, find the local extrema and determine whether they are minima, maxima, or neither.

23. $f(x) = x^3 - 4x^2 + 4x$

24. $s(t) = t^4 - 8t^2$

25. $f(x) = x^2(x + 2)^3$

26. $f(x) = x^{2/3}(1 - x)$

In Exercises 27–34, find the extreme values on the interval.

27. $f(x) = x(10 - x)$, $[-1, 3]$

28. $f(x) = 6x^4 - 4x^6$, $[-2, 2]$

29. $g(\theta) = \sin^2 \theta - \cos \theta$, $[0, 2\pi]$

30. $R(t) = \dfrac{t}{t^2 + t + 1}$, $[0, 3]$

31. $f(x) = x^{2/3} - 2x^{1/3}$, $[-1, 3]$

32. $f(x) = x - \tan x$, $[-1, 1]$

33. $f(x) = x - 12 \ln x$, $[5, 40]$

34. $f(x) = e^x - 20x - 1$, $[0, 5]$

35. Find the critical points and extreme values of $f(x) = |x - 1| + |2x - 6|$ in $[0, 8]$.

36. A function $f(x)$ has derivative $f'(x) = \dfrac{1}{x^4 + 1}$. Where on the interval $[1, 4]$ does $f(x)$ take on its maximum value?

In Exercises 37–42, find the points of inflection.

37. $y = x^3 - 4x^2 + 4x$

38. $y = x - 2 \cos x$

39. $y = \dfrac{x^2}{x^2 + 4}$

40. $y = \dfrac{x}{(x^2 - 4)^{1/3}}$

41. $f(x) = (x^2 - x)e^{-x}$

42. $f(x) = x(\ln x)^2$

43. Match the description of $f(x)$ with the graph of its derivative $f'(x)$ in Figure 1.

(a) $f(x)$ is increasing and concave up.

(b) $f(x)$ is decreasing and concave up.

(c) $f(x)$ is increasing and concave down.

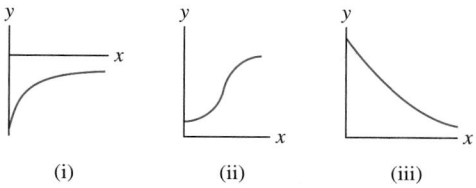

(i) (ii) (iii)

FIGURE 1 Graphs of the derivative.

44. Draw a curve $y = f(x)$ for which f' and f'' have signs as indicated in Figure 2.

FIGURE 2

In Exercises 45–60, evaluate the limit.

45. $\lim\limits_{x \to \infty} (9x^4 - 12x^3)$

46. $\lim\limits_{x \to -\infty} (7x^2 - 9x^5)$

47. $\lim\limits_{x \to \infty} \dfrac{x^3 + 2x}{4x^2 - 9}$

48. $\lim\limits_{x \to -\infty} \dfrac{x^3 + 2x}{4x^2 - 9}$

49. $\lim\limits_{x \to \infty} \dfrac{x^3 + 2x}{4x^3 - 9}$

50. $\lim\limits_{x \to -\infty} \dfrac{x^3 + 2x}{4x^5 - 40x^3}$

51. $\lim\limits_{x \to \infty} \dfrac{x^2 - 9x}{\sqrt{x^2 + 4x}}$

52. $\lim\limits_{x \to -\infty} \dfrac{12x + 1}{\sqrt{4x^2 + 4x}}$

53. $\lim\limits_{x \to \infty} \dfrac{x^{1/2}}{\sqrt{4x - 9}}$

54. $\lim\limits_{x \to \infty} \dfrac{x^{3/2}}{(16x^6 - 9x^4)^{1/4}}$

55. $\lim\limits_{x \to 3+} \dfrac{1 - 2x}{x - 3}$

56. $\lim\limits_{x \to 2-} \dfrac{1}{x^2 - 4}$

57. $\lim\limits_{x \to 2+} \dfrac{x - 5}{x - 2}$

58. $\lim\limits_{x \to -3+} \dfrac{1}{(x + 3)^3}$

59. $\lim\limits_{x \to \infty} x^2 e^{-x^2}$

60. $\lim\limits_{x \to 1} \dfrac{x^2 - 1}{e^{1-x^2} - 1}$

In Exercises 61–70, sketch the graph, noting the transition points and asymptotic behavior.

61. $y = 12x - 3x^2$

62. $y = 8x^2 - x^4$

63. $y = x^3 - 2x^2 + 3$

64. $y = 4x - x^{3/2}$

65. $y = \dfrac{x}{x^3 + 1}$

66. $y = \dfrac{x}{(x^2 - 4)^{2/3}}$

67. $y = \dfrac{1}{|x + 2| + 1}$

68. $y = \sqrt{2 - x^3}$

69. $y = 2 \sin x - \cos x$ on $[0, 2\pi]$

70. $y = 2x - \tan x$ on $[0, 2\pi]$

71. Find the maximum volume of a right-circular cone placed upside-down in a right-circular cone of radius R and height H (Figure 3). The volume of a cone of radius r and height h is $\frac{1}{3}\pi r^2 h$.

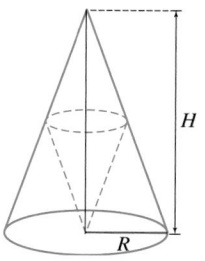

FIGURE 3

72. On a certain farm, the corn yield is

$$Y = -0.118x^2 + 8.5x + 12.9 \quad \text{(bushels per acre)}$$

where x is the number of corn plants per acre (in thousands). Assume that corn seed costs $1.25 (per thousand seeds) and that corn can be sold for $1.50/bushel.

(a) Find the value x_0 that maximizes yield Y. Then compute the profit (revenue minus the cost of seeds) at planting level x_0.

(b) Compute the profit $P(x)$ as a function of x and find the value x_1 that maximizes profit.

(c) Compare the profit at levels x_0 and x_1. Does a maximum yield lead to maximum profit?

73. Show that the maximum area of a parallelogram ADEF inscribed in a triangle $\triangle ABC$, as in Figure 4, is equal to one-half the area of $\triangle ABC$.

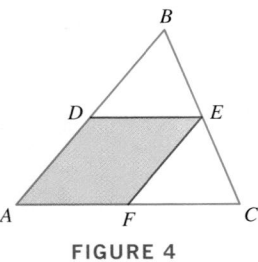

FIGURE 4

74. The area of the shaded region in Figure 5 is $\frac{1}{2}r^2(\theta - \sin\theta)$. What is the maximum possible area of this region if the length of the circular arc is 1?

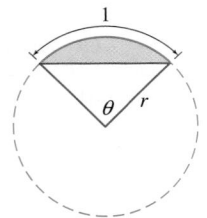

FIGURE 5

75. Let $f(x)$ be a function whose graph does not pass through the x-axis and let $Q = (a, 0)$. Let $P = (x_0, f(x_0))$ be the point on the graph closest to Q (Figure 6). Prove that $\overline{PQ}$ is perpendicular to the

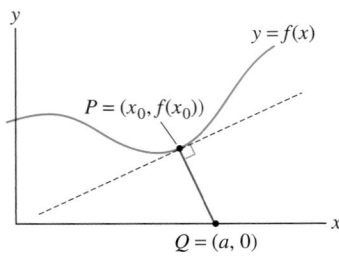

FIGURE 6

tangent line to the graph of x_0. *Hint:* Let $q(x)$ be the distance from $(x, f(x))$ to $(a, 0)$ and observe that x_0 is a critical point of $q(x)$.

76. Take a circular piece of paper of radius R, remove a sector of angle θ (Figure 7), and fold the remaining piece into a cone-shaped cup. Which angle θ produces the cup of largest volume?

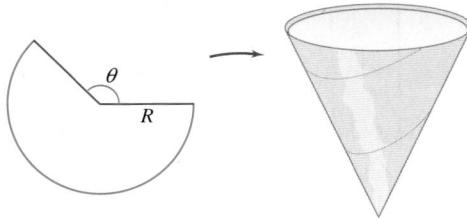

FIGURE 7

77. Use Newton's Method to estimate $\sqrt[3]{25}$ to four decimal places.

78. Use Newton's Method to find a root of $f(x) = x^2 - x - 1$ to four decimal places.

In Exercises 79–92, calculate the indefinite integral.

79. $\displaystyle\int \left(4x^3 - 2x^2\right) dx$

80. $\displaystyle\int x^{9/4}\, dx$

81. $\displaystyle\int \sin(\theta - 8)\, d\theta$

82. $\displaystyle\int \cos(5 - 7\theta)\, d\theta$

83. $\displaystyle\int \left(4t^{-3} - 12t^{-4}\right) dt$

84. $\displaystyle\int \left(9t^{-2/3} + 4t^{7/3}\right) dt$

85. $\displaystyle\int \sec^2 x\, dx$

86. $\displaystyle\int \tan 3\theta \sec 3\theta\, d\theta$

87. $\displaystyle\int (y + 2)^4\, dy$

88. $\displaystyle\int \frac{3x^3 - 9}{x^2}\, dx$

89. $\displaystyle\int (e^x - x)\, dx$

90. $\displaystyle\int e^{-4x}\, dx$

91. $\displaystyle\int 4x^{-1}\, dx$

92. $\displaystyle\int \sin(4x - 9)\, dx$

In Exercises 93–98, solve the differential equation with initial condition.

93. $\dfrac{dy}{dx} = 4x^3, \quad y(1) = 4$

94. $\dfrac{dy}{dt} = 3t^2 + \cos t, \quad y(0) = 12$

95. $\dfrac{dy}{dx} = x^{-1/2}, \quad y(1) = 1$

96. $\dfrac{dy}{dx} = \sec^2 x, \quad y(\frac{\pi}{4}) = 2$

97. $\dfrac{dy}{dx} = e^{-x}, \quad y(0) = 3$

98. $\dfrac{dy}{dx} = e^{4x}, \quad y(1) = 1$

99. Find $f(t)$, assuming that $f''(t) = 1 - 2t$, $f(0) = 2$, and $f'(0) = -1$.

100. The driver of an automobile applies the brakes at time $t = 0$ and comes to a halt after traveling 500 ft. Find the automobile's velocity at $t = 0$ assuming that the rate of deceleration was a constant -10 ft/s^2.

101. Find the local extrema of $f(x) = \dfrac{e^{2x} + 1}{e^{x+1}}$.

102. Find the points of inflection of $f(x) = \ln(x^2 + 1)$ and determine whether the concavity changes from up to down or vice versa.

In Exercises 103–106, find the local extrema and points of inflection, and sketch the graph. Use L'Hôpital's Rule to determine the limits as $x \to 0+$ or $x \to \pm\infty$ if necessary.

103. $y = x \ln x, \quad x > 0$

104. $y = e^{x - x^2}$

105. $y = x(\ln x)^2, \quad x > 0$

106. $y = \tan^{-1}\left(\dfrac{x^2}{4}\right)$

107. Explain why L'Hôpital's Rule gives no information about $\displaystyle\lim_{x \to \infty} \dfrac{2x - \sin x}{3x + \cos 2x}$. Evaluate the limit by another method.

108. Let $f(x)$ be a differentiable function with inverse $g(x)$ such that $f(0) = 0$ and $f'(0) \neq 0$. Prove that

$$\lim_{x \to 0} \dfrac{f(x)}{g(x)} = f'(0)^2$$

In Exercises 109–120, verify that L'Hôpital's Rule applies and evaluate the limit.

109. $\displaystyle\lim_{x \to 3} \dfrac{4x - 12}{x^2 - 5x + 6}$

110. $\displaystyle\lim_{x \to -2} \dfrac{x^3 + 2x^2 - x - 2}{x^4 + 2x^3 - 4x - 8}$

111. $\displaystyle\lim_{x \to 0+} x^{1/2} \ln x$

112. $\displaystyle\lim_{t \to \infty} \dfrac{\ln(e^t + 1)}{t}$

113. $\displaystyle\lim_{\theta \to 0} \dfrac{2\sin\theta - \sin 2\theta}{\sin\theta - \theta\cos\theta}$

114. $\displaystyle\lim_{x \to 0} \dfrac{\sqrt{4+x} - 2\sqrt[8]{1+x}}{x^2}$

115. $\displaystyle\lim_{t \to \infty} \dfrac{\ln(t+2)}{\log_2 t}$

116. $\displaystyle\lim_{x \to 0} \left(\dfrac{e^x}{e^x - 1} - \dfrac{1}{x}\right)$

117. $\displaystyle\lim_{y \to 0} \dfrac{\sin^{-1} y - y}{y^3}$

118. $\displaystyle\lim_{x \to 1} \dfrac{\sqrt{1-x^2}}{\cos^{-1} x}$

119. $\displaystyle\lim_{x \to 0} \dfrac{\sinh(x^2)}{\cosh x - 1}$

120. $\displaystyle\lim_{x \to 0} \dfrac{\tanh x - \sinh x}{\sin x - x}$

121. Let $f(x) = e^{-Ax^2/2}$, where $A > 0$. Given any n numbers $a_1, a_2, \ldots, a_n$, set

$$\Phi(x) = f(x - a_1)f(x - a_2)\cdots f(x - a_n)$$

(a) Assume $n = 2$ and prove that $\Phi(x)$ attains its maximum value at the average $x = \frac{1}{2}(a_1 + a_2)$. *Hint:* Show that $\dfrac{d}{dx}\ln(f(x)) = -Ax$ and calculate $\Phi'(x)$ using logarithmic differentiation.

(b) Show that for any n, $\Phi(x)$ attains its maximum value at $x = \frac{1}{n}(a_1 + a_2 + \cdots + a_n)$. This fact is related to the role of $f(x)$ (whose graph is a bell-shaped curve) in statistics.

5 | THE INTEGRAL

The starting point in integral calculus is the problem of finding the area under a curve. You may wonder why calculus is concerned with two seemingly unrelated topics such as areas and tangent lines. One reason is that both are computed using limits. However, another deeper connection between areas and tangent lines is revealed by the Fundamental Theorem of Calculus, discussed in Sections 5.3 and 5.4. This theorem expresses the fundamental "inverse" relationship between integration and differentiation that plays a key role in almost all applications of calculus, both theoretical and practical.

Integration solves an ancient mathematical problem—finding the area of an irregular region.

5.1 Approximating and Computing Area

We begin by discussing a basic example in which the area under a curve plays a role. If an object moves in a straight line with *constant velocity* v (assumed to be positive), then the distance traveled over a time interval $[t_1, t_2]$ is equal to $v(t_2 - t_1)$. This is the well-known formula

$$\boxed{\text{Distance traveled} = \text{velocity} \times \text{time}} \qquad \boxed{1}$$

We may interpret this formula in terms of area by observing that the graph of velocity is a horizontal line at height v, and $v(t_2 - t_1)$ is the area of the rectangular region under the graph lying over the interval $[t_1, t_2]$ (Figure 1). Thus,

$$\boxed{\text{Distance traveled during } [t_1, t_2] = \text{area under the graph of velocity over } [t_1, t_2]}$$

$$\boxed{2}$$

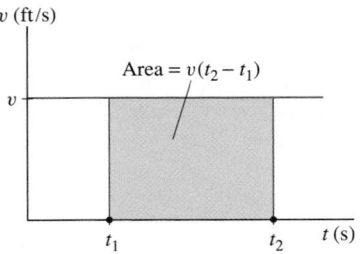

FIGURE 1 Distance traveled equals the area of the rectangle.

Although Eqs. (1) and (2) are equivalent, the advantage of Eq. (2) is that it remains true, even if the velocity is not constant (Figure 2). This is discussed in Section 5.5, but we can understand the main idea by considering the graph of velocity in Figure 3. In this case, the velocity changes over time but is constant on intervals (of course, this graph is not realistic because an object's velocity cannot change so abruptly). The distance traveled over each time interval is equal to the area of the rectangle above that interval, and the total distance traveled is the sum of the areas of the rectangles:

$$\text{Distance traveled over } [0, 8] = \underbrace{10 + 15 + 30 + 10}_{\text{Sum of areas of rectangles}} = 65 \text{ ft}$$

This reasoning shows more generally that distance traveled is equal to the area under the graph whenever the velocity is constant on intervals. This argument does not work when the velocity changes continuously as in Figure 2. Our strategy in this case is to *approximate* the area under the graph by sums of areas of rectangles and then pass to a limit.

Approximating Area by Rectangles

For the rest of this section, we assume that $f(x)$ is continuous and *positive*, so that the graph of $f(x)$ lies above the x-axis. Our goal is to compute the area "under the graph,"

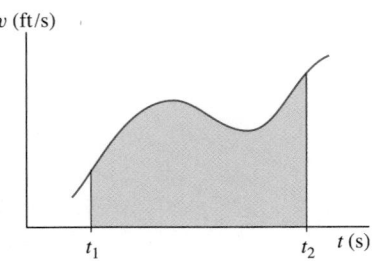

FIGURE 2 Distance traveled is the area under the graph of velocity.

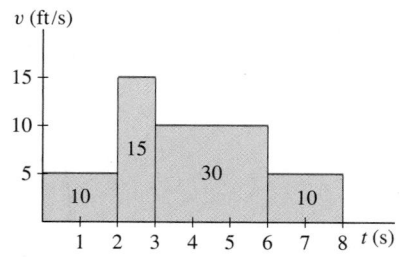

FIGURE 3 Distance traveled is the sum of the areas of the rectangles.

that is, the area between the graph and the x-axis. As a first step, we approximate the area using rectangles.

First, choose a whole number N and divide $[a, b]$ into N subintervals of equal width, as in Figure 4. Each subinterval has width $\Delta x = \dfrac{b - a}{N}$ since $[a, b]$ has width $b - a$. The right endpoints of the subintervals are

$$a + \Delta x, \ a + 2\Delta x, \ \ldots, \ a + (N - 1)\Delta x, \ a + N\Delta x$$

Notice that the last right endpoint is b because $a + N\Delta x = a + N\left(\dfrac{b - a}{N}\right) = b$. Next, above each subinterval, construct the rectangle whose height is the value of $f(x)$ at the *right endpoint* of the subinterval.

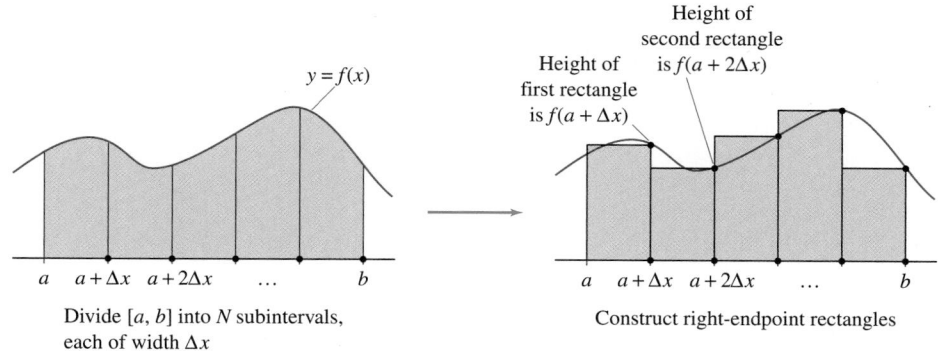

FIGURE 4 First subdivide the interval, then construct the rectangles.

Divide $[a, b]$ into N subintervals, each of width Δx

Construct right-endpoint rectangles

Each rectangle has width Δx. The height of the first rectangle is $f(a + \Delta x)$ and its area is $f(a + \Delta x)\Delta x$. Similarly, the second rectangle has height $f(a + 2\Delta x)$ and area $f(a + 2\Delta x)\,\Delta x$, etc. The sum of the areas of the rectangles is

$$f(a + \Delta x)\Delta x + f(a + 2\Delta x)\Delta x + \cdots + f(a + N\Delta x)\Delta x$$

This sum is called the *N***th right-endpoint approximation** and is denoted R_N. Factoring out Δx, we may write R_N as

$$R_N = \Delta x \left(f(a + \Delta x) + f(a + 2\Delta x) + \cdots + f(a + N\Delta x) \right)$$

In words, R_N is equal to Δx times the sum of the function values at the right endpoints of the subintervals.

To summarize,

   $a = $ *left endpoint of interval* $[a, b]$

   $b = $ *right endpoint of interval* $[a, b]$

   $N = $ *number of subintervals in* $[a, b]$

   $\Delta x = \dfrac{b - a}{N}$

■ **EXAMPLE 1** Computing Right-Endpoint Approximations Calculate R_4 and R_6 for $f(x) = x^2$ on the interval $[1, 3]$.

Solution

Step 1. **Determine Δx and the endpoints.**

To calculate R_4, we divide $[1, 3]$ into four subintervals of width $\Delta x = \frac{3-1}{4} = \frac{1}{2}$. The right endpoints are the numbers $a + j\Delta x = 1 + j(\frac{1}{2})$ for $j = 1, 2, 3, 4$. They are spaced at intervals of $\frac{1}{2}$ beginning at $\frac{3}{2}$, so as we see in Figure 5, the right endpoints are $\frac{3}{2}, \frac{4}{2}, \frac{5}{2}, \frac{6}{2}$.

Step 2. **Sum the function values.**

R_4 is Δx times the sum of the values of $f(x) = x^2$ at the right endpoints:

$$R_4 = \frac{1}{2}\left(f\left(\frac{3}{2}\right) + f\left(\frac{4}{2}\right) + f\left(\frac{5}{2}\right) + f\left(\frac{6}{2}\right)\right)$$

$$= \frac{1}{2}\left(\frac{9}{4} + \frac{16}{4} + \frac{25}{4} + \frac{36}{4}\right) = \frac{43}{4} = 10.75$$

The calculation of R_6 is similar: $\Delta x = \frac{3-1}{6} = \frac{1}{3}$, and the right endpoints are spaced at intervals of $\frac{1}{3}$ beginning at $\frac{4}{3}$ and ending at 3, as in Figure 5. Thus,

$$R_6 = \frac{1}{3}\left(f\left(\frac{4}{3}\right) + f\left(\frac{5}{3}\right) + f\left(\frac{6}{3}\right) + f\left(\frac{7}{3}\right) + f\left(\frac{8}{3}\right) + f\left(\frac{9}{3}\right)\right)$$

$$= \frac{1}{3}\left(\frac{16}{9} + \frac{25}{9} + \frac{36}{9} + \frac{49}{9} + \frac{64}{9} + \frac{81}{9}\right) = \frac{271}{27} \approx 10.037 \qquad \blacksquare$$

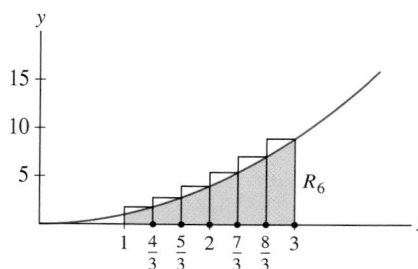

FIGURE 5 Approximations to the area under the graph of $f(x) = x^2$ over $[1, 3]$.

Summation Notation

Since approximations to area involve sums of function values, we introduce **summation notation**, which is a standard way of writing sums in compact form. The sum of numbers $a_m, \ldots, a_n$ $(m \le n)$ is denoted

$$\sum_{j=m}^{n} a_j = a_m + a_{m+1} + \cdots + a_n$$

The capital Greek letter $\sum$ (sigma) stands for "sum" and the notation $\displaystyle\sum_{j=m}^{n}$ tells us to start the summation at $j = m$ and end it at $j = n$. For example,

$$\sum_{j=1}^{5} j^2 = 1^2 + 2^2 + 3^2 + 4^2 + 5^2 = 55$$

In this summation, the jth term is given by the formula $a_j = j^2$. We call j^2 the **general term**. The letter j plays the role of index and is sometimes called a **dummy variable** because any other letter can be used instead. For example,

$$\sum_{j=4}^{6}(j^3 - 2j) = \sum_{k=4}^{6}(k^3 - 2k) = \left(4^3 - 2(4)\right) + \left(5^3 - 2(5)\right) + \left(6^3 - 2(6)\right) = 375$$

The following rules are used to manipulate summations. They are consequences of the usual commutative, associative, and distributive laws of addition.

Linearity of Summations

- $$\sum_{j=m}^{n}(a_j + b_j) = \sum_{j=m}^{n} a_j + \sum_{j=m}^{n} b_j$$

- $$\sum_{j=m}^{n} Ca_j = C \sum_{j=m}^{n} a_j \qquad (C \text{ any constant})$$

- $$\sum_{j=1}^{n} k = nk \qquad (k \text{ any constant and } n \geq 1)$$

For example,

$$\sum_{i=3}^{5}(i^2 + i) = \overbrace{(3^2 + 3)}^{i=3} + \overbrace{(4^2 + 4)}^{i=4} + \overbrace{(5^2 + 5)}^{i=5}$$

is equal to

$$\sum_{i=3}^{5} i^2 + \sum_{i=3}^{5} i = \left(3^2 + 4^2 + 5^2\right) + \left(3 + 4 + 5\right)$$

We often use linearity to write a single summation as a sum of several summations. For example,

$$\sum_{k=0}^{100}(7k^2 - 4k + 9) = \sum_{k=0}^{100} 7k^2 + \sum_{k=0}^{100}(-4k) + \sum_{k=0}^{100} 9$$

$$= 7\sum_{k=0}^{100} k^2 - 4\sum_{k=0}^{100} k + 9\sum_{k=0}^{100} 1$$

It is convenient to write R_N in summation notation. By definition,

$$R_N = \Delta x \big[f(a + \Delta x) + f(a + 2\Delta x) + \cdots + f(a + N\Delta x) \big]$$

The general term is $f(a + j\Delta x)$ and the summation extends from $j = 1$ to $j = N$, so

$$R_N = \Delta x \sum_{j=1}^{N} f(a + j\Delta x)$$

We now discuss two other approximations to area: the left-endpoint and the midpoint approximations. To calculate the **left-endpoint approximation**, we divide $[a, b]$ into N subintervals as before, but the heights of the rectangles are the values of $f(x)$ at the left endpoints [Figure 6(A)]. The left endpoints are

$$a, \ a + \Delta x, \ a + 2\Delta x, \ldots, \ a + (N - 1)\Delta x$$

and the sum of the areas of the left-endpoint rectangles is

$$L_N = \Delta x \big(f(a) + f(a + \Delta x) + f(a + 2\Delta x) + \cdots + f(a + (N - 1)\Delta x) \big)$$

Note that both R_N and L_N have general term $f(a + j\Delta x)$, but the sum for L_N runs from $j = 0$ to $j = N - 1$ rather than from $j = 1$ to $j = N$. Thus,

$$L_N = \Delta x \sum_{j=0}^{N-1} f(a + j\Delta x)$$

In the **midpoint approximation** M_N, the heights of the rectangles are the values of $f(x)$ at the midpoints of the subintervals rather than at the endpoints [Figure 6(B)]. The midpoints are

$$a + \frac{1}{2}\Delta x, \ a + \frac{3}{2}\Delta x, \ \ldots, \ a + \left(N - \frac{1}{2}\right)\Delta x$$

and the sum of the areas of the midpoint rectangles is

$$M_N = \Delta x \left(f\left(a + \frac{1}{2}\Delta x\right) + f\left(a + \frac{3}{2}\Delta x\right) + \cdots + f\left(a + \left(N - \frac{1}{2}\right)\Delta x\right) \right)$$

In summation notation,

$$M_N = \Delta x \sum_{j=1}^{N} f\left(a + \left(j - \frac{1}{2}\right)\Delta x\right)$$

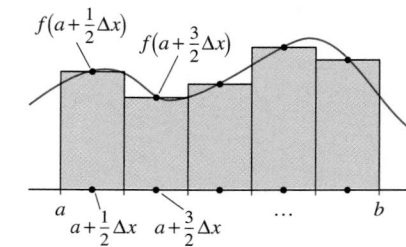

(A) Left-endpoint rectangles

(B) Midpoint rectangles

FIGURE 6 Left-endpoint and midpoint approximations.

■ **EXAMPLE 2** Calculate R_6, L_6, and M_6 for $f(x) = x^{-1}$ on [2, 4].

Solution In this case, $a = 2$, $b = 4$, and $\Delta x = \frac{4-2}{6} = \frac{1}{3}$. To compute R_6, let's evaluate the general term in the sum representing R_6:

$$f(a + j\Delta x) = f\left(2 + j\left(\frac{1}{3}\right)\right) = \frac{1}{2 + \frac{j}{3}}$$

Therefore,

$$R_6 = \frac{1}{3} \sum_{j=1}^{6} f\left(2 + \left(\frac{1}{3}\right)j\right) = \frac{1}{3} \sum_{j=1}^{6} \frac{1}{2 + \frac{j}{3}}$$

$$= \frac{1}{3}\left(\frac{1}{2 + \frac{1}{3}} + \frac{1}{2 + \frac{2}{3}} + \frac{1}{2 + \frac{3}{3}} + \frac{1}{2 + \frac{4}{3}} + \frac{1}{2 + \frac{5}{3}} + \frac{1}{2 + \frac{6}{3}}\right)$$

$$= \frac{1}{3}\left(\frac{3}{7} + \frac{3}{8} + \frac{3}{9} + \frac{3}{10} + \frac{3}{11} + \frac{3}{12}\right) \approx 0.653$$

The left-endpoint approximation sum is nearly the same, but the sum begins with $j = 0$ and ends at $j = 5$:

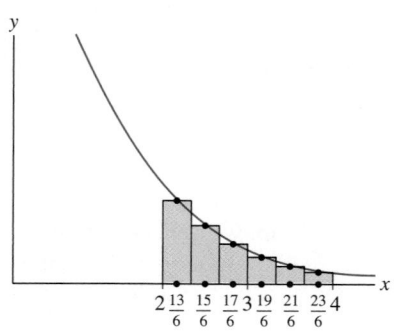

FIGURE 7 M_6 for $f(x) = x^{-1}$ on [2, 4].

$$L_6 = \frac{1}{3} \sum_{j=0}^{5} \frac{1}{2 + \frac{j}{3}} = \frac{1}{3}\left(\frac{3}{6} + \frac{3}{7} + \frac{3}{8} + \frac{3}{9} + \frac{3}{10} + \frac{3}{11}\right) \approx 0.737$$

To compute M_6, we note that the general term is

$$f\left(a + \left(j - \frac{1}{2}\right)\Delta x\right) = f\left(2 + \left(j - \frac{1}{2}\right)\frac{1}{3}\right) = \frac{1}{2 + \frac{j}{3} - \frac{1}{6}}$$

Summing up from $j = 1$ to 6, we obtain (Figure 7)

$$M_6 = \frac{1}{3} \sum_{j=1}^{6} f\left(2 + \left(j - \frac{1}{2}\right)\frac{1}{3}\right) = \frac{1}{3}\sum_{j=1}^{6} \frac{1}{2 + \frac{j}{3} - \frac{1}{6}}$$

$$= \frac{1}{3}\left(\frac{1}{2 + \frac{1}{6}} + \frac{1}{2 + \frac{3}{6}} + \frac{1}{2 + \frac{5}{6}} + \frac{1}{2 + \frac{7}{6}} + \frac{1}{2 + \frac{9}{6}} + \frac{1}{2 + \frac{11}{6}}\right)$$

$$= \frac{1}{3}\left(\frac{6}{13} + \frac{6}{15} + \frac{6}{17} + \frac{6}{19} + \frac{6}{21} + \frac{6}{23}\right) \approx 0.692 \qquad \blacksquare$$

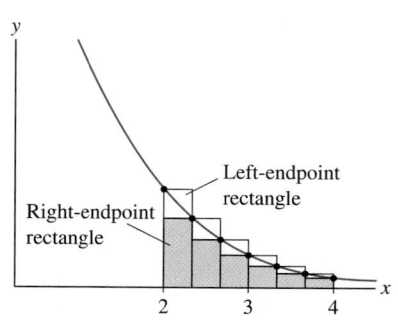

FIGURE 8 L_6 and R_6 for $f(x) = x^{-1}$ on $[2, 4]$.

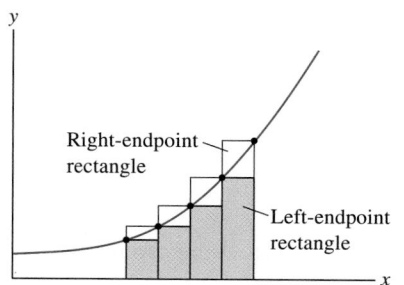

FIGURE 9 When $f(x)$ is increasing, the left-endpoint rectangles lie below the graph and right-endpoint rectangles lie above it.

GRAPHICAL INSIGHT **Monotonic Functions**   Observe in Figure 8 that the left-endpoint rectangles for $f(x) = x^{-1}$ extend above the graph and the right-endpoint rectangles lie below it. Therefore, the exact area A lies between R_6 and L_6 and so, according to the previous example, $0.65 \le A \le 0.74$. More generally, *when $f(x)$ is monotonic (increasing or decreasing), the exact area lies between R_N and L_N* (Figures 8 and 9):

- $f(x)$ increasing $\Rightarrow$ $L_N \le$ area under graph $\le R_N$
- $f(x)$ decreasing $\Rightarrow$ $R_N \le$ area under graph $\le L_N$

Computing Area as the Limit of Approximations

What can be said about the accuracy of our approximations to the area under a graph? Figure 10 shows the error for several right-endpoint approximations as the yellow region above the graph. We see that the error gets smaller as the number of rectangles increases. Furthermore, it appears that *we can achieve arbitrarily high accuracy by taking the number N of rectangles large enough.* If so, it makes sense to consider the limit as $N \to \infty$, which should give us the exact area under the curve. The following theorem guarantees that the limit exists (see Exercise 91 for a proof for monotonic functions).

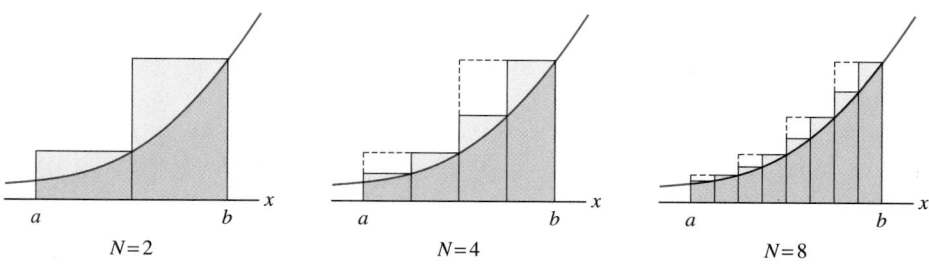

FIGURE 10 The error decreases as we use more rectangles.

THEOREM 1   If $f(x)$ is continuous on $[a, b]$, then the endpoint and midpoint approximations approach one and the same limit as $N \to \infty$. In other words, there is a value L such that

$$\lim_{N \to \infty} R_N = \lim_{N \to \infty} L_N = \lim_{N \to \infty} M_N = L$$

Theorem 1 gives us a precise definition of the area under a graph as a limit L. Recall that the tangent line is another basic concept which is defined as a limit. When we studied the derivative, we found that the slope of the tangent line is the limit of slopes of secant lines.

In the next two examples, we illustrate Theorem 1 by finding the area under the graphs of some simple polynomials. These calculations rely on the formulas for the **power sums**. The following are the first three power sum formulas.

DEFINITION Power Sums   The kth power sum is the sum of the kth powers of the first N integers.

$$\sum_{j=1}^{N} j = 1 + 2 + \cdots + N = \frac{N(N+1)}{2} = \frac{N^2}{2} + \frac{N}{2} \qquad \boxed{3}$$

$$\sum_{j=1}^{N} j^2 = 1^2 + 2^2 + \cdots + N^2 = \frac{N(N+1)(2N+1)}{6} = \frac{N^3}{3} + \frac{N^2}{2} + \frac{N}{6} \qquad \boxed{4}$$

$$\sum_{j=1}^{N} j^3 = 1^3 + 2^3 + \cdots + N^3 = \frac{N^2(N+1)^2}{4} = \frac{N^4}{4} + \frac{N^3}{2} + \frac{N^2}{4} \qquad \boxed{5}$$

For example, by Eq. (4),

$$\sum_{j=1}^{6} j^2 = 1^2 + 2^2 + 3^2 + 4^2 + 5^2 + 6^2 = \underbrace{\frac{6^3}{3} + \frac{6^2}{2} + \frac{6}{6}}_{\frac{N^3}{3} + \frac{N^2}{2} + \frac{N}{6} \text{ for } N=6} = 91$$

As a first illustration, we compute the area of a right triangle "the hard way."

■ **EXAMPLE 3**   Calculating Area as a Limit   Calculate the area under the graph of $f(x) = x$ over $[0, 4]$ in three ways:

(a) $\lim\limits_{N \to \infty} R_N$ **(b)** $\lim\limits_{N \to \infty} L_N$ **(c)** Using geometry

Solution   The interval is $[a, b] = [0, 4]$, so $\Delta x = \dfrac{b-a}{N} = \dfrac{4}{N}$, and since $f(x) = x$,

$$f(a + j\Delta x) = f\left(0 + \frac{4j}{N}\right) = \frac{4j}{N}$$

Therefore,

$$R_N = \Delta x \sum_{j=1}^{N} f(a + j\Delta x) = \frac{4}{N} \sum_{j=1}^{N} \frac{4j}{N} = \frac{16}{N^2} \sum_{j=1}^{N} j$$

In the last equality, we factor out $\dfrac{4}{N}$ from the sum. This is valid because $\dfrac{4}{N}$ is a constant that does not depend on j. Now use formula (3) for $\sum\limits_{j=1}^{N} j$ to obtain

$$R_N = \frac{16}{N^2} \sum_{j=1}^{N} j = \frac{16}{N^2} \underbrace{\left(\frac{N^2}{2} + \frac{N}{2}\right)}_{\text{Formula for power sum}} = 8 + \frac{8}{N} \qquad \boxed{6}$$

The second term $\dfrac{8}{N}$ tends to zero as N approaches ∞, so the limit is equal to

$$\text{Area} = \lim_{N \to \infty} R_N = \lim_{N \to \infty} \left(8 + \frac{8}{N}\right) = 8$$

The left-endpoint approximation is nearly the same as Eq. (6), but the sum begins at $j = 0$ and ends at $j = N - 1$ (we leave the algebra of the last step to you):

$$L_N = \frac{16}{N^2} \sum_{j=0}^{N-1} j = \frac{16}{N^2} \sum_{j=1}^{N-1} j = \frac{16}{N^2}\left(\frac{(N-1)^2}{2} + \frac{(N-1)}{2}\right) = 8 - \frac{8}{N}$$

Notice that in the second step, we replaced the sum beginning at $j = 0$ with a sum beginning at $j = 1$. This is valid because the general term j is zero for $j = 0$ and may be dropped. Again, we find that $\lim\limits_{N \to \infty} L_N = \lim\limits_{N \to \infty}(8 - 8/N) = 8$.

Finally, we note that the region under the graph is a right triangle with base 4 and height 4 (Figure 11), so its area is $(\frac{1}{2})(4)(4) = 8$. All three methods of computing the area yield the same value of 8. ∎

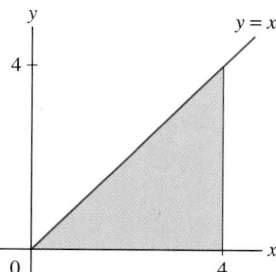

FIGURE 11 The right-endpoint approximations approach the area of the triangle.

In the next example, we compute the area under a curved graph. Unlike the previous example, it is not possible to compute the area directly using geometry.

■ **EXAMPLE 4** Let A be the area under the graph of $f(x) = 2x^2 - x + 3$ over $[2, 4]$ (Figure 12). Find a formula for R_N and compute A as the limit $\lim\limits_{N \to \infty} R_N$.

Solution For clarity, we break up the solution into steps.

Step 1. **Express R_N in terms of power sums.**

The interval is $[2, 4]$, so $\Delta x = \dfrac{4 - 2}{N} = \dfrac{2}{N}$ and

$$R_N = \Delta x \sum_{j=1}^{N} f(a + j\Delta x) = \frac{2}{N} \sum_{j=1}^{N} f\left(2 + \frac{2j}{N}\right)$$

Let's compute the general term in this sum. Since $f(x) = 2x^2 - x + 3$,

$$f\left(2 + \frac{2j}{N}\right) = 2\left(2 + \frac{2j}{N}\right)^2 - \left(2 + \frac{2j}{N}\right) + 3$$

$$= 2\left(4 + \frac{8j}{N} + \frac{4j^2}{N^2}\right) - \left(2 + \frac{2j}{N}\right) + 3 = \frac{8}{N^2}j^2 + \frac{14}{N}j + 9$$

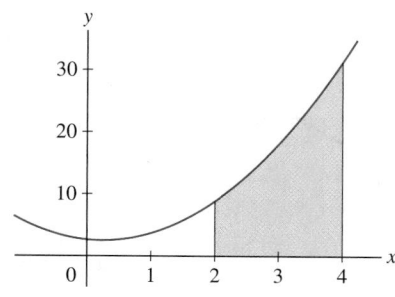

FIGURE 12 Area under the graph of $f(x) = 2x^2 - x + 3$ over $[2, 4]$.

Now we can express R_N in terms of power sums:

$$R_N = \frac{2}{N} \sum_{j=1}^{N} \left(\frac{8}{N^2} j^2 + \frac{14}{N} j + 9 \right) = \frac{2}{N} \sum_{j=1}^{N} \frac{8}{N^2} j^2 + \frac{2}{N} \sum_{j=1}^{N} \frac{14}{N} j + \frac{2}{N} \sum_{j=1}^{N} 9$$

$$= \frac{16}{N^3} \sum_{j=1}^{N} j^2 + \frac{28}{N^2} \sum_{j=1}^{N} j + \frac{18}{N} \sum_{j=1}^{N} 1 \qquad \boxed{7}$$

Step 2. Use the formulas for the power sums.

We use formulas (3) and (4) for the power sums in Eq. (7) to obtain

$$R_N = \frac{16}{N^3} \left(\frac{N^3}{3} + \frac{N^2}{2} + \frac{N}{6} \right) + \frac{28}{N^2} \left(\frac{N^2}{2} + \frac{N}{2} \right) + \frac{18}{N} (N)$$

$$= \left(\frac{16}{3} + \frac{8}{N} + \frac{8}{3N^2} \right) + \left(14 + \frac{14}{N} \right) + 18$$

$$= \frac{112}{3} + \frac{22}{N} + \frac{8}{3N^2}$$

Step 3. Calculate the limit.

$$\text{Area} = \lim_{N \to \infty} R_N = \lim_{N \to \infty} \left(\frac{112}{3} + \frac{22}{N} + \frac{8}{3N^2} \right) = \frac{112}{3} \qquad \blacksquare$$

We can calculate the area under the graph of a polynomial using the formulas for power sums as in the previous example. For more general functions, the area can be expressed as a limit, but the limit may be more complicated. Consider $f(x) = \sin x$ on the interval $[\frac{\pi}{4}, \frac{3\pi}{4}]$ (Figure 13). In this case,

$$\Delta x = \frac{3\pi/4 - \pi/4}{N} = \frac{\pi}{2N}$$

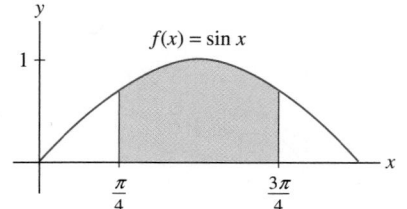

$f(x) = \sin x$

FIGURE 13 This area is more difficult to compute as a limit of endpoint approximations.

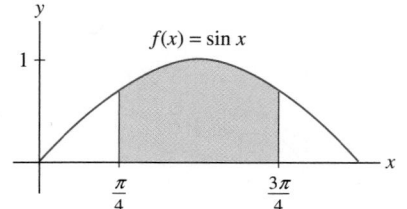

Jacob Bernoulli
(1654–1705)

HISTORICAL PERSPECTIVE

In the calculations above, we used the formulas for the kth power sums for $k = 1, 2, 3$. You may be curious whether or not similar formulas exist for all powers k. The problem of finding such formulas was studied in the seventeenth century and eventually solved around 1690 by the great Swiss mathematician Jacob Bernoulli (1654–1705). Of this discovery, he wrote

With the help of [these formulas] it took me less than half of a quarter of an hour to find that the 10th

powers of the first 1000 numbers being added together will yield the sum

9 1409924241424243424241924242500

Bernoulli's formula has the general form

$$\sum_{j=1}^{n} j^k = \frac{1}{k+1} n^{k+1} + \frac{1}{2} n^k + \frac{k}{12} n^{k-1} + \cdots$$

The dots indicate terms involving smaller powers of n whose coefficients are expressed in terms of the so-called Bernoulli numbers. For example,

$$\sum_{j=1}^{n} j^4 = \frac{1}{5} n^5 + \frac{1}{2} n^4 + \frac{1}{3} n^3 - \frac{1}{30} n$$

These formulas are available on most computer algebra systems.

and the area A under the graph is equal to

$$A = \lim_{N \to \infty} R_N = \lim_{N \to \infty} \Delta x \sum_{j=1}^{N} f(a + j\Delta x) = \lim_{N \to \infty} \frac{\pi}{2N} \sum_{j=1}^{N} \sin\left(\frac{\pi}{4} + \frac{\pi j}{2N}\right)$$

Although this limit can be evaluated with some work, we will see in Section 5.3 that it is much easier to use the Fundamental Theorem of Calculus, which reduces area computations to the problem of finding antiderivatives.

5.1 SUMMARY

- Approximations to the area under the graph of $f(x)$ over $[a, b]$ $\left(\Delta x = \dfrac{b-a}{N}\right)$:

$$R_N = \Delta x \sum_{j=1}^{N} f(a + j\Delta x) = \Delta x\big(f(a + \Delta x) + f(a + 2\Delta x) + \cdots + f(a + N\Delta x)\big)$$

$$L_N = \Delta x \sum_{j=0}^{N-1} f(a + j\Delta x) = \Delta x\big(f(a) + f(a + \Delta x) + \cdots + f(a + (N-1)\Delta x)\big)$$

$$M_N = \Delta x \sum_{j=1}^{N} f\left(a + \left(j - \frac{1}{2}\right)\Delta x\right)$$

$$= \Delta x \left(f\left(a + \frac{1}{2}\Delta x\right) + \cdots + f\left(a + \left(N - \frac{1}{2}\right)\Delta x\right)\right)$$

- If $f(x)$ is continuous on $[a, b]$, then the endpoint and midpoint approximations approach one and the same limit L:

$$\lim_{N \to \infty} R_N = \lim_{N \to \infty} L_N = \lim_{N \to \infty} M_N = L$$

- If $f(x) \geq 0$ on $[a, b]$, we take L as the definition of the area under the graph of $y = f(x)$ over $[a, b]$.

5.1 EXERCISES

Preliminary Questions

1. Suppose that $[2, 5]$ is divided into six subintervals. What are the right and left endpoints of the subintervals?

2. If $f(x) = x^{-2}$ on $[3, 7]$, which is larger: R_2 or L_2?

3. Which of the following pairs of sums are *not* equal?

(a) $\sum_{i=1}^{4} i$, $\sum_{\ell=1}^{4} \ell$

(b) $\sum_{j=1}^{4} j^2$, $\sum_{k=2}^{5} k^2$

(c) $\sum_{j=1}^{4} j$, $\sum_{i=2}^{5} (i - 1)$

(d) $\sum_{i=1}^{4} i(i + 1)$, $\sum_{j=2}^{5} (j - 1)j$

4. Explain why $\sum_{j=1}^{100} j$ is equal to $\sum_{j=0}^{100} j$ but $\sum_{j=1}^{100} 1$ is not equal to $\sum_{j=0}^{100} 1$.

5. We divide the interval $[1, 5]$ into 16 subintervals.
(a) What are the left endpoints of the first and last subintervals?
(b) What are the right endpoints of the first two subintervals?

6. Are the following statements true or false?
(a) The right-endpoint rectangles lie below the graph if $f(x)$ is increasing.

(b) If $f(x)$ is monotonic, then the area under the graph lies between R_N and L_N.

(c) If $f(x)$ is constant, then the right-endpoint rectangles all have the same height.

Exercises

1. An athlete runs with velocity 4 mph for half an hour, 6 mph for the next hour, and 5 mph for another half-hour. Compute the total distance traveled and indicate on a graph how this quantity can be interpreted as an area.

2. Figure 14 shows the velocity of an object over a 3-min interval. Determine the distance traveled over the intervals [0, 3] and [1, 2.5] (remember to convert from miles per hour to miles per minute).

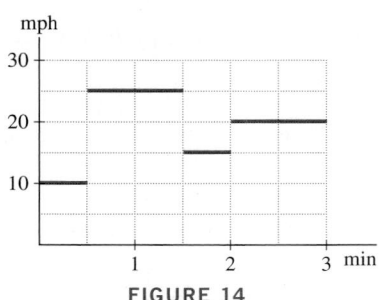

FIGURE 14

3. A rainstorm hit Portland, Maine, in October 1996, resulting in record rainfall. The rainfall rate $R(t)$ on October 21 is recorded, in inches per hour, in the following table, where t is the number of hours since midnight. Compute the total rainfall during this 24-hour period and indicate on a graph how this quantity can be interpreted as an area.

t	0–2	2–4	4–9	9–12	12–20	20–24
$R(t)$	0.2	0.1	0.4	1.0	0.6	0.25

4. The velocity of an object is $v(t) = 32t$ ft/s. Use Eq. (2) and geometry to find the distance traveled over the time intervals [0, 2] and [2, 5].

5. Compute R_6, L_6, and M_3 to estimate the distance traveled over [0, 3] if the velocity at half-second intervals is as follows:

t (s)	0	0.5	1	1.5	2	2.5	3
v (ft/s)	0	12	18	25	20	14	20

6. Use the following table of values to estimate the area under the graph of $f(x)$ over [0, 1] by computing the average of R_5 and L_5.

x	0	0.2	0.4	0.6	0.8	1
$f(x)$	50	48	46	44	42	40

7. Consider $f(x) = 2x + 3$ on [0, 3].

(a) Compute R_6 and L_6 over [0, 3].

(b) Find the error in these approximations by computing the area exactly using geometry.

8. Let $f(x) = x^2 + x - 2$.

(a) Calculate R_3 and L_3 over [2, 5].

(b) Sketch the graph of f and the rectangles that make up each approximation. Is the area under the graph larger or smaller than R_3? Than L_3?

9. Estimate R_6, L_6, and M_6 over [0, 1.5] for the function in Figure 15.

FIGURE 15

10. Estimate R_2, M_3, and L_6 for the graph in Figure 16.

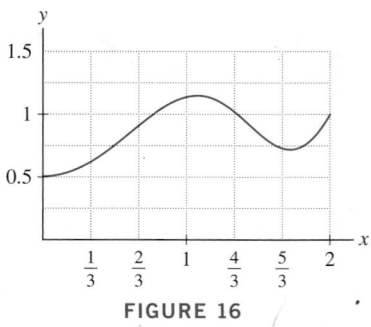

FIGURE 16

11. Let $f(x) = \sqrt{x^2 + 1}$ and $\Delta x = \frac{1}{3}$. Sketch the graph of $f(x)$ and draw the rectangles whose area is represented by the sum $\sum_{i=1}^{6} f(1 + i\Delta x)\Delta x$.

12. Calculate the area of the shaded rectangles in Figure 17. Which approximation do these rectangles represent?

$$y = \frac{4-x}{1+x^2}$$

FIGURE 17

In Exercises 13–24, calculate the approximation for the given function and interval.

13. R_8, $f(x) = 7 - x$, $[3, 5]$

14. M_4, $f(x) = 7 - x$, $[3, 5]$

15. M_4, $f(x) = x^2$, $[0, 1]$

16. M_6, $f(x) = \sqrt{x}$, $[2, 5]$

17. R_6, $f(x) = 2x^2 - x + 2$, $[1, 4]$

18. L_6, $f(x) = 2x^2 - x + 2$, $[1, 4]$

19. L_5, $f(x) = x^{-1}$, $[1, 2]$

20. M_4, $f(x) = x^{-2}$, $[1, 3]$

21. L_4, $f(x) = \cos x$, $[\frac{\pi}{4}, \frac{\pi}{2}]$

22. R_6, $f(x) = e^x$, $[0, 2]$

23. M_6, $f(x) = \ln x$, $[1, 2]$

24. L_5, $f(x) = x^2 + 3|x|$, $[-3, 2]$

In Exercises 25–28, use the Graphical Insight on page 303 to obtain bounds on the area.

25. Let A be the area under the graph of $f(x) = \sqrt{x}$ over $[0, 1]$. Prove that $0.51 \le A \le 0.77$ by computing R_4 and L_4. Explain your reasoning.

26. Use R_6 and L_6 to show that the area A under $y = x^{-2}$ over $[10, 12]$ satisfies $0.0161 \le A \le 0.0172$.

27. Use R_4 and L_4 to show that the area A under the graph of $y = \sin x$ over $[0, \pi/2]$ satisfies $0.79 \le A \le 1.19$.

28. Show that the area A under the graph of $f(x) = x^{-1}$ over $[1, 8]$ satisfies

$$\frac{1}{2} + \frac{1}{3} + \frac{1}{4} + \frac{1}{5} + \frac{1}{6} + \frac{1}{7} + \frac{1}{8} \le A \le 1 + \frac{1}{2} + \frac{1}{3} + \frac{1}{4} + \frac{1}{5} + \frac{1}{6} + \frac{1}{7}$$

29. *CAS* Show that the area A in Exercise 25 satisfies $L_N \le A \le R_N$ for all N. Then use a computer algebra system to calculate L_N and R_N for $N = 100$ and 150. Which of these calculations allows you to conclude that $A \approx 0.66$ to two decimal places?

30. *CAS* Show that the area A in Exercise 26 satisfies $R_N \le A \le L_N$ for all N. Use a computer algebra system to calculate L_N and R_N for N sufficiently large to determine A to within an error of at most 10^{-4}.

31. Calculate the following sums:

(a) $\sum_{i=1}^{5} 3$

(b) $\sum_{i=0}^{5} 3$

(c) $\sum_{k=2}^{4} k^3$

(d) $\sum_{j=3}^{4} \sin\left(j\frac{\pi}{2}\right)$

(e) $\sum_{k=2}^{4} \frac{1}{k-1}$

(f) $\sum_{j=0}^{3} 3^j$

32. Let $b_1 = 3$, $b_2 = 1$, $b_3 = 17$, and $b_4 = -17$. Calculate the sums.

(a) $\sum_{i=2}^{4} b_i$

(b) $\sum_{j=1}^{2} (b_j + 2b_j)$

(c) $\sum_{k=1}^{3} \frac{b_k}{b_{k+1}}$

33. Calculate $\sum_{j=101}^{200} j$ by writing it as a difference of two sums and using formula (3).

In Exercises 34–39, write the sum in summation notation.

34. $4^7 + 5^7 + 6^7 + 7^7 + 8^7$

35. $(2^2 + 2) + (3^2 + 3) + (4^2 + 4) + (5^2 + 5)$

36. $(2^2 + 2) + (2^3 + 2) + (2^4 + 2) + (2^5 + 2)$

37. $\sqrt{1 + 1^3} + \sqrt{2 + 2^3} + \cdots + \sqrt{n + n^3}$

38. $\frac{1}{2 \cdot 3} + \frac{2}{3 \cdot 4} + \cdots + \frac{n}{(n+1)(n+2)}$

39. $e^\pi + e^{\pi/2} + e^{\pi/3} + \cdots + e^{\pi/n}$

In Exercises 40–47, use linearity and formulas (3)–(5) to rewrite and evaluate the sums.

40. $\sum_{j=1}^{15} 12j^3$

41. $\sum_{k=1}^{20} (2k + 1)$

42. $\sum_{k=51}^{150} (2k + 1)$

43. $\sum_{k=100}^{200} k^3$

44. $\sum_{\ell=1}^{10} (\ell^3 - 2\ell^2)$

45. $\sum_{j=2}^{30} \left(6j + \frac{4j^2}{3}\right)$

46. $\sum_{j=0}^{50} j(j - 1)$

47. $\sum_{s=1}^{30} (3s^2 - 4s - 1)$

In Exercises 48–51, calculate the sum, assuming that $a_1 = -1$, $\sum_{i=1}^{10} a_i = 10$, and $\sum_{i=1}^{10} b_i = 7$.

48. $\sum_{i=1}^{10} 2a_i$

49. $\sum_{i=1}^{10} (a_i - b_i)$

50. $\displaystyle\sum_{\ell=1}^{10}(3a_\ell + 4b_\ell)$

51. $\displaystyle\sum_{i=2}^{10} a_i$

In Exercises 52–55, use formulas (3)–(5) to evaluate the limit.

52. $\displaystyle\lim_{N\to\infty}\sum_{i=1}^{N}\frac{i}{N^2}$

53. $\displaystyle\lim_{N\to\infty}\sum_{j=1}^{N}\frac{j^3}{N^4}$

54. $\displaystyle\lim_{N\to\infty}\sum_{i=1}^{N}\frac{i^2 - i + 1}{N^3}$

55. $\displaystyle\lim_{N\to\infty}\sum_{i=1}^{N}\left(\frac{i^3}{N^4} - \frac{20}{N}\right)$

In Exercises 56–59, calculate the limit for the given function and interval. Verify your answer by using geometry.

56. $\displaystyle\lim_{N\to\infty} R_N, \quad f(x) = 5x, \quad [0,3]$

57. $\displaystyle\lim_{N\to\infty} L_N, \quad f(x) = 5x, \quad [1,3]$

58. $\displaystyle\lim_{N\to\infty} L_N, \quad f(x) = 6 - 2x, \quad [0,2]$

59. $\displaystyle\lim_{N\to\infty} M_N, \quad f(x) = x, \quad [0,1]$

In Exercises 60–69, find a formula for R_N for the given function and interval. Then compute the area under the graph as a limit.

60. $f(x) = x^2, \quad [0,1]$

61. $f(x) = x^3, \quad [0,1]$

62. $f(x) = x^3 + 2x^2, \quad [0,3]$

63. $f(x) = 1 - x^3, \quad [0,1]$

64. $f(x) = 3x^2 - x + 4, \quad [0,1]$

65. $f(x) = 3x^2 - x + 4, \quad [1,5]$

66. $f(x) = 2x + 7, \quad [3,6]$

67. $f(x) = x^2, \quad [2,4]$

68. $f(x) = 2x + 1, \quad [a,b] \quad (a, b \text{ constants with } a < b)$

69. $f(x) = x^2, \quad [a,b] \quad (a, b \text{ constants with } a < b)$

70. Let A be the area under the graph of $y = e^x$ for $0 \le x \le 1$ [Figure 18(A)]. In this exercise, we evaluate A using the formula for a geometric sum (valid for $r \ne 1$):

$$1 + r + r^2 + \cdots + r^{N-1} = \sum_{j=0}^{N-1} r^j = \frac{r^N - 1}{r - 1} \qquad \boxed{8}$$

(a) Show that the left-endpoint approximation to A is

$$L_N = \frac{1}{N}\sum_{j=0}^{N-1} e^{j/N}$$

(b) Apply Eq. (8) with $r = e^{1/N}$ to prove that

$$A = (e - 1)\lim_{N\to\infty}\frac{1}{N(e^{1/N} - 1)}$$

(c) Evaluate the limit in (B) and calculate A. *Hint:* Show that L'Hôpital's Rule may be used after writing

$$\frac{1}{N(e^{1/N} - 1)} = \frac{N^{-1}}{e^{1/N} - 1}$$

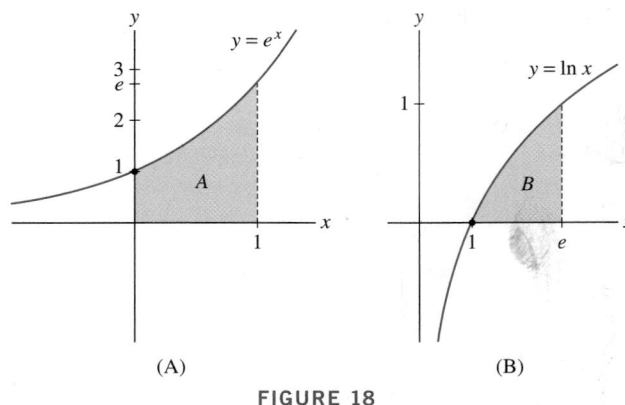

(A)   (B)

FIGURE 18

71. Use the result of Exercise 70 to show that the area B under the graph of $f(x) = \ln x$ over $[1, e]$ is equal to 1. *Hint:* Relate B to the area A computed in Exercise 70.

In Exercises 72–75, describe the area represented by the limits.

72. $\displaystyle\lim_{N\to\infty}\frac{1}{N}\sum_{j=1}^{N}\left(\frac{j}{N}\right)^4$

73. $\displaystyle\lim_{N\to\infty}\frac{3}{N}\sum_{j=1}^{N}\left(2 + \frac{3j}{N}\right)^4$

74. $\displaystyle\lim_{N\to\infty}\frac{5}{N}\sum_{j=0}^{N-1} e^{-2+5j/N}$

75. $\displaystyle\lim_{N\to\infty}\frac{\pi}{2N}\sum_{j=1}^{N}\sin\left(\frac{\pi}{3} + \frac{j\pi}{2N}\right)$

76. Evaluate $\displaystyle\lim_{N\to\infty}\frac{1}{N}\sum_{j=1}^{N}\sqrt{1 - \left(\frac{j}{N}\right)^2}$ by interpreting it as the area of part of a familiar geometric figure.

In Exercises 77–82, use the approximation indicated (in summation notation) to express the area under the graph as a limit but do not evaluate.

77. $R_N, \quad f(x) = \sin x \text{ over } [0, \pi]$

78. $R_N, \quad f(x) = x^{-1} \text{ over } [1, 7]$

79. $M_N, \quad f(x) = \tan x \text{ over } [\frac{1}{2}, 1]$

80. $M_N, \quad f(x) = x^{-2} \text{ over } [3, 5]$

81. L_N, $f(x) = \cos x$ over $[\frac{\pi}{8}, \pi]$

82. L_N, $f(x) = \cos x$ over $[\frac{\pi}{8}, \frac{\pi}{4}]$

In Exercises 83–85, let $f(x) = x^2$ and let R_N, L_N, and M_N be the approximations for the interval $[0, 1]$.

83. Show that $R_N = \dfrac{1}{3} + \dfrac{1}{2N} + \dfrac{1}{6N^2}$. Interpret the quantity $\dfrac{1}{2N} + \dfrac{1}{6N^2}$ as the area of a region.

84. Show that
$$L_N = \frac{1}{3} - \frac{1}{2N} + \frac{1}{6N^2}, \qquad M_N = \frac{1}{3} - \frac{1}{12N^2}$$
Then rank the three approximations R_N, L_N, and M_N in order of increasing accuracy (use the formula for R_N in Exercise 83).

85. For each of R_N, L_N, and M_N, find the smallest integer N for which the error is less than 0.001.

Further Insights and Challenges

86. Although the accuracy of R_N generally improves as N increases, this need not be true for small values of N. Draw the graph of a positive continuous function $f(x)$ on an interval such that R_1 is closer than R_2 to the exact area under the graph. Can such a function be monotonic?

87. Draw the graph of a positive continuous function on an interval such that R_2 and L_2 are both smaller than the exact area under the graph. Can such a function be monotonic?

88. Explain the following statement graphically: *The endpoint approximations are less accurate when $f'(x)$ is large.*

89. Assume that $f(x)$ is monotonic. Prove that M_N lies between R_N and L_N and that M_N is closer to the actual area under the graph than both R_N and L_N. *Hint:* Argue from Figure 19; the part of the error in R_N due to the ith rectangle is the sum of the areas $A + B + D$, and for M_N it is $|B - E|$.

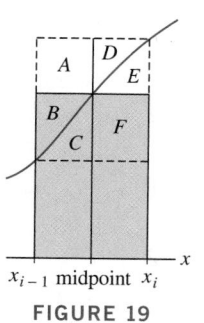

x_{i-1} midpoint x_i

FIGURE 19

90. Prove that for any function $f(x)$ on $[a, b]$,
$$R_N - L_N = \frac{b-a}{N}(f(b) - f(a)) \qquad \boxed{9}$$

91. In this exercise, we prove that the limits $\displaystyle\lim_{N\to\infty} R_N$ and $\displaystyle\lim_{N\to\infty} L_N$ exist and are equal if $f(x)$ is positive and increasing [the case of $f(x)$ decreasing is similar]. We use the concept of a least upper bound discussed in Appendix B.

(a) Explain with a graph why $L_N \le R_M$ for all $N, M \ge 1$.

(b) By part (a), the sequence $\{L_N\}$ is bounded by R_M for any M, so it has a least upper bound L. By definition, L is the smallest number such that $L_N \le L$ for all N. Show that $L \le R_M$ for all M.

(c) According to part (b), $L_N \le L \le R_N$ for all N. Use Eq. (9) to show that $\displaystyle\lim_{N\to\infty} L_N = L$ and $\displaystyle\lim_{N\to\infty} R_N = L$.

92. Assume that $f(x)$ is positive and monotonic, and let A be the area under its graph over $[a, b]$. Use Eq. (9) to show that
$$|R_N - A| \le \frac{b-a}{N}|f(b) - f(a)| \qquad \boxed{10}$$

In Exercises 93–94, use Eq. (10) to find a value of N such that $|R_N - A| < 10^{-4}$ for the given function and interval.

93. $f(x) = \sqrt{x}$, $[1, 4]$

94. $f(x) = \sqrt{9 - x^2}$, $[0, 3]$

5.2 The Definite Integral

In the previous section we saw that if $f(x)$ is continuous on an interval $[a, b]$, then the endpoint and midpoint approximations approach a common limit L as $N \to \infty$:

$$L = \lim_{N\to\infty} R_N = \lim_{N\to\infty} L_N = \lim_{N\to\infty} M_N \qquad \boxed{1}$$

Furthermore, when $f(x) \ge 0$, we interpret L as the area under the graph of $f(x)$ over $[a, b]$. In a moment, we will state formally that the limit L is the *definite integral* of $f(x)$ over $[a, b]$. Before doing so, we introduce **Riemann sums**, which are more general approximations to area.

Georg Friedrich Riemann (1826–1866)

One of the greatest mathematicians of the nineteenth century and perhaps second only to his teacher C. F. Gauss, Riemann transformed the fields of geometry, analysis, and number theory. In 1916, Albert Einstein based his General Theory of Relativity *on Riemann's geometry. The "Riemann hypothesis" dealing with prime numbers is one of the great unsolved problems in present-day mathematics. The Clay Foundation has offered a $1 million prize for its solution (http://www.claymath.org/millennium).*

In the endpoint and midpoint approximations, we approximate the area under the graph by rectangles of equal width. The heights of the rectangles are the values of $f(x)$ at the endpoints or midpoints of the subintervals. In Riemann sum approximations, we relax these requirements; the rectangles need not have equal width and the height of a rectangle may be *any* value of $f(x)$ within the subinterval. More formally, to define a Riemann sum, we choose a partition and a set of intermediate points. A **partition** P of length N is any choice of points in $[a, b]$ that divides the interval into N subintervals:

$$\text{Partition } P: \quad a = x_0 < x_1 < x_2 \quad \cdots \quad < x_N = b$$

$$\text{Subintervals:} \quad [x_0, x_1], \quad [x_1, x_2], \quad \ldots \quad [x_{N-1}, x_N]$$

We denote the length of the ith interval by

$$\Delta x_i = x_i - x_{i-1}$$

The maximum of the lengths Δx_i is called the **norm** of the partition P and is denoted $\|P\|$. A **set of intermediate points** is a set of points $C = \{c_1, \ldots, c_N\}$, where c_i belongs to $[x_{i-1}, x_i]$ [Figure 1(A)].

Now, over each subinterval $[x_{i-1}, x_i]$ we construct the rectangle of height $f(c_i)$ and base Δx_i, as in Figure 1(B). This rectangle has area $f(c_i)\Delta x_i$. If $f(c_i) < 0$, the rectangle extends below the x-axis and the "area" $f(c_i)\Delta x_i$ is negative. The Riemann sum is the sum of these positive or negative areas:

$$R(f, P, C) = \sum_{i=1}^{N} f(c_i)\Delta x_i = f(c_1)\Delta x_1 + f(c_2)\Delta x_2 + \cdots + f(c_N)\Delta x_N$$

2

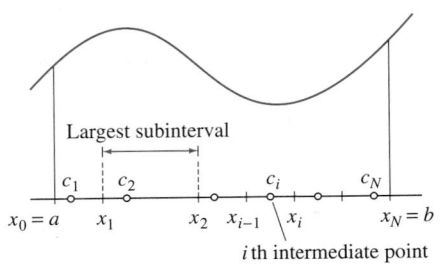

(A) Partition of $[a, b]$ into subintervals

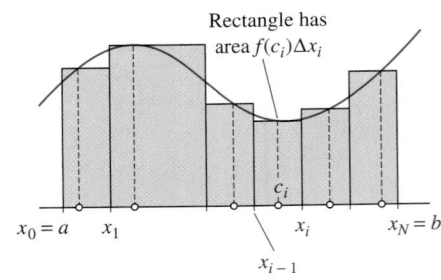

(B) Construct rectangle above each subinterval of height $f(c_i)$

FIGURE 1 The Riemann sum $R(f, P, C)$ is the sum of the areas of the rectangles.

The endpoint and midpoint approximations are examples of Riemann sums. They correspond to the partition of $[a, b]$ into N subintervals of equal length $\Delta x = \dfrac{b - a}{N}$ and the choice of right endpoints, left endpoints, or midpoints as intermediate points. In the next example, we consider a Riemann sum with intervals of unequal length.

■ **EXAMPLE 1** Let $f(x) = 8 + 12\sin x - 4x$. Calculate $R(f, P, C)$ for the partition P of $[0, 4]$:

$$P: \quad x_0 = 0 < x_1 = 1 < x_2 = 1.8 < x_3 = 2.9 < x_4 = 4$$

and intermediate points $C = \{0.4, 1.2, 2, 3.5\}$. What is the norm $\|P\|$?

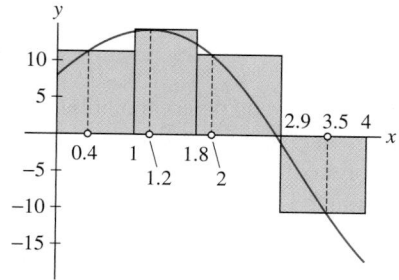

FIGURE 2 Rectangles defined by a Riemann sum for $f(x) = 8 + 12 \sin x - 4x$ on $[0, 4]$.

Solution The widths of the subintervals in the partition (Figure 2) are

$$\Delta x_1 = x_1 - x_0 = 1 - 0 = 1, \qquad \Delta x_2 = x_2 - x_1 = 1.8 - 1 = 0.8$$

$$\Delta x_3 = x_3 - x_2 = 2.9 - 1.8 = 1.1, \qquad \Delta x_4 = x_4 - x_3 = 4 - 2.9 = 1.1$$

The norm of the partition is $\|P\| = 1.1$ since the two longest subintervals have width 1.1. Using a calculator, we obtain:

$$R(f, P, C) = f(0.4)\Delta x_1 + f(1.2)\Delta x_2 + f(2)\Delta x_3 + f(3.5)\Delta x_4$$

$$\approx 11.07(1) + 14.38(0.8) + 10.91(1.1) - 10.2(1.1) \approx 23.35 \qquad \blacksquare$$

Having introduced Riemann sums, we are ready to make the following definition: A function is **integrable** over $[a, b]$ if *all* of the Riemann sums (not just the endpoint and midpoint approximations) approach one and the same limit L as the norm of the partition tends to zero. More formally, we write

$$L = \lim_{\|P\| \to 0} R(f, P, C) = \lim_{\|P\| \to 0} \sum_{i=1}^{N} f(c_i)\,\Delta x_i \qquad \boxed{3}$$

if $|R(f, P, C) - L|$ gets arbitrarily small as the norm $\|P\|$ tends to zero, no matter how we choose the partition and intermediate points. Note that as $\|P\| \to 0$, the number N of intervals tends to ∞. The limit L is called the **definite integral** of $f(x)$ over $[a, b]$.

The notation $\int f(x)\,dx$, introduced by Leibniz in 1686, is appropriate because it reminds us of the Riemann sums used to define the definite integral. The symbol $\int$ is an elongated S standing for summation, and dx corresponds to the lengths Δx_i along the x-axis.

DEFINITION Definite Integral The definite integral of $f(x)$ over $[a, b]$ is the limit of Riemann sums and is denoted by the integral sign:

$$\int_a^b f(x)\,dx = \lim_{\|P\| \to 0} R(f, P, C) = \lim_{\|P\| \to 0} \sum_{i=1}^{N} f(c_i)\,\Delta x_i$$

When this limit exists, we say that $f(x)$ is integrable over $[a, b]$.

Remarks

- We often refer to the definite integral more simply as the *integral* of f over $[a, b]$.
- The function $f(x)$ inside the integral sign is called the **integrand**.
- The numbers a and b representing the interval $[a, b]$ are called the **limits of integration**.
- Any variable may be used as a variable of integration (this is a "dummy" variable). Thus, the following three integrals all denote the same quantity:

$$\int_a^b f(x)\,dx, \qquad \int_a^b f(t)\,dt, \qquad \int_a^b f(u)\,du$$

- Recall that the *indefinite integral* $\int f(x)\,dx$ denotes the general antiderivative of $f(x)$ (Section 4.9). The connection between definite and indefinite integrals is explained in Section 5.3.

Most functions that arise in practice are integrable. In particular, every continuous function is integrable, as stated in the following theorem. A proof is given in Appendix D.

> **THEOREM 1**   If $f(x)$ is continuous on $[a, b]$, then $f(x)$ is integrable over $[a, b]$.

CONCEPTUAL INSIGHT   General Riemann sums are rarely used for computations. In practice, we compute definite integrals using either particular approximations such as the midpoint approximation or the Fundamental Theorem of Calculus discussed in the next section. If so, why bother introducing Riemann sums? The answer is that Riemann sums play a theoretical rather than a computational role. They are useful in proofs and for dealing rigorously with certain discontinuous functions. In later sections, Riemann sums are used to show that volumes and other quantities can be expressed as definite integrals.

Graphical Interpretation of the Definite Integral

When $f(x)$ is continuous and positive, we interpret the definite integral as the area under the graph. If $f(x)$ is not positive, then the definite integral is not equal to an area in the usual sense, but we may interpret it as the **signed area** between the graph and the x-axis. Signed area is defined as follows:

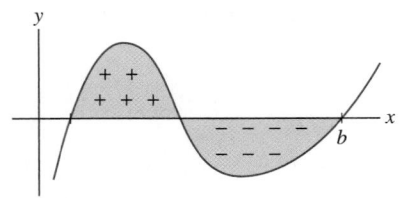

FIGURE 3 The signed area is the area above the x-axis minus the area below the x-axis.

> Signed area of a region = (area above x-axis) − (area below x-axis)

Thus, signed area treats regions under the x-axis as "negative area" (Figure 3).

To see why the definite integral gives us signed area, suppose first that $f(x)$ is negative on $[a, b]$ as in Figure 4(A) and consider a Riemann sum:

$$R(f, C, P) = f(c_1)\Delta x_1 + f(c_2)\Delta x_2 + \cdots + f(c_N)\Delta x_N$$

Since $f(c_i) < 0$, each term is equal to the negative of the area of a rectangle:

$$f(c_i)\Delta x_i = -\big(\text{area of rectangle of height } |f(c_i)|\big)$$

Therefore, the Riemann sums $R(f, C, P)$ converge to the negative of the area between the graph and the x-axis. If $f(x)$ takes on both positive and negative values as in Figure 4(B), then $f(c_i)\Delta x_i$ is positive or negative, depending on whether the corresponding rectangle lies above or below the x-axis. In this case, the Riemann sums converge to the signed area. In summary,

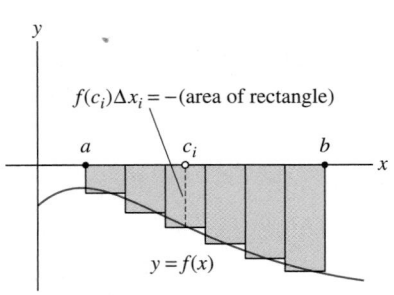

(A) Approximation to the definite integral of a negative function

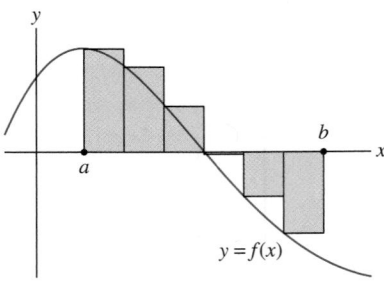

(B) Approximation to the definite integral of a function taking positive and negative values

FIGURE 4

$$\int_a^b f(x)\,dx = \text{signed area of region between the graph and } x\text{-axis over } [a, b]$$

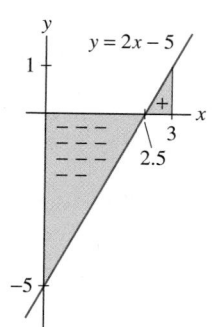

FIGURE 5 Signed area under the graph of $f(x) = 2x - 5$.

■ **EXAMPLE 2** Calculating Signed Area Calculate $\int_0^3 (2x - 5)\,dx$ using the interpretation as signed area.

Solution Figure 5 shows that the region between the graph and the x-axis consists of two triangles. The first triangle has area $(\frac{1}{2})(2.5)(5) = \frac{25}{4}$, but it lies below the x-axis, so its signed area is $-\frac{25}{4}$. The second triangle lies above the x-axis and has area $(\frac{1}{2})(0.5)(1) = \frac{1}{4}$. Therefore,

$$\int_0^3 (2x - 5)\,dx = -\frac{25}{4} + \frac{1}{4} = -6 \qquad ■$$

■ **EXAMPLE 3** Integral of an Absolute Value Calculate $\int_0^4 |3 - x|\,dx$.

Solution The graph of $y = |3 - x|$ in Figure 6(B) is derived from the graph of $y = 3 - x$ by reflecting it through the x-axis for $x \geq 3$. We see that the integral is equal to the sum of the areas of two triangles:

$$\int_0^4 |3 - x|\,dx = \frac{1}{2}(3)(3) + \frac{1}{2}(1)(1) = 5 \qquad ■$$

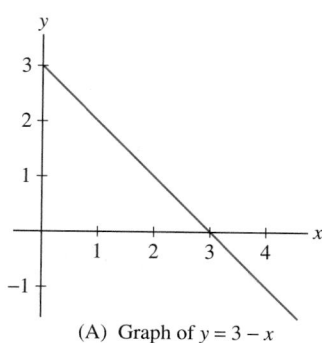

(A) Graph of $y = 3 - x$

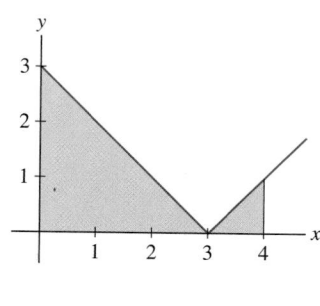

(B) Graph of $y = |3 - x|$

FIGURE 6

In the next example, we compute the integral of e^x as a limit of left-endpoint approximations (Figure 7). In the next section, we will derive this same result with less effort using the Fundamental Theorem of Calculus.

■ **EXAMPLE 4** The Integral of $f(x) = e^x$ Prove the formula

$$\int_0^b e^x\,dx = e^b - 1 \qquad \boxed{4}$$

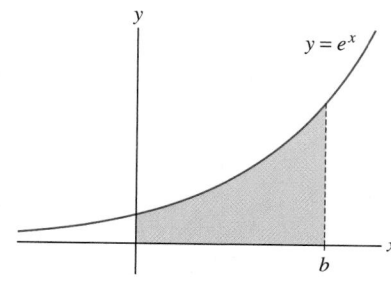

FIGURE 7 The area of the shaded region is equal to $e^b - 1$.

Solution We compute the integral as a limit of left-endpoint approximations L_N, using the formula for a geometric sum to evaluate L_N and then L'Hôpital's Rule.

The following formula for the geometric sum *is valid for all* $r \neq 1$:

$$1 + r + r^2 + \cdots + r^{N-1} = \frac{r^N - 1}{r - 1} \qquad \boxed{5}$$

Step 1. Evaluate L_N as a geometric sum.

We integrate over $[0, b]$, so $\Delta x = b/N$ and the left-endpoint approximation is

$$L_N = \frac{b}{N}\left(f(0) + f\left(\frac{b}{N}\right) + + f\left(\frac{2b}{N}\right) + \cdots + f\left(\frac{(N-1)b}{N}\right) \right)$$

$$= \frac{b}{N}\left(e^0 + e^{b/N} + e^{2b/N} + \cdots + e^{(N-1)b/N} \right)$$

Now let $r = e^{b/N}$. Then $r^j = e^{jb/N}$ and $r^N = e^b$. We may write L_N in terms of r and evaluate using Eq. (5) in the margin:

$$L_N = \frac{b}{N}\left(1 + r + r^2 + \cdots + r^{N-1} \right) = \frac{b}{N}\left(\frac{r^N - 1}{r - 1} \right) = \frac{b}{N}\left(\frac{e^b - 1}{e^{b/N} - 1} \right)$$

The integral is equal to the limit

$$\int_0^b e^x \, dx = \lim_{N \to \infty} L_N = \lim_{N \to \infty} \frac{b}{N}\left(\frac{e^b - 1}{e^{b/N} - 1} \right) = b(e^b - 1) \lim_{N \to \infty} \frac{1}{N(e^{b/N} - 1)}$$

Step 2. Use L'Hôpital's Rule to evaluate the limit.

We observe that

$$\frac{1}{N(e^{b/N} - 1)} = \frac{N^{-1}}{e^{b/N} - 1}$$

The fraction on the right yields the indeterminate form $0/0$ as $N \to \infty$ since $e^{b/N} - 1$ approaches $e^0 - 1 = 0$. Therefore, we may apply L'Hôpital's Rule (with respect to the variable N):

$$\lim_{N \to \infty} \frac{1}{N(e^{b/N} - 1)} = \lim_{N \to \infty} \frac{N^{-1}}{e^{b/N} - 1} \overset{\text{L'Hôpital's Rule}}{=} \lim_{N \to \infty} \left(\frac{-N^{-2}}{-bN^{-2}e^{b/N}} \right)$$

$$= \frac{1}{b} \lim_{N \to \infty} e^{-b/N} = \frac{1}{b}e^0 = \frac{1}{b}$$

Thus, we obtain the desired result:

$$\int_0^b e^x \, dx = b(e^b - 1) \lim_{N \to \infty} \frac{1}{N(e^{b/N} - 1)} = b(e^b - 1)\frac{1}{b} = e^b - 1 \qquad \blacksquare$$

Properties of the Definite Integral

In the rest of this section, we discuss some basic properties of definite integrals. First, we note that the integral of a constant function $f(x) = C$ over $[a, b]$ is the signed area $C(b - a)$ of a rectangle as shown in Figure 8.

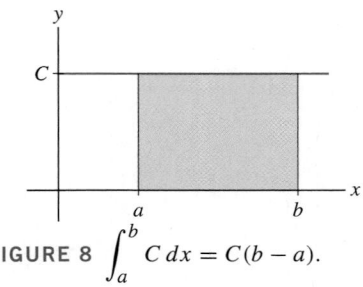

FIGURE 8 $\int_a^b C \, dx = C(b - a)$.

> **THEOREM 2 Integral of a Constant**   For any constant C, $\displaystyle\int_a^b C \, dx = C(b - a)$.

Next, we state the linearity properties of the definite integral.

THEOREM 3 Linearity of the Definite Integral If $f(x)$ and $g(x)$ are integrable over $[a, b]$, then

- $$\int_a^b \big(f(x) + g(x)\big)\,dx = \int_a^b f(x)\,dx + \int_a^b g(x)\,dx$$

- $$\int_a^b Cf(x)\,dx = C\int_a^b f(x)\,dx \quad \text{for any constant } C$$

Proof These properties follow from the corresponding linearity properties of sums and limits. To prove additivity, we observe that Riemann sums are additive:

$$R(f + g, P, C) = \sum_{i=1}^N \big(f(c_i) + g(c_i)\big)\Delta x_i = \sum_{i=1}^N f(c_i)\,\Delta x_i + \sum_{i=1}^N g(c_i)\,\Delta x_i$$

$$= R(f, P, C) + R(g, P, C)$$

By the additivity of limits,

$$\int_a^b \big(f(x) + g(x)\big)\,dx = \lim_{\|P\|\to 0} R(f + g, P, C)$$

$$= \lim_{\|P\|\to 0} R(f, P, C) + \lim_{\|P\|\to 0} R(g, P, C)$$

$$= \int_a^b f(x)\,dx + \int_a^b g(x)\,dx$$

The second property is proved similarly. ∎

In Example 5, the function $f(x) = x^2$ is continuous, hence integrable, so all Riemann sum approximations approach the same limit. We would obtain the same answer using L_N or M_N. The formula (6) is valid for all b (see Exercise 87).

■ **EXAMPLE 5** Use the limit of right-endpoint approximations to prove that for $b > 0$

$$\int_0^b x^2\,dx = \frac{b^3}{3}$$ **6**

Then use the Linearity Rules to calculate $\int_0^3 (2x^2 - 5)\,dx$.

Solution In the right-endpoint approximation for $f(x) = x^2$ on $[0, b]$, we have $\Delta x = b/N$. Using the formula for the power sum recalled in the margin, we obtain

◀·· **REMINDER**

$$\sum_{j=1}^N j^2 = \frac{N^3}{3} + \frac{N^2}{2} + \frac{N}{6}$$

$$R_N = \Delta x \sum_{j=1}^N f(0 + j\,\Delta x) = \frac{b}{N}\sum_{j=1}^N \left(\frac{jb}{N}\right)^2 = \frac{b^3}{N^3}\sum_{j=1}^N j^2$$

$$= \frac{b^3}{N^3}\left(\frac{N^3}{3} + \frac{N^2}{2} + \frac{N}{6}\right) = \frac{b^3}{3} + \frac{b^3}{2N} + \frac{b^3}{6N^2}$$

We prove Eq. (6) by taking the limit:

$$\int_0^b x^2\,dx = \lim_{N\to\infty}\left(\frac{b^3}{3} + \frac{b^3}{2N} + \frac{b^3}{6N^2}\right) = \frac{b^3}{3}$$

Next, by linearity and Theorem 2,

$$\int_0^3 (2x^2 - 5)\,dx = \int_0^3 2x^2\,dx + \int_0^3 (-5)\,dx = 2\int_0^3 x^2\,dx - \int_0^3 5\,dx$$

$$= 2\left(\frac{3^3}{3}\right) - 5(3 - 0) = 3 \qquad \blacksquare$$

So far we have used the symbol $\int_a^b f(x)\,dx$ with the understanding that $a < b$. It is convenient to define the definite integral for arbitrary a and b.

DEFINITION Reversing the Limits of Integration For $a < b$, we set

$$\int_b^a f(x)\,dx = -\int_a^b f(x)\,dx$$

For example, using the formula $\int_0^b x^2\,dx = b^3/3$ from Example 5 with $b = 5$, we have

$$\int_5^0 x^2\,dx = -\int_0^5 x^2\,dx = -\frac{5^3}{3} = -\frac{125}{3}$$

When $a = b$, the interval $[a, b] = [a, a]$ has length zero and we define the definite integral to be zero:

$$\int_a^a f(x)\,dx = 0$$

■ **EXAMPLE 6** Prove for all b (positive or negative)

$$\int_0^b x\,dx = \frac{1}{2}b^2 \qquad \boxed{7}$$

Solution If $b > 0$, $\int_0^b x\,dx$ is the area of the triangle of height b and base b in Figure 9(A). The area is $\frac{1}{2}b^2$. For $b < 0$, $\int_b^0 x\,dx$ is the signed area of the triangle in Figure 9(B), which equals $-\frac{1}{2}b^2$. Therefore,

$$\int_0^b x\,dx = -\int_b^0 x\,dx = -\left(-\frac{1}{2}b^2\right) = \frac{1}{2}b^2 \qquad \blacksquare$$

Definite integrals satisfy an important additivity property. If $f(x)$ is integrable and $a \le b \le c$ as in Figure 10, then the integral from a to c is equal to the integral from a to b plus the integral from b to c. We state this formally in the next theorem (proof omitted).

(A) Signed area $\frac{1}{2}b^2$

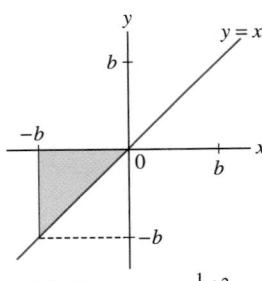

(B) Signed area $-\frac{1}{2}b^2$

FIGURE 9

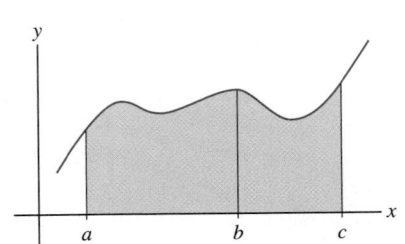

FIGURE 10 The area over $[a, c]$ is the sum of the areas over $[a, b]$ and $[b, c]$.

The property stated in Theorem 4 remains true as stated for all a, b, c even if the condition $a \leq b \leq c$ is not satisfied (Exercise 88).

> **THEOREM 4 Additivity for Adjacent Intervals**   For $a \leq b \leq c$,
> $$\int_a^c f(x)\,dx = \int_a^b f(x)\,dx + \int_b^c f(x)\,dx$$

■ **EXAMPLE 7** Calculate $\displaystyle\int_4^7 x^2\,dx$ using the formula $\displaystyle\int_0^b x^2 = \frac{b^3}{3}$ from Example 5.

Solution By additivity for adjacent intervals, we have

$$\underbrace{\int_0^4 x^2\,dx}_{\left(\frac{1}{3}\right)4^3} + \int_4^7 x^2\,dx = \underbrace{\int_0^7 x^2\,dx}_{\left(\frac{1}{3}\right)7^3}$$

Therefore,

$$\int_4^7 x^2\,dx = \int_0^7 x^2\,dx - \int_0^4 x^2\,dx = \left(\frac{1}{3}\right)7^3 - \left(\frac{1}{3}\right)4^3 = 93 \qquad ■$$

Sometimes, we wish to compare the integrals of two functions. If $f(x)$ and $g(x)$ are integrable and $g(x) \leq f(x)$, then the larger function $f(x)$ has a larger integral (Figure 11).

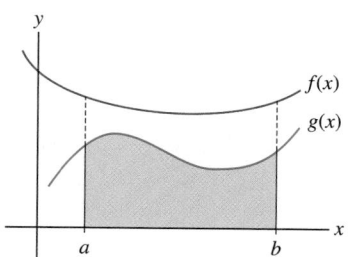

FIGURE 11 The integral of $f(x)$ is larger than the integral of $g(x)$.

> **THEOREM 5 Comparison Theorem**   If $g(x) \leq f(x)$ on an interval $[a, b]$, then
> $$\int_a^b g(x)\,dx \leq \int_a^b f(x)\,dx$$

Proof For any partition P of $[a, b]$ with endpoints $x_1, \ldots, x_N$ and choice of intermediate points $\{c_i\}$, we have $g(c_i)\,\Delta x_i \leq f(c_i)\,\Delta x_i$ for all i. Therefore, the Riemann sums satisfy

$$\sum_{i=1}^N g(c_i)\,\Delta x_i \leq \sum_{i=1}^N f(c_i)\,\Delta x_i$$

Taking the limit as the norm $\|P\|$ tends to zero, we obtain

$$\int_a^b g(x)\,dx = \lim_{\|P\|\to 0}\sum_{i=1}^N g(c_i)\,\Delta x_i \leq \lim_{\|P\|\to 0}\sum_{i=1}^N f(c_i)\,\Delta x_i = \int_a^b f(x)\,dx \qquad ■$$

Here is a simple but useful application of the Comparison Theorem: Suppose that $f(x)$ satisfies $m \leq f(x) \leq M$ for x in $[a, b]$. The numbers m and M are called lower and upper bounds for $f(x)$ on $[a, b]$. In this case, the integral lies between the areas of two rectangles (Figure 12):

$$\int_a^b m\,dx \leq \int_a^b f(x)\,dx \leq \int_a^b M\,dx$$

This gives us

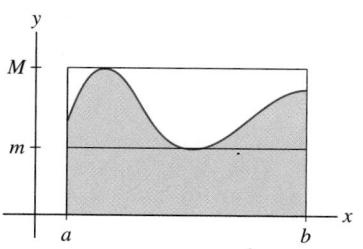

FIGURE 12 The integral $\displaystyle\int_a^b f(x)\,dx$ lies between the areas of the rectangles of heights m and M.

$$\boxed{m(b-a) \leq \int_a^b f(x)\,dx \leq M(b-a)}$$

$$\boxed{8}$$

EXAMPLE 8 Prove the inequalities:

(a) $\displaystyle\int_1^5 \frac{1}{x^2}\,dx \le \int_1^5 \frac{1}{x}\,dx$

(b) $\displaystyle\frac{3}{4} \le \int_{1/2}^2 \frac{1}{x}\,dx \le 3.$

Solution

(a) If $x \ge 1$, then $x^2 \ge x$, and hence by taking inverses, we obtain $x^{-2} \le x^{-1}$. By the Comparison Theorem,

$$\int_1^5 \frac{1}{x^2}\,dx \le \int_1^5 \frac{1}{x}\,dx$$

(b) Since $f(x) = x^{-1}$ is decreasing, its maximum on $\left[\frac{1}{2}, 2\right]$ is $M = f\left(\frac{1}{2}\right) = 2$ and its minimum is $m = f(2) = \frac{1}{2}$ (Figure 13). By Eq. (8),

$$\underbrace{\frac{1}{2}\left(2 - \frac{1}{2}\right)}_{m(b-a)} = \frac{3}{4} \le \int_{1/2}^2 \frac{1}{x}\,dx \le \underbrace{2\left(2 - \frac{1}{2}\right)}_{M(b-a)} = 3 \qquad \blacksquare$$

FIGURE 13

5.2 SUMMARY

- A Riemann sum $R(f, P, C)$ for the interval $[a, b]$ is defined by choosing a *partition*

$$P: \qquad a = x_0 < x_1 < x_2 < \cdots < x_N = b$$

and *intermediate points* $C = \{c_i\}$, where $c_i \in [x_{i-1}, x_i]$. Let $\Delta x_i = x_i - x_{i-1}$. Then

$$R(f, P, C) = \sum_{i=1}^N f(c_i)\,\Delta x_i$$

- The maximum of the widths Δx_i is called the norm $\|P\|$ of the partition.
- The *definite integral* is the limit of the Riemann sums (if it exists):

$$\int_a^b f(x)\,dx = \lim_{\|P\| \to 0} R(f, P, C)$$

We say that $f(x)$ is *integrable* over $[a, b]$ if the limit exists.
- Theorem: If $f(x)$ is continuous on $[a, b]$, then $f(x)$ is integrable over $[a, b]$.
- The definite integral $\displaystyle\int_a^b f(x)\,dx$ is equal to the signed area of the region between the graph of $f(x)$ and the x-axis.
- Properties of definite integrals [for $f(x)$ and $g(x)$ integrable]:

$$\int_a^b \left(f(x) + g(x)\right)dx = \int_a^b f(x)\,dx + \int_a^b g(x)\,dx$$

$$\int_a^b Cf(x)\,dx = C\int_a^b f(x)\,dx \quad \text{for any constant } C$$

$$\int_a^b f(x)\,dx + \int_b^c f(x)\,dx = \int_a^c f(x)\,dx \quad \text{for all } a, b, c$$

$$\int_a^b f(x)\,dx = -\int_b^a f(x)\,dx$$

$$\int_a^a f(x)\,dx = 0$$

- Formulas:

$$\int_a^b C\,dx = C(b-a) \quad (C \text{ any constant})$$

$$\int_0^b x\,dx = \frac{1}{2}b^2$$

$$\int_0^b x^2\,dx = \frac{1}{3}b^3$$

$$\int_0^b e^x\,dx = e^b - 1$$

- Comparison Theorem: If $f(x) \le g(x)$ on $[a, b]$, then

$$\int_a^b f(x)\,dx \le \int_a^b g(x)\,dx$$

If $m \le f(x) \le M$ on $[a, b]$, then

$$m(b-a) \le \int_a^b f(x)\,dx \le M(b-a)$$

5.2 EXERCISES

Preliminary Questions

1. What is $\int_a^b dx$ [here the function is $f(x) = 1$]?

2. Are the following statements true or false [assume that $f(x)$ is continuous]?

(a) $\int_a^b f(x)\,dx$ is the area between the graph and the x-axis over $[a, b]$.

(b) $\int_a^b f(x)\,dx$ is the area between the graph and the x-axis over $[a, b]$ if $f(x) \ge 0$.

(c) If $f(x) \le 0$, then $-\int_a^b f(x)\,dx$ is the area between the graph of $f(x)$ and the x-axis over $[a, b]$.

3. Explain graphically why $\int_0^\pi \cos x\,dx = 0$.

4. Is $\int_{-5}^{-1} 8\,dx$ negative?

5. What is the largest possible value of $\int_0^6 f(x)\,dx$ if $f(x) \le \frac{1}{3}$?

Exercises

In Exercises 1–10, draw a graph of the signed area represented by the integral and compute it using geometry.

1. $\int_{-3}^3 2x\,dx$

2. $\int_{-2}^3 (2x+4)\,dx$

3. $\int_{-2}^1 (3x+4)\,dx$

4. $\int_{-2}^1 4\,dx$

5. $\int_6^8 (7-x)\,dx$

6. $\int_{\pi/2}^{3\pi/2} \sin x\,dx$

7. $\int_0^5 \sqrt{25-x^2}\,dx$

8. $\int_{-2}^3 |x|\,dx$

9. $\int_{-2}^2 (2-|x|)\,dx$

10. $\int_{-1}^1 (2x-|x|)\,dx$

11. Calculate $\int_0^6 (4 - x)\, dx$ in two ways:

(a) As the limit $\lim\limits_{N \to \infty} R_N$

(b) By sketching the relevant signed area and using geometry

12. Calculate $\int_2^5 (2x + 1)\, dx$ in two ways: As the limit $\lim\limits_{N \to \infty} R_N$ and using geometry.

13. Evaluate the integrals for $f(x)$ shown in Figure 14.

(a) $\int_0^2 f(x)\, dx$ **(b)** $\int_0^6 f(x)\, dx$

(c) $\int_1^4 f(x)\, dx$ **(d)** $\int_1^6 |f(x)|\, dx$

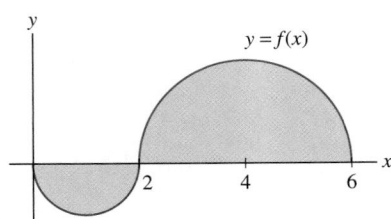

FIGURE 14 The two parts of the graph are semicircles.

In Exercises 14–15, refer to Figure 15.

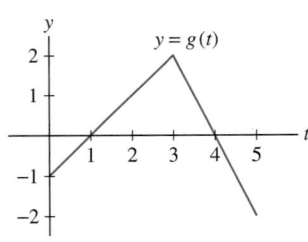

FIGURE 15

14. Evaluate $\int_0^3 g(t)\, dt$ and $\int_3^5 g(t)\, dt$.

15. Find a, b, and c such that $\int_0^a g(t)\, dt$ and $\int_b^c g(t)\, dt$ are as large as possible.

16. Describe the partition P and the set of intermediate points C for the Riemann sum shown in Figure 16. Compute the value of the Riemann sum.

In Exercises 17–22, sketch the signed area represented by the integral. Indicate the regions of positive and negative area.

17. $\int_0^2 (x - x^2)\, dx$ **18.** $\int_0^3 (2x - x^2)\, dx$

FIGURE 16

19. $\int_\pi^{2\pi} \sin x\, dx$ **20.** $\int_0^{3\pi} \sin x\, dx$

21. $\int_{1/2}^2 \ln x\, dx$ **22.** $\int_{-1}^1 \tan^{-1} x\, dx$

In Exercises 23–26, determine the sign of the integral without calculating it. Draw a graph if necessary.

23. $\int_{-2}^1 x^4\, dx$ **24.** $\int_{-2}^1 x^3\, dx$

25. GU $\int_0^{2\pi} x \sin x\, dx$ **26.** GU $\int_0^{2\pi} \dfrac{\sin x}{x}\, dx$

In Exercises 27–30, calculate the Riemann sum $R(f, P, C)$ for the given function, partition, and choice of intermediate points. Also, sketch the graph of f and the rectangles corresponding to $R(f, P, C)$.

27. $f(x) = x$, $P = \{1, 1.2, 1.5, 2\}$, $C = \{1.1, 1.4, 1.9\}$

28. $f(x) = x^2 + x$, $P = \{2, 3, 4.5, 5\}$, $C = \{2, 3.5, 5\}$

29. $f(x) = x + 1$, $P = \{-2, -1.6, -1.2, -0.8, -0.4, 0\}$, $C = \{-1.7, -1.3, -0.9, -0.5, 0\}$

30. $f(x) = \sin x$, $P = \{0, \frac{\pi}{6}, \frac{\pi}{3}, \frac{\pi}{2}\}$, $C = \{0.4, 0.7, 1.2\}$

In Exercises 31–40, use the basic properties of the integral and the formulas in the summary to calculate the integrals.

31. $\int_0^4 x^2\, dx$ **32.** $\int_1^4 x^2\, dx$

33. $\int_0^3 (3t + 4)\, dt$ **34.** $\int_{-2}^3 (3x + 4)\, dx$

35. $\int_0^1 (u^2 - 2u)\, du$ **36.** $\int_0^3 (6y^2 + 7y + 1)\, dy$

37. $\int_{-a}^1 (x^2 + x)\, dx$ **38.** $\int_a^{a^2} x^2\, dx$

39. $\int_0^4 e^x\, dx$ **40.** $\int_2^0 (x^2 - e^x)\, dx$

41. Prove by computing the limit of right-endpoint approximations:

$$\int_0^b x^3\, dx = \frac{b^4}{4}$$

9

In Exercises 42–49, use the formulas in the summary and Eq. (9) to evaluate the integral.

42. $\int_0^3 x^2 \, dx$

43. $\int_0^2 (x^2 + 2x) \, dx$

44. $\int_0^3 x^3 \, dx$

45. $\int_0^2 (x - x^3) \, dx$

46. $\int_0^1 (2x^3 - x + 4) \, dx$

47. $\int_{-3}^0 (2x - 5) \, dx$

48. $\int_1^3 x^3 \, dx$

49. $\int_1^2 (x - x^3) \, dx$

In Exercises 50–54, calculate the integral, assuming that

$$\int_0^5 f(x) \, dx = 5, \qquad \int_0^5 g(x) \, dx = 12$$

50. $\int_0^5 (f(x) + g(x)) \, dx$

51. $\int_0^5 (f(x) + 4g(x)) \, dx$

52. $\int_5^0 g(x) \, dx$

53. $\int_0^5 (3f(x) - 5g(x)) \, dx$

54. Is it possible to calculate $\int_0^5 g(x) f(x) \, dx$ from the information given?

In Exercises 55–58, calculate the integral, assuming that

$$\int_0^1 f(x) \, dx = 1, \qquad \int_0^2 f(x) \, dx = 4, \qquad \int_1^4 f(x) \, dx = 7$$

55. $\int_0^4 f(x) \, dx$

56. $\int_1^2 f(x) \, dx$

57. $\int_4^1 f(x) \, dx$

58. $\int_2^4 f(x) \, dx$

In Exercises 59–62, express each integral as a single integral.

59. $\int_0^3 f(x) \, dx + \int_3^7 f(x) \, dx$

60. $\int_2^9 f(x) \, dx - \int_4^9 f(x) \, dx$

61. $\int_2^9 f(x) \, dx - \int_2^5 f(x) \, dx$

62. $\int_7^3 f(x) \, dx + \int_3^9 f(x) \, dx$

In Exercises 63–66, calculate the integral, assuming that f is an integrable function such that $\int_1^b f(x) \, dx = 1 - b^{-1}$ for all $b > 0$.

63. $\int_1^3 f(x) \, dx$

64. $\int_2^4 f(x) \, dx$

65. $\int_1^4 (4f(x) - 2) \, dx$

66. $\int_{1/2}^1 f(x) \, dx$

67. Use the result of Example 4 and Theorem 4 to prove that for $b > a > 0$,

$$\int_a^b e^x \, dx = e^b - e^a$$

68. Use the result of Exercise 67 to evaluate $\int_2^4 (x - e^x) \, dx$.

69. ✏ Explain the difference in graphical interpretation between $\int_a^b f(x) \, dx$ and $\int_a^b |f(x)| \, dx$.

70. ✏ Let $I = \int_0^{2\pi} \sin^2 x \, dx$ and $J = \int_0^{2\pi} \cos^2 x \, dx$. Use the following trick to prove that $I = J = \pi$: First show with a graph that $I = J$ and then prove $I + J = \int_0^{2\pi} dx$.

In Exercises 71–74, calculate the integral.

71. $\int_0^6 |3 - x| \, dx$

72. $\int_1^3 |2x - 4| \, dx$

73. $\int_{-1}^1 |x^3| \, dx$

74. $\int_0^2 |x^2 - 1| \, dx$

75. Use the Comparison Theorem to show that

$$\int_0^1 x^5 \, dx \le \int_0^1 x^4 \, dx, \qquad \int_1^2 x^4 \, dx \le \int_1^2 x^5 \, dx$$

76. Prove that $\dfrac{1}{3} \le \int_4^6 \dfrac{1}{x} \, dx \le \dfrac{1}{2}$.

77. Prove that $0.0198 \le \int_{0.2}^{0.3} \sin x \, dx \le 0.0296$. *Hint:* Show that $0.198 \le \sin x \le 0.296$ for x in $[0.2, 0.3]$.

78. Prove that $0.277 \le \int_{\pi/8}^{\pi/4} \cos x \, dx \le 0.363$.

79. GU Prove that

$$\int_{\pi/4}^{\pi/2} \frac{\sin x}{x} \, dx \le \frac{\sqrt{2}}{2}$$

Hint: Graph $y = \dfrac{\sin x}{x}$ and observe that it is decreasing on $[\frac{\pi}{4}, \frac{\pi}{2}]$.

80. Find upper and lower bounds for $\int_0^1 \dfrac{dx}{\sqrt{x^3 + 4}}$.

81. ✏ Suppose that $f(x) \le g(x)$ on $[a, b]$. By the Comparison Theorem, $\int_a^b f(x) \, dx \le \int_a^b g(x) \, dx$. Is it also true that $f'(x) \le g'(x)$ for $x \in [a, b]$? If not, give a counterexample.

82. State whether true or false. If false, sketch the graph of a counterexample.

(a) If $f(x) > 0$, then $\int_a^b f(x)\,dx > 0$.

(b) If $\int_a^b f(x)\,dx > 0$, then $f(x) > 0$.

Further Insights and Challenges

83. Explain graphically: $\int_{-a}^{a} f(x)\,dx = 0$ if $f(x)$ is an odd function.

84. Compute $\int_{-1}^{1} \sin(\sin(x))(\sin^2(x) + 1)\,dx$.

85. Let k and b be positive. Show, by comparing the right-endpoint approximations, that

$$\int_0^b x^k\,dx = b^{k+1}\int_0^1 x^k\,dx$$

86. Verify by interpreting the integral as an area:

$$\int_0^b \sqrt{1-x^2}\,dx = \frac{1}{2}b\sqrt{1-b^2} + \frac{1}{2}\theta$$

Here, $0 \le b \le 1$ and θ is the angle between 0 and $\frac{\pi}{2}$ such that $\sin\theta = b$.

87. Show that Eq. (6) holds for $b \le 0$.

88. Theorem 4 remains true without the assumption $a \le b \le c$. Verify this for the cases $b < a < c$ and $c < a < b$.

5.3 The Fundamental Theorem of Calculus, Part I

The FTC was first stated clearly by Isaac Newton in 1666, although other mathematicians, including Newton's teacher Isaac Barrow, had come close to discovering it.

The Fundamental Theorem of Calculus (FTC) establishes a relationship between the two main operations of calculus: differentiation and integration. The theorem has two parts. Although they are closely related, we discuss them in separate sections to emphasize their different roles. To motivate the first part of the FTC, recall a formula proved in Example 5 of Section 5.2:

$$\int_0^b x^2\,dx = \frac{1}{3}b^3$$

Notice that $F(x) = \frac{1}{3}x^3$ is an *antiderivative* of x^2, and since $F(0) = 0$, this formula can be rewritten as

$$\int_0^b x^2\,dx = F(b) - F(0)$$

According to the FTC, Part I, this is no coincidence—such a relationship between the definite integral of $f(x)$ and its antiderivative holds in general.

THEOREM 1 The Fundamental Theorem of Calculus, Part I   Assume that $f(x)$ is continuous on $[a, b]$ and let $F(x)$ be an antiderivative of $f(x)$ on $[a, b]$. Then

$$\int_a^b f(x)\,dx = F(b) - F(a)$$

This result allows us to compute definite integrals without calculating any limits, provided that we can find an antiderivative F. It also reveals an unexpected connection between two thus far unrelated concepts: antiderivatives and areas under curves. Furthermore, the FTC shows that the notation $\int f(x)\,dx$ for the indefinite integral is appropriate because the antiderivative is used to evaluate the definite integral.

Notation: We denote $F(b) - F(a)$ by $F(x)\big|_a^b$. In this notation, the FTC reads

$$\int_a^b f(x)\, dx = F(x)\big|_a^b$$

Proof Consider an arbitrary partition P of $[a, b]$:

$$x_0 = a < x_1 < x_2 < \cdots < x_N = b$$

The first key step is to write the change $F(b) - F(a)$ as the sum of the changes of $F(x)$ over the subintervals $[x_{i-1}, x_i]$ (Figure 1):

$$F(b) - F(a) = F(x_N) - F(x_0)$$

$$= \big(F(x_1) - F(x_0)\big) + \big(F(x_2) - F(x_1)\big) + \cdots + \big(F(x_N) - F(x_{N-1})\big)$$

In this equality, $F(x_1)$ is cancelled by $-F(x_1)$ in the next term, $F(x_2)$ is cancelled by $-F(x_2)$, etc. In summation notation,

$$F(b) - F(a) = \sum_{i=1}^{N} \big(F(x_i) - F(x_{i-1})\big) \qquad \boxed{1}$$

In this proof, the MVT shows that for any partition P, there is a choice of intermediate points $C = \{c_i^*\}$ such that

$$F(b) - F(a) = R(f, P, C)$$

Once we know this, FTC I follows by taking a limit.

The second key step is to apply the Mean Value Theorem (MVT): In each subinterval $[x_{i-1}, x_i]$, there exists an intermediate point c_i^* such that

$$\frac{F(x_i) - F(x_{i-1})}{x_i - x_{i-1}} = F'(c_i^*) = f(c_i^*)$$

Therefore, $F(x_i) - F(x_{i-1}) = f(c_i^*)(x_i - x_{i-1})$, and Eq. (1) can be written

$$F(b) - F(a) = \sum_{i=1}^{N} f(c_i^*)(x_i - x_{i-1}) = \sum_{i=1}^{N} f(c_i^*)\, \Delta x_i$$

This is the Riemann sum approximation to the integral $\int_a^b f(x)\, dx$ with $C = \{c_i^*\}$, so

$$F(b) - F(a) = R(f, P, C)$$

Since $f(x)$ is integrable, $R(f, P, C)$ approaches $\int_a^b f(x)\, dx$ as the norm $\|P\|$ tends to zero. On the other hand, $R(f, P, C)$ is actually *equal* to $F(b) - F(a)$ by our choice of

FIGURE 1 The total change $F(b) - F(a)$ is the sum of terms in a Riemann sum for f.

intermediate points. This proves the desired result:

$$F(b) - F(a) = \lim_{||P|| \to 0} R(f, P, C) = \int_a^b f(x)\, dx \qquad \blacksquare$$

Area computations are much easier with the help of the FTC—but only in cases where an antiderivative can be found.

◄·· **REMINDER** *The Power Rule for Integrals (valid for $n \neq -1$) states that*

$$\int x^n\, dx = \frac{x^{n+1}}{n+1} + C$$

■ EXAMPLE 1   Calculate the area under the graph:

(a) $f(x) = x^3$ over [2, 4]   **(b)** $g(x) = x^{-3/4} + 3x^{5/3}$ over [1, 3]

Solution

(a) Since $F(x) = \frac{1}{4}x^4$ is an antiderivative of $f(x) = x^3$, the FTC I gives us

$$\int_2^4 x^3\, dx = F(4) - F(2) = \frac{1}{4}x^4 \Big|_2^4 = \frac{1}{4}4^4 - \frac{1}{4}2^4 = 60$$

(b) Since $G(x) = 4x^{1/4} + \frac{9}{8}x^{8/3}$ is an antiderivative of $g(x) = x^{-3/4} + 3x^{5/3}$,

$$\int_1^3 (x^{-3/4} + 3x^{5/3})\, dx = G(x) \Big|_1^3 = \left(4x^{1/4} + \frac{9}{8}x^{8/3}\right) \Big|_1^3$$

$$= \left(4 \cdot 3^{1/4} + \frac{9}{8} \cdot 3^{8/3}\right) - \left(4 \cdot 1^{1/4} + \frac{9}{8} \cdot 1^{8/3}\right)$$

$$\approx 26.325 - 5.125 = 21.2$$

This is the area of the region shown in Figure 2.   ■

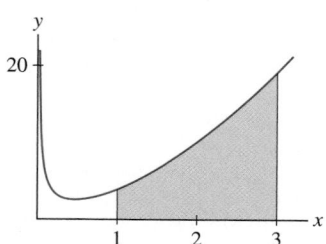

FIGURE 2 The area under the graph of $y = x^{-3/4} + 3x^{5/3}$ over [1, 3] is approximately 21.2.

■ EXAMPLE 2   Sketch the region under $y = \sec^2 x$ for $-\frac{\pi}{4} \leq x \leq \frac{\pi}{4}$ and find its area.

Solution   The region is shown in Figure 3. Recall that $(\tan x)' = \sec^2 x$. Therefore,

$$\int_{-\pi/4}^{\pi/4} \sec^2 x\, dx = \tan x \Big|_{-\pi/4}^{\pi/4} = \tan\left(\frac{\pi}{4}\right) - \tan\left(-\frac{\pi}{4}\right)$$

$$= 1 - (-1) = 2 \qquad \blacksquare$$

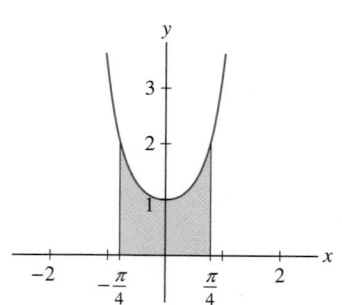

FIGURE 3 Graph of $y = \sec^2 x$.

In the previous section, we saw that the definite integral is equal to the *signed* area between the graph and the x-axis. Needless to say, the FTC "knows" this: When you evaluate an integral using the FTC, you obtain the signed area. This is the principal reason we consider signed area.

■ EXAMPLE 3   Evaluate **(a)** $\int_0^\pi \sin x\, dx$ and **(b)** $\int_0^{2\pi} \sin x\, dx$.

Solution

(a) Since $(-\cos x)' = \sin x$, the area of one "hump" (Figure 4) is

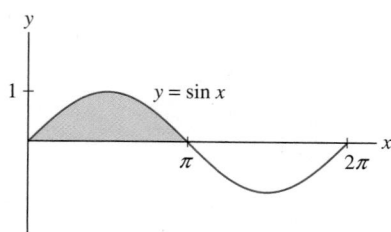

FIGURE 4 Area of one hump is 2. Signed area over $[0, 2\pi]$ is zero.

$$\int_0^\pi \sin x\, dx = -\cos x \Big|_0^\pi = -\cos \pi - (-\cos 0) = -(-1) - (-1) = 2$$

(b) We expect the signed area over $[0, 2\pi]$ to be zero since the second hump lies below the x-axis, and indeed,

$$\int_0^{2\pi} \sin x \, dx = -\cos x \Big|_0^{2\pi} = (-\cos(2\pi)) - (-\cos 0)) = -1 - (-1) = 0 \qquad \blacksquare$$

■ **EXAMPLE 4** Integral of the Exponential Function   Evaluate $\displaystyle\int_{-2}^{1} e^x \, dx$.

Solution The region is shown in Figure 5. Since $F(x) = e^x$ is an antiderivative of $f(x) = e^x$, we have

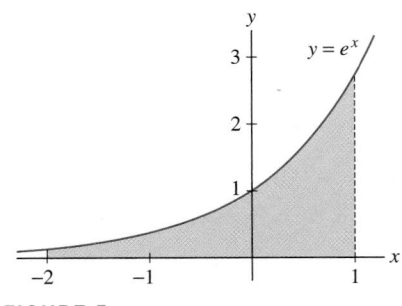

$y = e^x$

FIGURE 5

$$\int_{-2}^{1} e^x \, dx = F(1) - F(-2) = e - e^{-2} \approx 2.58 \qquad \blacksquare$$

Recall, from Section 4.9, that $F(x) = \ln|x|$ is an antiderivative of $f(x) = x^{-1}$ in the domain $\{x : x \neq 0\}$. Therefore, the FTC yields the following formula, valid if both a and b are positive or both are negative:

$$\int_a^b \frac{dx}{x} = \ln|b| - \ln|a| = \ln\frac{b}{a} \qquad \boxed{2}$$

In other words, the area under the hyperbola $y = x^{-1}$ over $[a, b]$ is equal to $\ln\dfrac{b}{a}$ (Figure 6).

■ **EXAMPLE 5** The Logarithm as an Antiderivative   Evaluate

(a) $\displaystyle\int_2^8 \frac{dx}{x}$ and   **(b)** $\displaystyle\int_{-4}^{-2} \frac{dx}{x}$.

Solution By Eq. (2),

$$\int_2^8 \frac{dx}{x} = \ln\frac{8}{2} = \ln 4 \approx 1.39$$

$$\int_{-4}^{-2} \frac{dx}{x} = \ln\left(\frac{-2}{-4}\right) = \ln\frac{1}{2} \approx -0.69$$

The area represented by these integrals is shown in Figures 7(A) and (B).   ■

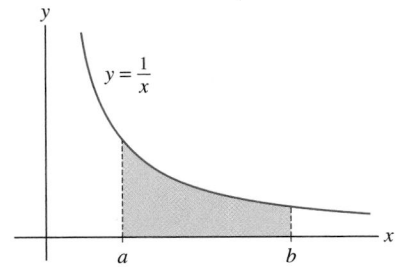

$y = \dfrac{1}{x}$

FIGURE 6 The area under $y = x^{-1}$ over $[a, b]$ is $\ln\dfrac{b}{a}$.

(A) Shaded region has area $\ln 4$.   (B) Shaded region has signed area $\ln\dfrac{1}{2}$.

FIGURE 7

328 CHAPTER 5 | THE INTEGRAL

CONCEPTUAL INSIGHT **Which Antiderivative?** We saw in Chapter 4 that the antiderivative of a function is unique only to within an additive constant. Does it matter which antiderivative is used in the FTC? The answer is no. If $F(x)$ and $G(x)$ are both antiderivatives of $f(x)$, then $F(x) = G(x) + C$ for some constant C. Therefore,

$$F(b) - F(a) = \underbrace{(G(b) + C) - (G(a) + C)}_{\text{The constant cancels}} = G(b) - G(a)$$

and

$$\int_a^b f(x)\,dx = F(b) - F(a) = G(b) - G(a)$$

For example, $F(x) = \frac{1}{3}x^3$ and $G(x) = \frac{1}{3}x^3 - 9$ are both antiderivatives of x^2. Either one may be used to evaluate the definite integral:

$$\int_1^3 x^2\,dx = F(x)\Big|_1^3 = \frac{1}{3}3^3 - \frac{1}{3}1^3 = \frac{26}{3}$$

$$\int_1^3 x^2\,dx = G(x)\Big|_1^3 = \left(\frac{1}{3}3^3 - 9\right) - \left(\frac{1}{3}1^3 - 9\right) = \frac{26}{3}$$

5.3 SUMMARY

- The Fundamental Theorem of Calculus, Part I, states that

$$\int_a^b f(x)\,dx = F(b) - F(a)$$

where $F(x)$ is an antiderivative of $f(x)$. The FTC I is used to evaluate definite integrals in cases where we can find an antiderivative of the integrand.
- Basic antiderivative formulas for evaluating definite integrals:

$$\int x^n\,dx = \frac{x^{n+1}}{n+1} + C \qquad \text{for } n \neq -1$$

$$\int \frac{dx}{x} = \ln|x| + C$$

$$\int e^x\,dx = e^x + C$$

$$\int \sin x\,dx = -\cos x + C, \qquad\qquad \int \cos x\,dx = \sin x + C$$

$$\int \sec^2 x\,dx = \tan x + C, \qquad\qquad \int \csc^2 x\,dx = -\cot x + C$$

$$\int \sec x \tan x\,dx = \sec x + C, \qquad\qquad \int \csc x \cot x\,dx = -\csc x + C$$

5.3 EXERCISES

Preliminary Questions

1. Assume that $f(x) \geq 0$. What is the area under the graph of $f(x)$ over $[0, 2]$ if $f(x)$ has an antiderivative $F(x)$ such that $F(0) = 3$ and $F(2) = 7$?

2. Suppose that $F(x)$ is an antiderivative of $f(x)$. What is the graphical interpretation of $F(4) - F(1)$ if $f(x)$ takes on both positive and negative values?

3. Evaluate $\int_0^7 f(x)\,dx$ and $\int_2^7 f(x)\,dx$, assuming that $f(x)$ has an antiderivative $F(x)$ with values from the following table:

x	0	2	7
$F(x)$	3	7	9

4. Are the following statements true or false? Explain.

(a) The FTC I is only valid for positive functions.

(b) To use the FTC I, you have to choose the right antiderivative.

(c) If you cannot find an antiderivative of $f(x)$, then the definite integral does not exist.

5. What is the value of $\int_2^9 f'(x)\,dx$ if $f(x)$ is differentiable and $f(2) = f(9) = 4$?

Exercises

In Exercises 1–4, sketch the region under the graph of the function and find its area using the FTC I.

1. $f(x) = x^2, \quad [0, 1]$

2. $f(x) = 2x - x^2, \quad [0, 2]$

3. $f(x) = \sin x, \quad [0, \pi/2]$

4. $f(x) = \cos x, \quad [0, \pi/2]$

In Exercises 5–40, evaluate the integral using the FTC I.

5. $\int_3^6 x\,dx$

6. $\int_0^9 2\,dx$

7. $\int_{-3}^2 u^2\,du$

8. $\int_0^1 (x - x^2)\,dx$

9. $\int_3^5 e^x\,dx$

10. $\int_1^4 \left(x + \frac{1}{x}\right)\,dx$

11. $\int_{-2}^0 (3x - 2e^x)\,dx$

12. $\int_{-12}^{-4} \frac{dx}{x}\,dx$

13. $\int_1^3 (t^3 - t^2)\,dt$

14. $\int_0^1 (4 - 5u^4)\,du$

15. $\int_{-3}^4 (x^2 + 2)\,dx$

16. $\int_0^4 (3x^5 + x^2 - 2x)\,dx$

17. $\int_{-2}^2 (10x^9 + 3x^5)\,dx$

18. $\int_{-1}^1 (5u^4 - 6u^2)\,du$

19. $\int_3^1 (4t^{3/2} + t^{7/2})\,dt$

20. $\int_1^2 (x^2 - x^{-2})\,dx$

21. $\int_1^4 \frac{1}{t^2}\,dt$

22. $\int_0^4 \sqrt{y}\,dy$

23. $\int_1^{27} x^{1/3}\,dx$

24. $\int_1^4 x^{-4}\,dx$

25. $\int_1^9 t^{-1/2}\,dt$

26. $\int_4^9 \frac{8}{x^3}\,dx$

27. $\int_{0.2}^{10} \frac{dx}{3x}$

28. $\int_0^1 (9e^x)\,dx$

29. $\int_{-2}^{-1} \frac{1}{x^3}\,dx$

30. $\int_2^4 \pi^2\,dx$

31. $\int_1^{27} \frac{t+1}{\sqrt{t}}\,dt$

32. $\int_0^{\pi/2} \cos\theta\,d\theta$

33. $\int_{-\pi/2}^{\pi/2} \cos x\,dx$

34. $\int_0^{2\pi} \cos t\,dt$

35. $\int_{\pi/4}^{3\pi/4} \sin\theta\,d\theta$

36. $\int_{2\pi}^{4\pi} \sin x\,dx$

37. $\int_0^{\pi/4} \sec^2 t\,dt$

38. $\int_0^{\pi/4} \sec\theta\tan\theta\,d\theta$

39. $\int_{\pi/6}^{\pi/3} \csc x \cot x\,dx$

40. $\int_{\pi/6}^{\pi/2} \csc^2 y\,dy$

In Exercises 41–46, write the integral as a sum of integrals without absolute values and evaluate.

41. $\int_{-2}^1 |x|\,dx$

42. $\int_0^5 |3 - x|\,dx$

43. $\int_{-2}^{3} |x^3|\, dx$

44. $\int_{0}^{3} |x^2 - 1|\, dx$

45. $\int_{0}^{\pi} |\cos x|\, dx$

46. $\int_{0}^{5} |x^2 - 4x + 3|\, dx$

In Exercises 47–52, evaluate the integral in terms of the constants.

47. $\int_{1}^{b} x^3\, dx$

48. $\int_{b}^{a} x^4\, dx$

49. $\int_{1}^{b} x^5\, dx$

50. $\int_{-x}^{x} (t^3 + t)\, dt$

51. $\int_{a}^{5a} \dfrac{dx}{x}$

52. $\int_{b}^{b^2} \dfrac{dx}{x}$

53. Use the FTC I to show that $\int_{-1}^{1} x^n\, dx = 0$ if n is an odd whole number. Explain graphically.

54. What is the area (a positive number) between the x-axis and the graph of $f(x)$ on $[1, 3]$ if $f(x)$ is a *negative* function whose antiderivative F has the values $F(1) = 7$ and $F(3) = 4$?

55. Show that the area of a parabolic arch (the shaded region in Figure 8) is equal to four-thirds the area of the triangle shown.

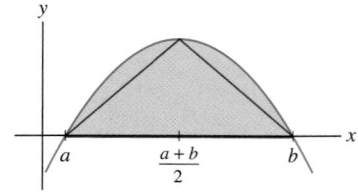

FIGURE 8 Graph of $y = (x - a)(b - x)$.

56. 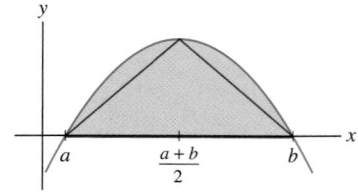 Does $\int_{0}^{1} x^n\, dx$ get larger or smaller as n increases? Explain graphically.

57. Calculate $\int_{-2}^{3} f(x)\, dx$, where

$$f(x) = \begin{cases} 12 - x^2 & \text{for } x \le 2 \\ x^3 & \text{for } x > 2 \end{cases}$$

58. *CAS* Plot the function $f(x) = \sin 3x - x$. Find the positive root of $f(x)$ to three places and use it to find the area under the graph of $f(x)$ in the first quadrant.

Further Insights and Challenges

59. In this exercise, we generalize the result of Exercise 55 by proving the famous result of Archimedes: For $r < s$, the area of the shaded region in Figure 9 is equal to four-thirds the area of triangle $\triangle ACE$, where C is the point on the parabola at which the tangent line is parallel to secant line $\overline{AE}$.

(a) Show that C has x-coordinate $(r + s)/2$.
(b) Show that $ABDE$ has area $(s - r)^3/4$ by viewing it as a parallelogram of height $s - r$ and base of length $\overline{CF}$.
(c) Show that $\triangle ACE$ has area $(s - r)^3/8$ by observing that it has the same base and height as the parallelogram.
(d) Compute the shaded area as the area under the graph minus the area of a trapezoid and prove Archimedes's result.

FIGURE 9 Graph of $f(x) = (x - a)(b - x)$.

60. (a) Apply the Comparison Theorem (Theorem 5 in Section 5.2) to the inequality $\sin x \le x$ (valid for $x \ge 0$) to prove

$$1 - \frac{x^2}{2} \le \cos x \le 1$$

(b) Apply it again to prove

$$x - \frac{x^3}{6} \le \sin x \le x \quad \text{(for } x \ge 0\text{)}$$

(c) Verify these inequalities for $x = 0.3$.

61. Use the method of Exercise 60 to prove that

$$1 - \frac{x^2}{2} \le \cos x \le 1 - \frac{x^2}{2} + \frac{x^4}{24}$$

$$x - \frac{x^3}{6} \le \sin x \le x - \frac{x^3}{6} + \frac{x^5}{120} \quad \text{(for } x \ge 0\text{)}$$

Verify these inequalities for $x = 0.1$. Why have we specified $x \ge 0$ for $\sin x$ but not $\cos x$?

62. Calculate the next pair of inequalities for $\sin x$ and $\cos x$ by integrating the results of Exercise 61. Can you guess the general pattern?

63. Assume that $|f'(x)| \le K$ for $x \in [a, b]$. Use FTC I to prove that $|f(x) - f(a)| \le K|x - a|$ for $x \in [a, b]$.

64. (a) Prove that $|\sin a - \sin b| \le |a - b|$ for all a, b (use Exercise 63).
(b) Let $f(x) = \sin(x + a) - \sin x$. Use part (a) to show that the graph of $f(x)$ lies between the horizontal lines $y = \pm a$.
(c) GU Produce a graph of $f(x)$ and verify part (b) for $a = 0.5$ and $a = 0.2$.

5.4 The Fundamental Theorem of Calculus, Part II

Part I of the Fundamental Theorem shows that we can compute definite integrals using antiderivatives. Part II turns this relationship around: It tells us that we can use the definite integral to *construct* antiderivatives.

To state Part II, we introduce the **area function** (or **cumulative area function**) associated to a function $f(x)$ and lower limit a (Figure 1):

$$A(x) = \int_a^x f(t)\, dt = \text{signed area from } a \text{ to } x$$

In essence, we turn the definite integral into a function by treating the upper limit x as a variable rather than a constant. Notice that $A(a) = 0$ because $A(a) = \int_a^a f(t)\, dt = 0$.

Although $A(x)$ is defined by an integral, in many cases we can find an explicit formula for it.

In the definition of $A(x)$, we use t as the variable of integration to avoid confusion with x, which is the upper limit of integration. In fact, t is a dummy variable and may be replaced by any other variable.

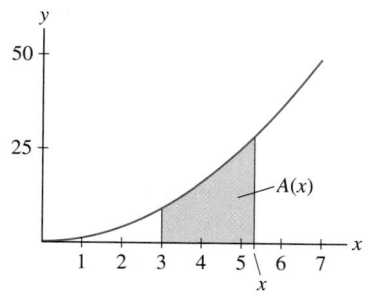

FIGURE 1 $A(x)$ is the area under the graph from a to x.

■ **EXAMPLE 1** Find a formula for the area function $A(x) = \int_3^x t^2\, dt$.

Solution The function $F(t) = \frac{1}{3}t^3$ is an antiderivative for $f(t) = t^2$. By FTC I (Figure 2):

$$A(x) = \int_3^x t^2\, dt = F(x) - F(3) = \frac{1}{3}x^3 - \frac{1}{3}\cdot 3^3 = \frac{1}{3}x^3 - 9 \qquad ■$$

Notice in the previous example that *the derivative of the area function is $f(x)$ itself:*

$$A'(x) = \frac{d}{dx}\left(\frac{1}{3}x^3 - 9\right) = x^2$$

FTC II states that this relation always holds: The derivative of the area function is equal to the original function.

FIGURE 2 The area under $y = x^2$ from 3 to x is $A(x) = \frac{1}{3}x^3 - 9$.

> **THEOREM 1 Fundamental Theorem of Calculus, Part II**   Let $f(x)$ be a continuous function on $[a, b]$. Then $A(x) = \int_a^x f(t)\, dt$ is an antiderivative of $f(x)$, that is, $A'(x) = f(x)$, or equivalently,
>
> $$\frac{d}{dx}\int_a^x f(t)\, dt = f(x)$$
>
> Furthermore, $A(x)$ satisfies the initial condition $A(a) = 0$.

Proof To simplify the proof, we assume that $f(x)$ is increasing (see Exercise 50 for the general case). We compute $A'(x)$ using the limit definition of the derivative. First, let's analyze the numerator of the difference quotient. By the additivity property of the definite integral,

$$A(x + h) - A(x) = \int_a^{x+h} f(t)\, dt - \int_a^x f(t)\, dt = \int_x^{x+h} f(t)\, dt$$

In this proof,

$$A(x) = \int_a^x f(t)\,dt$$

$$A(x+h) - A(x) = \int_x^{x+h} f(t)\,dt$$

$$A'(x) = \lim_{h\to 0} \frac{A(x+h) - A(x)}{h}$$

Assume first that $h > 0$. Then this numerator is equal to the signed area of the thin sliver between the graph and the x-axis from x to $x + h$ (Figure 3).

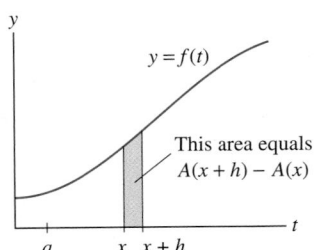

FIGURE 3   The area of the thin sliver equals $A(x+h) - A(x)$.

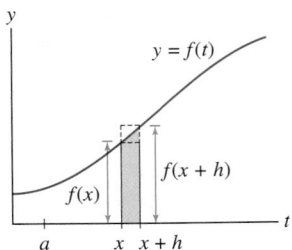

FIGURE 4   The shaded sliver lies between the rectangles of heights $f(x)$ and $f(x+h)$.

This thin sliver lies between the two rectangles of heights $f(x)$ and $f(x + h)$ in Figure 4. Since the rectangles have base h,

$$\underbrace{hf(x)}_{\text{Area of smaller rectangle}} \le \underbrace{A(x+h) - A(x)}_{\text{Area of sliver}} \le \underbrace{hf(x+h)}_{\text{Area of larger rectangle}}$$

Now divide by h to obtain the difference quotient:

$$f(x) \le \frac{A(x+h) - A(x)}{h} \le f(x+h)$$

and apply the Squeeze Theorem. The continuity of $f(x)$ implies that $\lim_{h\to 0} f(x+h) = f(x)$ and $\lim_{h\to 0} f(x) = f(x)$. We conclude that

$$\lim_{h\to 0+} \frac{A(x+h) - A(x)}{h} = f(x) \qquad \boxed{1}$$

If $h < 0$, then $x + h$ lies to the left of x. In this case, the sliver in Figure 5 has signed area $A(x) - A(x + h)$, and a comparison with the rectangles yields

$$-hf(x+h) \le A(x) - A(x+h) \le -hf(x)$$

Dividing by $-h$ (which is positive), we obtain

$$f(x+h) \le \frac{A(x+h) - A(x)}{h} \le f(x)$$

and again, the Squeeze Theorem gives us

$$\lim_{h\to 0-} \frac{A(x+h) - A(x)}{h} = f(x) \qquad \boxed{2}$$

Equations (1) and (2) show that $A'(x)$ exists and $A'(x) = f(x)$.   ∎

FIGURE 5   Here, $h < 0$ and the area of the thin sliver is $A(x) - A(x + h)$.

CONCEPTUAL INSIGHT   Although most elementary functions are defined by formulas or geometry (the trigonometric functions), many new and important functions are defined by integrals. We can graph such functions and compute their values numerically using a computer algebra system. Figure 7 shows a computer-generated graph of the antiderivative of $f(x) = \sin(x^2)$ obtained as an area function (Figure 6).

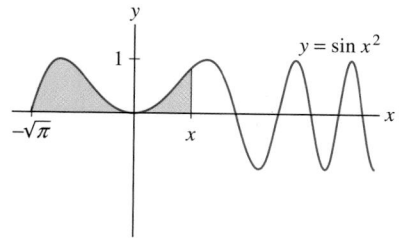

FIGURE 6 Graph of $f(x) = \sin(x^2)$.

FIGURE 7 Computer-generated graph of $F(x) = \displaystyle\int_{-\sqrt{\pi}}^{x} \sin(t^2)\,dt$.

■ **EXAMPLE 2** Expressing an Antiderivative as an Integral   There is no elementary formula for an antiderivative of $f(x) = \sin(x^2)$. Express the antiderivative $F(x)$ satisfying $F(-\sqrt{\pi}) = 0$ as an integral.

Solution   According to FTC II, the area function with lower limit $a = -\sqrt{\pi}$ is an antiderivative satisfying $F(-\sqrt{\pi}) = 0$:

$$F(x) = \int_{-\sqrt{\pi}}^{x} \sin(t^2)\,dt \qquad\blacksquare$$

■ **EXAMPLE 3** Differentiating an Integral   Find the derivative of

$$A(x) = \int_{2}^{x} \sqrt{1 + t^3}\,dt$$

and calculate $A'(2)$, $A'(3)$, and $A(2)$.

Solution   By FTC II, $A'(x) = \sqrt{1 + x^3}$. In particular,

$$A'(2) = \sqrt{1 + 2^3} = 3 \qquad \text{and} \qquad A'(3) = \sqrt{1 + 3^3} = \sqrt{28}$$

On the other hand, $A(2) = \displaystyle\int_{2}^{2} \sqrt{1 + t^3}\,dt = 0$.   ·   ■

CONCEPTUAL INSIGHT   The FTC shows that integration and differentiation are *inverse operations*. By FTC II, if you start with a continuous function $f(x)$ and form the integral $\displaystyle\int_{a}^{x} f(x)\,dx$, then you get back the original function $f(x)$ by differentiating:

$$f(x) \stackrel{\text{Integrate}}{\longrightarrow} \int_{a}^{x} f(t)\,dt \stackrel{\text{Differentiate}}{\longrightarrow} \frac{d}{dx} \int_{a}^{x} f(t)\,dt = f(x)$$

On the other hand, by FTC I, if you first differentiate $f(x)$ and then integrate, you also recover $f(x)$ [but only up to a constant $f(a)$]:

$$f(x) \stackrel{\text{Differentiate}}{\longrightarrow} f'(x) \stackrel{\text{Integrate}}{\longrightarrow} \int_{a}^{x} f'(t)\,dt = f(x) - f(a)$$

When the upper limit of the integral is a *function* of x rather than x itself, we use FTC II together with the Chain Rule to differentiate the integral.

■ **EXAMPLE 4** Combining the FTC and the Chain Rule   Find the derivative of

$$G(x) = \int_{-2}^{x^2} \sin t\,dt.$$

Solution   FTC II does not apply directly because the upper limit is x^2 rather than x. It is necessary to recognize that $G(x)$ is a *composite function*. To see this clearly, define the usual area function $A(x) = \displaystyle\int_{-2}^{x} \sin t\,dt$. Then

$$G(x) = A(x^2) = \int_{-2}^{x^2} \sin t\,dt$$

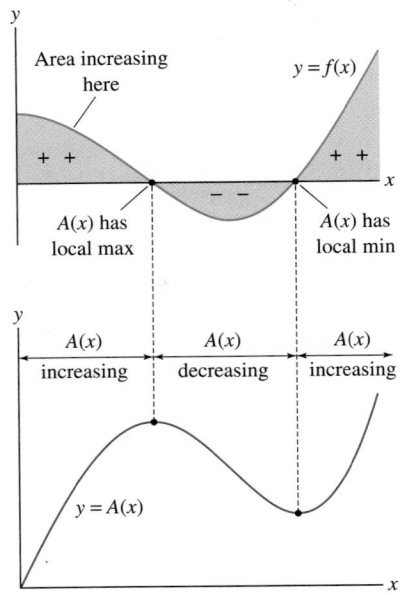

FIGURE 8 The change in area ΔA for a given Δx is larger when $f(x)$ is larger.

FIGURE 9 The sign of $f(x)$ determines the increasing/decreasing behavior of $A(x)$.

FTC II tells us that $A'(x) = \sin x$, so by the Chain Rule,

$$G'(x) = A'(x^2) \cdot (x^2)' = \sin(x^2) \cdot (2x) = 2x \sin(x^2)$$

Alternatively, we may set $u = x^2$ and use the Chain Rule as follows:

$$\frac{dG}{dx} = \frac{d}{dx} \int_{-2}^{x^2} \sin t \, dt = \left(\frac{d}{du} \int_{-2}^{u} \sin t \, dt \right) \frac{du}{dx} = (\sin u)2x = 2x \sin(x^2) \qquad \blacksquare$$

GRAPHICAL INSIGHT   FTC II claims that $A'(x) = f(x)$, which means $f(x)$ is the rate at which $A(x)$ increases or decreases. Figures 8 and 9 provide "evidence" for this claim. First, Figure 8 shows that for a given change Δx, the increase in area is larger at x_2 [where $f(x)$ is larger] than at x_1. Thus, the size of $f(x)$ indeed determines how quickly $A(x)$ changes.

Second, Figure 9 shows that the sign of $f(x)$ determines the increasing/decreasing behavior of $A(x)$. Indeed, $A(x)$ is increasing when $f(x) > 0$ because positive area is added as x moves to the right. When $f(x)$ turns negative, $A(x)$ begins to decrease because we start adding negative area. In particular, $A(x)$ has a local maximum at points where $f(x)$ changes sign from $+$ to $-$, in agreement with the First Derivative Test. Similarly, when $f(x)$ changes sign from $-$ to $+$, $A(x)$ reaches a local minimum. Thus, $f(x)$ "behaves" like the derivative of $A(x)$, as claimed by FTC II.

5.4 SUMMARY

- The *area function* with lower limit a is $A(x) = \int_a^x f(t) \, dt$ and satisfies $A(a) = 0$.

- The FTC II states that $A'(x) = f(x)$, or equivalently, $\dfrac{d}{dx} \int_a^x f(t) \, dt = f(x)$.

- The FTC II shows that every continuous function has an antiderivative, namely the associated area function (with any lower limit).

- To differentiate a function of the form $G(x) = \int_a^{g(x)} f(t) \, dt$, write it as a composite $G(x) = A(g(x))$, where $A(x) = \int_a^x f(t) \, dt$. Then use the Chain Rule:

$$G'(x) = A'(g(x))g'(x) = f(g(x))g'(x)$$

5.4 EXERCISES

Preliminary Questions

1. What is $A(-2)$, where $A(x) = \int_{-2}^x f(t) \, dt$?

2. Let $G(x) = \int_4^x \sqrt{t^3 + 1} \, dt$.

(a) Is the FTC needed to calculate $G(4)$?

(b) Is the FTC needed to calculate $G'(4)$?

3. Which of the following defines an antiderivative $F(x)$ of $f(x) = x^2$ satisfying $F(2) = 0$?

(a) $\int_2^x 2t \, dt$ **(b)** $\int_0^2 t^2 \, dt$ **(c)** $\int_2^x t^2 \, dt$

4. True or false? Some continuous functions do not have antiderivatives. Explain.

5. Let $G(x) = \int_4^{x^3} \sin t \, dt$. Which of the following statements are correct?

(a) $G(x)$ is the composite function $\sin(x^3)$.

(b) $G(x)$ is the composite function $A(x^3)$, where

$$A(x) = \int_4^x \sin(t)\,dt$$

(c) $G(x)$ is too complicated to differentiate.
(d) The Product Rule is used to differentiate $G(x)$.

(e) The Chain Rule is used to differentiate $G(x)$.

(f) $G'(x) = 3x^2 \sin(x^3)$.

6. Trick question: Find the derivative of $\int_1^3 t^3\,dt$ at $x = 2$.

Exercises

1. Write the area function of $f(x) = 2x + 4$ with lower limit $a = -2$ as an integral and find a formula for it.

2. Find a formula for the area function of $f(x) = 2x + 4$ with lower limit $a = 0$.

3. Let $G(x) = \int_1^x (t^2 - 2)\,dt$.

(a) What is $G(1)$?
(b) Use FTC II to find $G'(1)$ and $G'(2)$.
(c) Find a formula for $G(x)$ and use it to verify your answers to (a) and (b).

4. Find $F(0)$, $F'(0)$, and $F'(3)$, where $F(x) = \int_0^x \sqrt{t^2 + t}\,dt$.

5. Find $G(1)$, $G'(0)$, and $G'(\pi/4)$, where $G(x) = \int_1^x \tan t\,dt$.

6. Find $H(-2)$ and $H'(-2)$, where $H(x) = \int_{-2}^x \dfrac{du}{u^2 + 1}$.

In Exercises 7–14, find formulas for the functions represented by the integrals.

7. $\displaystyle\int_2^x u^3\,du$

8. $\displaystyle\int_0^x \sin u\,du$

9. $\displaystyle\int_1^{x^2} t\,dt$

10. $\displaystyle\int_2^x (t^2 - t)\,dt$

11. $\displaystyle\int_x^5 e^t\,dt$

12. $\displaystyle\int_{\pi/4}^x \cos u\,du$

13. $\displaystyle\int_{-\pi/4}^x \sec^2 \theta\,d\theta$

14. $\displaystyle\int_2^{\sqrt{x}} \dfrac{dt}{t}$

In Exercises 15–18, express the antiderivative $F(x)$ of $f(x)$ satisfying the given initial condition as an integral.

15. $f(x) = \sqrt{x^4 + 1}, \quad F(3) = 0$

16. $f(x) = \dfrac{x+1}{x^2 + 9}, \quad F(7) = 0$

17. $f(x) = \sec x, \quad F(0) = 0$

18. $f(x) = e^{-x^2}, \quad F(4) = 0$

In Exercises 19–22, calculate the derivative.

19. $\dfrac{d}{dx}\displaystyle\int_0^x (t^3 - t)\,dt$

20. $\dfrac{d}{dx}\displaystyle\int_1^x \sin(t^2)\,dt$

21. $\dfrac{d}{dt}\displaystyle\int_{100}^t \cos 5x\,dx$

22. $\dfrac{d}{ds}\displaystyle\int_{-2}^s \tan\left(\dfrac{1}{1 + u^2}\right)\,du$

23. Sketch the graph of $A(x) = \displaystyle\int_0^x f(t)\,dt$ for each of the functions shown in Figure 10.

(A)

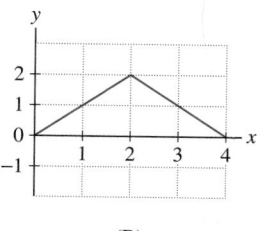

(B)

FIGURE 10

24. Let $A(x) = \displaystyle\int_0^x f(t)\,dt$ for $f(x)$ shown in Figure 11. Calculate $A(2)$, $A(3)$, $A'(2)$, and $A'(3)$. Then find a formula for $A(x)$ (actually two formulas, one for $0 \le x \le 2$ and one for $2 \le x \le 4$) and sketch the graph of $A(x)$.

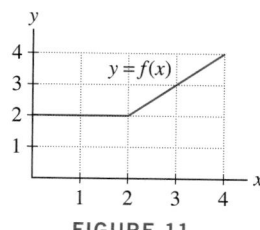

FIGURE 11

25. Make a rough sketch of the graph of the area function of $g(x)$ shown in Figure 12.

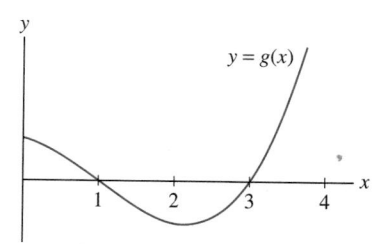

FIGURE 12

26. Show that $\int_0^x |t|\,dt$ is equal to $\frac{1}{2}x|x|$. *Hint:* Consider $x \geq 0$ and $x \leq 0$ separately.

27. Find $G'(x)$, where $G(x) = \int_3^{x^3} \tan t\,dt$.

28. Find $G'(1)$, where $G(x) = \int_0^{x^2} \sqrt{t^3 + 3}\,dt$.

In Exercises 29–36, calculate the derivative.

29. $\dfrac{d}{dx} \displaystyle\int_0^{x^2} \sin^2 t\,dt$

30. $\dfrac{d}{dx} \displaystyle\int_1^{1/x} \sin(t^2)\,dt$

31. $\dfrac{d}{ds} \displaystyle\int_{-6}^{\cos s} (u^4 - 3u)\,du$

32. $\dfrac{d}{dx} \displaystyle\int_x^0 \sin^2 t\,dt$

33. $\dfrac{d}{dx} \displaystyle\int_{x^3}^0 \sin^2 t\,dt$

34. $\dfrac{d}{dx} \displaystyle\int_{x^2}^{x^4} \sqrt{t}\,dt$

Hint for Exercise 34: $F(x) = A(x^4) - A(x^2)$, where

$$A(x) = \int_0^x \sqrt{t}\,dt$$

35. $\dfrac{d}{dx} \displaystyle\int_{\sqrt{x}}^{x^2} \tan t\,dt$

36. $\dfrac{d}{du} \displaystyle\int_{-u}^{3u+9} \sqrt{x^2 + 1}\,dx$

In Exercises 37–38, let $A(x) = \int_0^x f(t)\,dt$ and $B(x) = \int_2^x f(t)\,dt$, with $f(x)$ as in Figure 13.

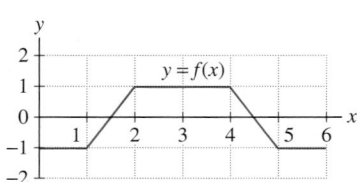

FIGURE 13

37. Find the min and max of $A(x)$ on $[0, 6]$.

38. Find formulas for $A(x)$ and $B(x)$ valid on $[2, 4]$.

39. Let $A(x) = \int_0^x f(t)\,dt$, with $f(x)$ as in Figure 14.

(a) Does $A(x)$ have a local maximum at P?

(b) Where does $A(x)$ have a local minimum?

(c) Where does $A(x)$ have a local maximum?

(d) True or false? $A(x) < 0$ for all x in the interval shown.

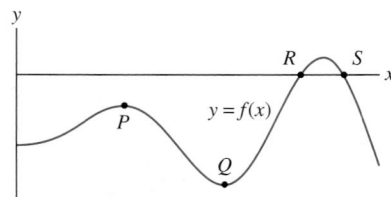

FIGURE 14 Graph of $f(x)$.

40. Find the smallest positive critical point of

$$F(x) = \int_0^x \cos(t^{3/2})\,dt$$

and determine whether it is a local min or max.

In Exercises 41–42, let $A(x) = \int_a^x f(t)\,dt$, where $f(x)$ is continuous.

41. ✎ **Area Functions and Concavity**   Explain why the following statements are true. Assume $f(x)$ is differentiable.

(a) If c is an inflection point of $A(x)$, then $f'(c) = 0$.

(b) $A(x)$ is concave up if $f(x)$ is increasing.

(c) $A(x)$ is concave down if $f(x)$ is decreasing.

42. Match the property of $A(x)$ with the corresponding property of the graph of $f(x)$. Assume $f(x)$ is differentiable.

Area function $A(x)$

(a) $A(x)$ is decreasing.

(b) $A(x)$ has a local maximum.

(c) $A(x)$ is concave up.

(d) $A(x)$ goes from concave up to concave down.

Graph of $f(x)$

(i) Lies below the x-axis.

(ii) Crosses the x-axis from positive to negative.

(iii) Has a local maximum.

(iv) $f(x)$ is increasing.

43. Let $A(x) = \int_0^x f(t)\,dt$, with $f(x)$ as in Figure 15. Determine:

(a) The intervals on which $A(x)$ is increasing and decreasing

(b) The values x where $A(x)$ has a local min or max

(c) The inflection points of $A(x)$

(d) The intervals where $A(x)$ is concave up or concave down

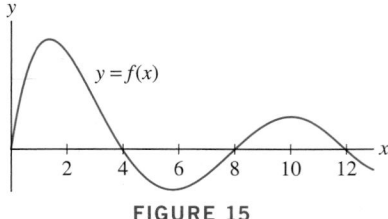

FIGURE 15

44. Let $f(x) = x^2 - 5x - 6$ and $F(x) = \int_0^x f(t)\,dt$.

(a) Find the critical points of $F(x)$ and determine whether they are local minima or maxima.

(b) Find the points of inflection of $F(x)$ and determine whether the concavity changes from up to down or vice versa.

(c) GU   Plot $f(x)$ and $F(x)$ on the same set of axes and confirm your answers to (a) and (b).

45. Sketch the graph of an increasing function $f(x)$ such that both $f'(x)$ and $A(x) = \int_0^x f(t)\,dt$ are decreasing.

46. 📖   Figure 16 shows the graph of $f(x) = x\sin x$. Let $F(x) = \int_0^x t\sin t\,dt$.

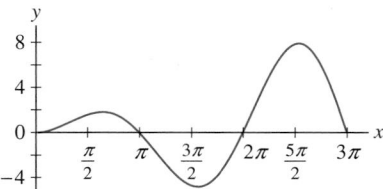

FIGURE 16 Graph of $y = x\sin x$.

(a) Locate the local maxima and absolute maximum of $F(x)$ on $[0, 3\pi]$.

(b) Justify graphically that $F(x)$ has precisely one zero in the interval $[\pi, 2\pi]$.

(c) How many zeros does $F(x)$ have in $[0, 3\pi]$?

(d) Find the inflection points of $F(x)$ on $[0, 3\pi]$ and, for each one, state whether the concavity changes from up to down or vice versa.

47. GU   Find the smallest positive inflection point of

$$F(x) = \int_0^x \cos(t^{3/2})\,dt$$

Use a graph of $y = \cos(x^{3/2})$ to determine whether the concavity changes from up to down or vice versa at this point of inflection.

48. Determine $f(x)$, assuming that $\int_0^x f(t)\,dt$ is equal to $x^2 + x$.

49. Determine the function $g(x)$ and all values of c such that

$$\int_c^x g(t)\,dt = x^2 + x - 6$$

Further Insights and Challenges

50. Proof of FTC II   The proof in the text assumes that $f(x)$ is increasing. To prove it for all continuous functions, let $m(h)$ and $M(h)$ denote the *minimum* and *maximum* of $f(x)$ on $[x, x+h]$ (Figure 17). The continuity of $f(x)$ implies that $\lim_{h\to 0} m(h) = \lim_{h\to 0} M(h) = f(x)$. Show that for $h > 0$,

$$hm(h) \le A(x+h) - A(x) \le hM(h)$$

For $h < 0$, the inequalities are reversed. Prove that $A'(x) = f(x)$.

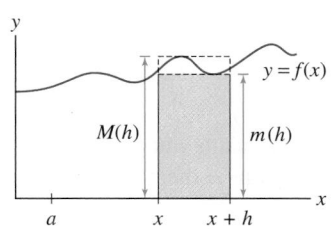

FIGURE 17 Graphical interpretation of $A(x+h) - A(x)$.

51. Proof of FTC I   FTC I asserts that $\int_a^b f(t)\,dt = F(b) - F(a)$ if $F'(x) = f(x)$. Assume FTC II and give a new proof of FTC I as follows. Set $A(x) = \int_a^x f(t)\,dt$.

(a) Show that $F(x) = A(x) + C$ for some constant.

(b) Show that $F(b) - F(a) = A(b) - A(a) = \int_a^b f(t)\,dt$.

52. Can Every Antiderivative Be Expressed as an Integral?   The area function $\int_a^x f(t)\,dt$ is an antiderivative of $f(x)$ for every value of a. However, not all antiderivatives are obtained in this way. The general antiderivative of $f(x) = x$ is $F(x) = \frac{1}{2}x^2 + C$. Show that $F(x)$ is an area function if $C \le 0$ but not if $C > 0$.

53. Find the values $a \le b$ such that $\int_a^b (x^2 - 9)\,dx$ has minimal value.

5.5 Net or Total Change as the Integral of a Rate

So far we have focused on the area interpretation of the integral. In this section, we study the integral as a tool for computing net or total change.

Consider the following problem: Water flows into an empty bucket at a rate of $r(t)$ gallons per second. How much water is in the bucket after 4 s? This would be easy to answer if the rate of water flow were *constant*, say, 0.3 gal/s. We would have

Total amount in bucket = flow rate × length of time = $(0.3)4 = 1.2$ gal

Suppose, however, that the flow rate $r(t)$ varies as in Figure 1. Then *the quantity of water is equal to the area under the graph of $r(t)$.* To prove this, let $s(t)$ be the amount of water

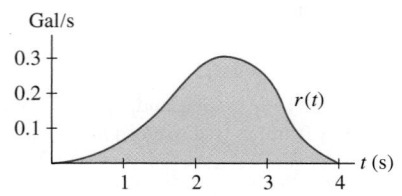

FIGURE 1 The amount of water in the bucket is equal to the area under the graph of the flow rate $r(t)$.

In Theorem 1, the variable t does not have to be a time variable.

in the bucket at time t. Then $s'(t) = r(t)$ because $s'(t)$ is the rate at which the quantity of water is changing, and $s(0) = 0$ because the bucket is initially empty. By FTC I,

$$\underbrace{\int_0^4 s'(t)\, dt}_{\substack{\text{Area under the graph} \\ \text{of the flow rate}}} = s(4) - s(0) = \underbrace{s(4)}_{\substack{\text{Water in bucket} \\ \text{at } t = 4}}$$

More generally, we call the quantity $s(t_2) - s(t_1)$ the **net change** in $s(t)$ over the interval $[t_1, t_2]$. The FTC I yields the following result.

THEOREM 1 Net Change as the Integral of a Rate The net change in $s(t)$ over an interval $[t_1, t_2]$ is given by the integral

$$\underbrace{\int_{t_1}^{t_2} s'(t)\, dt}_{\text{Integral of the rate of change}} = \underbrace{s(t_2) - s(t_1)}_{\text{Total change over } [t_1, t_2]}$$

■ **EXAMPLE 1** At 7 AM, water begins leaking from a tank at a rate of $2 + 0.25t$ gal/hour (t is the number of hours after 7 AM). How much water is lost between 9 and 11 AM?

Solution Let $s(t)$ be the quantity of water in the tank at time t. Then $s(t)$ is decreasing and thus $s'(t) = -(2 + 0.25t)$. Since 9 AM corresponds to $t = 2$, the net change in $s(t)$ between 9 and 11 AM is

$$s(4) - s(2) = \int_2^4 s'(t)\, dt = -\int_2^4 (2 + 0.25t)\, dt$$

$$= -\left(2t + \frac{1}{8}t^2 \right)\Big|_2^4 = -(10 - 4.5) = -5.5 \text{ gal}$$

Thus, the quantity of water lost between 9 and 11 AM is 5.5 gal. ■

In the next example, an integral is estimated using numerical data. We compute both the left- and right-endpoint approximations and take their average. This average, which is usually more accurate than either endpoint approximation alone, is called the **Trapezoidal Approximation;** we will study it in Chapter 7.

■ **EXAMPLE 2** Traffic Flow The number of cars per hour passing an observation point along a highway is called the rate of traffic flow $q(t)$ (in cars per hour).

(a) Which quantity is represented by the integral $\displaystyle\int_{t_1}^{t_2} q(t)\, dt$?

(b) The flow rate is recorded at 15-min intervals between 7:00 and 9:00 AM. Estimate the number of cars using the highway during this 2-hour period by taking the average of the left- and right-endpoint approximations.

t	7:00	7:15	7:30	7:45	8:00	8:15	8:30	8:45	9:00
$q(t)$	1,044	1,297	1,478	1,844	1,451	1,378	1,155	802	542

Solution

(a) The integral represents the total number of cars that passed the observation point during the time interval $[t_1, t_2]$.

(b) The data values are spaced at intervals of $\Delta t = 0.25$ hour. Thus,

$$L_N = 0.25\Big(1{,}044 + 1{,}297 + 1{,}478 + 1{,}844 + 1{,}451 + 1{,}378 + 1{,}155 + 802\Big)$$

$$\approx 2{,}612$$

$$R_N = 0.25\Big(1{,}297 + 1{,}478 + 1{,}844 + 1{,}451 + 1{,}378 + 1{,}155 + 802 + 542\Big)$$

$$\approx 2{,}487$$

We estimate the number of cars that passed the observation point between 7 and 9 AM by taking the average of R_N and L_N:

$$\frac{1}{2}(R_N + L_N) = \frac{1}{2}(2{,}612 + 2{,}487) \approx 2{,}550 \text{ cars passed between 7 and 9 AM} \quad \blacksquare$$

In the left margin:

In Example 2, L_N is the sum of the values of $q(t)$ at the left endpoints

7:00, 7:15, ..., 8:45

and R_N is the sum of the values of $q(t)$ at the right endpoints

7:15, ..., 8:45, 9:00

The Integral of Velocity

At the beginning of this chapter, we claimed that the area under the graph of velocity is equal to the distance traveled. We can now verify this claim, with the understanding that if the velocity is not positive, then the integral computes the *net change in position* or *displacement*. Indeed, if $s(t)$ is the position at time t, then $v(t) = s'(t)$ is the velocity and

$$\int_{t_1}^{t_2} v(t)\, dt = \int_{t_1}^{t_2} s'(t)\, dt = \underbrace{s(t_2) - s(t_1)}_{\text{Displacement or net change in position}}$$

To calculate the actual distance traveled rather than displacement, we must integrate the absolute value of velocity, which is the *speed* $|v(t)|$.

THEOREM 2 The Integral of Velocity   If an object travels in a straight line with velocity $v(t)$, then

$$\text{Displacement during } [t_1, t_2] = \int_{t_1}^{t_2} v(t)\, dt$$

$$\text{Total distance traveled during } [t_1, t_2] = \int_{t_1}^{t_2} |v(t)|\, dt$$

$v(t)$ ft/s

FIGURE 2 Graph of $v(t) = t^3 - 10t^2 + 24t$. Over $[4, 6]$, the broken curve is the graph of $|v(t)|$.

■ **EXAMPLE 3**   The velocity of a particle is $v(t) = t^3 - 10t^2 + 24t$ ft/s (Figure 2). Compute the

(a) Displacement over $[0, 4]$, $[4, 6]$, and $[0, 6]$

(b) Total distance traveled over $[0, 6]$

Indicate the particle's trajectory with a motion diagram.

Solution   We first compute an indefinite integral:

$$\int v(t)\, dt = \int (t^3 - 10t^2 + 24t)\, dt = \frac{1}{4}t^4 - \frac{10}{3}t^3 + 12t^2 + C$$

(a) The displacements over the given intervals are

$$[0, 4]: \quad \int_0^4 v(t)\, dt = \left(\frac{1}{4}t^4 - \frac{10}{3}t^3 + 12t^2\right)\Big|_0^4 = 42\frac{2}{3} \text{ ft}$$

$$[4, 6]: \quad \int_4^6 v(t)\, dt = \left(\frac{1}{4}t^4 - \frac{10}{3}t^3 + 12t^2\right)\Big|_4^6 = -6\frac{2}{3} \text{ ft}$$

$$[0, 6]: \quad \int_0^6 v(t)\, dt = \left(\frac{1}{4}t^4 - \frac{10}{3}t^3 + 12t^2\right)\Big|_0^6 = 36 \text{ ft}$$

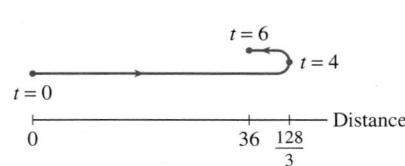

FIGURE 3 Path of the particle along a straight line.

Figure 3 is a motion diagram indicating the particle's trajectory. It travels $42\frac{2}{3}$ ft during the first 4 s and then backtracks $6\frac{2}{3}$ ft over the next 2 s.

(b) To integrate $|v(t)|$, we refer to Figure 2, which shows that $v(t)$ is negative on $[4, 6]$. Although the displacement over $[0, 6]$ is 36 ft, the total distance traveled is

$$\int_0^6 |v(t)|\, dt = \int_0^4 v(t)\, dt - \int_4^6 v(t)\, dt = 42\frac{2}{3} - \left(-6\frac{2}{3}\right) = 49\frac{1}{3} \text{ ft} \quad \blacksquare$$

Total Versus Marginal Cost

Economists and business managers often study how the cost of manufacturing a product varies with the number of units produced. The cost function $C(x)$ is defined as the cost of producing x units of a commodity and the **marginal cost** is the derivative $C'(x)$. We then have

$$\text{Cost of increasing production from } a \text{ to } b = \int_a^b C'(x)\, dx$$

■ **EXAMPLE 4**   The marginal cost of producing x computer chips (in units of 1,000) is $C'(x) = 150x^2 - 3{,}000x + 17{,}500$ (dollars per thousand chips).

(a) Find the cost of increasing production from 10,000 to 15,000 chips.

(b) Determine the total cost of producing 15,000 chips, assuming that $C(0) = 35{,}000$ (in other words, it costs \$35,000 to set up the manufacturing run).

Solution

(a) The cost of increasing production from 10,000 ($x = 10$) to 15,000 ($x = 15$) is

$$C(15) - C(10) = \int_{10}^{15} (150x^2 - 3{,}000x + 17{,}500)\, dx$$

$$= (50x^3 - 1{,}500x^2 + 17{,}500x)\Big|_{10}^{15}$$

$$= 93{,}750 - 75{,}000 = \$18{,}750$$

(b) We have

$$C(15) - C(0) = \int_0^{15} (150x^2 - 3{,}000x + 17{,}500)\, dx$$

$$= (50x^3 - 1{,}500x^2 + 17{,}500x)\Big|_0^{15}$$

$$= \$93{,}750$$

This is the increase in cost from 0 to 15,000 chips. Since setup costs are $C(0) = \$35,000$, the total cost of producing 15,000 chips is

$$C(15) = C(0) + 93,750 = 35,000 + 93,750 = \$128,750$$ ■

5.5 SUMMARY

- Many applications are based on the following principle: *The integral of the rate of change is the net change:*

$$\underbrace{s(t_2) - s(t_1)}_{\text{Net change over } [t_1, t_2]} = \int_{t_1}^{t_2} s'(t)\, dt$$

- For an object traveling in a straight line at velocity $v(t)$,

$$\text{Displacement during } [t_1, t_2] = \int_{t_1}^{t_2} v(t)\, dt$$

$$\text{Total distance traveled during } [t_1, t_2] = \int_{t_1}^{t_2} |v(t)|\, dt$$

- If $C(x)$ is the cost of producing x units of a commodity, then $C'(x)$ is the marginal cost and

$$\text{Cost of increasing production from } a \text{ to } b = \int_a^b C'(x)\, dx$$

5.5 EXERCISES

Preliminary Questions

1. An airplane makes the 350-mile trip from Los Angeles to San Francisco in 1 hour. Assuming that the plane's velocity at time t is $v(t)$ mph, what is the value of the integral $\int_0^1 v(t)\, dt$?

2. A hot metal object is submerged in cold water. The rate at which the object cools (in degrees per minute) is a function $f(t)$ of time. Which quantity is represented by the integral $\int_0^T f(t)\, dt$?

3. Which of the following quantities would be naturally represented as derivatives and which as integrals?

(a) Velocity of a train

(b) Rainfall during a 6-month period

(c) Mileage per gallon of an automobile

(d) Increase in the population of Los Angeles from 1970 to 1990

4. Two airplanes take off at $t = 0$ from the same place and in the same direction. Their velocities are $v_1(t)$ and $v_2(t)$, respectively. What is the physical interpretation of the area between the graphs of $v_1(t)$ and $v_2(t)$ over an interval $[0, T]$?

Exercises

1. Water flows into an empty reservoir at a rate of $3,000 + 5t$ gal/hour. What is the quantity of water in the reservoir after 5 hours?

2. Find the displacement of a particle moving in a straight line with velocity $v(t) = 4t - 3$ ft/s over the time interval $[2, 5]$.

3. A population of insects increases at a rate of $200 + 10t + 0.25t^2$ insects per day. Find the insect population after 3 days, assuming that there are 35 insects at $t = 0$.

4. A survey shows that a mayoral candidate is gaining votes at a rate of $2,000t + 1,000$ votes per day, where t is the number of days since she announced her candidacy. How many supporters will the candidate have after 60 days, assuming that she had no supporters at $t = 0$?

5. A factory produces bicycles at a rate of $95 + 0.1t^2 - t$ bicycles per week (t in weeks). How many bicycles were produced from day 8 to 21?

6. Find the displacement over the time interval $[1, 6]$ of a helicopter whose (vertical) velocity at time t is $v(t) = 0.02t^2 + t$ ft/s.

7. A cat falls from a tree (with zero initial velocity) at time $t = 0$. How far does the cat fall between $t = 0.5$ and $t = 1$ s? Use Galileo's formula $v(t) = -32t$ ft/s.

8. A projectile is released with initial (vertical) velocity 100 m/s. Use the formula $v(t) = 100 - 9.8t$ for velocity to determine the distance traveled during the first 15 s.

In Exercises 9–12, assume that a particle moves in a straight line with given velocity. Find the total displacement and total distance traveled over the time interval, and draw a motion diagram like Figure 3 (with distance and time labels).

9. $12 - 4t$ ft/s, $[0, 5]$

10. $32 - 2t^2$ ft/s, $[0, 6]$

11. $t^{-2} - 1$ m/s, $[0.5, 2]$

12. $\cos t$ m/s, $[0, 4\pi]$

13. The rate (in liters per minute) at which water drains from a tank is recorded at half-minute intervals. Use the average of the left- and right-endpoint approximations to estimate the total amount of water drained during the first 3 min.

t (min)	0	0.5	1	1.5	2	2.5	3
l/min	50	48	46	44	42	40	38

14. The velocity of a car is recorded at half-second intervals (in feet per second). Use the average of the left- and right-endpoint approximations to estimate the total distance traveled during the first 4 s.

t	0	0.5	1	1.5	2	2.5	3	3.5	4
$v(t)$	0	12	20	29	38	44	32	35	30

15. Let $a(t)$ be the acceleration of an object in linear motion at time t. Explain why $\int_{t_1}^{t_2} a(t)\,dt$ is the net change in velocity over $[t_1, t_2]$. Find the net change in velocity over $[1, 6]$ if $a(t) = 24t - 3t^2$ ft/s^2.

16. Show that if acceleration a is constant, then the change in velocity is proportional to the length of the time interval.

17. The traffic flow rate past a certain point on a highway is $q(t) = 3{,}000 + 2{,}000t - 300t^2$, where t is in hours and $t = 0$ is 8 AM. How many cars pass by during the time interval from 8 to 10 AM?

18. Suppose that the marginal cost of producing x video recorders is $0.001x^2 - 0.6x + 350$ dollars. What is the cost of producing 300 units if the setup cost is \$20,000 (see Example 4)? If production is set at 300 units, what is the cost of producing 20 additional units?

19. Carbon Tax   To encourage manufacturers to reduce pollution, a carbon tax on each ton of CO$_2$ released into the atmosphere has been proposed. To model the effects of such a tax, policymakers study the *marginal cost of abatement $B(x)$*, defined as the cost of increasing CO$_2$

reduction from x to $x + 1$ tons (in units of ten thousand tons—Figure 4). Which quantity is represented by $\int_0^3 B(t)\,dt$?

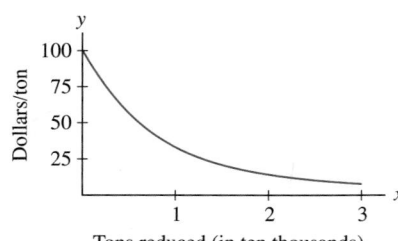

FIGURE 4   Marginal cost of abatement $B(x)$.

20. Power is the rate of energy consumption per unit time. A megawatt of power is 10^6 W or 3.6×10^9 J/hour. Figure 5 shows the power supplied by the California power grid over a typical 1-day period. Which quantity is represented by the area under the graph?

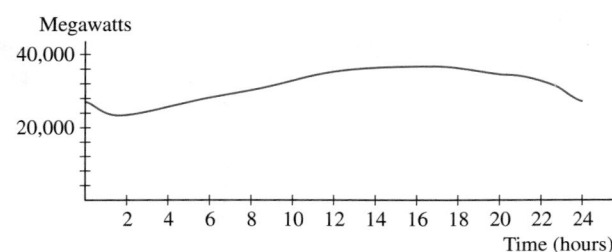

FIGURE 5   Power consumption over 1-day period in California.

21. 📖✏️   Figure 6 shows the migration rate $M(t)$ of Ireland during the period 1988–1998. This is the rate at which people (in thousands per year) move in or out of the country.

(a) What does $\int_{1988}^{1991} M(t)\,dt$ represent?

(b) Did migration over the 11-year period 1988–1998 result in a net influx or outflow of people from Ireland? Base your answer on a rough estimate of the positive and negative areas involved.

(c) During which year could the Irish prime minister announce, "We are still losing population but we've hit an inflection point—the trend is now improving."

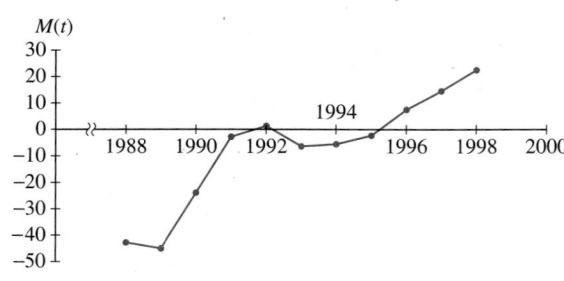

FIGURE 6   Irish migration rate (in thousands per year).

22. Figure 7 shows the graph of $Q(t)$, the rate of retail truck sales in the United States (in thousands sold per year).

(a) What does the area under the graph over the interval [1995, 1997] represent?

(b) Express the total number of trucks sold in the period 1994–1997 as an integral (but do not compute it).

(c) Use the following data to compute the average of the right- and left-endpoint approximations as an estimate for the total number of trucks sold during the 2-year period 1995–1996.

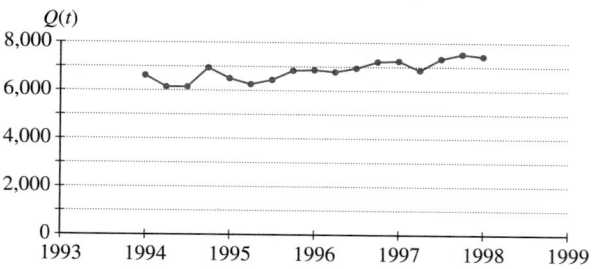

Year (qtr.)	$Q(t)$ ($)	Year (qtr.)	$Q(t)$ ($)
1995(1)	6,484	1996(1)	7,216
1995(2)	6,255	1996(2)	6,850
1995(3)	6,424	1996(3)	7,322
1995(4)	6,818	1996(4)	7,537

FIGURE 7 Quarterly retail sales rate of trucks in the United States (in thousands per year).

23. Heat Capacity The heat capacity $C(T)$ of a substance is the amount of energy (in joules) required to raise the temperature of 1 g by 1°C at temperature T.

(a) Explain why the energy required to raise the temperature from T_1 to T_2 is the area under the graph of $C(T)$ over $[T_1, T_2]$.

(b) How much energy is required to raise the temperature from 50 to 100°C if $C(T) = 6 + 0.2\sqrt{T}$?

In Exercises 24 and 25, consider the following. Paleobiologists have studied the extinction of marine animal families during the phanero-zoic period, which began 544 million years ago. A recent study suggests that the extinction rate $r(t)$ may be modeled by the function

$r(t) = 3{,}130/(t + 262)$ *for* $0 \le t \le 544$. *Here, t is time elapsed (in millions of years) since the beginning of the phanerozoic period. Thus, $t = 544$ refers to the present time, $t = 540$ is 4 million years ago, etc.*

24. Use R_N or L_N with $N = 10$ (or their average) to estimate the total number of families that became extinct in the periods $100 \le t \le 150$ and $350 \le t \le 400$.

25. *CAS* Estimate the total number of extinct families from $t = 0$ to the present, using M_N with $N = 544$.

26. Cardiac output is the rate R of volume of blood pumped by the heart per unit time (in liters per minute). Doctors measure R by injecting A mg of dye into a vein leading into the heart at $t = 0$ and recording the concentration $c(t)$ of dye (in milligrams per liter) pumped out at short regular time intervals (Figure 8).

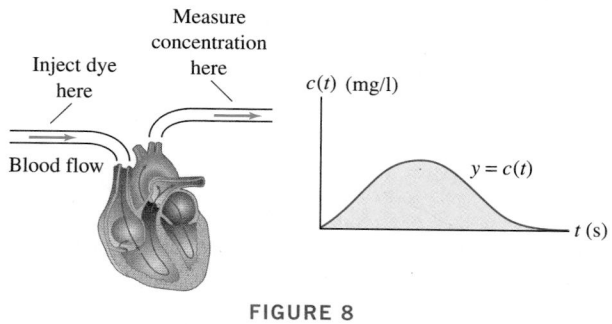

FIGURE 8

(a) The quantity of dye pumped out in a small time interval $[t, t + \Delta t]$ is approximately $Rc(t)\Delta t$. Explain why.

(b) Show that $A = R \displaystyle\int_0^T c(t)\, dt$, where T is large enough that all of the dye is pumped through the heart but not so large that the dye returns by recirculation.

(c) Use the following data to estimate R, assuming that $A = 5$ mg:

t (s)	0	1	2	3	4	5
$c(t)$	0	0.4	2.8	6.5	9.8	8.9
t (s)	6	7	8	9	10	
$c(t)$	6.1	4	2.3	1.1	0	

Further Insights and Challenges

27. A particle located at the origin at $t = 0$ moves along the x-axis with velocity $v(t) = (t + 1)^{-2}$. Show that the particle will never pass the point $x = 1$.

28. A particle located at the origin at $t = 0$ moves along the x-axis with velocity $v(t) = (t + 1)^{-1/2}$. Will the particle be at the point $x = 1$ at any time t? If so, find t.

5.6 Substitution Method

In calculus, the term "integration" is used in two ways. It denotes the process of finding signed area under a graph (computing a definite integral) and the process of finding an antiderivative (evaluating an indefinite integral).

Integration (antidifferentiation) is generally more difficult than differentiation. There are no sure-fire methods for finding an antiderivative, and, in some cases, there is no elementary formula for the antiderivative. However, there are several general techniques that are widely applicable. One such technique is the **Substitution Method**, which uses the Chain Rule "in reverse."

Consider the integral $\int 2x \cos(x^2)\, dx$. We can evaluate it if we remember the Chain Rule calculation

$$\frac{d}{dx} \sin(x^2) = 2x \cos(x^2)$$

This says that $\sin(x^2)$ is an antiderivative of $2x \cos(x^2)$. Therefore,

$$\int \underbrace{2x}_{\substack{\text{Derivative of} \\ \text{inside function}}} \underbrace{\cos(x^2)}_{\substack{\text{Inside} \\ \text{function}}} dx = \sin(x^2) + C$$

A similar Chain Rule calculation shows that

$$\int \underbrace{(1 + 3x^2)}_{\substack{\text{Derivative of} \\ \text{inside function}}} \underbrace{\cos(x + x^3)}_{\substack{\text{Inside} \\ \text{function}}} dx = \sin(x + x^3) + C$$

REMINDER A "composite function" is a function of the form $f(g(x))$. For convenience, we call $g(x)$ the "inside function" and $f(u)$ the "outside function."

In both cases, the integrand is the product of a composite function *and* the derivative of the inside function. The Chain Rule does not help if the derivative of the inside function is missing. For instance, we cannot use the Chain Rule to compute $\int \cos(x^2)\, dx$.

In general, this method works when the integrand is of the form $f(u(x))u'(x)$. If $F(u)$ is an antiderivative of $f(u)$, then by the Chain Rule,

$$\frac{d}{dx} F(u(x)) = F'(u(x))u'(x) = f(u(x))u'(x)$$

This translates into the following integration formula:

THEOREM 1 The Substitution Method   If $F'(x) = f(x)$, then

$$\int f(u(x))u'(x)\, dx = F(u(x)) + C$$

Substitution Using Differentials

Before proceeding to examples, we discuss the procedure for carrying out substitution using differentials. A **differential** is a symbol such as dx or du that occurs in Leibniz's notations du/dx and $\int f(x)\, dx$. We treat differentials as symbols that do not represent actual quantities, but we manipulate them algebraically as if they were related by an equation in which we may cancel dx:

$$du = \frac{du}{dx}\, dx$$

Equivalently,

$$du = u'(x)\, dx$$

For example,

If $u = x^2$, then $du = 2x\, dx$ [since $u'(x) = 2x$]

If $u = \cos(x^3)$, then $du = -3x^2 \sin(x^3)\, dx$ [since $u'(x) = -3x^2 \sin(x^3)$]

Now, when the integrand has the form $f(u(x))\, u'(x)$, we may use Eq. (1) to rewrite the entire integral (including the term dx) in terms of u and its differential du:

The symbolic calculus of substitution using differentials was invented by Leibniz and is considered one of his greatest achievements. It reduces the otherwise complicated process of transforming integrals to a convenient set of rules.

$$\int \underbrace{f(u(x))}_{f(u)} \underbrace{u'(x)\, dx}_{du} = \int f(u)\, du$$

This is often referred to as the **Change of Variables Formula**. It transforms an integral in the variable x into a (hopefully simpler) integral in the new variable u.

In substitution, the key step is to choose the appropriate inside function u.

■ **EXAMPLE 1** Evaluate $\displaystyle\int 3x^2 \sin(x^3)\, dx$ and check the answer.

Solution The integrand contains the composite function $\sin(x^3)$, so we set $u = x^3$. Since the differential $du = 3x^2 dx$ also appears in the integrand, we can carry out the substitution:

$$\int 3x^2 \sin(x^3)\, dx = \int \underbrace{\sin(x^3)}_{\sin u}\ \underbrace{3x^2\, dx}_{du} = \int \sin u\, du$$

Next, we evaluate the integral in the u variable and replace u by x^3 in the answer:

$$\int 3x^2 \sin(x^3)\, dx = \int \sin u\, du = -\cos u + C = -\cos(x^3) + C$$

We can check the answer by differentiating:

$$\frac{d}{dx}(-\cos(x^3)) = \sin(x^3)\frac{d}{dx}x^3 = 3x^2 \sin(x^3)$$

■

■ **EXAMPLE 2** Evaluate $\displaystyle\int 2x(x^2 + 9)^5\, dx$.

Solution For clarity, we break up the solution into three steps.

Step 1. **Choose the function u and compute du.**
 In this case, we let $u = x^2 + 9$. This is a good choice because both the composite $u^5 = (x^2 + 9)^5$ and the differential $du = 2x\, dx$ appear in the integral.

Step 2. **Rewrite the integral in terms of u and du, and evaluate.**

$$\int 2x(x^2 + 9)^5\, dx = \int \overbrace{(x^2 + 9)^5}^{u^5}\ \overbrace{2x\, dx}^{du} = \int u^5\, du = \frac{1}{6}u^6 + C$$

Step 3. Express the final answer in terms of x.
Rewrite the previous answer in terms of x by substituting $u = x^2 + 9$:

$$\int 2x(x^2 + 9)^5 \, dx = \frac{1}{6}u^6 + C = \frac{1}{6}(x^2 + 9)^6 + C$$

∎

■ **EXAMPLE 3** Integral of $\tan\theta$ Evaluate $\displaystyle\int \tan\theta \, d\theta$.

Solution Use the substitution $u = \cos x$, $du = -\sin x \, dx$:

$$\int \tan x \, dx = \int \frac{\sin x}{\cos x} \, dx = -\int \frac{du}{u} = -\ln|u| + C = -\ln|\cos x| + C$$

Now recall that $-\ln u = \ln \dfrac{1}{u}$. Thus, $-\ln|\cos x| = \ln \dfrac{1}{|\cos x|}$, and we obtain

$$\int \tan x \, dx = \ln \left| \frac{1}{\cos x} \right| + C = \ln|\sec x| + C$$

∎

Applying Substitution:

(1) Choose u and compute du.

(2) Rewrite the integral in terms of u and du, and evaluate.

(3) Express the final answer in terms of x.

■ **EXAMPLE 4** Multiplying du by a Constant Evaluate $\displaystyle\int \frac{x^2 + 2x}{(x^3 + 3x^2 + 9)^4} \, dx$.

Solution The composite function $(x^3 + 3x^2 + 9)^{-4}$ in the integrand suggests the substitution $u = x^3 + 3x^2 + 9$. However, the differential

$$du = (3x^2 + 6x) \, dx = 3(x^2 + 2x) \, dx \qquad \boxed{2}$$

does *not* appear in the integrand. What does appear is $(x^2 + 2x) \, dx$. To remedy the situation, we multiply Eq. (2) by $\frac{1}{3}$ to obtain

$$\frac{1}{3} \, du = (x^2 + 2x) \, dx$$

Now we can carry out the substitution:

$$\int \frac{x^2 + 2x}{(x^3 + 3x^2 + 9)^4} \, dx = \int \overbrace{(x^3 + 3x^2 + 9)^{-4}}^{u^{-4}} \overbrace{(x^2 + 2x) \, dx}^{\frac{1}{3} du}$$

$$= \frac{1}{3} \int u^{-4} \, du = -\frac{1}{9}u^{-3} + C$$

$$= -\frac{1}{9}(x^3 + 3x^2 + 9)^{-3} + C$$

∎

■ **EXAMPLE 5** Evaluate $\displaystyle\int \sin(7\theta + 5) \, d\theta$.

Solution Let $u = 7\theta + 5$. Then $du = 7 \, d\theta$ and $\frac{1}{7}du = d\theta$. We obtain

$$\int \sin(7\theta + 5) \, d\theta = \int (\sin u)\left(\frac{1}{7} du\right) = \frac{1}{7}\int \sin u \, du$$

$$= -\frac{1}{7}\cos u + C = -\frac{1}{7}\cos(7\theta + 5) + C$$

∎

■ **EXAMPLE 6** Evaluate $\int e^{-9x}\,dx$.

Solution We use the substitution $u = -9x$, $du = -9\,dx$:

$$\int e^{-9x}\,dx = \int e^u\left(-\frac{1}{9}\,du\right) = -\frac{1}{9}\int e^u\,du = -\frac{1}{9}e^u + C = -\frac{1}{9}e^{-9x} + C \qquad ■$$

■ **EXAMPLE 7** **Additional Step Necessary** Evaluate $\int x\sqrt{5x+1}\,dx$.

Solution We recognize that $\sqrt{5x+1}$ is a composite function, so it makes sense to try $u = 5x + 1$. Then $\frac{1}{5}\,du = dx$ and $\sqrt{5x+1} = u^{1/2}$, and

$$\sqrt{5x+1}\,dx = \frac{1}{5}u^{1/2}\,du$$

Unfortunately, the integrand is not $\sqrt{5x+1}$ but $x\sqrt{5x+1}$. To take care of the extra factor of x, we solve $u = 5x + 1$ to obtain $x = \frac{1}{5}(u-1)$. Then

$$x\sqrt{5x+1}\,dx = \left(\frac{1}{5}(u-1)\right)\frac{1}{5}u^{1/2}\,du = \frac{1}{25}(u-1)\,u^{1/2}\,du$$

and we evaluate as follows:

$$\int x\sqrt{5x+1}\,dx = \frac{1}{25}\int (u-1)\,u^{1/2}\,du = \frac{1}{25}\int (u^{3/2} - u^{1/2})\,du$$

$$= \frac{1}{25}\left(\frac{2}{5}u^{5/2} - \frac{2}{3}u^{3/2}\right) + C$$

$$= \frac{2}{125}(5x+1)^{5/2} - \frac{2}{75}(5x+1)^{3/2} + C \qquad ■$$

Change of Variables Formula for Definite Integrals

There are two ways of using substitution to compute definite integrals. One way is to carry out the Substitution Method fully to find an antiderivative in terms of the x-variable. However, it is often more efficient to evaluate the definite integral directly in terms of the u-variable. This can be done provided that the limits of integration are changed as indicated in the next theorem.

The new limits of integration with respect to the u-variable are $u(a)$ and $u(b)$. Think of it this way: as x varies from a to b, the variable $u = u(x)$ varies from $u(a)$ to $u(b)$.

Change of Variables Formula for Definite Integrals   If $u(x)$ is differentiable on $[a, b]$ and $f(x)$ is integrable on the range of $u(x)$, then

$$\int_a^b f(u(x))u'(x)\,dx = \int_{u(a)}^{u(b)} f(u)\,du$$

Proof If $F(x)$ is an antiderivative of $f(x)$, then $F(u(x))$ is an antiderivative of $f(u(x))u'(x)$. By FTC I,

$$\int_a^b f(u(x))u'(x)\,dx = F(u(b)) - F(u(a))$$

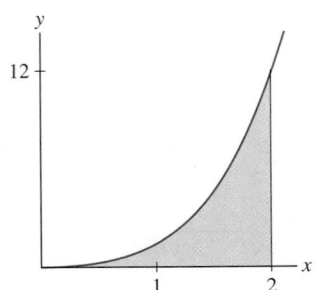

FIGURE 1 Region represented by $\int_0^2 x^2\sqrt{x^3+1}\,dx$.

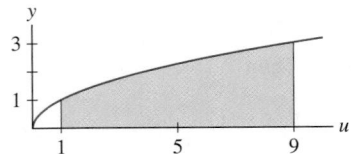

FIGURE 2 Region represented by $\frac{1}{3}\int_1^9 \sqrt{u}\,du$.

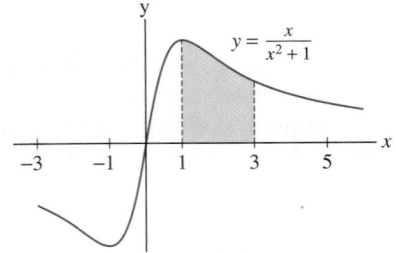

FIGURE 3 Area under the graph of $y=\dfrac{x}{x^2+1}$ over $[1,3]$.

The FTC applied to the u variable also yields

$$\int_{u(a)}^{u(b)} f(u)\,du = F(u(b)) - F(u(a))$$ ∎

■ **EXAMPLE 8** Evaluate $\int_0^2 x^2\sqrt{x^3+1}\,dx$.

Solution Use the substitution $u=x^3+1$, $du=3x^2\,dx$ to obtain

$$x^2\sqrt{x^3+1}\,dx = \frac{1}{3}\sqrt{u}\,du$$

According to the Change of Variables Formula, the new limits of integration are $u(0)=0^3+1=1$ and $u(2)=2^3+1=9$. We obtain

$$\int_0^2 x^2\sqrt{x^3+1}\,dx = \frac{1}{3}\int_1^9 \sqrt{u}\,du = \frac{2}{9}u^{3/2}\Big|_1^9 = \frac{52}{9}$$

This substitution shows that the regions in Figures 1 and 2 have equal area. ∎

■ **EXAMPLE 9** Evaluate $\int_0^{\pi/4} \tan^3 x \sec^2 x\,dx$.

Solution To carry out the substitution, we must remember that $\sec^2 x$ is the derivative of $\tan x$. This suggests setting $u=\tan x$, so that $u^3=\tan^3 x$ and $du=\sec^2 x\,dx$. The new limits of integration are $u(0)=\tan 0=0$ and $u(\frac{\pi}{4})=\tan(\frac{\pi}{4})=1$. Thus,

$$\int_0^{\pi/4}\tan^3 x\sec^2 x\,dx = \int_0^1 u^3\,du = \frac{u^4}{4}\Big|_0^1 = \frac{1}{4}$$ ∎

■ **EXAMPLE 10** Calculate the area under the graph of $y=\dfrac{x}{x^2+1}$ over $[1,3]$.

Solution The area in question (the shaded region in Figure 3) is equal to $\int_1^3 \dfrac{x}{x^2+1}\,dx$. We use the substitution

$$u=x^2+1,\qquad \frac{1}{2}du=x\,dx,\qquad \frac{1}{2}\frac{du}{u}=\frac{x\,dx}{x^2+1}$$

The new limits of integration are $u(1)=1^2+1=2$ and $u(3)=3^2+1=10$. By the Change of Variables Formula,

$$\int_1^3 \frac{x}{x^2+1}\,dx = \frac{1}{2}\int_2^{10}\frac{du}{u} = \frac{1}{2}\ln|u|\Big|_2^{10} = \frac{1}{2}\ln 10 - \frac{1}{2}\ln 2 \approx 0.805$$ ∎

5.6 SUMMARY

• Try the Substitution Method when the integrand has the form $f(u(x))\,u'(x)$. If F is an antiderivative of f, then:

$$\int f(u(x))\,u'(x)\,dx = F(u(x)) + C$$

- The differential of $u(x)$ is the symbol $du = u'(x)\,dx$. The Substitution Method is expressed by the Change of Variables Formula:

$$\int f(u(x))\,u'(x)\,dx = \int f(u)\,du$$

- Definite integrals may be evaluated using the Change of Variables Formula, provided that the limits of integration are changed appropriately:

$$\int_a^b f(u(x))\,u'(x)\,dx = \int_{u(a)}^{u(b)} f(u)\,du$$

5.6 EXERCISES

Preliminary Questions

1. Which of the following integrals is a candidate for the Substitution Method?

(a) $\int 5x^4 \sin(x^5)\,dx$

(b) $\int \sin^5 x \cos x\,dx$

(c) $\int x^5 \sin x\,dx$

2. Write each of the following functions in the form $cg(u(x))u'(x)$,

where c is a constant.

(a) $x(x^2+9)^4$

(b) $x^2 \sin(x^3)$

(c) $\sin x \cos^2 x$

3. Which of the following is equal to $\int_0^2 x^2(x^3+1)\,dx$ for a suitable substitution?

(a) $\dfrac{1}{3}\int_0^2 u\,du$

(b) $\int_0^9 u\,du$

(c) $\dfrac{1}{3}\int_1^9 u\,du$

Exercises

In Exercises 1–6, calculate du for the given function.

1. $u = 1 - x^2$

2. $u = \sin x$

3. $u = x^3 - 2$

4. $u = 2x^4 + 8x$

5. $u = \cos(x^2)$

6. $u = \tan x$

In Exercises 7–28, write the integral in terms of u and du. Then evaluate.

7. $\int (x-7)^3\,dx, \quad u = x - 7$

8. $\int 2x\sqrt{x^2+1}\,dx, \quad u = x^2 + 1$

9. $\int (x+1)^{-2}\,dx, \quad u = x + 1$

10. $\int x(x+1)^9\,dx, \quad u = x + 1$

11. $\int \sin(2x-4)\,dx, \quad u = 2x - 4$

12. $\int \dfrac{x^3}{(x^4+1)^4}\,dx, \quad u = x^4 + 1$

13. $\int \dfrac{x+1}{(x^2+2x)^3}\,dx, \quad u = x^2 + 2x$

14. $\int \dfrac{x}{(8x+5)^3}\,dx, \quad u = 8x + 5$

15. $\int \sqrt{4x-1}\,dx, \quad u = 4x - 1$

16. $\int x\sqrt{4x-1}\,dx, \quad u = 4x - 1$

17. $\int x^2\sqrt{4x-1}\,dx, \quad u = 4x - 1$

18. $\int x\cos(x^2)\,dx, \quad u = x^2$

19. $\int \sin^2 x \cos x\,dx, \quad u = \sin x$

20. $\int \sec^2 x \tan x\,dx, \quad u = \tan x$

21. $\int \tan 2x\,dx, \quad u = \cos 2x$

22. $\int \cot x\,dx, \quad u = \sin x$

23. $\int xe^{-x^2}\,dx, \quad u = -x^2$

24. $\int (\sec^2 \theta)\,e^{\tan \theta}\,d\theta, \quad u = \tan \theta$

25. $\int \dfrac{e^t\,dt}{e^{2t}+2e^t+1}, \quad u = e^t$

26. $\int \dfrac{(\ln x)^2 \, dx}{x}, \quad u = \ln x$

27. $\int \dfrac{dx}{x(\ln x)^2}, \quad u = \ln x$

28. $\int \dfrac{(\tan^{-1} x)^2 \, dx}{x^2 + 1}, \quad u = \tan^{-1} x$

In Exercises 29–32, show that each of the following integrals is equal to a multiple of $\sin(u(x)) + C$ for an appropriate choice of $u(x)$.

29. $\int x^3 \cos(x^4) \, dx$

30. $\int x^2 \cos(x^3 + 1) \, dx$

31. $\int x^{1/2} \cos(x^{3/2}) \, dx$

32. $\int \cos x \cos(\sin x) \, dx$

In Exercises 33–70, evaluate the indefinite integral.

33. $\int (4x + 3)^4 \, dx$

34. $\int x^2 (x^3 + 1)^3 \, dx$

35. $\int \dfrac{1}{\sqrt{x - 7}} \, dx$

36. $\int \sin(x - 7) \, dx$

37. $\int x\sqrt{x^2 - 4} \, dx$

38. $\int (2x + 1)(x^2 + x)^3 \, dx$

39. $\int \dfrac{dx}{(x + 9)^2}$

40. $\int \dfrac{x}{\sqrt{x^2 + 9}} \, dx$

41. $\int \dfrac{2x^2 + x}{(4x^3 + 3x^2)^2} \, dx$

42. $\int (3x^2 + 1)(x^3 + x)^2 \, dx$

43. $\int \dfrac{5x^4 + 2x}{(x^5 + x^2)^3} \, dx$

44. $\int x^2 (x^3 + 1)^4 \, dx$

45. $\int (3x + 9)^{10} \, dx$

46. $\int x(3x + 9)^{10} \, dx$

47. $\int x(x + 1)^{1/4} \, dx$

48. $\int x^2 (x + 1)^7 \, dx$

49. $\int x^3 (x^2 - 1)^{3/2} \, dx$

50. $\int x^2 \sin(x^3) \, dx$

51. $\int \sin^5 x \cos x \, dx$

52. $\int x^2 \sin(x^3 + 1) \, dx$

53. $\int \tan 3x \, dx$

54. $\int \dfrac{\tan(\ln x)}{x} \, dx$

55. $\int \sec^2(4x + 9) \, dx$

56. $\int \sec^2 x \tan^4 x \, dx$

57. $\int \dfrac{\cos 2x}{(1 + \sin 2x)^2} \, dx$

58. $\int \sin 4x \sqrt{\cos 4x + 1} \, dx$

59. $\int \cos x (3 \sin x - 1) \, dx$

60. $\int \dfrac{\cos \sqrt{x}}{\sqrt{x}} \, dx$

61. $\int \sec^2 x (4 \tan^3 x - 3 \tan^2 x) \, dx$

62. $\int e^{14x - 7} \, dx$

63. $\int (x + 1)e^{x^2 + 2x} \, dx$

64. $\int \dfrac{dx}{(x + 1)^4}$

65. $\int \dfrac{e^x \, dx}{(e^x + 1)^4}$

66. $\int \dfrac{\sec^2(\sqrt{x}) \, dx}{\sqrt{x}}$

67. $\int \dfrac{(\ln x)^4 \, dx}{x}$

68. $\int \dfrac{dx}{x\sqrt{\ln x}}$

69. $\int \dfrac{dx}{x \ln x}$

70. $\int (\cot x) \ln(\sin x) \, dx$

71. Evaluate $\int x^5 \sqrt{x^3 + 1} \, dx$ using $u = x^3 + 1$. *Hint:* $x^5 \, dx = x^3 \cdot x^2 \, dx$ and $x^3 = u - 1$.

72. Evaluate $\int (x^3 + 1)^{1/4} x^5 \, dx$.

73. Can They Both Be Right?   Hannah uses the substitution $u = \tan x$ and Akiva uses $u = \sec x$ to evaluate $\int \tan x \sec^2 x \, dx$. Show that they obtain different answers and explain the apparent contradiction.

74. Evaluate $\int \sin x \cos x \, dx$ using substitution in two different ways: first using $u = \sin x$ and then $u = \cos x$. Reconcile the two different answers.

75. Some Choices Are Better Than Others   Evaluate

$$\int \sin x \, \cos^2 x \, dx$$

twice. First use $u = \sin x$ to show that

$$\int \sin x \, \cos^2 x \, dx = \int u\sqrt{1 - u^2} \, du$$

and evaluate the integral on the right by a further substitution. Then show that $u = \cos x$ is a better choice.

76. What are the new limits of integration if we apply the substitution $u = 3x + \pi$ to the integral $\int_0^\pi \sin(3x + \pi) \, dx$?

77. Which of the following is the result of applying the substitution $u = 4x - 9$ to the integral $\int_2^8 (4x - 9)^{20} \, dx$?

(a) $\int_2^8 u^{20} \, du$

(b) $\dfrac{1}{4} \int_2^8 u^{20} \, du$

(c) $4 \int_{-1}^{23} u^{20} \, du$

(d) $\dfrac{1}{4} \int_{-1}^{23} u^{20} \, du$

In Exercises 78–91, use the Change of Variables Formula to evaluate the definite integral.

78. $\int_1^3 (x + 2)^3 \, dx$

79. $\int_1^6 \sqrt{x + 3} \, dx$

80. $\displaystyle\int_0^1 \frac{x}{(x^2+1)^3}\,dx$

81. $\displaystyle\int_{-1}^2 \sqrt{5x+6}\,dx$

82. $\displaystyle\int_0^4 x\sqrt{x^2+9}\,dx$

83. $\displaystyle\int_0^2 \frac{x+3}{(x^2+6x+1)^3}\,dx$

84. $\displaystyle\int_1^2 (x+1)(x^2+2x)^3\,dx$

85. $\displaystyle\int_{10}^{17} (x-9)^{-2/3}\,dx$

86. $\displaystyle\int_0^{\pi/4} \tan\theta\,d\theta$

87. $\displaystyle\int_0^1 \theta\tan(\theta^2)\,d\theta$

88. $\displaystyle\int_0^{\pi/2} \cos 3x\,dx$

89. $\displaystyle\int_0^{\pi/2} \cos\left(3x+\frac{\pi}{2}\right)dx$

90. $\displaystyle\int_0^{\pi/2} \cos^3 x\sin x\,dx$

91. $\displaystyle\int_0^{\pi/4} \tan^2 x\sec^2 x\,dx$

92. Evaluate $\displaystyle\int \frac{dx}{(2+\sqrt{x})^3}$ using $u=2+\sqrt{x}$.

93. Evaluate $\displaystyle\int_0^2 r\sqrt{5-\sqrt{4-r^2}}\,dr$.

In Exercises 94–95, use substitution to evaluate the integral in terms of $f(x)$.

94. $\displaystyle\int f(x)^3 f'(x)\,dx$

95. $\displaystyle\int \frac{f'(x)}{f(x)^2}\,dx$

96. Show that $\displaystyle\int_0^{\pi/6} f(\sin\theta)\,d\theta = \int_0^{1/2} f(u)\frac{1}{\sqrt{1-u^2}}\,du$.

97. Evaluate $\displaystyle\int_0^{\pi/2} \sin^n x\cos x\,dx$, where n is an integer, $n\neq -1$.

Further Insights and Challenges

98. Use the substitution $u=1+x^{1/n}$ to show that
$$\int \sqrt{1+x^{1/n}}\,dx = n\int u^{1/2}(u-1)^{n-1}\,du$$
Evaluate for $n=2,3$.

99. Evaluate $\displaystyle I = \int_0^{\pi/2} \frac{d\theta}{1+\tan^{6,000}\theta}$. *Hint:* Use substitution to show that I is equal to $\displaystyle J = \int_0^{\pi/2} \frac{d\theta}{1+\cot^{6,000}\theta}$ and then check that $\displaystyle I+J = \int_0^{\pi/2} d\theta$.

100. Show that $\displaystyle\int_{-a}^a f(x)\,dx = 0$ if f is an odd function.

101. **(a)** Use the substitution $u=x/a$ to prove that the hyperbola $y=x^{-1}$ (Figure 4) has the following special property: If $a,b>0$, then $\displaystyle\int_a^b \frac{1}{x}\,dx = \int_1^{b/a} \frac{1}{x}\,dx$.
(b) Show that the areas under the hyperbola over the intervals $[1,2]$, $[2,4]$, $[4,8]$, ... are all equal.

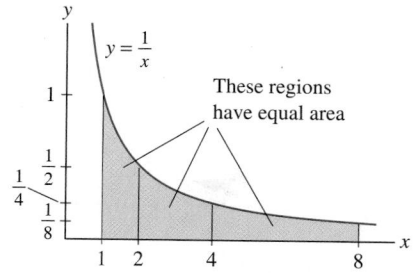

FIGURE 4 The area under $y=\frac{1}{x}$ over $[2^n, 2^{n+1}]$ is the same for all $n=0,1,2\ldots$.

102. Show that the two regions in Figure 5 have the same area. Then use the identity $\cos^2 u = \frac{1}{2}(1+\cos 2u)$ to compute the second area.

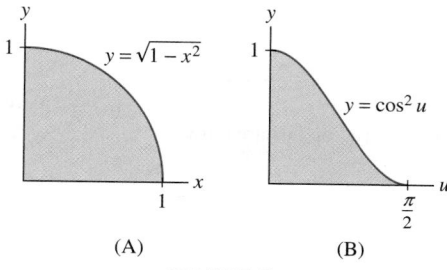

(A)   (B)

FIGURE 5

103. [image] **Area of a Circle**   The number π is defined as one-half the *circumference* of the unit circle. Prove that the area of a circle of radius r is $A=\pi r^2$. The case $r=1$ follows from Exercise 102. Prove it for all $r>0$ by showing that
$$\int_0^r \sqrt{r^2-x^2}\,dx = r^2\int_0^1 \sqrt{1-x^2}\,dx$$

104. Area of an Ellipse   Prove the formula $A=\pi ab$ for the area of the ellipse with equation
$$\frac{x^2}{a^2}+\frac{y^2}{b^2}=1$$
Hint: Show that $A = 2b\displaystyle\int_{-a}^a \sqrt{1-(x/a)^2}\,dx$, change variables, and use the formula for the area of a circle (Figure 6).

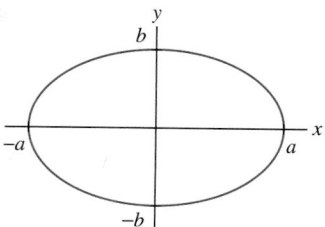

FIGURE 6 Graph of $\dfrac{x^2}{a^2}+\dfrac{y^2}{b^2}=1$.

5.7 Further Transcendental Functions

In Section 5.3, we used the FTC to derive the formula

$$\int_a^b \frac{dx}{x} = \ln \frac{b}{a}$$

Setting $a = 1$ and $b = x$, we obtain the following formula of $\ln x$ as a definite integral:

$$\boxed{\ln x = \int_1^x \frac{dt}{t} \qquad \text{for } x > 0} \qquad \boxed{1}$$

In other words, $\ln x$ is equal to the area under the hyperbola over $[1, x]$ as shown in Figure 1. In some treatments of logarithms, Eq. (1) is taken as the definition of the logarithm. All the properties of $\ln x$ may be derived from this integral expression (see Exercise 77).

In a similar fashion, the inverse trigonometric functions can be expressed as integrals. For example, we showed in Section 3.9 that

$$\frac{d}{dx} \sin^{-1} x = \frac{1}{\sqrt{1 - x^2}}$$

This yields the formula

$$\int \frac{dx}{\sqrt{1 - x^2}} = \sin^{-1} x + C$$

Since $\sin^{-1} 0 = 0$, we may express the arcsine as a definite integral:

$$\boxed{\sin^{-1} x = \int_0^x \frac{dt}{\sqrt{1 - t^2}} \qquad \text{for } -1 < x < 1}$$

Thus, $\sin^{-1} x$ is equal to the area under the graph of $y = (1 - x^2)^{-1/2}$ over $[0, x]$ as shown in Figure 2. The following integral formulas follow from the derivative formulas derived in Section 3.9.

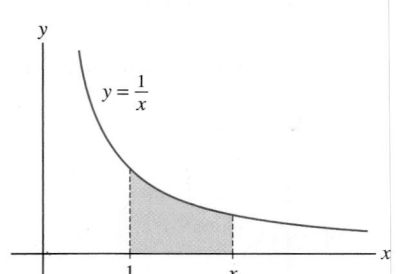

FIGURE 1 Shaded region has area $\ln x$.

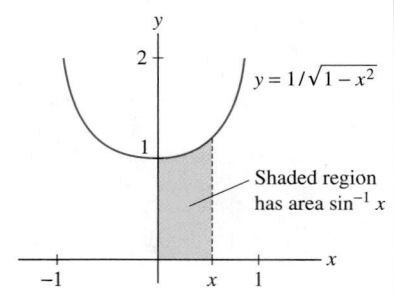

FIGURE 2

THEOREM 1 Inverse Trigonometric Functions

$$\frac{d}{dx} \sin^{-1} x = \frac{1}{\sqrt{1 - x^2}}, \qquad \int \frac{dx}{\sqrt{1 - x^2}} = \sin^{-1} x + C \qquad \boxed{2}$$

$$\frac{d}{dx} \tan^{-1} x = \frac{1}{x^2 + 1}, \qquad \int \frac{dx}{x^2 + 1} = \tan^{-1} x + C \qquad \boxed{3}$$

$$\frac{d}{dx} \sec^{-1} x = \frac{1}{|x|\sqrt{x^2 - 1}}, \qquad \int \frac{dx}{|x|\sqrt{x^2 - 1}} = \sec^{-1} x + C \qquad \boxed{4}$$

In this list, we omit the integral formulas corresponding to the derivatives of $\cos^{-1} x$, $\cot^{-1}$, and $\csc^{-1} x$. The resulting integrals differ only by a minus sign from those already on the list. For example,

$$\frac{d}{dx} \cos^{-1} x = -\frac{1}{\sqrt{1 - x^2}}, \qquad \int \frac{dx}{\sqrt{1 - x^2}} = -\cos^{-1} x + C$$

Thus, we may evaluate integrals of the type $\int \dfrac{dx}{\sqrt{1-x^2}}$ using $\sin^{-1} x$ or $\cos^{-1} x$.

■ **EXAMPLE 1** Evaluate **(a)** $\displaystyle\int_0^1 \dfrac{dx}{x^2+1}$ and **(b)** $\displaystyle\int_0^2 \dfrac{dx}{\sqrt{9-x^2}}$.

Solution

(a) Use Eq. (3):

$$\int_0^1 \frac{dx}{x^2+1} = \tan^{-1} x \Big|_0^1 = \tan^{-1} 1 - \tan^{-1} 0 = \frac{\pi}{4} - 0 = \frac{\pi}{4}$$

This integral represents the shaded area in Figure 3.

(b) Since the integral involves $\sqrt{9-x^2}$ instead of $\sqrt{1-x^2}$, Eq. (2) cannot be applied directly. To take care of the 9, we use the substitution

$$x = 3u, \qquad dx = 3\,du$$

$$\sqrt{9-x^2} = \sqrt{9-(3u)^2} = \sqrt{9(1-u^2)} = 3\sqrt{1-u^2}$$

Since $u = \frac{1}{3}x$, the new limits of integration are $u(0) = 0$ and $u(2) = \frac{2}{3}$:

$$\int_0^2 \frac{dx}{\sqrt{9-x^2}} = \int_0^{2/3} \frac{3\,du}{3\sqrt{1-u^2}} = \sin^{-1}\frac{2}{3} - \sin^{-1} 0 \approx 0.73 - 0 = 0.73 \qquad ■$$

5

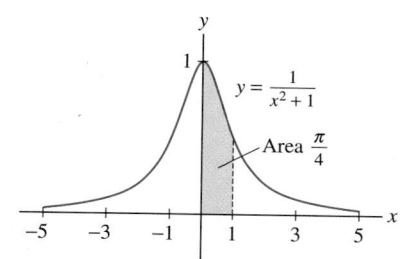

In substitution, we usually define u as a function of x. Sometimes, it is more convenient to define x as a function of u. We do this in Eq. (5), where we set $x = 3u$.

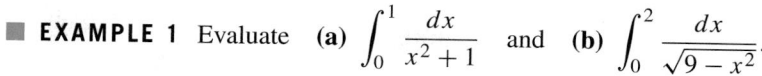

y

$y = \dfrac{1}{x^2+1}$

Area $\dfrac{\pi}{4}$

−5 −3 −1   1   3   5   *x*

FIGURE 3 The shaded region has an area equal to $\tan^{-1} 1 = \frac{\pi}{4}$.

■ **EXAMPLE 2** **Using Substitution** Evaluate $\displaystyle\int_{1/\sqrt{2}}^1 \dfrac{dx}{x\sqrt{4x^2-1}}$.

Solution To apply Eq. (4), we rewrite the integral using the substitution

$$u = 2x, \qquad du = 2\,dx$$

The new limits are $u\left(\frac{1}{\sqrt{2}}\right) = 2\left(\frac{1}{\sqrt{2}}\right) = \sqrt{2}$ and $u(1) = 2$. Thus, we obtain

$$\int_{1/\sqrt{2}}^1 \frac{dx}{x\sqrt{4x^2-1}} = \int_{\sqrt{2}}^2 \frac{\frac{1}{2}\,du}{(u/2)\sqrt{u^2-1}}$$

$$= \int_{\sqrt{2}}^2 \frac{du}{u\sqrt{u^2-1}}$$

$$= \sec^{-1} 2 - \sec^{-1}\sqrt{2} = \frac{\pi}{3} - \frac{\pi}{4} = \frac{\pi}{12} \qquad ■$$

Integrals Involving $f(x) = b^x$

We have seen that the exponential function to the base e is particularly convenient because $f(x) = e^x$ is both its own derivative and its own antiderivative. Thus,

$$\frac{d}{dx}e^x = e^x, \qquad \int e^x\,dx = e^x + C$$

We now consider an arbitrary base b, where $b > 0$ and $b \neq 1$. Recall that to differentiate $f(x) = b^x$, we write $b^x = (e^{\ln b})^x = e^{(\ln b)x}$ and use the Chain Rule:

$$\frac{d}{dx}b^x = \frac{d}{dx}e^{(\ln b)x} = (\ln b)e^{(\ln b)x} = (\ln b)b^x$$

Therefore, $\dfrac{d}{dx}\left(\dfrac{b^x}{\ln b}\right) = b^x$ and we obtain the integral formula

$$\boxed{\int b^x \, dx = \dfrac{b^x}{\ln b} + C}$$ $\boxed{6}$

■ **EXAMPLE 3** Evaluate **(a)** $\displaystyle\int_2^5 10^x \, dx$ and **(b)** $\displaystyle\int_0^{\pi/2} (\cos\theta) 10^{\sin\theta} \, d\theta$.

Solution We apply Eq. (6) with $b = 10$.

(a) $\displaystyle\int_2^5 10^x \, dx = \dfrac{10^x}{\ln 10}\bigg|_2^5 = \dfrac{10^5 - 10^2}{\ln 10} \approx 43{,}386.0$

(b) Use the substitution $u = \sin\theta$, $du = \cos\theta \, d\theta$. The new limits of integration become $u(0) = 0$ and $u(\pi/2) = 1$:

$$\int_0^{\pi/2} (\cos\theta) 10^{\sin\theta} \, d\theta = \int_0^1 10^u \, du = \dfrac{10^u}{\ln 10}\bigg|_0^1 = \dfrac{10^1 - 10^0}{\ln 10} \approx 3.9 \qquad ■$$

5.7 SUMMARY

- Formula for the logarithm as a definite integral:

$$\ln x = \int_1^x \dfrac{dt}{t}$$

- Integral formulas for inverse trigonometric functions:

$$\int \dfrac{dx}{\sqrt{1 - x^2}} = \sin^{-1} x + C$$

$$\int \dfrac{dx}{x^2 + 1} = \tan^{-1} x + C$$

$$\int \dfrac{dx}{|x|\sqrt{x^2 - 1}} = \sec^{-1} x + C$$

- Integrals of exponential functions ($b > 0$, $b \neq 1$):

$$\int e^x \, dx = e^x + C, \qquad \int b^x \, dx = \dfrac{b^x}{\ln b} + C$$

5.7 EXERCISES

Preliminary Questions

1. What is the general antiderivative of the function?

(a) $f(x) = 2^x$ **(b)** $f(x) = x^{-1}$

(c) $f(x) = (1 - x^2)^{-1/2}$

2. Find a value of b such that $\displaystyle\int_1^b \dfrac{dx}{x}$ is equal to

(a) $\ln 3$ **(b)** 3

3. For which value of b is $\displaystyle\int_0^b \dfrac{dx}{1 + x^2} = \dfrac{\pi}{3}$?

4. Which of the following integrals should be evaluated using substitution?

(a) $\displaystyle\int \dfrac{9 \, dx}{1 + x^2}$ **(b)** $\displaystyle\int \dfrac{dx}{1 + 9x^2}$

5. If we set $x = 3u$, then $\sqrt{9 - x^2} = 3\sqrt{1 - u^2}$. Which relation between x and u yields the equality $\sqrt{16 + x^2} = 4\sqrt{1 + u^2}$?

Exercises

In Exercises 1–10, evaluate the definite integral.

1. $\displaystyle\int_1^2 \frac{1}{x}\,dx$

2. $\displaystyle\int_4^{12} \frac{1}{x}\,dx$

3. $\displaystyle\int_1^e \frac{1}{x}\,dx$

4. $\displaystyle\int_2^4 \frac{dt}{3t+4}$

5. $\displaystyle\int_{-e^2}^{-e} \frac{1}{t}\,dt$

6. $\displaystyle\int_e^{e^2} \frac{1}{t\ln t}\,dt$

7. $\displaystyle\int_0^{1/2} \frac{dx}{\sqrt{1-x^2}}$

8. $\displaystyle\int_{\tan 1}^{\tan 8} \frac{dx}{x^2+1}$

9. $\displaystyle\int_{-2}^{-2/\sqrt{3}} \frac{dx}{|x|\sqrt{x^2-1}}$

10. $\displaystyle\int_{-1/2}^{\sqrt{3}/2} \frac{dx}{\sqrt{1-x^2}}$

11. Use the substitution $u = x/3$ to prove

$$\int \frac{dx}{9+x^2} = \frac{1}{3}\tan^{-1}\frac{x}{3} + C$$

12. Use the substitution $u = 2x$ to evaluate $\displaystyle\int \frac{dx}{4x^2+1}$.

In Exercises 13–32, calculate the integral.

13. $\displaystyle\int_0^2 \frac{dx}{x^2+4}$

14. $\displaystyle\int_{1/\sqrt{3}}^{1/\sqrt{2}} \frac{dx}{x\sqrt{x^2-4}}$

15. $\displaystyle\int \frac{dt}{\sqrt{16-t^2}}$

16. $\displaystyle\int \frac{dt}{\sqrt{1-16t^2}}$

17. $\displaystyle\int \frac{dt}{\sqrt{25-4t^2}}$

18. $\displaystyle\int \frac{dx}{x\sqrt{4x^2-1}}$

19. $\displaystyle\int \frac{dx}{\sqrt{1-4x^2}}$

20. $\displaystyle\int \frac{dx}{4+x^2}$

21. $\displaystyle\int \frac{(x+1)dx}{\sqrt{1-x^2}}$

22. $\displaystyle\int \frac{dx}{x\sqrt{x^4-1}}$

23. $\displaystyle\int \frac{e^x\,dx}{1+e^{2x}}$

24. $\displaystyle\int \frac{\ln(\cos^{-1}x)\,dx}{(\cos^{-1}x)\sqrt{1-x^2}}$

25. $\displaystyle\int \frac{\tan^{-1}x\,dx}{1+x^2}$

26. $\displaystyle\int \frac{dx}{(\tan^{-1}x)(1+x^2)}$

27. $\displaystyle\int_0^1 3^x\,dx$

28. $\displaystyle\int_0^1 3^{-x}\,dx$

29. $\displaystyle\int_0^{\log_4(3)} 4^x\,dx$

30. $\displaystyle\int_{-2}^2 x10^{x^2}\,dx$

31. $\displaystyle\int 9^x \sin(9^x)\,dx$

32. $\displaystyle\int \frac{dx}{\sqrt{5^{2x}-1}}$

In Exercises 33–70, evaluate the integral using the methods covered in the text so far.

33. $\displaystyle\int (e^x+2)\,dx$

34. $\displaystyle\int e^{4x}\,dx$

35. $\displaystyle\int 7^{-x}\,dx$

36. $\displaystyle\int ye^{y^2}\,dy$

37. $\displaystyle\int (e^{4x}+1)\,dx$

38. $\displaystyle\int \frac{4x\,dx}{x^2+1}$

39. $\displaystyle\int e^{-9t}\,dt$

40. $\displaystyle\int (e^x + e^{-x})\,dx$

41. $\displaystyle\int \frac{dx}{\sqrt{1-16x^2}}$

42. $\displaystyle\int \frac{dx}{\sqrt{9-16x^2}}$

43. $\displaystyle\int e^t\sqrt{e^t+1}\,dt$

44. $\displaystyle\int (e^{-x}-4x)\,dx$

45. $\displaystyle\int (7-e^{10x})\,dx$

46. $\displaystyle\int \frac{e^{2x}-e^{4x}}{e^x}\,dx$

47. $\displaystyle\int \frac{dx}{x\sqrt{25x^2-1}}$

48. $\displaystyle\int \frac{x\,dx}{\sqrt{4x^2+9}}$

49. $\displaystyle\int xe^{-4x^2}\,dx$

50. $\displaystyle\int e^x\cos(e^x)\,dx$

51. $\displaystyle\int \frac{e^x}{\sqrt{e^x+1}}\,dx$

52. $\displaystyle\int e^x(e^{2x}+1)^3\,dx$

53. $\displaystyle\int \frac{dx}{2x+4}$

54. $\displaystyle\int \frac{t\,dt}{t^2+4}$

55. $\displaystyle\int \frac{x^2\,dx}{x^3+2}$

56. $\displaystyle\int \frac{(3x-1)\,dx}{9-2x+3x^2}$

57. $\displaystyle\int \tan(4x+1)\,dx$

58. $\displaystyle\int \cot x\,dx$

59. $\displaystyle\int \frac{\cos x}{2\sin x+3}\,dx$

60. $\displaystyle\int \frac{\ln x}{x}\,dx$

61. $\displaystyle\int \frac{4\ln x+5}{x}\,dx$

62. $\displaystyle\int \frac{(\ln x)^2}{x}\,dx$

63. $\displaystyle\int \frac{dx}{x\ln x}$

64. $\displaystyle\int \frac{dx}{(4x-1)\ln(8x-2)}$

65. $\displaystyle\int \frac{\ln(\ln x)}{x\ln x}\,dx$

66. $\displaystyle\int \cot x\ln(\sin x)\,dx$

67. $\displaystyle\int 3^x\,dx$

68. $\displaystyle\int x3^{x^2}\,dx$

69. $\displaystyle\int \cos x\, 3^{\sin x}\,dx$

70. $\displaystyle\int \left(\frac{1}{2}\right)^{3x+2}\,dx$

71. Use Figure 4 on the following page to prove the formula

$$\int_0^x \sqrt{1-t^2}\,dt = \frac{1}{2}x\sqrt{1-x^2} + \frac{1}{2}\sin^{-1}x$$

Hint: The area represented by the integral is the sum of a triangle and a sector.

72. Show that $G(t) = \sqrt{1-t^2} + t\sin^{-1}t$ is an antiderivative of $\sin^{-1}t$.

73. Verify by differentiation:

$$\int_0^T te^{rt}\,dt = \frac{e^{rT}(rT-1)+1}{r^2}$$

Then use L'Hôpital's Rule to show that the limit of the right-hand side as $r \to 0$ is equal to the value of the integral for $r = 0$.

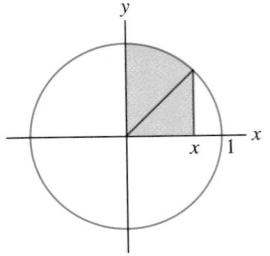

FIGURE 4

Further Insights and Challenges

74. Recall the following property of integrals: If $f(t) \geq g(t)$ for all $t \geq 0$, then for all $x \geq 0$,

$$\int_0^x f(t)\,dt \geq \int_0^x g(t)\,dt \qquad \boxed{7}$$

The inequality $e^t \geq 1$ holds for $t \geq 0$ because $e > 1$. Use (7) to prove that $e^x \geq 1 + x$ for $x \geq 0$. Then prove, by successive integration, the following inequalities (for $x \geq 0$):

$$e^x \geq 1 + x + \frac{1}{2}x^2, \qquad e^x \geq 1 + x + \frac{1}{2}x^2 + \frac{1}{6}x^3$$

75. Generalize Exercise 74; that is, use induction (if you are familiar with this method of proof) to prove that for all $n \geq 0$,

$$e^x \geq 1 + x + \frac{1}{2}x^2 + \frac{1}{6}x^3 + \cdots + \frac{1}{n!}x^n \quad (x \geq 0)$$

76. Use Exercise 74 to show that $\dfrac{e^x}{x^2} \geq \dfrac{x}{6}$ and conclude that $\lim\limits_{x\to\infty} \dfrac{e^x}{x^2} = \infty$. Then use Exercise 75 to prove more generally that $\lim\limits_{x\to\infty} \dfrac{e^x}{x^n} = \infty$ for all n.

77. Defining $\ln x$ as an Integral   Define a function $\varphi(x)$ in the domain $x > 0$:

$$\varphi(x) = \int_1^x \frac{1}{t}\,dt$$

This exercise proceeds as if we didn't know that $\varphi(x) = \ln x$ and shows directly that $\varphi(x)$ has all the basic properties of the logarithm. Prove the following statements:

(a) $\displaystyle\int_1^b \frac{1}{t}\,dt = \int_a^{ab} \frac{1}{t}\,dt$ for all $a, b > 0$. *Hint:* Use the substitution $u = t/a$.

(b) $\varphi(ab) = \varphi(a) + \varphi(b)$. *Hint:* Break up the integral from 1 to ab into two integrals and use (a).

(c) $\varphi(1) = 0$ and $\varphi(a^{-1}) = -\varphi(a)$ for $a > 0$.

(d) $\varphi(a^n) = n\varphi(a)$ for all $a > 0$ and integers n.

(e) $\varphi(a^{1/n}) = \dfrac{1}{n}\varphi(a)$ for all $a > 0$ and integers $n \neq 0$.

(f) $\varphi(a^r) = r\varphi(a)$ for all $a > 0$ and rational number r.

(g) There exists x such that $\varphi(x) > 1$. *Hint:* Show that $\varphi(a) > 0$ for every $a > 1$. Then take $x = a^m$ for $m > 1/\varphi(a)$.

(h) Show that $\varphi(t)$ is increasing and use the Intermediate Value Theorem to show that there exists a unique number e such that $\varphi(e) = 1$.

(i) $\varphi(e^r) = r$ for any rational number r.

78. Show that if $f(x)$ is increasing and satisfies $f(xy) = f(x) + f(y)$, then its inverse $g(x)$ satisfies $g(x+y) = g(x)g(y)$.

79. This is a continuation of the previous two exercises. Let $g(x)$ be the inverse of $\varphi(x)$. Show that

(a) $g(x)g(y) = g(x+y)$.

(b) $g(r) = e^r$ for any rational number.

(c) $g'(x) = g(x)$.

Exercises 77–79 provide a mathematically elegant approach to the exponential and logarithm functions, which avoids the problem of defining e^x for irrational x and of proving that e^x is differentiable.

80. The formula $\displaystyle\int x^n\,dx = \frac{x^{n+1}}{n+1} + C$ is valid for $n \neq -1$. Use L'Hôpital's Rule to show that the exceptional case $n = -1$ is a limit of the general case in the following sense: For fixed $x > 0$,

$$\lim_{n\to -1}\int_1^x t^n\,dt = \int_1^x t^{-1}\,dt$$

Note that the integral on the left is equal to $\dfrac{x^{n+1}-1}{n+1}$.

81. CAS   The integral on the left in Exercise 80 is equal to $f_n(x) = \dfrac{x^{n+1}-1}{n+1}$. Investigate the limit graphically by plotting $f_n(x)$ for $n = 0, -0.3, -0.6,$ and -0.9 together with $\ln x$ on a single plot.

82. Use the substitution $u = \tan x$ to evaluate $\displaystyle\int \frac{dx}{1+\sin^2 x}$. *Hint:* Show that

$$\frac{dx}{1+\sin^2 x} = \frac{du}{1+2u^2}$$

5.8 Exponential Growth and Decay

In this section, we explore some of the applications of the exponential function to biology, physics, and economics. Consider a quantity $P(t)$ that depends exponentially on time:

$$P(t) = P_0 e^{kt}$$

where P_0 and k are constants. If $k > 0$, then $P(t)$ *grows exponentially* with *growth constant* k, and if $k < 0$, then $P(t)$ *decays exponentially* with *decay constant* k. The coefficient P_0 is the initial size since $P(0) = P_0 e^{k \cdot 0} = P_0$. Exponential growth or decay may also be expressed by the formula $P(t) = P_0 b^t$ with $b = e^k$, since $P_0 b^t = P_0 (e^k)^t = P_0 e^{kt}$.

The constant k has units of "inverse time"; if t is measured in days, then k has units of $(\text{days})^{-1}$.

Population is a typical example of a quantity that grows exponentially, at least under suitable conditions. To understand why, consider a cell colony with initial population $P_0 = 100$ and assume that each cell divides into two cells after 1 hour. Then with each passing hour, the population $P(t)$ doubles:

$$P(0) \qquad\qquad = 100 \quad \text{(initial population)}$$
$$P(1) = 2(100) = 200 \quad \text{(population doubles)}$$
$$P(2) = 2(200) = 400 \quad \text{(population doubles again)}$$

After t hours, $P(t) = (100)2^t$.

FIGURE 1 *E. coli* bacteria.

Exponential growth cannot continue over long periods of time. A colony starting with one E. coli cell would grow to 5×10^{89} cells after 3 weeks, much more than the estimated number of atoms in the observable universe. In actual cell growth, the exponential phase is followed by a period in which growth slows and may decline.

■ **EXAMPLE 1** When the *Escherichia coli* bacteria (found in the human intestine; see Figure 1) is grown in the laboratory, it increases exponentially with a growth constant of $k = 0.41$ $(\text{hours})^{-1}$. Assume that 1,000 bacteria are present at time $t = 0$.

(a) Find the formula for the number of bacteria $P(t)$ at time t.

(b) How large is the population after 5 hours?

(c) When will the population reach 10,000?

Solution

(a) The initial size is $P_0 = 1,000$ and $k = 0.41$, so $P(t) = 1,000e^{0.41t}$ (t in hours).

(b) After 5 hours, $P(5) = 1,000e^{0.41 \cdot 5} = 1,000e^{2.05} \approx 7767.9$. Because the number of bacteria must be a whole number, we round off the answer to 7,768.

(c) The problem asks for the time t such that $P(t) = 10,000$, so we must solve the equation $1,000e^{0.41t} = 10,000$, or

$$e^{0.41t} = \frac{10,000}{1,000} = 10$$

Taking the logarithm of both sides, we obtain $\ln\left(e^{0.41t}\right) = \ln 10$, or

$$0.41t = \ln 10 \Rightarrow t = \frac{\ln 10}{0.41} \approx 5.62$$

Therefore, $P(t)$ reaches 10,000 after approximately 5 hours and 37 min (Figure 2). ■

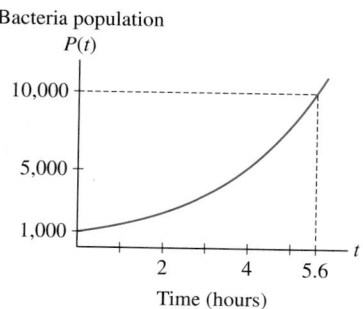

Bacteria population

$P(t)$

10,000

5,000

1,000

2   4   5.6 t

Time (hours)

FIGURE 2 Growth of *E. coli* population.

The important role played by exponential functions is best understood in terms of the differential equation $y' = ky$. The function $y = P_0 e^{kt}$ satisfies this differential equation, as we can check directly:

$$y' = \frac{d}{dt}\left(P_0 e^{kt}\right) = k P_0 e^{kt} = ky$$

Theorem 1 goes further and asserts that the exponential functions $y = P_0 e^{kt}$ are the *only* functions that satisfy this differential equation.

> **THEOREM 1**   If $y(t)$ is a differentiable function satisfying the differential equation
>
> $$y' = ky$$
>
> then $y(t) = P_0 e^{kt}$, where P_0 is the initial value $P_0 = y(0)$.

Proof Compute the derivative of ye^{-kt}. If $y' = ky$, then

$$\frac{d}{dt}\left(ye^{-kt}\right) = y'e^{-kt} - ke^{-kt}y = e^{-kt}(y' - ky) = 0 \quad \text{(since } y' = ky\text{)}$$

Because the derivative is zero, $y(t)e^{-kt} = P_0$ for some constant P_0 and $y(t) = P_0 e^{kt}$ as claimed. ∎

CONCEPTUAL INSIGHT   Theorem 1 tells us that a process obeys an exponential law precisely when *its rate of change is proportional to the amount present at time t*. This formulation helps us understand why certain quantities grow exponentially. A population grows exponentially because its growth rate is proportional to the size of the population (each organism contributes to the growth through reproduction). However, this is true only if the organisms do not interact. If they do interact, say, by competing for food or mates, then the growth rate may not be proportional to population size and we cannot expect exponential growth. Similarly, the experimental fact that radioactive substances decay exponentially is consistent with the hypothesis that atoms decay randomly and independently. Under this hypothesis, a fixed fraction of the atoms decay per unit time and the total number decaying per unit time is proportional to the total number of atoms present, leading to the observed exponential decay. If exponential decay were not observed, we might suspect that the decay is influenced by some interaction between the atoms.

■ **EXAMPLE 2** Find all solutions of $y' = 3y$. Which particular solution satisfies $y(0) = 9$?

Solution The solutions to $y' = 3y$ are the functions $y(t) = Ce^{3t}$, where C is the initial value $C = y(0)$. The particular solution satisfying $y(0) = 9$ is $y(t) = 9e^{3t}$. ■

■ **EXAMPLE 3** Modeling Penicillin   Pharmacologists have shown that the rate at which penicillin leaves a person's bloodstream is proportional to the amount of penicillin present.

(a) Express this statement as a differential equation.
(b) Find the decay constant if 50 mg of penicillin remain in the bloodstream 7 hours after an initial injection of 450 mg.
(c) Under the hypothesis of (b), at what time was 200 mg of penicillin present?

Solution

(a) Let's denote the quantity of penicillin present in the bloodstream at time t by $A(t)$. Then the rate at which penicillin leaves the bloodstream is $A'(t)$. If this rate is proportional to $A(t)$, then

$$A'(t) = -kA(t) \qquad \boxed{1}$$

for some constant k. Note that $k > 0$ because $A(t)$ is decreasing.

(b) Since $A(t)$ satisfies Eq. (1) and $A(0) = 450$, $A(t) = 450e^{-kt}$. We use the condition $A(7) = 50$ to solve for k:

$$A(7) = 450e^{-7k} = 50 \quad \Rightarrow \quad e^{-7k} = \frac{1}{9}$$

Thus, $-7k = \ln \frac{1}{9}$ and $k = -\ln(\frac{1}{9})/7 \approx 0.31$.

(c) To find the time at which 200 mg was present, we solve

$$A(t) = 450e^{-0.31t} = 200$$

$$e^{-0.31t} = \frac{4}{9}$$

$$t = -\frac{1}{0.31} \ln \left(\frac{4}{9} \right) \approx 2.62$$

This shows that 200 mg was present at time $t = 2.62$ hours (Figure 3). ∎

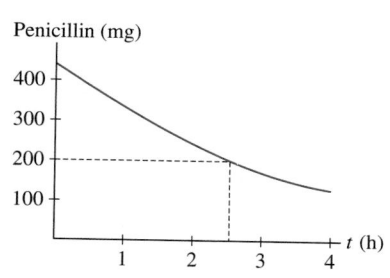

Penicillin (mg)

FIGURE 3

Quantities that grow exponentially possess an important property: There is a doubling time T such that $P(t)$ doubles in size over every time interval of length T. To prove this, let $P(t) = P_0e^{kt}$ and solve for T in the equation $P(t + T) = 2P(t)$:

$$P_0e^{k(t+T)} = 2P_0e^{kt}$$

$$e^{kt}e^{kT} = 2e^{kt}$$

$$e^{kT} = 2$$

We obtain $kT = \ln 2$ or $T = \ln 2/k$.

The constant k has units of time^{-1}, so the doubling time $T = (\ln 2)/k$ has units of time, as we should expect.

Doubling Time   If $P(t) = P_0e^{kt}$ with $k > 0$, then the doubling time of P is:

$$\text{Doubling time} = \frac{\ln 2}{k}$$

■ **EXAMPLE 4** Computing Doubling Time from k Some studies have suggested that from 1955 to 1970, the number of bachelor's degrees in physics awarded per year by U.S. universities grew exponentially, with growth constant $k = 0.1$ (approximately 2,500 degrees awarded in 1955). If this was true,

(a) What was the doubling time?

(b) How long would it take for the number of degrees awarded per year to increase 8-fold, assuming that exponential growth continued?

Solution

(a) The doubling time is $(\ln 2)/0.1 \approx 0.693/0.1 = 6.93$ (Figure 4). Since degrees are usually awarded once a year, we round off the doubling time to 7 years.

Degrees awarded

FIGURE 4 Doubling (from 2,500 to 5,000 to 10,000, etc.) occurs at equal time intervals.

(b) The number doubles after 7 years, quadruples after 14 years, and increases 8-fold after 21 years. ∎

■ **EXAMPLE 5** Calculating k from the Doubling Time   One of the world's smallest flowering plants, *Wolffia globosa* (Figure 5), has a doubling time of approximately 30 hours. Find the growth constant k and determine the initial population if the population grew to 1,000 after 48 hours.

Solution By the formula for the doubling time, $30 = \dfrac{\ln 2}{k}$. Therefore, $k = \dfrac{\ln 2}{30} \approx 0.023$. The plant population after t hours is $P(t) = P_0 e^{0.023t}$. If $P(48) = 1,000$, then

$$P_0 e^{(0.023)48} = 1,000 \quad \Rightarrow \quad P_0 = 1,000 e^{-(0.023)48} \approx 332 \qquad ∎$$

Scientists have observed that if a quantity R_0 of a radioactive isotope is present at time $t = 0$, then the quantity present at time t is $R(t) = R_0 e^{-kt}$, where $k > 0$. In this situation of exponential decay, we have a **half-life** rather than a doubling time. The half-life is the time it takes for the quantity to decrease by a factor of $\frac{1}{2}$. The calculation of doubling time above shows similarly that

$$\boxed{\text{Half-life} = \frac{\ln 2}{k}}$$

■ **EXAMPLE 6**   The isotope Radon-222 has a half-life of 3.825 days. Find the decay constant and determine how long will it take for 80% of the isotope to decay.

Solution By the equation for half-life, $k = \dfrac{\ln 2}{3.825} \approx 0.181$. Therefore, the quantity of Radon-222 at time t is $R(t) = R_0 e^{-0.181t}$, where R_0 is the amount present at $t = 0$ (Figure 6). To determine when 20% remains, we solve for t in the equation $P(t) = 0.2 R_0$:

$$R_0 e^{-0.181t} = 0.2 R_0$$

$$e^{-0.181t} = 0.2$$

$$-0.181t = \ln(0.2) \quad \Rightarrow \quad t = \frac{\ln(0.2)}{-0.181} \approx 8.9 \text{ days}$$

Therefore, the quantity of Radon-222 decreases by 80% after 8.9 days. ∎

Carbon Dating

Carbon dating relies on the fact that all living organisms contain carbon. The carbon, which enters the food chain through the carbon dioxide absorbed by plants from the atmosphere, is made up mostly of nonradioactive C^{12} and a much smaller amount of radioactive C^{14} that decays into nitrogen. The ratio R of C^{14} to C^{12} in the atmosphere is approximately $R = 10^{-12}$. The ratio of C^{14} to C^{12} in a living organism is also R because the organism's carbon originates in the atmosphere.

When the organism dies, its carbon is no longer replenished. The C^{14} begins to decay exponentially while the C^{12} remains unchanged. Therefore, the ratio of C^{14} to C^{12} in the organism decreases exponentially as a function of the time t since the organism died. By measuring this ratio, we can determine when the death occurred. The decay constant for

FIGURE 5 The tiny plants are *Wolffia*, with plant bodies smaller than the head of a pin.

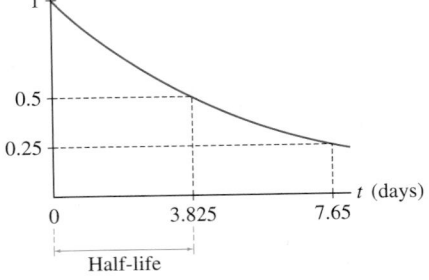

Fraction present

FIGURE 6 Fraction of Radon-222 present at time t.

FIGURE 7 American chemist Willard Libby (1908–1980) developed the technique of carbon dating in 1946 to determine the age of fossils and was awarded the Nobel Prize in Chemistry for this work in 1960. Since then the technique has been considerably refined.

C^{14} is $k = 0.000121$ years^{-1}, so we obtain

$$\text{Ratio of } C^{14} \text{ to } C^{12} \text{ after } t \text{ years} = Re^{-0.000121t}$$

■ **EXAMPLE 7** Lascaux Cave Paintings A remarkable gallery of prehistoric paintings of animals was discovered in the Lascaux cave in Dordogne, France, in 1940 (Figure 8). Scientists determined that a charcoal sample taken from the cave walls had a C^{14} to C^{12} ratio equal to 15% of that found in the atmosphere. What did they estimate the age of the paintings to be?

Solution Let t be the age of the paintings. Then the C^{14} to C^{12} ratio decreased by a factor of $e^{-0.000121t}$. If the ratio in the sample decreased by a factor of 0.15, then

$$e^{-0.000121t} = 0.15$$
$$-0.000121t = \ln(0.15) \approx -1.9$$

Thus $t = 1.9/0.000121 \approx 15{,}700$. The scientists concluded that the cave paintings are approximately 16,000 years old (Figure 9). ■

FIGURE 8 Detail of bison and other animals from a replica of the Lascaux cave mural.

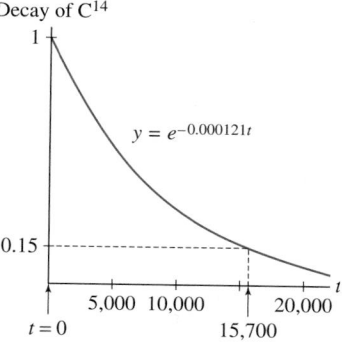

Decay of C^{14}

$y = e^{-0.000121t}$

$t = $ years since organism died

FIGURE 9 If only 15% of the C^{14} remains, the object is approximately 16,000 years old.

Convention: Time t is measured in years and interest rates are given as yearly rates, either as a decimal or as a percentage. Thus, $r = 0.05$ corresponds to an interest rate of 5% per year.

Compound Interest and Present Value

Exponential functions play an important role in financial calculations. For example, they are used to compute compound interest and present value.

The initial sum of money P_0 placed in an account or investment is called the **principal**. When a principal of P_0 dollars is deposited into a bank account, the amount or *balance* in the account at time t depends on two factors: the **interest rate** r and how often interest is **compounded**. If interest is paid out once a year at the end of the year, we say that the interest is *compounded annually*. If the principal and interest are retained in the account, then the balance grows exponentially:

$$\text{Balance after 1 year} = \overbrace{P_0}^{\text{Principal}} + \overbrace{rP_0}^{\text{Interest}} = P_0(1+r)$$
$$\text{Balance after 2 years} = P_0(1+r) + rP_0(1+r) = P_0(1+r)^2$$
$$\text{Balance after } t \text{ years} = P_0(1+r)^{t-1} + rP_0(1+r)^{t-1} = P_0(1+r)^t$$

When interest is paid out quarterly, the interest earned after 3 months (the first quarter) is $\frac{r}{4}P_0$ dollars. The principal increases by the factor $\left(1 + \frac{r}{4}\right)$ after each 3-month period, and

$$\text{Balance after 1 year} = P_0\left(1+\frac{r}{4}\right)\left(1+\frac{r}{4}\right)\left(1+\frac{r}{4}\right)\left(1+\frac{r}{4}\right) = P_0\left(1+\frac{r}{4}\right)^4$$
$$\text{Balance after } t \text{ years} = P_0\left(1+\frac{r}{4}\right)^{4t}$$

After t years, the principal increases by the factor
$$\left(1+\frac{r}{M}\right)^{Mt}$$
This factor is called the "multiplier."

In general, if interest is compounded M times a year, the principal increases by the factor $\left(1+\frac{r}{M}\right)^M$ after each year, and after t years, $P(t) = P_0\left(1+\frac{r}{M}\right)^{Mt}$.

TABLE 1   Compound Interest with Principal $P_0 = \$100$ and $r = 0.09$

	Principal after 1 Year
Annual	$100(1 + 0.09) = \$109$
Quarterly	$100\left(1 + \frac{0.09}{4}\right)^4 \approx \$109.31$
Monthly	$100\left(1 + \frac{0.09}{12}\right)^{12} \approx \$109.38$
Weekly	$100\left(1 + \frac{0.09}{52}\right)^{52} \approx \$109.41$
Daily	$100\left(1 + \frac{0.09}{365}\right)^{365} \approx \$109.42$

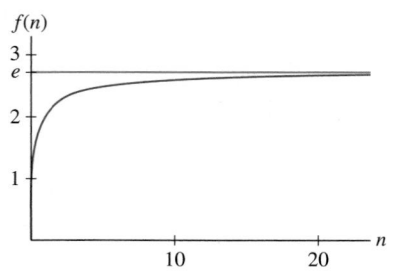

FIGURE 10 The function $f(n) = \left(1 + \frac{1}{n}\right)^n$ approaches e as $n \to \infty$.

Compound Interest   If P_0 dollars are deposited into an account earning interest at an annual rate r, compounded M times yearly, then the value of the account after t years is

$$P(t) = P_0\left(1 + \frac{r}{M}\right)^{Mt}$$

Table 1 shows that there is a benefit to more frequent compounding. What happens in the limit as M tends to infinity? The yearly multiplier approaches the limit

$$\lim_{M \to \infty}\left(1 + \frac{r}{M}\right)^M \qquad \boxed{2}$$

Theorem 2 relates this limit to the exponential function (a proof is given at the end of this section). Figure 10 illustrates the first limit in Theorem 2 graphically.

THEOREM 2   Limit Formula for e and e^x

$$e = \lim_{n \to \infty}\left(1 + \frac{1}{n}\right)^n \qquad \text{and} \qquad e^x = \lim_{n \to \infty}\left(1 + \frac{x}{n}\right)^n \qquad \text{for all } x$$

To evaluate (2), we apply Theorem 2 with $x = r$ and $n = M$:

$$\lim_{M \to \infty}\left(1 + \frac{r}{M}\right)^M = e^r$$

Thus, as we compound more and more frequently, the yearly multiplier approaches e^r and the multiplier for t years approaches $(e^r)^t = e^{rt}$. We use this result to define **continuously compounded interest**.

Continuously Compounded Interest   If P_0 dollars are deposited into an account earning interest at an annual rate r, compounded continuously, then the value of the account after t years is

$$P(t) = P_0 e^{rt}$$

■ **EXAMPLE 8**   A principal $P_0 = \$10,000$ is deposited into an account paying 6% interest. Find the balance after 3 years if interest is compounded quarterly and if interest is compounded continuously.

Solution   After 3 years, the balance is

$$\text{Quarterly compounding:} \qquad 10,000\left(1 + \frac{0.06}{4}\right)^{4(3)} \approx \$11,956.18$$

$$\text{Continuous compounding:} \qquad 10,000 e^{(0.06)3} \approx \$11,972.17 \qquad ■$$

Present Value

The concept of *present value* (PV) is used in business and finance to compare the values of payments made at different times. Assume that there is an interest rate r at which any investor can lend or borrow money and that interest is compounded continuously. Then the PV of P dollars to be received t years in the future is defined to be Pe^{-rt}:

The financial world has many different interest rates (federal funds rate, prime rate, etc.). We simplify the discussion by assuming that there is just one rate.

> The PV of P dollars received at time t is Pe^{-rt}

What is the reasoning behind this definition? When you invest at the interest rate r for t years, your principal increases by the factor e^{rt}, so if you invest Pe^{-rt} dollars, your principal grows to $Pe^{-rt}e^{rt} = P$ dollars at time t. Thus, the present value Pe^{-rt} is the amount you would have to invest *today* in order to have P dollars at time t.

■ **EXAMPLE 9** Is it better to receive $2,000 today or $2,200 in 2 years? Consider $r = 0.07$ and $r = 0.03$.

Solution We compare $2,000 today with the present value of $2,200 received in 2 years. If $r = 0.07$, the PV is $2,200e^{-(0.07)2} \approx \$1,912.59$. This PV is less than $2,000, so it is better to receive $2,000 today.

If $r = 0.03$, the PV is $2,200e^{-(0.03)2} \approx \$2,071.88$. In this case, a payment of $2,200 in 2 years is preferable to a $2,000 payment today. ■

■ **EXAMPLE 10** Deciding Whether to Invest   Chief Operating Officer Marjorie Bean must decide whether to invest $400,000 to upgrade her company's computer system. The upgrade will save $150,000 a year for each of the next 3 years. Is this a good investment if $r = 7\%$? What if $r = 4\%$?

Solution Marjorie must compare the PV of the money saved ($150,000 annually for 3 years) with the $400,000 cost of the upgrade. For simplicity, assume that the $150,000 in savings is received as a lump sum at the end of each year.

If $r = 0.07$, the PV of the savings over 3 years is

$$150,000e^{-(0.07)} + 150,000e^{-(0.07)2} + 150,000e^{-(0.07)3} \approx \$391,850$$

Since this is *less* than the initial cost, the upgrade is not a worthwhile investment.

If $r = 0.04$, then the present value of the savings is

$$150,000e^{-(0.04)} + 150,000e^{-(0.04)2} + 150,000e^{-(0.04)3} \approx \$415,624$$

In this "interest-rate environment," the upgrade is a good investment because it saves the company $15,624. ■

An **income stream** is a sequence of periodic payments that continue over an interval of T years. Consider an investment that produces income at a rate of $800/year for 5 years. A total of $4,000 is paid out over 5 years but the PV of the income stream is less. For instance, if $r = 0.06$ and payments are made at the end of the year, then the PV is

$$800e^{-0.06} + 800e^{-(0.06)2} + 800e^{-(0.06)3} + 800e^{-(0.06)4} + 800e^{-(0.06)5} \approx \$3,353.12$$

It is more convenient mathematically to assume that payments are made *continuously* at a rate of $R(t)$ dollars/year. We may then calculate PV as an integral. Divide the time interval $[0, T]$ into N subintervals of length $\Delta t = T/N$. If Δt is small, the amount paid out between time t and $t + \Delta t$ is approximately equal to

$$\underbrace{R(t)}_{\text{Rate}} \times \underbrace{\Delta t}_{\text{Time interval}} = R(t)\,\Delta t$$

The PV of this payment is approximately $e^{-rt}R(t)\Delta t$. If $t_0 = 0, t_1, \ldots, t_N$ are the endpoints of the subintervals, we obtain the approximation

$$\text{PV of income stream} \approx \sum_{i=1}^{N} e^{-rt_i} R(t_i)\,\Delta t$$

This is a Riemann sum whose value approaches $\displaystyle\int_0^T R(t)e^{-rt}\,dt$ as Δt tends to zero.

PV of an Income Stream   The present value at interest rate r of an income stream paying out $R(t)$ dollars/year continuously for T years is

$$PV = \int_0^T R(t)e^{-rt}\,dt \qquad \boxed{3}$$

■ **EXAMPLE 11**   An investment pays out $800/year continuously for 5 years. Find the PV of the investment for $r = 0.06$ and $r = 0.04$.

Solution   If $r = 0.06$, the PV is equal to

$$\int_0^5 800e^{-0.06t}\,dt = -800\frac{e^{-0.06t}}{0.06}\Bigg|_0^5 \approx -9{,}877.58 - (-13{,}333.33) = \$3{,}455.76$$

If $r = 0.04$, the PV is equal to

$$\int_0^5 800e^{-0.04t}\,dt = -800\frac{e^{-0.04t}}{0.04}\Bigg|_0^5 \approx -16{,}374.62 - (-20{,}000) = \$3{,}625.38 \qquad ■$$

Proof of Theorem 2   To verify the first limit, $e = \lim_{n\to\infty}(1 + 1/n)^n$, we use the formula $\ln b = \int_1^b t^{-1}\,dt$ with $b = 1 + 1/n$:

$$\ln\left(1 + \frac{1}{n}\right) = \int_1^{1+1/n} \frac{dt}{t}$$

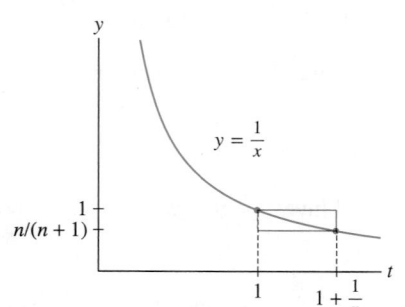

FIGURE 11

Figure 11 shows that the area represented by this integral lies between the areas of two rectangles of heights $n/(n+1)$ and 1, both of base $1/n$. These rectangles have areas $1/(n+1)$ and $1/n$, so

$$\frac{1}{n+1} \le \ln\left(1 + \frac{1}{n}\right) = \int_1^{1+1/n} \frac{dt}{t} \le \frac{1}{n}$$

Since e^x is increasing, these inequalities remain true when we apply e^x to each term:

$$e^{1/(n+1)} \le 1 + \frac{1}{n} \le e^{1/n} \qquad \boxed{4}$$

$$e^{n/(n+1)} \le \left(1 + \frac{1}{n}\right)^n \le e \qquad \text{(raise both sides to the nth power)} \qquad \boxed{5}$$

Now apply the Squeeze Theorem. Since e^x is continuous,

$$\lim_{n \to \infty} e^{n/(n+1)} = e^{\lim_{n \to \infty} (n/(n+1))} = e^1 = e$$

We conclude that $\lim_{n \to \infty} (1 + \frac{1}{n})^n = e$ as desired. See Exercise 58 for a proof of the more general formula $e^x = \lim_{n \to \infty} (1 + \frac{x}{n})^n$. ∎

5.8 SUMMARY

- $P(t)$ *grows exponentially* with growth constant $k > 0$ if $P(t) = P_0 e^{kt}$. The initial amount is P_0.
- The solutions of the differential equation $y' = ky$ are the exponential functions $y = Ce^{kt}$, where C is a constant.
- A quantity $P(t)$ grows exponentially if it grows at a rate proportional to its size, that is, if $P'(t) = kP(t)$.
- If $P(t)$ grows exponentially, its *doubling time* is $\dfrac{\ln 2}{k}$.
- $P(t)$ is said to *decay exponentially* with decay constant $k > 0$ if $P(t) = P_0 e^{-kt}$. If $P(t)$ decays exponentially, then $P'(t) = -kP(t)$. The *half-life* of $P(t)$ is $(\ln 2)/k$.
- If a principal P_0 is deposited in an account bearing interest at the rate r, compounded M times per year, the value of the account after t years is $P(t) = P_0(1 + r/M)^{Mt}$.
- When interest is compounded continuously, $P(t) = P_0 e^{rt}$.
- The *present value* (PV) of P dollars, to be paid t years in the future, is Pe^{-rt}.
- The present value of an income stream paying out $R(t)$ dollars/year continuously for T years is

$$PV = \int_0^T R(t)e^{-rt}\, dt$$

5.8 EXERCISES

Preliminary Questions

1. Two quantities increase exponentially with growth constants $k = 1.2$ and $k = 3.4$, respectively. Which quantity doubles more rapidly?

2. If you are given both the doubling time and the growth constant of a quantity that increases exponentially, can you determine the initial amount?

3. A cell population grows exponentially beginning with one cell. Does it take less time for the population to increase from one to two cells than from 10 million to 20 million cells?

4. Referring to his popular book *A Brief History of Time*, the renowned physicist Stephen Hawking said, "Someone told me that each equation I included in the book would halve its sales." If this is so, write a differential equation satisfied by the sales function $S(n)$, where n is the number of equations in the book.

5. Carbon dating is based on the assumption that the ratio R of C^{14} to C^{12} in the atmosphere has been constant over the past 50,000 years.

If R were actually smaller in the past than it is today, would the age estimates produced by carbon dating be too ancient or too recent?

6. Which is preferable: an interest rate of 12% compounded quarterly, or an interest rate of 11% compounded continuously?

7. Find the yearly multiplier if $r = 9\%$ and interest is compounded (a) continuously and (b) quarterly.

8. The PV of N dollars received at time T is (choose the correct answer):

(a) The value at time T of N dollars invested today

(b) The amount you would have to invest today in order to receive N dollars at time T

9. A year from now, $1 will be received. Will its PV increase or decrease if the interest rate goes up?

10. Xavier expects to receive a check for $1,000 1 year from today. Explain, using the concept of PV, whether he will be happy or sad to learn that the interest rate has just increased from 6% to 7%.

Exercises

1. A certain bacteria population P obeys the exponential growth law $P(t) = 2,000e^{1.3t}$ (t in hours).

(a) How many bacteria are present initially?

(b) At what time will there be 10,000 bacteria?

2. A quantity P obeys the exponential growth law $P(t) = e^{5t}$ (t in years).

(a) At what time t is $P = 10$?

(b) At what time t is $P = 20$?

(c) What is the doubling time for P?

3. A certain RNA molecule replicates every 3 minutes. Find the differential equation for the number $N(t)$ of molecules present at time t (in minutes). Starting with one molecule, how many will be present after 10 min?

4. A quantity P obeys the exponential growth law $P(t) = Ce^{kt}$ (t in years). Find the formula for $P(t)$, assuming that the doubling time is 7 years and $P(0) = 100$.

5. The decay constant of Cobalt-60 is 0.13 years^{-1}. What is its half-life?

6. Find the decay constant of Radium-226, given that its half-life is 1,622 years.

7. Find all solutions to the differential equation $y' = -5y$. Which solution satisfies the initial condition $y(0) = 3.4$?

8. Find the solution to $y' = \sqrt{2}y$ satisfying $y(0) = 20$.

9. Find the solution to $y' = 3y$ satisfying $y(2) = 4$.

10. Find the function $y = f(t)$ that satisfies the differential equation $y' = -0.7y$ and initial condition $y(0) = 10$.

11. The population of a city is $P(t) = 2 \cdot e^{0.06t}$ (in millions), where t is measured in years.

(a) Calculate the doubling time of the population.

(b) How long does it take for the population to triple in size?

(c) How long does it take for the population to quadruple in size?

12. The population of Washington state increased from 4.86 million in 1990 to 5.89 million in 2000. Assuming exponential growth,

(a) What will the population be in 2010?

(b) What is the doubling time?

13. Assuming that population growth is approximately exponential, which of the two sets of data is most likely to represent the population (in millions) of a city over a 5-year period?

Year	2000	2001	2002	2003	2004
Data I	3.14	3.36	3.60	3.85	4.11
Data II	3.14	3.24	3.54	4.04	4.74

14. Light Intensity The intensity of light passing through an absorbing medium decreases exponentially with the distance traveled. Suppose the decay constant for a certain plastic block is $k = 2$ when the distance is measured in feet. How thick must the block be to reduce the intensity by a factor of one-third?

15. The **Beer–Lambert Law** is used in spectroscopy to determine the molar absorptivity α or the concentration c of a compound dissolved in a solution at low concentrations (Figure 12). The law states that the intensity I of light as it passes through the solution satisfies $\ln(I/I_0) = \alpha cx$, where I_0 is the initial intensity and x is the distance traveled by the light. Show that I satisfies a differential equation $dI/dx = -kI$ for some constant k.

FIGURE 12 Light of intensity passing through a solution.

16. An insect population triples in size after 5 months. Assuming exponential growth, when will it quadruple in size?

17. A 10-kg quantity of a radioactive isotope decays to 3 kg after 17 years. Find the decay constant of the isotope.

18. Measurements showed that a sample of sheepskin parchment discovered by archaeologists had a C^{14} to C^{12} ratio equal to 40% of that found in the atmosphere. Approximately how old is the parchment?

19. Chauvet Caves In 1994, rock climbers in southern France stumbled on a cave containing prehistoric cave paintings. A C^{14}-analysis carried out by French archeologist Helene Valladas showed that the paintings are between 29,700 and 32,400 years old, much older than any previously known human art. Given that the C^{14} to C^{12} ratio of the atmosphere is $R = 10^{-12}$, what range of C^{14} to C^{12} ratios did Valladas find in the charcoal specimens?

20. A paleontologist has discovered the remains of animals that appear to have died at the onset of the Holocene ice age. She applies carbon dating to test her theory that the Holocene age started between 10,000 and 12,000 years ago. What range of C^{14} to C^{12} ratio would she expect to find in the animal remains?

21. Atmospheric Pressure The atmospheric pressure $P(h)$ (in pounds per square inch) at a height h (in miles) above sea level on earth satisfies a differential equation $P' = -kP$ for some positive constant k.

(a) Measurements with a barometer show that $P(0) = 14.7$ and $P(10) = 2.13$. What is the decay constant k?

(b) Determine the atmospheric pressure 15 miles above sea level.

22. Inversion of Sugar   When cane sugar is dissolved in water, it converts to invert sugar over a period of several hours. The percentage $f(t)$ of unconverted cane sugar at time t (in hours) decreases exponentially. Suppose that $f' = -0.2f$. What percentage of cane sugar remains after 5 hours? After 10 hours?

23. A quantity P increases exponentially with doubling time 6 hours. After how many hours has P increased by 50%?

24. Two bacteria colonies are cultivated in a laboratory. The first colony has a doubling time of 2 hours and the second a doubling time of 3 hours. Initially, the first colony contains 1,000 bacteria and the second colony 3,000 bacteria. At what time t will sizes of the colonies be equal?

25. Moore's Law   In 1965, Gordon Moore predicted that the number N of transistors on a microchip would increase exponentially.

(a) Does the table of data below confirm Moore's prediction for the period from 1971 to 2000? If so, estimate the growth constant k.

(b) $\mathcal{CAS}$ Plot the data in the table.

(c) Let $N(t)$ be the number of transistors t years after 1971. Find an approximate formula $N(t) \approx Ce^{kt}$, where t is the number of years after 1971.

(d) Estimate the doubling time in Moore's Law for the period from 1971 to 2000.

(e) If Moore's Law continues to hold until the end of the decade, how many transistors will a chip contain in 2010?

(f) Can Moore have expected his prediction to hold indefinitely?

Transistors	Year	No. Transistors
4004	1971	2,250
8008	1972	2,500
8080	1974	5,000
8086	1978	29,000
286	1982	120,000
386 processor	1985	275,000
486 DX processor	1989	1,180,000
Pentium processor	1993	3,100,000
Pentium II processor	1997	7,500,000
Pentium III processor	1999	24,000,000
Pentium 4 processor	2000	42,000,000

26. Assume that in a certain country, the rate at which jobs are created is proportional to the number of people who already have jobs. If there are 15 million jobs at $t = 0$ and 15.1 million jobs 3 months later, how many jobs will there be after two years?

*In Exercises 27–28, we consider the **Gompertz differential equation:***

$$\frac{dy}{dt} = ky \ln\left(\frac{y}{M}\right)$$

(where M and k are constants), introduced in 1825 by the English mathematician Benjamin Gompertz and still used today to model aging and mortality.

27. Show that $y = Me^{ae^{kt}}$ is a solution for any constant a.

28. To model mortality in a population of 200 laboratory rats, a scientist assumes that the number $P(t)$ of rats alive at time t (in months) satisfies the Gompertz equation with $M = 204$ and $k = 0.15$ months^{-1} (Figure 13). Find $P(t)$ [note that $P(0) = 200$] and determine the population after 20 months.

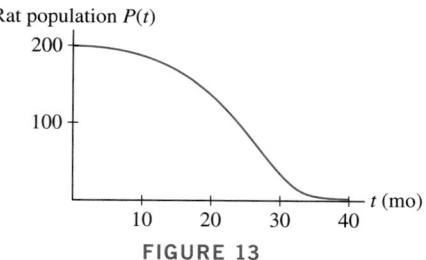

Rat population $P(t)$

FIGURE 13

29. [icon]   A certain quantity increases quadratically: $P(t) = P_0 t^2$.

(a) Starting at time $t_0 = 1$, how long will it take for P to double in size? How long will it take starting at $t_0 = 2$ or 3?

(b) In general, starting at time t_0, how long will it take for P to double in size?

30. Verify that the half-life of a quantity that decays exponentially with decay constant k is equal to $\ln 2/k$.

31. Compute the balance after 10 years if $2,000 is deposited in an account paying 9% interest and interest is compounded (a) quarterly, (b) monthly, and (c) continuously.

32. Suppose $500 is deposited into an account paying interest at a rate of 7%, continuously compounded. Find a formula for the value of the account at time t. What is the value of the account after 3 years?

33. A bank pays interest at a rate of 5%. What is the yearly multiplier if interest is compounded

(a) yearly?   **(b)** three times a year?

(c) continuously?

34. How long will it take for $4,000 to double in value if it is deposited in an account bearing 7% interest, continuously compounded?

35. Show that if interest is compounded continuously at a rate r, then an account doubles after $(\ln 2)/r$ years.

36. How much must be invested today in order to receive $20,000 after 5 years if interest is compounded continuously at the rate $r = 9\%$?

37. An investment increases in value at a continuously compounded rate of 9%. How large must the initial investment be in order to build up a value of $50,000 over a seven-year period?

38. Compute the PV of $5,000 received in 3 years if the interest rate is (a) 6% and (b) 11%. What is the PV in these two cases if the sum is instead received in 5 years?

39. Is it better to receive $1,000 today or $1,300 in 4 years? Consider $r = 0.08$ and $r = 0.03$.

40. Find the interest rate r if the PV of $8,000 to be received in 1 year is $7,300.

41. If a company invests $2 million to upgrade its factory, it will earn additional profits of $500,000/year for 5 years. Is the investment worthwhile, assuming an interest rate of 6% (assume that the savings are received as a lump sum at the end of each year)?

42. A new computer system costing $25,000 will reduce labor costs by $7,000/year for 5 years.
(a) Is it a good investment if $r = 8\%$?
(b) How much money will the company actually save?

43. After winning $25 million in the state lottery, Jessica learns that she will receive five yearly payments of $5 million beginning immediately.
(a) What is the PV of Jessica's prize if $r = 6\%$?
(b) How much more would the prize be worth if the entire amount were paid today?

44. An investment group purchased an office building in 1998 for $17 million and sold it 5 years later for $26 million. Calculate the annual (continuously compounded) rate of return on this investment.

45. Use Eq. (3) to compute the PV of an income stream paying out $R(t) = \$5,000$/year continuously for 10 years and $r = 0.05$.

46. Compute the PV of an income stream if income is paid out continuously at a rate $R(t) = \$5,000e^{0.1t}$/year for 5 years and $r = 0.05$.

47. Find the PV of an investment that produces income continuously at a rate of $800/year for 5 years, assuming an interest rate of $r = 0.08$.

48. The rate of yearly income generated by a commercial property is $50,000/year at $t = 0$ and increases at a continuously compounded rate of 5%. Find the PV of the income generated in the first four years if $r = 8\%$.

49. Show that the PV of an investment that pays out R dollars/year continuously for T years is $R(1 - e^{-rT})/r$, where r is the interest rate.

50. Explain this statement: If T is very large, then the PV of the income stream described in Exercise 49 is approximately R/r.

51. Suppose that $r = 0.06$. Use the result of Exercise 50 to estimate the payout rate R needed to produce an income stream whose PV is $20,000, assuming that the stream continues for a large number of years.

52. Verify by differentiation

$$\int te^{-rt}\,dt = -\frac{e^{-rt}(1+rt)}{r^2} + C \qquad \boxed{6}$$

Use Eq. (6) to compute the PV of an investment that pays out income continuously at a rate $R(t) = (5,000 + 1,000t)$ dollars/year for 5 years and $r = 0.05$.

53. Use Eq. (6) to compute the PV of an investment that pays out income continuously at a rate $R(t) = (5,000 + 1,000t)e^{0.02t}$ dollars/year for 10 years and $r = 0.08$.

54.   **Banker's Rule of 70**   Bankers have a rule of thumb that if you receive R percent interest, continuously compounded, then your money doubles after approximately $70/R$ years. For example, at $R = 5\%$, your money doubles after $70/5$ or 14 years. Use the concept of doubling time to justify the Banker's Rule. (*Note:* Sometimes, the approximation $72/R$ is used. It is less accurate but easier to apply because 72 is divisible by more numbers than 70.)

Further Insights and Challenges

55.   **Isotopes for Dating**   Which of the following isotopes would be most suitable for dating extremely old rocks: Carbon-14 (half-life 5,570 years), Lead-210 (half-life 22.26 years), and Potassium-49 (half-life 1.3 billion years)? Explain why.

56. Let $P = P(t)$ be a quantity that obeys an exponential growth law with growth constant k. Show that P increases m-fold after an interval of $(\ln m)/k$ years.

57.   **Average Time of Decay**   Physicists use the radioactive decay law $R = R_0 e^{-kt}$ to compute the average or *mean time M* until an atom decays. Let $F(t) = R/R_0 = e^{-kt}$ be the fraction of atoms that have survived to time t without decaying.
(a) Find the inverse function $t(F)$.
(b) The error in the following approximation tends to zero as $N \to \infty$:

$$M = \text{mean time to decay} \approx \frac{1}{N}\sum_{j=1}^{N} t\left(\frac{j}{N}\right)$$

Argue that $M = \displaystyle\int_0^1 t(F)\,dF$.

(c) Verify the formula $\displaystyle\int \ln x\,dx = x\ln x - x$ by differentiation and use it to show that for $c > 0$,

$$\int_c^1 t(F)\,dF = \frac{1}{k} + \frac{1}{k}(c\ln c - c)$$

(d) Verify numerically that $\displaystyle\lim_{c \to 0}(c - c\ln c) = 0$.

(e) The integral defining M is "improper" because $t(0)$ is infinite. Show that $M = 1/k$ by computing the limit

$$M = \lim_{c \to 0} \int_c^1 t(F)\,dF$$

(f) What is the mean time to decay for Radon (with a half-life of 3.825 days)?

58. The text proves that $e = \lim_{n \to \infty} (1 + \frac{1}{n})^n$. Use a change of variables to show that for any x,

$$\lim_{n \to \infty} \left(1 + \frac{x}{n}\right)^n = \lim_{n \to \infty} \left(1 + \frac{1}{n}\right)^{nx}$$

Use this to conclude that $e^x = \lim_{n \to \infty} (1 + \frac{x}{n})^n$.

59. Use Eq. (4) to prove that for $n > 0$,

$$\left(1 + \frac{1}{n}\right)^n \le e \le \left(1 + \frac{1}{n}\right)^{n+1}$$

60. A bank pays interest at the rate r, compounded M times yearly. The **effective interest rate** r_e is the rate at which interest, if compounded annually, would have to be paid to produce the same yearly return.

(a) Find r_e if $r = 9\%$ compounded monthly.

(b) Show that $r_e = (1 + r/M)^M - 1$ and that $r_e = e^r - 1$ if interest is compounded continuously.

(c) Find r_e if $r = 11\%$ compounded continuously.

(d) Find the rate r, compounded weekly, that would yield an effective rate of 20%.

CHAPTER REVIEW EXERCISES

In Exercises 1–4, refer to the function $f(x)$ whose graph is shown in Figure 1.

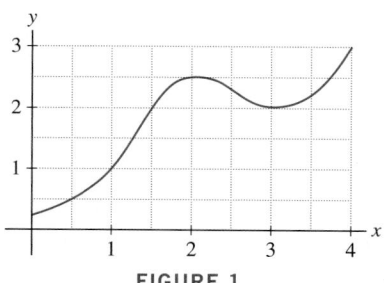

FIGURE 1

1. Estimate L_4 and M_4 on $[0, 4]$.

2. Estimate R_4, L_4, and M_4 on $[1, 3]$.

3. Find an interval $[a, b]$ on which R_4 is larger than $\int_a^b f(x)\,dx$. Do the same for L_4.

4. Justify $\frac{7}{4} \le \int_1^2 f(x)\,dx \le \frac{9}{4}$.

In Exercises 5–8, let $f(x) = x^2 + 4x$.

5. Calculate R_6, M_6, and L_6 for $f(x)$ on the interval $[1, 4]$. Sketch the graph of $f(x)$ and the corresponding rectangles for each approximation.

6. Find a formula for R_N for $f(x)$ on $[1, 4]$ and compute $\int_1^4 f(x)\,dx$ by taking the limit.

7. Find a formula for L_N for $f(x)$ on $[0, 2]$ and compute $\int_0^2 f(x)\,dx$ by taking the limit.

8. Use FTC I to evaluate $A(x) = \int_{-2}^x f(t)\,dt$.

9. Calculate R_6, M_6, and L_6 for $f(x) = (x^2 + 1)^{-1}$ on the interval $[0, 1]$.

10. Let R_N be the Nth right-endpoint approximation for $f(x) = x^3$ on $[0, 4]$ (Figure 2).

(a) Prove that $R_N = \dfrac{64(N+1)^2}{N^2}$.

(b) Prove that the area of the region below the right-endpoint rectangles and above the graph is equal to

$$\frac{64(2N + 1)}{N^2}$$

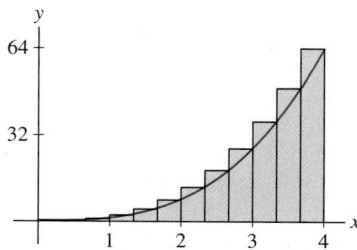

FIGURE 2 Approximation R_N for $f(x) = x^3$ on $[0, 4]$.

11. Which approximation to the area is represented by the shaded rectangles in Figure 3? Compute R_5 and L_5.

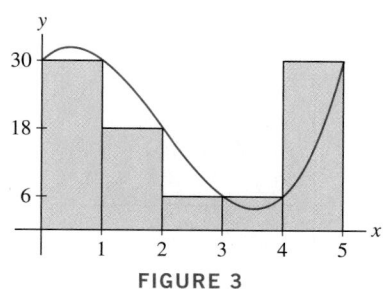

FIGURE 3

12. Calculate any two Riemann sums for $f(x) = x^2$ on the interval $[2, 5]$, but choose partitions with at least five subintervals of unequal widths and intermediate points that are neither endpoints nor midpoints.

In Exercises 13–34, evaluate the integral.

13. $\displaystyle\int (6x^3 - 9x^2 + 4x)\,dx$

14. $\displaystyle\int_0^1 (4x^3 - 2x^5)\,dx$

15. $\displaystyle\int (2x^3 - 1)^2\,dx$

16. $\displaystyle\int_1^4 (x^{5/2} - 2x^{-1/2})\,dx$

17. $\displaystyle\int \frac{x^4 + 1}{x^2}\,dx$

18. $\displaystyle\int_1^4 r^{-2}\,dr$

19. $\displaystyle\int_{-1}^4 |x^2 - 9|\,dx$

20. $\displaystyle\int_1^3 [t]\,dt$

21. $\displaystyle\int \csc^2\theta\,d\theta$

22. $\displaystyle\int_0^{\pi/4} \sec t \tan t\,dt$

23. $\displaystyle\int \sec^2(9t - 4)\,dt$

24. $\displaystyle\int_0^{\pi/3} \sin 4\theta\,d\theta$

25. $\displaystyle\int (9t - 4)^{11}\,dt$

26. $\displaystyle\int_6^2 \sqrt{4y + 1}\,dy$

27. $\displaystyle\int \sin^2(3\theta) \cos(3\theta)\,d\theta$

28. $\displaystyle\int_0^{\pi/2} \sec^2(\cos\theta) \sin\theta\,d\theta$

29. $\displaystyle\int \frac{(2x^3 + 3x)\,dx}{(3x^4 + 9x^2)^5}$

30. $\displaystyle\int_{-4}^{-2} \frac{12x\,dx}{(x^2 + 2)^3}$

31. $\displaystyle\int \sin\theta\sqrt{4 - \cos\theta}\,d\theta$

32. $\displaystyle\int_0^{\pi/3} \frac{\sin\theta}{\cos^{2/3}\theta}\,d\theta$

33. $\displaystyle\int y\sqrt{2y + 3}\,dy$

34. $\displaystyle\int_1^8 t^2\sqrt{t + 8}\,dt$

35. Combine to write as a single integral

$$\int_0^8 f(x)\,dx + \int_{-2}^0 f(x)\,dx + \int_8^6 f(x)\,dx$$

36. Let $A(x) = \displaystyle\int_0^x f(x)\,dx$, where $f(x)$ is the function shown in Figure 4. Indicate on the graph of f where the local minima, maxima, and points of inflection of $A(x)$ occur and identify the intervals where $A(x)$ is increasing, decreasing, concave up, or concave down.

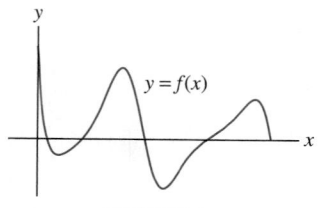

FIGURE 4

37. Find inflection points of $A(x) = \displaystyle\int_3^x \frac{t\,dt}{t^2 + 1}$.

38. A particle starts at the origin at time $t = 0$ and moves with velocity $v(t)$ as shown in Figure 5.

(a) How many times does the particle return to the origin in the first 12 s?

(b) Where is the particle located at time $t = 12$?

(c) At which time t is the particle's distance to the origin at a maximum?

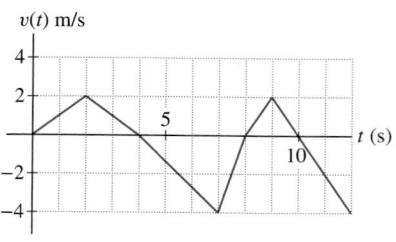

FIGURE 5

39. On a typical day, a city consumes water at the rate of $r(t) = 100 + 72t - 3t^2$ (in thousands of gallons per hour), where t is the number of hours past midnight. What is the daily water consumption? How much water is consumed between 6 PM and midnight?

40. The learning curve for producing bicycles in a certain factory is $L(x) = 12x^{-1/5}$ (in hours per bicycle), which means that it takes a bike mechanic $L(n)$ hours to assemble the nth bicycle. If 24 bicycles are produced, how long does it take to produce the second batch of 12?

41. Cost engineers at NASA have the task of projecting the cost P of major space projects. It has been found that the cost C of developing a projection increases with P at the rate $dC/dP \approx 21P^{-0.65}$, where C is in thousands of dollars and P in millions of dollars. What is the cost of developing a projection for a project whose cost turns out to be $P = \$35$ million?

42. The cost of jet fuel increased dramatically in 2005. Figure 6 displays Department of Transportation estimates for the rate of percentage price increase $R(t)$ (in units of percentage per year) during the first 6 months of the year. Express the total percentage price increase I during the first 6 months as an integral and calculate I. When determining the limits of integration, keep in mind that t is in years since $R(t)$ is a yearly rate.

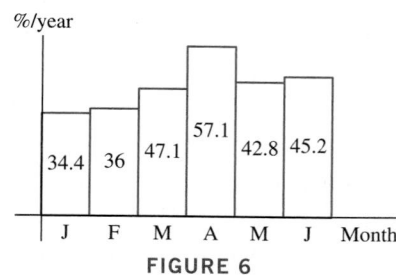

FIGURE 6

43. Let $f(x)$ be a positive increasing continuous function on $[a, b]$, where $0 \le a < b$ as in Figure 7. Show that the shaded region has area

$$I = bf(b) - af(a) - \int_a^b f(x)\, dx \qquad \boxed{1}$$

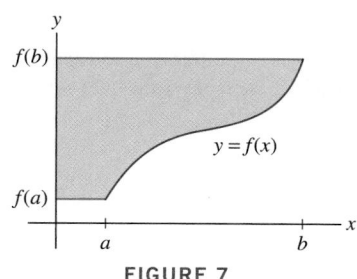

FIGURE 7

44. How can we interpret the quantity I in Eq. (1) if $a < b \le 0$? Explain with a graph.

In Exercises 45–49, express the limit as an integral (or multiple of an integral) and evaluate.

45. $\displaystyle \lim_{N\to\infty} \frac{2}{N} \sum_{j=1}^{N} \sin\left(\frac{2j}{N}\right)$

46. $\displaystyle \lim_{N\to\infty} \frac{4}{N} \sum_{k=1}^{N} \left(3 + \frac{4k}{N}\right)$

47. $\displaystyle \lim_{N\to\infty} \frac{\pi}{N} \sum_{j=0}^{N-1} \sin\left(\frac{\pi}{2} + \frac{\pi j}{N}\right)$

48. $\displaystyle \lim_{N\to\infty} \frac{4}{N} \sum_{k=1}^{N} \frac{1}{(3 + \frac{4k}{N})^2}$

49. $\displaystyle \lim_{N\to\infty} \frac{1^k + 2^k + \cdots N^k}{N^{k+1}} \qquad (k > 0)$

50. Evaluate $\displaystyle \int_{-\pi/4}^{\pi/4} \frac{x^9\, dx}{\cos^2 x}$, using the properties of odd functions.

51. Evaluate $\displaystyle \int_0^1 f(x)\, dx$, assuming that $f(x)$ is an even continuous function such that

$$\int_1^2 f(x)\, dx = 5, \qquad \int_{-2}^1 f(x)\, dx = 8$$

52. GU Plot the graph of $f(x) = \sin mx \sin nx$ on $[0, \pi]$ for the pairs $(m, n) = (2, 4),\ (3, 5)$ and in each case guess the value of $I = \int_0^\pi f(x)\, dx$. Experiment with a few more values (including two cases with $m = n$) and formulate a conjecture for when I is zero.

53. Show that

$$\int x\, f(x)\, dx = xF(x) - G(x)$$

where $F'(x) = f(x)$ and $G'(x) = F(x)$. Use this to evaluate $\int x \cos x\, dx$.

54. Prove

$$2 \le \int_1^2 2^x\, dx \le 4 \qquad \text{and} \qquad \frac{1}{9} \le \int_1^2 3^{-x}\, dx \le \frac{1}{3}$$

55. GU Plot the graph of $f(x) = x^{-2} \sin x$ and show that

$$0.2 \le \int_1^2 f(x)\, dx \le 0.9.$$

56. Find upper and lower bounds for $\displaystyle \int_0^1 f(x)\, dx$, where $f(x)$ has the graph shown in Figure 8.

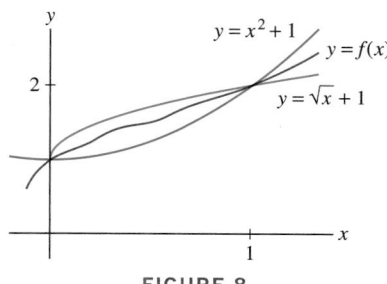

FIGURE 8

In Exercises 57–62, find the derivative.

57. $A'(x)$, where $\displaystyle A(x) = \int_3^x \sin(t^3)\, dt$

58. $A'(\pi)$, where $\displaystyle A(x) = \int_2^x \frac{\cos t}{1 + t}\, dt$

59. $\displaystyle \frac{d}{dy} \int_{-2}^y 3^x\, dx$

60. $G'(x)$, where $\displaystyle G(x) = \int_{-2}^{\sin x} t^3\, dt$

61. $G'(2)$, where $\displaystyle G(x) = \int_0^{x^3} \sqrt{t + 1}\, dt$

62. $H'(1)$, where $\displaystyle H(x) = \int_{4x^2}^9 \frac{1}{t}\, dt$

63. Explain with a graph: If $f(x)$ is increasing and concave up on $[a, b]$, then L_N is more accurate than R_N. Which is more accurate if $f(x)$ is increasing and concave down?

64. Explain with a graph: If $f(x)$ is linear on $[a, b]$, then the $\displaystyle \int_a^b f(x)\, dx = \frac{1}{2}(R_N + L_N)$ for all N.

In Exercises 65–70, use the given substitution to evaluate the integral.

65. $\displaystyle \int \frac{(\ln x)^2 dx}{x}, \quad u = \ln x$

66. $\displaystyle \int \frac{dx}{4x^2 + 9}, \quad u = \frac{2x}{3}$

67. $\displaystyle \int \frac{dx}{\sqrt{e^{2x} - 1}}, \quad u = e^x$

68. $\displaystyle \int \frac{\cos^{-1} t \, dt}{\sqrt{1 - t^2}}, \quad u = \cos^{-1} t$

69. $\displaystyle \int \frac{dt}{t(1 + (\ln t)^2)}, \quad u = \ln t$

70. $\displaystyle \int \sec^2(2\theta) \tan(2\theta) \, d\theta, \quad u = \tan(2\theta)$

In Exercises 71–92, calculate the integral.

71. $\displaystyle \int e^{9-2x} \, dx$

72. $\displaystyle \int x^2 e^{x^3} \, dx$

73. $\displaystyle \int e^{-2x} \sin(e^{-2x}) \, dx$

74. $\displaystyle \int \frac{\cos(\ln x) \, dx}{x}$

75. $\displaystyle \int_1^e \frac{\ln x \, dx}{x}$

76. $\displaystyle \int_0^{\ln 3} e^{x - e^x} \, dx$

77. $\displaystyle \int_{1/3}^{2/3} \frac{dx}{\sqrt{1 - x^2}}$

78. $\displaystyle \int_4^{12} \frac{dx}{x\sqrt{x^2 - 1}}$

79. $\displaystyle \int_0^{\pi/3} \tan\theta \, d\theta$

80. $\displaystyle \int_{\pi/6}^{2\pi/3} \cot\left(\frac{1}{2}\theta\right) d\theta$

81. $\displaystyle \int_0^1 \cos 2t \, dt$

82. $\displaystyle \int_0^2 \frac{dt}{4t + 12}$

83. $\displaystyle \int_0^3 \frac{x \, dx}{x^2 + 9}$

84. $\displaystyle \int_0^3 \frac{dx}{x^2 + 9}$

85. $\displaystyle \int \frac{x \, dx}{\sqrt{1 - x^4}}$

86. $\displaystyle \int e^x 10^x \, dx$

87. $\displaystyle \int \frac{\sin^{-1} x \, dx}{\sqrt{1 - x^2}}$

88. $\displaystyle \int \tan 5x \, dx$

89. $\displaystyle \int \sin x \cos^3 x \, dx$

90. $\displaystyle \int_0^1 \frac{dx}{25 + x^2}$

91. $\displaystyle \int_0^4 \frac{dx}{2x^2 + 1}$

92. $\displaystyle \int_5^8 \frac{dx}{x\sqrt{x^2 - 16}}$

93. In this exercise, we prove that for all $x > 0$,

$$x - \frac{x^2}{2} \leq \ln(1 + x) \leq x \qquad \boxed{2}$$

(a) Show that $\ln(1 + x) = \displaystyle\int_0^x \frac{dt}{1 + t}$ for $x > 0$.

(b) Verify that $1 - t \leq \dfrac{1}{1 + t} \leq 1$ for all $t > 0$.

(c) Use (b) to prove Eq. (2).

(d) Verify Eq. (2) for $x = 0.5, 0.1$, and 0.01.

94. Let

$$F(x) = x\sqrt{x^2 - 1} - 2\int_1^x \sqrt{t^2 - 1} \, dt$$

Prove that $F(x)$ and $\cosh^{-1} x$ differ by a constant by showing that they have the same derivative. Then prove they are equal by evaluating both at $x = 1$.

95. Let

$$F(x) = \int_2^x \frac{dt}{\ln t} \qquad \text{and} \qquad G(x) = \frac{x}{\ln x}$$

Verify that L'Hôpital's Rule may be applied to the limit $L = \displaystyle\lim_{x \to \infty} \frac{F(x)}{G(x)}$ and evaluate L.

96. The isotope Thorium-234 has a half-life of 24.5 days.

(a) Find the differential equation satisfied by the amount $y(t)$ of Thorium-234 in a sample at time t.

(b) At $t = 0$, a sample contains 2 kg of Thorium-234. How much remains after 1 year?

97. The Oldest Snack Food In Bat Cave, New Mexico, archaeologists found ancient human remains, including cobs of popping corn, that had a C^{14} to C^{12} ratio equal to around 48% of that found in living matter. Estimate the age of the corn cobs.

98. The C^{14} to C^{12} ratio of a sample is proportional to the disintegration rate (number of beta particles emitted per minute) that is measured directly with a Geiger counter. The disintegration rate of carbon in a living organism is 15.3 beta particles/min per gram. Find the age of a sample that emits 9.5 beta particles/min per gram.

99. An investment pays out $5,000 at the end of the year for 3 years. Compute the PV, assuming an interest rate of 8%.

100. Use Eq. (3) of Section 5.8 to show that the PV of an investment which pays out income continuously at a constant rate of R dollars/year for T years is PV $= R\dfrac{1 - e^{-rT}}{r}$, where r is the interest rate. Use L'Hôpital's Rule to prove that the PV approaches RT as $r \to 0$.

101. In a first-order chemical reaction, the quantity $y(t)$ of reactant at time t satisfies $y' = -ky$, where $k > 0$. The dependence of k on temperature T (in kelvins) is given by the **Arrhenius equation** $k = Ae^{-E_a/(RT)}$, where E_a is the activation energy (J-mol^{-1}), $R = 8.314$ J-mol^{-1}-K^{-1}, and A is a constant. Assume that $A = 72 \times 10^{12}$ hour^{-1} and $E_a = 1.1 \times 10^5$. Calculate $\dfrac{dk}{dT}$ for $T = 500$ and use the Linear Approximation to estimate the change in k if T is raised from 500 to 510 K.

102. Find the interest rate if the PV of $50,000 to be received in 3 years is 43,000.

103. An equipment upgrade costing $1 million will save a company $320,000 per year for 4 years. Is this a good investment if the interest rate is $r = 5\%$? What is the largest interest rate that would make the investment worthwhile? Assume that the savings are received as a lump sum at the end of each year.

104. Calculate the limit:

(a) $\displaystyle\lim_{n \to \infty} \left(1 + \frac{4}{n}\right)^n$

(b) $\displaystyle\lim_{n \to \infty} \left(1 + \frac{1}{n}\right)^{4n}$

(c) $\displaystyle\lim_{n \to \infty} \left(1 + \frac{4}{n}\right)^{3n}$

6 | APPLICATIONS OF THE INTEGRAL

The integral, like the derivative, has a wide variety of applications. In the previous chapter, we used the integral to compute areas under curves and net change. In this chapter, we discuss some of the other quantities that are represented by integrals, including volume, average value, work, total mass, population, and fluid flow.

The CAT scan is based on tomography, a mathematical technique for combining a large series of X-rays of the body taken at different angles into a single cross-sectional image.

6.1 Area Between Two Curves

In Section 5.2, we learned that the definite integral $\int_a^b f(x)\,dx$ represents the signed area between the graph of $f(x)$ and the x-axis over the interval $[a, b]$. Sometimes, we are interested in computing the area between two graphs (Figure 1). If $f(x) \geq g(x) \geq 0$ for $x \in [a, b]$, then the graph of $f(x)$ lies above the graph of $g(x)$, and the region between the two graphs is obtained by removing the region under $y = g(x)$ from the region under $y = f(x)$. Therefore,

$$\text{Area between the graphs} = \int_a^b f(x)\,dx - \int_a^b g(x)\,dx$$

$$= \int_a^b (f(x) - g(x))\,dx \qquad \boxed{1}$$

 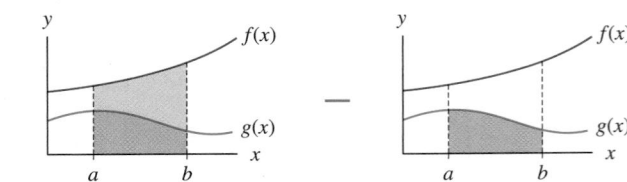

Region between the graphs

FIGURE 1 The area between the graphs is a difference of two areas.

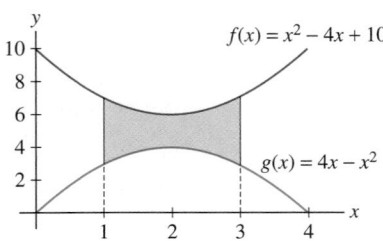

FIGURE 2 Region between the graphs of $f(x) = x^2 - 4x + 10$ and $g(x) = 4x - x^2$ over $[1, 3]$.

■ **EXAMPLE 1** Calculate the area of the region between the graphs of

$$f(x) = x^2 - 4x + 10 \qquad \text{and} \qquad g(x) = 4x - x^2$$

over $[1, 3]$.

Solution To calculate the area between the graphs, we must first determine which graph lies on top. Figure 2 shows that $f(x) \geq g(x)$. We can verify this without appealing to the graph by completing the square:

$$f(x) - g(x) = (x^2 - 4x + 10) - (4x - x^2) = 2x^2 - 8x + 10 = 2(x - 2)^2 + 2 > 0$$

By Eq. (1), the area between the graphs is

$$\int_1^3 \big(f(x) - g(x)\big)\,dx = \int_1^3 \big((x^2 - 4x + 10) - (4x - x^2)\big)\,dx$$

$$= \int_1^3 (2x^2 - 8x + 10)\,dx = \left(\frac{2}{3}x^3 - 4x^2 + 10x\right)\Bigg|_1^3 = 12 - \frac{20}{3} = \frac{16}{3} \quad \blacksquare$$

Before continuing with more examples, we use Riemann sums to explain why Eq. (1) remains valid if $f(x) \geq g(x)$ but $f(x)$ and $g(x)$ are not assumed to be positive:

$$\int_a^b \big(f(x) - g(x)\big)\,dx = \lim_{\|P\| \to 0} R(f - g, P, C) = \lim_{N \to \infty} \sum_{i=1}^N \big(f(c_i) - g(c_i)\big)\Delta x_i$$

Recall that P denotes a partition of $[a, b]$:

$$\text{Partition } P: \quad a = x_0 < x_1 < x_2 < \cdots < x_N = b$$

$C = \{c_1, \ldots, c_N\}$ is a choice of intermediate points where $c_i \in [x_{i-1}, x_i]$, and $\Delta x_i = x_i - x_{i-1}$. The ith term in the Riemann sum is equal to the area of a thin vertical rectangle of height $(f(c_i) - g(c_i))$ and width Δx_i (Figure 3):

$$\big(f(c_i) - g(c_i)\big)\Delta x_i = \text{height} \times \text{width}$$

Therefore, $R(f - g, P, C)$ is an approximation to the area between the graphs using thin vertical rectangles. As the norm $\|P\|$ (the maximum width of the rectangles) approaches zero, the Riemann sum converges to the area between the graphs and we obtain Eq. (1).

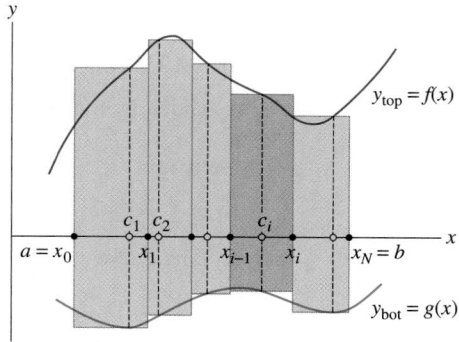

FIGURE 3 The ith rectangle has width $\Delta x_i = x_i - x_{i-1}$ and height $f(c_i) - g(c_i)$.

To help identify the functions, we sometimes denote the upper graph by $y_{\text{top}} = f(x)$ and the lower graph by $y_{\text{bot}} = g(x)$:

Keep in mind that $(y_{\text{top}} - y_{\text{bot}})$ is the height of a thin vertical slice of the region.

$$\boxed{\text{Area between the graphs} = \int_a^b (y_{\text{top}} - y_{\text{bot}})\,dx = \int_a^b \big(f(x) - g(x)\big)\,dx}$$

■ **EXAMPLE 2** Find the area between the graphs of $f(x) = x^2 - 5x - 7$ and $g(x) = x - 12$ over $[-2, 5]$.

Solution To calculate the area, we must first determine which graph lies on top.

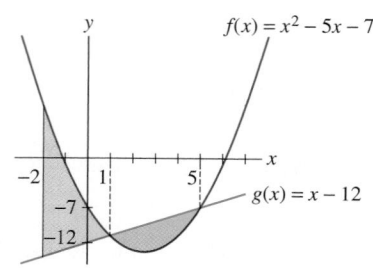

$f(x) = x^2 - 5x - 7$

$g(x) = x - 12$

FIGURE 4

In Example 2, we found the intersection points of $y = f(x)$ and $y = g(x)$ algebraically. For more complicated functions, it may be necessary to use a computer algebra system.

Step 1. Sketch the region (especially, find any points of intersection).
We know that $y = f(x)$ is a parabola with y-intercept -7 and $y = g(x)$ is a line with y-intercept -12. To determine where the graphs intersect, we solve $f(x) = g(x)$:

$$x^2 - 5x - 7 = x - 12 \quad \text{or} \quad x^2 - 6x + 5 = (x-1)(x-5) = 0$$

Thus, the points of intersection are $x = 1, 5$ (Figure 4).

Step 2. Set up the integrals and evaluate.
Figure 4 shows that

$$f(x) \geq g(x) \text{ on } [-2, 1] \quad \text{and} \quad g(x) \geq f(x) \text{ on } [1, 5]$$

Therefore, we write the area as a sum of integrals over the two intervals:

$$\int_{-2}^{5} (y_{\text{top}} - y_{\text{bot}}) \, dx$$

$$= \int_{-2}^{1} \left(f(x) - g(x) \right) dx + \int_{1}^{5} \left(g(x) - f(x) \right) dx$$

$$= \int_{-2}^{1} \left((x^2 - 5x - 7) - (x - 12) \right) dx + \int_{1}^{5} \left((x - 12) - (x^2 - 5x - 7) \right) dx$$

$$= \int_{-2}^{1} (x^2 - 6x + 5) \, dx + \int_{1}^{5} (-x^2 + 6x - 5) \, dx$$

$$= \left(\frac{1}{3}x^3 - 3x^2 + 5x \right) \Big|_{-2}^{1} + \left(-\frac{1}{3}x^3 + 3x^2 - 5x \right) \Big|_{1}^{5}$$

$$= \left(\frac{7}{3} - \frac{(-74)}{3} \right) + \left(\frac{25}{3} - \frac{(-7)}{3} \right) = \frac{113}{3} \qquad \blacksquare$$

■ **EXAMPLE 3** Calculating Area by Dividing the Region Find the area of the region bounded by the graphs of $y = \dfrac{8}{x^2}$, $y = 8x$, and $y = x$.

Solution

Step 1. Sketch the region (especially, find any points of intersection).
The curve $y = \dfrac{8}{x^2}$ cuts off a region in the sector between the two lines $y = 8x$ and $y = x$ (Figure 5). To find the intersection of $y = \dfrac{8}{x^2}$ and $y = 8x$, we solve

$$\frac{8}{x^2} = 8x \Rightarrow x^3 = 1 \Rightarrow x = 1$$

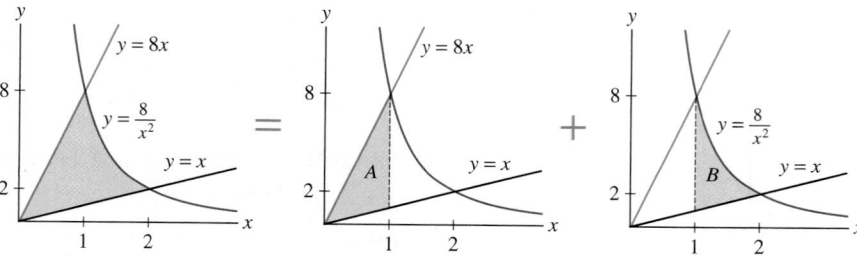

FIGURE 5 Area bounded by $y = \dfrac{8}{x^2}$, $y = 8x$, and $y = x$ as a sum of two areas.

To find the intersection of $y = \dfrac{8}{x^2}$ and $y = x$, we solve

$$\frac{8}{x^2} = x \Rightarrow x^3 = 8 \Rightarrow x = 2$$

Step 2. **Set up the integrals and evaluate.**

Figure 5 shows that $y_{\text{bot}} = x$, but y_{top} changes at $x = 1$ from $y_{\text{top}} = 8x$ to $y_{\text{top}} = \dfrac{8}{x^2}$. Therefore, we break up the regions into two parts, A and B, and compute their areas separately:

$$\text{Area of } A = \int_0^1 \left(y_{\text{top}} - y_{\text{bot}}\right) dx = \int_0^1 (8x - x)\, dx = \int_0^1 7x\, dx = \frac{7}{2} x^2 \bigg|_0^1 = \frac{7}{2}$$

$$\text{Area of } B = \int_1^2 \left(y_{\text{top}} - y_{\text{bot}}\right) dx = \int_1^2 \left(\frac{8}{x^2} - x\right) dx = \left(-\frac{8}{x} - \frac{1}{2} x^2\right) \bigg|_1^2 = \frac{5}{2}$$

The total area bounded by the curves is the sum $\frac{7}{2} + \frac{5}{2} = 6$. ∎

Integration Along the y-Axis

Suppose we are given x as a function of y, say, $x = g(y)$. What is the meaning of the integral $\displaystyle\int_c^d g(y)\, dy$? This integral may be interpreted as *signed area*, where regions to the *right* of the y-axis have positive area and regions to the *left* have negative area:

$$\int_c^d g(y)\, dy = \text{signed area between graph and } y\text{-axis for } c \le y \le d$$

Figure 6(A) shows the graph of $g(y) = y^2 - 1$. The region to the left of the y-axis has negative signed area. The integral is equal to the signed area:

$$\underbrace{\int_{-2}^2 (y^2 - 1)\, dy}_{\substack{\text{Area to the right of } y\text{-axis minus} \\ \text{area to the left of } y\text{-axis}}} = \left(\frac{1}{3} y^3 - y\right) \bigg|_{-2}^2 = \frac{4}{3}$$

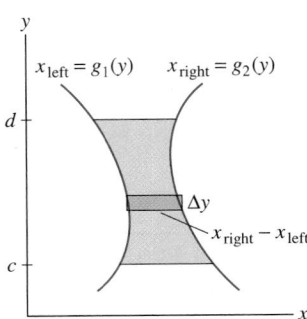

(A) Region between $x = y^2 - 1$ and the y-axis

(B) Region between $x = g_2(y)$ and $x = g_1(y)$

FIGURE 6

More generally, if $g_2(y) \geq g_1(y)$ as in Figure 6(B), then the graph of $x = g_2(y)$ lies to the right of the graph of $x = g_1(y)$. As a reminder, we write $x_{\text{right}} = g_2(y)$ and $x_{\text{left}} = g_1(y)$. The area between the two graphs for $c \leq y \leq d$ is equal to

$$\text{Area between the graphs} = \int_c^d \big(g_2(y) - g_1(y)\big)\, dy = \int_c^d (x_{\text{right}} - x_{\text{left}})\, dy \qquad \boxed{2}$$

In this case, the Riemann sums approximate the area by thin horizontal rectangles of width $x_{\text{right}} - x_{\text{left}}$ and height Δy.

■ EXAMPLE 4 Calculate the area between the graphs of $g_1(y) = y^3 - 4y$ and $g_2(y) = y^3 + y^2 + 8$ for $-2 \leq y \leq 2$.

Solution We confirm that $g_2(y) \geq g_1(y)$ as shown in Figure 7:

$$g_2(y) - g_1(y) = (y^3 + y^2 + 8) - (y^3 - 4y) = y^2 + 4y + 8 = (y+2)^2 + 4 > 0$$

Therefore, $x_{\text{right}} = g_2(y)$ and $x_{\text{left}} = g_1(y)$, and the area is

$$\int_{-2}^{2} (x_{\text{right}} - x_{\text{left}})\, dy = \int_{-2}^{2} (y^2 + 4y + 8)\, dy = \left(\frac{1}{3} y^3 + 2y^2 + 8y \right) \Bigg|_{-2}^{2}$$

$$= \frac{80}{3} - \frac{-32}{3} = \frac{112}{3} \qquad ■$$

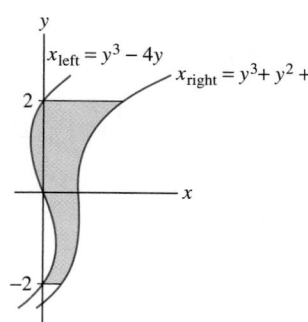

FIGURE 7 Region between $g_1(x) = y^3 - 4y$ and $g_2(x) = y^3 + y^2 + 8$ for $-2 \leq y \leq 2$.

It would be more difficult to calculate the area of the region in Figure 7 as an integral with respect to x because the curves are not graphs of functions of x.

6.1 SUMMARY

- If $f(x) \geq g(x)$ on $[a, b]$, then the area between the graphs of f and g over $[a, b]$ is

$$\text{Area between graphs} = \int_a^b \big(f(x) - g(x)\big)\, dx = \int_a^b \big(y_{\text{top}} - y_{\text{bot}}\big)\, dx$$

- To calculate the area between two graphs $y = f(x)$ and $y = g(x)$, sketch the region to find y_{top}. If necessary, find points of intersection by solving $f(x) = g(x)$.

- The integral along the y-axis, $\int_c^d g(y)\, dy$, is equal to the signed area between the graph and the y-axis for $c \leq y \leq d$, where area to the right of the y-axis is positive and area to the left is negative.

- If $g_2(y) \geq g_1(y)$, then the graph of $x = g_2(y)$ lies to the right of the graph of $x = g_1(y)$ and the area between the graphs for $c \leq y \leq d$ is

$$\text{Area between graphs} = \int_c^d \big(g_2(y) - g_1(y)\big)\, dy = \int_c^d \big(x_{\text{right}} - x_{\text{left}}\big)\, dy$$

6.1 EXERCISES

Preliminary Questions

1. What is the area interpretation of $\int_a^b \big(f(x) - g(x)\big)\, dx$ if $f(x) \geq g(x)$?

2. Is $\int_a^b \big(f(x) - g(x)\big)\, dx$ still equal to the area between the graphs of f and g if $f(x) \geq 0$ but $g(x) \leq 0$?

3. Suppose that $f(x) \geq g(x)$ on $[0, 3]$ and $g(x) \geq f(x)$ on $[3, 5]$. Express the area between the graphs over $[0, 5]$ as a sum of integrals.

4. Suppose that the graph of $x = f(y)$ lies to the left of the y-axis. Is $\int_a^b f(y)\,dy$ positive or negative?

Exercises

1. Find the area of the region between $y = 3x^2 + 12$ and $y = 4x + 4$ over $[-3, 3]$ (Figure 8).

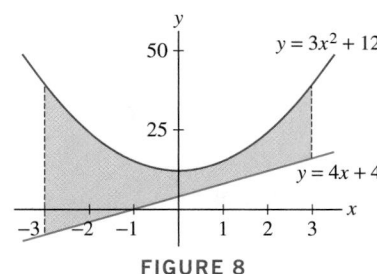

FIGURE 8

2. Compute the area of the region in Figure 9(A), which lies between $y = 2 - x^2$ and $y = -2$ over $[-2, 2]$.

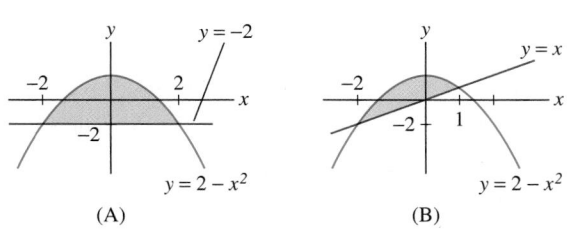

(A)   (B)

FIGURE 9

3. Let $f(x) = x$ and $g(x) = 2 - x^2$ [Figure 9(B)].
(a) Find the points of intersection of the graphs.
(b) Find the area enclosed by the graphs of f and g.

4. Let $f(x) = 8x - 10$ and $g(x) = x^2 - 4x + 10$.
(a) Find the points of intersection of the graphs.
(b) Compute the area of the region *below* the graph of f and *above* the graph of g.

In Exercises 5–7, find the area between $y = \sin x$ and $y = \cos x$ over the interval. Sketch the curves if necessary.

5. $\left[0, \dfrac{\pi}{4}\right]$ **6.** $\left[\dfrac{\pi}{4}, \dfrac{\pi}{2}\right]$ **7.** $[0, \pi]$

In Exercises 8–10, let $f(x) = 20 + x - x^2$ and $g(x) = x^2 - 5x$.

8. Find the area between the graphs of f and g over $[1, 3]$.

9. Find the area of the region enclosed by the two graphs.

10. Compute the area of the region between the two graphs over $[4, 8]$ as a sum of two integrals.

11. Find the area between $y = e^x$ and $y = e^{2x}$ over $[0, 1]$.

12. Find the area of the region bounded by $y = e^x$ and $y = 12 - e^x$ and the y-axis.

13. Sketch the region bounded by $y = \dfrac{1}{\sqrt{1 - x^2}}$ and $y = -\dfrac{1}{\sqrt{1 - x^2}}$ for $-\dfrac{1}{2} \leq x \leq \dfrac{1}{2}$ and find its area.

14. Sketch the region bounded by $y = \sec^2 x$ and $y = 2$ and find its area.

In Exercises 15–18, find the area of the shaded region in the figure.

15.

FIGURE 10

16.

FIGURE 11

17.

FIGURE 12

18.

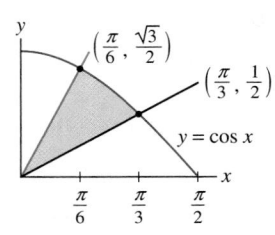

FIGURE 13

19. Find the area of the region enclosed by the curves $y = x^3 - 6x$ and $y = 8 - 3x^2$.

20. Find the area of the region enclosed by the *semicubical parabola* $y^2 = x^3$ and the line $x = 1$.

In Exercises 21–22, find the area between the graphs of $x = \sin y$ and $x = 1 - \cos y$ over the given interval (Figure 14).

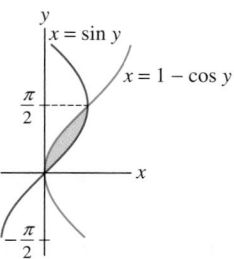

FIGURE 14

21. $0 \le y \le \dfrac{\pi}{2}$

22. $-\dfrac{\pi}{2} \le y \le \dfrac{\pi}{2}$

23. Find the area of the region lying to the right of $x = y^2 + 4y - 22$ and the left of $x = 3y + 8$.

24. Find the area of the region lying to the right of $x = y^2 - 5$ and the left of $x = 3 - y^2$.

25. Calculate the area enclosed by $x = 9 - y^2$ and $x = 5$ in two ways: as an integral along the y-axis and as an integral along the x-axis.

26. Figure 15 shows the graphs of $x = y^3 - 26y + 10$ and $x = 40 - 6y^2 - y^3$. Match the equations with the curve and compute the area of the shaded region.

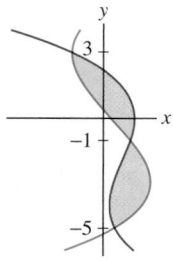

FIGURE 15

In Exercises 27–28, find the area of the region using the method (integration along either the x- or y-axis) that requires you to evaluate just one integral.

27. Region between $y^2 = x + 5$ and $y^2 = 3 - x$

28. Region between $y = x$ and $x + y = 8$ over [2, 3]

In Exercises 29–45, sketch the region enclosed by the curves and compute its area as an integral along the x- or y-axis.

29. $y = 4 - x^2$, $y = x^2 - 4$

30. $y = x^2 - 6$, $y = 6 - x^3$, y-axis

31. $x = \sin y$, $x = \dfrac{2}{\pi} y$

32. $x + y = 4$, $x - y = 0$, $y + 3x = 4$

33. $y = 3x^{-3}$, $y = 4 - x$, $y = \dfrac{x}{3}$

34. $y = 2 - \sqrt{x}$, $y = \sqrt{x}$, $x = 0$

35. $y = x\sqrt{x - 2}$, $y = -x\sqrt{x - 2}$, $x = 4$

36. $y = |x|$, $y = x^2 - 6$

37. $x = |y|$, $x = 6 - y^2$

38. $x = |y|$, $x = 1 - |y|$

39. $x = 12 - y$, $x = y$, $x = 2y$

40. $x = y^3 - 18y$, $y + 2x = 0$

41. $x = 2y$, $x + 1 = (y - 1)^2$

42. $x + y = 1$, $x^{1/2} + y^{1/2} = 1$

43. $y = 6$, $y = x^{-2} + x^2$ (in the region $x > 0$)

44. $y = \cos x$, $y = \cos(2x)$, $x = 0$, $x = \dfrac{2\pi}{3}$

45. $y = \sin x$, $y = \csc^2 x$, $x = \dfrac{\pi}{4}$, $x = \dfrac{3\pi}{4}$

46. ⌨ **CAS** Plot $y = \dfrac{x}{\sqrt{x^2 + 1}}$ and $y = (x - 1)^2$ on the same set of axes. Use a computer algebra system to find the points of intersection numerically and compute the area between the curves.

47. Sketch a region whose area is represented by

$$\int_{-\sqrt{2}/2}^{\sqrt{2}/2} \left(\sqrt{1 - x^2} - |x| \right) dx$$

and evaluate using geometry.

48. ✍ Two athletes run in the same direction along a straight track with velocities $v_1(t)$ and $v_2(t)$ (in ft/s). Assume that

$$\int_0^5 (v_1(t) - v_2(t))\, dt = 2, \qquad \int_0^{20} (v_1(t) - v_2(t))\, dt = 5$$

$$\int_{30}^{35} (v_1(t) - v_2(t))\, dt = -2$$

(a) Give a verbal interpretation of the integral $\displaystyle\int_{t_1}^{t_2} \left(v_1(t) - v_2(t) \right) dt$.

(b) Is enough information given to determine the distance between the two runners at time $t = 5$ s?

(c) If the runners begin at the same time and place, how far ahead is runner 1 at time $t = 20$ s?

(d) Suppose that runner 1 is 8 ft ahead at $t = 30$ s. How far is she ahead or behind at $t = 35$ s?

49. Express the area (not signed) of the shaded region in Figure 16 as a sum of three integrals involving the functions f and g.

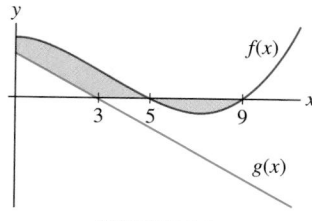

FIGURE 16

50. Find the area enclosed by the curves $y = c - x^2$ and $y = x^2 - c$ as a function of c. Find the value of c for which this area is equal to 1.

51. Set up (but do not evaluate) an integral that expresses the area between the circles $x^2 + y^2 = 2$ and $x^2 + (y - 1)^2 = 1$.

52. Set up (but do not evaluate) an integral that expresses the area between the graphs of $y = (1 + x^2)^{-1}$ and $y = x^2$.

53. *CAS* Find a numerical approximation to the area above $y = 1 - (x/\pi)$ and below $y = \sin x$ (find the points of intersection numerically).

54. *CAS* Find a numerical approximation to the area above $y = |x|$ and below $y = \cos x$.

55. *CAS* Use a computer algebra system to find a numerical approximation to the number c (besides zero) in $[0, \frac{\pi}{2}]$, where the curves $y = \sin x$ and $y = \tan^2 x$ intersect. Then find the area enclosed by the graphs over $[0, c]$.

56. The back of Jon's guitar (Figure 17) has a length 19 in. He measured the widths at 1-in. intervals, beginning and ending $\frac{1}{2}$ in. from the ends, obtaining the results

6, 9, 10.25, 10.75, 10.75, 10.25, 9.75, 9.5, 10, 11.25,

12.75, 13.75, 14.25, 14.5, 14.5, 14, 13.25, 11.25, 9

Use the midpoint rule to estimate the area of the back.

FIGURE 17 Back of guitar.

Further Insights and Challenges

57. Find the line $y = mx$ that divides the area under the curve $y = x(1 - x)$ over $[0, 1]$ into two regions of equal area.

58. *CAS* Let c be the number such that the area under $y = \sin x$ over $[0, \pi]$ is divided in half by the line $y = cx$ (Figure 18). Find an

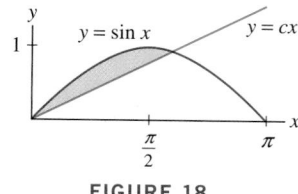

FIGURE 18

equation for c and solve this equation *numerically* using a computer algebra system.

59. Explain geometrically (without calculation) why the following holds for any $n > 0$:

$$\int_0^1 x^n \, dx + \int_0^1 x^{1/n} \, dx = 1$$

60. Let $f(x)$ be a strictly increasing function with inverse $g(x)$. Explain the equality geometrically:

$$\int_0^a f(x) \, dx + \int_{f(0)}^{f(a)} g(x) \, dx = af(a)$$

6.2 Setting Up Integrals: Volume, Density, Average Value

In this section, we use the integral to compute quantities such as volume, total mass, and fluid flow. The common thread in these diverse applications is that we approximate the relevant quantity by a Riemann sum and then pass to the limit to obtain an exact value.

Volume

The term "solid" or "solid body" refers to a solid three-dimensional object.

We begin by showing how integration can be used to compute the **volume** of a solid body. Before proceeding, let's recall that the volume of a *right cylinder* (Figure 1) is Ah, where A is the area of the base and h is the height, measured perpendicular to the base. Here we use the term "cylinder" in a general sense; the base does not have to be circular.

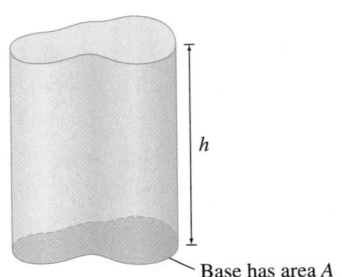

FIGURE 1 The volume of a right cylinder of height h is Ah.

Now let V be the volume of a solid body that extends from height $y = a$ to $y = b$ along the y-axis as in Figure 2. The intersection of the solid with the horizontal plane at height y is called the **horizontal cross section** at height y. Let $A(y)$ be its area.

To compute V, we divide the solid into N horizontal slices of thickness $\Delta y = \dfrac{b-a}{N}$. The ith slice extends from y_{i-1} to y_i, where $y_i = a + i\,\Delta y$. Let V_i be the volume of the slice. If N is very large, then Δy is very small and the slices are very thin. In this case, the ith slice is nearly a right cylinder of base $A(y_{i-1})$ and height Δy, and therefore $V_i \approx A(y_{i-1})\Delta y$. Summing up, we obtain

$$V = \sum_{i=1}^{N} V_i \approx \sum_{i=1}^{N} A(y_{i-1})\Delta y$$

FIGURE 2 Divide the solid into thin horizontal slices. Each slice is nearly a right cylinder whose volume can be approximated.

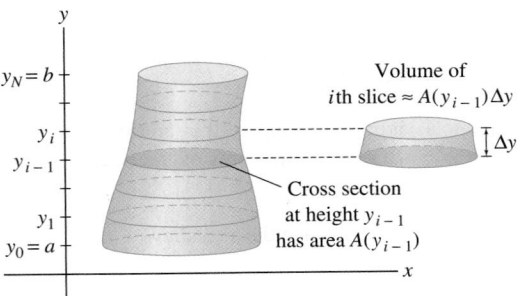

The sum on the right is a left-endpoint approximation to the integral $\displaystyle\int_a^b A(y)\,dy$. We assume that $A(y)$ is a continuous function. Then the approximation improves in accuracy and converges to the integral as $N \to \infty$. We conclude that *the volume of the solid is equal to the integral of its cross-sectional area.*

Volume as the Integral of Cross-Sectional Area   Suppose that a solid body extends from height $y = a$ to $y = b$. Let $A(y)$ be the area of the horizontal cross section at height y. Then

$$\text{Volume of the solid body} = \int_a^b A(y)\,dy \qquad \boxed{1}$$

■ **EXAMPLE 1   Volume of a Pyramid: Horizontal Cross Sections**   Calculate the volume V of a pyramid of height 12 ft whose base is a square of side 4 ft.

Solution   To use Eq. (1), we need a formula for $A(y)$.

Step 1. Find a formula for $A(y)$.
We see in Figure 3 that the horizontal cross section at height y is a square. Let $\ell(y)$ be the length of its sides. We apply the law of similar triangles to $\triangle ABC$ and the triangle of height $12 - y$ whose base of length $\frac{1}{2}\ell(y)$ lies on the cross section:

$$\frac{\text{Base}}{\text{Height}} = \frac{2}{12} = \frac{\frac{1}{2}\ell(y)}{12 - y} \quad \Rightarrow \quad 2(12 - y) = 6\ell(y)$$

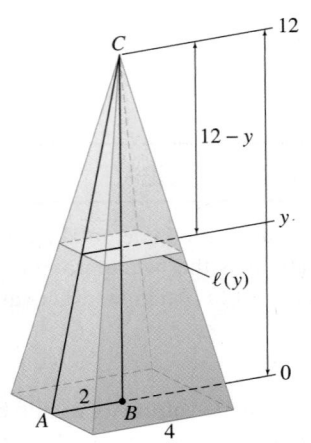

FIGURE 3 A horizontal cross section of the pyramid is a square.

We find that $\ell(y) = \frac{1}{3}(12 - y)$ and therefore $A(y) = \ell(y)^2 = \frac{1}{9}(12 - y)^2$.

Step 2. **Compute V as the integral of $A(y)$.**

$$V = \int_0^{12} A(y)\,dy = \int_0^{12} \frac{1}{9}(12 - y)^2 dy = -\frac{1}{27}(12 - y)^3 \Big|_0^{12} = 64\ \text{ft}^3$$

This is the same result we would obtain using the formula $V = \frac{1}{3}Ah$ for the volume of a pyramid of base A and height h, since $\frac{1}{3}Ah = \frac{1}{3}(4^2)(12) = 64$. ■

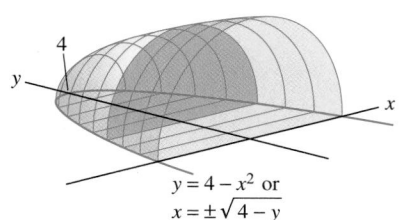

FIGURE 4 A solid whose base is the region between $y = 4 - x^2$ and the x-axis.

■ **EXAMPLE 2** The base of a solid is the region between the x-axis and the inverted parabola $y = 4 - x^2$. The vertical cross sections of the solid perpendicular to the y-axis are semicircles (Figure 4). Compute the volume of the solid.

Solution First, we find a formula for the cross section at height y. Note that $y = 4 - x^2$ can be written $x = \pm\sqrt{4 - y}$. We see in Figure 4 that the cross section at height y is a semicircle of radius $r = \sqrt{4 - y}$. The area of the semicircle is $A(y) = \frac{1}{2}\pi r^2 = \frac{\pi}{2}(4 - y)$, and the volume of the solid is

$$V = \int_0^4 A(y)\,dy = \frac{\pi}{2} \int_0^4 (4 - y)\,dy = \frac{\pi}{2}\left(4y - \frac{1}{2}y^2\right)\Big|_0^4 = 4\pi$$ ■

The volume of a solid body may be computed using vertical rather than horizontal cross sections. We then obtain an integral with respect to x rather than y.

■ **EXAMPLE 3** Volume of a Sphere: Vertical Cross Sections Compute the volume of a sphere of radius R as an integral of cross-sectional area.

Solution As we see in Figure 5, the vertical cross section of the sphere at x is a circle whose radius r satisfies $x^2 + r^2 = R^2$ or $r = \sqrt{R^2 - x^2}$. The area of the cross section is $A(x) = \pi r^2 = \pi(R^2 - x^2)$. Therefore, the volume of the sphere is

$$\int_{-R}^R \pi(R^2 - x^2)\,dx = \pi\left(R^2 x - \frac{x^3}{3}\right)\Big|_{-R}^R = 2\left(\pi R^3 - \pi\frac{R^3}{3}\right) = \frac{4}{3}\pi R^3$$ ■

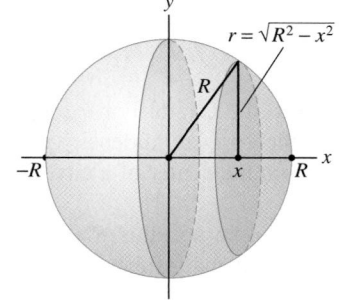

FIGURE 5 Vertical cross section is a circle of radius $\sqrt{R^2 - x^2}$.

Density

Next, we show how an integral may be used to compute the total mass of an object given its mass density. Consider a rod of length ℓ. The rod's **linear mass density** ρ is defined as the mass per unit length. If the density ρ is constant, then

$$\text{Total mass} = \text{linear mass density} \times \text{length} = \rho \cdot \ell \qquad \boxed{2}$$

For example, if $\ell = 10$ cm and $\rho = 9$ g/cm, then the total mass is $\rho\ell = 9 \cdot 10 = 90$ g.

Integration is needed to compute total mass when the density is not constant. Consider a rod extending along the x-axis from $x = a$ to $x = b$ whose density $\rho(x)$ depends on x, as in Figure 6. To compute the total mass M, we decompose the rod into N small segments of length $\Delta x = \dfrac{b - a}{N}$. Then $M = \displaystyle\sum_{i=1}^{N} M_i$, where M_i is the mass of the ith segment. Although Eq. (2) cannot be used when $\rho(x)$ is not constant, we can argue that if Δx is small, then $\rho(x)$ is nearly constant along the ith segment. Therefore, if the ith segment extends from x_{i-1} to x_i and if c_i is any sample point in $[x_{i-1}, x_i]$, then $\Delta x = x_i - x_{i-1}$ and $M_i \approx \rho(c_i)\Delta x$. We obtain

$$\text{Total mass } M = \sum_{i=1}^{N} M_i \approx \sum_{i=1}^{N} \rho(c_i)\Delta x$$

Do you see the similarity in the way we use thin slices to compute volume and small pieces to compute total mass?

As $N \to \infty$, the accuracy of the approximation improves. However, the sum on the right is a Riemann sum whose value approaches $\int_a^b \rho(x)\, dx$. We conclude that *the total mass of a rod is equal to the integral of its linear mass density*:

$$\text{Total mass } M = \int_a^b \rho(x)\, dx$$

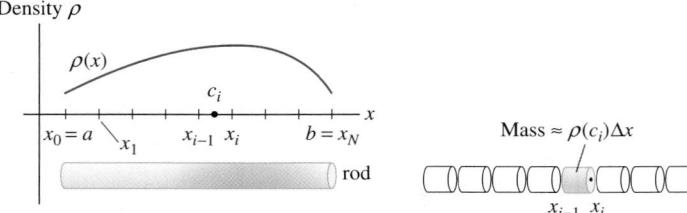

FIGURE 6 The total mass of the rod is equal to the area under the graph of mass density ρ.

■ **EXAMPLE 4** Total Mass Find the total mass of a 2-m rod of linear density $\rho(x) = 1 + x(2 - x)$ kg/m, where x is the distance from one end of the rod.

Solution The total mass is

$$\int_0^2 \rho(x)\, dx = \int_0^2 \left(1 + x(2 - x)\right) dx = \left(x + x^2 - \frac{1}{3}x^3\right)\Big|_0^2 = \frac{10}{3}\text{ kg} \qquad ■$$

In some situations, density is a function of distance to the origin. For example, in the study of urban populations, it might be assumed that the population density $\rho(r)$ (in people per square mile) depends only on the distance r from the center of a city. Such a density function is called a **radial density function.**

More general density functions depend on two variables $\rho(x, y)$. In this case, total mass or population is computed using double integration, a topic in multivariable calculus.

We now derive a formula for the total population P within a radius R of the city center, assuming that the population density $\rho(r)$ is a radial density function. To compute P, it makes sense to divide the circle of radius R into N thin rings of equal width $\Delta r = R/N$ as in Figure 7.

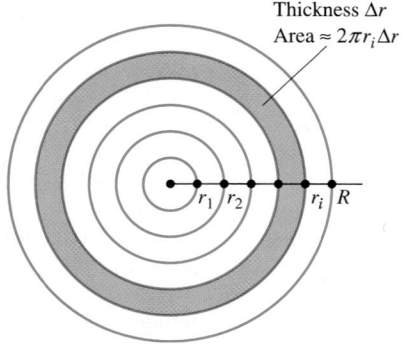

FIGURE 7 Dividing the circle of radius R into N thin rings of thickness $\Delta r = \dfrac{R}{N}$.

Let P_i be the population within the ith ring, so that $P = \sum_{i=1}^{N} P_i$. If the outer radius of the ith ring is r_i, then circumference is $2\pi r_i$, and if Δr is small, the area of this ring is *approximately* $2\pi r_i \Delta r$ (outer circumference times thickness). Furthermore, the population density within the thin ring is nearly constant with value $\rho(r_i)$. With these approximations,

$$P_i \approx \underbrace{2\pi r_i \Delta r}_{\text{Area}} \times \underbrace{\rho(r_i)}_{\substack{\text{Population} \\ \text{density}}} = 2\pi r_i \rho(r_i) \Delta r$$

and

$$P = \sum_{i=1}^{N} P_i \approx 2\pi \sum_{i=1}^{N} r_i \rho(r_i) \Delta r$$

This sum is a right-endpoint approximation to the integral $2\pi \int_0^R r\rho(r)\,dr$. As N tends to ∞, the approximation improves in accuracy and the sum converges to the integral. We conclude that for a population with a radial density function $\rho(r)$,

Remember that for a radial density function, the total population is obtained by integrating $2\pi r \rho(r)$ rather than $\rho(r)$.

> Population P within a radius $R = 2\pi \int_0^R r\rho(r)\,dr$ $\boxed{3}$

■ **EXAMPLE 5** **Computing Total Population from Population Density** The density function for the population in a certain city is $\rho(r) = 13.2(1 + r^2)^{-1/2}$, where r is the distance from the center in miles and ρ has units of thousands per square mile. How many people live within a 20-mile radius of the city center?

Solution The population (in thousands) within a 20-mile radius is

$$2\pi \int_0^{20} r\left(13.2(1 + r^2)^{-1/2}\right) dr = 2\pi(13.2) \int_0^{20} \frac{r}{(1 + r^2)^{1/2}}\,dr$$

To evaluate the integral, use the substitution $u = 1 + r^2$, $du = 2r\,dr$. The limits of integration become $u(0) = 1$ and $u(20) = 401$, and we obtain

$$13.2\pi \int_1^{401} u^{-1/2}\,du = 26.4\pi\, u^{1/2}\Big|_1^{401} \approx 1{,}577.9 \text{ thousand}$$

In other words, the population is nearly 1.6 million people.   ■

Flow Rate

When liquid flows through a tube, the **flow rate** Q is the *volume per unit time* of fluid passing through the tube. The flow rate depends on the velocity of the fluid particles. If all particles of the liquid travel with the same velocity v (say, in units of centimeters per minute), then the flow rate through a tube of radius R is

$$\underbrace{\text{Flow rate } Q}_{\text{Volume per unit time}} = \text{cross-sectional area} \times \text{velocity} = \pi R^2 v \text{ cm}^3/\text{min}$$

How do we obtain this formula? Fix an observation point P in the tube and ask the following question: Which liquid particles flow past P in a 1-min interval? The particles passing P during this minute are located at most v centimeters to the left of P since each particle travels v centimeters per minute (assuming the liquid flows from left to right).

Therefore, the column of liquid flowing past P in a 1-min interval is a cylinder of radius R and length v, which has volume $\pi R^2 v$ (Figure 8).

FIGURE 8 The column of fluid flowing past P in one unit of time is a cylinder of volume $\pi R^2 v$.

In reality, the particles of liquid do not all travel at the same velocity because of friction. However, for a slowly moving liquid, the flow is **laminar**, by which we mean that the velocity $v(r)$ depends only on the distance r from the center of the tube. The particles traveling along the center of the tube travel most quickly and the velocity tapers off to zero near the walls of the tube (Figure 9).

FIGURE 9 Laminar flow: Velocity of liquid increases toward the center of the tube.

If the flow is laminar, we can express the flow rate Q as an integral. The computation is similar to that of population with a radial density function. We divide the tube into N thin concentric cylindrical shells of width $\Delta r = R/N$ (Figure 10). The cross section of the cylindrical shell is a circular band. If r_i is the outer radius of the ith shell, then the area of this cross section is approximately $2\pi r_i \Delta r$. Furthermore, the fluid velocity within a shell is nearly constant with value $v(r_i)$, so we can approximate flow rate Q_i through the ith cylindrical shell by

$$Q_i \approx \text{cross-sectional area} \times \text{velocity} \approx 2\pi r_i \Delta r v(r_i)$$

We obtain

$$Q \approx \sum_{i=1}^{N} Q_i = 2\pi \sum_{i=1}^{N} r_i \, v(r_i) \Delta r$$

FIGURE 10 In a laminar flow, the fluid particles in a thin cylindrical shell all travel at nearly the same velocity.

The sum on the right is a right-endpoint approximation to the integral $2\pi \displaystyle\int_0^R rv(r)\,dr$. Once again, we let N tend to ∞ to obtain the formula

$$\boxed{\text{Flow rate } Q = 2\pi \int_0^R rv(r)\,dr}$$

$\boxed{4}$

The French physician Jean Poiseuille (1799–1869) discovered the law of laminar flow that cardiologists use to study blood flow in humans. Poiseuille's Law highlights the danger of cholesterol buildup in blood vessels: The flow rate through a blood vessel of radius R is proportional to R^4, so if R is reduced by one-half, the flow is reduced by a factor of 16.

■ EXAMPLE 6 Poiseuille's Law of Laminar Flow   According to Poiseuille's Law, the velocity of blood flowing in a blood vessel of radius R cm is $v(r) = k(R^2 - r^2)$, where r is the distance from the center of the vessel (in centimeters) and k is a constant. Calculate the flow rate Q as function of R, assuming that $k = 0.5$ (cm-s)$^{-1}$.

Solution By Eq. (4),

$$Q = 2\pi \int_0^R (0.5)r(R^2 - r^2)\,dr = \pi \left(R^2 \frac{r^2}{2} - \frac{r^4}{4} \right)\Bigg|_0^R = \frac{\pi}{4} R^4 \text{ cm}^3/\text{s}$$

This shows that the flow rate is proportional to R^4 (this is true for any value of k).   ■

Average Value

As a final example, we discuss the *average value* of a function. Recall that the average of N numbers $a_1, a_2, \ldots, a_N$ is the sum divided by N:

$$\frac{a_1 + a_2 + \cdots + a_N}{N} = \frac{1}{N} \sum_{j=1}^{N} a_j$$

For example, the average of 18, 25, 22, and 31, is $\frac{1}{4}(18 + 25 + 22 + 31) = 24$.

We cannot define the average value of a function $f(x)$ on an interval $[a, b]$ as a sum because there are infinitely many values of x to consider. However, the right-endpoint approximation may be interpreted as an average value (Figure 11):

$$R_N = \frac{b-a}{N}\big(f(x_1) + f(x_2) + \cdots + f(x_N)\big)$$

where $x_i = a + i\left(\dfrac{b-a}{N}\right)$. Dividing by $(b-a)$, we obtain the average of the function values $f(x_i)$:

$$\frac{1}{b-a} R_N = \underbrace{\frac{f(x_1) + f(x_2) + \cdots + f(x_N)}{N}}_{\text{Average of the function values}}$$

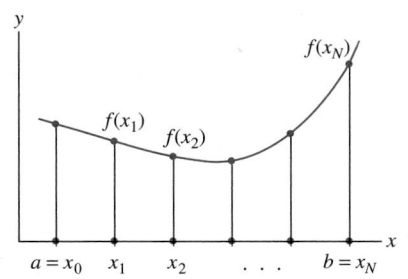

FIGURE 11 $\dfrac{R_N}{b-a}$ is the average of the values of $f(x)$ at the right endpoints.

If N is large, it is reasonable to think of this quantity as an *approximation* to the average of $f(x)$ on $[a, b]$. Therefore, we define the average value itself as the limit:

$$\text{Average value} = \lim_{N \to \infty} \frac{1}{b-a} R_N(f) = \frac{1}{b-a} \int_a^b f(x)\, dx$$

The average value is also called the **mean value**.

> **DEFINITION Average Value**   The **average value** of an integrable function $f(x)$ on $[a, b]$ is the quantity
>
> $$\text{Average value} = \frac{1}{b-a} \int_a^b f(x)\, dx$$

■ **EXAMPLE 7** Find the average value of $f(x) = \sin x$ on **(a)** $[0, \pi]$ and **(b)** $[0, 2\pi]$.

Solution

(a) The average value of $\sin x$ on $[0, \pi]$ is

$$\frac{1}{\pi} \int_0^\pi \sin x\, dx = -\frac{1}{\pi} \cos x \Big|_0^\pi = \frac{1}{\pi}\big(-(-1) - (-1)\big) = \frac{2}{\pi} \approx 0.637$$

This answer is reasonable because $\sin x$ varies from 0 to 1 on the interval $[0, \pi]$ and the average 0.637 lies somewhere between the two extremes (Figure 12).

(b) The average value over $[0, 2\pi]$ is

$$\frac{1}{2\pi} \int_0^{2\pi} \sin x\, dx = -\frac{1}{2\pi} \cos x \Big|_0^{2\pi} = -\frac{1}{2\pi}\big(1 - (1)\big) = 0$$

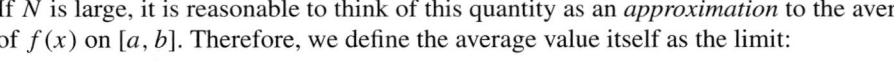

FIGURE 12 The area under the graph is equal to the area of the rectangle whose height is the average value.

This answer is also reasonable: The positive and negative values of $\sin x$ on $[0, 2\pi]$ cancel each other out, yielding an average of zero.   ∎

GRAPHICAL INSIGHT   The average value M of a function $f(x)$ on $[a, b]$ is the average height of its graph over $[a, b]$. Furthermore, the signed area of the rectangle of height M over $[a, b]$, $M(b - a)$, is equal to $\displaystyle\int_a^b f(x)\, dx$ (Figure 15).

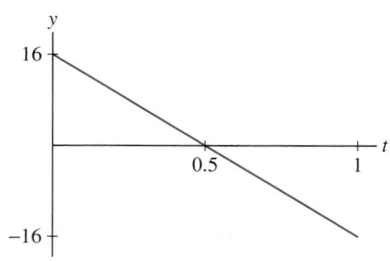

FIGURE 13 The frog's velocity is $h'(t) = 16 - 32t$.

■ **EXAMPLE 8**   A frog hops straight up from the ground with an initial velocity of $v_0 = 16$ ft/s. Use Galileo's formula for the height $h(t) = v_0 t - 16t^2$ to find the frog's average speed.

Solution   The first step is to determine the time interval of the jump. Since $v_0 = 16$, the frog's height at time t is $h(t) = 16t - 16t^2 = 16t(1 - t)$. We have $h(0) = h(1) = 0$. Therefore, the jump begins at $t = 0$ s and ends at $t = 1$ s.

The frog's velocity is $h'(t) = 16 - 32t$ (Figure 13). The frog's average speed is the average of $|h'(t)| = |16 - 32t|$, which we evaluate as the sum of two integrals since $h'(t)$ is positive on $[0, \frac{1}{2}]$ and negative on $[\frac{1}{2}, 1]$ (Figure 14):

$$\frac{1}{1 - 0}\int_0^1 |16 - 32t|\, dt = \int_0^{0.5} (16 - 32t)\, dt + \int_{0.5}^1 (32t - 16)\, dt$$

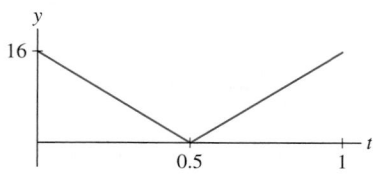

FIGURE 14 The frog's speed is $|h'(t)| = |16 - 32t|$.

$$= (16t - 16t^2)\Big|_0^{0.5} + (16t^2 - 16t)\Big|_{0.5}^1$$

$$= 4 + 4 = 8 \text{ ft/s}$$ ∎

Our next result, the Mean Value Theorem (MVT) for Integrals, highlights an important difference between the average of a list of numbers and the average value of a continuous function. If the average score on an exam is 84, then 84 lies between the highest and lowest scores, but it is possible that no student received a score of 84. By contrast, the MVT for Integrals asserts that a continuous function always takes on its average value at some point in the interval (Figure 15).

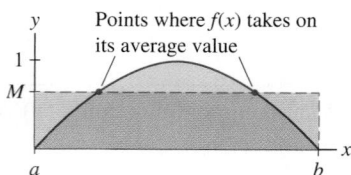

FIGURE 15 The function $f(x)$ takes on its average value M at the points where the upper edge of the rectangle intersects the graph.

THEOREM 1   Mean Value Theorem for Integrals   If $f(x)$ is continuous on $[a, b]$, then there exists a value $c \in [a, b]$ such that

$$f(c) = \frac{1}{b - a}\int_a^b f(x)\, dx$$

Proof   Because $f(x)$ is continuous and $[a, b]$ is closed, there exist points $c_{\min}$ and $c_{\max}$ in $[a, b]$ such that $f(c_{\min})$ and $f(c_{\max})$ are the minimum and maximum values of $f(x)$ on $[a, b]$. Thus, $f(c_{\min}) \leq f(x) \leq f(c_{\max})$ for all $x \in [a, b]$ and therefore,

$$\int_a^b f(c_{\min})\, dx \leq \int_a^b f(x)\, dx \leq \int_a^b f(c_{\max})\, dx$$

$$f(c_{\min})(b - a) \leq \int_a^b f(x)\, dx \leq f(c_{\max})(b - a)$$

Notice how the proof of Theorem 1 uses important parts of the theory we have developed so far: the existence of extreme values on a closed interval, the IVT, and the basic property that if $f(x) \leq g(x)$, then

$$\int_a^b f(x)\,dx \leq \int_a^b g(x)\,dx$$

Now divide by $(b - a)$:

$$f(c_{\min}) \leq \underbrace{\frac{1}{b-a}\int_a^b f(x)\,dx}_{\text{Average value } M} \leq f(c_{\max}) \qquad \boxed{5}$$

The expression in the middle is the average value M of $f(x)$ on $[a, b]$. We see that M lies between the min and max of $f(x)$ on $[a, b]$. Because $f(x)$ is continuous, the Intermediate Value Theorem (IVT) guarantees that $f(c) = M$ for some $c \in [a, b]$ as claimed. ∎

6.2 SUMMARY

- The volume V of a solid body is equal to the integral of the area of the horizontal (or vertical) cross sections $A(y)$:

$$V = \int_a^b A(y)\,dy$$

- *Linear mass density* $\rho(x)$ is defined as mass per unit length. If a rod (or other object) with density ρ extends from $x = a$ to $x = b$, then its total mass is $M = \int_a^b \rho(x)\,dx$.
- If the density function $\rho(r)$ depends only on the distance r from the origin (*radial density function*), then the total amount (of population, mass, etc.) within a radius R of the center is equal to $2\pi \int_0^R r\rho(r)\,dr$.
- The volume of fluid passing through a tube of radius R per unit time is called the *flow rate* Q. The flow is *laminar* if the velocity $v(r)$ of a fluid particle depends only on its distance r from the center of the tube. For a laminar flow, $Q = 2\pi \int_0^R rv(r)\,dr$.
- The *average* (or *mean*) value on $[a, b]$: $M = \frac{1}{b-a}\int_a^b f(x)\,dx$.
- The MVT for Integrals: If $f(x)$ is continuous on $[a, b]$ with average value M, then $f(c) = M$ for some $c \in [a, b]$.

6.2 EXERCISES

Preliminary Questions

1. What is the average value of $f(x)$ on $[1, 4]$ if the area between the graph of $f(x)$ and the x-axis is equal to 9?

2. Find the volume of a solid extending from $y = 2$ to $y = 5$ if the cross section at y has area $A(y) = 5$ for all y.

3. Describe the horizontal cross sections of an ice cream cone and the vertical cross sections of a football (when it is held horizontally).

4. What is the formula for the total population within a circle of radius R around a city center if the population has a radial function?

5. What is the definition of flow rate?

6. Which assumption about fluid velocity did we use to compute the flow rate as an integral?

Exercises

1. Let V be the volume of a pyramid of height 20 whose base is a square of side 8.

(a) Use similar triangles as in Example 1 to find the area of the horizontal cross section at a height y.

(b) Calculate V by integrating the cross-sectional area.

2. Let V be the volume of a right circular cone of height 10 whose base is a circle of radius 4 (Figure 16).

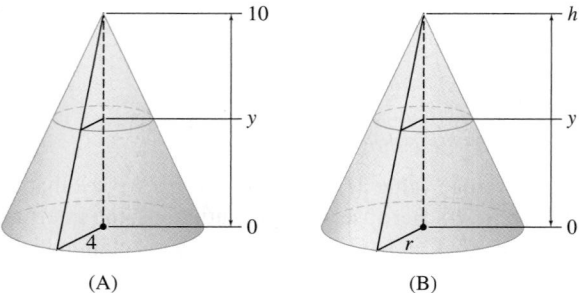

(A)   (B)

FIGURE 16 Right circular cones.

(a) Use similar triangles to find the area of a horizontal cross section at a height y.

(b) Calculate V by integrating the cross-sectional area.

3. Use the method of Exercise 2 to find the formula for the volume of a right circular cone of height h whose base is a circle of radius r (Figure 16).

4. Calculate the volume of the ramp in Figure 17 in three ways by integrating the area of the cross sections:

(a) Perpendicular to the x-axis (rectangles)

(b) Perpendicular to the y-axis (triangles)

(c) Perpendicular to the z-axis (rectangles)

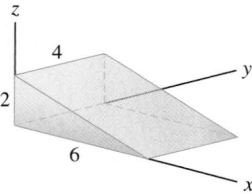

FIGURE 17 Ramp of length 6, width 4, and height 2.

5. Find the volume of liquid needed to fill a sphere of radius R to height h (Figure 18).

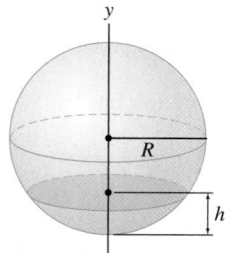

FIGURE 18 Sphere filled with liquid to height h.

6. Find the volume of the wedge in Figure 19(A) by integrating the area of vertical cross sections.

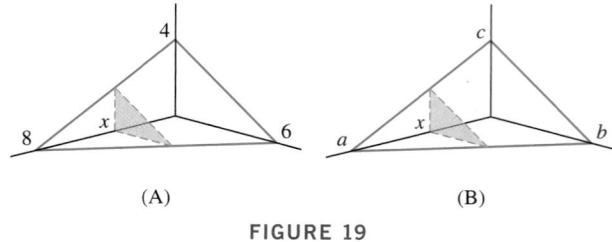

(A)   (B)

FIGURE 19

7. Derive a formula for the volume of the wedge in Figure 19(B) in terms of the constants a, b, and c.

8. Let B be the solid whose base is the unit circle $x^2 + y^2 = 1$ and whose vertical cross sections perpendicular to the x-axis are equilateral triangles. Show that the vertical cross sections have area $A(x) = \sqrt{3}(1 - x^2)$ and compute the volume of B.

In Exercises 9–14, find the volume of the solid with given base and cross sections.

9. The base is the unit circle $x^2 + y^2 = 1$ and the cross sections perpendicular to the x-axis are triangles whose height and base are equal.

10. The base is the triangle enclosed by $x + y = 1$, the x-axis, and the y-axis. The cross sections perpendicular to the y-axis are semicircles.

11. The base is the semicircle $y = \sqrt{9 - x^2}$, where $-3 \le x \le 3$. The cross sections perpendicular to the x-axis are squares.

12. The base is a square, one of whose sides is the interval $[0, \ell]$ along the x-axis. The cross sections perpendicular to the x-axis are rectangles of height $f(x) = x^2$.

13. The base is the region enclosed by $y = x^2$ and $y = 3$. The cross sections perpendicular to the y-axis are squares.

14. The base is the region enclosed by $y = x^2$ and $y = 3$. The cross sections perpendicular to the y-axis are rectangles of height y^3.

15. Find the volume of the solid whose base is the region $|x| + |y| \le 1$ and whose vertical cross sections perpendicular to the y-axis are semicircles (with diameter along the base).

16. Show that the volume of a pyramid of height h whose base is an equilateral triangle of side s is equal to $\dfrac{\sqrt{3}}{12} hs^2$.

17. Find the volume V of a *regular* tetrahedron whose face is an equilateral triangle of side s (Figure 20).

FIGURE 20 Regular tetrahedron.

18. The area of an ellipse is πab, where a and b are the lengths of the semimajor and semiminor axes (Figure 21). Compute the volume of a cone of height 12 whose base is an ellipse with semimajor axis $a = 6$ and semiminor axis $b = 4$.

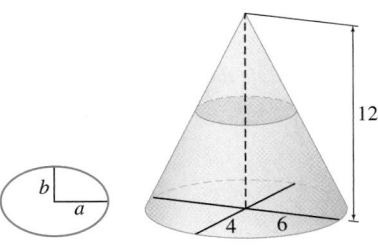

FIGURE 21

19. A frustum of a pyramid is a pyramid with its top cut off [Figure 22(A)]. Let V be the volume of a frustum of height h whose base is a square of side a and top is a square of side b with $a > b \geq 0$.

(a) Show that if the frustum were continued to a full pyramid, it would have height $\dfrac{ha}{a - b}$ [Figure 22(B)].

(b) Show that the cross section at height x is a square of side $(1/h)(a(h - x) + bx)$.

(c) Show that $V = \frac{1}{3}h(a^2 + ab + b^2)$. A papyrus dating to the year 1850 BCE indicates that Egyptian mathematicians had discovered this formula almost 4,000 years ago.

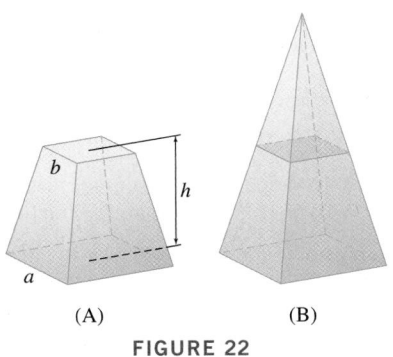

(A)   (B)

FIGURE 22

20. A plane inclined at an angle of $45°$ passes through a diameter of the base of a cylinder of radius r. Find the volume of the region within the cylinder and below the plane (Figure 23).

FIGURE 23

21. Figure 24 shows the solid S obtained by intersecting two cylinders of radius r whose axes are perpendicular.

(a) The horizontal cross section of each cylinder at distance y from the central axis is a rectangular strip. Find the strip's width.

(b) Find the area of the horizontal cross section of S at distance y.

(c) Find the volume of S as a function of r.

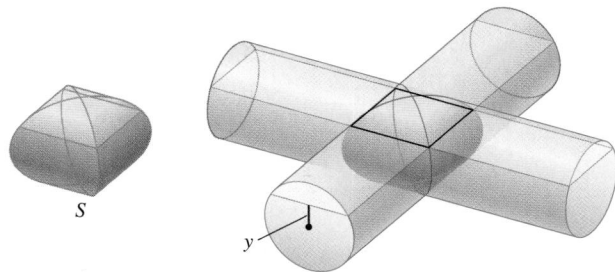

FIGURE 24 Intersection of two cylinders intersecting at right angles.

22. Let S be the solid obtained by intersecting two cylinders of radius r whose axes intersect at an angle θ. Find the volume of S as a function of r and θ.

23. Calculate the volume of a cylinder inclined at an angle $\theta = 30°$ whose height is 10 and whose base is a circle of radius 4 (Figure 25).

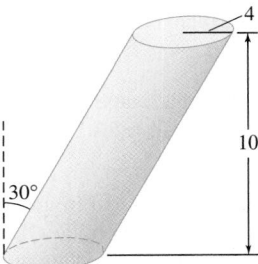

FIGURE 25 Cylinder inclined at an angle $\theta = 30°$.

24. Find the total mass of a 1-m rod whose linear density function is $\rho(x) = 10(x + 1)^{-2}$ kg/m for $0 \leq x \leq 1$.

25. Find the total mass of a 2-m rod whose linear density function is $\rho(x) = 1 + 0.5 \sin(\pi x)$ kg/m for $0 \leq x \leq 2$.

26. A mineral deposit along a strip of length 6 cm has density $s(x) = 0.01x(6 - x)$ g/cm for $0 \leq x \leq 6$. Calculate the total mass of the deposit.

27. Calculate the population within a 10-mile radius of the city center if the radial population density is $\rho(r) = 4(1 + r^2)^{1/3}$ (in thousands per square mile).

28. Odzala National Park in the Congo has a high density of gorillas. Suppose that the radial population density is $\rho(r) = 52(1 + r^2)^{-2}$ gorillas per square kilometer, where r is the distance from a large grassy clearing with a source of food and water. Calculate the number of gorillas within a 5-km radius of the clearing.

29. Table 1 lists the population density (in people per squared kilometer) as a function of distance r (in kilometers) from the center of a rural town. Estimate the total population within a 2-km radius of the center by taking the average of the left- and right-endpoint approximations.

TABLE 1   Population Density

r	$\rho(r)$	r	$\rho(r)$
0.0	125.0	1.2	37.6
0.2	102.3	1.4	30.8
0.4	83.8	1.6	25.2
0.6	68.6	1.8	20.7
0.8	56.2	2.0	16.9
1.0	46.0		

30. Find the total mass of a circular plate of radius 20 cm whose mass density is the radial function $\rho(r) = 0.03 + 0.01 \cos(\pi r^2)$ g/cm^2.

31. The density of deer in a forest is the radial function $\rho(r) = 150(r^2 + 2)^{-2}$ deer per km^2, where r is the distance (in kilometers) to a small meadow. Calculate the number of deer in the region $2 \leq r \leq 5$ km.

32. Show that a circular plate of radius 2 cm with radial mass density $\rho(r) = \dfrac{4}{r}$ g/cm has finite total mass, even though the density becomes infinite at the origin.

33. Find the flow rate through a tube of radius 4 cm, assuming that the velocity of fluid particles at a distance r cm from the center is $v(r) = 16 - r^2$ cm/s.

34. Let $v(r)$ be the velocity of blood in an arterial capillary of radius $R = 4 \times 10^{-5}$ m. Use Poiseuille's Law (Example 6) with $k = 10^6$ (m-s)$^{-1}$ to determine the velocity at the center of the capillary and the flow rate (use correct units).

35. A solid rod of radius 1 cm is placed in a pipe of radius 3 cm so that their axes are aligned. Water flows through the pipe and around the rod. Find the flow rate if the velocity of the water is given by the radial function $v(r) = 0.5(r - 1)(3 - r)$ cm/s.

36. To estimate the volume V of Lake Nogebow, the Minnesota Bureau of Fisheries created the depth contour map in Figure 26 and determined the area of the cross section of the lake at the depths recorded

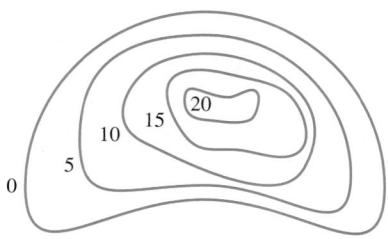

FIGURE 26 Depth contour map of Lake Nogebow.

in the table below. Estimate V by taking the average of the right- and left-endpoint approximations to the integral of cross-sectional area.

Depth (ft)	0	5	10	15	20
Area (million ft^2)	2.1	1.5	1.1	0.835	0.217

In Exercises 37–46, calculate the average over the given interval.

37. $f(x) = x^3$, $[0, 1]$

38. $f(x) = x^3$, $[-1, 1]$

39. $f(x) = \cos x$, $[0, \frac{\pi}{2}]$

40. $f(x) = \sec^2 x$, $[0, \frac{\pi}{4}]$

41. $f(s) = s^{-2}$, $[2, 5]$

42. $f(x) = \dfrac{\sin(\pi/x)}{x^2}$, $[1, 2]$

43. $f(x) = 2x^3 - 3x^2$, $[-1, 3]$ **44.** $f(x) = x^n$, $[0, 1]$

45. $f(x) = \dfrac{1}{x^2 + 1}$, $[-1, 1]$ **46.** $f(x) = e^{-nx}$, $[-1, 1]$

47. Let M be the average value of $f(x) = x^3$ on $[0, A]$, where $A > 0$. Which theorem guarantees that $f(c) = M$ has a solution c in $[0, A]$? Find c.

48. CAS Let $f(x) = 2 \sin x - x$. Use a computer algebra system to plot $f(x)$ and estimate:
(a) The positive root α of $f(x)$.
(b) The average value M of $f(x)$ on $[0, \alpha]$.
(c) A value $c \in [0, \alpha]$ such that $f(c) = M$.

49. Which of $f(x) = x \sin^2 x$ and $g(x) = x^2 \sin^2 x$ has a larger average value over $[0, 1]$? Over $[1, 2]$?

50. Show that the average value of $f(x) = \dfrac{\sin x}{x}$ over $[\frac{\pi}{2}, \pi]$ is less than 0.41. Sketch the graph if necessary.

51. 🖵   Sketch the graph of a function $f(x)$ such that $f(x) \geq 0$ on $[0, 1]$ and $f(x) \leq 0$ on $[1, 2]$, whose average on $[0, 2]$ is negative.

52. Find the average of $f(x) = ax + b$ over the interval $[-M, M]$, where a, b, and M are arbitrary constants.

53. The temperature $T(t)$ at time t (in hours) in an art museum varies according to $T(t) = 70 + 5 \cos\left(\dfrac{\pi}{12}t\right)$. Find the average over the time periods $[0, 24]$ and $[2, 6]$.

54. A ball is thrown in the air vertically from ground level with initial velocity 64 ft/s. Find the average height of the ball over the time interval extending from the time of the ball's release to its return to ground level. Recall that the height at time t is $h(t) = 64t - 16t^2$.

55. What is the average area of the circles whose radii vary from 0 to 1?

56. An object with zero initial velocity accelerates at a constant rate of 10 m/s^2. Find its average velocity during the first 15 s.

57. The acceleration of a particle is $a(t) = t - t^3$ m/s^2 for $0 \leq t \leq 1$. Compute the average acceleration and average velocity over the time interval $[0, 1]$, assuming that the particle's initial velocity is zero.

58. Let M be the average value of $f(x) = x^4$ on $[0, 3]$. Find a value of c in $[0, 3]$ such that $f(c) = M$.

59. Let $f(x) = \sqrt{x}$. Find a value of c in $[4, 9]$ such that $f(c)$ is equal to the average of f on $[4, 9]$.

60. Give an example of a function (necessarily discontinuous) that does not satisfy the conclusion of the MVT for Integrals.

Further Insights and Challenges

61. An object is tossed in the air vertically from ground level with initial velocity v_0 ft/s at time $t = 0$. Find the average speed of the object over the time interval $[0, T]$, where T is the time the object returns to earth.

62. ✎ Review the MVT stated in Section 4.3 (Theorem 1, p. 230) and show how it can be used, together with the Fundamental Theorem of Calculus, to prove the MVT for integrals.

6.3 Volumes of Revolution

We use the terms "revolve" and "rotate" interchangeably.

A **solid of revolution** is a solid obtained by rotating a region in the plane about an axis. The sphere and right circular cone are familiar examples of such solids. Each of these is "swept out" as a plane region revolves around an axis (Figure 1).

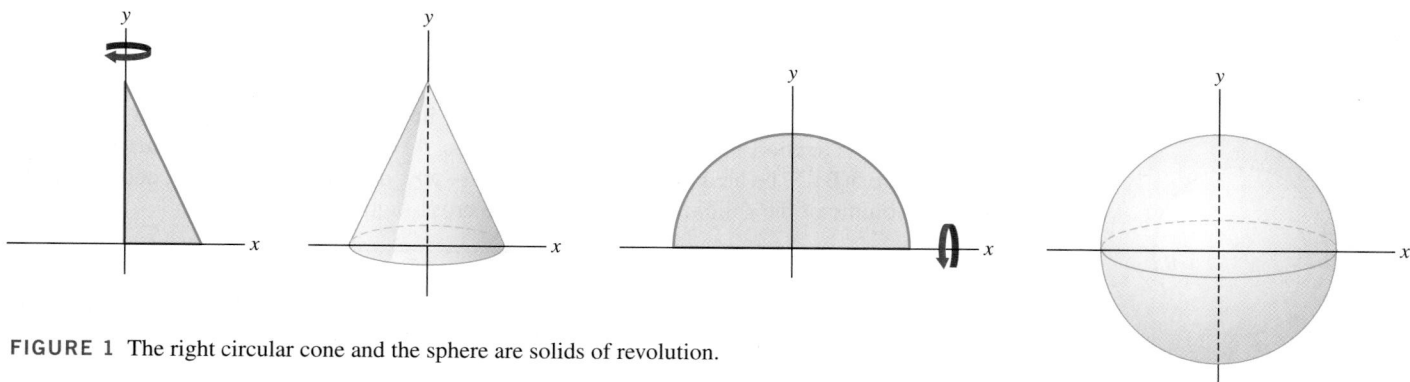

FIGURE 1 The right circular cone and the sphere are solids of revolution.

This method for computing the volume is often referred to as the "disk method" because the vertical slices of the solid are circular disks.

In general, if $f(x) \geq 0$ for $a \leq x \leq b$, then the region under the graph lies above the x-axis. Rotating this region around the x-axis produces a solid with a special feature: All vertical cross sections are circles (Figure 2). In fact, the vertical cross section at location x is a circle of radius $R = f(x)$ and has area

$$\text{Area of the vertical cross section} = \pi R^2 = \pi f(x)^2$$

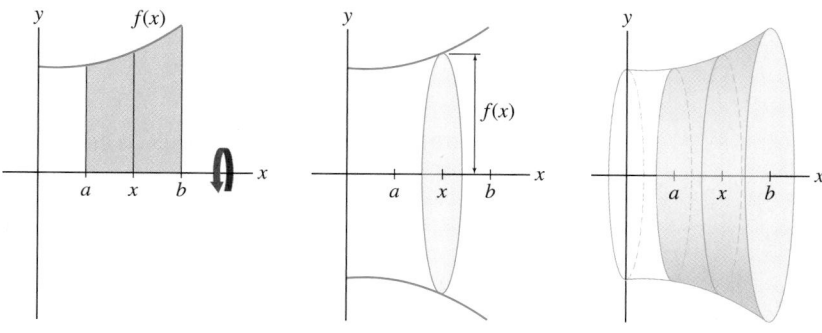

FIGURE 2 The cross section of a solid of revolution is a circle of radius $f(x)$ and area $\pi f(x)^2$.

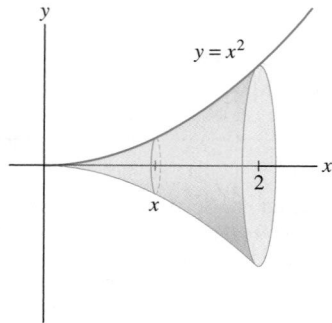

As we saw in Section 6.2, the total volume V is equal to the integral of cross-sectional area. Therefore, $V = \int_a^b \pi f(x)^2 \, dx$.

> **Volume of a Solid of Revolution: Disk Method** If $f(x)$ is continuous and $f(x) \geq 0$ on $[a, b]$, then the volume V obtained by rotating the region under the graph about the x-axis is [with $R = f(x)$]:
>
> $$V = \pi \int_a^b R^2 \, dx = \pi \int_a^b f(x)^2 \, dx \qquad \boxed{1}$$

▮ **EXAMPLE 1** Calculate the volume V of the solid obtained by rotating the region under $y = x^2$ about the x-axis for $0 \leq x \leq 2$.

Solution In this case, $f(x) = x^2$, and by Eq. (1), the volume is (Figure 3)

$$V = \pi \int_0^2 R^2 \, dx = \pi \int_0^2 (x^2)^2 \, dx = \pi \int_0^2 x^4 \, dx = \pi \left.\frac{x^5}{5}\right|_0^2 = \pi \frac{2^5}{5} = \frac{32}{5} \pi \qquad ▮$$

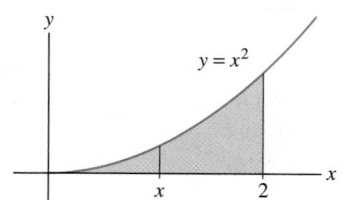

FIGURE 3 Region under $y = x^2$ rotated about the x-axis.

We now consider some variations on the formula for a volume of revolution. First, suppose that the region *between* two curves $y = f(x)$ and $y = g(x)$, where $f(x) \geq g(x) \geq 0$, is rotated about the x-axis [Figure 5(C)]. Then the vertical cross section of the solid at x is generated as the segment $\overline{AB}$ in Figure 5 revolves around the x-axis. This cross section is a **washer** of outer radius $R = f(x)$ and inner radius $r = g(x)$ [Figure 5(B)]. The area of this washer is $\pi R^2 - \pi r^2$ or $\pi(f(x)^2 - g(x)^2)$, and we obtain the volume of the solid as the integral of the cross-sectional area:

$$V = \pi \int_a^b \left(R^2 - r^2 \right) dx = \pi \int_a^b \left(f(x)^2 - g(x)^2 \right) dx \qquad \boxed{2}$$

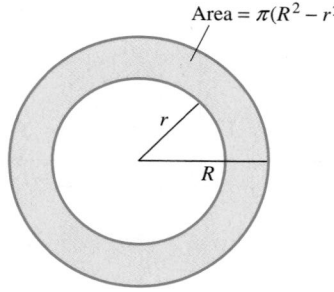

FIGURE 4 The region between two concentric circles is called an "annulus," or more informally, a "washer."

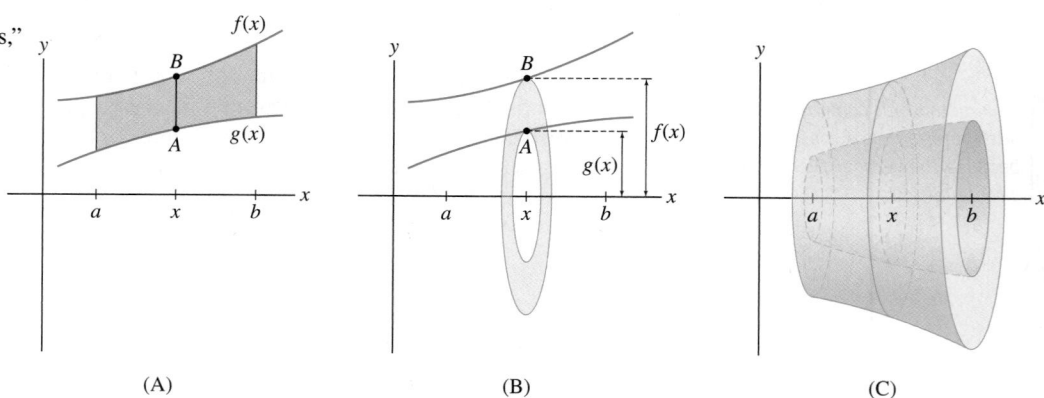

(A)   (B)   (C)

FIGURE 5 The vertical cross section is the washer generated when $\overline{AB}$ is rotated about the x-axis.

▮ **EXAMPLE 2** *Rotating the Area Between Two Curves* Find the volume V of the solid obtained by rotating the region between $y = x^2 + 4$ and $y = 2$ about the x-axis for $1 \leq x \leq 3$.

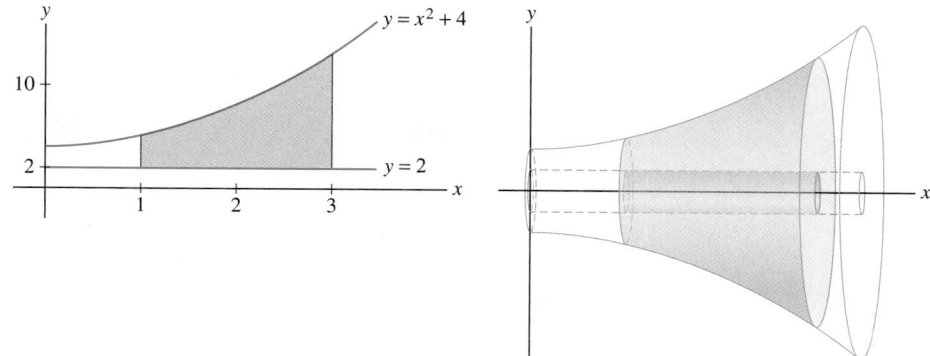

FIGURE 6 The area between $y = x^2 + 4$ and $y = 2$ over $[1, 3]$ rotated about the x-axis.

Solution The graph of $y = x^2 + 4$ lies above the graph of $y = 2$ (Figure 6). Therefore, the outer radius is $R = x^2 + 4$ and the inner radius is $r = 2$. By Eq. (2),

$$V = \pi \int_1^3 (R^2 - r^2)\, dx = \pi \int_1^3 \left((x^2 + 4)^2 - 2^2\right) dx$$

$$= \pi \int_1^3 \left(x^4 + 8x^2 + 12\right) dx = \pi \left(\frac{1}{5}x^5 + \frac{8}{3}x^3 + 12x\right)\Big|_1^3 = \frac{2{,}126}{15}\pi \qquad \blacksquare$$

Equation (2) can be modified to compute volumes of revolution about horizontal lines parallel to the x-axis.

■ **EXAMPLE 3** Revolving About a Horizontal Axis Find the volume V of the "wedding band" in Figure 7(C), obtained by rotating the region between the graphs of $f(x) = x^2 + 2$ and $g(x) = 4 - x^2$ about the horizontal line $y = -3$.

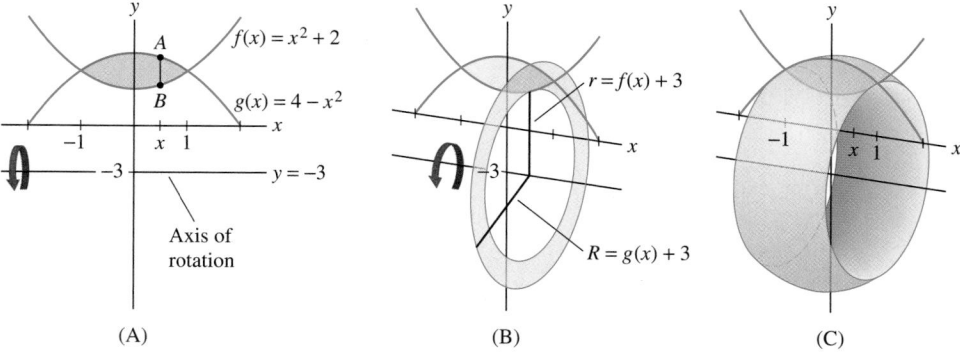

FIGURE 7 Segment $\overline{AB}$ generates a washer when it is rotated about the axis $y = -3$.

(A)   (B)   (C)

Solution We first determine the points of intersection of the two graphs by solving

$$x^2 + 2 = 4 - x^2 \qquad \text{or} \qquad x^2 = 1$$

The graphs intersect at $x = \pm 1$. Figure 7(A) shows that the graph of $g(x) = 4 - x^2$ lies above the graph of $f(x) = x^2 + 2$ on $[-1, 1]$. *If we revolved the region between the graphs about the x-axis*, the volume would be given by Eq. (2):

$$V \text{ (about x-axis)} = \pi \int_{-1}^1 \left(g(x)^2 - f(x)^2\right) dx = \pi \int_{-1}^1 \left((4 - x^2)^2 - (x^2 + 2)^2\right) dx$$

When you set up the integral for a volume of revolution, visualize the cross sections. These cross sections are washers (or disks) whose inner and outer radii depend on the axis of rotation.

The formula is similar when we rotate about a horizontal line other than the x-axis, but we must use the appropriate outer and inner radii. When we rotate about $y = -3$, $\overline{AB}$ generates a washer whose outer and inner radii are both three units longer [Figure 7(B)]:

- Outer radius extends from $y = -3$ to $y = g(x)$, so $R = g(x) + 3 = 7 - x^2$.
- Inner radius extends from $y = -3$ to $y = f(x)$, so $r = f(x) + 3 = x^2 + 5$.

The volume of revolution is obtained by integrating the area of this washer:

$$V \text{ (about } y = -3) = \pi \int_{-1}^{1} (R^2 - r^2)\, dx = \pi \int_{-1}^{1} \left((g(x) + 3)^2 - (f(x) + 3)^2 \right) dx$$

$$= \pi \int_{-1}^{1} \left((7 - x^2)^2 - (x^2 + 5)^2 \right) dx$$

$$= \pi \int_{-1}^{1} \left((49 - 14x^2 + x^4) - (25 + 10x^2 + x^4) \right) dx$$

$$= \pi \int_{-1}^{1} (24 - 24x^2)\, dx = \pi (24x - 8x^3) \Big|_{-1}^{1} = 32\pi \qquad \blacksquare$$

■ **EXAMPLE 4** Find the volume of the solid obtained by rotating the region between the graph of $f(x) = 9 - x^2$ and the line $y = 12$ for $0 \le x \le 3$ about **(a)** the line $y = 12$ and **(b)** the line $y = 15$.

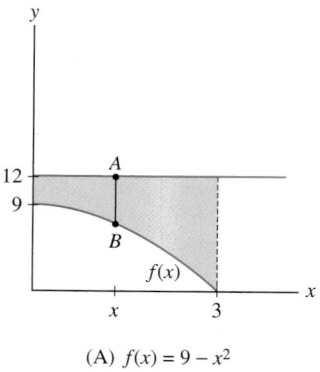

(A) $f(x) = 9 - x^2$

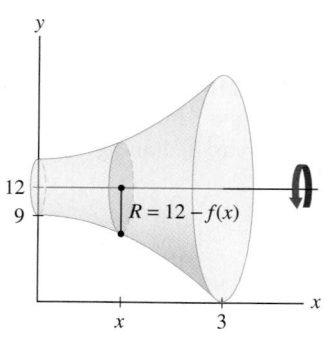

(B) Rotation about the line $y = 12$

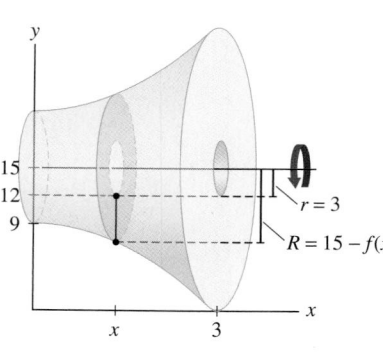

(C) Rotation about the line $y = 15$

FIGURE 8 Segment $\overline{AB}$ generates a disk when rotated about $y = 12$, but it generates a washer when rotated about $y = 15$.

Solution To set up the integrals, we must first visualize whether the cross section is a disk or washer.

In Figure 8, the length of $\overline{AB}$ is $12 - f(x)$ rather than $f(x) - 12$ because the line $y = 12$ lies above the graph of $f(x)$.

(a) Figure 8(B) shows that when $\overline{AB}$ is rotated about $y = 12$, it generates a disk of radius

$$R = \text{length of } \overline{AB} = 12 - f(x) = 12 - (9 - x^2) = 3 + x^2$$

The volume of the solid of revolution about $y = 12$ is

$$\pi \int_{0}^{3} R^2\, dx = \pi \int_{0}^{3} (3 + x^2)^2\, dx = \pi \int_{0}^{3} (9 + 6x^2 + x^4)\, dx$$

$$= \pi \left(9x + 2x^3 + \frac{1}{5}x^5 \right) \Big|_{0}^{3} = \frac{648}{5}\pi$$

(b) Figure 8(C) shows that $\overline{AB}$ generates a washer when rotated about $y = 15$. The outer radius R of this washer is the distance from B to the line $y = 15$:

$$R = 15 - f(x) = 15 - (9 - x^2) = 6 + x^2$$

The inner radius is $r = 3$, so the volume of revolution about $y = 15$ is

$$\pi \int_0^3 (R^2 - r^2)\, dx = \pi \int_0^3 \left((6 + x^2)^2 - 3^2 \right) dx = \pi \int_0^3 \left(36 + 12x^2 + x^4 - 9 \right) dx$$

$$= \pi \left(27x + 4x^3 + \frac{1}{5}x^5 \right) \Big|_0^3 = \frac{1{,}188}{5} \pi \qquad \blacksquare$$

To use the disk and washer methods for solids of revolution about vertical axes, the graph must be described as a function of y rather than x.

■ **EXAMPLE 5** Revolving About a Vertical Axis Find the volume of the solid obtained by rotating the region under the graph of $f(x) = 9 - x^2$ for $0 \le x \le 3$ about the vertical axis $x = -2$.

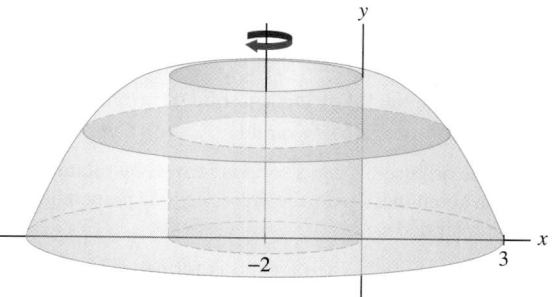

FIGURE 9

Solution In Figure 9, the horizontal segment $\overline{AB}$ at height y sweeps out a horizontal washer when it is rotated about the vertical line $x = -2$. To find the area of this washer as a function of y, we first solve for x in $y = 9 - x^2$ to obtain $x^2 = 9 - y$ or $x = \sqrt{9 - y}$. Thus

$$\text{Outer radius } R = \sqrt{9 - y} + 2, \qquad \text{inner radius } r = 2$$

Since the region extends from $y = 0$ to $y = 9$ along the y-axis, the volume is obtained by integrating from $y = 0$ to $y = 9$:

$$\pi \int_0^9 (R^2 - r^2)\, dy = \pi \int_0^9 \left((\sqrt{9 - y} + 2)^2 - 2^2 \right) dy$$

$$= \pi \int_0^9 \left(9 - y + 4\sqrt{9 - y} \right) dy$$

$$= \pi \left(9y - \frac{1}{2}y^2 - \frac{8}{3}(9 - y)^{3/2} \right) \Big|_0^9 = \frac{225}{2} \pi \qquad \blacksquare$$

6.3 SUMMARY

- *Disk method:* When we rotate the region under the graph of $f(x)$ about the x-axis for $a \leq x \leq b$, we obtain a solid whose vertical cross section is a circle of radius $R = f(x)$ and area $\pi R^2 = \pi f(x)^2$. The volume V of the solid is

$$V = \pi \int_a^b R^2 \, dx = \pi \int_a^b f(x)^2 \, dx$$

- *Washer method:* Assume that $f(x) \geq g(x) \geq 0$ for $a \leq x \leq b$. When we rotate the region between the graphs of $f(x)$ and $g(x)$ about the x-axis, we obtain a solid whose vertical cross section is a washer of outer radius $R = f(x)$ and inner radius $r = g(x)$. The volume V of the solid is

$$V = \pi \int_a^b (R^2 - r^2) \, dx = \pi \int_a^b \left(f(x)^2 - g(x)^2 \right) dx$$

- These formulas apply when we rotate about a horizontal line $y = c$, but the radii R and r must be modified appropriately.

6.3 EXERCISES

Preliminary Questions

1. Which of the following is a solid of revolution?
(a) Sphere   **(b)** Pyramid   **(c)** Cylinder   **(d)** Cube

2. True or false? When a solid is formed by rotating the region under a graph about the x-axis, the cross sections perpendicular to the x-axis are circular disks.

3. True or false? When a solid is formed by rotating the region between two graphs about the x-axis, the cross sections perpendicular to the x-axis are circular disks.

4. Which of the following integrals expresses the volume of the solid obtained by rotating the area between $y = f(x)$ and $y = g(x)$ over $[a, b]$ around the x-axis [assume $f(x) \geq g(x) \geq 0$]?

(a) $\pi \int_a^b \left(f(x) - g(x) \right)^2 dx$

(b) $\pi \int_a^b \left(f(x)^2 - g(x)^2 \right) dx$

Exercises

In Exercises 1–4, (a) sketch the solid obtained by revolving the region under the graph of $f(x)$ about the x-axis over the given interval, (b) describe the cross section perpendicular to the x-axis located at x, and (c) calculate the volume of the solid.

1. $f(x) = x + 1$, $[0, 3]$ **2.** $f(x) = x^2$, $[1, 3]$

3. $f(x) = \sqrt{x+1}$, $[1, 4]$ **4.** $f(x) = x^{-1}$, $[1, 2]$

In Exercises 5–12, find the volume of the solid obtained by rotating the region under the graph of the function about the x-axis over the given interval.

5. $f(x) = x^2 - 3x$, $[0, 3]$ **6.** $f(x) = \dfrac{1}{x^2}$, $[1, 4]$

7. $f(x) = x^{5/3}$, $[1, 8]$ **8.** $f(x) = 4 - x^2$, $[0, 2]$

9. $f(x) = \dfrac{2}{x+1}$, $[1, 3]$

10. $f(x) = \sqrt{x^4 + 1}$, $[1, 3]$

11. $f(x) = e^x$, $[0, 1]$

12. $f(x) = \sqrt{\cos x \sin x}$, $[0, \pi/2]$

In Exercises 13–18, (a) sketch the region enclosed by the curves, (b) describe the cross section perpendicular to the x-axis located at x, and (c) find the volume of the solid obtained by rotating the region about the x-axis.

13. $y = x^2 + 2$, $y = 10 - x^2$

14. $y = x^2$, $y = 2x + 3$

15. $y = 16 - x$, $y = 3x + 12$, $x = -1$

16. $y = \dfrac{1}{x}$, $y = \dfrac{5}{2} - x$

17. $y = \sec x$, $y = 0$, $x = -\dfrac{\pi}{4}$, $x = \dfrac{\pi}{4}$

18. $y = \sec x$, $y = \csc x$, $y = 0$, $x = 0$,   and $x = \dfrac{\pi}{2}$.

In Exercises 19–22, find the volume of the solid obtained by rotating the region enclosed by the graphs about the y-axis over the given interval.

19. $x = \sqrt{y}$, $x = 0$; $1 \le y \le 4$

20. $x = \sqrt{\sin y}$, $x = 0$; $0 \le y \le \pi$

21. $x = y^2$, $x = \sqrt{y}$; $0 \le y \le 1$

22. $x = 4 - y$, $x = 16 - y^2$; $-3 \le y \le 4$

In Exercises 23–28, find the volume of the solid obtained by rotating region A in Figure 10 about the given axis.

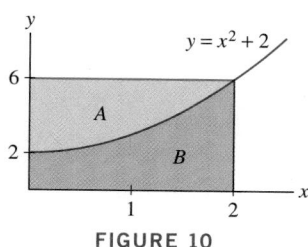

FIGURE 10

23. x-axis

24. $y = -2$

25. $y = 2$

26. y-axis

27. $x = -3$

28. $x = 2$

In Exercises 29–34, find the volume of the solid obtained by rotating region B in Figure 10 about the given axis.

29. x-axis

30. $y = -2$

31. $y = 6$

32. y-axis

Hint for Exercise 32: Express the volume as a sum of two integrals along the y-axis, or use Exercise 26.

33. $x = 2$

34. $x = -3$

In Exercises 35–48, find the volume of the solid obtained by rotating the region enclosed by the graphs about the given axis.

35. $y = x^2$, $y = 12 - x$, $x = 0$,   about $y = -2$

36. $y = x^2$, $y = 12 - x$, $x = 0$,   about $y = 15$

37. $y = 16 - x$, $y = 3x + 12$, $x = 0$,   about y-axis

38. $y = 16 - x$, $y = 3x + 12$, $x = 0$,   about $x = 2$

39. $y = \dfrac{9}{x^2}$, $y = 10 - x^2$,   about x-axis

40. $y = \dfrac{9}{x^2}$, $y = 10 - x^2$,   about $y = 12$

41. $y = \dfrac{1}{x}$, $y = \dfrac{5}{2} - x$,   about y-axis

42. $x = 2$, $x = 3$, $y = 16 - x^4$, $y = 0$,   about y-axis

43. $y = x^3$, $y = x^{1/3}$,   about y-axis

44. $y = x^3$, $y = x^{1/3}$,   about $x = -2$

45. $y = e^{-x}$, $y = 1 - e^{-x}$, $x = 0$,   about $y = 4$

46. $y = \cosh x$, $x = \pm 2$,   about x-axis

47. $y^2 = 4x$, $y = x$, $y = 0$,   about x-axis

48. $y^2 = 4x$, $y = x$,   about $y = 8$

49. $\boxed{\text{GU}}$ Sketch the hypocycloid $x^{2/3} + y^{2/3} = 1$ and find the volume of the solid obtained by revolving it about the x-axis.

50. The solid generated by rotating the region between the branches of the hyperbola $y^2 - x^2 = 1$ about the x-axis is called a **hyperboloid** (Figure 11). Find the volume of the hyperboloid for $-a \le x \le a$.

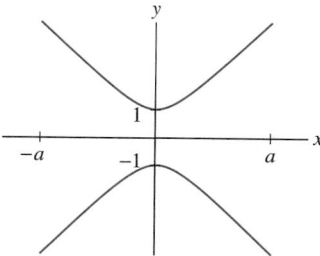

FIGURE 11 The hyperbola with equation $y^2 - x^2 = 1$.

51. A "bead" is formed by removing a cylinder of radius r from the center of a sphere of radius R (Figure 12). Find the volume of the bead with $r = 1$ and $R = 2$.

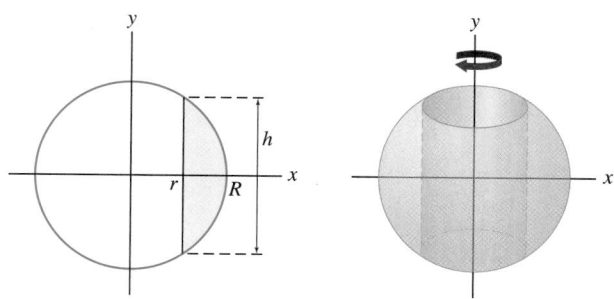

FIGURE 12 A bead is a sphere with a cylinder removed.

Further Insights and Challenges

52. Find the volume V of the bead (Figure 12) in terms of r and R. Then show that $V = \frac{\pi}{6}h^3$, where h is the height of the bead. This formula has a surprising consequence: Since V can be expressed in terms of h alone, it follows that two beads of height 2 in., one formed from a sphere the size of an orange and the other the size of the earth would have the same volume! Can you explain intuitively how this is possible?

53. The solid generated by rotating the region inside the ellipse with equation $\left(\frac{x}{a}\right)^2 + \left(\frac{y}{b}\right)^2 = 1$ around the x-axis is called an **ellipsoid**. Show that the ellipsoid has volume $\frac{4}{3}\pi ab^2$. What is the volume if the ellipse is rotated around the y-axis?

54. A doughnut-shaped solid is called a **torus** (Figure 13). Use the washer method to calculate the volume of the torus obtained by rotating the region inside the circle with equation $(x - a)^2 + y^2 = b^2$ around the y-axis (assume that $a > b$). *Hint:* Evaluate the integral by interpreting it as the area of a circle.

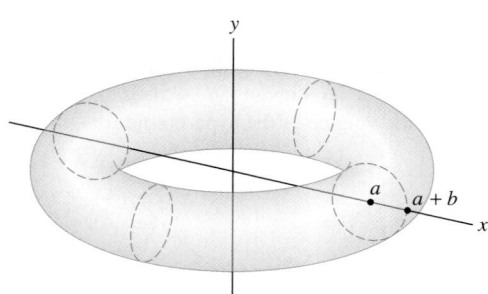

FIGURE 13 Torus obtained by rotating a circle about the y-axis.

55. The curve $y = f(x)$ in Figure 14, called a **tractrix**, has the following property: the tangent line at each point (x, y) on the curve has slope

$$\frac{dy}{dx} = \frac{-y}{\sqrt{1 - y^2}}.$$

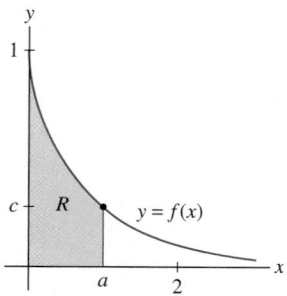

FIGURE 14 The tractrix.

Let R be the shaded region under the graph of $0 \le x \le a$ in Figure 14. Compute the volume V of the solid obtained by revolving R around the x-axis in terms of the constant $c = f(a)$. *Hint:* Use the disk method and the substitution $u = f(x)$ to show that

$$V = \pi \int_c^1 u\sqrt{1 - u^2}\, du$$

56. Verify the formula

$$\int_{x_1}^{x_2} (x - x_1)(x - x_2)\, dx = \frac{1}{6}(x_1 - x_2)^3 \qquad \boxed{3}$$

Then prove that the solid obtained by rotating the shaded region in Figure 15 about the x-axis has volume $V = \frac{\pi}{6}BH^2$, with B and H as in the figure. *Hint:* Let x_1 and x_2 be the roots of $f(x) = ax + b - (mx + c)^2$, where $x_1 < x_2$. Show that

$$V = \pi \int_{x_1}^{x_2} f(x)\, dx$$

and use Eq. (3).

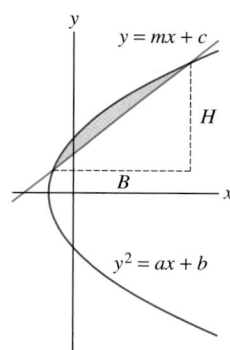

FIGURE 15 The line $y = mx + c$ intersects the parabola $y^2 = ax + b$ at two points above the x-axis.

57. Let R be the region in the unit circle lying above the cut with the line $y = mx + b$ (Figure 16). Assume the points where the line intersects the circle lie above the x-axis. Use the method of Exercise 56 to show that the solid obtained by rotating R about the x-axis has volume $V = \frac{\pi}{6}hd^2$, with h and d as in the figure.

FIGURE 16

FIGURE 1 The volume of the cylindrical shell is *approximately* $2\pi R h \,\Delta r$, where $\Delta r = R - r$.

Observe that the outer circumference of the cylindrical shell is $2\pi R$. So Eq. (2) says that the volume is approximately equal to

$(2\pi R)(h)(\Delta r)$

$= (circumference) \times (height) \times (thickness)$

6.4 The Method of Cylindrical Shells

In the previous two sections, we computed volumes by integrating cross-sectional area. The Shell Method is based on a different idea and is more convenient in some cases.

In the **Shell Method**, we use cylindrical shells like the one in Figure 1 to approximate a volume of revolution. To this end, let us first derive an approximation to the volume of a cylindrical shell of height h, outer radius R, and inner radius r. Since the shell is obtained by removing a cylinder of radius r from the wider cylinder of radius R, its volume is equal to

$$\pi R^2 h - \pi r^2 h = \pi h (R^2 - r^2) = \pi h (R + r)(R - r) = \pi h (R + r)\Delta r \qquad \boxed{1}$$

where $\Delta r = R - r$ is the shell's thickness. If the shell is very thin, then R and r are nearly equal and we may replace $(R + r)$ by $2R$ in Eq. (1) to obtain the approximation

$$\text{Volume of shell} \approx 2\pi R h \,\Delta r \qquad \boxed{2}$$

Now consider a solid obtained by rotating the region under $y = f(x)$ from $x = a$ to $x = b$ around the y-axis as in Figure 2. The idea is to divide the solid into thin concentric shells. More precisely, we divide $[a, b]$ into N subintervals of length $\Delta x = \dfrac{b - a}{N}$ with endpoints $x_0, x_1, \ldots, x_N$. When we rotate the thin strip of area above $[x_{i-1}, x_i]$ about the y-axis, we obtain a thin shell whose volume we denote by V_i. The total volume V of the solid is equal to $V = \displaystyle\sum_{i=1}^{N} V_i$.

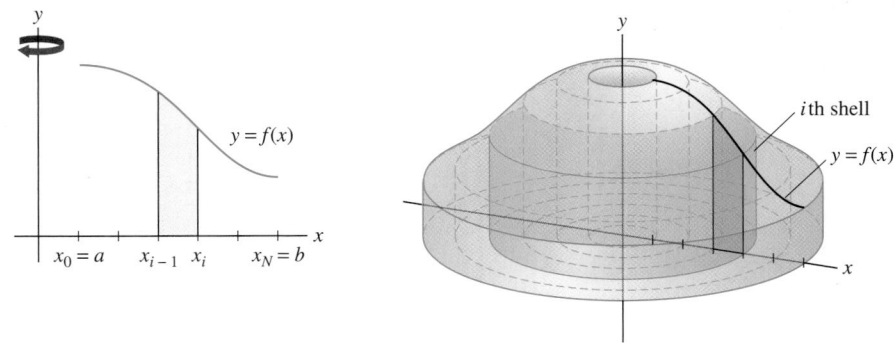

FIGURE 2 The shaded strip, when rotated about the y-axis, generates a "thin shell."

The top rim of the ith thin shell in Figure 2 is curved. However, when Δx is small, we may approximate this thin shell by the cylindrical shell (with a flat rim) of height $f(x_i)$ and use (2) to obtain

$$V_i \approx (\text{circumference})(\text{height})(\text{thickness}) = 2\pi x_i \, f(x_i) \,\Delta x$$

Therefore,

$$V = \sum_{i=1}^{N} V_i \approx \sum_{i=1}^{N} 2\pi x_i \, f(x_i) \,\Delta x$$

The sum on the right is the volume of the cylindrical approximation to V illustrated in Figure 3. We complete the argument in the usual way. As $N \to \infty$, the accuracy of the approximation improves, and the sum on the right is a right-endpoint approximation that converges to $V = \displaystyle\int_a^b 2\pi x f(x)\, dx$.

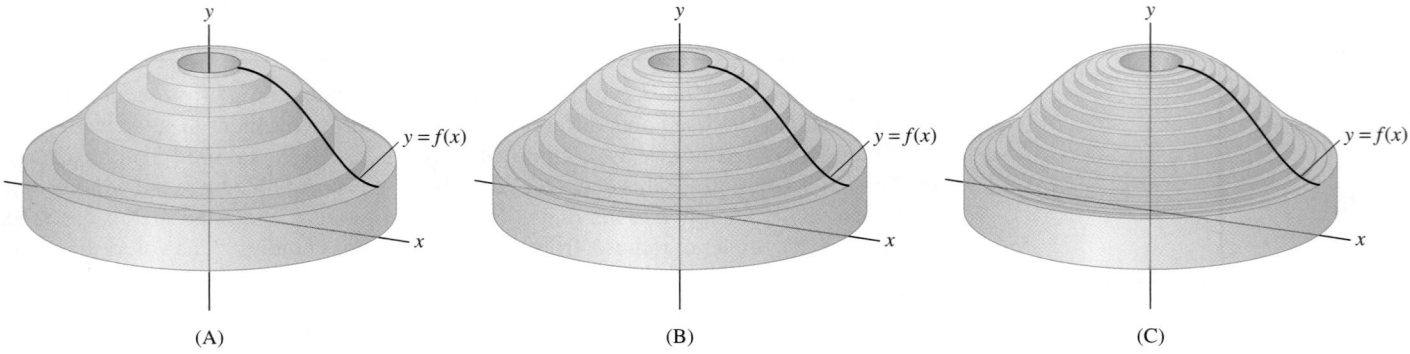

(A)   (B)   (C)

FIGURE 3 The volume is approximated by the sum of volumes of thin cylinders.

In the Shell Method, we integrate with respect to x even though we are rotating about the y-axis.

Volume of a Solid of Revolution: The Shell Method   The volume V of the solid obtained by rotating the region under the graph of $y = f(x)$ over the interval $[a, b]$ about the y-axis is equal to

$$V = 2\pi \int_a^b x f(x)\, dx \qquad \boxed{3}$$

It would be hard to use the disk method in Example 1. Since the axis of revolution is the y-axis, we would have to integrate with respect to y. This would require finding the inverse function $g(y) = f^{-1}(y)$.

■ **EXAMPLE 1** Find the volume V of the solid obtained by rotating the area under the graph of $f(x) = 1 - 2x + 3x^2 - 2x^3$ over $[0, 1]$ about the y-axis.

Solution The solid is shown in Figure 4. By Eq. (3),

$$V = 2\pi \int_0^1 x f(x)\, dx = 2\pi \int_0^1 x(1 - 2x + 3x^2 - 2x^3)\, dx$$

$$= 2\pi \left(\frac{1}{2}x^2 - \frac{2}{3}x^3 + \frac{3}{4}x^4 - \frac{2}{5}x^5 \right) \Big|_0^1 = \frac{11}{30}\pi \qquad ■$$

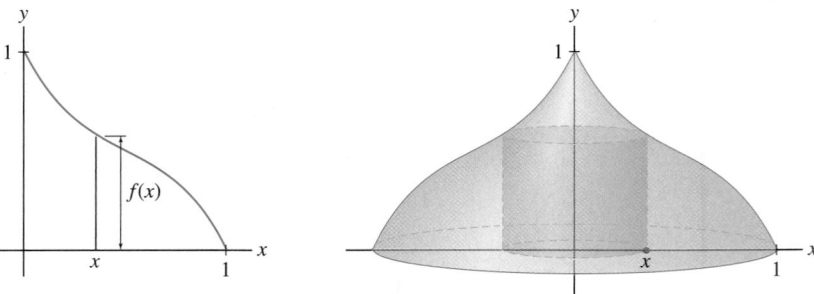

FIGURE 4 The graph of $f(x) = 1 - 2x + 3x^2 - 2x^3$ rotated about the y-axis.

GRAPHICAL INSIGHT   When we rotate the vertical segment at location x about the y-axis, we obtain a cylinder of radius x and height $f(x)$. Thus, we have the interpretation

$$V = 2\pi \int_a^b x f(x)\, dx = 2\pi \int_a^b (\text{radius})(\text{height of shell})\, dx \qquad \boxed{4}$$

For some solids, it is necessary to modify Eq. (4) in various ways. As a first example, let us rotate the region between the graphs of functions $f(x)$ and $g(x)$ over $[a, b]$ around the y-axis, where $f(x) \geq g(x)$. The vertical segment at location x generates a cylinder of radius x and height $f(x) - g(x)$ (Figure 5), so the volume V is

$$V = 2\pi \int_a^b (\text{radius})(\text{height of shell}) \, dx = 2\pi \int_a^b x\big(f(x) - g(x)\big) \, dx \qquad \boxed{5}$$

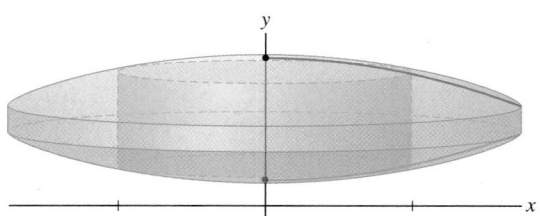

FIGURE 5 The vertical segment at location x generates a cylinder of radius x and height $f(x) - g(x)$.

■ **EXAMPLE 2** Rotating the Area Between Two Curves Use the Shell Method to calculate the volume V of the solid obtained by rotating the area enclosed by the graphs of $f(x) = 9 - x^2$ and $g(x) = 9 - 3x$ about the y-axis.

Solution We solve $9 - x^2 = 9 - 3x$ to find the points of intersection $x = 0, 3$. Figure 6 shows that $9 - x^2 \geq 9 - 3x$ for x in $[0, 3]$, so $f(x) = 9 - x^2$ is the upper curve and $g(x) = 9 - 3x$ is the lower curve. We have

$$\text{Height} = f(x) - g(x) = (9 - x^2) - (9 - 3x) = 3x - x^2$$

$$V = 2\pi \int_0^3 (\text{radius})(\text{height}) \, dx = 2\pi \int_0^3 x\big(f(x) - g(x)\big) \, dx = 2\pi \int_0^3 (3x^2 - x^3) \, dx$$

$$= 2\pi \left(x^3 - \frac{1}{4}x^4 \right) \Big|_0^3 = \frac{27}{2}\pi \qquad\blacksquare$$

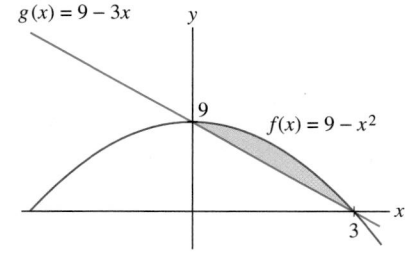

FIGURE 6

■ **EXAMPLE 3** Rotating Around a Vertical Axis Use the Shell Method to calculate the volume V of the solid obtained by rotating the region under the graph of $f(x) = x^{-1/2}$ over $[1, 4]$ about $x = -3$.

The reasoning in Example 3 shows that if we rotate the region under $y = f(x)$ over $[a, b]$ about the vertical line $x = c$ with $c \leq a$, then the volume is

$$V = 2\pi \int_a^b (x - c) f(x) \, dx$$

Solution If we revolved around the y-axis, the volume would be

$$V \text{ (around } y\text{-axis)} = 2\pi \int_1^4 (\text{radius})(\text{height}) \, dx = 2\pi \int_1^4 x \cdot x^{-1/2} \, dx$$

The formula is similar when we rotate around $x = -3$, but we must use the correct radius (distance to the axis of rotation). The height is still $f(x) = x^{-1/2}$, but the radius is now three units longer, that is, $x + 3$ (Figure 7), so

$$V \text{ (around } x = -3) = 2\pi \int_1^4 (\text{radius})(\text{height}) \, dx$$

$$= 2\pi \int_1^4 (x + 3)x^{-1/2} \, dx = 2\pi \left(\frac{2}{3}x^{3/2} + 6x^{1/2} \right) \Big|_1^4 = \frac{64\pi}{3} \qquad\blacksquare$$

FIGURE 7 Region under the graph of $y = x^{-1/2}$ over $[1, 4]$ rotated about the axis $x = -3$.

The method of cylindrical shells can be applied to rotations about horizontal axes, but in this case, the graph must be described in the form $x = g(y)$.

■ **EXAMPLE 4** *Rotating About the x-Axis* Use the Shell Method to compute the volume V of the solid obtained by rotating the area under $y = 9 - x^2$ over $[0, 3]$ about the x-axis.

Solution Since we are rotating about the x-axis rather than the y-axis, the Shell Method gives us an integral with respect to y. Therefore, we solve $y = 9 - x^2$ to obtain $x^2 = 9 - y$ or $x = \sqrt{9 - y}$.

The cylindrical shells are generated by *horizontal* segments. The segment $\overline{AB}$ in Figure 8 generates a cylindrical shell of radius y and height $\sqrt{9 - y}$ (we still use the term "height" although the cylinder is horizontal). Using the substitution $u = 9 - y$, $du = -dy$ in the resulting integral, we obtain

$$V = 2\pi \int_0^9 (\text{radius})(\text{height})\, dy = 2\pi \int_0^9 y \sqrt{9 - y}\, dy = -2\pi \int_9^0 (9 - u) \sqrt{u}\, du$$

$$= 2\pi \int_0^9 (9u^{1/2} - u^{3/2})\, du = 2\pi \left(6u^{3/2} - \frac{2}{5} u^{5/2} \right) \Big|_0^9 = \frac{648}{5} \pi \qquad ■$$

◄·· REMINDER After making the substitution $u = 9 - y$, the limits of integration must be changed. Since $u(0) = 9$ and $u(9) = 0$, we change $\int_0^9$ to $\int_9^0$.

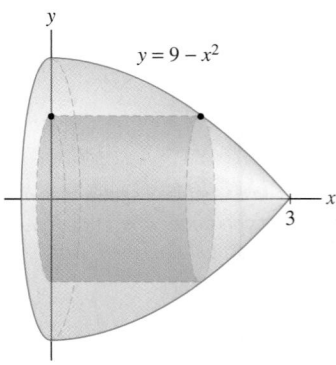

FIGURE 8 Shell generated by a horizontal segment in the region under the graph of $y = 9 - x^2$.

6.4 SUMMARY

• If $f(x) \geq 0$, then the volume V of the solid obtained by rotating the region underneath the graph of $y = f(x)$ over $[a, b]$ about the y-axis is

$$V = 2\pi \int_a^b (\text{radius})(\text{height}) \, dx = \boxed{2\pi \int_a^b x f(x) \, dx}$$

- If we revolve the region about the vertical axis $x = c$ rather than the y-axis, then the radius of the shell (distance to the axis of rotation) is no longer x. For example, if $c < a$, the radius is $(x - c)$ and

$$V = 2\pi \int_a^b (\text{radius})(\text{height}) \, dx = \boxed{2\pi \int_a^b (x - c) f(x) \, dx}$$

- We can use the Shell Method to compute volumes of revolution about the x-axis. It is necessary to express the curve in the form $x = g(y)$:

$$V = 2\pi \int_c^d (\text{radius})(\text{height}) \, dy = \boxed{2\pi \int_c^d y \, g(y) \, dy}$$

6.4 EXERCISES

Preliminary Questions

1. Consider the region $\mathcal{R}$ under the graph of the constant function $f(x) = h$ over the interval $[0, r]$. What are the height and radius of the cylinder generated when $\mathcal{R}$ is rotated about:

(a) the x-axis

(b) the y-axis

2. Let V be the volume of a solid of revolution about the y-axis.

(a) Does the Shell Method for computing V lead to an integral with respect to x or y?

(b) Does the Disk or Washer Method for computing V lead to an integral with respect to x or y?

Exercises

In Exercises 1–10, sketch the solid obtained by rotating the region underneath the graph of the function over the given interval about the y-axis and find its volume.

1. $f(x) = x^3$, $[0, 1]$

2. $f(x) = \sqrt{x}$, $[0, 4]$

3. $f(x) = 3x + 2$, $[2, 4]$

4. $f(x) = 1 + x^2$, $[1, 3]$

5. $f(x) = 4 - x^2$, $[0, 2]$

6. $f(x) = \sqrt{x^2 + 9}$, $[0, 3]$

7. $f(x) = \sin(x^2)$, $[0, \sqrt{\pi}]$

8. $f(x) = x^{-1}$, $[1, 3]$

9. $f(x) = x + 1 - 2x^2$, $[0, 1]$

10. $f(x) = \dfrac{x}{\sqrt{1 + x^3}}$, $[1, 4]$

In Exercises 11–14, use the Shell Method to compute the volume of the solids obtained by rotating the region enclosed by the graphs of the functions about the y-axis.

11. $y = x^2$, $y = 8 - x^2$, $x = 0$

12. $y = 8 - x^3$, $y = 8 - 4x$

13. $y = \sqrt{x}$, $y = x^2$

14. $y = 1 - |x - 1|$, $y = 0$

GU *In Exercises 15–16, use the Shell Method to compute the volume of rotation of the region enclosed by the curves about the y-axis. Use a computer algebra system or graphing utility to find the points of intersection numerically.*

15. $y = \frac{1}{2}x^2$, $y = \sin(x^2)$

16. $y = e^{-x^2/2}$, $y = x$, $x = 0$

In Exercises 17–22, sketch the solid obtained by rotating the region underneath the graph of the function over the interval about the given axis and calculate its volume using the Shell Method.

17. $f(x) = x^3$, $[0, 1]$, $x = 2$

18. $f(x) = x^3$, $[0, 1]$, $x = -2$

19. $f(x) = x^{-4}$, $[-3, -1]$, $x = 4$

20. $f(x) = \dfrac{1}{\sqrt{x^2 + 1}}$, $[0, 2]$, $x = 0$

21. $f(x) = a - bx$, $[0, a/b]$, $x = -1$, $a, b > 0$

22. $f(x) = 1 - x^2$, $[-1, 1]$, $x = c$ (with $c > 1$)

In Exercises 23–28, use the Shell Method to calculate the volume of rotation about the x-axis for the region underneath the graph.

23. $y = x$, $0 \le x \le 1$

24. $y = 4 - x^2$, $0 \le x \le 2$

25. $y = x^{1/3} - 2$, $8 \le x \le 27$

26. $y = x^{-1}$, $1 \le x \le 4$. Sketch the region and express the volume as a sum of two integrals.

27. $y = x^{-2}$, $2 \le x \le 4$

28. $y = \sqrt{x}$, $1 \le x \le 4$

29. Use both the Shell and Disk Methods to calculate the volume of the solid obtained by rotating the region under the graph of $f(x) = 8 - x^3$ for $0 \le x \le 2$ about:

(a) the x-axis

(b) the y-axis

30. Sketch the solid of rotation about the y-axis for the region under the graph of the constant function $f(x) = c$ (where $c > 0$) for $0 \le x \le r$.

(a) Find the volume without using integration.

(b) Use the Shell Method to compute the volume.

31. Assume that the graph in Figure 9(A) can be described by both $y = f(x)$ and $x = h(y)$. Let V be the volume of the solid obtained by rotating the region under the curve about the y-axis.

(a) Describe the figures generated by rotating segments $\overline{AB}$ and $\overline{CB}$ about the y-axis.

(b) Set up integrals that compute V by the Shell and Disk Methods.

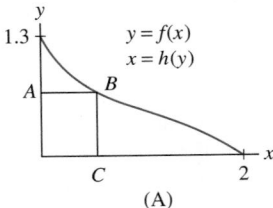

FIGURE 9

32. 📖 Let W be the volume of the solid obtained by rotating the region under the curve in Figure 9(B) about the y-axis.

(a) Describe the figures generated by rotating segments $\overline{A'B'}$ and $\overline{A'C'}$ about the y-axis.

(b) Set up an integral that computes W by the Shell Method.

(c) Explain the difficulty in computing W by the Washer Method.

In Exercises 33–38, use the Shell Method to find the volume of the solid obtained by rotating region A in Figure 10 about the given axis.

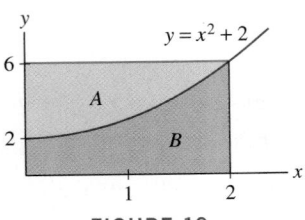

FIGURE 10

33. y-axis

34. $x = -3$

35. $x = 2$

36. x-axis

37. $y = -2$

38. $y = 6$

In Exercises 39–44, use the Shell Method to find the volumes of the solids obtained by rotating region B in Figure 10 about the given axis.

39. y-axis

40. $x = -3$

41. $x = 2$

42. x-axis

43. $y = -2$

44. $y = 8$

45. Use the Shell Method to compute the volume of a sphere of radius r.

46. Use the Shell Method to calculate the volume V of the "bead" formed by removing a cylinder of radius r from the center of a sphere of radius R (compare with Exercise 51 in Section 6.3).

47. Use the Shell Method to compute the volume of the torus obtained by rotating the interior of the circle $(x - a)^2 + y^2 = r^2$ about the y-axis, where $a > r$. *Hint:* Evaluate the integral by interpreting part of it as the area of a circle.

48. Use the Shell or Disk Method (whichever is easier) to compute the volume of the solid obtained by rotating the region in Figure 11 about:

(a) the x-axis

(b) the y-axis

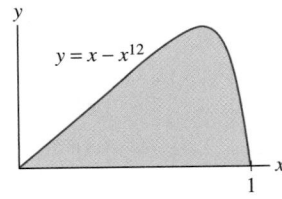

FIGURE 11

49. Use the most convenient method to compute the volume of the solid obtained by rotating the region in Figure 12 about the axis:

(a) $x = 4$

(b) $y = -2$

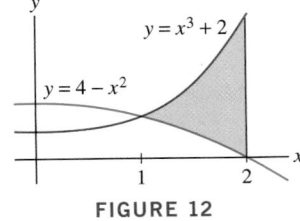

FIGURE 12

Further Insights and Challenges

50. 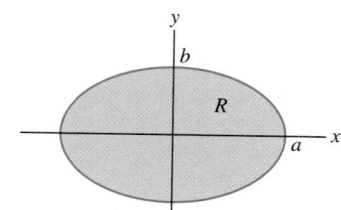 The surface area of a sphere of radius r is $4\pi r^2$. Use this to derive the formula for the volume V of a sphere of radius R in a new way.

(a) Show that the volume of a thin spherical shell of inner radius r and thickness Δx is approximately $4\pi r^2 \Delta x$.

(b) Approximate V by decomposing the sphere of radius R into N thin spherical shells of thickness $\Delta x = R/N$.

(c) Show that the approximation is a Riemann sum which converges to an integral. Evaluate the integral.

51. Let R be the region bounded by the ellipse $\left(\dfrac{x}{a}\right)^2 + \left(\dfrac{y}{b}\right)^2 = 1$ (Figure 13). Show that the solid obtained by rotating R about the y-axis (called an **ellipsoid**) has volume $\frac{4}{3}\pi a^2 b$.

52. The bell-shaped curve in Figure 14 is the graph of a certain function $y = f(x)$ with the following property: The tangent line at a point (x, y) on the graph has slope $dy/dx = -xy$. Let R be the shaded region under the graph for $0 \le x \le a$ in Figure 14. Use the Shell Method and the substitution $u = f(x)$ to show that the solid obtained by revolving R around the y-axis has volume $V = 2\pi(1 - c)$, where $c = f(a)$. Observe that as $c \to 0$, the region R becomes infinite but the volume V approaches 2π.

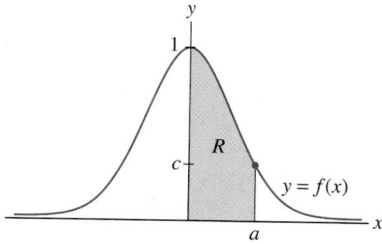

FIGURE 14 The bell-shaped curve.

FIGURE 13 The ellipse $\left(\dfrac{x}{a}\right)^2 + \left(\dfrac{y}{b}\right)^2 = 1.$

6.5 Work and Energy

> "For those who want some proof that physicists are human, the proof is in the idiocy of all the different units which they use for measuring energy."
>
> —Richard Feynman,
> The Character of Physical Law

All physical tasks, such as boiling water or lighting a fire, require an expenditure of energy. When a force is applied to an object to move it, the energy expended is called **work**. If a *constant* force F is applied through a distance d, then the work W is defined as "force times distance" (Figure 1)

$$W = F \cdot d \qquad \boxed{1}$$

FIGURE 1 The work expended to move the object from A to B is $W = F \cdot d$.

Force F

A Distance d B

In the metric system, the unit of force is the *newton* (abbreviated N), defined as 1 kg-m/s^2. Energy and work are both measured in units of the *joule* (J), equal to 1 N-m. In the British system, the unit of force is the pound, and both energy and work are measured in foot-pounds (ft-lb). Another unit of energy is the *calorie*. One ft-lb is approximately 0.738 J or 3.088 calories.

To become familiar with the units, let's calculate the work W required to lift a 2-kg stone 3 m above the ground. Gravity pulls down on the stone of mass m with a force equal to $-mg$, where $g = 9.8$ m/s^2. Therefore, lifting the stone requires an upward vertical force $F = mg$, and the work expended is

$$W = \underbrace{(mg)h}_{F \cdot d} = (2 \text{ kg})(9.8 \text{ m/s}^2)(3 \text{ m}) = 58.8 \text{ J}$$

While the kilogram is a unit of mass, the pound is a unit of force rather than mass, so the factor g does not appear when computing work against gravity in the British system. The work required to lift a 2-lb stone 3 ft above ground is

$$W = \underbrace{(2 \text{ lb})(3 \text{ ft})}_{F \cdot d} = 6 \text{ ft-lb}$$

We use integration to calculate work when the force is not constant. Suppose that the force $F(x)$ varies as the object moves from a to b along the x-axis. Then Eq. (1) does not apply directly, but we may break up the task into a large number of smaller tasks where Eq. (1) gives a good approximation. Divide $[a, b]$ into N subintervals of length $\Delta x = \dfrac{b - a}{N}$, with endpoints: $a = x_0, x_1, x_2, \ldots, x_{N-1}, x_N = b$. Let W_i be the work required to move the object from x_{i-1} to x_i (Figure 2). If Δx is small, then the force $F(x)$ is nearly constant on the interval $[x_{i-1}, x_i]$ with value $F(x_i)$, so $W_i \approx F(x_i)\,\Delta x$. Summing the contributions, we obtain

$$W = \sum_{i=1}^{N} W_i \approx \underbrace{\sum_{i=1}^{N} F(x_i)\,\Delta x}_{\text{Right-endpoint approximation}}$$

The sum on the right is a right-endpoint approximation converging to $\displaystyle\int_a^b F(x)\,dx$. This leads to the following definition.

FIGURE 2 The work to move an object from x_{i-1} to x_i is approximately $F(x_i)\,\Delta x$.

> **DEFINITION Work**   The work performed in moving an object along the x-axis from a to b by applying a force of magnitude $F(x)$ is
>
> $$W = \int_a^b F(x)\,dx \qquad \boxed{2}$$

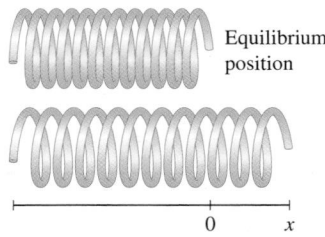

FIGURE 3 According to Hooke's Law, a spring stretched x units past equilibrium exerts a restoring force $-kx$ in the opposite direction.

One typical calculation involves finding the work required to stretch a spring. Assume that the end of the spring has position $x = 0$ at equilibrium, when no force is acting (Figure 3). The spring may be stretched x units or compressed (if $x < 0$). **Hooke's Law** states that the spring exerts a restoring force $-kx$ in the opposite direction, where k is the **spring constant**, measured in units of kilograms per second squared.

■ **EXAMPLE 1** Hooke's Law   Assuming a spring constant of $k = 400 \text{ kg/s}^2$, find the work (in joules) required to **(a)** stretch the spring 10 cm beyond equilibrium and **(b)** compress the spring 2 additional cm when it is already compressed 3 cm.

Solution   By Hooke's Law, the spring exerts a restoring force of $-400x$ N when it is stretched x units. Therefore, we must apply a force $F(x) = 400x$ N to stretch the spring further. To compute the work in joules, we must convert from centimeters to meters since 1 J is equal to a Newton-meter. Therefore,

(a) The work required to stretch the spring 10 cm (0.1 m) beyond equilibrium is

$$W = \int_0^{0.1} 400x\,dx = 200x^2 \Big|_0^{0.1} = 2 \text{ J}$$

(b) If the spring is at position $x = -0.03$ m, then the work W required to compress it further to $x = -0.05$ m is

$$W = \int_{-0.03}^{-0.05} 400x\,dx = 200x^2 \Big|_{-0.03}^{-0.05} = 0.5 - 0.18 = 0.32 \text{ J}$$

Hooke's Law is named after the English scientist, inventor, and architect Robert Hooke (1635–1703) who made important discoveries in physics, astronomy, chemistry, and biology. He was a pioneer in the use of the microscope to study organisms. Unfortunately, Hooke was involved in several bitter disputes with other scientists, most notably with his contemporary Isaac Newton. Newton was furious when Hooke criticized his work on optics. Later, Hooke told Newton that he believed Kepler's Laws would follow from an inverse square law of gravitation. Newton refused to acknowledge Hooke's contributions in his masterwork Principia. It was in a letter to Hooke that Newton made his famous remark "If I have seen further it is by standing on the shoulders of giants."

Note that we integrate from right to left (the lower limit -0.03 is *larger* than the upper limit -0.05) because we're compressing the spring to the left. ∎

In the next two examples, we compute work in a different way. In these examples, the formula $W = \int_a^b F(x)\,dx$ cannot be used because we are not moving a single object through a fixed distance. Rather, each thin layer of the object is moved through a different distance. We compute total work by "summing" (i.e., *integrating*) the work performed on each thin layer.

■ **EXAMPLE 2** Building a Cement Column Compute the work (against gravity) required to build a cement column of height 5 m and square base of side 2 m. Assume that cement has density $1{,}500\ \mathrm{kg/m^3}$.

Solution Think of the column as a stack of N thin layers of thickness $\Delta y = 5/N$. The work consists of lifting up these layers and placing them on the stack (Figure 4), but the work performed on a given layer depends on how high we lift it. To find the work required to lift a thin layer of thickness Δy to height y, we compute the volume, mass, and gravitational force on the layer:

On the earth's surface, work against gravity is equal to the force mg times the vertical distance through which the object is moved. No work against gravity is done when an object is moved sideways.

$$\text{Volume of layer} = \text{area} \times \text{thickness} = 4\,\Delta y\ \mathrm{m^3}$$

$$\text{Mass of layer} = \text{density} \times \text{volume} = 1{,}500 \cdot 4\,\Delta y\ \mathrm{kg}$$

$$\text{Force on layer} = g \times \text{mass} = 9.8 \cdot 1{,}500 \cdot 4\,\Delta y = 58{,}800\,\Delta y\ \mathrm{N}$$

The work required to raise this layer to height y is approximately equal to force times the distance y, that is, $58{,}800 y\,\Delta y$. We set $W(y) = 58{,}800 y$ and write

FIGURE 4 Total work is the sum of the work performed on each layer of the column.

$$\boxed{\text{Work performed lifting layer to height } y \approx W(y)\,\Delta y}$$

This is only an approximation because the layer has a nonzero thickness and the cement particles at the top have been lifted a little bit higher than those at the bottom.

The ith layer is lifted to height $y_i = i\,\Delta y$, and the total work performed is

$$W \approx \sum_{i=1}^{N} W(y_i)\,\Delta y$$

The sum on the right is a right-endpoint approximation to $\int_0^5 W(y)\,dy$. Letting N tend to ∞, we obtain

$$W = \int_0^5 W(y)\,dy = \int_0^5 58{,}800 y\,dy = 58{,}800\frac{y^2}{2}\Big|_0^5 = 735{,}000\ \mathrm{J} \quad ■$$

■ **EXAMPLE 3** Pumping Water out of a Tank A spherical tank of radius R meters with a small hole at the top is filled with water. How much work (against gravity) is done pumping the water out through the hole? The density of water is $1{,}000\ \mathrm{kg/m^3}$.

Solution This is similar to the previous example. We divide the sphere into N thin layers of thickness Δy (Figure 5) and show that the work performed on the layer at height y is approximately $W(y)\Delta y$ for some function $W(y)$.

Step 1. **Approximate the work performed on a single layer.**

We place the origin at the center of the sphere so that y ranges from $-R$ to R (Figure 5). The ith layer at height $y = y_i$ is nearly cylindrical of thickness $\Delta y = \dfrac{2R}{N}$ and some radius which we call x. By the Pythagorean Theorem, $x^2 + y^2 = R^2$. Therefore, $x^2 = R^2 - y^2$ and the volume of the ith layer is approximately

$$\text{Volume of } i\text{th layer} \approx \pi x^2 \,\Delta y = \pi(R^2 - y^2)\,\Delta y \text{ m}^3$$

> ◀⋯ *REMINDER In the MKS system of units, the force due to gravity on an object of mass m kg is 9.8m N.*

Furthermore,

$$\text{Mass of } i\text{th layer} = \text{density} \times \text{volume} \approx 1{,}000\pi(R^2 - y^2)\,\Delta y \text{ kg}$$

$$\text{Force on } i\text{th layer} = g \times \text{mass} \qquad \approx (9.8)1{,}000\pi(R^2 - y^2)\,\Delta y \text{ N}$$

We must raise up this thin layer a vertical distance of approximately $R - y$ (no work against gravity is required to move an object sideways), so the work performed on the layer is approximately

$$\underbrace{9{,}800\pi(R^2 - y^2)\,\Delta y}_{\text{Force against gravity}} \cdot \underbrace{(R - y)}_{\text{Vertical distance moved}} = \underbrace{9{,}800\pi(R^3 - R^2 y - R y^2 + y^3)\,\Delta y}_{\text{Call this } W(y)}$$

With $W(y)$ as indicated and $y = y_i$,

$$\boxed{\text{Work performed on } i\text{th layer} \approx W(y_i)\,\Delta y}$$

Step 2. **Integrate the work performed on the layers.**

We complete the argument as in Example 2. The total work W is the sum of the work performed on the N layers. Thus $W \approx \displaystyle\sum_{i=1}^{N} W(y_i)\,\Delta y$ and in the limit as $N \to \infty$, we obtain

> *The limits of integration in Eq. (3) are $-R$ to R because the y-coordinate along the sphere varies from $-R$ to R.*

$$W = \int_{-R}^{R} W(y)\,dy = 9{,}800\pi \int_{-R}^{R} (R^3 - R^2 y - R y^2 + y^3)\,dy \qquad \boxed{3}$$

$$= 9{,}800\pi \left(R^3 y - \frac{1}{2}R^2 y^2 - \frac{1}{3}R y^3 + \frac{1}{4}y^4 \right)\Bigg|_{-R}^{R} = \frac{39{,}200\pi}{3}R^4 \text{ J} \qquad ■$$

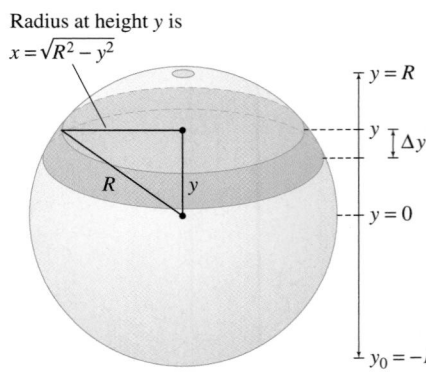

FIGURE 5 The radius of a thin layer at height y is $x = \sqrt{R^2 - y^2}$.

6.5 SUMMARY

- The work W performed when a force F is applied to move an object along a straight line:

$$\text{Constant force:} \quad W = F \cdot d, \qquad \text{Variable force:} \quad W = \int_a^b F(x)\, dx$$

- In some cases, the work is computed by decomposing an object into N thin layers of thickness $\Delta y = \dfrac{b-a}{N}$ (where the object extends from $y = a$ to $y = b$). We approximate the work W_i performed on the ith layer as $W_i \approx W(y_i)\, \Delta y$ for some function $W(y)$. The total work is equal to $W = \int_a^b W(y)\, dy$.

6.5 EXERCISES

Preliminary Questions

1. Why is integration needed to compute the work performed in stretching a spring?

2. Why is integration needed to compute the work performed in pumping water out of a tank but not to compute the work performed in lifting up the tank?

3. Which of the following represents the work required to stretch a spring (with spring constant k) a distance x beyond its equilibrium position: kx, $-kx$, $\frac{1}{2}mk^2$, $\frac{1}{2}kx^2$, or $\frac{1}{2}mx^2$?

Exercises

1. How much work is done raising a 4-kg mass to a height of 16 m above ground?

2. How much work is done raising a 4-lb mass to a height of 16 ft above ground?

In Exercises 3–6, compute the work (in joules) required to stretch or compress a spring as indicated, assuming that the spring constant is $k = 150$ kg/s^2.

3. Stretching from equilibrium to 12 cm past equilibrium

4. Compressing from equilibrium to 4 cm past equilibrium

5. Stretching from 5 to 15 cm past equilibrium

6. Compressing the spring 4 more cm when it is already compressed 5 cm

7. If 5 J of work are needed to stretch a spring 10 cm beyond equilibrium, how much work is required to stretch it 15 cm beyond equilibrium?

8. If 5 J of work are needed to stretch a spring 10 cm beyond equilibrium, how much work is required to compress it 5 cm beyond equilibrium?

9. If 10 ft-lb of work are needed to stretch a spring 1 ft beyond equilibrium, how far will the spring stretch if a 10-lb weight is attached to its end?

10. Show that the work required to stretch a spring from position a to position b is $\frac{1}{2}k(b^2 - a^2)$, where k is the spring constant. How do you interpret the negative work obtained when $|b| < |a|$?

In Exercises 11–14, calculate the work against gravity required to build the structure out of brick using the method of Examples 2 and 3. Assume that brick has density 80 lb/ft^3.

11. A tower of height 20 ft and square base of side 10 ft

12. A cylindrical tower of height 20 ft and radius 10 ft

13. A 20-ft-high tower in the shape of a right circular cone with base of radius 4 ft

14. A structure in the shape of a hemisphere of radius 4 ft

15. Built around 2600 BCE, the Great Pyramid of Giza in Egypt is 485 ft high (due to erosion, its current height is slightly less) and has a square base of side 755.5 ft (Figure 6). Find the work needed to build the pyramid if the density of the stone is estimated at 125 lb/ft^3.

FIGURE 6 The Great Pyramid in Giza, Egypt.

In Exercises 16–20, calculate the work (in joules) required to pump all of the water out of the tank. Assume that the tank is full, distances are measured in meters, and the density of water is 1,000 kg/m³.

16. The box in Figure 7; water exits from a small hole at the top.

Water exits here

5

4

8

FIGURE 7

17. The hemisphere in Figure 8; water exits from the spout as shown.

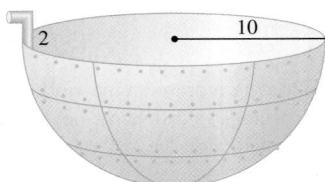

2   10

FIGURE 8

18. The conical tank in Figure 9; water exits through the spout as shown.

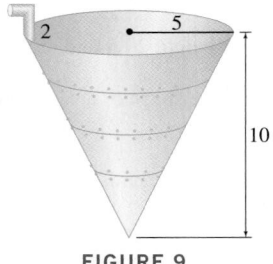

2   5

10

FIGURE 9

19. The horizontal cylinder in Figure 10; water exits from a small hole at the top. *Hint:* Evaluate the integral by interpreting part of it as the area of a circle.

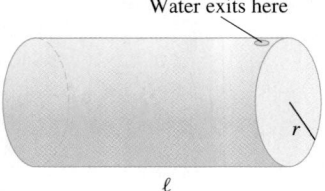

Water exits here

r

ℓ

FIGURE 10

20. The trough in Figure 11; water exits by pouring over the sides.

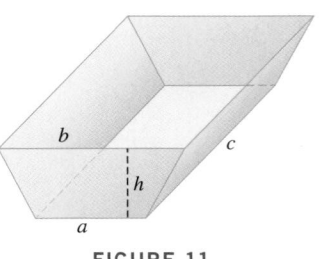

b   c

h

a

FIGURE 11

21. Find the work W required to empty the tank in Figure 7 if it is half full of water.

22. ✎   Assume the tank in Figure 7 is full of water and let W be the work required to pump out half of the water. Do you expect W to equal the work computed in Exercise 21? Explain and then compute W.

23. Find the work required to empty the tank in Figure 9 if it is half full of water.

24. Assume the tank in Figure 9 is full of water and find the work required to pump out half of the water.

25. Assume that the tank in Figure 9 is full.
(a) Calculate the work $F(y)$ required to pump out water until the water level has reached level y.
(b) *CAS*   Plot $F(y)$.
(c) ✎   What is the significance of $F'(y)$ as a rate of change?
(d) *CAS*   If your goal is to pump out all of the water, at which water level y_0 will half of the work be done?

26. How much work is done lifting a 25-ft chain over the side of a building (Figure 12)? Assume that the chain has a density of 4 lb/ft. *Hint:* Break up the chain into N segments, estimate the work performed on a segment, and compute the limit as $N \to \infty$ as an integral.

y

Segment of length Δy

FIGURE 12 The small segment of the chain of length Δy located y feet from the top is lifted through a vertical distance y.

27. How much work is done lifting a 3-m chain over the side of a building if the chain has mass density 4 kg/m?

28. An 8-ft chain weighs 16 lb. Find the work required to lift the chain over the side of a building.

29. A 20-ft chain with mass density 3 lb/ft is initially coiled on the ground. How much work is performed in lifting the chain so that it is fully extended (and one end touches the ground)?

30. How much work is done lifting a 20-ft chain with mass density 3 lb/ft (initially coiled on the ground) so that its top end is 30 ft above the ground?

31. A 1,000-lb wrecking ball hangs from a 30-ft cable of density 10 lb/ft attached to a crane. Calculate the work done if the crane lifts the ball from ground level to 30 ft in the air by drawing in the cable.

In Exercises 32–34, use Newton's Universal Law of Gravity, according to which the gravitational force between two objects of mass m and M separated by a distance r has magnitude GMm/r^2, where $G = 6.67 \times 10^{-11} \ \mathrm{m^3 kg^{-1} s^{-1}}$. Although the Universal Law refers to point masses, Newton proved that it also holds for uniform spherical objects, where r is the distance between their centers.

32. Two spheres of mass M and m are separated by a distance r_1. Show that the work required to increase the separation to a distance r_2 is equal to $W = GMm(r_1^{-1} - r_2^{-1})$.

33. Use the result of Exercise 32 to calculate the work required to place a 2,000-kg satellite in an orbit 1,200 km above the surface of the

earth. Assume that the earth is a sphere of mass $M_e = 5.98 \times 10^{24}$ kg and radius $r_e = 6.37 \times 10^6$ m. Treat the satellite as a point mass.

34. Use the result of Exercise 32 to compute the work required to move a 1,500-kg satellite from an orbit 1,000 to 1,500 km above the surface of the earth.

35. Assume that the pressure P and volume V of the gas in a 30-in. cylinder of radius 3 in. with a movable piston are related by $PV^{1.4} = k$, where k is a constant (Figure 13). When the cylinder is full, the gas pressure is 200 lb/in.^{2}.

(a) Calculate k.

(b) Calculate the force on the piston as a function of the length x of the column of gas (the force is PA, where A is the piston's area).

(c) Calculate the work required to compress the gas column from 30 to 20 in.

FIGURE 13 Gas in a cylinder with a piston.

Further Insights and Challenges

36. A 20-ft chain with linear mass density

$$\rho(x) = 0.02x(20 - x) \ \mathrm{lb/ft}$$

lies on the ground.

(a) How much work is done lifting the chain so that it is fully extended (and one end touches the ground)?

(b) How much work is done lifting the chain so that its top end has a height of 30 ft?

37. Work-Kinetic Energy Theorem The **kinetic energy** of an object of mass m moving with velocity v is KE $= \frac{1}{2}mv^2$.

(a) Suppose that the object moves from x_1 to x_2 during the time interval $[t_1, t_2]$ due to a net force $F(x)$ acting along the interval $[x_1, x_2]$. Let $x(t)$ be the position of the object at time t. Use the Change of Variables formula to show that the work performed is equal to

$$W = \int_{x_1}^{x_2} F(x)\, dx = \int_{t_1}^{t_2} F(x(t))v(t)\, dt$$

(b) By Newton's Second Law, $F(x(t)) = ma(t)$, where $a(t)$ is the acceleration at time t. Show that

$$\frac{d}{dt}\left(\frac{1}{2}mv(t)^2\right) = F(x(t))v(t)$$

(c) Use the FTC to show that the change in kinetic energy during the time interval $[t_1, t_2]$ is equal to

$$\int_{t_1}^{t_2} F(x(t))v(t)\, dt.$$

(d) Prove the Work-Kinetic Energy Theorem: The change in KE is equal to the work W performed.

38. A model train of mass 0.5 kg is placed at one end of a straight 3-m electric track. Assume that a force $F(x) = 3x - x^2$ N acts on the train at distance x along the track. Use the Work-Kinetic Energy Theorem (Exercise 37) to determine the velocity of the train when it reaches the end of the track.

39. With what initial velocity v_0 must we fire a rocket so it attains a maximum height r above the earth? *Hint:* Use the results of Exercises 32 and 37. As the rocket reaches its maximum height, its KE decreases from $\frac{1}{2}mv_0^2$ to zero.

40. With what initial velocity must we fire a rocket so it attains a maximum height of $r = 20$ km above the surface of the earth?

41. Calculate **escape velocity,** the minimum initial velocity of an object to ensure that it will continue traveling into space and never fall back to earth (assuming that no force is applied after takeoff). *Hint:* Take the limit as $r \to \infty$ in Exercise 39.

CHAPTER REVIEW EXERCISES

In Exercises 1–6, find the area of the region bounded by the graphs of the functions.

1. $y = \sin x$, $y = \cos x$, $0 \le x \le \dfrac{5\pi}{4}$

2. $f(x) = x^3 - 2x^2 + x$, $g(x) = x^2 - x$

3. $f(x) = x^2 + 2x$, $g(x) = x^2 - 1$, $h(x) = x^2 + x - 2$

4. $f(x) = \sin x$, $g(x) = \sin 2x$, $\dfrac{\pi}{3} \le x \le \pi$

5. $y = e^x$, $y = 1 - x$, $x = 1$

6. $y = \cosh 1 - \cosh x$, $y = \cosh x - \cosh 1$

In Exercises 7–10, sketch the region bounded by the graphs of the functions and find its area.

7. $f(x) = x^3 - x^2 - x + 1$, $g(x) = \sqrt{1 - x^2}$, $0 \le x \le 1$
Hint: Use geometry to evaluate the integral.

8. $x = \dfrac{1}{2}y$, $x = y\sqrt{1 - y^2}$, $0 \le y \le 1$

9. $y = 4 - x^2$, $y = 3x$, $y = 4$

10. $x = y^3 - 2y^2 + y$, $x = y^2 - y$

11. GU   Use a graphing utility to locate the points of intersection of $y = e^{-x}$ and $y = 1 - x^2$ and find the area between the two curves (approximately).

12. Figure 1 shows a solid whose horizontal cross section at height y is a circle of radius $(1 + y)^{-2}$ for $0 \le y \le H$. Find the volume of the solid.

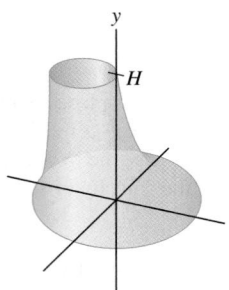

FIGURE 1

13. Find the total weight of a 3-ft metal rod of linear density

$$\rho(x) = 1 + 2x + \frac{2}{9}x^3 \text{ lb/ft.}$$

14. Find the flow rate (in the correct units) through a pipe of diameter 6 cm if the velocity of fluid particles at a distance r from the center of the pipe is $v(r) = (3 - r)$ cm/s.

In Exercises 15–20, find the average value of the function over the interval.

15. $f(x) = x^3 - 2x + 2$, $[-1, 2]$

16. $f(x) = \sqrt{9 - x^2}$, $[0, 3]$ *Hint:* Use geometry to evaluate the integral.

17. $f(x) = |x|$, $[-4, 4]$ **18.** $f(x) = x[x]$, $[0, 3]$

19. $f(x) = x \cosh(x^2)$, $[0, 1]$ **20.** $f(x) = \dfrac{e^x}{1 + e^{2x}}$, $\left[0, \dfrac{1}{2}\right]$

21. The average value of $g(t)$ on $[2, 5]$ is 9. Find $\displaystyle\int_2^5 g(t)\, dt$.

22. For all $x \ge 0$, the average value of $R(x)$ over $[0, x]$ is equal to x. Find $R(x)$.

23. Use the Shell Method to find the volume of the solid obtained by revolving the region between $y = x^2$ and $y = mx$ about the x-axis (Figure 2).

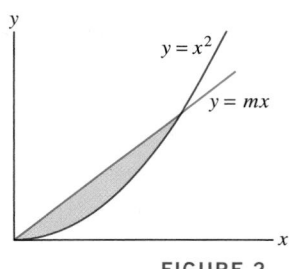

FIGURE 2

24. Use the washer method to find the volume of the solid obtained by revolving the region between $y = x^2$ and $y = mx$ about the y-axis (Figure 2).

25. Let R be the intersection of the circles of radius 1 centered at $(1, 0)$ and $(0, 1)$. Express as an integral (but do not evaluate): **(a)** the area of R and **(b)** the volume of revolution of R about the x-axis.

26. Use the Shell Method to set up an integral (but do not evaluate) expressing the volume of the solid obtained by rotating the region under $y = \cos x$ over $[0, \pi/2]$ about the line $x = \pi$.

In Exercises 27–35, find the volume of the solid obtained by rotating the region enclosed by the curves about the given axis.

27. $y = 2x$, $y = 0$, $x = 8$; x-axis

28. $y = 2x$, $y = 0$, $x = 8$;   axis $x = -3$

29. $y = x^2 - 1$, $y = 2x - 1$,   axis $x = -2$

30. $y = x^2 - 1$, $y = 2x - 1$,   axis $y = 4$

31. $y^2 = x^3$, $y = x$, $x = 8$;   axis $x = -1$

32. $y^2 = x^{-1}$, $x = 1$, $x = 3$;   axis $y = -3$

33. $y = -x^2 + 4x - 3$, $y = 0$;   axis $y = -1$

34. $x = 4y - y^3$, $y = 0$, $y = 2$;   y-axis

35. $y^2 = x^{-1}$, $x = 1$, $x = 3$;   axis $x = -3$

In Exercises 36–38, the regions refer to the graph of the hyperbola $y^2 - x^2 = 1$ in Figure 3. Calculate the volume of revolution about both the x- and y-axes.

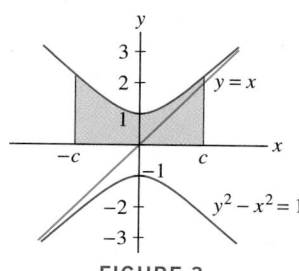

FIGURE 3

36. The shaded region between the upper branch of the hyperbola and the x-axis for $-c \leq x \leq c$.

37. The region between the upper branch of the hyperbola and the line $y = x$ for $0 \leq x \leq c$.

38. The region between the upper branch of the hyperbola and $y = 2$.

39. Let $a > 0$. Show that when the region between $y = a\sqrt{x - ax^2}$ and the x-axis is rotated about the x-axis, the resulting volume is independent of the constant a.

40. A spring whose equilibrium length is 15 cm exerts a force of 50 N when it is stretched to 20 cm. Find the work required to stretch the spring from 22 to 24 cm.

In Exercises 41–42, water is pumped into a spherical tank of radius 5 ft from a source located 2 ft below a hole at the bottom (Figure 4). The density of water is 64.2 lb/ft³.

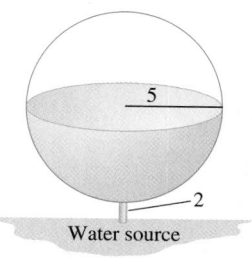

Water source

FIGURE 4

41. Calculate the work required to fill the tank.

42. Calculate the work $F(h)$ required to fill the tank to height h ft from the bottom of the sphere.

43. A container weighing 50 lb is filled with 20 ft³ of water. The container is raised vertically at a constant speed of 2 ft/s for 1 min, during which time it leaks water at a rate of $\frac{1}{3}$ ft³/s. Calculate the total work performed in raising the container. The density of water is 64.2 lb/ft³.

44. Let W be the work (against the sun's gravitational force) required to transport an 80-kg person from Earth to Mars when the two planets are aligned with the sun at their minimal distance of 55.7×10^6 km. Use Newton's Universal Law of Gravity (see Exercises 32–34 in Section 6.5) to express W as an integral and evaluate it. The sun has mass $M_s = 1.99 \times 10^{30}$ kg, and the distance from the sun to the earth is 149.6×10^6 km.

In this chapter, we develop some basic techniques of integration. The two most important general techniques are integration by parts and substitution (which was covered in Chapter 5). We also have specialized techniques for treating particular classes of functions such as trigonometric or rational functions. However, there is no surefire method for finding antiderivatives; in fact, it is not always possible to express an antiderivative in elementary terms. Therefore, we begin with a discussion of numerical integration. Every definite integral can be approximated numerically to any desired degree of accuracy.

Computer simulation of the 1996 Kamchatka tsunami by the National Oceanic and Atmospheric Administration, using models of wave motion based on advanced calculus.

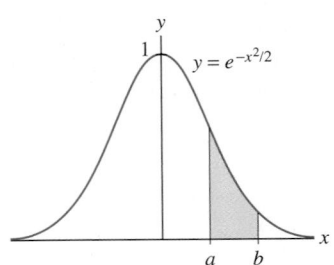

FIGURE 1 There is no explicit formula for areas under the bell-shaped curve.

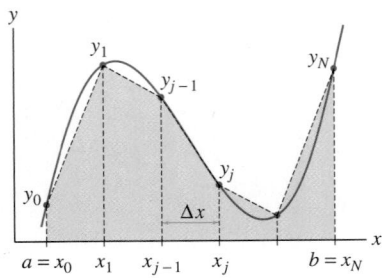

FIGURE 2 T_N approximates the area under the graph by trapezoids.

7.1 Numerical Integration

Numerical integration is the process of approximating a definite integral using well-chosen sums of function values. We use numerical integration when we cannot find a formula for an antiderivative and thus cannot apply the FTC. For instance, the Gaussian function $f(x) = e^{-x^2/2}$, whose graph is a bell-shaped curve (Figure 1), does not have an elementary antiderivative. Numerical methods are needed to approximate $\int_a^b e^{-x^2/2}\,dx$.

Let N be a whole number. The **Trapezoidal Rule** T_N consists of approximating $\int_a^b f(x)\,dx$ by the average of the left- and right-endpoint approximations:

$$T_N = \frac{1}{2}(R_N + L_N)$$

This approximation was used informally in previous chapters, where we observed that the average was likely to give a better approximation than either of R_N or L_N alone. To derive a formula for T_N, let's recall the formulas for the endpoint approximations. We divide $[a, b]$ into N subintervals of length $\Delta x = \dfrac{b-a}{N}$ (Figure 2) with endpoints

$$x_0 = a, \qquad x_1 = a + \Delta x, \qquad x_2 = a + 2\Delta x, \qquad \ldots, \qquad x_N = b$$

For convenience, we denote the values of $f(x)$ at the endpoints by y_j:

$$y_j = f(x_j) = f(a + j\Delta x) \quad (j = 0, 1, 2, \ldots, N)$$

In particular, $y_0 = f(a)$ and $y_N = f(b)$. Then

$$R_N = \Delta x \sum_{j=1}^{N} y_j, \qquad L_N = \Delta x \sum_{j=0}^{N-1} y_j$$

When we form the sum $R_N + L_N$, each value y_j occurs twice except for the first y_0 (which only occurs in L_N) and the last y_N (which only occurs in R_N). Therefore,

$$R_N + L_N = \Delta x \big(y_0 + 2y_1 + 2y_2 + \cdots + 2y_{N-1} + y_N\big)$$

Dividing by 2, we obtain the formula for T_N.

Trapezoidal Rule   The Nth trapezoidal approximation to $\displaystyle\int_a^b f(x)\,dx$ is

$$T_N = \frac{1}{2}\Delta x\big(y_0 + 2y_1 + \cdots + 2y_{N-1} + y_N\big)$$

where $\Delta x = \dfrac{b-a}{N}$ and $y_j = f(a + j\,\Delta x)$.

The name "Trapezoidal Rule" is appropriate because T_N is equal to the sum of the areas of the trapezoids obtained by joining the points (x_0, y_0), (x_1, y_1), ..., (x_N, y_N) with line segments (Figure 2). To verify this, we note that the area of a typical trapezoid is equal to the average of the areas of the left- and right-hand rectangles (Figure 3).

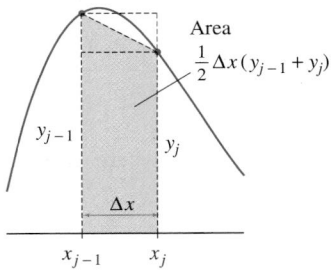

FIGURE 3 The area of a trapezoid is equal to the average of the areas of the left- and right-endpoint rectangles.

■ **EXAMPLE 1** ⌐⌐⌐⌐ Calculate the tenth trapezoidal approximation to $\displaystyle\int_0^1 \sin(x^2)\,dx$. Then use a computer algebra system to calculate T_N for $N = 20, 50, 100, 500, 1{,}000,$ and $5{,}000$.

Solution   Divide $[0, 1]$ into 10 subintervals of length $\Delta x = \frac{1}{10} = 0.1$. To obtain T_{10}, sum the function values at the endpoints $0, 0.1, 0.2, \ldots, 1$ with the appropriate coefficients:

$$T_{10} = \frac{1}{2}(0.1)\big[\sin(0^2) + 2\sin(0.1^2) + 2\sin(0.2^2) + 2\sin(0.3^2) + 2\sin(0.4^2)$$
$$+ 2\sin(0.5^2) + 2\sin(0.6^2) + 2\sin(0.7^2) + 2\sin(0.8^2) + 2\sin(0.9^2) + \sin(1^2)\big]$$
$$\approx 0.311$$

For general N, $\Delta x = \dfrac{1-0}{N} = \dfrac{1}{N}$ and

$$T_N = \frac{1}{2N}\left[\sin(0^2) + 2\underbrace{\left(\sin\!\left(\left(\tfrac{1}{N}\right)^2\right) + \sin\!\left(\left(\tfrac{2}{N}\right)^2\right) + \cdots + \sin\!\left(\left(\tfrac{N-1}{N}\right)^2\right)\right)}_{\text{Inner sum}} + \sin(1^2)\right]$$

The inner sum may be evaluated on a computer algebra system using a command such as

```
Sum[Sin[(i/N)^2], {i, 1, N − 1}]
```

We obtain the results in Table 1. Thus $\displaystyle\int_0^1 \sin(x^2)\,dx$ is approximately 0.31027. ■

TABLE 1	
N	T_N
20	0.3104936
50	0.3103043
100	0.3102773
500	0.3102687
1000	0.3102684
5000	**0.3102683**

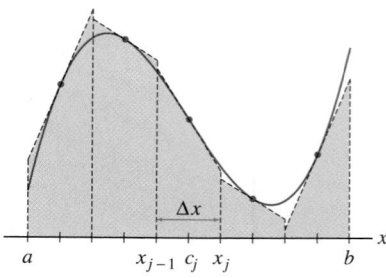

(A) M_N is the sum of the areas
of the midpoint rectangles.

(B) M_N is also the sum of the areas
of the tangential trapezoids.

FIGURE 4 Two interpretations of M_N.

We introduced the Midpoint Rule M_N in Section 5.1. Recall that M_N is the sum of the areas of the rectangles of height $f(c_j)$ and base Δx, where c_j is the midpoint of the interval $[x_{j-1}, x_j]$ [Figure 4(A)].

Midpoint Rule   The Nth Midpoint Approximation to $\displaystyle\int_a^b f(x)\,dx$ is

$$M_N = \Delta x \big(f(c_1) + f(c_2) + \cdots + f(c_N)\big)$$

where $\Delta x = \dfrac{b-a}{N}$ and $c_j = a + (j - \frac{1}{2})\,\Delta x$ is the midpoint of the jth interval $[x_{j-1}, x_j]$.

GRAPHICAL INSIGHT   Although M_N is defined using midpoint rectangles [Figure 4(A)], there is a second interpretation of M_N as the sum of the areas of the trapezoids whose top edges are tangent to the graph of $y = f(x)$ at the midpoints c_j [Figure 4(B)]. What allows us to replace the rectangles by trapezoids? The following general fact: A rectangle has the same area as any trapezoid of the same base whose top edge passes through the midpoint of the top of the rectangle (Figure 5). Because of this interpretation, the Midpoint Rule is sometimes called the Tangent Rule.

FIGURE 5 The rectangle and the trapezoid have the same area.

Error Bounds

In applications, it is important to know the accuracy of a numerical approximation. We define the error in T_N and M_N by

$$\text{Error}(T_N) = \left| T_N - \int_a^b f(x)\,dx \right|, \qquad \text{Error}(M_N) = \left| M_N - \int_a^b f(x)\,dx \right|$$

According to the following theorem, the magnitudes of these errors are related to the size of the *second* derivative $f''(x)$—here we assume $f''(x)$ exists and is continuous.

In the Error Bound, you can let K_2 be the maximum of $|f''(x)|$ on $[a, b]$, but if it is inconvenient to find this maximum exactly, take K_2 to be any number that is definitely larger than the maximum.

THEOREM 1 Error Bound for T_N and M_N   Let K_2 be a number such that $|f''(x)| \leq K_2$ for all $x \in [a, b]$. Then

$$\text{Error}(T_N) \leq \frac{K_2(b-a)^3}{12N^2}, \qquad \text{Error}(M_N) \leq \frac{K_2(b-a)^3}{24N^2}$$

GRAPHICAL INSIGHT   The proof of the Error Bound is technical and we omit it. Intuitively, however, we can see why Error(T_N) is related to the size of $f''(x)$. The second derivative measures concavity: If $|f''(x)|$ is large, then the graph of f bends a lot (either up or down) and the trapezoids do not provide a good approximation to the area under the graph (Figure 6). Since the Midpoint Rule can be interpreted in terms of tangential trapezoids, Error(M_N) is also related to the size of $f''(x)$.

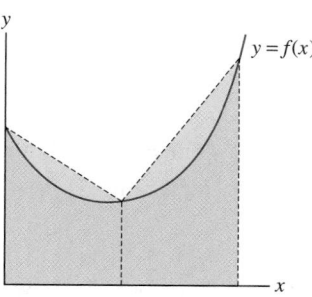
(A) Larger second derivative $f''(x)$

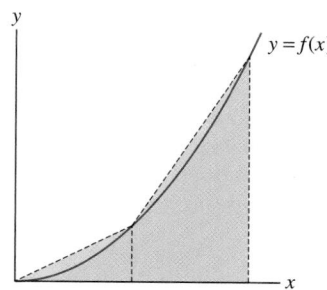
(B) Smaller second derivative $f''(x)$

FIGURE 6 The trapezoidal approximation is more accurate when $|f''(x)|$ is small.

■ **EXAMPLE 2** Checking the Error Bound   Calculate T_6 and M_6 for $\int_1^4 \sqrt{x}\,dx$. Then

(a) Use the Error Bound to find bounds for the errors.

(b) Calculate the integral exactly and verify that the Error Bounds are satisfied.

Solution We divide $[1, 4]$ into six subintervals of width $\Delta x = \frac{4-1}{6} = \frac{1}{2}$. Using the endpoints and midpoints shown in Figure 7, we obtain

$$T_6 = \frac{1}{2}\left(\frac{1}{2}\right)\left(\sqrt{1} + 2\sqrt{1.5} + 2\sqrt{2} + 2\sqrt{2.5} + 2\sqrt{3} + 2\sqrt{3.5} + \sqrt{4}\right)$$

$$\approx 4.661488$$

$$M_6 = \frac{1}{2}\left(\sqrt{1.25} + \sqrt{1.75} + \sqrt{2.25} + \sqrt{2.75} + \sqrt{3.25} + \sqrt{3.75}\right)$$

$$\approx 4.669245$$

FIGURE 7 Interval $[1, 4]$ divided into $N = 6$ subintervals.

(a) Let $f(x) = \sqrt{x}$. To use the Error Bound, we must find a number K_2 such that $|f''(x)| \leq K_2$ for $1 \leq x \leq 4$. We have $f'(x) = \frac{1}{2}x^{-1/2}$ and $f''(x) = -\frac{1}{4}x^{-3/2}$. The absolute value $|f''(x)| = \frac{1}{4}x^{-3/2}$ is decreasing on $[1, 4]$, so its maximum occurs at $x = 1$ (Figure 8). Thus, we may take $K_2 = |f''(1)| = \frac{1}{4}$. By Theorem 1,

$$\text{Error}(T_6) \leq \frac{K_2(b-a)^3}{12N^2} = \frac{\frac{1}{4}(4-1)^3}{12(6)^2} = \frac{1}{64} \approx 0.0156 \qquad \boxed{1}$$

$$\text{Error}(M_6) \leq \frac{K_2(b-a)^3}{24N^2} = \frac{\frac{1}{4}(4-1)^3}{24(6)^2} = \frac{1}{128} \approx 0.0078 \qquad \boxed{2}$$

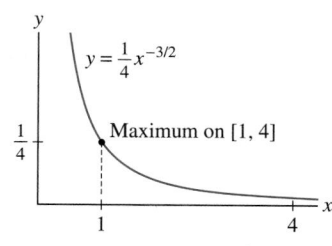

FIGURE 8 Graph of $y = |f''(x)|$ for $f(x) = \sqrt{x}$.

(b) We use the exact value $\int_1^4 \sqrt{x}\,dx = \frac{2}{3}x^{3/2}\Big|_1^4 = \frac{14}{3}$ to compute the actual errors:

In Example 2, the error in T_6 is approximately twice as large as the error in M_6. In practice, this is often the case. Note that the Error Bound for T_N in Theorem 1 is twice as large as the Error Bound for M_N.

$$\text{Error}(T_6) = \left| T_6 - \frac{14}{3} \right| \approx |4.661488 - 4.666667| \approx 0.00518$$

$$\text{Error}(M_6) = \left| M_6 - \frac{14}{3} \right| \approx |4.669245 - 4.666667| \approx 0.00258$$

Both of these errors lie well within the error bounds (1) and (2). ∎

The Error Bound can be used to determine values of N that provide a given accuracy.

■ **EXAMPLE 3** Obtaining the Desired Accuracy Find N such that M_N approximates $\int_0^3 e^{-x^2}\, dx$ with an error of at most 0.0001.

A quick way to find a value for K_2 is to plot $f''(x)$ using a graphing utility and find a bound for $|f''(x)|$ visually, as we do in Example 3.

Solution Let $f(x) = e^{-x^2}$. To apply the Error Bound, we must find a number K_2 such that $|f''(x)| \le K_2$ for all $x \in [0, 3]$. We have $f'(x) = -2xe^{-x^2}$ and

$$f''(x) = (4x^2 - 2)e^{-x^2}$$

To find a bound for $|f''(x)|$, we use a graphing utility to plot $f''(x)$ over $[0, 3]$ (Figure 9). The graph shows that the maximum value of $|f''(x)|$ is $|f''(0)| = |-2| = 2$, so we take $K_2 = 2$ in the Error Bound:

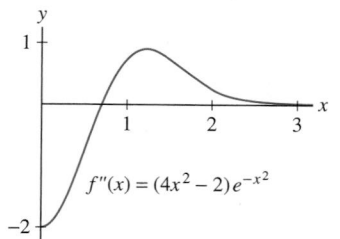

$$\text{Error}(M_N) \le \frac{K_2(b-a)^3}{24N^2} = \frac{2(3-0)^3}{24N^2} = \frac{9}{4N^2}$$

The error is at most 0.0001 if

$$\frac{9}{4N^2} \le 0.0001 \qquad \text{or} \qquad N^2 \ge \frac{9 \times 10^4}{4}$$

FIGURE 9 Graph of the second derivative of $f(x) = e^{-x^2}$.

Thus, $N \ge 150$. We conclude that M_{150} has error at most 0.0001. Computing M_{150} with a computer algebra system, we find that $M_{150} \approx 0.886207$, so the value of the integral lies between 0.886107 and 0.886307 (in fact, the value to nine places is 0.886207348). ∎

Can we improve on the Trapezoidal and Midpoint Rules? One clue is that when $f(x)$ is concave (up or down), then the exact value of the integral lies between T_N and M_N. In fact, we see geometrically (Figure 10) that

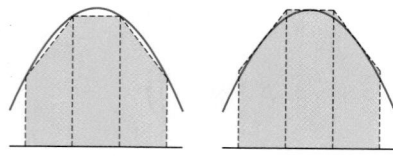

(A) Trapezoids used to compute T_N (B) Trapezoids used to compute M_N

FIGURE 10 If $f(x)$ is concave down, then T_N is smaller and M_N is larger than the integral.

- If $f(x)$ is concave down $\Rightarrow$ $T_N \le \int_a^b f(x)\, dx \le M_N$.

- If $f(x)$ is concave up $\Rightarrow$ $M_N \le \int_a^b f(x)\, dx \le T_N$

This suggests that the average of T_N and M_N may be more accurate than either T_N or M_N alone.

Simpson's Rule S_N exploits this idea, but it takes into account that M_N is roughly twice as accurate as T_N. Therefore, S_N uses a **weighted average**, which is more accurate than the ordinary average. For N even, we define

$$S_N = \frac{1}{3} T_{N/2} + \frac{2}{3} M_{N/2} \qquad \boxed{3}$$

To derive a formula for S_N, let $\Delta x = \dfrac{b-a}{N}$ and, as before, set $x_j = a + j\,\Delta x$ and $y_j = f(a + j\,\Delta x)$. If we use just the even-numbered endpoints, we obtain a partition of

$[a, b]$ into $N/2$ subintervals (keep in mind that N is even):

$$[x_0, x_2], [x_2, x_4], \ldots, [x_{N-2}, x_N] \qquad \boxed{4}$$

The *endpoints* of these intervals are used to compute $T_{N/2}$, and we use the *midpoints* $x_1, x_3, \ldots, x_{N-1}$ to compute $M_{N/2}$ (see Figure 11 for the case $N = 8$):

$$T_{N/2} = \left(\frac{1}{2}\right) \frac{b - a}{N/2} \left(y_0 + 2y_2 + 2y_4 + \cdots + 2y_{N-2} + y_N\right)$$

$$M_{N/2} = \frac{b - a}{N/2}\left(y_1 + y_3 + y_5 + \cdots + y_{N-1}\right) = \frac{b - a}{N}\left(2y_1 + 2y_3 + 2y_5 + \cdots + 2y_{N-1}\right)$$

Thus,

$$S_N = \frac{1}{3}T_{N/2} + \frac{2}{3}M_{N/2} = \frac{1}{3}\Delta x \left(y_0 + 2y_2 + 2y_4 + \cdots + 2y_{N-2} + y_N\right)$$
$$+ \frac{1}{3}\Delta x \left(4y_1 + 4y_3 + 4y_5 + \cdots + 4y_{N-1}\right)$$

FIGURE 11 We compute S_8 using eight subintervals. The even endpoints are used for T_4, the odd endpoints for M_4, and $S_8 = \frac{1}{3}T_4 + \frac{2}{3}M_4$.

Here is the pattern of coefficients in S_N:

$$1, 4, 2, 4, 2, 4, \ldots, 4, 2, 4, 1$$

The intermediate coefficients alternate $4, 2, 4, 2, \ldots, 2, 4$ (beginning and ending with 4).

Simpson's Rule Assume that N is even. Let $\Delta x = \dfrac{b - a}{N}$ and $y_j = f(a + j\,\Delta x)$. The Nth approximation to $\displaystyle\int_a^b f(x)\,dx$ by Simpson's Rule is the quantity

$$S_N = \frac{1}{3}\Delta x\left[y_0 + 4y_1 + 2y_2 + \cdots + 4y_{N-3} + 2y_{N-2} + 4y_{N-1} + y_N\right] \qquad \boxed{5}$$

CONCEPTUAL INSIGHT Both T_N and M_N (for all $N \geq 1$) give the exact value of the integral when $f(x)$ is a linear function (Exercise 57). However, of all possible combinations of $T_{N/2}$ and $M_{N/2}$, only the particular combination $S_N = \frac{1}{3}T_{N/2} + \frac{2}{3}M_{N/2}$ gives the exact value of the integral for all quadratic polynomials for all N (Exercises 58 and 59). In fact, S_N is exact for all cubic polynomials (Exercise 60).

■ **EXAMPLE 4** Use Simpson's Rule with $N = 8$ to approximate $\displaystyle\int_2^4 \sqrt{1 + x^3}\,dx$.

Solution We have $\Delta x = \dfrac{4 - 2}{8} = 0.25$. Figure 12 shows the endpoints and coefficients needed to compute S_8 using Eq. (5):

FIGURE 12 Pattern of coefficients for S_8 on $[2, 4]$.

$$\frac{1}{3}(0.25)\left[\sqrt{1 + 2^3} + 4\sqrt{1 + 2.25^3} + 2\sqrt{1 + 2.5^3} + 4\sqrt{1 + 2.75^3} + 2\sqrt{1 + 3^3}\right.$$
$$\left. + 4\sqrt{1 + 3.25^3} + 2\sqrt{1 + 3.5^3} + 4\sqrt{1 + 3.75^3} + \sqrt{1 + 4^3}\right]$$
$$\approx \frac{1}{12}\left[3 + 4(3.52003) + 2(4.07738) + 4(4.66871) + 2(5.2915)\right.$$
$$\left. + 4(5.94375) + 2(6.62382) + 4(7.33037) + 8.06226\right] \approx 10.74159 \qquad ■$$

Simpson's Rule yields impressive accuracy. Using a computer algebra system, we find that the approximation in Example 4 has an error of less than 3×10^{-6}.

We now state (without proof) a bound for the error in Simpson's Rule, where

$$\text{Error}(S_N) = \left|S_N(f) - \int_a^b f(x)\,dx\right|$$

The error involves the fourth derivative, which we assume exists and is continuous.

Although Simpson's Rule provides quite good approximations, more sophisticated techniques are implemented in computer algebra systems. These techniques are studied in the area of mathematics called "numerical analysis."

THEOREM 2 Error Bound for S_N   Let K_4 be a number such that $|f^{(4)}(x)| \le K_4$ for all $x \in [a, b]$. Then

$$\text{Error}(S_N) \le \frac{K_4(b-a)^5}{180N^4}$$

■ **EXAMPLE 5**   Calculate S_8 for $\displaystyle\int_1^3 \frac{1}{x}\,dx$. Then

(a) Find a bound for the error in S_8.

(b) Find N such that S_N has an error of at most 10^{-6}.

Solution We have $\Delta x = \dfrac{3-1}{8} = 0.25$ and the endpoints in the partition of $[1, 3]$ are $1, 1.25, 1.5, \ldots, 2.75,$ and 3. Therefore, using Eq. (5),

$$S_8 = \frac{1}{3}(0.25)\left[\frac{1}{1} + \frac{4}{1.25} + \frac{2}{1.5} + \frac{4}{1.75} + \frac{2}{2} + \frac{4}{2.25} + \frac{2}{2.5} + \frac{4}{2.75} + \frac{1}{3}\right]$$
$$\approx 1.09873$$

(a) Let $f(x) = x^{-1}$. The fourth derivative $f^{(4)}(x) = 24x^{-5}$ is decreasing, so its maximum on $[1, 3]$ is $f^{(4)}(1) = 24$. Therefore, we may set $K_4 = 24$, and

$$\text{Error}(S_N) \le \frac{K_4(b-a)^5}{180N^4} = \frac{24(3-1)^5}{180N^4} = \frac{64}{15N^4}$$

$$\text{Error}(S_8) \le \frac{K_4(b-a)^5}{180(8)^4} = \frac{24(3-1)^5}{180(8^4)} \approx 0.001$$

Using a CAS, we find that

$$S_{46} \approx 1.09861241$$

$$\int_1^3 \frac{1}{x}\,dx = \ln 3 \approx 1.09861229$$

The error is indeed less than 10^{-6}.

(b) To ensure that the error is at most 10^{-6}, we choose N so that

$$\text{Error}(S_N) = \frac{64}{15N^4} \le 10^{-6}$$

This gives

$$N^4 \ge 10^6\left(\frac{64}{15}\right) \quad \text{or} \quad N \ge \left(\frac{10^6 \cdot 64}{15}\right)^{1/4} \approx 45.45$$

Thus, we may take $N = 46$ (see the comment in the margin). ■

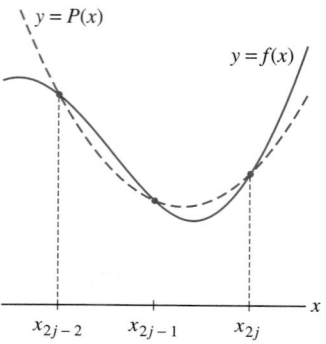

FIGURE 13 There is a unique quadratic function $P(x)$ that agrees with $f(x)$ at $x_{2j-2}, x_{2j-1}, x_{2j}$.

GRAPHICAL INSIGHT   We defined Simpson's Rule in terms of T_N and M_N, but it has another interpretation as an approximation to area under a graph using parabolas rather than rectangles. If we take three consecutive points $x_{2j-2}, x_{2j-1}, x_{2j}$, there is a unique quadratic function $P(x)$ (whose graph is a parabola) taking the same values as $f(x)$ at $x_{2j-2}, x_{2j-1}, x_{2j}$ (Figure 13). The area under the parabola over $[x_{2j-2}, x_{2j}]$ is an approximation to the area under the graph of $f(x)$ on this interval. Simpson's Rule S_N is equal to the sum of these parabolic approximations (Figure 14). This interpretation follows from the result of Exercise 58.

Graph of $y = f(x)$

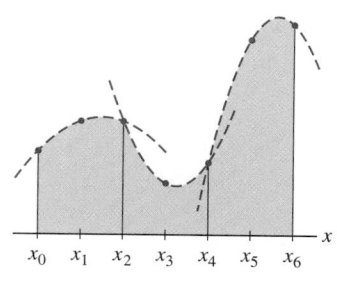

Parabolic arcs used in Simpson's Rule.

FIGURE 14 Simpson's Rule approximates the graph of $y = f(x)$ by parabolic arcs.

■ **EXAMPLE 6** Estimating Integrals from Numerical Data The velocity (in miles per hour) of a Piper Cub aircraft traveling due west is recorded every minute during the first 10 min after takeoff. Use the Trapezoidal Rule and Simpson's Rule to estimate the distance traveled after 10 min.

t	0	1	2	3	4	5	6	7	8	9	10
$v(t)$	0	50	60	80	90	100	95	85	80	75	85

Solution Since the velocity is given in miles per hour, we use hours as our unit of time. The 10-min interval has length $\frac{1}{6}$ hours. We divide this interval into $N = 10$ subintervals of length $\Delta t = \frac{1}{60}$ hours. Then

$$T_{10} = \frac{1}{120}\big(0 + 2(50 + 60 + 80 + 90 + 100 + 95 + 85 + 80 + 75) + 85\big) = 12.625$$

$$S_{10} = \frac{1}{180}\big(0 + 4(50) + 2(60) + 4(80) + 2(90) + 4(100)$$
$$+ 2(95) + 4(85) + 2(80) + 4(75) + 85\big) = 12.75$$

The plane has traveled approximately 12.7 miles (Figure 15).   ■

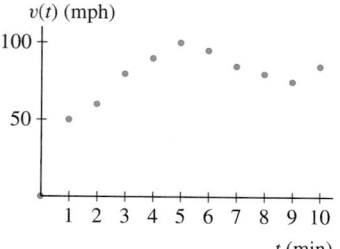

FIGURE 15 Velocity of a Piper Cub. The distance traveled is estimated by approximating the area under the curve through these data points.

7.1 SUMMARY

- We consider three numerical approximations to $\displaystyle\int_a^b f(x)\,dx$: the *Trapezoidal Rule* T_N, the *Midpoint Rule* M_N, and *Simpson's Rule* S_N (if N is even).

$$T_N = \frac{1}{2}\,\Delta x\big(y_0 + 2y_1 + 2y_2 + \cdots + 2y_{N-1} + y_N\big)$$

$$M_N = \Delta x\big(f(c_1) + f(c_2) + \cdots + f(c_N)\big) \qquad \left(c_j = a + \left(j - \frac{1}{2}\right)\Delta x\right)$$

$$S_N = \frac{1}{3}\Delta x\big[y_0 + 4y_1 + 2y_2 + \cdots + 4y_{N-3} + 2y_{N-2} + 4y_{N-1} + y_N\big]$$

where $\Delta x = \dfrac{b-a}{N}$ and $y_j = f(a + j\,\Delta x)$.

- T_N is equal to the sum of the areas of the trapezoids obtained by connecting the points $(x_0, y_0), (x_1, y_1), \ldots, (x_N, y_N)$ with line segments.
- M_N has two geometric interpretations, either as the sum of the areas of the midpoint rectangles or as the sum of the areas of the tangential trapezoids.
- S_N is equal to $\frac{1}{3}T_{N/2} + \frac{2}{3}M_{N/2}$.
- Error Bounds:

$$\text{Error}(T_N) \le \frac{K_2(b-a)^3}{12N^2}, \qquad \text{Error}(M_N) \le \frac{K_2(b-a)^3}{24N^2}, \qquad \text{Error}(S_N) \le \frac{K_4(b-a)^5}{180N^4}$$

where K_2 is any number such that $|f''(x)| \le K_2$ for all $x \in [a, b]$ and K_4 is any number such that $|f^{(4)}(x)| \le K_4$ for all $x \in [a, b]$.

7.1 EXERCISES

Preliminary Questions

1. What are T_1 and T_2 for a function on $[0, 2]$ such that $f(0) = 3$, $f(1) = 4$, and $f(2) = 3$?

2. For which graph in Figure 16 will T_N *overestimate* the integral? What about M_N?

FIGURE 16

3. How large is the error when the Trapezoidal Rule is applied to a linear function? Explain graphically.

4. Suppose T_4 is used to approximate $\int_0^3 f(x)\,dx$, where $|f''(x)| \le 2$ for all x. What is the maximum possible error?

5. What are the two graphical interpretations of the Midpoint Rule?

Exercises

In Exercises 1–12, calculate T_N and M_N for the value of N indicated.

1. $\displaystyle\int_0^2 x^2\,dx, \quad N = 4$

2. $\displaystyle\int_0^4 \sqrt{x}\,dx, \quad N = 4$

3. $\displaystyle\int_1^4 x^3\,dx, \quad N = 6$

4. $\displaystyle\int_0^{\pi/2} \sqrt{\sin x}\,dx, \quad N = 8$

5. $\displaystyle\int_1^4 \frac{dx}{x}, \quad N = 6$

6. $\displaystyle\int_{-2}^{-1} \frac{dx}{x}, \quad N = 5$

7. $\displaystyle\int_0^{\pi/4} \sec x\,dx, \quad N = 6$

8. $\displaystyle\int_1^2 \ln x, \quad N = 5$

9. $\displaystyle\int_2^3 \frac{dx}{\ln x}, \quad N = 5$

10. $\displaystyle\int_0^1 e^{-x^2}\,dx, \quad N = 6$

11. $\displaystyle\int_0^2 \frac{e^x}{x+1}\,dx, \quad N = 8$

12. $\displaystyle\int_{-2}^1 e^{x^2}\,dx, \quad N = 6$

In Exercises 13–22, calculate S_N given by Simpson's Rule for the value of N indicated.

13. $\displaystyle\int_0^4 \sqrt{x}\,dx, \quad N = 4$

14. $\displaystyle\int_3^5 (9 - x^2)\,dx, \quad N = 4$

15. $\displaystyle\int_0^2 \frac{dx}{x^4+1}, \quad N = 4$

16. $\displaystyle\int_0^1 \cos(x^2)\,dx, \quad N = 6$

17. $\displaystyle\int_0^1 e^{-x^2}\,dx, \quad N = 6$

18. $\displaystyle\int_1^2 e^{-x}\,dx, \quad N = 6$

19. $\displaystyle\int_1^4 \ln x\,dx, \quad N = 8$

20. $\displaystyle\int_2^4 \sqrt{x^4+1}\,dx, \quad N = 8$

21. $\displaystyle\int_0^{\pi/4} \sec x\,dx, \quad N = 10$

22. $\displaystyle\int_2^4 \sqrt{x^4+1}\,dx, \quad N = 10$

In Exercises 23–26, calculate the approximation to the volume of the solid obtained by rotating the graph around the given axis.

23. $y = \cos x; \quad [0, \frac{\pi}{2}]; \quad x\text{-axis}; \quad M_8$

24. $y = \cos x; \quad [0, \frac{\pi}{2}]; \quad y\text{-axis}; \quad S_8$

25. $y = e^{-x^2}; \quad [0, 1]; \quad x\text{-axis}; \quad T_8$

26. $y = e^{-x^2}; \quad [0, 1]; \quad y\text{-axis}; \quad S_8$

27. Use S_8 to estimate $\int_0^{\pi/2} \frac{\sin x}{x} \, dx$, taking the value of $(\sin x)/x$ at $x = 0$ to be 1.

28. Calculate T_6 for the integral $I = \int_0^2 x^3 \, dx$.

(a) Is T_6 too large or too small? Explain graphically.

(b) Show that $K_2 = |f''(2)|$ may be used in the Error Bound and find a bound for the error.

(c) Evaluate I and check that the actual error is less than the bound computed in (b).

29. Calculate M_6 for the integral $I = \int_0^{\pi/2} \cos x \, dx$.

(a) Is M_6 too large or too small? Explain graphically.

(b) Show that $K_2 = 1$ may be used in the Error Bound and find a bound for the error.

(c) Evaluate I and check that the actual error is less than the bound computed in (b).

In Exercises 30–33, state whether T_N or M_N underestimates or overestimates the integral and find a bound for the error (but do not calculate T_N or M_N).

30. $\int_1^4 \frac{1}{x} \, dx, \quad T_{10}$

31. $\int_0^2 e^{-x/4} \, dx, \quad T_{20}$

32. $\int_1^4 \ln x \, dx, \quad M_{10}$

33. $\int_0^{\pi/4} \cos x, \quad M_{20}$

CAS *In Exercises 34–37, use the Error Bound to find a value of N for which $Error(T_N) \le 10^{-6}$. If you have a computer algebra system, calculate the corresponding approximation and confirm that the error satisfies the required bound.*

34. $\int_0^1 x^4 \, dx$

35. $\int_0^{\pi/6} \cos x \, dx$

36. $\int_2^5 \frac{1}{x} \, dx$

37. $\int_0^3 e^{-x} \, dx$

38. Find a bound for the error in the approximations T_{10} and M_{10} to $\int_0^5 (x^3 + 1)^{-1} \, dx$ (use Figure 17 to determine a value of K_2). Then find a value of N such that the error in M_N is at most 10^{-6}.

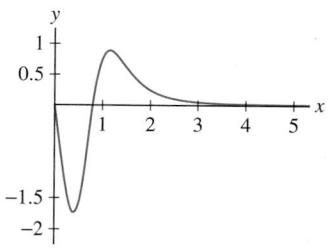

FIGURE 17 Graph of $f''(x)$, where $f(x) = (x^3 + 1)^{-1}$.

39. (a) Compute S_6 for the integral $I = \int_0^1 e^{-2x} \, dx$.

(b) Show that $K_4 = 16$ may be used in the Error Bound and find a bound for the error.

(c) Evaluate I and check that the actual error is less than the bound for the error computed in (b).

40. Calculate S_8 for the integral $\int_0^1 x^4 \, dx$. Use the Error Bound to find a bound for the error and verify that the actual error satisfies this bound.

41. Calculate S_8 for $\int_1^5 \ln x \, dx$ and find a bound for the error. Then find a value of N such that S_N has an error of at most 10^{-6}.

42. Find a bound for the error in the approximation S_{10} to $\int_0^3 e^{-x^2} \, dx$ (use Figure 18 to determine a value of K_4). Then find a value of N such that S_N has an error of at most 10^{-6}.

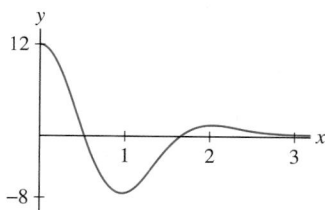

FIGURE 18 Graph of $f^{(4)}(x)$, where $f(x) = e^{-x^2}$.

43. CAS Use a computer algebra system to compute and graph $f^{(4)}(x)$ for $f(x) = \sqrt{1 + x^4}$ and find a bound for the error in the approximation S_{40} to $\int_0^5 f(x) \, dx$.

44. CAS Use a computer algebra system to compute and graph $f^{(4)}(x)$ for $f(x) = \tan x - \sec x$ and find a bound for the error in the approximation S_{40} to $\int_0^{\pi/4} f(x) \, dx$.

In Exercises 45–48, use the Error Bound to find a value of N for which $Error(S_N) \le 10^{-9}$.

45. $\int_1^7 x^{3/2} \, dx$

46. $\int_0^4 xe^x \, dx$

47. $\int_0^1 e^{x^2} \, dx$

48. $\int_1^4 \sin(\ln x) \, dx$

49. CAS Show that $\int_0^1 \frac{dx}{1 + x^2} = \frac{\pi}{4}$ [use Eq. (3) in Section 5.7].

(a) Use a computer algebra system to graph $f^{(4)}(x)$ for $f(x) = (1 + x^2)^{-1}$ and find its maximum on $[0, 1]$.

(b) Find a value of N such that S_N approximates the integral with an error of at most 10^{-6}. Calculate the corresponding approximation and confirm that you have computed $\frac{\pi}{4}$ to at least four places.

50. Let $f(x) = \sin(x^2)$ and $I = \int_0^1 f(x)\,dx$.

(a) Check that $f''(x) = 2\cos(x^2) - 4x^2\sin(x^2)$. Then show that $|f''(x)| \le 6$ for $x \in [0, 1]$. *Hint:* Note that $|2\cos(x^2)| \le 2$ and $|4x^2\sin(x^2)| \le 4$ for $x \in [0, 1]$.

(b) Show that Error(M_N) is at most $\dfrac{1}{4N^2}$.

(c) Find an N such that $|I - M_N| \le 10^{-3}$.

51. *CAS* 📖 The Error Bound for M_N is proportional to $1/N^2$, so the Error Bound decreases by $\frac{1}{4}$ if N is increased to $2N$. Compute the actual error in M_N for $\int_0^\pi \sin x\,dx$ for $N = 4, 8, 16, 32$, and 64. Does the actual error seem to decrease by $\frac{1}{4}$ as N is doubled?

52. *CAS* 📖 Observe that the Error Bound for T_N (which has 12 in the denominator) is twice as large as the Error Bound for M_N (which has 24 in the denominator). Compute the actual error in T_N for $\int_0^\pi \sin x\,dx$ for $N = 4, 8, 16, 32$, and 64 and compare with the calculations of Exercise 51. Does the actual error in T_N seem to be roughly twice as large as the error in M_N in this case?

53. *CAS* 📖 Explain why the Error Bound for S_N decreases by $\frac{1}{16}$ if N is increased to $2N$. Compute the actual error in S_N for $\int_0^\pi \sin x\,dx$ for $N = 4, 8, 16, 32$, and 64. Does the actual error seem to decrease by $\frac{1}{16}$ as N is doubled?

54. An airplane's velocity is recorded at 5-min intervals during a 1-hour period with the following results, in mph:

$$550, 575, 600, 580, 610, 640, 625, 595, 590, 620, 640, 640, 630$$

Use Simpson's Rule to estimate the distance traveled during the hour.

55. Use Simpson's Rule to determine the average temperature in a museum over a 3-hour period, if the temperatures (in degrees Celsius), recorded at 15-min intervals, are

$$21, 21.3, 21.5, 21.8, 21.6, 21.2, 20.8,$$
$$20.6, 20.9, 21.2, 21.1, 21.3, 21.2$$

56. 📖 **Tsunami Arrival Times** Scientists estimate the arrival times of tsunamis (seismic ocean waves) based on the point of origin P and ocean depths. The speed s of a tsunami in miles per hour is approximately $s = \sqrt{15d}$, where d is the ocean depth in feet.

(a) Let $f(x)$ be the ocean depth x miles from P (in the direction of the coast). Argue using Riemann sums that the time T required for the tsunami to travel M miles toward the coast is

$$T = \int_0^M \frac{dx}{\sqrt{15f(x)}}$$

(b) Use Simpson's Rule to estimate T if $M = 1{,}000$ and the ocean depths (in feet), measured at 100-mile intervals starting from P, are

$$13{,}000,\ 11{,}500,\ 10{,}500,\ 9{,}000,\ 8{,}500,\ 7{,}000,$$
$$6{,}000,\ 4{,}400,\ 3{,}800,\ 3{,}200,\ 2{,}000$$

Further Insights and Challenges

57. Show that $T_N = \int_a^b f(x)\,dx$ for all N and all endpoints a, b if $f(x) = rx + s$ is a linear function (r, s constants).

58. Show that if $f(x) = px^2 + qx + r$ is a quadratic polynomial, then $S_2 = \int_a^b f(x)\,dx$. In other words, show that

$$\int_a^b f(x)\,dx = \frac{b-a}{6}(y_0 + 4y_1 + y_2)$$

where $y_0 = f(a)$, $y_1 = f\left(\dfrac{a+b}{2}\right)$, and $y_2 = f(b)$. *Hint:* Show this first for $f(x) = 1, x, x^2$ and use linearity.

59. For N even, divide $[a, b]$ into N subintervals of width $\Delta x = \dfrac{b-a}{N}$. Set $x_j = a + j\,\Delta x$, $y_j = f(x_j)$, and

$$S_2^{2j} = \frac{b-a}{3N}(y_{2j} + 4y_{2j+1} + y_{2j+2})$$

(a) Show that S_N is the sum of the approximations on the intervals $[x_{2j}, x_{2j+2}]$, that is, $S_N = S_2^0 + S_2^2 + \cdots + S_2^{N-2}$.

(b) By Exercise 58, $S_2^{2j} = \int_{x_{2j}}^{x_{2j+2}} f(x)\,dx$ if $f(x)$ is a quadratic polynomial. Use (a) to show that S_N is exact *for all N* if $f(x)$ is a quadratic polynomial.

60. Show that S_2 also gives the exact value for $\int_a^b x^3\,dx$ and conclude, as in Exercise 59, that S_N is exact for all cubic polynomials. Show by counterexample that S_2 is not exact for integrals of x^4.

61. Use the Error Bound for S_N to obtain another proof that Simpson's Rule is exact for all cubic polynomials.

62. 📖 **Sometimes, Simpson's Rule Performs Poorly** Calculate M_{10} and S_{10} for the integral $\int_0^1 \sqrt{1 - x^2}\,dx$, whose value we know to be $\frac{\pi}{4}$ (one-quarter of the area of the unit circle).

(a) We usually expect S_N to be more accurate than M_N. Which of M_{10} and S_{10} is more accurate in this case?

(b) How do you explain the result of part (a)? *Hint:* The Error Bounds are not valid because $|f''(x)|$ and $|f^{(4)}(x)|$ tend to ∞ as $x \to 1$, but $|f^{(4)}(x)|$ goes to infinity faster.

7.2 Integration by Parts

The Integration by Parts formula is derived from the Product Rule:

$$\big(u(x)v(x)\big)' = u(x)v'(x) + u'(x)v(x)$$

This formula states that $u(x)v(x)$ is an antiderivative of the right-hand side, so

$$u(x)v(x) = \int u(x)v'(x)\,dx + \int u'(x)v(x)\,dx$$

Combining the term on the left with the second integral on the right, we obtain

> *The Integration by Parts formula is often written using differentials:*
>
> $$\int u\,dv = uv - \int v\,du$$
>
> *Recall that $dv = v'(x)\,dx$ and $du = u'(x)\,dx$.*

Integration by Parts Formula

$$\int u(x)v'(x)\,dx = u(x)v(x) - \int u'(x)v(x)\,dx \qquad \boxed{1}$$

Note that the Integration by Parts formula applies to a product $u(x)v'(x)$, so we should consider using it when the integrand is a product of two functions.

■ **EXAMPLE 1** Evaluate $\displaystyle\int x \cos x\,dx$.

Solution The integrand is the product $x \cos x$, so we try $u(x) = x$ and $v'(x) = \cos x$. With this choice, the integral has the required form:

$$\int x \cos x\,dx = \int u(x)v'(x)\,dx$$

> *In applying Eq. (1), any antiderivative $v(x)$ of $v'(x)$ may be used.*

We apply Integration by Parts, observing that $u'(x) = 1$ and $v(x) = \sin x$ (since $v'(x) = \cos x$):

$$\int \underbrace{x \cos x}_{uv'}\,dx = \underbrace{x \sin x}_{uv} - \int \underbrace{\sin x}_{u'v}\,dx = x \sin x + \cos x + C$$

Let's check the answer by taking the derivative:

$$\frac{d}{dx}(x \sin x + \cos x + C) = x \cos x + \sin x - \sin x = x \cos x \qquad ■$$

The key step in Integration by Parts is deciding how to write the integrand as a product uv'. Keep in mind that Integration by Parts expresses $\displaystyle\int uv'\,dx$ in terms of $\displaystyle\int u'v\,dx$. *This is useful if $u'v$ is easier to integrate than uv'.* Here are two guidelines:

- Choose u so that u' is "simpler" than u itself.
- Choose v' so that $v = \displaystyle\int v'\,dx$ can be evaluated.

■ **EXAMPLE 2** Good Versus Bad Choices of u and v' Evaluate $\int xe^x\,dx$.

Solution Based on our guidelines, it makes sense to let

- $u = x$ (since $u' = 1$ is simpler)
- $v' = e^x$ (since we can evaluate $v = \int e^x\,dx = e^x + C$)

Integration by Parts gives us

$$\int xe^x\,dx = u(x)v(x) - \int u'(x)v(x)\,dx = xe^x - \int e^x\,dx = xe^x - e^x + C$$

Let's see what would have happened if we had chosen $u = e^x$, $v' = x$. Then

$$u'(x) = e^x, \qquad v(x) = \int x\,dx = \frac{1}{2}x^2 + C$$

$$\int \underbrace{xe^x}_{uv'}\,dx = \underbrace{\tfrac{1}{2}x^2e^x}_{uv} - \boxed{\int \underbrace{\tfrac{1}{2}x^2e^x}_{u'v}\,dx}$$

This is a poor choice because the integral on the right is more complicated than our original integral. ■

The Integration by Parts formula applies to *definite integrals*:

$$\int_a^b u(x)v'(x)\,dx = u(x)v(x)\Big|_a^b - \int_a^b u'(x)v(x)\,dx$$

■ **EXAMPLE 3** Taking $v' = 1$ Evaluate $\int_1^3 \ln x\,dx$.

Surprisingly, the choice $v' = 1$ is effective in some cases. Using it as in Example 3, we find that

$$\int \ln x\,dx = x\ln x - x + C$$

This trick also works for the inverse trigonometric functions (see Exercise 6).

Solution At first glance, this integral does not appear to be a candidate for Integration by Parts because the integrand is not a product. However, we are free to add a factor of 1 and write $\ln x = (\ln x)\cdot 1 = uv'$. Then

$$u = \ln x, \qquad v' = 1$$
$$u' = x^{-1}, \qquad v = x$$

Noting that $u'v = 1$, we obtain

$$\int_1^3 \ln x\,dx = \int_1^3 \underbrace{(\ln x)\cdot 1}_{uv'}\,dx = x\ln x\Big|_1^3 - \int_1^3 1\,dx = (3\ln 3 - 0) - 2 = 3\ln 3 - 2 \ ■$$

■ **EXAMPLE 4** Integrating by Parts More Than Once Evaluate $\int x^2\cos x\,dx$.

In Example 4, it makes sense to take $u = x^2$ because Integration by Parts reduces the integration of $x^2\cos x$ to the integration of $2x\sin x$, which is easier.

Solution We use Integration by Parts with $u = x^2$ and $v' = \cos x$:

$$\int \underbrace{x^2\cos x}_{uv'}\,dx = \underbrace{x^2\sin x}_{uv} - \int \underbrace{(2x)(\sin x)}_{u'v}\,dx = x^2\sin x - 2\int x\sin x\,dx \qquad \boxed{2}$$

We apply Integration by Parts a second time to evaluate $\int x \sin x \, dx$. Taking $u = x$ and $v' = \sin x$, we obtain

$$\int \underbrace{x \sin x}_{uv'} \, dx = \underbrace{-x \cos x}_{uv} - \int \underbrace{(-\cos x)}_{u'v} \, dx = -x \cos x + \sin x + C$$

Now use this result in Eq. (2):

$$\int x^2 \cos x \, dx = x^2 \sin x - 2(-x \cos x + \sin x) + C$$

$$= x^2 \sin x + 2x \cos x - 2 \sin x + C$$

■ **EXAMPLE 5** Going in a Circle? Evaluate $\int e^x \cos x \, dx$.

We can also carry out the computation in Example 5 using the choice $u = e^x$, $v' = \cos x$.

Solution There are two natural ways of writing $e^x \cos x$ as uv'. Let's try the following:

$$u = \cos x, \qquad v' = e^x$$
$$u' = -\sin x, \qquad v = e^x$$

Integration by Parts yields

$$\int \underbrace{e^x \cos x}_{uv'} \, dx = \underbrace{e^x \cos x}_{uv} + \int \underbrace{e^x \sin x}_{-u'v} \, dx \qquad \boxed{3}$$

Now use Integration by Parts again to evaluate $\int e^x \sin x \, dx$:

$$u = \sin x, \qquad v' = e^x$$
$$u' = \cos x, \qquad v = e^x$$

$$\int e^x \sin x \, dx = e^x \sin x - \int e^x \cos x \, dx \qquad \boxed{4}$$

Equation (4) brings us back to our original integral of $e^x \cos x$, so it may seem that we're going in a circle. Fortunately, we can combine Eq. (4) with Eq. (3) and solve

$$\int e^x \cos x \, dx = e^x \cos x + e^x \sin x - \int e^x \cos x \, dx$$

$$2 \int e^x \cos x \, dx = e^x \cos x + e^x \sin x + C$$

$$\int e^x \cos x \, dx = \frac{1}{2} e^x (\cos x + \sin x) + C$$

Integration by Parts can be used to derive **reduction formulas** for integrals that depend on a positive integer n such as $\int x^n e^x \, dx$ or $\int \ln^n x \, dx$. A reduction formula (also called a **recursive formula**) expresses the integral for a given value of n in terms of a similar integral for a smaller value of n. The desired integral is evaluated by applying the reduction formula repeatedly.

■ **EXAMPLE 6** **A Reduction Formula** Verify the reduction formula

$$\boxed{\int x^n e^x \, dx = x^n e^x - n \int x^{n-1} e^x \, dx} \qquad \boxed{5}$$

Then use the reduction formula to evaluate $\int x^3 e^x \, dx$.

Solution We derive the reduction formula using Integration by Parts once:

$$u = x^n, \qquad v' = e^x$$

$$u' = nx^{n-1}, \qquad v = e^x$$

$$\int x^n e^x \, dx = uv - \int u'v \, dx = x^n e^x - n \int x^{n-1} e^x \, dx$$

To evaluate $\int x^3 e^x \, dx$, we'll need to use the reduction formula for $n = 3, 2, 1$:

Notice that $\int x^3 e^x \, dx$ is equal to $P(x)e^x + C$, where $P(x)$ is a polynomial of degree 3. In general, $\int x^n e^x \, dx = Q(x)e^x + C$, where $Q(x)$ is a polynomial of degree n.

$$\int x^3 e^x \, dx = x^3 e^x - 3 \int x^2 e^x \, dx$$

$$= x^3 e^x - 3 \left(x^2 e^x - 2 \int x e^x \, dx \right) = x^3 e^x - 3x^2 e^x + 6 \int x e^x \, dx$$

$$= x^3 e^x - 3x^2 e^x + 6 \left(x e^x - \int e^x \, dx \right) = x^3 e^x - 3x^2 e^x + 6x e^x - 6e^x + C$$

$$= (x^3 - 3x^2 + 6x - 6)e^x + C \qquad ■$$

7.2 SUMMARY

- *Integration by Parts* formula: $\int u(x)v'(x) \, dx = u(x)v(x) - \int u'(x)v(x) \, dx$.
- The key step in applying Integration by Parts is deciding how to write the integrand as a product uv'. Keep in mind that Integration by Parts is useful when $u'v$ is easier (or, at least, not more difficult) to integrate than uv'. Here are some guidelines:
  - Choose u so that u' is simpler than u itself.
  - Choose v' so that $v = \int v' \, dx$ can be evaluated.
  - Sometimes, $v' = 1$ is a good choice.

7.2 EXERCISES

Preliminary Questions

1. Which derivative rule is used to derive the Integration by Parts formula?

2. For each of the following integrals, state whether substitution or Integration by Parts should be used:

$$\int x \cos(x^2) \, dx, \qquad \int x \cos x \, dx, \qquad \int x^2 e^x \, dx, \qquad \int x e^{x^2} \, dx$$

3. Why is $u = \cos x$, $v' = x$ a poor choice for evaluating $\int x \cos x \, dx$?

Exercises

In Exercises 1–6, evaluate the integral using the Integration by Parts formula with the given choice of u and v′.

1. $\int x \sin x \, dx; \quad u = x, \ v' = \sin x$

2. $\int x e^{2x} \, dx; \quad u = x, \ v' = e^{2x}$

3. $\int (2x + 9)e^x \, dx; \quad u = 2x + 9, \ v' = e^x$

4. $\int x \cos(4x) \, dx; \quad u = x, \ v' = \cos(4x)$

5. $\int x^3 \ln x \, dx; \quad u = \ln x, \ v' = x^3$

6. $\int \tan^{-1} x \, dx; \quad u = \tan^{-1} x, \ v' = 1$

In Exercises 7–32, use Integration by Parts to evaluate the integral.

7. $\int (3x - 1)e^{-x} \, dx$

8. $\int x e^{-x} \, dx$

9. $\int x^2 e^x \, dx$

10. $\int x^2 \sin x \, dx$

11. $\int x \cos 2x \, dx$

12. $\int x^2 \sin(3x + 1) \, dx$

13. $\int e^{-x} \sin x \, dx$

14. $\int e^{4x} \cos 3x \, dx$

15. $\int x \ln x \, dx$

16. $\int \frac{\ln x}{x^2} \, dx$

17. $\int x^{-9} \ln x \, dx$

18. $\int e^x \sin(2x) \, dx$

19. $\int x \cos(2 - x) \, dx$

20. $\int x^2 \ln x \, dx$

21. $\int x \, 2^x \, dx$

22. $\int x \sec^2 x \, dx$

23. $\int (\ln x)^2 \, dx$

24. $\int \cos^{-1} x \, dx$

25. $\int \sin^{-1} x \, dx$

26. $\int \sec^{-1} x \, dx$

27. $\int x \, 5^x \, dx$

28. $\int (\sin x) 5^x \, dx$

29. $\int x \cosh 2x \, dx$

30. $\int \tanh^{-1}(4x) \, dx$

31. $\int \sinh^{-1} x \, dx$

32. $\int (\cos x)(\cosh x) \, dx$

33. Use the substitution $u = x^{1/2}$ and then Integration by Parts to evaluate $\int e^{\sqrt{x}} \, dx$.

34. Use substitution and then Integration by Parts to evaluate $\int x^3 e^{x^2} \, dx$.

In Exercises 35–44, evaluate using Integration by Parts, substitution, or both if necessary.

35. $\int x \cos 4x \, dx$

36. $\int \frac{\ln(\ln x) \, dx}{x}$

37. $\int \frac{x \, dx}{\sqrt{x + 1}}$

38. $\int \sin(\ln x) \, dx$

39. $\int \cos x \ln(\sin x) \, dx$

40. $\int x^3 (x^2 + 1)^{12} \, dx$

41. $\int \sin \sqrt{x} \, dx$

42. $\int \sqrt{x} e^{\sqrt{x}} \, dx$

43. $\int \frac{\ln(\ln x) \ln x \, dx}{x}$

44. $\int x \tan^{-1} x \, dx$

In Exercises 45–50, compute the definite integral.

45. $\int_0^2 x e^{9x} \, dx$

46. $\int_1^3 \ln x \, dx$

47. $\int_0^4 x \sqrt{4 - x} \, dx$

48. $\int_0^{\pi/4} x \sin(2x) \, dx$

49. $\int_1^4 \sqrt{x} \ln x \, dx$

50. $\int_0^1 \tan^{-1} x \, dx$

51. Use Eq. (5) to evaluate $\int x^4 e^x \, dx$.

52. Use substitution and then Eq. (5) to evaluate $\int x^4 e^{7x} \, dx$.

53. Find a reduction formula for $\int x^n e^{-x} \, dx$ similar to Eq. (5).

54. Evaluate $\int x^n \ln x \, dx$ for $n \neq -1$. Which method should be used to evaluate $\int x^{-1} \ln x \, dx$?

In Exercises 55–62, indicate a good method for evaluating the integral (but do not evaluate). Your choices are algebraic manipulation, substitution (specify u and du), and Integration by Parts (specify u and v′). If it appears that the techniques you have learned thus far are not sufficient, state this.

55. $\int \sqrt{x} \ln x \, dx$

56. $\int \frac{x^2 - \sqrt{x}}{2x} \, dx$

57. $\int \frac{x^3}{\sqrt{4 - x^2}} \, dx$

58. $\int \frac{dx}{\sqrt{4 - x^2}}$

59. $\int \frac{2x + 3}{x^2 + 3x + 6} \, dx$

60. $\int \frac{1}{x^2 + 3x + 6} \, dx$

61. $\int x \sin(3x + 4) \, dx$

62. $\int x \cos 3x \, dx$

63. Evaluate $\int (\sin^{-1} x)^2 \, dx$. *Hint:* First use Integration by Parts and then substitution.

64. Evaluate $\int \dfrac{(\ln x)^2 \, dx}{x^2}$. *Hint:* First use substitution and then Integration by Parts.

65. Evaluate $\int x^7 \cos(x^4) \, dx$.

66. Find $f(x)$, assuming that

$$\int f(x)e^x \, dx = f(x)e^x - \int x^{-1}e^x \, dx$$

67. Find the volume of the solid obtained by revolving $y = e^x$ for $0 \le x \le 2$ about the y-axis.

68. Find the volume of the solid obtained by revolving $y = \cos x$ for $0 \le x \le \frac{\pi}{2}$ about the y-axis.

69. Recall that the *present value* of an investment which pays out income continuously at a rate $R(t)$ for T years is $\int_0^T R(t)e^{-rt} \, dt$, where

r is the interest rate. Find the present value if income is produced at a rate $R(t) = 5,000 + 100t$ dollars/year for 10 years.

70. Prove the reduction formula

$$\int (\ln x)^k \, dx = x(\ln x)^k - k \int (\ln x)^{k-1} \, dx \qquad \boxed{6}$$

71. Use Eq. (6) to calculate $\int (\ln x)^k \, dx$ for $k = 2, 3$.

72. Prove the reduction formulas

$$\int x^n \cos x \, dx = x^n \sin x - n \int x^{n-1} \sin x \, dx$$

$$\int x^n \sin x \, dx = -x^n \cos x + n \int x^{n-1} \cos x \, dx$$

73. Prove $\int x b^x \, dx = b^x \left(\dfrac{x}{\ln b} - \dfrac{1}{\ln^2 b} \right) + C$.

Further Insights and Challenges

74. The Integration by Parts formula can be written

$$\int u(x)v(x) \, dx = u(x)V(x) - \int u'(x)V(x) \, dx \qquad \boxed{7}$$

where $V(x)$ satisfies $V'(x) = v(x)$.

(a) Show directly that the right-hand side of Eq. (7) does not change if $V(x)$ is replaced by $V(x) + C$, where C is a constant.

(b) Use $u = \tan^{-1} x$ and $v = x$ in Eq. (7) to calculate $\int x \tan^{-1} x \, dx$, but carry out the calculation twice: first with $V(x) = \frac{1}{2}x^2$ and then with $V(x) = \frac{1}{2}x^2 + \frac{1}{2}$. Which choice of $V(x)$ results in a simpler calculation?

75. Prove in two ways

$$\int_0^a f(x) \, dx = af(a) - \int_0^a x f'(x) \, dx \qquad \boxed{8}$$

First use Integration by Parts. Then assume $f(x)$ is increasing. Use the substitution $u = f(x)$ to prove that $\int_0^a x f'(x) \, dx$ is equal to the area of the shaded region in Figure 1 and derive Eq. (8) a second time.

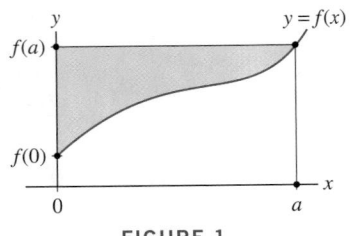

FIGURE 1

76. Assume that $f(0) = f(1) = 0$ and that f'' exists. Prove

$$\int_0^1 f''(x) f(x) \, dx = - \int_0^1 f'(x)^2 \, dx \qquad \boxed{9}$$

Use this to prove that if $f(0) = f(1) = 0$ and $f''(x) = \lambda f(x)$ for some constant λ, then $\lambda < 0$. Can you think of a function satisfying these conditions for some λ?

77. Set $I(a, b) = \int_0^1 x^a (1 - x)^b \, dx$, where a, b are whole numbers.

(a) Use substitution to show that $I(a, b) = I(b, a)$.

(b) Show that $I(a, 0) = I(0, a) = \dfrac{1}{a + 1}$.

(c) Prove that for $a \ge 1$ and $b \ge 0$,

$$I(a, b) = \dfrac{a}{b + 1} I(a - 1, b + 1)$$

(d) Use (b) and (c) to calculate $I(1, 1)$ and $I(3, 2)$.

(e) Show that $I(a, b) = \dfrac{a! \, b!}{(a + b + 1)!}$.

78. Show by differentiation that if $P_n(x)$ is a polynomial of degree n satisfying $P_n(x) + P_n'(x) = x^n$, then

$$\int x^n e^x \, dx = P_n(x) e^x + C$$

Find $P_n(x)$ for $n = 2, 3, 4$.

79. Let $I_n = \int x^n \cos(x^2) \, dx$ and $J_n = \int x^n \sin(x^2) \, dx$.

(a) Find a reduction formula that expresses I_n in terms of J_{n-2}. *Hint:* Write $x^n \cos(x^2)$ as $x^{n-1}(x \cos(x^2))$.

(b) 🖉 Use the result of (a) to show that I_n can be evaluated explicitly if n is odd.

(c) Evaluate I_3.

7.3 Trigonometric Integrals

In this section, we combine substitution and Integration by Parts with the appropriate identities to integrate various trigonometric functions. First, consider

$$\int \sin^m x \cos^n x \, dx$$

where n, m are positive integers. The easier case occurs when one or both of the exponents m, n is *odd*.

■ **EXAMPLE 1** Odd Power of $\sin x$ Evaluate $\int \sin^3 x \, dx$.

Solution First use the identity $\sin^2 x = 1 - \cos^2 x$ to rewrite the integrand:

$$\sin^3 x = \sin^2 x \sin x = (1 - \cos^2 x) \sin x$$

$$\int \sin^3 x \, dx = \int (1 - \cos^2 x) \sin x \, dx$$

Then use the substitution $u = \cos x$, $du = -\sin x \, dx$:

$$\int \sin^3 x \, dx = \int (1 - \cos^2 x) \sin x \, dx = -\int (1 - u^2) \, du \qquad \boxed{1}$$

$$= \frac{u^3}{3} - u + C = \frac{\cos^3 x}{3} - \cos x + C \qquad \boxed{2}$$

■

Sine or Cosine to Odd Power

If $\cos x$ occurs to an odd power $2k + 1$, write $\cos^{2k+1} x$ as $(1 - \sin^2 x)^k \cos x$. Then

$$\int \sin^m x \cos^{2k+1} x \, dx \text{ becomes}$$

$$\int (\sin^m x)(1 - \sin^2 x)^k \cos x \, dx$$

Now substitute $u = \sin x$, $du = \cos x \, dx$. If $\sin x$ occurs to an odd power $2k + 1$, write $\sin^{2k+1} x$ as $(1 - \cos^2 x)^k \sin x$ and substitute $u = \cos x$, $du = -\sin x \, dx$.

In general, if n is odd, we may use the identity $\cos^2 x = 1 - \sin^2 x$ to write $\cos^n x$ as a power of $(1 - \sin^2 x)$ times $\cos x$ and substitute $u = \sin x$, $du = \cos x \, dx$. Alternatively, if m is odd, we write $\sin^m x$ as a power of $(1 - \cos^2 x)$ times $\sin x$.

■ **EXAMPLE 2** Odd Power of $\sin x$ or $\cos x$ Evaluate $\int \sin^4 x \cos^5 x \, dx$.

Solution Write $\cos^5 x = \cos^4 x \cdot \cos x = (1 - \sin^2 x)^2 \cos x$. Then

$$\int \sin^4 x \cos^5 x \, dx = \int (\sin^4 x)(1 - \sin^2 x)^2 \cos x \, dx$$

The factor of $\cos x$ in the integrand allows us to use the substitution $u = \sin x$. We have $(\sin^4 x)(1 - \sin^2 x)^2 = u^4 (1 - u^2)^2$ and the leftover factor $\cos x \, dx$ is equal to du:

$$\int \sin^4 x \cos^5 x \, dx = \int u^4 (1 - u^2)^2 \, du = \int (u^4 - 2u^6 + u^8) \, du$$

$$= \frac{u^5}{5} - \frac{2u^7}{7} + \frac{u^9}{9} + C = \frac{\sin^5 x}{5} - \frac{2 \sin^7 x}{7} + \frac{\sin^9 x}{9} + C \qquad ■$$

Since the method of previous examples require an odd exponent, we state the following reduction formulas that can be used for any exponent, even or odd. They are derived using Integration by Parts (see Exercise 62).

Reduction Formulas for Sine and Cosine

$$\int \sin^n x \, dx = -\frac{1}{n} \sin^{n-1} x \cos x + \frac{n-1}{n} \int \sin^{n-2} x \, dx \qquad \boxed{3}$$

$$\int \cos^n x \, dx = \frac{1}{n} \cos^{n-1} x \sin x + \frac{n-1}{n} \int \cos^{n-2} x \, dx \qquad \boxed{4}$$

■ **EXAMPLE 3** Evaluate $\int \sin^4 x \, dx$.

Solution By reduction formula (3) with $n = 4$, we obtain

$$\int \sin^4 x \, dx = -\frac{1}{4} \sin^3 x \cos x + \frac{3}{4} \int \sin^2 x \, dx \qquad \boxed{5}$$

To evaluate the integral on the right, we apply the reduction formula again with $n = 2$:

$$\int \sin^2 x \, dx = -\frac{1}{2} \sin x \cos x + \frac{1}{2} \int dx = -\frac{1}{2} \sin x \cos x + \frac{1}{2} x + C \qquad \boxed{6}$$

Using Eq. (6) in Eq. (5), we obtain

$$\int \sin^4 x \, dx = -\frac{1}{4} \sin^3 x \cos x - \frac{3}{8} \sin x \cos x + \frac{3}{8} x + C \qquad ■$$

Trigonometric integrals can be expressed in more than one way because of the large number of trigonometric identities. For example, a computer algebra system may yield the following evaluation of the integral in the previous example:

$$\int \sin^4 x \, dx = \frac{1}{32}(x - 8 \sin 2x + \sin 4x) + C$$

You can check that this agrees with the result in Example 3 (Exercise 55). The following formulas are verified using the identities recalled in the margin.

←·· REMINDER *Useful Identities:*

$$\sin^2 x = \frac{1}{2}(1 - \cos 2x)$$

$$\cos^2 x = \frac{1}{2}(1 + \cos 2x)$$

$$\sin 2x = 2 \sin x \cos x$$

$$\cos 2x = \cos^2 x - \sin^2 x$$

$$\int \sin^2 x \, dx = \frac{x}{2} - \frac{\sin 2x}{4} + C = \frac{x}{2} - \frac{1}{2} \sin x \cos x + C$$

$$\int \cos^2 x \, dx = \frac{x}{2} + \frac{\sin 2x}{4} + C = \frac{x}{2} + \frac{1}{2} \sin x \cos x + C$$

More work is required to integrate $\sin^m x \cos^n x$ when both n and m are even:

• If $m \le n$, use the identity $\sin^2 x = 1 - \cos^2 x$ to replace $\sin^m x$ by $(1 - \cos^2 x)^{m/2}$:

$$\int \sin^m x \cos^n x \, dx = \int (1 - \cos^2 x)^{m/2} \cos^n x \, dx$$

Expand the integral on the right to obtain a sum of integrals of powers of $\cos x$ and use reduction formula (4).

- If $m \geq n$, replace $\cos^n x$ by $(1 - \sin^2 x)^{n/2}$:

$$\int \sin^m x \cos^n x \, dx = \int (\sin^m x)(1 - \sin^2 x)^{n/2} \, dx$$

Expand the integral on the right to obtain a sum of integrals of powers of $\sin x$ and again evaluate using reduction formula (3).

Another method for integrating $\sin^m x \cos^n x$ when m and n are even is discussed in Exercise 63.

■ **EXAMPLE 4** Even Powers of $\sin x$ and $\cos x$ Evaluate $\displaystyle\int \sin^2 x \cos^4 x \, dx$.

Solution Since $m < n$, we replace $\sin^2 x$ by $1 - \cos^2 x$:

$$\int \sin^2 x \cos^4 x \, dx = \int (1 - \cos^2 x) \cos^4 x \, dx = \int \cos^4 x \, dx - \int \cos^6 x \, dx \quad \boxed{7}$$

The reduction formula for $n = 6$ gives

$$\int \cos^6 x \, dx = \frac{1}{6} \cos^5 x \sin x + \frac{5}{6} \int \cos^4 x \, dx$$

Using this result in Eq. (7), we obtain

$$\int \sin^2 x \cos^4 x \, dx = -\frac{1}{6} \cos^5 x \sin x + \frac{1}{6} \int \cos^4 x \, dx$$

Next evaluate $\displaystyle\int \cos^4 x \, dx$ using the reduction formulas for $n = 4$ and $n = 2$:

$$\int \cos^4 x \, dx = \frac{1}{4} \cos^3 x \sin x + \frac{3}{4} \int \cos^2 x \, dx$$

$$= \frac{1}{4} \cos^3 x \sin x + \frac{3}{4} \left(\frac{1}{2} \cos x \sin x + \frac{1}{2} x \right) + C$$

$$= \frac{1}{4} \cos^3 x \sin x + \frac{3}{8} \cos x \sin x + \frac{3}{8} x + C$$

As noted above, trigonometric integrals can be expressed in more than one way. According to Mathematica,

$$\int \sin^2 x \cos^4 x \, dx$$

$$= \tfrac{1}{16} x + \tfrac{1}{64} \sin 2x - \tfrac{1}{64} \sin 4x - \tfrac{1}{192} \sin 6x$$

Trigonometric identities can be used to show that this agrees with Eq. (8).

Altogether,

$$\int \sin^2 x \cos^4 x \, dx = -\frac{1}{6} \cos^5 x \sin x + \frac{1}{6} \left(\frac{1}{4} \cos^3 x \sin x + \frac{3}{8} \cos x \sin x + \frac{3}{8} x \right) + C$$

$$= -\frac{1}{6} \cos^5 x \sin x + \frac{1}{24} \cos^3 x \sin x + \frac{1}{16} \cos x \sin x + \frac{1}{16} x + C$$

$$\boxed{8}$$

■

Many other trigonometric integrals can be evaluated using the appropriate trigonometric identities combined with substitution or Integration by Parts.

■ **EXAMPLE 5** Integral of the Tangent and Secant Derive the formulas

$$\int \tan x \, dx = \ln |\sec x| + C, \qquad \int \sec x \, dx = \ln \left| \sec x + \tan x \right| + C$$

The integral $\int \sec x \, dx$ was first computed numerically in the 1590s by the English mathematician Edward Wright, decades before the invention of calculus. Although he did not invent the concept of an integral, Wright realized that the sums which approximate the integral hold the key to understanding the Mercator map projection, of great importance in sea navigation because it allowed sailors to reach their destinations along lines of fixed compass direction. The formula for the integral was first proved by James Gregory in 1668.

Solution To integrate $\tan x$, use the substitution $u = \cos x$, $du = -\sin x \, dx$:

$$\int \tan x \, dx = \int \frac{\sin x}{\cos x} \, dx = -\int \frac{du}{u} = -\ln |u| + C = -\ln |\cos x| + C$$

$$= \ln \frac{1}{|\cos x|} + C = \ln |\sec x| + C$$

To integrate $\sec x$, we employ a clever substitution $u = \sec x + \tan x$. Then

$$du = (\sec x \tan x + \sec^2 x) \, dx = (\sec x) \underbrace{(\tan x + \sec x)}_{u} \, dx$$

Dividing both sides by u, we obtain $\dfrac{du}{u} = \sec x \, dx$ and thus

$$\int \sec x \, dx = \int \frac{du}{u} = \ln |u| + C = \ln \left| \sec x + \tan x \right| + C \qquad \blacksquare$$

The table of integrals at the end of this section (page 438) contains a list of additional trigonometric integrals and reduction formulas.

■ **EXAMPLE 6** **Power of $\tan x$** Evaluate $\displaystyle\int_0^{\pi/4} \tan^3 x \, dx$.

Solution We use reduction formula (18) in the table with $k = 3$:

$$\int_0^{\pi/4} \tan^3 x \, dx = \frac{\tan^2 x}{2} \bigg|_0^{\pi/4} - \int_0^{\pi/4} \tan x \, dx = \left(\frac{1}{2} \tan^2 x - \ln |\sec x| \right) \bigg|_0^{\pi/4}$$

$$= \left(\frac{1}{2} \tan^2 \frac{\pi}{4} - \ln \left| \sec \frac{\pi}{4} \right| \right) - \left(\frac{1}{2} \tan^2 0 - \ln |\sec 0| \right)$$

$$= \left(\frac{1}{2} (1)^2 - \ln \sqrt{2} \right) - \left(\frac{1}{2} 0^2 - \ln |1| \right) = \frac{1}{2} - \ln \sqrt{2} \qquad \blacksquare$$

In the margin, we describe a method for integrating $\tan^m x \sec^n x$.

■ **EXAMPLE 7** **Using Reduction Formulas and the Table of Integrals** Evaluate

$$\int \tan^2 x \sec^3 x \, dx.$$

Solution Referring to the cases described in the marginal note, we see that this integral is covered by Case 3 since the integrand is $\tan^m x \sec^n x$, with $m = 2$ and $n = 3$. The first step is to use the identity $\tan^2 x = \sec^2 x - 1$:

$$\int \tan^2 x \sec^3 x \, dx = \int (\sec^2 x - 1) \sec^3 x \, dx = \int \sec^5 x \, dx - \int \sec^3 x \, dx \qquad \boxed{9}$$

Next we use the reduction formula (22) in the table with $m = 5$:

$$\int \sec^5 x \, dx = \frac{\tan x \sec^3 x}{4} + \frac{3}{4} \int \sec^3 x \, dx$$

$$\int \tan^2 x \sec^3 x \, dx = \left(\frac{\tan x \sec^3 x}{4} + \frac{3}{4} \int \sec^3 x \, dx \right) - \int \sec^3 x \, dx$$

$$= \frac{1}{4} \tan x \sec^3 x - \frac{1}{4} \int \sec^3 x \, dx \qquad \boxed{10}$$

Integrating $\tan^m x \sec^n x$

Case 1 $m = 2k + 1$ odd

Use the identity $\tan^2 x = \sec^2 x - 1$ to write $\tan^{2k+1} x \sec^n x$ as

$$(\sec^2 x - 1)^k (\sec^{n-1} x)(\sec x \tan x)$$

Then substitute $u = \sec x$, $du = \sec x \tan x \, dx$ to obtain an integral involving only powers of u.

Case 2 $n = 2k$ even

Use the identity $\sec^2 x = 1 + \tan^2 x$ to write $\tan^m x \sec^n x$ as

$$(\tan^m x)(1 + \tan^2 x)^{k-1} \sec^2 x$$

Then substitute $u = \tan x$, $du = \sec^2 x \, dx$ to obtain an integral involving only powers of u.

Case 3 m even and n odd

Use the identity $\tan^2 x = \sec^2 x - 1$ to write $\tan^m x \sec^n x$ as

$$(\sec^2 x - 1)^{m/2} \sec^n x$$

Expand to obtain an integral involving only powers of $\sec x$ and use the reduction formula (22).

Finally, we use the reduction formula (22) again with $m = 3$ and formula (21):

$$\int \sec^3 x\, dx = \frac{\tan x \sec x}{2} + \frac{1}{2}\int \sec x\, dx = \frac{1}{2}\tan x \sec x + \frac{1}{2}\ln|\sec x + \tan x| + C$$

Then Eq. (10) becomes

$$\int \tan^2 x \sec^3 x\, dx = \frac{1}{4}\tan x \sec^3 x - \frac{1}{4}\left(\frac{1}{2}\tan x \sec x + \frac{1}{2}\ln|\sec x + \tan x|\right) + C$$

$$= \frac{1}{4}\tan x \sec^3 x - \frac{1}{8}\tan x \sec x - \frac{1}{8}\ln|\sec x + \tan x| + C \quad \blacksquare$$

Formulas (25)–(27) in the table describe the integrals of products of the form $\sin mx \sin nx$, $\cos mx \cos nx$, and $\sin mx \cos nx$. These integrals play an important role in the theory of Fourier Series, a more advanced technique used extensively in engineering and physics.

■ **EXAMPLE 8** Integral of $\sin mx \cos nx$ Evaluate $\int_0^\pi \sin 4x \cos 3x\, dx$.

Solution   We apply (26), with $m = 4$ and $n = 3$:

$$\int_0^\pi \sin 4x \cos 3x\, dx = \left(-\frac{\cos(4-3)x}{2(4-3)} - \frac{\cos(4+3)x}{2(4+3)}\right)\Bigg|_0^\pi$$

$$= \left(-\frac{\cos x}{2} - \frac{\cos 7x}{14}\right)\Bigg|_0^\pi$$

$$= \left(\frac{1}{2} + \frac{1}{14}\right) - \left(-\frac{1}{2} - \frac{1}{14}\right) = \frac{8}{7} \quad \blacksquare$$

7.3 SUMMARY

- To integrate an odd power of $\sin x$ times $\cos^n x$, write

$$\int \sin^{2k+1} x \cos^n x\, dx = \int (1 - \cos^2 x)^k \cos^n x \sin x\, dx$$

Then use the substitution $u = \cos x$, $du = -\sin x\, dx$.
- To integrate an odd power of $\cos x$ times $\sin^m x$, write

$$\int \sin^m x \cos^{2k+1} x\, dx = \int (\sin^m x)(1 - \sin^2 x)^k \cos x\, dx$$

Then use the substitution $u = \sin x$, $du = \cos x\, dx$.
- If both $\sin x$ and $\cos x$ occur to an even power, write

$$\int \sin^m x \cos^n x\, dx = \int (1 - \cos^2 x)^{m/2} \cos^n x\, dx \quad (\text{if } m \le n)$$

$$\int \sin^m x \cos^n x\, dx = \int \sin^m x(1 - \sin^2 x)^{n/2}\, dx \quad (\text{if } m \ge n)$$

Expand the right-hand side to obtain a sum of powers of $\cos x$ or powers of $\sin x$. Then use the reduction formulas

$$\int \sin^n x \, dx = -\frac{1}{n} \sin^{n-1} x \cos x + \frac{n-1}{n} \int \sin^{n-2} x \, dx$$

$$\int \cos^n x \, dx = \frac{1}{n} \cos^{n-1} x \sin x + \frac{n-1}{n} \int \cos^{n-2} x \, dx$$

- The integral $\int \tan^m x \sec^n x \, dx$ can be evaluated by a substitution that depends on three cases: m odd, n even, or m even and n odd. See the marginal note on page 436.

TABLE OF TRIGONOMETRIC INTEGRALS

$$\int \sin^2 x \, dx = \frac{x}{2} - \frac{\sin 2x}{4} + C = \frac{x}{2} - \frac{1}{2} \sin x \cos x + C \qquad \boxed{11}$$

$$\int \cos^2 x \, dx = \frac{x}{2} + \frac{\sin 2x}{4} + C = \frac{x}{2} + \frac{1}{2} \sin x \cos x + C \qquad \boxed{12}$$

$$\int \sin^n x \, dx = -\frac{\sin^{n-1} x \cos x}{n} + \frac{n-1}{n} \int \sin^{n-2} x \, dx \qquad \boxed{13}$$

$$\int \cos^n x \, dx = \frac{\cos^{n-1} x \sin x}{n} + \frac{n-1}{n} \int \cos^{n-2} x \, dx \qquad \boxed{14}$$

$$\int \sin^m x \cos^n x \, dx = \frac{\sin^{m+1} x \cos^{n-1} x}{m+n} + \frac{n-1}{m+n} \int \sin^m x \cos^{n-2} x \, dx \qquad \boxed{15}$$

$$\int \sin^m x \cos^n x \, dx = -\frac{\sin^{m-1} x \cos^{n+1} x}{m+n} + \frac{m-1}{m+n} \int \sin^{m-2} x \cos^n x \, dx \qquad \boxed{16}$$

$$\int \tan x \, dx = \ln |\sec x| + C = -\ln |\cos x| + C \qquad \boxed{17}$$

$$\int \tan^m x \, dx = \frac{\tan^{m-1} x}{m-1} - \int \tan^{m-2} x \, dx \qquad \boxed{18}$$

$$\int \cot x \, dx = -\ln |\csc x| + C = \ln |\sin x| + C \qquad \boxed{19}$$

$$\int \cot^m x \, dx = -\frac{\cot^{m-1} x}{m-1} - \int \cot^{m-2} x \, dx \qquad \boxed{20}$$

$$\int \sec x \, dx = \ln|\sec x + \tan x| + C \qquad \boxed{21}$$

$$\int \sec^m x \, dx = \frac{\tan x \sec^{m-2} x}{m-1} + \frac{m-2}{m-1} \int \sec^{m-2} x \, dx \qquad \boxed{22}$$

$$\int \csc x \, dx = \ln|\csc x - \cot x| + C \qquad \boxed{23}$$

$$\int \csc^m x \, dx = -\frac{\cot x \csc^{m-2} x}{m-1} + \frac{m-2}{m-1} \int \csc^{m-2} x \, dx \qquad \boxed{24}$$

$$\int \sin mx \sin nx \, dx = \frac{\sin(m-n)x}{2(m-n)} - \frac{\sin(m+n)x}{2(m+n)} + C \quad (m \neq \pm n) \qquad \boxed{25}$$

$$\int \sin mx \cos nx \, dx = -\frac{\cos(m-n)x}{2(m-n)} - \frac{\cos(m+n)x}{2(m+n)} + C \quad (m \neq \pm n) \qquad \boxed{26}$$

$$\int \cos mx \cos nx \, dx = \frac{\sin(m-n)x}{2(m-n)} + \frac{\sin(m+n)x}{2(m+n)} + C \quad (m \neq \pm n) \qquad \boxed{27}$$

7.3 EXERCISES

Preliminary Questions

1. Describe the technique used to evaluate $\int \sin^5 x \, dx$.

2. Describe a way of evaluating $\int \sin^6 x \, dx$.

3. Are reduction formulas needed to evaluate $\int \sin^7 x \cos^2 x \, dx$? Why or why not?

4. Describe a way of evaluating $\int \sin^6 x \cos^2 x \, dx$.

5. Which integral requires more work to evaluate?

$$\int \sin^{798} x \cos x \, dx \qquad \text{or} \qquad \int \sin^4 x \cos^4 x \, dx$$

Explain your answer.

Exercises

In Exercises 1–6, use the method for odd powers to evaluate the integral.

1. $\int \cos^3 x \, dx$.

2. $\int \sin^5 x \, dx$

3. $\int \sin^3 \theta \cos^2 \theta \, d\theta$

4. $\int \cos x \sin^5 x \, dx$

5. $\int \sin^3 t \cos^3 t \, dt$

6. $\int \sin^2 x \cos^5 x \, dx$

7. Find the area of the shaded region in Figure 1.

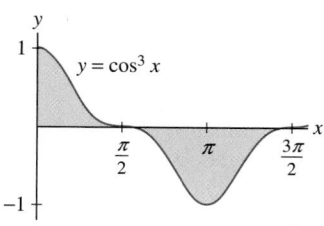

FIGURE 1 Graph of $y = \cos^3 x$.

8. Use the identity $\sin^2 x = 1 - \cos^2 x$ to write $\int \sin^2 x \cos^2 x \, dx$ as a sum of two integrals and then evaluate using the reduction formula.

In Exercises 9–12, evaluate the integral using methods employed in Examples 3 and 4.

9. $\int \cos^4 y \, dy$

10. $\int \cos^2 \theta \sin^2 \theta \, d\theta$

11. $\int \sin^4 x \cos^2 x \, dx$

12. $\int \sin^4 x \cos^6 x \, dx$

In Exercises 13–20, evaluate the integral using the reduction formulas on page 438 as necessary.

13. $\int \tan^2 x \sec^2 x \, dx$

14. $\int \tan^3 x \sec^2 x \, dx$

15. $\int \tan^4 \theta \sec^2 \theta \, d\theta$

16. $\int \tan^2 x \sec x \, dx$

17. $\int \tan^2 t \, dt$

18. $\int \cot^3 x \, dx$

19. $\int \sec^3 x \, dx$

20. $\int \csc^2 x \, dx$

In Exercises 21–54, use the techniques and reduction formulas necessary to evaluate the integral.

21. $\int \cos^5 x \sin x \, dx$

22. $\int \cos^3 2x \sin 2x \, dx$

23. $\int \cos^4 (3x) \, dx$

24. $\int \cos^7 3x \, dx$

25. $\int \cos^3 (\pi\theta) \sin^4 (\pi\theta) \, d\theta$

26. $\int \cos^{498} y \sin^3 y \, dy$

27. $\int \sin^4 (3x) \, dx$

28. $\int \sin^2 x \cos^6 x \, dx$

29. $\int \sec 7t \, dt$

30. $\int \csc^3 x \, dx$

31. $\int \tan x \sec^2 x \, dx$

32. $\int \tan^3 \theta \sec^3 \theta \, d\theta$

33. $\int \tan^5 x \sec^4 x \, dx$

34. $\int \tan^2 x \sec^4 x \, dx$

35. $\int \tan^4 x \sec x \, dx$

36. $\int \tan^6 x \sec^4 x \, dx$

37. $\int \tan^4 x \sec^3 x \, dx$

38. $\int \tan^2 x \csc^2 x \, dx$

39. $\int \sin 2x \cos 2x \, dx$

40. $\int \cos 4x \cos 6x \, dx$

41. $\int \sin 2x \cos 4x \, dx$

42. $\int t \cos^3 (t^2) \, dt$

43. $\int \dfrac{\tan^3 (\ln t)}{t} \, dt$

44. $\int \cos^2 (\sin t) \cos t \, dt$

45. $\int_0^{2\pi} \sin^2 x \, dx$

46. $\int_0^{\pi/2} \cos^3 x \, dx$

47. $\int_0^{\pi/3} \sin^3 x \, dx$

48. $\int_0^{\pi/4} \dfrac{dx}{\cos x}$

49. $\int_{\pi/4}^{\pi/2} \dfrac{dx}{\sin x}$

50. $\int_{0}^{\pi/3} \tan x \, dx$

51. $\int_{0}^{\pi/4} \tan^5 x \, dx$

52. $\int_{-\pi/4}^{\pi/4} \sec^4 x \, dx$

53. $\int_{0}^{\pi} \sin 3x \cos 4x \, dx$

54. $\int_{0}^{\pi} \sin x \sin 3x \, dx$

55. Use the identities for $\sin 2x$ and $\cos 2x$ listed on page 434 to verify that the first of the following formulas is equivalent to the second:

$$\int \sin^4 x \, dx = \frac{1}{32}(12x - 8\sin 2x + \sin 4x) + C$$

$$\int \sin^4 x \, dx = -\frac{1}{4}\sin^3 x \cos x - \frac{3}{8}\sin x \cos x + \frac{3}{8}x + C$$

56. Evaluate $\int \sin^2 x \cos^3 x \, dx$ using the method described in the text and verify that your result is equivalent to the following result produced by a computer algebra system:

$$\int \sin^2 x \cos^3 x \, dx = \frac{1}{30}(7 + 3\cos 2x)\sin^3 x + C$$

In Exercises 57–60, evaluate using the identity $\cot^2 x + 1 = \csc^2 x$ and methods similar to those for integrating $\tan^m x \sec^n x$.

57. $\int \cot^3 x \csc x \, dx$

58. $\int \cot^4 x \csc^2 x \, dx$

59. $\int \cot^2 x \csc^2 x \, dx$

60. $\int \cot^2 x \csc^3 x \, dx$

61. Find the volume of the solid obtained by revolving $y = \sin x$ for $0 \le x \le \pi$ about the x-axis.

62. Use Integration by Parts to prove the reductions formulas Eqs. (3) and (4).

63. Here is another reduction method for evaluating the integral $J = \int \sin^m x \cos^n x \, dx$ when m and n are even. Use the identities

$$\sin^2 x = \frac{1}{2}(1 - \cos 2x), \qquad \cos^2 x = \frac{1}{2}(1 + \cos 2x)$$

to write $J = \frac{1}{4}\int (1 - \cos 2x)^{m/2}(1 + \cos 2x)^{n/2}\, dx$. Then expand the right-hand side as a sum of integrals involving smaller powers of sine and cosine in the variable $2x$. Use this method to evaluate $J = \int \sin^2 x \cos^2 x \, dx$.

64. Use the method of Exercise 63 to evaluate $\int \cos^4 x \, dx$.

65. Use the method of Exercise 63 to evaluate $\int \sin^4 x \cos^2 x \, dx$.

66. Show that for $n \ge 2$,

$$\int_{0}^{\pi/2} \sin^n x \, dx = \frac{n-1}{n} \int_{0}^{\pi/2} \sin^{n-2} x \, dx \qquad \boxed{28}$$

67. Prove the reduction formula

$$\int \tan^k x \, dx = \frac{\tan^{k-1} x}{k-1} - \int \tan^{k-2} x \, dx$$

Hint: Use the identity $\tan^2 x = (\sec^2 x - 1)$ to write

$$\tan^k x = (\sec^2 x - 1)\tan^{k-2} x.$$

68. Evaluate $I = \int \sin^2 x \cos^4 x \, dx$ using (15). Show:

(a) $I = \frac{1}{6}\sin^3 x \cos^3 x + \frac{1}{2}\int \sin^2 x \cos^2 x \, dx$

(b) $\int \sin^2 x \cos^2 x \, dx = \frac{1}{4}\sin^3 x \cos x + \frac{1}{4}\int \sin^2 x \, dx$

69. Use the substitution $u = \csc x - \cot x$ to evaluate $\int \csc x \, dx$ (see Example 5).

70. Total Energy A 100-W light bulb has resistance $R = 144 \; \Omega$ (ohms) when attached to household current, where the voltage varies as $V = V_0 \sin(2\pi ft)$ ($V_0 = 110$ V, $f = 60$ Hz). The power supplied to the bulb is $P = V^2/R$ (joules per second) and the total energy expended over a time period $[0, T]$ (in seconds) is $U = \int_{0}^{T} P(t)\, dt$. Compute U if the bulb remains on for 5 hours.

71. Let m, n be integers with $m \ne \pm n$. Use formulas (25)–(27) in the table of trigonometric integrals to prove that

$$\int_{0}^{\pi} \sin mx \sin nx \, dx = 0, \qquad \int_{0}^{\pi} \cos mx \cos nx \, dx = 0$$

$$\int_{0}^{2\pi} \sin mx \cos nx \, dx = 0$$

These formulas, known as the **orthogonality relations**, play a basic role in the theory of Fourier Series (Figure 2).

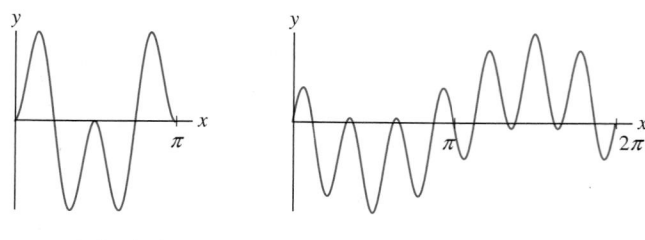

$y = \sin 2x \sin 4x$ $y = \sin 3x \cos 4x$

FIGURE 2 By the orthogonality relations, the signed area under these graphs is zero.

Further Insights and Challenges

72. Use the trigonometric identity

$$\sin mx \cos nx = \frac{1}{2}\big(\sin(m-n)x + \sin(m+n)x\big)$$

to prove Eq. (26) in the table of integrals on page 438.

73. Evaluate $\displaystyle\int_0^\pi \sin^2 mx\, dx$ for m an arbitrary integer.

74. Use Integration by Parts to prove (for $m \neq 1$)

$$\int \sec^m x\, dx = \frac{\tan x \sec^{m-2} x}{m-1} + \frac{m-2}{m-1}\int \sec^{m-2} x\, dx$$

75. Set $\displaystyle I_m = \int_0^{\pi/2} \sin^m x\, dx.$

(a) Show that $I_1 = 1$, $I_2 = \left(\frac{1}{2}\right)\left(\frac{\pi}{2}\right)$ and use Eq. (28) to prove that for $m > 1$, $I_m = \left(\dfrac{m-1}{m}\right) I_{m-2}.$

(b) Show that $I_3 = \frac{2}{3}$ and $I_4 = \left(\frac{3}{4}\right)\left(\frac{1}{2}\right)\left(\frac{\pi}{2}\right).$

(c) Show more generally:

$$I_{2m} = \frac{2m-1}{2m}\frac{2m-3}{2m-2}\cdots\frac{1}{2}\cdot\frac{\pi}{2}$$

$$I_{2m+1} = \frac{2m}{2m+1}\frac{2m-2}{2m-1}\cdots\frac{2}{3}$$

(d) Conclude that

$$\frac{\pi}{2} = \frac{2\cdot 2}{1\cdot 3}\cdot\frac{4\cdot 4}{3\cdot 5}\cdots\frac{2m\cdot 2m}{(2m-1)(2m+1)}\frac{I_{2m}}{I_{2m+1}}$$

76. This is a continuation of Exercise 75.

(a) Prove that $I_{2m+1} \leq I_{2m} \leq I_{2m-1}$. *Hint:* Observe that

$$\sin^{2m+1} x \leq \sin^{2m} x \leq \sin^{2m-1} x \quad \text{for} \quad 0 \leq x \leq \frac{\pi}{2}$$

(b) Show that $\dfrac{I_{2m-1}}{I_{2m+1}} = 1 + \dfrac{1}{2m}.$

(c) Show that

$$1 \leq \frac{I_{2m}}{I_{2m+1}} \leq 1 + \frac{1}{2m}$$

(d) Prove that $\displaystyle\lim_{m\to\infty}\frac{I_{2m}}{I_{2m+1}} = 1.$

(e) Finally, deduce the infinite product for $\frac{\pi}{2}$ discovered by English mathematician John Wallis (1616–1703):

$$\frac{\pi}{2} = \lim_{m\to\infty}\frac{2}{1}\cdot\frac{2}{3}\cdot\frac{4}{3}\cdot\frac{4}{5}\cdots\frac{2m\cdot 2m}{(2m-1)(2m+1)}$$

7.4 Trigonometric Substitution

Our next goal is to integrate functions involving one of the square root expressions:

$$\sqrt{a^2 - x^2}, \qquad \sqrt{x^2 + a^2}, \qquad \sqrt{x^2 - a^2}$$

In each case, a substitution transforms the integral into a trigonometric integral. For example, the substitution $x = a\sin\theta$ may be used when the integrand involves $\sqrt{a^2 - x^2}$.

■ **EXAMPLE 1** Evaluate $\displaystyle\int \sqrt{1 - x^2}\, dx.$

Solution

When making a trigonometric substitution such as $x = \sin\theta$ or $x = \tan\theta$, we choose the angle θ in the domain of the relevant inverse trigonometric function, i.e., $\sin^{-1} x$ or $\tan^{-1} x$.

Step 1. Substitute to eliminate the square root.

In this integral, $-1 \leq x \leq 1$. Therefore, we may use the substitution $x = \sin\theta$, $dx = \cos\theta\, d\theta$ where $-\frac{\pi}{2} \leq \theta \leq \frac{\pi}{2}$. For such θ, $\cos\theta \geq 0$, and we obtain

$$\sqrt{1 - x^2} = \sqrt{1 - \sin^2\theta} = \cos\theta \qquad \boxed{1}$$

$$\int \sqrt{1 - x^2}\, dx = \int \cos\theta \overbrace{\cos\theta\, d\theta}^{dx} = \int \cos^2\theta\, d\theta$$

Step 2. Evaluate the trigonometric integral.

Use the formula recalled in the margin:

··◄ REMINDER
$$\int \cos^2\theta\, d\theta = \frac{1}{2}\theta + \frac{1}{4}\sin 2\theta + C$$

$$\int \sqrt{1 - x^2}\, dx = \int \cos^2\theta\, d\theta = \frac{1}{2}\theta + \frac{1}{4}\sin 2\theta + C \qquad \boxed{2}$$

Step 3. Convert back to original variable.
We need to express our answer in terms of the original variable x. We use the trigonometric identity $\sin 2\theta = 2 \sin \theta \cos \theta$ and Eq. (1) to write

$$\sin 2\theta = 2 \sin \theta \cos \theta = 2x\sqrt{1 - x^2}$$

Furthermore, $\theta = \sin^{-1} x$, so Eq. (2) becomes

$$\int \sqrt{1 - x^2}\, dx = \frac{1}{2}\theta + \frac{1}{4}\sin 2\theta + C = \frac{1}{2}\sin^{-1} x + \frac{1}{2}x\sqrt{1 - x^2} + C \qquad \blacksquare$$

Integrals Involving $\sqrt{a^2 - x^2}$ If $\sqrt{a^2 - x^2}$ occurs in an integral where $a > 0$, try the substitution

$$x = a \sin \theta, \qquad dx = a \cos \theta\, d\theta, \qquad \sqrt{a^2 - x^2} = a \cos \theta$$

Note that if $x = a \sin \theta$ and $a > 0$, then

$$a^2 - x^2 = a^2(1 - \sin^2 \theta) = a^2 \cos^2 \theta$$

For $-\frac{\pi}{2} \le \theta \le \frac{\pi}{2}$, $\cos \theta \ge 0$ and

$$\sqrt{a^2 - x^2} = a \cos \theta$$

■ **EXAMPLE 2** Integrand Involving $(a^2 - x^2)^{3/2}$ Evaluate $\displaystyle\int \frac{x^2}{(4 - x^2)^{3/2}}\, dx$.

Solution

Step 1. Substitute to eliminate the square root.
In this case, $a = 2$ since $\sqrt{4 - x^2} = \sqrt{2^2 - x^2}$. Therefore, we use

$$x = 2 \sin \theta, \qquad dx = 2 \cos \theta\, d\theta, \qquad \sqrt{4 - x^2} = 2 \cos \theta$$

$$\int \frac{x^2}{(4 - x^2)^{3/2}}\, dx = \int \frac{4 \sin^2 \theta}{2^3 \cos^3 \theta}\, 2 \cos \theta\, d\theta = \int \frac{\sin^2 \theta}{\cos^2 \theta}\, d\theta = \int \tan^2 \theta\, d\theta$$

Example 2 shows that trigonometric substitution can be used with integrands involving $(a^2 - x^2)^{n/2}$, where n is any integer.

Step 2. Evaluate the trigonometric integral.
Use the reduction formula in the marginal note with $m = 2$:

$$\int \tan^2 \theta\, d\theta = \tan \theta - \int d\theta = \tan \theta - \theta + C$$

◄·· REMINDER

$$\int \tan^m x\, dx = \frac{\tan^{m-1} x}{m - 1} - \int \tan^{m-2} x\, dx$$

Step 3. Convert back to original variable.
We must express $\tan \theta$ and θ in terms of x. By definition, $x = 2 \sin \theta$, so

$$\sin \theta = \frac{x}{2}, \qquad \theta = \sin^{-1} \frac{x}{2}$$

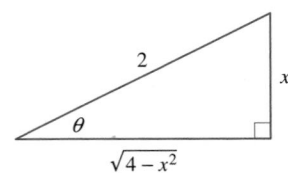

FIGURE 1 Right triangle with $\sin \theta = x/2$.

How do we express $\tan \theta$ in terms of x? We construct a right triangle with angle θ such that $\sin \theta = x/2$ as in Figure 1. By the Pythagorean Theorem, the adjacent side has length $\sqrt{4 - x^2}$, and thus

$$\tan \theta = \frac{\text{opposite}}{\text{adjacent}} = \frac{x}{\sqrt{4 - x^2}}$$

$$\int \frac{x^2}{(4 - x^2)^{3/2}}\, dx = \tan \theta - \theta + C = \frac{x}{\sqrt{4 - x^2}} - \sin^{-1} \frac{x}{2} + C \qquad \blacksquare$$

When the integrand involves $\sqrt{a^2 + x^2}$, try the substitution $x = a \tan \theta$. Then

$$x^2 + a^2 = a^2 \tan^2 \theta + a^2 = a^2(1 + \tan^2 \theta) = a^2 \sec^2 \theta$$

and thus $\sqrt{a^2 + x^2} = a \sec \theta$.

In the substitution $x = a \tan\theta$, we choose $-\frac{\pi}{2} < \theta < \frac{\pi}{2}$. Therefore $a \sec\theta$ is the positive square root $\sqrt{a^2 + x^2}$.

> **Integrals Involving $\sqrt{a^2 + x^2}$** If $\sqrt{a^2 + x^2}$ occurs in an integral where $a > 0$, try the substitution
>
> $$x = a \tan\theta, \qquad dx = a \sec^2\theta\, d\theta, \qquad \sqrt{a^2 + x^2} = a \sec\theta$$

■ **EXAMPLE 3** Evaluate $\displaystyle\int \sqrt{4x^2 + 20}\, dx$.

Solution To obtain an integrand of the form $\sqrt{a^2 + x^2}$, we factor out a constant:

$$\int \sqrt{4x^2 + 20}\, dx = \int \sqrt{4(x^2 + 5)}\, dx = 2\int \sqrt{x^2 + 5}\, dx$$

Next, let $a = \sqrt{5}$ and use the substitution

$$x = \sqrt{5}\tan\theta, \qquad dx = \sqrt{5}\sec^2\theta\, d\theta, \qquad \sqrt{x^2 + 5} = \sqrt{5}\sec\theta$$

$$2\int \sqrt{x^2 + 5}\, dx = 2\int \left(\sqrt{5}\sec\theta\right)\sqrt{5}\sec^2\theta\, d\theta = 10\int \sec^3\theta\, d\theta$$

Now use the reduction formula for $\sec^m x$ in the marginal note with $m = 3$:

$$\int \sqrt{4x^2 + 20}\, dx = 10\int \sec^3\theta\, d\theta = 10\frac{\tan\theta\sec\theta}{2} + 10\left(\frac{1}{2}\right)\int \sec\theta\, dx$$

$$= 5\tan\theta\sec\theta + 5\ln\left|\sec\theta + \tan\theta\right| + C \qquad \boxed{3}$$

◀·· REMINDER

$$\int \sec^m x\, dx = \frac{\tan x \sec^{m-2} x}{m - 1}$$
$$+ \frac{m - 2}{m - 1}\int \sec^{m-2} x\, dx$$

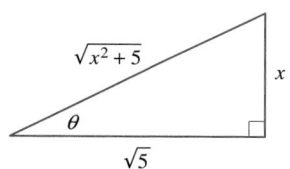

FIGURE 2

It remains to express the result in terms of x. Since $x = \sqrt{5}\tan\theta$, we refer to the right triangle in Figure 2 to obtain

$$\tan\theta = \frac{\text{opposite}}{\text{adjacent}} = \frac{x}{\sqrt{5}}, \qquad \sec\theta = \frac{\text{hypotenuse}}{\text{adjacent}} = \frac{\sqrt{x^2 + 5}}{\sqrt{5}}$$

Thus Eq. (3) yields

$$\int \sqrt{4x^2 + 20}\, dx = 5\frac{x}{\sqrt{5}}\frac{\sqrt{x^2 + 5}}{\sqrt{5}} + 5\ln\left|\frac{\sqrt{x^2 + 5}}{\sqrt{5}} + \frac{x}{\sqrt{5}}\right| + C$$

$$= x\sqrt{x^2 + 5} + 5\ln\left|\frac{\sqrt{x^2 + 5} + x}{\sqrt{5}}\right| + C$$

The logarithmic term can be rewritten as

$$5\ln\left|\frac{\sqrt{x^2 + 5} + x}{\sqrt{5}}\right| + C = 5\ln\left|\sqrt{x^2 + 5} + x\right| + \underbrace{5\ln\left|\frac{1}{\sqrt{5}}\right| + C}_{\text{Constant}}$$

Since C is an arbitrary constant, we may absorb $5\ln\left|\dfrac{1}{\sqrt{5}}\right|$ into C and write

$$\int \sqrt{4x^2 + 20}\, dx = x\sqrt{x^2 + 5} + 5\ln\left|\sqrt{x^2 + 5} + x\right| + C$$

■

Our last trigonometric substitution is $x = a \sec \theta$, which transforms $\sqrt{x^2 - a^2}$ into $a \tan \theta$ because

$$x^2 - a^2 = a^2 \sec^2 \theta - a^2 = a^2(\sec^2 \theta - 1) = a^2 \tan^2 \theta$$

In the substitution $x = a \sec \theta$, we choose $0 \le \theta < \frac{\pi}{2}$ if $x \ge a$ and $\pi \le \theta < \frac{3\pi}{2}$ if $x \le -a$. With these choices, $a \tan \theta$ is the positive square root $\sqrt{x^2 - a^2}$.

> **Integrals Involving $\sqrt{x^2 - a^2}$**   If $\sqrt{x^2 - a^2}$ occurs in an integral where $a > 0$, try the substitution
> $$x = a \sec \theta, \qquad dx = a \sec \theta \tan \theta \, d\theta, \qquad \sqrt{x^2 - a^2} = a \tan \theta$$

■ **EXAMPLE 4** Evaluate $\displaystyle\int \frac{dx}{x^2 \sqrt{x^2 - 9}}$.

Solution In this case, make the substitution

$$x = 3 \sec \theta, \qquad dx = 3 \sec \theta \tan \theta \, d\theta, \qquad \sqrt{x^2 - 9} = 3 \tan \theta$$

Then

$$\int \frac{dx}{x^2 \sqrt{x^2 - 9}} = \int \frac{3 \sec \theta \tan \theta \, d\theta}{(9 \sec^2 \theta)(3 \tan \theta)} = \frac{1}{9} \int \cos \theta \, d\theta = \frac{1}{9} \sin \theta + C$$

Since $x = 3 \sec \theta$, we construct a right triangle as in Figure 3 for which

$$\sec \theta = \frac{\text{hypotenuse}}{\text{adjacent}} = \frac{x}{3}$$

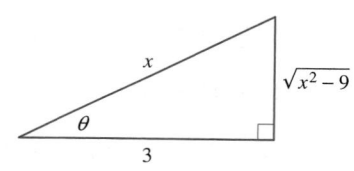

FIGURE 3

The right triangle shows that

$$\sin \theta = \frac{\text{opposite}}{\text{hypotenuse}} = \frac{\sqrt{x^2 - 9}}{x}$$

Therefore,

$$\int \frac{dx}{x^2 \sqrt{x^2 - 9}} = \frac{1}{9} \sin \theta + C = \frac{\sqrt{x^2 - 9}}{9x} + C \qquad ■$$

So far we have used trigonometric substitution to deal with integrands involving the square roots $\sqrt{x^2 \pm a^2}$ or $\sqrt{a^2 - x^2}$. We treat more general square roots of the form $\sqrt{ax^2 + bx + c}$ by completing the square (Section 1.2).

■ **EXAMPLE 5** Completing the Square Evaluate $\displaystyle\int \frac{dx}{(x^2 - 6x + 11)^2}$.

Solution

Step 1. **Complete the square.**

$$x^2 - 6x + 11 = (x^2 - 6x + 9) + 2 = (x - 3)^2 + 2$$

Step 2. **Substitution.**

Let $u = x - 3$. Then $x^2 - 6x + 11 = u^2 + 2$ and

$$\int \frac{dx}{(x^2 - 6x + 11)^2} = \int \frac{du}{(u^2 + 2)^2}$$

$\boxed{4}$

Step 3. **Trigonometric substitution.**

Evaluate the u-integral using trigonometric substitution:

$$u = \sqrt{2}\tan\theta, \qquad \sqrt{u^2 + 2} = \sqrt{2}\sec\theta, \qquad du = \sqrt{2}\sec^2\theta\,d\theta$$

$$\int \frac{du}{(u^2 + 2)^2} = \int \frac{\sqrt{2}\sec^2\theta\,d\theta}{4\sec^4\theta} = \frac{1}{2\sqrt{2}}\int\cos^2\theta\,d\theta$$

$$= \frac{1}{2\sqrt{2}}\left(\frac{\theta}{2} + \frac{\sin\theta\cos\theta}{2}\right) + C \qquad \boxed{5}$$

To write this result in terms of u, note that $\theta = \tan^{-1}\dfrac{u}{\sqrt{2}}$. From the right triangle in Figure 4, we obtain

$$\sin\theta\cos\theta = \left(\frac{\text{opposite}}{\text{hypotenuse}}\right)\left(\frac{\text{adjacent}}{\text{hypotenuse}}\right) = \frac{u}{\sqrt{u^2 + 2}}\cdot\frac{\sqrt{2}}{\sqrt{u^2 + 2}} = \frac{\sqrt{2}\,u}{u^2 + 2}$$

Thus, Eq. (5) becomes

$$\int \frac{du}{(u^2 + 2)^2} = \frac{1}{4\sqrt{2}}\left(\tan^{-1}\frac{u}{\sqrt{2}} + \frac{\sqrt{2}\,u}{u^2 + 2}\right) + C$$

$$= \frac{1}{4\sqrt{2}}\tan^{-1}\frac{u}{\sqrt{2}} + \frac{u}{4(u^2 + 2)} + C \qquad \boxed{6}$$

Step 4. **Convert to original variable.**

Since $u = x - 3$ and $u^2 + 2 = x^2 - 6x + 11$, Eq. (6) becomes

$$\int \frac{du}{(u^2 + 2)^2} = \frac{1}{4\sqrt{2}}\tan^{-1}\frac{x - 3}{\sqrt{2}} + \frac{x - 3}{4(x^2 - 6x + 11)} + C$$

This is our final answer by Eq. (4):

$$\int \frac{dx}{(x^2 - 6x + 11)^2} = \frac{1}{4\sqrt{2}}\tan^{-1}\frac{x - 3}{\sqrt{2}} + \frac{x - 3}{4(x^2 - 6x + 11)} + C \qquad \blacksquare$$

◄⋯ **REMINDER**

$$\int\cos^2\theta\,d\theta = \frac{\theta}{2} + \frac{\sin\theta\cos\theta}{2} + C$$

FIGURE 4

7.4 SUMMARY

- Trigonometric substitution:

Square Root Form in Integrand	Trigonometric Substitution
$\sqrt{a^2 - x^2}$	$x = a\sin\theta, \quad dx = a\cos\theta\,d\theta, \quad \sqrt{a^2 - x^2} = a\cos\theta$
$\sqrt{a^2 + x^2}$	$x = a\tan\theta, \quad dx = a\sec^2\theta\,d\theta, \quad \sqrt{a^2 + x^2} = a\sec\theta$
$\sqrt{x^2 - a^2}$	$x = a\sec\theta, \quad dx = a\sec\theta\tan\theta\,d\theta, \quad \sqrt{x^2 - a^2} = a\tan\theta$

Step 1. Substitute to eliminate the square root.

Step 2. Evaluate the trigonometric integral.

Step 3. Convert back to original variable.

- The three trigonometric substitutions correspond to three right triangles (Figure 5) that we use to express the trigonometric functions of θ in terms of x.

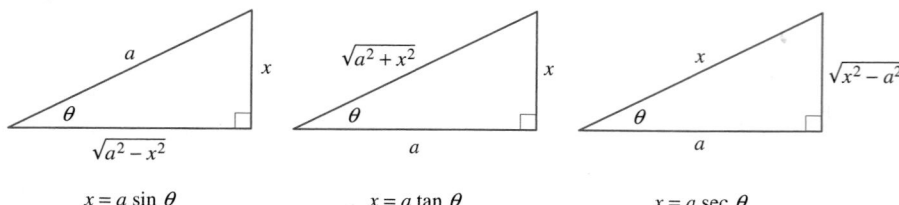

FIGURE 5

- Integrands involving $\sqrt{x^2 + bx + c}$ are treated by completing the square (see Example 5).

7.4 EXERCISES

Preliminary Questions

1. Explain why trigonometric substitution is *not needed* to evaluate $\int x\sqrt{9 - x^2}\, dx$.

2. State the trigonometric substitution appropriate to the given integral:

(a) $\int \sqrt{9 - x^2}\, dx$

(b) $\int x^2(x^2 - 16)^{3/2}\, dx$

(c) $\int x^2(x^2 + 16)^{3/2}\, dx$

(d) $\int (x^2 - 5)^{-2}\, dx$

3. Which of the triangles in Figure 6 would be used together with the substitution $x = 3\sin\theta$?

4. Express $\tan\theta$ in terms of x for the angle in Figure 6(A).

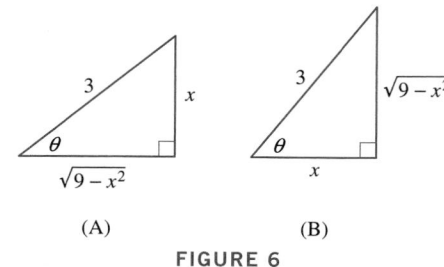

FIGURE 6

5. Express $\sec\theta$ in terms of x for the angle in Figure 6(B).

6. Express $\sin 2\theta$ in terms of x, where $x = \sin\theta$.

Exercises

In Exercises 1–4, evaluate the integral by following the steps.

1. $I = \displaystyle\int \frac{dx}{\sqrt{9 - x^2}}$

(a) Show that the substitution $x = 3\sin\theta$ transforms I into $\int d\theta$ and evaluate I in terms of θ.

(b) Evaluate I in terms of x.

2. $I = \displaystyle\int \frac{dx}{x^2\sqrt{x^2 - 2}}$

(a) Show that the substitution $x = \sqrt{2}\sec\theta$ transforms the integral I into $\frac{1}{2}\int \cos\theta\, d\theta$ and evaluate I in terms of θ.

(b) Use a right triangle to show that with the above substitution, $\sin\theta = \sqrt{x^2 - 2}/x$.

(c) Evaluate I in terms of x.

3. $I = \displaystyle\int \frac{dx}{\sqrt{x^2 + 9}}$

(a) Show that the substitution $x = 3\tan\theta$ transforms I into $\int \sec\theta\, d\theta$ and evaluate I in terms of θ (refer to the table of integrals in Section 7.3 if necessary).

(b) Show that if $x = 3\tan\theta$, then $\sec\theta = \frac{1}{3}\sqrt{x^2 + 9}$.

(c) Express I in terms of x.

4. $I = \displaystyle\int \frac{dx}{(x^2 + 4)^2}$

(a) Show that the substitution $x = 2\tan\theta$ transforms the integral I into $\frac{1}{8}\int \cos^2\theta\, d\theta$.

(b) Use the formula $\int \cos^2\theta\, d\theta = \frac{1}{2}\theta + \frac{1}{2}\sin\theta\cos\theta$ to evaluate I in terms of θ.

(c) Show that $\sin\theta = \dfrac{x}{\sqrt{x^2 + 4}}$ and $\cos\theta = \dfrac{2}{\sqrt{x^2 + 4}}$.

(d) Express I in terms of x.

In Exercises 5–10, use the indicated substitution to evaluate the integral.

5. $\displaystyle\int \sqrt{4 - x^2}\, dx, \quad x = 2\sin\theta$

6. $\displaystyle\int_0^{1/2} \frac{x^2}{\sqrt{1 - x^2}}\, dx, \quad x = \sin\theta$

7. $\displaystyle\int \frac{dx}{x\sqrt{x^2-9}}$, $x = 3 \sec\theta$

8. $\displaystyle\int_{1/2}^{1} \frac{dx}{x^2\sqrt{x^2+4}}$, $x = 2 \tan\theta$

9. $\displaystyle\int \frac{dx}{(x^2-4)^{3/2}}$, $x = 2 \sec\theta$

10. $\displaystyle\int_{0}^{1} \frac{dx}{(16+x^2)^2}$, $x = 4 \tan\theta$

11. Is the substitution $u = x^2 - 4$ effective for evaluating the integral $\displaystyle\int \frac{x^2\,dx}{\sqrt{x^2-4}}$? If not, evaluate using trigonometric substitution.

12. Evaluate both $\displaystyle\int \frac{x\,dx}{\sqrt{x^2-4}}$ and $\displaystyle\int \frac{x^3\,dx}{\sqrt{x^2-4}}$ in two ways: using trigonometric substitution and using the direct substitution $u = x^2 - 4$.

In Exercises 13–30, evaluate the integral using trigonometric substitution. Refer to the table of trigonometric integrals as necessary.

13. $\displaystyle\int \frac{x^2\,dx}{\sqrt{9-x^2}}$

14. $\displaystyle\int \frac{dx}{\sqrt{x^2-9}}$

15. $\displaystyle\int \sqrt{12+4x^2}\,dx$

16. $\displaystyle\int \frac{dx}{x^2\sqrt{x^2-25}}$

17. $\displaystyle\int \frac{dt}{(4-t^2)^{3/2}}$

18. $\displaystyle\int \frac{dt}{\sqrt{t^2-5}}$

19. $\displaystyle\int \frac{dy}{y^2\sqrt{5-y^2}}$

20. $\displaystyle\int \frac{dt}{t\sqrt{t^2-4}}$

21. $\displaystyle\int \frac{dz}{z^3\sqrt{z^2-4}}$

22. $\displaystyle\int \frac{dx}{\sqrt{25+x^2}}$

23. $\displaystyle\int \frac{x^2\,dx}{\sqrt{x^2+1}}$

24. $\displaystyle\int \frac{x^2\,dx}{(x^2-4)^{3/2}}$

25. $\displaystyle\int \frac{dx}{(x^2+9)^2}$

26. $\displaystyle\int \frac{dx}{(9x^2+4)^2}$

27. $\displaystyle\int \frac{dx}{(x^2-4)^2}$

28. $\displaystyle\int \frac{dx}{(x^2+1)^3}$

29. $\displaystyle\int x^3\sqrt{9-x^2}\,dx$

30. $\displaystyle\int \frac{x^2\,dx}{(x^2+1)^{3/2}}$

31. Prove the following for $a > 0$:
$$\int \frac{dx}{x^2+a} = \frac{1}{\sqrt{a}}\tan^{-1}\frac{x}{\sqrt{a}} + C$$

32. Prove the following for $a > 0$:
$$\int \frac{dx}{(x^2+a)^2} = \frac{1}{2a}\left(\frac{x}{x^2+a} + \frac{1}{\sqrt{a}}\tan^{-1}\frac{x}{\sqrt{a}}\right) + C$$

33. Let $I = \displaystyle\int \frac{dx}{\sqrt{x^2-4x+8}}$.

(a) Complete the square to show that $x^2 - 4x + 8 = (x-2)^2 + 4$.

(b) Use the substitution $u = x - 2$ to show that $I = \displaystyle\int \frac{du}{\sqrt{u^2+2^2}}$. Evaluate I.

34. Evaluate $\displaystyle\int \frac{dx}{\sqrt{12x-x^2}}$. First complete the square to write $12x - x^2 = 36 - (x-6)^2$.

In Exercises 35–40, evaluate the integral by completing the square and using trigonometric substitution.

35. $\displaystyle\int \frac{dx}{\sqrt{x^2+4x+13}}$

36. $\displaystyle\int \frac{dx}{\sqrt{2+x-x^2}}$

37. $\displaystyle\int \frac{dx}{\sqrt{x+x^2}}$

38. $\displaystyle\int \sqrt{x^2-4x+7}\,dx$

39. $\displaystyle\int \sqrt{x^2-4x+3}\,dx$

40. $\displaystyle\int \frac{dx}{(x^2+6x+6)^2}$

41. Evaluate $\displaystyle\int \sec^{-1}x\,dx$. *Hint:* First use Integration by Parts.

42. Evaluate $\displaystyle\int \frac{\sin^{-1}x}{x^2}\,dx$. *Hint:* First use Integration by Parts.

In Exercises 43–52, indicate a good method for evaluating the integral (but do not evaluate). Your choices are recognizing a basic integration formula, algebraic manipulation, substitution (specify u and du), Integration by Parts (specify u and v'), a trigonometric method, or trigonometric substitution (specify). If it appears that the techniques you have learned thus far are not sufficient, state this.

43. $\displaystyle\int \frac{dx}{\sqrt{12-6x-x^2}}$

44. $\displaystyle\int \sin^3 x \cos^3 x\,dx$

45. $\displaystyle\int x \sec^2 x\,dx$

46. $\displaystyle\int \sqrt{4x^2-1}\,dx$

47. $\displaystyle\int \frac{e^{2x}}{e^{4x}+1}\,dx$

48. $\displaystyle\int \frac{dx}{\sqrt{9-x^2}}$

49. $\displaystyle\int \cot x \csc x\,dx$

50. $\displaystyle\int \ln x\,dx$

51. $\displaystyle\int \frac{dx}{(x+2)^3}$

52. $\displaystyle\int \frac{dx}{(x+1)(x+2)^3}$

53. Which of the following integrals can be evaluated using the substitution $u = 1 - x^2$ and which require trigonometric substitution? Determine the integral obtained after substitution in each case.

(a) $\displaystyle\int x^3\sqrt{1-x^2}\,dx$

(b) $\displaystyle\int x^2\sqrt{1-x^2}\,dx$

(c) $\displaystyle\int \frac{x^4}{\sqrt{1-x^2}}\,dx$

(d) $\displaystyle\int \frac{x}{\sqrt{1-x^2}}\,dx$

54. Find the average height of a point on the semicircle $y = \sqrt{1 - x^2}$ for $-1 \le x \le 1$.

55. Find the volume of the solid obtained by revolving the graph of $y = x\sqrt{1 - x^2}$ over $[0, 1]$ about the y-axis.

56. Find the volume of the solid obtained by revolving the region between the graph of $y^2 - x^2 = 1$ and the line $y = 2$ about the line $y = 2$.

57. Find the volume of revolution for the region in Exercise 56, but revolve around $y = 3$.

58. A charged wire creates an electric field at a point P located at a distance D from the wire (Figure 7). The component $E_\perp$ of the field perpendicular to the wire (in N/C) is

$$E_\perp = \int_{x_1}^{x_2} \frac{k\lambda D}{(x^2 + D^2)^{3/2}}\, dx$$

where $k = 8.99 \times 10^9$ N $\cdot$ m^2/C^2 (Coulomb constant), λ is the charge density (coulombs per meter), and x_1, x_2 are shown in the figure. Suppose that $\lambda = 6 \times 10^{-4}$ C/m, and $D = 3$ m. Find $E_\perp$ if (a) $x_1 = 0$ and $x_2 = 30$ m, and (b) $x_1 = -15$ m and $x_2 = 15$ m.

59. $\boxed{CAS}$ Having ordered an 18-in. pizza for yourself and two friends, you want to divide it up using vertical slices as in Figure 8. Use Eq. (7) in Exercise 63 below and a computer algebra system to find the value of x that divides the pizza into equal parts.

FIGURE 7

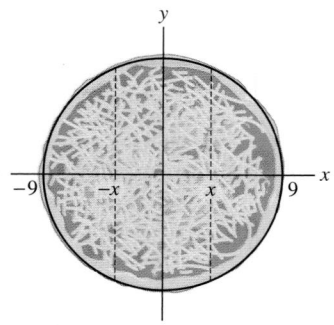

FIGURE 8 Dividing a pizza into three equal parts.

Further Insights and Challenges

60. Let $J_n = \int \dfrac{dx}{(x^2 + 1)^n}$. Prove

$$J_{n+1} = \left(1 - \frac{1}{2n}\right) J_n + \left(\frac{1}{2n}\right) \frac{x}{(x^2 + 1)^n}$$

Hint: Compute J_n using Integration by Parts with $v' = 1$. Use this recursion relation to calculate J_2 and J_3.

61. Hyperbolic Substitution Hyperbolic functions can be used instead of trigonometric substitution to treat integrals involving $\sqrt{x^2 \pm a^2}$. Let $I = \int \dfrac{dx}{\sqrt{x^2 - 1}}$.

(a) Show that the substitution $x = \cosh t$ transforms I into the integral $\int dt = t + C$.

(b) Show that $I = \cosh^{-1} x + C$.

(c) Trigonometric substitution with $x = \sec \theta$ leads to

$$I = \ln |x + \sqrt{x^2 - 1}| + C$$

Show that the two answers coincide.

62. Let $I = \int \sqrt{x^2 - 9}\, dx$.

(a) Show that the substitution $x = 3 \cosh t$ transforms I into the integral $9 \int \sinh^2 t\, dt$.

(b) Evaluate the hyperbolic integral using the identity

$$\sinh^2 t = \frac{1}{2}(\cosh 2t - 1)$$

(c) Express the result in terms of x. Note that $\sinh 2t = 2 \sinh t \cosh t$.

63. In Example 1, we proved the formula

$$\int \sqrt{1 - x^2}\, dx = \frac{1}{2} \sin^{-1} x + \frac{1}{2} x \sqrt{1 - x^2} + C \qquad \boxed{7}$$

Derive this formula using geometry rather than calculus by interpreting the integral as the area of part of the unit circle.

64. Compute $\int \dfrac{dx}{x^2 - 1}$ in two ways and verify that the answers agree: first via trigonometric substitution and then using the identity

$$\frac{1}{x^2 - 1} = \frac{1}{2}\left(\frac{1}{x - 1} - \frac{1}{x + 1}\right)$$

7.5 Integrals of Hyperbolic and Inverse Hyperbolic Functions

As we noted in Section 1.6, hyperbolic and trigonometric functions are similar in many ways (see the Excursion at the end of this section for an explanation of these similarities). We also saw in Section 3.10 that the formulas for their derivatives resemble each other, differing in at most a sign. The derivative formulas for the hyperbolic functions are equivalent to the following integral formulas.

◄·· REMINDER

$$\sinh x = \frac{e^x - e^{-x}}{2} \qquad \cosh x = \frac{e^x + e^{-x}}{2}$$

$$\frac{d}{dx} \sinh x = \cosh x \qquad \frac{d}{dx} \cosh x = \sinh x$$

$$\frac{d}{dx} \tanh x = \operatorname{sech}^2 x \qquad \frac{d}{dx} \coth x = -\operatorname{csch}^2 x$$

$$\frac{d}{dx} \operatorname{sech} x = -\operatorname{sech} x \tanh x$$

$$\frac{d}{dx} \operatorname{csch} x = -\operatorname{csch} x \coth x$$

Hyperbolic Integral Formulas

$$\int \sinh x \, dx = \cosh x + C, \qquad \int \cosh x \, dx = \sinh x + C$$

$$\int \operatorname{sech}^2 x \, dx = \tanh x + C, \qquad \int \operatorname{csch}^2 x \, dx = -\coth x + C$$

$$\int \operatorname{sech} x \tanh x \, dx = -\operatorname{sech} x + C, \qquad \int \operatorname{csch} x \coth x \, dx = -\operatorname{csch} x + C$$

■ **EXAMPLE 1** Calculate $\int x \cosh(x^2) \, dx$.

Solution The substitution $u = x^2$, $du = 2x \, dx$ yields

$$\int x \cosh(x^2) \, dx = \frac{1}{2} \int \cosh u \, du = \frac{1}{2} \sinh u + C = \frac{1}{2} \sinh(x^2) + C \qquad ■$$

The techniques for computing trigonometric integrals discussed in Section 7.3 apply with little change to hyperbolic integrals. In place of trigonometric identities, we use the corresponding hyperbolic identities. The following identities are useful.

$$\cosh^2 x - \sinh^2 x = 1, \qquad \cosh^2 x = 1 + \sinh^2 x$$

$$\cosh^2 x = \frac{1}{2}(\cosh 2x + 1), \qquad \sinh^2 x = \frac{1}{2}(\cosh 2x - 1)$$

$$\sinh 2x = 2 \sinh x \cosh x, \qquad \cosh 2x = \cosh^2 x + \sinh^2 x$$

■ **EXAMPLE 2** Powers of $\sinh x$ or $\cosh x$ Calculate (a) $\int \sinh^4 x \cosh^5 x \, dx$ and
(b) $\int \cosh^2 x \, dx$.

Solution

(a) Since $\cosh x$ appears to an odd power, use the identity $\cosh^2 x = 1 + \sinh^2 x$ to write

$$\cosh^5 x = \cosh^4 x \cdot \cosh x = (\sinh^2 x + 1)^2 \cosh x$$

Then

$$\int \sinh^4 x \cosh^5 x \, dx = \int (\sinh^4 x)(\sinh^2 x + 1)^2 \cosh x \, dx$$

The extra factor of $\cosh x$ in the integrand allows us to use the substitution $u = \sinh x$, $du = \cosh x \, dx$. We have

$$(\sinh^4 x)(\sinh^2 x + 1)^2 \cosh x \, dx = u^4 (u^2 + 1)^2 \, du$$

$$\int \sinh^4 x \cosh^5 x \, dx = \int u^4 (u^2 + 1)^2 \, du = \int (u^8 + 2u^6 + u^4) \, du$$

$$= \frac{u^9}{9} + \frac{2u^7}{7} + \frac{u^5}{5} + C = \frac{\sinh^9 x}{9} + \frac{2 \sinh^7 x}{7} + \frac{\sinh^5 x}{5} + C$$

(b) Use the identity $\cosh^2 x = \frac{1}{2}(\cosh 2x + 1)$:

$$\int \cosh^2 x \, dx = \frac{1}{2} \int (\cosh 2x + 1) \, dx = \frac{1}{2} \left(\frac{\sinh 2x}{2} + x \right) + C$$

$$= \frac{1}{4} \sinh 2x + \frac{1}{2}x + C \qquad \blacksquare$$

Hyperbolic substitution may be used as an alternative to trigonometric substitution to integrate functions involving the following square root expressions:

In trigonometric substitution, we treat $\sqrt{x^2 + a^2}$ using the substitution $x = a \tan \theta$ and $\sqrt{x^2 - a^2}$ using $x = a \sec \theta$. Identities may be used to show that the results coincide with those obtained from hyperbolic substitution (see Exercises 31–35).

Square Root Form	Hyperbolic Substitution
$\sqrt{x^2 + a^2}$	$x = a \sinh u, \, dx = a \cosh u, \, \sqrt{x^2 + a^2} = a \cosh u$
$\sqrt{x^2 - a^2}$	$x = a \cosh u, \, dx = a \sinh u, \, \sqrt{x^2 - a^2} = a \sinh u$

When hyperbolic substitution is applied, the answer may involve inverse hyperbolic functions just as trigonometric substitution may yield inverse trigonometric functions.

■ **EXAMPLE 3** Hyperbolic Substitution Calculate $\int \sqrt{x^2 + 16} \, dx$.

Solution

Step 1. **Substitute to eliminate the square root.**
Use the hyperbolic substitution $x = 4 \sinh u$, $dx = 4 \cosh u \, du$. Then

$$x^2 + 16 = 16(\sinh^2 u + 1) = (4 \cosh u)^2$$

$$\int \sqrt{x^2 + 16} \, dx = \int (4 \cosh u) 4 \cosh u \, du = 16 \int \cosh^2 u \, du$$

Step 2. **Evaluate the hyperbolic integral.**
We evaluated the integral of $y = \cosh^2 u$ in Example 2(b):

$$\int \sqrt{x^2 + 16} \, dx = 16 \int \cosh^2 u \, du = 16 \left(\frac{1}{4} \sinh 2u + \frac{1}{2}u + C \right)$$

$$= 4 \sinh 2u + 8u + C \qquad \boxed{1}$$

Step 3. **Convert back to original variable.**
To write the answer in terms of the original variable x, we note that

$$\sinh u = \frac{x}{4}, \qquad u = \sinh^{-1} \frac{x}{4}$$

However, in Eq. (1) we must express $\sinh 2u$ in terms of x, so we use the identities recalled in the margin to write

← REMINDER

$$\sinh 2u = 2\sinh u \cosh u$$

$$\cosh u = \sqrt{\sinh^2 u + 1}$$

$$4\sinh 2u = 4(2\sinh u \cosh u) = 8\sinh u\sqrt{\sinh^2 u + 1}$$

$$= 8\left(\frac{x}{4}\right)\sqrt{\left(\frac{x}{4}\right)^2 + 1} = 2x\sqrt{\frac{x^2}{16} + 1} = \frac{1}{2}x\sqrt{x^2 + 16}$$

Then Eq. (1) becomes

$$\int \sqrt{x^2 + 16}\, dx = 4\sinh 2u + 8u + C = \frac{1}{2}x\sqrt{x^2 + 16} + 8\sinh^{-1}\frac{x}{4} + C \quad \blacksquare$$

In the next theorem, we state the integral formulas corresponding to the derivative formulas for the inverse hyperbolic functions recorded in Section 3.10. Each formula is valid on the domain where the integrand and inverse hyperbolic function are defined.

THEOREM 1 Integrals Involving Inverse Hyperbolic Functions

$$\int \frac{dx}{\sqrt{x^2 + 1}} = \sinh^{-1} x + C$$

$$\int \frac{dx}{\sqrt{x^2 - 1}} = \cosh^{-1} x + C \qquad (\text{for } x > 1)$$

$$\int \frac{dx}{1 - x^2} = \tanh^{-1} x + C \qquad (\text{for } |x| < 1)$$

$$\int \frac{dx}{1 - x^2} = \coth^{-1} x + C \qquad (\text{for } |x| > 1)$$

$$\int \frac{dx}{x\sqrt{1 - x^2}} = -\operatorname{sech}^{-1} x + C \qquad (\text{for } 0 < x < 1)$$

$$\int \frac{dx}{|x|\sqrt{1 + x^2}} = -\operatorname{csch}^{-1} x + C \qquad (\text{for } x \neq 0)$$

■ **EXAMPLE 4** Evaluate **(a)** $\displaystyle\int_2^4 \frac{dx}{\sqrt{x^2 - 1}}$ and **(b)** $\displaystyle\int_{0.2}^{0.6} \frac{x\, dx}{1 - x^4}$.

Solution

(a) By Theorem 1,

$$\int_2^4 \frac{dx}{\sqrt{x^2 - 1}} = \cosh^{-1} x\Big|_2^4 = \cosh^{-1} 4 - \cosh^{-1} 2 \approx 0.75$$

(b) First use the substitution $u = x^2$, $du = 2x\, du$. The new limits of integration become $u = (0.2)^2 = 0.04$ and $u = (0.6)^2 = 0.36$. We obtain

$$\int_{0.2}^{0.6} \frac{x\, dx}{1 - x^4} = \int_{0.04}^{0.36} \frac{\frac{1}{2} du}{1 - u^2} = \frac{1}{2}\int_{0.04}^{0.36} \frac{du}{1 - u^2}$$

By Theorem 1, both $\tanh^{-1} u$ and $\coth^{-1} u$ are antiderivatives of $y = (1 - u^2)^{-1}$. We use $\tanh^{-1} u$ because the interval of integration $[0.04, 0.36]$ is contained in the domain $\{u : |u| < 1\}$ of $\tanh^{-1} u$. If the limits of integration were contained in $\{u : |u| > 1\}$, we would use $\coth^{-1} u$. The result is

$$\frac{1}{2} \int_{0.04}^{0.36} \frac{du}{1 - u^2} = \frac{1}{2}\left(\tanh^{-1}(0.36) - \tanh^{-1}(0.04)\right) \approx 0.1684 \qquad \blacksquare$$

Excursion: A Leap of Imagination

The terms "hyperbolic sine" and "hyperbolic cosine" suggest a connection between the hyperbolic and trigonometric functions. This excursion explores the source of this connection, which leads us to **complex numbers** and Euler's famous formula.

Recall that $y = e^t$ satisfies the differential equation $y' = y$ and *every* solution is of the form $y = Ce^t$ for some constant C. On the other hand, we can check that both $y = e^t$ and $y = e^{-t}$ satisfy the **second-order differential equation**

This differential equation is called "second-order" because it involves the second derivative y''.

$$\boxed{y'' = y} \qquad \boxed{2}$$

Indeed, $(e^t)'' = e^t$ and $(e^{-t})'' = (-e^{-t})' = e^{-t}$. It can be shown that every solution of Eq. (2) has the form $y = Ae^t + Be^{-t}$ for some constants A and B.

Now let's examine what happens when we change Eq. (2) by a minus sign:

$$\boxed{y'' = -y} \qquad \boxed{3}$$

In this case, $y = \sin t$ and $y = \cos t$ are solutions because

$$(\sin t)'' = (\cos t)' = -\sin t, \qquad (\cos t)'' = (-\sin t)' = -\cos t$$

Furthermore, it can be proved as before that every solution of Eq. (3) has the form

$$y = A \cos t + B \sin t$$

This might seem to be the end of the story. However, there is another way to write down solutions to Eq. (3). Consider the exponential functions $y = e^{it}$ and $y = e^{-it}$, where

$$i = \sqrt{-1}$$

The number i is an *imaginary* complex number satisfying $i^2 = -1$. Since i is not a real number, the exponential e^{it} is not defined without further explanation. But if we accept that e^{it} *can* be defined and that it is legitimate to apply the usual rules of calculus to it, we obtain

$$\frac{d}{dt}e^{it} = ie^{it}$$

Differentiating a second time yields

$$(e^{it})'' = (ie^{it})' = i^2 e^{it} = -e^{it}$$

In other words, $y = e^{it}$ is a solution to the equation $y'' = -y$ and must therefore be expressible in terms of $\sin t$ and $\cos t$. That is, for some constants A and B,

$$e^{it} = A \cos t + B \sin t \qquad \boxed{4}$$

We determine A and B by considering initial conditions. First, set $t = 0$ in Eq. (4):

$$1 = e^{i0} = A \cos 0 + B \sin 0 = A$$

FIGURE 1 Leonhard Euler (1707–1783). Euler (pronounced "oil-er") ranks among the greatest mathematicians of all time. His work (printed in more than 70 volumes) contains fundamental contributions to almost every aspect of the mathematics and physics of his time. The French mathematician Pierre Simon de Laplace once declared: "Read Euler, he is our master in everything."

Then take the derivative of Eq. (4) and set $t = 0$:

$$ie^{it} = \frac{d}{dt}e^{it} = A\cos' t + B\sin' t = -A\sin t + B\cos t$$

$$i = ie^{i0} = -A\sin 0 + B\cos 0 = B$$

Thus, $A = 1$ and $B = i$, and Eq. (4) yields **Euler's Formula**:

$$\boxed{e^{it} = \cos t + i\sin t}$$

Euler derived this result using power series, which allow us to define e^{it} in a rigorous fashion.

At $t = \pi$, Euler's Formula yields

$$\boxed{e^{i\pi} = -1}$$

Here we have a simple but surprising relation between the three important numbers e, π, and i.

Euler's Formula also reveals the source of the analogy between hyperbolic and trigonometric functions. Let us calculate the hyperbolic cosine at $x = it$:

$$\cosh(it) = \frac{e^{it} + e^{-it}}{2} = \frac{\cos t + i\sin t}{2} + \frac{\cos(-t) + i\sin(-t)}{2} = \cos t$$

A similar calculation shows that $\sinh(it) = i\sin(t)$. In other words, the hyperbolic and trigonometric functions are not merely analogous—once we introduce complex numbers, we see that they are very nearly the same functions.

7.5 SUMMARY

- Integrals of hyperbolic functions:

$$\int \sinh x \, dx = \cosh x + C, \qquad \int \cosh x \, dx = \sinh x + C$$

$$\int \operatorname{sech}^2 x \, dx = \tanh x + C, \qquad \int \operatorname{csch}^2 x \, dx = -\coth x + C$$

$$\int \operatorname{sech} x \tanh x \, dx = -\operatorname{sech} x + C, \qquad \int \operatorname{csch} x \coth x \, dx = -\operatorname{csch} x + C$$

- Integrals involving inverse hyperbolic functions:

$$\int \frac{dx}{\sqrt{x^2 + 1}} = \sinh^{-1} x + C$$

$$\int \frac{dx}{\sqrt{x^2 - 1}} = \cosh^{-1} x + C \qquad \text{(for } x > 1\text{)}$$

$$\int \frac{dx}{1 - x^2} = \tanh^{-1} x + C \qquad \text{(for } |x| < 1\text{)}$$

$$\int \frac{dx}{1 - x^2} = \coth^{-1} x + C \qquad \text{(for } |x| > 1\text{)}$$

$$\int \frac{dx}{x\sqrt{1-x^2}} = -\operatorname{sech}^{-1} x + C \qquad (\text{for } 0 < x < 1)$$

$$\int \frac{dx}{|x|\sqrt{1+x^2}} = -\operatorname{csch}^{-1} x + C \quad (\text{for } x \neq 0)$$

7.5 EXERCISES

Preliminary Questions

1. Which hyperbolic substitution can be used to evaluate the following integrals?

(a) $\displaystyle\int \frac{dx}{\sqrt{x^2+1}}$ **(b)** $\displaystyle\int \frac{dx}{\sqrt{x^2+9}}$ **(c)** $\displaystyle\int \frac{dx}{\sqrt{9x^2+1}}$

2. Which two hyperbolic integration formulas differ from their trigonometric counterparts by a minus sign?

3. Which antiderivative of $y = (1-x^2)^{-1}$ should be used to evaluate $\displaystyle\int_3^5 (1-x^2)^{-1}\,dx$?

Exercises

In Exercises 1–16, calculate the integral.

1. $\displaystyle\int \cosh(3x)\,dx$

2. $\displaystyle\int \sinh(x+1)\,dx$

3. $\displaystyle\int x\sinh(x^2+1)\,dx$

4. $\displaystyle\int \sinh^2 x \cosh x\,dx$

5. $\displaystyle\int \operatorname{sech}^2(1-2x)\,dx$

6. $\displaystyle\int \tanh(3x)\operatorname{sech}(3x)\,dx$

7. $\displaystyle\int \tanh x \operatorname{sech}^2 x\,dx$

8. $\displaystyle\int \frac{\cosh x}{3\sinh x + 4}\,dx$

9. $\displaystyle\int \tanh x\,dx$

10. $\displaystyle\int x\operatorname{csch}(x^2)\coth(x^2)\,dx$

11. $\displaystyle\int \frac{\cosh x}{\sinh x}\,dx$

12. $\displaystyle\int \frac{\cosh x}{\sinh^2 x}\,dx$

13. $\displaystyle\int \sinh^2(4x-9)\,dx$

14. $\displaystyle\int \sinh^3 x \cosh^6 x\,dx$

15. $\displaystyle\int \sinh^2 x \cosh^2 x\,dx$

16. $\displaystyle\int \tanh^3 x\,dx$

In Exercises 17–30, calculate the integral in terms of the inverse hyperbolic functions.

17. $\displaystyle\int \frac{dx}{\sqrt{x^2-1}}$

18. $\displaystyle\int \frac{dx}{\sqrt{x^2-4}}$

19. $\displaystyle\int \frac{dx}{\sqrt{16+x^2}}$

20. $\displaystyle\int \frac{dx}{\sqrt{1+3x^2}}$

21. $\displaystyle\int \sqrt{x^2-1}\,dx$

22. $\displaystyle\int \frac{x^2\,dx}{\sqrt{x^2+1}}$

23. $\displaystyle\int_{-1/2}^{1/2} \frac{dx}{1-x^2}$

24. $\displaystyle\int_4^5 \frac{dx}{1-x^2}$

25. $\displaystyle\int_0^1 \frac{dx}{\sqrt{1+x^2}}$

26. $\displaystyle\int_2^{10} \frac{dx}{4x^2-1}$

27. $\displaystyle\int_{-3}^{-1} \frac{dx}{x\sqrt{x^2+16}}$

28. $\displaystyle\int_{0.2}^{0.8} \frac{dx}{x\sqrt{1-x^2}}$

29. $\displaystyle\int \frac{\sqrt{x^2-1}\,dx}{x^2}$

30. $\displaystyle\int_1^9 \frac{dx}{x\sqrt{x^4+1}}$

31. Verify the formulas

$$\sinh^{-1} x = \ln|x + \sqrt{x^2+1}|$$

$$\cosh^{-1} x = \ln|x + \sqrt{x^2-1}| \quad \text{for } x \geq 1$$

32. Verify that for $|x| < 1$,

$$\tanh^{-1} x = \frac{1}{2}\ln\left|\frac{1+x}{1-x}\right|$$

33. Evaluate $\displaystyle\int \sqrt{x^2+16}\,dx$ using trigonometric substitution. Then use Exercise 31 to verify that your answer agrees with the answer in Example 3.

34. Evaluate $\displaystyle\int \sqrt{x^2-9}\,dx$ in two ways: using trigonometric substitution and using hyperbolic substitution. Then use Exercise 31 to verify that the two answers agree.

35. Evaluate $\displaystyle\int \frac{dx}{1-x^2}$ in terms of the logarithm by writing $\dfrac{1}{1-x^2} = \dfrac{1}{2}\left(\dfrac{1}{1+x} + \dfrac{1}{1-x}\right)$. Then use Exercise 32 to verify that your result agrees with Theorem 1.

36. [GU] Find an equation for the tangent line to the graph of $f(x) = \tanh^{-1} x$ at $x = 0.8$. Using a graphing utility, plot the graphs of $f(x)$ and the tangent line on the same pair of axes.

37. [GU] Repeat Exercise 36 with $f(x) = \operatorname{csch}^{-1} x$.

38. Prove the reduction formula for $n \geq 2$:

$$\int \cosh^n x \, dx = \frac{1}{n} \cosh^{n-1} x \sinh x$$

$$+ \frac{n-1}{n} \int \cosh^{n-2} x \, dx \qquad \boxed{5}$$

39. Use Eq. (5) to compute $\int \cosh^4 x \, dx$.

40. Compute $\int \dfrac{\tanh^{-1} x \, dx}{x^2 - 1}$.

41. Compute $\int \sinh^{-1} x \, dx$. *Hint:* Use Integration by Parts with $v' = 1$.

42. Compute $\int \tanh^{-1} x \, dx$.

43. Compute $\int x \tanh^{-1} x \, dx$.

Further Insights and Challenges

44. Show that if $u = \tanh(x/2)$, then

$$\cosh x = \frac{1 + u^2}{1 - u^2}, \quad \sinh x = \frac{2u}{1 - u^2}, \quad dx = \frac{2 \, du}{1 - u^2}$$

Hint: For the first relation, use the identities

$$\sinh^2 \left(\frac{x}{2} \right) = \frac{1}{2} (\cosh x - 1), \quad \cosh^2 \left(\frac{x}{2} \right) = \frac{1}{2} (\cosh x + 1)$$

45. Use the substitution of Exercise 44 to evaluate $\int \operatorname{sech} x \, dx$.

46. Use the substitution of Exercise 44 to evaluate $\int \dfrac{dx}{1 + \cosh x}$.

47. The relations $\cosh(it) = \cos t$ and $\sinh(it) = i \sin t$ were discussed in the Excursion. Use these relations to show that the identity $\cos^2 t + \sin^2 t = 1$ results from setting $x = it$ in the identity $\cosh^2 x - \sinh^2 x = 1$.

7.6 The Method of Partial Fractions

◄-- *REMINDER A "rational function" is a quotient of two polynomials.*

In this section, we introduce the Method of Partial Fractions, used to integrate rational functions $\dfrac{P(x)}{Q(x)}$, where $P(x)$ and $Q(x)$ are polynomials. As we will see, the integral of a rational function can be expressed as a sum of three types of terms: rational functions, arctangents of linear or quadratic functions, and logarithms of polynomials. Here is a typical example:

$$\int \frac{(2x^3 + x^2 - 2x + 2) \, dx}{(x^2 + 1)^2} = \frac{x + 4}{2x^2 + 2} + \frac{3}{2} \tan^{-1} x + \ln(x^2 + 1) + C$$

It is a fact from algebra that every polynomial $Q(x)$ with real coefficients can be written as a product of linear and quadratic factors with real coefficients. However, it is not always possible to find these factors explicitly.

The Method of Partial Fractions yields an explicit antiderivative of this type whenever we can factor the denominator $Q(x)$ into a product of linear and quadratic factors.

A rational function $\dfrac{P(x)}{Q(x)}$ is called **proper** if the degree of $P(x)$ [denoted $\deg(P)$] is *less than* the degree of $Q(x)$. For example,

$$\underbrace{\frac{x^2 - 3x + 7}{x^4 - 16}}_{\text{Proper}}, \qquad \underbrace{\frac{2x^2 + 7}{x - 5}, \quad \frac{x - 2}{x - 5}}_{\text{Not proper}}$$

Suppose first that $\dfrac{P(x)}{Q(x)}$ is proper and that the denominator $Q(x)$ factors as a product of *distinct linear factors*. In other words,

$$\frac{P(x)}{Q(x)} = \frac{P(x)}{(x - a_1)(x - a_2) \cdots (x - a_n)}$$

Each distinct linear factor $(x - a)$ in the denominator contributes a term of the form

$$\frac{A}{x - a}$$

to the partial fraction decomposition.

where the roots $a_1, a_2, \ldots, a_n$ are all distinct and $\deg(P) < n$. Then there is a **partial fraction decomposition**:

$$\frac{P(x)}{Q(x)} = \frac{A_1}{(x - a_1)} + \frac{A_2}{(x - a_2)} + \cdots + \frac{A_n}{(x - a_n)}$$

for suitable constants $A_1, \ldots, A_n$. For example,

$$\frac{-5x^2 - x + 28}{(x + 1)(x - 2)(x - 3)} = \frac{2}{x + 1} - \frac{2}{x - 2} - \frac{5}{x - 3}$$

Once we have found the partial fraction decomposition, we can integrate the individual terms.

■ **EXAMPLE 1** Finding the Constants $A_1, A_2, \ldots, A_n$ Evaluate $\displaystyle\int \frac{dx}{x^2 - 7x + 10}$.

Solution The denominator factors as $x^2 - 7x + 10 = (x - 2)(x - 5)$, so we look for a partial fraction decomposition:

$$\frac{1}{(x - 2)(x - 5)} = \frac{A}{x - 2} + \frac{B}{x - 5}$$

To find the constants A and B, first multiply by $(x - 2)(x - 5)$ to clear denominators:

$$1 = (x - 2)(x - 5)\left(\frac{A}{x - 2} + \frac{B}{x - 5}\right)$$

$$1 = A(x - 5) + B(x - 2) \qquad \boxed{1}$$

This equation holds for all values of x (including $x = 2$ and $x = 5$, by continuity). We determine A by setting $x = 2$ (which makes the second term disappear):

$$1 = A(2 - 5) + \underbrace{B(2 - 2)}_{\text{This is zero}} = -3A \quad \Rightarrow \quad \boxed{A = -\frac{1}{3}}$$

Similarly, to calculate B, set $x = 5$ in Eq. (1):

$$1 = A(5 - 5) + B(5 - 2) = 3B \quad \Rightarrow \quad \boxed{B = \frac{1}{3}}$$

The resulting partial fraction decomposition is

$$\boxed{\frac{1}{(x - 2)(x - 5)} = \frac{-1/3}{x - 2} + \frac{1/3}{x - 5}}$$

The integration can now be carried out:

$$\int \frac{dx}{(x - 2)(x - 5)} = -\frac{1}{3}\int \frac{dx}{x - 2} + \frac{1}{3}\int \frac{dx}{x - 5}$$

$$= -\frac{1}{3}\ln|x - 2| + \frac{1}{3}\ln|x - 5| + C \qquad ■$$

■ **EXAMPLE 2** Evaluate $\displaystyle\int \frac{x^2 + 2}{(x - 1)(2x - 8)(x + 2)}\, dx$.

Solution

Step 1. Find the partial fraction decomposition.

The decomposition has the form

In Eq. (2), the linear factor $2x - 8$ does not have the form $(x - a)$ used previously, but the partial fraction decomposition can be carried out in the same way.

$$\frac{x^2 + 2}{(x - 1)(2x - 8)(x + 2)} = \frac{A}{x - 1} + \frac{B}{2x - 8} + \frac{C}{x + 2} \qquad \boxed{2}$$

As before, multiply by $(x - 1)(2x - 8)(x + 2)$ to clear denominators:

$$x^2 + 2 = A(2x - 8)(x + 2) + B(x - 1)(x + 2) + C(x - 1)(2x - 8) \qquad \boxed{3}$$

Since A goes with the factor $(x - 1)$, we set $x = 1$ in Eq. (3) to compute A:

$$1^2 + 2 = A(2 - 8)(1 + 2) + \overbrace{B(1 - 1)(1 + 2)}^{\text{Zero}} + C(1 - 1)(2 - 8)$$

$$3 = -18A \quad \Rightarrow \quad \boxed{A = -\frac{1}{6}}$$

Similarly, 4 is the root of $2x - 8$, so we compute B by setting $x = 4$ in Eq. (3):

$$4^2 + 2 = A(8 - 8)(4 + 2) + B(4 - 1)(4 + 2) + C(4 - 1)(8 - 8)$$

$$18 = 18B \quad \Rightarrow \quad \boxed{B = 1}$$

Finally, C is determined by setting $x = -2$ in Eq. (3):

$$(-2)^2 + 2 = A(-4 - 8)(-2 + 2) + B(-2 - 1)(-2 + 2) + C(-2 - 1)(-4 - 8)$$

$$6 = 36C \quad \Rightarrow \quad \boxed{C = \frac{1}{6}}$$

The result is

$$\boxed{\frac{x^2 + 2}{(x - 1)(2x - 8)(x + 2)} = -\frac{1/6}{x - 1} + \frac{1}{2x - 8} + \frac{1/6}{x + 2}}$$

Step 2. Carry out the integration.

$$\int \frac{x^2 + 2}{(x - 1)(2x - 8)(x + 2)} \, dx = -\frac{1}{6} \int \frac{dx}{x - 1} + \int \frac{dx}{2x - 8} + \frac{1}{6} \int \frac{dx}{x + 2}$$

$$= -\frac{1}{6} \ln|x - 1| + \frac{1}{2} \ln|2x - 8| + \frac{1}{6} \ln|x + 2| + C$$

∎

If $\dfrac{P(x)}{Q(x)}$ is not proper, that is, if $\deg(P) \geq \deg(Q)$, we use long division to write

$$\frac{P(x)}{Q(x)} = g(x) + \frac{R(x)}{Q(x)}$$

where $g(x)$ is a polynomial and $\dfrac{R(x)}{Q(x)}$ is proper. We may then integrate $\dfrac{P(x)}{Q(x)}$ using the partial fraction decomposition of $\dfrac{R(x)}{Q(x)}$.

■ **EXAMPLE 3** Long Division Necessary Evaluate $\displaystyle\int \dfrac{x^3+1}{x^2-4}\,dx$.

Solution Using long division, we write

$$\frac{x^3+1}{x^2-4} = x + \frac{4x+1}{x^2-4} = x + \frac{4x+1}{(x-2)(x+2)}$$

The second term has a partial fraction decomposition:

$$\frac{4x+1}{(x-2)(x+2)} = \frac{A}{x-2} + \frac{B}{x+2}$$

To find A and B, clear denominators:

$$4x+1 = A(x+2) + B(x-2)$$

Setting $x = 2$, we get $A = 9/4$, and setting $x = -2$, we get $B = 7/4$. Therefore,

$$\frac{x^3+1}{x^2-4} = x + \frac{9/4}{x-2} + \frac{7/4}{x+2}$$

$$\int \frac{(x^3+1)\,dx}{x^2-4} = \int x\,dx + \frac{9}{4}\int \frac{dx}{x-2} + \frac{7}{4}\int \frac{dx}{x+2}$$

$$= \frac{1}{2}x^2 + \frac{9}{4}\ln|x-2| + \frac{7}{4}\ln|x+2| + C \qquad ■$$

Now suppose we have a rational function $\dfrac{P(x)}{Q(x)}$ whose denominator has repeated linear factors:

$$\frac{P(x)}{Q(x)} = \frac{P(x)}{(x-a_1)^{M_1}(x-a_2)^{M_2}\cdots(x-a_n)^{M_n}}$$

Each factor $(x-a_i)^{M_i}$ contributes the following sum of terms to the partial fraction decomposition:

$$\frac{B_1}{(x-a_i)} + \frac{B_2}{(x-a_i)^2} + \cdots + \frac{B_{M_i}}{(x-a_i)^{M_i}}$$

■ **EXAMPLE 4** Repeated Linear Factors Evaluate $\displaystyle\int \dfrac{3x-9}{(x-1)(x+2)^2}\,dx$.

Solution We are looking for a partial fraction decomposition of the form

$$\frac{3x-9}{(x-1)(x+2)^2} = \frac{A}{x-1} + \frac{B_1}{x+2} + \frac{B_2}{(x+2)^2}$$

Let's clear denominators to obtain

$$3x - 9 = A(x+2)^2 + B_1(x-1)(x+2) + B_2(x-1) \qquad \boxed{4}$$

Long division:

$$\begin{array}{r} x \\ x^2-4\,\overline{)\,x^3\ +\ 1} \\ \underline{x^3\ -\ 4x} \\ 4x\ +\ 1 \end{array}$$

The quotient $\dfrac{x^3+1}{x^2-4}$ is equal to x with remainder $4x+1$.

We compute A and B_2 by substituting in Eq. (4) in the usual way:

- Set $x = 1$: This gives $-6 = 9A$ or $\boxed{A = -\dfrac{2}{3}}$.

- Set $x = -2$: This gives $-15 = -3B_2$ or $\boxed{B_2 = 5}$.

With these constants, Eq. (4) becomes

$$3x - 9 = -\frac{2}{3}(x + 2)^2 + B_1(x - 1)(x + 2) + 5(x - 1) \qquad \boxed{5}$$

We cannot determine B_1 in the same way as A and B_2. Here are two ways to proceed.

- **First method (substitution):** If we substitute $x = 1$ or $x = -2$ in Eq. (5), then the term involving B_1 drops out. But Eq. (5) is valid for all values of x, so let's plug in a value different from $x = 1, -2$, such as $x = 2$. Then Eq. (5) gives us

$$3(2) - 9 = -\frac{2}{3}(2 + 2)^2 + B_1(2 - 1)(2 + 2) + 5(2 - 1)$$

$$-3 = -\frac{32}{3} + 4B_1 + 5$$

$$B_1 = \frac{1}{4}\left(-8 + \frac{32}{3}\right) = \frac{2}{3}$$

- **Second method (undetermined coefficients):** Expand the terms in Eq. (5):

$$3x - 9 = -\frac{2}{3}(x^2 + 4x + 4) + B_1(x^2 + x - 2) + 5(x - 1)$$

$$3x - 9 = \left(B_1 - \frac{2}{3}\right)x^2 + \left(B_1 + \frac{7}{3}\right)x - \left(2B_1 + \frac{23}{3}\right)$$

The coefficients of the powers of x on each side of the equation must be equal. Since x^2 does not occur on the left-hand side, $0 = B_1 - \frac{2}{3}$ or $B_1 = \frac{2}{3}$.

Either way, we have shown that

$$\boxed{\frac{3x - 9}{(x - 1)(x + 2)^2} = -\frac{2/3}{x - 1} + \frac{2/3}{x + 2} + \frac{5}{(x + 2)^2}}$$

$$\int \frac{3x - 9}{(x - 1)(x + 2)^2}\, dx = -\frac{2}{3} \int \frac{dx}{x - 1} + \frac{2}{3} \int \frac{dx}{x + 2} + 5 \int \frac{dx}{(x + 2)^2}$$

$$= -\frac{2}{3} \ln |x - 1| + \frac{2}{3} \ln |x + 2| - \frac{5}{x + 2} + C \qquad \blacksquare$$

Quadratic Factors

A quadratic polynomial $ax^2 + bx + c$ is called **irreducible** if it cannot be written as a product of two linear factors (without using complex numbers). A power of an irreducible quadratic factor $(ax^2 + bx + c)^M$ contributes a sum of the following type to a partial fraction decomposition:

$$\frac{A_1x + B_1}{ax^2 + bx + c} + \frac{A_2x + B_2}{(ax^2 + bx + c)^2} + \cdots + \frac{A_Mx + B_M}{(ax^2 + bx + c)^M}$$

For example,

$$\frac{4 - x}{x(x^2 + 4x + 2)^2} = \frac{1}{x} - \frac{x + 4}{x^2 + 4x + 2} - \frac{2x + 9}{(x^2 + 4x + 2)^2}$$

You may need to use trigonometric substitution to integrate these terms. In particular, the following result may be useful (see Exercise 31 in Section 7.4).

$$\int \frac{dx}{x^2 + a} = \frac{1}{\sqrt{a}} \tan^{-1}\left(\frac{x}{\sqrt{a}}\right) + C \quad \text{(for } a > 0\text{)}$$ **6**

◄·· REMINDER *If $b > 0$, then $x^2 + b$ is irreducible, but $x^2 - b$ is reducible because*

$$x^2 - b = (x + \sqrt{b})(x - \sqrt{b})$$

■ **EXAMPLE 5**   Irreducible versus Reducible Quadratic Factors   Evaluate

(a) $\displaystyle\int \frac{18}{(x + 3)(x^2 + 9)}\, dx$ **(b)** $\displaystyle\int \frac{18}{(x + 3)(x^2 - 9)}\, dx$

Solution

(a) The quadratic factor $x^2 + 9$ is irreducible, so the partial fraction decomposition has the form

$$\frac{18}{(x + 3)(x^2 + 9)} = \frac{A}{x + 3} + \frac{Bx + C}{x^2 + 9}$$

Clear denominators to obtain

$$18 = A(x^2 + 9) + (Bx + C)(x + 3)$$ **7**

To find A, set $x = -3$:

$$18 = A((-3)^2 + 9) + 0 \quad \Rightarrow \quad A = 1$$

Then substitute $A = 1$ in (7) to obtain

$$18 = (x^2 + 9) + (Bx + C)(x + 3) = (B + 1)x^2 + (C + 3B)x + (9 + 3C)$$

Equating coefficients, we get $B + 1 = 0$ or $B = -1$ and $9 + 3C = 18$ or $C = 3$. Thus,

$$\int \frac{18\, dx}{(x + 3)(x^2 + 9)} = \int \frac{dx}{x + 3} - \int \frac{(x - 3)\, dx}{x^2 + 9}$$

$$= \int \frac{dx}{x + 3} - \int \frac{x\, dx}{x^2 + 9} + \int \frac{3\, dx}{x^2 + 9}$$

$$= \ln|x + 3| - \frac{1}{2}\ln(x^2 + 9) + \tan^{-1}\frac{x}{3} + C$$

Here we have used

$$\int \frac{x}{x^2 + 9}\, dx = \frac{1}{2}\int \frac{du}{u} = \frac{1}{2}\ln(x^2 + 9) + C \quad (u = x^2 + 9,\ du = 2x\, dx)$$

$$\int \frac{dx}{x^2 + 9} = \frac{1}{3}\tan^{-1}\frac{x}{3} + C \quad\quad\quad\quad \text{[by Eq. (6)]}$$

(b) The polynomial $x^2 - 9$ is not irreducible because $x^2 - 9 = (x - 3)(x + 3)$. Therefore, the partial fraction decomposition has the form

$$\frac{18}{(x+3)(x^2-9)} = \frac{18}{(x+3)^2(x-3)} = \frac{A}{x-3} + \frac{B}{x+3} + \frac{C}{(x+3)^2}$$

Clear denominators:

$$18 = A(x+3)^2 + B(x+3)(x-3) + C(x-3)$$

Setting $x = 3$, we obtain $18 = (6^2)A$ or $A = \frac{1}{2}$, and setting $x = -3$, we obtain $18 = -6C$ or $C = -3$. Therefore,

$$18 = \frac{1}{2}(x+3)^2 + B(x+3)(x-3) - 3(x-3)$$

To find B, we plug in any value other than ± 3, such as $x = 2$. Then $18 = \frac{1}{2}(25) - 5B + 3$ or $B = -\frac{1}{2}$, and

$$\int \frac{18}{(x+3)(x^2-9)}\,dx = \frac{1}{2}\int \frac{dx}{x-3} - \frac{1}{2}\int \frac{dx}{x+3} - 3\int \frac{dx}{(x+3)^2}$$

$$= \frac{1}{2}\ln|x-3| - \frac{1}{2}\ln|x+3| + 3(x+3)^{-1} + C \qquad \blacksquare$$

■ **EXAMPLE 6** Repeated Quadratic Factor Evaluate $\displaystyle\int \frac{4-x}{x(x^2+2)^2}\,dx$.

Solution The partial fraction decomposition has the form

$$\frac{4-x}{x(x^2+2)^2} = \frac{A}{x} + \frac{Bx+C}{x^2+2} + \frac{Dx+E}{(x^2+2)^2}$$

Clear denominators by multiplying through by $x(x^2+2)^2$:

$$4 - x = A(x^2+2)^2 + (Bx+C)(x(x^2+2)) + (Dx+E)x \qquad \boxed{8}$$

We compute A directly by setting $x = 0$. Then Eq. (8) reduces to $4 = 4A$ or $A = 1$. We find the remaining coefficients by the method of undetermined coefficients. Set $A = 1$ in Eq. (8) and expand:

$$4 - x = (x^4 + 4x^2 + 4) + (Bx^4 + 2Bx^2 + Cx^3 + 2C) + (Dx^2 + Ex)$$

$$= (1+B)x^4 + Cx^3 + (4+2B+D)x^2 + Ex + 2C + 4$$

Now equate the coefficients on the two sides of the equation:

$$1 + B = 0 \qquad \text{(Coefficient of } x^4)$$

$$C = 0 \qquad \text{(Coefficient of } x^3)$$

$$4 + 2B + D = 0 \qquad \text{(Coefficient of } x^2)$$

$$E = -1 \qquad \text{(Coefficient of } x)$$

$$2C + 4 = 4 \qquad \text{(Constant term)}$$

By the first equation, $B = -1$. Setting $B = -1$ in the third equation gives $D = -2$. Thus,

$$\int \frac{(4-x)\,dx}{x(x^2+2)^2} = \int \frac{dx}{x} - \int \frac{x\,dx}{x^2+2} - \int \frac{(2x+1)\,dx}{(x^2+2)^2}$$

$$= \ln|x| - \frac{1}{2}\ln(x^2+2) - \int \frac{(2x+1)dx}{(x^2+2)^2}$$

The middle integral was evaluated using the substitution $u = x^2 + 2$, $du = 2x\,dx$. The third integral breaks up as a sum:

$$\int \frac{(2x+1)\,dx}{(x^2+2)^2} = \overbrace{\int \frac{2x\,dx}{(x^2+2)^2}}^{\text{Use substitution } u = x^2+2} + \int \frac{dx}{(x^2+2)^2}$$

$$= -(x^2+2)^{-1} + \int \frac{dx}{(x^2+2)^2} \qquad \boxed{9}$$

To evaluate the integral in Eq. (9), we use the trigonometric substitution

$$x = \sqrt{2}\tan\theta, \qquad dx = \sqrt{2}\sec^2\theta\,d\theta, \qquad x^2 + 2 = 2\tan^2\theta + 2 = 2\sec^2\theta$$

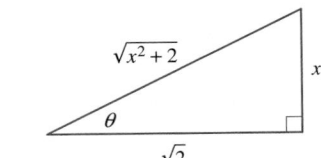

FIGURE 1

Referring to Figure 1, we obtain

$$\int \frac{dx}{(x^2+2)^2} = \int \frac{\sqrt{2}\sec^2\theta\,d\theta}{(2\tan^2\theta+2)^2} = \int \frac{\sqrt{2}\sec^2\theta\,d\theta}{4\sec^4\theta}$$

$$= \frac{\sqrt{2}}{4}\int \cos^2\theta\,d\theta = \frac{\sqrt{2}}{4}\left(\frac{1}{2}\theta + \frac{1}{2}\sin\theta\cos\theta\right) + C$$

$$= \frac{\sqrt{2}}{8}\tan^{-1}\frac{x}{\sqrt{2}} + \frac{\sqrt{2}}{8}\frac{x}{\sqrt{x^2+2}}\frac{\sqrt{2}}{\sqrt{x^2+2}} + C$$

$$= \frac{1}{4\sqrt{2}}\tan^{-1}\frac{x}{\sqrt{2}} + \frac{1}{4}\frac{x}{x^2+2} + C$$

Finally, collecting all the terms, we have

$$\int \frac{4-x}{x(x^2+2)^2}\,dx = \ln|x| - \frac{1}{2}\ln(x^2+2) + \frac{1}{x^2+2} - \frac{1}{4\sqrt{2}}\arctan\frac{x}{\sqrt{2}}$$

$$-\frac{1}{4}\frac{x}{x^2+2} + C$$

$$= \ln|x| - \frac{1}{2}\ln(x^2+2) + \frac{1-\frac{1}{4}x}{x^2+2} - \frac{1}{4\sqrt{2}}\arctan\frac{x}{\sqrt{2}} + C \quad \blacksquare$$

Using a Computer Algebra System

Finding partial fraction decompositions often requires laborious computation. Fortunately, most computer algebra systems have a command (with names such as "Apart" or "parfrac") that produce partial fraction decompositions. For example, the command

```
Apart[(x^2 − 2)/((x + 2)(x^2 + 4)^3)]
```

The polynomial $x^5 + 2x + 2$ cannot be factored explicitly, so the command

```
Apart[1/(x^5 + 2x + 2)]
```

returns the useless response

$$\frac{1}{x^5 + 2x + 2}$$

produces the partial fraction decomposition

$$\frac{x^2-2}{(x+2)(x^2+4)^3} = \frac{1}{256(2+x)} + \frac{3(x-2)}{4(4+x^2)^3} + \frac{2-x}{32(4+x^2)^2} + \frac{2-x}{256(4+x^2)}$$

However, a computer algebra system cannot produce a partial fraction decomposition in cases where $Q(x)$ cannot be factored explicitly.

7.6 SUMMARY

The integral of a rational function $\int \dfrac{P(x)\,dx}{Q(x)}$ is evaluated using the Method of Partial Fractions. Assume that $P(x)/Q(x)$ is *proper* [i.e., $\deg(P) < \deg(Q)$] and that $Q(x)$ can be factored explicitly as a product of linear and irreducible quadratic terms.

- If $Q(x) = (x - a_1)(x - a_2) \cdots (x - a_n)$ where the roots a_j are distinct, then

$$\frac{P(x)}{(x - a_1)(x - a_2) \cdots (x - a_n)} = \frac{A_1}{x - a_1} + \frac{A_2}{x - a_2} + \cdots + \frac{A_n}{x - a_n}$$

To calculate the constants, clear denominators and substitute, in turn, the values $x = a_1,$ $a_2, \ldots, a_n$.

- If $Q(x)$ is equal to a product of powers of linear factors $(x - a)^M$ and irreducible quadratic factors $(x^2 + b)^N$ with $b > 0$, then the partial fraction decomposition of $\dfrac{P(x)}{Q(x)}$ is a sum of terms of the following type:

$$(x - a)^M \quad \text{contributes} \quad \frac{A_1}{x - a} + \frac{A_2}{(x - a)^2} + \cdots + \frac{A_M}{(x - a)^M}$$

$$(x^2 + b)^N \quad \text{contributes} \quad \frac{A_1 x + B_1}{x^2 + b} + \frac{A_2 x + B_2}{(x^2 + b)^2} + \cdots + \frac{A_N x + B_N}{(x^2 + b)^N}$$

Substitution and trigonometric substitution may be needed to integrate the terms corresponding to $(x^2 + b)^N$ (see Example 6).

- If $P(x)/Q(x)$ is improper, use long division (see Example 3).

7.6 EXERCISES

Preliminary Questions

1. Suppose that $\int f(x)\,dx = \ln x + \sqrt{x + 1} + C$. Can $f(x)$ be a rational function? Explain.

2. Which of the following are *proper* rational functions?

(a) $\dfrac{x}{x - 3}$

(b) $\dfrac{4}{9 - x}$

(c) $\dfrac{x^2 + 12}{(x + 2)(x + 1)(x - 3)}$

(d) $\dfrac{4x^3 - 7x}{(x - 3)(2x + 5)(9 - x)}$

3. Which of the following quadratic polynomials are irreducible? To check, complete the square if necessary.

(a) $x^2 + 5$

(b) $x^2 - 5$

(c) $x^2 + 4x + 6$

(d) $x^2 + 4x + 2$

4. Let $P(x)/Q(x)$ be a proper rational function where $Q(x)$ factors as a product of distinct linear factors $(x - a_i)$. Then

$$\int \frac{P(x)\,dx}{Q(x)}$$

(choose correct answer):

(a) is a sum of logarithmic terms $A_i \ln(x - a_i)$ for some constants A_i.

(b) may contain a term involving the arctangent.

Exercises

1. Match the rational function (a)–(d) with the corresponding partial fraction decomposition (i)–(iv).

(a) $\dfrac{x^2 + 4x + 12}{(x + 2)(x^2 + 4)}$

(b) $\dfrac{2x^2 + 8x + 24}{(x + 2)^2(x^2 + 4)}$

(c) $\dfrac{x^2 - 4x + 8}{(x - 1)^2(x - 2)^2}$

(d) $\dfrac{x^4 - 4x + 8}{(x + 2)(x^2 + 4)}$

(i) $x - 2 + \dfrac{4}{x + 2} - \dfrac{4x - 4}{x^2 + 4}$

(ii) $\dfrac{-8}{x - 2} + \dfrac{4}{(x - 2)^2} + \dfrac{8}{x - 1} + \dfrac{5}{(x - 1)^2}$

(iii) $\dfrac{1}{x + 2} + \dfrac{2}{(x + 2)^2} + \dfrac{-x + 2}{x^2 + 4}$

(iv) $\dfrac{1}{x + 2} + \dfrac{4}{x^2 + 4}$

2. Determine the constants A, B:

$$\frac{2x-3}{(x-3)(x-4)} = \frac{A}{x-3} + \frac{B}{x-4}$$

3. Clear denominators in the following partial fraction decomposition and determine the constant B (substitute a value of x or use the method of undetermined coefficients):

$$\frac{3x^2+11x+12}{(x+1)(x+3)^2} = \frac{1}{x+1} - \frac{B}{x+3} - \frac{3}{(x+3)^2}$$

4. Find the constants in the partial fraction decomposition

$$\frac{2x+4}{(x-2)(x^2+4)} = \frac{A}{x-2} + \frac{Bx+C}{x^2+4}$$

In Exercises 5–8, use long division to write $f(x)$ as the sum of a polynomial and a proper rational function. Then calculate $\int f(x)\,dx$.

5. $f(x) = \dfrac{x}{3x-9}$

6. $f(x) = \dfrac{x^2+2}{x+3}$

7. $f(x) = \dfrac{x^3+x+1}{x-2}$

8. $f(x) = \dfrac{x^3-1}{x^2-x}$

In Exercises 9–46, evaluate the integral.

9. $\displaystyle\int \frac{dx}{(x-2)(x-4)}$

10. $\displaystyle\int \frac{dx}{(x-3)(x+7)}$

11. $\displaystyle\int \frac{dx}{x(2x+1)}$

12. $\displaystyle\int \frac{(3x+5)\,dx}{x^2-4x-5}$

13. $\displaystyle\int \frac{(2x-1)\,dx}{x^2-5x+6}$

14. $\displaystyle\int \frac{dx}{(x-2)(x-3)(x+2)}$

15. $\displaystyle\int \frac{(x^2+3x-44)\,dx}{(x+3)(x+5)(3x-2)}$

16. $\displaystyle\int \frac{3\,dx}{(x+1)(x^2+x)}$

17. $\displaystyle\int \frac{(x^2+11x)\,dx}{(x-1)(x+1)^2}$

18. $\displaystyle\int \frac{(4x^2-21x)\,dx}{(x-3)^2(2x+3)}$

19. $\displaystyle\int \frac{dx}{(x-1)^2(x-2)^2}$

20. $\displaystyle\int \frac{dx}{(x+4)^3}$

21. $\displaystyle\int \frac{48\,dx}{x(x+4)^2}$

22. $\displaystyle\int \frac{(x^2+x+3)\,dx}{(x-1)^3}$

23. $\displaystyle\int \frac{dx}{(x-4)^2(x-1)}$

24. $\displaystyle\int \frac{x}{(3x+7)^3}\,dx$

25. $\displaystyle\int \frac{3x+6}{x^2(x-1)(x-3)}\,dx$

26. $\displaystyle\int \frac{dx}{x(x-1)^3}$

27. $\displaystyle\int \frac{(3x^2-2)\,dx}{x-4}$

28. $\displaystyle\int \frac{(x^2-x+1)\,dx}{x^2+x}$

29. $\displaystyle\int \frac{dx}{x(x^2+1)}$

30. $\displaystyle\int \frac{(3x^2-4x+5)\,dx}{(x-1)(x^2+1)}$

31. $\displaystyle\int \frac{x^2\,dx}{x^2+3}$

32. $\displaystyle\int \frac{dx}{2x^2-3}$

33. $\displaystyle\int \frac{x^2}{(x+1)(x^2+1)}\,dx$

34. $\displaystyle\int \frac{6x^2+7x-6}{(x^2-4)(x+2)}\,dx$

35. $\displaystyle\int \frac{x^2\,dx}{(3x+7)^3}$

36. $\displaystyle\int \frac{dx}{x(x^2+25)}$

37. $\displaystyle\int \frac{dx}{x^2(x^2+25)}$

38. $\displaystyle\int \frac{10\,dx}{(x-1)^2(x^2+9)}$

39. $\displaystyle\int \frac{10\,dx}{(x+1)(x^2+9)^2}$

40. $\displaystyle\int \frac{dx}{x(x^2+8)^2}$

41. $\displaystyle\int \frac{100x\,dx}{(x-3)(x^2+1)^2}$

42. $\displaystyle\int \frac{dx}{(x+2)(x^2+4x+10)}$

43. $\displaystyle\int \frac{9\,dx}{(x+1)(x^2-2x+6)}$

44. $\displaystyle\int \frac{25\,dx}{x(x^2+2x+5)^2}$

45. $\displaystyle\int \frac{(x^2+3)\,dx}{(x^2+2x+3)^2}$

46. $\displaystyle\int \frac{dx}{x^4-1}$

47. Evaluate $\displaystyle\int \frac{dx}{x^2-1}$ in two ways: using partial fractions and trigonometric substitution. Verify that the two answers agree.

48. [GU] Graph the equation $(x-40)y^2 = 10x(x-30)$ and find the volume of the solid obtained by revolving the region between the graph and the x-axis for $0 \le x \le 30$ around the x-axis.

49. Evaluate $\displaystyle\int \frac{\sqrt{x}\,dx}{x-1}$. *Hint:* Use the substitution $u = \sqrt{x}$ (sometimes called a **rationalizing substitution**).

50. Evaluate $\displaystyle\int \frac{x\,dx}{x^{1/2}-x^{1/3}}$.

In Exercises 51–66, evaluate the integral using the appropriate method or combination of methods covered thus far in the text.

51. $\displaystyle\int \frac{dx}{x^2\sqrt{4-x^2}}$

52. $\displaystyle\int \cos^2 4x\,dx$

53. $\displaystyle\int \frac{dx}{x(x-1)^2}$

54. $\displaystyle\int x\sec^2 x\,dx$

55. $\displaystyle\int \frac{dx}{(x^2+9)^2}$

56. $\displaystyle\int \theta \sec^{-1}\theta\,d\theta$

57. $\displaystyle\int \tan^5 x\sec x\,dx$

58. $\displaystyle\int \frac{x\,dx}{(x^2-1)^{3/2}}$

59. $\displaystyle\int \frac{x^2\,dx}{(x^2-1)^{3/2}}$

60. $\displaystyle\int \frac{dx}{x(x^2+x)}$

61. $\displaystyle\int \frac{dx}{x(x^2-1)}$

62. $\displaystyle\int \frac{(x+1)\,dx}{(x^2+4x+8)^2}$

63. $\displaystyle\int \frac{\sqrt{x}\,dx}{x^3+1}$

64. $\displaystyle\int \tan^{-1} 3\theta\,d\theta$

65. $\displaystyle\int \frac{dx}{x^4(x^3+1)}$

66. $\displaystyle\int \frac{x^{1/2}dx}{x^{1/3}+1}$

67. Show that the substitution $\theta = 2\tan^{-1} t$ (Figure 2) yields the formulas

$$\cos\theta = \frac{1-t^2}{1+t^2}, \qquad \sin\theta = \frac{2t}{1+t^2}, \qquad d\theta = \frac{2\,dt}{1+t^2} \qquad \boxed{10}$$

This substitution transforms the integral of any rational function of $\cos\theta$ and $\sin\theta$ into an integral of a rational function of t (which

can then be evaluated using partial fractions). Use it to evaluate

$$\int \frac{d\theta}{\cos\theta + (3/4)\sin\theta}.$$

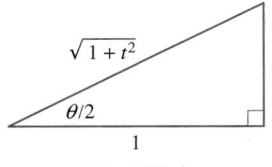

FIGURE 2

68. Use the substitution of Exercise 67 to evaluate $\displaystyle\int \frac{d\theta}{\cos\theta + \sin\theta}$.

Further Insights and Challenges

69. Prove the general formula

$$\int \frac{dx}{(x-a)(x-b)} = \frac{1}{a-b}\ln\frac{x-a}{x-b} + C$$

where a, b are constants such that $a \neq b$.

70. The method of partial fractions shows that

$$\int \frac{dx}{x^2-1} = \frac{1}{2}\ln|x-1| - \frac{1}{2}\ln|x+1| + C$$

The computer algebra system Mathematica evaluates this integral as $-\tanh^{-1} x$, where $\tanh^{-1} x$ is the inverse hyperbolic tangent function. Can you reconcile the two answers?

71. Suppose that $Q(x) = (x-a)(x-b)$, where $a \neq b$, and let $\dfrac{P(x)}{Q(x)}$ be a proper rational function so that

$$\frac{P(x)}{Q(x)} = \frac{A}{(x-a)} + \frac{B}{(x-b)}$$

(a) Show that $A = \dfrac{P(a)}{Q'(a)}$ and $B = \dfrac{P(b)}{Q'(b)}$.

(b) Use this result to find the partial fraction decomposition for $P(x) = 3x - 2$ and $Q(x) = x^2 - 4x - 12$.

72. Suppose that $Q(x) = (x-a_1)(x-a_2)\cdots(x-a_n)$ where the roots a_j are all distinct. Let $\dfrac{P(x)}{Q(x)}$ be a proper rational function so that

$$\frac{P(x)}{Q(x)} = \frac{A_1}{(x-a_1)} + \frac{A_2}{(x-a_2)} + \cdots + \frac{A_n}{(x-a_n)}$$

(a) Show that $A_j = \dfrac{P(a_j)}{Q'(a_j)}$ for $j = 1, \ldots, n$.

(b) Use this result to find the partial fraction decomposition for $P(x) = 2x^2 - 1$, $Q(x) = x^3 - 4x^2 + x + 6 = (x+1)(x-2)(x-3)$.

7.7 Improper Integrals

Sometimes, we are interested in integrating a function over an infinite interval. For example, in statistics, the probability integral

$$\int_{-\infty}^{\infty} e^{-x^2/2}\,dx$$

represents the area under the bell-shaped curve for $-\infty < x < \infty$ (Figure 1). This is an example of an **improper integral**. Although the region under the curve extends infinitely to the left and right, the total area is finite (see Exercise 62).

We consider three types of improper integrals, depending on whether one or both endpoints are infinite:

$$\int_{-\infty}^{b} f(x)\,dx, \qquad \int_{a}^{\infty} f(x)\,dx, \qquad \int_{-\infty}^{\infty} f(x)\,dx$$

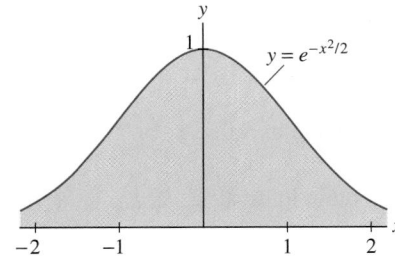

FIGURE 1 Total area under the bell-shaped curve is finite.

In what sense might the area of an infinite region be finite? Consider the function $f(x) = e^{-x}$. The area under the graph over the finite interval $[0, R]$ is [Figure 2(A)]

$$\int_0^R e^{-x}\, dx = -e^{-x}\Big|_0^R = -e^{-R} + e^0 = 1 - e^{-R}$$

Now observe that this area tends to a limit as $R \to \infty$:

$$\int_0^\infty e^{-x}\, dx = \lim_{R\to\infty} \int_0^R e^{-x}\, dx = \lim_{R\to\infty} \left(1 - e^{-R}\right) = 1 \qquad \boxed{1}$$

We take this limit as the definition of the integral over the infinite interval $[0, \infty)$, and thus the infinite region in Figure 2(B) has area 1.

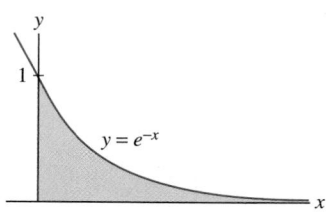

(A) Finite region has area $1 - e^{-R}$.   (B) Infinite region has area 1.

FIGURE 2 The area under $y = e^{-x}$ for $0 \le x \le R$ approaches the limit 1 as $R \to \infty$.

DEFINITION Improper Integral   Fix a number a and assume that $f(x)$ is integrable over $[a, b]$ for all $b \ge a$. The *improper integral* of $f(x)$ over $[a, \infty)$ is defined as the limit (if it exists)

$$\int_a^\infty f(x)\, dx = \lim_{R\to\infty} \int_a^R f(x)\, dx$$

When the limit exists, we say that the improper integral *converges*. In a similar fashion, we define $\displaystyle\int_{-\infty}^a f(x)\, dx = \lim_{R\to-\infty} \int_R^a f(x)\, dx$.

A doubly infinite improper integral is defined as a sum (provided that both integrals on the right converge):

$$\int_{-\infty}^\infty f(x)\, dx = \int_{-\infty}^0 f(x)\, dx + \int_0^\infty f(x)\, dx \qquad \boxed{2}$$

■ **EXAMPLE 1** Show that $\displaystyle\int_2^\infty \frac{dx}{x^2}$ converges and compute its value.

Solution First we evaluate the definite integral over a finite interval $[2, R]$:

$$\int_2^R \frac{dx}{x^2} = -x^{-1}\Big|_2^R = \frac{1}{2} - \frac{1}{R}$$

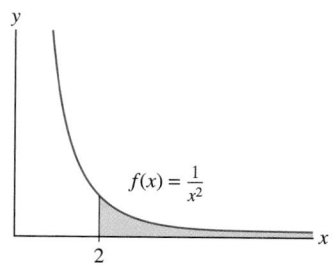

FIGURE 3 The area over $[2, \infty)$ is equal to $\frac{1}{2}$.

The limit as $R \to \infty$ exists, so the improper integral converges (Figure 3):

$$\int_2^\infty \frac{dx}{x^2} = \lim_{R \to \infty} \int_2^R \frac{dx}{x^2} = \lim_{R \to \infty} \left(\frac{1}{2} - \frac{1}{R} \right) = \frac{1}{2}$$

When the limit defining an improper integral is either infinite or does not exist, we say that the improper integral *diverges*. For instance, the area under the parabola $y = x^2$ appears infinite (Figure 4) and indeed, the following integral diverges:

$$\int_0^\infty x^2 \, dx = \lim_{R \to \infty} \int_0^R x^2 \, dx = \lim_{R \to \infty} \frac{1}{3} R^3 = \infty$$

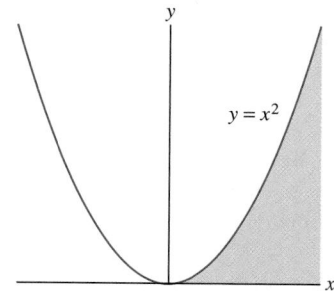

FIGURE 4 Total area under the graph of $y = x^2$ over $[0, \infty)$ is infinite.

In general, however, we cannot tell if an improper integral converges or diverges merely by looking at the graph of the function. The functions $f(x) = x^{-2}$ and $g(x) = x^{-1}$ are both decreasing, but $f(x) = x^{-2}$ decreases more rapidly as $x \to \infty$, and the area under its graph over $[2, \infty)$ is finite, as we saw in the previous example. The next example shows that $g(x) = x^{-1}$ does not tend to zero quickly enough for the improper integral to converge.

■ **EXAMPLE 2** Determine if $\displaystyle\int_{-\infty}^{-1} \frac{dx}{x}$ converges.

Solution The integral over a finite interval $[R, -1]$ for $R < -1$ is

$$\int_R^{-1} \frac{dx}{x} = \ln|x| \Big|_R^{-1} = \ln|-1| - \ln|R| = -\ln|R|$$

The improper integral diverges because the limit as $R \to -\infty$ is infinite (Figure 5):

$$\lim_{R \to -\infty} \int_R^{-1} \frac{dx}{x} = \lim_{R \to -\infty} (-\ln|R|) = -\infty$$

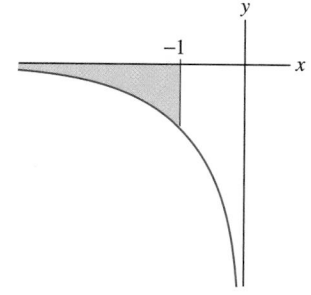

FIGURE 5 The integral of $f(x) = x^{-1}$ over $(-\infty, -1]$ is infinite.

■ **EXAMPLE 3** Using L'Hôpital's Rule Determine if $\displaystyle\int_0^\infty x e^{-x} \, dx$ converges and, if so, find its value.

Solution First, we use Integration by Parts with $u = x$ and $v' = e^{-x}$:

$$\int x e^{-x} \, dx = -x e^{-x} + \int e^{-x} \, dx = -x e^{-x} - e^{-x} = -(x+1)e^{-x} + C$$

$$\int_0^R x e^{-x} \, dx = -(x+1)e^{-x} \Big|_0^R = -(R+1)e^{-R} + 1 = 1 - \frac{R+1}{e^R}$$

Now we can compute the improper integral as a limit using L'Hôpital's Rule:

$$\int_0^\infty x e^{-x} \, dx = 1 - \lim_{R \to \infty} \frac{R+1}{e^R} = 1 - \lim_{R \to \infty} \frac{1}{e^R} = 1 - 0 = 1$$

For which exponents p does the integral $\displaystyle\int_1^\infty \frac{dx}{x^p}$ converge? The answer is given in the following theorem.

The integral $\int_1^\infty \frac{dx}{x^p}$ is called the "p-integral." It is often used in the Comparison Test to determine convergence or divergence of more complicated improper integrals (see Example 8).

THEOREM 1 Improper Integral of x^{-p} over $[1, \infty)$

$$\int_1^\infty \frac{dx}{x^p} = \begin{cases} \dfrac{1}{p-1} & \text{if } p > 1 \\ \infty & \text{if } p \leq 1 \end{cases}$$

Proof Let $J = \int_1^\infty \frac{dx}{x^p}$. If $p \neq 1$, then

$$J = \lim_{R \to \infty} \int_1^R x^{-p}\,dx = \lim_{R \to \infty} \left. \frac{x^{1-p}}{1-p} \right|_1^R = \lim_{R \to \infty} \left(\frac{R^{1-p}}{1-p} - \frac{1}{1-p} \right)$$

If $p > 1$, then R^{1-p} tends to zero as $R \to \infty$. In this case, J converges and $J = 1/(p-1)$. If $p < 1$, then R^{1-p} tends to ∞ and J diverges. Finally, if $p = 1$, then J also diverges because $\int_1^R x^{-1}\,dx = \ln R$ and $\lim_{R \to \infty} \ln R = \infty$. ∎

Improper integrals arise in physical applications when it makes sense to treat a very large quantity as if it were infinite. A typical example is the calculation of escape velocity, the minimum initial velocity needed to launch an object into space.

In physics, we speak of moving an object "infinitely far away." In practice this means "very far way," but it is more convenient to work with an improper integral.

■ **EXAMPLE 4 Escape Velocity** The earth exerts a gravitational force of magnitude $F(r) = \dfrac{GM_em}{r^2}$ on an object of mass m, where r is the distance to the center of the earth.

(a) Find the work required to move an object of mass m infinitely far from the earth.

(b) Calculate the escape velocity on the earth's surface.

Solution Recall that work is the integral of force as a function of distance (Section 6.5).

(a) The work required to move an object from the earth's surface ($r = r_e$) to a distance R from the center is

$$\int_{r_e}^R \frac{GM_em}{r^2}\,dr = \left. -\frac{GM_em}{r} \right|_{r_e}^R = GM_em \left(\frac{1}{r_e} - \frac{1}{R} \right) \text{ J}$$

The work moving the object "infinitely far away" is the improper integral

$$GM_em \int_{r_e}^\infty \frac{dr}{r^2} = \lim_{R \to \infty} GM_em \left(\frac{1}{r_e} - \frac{1}{R} \right) = \frac{GM_em}{r_e} \text{ J}$$

◄⋯ **REMINDER** The mass of the earth is

$$M_e \approx 5.98 \cdot 10^{24} \text{ kg}$$

The radius of the earth is

$$r_e \approx 6.37 \cdot 10^6 \text{ m}$$

The universal gravitational constant is

$$G \approx 6.67 \cdot 10^{-11} \text{ N-m}^2/\text{kg}^2$$

where a newton is 1 kg-m/s^2. A joule is 1 N-m.

(b) Using conservation of energy in physics, we can show that an object launched with initial velocity v_0 will escape the earth's gravitational field if its kinetic energy $\frac{1}{2}mv_0^2$ is at least as large as the work required to move the object to infinity, that is, if

$$\frac{1}{2}mv_0^2 \geq \frac{GM_em}{r_e} \qquad \text{or} \qquad v_0 \geq \left(\frac{2GM_e}{r_e} \right)^{1/2}$$

Escape velocity in miles per hour is approximately 25,000 mph.

Using the values of the constants recalled in the marginal note, we find that the escape velocity v_{esc} is

$$v_{esc} = \left(\frac{2GM_e}{r_e} \right)^{1/2} \approx \sqrt{\frac{2(6.67 \cdot 10^{-11})(5.98 \cdot 10^{24})}{6.37 \cdot 10^6}} \approx 11{,}200 \text{ m/s} \qquad ■$$

In the next example, we calculate the present value of an investment that pays out a dividend "forever."

In practice, the word "forever" means a "long but unspecified length of time." For example, if the investment pays out dividends for 100 years, then its present value is

$$\int_0^{100} 5{,}000e^{-0.5t}\, dt \approx \$99{,}326$$

The improper integral ($100,000) gives a useful approximation to this value.

EXAMPLE 5 Perpetual Annuity An investment pays a dividend continuously at a rate of $5,000/year. Compute the present value of the income stream if the interest rate is 5% and the dividends continue forever.

Solution Recall from Section 5.8 that the present value (PV) of the income stream after T years is $\int_0^T 5{,}000e^{-0.05t}\, dt$. The PV over an infinite time interval is the improper integral

$$\int_0^\infty 5{,}000e^{-0.05t}\, dt = \lim_{T\to\infty} \left.\frac{5{,}000e^{-0.05t}}{-0.05}\right|_0^T = \frac{5{,}000}{0.05} = \$100{,}000$$

An infinite number of dollars are paid out during the infinite time interval, but their total present value is finite. ∎

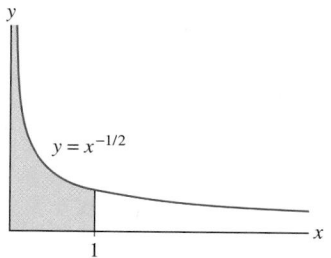

FIGURE 6 Region under the graph over $(0, 1]$ has total area 2 by Example 7.

Infinite Discontinuities at the Endpoints

Another type of improper integral occurs when the integrand $f(x)$ becomes infinite at one or both of the endpoints of the interval of integration. For example, $\int_0^1 \frac{dx}{\sqrt{x}}$ is an improper integral because the integrand $x^{-1/2}$ tends to ∞ as $x \to 0+$ (Figure 6). This integral represents the area of a region that is infinite in the vertical direction. Improper integrals of this type are defined as one-sided limits.

DEFINITION Integrands with Infinite Discontinuities If $f(x)$ is continuous on $[a, b)$ but discontinuous at $x = b$, we define

$$\int_a^b f(x)\, dx = \lim_{R\to b-} \int_a^R f(x)\, dx$$

Similarly, if $f(x)$ is continuous on $(a, b]$ but discontinuous at $x = a$,

$$\int_a^b f(x)\, dx = \lim_{R\to a+} \int_R^b f(x)\, dx$$

If, in either case, the limit does not exist, the integral is said to diverge.

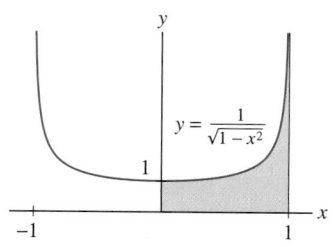

FIGURE 7 The infinite shaded region under the graph of $y = (1 - x^2)^{-1/2}$ has area $\frac{\pi}{2}$.

EXAMPLE 6 Evaluate $\int_0^1 \frac{dx}{\sqrt{1 - x^2}}$.

Solution This is an improper integral because the integrand has an infinite discontinuity at $x = 1$ (Figure 7). Recall that $\int \frac{dx}{\sqrt{1 - x^2}} = \sin^{-1} x + C$. Thus,

$$\int_0^1 \frac{dx}{\sqrt{1 - x^2}} = \lim_{R\to 1-} \int_0^R \frac{dx}{\sqrt{1 - x^2}} = \lim_{R\to 1-} (\sin^{-1} R - \sin^{-1} 0)$$

$$= \sin^{-1} 1 - \sin^{-1} 0 = \frac{\pi}{2} - 0 = \frac{\pi}{2} \qquad ∎$$

■ EXAMPLE 7 Determine the convergence of   **(a)** $\displaystyle\int_0^1 \frac{dx}{\sqrt{x}}$ and **(b)** $\displaystyle\int_0^1 \frac{dx}{x}$.

Solution Both integrals are improper because the integrands have infinite discontinuities at $x = 0$. The first integral converges:

$$\int_0^1 \frac{dx}{\sqrt{x}} = \lim_{R\to 0+} \int_R^1 \frac{dx}{\sqrt{x}} = \lim_{R\to 0+} 2x^{1/2}\Big|_R^1 = \lim_{R\to 0+}(2 - 2R^{1/2}) = 2$$

The second integral diverges (recall that $\ln R$ approaches $-\infty$ as $R \to 0+$):

$$\int_0^1 \frac{dx}{x} = \lim_{R\to 0+} \int_R^1 \frac{dx}{x} = \lim_{R\to 0+}(\ln 1 - \ln R) = -\lim_{R\to 0+}\ln R = \infty \qquad ■$$

The integrals in the previous example are both of the form $\displaystyle\int_0^1 x^{-p}\,dx$. According to the next theorem, this integral converges if $p < 1$ and diverges otherwise. The proof is similar to the proof of Theorem 1 (see Exercise 56).

The result of Theorem 2 is valid for all exponents p. However, the integral is not improper if $p < 0$.

THEOREM 2 Improper Integral of x^{-p} over $[0, 1]$

$$\int_0^1 \frac{dx}{x^p} = \begin{cases} \dfrac{1}{1-p} & \text{if } p < 1 \\[2mm] \infty & \text{if } p \geq 1 \end{cases}$$

Comparing Theorems 1 and 2, we see that the integrals $\displaystyle\int_1^\infty \frac{dx}{x^p}$ and $\displaystyle\int_0^1 \frac{dx}{x^p}$ have opposite behavior for $p \neq 1$. The first integral converges only for $p > 1$ and the second integral converges only for $p < 1$ (Figure 8). They both diverge for $p = 1$.

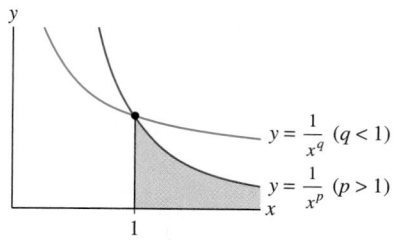

$$\int_1^\infty \frac{dx}{x^p} < \infty \quad \text{but} \quad \int_1^\infty \frac{dx}{x^q} = \infty$$

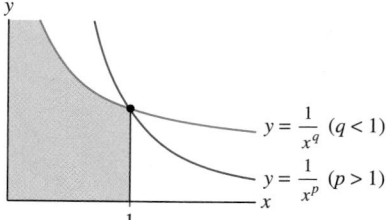

$$\int_0^1 \frac{dx}{x^p} = \infty \quad \text{but} \quad \int_0^1 \frac{dx}{x^q} < \infty$$

FIGURE 8 Comparison of $y = \dfrac{1}{x^p}$ and $y = \dfrac{1}{x^q}$ for $p > 1$ and $q < 1$.

Comparison of Integrals

Sometimes we are interested in determining whether an improper integral converges, even if we cannot find its exact value. For instance, the following integral cannot be evaluated explicitly:

$$\int_1^\infty \frac{e^{-x}}{x}\,dx$$

However, we have the inequalities

$$0 \le \frac{e^{-x}}{x} \le e^{-x} \qquad \text{for } x \ge 1$$

Thus, the graph of e^{-x}/x lies *underneath* the graph of e^{-x} to the right of $x = 1$ (Figure 9) and we should be able to conclude that our integral converges:

$$\underbrace{\int_1^\infty \frac{e^{-x}}{x} \, dx} \le \underbrace{\int_1^\infty e^{-x} \, dx = e^{-1}}$$

This converges by comparison $\qquad$ Converges by direct computation

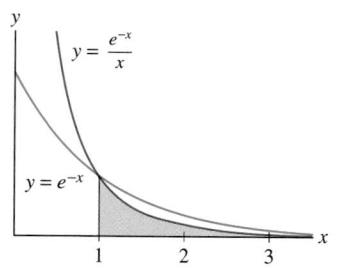

FIGURE 9 There is less area under $y = \dfrac{e^{-x}}{x}$ than under $y = e^{-x}$ in the interval $[1, \infty)$.

The next theorem (whose proof is omitted) summarizes this type of argument.

THEOREM 3 Comparison Test for Improper Integrals
Assume that $f(x) \ge g(x) \ge 0$ for $x \ge a$.

- If $\displaystyle\int_a^\infty f(x) \, dx$ converges, then $\displaystyle\int_a^\infty g(x) \, dx$ also converges.

- If $\displaystyle\int_a^\infty g(x) \, dx$ diverges, then $\displaystyle\int_a^\infty f(x) \, dx$ also diverges.

What the Comparison Test says (for nonnegative functions):

- *If the integral of the bigger function converges, then the integral of the smaller function also converges.*
- *If the integral of the smaller function diverges, then the integral of the larger function also diverges.*
- *The Comparison Test is also valid for improper integrals with infinite discontinuities at endpoints.*

■ **EXAMPLE 8** Show that $\displaystyle\int_1^\infty \frac{dx}{\sqrt{x^3 + 1}}$ converges.

Solution To show that the integral converges, we compare the integrand $(x^3 + 1)^{-1/2}$ with a *larger* function whose integral we can compute. It makes sense to compare with $x^{-3/2}$ because, for $x \ge 1$,

$$\sqrt{x^3 + 1} \ge \sqrt{x^3} = x^{3/2} \qquad \Rightarrow \qquad \frac{1}{x^{3/2}} \ge \frac{1}{\sqrt{x^3 + 1}}$$

The integral $\displaystyle\int_1^\infty \frac{dx}{x^{3/2}}$ converges by Theorem 1 (the case $p = \frac{3}{2}$). Therefore, we may apply the Comparison Test:

$$\int_1^\infty \frac{dx}{x^{3/2}} \text{ converges} \qquad \Rightarrow \qquad \int_1^\infty \frac{dx}{\sqrt{x^3 + 1}} \text{ also converges} \qquad ■$$

If $f(x)$ is continuous, the convergence or divergence of $\displaystyle\int_a^\infty f(x) \, dx$ is not affected by a change in the lower limit a. If we change the lower limit to b, the integral changes by a finite amount:

$$\int_b^\infty f(x) \, dx = \underbrace{\int_b^a f(x) \, dx} + \int_a^\infty f(x) \, dx$$

Ordinary integral whose value is finite

Therefore, the integral over $[b, \infty)$ converges if and only if the integral over $[a, \infty)$ converges:

$$\int_b^\infty f(x) \, dx \text{ converges} \quad \Leftrightarrow \quad \int_a^\infty f(x) \, dx \text{ converges} \qquad \boxed{3}$$

■ **EXAMPLE 9** Determine whether $\displaystyle\int_0^\infty \frac{dx}{\sqrt{x^2+1}}$ converges or diverges.

Solution To decide whether the integral converges or diverges, we might start by observing that if x is large, then

$$\frac{1}{\sqrt{x^2+1}} \approx \frac{1}{\sqrt{x^2}} = \frac{1}{x}$$

We know that $\displaystyle\int_1^\infty \frac{dx}{x}$ diverges (by Theorem 1 with $p=1$), so this suggests that our integral also diverges. However, $\dfrac{1}{\sqrt{x^2+1}} \le \dfrac{1}{x}$, and we need an inequality in the other direction to prove divergence. But note that if $x \ge 1$, then $x^2 + 1 \le x^2 + x^2 = 2x^2$ and thus $\sqrt{x^2+1} \le \sqrt{2}\,x$. This gives us the inequality

$$\frac{1}{\sqrt{x^2+1}} \ge \frac{1}{\sqrt{2}\,x} \qquad \text{(for } x \ge 1)$$

Now we may apply the Comparison Test and (3):

$$\frac{1}{\sqrt{2}}\int_1^\infty \frac{dx}{x} \text{ diverges} \Rightarrow \int_1^\infty \frac{dx}{\sqrt{x^2+1}} \text{ diverges} \Rightarrow \int_0^\infty \frac{dx}{\sqrt{x^2+1}} \text{ diverges} \qquad ■$$

■ **EXAMPLE 10** **Assumptions Matter: Choosing the Right Comparison** Determine whether $\displaystyle\int_1^\infty \frac{dx}{\sqrt{x}+e^{3x}}$ converges or diverges.

Solution Let's try using the inequality

$$\frac{1}{\sqrt{x}+e^{3x}} \le \frac{1}{e^{3x}} \qquad (x \ge 0)$$

$$\int_1^\infty \frac{dx}{e^{3x}} = \lim_{R\to\infty} -\frac{1}{3}e^{-3x}\Big|_1^R = \lim_{R\to\infty}\frac{1}{3}(e^{-3} - e^{-3R}) = e^{-3} \quad \text{(converges)}$$

The Comparison Test tells us that our integral converges:

$$\int_1^\infty \frac{dx}{e^{3x}} \text{ converges} \quad \Rightarrow \quad \int_1^\infty \frac{dx}{\sqrt{x}+e^{3x}} \text{ also converges}$$

Had we not been thinking, we might have tried to use the inequality

$$\frac{1}{\sqrt{x}+e^{3x}} \le \frac{1}{\sqrt{x}} \qquad \boxed{4}$$

The integral $\displaystyle\int_1^\infty \frac{dx}{\sqrt{x}}$ diverges by Theorem 1 (with $p=\frac{1}{2}$), but this tells us nothing about our integral which, by (4), is smaller (Figure 10). ■

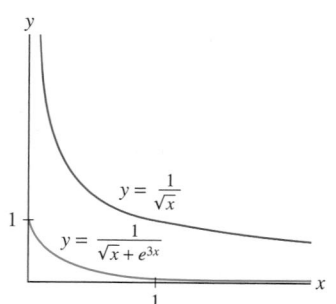

FIGURE 10 The divergence of $\displaystyle\int_1^\infty \frac{dx}{\sqrt{x}}$ says nothing about the divergence or convergence of the smaller integral $\displaystyle\int_1^\infty \frac{dx}{\sqrt{x}+e^{3x}}$.

7.7 SUMMARY

• An *improper integral* is defined as the limit of ordinary integrals. For example,

$$\int_a^\infty f(x)\,dx = \lim_{R\to\infty} \int_a^R f(x)\,dx$$

- If this limit exists, we say that the improper integral *converges*. Otherwise, we say that it *diverges*.

-
$$p > 1: \quad \int_1^\infty \frac{dx}{x^p} \ \text{converges} \quad \text{and} \quad \int_0^1 \frac{dx}{x^p} \ \text{diverges}$$

$$p < 1: \quad \int_1^\infty \frac{dx}{x^p} \ \text{diverges} \quad \text{and} \quad \int_0^1 \frac{dx}{x^p} \ \text{converges}$$

$$p = 1: \quad \int_1^\infty \frac{dx}{x} \quad \text{and} \quad \int_0^1 \frac{dx}{x} \ \text{both diverge}$$

- The Comparison Test: Assume that $f(x) \geq g(x) \geq 0$ for $x \geq a$. Then

$$\text{If } \int_a^\infty f(x)\,dx \ \text{converges,} \quad \text{then} \quad \int_a^\infty g(x)\,dx \ \text{converges.}$$

$$\text{If } \int_a^\infty g(x)\,dx \ \text{diverges,} \quad \text{then} \quad \int_a^\infty f(x)\,dx \ \text{diverges.}$$

Remember: The Comparison Test provides no information if the larger integral $\int_a^\infty f(x)\,dx$ diverges or the smaller integral $\int_a^\infty g(x)\,dx$ converges.

- The Comparison Test is also valid for improper integrals with infinite discontinuities at endpoints.

7.7 EXERCISES

Preliminary Questions

1. State whether the integral converges or diverges:

(a) $\int_1^\infty x^{-3}\,dx$

(b) $\int_0^1 x^{-3}\,dx$

(c) $\int_1^\infty x^{-2/3}\,dx$

(d) $\int_0^1 x^{-2/3}\,dx$

2. Is $\int_0^{\pi/2} \cot x\,dx$ an improper integral? Explain.

3. Find a value of $b > 0$ that makes $\int_0^b \frac{1}{x^2 - 4}\,dx$ an improper integral.

4. Which comparison would show that $\int_0^\infty \frac{dx}{x + e^x}$ converges?

5. Explain why it is not possible to draw any conclusions about the convergence of $\int_1^\infty \frac{e^{-x}}{x}\,dx$ by comparing with the integral $\int_1^\infty \frac{dx}{x}$.

Exercises

1. Which of the following integrals is improper? Explain your answer, but do not evaluate the integral.

(a) $\int_0^2 \frac{dx}{x^{1/3}}\,dx$

(b) $\int_1^\infty \frac{dx}{x^{0.2}}$

(c) $\int_{-1}^\infty e^{-x}\,dx$

(d) $\int_0^1 e^{-x}$

(e) $\int_0^{\pi/2} \sec x\,dx$

(f) $\int_0^\infty \sin x\,dx$

(g) $\int_0^1 \sin x\,dx$

(h) $\int_0^1 \frac{dx}{\sqrt{3 - x^2}}$

(i) $\int_1^\infty \ln x\,dx$

(j) $\int_0^3 \ln x\,dx$

2. Let $f(x) = x^{-4/3}$.

(a) Evaluate $\int_1^R f(x)\,dx$.

(b) Evaluate $\int_1^\infty f(x)\,dx$ by computing the limit

$$\lim_{R \to \infty} \int_1^R f(x)\,dx$$

3. Prove that $\int_1^\infty x^{-2/3}\,dx$ diverges by showing that

$$\lim_{R\to\infty} \int_1^R x^{-2/3}\,dx = \infty$$

4. Determine if $\int_0^3 \dfrac{dx}{(3-x)^{3/2}}$ converges by computing

$$\lim_{R\to 3^-} \int_0^R \dfrac{dx}{(3-x)^{3/2}}$$

In Exercises 5–46, determine whether the improper integral converges and, if so, evaluate it.

5. $\displaystyle\int_1^\infty \dfrac{dx}{x^{19/20}}$

6. $\displaystyle\int_1^\infty \dfrac{dx}{x^{20/19}}$

7. $\displaystyle\int_{-\infty}^4 e^{0.0001t}\,dt$

8. $\displaystyle\int_{20}^\infty \dfrac{dt}{t}$

9. $\displaystyle\int_0^5 \dfrac{dx}{x^{20/19}}$

10. $\displaystyle\int_0^5 \dfrac{dx}{x^{19/20}}$

11. $\displaystyle\int_0^4 \dfrac{dx}{\sqrt{4-x}}$

12. $\displaystyle\int_5^6 \dfrac{dx}{(x-5)^{3/2}}$

13. $\displaystyle\int_2^\infty x^{-3}\,dx$

14. $\displaystyle\int_0^\infty \dfrac{dx}{(x+1)^3}$

15. $\displaystyle\int_{-3}^\infty \dfrac{dx}{(x+4)^{3/2}}$

16. $\displaystyle\int_2^\infty e^{-2x}\,dx$

17. $\displaystyle\int_0^1 \dfrac{dx}{x^{0.2}}$

18. $\displaystyle\int_2^\infty x^{-1/3}\,dx$

19. $\displaystyle\int_4^\infty e^{-3x}\,dx$

20. $\displaystyle\int_4^\infty e^{3x}\,dx$

21. $\displaystyle\int_{-\infty}^0 e^{3x}\,dx$

22. $\displaystyle\int_0^1 \dfrac{dx}{x^{3.7}}$

23. $\displaystyle\int_2^\infty \dfrac{dx}{(x+3)^4}$

24. $\displaystyle\int_1^2 \dfrac{dx}{(x-1)^2}$

25. $\displaystyle\int_1^3 \dfrac{dx}{\sqrt{3-x}}$

26. $\displaystyle\int_{-2}^4 \dfrac{dx}{(x+2)^{1/3}}$

27. $\displaystyle\int_0^\infty \dfrac{dx}{1+x}$

28. $\displaystyle\int_{-\infty}^0 xe^{-x^2}\,dx$

29. $\displaystyle\int_0^\infty \dfrac{dx}{(1+x^2)^2}$

30. $\displaystyle\int_0^3 \dfrac{x\,dx}{\sqrt{x-3}}$

31. $\displaystyle\int_0^\infty e^{-x}\cos x\,dx$

32. $\displaystyle\int_{-\infty}^0 x^2 e^{-x}\,dx$

33. $\displaystyle\int_0^1 \dfrac{dx}{\sqrt{1-x^2}}$

34. $\displaystyle\int_0^1 \dfrac{e^{\sqrt{x}}\,dx}{\sqrt{x}}$

35. $\displaystyle\int_1^\infty \dfrac{e^{\sqrt{x}}\,dx}{\sqrt{x}}$

36. $\displaystyle\int_1^\infty xe^{-2x}\,dx$

37. $\displaystyle\int_0^\infty \dfrac{x\,dx}{x^2+1}$

38. $\displaystyle\int_0^{\pi/2} \sec\theta\,d\theta$

39. $\displaystyle\int_0^\infty \sin x\,dx$

40. $\displaystyle\int_0^{\pi/2} \tan x\,dx$

41. $\displaystyle\int_0^{\pi/2} \tan x\sec x\,dx$

42. $\displaystyle\int_0^1 \ln x\,dx$

43. $\displaystyle\int_0^1 x\ln x\,dx$

44. $\displaystyle\int_1^2 \dfrac{dx}{x\ln x}$

45. $\displaystyle\int_0^1 \dfrac{\ln x}{x^2}\,dx$

46. $\displaystyle\int_1^\infty \dfrac{\ln x}{x^2}\,dx$

47. Let $I = \displaystyle\int_4^\infty \dfrac{dx}{(x-2)(x-3)}$.

(a) Show that for $R > 4$,

$$\int_4^R \dfrac{dx}{(x-2)(x-3)} = \ln\left|\dfrac{R-3}{R-2}\right| - \ln\dfrac{1}{2}$$

(b) Then show that $I = \ln 2$.

48. Evaluate the integral $I = \displaystyle\int_1^\infty \dfrac{dx}{x(2x+5)}$.

49. Determine whether $I = \displaystyle\int_0^1 \dfrac{dx}{x(2x+5)}$ converges and, if so, evaluate.

50. Determine whether $I = \displaystyle\int_2^\infty \dfrac{dx}{(x+3)(x+1)^2}$ converges and, if so, evaluate.

In Exercises 51–54, determine if the doubly infinite improper integral converges and, if so, evaluate it. Use definition (2).

51. $\displaystyle\int_{-\infty}^\infty \dfrac{dx}{1+x^2}$

52. $\displaystyle\int_{-\infty}^\infty e^{-|x|}\,dx$

53. $\displaystyle\int_{-\infty}^\infty xe^{-x^2}\,dx$

54. $\displaystyle\int_{-\infty}^\infty \dfrac{dx}{(x^2+1)^{3/2}}$

55. For which values of a does $\displaystyle\int_0^\infty e^{ax}\,dx$ converge?

56. Show that $\displaystyle\int_0^1 \dfrac{dx}{x^p}$ converges if $p < 1$ and diverges if $p \geq 1$.

57. Sketch the region under the graph of $f(x) = \dfrac{1}{1+x^2}$ for $-\infty < x < \infty$ and show that its area is π.

58. Show that $\dfrac{1}{\sqrt{x^4+1}} \leq \dfrac{1}{x^2}$ for all x and use this to prove that $\displaystyle\int_1^\infty \dfrac{dx}{\sqrt{x^4+1}}$ converges.

59. Show that $\displaystyle\int_1^\infty \frac{dx}{x^3+4}$ converges by comparing with $\displaystyle\int_1^\infty x^{-3}\,dx$.

60. Show that $\displaystyle\int_2^\infty \frac{dx}{x^3-4}$ converges by comparing with $\displaystyle\int_2^\infty 2x^{-3}\,dx$.

61. Show that $0 \le e^{-x^2} \le e^{-x}$ for $x \ge 1$ (Figure 11). Then use the Comparison Test and (3) to show that $\displaystyle\int_0^\infty e^{-x^2}\,dx$ converges.

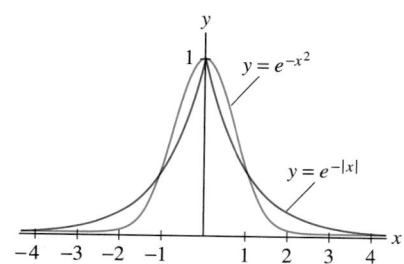

FIGURE 11 Comparison of $y = e^{-|x|}$ and $y = e^{-x^2}$.

62. Prove that $\displaystyle\int_{-\infty}^\infty e^{-x^2}\,dx$ converges by comparing with $\displaystyle\int_{-\infty}^\infty e^{-|x|}\,dx$ (Figure 11).

63. Show that $\displaystyle\int_1^\infty \frac{1-\sin x}{x^2}\,dx$ converges.

64. Let $a > 0$. Recall that $\displaystyle\lim_{x\to\infty}\frac{x^a}{\ln x} = \infty$ (by Exercise 60 in Section 4.7).
(a) Show that $x^a > 2\ln x$ for all x sufficiently large.
(b) Show that $e^{-x^a} < x^{-2}$ for all x sufficiently large.
(c) Show that $\displaystyle\int_1^\infty e^{-x^a}\,dx$ converges.

In Exercises 65–74, use the Comparison Test to determine whether or not the integral converges.

65. $\displaystyle\int_1^\infty \frac{1}{\sqrt{x^5+2}}\,dx$

66. $\displaystyle\int_1^\infty \frac{dx}{(x^3+2x+4)^{1/2}}$

67. $\displaystyle\int_3^\infty \frac{dx}{\sqrt{x-1}}$

68. $\displaystyle\int_0^5 \frac{dx}{x^{1/3}+x^3}$

69. $\displaystyle\int_1^\infty e^{-(x+x^{-1})}\,dx$

70. $\displaystyle\int_0^1 \frac{\sin x}{\sqrt{x}}\,dx$

71. $\displaystyle\int_0^1 \frac{e^x}{x^2}\,dx$

72. $\displaystyle\int_1^\infty \frac{1}{x^4+e^x}\,dx$

73. $\displaystyle\int_0^1 \frac{1}{x^4+\sqrt{x}}\,dx$

74. $\displaystyle\int_1^\infty \frac{\ln x}{\sinh x}\,dx$

75. An investment pays a dividend of \$250/year continuously forever. If the interest rate is 7%, what is the present value of the entire income stream generated by the investment?

76. An investment is expected to earn profits at a rate of $10,000e^{0.01t}$ dollars/year forever. Find the present value of the income stream if the interest rate is 4%.

77. Compute the present value of an investment that generates income at a rate of $5,000te^{0.01t}$ dollars/year forever, assuming an interest rate of 6%.

78. Find the volume of the solid obtained by rotating the region below the graph of $y = e^{-x}$ about the x-axis for $0 \le x < \infty$.

79. Let S be the solid obtained by rotating the region below the graph of $y = x^{-1}$ about the x-axis for $1 \le x < \infty$ (Figure 12).
(a) Use the Disk Method (Section 6.3) to compute the volume of S. Note that the volume is finite even though S is an infinite region.
(b) It can be shown that the surface area of S is

$$A = 2\pi \int_1^\infty x^{-1}\sqrt{1+x^{-4}}\,dx.$$

Show that A is infinite.

If S were a container, you could fill its interior with a finite amount of paint, but you could not paint its surface with a finite amount of paint.

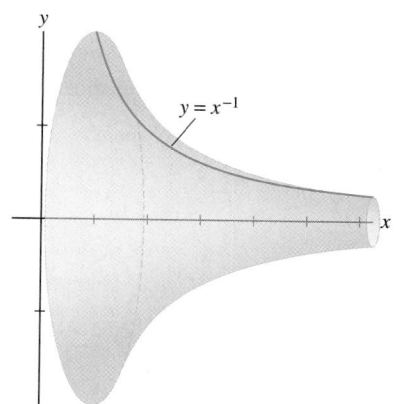

FIGURE 12

80. Compute the volume of the solid obtained by rotating the region below the graph of $y = (x^2+1)^{-1}$ about the x-axis for $-\infty < x < \infty$.

81. When a capacitor of capacitance C is charged by a source of voltage V, the power expended at time t is

$$P(t) = \frac{V^2}{R}(e^{-t/RC} - e^{-2t/RC}),$$

where R is the resistance in the circuit. The total energy stored in the capacitor is

$$W = \int_0^\infty P(t)\, dt$$

Show that $W = \frac{1}{2}CV^2$.

82. Conservation of Energy can be used to show that when a mass m oscillates at the end of a spring with spring constant k, the period of oscillation is

$$T = 4\sqrt{m} \int_0^{\sqrt{2E/k}} \frac{dx}{\sqrt{2E - kx^2}}$$

where E is the total energy of the mass. Show that this is an improper integral with value $T = 2\pi\sqrt{m/k}$.

83. When a radioactive substance decays, the fraction of atoms present at time t is $f(t) = e^{-kt}$, where $k > 0$ is the decay constant. It can be shown that the *average* life of an atom (until it decays) is $A = -\int_0^\infty t f'(t)\, dt$. Use Integration by Parts to show that $A = \int_0^\infty f(t)\, dt$ and compute A. What is the average decay time of Radon-222, whose half-life is 3.825 days?

84. According to **Maxwell's Law of velocity distribution** for a gas at temperature T, the fraction of gas molecules with speed between v and $v + \Delta v$ is approximately $f(v)\, \Delta v$, where

$$f(v) = 4\pi \left(\frac{m}{2\pi kT}\right)^{3/2} v^2 e^{-mv^2/(2kT)}$$

(m is the molecular mass of the gas and k Boltzmann's constant). Show that the average speed $\overline{v}$ is $\left(\dfrac{8kT}{\pi m}\right)^{1/2}$, where

$$\overline{v} = \int_0^\infty v f(v)\, dv$$

85. Let $J_n = \int_0^\infty x^n e^{-\alpha x}\, dx$, where $n \geq 1$ is an integer and $\alpha > 0$. Prove that $J_n = (n/\alpha)J_{n-1}$ and $J_0 = 1/\alpha$. Use this to compute J_4. Show that $J_n = \dfrac{n!}{\alpha^{n+1}}$.

86. Let $a > 0$ and $n > 1$. Define $f(x) = \dfrac{x^n}{e^{ax} - 1}$ for $x \neq 0$ and $f(0) = 0$.
(a) Use L'Hôpital's Rule to show that $f(x)$ is continuous at $x = 0$.

(b) Show that $\int_0^\infty f(x)\, dx$ converges. *Hint:* Show that $f(x) \leq 2x^n e^{-ax}$ if x is large enough. Then use the Comparison Test and Exercise 85.

87. There is a function $F(v)$ such that the amount of electromagnetic energy with frequency between v and $v + \Delta v$ radiated by a so-called black body at temperature T is proportional to $F(v)\, \Delta v$. The total radiated energy is $E = \int_0^\infty F(v)\, dv$. According to **Planck's Radiation Law**,

$$F(v) = \left(\frac{8\pi h}{c^3}\right) \frac{v^3}{e^{hv/kT} - 1}$$

where c, h, k are physical constants. To derive this law, Planck introduced the quantum hypothesis in 1900, which thus marked the birth of quantum mechanics. Show that E is finite (use Exercise 86).

88. Let $J = \int_0^\infty e^{-x^2}\, dx$ and $J_N = \int_0^N e^{-x^2}\, dx$. Although e^{-x^2} has no elementary antiderivative, it is known that $J = \sqrt{\pi}/2$. Let T_N be the Nth trapezoidal approximation to J_N. Calculate T_4 and show that T_4 approximates J to three decimal places.

89. A **probability density** function on $[0, \infty)$ is a function $f(x)$ defined for $x \geq 0$ such that $f(x) \geq 0$ and $\int_0^\infty f(x)\, dx = 1$. The *mean value* of $f(x)$ is the quantity $\mu = \int_0^\infty x f(x)\, dx$. For $k > 0$, find a constant C such that Ce^{-kx} is a probability density and compute μ.

*In Exercises 90–93, the **Laplace transform** of a function $f(x)$ is the function $Lf(s)$ of the variable s defined by the improper integral (if it converges):*

$$Lf(s) = \int_0^\infty f(x)e^{-sx}\, dx$$

Laplace transforms are widely used in physics and engineering.

90. Show that if $f(x) = C$ where C a constant, then $Lf(s) = C/s$ for $s > 0$.

91. Show that if $f(x) = \sin \alpha x$, then $Lf(s) = \dfrac{\alpha}{s^2 + \alpha^2}$.

92. Compute $Lf(s)$, where $f(x) = e^{\alpha x}$ and $s > \alpha$.

93. Compute $Lf(s)$, where $f(x) = \cos \alpha x$ and $s > 0$.

Further Insights and Challenges

94. Let p be an integer. Show that $\displaystyle\int_0^{1/2} \frac{dx}{x(\ln x)^p}$ converges if and only if $p > 1$.

95. Let $I = \displaystyle\int_0^1 x^p \ln x\, dx$.
(a) Show that I diverges for $p = -1$.

(b) Show that if $p \neq -1$, then

$$\int x^p \ln x\, dx = \frac{x^{p+1}}{p+1}\left(\ln x - \frac{1}{p+1}\right) + C$$

(c) Use L'Hôpital's Rule to show that I converges if $p > -1$ and diverges if $p < -1$.

In Exercises 96–98, an improper integral $I = \int_a^\infty f(x)\,dx$ is called **absolutely convergent** if $\int_a^\infty |f(x)|\,dx$ converges. It can be shown that if I is absolutely convergent, then it is convergent.

96. Show that $\int_1^\infty \dfrac{\sin x}{x^2}\,dx$ is absolutely convergent.

97. Show that $\int_1^\infty e^{-x^2}\cos x\,dx$ is absolutely convergent.

98. Let $f(x) = \dfrac{\sin x}{x}$ and $I = \int_0^\infty f(x)\,dx$. We define $f(0) = 1$. Then $f(x)$ is continuous and I is not improper at $x = 0$.

(a) Show that

$$\int_1^R \frac{\sin x}{x}\,dx = -\left.\frac{\cos x}{x}\right|_1^R - \int_1^R \frac{\cos x}{x^2}\,dx$$

(b) Show that $\int_1^\infty \dfrac{\cos x}{x^2}\,dx$ converges. Conclude that the limit as $R \to \infty$ of the integral in (a) exists and is finite.

(c) Show that I converges.

It is known that $I = \frac{\pi}{2}$. However, I is *not* absolutely convergent. The convergence depends on cancellation, as shown in Figure 13.

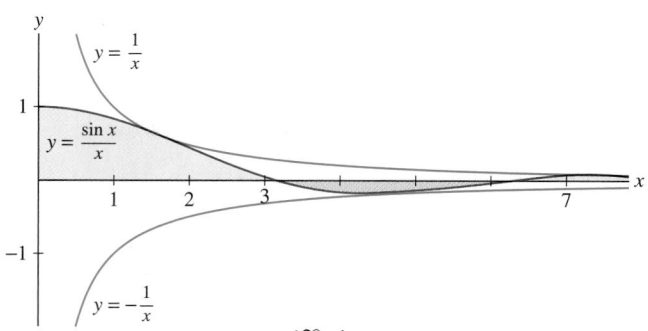

FIGURE 13 Convergence of $\int_1^\infty \dfrac{\sin x}{x}\,dx$ is due to the cancellation arising from the periodic change of sign.

99. The **gamma function**, which plays an important role in advanced applications, is defined for $n \geq 1$ by

$$\Gamma(n) = \int_0^\infty t^{n-1}e^{-t}\,dt$$

(a) Show that the integral defining $\Gamma(n)$ converges for $n \geq 1$ (it actually converges for all $n > 0$). *Hint:* Show that $t^{n-1}e^{-t} < t^{-2}$ for t sufficiently large.

(b) Show that $\Gamma(n+1) = n\Gamma(n)$ using Integration by Parts.

(c) Show that $\Gamma(n+1) = n!$ if $n \geq 1$ is an integer. *Hint:* Use (a) repeatedly. Thus, $\Gamma(n)$ provides a way of defining n-factorial when n is not an integer.

100. Use the results of Exercise 99 to show that the Laplace transform (see Exercises 90–93 above) of x^n is $\dfrac{n!}{s^{n+1}}$.

CHAPTER REVIEW EXERCISES

1. Estimate $\int_2^5 f(x)\,dx$ by computing T_2, M_3, T_6, and S_6 for a function $f(x)$ taking on the values in the table below:

x	2	2.5	3	3.5	4	4.5	5
$f(x)$	$\frac{1}{2}$	2	1	0	$-\frac{3}{2}$	-4	-2

2. State whether the approximation M_N or T_N is larger or smaller than the integral.

(a) $\int_0^\pi \sin x\,dx$

(b) $\int_\pi^{2\pi} \sin x\,dx$

(c) $\int_1^8 \dfrac{dx}{x^2}$

(d) $\int_2^5 \ln x\,dx$

3. The rainfall rate (in inches per hour) was measured hourly during a 10-hour thunderstorm with the following results:

0, 0.41, 0.49, 0.32, 0.3, 0.23, 0.09, 0.08, 0.05, 0.11, 0.12

Use Simpson's Rule to estimate the total rainfall during the 10-hour period.

In Exercises 4–9, compute the given approximation to the integral.

4. $\int_0^1 e^{-x^2}\,dx$, M_5

5. $\int_2^4 \sqrt{6t^3 + 1}\,dt$, T_3

6. $\int_{\pi/4}^{\pi/2} \sqrt{\sin\theta}\,d\theta$, M_4

7. $\int_1^4 \dfrac{dx}{x^3 + 1}$, T_6

8. $\int_0^1 e^{-x^2}\,dx$, S_4

9. $\int_5^9 \cos(x^2)\,dx$, S_8

10. The following table gives the area $A(h)$ of a horizontal cross section of a pond at depth h. Use the Trapezoidal Rule to estimate the volume V of the pond (Figure 1).

h (ft)	$A(h)$ (acres)	h (ft)	$A(h)$ (acres)
0	2.8	10	0.8
2	2.4	12	0.6
4	1.8	14	0.2
6	1.5	16	0.1
8	1.2	18	0

Area of horizontal
cross section is $A(h)$

FIGURE 1

11. Suppose that the second derivative of the function $A(h)$ in Exercise 10 satisfies $|A''(h)| \leq 1.5$. Use the Error Bound to find the maximum possible error in your estimate of the volume V of the pond.

12. Find a bound for the error $\left| M_{16} - \int_1^3 x^3 \, dx \right|$.

13. [GU]   Let $f(x) = \sin(x^3)$. Find a bound for the error

$$\left| T_{24} - \int_0^{\pi/2} f(x) \, dx \right|.$$

Hint: Find a bound K_2 for $|f''(x)|$ by plotting $f''(x)$ with a graphing utility.

14. Find a value of N such that

$$\left| M_N - \int_0^{\pi/4} \tan x \, dx \right| \leq 10^{-4}.$$

15. Find a value of N such that S_N approximates $\int_2^5 x^{-1/4} \, dx$ with an error of at most 10^{-2} (but do not calculate S_N).

16. Match the integrals (a)–(e) with their antiderivatives (i)–(v) on the basis of the general form (do not evaluate the integrals).

(a) $\int \dfrac{x \, dx}{x^2 - 4}$

(b) $\int \dfrac{(2x + 9) \, dx}{x^2 + 4}$

(c) $\int \sin^3 x \cos^2 x \, dx$

(d) $\int \dfrac{dx}{x\sqrt{16x^2 - 1}}$

(e) $\int \dfrac{16 \, dx}{x(x - 4)^2}$

(i) $\sec^{-1} 4x + C$

(ii) $\log|x| - \log|x - 4| - \dfrac{4}{x - 4} + C$

(iii) $\dfrac{1}{30}(3\cos^5 x - 3\cos^3 x \sin^2 x - 7\cos^3 x) + C$

(iv) $\dfrac{9}{2}\tan^{-1}\dfrac{x}{2} + \ln(x^2 + 4) + C$

(v) $\sqrt{x^2 - 4} + C$

In Exercises 17–25, compute the integral using the suggested method.

17. $\displaystyle\int \cos^3 \theta \sin^8 \theta \, d\theta$ [write $\cos^3 \theta$ as $\cos\theta(1 - \sin^2 \theta)$]

18. $\displaystyle\int xe^{-12x} \, dx$ (Integration by Parts)

19. $\displaystyle\int \sec^3 \theta \tan^4 \theta \, d\theta$ (trigonometric identity, reduction formula)

20. $\displaystyle\int \dfrac{4x + 4}{(x - 5)(x + 3)} \, dx$ (partial fractions)

21. $\displaystyle\int \dfrac{dx}{x(x^2 - 1)^{3/2}} \, dx$ (trigonometric substitution)

22. $\displaystyle\int \dfrac{dx}{x^{3/2} + x^{1/2}}$ (substitution)

23. $\displaystyle\int \dfrac{dx}{x + x^{-1}}$ (rewrite integrand)

24. $\displaystyle\int x^{-2} \tan^{-1} x \, dx$ (Integration by Parts)

25. $\displaystyle\int \dfrac{dx}{x^2 + 4x - 5}$ (complete square, substitution, partial fractions)

In Exercises 26–69, compute the integral using the appropriate method or combination of methods.

26. $\displaystyle\int x^2 e^{4x} \, dx$

27. $\displaystyle\int \dfrac{x^2}{\sqrt{9 - x^2}} \, dx$

28. $\displaystyle\int \cos^9 6\theta \sin^3 6\theta \, d\theta$

29. $\displaystyle\int \sec^2 \theta \tan^4 \theta \, d\theta$

30. $\displaystyle\int \dfrac{(6x + 4) \, dx}{x^2 - 1}$

31. $\displaystyle\int \dfrac{dt}{(t^2 - 1)^2}$

32. $\displaystyle\int \dfrac{d\theta}{\cos^4 \theta}$

33. $\displaystyle\int \sin 2\theta \sin^2 \theta \, d\theta$

34. $\displaystyle\int \ln(9 - 2x) \, dx$

35. $\displaystyle\int (\ln(x + 1))^2 \, dx$

36. $\displaystyle\int \sin^5 \theta \, d\theta$

37. $\displaystyle\int \cos^4(9x - 2) \, dx$

38. $\int \sin 3x \cos 5x \, dx$

39. $\int \sin 2x \sec^2 x \, dx$

40. $\int \sqrt{\tan x} \, \sec^2 x \, dx$

41. $\int (\sec x + \tan x)^2 \, dx$

42. $\int \sin^5 \theta \cos^3 \theta \, d\theta$

43. $\int \cot^2 \frac{\theta}{2} \, d\theta$

44. $\int \frac{dt}{(t-3)(t+4)}$

45. $\int \frac{dt}{(t-3)^2(t+4)}$

46. $\int \sqrt{x^2+9} \, dx$

47. $\int \frac{dx}{x\sqrt{x^2-4}}$

48. $\int \frac{dx}{x+x^{2/3}}$

49. $\int \frac{dx}{x^{3/2}+ax^{1/2}}$

50. $\int \frac{dx}{(x-b)^2+4}$

51. $\int \frac{(x^2-x)\,dx}{(x+2)^3}$

52. $\int \frac{(7x^2+x)\,dx}{(x-2)(2x+1)(x+1)}$

53. $\int \frac{16\,dx}{(x-2)^2(x^2+4)}$

54. $\int \frac{dx}{(x^2+25)^2}$

55. $\int \frac{dx}{x^2+8x+25}$

56. $\int \frac{dx}{x^2+8x+4}$

57. $\int \frac{(x^2-x)\,dx}{(x+2)^3}$

58. $\int t^2\sqrt{1-t^2}\,dt$

59. $\int \frac{dx}{x^4\sqrt{x^2+4}}$

60. $\int \frac{dx}{(x^2+5)^{3/2}}$

61. $\int (x+1)e^{4-3x}\,dx$

62. $\int \frac{dx}{x^2+x^{-2}}$

63. $\int x^{-3}\sin^{-1}x\,dx$

64. $\int x^3\cos(x^2)\,dx$

65. $\int x^2(\ln x)^2\,dx$

66. $\int x\tanh^{-1}x\,dx$

67. $\int \frac{\tan^{-1}t\,dt}{1+t^2}$

68. $\int \ln(x^2+9)\,dx$

69. $\int (\sin x)(\cosh x)\,dx$

In Exercises 70–77, evaluate the integral.

70. $\int_0^1 \cosh 2t\,dt$

71. $\int \tanh 5x\,dx$

72. $\int \sinh^3 x \cosh x\,dx$

73. $\int \coth^2(1-4t)\,dt$

74. $\int_{-0.3}^{0.3} \frac{dx}{1-x^2}$

75. $\int_0^{3\sqrt{3}/2} \frac{dx}{\sqrt{9-x^2}}$

76. $\int \frac{\sqrt{x^2+1}\,dx}{x^2}$

77. $\int \frac{dt}{\cosh^2 t+\sinh^2 t}$ *Hint:* Use the substitution $u=\tanh t$.

In Exercises 78–81, let $gd(y)=\tan^{-1}(\sinh y)$ *be the so-called* **gudermannian**, *which arises in cartography. In a map of the earth constructed by Mercator projection, points located y radial units from the equator correspond to points on the globe of latitude $gd(y)$.*

78. Prove that $\frac{d}{dy}gd(y)=\operatorname{sech} y$.

79. Let $f(y)=2\tan^{-1}(e^y)-\pi/2$. Prove that $gd(y)=f(y)$. *Hint:* Show that $gd'(y)=f'(y)$ and $f(0)=g(0)$.

80. Show that $t(y)=\sinh^{-1}(\tan y)$ is the inverse of $gd(y)$ for $0\le y<\pi/2$.

81. Verify that $t(y)$ in Exercise 80 satisfies $t'(y)=\sec y$ and find a value of a such that

$$t(y)=\int_a^y \frac{dt}{\cos t}$$

In Exercises 82–91, determine whether the improper integral converges and, if so, evaluate it.

82. $\int_0^\infty \frac{dx}{(x+2)^2}$

83. $\int_4^\infty \frac{dx}{x^{2/3}}$

84. $\int_0^4 \frac{dx}{x^{2/3}}$

85. $\int_9^\infty \frac{dx}{x^{12/5}}$

86. $\int_{-\infty}^0 \frac{dx}{x^2+1}$

87. $\int_{-\infty}^9 e^{4x}\,dx$

88. $\int_0^{\pi/2} \cot\theta\,d\theta$

89. $\int_1^\infty \frac{dx}{(x+2)(2x+3)}$

90. $\int_0^\infty (5+x)^{-1/3}\,dx$

91. $\int_2^5 (5-x)^{-1/3}\,dx$

In Exercises 92–97, use the Comparison Test to determine if the improper integral converges or diverges.

92. $\int_8^\infty \frac{dx}{x^2-4}$

93. $\int_8^\infty (\sin^2 x)e^{-x}\,dx$

94. $\int_3^\infty \frac{dx}{x^4+\cos^2 x}$

95. $\int_1^\infty \frac{dx}{x^{1/3}+x^{2/3}}$

96. $\int_0^1 \frac{dx}{x^{1/3}+x^{2/3}}$

97. $\int_0^\infty e^{-x^3}\,dx$

98. Calculate the volume of the infinite solid obtained by rotating the region under $y=(x^2+1)^{-2}$ for $0\le x<\infty$ about the y-axis.

99. Let R be the region under the graph of $y=(x+1)^{-1}$ for $0\le x<\infty$. Which of the following quantities is finite?
(a) The area of R
(b) The volume of the solid obtained by rotating R about the x-axis
(c) The volume of the solid obtained by rotating R about the y-axis

100. Show that $\displaystyle\int_0^\infty x^n e^{-x^2}\,dx$ converges for all $n > 0$. *Hint:* First observe that $x^n e^{-x^2} < x^n e^{-x}$ for $x > 1$. Then show that $x^n e^{-x} < x^{-2}$ for x sufficiently large.

101. According to kinetic theory, the molecules of ordinary matter are in constant random motion. The probability that a molecule has kinetic energy in a small interval $[E, E + \Delta E]$ is approximately $\dfrac{1}{kT} e^{-E/(kT)}$, where T is the temperature (in kelvins) and k Boltzmann's constant. Compute the *average* kinetic energy $\overline{E}$ in terms of k and T, where

$$\overline{E} = \frac{1}{kT}\int_0^\infty E e^{-E/kT}\,dE$$

102. Compute the Laplace transform $Lf(s)$ of the function $f(x) = x$ for $s > 0$. See Exercises 90–93 in Section 7.7 for the definition of $Lf(s)$.

103. Compute the Laplace transform $Lf(s)$ of the function $f(x) = x^2 e^{\alpha x}$ for $s > \alpha$.

104. Let $I_n = \displaystyle\int \frac{x^n\,dx}{x^2 + 1}$.

(a) Prove that $I_n = \dfrac{x^{n-1}}{n-1} - I_{n-2}$.

(b) Use (a) to calculate I_n for $0 \leq n \leq 5$.

(c) Show that, in general,

$$I_{2n+1} = \frac{x^{2n}}{2n} - \frac{x^{2n-2}}{2n-2} + \cdots$$
$$+ (-1)^{n-1}\frac{x^2}{2} + (-1)^n\frac{1}{2}\ln(x^2+1) + C$$

$$I_{2n} = \frac{x^{2n-1}}{2n-1} - \frac{x^{2n-3}}{2n-3} + \cdots$$
$$+ (-1)^{n-1}x + (-1)^n \tan^{-1} x + C$$

105. Let $J_n = \displaystyle\int x^n e^{-x^2/2}\,dx$.

(a) Show that $J_1 = -e^{-x^2/2}$.

(b) Prove $J_n = -x^{n-1}e^{-x^2/2} + (n-1)J_{n-2}$.

(c) Use (a) and (b) to compute J_3 and J_5.

8 | FURTHER APPLICATIONS OF THE INTEGRAL AND TAYLOR POLYNOMIALS

In the first three sections of this chapter, we discuss some additional uses of integration, including two applications to physics. The last section introduces Taylor polynomials, which are higher-order generalizations of the linear approximation used to approximate algebraic and transcendental functions to arbitrary accuracy. Taylor polynomials are a high point of the theory we have developed so far and provide a beautiful illustration of the power and utility of calculus.

This NASA simulation, depicting streamlines of hot gas from the nozzles of a Harrier Jet during vertical takeoff, is based on a branch of mathematics called computational fluid dynamics.

8.1 Arc Length and Surface Area

One of the first problems we solved using integration was finding the area under the parabola $y = x^2$. What about the length of the parabola (Figure 1) or more generally, the length of a curve $y = f(x)$? In this section, we show that the length of a curve (often called **arc length**) can also be computed as an integral.

The arc length of $y = f(x)$ over an interval $[a, b]$ is defined using **polygonal approximations** to the graph. To construct a polygonal approximation L, choose a partition of $[a, b]$ into N subintervals with endpoints

$$a = x_0 < x_1 < \cdots < x_N = b$$

and join the corresponding points $P_0, P_1, \ldots, P_N$ on the graph by line segments (Figure 2). Let $|L|$ be the length of L.

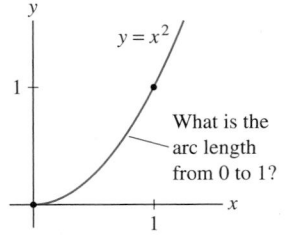

FIGURE 1 The arc length of $y = x^2$ over $[0, 1]$ is

$$\tfrac{1}{2}\sqrt{5} + \tfrac{1}{4}\ln(2 + \sqrt{5}) \approx 1.148$$

(see Exercise 22).

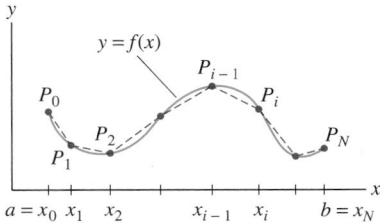

FIGURE 2 A polygonal approximation L.

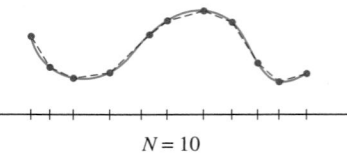

$N = 3$ $N = 5$ $N = 10$

FIGURE 3 The polygonal approximations L improve as the widths of the subintervals decrease.

Figure 3 suggests that L approximates the graph more closely as the width of the partition decreases. We define the arc length of the graph to be the limit $\lim |L|$ as the width of the partition tends to zero, provided that the limit exists.

Now assume that $f(x)$ is differentiable and that $f'(x)$ is continuous on $[a, b]$. In this case, the limit $\lim |L|$ exists and we can compute it as an integral. Let $|L_i|$ be the length of the ith segment $L_i = \overline{P_{i-1}P_i}$. Figure 4 shows that $|L_i|$ is the hypotenuse of a right triangle of base $\Delta x_i = x_i - x_{i-1}$ and height $|f(x_i) - f(x_{i-1})|$. By the Pythagorean Theorem,

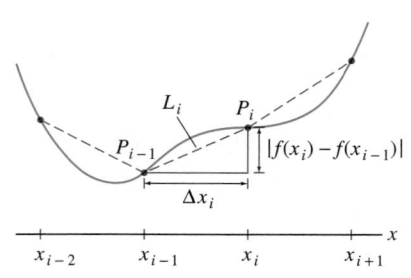

FIGURE 4

$$|L_i| = \sqrt{\Delta x_i^2 + (f(x_i) - f(x_{i-1}))^2}$$

The Mean Value Theorem tells us that for some point c_i in $[x_{i-1}, x_i]$, we have

$$f(x_i) - f(x_{i-1}) = f'(c_i)(x_i - x_{i-1}) = f'(c_i)\Delta x_i$$

Therefore,

$$|L_i| = \sqrt{(\Delta x_i)^2 + (f'(c_i)\Delta x_i)^2} = \sqrt{(\Delta x_i)^2(1 + f'(c_i)^2)} = \sqrt{1 + f'(c_i)^2}\,\Delta x_i$$

The total length $|L|$ is the sum of the lengths of the segments L_i:

$$|L| = |L_1| + |L_2| + \cdots + |L_N| = \sum_{i=1}^{N} \sqrt{1 + f'(c_i)^2}\,\Delta x_i \qquad \boxed{1}$$

◄··· REMINDER *A Riemann sum for the integral* $\int_a^b g(x)\,dx$ *is a sum of the form*

$$\sum_{i=1}^{N} g(c_i)\,\Delta x_i$$

where $x_0, x_1, \ldots, x_N$ *is any partition of* $[a, b]$, $\Delta x_i = x_i - x_{i-1}$, *and* c_i *is any number in* $[x_{i-1}, x_i]$.

We recognize the sum in Eq. (1) as a Riemann sum for the integral $\int_a^b \sqrt{1 + f'(x)^2}\,dx$. Since $f'(x)$ is continuous, the integrand $\sqrt{1 + f'(x)^2}$ is also continuous and hence integrable, so the limit $\lim |L|$ exists and is equal to the integral.

In Exercises 19–21, we verify that Eq. (2) gives the correct result for the length of a line and circumference of a circle.

THEOREM 1 Formula for Arc Length   If $f'(x)$ exists and is continuous on $[a, b]$, then the arc length of the graph of $y = f(x)$ over $[a, b]$ is equal to

$$\text{Arc length over } [a, b] = \int_a^b \sqrt{1 + f'(x)^2}\,dx \qquad \boxed{2}$$

■ **EXAMPLE 1** Calculate the arc length of the graph of $f(x) = \dfrac{1}{12}x^3 + x^{-1}$ over $[1, 3]$.

Solution First, let's calculate $1 + f'(x)^2$. Since $f'(x) = \frac{1}{4}x^2 - x^{-2}$,

$$1 + f'(x)^2 = 1 + \left(\frac{1}{4}x^2 - x^{-2}\right)^2 = 1 + \left(\frac{1}{16}x^4 - \frac{1}{2} + x^{-4}\right)$$

$$= \frac{1}{16}x^4 + \frac{1}{2} + x^{-4} = \left(\frac{1}{4}x^2 + x^{-2}\right)^2$$

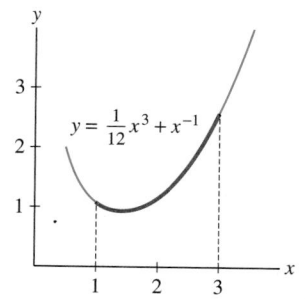

FIGURE 5 The arc length over [1, 3] is $\frac{17}{6}$.

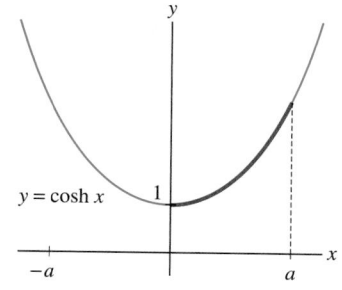

FIGURE 6

$$\cosh x = \frac{1}{2}(e^x + e^{-x})$$

$$\sinh x = \frac{1}{2}(e^x - e^{-x})$$

$$\cosh^2 x - \sinh^2 x = 1 \qquad \boxed{3}$$

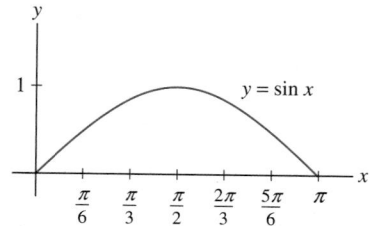

FIGURE 7 The arc length from 0 to π is approximately 3.82.

Fortunately, $1 + f'(x)^2$ is a square, so we can easily compute the arc length (Figure 5)

$$\int_1^3 \sqrt{1 + f'(x)^2}\, dx = \int_1^3 \left(\frac{1}{4}x^2 + x^{-2}\right) dx = \left(\frac{1}{12}x^3 - x^{-1}\right)\Big|_1^3$$

$$= \left(\frac{9}{4} - \frac{1}{3}\right) - \left(\frac{1}{12} - 1\right) = \frac{17}{6} \qquad \blacksquare$$

■ **EXAMPLE 2** **Arc Length as a Function of the Upper Limit** Find the arc length of $y = \cosh x$ over $[0, a]$ (Figure 6). Then find the arc length over $[0, 2]$.

Solution Recall that $y' = (\cosh x)' = \sinh x$. Using Eq. (3) in the margin, we obtain

$$1 + (y')^2 = 1 + \sinh^2 x = \cosh^2 x$$

$$\int_0^a \sqrt{1 + (y')^2}\, dx = \int_0^a \cosh x\, dx = \sinh x\Big|_0^a = \sinh a$$

Therefore, the arc length over $[0, a]$ is $\sinh a$. Setting $a = 2$, we find that the arc length over $[0, 2]$ is $\sinh 2 \approx 3.63$. ■

In Examples 1 and 2, the quantity $1 + f'(x)^2$ turned out to be a perfect square and we were able to evaluate the arc length integral. When $\sqrt{1 + f'(x)^2}$ does not have an elementary antiderivative, as is often the case, we may approximate the arc length using numerical integration.

■ **EXAMPLE 3** **When Arc Length Cannot Be Calculated Exactly** *CAS* Express the arc length L of $y = \sin x$ over $[0, \pi]$ as an integral. Then approximate L using Simpson's Rule S_N with $N = 6$ and using a computer algebra system.

Solution We have $y' = \cos x$. Setting $g(x) = \sqrt{1 + y'^2} = \sqrt{1 + \cos^2 x}$,

$$\text{Arc length} = \int_0^\pi \sqrt{1 + \cos^2 x}\, dx = \int_0^\pi g(x)\, dx$$

This integral cannot be evaluated explicitly, so we approximate it using Simpson's Rule (Section 7.1). We divide $[0, \pi]$ into $N = 6$ subintervals of width $\Delta x = \pi/6$ and endpoints $j\,\Delta x = j\pi/6$ for $j = 0, 1, \ldots, 6$. Then

$$S_6 = \frac{\Delta x}{3}\left(g(0) + 4g\left(\frac{\pi}{6}\right) + 2g\left(\frac{2\pi}{6}\right) + 4g\left(\frac{3\pi}{6}\right) + 2g\left(\frac{4\pi}{6}\right) + 4g\left(\frac{5\pi}{6}\right) + g(\pi)\right)$$

$$\approx \frac{\pi}{18}(1.4142 + 5.2915 + 2.2361 + 4 + 2.2361 + 5.2915 + 1.4142)$$

$$\approx 3.82$$

Using a computer algebra system to evaluate the integral numerically, we obtain the approximation 3.820198 (Figure 7). ■

The surface area of a surface of revolution can be computed by an integral that is closely related to the arc length integral (Figure 8). This formula can be justified using polygonal approximations to $y = f(x)$ as discussed above for arc length.

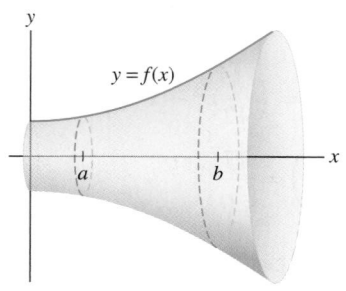

FIGURE 8 Surface obtained by revolving $y = f(x)$ about the x-axis.

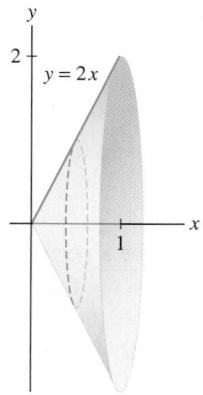

FIGURE 9 Surface obtained by revolving $y = 2x$ about the x-axis.

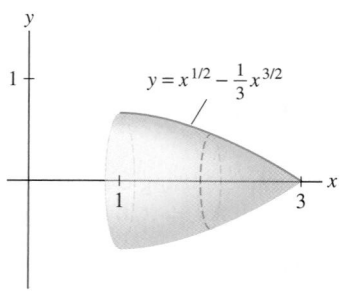

FIGURE 10 Surface obtained by revolving $y = x^{1/2} - \frac{1}{3}x^{3/2}$ about the x-axis for $1 \leq x \leq 3$.

Area of a Surface of Revolution   If $f'(x)$ exists and is continuous on $[a, b]$, then the area of the surface obtained by rotating the graph of $f(x)$ about the x-axis for $a \leq x \leq b$ is equal to

$$\text{Surface area over } [a, b] = 2\pi \int_a^b f(x)\sqrt{1 + f'(x)^2}\,dx$$

■ **EXAMPLE 4**   Find the surface area S of the cone obtained by rotating the line $y = 2x$ about the x-axis for $0 \leq x \leq 4$.

Solution   We have $f(x) = 2x$ and $\sqrt{1 + f'(x)^2} = \sqrt{1 + 4} = \sqrt{5}$, so the surface area is equal to (Figure 9)

$$S = 2\pi \int_0^4 f(x)\sqrt{1 + f'(x)^2}\,dx = 2\pi \int_0^4 2x\sqrt{5}\,dx = 2\pi\sqrt{5}\,x^2 \Big|_0^4 = 32\pi\sqrt{5} \quad ■$$

■ **EXAMPLE 5**   Find the area S of the surface obtained by rotating $y = x^{1/2} - \frac{1}{3}x^{3/2}$ about the x-axis for $1 \leq x \leq 3$.

Solution   Let $f(x) = x^{1/2} - \frac{1}{3}x^{3/2}$. Then $f'(x) = \frac{1}{2}(x^{-1/2} - x^{1/2})$ and

$$1 + f'(x)^2 = 1 + \left(\frac{x^{-1/2} - x^{1/2}}{2}\right)^2 = 1 + \frac{x^{-1} - 2 + x}{4}$$

$$= \frac{x^{-1} + 2 + x}{4} = \left(\frac{x^{1/2} + x^{-1/2}}{2}\right)^2$$

The surface area (Figure 10) is equal to

$$S = 2\pi \int_1^3 f(x)\sqrt{1 + f'(x)^2}\,dx$$

$$= 2\pi \int_1^3 \left(x^{1/2} - \frac{1}{3}x^{3/2}\right)\left(\frac{x^{1/2} + x^{-1/2}}{2}\right)dx$$

$$= \pi \int_1^3 \left(1 + \frac{2}{3}x - \frac{1}{3}x^2\right)dx = \pi\left(x + \frac{1}{3}x^2 - \frac{1}{9}x^3\right)\Big|_1^3 = \frac{16\pi}{9} \quad ■$$

8.1 SUMMARY

- The *arc length* of the curve $y = f(x)$ for $a \leq x \leq b$ is $\int_a^b \sqrt{1 + f'(x)^2}\,dx$.
- When the arc length integral cannot be evaluated directly, we may use numerical integration to obtain an approximation.
- The *surface area* of the surface obtained by rotating the graph of $f(x)$ about the x-axis for $a \leq x \leq b$ is

$$\text{Surface area} = 2\pi \int_a^b f(x)\sqrt{1 + f'(x)^2}\,dx$$

8.1 EXERCISES

Preliminary Questions

1. Which integral represents the length of the curve $y = \cos x$ between 0 and π?

$$\int_0^\pi \sqrt{1 + \cos^2 x}\, dx, \qquad \int_0^\pi \sqrt{1 + \sin^2 x}\, dx$$

2. How do the arc lengths of the curves $y = f(x)$ and $y = f(x) + C$ over an interval $[a, b]$ differ (C is a constant)? Explain geometrically and then justify using the arc length formula.

Exercises

1. Express the arc length of the curve $y = x^4$ between $x = 2$ and $x = 6$ as an integral (but do not evaluate).

2. Express the arc length of the curve $y = \tan x$ for $0 \le x \le \frac{\pi}{4}$ as an integral (but do not evaluate).

3. Find the arc length of $y = \frac{1}{12}x^3 + x^{-1}$ for $1 \le x \le 2$. *Hint:* Show that $1 + (y')^2 = \left(\frac{1}{4}x^2 + x^{-2}\right)^2$.

4. Find the arc length of $y = \left(\frac{x}{2}\right)^4 + \frac{1}{2x^2}$ over $[1, 4]$. *Hint:* Show that $1 + (y')^2$ is a perfect square.

In Exercises 5–10, calculate the arc length over the given interval.

5. $y = 3x + 1$, $[0, 3]$

6. $y = 9 - 3x$, $[1, 3]$

7. $y = x^{3/2}$, $[1, 2]$

8. $y = \frac{1}{3}x^{3/2} - x^{1/2}$, $[2, 8]$

9. $y = \frac{1}{4}x^2 - \frac{1}{2}\ln x$, $[1, 2e]$

10. $y = \ln(\cos x)$, $[0, \frac{\pi}{4}]$

In Exercises 11–14, approximate the arc length of the curve over the interval using the Trapezoidal Rule T_N, the Midpoint Rule M_N, or Simpson's Rule S_N as indicated.

11. $y = \frac{1}{4}x^4$, $[1, 2]$, T_5

12. $y = \sin x$, $[0, \frac{\pi}{2}]$, M_8

13. $y = x^{-1}$, $[1, 2]$, S_8

14. $y = e^{-x^2}$, $[0, 2]$, S_8

15. Calculate the length of the astroid $x^{2/3} + y^{2/3} = 1$ (Figure 11).

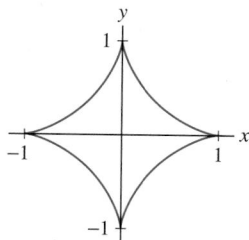

FIGURE 11 Graph of $x^{2/3} + y^{2/3} = 1$.

16. Show that the arc length of the astroid $x^{2/3} + y^{2/3} = a^{2/3}$ (for $a > 0$) is proportional to a.

17. Find the arc length of the curve shown in Figure 12.

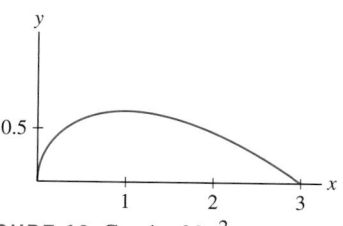

FIGURE 12 Graph of $9y^2 = x(x - 3)^2$.

18. Find the value of a such that the arc length of the *catenary* $y = \cosh x$ for $-a \le x \le a$ equals 10.

19. Let $f(x) = mx + r$ be a linear function (Figure 13). Use the Pythagorean Theorem to verify that the arc length over $[a, b]$ is equal to

$$\int_a^b \sqrt{1 + f'(x)^2}\, dx$$

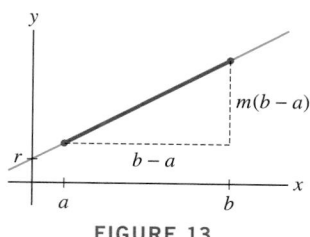

FIGURE 13

20. Show that the circumference of the unit circle is equal to

$$2\int_{-1}^1 \frac{dx}{\sqrt{1 - x^2}} \quad \text{(an improper integral)}.$$ Evaluate, thus verifying that the circumference is 2π.

21. Show that the circumference C of the circle of radius r is $C = 2\int_{-r}^r \frac{dx}{\sqrt{1 - x^2/r^2}}$. Use a substitution to show that C is proportional to r.

22. Calculate the arc length of $y = x^2$ over $[0, a]$. *Hint:* Use trigonometric substitution. Evaluate for $a = 1$.

23. 📖 Express the arc length of $g(x) = \sqrt{x}$ over $[0, 1]$ as a definite integral. Then use the substitution $u = \sqrt{x}$ to show that this arc length is equal to the arc length of x^2 over $[0, 1]$ (but do not evaluate the integrals). Explain this result graphically.

24. [GU] Use a graphing utility to sketch the graphs of $f(x) = \sqrt{x}$ and $g(x) = \ln x$ over $[1, 4]$. Then, by applying the Comparison Test to the arc length integrals, show that the arc length of $y = f(x)$ is:
(a) Less than the arc length of $y = g(x)$ over $[1, 4]$.
(b) Greater than the arc length of $y = g(x)$ over $[4, 8]$.

25. Find the arc length of $y = e^x$ over $[0, a]$. *Hint:* Try the substitution $u = \sqrt{1 + e^{2x}}$ followed by partial fractions.

26. Show that the arc length of $y = \ln(f(x))$ for $a \le x \le b$ is

$$\int_a^b \frac{\sqrt{f(x)^2 + f'(x)^2}}{f(x)} \, dx \qquad \boxed{4}$$

27. Use Eq. (4) to compute the arc length of $y = \ln(\sin x)$ for $\frac{\pi}{4} \le x \le \frac{\pi}{2}$.

28. Use Eq. (4) to compute the arc length of $y = \ln\left(\dfrac{e^x + 1}{e^x - 1}\right)$ over $[1, 3]$.

29. Show that if $0 \le f'(x) \le 1$ for all x, then the arc length of $y = f(x)$ over $[a, b]$ is at most $\sqrt{2}(b - a)$. Show that for $f(x) = x$, the arc length equals $\sqrt{2}(b - a)$.

30. Let $a, r > 0$. Show that the arc length of the curve $x^r + y^r = a^r$ for $0 \le x \le a$ is proportional to a.

31. Approximate the arc length of one-quarter of the unit circle (which we know is $\frac{\pi}{2}$) by computing the length of the polygonal approximation with $N = 4$ segments (Figure 14).

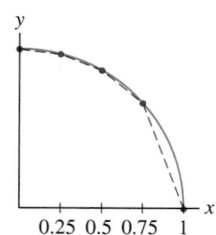

FIGURE 14 One-quarter of the unit circle

In Exercises 32–39, compute the surface area of revolution about the x-axis over the interval.

32. $y = 4x + 3$, $[0, 1]$

33. $y = 4x + 3$, $[2, 4]$

34. $y = \frac{1}{12}x^3 + x^{-1}$, $[1, 4]$

35. $y = (4 - x^{2/3})^{3/2}$, $[0, 8]$

36. $y = e^{-x}$, $[0, 1]$

37. $y = e^x$, $[0, 1]$

38. $y = x^2$, $[0, 4]$

39. $y = \sin x$, $[0, \pi]$

40. Find the area of the surface obtained by rotating $y = \cosh x$ over $[-\ln 2, \ln 2]$ around the x-axis.

41. Prove that the surface area of a sphere of radius r is $4\pi r^2$ by rotating the top half of the unit circle $x^2 + y^2 = r^2$ about the x-axis.

CRS *In Exercises 42–45, use a computer algebra system to find the exact or approximate surface area of the solid generated by rotating the curve about the x-axis.*

42. $y = \frac{1}{4}x^2 - \frac{1}{2}\ln x$ for $1 \le x \le e$, exact area

43. $y = x^2$ for $0 \le x \le 4$, exact area

44. $y = x^3$ for $0 \le x \le 4$, approximate area

45. $y = \tan x$ for $0 \le x \le \frac{\pi}{4}$, approximate area

46. Find the surface area of the torus obtained by rotating the circle $x^2 + (y - b)^2 = r^2$ about the x-axis.

47. **CRS** A merchant intends to produce specialty carpets in the shape of the region in Figure 15, bounded by the axes and graph of $y = 1 - x^n$ (units in yards). Assume that material costs \$50/yd^2 and that it costs $50L$ dollars to cut the carpet, where L is the length of the curved side of the carpet. The carpet can be sold for $150A$ dollars, where A is the carpet's area. Using numerical integration with a computer algebra system, find the whole number n for which the merchant's profits are maximal.

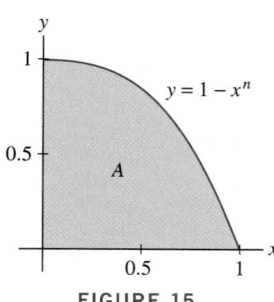

FIGURE 15

48. Prove that the portion of a sphere of radius R seen by an observer located at a distance d above the North Pole has area $A = \dfrac{2\pi d R^2}{d + R}$.
Hint: First show that if the cap has height h, then its surface area is $2\pi R h$. Then show that $h = \dfrac{dR}{d + R}$ by applying the Pythagorean Theorem to three right triangles in Figure 16.

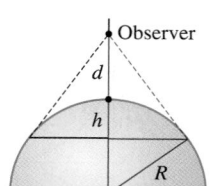

FIGURE 16 Spherical cap observed from a distance d above the North Pole.

49. [notebook icon] Suppose that the observer in Exercise 48 moves off to infinity, that is, $d \to \infty$. What do you expect the limiting value of the observed area to be? Check your guess by calculating the limit using the formula for the area in the previous exercise.

Further Insights and Challenges

50. Show that the surface area of a right circular cone of radius r and height h is $\pi r\sqrt{r^2+h^2}$. *Hint:* Rotate a line $y=mx$ about the x-axis for $0 \le x \le h$, where m is determined suitably by the radius r.

51. Find the surface area of the ellipsoid obtained by rotating the ellipse $\left(\dfrac{x}{a}\right)^2 + \left(\dfrac{y}{b}\right)^2 = 1$ about the x-axis.

52. Show that if the arc length of $f(x)$ over $[0, a]$ is proportional to a, then $f(x)$ must be a linear function

53. *CAS*   Let L be the arc length of the upper half of the ellipse with equation $y = \left(\dfrac{b}{a}\right)\sqrt{a^2-x^2}$ (Figure 17) and let $\eta = \sqrt{1-\dfrac{b^2}{a^2}}$.

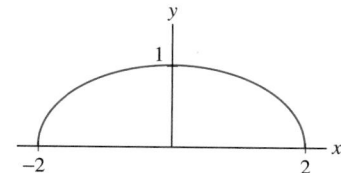

FIGURE 17 Graph of the ellipse $y = \frac{1}{2}\sqrt{4-x^2}$.

Use substitution to show that

$$L = a \int_{-\pi/2}^{\pi/2} \sqrt{1 - \eta^2 \sin^2 \theta}\, d\theta$$

Use a computer algebra system to approximate L for $a=2, b=1$.

54. Let M be the total mass of a metal rod in the shape of the curve $y=f(x)$ over $[a, b]$ whose mass density $\rho(x)$ varies as a function of x. Use Riemann sums to justify the formula

$$M = \int_a^b \rho(x)\sqrt{1 + f'(x)^2}\, dx$$

55. Let $f(x)$ be an increasing function on $[a, b]$ and let $g(x)$ be its inverse. Argue on the basis of arc length that the following equality holds:

$$\int_a^b \sqrt{1 + f'(x)^2}\, dx = \int_{f(a)}^{f(b)} \sqrt{1 + g'(y)^2}\, dy \qquad \boxed{5}$$

Then use the substitution $u = f(x)$ to prove Eq. (5).

8.2 Fluid Pressure and Force

In this section, we use integration to compute the force exerted on an object submerged in a fluid. This is the force you feel when you try to push a basketball underwater in a swimming pool. Our computations are based on a law from physics which states that for a fluid at rest (such as water in a tank as opposed to water flowing in a stream), fluid pressure p is proportional to depth h.

Pressure is defined as force per unit area. However, there is an important difference between fluid pressure and the pressure exerted by one solid object on another. Fluid pressure does not act in any specific direction. Instead, a fluid exerts pressure on each side of an object in the direction perpendicular to that side (Figure 2). Recall that the density of a fluid is measured in units of mass per volume or weight per volume.

FIGURE 1 Since water pressure is proportional to depth, a diver's lungs are compressed and may be injured during a deep dive.

Pressure as a Function of Depth   The fluid pressure at depth h in a fluid of density w (in weight per volume) is

$$p = \text{fluid pressure} = wh \qquad \boxed{1}$$

The pressure acts at each point on an object in the direction perpendicular to the object's surface at that point.

In our first example, we calculate the fluid force (also called the "total" or "resultant" force) in a situation where the pressure p is constant. In this case, the total force acting

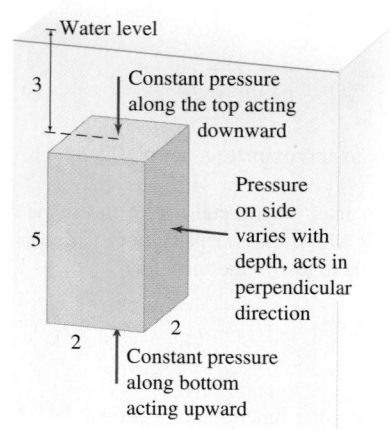

FIGURE 2 The pressure acts on each side in the perpendicular direction.

on a surface of area A is

$$\boxed{\text{Force} = \text{pressure} \times \text{area} = pA}$$

The density of water is $w = 62.5 \text{ lb/ft}^3$.

■ **EXAMPLE 1** Calculate the force on the top and bottom of a box of dimensions $2 \times 2 \times 5$ ft, submerged in a pool of water with its top 3 ft below the water surface (Figure 2).

Solution The top of the box is located at depth $h = 3$ ft, so by Eq. (1),

$$\text{Pressure on top} = wh = 62.5(3) = 187.5 \text{ lb/ft}^2$$

Since the pressure along the top is constant and the top has area $A = 4 \text{ ft}^2$, the total force on the top, acting downward, is

$$\text{Force on top} = \text{pressure} \times \text{area} = 187.5 \times 4 = 750 \text{ lb}$$

The bottom of the box is at depth $h = 8$ ft, so the total force on the bottom is

$$\text{Force on bottom} = \text{pressure} \times \text{area} = 62.5(8) \times 4 = 2{,}000 \text{ lb}$$

The force on the bottom acts in the upward direction.   ■

In the next example, we calculate the total force on the side of a box. Since the pressure varies with depth, we use integration to find the total force.

■ **EXAMPLE 2** Calculating Force on a Side Using Integration Calculate the force F on the side of a box submerged with its top 3 ft below the surface of the water. As before, the box has height 5 and a square base with a side of 2 ft.

Solution Since the pressure varies with depth, we divide the side into N thin horizontal strips (Figure 3). If F_j is the force on the jth strip, then the total force F is

$$F = F_1 + F_2 + \cdots + F_N$$

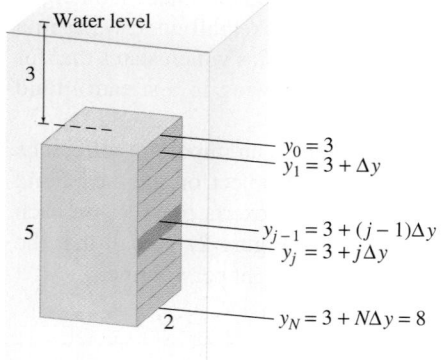

FIGURE 3 Each strip has area $2\Delta y$.

Step 1. **Approximate the force on a strip.**
We'll use the variable y to denote depth. Thus, a larger value of y denotes greater depth. Each strip has height $\Delta y = 5/N$ and width 2, so the area of a strip is $2\Delta y$. The bottom edge of the jth strip has depth $y_j = 3 + j\Delta y$ (Figure 3).

If Δy is small, then the pressure on the jth strip is nearly constant with value wy_j (since the strip is thin, all points on it lie at nearly the same depth y_j), so we can approximate the force on the jth strip:

$$F_j = \text{force on the } j\text{th strip at depth } y_j \approx \underbrace{wy_j}_{\text{Pressure}} \times \underbrace{(2\Delta y)}_{\text{Area}} = 2wy_j\,\Delta y$$

Step 2. **Approximate total force as a Riemann sum.**
The total force on the side is

$$F = F_1 + F_2 + \cdots + F_N \approx \sum_{j=1}^{N} 2wy_j\,\Delta y$$

We recognize the sum on the right as a Riemann sum for the integral $w \displaystyle\int_3^8 2y\,dy$. The interval of integration is $[3, 8]$ because the box extends from $y = 3$ to $y = 8$ (the Riemann sum has been set up with $y_0 = 3$ and $y_N = 8$).

The computations of fluid force and work (Example 2 in Section 6.5) are similar. In both cases, we estimate the force or work as a sum (corresponding to thin strips, slices, etc.), and express the quantity precisely as an integral.

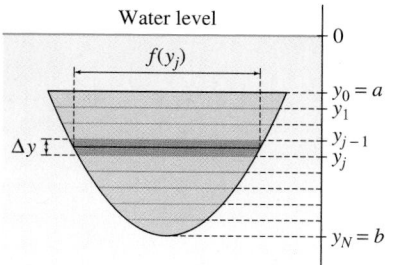

FIGURE 4 The area of the shaded strip is approximately $f(y_j)\,\Delta y$.

Step 3. Evaluate total force as an integral.

As Δy tends to zero, the Riemann sum approaches the integral and we obtain

$$F = w \int_3^8 2y\,dy = (62.5)y^2 \Big|_3^8 = 62.5(8^2 - 3^2) = 3{,}437.5 \text{ lb} \qquad ▮$$

The method of the previous example can be used to find the fluid force on a flat side of any object submerged vertically (not at an incline) in a fluid. Assume that the object extends from depth $y = a$ to $y = b$ and, as before, divide the side into N horizontal strips of thickness $\Delta y = (b - a)/N$ (Figure 4). Let

$$f(y) = \text{width of the side at depth } y$$

If Δy is small, then the jth strip is nearly rectangular with height Δy, width $f(y)$ for some continuous function $f(y)$, and area $f(y)\Delta y$. Since the strip lies at depth $y_j = a + j\Delta y$, the force F_j on the jth strip can be approximated:

$$F_j = \text{force on } j\text{th strip at depth } y_j \approx \underbrace{wy_j}_{\text{Pressure}} \times \underbrace{f(y_j)\Delta y}_{\text{Area}} = wy_j f(y_j)\,\Delta y$$

The total force F on the side is approximated by a Riemann sum that converges to an integral as $N \to \infty$:

$$F = F_1 + \cdots + F_N = \sum_{j=1}^{N} wy_j f(y_j)\,\Delta y \quad \to \quad w \int_a^b y f(y)\,dy$$

THEOREM 1 Fluid Force on a Flat Surface Submerged Vertically   The force on a flat side of an object submerged vertically in a fluid is

$$\text{Force on the side} = w \int_a^b \big(\text{depth} \times \text{width}\big)\,d(\text{depth}) = w \int_a^b y f(y)\,dy \qquad \boxed{2}$$

where $f(y)$ is the horizontal width of the side at depth y, w is the fluid density (in weight per unit volume), and the object extends from depth $y = a$ to depth $y = b$.

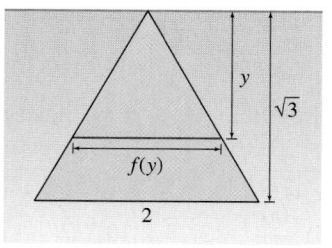

FIGURE 5 Triangular plate submerged in a tank of oil.

▮ **EXAMPLE 3** A plate in the shape of an equilateral triangle of side 2 m is submerged vertically in a tank of oil of mass density $\rho = 900$ kg/m^3 as in Figure 5. Calculate the total force F on one side of the plate.

Solution To use Eq. (2), we calculate the horizontal width $f(y)$ of the plate at depth y. Recall that an equilateral triangle of side s has height $\dfrac{\sqrt{3}}{2}s$. Since $s = 2$, our triangle has height $\sqrt{3}$. By similar triangles (Figure 5), $y/f(y) = \sqrt{3}/2$ and thus $f(y) = 2y/\sqrt{3}$. The fluid density in weight per volume is $w = \rho g = (900)(9.8) = 8{,}820$ N/m^3. By Eq. (2),

$$F = w \int_0^{\sqrt{3}} y\, f(y)\,dy = 8{,}820 \int_0^{\sqrt{3}} \frac{2}{\sqrt{3}} y^2\,dy = \left(\frac{17{,}640}{\sqrt{3}}\right) \frac{y^3}{3} \Big|_0^{\sqrt{3}} = 17{,}640 \text{ N} \quad ▮$$

If the side of the submerged object is inclined at an angle, we must modify the force calculation, as explained in the next example.

Hoover Dam, located on the Nevada–Arizona border.

FIGURE 6 Water exerts pressure on the side of a dam.

■ **EXAMPLE 4** Force on an Inclined Surface Calculate the fluid force F on the side of a dam, which is inclined at an angle of 45°. The height of the dam is 700 ft and the width of the base is 1,500 ft as in Figure 6. Assume the reservoir is filled so that water pressure acts along the full slanted side of the dam.

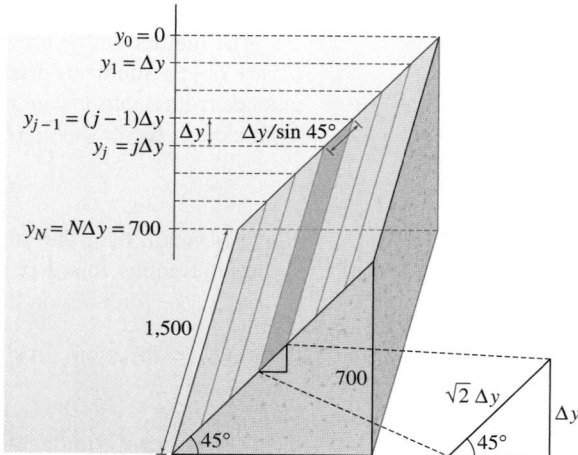

Solution The vertical height of the dam is 700 ft. We divide the vertical axis from 0 to 700 into N subintervals of length $\Delta y = 700/N$. This divides the face of the dam into N strips as in Figure 6. Note that the width of each strip is the length of the hypotenuse of a triangle that, by trigonometry, has length $\Delta y / \sin(45°) = \sqrt{2}\,\Delta y$. Therefore,

$$\text{Area of each strip} = \text{length} \times \text{width} = 1,500(\sqrt{2}\,\Delta y)$$

Now we may proceed as in the previous example. We approximate the force F_j on the jth strip and write the total force F as the sum of the forces on the strips:

$$F_j \approx \underbrace{wy_j}_{\text{Pressure}} \times \underbrace{1,500\sqrt{2}\Delta y}_{\text{Area of strip}} = wy_j \times 1,500\sqrt{2}\,\Delta y$$

$$F = \sum_{j=1}^{N} F_j \approx \sum_{j=1}^{N} wy_j\left(1,500\sqrt{2}\,\Delta y\right) = 1,500\sqrt{2}\,w \sum_{j=1}^{N} y_j\,\Delta y$$

The sum on the right is a Riemann sum for the integral $1,500\sqrt{2}w \displaystyle\int_0^{700} y\,dy$. Since water has density $w = 62.5$ lb/ft^3,

$$F = 1,500\sqrt{2}w \int_0^{700} y\,dy = 1,500\sqrt{2}(62.5)\frac{700^2}{2} \approx 3.25 \times 10^{10}\text{ lb} \qquad ■$$

8.2 SUMMARY

- The fluid pressure at depth h is equal to wh, where w is the density of the fluid (weight per volume). Fluid pressure acts on a surface in the direction perpendicular to the surface. Water has density 62.5 lb/ft^3 and mass density 1,000 kg/m^3.

• If an object is submerged vertically in a fluid and extends from depth $y = a$ to $y = b$, then the total force on a side of the object is

$$F = w \int_a^b yf(y)\, dy$$

where $f(y)$ is the horizontal width of the side at depth y.

8.2 EXERCISES

Preliminary Questions

1. How is pressure defined?

2. Fluid pressure is proportional to depth. What is the factor of proportionality?

3. When fluid force acts on the side of a submerged object, in which direction does it act?

4. Why is fluid pressure on a surface calculated using thin horizontal strips rather than thin vertical strips?

5. If a thin plate is submerged horizontally, then the fluid force on one side of the plate is equal to pressure times area. Is this true if the plate is submerged vertically?

Exercises

1. A box of height 6 ft and square base of side 3 ft is submerged in a pool of water. The top of the box is 2 ft below the surface of the water.

(a) Calculate the fluid force on the top and bottom of the box.

(b) Write a Riemann sum that approximates the fluid force on a side of the box by dividing the side into N horizontal strips of thickness $\Delta y = 6/N$.

(c) To which integral does the Riemann sum converge?

(d) Compute the fluid force on a side of the box.

2. A plate in the shape of an isosceles triangle with base 1 ft and height 2 ft is submerged vertically in a tank of water so that its vertex touches the surface of the water (Figure 7).

(a) Show that the width of the triangle at depth y is $f(y) = \frac{1}{2}y$.

(b) Consider a thin strip of thickness Δy at depth y. Explain why the fluid force on a side of this strip is approximately equal to $w\frac{1}{2}y^2\Delta y$, where $w = 62.5 \text{ lb/ft}^3$.

(c) Write an approximation for the total fluid force F on a side of the plate as a Riemann sum and indicate the integral to which it converges.

(d) Calculate F.

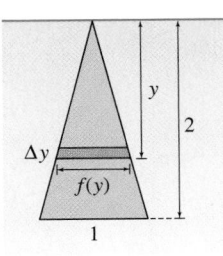

FIGURE 7

3. Repeat Exercise 2, but assume that the top of the triangle is located 3 ft below the surface of the water.

4. The thin plate R in Figure 8, bounded by the parabola $y = x^2$ and $y = 1$, is submerged vertically in water. Let F be the fluid force on one side of R.

(a) Show that the width of R at height y is $f(y) = 2\sqrt{y}$ and the fluid force on a side of a horizontal strip of thickness Δy at height y is approximately $2wy^{1/2}(1 - y)\Delta y$.

(b) Write a Riemann sum that approximates F and use it to explain why $F = 2w \int_0^1 y^{1/2}(1 - y)\, dy$.

(c) Calculate F.

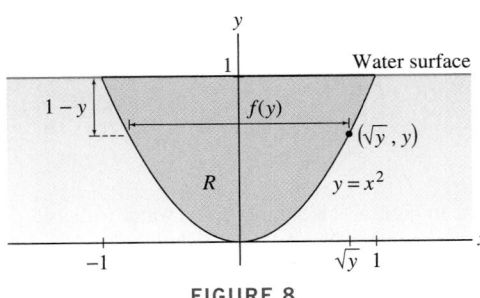

FIGURE 8

5. Let F be the fluid force (in Newtons) on a side of a semicircular plate of radius r meters, submerged in water so that its diameter is level with the water's surface (Figure 9).

FIGURE 9

(a) Show that the width of the plate at depth y is $2\sqrt{r^2 - y^2}$.

(b) Calculate F using Eq. (2).

6. Calculate the force on one side of a circular plate with radius 2 ft, submerged vertically in a tank of water so that the top of the circle is tangent to the water surface.

7. A semicircular plate of radius r, oriented as in Figure 9, is submerged in water so that its diameter is located at a depth of m feet. Calculate the force on one side of the plate in terms of m and r.

8. Figure 10 shows the wall of a dam on a water reservoir. Use the Trapezoidal Rule and the width and depth measurements in the figure to estimate the total force on the wall.

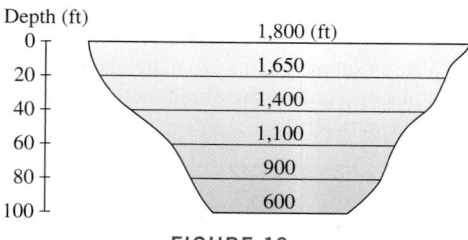

FIGURE 10

9. Calculate the total force (in Newtons) on a side of the plate in Figure 11(A), submerged in water.

10. Calculate the total force (in Newtons) on a side of the plate in Figure 11(B), submerged in a fluid of mass density $\rho = 800$ kg/m^3.

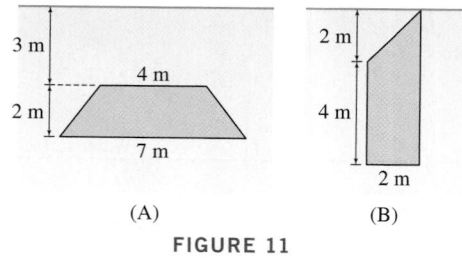

(A)   (B)

FIGURE 11

11. The plate in Figure 12 is submerged in water with its top level with the surface of the water. The left and right edges of the plate are the curves $y = x^{1/3}$ and $y = -x^{1/3}$. Find the fluid force on a side of the plate.

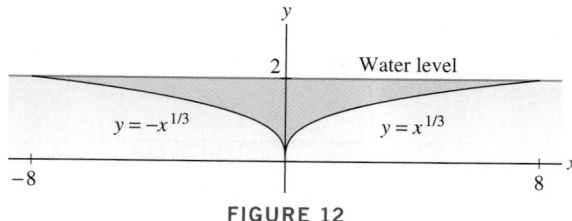

FIGURE 12

12. Let R be the plate in the shape of the region under $y = \sin x$ for $0 \le x \le \frac{\pi}{2}$ in Figure 13(A). Find the fluid force on a side of R if it is rotated counterclockwise by 90° and submerged in a fluid of density 140 lb/ft^3 with its top edge level with the surface of the fluid as in (B).

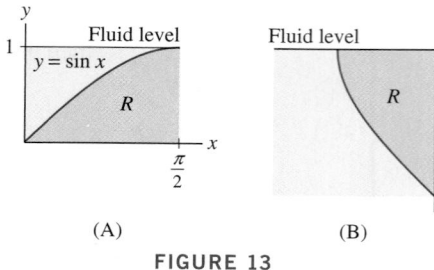

(A)   (B)

FIGURE 13

13. In the notation of Exercise 12, calculate the fluid force on a side of the plate R if it is oriented as in Figure 13(A). You may need to use Integration by Parts and trigonometric substitution.

14. Let A be the region under the graph of $y = \ln x$ for $1 \le x \le e$ (Figure 14). Calculate the fluid force on one side of a plate in the shape of region A if the water surface is at $y = 1$.

15. Calculate the fluid force on one side of the "infinite" plate B in Figure 14.

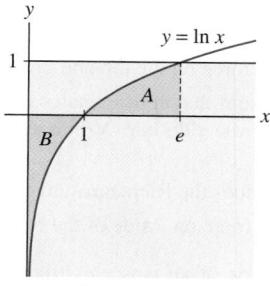

FIGURE 14

16. A square plate of side 3 m is submerged in water at an incline of 30° with the horizontal. Its top edge is located at the surface of the water. Calculate the fluid force (in Newtons) on one side of the plate.

17. Repeat Exercise 16, but assume that the top edge of the plate lies at a depth of 6 m.

18. Figure 15(A) shows a ramp inclined at 30° leading into a swimming pool. Calculate the fluid force on the ramp.

19. Calculate the fluid force on one side of the plate (an isosceles triangle) shown in Figure 15(B).

(A)   (B)

FIGURE 15

20. The trough in Figure 16 is filled with corn syrup, whose density is 90 lb/ft^3. Calculate the force on the front side of the trough.

21. Calculate the fluid pressure on one of the slanted sides of the trough in Figure 16, filled with corn syrup as in Exercise 20.

22. Figure 17 shows an object whose face is an equilateral triangle with 5-ft sides. The object is 2 ft thick and is submerged in water with its vertex 3 ft below the water surface. Calculate the fluid force on both a triangular face and a slanted rectangular edge of the object.

FIGURE 16

FIGURE 17

Further Insights and Challenges

23. The end of the trough in Figure 18 is an equilateral triangle of side 3. Assume that the trough is filled with water to height y. Calculate the fluid force on each side of the trough as a function of the level y and the length l of the trough.

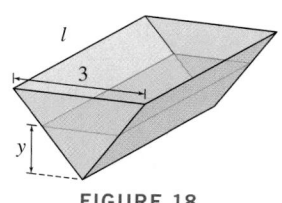

FIGURE 18

24. A rectangular plate of side ℓ is submerged vertically in a fluid of density w, with its top edge at depth h. Show that if the depth is increased by an amount Δh, then the force on a side of the plate increases by $wA\Delta h$, where A is the area of the plate.

25. Prove that the force on the side of a rectangular plate of area A submerged vertically in a fluid is equal to $p_0 A$, where p_0 is the fluid pressure at the center point of the rectangle.

26. If the density of a fluid varies with depth, then the pressure at depth y is a function $p(y)$ (which need not equal wy as in the case of constant density). Use Riemann sums to argue that the total force F on the flat side of a submerged object submerged vertically is

$$F = \int_a^b f(y)p(y)\,dy,$$ where $f(y)$ is the width of the side at depth y.

8.3 Center of Mass

Every object has a balance point called the *center of mass* (COM) (Figure 1). You can balance an object such as a tray or ball on the tip of one finger if you place your finger at a point directly below the COM. More generally, an object does not rotate when a force is applied along a line passing through the COM.

In this section, we review the formula for the COM of a system of particles (or "point masses") and then show how integration may be used to compute the centers of mass of thin plates of uniform density.

The center of mass is expressed in terms of quantities called **moments**. For a single particle of mass m, the moment with respect to a line L is defined as m times the directed distance (positive or negative) from the particle to the line:

Moment with respect to line $L = m \times$ directed distance to L

The moments with respect to the x- and y-axes, denoted M_x and M_y, are called the x- and y-moments. If the mass m is located at the point (x, y) in the plane (Figure 2),

FIGURE 1 An "extended object" such as a dancer may rotate in a complicated fashion as it moves through the air under the influence of gravity, but its center of mass traces a simple parabolic trajectory (ignoring air friction).

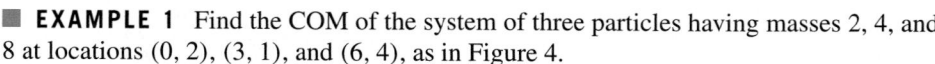

FIGURE 2   FIGURE 3

$$M_x = \text{moment with respect to } x\text{-axis} = \underbrace{m}_{\text{Mass}} \times \underbrace{y}_{\text{Directed distance to } x\text{-axis}}$$

$$M_y = \text{moment with respect to } y\text{-axis} = \underbrace{m}_{\text{Mass}} \times \underbrace{x}_{\text{Directed distance to } y\text{-axis}}$$

Given n particles with coordinates (x_j, y_j) and mass m_j, we define the moment of the system as the sum of the moments of the individual particles (Figure 3):

$$M_x = m_1 y_1 + m_2 y_2 + \cdots + m_n y_n$$

$$M_y = m_1 x_1 + m_2 x_2 + \cdots + m_n x_n$$

The **center of mass** of the system is the point $P_{\text{CM}} = (x_{\text{CM}}, y_{\text{CM}})$ with coordinates

$$x_{\text{CM}} = \frac{M_y}{M}, \qquad y_{\text{CM}} = \frac{M_x}{M}$$

| |
CAUTION *The notation is potentially confusing: M_x is defined in terms of y-coordinates and M_y in terms of x-coordinates. However, $x_{CM} = M_y/M$ and $y_{CM} = M_x/M$.*

where $M = m_1 + m_2 + \cdots + m_n$ is the total mass of the system.

■ **EXAMPLE 1** Find the COM of the system of three particles having masses 2, 4, and 8 at locations $(0, 2)$, $(3, 1)$, and $(6, 4)$, as in Figure 4.

Solution The total mass is $M = 2 + 4 + 8 = 14$ and the moments are

$$M_x = m_1 y_1 + m_2 y_2 + m_3 y_3 = 2 \cdot 2 + 4 \cdot 1 + 8 \cdot 4 = 40$$

$$M_y = m_1 x_1 + m_2 x_2 + m_3 x_3 = 2 \cdot 0 + 4 \cdot 3 + 8 \cdot 6 = 60$$

Therefore, $x_{\text{CM}} = \frac{60}{14} = \frac{30}{7}$ and $y_{\text{CM}} = \frac{40}{14} = \frac{20}{7}$. The COM is $\left(\frac{30}{7}, \frac{20}{7}\right)$.   ■

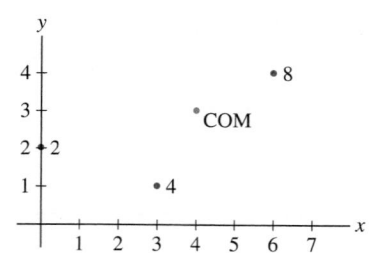

FIGURE 4   Centers of mass for Example 1.

Laminas (Thin Plates)

A lamina is a thin plate whose mass is distributed continuously throughout a region in the plane. We shall compute the moments and the center of mass of a lamina assuming that the mass density ρ is constant. Suppose first that the lamina occupies a region in the plane under the graph of a continuous function $f(x)$ over an interval $[a, b]$, where $f(x) \geq 0$ (Figure 5).

We follow our usual procedure to compute M_y. That is, we divide $[a, b]$ into N subintervals of width $\Delta x = \dfrac{b - a}{N}$ with endpoints $x_j = a + j\Delta x$. This divides the lamina into N vertical strips (Figure 6). If Δx is small, the jth strip is nearly rectangular with base Δx and height $f(x_j)$. Hence, the strip has area $f(x_j)\, \Delta x$ and mass $\rho f(x_j)\, \Delta x$. Since the strip is thin, all points in the strip lie at approximately the same distance x_j from the y-axis (Figure 6). Therefore, we can approximate the moment of the jth strip, which

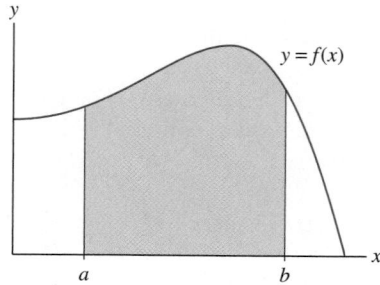

FIGURE 5 Lamina occupying the region under the graph of $f(x)$ over $[a, b]$.

In this section, we restrict our attention to thin plates of constant mass density (also called "uniform density"). COM computations when mass density is not constant require multiple integration, covered in Chapter 15.

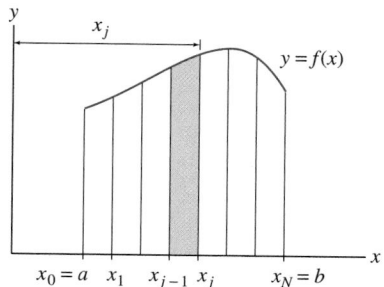

FIGURE 6 The shaded strip is nearly rectangular of area $f(x_j)\,\Delta x$.

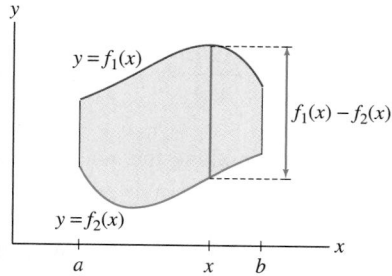

FIGURE 7 The vertical cut at x has length $f_1(x) - f_2(x)$.

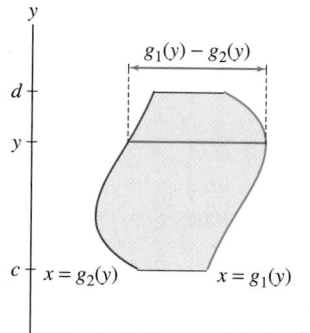

FIGURE 8 The horizontal cut at y has length $g_1(y) - g_2(y)$.

we call $M_{y,j}$:

$$M_{y,j} \approx (\text{mass}) \times (\text{directed distance to } y\text{-axis}) = (\rho f(x_j)\,\Delta x)x_j$$

The moment M_y is the sum of the $M_{y,j}$ and thus may be approximated:

$$M_y = \sum_{j=1}^{N} M_{y,j} \approx \rho \sum_{j=1}^{N} x_j f(x_j)\,\Delta x$$

The sum on the right is a Riemann sum whose value approaches $\int_a^b \rho x f(x)\,dx$ as $N \to \infty$, and the moment itself is given by the integral:

$$M_y = \rho \int_a^b x f(x)\,dx$$

We argue similarly if the lamina occupies the region *between* the graphs of two functions $f_1(x)$ and $f_2(x)$ over $[a, b]$, where $f_1(x) \geq f_2(x)$. The length of the ith vertical strip is then $f_1(x_i) - f_2(x_i)$, so we replace $f(x)$ by $f_1(x) - f_2(x)$ in the formula for M_y. We may think of $f_1(x) - f_2(x)$ as the length of a *vertical* cut or strip of "width zero" (Figure 7):

$$M_y = \rho \int_a^b x\,(\text{length of vertical cut})\,dx = \rho \int_a^b x\big(f_1(x) - f_2(x)\big)\,dx \qquad \boxed{1}$$

We may compute the x-moment in a similar way by dividing the lamina into thick *horizontal* strips. To do so, we must describe the lamina as a region between two curves $x = g_1(y)$ and $x = g_2(y)$ with $g_1(y) \geq g_2(y)$ over an interval $[c, d]$ along the y-axis (Figure 8). We obtain the formula

$$M_x = \rho \int_c^d y\,(\text{length of horizontal cut})\,dy = \rho \int_c^d y\big(g_1(y) - g_2(y)\big)\,dy \qquad \boxed{2}$$

The coordinates of the center of mass are the moments divided by the total mass M of the lamina:

$$x_{\mathrm{CM}} = \frac{M_y}{M}, \qquad y_{\mathrm{CM}} = \frac{M_x}{M}$$

The total mass of the lamina is $M = \rho A$ where A is the area of the lamina:

$$M = \rho A = \rho \int_a^b \big(f_1(x) - f_2(x)\big)\,dx \quad \text{or} \quad \rho \int_c^d \big(g_1(y) - g_2(y)\big)\,dy$$

The COM of a lamina of constant mass density ρ occupying a given region is also called the **centroid** of the lamina or region. Notice that the moments M_x and M_y and the total mass M all contain the factor ρ. Since this factor cancels in the ratios M_y/M and M_x/M, the centroid depends on only the shape of the lamina, not on its mass density. Therefore, *in calculating the centroid, we are free to take $\rho = 1$.* In the more general situation where the mass density is not constant (not covered here), the COM depends on both shape and mass density.

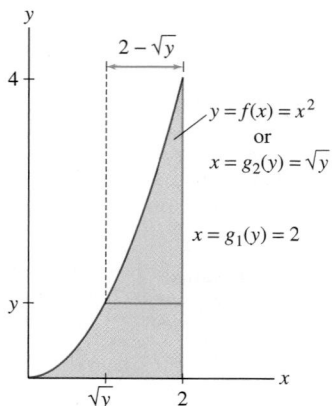

FIGURE 9 Lamina occupying the region under the graph of $f(x) = x^2$ over [0, 2].

■ **EXAMPLE 2** Find the moments and COM of the lamina of uniform density ρ occupying the region underneath the curve $y = x^2$ for $0 \le x \le 2$.

Solution First, we compute M_y using Eq. (1):

$$M_y = \rho \int_0^2 x f(x)\, dx = \rho \int_0^2 x(x^2)\, dx = \rho \left. \frac{x^4}{4} \right|_0^2 = 4\rho$$

To compute M_x using Eq. (2), we must describe our lamina as the region between the graph of $x = g_2(y) = \sqrt{y}$ and the vertical line $x = g_1(y) = 2$ over the interval [0, 4] along the y-axis (Figure 9). Then we obtain

$$M_x = \rho \int_0^4 y\big(g_1(y) - g_2(y)\big)\, dy = \rho \int_0^4 y(2 - \sqrt{y})\, dy$$

$$= \rho \left(y^2 - \frac{2}{5} y^{5/2} \right) \Big|_0^4 = \rho \left(16 - \frac{2}{5} \cdot 32 \right) = \frac{16}{5}\rho$$

The area of the plate is $A = \int_0^2 x^2\, dx = \frac{8}{3}$ and its total mass is $M = \frac{8}{3}\rho$. Therefore, the coordinates of the COM are

$$x_{\text{CM}} = \frac{M_y}{M} = \frac{4\rho}{(8/3)\rho} = \frac{3}{2}, \qquad y_{\text{CM}} = \frac{M_x}{M} = \frac{(16/5)\rho}{(8/3)\rho} = \frac{6}{5} \qquad ■$$

A drawback of Eq. (2) for M_x is that it requires integration along the y-axis. Fortunately, there is a second formula for M_x as an integral along the x-axis. As before, divide the region into N thin vertical strips of width Δx (see Figure 10). Let $M_{x,j}$ be the x-moment of the jth strip and, in addition, let m_j be its mass and y_j the y-coordinate of its COM. Then $y_j = M_{x,j}/m_j$ and we can use the following trick to approximate $M_{x,j}$. The strip is nearly rectangular with height $f(x_j)$ and width Δx, so $m_j \approx \rho f(x_j) \Delta x$. Furthermore, $y_j \approx \frac{1}{2} f(x_j)$ because the COM of a rectangle is located at its center. Thus,

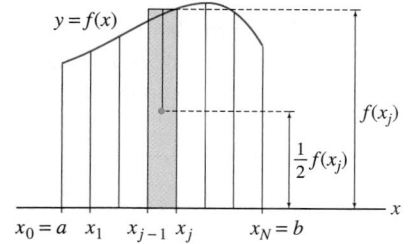

FIGURE 10 The shaded strip is nearly rectangular and its COM has an approximate height of $\frac{1}{2} f(x_j)$.

$$M_{x,j} = m_j y_j \approx \rho f(x_j) \Delta x \cdot \frac{1}{2} f(x_j) = \frac{1}{2} \rho f(x_j)^2 \Delta x$$

$$M_x = \sum_{j=1}^N M_{x,j} \approx \frac{1}{2}\rho \sum_{j=1}^N f(x_j)^2 \Delta x$$

The sum on the right is a Riemann sum whose value approaches $\frac{1}{2}\rho \int_a^b f(x)^2\, dx$ as $N \to \infty$. The region *between* the graphs of two functions $f_1(x)$ and $f_2(x)$ where $f_1(x) \ge f_2(x) \ge 0$ is treated similarly, so we obtain the following alternate formulas for M_x:

$$\boxed{M_x = \frac{1}{2}\rho \int_a^b f(x)^2\, dx \qquad \text{or} \qquad \frac{1}{2}\rho \int_a^b \big(f_1(x)^2 - f_2(x)^2\big)\, dx} \qquad \boxed{3}$$

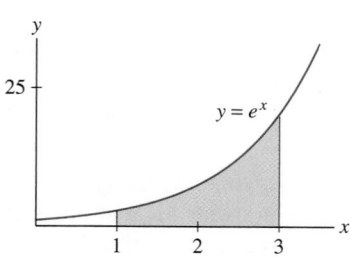

FIGURE 11 Region under the curve $y = e^x$ between $x = 1$ and $x = 3$.

■ **EXAMPLE 3** Find the centroid of the region under the curve $y = e^x$ over [1, 3] (Figure 11).

Solution The centroid does not depend on ρ, so we may set $\rho = 1$. By Eqs. (1) and (3),

$$M_x = \frac{1}{2}\int_1^3 f(x)^2\,dx = \frac{1}{2}\int_1^3 e^{2x}\,dx = \frac{1}{4}e^{2x}\Big|_1^3 = \frac{e^6 - e^2}{4}$$

$$M_y = \int_1^3 xf(x)\,dx = \int_1^3 xe^x\,dx = (x-1)e^x\Big|_1^3 = 2e^3$$

The total mass is $M = \int_1^3 e^x\,dx = (e^3 - e)$. Therefore, the centroid has coordinates

$$x_{CM} = \frac{M_y}{M} = \frac{2e^3}{e^3 - e} \approx 2.313, \qquad y_{CM} = \frac{M_x}{M} = \frac{e^6 - e^2}{4(e^3 - e)} \approx 5.701 \quad \blacksquare$$

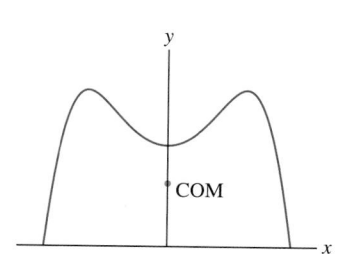

FIGURE 12 The COM of a symmetrical plate lies on the axis of symmetry.

The symmetry properties of an object give information about its centroid (Figure 12). For instance, the centroid of a square or circular plate is located at the center of the plate. Here is a precise formulation of the Symmetry Principle (see Exercise 42).

> **THEOREM 1 Symmetry Principle**   If a lamina of constant density is symmetrical with respect to a line, then its centroid lies on that line.

■ **EXAMPLE 4** Using Symmetry   Find the centroid of a semicircle of radius 3.

Solution We use symmetry to cut our work in half. Set $\rho = 1$. Since the semicircle is symmetrical with respect to y-axis, the centroid lies on the y-axis and hence $x_{CM} = 0$. It remains to calculate M_x and y_{CM}. The semicircle is the graph of $f(x) = \sqrt{9 - x^2}$ (Figure 13). By Eq. (3),

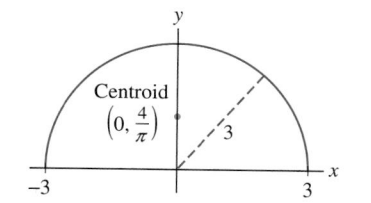

FIGURE 13 The semicircle $y = \sqrt{9 - x^2}$.

$$M_x = \frac{1}{2}\int_{-3}^3 f(x)^2\,dx = \frac{1}{2}\int_{-3}^3 (9 - x^2)\,dx = \frac{1}{2}\left(9x - \frac{1}{3}x^3\right)\Big|_{-3}^3 = 9 - (-9) = 18$$

Furthermore, the semicircle has area (and mass) equal to $A = \frac{1}{2}\pi(3^2) = 9\pi/2$, so

$$y_{CM} = \frac{M_x}{M} = \frac{18}{9\pi/2} = \frac{4}{\pi} \approx 1.27 \quad \blacksquare$$

Moments satisfy a key additivity property: If a region S consists of two or more disjoint smaller regions, then the moment of S (with respect to any axis or line) is the *sum* of the moments of the smaller regions. This follows from the additivity of the integral. The next example shows how the additivity property can be used to simplify COM calculations.

■ **EXAMPLE 5** Using Additivity of Moments   Find the centroid of the region R in Figure 14.

Solution We set $\rho = 1$ because we are computing a centroid. The region R is symmetrical with respect to the y-axis, so we know that the centroid lies on the y-axis. Thus $x_{CM} = 0$. It remains to compute y_{CM}.

Step 1. Use additivity of moments.
  To find y_{CM}, we use additivity of the x-moment:

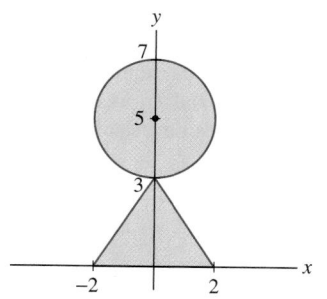

FIGURE 14 The moment of region R is the sum of the moments of the triangle and circle.

$$M_x^R = M_x^{\text{triangle}} + M_x^{\text{circle}}$$

where M_x^R, M_x^{triangle}, and M_x^{circle} are the x-moments of R, the triangle, and circle, respectively.

Step 2. Moment of a triangle.

Let's compute M_x^{triangle} for an arbitrary triangle of height h and base b. If $\ell(y)$ is the width of the triangle at height y (Figure 15), then by similar triangles,

$$\frac{\ell(y)}{h-y} = \frac{b}{h} \quad \Rightarrow \quad \ell(y) = b - \frac{b}{h}y$$

Therefore,

$$M_x^{\text{triangle}} = \int_0^h y\,\ell(y)\,dy = \int_0^h y\left(b - \frac{b}{h}y\right)dy = \left(\frac{by^2}{2} - \frac{by^3}{3h}\right)\Bigg|_0^h = \frac{bh^2}{6}$$

In our case, $b = 4$, $h = 3$, and $M_x^{\text{triangle}} = \dfrac{4 \cdot 3^2}{6} = 6$.

Step 3. Moment of the circle.

To compute M_x^{circle}, we exploit the fact that, by symmetry, the centroid of the circle is located at the center $(0, 5)$. Thus $y_{\text{CM}}^{\text{circle}} = 5$. The circle has radius 2, hence it has area 4π and also mass $M^{\text{circle}} = 4\pi$. Now we can solve for M_x^{circle}:

$$y_{\text{CM}}^{\text{circle}} = \frac{M_x^{\text{circle}}}{M^{\text{circle}}} = \frac{M_x^{\text{circle}}}{4\pi} = 5 \quad \Rightarrow \quad M_x^{\text{circle}} = 20\pi$$

Step 4. Computation of y_{CM}.

We may now conclude that

$$M_x^R = M_x^{\text{triangle}} + M_x^{\text{circle}} = 6 + 20\pi$$

The triangle has mass $\frac{1}{2} \cdot 4 \cdot 3 = 6$ and the circle has mass 4π, so R has mass $M = 6 + 4\pi$, and

$$y_{\text{CM}} = \frac{M_x^R}{M} = \frac{6 + 20\pi}{6 + 4\pi} \approx 3.71 \qquad \blacksquare$$

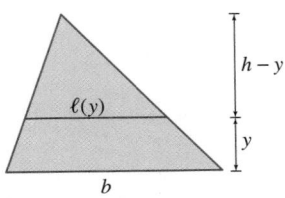

FIGURE 15 By similar triangles, $\dfrac{\ell(y)}{h-y} = \dfrac{b}{h}$.

HISTORICAL PERSPECTIVE

We take it for granted that the basic laws of physics are best expressed as mathematical relationships. Think of $F = ma$ or the universal law of gravitation. However, the fundamental insight that mathematics could be used to formulate laws of nature (and not just for counting or measuring) developed gradually, beginning with the philosophers of ancient Greece and culminating some 2,000 years later in the discoveries of Galileo and Newton. Archimedes

(287–212 BCE) was one of the first scientists (perhaps the first) to formulate a precise physical law. He considered the following question: If weights of mass m_1 and m_2 are placed at the ends of a weightless lever, where should the fulcrum P be located so that the lever does not tip to one side? Suppose that the distance from P to m_1 and m_2 is L_1 and L_2, respectively (Figure 16). Archimedes's Law states that the lever will balance if

$$\boxed{L_1 m_1 = L_2 m_2}$$

In our terminology, what Archimedes had discovered was the center of mass P of the system of weights (see Exercises 40 and 41).

FIGURE 16 Archimedes's Law of the Lever: $L_1 m_1 = L_2 m_2$.

8.3 SUMMARY

- The *moments* of a system of particles of mass m_j located at (x_j, y_j) are

$$M_x = m_1 y_1 + \cdots + m_n y_n, \qquad M_y = m_1 x_1 + \cdots + m_n x_n$$

The *center of mass* (COM) has coordinates

$$x_{\mathrm{CM}} = \frac{M_y}{M} \quad \text{and} \quad y_{\mathrm{CM}} = \frac{M_x}{M},$$

where $M = m_1 + \cdots + m_n$.

- The y-moment M_y of a lamina (thin plate) of constant mass density ρ occupying the region under the graph of $y = f(x)$ where $f(x) \geq 0$, or *between* the graphs of $f_1(x)$ and $f_2(x)$ where $f_1(x) \geq f_2(x)$ is equal to

$$M_y = \rho \int_a^b x f(x)\, dx \qquad \text{or} \qquad \rho \int_a^b x\big(f_1(x) - f_2(x)\big)\, dx$$

- There are two ways to compute the x-moment M_x. If the lamina occupies the region between the graph of $x = g(y)$ and the y-axis where $g(y) \geq 0$, or *between* the graphs of $g_1(y)$ and $g_2(y)$ where $g_1(y) \geq g_2(y)$, then

$$M_x = \rho \int_c^d y g(y)\, dy \qquad \text{or} \qquad \rho \int_c^d y\big(g_1(y) - g_2(y)\big)\, dy$$

- Alternate (often more convenient) formula for M_x:

$$M_x = \frac{1}{2}\rho \int_a^b f(x)^2\, dx \qquad \text{or} \qquad \frac{1}{2}\rho \int_a^b \big(f_1(x)^2 - f_2(x)^2\big)\, dx$$

- The total mass of the lamina is $M = \rho \int_a^b \big(f_1(x) - f_2(x)\big)\, dx$. The coordinates of the center of mass (also called the *centroid*) are

$$x_{\mathrm{CM}} = \frac{M_y}{M}, \qquad y_{\mathrm{CM}} = \frac{M_x}{M}$$

- Symmetry Principle: If a lamina of constant mass density is symmetrical with respect to a given line, then the center of mass (centroid) lies on that line.
- Additivity property: If a region S consists of two or more disjoint smaller regions, then each moment of S is the sum of the corresponding moments of the smaller regions.

8.3 EXERCISES

Preliminary Questions

1. What are the x- and y-moments of a lamina whose center of mass is located at the origin?

2. A thin plate has mass 3. What is the x-moment of the plate if its center of mass has coordinates $(2, 7)$?

3. The center of mass of a lamina of total mass 5 has coordinates $(2, 1)$. What are the lamina's x- and y-moments?

4. Explain how the Symmetry Principle is used to conclude that the centroid of a rectangle is the center of the rectangle.

Exercises

1. Four particles are located at points

$$(1, 1) \quad (1, 2) \quad (4, 0) \quad (3, 1)$$

(a) Find the moments M_x and M_y and the center of mass of the system, assuming that the particles have equal mass m.

(b) Find the center of mass of the system, assuming the particles have mass 3, 2, 5, and 7, respectively.

2. Find the center of mass for the system of particles of mass 4, 2, 5, 1 located at $(1, 2)$, $(-3, 2)$, $(2, -1)$, $(4, 0)$.

3. Point masses of equal size are placed at the vertices of the triangle with coordinates $(a, 0)$, $(b, 0)$, and $(0, c)$. Show that the center of mass of the system of masses has coordinates $(\frac{1}{3}(a + b), \frac{1}{3}c)$.

4. Point masses of mass m_1, m_2, and m_3 are placed at the points $(-1, 0)$, $(3, 0)$, and $(0, 4)$.
(a) Suppose that $m_1 = 6$. Show that there is a unique value of m_2 such that the center of mass lies on the y-axis.
(b) Suppose that $m_1 = 6$ and $m_2 = 4$. Find the value of m_3 such that $y_{CM} = 2$.

5. Sketch the lamina S of constant density $\rho = 3$ g/cm² occupying the region beneath the graph of $y = x^2$ for $0 \le x \le 3$.
(a) Use formulas (1) and (2) to compute M_x and M_y.
(b) Find the area and the center of mass of S.

6. Use Eqs. (1) and (3) to find the moments and center of mass of the lamina S of constant density $\rho = 2$ g/cm² occupying the region between $y = x^2$ and $y = 9x$ over $[0, 3]$. Sketch S, indicating the location of the center of mass.

7. Find the moments and center of mass of the lamina of uniform density ρ occupying the region underneath $y = x^3$ for $0 \le x \le 2$.

8. Calculate M_x (assuming $\rho = 1$) for the region underneath the graph of $y = 1 - x^2$ for $0 \le x \le 1$ in two ways, first using Eq. (2) and then using Eq. (3).

9. Let T be the triangular lamina in Figure 17.
(a) Show that the horizontal cut at height y has length $4 - \frac{2}{3}y$ and use Eq. (2) to compute M_x (with $\rho = 1$).
(b) Use the Symmetry Principle to show that $M_y = 0$ and find the center of mass.

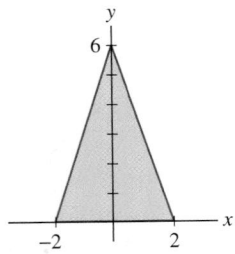

FIGURE 17 Isosceles triangle.

In Exercises 10–17, find the centroid of the region lying underneath the graph of the function over the given interval.

10. $f(x) = 6 - 2x$, $[0, 3]$ **11.** $f(x) = \sqrt{x}$, $[4, 9]$

12. $f(x) = x^3$, $[0, 1]$ **13.** $f(x) = 9 - x^2$, $[0, 3]$

14. $f(x) = (1 + x^2)^{-1/2}$, $[0, 1]$ **15.** $f(x) = e^{-x}$, $[0, 4]$

16. $f(x) = \ln x$, $[1, 2]$ **17.** $f(x) = \sin x$, $[0, \pi]$

18. Calculate the moments and center of mass of the lamina occupying the region between the curves $y = x$ and $y = x^2$ for $0 \le x \le 1$.

19. Sketch the region between $y = x + 4$ and $y = 2 - x$ for $0 \le x \le 2$. Using symmetry, explain why the centroid of the region lies on the line $y = 3$. Verify this by computing the moments and the centroid.

In Exercises 20–25, find the centroid of the region lying between the graphs of the functions over the given interval.

20. $y = x$, $y = \sqrt{x}$, $[0, 1]$

21. $y = x^2$, $y = \sqrt{x}$, $[0, 1]$

22. $y = x^{-1}$, $y = 2 - x$, $[1, 2]$

23. $y = e^x$, $y = 1$, $[0, 1]$

24. $y = \ln x$, $y = x - 1$, $[1, 3]$

25. $y = \sin x$, $y = \cos x$, $[0, \pi/4]$

26. Sketch the region lying above $y = 0$ and below $y = x + 1$ and $y = (x - 1)^2$, and find its centroid.

27. Sketch the region enclosed by $y = 0$, $y = (x + 1)^3$, and $y = (1 - x)^3$ and find its centroid.

In Exercises 28–32, find the centroid of the region.

28. Top half of the ellipse $\left(\frac{x}{2}\right)^2 + \left(\frac{y}{4}\right)^2 = 1$

29. Top half of the ellipse $\left(\frac{x}{a}\right)^2 + \left(\frac{y}{b}\right)^2 = 1$ for arbitrary $a, b > 0$

30. Semicircle of radius r with center at the origin

31. Quarter of the unit circle lying in the first quadrant

32. Triangular plate with vertices $(-c, 0)$, $(0, c)$, (a, b), where $a, b, c > 0$, and $b < c$

33. Find the centroid for the shaded region of the semicircle of radius r in Figure 18. What is the centroid when $r = 1$ and $h = \frac{1}{2}$? *Hint:* Use geometry rather than integration to show that the *area* of the region is $r^2 \sin^{-1}(\sqrt{1 - h^2/r^2}) - h\sqrt{r^2 - h^2}$.

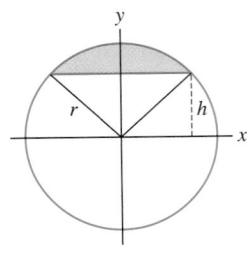

FIGURE 18

In Exercises 34–36, use the additivity of moments to find the COM of the region.

34. Isosceles triangle of height 2 on top of a rectangle of base 4 and height 3 (Figure 19)

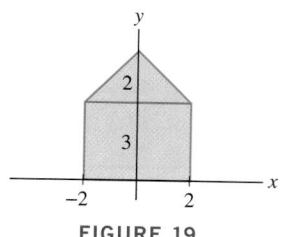

FIGURE 19

35. An ice cream cone consisting of a semicircle on top of an equilateral triangle of side 6 (Figure 20)

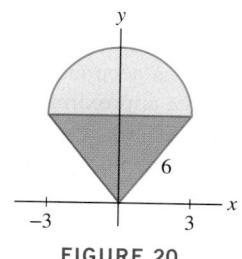

FIGURE 20

36. Three-quarters of the unit circle (remove the part in the fourth quadrant)

37. Let S be the lamina of mass density $\rho = 1$ obtained by removing a circle of radius r from the circle of radius $2r$ shown in Figure 21. Let

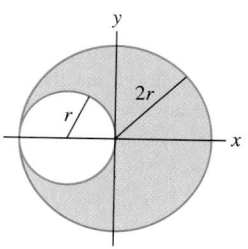

FIGURE 21

M_x^S and M_y^S denote the moments of S. Similarly, let M_y^{big} and M_y^{small} be the y-moments of the larger and smaller circles.

(a) Use the Symmetry Principle to show that $M_x^S = 0$.

(b) Show that $M_y^S = M_y^{\text{big}} - M_y^{\text{small}}$ using the additivity of moments.

(c) Find M_y^{big} and M_y^{small} using the fact that the COM of a circle is its center. Then compute M_y^S using (b).

(d) Determine the COM of S.

38. Find the COM of the laminas in Figure 22 obtained by removing squares of side 2 from a square of side 8.

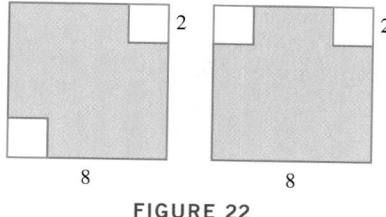

FIGURE 22

Further Insights and Challenges

39. A **median** of a triangle is a segment joining a vertex to the midpoint of the opposite side. Show that the centroid of a triangle lies on each of its medians, at a distance two-thirds down from the vertex. Then use this fact to prove that the three medians intersect at a single point. *Hint:* Simplify the calculation by assuming that one vertex lies at the origin and another on the x-axis.

40. Let P be the COM of a system of two weights with masses m_1 and m_2 separated by a distance d. Prove Archimedes's Law of the (weightless) Lever: P is the point on line between the two weights such that $\ell_1 m_1 = \ell_2 m_2$, where ℓ_j is the distance from mass j to P.

41. Find the COM of a system of two weights of masses m_1 and m_2 connected by a lever of length d whose mass density ρ is uniform. *Hint:* The moment of the system is the sum of the moments of the weights and the lever.

42. **Symmetry Principle** Let $\mathcal{R}$ be the region under the graph of $f(x)$ over the interval $[-a, a]$, where $f(x) \geq 0$. Assume that $\mathcal{R}$ is symmetrical with respect to the y-axis.

(a) Explain why $f(x)$ is even, that is, $f(x) = f(-x)$.

(b) Show that $xf(x)$ is an *odd* function.

(c) Use (b) to prove that $M_y = 0$.

(d) Prove that the COM of $\mathcal{R}$ lies on the y-axis (a similar argument applies to symmetry with respect to the x-axis).

43. Prove directly that Eqs. (2) and (3) are equivalent in the following situation. Let $f(x)$ be a positive decreasing function on $[0, b]$ such that $f(b) = 0$. Set $d = f(0)$ and $g(y) = f^{-1}(y)$. Show that

$$\frac{1}{2}\int_0^b f(x)^2\,dx = \int_0^d yg(y)\,dy$$

Hint: First apply the substitution $y = f(x)$ to the integral on the left and observe that $dx = g'(y)\,dy$. Then apply Integration by Parts.

44. Let R be a lamina of uniform density submerged in a fluid of density w (Figure 23). Prove the following law: The fluid force on one side of R is equal to the area of R times the fluid pressure on the centroid. *Hint:* Let $g(y)$ be the horizontal width of R at depth y. Express both the fluid pressure [Eq. (2) in Section 8.2] and y-coordinate of the centroid in terms of $g(y)$.

FIGURE 23

8.4 Taylor Polynomials

We have seen that approximation is a basic theme in calculus. For example, we may approximate a differentiable function $f(x)$ at $x = a$ by the linear function (called the linearization)

$$L(x) = f(a) + f'(a)(x - a)$$

However, a drawback of the linearization is that it is accurate only in a small interval around $x = a$. Taylor polynomials are higher-degree approximations that generalize the linearization using the higher derivatives $f^{(k)}(a)$. They are useful because, by taking sufficiently high degree, we can approximate transcendental functions such as $\sin x$ and e^x to arbitrary accuracy on any given interval.

Assume that $f(x)$ is defined on an open interval I and that all higher derivatives $f^{(k)}(x)$ exist on I. Fix a number $a \in I$. The nth **Taylor polynomial** for f centered at $x = a$ is the polynomial

$$T_n(x) = f(a) + \frac{f'(a)}{1!}(x - a) + \frac{f''(a)}{2!}(x - a)^2 + \cdots + \frac{f^{(n)}(a)}{n!}(x - a)^n$$

It is convenient to regard $f(x)$ itself as the *zeroth* derivative $f^{(0)}(x)$. Then we may write the Taylor polynomial in summation notation,

$$T_n(x) = \sum_{j=0}^{n} \frac{f^{(j)}(a)}{j!}(x - a)^j$$

When $a = 0$, $T_n(x)$ is also called the nth **Maclaurin polynomial**. The first few Taylor polynomials are

$$T_0(x) = f(a)$$

$$T_1(x) = f(a) + f'(a)(x - a)$$

$$T_2(x) = f(a) + f'(a)(x - a) + \frac{1}{2}f''(a)(x - a)^2$$

$$T_3(x) = f(a) + f'(a)(x - a) + \frac{1}{2}f''(a)(x - a)^2 + \frac{1}{6}f'''(a)(x - a)^3$$

Note that $T_1(x)$ is the linearization of $f(x)$ at a. In most cases, the higher-degree Taylor polynomials provide increasingly better approximations to $f(x)$. Before computing some Taylor polynomials, we record two important properties that follow from the definition:

English mathematician Brook Taylor (1685–1731) made important contributions to calculus and physics, as well as to the theory of linear perspective used in drawing.

◄‧‧ *REMINDER* k-factorial *is the number* $k! = k(k-1)(k-2)\cdots(2)(1)$. *Thus,*

$$1! = 1, \quad 2! = (2)1 = 2, \quad 3! = (3)(2)1 = 6$$

By convention, we define $0! = 1$.

Scottish mathematician Colin Maclaurin (1698–1746) was a professor in Edinburgh. Newton was so impressed by his work that he once offered to pay part of Maclaurin's salary.

- $T_n(a) = f(a)$ [since all terms in $T_n(x)$ after the first are zero at $x = a$].
- $T_n(x)$ is obtained from $T_{n-1}(x)$ by adding on a term of degree n:

$$T_n(x) = T_{n-1}(x) + \frac{f^{(n)}(a)}{n!}(x-a)^n$$

■ **EXAMPLE 1** Computing Taylor Polynomials Let $f(x) = \sqrt{x+1}$. Compute $T_n(x)$ at $a = 3$ for $n = 0, 1, 2, 3,$ and 4.

Solution First evaluate the derivatives $f^{(j)}(3)$:

$$f(x) = (x+1)^{1/2}, \qquad\qquad f(3) = 2$$

$$f'(x) = \frac{1}{2}(x+1)^{-1/2} \qquad\qquad f'(3) = \frac{1}{4}$$

$$f''(x) = -\frac{1}{4}(x+1)^{-3/2} \qquad\qquad f''(3) = -\frac{1}{32}$$

$$f'''(x) = \frac{3}{8}(x+1)^{-5/2} \qquad\qquad f'''(3) = \frac{3}{256}$$

$$f^{(4)}(x) = -\frac{15}{16}(x+1)^{-7/2} \quad f^{(4)}(3) = -\frac{15}{2,048}$$

Then compute the coefficients $\dfrac{f^{(j)}(3)}{j!}$:

Constant term $\qquad\qquad = f(3) = 2$

Coefficient of $(x-3) \;\;= f'(3) = \dfrac{1}{4}$

Coefficient of $(x-3)^2 = \dfrac{f''(3)}{2!} = -\dfrac{1}{32}\cdot\dfrac{1}{2!} = -\dfrac{1}{64}$

Coefficient of $(x-3)^3 = \dfrac{f'''(3)}{3!} = \dfrac{3}{256}\cdot\dfrac{1}{3!} = \dfrac{1}{512}$

Coefficient of $(x-3)^4 = \dfrac{f^{(4)}(3)}{4!} = -\dfrac{15}{2,048}\cdot\dfrac{1}{4!} = -\dfrac{5}{16,384}$

The first four Taylor polynomials centered at $a = 3$ are (see Figure 1):

$$T_0(x) = 2$$

$$T_1(x) = 2 + \frac{1}{4}(x-3)$$

$$T_2(x) = 2 + \frac{1}{4}(x-3) - \frac{1}{64}(x-3)^2$$

$$T_3(x) = 2 + \frac{1}{4}(x-3) - \frac{1}{64}(x-3)^2 + \frac{1}{512}(x-3)^3$$

$$T_4(x) = 2 + \frac{1}{4}(x-3) - \frac{1}{64}(x-3)^2 + \frac{1}{512}(x-3)^3 - \frac{5}{16,384}(x-3)^4 \qquad ■$$

The first term $f(a)$ in the Taylor polynomial $T_n(x)$ is called the constant term.

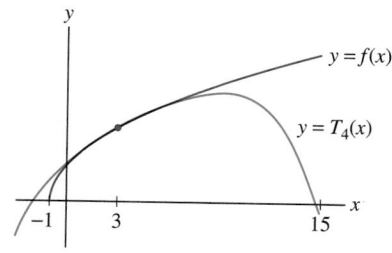

FIGURE 1 Graph of $f(x) = \sqrt{x+1}$ and $T_4(x)$ centered at $x = 3$.

■ **EXAMPLE 2** Finding a General Formula for T_n Find the Taylor polynomials $T_n(x)$ of $f(x) = \ln x$ at $a = 1$.

After computing several derivatives of $f(x) = \ln x$, we begin to discern the pattern. Often, however, the derivatives follow no simple pattern and there is no simple formula for the general Taylor polynomial.

Taylor polynomials for $\ln x$ at $a = 1$:

$T_1(x) = (x - 1)$

$T_2(x) = (x - 1) - \dfrac{1}{2}(x - 1)^2$

$T_3(x) = (x - 1) - \dfrac{1}{2}(x - 1)^2 + \dfrac{1}{3}(x - 1)^3$

Solution For $f(x) = \ln x$, the constant term of $T_n(x)$ at $a = 1$ is zero since $f(1) = \ln 1 = 0$. Next, we compute the derivatives:

$$f'(x) = x^{-1}, \qquad f''(x) = -x^{-2}, \qquad f'''(x) = 2x^{-3}, \qquad f^{(4)}(x) = -3 \cdot 2x^{-4}$$

Similarly, $f^{(5)}(x) = 4 \cdot 3 \cdot 2x^{-5}$. The general pattern is that $f^{(k)}(x)$ is a multiple of x^{-k}, with a coefficient $(k - 1)!$ that alternates in sign:

$$f^{(k)}(x) = (-1)^{k-1}(k - 1)! \, x^{-k} \qquad \boxed{1}$$

Therefore, the coefficient of $(x - 1)^k$ in $T_n(x)$ is

$$\frac{f^{(k)}(1)}{k!} = \frac{(-1)^{k-1}(k - 1)!}{k!} = \frac{(-1)^{k-1}}{k} \qquad \text{(for } k \geq 1\text{)}$$

In other words, the coefficients for $k \geq 1$ are $1, -\frac{1}{2}, \frac{1}{3}, -\frac{1}{4}, \ldots$, and

$$T_n(x) = (x - 1) - \frac{1}{2}(x - 1)^2 + \frac{1}{3}(x - 1)^3 - \cdots + (-1)^{n-1}\frac{1}{n}(x - 1)^n \qquad \blacksquare$$

■ **EXAMPLE 3** Maclaurin Polynomials for $f(x) = \cos x$ Find the Maclaurin polynomials of $f(x) = \cos x$.

Solution Recall that the Maclaurin polynomials are the Taylor polynomials centered at $a = 0$. The key observation is that the derivatives of $f(x) = \cos x$ form a pattern that repeats with period 4:

$$f'(x) = -\sin x, \qquad f''(x) = -\cos x, \qquad f'''(x) = \sin x, \qquad f^{(4)}(x) = \cos x,$$

and, in general, $f^{(j+4)}(x) = f^{(j)}(x)$. At $x = 0$, the derivatives form the repeating pattern $1, 0, -1$, and 0:

$f(0)$	$f'(0)$	$f''(0)$	$f'''(0)$	$f^{(4)}(0)$	$f^{(5)}(0)$	$f^{(6)}(0)$	$f^{(7)}(0)$	$\cdots$
1	0	-1	0	1	0	-1	0	$\cdots$

In other words, the even derivatives are $f^{(2k)}(0) = (-1)^k$ and the odd derivatives are zero: $f^{(2k+1)}(0) = 0$. Therefore, the coefficient of x^{2k} is $(-1)^k/(2k)!$ and the coefficient of x^{2k+1} is zero. We have

$$T_0(x) = T_1(x) = 1$$

$$T_2(x) = T_3(x) = 1 - \frac{1}{2!}x^2$$

$$T_4(x) = T_5(x) = 1 - \frac{x^2}{2} + \frac{x^4}{4!}$$

and, in general,

$$T_{2n}(x) = T_{2n+1}(x) = 1 - \frac{1}{2}x^2 + \frac{1}{4!}x^4 - \frac{1}{6!}x^6 + \cdots + (-1)^n \frac{1}{(2n)!}x^{2n} \qquad \blacksquare$$

Taylor polynomials $T_n(x)$ are designed to approximate $f(x)$ in an interval around $x = a$. Figure 2 shows the first few Maclaurin polynomials for $f(x) = \cos x$. Observe that as n gets larger, $T_n(x)$ approximates $f(x) = \cos x$ well over larger and larger intervals. Outside this interval, the approximation fails.

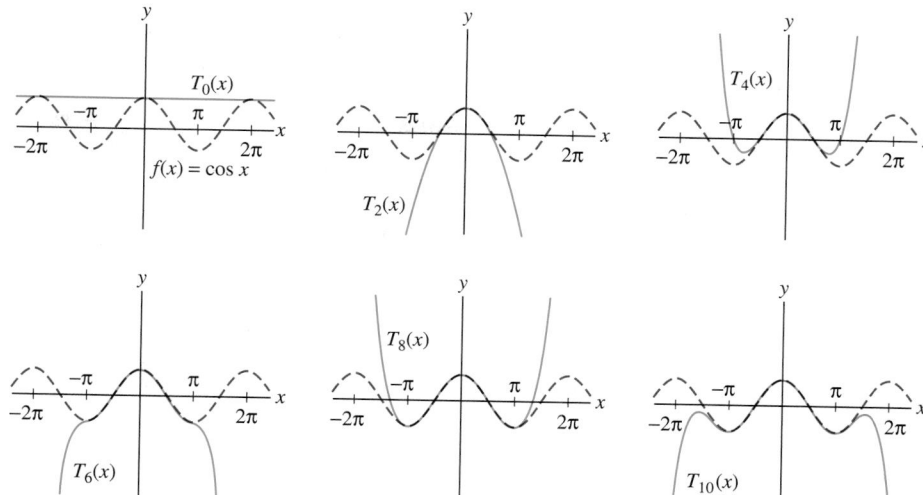

FIGURE 2 Graphs of Taylor polynomials for $f(x) = \cos x$. The graph of $f(x)$ is shown as a dashed curve.

CONCEPTUAL INSIGHT To understand how the formula for T_n arises, it is useful to introduce a notion of closeness for two functions. We say that f and g *agree to order n* at $x = a$ if

$$f(a) = g(a), \quad f'(a) = g'(a), \quad f''(a) = g''(a), \ldots, \quad f^{(n)}(a) = g^{(n)}(a)$$

Taylor polynomials are defined so that $T_n(x)$ and $f(x)$ agree to order n at $x = a$ (see Exercise 59) and, in fact, $T_n(x)$ is the only polynomial of degree at most n with this property. For example, we can check that $T_2(x)$ agrees with $f(x)$ to order $n = 2$:

$$T_2(x) = f(a) + f'(a)(x - a) + \frac{1}{2}f''(a)(x-a)^2, \quad T_2(a) = f(a)$$

$$T_2'(x) = f'(a) + f''(a)(x - a), \qquad\qquad\qquad T_2'(a) = f'(a)$$

$$T_2''(x) = f''(a), \qquad\qquad\qquad\qquad\qquad T_2''(a) = f''(a)$$

The Remainder Term

Our next goal is to study the error $|T_n(x) - f(x)|$ in the approximation provided by the nth Taylor polynomial centered at $x = a$. Define the **nth remainder** for $f(x)$ at $x = a$ by

$$R_n(x) = f(x) - T_n(x)$$

The error is the absolute value of the remainder. Also, $f(x) = T_n(x) + R_n(x)$, so

$$f(x) = f(a) + \frac{f'(a)}{1!}(x - a)^1 + \frac{f''(a)}{2!}(x - a)^2 + \cdots + \frac{f^{(n)}(a)}{n!}(x - a)^n + R_n(x)$$

Taylor's Theorem Assume that $f^{(n+1)}(x)$ exists and is continuous. Then

$$R_n(x) = \frac{1}{n!}\int_a^x (x - u)^n f^{(n+1)}(u)\,du \qquad \boxed{2}$$

Proof Set

$$I_n(x) = \frac{1}{n!} \int_a^x (x-u)^n f^{(n+1)}(u) \, du$$

Our goal is to show that $R_n(x) = I_n(x)$. For $n = 0$, $R_0(x) = f(x) - f(a)$ and the desired result is just a restatement of the Fundamental Theorem of Calculus:

$$I_0(x) = \int_a^x f'(u) \, du = f(x) - f(a) = R_0(x)$$

Exercise 55 reviews this proof for the special case $n = 2$.

To prove the formula for $n > 0$, we apply Integration by Parts to $I_n(x)$ with

$$h(u) = \frac{(x-u)^n}{n!}, \qquad g(u) = f^{(n)}(u)$$

$$I_n(x) = \int_a^x h(u) \, g'(u) \, du = h(u)g(u) \Big|_a^x - \int_a^x h'(u)g(u) \, du$$

$$= \frac{1}{n!}(x-u)^n f^{(n)}(u) \Big|_a^x - \frac{1}{n!} \int_a^x (-n)(x-u)^{n-1} f^{(n)}(u) \, du$$

$$= -\frac{1}{n!}(x-a)^n f^{(n)}(a) + I_{n-1}(x)$$

This result can be rewritten as

$$I_{n-1}(x) = \frac{f^{(n)}(a)}{n!}(x-a)^n + I_n(x)$$

Now apply this relation n times:

$$f(x) = f(a) + I_0(x)$$

$$= f(a) + \frac{f'(a)}{1!}(x-a) + I_1(x)$$

$$= f(a) + \frac{f'(a)}{1!}(x-a) + \frac{f''(a)}{2!}(x-a)^2 + I_2(x)$$

$$\vdots$$

$$= f(a) + \frac{f'(a)}{1!}(x-a) + \cdots + \frac{f^{(n)}(a)}{n!}(x-a)^n + I_n(x)$$

This shows that $f(x) = T_n(x) + I_n(x)$ and hence $I_n(x) = R_n(x)$ as desired.   ∎

Although Taylor's Theorem gives us an explicit formula for the remainder, we will not use this formula directly. Instead, we will use it to *estimate* the size of the error.

THEOREM 1 Error Bound   Assume that $f^{(n+1)}(x)$ exists and is continuous. Let K be a number such that $|f^{(n+1)}(u)| \leq K$ for all u between a and x. Then

$$|T_n(x) - f(x)| \leq K \frac{|x-a|^{n+1}}{(n+1)!}$$

Proof Assume that $x \geq a$ (the case $x \leq a$ is similar). Then, since $|f^{(n+1)}(u)| \leq K$ for $a \leq u \leq x$,

$$|T_n(x) - f(x)| = |R_n(x)| = \left| \frac{1}{n!} \int_a^x (x-u)^n f^{(n+1)}(u) \, du \right|$$

$$\leq \frac{1}{n!} \int_a^x \left| (x-u)^n f^{(n+1)}(u) \right| du \qquad \boxed{3}$$

$$\leq \frac{K}{n!} \int_a^x (x-u)^n \, du$$

$$= \frac{K}{n!} \left. \frac{-(x-u)^{n+1}}{n+1} \right|_{u=a}^x = K \frac{|x-a|^{n+1}}{(n+1)!} \qquad \blacksquare$$

In (3), we use the inequality

$$\left| \int_a^b f(x) \, dx \right| \leq \int_a^b |f(x)| \, dx$$

which is valid for all integrable functions.

EXAMPLE 4 Using the Error Bound Use the Error Bound to find a bound for the error $|T_3(1.2) - \ln 1.2|$, where $T_3(x)$ is the third Taylor polynomial for $f(x) = \ln x$ at $a = 1$. Check your result using a calculator.

Solution To use the Error Bound with $n = 3$, we must find a value of K such that $|f^{(4)}(u)| \leq K$ for $1 \leq u \leq 1.2$. By Eq. (1) in Example 2, $|f^{(4)}(x)| = 6x^{-4}$. Since $|f^{(4)}(x)|$ is decreasing for $x \geq 1$, its maximum value on $[1, 1.2]$ is $|f^{(4)}(1)| = 6$. Therefore, we may apply the Error Bound with $K = 6$ to obtain

$$|T_3(1.2) - \ln 1.2| \leq K \frac{|x-a|^{n+1}}{(n+1)!} = 6 \frac{|1.2-1|^4}{4!} \approx 0.0004$$

By Example 2,

$$T_3(x) = (x-1) - \frac{1}{2}(x-1)^2 + \frac{1}{3}(x-1)^3$$

The following values from a calculator confirm that the error is at most 0.0004:

$$|T_3(1.2) - \ln 1.2| \approx |0.182667 - 0.182322| \approx 0.00035 < 0.0004 \qquad \blacksquare$$

EXAMPLE 5 Approximating with a Given Accuracy Let $T_n(x)$ be the nth Maclaurin polynomial for $f(x) = \cos x$. Find a value of n such that

$$|T_n(0.2) - \cos(0.2)| < 10^{-5}$$

Use a calculator to verify that this value of n works.

To use the Error Bound, it is not necessary to find the smallest possible value of K. In this example, we take $K = 1$. This works for all n, but for odd n we could have used the smaller value $K = \sin(0.2) \approx 0.2$.

Solution Since $|f^{(n)}(x)|$ is $|\cos x|$ or $|\sin x|$, depending on whether n is even or odd, we have $|f^{(n)}(u)| \leq 1$ for all u. Thus, we may apply the Error Bound with $K = 1$:

$$|T_n(0.2) - \cos(0.2)| \leq K \frac{|0.2 - 0|^{n+1}}{(n+1)!} = \frac{|0.2|^{n+1}}{(n+1)!}$$

To make the error less than 10^{-5}, we must choose n so that

$$\frac{|0.2|^{n+1}}{(n+1)!} < 10^{-5}$$

We find a suitable n by checking several values:

n	2	3	4
$\dfrac{\|0.2\|^{n+1}}{(n+1)!}$	$\dfrac{0.2^3}{3!} \approx 0.0013$	$\dfrac{0.2^4}{4!} \approx 6.67 \times 10^{-5}$	$\dfrac{0.2^5}{5!} \approx 2.67 \times 10^{-6} < 10^{-5}$

We see that the error is less than 10^{-5} for $n = 4$. To verify this, recall that $T_4(x) = 1 - \frac{1}{2}x^2 + \frac{1}{4!}x^4$ by Example 3. The following values from a calculator confirm that the error is significantly less than 10^{-5} as required:

$$\text{Actual error} = |T_4(0.2) - \cos(0.2)| \approx |0.98006667 - 0.98006657| = 10^{-7} \qquad \blacksquare$$

CONCEPTUAL INSIGHT   Recall that functions $g(x)$ and $f(x)$ agree at $x = a$ to order n if $g^{(k)}(a) = f^{(k)}(a)$ for $0 \le k \le n$. The defining property of the Taylor polynomial $T_n(x)$ is that it has at most degree n and agrees with $f(x)$ at $x = a$ to order n. Now, let us say that a function $g(x)$ *approximates* $f(x)$ at $x = a$ *to order* n if the error $E(x) = |f(x) - g(x)|$ satisfies

$$\lim_{x \to a} \frac{E(x)}{|x - a|^n} = 0 \qquad \boxed{4}$$

Thus, in an nth-order approximation, the error tends to zero faster than $|x - a|^n$. The Error Bound tells us that $T_n(x)$ is an nth-order approximation to $f(x)$ at $x = a$, provided we assume that $f^{(n+1)}(x)$ exists and is continuous and that $|f^{(n+1)}(u)| \le K$ for u in some open interval I around a. In fact, the Error Bound gives

$$|f(x) - T_n(x)| \le C|x - a|^{n+1} \qquad \text{for } x \in I$$

where $C = \dfrac{K}{(n+1)!}$, so

$$\frac{|f(x) - T_n(x)|}{|x - a|^n} \le C|x - a| \to 0 \qquad (\text{as } x \to a)$$

If $f^{(k)}(x)$ exists for all k [as is the case for transcendental functions such as $f(x) = \sin x$ or $f(x) = e^x$], then we can approximate $f(x)$ to arbitrarily high order using Taylor polynomials.

8.4 SUMMARY

- The nth *Taylor polynomial* centered at $x = a$ for the function $f(x)$ is

$$T_n(x) = f(a) + \frac{f'(a)}{1!}(x - a)^1 + \frac{f''(a)}{2!}(x - a)^2 + \cdots + \frac{f^{(n)}(a)}{n!}(x - a)^n$$

When $a = 0$, $T_n(x)$ is also called the nth *Maclaurin polynomial*.
- If $f^{(n+1)}(x)$ exists and is continuous, then we have the *Error Bound*

$$|T_n(x) - f(x)| \le K \frac{|x - a|^{n+1}}{(n+1)!}$$

where K is a number such that $|f^{(n+1)}(u)| \le K$ for all u between a and x.

• For reference, we include a table of standard Maclaurin and Taylor polynomials.

$f(x)$	a	Maclaurin or Taylor Polynomial
e^x	0	$T_n(x) = 1 + x + \dfrac{x^2}{2!} + \dfrac{x^3}{3!} + \cdots + \dfrac{x^n}{n!}$
$\sin x$	0	$T_{2n+1}(x) = T_{2n+2}(x) = x - \dfrac{x^3}{3!} + \cdots + (-1)^n \dfrac{x^{2n+1}}{(2n+1)!}$
$\cos x$	0	$T_{2n}(x) = T_{2n+1}(x) = 1 - \dfrac{x^2}{2!} + \dfrac{x^4}{4!} - \cdots + (-1)^n \dfrac{x^{2n}}{(2n)!}$
$\ln x$	1	$T_n(x) = (x-1) - \dfrac{1}{2}(x-1)^2 + \cdots + \dfrac{(-1)^{n-1}}{n}(x-1)^n$
$\dfrac{1}{1-x}$	0	$T_n(x) = 1 + x + x^2 + \cdots + x^n$

8.4 EXERCISES

Preliminary Questions

1. What is $T_3(x)$ centered at $a = 3$ for a function $f(x)$ such that $f(3) = 9$, $f'(3) = 8$, $f''(3) = 4$, and $f'''(3) = 12$.

2. The dashed graphs in Figure 3 are Taylor polynomials for a function $f(x)$. Which of the two is a Maclaurin polynomial?

3. For which value of x does the Maclaurin polynomial $T_n(x)$ satisfy $T_n(x) = f(x)$, no matter what $f(x)$ is?

4. Let $T_n(x)$ be the Maclaurin polynomial of a function $f(x)$ satisfying $|f^{(4)}(x)| \leq 1$ for all x. Which of the following statements follow from the Error Bound?

(a) $|T_4(2) - f(2)| \leq 2^4/24$

(b) $|T_3(2) - f(2)| \leq 2^3/6$

(c) $|T_3(2) - f(2)| \leq 1/3$

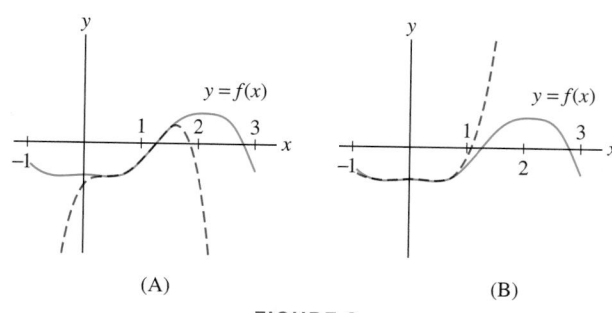

(A)   (B)

FIGURE 3

Exercises

In Exercises 1–14, calculate the Taylor polynomials $T_2(x)$ and $T_3(x)$ centered at $x = a$ for the given function and value of a.

1. $f(x) = \sin x$, $a = 0$

2. $f(x) = \sin x$, $a = \pi/2$

3. $f(x) = \dfrac{1}{1+x}$, $a = 0$

4. $f(x) = \dfrac{1}{1+x}$, $a = 1$

5. $f(x) = \tan x$, $a = 0$

6. $f(x) = \tan x$, $a = \frac{\pi}{4}$

7. $f(x) = \dfrac{1}{1+x^2}$, $a = 0$

8. $f(x) = \dfrac{1}{1+x^2}$, $a = -1$

9. $f(x) = e^x$, $a = 0$

10. $f(x) = e^x$, $a = \ln 2$

11. $f(x) = e^{-x} + e^{-2x}$, $a = 0$

12. $f(x) = x^2 e^{-x}$, $a = 1$

13. $f(x) = \ln(x+1)$, $a = 0$

14. $f(x) = \cosh x$, $a = 0$

In Exercises 15–18, compute $T_2(x)$ at $x = a$ and use a calculator to compute the error $|f(x) - T_2(x)|$ at the given value of x.

15. $y = e^x$, $x = -0.5$, $a = 0$

16. $y = \cos x$, $x = \frac{\pi}{12}$, $a = 0$

17. $y = x^{-3/2}$, $x = 0.3$, $a = 1$

18. $y = e^{\sin x}$, $x = 1.5$, $a = \frac{\pi}{2}$

19. Show that the nth Maclaurin polynomial for $f(x) = e^x$ is

$$T_n(x) = 1 + \frac{x}{1!} + \frac{x^2}{2!} + \cdots + \frac{x^n}{n!}$$

20. Show that the nth Taylor polynomial for $\dfrac{1}{x+1}$ at $a = 1$ is

$$T_n(x) = \frac{1}{2} - \frac{x-1}{4} + \frac{(x-1)^2}{8} + \cdots + (-1)^n \frac{(x-1)^n}{2^{n+1}}$$

In Exercises 21–26, find $T_n(x)$ at $x = a$ for all n.

21. $f(x) = \dfrac{1}{1-x}$, $a = 0$

22. $f(x) = \dfrac{1}{x-1}$, $a = 4$

23. $f(x) = e^x$, $a = 1$

24. $f(x) = \sqrt{x}$, $a = 1$

25. $f(x) = x^{5/2}$, $a = 2$

26. $f(x) = \cos x$, $a = \pi/4$

27. *CAS* Plot $y = e^x$ together with the Maclaurin polynomials $T_n(x)$ for $n = 1, 3, 5$ and then for $n = 2, 4, 6$ on the interval $[-3, 3]$. What difference do you notice between the even and odd Maclaurin polynomials?

28. *CAS* Plot $f(x) = \dfrac{1}{1+x}$ together with the Taylor polynomials $T_n(x)$ at $a = 1$ for $1 \le n \le 4$ on the interval $[-2, 8]$ (be sure to limit the upper plot range).

(a) Over which interval does T_4 appear to give a close approximation to $f(x)$?

(b) What happens with $x < -1$?

(c) Use your computer algebra system to produce T_{30} and plot it together with $f(x)$ on $[-2, 8]$. Over which interval does T_{30} appear to give a close approximation?

29. Use the Error Bound to find the maximum possible size of $|\cos 0.3 - T_5(0.3)|$, where $T_5(x)$ is the Maclaurin polynomial. Verify your result with a calculator.

30. Calculate $T_4(x)$ at $a = 1$ for $f(x) = x^{11/2}$ and use the Error Bound to find the maximum possible size of $|T_4(1.2) - f(1.2)|$. *Hint:* Show that the max $|f^{(5)}(x)|$ on $[1, c]$ with $c > 1$ is $f^{(5)}(c)$.

31. Let T_n be the Maclaurin polynomial of $f(x) = e^x$ and let $c > 0$. Show that we may take $K = e^c$ in the Error Bound for $|T_n(c) - f(c)|$. Then use the Error Bound to determine a number n such that $|T_n(0.1) - e^{0.1}| \le 10^{-5}$.

32. Let $f(x) = \sqrt{1+x}$ and let $T_n(x)$ be the Taylor polynomial centered at $a = 8$.

(a) Find $T_3(x)$ and calculate $T_3(8.02)$.

(b) Use the Error Bound to find a bound for $|T_3(8.02) - \sqrt{9.02}|$.

33. Calculate $T_3(x)$ at $a = 0$ for $f(x) = \tan^{-1} x$. Then compute $T_3(\frac{1}{2})$ and use the Error Bound to find a bound for $|T_3(\frac{1}{2}) - \tan^{-1}(\frac{1}{2})|$. Refer to the graph in Figure 4 to find an acceptable value of K.

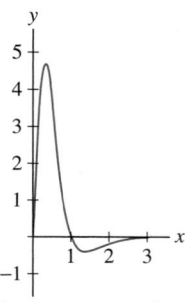

FIGURE 4 Graph of $f^{(4)}(x) = -24x(x^2 - 1)/(x^2 + 1)^4$, where $f(x) = \tan^{-1} x$.

34. *GU* Calculate $T_2(x)$ for $f(x) = \operatorname{sech} x$ at $a = 0$. Then compute $T_2(\frac{1}{2})$ and use the Error Bound to find a bound for

$$\left| T_2\left(\frac{1}{2}\right) - f\left(\frac{1}{2}\right) \right|.$$

Hint: Plot $f'''(x)$ to find an acceptable value of K.

35. Show that the Maclaurin polynomials for $f(x) = \sin x$ are

$$T_{2n-1}(x) = T_{2n} = x - \frac{x^3}{3!} + \frac{x^5}{5!} - \cdots + (-1)^{n-1} \frac{x^{2n-1}}{(2n-1)!}$$

Use the Error Bound with $n = 4$ to show that

$$\left| \sin x - \left(x - \frac{x^3}{6} \right) \right| \le \frac{|x|^5}{120} \quad \text{(for all } x\text{)}$$

36. Find n such that $|T_n(0.1) - \cos(0.1)| \le 10^{-7}$, where T_n is the Maclaurin polynomial for $f(x) = \cos x$ (see Example 5). Calculate $|T_n(0.1) - \cos(0.1)|$ for this value of n and verify the bound on the error.

37. Find n such that $|T_n(1.3) - \ln(1.3)| \le 10^{-4}$, where T_n is the Taylor polynomial for $f(x) = \ln x$ at $a = 1$.

38. Find n such that $|T_n(1) - e| \le 10^{-6}$, where T_n is the Maclaurin polynomial for $f(x) = e^x$.

39. Find n such that $|T_n(1.3) - \sqrt{1.3}| \le 10^{-6}$, where $T_n(x)$ is the Taylor polynomial for $f(x) = \sqrt{x}$ at $a = 1$.

40. Let $T_n(x)$ be the Taylor polynomial for $f(x) = \ln x$ at $a = 1$ and let $c > 1$.

(a) Show that the maximum of $f^{(k+1)}(x)$ on $[1, c]$ is $f^{(k+1)}(1)$ (see Example 4).

(b) Prove $|T_n(c) - \ln c| \le \dfrac{|c-1|^{n+1}}{n+1}$.

(c) Find n such that $|T_n(1.5) - \ln 1.5| \le 10^{-2}$.

41. Let $n \ge 1$. Show that if $|x|$ is small, then

$$(x+1)^{1/n} \approx 1 + \frac{x}{n} + \frac{1-n}{2n^2} x^2$$

Use this approximation with $n = 6$ to estimate $1.5^{1/6}$.

42. Verify that the third Maclaurin polynomial for $f(x) = e^x \sin x$ is equal to the product of the third Maclaurin polynomials of e^x and $\sin x$ (after discarding terms of degree greater than 3 in the product).

43. Find the fourth Maclaurin polynomial for $f(x) = \sin x \cos x$ by multiplying the fourth Maclaurin polynomials for $f(x) = \sin x$ and $f(x) = \cos x$.

44. Find the Maclaurin polynomials $T_n(x)$ for $f(x) = \cos(x^2)$. You may use the fact that $T_n(x)$ is equal to the sum of the terms up to degree n obtained by substituting x^2 for x in the nth Maclaurin polynomial of $\cos x$.

45. Find the Maclaurin polynomials of $\dfrac{1}{1+x^2}$ by substituting $-x^2$ for x in the Maclaurin polynomials of $\dfrac{1}{1-x}$ (see Exercise 21).

46. The seventeenth-century Dutch scientist Christian Huygens used the approximation $\theta \approx \dfrac{8b - a}{3}$ for the length θ of a circular arc of the unit circle, where a is the length of the chord $\overline{AC}$ of angle θ and b is length of the chord $\overline{AB}$ of angle $\theta/2$ (Figure 5).

(a) Prove that $a = 2\sin(\theta/2)$ and $b = 2\sin(\theta/4)$, and show that the Huygens approximation amounts to the approximation

$$\theta \approx \frac{16}{3}\sin\frac{\theta}{4} - \frac{2}{3}\sin\frac{\theta}{2}$$

(b) Compute the fifth MacLaurin polynomial of the function on the right.

(c) Use the Error Bound to show that the error in the Huygens approximation is less than $0.00022|\theta|^5$.

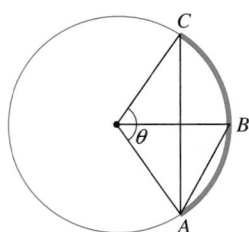

FIGURE 5 Unit circle.

47. Let $f(x) = 3x^3 + 2x^2 - x - 4$. Calculate $T_j(x)$ for $j = 1, 2, 3, 4, 5$ at both $a = 0$ and $a = 1$. Show that $T_3(x) = f(x)$ in both cases.

48. Let $T_n(x)$ be the nth Taylor polynomial at $x = a$ for a polynomial $f(x)$ of degree n. Based on the result of Exercise 47, guess the value of $|f(x) - T_n(x)|$. Prove that your guess is correct using the Error Bound.

49. Taylor polynomials can be used instead of L'Hôpital's Rule to evaluate limits. Consider $L = \lim\limits_{x \to 0} \dfrac{e^x + e^{-x} - 2}{1 - \cos x}$.

(a) Show that the second Maclaurin polynomial for

$$f(x) = e^x + e^{-x} - 2$$

is $T_2(x) = x^2$. Use the Error Bound with $n = 2$ to show that

$$e^x + e^{-x} - 2 = x^2 + g_1(x)$$

where $\lim\limits_{x \to 0} \dfrac{g_1(x)}{x^2} = 0$. Similarly, prove that

$$1 - \cos x = \frac{1}{2}x^2 + g_2(x)$$

where $\lim\limits_{x \to 0} \dfrac{g_2(x)}{x^2} = 0$.

(b) Evaluate L by using (a) to show that $L = \lim\limits_{x \to 0} \dfrac{1 + \dfrac{g_1(x)}{x^2}}{\dfrac{1}{2} + \dfrac{g_2(x)}{x^2}}$.

(c) Evaluate L again using L'Hôpital's Rule.

50. Use the method of Exercise 49 to evaluate

$$\lim_{x \to 0} \frac{1 - \cos x}{x^2} \qquad \text{and} \qquad \lim_{x \to 0} \frac{\sin x - x}{x^3}.$$

51. A light wave of wavelength λ travels from A to B by passing through an aperture (see Figure 6 for the notation). The aperture (circular region) is located in a plane that is perpendicular to $\overline{AB}$. Let $f(r) = d' + h'$, that is, $f(r)$ is the distance $AC + CB$ as a function of r. The **Fresnel zones**, used to determine the optical disturbance at B, are the concentric bands bounded by the circles of radius R_n such that $f(R_n) = AB + n\lambda/2 = d + h + n\lambda/2$.

(a) Show that $f(r) = \sqrt{d^2 + r^2} + \sqrt{h^2 + r^2}$, and use the Maclaurin polynomial of order 2 to show

$$f(r) \approx d + h + \frac{1}{2}\left(\frac{1}{d} + \frac{1}{h}\right)r^2$$

(b) Deduce that $R_n \approx \sqrt{n\lambda L}$, where $L = (d^{-1} + h^{-1})^{-1}$.

(c) Estimate the radii R_1 and R_{100} for blue light ($\lambda = 475 \times 10^{-7}$ cm) if $d = h = 100$ cm.

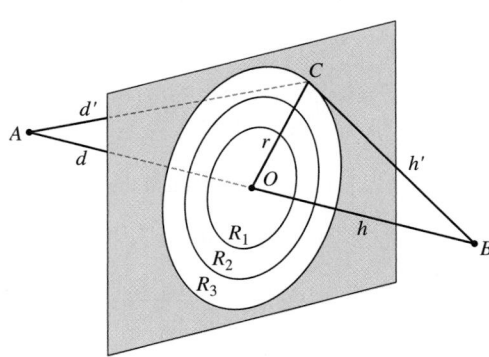

FIGURE 6 The Fresnel zones are the regions between the circles of radius R_n.

Further Insights and Challenges

52. Show that the nth Maclaurin polynomial of $f(x) = \arcsin x$ for n odd is

$$T_n(x) = x + \frac{1}{2}\frac{x^3}{3} + \frac{1 \cdot 3}{2 \cdot 4}\frac{x^5}{5} + \frac{1 \cdot 3 \cdot 5}{2 \cdot 4 \cdot 6}\frac{x^7}{7} + \cdots$$
$$+ \frac{1 \cdot 3 \cdot 5 \cdots (n-2)}{2 \cdot 4 \cdot 6 \cdots (n-1)}\frac{x^n}{n}$$

53. Use Taylor's Theorem to show that if $f^{(n+1)}(t) \geq 0$ for all t, then the nth Maclaurin polynomial $T_n(x)$ satisfies $T_n(x) \leq f(x)$ for all $x \geq 0$.

54. Use Exercise 53 to show that for $x \geq 0$ and all n,

$$e^x \geq 1 + x + \frac{x^2}{2!} + \cdots + \frac{x^n}{n!}$$

Sketch the graphs of e^x, $T_1(x)$, and $T_2(x)$ on the same coordinate axes. Does this inequality remain true for $x < 0$?

55. This exercise is intended to reinforce the proof of Taylor's Theorem.

(a) Show that $f(x) = T_0(x) + \int_a^x f'(u)\,du$.

(b) Use Integration by Parts to prove the formula

$$\int_a^x (x-u) f^{(2)}(u)\,du = -f'(a)(x-a) + \int_a^x f'(u)\,du$$

(c) Prove the case $n = 2$ of Taylor's Theorem:

$$f(x) = T_1(x) + \int_a^x (x-u) f^{(2)}(u)\,du.$$

56. Approximating Integrals Using Taylor Polynomials Taylor polynomials can be used to obtain numerical approximations to integrals.

(a) Let $L > 0$. Show that if two functions $f(x)$ and $g(x)$ satisfy $|f(x) - g(x)| < L$ for all $x \in [a, b]$, then

$$\int_a^b \big(f(x) - g(x)\big)\,dx < L(b - a)$$

(b) Show that $|T_3(x) - \sin x| \leq (\frac{1}{2})^5/5!$ for all $x \in [0, \frac{1}{2}]$.

(c) Evaluate $\int_0^{1/2} T_3(x)\,dx$ as an approximation to $\int_0^{1/2} \sin x\,dx$. Use (a) and (b) to find a bound for the size of the error.

57. Use the fourth Maclaurin polynomial and the method of Exercise 56 to approximate $\int_0^{1/2} \sin(x^2)\,dx$. Find a bound for the error.

58. The Second Derivative Test for local extrema fails when $f''(a) = 0$. This exercise shows us how to extend the test under the following assumption:

$$f'(a) = f''(a) = f'''(a) = 0 \quad \text{but} \quad f^{(4)}(a) > 0$$

(a) Show that $T_4(x) = f(a) + \frac{1}{24} f^{(4)}(a)(x-a)^4$.

(b) Show that $T_4(x) > f(a)$ for $x \neq a$.

(c) By the Error Bound, there is a constant C such that $|T_4(x) - f(x)| \leq C|x-a|^5$ for all x in an interval containing a. Use this and (b) to show that $f(x) > f(a)$ for x sufficiently close to but not equal to a.

(d) Conclude that $f(a)$ is a local minimum. A similar argument shows that $f(a)$ is a local maximum if $f^{(4)}(a) < 0$.

59. Prove by induction that for all k,

$$\frac{d^j}{dx^j}\left(\frac{(x-a)^k}{k!}\right) = \frac{k(k-1)\cdots(k-j+1)(x-a)^{k-j}}{k!}$$

$$\frac{d^j}{dx^j}\left(\frac{(x-a)^k}{k!}\right)\bigg|_{x=a} = \begin{cases} 1 & \text{for } k = j \\ 0 & \text{for } k \neq j \end{cases}$$

Use this to prove that $T_n(x)$ agrees with $f(x)$ at $x = a$ to order n.

60. The following equation arises in the description of Bose–Einstein condensation (the quantum theory of gases cooled to near absolute zero):

$$A_0 = \frac{4A}{\sqrt{\pi}} \int_0^\infty \frac{x^2 e^{-x^2}\,dx}{1 - Ae^{-x^2}}$$

It is necessary to derive an approximate expression for A_0 in terms of A for $|A|$ small.

(a) Show that the second Maclaurin polynomial for the function

$$f(A) = \frac{x^2 e^{-x^2}}{1 - Ae^{-x^2}} \quad \text{(where } A \text{ is the variable and } x \text{ is treated as a constant) is}$$

$$T_2(A) = x^2 e^{-x^2} + x^2 e^{-2x^2} A + x^2 e^{-3x^2} A^2$$

(b) Use the approximation $A_0 \approx \frac{4A}{\sqrt{\pi}} \int_0^\infty T_2(A)\,dx$ to show that

$$A_0 \approx A + \frac{1}{2\sqrt{2}} A^2 + \frac{1}{3\sqrt{3}} A^3$$

You may use the formula (valid for $\lambda > 0$)

$$\int_0^\infty x^2 e^{-\lambda x^2}\,dx = \frac{1}{4}\sqrt{\frac{\pi}{\lambda^3}}$$

CHAPTER REVIEW EXERCISES

In Exercises 1–4, calculate the arc length over the given interval.

1. $y = \dfrac{x^5}{10} + \dfrac{x^{-3}}{6}$,   [3, 5]

2. $y = e^{x/2} + e^{-x/2}$,   [0, 2]

3. $y = 4x - 2$,   [-2, 2]

4. $y = x^{2/3}$,   [1, 2]

5. Show that the arc length of $y = 2\sqrt{x}$ over $[0, a]$ is equal to $\sqrt{a(a+1)} + \ln(\sqrt{a} + \sqrt{a+1})$. *Hint:* Apply the substitution $x = \tan^2 \theta$ to the arc length integral.

6. *CAS* Find a numerical approximation to the arc length of $y = \tan x$ over $[0, \pi/4]$.

In Exercises 7–10, calculate the surface area of the solid obtained by rotating the curve over the given interval about the x-axis.

7. $y = x + 1$,   [0, 4]

8. $y = \dfrac{2}{3}x^{3/4} - \dfrac{2}{5}x^{5/4}$,   [0, 1]

9. $y = \dfrac{2}{3}x^{3/2} - \dfrac{1}{2}x^{1/2}$,   [1, 2]

10. $y = \dfrac{1}{2}x^2$,   [0, 2]

11. Compute the total surface area of the coin obtained by rotating the region in Figure 1 about the x-axis. The top and bottom parts of the region are semicircles with a radius of 1 mm.

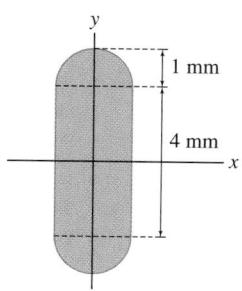

FIGURE 1

12. Calculate the fluid force on the side of a right triangle of height 3 ft and base 2 ft submerged in water vertically, with its upper vertex at the surface of the water.

13. Calculate the fluid force on the side of a right triangle of height 3 ft and base 2 ft submerged in water vertically, with its upper vertex located at a depth of 4 ft.

14. The end of a horizontal oil tank is an ellipse (Figure 2) with equation $(x/4)^2 + (y/3)^2 = 1$ (length in feet). Assume that the tank is filled with oil of density 55 lb/ft^3.

(a) Calculate the total force F on the end of the tank when the tank is full.

(b) [image] Would you expect the total force on the lower half of the tank to be greater than, less than, or equal to $\frac{1}{2}F$? Explain. Then compute the force on the lower half exactly and confirm (or refute) your expectation.

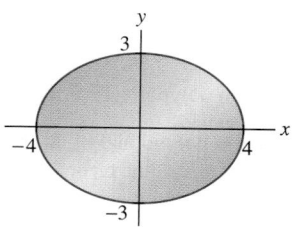

FIGURE 2

15. Calculate the moments and COM of the lamina occupying the region under $y = x(4 - x)$ for $0 \le x \le 4$, assuming a density of $\rho = 5$ lb/ft^2.

16. Sketch the region between $y = 4(x + 1)^{-1}$ and $y = 1$ for $0 \le x \le 3$ and find its centroid.

17. Find the centroid of the region between the semicircle $y = \sqrt{1 - x^2}$ and the top half of the ellipse $y = \frac{1}{2}\sqrt{1 - x^2}$ (Figure 3).

18. A plate in the shape of the shaded region in Figure 3 is submerged in water. Calculate the fluid pressure on a side of the plate if the water surface is $y = 1$.

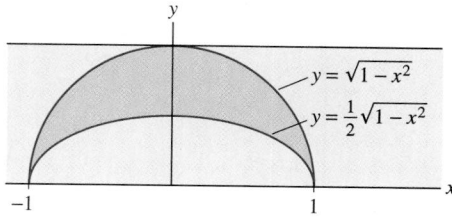

FIGURE 3

19. Find the centroid of the shaded region in Figure 4 bounded on the left by $x = 2y^2 - 2$ and on the right by a semicircle of radius 1. *Hint:* Use symmetry and additivity of moments.

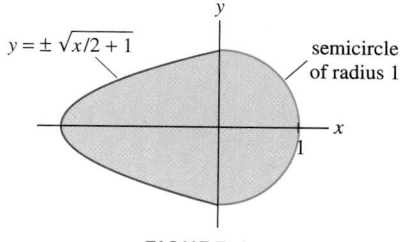

FIGURE 4

In Exercises 20–25, find the Taylor polynomial at $x = a$ for the given function.

20. $f(x) = x^3$, $T_3(x)$, $a = 1$

21. $f(x) = 3(x + 2)^3 - 5(x + 2)$, $T_3(x)$, $a = -2$

22. $f(x) = x \ln(x)$, $T_4(x)$, $a = 1$

23. $f(x) = (3x + 2)^{1/3}$, $T_3(x)$, $a = 2$

24. $f(x) = xe^{-x^2}$, $T_4(x)$, $a = 0$

25. $f(x) = \ln(\cos x)$, $T_3(x)$, $a = 0$

26. Find the nth Maclaurin polynomial for $f(x) = e^{3x}$.

27. Use the fifth Maclaurin polynomial of $f(x) = e^x$ to approximate $\sqrt{e}$. Use a calculator to determine the error.

28. Use the third Taylor polynomial of $f(x) = \tan^{-1} x$ at $a = 1$ to approximate $f(1.1)$. Use a calculator to determine the error.

29. Let $T_4(x)$ be the Taylor polynomial for $f(x) = \sqrt{x}$ at $a = 16$. Use the Error Bound to find the maximum possible size of $|f(17) - T_4(17)|$.

30. Find n such that $|T_n(1) - e| < 10^{-8}$, where $T_n(x)$ is the nth Maclaurin polynomial for $f(x) = e^x$.

31. Let $T_4(x)$ be the Taylor polynomial for $f(x) = x \ln x$ at $x = 1$ computed in Exercise 22. Use the Error Bound to find a bound for $|f(1.2) - T_4(1.2)|$.

32. Let $T_n(x)$ be the nth Maclaurin polynomial for $f(x) = \sin x + \sinh x$.

(a) Show that $T_5(x) = T_6(x) = T_7(x) = T_8(x)$. *Hint:* $T_n(x)$ is the sum of the Maclaurin polynomials of $\sin x$ and $\sinh x$.

(b) Show that $|f^n(x)| \le 1 + \cosh x$ for all n. *Hint:* Note that $|\sinh x| \le |\cosh x|$ for all x.

(c) Show that $|T_8(x) - f(x)| \le \dfrac{2.6}{9!}|x|^9$ for $-1 \le x \le 1$.

33. Show that the nth Maclaurin polynomial for $f(x) = \dfrac{1}{1 - x}$ is $T_n(x) = 1 + x + x^2 + \cdots + x^n$. Conclude by substituting $x/4$ for x so that the nth Maclaurin polynomial for $f(x) = \dfrac{1}{1 + x/4}$ is

$$T_n(x) = 1 + \frac{1}{4}x + \frac{1}{4^2}x^2 + \cdots + \frac{1}{4^n}x^n$$

What is the nth Maclaurin polynomial for $g(x) = \dfrac{1}{1 + x}$? *Hint:* $g(x) = f(-x)$.

34. Let $T_n(x)$ be the Maclaurin polynomial of $f(x) = \dfrac{5}{4 + 3x - x^2}$.

(a) Show that $f(x) = 5\left(\dfrac{1/4}{1 - x/4} + \dfrac{1}{1 + x}\right)$.

(b) Let a_k be the coefficient of x^k in $T_n(x)$ for $k \le n$. Use the result of Exercise 33 to show that

$$a_k = \frac{1}{4^{k+1}} + (-1)^k$$

9 | INTRODUCTION TO DIFFERENTIAL EQUATIONS

NASA technicians preparing a ground vibration test of an F/A-18 aircraft with aeroelastic wings.

Will this airplane fly? . . . How can we create an image of the interior of the human body using very weak X-rays? . . . What is a design of a bicycle frame that combines low weight with rigidity? . . . How much would the mean temperature of the earth increase if the amount of carbon dioxide in the atmosphere increased by 20 percent?

—An overview of applications of differential equations in *Computational Differential Equations*, K. Eriksson, D. Estep, P. Hansbo, and C. Johnson, Cambridge University Press, New York, 1996

Differential equations are among the most powerful tools we have for analyzing the world mathematically. They are used to formulate the fundamental laws of nature (from Newton's Laws to Maxwell's equations and the laws of quantum mechanics) and to model the most diverse physical phenomena. The above quote lists just a few of the myriad applications. In this chapter, we study some elementary techniques and basic applications of differential equations and give a few indications of where the more general theory leads.

9.1 Solving Differential Equations

A differential equation is an equation that involves an unknown function $y = y(x)$ and its first or higher derivatives. A **solution** is a function $y = f(x)$ satisfying the given equation. As we have seen in previous chapters, solutions usually depend on one or more arbitrary constants (denoted A, B, and C in the following examples):

Differential Equation	General Solution
$y' = -2y$	$y = Ce^{-2x}$
$\dfrac{dy}{dt} = t$	$y = \dfrac{1}{2}t^2 + C$
$y'' + y = 0$	$y = A\sin x + B\cos x$

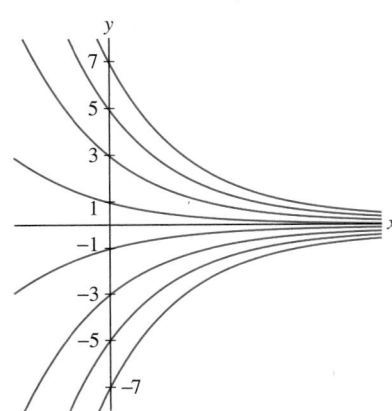

FIGURE 1 Family of solutions of $y' = -2y$.

An expression such as $y = Ce^{-2x}$ is called a **general solution**. For each value of C, we obtain a particular solution. The graphs of the solutions as C varies form a family of curves in the xy-plane (Figure 1).

The first step in any study of differential equations is to classify the equations according to various properties. The most important attributes of a differential equation are its order and whether or not it is linear. The **order** of a differential equation is the order of the highest derivative appearing in the equation. The general solution of an equation of order n usually involves n arbitrary constants. For example, $y'' + y = 0$ has order two and its general solution has two arbitrary constants A and B as listed above.

A differential equation is called **linear** if it can be written in the form

$$a_n(x)y^{(n)} + a_{n-1}(x)y^{(n-1)} + \cdots + a_1(x)y' + a_0(x)y = b(x)$$

The coefficients $a_j(x)$ and $b(x)$ are functions of x alone. A linear equation cannot have terms such as y^3, yy', or $\sin y$, but the coefficients $a_j(x)$ may be arbitrary.

Differential Equation	Type
$x^2 y' + e^x y = 4$	First-order, linear
$x(y')^2 = y + x$	First-order, nonlinear [since $(y')^2$ appears]
$y'' = (\sin x)y'$	Second-order, linear
$y'' = x(\sin y)$	Second-order, nonlinear (since $\sin y$ appears)

Separation of Variables

The simplest differential equations are those of the form $y' = f(x)$. A solution is an antiderivative of $f(x)$, and thus we may write the general solution as $y = \int f(x)\,dx$. A more general class of first-order equations that can be solved directly by integration are the **separable equations**, which have the form

$$\frac{dy}{dx} = f(x)g(y) \qquad \boxed{1}$$

For example,

- $\dfrac{dy}{dx} = (\sin x)y$ is separable.

- $\dfrac{dy}{dx} = x + y$ is not separable because $x + y$ is not a *product* $f(x)g(y)$.

In separation of variables, we manipulate dx and dy symbolically, just as in the Substitution Rule.

To solve a separable equation, we use the method of **separation of variables:** Move the terms involving y and dy to the left and those involving x and dx to the right. Then integrate both sides:

$$\frac{dy}{dx} = f(x)g(y) \qquad \text{(separable equation)}$$

$$\frac{dy}{g(y)} = f(x)\,dx \qquad \text{(separate the variables)}$$

$$\int \frac{dy}{g(y)} = \int f(x)\,dx \qquad \text{(integrate)}$$

If these integrals can be evaluated, we can try to solve for y as a function of x.

■ **EXAMPLE 1** Show that $y\dfrac{dy}{dx} - x = 0$ is separable but not linear. Then find the general solution and plot the family of solutions.

Solution This differential equation is nonlinear because it contains the term yy'. To show that it is separable, rewrite the equation:

$$y\frac{dy}{dx} - x = 0 \quad \Rightarrow \quad \frac{dy}{dx} = xy^{-1} \quad \text{(separable)}$$

Now use separation of variables:

$$y\,dy = x\,dx \qquad \text{(separate variables)}$$

$$\int y\,dy = \int x\,dx \qquad \text{(integrate)}$$

$$\frac{1}{2}y^2 = \frac{1}{2}x^2 + C \qquad \boxed{2}$$

$$y = \pm\sqrt{x^2 + 2C} \quad \text{(solve for } y\text{)}$$

Note that one constant of integration is sufficient in Eq. (2). An additional constant for the integral on the left is not needed.

Since C is arbitrary, we may replace $2C$ by C to obtain (Figure 2)

$$\boxed{y = \pm\sqrt{x^2 + C}}$$

Each choice of sign yields a solution. It is a good idea to check the solution by verifying that it satisfies the differential equation. For the positive square root (the negative square root is similar),

$$\frac{dy}{dx} = \frac{d}{dx}\sqrt{x^2 + C} = \frac{x}{\sqrt{x^2 + C}}$$

$$y\frac{dy}{dx} = \sqrt{x^2 + C}\left(\frac{x}{\sqrt{x^2 + C}}\right) = x$$

Thus, $y\dfrac{dy}{dx} - x = 0$ as required. ∎

FIGURE 2 Solutions $y = \sqrt{x^2 + C}$ to $y\dfrac{dy}{dx} - x = 0$.

Most of the differential equations arising in applications have an existence and uniqueness property: There exists one and only one solution satisfying a given initial condition. General existence and uniqueness theorems are discussed in more advanced treatments of differential equations.

Although it is useful to find general solutions, in applications we are usually interested in the solution that describes a particular physical situation. The general solution to a first-order equation generally depends on one arbitrary constant, so we can pick out a particular solution $y(x)$ by specifying the value $y(x_0)$ for some fixed x_0 (Figure 3). This specification is called an **initial condition.** A differential equation together with an initial condition is called an **initial value problem.**

■ **EXAMPLE 2** Initial Value Problem Solve the initial value problem $y' = -ty$ with $y(0) = 3$.

Solution First we find the general solution by separation of variables:

$$\frac{dy}{dt} = -ty$$

$$\frac{dy}{y} = -t\,dt$$

$$\int \frac{dy}{y} = -\int t\,dt$$

$$\ln|y| = -\frac{1}{2}t^2 + C$$

$$|y| = e^{-t^2/2+C} = e^C e^{-t^2/2}$$

Thus, $y = \pm e^C e^{-t^2/2}$. Since C is arbitrary, e^C represents an arbitrary *positive number* and $\pm e^C$ an arbitrary nonzero number. We replace $\pm e^C$ by C and write the general solution as

$$\boxed{y = Ce^{-t^2/2}} \qquad\qquad \boxed{3}$$

If we set $C = 0$ in Eq. (3), we obtain the solution $y = 0$. The separation of variables procedure did not directly yield this solution because we divided by y (and thus assumed implicitly that $y \neq 0$).

Now use the initial condition $y(0) = Ce^{-0^2/2} = 3$. Thus, $C = 3$ and $y = 3e^{-t^2/2}$ is the solution to the initial value problem (Figure 3). ■

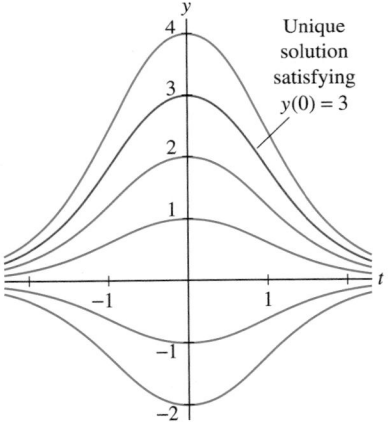

FIGURE 3 The initial condition $y(0) = 3$ determines one curve in the family of solutions to $y' = -ty$.

In the context of differential equations, the term "modeling" means finding a differential equation that describes a given physical situation. As an example, we consider water leaking through a hole at the bottom of a tank (Figure 4). The problem is to find the water level $y(t)$ at time t.

We solve this problem by showing that $y(t)$ satisfies a differential equation. The key observation is that the water lost during the interval from t to $t + \Delta t$ can be computed in two ways. Let

$$v(y) = \text{velocity of the water flowing through the hole}$$
$$\text{when the tank is filled to height } y$$

$$B = \text{area of the hole}$$

$$A(y) = \text{area of horizontal cross section of the tank at height } y$$

FIGURE 4 Water leaks out of a tank through a hole of area B at the bottom.

Like most if not all mathematical models, our model of water draining from a tank is at best an approximation. The differential equation (4) does not take into account viscosity (resistance of a fluid to flow). This can be remedied by using the differential equation

$$\frac{dy}{dt} = k\frac{Bv(y)}{A(y)}$$

where $k < 1$ is a viscosity constant. Furthermore, Torricelli's Law is valid only when the hole size B is small relative to the cross-sectional areas $A(y)$.

First we observe that the water exiting through the hole during a time interval Δt forms a cylinder of base B and height $v(y)\,\Delta t$ (because the water travels a distance $v(y)\,\Delta t$—see Figure 4). The volume of this cylinder is approximately $Bv(y)\,\Delta t$ [approximately but not exactly, because $v(y)$ may not be constant]. Thus,

$$\text{Water lost between } t \text{ and } t + \Delta t \approx Bv(y)\,\Delta t$$

Second, we note that if the water level drops by an amount Δy during the interval Δt, then the volume of water lost is approximately $A(y)\,\Delta y$ (Figure 4). Therefore,

$$\text{Water lost between } t \text{ and } t + \Delta t \approx A(y)\,\Delta y$$

This is also an approximation because the cross-sectional area may not be constant. Comparing the two results, we obtain $A(y)\Delta y \approx Bv(y)\,\Delta t$, or

$$\frac{\Delta y}{\Delta t} \approx \frac{Bv(y)}{A(y)}$$

Now take the limit as $\Delta t \to 0$ to obtain the differential equation

$$\boxed{\frac{dy}{dt} = \frac{Bv(y)}{A(y)}} \qquad \boxed{4}$$

To use Eq. (4), we need to know the velocity of the water leaving the hole. This is given by **Torricelli's Law**, which states that

$$v(y) = -\sqrt{2gy} = -\sqrt{2(32)y} = -8\sqrt{y} \text{ ft/s} \qquad \boxed{5}$$

where $g = 32$ ft/s^2 is the acceleration due to gravity on the earth's surface.

■ **EXAMPLE 3** Application of Torricelli's Law   A cylindrical tank of height 9 ft and radius 2 ft is filled with water. Water drains through a square hole of side 1 inch in the bottom. Determine the water level $y(t)$ at time t (seconds). How long does it take for the tank to go from full to empty?

Solution The horizontal cross section of the cylinder is a circle of radius $r = 2$ ft and area $A(y) = \pi r^2 = 4\pi$ ft^2 (Figure 5). The hole is a square of side $\frac{1}{12}$ ft and area $B = \frac{1}{144}$ ft^2. By Torricelli's Law [Eq. (5)], $v(y) = -8\sqrt{y}$, so Eq. (4) becomes

$$\frac{dy}{dt} = \frac{Bv(y)}{A(y)} = -\frac{\frac{1}{144}(8\sqrt{y})}{4\pi} = -\frac{\sqrt{y}}{72\pi} \qquad \boxed{6}$$

We solve this equation using separation of variables:

$$\int \frac{dy}{\sqrt{y}} = -\int \frac{dt}{72\pi}$$

$$2y^{1/2} = -\frac{1}{72\pi}t + C \qquad \boxed{7}$$

$$y = \left(-\frac{1}{144\pi}t + \frac{1}{2}C\right)^2$$

Hole = square of side 1 in.

FIGURE 5

Since C is arbitrary, we may replace $\frac{1}{2}C$ by C and write the general solution as

$$y = \left(-\frac{t}{144\pi} + C \right)^2$$

The tank is full at $t = 0$ so we have the initial condition $y(0) = 9$. Thus

$$y(0) = \left(-\frac{0}{144\pi} + C \right)^2 = C^2 = 9 \quad \Rightarrow \quad C = \pm 3$$

Which sign is correct? You might think that both sign choices are possible, but notice that the *positive* square root $y^{1/2}$ appears on the left in Eq. (7). Therefore, the right-hand side of Eq. (7) must be nonnegative for $t \geq 0$ and this occurs only if $C \geq 0$. Thus, $C = 3$ is the only possible choice (Figure 6):

$$y(t) = \left(3 - \frac{t}{144\pi} \right)^2$$

Suppose that the tank empties at time t_e. Then $y(t_e) = 0$ and we may solve

$$y(t_e) = \left(3 - \frac{t_e}{144\pi} \right)^2 = 0 \quad \Rightarrow \quad t_e = (3)144\pi \approx 1{,}357 \text{ s}$$

Thus, the tank is empty after $t_e = 1{,}357$ s, or approximately 23 min. ∎

Water level (ft)

(A) Water level $y(t) = \left(3 - \frac{t}{144\pi}\right)^2$ $(C = 3)$

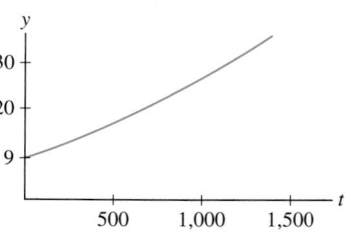

(B) Graph of $y(t) = \left(3 + \frac{t}{144\pi}\right)^2$ $(C = -3)$

FIGURE 6

CONCEPTUAL INSIGHT   The previous example highlights the need to check and analyze solutions to a differential equation rather than rely on algebra alone. The algebra seemed to suggest that $C = \pm 3$, but further analysis of the differential equation showed that $C = -3$ does not yield a solution for $t \geq 0$. Furthermore, our formula for the solution is valid only for $t \leq t_e$. The function

$$y(t) = \left(3 - \frac{t}{144\pi} \right)^2$$

does not satisfy Eq. (6) for $t > t_e$ because $y'(t)$ is positive for $t > t_e$, but by Eq. (6) the derivative of a solution is ≤ 0. In fact, the tank is empty for $t \geq t_e$ so the full solution, valid for all $t \geq 0$, is

$$y(t) = \begin{cases} \left(3 - \dfrac{t}{144\pi}\right)^2 & \text{for } 0 \leq t \leq t_e \\ 0 & \text{for } t > t_e \end{cases} \qquad \boxed{8}$$

9.1 SUMMARY

- A differential equation has order n if $y^{(n)}$ is the highest-order derivative appearing in the equation.
- A differential equation is *linear* if it can be written as

$$a_n(x)y^{(n)} + a_{n-1}(x)y^{(n-1)} + \cdots + a_1(x)y' + a_0(x)y = b(x)$$

- A first-order equation is called *separable* if it can be written in the form

$$\frac{dy}{dx} = f(x)g(y)$$

To solve a separable equation, move all terms involving y to the left and all terms involving x to the right and integrate:

$$\frac{dy}{g(y)} = f(x)\,dx$$

$$\int \frac{dy}{g(y)} = \int f(x)\,dx$$

9.1 EXERCISES

Preliminary Questions

1. Determine the order of the following differential equations:

(a) $x^5 y' = 1$

(b) $(y')^3 + x = 1$

(c) $y''' + x^4 y' = 2$

2. Is $y' = \sin x$ a linear differential equation?

3. Give an example of a nonlinear differential equation of the form $y' = f(y)$.

4. Can a nonlinear differential equation be separable? If so, give an example.

5. Give an example of a linear, nonseparable differential equation.

Exercises

1. Which of the following differential equations are first-order?

(a) $y' = x^2$

(b) $y'' = y^2$

(c) $(y')^3 + yy' = \sin x$

(d) $x^2 y' - e^x y = \sin y$

(e) $y'' + 3y' = \dfrac{y}{x}$

(f) $yy' + x + y = 0$

2. Which of the equations in Exercise 1 are linear?

In Exercises 3–9, verify that the given function is a solution of the differential equation.

3. $y' - 8x = 0, \quad y = 4x^2$

4. $y' + 8y = 0, \quad y = 4e^{-8x}$

5. $yy' + 4x = 0, \quad y = \sqrt{12 - 4x^2}$

6. $y' + 4xy = 0, \quad y = 3e^{-2x^2}$

7. $(x^2 - 1)y' + xy = 0, \quad y = 4(x^2 - 1)^{-1/2}$

8. $y'' - 2xy' + 8y = 0, \quad y = 4x^4 - 12x^2 + 3$

9. $y'' - 2y' + 5y = 0, \quad y = e^x \sin 2x$

10. Which of the following equations are separable? Write those that are separable in the form $y' = f(x)g(y)$ (but do not solve).

(a) $xy' + y^2 = 0$ (b) $\sqrt{1 - x^2}\,y' = e^y \sin x$

(c) $y' = x^2 + y^2$ (d) $y' = 9 - y^2$

11. Consider the differential equation $y' = e^y \cos x$.

(a) Write it as $e^{-y}\,dy = \cos x\,dx$.

(b) Integrate both sides of $\displaystyle\int e^{-y}\,dy = \int \cos x\,dx$.

(c) Show that $y = -\ln|C - \sin x|$ is the general solution.

(d) Find the particular solution satisfying $y(0) = 0$.

12. Verify that $x^2 y' + e^y = 0$ is separable by rewriting it in the form $y' = f(x)g(y)$.

(a) Show that $e^{-y}\,dy = -x^{-2}\,dx$.

(b) Show that $e^{-y} = x^{-1} + C$.

(c) Verify that $y = -\ln(C - x^{-1})$ is the general solution.

In Exercises 13–28, solve using separation of variables.

13. $y' = xy^2$

14. $y' = \dfrac{1}{2}xy$

15. $y' = 9y$

16. $y' = 2(4 - y)$

17. $2\dfrac{dy}{dx} + 6y + 4 = 0$

18. $\dfrac{dy}{dt} = 2\sqrt{y}$

19. $\dfrac{dy}{dt} - te^y = 0$

20. $\sqrt{1 - x^2}\,y' = xy$

21. $y' = y^2(1 - x^2)$

22. $yy' = x$

23. $(t^2 + 1)\dfrac{dx}{dt} = x^2 + 1$

24. $(1 + x^2)y' = x^3 y$

25. $y' = x \sec y$

26. $\dfrac{dy}{dt} = \tan y$

27. $\dfrac{dy}{dt} = y \tan t$

28. $\dfrac{dx}{dt} = t \tan x$

In Exercises 29–41, solve the initial value problem.

29. $y' + 2y = 0$, $y(\ln 2) = 3$

30. $y' - 2y + 4 = 0$, $y(1) = 4$

31. $yy' = xe^{-y^2}$, $y(0) = -1$

32. $y^2 \dfrac{dy}{dx} = x^{-3}$, $y(2) = 0$

33. $y' = (x - 1)(y - 2)$, $y(0) = 3$

34. $(1 - t)\dfrac{dy}{dt} - y = 0$, $y(2) = -4$

35. $\dfrac{dy}{dt} = ye^{-t}$, $y(0) = 1$

36. $\dfrac{dy}{dt} = te^{-y}$, $y(1) = 0$

37. $t^2 \dfrac{dy}{dt} - t = 1 + y + ty$, $y(1) = 0$

38. $\sqrt{1 - x^2}\,y' = y^2 + 1$, $y(0) = 0$

39. $\sqrt{1 - x^2}\,y' = y^2$, $y(0) = 1$

40. $y' = \tan y$, $y(\ln 2) = \dfrac{\pi}{2}$

41. $y' = y^2 \sin x$, $y(0) = 3$

42. Find all values of a such that $y = x^a$ is a solution of

$$y'' - 6x^{-2}y = 0$$

43. Find all values of a such that $y = e^{ax}$ is a solution of

$$y'' + 2y' - 8y = 0$$

44. Show that if $y(t)$ is a solution of $t(y - 1)y' = 2y$, then $t^2 y = Ce^y$ for some constant C [we cannot solve for $y(t)$ explicitly]. Find t such that $y(t) = 2$, assuming that $y(1) = 1$.

In Exercises 45–48, use Eq. (4) and Torricelli's Law [Eq. (5)].

45. A cylindrical tank filled with water has height 10 ft and a base of area 30 ft^2. Water leaks through a hole in the bottom of area $\frac{1}{3}$ ft^2. How long does it take (a) for half of the water to leak out and (b) for the tank to empty?

46. A conical tank filled with water has height 12 ft [Figure 7(A)]. Assume that the top is a circle of radius 4 ft and that water leaks through a hole in the bottom of area 2 in.^{2}. Let $y(t)$ be the water level at time t.

(A) Conical tank   (B) Horizontal tank

FIGURE 7

(a) Show that the cross-sectional area of the tank at height y is $A(y) = \frac{\pi}{9}y^2$.

(b) Find the differential equation satisfied by $y(t)$ and solve for $y(t)$. Use the initial condition $y(0) = 12$.

(c) How long does it take for the tank to empty?

47. The tank in Figure 7(B) is a cylinder of radius 10 ft and length 40 ft. Assume that the tank is half-filled with water and that water leaks through a hole in the bottom of area $B = 3$ in.^{2}. Determine the water level $y(t)$ and the time t_e when the tank is empty.

48. A cylindrical tank filled with water has height h ft and a base of area A ft^2. Water leaks through a hole in the bottom of area B ft^2.

(a) Show that the time required for the tank to empty is proportional to $A\sqrt{h}/B$.

(b) Show that the emptying time is proportional to $Vh^{-1/2}$, where V is the volume of the tank.

(c) Two tanks have the same volume and same-sized hole, but different heights and bases. Which tank empties first: the taller or shorter tank?

49. Figure 8 shows a circuit consisting of a resistor of R ohms, a capacitor of C farads, and a battery of voltage V. When the circuit is completed, the amount of charge $q(t)$ (in coulombs) on the plates of the capacitor varies according to the differential equation (t in seconds)

$$R\frac{dq}{dt} + \frac{1}{C}q = V$$

(a) Solve for $q(t)$.

(b) Show that $\lim_{t\to\infty} q(t) = CV$.

(c) Find $q(t)$, assuming that $q(0) = 0$. Show that the capacitor charges to approximately 63% of its final value CV after a time period of length $\tau = RC$ (τ is called the time constant of the capacitor).

FIGURE 8 An RC circuit.

50. Assume in the circuit of Figure 8 that $R = 100\ \Omega$, $C = 0.01$ F, and $V = 10$ V. How many seconds does it take for the charge on the capacitor plates to reach half of its limiting value?

51. One hypothesis for the growth rate of the volume V of a cell is that $\dfrac{dV}{dt}$ is proportional to the cell's surface area A. Since V has cubic units such as cm^3 and A has square units such as cm^2, we may assume roughly that $A \propto V^{2/3}$, and hence $\dfrac{dV}{dt} = kV^{2/3}$ for some constant k. If this hypothesis is correct, which dependence of volume on time would we expect to see (again, roughly speaking) in the laboratory?

(a) Linear   **(b)** Quadratic   **(c)** Cubic

52. We might also guess that the rate at which a snowball melts is proportional to its surface area. What is the differential equation satisfied by the volume V of a spherical snowball at time t? Suppose the snowball has radius 4 cm and that it loses half of its volume after 10 min. According to this model, when will the snowball disappear?

53. In general, $(fg)'$ is not equal to $f'g'$, but let $f(x) = e^{2x}$ and find a function $g(x)$ such that $(fg)' = f'g'$. Do the same for $f(x) = x$.

54. A boy standing at point B on a dock holds a rope of length ℓ attached to a boat at point A [Figure 9(A)]. As the boy walks along the dock, holding the rope taut, the boat moves along a curve called a **tractrix** (from the Latin *tractus* meaning "to pull"). The segment from a point P on the curve to the x-axis along the tangent line has constant length ℓ. Let $y = f(x)$ be the equation of the tractrix.

(a) Show that $y^2 + (y/y')^2 = \ell^2$ and conclude $y' = -\dfrac{y}{\sqrt{\ell^2 - y^2}}$. Why must we choose the negative square root?

(b) Prove that the tractrix is the graph of

$$x = \ell \ln\left(\frac{\ell + \sqrt{\ell^2 - y^2}}{y}\right) - \sqrt{\ell^2 - y^2}$$

 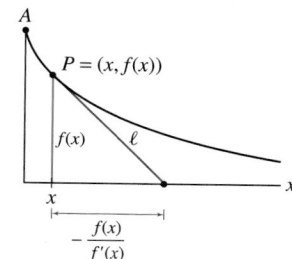

FIGURE 9

55. If a bucket of water spins about a vertical axis with constant angular velocity ω (in radians per second), the water climbs up the side of the bucket until it reaches an equilibrium position (Figure 10). Two forces act on a particle located at a distance x from the vertical axis: the

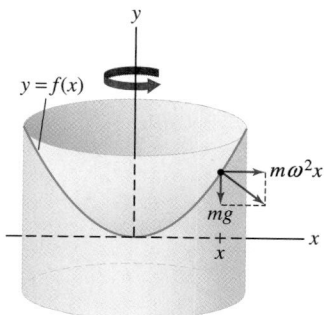

FIGURE 10

gravitational force $-mg$ acting downward and the force of the bucket on the particle (transmitted indirectly through the liquid) in the direction perpendicular to the surface of the water. These two forces must combine to supply a centripetal force $m\omega^2 x$, and this occurs if the diagonal of the rectangle in Figure 10 is normal to the water's surface (that is, perpendicular to the tangent line). Prove that if $y = f(x)$ is the equation of the curve obtained by taking a vertical cross section through the axis, then $-1/y' = -g/(\omega^2 x)$. Show that $y = f(x)$ is a parabola.

56. Show that the differential equations $y' = \dfrac{3y}{x}$ and $y' = -\dfrac{x}{3y}$ define **orthogonal families** of curves; that is, the graphs of solutions to the first equation intersect the graphs of the solutions to the second equation in right angles (Figure 11). Find these curves explicitly.

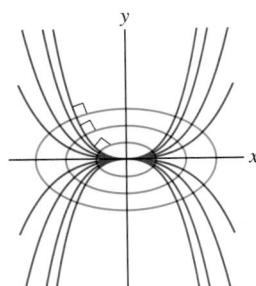

FIGURE 11 Two orthogonal families of curves.

57. Find the family of curves satisfying $y' = x/y$ and sketch several members of the family. Then find the differential equation for the orthogonal family (see Exercise 56), find its general solution, and add some members of this orthogonal family to your plot.

58. A 50-kg model rocket lifts off by expelling fuel at a rate of $k = 4.75$ kg/s for 10 s. The fuel leaves the end of the rocket with an exhaust velocity of $b = 100$ m/s. Let $m(t)$ be the mass of the rocket at time t. From the law of conservation of momentum, we find the following differential equation for the rocket's velocity $v(t)$ (in meters per second):

$$m(t)v'(t) = -9.8m(t) + b\frac{dm}{dt}$$

(a) Show that $m(t) = 50 - 4.75t$ kg.

(b) Solve for $v(t)$ and compute the rocket's velocity at rocket burnout (after 10 s).

59. Let $v(t)$ be the velocity of an object of mass m in free fall near the earth's surface. If we assume that air resistance is proportional to v^2, then v satisfies the differential equation $m\dfrac{dv}{dt} = -mg + kv^2$ for some constant $k > 0$.

(a) Set $\alpha = (g/k)^{1/2}$ and rewrite the differential equation as

$$\frac{dv}{dt} = -\frac{k}{m}(\alpha^2 - v^2)$$

Then solve using separation of variables with initial condition $v(0) = 0$.

(b) Show that the terminal velocity $\displaystyle\lim_{t \to \infty} v(t)$ is equal to $-\alpha$.

Further Insights and Challenges

60. In Section 6.2, we computed the volume V of a solid as the integral of cross-sectional area. Explain this formula in terms of differential equations. Let $V(y)$ be the volume of the solid up to height y and let $A(y)$ be the cross-sectional area at height y as in Figure 12.

(a) Explain the following approximation for small Δy:

$$V(y + \Delta y) - V(y) \approx A(y)\,\Delta y \qquad \boxed{9}$$

(b) Use Eq. (9) to justify the differential equation $\dfrac{dV}{dy} = A(y)$. Then derive the formula $V = \displaystyle\int_a^b A(y)\,dy$.

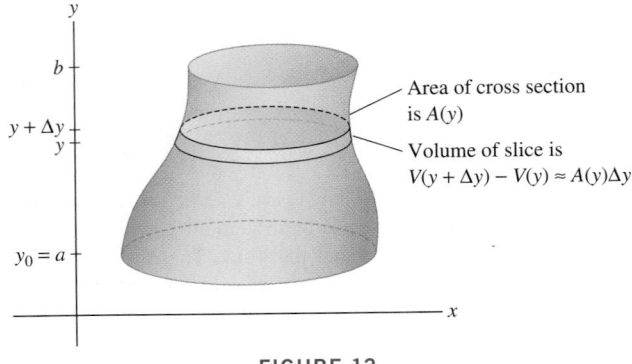

FIGURE 12

61. In most cases of interest, the general solution of a differential equation of order n depends on n arbitrary constants. This exercise shows there are exceptions.

(a) Show that $(y')^2 + y^2 = 0$ is a first-order equation with only one solution $y = 0$.

(b) Show that $(y')^2 + y^2 + 1 = 0$ is a first-order equation with no solutions.

62. **(a)** Let $P(x) = x^2 + ax + b$ with a, b constants. Show that $y = Ce^{\lambda x}$ (for any constant C) is a solution of $y'' + ay' + by = 0$ if and only if $P(\lambda) = 0$.

(b) Show that $y = C_1 e^{3x} + C_2 e^{-x}$ is a solution of $y'' - 2y' - 3y = 0$ for any constants C_1, C_2.

63. A spherical tank of radius R is half-filled with water. Suppose that water leaks through a hole in the bottom of area B ft^2. Let $y(t)$ be the water level at time t (seconds).

(a) Show that $\dfrac{dy}{dt} = \dfrac{-8B\sqrt{y}}{\pi(2Ry - y^2)}$.

(b) Show that for some constant C,

$$\frac{\pi}{4B}\left(\frac{2}{3}Ry^{3/2} - \frac{1}{5}y^{5/2}\right) = C - t$$

(c) Use the initial condition $y(0) = R$ to compute C and show that $C = t_e$, the time at which the tank is empty.

(d) Show that t_e is proportional to $R^{5/2}$ and inversely proportional to B.

9.2 Models Involving $y' = k(y - b)$

We have seen that a quantity grows or decays exponentially if its *rate of change* is proportional to the amount present. This characteristic property is expressed by the differential equation $y' = ky$. We now study a closely related differential equation:

$$\frac{dy}{dt} = k(y - b) \qquad \boxed{1}$$

Every first-order, linear differential equation with constant coefficients may be written in the form of Eq. (1). This equation is used to model a variety of phenomena such as the cooling process, free-fall with air resistance, and current in a circuit.

where k and b are constants and $k \neq 0$. This differential equation describes a quantity y whose *rate of change is proportional to the size of the difference $y - b$*. The general solution is

$$y(t) = b + Ce^{kt} \qquad \boxed{2}$$

where C is a constant determined by the initial condition. To verify this formula, we apply separation of variables to Eq. (1). Alternatively, we may observe that $(y - b)' = y'$ since b is a constant, so Eq. (1) may be rewritten

$$\frac{d}{dt}(y - b) = k(y - b)$$

In other words, $y - b$ satisfies the differential equation of an exponential function. Thus, $y - b = Ce^{kt}$ or $y = b + Ce^{kt}$, as claimed.

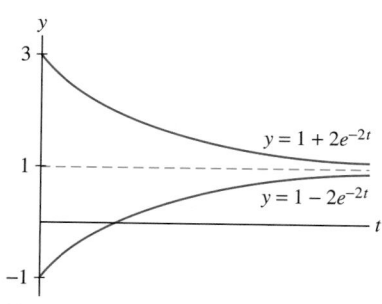

FIGURE 1 Two solutions to $y' = -2(y - 1)$ corresponding to $C = 2$ and $C = -2$.

> **GRAPHICAL INSIGHT**   The behavior of the solution $y(t)$ depends on whether C and k are positive or negative. When $k < 0$, we usually rewrite the differential equation as $y' = -k(y - b)$ with $k > 0$. In this case, $y(t) = b + Ce^{-kt}$ and $y(t)$ approaches the horizontal asymptote $y = b$ since Ce^{-kt} tends to zero as $t \to \infty$ (Figure 1). However, $y(t)$ approaches the asymptote from above or below, depending on whether $C > 0$ or $C < 0$.

We now consider several applications of Eq. (1), beginning with Newton's Law of Cooling. Let $y(t)$ be the temperature of a hot object cooling down in an environment where the ambient temperature is T_0. Newton assumed that the *rate of cooling* is proportional to the temperature difference $y - T_0$. We express this hypothesis in a precise way by the differential equation

Newton's Law of Cooling implies that the object cools quickly when it is very hot (when $y - T_0$ is large). The rate of cooling slows as y approaches T_0. When the object's initial temperature is less than T_0, y' is positive and Newton's Law models warming.

$$y' = -k(y - T_0)$$

The constant k [in units of $(\text{time})^{-1}$] is called the **cooling constant** and depends on the physical properties of the object.

■ **EXAMPLE 1**   Newton's Law of Cooling   At $t = 0$, we submerge a hot metal bar with cooling constant $k = 2.1 \text{ min}^{-1}$ in a large tank of water at temperature $T_0 = 55°\text{F}$. Let $y(t)$ be the bar's temperature at time t.

(a) Find the differential equation satisfied by $y(t)$ and determine its general solution.
(b) What is the bar's temperature after 1 min if its initial temperature is $400°\text{F}$?
(c) What was the bar's initial temperature if, after half a minute, it had cooled to $120°\text{F}$?

Solution

(a) Since $k = 2.1 \text{ min}^{-1}$, $y(t)$ (with t in minutes) satisfies

$$y' = -2.1(y - 55)$$

By Eq. (2), the general solution is $y(t) = 55 + Ce^{-2.1t}$ for some constant C.

(b) If the initial temperature is $400°F$, then $y(0) = 55 + C = 400$. Thus, $C = 345$ and $y(t) = 55 + 345e^{-2.1t}$ (Figure 2). After 1 min,

$$y(1) = 55 + 345e^{-2.1(1)} \approx 97.2°F$$

(c) If the temperature after half a minute is $120°F$, then $y(t) = 55 + Ce^{-2.1t}$ where $y(0.5) = 120$. We solve for C:

$$55 + Ce^{-2.1(0.5)} = 120$$

$$Ce^{-1.05} = 65 \quad \Rightarrow \quad C = 65e^{1.05} \approx 185.7$$

It follows that $y(t) = 55 + 185.7e^{-2.1t}$ and the initial temperature is

$$y(0) = 55 + 185.7e^{-2.1(0)} = 55 + 185.7 = 240.7°F$$ ∎

FIGURE 2 Temperature of metal bar as it cools.

The differential equation $y' = k(y - b)$ is also used to model free fall when air resistance is taken into account. Assume that the force due to air resistance is proportional to the velocity v and acts opposite to the direction of the fall. We write this force as $-kv$ where $k > 0$. The force due to gravity on a falling object of mass m is $-mg$, where g is the acceleration due to gravity, so the total force is $F = -mg - kv$. By Newton's Law, $F = ma = mv'$ ($a = v'$ is the acceleration). Thus $mv' = -mg - kv$, which can be written

> The effect of air resistance depends on the physical situation. A high-speed bullet is affected differently than a skydiver. Our model is fairly realistic for a large object such as a skydiver falling from high altitudes.

> In this model of free fall, k has units of mass per time, such as kg/s.

$$v' = -\frac{k}{m}\left(v + \frac{gm}{k}\right)$$

This equation has the form $v' = -k(v - b)$ with k/m instead of k and $b = -gm/k$. By Eq. (2) the general solution is

$$v(t) = -\frac{gm}{k} + Ce^{-(k/m)t} \qquad \boxed{3}$$

Since $Ce^{-(k/m)t}$ tends to zero as $t \to \infty$, $v(t)$ tends to a limiting terminal velocity:

$$\text{Terminal velocity} = \lim_{t \to \infty} v(t) = -\frac{gm}{k} \qquad \boxed{4}$$

Without air resistance the velocity would increase indefinitely.

■ **EXAMPLE 2** An 80-kg skydiver jumps out of an airplane (with zero initial velocity).

(a) What is his terminal velocity if $k = 8$ kg/s?

(b) What is his velocity after 30 s?

Solution

(a) By Eq. (4), with $k = 8$ kg/s and $g = 9.8$ m/s^2, the terminal velocity is

$$-\frac{gm}{k} = -\frac{9.8(80)}{8} = -98 \text{ m/s}$$

(b) By Eq. (3),

$$v(t) = -98 + Ce^{-(k/m)t} = -98 + Ce^{-(8/80)t} = -98 + Ce^{-0.1t}$$

Since the initial velocity is zero, $v(0) = -98 + C = 0$ or $C = 98$. Therefore, $v(t) = -98(1 - e^{-0.1t})$ (Figure 3). The skydiver's velocity after 30 s is

$$v(30) = -98(1 - e^{-0.1(30)}) \approx -93.1 \text{ m/s}$$ ■

FIGURE 3 Velocity of 80-kg skydiver in free fall with air resistance ($k = 8$).

An **annuity** is an investment in which a principal P_0 is placed in an account earning interest at a rate r and money is withdrawn at regular intervals. To model an annuity by a differential equation, we assume that the money is withdrawn continuously at a rate of N dollars/year. The balance $P(t)$ in the annuity at time t satisfies

Notice in Eq. (5) that $P'(t)$ is determined by the growth rate and the withdrawal rate. If no withdrawals occurred, $P(t)$ would grow with compound interest and would satisfy $P'(t) = rP(t)$.

$$\underbrace{P'(t)}_{\substack{\text{Rate of} \\ \text{change}}} = \underbrace{rP(t)}_{\substack{\text{Growth due} \\ \text{to interest}}} - \underbrace{N}_{\substack{\text{Withdrawal} \\ \text{rate}}} = r\left(P(t) - \frac{N}{r}\right) \qquad \boxed{5}$$

This equation has the form $y' = k(y - b)$ with $k = r$ and $b = N/r$, so by Eq. (2), the general solution is

$$P(t) = \frac{N}{r} + Ce^{rt} \qquad \boxed{6}$$

Since e^{rt} tends to infinity as $t \to \infty$, the balance $P(t)$ tends to ∞ if $C > 0$. If $C < 0$, then $P(t)$ tends to $-\infty$ (i.e., the annuity eventually runs out of money).

■ **EXAMPLE 3** Does an Annuity Pay Out Forever? Let $P(t)$ be the balance in an annuity earning interest at the rate $r = 0.07$, with withdrawals made continuously at a rate of $N = \$500$/year.

(a) Assume that $P(0) = \$5,000$. Find $P(t)$ and determine when the annuity runs out of money.

(b) Assume that $P(0) = \$9,000$. Find $P(t)$ and show that the balance increases indefinitely.

Solution We have $\dfrac{N}{r} = \dfrac{500}{0.07} \approx 7,143$, so $P(t) = 7,143 + Ce^{0.07t}$ by Eq. (6).

(a) If $P(0) = 5,000 = 7,143 + Ce^0$, then $C = -2,143$ and

$$P(t) = 7,143 - 2,143e^{0.07t}$$

The account runs out of money when $P(t) = 7,143 - 2,143e^{0.07t} = 0$. We solve for t:

$$e^{0.07t} = \frac{7,143}{2,143} \quad \Longrightarrow \quad 0.07t = \ln\left(\frac{7,143}{2,143}\right) \approx 1.2$$

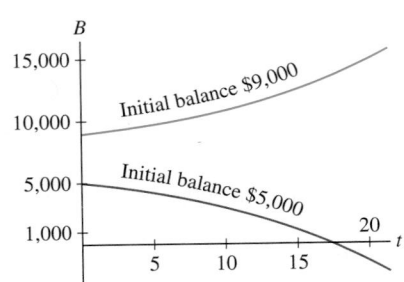

FIGURE 4 The balance in an annuity may increase indefinitely or decrease to zero (eventually becoming negative), depending on the size of initial deposit P_0.

The money runs out at time $t = \dfrac{1.2}{0.07} \approx 17$ years.

(b) If $P(0) = 9{,}000 = 7{,}143 + Ce^0$, then $C = 1{,}857$ and

$$P(t) = 7{,}143 + 1{,}857e^{0.07t}$$

Since the coefficient $C = 1{,}857$ is positive, the account never runs out of money. In fact, $P(t)$ increases indefinitely as $t \to \infty$. Figure 4 illustrates the two cases. ∎

9.2 SUMMARY

- The general solution of $y' = k(y - b)$ is $y = b + Ce^{kt}$, where C is a constant.

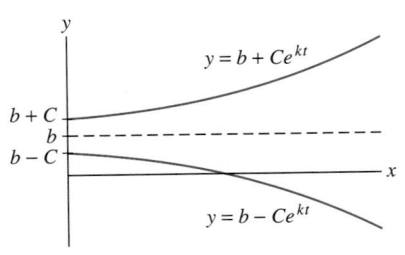

Solutions to $y' = k(y - b)$ with k, $C > 0$

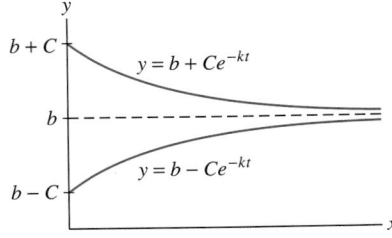

Solutions to $y' = -k(y - b)$ with k, $C > 0$

FIGURE 5

- The following tables describe the solutions to $y' = k(y - b)$ (see Figure 5):

Equation ($k > 0$)	Solution	Behavior as $t \to \infty$
$y' = k(y - b)$	$y(t) = b + Ce^{kt}$	$\displaystyle\lim_{t \to \infty} y(t) = \begin{cases} \infty & \text{if } C > 0 \\ -\infty & \text{if } C < 0 \end{cases}$
$y' = -k(y - b)$	$y(t) = b + Ce^{-kt}$	$\displaystyle\lim_{t \to \infty} y(t) = b$

- The equation $P' = r(P - N/r)$ models the balance $P(t)$ of a continuous annuity with interest rate r and withdrawal rate N.

9.2 EXERCISES

Preliminary Questions

1. What is the general solution to $y' = -k(y - b)$?

2. Write down a solution to $y' = 4(y - 5)$ that tends to $-\infty$ as $t \to \infty$.

3. Does there exist a solution of $y' = -4(y - 5)$ that tends to ∞ as $t \to \infty$?

4. True or false? If $k > 0$, then all solutions of $y' = -k(y - b)$ approach the same limit as $t \to \infty$.

5. Suppose that material A cools more rapidly than material B. Which material has a larger cooling constant k in Newton's Law of Cooling, $y' = -k(y - T_0)$?

Exercises

1. Find the general solution of $y' = 2(y - 10)$. Then find the two solutions satisfying $y(0) = 25$ and $y(0) = 5$, and sketch their graphs.

2. Find the general solution of $y' = -3(y - 12)$. Then find the two solutions satisfying $y(0) = 20$ and $y(0) = 0$, and sketch their graphs.

3. Verify directly that $y = b + Ce^{kt}$ satisfies $y' = k(y - b)$ for any constant C.

4. Let $F(t)$ be the temperature of a hot object submerged in a large pool of water whose temperature is 70°F.
(a) What is the differential equation satisfied by $F(t)$ if the cooling constant is $k = 1.5$?
(b) Find a formula for $F(t)$ if the object's initial temperature is 250°F.

5. A hot metal bar is submerged in a large reservoir of water whose temperature is 60°F. The temperature of the bar 20 s after submersion is 100°F. After 1 min, the temperature has cooled to 80°F.
(a) Determine the cooling constant k.
(b) What is the differential equation satisfied by the temperature $F(t)$ of the bar?
(c) What is the formula for $F(t)$?
(d) Determine the temperature of the bar at the moment it is submerged.

6. A hot metal rod is placed in a water bath whose temperature is 40°F. The rod cools from 300 to 200°F in 1 min. How long will it take for the rod to cool to 150°F?

7. When a hot object is placed in a water bath whose temperature is 25°C, it cools from 100 to 50°C in 150 s. In another bath, the same cooling occurs in 120 s. Find the temperature of the second bath.

8. A cold metal bar at -30°C is submerged in a pool maintained at a temperature of 40°C. Half a minute later, the temperature of the bar is 20°C. How long will it take for the bar to attain a temperature of 30°C?

9. A cup of coffee, cooling off in a room at temperature 20°C, has cooling constant $k = 0.09$ min^{-1}.
(a) How fast is the coffee cooling (in degrees per minute) when its temperature is $T = 80$°C?
(b) Use the Linear Approximation to estimate the change in temperature over the next 6 s when $T = 80$°C.
(c) The coffee is served at a temperature of 90°C. How long should you wait before drinking it if the optimal temperature is 65°C?

10. Two identical objects are heated to different temperatures T_1 and T_2. Both are submerged in a cold bath of temperature $T_0 = 40$°C at $t = 0$. Measurements show that $T_2 = 400$°C and the cooling rate of object 1 is twice as large as the cooling rate of object 2 (at each time t). Find T_1.

In Exercises 11–14, use the model for free-fall with air resistance discussed in this section. If the weight w of an object (in pounds) is given rather than its mass (as in Example 2), then the differential equation for free fall with air resistance is $v' = -(kg/w)(v + w/k)$, where $g = 32$ ft/s^2.

11. A 60-kg skydiver jumps out of an airplane. What is her terminal velocity in miles per hour, assuming that $k = 10$ kg/s for free-fall (no parachute)?

12. Find the terminal velocity of a 192-lb skydiver if $k = 1.2$ lb-s/ft. How long does it take him to reach half of his terminal velocity if his initial velocity is zero?

13. A 175-lb skydiver jumps out of an airplane (with zero initial velocity). Assume that $k = 0.7$ lb-s/ft with a closed parachute and $k = 5$ lb-s/ft with an open parachute. What is the skydiver's velocity at $t = 25$ s if the parachute opens after 20 seconds of free fall?

14. Does a heavier or lighter skydiver reach terminal velocity faster?

15. A continuous annuity with withdrawal rate $N = \$1,000$/year and interest rate $r = 5\%$ is funded by an initial deposit P_0.
(a) When will the annuity run out of funds if $P_0 = \$15,000$?
(b) Which initial deposit P_0 yields a constant balance?

16. Show that a continuous annuity with withdrawal rate $N = \$5,000$/year and interest rate $r = 8\%$, funded by an initial deposit of $P_0 = \$75,000$, never runs out of money.

17. Find the minimum initial deposit that will allow an annuity to pay out $\$500$/year indefinitely if it earns interest at a rate of 5%.

18. What is the minimum initial deposit necessary to fund an annuity for 30 years if withdrawals are made at a rate of $\$2,000$/year at an interest rate of 7%?

19. An initial deposit of $\$5,000$ is placed in a bank account. What is the minimum interest rate the bank must pay to allow continuous withdrawals at a rate of $\$500$/year to continue indefinitely?

20. Show that a continuous annuity never runs out of money if the initial balance is greater than or equal to N/r, where N is the withdrawal rate and r the interest rate.

21. Sam borrows $\$10,000$ from a bank at an interest rate of 9% and pays back the loan continuously at a rate of N dollars/year. Let $P(t)$ denote the amount still owed at time t.
(a) Explain why $P(t)$ satisfies the differential equation

$$y' = 0.09y - N$$

(b) How long will it take Sam to pay back the loan if $N = \$1,200$?
(c) Will he ever be able to pay back the loan if $N = \$800$?

22. Let $N(t)$ be the fraction of the population who have heard a given piece of news t hours after its initial release. According to one model, the rate $N'(t)$ at which the news spreads is equal to k times the fraction of the population that has not yet heard the news, for some constant k.
(a) Determine the differential equation satisfied by $N(t)$.
(b) Find the solution of this differential equation with the initial condition $N(0) = 0$ in terms of k.

(c) Suppose that half of the population is aware of an earthquake 8 hours after it occurs. Use the model to calculate k and estimate the percentage that will know about the earthquake 12 hours after it occurs.

23. Current in a Circuit The electric current flowing in the circuit in Figure 6 (consisting of a battery of V volts, a resistor of R ohms, and an inductor) satisfies

$$\frac{dI}{dt} = -k(I - b)$$

for some constants k and b with $k > 0$. Initially, $I(0) = 0$ and $I(t)$ approaches a maximum level V/R as $t \to \infty$.

(a) What is the value of b?

(b) Find a formula for $I(t)$ in terms of k, V, and R.

(c) Show that $I(t)$ reaches approximately 63% of its maximum value at time $t = 1/k$.

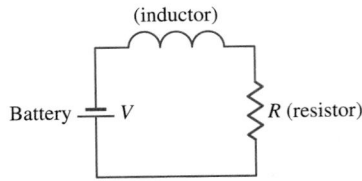

FIGURE 6 Current flow approaches the level $I_{max} = V/R$.

Further Insights and Challenges

24. Show that the cooling constant of an object can be determined from two temperature readings $y(t_1)$ and $y(t_2)$ at times $t_1 \neq t_2$ by the formula

$$k = \frac{1}{t_1 - t_2} \ln\left(\frac{y(t_2) - T_0}{y(t_1) - T_0}\right)$$

25. Show that by Newton's Law of Cooling, the time required to cool an object from temperature A to temperature B is

$$t = \frac{1}{k} \ln\left(\frac{A - T_0}{B - T_0}\right)$$

where T_0 is the ambient temperature.

26. Air Resistance A projectile of mass $m = 1$ travels straight up from ground level with initial velocity v_0. Suppose that the velocity v satisfies $v' = -g - kv$.

(a) Find a formula for $v(t)$.

(b) Show that the projectile's height $h(t)$ is given by

$$h(t) = C(1 - e^{-kt}) - \frac{g}{k}t$$

where $C = k^{-2}(g + kv_0)$.

(c) Show that the projectile reaches its maximum height at time $t_{max} = k^{-1} \ln(1 + kv_0/g)$.

(d) In the absence of air resistance, the maximum height is reached at time $t = v_0/g$. In view of this, explain why we should expect that

$$\lim_{k \to 0} \frac{\ln(1 + \frac{kv_0}{g})}{k} = \frac{v_0}{g} \qquad \boxed{7}$$

(e) Verify Eq. (7). *Hint:* Use Theorem 2 in Section 5.8 to show that

$$\lim_{k \to 0} \left(1 + \frac{kv_0}{g}\right)^{1/k} = e^{v_0/g} \text{ or use L'Hôpital's Rule.}$$

9.3 Graphical and Numerical Methods

"To imagine yourself subject to a differential equation, start somewhere. There you are tugged in some direction, so you move that way ... as you move, the tugging forces change, pulling you in a new direction; for your motion to solve the differential equation you must keep drifting with and responding to the ambient forces."

——From the introduction to *Differential Equations*, J. H. Hubbard and Beverly West, Springer-Verlag, New York, 1991

In the previous two sections, we focused on finding solutions to differential equations. However, most differential equations cannot be solved explicitly. Fortunately, there are techniques for analyzing the solutions that do not rely on explicit formulas. In this section, we discuss the method of slope fields, which provide us with a good visual understanding of first-order equations. We also discuss Euler's Method for finding numerical approximations to solutions.

We use t as the independent variable and write $\dot{y}$ for $\dfrac{dy}{dt}$. The notation $\dot{y}$, often used for time derivatives in physics and engineering, was introduced by Isaac Newton. A first-order differential equation can then be written in the form

$$\boxed{\dot{y} = F(t, y)} \qquad \boxed{1}$$

where $F(t, y)$ is a function of t and y. For example, $\dfrac{dy}{dt} = ty$ becomes $\dot{y} = ty$.

It is useful to think of Eq. (1) as a set of instructions that "tells a solution" which direction to go in. Thus, a solution passing through a point (t, y) is "instructed" to continue in the direction of slope $F(t, y)$. To visualize this set of instructions, we draw a **slope field**, which is an array of small segments of slope $F(t, y)$ at points (t, y) lying on a rectangular grid in the plane.

To illustrate, let's return to the differential equation:

$$\dot{y} = -ty$$

In this case, $F(t, y) = -ty$. According to Example 2 of Section 9.1, the general solution is $y = Ce^{-t^2/2}$. Figure 1(A) shows segments of slope $-ty$ at points (t, y) along the graph of a particular solution $y(t)$. This particular solution passes through $(-1, 3)$ and, according to the differential equation, $\dot{y}(-1) = -ty = -(-1)3 = 3$. Thus, the segment located at the point $(-1, 3)$ has slope 3. The graph of the solution is tangent to each segment [Figure 1(B)].

FIGURE 1 The solution of $\dot{y} = -ty$ satisfying $y(-1) = 3$.

(A) Slope segments

(B) Solution is tangent to each slope segment

To sketch the slope field for $\dot{y} = -ty$, we draw small segments of slope $-ty$ at an array of points in the plane, as in Figure 2(A). *The slope field allows us to visualize all*

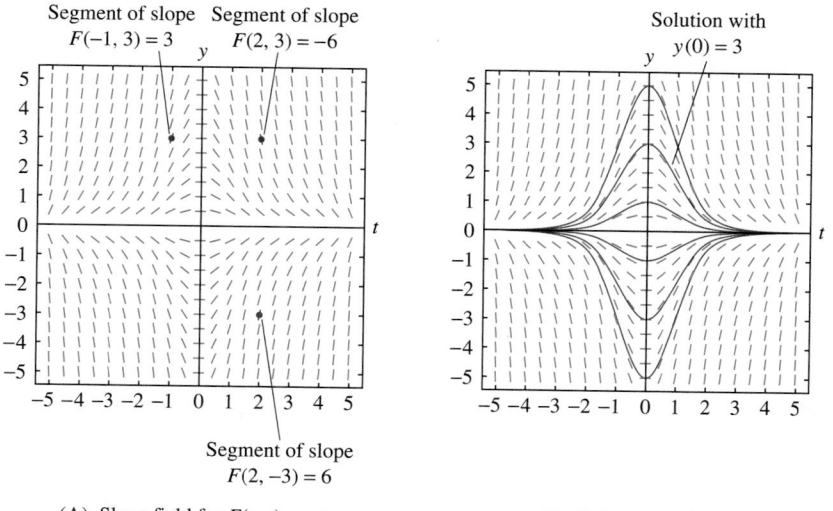

FIGURE 2 Slope field for $F(t, y) = -ty$.

(A) Slope field for $F(t, y) = -ty$

(B) Solutions of $\dot{y} = -ty$

of the solutions at a glance. Starting at any point, we may sketch a solution by drawing a curve that runs tangent to the slope segments at each point [Figure 2(B)]. The graph of a solution is also called an **integral curve**.

■ **EXAMPLE 1** Using Isoclines Draw the slope field for $\dot{y} = y - t$ and sketch the integral curves satisfying $y(0) = 1$ and $y(1) = -2$.

Solution A good way to sketch the slope field of $\dot{y} = F(t, y)$ is to choose several values c and identify the points (t, y) where the slope has the given value c, that is, where $F(t, y) = c$. The points where $F(t, y) = c$ form a curve called the **isocline** of slope c (Figure 3).

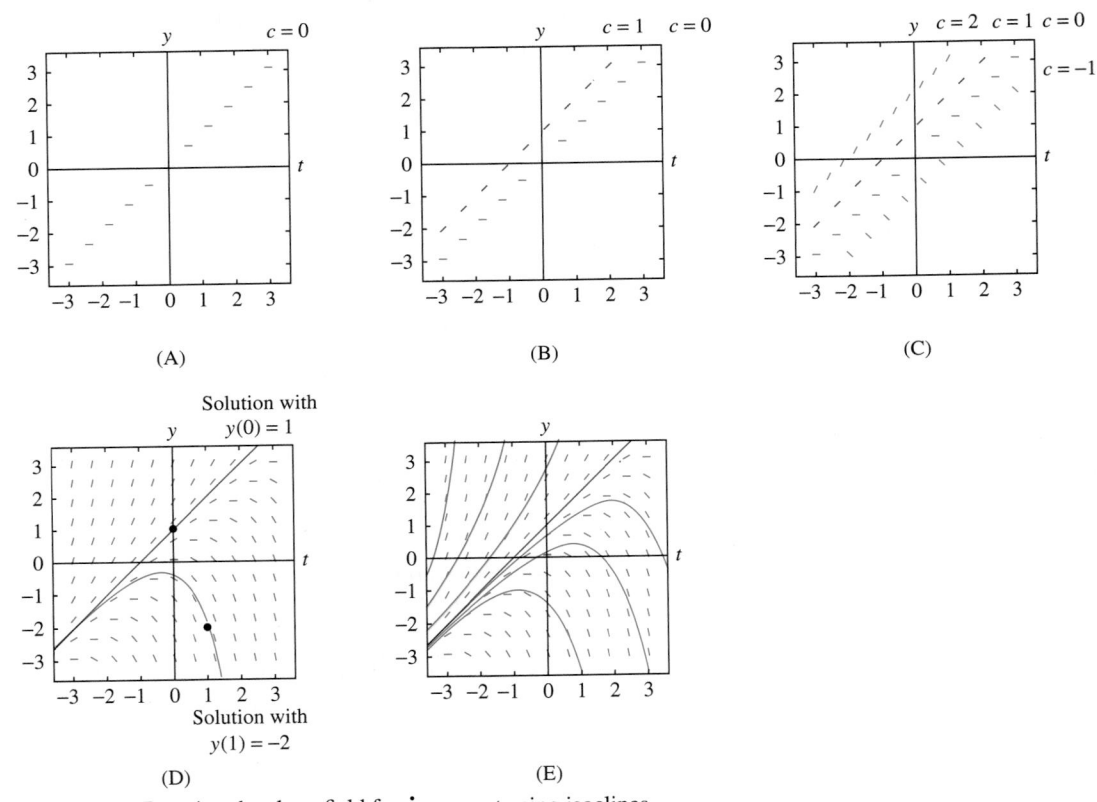

FIGURE 3 Drawing the slope field for $\dot{y} = y - t$ using isoclines.

In our case, $F(t, y) = y - t$, so the isocline of slope c has equation $y - t = c$ or $y = t + c$, which is a line. Consider the following values:

- $c = 0$: This isocline is $y - t = 0$ or $y = t$. We draw segments of slope $c = 0$ at points along the line $y = t$, as in Figure 3(A).
- $c = 1$: This isocline is $y - t = 1$ or $y = t + 1$. We draw segments of slope 1 at points along $y = t + 1$, as in Figure 3(B).
- $c = 2$: This isocline is $y - t = 2$ or $y = t + 2$. We draw segments of slope 2 at points along $y = t + 2$, as in Figure 3(C).
- $c = -1$: This isocline is $y - t = -1$ or $y = t - 1$ [Figure 3(C)].

A more detailed slope field is shown in Figure 3(D). To sketch the solution satisfying $y(0) = 1$, begin at the point $(t_0, y_0) = (0, 1)$ and draw the integral curve that follows the directions indicated by the slope field. Similarly, the graph of the solution satisfying $y(1) = -2$ is the integral curve obtained by starting at $(t_0, y_0) = (1, -2)$ and moving along the slope field. ■

■ **EXAMPLE 2** Newton's Law of Cooling Revisited   The temperature $y(t)$ (°F) of an object placed in a refrigerator satisfies $\dot{y} = -3(y - 40)$. Draw the slope field and describe the behavior of the solutions.

Solution The function $F(t, y) = -3(y - 40)$ depends only on y, so slopes of the segments in the slope field do not vary in the t-direction. The slope $F(t, y)$ is positive for $y < 40$ and negative for $y > 40$. Furthermore, the slope at height y is equal to $-3(y - 40) = -3y + 120$, so the segments grow steeper as y tends to ∞ or $-\infty$ (Figure 4).

The slope field shows that if the initial temperature satisfies $y_0 > 40$, then $y(t)$ decreases to 40 as $t \to \infty$. The object cools down to 40° when placed in the refrigerator. If $y_0 < 40$, then $y(t)$ increases to 40 as $t \to \infty$. In this case, the object warms up when placed in the refrigerator. If $y_0 = 40$, then y remains 40 for all time t. ■

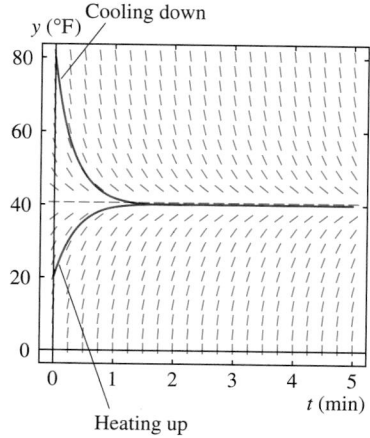

FIGURE 4 Slope field for $\dot{y} = -3(y - 40)$.

curve (solution) passes through the point (t_0, y_0). Thus, when uniqueness holds, distinct integral curves never cross as in Figure 3(E). Figure 5 shows the slope field of $\dot{y} = -\sqrt{|y|}$, where uniqueness fails. We can show that when a solution touches the t-axis, it can either remain on the t-axis or continue to the region below the t-axis. However, the slope field does not show this clearly, and we see again that it is necessary to analyze solutions rather than rely on visual impressions alone.

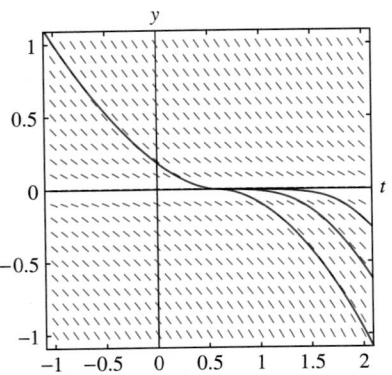

FIGURE 5 Overlapping integral curves for $\dot{y} = -\sqrt{|y|}$ (uniqueness fails for this differential equation).

Euler's Method

Euler's Method is the simplest method for solving initial value problems numerically, but it is not very efficient. More sophisticated versions are implemented on computer systems to plot and analyze solutions to the complex systems of differential equations arising in areas such as weather prediction, aerodynamic modeling, and economic forecasting.

Euler's Method produces numerical approximations to the solution of a first-order initial value problem:

$$\dot{y} = F(t, y), \qquad y(t_0) = y_0 \qquad \boxed{2}$$

We begin by choosing a small number h, called the **time step**, and consider the sequence of times spaced at intervals of size h:

$$t_0, \qquad t_1 = t_0 + h, \qquad t_2 = t_0 + 2h, \qquad t_3 = t_0 + 3h, \qquad \cdots$$

In general, $t_k = t_0 + kh$. Euler's Method consists of computing a sequence of values $y_1, y_2, y_3, \ldots, y_n$ successively using the formula

$$\boxed{y_k = y_{k-1} + hF(t_{k-1}, y_{k-1})} \qquad \boxed{3}$$

Starting with the initial value $y_0 = y(t_0)$, we compute $y_1 = y_0 + hF(t_0, y_0)$, etc. The value y_k is the Euler approximation to $y(t_k)$. We connect the points $P_k = (t_k, y_k)$ by segments to obtain an approximation to the graph of $y(t)$ (Figure 6).

GRAPHICAL INSIGHT The values y_k are defined so that the segment joining P_{k-1} to P_k has slope

$$\frac{y_k - y_{k-1}}{t_k - t_{k-1}} = \frac{(y_{k-1} + hF(t_{k-1}, y_{k-1})) - y_{k-1}}{h} = F(t_{k-1}, y_{k-1})$$

Thus, in Euler's method we move from P_{k-1} to P_k by traveling in the direction specified by the slope field at P_{k-1} for a time interval of length h (Figure 6).

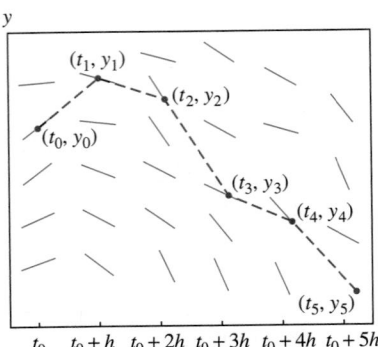

FIGURE 6 In Euler's Method, we move from one point to the next by traveling along the line indicated by the slope field.

■ **EXAMPLE 3** Use Euler's Method with time step $h = 0.2$ and $n = 4$ steps to approximate the solution of $\dot{y} = y - t^2$, $y(0) = 3$.

Solution Our initial value at $t_0 = 0$ is $y_0 = 3$. Since $h = 0.2$, the time values are $t_1 = 0.2$, $t_2 = 0.4$, $t_3 = 0.6$, and $t_4 = 0.8$. We use Eq. (3) with $F(t, y) = y - t^2$ to calculate

$$y_1 = y_0 + hF(t_0, y_0) = 3 + 0.2(3 - (0)^2) = 3.6$$

$$y_2 = y_1 + hF(t_1, y_1) = 3.6 + 0.2(3.6 - (0.2)^2) \approx 4.3$$

$$y_3 = y_2 + hF(t_2, y_2) = 4.3 + 0.2(4.3 - (0.4)^2) \approx 5.14$$

$$y_4 = y_3 + hF(t_3, y_3) = 5.14 + 0.2(5.14 - (0.6)^2) \approx 6.1$$

Figure 7(A) shows the exact solution $y(t) = 2 + 2t + t^2 + e^t$ together with a plot of the points (t_k, y_k) for $k = 0, 1, 2, 3, 4$ connected by line segments.   ■

FIGURE 7 Euler's Method applied to $\dot{y} = y - t^2$, $y(0) = 3$.

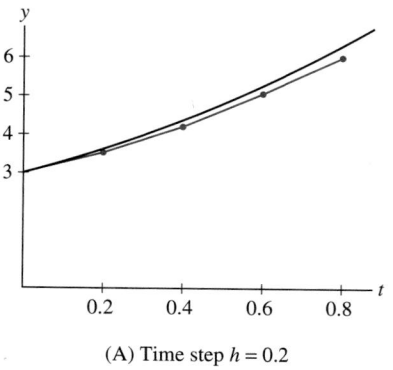

(A) Time step $h = 0.2$

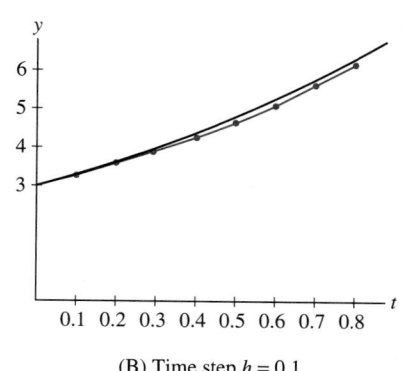

(B) Time step $h = 0.1$

CONCEPTUAL INSIGHT   Figure 7(B) shows that the timestep $h = 0.1$ gives a better approximation than $h = 0.2$. In general, the smaller the time step, the better the approximation. In fact, if we start at a point $(a, y(a))$ and use Euler's Method to approximate $(b, y(b))$ using N steps (i.e., $h = \dfrac{b - a}{N}$), then the error is roughly proportional to $1/N$ [provided that $F(t, y)$ is a well-behaved function]. This is similar to the error size in the Nth left- and right-endpoint approximations to an integral. What this means, however, is that Euler's Method is quite inefficient; to cut the error in half, it is necessary to double the number of steps and to achieve n-digit accuracy requires roughly 10^n steps. Fortunately, there are several methods that improve on Euler's Method in much the same way that the Midpoint Rule and Simpson's Rule improve on the endpoint approximations (see Exercises 21–26).

■ **EXAMPLE 4** *CAS*   Let $y(t)$ be the solution of $\dot{y} = \sin t \cos y$, $y(0) = 0$.

(a) Use Euler's Method with time step $h = 0.1$ to approximate $y(0.5)$.

(b) Use a computer algebra system to implement Euler's Method with time steps $h = 0.01, 0.001$, and 0.0001 to approximate $y(0.5)$.

Solution

(a) When $h = 0.1$, y_k is an approximation to $y(0 + k(0.1)) = y(0.1k)$. So y_5 is an approximation to $y(0.5)$. It is convenient to organize calculations in the following table. Note that the value y_{k+1} computed in the last column of each line is used in the next line to continue the process.

Euler's Method:

$$y_k = y_{k-1} + hF(t_{k-1}, y_{k-1})$$

t_k	y_k	$F(t_k, y_k) = \sin t_k \cos y_k$	$y_{k+1} = y_k + hF(t_k, y_k)$
$t_0 = 0$	$y_0 = 0$	$(\sin 0)\cos 0 = 0$	$y_1 = 0 + 0.1(0) = 0$
$t_1 = 0.1$	$y_1 = 0$	$(\sin 0.1)\cos 0 \approx 0.1$	$y_2 \approx 0 + 0.1(0.1) = 0.01$
$t_2 = 0.2$	$y_2 \approx 0.01$	$(\sin 0.2)\cos(0.01) \approx 0.2$	$y_3 \approx 0.01 + 0.1(0.2) = 0.03$
$t_3 = 0.3$	$y_3 \approx 0.03$	$(\sin 0.3)\cos(0.03) \approx 0.3$	$y_4 \approx 0.03 + 0.1(0.3) = 0.06$
$t_4 = 0.4$	$y_4 \approx 0.06$	$(\sin 0.4)\cos(0.06) \approx 0.4$	$y_5 \approx 0.06 + 0.1(0.4) = 0.10$

Thus, Euler's Method yields the approximation $y(0.5) \approx y_5 \approx 0.1$.

(b) When the number of steps is large, the calculations are too lengthy to do by hand, but they are easily carried out using a CAS. Note that for $h = 0.01$, the kth value y_k is an approximation to $y(0 + k(0.01)) = y(0.01k)$, and y_{50} gives an approximation to $y(0.5)$. Similarly, when $h = 0.001$, y_{500} is an approximation to $y(0.5)$, and when $h = 0.0001$, $y_{5,000}$ is an approximation to $y(0.5)$. Here are the results obtained using a CAS:

A typical CAS command to implement Euler's Method with time step $h = 0.01$ reads:

```
>> For[n = 0; y = 0, n < 50, n++,
>> y = y + (.01) * (Sin[.01 * n] * Cos[y])]
>> y
>> 0.119746
```

The command For[...] updates the variable y successively through the values $y_1, y_2, \ldots, y_{50}$ according to Euler's Method.

Time step $h = 0.01$	$y_{50} \approx 0.120$
Time step $h = 0.001$	$y_{500} \approx 0.1219$
Time step $h = 0.0001$	$y_{5,000} \approx 0.1221$

The values appear to converge and we may assume that $y(0.5) \approx 0.12$. However, we see here that Euler's Method converges quite slowly. ∎

9.3 SUMMARY

- The *slope field* for a first-order differential equation $\dot y = F(t, y)$ is obtained by drawing small segments of slope $F(t, y)$ at points (t, y) lying on a rectangular grid in the plane.
- The graph of a solution (also called an *integral curve*) satisfying $y(t_0) = y_0$ is a curve through (t_0, y_0) that runs tangent to the segments of the slope field at each point.
- *Euler's Method* is used to approximate a solution to $\dot y = F(t, y)$ with initial condition $y(t_0) = y_0$. Fix a time step h and set $t_k = t_0 + kh$. Define $y_1, y_2, \ldots$ successively by the formula

$$y_k = y_{k-1} + hF(t_{k-1}, y_{k-1})$$ 4

The values $y_0, y_1, y_2, \ldots$ are approximations to the values $y(t_0), y(t_1), y(t_2), \ldots$.

9.3 EXERCISES

Preliminary Questions

1. What is the slope of the segment in the slope field for $\dot y = ty + 1$ at the point $(2, 3)$?

2. What is the equation of the isocline of slope $c = 1$ for $\dot y = y^2 - t$?

3. True or false? In the slope field for $\dot{y} = \ln y$, the slopes at points on a vertical line $t = C$ are all equal.

4. What about the slope field for $\dot{y} = \ln t$? Are the slopes at points on a vertical line $t = C$ all equal?

5. Let $y(t)$ be the solution to $\dot{y} = F(t, y)$ with $y(1) = 3$. How many iterations of Euler's Method are required to approximate $y(3)$ if the time step is $h = 0.1$?

Exercises

1. Figure 8 shows the slope field for $\dot{y} = \sin y \sin t$. Sketch the graphs of the solutions with initial conditions $y(0) = 1$ and $y(0) = -1$. Show that $y(t) = 0$ is a solution and add its graph to the plot.

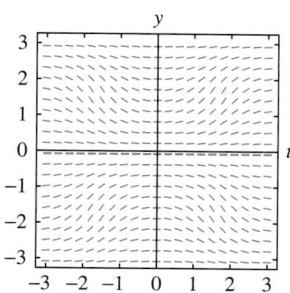

FIGURE 8 Slope field for $\dot{y} = \sin y \sin t$.

2. Figure 9 shows the slope field of $\dot{y} = y^2 - t^2$. Sketch the integral curve passing through the point $(0, -1)$, the curve through $(0, 0)$, and the curve through $(0, 2)$. Is $y(t) = 0$ a solution?

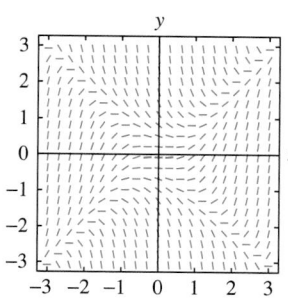

FIGURE 9 Slope field for $\dot{y} = y^2 - t^2$.

3. Show that $f(t) = \frac{1}{2}(t - \frac{1}{2})$ is a solution to $\dot{y} = t - 2y$. Sketch the four solutions with $y(0) = \pm 0.5, \pm 1$ on the slope field in Fig-

ure 10. The slope field suggests that every solution approaches $f(t)$ as $t \to \infty$. Confirm this by showing that $y = f(t) + Ce^{-2t}$ is the general solution.

4. Consider the differential equation $\dot{y} = t - y$.

(a) Sketch the slope field of the differential equation $\dot{y} = t - y$ in the range $-1 \le t \le 3$, $-1 \le y \le 3$. As an aid, observe that the isocline of slope c is the line $t - y = c$, so the segments have slope c at points on the line $y = t - c$.

(b) Show that $y = t - 1 + Ce^{-t}$ is a solution for all C. Since $\lim_{t \to \infty} e^{-t} = 0$, these solutions approach the particular solution $y = t - 1$ as $t \to \infty$. Explain how this behavior is reflected in your slope field.

5. Show that the isoclines of $\dot{y} = 1/y$ are horizontal lines. Sketch the slope field for $-2 \le t, y \le 2$ and plot the solutions with initial conditions $y(0) = 0$ and $y(0) = 1$.

6. Show that the isoclines of $\dot{y} = t$ are vertical lines. Sketch the slope field for $-2 \le t, y \le 2$ and plot the integral curves passing through $(0, -1)$ and $(0, 1)$.

7. Sketch the slope field of $\dot{y} = ty$ for $-2 \le t, y \le 2$. Based on the sketch, determine $\lim_{t \to \infty} y(t)$, where $y(t)$ is a solution with $y(0) > 0$. What is $\lim_{t \to \infty} y(t)$ if $y(0) < 0$?

8. Match the differential equation with its slope field in Figures 11(A)–(F).

(i) $\dot{y} = -1$

(ii) $\dot{y} = \dfrac{y}{t}$

(iii) $\dot{y} = t^2 y$

(iv) $\dot{y} = ty^2$

(v) $\dot{y} = t^2 + y^2$

(vi) $\dot{y} = t$

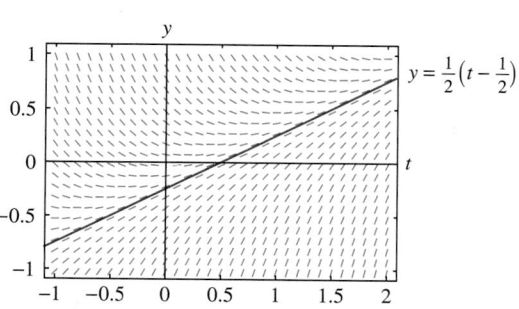

FIGURE 10 Slope field for $\dot{y} = t - 2y$.

FIGURE 11(A)

FIGURE 11(B)

FIGURE 11(C)

FIGURE 11(D)

FIGURE 11(E)

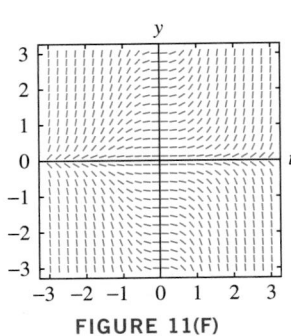

FIGURE 11(F)

9. One of the slope fields in Figures 12(A) and (B) is the slope field for $\dot{y} = t^2$. The other is for $\dot{y} = y^2$. Identify which is which. In each case, sketch the solutions with initial conditions $y(0) = 1$, $y(0) = 0$, and $y(0) = -1$.

FIGURE 12(A)

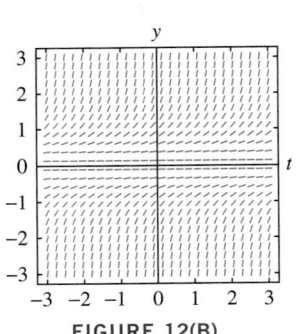

FIGURE 12(B)

10. (a) Sketch the slope field of $\dot{y} = t/y$ in the region $-2 \le t \le 2$, $-2 \le y \le 2$.

(b) Check that $y = \pm\sqrt{t^2 + C}$ is the general solution.

(c) Sketch the solutions on the slope field with initial conditions $y(0) = 1$ and $y(0) = -1$.

11. Sketch the slope field of $\dot{y} = t^2 - y$ in the region $-3 \le t \le 3$, $-3 \le y \le 3$ and sketch the solutions satisfying $y(1) = 0$, $y(1) = 1$, and $y(1) = -1$.

12. Let $F(t, y) = t^2 - y$ and let $y(t)$ be the solution of $\dot{y} = F(t, y)$ satisfying $y(2) = 3$. Let $h = 0.1$ be the time step in Euler's Method and set $y_0 = y(2) = 3$.

(a) Calculate $y_1 = y_0 + hF(2, 3)$.

(b) Calculate $y_2 = y_1 + hF(2.1, y_1)$.

(c) Calculate $y_3 = y_2 + hF(2.2, y_2)$ and continue computing y_4, y_5, and y_6.

(d) Find approximations to $y(2.2)$ and $y(2.5)$.

13. Let $y(t)$ be the solution to $\dot{y} = te^{-y}$ satisfying $y(0) = 0$.

(a) Use Euler's Method with time step $h = 0.1$ to approximate $y(0.1), y(0.2), \ldots , y(1)$.

(b) Use separation of variables to find $y(t)$ exactly.

(c) Compute the error in the approximations to $y(0.1)$, $y(0.5)$, and $y(1)$.

In Exercises 14–19, use Euler's Method with $h = 0.1$ to approximate the given value of $y(t)$.

14. $y(2.2)$; $\dot{y} = ty - 1$, $y(2) = 0$

15. $y(1)$; $\dot{y} = t - y^2$, $y(1) = -2$

16. $y(0.7)$; $\dot{y} = -yt$, $y(0) = 1$

17. $y(1.5)$; $\dot{y} = t \sin y$, $y(1) = 2$

18. $y(2.6)$; $\dot{y} = t/y$, $y(0) = 2$

19. $y(0.5)$; $\dot{y} = t - y$, $y(0) = 1$

Further Insights and Challenges

20. If $f(t)$ is continuous on $[a, b]$, then the solution to $\dot{y} = f(t)$ with initial condition $y(a) = 0$ is $y(t) = \int_a^t f(u)\, du$. Show that Euler's Method with time step $h = \dfrac{b - a}{N}$ for N steps yields the Nth left-endpoint approximation to $y(b) = \int_a^b f(u)\, du$.

In Exercises 21–22, use a modification of Euler's Method, Euler's Midpoint Method, that gives a significant improvement in accuracy. With

time step h and initial value $y_0 = y(t_0)$, the values y_k are defined successively by the equation

$$y_k = y_{k-1} + hm_{k-1}$$

where $m_{k-1} = F\left(t_{k-1} + \dfrac{h}{2}, y_{k-1} + \dfrac{h}{2}F(t_{k-1}, y_{k-1})\right)$.

21. Apply both Euler's Method and the Euler Midpoint Method with $h = 0.1$ to estimate $y(1.5)$, where $y(t)$ satisfies $\dot{y} = y$ with $y(0) = 1$. Find $y(t)$ exactly and compute the errors in these two approximations.

22. If $f(t)$ is continuous on $[a, b]$, then the solution to $\dot{y} = f(t)$ with initial condition $y(a) = 0$ is $y(t) = \int_a^t f(u) \, du$. Show that the Euler Midpoint Method with time step $h = \dfrac{b-a}{N}$ for N steps yields the Nth midpoint approximation to $y(b) = \int_a^b f(u) \, du$.

In Exercises 23–26, use Euler's Midpoint Method with $h = 0.1$ to approximate the given value of $y(t)$.

23. $y(0.5)$; $\dot{y} = y^2$, $y(0) = 0$

24. $y(1)$; $\dot{y} = y^2$, $y(0) = 0$

25. $y(1)$; $\dot{y} = t + y$, $y(0) = 0$

26. $y(0.5)$; $\dot{y} = y^2 - t$, $y(0) = 1$

9.4 The Logistic Equation

The logistic equation was first introduced in 1838 by the Belgian mathematician Pierre-François Verhulst (1804–1849). Based on the population of Belgium for three years (1815, 1830, and 1845), which was then between 4 and 4.5 million, Verhulst predicted that the population would never exceed 9.4 million. This prediction has held up reasonably well. Belgium's current population is around 10.4 million.

The first and simplest model of population growth is $dy/dt = ky$, which predicts that populations grow exponentially. This may be true over short periods of time, but it is obvious that no population can increase without limit. Therefore, population biologists use a variety of other differential equations that take into account factors which limit growth, such as food scarcity, competition between species, and environmental limitations. One widely used model is based on the **logistic differential equation**:

$$\frac{dy}{dt} = ky\left(1 - \frac{y}{A}\right) \qquad \boxed{1}$$

Here $k > 0$ is the growth constant (in units of time^{-1}) and A is a nonzero constant called the **carrying capacity**. Figure 1 shows a typical S-shaped solution of Eq. (1).

As in the previous section, we also denote dy/dt by $\dot{y}$.

CONCEPTUAL INSIGHT   The logistic equation $\dot{y} = ky(1 - y/A)$ is obtained by inserting the factor $(1 - y/A)$ in the exponential differential equation $\dot{y} = ky$. As long as $y = y(t)$ is small relative to A, $(1 - y/A)$ is close to 1 and can be ignored, yielding $\dot{y} \approx ky$. Thus, $y(t)$ grows nearly exponentially when the population is small (Figure 1). As $y(t)$ approaches A, the factor $(1 - y/A)$ tends to zero. This causes the growth rate to decrease to zero and prevents $y(t)$ from exceeding the carrying capacity A.

The slope field in Figure 2 provides a more complete picture of the solutions. Let's write Eq. (1) as $\dot{y} = F(y)$ with $F(y) = ky(1 - y/A)$. Since $F(y)$ does not depend on

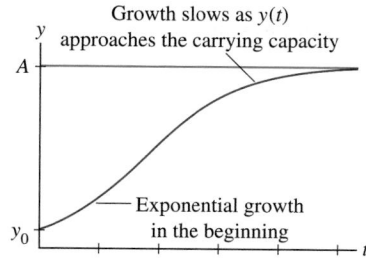

FIGURE 1 Solution of the logistic equation.

FIGURE 2 Slope field for $\dfrac{dy}{dt} = ky\left(1 - \dfrac{y}{A}\right)$.

Solutions of the logistic equation with $y_0 < 0$ are not relevant to populations because a population cannot be negative (see Exercise 18).

t, the slope field varies only in the y-direction. We see that the solutions come in three families plus two special cases, depending on the initial value $y_0 = y(0)$:

- If $y_0 > A$, then $y(t)$ is decreasing and approaches A as $t \to \infty$.
- If $0 < y_0 < A$, then $y(t)$ is increasing and approaches A as $t \to \infty$.
- If $y_0 < 0$, then $y(t)$ is decreasing and $\lim_{t \to t_b-} y(t) = -\infty$ for some time t_b.

Equation (1) also has two constant solutions, namely $y = 0$ and $y = A$ [because both sides of $\dot{y} = F(y)$ are zero for $y = 0$ and $y = A$]. These are called **equilibrium** or **steady-state** solutions. The equilibrium solution $y = A$ is called a **stable equilibrium** because every solution with initial condition $y_0 > 0$ approaches the equilibrium $y = A$ as $t \to \infty$. By contrast, $y = 0$ is an **unstable equilibrium**. Nonequilibrium solutions that start near $y = 0$ move away from $y = 0$, either increasing to A or decreasing to $-\infty$.

Having described the solutions of the logistic equation qualitatively, we now find the nonequilibrium solutions explicitly using separation of variables (assume $y \neq 0, A$):

$$\frac{dy}{dt} = ky\left(1 - \frac{y}{A}\right)$$

$$\frac{dy}{y(1 - y/A)} = k\,dt$$

In Eq. (2), we use the the partial fraction decomposition

$$\frac{1}{y(1-y/A)} = \frac{1}{y} - \frac{1}{y-A}$$

$$\int\left(\frac{1}{y} - \frac{1}{y-A}\right)dy = \int k\,dt \qquad \boxed{2}$$

$$\ln|y| - \ln|y-A| = kt + C$$

$$\left|\frac{y}{y-A}\right| = e^{kt+C} \quad \Rightarrow \quad \frac{y}{y-A} = \pm e^C e^{kt}$$

Since $\pm e^C$ takes on arbitrary nonzero values, we replace $\pm e^C$ with C (nonzero):

$$\frac{y}{y-A} = Ce^{kt} \qquad \boxed{3}$$

For $t = 0$, this gives a useful relation between C and the initial value $y_0 = y(0)$:

$$\frac{y_0}{y_0 - A} = C \qquad \boxed{4}$$

To solve for y, multiply each side of Eq. (3) by $(y - A)$:

$$y = (y - A)Ce^{kt}$$

$$y(1 - Ce^{kt}) = -ACe^{kt}$$

$$y = \frac{ACe^{kt}}{Ce^{kt} - 1}$$

As $C \neq 0$, we may divide by Ce^{kt} to obtain the general nonequilibrium solution:

$$\boxed{\frac{dy}{dt} = ky\left(1 - \frac{y}{A}\right), \qquad y = \frac{A}{1 - e^{-kt}/C}} \qquad \boxed{5}$$

■ EXAMPLE 1 Solve $\dot{y} = 0.3y(4 - y)$ with initial condition $y(0) = 1$.

Solution To apply (5), we must rewrite the equation in the form

$$\dot{y} = 1.2y \left(1 - \frac{y}{4}\right)$$

Thus, $k = 1.2$ and $A = 4$, and the general solution is

$$y = \frac{4}{1 - e^{-1.2t}/C}$$

There are two ways to find C. One way is to solve $y(0) = 1$ for C directly. An easier way is to use Eq. (4):

$$C = \frac{y_0}{y_0 - A} = \frac{1}{1 - 4} = -\frac{1}{3}$$

We find that the particular solution is $y = \dfrac{4}{1 + 3e^{-1.2t}}$ (Figure 3). ■

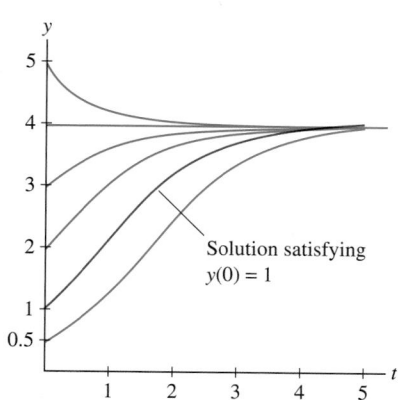

FIGURE 3 Several solutions of $\dot{y} = 0.3y(4 - y)$.

The logistic equation may be too simple to describe a real deer population accurately, but it serves as a starting point for more sophisticated models used by ecologists, population biologists, and forestry professionals.

■ **EXAMPLE 2** A 10,000-acre forest has a carrying capacity of 1,000 deer. Assume that the deer population grows logistically with growth constant $k = 0.4 \ \text{yr}^{-1}$.

(a) Find the deer population $P(t)$ if the initial population is $P_0 = 100$.

(b) How long does it take for the deer population to reach 500?

Solution

(a) Since $k = 0.4$ and $A = 1,000$, $P(t)$ satisfies the differential equation

$$\frac{dP}{dt} = 0.4P \left(1 - \frac{P}{1,000}\right)$$

The general solution is given by Eq. (5):

$$P(t) = \frac{1,000}{1 - e^{-0.4t}/C} \qquad \boxed{6}$$

According to Eq. (4) (see Figure 4),

$$C = \frac{P_0}{P_0 - A} = \frac{100}{100 - 1,000} = -\frac{1}{9} \quad \Rightarrow \quad P(t) = \frac{1,000}{1 + 9e^{-0.4t}}$$

Logistic Growth of Deer Population			
Year	Deer Population	Year	Deer Population
0	100	8	732
1	142	9	803
2	198	10	858
3	269	11	900
4	355	12	931
5	451	13	953
6	551	14	968
7	646	15	978

(b) To find the time t when $P(t) = 500$, we could solve the equation

$$P(t) = \frac{1{,}000}{1 + 9e^{-0.4t}} = 500$$

But it is easier to use Eq. (3):

$$\frac{P}{P - A} = Ce^{kt}$$

$$\frac{P}{P - 1{,}000} = -\frac{1}{9}e^{0.4t}$$

Set $P = 500$ and solve for t:

$$-\frac{1}{9}e^{0.4t} = \frac{500}{500 - 1{,}000} = -1 \quad \Rightarrow \quad e^{0.4t} = 9 \quad \Rightarrow \quad 0.4t = \ln 9$$

This gives $t = (\ln 9)/0.4 \approx 5.5$ years. ∎

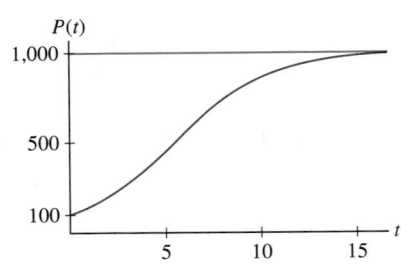

FIGURE 4 Deer population as a function of t (in years).

9.4 SUMMARY

- The *logistic equation* and its general nonequilibrium solution are

$$\frac{dy}{dt} = ky\left(1 - \frac{y}{A}\right), \qquad y = \frac{A}{1 - e^{-kt}/C}$$

- The logistic equation has two equilibrium (constant) solutions: $y = 0$ is an unstable equilibrium and $y = A$ is a stable equilibrium.
- Assume that $k > 0$. If the initial value $y_0 = y(0)$ satisfies $y_0 > 0$, then $y(t)$ approaches the stable equilibrium $y = A$, that is,

$$\lim_{t \to \infty} y(t) = A$$

9.4 EXERCISES

Preliminary Questions

1. Which of the following is a logistic differential equation?

(a) $\dot{y} = 2y(1 - y^2)$

(b) $\dot{y} = 2y\left(1 - \frac{y}{3}\right)$

(c) $\dot{y} = 2y\left(1 - \frac{x}{4}\right)$

(d) $\dot{y} = 2y(1 - 3y)$

2. True or false? The logistic equation is linear.

3. True or false? The logistic equation is separable.

4. Let $y(t)$ be a solution to $\dot{y} = 4y(3 - y)$. What is $\lim_{t \to \infty} y(t)$ in the following three cases:

(a) $y(0) = 3$ **(b)** $y(0) = 4$ **(c)** $y(0) = -2$

Exercises

1. Find the general solution of the logistic equation

$$\dot{y} = 3y(1 - y/5)$$

Then find the particular solution satisfying $y(0) = 2$.

2. Find the solution of $\dot{y} = 2y(3 - y)$, $y(0) = 10$.

3. Let $y(t)$ be a solution of $\dot{y} = 0.5y(1 - 0.5y)$ such that $y(0) = 4$. Determine $\lim_{t \to \infty} y(t)$ without finding $y(t)$ explicitly.

4. Let $y(t)$ be a solution of $\dot{y} = 5y(1 - y/A)$ satisfying $y(0) = 10$. Find $\lim_{t \to \infty} y(t)$, assuming that

(a) $A = 15$ **(b)** $A = 5$ **(c)** $A = 10$

5. A population of squirrels lives in a forest with a carrying capacity of 2,000. Assume logistic growth with growth constant $k = 0.6$ yr^{-1}.

(a) Find a formula for the squirrel population $P(t)$, assuming an initial population of 500 squirrels.

(b) How long will it take for the squirrel population to double?

6. The population $P(t)$ of mosquito larvae growing in a tree hole increases according to the logistic equation with growth constant $k = 0.3$ $days^{-1}$ and carrying capacity $A = 500$.

(a) Find a formula for the larvae population $P(t)$, assuming an initial population of $P_0 = 50$ larvae.

(b) After how many days will the larvae population reach 200?

7. Sunset Lake is stocked with 2,000 rainbow trout and after 1 year the population has grown to 4,500. Assuming logistic growth with a carrying capacity of 20,000, find the growth constant k (specify the units) and determine when the population will increase to 10,000.

8. Spread of a Rumor A rumor spreads through a small town. Let $y(t)$ be the fraction of the population that has heard the rumor at time t and assume that the rate at which the rumor spreads is proportional to the product of the fraction y of the population that has heard the rumor and the fraction $1 - y$ that has not yet heard the rumor.

(a) Write down the differential equation satisfied by y in terms of a proportionality factor k.

(b) Find k (in units of $days^{-1}$), assuming that 10% of the population knows the rumor at $t = 0$ and 40% knows it at $t = 2$ days.

(c) Using the assumptions of part (b), determine when 75% of the population will know the rumor.

9. A rumor spreads through a school with 1,000 students. At 8 AM, 80 students have heard the rumor, and by noon, half the school has heard it. Using the logistic model of Exercise 8, determine when 90% of the students will have heard the rumor.

10. GU A simpler model for the spread of a rumor assumes that the rate at which the rumor spreads is proportional (with factor k) to the fraction of the population that has not yet heard the rumor.

(a) Compute the solutions to this model and the model of Exercise 8 with the values $k = 0.9$ and $y_0 = 0.1$.

(b) Graph the two solutions on the same axis.

(c) Which model seems more realistic? Why?

11. Let $k = 1$ and $A = 1$ in the logistic equation.

(a) Find the solutions satisfying $y_1(0) = 10$ and $y_2(0) = -1$.

(b) Find the time t when $y_1(t) = 5$.

(c) When does $y_2(t)$ become infinite?

12. **Reverse Logistic Equation** Consider the logistic equation (with $k, B > 0$)

$$\frac{dP}{dt} = -kP\left(1 - \frac{P}{B}\right) \qquad \boxed{7}$$

(a) Sketch the slope field of this equation.

(b) The general solution is $P(t) = B/(1 - e^{kt}/C)$, where C is a nonzero constant. Show that $P(0) > B$ if $C > 1$ and $0 < P(0) < B$ if $C < 0$.

(c) Show that Eq. (7) models an "extinction–explosion" population. That is, $P(t)$ tends to zero if the initial population satisfies $0 < P(0) < B$ and it tends to ∞ after a finite amount of time if $P(0) > B$.

(d) Show that $P = 0$ is a stable equilibrium and $P = B$ an unstable equilibrium.

13. A tissue culture grows until it has a maximum area of M cm^2. The area $A(t)$ of the culture at time t may be modeled by the differential equation

$$\dot{A} = k\sqrt{A}\left(1 - \frac{A}{M}\right) \qquad \boxed{8}$$

where k is a growth constant.

(a) By setting $A = u^2$, show that the equation can be rewritten

$$\dot{u} = \frac{1}{2}k\left(1 - \frac{u^2}{M}\right)$$

Then find the general solution using separation of variables.

(b) Show that the general solution to Eq. (8) is

$$A(t) = M\left(\frac{Ce^{(k/\sqrt{M})t} - 1}{Ce^{(k/\sqrt{M})t} + 1}\right)^2 \qquad \boxed{9}$$

14. GU Use the model of Exercise 13 to determine the area $A(t)$ (t in hours) of a tissue culture with initial size $A(0) = 0.2$ cm^2, assuming that the maximum area is $M = 5$ cm^2 and the growth constant is $k = 0.06$. Graph the solution using a graphing utility.

15. Show that if a tissue culture grows according to Eq. (8), then the growth rate reaches a maximum when $A = M/3$.

16. In 1751, Benjamin Franklin predicted that the U.S. population $P(t)$ would increase with growth constant $k = 0.028$ yr^{-1}. According to the census, the U.S. population was 5 million in 1800 and 76 million in 1900. Assuming logistic growth with $k = 0.028$, what is the predicted carrying capacity for the U.S. population? *Hint:* Use Eqs. (3) and (4) to show that

$$\frac{P(t)}{P(t) - A} = \frac{P_0}{P_0 - A}e^{kt}$$

Further Insights and Challenges

17. Let $y(t)$ be a solution of the logistic equation

$$\frac{dy}{dt} = ky\left(1 - \frac{y}{A}\right) \qquad \boxed{10}$$

(a) Differentiate Eq. (10) with respect to t and use the Chain Rule to show that

$$\frac{d^2y}{dt^2} = k^2y\left(1 - \frac{y}{A}\right)\left(1 - \frac{2y}{A}\right)$$

(b) Show that $y(t)$ is concave up if $0 < y < A/2$ and concave down if $A/2 < y < A$.

(c) Show that if $0 < y(0) < A/2$, then $y(t)$ has a point of inflection at $y = A/2$ (Figure 5).

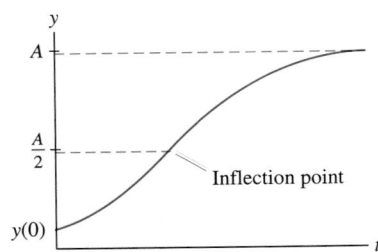

FIGURE 5 Inflection point in a logistic curve occurs at $y = A/2$.

(d) Assume that $0 < y(0) < A/2$. Find the time t when $y(t)$ reaches the inflection point.

18. Let

$$y = \frac{A}{1 - e^{-kt}/C}$$

be the general nonequilibrium solution of $\dot{y} = ky(1 - y/A)$ with $k > 0$. If $y(t)$ has a vertical asymptote at $t = t_b$, that is, if $\lim_{t \to t_b -} y(t) = \pm\infty$, we say that the solution "blows up" at $t = t_b$.

(a) Show that if $0 < y(0) < A$, then y does not blow up at any time t_b.

(b) Show that if $y(0) > A$, then y blows up at a time t_b, which is negative (and hence does not correspond to a real time).

(c) Show that y blows up at some positive time t_b if and only if $y(0) < 0$ (and hence does not correspond to a real population).

9.5 First-Order Linear Equations

The separation of variables method applies to all separable first-order equations, whether linear or not. In this section, we introduce a method for solving all linear first-order equations whether separable or not.

A first-order linear equation has the form $a(x)y' + b(x)y = c(x)$, where $a(x)$ is not identically equal to zero. We divide by $a(x)$ and write the equation in the standard form

$$y' + A(x)y = B(x) \qquad \boxed{1}$$

The idea in solving Eq. (1) is to multiply through by a function $\alpha(x)$, called an **integrating factor**, which turns the left-hand side into the derivative of $\alpha(x)y$:

$$\alpha(x)\big(y' + A(x)y\big) = \big(\alpha(x)y\big)' \qquad \boxed{2}$$

Suppose we can find a function $\alpha(x)$ satisfying Eq. (2). Then Eq. (1) yields

$$\alpha(x)\big(y' + A(x)y\big) = \alpha(x)B(x)$$
$$\big(\alpha(x)y\big)' = \alpha(x)B(x)$$

We can solve this equation by integration:

$$\alpha(x)y = \int \alpha(x)B(x)\,dx + C \qquad \text{or} \qquad y = \frac{1}{\alpha(x)}\left(\int \alpha(x)B(x)\,dx + C\right)$$

To find $\alpha(x)$, expand Eq. (2), using the Product Rule on the right-hand side:

$$\alpha(x)y' + \alpha(x)A(x)y = \alpha(x)y' + \alpha'(x)y$$

Then subtract $\alpha(x)y'$ from both sides and divide by y to obtain $\alpha'(x) = \alpha(x)A(x)$, or

$$\frac{d\alpha}{dx} = \alpha(x)A(x) \qquad \boxed{3}$$

We solve this equation using separation of variables:

$$\frac{d\alpha}{\alpha} = A(x)\,dx \quad \Rightarrow \quad \int \frac{d\alpha}{\alpha} = \int A(x)\,dx$$

Therefore, $\ln|\alpha(x)| = \displaystyle\int A(x)\,dx$, and by exponentiation, $\alpha(x) = \pm e^{\int A(x)\,dx}$. Since we need just one solution of Eq. (3), we choose the positive solution.

> *In the formula for the integrating factor $\alpha(x)$, the integral $\displaystyle\int A(x)\,dx$ denotes any antiderivative of $A(x)$.*

THEOREM 1 The General Solution   The general solution of $y' + A(x)y = B(x)$ is

$$y = \alpha(x)^{-1}\left(\int \alpha(x)B(x)\,dx + C \right) \qquad \boxed{4}$$

where $\alpha(x)$ is an integrating factor:

$$\alpha(x) = e^{\int A(x)\,dx} \qquad \boxed{5}$$

■ **EXAMPLE 1** Solve $xy' - 3y = x^2$, $y(1) = 2$.

Solution First put the equation in the form $y' + A(x)y = B(x)$ by dividing by x:

$$y' - \frac{3}{x}y = x \qquad \boxed{6}$$

Step 1. **Find an integrating factor.**
In our case, $A(x) = -\dfrac{3}{x}$ and by Eq. (5)

$$\alpha(x) = e^{\int A(x)\,dx} = e^{\int(-3/x)\,dx} = e^{-3\ln x} = e^{\ln(x^{-3})} = x^{-3}$$

Step 2. **Find the general solution.**
We have found $\alpha(x)$, so we can use Eq. (4) to write down the general solution. However, to illustrate the ideas, let's go through the steps leading to Eq. (4). First, multiply Eq. (6) by $\alpha(x) = x^{-3}$ to rewrite the equation:

$$x^{-3}\left(y' - \frac{3}{x}y \right) = x^{-3}x \quad \Rightarrow \quad x^{-3}y' - 3x^{-4}y = x^{-2} \quad \Rightarrow \quad \frac{d}{dx}\left(x^{-3}y \right) = x^{-2}$$

Thus, $x^{-3}y$ is an antiderivative of x^{-2}, and we obtain the general solution (see Figure 1)

$$x^{-3}y = \int x^{-2}\,dx = -x^{-1} + C \qquad \boxed{7}$$

$$\boxed{y = -x^2 + Cx^3}$$

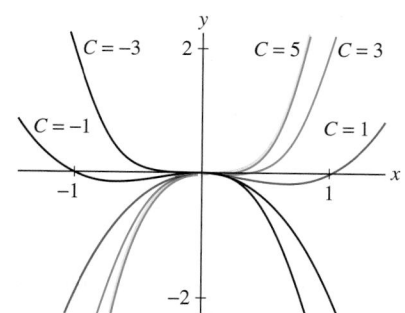

FIGURE 1 Family of solutions to $xy' - 3y = x^2$.

> **CAUTION** *The constant of integration C in Eq. (7) is essential, but note that in the general solution, C does not appear as an additive constant. The general solution is $y = -x^2 + Cx^3$, not $-x^2 + C$, a common mistake.*

Step 3. Solve the initial value problem.

The initial condition gives

$$y(1) = -1^2 + C \cdot 1^3 = 2 \qquad \text{or} \qquad C = 3$$

Therefore, the solution of the initial value problem is $y = -x^2 + 3x^3$.

Finally, let's check that $y = -x^2 + 3x^3$ satisfies our equation $xy' - 3y = x^2$:

$$xy' - 3y = x(-2x + 9x^2) - 3(-x^2 + 3x^3)$$

$$= (-2x^2 + 9x^3) + (3x^2 - 9x^3) = x^2 \qquad \blacksquare$$

■ **EXAMPLE 2** Solve the initial value problem: $y' + (1 - x^{-1})y = x^2$, $y(1) = 2$.

Solution This equation has the form $y' + A(x)y = B(x)$ with $A(x) = (1 - x^{-1})$. By Eq. (5), an integrating factor is

$$\alpha(x) = e^{\int (1 - x^{-1}) \, dx} = e^{x - \ln x} = e^x e^{\ln x^{-1}} = x^{-1} e^x$$

Using Eq. (4) with $B(x) = x^2$, we obtain the general solution:

$$y = \alpha(x)^{-1} \left(\int \alpha(x) B(x) \, dx + C \right) = xe^{-x} \left(\int (x^{-1} e^x) x^2 \, dx + C \right)$$

$$= xe^{-x} \left(\int xe^x \, dx + C \right)$$

Integration by Parts shows that $\int xe^x \, dx = (x - 1)e^x + C$, so we obtain

$$y = xe^{-x} \left((x - 1)e^x + C \right) = x(x - 1) + Cxe^{-x}$$

The initial condition $y(1) = 2$ gives

$$y(1) = 1(1 - 1) + Ce^{-1} = Ce^{-1} = 2 \quad \Rightarrow \quad C = 2e$$

The desired particular solution is

$$y = x(x - 1) + (2e)xe^{-x} = x(x - 1) + 2xe^{1-x} \qquad \blacksquare$$

Summary: The general solution of $y' + A(x)y = B(x)$ is

$$y = \alpha(x)^{-1} \left(\int \alpha(x)B(x) + C \right)$$

where

$$\alpha(x) = e^{\int A(x) \, dx}$$

CONCEPTUAL INSIGHT Our method expresses the general solution in terms of the integrals in Eqs. (4) and (5). In this sense, we have an explicit general solution of any first-order linear equation. However, it is not always possible to evaluate the integrals in elementary terms. For example, the general solution of $y' + xy = 1$ is

$$y = e^{-x^2/2} \left(\int e^{x^2/2} \, dx + C \right)$$

and $\int e^{x^2/2} \, dx$ cannot be evaluated in elementary terms. However, we may approximate the integral numerically and plot the solutions by computer (Figure 2).

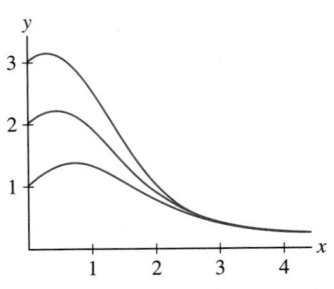

FIGURE 2 Solutions to $y' + xy = 1$ solved numerically and plotted by computer.

In the next example, we use a differential equation to model the "mixing problem," which has applications in biology, chemistry, and medicine.

R_{in} (gal/min)

Water level

R_{out} (gal/min)

FIGURE 3

■ **EXAMPLE 3**   **A Mixing Problem**   A tank contains 100 gal of water with a sucrose concentration of 0.2 lb/gal. We begin adding water with a sucrose concentration of 0.4 lb/gal at a rate of $R_{in} = 10$ gal/min (Figure 3). The water is mixed instantaneously and exits the bottom of the tank at a rate of $R_{out} = 5$ gal/min. Let $y(t)$ be the quantity of sucrose in the tank at time t. Set up a differential equation for $y(t)$ and solve for $y(t)$.

Solution

Step 1. Set up the differential equation.
The derivative dy/dt is the difference of two rates of change, namely the rate at which sucrose enters the tank and the rate at which it leaves:

$$\frac{dy}{dt} = \text{sucrose rate in} - \text{sucrose rate out} \qquad \boxed{8}$$

The rate at which sucrose enters the tank is

$$\text{Sucrose rate in} = \underbrace{(0.4 \text{ lb/gal})(10 \text{ gal/min})}_{\text{Concentration times water rate in}} = 4 \text{ lb/min}$$

Next, we compute the sucrose concentration in the tank at time t. Water flows in at 10 gal/min and out at 5 gal/min, so there is a net inflow of 5 gal/min. Since the tank has 100 gal at time $t = 0$, it has $100 + 5t$ gallons at time t minutes and

$$\text{Sucrose concentration at time } t = \frac{\text{lb of sucrose in tank}}{\text{gal of water in tank}} = \frac{y(t)}{100 + 5t} \text{ lb/gal}$$

The rate at which sucrose leaves the tank is the product of the concentration and the rate at which water flows out:

$$\text{Sucrose rate out} = \underbrace{\left(\frac{y}{100 + 5t} \frac{\text{lb}}{\text{gal}} \right) \left(5 \frac{\text{gal}}{\text{min}} \right)}_{\text{Concentration times water rate out}} = \frac{5y}{100 + 5t} = \frac{y}{t + 20} \text{ lb/min}$$

Now Eq. (8) gives us the differential equation

$$\frac{dy}{dt} = 4 - \frac{y}{t + 20} \qquad \boxed{9}$$

Step 2. Find the general solution.
We write Eq. (9) in standard form:

$$\frac{dy}{dt} + \underbrace{\frac{1}{t + 20}}_{A(t)} y = \underbrace{4}_{B(t)} \qquad \boxed{10}$$

An integrating factor is

$$\alpha(t) = e^{\int A(t)\, dt} = e^{\int dt/(t+20)} = e^{\ln(t+20)} = t + 20$$

The general solution is

$$y(t) = \alpha(t)^{-1} \left(\int \alpha(t) B(t)\, dt + C \right) = \frac{1}{t + 20} \left(\int (t + 20)(4)\, dt + C \right)$$

$$= \frac{1}{t + 20} \left(2(t + 20)^2 + C \right) = 2t + 40 + \frac{C}{t + 20}$$

Summary:

sucrose rate in $= 4$ lb/min

sucrose rate out $= \dfrac{y}{t + 20}$ lb/min

$\dfrac{dy}{dt} = 4 - \dfrac{y}{t + 20}$

$y(t) = 2t + 40 + \dfrac{C}{t + 20}$

Step 3. **Solve the initial value problem.**

At $t = 0$, the tank contains 100 gal of water with a sucrose concentration of 0.2 lb/gal. Thus, the total sucrose at $t = 0$ is $y(0) = (100)(0.2) = 20$ lb, and

$$y(0) = 2(0) + 40 + \frac{C}{0 + 20} = 40 + \frac{C}{20} = 20 \quad \Rightarrow \quad C = -400$$

We obtain the following formula (for t in minutes), valid until the tank overflows:

$$y(t) = 2t + 40 - \frac{400}{t + 20} \text{ lb sucrose}$$

∎

9.5 SUMMARY

- A *first-order linear differential equation* can always be written in the form

$$y' + A(x)y = B(x) \qquad \boxed{11}$$

- The general solution of Eq. (11) is

$$y = \alpha(x)^{-1} \left(\int \alpha(x)B(x)\, dx + C \right)$$

where $\alpha(x)$ is an *integrating factor*: $\alpha(x) = e^{\int A(x)\, dx}$.

9.5 EXERCISES

Preliminary Questions

1. Which of the following are first-order linear equations?

(a) $y' + x^2 y = 1$

(b) $y' + xy^2 = 1$

(c) $x^5 y' + y = e^x$

(d) $x^5 y' + y = e^y$

2. If $\alpha(x)$ is an integrating factor for $y' + A(x)y = B(x)$, then $\alpha'(x)$ is equal to (choose the correct answer):

(a) $B(x)$

(b) $\alpha(x)A(x)$

(c) $\alpha(x)A'(x)$

(d) $\alpha(x)B(x)$

Exercises

1. Consider $y' + x^{-1}y = x^3$.

(a) Verify that $\alpha(x) = x$ is an integrating factor.

(b) Show that when multiplied by $\alpha(x)$, the differential equation can be written $(xy)' = x^4$.

(c) Conclude that xy is an antiderivative of x^4 and use this information to find the general solution.

(d) Find the particular solution satisfying $y(1) = 0$.

2. Consider $\dfrac{dy}{dt} + 2y = e^{-3t}$.

(a) Verify that $\alpha(t) = e^{2t}$ is an integrating factor.

(b) Use Eq. (4) to find the general solution.

(c) Find the particular solution with initial condition $y(0) = 1$.

3. Let $\alpha(x) = e^{x^2}$. Verify the identity

$$(\alpha(x)y)' = \alpha(x)(y' + 2xy)$$

and explain how it is used to find the general solution of

$$y' + 2xy = x.$$

4. Find the solution of $y' - y = e^{2x}$, $y(0) = 1$.

In Exercises 5–18, find the general solution of the first-order linear differential equation.

5. $xy' + y = x$

6. $xy' - y = x^2 - x$

7. $3xy' - y = x^{-1}$

8. $y' + xy = x$

9. $y' + 3x^{-1}y = x + x^{-1}$

10. $y' + x^{-1}y = \cos(x^2)$

11. $xy' = y - x$

12. $xy' = x^3 - \dfrac{3y}{x}$

13. $y' + y = e^x$

14. $y' + (\sec x)y = \cos x$

15. $y' + (\tan x)y = \cos x$

16. $e^{2x}y' = 1 - e^x y$

17. $y' - (\ln x)y = x^x$

18. $y' + y = \cos x$

In Exercises 19–26, solve the initial value problem.

19. $y' + 3y = e^{2x}, \quad y(0) = -1$

20. $xy' + y = e^x, \quad y(1) = 3$

21. $y' + \dfrac{1}{x+1}y = x^{-2}, \quad y(1) = 2$

22. $y' + y = \sin x, \quad y(0) = 1$

23. $(\sin x)y' = (\cos x)y + 1, \quad y\left(\tfrac{\pi}{4}\right) = 0$

24. $y' + (\sec t)y = \sec t, \quad y(0) = 1$ and $y\left(\tfrac{\pi}{4}\right) = 1$

25. $y' + (\tanh x)y = 1, \quad y(0) = 3$

26. $y' + \dfrac{x}{1+x^2}y = \dfrac{1}{(1+x^2)^{3/2}}, \quad y(1) = 0$

27. Find the general solution of $y' + ny = e^{mx}$ for all m, n. *Note:* The case $m = -n$ must be treated separately.

28. A 200-gal tank contains 100 gal of water with a salt concentration of 0.1 lb/gal. Water with a salt concentration of 0.4 lb/gal flows into the tank at a rate of 20 gal/min. The fluid is mixed instantaneously, and water is pumped out at a rate of 10 gal/min. Let $y(t)$ be the amount of salt in the tank at time t.

(a) Set up and solve the differential equation for $y(t)$.

(b) What is the salt concentration when the tank overflows?

29. Repeat Exercise 28(a), assuming that water is pumped out at a rate of 20 gal/min. What is the limiting salt concentration for large t?

30. Repeat Exercise 28(a), assuming that water is pumped out at a rate of 25 gal/min. What is the limiting salt concentration for t large?

31. Water flows into a tank at the variable rate $R_{in} = \dfrac{20}{1+t}$ gal/min and out at the constant rate $R_{out} = 5$ gal/min. Let $V(t)$ be the volume of water in the tank at time t.

(a) Set up a differential equation for $V(t)$ and solve it with the initial condition $V(0) = 100$.

(b) Find the maximum value of V.

(c) *CAS* Plot $V(t)$ and estimate the time t when the tank is empty.

32. A stream feeds into a lake at a rate of 1,000 m^3/day. The stream is polluted with a toxin whose concentration is 5 g/m^3. Assume that the lake has volume 10^6 m^3 and that water flows out of the lake at the same rate of 1,000 m^3/day. Set up a differential equation for the concentration $c(t)$ of toxin in the lake and solve for $c(t)$, assuming that $c(0) = 0$. What is the limiting concentration for large t?

In Exercises 33–35, consider a series circuit (Figure 4) consisting of a resistor of R ohms, an inductor of L henries, and variable voltage source of V(t) volts (time t in seconds). The current through the circuit I(t) (in amperes) satisfies the differential equation:

$$\frac{dI}{dt} + \frac{R}{L}I = \frac{1}{L}V(t) \qquad \boxed{12}$$

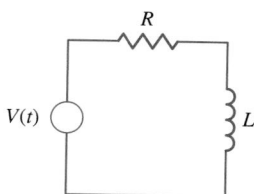

FIGURE 4 RL circuit.

33. Find the solution to Eq. (12) with initial condition $I(0) = 0$, assuming that $R = 100 \ \Omega$, $L = 5$ H, and $V(t)$ is constant with $V(t) = 10$ V.

34. $\boxed{\text{GU}}$ Assume that $R = 110 \ \Omega$, $L = 10$ H, and $V(t) = e^{-t}$.

(a) Solve Eq. (12) with initial condition $I(0) = 0$.

(b) Use a computer algebra system to sketch the graph of the solution for $0 \le t \le 3$.

(c) Calculate t_m and $I(t_m)$, where t_m is the time at which $I(t)$ has a maximum value.

35. Assume that $V(t) = V$ is constant and $I(0) = 0$.

(a) Solve for $I(t)$.

(b) Show that $\lim\limits_{t \to \infty} I(t) = V/R$ and that $I(t)$ reaches approximately 63% of its limiting value after L/R seconds.

(c) How long does it take for $I(t)$ to reach 90% of its limiting value if $R = 500 \ \Omega$, $L = 4$ H, and $V = 20$ V?

36. Tank 1 in Figure 5 is filled with V_1 gallons of water. Water flows into the tank at a rate of R gal/min and out through the bottom at the same rate R. Suppose that I gallons of blue ink are dumped into the tank at time $t = 0$ and mixed instantaneously. Let $y_1(t)$ be the quantity of ink in the tank at time t.

(a) Explain why y_1 satisfies the differential equation $\dfrac{dy_1}{dt} = -\dfrac{R}{V_1}y_1$.

(b) Solve for $y_1(t)$ with $V_1 = 100$, $R = 10$, and $I = 2$.

37. Continuing with the previous exercise, let Tank 2 be another tank filled with V_2 gallons of water. Assume that the inky water from Tank 1 empties into Tank 2 as in Figure 5, mixes instantaneously, and leaves Tank 2 at the same rate R. Let $y_2(t)$ be the amount of ink in Tank 2 at time t.

R(gal/min)

Tank 1

R(gal/min)

Tank 2

R(gal/min)

FIGURE 5

(a) Explain why y_2 satisfies the differential equation

$$\frac{dy_2}{dt} = R\left(\frac{y_1}{V_1} - \frac{y_2}{V_2}\right)$$

(b) Use the solution to Exercise 36 to solve for $y_2(t)$ if $V_1 = 100$, $V_2 = 200$, $R = 10$, $I = 2$, and $y_2(0) = 0$.

(c) [GU] Plot the solution for $0 \le t \le 120$.

(d) Find the maximum ink concentration in Tank 2.

Further Insights and Challenges

38. Let $\alpha(x)$ be an integrating factor for $y' + A(x)y = B(x)$. The differential equation $y' + A(x)y = 0$ is called the associated **homogeneous equation**.

(a) Show that $\dfrac{1}{\alpha(x)}$ is a solution of the associated homogeneous equation.

(b) Show that if $y = f(x)$ is a particular solution of $y' + A(x)y = B(x)$, then $f(x) + \dfrac{C}{\alpha(x)}$ is also a solution for any constant C.

39. Use the Fundamental Theorem of Calculus and the Product Rule to verify directly that for any x_0, the function

$$f(x) = \alpha(x)^{-1} \int_{x_0}^{x} \alpha(t)B(t)\,dt$$

is a solution of the initial value problem

$$y' + A(x)y = B(x), \qquad y(x_0) = 0,$$

where $\alpha(x)$ is an integrating factor [a solution to Eq. (3)].

40. Transient Currents Suppose the circuit described by Eq. (12) is driven by a sinusoidal voltage source $V(t) = V \sin \omega t$ (where V and ω are constant).

(a) Show that

$$I(t) = \frac{V}{R^2 + L^2\omega^2}(R \sin \omega t - L\omega \cos \omega t) + Ce^{-(R/L)t}$$

(b) Let $Z = \sqrt{R^2 + L^2\omega^2}$. Choose θ so that $Z \cos \theta = R$ and $Z \sin \theta = L\omega$. Use the addition formula for the sine function to show that

$$I(t) = \frac{V}{Z}\sin(\omega t - \theta) + +Ce^{-(R/L)t}$$

This shows that the current in the circuit varies sinusoidally apart from a DC term (called the **transient current** in electronics) that decreases exponentially.

CHAPTER REVIEW EXERCISES

1. Which of the following differential equations are linear? Determine the order of each equation.

(a) $y' = y^5 - 3x^4y$

(b) $y' = x^5 - 3x^4y$

(c) $y = y''' - 3x\sqrt{y}$

(d) $\sin x \cdot y'' = y - 1$

2. Find a value of c such that $y = x - 2 + e^{cx}$ is a solution of $2y' + y = x$.

In Exercises 3–6, solve using separation of variables.

3. $\dfrac{dy}{dt} = t^2 y^{-3}$

4. $xyy' = 1 - x^2$

5. $x\dfrac{dy}{dx} - y = 1$

6. $y' = \dfrac{xy^2}{x^2 + 1}$

In Exercises 7–10, solve the initial value problem using separation of variables.

7. $y' = \cos^2 x$, $\quad y(0) = \dfrac{\pi}{4}$

8. $y' = \cos^2 y$, $\quad y(0) = \dfrac{\pi}{4}$

9. $y' = xy^2$, $\quad y(1) = 2$

10. $xyy' = 1$, $\quad y(3) = 2$

11. Figure 1 shows the slope field for $\dot{y} = \sin y + ty$. Sketch the graphs of the solutions with the initial conditions $y(0) = 1$, $y(0) = 0$, and $y(0) = -1$.

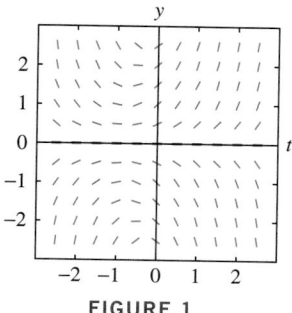

FIGURE 1

12. Which of the equations (i)–(iii) corresponds to the slope field in Figure 2?

(i) $\dot{y} = 1 - y^2$ **(ii)** $\dot{y} = 1 + y^2$ **(iii)** $\dot{y} = y^2$

13. Let $y(t)$ be the solution to the differential equation with slope field as shown in Figure 2, satisfying $y(0) = 0$. Sketch the graph of $y(t)$. Then use your answer to Exercise 12 to solve for $y(t)$.

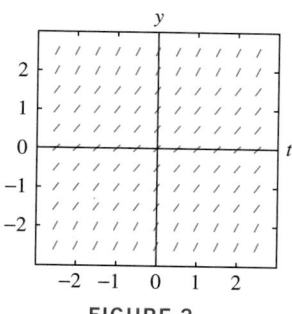

FIGURE 2

14. Let $y(t)$ be the solution of $4\dot{y} = y^2 + t$ satisfying $y(2) = 1$. Carry out Euler's Method with time step $h = 0.05$ for $n = 6$ steps.

15. Let $y(t)$ be the solution of $(x^3 + 1)\dot{y} = y$ satisfying $y(0) = 1$. Compute approximations to $y(0.1)$, $y(0.2)$, and $y(0.3)$ using Euler's Method with time step $h = 0.1$.

In Exercises 16–19, solve using the method of integrating factors.

16. $\dfrac{dy}{dt} = y + t^2$, $y(0) = 4$

17. $\dfrac{dy}{dx} = \dfrac{y}{x} + x$, $y(1) = 3$

18. $\dfrac{dy}{dt} = y - 3t$, $y(-1) = 2$

19. $y' + 2y = 1 + e^{-x}$, $y(0) = -4$

In Exercises 20–27, solve using the appropriate method.

20. $x^2 y' = x^2 + 1$, $y(1) = 10$

21. $y' + (\tan x)y = \cos^2 x$, $y(\pi) = 2$

22. $xy' = 2y + x - 1$, $y(\tfrac{3}{2}) = 9$

23. $(y - 1)y' = t$, $y(1) = -3$

24. $(\sqrt{y} + 1)y' = yte^{t^2}$, $y(0) = 1$.

25. $\dfrac{dw}{dx} = k\dfrac{1 + w^2}{x}$, $y(1) = 1$

26. $y' + \dfrac{3y - 1}{t} = t + 2$

27. $y' + \dfrac{y}{x} = \sin x$

28. Find the solutions to $y' = 4(y - 12)$ satisfying $y(0) = 20$ and $y(0) = 0$, and sketch their graphs.

29. Find the solutions to $y' = -2y + 8$ satisfying $y(0) = 3$ and $y(0) = 4$, and sketch their graphs.

30. Show that $y = \sin^{-1} x$ satisfies the differential equation $y' = \sec y$ with initial condition $y(0) = 0$.

31. What is the limit $\lim\limits_{t \to \infty} y(t)$ if $y(t)$ is a solution of:

(a) $\dfrac{dy}{dt} = -4(y - 12)$?

(b) $\dfrac{dy}{dt} = 4(y - 12)$?

(c) $\dfrac{dy}{dt} = -4y - 12$?

In Exercises 32–35, let $P(t)$ denote the balance at time t (years) of an annuity that earns 5% interest continuously compounded and pays out $2000/year continuously.

32. Find the differential equation satisfied by $P(t)$.

33. Determine $P(2)$ if $P(0) = \$5{,}000$.

34. When does the annuity run out of money if $P(0) = \$2{,}000$?

35. What is the minimum initial balance that will allow the annuity to make payments indefinitely?

36. State whether the differential equation can be solved using separation of variables, the method of integrating factors, both, or neither.

(a) $y' = y + x^2$ **(b)** $xy' = y + 1$

(c) $y' = y^2 + x^2$ **(d)** $xy' = y^2$

37. Let A and B be constants. Prove that if $A > 0$, then all solution of $\dfrac{dy}{dt} + Ay = B$ approach the same limit as $t \to \infty$.

38. The trough in Figure 3 is filled with water. At time $t = 0$ (in seconds), water begins leaking through a circular hole at the bottom of radius 3 in. Let $y(t)$ be the water height at time t. Find a differential

FIGURE 3

equation for $y(t)$ and solve it to determine when the trough will be half empty.

39. Find the solution of the logistic equation $\dot{y} = 0.4y(4 - y)$ satisfying $y(0) = 8$.

40. Let $y(t)$ be the solution of $\dot{y} = 0.3y(2 - y)$ with $y(0) = 1$. Determine $\lim_{t \to \infty} y(t)$ without solving for y explicitly.

41. Suppose that $y = ky(1 - y/8)$ has a solution satisfying $y(0) = 12$ and $y(10) = 24$. Find k.

42. A lake has a carrying capacity of 1,000 fish. Assume that the fish population grows logistically with growth constant $k = 0.2$ days^{-1}. How many days will it take for the population to reach 900 fish if the initial population is 20 fish?

43. A rabbit population on a deserted island increases exponentially with growth rate $k = 0.12$ months^{-1}. When the population reaches 150 rabbits (say, at time $t = 0$), hunters begin killing the rabbits at a rate of r rabbits per month.

(a) Find a differential equation satisfied by the rabbit population $P(t)$.

(b) How large can r be without the rabbit population becoming extinct?

44. Show that $y = \sin(\tan^{-1} x + C)$ is the general solution of $y' = \dfrac{\sqrt{1 - y^2}}{1 + x^2}$. Then use the addition formula for the sine function to show that the general solution may be written $y = \dfrac{(\cos C)x + \sin C}{\sqrt{1 + x^2}}$.

45. A tank contains 100 gal of pure water. Water is pumped out at a rate of 20 gal/min, and polluted water with a toxin concentration of 0.1 lb/gal is pumped in at a rate of 15 gal/min. Let $y(t)$ be the amount of toxin present in the tank at time t.

(a) Find a differential equation satisfied by $y(t)$.

(b) Solve for $y(t)$.

(c) CAS Plot $y(t)$ and find the time t at which the amount of toxin is maximal.

10 | INFINITE SERIES

Our knowledge of what stars are made of is based on the study of absorption spectra, the sequences of wavelengths absorbed by gases in the star's atmosphere.

The theory of infinite series is a third branch of calculus, in addition to differential and integral calculus. Infinite series provide us with a new perspective on functions and on many interesting numbers. Two examples are the Gregory–Leibniz series

$$\frac{\pi}{4} = 1 - \frac{1}{3} + \frac{1}{5} - \frac{1}{7} + \frac{1}{9} - \cdots$$

and the infinite series for the exponential function

$$e^x = 1 + x + \frac{x^2}{2!} + \frac{x^3}{3!} + \frac{x^4}{4!} + \cdots$$

The first reveals that π is related to the reciprocals of the odd integers in an unexpected way, whereas the second shows that e^x can be expressed as an "infinite polynomial." Series of this type are widely used in applications, both in computations and in the analysis of functions. To make sense of infinite series, we need to define precisely what it means to add up infinitely many terms. Limits play a key role here, just as they do in differential and integral calculus.

10.1 Sequences

Sequences of numbers appear in diverse situations. If you divide a cake in half, and then divide the remaining half in half, and continue dividing in half indefinitely (Figure 1), then the fraction of cake left at each step forms the sequence

$$1, \quad \frac{1}{2}, \quad \frac{1}{4}, \quad \frac{1}{8}, \quad \cdots$$

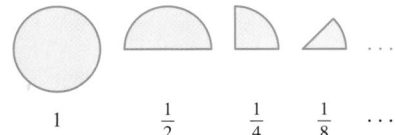

FIGURE 1

This is the sequence of values of $f(n) = \dfrac{1}{2^n}$ for $n = 0, 1, 2, \ldots$.

Formally, a **sequence** is a function $f(n)$ whose domain is a subset of the integers. The values $a_n = f(n)$ are called the **terms** of the sequence and n is called the **index**. We usually think of a sequence informally as a collection of values $\{a_n\}$, or a list of terms:

$$a_1, \quad a_2, \quad a_3, \quad a_4, \quad \cdots$$

When a_n is given by a formula, we refer to a_n as the **general term**.

General Term	Domain	Sequence
$a_n = 1 - \dfrac{1}{n}$	$n \geq 1$	$0, \ \dfrac{1}{2}, \ \dfrac{2}{3}, \ \dfrac{3}{4}, \ \dfrac{4}{5}, \ \cdots$
$a_n = (-1)^n n$	$n \geq 0$	$0, \ -1, \ 2, \ -3, \ 4, \ \ldots$
$b_n = \dfrac{364.5n^2}{n^2 - 4}$	$n \geq 3$	$656.1, \ 486, \ 433.9, \ 410.1, \ 396.9, \ \ldots$

The sequence $b_n = \dfrac{364.5n^2}{n^2 - 4}$, known as the "Balmer series" in physics and chemistry, plays a key role in spectroscopy. The terms of this sequence are the absorption wavelengths of the hydrogen atom in nanometers.

In the following example, we consider a sequence whose terms are defined *recursively*. The first term is given and the nth term a_n is computed in terms of the preceding term a_{n-1}.

■ **EXAMPLE 1** **Recursively Defined Sequence**   Compute a_2, a_3, a_4 for the sequence defined recursively by

$$a_1 = 1, \qquad a_n = \frac{1}{2}\left(a_{n-1} + \frac{2}{a_{n-1}}\right)$$

Solution

You may recognize the sequence in Example 1 as the sequence of approximations to $\sqrt{2} \approx 1.4142136$ produced by Newton's method with starting value $a_1 = 1$. As n tends to infinity, a_n approaches $\sqrt{2}$.

$$a_2 = \frac{1}{2}\left(a_1 + \frac{2}{a_1}\right) = \frac{1}{2}\left(1 + \frac{2}{1}\right) = \frac{3}{2} = 1.5$$

$$a_3 = \frac{1}{2}\left(a_2 + \frac{2}{a_2}\right) = \frac{1}{2}\left(\frac{3}{2} + \frac{2}{3/2}\right) = \frac{17}{12} \approx 1.4167$$

$$a_4 = \frac{1}{2}\left(a_3 + \frac{2}{a_3}\right) = \frac{1}{2}\left(\frac{17}{12} + \frac{2}{17/12}\right) = \frac{577}{408} \approx 1.414216 \qquad ■$$

Our main goal is to study convergence of sequences. A sequence $\{a_n\}$ converges to a limit L if the terms a_n get closer and closer to L as $n \to \infty$.

DEFINITION **Limit of a Sequence**   A sequence $\{a_n\}$ **converges** to a limit L, and we write

$$\lim_{n\to\infty} a_n = L \qquad \text{or} \qquad a_n \to L$$

if, for every $\epsilon > 0$, there is a number M such that $|a_n - L| < \epsilon$ for all $n > M$. If no limit exists, we say that $\{a_n\}$ **diverges**.

■ **EXAMPLE 2** **Proving Convergence of a Sequence**   Let $a_n = \dfrac{n+4}{n+1}$. Prove formally that $\lim\limits_{n\to\infty} a_n = 1$.

Solution   The definition requires us to find, for every $\epsilon > 0$, a number M such that

$$|a_n - 1| < \epsilon \qquad \text{for all } n > M \qquad \boxed{1}$$

We have

$$|a_n - 1| = \left|\frac{n+4}{n+1} - 1\right| = \frac{3}{n+1}$$

Therefore, $|a_n - 1| < \epsilon$ if

$$\frac{3}{n+1} < \epsilon \qquad \text{or} \qquad n > \frac{3}{\epsilon} - 1$$

It follows that (1) is valid with $M = \dfrac{3}{\epsilon} - 1$. For example, if $\epsilon = 0.01$, then we may take

$$M = \frac{3}{0.01} - 1 = 299. \text{ Thus, } |a_n - 1| < 0.01 \text{ for } n = 300, 301, 302, \ldots. \qquad ■$$

FIGURE 2 Plot of a sequence with limit L. For any ϵ, the dots eventually remain within an ϵ-band around L.

We may visualize a sequence by plotting its "graph," that is, by plotting the points $(1, a_1), (2, a_2), (3, a_3), \ldots$ (Figure 2). The sequence converges to a limit L if, for every $\epsilon > 0$, the plotted points eventually remain within an ϵ-band around the horizontal line $y = L$ (Figure 2). Figure 3 shows the plot of a sequence converging to $L = 1$. On the other hand, it can be shown that the sequence $a_n = \cos n$ shown in Figure 4 has no limit.

FIGURE 3 The sequence $a_n = \dfrac{n+4}{n+1}$.

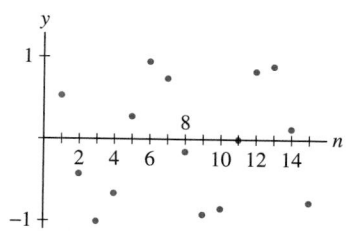

FIGURE 4 The sequence $a_n = \cos n$ has no limit.

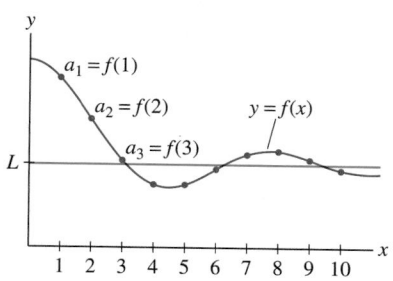

FIGURE 5 If $f(x)$ converges to L, then the sequence $a_n = f(n)$ also converges to L.

We note the following:

- The limit does not change if we modify or drop finitely many terms of the sequence.
- If C is a constant and $a_n = C$ for all n sufficiently large, then $\lim\limits_{n\to\infty} a_n = C$.

Suppose that $f(x)$ is a function and that $f(x)$ approaches a limit L as $x \to \infty$. In this case, the sequence $a_n = f(n)$ approaches the same limit L (Figure 5). Indeed, in this case, for all $\epsilon > 0$, we can find M so that $|f(x) - L| < \epsilon$ for all $x > M$. It follows automatically that $|f(n) - L| < \epsilon$ for all integers $n > M$.

THEOREM 1 Sequence Defined by a Function Let $f(x)$ be a function defined on $[c, \infty)$ for some constant c. If $\lim\limits_{x\to\infty} f(x)$ exists, then the sequence $a_n = f(n)$, defined for $n \geq c$, converges and

$$\lim_{n\to\infty} a_n = \lim_{x\to\infty} f(x)$$

■ **EXAMPLE 3** Find the limit of the sequence $0, \dfrac{1}{2}, \dfrac{2}{3}, \dfrac{3}{4}, \dfrac{4}{5} \ldots$.

Solution This is the sequence with general term

$$a_n = \frac{n-1}{n} = 1 - \frac{1}{n}$$

Let $f(x) = 1 - \dfrac{1}{x}$. Then $a_n = f(n)$ and by Theorem 1,

$$\lim_{n\to\infty} a_n = \lim_{x\to\infty} f(x) = \lim_{x\to\infty} \left(1 - \frac{1}{x}\right) = 1 - \lim_{x\to\infty} \frac{1}{x} = 1 \qquad ■$$

■ **EXAMPLE 4** Calculate $\lim\limits_{n\to\infty} a_n$, where $a_n = \dfrac{n + \ln n}{n^2}$.

Solution The limit of the sequence is equal to the limit of the function $f(x) = \dfrac{x + \ln x}{x^2}$, which we calculate using L'Hôpital's Rule:

$$\lim_{n\to\infty} a_n = \lim_{x\to\infty} f(x) = \lim_{x\to\infty} \frac{x + \ln x}{x^2} = \lim_{x\to\infty} \frac{1 + (1/x)}{2x} = 0 \qquad ■$$

The limit of the Balmer wavelengths b_n defined in the margin on page 553 is of interest in physics and chemistry because it determines the ionization energy of the hydrogen

TABLE 1   The Wavelengths in the Balmer Series Approach the Limit $L = 364.5$

n	b_n
3	656.1
4	486
5	433.9
6	410.1
7	396.9
10	379.7
20	368.2
40	365.4
60	364.9
80	364.7
100	364.6

atom. Table 1 suggests that b_n approaches 364.5 as $n \to \infty$. Figure 6 shows the graph of b_n, and in Figure 7, the wavelengths are shown "crowding in" toward their limiting value.

■ **EXAMPLE 5 Limit of Balmer Wavelengths** Calculate the limit of the Balmer wavelengths $b_n = \dfrac{364.5n^2}{n^2 - 4}$, where $n \geq 3$.

Solution Observe that $b_n = f(n)$, where $f(x) = \dfrac{364.5x^2}{x^2 - 4}$. We compute the limit by dividing the numerator and denominator by x^2:

$$\lim_{n \to \infty} b_n = \lim_{x \to \infty} f(x) = 364.5 \lim_{x \to \infty} \frac{x^2}{x^2 - 4} = 364.5 \lim_{x \to \infty} \frac{1}{1 - 4/x^2}$$

$$= \frac{364.5}{\lim_{x \to \infty} (1 - 4/x^2)} = 364.5 \qquad ■$$

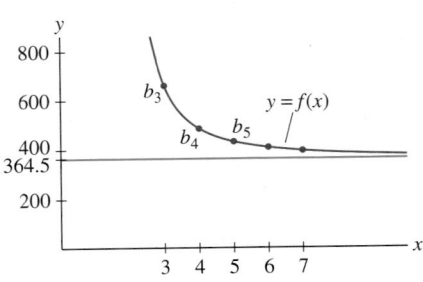

FIGURE 6 The sequence b_n and the function approach the same limit.

FIGURE 7

The geometric sequence $a_n = cr^n$ is the sequence defined by the exponential function $f(x) = cr^x$ whose base r is the common ratio.

A **geometric sequence** is a sequence of the form $a_n = cr^n$, where c and r are nonzero constants. For instance, if $c = 2$ and $r = 3$, we obtain the geometric sequence

$$2, \quad 2 \cdot 3, \quad 2 \cdot 3^2, \quad 2 \cdot 3^3, \quad 2 \cdot 3^4, \quad 2 \cdot 3^5, \quad \dots$$

The number r is called the **common ratio**. Each term a_n is r times the previous term a_{n-1}, that is, $a_n / a_{n-1} = r$.

We say that $\{a_n\}$ *diverges to* ∞, and we write $\lim_{n \to \infty} a_n = \infty$, if the terms a_n increase beyond all bounds, that is, if, for every $N > 0$, we have $a_n > N$ for all sufficiently large n (Figure 8).

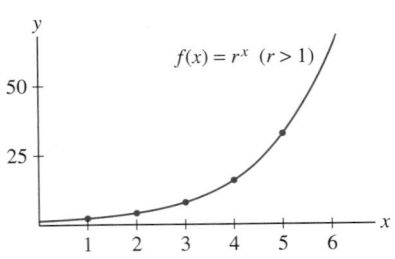

FIGURE 8 If $r > 1$, the geometric sequence $a_n = r^n$ diverges to ∞.

■ **EXAMPLE 6 Limit of a Geometric Sequence** Prove that:

$$\lim_{n \to \infty} r^n = \begin{cases} 0 & \text{if} & 0 < r < 1 \\ 1 & \text{if} & r = 1 \\ \text{diverges to } \infty & \text{if} & r > 1 \end{cases}$$

Solution We apply Theorem 1 to the exponential function $f(x) = r^x$. If $0 < r < 1$, then (Figure 9)

$$\lim_{n \to \infty} r^n = \lim_{x \to \infty} r^x = 0$$

Similarly, if $r > 1$, then $f(x)$ diverges to ∞ as $x \to \infty$, so $\{r^n\}$ also diverges to ∞ (Figure 8). If $r = 1$, $r^n = 1$ for all n and the limit is 1.   ■

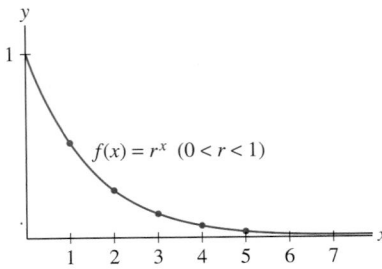

FIGURE 9 If $0 < r < 1$, the geometric sequence $a_n = r^n$ converges to 0.

Most of the limit laws for functions also apply to sequences. The proofs are similar and are omitted.

THEOREM 2 Limit Laws for Sequences   Assume that $\{a_n\}$ and $\{b_n\}$ are convergent sequences with

$$\lim_{n \to \infty} a_n = L, \qquad \lim_{n \to \infty} b_n = M$$

Then:

(i) $\displaystyle \lim_{n \to \infty} (a_n \pm b_n) = \lim_{n \to \infty} a_n \pm \lim_{n \to \infty} b_n = L \pm M$

(ii) $\displaystyle \lim_{n \to \infty} a_n b_n = \left(\lim_{n \to \infty} a_n \right)\left(\lim_{n \to \infty} b_n \right) = LM$

(iii) $\displaystyle \lim_{n \to \infty} \frac{a_n}{b_n} = \frac{\lim_{n \to \infty} a_n}{\lim_{n \to \infty} b_n} = \frac{L}{M} \quad$ if $M \neq 0$

(iv) $\displaystyle \lim_{n \to \infty} c a_n = c \lim_{n \to \infty} a_n = cL \quad$ for any constant c

---

THEOREM 3 Squeeze Theorem for Sequences   Let $\{a_n\}$, $\{b_n\}$, $\{c_n\}$ be sequences such that for some number M,

$$b_n \leq a_n \leq c_n \quad \text{for } n > M \qquad \text{and} \qquad \lim_{n \to \infty} b_n = \lim_{n \to \infty} c_n = L$$

Then $\displaystyle \lim_{n \to \infty} a_n = L$.

■ **EXAMPLE 7**   Show that if $\displaystyle \lim_{n \to \infty} |a_n| = 0$, then $\displaystyle \lim_{n \to \infty} a_n = 0$.

Solution   We have

$$-|a_n| \leq a_n \leq |a_n|$$

Since $|a_n|$ tends to zero, $-|a_n|$ also tends to zero and the Squeeze Theorem implies that $\displaystyle \lim_{n \to \infty} a_n = 0$. ■

As another application of the Squeeze Theorem, consider the sequence

$$a_n = \frac{5^n}{n!}$$

Both the numerator and denominator tend to infinity, so it is not clear in advance whether $\{a_n\}$ converges. Figure 10 and Table 2 suggest that a_n increases initially and then tends to zero. In the next example, we prove that, indeed, $a_n = \dfrac{R^n}{n!}$ converges to zero for all R. We will use this fact in the discussion of Taylor series in Section 10.7.

■ **EXAMPLE 8**   Prove that $\displaystyle \lim_{n \to \infty} \frac{R^n}{n!} = 0$ for all R.

Solution   We may assume without loss of generality that $R > 0$ by the result of Example 7. Then there is a unique integer $M \geq 0$ such that

$$M \leq R < M + 1$$

◀·· **REMINDER** $n!$ (*n-factorial*) is the number

$$n! = n(n-1)(n-2) \cdots 2 \cdot 1$$

For example, $4! = 4 \cdot 3 \cdot 2 \cdot 1 = 24$.

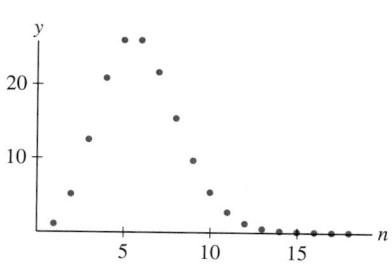

FIGURE 10 Graph of the sequence $a_n = \dfrac{5^n}{n!}$.

TABLE 2

n	$a_n = \dfrac{5^n}{n!}$
1	5
2	12.5
3	20.83
4	26.04
10	2.69
15	0.023
20	0.00004

For $n > M$, we write $R^n/n!$ as a product of n factors:

$$\frac{R^n}{n!} = \underbrace{\left(\frac{R}{1}\frac{R}{2}\cdots\frac{R}{M}\right)}_{\text{Call this constant }C}\underbrace{\left(\frac{R}{M+1}\right)\left(\frac{R}{M+2}\right)\cdots\left(\frac{R}{n}\right)}_{\text{Each factor is less than 1}} \leq C\left(\frac{R}{n}\right)$$

The first M factors are ≥ 1 and the last $n - M$ factors are < 1. If we lump together the first M factors and call the product C, and drop all the remaining factors except the last factor R/n, we obtain

$$0 \leq \frac{R^n}{n!} \leq \frac{CR}{n}$$

Since $\dfrac{CR}{n} \to 0$, the Squeeze Theorem implies that $\displaystyle\lim_{n\to\infty} \frac{R^n}{n!} = 0$. ∎

We can apply a function $f(x)$ to a sequence $\{a_n\}$ to obtain a new sequence $\{f(a_n)\}$. It is useful to know that if $f(x)$ is continuous and $a_n \to L$, then $f(a_n) \to f(L)$. We state this result in the next theorem. See Appendix D for a proof.

THEOREM 4   If $f(x)$ is continuous and the limit $\displaystyle\lim_{n\to\infty} a_n = L$ exists, then

$$\lim_{n\to\infty} f(a_n) = f\left(\lim_{n\to\infty} a_n\right) = f(L)$$

■ **EXAMPLE 9** Calculate $\displaystyle\lim_{n\to\infty} e^{3n/(n+1)}$.

Solution We have $e^{3n/(n+1)} = f(a_n)$, where $a_n = \dfrac{3n}{n+1}$ and $f(x) = e^x$. Furthermore,

$$\lim_{n\to\infty} a_n = \lim_{n\to\infty} \frac{3n}{n+1} = 3$$

By Theorem 4, $\displaystyle\lim_{n\to\infty} f(a_n) = f\left(\lim_{n\to\infty} a_n\right) = f(3)$, that is,

$$\lim_{n\to\infty} e^{3n/(n+1)} = e^{\lim_{n\to\infty} 3n/(n+1)} = e^3$$ ■

We now introduce two concepts that are important for understanding convergence: the concepts of a bounded sequence and a monotonic sequence.

DEFINITION Bounded Sequences   A sequence $\{a_n\}$ is:

- **Bounded from above** if there is a number M such that $a_n \leq M$ for all n. The number M is called an *upper bound*.
- **Bounded from below** if there exists m such that $a_n \geq m$ for all n. The number m is called a *lower bound*.

If $\{a_n\}$ is bounded from above and below, we say that $\{a_n\}$ is *bounded*. If $\{a_n\}$ is not bounded, we call $\{a_n\}$ an *unbounded sequence*.

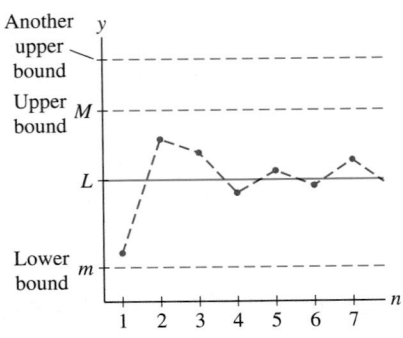

FIGURE 11 A convergent sequence is bounded.

Upper and lower bounds are not unique. If M is an upper bound, then any number larger than M is also an upper bound (Figure 11). Similarly, if m is a lower bound, then any number smaller than m is also a lower bound.

It would seem that a convergent sequence $\{a_n\}$ must be bounded because the terms a_n get closer and closer to the limit (Figure 11). This leads to the next theorem.

THEOREM 5 Convergent Sequences Are Bounded If $\{a_n\}$ converges, then $\{a_n\}$ is bounded.

Proof Let $L = \lim\limits_{n \to \infty} a_n$. Then there exists $N > 0$ such that $|a_n - L| < 1$ for $n > N$. In other words,

$$L - 1 < a_n < L + 1 \qquad \text{for } n > N$$

If M is any number larger than $L + 1$ and also larger than the numbers $a_1, a_2, \dots, a_N$, then $a_n < M$ for all n. Thus, M is an upper bound. Similarly, any number m smaller than $L - 1$ and $a_1, a_2, \dots, a_N$ is a lower bound. ∎

There are two ways for a sequence $\{a_n\}$ to be divergent. First, if $\{a_n\}$ is unbounded, then it certainly diverges by Theorem 5. For example, the following sequence diverges:

$$-1, \quad 2, \quad -3, \quad 4, \quad -5, \quad 6, \quad \dots$$

On the other hand, a sequence may diverge even if it is bounded if the terms a_n bounce around and never settle down to approach a limit. For example, the sequence $a_n = (-1)^n$ is bounded but does not converge:

$$1, \quad -1, \quad 1, \quad -1, \quad 1, \quad -1, \quad \dots$$

When can we be sure that a sequence converges? One situation is when $\{a_n\}$ is both bounded and **monotonic** increasing or decreasing. The reason, intuitively, is that if $\{a_n\}$ is increasing and bounded above by M, then the terms must eventually bunch up near some limiting value L that is not greater than M (Figure 12). We state this formally in the next theorem, whose proof is given in Appendix B.

A sequence $\{a_n\}$ is monotonic

- *Increasing if $a_j \le a_{j+1}$ for all j*
- *Decreasing if $a_j \ge a_{j+1}$ for all j.*

FIGURE 12 An increasing sequence with upper bound M approaches a limit L.

THEOREM 6 Bounded Monotonic Sequences Converge

- If $\{a_n\}$ is increasing and $a_n \le M$ for all n, then $\{a_n\}$ converges and $\lim\limits_{n \to \infty} a_n \le M$.
- If $\{a_n\}$ is decreasing and $a_n \ge m$ for all n, then $\{a_n\}$ converges and $\lim\limits_{n \to \infty} a_n \ge m$.

■ **EXAMPLE 10** Verify that $a_n = \sqrt{n + 1} - \sqrt{n}$ is decreasing and bounded below. Does $\lim\limits_{n \to \infty} a_n$ exist?

Solution The function $f(x) = \sqrt{x + 1} - \sqrt{x}$ is decreasing because its derivative is negative:

$$f'(x) = \frac{1}{2\sqrt{x + 1}} - \frac{1}{2\sqrt{x}} < 0 \qquad \text{for } x > 0$$

It follows that $a_n = f(n)$ is also decreasing (see Table 3). The sequence is bounded below by $m = 0$ since $a_n > 0$ for all n. Theorem 6 guarantees that the limit $L = \lim\limits_{n \to \infty} a_n$ exists and $L \ge 0$ (it can be shown that $L = 0$). ■

TABLE 3   The Sequence $a_n = \sqrt{n+1} - \sqrt{n}$ Is Decreasing

$a_n = \sqrt{n+1} - \sqrt{n}$
$a_1 \approx 0.414$
$a_2 \approx 0.318$
$a_3 \approx 0.268$
$a_4 \approx 0.213$
$a_5 \approx 0.196$

■ **EXAMPLE 11** Show that the following sequence is bounded and increasing:

$$a_0 = 0, \quad a_1 = \sqrt{2}, \quad a_2 = \sqrt{2 + \sqrt{2}}, \quad a_3 = \sqrt{2 + \sqrt{2 + \sqrt{2}}}, \quad \ldots$$

Prove that $L = \lim_{n \to \infty} a_n$ exists and compute its value.

Solution This sequence is defined recursively by

$$a_0 = 0, \qquad a_{n+1} = \sqrt{2 + a_n}$$

It would not be hard to find the limit L if we knew in advance that the limit exists. We could proceed as follows. The sequence $b_n = a_{n+1}$ (the same sequence $\{a_n\}$ but starting with a_1) would converge to the same limit L and we would have, using Theorem 4,

$$L = \lim_{n \to \infty} a_{n+1} = \lim_{n \to \infty} \sqrt{2 + a_n} = \sqrt{2 + \lim_{n \to \infty} a_n} = \sqrt{2 + L}$$

Thus, $L = \sqrt{2 + L}$ and hence

$$L^2 = 2 + L \quad \Rightarrow \quad L^2 - L - 2 = (L - 2)(L + 1) = 0$$

It follows that $L = -1$ or $L = 2$, and since $L \geq 0$, we conclude that $L = 2$ (see Table 4). To justify this conclusion, we must prove that the limit L exists. By Theorem 6, it suffices to show that $\{a_n\}$ is bounded above and increasing.

Step 1. Show that $\{a_n\}$ is bounded above by $M = 2$.
First, we observe that

$$\text{if } a_n < 2, \quad \text{then} \quad a_{n+1} = \sqrt{a_n + 2} < \sqrt{2 + 2} = 2 \qquad \boxed{2}$$

Now we can prove that $a_n < 2$ for all n. Since $a_0 = 0$, (2) implies that $a_1 < 2$. But then $a_1 < 2$ implies that $a_2 < 2$ by (2), and $a_2 < 2$ implies $a_3 < 2$, etc. for all n (formally speaking, this is a proof by induction).

Step 2. Show that $\{a_n\}$ is increasing.
Since a_n is positive and $a_n < 2$,

$$a_{n+1} = \sqrt{a_n + 2} > \sqrt{a_n + a_n} = \sqrt{2a_n} > \sqrt{a_n \cdot a_n} = a_n$$

Thus, $a_{n+1} > a_n$ for all n and $\{a_n\}$ is increasing. ■

TABLE 4   Terms of the Recursive Sequence $a_{n+1} = \sqrt{2 + a_n}$

a_0	0
a_1	1.4142
a_2	1.8478
a_3	1.9616
a_4	1.9904
a_5	1.9976

10.1 SUMMARY

- A *sequence* is a function $f(n)$ whose domain is a subset of the integers. We write $a_n = f(n)$ for the nth term and denote the sequence itself by $\{a_n\}$ or simply a_n.
- We say that $\{a_n\}$ *converges* to a limit L, and we write $\lim_{n \to \infty} a_n = L$ or $a_n \to L$ if, for every $\epsilon > 0$, there is a number M such that

$$|a_n - L| < \epsilon \qquad \text{for all } n > M$$

If no limit exists, we say that $\{a_n\}$ *diverges*.
- Let $f(x)$ be a function on $[c, \infty)$ for some number c and let $a_n = f(n)$ for $n \geq c$. If $\lim_{x \to \infty} f(x) = L$, then $\lim_{n \to \infty} a_n = L$.

- A *geometric sequence* is a sequence of the form $a_n = cr^n$, where c and r are nonzero.
- The basic Limit Laws and the Squeeze Theorem apply to sequences.
- If $f(x)$ is continuous and $\lim_{n\to\infty} a_n = L$, then $\lim_{n\to\infty} f(a_n) = f(L)$.
- We say that $\{a_n\}$ is *bounded above* by M if $a_n \leq M$ for all n and *bounded below* by m if $a_n \geq m$ for all n. If $\{a_n\}$ is bounded above and below, $\{a_n\}$ is called *bounded*.
- A sequence $\{a_n\}$ is *monotonic* if it is increasing $(a_j \leq a_{j+1})$ or decreasing $(a_j \geq a_{j+1})$ for all j.
- Theorem 6 states that every increasing sequence that is bounded above and every decreasing sequence that is bounded below converges.

10.1 EXERCISES

Preliminary Questions

1. What is a_4 for the sequence $a_n = n^2 - n$?

2. Which of the following sequences converge to zero?

(a) $\dfrac{n^2}{n^2 + 1}$
(b) 2^n
(c) $\left(\dfrac{-1}{2}\right)^n$

3. Let a_n be the nth decimal approximation to $\sqrt{2}$. That is, $a_1 = 1$, $a_2 = 1.4$, $a_3 = 1.41$, etc. What is $\lim_{n\to\infty} a_n$?

4. Which sequence is defined recursively?

(a) $a_n = \sqrt{2 + n^{-1}}$
(b) $b_n = \sqrt{4 + b_{n-1}}$

5. Theorem 5 says that every convergent sequence is bounded. Which of the following statements follow from Theorem 5 and which are false? If false, give a counterexample.

(a) If $\{a_n\}$ is bounded, then it converges.

(b) If $\{a_n\}$ is not bounded, then it diverges.

(c) If $\{a_n\}$ diverges, then it is not bounded.

Exercises

1. Match the sequence with the general term:

$a_1, a_2, a_3, a_4, \ldots$	General term
(a) $\frac{1}{2}, \frac{2}{3}, \frac{3}{4}, \frac{4}{5}, \ldots$	(i) $\cos \pi n$
(b) $-1, 1, -1, 1, \ldots$	(ii) $\dfrac{n!}{2^n}$
(c) $1, -1, 1, -1, \ldots$	(iii) $(-1)^{n+1}$
(d) $\frac{1}{2}, \frac{2}{4}, \frac{6}{8}, \frac{24}{16} \cdots$	(iv) $\dfrac{n}{n+1}$

2. Let $a_n = \dfrac{1}{2n-1}$ for $n = 1, 2, 3, \ldots$. Write out the first three terms of the following sequences.

(a) $b_n = a_{n+1}$
(b) $c_n = a_{n+3}$
(c) $d_n = a_n^2$
(d) $e_n = 2a_n - a_{n+1}$

In Exercises 3–10, calculate the first four terms of the following sequences, starting with $n = 1$.

3. $c_n = \dfrac{2^n}{n!}$

4. $b_n = \cos \pi n$

5. $a_1 = 3$, $a_{n+1} = 1 + a_n^2$

6. $b_n = 2 + (-1)^n$

7. $c_n = 1 + \dfrac{1}{2} + \dfrac{1}{3} + \cdots + \dfrac{1}{n}$

8. $a_n = n + (n+1) + (n+2) + \cdots + (2n)$

9. $b_1 = 2$, $b_2 = 5$, $b_n = b_{n-1} + 2b_{n-2}$

10. $c_n = n$th decimal approximation to e^{-1}

11. Find a formula for the nth term of the following sequence:

(a) $\dfrac{1}{1}, \dfrac{-1}{8}, \dfrac{1}{27}, \ldots$
(b) $\dfrac{2}{6}, \dfrac{3}{7}, \dfrac{4}{8}, \ldots$

12. Suppose that $\lim_{n\to\infty} a_n = 4$ and $\lim_{n\to\infty} b_n = 7$. Determine:

(a) $\lim_{n\to\infty} (a_n + b_n)$
(b) $\lim_{n\to\infty} a_n^3$
(c) $\lim_{n\to\infty} 4b_n$
(d) $\lim_{n\to\infty} (a_n^2 - 2a_n b_n)$

In Exercises 13–26, use Theorem 1 to determine the limit of the sequence or state that the sequence diverges.

13. $a_n = 4$

14. $b_n = \dfrac{3n+1}{2n+4}$

15. $a_n = 5 - \dfrac{9}{n^2}$

16. $b_n = (-1)^{2n+1}$

17. $c_n = -2^{-n}$

18. $c_n = 4(2^n)$

19. $z_n = \left(\dfrac{1}{3}\right)^n$

20. $z_n = (0.1)^{-1/n}$

21. $a_n = \dfrac{(-1)^n n^2 + n}{4n^2 + 1}$

22. $a_n = \dfrac{n}{\sqrt{n^2 + 1}}$

23. $a_n = \dfrac{n}{\sqrt{n^3 + 1}}$

24. $a_n = \sin \pi n$

25. $a_n = \cos \pi n$

26. $a_n = n((1 + n^{-1})^2 - 1)$

27. Let $a_n = \dfrac{n}{n+1}$. Find a number M such that:

(a) $|a_n - 1| \le 0.001$ for $n \ge M$.

(b) $|a_n - 1| \le 0.00001$ for $n \ge M$.

Then use the limit definition to prove that $\lim\limits_{n \to \infty} a_n = 1$.

28. Let $b_n = (\frac{1}{3})^n$.

(a) Find a value of M such that $|b_n| \le 10^{-5}$ for $n \ge M$.

(b) Use the limit definition to prove that $\lim\limits_{n \to \infty} b_n = 0$.

29. Use the limit definition to prove that $\lim\limits_{n \to \infty} n^{-2} = 0$.

30. Find the limit of $d_n = \sqrt{n+3} - \sqrt{n}$.

31. Find $\lim\limits_{n \to \infty} 2^{1/n}$.

32. Show that $\lim\limits_{n \to \infty} b^{1/n}$ is independent of b for $b > 0$.

33. Find $\lim\limits_{n \to \infty} n^{1/n}$.

34. Find the limit of $a_n = n^2(\sqrt[3]{n^3 + 1} - n)$. *Hint:* Write $a_n = \dfrac{(1 + n^{-3})^{1/3} - 1}{n^{-3}}$ and apply L'Hôpital's Rule.

35. Find $\lim\limits_{n \to \infty} \left(1 + \dfrac{1}{n}\right)^n$.

36. Find $\lim\limits_{n \to \infty} \left(1 + \dfrac{1}{n^2}\right)^n$.

37. Use the Squeeze Theorem to find $\lim\limits_{n \to \infty} a_n$, where $a_n = \dfrac{1}{\sqrt{n^4 + n^8}}$ by proving that

$$\frac{1}{\sqrt{2n^4}} \le a_n \le \frac{1}{\sqrt{2n^2}}$$

38. Evaluate $\lim\limits_{n \to \infty} \dfrac{\cos n}{n}$.

39. Evaluate $\lim\limits_{n \to \infty} n \sin \dfrac{1}{n}$.

40. Evaluate $\lim\limits_{n \to \infty} (2^n + 3^n)^{1/n}$. *Hint:* Show that

$$3 \le a_n \le (2 \cdot 3^n)^{1/n}$$

41. [image] Which statement is equivalent to the assertion $\lim\limits_{n \to \infty} a_n = L$? Explain.

(a) For every $\epsilon > 0$, the interval $(L - \epsilon, L + \epsilon)$ contains at least one element of the sequence $\{a_n\}$.

(b) For every $\epsilon > 0$, the interval $(L - \epsilon, L + \epsilon)$ contains all but at most finitely many elements of the sequence $\{a_n\}$.

42. [image] Which statement is equivalent to the assertion that $\{a_n\}$ is bounded? Explain.

(a) There exists a finite interval $[m, M]$ containing every element of the sequence $\{a_n\}$.

(b) There exists a finite interval $[m, M]$ containing an element of the sequence $\{a_n\}$.

In Exercises 43–63, determine the limit of the sequence or show that the sequence diverges by using the appropriate Limit Laws or theorems.

43. $a_n = \dfrac{3n^2 + n + 2}{2n^2 - 3}$

44. $a_n = \dfrac{\sqrt{n}}{\sqrt{n} + 4}$

45. $a_n = 3 + \left(-\dfrac{1}{2}\right)^n$

46. $a_n = \left(2 + \dfrac{4}{n^2}\right)^{1/3}$

47. $b_n = \tan^{-1}\left(1 - \dfrac{2}{n}\right)$

48. $b_n = e^{n^2 - n}$

49. $c_n = \ln\left(\dfrac{2n+1}{3n+4}\right)$

50. $c_n = \dfrac{n}{n + n^{1/n}}$

51. $y_n = \dfrac{e^n + 3^n}{5^n}$

52. $y_n = \dfrac{e^n}{2^n}$

53. $a_n = \dfrac{e^n}{2^{n^2}}$

54. $a_n = \dfrac{n}{2^n}$

55. $b_n = \dfrac{n^3 + 2e^{-n}}{3n^3 + 4e^{-n}}$

56. $b_n = \dfrac{3 - 4^n}{2 + 7 \cdot 4^n}$

57. $a_n = \dfrac{3 - 4^n}{2 + 7 \cdot 3^n}$

58. $b_n = \dfrac{10^n}{n!}$

59. $a_n = \dfrac{(-4{,}000)^n}{n!}$

60. $b_n = \dfrac{n!}{2^n}$

61. $a_n = \cos \dfrac{\pi}{n}$

62. $a_n = n \sin \dfrac{\pi}{n}$

63. $a_n = \sqrt[n]{n}$

64. Show that $a_n = \dfrac{1}{2n+1}$ is strictly decreasing.

65. Show that $a_n = \dfrac{3n^2}{n^2 + 2}$ is strictly increasing. Find an upper bound.

66. Show that $a_n = \sqrt[3]{n+1} - n$ is decreasing.

67. Use the limit definition to prove that the limit does not change if a finite number of terms are added or removed from a convergent sequence.

68. Let $b_n = a_{n+1}$. Prove that if $\{a_n\}$ converges, then $\{b_n\}$ also converges and $\lim\limits_{n \to \infty} a_n = \lim\limits_{n \to \infty} b_n$.

69. Let $\{a_n\}$ be a sequence such that $\lim\limits_{n\to\infty} |a_n|$ exists and is nonzero. Show that $\lim\limits_{n\to\infty} a_n$ exists if and only if there exists an integer M such that the sign of a_n does not change for $n > M$.

70. Give an example of a divergent sequence $\{a_n\}$ such that $\lim\limits_{n\to\infty} |a_n|$ converges.

71. Show, by giving an example, that there exist *divergent* sequences $\{a_n\}$ and $\{b_n\}$ such that $\{a_n + b_n\}$ converges.

72. Using the limit definition, prove that if $\{a_n\}$ converges and $\{b_n\}$ diverges, then $\{a_n + b_n\}$ diverges.

73. Use the limit definition to prove that if $\{a_n\}$ is a convergent sequence of integers with limit L, then there exists a number M such that $a_n = L$ for all $n \ge M$.

74. Theorem 1 states that if $\lim\limits_{x\to\infty} f(x) = L$, then the sequence $a_n = f(n)$ converges and $\lim\limits_{n\to\infty} a_n = L$. Show that the *converse* is false. In other words, find a function $f(x)$ such that $a_n = f(n)$ converges but $\lim\limits_{x\to\infty} f(x)$ does not exist.

75. Prove that the following sequence is bounded and increasing. Then find its limit:

$$a_1 = \sqrt{5}, \quad a_2 = \sqrt{5 + \sqrt{5}}, \quad a_3 = \sqrt{5 + \sqrt{5 + \sqrt{5}}}, \dots$$

76. Let $\{a_n\}$ be the sequence

$$\sqrt{2}, \quad \sqrt{2\sqrt{2}}, \quad \sqrt{2\sqrt{2\sqrt{2}}}, \dots$$

Show that $\{a_n\}$ is increasing and $0 \le a_n \le 2$. Then prove that $\{a_n\}$ converges and find the limit.

77. Find the limit of the sequence

$$c_n = \frac{1}{\sqrt{n^2 + 1}} + \frac{1}{\sqrt{n^2 + 2}} + \cdots + \frac{1}{\sqrt{n^2 + n}}$$

Hint: Show that

$$\frac{n}{\sqrt{n^2 + n}} \le c_n \le \frac{n}{\sqrt{n^2 + 1}}$$

Further Insights and Challenges

78. Show that $a_n = \sqrt[n]{n!}$ diverges. *Hint:* Show that $n! \ge (n/2)^{n/2}$ by observing that half of the factors of $n!$ are greater than or equal to $n/2$.

79. Let $b_n = \dfrac{\sqrt[n]{n!}}{n}$.

(a) Show that $\ln b_n = \dfrac{\ln(n!) - n \ln n}{n} = \dfrac{1}{n} \sum\limits_{k=1}^{n} \ln \dfrac{k}{n}$.

(b) Show that $\ln b_n$ converges to $\displaystyle\int_0^1 \ln x \, dx$ and conclude that $b_n \to e^{-1}$.

80. Given positive numbers $a_1 < b_1$, define two sequences recursively by

$$a_{n+1} = \sqrt{a_n b_n}, \qquad b_{n+1} = \frac{a_n + b_n}{2}$$

(a) Show that $a_n \le b_n$ for all n (Figure 13).

(b) Show that $\{a_n\}$ is increasing and $\{b_n\}$ is decreasing.

(c) Show that

$$b_{n+1} - a_{n+1} \le \frac{b_n - a_n}{2}$$

Prove that both $\{a_n\}$ and $\{b_n\}$ converge and have the same limit. This limit, denoted $\text{AGM}(a_1, b_1)$, is called the **arithmetic-geometric mean** of a_1 and b_1. See Figure 13.

(d) Estimate $\text{AGM}(1, \sqrt{2})$ to three decimal places.

81. Let $c_n = \dfrac{1}{n} + \dfrac{1}{n+1} + \dfrac{1}{n+2} + \cdots + \dfrac{1}{2n}$.

Geometric Arithmetic
mean mean

FIGURE 13

(a) Calculate c_1, c_2, c_3, c_4.

(b) Use a comparison of rectangles with the area under $y = x^{-1}$ over the interval $[n, 2n]$ to prove that

$$\int_n^{2n} \frac{dx}{x} + \frac{1}{2n} \le c_n \le \int_n^{2n} \frac{dx}{x} + \frac{1}{n}$$

(c) Use the Squeeze Theorem to determine $\lim\limits_{n\to\infty} c_n$.

82. The nth harmonic number is the number

$$H_n = 1 + \frac{1}{2} + \frac{1}{3} + \cdots + \frac{1}{n}$$

Let $a_n = H_n - \ln n$.

(a) Show that $a_n \ge 0$ for $n \ge 1$. *Hint:* Show that $H_n \ge \displaystyle\int_1^{n+1} \frac{dx}{x}$.

(b) Show that $\{a_n\}$ is decreasing by interpreting $a_n - a_{n+1}$ as an area.

(c) Prove that $\lim\limits_{n\to\infty} a_n$ exists. This limit, denoted γ and known as *Euler's Constant*, appears in many areas of mathematics, including analysis and number theory. It has been calculated to more than 100 million decimal places, but it is still not known if γ is an irrational number. The first 10 digits are $\gamma \approx 0.5772156649$.

10.2 Summing an Infinite Series

Quantities that arise in applications often cannot be computed exactly. We cannot write down an exact decimal expression for the number π or for most values of the sine function such as $\sin 1$. Sometimes these quantities can be represented as infinite sums. For example,

$$\sin 1 = 1 - \frac{1}{3!} + \frac{1}{5!} - \frac{1}{7!} + \frac{1}{9!} - \frac{1}{11!} + \cdots \qquad \boxed{1}$$

Infinite sums of this type are called **infinite series**.

What precisely does Eq. (1) mean? Although it is impossible to add up infinitely many numbers, we can compute the **partial sums** S_N, defined as the first N terms of the infinite series. Let's compare the first few partial sums of the series above with $\sin 1$:

$$S_1 = 1$$

$$S_2 = 1 - \frac{1}{3!} = 1 - \frac{1}{6} \qquad \approx 0.833$$

$$S_3 = 1 - \frac{1}{3!} + \frac{1}{5!} = 1 - \frac{1}{6} + \frac{1}{120} \qquad \approx 0.841667$$

$$S_4 = 1 - \frac{1}{6} + \frac{1}{120} - \frac{1}{5{,}040} \qquad \approx 0.841468$$

$$S_5 = 1 - \frac{1}{6} + \frac{1}{120} - \frac{1}{5{,}040} + \frac{1}{362{,}880} \approx \mathbf{0.8414709846}$$

$$\sin 1 \approx \mathbf{0.8414709848079}$$

The partial sums appear to converge to $\sin 1$ and, indeed, we will prove that $\lim\limits_{N \to \infty} S_N = \sin 1$ in Section 10.7. This is the precise meaning of Eq. (1).

In general, an infinite series is an expression of the form

$$\sum_{n=1}^{\infty} a_n = a_1 + a_2 + a_3 + \cdots$$

where $\{a_n\}$ is any sequence. For example,

Infinite series may begin with any index. For example,

$$\sum_{n=3}^{\infty} \frac{1}{n} = \frac{1}{3} + \frac{1}{4} + \frac{1}{5} + \cdots$$

When it is not necessary to specify the starting point, we write simply $\sum a_n$. Any letter may be used for the index. Thus, we may write a_m, a_k, a_i, etc.

Sequence	General Term	Infinite Series
$\dfrac{1}{3}, \dfrac{1}{9}, \dfrac{1}{27}, \ldots$	$a_n = \dfrac{1}{3^n}$	$\displaystyle\sum_{n=1}^{\infty} \frac{1}{3^n} = \frac{1}{3} + \frac{1}{9} + \frac{1}{27} + \frac{1}{81} + \cdots$
$\dfrac{1}{1}, \dfrac{1}{4}, \dfrac{1}{9}, \dfrac{1}{16}, \ldots$	$a_n = \dfrac{1}{n^2}$	$\displaystyle\sum_{n=1}^{\infty} \frac{1}{n^2} = \frac{1}{1^2} + \frac{1}{2^2} + \frac{1}{3^2} + \frac{1}{4^2} + \cdots$

The Nth partial sum S_N is the finite sum of the first N terms of the series:

$$S_N = \sum_{n=1}^{N} a_n = a_1 + a_2 + a_3 + \cdots + a_N$$

The sum of the infinite series $\displaystyle\sum_{n=1}^{\infty} a_n$ is defined as the limit of the sequence of partial sums S_N, if this limit exists.

> **DEFINITION Convergence of an Infinite Series** An infinite series $\displaystyle\sum_{n=1}^{\infty} a_n$ converges to S if $\displaystyle\lim_{N\to\infty} S_N = S$. The limit S is called the *sum* of the infinite series, and we write $S = \displaystyle\sum_{n=1}^{\infty} a_n$. If the limit does not exist, the infinite series is said to *diverge*.

It is easy to give examples of series that diverge. For example, $\displaystyle\sum_{n=1}^{\infty} 1$ diverges because the partial sums S_N diverge to ∞:

$$S_N = \sum_{n=1}^{N} 1 = \overbrace{1 + 1 + 1 + 1 + \cdots + 1}^{N \text{ times}} = N$$

Similarly, $\displaystyle\sum_{n=1}^{\infty} (-1)^{n-1}$ diverges because the partial sums jump back and forth between 1 and 0:

$$S_1 = 1, \quad S_2 = 1 - 1 = 0, \quad S_3 = 1 - 1 + 1 = 1, \quad S_4 = 1 - 1 + 1 - 1 = 0, \quad \ldots$$

We may investigate series numerically by computing several partial sums S_N. If the partial sums show a trend of convergence to some number S, then we have evidence (but not proof) that the series converges to S. The next example treats a convergent **telescoping series**, where the partial sums are particularly easy to evaluate.

Although there is a simple formula for the partial sums in Example 1, this is the exception rather than the rule. Apart from telescoping series and the geometric series introduced below, there is usually no formula for S_N, and we must develop techniques for studying infinite series that do not rely on formulas.

■ EXAMPLE 1 Telescoping Series Investigate the following series numerically:

$$S = \sum_{n=1}^{\infty} \frac{1}{n(n+1)} = \frac{1}{1(2)} + \frac{1}{2(3)} + \frac{1}{3(4)} + \frac{1}{4(5)} + \cdots$$

Then compute the sum S using the identity:

$$\frac{1}{n(n+1)} = \frac{1}{n} - \frac{1}{n+1}$$

Solution Table 1 lists some partial sums, computed with the help of a CAS. These numerics suggest convergence to $S = 1$. To evaluate the limit exactly, we use the identity to rewrite the terms of the series. We find that each partial sum collapses down to two terms due to cancellation:

$$S_1 = \frac{1}{1(2)} = \frac{1}{1} - \frac{1}{2}$$

$$S_2 = \frac{1}{1(2)} + \frac{1}{2(3)} = \left(\frac{1}{1} - \frac{1}{2}\right) + \left(\frac{1}{2} - \frac{1}{3}\right) = 1 - \frac{1}{3}$$

$$S_3 = \frac{1}{1(2)} + \frac{1}{2(3)} + \frac{1}{3(4)} = \left(\frac{1}{1} - \frac{1}{2}\right) + \left(\frac{1}{2} - \frac{1}{3}\right) + \left(\frac{1}{3} - \frac{1}{4}\right) = 1 - \frac{1}{4}$$

In general,

$$S_N = \left(\frac{1}{1} - \frac{1}{2}\right) + \left(\frac{1}{2} - \frac{1}{3}\right) + \cdots + \left(\frac{1}{N} - \frac{1}{N+1}\right) = 1 - \frac{1}{N+1}$$

TABLE 1 Partial Sums

for $\displaystyle\sum_{n=1}^{\infty} \frac{1}{n(n+1)}$

N	S_N
10	0.90909
50	0.98039
100	0.990099
200	0.995025
300	0.996678

Now we may compute the sum S as the limit of the partial sums:

$$S = \lim_{N \to \infty} S_N = \lim_{N \to \infty} \left(1 - \frac{1}{N+1}\right) = 1 \qquad \blacksquare$$

It is important to keep in mind the difference between a sequence $\{a_n\}$ and an infinite series $\sum_{n=1}^{\infty} a_n$, which is a sum of terms of a sequence.

■ **EXAMPLE 2** Difference between a Sequence and a Series Discuss the difference between $\{a_n\}$ and $\sum_{n=0}^{\infty} a_n$, where $a_n = 3^{-n}$.

Solution The sequence $a_n = 3^{-n}$ converges to zero:

$$1, \quad \frac{1}{3}, \quad \frac{1}{3^2}, \quad \frac{1}{3^3}, \quad \to 0$$

The infinite series defined by this sequence is an infinite sum:

$$\sum_{n=0}^{\infty} 3^{-n} = 1 + \frac{1}{3} + \frac{1}{3^2} + \cdots$$

The value of this sum is nonzero. In fact, the partial sum S_5 gives an approximation to this sum:

$$S_5 = 1 + \frac{1}{3} + \frac{1}{9} + \frac{1}{27} + \frac{1}{81} \approx 1.494 \qquad \blacksquare$$

One of the most important types of infinite series is the **geometric series**, defined as the sum of terms cr^n, where c and r are fixed nonzero numbers:

$$\sum_{n=0}^{\infty} cr^n = c + cr + cr^2 + cr^3 + cr^4 + cr^5 + \cdots$$

The nonzero number r is called the **common ratio**.

For $r = \frac{1}{2}$, we can visualize the sum of the geometric series (Figure 1):

$$\sum_{n=1}^{\infty} \frac{1}{2^n} = \frac{1}{2} + \frac{1}{4} + \frac{1}{8} + \frac{1}{16} + \cdots = 1$$

The sum is 1 because adding terms in the series corresponds to moving stepwise from 0 to 1, where each step consists of a move to the right by half of the remaining distance. A simple device exists for computing the partial sums of a geometric series:

$$S_N = c + cr + cr^2 + cr^3 + \cdots + cr^N$$
$$rS_N = cr + cr^2 + cr^3 + \cdots + cr^N + cr^{N+1}$$
$$S_N - rS_N = S_N(1 - r) = c - cr^{N+1} = c(1 - r^{N+1})$$

If $r \neq 1$, we may divide by $(1 - r)$ to obtain

$$\boxed{S_N = c + cr + cr^2 + cr^3 + \cdots + cr^N = \frac{c(1 - r^{N+1})}{1 - r}} \qquad \boxed{2}$$

FIGURE 1 Partial sums of $\sum_{n=1}^{\infty} \frac{1}{2^n}$.

Geometric series are important because they

• Arise often in applications.
• Can be evaluated explicitly.
• Are used to study other nongeometric series (by comparison).

> **THEOREM 1 Sum of a Geometric Series**   A geometric series with common ratio r converges if $|r| < 1$ and diverges if $|r| \geq 1$. Furthermore,
>
> $$\sum_{n=0}^{\infty} cr^n = \frac{c}{1-r}, \qquad |r| < 1 \qquad \boxed{3}$$
>
> $$\sum_{n=M}^{\infty} cr^n = \frac{cr^M}{1-r}, \qquad |r| < 1 \qquad \boxed{4}$$

Proof If $r \neq 1$, then by Eq. (2),

$$\lim_{N\to\infty} S_N = \lim_{N\to\infty} \frac{c(1-r^{N+1})}{1-r} = \frac{c}{1-r}\left(1 - \lim_{N\to\infty} r^{N+1}\right)$$

If $|r| < 1$, then $\lim_{N\to\infty} r^{N+1} = 0$ and we obtain Eq. (3):

$$\sum_{n=0}^{\infty} cr^n = \lim_{N\to\infty} S_N = \frac{c}{1-r}$$

If $|r| > 1$, $\lim_{N\to\infty} r^{N+1}$ diverges and thus the geometric series diverges. It also diverges in the borderline cases $r = \pm 1$, as we saw in the discussion before Example 1. If the geometric series starts with the term cr^M rather than cr^0, then

$$\sum_{n=M}^{\infty} cr^n = \sum_{n=0}^{\infty} cr^{M+n} = \sum_{n=0}^{\infty} cr^M r^n = r^M \sum_{n=0}^{\infty} cr^n = \frac{cr^M}{1-r} \qquad \blacksquare$$

■ **EXAMPLE 3** Evaluate $\displaystyle\sum_{n=3}^{\infty} 7\left(-\frac{3}{4}\right)^n$.

Solution This is the geometric series with $c = 7$ and $r = -\frac{3}{4}$. The general term is $cr^n = 7\left(-\frac{3}{4}\right)^n$ and the sum starts at $n = 3$. By Eq. (4), the sum is

$$S = \sum_{n=3}^{\infty} 7\left(-\frac{3}{4}\right)^n = \frac{7\left(-\frac{3}{4}\right)^3}{1-\left(-\frac{3}{4}\right)} = -\frac{27}{16} \qquad \blacksquare$$

A main goal in this chapter is to develop techniques for determining whether a series converges or diverges. Sometimes, it is obvious that an infinite series diverges. For example, $\displaystyle\sum_{k=1}^{\infty} 1$ diverges because its Nth partial sum is $S_N = N$. It is less clear if the following series converges or diverges:

$$\sum_{n=1}^{\infty} \frac{(-1)^{n+1}n}{n+1} = \frac{1}{2} - \frac{2}{3} + \frac{3}{4} - \frac{4}{5} + \frac{5}{6} - \cdots$$

We show that this series diverges in Example 4, using the next theorem.

THEOREM 2 Divergence Test If $\{a_n\}$ does not converge to zero, then $\displaystyle\sum_{n=1}^{\infty} a_n$ diverges.

Proof We use the relation

$$S_n = a_1 + a_2 + \cdots + a_{n-1} + a_n = S_{n-1} + a_n$$

to write $a_n = S_n - S_{n-1}$. If $\displaystyle\sum_{n=1}^{\infty} a_n$ is convergent with sum S, then

$$\lim_{n\to\infty} a_n = \lim_{n\to\infty} (S_n - S_{n-1}) = \lim_{n\to\infty} S_n - \lim_{n\to\infty} S_{n-1} = S - S = 0$$

Therefore, if $\{a_n\}$ does not converge to zero, $\displaystyle\sum_{n=1}^{\infty} a_n$ must diverge. ∎

■ **EXAMPLE 4 Using the Divergence Test** Does

$$\sum_{n=1}^{\infty} (-1)^n \frac{n}{n+1} = -\frac{1}{2} + \frac{2}{3} - \frac{3}{4} + \frac{4}{5} - \cdots$$

converge?

Solution The general term $a_n = (-1)^n \dfrac{n}{n+1}$ does not tend to zero. In fact, $\dfrac{n}{n+1}$ tends to 1, so the even terms a_{2n} tend to 1 and the odd terms a_{2n+1} to -1. Therefore, the infinite series diverges by Theorem 2. ∎

The Divergence Test only tells part of the story. If $\{a_n\}$ does not tend to zero, then $\sum a_n$ certainly diverges. But what if $\{a_n\}$ does tend to zero? In this case, the series may or may not converge. Here is an example of a series that diverges even though its terms tend to zero.

■ **EXAMPLE 5 A Divergent Series Where $\{a_n\}$ Tends to Zero** Show that

$$\sum_{n=1}^{\infty} \frac{1}{\sqrt{n}} = \frac{1}{\sqrt{1}} + \frac{1}{\sqrt{2}} + \frac{1}{\sqrt{3}} + \cdots$$

is divergent.

Solution Each term in the Nth partial sum is greater than or equal to $1/\sqrt{N}$:

$$S_N = \overbrace{\frac{1}{\sqrt{1}} + \frac{1}{\sqrt{2}} + \cdots + \frac{1}{\sqrt{N}}}^{N \text{ terms}} \geq \frac{1}{\sqrt{N}} + \frac{1}{\sqrt{N}} + \cdots + \frac{1}{\sqrt{N}} = N\left(\frac{1}{\sqrt{N}}\right)$$

Therefore,

$$S_N \geq N\left(\frac{1}{\sqrt{N}}\right) = \sqrt{N}$$

Since $S_N \geq \sqrt{N}$, we have $\displaystyle\lim_{N\to\infty} S_N = \infty$ and the series diverges. ∎

The next theorem shows that infinite series may be added or subtracted like ordinary sums, *provided that the series are convergent*.

THEOREM 3 Linearity of Infinite Series   If $\sum a_n$ and $\sum b_n$ are both convergent, then $\sum (a_n \pm b_n)$ and $\sum c a_n$ are convergent (c any constant), and

$$\sum a_n + \sum b_n = \sum (a_n + b_n)$$

$$\sum a_n - \sum b_n = \sum (a_n - b_n)$$

$$\sum c a_n = c \sum a_n \qquad \text{(c any constant)}$$

Proof   These rules follow from the corresponding linearity rules for limits. For the first rule, we have

$$\sum_{n=1}^{\infty} (a_n + b_n) = \lim_{N \to \infty} \sum_{n=1}^{N} (a_n + b_n) = \lim_{N \to \infty} \left(\sum_{n=1}^{N} a_n + \sum_{n=1}^{N} b_n \right)$$

$$= \lim_{N \to \infty} \sum_{n=1}^{N} a_n + \lim_{N \to \infty} \sum_{n=1}^{\infty} b_n = \sum_{n=1}^{\infty} a_n + \sum_{n=1}^{\infty} b_n$$

The remaining statements are proved similarly.   ■

■ **EXAMPLE 6**   Evaluate $S = \displaystyle\sum_{n=0}^{\infty} \frac{2 + 3^n}{5^n}$.

Solution   We write the series as a sum of two geometric series. This is valid by Theorem 3 because both geometric series are convergent:

$$\sum_{n=0}^{\infty} \frac{2+3^n}{5^n} = \sum_{n=0}^{\infty} \frac{2}{5^n} + \sum_{n=0}^{\infty} \frac{3^n}{5^n} = \overbrace{2 \sum_{n=0}^{\infty} \frac{1}{5^n} + \sum_{n=0}^{\infty} \left(\frac{3}{5} \right)^n}^{\text{Both convergent geometric series}}$$

$$= 2 \cdot \frac{1}{1 - \frac{1}{5}} + \frac{1}{1 - \frac{3}{5}} = 5$$ ■

CONCEPTUAL INSIGHT   Sometimes, the following *incorrect argument* is given for summing a geometric series:

$$S = \qquad \frac{1}{2} + \frac{1}{4} + \frac{1}{8} + \cdots$$

$$2S = 1 + \frac{1}{2} + \frac{1}{4} + \frac{1}{8} + \cdots = 1 + S$$

Thus, $2S = 1 + S$ or $S = 1$. The answer is correct, so why is the argument wrong? It is wrong because we do not know in advance that the geometric series converges. Observe what happens when this argument is applied to a divergent series:

$$S = 1 - 1 + 1 - 1 + \cdots$$

$$-S = -1 + (1 - 1 + \cdots) = -1 + S$$

This would yield $-S = -1 + S$ or $S = \frac{1}{2}$, which is clearly incorrect since S diverges. Mathematicians developed the formal definition of the sum of an infinite series as the limit of partial sums in order to avoid erroneous conclusions of this type.

Archimedes (287 BC–212 BC), who discovered the law of the lever, said "Give me a place to stand on, and I can move the earth" (quoted by Pappus of Alexandria c. AD 340).

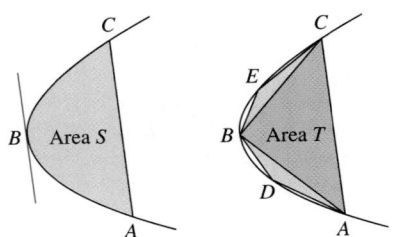

FIGURE 2 Archimedes showed that the area S of the parabolic segment is $\frac{4}{3}T$, where T is the area of $\triangle ABC$.

Infinite series have been a part of calculus since the beginning of the subject and they have remained an indispensable tool of mathematical analysis ever since. Geometric series were used as early as the third century BC by Archimedes in a brilliant argument for determining the area S of a parabolic segment (shaded region in Figure 2). Archimedes's result is equivalent to our formula for the integral of $f(x) = x^2$, but he discovered it 2,000 years before the invention of calculus. Archimedes expressed his result geometrically rather than in terms of functions (which had not yet been invented). Given any two points A and C on a parabola, we may choose B between A and C so that the tangent at B is parallel to $\overline{AC}$. Let T be the area of triangle $\triangle ABC$. Archimedes proved that if D is chosen in a similar fashion relative to $\overline{AB}$ and E relative to $\overline{BC}$, then

$$\frac{1}{4}T = \frac{1}{4}\text{Area}(\triangle ABC)$$

$$= \text{Area}(\triangle ADB) + \text{Area}(\triangle BEC) \quad \boxed{5}$$

This construction of triangles can be continued. The next step would be to construct the four triangles on the segments $\overline{AD}, \overline{DB}, \overline{BE}, \overline{EC}$, of total area $(1/4)^2 T$, etc. In this way, we obtain infinitely many triangles that completely fill up the parabolic segment. By Eq. (5) and the formula for the sum of a geometric series,

$$S = T + \frac{1}{4}T + \frac{1}{16}T + \cdots = T\sum_{n=0}^{\infty}\frac{1}{4^n} = \frac{4}{3}T$$

For this and many other achievements, Archimedes is ranked as one of the greatest scientists of all time, in the same league as Newton and Gauss.

The modern study of infinite series began in the seventeenth century with Newton, Leibniz, and their contemporaries. The divergence of $\sum_{n=1}^{\infty}\frac{1}{n}$ (called the **harmonic series**) was known to the medieval scholar Nicole d'Oresme (1323–1382), but his proof was lost for centuries and the result was rediscovered on more than one occasion. It was also known that the sum of the reciprocal squares $\sum_{n=1}^{\infty}\frac{1}{n^2}$ converges and, in the 1640s, the Italian Pietro Mengoli put forward the challenge of finding its sum. Despite the efforts of the best mathematicians of the day, including Leibniz and the Bernoulli brothers Jakob and Johann, the problem resisted solution for nearly a century. In 1735, the great master Leonhard Euler astonished his contemporaries by proving that

$$\frac{1}{1^2} + \frac{1}{2^2} + \frac{1}{3^2} + \frac{1}{4^2} + \frac{1}{5^2} + \frac{1}{6^2} + \cdots = \frac{\pi^2}{6}$$

$\boxed{6}$

This formula is used in a variety of ways in number theory. For example, the probability p that two whole numbers, chosen randomly, have no common factor is $p = 6/\pi^2 \approx 0.6$ (the reciprocal of Euler's result). This application and others like it lie in the realm of "pure mathematics," and for hundreds of years, it seemed that Euler's result had no real-world applications. Surprisingly, there is now evidence that this result and its generalizations may play a role in the area of advanced physics called quantum field theory. History seems to show that even the "purest" branches of mathematics are connected to the real world.

10.2 SUMMARY

- An *infinite series* is an expression

$$\sum_{n=1}^{\infty} a_n = a_1 + a_2 + a_3 + a_4 + \cdots$$

We call a_n the *general term* of the series.

- The Nth *partial sum* is the finite sum:

$$S_N = \sum_{n=1}^{N} a_n = a_1 + a_2 + a_3 + \cdots + a_N$$

If the limit $S = \lim_{N \to \infty} S_N$ exists, we say that the infinite series is *convergent* or *converges* to the sum S. If the limit does not exist, the infinite series is called *divergent*.

- *Divergence Test:* If $\{a_n\}$ does not tend to zero, then $\sum_{n=1}^{\infty} a_n$ diverges. However, a series may diverge, even if its general term $\{a_n\}$ tends to zero.
- A *geometric series* with common ratio r satisfying $|r| < 1$ converges:

$$\sum_{n=M}^{\infty} cr^n = cr^M + cr^{M+1} + cr^{M+2} + \cdots = \frac{cr^M}{1-r}$$

The geometric series diverges for $|r| \geq 1$. There is a formula for the partial sum:

$$c + cr + cr^2 + cr^3 + \cdots + cr^N = \frac{c(1 - r^{N+1})}{1 - r}$$

10.2 EXERCISES

Preliminary Questions

1. What role do partial sums play in defining the sum of an infinite series?

2. What is the sum of the following infinite series?

$$\frac{1}{4} + \frac{1}{8} + \frac{1}{16} + \frac{1}{32} + \frac{1}{64} + \cdots$$

3. What happens if you apply the formula for the sum of a geometric series to the following series? Is the formula valid?

$$1 + 3 + 3^2 + 3^3 + 3^4 + \cdots$$

4. Arvind asserts that $\sum_{n=1}^{\infty} \frac{1}{n^2} = 0$ because $\frac{1}{n^2}$ tends to zero. Is this valid reasoning?

5. Colleen claims that $\sum_{n=1}^{\infty} \frac{1}{\sqrt{n}}$ converges because $\lim_{n \to \infty} \frac{1}{\sqrt{n}} = 0$. Is this valid reasoning?

6. Find an N such that $S_N > 25$ for the series $\sum_{n=1}^{\infty} 2$.

7. Does there exist an N such that $S_N > 25$ for the series $\sum_{n=1}^{\infty} 2^{-n}$? Explain.

8. Give an example of a divergent infinite series whose general term tends to zero.

Exercises

1. Find a formula for the general term a_n (not the partial sum) of the infinite series.

(a) $\dfrac{1}{3} + \dfrac{1}{9} + \dfrac{1}{27} + \dfrac{1}{81} + \cdots$

(b) $\dfrac{1}{1} + \dfrac{5}{2} + \dfrac{25}{4} + \dfrac{125}{8} + \cdots$

(c) $\dfrac{1}{1} - \dfrac{2^2}{2 \cdot 1} + \dfrac{3^3}{3 \cdot 2 \cdot 1} - \dfrac{4^4}{4 \cdot 3 \cdot 2 \cdot 1} + \cdots$

(d) $\dfrac{2}{1^2 + 1} + \dfrac{1}{2^2 + 1} + \dfrac{2}{3^2 + 1} + \dfrac{1}{4^2 + 1} + \cdots$

2. Write in summation notation:

(a) $1 + \dfrac{1}{4} + \dfrac{1}{9} + \dfrac{1}{16} + \cdots$

(b) $\dfrac{1}{9} + \dfrac{1}{16} + \dfrac{1}{25} + \dfrac{1}{36} + \cdots$

(c) $1 - \dfrac{1}{3} + \dfrac{1}{5} - \dfrac{1}{7} + \cdots$

(d) $\dfrac{125}{9} + \dfrac{625}{16} + \dfrac{3,125}{25} + \dfrac{15,625}{36} + \cdots$

In Exercises 3–6, compute the partial sums S_2, S_4, and S_6.

3. $1 + \dfrac{1}{2^2} + \dfrac{1}{3^2} + \dfrac{1}{4^2} + \cdots$

4. $\displaystyle\sum_{k=1}^{\infty} (-1)^k k^{-1}$

5. $\dfrac{1}{1 \cdot 2} + \dfrac{1}{2 \cdot 3} + \dfrac{1}{3 \cdot 4} + \cdots$

6. $\displaystyle\sum_{j=1}^{\infty} \dfrac{1}{j!}$

7. Compute S_5, S_{10}, and S_{15} for the series

$$S = \dfrac{1}{2 \cdot 3 \cdot 4} - \dfrac{1}{4 \cdot 5 \cdot 6} + \dfrac{1}{6 \cdot 7 \cdot 8} - \dfrac{1}{8 \cdot 9 \cdot 10} + \cdots$$

This series S is known to converge to $\dfrac{\pi - 3}{4}$. Do your calculations support this conclusion?

8. The series

$$S = \dfrac{1}{0!} - \dfrac{1}{1!} + \dfrac{1}{2!} - \dfrac{1}{3!} + \cdots$$

is known to converge to e^{-1} (recall that $0! = 1$). Find a partial sum that approximates e^{-1} with an error at most 10^{-3}.

9. Calculate S_3, S_4, and S_5 and then find the sum of the telescoping series

$$S = \sum_{n=1}^{\infty} \left(\dfrac{1}{n+1} - \dfrac{1}{n+2} \right)$$

10. Calculate S_3, S_4, and S_5 and then find the sum $S = \displaystyle\sum_{n=1}^{\infty} \dfrac{1}{4n^2 - 1}$ using the identity

$$\dfrac{1}{4n^2 - 1} = \dfrac{1}{2} \left(\dfrac{1}{2n - 1} - \dfrac{1}{2n + 1} \right)$$

11. Write $\displaystyle\sum_{n=3}^{\infty} \dfrac{1}{n(n-1)}$ as a telescoping series and find its sum.

12. Find a formula for the partial sums S_N of $\displaystyle\sum_{n=1}^{\infty} (-1)^{n-1}$ and show that the series diverges.

In Exercises 13–16, use Theorem 2 to prove that the following series diverge.

13. $\displaystyle\sum_{n=1}^{\infty} (-1)^n n^2$

14. $\dfrac{0}{1} - \dfrac{1}{2} + \dfrac{2}{3} - \dfrac{3}{4} + \cdots$

15. $\displaystyle\sum_{n=0}^{\infty} \left(\sqrt{4n^2 + 1} - n \right)$

16. $\cos \dfrac{1}{2} + \cos \dfrac{1}{3} + \cos \dfrac{1}{4} + \cdots$

17. Which of these series converge?

(a) $\displaystyle\sum_{n=1}^{\infty} \left(\dfrac{1}{\sqrt{n}} - \dfrac{1}{\sqrt{n+1}} \right)$

(b) $\displaystyle\sum_{n=1}^{\infty} \left(\ln n - \ln(n+1) \right)$

In Exercises 18–31, use the formula for the sum of a geometric series to find the sum or state that the series diverges.

18. $1 + \dfrac{1}{5} + \dfrac{1}{5^2} + \dfrac{1}{5^3} + \cdots$

19. $\dfrac{1}{3^3} + \dfrac{1}{3^4} + \dfrac{1}{3^5} + \cdots$

20. $\displaystyle\sum_{n=0}^{\infty} \dfrac{3^n}{11^n}$

21. $\displaystyle\sum_{n=3}^{\infty} \dfrac{3^n}{11^n}$

22. $\displaystyle\sum_{n=0}^{\infty} \dfrac{7 \cdot 3^n}{11^n}$

23. $1 + \dfrac{2}{7} + \dfrac{2^2}{7^2} + \dfrac{2^3}{7^3} + \cdots$

24. $\displaystyle\sum_{n=1}^{\infty} e^{-n}$

25. $\displaystyle\sum_{n=2}^{\infty} e^{3-2n}$

26. $\displaystyle\sum_{n=0}^{\infty} \dfrac{8 + 2^n}{5^n}$

27. $\displaystyle\sum_{n=0}^{\infty} \dfrac{93^n + 4^{n-2}}{5^n}$

28. $5 - \dfrac{5}{4} + \dfrac{5}{4^2} - \dfrac{5}{4^3} + \cdots$

29. $\dfrac{2^3}{7} + \dfrac{2^4}{7^2} + \dfrac{2^5}{7^3} + \dfrac{2^6}{7^4} + \cdots$

30. $\dfrac{7}{8} - \dfrac{49}{64} + \dfrac{343}{512} - \dfrac{2{,}401}{4{,}096} + \cdots$

31. $\dfrac{64}{49} + \dfrac{8}{7} + 1 + \dfrac{7}{8} + \dfrac{49}{64} + \dfrac{343}{512} + \cdots$

32. Which of the following are *not* geometric series?

(a) $\displaystyle\sum_{n=0}^{\infty} \dfrac{7^n}{29^n}$

(b) $\displaystyle\sum_{n=3}^{\infty} \dfrac{1}{n^4}$

(c) $\displaystyle\sum_{n=0}^{\infty} \dfrac{n^2}{2^n}$

(d) $\displaystyle\sum_{n=5}^{\infty} \pi^{-n}$

33. Which of the following series are divergent?

(a) $\displaystyle\sum_{n=0}^{\infty} \dfrac{2^n}{5^n}$

(b) $\displaystyle\sum_{n=3}^{\infty} 1.5^n$

(c) $\displaystyle\sum_{n=0}^{\infty} \dfrac{5^n}{2^n}$

(d) $\displaystyle\sum_{n=0}^{\infty} (0.4)^n$

34. ✍ Explain why each of the following statements is incorrect.

(a) If the general term a_n tends to zero, $\displaystyle\sum_{n=1}^{\infty} a_n = 0$.

(b) The Nth partial sum of the infinite series defined by $\{a_n\}$ is a_N.

(c) If a_n tends to zero, then $\displaystyle\sum_{n=1}^{\infty} a_n$ converges.

(d) If a_n tends to L, then $\displaystyle\sum_{n=1}^{\infty} a_n = L$.

35. Let $S = \displaystyle\sum_{n=1}^{\infty} a_n$ be an infinite series such that $S_N = 5 - \dfrac{2}{N^2}$.

(a) What are the values of $\displaystyle\sum_{n=1}^{10} a_n$ and $\displaystyle\sum_{n=4}^{16} a_n$?

(b) What is the value of a_3?

(c) Find a general formula for a_n.

(d) Find the sum $\displaystyle\sum_{n=1}^{\infty} a_n$.

36. Compute the total area of the (infinitely many) triangles in Figure 3.

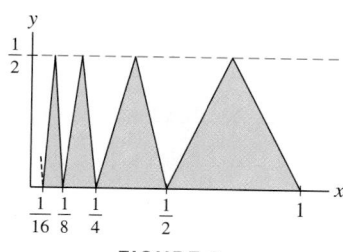

FIGURE 3

37. Use the method of Example 5 to show that $\displaystyle\sum_{k=1}^{\infty} \frac{1}{k^{1/3}}$ diverges.

38. Let S_N be the Nth partial sum of the **harmonic series** $S = \displaystyle\sum_{n=1}^{\infty} \frac{1}{n}$.

(a) Verify the following inequality for $n = 1, 2, 3$. Then prove it for general n.

$$\frac{1}{2^{n-1}+1} + \frac{1}{2^{n-1}+2} + \frac{1}{2^{n-1}+3} + \cdots + \frac{1}{2^n} \geq \frac{1}{2}$$

(b) Prove that S diverges by showing that $S_N \geq 1 + \dfrac{n}{2}$ for $N = 2^n$.

Hint: Break up S_N into $n+1$ sums of length $1, 2, 4, 8 \ldots$, as in the following:

$$S_{2^3} = 1 + \left(\frac{1}{2}\right) + \left(\frac{1}{3} + \frac{1}{4}\right) + \left(\frac{1}{5} + \frac{1}{6} + \frac{1}{7} + \frac{1}{8}\right)$$

39. A ball dropped from a height of 10 ft begins to bounce. Each time it strikes the ground, it returns to two-thirds of its previous height. What is the total distance traveled by the ball if it bounces infinitely many times?

40. Use partial fractions to rewrite $\displaystyle\sum_{n=1}^{\infty} \frac{1}{n(n+3)}$ as a telescoping series and find its sum.

41. Find the sum of $\dfrac{1}{1 \cdot 3} + \dfrac{1}{3 \cdot 5} + \dfrac{1}{5 \cdot 7} + \cdots$.

42. Let $S = \displaystyle\sum_{n=1}^{\infty} \left(\frac{1}{n} - \frac{1}{n+2}\right)$. Compute S_N for $N = 1, 2, 3, 4$. Find S by showing that

$$S_N = \frac{3}{2} - \frac{1}{N+1} - \frac{1}{N+2}$$

43. Let $\{b_n\}$ be a sequence and let $a_n = b_n - b_{n-1}$. Show that $\displaystyle\sum_{n=1}^{\infty} a_n$ converges if and only if $\displaystyle\lim_{n\to\infty} b_n$ exists.

44. The winner of a lottery receives m dollars at the end of each year for N years. The present value (PV) of this prize in today's dollars is PV $= \displaystyle\sum_{i=1}^{N} m(1+r)^{-i}$, where r is the interest rate. Calculate PV if $m = \$50,000$, $r = 0.06$, and $N = 20$. What is the PV if $N = \infty$?

45. Find the total length of the infinite zigzag path in Figure 4 (each zag occurs at an angle of $\frac{\pi}{4}$).

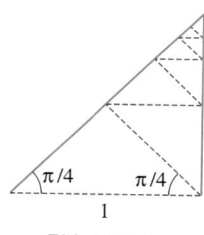

FIGURE 4

46. Evaluate $\displaystyle\sum_{n=1}^{\infty} \frac{1}{n(n+1)(n+2)}$. *Hint:* Find constants A, B, and C such that

$$\frac{1}{n(n+1)(n+2)} = \frac{A}{n} + \frac{B}{(n+1)} + \frac{C}{n+2}$$

47. Show that if a is a positive integer, then

$$\sum_{n=1}^{\infty} \frac{1}{n(n+a)} = \frac{1}{a}\left(1 + \frac{1}{2} + \cdots + \frac{1}{a}\right)$$

48. Assumptions Matter Show, by giving counterexamples, that the assertions of Theorem 3 is not valid if the series $\displaystyle\sum_{n=0}^{\infty} a_n$ and $\displaystyle\sum_{n=0}^{\infty} b_n$ are not convergent.

Further Insights and Challenges

49. Professor George Andrews of Pennsylvania State University observed that geometric sums can be used to calculate the derivative of $f(x) = x^N$ in a new way. By Eq. (2),

$$1 + r + r^2 + \cdots + r^{N-1} = \frac{1 - r^N}{1 - r} \qquad \boxed{7}$$

Assume that $a \neq 0$ and let $x = ra$. Show that

$$f'(a) = \lim_{x \to a} \frac{x^N - a^N}{x - a} = a^{N-1} \lim_{r \to 1} \frac{r^N - 1}{r - 1}$$

Then use Eq. (7) to evaluate the limit on the right.

50. Pierre de Fermat used geometric series to compute the area under the graph of $f(x) = x^N$ over $[0, A]$. For $0 < r < 1$, let $F(r)$ be the sum of the areas of the infinitely many right-endpoint rectangles with endpoints Ar^n, as in Figure 5. As r tends to 1, the rectangles become narrower and $F(r)$ tends to the area under the graph.

(a) Show that $F(r) = A^{N+1} \dfrac{1 - r}{1 - r^{N+1}}$.

(b) Use Eq. (7) to evaluate $\displaystyle\int_0^A x^N \, dx = \lim_{r \to 1} F(r)$.

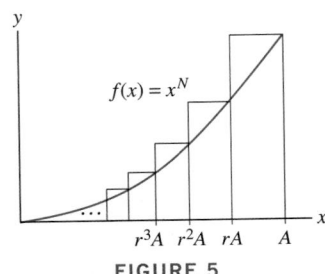

FIGURE 5

51. Cantor's Disappearing Table (following Larry Knop of Hamilton College)   Take a table of length L (Figure 6). At stage 1, remove the section of length $L/4$ centered at the midpoint. Two sections remain, each with a length less than $L/2$. At stage 2, remove sections of length $L/4^2$ from each of these two sections (this stage removes $L/8$ of the table). Now four sections remain, each of length less than $L/4$. At stage 3, remove the four central sections of length $L/4^3$, etc.

(a) Show that at the Nth stage, each remaining section has length less than $L/2^N$ and that the total amount of table removed is

$$L\left(\frac{1}{4} + \frac{1}{8} + \frac{1}{16} + \cdots + \frac{1}{2^{N+1}}\right)$$

(b) Show that in the limit as $N \to \infty$, precisely one-half of the table remains.

This result is curious, because there are no nonzero intervals of table left (at each stage, the remaining sections have a length less than $L/2^N$). So the table has "disappeared." However, we can place any object longer than $L/4$ on the table and it will not fall through since it will not fit through any of the removed sections.

FIGURE 6

52. The **Koch snowflake** (described in 1904 by Swedish mathematician Helge von Koch) is an infinitely jagged "fractal" curve obtained as a limit of polygonal curves (it is continuous, but has no tangent line at any point). Begin with an equilateral triangle (stage 0) and produce stage 1 by replacing each edge with four edges of one-third the length, arranged as in Figure 7. Continue the process: At the nth stage, replace each edge with four edges of one-third the length.

(a) Show that the perimeter P_n of the polygon at the nth stage satisfies $P_n = \frac{4}{3} P_{n-1}$. Prove that $\lim_{n \to \infty} P_n = \infty$. The snowflake has infinite length.

(b) Let A_0 be the area of the original equilateral triangle. Show that $(3)4^{n-1}$ new triangles are added at the nth stage, each with area $A_0/9^n$ (for $n \geq 1$). Show that the total area of the Koch snowflake is $\frac{8}{5}A_0$.

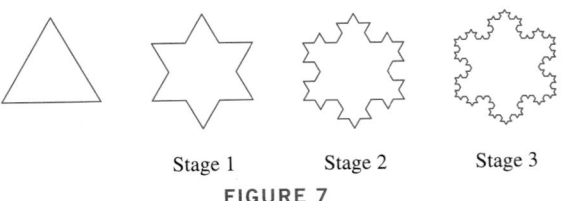

Stage 1   Stage 2   Stage 3

FIGURE 7

10.3 Convergence of Series with Positive Terms

In the next three sections, we focus on the problem of determining whether an infinite series converges or diverges. This is easier than finding the sum of an infinite series, which is possible only in special cases.

In this section, we consider **positive series** $\sum a_n$, defined as a series such that $a_n \geq 0$ for all n (thus the terms of the series are nonnegative). The terms of a positive series may be visualized as rectangles of width 1 and height a_n (Figure 1). The partial sum

$$S_N = a_1 + a_2 + \cdots + a_N$$

is equal to the area of the first N rectangles.

There are powerful numerical methods for finding approximations to infinite series. When implemented on a computer, these methods can be used to compute sums to millions of digits (see Exercises 75–77).

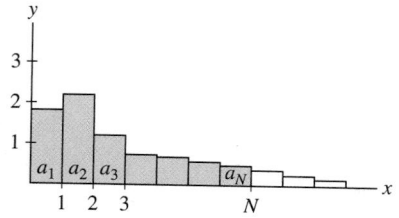

FIGURE 1 The partial sum S_N is the sum of the areas of the N shaded rectangles.

A key property of positive series is that the partial sums S_N form an increasing sequence. Each partial sum is obtained from the previous one by adding a nonnegative number:

$$S_{N+1} = \left(a_1 + a_2 + \cdots + a_N\right) + a_{N+1} = S_N + \underbrace{a_{N+1}}_{\text{Nonnegative}}$$

and thus $S_N \le S_{N+1}$. Recall that an increasing sequence converges if it is bounded above and diverges otherwise (Theorem 6, Section 10.1). It follows that there are just two ways a positive series can behave (we refer to this as the "dichotomy").

THEOREM 1 Dichotomy Theorem for Positive Series   If $S = \displaystyle\sum_{n=1}^{\infty} a_n$ is a positive series, then there are two possibilities:

(i) The partial sums S_N are bounded above. In this case, S converges.
(ii) The partial sums S_N are not bounded above. In this case, S diverges.

Assumptions Matter   This dichotomy does not hold in general for nonpositive series. The partial sums of the nonpositive series

$$S = \sum_{n=1}^{\infty} (-1)^n = 1 - 1 + 1 - 1 + 1 - 1 + \cdots$$

are bounded since $S_N = 1$ or 0, but S diverges.

One of the most important applications of Theorem 1 is the following Integral Test, which is useful because integrals are often easier to evaluate than series.

The Integral Test is valid for any series $\displaystyle\sum_{n=k}^{\infty} f(n)$, provided that $f(x)$ is positive, decreasing, and continuous for $x \ge M$ for some M. The convergence of the series is determined by the convergence of

$$\int_{M}^{\infty} f(x)\,dx$$

THEOREM 2 Integral Test   Let $a_n = f(n)$, where $f(x)$ is positive, decreasing, and continuous for $x \ge 1$.

(i) If $\displaystyle\int_{1}^{\infty} f(x)\,dx$ converges, then $\displaystyle\sum_{n=1}^{\infty} a_n$ converges.

(ii) If $\displaystyle\int_{1}^{\infty} f(x)\,dx$ diverges, then $\displaystyle\sum_{n=1}^{\infty} a_n$ diverges.

Proof   We compare S_N with the area under the graph of $f(x)$ over the interval $[1, N]$. Since $f(x)$ is decreasing (Figure 2),

$$\underbrace{a_2 + \cdots + a_N}_{\text{Area of shaded rectangles in Figure 2}} \le \int_{1}^{N} f(x)\,dx \le \int_{1}^{\infty} f(x)\,dx$$

If the improper integral on the right converges, then the sums $a_2 + \cdots + a_N$ remain bounded. In this case, S_N also remains bounded and the infinite series converges by the Dichotomy Theorem (Theorem 1). This proves (i).

On the other hand (Figure 3),

$$\int_{1}^{N} f(x)\,dx \le \underbrace{a_1 + a_2 + \cdots + a_{N-1}}_{\text{Area of shaded rectangles in Figure 3}} \qquad \boxed{1}$$

FIGURE 2

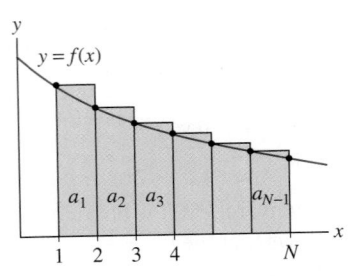

$y = f(x)$

a_1 a_2 a_3 a_{N-1}

1   2   3   4 N

FIGURE 3

The infinite series

$$\sum_{n=1}^{\infty} \frac{1}{n}$$

is called the "harmonic series."

If $\int_{1}^{\infty} f(x)\,dx$ diverges, then $\int_{1}^{N} f(x)\,dx$ tends to ∞ and (1) shows that S_N also tends to ∞. This proves (ii). ∎

■ **EXAMPLE 1**   **Divergence of the Harmonic Series**   Show that $\displaystyle\sum_{n=1}^{\infty} \frac{1}{n}$ diverges.

Solution   The function $f(x) = \dfrac{1}{x}$ is positive, decreasing, and continuous for $x \geq 1$, so we may apply the Integral Test:

$$\int_{1}^{\infty} \frac{dx}{x} = \lim_{R\to\infty} \int_{1}^{R} \frac{dx}{x} = \lim_{R\to\infty} \ln R = \infty$$

The integral diverges, and hence the sum $\displaystyle\sum_{n=1}^{\infty} \frac{1}{n}$ also diverges. ∎

■ **EXAMPLE 2**   Determine whether $\displaystyle\sum_{n=1}^{\infty} \frac{n}{(n^2+1)^2} = \frac{1}{2^2} + \frac{2}{5^2} + \frac{3}{10^2} + \cdots$ converges.

Solution   The function $f(x) = \dfrac{x}{(x^2+1)^2}$ is positive and continuous for $x \geq 1$, and it is decreasing since $f'(x)$ is negative:

$$f'(x) = \frac{1 - 3x^2}{(x^2+1)^3} < 0 \qquad \text{for } x \geq 1$$

Therefore, we may apply the Integral Test. We use the substitution $u = x^2 + 1$, $du = 2x\,dx$ to evaluate the improper integral:

$$\int_{1}^{\infty} \frac{x}{(x^2+1)^2}\,dx = \lim_{R\to\infty} \int_{1}^{R} \frac{x}{(x^2+1)^2}\,dx = \lim_{R\to\infty} \frac{1}{2} \int_{2}^{R} \frac{du}{u^2}$$

$$= \lim_{R\to\infty} \frac{-1}{2u}\Big|_{2}^{R} = \lim_{R\to\infty} \left(\frac{1}{4} - \frac{1}{2R} \right) = \frac{1}{4}$$

The integral converges and hence $\displaystyle\sum_{n=1}^{\infty} \frac{n}{(n^2+1)^2}$ also converges. ∎

The Integral Test applies to the sums of reciprocal powers n^{-p}, called **p-series**.

THEOREM 3 Convergence of p-Series   The infinite series $\displaystyle\sum_{n=1}^{\infty} \frac{1}{n^p}$ converges if $p > 1$ and diverges otherwise.

Proof   If $p \neq 1$, we have

$$\int_{1}^{\infty} \frac{1}{x^p}\,dx = \lim_{R\to\infty} \int_{1}^{R} \frac{1}{x^p}\,dx = \lim_{R\to\infty} \frac{x^{1-p}}{1-p}\Big|_{1}^{R} = \lim_{R\to\infty} \frac{R^{1-p} - 1}{1-p}$$

Since R^{1-p} tends to zero if $p > 1$ and to ∞ if $p < 1$, the improper integral converges for $p > 1$ and diverges for $p < 1$. The same is true of the p-series by the Integral Test. For $p = 1$, the series diverges, as shown in Example 1. ∎

Here are two examples of p-series:

$$p = \frac{1}{2}: \quad \sum_{n=1}^{\infty} \frac{1}{\sqrt{n}} = \frac{1}{\sqrt{1}} + \frac{1}{\sqrt{2}} + \frac{1}{\sqrt{3}} + \frac{1}{\sqrt{4}} + \cdots = \infty \quad \text{diverges}$$

$$p = 2: \quad \sum_{n=1}^{\infty} \frac{1}{n^2} = \frac{1}{1} + \frac{1}{2^2} + \frac{1}{3^2} + \frac{1}{4^2} + \cdots \qquad \text{converges}$$

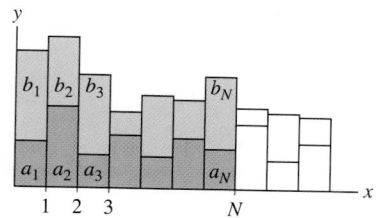

FIGURE 4 The series $\sum a_n$ is dominated by the series $\sum b_n$.

Another powerful method for determining convergence of positive series is comparison. Suppose that $0 \le a_n \le b_n$. Figure 4 suggests that if the larger sum $\sum b_n$ *converges*, then the smaller sum $\sum a_n$ also converges and, similarly, if the smaller sum *diverges*, then the larger sum also diverges.

THEOREM 4 Comparison Test
Assume that there exists $M > 0$ such that $0 \le a_n \le b_n$ for $n \ge M$.

(i) If $\displaystyle\sum_{n=1}^{\infty} b_n$ converges, then $\displaystyle\sum_{n=1}^{\infty} a_n$ also converges.

(ii) If $\displaystyle\sum_{n=1}^{\infty} a_n$ diverges, then $\displaystyle\sum_{n=1}^{\infty} b_n$ also diverges.

The convergence of an infinite series does not depend on where the series begins. Therefore, the Comparison Test remains valid even if the series does not begin with $n = 1$.

Proof Assume without loss of generality that $M = 1$. If $S = \displaystyle\sum_{n=1}^{\infty} b_n$ converges, then

$$a_1 + a_2 + \cdots + a_N \le b_1 + b_2 + \cdots + b_N \le \sum_{n=1}^{\infty} b_n = S \qquad \boxed{2}$$

Thus, the partial sums of $\displaystyle\sum_{n=1}^{\infty} a_n$ are bounded above by S, and $\displaystyle\sum_{n=1}^{\infty} a_n$ converges by the Dichotomy Theorem (Theorem 1). On the other hand, if $\displaystyle\sum_{n=1}^{\infty} a_n$ diverges, then its partial sums increase beyond bound and (2) shows that $\displaystyle\sum_{n=1}^{\infty} b_n$ also diverges. ∎

In words, the Comparison Test states that for positive series:

- *Convergence of larger series forces convergence of smaller series.*
- *Divergence of smaller series forces divergence of larger series.*

■ **EXAMPLE 3** Show that $S = \displaystyle\sum_{n=1}^{\infty} 2^{-n!} = 2^{-1} + 2^{-2} + 2^{-6} + 2^{-24} + \cdots$ converges.

Solution We apply the Comparison Test with $a_n = 2^{-n!}$ and $b_n = 2^{-n}$. This is valid because $n! \ge n$ and thus $0 \le 2^{-n!} \le 2^{-n}$ for $n \ge 1$. The geometric series converges:

$$\sum_{n=1}^{\infty} b_n = \sum_{n=1}^{\infty} 2^{-n} = 1 \quad \text{(converges)}$$

Therefore, the smaller series $\displaystyle\sum_{n=1}^{\infty} a_n = \sum_{n=1}^{\infty} 2^{-n!}$ converges. ∎

■ **EXAMPLE 4** Show that $\displaystyle\sum_{n=1}^{\infty} \frac{1}{\sqrt{n}\, 3^n}$ converges.

Solution We compare $\sum_{n=1}^{\infty} \dfrac{1}{\sqrt{n}\,3^n}$ with the geometric series $\sum_{n=1}^{\infty} \dfrac{1}{3^n}$. For $n \geq 1$,

$$\frac{1}{\sqrt{n}\,3^n} \leq \frac{1}{3^n}$$

The geometric series $\sum_{n=1}^{\infty} \dfrac{1}{3^n}$ converges, so $\sum_{n=1}^{\infty} \dfrac{1}{\sqrt{n}\,3^n}$ also converges. ■

In Example 5, the series begins with $n = 2$ because $1/\ln n$ is undefined for $n = 1$.

■ **EXAMPLE 5** Determine whether $\sum_{n=2}^{\infty} \dfrac{1}{\ln n}$ converges.

Solution We compare $\sum_{n=2}^{\infty} \dfrac{1}{\ln n}$ with the harmonic series $\sum_{n=2}^{\infty} \dfrac{1}{n}$ by showing that for $n \geq 2$,

$$\frac{1}{\ln n} \geq \frac{1}{n} \qquad \text{or equivalently,} \qquad n \geq \ln n$$

It suffices to show that $f(x) = x - \ln x$ is positive for $x \geq 2$. However, $f(1) = 1$ and $f(x)$ is increasing since $f'(x) = 1 - x^{-1} > 0$ for $x > 1$. Therefore, $f(x) > 1$ for $x > 1$ as required. Since the harmonic series diverges, the larger series $\sum_{n=2}^{\infty} \dfrac{1}{\ln n}$ also diverges. ■

■ **EXAMPLE 6** **Using the Comparison Test Correctly** Study the convergence of

$$\sum_{n=2}^{\infty} \frac{1}{n(\ln n)^2}$$

Solution We might be tempted to compare $\sum_{n=2}^{\infty} \dfrac{1}{n(\ln n)^2}$ to the harmonic series $\sum_{n=2}^{\infty} \dfrac{1}{n}$ using the inequality (valid for $n \geq 3$)

$$\frac{1}{n(\ln n)^2} \leq \frac{1}{n}$$

However, $\sum_{n=2}^{\infty} \dfrac{1}{n}$ diverges, so this inequality gives us no information about the *smaller* series $\sum \dfrac{1}{n(\ln n)^2}$. Fortunately, in this case we can use the Integral Test. The substitution $u = \ln x$ yields

$$\int_2^{\infty} \frac{dx}{x(\ln x)^2} = \int_{\ln 2}^{\infty} \frac{du}{u^2} = \lim_{R \to \infty} \left(\frac{1}{\ln 2} - \frac{1}{\ln R} \right) = \frac{1}{\ln 2} < \infty$$

The Integral Test shows that $\sum_{n=2}^{\infty} \dfrac{1}{n(\ln n)^2}$ converges. ■

Suppose we wish to study the convergence of

$$S = \sum_{n=2}^{\infty} \frac{n^2}{n^4 - n}$$

3

For large n, the general term is very close to $\dfrac{1}{n^2}$:

$$\frac{n^2}{n^4 - n} = \frac{1}{n^2 - n^{-1}} \approx \frac{1}{n^2}$$

so we might try to compare S with the convergent series $\displaystyle\sum_{n=2}^{\infty} \frac{1}{n^2}$. However, the Comparison Test cannot be used directly because the precise inequality goes in the wrong direction:

$$\frac{n^2}{n^4 - n} > \frac{n^2}{n^4} = \frac{1}{n^2}$$

In this case, we may apply the following variation of the Comparison Test.

CAUTION The Limit Comparison Test may not be applied if the series are not positive. See Exercise 38 in Section 10.4.

THEOREM 5 Limit Comparison Test Let $\{a_n\}$ and $\{b_n\}$ be *positive* sequences. Assume that the following limit exists:

$$L = \lim_{n \to \infty} \frac{a_n}{b_n}$$

- If $L > 0$, then $\displaystyle\sum_{n=1}^{\infty} a_n$ converges if and only if $\displaystyle\sum_{n=1}^{\infty} b_n$ converges.
- If $L = 0$ and $\displaystyle\sum_{n=1}^{\infty} b_n$ converges, then $\displaystyle\sum_{n=1}^{\infty} a_n$ converges.

Proof First we show that if $\displaystyle\sum_{n=1}^{\infty} b_n$ converges, then $\displaystyle\sum_{n=1}^{\infty} a_n$ converges for $L > 0$ or $L = 0$. Choose a positive number $R > L$. Since the sequences are positive and a_n/b_n approaches L, we have $0 \le a_n/b_n \le R$ and thus $a_n \le Rb_n$ for all n sufficiently large. Since $\displaystyle\sum_{n=1}^{\infty} Rb_n$ also converges, $\displaystyle\sum_{n=1}^{\infty} a_n$ converges by the Comparison Test.

Now suppose that $L > 0$ and $\displaystyle\sum_{n=1}^{\infty} a_n$ converges. We may choose r such that $0 < r < L$. Since a_n/b_n approaches L, we have $a_n/b_n \ge r$ and thus $a_n \ge rb_n$ for all n sufficiently large. Therefore, $\displaystyle\sum_{n=1}^{\infty} rb_n$ converges by the Comparison Test and hence $\displaystyle\sum_{n=1}^{\infty} b_n$ converges. ∎

■ **EXAMPLE 7** Show that $\displaystyle\sum_{n=2}^{\infty} \frac{n^2}{n^4 - n}$ converges.

Solution Let $a_n = \dfrac{n^2}{n^4 - n}$. We observed above that $a_n \approx n^{-2}$ for large n, so it makes sense to apply the Limit Comparison Test with $b_n = n^{-2}$:

$$\frac{a_n}{b_n} = \left(\frac{n^2}{n^4 - n} \right) n^2 = \frac{n^4}{n^4 - n} = \frac{1}{1 - n^{-3}}$$

$$L = \lim_{n \to \infty} \frac{a_n}{b_n} = \lim_{n \to \infty} \frac{1}{1 - n^{-3}} = 1$$

Since L exists and $\displaystyle\sum_{n=2}^{\infty} b_n = \sum_{n=2}^{\infty} n^{-2}$ converges, $\displaystyle\sum_{n=2}^{\infty} a_n$ also converges. ■

■ **EXAMPLE 8** Determine whether $\displaystyle\sum_{n=3}^{\infty} \frac{1}{\sqrt{n^2-4}}$ converges.

Solution Let $a_n = \dfrac{1}{\sqrt{n^2-4}}$ and $b_n = \dfrac{1}{n}$. Then

$$L = \lim_{n\to\infty} \frac{a_n}{b_n} = \lim_{n\to\infty} \frac{n}{\sqrt{n^2-4}} = \lim_{n\to\infty} \frac{1}{\sqrt{1-4/n^2}} = 1$$

Since $\displaystyle\sum_{n=3}^{\infty} \frac{1}{n}$ diverges and $L > 0$, the series $\displaystyle\sum_{n=3}^{\infty} \frac{1}{\sqrt{n^2-4}}$ also diverges. ■

10.3 SUMMARY

- The partial sums S_N of a positive series $S = \sum a_n$ form an increasing sequence.
- *Dichotomy Theorem:* A positive series S converges if its partial sums S_N remain bounded. Otherwise, it diverges.
- *Integral Test:* If f is positive, decreasing, and continuous, then $S = \sum f(n)$ converges (or diverges) if $\displaystyle\int_{M}^{\infty} f(x)\,dx$ converges (or diverges) for some $M > 0$.
- *p-Series:* The series $\displaystyle\sum_{n=1}^{\infty} \frac{1}{n^p}$ converges if $p > 1$ and diverges if $p \le 1$.
- *Comparison Test:* Assume that there exists $M > 0$ such that $0 \le a_n \le b_n$ for $n \ge M$. If $\displaystyle\sum_{n=1}^{\infty} b_n$ converges, then $\displaystyle\sum_{n=1}^{\infty} a_n$ converges; if $\displaystyle\sum_{n=1}^{\infty} a_n$ diverges, then $\displaystyle\sum_{n=1}^{\infty} b_n$ diverges.
- *Limit Comparison Test:* Let $\{a_n\}$ and $\{b_n\}$ be positive sequences and assume that the following limit exists:

$$L = \lim_{n\to\infty} \frac{a_n}{b_n}$$

  - If $L > 0$, then $\displaystyle\sum_{n=1}^{\infty} a_n$ converges if and only if $\displaystyle\sum_{n=1}^{\infty} b_n$ converges.

  - If $L = 0$ and $\displaystyle\sum_{n=1}^{\infty} b_n$ converges, then $\displaystyle\sum_{n=1}^{\infty} a_n$ converges.

10.3 EXERCISES

Preliminary Questions

1. Let $S = \displaystyle\sum_{n=1}^{\infty} a_n$. If the partial sums S_N are increasing, then (choose correct conclusion)

(a) $\{a_n\}$ is an increasing sequence.

(b) $\{a_n\}$ is a positive sequence.

2. What are the hypotheses of the Integral Test?

3. Which test would you use to determine whether $\displaystyle\sum_{n=1}^{\infty} n^{-3.2}$ converges?

4. Which test would you use to determine whether $\displaystyle\sum_{n=1}^{\infty} \frac{1}{2^n + \sqrt{n}}$ converges?

5. Ralph hopes to investigate the convergence of $\displaystyle\sum_{n=1}^{\infty} \frac{e^{-n}}{n}$ by comparing it with $\displaystyle\sum_{n=1}^{\infty} \frac{1}{n}$. Is Ralph on the right track?

Exercises

In Exercises 1–14, use the Integral Test to determine whether the infinite series is convergent.

1. $\displaystyle\sum_{n=1}^{\infty} \frac{1}{n^4}$

2. $\displaystyle\sum_{n=1}^{\infty} \frac{1}{n+3}$

3. $\displaystyle\sum_{n=1}^{\infty} n^{-1/3}$

4. $\displaystyle\sum_{n=5}^{\infty} \frac{1}{\sqrt{n-4}}$

5. $\displaystyle\sum_{n=25}^{\infty} \frac{n^2}{(n^3+9)^{5/2}}$

6. $\displaystyle\sum_{n=1}^{\infty} \frac{n}{(n^2+1)^{3/5}}$

7. $\displaystyle\sum_{n=1}^{\infty} \frac{1}{n^2+1}$

8. $\displaystyle\sum_{n=1}^{\infty} \frac{1}{n(n+1)}$

9. $\displaystyle\sum_{n=1}^{\infty} ne^{-n^2}$

10. $\displaystyle\sum_{n=2}^{\infty} \frac{1}{n(\ln n)^2}$

11. $\displaystyle\sum_{n=1}^{\infty} \frac{1}{2^{\ln n}}$

12. $\displaystyle\sum_{n=4}^{\infty} \frac{1}{n^2-1}$

13. $\displaystyle\sum_{n=1}^{\infty} \frac{\ln n}{n^2}$

14. $\displaystyle\sum_{n=1}^{\infty} \frac{n}{2^n}$

15. Use the Comparison Test to show that $\displaystyle\sum_{n=1}^{\infty} \frac{1}{n^3+8n}$ converges.

Hint: Compare with $\displaystyle\sum_{n=1}^{\infty} n^{-3}$.

16. Show that $\displaystyle\sum_{n=2}^{\infty} \frac{1}{\sqrt{n^2-3}}$ diverges by comparing with $\displaystyle\sum_{n=2}^{\infty} n^{-1}$.

17. Let $S = \displaystyle\sum_{n=1}^{\infty} \frac{1}{n+\sqrt{n}}$. Verify that for $n \geq 1$

$$\frac{1}{n+\sqrt{n}} \leq \frac{1}{n}, \qquad \frac{1}{n+\sqrt{n}} \leq \frac{1}{\sqrt{n}}$$

Can either inequality be used to show that S diverges? Show that $\dfrac{1}{n+\sqrt{n}} \geq \dfrac{1}{2n}$ and conclude that S diverges.

18. Which of the following inequalities can be used to study the convergence of $\displaystyle\sum_{n=2}^{\infty} \frac{1}{n^2+\sqrt{n}}$? Explain.

$$\frac{1}{n^2+\sqrt{n}} \leq \frac{1}{\sqrt{n}}, \qquad \frac{1}{n^2+\sqrt{n}} \leq \frac{1}{n^2}$$

In Exercises 19–31, use the Comparison Test to determine whether the infinite series is convergent.

19. $\displaystyle\sum_{n=1}^{\infty} \frac{1}{n2^n}$

20. $\displaystyle\sum_{n=1}^{\infty} \frac{1}{\sqrt{n}+2^n}$

21. $\displaystyle\sum_{k=1}^{\infty} \frac{k^{1/3}}{k^2+k}$

22. $\displaystyle\sum_{n=4}^{\infty} \frac{\sqrt{n}}{n-3}$

23. $\displaystyle\sum_{n=1}^{\infty} \frac{1}{\sqrt{n^3+1}}$

24. $\displaystyle\sum_{n=1}^{\infty} \frac{n^3}{n^5+4n+1}$

25. $\displaystyle\sum_{k=1}^{\infty} \frac{\sin^2 k}{k^2}$

26. $\displaystyle\sum_{n=2}^{\infty} \frac{1}{(\ln n)2^n}$

27. $\displaystyle\sum_{m=1}^{\infty} \frac{4}{m!+4^m}$

28. $\displaystyle\sum_{n=1}^{\infty} \frac{2}{3^n+3^{-n}}$

29. $\displaystyle\sum_{k=1}^{\infty} 2^{-k^2}$

30. $\displaystyle\sum_{n=1}^{\infty} \frac{\ln n}{n^3}$

31. $\displaystyle\sum_{n=1}^{\infty} \frac{\ln n}{n^3+3\ln n}$

32. Show that $\displaystyle\sum_{n=1}^{\infty} \sin\frac{1}{n^2}$ is a positive, convergent series. *Hint:* Use the inequality $\sin x \leq x$ for $x \geq 0$.

33. Does $\displaystyle\sum_{n=1}^{\infty} \frac{n}{\sqrt{n^2+c}}$ converge for any c?

In Exercises 34–42, use the Limit Comparison Test to prove convergence or divergence of the infinite series.

34. $\displaystyle\sum_{n=2}^{\infty} \frac{n^2}{n^4-1}$

35. $\displaystyle\sum_{n=2}^{\infty} \frac{1}{n^2-\sqrt{n}}$

36. $\displaystyle\sum_{n=2}^{\infty} \frac{n}{\sqrt{n^3-1}}$

37. $\displaystyle\sum_{n=2}^{\infty} \frac{n^3}{\sqrt{n^7-2n^2+1}}$

38. $\displaystyle\sum_{n=3}^{\infty} \frac{3n+5}{n(n-1)(n-2)}$

39. $\displaystyle\sum_{n=1}^{\infty} \frac{e^n+n}{e^{2n}-n^2}$

40. $\displaystyle\sum_{n=1}^{\infty} \frac{\ln n}{n^2}$. *Hint:* Use L'Hôpital's Rule to compare with $\displaystyle\sum_{n=1}^{\infty} \frac{1}{n^{3/2}}$.

41. $\displaystyle\sum_{n=1}^{\infty} \left(1-\cos\frac{1}{n}\right)$ *Hint:* Compare with $\displaystyle\sum_{n=1}^{\infty} n^{-2}$.

42. $\displaystyle\sum_{n=1}^{\infty} (1-2^{-1/n})$ *Hint:* Compare with the harmonic series.

43. Show that if $a_n \geq 0$ and $\lim_{n\to\infty} n^2 a_n$ exists, then $\displaystyle\sum_{n=1}^{\infty} a_n$ converges. *Hint:* Show that if M is larger than $\lim_{n\to\infty} n^2 a_n$, then $a_n \leq M/n^2$ for n sufficiently large.

44. Show that $\displaystyle\sum_{n=2}^{\infty} \frac{1}{n^{\ln n}}$ converges. *Hint:* Show that $n^{\ln n} \geq n^2$ for $n > e^2$.

45. Show that $\displaystyle\sum_{n=2}^{\infty}(\ln n)^{-2}$ diverges. *Hint:* Show that for x sufficiently large, $\ln x < x^{1/2}$.

46. For which a does $\displaystyle\sum_{n=2}^{\infty}\frac{1}{n(\ln n)^a}$ converge?

47. For which a does $\displaystyle\sum_{n=2}^{\infty}\frac{1}{n^a \ln n}$ converge?

48. Use the Integral Test to show that $\displaystyle\sum_{n=2}^{\infty}\frac{(\ln n)^k}{n^2}$ converges for all exponents k. You may use that $\displaystyle\int_{1}^{\infty}u^k e^{-u}\,du$ converges for all k.

In Exercises 49–74, determine convergence or divergence using any method covered so far.

49. $\displaystyle\sum_{n=4}^{\infty}\frac{1}{n^2-9}$

50. $\displaystyle\sum_{n=1}^{\infty}\frac{\cos^2 n}{n^2}$

51. $\displaystyle\sum_{n=1}^{\infty}\frac{\sqrt{n}}{4n+9}$

52. $\displaystyle\sum_{n=1}^{\infty}e^{-n}$

53. $\displaystyle\sum_{n=1}^{\infty}\frac{1}{3^{n^2}}$

54. $\displaystyle\sum_{n=1}^{\infty}\frac{1}{n^2+\sin n}$

55. $\displaystyle\sum_{n=2}^{\infty}\frac{1}{n^{3/2}\ln n}$

56. $\displaystyle\sum_{k=1}^{\infty}2^{1/k}$

57. $\displaystyle\sum_{n=2}^{\infty}\frac{1}{n^{1/2}\ln n}$

58. $\displaystyle\sum_{n=1}^{\infty}\frac{4^n}{5^n-2n}$

59. $\displaystyle\sum_{n=2}^{\infty}\frac{n}{e^{n^2}}$

60. $\displaystyle\sum_{n=1}^{\infty}\frac{n-\sin n}{n^3}$

61. $\displaystyle\sum_{n=1}^{\infty}\frac{2^n}{3^n-n}$

62. $\displaystyle\sum_{n=2}^{\infty}\frac{1}{n\ln n-n}$

63. $\displaystyle\sum_{n=1}^{\infty}\frac{\tan^{-1}n}{n^2}$

64. $\displaystyle\sum_{n=1}^{\infty}\frac{1}{n^n}$

65. $\displaystyle\sum_{n=1}^{\infty}\frac{\ln n}{n^3}$

66. $\displaystyle\sum_{n=1}^{\infty}\frac{2+(-1)^n}{n}$

67. $\displaystyle\sum_{n=1}^{\infty}\frac{2+(-1)^n}{n^{3/2}}$

68. $\displaystyle\sum_{n=1}^{\infty}\sin\frac{1}{n}$

69. $\displaystyle\sum_{n=1}^{\infty}\frac{2n+1}{4^n}$

70. $\displaystyle\sum_{n=3}^{\infty}\frac{1}{n^4-\sqrt{n}}$

71. $\displaystyle\sum_{n=1}^{\infty}\frac{n^2-n}{n^5+n}$

72. $\displaystyle\sum_{n=2}^{\infty}\frac{n^2+n}{n^5-n}$

73. $\displaystyle\sum_{n=2}^{\infty}\frac{1}{n^{1.2}\ln n}$

74. $\displaystyle\sum_{n=1}^{\infty}\frac{\ln n}{n^{1.2}}$

Approximating Infinite Sums *In Exercises 75–77, let $a_n = f(n)$, where $f(x)$ is a continuous, decreasing function such that*

$$\int_{1}^{\infty}f(x)\,dx$$

converges.

75. Show that

$$\int_{1}^{\infty}f(x)\,dx \le \sum_{n=1}^{\infty}a_n \le a_1 + \int_{1}^{\infty}f(x)\,dx \qquad \boxed{4}$$

76. ⊂🄰🄂 Using Eq. (4), show that

$$5 \le \sum_{n=1}^{\infty}\frac{1}{n^{1.2}} \le 6$$

This series converges slowly. Use a computer algebra system to verify that $S_N < 5$ for $N \le 43{,}128$ and $S_{43{,}129} \approx 5.00000021$.

77. Let $S = \displaystyle\sum_{n=1}^{\infty}a_n$. Arguing as in Exercise 75, show that

$$\sum_{n=1}^{M}a_n + \int_{M+1}^{\infty}f(x)\,dx \le S \le \sum_{n=1}^{M+1}a_n + \int_{M+1}^{\infty}f(x)\,dx \qquad \boxed{5}$$

Conclude that

$$0 \le S - \left(\sum_{n=1}^{M}a_n + \int_{M+1}^{\infty}f(x)\,dx\right) \le a_{M+1} \qquad \boxed{6}$$

This yields a method for approximating S with an error of at most a_{M+1}.

78. ⊂🄰🄂 Use Eq. (5) with $M = 43{,}129$ to prove that

$$5.5915810 \le \sum_{n=1}^{\infty}\frac{1}{n^{1.2}} \le 5.5915839$$

79. ⊂🄰🄂 Apply Eq. (5) with $M = 40{,}000$ to show that

$$1.644934066 \le \sum_{n=1}^{\infty}\frac{1}{n^2} \le 1.644934068$$

Is this consistent with Euler's result, according to which this infinite series has sum $\pi^2/6$?

80. ⊂🄰🄂 Using a CAS and Eq. (6), determine the value of $\displaystyle\sum_{n=1}^{\infty}n^{-4}$ to within an error less than 10^{-4}. Check that your result is consistent with that of Euler, who proved that the sum is equal to $\pi^4/90$.

81. ⊂🄰🄂 Using a CAS and Eq. (6), determine the value of $\displaystyle\sum_{n=1}^{\infty}n^{-5}$ to within an error less than 10^{-4}.

82. The harmonic series diverges, but it does so very slowly. Show that the partial sum S_N satisfies

$$\ln N \le 1 + \frac{1}{2} + \frac{1}{3} + \cdots + \frac{1}{N} \le 1 + \ln N$$

Verify that $S_N \le 10$ for $N \le 8{,}200$. Find an N such that $S_N \ge 100$.

83. Let p_n denote the nth prime number ($p_1 = 2$, $p_2 = 3$, etc.). It is known that there is a constant C such that $p_n \le C n \ln n$. Prove the divergence of

$$\sum_{n=1}^{\infty} \frac{1}{p_n} = \frac{1}{2} + \frac{1}{3} + \frac{1}{5} + \frac{1}{7} + \frac{1}{11} + \cdots$$

84. How far can a stack of identical books (each of unit length) extend without tipping over? The stack will not tip over if the $(n+1)$st book is placed at the bottom of the stack with its left edge located at the center of mass of the first n books (Figure 5). Let c_n be the center of mass of the first n books, measured along the x-axis.

(a) Prove that $c_{n+1} = c_n + \frac{1}{2(n+1)}$. Recall that if objects of mass $m_1, \ldots, m_n$ are placed along the x-axis with their centers of mass at $x_1, \ldots, x_n$, then the center of mass of the system is located at

$$\frac{m_1 x_1 + \cdots + m_n x_n}{x_1 + \cdots + x_n}$$

(b) Prove that $\lim_{n \to \infty} c_n = \infty$. Thus, by using enough books, the stack can be extended as far as desired without tipping over.

FIGURE 5

Further Insights and Challenges

85. Use the Integral Test to prove again that the geometric series $\sum_{n=1}^{\infty} r^n$ converges if $0 < r < 1$ and diverges if $r > 1$.

86. Let $S = \sum_{n=2}^{\infty} a_n$, where $a_n = (\ln(\ln n))^{-\ln n}$.

(a) Show, by taking logarithms, that $a_n = n^{-\ln(\ln(\ln n))}$.

(b) Show that $\ln(\ln(\ln n)) \ge 2$ if $n > C$, where $C = e^{e^{e^2}}$.

(c) Show that S converges.

87. Kummer's Acceleration Method Suppose we wish to approximate $S = \sum_{n=1}^{\infty} \frac{1}{n^2}$. There is a similar telescoping series whose value can be computed exactly (see Example 1 in Section 10.2):

$$\sum_{n=1}^{\infty} \frac{1}{n(n+1)} = 1$$

(a) Verify that

$$S = \sum_{n=1}^{\infty} \frac{1}{n(n+1)} + \sum_{n=1}^{\infty} \left(\frac{1}{n^2} - \frac{1}{n(n+1)} \right)$$

Thus for M large,

$$S \approx 1 + \sum_{n=1}^{M} \frac{1}{n^2(n+1)} \qquad \boxed{7}$$

(b) Explain what has been gained. Why is (7) a better approximation to S than $\sum_{n=1}^{M} \frac{1}{n^2}$?

(c) ⌂AS Compute

$$\sum_{n=1}^{1{,}000} \frac{1}{n^2}, \qquad 1 + \sum_{n=1}^{100} \frac{1}{n^2(n+1)}$$

Which is a better approximation to S, whose exact value is $\pi^2/6$?

88. ⌂AS The series $S = \sum_{k=1}^{\infty} k^{-3}$ has been computed to more than 100 million digits. The first 30 digits are

$$S = 1.202056903159594285399738161511$$

Approximate S using the Acceleration Method of Exercise 87 with $M = 100$ and auxiliary series $R = \sum_{n=1}^{\infty} (n(n+1)(n+2))^{-3}$. According to Exercise 46 in Section 10.2, R is a telescoping series with the sum $R = \frac{1}{4}$.

10.4 Absolute and Conditional Convergence

In the previous section, we studied the convergence of positive series, but we still lack the tools to analyze series with both positive and negative terms such as

$$\frac{1}{1^2} - \frac{1}{2^2} + \frac{1}{3^2} - \frac{1}{4^2} + \cdots \qquad \boxed{1}$$

One of the keys to understanding these more general series is the concept of absolute convergence.

DEFINITION Absolute Convergence $\sum a_n$ is called **absolutely convergent** if $\sum |a_n|$ converges.

As an example, consider the series (1). It is absolutely convergent because the sequence of absolute values is a convergent p-series:

$$\frac{1}{1^2} + \frac{1}{2^2} + \frac{1}{3^2} + \frac{1}{4^2} + \cdots$$

The following theorem states that if the series of absolute values converges, then the original series also converges.

THEOREM 1 Absolute Convergence Implies Convergence If $\sum a_n$ is absolutely convergent, then $\sum a_n$ converges.

Proof We have

$$0 \le a_n + |a_n| \le 2|a_n|$$

and thus $\sum (a_n + |a_n|)$ is a positive series. If $\sum |a_n|$ converges, then $\sum 2|a_n|$ also converges and hence $\sum (a_n + |a_n|)$ converges by the Comparison Test. Our original series is the difference of two convergent series and hence it converges:

$$\sum a_n = \sum (a_n + |a_n|) - \sum |a_n| \qquad \blacksquare$$

■ **EXAMPLE 1** Verify the convergence of $S = \dfrac{1}{1^2} - \dfrac{1}{2^2} + \dfrac{1}{3^2} - \dfrac{1}{4^2} + \cdots$.

Solution The positive series $\displaystyle\sum_{n=1}^{\infty} \frac{1}{n^2}$ converges because it is a p-series with $p = 2$. Thus, S converges absolutely. By Theorem 1, S converges. ■

■ **EXAMPLE 2** Does $S = \displaystyle\sum_{n=2}^{\infty} \frac{(-1)^n}{n \ln n}$ converge absolutely?

Solution We apply the Integral Test to the positive series $\displaystyle\sum_{n=2}^{\infty} \frac{1}{n \ln n}$. Using the substitution $u = \ln x$, $du = x^{-1} \, dx$, we find that the improper integral diverges:

$$\int_2^{\infty} \frac{dx}{x \ln x} = \int_{\ln 2}^{\infty} \frac{du}{u} = \lim_{R \to \infty} \int_{\ln 2}^{R} \frac{du}{u} = \lim_{R \to \infty} \left(\ln R - \ln(\ln 2) \right) = \infty$$

Therefore, $\displaystyle\sum_{n=2}^{\infty} \frac{1}{n \ln n}$ diverges and S does not converge absolutely. ■

In the previous example, we showed that the series $\displaystyle\sum_{n=2}^{\infty} \frac{(-1)^n}{n \ln n}$ does not converge *absolutely*, but we still do not know whether or not it converges. A series $\sum a_n$ may converge without converging absolutely. In this case, we say that $\sum a_n$ is conditionally convergent.

> **DEFINITION** **Conditional Convergence** An infinite series $\sum a_n$ is called **conditionally convergent** if it converges but $\sum |a_n|$ diverges.

If we encounter a series that is not absolutely convergent, how can we determine whether it is conditionally convergent? The Integral Test and Comparison Test cannot be used because they apply only to positive series. Our next test applies to **alternating series**, that is, series in which the terms alternate in sign. Such a series has the following form, where $\{a_n\}$ is a *positive* sequence (Figure 1):

$$\sum_{n=1}^{\infty} (-1)^{n-1} a_n = a_1 - a_2 + a_3 - a_4 + a_5 + \cdots$$

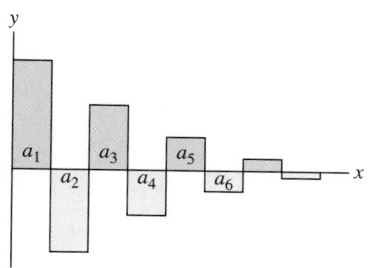

FIGURE 1 An alternating series with decreasing terms. The sum of the signed areas is positive and at most a_1.

Assumptions Matter The Leibniz Test is not valid if we drop the assumption that a_n is decreasing (see Exercise 35).

> **THEOREM 2** **Leibniz Test for Alternating Series** Let $\{a_n\}$ be a decreasing positive sequence that converges to 0:
>
> $$a_1 \geq a_2 \geq a_3 \geq a_4 \geq \cdots \geq 0, \qquad \lim_{n \to \infty} a_n = 0$$
>
> Then the following alternating series converges:
>
> $$S = \sum_{n=1}^{\infty} (-1)^{n-1} a_n = a_1 - a_2 + a_3 - a_4 + \cdots$$
>
> Furthermore, $0 \leq S \leq a_1$ and $S_{2N} \leq S \leq S_{2N+1}$ for all N.

Proof We analyze the even and odd partial sums separately. First, we observe that the even partial sums form an increasing sequence:

$$S_{2N+2} = \underbrace{(a_1 - a_2) + \cdots + (a_{2N-1} - a_{2N})}_{S_{2N}} + (a_{2N+1} - a_{2N+2})$$

$$= S_{2N} + (a_{2N+1} - a_{2N+2})$$

Since $\{a_n\}$ is decreasing, the quantity $a_{2N+1} - a_{2N+2}$ is positive and thus S_{2N+2} is obtained by adding the positive quantity $(a_{2N+1} - a_{2N+2})$ to S_{2N}. On the other hand, we may regroup the terms:

$$S_{2N} = a_1 - (a_2 - a_3) - (a_4 - a_5) - \cdots - (a_{2N-2} - a_{2N-1}) - a_{2N}$$

We conclude that $S_{2N} \leq a_1$ because each term $(a_{2j} - a_{2j+1})$ is nonnegative. This shows that $\{S_{2N}\}$ is a positive increasing sequence with upper bound a_1. Consequently, the limit $S = \lim S_{2N}$ exists and $0 \leq S_{2N} \leq S \leq a_1$ by Theorem 6 in Section 10.1.

The odd partial sums form a decreasing sequence:

$$S_{2N+1} = S_{2N-1} - (a_{2N} - a_{2N+1}) \leq S_{2N-1}$$

The Leibniz Test is the only test for conditional convergence developed in this text. More sophisticated tests such as Abel's Criterion are treated in analysis textbooks.

Furthermore, the odd partial sums converge to the same limit:

$$\lim_{N\to\infty} S_{2N+1} = \lim_{N\to\infty} (S_{2N} + a_{2N+1}) = \lim_{N\to\infty} S_{2N} + \lim_{N\to\infty} a_{2N+1} = S + 0 = S$$

This proves that the alternating series converges to S and that $S_{2N+1} \geq S$ for all N. ■

■ **EXAMPLE 3** Show that $S = \displaystyle\sum_{n=1}^{\infty} \frac{(-1)^{n-1}}{\sqrt{n}}$ is conditionally convergent and that $0 \leq S \leq 1$.

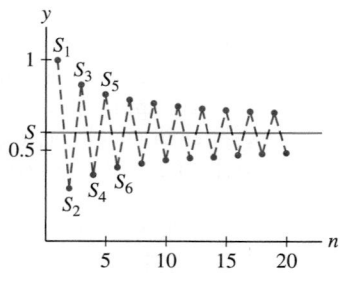

(A) Partial sums of $S = \displaystyle\sum_{n=1}^{\infty} (-1)^{n-1} \frac{1}{\sqrt{n}}$

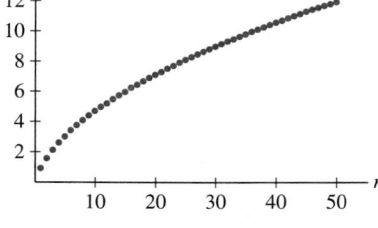

(B) Partial sums of $\displaystyle\sum_{n=1}^{\infty} \frac{1}{\sqrt{n}}$

FIGURE 2 The series $S = \displaystyle\sum_{n=1}^{\infty} \frac{(-1)^{n-1}}{\sqrt{n}}$ converges conditionally but not absolutely.

Solution The terms $a_n = \dfrac{1}{\sqrt{n}}$ form a decreasing sequence that tends to zero. The Leibniz Test implies that $S = \displaystyle\sum_{n=1}^{\infty} \frac{(-1)^{n-1}}{\sqrt{n}}$ converges and that $0 \leq S \leq a_1 = 1$ [Figure 2(A)]. However, S is only conditionally convergent because the positive series $\displaystyle\sum_{n=1}^{\infty} \frac{1}{\sqrt{n}}$ is a divergent p-series [Figure 2(B)]. ■

According to Theorem 2, if $\{a_n\}$ is decreasing and positive, then the partial sums S_N of the alternating series satisfy

$$S_{2N} \leq S \leq S_{2N+1}$$

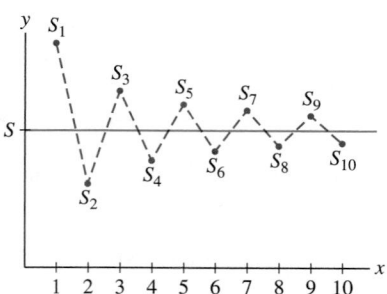

FIGURE 3 The partial sums of an alternating series zigzag, above and below the limit.

Therefore, the partial sums zigzag above and below the limit (Figure 3). It follows that the error $|S_N - S|$ is not greater than $|S_N - S_{N+1}| = a_{N+1}$:

$$\boxed{|S_N - S| \leq a_{N+1}}$$ 2

In other words, for an alternating series, *the error committed when we approximate S by S_N is at most the size of the first omitted term a_{N+1}.*

Leibniz proved that

$$\sum_{n=1}^{\infty} \frac{(-1)^{n+1}}{n} = \ln 2 \approx 0.6931$$

We will prove this in Exercise 83 in Section 10.7. The result of Example 4 is consistent with this fact.

■ **EXAMPLE 4** **The Alternating Harmonic Series** Show that $S = \displaystyle\sum_{n=1}^{\infty} \frac{(-1)^{n+1}}{n}$ converges conditionally. Then

(a) Show that $|S_6 - S| \leq \frac{1}{7}$.

(b) Find an N such that S_N approximates S with an error less than 10^{-3}.

Solution The harmonic series $\displaystyle\sum_{n=1}^{\infty} n^{-1}$ diverges, but S converges conditionally by the Leibniz Test, since $a_n = n^{-1}$ is a positive, decreasing sequence.

(a) By Eq. (2), $|S_6 - S| \leq a_7 = \frac{1}{7}$.

(b) The error in the Nth partial sum is at most a_{N+1}:

$$\left| \sum_{n=1}^{N} \frac{(-1)^{n+1}}{n} - S \right| \leq a_{N+1} = \frac{1}{N+1}$$

To make the error less than 10^{-3}, we choose N so that $\dfrac{1}{N+1} < 10^{-3}$ or $N + 1 > 10^3$. This gives $N > 999$. Using a computer algebra system, we find that $S_{1,000} \approx 0.6926$. ∎

CONCEPTUAL INSIGHT   The convergence of an infinite series $\sum a_n$ depends on two factors: (1) how quickly a_n tends to zero, and (2) how much cancellation takes place among the terms. Consider the series

Harmonic series (diverges):	$1 + \dfrac{1}{2} + \dfrac{1}{3} + \dfrac{1}{4} + \dfrac{1}{5} + \cdots$
p-Series with $p = 2$ (converges):	$1 + \dfrac{1}{2^2} + \dfrac{1}{3^2} + \dfrac{1}{4^2} + \dfrac{1}{5^2} + \cdots$
Alternating harmonic series (converges):	$1 - \dfrac{1}{2} + \dfrac{1}{3} - \dfrac{1}{4} + \dfrac{1}{5} - \cdots$

The harmonic series diverges because reciprocals $\dfrac{1}{n}$ do not tend to zero quickly enough. By contrast, the reciprocal squares $\dfrac{1}{n^2}$ tend to zero quickly enough for the series $\displaystyle\sum_{n=1}^{\infty} \dfrac{1}{n^2}$ to converge. The alternating harmonic series converges as a result of cancellation among the terms.

10.4 SUMMARY

- An infinite series $\sum a_n$ is called *absolutely convergent* if the positive series $\sum |a_n|$ converges.
- Theorem: Absolute convergence implies convergence. Namely, if $\sum |a_n|$ converges, then $\sum a_n$ also converges.
- An infinite series $\sum a_n$ is *conditionally convergent* if it converges but $\sum |a_n|$ diverges.
- *Leibniz Test:* If $\{a_n\}$ is a positive decreasing sequence such that $\lim_{n\to\infty} a_n = 0$, then the following alternating series converges:

$$S = \sum_{n=1}^{\infty} (-1)^{n-1} a_n = a_1 - a_2 + a_3 - a_4 + a_5 - \cdots$$

Furthermore, $|S - S_N| \leq a_{N+1}$.

- We have developed two ways to handle nonpositive series: Either show absolute convergence or use the Leibniz Test, if applicable.

10.4 EXERCISES

Preliminary Questions

1. Suppose that $S = \sum_{n=0}^{\infty} a_n$ is conditionally convergent. Which of the following statements are correct?

(a) $\sum_{n=0}^{\infty} |a_n|$ may or may not converge.

(b) S may or may not converge.

(c) $\sum_{n=0}^{\infty} |a_n|$ diverges.

2. Which of the following statements is equivalent to Theorem 1?

(a) If $\sum_{n=0}^{\infty} |a_n|$ diverges, then $\sum_{n=0}^{\infty} a_n$ also diverges.

(b) If $\sum_{n=0}^{\infty} a_n$ diverges, then $\sum_{n=0}^{\infty} |a_n|$ also diverges.

(c) If $\sum_{n=0}^{\infty} a_n$ converges, then $\sum_{n=0}^{\infty} |a_n|$ also converges.

3. Lathika argues that $\sum_{n=1}^{\infty} (-1)^n \sqrt{n}$ is an alternating series and therefore converges. Is Lathika right?

4. Give an example of a series such that $\sum a_n$ converges but $\sum |a_n|$ diverges.

Exercises

1. Show that $\displaystyle\sum_{n=0}^{\infty} \frac{(-1)^n}{2^n}$ converges absolutely.

2. Show that the following series converges conditionally:

$$\sum_{n=1}^{\infty} (-1)^{n-1} \frac{1}{n^{2/3}} = \frac{1}{1^{2/3}} - \frac{1}{2^{2/3}} + \frac{1}{3^{2/3}} - \frac{1}{4^{2/3}} + \cdots$$

In Exercises 3–12, determine whether the series converges absolutely, conditionally, or not at all.

3. $\displaystyle\sum_{n=1}^{\infty} \frac{(-1)^n}{\sqrt{n}}$

4. $\displaystyle\sum_{n=1}^{\infty} \frac{(-1)^n n^4}{n^3 + 1}$

5. $\displaystyle\sum_{n=1}^{\infty} \frac{(-1)^{n-1}}{(1.1)^n}$

6. $\displaystyle\sum_{n=1}^{\infty} \frac{\sin n}{n^2}$

7. $\displaystyle\sum_{n=2}^{\infty} \frac{(-1)^{n+1}}{n \ln n}$

8. $\displaystyle\sum_{n=1}^{\infty} \frac{(-1)^n}{1 + \frac{1}{n}}$

9. $\displaystyle\sum_{n=1}^{\infty} \frac{\sin n\pi}{\sqrt{n}}$

10. $\displaystyle\sum_{n=4}^{\infty} (-1)^n \tan \frac{1}{n}$

11. $\displaystyle\sum_{n=1}^{\infty} \frac{\cos \frac{1}{n}}{n^2}$

12. $\displaystyle\sum_{n=1}^{\infty} \frac{\cos \frac{1}{n}}{n}$

13. Let $S = \displaystyle\sum_{n=1}^{\infty} (-1)^{n+1} \frac{1}{n^3}$.

(a) Calculate S_n for $1 \leq n \leq 10$.

(b) Use Eq. (2) to show that $0.9 \leq S \leq 0.902$.

14. Use Eq. (2) to approximate $\displaystyle\sum_{n=1}^{\infty} \frac{(-1)^{n+1}}{n!}$ to four decimal places.

15. Approximate $\displaystyle\sum_{n=1}^{\infty} \frac{(-1)^{n+1}}{n^4}$ to three decimal places.

16. CAS Let $S = \displaystyle\sum_{n=1}^{\infty} (-1)^{n-1} \frac{n}{n^2 + 1}$. Use a computer algebra system to calculate and plot the partial sums S_n for $1 \leq n \leq 100$. Observe that the partial sums zigzag above and below the limit.

In Exercises 17–18, use Eq. (2) to approximate the value of the series to within an error of at most 10^{-5}.

17. $\displaystyle\sum_{n=1}^{\infty} \frac{(-1)^{n+1}}{n(n+2)(n+3)}$

18. $\displaystyle\sum_{n=1}^{\infty} \frac{(-1)^{n+1} \ln n}{n!}$

In Exercises 19–26, determine convergence or divergence by any method.

19. $\displaystyle\sum_{n=1}^{\infty} \frac{1}{3^n + 5^n}$

20. $\displaystyle\sum_{n=2}^{\infty} \frac{n}{n^2 - n}$

21. $\displaystyle\sum_{n=1}^{\infty} \frac{(-1)^n}{\sqrt{n^2 + 1}}$

22. $\displaystyle\sum_{n=1}^{\infty} \frac{1}{\sqrt{n^2 + 1}}$

23. $\displaystyle\sum_{n=1}^{\infty} \frac{3^n + (-1)^n 2^n}{5^n}$

24. $\displaystyle\sum_{n=1}^{\infty} \frac{(-1)^{n+1}}{(2n+1)!}$

25. $\displaystyle\sum_{n=1}^{\infty} (-1)^n n e^{-n}$

26. $\displaystyle\sum_{n=1}^{\infty} (-1)^n n^4 2^{-n}$

27. Show that

$$S = \frac{1}{2} - \frac{1}{2} + \frac{1}{3} - \frac{1}{3} + \frac{1}{4} - \frac{1}{4}$$

converges by computing the partial sums. Does it converge absolutely?

28. The Leibniz Test cannot be applied to

$$\frac{1}{2} - \frac{1}{3} + \frac{1}{2^2} - \frac{1}{3^2} + \frac{1}{2^3} - \frac{1}{3^3} + \cdots$$

Why not? Show that it converges by another method.

29. Determine whether the following series converges conditionally:

$$1 - \frac{1}{3} + \frac{1}{2} - \frac{1}{5} + \frac{1}{3} - \frac{1}{7} + \frac{1}{4} - \frac{1}{9} + \frac{1}{5} - \frac{1}{11} + \cdots$$

30. Prove that if $\sum a_n$ converges absolutely, then $\sum a_n^2$ also converges. Then show by giving a counterexample that $\sum a_n^2$ need not converge if $\sum a_n$ is only conditionally convergent.

Further Insights and Challenges

31. Prove the following variant of the Leibniz Test: If $\{a_n\}$ is a positive, decreasing sequence with $\lim_{n\to\infty} a_n = 0$, then the series

$$a_1 + a_2 - 2a_3 + a_4 + a_5 - 2a_6 + \cdots$$

converges. *Hint:* Show that S_{3N} is increasing and bounded by $a_1 + a_2$, and continue as in the proof of the Leibniz Test.

32. Use Exercise 31 to show that the following series converges:

$$S = \frac{1}{\ln 2} + \frac{1}{\ln 3} - \frac{2}{\ln 4} + \frac{1}{\ln 5} + \frac{1}{\ln 6} - \frac{2}{\ln 7} + \cdots$$

33. Prove the conditional convergence of

$$R = 1 + \frac{1}{2} + \frac{1}{3} - \frac{3}{4} + \frac{1}{5} + \frac{1}{6} + \frac{1}{7} - \frac{3}{8} + \cdots$$

34. Show that the following series diverges:

$$S = 1 + \frac{1}{2} + \frac{1}{3} - \frac{2}{4} + \frac{1}{5} + \frac{1}{6} + \frac{1}{7} - \frac{2}{8} + \cdots$$

Hint: Use the result of Exercise 33 to write S as the sum of a convergent and a divergent series.

35. **Assumptions Matter** Show by counterexample that the Leibniz Test does not remain true if $\{a_n\}$ tends to zero but we drop the assumption that the sequence a_n is decreasing. *Hint:* Consider

$$R = \frac{1}{2} - \frac{1}{4} + \frac{1}{3} - \frac{1}{8} + \frac{1}{4} - \frac{1}{16} + \cdots + \left(\frac{1}{n} - \frac{1}{2^n}\right) + \cdots$$

36. Prove that $\displaystyle\sum_{n=1}^{\infty} (-1)^{n+1} \frac{(\ln n)^a}{n}$ converges for all exponents a.

Hint: Show that $f(x) = \dfrac{(\ln x)^a}{x}$ is decreasing for x sufficiently large.

37. We say that $\{b_n\}$ is a rearrangement of $\{a_n\}$ if $\{b_n\}$ has the same terms as $\{a_n\}$ but occurring in a different order. Show that if $\{b_n\}$ is a rearrangement of $\{a_n\}$ and $S = \displaystyle\sum_{n=1}^{\infty} a_n$ converges absolutely, then

$$T = \sum_{n=1}^{\infty} b_n$$ also converges absolutely. (This result does not hold if S is only conditionally convergent.) *Hint:* Prove that the partial sums $\displaystyle\sum_{n=1}^{N} |b_n|$ are bounded. It can be shown further that $S = T$.

38. **Assumptions Matter** In 1829, Lejeune Dirichlet pointed out that the great French mathematician Augustin Louis Cauchy made a mistake in a published paper by improperly assuming the Limit Comparison Test to be valid for nonpositive series. Here are Dirichlet's two series:

$$\sum_{n=1}^{\infty} \frac{(-1)^n}{\sqrt{n}}, \qquad \sum_{n=1}^{\infty} \frac{(-1)^n}{\sqrt{n}}\left(1 + \frac{(-1)^n}{\sqrt{n}}\right)$$

Explain how they provide a counterexample to the Limit Comparison Test when the series are not assumed to be positive.

10.5 The Ratio and Root Tests

As we will show in Section 10.7, the number e has a well-known expression as an infinite series:

$$e = 1 + \frac{1}{1!} + \frac{1}{2!} + \frac{1}{3!} + \cdots$$

However, the convergence tests developed so far cannot be easily applied to this series. This points to the need for the following test, which is also of key importance in the study of power series (Section 10.6).

The symbol ρ, pronounced "rho," is the seventeenth letter of the Greek alphabet.

THEOREM 1 Ratio Test Let $\{a_n\}$ be a sequence and assume that the following limit exists:

$$\rho = \lim_{n \to \infty} \left| \frac{a_{n+1}}{a_n} \right|$$

(i) If $\rho < 1$, then $\sum a_n$ converges absolutely.

(ii) If $\rho > 1$, then $\sum a_n$ diverges.

(iii) If $\rho = 1$, the Ratio Test is inconclusive (the series may converge or diverge).

Proof If $\rho < 1$, we may choose a number r such that $\rho < r < 1$. Since $\left| \frac{a_{n+1}}{a_n} \right|$ converges to ρ, there exists a number M such that $\left| \frac{a_{n+1}}{a_n} \right| < r$ for $n \geq M$. Therefore,

$$|a_{M+1}| < r|a_M|$$

$$|a_{M+2}| < r|a_{M+1}| < r^2|a_M|$$

$$|a_{M+3}| < r|a_{M+2}| < r^3|a_M|$$

In general, $|a_{M+n}| < r^n|a_M|$ and therefore

$$\sum_{n=M}^{\infty} |a_n| = \sum_{n=0}^{\infty} |a_{M+n}| \leq \sum_{n=0}^{\infty} |a_M| r^n = |a_M| \sum_{n=0}^{\infty} r^n$$

The geometric series on the right converges because $0 < r < 1$, so $\sum_{n=M}^{\infty} |a_n|$ converges by the Comparison Test. Thus, $\sum_{n=1}^{\infty} a_n$ converges absolutely.

If $\rho > 1$, choose a number r such that $1 < r < \rho$. Since $\left| \frac{a_{n+1}}{a_n} \right|$ converges to ρ, there exists a number M such that $\left| \frac{a_{n+1}}{a_n} \right| > r$ for $n \geq M$. Arguing as before with the inequalities reversed, we find that $|a_{M+n}| \geq r^n|a_M|$. Since r^n tends to ∞, we see that the terms a_{M+n} do not tend to zero and, consequently, $\sum_{n=1}^{\infty} a_n$ diverges. Finally, Example 4 below shows that when $\rho = 1$, both convergence and divergence are possible, so the test is inconclusive. ∎

EXAMPLE 1 Prove that $S = \sum_{n=1}^{\infty} \frac{1}{n!}$ converges.

Solution We compute the limit ρ. Let $a_n = \frac{1}{n!}$. Then

$$\frac{a_{n+1}}{a_n} = \frac{1}{(n+1)!} \frac{n!}{1} = \frac{n!}{(n+1)!} = \frac{1}{n+1}$$

$$\rho = \lim_{n \to \infty} \left| \frac{a_{n+1}}{a_n} \right| = \lim_{n \to \infty} \frac{1}{n+1} = 0$$

Since $\rho < 1$, S converges by the Ratio Test. ∎

■ **EXAMPLE 2** Apply the Ratio Test to determine if $\sum_{n=1}^{\infty} \dfrac{n^2}{2^n}$ converges.

Solution Let $a_n = \dfrac{n^2}{2^n}$. We have

$$\left|\frac{a_{n+1}}{a_n}\right| = \frac{(n+1)^2}{2^{n+1}}\frac{2^n}{n^2} = \frac{1}{2}\left(\frac{n^2+2n+1}{n^2}\right) = \frac{1}{2}\left(1+\frac{2}{n}+\frac{1}{n^2}\right)$$

$$\rho = \lim_{n\to\infty}\left|\frac{a_{n+1}}{a_n}\right| = \frac{1}{2}\lim_{n\to\infty}\left(1+\frac{2}{n}+\frac{1}{n^2}\right) = \frac{1}{2}$$

Since $\rho < 1$, the series converges by the Ratio Test. ■

■ **EXAMPLE 3** Determine whether $\sum_{n=0}^{\infty}(-1)^n\dfrac{n!}{100^n}$ converges.

Solution Let $a_n = (-1)^n\dfrac{n!}{100^n}$. Then

$$\left|\frac{a_{n+1}}{a_n}\right| = \frac{(n+1)!}{100^{n+1}}\frac{100^n}{n!} = \frac{n+1}{100}$$

We see that the ratio of the coefficients tends to infinity:

$$\rho = \lim_{n\to\infty}\left|\frac{a_{n+1}}{a_n}\right| = \lim_{n\to\infty}\frac{n+1}{100} = \infty$$

Since $\rho > 1$, $\sum_{n=0}^{\infty}(-1)^n\dfrac{n!}{100^n}$ diverges by the Ratio Test. ■

■ **EXAMPLE 4** Ratio Test Inconclusive Show that $\rho = 1$ for both $\sum_{n=1}^{\infty}n^2$ and $\sum_{n=1}^{\infty}n^{-2}$.
Conclude that the Ratio Test is inconclusive when $\rho = 1$.

Solution For $a_n = n^2$, we have

$$\rho = \lim_{n\to\infty}\left|\frac{a_{n+1}}{a_n}\right| = \lim_{n\to\infty}\frac{(n+1)^2}{n^2} = \lim_{n\to\infty}\frac{n^2+2n+1}{n^2} = \lim_{n\to\infty}\left(1+\frac{2}{n}+\frac{1}{n^2}\right) = 1$$

On the other hand, for $b_n = n^{-2}$,

$$\rho = \lim_{n\to\infty}\left|\frac{b_{n+1}}{b_n}\right| = \lim_{n\to\infty}\left|\frac{a_n}{a_{n+1}}\right| = \frac{1}{\lim_{n\to\infty}\frac{a_{n+1}}{a_n}} = 1$$

Thus, $\rho = 1$ in both cases, but $\sum_{n=1}^{\infty}n^2$ diverges and $\sum_{n=1}^{\infty}n^{-2}$ converges (a p-series with $p = 2$). This shows that both convergence and divergence are possible when $\rho = 1$. ■

For some series, it is more convenient to use the following Root Test, based on the limit of the nth roots $\sqrt[n]{a_n}$ rather than the ratios a_{n+1}/a_n. The proof of the Root Test, like that of the Ratio Test, is based on a comparison with geometric series (see Exercise 53).

> **THEOREM 2 Root Test** Let $\{a_n\}$ be a sequence and assume that the following limit exists:
>
> $$L = \lim_{n \to \infty} \sqrt[n]{|a_n|}$$
>
> **(i)** If $L < 1$, then $\sum a_n$ converges absolutely.
> **(ii)** If $L > 1$, then $\sum a_n$ diverges.
> **(iii)** If $L = 1$, the Root Test is inconclusive: The series may converge or diverge.

■ **EXAMPLE 5** Determine whether $\displaystyle\sum_{n=1}^{\infty} \left(\frac{n}{2n+3}\right)^n$ converges.

Solution Let $a_n = \left(\dfrac{n}{2n+3}\right)^n$. Then

$$L = \lim_{n \to \infty} \sqrt[n]{a_n} = \lim_{n \to \infty} \frac{n}{2n+3} = \frac{1}{2}$$

Since $L < \frac{1}{2}$, the series $\displaystyle\sum_{n=1}^{\infty} \left(\frac{n}{2n+3}\right)^n$ converges. ■

10.5 SUMMARY

- *Ratio Test:* Assume that the following limit exists:

$$\rho = \lim_{n \to \infty} \left| \frac{a_{n+1}}{a_n} \right|$$

Then $\sum a_n$ converges absolutely if $\rho < 1$ and it diverges if $\rho > 1$. The test is inconclusive if $\rho = 1$.

- *Root Test:* Assume that the following limit exists: $L = \lim_{n \to \infty} \sqrt[n]{|a_n|}$. Then $\sum a_n$ converges if $L < 1$ and it diverges if $L > 1$. The test is inconclusive if $L = 1$.

10.5 EXERCISES

Preliminary Questions

1. In the Ratio Test, is ρ equal to $\displaystyle\lim_{n \to \infty}\left|\frac{a_{n+1}}{a_n}\right|$ or $\displaystyle\lim_{n \to \infty}\left|\frac{a_n}{a_{n+1}}\right|$?

2. Is the Ratio Test conclusive for $\displaystyle\sum_{n=1}^{\infty}\frac{1}{2^n}$? Is it conclusive for $\displaystyle\sum_{n=1}^{\infty}\frac{1}{n}$?

3. Can the Ratio Test be used to show convergence if the series is only conditionally convergent?

Exercises

In Exercises 1–18, apply the Ratio Test to determine convergence or divergence, or state that the Ratio Test is inconclusive.

1. $\displaystyle\sum_{n=1}^{\infty}\frac{1}{5^n}$

2. $\displaystyle\sum_{n=1}^{\infty}\frac{(-1)^{n-1}n}{5^n}$

3. $\displaystyle\sum_{n=1}^{\infty}\frac{(-1)^{n-1}}{n^n}$

4. $\displaystyle\sum_{n=0}^{\infty}\frac{3n+2}{5n^3+1}$

5. $\displaystyle\sum_{n=1}^{\infty}\frac{n}{n^2+1}$

6. $\displaystyle\sum_{n=1}^{\infty}\frac{2^n}{n}$

7. $\displaystyle\sum_{n=1}^{\infty} \frac{2^n}{n^{100}}$

8. $\displaystyle\sum_{n=1}^{\infty} \frac{n^3}{3^{n^2}}$

9. $\displaystyle\sum_{n=1}^{\infty} \frac{10^n}{2^{n^2}}$

10. $\displaystyle\sum_{n=1}^{\infty} \frac{e^n}{n!}$

11. $\displaystyle\sum_{n=1}^{\infty} \frac{e^n}{n^n}$

12. $\displaystyle\sum_{n=1}^{\infty} \frac{n^{50}}{n!}$

13. $\displaystyle\sum_{n=0}^{\infty} (-1)^n \frac{n!}{4^n}$

14. $\displaystyle\sum_{n=1}^{\infty} \frac{n!}{n^4}$

15. $\displaystyle\sum_{n=2}^{\infty} \frac{1}{n \ln n}$

16. $\displaystyle\sum_{n=1}^{\infty} \frac{1}{(2n)!}$

17. $\displaystyle\sum_{n=1}^{\infty} \frac{n^2}{(2n+1)!}$

18. $\displaystyle\sum_{n=1}^{\infty} \frac{(n!)^2}{(2n)!}$

19. Show that $\displaystyle\sum_{n=1}^{\infty} n^k 3^{-n}$ converges for all exponents k.

20. Show that $\displaystyle\sum_{n=1}^{\infty} n^2 x^n$ converges if $|x| < 1$.

21. Show that $\displaystyle\sum_{n=1}^{\infty} 2^n x^n$ converges if $|x| < \frac{1}{2}$.

22. Show that $\displaystyle\sum_{n=1}^{\infty} \frac{r^n}{n!}$ converges for all r.

23. Show that $\displaystyle\sum_{n=1}^{\infty} \frac{r^n}{n}$ converges if $|r| < 1$.

24. Is there any value of k such that $\displaystyle\sum_{n=1}^{\infty} \frac{2^n}{n^k}$ converges?

25. Show that $\displaystyle\sum_{n=1}^{\infty} \frac{n!}{n^n}$ converges. *Hint:* Use $\displaystyle\lim_{n\to\infty} \left(1 + \frac{1}{n}\right)^n = e$.

In Exercises 26–31, assume that $|a_{n+1}/a_n|$ converges to $\rho = \frac{1}{3}$. What can you say about the convergence of the given series?

26. $\displaystyle\sum_{n=1}^{\infty} n a_n$

27. $\displaystyle\sum_{n=1}^{\infty} n^3 a_n$

28. $\displaystyle\sum_{n=1}^{\infty} 2^n a_n$

29. $\displaystyle\sum_{n=1}^{\infty} 3^n a_n$

30. $\displaystyle\sum_{n=1}^{\infty} 4^n a_n$

31. $\displaystyle\sum_{n=1}^{\infty} a_n^2$

32. Assume that $\left|\dfrac{a_{n+1}}{a_n}\right|$ converges to $\rho = 4$. Does $\displaystyle\sum_{n=1}^{\infty} a_n^{-1}$ converge (assume that $a_n \neq 0$ for all n)?

33. Is the Ratio Test conclusive for the p-series $\displaystyle\sum_{n=1}^{\infty} \frac{1}{n^p}$?

In Exercises 34–39, use the Root Test to determine convergence or divergence (or state that the test is inconclusive).

34. $\displaystyle\sum_{n=0}^{\infty} \frac{1}{10^n}$

35. $\displaystyle\sum_{n=1}^{\infty} \frac{1}{n^n}$

36. $\displaystyle\sum_{k=0}^{\infty} \left(\frac{k}{k+10}\right)^k$

37. $\displaystyle\sum_{k=0}^{\infty} \left(\frac{k}{3k+1}\right)^k$

38. $\displaystyle\sum_{n=1}^{\infty} \left(1 + \frac{1}{n}\right)^{-n}$

39. $\displaystyle\sum_{n=4}^{\infty} \left(1 + \frac{1}{n}\right)^{-n^2}$

40. Prove that $\displaystyle\sum_{n=1}^{\infty} \frac{2^{n^2}}{n!}$ diverges. *Hint:* Use $2^{n^2} = (2^n)^n$ and $n! \leq n^n$.

In Exercises 41–52, determine convergence or divergence using any method covered in the text so far.

41. $\displaystyle\sum_{n=1}^{\infty} \frac{2^n + 4^n}{7^n}$

42. $\displaystyle\sum_{n=1}^{\infty} \frac{n^3}{n!}$

43. $\displaystyle\sum_{n=1}^{\infty} \frac{n^3}{5^n}$

44. $\displaystyle\sum_{n=2}^{\infty} \frac{1}{n(\ln n)^3}$

45. $\displaystyle\sum_{n=2}^{\infty} \frac{1}{\sqrt{n^3 - n^2}}$

46. $\displaystyle\sum_{k=1}^{\infty} 4^{-2k+1}$

47. $\displaystyle\sum_{n=1}^{\infty} \frac{n^2 + 4n}{3n^4 + 9}$

48. $\displaystyle\sum_{n=1}^{\infty} (-1)^n \cos \frac{1}{n}$

49. $\displaystyle\sum_{n=1}^{\infty} \sin \frac{1}{n^2}$

50. $\displaystyle\sum_{n=1}^{\infty} \frac{(-1)^{n-1}}{\sqrt{n}}$

51. $\displaystyle\sum_{n=1}^{\infty} \left(\frac{n}{n+12}\right)^n$

52. $\displaystyle\sum_{n=1}^{\infty} \frac{(-2)^n}{\sqrt{n}}$

Further Insights and Challenges

53. ✏️ **Proof of the Root Test** Let $S = \displaystyle\sum_{n=0}^{\infty} a_n$ be a positive series and assume that $L = \displaystyle\lim_{n\to\infty} \sqrt[n]{a_n}$ exists.

(a) Show that S converges if $L < 1$. *Hint:* Choose R with $\rho < R < 1$ and show that $a_n \leq R^n$ for n sufficiently large. Then compare with the geometric series $\sum R^n$.

(b) Show that S diverges if $L > 1$.

54. Show that the Ratio Test is inconclusive but verify convergence using the comparison test for the series

$$\frac{1}{2} + \frac{1}{3^2} + \frac{1}{2^3} + \frac{1}{3^4} + \frac{1}{2^5} + \cdots$$

55. Let $S = \sum_{n=1}^{\infty} \dfrac{c^n n!}{n^n}$, where c is a constant.

(a) Prove that S converges absolutely if $|c| < e$ and diverges if $|c| > e$.

(b) It is known that $\displaystyle\lim_{n\to\infty} \dfrac{e^n n!}{n^{n+1/2}} = \sqrt{2\pi}$. Verify this numerically.

(c) Use the Limit Comparison Test to prove that S diverges for $c = e$.

10.6 Power Series

Most functions that arise in applications can be represented as power series. This includes not only the familiar trigonometric, exponential, logarithm, and root functions, but also the host of more advanced "special functions" of physics and engineering such as Bessel functions and elliptic functions.

In the introduction to this chapter, we mentioned that e^x can be expressed as an "infinite polynomial" called a power series:

$$e^x = 1 + \frac{x}{1!} + \frac{x^2}{2!} + \frac{x^3}{3!} + \frac{x^4}{4!} + \cdots$$

In this section, we develop the basic properties of power series, especially the key concept of *radius of convergence*.

A **power series** centered at the point $x = c$ is an infinite series of the form

$$F(x) = \sum_{n=0}^{\infty} a_n(x - c)^n = a_0 + a_1(x - c) + a_2(x - c)^2 + a_3(x - c)^3 + \cdots$$

To make use of a power series, we must determine the values of x for which the series converges. It certainly converges at its center $x = c$:

$$F(c) = a_0 + a_1(c - c) + a_2(c - c)^2 + a_3(c - c)^3 + \cdots = a_0$$

Where else does it converge? The following basic theorem states that every power series converges absolutely on an interval that is symmetric around the center $x = c$ (the interval may be infinite or possibly reduced to the single point c).

THEOREM 1   Radius of Convergence   Let $F(x) = \sum_{n=0}^{\infty} a_n(x - c)^n$. There are three possibilities:

(i) $F(x)$ converges only for $x = c$, or

(ii) $F(x)$ converges for all x, or

(iii) There is a number $R > 0$ such that $F(x)$ converges absolutely if $|x - c| < R$ and diverges if $|x - c| > R$. It may or may not converge at the endpoints $|x - c| = R$.

In Case (i), set $R = 0$, and in Case (ii), set $R = \infty$. We call R the **radius of convergence** of $F(x)$.

Proof   We assume that $c = 0$ to simplify the notation. The key observation is that if $F(x)$ converges for some nonzero value $x = B$, then it converges absolutely for all $|x| < |B|$. To prove this, note that if $F(B) = \sum_{n=0}^{\infty} a_n B^n$ converges, then the general term $a_n B^n$ must tend to zero. In particular, there exists $M > 0$ such that $|a_n B^n| < M$ for all n, and therefore,

$$\sum_{n=0}^{\infty} |a_n x^n| = \sum_{n=0}^{\infty} |a_n B^n| \left| \frac{x}{B} \right|^n < M \sum_{n=0}^{\infty} \left| \frac{x}{B} \right|^n$$

If $|x| < |B|$, then $|x/B| < 1$ and the series on the right is a convergent geometric series. By the Comparison Test, the series on the left also converges and thus $F(x)$ converges absolutely if $|x| < |B|$.

Least Upper Bound Property: If S is a set of real numbers with an upper bound M (that is, $x \le M$ for all $x \in S$), then S has a least upper bound L. See Appendix B.

Let S be the set of numbers x such that $F(x)$ converges. Then S contains 0. If $S = \{0\}$, then $F(x)$ converges only for $x = 0$ and Case (i) holds. Otherwise, S contains a number $B \ne 0$. In this case, S contains the open interval $(-|B|, |B|)$ by the previous paragraph. If S is bounded, then S has a least upper bound $L > 0$ (see marginal note). Since there exist numbers $B \in S$ smaller than but arbitrarily close to L, S contains $(-B, B)$ for all $0 < B < L$. It follows that S contains the open interval $(-L, L)$. S cannot contain any number x with $|x| > L$, but S may contain one or both of the endpoints $x = \pm L$. This is Case (iii). If S is not bounded, then S contains intervals $(-B, B)$ for B arbitrarily large. Thus $S = \mathbf{R}$ and we are in Case (ii). ∎

According to Theorem 1, if the radius of convergence R is nonzero and finite, then $F(x)$ converges absolutely on an open interval around c of radius R (Figure 1). *It is necessary to check convergence at the endpoints of the interval separately.*

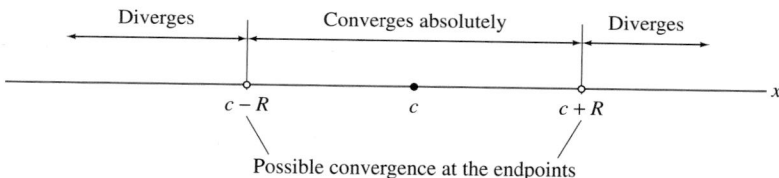

FIGURE 1 Interval of convergence of a power series.

■ **EXAMPLE 1** Using the Ratio Test For which values of x does $F(x) = \sum_{n=0}^{\infty} \dfrac{x^n}{2^n}$ converge?

Solution Let $b_n = \dfrac{x^n}{2^n}$ and let us compute the ratio ρ of the Ratio Test:

$$\rho = \lim_{n\to\infty} \left| \frac{b_{n+1}}{b_n} \right| = \lim_{n\to\infty} \left| \frac{x^{n+1}}{2^{n+1}} \right| \cdot \left| \frac{2^n}{x^n} \right| = \lim_{n\to\infty} \left| \frac{2^{-(n+1)} x^{n+1}}{2^{-n} x^n} \right| = \lim_{n\to\infty} \frac{1}{2}|x| = \frac{1}{2}|x|$$

By the Ratio Test, $F(x)$ converges if $\rho = \frac{1}{2}|x| < 1$, that is, for $|x| < 2$. Similarly, $F(x)$ diverges if $\rho = \frac{1}{2}|x| > 1$ or $|x| > 2$. Therefore, the radius of convergence is $R = 2$.

What about the endpoints? The Ratio Test is inconclusive for $x = \pm 2$, so we must check these cases directly. Both series diverge:

$$F(2) = \sum_{n=0}^{\infty} \frac{2^n}{2^n} = 1 + 1 + 1 + 1 + 1 + 1 \cdots$$

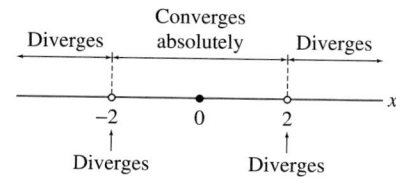

FIGURE 2 Interval of convergence of $\sum_{n=0}^{\infty} \dfrac{x^n}{2^n}$.

$$F(-2) = \sum_{n=0}^{\infty} \frac{(-2)^n}{2^n} = 1 - 1 + 1 - 1 + 1 - 1 \cdots$$

Therefore, $F(x)$ converges only for $|x| < 2$ (Figure 2). ■

The method of the previous example may be applied more generally to any power series $F(x) = \sum b_n(x - c)^n$ for which the following limit exists:

$$r = \lim_{n \to \infty} \left| \frac{b_{n+1}}{b_n} \right|$$

We compute the quantity ρ of the Ratio Test applied to $F(x)$:

$$\rho = \lim_{n \to \infty} \left| \frac{b_{n+1}(x - c)^{n+1}}{b_n(x - c)^n} \right| = |x - c| \left(\lim_{n \to \infty} \left| \frac{b_{n+1}}{b_n} \right| \right) = r|x - c|$$

By the Ratio Test, $F(x)$ converges if $\rho = r|x - c| < 1$ and diverges if $\rho = r|x - c| > 1$. Thus, if r is finite and nonzero, then $F(x)$ converges if $|x - c| < r^{-1}$ and the radius of convergence is $R = r^{-1}$. If $r = 0$, then $F(x)$ converges for all x and the radius of convergence is $R = \infty$. If $r = \infty$, then $F(x)$ diverges for all $x \neq c$ and $R = 0$.

THEOREM 2 Finding the Radius of Convergence Let $F(x) = \sum b_n(x - c)^n$ and assume that the following limit exists:

$$r = \lim_{n \to \infty} \left| \frac{b_{n+1}}{b_n} \right|$$

Then $F(x)$ has radius of convergence $R = r^{-1}$ (where we set $R = \infty$ if $r = 0$ and $R = 0$ if $r = \infty$).

■ **EXAMPLE 2** Determine the convergence of $F(x) = \sum_{n=1}^{\infty} \frac{(-1)^n}{n}(x - 5)^n$.

Solution Let $b_n = \dfrac{(-1)^n}{n}$. Then

$$r = \lim_{n \to \infty} \left| \frac{b_{n+1}}{b_n} \right| = \lim_{n \to \infty} \left| \frac{1}{n + 1} \frac{n}{1} \right| = \lim_{n \to \infty} \left| \frac{n}{n + 1} \right| = 1$$

The radius of convergence is $R = r^{-1} = 1$. Therefore, the power series converges absolutely if $|x - 5| < 1$ and diverges if $|x - 5| > 1$. In other words, $F(x)$ converges absolutely on the open interval $(4, 6)$. At the endpoints we have

$$x = 6: \quad \sum_{n=1}^{\infty} \frac{(-1)^n}{n}(6 - 5)^n = \sum_{n=1}^{\infty} \frac{(-1)^n}{n} \qquad \text{convergent by the Leibniz Test}$$

$$x = 4: \quad \sum_{n=1}^{\infty} \frac{(-1)^n}{n}(4 - 5)^n = \sum_{n=1}^{\infty} \frac{1}{n} \qquad \text{divergent (harmonic series)}$$

Therefore, the power series converges on the half-open interval $(4, 6]$. ■

■ **EXAMPLE 3** Infinite Radius of Convergence Show that $\sum_{n=0}^{\infty} \frac{x^n}{n!}$ converges for all x.

Solution Let $b_n = \dfrac{1}{n!}$. Then

$$r = \lim_{n\to\infty} \left| \frac{b_{n+1}}{b_n} \right| = \lim_{n\to\infty} \frac{1}{(n+1)!} \frac{n!}{1} = \lim_{n\to\infty} \frac{1}{n+1} = 0$$

Thus, $R = r^{-1} = \infty$ and the series converges for all x by Theorem 2. ∎

When a function $f(x)$ is represented by a power series on an interval I, we refer to the power series as the "power series expansion" of $f(x)$ on I. In the next section, we show that a function has at most one power series expansion with a given center c on an interval.

An important example of a power series is provided by the geometric series. Recall that $\displaystyle\sum_{n=0}^{\infty} r^n = \dfrac{1}{1-r}$ for $|r| < 1$. Writing x instead of r, we view this formula as a power series expansion:

$$\boxed{\frac{1}{1-x} = \sum_{n=0}^{\infty} x^n \qquad \text{for } |x| < 1} \qquad \boxed{1}$$

The next two examples show how this formula may be adapted to find the power series representations of other functions.

■ **EXAMPLE 4** Using the Formula for Geometric Series Prove that $\dfrac{1}{1-2x} = \displaystyle\sum_{n=0}^{\infty} 2^n x^n$ for $|x| < \frac{1}{2}$.

Solution Substitute $2x$ for x in Eq. (1):

$$\frac{1}{1-2x} = \sum_{n=0}^{\infty} (2x)^n = \sum_{n=0}^{\infty} 2^n x^n \qquad \boxed{2}$$

Expansion (1) is valid for $|x| < 1$ and thus expansion (2) is valid for $|2x| < 1$ or $|x| < \frac{1}{2}$. ∎

■ **EXAMPLE 5** Prove that $\dfrac{1}{2+x^2} = \displaystyle\sum_{n=0}^{\infty} \frac{(-1)^n x^{2n}}{2^{n+1}}$. For which x is this formula valid?

Solution We first rewrite $\dfrac{1}{2+x^2}$ in the form $\dfrac{1}{1-u}$ so we can use Eq. (1):

$$\frac{1}{2+x^2} = \frac{1}{2} \frac{1}{1 + \frac{1}{2}x^2} = \frac{1}{2} \frac{1}{1 - (-\frac{1}{2}x^2)}$$

We may substitute $-\frac{1}{2}x^2$ for x in Eq. (1), provided that $\left|\frac{1}{2}x^2\right| < 1$, to obtain

$$\frac{1}{2+x^2} = \frac{1}{2} \sum_{n=0}^{\infty} \left(-\frac{x^2}{2}\right)^n = \sum_{n=0}^{\infty} \frac{(-1)^n x^{2n}}{2^{n+1}}$$

This expansion is valid if $\left|-x^2/2\right| < 1$ or $|x| < \sqrt{2}$. ∎

Our next theorem states, in essence, that power series are well-behaved functions in the following sense: A power series $F(x)$ is differentiable within its interval of convergence and we may differentiate and integrate $F(x)$ as if it were a polynomial.

THEOREM 3 Term-by-Term Differentiation and Integration Suppose that

$$F(x) = \sum_{n=0}^{\infty} a_n(x - c)^n$$

has radius of convergence $R > 0$. Then $F(x)$ is differentiable on $(c - R, c + R)$ and its derivative and antiderivative may be computed term by term. More precisely, for $x \in (c - R, c + R)$ we have

$$F'(x) = \sum_{n=1}^{\infty} na_n(x - c)^{n-1}$$

$$\int F(x)\, dx = A + \sum_{n=0}^{\infty} \frac{a_n}{n + 1}(x - c)^{n+1} \quad (A \text{ any constant})$$

These series have the same radius of convergence R.

See Exercise 58 for a proof that $F(x)$ is continuous. The proofs of the remaining statements are omitted.

■ **EXAMPLE 6** Differentiating a Power Series Prove that

$$\frac{1}{(1 - x)^2} = 1 + 2x + 3x^2 + 4x^3 + 5x^4 + \cdots$$

for $-1 < x < 1$.

Solution Noting that

$$\frac{d}{dx} \frac{1}{1 - x} = \frac{1}{(1 - x)^2}$$

we obtain the result by differentiating the geometric series term by term for $|x| < 1$:

$$\frac{d}{dx} \frac{1}{1 - x} = \frac{d}{dx}\left(1 + x + x^2 + x^3 + x^4 + \cdots\right)$$

$$\frac{1}{(1 - x)^2} = 1 + 2x + 3x^2 + 4x^3 + 5x^4 + \cdots \qquad \boxed{3}$$

Expansion (3) is valid for $|x| < 1$ because the geometric series has radius of convergence $R = 1$. ■

■ **EXAMPLE 7** The Power Series for $f(x) = \tan^{-1} x$ via Integration Prove that for $-1 < x < 1$,

$$\tan^{-1} x = \sum_{n=0}^{\infty} \frac{(-1)^n x^{2n+1}}{2n + 1} = x - \frac{x^3}{3} + \frac{x^5}{5} - \frac{x^7}{7} + \cdots \qquad \boxed{4}$$

Solution First, substitute $-x^2$ for x in (1) to obtain

$$\frac{1}{1 + x^2} = 1 - x^2 + x^4 - x^6 + \cdots$$

Since the geometric series has radius of convergence $R = 1$, this expansion is valid for $|x^2| < 1$, that is, $|x| < 1$. Now apply Theorem 3 to integrate this series term by term, recalling that $\tan^{-1} x$ is an antiderivative $(1 + x^2)^{-1}$:

$$\tan^{-1} x = \int \frac{dx}{1+x^2} = \int \left(1 - x^2 + x^4 - x^6 + \cdots\right) dx = A + x - \frac{x^3}{3} + \frac{x^5}{5} - \frac{x^7}{7} + \cdots$$

To determine the constant A, set $x = 0$. We obtain $\tan^{-1} 0 = 0 = A$ and therefore $A = 0$. This proves Eq. (4) for $-1 < x < 1$.   ∎

GRAPHICAL INSIGHT   Let's examine the expansion of the previous example graphically. The partial sums of the power series for $f(x) = \tan^{-1} x$ are

$$S_N(x) = \sum_{n=0}^{N} (-1)^n \frac{x^{2n+1}}{2n+1} = x - \frac{x^3}{3} + \frac{x^5}{5} - \cdots + (-1)^N \frac{x^{2N+1}}{2N+1}$$

We can expect $S_N(x)$ to provide a good approximation to $f(x) = \tan^{-1} x$ on the interval $(-1, 1)$, where the power series expansion is valid. Figure 3 confirms this: The graphs of the partial sums $S_{50}(x)$ and $S_{51}(x)$ are nearly indistinguishable from the graph of $\tan^{-1} x$ on $(-1, 1)$. Thus we may use the partial sums to approximate values of the arctangent. For example, an approximation to $\tan^{-1}(0.3)$ is given by

$$S_4(0.3) = 0.3 - \frac{(0.3)^3}{3} + \frac{(0.3)^5}{5} - \frac{(0.3)^7}{7} + \frac{(0.3)^9}{9} \approx 0.2914569$$

Since the power series is an alternating series, the error is not greater than the first omitted term:

$$|\tan^{-1}(0.3) - S_4(0.3)| \leq \frac{(0.3)^{11}}{11} \approx 1.61 \times 10^{-7}$$

The situation changes drastically in the region $|x| > 1$, where the power series diverges. The partial sums $S_N(x)$ deviate sharply from $\tan^{-1} x$ outside $(-1, 1)$.

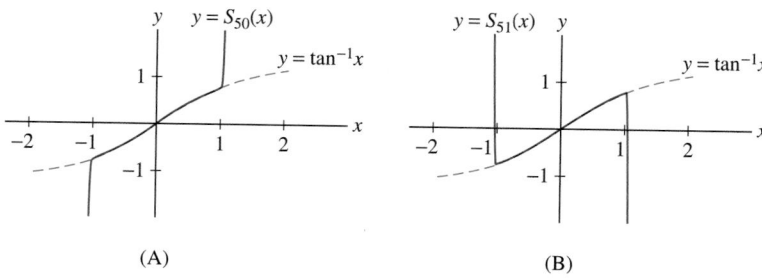

FIGURE 3 $S_{50}(x)$ and $S_{51}(x)$ are nearly indistinguishable from $\tan^{-1} x$ on $(-1, 1)$.

(A)   (B)

Power Series Solutions of Differential Equations

In the next section, we use the theory of Taylor series to prove that the exponential function $f(x) = e^x$ is represented by a power series. However, we can already show this with the tools at our disposal by making use of the differential equation satisfied by $f(x) = e^x$. Recall that by Theorem 1 in Section 5.8, $y = e^x$ is the unique function satisfying the differential equation $y' = y$ with initial condition $y(0) = 1$. Let's try to find a power series

$$P(x) = \sum_{n=0}^{\infty} a_n x^n$$ that also satisfies $P'(x) = P(x)$ and $P(0) = 1$.

We have

$$P(x) = \sum_{n=0}^{\infty} a_n x^n = a_0 + a_1 x + a_2 x^2 + a_3 x^3 + \cdots$$

$$P'(x) = \sum_{n=0}^{\infty} n a_n x^{n-1} = a_1 + 2a_2 x + 3a_3 x^2 + 4a_4 x^3 + \cdots$$

We see that $P(x)$ satisfies $P'(x) = P(x)$ if

$$a_0 = a_1, \quad a_1 = 2a_2, \quad a_2 = 3a_3, \quad a_3 = 4a_4, \quad \cdots$$

In general, $a_{n-1} = n a_n$, or

$$a_n = \frac{a_{n-1}}{n}$$

This equation is called a *recursion relation*. It allows us to successively determine all of the coefficients a_n from the first coefficient a_0, which may be chosen arbitrarily. For example, the recursion relation yields

$$n = 1: \quad a_1 = \frac{a_0}{1}$$

$$n = 2: \quad a_2 = \frac{a_1}{2} = \frac{a_0}{2 \cdot 1} = \frac{a_0}{2!}$$

$$n = 3: \quad a_3 = \frac{a_2}{3} = \frac{a_1}{3 \cdot 2} = \frac{a_0}{3 \cdot 2 \cdot 1} = \frac{a_0}{3!}$$

To obtain a general formula for a_n, apply the recursion relation n times:

$$a_n = \frac{a_{n-1}}{n} = \frac{a_{n-2}}{n(n-1)} = \frac{a_{n-3}}{n(n-1)(n-2)} = \cdots = \frac{a_0}{n!}$$

We conclude that $P(x) = a_0 \sum_{n=0}^{\infty} \frac{x^n}{n!}$. As we showed in Example 3, this power series has an infinite radius of convergence and thus $P(x)$ is a solution of $y' = y$ for all x.

Now observe that $P(0) = a_0$, so we set $a_0 = 1$ to obtain a solution satisfying the initial condition $y(0) = 1$. Now, since $f(x) = e^x$ and $P(x)$ both satisfy the differential condition with initial condition, they are equal. Thus we have proven that for all x,

$$e^x = \sum_{n=0}^{\infty} \frac{x^n}{n!} = 1 + x + \frac{x^2}{2!} + \frac{x^3}{3!} + \frac{x^4}{4!} + \cdots \qquad \boxed{5}$$

The method just employed is a powerful tool in the study of differential equations. We knew in advance that $y = e^x$ is a solution of $y' = y$, but suppose we are given a differential equation whose solution is unknown. We may try to find a solution in the form of a power series $P(x) = \sum_{n=0}^{\infty} a_n x^n$. In favorable cases, the differential equation leads to a recursion relation that enables us to determine the coefficients a_n.

The solution of Eq. (6) satisfying $y'(0) = 1$ is called the "Bessel function of order one." The Bessel function of order n is a solution of

$$x^2 y'' + xy' + (x^2 - n^2)y = 0$$

Bessel functions appear in many areas of physics and engineering.

■ **EXAMPLE 8** Find a power series solution to the differential equation

$$x^2 y'' + xy' + (x^2 - 1)y = 0 \qquad \boxed{6}$$

and the initial condition $y'(0) = 1$.

Solution Assume that Eq. (6) has a power series solution $P(x) = \sum_{n=0}^{\infty} a_n x^n$. Then

$$y' = P'(x) = \sum_{n=0}^{\infty} n a_n x^{n-1} = a_1 + 2a_2 x + 3a_3 x^2 + \cdots$$

$$y'' = P''(x) = \sum_{n=0}^{\infty} n(n-1) a_n x^{n-2} = 2a_2 + 6a_3 x + 12a_4 x^2 + \cdots$$

Now substitute the series for y, y', and y'' into the differential equation (6) to determine the recursion relation satisfied by the coefficients a_n:

$$x^2 y'' + xy' + (x^2 - 1)y$$

$$= x^2 \sum_{n=0}^{\infty} n(n-1) a_n x^{n-2} + x \sum_{n=0}^{\infty} n a_n x^{n-1} + (x^2 - 1) \sum_{n=0}^{\infty} a_n x^n$$

$$= \sum_{n=0}^{\infty} n(n-1) a_n x^n + \sum_{n=0}^{\infty} n a_n x^n - \sum_{n=0}^{\infty} a_n x^n + \sum_{n=0}^{\infty} a_n x^{n+2}$$

$$= \sum_{n=0}^{\infty} (n^2 - 1) a_n x^n + \sum_{n=2}^{\infty} a_{n-2} x^n$$

We see that the equation $x^2 y'' + xy' + (x^2 - 1)y = 0$ is satisfied if

$$\sum_{n=0}^{\infty} (n^2 - 1) a_n x^n = - \sum_{n=2}^{\infty} a_{n-2} x^n \qquad \boxed{7}$$

The first few terms on each side of this equation are

$$-a_0 + 0 \cdot x + 3a_2 x^2 + 8a_3 x^3 + 15a_4 x^4 + \cdots = 0 + 0 \cdot x - a_0 x^2 - a_1 x^3 - a_2 x^4 - \cdots$$

Matching up the coefficients of x^n, we find that

$$-a_0 = 0, \qquad 3a_2 = -a_0, \qquad 8a_3 = -a_1, \qquad 15a_4 = -a_2 \qquad \boxed{8}$$

In general, $(n^2 - 1)a_n = -a_{n-2}$, and this yields the recursion relation

$$\boxed{a_n = -\frac{a_{n-2}}{n^2 - 1} \qquad \text{for } n \geq 2} \qquad \boxed{9}$$

Note that $a_0 = 0$ by (8). The recursion relation implies that all of the even coefficients $a_2, a_4, a_6, \ldots$ are zero:

$$a_2 = \frac{a_0}{2^2 - 1} \text{ so } a_2 = 0, \qquad \text{and then} \qquad a_4 = \frac{a_2}{4^2 - 1} = 0 \text{ so } a_4 = 0, \qquad \text{etc.}$$

As for the odd coefficients, note that we may choose a_1 arbitrarily. Observe that $P'(0) = a_1$. Thus, we set $a_1 = 1$ so that $y = P(x)$ satisfies $y'(0) = 1$. Now apply Eq. (9):

$$n = 3 \qquad a_3 = -\frac{a_1}{3^2 - 1} = -\frac{1}{3^2 - 1}$$

$$n = 5: \qquad a_5 = -\frac{a_3}{5^2 - 1} = \frac{1}{(5^2 - 1)(3^2 - 1)}$$

$$n = 7: \qquad a_7 = -\frac{a_5}{7^2 - 1} = -\frac{1}{(7^2 - 1)(3^2 - 1)(5^2 - 1)}$$

This shows the general pattern of coefficients. To express the coefficients in a compact form, let $n = 2k + 1$. Then

$$n^2 - 1 = (2k + 1)^2 - 1 = 4k^2 + 4k = 4k(k + 1)$$

and the recursion relation may be written

$$a_{2k+1} = -\frac{a_{2k-1}}{4k(k + 1)}$$

Applying this recursion relation k times, we obtain the closed formula

$$a_{2k+1} = (-1)^k \left(\frac{1}{4k(k + 1)}\right)\left(\frac{1}{4(k - 1)k}\right)\cdots\left(\frac{1}{4(1)(2)}\right) = \frac{(-1)^k}{4^k\, k!\, (k + 1)!}$$

Thus we obtain a power series representation of our solution:

$$P(x) = \sum_{k=0}^{\infty} \frac{(-1)^k}{4^k k!(k + 1)!} x^{2k+1}$$

A straightforward application of the Ratio Test shows that $P(x)$ has an infinite radius of convergence. Therefore $P(x)$ is a solution of the initial value problem for all x.   ∎

10.6 SUMMARY

- A *power series* is an infinite series of the form

$$F(x) = \sum_{n=0}^{\infty} a_n (x - c)^n$$

The constant c is called the *center* of $F(x)$.
- A power series behaves in one of three ways:

   (i) $F(x)$ converges only for $x = c$, or
   (ii) $F(x)$ converges for all x, or
   (iii) There exists $R > 0$ such that $F(x)$ converges absolutely for $|x - c| < R$ and diverges for $|x - c| > R$.

The number R is called the *radius of convergence* of $F(x)$. Convergence at the endpoints $x = c \pm R$ must be checked separately. We set $R = 0$ in Case (i), and $R = \infty$ in Case (ii).
- If $r = \lim_{n \to \infty} \left|\frac{a_{n+1}}{a_n}\right|$ exists, then $F(x)$ has radius of convergence $R = r^{-1}$ (where $R = 0$ if $r = \infty$ and $R = \infty$ if $r = 0$).

• If $R > 0$, then $F(x)$ is differentiable on $(c - R, c + R)$ and may be differentiated and integrated term by term:

$$F'(x) = \sum_{n=1}^{\infty} n a_n (x - c)^{n-1}, \qquad \int F(x)\, dx = A + \sum_{n=0}^{\infty} \frac{a_n}{n+1} (x - c)^{n+1}$$

(A is any constant). The power series for $F'(x)$ and $\int F(x)\, dx$ have the same radius of convergence R.

• The power series expansion $\dfrac{1}{1-x} = \displaystyle\sum_{n=0}^{\infty} x^n$ is valid for $|x| < 1$. It may be used to derive expansions of other related functions by substitution, integration, or differentiation.

10.6 EXERCISES

Preliminary Questions

1. Suppose that $\sum a_n x^n$ converges for $x = 5$. Must it also converge for $x = 4$? What about $x = -3$?

2. Suppose that $\sum a_n (x - 6)^n$ converges for $x = 10$. At which of the points (a)–(d) must it also converge?

(a) $x = 8$ **(b)** $x = 12$ **(c)** $x = 2$ **(d)** $x = 0$

3. Suppose that $F(x)$ is a power series with radius of convergence $R = 12$. What is the radius of convergence of $F(3x)$?

4. The power series $F(x) = \displaystyle\sum_{n=1}^{\infty} n x^n$ has radius of convergence $R = 1$. What is the power series expansion of $F'(x)$ and what is its radius of convergence?

Exercises

1. Use the Ratio Test to determine the radius of convergence of

$$\sum_{n=0}^{\infty} \frac{x^n}{2^n}.$$

2. Use the Ratio Test to show that $\displaystyle\sum_{n=1}^{\infty} \frac{x^n}{\sqrt{n}\, 2^n}$ has radius of convergence $R = 2$. Then determine whether it converges absolutely or conditionally at the endpoints $R = \pm 2$.

3. Show that the following three power series have the same radius of convergence. Then show that (a) diverges at both endpoints, (b) converges at one endpoint but diverges at the other, and (c) converges at both endpoints.

(a) $\displaystyle\sum_{n=1}^{\infty} \frac{x^n}{3^n}$ **(b)** $\displaystyle\sum_{n=1}^{\infty} \frac{x^n}{n 3^n}$ **(c)** $\displaystyle\sum_{n=1}^{\infty} \frac{x^n}{n^2 3^n}$

4. Repeat Exercise 3 for the following series:

(a) $\displaystyle\sum_{n=1}^{\infty} \frac{(x-5)^n}{9^n}$ **(b)** $\displaystyle\sum_{n=1}^{\infty} \frac{(x-5)^n}{n 9^n}$ **(c)** $\displaystyle\sum_{n=1}^{\infty} \frac{(x-5)^n}{n^2 9^n}$

5. Show that $\displaystyle\sum_{n=0}^{\infty} n^n x^n$ diverges for all $x \neq 0$.

6. (a) Find the radius of convergence of $\displaystyle\sum_{n=1}^{\infty} \frac{x^n}{n^2}$.

(b) Determine whether the series converges at the endpoints of the interval of convergence.

In Exercises 7–26, find the values of x for which the following power series converge.

7. $\displaystyle\sum_{n=1}^{\infty} n x^n$

8. $\displaystyle\sum_{n=1}^{\infty} n (x - 3)^n$

9. $\displaystyle\sum_{n=1}^{\infty} \frac{2^n x^n}{n}$

10. $\displaystyle\sum_{n=1}^{\infty} \frac{(-5)^n (x - 3)^n}{n^2}$

11. $\displaystyle\sum_{n=2}^{\infty} \frac{x^n}{\ln n}$

12. $\displaystyle\sum_{n=1}^{\infty} \frac{x^n}{n 2^n}$

13. $\displaystyle\sum_{n=1}^{\infty} \frac{x^n}{(n!)^2}$

14. $\displaystyle\sum_{n=4}^{\infty} \frac{x^n}{n^5}$

15. $\displaystyle\sum_{n=1}^{\infty} (-1)^n n^4 (x + 4)^n$

16. $\displaystyle\sum_{n=0}^{\infty} \frac{(-1)^n x^n}{\sqrt{n^2 + 1}}$

17. $\displaystyle\sum_{n=0}^{\infty} \frac{n}{2^n} x^n$

18. $\displaystyle\sum_{n=0}^{\infty} 4^n x^n$

19. $\displaystyle\sum_{n=1}^{\infty} \frac{(x - 4)^n}{n^4}$

20. $\displaystyle\sum_{n=1}^{\infty} \frac{2^n}{3n} (x + 3)^n$

21. $\displaystyle\sum_{n=10}^{\infty} n!\,(x+5)^n$

22. $\displaystyle\sum_{n=15}^{\infty} \frac{x^{2n+1}}{3n+1}$

23. $\displaystyle\sum_{n=12}^{\infty} e^n (x-2)^n$

24. $\displaystyle\sum_{n=0}^{\infty} \frac{x^n}{n^4+2}$

25. $\displaystyle\sum_{n=1}^{\infty} \frac{x^n}{n-4\ln n}$

26. $\displaystyle\sum_{n=2}^{\infty} \frac{(x-2)^n}{(n\ln n)^2}$

In Exercises 27–34, use Eq. (1) to expand the function in a power series with center $c = 0$ and determine the set of x for which the expansion is valid.

27. $f(x) = \dfrac{1}{1-3x}$

28. $f(x) = \dfrac{1}{1+3x}$

29. $f(x) = \dfrac{1}{3-x}$

30. $f(x) = \dfrac{1}{4+3x}$

31. $f(x) = \dfrac{1}{1+x^9}$

32. $f(x) = \dfrac{1}{5-x^2}$

33. $f(x) = \dfrac{1}{1+3x^7}$

34. $f(x) = \dfrac{1}{4-2x^3}$

35. Use the equalities

$$\frac{1}{1-x} = \frac{1}{-3-(x-4)} = \frac{-\frac{1}{3}}{1+\left(\frac{x-4}{3}\right)}$$

to show that for $|x-4| < 3$

$$\frac{1}{1-x} = \sum_{n=0}^{\infty} (-1)^{n+1} \frac{(x-4)^n}{3^{n+1}}$$

36. Use the method of Exercise 35 to expand $\dfrac{1}{1-x}$ in power series with centers $c = 2$ and $c = -2$. Determine the set of x for which the expansions are valid.

37. Use the method of Exercise 35 to expand $\dfrac{1}{4-x}$ in a power series with center $c = 5$. Determine the set of x for which the expansion is valid.

38. Evaluate $\displaystyle\sum_{n=1}^{\infty} \frac{n}{2^n}$. *Hint:* Show that

$$(1-x)^{-2} = \sum_{n=1}^{\infty} nx^{n-1} \quad \text{(for } |x| < 1)$$

39. Give an example of a power series that converges for x in $[2, 6)$.

40. Prove that for $-1 < x < 1$

$$\frac{1}{1+x} = \sum_{n=0}^{\infty} (-1)^n x^n = 1 - x + x^2 - x^3 + \cdots$$

$$\ln(1+x) = \sum_{n=1}^{\infty} \frac{(-1)^{n-1} x^n}{n} = x - \frac{x^2}{2} + \frac{x^3}{3} - \frac{x^4}{4} + \cdots$$

41. Use Exercise 40 to prove that

$$\ln\frac{3}{2} = \frac{1}{2} - \frac{1}{2\cdot 2^2} + \frac{1}{3\cdot 2^3} - \frac{1}{4\cdot 2^4} + \cdots$$

Use your knowledge of alternating series to find an N such that the partial sum S_N approximates $\ln\frac{3}{2}$ to within an error of at most 10^{-3}. Confirm this using a calculator to compute both S_N and $\ln\frac{3}{2}$.

42. Show that the following series converges absolutely for $|x| < 1$ and compute its sum:

$$F(x) = 1 - x - x^2 + x^3 - x^4 - x^5 + x^6 - x^7 - x^8 + \cdots$$

Hint: Write $F(x)$ as a sum of three geometric series with common ratio x^3.

43. Show that for $|x| < 1$

$$\frac{1+2x}{1+x+x^2} = 1 + x - 2x^2 + x^3 + x^4 - 2x^5 + x^6 + x^7 - 2x^8 + \cdots$$

Hint: Use the hint from Exercise 42.

44. Find a power series $P(x) = \displaystyle\sum_{n=0}^{\infty} a_n x^n$ satisfying the differential equation $y' = -y$ with initial condition $y(0) = 1$. Then use Theorem 1 of Section 5.8 to conclude that $P(x) = e^{-x}$.

45. Use the power series for $y = e^x$ to show that

$$\frac{1}{e} = \frac{1}{2!} - \frac{1}{3!} + \frac{1}{4!} - \cdots$$

Use your knowledge of alternating series to find an N such that the partial sum S_N approximates e^{-1} to within an error of at most 10^{-3}. Confirm this using a calculator to compute both S_N and e^{-1}.

46. Let $P(x) = \displaystyle\sum_{n=0}^{\infty} a_n x^n$ be a power series solution to $y' = 2xy$ with initial condition $y(0) = 1$.

(a) Show that the odd coefficients a_{2k+1} are all zero.

(b) Prove that $a_{2k} = \dfrac{a_{2k-2}}{k}$ and use this result to determine the coefficients a_{2k}.

47. Find a power series $P(x)$ satisfying the differential equation:

$$y'' - xy' + y = 0 \qquad \boxed{10}$$

with initial condition $y(0) = 1$, $y'(0) = 0$. What is the radius of convergence of the power series?

48. Find a power series satisfying Eq. (10) with initial condition $y(0) = 0$, $y'(0) = 1$.

49. Prove that $J_2(x) = \displaystyle\sum_{k=0}^{\infty} \frac{(-1)^k}{2^{2k+2}\,k!\,(k+3)!} x^{2k+2}$ is a solution of the Bessel differential equation of order two:

$$x^2 y'' + xy' + (x^2 - 4)y = 0$$

50. Use Eq. (4) to approximate $\tan^{-1}(0.5)$ to three decimal places.

51. Let $C(x) = 1 - \dfrac{x^2}{2!} + \dfrac{x^4}{4!} - \dfrac{x^6}{6!} + \cdots$.

(a) Show that $C(x)$ has an infinite radius of convergence.

(b) Prove that $C(x)$ and $f(x) = \cos x$ are both solutions of $y'' = -y$ with initial conditions $y(0) = 1$, $y'(0) = 0$. This initial value problem has a unique solution, so it follows that $C(x) = \cos x$ for all x.

52. Find all values of x such that $\displaystyle\sum_{n=1}^{\infty} \frac{x^{n^2}}{n!}$ converges.

53. Find all values of x such that the following series converges:

$$F(x) = 1 + 3x + x^2 + 27x^3 + x^4 + 243x^5 + \cdots$$

54. Explain why Theorem 2 cannot be applied directly to find the radius of convergence of $\displaystyle\sum_{n=1}^{\infty} \frac{x^{3n}}{5^n}$. What is the radius of convergence of this series?

55. Why is it impossible to expand $f(x) = |x|$ as a power series that converges in an interval around $x = 0$? Explain this using Theorem 3.

Further Insights and Challenges

56. Prove that for $b \neq 0$

$$\sum_{n=1}^{\infty} \frac{1}{n(n+b)} = \frac{1}{b} \int_0^1 \frac{1 - x^b}{1 - x} \, dx$$

Conclude that the sum on the left has the value

$$\frac{1}{b}\left(1 + \frac{1}{2} + \frac{1}{3} + \cdots + \frac{1}{b}\right)$$

57. Suppose that the coefficients of $F(x) = \displaystyle\sum_{n=0}^{\infty} a_n x^n$ are *periodic*, that is, for some whole number $M > 0$, we have $a_{M+n} = a_n$. Prove that $F(x)$ converges absolutely for $|x| < 1$ and that

$$F(x) = \frac{a_0 + a_1 x + \cdots + a_{M-1} x^{M-1}}{1 - x^M}$$

Hint: Use the hint for Exercise 42.

58. Continuity of Power Series   Let $F(x) = \displaystyle\sum_{n=0}^{\infty} a_n x^n$ be a power series with radius of convergence $R > 0$.

(a) Prove the inequality

$$|x^n - y^n| \le n|x - y|(|x|^{n-1} + |y|^{n-1}) \qquad \boxed{11}$$

Hint: $x^n - y^n = (x - y)(x^{n-1} + x^{n-2}y + \cdots + y^{n-1})$.

(b) Choose R_1 with $0 < R_1 < R$. Show that the infinite series $M = \displaystyle\sum_{n=0}^{\infty} 2n|a_n|R_1^n$ converges. *Hint:* Show that $n|a_n|R_1^n < |a_n|x^n$ for all n sufficiently large if $R_1 < x < R$.

(c) Use Eq. (11) to show that if $|x| < R_1$ and $|y| < R_1$, then $|F(x) - F(y)| \le M|x - y|$.

(d) Prove that if $|x| < R$, then $F(x)$ is continuous at x. *Hint:* Choose R_1 such that $|x| < R_1 < R$. Show that if $\epsilon > 0$ is given, then $|F(x) - F(y)| \le \epsilon$ for all y such that $|x - y| < \delta$, where δ is any positive number that is less than ϵ/M and $R_1 - |x|$ (see Figure 4).

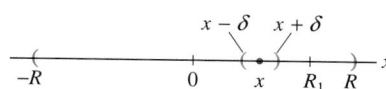

FIGURE 4 If $x > 0$, choose $\delta > 0$ less than ϵ/M and $R_1 - x$.

10.7 Taylor Series

We saw in the previous section that functions such as $f(x) = e^x$ and $f(x) = \tan^{-1} x$ can be represented as power series. These power series give us a certain tangible insight into the function represented and they allow us to approximate the values of $f(x)$ to any desired degree of accuracy. Thus, it is desirable to develop general methods for finding power series representations.

Suppose that $f(x)$ has a power series expansion centered at $x = c$ that is valid for all x in an interval $(c - R, c + R)$ with $R > 0$:

$$f(x) = \sum_{n=0}^{\infty} a_n (x - c)^n = a_0 + a_1(x - c) + a_2(x - c)^2 + \cdots$$

Then we may differentiate the series term by term (Theorem 3 in Section 10.6) to obtain

$$f(x) = a_0 + a_1(x - c) + a_2(x - c)^2 + a_3(x - c)^3 + \cdots$$

$$f'(x) = a_1 + 2a_2(x - c) + 3a_3(x - c)^2 + 4a_4(x - c)^3 + \cdots$$

$$f''(x) = 2a_2 + 2 \cdot 3a_3(x - c) + 3 \cdot 4a_4(x - c)^2 + \cdots$$

$$\vdots$$

$$f^{(k)}(x) = k!a_k + \left(2 \cdot 3 \cdots (k+1)\right)a_{k+1}(x - c) + \cdots$$

Setting $x = c$ in each of these series, we find that

$$f(c) = a_0, \quad f'(c) = a_1, \quad f''(c) = 2a_2, \quad \ldots, \quad f^{(k)}(c) = k!a_k$$

This shows that the coefficients are given by the formula (and proves Theorem 1 below):

$$\boxed{a_k = \frac{f^{(k)}(c)}{k!}}$$

$\boxed{1}$

Recall that these are the coefficients of the Taylor polynomials. In summary:

$$f(x) = \sum_{n=0}^{\infty} \frac{f^{(n)}(c)}{n!}(x - c)^n$$

$$= f(c) + f'(c)(x - c) + \frac{f''(c)}{2!}(x - c)^2 + \frac{f'''(c)}{3!}(x - c)^3 + \cdots$$

The power series on the right is called the **Taylor series** of $f(x)$ centered at $x = c$. In the special case $c = 0$, the Taylor series is also called the **Maclaurin series**:

$$f(x) = f(0) + f'(0)x + \frac{f''(0)}{2!}x^2 + \frac{f'''(0)}{3!}x^3 + \frac{f^{(4)}(0)}{4!}x^4 + \cdots$$

THEOREM 1 Uniqueness of the Power Series Expansion If $f(x)$ is represented by a power series $F(x)$ centered at c on an interval $(c - R, c + R)$ with $R > 0$, then $F(x)$ is the Taylor series of $f(x)$ centered at $x = c$.

■ **EXAMPLE 1** Find the Maclaurin series for $f(x) = e^x$.

Solution The nth derivative $f^{(n)}(x)$ is $f^{(n)}(x) = e^x$ for all n and thus

$$f(0) = f'(0) = f''(0) = \cdots = e^0 = 1$$

Therefore, the coefficients of the Maclaurin series are $a_k = \dfrac{f^{(k)}(0)}{k!} = \dfrac{1}{k!}$ and the Maclaurin series is

$$1 + \frac{x}{1} + \frac{x^2}{2!} + \frac{x^3}{3!} + \cdots$$

■

Theorem 1 tells us that if we want to represent $f(x)$ by a power series centered at c, the only candidate for the job is the Taylor series:

$$T(x) = \sum_{n=0}^{\infty} \frac{f^{(n)}(c)}{n!}(x-c)^n$$

However, there is no guarantee that $T(x)$ converges to $f(x)$. To study convergence, we consider the kth partial sum, which is the Taylor polynomial of degree k:

$$T_k(x) = f(c) + f'(c)(x-c) + \frac{f''(c)}{2!}(x-c)^2 + \cdots + \frac{f^{(k)}(c)}{k!}(x-c)^k$$

Recall that the remainder is defined by

$$R_k(x) = f(x) - T_k(x)$$

Since $T(x)$ is the limit of the partial sums $T_k(x)$, we see that

The Taylor series converges to $f(x)$ if and only if $\lim_{k \to \infty} R_k(x) = 0$

Although there is no general method for determining whether $R_k(x)$ tends to zero, the following theorem can often be applied.

◄·· **REMINDER** $f(x)$ is called "infinitely differentiable" if $f^{(n)}(x)$ exists for all n.

THEOREM 2 Let $f(x)$ be an infinitely differentiable function on the open interval $I = (c - R, c + R)$ with $R > 0$. Assume there exists $K \geq 0$ such that for all $k \geq 0$,

$$|f^{(k)}(x)| \leq K \qquad \text{for all} \quad x \in I$$

Then $f(x)$ is represented by its Taylor series in I:

$$f(x) = \sum_{n=0}^{\infty} \frac{f^{(n)}(c)}{n!}(x-c)^n \qquad \text{for all} \quad x \in I$$

Proof We apply the Error Bound for Taylor polynomials (Theorem 1 in Section 8.4):

$$|R_k(x)| = |f(x) - T_k(x)| \leq K \frac{|x-c|^{k+1}}{(k+1)!}$$

If $x \in I$, then $|x - c| < R$ and

$$|R_k(x)| \leq \underbrace{K \frac{R^{k+1}}{(k+1)!}}_{\text{This tends to zero as } k \to \infty}$$

As shown in Example 8 of Section 10.1, the quantity $R^k / k!$ tends to zero as $k \to \infty$ for every number R. We conclude that $\lim_{k \to \infty} R_k(x) = 0$ for all $x \in (c - R, c + R)$ and Theorem 2 follows. ∎

Taylor expansions were studied throughout the seventeenth and eighteenth centuries by Euler, Gregory, Leibniz, Maclaurin, Newton, Taylor, and others. These developments in Europe and England were anticipated by the great Indian mathematician Madhava (c. 1340–1425), who discovered the expansions of sine and cosine and many other results two centuries earlier.

■ **EXAMPLE 2** Maclaurin Expansions of Sine and Cosine Show that the following Taylor expansions are valid for all x:

$$\sin x = \sum_{n=0}^{\infty} (-1)^n \frac{x^{2n+1}}{(2n+1)!} = x - \frac{x^3}{3!} + \frac{x^5}{5!} - \frac{x^7}{7!} + \cdots$$

$$\cos x = \sum_{n=0}^{\infty} (-1)^n \frac{x^{2n}}{(2n)!} = 1 - \frac{x^2}{2!} + \frac{x^4}{4!} - \frac{x^6}{6!} + \cdots$$

Solution For $f(x) = \sin x$, we have

$$f^{(2n)}(x) = (-1)^n \sin x, \qquad f^{(2n+1)}(x) = (-1)^n \cos x$$

Therefore, $f^{(2n)}(0) = 0$ and $f^{(2n+1)}(0) = (-1)^n$. The nonzero Taylor coefficients for $\sin x$ are $a_{2n+1} = \dfrac{(-1)^n}{(2n+1)!}$. Similarly, for $f(x) = \cos x$,

$$f^{(2n)}(x) = (-1)^n \cos x, \qquad f^{(2n+1)}(x) = (-1)^{n+1} \sin x$$

Therefore, $f^{(2n)}(0) = (-1)^n$ and $f^{(2n+1)}(0) = 0$. The nonzero Taylor coefficients for $\cos x$ are $a_{2n} = \dfrac{(-1)^n}{(2n)!}$.

In both cases, $|f^{(n)}(x)| \le 1$ for all x and n. Thus, we may apply Theorem 2 with $M = 1$ and any R to conclude that the Taylor series converges to $f(x)$ for $|x| < R$. Since R is arbitrary, the Taylor expansions hold for all x. ∎

■ **EXAMPLE 3** Taylor Expansion $f(x) = e^x$ at $x = c$ Find the Taylor series $T(x)$ of $f(x) = e^x$ at $x = c$.

Solution We have $f^{(n)}(c) = e^c$ for all x and thus

$$T(x) = \sum_{n=0}^{\infty} \frac{e^c}{n!}(x-c)^n$$

To prove convergence, we note that e^x is increasing and therefore, for any R, $|f^{(k)}(x)| \le e^{c+R}$ for $x \in (c - R, c + R)$. Applying Theorem 2 with $M = e^{c+R}$, we conclude that $T(x)$ converges to $f(x)$ for all $x \in (c - R, c + R)$. Since R is arbitrary, the Taylor expansion holds for all x. For $c = 0$ we obtain the standard Taylor series

$$e^x = 1 + x + \frac{x^2}{2!} + \frac{x^3}{3!} + \cdots$$

∎

Shortcuts to Finding Taylor Series

Since a Taylor series is a power series, we may differentiate and integrate a Taylor series term by term within its interval of convergence by Theorem 3 in Section 10.6. We may also multiply two Taylor series or substitute one Taylor series into another (we omit the proofs of these facts). This leads to shortcuts for generating new Taylor series from known ones.

■ **EXAMPLE 4** Find the Maclaurin series for $f(x) = x^2 e^x$.

Solution We obtain the Maclaurin series of $f(x)$ by multiplying the known Maclaurin series for e^x by x^2:

$$x^2 e^x = x^2 \left(1 + x + \frac{x^2}{2!} + \frac{x^3}{3!} + \frac{x^4}{4!} + \frac{x^5}{5!} + \cdots \right)$$

$$= x^2 + x^3 + \frac{x^4}{2!} + \frac{x^5}{3!} + \frac{x^6}{4!} + \frac{x^7}{5!} + \cdots = \sum_{n=2}^{\infty} \frac{x^n}{(n-2)!}$$

∎

In some cases, there is no convenient formula for the Taylor coefficients of a product, but we can compute as many coefficients as desired numerically.

■ **EXAMPLE 5** Multiplying Taylor Series   Write out the first five terms in the Maclaurin series for $f(x) = e^x \cos x$.

Solution We multiply the fifth-order Taylor polynomials of e^x and $\cos x$ together, dropping the terms of degree greater than 5:

$$\left(1 + x + \frac{x^2}{2} + \frac{x^3}{6} + \frac{x^4}{24} + \frac{x^5}{120}\right)\left(1 - \frac{x^2}{2} + \frac{x^4}{24}\right)$$

Distributing the term on the left (and ignoring terms of degree greater than 5), we obtain

$$\left(1 + x + \frac{x^2}{2} + \frac{x^3}{6} + \frac{x^4}{24} + \frac{x^5}{120}\right) - \left(\frac{x^2}{2}\right)\left(1 + x + \frac{x^2}{2} + \frac{x^3}{6}\right) + \left(\frac{x^4}{24}\right)(1 + x)$$

$$= \underbrace{1 + x - \frac{x^3}{3} - \frac{x^4}{6} - \frac{x^5}{30}}_{\text{Retain terms of degree} \leq 5} + \text{ higher-order terms}$$

We conclude that the first five terms of the Taylor series for $f(x) = e^x \cos x$ are

$$T_5(x) = 1 + x - \frac{x^3}{3} - \frac{x^4}{6} - \frac{x^5}{30} \qquad \blacksquare$$

■ **EXAMPLE 6** Substitution   Use substitution to determine the Maclaurin series for e^{-x^2}.

Solution The Maclaurin series for e^{-x^2} is obtained by substituting $-x^2$ in the Maclaurin series for e^x:

$$e^{-x^2} = \sum_{n=0}^{\infty} \frac{(-1)^n x^{2n}}{n!} = 1 - x^2 + \frac{x^4}{2!} - \frac{x^6}{3!} + \frac{x^8}{4!} - \cdots \qquad \boxed{2}$$

Since the Taylor expansion of e^x is valid for all x, this expansion is also valid for all x.

$\blacksquare$

■ **EXAMPLE 7** Integrating a Taylor Series   Find the Maclaurin series for $f(x) = \ln(1 + x)$.

Solution In our study of Taylor polynomials, we computed the Maclaurin polynomials of $\ln(1 + x)$ directly. Here we obtain the same result by integrating the geometric series with common ratio $-x$:

$$\frac{1}{1+x} = 1 - x + x^2 - x^3 + \cdots \qquad \boxed{3}$$

$$\ln(1 + x) = \int \frac{dx}{1+x} = x - \frac{x^2}{2} + \frac{x^3}{3} - \frac{x^4}{4} + \cdots \qquad \boxed{4}$$

In principle, we might need a constant of integration on the right-hand side of Eq. (4). However, the constant of integration is zero because both $\ln(1 + x)$ and the power series take the value 0 at $x = 0$.

Furthermore, (3) is valid for $|x| < 1$, so the expansion of $\ln(1 + x)$ is also valid for $|x| < 1$. It also holds for $x = 1$ (see Exercise 83). $\blacksquare$

Taylor series may be used to express definite integrals as infinite series. This is useful when the integrand does not have an explicit antiderivative and the FTC cannot be applied. To justify this use of Taylor series, we appeal to Theorem 3 in Section 10.6, which implies the following: If a power series $P(x)$ centered at c has radius of convergence R, then the definite integral $\int_a^b P(x)\,dx$ over an interval $[a, b]$ contained in $(c - R, c + R)$ may be evaluated term by term.

■ **EXAMPLE 8** Let $J = \int_0^1 \sin(x^2)\,dx$.

(a) Express J as an infinite series.

(b) Determine J to within an error less than 10^{-4}.

Solution

(a) The Maclaurin expansion for $\sin x$ is valid for all x, so we have

$$\sin x = \sum_{n=0}^{\infty} \frac{(-1)^n}{(2n+1)!} x^{2n+1} \quad \Rightarrow \quad \sin(x^2) = \sum_{n=0}^{\infty} \frac{(-1)^n}{(2n+1)!} x^{4n+2}$$

We obtain an infinite series for J by integration:

$$J = \int_0^1 \sin(x^2)\,dx = \sum_{n=0}^{\infty} \frac{(-1)^n}{(2n+1)!} \int_0^1 x^{4n+2}\,dx$$

$$= \sum_{n=0}^{\infty} \frac{(-1)^n}{(2n+1)!} \left(\frac{1}{4n+3} \right)$$

$$= \frac{1}{3} - \frac{1}{42} + \frac{1}{1,320} - \frac{1}{75,600} + \cdots \qquad \boxed{5}$$

(b) The infinite series for J is an alternating series with decreasing terms, so the sum of the first N terms is accurate to within an error not greater than the $(N+1)$st term. In other words,

$$\left| J - \sum_{n=0}^{N-1} \frac{(-1)^n}{(4n+3)(2n+1)!} \right| \le \frac{1}{(4N+3)(2N+1)!}$$

For $N = 3$, we obtain

$$J \approx \frac{1}{3} - \frac{1}{42} + \frac{1}{1,320} \approx 0.31028$$

with an error

$$\left| J - \left(\frac{1}{3} - \frac{1}{42} + \frac{1}{1,320} \right) \right| \le \frac{1}{(4(3)+3)(2(3)+1)!} = \frac{1}{75,600} \approx 1.3 \times 10^{-5}$$

We see that three terms of the series suffice to compute the integral with an error less than 10^{-4}.   ■

Binomial Series

Taylor series yield a generalization of the Binomial Theorem that was first discovered by Isaac Newton around 1665. For any number a (integer or not) and integer $n \ge 0$, we define the **binomial coefficient**:

$$\binom{a}{n} = \frac{a(a-1)(a-2)\cdots(a-n+1)}{n!}, \qquad \binom{a}{0} = 1$$

For example,

$$\binom{6}{3} = \frac{6\cdot5\cdot4}{3\cdot2\cdot1} = 20, \qquad \binom{\frac{4}{3}}{3} = \frac{\frac{4}{3}\cdot\frac{1}{3}\cdot(-\frac{2}{3})}{3\cdot2\cdot1} = -\frac{4}{81}$$

The **Binomial Theorem** of algebra (see Appendix C) states that for any whole number a,

$$(r+s)^a = r^a + \binom{a}{1}r^{a-1}s + \binom{a}{2}r^{a-2}s^2 + \cdots + \binom{a}{a-1}rs^{a-1} + s^a$$

Setting $r = 1$ and $s = x$, we obtain an expansion of $f(x) = (1+x)^a$:

$$f(x) = (1+x)^a = 1 + \binom{a}{1}x + \binom{a}{2}x^2 + \cdots + \binom{a}{a-1}x^{a-1} + x^a$$

To derive Newton's generalization, we compute the Taylor series of $f(x) = (1+x)^a$ without assuming that a is a whole number. Observe that the derivatives $f^{(k)}(0)$ follow a pattern:

$$f(x) = (1+x)^a \qquad\qquad f(0) = 1$$
$$f'(x) = a(1+x)^{a-1} \qquad\qquad f'(0) = a$$
$$f''(x) = a(a-1)(1+x)^{a-2} \qquad f''(0) = a(a-1)$$
$$f'''(x) = a(a-1)(a-2)(1+x)^{a-3} \quad f'''(0) = a(a-1)(a-2)$$

In general, $f^{(n)}(0) = a(a-1)(a-2)\cdots(a-n+1)$ and

$$\frac{f^{(n)}(0)}{n!} = \frac{a(a-1)(a-2)\cdots(a-n+1)}{n!} = \binom{a}{n}$$

Hence the Taylor series for $(1+x)^a$ is the binomial series

When a is a whole number, $\binom{a}{n}$ is zero for $n > a$, and in this case, the binomial series breaks off at degree n. The binomial series is infinite when a is not a whole number.

$$\sum_{n=0}^{\infty}\binom{a}{n}x^n = 1 + ax + \frac{a(a-1)}{2!}x^2 + \frac{a(a-1)(a-2)}{3!}x^3 + \cdots + \binom{a}{n}x^n + \cdots$$

The Ratio Test shows that this series has radius of convergence $R = 1$ (see Exercise 83). Furthermore, the binomial series converges to $(1+x)^a$ for $|x| < 1$ (see Exercise 84).

THEOREM 3 The Binomial Series   For any exponent a, the following expansion is valid for $|x| < 1$:

$$(1+x)^a = 1 + ax + \frac{a(a-1)}{2!}x^2 + \frac{a(a-1)(a-2)}{3!}x^3 + \cdots + \binom{a}{n}x^n + \cdots \qquad \boxed{6}$$

■ **EXAMPLE 9**   Find the first five terms in the Maclaurin expansion of

$$f(x) = (1+x)^{1/3}$$

Solution The binomial coefficients $\binom{a}{n}$ for $a = \dfrac{1}{3}$ are

$$1, \quad \frac{1}{3}, \quad \frac{\frac{1}{3}(-\frac{2}{3})}{2!} = -\frac{1}{9}, \quad \frac{\frac{1}{3}(-\frac{2}{3})(-\frac{5}{3})}{3!} = \frac{5}{81}, \quad \frac{\frac{1}{3}(-\frac{2}{3})(-\frac{5}{3})(-\frac{8}{3})}{4!} = -\frac{10}{243}$$

Therefore, $(1 + x)^{1/3} \approx 1 + \dfrac{1}{3}x - \dfrac{1}{9}x^2 + \dfrac{5}{81}x^3 - \dfrac{10}{243}x^4 + \cdots$. ∎

■ **EXAMPLE 10** Find the Maclaurin series for $f(x) = \dfrac{1}{\sqrt{1 - x^2}}$.

Solution First, we compute the coefficients $\binom{-\frac{1}{2}}{n}$ in the binomial series for $(1 + x)^{-1/2}$. The coefficients for $n = 0, 1, 2, 3$ are

$$1, \quad \frac{(-\frac{1}{2})}{1} = -\frac{1}{2}, \quad \frac{(-\frac{1}{2})(-\frac{3}{2})}{2!} = \frac{1 \cdot 3}{2 \cdot 4}, \quad \frac{(-\frac{1}{2})(-\frac{3}{2})(-\frac{5}{2})}{3!} = -\frac{1 \cdot 3 \cdot 5}{2 \cdot 4 \cdot 6}$$

The general pattern is

$$\binom{-\frac{1}{2}}{n} = (-1)^n \frac{1 \cdot 3 \cdot 5 \cdots (2n - 1)}{2^n \, n!} = (-1)^n \frac{1 \cdot 3 \cdot 5 \cdots (2n - 1)}{2 \cdot 4 \cdot 6 \cdots 2n}$$

Thus, the following binomial expansion is valid for $|x| < 1$:

$$\frac{1}{\sqrt{1 + x}} = 1 + \sum_{n=1}^{\infty} (-1)^n \frac{1 \cdot 3 \cdot 5 \cdots (2n - 1)}{2 \cdot 4 \cdot 6 \cdots (2n)} x^n = 1 - \frac{1}{2}x + \frac{1 \cdot 3}{2 \cdot 4} x^2 - \cdots$$

If $|x| < 1$, then $|x|^2 < 1$ and we may substitute $-x^2$ for x. We obtain for $|x| < 1$,

$$\boxed{\frac{1}{\sqrt{1 - x^2}} = 1 + \sum_{n=1}^{\infty} \frac{1 \cdot 3 \cdot 5 \cdots (2n - 1)}{2 \cdot 4 \cdot 6 \cdots 2n} x^{2n} = 1 + \frac{1}{2}x^2 + \frac{1 \cdot 3}{2 \cdot 4} x^4 + \cdots} \qquad \boxed{7}$$

∎

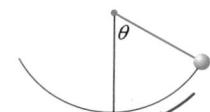

FIGURE 1 Pendulum released at an angle θ.

Period T

(graph: curve labeled "Small angle approximation", vertical axis values 8, 6, 4, 2, horizontal axis "Angle θ" with marks at $\dfrac{\pi}{2}$ and π)

FIGURE 2 The period T of a 1-m pendulum as a function of the angle θ at which it is released.

Taylor series are used to study the transcendental functions occurring in physics and engineering, such as Bessel functions and hypergeometric functions. Though less familiar than the trigonometric and exponential functions, these so-called *special functions* appear in a wide range of applications. An example is the following **elliptic function of the first kind**, defined for $|k| < 1$:

$$E(k) = \int_0^{\pi/2} \frac{dt}{\sqrt{1 - k^2 \sin^2 t}}$$

In physics, one shows that the period T of pendulum of length L released from an angle θ is equal to $T = 4\sqrt{L/g}\, E(k)$, where $k = \sin \frac{1}{2}\theta$ (Figure 1). For small θ, we have the "small angle approximation" $T \approx 2\pi\sqrt{L/g}$, but this approximation breaks down for large angles (as we see in Figure 2).

■ **EXAMPLE 11 Elliptic Functions** Find the Maclaurin series for $E(k)$ and estimate $E(k)$ for $k = \sin \dfrac{\pi}{6}$.

Solution Assume that $|k| < 1$ and substitute $x = k \sin t$ in the Taylor expansion (7):

$$\frac{1}{\sqrt{1 - k^2 \sin^2 t}} = 1 + \frac{1}{2}k^2 \sin^2 t + \frac{1 \cdot 3}{2 \cdot 4}k^4 \sin^4 t + \frac{1 \cdot 3 \cdot 5}{2 \cdot 4 \cdot 6}k^6 \sin^6 t + \cdots$$

This expansion is valid since $|x| = |k \sin t| < 1$. Thus $E(k)$ is equal to

$$\int_0^{\pi/2} \frac{dt}{\sqrt{1 - k^2 \sin^2 t}} = \int_0^{\pi/2} dt + \sum_{n=1}^{\infty} \frac{1 \cdot 3 \cdots (2n-1)}{2 \cdot 4 \cdot (2n)} \left(\int_0^{\pi/2} \sin^{2n} t \, dt \right) k^{2n}$$

According to Exercise 75 in Section 7.3,

$$\int_0^{\pi/2} \sin^{2n} t \, dt = \left(\frac{1 \cdot 3 \cdots (2n-1)}{2 \cdot 4 \cdot (2n)} \right) \frac{\pi}{2}$$

This yields

$$E(k) = \frac{\pi}{2} + \frac{\pi}{2} \sum_{n=1}^{\infty} \left(\frac{1 \cdot 3 \cdots (2n-1)^2}{2 \cdot 4 \cdots (2n)} \right)^2 k^{2n}$$

We approximate $E(k)$ for $k = \sin(\frac{\pi}{6}) = \frac{1}{2}$ using the first five terms:

$$F\left(\frac{1}{2}\right) \approx \frac{\pi}{2}\left(1 + \left(\frac{1}{2}\right)^2 \left(\frac{1}{2}\right)^2 + \left(\frac{1 \cdot 3}{2 \cdot 4}\right)^2 \left(\frac{1}{2}\right)^4 \right.$$

$$\left. + \left(\frac{1 \cdot 3 \cdot 5}{2 \cdot 4 \cdot 6}\right)^2 \left(\frac{1}{2}\right)^6 + \left(\frac{1 \cdot 3 \cdot 5 \cdot 7}{2 \cdot 4 \cdot 6 \cdot 8}\right)^2 \left(\frac{1}{2}\right)^8 \right)$$

$$\approx 1.68517$$

The value given by a computer algebra system to seven places is $F(\frac{1}{2}) \approx 1.6856325$. ∎

10.7 SUMMARY

- The *Taylor series* of $f(x)$ centered at $x = c$ is

$$T(x) = \sum_{n=0}^{\infty} \frac{f^{(n)}(c)}{n!}(x - c)^n$$

The partial sum $T_k(x)$ of $T(x)$ is the kth Taylor polynomial.
- If $f(x)$ is represented as a power series $\sum_{n=0}^{\infty} a_n(x - c)^n$ on an interval $(c - R, c + R)$ with $R > 0$, then this power series is the Taylor series centered at $x = c$.
- When $c = 0$, $T(x)$ is called the *Maclaurin series* of $f(x)$.
- The equality $f(x) = T(x)$ holds if and only if the remainder, defined as $R_k(x) = f(x) - T_k(x)$, tends to zero as $k \to \infty$.
- Suppose that $f(x)$ is infinitely differentiable on an interval $I = (c - R, c + R)$ with $R > 0$ and assume there exists $K > 0$ such that $|f^{(k)}(x)| < K$ for $x \in I$. Then $f(x)$ is represented by its Taylor series on I, that is, $f(x) = T(x)$ for $x \in I$.

TABLE 1

Function $f(x)$	Maclaurin Series	Converges to $f(x)$ for
e^x	$\displaystyle\sum_{n=0}^{\infty} \frac{x^n}{n!} = 1 + x + \frac{x^2}{2!} + \frac{x^3}{3!} + \frac{x^4}{4!} + \cdots$	All x
$\sin x$	$\displaystyle\sum_{n=0}^{\infty} \frac{(-1)^n x^{2n+1}}{(2n+1)!} = x - \frac{x^3}{3!} + \frac{x^5}{5!} - \frac{x^7}{7!} + \cdots$	All x
$\cos x$	$\displaystyle\sum_{n=0}^{\infty} \frac{(-1)^n x^{2n}}{(2n)!} = 1 - \frac{x^2}{2!} + \frac{x^4}{4!} - \frac{x^6}{6!} + \cdots$	All x
$\dfrac{1}{1-x}$	$\displaystyle\sum_{n=0}^{\infty} x^n = 1 + x + x^2 + x^3 + x^4 + \cdots$	$|x| < 1$
$\dfrac{1}{1+x}$	$\displaystyle\sum_{n=0}^{\infty} (-1)^n x^n = 1 - x + x^2 - x^3 + x^4 - \cdots$	$|x| < 1$
$\ln(1+x)$	$\displaystyle\sum_{n=1}^{\infty} \frac{(-1)^{n-1} x^n}{n} = x - \frac{x^2}{2} + \frac{x^3}{3} - \frac{x^4}{4} + \cdots$	$|x| < 1$ and $x = 1$
$\tan^{-1} x$	$\displaystyle\sum_{n=0}^{\infty} \frac{(-1)^n x^{2n+1}}{2n+1} = x - \frac{x^3}{3} + \frac{x^5}{5} - \frac{x^7}{7} + \cdots$	$|x| \le 1$
$(1+x)^a$	$\displaystyle\sum_{n=0}^{\infty} \binom{a}{n} x^n = 1 + ax + \frac{a(a-1)}{2!} x^2 + \frac{a(a-1)(a-2)}{3!} x^3 + \cdots$	$|x| < 1$

- A good way to find the Taylor series of a function is to start with known Taylor expansions and apply one of the following operations: multiplication, substitution, differentiation, or integration.

10.7 EXERCISES

Preliminary Questions

1. Determine $f(0)$ and $f'''(0)$ for a function $f(x)$ with Maclaurin series

$$T(x) = 3 + 2x + 12x^2 + 5x^3 + \cdots$$

2. Determine $f(-2)$ and $f^{(4)}(-2)$ for a function with Taylor series

$$T(x) = 3(x+2) + (x+2)^2 - 4(x+2)^3 + 2(x+2)^4 + \cdots$$

3. What is the easiest way to find the Maclaurin series for the function $f(x) = \sin(x^2)$?

4. What is the Taylor series for $f(x)$ centered at $c = 3$ if $f(3) = 4$ and $f'(x)$ has a Taylor expansion

$$f'(x) = \sum_{n=1}^{\infty} \frac{(x-3)^n}{n}$$

5. Let $T(x)$ be the Maclaurin series of $f(x)$. Which of the following guarantees that $f(2) = T(2)$?

(a) $T(x)$ converges for $x = 2$.

(b) The remainder $R_k(2)$ approaches a limit as $k \to \infty$.

(c) The remainder $R_k(2)$ approaches zero as $k \to \infty$.

Exercises

1. Write out the first four terms of the Maclaurin series of $f(x)$ if

$$f(0) = 2, \quad f'(0) = 3, \quad f''(0) = 4, \quad f'''(0) = 12$$

2. Write out the first four terms of the Taylor series of $f(x)$ centered at $c = 3$ if

$$f(3) = 1, \quad f'(3) = 2, \quad f''(3) = 12, \quad f'''(3) = 3$$

In Exercises 3–20, find the Maclaurin series.

3. $f(x) = \dfrac{1}{1 - 2x}$

4. $f(x) = \dfrac{x}{1 - x^4}$

5. $f(x) = \cos 3x$

6. $f(x) = \sin(2x)$

7. $f(x) = \sin(x^2)$

8. $f(x) = e^{4x}$

9. $f(x) = \ln(1 - x^2)$

10. $f(x) = (1 - x)^{-1/2}$

11. $f(x) = \tan^{-1}(x^2)$

12. $f(x) = x^2 e^{x^2}$

13. $f(x) = e^{x-2}$

14. $f(x) = \cos \sqrt{x}$

15. $f(x) = \ln(1 - 5x)$

16. $f(x) = (x^2 + 2x)e^x$

17. $f(x) = \sinh x$

18. $f(x) = \cosh x$

19. $f(x) = \dfrac{1 - \cos(x^2)}{x}$

20. $f(x) = \dfrac{e^x - \cos x}{x}$

21. Use multiplication to find the first four terms in the Maclaurin series for $f(x) = e^x \sin x$.

22. Find the first five terms of the Maclaurin series for $f(x) = \dfrac{\sin x}{1 - x}$.

23. Find the first four terms of the Maclaurin series for $f(x) = e^x \ln(1 - x)$.

24. Write out the first five terms of the binomial series for $f(x) = (1 + x)^{1/4}$.

25. Write out the first five terms of the binomial series for $f(x) = (1 + x)^{-3/2}$.

26. Differentiate the Maclaurin series for $\dfrac{1}{1 - x}$ twice to find the Maclaurin series of $\dfrac{1}{(1 - x)^3}$.

27. Find the first four terms of the Maclaurin for $f(x) = e^{(e^x)}$.

28. Find the first three terms of the Maclaurin series for $f(x) = \dfrac{1}{1 + \sin x}$. *Hint:* First expand $f(x)$ as a geometric series.

29. Find the Taylor series for $\sin x$ at $c = \dfrac{\pi}{2}$.

30. What is the Maclaurin series for $f(x) = x^4 - 2x^2 + 3$? What is the Taylor series centered at $c = 2$?

In Exercises 31–40, find the Taylor series centered at c.

31. $f(x) = \dfrac{1}{x}$, $c = 1$

32. $f(x) = \sqrt{x}$, $c = 4$

33. $f(x) = \dfrac{1}{1 - x}$, $c = 5$

34. $f(x) = x^4 + 3x - 1$, $c = 0$

35. $f(x) = x^4 + 3x - 1$, $c = 2$

36. $f(x) = \dfrac{1}{x^2}$, $c = 4$

37. $f(x) = e^{3x}$, $c = -1$

38. $f(x) = \dfrac{1}{1 - 4x}$, $c = -2$

39. $f(x) = \dfrac{1}{1 - x^2}$, $c = 3$

40. $f(x) = \dfrac{1}{3x - 2}$, $c = -1$

41. Find the Maclaurin series for $f(x) = \dfrac{1}{\sqrt{1 - 9x^2}}$ (see Example 10).

42. Show, by integrating the Maclaurin series for $f(x) = \dfrac{1}{\sqrt{1 - x^2}}$, that for $|x| < 1$,

$$\sin^{-1} x = x + \sum_{n=1}^{\infty} \frac{1 \cdot 3 \cdot 5 \cdots (2n - 1)}{2 \cdot 4 \cdot 6 \cdots (2n)} \frac{x^{2n+1}}{2n + 1}$$

43. Use the first five terms of the Maclaurin series in Exercise 42 to approximate $\sin^{-1} \frac{1}{2}$. Compare the result with the calculator value.

44. Show that for $|x| < 1$

$$\tanh^{-1} x = x + \frac{x^3}{3} + \frac{x^5}{5} + \cdots$$

Hint: Recall that $\dfrac{d}{dx} \tanh^{-1} x = \dfrac{1}{1 - x^2}$.

45. Use the Maclaurin series for $\ln(1 + x)$ and $\ln(1 - x)$ to show that

$$\frac{1}{2} \ln \left(\frac{1 + x}{1 - x} \right) = x + \frac{x^3}{3} + \frac{x^5}{5} + \cdots$$

What can you conclude by comparing this result with that of Exercise 44?

46. Use the Taylor series for $\cos x$ to compute $\cos 1$ to within an error of at most 10^{-6}. Use the fact that $\cos x$ is an alternating series with decreasing terms to estimate the error.

47. Use the Maclaurin expansion for e^{-t^2} to express $\displaystyle\int_0^x e^{-t^2}\, dt$ as an alternating power series in t.

(a) How many terms of the infinite series are needed to approximate the integral for $x = 1$ to within an error of at most 0.001?

(b) *CAS* Carry out the computation and check your answer using a computer algebra system.

48. Let $F(x) = \displaystyle\int_0^x \dfrac{\sin t\, dt}{t}$. Show that

$$F(x) = x - \frac{x^3}{3 \cdot 3!} + \frac{x^5}{5 \cdot 5!} - \frac{x^7}{7 \cdot 7!} + \cdots$$

Evaluate $F(1)$ to three decimal places.

In Exercises 49–52, express the definite integral as an infinite series and find its value to within an error of at most 10^{-4}.

49. $\int_0^1 \cos(x^2)\, dx$

50. $\int_0^1 \tan^{-1}(x^2)\, dx$

51. $\int_0^1 e^{-x^3}\, dx$

52. $\int_0^1 \dfrac{dx}{\sqrt{x^4+1}}$

In Exercises 53–56, express the integral as an infinite series.

53. $\int_0^x \dfrac{1-\cos(t)}{t}\, dt,$ for all x

54. $\int_0^x \dfrac{t-\sin t}{t}\, dt,$ for all x

55. $\int_0^x \ln(1+t^2)\, dt,$ for $|x|<1$

56. $\int_0^x \dfrac{dt}{\sqrt{1-t^4}},$ for $|x|<1$

57. Which function has Maclaurin series $\sum_{n=0}^{\infty}(-1)^n 2^n x^n$?

58. Which function has Maclaurin series

$$\sum_{k=0}^{\infty}\frac{(-1)^k}{3^{k+1}}(x-3)^k?$$

For which values of x is the expansion valid?

In Exercises 59–62, find the first four terms of the Taylor series.

59. $f(x)=\sin(x^2)\cos(x^2)$

60. $f(x)=e^x \tan^{-1}x$

61. $f(x)=e^{\sin x}$

62. $f(x)=\sin(x^3+2x)$

In Exercises 63–66, find the functions with the following Maclaurin series (refer to Table 1).

63. $1+x^3+\dfrac{x^6}{2!}+\dfrac{x^9}{3!}+\dfrac{x^{12}}{4!}+\cdots$

64. $1-4x+4^2x^2-4^3x^3+4^4x^4-4^5x^5+\cdots$

65. $1-\dfrac{5^3x^3}{3!}+\dfrac{5^5x^5}{5!}-\dfrac{5^7x^7}{7!}+\cdots$

66. $x^4-\dfrac{x^{12}}{3}+\dfrac{x^{20}}{5}-\dfrac{x^{28}}{7}+\cdots$

67. When a voltage V is applied to a series circuit consisting of a resistor R and an inductor L, the current at time t is

$$I(t)=\left(\frac{V}{R}\right)\left(1-e^{-Rt/L}\right)$$

Expand $I(t)$ in a Maclaurin series. Show that $I(t)\approx Vt/L$ if R is small.

68. Use substitution to write out the first three terms of the Maclaurin series for $f(x)=e^{x^{20}}$. Explain how the result implies that $f^{(k)}(0)=0$ for $1\le k\le 19$.

69. Find the Maclaurin series for $f(x)=\cos(\sqrt{x})$ and use it to determine $f^{(5)}(0)$.

70. Find $f^{(7)}(0)$ and $f^{(8)}(0)$ for $f(x)=\tan^{-1}x$.

71. Use the binomial series to find $f^{(8)}(0)$ for $f(x)=\sqrt{1-x^2}$.

72. Show that $\pi-\dfrac{\pi^3}{3!}+\dfrac{\pi^5}{5!}-\dfrac{\pi^7}{7!}+\cdots$ converges to zero. How many terms must be computed to get within 0.01 of zero?

73. Does the Taylor series for $f(x)=(1+x)^{3/4}$ converge to $f(x)$ at $x=2$? Give numerical evidence to support your answer.

74. Explain the steps required to verify that the Maclaurin series for $f(x)=\sin x$ converges to $f(x)$ at $x=1$.

75. Explain the steps required to verify that the Maclaurin series for $f(x)=\tan^{-1}x$ converges to $f(x)$ at $x=0.5$.

76. GU Let $f(x)=\sqrt{1+x}$.
(a) Use a graphing calculator to compare the graph of f with the graphs of the first five Taylor polynomials for f. What do they suggest about the interval of convergence of the Taylor series?
(b) Investigate numerically whether or not the Taylor expansion for f is valid for $x=1$ and $x=-1$.

77. How many terms of the Maclaurin series of $f(x)=\ln(1+x)$ are needed to compute $\ln 1.2$ to within an error of at most 0.0001? Make the computation and compare the result with the calculator value.

In Exercises 78–79, let

$$f(x)=\frac{1}{(1-x)(1-2x)}$$

78. Find the Maclaurin series of $f(x)$ using the identity

$$f(x)=\frac{2}{1-2x}-\frac{1}{1-x}$$

79. Find the Taylor series for $f(x)$ at $c=2$. *Hint:* Rewrite the identity of Exercise 78 as

$$f(x)=\frac{2}{-3-2(x-2)}-\frac{1}{-1-(x-2)}$$

80. Use the first five terms of the Maclaurin series for the elliptic function $E(k)$ to estimate the period T of a 1-m pendulum released at an angle $\theta=\frac{\pi}{4}$ (see Example 11).

81. Use Example 11 and the approximation $\sin x\approx x$ to show that the period T of a pendulum released at an angle θ has the following second-order approximation:

$$T\approx 2\pi\sqrt{\frac{L}{g}}\left(1+\frac{\theta^2}{16}\right)$$

Further Insights and Challenges

82. In this exercise we show that the Maclaurin expansion of the function $f(x) = \ln(1 + x)$ is valid for $x = 1$.

(a) Show that for all $x \neq -1$,

$$\frac{1}{1 + x} = \sum_{n=0}^{N} (-1)^n x^n + \frac{(-1)^{N+1} x^{N+1}}{1 + x}$$

(b) Integrate from 0 to 1 to obtain

$$\ln 2 = \sum_{n=1}^{N} \frac{(-1)^{n-1}}{n} + (-1)^{N+1} \int_0^1 \frac{x^{N+1}\,dx}{1 + x}$$

(c) Verify that the integral on the right tends to zero as $N \to \infty$ by showing that it is smaller than $\int_0^1 x^{N+1}\,dx$.

(d) Prove Leibniz's formula

$$\ln 2 = 1 - \frac{1}{2} + \frac{1}{3} - \frac{1}{4} + \cdots$$

In Exercises 83–84, we investigate the convergence of the binomial series

$$T_a(x) = \sum_{n=0}^{\infty} \binom{a}{n} x^n$$

83. Prove that $T_a(x)$ has radius of convergence $R = 1$ if a is not a whole number. What is the radius of convergence if a is a whole number?

84. By Exercise 83, $T_a(x)$ converges for $|x| < 1$, but we do not yet know whether $T_a(x) = (1 + x)^a$.

(a) Verify the identity

$$a \binom{a}{n} = n \binom{a}{n} + (n + 1) \binom{a}{n + 1}$$

(b) Use (a) to show that $y = T_a(x)$ satisfies the differential equation $(1 + x)y' = ay$ with initial condition $y(0) = 1$.

(c) Prove that $T_a(x) = (1 + x)^a$ for $|x| < 1$ by showing that the derivative of the ratio $\dfrac{T_a(x)}{(1 + x)^a}$ is constant.

85. The function $G(k) = \displaystyle\int_0^{\pi/2} \sqrt{1 - k^2 \sin^2 t}\, dt$ is called an **elliptic function of the second kind**. Prove that for $|k| < 1$

$$G(k) = \frac{\pi}{2} - \frac{\pi}{2} \sum_{n=1}^{\infty} \left(\frac{1 \cdot 3 \cdots (2n - 1)}{2 \cdots 4 \cdot (2n)} \right)^2 \frac{k^{2n}}{2n - 1}$$

86. Assume that $a < b$ and let L be the arc length (circumference) of the ellipse $\left(\dfrac{x}{a}\right)^2 + \left(\dfrac{y}{b}\right)^2 = 1$ shown in Figure 3. There is no explicit formula for L, but it is known that $L = 4bG(k)$, where $k = \sqrt{1 - a^2/b^2}$. Use the first three terms of the expansion of Exercise 85 to estimate the arc length when $a = 4$ and $b = 5$.

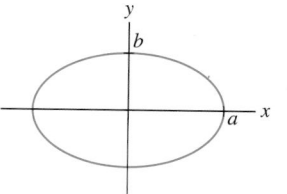

FIGURE 3 The ellipse $\left(\dfrac{x}{a}\right)^2 + \left(\dfrac{y}{b}\right)^2 = 1$

87. Use Exercise 85 to prove that if $a < b$ and a/b is near 1 (a nearly circular ellipse), then

$$L \approx \frac{\pi}{2}\left(3b + \frac{a^2}{b}\right)$$

Hint: Use the first two terms of the series for $G(k)$.

88. Irrationality of e   Prove that e is an irrational number using the following argument by contradiction. Suppose that $e = M/N$, where M, N are nonzero integers.

(a) Show that $M! e^{-1}$ is a whole number.

(b) Use the power series for e^x at $x = -1$ to show that there is an integer B such that $M! e^{-1}$ equals

$$B + (-1)^{M+1} \left(\frac{1}{M + 1} - \frac{1}{(M + 1)(M + 2)} + \cdots \right)$$

(c) Use your knowledge of alternating series with decreasing terms to conclude that $0 < |M! e^{-1} - B| < 1$. However, this contradicts (a). Hence, e is not equal to M/N.

CHAPTER REVIEW EXERCISES

1. Let $a_n = \dfrac{n - 3}{n!}$ and $b_n = a_{n+3}$. Calculate the first three terms in the sequence:

(a) a_n^2

(b) b_n

(c) $a_n b_n$

(d) $2a_{n+1} - 3a_n$

2. Prove that $\displaystyle\lim_{n \to \infty} \frac{2n - 1}{3n + 2} = \frac{2}{3}$ using the limit definition.

In Exercises 3–8, compute the limit (or state that it does not exist) assuming that $\lim_{n\to\infty} a_n = 2$.

3. $\lim_{n\to\infty} (5a_n - 2a_n^2)$

4. $\lim_{n\to\infty} \dfrac{1}{a_n}$

5. $\lim_{n\to\infty} e^{a_n}$

6. $\lim_{n\to\infty} \cos(\pi a_n)$

7. $\lim_{n\to\infty} (-1)^n a_n$

8. $\lim_{n\to\infty} \dfrac{a_n + n}{a_n + n^2}$

In Exercises 9–22, determine the limit of the sequence or show that the sequence diverges.

9. $a_n = \sqrt{n+5} - \sqrt{n+2}$

10. $a_n = \dfrac{3n^3 - n}{1 - 2n^3}$

11. $a_n = 2^{1/n^2}$

12. $a_n = \dfrac{10^n}{n!}$

13. $b_m = 1 + (-1)^m$

14. $b_m = \dfrac{1 + (-1)^m}{m}$

15. $b_n = \tan^{-1}\left(\dfrac{n+2}{n+5}\right)$

16. $a_n = \ln(n+1) - \ln n$

17. $b_n = \sqrt{n^2+n} - \sqrt{n^2+1}$

18. $c_n = \sqrt{n^2+n} - \sqrt{n^2-n}$

19. $a_n = \dfrac{100^n}{n!} - \dfrac{3+\pi^n}{5^n}$

20. $b_m = \left(1 + \dfrac{1}{m}\right)^m$

21. $c_n = \left(1 + \dfrac{3}{n}\right)^n$

22. $b_n = n(\ln(n+1) - \ln n)$

23. Use the Squeeze Theorem to show that $\lim_{n\to\infty} \dfrac{\arctan(n^2)}{\sqrt{n}} = 0$.

24. Give an example of a divergent sequence $\{a_n\}$ such that $\{\sin a_n\}$ is convergent.

25. Given $a_n = \dfrac{1}{2}3^n - \dfrac{1}{3}2^n$,
(a) Calculate $\lim_{n\to\infty} a_n$.
(b) Calculate $\lim_{n\to\infty} \dfrac{a_{n+1}}{a_n}$.

26. Define $a_{n+1} = \sqrt{a_n + 6}$ with $a_1 = 2$.
(a) Compute a_n for $n = 2, 3, 4, 5$.

(b) Show that $\{a_n\}$ is increasing and bounded by 3.
(c) Prove that $\lim_{n\to\infty} a_n$ exists and find its value.

27. Calculate the partial sums S_4 and S_7 of the series $\displaystyle\sum_{n=1}^{\infty} \dfrac{n-2}{n^2+2n}$.

28. Find the sum $1 - \dfrac{1}{4} + \dfrac{1}{4^2} - \dfrac{1}{4^3} + \cdots$.

29. Find the sum $\dfrac{4}{9} + \dfrac{8}{27} + \dfrac{16}{81} + \dfrac{32}{243} + \cdots$.

30. Find the sum $\displaystyle\sum_{n=2}^{\infty} \left(\dfrac{2}{e}\right)^n$.

31. Find the sum $\displaystyle\sum_{n=0}^{\infty} \dfrac{2^{n+1}}{3^n}$.

32. Show that $\displaystyle\sum_{n=1}^{\infty} (b - \tan^{-1} n^2)$ diverges if $b \neq \dfrac{\pi}{2}$.

33. Give an example of divergent series $\displaystyle\sum_{n=1}^{\infty} a_n, \sum_{n=1}^{\infty} b_n$ such that $\displaystyle\sum_{n=1}^{\infty}(a_n + b_n) = 1$.

34. Find the total area of the infinitely many circles on the interval $[0, 1]$ in Figure 1.

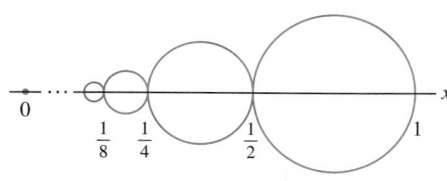

FIGURE 1

In Exercises 35–38, use the Integral Test to determine if the infinite series converges.

35. $\displaystyle\sum_{n=1}^{\infty} \dfrac{n^2}{n^3+1}$

36. $\displaystyle\sum_{n=1}^{\infty} \dfrac{n^2}{(n^3+1)^{1.01}}$

37. $\displaystyle\sum_{n=1}^{\infty} \dfrac{n^3}{e^{n^4}}$

38. $\displaystyle\sum_{n=1}^{\infty} \dfrac{1}{(2n+1)(\ln(2n+1))^2}$

In Exercises 39–46, use the Comparison or Limit Comparison Test to determine whether the infinite series converges.

39. $\displaystyle\sum_{n=1}^{\infty} \frac{1}{(n+1)^2}$

40. $\displaystyle\sum_{n=1}^{\infty} \frac{1}{\sqrt{n}+n}$

41. $\displaystyle\sum_{n=2}^{\infty} \frac{n^2+1}{n^{3.5}-2}$

42. $\displaystyle\sum_{n=1}^{\infty} \frac{1}{n-\ln n}$

43. $\displaystyle\sum_{n=2}^{\infty} \frac{\ln n}{1.5^n}$

44. $\displaystyle\sum_{n=2}^{\infty} \frac{n}{\sqrt{n^5+5}}$

45. $\displaystyle\sum_{n=1}^{\infty} \frac{1}{3^n-2^n}$

46. $\displaystyle\sum_{n=1}^{\infty} \frac{n^{10}+10^n}{n^{11}+11^n}$

47. Show that $\displaystyle\sum_{n=2}^{\infty}\left(1-\sqrt{1-\frac{1}{n}}\right)$ diverges. *Hint:* Show that

$$1-\sqrt{1-\frac{1}{n}} \ge \frac{1}{2n}$$

48. Determine whether $\displaystyle\sum_{n=2}^{\infty}\left(1-\sqrt{1-\frac{1}{n^2}}\right)$ converges.

49. Let $S = \displaystyle\sum_{n=1}^{\infty} \frac{n}{(n^2+1)^2}$.

(a) Show that S converges.

(b) $\mathcal{CAS}$ Use Eq. (5) in Exercise 77 of Section 10.3 with $M=99$ to approximate S. What is the maximum size of the error?

In Exercises 50–53, determine whether the series converges absolutely. If not, determine whether it converges conditionally.

50. $\displaystyle\sum_{n=1}^{\infty} \frac{(-1)^n}{\sqrt[3]{n}+2n}$

51. $\displaystyle\sum_{n=1}^{\infty} \frac{(-1)^n}{n^{1.1}\ln(n+1)}$

52. $\displaystyle\sum_{n=1}^{\infty} \frac{\cos\left(\frac{\pi}{4}+\pi n\right)}{\sqrt{n}}$

53. $\displaystyle\sum_{n=1}^{\infty} \frac{\cos\left(\frac{\pi}{4}+2\pi n\right)}{\sqrt{n}}$

54. $\mathcal{CAS}$ Use a computer algebra system to approximate $\displaystyle\sum_{n=1}^{\infty} \frac{(-1)^n}{n^3+\sqrt{n}}$ to within an error of at most 10^{-5}.

55. How many terms of the series are needed to calculate Catalan's constant $K = \displaystyle\sum_{k=0}^{\infty} \frac{(-1)^k}{(2k+1)^2}$ to three decimal places? Carry out the calculation.

56. Give an example of conditionally convergent series $\displaystyle\sum_{n=1}^{\infty} a_n$, $\displaystyle\sum_{n=1}^{\infty} b_n$ such that $\displaystyle\sum_{n=1}^{\infty}(a_n+b_n)$ converges absolutely.

57. Let $\displaystyle\sum_{n=1}^{\infty} a_n$ be an absolutely convergent series. Determine whether the following series are convergent or divergent:

(a) $\displaystyle\sum_{n=1}^{\infty}\left(a_n+\frac{1}{n^2}\right)$

(b) $\displaystyle\sum_{n=1}^{\infty}(-1)^n a_n$

(c) $\displaystyle\sum_{n=1}^{\infty} \frac{1}{1+a_n^2}$

(d) $\displaystyle\sum_{n=1}^{\infty} \frac{|a_n|}{n}$

In Exercises 58–65, apply the Ratio Test to determine convergence or divergence, or state that the Ratio Test is inconclusive.

58. $\displaystyle\sum_{n=1}^{\infty} \frac{n^5}{5^n}$

59. $\displaystyle\sum_{n=1}^{\infty} \frac{\sqrt{n+1}}{n^8}$

60. $\displaystyle\sum_{n=1}^{\infty} \frac{1}{n2^n+n^3}$

61. $\displaystyle\sum_{n=1}^{\infty} \frac{n^4}{n!}$

62. $\displaystyle\sum_{n=1}^{\infty} \frac{2^{n^2}}{n!}$

63. $\displaystyle\sum_{n=4}^{\infty} \frac{\ln n}{n^{3/2}}$

64. $\displaystyle\sum_{n=1}^{\infty} \left(\frac{n}{2}\right)^n \frac{1}{n!}$

65. $\displaystyle\sum_{n=1}^{\infty} \left(\frac{n}{4}\right)^n \frac{1}{n!}$

In Exercises 66–69, apply the Root Test to determine convergence or divergence, or state that the Root Test is inconclusive.

66. $\displaystyle\sum_{n=1}^{\infty} \frac{1}{4^n}$

67. $\displaystyle\sum_{n=1}^{\infty} \left(\frac{2}{n}\right)^n$

68. $\displaystyle\sum_{n=1}^{\infty} \left(\frac{3}{4n}\right)^n$

69. $\displaystyle\sum_{n=1}^{\infty} \left(\cos\frac{1}{n}\right)^{n^3}$

70. Let $\{a_n\}$ be a positive sequence such that $\displaystyle\lim_{n\to\infty} \sqrt[n]{a_n} = \frac{1}{2}$. Determine whether the following series converge or diverge:

(a) $\displaystyle\sum_{n=1}^{\infty} 2a_n$

(b) $\displaystyle\sum_{n=1}^{\infty} 3^n a_n$

(c) $\displaystyle\sum_{n=1}^{\infty} \sqrt{a_n}$

In Exercises 71–84, determine convergence or divergence using any method covered in the text.

71. $\displaystyle\sum_{n=1}^{\infty} \left(\frac{2}{3}\right)^n$

72. $\displaystyle\sum_{n=1}^{\infty} \frac{n+2^n}{3^n-1}$

73. $\displaystyle\sum_{n=1}^{\infty} e^{-0.02n}$

74. $\displaystyle\sum_{n=1}^{\infty} ne^{-0.02n}$

75. $\displaystyle\sum_{n=1}^{\infty} \frac{(-1)^{n-1}}{\sqrt{n}+\sqrt{n+1}}$

76. $\displaystyle\sum_{n=10}^{\infty} \frac{1}{n(\log n)^{3/2}}$

77. $\displaystyle\sum_{n=10}^{\infty} \frac{(-1)^n}{\log n}$

78. $\displaystyle\sum_{n=1}^{\infty} \frac{1}{n\sqrt{n}+\ln n}$

79. $\displaystyle\sum_{n=1}^{\infty} \frac{e^n}{n!}$

80. $\displaystyle\sum_{n=1}^{\infty} \frac{1}{\sqrt[3]{n}(1+\sqrt{n})}$

81. $\displaystyle\sum_{n=1}^{\infty} \frac{1}{n - 100.1}$

82. $\displaystyle\sum_{n=2}^{\infty} \frac{\cos(\pi n)}{n^{2/3}}$

83. $\displaystyle\sum_{n=1}^{\infty} \sin^2 \frac{\pi}{n}$

84. $\displaystyle\sum_{n=0}^{\infty} \frac{2^{2n}}{n!}$

In Exercises 85–90, find the values of x for which the power series converges.

85. $\displaystyle\sum_{n=0}^{\infty} \frac{2^n x^n}{n!}$

86. $\displaystyle\sum_{n=0}^{\infty} \frac{x^n}{n+1}$

87. $\displaystyle\sum_{n=0}^{\infty} \frac{n^6 (x-3)^n}{n^8 + 1}$

88. $\displaystyle\sum_{n=0}^{\infty} nx^n$

89. $\displaystyle\sum_{n=0}^{\infty} (nx)^n$

90. $\displaystyle\sum_{n=0}^{\infty} \frac{(2x-3)^n}{n \ln n}$

91. Expand the function $f(x) = \dfrac{2}{4 - 3x}$ as a power series centered at $c = 0$. Determine the values of x for which the series converges.

92. Prove that $\displaystyle\sum_{n=0}^{\infty} ne^{-nx} = \frac{e^{-x}}{(1 - e^{-x})^2}$. *Hint:* Express the left-hand side as the derivative of a geometric series.

93. Let $F(x) = \displaystyle\sum_{k=0}^{\infty} \frac{x^{2k}}{2^k \cdot k!}$.

(a) Show that $F(x)$ has infinite radius of convergence.

(b) Show that $y = F(x)$ is a solution to the differential equation

$$y'' = xy' + y$$

satisfying $y(0) = 1$, $y'(0) = 0$.

(c) *CAS*   Plot the partial sums S_N for $N = 1, 3, 5, 7$ on the same set of axes.

94. Find a power series solution $P(x) = \displaystyle\sum_{n=0}^{\infty} a_n x^n$ to the Laguerre differential equation

$$xy'' + (1 - x)y' - y = 0$$

satisfying $P(0) = 1$.

95. Use the Maclaurin series for $f(x) = \cos x$ to calculate the limit

$$\lim_{x \to 0} \frac{1 - \frac{x^2}{2} - \cos x}{x^4}.$$

In Exercises 96–103, find the Taylor series centered at c.

96. $f(x) = e^{4x}$, $c = 0$

97. $f(x) = e^{2x}$, $c = -1$

98. $f(x) = x \sin x$, $c = \pi$

99. $f(x) = \ln \dfrac{x}{2}$, $c = 2$

100. $f(x) = x \ln \left(1 + \dfrac{x}{2}\right)$, $c = 0$

101. $f(x) = \sqrt{x} \arctan \sqrt{x}$, $c = 0$

102. $f(x) = \dfrac{1}{1 - 2x}$, $c = -2$

103. $f(x) = e^{x-1}$, $c = -1$

104. Find the Maclaurin series of $f(x) = x^5 - 4x^2 + 10$. Determine its Taylor series centered at $c = 2$.

105. Use the Maclaurin series of $\sin x$ and $\sqrt{1 + x}$ to calculate $f^{(4)}(0)$, where $f(x) = (\sin x)\sqrt{1 + x}$.

106. Calculate $\dfrac{\pi}{2} - \dfrac{\pi^3}{2^3 3!} + \dfrac{\pi^5}{2^5 5!} - \dfrac{\pi^7}{2^7 7!} + \cdots$.

107. Find the Maclaurin series of the function $F(x) = \displaystyle\int_0^x \frac{e^t - 1}{t}\, dt$.

11 | PARAMETRIC EQUATIONS, POLAR COORDINATES, AND CONIC SECTIONS

The beautiful shell of a chambered nautilus grows in the shape of an equiangular spiral, a curve described in polar coordinates by an equation $r = e^{a\theta}$.

Parametric equations are useful in multivariable calculus, especially in three dimensions, where it is no longer possible to describe a curve as a graph of a function (the graph of a function of two variables is a surface in $\mathbf{R}^3$ rather than a curve).

This chapter develops several new tools of calculus. First, we introduce parametric equations, which are used to analyze motion and to describe curves that cannot be represented as graphs of functions or equations. We then discuss polar coordinates, which are an alternative to rectangular coordinates and are useful in many applications. The chapter closes with a discussion of conic sections (ellipses, hyperbolas, and parabolas). The conic sections were first defined and studied by the ancient Greek mathematicians Menaechmus (c. 380–320 BC) and Apollonius (c. 262–190 BC) and they have played an important role in mathematics ever since.

11.1 Parametric Equations

In previous chapters, we have studied curves that are graphs of functions or equations. In this section, we introduce a new and important way of describing a curve, via parametric equations. Imagine a particle P moving along a path $\mathcal{C}$ as a function of time t (Figure 1). Then the coordinates of P are functions of t:

$$\boxed{x = f(t), \qquad y = g(t)} \qquad \boxed{1}$$

The equations (1) are called **parametric equations** of the path $\mathcal{C}$ with **parameter** t. We also write

$$c(t) = (f(t), g(t))$$

As t varies in a domain such as an interval or the real line, $c(t)$ represents a point moving along the path $\mathcal{C}$. We refer to $\mathcal{C}$ as a **parametrized** or **parametric** curve. The direction of motion is often indicated by an arrow next to the plot of $c(t)$ as in Figure 1. Finally, since x and y are functions of t, we often write $c(t) = (x(t), y(t))$ instead of $(f(t), g(t))$.

In physical problems, t often represents time, but we are free to use other variables such as s or θ for the parameter.

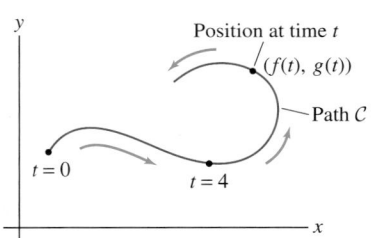

FIGURE 1 Path of a particle moving along a path $\mathcal{C}$ in the plane.

t	$x = 2t - 4$	$y = 3 + t^2$
-2	-8	7
0	-4	3
2	0	7
4	4	19

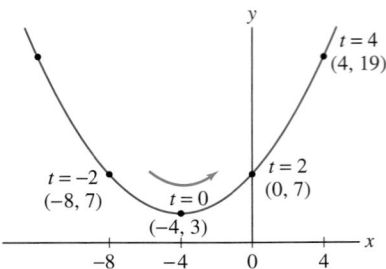

FIGURE 2 The parametric curve
$x = 2t - 4$, $y = 3 + t^2$.

Consider the parametric equations

$$x = 2t - 4, \qquad y = 3 + t^2 \qquad \boxed{2}$$

The table above gives the x and y coordinates for several values of t. In Figure 2, we plot these points and join them by a smooth curve, indicating the direction of motion with an arrow.

In some cases, a parametric curve can be expressed as the graph of a function.

■ **EXAMPLE 1** Eliminating the Parameter Describe the parametric curve

$$c(t) = (2t - 4, 3 + t^2)$$

in the form $y = f(x)$.

Solution We "eliminate the parameter" by solving for y as a function of x. First, express t in terms of x: Since $x = 2t - 4$, we have $t = \frac{1}{2}x + 2$. Then substitute

$$y = 3 + t^2 = 3 + \left(\frac{1}{2}x + 2\right)^2 = 7 + 2x + \frac{1}{4}x^2$$

Thus, $c(t)$ traces out the parabola $y = 7 + 2x + \frac{1}{4}x^2$ (Figure 2). ■

FIGURE 3 The parametric curve:
$x = 5\cos(3t)\cos\left(\frac{2}{3}\sin(5t)\right)$
$y = 4\sin(3t)\cos\left(\frac{2}{3}\sin(5t)\right)$.

> **CONCEPTUAL INSIGHT** The graph of a function $y = f(x)$ can always be parametrized as $c(t) = (t, f(t))$. For example, the parabola $y = x^2$ is parametrized by $c(t) = (t, t^2)$ and $y = e^t$ by $c(t) = (t, e^t)$. An advantage of parametric equations is that they allow us to describe curves that are not graphs of functions, such as the curve in Figure 3.

■ **EXAMPLE 2** A bullet follows the trajectory (t in seconds, distance in feet):

$$x(t) = 200t, \qquad y(t) = 400t - 16t^2 \qquad 0 \le t \le 25$$

Find the bullet's height at $t = 5$ and its maximum height.

Solution The bullet's height is $y(t) = 400t - 16t^2$. At $t = 5$, the height is

$$y(5) = 400(5) - 16(5^2) = 1{,}600 \text{ ft}$$

The function $y(t)$ is quadratic and takes on a maximum value where $y'(t) = 0$:

$$y'(t) = \frac{d}{dt}(400t - 16t^2) = 400 - 32t = 0 \qquad \Rightarrow \qquad t = \frac{400}{32} = 12.5$$

Thus, the maximum height is $y(12.5) = 400(12.5) - 16(12.5)^2 = 2{,}500$ ft (Figure 4). ■

FIGURE 4 Parabolic path of a bullet.

We now discuss parametrizations of straight lines and circles. These are used frequently in multivariable calculus.

■ **EXAMPLE 3** **Parametric Representation of a Line** Show that the line through the point $P = (a, b)$ with slope m has parametrization

$$c(t) = (a + t, b + mt) \qquad -\infty < t < \infty \qquad \boxed{3}$$

Find parametric equations for the line through $P = (3, -1)$ of slope $m = 4$.

Solution We eliminate the parameter $x = a + t$. Thus, $t = x - a$ and

$$y = b + mt = b + m(x - a) \qquad \text{or} \qquad y - b = m(x - a)$$

This is the equation of the line through $P = (a, b)$ of slope m as claimed (Figure 5). The line through $P = (3, -1)$ of slope $m = 4$ has parametrization

$$c(t) = (3 + t, -1 + 4t) \qquad ■$$

A circle of radius R centered at the origin has the parametrization

$$c(\theta) = (R \cos \theta, R \sin \theta)$$

The parameter θ represents the angle corresponding to the point (x, y) on the circle (Figure 6). The circle is traversed once in the counterclockwise direction as θ varies from 0 to 2π.

To move or "translate" a parametric curve a units horizontally and b units vertically, we replace $c(t) = (x(t), y(t))$ by $c(t) = (a + x(t), b + y(t))$. In particular, the circle of radius R with center (a, b) has parametrization (Figure 6)

$$x = a + R \cos \theta, \qquad y = b + R \sin \theta \qquad \boxed{4}$$

As a check, let's verify that the point (x, y) satisfies the equation of the circle of radius R centered at (a, b):

$$(x - a)^2 + (y - b)^2 = (a + R \cos \theta - a)^2 + (b + R \sin \theta - b)^2$$
$$= R^2 \cos^2 \theta + R^2 \sin^2 \theta = R^2$$

A modification of these equations may be used to parametrize ellipses.

■ **EXAMPLE 4** **Parametrization of an Ellipse** Show that the ellipse $\left(\dfrac{x}{a}\right)^2 + \left(\dfrac{y}{b}\right)^2 = 1$ is parametrized by

$$c(t) = (a \cos t, \ b \sin t) \qquad 0 \le t < 2\pi$$

Graph the case $a = 4$, $b = 2$, and indicate the points corresponding to $t = 0, \dfrac{\pi}{6}, \dfrac{\pi}{3}, \dfrac{\pi}{2}$.

Solution We check that $x = a \cos t$, $y = b \sin t$ satisfies the equation of the ellipse:

$$\left(\frac{x}{a}\right)^2 + \left(\frac{y}{b}\right)^2 = \left(\frac{a \cos t}{a}\right)^2 + \left(\frac{b \sin t}{b}\right)^2 = \cos^2 t + \sin^2 t = 1$$

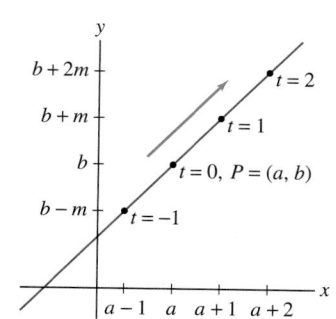

FIGURE 5 The parametric line $c(t) = (a + t, b + mt)$.

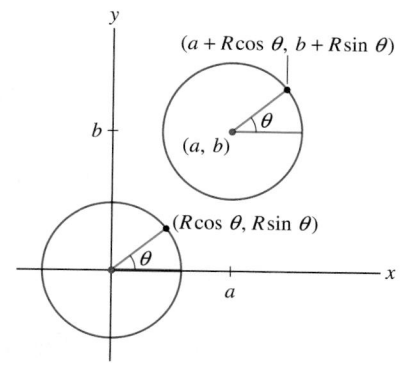

FIGURE 6 Parametrization of a circle of radius R with center (a, b).

As t varies from 0 to π, $c(t)$ traces the top half of the ellipse because x decreases from a to $-a$ and y varies from 0 to b and back again to 0. Similarly, $c(t)$ traces the bottom half as t varies from π to 2π. For $a = 4$, $b = 2$, we obtain $c(t) = (4\cos t, 2\sin t)$, which parametrizes the ellipse $\left(\dfrac{x}{4}\right)^2 + \left(\dfrac{y}{2}\right)^2 = 1$. Figure 7 shows the ellipse and points indicated.   ∎

t	$x(t) = 4\cos t$	$y(t) = 2\sin t$
0	4	0
$\dfrac{\pi}{6}$	$2\sqrt{3}$	1
$\dfrac{\pi}{3}$	2	$\sqrt{3}$
$\dfrac{\pi}{2}$	0	2

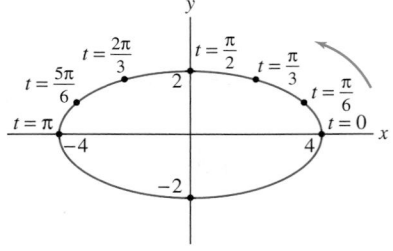

FIGURE 7 Ellipse with parametric equations $x = 4\cos t$, $y = 2\sin t$.

Before continuing, we make two important remarks:

- There is a difference between a path $c(t)$—such as the orbit of a moon around a planet—and the underlying curve $\mathcal{C}$. The curve $\mathcal{C}$ is a set of points in the plane. The path $c(t)$ (the orbit) describes not just the curve, but also the location of a point along the curve as a function of the parameter. The path may traverse all or part of $\mathcal{C}$ several times. For example, $c(t) = (\cos t, \sin t)$ describes a *path* that goes around the unit circle twice as t varies from 0 to 4π.
- Parametrizations are not unique, and in fact, every curve may be parametrized in infinitely many different ways. For instance, the parabola $y = x^2$ is parametrized by (t, t^2), but also (t^3, t^6), or (t^5, t^{10}), etc.

■ **EXAMPLE 5** Parametrizations and Paths Versus Curves Describe the motion of a particle moving along the paths (for $-\infty < t < \infty$):

(a) $c_1(t) = (t^3, t^6)$ **(b)** $c_2(t) = (t^2, t^4)$ **(c)** $c_3(t) = (\cos t, \cos^2 t)$

Solution The relation $y = x^2$ holds for each of these parametrizations, so all three parametrize portions of the parabola $y = x^2$.

(a) As t varies from $-\infty$ to ∞, the function t^3 also varies from $-\infty$ to ∞. Therefore, a particle following the path $c_1(t) = (t^3, t^6)$ traces the entire parabola $y = x^2$, moving from left to right and passing through each point once [Figure 8(A)].

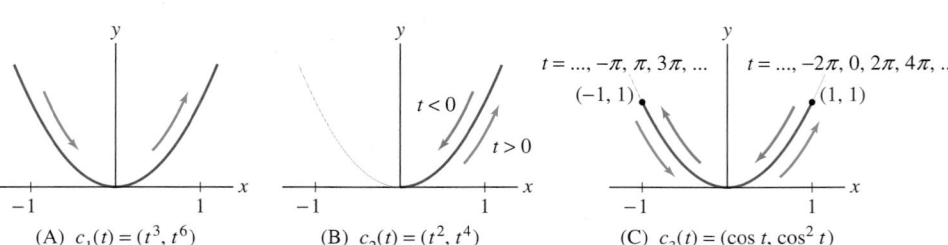

FIGURE 8 Three parametrizations of portions of the parabola.

(A) $c_1(t) = (t^3, t^6)$ (B) $c_2(t) = (t^2, t^4)$ (C) $c_3(t) = (\cos t, \cos^2 t)$

(b) In this parametrization, $x = t^2 \geq 0$, so the path $c_2(t) = (t^2, t^4)$ only traces the right half of the parabola. The particle comes in toward the origin as t varies from $-\infty$ to 0, and goes back out to the right as t varies from 0 to ∞ [Figure 8(B)].

(c) The function $\cos t$ oscillates between 1 to -1. Therefore, a particle following the path $c_3(t) = (\cos t, \cos^2 t)$ oscillates back and forth between the points $(1, 1)$ and $(-1, 1)$ on the parabola [Figure 8(C)]. The motion is repeated as t varies on intervals of length 2π.

■

■ **EXAMPLE 6** Using Symmetry to Sketch a Loop Sketch the curve parametrized by

$$c(t) = (t^2 + 1, t^3 - 4t)$$

Label the points corresponding to $t = 0, \pm 1, \pm 2, \pm 2.5$.

Solution We note that $x(t) = t^2 + 1$ is an even function and $y(t) = t^3 - 4t$ is an odd function:

$$x(-t) = x(t), \qquad y(-t) = -y(t)$$

It follows that if a point $P = c(t)$ lies on the curve, then its reflection across the x-axis also lies on the curve and corresponds to "time" $-t$:

$$c(-t) = (x(t), -y(t)) = \text{reflection of } c(t) \text{ across the } x\text{-axis}$$

In other words, the curve is symmetric with respect to the x-axis. Therefore, it suffices to sketch the curve for $t \geq 0$ and then use symmetry.

Next, we observe that the coordinate $x(t) = t^2 + 1$ tends to ∞ as $t \to \infty$. To analyze the y-coordinate, we graph $y(t) = t^3 - 4t$ as a function of t (*not* as a function of x). Figure 9(A) shows that

$$\begin{aligned} y(t) &< 0 && \text{for} && 0 < t < 2, \\ y(t) &> 0 && \text{for} && t > 2, \\ y(t) &\to \infty && \text{as} && t \to \infty \end{aligned}$$

Therefore, starting at $c(0) = (1, 0)$, the curve first dips below the x-axis. Then it turns up and tends to ∞. The points $c(0)$, $c(1)$, $c(2)$, and $c(2.5)$ are tabulated in Table 1 and plotted in Figure 9(B). The part of the path for $t \leq 0$ is obtained by reflecting across the x-axis as shown in Figure 9(C).

■

TABLE 1

t	$x = t^2 + 1$	$y = t^3 - 4t$
0	1	0
1	2	-3
2	5	0
2.5	7.25	5.625

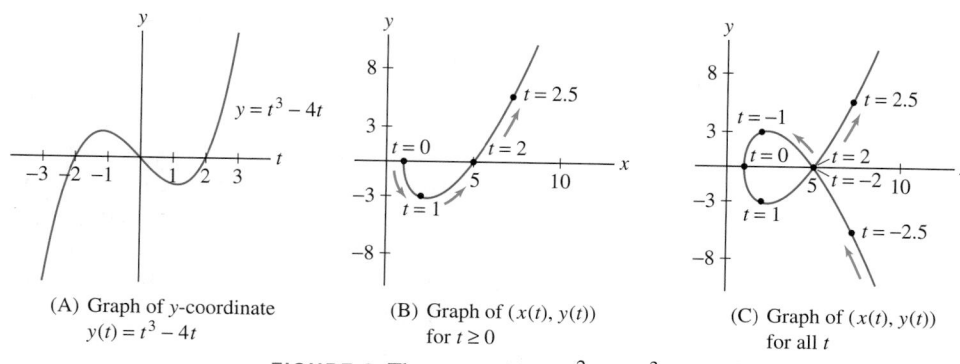

(A) Graph of y-coordinate $y(t) = t^3 - 4t$

(B) Graph of $(x(t), y(t))$ for $t \geq 0$

(C) Graph of $(x(t), y(t))$ for all t

FIGURE 9 The curve $c(t) = (t^2 + 1, t^3 - 4t)$.

The **cycloid** is the curve traced by a point on the circumference of a rolling wheel (Figure 10).

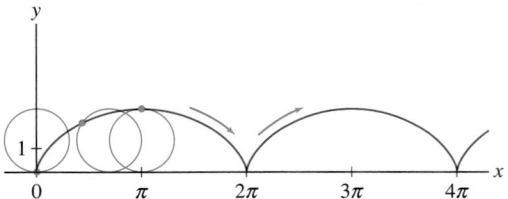

FIGURE 10 A cycloid.

The cycloid was studied intensively in the sixteenth and seventeenth centuries by a stellar cast of mathematicians (including Galileo, Pascal, Newton, Leibniz, Huygens, and Bernoulli) who discovered many of its remarkable properties. For example, a slide with the property that an object sliding down (without friction) reaches the bottom in the least amount of time must have the shape of an inverted cycloid. This is called the "brachistochrone property" from the Greek brachistos, "the shortest," and chronos, "time."

■ **EXAMPLE 7** Parametrizing the Cycloid   Find parametric equations for the cycloid generated by a point P on the unit circle.

Solution We refer to Figure 11. Let P be the point on the rolling unit circle located at the origin at $t = 0$. Now let the circle roll a distance t along the x-axis. Then the length of the circular arc $\overset{\frown}{QP}$ is also t and the angle $\angle PCQ$ has radian measure t.

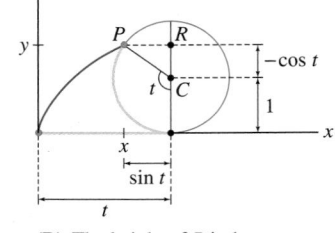

(A) The segment $\overline{OQ}$ and the arc $\overset{\frown}{QP}$ have the same length.

(B) The height of P is $1 - \cos t$. In this diagram, $\cos t$ is negative since t is obtuse.

FIGURE 11 Parametric representation of the cycloid.

To calculate the coordinates of P, observe in Figure 11(B) that the height y of P is $y = 1 - \cos t$. Note that in the figure, $\cos t$ is negative since t is obtuse and the segment $\overline{CR}$ has length $-\cos t$. The x-coordinate of P is $x = t - PR = t - \sin t$, so we obtain

$$x(t) = t - \sin t, \qquad y(t) = 1 - \cos t \qquad \boxed{5}$$

■

The argument in Example 7 may be modified to show that the cycloid generated by a circle of radius R has parametric equations

$$x = Rt - R\sin t, \qquad y = R - R\cos t \qquad \boxed{6}$$

Next, we address the problem of finding the tangent line for a curve in parametric form. For any curve in the xy-plane, the slope of the tangent line is the derivative dy/dx (whether described parametrically or not), but for a parametric curve, we must use the Chain Rule to compute dy/dx because y is not given explicitly as a function of x. If $x = f(t)$, $y = g(t)$, then

$$g'(t) = \frac{dy}{dt} = \frac{dy}{dx}\frac{dx}{dt} = \frac{dy}{dx}f'(t)$$

If $f'(t) \neq 0$, we may divide by $f'(t)$ to obtain

$$\frac{dy}{dx} = \frac{g'(t)}{f'(t)}$$

This calculation is valid on an interval I where $f(t)$ and $g(t)$ are differentiable, $f'(t)$ is continuous, and $f'(t) \neq 0$. In this case, $t = f^{-1}(x)$ exists and the composite $y = g(f^{-1}(x))$ is a differentiable function of x.

THEOREM 1   Slope of the Tangent Line   Let $c(t) = (x(t), y(t))$, where $x(t)$ and $y(t)$ are differentiable. Assume that $x'(t)$ is continuous and $x'(t) \neq 0$. Then

$$\frac{dy}{dx} = \frac{dy/dt}{dx/dt} = \frac{y'(t)}{x'(t)} \qquad \boxed{7}$$

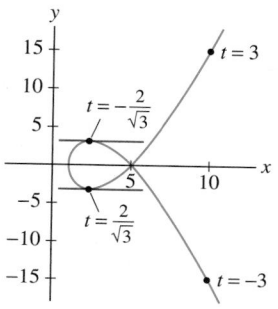

FIGURE 12 Horizontal tangent lines on $c(t) = (t^2 + 1, t^3 - 4t)$.

■ **EXAMPLE 8**   Let $c(t) = (t^2 + 1, t^3 - 4t)$. Find the equation of the tangent line at $t = 3$ and find the points where the tangent is horizontal.

Solution   We have

$$\frac{dy}{dx} = \frac{y'(t)}{x'(t)} = \frac{(t^3 - 4t)'}{(t^2 + 1)'} = \frac{3t^2 - 4}{2t}$$

The slope at $t = 3$ is

$$\frac{dy}{dx} = \frac{3t^2 - 4}{2t}\bigg|_{t=3} = \frac{3(3)^2 - 4}{2(3)} = \frac{23}{6}$$

Since $c(3) = (10, 15)$, the equation of the tangent line in point-slope form is

$$y - 15 = \frac{23}{6}(x - 10)$$

The slope $\dfrac{dy}{dx}$ is zero if $3t^2 - 4 = 0$ and $2t \neq 0$. This gives $t = \pm 2/\sqrt{3}$ (Figure 12). Therefore, the tangent line is horizontal at the points

$$c\left(-\frac{2}{\sqrt{3}}\right) = \left(\frac{7}{3}, \frac{16}{3\sqrt{3}}\right), \qquad c\left(\frac{2}{\sqrt{3}}\right) = \left(\frac{7}{3}, -\frac{16}{3\sqrt{3}}\right) \qquad ■$$

Parametric curves are widely used in the field of computer graphics. A particularly important class of curves are **Bézier curves**, which we discuss here briefly in the cubic case. Given four "control points" (Figure 13):

$$P_0 = (a_0, b_0), \qquad P_1 = (a_1, b_1), \qquad P_2 = (a_2, b_2), \qquad P_3 = (a_3, b_3)$$

we define the Bézier curve $c(t) = (x(t), y(t))$ (for $0 \leq t \leq 1$), where

$$x(t) = a_0(1 - t)^3 + 3a_1 t(1 - t)^2 + 3a_2 t^2(1 - t) + a_3 t^3 \qquad \boxed{8}$$

$$y(t) = b_0(1 - t)^3 + 3b_1 t(1 - t)^2 + 3b_2 t^2(1 - t) + b_3 t^3 \qquad \boxed{9}$$

Note that $c(0) = (a_0, b_0)$ and $c(1) = (a_3, b_3)$, so the Bézier curve begins at P_0 and ends at P_3 (Figure 13). It can also be shown that the Bézier curve is contained within the quadrilateral (shown in yellow) with vertices P_0, P_1, P_2, P_3. However, $c(t)$ does not pass

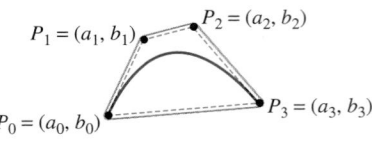

FIGURE 13 Cubic Bézier curves specified by four control points.

Hand sketch by Pierre Bézier for the French automobile manufacturer Renault, made in 1964.

through P_1 and P_2. Instead, these intermediate control points determine the slopes of the tangent lines at P_0 and P_3, as we show in the next example (also, see Exercises 58–60).

■ **EXAMPLE 9** Show that the Bézier curve is tangent to the segment $\overline{P_0 P_1}$ at P_0.

Solution The Bézier curve passes through P_0 at $t = 0$, so we must show that the slope of the tangent line at $c(0) = P_0 = (a_0, b_0)$ is equal to the slope of $\overline{P_0 P_1}$. To find the slope, we compute the derivatives:

$$x'(t) = -3a_0(1-t)^2 + 3a_1(1 - 4t + 3t^2) + a_2(2t - 3t^2) + 3a_3 t^2$$

$$y'(t) = -3b_0(1-t)^2 + 3b_1(1 - 4t + 3t^2) + b_2(2t - 3t^2) + 3b_3 t^2$$

Evaluating at $t = 0$, we obtain $x'(0) = 3(a_1 - a_0)$, $y'(0) = 3(b_1 - b_0)$, and

$$\left.\frac{dy}{dx}\right|_{t=0} = \frac{y'(0)}{x'(0)} = \frac{3(b_1 - b_0)}{3(a_1 - a_0)} = \frac{b_1 - b_0}{a_1 - a_0}$$

This is equal to the slope of the line segment through $P_0 = (a_0, b_0)$ and $P_1 = (a_1, b_1)$ as claimed (provided that $a_1 \neq a_0$). ■

11.1 SUMMARY

- A path traced by a point $P = (x, y)$, where x and y are functions of a parameter t, is called a *parametric* or *parametrized* path or curve. We write $c(t) = (f(t), g(t))$ or $c(t) = (x(t), y(t))$.
- Keep in mind that the path $c(t) = (x(t), y(t))$ and the curve that it traces are different. The path $c(t)$ describes the location of a point as a function of t, whereas the underlying curve is a set of points in the plane. For example, the path $(\cos t, \sin t)$ moves around the unit circle infinitely many times as t varies from 0 to ∞.
- Parametrizations are not unique: Every path may be parametrized in infinitely many ways.
- Let $c(t) = (x(t), y(t))$, where $x(t)$ and $y(t)$ are differentiable and $x'(t)$ is continuous. Then the slope of the tangent line at $c(t)$ is the derivative

$$\frac{dy}{dx} = \frac{dy/dt}{dx/dt} = \frac{y'(t)}{x'(t)} \quad \text{[valid if } x'(t) \neq 0\text{]}$$

- Do not confuse the derivatives with respect to t, dy/dt, and dx/dt, with the derivative dy/dx (the slope of the tangent line).

11.1 EXERCISES

Preliminary Questions

1. Describe the shape of the curve $x = 3 \cos t$, $y = 3 \sin t$.

2. How does $x = 4 + 3 \cos t$, $y = 5 + 3 \sin t$ differ from the curve in the previous question?

3. What is the maximum height of a particle whose path has parametric equations $x = t^9$, $y = 4 - t^2$?

4. Can the parametric curve $(t, \sin t)$ be represented as a graph $y = f(x)$? What about $(\sin t, t)$?

5. Match the derivatives with a verbal description:

(a) $\dfrac{dx}{dt}$ (b) $\dfrac{dy}{dt}$ (c) $\dfrac{dy}{dx}$

(i) Slope of the tangent line to the curve

(ii) Vertical rate of change with respect to time

(iii) Horizontal rate of change with respect to time

Exercises

1. Find the coordinates at times $t = 0, 2, 4$ of a particle following the path $x = 1 + t^3$, $y = 9 - 3t^2$.

2. Find the coordinates at $t = 0, \frac{\pi}{4}, \pi$ of a particle moving along the path $c(t) = (\cos 2t, \sin^2 t)$.

3. Show that the path traced by the bullet in Example 2 is a parabola by eliminating the parameter.

4. Use the table of values to sketch the parametric curve $(x(t), y(t))$, indicating the direction of motion.

t	-3	-2	-1	0	1	2	3
x	-15	0	3	0	-3	0	15
y	5	0	-3	-4	-3	0	5

5. Graph the parametric curves. Include arrows indicating the direction of motion.

(a) (t, t), $-\infty < t < \infty$ (b) $(\sin t, \sin t)$, $0 \le t \le 2\pi$
(c) (e^t, e^t), $-\infty < t < \infty$ (d) (t^3, t^3), $-1 \le t \le 1$

6. Give two different parametrizations of the line through $(4, 1)$ with slope 2.

In Exercises 7–14, express in the form $y = f(x)$ by eliminating the parameter.

7. $x = t + 3$, $y = 4t$ **8.** $x = t^{-1}$, $y = t^{-2}$

9. $x = t$, $y = \tan^{-1}(t^3 + e^t)$ **10.** $x = t^2$, $y = t^3 + 1$

11. $x = e^{-2t}$, $y = 6e^{4t}$ **12.** $x = 1 + t^{-1}$, $y = t^2$

13. $x = \ln t$, $y = 2 - t$ **14.** $x = \cos t$, $y = \tan t$

In Exercises 15–18, graph the curve and draw an arrow specifying the direction corresponding to motion.

15. $x = \frac{1}{2}t$, $y = 2t^2$ **16.** $x = 2 + 4t$, $y = 3 + 2t$

17. $x = \pi t$, $y = \sin t$ **18.** $x = t^2$, $y = t^3$

19. Match the parametrizations (a)–(d) below with their plots in Figure 14 and draw an arrow indicating the direction of motion.

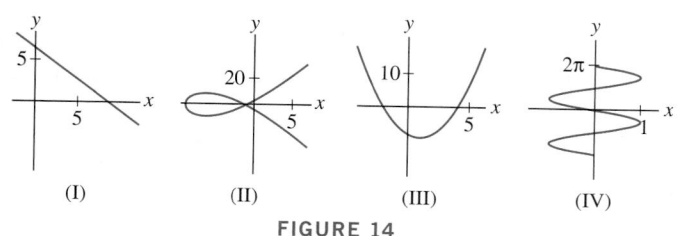

(I)   (II)   (III)   (IV)

FIGURE 14

(a) $c(t) = (\sin t, -t)$ (b) $c(t) = (t^2 - 9, 8t - t^3)$
(c) $c(t) = (1 - t, t^2 - 9)$ (d) $c(t) = (4t + 2, 5 - 3t)$

20. The graphs of $x(t)$ and $y(t)$ as functions of t are shown in Figure 15(A). Which of (I)–(III) is the plot of $c(t) = (x(t), y(t))$? Explain.

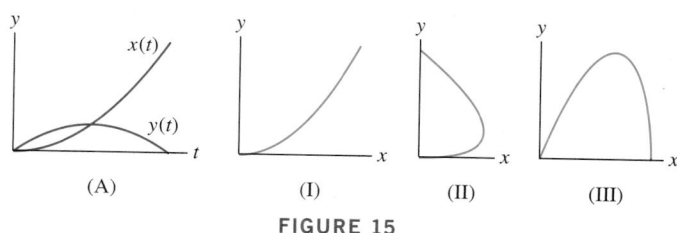

(A)   (I)   (II)   (III)

FIGURE 15

21. Find an interval of t-values such that $c(t) = (\cos t, \sin t)$ traces the lower half of the unit circle.

22. Find an interval of t-values such that $c(t) = (2t + 1, 4t - 5)$ parametrizes the segment from $(0, -7)$ to $(7, 7)$.

In Exercises 23–34, find parametric equations for the given curve.

23. $y = 9 - 4x$ **24.** $y = 8x^2 - 3x$

25. $4x - y^2 = 5$ **26.** $x^2 + y^2 = 49$

27. $(x + 9)^2 + (y - 4)^2 = 49$

28. Line of slope 8 through $(-4, 9)$

29. Line through $(2, 5)$ perpendicular to $y = 3x$

30. Circle of radius 4 with center $(3, 9)$

31. $\left(\dfrac{x}{4}\right)^2 + \left(\dfrac{y}{9}\right)^2 = 1$

32. Ellipse of Exercise 31, with its center translated to $(2, 11)$

33. The parabola $y = x^2$ translated so that its minimum occurs at $(2, 3)$

34. The curve $y = \cos x$ translated so that a maximum occurs at $(3, 5)$

35. Describe the parametrized curve $c(t) = (\sin^2 t, \cos^2 t)$ for $0 \le t \le \pi$.

36. Find the graph $y = f(x)$ traced by the path $x = \sec t$, $y = \tan t$. Which intervals of t-values trace the graph exactly once?

37. Find a parametrization $c(t)$ of the line $y = 3x - 4$ such that $c(0) = (2, 2)$.

38. Find a parametrization $c(t)$ of $y = x^2$ such that $c(0) = (3, 9)$.

39. Show that $x = \cosh t$, $y = \sinh t$ parametrizes the hyperbola $x^2 - y^2 = 1$. Calculate $\dfrac{dy}{dx}$ as a function of t. Generalize to obtain a parametrization of $\left(\dfrac{x}{a}\right)^2 - \left(\dfrac{y}{b}\right)^2 = 1$.

40. A particle moves along the path $x(t) = \dfrac{1}{4}t^3 + 2t$, $y(t) = 20t - t^2$ (in centimeters).
(a) What is the maximum height attained by the object?
(b) At what time does the object hit the ground?
(c) How far is the object from the origin when it hits the ground?

41. Which of (I) or (II) is the graph of $x(t)$ for the parametric curve in Figure 16(A)? Which represents $y(t)$?

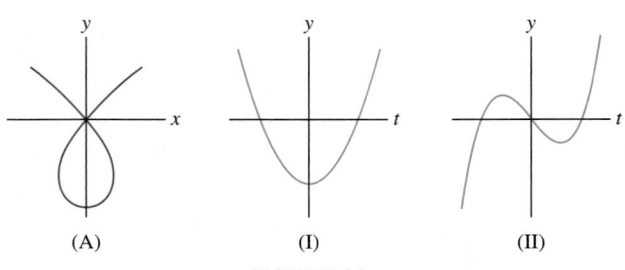

(A)   (I)   (II)

FIGURE 16

42. Sketch the graph of $c(t) = (t^3 - 4t, t^2)$ following the steps in Example 6.

43. Sketch $c(t) = (t^2, \sin t)$ for $-2\pi \le t \le 2\pi$.

44. Sketch $c(t) = (t^2 - 4t, 9 - t^2)$ for $-4 \le t \le 10$.

In Exercises 45–48, use Eq. (7) to find $\dfrac{dy}{dx}$ at the given point.

45. $(t^3, t^2 - 1)$, $t = -4$

46. $(2t + 9, 7t - 9)$, $t = 1$

47. $(s^{-1} - 3s, s^3)$, $s = -1$

48. $(\sin 2\theta, \cos 3\theta)$, $\theta = \frac{\pi}{4}$

In Exercises 49–52, find an equation $y = f(x)$ for the parametric curve and compute $\dfrac{dy}{dx}$ in two ways: using Eq. (7) and by differentiating $f(x)$.

49. $c(t) = (2t + 1, 1 - 9t)$

50. $c(t) = \left(\frac{1}{2}t, \frac{1}{4}t^2 - t\right)$

51. $x = s^3$, $y = s^6 + s^{-3}$

52. $x = \cos\theta$, $y = \cos\theta + \sin^2\theta$

In Exercises 53–56, let $c(t) = (t^2 - 9, t^2 - 8t)$ (see Figure 17).

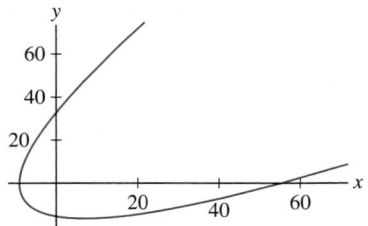

FIGURE 17 Plot of $c(t) = (t^2 - 9, t^2 - 8t)$.

53. Draw an arrow indicating the direction of motion and determine the interval of t-values corresponding to the portion of the curve in each of the four quadrants.

54. Find the equation of the tangent line at $t = 4$.

55. Find the points where the tangent has slope $\frac{1}{2}$.

56. Find the points where the tangent is horizontal or vertical.

57. Find the equation of the ellipse represented parametrically by $x = 4\cos t$, $y = 7\sin t$. Calculate the slope of the tangent line at the point $(2\sqrt{2}, 7\sqrt{2}/2)$.

In Exercises 58–60, refer to the Bézier curve defined by Eqs. (8) and (9).

58. Show that the Bézier curve with control points
$$P_0 = (1, 4), \quad P_1 = (3, 12), \quad P_2 = (6, 15), \quad P_3 = (7, 4)$$
has parametrization
$$c(t) = (1 + 6t + 3t^2 - 3t^3, 4 + 24t - 15t^2 - 9t^3)$$
Verify that the slope at $t = 0$ is equal to the slope of the segment $\overline{P_0 P_1}$.

59. Find and plot the Bézier curve $c(t)$ passing through the control points
$$P_0 = (3, 2), \quad P_1 = (0, 2), \quad P_2 = (5, 4), \quad P_3 = (2, 4)$$

60. Show that a cubic Bézier curve is tangent to the segment $\overline{P_2 P_3}$ at P_3.

61. A bullet fired from a gun follows the trajectory
$$x = at, \quad y = bt - 16t^2 \quad (a, b > 0)$$
Show that the bullet leaves the gun at an angle $\theta = \tan^{-1}\left(\dfrac{b}{a}\right)$ and lands at a distance $\dfrac{ab}{16}$ from the origin.

62. CAS Plot $c(t) = (t^3 - 4t, t^4 - 12t^2 + 48)$ for $-3 \le t \le 3$. Find the points where the tangent line is horizontal and vertical.

63. *CAS* Plot the astroid $x = \cos^3 \theta$, $y = \sin^3 \theta$ and find the equation of the tangent line at $\theta = \frac{\pi}{3}$.

64. Find the equation of the tangent line at $t = \frac{\pi}{4}$ to the cycloid generated by the unit circle with parametric equation (5).

65. Find the points with horizontal tangent line on the cycloid with parametric equation (5).

66. Property of the Cycloid Prove that the tangent line at a point P on the cycloid always passes through the top point on the rolling circle as indicated in Figure 18. Assume the circle has radius 1 and use (5).

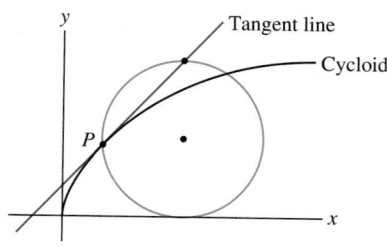

FIGURE 18

67. A *curtate cycloid* (Figure 19) is the curve traced by a point at a distance h from the center of a circle of radius R rolling along the x-axis where $h < R$. Show that this curve has parametric equations $x = Rt - h \sin t$, $y = R - h \cos t$.

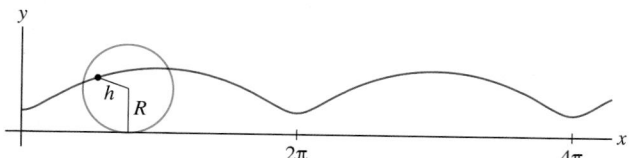

FIGURE 19 Curtate cycloid.

68. *CAS* Use a computer algebra system to explore what happens when $h > R$ in the parametric equations of Exercise 67. Describe the result.

69. Show that the line of slope t through $(-1, 0)$ intersects the unit circle in the point with coordinates

$$x = \frac{1 - t^2}{t^2 + 1}, \qquad y = \frac{2t}{t^2 + 1} \qquad \boxed{10}$$

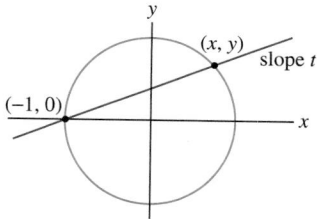

FIGURE 20 Unit circle.

Conclude that these equations parametrize the unit circle with the point $(-1, 0)$ excluded (Figure 20). Show further that $t = \dfrac{y}{x + 1}$.

70. The **folium of Descartes** is the curve with equation $x^3 + y^3 = 3axy$, where $a \neq 0$ is a constant (Figure 21).

(a) Show that for $t \neq -1, 0$, the line $y = tx$ intersects the folium at the origin and at one other point P. Express the coordinates of P in terms of t to obtain a parametrization of the folium. Indicate the direction of the parametrization on the graph.

(b) Describe the interval of t values parametrizing the parts of the curve in quadrants I, II, and IV. Note that $t = -1$ is a point of discontinuity of the parametrization.

(c) Calculate dy/dx as a function of t and find the points with horizontal or vertical tangent.

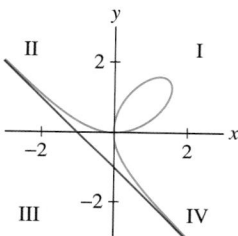

FIGURE 21 Folium $x^3 + y^3 = 3axy$.

71. Use the results of Exercise 70 to show that the asymptote of the folium is the line $x + y = -a$. Hint: show that $\lim\limits_{t \to -1} (x + y) = -a$.

72. Find a parametrization of $x^{2n+1} + y^{2n+1} = ax^n y^n$, where a and n are constants.

73. Verify that the **tractrix** curve ($\ell > 0$)

$$c(t) = \left(t - \ell \tanh \frac{t}{\ell}, \, \ell \operatorname{sech} \frac{t}{\ell} \right)$$

has the following property: For all t, the segment from $c(t)$ to $(t, 0)$ is tangent to the curve and has length ℓ (Figure 22).

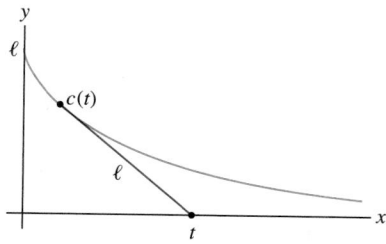

FIGURE 22 The tractrix $c(t) = \left(t - \ell \tanh \dfrac{t}{\ell}, \, \ell \operatorname{sech} \dfrac{t}{\ell} \right)$.

74. In Exercise 54 of Section 9.1, we described the tractrix by the differential equation

$$\frac{dy}{dx} = -\frac{y}{\sqrt{\ell^2 - y^2}}$$

Show that the curve $c(t)$ identified as the tractrix in Exercise 73 satisfies this differential equation. Note that the derivative on the left is taken with respect to x, not t.

75. Let A and B be the points where the ray of angle θ intersects the two concentric circles of radii $r < R$ centered at the origin (Figure 23). Let P be the point of intersection of the horizontal line through A and the vertical line through B. Express the coordinates of P as a function of θ and describe the curve traced by P for $0 \le \theta \le 2\pi$.

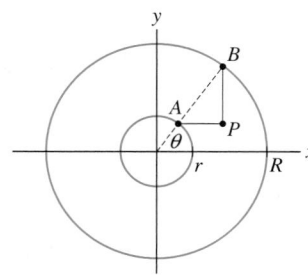

FIGURE 23

76. A 10-ft ladder slides down a wall as its bottom B is pulled away from the wall (Figure 24). Using the angle θ as parameter, find the parametric equations for the path followed by (a) the top of the ladder A, (b) the bottom of the ladder B, and (c) the point P located 4 ft from the top of the ladder. Show that P describes an ellipse.

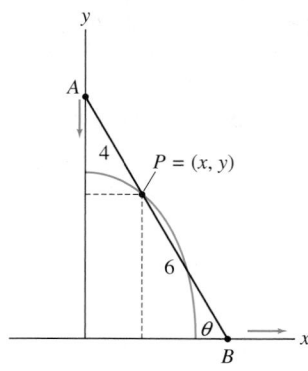

FIGURE 24

Further Insights and Challenges

77. 🖉 Derive the formula for the slope of the tangent line to a parametric curve $c(t) = (x(t), y(t))$ using a different method than that presented in the text. Assume that $x'(t_0)$ and $y'(t_0)$ exist and that $x'(t_0) \ne 0$. Show that

$$\lim_{h \to 0} \frac{y(t_0 + h) - y(t_0)}{x(t_0 + h) - x(t_0)} = \frac{y'(t_0)}{x'(t_0)}$$

Then explain why this limit is equal to the slope dy/dx. Draw a diagram showing that the ratio in the limit is the slope of a secant line.

78. Second Derivative for a Parametrized Curve   Given a parametrized curve $c(t) = (x(t), y(t))$, show that

$$\frac{d}{dt}\left(\frac{dy}{dx}\right) = \frac{x'(t)y''(t) - y'(t)x''(t)}{x'(t)^2}$$

Use this to prove the formula

$$\boxed{\frac{d^2y}{dx^2} = \frac{x'(t)y''(t) - y'(t)x''(t)}{x'(t)^3}} \qquad \boxed{11}$$

In Exercises 79–82, use Eq. (11) to find $\dfrac{d^2y}{dx^2}$.

79. $x = t^3 + t^2$, $y = 7t^2 - 4$, $t = 2$

80. $x = s^{-1} + s$, $y = 4 - s^{-2}$, $s = 1$

81. $x = 8t + 9$, $y = 1 - 4t$, $t = -3$

82. $x = \cos\theta$, $y = \sin\theta$, $\theta = \frac{\pi}{4}$

83. Use Eq. (11) to find the t-intervals on which $c(t) = (t^2, t^3 - 4t)$ is concave up.

84. Use Eq. (11) to find the t-intervals on which $c(t) = (t^2, t^4 - 4t)$ is concave up.

85. Area under a Parametrized Curve   Let $c(t) = (x(t), y(t))$ be a parametrized curve such that $x'(t) > 0$ and $y(t) > 0$ (Figure 25). Show that the area A under $c(t)$ for $t_0 \le t \le t_1$ is

$$A = \int_{t_0}^{t_1} y(t)x'(t)\, dt \qquad \boxed{12}$$

Hint: $x(t)$ is increasing and therefore has an inverse, say, $t = g(x)$. Observe that $c(t)$ is the graph of the function $y(g(x))$ and apply the Change of Variables formula to $A = \displaystyle\int_{x(t_0)}^{x(t_1)} y(g(x))\, dx$.

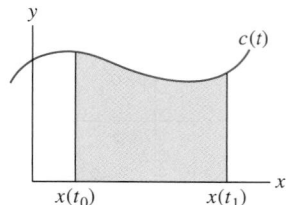

FIGURE 25

86. Calculate the area under $y = x^2$ over $[0, 1]$ using Eq. (12) with the parametrizations (t^3, t^6) and (t^2, t^4).

87. What does Eq. (12) say if $c(t) = (t, f(t))$?

88. Sketch the graph of $c(t) = (\ln t, 2 - t)$ for $1 \le t \le 2$ and compute the area under the graph using Eq. (12).

89. Use Eq. (12) to show that the area under one arch of the cycloid $c(t)$ (Figure 26) generated by a circle of radius R is equal to three times the area of the circle. Recall that

$$c(t) = (Rt - R\sin t, R - R\cos t)$$

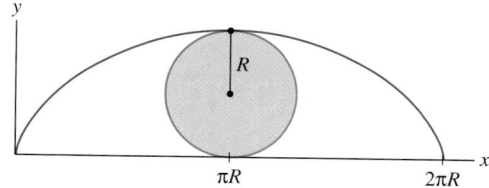

FIGURE 26 The area of the generating circle is one-third the area of one arch of the cycloid.

In Exercises 90–91, refer to Figure 27.

90. The parameter t in the standard parametrization of the ellipse $c(t) = (a\cos t, b\sin t)$ is *not* an angular parameter (unless $a = b$), but it can be interpreted as an area parameter. Show that if $c(t) = (x, y)$, then $t = 2A/ab$, where A is the area of the shaded region in Figure 27. *Hint:* Use Eq. (12).

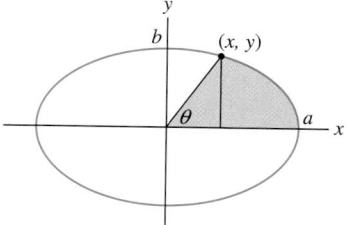

FIGURE 27 The parameter θ on the ellipse $\left(\frac{x}{a}\right)^2 + \left(\frac{y}{b}\right)^2 = 1$.

91. Show that the parametrization of the ellipse by the angle θ is

$$x = \frac{ab\cos\theta}{\sqrt{a^2\sin^2\theta + b^2\cos^2\theta}}$$

$$y = \frac{ab\sin\theta}{\sqrt{a^2\sin^2\theta + b^2\cos^2\theta}}$$

11.2 Arc Length and Speed

In Section 8.1, the length (or "arc length") s of a curve was defined as the limit of the lengths of polygonal approximations. We showed that if $f(x)$ has a continuous derivative, then the arc length of the graph of $y = f(x)$ over $[a, b]$ is equal to

$$s = \text{arc length} = \int_a^b \sqrt{1 + f'(x)^2}\, dx \qquad \boxed{1}$$

There is a more general formula for the arc length of a path $c(t) = (x(t), y(t))$ for $a \le t \le b$. To derive this formula, choose a partition P of $[a, b]$ by an increasing sequence of parameter values:

$$t_0 = a < t_1 < t_2 < \cdots < t_N = b$$

and let L be the polygonal path obtained by joining the points

$$P_0 = c(t_0), \quad P_1 = c(t_1), \ldots, P_N = c(t_N)$$

by segments as in Figure 1. We define the length s of the path to be the limit of the lengths $|L|$ of the polygonal approximations L as the norm $\|P\|$ tends to zero (recall that $\|P\|$ is the maximum of the widths $\Delta t_i = t_i - t_{i-1}$). Note that N tends to ∞ as $\|P\| \to 0$.

By the distance formula, the length of the ith segment in a polygonal approximation L is

$$P_{i-1}P_i = \sqrt{(x(t_i) - x(t_{i-1}))^2 + (y(t_i) - y(t_{i-1}))^2} \qquad \boxed{2}$$

We rewrite this expression by applying the Mean Value Theorem (MVT) to both $x(t)$ and $y(t)$. Assuming that $x(t)$ and $y(t)$ are differentiable, the MVT states that there exist

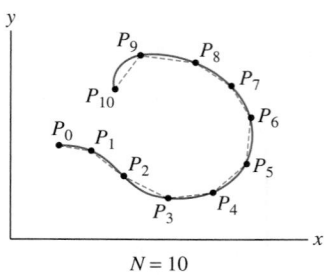

FIGURE 1 Polygonal approximations for $N = 5$ and $N = 10$.

intermediate values t_i^* and t_i^{**} in the interval $[t_{i-1}, t_i]$ such that

$$x(t_i) - x(t_{i-1}) = x'(t_i^*)\Delta t_i, \qquad y(t_i) - y(t_{i-1}) = y'(t_i^{**})\Delta t_i$$

where $\Delta t_i = t_i - t_{i-1}$. Therefore,

$$P_{i-1}P_i = \sqrt{x'(t_i^*)^2\Delta t_i^2 + y'(t_i^{**})^2\Delta t_i^2} = \sqrt{x'(t_i^*)^2 + y'(t_i^{**})^2}\,\Delta t_i$$

The length $|L|$ of L is the sum of the lengths of the segments:

$$|L| = \sum_{i=1}^{N} P_{i-1}P_i = \sum_{i=1}^{N} \sqrt{x'(t_i^*)^2 + y'(t_i^{**})^2}\,\Delta t_i \qquad \boxed{3}$$

This is *nearly* a Riemann sum for the function $\sqrt{x'(t)^2 + y'(t)^2}$. It would be a true Riemann sum if the two intermediate values t_i^* and t_i^{**} were equal. Although they need not be equal, it can be shown (and we will take it for granted) that if $x'(t)$ and $y'(t)$ are continuous, then the sum in Eq. (3) still approaches the integral as $\|P\| \to 0$. Thus,

$$\lim_{\|P\|\to 0} \sum_{i=1}^{N} P_{i-1}P_i = \int_a^b \sqrt{x'(t)^2 + y'(t)^2}\,dt$$

THEOREM 1 Arc Length Let $c(t) = (x(t), y(t))$ and assume that $x'(t)$ and $y'(t)$ exist and are continuous. Then the length s of $c(t)$ for $a \le t \le b$ is equal to

$$s = \int_a^b \sqrt{x'(t)^2 + y'(t)^2}\,dt \qquad \boxed{4}$$

It is often impossible to evaluate the arc length integral (4) explicitly, but we can always approximate it numerically.

Keep in mind that the length of a path $c(t)$ is the distance traveled by a particle following the path. This is greater than the length of the underlying curve if parts of the curve are traversed more than once.

■ **EXAMPLE 1 Length of a Circular Arc** Calculate the length s of the arc $0 \le \theta \le \theta_0$ of a circle of radius R.

Solution We use the parametrization $x = R\cos\theta$, $y = R\sin\theta$:

$$x'(\theta)^2 + y'(\theta)^2 = (-R\sin\theta)^2 + (R\cos\theta)^2 = R^2(\sin^2\theta + \cos^2\theta) = R^2$$

$$s = \int_0^{\theta_0} \sqrt{x'(\theta)^2 + y'(\theta)^2}\,d\theta = \int_0^{\theta_0} R\,d\theta = R\theta_0$$

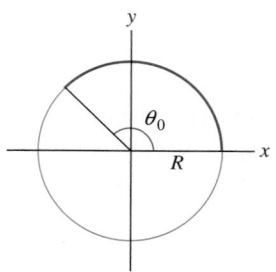

FIGURE 2 An arc of angle θ_0 on a circle of radius R has length $R\theta_0$.

This agrees with the standard formula for the length of a circular arc (Figure 2). ■

It turns out that the cycloid is a curve whose arc length can be computed explicitly.

■ **EXAMPLE 2** Length of the Cycloid Calculate the length s of one arch of the cycloid generated by a circle of radius $R = 2$ (Figure 3).

Solution By Eq. (6) in Section 1, the cycloid generated by a circle of radius $R = 2$ has parametric equations

$$x = 2(t - \sin t), \qquad y = 2(1 - \cos t)$$

Using the identity $\dfrac{1 - \cos t}{2} = \sin^2 \dfrac{t}{2}$, we find

$$x'(t)^2 + y'(t)^2 = 2^2(1 - \cos t)^2 + 2^2 \sin^2 t$$

$$= 4 - 8\cos t + 4\cos^2 t + 4\sin^2 t$$

$$= 8 - 8\cos t = 16\left(\frac{1 - \cos t}{2}\right) = 16\sin^2 \frac{t}{2}$$

and thus,

$$\sqrt{x'(t)^2 + y'(t)^2} = 4\left|\sin \frac{t}{2}\right|$$

One arch of the cycloid is traced as t varies from 0 to 2π, so

$$s = \int_0^{2\pi} \sqrt{x'(t)^2 + y'(t)^2}\, dt = 4\int_0^{2\pi} \sin \frac{t}{2}\, dt = -8\cos \frac{t}{2}\Big|_0^{2\pi} = -8(-1) + 8 = 16$$

Note that we may drop the absolute value in the integral over $[0, 2\pi]$ because $\sin \dfrac{t}{2} \geq 0$ for $0 \leq t \leq 2\pi$. ■

The arc length integral leads to an expression for the **speed** of a particle moving along a path $c(t) = (x(t), y(t))$. By definition, speed is the rate of change of distance traveled with respect to time. The distance traveled over the time interval $[t_0, t]$ is given by the arc length integral:

$$s(t) = \text{distance traveled} = \int_{t_0}^{t} \sqrt{x'(u)^2 + y'(u)^2}\, du$$

Therefore, by the Fundamental Theorem of Calculus,

$$\text{Speed} = \frac{ds}{dt} = \frac{d}{dt}\int_0^t \sqrt{x'(u)^2 + y'(u)^2}\, du = \sqrt{x'(t)^2 + y'(t)^2}$$

In Chapter 13, we will discuss not just the speed but also the velocity of a particle moving along a curved path. Velocity is "speed plus direction" and is represented by a "vector."

THEOREM 2 Speed along a Parametrized Path The speed at time t of a particle with trajectory $c(t) = (x(t), y(t))$ is the derivative of the arc length integral

$$s(t) = \int_{t_0}^{t} \sqrt{x'(u)^2 + y'(u)^2}\, du$$

Namely,

$$\text{Speed} = \frac{ds}{dt} = \sqrt{x'(t)^2 + y'(t)^2}$$

FIGURE 3 One arch of the cycloid generated by a circle of radius 2.

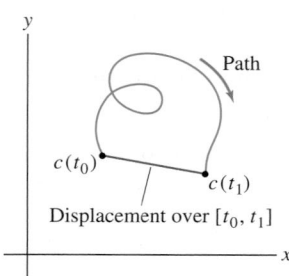

FIGURE 4 The distance traveled is greater than or equal to the displacement.

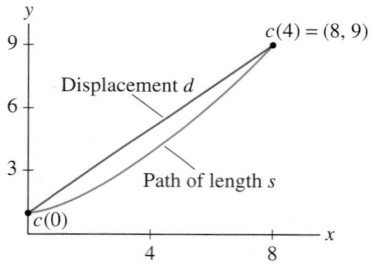

FIGURE 5 The path $c(t) = (2t, 1 + t^{3/2})$ for $0 \le t \le 4$.

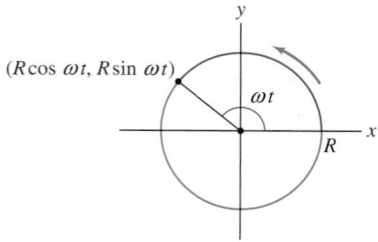

FIGURE 6 A particle rotating on a circular path with angular velocity ω has speed $|\omega R|$.

Before continuing, let us emphasize the distinction between distance traveled and **displacement** (also called net change in position). The displacement over a time interval $[t_0, t_1]$ is the distance between the initial point $c(t_0)$ and the endpoint $c(t_1)$. The distance traveled is greater than the displacement unless the particle happens to move in a straight line (Figure 4).

■ **EXAMPLE 3** A particle travels along the path $c(t) = (2t, 1 + t^{3/2})$ (t in minutes, distance in feet).

(a) Find the speed at $t = 1$.

(b) Compute distance traveled s and displacement d during the first 4 min.

Solution We have $x'(t) = 2$, $y'(t) = \frac{3}{2}t^{1/2}$, so the particle's speed at time t is

$$s'(t) = \sqrt{x'(t)^2 + y'(t)^2} = \sqrt{4 + \frac{9}{4}t} \quad \text{ft/min}$$

The speed at $t = 1$ is $s'(1) = \sqrt{4 + \frac{9}{4}} = 2.5$ ft/min. The distance traveled during the first 4 min is equal to

$$s = \int_0^4 \sqrt{4 + \frac{9}{4}t}\, dt = \frac{8}{27}\left(4 + \frac{9}{4}t\right)^{3/2}\Big|_0^4 = \frac{8}{27}(13^{3/2} - 8) \approx 11.52 \text{ ft}$$

The displacement d is the distance from the initial point $c(0) = (0, 1)$ to the endpoint $c(4) = (8, 1 + 4^{3/2}) = (8, 9)$ (see Figure 5):

$$d = \sqrt{(8 - 0)^2 + (9 - 1)^2} = 8\sqrt{2} \approx 11.31 \text{ ft} \qquad ■$$

■ **EXAMPLE 4 Angular Velocity** Let $c(t) = (R \cos \omega t, R \sin \omega t)$ be a circular path, where ω is a constant (called the **angular velocity**, in units of radians per unit time). Show that $c(t)$ is a path of constant speed. Find the angular velocity ω of a counterclockwise path if $R = 3$ m and the speed is 12 m/s.

Solution We have $x = R \cos \omega t$ and $y = R \sin \omega t$, and

$$x'(t) = -\omega R \sin \omega t, \qquad y'(t) = \omega R \cos \omega t$$

We see that the speed is a constant independent of t:

$$\frac{ds}{dt} = \sqrt{x'(t)^2 + y'(t)^2} = \sqrt{(-\omega R \sin \omega t)^2 + (\omega R \cos \omega t)^2}$$

$$= \sqrt{\omega^2 R^2 (\sin^2 \omega t + \cos^2 \omega t)} = |\omega| R$$

If $R = 3$ and the speed is 12 m/s, then $|\omega| R = 3|\omega| = 12$. Therefore $|\omega| = 4$ and $\omega = 4$ since the path is counterclockwise (Figure 6). ■

The surface area S of a surface of revolution obtained by rotating a parametric curve $c(t) = (x(t), y(t))$ about the x-axis may be computed by an integral that is closely related to the arc length integral. The formula, stated in the next theorem (without proof), may be justified using polygonal approximations as in the case of the arc length formula. Note that we assume $y(t) \ge 0$ so that the curve $c(t)$ lies above the x-axis.

> **THEOREM 3 Surface Area**   Let $c(t) = (x(t), y(t))$ and assume that $x'(t)$ and $y'(t)$ exist and are continuous. Assume further that $y(t) \geq 0$. Then the surface area S of the surface obtained by rotating $c(t)$ about the x-axis for $a \leq t \leq b$ is equal to
>
> $$S = 2\pi \int_a^b y(t)\sqrt{x'(t)^2 + y'(t)^2}\, dt \qquad \boxed{5}$$

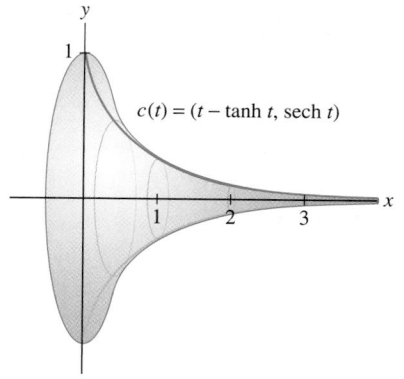

$c(t) = (t - \tanh t,\ \mathrm{sech}\, t)$

FIGURE 7 Surface generated by revolving the tractrix about the x-axis.

▪ **EXAMPLE 5**   The curve $c(t) = (t - \tanh t, \ \mathrm{sech}\, t)$ is called a *tractrix*. Calculate the surface area of the infinite surface generated by revolving the tractrix about the x-axis for $0 \leq t < \infty$ (Figure 7).

Solution   We have

$$x'(t) = \frac{d}{dt}(t - \tanh t) = 1 - \mathrm{sech}^2 t, \quad y'(t) = \frac{d}{dt}\mathrm{sech}\, t = -\mathrm{sech}\, t \tanh t$$

Using the identities $1 - \mathrm{sech}^2 t = \tanh^2 t$ and $\mathrm{sech}^2 t = 1 - \tanh^2 t$, we obtain

$$x'(t)^2 + y'(t)^2 = (1 - \mathrm{sech}^2 t)^2 + (-\mathrm{sech}\, t \tanh t)^2$$

$$= (\tanh^2 t)^2 + (1 - \tanh^2 t)\tanh^2 t = \tanh^2 t$$

The surface area is given by an improper integral, which we evaluate using the integral formula recalled in the margin:

$$S = 2\pi \int_0^\infty \mathrm{sech}\, t \sqrt{\tanh^2 t}\, dt = 2\pi \int_0^\infty \mathrm{sech}\, t \tanh t\, dt = 2\pi \lim_{R \to \infty} \int_0^R \mathrm{sech}\, t \tanh t\, dt$$

$$= 2\pi \lim_{R \to \infty} (-\mathrm{sech}\, t)\Big|_0^R = 2\pi \lim_{R \to \infty} (\mathrm{sech}\, 0 - \mathrm{sech}\, R) = 2\pi\, \mathrm{sech}\, 0 = 2\pi$$

Here we use that $\mathrm{sech}\, t = \dfrac{2}{e^t + e^{-t}}$ tends to 0 as $t \to \infty$ because the term e^t in the denominator tends to ∞. ▪

◀·· REMINDER

$$\mathrm{sech}\, t = \frac{1}{\cosh t} = \frac{2}{e^t + e^{-t}}$$

$$1 - \mathrm{sech}^2 t = \tanh^2 t$$

$$\frac{d}{dt}\tanh t = \mathrm{sech}^2 t$$

$$\frac{d}{dt}\mathrm{sech}\, t = -\mathrm{sech}\, t \tanh t$$

$$\int \mathrm{sech}\, t \tanh t\, dt = -\mathrm{sech}\, t + C$$

11.2 SUMMARY

- The *arc length* s of a path $c(t) = (x(t), y(t))$ for $a \leq t \leq b$ is

$$s = \text{arc length} = \int_a^b \sqrt{x'(t)^2 + y'(t)^2}\, dt$$

- The *arc length integral* $s(t) = \displaystyle\int_{t_0}^t \sqrt{x'(u)^2 + y'(u)^2}\, du$ is equal to the *distance traveled* over the time interval $[t_0, t]$. The *displacement* is the distance between the starting point $c(t_0)$ and endpoint $c(t)$.
- The speed at time t of a particle with trajectory $c(t) = (x(t), y(t))$ is

$$\frac{ds}{dt} = \sqrt{x'(t)^2 + y'(t)^2}$$

11.2 EXERCISES

Preliminary Questions

1. What is the definition of arc length?

2. What is the interpretation of $\sqrt{x'(t)^2 + y'(t)^2}$ for a particle following the trajectory $(x(t), y(t))$?

3. A particle travels along a path from $(0, 0)$ to $(3, 4)$. What is the displacement? Can the distance traveled be determined from the information given?

4. A particle traverses the parabola $y = x^2$ with constant speed 3 cm/s. What is the distance traveled during the first minute? *Hint:* No computation is necessary.

Exercises

1. Use Eq. (4) to calculate the length of the semicircle

$$x = 3 \sin t, \quad y = 3 \cos t, \qquad 0 \le t \le \pi$$

2. Find the speed at $t = 4$ s of a particle whose position at time t seconds is $c(t) = (4 - t^2, t^3)$.

In Exercises 3–12, use Eq. (4) to find the length of the path over the given interval.

3. $(3t + 1, 9 - 4t), \quad 0 \le t \le 2$

4. $(1 + 2t, 2 + 4t), \quad 1 \le t \le 4$

5. $(2t^2, 3t^2 - 1), \quad 0 \le t \le 4$

6. $(3t, 4t^{3/2}), \quad 0 \le t \le 1$

7. $(3t^2, 4t^3), \quad 1 \le t \le 4$

8. $(t^3 + 1, t^2 - 3), \quad 0 \le t \le 1$

9. $(\sin 3t, \cos 3t), \quad 0 \le t \le \pi$

10. $(2 \cos t - \cos 2t, 2 \sin t - \sin 2t), \quad 0 \le t \le \frac{\pi}{2}$

11. $(\sin \theta - \theta \cos \theta, \cos \theta + \theta \sin \theta), \quad 0 \le \theta \le 2$

12. $(5(\theta - \sin \theta), 5(1 - \cos \theta)), \quad 0 \le \theta \le 2\pi$

13. Show that one arch of a cycloid generated by a circle of radius R has length $8R$.

14. Find the length of the tractrix (Figure 22 in Section 11.1) with parametrization

$$c(t) = (t - \tanh(t), \operatorname{sech}(t)), \qquad 0 \le t \le A$$

15. Find the length of the spiral $c(t) = (t \cos t, t \sin t)$ for $0 \le t \le 2\pi$ to three decimal places (Figure 8). *Hint:* Use the formula

$$\int \sqrt{1 + t^2}\, dt = \frac{1}{2} t \sqrt{1 + t^2} + \frac{1}{2} \ln\left(t + \sqrt{1 + t^2}\right)$$

In Exercises 16–19, determine the speed $s(t)$ of a particle with a given trajectory at time t_0 (in units of meters and seconds).

16. $(t^3, t^2), \quad t = 2$

17. $(3 \sin 5t, 8 \cos 5t), \quad t = \frac{\pi}{4}$

18. $(5t + 1, 4t - 3), \quad t = 9$

19. $(\ln(t^2 + 1), t^3), \quad t = 1$

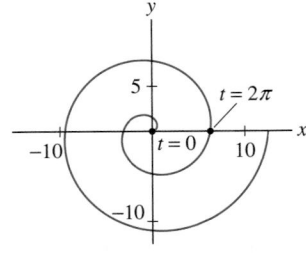

FIGURE 8 The spiral $c(t) = (t \cos t, t \sin t)$.

20. Find the speed of a particle whose path around a circle of radius r is described by the parametric curve $c(t) = (r \cos \omega t, r \sin \omega t)$.

21. Find the minimum speed of a particle with trajectory $c(t) = (t^3 - 4t, t^2 + 1)$ for $t \ge 0$. *Hint:* It is easier to find the minimum of the square of the speed.

22. Find the minimum speed of a particle with trajectory $c(t) = (t^3, t^{-2})$ for $t \ge 0.5$.

23. Find the speed of the cycloid $c(t) = (4t - 4 \sin t, 4 - 4 \cos t)$ at points where the tangent line is horizontal.

24. If you unwind thread from a stationary circular spool, keeping the thread taut at all times, then the endpoint traces a curve $\mathcal{C}$ called the **involute** of the circle (Figure 9). Observe that $\overline{PQ}$ has length $R\theta$. Show that $\mathcal{C}$ is parametrized by

$$c(\theta) = \left(R(\cos \theta + \theta \sin \theta), R(\sin \theta - \theta \cos \theta)\right)$$

Then find the length of the involute for $0 \le \theta \le 2\pi$.

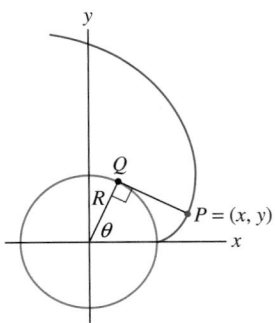

FIGURE 9 Involute of a circle.

CAS   *In Exercises 25–28, plot the curve and use the Midpoint Rule with N = 10, 20, 30, and 50 to approximate its length.*

25. $c(t) = (\cos t, e^{\sin t})$ for $0 \le t \le 2\pi$

26. $c(t) = (t - \sin 2t, 1 - \cos 2t)$ for $0 \le t \le 2\pi$

27. The ellipse $\left(\dfrac{x}{5}\right)^2 + \left(\dfrac{y}{3}\right)^2 = 1$

28. $x = \sin 2t, \quad y = \sin 3t$ for $0 \le t \le 2\pi$

29. Let $a > b$ and set $k = \sqrt{1 - \dfrac{b^2}{a^2}}$. Use a parametric representation to show that the ellipse $\left(\dfrac{x}{a}\right)^2 + \left(\dfrac{y}{b}\right)^2 = 1$ has length $L = 4aG\left(\dfrac{\pi}{2}, k\right)$, where

$$G(\theta, k) = \int_0^\theta \sqrt{1 - k^2 \sin^2 t}\, dt$$

is the elliptic integral of the second kind.

In Exercises 30–33, use Eq. (5) to compute the surface area of the given surface.

30. The cone generated by revolving $c(t) = (t, mt)$ about the x-axis for $0 \le t \le A$

31. The surface generated by revolving the astroid with parametrization $c(t) = (\cos^3 t, \sin^3 t)$ about the x-axis for $0 \le t \le \frac{\pi}{2}$

32. A sphere of radius R

33. The surface generated by revolving one arch of the cycloid $c(t) = (t - \sin t, 1 - \cos t)$ about the x-axis

Further Insights and Challenges

34. CAS   Let $b(t)$ be the "Butterfly Curve":

$$x(t) = \sin t \left(e^{\cos t} - 2\cos 4t - \sin\left(\frac{t}{12}\right)^5\right)$$

$$y(t) = \cos t \left(e^{\cos t} - 2\cos 4t - \sin\left(\frac{t}{12}\right)^5\right)$$

(a) Use a computer algebra system to plot $b(t)$ and the speed $s(t)$ for $0 \le t \le 12\pi$.

(b) Approximate the length $b(t)$ for $0 \le t \le 10\pi$.

35. CAS   Let $a \ge b > 0$ and set $k = \dfrac{2\sqrt{ab}}{a - b}$. Show that the **trochoid**

$$x = at - b\sin t, \qquad y = a - b\cos t, \qquad 0 \le t \le T$$

has length $2(a - b)G\left(\dfrac{T}{2}, k\right)$ with $G(\theta, k)$ as in Exercise 29.

36. The path of a satellite orbiting at a distance R from the center of the earth is parametrized by $x = R\cos \omega t, y = R\sin \omega t$.

(a) Show that the period T (the time of one revolution) is $T = 2\pi/\omega$.

(b) According to Newton's laws of motion and gravity,

$$x''(t) = -Gm_e \frac{x}{R^3}, \qquad y''(t) = -Gm_e \frac{y}{R^3}$$

where G is the universal gravitational constant and m_e is the mass of the earth. Prove that $\dfrac{R^3}{T^2} = \dfrac{Gm_e}{4\pi^2}$. Thus, $\dfrac{R^3}{T^2}$ has the same value for all orbits (a special case of Kepler's Third Law).

37. The acceleration due to gravity on the surface of the earth is $g = \dfrac{Gm_e}{R_e^2} = 9.8 \text{ m/s}^2$, where $R_e = 6{,}378$ km. Use Exercise 36(b) to show that a satellite orbiting at the earth's surface would have period $T_e = 2\pi\sqrt{R_e/g} \approx 84.5$ min. Then estimate the distance R_m from the moon to the center of the earth. Assume that the period of the moon (sidereal month) is $T_m \approx 27.43$ days.

11.3 Polar Coordinates

Polar coordinates are an alternative system of labeling points P in the plane. Instead of specifying the x- and y-coordinates, we specify coordinates (r, θ), where r is the distance from P to the origin O and θ is the angle between $\overline{OP}$ and the positive x-axis (Figure 1). By convention, an angle is positive if the corresponding rotation is counterclockwise. We call r the **radial coordinate** and θ the **angular coordinate**.

For example, the point P in Figure 2 has polar coordinates $(4, \frac{2\pi}{3})$. It is located at a distance 4 from the origin (so it lies on the circle of radius 4) and it lies on the ray of angle $\frac{2\pi}{3}$.

To relate polar and rectangular coordinates, we refer to Figure 1. We see that if P has polar coordinates (r, θ), then its rectangular coordinates are $x = r\cos\theta$ and $y = r\sin\theta$. On the other hand, $r^2 = x^2 + y^2$ by the distance formula and $\tan\theta = \dfrac{y}{x}$ if $x \ne 0$.

Polar coordinates are useful for problems in which the distance from the origin or angle plays a role, especially if the problem involves rotational symmetry. They are a natural choice for describing the gravitational force exerted by an object such as the sun, since the magnitude of the force depends only on distance from the sun.

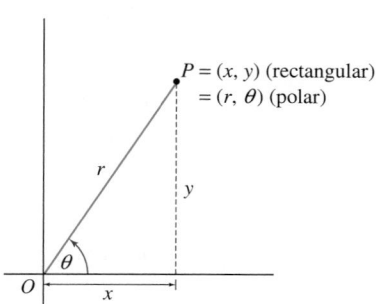

FIGURE 1 The polar coordinates of P satisfy $r = \sqrt{x^2 + y^2}$, $\tan \theta = y/x$.

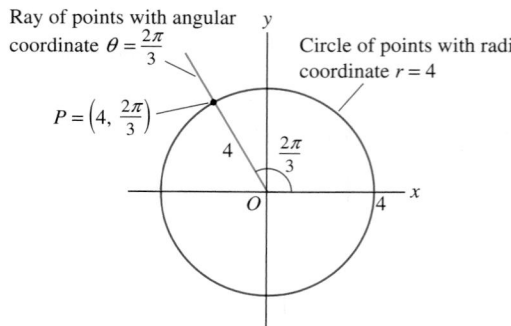

FIGURE 2 The point P lies at the intersection of the circle $r = 4$ and the ray $\theta = \frac{2\pi}{3}$.

Polar to Rectangular	Rectangular to Polar
$x = r \cos \theta$	$r = \sqrt{x^2 + y^2}$
$y = r \sin \theta$	$\tan \theta = \dfrac{y}{x} \quad (x \neq 0)$

A few remarks are in order before proceeding.

- The angular coordinate θ is not uniquely determined because the polar coordinates (r, θ) and $(r, \theta + 2\pi n)$ *label the same point* for any integer n. For instance, the point $(x, y) = (0, 2)$ has polar coordinates $\left(2, \frac{\pi}{2}\right)$ and $\left(2, \frac{5\pi}{2}\right)$ and $\left(2, \frac{9\pi}{2}\right)$, etc. (see Figure 3).
- The origin O does not have a well-defined angular coordinate, so we assign to O the polar coordinates $(0, \theta)$ for any angle θ.
- By convention, we also allow *negative* radial coordinates. For $r > 0$, the point $(-r, \theta)$ is defined to be the reflection of (r, θ) through the origin (see Figure 4). With this convention, $(-r, \theta)$ and $(r, \theta + \pi)$ represent the same point.
- We may specify unique polar coordinates for points other than the origin by placing restrictions on r and θ. We commonly choose $r > 0$ and $0 \leq \theta < 2\pi$.

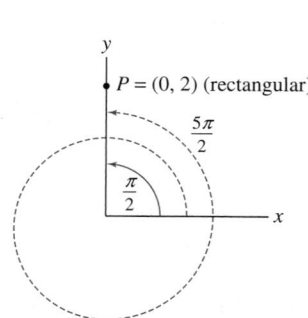

FIGURE 3 The angular coordinate of P can take on the value $\frac{\pi}{2} + 2\pi n$ for any integer n.

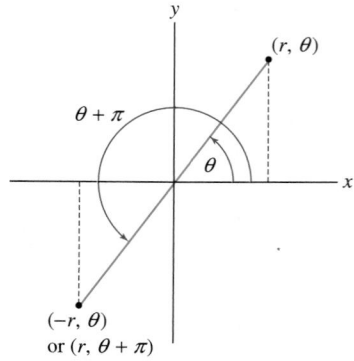

FIGURE 4 Relation between (r, θ) and $(-r, \theta)$.

FIGURE 5 Grid lines in polar coordinates.

Figure 5 shows the two families of **grid lines** in polar coordinates:

$$\text{Circle centered at } O \quad \longleftrightarrow \quad r = \text{constant}$$

$$\text{Ray starting at } O \quad \longleftrightarrow \quad \theta = \text{constant}$$

Every point in the plane other than the origin lies at the intersection of two grid lines that determine its polar coordinates. For example, the point Q in Figure 5 lies on the circle $r = 3$ and the ray $\theta = \frac{5\pi}{6}$ so $Q = (3, \frac{5\pi}{6})$ in polar coordinates.

■ **EXAMPLE 1** From Polar to Rectangular Coordinates   Find the rectangular coordinates of the point Q with polar coordinates $(r, \theta) = (3, \frac{5\pi}{6})$.

Solution The point $Q = (r, \theta) = (3, \frac{5\pi}{6})$ has rectangular coordinates:

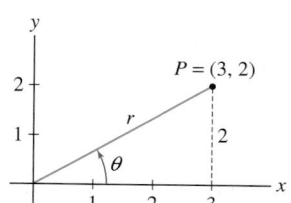

FIGURE 6 The polar coordinates of P satisfy $r = \sqrt{3^2 + 2^2}$ and $\tan \theta = \frac{2}{3}$.

$$x = r \cos \theta = 3 \cos\left(\frac{5\pi}{6}\right) = 3\left(-\frac{\sqrt{3}}{2}\right) = -\frac{3\sqrt{3}}{2}$$

$$y = r \sin \theta = 3 \sin\left(\frac{5\pi}{6}\right) = 3\left(\frac{1}{2}\right) = \frac{3}{2}$$

■ **EXAMPLE 2** From Rectangular to Polar Coordinates   Find polar coordinates of the point P with rectangular coordinates $(x, y) = (3, 2)$.

Solution Since $P = (x, y) = (3, 2)$,

$$r = \sqrt{x^2 + y^2} = \sqrt{3^2 + 2^2} = \sqrt{13} \approx 3.6$$

$$\tan \theta = \frac{y}{x} = \frac{2}{3}$$

$$\theta = \tan^{-1} \frac{y}{x} = \tan^{-1} \frac{2}{3} \approx 0.588$$

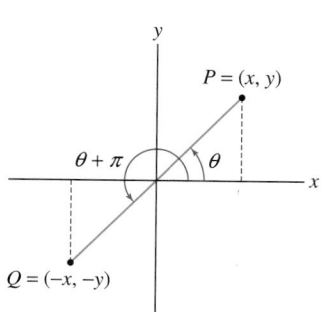

FIGURE 7 $\tan(\theta + \pi) = \tan \theta$.

By definition, the arctangent satisfies

$$-\frac{\pi}{2} < \tan^{-1} x < \frac{\pi}{2}$$

An angular coordinate θ of $P = (x, y)$ is

$$\theta = \begin{cases} \tan^{-1} \dfrac{y}{x} & \text{if } x > 0 \\ \tan^{-1} \dfrac{y}{x} + \pi & \text{if } x < 0 \\ \pm\dfrac{\pi}{2} & \text{if } x = 0 \end{cases}$$

Thus, P has polar coordinates $(r, \theta) \approx (3.6, 0.588)$ (see Figure 6). ■

The next example calls attention to a subtlety in determining the angular coordinate θ of a point $P = (x, y)$. There are two angles between 0 and 2π satisfying $\tan \theta = \frac{y}{x}$ because $\tan(\theta + \pi) = \tan \theta$. We must choose θ so that (r, θ) lies in the quadrant containing P and not its reflection through the origin Q (Figure 7).

FIGURE 8

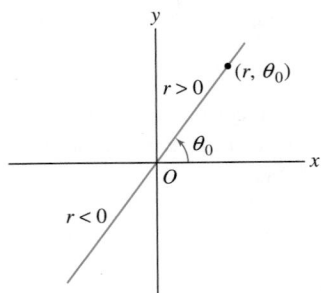

FIGURE 9 Lines through O with polar equation $\theta = \theta_0$.

■ **EXAMPLE 3** Choosing θ Correctly Find two polar representations of $P = (-1, 1)$, one using a positive radial coordinate and one using a negative radial coordinate.

Solution The positive radial coordinate of $P = (x, y) = (-1, 1)$ is

$$r = \sqrt{(-1)^2 + 1^2} = \sqrt{2}$$

To find the angular coordinate, we solve

$$\tan \theta = \frac{y}{x} = -1 \quad \Rightarrow \quad \theta = \frac{3\pi}{4}, \ \frac{3\pi}{4} + \pi = \frac{7\pi}{4}$$

Since P lies in the second quadrant, the appropriate angular coordinate of P is $\theta = \frac{3\pi}{4}$ (Figure 8). Alternatively, P may be represented with the negative radial coordinate $r = -\sqrt{2}$ and angle $\theta = \frac{7\pi}{4}$. Thus,

$$P = \left(\sqrt{2}, \frac{3\pi}{4} \right) \qquad \text{or} \qquad \left(-\sqrt{2}, \frac{7\pi}{4} \right) \qquad ■$$

Curves are described in polar coordinates by equations relating r and θ. We refer to an equation in polar coordinates as a **polar equation** to avoid confusion with the usual equation in rectangular coordinates. By convention, we allow solutions with negative r when considering equations in polar coordinates.

The polar equation of a line through the origin O has the simple form $\theta = \theta_0$, where θ_0 is the angle between the line and the x-axis (Figure 9). Indeed, the solutions of $\theta = \theta_0$ are (r, θ_0), where r is arbitrary (positive, negative, or zero).

■ **EXAMPLE 4** Line Through the Origin Find the polar equation of the line through the origin of slope $\frac{3}{2}$.

Solution A line of slope m makes an angle θ_0 with the x-axis, where $m = \tan \theta_0$. In our case, Figure 10 shows that $\tan \theta_0 = \frac{3}{2}$ and so $\theta_0 = \tan^{-1} \frac{3}{2} \approx 0.98$. The equation of the line is $\theta = \tan^{-1} \frac{3}{2}$. ■

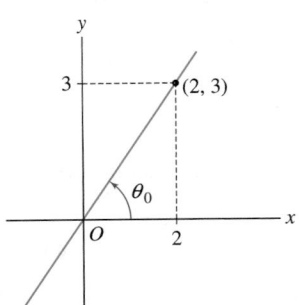

FIGURE 10 Line of slope $\frac{3}{2}$ through the origin.

FIGURE 11 Line $\mathcal{L}$ not passing through the origin.

To describe a line not passing through the origin in polar coordinates, the key is to consider the point P_0 on the line that is closest to the origin (Figure 11).

■ **EXAMPLE 5** Line Not Passing Through the Origin Let $P_0 = (d, \alpha)$ be polar coordinates of the point on a line $\mathcal{L}$ closest to the origin. Show that $\mathcal{L}$ has polar equation

$$r = d \sec(\theta - \alpha)$$ [1]

Solution The point P_0 on $\mathcal{L}$ closest to the origin is obtained by dropping a perpendicular from the origin to $\mathcal{L}$ (Figure 11). Therefore, if $P = (r, \theta)$ is any point on $\mathcal{L}$ other than P_0, then $\triangle OPP_0$ is a right triangle and as we see in Figure 11, $\dfrac{d}{r} = \cos(\theta - \alpha)$. We obtain the equation $r = d \sec(\theta - \alpha)$ as claimed. ■

■ **EXAMPLE 6** Find the polar equation of the line $\mathcal{L}$ tangent to the circle $r = 4$ at the point with polar coordinates $P_0 = (4, \frac{\pi}{3})$.

Solution Since $\mathcal{L}$ is tangent to the circle at P_0, all points $P \neq P_0$ on $\mathcal{L}$ lie outside the circle. Therefore, P_0 is the point on $\mathcal{L}$ closest to the origin (Figure 12). We take $(d, \alpha) = (4, \frac{\pi}{3})$ in Eq. (1) to obtain the equation $r = 4 \sec(\theta - \frac{\pi}{3})$. ■

It is often difficult to guess the shape of a graph of a polar equation. One way of sketching the graph of an equation $r = f(\theta)$ is to rewrite the equation in rectangular coordinates.

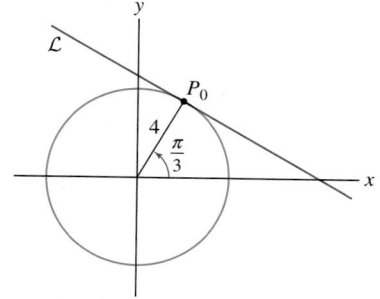

FIGURE 12 The tangent line has equation $r = 4 \sec\left(\theta - \dfrac{\pi}{3}\right)$.

■ **EXAMPLE 7** Converting to Rectangular Coordinates Identify the curve with polar equation $r = 2a \cos \theta$ (a a constant).

Solution We have $r = \pm\sqrt{x^2 + y^2}$ and $\cos \theta = \dfrac{x}{r} = \dfrac{x}{\pm\sqrt{x^2 + y^2}}$. Thus

$$r = 2a \cos \theta \quad \Rightarrow \quad \pm\sqrt{x^2 + y^2} = 2a \frac{x}{\pm\sqrt{x^2 + y^2}} \quad \Rightarrow \quad x^2 + y^2 = 2ax$$

Now complete the square:

$$x^2 - 2ax + y^2 = 0$$

$$(x - a)^2 - a^2 + y^2 = 0 \qquad \text{or} \qquad \boxed{(x - a)^2 + y^2 = a^2}$$

Thus, $r = 2a \cos \theta$ is the circle of radius a and center $(a, 0)$ (Figure 13). ■

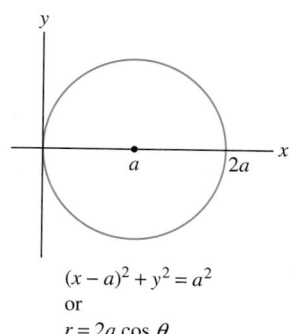

$(x - a)^2 + y^2 = a^2$
or
$r = 2a \cos \theta$

FIGURE 13 Circle with polar equation $r = 2a \cos \theta$.

A similar calculation shows that $x^2 + (y - a)^2 = a^2$ has polar equation $r = 2a \sin \theta$. In the next example, we make use of symmetry. Note that the points (r, θ) and $(r, -\theta)$ are symmetric about the x-axis (Figure 14).

■ **EXAMPLE 8** The *Limaçon*: Symmetry About the *x*-Axis Sketch the graph of $r = 2\cos\theta - 1$.

Solution To get started, we plot the points A–G on a grid and join them by a smooth curve (Figure 15). However, for a better understanding, it is helpful to graph r as a function of θ in rectangular coordinates. Figure 16(A) shows that

As θ varies from 0 to $\frac{\pi}{3}$, r varies from 1 to 0

As θ varies from $\frac{\pi}{3}$ to π, r is *negative* and varies from 0 to -3

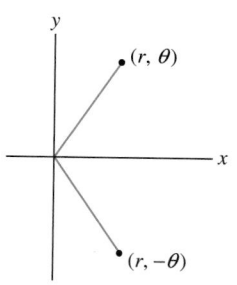

FIGURE 14 The points (r, θ) and $(r, -\theta)$ are symmetric with respect to the x-axis.

	A	B	C	D	E	F	G
θ	0	$\pi/6$	$\pi/3$	$\pi/2$	$2\pi/3$	$5\pi/6$	π
r	1	0.73	0	-1	-2	-2.73	-3

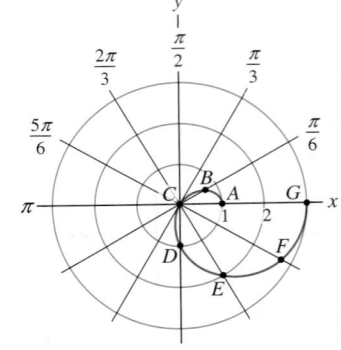

FIGURE 15 Plotting $r = 2\cos\theta - 1$ using a grid.

Since $r(\theta) = 2\cos\theta - 1$ is periodic, it suffices to plot points for $-\pi \le \theta \le \pi$.

- The graph begins at point A with polar coordinates $(r, \theta) = (1, 0)$ and moves in toward the origin as θ varies from 0 to $\pi/3$ [Figure 16(B)].
- For $\pi/3 \le \theta \le \pi$, r is negative. Therefore the curve continues into the third and fourth quadrants (rather than into the first and second quadrants), moving toward the point G with polar coordinates $(r, \theta) = (-3, \pi)$ as in Figure 16(C).
- Since $r(\theta) = r(-\theta)$, the curve is symmetric with respect to the x-axis. As θ varies from 0 to $-\pi$, we obtain the reflection through the x-axis of the first half of the curve, where $0 \le \theta \le \pi$, as in Figure 16(D).   ∎

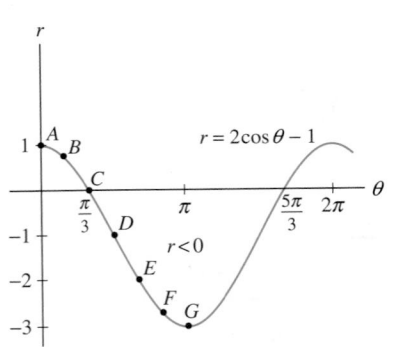

(A) Variation of r as a function of θ.

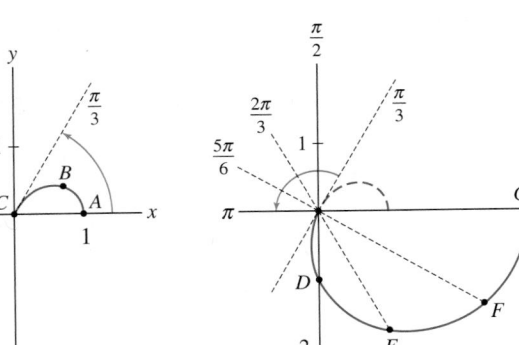

(B) As θ varies from 0 to $\pi/3$, r varies from 1 to 0.

(C) As θ varies from $\pi/3$ to π, r is negative and varies from 0 to -3.

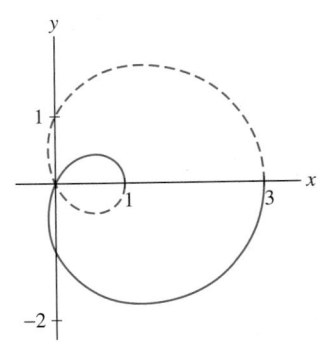

(D) The entire limaçon.

FIGURE 16 The curve $r = 2\cos\theta - 1$ is called the *limaçon*, from the Latin word for "snail." It was first described in 1525 by the German artist Albrecht Dürer.

11.3 SUMMARY

- A point P in the plane has polar coordinates (r, θ), where r is the distance from P to the origin and θ is the angle between the positive x-axis and the segment $\overline{OP}$, measured in the counterclockwise direction. The rectangular coordinates of P are

$$x = r\cos\theta, \qquad y = r\sin\theta$$

- If P has rectangular coordinates (x, y), then

$$r = \sqrt{x^2 + y^2}, \qquad \tan \theta = \frac{y}{x}$$

The angular coordinate θ must be chosen so that (r, θ) lies in the proper quadrant. We have

$$\theta = \begin{cases} \tan^{-1} \dfrac{y}{x} & \text{if } x > 0 \\[2mm] \tan^{-1} \dfrac{y}{x} + \pi & \text{if } x < 0 \\[2mm] \pm \dfrac{\pi}{2} & \text{if } x = 0 \end{cases}$$

- Nonuniqueness: (r, θ) and $(r, \theta + 2n\pi)$ represent the same point for all integers n. The origin O has polar coordinates $(0, \theta)$ for any θ.
- Negative radial coordinates: $(-r, \theta)$ and $(r, \theta + \pi)$ represent the same point.
- Polar equations:

Curve	Polar Equation
Circle of radius R, center at the origin	$r = R$
Line through origin of slope $m = \tan \theta_0$	$\theta = \theta_0$
Line on which $P_0 = (d, \alpha)$ is the point closest to the origin	$r = d \sec(\theta - \alpha)$
Circle of radius a, center at $(a, 0)$ $(x - a)^2 + y^2 = a^2$	$r = 2a \cos \theta$
Circle of radius a, center at $(0, a)$ $x^2 + (y - a)^2 = a^2$	$r = 2a \sin \theta$

11.3 EXERCISES

Preliminary Questions

1. If P and Q have the same radial coordinate, then (choose the correct answer):

(a) P and Q lie on the same circle with the center at the origin.

(b) P and Q lie on the same ray based at the origin.

2. Give two polar coordinate representations for the point $(x, y) = (0, 1)$, one with negative r and one with positive r.

3. Does a point (r, θ) have more than one representation in rectangular coordinates?

4. Describe the curves with polar equations

(a) $r = 2$ **(b)** $r^2 = 2$ **(c)** $r \cos \theta = 2$

5. If $f(-\theta) = f(\theta)$, then the curve $r = f(\theta)$ is symmetric with respect to the (choose the correct answer):

(a) x-axis   **(b)** y-axis   **(c)** origin

Exercises

1. Find polar coordinates for each of the seven points plotted in Figure 17.

2. Plot the points with polar coordinates

(a) $(2, \frac{\pi}{6})$ **(b)** $(4, \frac{3\pi}{4})$ **(c)** $(3, -\frac{\pi}{2})$ **(d)** $(0, \frac{\pi}{6})$

3. Convert from rectangular to polar coordinates:

(a) $(1, 0)$ **(b)** $(3, \sqrt{3})$

(c) $(-2, 2)$ **(d)** $(-1, \sqrt{3})$

FIGURE 17

$(x, y) = (2\sqrt{3}, 2)$

4. Use a calculator to convert from rectangular to polar coordinates (make sure your choice of θ gives the correct quadrant):

(a) $(2, 3)$ **(b)** $(4, -7)$

(c) $(-3, -8)$ **(d)** $(-5, 2)$

5. Convert from polar to rectangular coordinates:

(a) $(3, \frac{\pi}{6})$ **(b)** $(6, \frac{3\pi}{4})$ **(c)** $(5, -\frac{\pi}{2})$

6. Convert from polar to rectangular coordinates:

(a) $(0, 0)$ **(b)** $(-4, \frac{\pi}{3})$ **(c)** $(0, \frac{\pi}{6})$

7. Which of the following are possible polar coordinates for the point P with rectangular coordinates $(0, -2)$?

(a) $\left(2, \frac{\pi}{2}\right)$ **(b)** $\left(2, \frac{7\pi}{2}\right)$

(c) $\left(-2, -\frac{3\pi}{2}\right)$ **(d)** $\left(-2, \frac{7\pi}{2}\right)$

(e) $\left(-2, -\frac{\pi}{2}\right)$ **(f)** $\left(2, -\frac{7\pi}{2}\right)$

8. Describe each shaded sector in Figure 18 by inequalities in r and θ.

  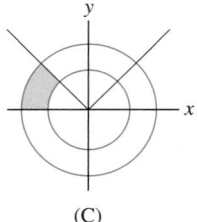

(A)   (B)   (C)

FIGURE 18

9. Find the equation in polar coordinates of the line through the origin with slope $\frac{1}{2}$.

10. What is the slope of the line $\theta = \frac{3\pi}{5}$?

11. Which of the two equations, $r = 2 \sec \theta$ and $r = 2 \csc \theta$, defines a horizontal line?

In Exercises 12–17, convert to an equation in rectangular coordinates.

12. $r = 7$ **13.** $r = \sin \theta$

14. $r = 2 \sin \theta$ **15.** $r = 2 \csc \theta$

16. $r = \dfrac{1}{\cos \theta - \sin \theta}$ **17.** $r = \dfrac{1}{2 - \cos \theta}$

In Exercises 18–21, convert to an equation in polar coordinates.

18. $x^2 + y^2 = 5$ **19.** $x = 5$

20. $y = x^2$ **21.** $xy = 1$

22. Match the equation with its description:

(a) $r = 2$ **(i)** Vertical line
(b) $\theta = 2$ **(ii)** Horizontal line
(c) $r = 2 \sec \theta$ **(iii)** Circle
(d) $r = 2 \csc \theta$ **(iv)** Line through origin

23. Find the values of θ in the plot of $r = 4 \cos \theta$ corresponding to points A, B, C, D in Figure 19. Then indicate the portion of the graph traced out as θ varies in the following intervals:

(a) $0 \leq \theta \leq \frac{\pi}{2}$ **(b)** $\frac{\pi}{2} \leq \theta \leq \pi$ **(c)** $\pi \leq \theta \leq \frac{3\pi}{2}$

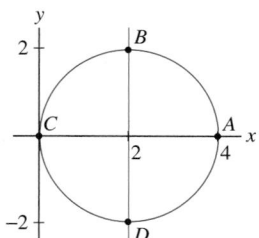

FIGURE 19 Plot of $r = 4 \cos \theta$.

24. Suppose that (x, y) has polar coordinates (r, θ). Find the polar coordinates of the following:

(a) $(x, -y)$ **(b)** $(-x, -y)$ **(c)** $(-x, y)$ **(d)** (y, x)

25. Match each equation in rectangular coordinates with its equation in polar coordinates.

(a) $x^2 + y^2 = 4$ **(i)** $r^2(1 - 2 \sin^2 \theta) = 4$
(b) $x^2 + (y - 1)^2 = 1$ **(ii)** $r(\cos \theta + \sin \theta) = 4$
(c) $x^2 - y^2 = 4$ **(iii)** $r = 2 \sin \theta$
(d) $x + y = 4$ **(iv)** $r = 2$

26. What are the polar equations of the lines parallel to the line $r \cos(\theta - \frac{\pi}{3}) = 1$?

27. Show that $r = \sin \theta + \cos \theta$ is the equation of the circle of radius $1/\sqrt{2}$ whose center in rectangular coordinates is $(\frac{1}{2}, \frac{1}{2})$. Then find the values of θ between 0 and π such that $(\theta, r(\theta))$ yields the points $A, B, C,$ and D in Figure 20.

28. Sketch the curve $r = \frac{1}{2}\theta$ (the spiral of Archimedes) for θ between 0 and 2π by plotting the points for $\theta = 0, \frac{\pi}{4}, \frac{\pi}{2}, \ldots, 2\pi$.

29. Sketch the graph of $r = 3 \cos \theta - 1$ (see Example 8).

30. Sketch the graph of $r = \cos \theta - 1$.

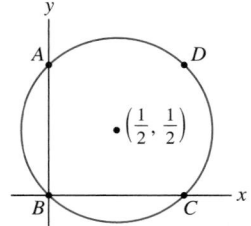

FIGURE 20 Plot of $r = \sin\theta + \cos\theta$.

31. Figure 21 displays the graphs of $r = \sin 2\theta$ in rectangular coordinates and in polar coordinates, where it is a "rose with four petals." Identify (a) the points in (B) corresponding to the points labeled A–I in (A), and (b) the parts of the curve in (B) corresponding to the angle intervals $[0, \frac{\pi}{2}]$, $[\frac{\pi}{2}, \pi]$, $[\pi, \frac{3\pi}{2}]$, and $[\frac{3\pi}{2}, 2\pi]$.

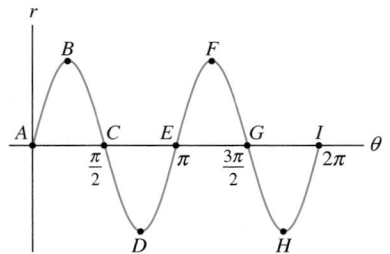

(A) Graph of r as a function of θ, where $r = \sin 2\theta$.

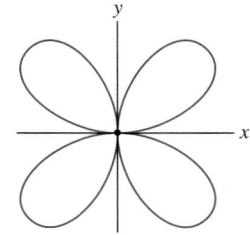

(B) Graph of $r = \sin 2\theta$ in polar coordinates.

FIGURE 21 Rose with four petals.

32. Sketch the curve $r = \sin 3\theta$. First fill in the table of r-values below and plot the corresponding points of the curve. Notice that the three petals of the curve correspond to the angle intervals $[0, \frac{\pi}{3}]$, $[\frac{\pi}{3}, \frac{2\pi}{3}]$, and $[\frac{\pi}{3}, \pi]$. Then plot $r = \sin 3\theta$ in rectangular coordinates and label the points on this graph corresponding to (r, θ) in the table.

θ	0	$\frac{\pi}{12}$	$\frac{\pi}{6}$	$\frac{\pi}{4}$	$\frac{\pi}{3}$	$\frac{5\pi}{12}$	$\cdots$	$\frac{11\pi}{12}$	π
r									

33. CAS Plot the **cissoid** $r = 2\sin\theta\tan\theta$ and show that its equation in rectangular coordinates is $y^2 = \dfrac{x^3}{2 - x}$.

34. Prove that $r = 2a\cos\theta$ is the equation of the circle in Figure 22 using only the fact that a triangle inscribed in a circle with one side a diameter is a right triangle.

35. Show that $r = a\cos\theta + b\sin\theta$ is the equation of a circle passing through the origin. Express the radius and center (in rectangular coordinates) in terms of a and b.

36. Use the previous exercise to write the equation of the circle of radius 5 and center $(3, 4)$ in the form $r = a\cos\theta + b\sin\theta$.

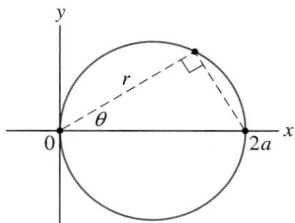

FIGURE 22

37. Use the identity $\cos 2\theta = \cos^2\theta - \sin^2\theta$ to find a polar equation of the hyperbola $x^2 - y^2 = 1$.

38. Find an equation in rectangular coordinates for the curve $r^2 = \cos 2\theta$.

39. Show that $\cos 3\theta = \cos^3\theta - 3\cos\theta\sin^2\theta$ and use this identity to find an equation in rectangular coordinates for the curve $r = \cos 3\theta$.

40. Use the addition formula for the cosine to show that the line $\mathcal{L}$ with polar equation $r\cos(\theta - \alpha) = d$ has the equation in rectangular coordinates $(\cos\alpha)x + (\sin\alpha)y = d$. Show that $\mathcal{L}$ has slope $m = -\cot\alpha$ and y-intercept $\dfrac{d}{\sin\alpha}$.

In Exercises 41–45, find an equation in polar coordinates of the line $\mathcal{L}$ with given description.

41. The point on $\mathcal{L}$ closest to the origin has polar coordinates $\left(2, \frac{\pi}{9}\right)$.

42. The point on $\mathcal{L}$ closest to the origin has rectangular coordinates $(-2, 2)$.

43. $\mathcal{L}$ is tangent to the circle $r = 2\sqrt{10}$ at the point with rectangular coordinates $(-2, -6)$.

44. $\mathcal{L}$ has slope 3 and is tangent to the unit circle in the fourth quadrant.

45. $y = 4x - 9$.

46. Show that the polar equation of the line $y = ax + b$ can be written in the form

$$r = \frac{b}{\sin\theta - a\cos\theta}$$

47. Distance Formula Use the Law of Cosines (Figure 23) to show that the distance d between two points with polar coordinates (r, θ) and (r_0, θ_0) is

$$d^2 = r^2 + r_0^2 - 2rr_0\cos(\theta - \theta_0) \qquad \boxed{2}$$

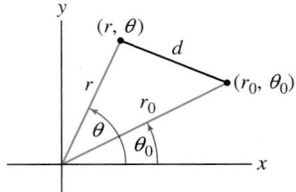

FIGURE 23

48. Use the distance formula (2) to show that the circle of radius 9 whose center has polar coordinates $\left(5, \frac{\pi}{4}\right)$ has equation:

$$r^2 - 10r \cos\left(\theta - \frac{\pi}{4}\right) = 56$$

49. Show that the cardiod $r = 1 + \sin\theta$ has equation

$$x^2 + y^2 = (x^2 + y^2 - x)^2$$

50. For $a > 0$, a **lemniscate curve** is the set of points P such that the product of the distances from P to $(a, 0)$ and $(-a, 0)$ is a^2. Show that the equation of the lemniscate is:

$$(x^2 + y^2)^2 = 2a^2(x^2 - y^2)$$

Then find the equation in polar coordinates. To obtain the simplest form of the equation, use the identity $\cos 2\theta = \cos^2\theta - \sin^2\theta$. Plot the lemniscate for $a = 2$ if you have a computer algebra system.

51. The Derivative in Polar Coordinates   A polar curve $r = f(\theta)$ has parametric equations (since $x = r\cos\theta$ and $y = r\sin\theta$):

$$x = f(\theta)\cos\theta, \quad y = f(\theta)\sin\theta$$

Apply Theorem 1 of Section 11.1 to prove the formula

$$\frac{dy}{dx} = \frac{f(\theta)\cos\theta + f'(\theta)\sin\theta}{-f(\theta)\sin\theta + f'(\theta)\cos\theta} \qquad \boxed{3}$$

where $f'(\theta) = df/d\theta$.

52. Use Eq. (3) to find the slope of the tangent line to $r = \theta$ at $\theta = \frac{\pi}{2}$ and $\theta = \pi$.

53. Find the equation in rectangular coordinates of the tangent line to $r = 4\cos 3\theta$ at $\theta = \frac{\pi}{6}$.

54. Show that for the circle $r = \sin\theta + \cos\theta$,

$$\frac{dy}{dx} = \frac{\cos 2\theta + \sin 2\theta}{\cos 2\theta - \sin 2\theta}$$

Calculate the slopes of the tangent lines at points A, B, C in Figure 20 and find the polar coordinates of the points at which the tangent line is horizontal.

Further Insights and Challenges

55. ✏️   Let c be a fixed constant. Explain the relationship between the graphs of:

(a) $y = f(x + c)$ and $y = f(x)$ (rectangular)

(b) $r = f(\theta + c)$ and $r = f(\theta)$ (polar)

(c) $y = f(x) + c$ and $y = f(x)$ (rectangular)

(d) $r = f(\theta) + c$ and $r = f(\theta)$ (polar)

56. ✏️   Let $f(x)$ be a periodic function of period 2π, that is, $f(x) = f(x + 2\pi)$. Explain how this periodicity is reflected in the graph of:

(a) $y = f(x)$ in rectangular coordinates

(b) $r = f(\theta)$ in polar coordinates

57. [GU]   Use a graphing utility to convince yourself that graphs of the polar equations $r = f_1(\theta) = 2\cos\theta - 1$ and $r = f_2(\theta) = 2\cos\theta + 1$ have the same graph. Then explain why. *Hint:* Show that the points $(f_1(\theta + \pi), \theta + \pi)$ and $(f_2(\theta), \theta)$ coincide.

58. 𝄢𝗔𝗦   Plot the limaçon curves $r = a + \cos\theta$ for several values of a.

(a) Describe how the shape changes as a increases. What happens for $a < 0$?

(b) A closed curve is called convex if, for any two points P and Q in the interior of the curve, the segment $\overline{PQ}$ is also contained in the interior. Figure 24 shows that the limaçon is convex if a is large and is not convex if a is small. Experiment with a computer algebra system to estimate the smallest value a for which it is convex.

(c) Use Eq. (3) of Exercise 51 to show that if the tangent line is vertical, then either $\theta = 0$, π or $\cos\theta = -a/2$.

(d) Show that the limaçon has three vertical tangent lines to the left of the y-axis if $1 < a < 2$ and a unique vertical tangent line to the left of the y-axis there if $a \geq 2$. What does this suggest about the convexity of the limaçon for $a > 2$?

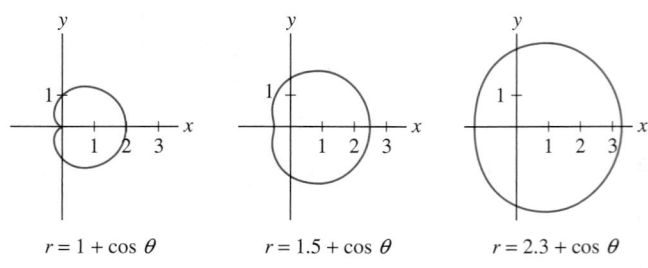

$r = 1 + \cos\theta$ $r = 1.5 + \cos\theta$ $r = 2.3 + \cos\theta$

FIGURE 24

11.4 Area and Arc Length in Polar Coordinates

In this section, we derive the formulas for area and arc length in polar coordinates. First, we compute the area of a sector bounded by a curve $r = f(\theta)$ and two rays $\theta = \alpha$ and $\theta = \beta$ [shaded region in Figure 1(A)]. Divide the region into N narrow sectors of angle

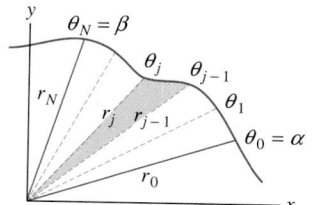

(A) Region defined by $\alpha \le \theta \le \beta$

(B) Region divided into narrow sectors

FIGURE 1 Area bounded by the curve $r = f(\theta)$ and the two rays $\theta = \alpha$ and $\theta = \beta$.

Rectangular and polar coordinates are suited for different kinds of area calculations. Rectangular coordinates should be used to compute the signed area under the graph $y = f(x)$ over an interval $[a, b]$. Polar coordinates should be used to calculate the area bounded by a curve in an angular sector $\alpha \le \theta \le \beta$.

$\Delta\theta = \frac{\beta - \alpha}{N}$ corresponding to a partition of the interval $[\alpha, \beta]$:

$$\theta_0 = \alpha \le \theta_1 \le \theta_2 \cdots \le \theta_N = \beta$$

Recall that a circular sector of width $\Delta\theta$ and radius r has area $\frac{1}{2}r^2 \Delta\theta$ (Figure 2). Each narrow sector in our region is nearly circular, of radius $r_j = f(\theta_j)$, and therefore has area approximately equal to $\frac{1}{2}r_j^2 \Delta\theta$ (Figure 3). The total area is approximated by the sum:

$$\text{Area of region} \approx \sum_{j=1}^{N} \frac{1}{2}r_j^2 \Delta\theta = \sum_{j=1}^{N} \frac{1}{2}f(\theta_j)^2 \Delta\theta \qquad \boxed{1}$$

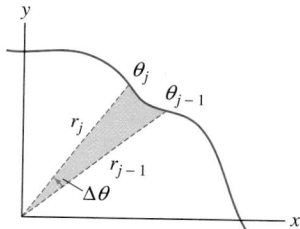

FIGURE 2 The area of a circular sector is *exactly* $\frac{1}{2}r^2 \Delta\theta$.

FIGURE 3 The area of the jth sector is *approximately* $\frac{1}{2}r_j^2 \Delta\theta$.

This is a Riemann sum for the integral $\int_{\alpha}^{\beta} \frac{1}{2}f(\theta)^2 \, d\theta$. If $f(\theta)$ is continuous, the sum approaches the integral as $N \to \infty$, and we obtain the following formula.

CAUTION In rectangular coordinates, the integral computes signed area, but in polar coordinates, formula (2) gives the actual (positive) area, not signed area. The formula remains valid if $f(\theta)$ takes on negative values.

> **THEOREM 1 Area in Polar Coordinates**   If $f(\theta)$ is continuous function, then the area bounded by a curve in polar form $r = f(\theta)$ and the rays $\theta = \alpha$ and $\theta = \beta$ is equal to
>
> $$\frac{1}{2}\int_{\alpha}^{\beta} r^2 \, d\theta = \frac{1}{2}\int_{\alpha}^{\beta} f(\theta)^2 \, d\theta \qquad \boxed{2}$$

As a check on this formula, note that the graph of $f(r) = R$ is a circle of radius R. According to Eq. (2), the area of the circle is $\frac{1}{2}\int_{0}^{2\pi} R^2 \, d\theta = \frac{1}{2}R^2(2\pi) = \pi R^2$, as expected.

■ **EXAMPLE 1** Find the area of the right semicircle with equation $r = 4\sin\theta$.

◀┄ **REMINDER** In Eq. (4), we use the identity

$$\sin^2\theta = \frac{1}{2}(1 - \cos 2\theta) \qquad \boxed{3}$$

to integrate $y = \sin^2\theta$.

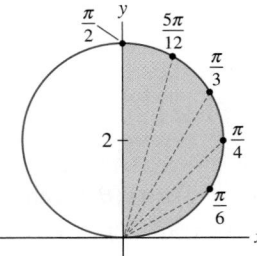

FIGURE 4 The circle with equation $r = 4\sin\theta$.

Solution The equation $r = 4\sin\theta$ defines a circle of radius 2 tangent to the x-axis at the origin (Figure 4). The right semicircle is "swept out" as θ varies from 0 to $\frac{\pi}{2}$. By Eq. (2), the area of the semicircle is

$$\frac{1}{2}\int_0^{\pi/2} r^2\,d\theta = \frac{1}{2}\int_0^{\pi/2}(4\sin\theta)^2\,d\theta = 8\int_0^{\pi/2}\sin^2\theta\,d\theta \qquad \boxed{4}$$

$$= 8\int_0^{\pi/2}\frac{1}{2}(1 - \cos 2\theta)\,d\theta$$

$$= (4\theta - 2\sin 2\theta)\Big|_0^{\pi/2} = 4\left(\frac{\pi}{2}\right) - 0 = 2\pi$$

This is the expected result since a circle of radius $R = 2$ has total area 4π. ∎

CONCEPTUAL INSIGHT Keep in mind that the integral $\dfrac{1}{2}\displaystyle\int_\alpha^\beta r^2\,d\theta$ does not compute the area *under* a curve, but rather the area "swept out" by a radial segment as θ varies from α to β. For example, integral $\dfrac{1}{2}\displaystyle\int_0^{\pi/3}(4\sin\theta)^2\,d\theta$ is equal to the area of the shaded region in Figure 5(A), not the area under the curve in Figure 5(B).

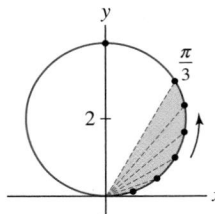

(A) Region swept out by radial segment has area
$$\frac{1}{2}\int_0^{\pi/3} r^2\,d\theta$$

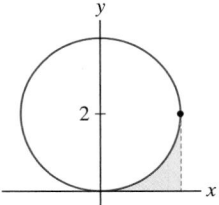

(B) Region under the curve.

FIGURE 5

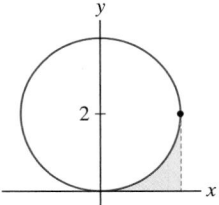

FIGURE 6 Graph of $r = \sin 3\theta$ as a function of θ.

■ **EXAMPLE 2** Sketch the graph of $r = \sin 3\theta$ and compute the area of one "petal."

Solution To sketch the curve, we first graph $r = \sin 3\theta$ in rectangular coordinates. Figure 6 shows that as θ varies from 0 to $\frac{\pi}{3}$, the radius r varies from 0 to 1, and then back to 0. This gives petal A (Figure 7). Petal B is traced as θ varies from $\frac{\pi}{3}$ to $\frac{2\pi}{3}$ (with $r \leq 0$), and petal C for $\frac{2\pi}{3} \leq \theta \leq \pi$. We compute the area of petal A using identity (3):

$$\frac{1}{2}\int_0^{\pi/3}(\sin 3\theta)^2\,d\theta = \frac{1}{4}\int_0^{\pi/3}(1 - \cos 6\theta)\,d\theta = \left(\frac{1}{4}\theta - \frac{1}{24}\sin 6\theta\right)\Big|_0^{\pi/3} = \frac{\pi}{12} \quad ∎$$

If $r = f_1(\theta)$ and $r = f_2(\theta)$ are two polar curves with $f_2(\theta) \geq f_1(\theta)$, then we may compute the area of the region between the two curves within the sector $\alpha \leq \theta \leq \beta$ by the integral (Figure 8):

$$\boxed{\text{Area between two curves} = \frac{1}{2}\int_\alpha^\beta \left(f_2(\theta)^2 - f_1(\theta)^2\right)d\theta \qquad \boxed{5}}$$

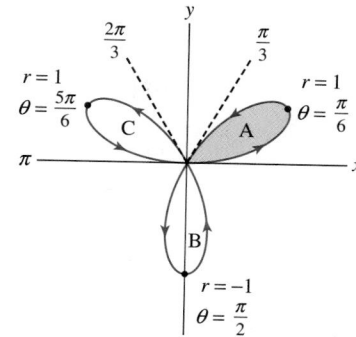

FIGURE 7 The "rose with three petals" $r = \sin 3\theta$.

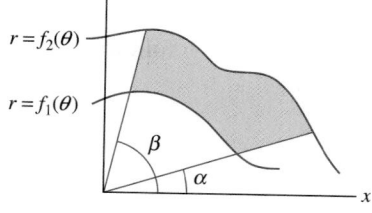

FIGURE 8 Area between two polar graphs in a sector.

■ **EXAMPLE 3** Area Between Two Curves   Find the area of the region inside the circle $r = 2\cos\theta$ but outside the circle $r = 1$ [Figure 9A].

Solution Recall that $r = 2a\cos\theta$ is the polar equation of the circle of radius a with center $(a, 0)$ (Section 11.3, Example 7). Thus $r = 2\cos\theta$ is a circle of radius 1 with center $(1, 0)$. We wish to compute the area of the region within this circle and outside the circle $r = 1$ of radius 1 centered at the origin. This is the region labeled (I) in Figure 9(A). The two circles intersect at the points where $r = 2\cos\theta = 1$ or $\cos\theta = \frac{1}{2}$. Thus the intersection points occur at $\theta = \pm\frac{\pi}{3}$. The area of region (I) is equal to

$$\text{Area of (I)} = \text{area of (II)} - \text{area of (III)}$$

←·· **REMINDER** In (6), we use the identity
$$\cos^2\theta = \frac{1}{2}(1 + \cos 2\theta)$$

$$= \frac{1}{2}\int_{-\pi/3}^{\pi/3}(2\cos\theta)^2\,d\theta - \frac{1}{2}\int_{-\pi/3}^{\pi/3}(1)^2\,d\theta$$

$$= \frac{1}{2}\int_{-\pi/3}^{\pi/3}(4\cos^2\theta - 1)\,d\theta = \frac{1}{2}\int_{-\pi/3}^{\pi/3}(2\cos 2\theta + 1)\,d\theta \qquad \boxed{6}$$

$$= \frac{1}{2}(\sin 2\theta + \theta)\Big|_{-\pi/3}^{\pi/3} = \frac{\sqrt{3}}{2} + \frac{\pi}{3} \approx 1.91 \qquad ■$$

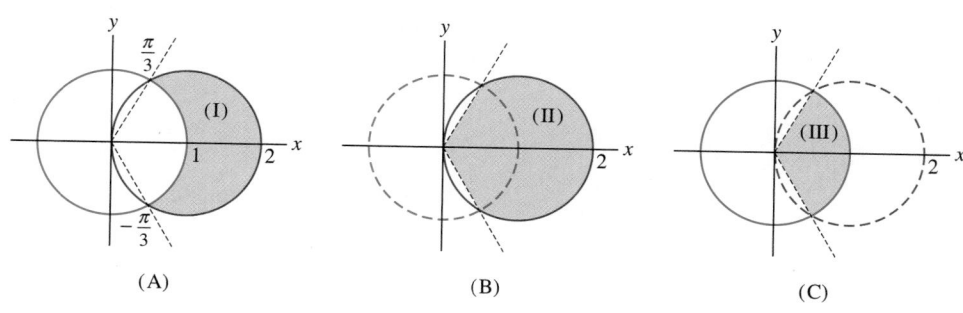

FIGURE 9 Region I is the difference of regions II and III.

(A)   (B)   (C)

To derive a formula for arc length in polar coordinates, we observe that a polar curve $r = f(\theta)$ has a natural parametrization with θ as a parameter:

$$x = r \cos \theta = f(\theta) \cos \theta, \qquad y = r \sin \theta = f(\theta) \sin \theta$$

Using a prime to denote the derivative with respect to θ, we have

$$x'(\theta) = \frac{dx}{d\theta} = -f(\theta) \sin \theta + f'(\theta) \cos \theta$$

$$y'(\theta) = \frac{dy}{d\theta} = f(\theta) \cos \theta + f'(\theta) \sin \theta$$

We check using algebra that $x'(\theta)^2 + y'(\theta)^2 = f(\theta)^2 + f'(\theta)^2$. Now recall from Section 11.2 that the length s of the arc traced out for $\alpha \le \theta \le \beta$ is $\int_\alpha^\beta \sqrt{x'(\theta)^2 + y'(\theta)^2}\, d\theta$. It follows that

$$\boxed{\text{Arc length } s = \int_\alpha^\beta \sqrt{f(\theta)^2 + f'(\theta)^2}\, d\theta} \qquad \boxed{7}$$

■ **EXAMPLE 4** Find the total length of the circle with equation $r = 2a \cos \theta$ for $a > 0$.

Solution In this case, $f(\theta) = 2a \cos \theta$ and

$$f(\theta)^2 + f'(\theta)^2 = 4a^2 \cos^2 \theta + 4a^2 \sin^2 \theta = 4a^2$$

The total length of this circle of radius a has the expected value:

$$\int_0^\pi \sqrt{f(\theta)^2 + f'(\theta)^2}\, d\theta = \int_0^\pi (2a)\, d\theta = 2\pi a$$

Note that the upper limit of integration is π rather than 2π because the entire circle is traced out as θ varies from 0 to π (see Figure 10). ■

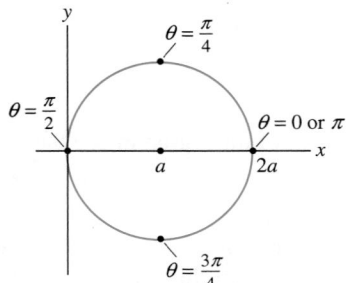

FIGURE 10 Graph of $r = 2a \cos \theta$.

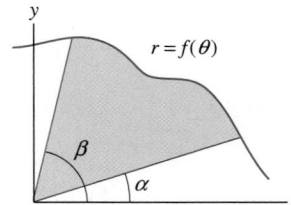

FIGURE 11 Region bounded by the polar curve $r = f(\theta)$ and the rays $\theta = \alpha$, $\theta = \beta$.

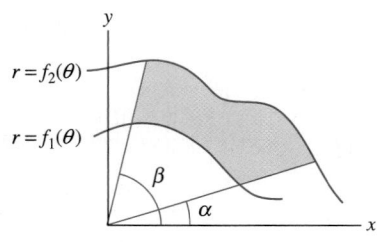

FIGURE 12 Region between two polar curves.

11.4 SUMMARY

- Polar coordinates are suited to calculating areas bounded by a polar curve $r = f(\theta)$ and two rays $\theta = \alpha$ and $\theta = \beta$ (Figure 11):

$$\text{Area} = \frac{1}{2} \int_\alpha^\beta f(\theta)^2\, d\theta$$

- The area between the polar curves $r = f_1(\theta)$ and $r = f_2(\theta)$, where $f_2(\theta) \ge f_1(\theta)$, is equal to (Figure 12)

$$\text{Area} = \frac{1}{2} \int_\alpha^\beta \left(f_2(\theta)^2 - f_1(\theta)^2 \right) d\theta$$

- The arc length of the polar curve $r = f(\theta)$ for $\alpha \le \theta \le \beta$ is

$$\text{Arc length} = \int_\alpha^\beta \sqrt{f(\theta)^2 + f'(\theta)^2}\, d\theta$$

11.4 EXERCISES

Preliminary Questions

1. True or False: The area under the curve with polar equation $r = f(\theta)$ is equal to the integral of $f(\theta)$.

2. Polar coordinates are best suited to finding the area (choose one):
(a) Under a curve between $x = a$ and $x = b$.
(b) Bounded by a curve and two rays through the origin.

3. True or False: The formula for area in polar coordinates is valid only if $f(\theta) \geq 0$.

4. The horizontal line $y = 1$ has polar equation $r = \csc \theta$. Which area is represented by the integral $\dfrac{1}{2} \displaystyle\int_{\pi/6}^{\pi/2} \csc^2 \theta \, d\theta$ (Figure 13)?

(a) $\square ABCD$ **(b)** $\triangle ABC$ **(c)** $\triangle ACD$

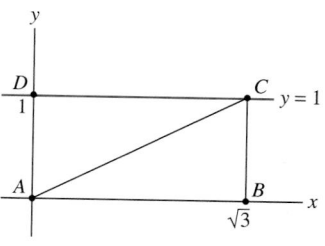

FIGURE 13

Exercises

1. Sketch the area bounded by the circle $r = 5$ and the rays $\theta = \frac{\pi}{2}$ and $\theta = \pi$, and compute its area as an integral in polar coordinates.

2. Sketch the region bounded by the line $r = \sec \theta$ and the rays $\theta = 0$ and $\theta = \frac{\pi}{3}$. Compute its area in two ways: as an integral in polar coordinates and using geometry.

3. Calculate the area of the circle $r = 4 \sin \theta$ as an integral in polar coordinates (see Figure 4). Be careful to choose the correct limits of integration.

4. Compute the area of the shaded region in Figure 14 as an integral in polar coordinates.

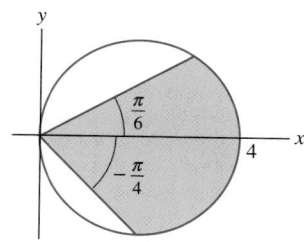

FIGURE 14 Graph of $r = 4 \cos \theta$.

5. Find the total area enclosed by the cardioid $r = 1 - \cos \theta$ (Figure 15).

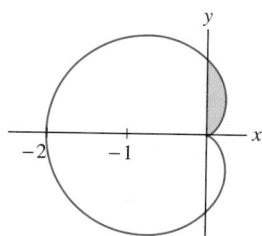

FIGURE 15 The cardioid $r = 1 - \cos \theta$.

6. Find the area of the shaded region in Figure 15.

7. Find the area of one leaf of the "four-petaled rose" $r = \sin 2\theta$ (Figure 16).

8. Prove that the total area of the four-petaled rose $r = \sin 2\theta$ is equal to one-half the area of the circumscribed circle (Figure 16).

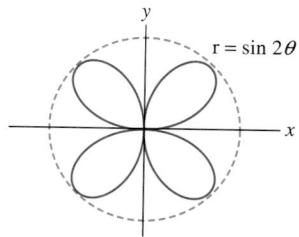

FIGURE 16 Graph of four-petaled rose $r = \sin 2\theta$.

9. Find the area enclosed by one loop of the lemniscate with equation $r^2 = \cos 2\theta$ (Figure 17). Choose your limits of integration carefully.

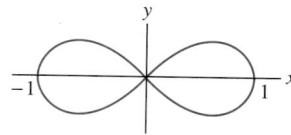

FIGURE 17 The lemniscate $r^2 = \cos 2\theta$.

10. Sketch the spiral $r = \theta$ for $0 \leq \theta \leq 2\pi$ and find the area bounded by the curve and the first quadrant.

11. Find the area enclosed by the cardioid $r = a(1 + \cos \theta)$, where $a > 0$.

12. Find the area of the intersection of the circles $r = \sin \theta$ and $r = \cos \theta$.

13. Find the area of region A in Figure 18.

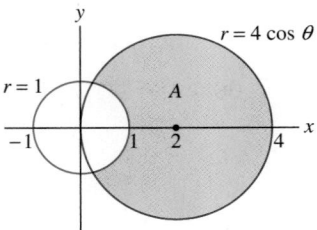

FIGURE 18

14. Find the area of the shaded region in Figure 19, enclosed by the circle $r = \frac{1}{2}$ and a petal of the curve $r = \cos 3\theta$. *Hint:* Compute the area of both the petal and the region inside the petal and outside the circle.

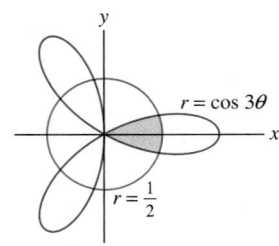

FIGURE 19

15. Find the area of the inner loop of the limaçon with polar equation $r = 2\cos\theta - 1$ (Figure 20).

16. Find the area of the region between the inner and outer loop of the limaçon $r = 2\cos\theta - 1$.

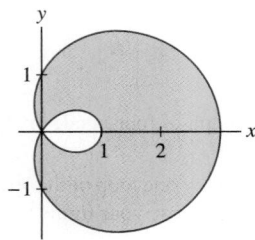

FIGURE 20 The limaçon $r = 2\cos\theta - 1$.

17. Find the area of the part of the circle $r = \sin\theta + \cos\theta$ in the fourth quadrant (see Exercise 27 in Section 11.3).

18. Compute the area of the shaded region in Figure 21.

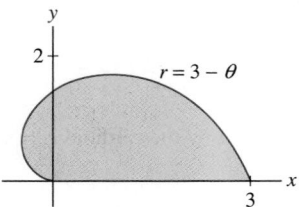

FIGURE 21

19. Find the area between the two curves in Figure 22(A).

20. Find the area between the two curves in Figure 22(B).

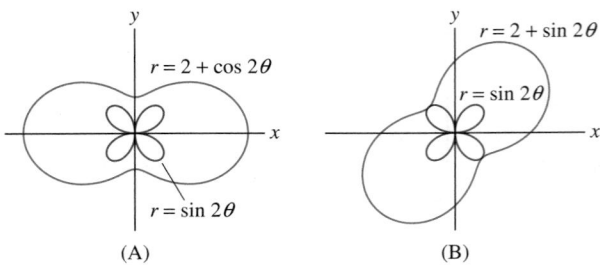

(A)   (B)

FIGURE 22

21. Find the area inside both curves in Figure 23.

22. Find the area of the region that lies inside one but not both of the curves in Figure 23.

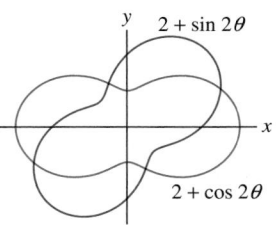

FIGURE 23

23. Figure 24 suggests that the circle $r = \sin\theta$ lies inside the spiral $r = \theta$. Which inequality from Chapter 2 assures us that this is the case? Find the area between the curves $r = \theta$ and $r = \sin\theta$ in the first quadrant.

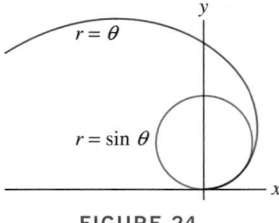

FIGURE 24

24. Calculate the total length of the circle $r = 4\sin\theta$ as an integral in polar coordinates.

25. Find the length of the spiral $r = \theta$ for $0 \le \theta \le A$.

26. Find the length of $r = \theta^2$ for $0 \le \theta \le \pi$.

27. Sketch the segment $r = \sec\theta$ for $0 \le \theta \le A$. Then compute its length in two ways: as an integral in polar coordinates and using trigonometry.

28. Sketch the circle $r = \sin\theta + \cos\theta$. Then compute its length in two ways: as an integral in polar coordinates and using trigonometry.

29. Find the length of the cardioid $r = 1 + \cos\theta$. *Hint:* Use the identity $1 + \cos\theta = 2\cos^2\left(\dfrac{\theta}{2}\right)$ to evaluate the arc length integral.

30. Find the length of the cardioid with equation $r = 1 + \cos\theta$ located in the first quadrant.

31. Find the length of the *equiangular spiral* $r = e^\theta$ for $0 \le \theta \le 2\pi$.

32. Find the length of the curve $r = \cos^2\theta$.

In Exercises 33–36, express the length of the curve as an integral but do not evaluate it.

33. $r = e^{a\theta}, \quad 0 \le \theta \le \pi$

34. $r = \sin 2\theta, \quad 0 \le \theta \le \pi$

35. $r = (2 - \cos\theta)^{-1}, \quad 0 \le \theta \le 2\pi$

36. The inner loop of $r = 2\cos\theta - 1$ (see Exercise 15)

Further Insights and Challenges

37. Suppose that the polar coordinates of a moving particle at time t are $(r(t), \theta(t))$. Prove that the particle's speed is equal to
$$\sqrt{(dr/dt)^2 + r^2(d\theta/dt)^2}.$$

38. Compute the speed at time $t = 1$ of a particle whose polar coordinates at time t are $r = t$, $\theta = t$ (use Exercise 37). What would the speed be if the particle's rectangular coordinates are $x = t$, $y = t$? Why is the speed increasing in one case and constant in the other?

11.5 Conic Sections

The three familiar families of curves—ellipses, hyperbolas, and parabolas—appear throughout mathematics and its applications. They were first studied by the ancient Greek mathematicians, who recognized that they are obtained by intersecting a plane with a cone (Figure 1) and hence referred to them as "conic sections." Our main goal in this section is to derive the equations of the conic sections from their geometric definitions as curves in the plane.

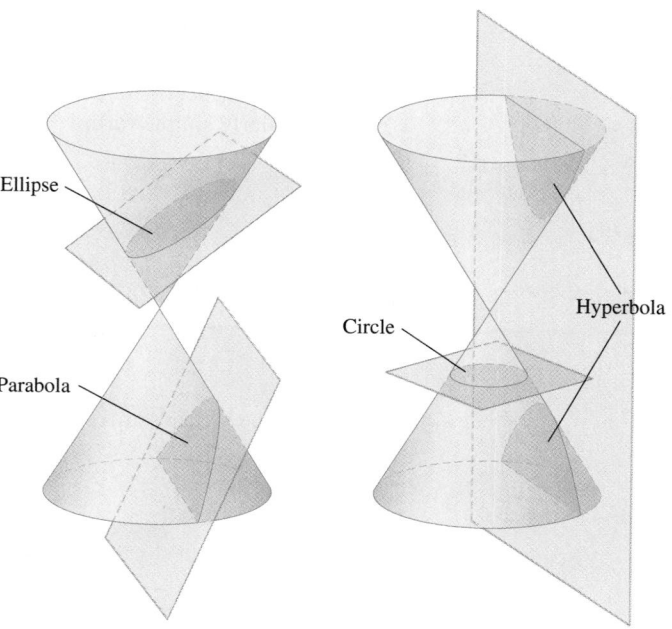

FIGURE 1 The conic sections are obtained by intersecting a plane and a cone.

A circle is an ellipse whose foci coincide. If $F_1 = F_2$, Eq. (1) reduces to $PF_1 = K/2$, which defines a circle of radius $K/2$ and center F_1.

An ellipse is the set of all points P such that the sum of the distances to two fixed points F_1 and F_2 is a constant $K > 0$:

$$PF_1 + PF_2 = K \qquad \boxed{1}$$

The points F_1 and F_2 are called the **foci** (plural of "focus") of the ellipse. We assume that K is greater than the distance $\overline{F_1 F_2}$ between the foci (if $K = F_1 F_2$, the ellipse reduces to the line segment $\overline{F_1 F_2}$; if $K < F_1 F_2$, the ellipse has no points). Figure 2(A) shows an ellipse with foci at F_1 and F_2.

We use the following terminology:

- The midpoint of $\overline{F_1 F_2}$ is the **center** of the ellipse.
- The line through the foci is the **focal axis**.
- The line through the center perpendicular to the focal axis is the **conjugate axis**.

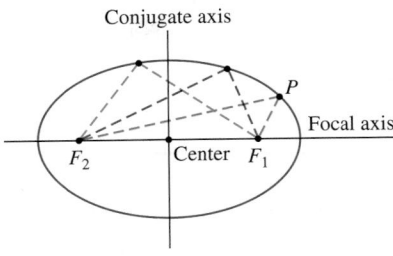

(A) The ellipse consists of all points P such that $PF_1 + PF_2 =$ constant.

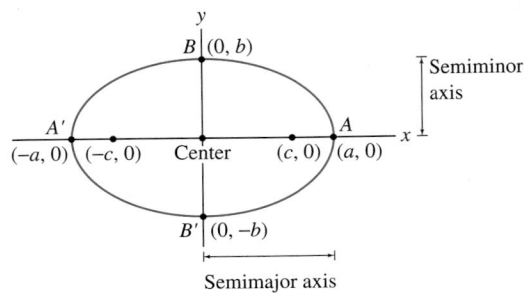

(B) An ellipse in standard position with equation
$$\left(\frac{x}{a}\right)^2 + \left(\frac{y}{b}\right)^2 = 1.$$

FIGURE 2

The ellipse is said to be in **standard position** if the focal and conjugate axes are the x- and y-axes, as shown in Figure 2(B). In this case, the foci have coordinates $F_1 = (c, 0)$ and $F_2 = (-c, 0)$ for some $c > 0$. Let us prove that the equation of this ellipse has the particularly simple form

$$\left(\frac{x}{a}\right)^2 + \left(\frac{x}{b}\right)^2 = 1 \qquad \boxed{2}$$

where $a = K/2$ and $b = \sqrt{a^2 - c^2}$. By definition, a point $P = (x, y)$ lies on the ellipse if $PF_1 + PF_2 = 2a$ and therefore, by the distance formula,

$$\sqrt{(x + c)^2 + y^2} + \sqrt{(x - c)^2 + y^2} = 2a \qquad \boxed{3}$$

Move the second term on the left over to the right and square both sides:

$$(x + c)^2 + y^2 = 4a^2 - 4a\sqrt{(x - c)^2 + y^2} + (x - c)^2 + y^2$$

$$4a\sqrt{(x - c)^2 + y^2} = 4a^2 + (x - c)^2 - (x + c)^2 = 4a^2 - 4cx$$

Now divide by 4, square, and simplify:

Strictly speaking, we must also show that every point (x, y) satisfying (4) also satisfies (3). When we begin with (4) and reverse the algebraic steps, the process of taking square roots leads to the relation

$$\sqrt{(x-c)^2 + y^2} \pm \sqrt{(x+c)^2 + y^2} = \pm 2a$$

However, since $a > c$, this equation cannot hold unless both signs are positive.

$$a^2(x^2 - 2cx + c^2 + y^2) = a^4 - 2a^2cx + c^2x^2$$

$$(a^2 - c^2)x^2 + a^2y^2 = a^4 - a^2c^2 = a^2(a^2 - c^2)$$

$$\frac{x^2}{a^2} + \frac{y^2}{a^2 - c^2} = 1 \qquad \boxed{4}$$

This is Eq. (2) with $b^2 = a^2 - c^2$ as claimed.

The axes intersect the ellipse in four points called **vertices** [labeled A, A', B, and B' in Figure 2(B)]. The segments $\overline{AA'}$ and $\overline{BB'}$ are called the major and minor axes of the ellipse. Following common usage, the numbers a and b are referred to as the **semimajor axis** and **semiminor axis** (even though they are numbers rather than axes).

THEOREM 1   Equation of an Ellipse in Standard Position   Let $a, b > 0$ be constants with $a > b > 0$, and set $c = \sqrt{a^2 - b^2}$. Then

$$\left(\frac{x}{a}\right)^2 + \left(\frac{y}{b}\right)^2 = 1 \qquad \boxed{5}$$

is an equation of the ellipse

$$PF_1 + PF_2 = 2a$$

with foci $F_1 = (c, 0)$ and $F_2 = (-c, 0)$. Furthermore, the ellipse has:

- Semimajor axis a, semiminor axis b.
- Vertices $(\pm a, 0)$, $(0, \pm b)$.

If $b > a > 0$, then (5) defines an ellipse with foci $(0, \pm c)$, where $c = \sqrt{b^2 - a^2}$.

■ **EXAMPLE 1** Find the equation of the ellipse with foci $(\pm\sqrt{11}, 0)$ and semimajor axis $a = 6$. Then find the semiminor axis and sketch the graph.

Solution The foci are $(\pm c, 0)$ with $c = \sqrt{11}$ and the semimajor axis is $a = 6$, so we may use the relation $c = \sqrt{a^2 - b^2}$ to find b:

$$b^2 = a^2 - c^2 = 6^2 - (\sqrt{11})^2 = 25$$

Thus, the semiminor axis is $b = 5$ and the ellipse has equation $\left(\frac{x}{6}\right)^2 + \left(\frac{y}{5}\right)^2 = 1$. To sketch an ellipse, plot the vertices $(\pm 6, 0)$ and $(0, \pm 5)$ and connect them as in Figure 3. ■

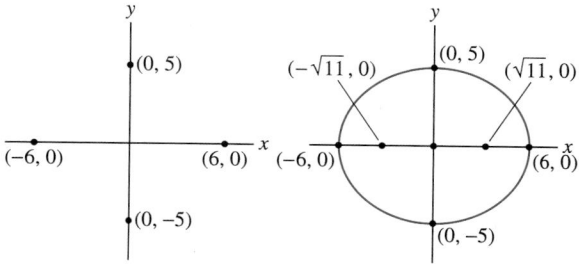

FIGURE 3 To sketch an ellipse, first draw the vertices.

We can easily write down the equation of an ellipse with axes parallel to the x- and y-axes and center translated to the point $C = (x_0, y_0)$ (Figure 4). The equation of the

$$\left(\frac{x-6}{3}\right)^2 + \left(\frac{y-7}{5}\right)^2 = 1$$

(6, 12) (6, 11)

$C = (6, 7)$

(6, 3)

(6, 2)

(0, 4)

$$\left(\frac{x}{3}\right)^2 + \left(\frac{y}{5}\right)^2 = 1$$

-5 (0, −4)

FIGURE 4 An ellipse with vertical major axis and its translate with center $C = (6, 7)$.

translated ellipse is

$$\left(\frac{x - x_0}{a}\right)^2 + \left(\frac{y - y_0}{b}\right)^2 = 1$$

■ **EXAMPLE 2** **Translating the Center of an Ellipse** Find the equation of the ellipse with vertical focal axis, center at $C = (6, 7)$, semimajor axis 5, and semiminor axis 3. Where are the foci located?

Solution First, we find the equation of the ellipse centered at the origin. Since the focal axis is vertical, we use Eq. (5) with $a < b$. We set $a = 3$ and $b = 5$ to obtain $\left(\frac{x}{3}\right)^2 + \left(\frac{y}{5}\right)^2 = 1$. Furthermore, $c = \sqrt{b^2 - a^2} = \sqrt{5^2 - 3^2} = 4$, so the foci are $F_1 = (0, 4)$ and $F_2 = (0, -4)$ on the y-axis. When we translate the ellipse so that its center is $(6, 7)$, the equation becomes

$$\left(\frac{x - 6}{3}\right)^2 + \left(\frac{y - 7}{5}\right)^2 = 1$$

The foci are still ± 4 vertical units from the center, so the foci of the translated ellipse are $F_1 = (6, 11)$ and $F_2 = (6, 3)$ (Figure 4). ■

A hyperbola is the set of all points P such that the difference of the distances to two fixed points F_1 and F_2 is $\pm K$, where $K > 0$ is a constant:

$$PF_1 - PF_2 = \pm K \qquad \boxed{6}$$

The points F_1 and F_2 are called the foci of the hyperbola. We assume that K is less than the distance $F_1 F_2$ between the foci. Note that a hyperbola consists of two branches corresponding to the choices of sign $\pm$ (Figure 5).

As before, the midpoint of $\overline{F_1 F_2}$ is the center of the hyperbola, the line through F_1 and F_2 is the focal axis, and the line through the center perpendicular to the focal axis is the conjugate axis. The vertices are the points where the focal axis intersects the hyperbola, labeled A and A' in Figure 5. The hyperbola is said to be in standard position

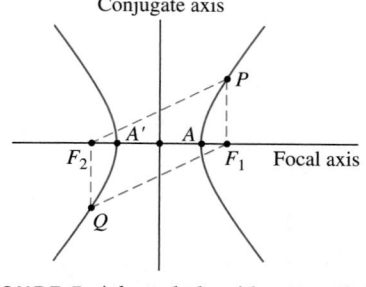

Conjugate axis

P

A' A

F_2 F_1 Focal axis

Q

FIGURE 5 A hyperbola with center $(0, 0)$.

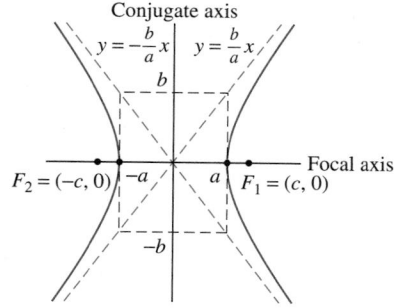

Conjugate axis

$y = -\dfrac{b}{a}x$ $y = \dfrac{b}{a}x$

b

$F_2 = (-c, 0)$ $-a$ a $F_1 = (c, 0)$ Focal axis

$-b$

FIGURE 6 Hyperbola in standard position.

when the focal and conjugate axes are the x- and y-axes as in Figure 6. In this case, the foci are $(\pm c, 0)$ for some $c > 0$ and we can verify that the equation of the hyperbola has the simple form (7) below, with $a = K/2$ and $b = \sqrt{c^2 - a^2}$. The vertices are $A = (a, 0)$ and $A' = (-a, 0)$ (see Exercise 64).

THEOREM 2 Equation of a Hyperbola in Standard Position Let a, b be positive constants and set $c = \sqrt{a^2 + b^2}$. Then

$$\left(\frac{x}{a}\right)^2 - \left(\frac{y}{b}\right)^2 = 1 \qquad \boxed{7}$$

is the equation of the hyperbola

$$PF_1 - PF_2 = \pm 2a$$

with foci $F_1 = (c, 0)$ and $F_2 = (-c, 0)$.

A hyperbola has two **asymptotes** that may be determined by drawing the rectangle whose sides pass through $(\pm a, 0)$ and $(0, \pm b)$ as in Figure 6. Let us prove that the asymptotes are the diagonals of this rectangle, that is, the lines with equations $y = \pm \dfrac{b}{a}x$. Consider a point (x, y) on the hyperbola in the first quadrant. Thus we assume that $x > 0$ and $y > 0$. Equation (7) yields

$$y = \sqrt{\frac{b^2}{a^2}x^2 - b^2} = \frac{b}{a}\sqrt{x^2 - a^2}$$

To show that the point (x, y) on the hyperbola approaches the line $y = \dfrac{b}{a}x$ as $x \to \infty$, we verify that the following limit is zero:

$$\lim_{x \to \infty}\left(y - \frac{b}{a}x\right) = \frac{b}{a}\lim_{x \to \infty}\left(\sqrt{x^2 - a^2} - x\right)$$

$$= \frac{b}{a}\lim_{x \to \infty}\left(\sqrt{x^2 - a^2} - x\right)\left(\frac{\sqrt{x^2 - a^2} + x}{\sqrt{x^2 - a^2} + x}\right)$$

$$= \frac{b}{a}\lim_{x \to \infty}\left(\frac{-a^2}{\sqrt{x^2 - a^2} + x}\right) = 0$$

The asymptotic behavior in the remaining quadrants may be verified by a similar computation.

■ **EXAMPLE 3** Find the foci of the hyperbola $9x^2 - 4y^2 = 36$. Sketch its graph and asymptotes.

Solution First divide by 36 to write the equation in standard form:

$$\frac{x^2}{4} - \frac{y^2}{9} = 1 \qquad \text{or} \qquad \left(\frac{x}{2}\right)^2 - \left(\frac{y}{3}\right)^2 = 1$$

Thus, $a = 2$, $b = 3$, and $c = \sqrt{a^2 + b^2} = \sqrt{4 + 9} = \sqrt{13}$. The foci are $F_1 = (\sqrt{13}, 0)$ and $F_2 = (-\sqrt{13}, 0)$. To sketch the graph, we draw the rectangle through the points

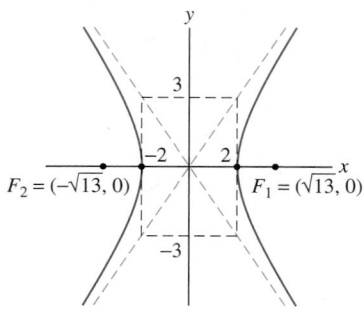

FIGURE 7 The hyperbola $9x^2 - 4y^2 = 36$.

$(\pm 2, 0)$ and $(0, \pm 3)$ as in Figure 7. The diagonals of the rectangle are the asymptotes $y = \pm \frac{3}{2}x$. We sketch the hyperbola so that it passes through the vertices $(\pm 2, 0)$ and approaches the asymptotes. ∎

Finally, we consider the parabola. Unlike the ellipse and hyperbola, which are defined in terms of two foci, a parabola is the set of points P equidistant from a focus F and a line $\mathcal{D}$ called the **directrix**:

$$PF = PD \qquad \boxed{8}$$

Here, when we speak of the *distance* from a point P to a line $\mathcal{D}$, we mean the distance from P to the point Q on $\mathcal{D}$ closest to P, obtained by dropping a perpendicular from P to $\mathcal{D}$ (Figure 8). We denote this distance by PD.

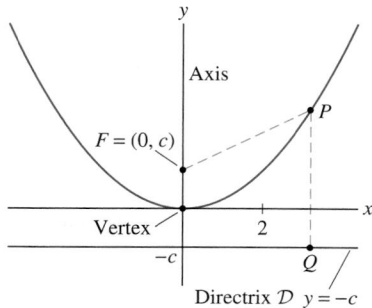

FIGURE 8 Parabola with focus $(0, c)$ and directrix $y = -c$.

The line through the focus F perpendicular to $\mathcal{D}$ is an axis of symmetry and is called the axis of the parabola. It intersects the parabola at a unique point called the vertex. We say that the parabola is in standard position if, for some c, the focus is $F = (0, c)$ and the directrix is $y = -c$, as shown in Figure 8. As we verify in Exercise 71, the vertex is then located at the origin and the equation of the parabola is $y = \dfrac{x^2}{4c}$. If $c < 0$, then the parabola opens downward.

> **THEOREM 3 Equation of a Parabola in Standard Position** Let $c \neq 0$. The parabola with focus $F = (0, c)$ and directrix $y = -c$ has equation
>
> $$y = \frac{1}{4c}x^2 \qquad \boxed{9}$$
>
> The vertex is located at the origin. If $c < 0$, then the parabola opens downward.

■ **EXAMPLE 4** Find the equation of the standard parabola with directrix $y = -2$. Where are its focus and vertex located? What is the equation if this parabola is translated so that its vertex is located at $(2, 4)$, and what are the directrix and focus in this case?

Solution We apply Eq. (9) with $c = 2$. The standard parabola with directrix $y = -2$ has equation $y = \dfrac{x^2}{8}$ (Figure 9). The focus is $(0, c) = (0, 2)$ and the vertex is the origin $(0, 0)$. Note that the focus is two units above the vertex.

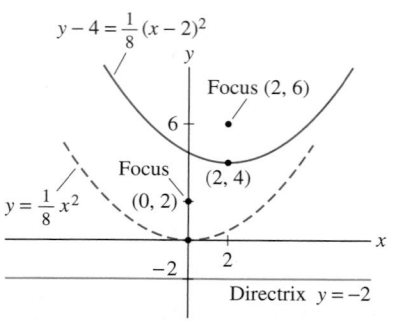

FIGURE 9 A parabola and its translate.

When the vertex is translated to $(2, 4)$, the equation becomes

$$y - 4 = \frac{(x-2)^2}{8} \qquad \text{or} \qquad y = \frac{1}{8}x^2 - \frac{1}{2}x + \frac{9}{2}$$

The focus F is still two units above the vertex, so now $F = (2, 6)$ (see Figure 9). The directrix is the horizontal line located two units below the vertex, namely $y = 2$. ∎

Eccentricity

Some ellipses are flatter than others, just as some hyperbolas are steeper. The "shape" of a conic section is measured by a quantity e called the **eccentricity**. For an ellipse or hyperbola,

$$e = \frac{\text{distance betweeen foci}}{\text{distance between vertices on focal axis}}$$

A parabola is defined to have eccentricity $e = 1$.

The eccentricity of an ellipse or hyperbola in standard position is given by the formula $e = c/a$ in the notation used above. To check this, recall that for a standard ellipse with equation $\left(\frac{x}{a}\right)^2 + \left(\frac{y}{b}\right)^2 = 1$, where $a > b > 0$, we set $c = \sqrt{a^2 - b^2}$. The foci are located at $(\pm c, 0)$ and the vertices on the focal axis at $(\pm a, 0)$. Therefore,

$$e = \frac{\text{distance between foci}}{\text{distance between vertices on focal axis}} = \frac{2c}{2a} = \frac{c}{a}$$

Similarly, the standard hyperbola $\left(\frac{x}{a}\right)^2 - \left(\frac{y}{b}\right)^2 = 1$ has foci at $(\pm c, 0)$, where $c = \sqrt{a^2 + b^2}$, and the vertices on the focal axis at $(\pm a, 0)$. Again, we obtain $e = c/a$. In summary:

$$e = \frac{c}{a} \qquad \text{where} \quad c = \begin{cases} \sqrt{a^2 - b^2} & \text{for an ellipse} \\ \sqrt{a^2 + b^2} & \text{for a hyperbola} \end{cases} \qquad \boxed{10}$$

The eccentricity of an ellipse satisfies $0 \le e < 1$ since $c = \sqrt{a^2 - b^2} < a$. If $e = 0$, the ellipse is a circle because in this case, $a = b$. Circles are the "roundest" possible ellipses and, as we see in Figure 10(A), the larger the eccentricity, the flatter the ellipse. In fact, it is straightforward to check that, for an ellipse,

$$\frac{b}{a} = \sqrt{1 - e^2}$$

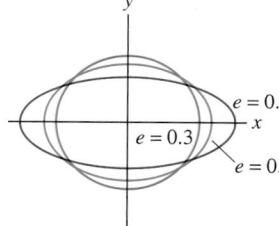

(A) Ellipse flattens as $e \to 1$.

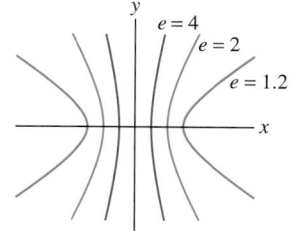

(B) Asymptotes of the hyperbola become more vertical as $e \to \infty$.

FIGURE 10 Effect of changing eccentricity.

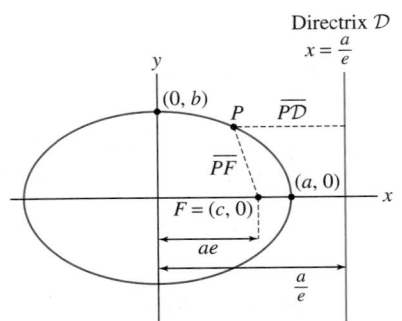

FIGURE 11 The ellipse consists of points P such that $PF = e\,PD$.

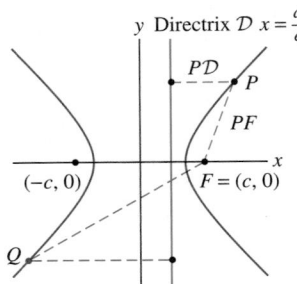

FIGURE 12 The hyperbola consists of points P such that $PF = e\,PD$.

As $e \to 1$, the ratio b/a tends to zero. But b/a is the ratio of the semiminor axis to the semimajor axis, and thus a smaller value of b/a results in a flatter ellipse.

A hyperbola has eccentricity $e > 1$ since $c = \sqrt{a^2 + b^2} < a$. As e increases, the asymptotes of the hyperbola become steeper. We check this by observing that $b/a = \sqrt{1 + e^2}$. The asymptotes of the hyperbola have slopes $\pm b/a$ and thus the slopes of the asymptotes approach infinity as $e \to \infty$.

The notion of eccentricity may be used to give a unified focus-directrix definition of all three types of conics. Given a point F (the focus), a line $\mathcal{D}$ (the directrix), and a number $e > 0$, we consider the set of all points P such that

$$\boxed{PF = e\,PD} \qquad \textbf{11}$$

For $e = 1$, this is precisely our definition of a parabola. According to the next theorem, (11) defines a conic section of eccentricity e for all $e > 0$ (Figures 11 and 12). Note, however, that there is no focus-directrix definition for circles ($e = 0$).

THEOREM 4 Focus-Directrix Definition   Let F be a point, $\mathcal{D}$ a line, and let $e > 0$. Then the set of points satisfying (11) is a conic section of eccentricity e. Furthermore,

- **Ellipse:** Let $a > b > 0$ and $c = \sqrt{a^2 - b^2}$. The ellipse

$$\left(\frac{x}{a}\right)^2 + \left(\frac{y}{b}\right)^2 = 1$$

   satisfies the focus-directrix definition (11) with focus $F = (c, 0)$, eccentricity $e = \dfrac{c}{a}$, and vertical directrix $x = \dfrac{a}{e}$.

- **Hyperbola:** Let $a, b > 0$ and $c = \sqrt{a^2 + b^2}$. The hyperbola

$$\left(\frac{x}{a}\right)^2 - \left(\frac{y}{b}\right)^2 = 1$$

   satisfies the focus-directrix definition (11) with focus $F = (c, 0)$, eccentricity $e = \dfrac{c}{a}$, and vertical directrix $x = \dfrac{a}{e}$.

Proof We assume that $e > 1$ and prove that $PF = e\,PD$ defines a hyperbola (the case $e < 1$ is similar, see Exercise 65). We may choose our coordinate axes so that the focus F lies on the x-axis and the directrix is vertical, lying to the left of F as in Figure 13. Anticipating the final result, we let d be the distance from the focus F to the directrix $\mathcal{D}$ and set

$$c = \frac{d}{1 - e^{-2}}, \qquad a = \frac{c}{e}, \qquad b = \sqrt{c^2 - a^2}$$

Since we are free to shift the y-axis, let us choose the y-axis so that the focus has coordinates $F = (c, 0)$. Then the directrix is the line

$$x = c - d = c - c(1 - e^{-2}) = c\,e^{-2} = \frac{a}{e}$$

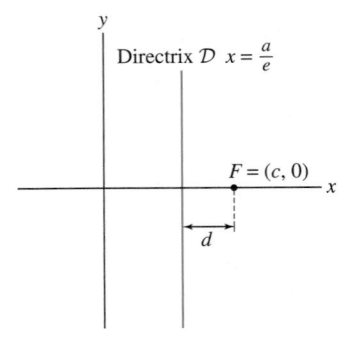

FIGURE 13

Now, the equation $PF = e\,PD$ for a point $P = (x, y)$ may be written

$$\underbrace{\sqrt{(x - c)^2 + y^2}}_{PF} = e\underbrace{\sqrt{(x - (a/e))^2}}_{PD}$$

Algebraic manipulation yields

$$(x - c)^2 + y^2 = e^2(x - (a/e))^2 \qquad \text{(square)}$$

$$x^2 - 2cx + c^2 + y^2 = e^2x^2 - 2aex + a^2 \qquad \text{(note: } c = ae\text{)}$$

$$(e^2 - 1)x^2 - y^2 = c^2 - a^2 = a^2(e^2 - 1) \qquad \text{(rearrange)}$$

$$\frac{x^2}{a^2} - \frac{y^2}{a^2(e^2 - 1)} = 1 \qquad \text{(divide)}$$

Because $a^2(e^2 - 1) = c^2 - a^2 = b^2$, we obtain the equation of the hyperbola

$$\left(\frac{x}{a}\right)^2 - \left(\frac{y}{b}\right)^2 = 1$$

as claimed. ∎

■ **EXAMPLE 5** Find the equation of the standard ellipse with eccentricity $e = 0.8$ and vertices $(\pm 10, 0)$. What are the foci and directrix?

Solution The vertices are $(\pm a, 0)$ with $a = 10$, so the ellipse has equation

$$\left(\frac{x}{10}\right)^2 + \left(\frac{y}{b}\right)^2 = 1$$

(Figure 14). By Theorem 4, $c = ae = 10(0.8) = 8$, and since $c = \sqrt{a^2 - b^2}$, we obtain $8 = \sqrt{10^2 - b^2}$ or $b = 6$. Thus, our ellipse has equation

$$\left(\frac{x}{10}\right)^2 + \left(\frac{y}{6}\right)^2 = 1$$

The foci are $(\pm c, 0) = (\pm 8, 0)$ and the directrix is $x = \dfrac{a}{e} = \dfrac{10}{0.8} = 12.5$. ∎

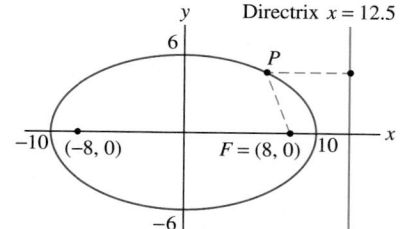

FIGURE 14 Ellipse of eccentricity $e = 0.8$ with focus at $(8, 0)$.

CONCEPTUAL INSIGHT We can prove that if two conic sections $\mathcal{C}_1$ and $\mathcal{C}_2$ have the same eccentricity e, then they have the same shape in the following precise sense: It is possible to scale $\mathcal{C}_1$ so that $\mathcal{C}_1$ and $\mathcal{C}_2$ become congruent. By scaling, we mean changing the units along the x- and y-axes by a common positive factor. A curve scaled by a factor of 10 would have the same shape but would be ten times as large. This would correspond, for example, to changing units from centimeters to millimeters (smaller units make for a larger figure). On the other hand, by "congruent" we mean that after scaling, it is possible to move $\mathcal{C}_1$ by a rigid motion (that is, without stretching or bending) so that it lies directly on top of $\mathcal{C}_2$.

All circles ($e = 0$) have the same shape because scaling by a factor $r > 0$ transforms a circle of radius R into a circle of radius rR. Similarly, any two parabolas ($e = 1$) become congruent after suitable scaling. However, an ellipse of eccentricity $e = 0.5$ cannot be made congruent to an ellipse of eccentricity $e = 0.8$ by scaling (see Exercise 72).

In Section 13.6, we discuss the famous law of Johannes Kepler stating that the orbit of a planet around the sun is an ellipse with one focus at the sun. In this discussion, we will need to write the equation of an ellipse in polar coordinates. To derive the polar equations of the conic sections, it is convenient to use the focus-directrix definition with focus F at the origin O, and vertical line $x = d$ as directrix $\mathcal{D}$. By Eq. (11), the equation

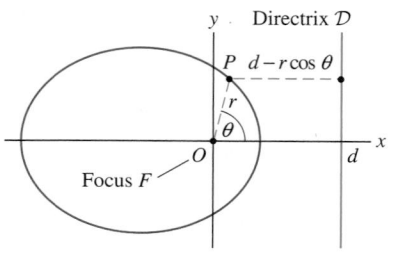

FIGURE 15 Focus-directrix definition of the ellipse in polar coordinates.

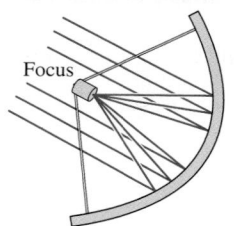

FIGURE 16 The paraboloid shape of this radio telescope directs the incoming signal to the focus.

of the ellipse is $PF = e\,PD$. Referring to Figure 15, we see that if $P = (r, \theta)$, then

$$PF = r, \qquad PD = d - r\cos\theta$$

Thus, the equation of the ellipse $PF = e\,PD$ becomes $r = e(d - r\cos\theta)$ or $r(1 + e\cos\theta) = ed$. This proves the following result, which is also valid for the hyperbola and parabola (see Exercise 66).

Polar Equation of a Conic Section The conic section of eccentricity $e > 0$ with focus at the origin and directrix at $x = d$ has polar equation

$$r = \frac{ed}{1 + e\cos\theta} \qquad \boxed{12}$$

Reflective Properties of Conic Sections

The conic sections have numerous geometric properties that are of importance in applications, especially in optics and communications (for example, in antenna and telescope design; Figure 16). Here, we limit ourselves to mentioning (without proof) the *reflective properties* of each family of conics. See Exercises 67–69 for a proof of the reflective property of an ellipse.

- **Ellipse:** The segments $F_1 P$ and $F_2 P$ make equal angles with the tangent line at a point P on the ellipse. Therefore, a beam of light originating at focus F_1 is reflected off the ellipse toward the second focus F_2 [Figure 17(A)].

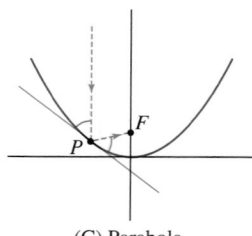

(A) Ellipse   (B) Hyperbola   (C) Parabola

FIGURE 17

- **Hyperbola:** The tangent line at a point P on the hyperbola bisects the angle formed by the segments $F_1 P$ and $F_2 P$. Therefore, a beam of light directed toward F_2 is reflected off the hyperbola toward the second focus F_1 [Figure 17(B)].
- **Parabola:** The segment FP and the line through P perpendicular to the directrix make equal angles with the tangent line at a point P on the parabola. Therefore, a beam of light approaching P from above, in the direction perpendicular to the directrix, is reflected off the parabola toward the focus F [Figure 17(C)].

General Equations of Degree Two

We now consider briefly the general equation of degree two in x and y:

$$Ax^2 + Bxy + Cy^2 + Dx + Ey + F = 0 \qquad \boxed{13}$$

where A, B, C, D, E, F are constants with A, B, C not all zero. The equations of the conic sections in standard position are special cases of this general quadratic equation. For example, the equation of a standard ellipse may be written in the form

$$Ax^2 + Cy^2 - 1 = 0$$

It turns out that the general quadratic equation does not give rise to any new types of curves. Apart from certain "degenerate cases," Eq. (13) defines a conic section with arbitrary center whose focal and conjugate axes may be rotated relative to the coordinate axes. For example, the equation

$$6x^2 - 8xy + 8y^2 - 12x - 24y + 38 = 0$$

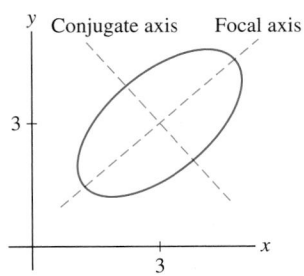

FIGURE 18 The ellipse with equation $6x^2 - 8xy + 8y^2 - 12x - 24y + 38 = 0$.

defines an ellipse with center at $(3, 3)$ whose axes are rotated (Figure 18).

We say that Eq. (13) is degenerate if the set of solutions is a pair of intersecting lines, a pair of parallel lines, a single line, a point, or the empty set. For example:

- $x^2 - y^2 = 0$ defines a pair of intersecting lines $y = x$ and $y = -x$.
- $x^2 - x = 0$ defines a pair of parallel lines $x = 0$ and $x = 1$.
- $x^2 = 0$ defines a single line (the y-axis).
- $x^2 + y^2 = 0$ has just one solution $(0, 0)$.
- $x^2 + y^2 = -1$ has no solutions.

Now assume that Eq. (13) is nondegenerate. The term Bxy is called the *cross term*. When the cross term is zero (i.e., $B = 0$), we can "complete the square" to show that Eq. (13) defines a translate of conic in standard position. In other words, the axes of the conic are parallel to the coordinate axes. This is illustrated in the next example.

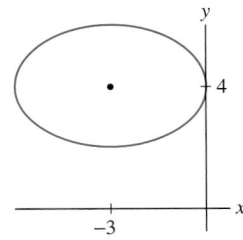

FIGURE 19 The ellipse with equation $4x^2 + 9y^2 + 24x - 72y + 144 = 0$.

■ **EXAMPLE 6** Completing the Square Show that $4x^2 + 9y^2 + 24x - 72y + 144 = 0$ defines a translate of a conic section in standard position (Figure 19).

Solution Since there is no cross term, we may complete the square of the terms involving x and y terms separately:

$$4x^2 + 9y^2 + 24x - 72y + 144 = 4(x^2 + 6x) + 9(y^2 - 8y) + 144 = 0$$

$$4(x + 3)^2 - 36 + 9(y - 4)^2 - 144 + 144 = 0$$

$$4(x + 3)^2 + 9(y - 4)^2 = 36$$

Therefore, this quadratic equation can be rewritten in the form

$$\left(\frac{x + 3}{3}\right)^2 + \left(\frac{y - 4}{2}\right)^2 = 1$$

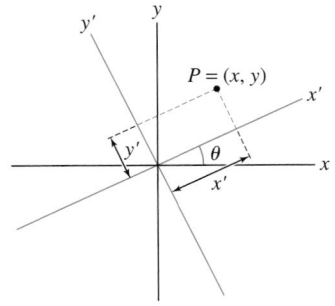

FIGURE 20

When the cross-term Bxy is nonzero, Eq. (13) defines a conic whose axes are rotated relative to the coordinate axes. The marginal note describes how this may be verified in general. We illustrate with the following example.

■ **EXAMPLE 7** Show that $2xy = 1$ defines a conic section whose focal and conjugate axes are rotated relative to the coordinate axes.

More generally, if (x', y') are coordinates relative to axes rotated by an angle θ as in Figure 20, then

$$x = x' \cos \theta - y' \sin \theta \qquad \boxed{14}$$

$$y = x' \sin \theta + y' \cos \theta \qquad \boxed{15}$$

In Exercise 73, we show that the cross term disappears when Eq. (13) is rewritten in terms of x' and y' for the angle

$$\theta = \frac{1}{2} \cot^{-1} \frac{A - C}{B} \qquad \boxed{16}$$

Solution Figure 21(A) shows axes labeled x' and y' that are rotated by $45°$ relative to the standard coordinate axes. A point P with coordinates (x, y) may also be described by coordinates (x', y') relative to these rotated axes. Applying (14) and (15) with $\theta = \frac{\pi}{4}$, we find that (x, y) and (x', y') are related by the formulas

$$x = \frac{x' - y'}{\sqrt{2}}, \qquad y = \frac{x' + y'}{\sqrt{2}}$$

Therefore, if $P = (x, y)$ lies on the hyperbola, that is, if $2xy = 1$, then

$$2xy = 2\left(\frac{x' - y'}{\sqrt{2}}\right)\left(\frac{x' + y'}{\sqrt{2}}\right) = x'^2 - y'^2 = 1$$

Thus, the coordinates (x', y') satisfy the equation of the standard hyperbola $x'^2 - y'^2 = 1$ whose focal and conjugate axes are the x'- and y'-axes, respectively. ∎

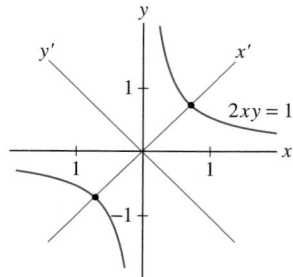

FIGURE 21 The x'- and y'-axes are rotated at a $45°$ angle relative to the x- and y-axes.

(A) The point $P = (x, y)$ may also be described by coordinates (x', y') relative to the rotated axis.

(B) The hyperbola $2xy = 1$ is in standard position relative to the x', y' axes.

We conclude our discussion of conics by stating the Discriminant Test. Suppose that the equation

$$Ax^2 + Bxy + Cy^2 + Dx + Ey + F = 0$$

is nondegenerate and thus defines a conic section. The type of conic is determined by the **discriminant** D (not to be confused with the coefficient D in the equation), defined by

$$D = B^2 - 4AC$$

According to the Discriminant Test (which we state without proof), the curve is

- an ellipse if $D < 0$.
- a hyperbola if $D > 0$.
- a parabola if $D = 0$.

For example, the discriminant of the equation $2xy = 1$ is

$$D = B^2 - 4AC = 2^2 - 0 = 4 > 0$$

According to the Discriminant Test, $2xy = 1$ defines a hyperbola. This agrees with our conclusion in Example 7.

11.5 SUMMARY

- An *ellipse* with foci F_1 and F_2 is the set of points P such that $PF_1 + PF_2 = K$, where $K > 0$ is a constant. The equation of an ellipse in standard position is

$$\left(\frac{x}{a}\right)^2 + \left(\frac{y}{b}\right)^2 = 1$$

	Focal Axis	Foci	Vertices
$a > b$	x-axis	$(\pm c, 0)$ with $c = \sqrt{a^2 - b^2}$	$(\pm a, 0), (0, \pm b)$
$a < b$	y-axis	$(0, \pm c)$ with $c = \sqrt{b^2 - a^2}$	$(\pm a, 0), (0, \pm b)$

- A *hyperbola* with foci F_1 and F_2 is the set of points P such that

$$PF_1 - PF_2 = \pm K$$

where $K > 0$ is a constant. The equation of a hyperbola in standard position is

$$\left(\frac{x}{a}\right)^2 - \left(\frac{y}{b}\right)^2 = 1$$

The focal axis is the x-axis and the foci are $(\pm c, 0)$ with $c = \sqrt{a^2 + b^2}$. The vertices are $(\pm a, 0)$ and the asymptotes are the lines $y = \pm \frac{b}{a}x$.

- A *parabola* with focus F and directrix $\mathcal{D}$ is the set of points P such that $PF = PD$. A parabola in standard position, with focus $F = (0, c)$ and directrix $y = -c$ with $c \neq 0$, has equation $y = \frac{1}{4c}x^2$. The vertex of a parabola in standard position is the origin $(0, 0)$.

- To translate a conic so that its center is located at (x_0, y_0) (while keeping its axes parallel to the x- and y-axes), replace x and y by $x - x_0$ and $y - y_0$, respectively, in the equation of the conic.

- The *eccentricity* e of an ellipse or a hyperbola is the quantity

$$e = \frac{\text{distance betweeen foci}}{\text{distance between vertices on focal axis}}$$

In standard position, $e = \frac{c}{a}$. For an ellipse, $0 \le e < 1$ (with $e = 0$ a circle); for a hyperbola, $e > 1$. By definition, a parabola has eccentricity $e = 1$.

- *Focus-directrix definition of conic sections:* Given a point P (a focus), a line $\mathcal{D}$ (a directrix), and a number $e > 0$, the set of points P such that $PF = ePD$ is a conic section of eccentricity e.

- The conic section of eccentricity $e > 0$ with focus at the origin and directrix $x = d$ has polar equation

$$r = \frac{ed}{1 + e \cos \theta}$$

11.5 EXERCISES

Preliminary Questions

1. Which of the following equations defines an ellipse? Which does not define a conic section?

(a) $4x^2 - 9y^2 = 12$

(b) $-4x + 9y^2 = 0$

(c) $4y^2 + 9x^2 = 12$

(d) $4x^3 + 9y^3 = 12$

2. For which conic sections do the vertices lie between the foci?

3. What are the foci of $\left(\frac{x}{a}\right)^2 + \left(\frac{y}{b}\right)^2 = 1$ if $a < b$?

4. For a hyperbola in standard position, the set of points equidistant from the foci is the y-axis. Use the definition $PF_1 - PF_2 = \pm K$ to explain why the hyperbola does not intersect the y-axis.

5. What is the geometric interpretation of the quantity $\frac{b}{a}$ in the equation of a hyperbola in standard position?

Exercises

In Exercises 1–8, find the vertices and foci of the conic section.

1. $\left(\frac{x}{9}\right)^2 + \left(\frac{y}{4}\right)^2 = 1$

2. $\left(\frac{x}{4}\right)^2 + \left(\frac{y}{9}\right)^2 = 1$

3. $\frac{x^2}{9} + \frac{y^2}{4} = 1$

4. $\left(\frac{x}{9}\right)^2 - \left(\frac{y}{4}\right)^2 = 1$

5. $\left(\frac{x}{4}\right)^2 - \left(\frac{y}{9}\right)^2 = 1$

6. $\frac{x^2}{9} - \frac{y^2}{4} = 1$

7. $\left(\frac{x-3}{7}\right)^2 - \left(\frac{y+1}{4}\right)^2 = 1$

8. $\left(\frac{x-3}{4}\right)^2 + \left(\frac{y+1}{7}\right)^2 = 1$

In Exercises 9–12, consider the ellipse

$$\left(\frac{x-12}{5}\right)^2 + \left(\frac{y-9}{7}\right)^2 = 1$$

Find the equation of the translated ellipse.

9. Translated so that its center is at the origin

10. Translated four units to the right

11. Translated three units down

12. Translated so its center is $(-4, -9)$

In Exercises 13–16, find the equation of the ellipse with the given properties.

13. Vertices at $(\pm 9, 0)$ and $(0, \pm 16)$

14. Foci $(\pm 6, 0)$ and two vertices at $(\pm 10, 0)$

15. Foci $(0, \pm 6)$ and two vertices at $(\pm 4, 0)$

16. Foci $(0, \pm 3)$ and eccentricity $\frac{3}{4}$

In Exercises 17–22, find the equation of the hyperbola with the given properties.

17. Vertices $(\pm 3, 0)$ and foci at $(\pm 5, 0)$

18. Vertices $(0, \pm 5)$ and foci $(0, \pm 8)$

19. Vertices $(\pm 4, 0)$ and asymptotes $y = \pm 3x$

20. Vertices $(\pm 3, 0)$ and asymptotes $y = \pm \frac{1}{2}x$

21. Vertices $(0, -5)$, $(0, 4)$ and foci $(0, -8)$, $(0, 7)$

22. Foci $(0, \pm 3)$ and eccentricity $\sqrt{2}$

In Exercises 23–30, find the equation of the parabola with the given properties.

23. Vertex $(0, 0)$, focus $(2, 0)$

24. Vertex $(0, 0)$, focus $(0, \frac{1}{2})$

25. Vertex $(0, 0)$, directrix $y = -5$

26. Vertex $(0, 0)$, directrix $y = -\frac{1}{8}$

27. Focus $(0, 4)$, directrix $y = -4$

28. Focus $(0, -4)$, directrix $y = 4$

29. Focus $(2, 0)$, directrix $x = -2$

30. Focus $(-2, 0)$, directrix $x = 2$

In Exercises 31–40, find the vertices, foci, axes, center (if an ellipse or a hyperbola) and asymptotes (if a hyperbola) of the conic section.

31. $x^2 + 4y^2 = 16$

32. $\left(\frac{x-3}{16}\right)^2 - \left(\frac{y+5}{49}\right)^2 = 1$

33. $4x^2 + y^2 = 16$

34. $3x^2 - 27y^2 = 12$

35. $4x^2 - 3y^2 + 8x + 30y = 215$

36. $y = 4x^2$

37. $y = 4(x-4)^2$

38. $8y^2 + 6x^2 - 36x - 64y + 134 = 0$

39. $4x^2 + 25y^2 - 8x - 10y = 20$

40. $y^2 - 2x^2 + 8x - 9y - 12 = 0$

In Exercises 41–44, use the Discriminant Test to determine the type of the conic section defined by the equation. You may assume that the equation is nondegenerate. Plot the curve if you have a computer algebra system.

41. $4x^2 + 5xy + 7y^2 = 24$

42. $2x^2 - 8xy + 3y^2 - 4 = 0$

43. $2x^2 - 8xy - 3y^2 - 4 = 0$

44. $x^2 - 2xy + y^2 + 24x - 8 = 0$

45. Show that $\dfrac{b}{a} = \sqrt{1 - e^2}$ for a standard ellipse of eccentricity e.

46. Show that the eccentricity of a hyperbola in standard position is $e = \sqrt{1 + m^2}$, where $\pm m$ are the slopes of the asymptotes.

47. Explain why the dots in Figure 22 lie on a parabola. Where are the focus and directrix located?

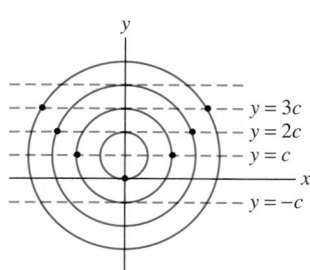

FIGURE 22

48. Show that the equation of the tangent line to the hyperbola $\left(\dfrac{x}{a}\right)^2 - \left(\dfrac{y}{b}\right)^2 = 1$ at a point (x_0, y_0) is

$$Ax - By = 1$$

where $A = \dfrac{x_0}{a^2}$ and $B = \dfrac{y_0}{b^2}$.

49. Kepler's First Law states that the orbits of the planets around the sun are ellipses with the sun at one focus. The orbit of Pluto has an eccentricity of approximately $e = 0.25$ and the **perihelion** (closest distance to the sun) of Pluto's orbit is approximately 2.7 billion miles. Find the **aphelion** (farthest distance from the sun).

50. Kepler's Third Law states that the ratio $T/a^{3/2}$ is equal to a constant C for all planetary orbits around the sun, where T is the period (time for a complete orbit) and a is the semimajor axis.

(a) Compute C in units of days and kilometers, given that the semimajor axis of the earth's orbit is 150×10^6 km.

(b) Compute the period of Saturn's orbit, given that its semimajor axis is approximately 1.43×10^9 km.

(c) Saturn's orbit has eccentricity $e = 0.056$. Find the perihelion and aphelion of Saturn.

In Exercises 51–54, find the polar equation of the conic with given eccentricity and directrix.

51. $e = \frac{1}{2}, \quad x = 3$

52. $e = \frac{1}{2}, \quad x = -3$

53. $e = 1, \quad x = 4$

54. $e = \frac{3}{2}, \quad x = -4$

In Exercises 55–58, identify the type of conic, the eccentricity, and the equation of the directrix.

55. $r = \dfrac{8}{1 + 4\cos\theta}$

56. $r = \dfrac{8}{4 + \cos\theta}$

57. $r = \dfrac{8}{4 + 3\cos\theta}$

58. $r = \dfrac{12}{4 + 3\cos\theta}$

59. Show that $r = f_1(\theta)$ and $r = f_2(\theta)$ define the same curves in polar coordinates if $f_1(\theta) = -f_2(\theta + \pi)$, and use this to show that the following define the same conic section:

$$r = \dfrac{de}{1 - e\cos\theta}, \qquad r = \dfrac{-de}{1 + e\cos\theta}$$

60. Find a polar equation for the hyperbola with focus at the origin, directrix $x = -2$, and eccentricity $e = 1.2$.

61. Find the equation of the ellipse $r = \dfrac{4}{2 + \cos\theta}$ in rectangular coordinates.

62. Let $\mathcal{C}$ be the ellipse $r = \dfrac{de}{1 + e\cos\theta}$, where $e < 1$. Express the x-coordinates of the vertices A, A', the center C, and the second focus F_2 in terms of d and e (Figure 23).

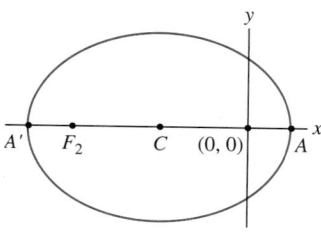

FIGURE 23

63. Let $e > 1$. Show that the vertices of the hyperbola $r = \dfrac{de}{1 + e\cos\theta}$ have x-coordinates $\dfrac{ed}{e + 1}$ and $\dfrac{ed}{e - 1}$.

Further Insights and Challenges

64. Verify Theorem 2.

65. Verify Theorem 4 in the case $0 < e < 1$. *Hint:* Repeat the proof of Theorem 4, but set $c = d/(e^{-2} - 1)$.

66. Verify that if $e > 1$, then Eq. (12) defines a hyperbola of eccentricity e, with its focus at the origin and directrix at $x = d$.

Reflective Property of the Ellipse *In Exercises 67–69, we prove that the focal radii at a point on an ellipse make equal angles with the tangent line. Let $P = (x_0, y_0)$ be a point on the ellipse $\left(\dfrac{x}{a}\right)^2 + \left(\dfrac{y}{b}\right)^2 = 1$ $(a > b)$ with foci $F_1 = (-c, 0)$ and $F_2 = (c, 0)$, and eccentricity e (Figure 24).*

67. Show that $PF_1 = a + x_0 e$ and $PF_2 = a - x_0 e$. *Hints:*

(a) Show that $PF_1{}^2 - PF_2{}^2 = 4x_0 c$.

(b) Divide the previous relation by $PF_1 + PF_2 = 2a$, and conclude that $PF_1 - PF_2 = 2x_0 e$.

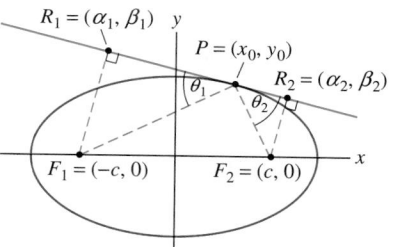

FIGURE 24 The ellipse $\left(\dfrac{x}{a}\right)^2 + \left(\dfrac{y}{b}\right)^2 = 1$.

68. Show that the equation of the tangent line at P is $Ax + By = 1$, where $A = \dfrac{x_0}{a^2}$ and $B = \dfrac{y_0}{b^2}$.

69. Define R_1 and R_2 as in the figure, so that $\overline{F_1 R_1}$ and $\overline{F_2 R_2}$ are perpendicular to the tangent line.

(a) Show that $\dfrac{\alpha_1 + c}{\beta_1} = \dfrac{\alpha_2 - c}{\beta_2} = \dfrac{A}{B}$.

(b) Use (a) and the distance formula to show that

$$\frac{F_1 R_1}{F_2 R_2} = \frac{\beta_1}{\beta_2}$$

(c) Solve for β_1 and β_2:

$$\beta_1 = \frac{B(1 + Ac)}{A^2 + B^2}, \qquad \beta_2 = \frac{B(1 - Ac)}{A^2 + B^2}$$

(d) Show that $\dfrac{F_1 R_1}{F_2 R_2} = \dfrac{P F_1}{P F_2}$. Conclude that $\theta_1 = \theta_2$.

70. Show that the length QR in Figure 25 is independent of the point P.

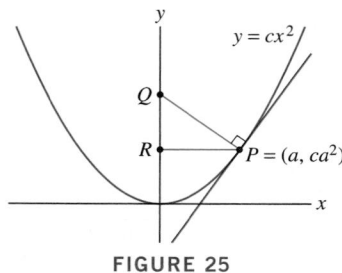

FIGURE 25

71. Show that $y = \dfrac{x^2}{4c}$ is the equation of a parabola with directrix $y = -c$, focus $(0, c)$, and the vertex at the origin, as stated in Theorem 3.

72. Consider two ellipses in standard position:

$$E_1 : \quad \left(\frac{x}{a_1}\right)^2 + \left(\frac{y}{b_1}\right)^2 = 1$$

$$E_2 : \quad \left(\frac{x}{a_2}\right)^2 + \left(\frac{y}{b_2}\right)^2 = 1$$

We say that E_1 is similar to E_2 under scaling if there exists a factor $r > 0$ such that for all (x, y) on E_1, the point (rx, ry) lies on E_2. Show that E_1 and E_2 are similar under scaling if and only if they have the same eccentricity. Show that any two circles are similar under scaling.

73. If we rewrite the general equation of degree two (13) in terms of variables x' and y' that are related to x and y by equations (14) and (15), we obtain a new equation of degree two in x' and y' of the same form but with different coefficients:

$$A'x^2 + B'xy + C'y^2 + D'x + E'y + F' = 0$$

(a) Show that $B' = B \cos 2\theta + (C - A) \sin 2\theta$.

(b) Show that if $B \neq 0$, then we obtain $B' = 0$ for

$$\theta = \frac{1}{2} \cot^{-1} \frac{A - C}{B}$$

This proves that it is always possible to eliminate the cross term Bxy by rotating the axes through a suitable angle.

CHAPTER REVIEW EXERCISES

1. Which of the following curves pass through the point $(1, 4)$?
(a) $c(t) = (t^2, t + 3)$
(b) $c(t) = (t^2, t - 3)$
(c) $c(t) = (t^2, 3 - t)$
(d) $c(t) = (t - 3, t^2)$

2. Find parametric equations for the line through $P = (2, 5)$ perpendicular to the line $y = 4x - 3$.

3. Find parametric equations for the circle of radius 2 with center $(1, 1)$. Use the equations to find the points of intersection of the circle with the x- and the y-axes.

4. Find a parametrization $c(t)$ of the line $y = 5 - 2x$ such that $c(0) = (2, 1)$.

5. Find a parametrization $c(\theta)$ of the unit circle such that $c(0) = (-1, 0)$.

6. Find a path $c(t)$ that traces the parabolic arc $y = x^2$ from $(0, 0)$ to $(3, 9)$ for $0 \le t \le 1$.

7. Find a path $c(t)$ that traces the line $y = 2x + 1$ from $(1, 3)$ to $(3, 7)$ for $0 \le t \le 1$.

8. Sketch the graph $c(t) = (1 + \cos t, \sin 2t)$ for $0 \le t \le 2\pi$ and draw arrows specifying the direction of motion.

In Exercises 9–12, express the parametric curve in the form $y = f(x)$.

9. $c(t) = (4t - 3, 10 - t)$
10. $c(t) = (t^3 + 1, t^2 - 4)$

11. $c(t) = \left(3 - \dfrac{2}{t}, t^3 + \dfrac{1}{t}\right)$
12. $x = \tan t, \quad y = \sec t$

13. Find all points visited twice by the path $c(t) = (t^2, \sin t)$. Plot $c(t)$ with a graphing utility.

In Exercises 14–17, calculate $\dfrac{dy}{dx}$ at the point indicated.

14. $c(t) = (t^3 + t, t^2 - 1), \quad t = 3$

15. $c(\theta) = (\tan^2 \theta, \cos \theta), \quad \theta = \frac{\pi}{4}$

16. $c(t) = (e^t - 1, \sin t), \quad t = 20$

17. $c(t) = (\ln t, 3t^2 - t), \quad P = (0, 2)$

18. $\boxed{C A S}$ Find the point on the cycloid $c(t) = (t - \sin t, 1 - \cos t)$ where the tangent line has slope $\frac{1}{2}$.

19. Find the points on $(t + \sin t, t - 2 \sin t)$ where the tangent is vertical or horizontal.

20. Find the equation of the Bézier curve with control points

$$P_0 = (-1, -1), \quad P_1 = (-1, 1), \quad P_2 = (1, 1), \quad P_3(1, -1)$$

21. Find the speed at $t = \frac{\pi}{4}$ of a particle whose position at time t seconds is $c(t) = (\sin 4t, \cos 3t)$.

22. Find the speed (as a function of t) of a particle whose position at time t seconds is $c(t) = (\sin t + t, \cos t + t)$. What is the particle's maximal speed?

23. Find the length of $(3e^t - 3, 4e^t + 7)$ for $0 \le t \le 1$.

In Exercises 24–25, let $c(t) = (e^{-t} \cos t, e^{-t} \sin t)$.

24. Show that $c(t)$ for $0 \le t < \infty$ has finite length and calculate its value.

25. Find the first positive value of t_0 such that the tangent line to $c(t_0)$ is vertical and calculate the speed at $t = t_0$.

26. $\boxed{C A S}$ Plot $c(t) = (\sin 2t, 2 \cos t)$ for $0 \le t \le \pi$. Express the length of the curve as a definite integral and approximate it using a computer algebra system.

27. Convert the points $(x, y) = (1, -3), (3, -1)$ from rectangular to polar coordinates.

28. Convert the points $(r, \theta) = \left(1, \frac{\pi}{6}\right), \left(3, \frac{5\pi}{4}\right)$ from polar to rectangular coordinates.

29. Write $(x + y)^2 = xy + 6$ as an equation in polar coordinates.

30. Write $r = \dfrac{2 \cos \theta}{\cos \theta - \sin \theta}$ as an equation in rectangular coordinates.

31. Show that $r = \dfrac{4}{7 \cos \theta - \sin \theta}$ is the polar equation of a line.

32. $\boxed{GU}$ Convert the equation

$$9(x^2 + y^2) = (x^2 + y^2 - 2y)^2$$

to polar coordinates and plot with a graphing utility.

33. Calculate the area of the circle $r = 3 \sin \theta$ bounded by the rays $\theta = \frac{\pi}{3}$ and $\theta = \frac{2\pi}{3}$.

34. $\boxed{GU}$ Plot the graph of $r = \sin 4\theta$ and calculate the area of one petal.

35. The equation $r = \sin(n\theta)$, where $n \ge 2$ is even, is a "rose" of $2n$ petals (Figure 1). Compute the total area of the flower and show that it does not depend on n.

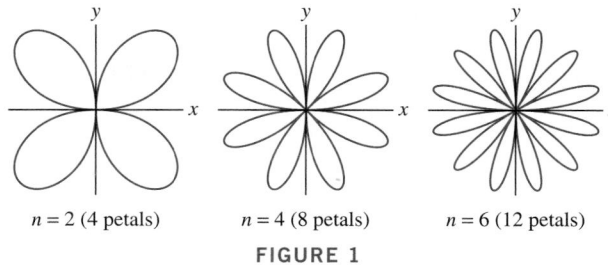

$n = 2$ (4 petals) $n = 4$ (8 petals) $n = 6$ (12 petals)

FIGURE 1

36. Calculate the total area enclosed by the curve $r^2 = \cos \theta e^{\sin \theta}$ (Figure 2).

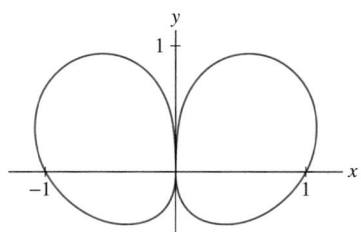

FIGURE 2 Graph of $r^2 = \cos \theta e^{\sin \theta}$.

37. Find the shaded area in Figure 3.

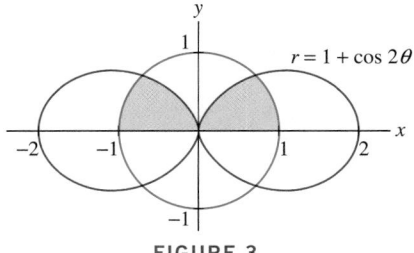

$r = 1 + \cos 2\theta$

FIGURE 3

38. Calculate the length of the curve with polar equation $r = \theta$ in Figure 4.

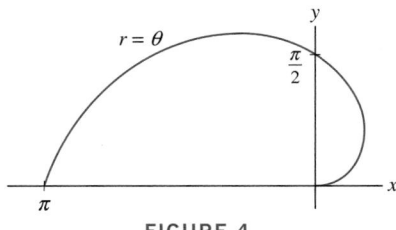

$r = \theta$

FIGURE 4

39. CAS Figure 5 shows the graph of $r = e^{0.5\theta} \sin \theta$ for $0 \le \theta \le 2\pi$. Use a CAS to approximate the difference in length between the outer and inner loops.

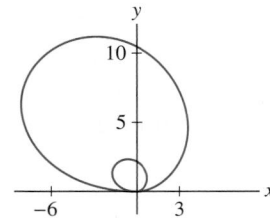

FIGURE 5

In Exercises 40–43, identify the conic section. Find the vertices and foci.

40. $\left(\dfrac{x}{3}\right)^2 + \left(\dfrac{y}{2}\right)^2 = 1$

41. $x^2 - 2y^2 = 4$

42. $(2x + \frac{1}{2}y)^2 = 4 - (x - y)^2$

43. $(y - 3)^2 = 2x^2 - 1$

44. Find the equation of a standard ellipse with two vertices at $(\pm 8, 0)$ and foci $(\pm\sqrt{3}, 0)$.

45. Find the equation of a standard hyperbola with vertices at $(\pm 8, 0)$ and asymptotes $y = \pm\dfrac{3}{4}x$.

46. Find the equation of a standard parabola with focus $(8, 0)$ and directrix $x = -8$.

47. Find the equation of a standard ellipse with foci at $(\pm 8, 0)$ and eccentricity $\frac{1}{8}$.

48. Find the asymptotes of the hyperbola $3x^2 + 6x - y^2 - 10y = 1$.

49. Show that the "conic section" with equation $x^2 - 4x + y^2 + 5 = 0$ has no points.

50. Show that the relation $\dfrac{dy}{dx} = (e^2 - 1)\dfrac{x}{y}$ holds on a standard ellipse or hyperbola of eccentricity e.

51. The orbit of Jupiter is an ellipse with the sun at a focus. Find the eccentricity of the orbit if the perihelion (closest distance to the sun) equals 740×10^6 km and the aphelion (farthest distance to the sun) equals 816×10^6 km.

52. Refer to Figure 24 in Section 11.5. Prove that the product of the perpendicular distances $F_1 R_1$ and $F_2 R_2$ from the foci to a tangent line of an ellipse is equal to the square b^2 of the semiminor axes.

12 | VECTOR GEOMETRY

The tension in the cables of the Tri-City Bridge over the Columbia River in Washington State is described using vectors.

Vectors play a role in nearly all areas of mathematics and its applications. In physics and engineering, vectors are used to represent quantities that have direction as well as magnitude, such as velocity or force. More advanced physical applications of vectors include aerodynamics, electromagnetic theory, quantum theory, and more recent fields such as computer graphics, image processing, and robotics. This chapter develops the basic geometric and algebraic properties of vectors. Although this chapter does not use any calculus, the concepts developed will be used throughout the remainder of the text.

The term "vector" was first used in 1844 by the Irish mathematician William Rowan Hamilton (1805–1865) who made fundamental contributions to both mathematics and physics. In his work Lectures on Quaternions, he wrote

> But in the new mode of speaking which it is here proposed to introduce, ... the vector of the sun has (itself) direction, as well as length. It is, therefore not sufficiently characterized by any single number, but requires, for its complete numerical expression, a system of three numbers.

NOTATION In this text, vectors are represented by boldface lowercase letters such as $\mathbf{v}$, $\mathbf{w}$, $\mathbf{a}$, $\mathbf{b}$, etc.

When referring to vectors, we use the terms "length" and "magnitude" interchangeably. The term "norm" is also commonly used.

12.1 Vectors in the Plane

A two-dimensional **vector v** is determined by two points in the plane: an initial point P (also called the "tail" or basepoint) and a terminal point Q (also called the "head"). We write $\mathbf{v} = \overrightarrow{PQ}$ and we draw $\mathbf{v}$ as an arrow pointing from P to Q [Figure 1(A)]. The vector $\mathbf{v} = \overrightarrow{PQ}$ is said to be based at P. The **position vector** of a point R is the vector $\mathbf{v} = \overrightarrow{OR}$, based at the origin with endpoint R [Figure 1(B)].

(A) The vector $\overrightarrow{PQ}$

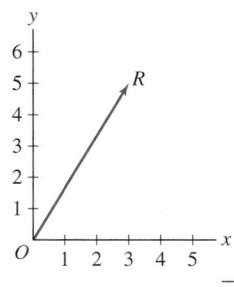
(B) The position vector $\overrightarrow{OR}$

FIGURE 1

The **length** or **magnitude** of $\mathbf{v} = \overrightarrow{PQ}$, denoted $\|\mathbf{v}\|$, is defined as the distance from P to Q. We denote the distance from P to Q by $|P - Q|$. If $P = (a_1, b_1)$ and $Q = (a_2, b_2)$, then by the distance formula,

$$\|\mathbf{v}\| = \|\overrightarrow{PQ}\| = |P - Q| = \sqrt{(a_2 - a_1)^2 + (b_2 - b_1)^2}$$

■ **EXAMPLE 1** Find the length of the vector $\mathbf{v} = \overrightarrow{PQ}$ in Figure 1(A).

Solution Observe in Figure 1(A) that the initial point is $P = (2, 2)$ and the terminal point is $Q = (7, 5)$. By the distance formula,

$$\|\mathbf{v}\| = \text{Distance from } P \text{ to } Q = \sqrt{(7 - 2)^2 + (5 - 2)^2} = \sqrt{5^2 + 3^2} = \sqrt{34} \quad ■$$

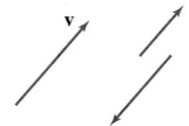

(A) Vectors parallel to **v**

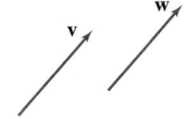

(B) **w** is a translate of **v**

FIGURE 2

Two vectors **v** and **w** of nonzero length are called **parallel** if the lines through **v** and **w** are parallel. Parallel vectors point either in the same or in opposite directions [Figure 2(A)]. A vector **v** is said to undergo a **translation** when it is moved parallel to itself without changing in length or direction. The resulting vector **w** is called a translate of **v** [Figure 2(B)]. Translates have the same length and direction but different basepoints.

As our study of vectors progresses, we will see that it is convenient to treat two vectors with the same length and direction as equivalent, even if they have different basepoints. With this in mind, we call vectors **v** and **w** **equivalent** if **w** is a translate of **v** [Figure 3(A)]. Every vector can be translated so that its tail is at the origin [Figure 3(C)]. Therefore,

*Every vector **v** is equivalent to a unique vector **v**$_0$ based at the origin.*

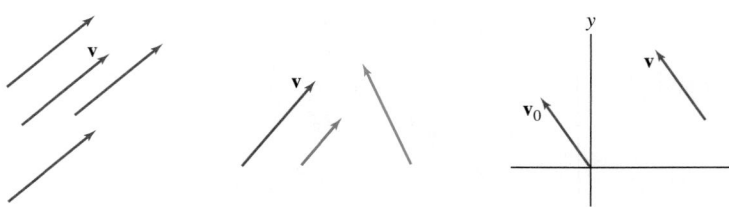

(A) Vectors equivalent   (B) Inequivalent vectors   (C) **v**$_0$ is the unique vector based
    to **v** (translates of **v**)   at the origin equivalent to **v**

FIGURE 3 Equivalent vectors have the same length and direction.

To work with vectors algebraically, we define the components of a vector.

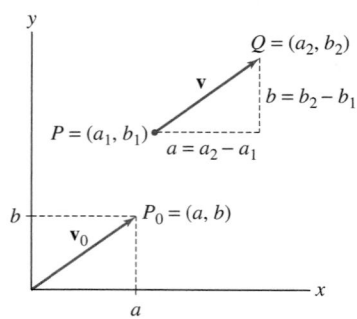

FIGURE 4 The vectors **v** and **v**$_0$ have components $\langle a, b \rangle$.

NOTATION *In this text, vectors in component form are written using "angle brackets." This allows us to clearly distinguish between the vector **v** $= \langle a, b \rangle$ and the point $P = (a, b)$. Some textbooks denote both **v** and P by the same symbol (a, b).*

> **DEFINITION Components of a Vector**   Let $\mathbf{v} = \overrightarrow{PQ}$, where $P = (a_1, b_1)$ and $Q = (a_2, b_2)$. The **components** of **v** are the quantities
>
> $$a = a_2 - a_1 \quad \text{(the } x\text{-component)}, \qquad b = b_2 - b_1 \quad \text{(the } y\text{-component)}$$
>
> The pair of components is denoted $\langle a, b \rangle$.

As we see in Figure 4, the components $\langle a, b \rangle$ of **v** are the horizontal and vertical displacements from P to Q. By the Pythagorean Theorem,

$$\|\mathbf{v}\| = \sqrt{a^2 + b^2}$$

Note that the components $\langle a, b \rangle$ determine the length and direction of **v**, but not its basepoint. Therefore, two vectors are equivalent if and only if they have the same components. Furthermore, **v** is equivalent to the position vector $\mathbf{v}_0 = \overrightarrow{OP_0}$ with terminal point $P_0 = (a, b)$ (Figure 4).

We write $\mathbf{v} = \langle a, b \rangle$ if **v** has components a and b. This notation is ambiguous because it does not indicate the basepoint of **v**. However, in practice this rarely causes confusion. To further avoid confusion, the following convention will be in force for the remainder of the text:

Convention: *We assume all vectors are based at the origin unless otherwise stated.*

The **zero vector** (whose head and tail coincide) is the vector $\mathbf{0} = \langle 0, 0 \rangle$ of length zero.

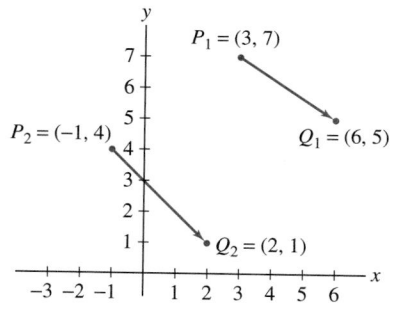

FIGURE 5

■ **EXAMPLE 2** Find the components of $\mathbf{v}_1 = \overrightarrow{P_1 Q_1}$ and $\mathbf{v}_2 = \overrightarrow{P_2 Q_2}$, where

$$P_1 = (3, 7), \quad Q_1 = (6, 5) \qquad \text{and} \qquad P_2 = (-1, 4), \quad Q_2 = (2, 1)$$

Are $\mathbf{v}_1$ and $\mathbf{v}_2$ equivalent?

Solution We compute the components of $\mathbf{v}_1$ and $\mathbf{v}_2$:

$$\mathbf{v}_1 = \langle 6 - 3, 5 - 7 \rangle = \langle 3, -2 \rangle, \qquad \mathbf{v}_2 = \langle 2 - (-1), 1 - 4 \rangle = \langle 3, -3 \rangle$$

The vectors $\mathbf{v}_1$ and $\mathbf{v}_2$ do not have the same components and hence are not equivalent (Figure 5). ■

■ **EXAMPLE 3** Sketch the vector $\mathbf{v} = \langle 2, -3 \rangle$ based at $P = (1, 4)$ and the vector $\mathbf{v}_0$ equivalent to $\mathbf{v}$ based at the origin.

Solution The vector $\mathbf{v} = \langle 2, -3 \rangle$ based at P has terminal point Q located two units to the right and three units down from P as shown in Figure 6. Therefore, $Q = (1 + 2, 4 - 3) = (3, 1)$. The vector $\mathbf{v}_0$ equivalent to $\mathbf{v}$ based at O has terminal point $(2, -3)$. ■

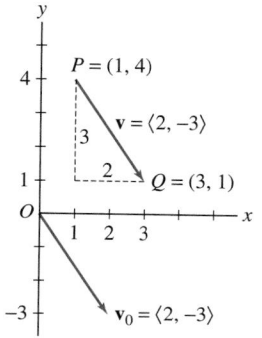

FIGURE 6 The vectors $\mathbf{v}$ and $\mathbf{v}_0$ have the same components $\langle 2, -3 \rangle$ but different basepoints.

NOTATION λ is the eleventh letter in the Greek alphabet. We often (but not exclusively) use the symbol λ to denote a scalar.

Vector Algebra

We now define two basic vector operations: vector addition and scalar multiplication. The term **scalar** is another word for "real number," and we often speak of scalar versus vector quantities. Thus, the number 8 is a scalar, while $\langle 8, 2 \rangle$ is a vector.

If λ is a scalar and $\mathbf{v}$ is a nonzero vector, the scalar multiple $\lambda \mathbf{v}$ is defined as the vector of length $|\lambda| \, \|\mathbf{v}\|$ pointing in the same direction as $\mathbf{v}$ if $\lambda > 0$ and in the opposite direction if $\lambda < 0$ (Figure 7). For $\lambda = 0$, we set $0\mathbf{v} = \mathbf{0}$. In all cases,

$$\boxed{\|\lambda \mathbf{v}\| = |\lambda| \, \|\mathbf{v}\|}$$

In particular, $-\mathbf{v}$ has the same length as $\mathbf{v}$ but points in the opposite direction. A vector $\mathbf{w}$ is parallel to $\mathbf{v}$ if and only if $\mathbf{w} = \lambda \mathbf{v}$ for some scalar λ.

The vector sum $\mathbf{v} + \mathbf{w}$ is defined when $\mathbf{v}$ and $\mathbf{w}$ have the same basepoint. The sum can be described in two ways. First, translate $\mathbf{w}$ to the equivalent vector $\mathbf{w}'$ whose tail coincides with the head of $\mathbf{v}$. Then $\mathbf{v} + \mathbf{w}$ is the vector pointing from the tail of $\mathbf{v}$ to the head of $\mathbf{w}'$ [Figure 8(A)]. Alternatively, we may use the **Parallelogram Law**: $\mathbf{v} + \mathbf{w}$ is the vector pointing from the basepoint to the opposite vertex of the parallelogram formed by $\mathbf{v}$ and $\mathbf{w}$ [Figure 8(B)].

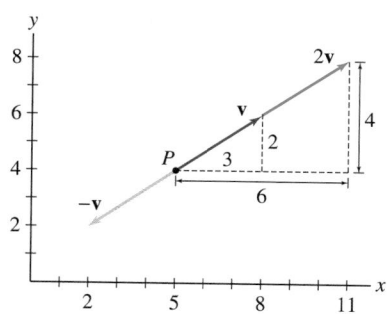

FIGURE 7 The vector $\mathbf{v} = \langle 3, 2 \rangle$ based at $P = (5, 4)$. The vector $2\mathbf{v}$ is also based at P and is twice as long as $\mathbf{v}$.

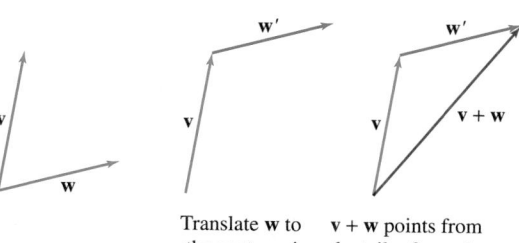

Translate $\mathbf{w}$ to the vector $\mathbf{w}'$

$\mathbf{v} + \mathbf{w}$ points from the tail of $\mathbf{v}$ to the head of $\mathbf{w}'$

(A) The vector sum $\mathbf{v} + \mathbf{w}$.

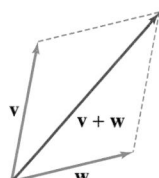

(B) $\mathbf{v} + \mathbf{w}$ can also be formed using the Parallelogram Law.

FIGURE 8

To add several vectors $\mathbf{v}_1, \mathbf{v}_2, \ldots, \mathbf{v}_n$, translate the vectors to $\mathbf{v}_1 = \mathbf{v}'_1, \mathbf{v}'_2, \ldots, \mathbf{v}'_n$ so that they lie head to tail as in Figure 9. The vector sum $\mathbf{v} = \mathbf{v}_1 + \mathbf{v}_2 + \cdots + \mathbf{v}_n$ is the vector whose terminal point is the terminal point of $\mathbf{v}'_n$.

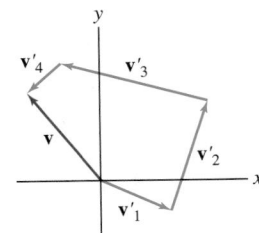

FIGURE 9 The sum
$\mathbf{v} = \mathbf{v}_1 + \mathbf{v}_2 + \mathbf{v}_3 + \mathbf{v}_4$.

Vector subtraction $\mathbf{v} - \mathbf{w}$ is carried out by adding $-\mathbf{w}$ to $\mathbf{v}$ as in Figure 10(A). More simply, draw the vector pointing from $\mathbf{w}$ to $\mathbf{v}$ as in Figure 10(B), and translate it back to the basepoint to obtain $\mathbf{v} - \mathbf{w}$.

| *Remember that the vector $\mathbf{v} - \mathbf{w}$ points in the direction from $\mathbf{w}$ to $\mathbf{v}$ (not from $\mathbf{v}$ to $\mathbf{w}$).*

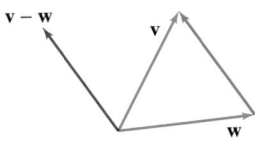

(A) The difference $\mathbf{v} - \mathbf{w}$ is obtained by adding $\mathbf{v}$ and $-\mathbf{w}$.

(B) More simply, to form $\mathbf{v} - \mathbf{w}$, draw the vector pointing from $\mathbf{w}$ to $\mathbf{v}$ and translate it back to the basepoint.

FIGURE 10 Vector subtraction.

The vector operations are easily performed using components. To add (or subtract) two vectors, we add (or subtract) their components. This follows from the parallelogram law as indicated in Figure 11(A). Similarly, if $\mathbf{v} = \langle a, b \rangle$, then $\lambda \mathbf{v} = \langle \lambda a, \lambda b \rangle$ because $\langle \lambda a, \lambda b \rangle$ is a vector of length $|\lambda| \, \|\mathbf{v}\|$ that, if λ and $\mathbf{v}$ are nonzero, points in the same or opposite direction as $\langle a, b \rangle$ depending on whether $\lambda > 0$ or $\lambda < 0$ [Figures 11(B) and (C)]. Therefore,

---

Vector Operations Using Components   If $\mathbf{v} = \langle a, b \rangle$ and $\mathbf{w} = \langle c, d \rangle$, then:

(i) $\mathbf{v} + \mathbf{w} = \langle a + c, b + d \rangle$

(ii) $\mathbf{v} - \mathbf{w} = \langle a - c, b - d \rangle$

(iii) $\lambda \mathbf{v} = \langle \lambda a, \lambda b \rangle$

(iv) $\mathbf{v} + \mathbf{0} = \mathbf{0} + \mathbf{v} = \mathbf{v}$ and $\mathbf{v} - \mathbf{v} = \mathbf{0}$

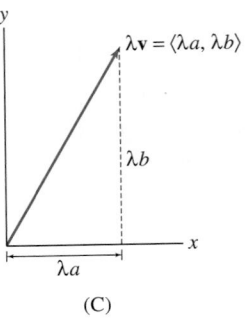

FIGURE 11 Vector operations using components.

(A)

(B)

(C)

We also note that if $P = (a_1, b_1)$ and $Q = (a_2, b_2)$, then components of the vector $\mathbf{v} = \overrightarrow{PQ}$ are conveniently computed as the difference:

$$\overrightarrow{PQ} = \overrightarrow{OQ} - \overrightarrow{OP} = \langle a_2, b_2 \rangle - \langle a_1, b_1 \rangle = \langle a_2 - a_1, b_2 - b_1 \rangle$$

■ EXAMPLE 4 Let $\mathbf{v} = \langle 1, 4 \rangle$ and $\mathbf{w} = \langle 3, 2 \rangle$. Calculate **(a)** $\mathbf{v} + \mathbf{w}$ and **(b)** $5\mathbf{v}$.

Solution

$$\mathbf{v} + \mathbf{w} = \langle 1, 4 \rangle + \langle 3, 2 \rangle = \langle 1 + 3, 4 + 2 \rangle = \langle 4, 6 \rangle$$
$$5\mathbf{v} = 5 \langle 1, 4 \rangle = \langle 5, 20 \rangle$$

The vector sum is illustrated in Figure 12. ■

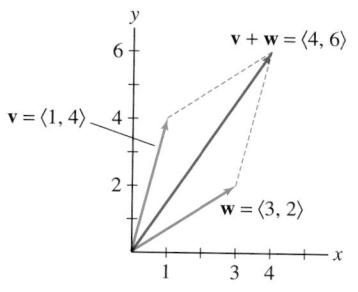

FIGURE 12

The vector operations obey the usual laws of algebra.

THEOREM 1 Basic Properties of Vector Algebra For all vectors $\mathbf{u}$, $\mathbf{v}$, $\mathbf{w}$ and scalars λ,

Commutative Law:	$\mathbf{v} + \mathbf{w} = \mathbf{w} + \mathbf{v}$
Associative Law:	$\mathbf{u} + (\mathbf{v} + \mathbf{w}) = (\mathbf{u} + \mathbf{v}) + \mathbf{w}$
Distributive Law for Scalars:	$\lambda(\mathbf{v} + \mathbf{w}) = \lambda\mathbf{v} + \lambda\mathbf{w}$

These properties are easily verified using components. For example, we may check that vector addition is commutative as follows:

$$\langle a, b \rangle + \langle c, d \rangle = \underbrace{\langle a + c, b + d \rangle = \langle c + a, d + b \rangle}_{\text{Commutativity of ordinary addition}} = \langle c, d \rangle + \langle a, b \rangle$$

A **linear combination** of vectors $\mathbf{v}$ and $\mathbf{w}$ is a vector of the form $r\mathbf{v} + s\mathbf{w}$, where r and s are scalars. If $\mathbf{v}$ and $\mathbf{w}$ are not parallel, then every vector $\mathbf{u}$ can be expressed as a linear combination $\mathbf{u} = r\mathbf{v} + s\mathbf{w}$ [Figure 13(A)]. We refer to the parallelogram $\mathcal{P}$ whose vertices are the origin and the terminal points of $\mathbf{v}$, $\mathbf{w}$ and $\mathbf{v} + \mathbf{w}$ as the parallelogram **spanned** by $\mathbf{v}$ and $\mathbf{w}$ [Figure 13(B)]. Observe that $\mathcal{P}$ consists of the linear combinations $r\mathbf{v} + s\mathbf{w}$ with $0 \le r, s \le 1$.

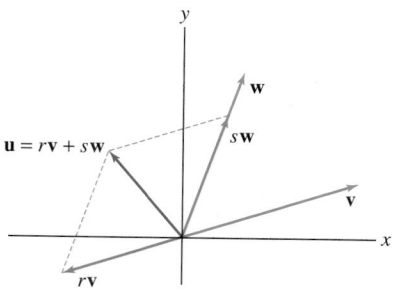

(A) The vector $\mathbf{u}$ can be expressed as a linear combination $\mathbf{u} = r\mathbf{v} + s\mathbf{w}$. In this figure, $r < 0$.

(B) The parallelogram $\mathcal{P}$ spanned by $\mathbf{v}$ and $\mathbf{w}$ consists of all linear combinations $r\mathbf{v} + s\mathbf{w}$ with $0 \le r, s \le 1$.

FIGURE 13

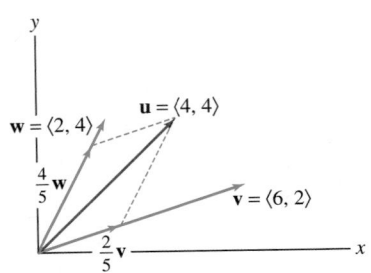

FIGURE 14

EXAMPLE 5 Linear Combinations Express $\mathbf{u} = \langle 4, 4 \rangle$ as a linear combination of $\mathbf{v} = \langle 6, 2 \rangle$ and $\mathbf{w} = \langle 2, 4 \rangle$.

Solution We must find r and s such that $r\mathbf{v} + s\mathbf{w} = \langle 4, 4 \rangle$, or

$$r\langle 6, 2 \rangle + s\langle 2, 4 \rangle = \langle 6r + 2s, 2r + 4s \rangle = \langle 4, 4 \rangle$$

This gives $6r + 2s = 4$ and $2r + 4s = 4$. Subtracting the equations, we obtain $4r - 2s = 0$ or $s = 2r$. Substitution in the first equation yields $6r + 4r = 4$ or $r = \frac{2}{5}$ and $s = \frac{4}{5}$. Thus (Figure 14),

$$\mathbf{u} = \langle 4, 4 \rangle = \frac{2}{5}\langle 6, 2 \rangle + \frac{4}{5}\langle 2, 4 \rangle$$ ■

A vector of length one is called a **unit vector**. Unit vectors are often used to indicate direction, when it is not necessary to specify length. The head of a unit vector $\mathbf{e}$ based at the origin lies on the unit circle and has components $\mathbf{e} = \langle \cos\theta, \sin\theta \rangle$, where θ is the angle between the x-axis and $\mathbf{e}$, measured counterclockwise (Figure 15).

If $\mathbf{v} = \langle a, b \rangle$ is a nonzero vector, then the vector

$$\boxed{\mathbf{e_v} = \frac{1}{\|\mathbf{v}\|}\mathbf{v}}$$

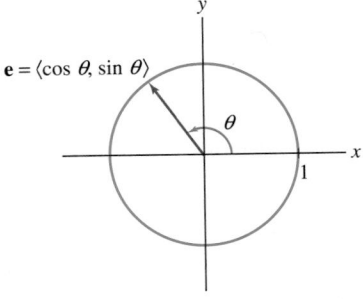

FIGURE 15 The head of a unit vector lies on the unit circle.

is the unit vector pointing in the direction of $\mathbf{v}$ (Figure 16). Indeed,

$$\|\mathbf{e_v}\| = \left\|\frac{1}{\|\mathbf{v}\|}\mathbf{v}\right\| = \frac{1}{\|\mathbf{v}\|}\|\mathbf{v}\| = 1$$

and $\mathbf{e_v}$ is a positive scalar multiple of $\mathbf{v}$, so $\mathbf{e_v}$ and $\mathbf{v}$ point in the same direction. If $\mathbf{v}$ makes an angle θ with the positive x-axis, then (Figure 16)

$$\mathbf{e_v} = \langle \cos\theta, \sin\theta \rangle$$

Furthermore, the components of $\mathbf{v}$ are $a = \|\mathbf{v}\|\cos\theta$, $b = \|\mathbf{v}\|\sin\theta$. Thus

$$\mathbf{v} = \langle a, b \rangle = \|\mathbf{v}\|\mathbf{e_v} = \|\mathbf{v}\|\langle \cos\theta, \sin\theta \rangle \qquad \boxed{1}$$

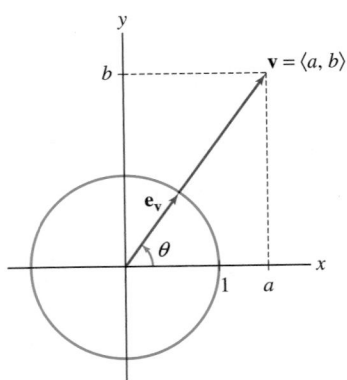

FIGURE 16 Unit vector in the direction of $\mathbf{v}$.

EXAMPLE 6 Find the unit vector in the direction of $\mathbf{v} = \langle 3, 5 \rangle$.

Solution $\|\mathbf{v}\| = \sqrt{3^2 + 5^2} = \sqrt{34}$ and $\mathbf{e_v} = \frac{1}{\sqrt{34}}\mathbf{v} = \left\langle \frac{3}{\sqrt{34}}, \frac{5}{\sqrt{34}} \right\rangle$. ■

It is customary to introduce a special notation for the unit vectors in the direction of the positive x- and y-axes (Figure 17):

$$\boxed{\mathbf{i} = \langle 1, 0 \rangle, \qquad \mathbf{j} = \langle 0, 1 \rangle}$$

The vectors $\mathbf{i}$ and $\mathbf{j}$ are called the **standard basis vectors**. Every vector can be written as a linear combination of these basis vectors (Figure 17):

$$\langle a, b \rangle = a\mathbf{i} + b\mathbf{j}$$

For example, $\langle 4, -2 \rangle = 4\mathbf{i} - 2\mathbf{j}$. Vector addition is performed by adding the $\mathbf{i}$ and $\mathbf{j}$ coefficients. For example,

$$(4\mathbf{i} - 2\mathbf{j}) + (5\mathbf{i} + 7\mathbf{j}) = (4 + 5)\mathbf{i} + (-2 + 7)\mathbf{j} = 9\mathbf{i} + 5\mathbf{j}$$

FIGURE 17

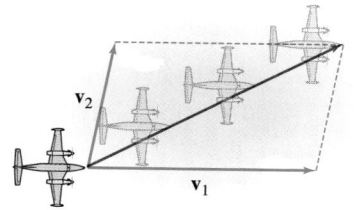

FIGURE 18 The resultant velocity of an airplane traveling with velocity $\mathbf{v}_1$ when it encounters a wind of velocity $\mathbf{v}_2$ is the vector sum $\mathbf{v}_1 + \mathbf{v}_2$.

CONCEPTUAL INSIGHT We commonly say that quantities such as force or velocity are vectors because they have both magnitude and direction, but there is more to this statement than meets the eye. A vector quantity must obey the law of vector addition (Figure 18), so when we say that force is a vector quantity, we are implicitly claiming that forces add according to the Parallelogram Law. In other words, if two forces $\mathbf{F}_1$ and $\mathbf{F}_2$ act on an object, then the resultant force is the **vector sum** $\mathbf{F}_1 + \mathbf{F}_2$. This is a physical fact that must be verified experimentally. It was well known to physicists in the time of Galileo and Newton, who used the Parallelogram Law long before the formal introduction of vectors by Hamilton and others in the 1800s.

■ **EXAMPLE 7** Calculate the magnitudes of the forces on cables 1 and 2 in Figure 19(A).

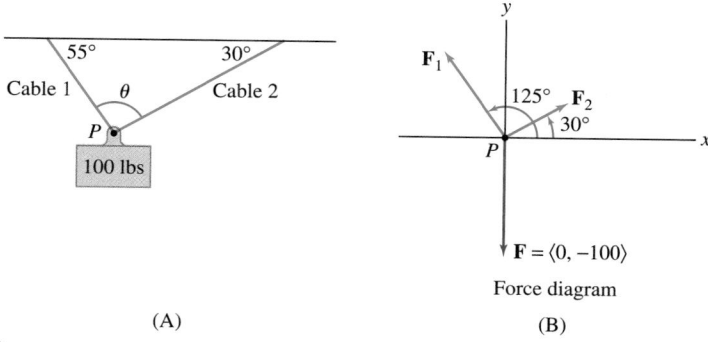

(A)   (B)

FIGURE 19

Solution Three forces act on the point P in Figure 19(A). A 100-lb force $\mathbf{F}$ due to gravity acts vertically downward and two unknown forces $\mathbf{F}_1$ and $\mathbf{F}_2$ act through cables 1 and 2. The point P is not in motion, so the net force acting on P must be zero, that is,

$$\mathbf{F}_1 + \mathbf{F}_2 + \mathbf{F} = \mathbf{0} \qquad \boxed{2}$$

The three forces are indicated in the force diagram in Figure 19(B). Notice that the angle between $\mathbf{F}_1$ and $\mathbf{F}_2$ is $\theta = 180 - 55 - 30 = 95°$. The vector $\mathbf{F}_2$ makes an angle of $30°$ with the positive x-axis and hence $\mathbf{F}_1$ makes an angle of $95 + 30 = 125°$. Let $f_1 = \|\mathbf{F}_1\|$ and $f_2 = \|\mathbf{F}_2\|$ be the magnitudes of $\mathbf{F}_1$ and $\mathbf{F}_2$. We use Eq. (1) and the table in the margin to write the force vectors in component form:

$$\mathbf{F}_1 = f_1 \langle \cos 125°, \sin 125° \rangle \approx f_1 \langle -0.573, 0.819 \rangle$$
$$\mathbf{F}_2 = f_2 \langle \cos 30°, \sin 30° \rangle \approx f_2 \langle 0.866, 0.5 \rangle$$
$$\mathbf{F} = \langle 0, -100 \rangle$$

θ	$\cos \theta$	$\sin \theta$
125°	−0.573	0.819
30°	0.866	0.5

Therefore, Eq. (2) becomes

$$\mathbf{F}_1 + \mathbf{F}_2 + \mathbf{F} = f_1 \langle -0.573, 0.819 \rangle + f_2 \langle 0.866, 0.5 \rangle + \langle 0, -100 \rangle = \langle 0, 0 \rangle$$

This gives us two equations:

$$-0.573 f_1 + 0.866 f_2 = 0, \qquad 0.819 f_1 + 0.5 f_2 - 100 = 0$$

By the first equation, $f_2 = \left(\dfrac{0.573}{0.866}\right) f_1$. The second equation then gives

$$0.819 f_1 + 0.5 f_2 - 100 = 0.819 f_1 + 0.5 \left(\frac{0.573}{0.866}\right) f_1 - 100 \approx 1.15 f_1 - 100 = 0$$

Therefore,

$$f_1 \approx \frac{100}{1.15} \approx 87 \text{ lb} \qquad \text{and} \qquad f_2 \approx \left(\frac{0.573}{0.866}\right) 87 \approx 58 \text{ lb} \qquad \blacksquare$$

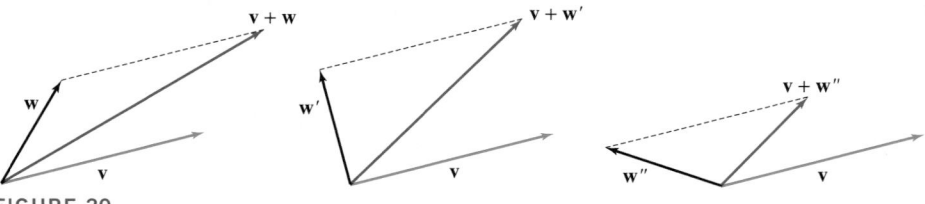

FIGURE 20

We close this section by stating the Triangle Inequality. As we see in Figure 20, the magnitude of a sum $\|\mathbf{v} + \mathbf{w}\|$ depends on the angle between $\mathbf{v}$ and $\mathbf{w}$. So in general, $\|\mathbf{v} + \mathbf{w}\|$ is not equal to the sum $\|\mathbf{v}\| + \|\mathbf{w}\|$. What we can say is that $\|\mathbf{v} + \mathbf{w}\|$ is *at most* equal to the sum $\|\mathbf{v}\| + \|\mathbf{w}\|$. This corresponds to the fact that the length of one side of a triangle is at most the sum of the lengths of the other two sides. A formal proof may be given using the dot product (see Exercise 81 in Section 12.3).

THEOREM 2 Triangle Inequality   For any two vectors $\mathbf{v}$ and $\mathbf{w}$,

$$\boxed{\|\mathbf{v} + \mathbf{w}\| \le \|\mathbf{v}\| + \|\mathbf{w}\|}$$

Equality holds if and only if $\mathbf{w} = \lambda \mathbf{v}$, where $\lambda \ge 0$, or $\mathbf{v} = \mathbf{0}$.

12.1 SUMMARY

- A *vector* $\mathbf{v} = \overrightarrow{PQ}$ is determined by a basepoint P (the "tail") and a terminal point Q (the "head"). The *length* of $\mathbf{v} = \overrightarrow{PQ}$, denoted $\|\mathbf{v}\|$, is the distance from P to Q.
- The vector $\mathbf{v} = \overrightarrow{OP_0}$ based at the origin $O = (0, 0)$, with head $P_0 = (a, b)$ is called the *position vector* of P_0.
- Two vectors $\mathbf{v}$ and $\mathbf{w}$ are *equivalent* if they are translates of each other, that is, if they have the same magnitude and direction, but possibly different basepoints.
- The vector $\mathbf{v} = \overrightarrow{PQ}$ with $P = (a_1, b_1)$ and $Q = (a_2, b_2)$ has *components* $a = a_2 - a_1$, $b = b_2 - b_1$.
- Two vectors are equivalent if and only if they have the same components. If $\mathbf{v}$ has components a and b, we write $\mathbf{v} = \langle a, b \rangle$. Although this notation is ambiguous (because equivalent vectors have the same components), it rarely causes confusion. Our convention is that all vectors are assumed to be based at the origin unless otherwise indicated.
- If $\mathbf{v} = \langle a, b \rangle$, then $\|\mathbf{v}\| = \sqrt{a^2 + b^2}$. The *zero vector* is the vector $\mathbf{0} = \langle 0, 0 \rangle$ of length 0.

- *Vector addition* is defined geometrically by the *Parallelogram Law*. If λ is a scalar, then the *scalar multiple* $\lambda \mathbf{v}$ is the vector of length $|\lambda| \, \|\mathbf{v}\|$ in the same direction as $\mathbf{v}$ if $\lambda > 0$ and in the opposite direction if $\lambda < 0$. For $\lambda = 0$, we set $0\mathbf{v} = \mathbf{0}$. In components,

$$\langle a_1, b_1 \rangle + \langle a_2, b_2 \rangle = \langle a_1 + a_2, b_1 + b_2 \rangle$$

$$\lambda \langle a, b \rangle = \langle \lambda a, \lambda b \rangle$$

- Nonzero vectors $\mathbf{v}$ and $\mathbf{w}$ are *parallel* if $\mathbf{w} = \lambda \mathbf{v}$ for some scalar λ (they point in the same direction if $\lambda > 0$).
- A *unit vector* is a vector of length one. If $\mathbf{v} \neq \mathbf{0}$, the unit vector pointing in the direction $\mathbf{v}$ is $\mathbf{e_v} = \dfrac{1}{\|\mathbf{v}\|} \mathbf{v}$.
- Every vector can be expressed as a linear combination of the *standard basis vectors* $\mathbf{i} = \langle 1, 0 \rangle$ and $\mathbf{j} = \langle 0, 1 \rangle$. If $\mathbf{v} = \langle a, b \rangle$, then $\mathbf{v} = a\mathbf{i} + b\mathbf{j}$.
- If the vector $\mathbf{v} = \langle a, b \rangle$ is nonzero and makes an angle θ with the positive x-axis, then $a = \|\mathbf{v}\| \cos \theta$ and $b = \|\mathbf{v}\| \sin \theta$, and $\mathbf{e_v} = \langle \cos \theta, \sin \theta \rangle$.
- The *Triangle Inequality:* For any two vectors $\mathbf{v}$ and $\mathbf{w}$,

$$\|\mathbf{v} + \mathbf{w}\| \leq \|\mathbf{v}\| + \|\mathbf{w}\|$$

Equality holds if and only if $\mathbf{w} = \lambda \mathbf{v}$, where $\lambda \geq 0$ or $\mathbf{v} = \mathbf{0}$.

12.1 EXERCISES

Preliminary Questions

1. Answer true or false. Every nonzero vector is:
(a) Equivalent to a vector based at the origin.
(b) Equivalent to a unit vector based at the origin.
(c) Parallel to a vector based at the origin.
(d) Parallel to a unit vector based at the origin.

2. What is the length of $-3\mathbf{a}$ if $\|\mathbf{a}\| = 5$?

3. Suppose that $\mathbf{v}$ has components $\langle 3, 1 \rangle$. How, if at all, do the components change if you translate $\mathbf{v}$ horizontally two units to the left?

4. What are the components of the zero vector based at $P = \langle 3, 5 \rangle$?

5. Are the following true or false?
(a) The vectors $\mathbf{v}$ and $-2\mathbf{v}$ are parallel.
(b) The vectors $\mathbf{v}$ and $-2\mathbf{v}$ point in the same direction.

6. Explain the commutativity of vector addition in terms of the Parallelogram Law.

Exercises

1. Sketch the vectors $\mathbf{v}_1, \mathbf{v}_2, \mathbf{v}_3, \mathbf{v}_4$ with tail P and head Q, and compute their lengths. Are any two of these vectors equivalent?

	$\mathbf{v}_1$	$\mathbf{v}_2$	$\mathbf{v}_3$	$\mathbf{v}_4$
P	$(2, 4)$	$(-1, 3)$	$(-1, 3)$	$(4, 1)$
Q	$(4, 4)$	$(1, 3)$	$(2, 4)$	$(6, 3)$

2. Sketch the vector $\mathbf{b} = \langle 3, 4 \rangle$ based at $P = (-2, -1)$.

3. What is the terminal point of the vector $\mathbf{a} = \langle 1, 3 \rangle$ based at $P = (2, 2)$? Sketch $\mathbf{a}$ and the vector $\mathbf{a}_0$ based at the origin equivalent to $\mathbf{a}$.

4. Let $\mathbf{v} = \overrightarrow{PQ}$, where $P = (1, 1)$ and $Q = (2, 2)$. What is the head of the vector $\mathbf{v}'$ equivalent to $\mathbf{v}$ based at $(2, 4)$? What is the head of the vector $\mathbf{v}_0$ equivalent to $\mathbf{v}$ based at the origin? Sketch $\mathbf{v}$, $\mathbf{v}_0$, and $\mathbf{v}'$.

In Exercises 5–8, find the components of $\overrightarrow{PQ}$.

5. $P = (3, 2)$, $Q = (2, 7)$
6. $P = (1, -4)$, $Q = (3, 5)$
7. $P = (3, 5)$, $Q = (1, -4)$
8. $P = (0, 2)$, $Q = (5, 0)$

In Exercises 9–14, calculate.

9. $\langle 2, 1 \rangle + \langle 3, 4 \rangle$
10. $\langle -4, 6 \rangle - \langle 3, -2 \rangle$
11. $5 \langle 6, 2 \rangle$
12. $4(\langle 1, 1 \rangle + \langle 3, 2 \rangle)$
13. $\left\langle -\frac{1}{2}, \frac{5}{3} \right\rangle + \left\langle 3, \frac{10}{3} \right\rangle$
14. $\langle \ln 2, e \rangle + \langle \ln 3, \pi \rangle$

15. Which of the vectors (A)–(C) in Figure 21 is equivalent to $\mathbf{v} - \mathbf{w}$?

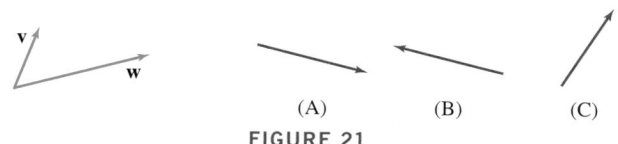

(A)   (B)   (C)

FIGURE 21

16. Sketch $\mathbf{v} + \mathbf{w}$ and $\mathbf{v} - \mathbf{w}$ for the vectors in Figure 22.

FIGURE 22

17. Sketch $2\mathbf{v}$, $-\mathbf{w}$, $\mathbf{v} + \mathbf{w}$, and $2\mathbf{v} - \mathbf{w}$ for the vectors in Figure 23.

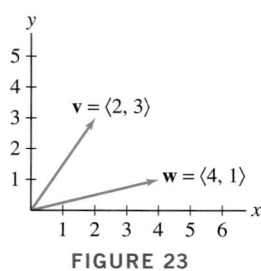

FIGURE 23

18. Sketch $\mathbf{v} = \langle 1, 3 \rangle$, $\mathbf{w} = \langle 2, -2 \rangle$, $\mathbf{v} + \mathbf{w}$, $\mathbf{v} - \mathbf{w}$.

19. Sketch $\mathbf{v} = \langle 0, 2 \rangle$, $\mathbf{w} = \langle -2, 4 \rangle$, $3\mathbf{v} + \mathbf{w}$, $2\mathbf{v} - 2\mathbf{w}$.

20. Sketch $\mathbf{v} = \langle -2, 1 \rangle$, $\mathbf{w} = \langle 2, 2 \rangle$, $\mathbf{v} + 2\mathbf{w}$, $\mathbf{v} - 2\mathbf{w}$.

21. Sketch the vector $\mathbf{v}$ such that $\mathbf{v} + \mathbf{v}_1 + \mathbf{v}_2 = \mathbf{0}$ for $\mathbf{v}_1$ and $\mathbf{v}_2$ in Figure 24(A).

22. Sketch the vector sum $\mathbf{v} = \mathbf{v}_1 + \mathbf{v}_2 + \mathbf{v}_3 + \mathbf{v}_4$ in Figure 24(B).

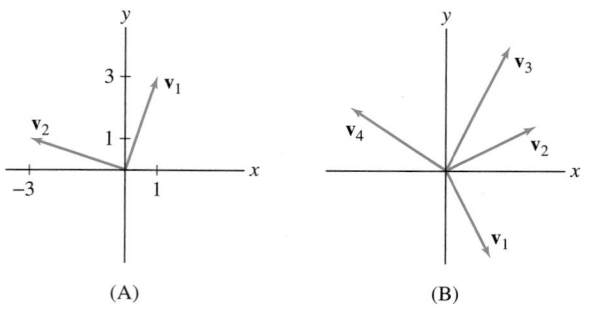

(A) (B)

FIGURE 24

23. Let $\mathbf{v} = \overrightarrow{PQ}$, where $P = (-2, 5)$, $Q = (1, -2)$. Which of the vectors with the following given tails and heads are equivalent to $\mathbf{v}$?
(a) $(-3, 3)$, $(0, 4)$ **(b)** $(0, 0)$, $(3, -7)$
(c) $(-1, 2)$, $(2, -5)$ **(d)** $(4, -5)$, $(1, 4)$

24. Let $\mathbf{v} = \langle 6, 9 \rangle$. Which of the following vectors are parallel to $\mathbf{v}$ and which point in the same direction?
(a) $\langle 12, 18 \rangle$ **(b)** $\langle 3, 2 \rangle$ **(c)** $\langle 2, 3 \rangle$
(d) $\langle -6, -9 \rangle$ **(e)** $\langle -24, -27 \rangle$ **(f)** $\langle -24, -36 \rangle$

In Exercises 25–28, sketch the vectors $\overrightarrow{AB}$ and $\overrightarrow{PQ}$, and determine whether they are equivalent.

25. $A = (1, 1)$, $B = (3, 7)$, $P = (4, -1)$, $Q = (6, 5)$

26. $A = (1, 4)$, $B = (-6, 3)$, $P = (1, 4)$, $Q = (6, 3)$

27. $A = (-3, 2)$, $B = (0, 0)$, $P = (0, 0)$, $Q = (3, -2)$

28. $A = (5, 8)$, $B = (1, 8)$, $P = (1, 8)$, $Q = (-3, 8)$

In Exercises 29–32, are $\overrightarrow{AB}$ and $\overrightarrow{PQ}$ parallel (and if so, do they point in the same direction)?

29. $A = (1, 1)$, $B = (3, 4)$, $P = (1, 1)$, $Q = (7, 10)$

30. $A = (-3, 2)$, $B = (0, 0)$, $P = (0, 0)$, $Q = (3, 2)$

31. $A = (2, 2)$, $B = (-6, 3)$, $P = (9, 5)$, $Q = (17, 4)$

32. $A = (5, 8)$, $B = (2, 2)$, $P = (2, 2)$, $Q = (-3, 8)$

In Exercises 33–36, let $R = (-2, 7)$. Calculate the following.

33. The length of $\overrightarrow{OR}$

34. The components of $\mathbf{u} = \overrightarrow{PR}$, where $P = (1, 2)$

35. The point P such that $\overrightarrow{PR}$ has components $\langle -2, 7 \rangle$

36. The point Q such that $\overrightarrow{RQ}$ has components $\langle 8, -3 \rangle$

In Exercises 37–44, find the given vector.

37. Unit vector $\mathbf{e}_\mathbf{v}$ where $\mathbf{v} = \langle 3, 4 \rangle$

38. Unit vector $\mathbf{e}_\mathbf{w}$ where $\mathbf{w} = \langle 21, -20 \rangle$

39. Unit vector in the direction of $\mathbf{u} = \langle -1, -1 \rangle$

40. Vector of length 4 in the direction of $\mathbf{u} = \langle -1, -1 \rangle$

41. Unit vector in the direction opposite to $\mathbf{v} = \langle -2, 4 \rangle$

42. Unit vector $\mathbf{e}$ making an angle of $\frac{4\pi}{7}$ with the x-axis

43. Vector of length 2 making an angle of $30°$ with the x-axis

44. Vector of length 4 making an angle of $\frac{7\pi}{3}$ with the y-axis

45. Find all scalars λ such that $\lambda \langle 2, 3 \rangle$ has length one.

46. What are the coordinates of the point P in the parallelogram in Figure 25(A)?

47. What are the coordinates a and b in the parallelogram in Figure 25(B)?

(A) (B)

FIGURE 25

48. Let $\mathbf{v} = \overrightarrow{AB}$ and $\mathbf{w} = \overrightarrow{AC}$, where A, B, C are three distinct points in the plane. Match (a)–(d) with (i)–(iv). (*Hint:* Draw a picture.)

(a) $-\mathbf{w}$ (b) $-\mathbf{v}$ (c) $\mathbf{w} - \mathbf{v}$ (d) $\mathbf{v} - \mathbf{w}$

(i) $\overrightarrow{CB}$ (ii) $\overrightarrow{CA}$ (iii) $\overrightarrow{BC}$ (iv) $\overrightarrow{BA}$

49. Find the components and length of the following vectors:

(a) $4\mathbf{i} + 3\mathbf{j}$ (b) $2\mathbf{i} - 3\mathbf{j}$ (c) $\mathbf{i} + \mathbf{j}$ (d) $\mathbf{i} - 3\mathbf{j}$

In Exercises 50–53, calculate the linear combination.

50. $(-2\mathbf{i} + 9\mathbf{j}) + (3\mathbf{i} - 4\mathbf{j})$ **51.** $6(\mathbf{i} + 9\mathbf{j}) + 2(\mathbf{i} - 4\mathbf{j})$

52. $(\mathbf{i} + \mathbf{j}) + (\mathbf{i} - 2\mathbf{j})$ **53.** $3(\mathbf{i} + \mathbf{j}) + 7(3\mathbf{i} - \mathbf{j})$

54. For each of the position vectors $\mathbf{u}$ with endpoints A, B, and C in Figure 26, indicate with a diagram the multiples $r\mathbf{v}$ and $s\mathbf{w}$ such that $\mathbf{u} = r\mathbf{v} + s\mathbf{w}$. A sample is shown for $\mathbf{u} = \overrightarrow{OQ}$.

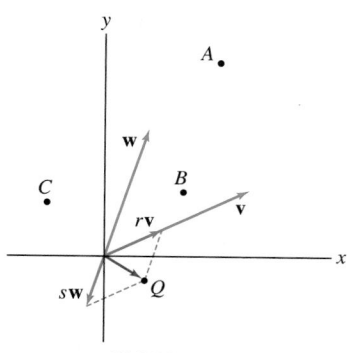

FIGURE 26

55. Express $\mathbf{u} = \langle 3, -1 \rangle$ as a linear combination $\mathbf{u} = r\mathbf{v} + s\mathbf{w}$, where $\mathbf{v} = \langle 2, 1 \rangle$ and $\mathbf{w} = \langle 1, 3 \rangle$. Sketch the vectors $\mathbf{u}$, $\mathbf{v}$, $\mathbf{w}$ and the parallelogram formed by $r\mathbf{v}$ and $s\mathbf{w}$.

56. Express $\mathbf{u} = \langle 6, -2 \rangle$ as a linear combination $\mathbf{u} = r\mathbf{v} + s\mathbf{w}$, where $\mathbf{v} = \langle 1, 1 \rangle$ and $\mathbf{w} = \langle 1, -1 \rangle$. Sketch the vectors $\mathbf{u}$, $\mathbf{v}$, $\mathbf{w}$ and the parallelogram formed by $r\mathbf{v}$ and $s\mathbf{w}$.

57. Sketch the parallelogram spanned by $\mathbf{v} = \langle 1, 4 \rangle$ and $\mathbf{w} = \langle 5, 2 \rangle$. Add the vector $\mathbf{u} = \langle 2, 3 \rangle$ to the sketch and express $\mathbf{u}$ as a linear combination of $\mathbf{v}$ and $\mathbf{w}$.

58. Calculate the magnitude of the force on cables 1 and 2 in Figure 27.

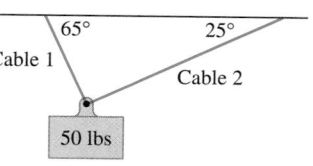

FIGURE 27

59. Determine the magnitude of the forces $\mathbf{F}_1$ and $\mathbf{F}_2$ in Figure 28, assuming that there is no net force on the object.

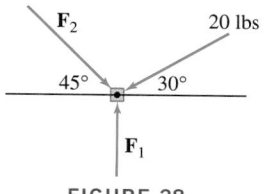

FIGURE 28

60. A plane flying due east at 200 km/hour encounters a 40 km/hour north-easterly wind. The resultant velocity of the plane is the vector sum $\mathbf{v} = \mathbf{v}_1 + \mathbf{v}_2$, where $\mathbf{v}_1$ is the velocity vector of the plane and $\mathbf{v}_2$ is the velocity vector of the wind (Figure 29). The angle between $\mathbf{v}_1$ and $\mathbf{v}_2$ is $\frac{\pi}{4}$. Determine the resultant *speed* of the plane (the length of the $\mathbf{v}$).

FIGURE 29

Further Insights and Challenges

In Exercises 61–63, refer to Figure 30, which shows a robotic arm consisting of two segments of lengths L_1 and L_2.

FIGURE 30

61. Find the components of the vector $\mathbf{r} = \overrightarrow{OP}$ in terms of θ_1 and θ_2.

62. Let $L_1 = 5$ and $L_2 = 3$. Find $\mathbf{r}$ for $\theta_1 = \frac{\pi}{3}$, $\theta_2 = \frac{\pi}{4}$.

63. Let $L_1 = 5$ and $L_2 = 3$. Show that the set of points reachable by the robotic arm with $\theta_1 = \theta_2$ is an ellipse.

64. Use vectors to prove that the diagonals $\overline{AC}$ and $\overline{BD}$ of a parallelogram bisect each other (Figure 31). *Hint:* Observe that the midpoint of $\overline{BD}$ is the terminal point of $\mathbf{w} + \frac{1}{2}(\mathbf{v} - \mathbf{w})$.

65. Use vectors to prove that the segments joining the midpoints of opposite sides of a quadrilateral bisect each other (Figure 32). *Hint:*

FIGURE 31

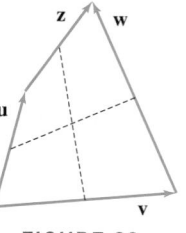

FIGURE 32

Show that the midpoints of these segments are the terminal points of $\frac{1}{4}(2\mathbf{u} + \mathbf{v} + \mathbf{z})$ and $\frac{1}{4}(2\mathbf{v} + \mathbf{w} + \mathbf{u})$.

66. Prove that two vectors $\mathbf{v} = \langle a, b \rangle$ and $\mathbf{w} = \langle c, d \rangle$ are perpendicular if and only if $ac + bd = 0$.

12.2 Vectors in Three Dimensions

This section extends the vector concepts introduced in the previous section to three-dimensional space. We begin with some introductory remarks about the three-dimensional coordinate system.

By convention, we label the x-, y-, and z-axes as in Figure 1(A). This labeling satisfies the **right-hand rule**, which means that when you position your right hand so that your fingers curl from the positive x-axis toward the positive y-axis, your thumb points in the positive z-direction. The axes in Figure 1(B) are not labeled according to the right-hand rule.

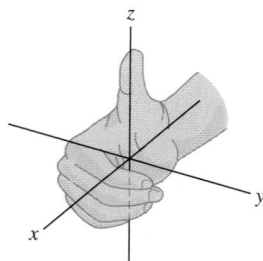

(A) Standard coordinate system (satisfies the right-hand rule)

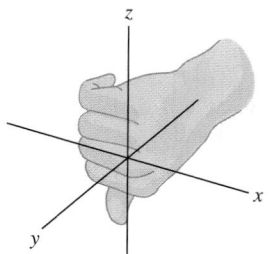

(B) This coordinate system does not satisfy the right-hand rule because the thumb points in the negative z-direction.

FIGURE 1 The fingers of the right hand curl from the positive x-axis to the positive y-axis.

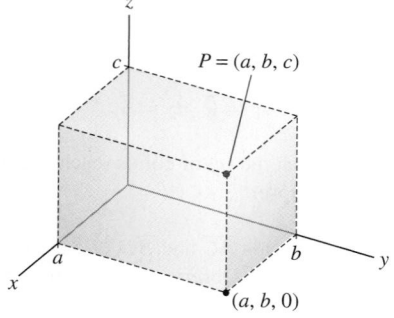

FIGURE 2

Each point in space has unique coordinates (a, b, c) relative to the axes (Figure 2). We denote the set of all triples (a, b, c) by $\mathbf{R}^3$. The **coordinate planes** in $\mathbf{R}^3$ are defined by setting one of the coordinates equal to zero (Figure 3). The xy-plane consists of the points $(a, b, 0)$ and is defined by the equation $z = 0$. Similarly, $x = 0$ defines the yz-plane consisting of the points $(0, b, c)$ and $y = 0$ defines the xz-plane consisting of the points $(a, 0, c)$. The coordinate planes divide $\mathbf{R}^3$ into eight **octants** (analogous to the four quadrants in the plane). Each octant corresponds to a possible combination of signs of the coordinates. The set of points (a, b, c) with $a, b, c > 0$ is called the **first octant**.

As in two dimensions, we derive the distance formula in $\mathbf{R}^3$ from the Pythagorean Theorem.

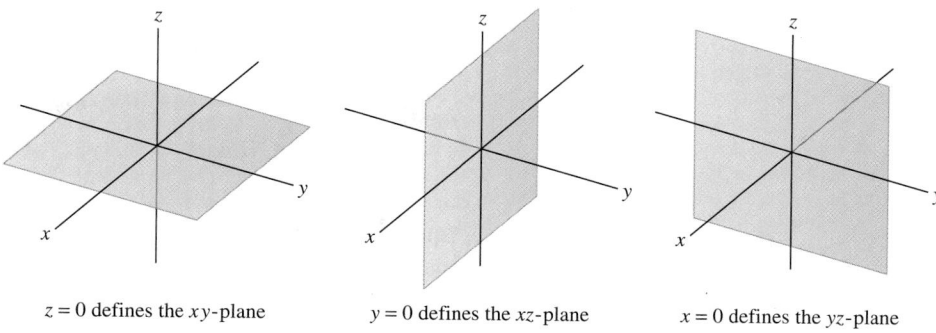

$z = 0$ defines the xy-plane $y = 0$ defines the xz-plane $x = 0$ defines the yz-plane

FIGURE 3

THEOREM 1 Distance Formula in R³ The distance $|P - Q|$ between the points $P = (a_1, b_1, c_1)$ and $Q = (a_2, b_2, c_2)$ is

$$|P - Q| = \sqrt{(a_2 - a_1)^2 + (b_2 - b_1)^2 + (c_2 - c_1)^2}$$ **1**

Proof First apply the distance formula in the plane to the points P and R (Figure 4):

$$|P - R|^2 = (a_2 - a_1)^2 + (b_2 - b_1)^2$$

Then observe that $\triangle PRQ$ is a right triangle [Figure 4(B)] and use the Pythagorean Theorem:

$$|P - Q|^2 = |P - R|^2 + |R - Q|^2 = (a_2 - a_1)^2 + (b_2 - b_1)^2 + (c_2 - c_1)^2 \quad \blacksquare$$

FIGURE 4 Compute $|P - Q|$ using the right triangle $\triangle PQR$.

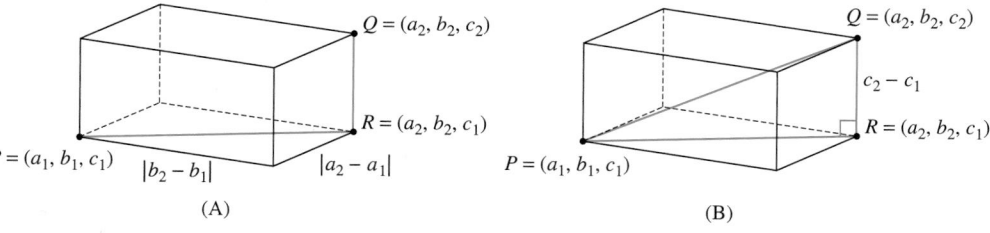

The sphere of radius R with center $Q = (a, b, c)$ consists of all points $P = (x, y, z)$ located a distance R from Q (Figure 5). By the distance formula, the coordinates of $P = (x, y, z)$ must satisfy

$$\sqrt{(x - a)^2 + (y - b)^2 + (z - c)^2} = R$$

On squaring both sides, we obtain the standard equation of the sphere [Eq. (3) below].

Now consider the equation

$$(x - a)^2 + (y - b)^2 = R^2$$ **2**

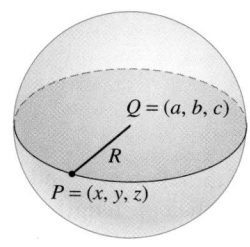

FIGURE 5 Sphere of radius R centered at (a, b, c).

In the xy-plane, Eq. (2) defines the circle of radius R with center (a, b). However, as an equation in $\mathbf{R}^3$, it defines the cylinder of radius R whose central axis is the vertical line through $(a, b, 0)$ (Figure 6). Indeed, a point (x, y, z) satisfies Eq. (2) for any value of z

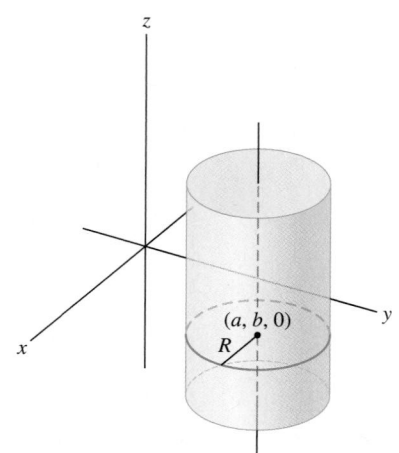

FIGURE 6 Cylinder of radius R centered at $(a, b, 0)$.

if (x, y) lies on the circle. It is usually clear from the context which of the following is intended:

$$\text{Circle} = \{(x, y) : (x - a)^2 + (y - b)^2 = R^2\}$$

$$\text{Cylinder} = \{(x, y, z) : (x - a)^2 + (y - b)^2 = R^2\}$$

Equations of Spheres and Cylinders   An equation of the sphere in $\mathbf{R}^3$ of radius R centered at $Q = (a, b, c)$ is

$$(x - a)^2 + (y - b)^2 + (z - c)^2 = R^2 \qquad \boxed{3}$$

An equation of the cylinder in $\mathbf{R}^3$ of radius R whose central axis is the vertical line through $(a, b, 0)$ is

$$(x - a)^2 + (y - b)^2 = R^2 \qquad \boxed{4}$$

■ **EXAMPLE 1**   Describe the sets of points defined by the following conditions:

(a) $x^2 + y^2 + z^2 = 4, \quad y \geq 0$ **(b)** $(x - 3)^2 + (y - 2)^2 = 1, \quad z \geq -1$

Solution

(a) The equation $x^2 + y^2 + z^2 = 4$ defines a sphere of radius 2 centered at the origin. The inequality $y \geq 0$ holds for points lying on the positive side of the xz-plane. We obtain the right hemisphere of radius 2 illustrated in Figure 7(A).

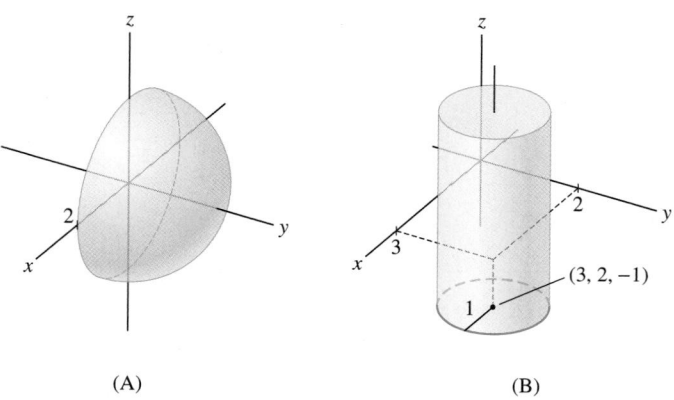

(A)   (B)

FIGURE 7 Hemisphere and upper cylinder.

(b) The equation $(x - 3)^2 + (y - 2)^2 = 1$ defines a cylinder of radius 1 whose central axis is the vertical line through $(3, 2, 0)$. The part of the cylinder where $z \geq -1$ is illustrated in Figure 7(B).

■

Vector Concepts

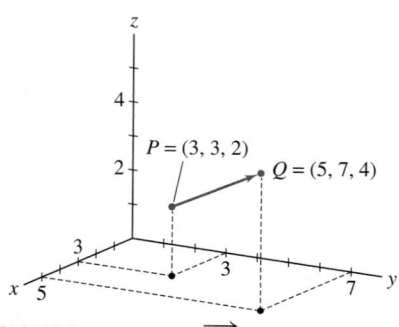

FIGURE 8 A vector $\overrightarrow{PQ}$ in 3-space.

As in the plane, a vector $\mathbf{v} = \overrightarrow{PQ}$ in $\mathbf{R}^3$ is determined by an initial point P and a terminal point Q (Figure 8). If $P = (a_1, b_1, c_1)$ and $Q = (a_2, b_2, c_2)$, then the **length** of $\mathbf{v} = \overrightarrow{PQ}$, denoted $\|\mathbf{v}\|$, is the distance from P to Q:

$$\|\mathbf{v}\| = \|\overrightarrow{PQ}\| = \sqrt{(a_2 - a_1)^2 + (b_2 - b_1)^2 + (c_2 - c_1)^2}$$

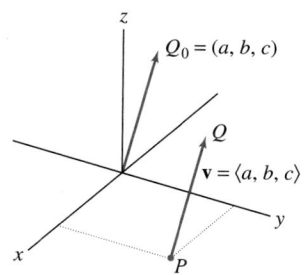

FIGURE 9 A vector **v** and its translate based at the origin.

Our convention about basepoints remains in force: All vectors are assumed to be based at the origin unless otherwise indicated.

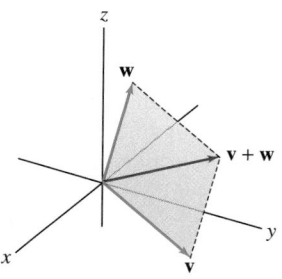

FIGURE 10 Vector addition in $\mathbf{R}^3$ is defined using the Parallelogram Law as in $\mathbf{R}^2$.

The terminology of the previous section carries to $\mathbf{R}^3$ with little change.

- A vector **v** is said to undergo a **translation** if it is moved without changing direction or magnitude.
- Two vectors **v** and **w** are called **equivalent** if **w** is a translate of **v** or, equivalently, if **v** and **w** have the same length and direction.
- The **position vector** of a point Q_0 is the vector $\mathbf{v} = \overrightarrow{OQ_0}$ based at the origin (Figure 9).
- If $P = (a_1, b_1, c_1)$ and $Q = (a_2, b_2, c_2)$, then the **components** of $\mathbf{v} = \overrightarrow{PQ}$ are the differences $a = a_2 - a_1, b = b_2 - b_1, c = c_2 - c_1$, that is,

$$\mathbf{v} = \overrightarrow{PQ} = \overrightarrow{OQ} - \overrightarrow{OP} = \langle a_2, b_2, c_2 \rangle - \langle a_1, b_1, c_1 \rangle$$

For example, if $P = (3, -4, -4)$ and $Q = (2, 5, -1)$, then

$$\mathbf{v} = \overrightarrow{PQ} = \langle 2, 5, -1 \rangle - \langle 3, -4, -4 \rangle = \langle -1, 9, 3 \rangle$$

- A vector $\mathbf{v} = \overrightarrow{PQ}$ with components $\langle a, b, c \rangle$ is equivalent to the vector $\mathbf{v}_0 = \overrightarrow{OQ_0}$ based at the origin with $Q_0 = (a, b, c)$ (Figure 9).
- The operations of vector addition and scalar multiplication are defined as in the two-dimensional case. Vector addition is commutative and associative, and satisfies the distributive property with respect to scalar multiplication (Theorem 1 in Section 12.1).
- Vector addition is visualized geometrically using the Parallelogram Law (Figure 10). In terms of components, if $\mathbf{v} = \langle a_1, b_1, c_1 \rangle$ and $\mathbf{w} = \langle a_2, b_2, c_2 \rangle$, then

$$\lambda \mathbf{v} = \lambda \langle a_1, b_1, c_1 \rangle = \langle \lambda a_1, \lambda b_1, \lambda c_1 \rangle$$

$$\mathbf{v} + \mathbf{w} = \langle a_1, b_1, c_1 \rangle + \langle a_2, b_2, c_2 \rangle = \langle a_1 + a_2, b_1 + b_2, c_1 + c_2 \rangle$$

■ **EXAMPLE 2** Vector Calculations Calculate $\|\mathbf{v}\|$ and $6\mathbf{v} - \frac{1}{2}\mathbf{w}$, where $\mathbf{v} = \langle 3, -1, 2 \rangle$ and $\mathbf{w} = \langle 4, 6, -8 \rangle$.

Solution

$$\|\mathbf{v}\| = \sqrt{3^2 + (-1)^2 + 2^2} = \sqrt{14}$$

$$6\mathbf{v} - \frac{1}{2}\mathbf{w} = 6\langle 3, -1, 2 \rangle - \frac{1}{2}\langle 4, 6, -8 \rangle$$

$$= \langle 18, -6, 12 \rangle - \langle 2, 3, -4 \rangle = \langle 16, -9, 16 \rangle \qquad ■$$

The **standard basis vectors** in $\mathbf{R}^3$ are

$$\mathbf{i} = \langle 1, 0, 0 \rangle, \qquad \mathbf{j} = \langle 0, 1, 0 \rangle, \qquad \mathbf{k} = \langle 0, 0, 1 \rangle$$

Every vector can be written as a **linear combination** of the standard basis vectors (Figure 11):

$$\langle a, b, c \rangle = a\langle 1, 0, 0 \rangle + b\langle 0, 1, 0 \rangle + c\langle 0, 0, 1 \rangle = a\mathbf{i} + b\mathbf{j} + c\mathbf{k}$$

For example, $\langle -9, -4, 17 \rangle = -9\mathbf{i} - 4\mathbf{j} + 17\mathbf{k}$.

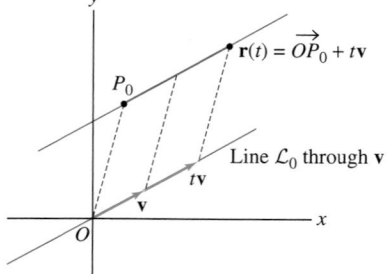

FIGURE 11 Writing $\mathbf{v} = \langle a, b, c \rangle$ as the sum $a\mathbf{i} + b\mathbf{j} + c\mathbf{k}$.

■ **EXAMPLE 3** Find the unit vector $\mathbf{e_v}$ in the direction of $\mathbf{v} = 3\mathbf{i} + 2\mathbf{j} - 4\mathbf{k}$.

Solution Since $\|\mathbf{v}\| = \sqrt{3^2 + 2^2 + (-4)^2} = \sqrt{29}$,

$$\mathbf{e_v} = \frac{1}{\|\mathbf{v}\|}\mathbf{v} = \frac{1}{\sqrt{29}}(3\mathbf{i} + 2\mathbf{j} - 4\mathbf{k}) = \left\langle \frac{3}{\sqrt{29}}, \frac{2}{\sqrt{29}}, \frac{-4}{\sqrt{29}} \right\rangle$$ ■

Parametric Equations of a Line

Although the basic vector concepts are essentially the same in two and three dimensions, there is an important difference in the way lines are described. A line in $\mathbf{R}^2$ can be described by a single linear equation such as $y = mx + b$. This is not so in $\mathbf{R}^3$ because a linear equation in $\mathbf{R}^3$ defines a plane rather than a line (as we will see in Section 12.5). Therefore, we describe lines in $\mathbf{R}^3$ parametrically.

We may specify a line $\mathcal{L}$ by stating that $\mathcal{L}$ passes through a given point $P_0 = (x_0, y_0, z_0)$ in the direction of a vector $\mathbf{v} = \langle a, b, c \rangle$ (Figure 12). The vector $\mathbf{v}$ is called a **direction vector** for $\mathcal{L}$. Figure 12 shows that a point on this line is the terminal point of the vector obtained by adding a multiple $t\mathbf{v}$ to the position vector $\overrightarrow{OP_0}$. Thus, $\mathcal{L}$ has the following parametrization:

$$\mathbf{r}(t) = \overrightarrow{OP_0} + t\mathbf{v} = \langle x_0, y_0, z_0 \rangle + t \langle a, b, c \rangle$$

FIGURE 12 The line through P_0 in the direction $\mathbf{v}$.

In this vector parametrization, the terminal point of $\mathbf{r}(t)$ traces the line as t varies from $-\infty$ to ∞. The components of $\mathbf{r}(t)$ are given by the parametric equations

$$x = x_0 + at, \quad y = y_0 + bt, \quad z = z_0 + ct$$

Equation of a Line (Point-Direction Form) The line $\mathcal{L}$ through $P_0 = (x_0, y_0, z_0)$ in the direction $\mathbf{v} = \langle a, b, c \rangle$ is described in vector or parametric form by

Vector parametrization: $\mathbf{r}(t) = \overrightarrow{OP_0} + t\mathbf{v} = \langle x_0, y_0, z_0 \rangle + t \langle a, b, c \rangle$ $\boxed{5}$

Parametric equations: $x = x_0 + at, \quad y = y_0 + bt, \quad z = z_0 + ct$ $\boxed{6}$

FIGURE 13 The line $\mathbf{r}(t)$ is obtained by translating the line $\mathcal{L}_0$ through $\mathbf{v}$ so that it passes through P_0.

A direction vector $\mathbf{v}$ for a line is a substitute for the slope (which only makes sense in the plane). However, the direction vector is not unique—any nonzero scalar multiple of $\mathbf{v}$ is also a direction vector for $\mathcal{L}$.

GRAPHICAL INSIGHT   For any nonzero vector $\mathbf{v}$, the line $\mathcal{L}_0$ through $\mathbf{v}$ consists of the multiples $\mathbf{r}_0(t) = t\mathbf{v}$. Observe in Figure 13 (illustrating the case of a line in the plane) that the line through P_0 in the direction of $\mathbf{v}$ is obtained by translating $\mathcal{L}_0$ in a parallel fashion so that it passes through P_0. This is the graphical interpretation of the parameterization $\mathbf{r}(t) = \overrightarrow{OP_0} + t\mathbf{v}$.

■ **EXAMPLE 4** Find a parametrization of the line through $P_0 = (3, -1, 4)$ with direction vector $\mathbf{v} = \langle 2, 1, 7 \rangle$.

Vector parametrizations are not unique. Every line $\mathcal{L}$ can be parametrized in infinitely many ways because we are free to choose any point P_0 on $\mathcal{L}$ and we may replace the direction vector by any nonzero scalar multiple.

Solution By Eq. (5),

$$\mathbf{r}(t) = \underbrace{\langle 3, -1, 4 \rangle}_{\text{Coordinates of } P_0} + \; t \underbrace{\langle 2, 1, 7 \rangle}_{\text{Direction vector}} = \langle 3 + 2t, -1 + t, 4 + 7t \rangle$$

The parametric equations are $x = 3 + 2t, \quad y = -1 + t, \quad z = 4 + 7t$.   ■

■ **EXAMPLE 5** Intersection of Two Lines Determine if the following two lines intersect:

$$\mathbf{r}_1(t) = \langle 1, 0, 1 \rangle + t \langle 3, 3, 5 \rangle, \qquad \mathbf{r}_2(t) = \langle 3, 6, 1 \rangle + t \langle 4, -2, 7 \rangle$$

CAUTION We cannot assume in Eq. (7) that the parameter values t_1 and t_2 are equal. The point of intersection may correspond to different parameter values on the two lines.

Solution The two lines intersect if there exist parameter values t_1 and t_2 such that $\mathbf{r}_1(t_1) = \mathbf{r}_2(t_2)$, that is,

$$\langle 1, 0, 1 \rangle + t_1 \langle 3, 3, 5 \rangle = \langle 3, 6, 1 \rangle + t_2 \langle 4, -2, 7 \rangle \qquad \boxed{7}$$

This is equivalent to the three equations for the components:

$$x = 1 + 3t_1 = 3 + 4t_2, \qquad y = 3t_1 = 6 - 2t_2, \qquad z = 1 + 5t_1 = 1 + 7t_2 \qquad \boxed{8}$$

Let's solve the first two equations for t_1 and t_2. Subtracting the second equation from the first, we get $1 = 6t_2 - 3$ or $t_2 = \frac{2}{3}$. Using this value in the second equation, we get $t_1 = 2 - \frac{2}{3}t_2 = \frac{14}{9}$. The values $t_1 = \frac{14}{9}$ and $t_2 = \frac{2}{3}$ satisfy the first two equations and thus $\mathbf{r}_1(t_1)$ and $\mathbf{r}_2(t_2)$ have the same x- and y-coordinates (Figure 14). However, t_1 and t_2 do not satisfy the third equation in (8):

$$1 + 5 \left(\frac{14}{9} \right) \neq 1 + 7 \left(\frac{2}{3} \right)$$

Therefore, Eq. (7) has no solution and the lines do not intersect.   ■

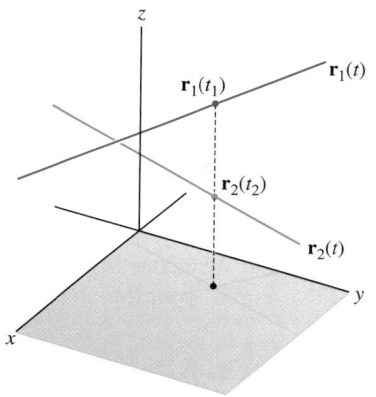

FIGURE 14 The lines $\mathbf{r}_1(t)$ and $\mathbf{r}_2(t)$ do not intersect, but $\mathbf{r}_1(t_1)$ and $\mathbf{r}_2(t_2)$ have the same x- and y-coordinates.

Suppose we wish to parametrize the line $\mathcal{L}$ passing through two points:

$$P = (a_1, b_1, c_1), \qquad Q = (a_2, b_2, c_2)$$

The vector

$$\mathbf{v} = \overrightarrow{PQ} = \langle a_2 - a_1, b_2 - b_1, c_2 - c_1 \rangle$$

points from P to Q, so it is a direction vector for $\mathcal{L}$ (Figure 15). Thus, a possible vector parametrization is

$$\mathbf{r}(t) = \overrightarrow{OP} + t\mathbf{v} = \langle a_1, b_1, c_1 \rangle + t \langle a_2 - a_1, b_2 - b_1, c_2 - c_1 \rangle$$

$$= (1 - t) \langle a_1, b_1, c_1 \rangle + t \langle a_2, b_2, c_2 \rangle$$

In this parametrization,

$$\mathbf{r}(0) = \langle a_1, b_1, c_1 \rangle = \overrightarrow{OP}, \qquad \mathbf{r}(1) = \langle a_2, b_2, c_2 \rangle = \overrightarrow{OQ}$$

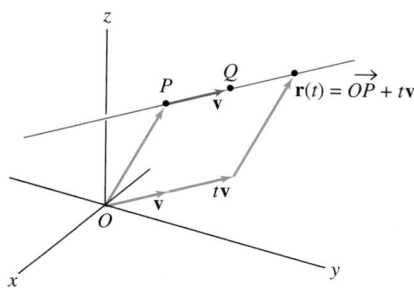

FIGURE 15 Line through two points P and Q.

Thus, $\mathbf{r}(t)$ begins at P and traces the segment $\overline{PQ}$ joining P and Q for $0 \leq t \leq 1$. For $t = \frac{1}{2}$, we obtain the **midpoint**:

$$\text{Midpoint of } \overline{PQ} = \left\langle \frac{a_1 + a_2}{2}, \frac{b_1 + b_2}{2}, \frac{c_1 + c_2}{2} \right\rangle$$

> **Equation of Line Through Two Points** The line through the points $P = (a_1, b_1, c_1)$ and $Q = (a_2, b_2, c_2)$ has **vector parametrization**
>
> $$\mathbf{r}(t) = (1 - t)\overrightarrow{OP} + t\overrightarrow{OQ} = (1 - t)\langle a_1, b_1, c_1 \rangle + t\langle a_2, b_2, c_2 \rangle \qquad \boxed{9}$$
>
> and **parametric equations:**
>
> $$x = a_1 + (a_2 - a_1)t, \qquad y = b_1 + (b_2 - b_1)t, \qquad z = c_1 + (c_2 - c_1)t \qquad \boxed{10}$$
>
> In this parametrization, $\mathbf{r}(0) = \overrightarrow{OP}$, $\mathbf{r}(1) = \overrightarrow{OQ}$, and $\mathbf{r}(t)$ traces the segment $\overline{PQ}$ from P to Q for $0 \le t \le 1$.

■ **EXAMPLE 6** Parametrize the line through $P = (1, 0, 4)$ and $Q = (3, 2, 1)$, and find the midpoint of $\overline{PQ}$.

Solution The line through $P = (1, 0, 4)$ and $Q = (3, 2, 1)$ has the parametrization

$$\mathbf{r}(t) = (1 - t)\langle 1, 0, 4 \rangle + t\langle 3, 2, 1 \rangle = \langle 1 + 2t, 2t, 4 - 3t \rangle$$

The midpoint of $\overline{PQ}$ is the endpoint of $\mathbf{r}\left(\tfrac{1}{2}\right)$:

$$\mathbf{r}\left(\frac{1}{2}\right) = \frac{1}{2}\langle 1, 0, 4 \rangle + \frac{1}{2}\langle 3, 2, 1 \rangle = \left\langle 2, 1, \frac{5}{2} \right\rangle \qquad ■$$

12.2 SUMMARY

- The axes in $\mathbf{R}^3$ are labeled so as to satisfy the *right-hand rule*: When the fingers of your right hand curl from the positive x-axis toward the positive y-axis, your thumb points in the positive z-direction (Figure 16).

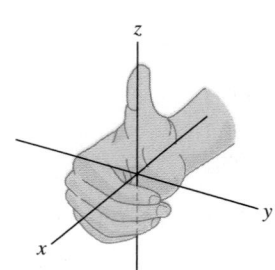

FIGURE 16

Sphere of radius R and center (a, b, c)	$(x - a)^2 + (y - b)^2 + (z - c)^2 = R^2$
Cylinder of radius R with vertical axis through $(a, b, 0)$	$(x - a)^2 + (y - b)^2 = R^2$

- The notation and terminology for vectors in the plane carry over to vectors in $\mathbf{R}^3$.
- If $P = (a_1, b_1, c_1)$ and $Q = (a_2, b_2, c_2)$, then the length of $\mathbf{v} = \overrightarrow{PQ}$ is

$$\|\mathbf{v}\| = \|\overrightarrow{PQ}\| = \sqrt{(a_2 - a_1)^2 + (b_2 - b_1)^2 + (c_2 - c_1)^2}$$

- Equations for the line through $P_0 = (x_0, y_0, z_0)$ with direction vector $\mathbf{v} = \langle a, b, c \rangle$:

$$\text{Vector parametrization:} \quad \mathbf{r}(t) = P_0 + t\mathbf{v} = \langle x_0, y_0, z_0 \rangle + t\langle a, b, c \rangle$$

$$\text{Parametric equations:} \quad x = x_0 + at, \quad y = y_0 + bt, \quad z = z_0 + ct$$

- Equation of the line through $P = (a_1, b_1, c_1)$ and $Q = (a_2, b_2, c_2)$:

$$\text{Vector parametrization:} \quad \mathbf{r}(t) = (1 - t)\langle a_1, b_1, c_1 \rangle + t\langle a_2, b_2, c_2 \rangle$$

$$\text{Parametric equations:} \quad x = a_1 + (a_2 - a_1)t, \quad y = b_1 + (b_2 - b_1)t,$$

$$z = c_1 + (c_2 - c_1)t$$

The segment $\overline{PQ}$ is parametrized by $\mathbf{r}(t)$ for $0 \leq t \leq 1$ and the midpoint of $\overline{PQ}$ is the terminal point of $\mathbf{r}(\frac{1}{2})$.

12.2 EXERCISES

Preliminary Questions

1. What is the terminal point of the vector $\mathbf{v} = \langle 3, 2, 1 \rangle$ based at the point $P = (1, 1, 1)$?

2. What are the components of the vector $\mathbf{v} = \langle 3, 2, 1 \rangle$ based at the point $P = (1, 1, 1)$?

3. If $\mathbf{v} = -3\mathbf{w}$, then (choose the correct answer):
(a) $\mathbf{v}$ and $\mathbf{w}$ are parallel.
(b) $\mathbf{v}$ and $\mathbf{w}$ point in the same direction.

4. Which of the following is a direction vector for the line through $P = (3, 2, 1)$ and $Q = (1, 1, 1)$?
(a) $\langle 3, 2, 1 \rangle$ **(b)** $\langle 1, 1, 1 \rangle$ **(c)** $\langle 2, 1, 0 \rangle$

5. How many different direction vectors does a line have?

6. True or false: If $\mathbf{v}$ is a direction vector for a line $\mathcal{L}$, then $-\mathbf{v}$ is also a direction vector for $\mathcal{L}$.

Exercises

1. Sketch the vector $\mathbf{v} = \langle 1, 3, 2 \rangle$ and compute its length.

2. Let $\mathbf{v} = \overrightarrow{P_0 Q_0}$, where $P_0 = (1, -2, 5)$ and $Q_0 = (0, 1, -4)$. Which of the following vectors (with tail P and head Q) are equivalent to $\mathbf{v}$?

	$\mathbf{v}_1$	$\mathbf{v}_2$	$\mathbf{v}_3$	$\mathbf{v}_4$
P	$(1, 2, 4)$	$(1, 5, 4)$	$(0, 0, 0)$	$(2, 4, 5)$
Q	$(0, 5, -5)$	$(0, -8, 13)$	$(-1, 3, -9)$	$(1, 7, 4)$

3. Sketch the vector $\mathbf{v} = \langle 1, 1, 0 \rangle$ based at $P = (0, 1, 1)$. Describe this vector in the form $\overrightarrow{PQ}$ for some point Q and sketch the vector $\mathbf{v}_0$ based at the origin equivalent to $\mathbf{v}$.

4. Determine if the coordinate systems (A)–(C) in Figure 17 satisfy the right-hand rule.

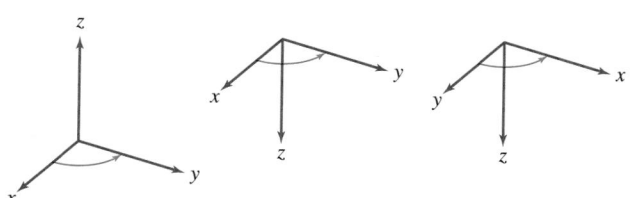

FIGURE 17

In Exercises 5–8, find the components of the vector $\overrightarrow{PQ}$.

5. $P = (1, 0, 1)$, $Q = (2, 1, 0)$

6. $P = (-3, -4, 2)$, $Q = (1, -4, 3)$

7. $P = (4, 6, 0)$, $Q = (-\frac{1}{2}, \frac{9}{2}, 1)$

8. $P = (-\frac{1}{2}, \frac{9}{2}, 1)$, $Q = (4, 6, 0)$

In Exercises 9–12, let $R = (1, 4, 3)$.

9. Calculate the length of $\overrightarrow{OR}$.

10. Find the point Q such that $\mathbf{v} = \overrightarrow{RQ}$ has components $\langle 4, 1, 1 \rangle$ and sketch $\mathbf{v}$.

11. Find the point P such that $\mathbf{w} = \overrightarrow{PR}$ has components $\langle 3, -2, 3 \rangle$ and sketch $\mathbf{w}$.

12. Find the components of $\mathbf{u} = \overrightarrow{PR}$, where $P = (1, 2, 2)$.

13. Let $\mathbf{v} = \langle 4, 8, 12 \rangle$. Which of the following vectors is parallel to $\mathbf{v}$? Which point in the same direction?
(a) $\langle 2, 4, 6 \rangle$ **(b)** $\langle -1, -2, 3 \rangle$
(c) $\langle -7, -14, -21 \rangle$ **(d)** $\langle 6, 10, 14 \rangle$

In Exercises 14–17, determine whether $\overrightarrow{AB}$ is equivalent to $\overrightarrow{PQ}$.

14. $A = (1, 1, 1)$, $B = (3, 3, 3)$, $P = (1, 4, 5)$, $Q = (3, 6, 7)$

15. $A = (1, 4, 1)$, $B = (-2, 2, 0)$, $P = (2, 5, 7)$, $Q = (-3, 2, 1)$

16. $A = (0, 0, 0)$, $B = (-4, 2, 3)$, $P = (4, -2, -3)$, $Q = (0, 0, 0)$

17. $A = (1, 1, 0)$, $B = (3, 3, 5)$, $P = (2, -9, 7)$, $Q = (4, -7, 13)$

In Exercises 18–21, calculate the linear combinations.

18. $5 \langle 2, 2, -3 \rangle + 3 \langle 1, 7, 2 \rangle$

19. $-2 \langle 8, 11, 3 \rangle + 4 \langle 2, 1, 1 \rangle$

20. $6 \langle 2, 0, -1 \rangle - 3 \langle 8, 6, 9 \rangle$

21. $- \langle 4, 3, 8 \rangle + \langle 8, 3, 3 \rangle$

In Exercises 22–25, find the given vector.

22. $\mathbf{e_v}$, where $\mathbf{v} = \langle 1, 1, 2 \rangle$

23. $\mathbf{e_w}$, where $\mathbf{w} = \langle 4, -2, -1 \rangle$

24. Unit vector in the direction of $\mathbf{u} = \langle 1, 0, 7 \rangle$

25. Unit vector in the direction opposite to $\mathbf{v} = \langle -4, 4, 2 \rangle$

26. Sketch the following vectors, and find their components and lengths.

(a) $4\mathbf{i} + 3\mathbf{j} - 2\mathbf{k}$ (b) $\mathbf{i} + \mathbf{j} + \mathbf{k}$
(c) $4\mathbf{j} + 3\mathbf{k}$ (d) $12\mathbf{i} + 8\mathbf{j} - \mathbf{k}$

In Exercises 27–34, find a vector parametrization for the line with the given description.

27. Passes through $P = (1, 2, -8)$, direction vector $\mathbf{v} = \langle 2, 1, 3 \rangle$

28. Passes through $P = (4, 0, 8)$, direction vector $\mathbf{v} = \langle 1, 0, 1 \rangle$

29. Passes through $P = (4, 0, 8)$, direction vector $\mathbf{v} = 7\mathbf{i} + 4\mathbf{k}$

30. Passes through O, direction vector $\mathbf{v} = \langle 4, 2, -1 \rangle$

31. Passes through $(1, 1, 1)$ and $(3, -5, 2)$

32. Passes through $(-2, 0, -2)$ and $(4, 3, 7)$

33. Passes through O and $(4, 1, 1)$

34. Passes through $(1, 1, 1)$ parallel to the line through $(2, 0, -1)$ and $(4, 1, 3)$

In Exercises 35–40, find parametric equations of the line with the given description.

35. Perpendicular to the xy-plane passing through the origin

36. Perpendicular to the xz-plane passing through the point $(1, -1, 2)$

37. Perpendicular to the yz-plane passing through the point $(0, 0, 2)$

38. Passes through the origin perpendicular to the yz-plane

39. Passes through $(2, 3, 1)$ in the direction $\mathbf{v} = \langle 2, 1, -2 \rangle$

40. Passes through the origin with direction vector $\mathbf{v} = \langle 3, 4, -2 \rangle$

In Exercises 41–44, let $P = (2, 1, -1)$ and $Q = (4, 7, 7)$. Find the coordinates of each of the following.

41. The midpoint of $\overline{PQ}$

42. The point on $\overline{PQ}$ lying two-thirds of the way from P to Q

43. The point R such that Q is the midpoint of $\overline{PR}$

44. All points on the line through $\overline{PQ}$ whose distance from P is twice its distance from Q

45. Show that $\mathbf{r}_1(t)$ and $\mathbf{r}_2(t)$ define the same line, where

$$\mathbf{r}_1(t) = \langle 3, -1, 4 \rangle + t \langle 8, 12, -6 \rangle$$
$$\mathbf{r}_2(t) = \langle 11, 11, -2 \rangle + t \langle 4, 6, -3 \rangle$$

Hint: Show that $\mathbf{r}_2$ passes through $(3, -1, 4)$ and that the direction vectors for $\mathbf{r}_1$ and $\mathbf{r}_2$ are parallel.

46. Show that $\mathbf{r}_1(t)$ and $\mathbf{r}_2(t)$ define the same line, where

$$\mathbf{r}_1(t) = t \langle 2, 1, 3 \rangle, \qquad \mathbf{r}_2(t) = \langle -6, -3, -9 \rangle + t \langle 8, 4, 12 \rangle$$

47. Find two different vector parametrizations of the line through $P = (5, 5, 2)$ with direction vector $\mathbf{v} = \langle 0, -2, 1 \rangle$.

48. Find the point of intersection of the lines $\mathbf{r}(t) = \langle 1, 0, 0 \rangle + t \langle -3, 1, 0 \rangle$ and $\langle 0, 1, 1 \rangle + t \langle 2, 0, 1 \rangle$.

49. Show that the lines $\mathbf{r}_1(t) = \langle -1, 2, 2 \rangle + t \langle 4, -2, 1 \rangle$ and $\mathbf{r}_2(t) = \langle 0, 1, 1 \rangle + t \langle 2, 0, 1 \rangle$ do not intersect.

50. Determine if the lines $\mathbf{r}_1(t) = \langle 2, 1, 1 \rangle + t \langle -4, 0, 1 \rangle$ and $\mathbf{r}_2(s) = \langle -4, 1, 5 \rangle + s \langle 2, 1, -2 \rangle$ intersect and, if so, find the point of intersection.

51. Determine if the lines $\mathbf{r}_1(t) = \langle 0, 1, 1 \rangle + t \langle 1, 1, 2 \rangle$ and $\mathbf{r}_2(s) = \langle 2, 0, 3 \rangle + s \langle 1, 4, 4 \rangle$ intersect and, if so, find the point of intersection.

52. Find the intersection of the lines $\mathbf{r}_1(t) = \langle -1, 1 \rangle + t \langle 2, 4 \rangle$ and $\mathbf{r}_2(s) = \langle 2, 1 \rangle + s \langle -1, 6 \rangle$ in $\mathbf{R}^2$.

53. Find the components of the vector $\mathbf{v}$ whose tail and head are the midpoints of segments $\overline{AC}$ and $\overline{BC}$ in Figure 18.

54. Find the components of the vector $\mathbf{w}$ whose tail is C and head is the midpoint of $\overline{AB}$ in Figure 18.

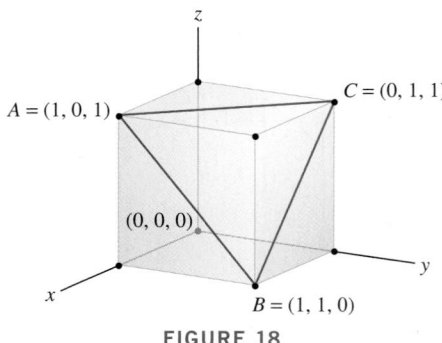

FIGURE 18

Further Insights and Challenges

*In Exercises 55–59, we consider the equations of a line in **symmetric form**.*

$$\frac{x - x_0}{a} = \frac{y - y_0}{b} = \frac{z - z_0}{c} \qquad \boxed{11}$$

55. Let $\mathcal{L}$ be the line through $P_0 = (x_0, y_0, c_0)$ with direction vector $\mathbf{v} = \langle a, b, c \rangle$. Show that $\mathcal{L}$ is defined by the symmetric equations (11). *Hint:* Use the vector parametrization to show that every point on $\mathcal{L}$ satisfies (11).

56. Find the symmetric equations of the line through $P_0 = (-2, 3, 3)$ with direction vector $\mathbf{v} = \langle 2, 4, 3 \rangle$.

57. Find a vector parametrization for the line with symmetric equations

$$\frac{x-5}{9} = \frac{y+3}{7} = z - 10$$

58. Find the symmetric equations of the line through $P = (1, 1, 2)$ and $Q = (-2, 4, 0)$.

59. Show that the line in the plane through (x_0, y_0) of slope m has symmetric equations

$$\frac{x - x_0}{1} = \frac{y - y_0}{m}$$

60. A median of a triangle is a segment joining a vertex to the midpoint of the opposite side. It is known that the three medians intersect at a point P called the *centroid*. Prove this for a triangle whose vertices

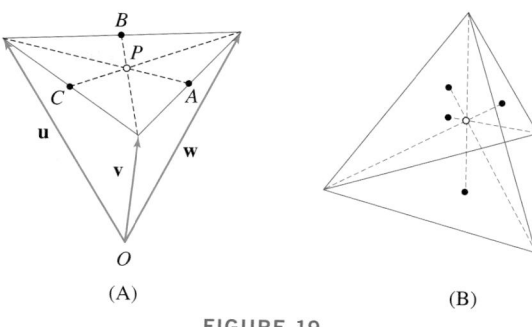

FIGURE 19

are the terminal points of vectors $\mathbf{u}$, $\mathbf{v}$, and $\mathbf{w}$ as in Figure 19(A). More precisely, show that the centroid is the terminal point of $\frac{1}{3}(\mathbf{u} + \mathbf{v} + \mathbf{w})$.

61. A median of a tetrahedron is a segment joining a vertex to the centroid of the opposite face. Show that the medians of the tetrahedron with vertices at the terminal points of vectors $\mathbf{u}, \mathbf{v}, \mathbf{w}, \mathbf{z}$ intersect at the terminal point of $\frac{1}{4}(\mathbf{u} + \mathbf{v} + \mathbf{w} + \mathbf{z})$ [Figure 19(B)].

12.3 Dot Product and the Angle Between Two Vectors

We now introduce one of the most important vector operations, the dot product. It plays a fundamental role in nearly all aspects of multivariable calculus.

Important concepts in mathematics often have multiple names or notations either for historical reasons or because they arise in more than one context. The dot product is also called the "scalar product" or "inner product" and in many texts, $\mathbf{v} \cdot \mathbf{w}$ is denoted $(\mathbf{v}, \mathbf{w})$ or $\langle \mathbf{v}, \mathbf{w} \rangle$.

DEFINITION Dot Product The **dot product** $\mathbf{v} \cdot \mathbf{w}$ of two vectors

$$\mathbf{v} = \langle a_1, b_1, c_1 \rangle, \qquad \mathbf{w} = \langle a_2, b_2, c_2 \rangle$$

is the scalar defined by

$$\mathbf{v} \cdot \mathbf{w} = a_1 a_2 + b_1 b_2 + c_1 c_2$$

The dot product of vectors $\mathbf{v} = \langle a_1, b_1 \rangle$ and $\mathbf{w} = \langle a_2, b_2 \rangle$ in $\mathbf{R}^2$ is defined similarly:

$$\mathbf{v} \cdot \mathbf{w} = a_1 a_2 + b_1 b_2$$

In words, to compute the dot product, *multiply the corresponding components and add.* For example,

$$\langle 2, 3, 1 \rangle \cdot \langle -4, 2, 5 \rangle = 2(-4) + 3(2) + 1(5) = -8 + 6 + 5 = 3$$

As we will see in a moment, the dot product is closely related to the angle between $\mathbf{v}$ and $\mathbf{w}$. Before getting to this, we describe some elementary properties of dot products.

First, the dot product is *commutative*: $\mathbf{v} \cdot \mathbf{w} = \mathbf{w} \cdot \mathbf{v}$. This follows because we may multiply the components in either order. Second, the dot product of a vector with itself is the square of the length: If $\mathbf{v} = \langle a, b, c \rangle$, then

$$\mathbf{v} \cdot \mathbf{v} = a \cdot a + b \cdot b + c \cdot c = a^2 + b^2 + c^2 = \|\mathbf{v}\|^2$$

The dot product also satisfies a Distributive Law and a scalar property as summarized in the next theorem (see Exercises 77–78).

THEOREM 1 Properties of the Dot Product

(i) $\mathbf{0} \cdot \mathbf{v} = \mathbf{v} \cdot \mathbf{0} = 0$

(ii) **Commutativity:** $\mathbf{v} \cdot \mathbf{w} = \mathbf{w} \cdot \mathbf{v}$

(iii) **Pulling out scalars:** $(\lambda \mathbf{v}) \cdot \mathbf{w} = \mathbf{v} \cdot (\lambda \mathbf{w}) = \lambda (\mathbf{v} \cdot \mathbf{w})$

(iv) **Distributive Law:** $\mathbf{u} \cdot (\mathbf{v} + \mathbf{w}) = \mathbf{u} \cdot \mathbf{v} + \mathbf{u} \cdot \mathbf{w}$

$\qquad\qquad\qquad\qquad (\mathbf{v} + \mathbf{w}) \cdot \mathbf{u} = \mathbf{v} \cdot \mathbf{u} + \mathbf{w} \cdot \mathbf{u}$

(v) **Relation with length:** $\mathbf{v} \cdot \mathbf{v} = \|\mathbf{v}\|^2$

■ **EXAMPLE 1**   Verify the Distributive Law $\mathbf{u} \cdot (\mathbf{v} + \mathbf{w}) = \mathbf{u} \cdot \mathbf{v} + \mathbf{u} \cdot \mathbf{w}$ for the vectors

$$\mathbf{u} = \langle 4, 3, 3 \rangle, \quad \mathbf{v} = \langle 1, 2, 2 \rangle, \quad \mathbf{w} = \langle 3, -2, 5 \rangle$$

Solution   We compute both sides and check that they are equal:

$$\mathbf{u} \cdot (\mathbf{v} + \mathbf{w}) = \langle 4, 3, 3 \rangle \cdot (\langle 1, 2, 2 \rangle + \langle 3, -2, 5 \rangle)$$

$$= \langle 4, 3, 3 \rangle \cdot \langle 4, 0, 7 \rangle = 4(4) + 3(0) + 3(7) = 37$$

$$\mathbf{u} \cdot \mathbf{v} + \mathbf{u} \cdot \mathbf{w} = \langle 4, 3, 3 \rangle \cdot \langle 1, 2, 2 \rangle + \langle 4, 3, 3 \rangle \langle 3, -2, 5 \rangle$$

$$= \big(4(1) + 3(2) + 3(2)\big) + \big(4(3) + 3(-2) + 3(5)\big)$$

$$= 16 + 21 = 37 \qquad\blacksquare$$

As mentioned above, the dot product $\mathbf{v} \cdot \mathbf{w}$ is related to the angle θ between $\mathbf{v}$ and $\mathbf{w}$. This angle θ is not uniquely defined because, as we see in Figure 1, both θ and $2\pi - \theta$ can serve as an angle between $\mathbf{v}$ and $\mathbf{w}$. Furthermore, any multiple of 2π may be added to θ. We adopt the convention:

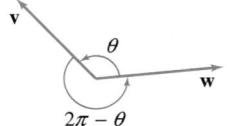

FIGURE 1 By convention, the angle θ between two vectors is chosen so that $0 \le \theta \le \pi$.

> *The angle between two vectors is chosen to satisfy $0 \le \theta \le \pi$.*

The next theorem establishes the fundamental relation between the dot product and the angle between two vectors.

THEOREM 2 Dot Product and the Angle   Let θ be the angle between two nonzero vectors $\mathbf{v}$ and $\mathbf{w}$. Then

$$\mathbf{v} \cdot \mathbf{w} = \|\mathbf{v}\| \, \|\mathbf{w}\| \cos \theta \quad \text{or} \quad \cos \theta = \frac{\mathbf{v} \cdot \mathbf{w}}{\|\mathbf{v}\| \, \|\mathbf{w}\|} \qquad \boxed{1}$$

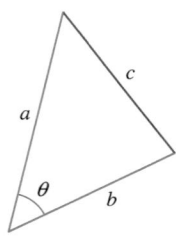

Proof   According to the Law of Cosines, the three sides of a triangle satisfy (Figure 2)

$$c^2 = a^2 + b^2 - 2ab \cos \theta$$

If two sides of the triangle are $\mathbf{v}$ and $\mathbf{w}$, then the third side is $\mathbf{v} - \mathbf{w}$, as in the figure, and the Law of Cosines gives

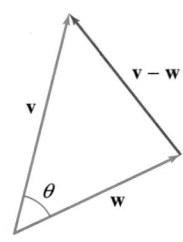

FIGURE 2

$$\|\mathbf{v} - \mathbf{w}\|^2 = \|\mathbf{v}\|^2 + \|\mathbf{w}\|^2 - 2 \cos \theta \|\mathbf{v}\| \, \|\mathbf{w}\| \qquad \boxed{2}$$

Now, by property (v) of Theorem 1 and the Distributive Law,

$$\|\mathbf{v} - \mathbf{w}\|^2 = (\mathbf{v} - \mathbf{w}) \cdot (\mathbf{v} - \mathbf{w}) = \mathbf{v} \cdot \mathbf{v} - 2\mathbf{v} \cdot \mathbf{w} + \mathbf{w} \cdot \mathbf{w}$$

$$= \|\mathbf{v}\|^2 + \|\mathbf{w}\|^2 - 2\mathbf{v} \cdot \mathbf{w} \qquad \boxed{3}$$

Comparing (2) and (3), we obtain $-2 \cos \theta \|\mathbf{v}\| \|\mathbf{w}\| = -2\mathbf{v} \cdot \mathbf{w}$, and Eq. (1) follows. ■

By definition of the arccosine, the angle $\theta = \cos^{-1} x$ is the angle in the interval $[0, \pi]$ satisfying $\cos \theta = x$. Thus, for nonzero vectors $\mathbf{v}$ and $\mathbf{w}$, we have

$$\boxed{\theta = \cos^{-1}\left(\frac{\mathbf{v} \cdot \mathbf{w}}{\|\mathbf{v}\| \|\mathbf{w}\|}\right)}$$

■ **EXAMPLE 2** Find the angle θ between $\mathbf{v} = \langle 3, 6, 2 \rangle$ and $\mathbf{w} = \langle 4, 2, 4 \rangle$.

Solution Compute $\cos \theta$ using the dot product:

$$\|\mathbf{v}\| = \sqrt{3^2 + 6^2 + 2^2} = \sqrt{49} = 7, \qquad \|\mathbf{w}\| = \sqrt{4^2 + 2^2 + 4^2} = \sqrt{36} = 6$$

$$\cos \theta = \frac{\mathbf{v} \cdot \mathbf{w}}{\|\mathbf{v}\| \|\mathbf{w}\|} = \frac{\langle 3, 6, 2 \rangle \cdot \langle 4, 2, 4 \rangle}{7 \cdot 6} = \frac{3 \cdot 4 + 6 \cdot 2 + 2 \cdot 4}{42} = \frac{32}{42} = \frac{16}{21}$$

The angle itself is $\theta = \cos^{-1}\left(\frac{16}{21}\right) \approx 0.705$ rad (Figure 3). ■

w = ⟨4, 2, 4⟩

v = ⟨3, 6, 2⟩

FIGURE 3

The terms "orthogonal" and "perpendicular" are synonymous and are used interchangeably, although orthogonal is commonly used when dealing with vectors.

Two nonzero vectors $\mathbf{v}$ and $\mathbf{w}$ are called **perpendicular** or **orthogonal** if the angle between them is $\frac{\pi}{2}$. (The zero vector is considered orthogonal to every vector.) The dot product gives us a simple way to test whether $\mathbf{v}$, $\mathbf{w}$ are orthogonal. An angle between 0 and π satisfies $\cos \theta = 0$ if and only if $\theta = \frac{\pi}{2}$, and therefore,

$$\mathbf{v} \cdot \mathbf{w} = \|\mathbf{v}\| \|\mathbf{w}\| \cos \theta = 0 \quad \Leftrightarrow \quad \theta = \frac{\pi}{2}$$

We write $\mathbf{v} \perp \mathbf{w}$ if $\mathbf{v}$ and $\mathbf{w}$ are orthogonal. Thus, we have

$$\boxed{\mathbf{v} \perp \mathbf{w} \text{ if and only if } \mathbf{v} \cdot \mathbf{w} = 0}$$

■ **EXAMPLE 3** Testing for Orthogonality Determine if $\mathbf{v} = \langle 2, 6, 1 \rangle$ is orthogonal to $\mathbf{u} = \langle 2, -1, 1 \rangle$ or $\mathbf{w} = \langle -4, 1, 2 \rangle$.

Solution We test for orthogonality by computing the dot products (Figure 4):

$$\mathbf{v} \cdot \mathbf{u} = \langle 2, 6, 1 \rangle \cdot \langle 2, -1, 1 \rangle = 2(2) + 6(-1) + 1(1) = -1 \quad \text{(not orthogonal)}$$

$$\mathbf{v} \cdot \mathbf{w} = \langle 2, 6, 1 \rangle \cdot \langle -4, 1, 2 \rangle = 2(-4) + 6(1) + 1(2) = 0 \quad \text{(orthogonal)} \qquad ■$$

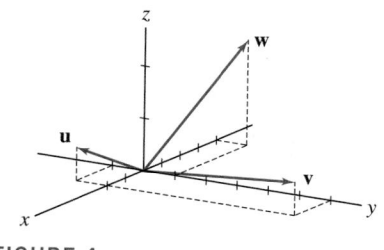

FIGURE 4

■ **EXAMPLE 4** Testing for Obtuseness Determine if the angles between $\mathbf{v} = \langle 3, 1, -2 \rangle$ and the vectors $\mathbf{u} = \left\langle \frac{1}{2}, \frac{1}{2}, 5 \right\rangle$ and $\mathbf{w} = \langle 4, -3, 0 \rangle$ are obtuse.

Solution The angle θ between two vectors is obtuse if $\frac{\pi}{2} < \theta \leq \pi$, and this is the case if $\cos \theta < 0$. Since $\mathbf{v} \cdot \mathbf{u} = \|\mathbf{v}\| \|\mathbf{u}\| \cos \theta$ and the lengths $\|\mathbf{v}\|$ and $\|\mathbf{u}\|$ are positive, we see that

$$\boxed{\text{The angle between } \mathbf{v} \text{ and } \mathbf{w} \text{ is obtuse if } \mathbf{v} \cdot \mathbf{u} < 0.}$$

We have

$$\mathbf{v} \cdot \mathbf{u} = \langle 3, 1, -2 \rangle \cdot \left\langle \frac{1}{2}, \frac{1}{2}, 5 \right\rangle = \frac{3}{2} + \frac{1}{2} - 10 = -8 < 0 \quad \text{(angle is obtuse)}$$

$$\mathbf{v} \cdot \mathbf{w} = \langle 3, 1, -2 \rangle \cdot \langle 4, -3, 0 \rangle = 12 - 3 + 0 = 9 > 0 \quad \text{(angle is acute)} \quad \blacksquare$$

The standard basis vectors are mutually orthogonal and have length one (Figure 5). In terms of dot products,

$$\mathbf{i} \cdot \mathbf{j} = \mathbf{i} \cdot \mathbf{k} = \mathbf{j} \cdot \mathbf{k} = 0, \qquad \mathbf{i} \cdot \mathbf{i} = \mathbf{j} \cdot \mathbf{j} = \mathbf{k} \cdot \mathbf{k} = 1$$

Furthermore, the components of a vector $\mathbf{v} = \langle a, b, c \rangle$ are simply the dot products with the standard basis vectors:

$$\mathbf{v} \cdot \mathbf{i} = \langle a, b, c \rangle \cdot \langle 1, 0, 0 \rangle = a$$

$$\mathbf{v} \cdot \mathbf{j} = \langle a, b, c \rangle \cdot \langle 0, 1, 0 \rangle = b$$

$$\mathbf{v} \cdot \mathbf{k} = \langle a, b, c \rangle \cdot \langle 0, 0, 1 \rangle = c$$

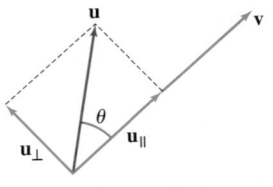

FIGURE 5 The standard basis vectors are mutually orthogonal and have length one.

■ **EXAMPLE 5** Using the Distributive Law Calculate $\mathbf{v} \cdot \mathbf{w}$, where $\mathbf{v} = 4\mathbf{i} - 3\mathbf{j}$ and $\mathbf{w} = \mathbf{i} + 2\mathbf{j} + \mathbf{k}$.

Solution Using the Distributive Law and the orthogonality of $\mathbf{i}$, $\mathbf{j}$, and $\mathbf{k}$, we obtain

$$\mathbf{v} \cdot \mathbf{w} = (4\mathbf{i} - 3\mathbf{j}) \cdot (\mathbf{i} + 2\mathbf{j} + \mathbf{k})$$

$$= 4\mathbf{i} \cdot (\mathbf{i} + 2\mathbf{j} + \mathbf{k}) - 3\mathbf{j} \cdot (\mathbf{i} + 2\mathbf{j} + \mathbf{k})$$

$$= 4\mathbf{i} \cdot \mathbf{i} - 3\mathbf{j} \cdot (2\mathbf{j}) = 4 - 6 = -2 \quad \blacksquare$$

Given vectors $\mathbf{u}$ and $\mathbf{v}$ with $\mathbf{v} \neq \mathbf{0}$, it is possible to decompose $\mathbf{u}$ as a sum:

$$\mathbf{u} = \mathbf{u}_{\parallel} + \mathbf{u}_{\perp}$$

where $\mathbf{u}_{\parallel}$ is parallel to $\mathbf{v}$ and $\mathbf{u}_{\perp}$ is orthogonal to $\mathbf{v}$ (Figure 6). The vector $\mathbf{u}_{\parallel}$ is called the **projection** of $\mathbf{u}$ along $\mathbf{v}$ and is also denoted $\text{proj}_{\mathbf{v}}(\mathbf{u})$.

To determine the projection $\mathbf{u}_{\parallel}$, we note that $\mathbf{u}_{\parallel}$ is parallel to $\mathbf{v}$, and therefore, $\mathbf{u}_{\parallel}$ is a multiple of the unit vector $\mathbf{e}_{\mathbf{v}} = \dfrac{\mathbf{v}}{\|\mathbf{v}\|}$. Let $\mathbf{u}_{\parallel} = c\mathbf{e}_{\mathbf{v}}$ for some constant c. Since $\mathbf{u}_{\perp} \cdot \mathbf{e}_{\mathbf{v}} = 0$,

$$\mathbf{u} \cdot \mathbf{e}_{\mathbf{v}} = (\mathbf{u}_{\parallel} + \mathbf{u}_{\perp}) \cdot \mathbf{e}_{\mathbf{v}} = \mathbf{u}_{\parallel} \cdot \mathbf{e}_{\mathbf{v}} + \mathbf{u}_{\perp} \cdot \mathbf{e}_{\mathbf{v}} = \mathbf{u}_{\parallel} \cdot \mathbf{e}_{\mathbf{v}} = c\mathbf{e}_{\mathbf{v}} \cdot \mathbf{e}_{\mathbf{v}} = c$$

Thus, $c = \mathbf{u} \cdot \mathbf{e}_{\mathbf{v}}$ and $\mathbf{u}_{\parallel} = (\mathbf{u} \cdot \mathbf{e}_{\mathbf{v}})\mathbf{e}_{\mathbf{v}}$. We can also write $\mathbf{u}_{\parallel}$ directly in terms of $\mathbf{v}$:

$$\mathbf{u}_{\parallel} = (\mathbf{u} \cdot \mathbf{e}_{\mathbf{v}})\mathbf{e}_{\mathbf{v}} = \left(\mathbf{u} \cdot \frac{\mathbf{v}}{\|\mathbf{v}\|} \right) \frac{\mathbf{v}}{\|\mathbf{v}\|} = \left(\frac{\mathbf{u} \cdot \mathbf{v}}{\|\mathbf{v}\|^2} \right) \mathbf{v} = \left(\frac{\mathbf{u} \cdot \mathbf{v}}{\mathbf{v} \cdot \mathbf{v}} \right) \mathbf{v}$$

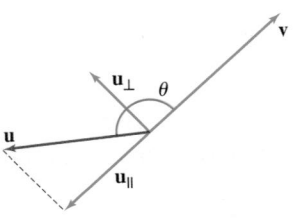

FIGURE 6 Resolution of $\mathbf{u}$ into parallel and orthogonal components with respect to $\mathbf{v}$.

In summary,

$$\boxed{\; \mathbf{u}_{\parallel} = \text{proj}_{\mathbf{v}}(\mathbf{u}) = (\mathbf{u} \cdot \mathbf{e}_{\mathbf{v}})\mathbf{e}_{\mathbf{v}} = \left(\frac{\mathbf{u} \cdot \mathbf{v}}{\mathbf{v} \cdot \mathbf{v}} \right) \mathbf{v} \;}$$ **4**

Note that if $\mathbf{u} \cdot \mathbf{e}_{\mathbf{v}} < 0$, then the angle θ between $\mathbf{u}$ and $\mathbf{v}$ is obtuse and the vectors $\mathbf{u}_{\parallel}$ and $\mathbf{v}$ point in opposite directions (Figure 7). Having found $\mathbf{u}_{\parallel}$, we obtain the vector $\mathbf{u}_{\perp}$

FIGURE 7 When θ is obtuse, $\mathbf{u}_{\parallel}$ and $\mathbf{v}$ point in opposite directions.

as the difference

$$\boxed{\mathbf{u}_\perp = \mathbf{u} - \mathbf{u}_{||}}$$

The vector $\mathbf{u}_\perp$ is orthogonal to $\mathbf{v}$ since

$$\mathbf{u}_\perp \cdot \mathbf{v} = \left(\mathbf{u} - \left(\frac{\mathbf{u} \cdot \mathbf{v}}{\mathbf{v} \cdot \mathbf{v}}\right)\mathbf{v}\right) \cdot \mathbf{v} = \mathbf{u} \cdot \mathbf{v} - \left(\frac{\mathbf{u} \cdot \mathbf{v}}{\mathbf{v} \cdot \mathbf{v}}\right)\mathbf{v} \cdot \mathbf{v} = 0$$

■ **EXAMPLE 6** Find the decomposition $\mathbf{u} = \mathbf{u}_{||} + \mathbf{u}_\perp$ of $\mathbf{u} = \langle 5, 1, -3\rangle$ with respect to $\mathbf{v} = \langle 4, 4, 2\rangle$.

Solution We compute $\mathbf{u}_{||}$ in two steps:

Step 1. **Compute $\mathbf{u} \cdot \mathbf{v}$ and $\mathbf{v} \cdot \mathbf{v}$.**

$$\mathbf{u} \cdot \mathbf{v} = \langle 5, 1, -3\rangle \cdot \langle 4, 4, 2\rangle = 20 + 4 - 6 = 18, \qquad \mathbf{v} \cdot \mathbf{v} = 4^2 + 4^2 + 2^2 = 36$$

Step 2. **Use formulas (4) and (5).**

$$\mathbf{u}_{||} = \text{proj}_\mathbf{v}(\mathbf{u}) = \left(\frac{\mathbf{u} \cdot \mathbf{v}}{\mathbf{v} \cdot \mathbf{v}}\right)\mathbf{v} = \left(\frac{18}{36}\right)\langle 4, 4, 2\rangle = \langle 2, 2, 1\rangle$$

$$\mathbf{u}_\perp = \mathbf{u} - \mathbf{u}_{||} = \langle 5, 1, -3\rangle - \langle 2, 2, 1\rangle = \langle 3, -1, -4\rangle$$

We obtain the decomposition

$$\mathbf{u} = \langle 5, 1, -3\rangle = \mathbf{u}_{||} + \mathbf{u}_\perp = \underbrace{\langle 2, 2, 1\rangle}_{\text{Projection along } \mathbf{v}} + \underbrace{\langle 3, -1, -4\rangle}_{\text{Orthogonal to } \mathbf{v}}$$

■

The coefficient of $\mathbf{e}_\mathbf{v}$ in Eq. (4) is called the **component** of $\mathbf{u}$ along $\mathbf{v}$:

$$\boxed{\text{Component of } \mathbf{u} \text{ along } \mathbf{v} = \mathbf{u} \cdot \mathbf{e}_\mathbf{v} = \frac{\mathbf{u} \cdot \mathbf{v}}{\|\mathbf{v}\|} = \|\mathbf{u}\|\cos\theta}$$

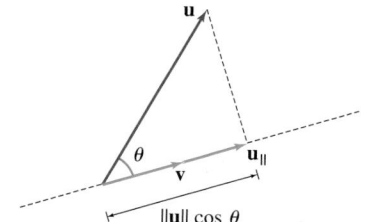

FIGURE 8 The component of $\mathbf{u}$ along $\mathbf{v}$ is $\mathbf{u} \cdot \mathbf{e}_\mathbf{v} = \|\mathbf{u}\|\cos\theta$.

The component $\mathbf{u} \cdot \mathbf{e}_\mathbf{v}$ tells us "how much" of $\mathbf{u}$ points in the direction $\mathbf{v}$.

where θ is the angle between $\mathbf{u}$ and $\mathbf{v}$ (Figure 8). Notice that by Eq. (4), the absolute value $|\mathbf{u} \cdot \mathbf{e}_\mathbf{v}|$ is the length of $\text{proj}_\mathbf{v}(\mathbf{u})$.

■ **EXAMPLE 7** What is the minimum force you must apply to pull a 35-lb wagon up a frictionless ramp, inclined at an angle $\theta = 15°$?

Solution Referring to Figure 9, we resolve the force due to gravity $\mathbf{F}_g$ into a sum $\mathbf{F}_g = \mathbf{F}_{||} + \mathbf{F}_\perp$, where $\mathbf{F}_{||}$ is the force along the ramp and $\mathbf{F}_\perp$ is the force orthogonal to the ramp (the **normal force**). The force $\mathbf{F}_\perp$ is canceled by the force of the ramp pushing against the wagon in the normal direction. To move the wagon, you need only pull against the projection $\mathbf{F}_{||}$ of the force along the ramp. The angle between $\mathbf{F}_g$ and the ramp is $90° - \theta = 75°$, so by Eq. (6),

$$\text{Component along ramp} = \|\mathbf{F}_{||}\| = \|\mathbf{F}_g\|\cos 75° \approx 35(0.26) = 9.1 \text{ lb}$$

Therefore, a minimum force of 9.1 lb is required to pull the wagon up the ramp. ■

Man pulls against component of force parallel to the ramp

Ramp pushes against wagon in normal direction

FIGURE 9

12.3 SUMMARY

- The *dot product* of $\mathbf{v} = \langle a_1, b_1, c_1 \rangle$ and $\mathbf{w} = \langle a_2, b_2, c_2 \rangle$ is

$$\mathbf{v} \cdot \mathbf{w} = a_1 a_2 + b_1 b_2 + c_1 c_2$$

- By convention, the angle θ between two vectors is chosen to satisfy $0 \le \theta \le \pi$.
- The dot product and the angle θ between nonzero vectors $\mathbf{v}$ and $\mathbf{w}$ are related by

$$\mathbf{v} \cdot \mathbf{w} = \|\mathbf{v}\| \, \|\mathbf{w}\| \cos \theta, \qquad \cos \theta = \frac{\mathbf{v} \cdot \mathbf{w}}{\|\mathbf{v}\| \, \|\mathbf{w}\|}, \qquad \theta = \cos^{-1}\left(\frac{\mathbf{v} \cdot \mathbf{w}}{\|\mathbf{v}\| \, \|\mathbf{w}\|} \right)$$

- Test for orthogonality: $\mathbf{v} \perp \mathbf{w}$ if and only if $\mathbf{v} \cdot \mathbf{w} = 0$.
- The angle between $\mathbf{v}$ and $\mathbf{w}$ is acute if $\mathbf{v} \cdot \mathbf{w} > 0$ and obtuse if $\mathbf{v} \cdot \mathbf{w} < 0$.
- Assume $\mathbf{v} \ne \mathbf{0}$. Then every vector $\mathbf{u}$ has a decomposition $\mathbf{u} = \mathbf{u}_{\|} + \mathbf{u}_{\perp}$, where $\mathbf{u}_{\|}$ is parallel to $\mathbf{v}$ and $\mathbf{u}_{\perp}$ is orthogonal to $\mathbf{v}$. The vector $\mathbf{u}_{\|}$ is called the *projection of* $\mathbf{u}$ along $\mathbf{v}$ and is also denoted $\text{proj}_{\mathbf{v}}(\mathbf{u})$.
- Let $\mathbf{e}_{\mathbf{v}} = \dfrac{\mathbf{v}}{\|\mathbf{v}\|}$ for some nonzero vector $\mathbf{v}$. Then

$$\mathbf{u}_{\|} = \text{proj}_{\mathbf{v}}(\mathbf{u}) = (\mathbf{u} \cdot \mathbf{e}_{\mathbf{v}}) \mathbf{e}_{\mathbf{v}} = \left(\frac{\mathbf{u} \cdot \mathbf{v}}{\mathbf{v} \cdot \mathbf{v}} \right) \mathbf{v}, \qquad \mathbf{u}_{\perp} = \mathbf{u} - \mathbf{u}_{\|}$$

- The *component* of $\mathbf{u}$ along $\mathbf{v}$ is the quantity

$$\text{Component of } \mathbf{u} \text{ along } \mathbf{v} = \mathbf{u} \cdot \mathbf{e}_{\mathbf{v}} = \frac{\mathbf{u} \cdot \mathbf{v}}{\|\mathbf{v}\|} = \|\mathbf{u}\| \cos \theta$$

12.3 EXERCISES

Preliminary Questions

1. Is the dot product of two vectors a scalar or a vector?

2. What can you say about the angle between $\mathbf{a}$ and $\mathbf{b}$ if $\mathbf{a} \cdot \mathbf{b} < 0$?

3. Suppose that $\mathbf{v}$ is orthogonal to both $\mathbf{u}$ and $\mathbf{w}$. Which property of dot products allows us to conclude that $\mathbf{v}$ is orthogonal to $\mathbf{u} + \mathbf{w}$?

4. What is $\text{proj}_{\mathbf{v}}(\mathbf{v})$?

5. What is the difference, if any, between the projection of $\mathbf{u}$ along $\mathbf{v}$ and the projection along the unit vector $\mathbf{e}_{\mathbf{v}}$?

6. Suppose that $\text{proj}_{\mathbf{v}}(\mathbf{u}) = \mathbf{v}$. Determine:
(a) $\text{proj}_{2\mathbf{v}}(\mathbf{u})$
(b) $\text{proj}_{\mathbf{v}}(2\mathbf{u})$

7. Let $\mathbf{u}_{\|}$ be the projection of $\mathbf{u}$ along $\mathbf{v}$. Which of the following is the projection $\mathbf{u}$ along the vector $2\mathbf{v}$?
(a) $\frac{1}{2}\mathbf{u}_{\|}$
(b) $\mathbf{u}_{\|}$
(c) $2\mathbf{u}_{\|}$

8. Let θ be the angle between $\mathbf{u}$ and $\mathbf{v}$. Which of the following is equal to the $\cos \theta$?
(a) $\mathbf{u} \cdot \mathbf{v}$
(b) $\mathbf{u} \cdot \mathbf{e}_{\mathbf{v}}$
(c) $\mathbf{e}_{\mathbf{u}} \cdot \mathbf{e}_{\mathbf{v}}$

Exercises

In Exercises 1–12, compute the dot product.

1. $\langle 1, 2, 1 \rangle \cdot \langle 4, 3, 5 \rangle$

2. $\langle 3, -2, 2 \rangle \cdot \langle 1, 0, 1 \rangle$

3. $\langle 0, 1, 0 \rangle \cdot \langle 7, 41, -3 \rangle$

4. $\langle 1, 1, 1 \rangle \cdot \langle 6, 4, 2 \rangle$

5. $\langle 3, 1 \rangle \cdot \langle 4, -7 \rangle$

6. $\left\langle \frac{1}{6}, \frac{1}{2} \right\rangle \cdot \left\langle 3, \frac{1}{2} \right\rangle$

7. $\mathbf{i} \cdot \mathbf{k}$

8. $\mathbf{j} \cdot \mathbf{j}$

9. $(\mathbf{i} + \mathbf{j}) \cdot (\mathbf{i} + \mathbf{k})$

10. $(3\mathbf{j} + 2\mathbf{k}) \cdot (\mathbf{i} - 4\mathbf{k})$

11. $(\mathbf{i} + \mathbf{j} + \mathbf{k}) \cdot (3\mathbf{i} + 2\mathbf{j} - 5\mathbf{k})$ **12.** $(-\mathbf{k}) \cdot (\mathbf{i} - 2\mathbf{j} + 7\mathbf{k})$

In Exercises 13–18, determine whether the two vectors are orthogonal and if not, whether the angle between them is acute or obtuse.

13. $\langle 1, 1, 1 \rangle, \quad \langle 1, -2, -2 \rangle$ **14.** $\langle 1, 1, 1 \rangle, \quad \langle 3, -2, -1 \rangle$

15. $\langle 0, 2, 4 \rangle, \quad \langle 3, 1, 0 \rangle$ **16.** $\langle 0, 2, 4 \rangle, \quad \langle -5, 0, 0 \rangle$

17. $\langle 4, 3 \rangle, \quad \langle 2, -4 \rangle$ **18.** $\langle 12, 6 \rangle, \quad \langle 2, -4 \rangle$

In Exercises 19–22, find the cosine of the angle between the vectors.

19. $\langle 0, 3, 1 \rangle, \quad \langle 4, 0, 0 \rangle$ **20.** $\langle 1, 1, 1 \rangle, \quad \langle 2, -1, 2 \rangle$

21. $\mathbf{i} + \mathbf{j}, \quad \mathbf{j} + 2\mathbf{k}$ **22.** $3\mathbf{i} + \mathbf{k}, \quad \mathbf{i} + \mathbf{j} + \mathbf{k}$

In Exercises 23–28, find the angle between the vectors.

23. $\langle 1, 2 \rangle, \quad \langle 3, 5 \rangle$ **24.** $\langle 0, -4 \rangle, \quad \langle 1, 3 \rangle$

25. $\langle 1, 1, 1 \rangle, \quad \langle 1, -1, 1 \rangle$ **26.** $\langle 2, 4, 1 \rangle, \quad \langle 1, -3, 5 \rangle$

27. $\langle \pi, e, 3 \rangle, \quad \langle \cos 1, \tan 1, e^2 \rangle$ **28.** $\langle 1, 1, 5 \rangle, \quad \langle 1, -1, 5 \rangle$

29. Find all values of b for which the vectors are orthogonal.
(a) $\langle b, 3, 2 \rangle, \quad \langle 1, b, 1 \rangle$ **(b)** $\langle 4, -2, 7 \rangle, \quad \langle b^2, b, 0 \rangle$

30. Find a vector that is orthogonal to $\langle -1, 2, 2 \rangle$.

31. Find two vectors (which are not multiples of each other) that are both orthogonal to $\langle 2, 0, -3 \rangle$.

32. The angle between two unit vectors $\mathbf{e}$ and $\mathbf{f}$ is $50°$. Find $\mathbf{e} \cdot \mathbf{f}$.

In Exercises 33–36, assume that $\mathbf{u} \cdot \mathbf{v} = 2$, $\|\mathbf{u}\| = 1$, and $\|\mathbf{v}\| = 3$ and evaluate the expression.

33. $\mathbf{u} \cdot (4\mathbf{v})$ **34.** $(\mathbf{u} + \mathbf{v}) \cdot \mathbf{v}$

35. $2\mathbf{u} \cdot (3\mathbf{u} - \mathbf{v})$ **36.** $(\mathbf{u} + \mathbf{v}) \cdot (\mathbf{u} - \mathbf{v})$

In Exercises 37–40, simplify the expression.

37. $(\mathbf{v} - \mathbf{w}) \cdot \mathbf{v} + \mathbf{v} \cdot \mathbf{w}$

38. $(\mathbf{v} + \mathbf{w}) \cdot (\mathbf{v} + \mathbf{w}) - 2\mathbf{v} \cdot \mathbf{w}$

39. $(\mathbf{v} + \mathbf{w}) \cdot \mathbf{v} - (\mathbf{v} + \mathbf{w}) \cdot \mathbf{w}$

40. $(\mathbf{v} + \mathbf{w}) \cdot \mathbf{v} - (\mathbf{v} - \mathbf{w}) \cdot \mathbf{w}$

In Exercises 41–48, find the projection $\text{proj}_{\mathbf{v}}(\mathbf{u})$.

41. $\mathbf{u} = \langle -1, 2, 0 \rangle, \quad \mathbf{v} = \langle 2, 0, 1 \rangle$

42. $\mathbf{u} = \langle 2, 0 \rangle, \quad \mathbf{v} = \langle 4, 3 \rangle$

43. $\mathbf{u} = \langle 1, 1, 1 \rangle, \quad \mathbf{v} = \langle 1, 1, 0 \rangle$

44. $\mathbf{u} = \langle 3, 2, 1 \rangle, \quad \mathbf{v} = \langle 6, 4, 2 \rangle$

45. $\mathbf{u} = 5\mathbf{i} + 7\mathbf{j} - 4\mathbf{k}, \quad \mathbf{v} = \mathbf{k}$

46. $\mathbf{u} = \mathbf{i} + 29\mathbf{k}, \quad \mathbf{v} = \mathbf{j}$

47. $\mathbf{u} = \langle a, b, c \rangle, \quad \mathbf{v} = \mathbf{i}$

48. $\mathbf{u} = \langle a, a, b \rangle, \quad \mathbf{v} = \mathbf{i} - \mathbf{j}$

In Exercises 49–54, find the decomposition $\mathbf{a} = \mathbf{a}_{\|} + \mathbf{a}_{\perp}$ with respect to $\mathbf{b}$.

49. $\mathbf{a} = \langle -2, 1, 1 \rangle, \quad \mathbf{b} = \langle 1, 0, 0 \rangle$

50. $\mathbf{a} = \langle 4, -1, 0 \rangle, \quad \mathbf{b} = \langle 0, 1, 1 \rangle$

51. $\mathbf{a} = \langle 1, 0 \rangle, \quad \mathbf{b} = \langle 1, 1 \rangle$

52. $\mathbf{a} = \langle 4, 4, -5 \rangle, \quad \mathbf{b} = \langle 2, -1, 1 \rangle$

53. $\mathbf{a} = \langle 4, 2 \rangle, \quad \mathbf{b} = \langle -1, 2 \rangle$

54. $\mathbf{a} = \langle 3, 5, 5 \rangle, \quad \mathbf{b} = \langle 5, 2, 2 \rangle$

55. Let $\mathbf{e}_\theta = \langle \cos \theta, \sin \theta \rangle$. Show that $\mathbf{e}_\theta$ is a unit vector making an angle θ with the x-axis. Show that $\mathbf{e}_\theta \cdot \mathbf{e}_\psi = \cos(\theta - \psi)$ for any two angles θ and ψ.

56. [icon] Let $\mathbf{v}$ and $\mathbf{w}$ be vectors in the plane.
(a) Use Theorem 2 to explain why the dot product $\mathbf{v} \cdot \mathbf{w}$ does not change if both $\mathbf{v}$ and $\mathbf{w}$ are rotated by the same angle θ.

(b) Sketch the vectors $\mathbf{e}_1 = \langle 1, 0 \rangle$ and $\mathbf{e}_2 = \left\langle \dfrac{\sqrt{2}}{2}, \dfrac{\sqrt{2}}{2} \right\rangle$, and determine the vectors $\mathbf{e}_1', \mathbf{e}_2'$ obtained by rotating $\mathbf{e}_1, \mathbf{e}_2$ through an angle $\dfrac{\pi}{4}$. Verify that $\mathbf{e}_1 \cdot \mathbf{e}_2 = \mathbf{e}_1' \cdot \mathbf{e}_2'$.

57. Determine $\|\mathbf{v} + \mathbf{w}\|$ if $\mathbf{v}$ and $\mathbf{w}$ are unit vectors separated by an angle of $30°$.

58. What is the angle between $\mathbf{v}$ and $\mathbf{w}$ if:
(a) $\mathbf{v} \cdot \mathbf{w} = -\|\mathbf{v}\| \, \|\mathbf{w}\|$ **(b)** $\mathbf{v} \cdot \mathbf{w} = \frac{1}{2}\|\mathbf{v}\| \, \|\mathbf{w}\|$

59. Suppose that $\|\mathbf{v}\| = 2$ and $\|\mathbf{w}\| = 3$, and the angle between $\mathbf{v}$ and $\mathbf{w}$ is $120°$. Determine:
(a) $\mathbf{v} \cdot \mathbf{w}$ **(b)** $\|2\mathbf{v} + \mathbf{w}\|$ **(c)** $\|2\mathbf{v} - 3\mathbf{w}\|$

60. Find the angle θ in the triangle in Figure 10.

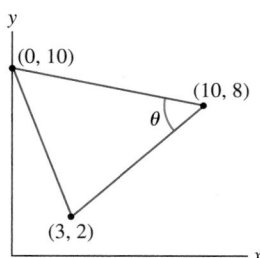

FIGURE 10

61. Find all three angles in the triangle in Figure 11.

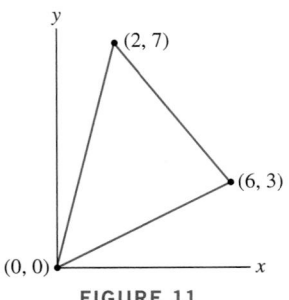

FIGURE 11

In Exercises 62–65, refer to Figure 12.

62. Find the angle between $\overrightarrow{AB}$ and $\overrightarrow{AC}$.

63. Find the angle between $\overrightarrow{AB}$ and $\overrightarrow{AD}$.

64. Calculate the projection of $\overrightarrow{AC}$ along $\overrightarrow{AD}$.

65. Calculate the projection of $\overrightarrow{AD}$ along $\overrightarrow{AB}$.

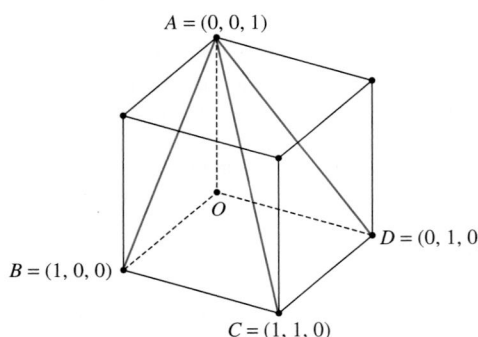

FIGURE 12 Unit cube in $\mathbf{R}^3$.

66. Let $\mathbf{v}$ and $\mathbf{w}$ be nonzero vectors and set $\mathbf{u} = \mathbf{e_v} + \mathbf{e_w}$. Use the dot product to show that the angle between $\mathbf{u}$ and $\mathbf{v}$ is equal to the angle between $\mathbf{u}$ and $\mathbf{w}$. Explain this result geometrically with a diagram.

67. Let $\mathbf{v}$, $\mathbf{w}$, and $\mathbf{a}$ be nonzero vectors such that $\mathbf{v} \cdot \mathbf{a} = \mathbf{w} \cdot \mathbf{a}$. Is it true that $\mathbf{v} = \mathbf{w}$? Either prove this or give a counterexample.

68. Prove that if $\mathbf{v}$ and $\mathbf{w}$ are nonzero and orthogonal, and if $c_1\mathbf{v} + c_2\mathbf{w} = \mathbf{0}$, then $c_1 = c_2 = 0$. *Hint:* Take the dot product of the equation with $\mathbf{v}$ and $\mathbf{w}$.

69. Calculate the force (in Newtons) required to push a 40-kg wagon up a 10° incline (Figure 13). One N is equal to 1 kg-m/s², and the force due to gravity on the wagon is $\mathbf{F} = 40g$ N, where $g = 9.8$.

FIGURE 13

70. A force $\mathbf{F}$ is applied to each of two ropes (of negligible weight) attached to opposite ends of a 75-lb wagon and making an angle of 35° with the horizontal (Figure 14). What is the maximum magnitude of $\mathbf{F}$ that can be applied without lifting the wagon off the ground?

FIGURE 14

71. A 10-kg mass hangs from two ropes (of negligible weight) as in Figure 15. Rope 1 exerts a force of magnitude $\mathbf{F_1}$ acting in the direction $\overrightarrow{PA}$ and rope 2 exerts a force $\mathbf{F_2}$ in the direction $\overrightarrow{PB}$. The sum of these forces balances the force of gravity $\mathbf{F_g} = 10g$ N acting downward at P (where $g = 9.8$ m/s²). Determine $\mathbf{F_1}$ and $\mathbf{F_2}$.

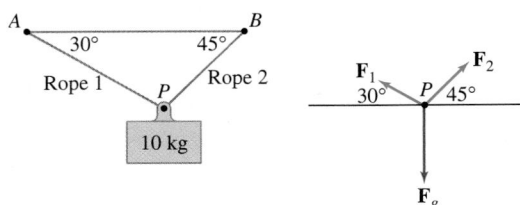

FIGURE 15

72. A light beam travels along the ray determined by a unit vector $\mathbf{L}$, strikes a flat surface at point P, and is reflected along the ray determined by a unit vector $\mathbf{R}$ where $\theta_1 = \theta_2$ (Figure 16). Show that if $\mathbf{N}$ is the unit vector orthogonal to the surface, then $\mathbf{R} = 2(\mathbf{L} \cdot \mathbf{N})\mathbf{N} - \mathbf{L}$.

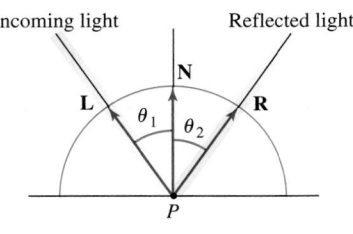

FIGURE 16

73. Let P and Q be antipodal (opposite) points on a sphere of radius r centered at the origin and let R be a third point on the sphere (Figure 17). Prove that $\overrightarrow{PR}$ and $\overrightarrow{QR}$ are orthogonal.

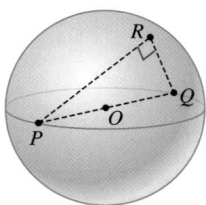

FIGURE 17

74. Prove that $\|\mathbf{v} + \mathbf{w}\|^2 - \|\mathbf{v} - \mathbf{w}\|^2 = 4\mathbf{v} \cdot \mathbf{w}$. *Hint:* Expand the dot products $(\mathbf{v} \pm \mathbf{w}) \cdot (\mathbf{v} \pm \mathbf{w})$.

75. Use Exercise 74 to show that $\mathbf{v}$ and $\mathbf{w}$ are orthogonal if and only if $\|\mathbf{v} - \mathbf{w}\| = \|\mathbf{v} + \mathbf{w}\|$.

76. Show that the two diagonals of a parallelogram are perpendicular if and only if its sides have equal length. *Hint:* Use Exercise 75 to show

that $\mathbf{v} - \mathbf{w}$ and $\mathbf{v} + \mathbf{w}$ are orthogonal if and only if $\|\mathbf{v}\| = \|\mathbf{w}\|$.

77. Verify the Distributive Law:

$$\mathbf{u} \cdot (\mathbf{v} + \mathbf{w}) = \mathbf{u} \cdot \mathbf{v} + \mathbf{u} \cdot \mathbf{w}$$

78. Verify that $(\lambda \mathbf{v}) \cdot \mathbf{w} = \lambda (\mathbf{v} \cdot \mathbf{w})$ for any scalar λ.

Further Insights and Challenges

79. Prove the Law of Cosines: $c^2 = a^2 + b^2 - 2ab \cos \theta$ by referring to Figure 18. *Hint:* Consider the right triangle $\triangle PQR$.

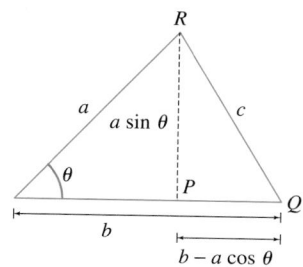

FIGURE 18

80. In this exercise, we prove the Cauchy–Schwarz inequality: If $\mathbf{v}$ and $\mathbf{w}$ are any two vectors, then

$$|\mathbf{v} \cdot \mathbf{w}| \le \|\mathbf{v}\| \, \|\mathbf{w}\| \qquad \boxed{7}$$

(a) Let $f(x) = \|x\mathbf{v} + \mathbf{w}\|^2$ for x a scalar. Show that $f(x) = ax^2 + bx + c$, where $a = \|\mathbf{v}\|^2$, $b = 2\mathbf{v} \cdot \mathbf{w}$, and $c = \|\mathbf{w}\|^2$.
(b) Conclude that $b^2 - 4ac \le 0$. *Hint:* Observe that $f(x) \ge 0$ for all x.

81. Use (7) to prove the Triangle Inequality

$$\|\mathbf{v} + \mathbf{w}\| \le \|\mathbf{v}\| + \|\mathbf{w}\|$$

Hint: First use the Triangle Inequality for numbers to prove

$$|(\mathbf{v} + \mathbf{w}) \cdot (\mathbf{v} + \mathbf{w})| \le |(\mathbf{v} + \mathbf{w}) \cdot \mathbf{v}| + |(\mathbf{v} + \mathbf{w}) \cdot \mathbf{w}|$$

82. This exercise gives another proof of the relation between the dot product and the angle θ between two vectors $\mathbf{v} = \langle a_1, b_1 \rangle$ and $\mathbf{w} = \langle a_2, b_2 \rangle$ in the plane. Let θ_1 and θ_2 be the angles between $\mathbf{v}$ and $\mathbf{w}$ and the positive x-axis as in Figure 19. Use the addition formula $\cos(\theta_1 - \theta_2) = \cos \theta_1 \cos \theta_2 + \sin \theta_1 \sin \theta_2$ to prove that

$\mathbf{v} \cdot \mathbf{w} = \|\mathbf{v}\| \, \|\mathbf{w}\| \cos \theta$. *Hint:* $\mathbf{v} = \|\mathbf{v}\| \langle \cos \theta_1, \sin \theta_1 \rangle$ and $\mathbf{w} = \|\mathbf{w}\| \langle \cos \theta_2, \sin \theta_2 \rangle$.

83. Let $\mathbf{v} = \langle x, y \rangle$ and

$$\mathbf{v}_\theta = \langle x \cos \theta + y \sin \theta, -x \sin \theta + y \cos \theta \rangle$$

Prove that the angle between $\mathbf{v}$ and $\mathbf{v}_\theta$ is θ.

84. Let $\mathbf{v}$ be a nonzero vector. The angles α, β, γ between $\mathbf{v}$ and the unit vectors $\mathbf{i}$, $\mathbf{j}$, $\mathbf{k}$ are called the direction angles of $\mathbf{v}$ (Figure 20). The cosines of these angles are called the **direction cosines** of $\mathbf{v}$. Prove that $\cos^2 \alpha + \cos^2 \beta + \cos^2 \gamma = 1$.

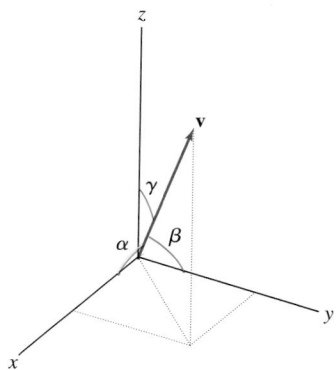

FIGURE 20 Direction angles of $\mathbf{v}$.

85. Find the direction cosines of $\mathbf{v} = \langle 3, 6, -2 \rangle$.

86. The set of all points $X = (x, y, z)$ equidistant from two points P, Q in $\mathbf{R}^3$ is a plane (Figure 21). Show that X lies on this plane if

$$\overrightarrow{PQ} \cdot \overrightarrow{OX} = \frac{1}{2}(\|\overrightarrow{OQ}\|^2 - \|\overrightarrow{OP}\|^2) \qquad \boxed{8}$$

FIGURE 19

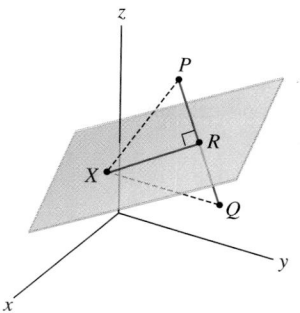

FIGURE 21

Hint: If R is the midpoint of $\overline{PQ}$, then X is equidistant from P and Q if and only if $\overrightarrow{XR}$ is orthogonal to $\overrightarrow{PQ}$.

87. Sketch the plane consisting of all points $X = (x, y, z)$ equidistant from the points $P = (0, 1, 0)$ and $Q = (0, 0, 1)$. Use Eq. (8) to show

that X lies on this plane if and only if $y = z$.

88. Use Eq. (8) to find the equation of the plane consisting of all points $X = (x, y, z)$ equidistant from $P = (2, 1, 1)$ and $Q = (1, 0, 2)$.

12.4 The Cross Product

This section introduces the **cross product** $\mathbf{v} \times \mathbf{w}$ of two vectors $\mathbf{v}$ and $\mathbf{w}$. The cross product (sometimes called the **vector product**) is used in physics and engineering to describe quantities involving rotation such as torque and angular momentum. In electromagnetic theory, magnetic forces are described using cross products (Figures 1 and 2).

Unlike the dot product $\mathbf{v} \cdot \mathbf{w}$ (which is a scalar), the cross product $\mathbf{v} \times \mathbf{w}$ is again a vector. It is defined using determinants, which we now define in the 2×2 and 3×3 cases. A 2×2 determinant is a number formed from an array A of numbers with two rows and two columns (called a **matrix**) according to the formula

$$\det(A) = \begin{vmatrix} a_{11} & a_{12} \\ a_{21} & a_{22} \end{vmatrix} = a_{11}a_{22} - a_{12}a_{21} \qquad \boxed{1}$$

FIGURE 1 The spiral paths of charged particles in a bubble chamber in the presence of a magnetic field are described using cross products.

For example,

$$\begin{vmatrix} 3 & 2 \\ \frac{1}{2} & 4 \end{vmatrix} = 3 \cdot 4 - 2 \cdot \frac{1}{2} = 11$$

The determinant of a 3×3 matrix is defined by the formula

$$\begin{vmatrix} a_{11} & a_{12} & a_{13} \\ a_{21} & a_{22} & a_{23} \\ a_{31} & a_{32} & a_{33} \end{vmatrix} = a_{11}\underbrace{\begin{vmatrix} a_{22} & a_{23} \\ a_{32} & a_{33} \end{vmatrix}}_{(1,1)\text{-minor}} - a_{12}\underbrace{\begin{vmatrix} a_{21} & a_{23} \\ a_{31} & a_{33} \end{vmatrix}}_{(1,2)\text{-minor}} + a_{13}\underbrace{\begin{vmatrix} a_{21} & a_{22} \\ a_{31} & a_{32} \end{vmatrix}}_{(1,3)\text{-minor}} \qquad \boxed{2}$$

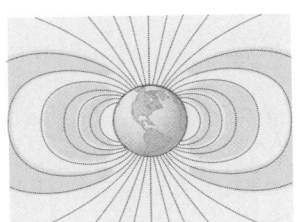

FIGURE 2 The Van Allen radiation belts, located thousands of miles above the earth's surface, are made up of streams of protons and electrons that oscillate back and forth in helical paths between two "magnetic mirrors" set up by the earth's magnetic field. This helical motion is explained by the "cross-product" nature of magnetic forces.

This formula expresses the 3×3 determinant in terms of 2×2 determinants called **minors**. The minors are obtained by crossing out the first row and one of the three columns of the 3×3 matrix. For example, the minor labeled $(1, 2)$ above is obtained as follows:

The theory of matrices and determinants is part of linear algebra, a subject of great importance throughout mathematics. In this section, we discuss just a few basic definitions and facts needed for our treatment of multivariable calculus.

$$\begin{vmatrix} a_{11} & \cancel{a_{12}} & a_{13} \\ a_{21} & a_{22} & a_{23} \\ a_{31} & a_{32} & a_{33} \end{vmatrix} \qquad \text{to obtain the } (1, 2)\text{-minor} \qquad \begin{vmatrix} a_{21} & a_{23} \\ a_{31} & a_{33} \end{vmatrix}$$

Cross out row 1 and column 2 $\qquad\qquad\qquad\qquad\qquad\qquad\qquad\qquad$ (1,2)-minor

■ **EXAMPLE 1** A 3×3 Determinant   Calculate $\begin{vmatrix} 2 & 4 & 3 \\ 0 & 1 & -7 \\ -1 & 5 & 3 \end{vmatrix}$.

Solution

$$\begin{vmatrix} ② & ④ & ③ \\ 0 & 1 & -7 \\ -1 & 5 & 3 \end{vmatrix} = ②\begin{vmatrix} 1 & -7 \\ 5 & 3 \end{vmatrix} - ④\begin{vmatrix} 0 & -7 \\ -1 & 3 \end{vmatrix} + ③\begin{vmatrix} 0 & 1 \\ -1 & 5 \end{vmatrix}$$

$$= 2(38) - 4(-7) + 3(1) = 107$$

We will see below that determinants are related to area and volume. Our immediate goal is to introduce the cross product, which is defined as a "symbolic" determinant whose first row has the vector entries $\mathbf{i}, \mathbf{j}, \mathbf{k}$.

DEFINITION The Cross Product   The cross product of vectors $\mathbf{v} = \langle a_1, b_1, c_1 \rangle$ and $\mathbf{w} = \langle a_2, b_2, c_2 \rangle$ is the vector

$$\mathbf{v} \times \mathbf{w} = \begin{vmatrix} \mathbf{i} & \mathbf{j} & \mathbf{k} \\ a_1 & b_1 & c_1 \\ a_2 & b_2 & c_2 \end{vmatrix} = \begin{vmatrix} b_1 & c_1 \\ b_2 & c_2 \end{vmatrix} \mathbf{i} - \begin{vmatrix} a_1 & c_1 \\ a_2 & c_2 \end{vmatrix} \mathbf{j} + \begin{vmatrix} a_1 & b_1 \\ a_2 & b_2 \end{vmatrix} \mathbf{k} \qquad \boxed{3}$$

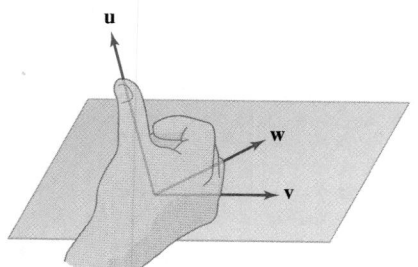

FIGURE 3 $\{\mathbf{v}, \mathbf{w}, \mathbf{u}\}$ forms a right-handed system.

■ **EXAMPLE 2** Calculate $\mathbf{v} \times \mathbf{w}$, where $\mathbf{v} = \langle -2, 1, 4 \rangle$ and $\mathbf{w} = \langle 3, 2, 5 \rangle$.

Solution

$$\mathbf{v} \times \mathbf{w} = \begin{vmatrix} \mathbf{i} & \mathbf{j} & \mathbf{k} \\ -2 & 1 & 4 \\ 3 & 2 & 5 \end{vmatrix} = \begin{vmatrix} 1 & 4 \\ 2 & 5 \end{vmatrix} \mathbf{i} - \begin{vmatrix} -2 & 4 \\ 3 & 5 \end{vmatrix} \mathbf{j} + \begin{vmatrix} -2 & 1 \\ 3 & 2 \end{vmatrix} \mathbf{k}$$

$$= (-3)\mathbf{i} - (-22)\mathbf{j} + (-7)\mathbf{k} = \langle -3, 22, -7 \rangle \qquad ■$$

Formula (3) gives no hint of the geometric meaning of the cross product. However, there is a simple way to visualize the vector $\mathbf{v} \times \mathbf{w}$ using the **right-hand rule**. Suppose that $\mathbf{v}, \mathbf{w}$, and $\mathbf{u}$ are nonzero vectors that do not all lie in a plane. We say that $\{\mathbf{v}, \mathbf{w}, \mathbf{u}\}$ forms a **right-handed system** if the direction of $\mathbf{u}$ is determined by the right-hand rule: *When the fingers of your right hand curl from $\mathbf{v}$ to $\mathbf{w}$, your thumb points to the same side of the plane spanned by $\mathbf{v}$ and $\mathbf{w}$ as $\mathbf{u}$* (Figure 3). The following theorem is proved at the end of this section.

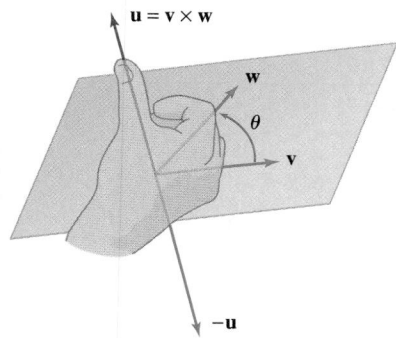

FIGURE 4 There are two vectors orthogonal to $\mathbf{v}$ and $\mathbf{w}$ with length $\|\mathbf{v}\| \, \|\mathbf{w}\| \sin \theta$. The right-hand rule determines which is $\mathbf{v} \times \mathbf{w}$.

THEOREM 1 Geometric Description of the Cross Product   The cross product $\mathbf{v} \times \mathbf{w}$ is the unique vector with the following three properties:

(i) $\mathbf{v} \times \mathbf{w}$ is orthogonal to $\mathbf{v}$ and $\mathbf{w}$.
(ii) $\mathbf{v} \times \mathbf{w}$ has length $\|\mathbf{v}\| \, \|\mathbf{w}\| \, |\sin \theta|$ (θ = angle between $\mathbf{v}$ and $\mathbf{w}$).
(iii) $\{\mathbf{v}, \mathbf{w}, \mathbf{v} \times \mathbf{w}\}$ forms a right-handed system.

How do the three properties in Theorem 1 determine $\mathbf{v} \times \mathbf{w}$? By property (i), $\mathbf{v} \times \mathbf{w}$ lies on the line orthogonal to $\mathbf{v}$ and $\mathbf{w}$. By property (ii), $\mathbf{v} \times \mathbf{w}$ is one of the two vectors on this line of length $\|\mathbf{v}\| \, \|\mathbf{w}\| \, |\sin \theta|$. Finally, property (iii) tells us which of these two vectors is $\mathbf{v} \times \mathbf{w}$, namely, the vector for which $\{\mathbf{v}, \mathbf{w}, \mathbf{u}\}$ is right-handed (Figure 4).

■ **EXAMPLE 3** Let $\mathbf{v} = \langle 2, 0, 0 \rangle$ and $\mathbf{w} = \langle 0, 1, 1 \rangle$. Determine $\mathbf{u} = \mathbf{v} \times \mathbf{w}$ using the geometric properties of the cross product rather than the formula.

Solution We use Theorem 1. By (i), $\mathbf{u} = \mathbf{v} \times \mathbf{w}$ is orthogonal to $\mathbf{v}$ and $\mathbf{w}$. Since $\mathbf{v}$ lies along the x-axis, $\mathbf{u}$ must lie in the yz-plane (Figure 5). In other words, $\mathbf{u} = \langle 0, b, c \rangle$. But $\mathbf{u}$ is also orthogonal to $\mathbf{w} = \langle 0, 1, 1 \rangle$, so $\mathbf{u} \cdot \mathbf{w} = b + c = 0$ and thus $\mathbf{u} = \langle 0, b, -b \rangle$.

Direct computation shows that $\|\mathbf{v}\| = 2$ and $\|\mathbf{w}\| = \sqrt{2}$. Furthermore, the angle between $\mathbf{v}$ and $\mathbf{w}$ is $\theta = \frac{\pi}{2}$ since $\mathbf{v} \cdot \mathbf{w} = 0$. By property (ii),

$$\|\mathbf{u}\| = \sqrt{b^2 + (-b)^2} = |b| \sqrt{2} \qquad \text{is equal to} \qquad \|\mathbf{v}\| \, \|\mathbf{w}\| \sin \frac{\pi}{2} = 2\sqrt{2}$$

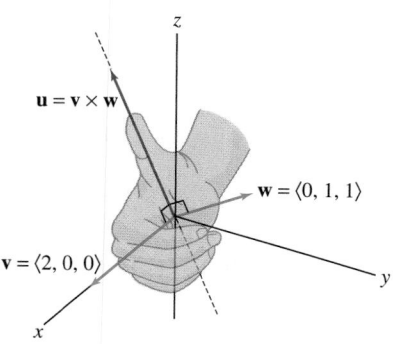

FIGURE 5 The direction of $\mathbf{u} = \mathbf{v} \times \mathbf{w}$ is determined by the right-hand rule. Thus, $\mathbf{u}$ has a positive z-component.

Therefore, $|b| = 2$ and $b = \pm 2$. Finally, property (iii) tells us that $\mathbf{u}$ points in the positive z-direction (Figure 5). Thus, $b = -2$ and $\mathbf{u} = \langle 0, -2, 2 \rangle$. You can verify that the formula for the cross product yields the same answer.   ∎

One of the most striking properties of the cross product is that it is *anticommutative*. Reversing the order changes the sign:

$$\boxed{\mathbf{w} \times \mathbf{v} = -\mathbf{v} \times \mathbf{w}}$$ **4**

This follows from Eq. (3) and the fact that a 2×2 determinant changes sign when the rows are interchanged:

$$\begin{vmatrix} a_1 & b_1 \\ a_2 & b_2 \end{vmatrix} = a_1 b_2 - b_1 a_2 = -(b_1 a_2 - a_1 b_2) = - \begin{vmatrix} a_2 & b_2 \\ a_1 & b_1 \end{vmatrix}$$

Anticommutativity also follows from the geometric description of the cross product. By properties (i) and (ii) in Theorem 1, $\mathbf{v} \times \mathbf{w}$ and $\mathbf{w} \times \mathbf{v}$ are both orthogonal to $\mathbf{v}$ and $\mathbf{w}$ and have the same length. However, $\mathbf{v} \times \mathbf{w}$ and $\mathbf{w} \times \mathbf{v}$ point in opposite directions by the right-hand rule and thus $\mathbf{v} \times \mathbf{w} = -\mathbf{w} \times \mathbf{v}$ (Figure 6). In particular, $\mathbf{v} \times \mathbf{v} = -\mathbf{v} \times \mathbf{v}$ and hence $\mathbf{v} \times \mathbf{v} = \mathbf{0}$.

The next theorem lists some further properties of cross products (the proofs are given as Exercises 49–51).

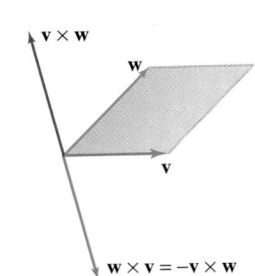

FIGURE 6

Note an important distinction between the dot and cross products of a vector with itself:

$$\mathbf{v} \times \mathbf{v} = \mathbf{0}$$

$$\mathbf{v} \cdot \mathbf{v} = \|\mathbf{v}\|^2$$

THEOREM 2 Basic Properties of the Cross Product

(i) $\mathbf{w} \times \mathbf{v} = -\mathbf{v} \times \mathbf{w}$

(ii) $\mathbf{v} \times \mathbf{v} = \mathbf{0}$

(iii) $\mathbf{v} \times \mathbf{w} = \mathbf{0}$ if and only if $\mathbf{w} = \lambda \mathbf{v}$ for some scalar λ or $\mathbf{v} = \mathbf{0}$.

(iv) $(\lambda \mathbf{v}) \times \mathbf{w} = \mathbf{v} \times (\lambda \mathbf{w}) = \lambda (\mathbf{v} \times \mathbf{w})$

(v) $(\mathbf{u} + \mathbf{v}) \times \mathbf{w} = \mathbf{u} \times \mathbf{w} + \mathbf{v} \times \mathbf{w}$

$\mathbf{v} \times (\mathbf{u} + \mathbf{w}) = \mathbf{v} \times \mathbf{u} + \mathbf{v} \times \mathbf{w}$

The cross product of any two of the standard basis vectors $\mathbf{i}$, $\mathbf{j}$, and $\mathbf{k}$ is equal to the third, possibly with a minus sign. More precisely (see Exercise 52),

$$\mathbf{i} \times \mathbf{j} = \mathbf{k}, \qquad \mathbf{j} \times \mathbf{k} = \mathbf{i}, \qquad \mathbf{k} \times \mathbf{i} = \mathbf{j}$$ **5**

$$\mathbf{i} \times \mathbf{i} = \mathbf{j} \times \mathbf{j} = \mathbf{k} \times \mathbf{k} = \mathbf{0}$$

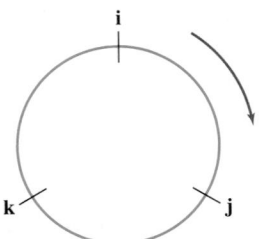

FIGURE 7 Circle for computing the cross products of the basic vectors.

Since the cross product is anticommutative, minus signs occur when the cross products are taken in the opposite order. An easy way to remember these relations is to draw $\mathbf{i}$, $\mathbf{j}$, and $\mathbf{k}$ in a circle as in Figure 7. Go around the circle in the clockwise direction (starting at any point) and you obtain one of the relations (5). For example, starting at $\mathbf{i}$ and moving clockwise yields $\mathbf{i} \times \mathbf{j} = \mathbf{k}$. If you go around in the counterclockwise direction, you obtain the relations with a minus sign. Thus, starting at $\mathbf{k}$ and going counterclockwise gives the relation $\mathbf{k} \times \mathbf{j} = -\mathbf{i}$.

■ EXAMPLE 4 Using the ijk Relations   Compute $(2\mathbf{i} + \mathbf{k}) \times (3\mathbf{j} + 5\mathbf{k})$.

Solution We use the Distributive Law for cross products:

$$(2\mathbf{i} + \mathbf{k}) \times (3\mathbf{j} + 5\mathbf{k}) = (2\mathbf{i}) \times (3\mathbf{j}) + (2\mathbf{i}) \times (5\mathbf{k}) + \mathbf{k} \times (3\mathbf{j}) + \mathbf{k} \times (5\mathbf{k})$$
$$= 6(\mathbf{i} \times \mathbf{j}) + 10(\mathbf{i} \times \mathbf{k}) + 3(\mathbf{k} \times \mathbf{j}) + 5(\mathbf{k} \times \mathbf{k})$$
$$= 6\mathbf{k} - 10\mathbf{j} - 3\mathbf{i} + 5(\mathbf{0}) = -3\mathbf{i} - 10\mathbf{j} + 6\mathbf{k} \qquad \blacksquare$$

■ **EXAMPLE 5** **Velocity in a Magnetic Field** The force $\mathbf{F}$ on a proton moving at velocity $\mathbf{v}$ m/s in a uniform magnetic field $\mathbf{B}$ (in teslas) is $\mathbf{F} = q(\mathbf{v} \times \mathbf{B})$ in newtons, where $q = 1.6 \times 10^{-19}$ Coulombs (Figure 8). Calculate $\mathbf{F}$ if $\mathbf{B} = 0.0004\mathbf{k}$ T and $\mathbf{v}$ has magnitude 10^6 m/s in the direction $-\mathbf{j} + \mathbf{k}$.

Solution The vector $-\mathbf{j} + \mathbf{k}$ has length $\sqrt{2}$, and since $\mathbf{v}$ has magnitude 10^6,

$$\mathbf{v} = 10^6 \left(\frac{-\mathbf{j} + \mathbf{k}}{\sqrt{2}} \right)$$

Therefore, the force (in newtons) is

$$\mathbf{F} = q(\mathbf{v} \times \mathbf{B}) = 10^6 q \left(\frac{-\mathbf{j} + \mathbf{k}}{\sqrt{2}} \right) \times (0.0004\mathbf{k}) = \frac{400q}{\sqrt{2}} ((-\mathbf{j} + \mathbf{k}) \times \mathbf{k})$$

$$= -\frac{400q}{\sqrt{2}} \mathbf{i} = \frac{-400(1.6 \times 10^{-19})}{\sqrt{2}} \mathbf{i} \approx -(4.5 \times 10^{-17})\mathbf{i} \qquad \blacksquare$$

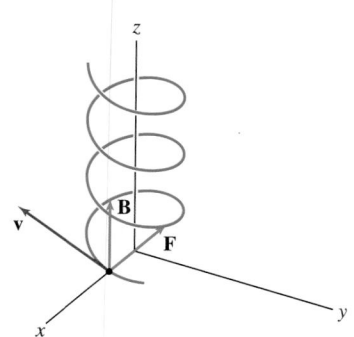

FIGURE 8 A proton in a uniform magnetic field travels in a helical path.

Relation Between the Cross Product, Area, and Volume

The cross product and determinants are closely related to area and volume. Consider the parallelogram $\mathcal{P}$ spanned by nonzero vectors $\mathbf{v}$ and $\mathbf{w}$ with a common basepoint [Figure 9(A)]. Recall that $\mathcal{P}$ is the parallelogram whose vertices are the basepoint and the terminal points of $\mathbf{v}$, $\mathbf{w}$, and $\mathbf{v} + \mathbf{w}$. This parallelogram has base $b = \|\mathbf{v}\|$ and height $h = \|\mathbf{w}\| \sin \theta$, where θ is the angle between $\mathbf{v}$ and $\mathbf{w}$. Therefore, the area of $\mathcal{P}$ is equal to $A = bh = \|\mathbf{v}\| \|\mathbf{w}\| \sin \theta = \|\mathbf{v} \times \mathbf{w}\|$.

 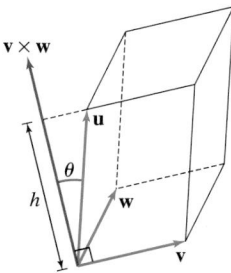

(A) The area of the parallelogram $\mathcal{P}$ is $\|\mathbf{v} \times \mathbf{w}\| = \|\mathbf{v}\| \|\mathbf{w}\| \sin \theta$.

(B) The volume of the parallelpiped $\mathbf{P}$ is $|\mathbf{u} \cdot (\mathbf{v} \times \mathbf{w})|$.

FIGURE 9

A *"parallelepiped"* is the solid spanned by three vectors. Each face is a parallelogram.

Next, consider the **parallelepiped P** spanned by three nonzero vectors $\mathbf{u}, \mathbf{v}, \mathbf{w}$ in $\mathbf{R}^3$ [the three-dimensional prism in Figure 9(B)]. The base of $\mathbf{P}$ is the parallelogram spanned by $\mathbf{v}$ and $\mathbf{w}$, so the area of the base is $\|\mathbf{v} \times \mathbf{w}\|$. The height of $\mathbf{P}$ is $h = \|\mathbf{u}\| \cdot |\cos \theta|$, where θ is the angle between $\mathbf{u}$ and $\mathbf{v} \times \mathbf{w}$. Therefore,

$$\text{Volume of } \mathbf{P} = (\text{area of base})(\text{height}) = \|\mathbf{v} \times \mathbf{w}\| \cdot \|\mathbf{u}\| \cdot |\cos \theta| = |\mathbf{u} \cdot (\mathbf{v} \times \mathbf{w})|$$

The quantity $\mathbf{u} \cdot (\mathbf{v} \times \mathbf{w})$, called the **vector triple product**, can be expressed as a determinant. Let

$$\mathbf{u} = \langle a_1, b_1, c_1 \rangle, \qquad \mathbf{v} = \langle a_2, b_2, c_2 \rangle, \qquad \mathbf{w} = \langle a_3, b_3, c_3 \rangle$$

We use the following notation for the determinant of the matrix whose rows are the vectors $\mathbf{v}, \mathbf{w}, \mathbf{u}$:

$$\det \begin{pmatrix} \mathbf{u} \\ \mathbf{v} \\ \mathbf{w} \end{pmatrix} = \begin{vmatrix} a_1 & b_1 & c_1 \\ a_2 & b_2 & c_2 \\ a_3 & b_3 & c_3 \end{vmatrix}$$

It is awkward to write the absolute value of a determinant in the notation on the right, but we may denote it

$$\left| \det \begin{pmatrix} \mathbf{u} \\ \mathbf{v} \\ \mathbf{w} \end{pmatrix} \right|$$

Then

$$\mathbf{u} \cdot (\mathbf{v} \times \mathbf{w}) = \mathbf{u} \cdot \left(\begin{vmatrix} b_2 & c_2 \\ b_3 & c_3 \end{vmatrix} \mathbf{i} - \begin{vmatrix} a_2 & c_2 \\ a_3 & c_3 \end{vmatrix} \mathbf{j} + \begin{vmatrix} a_2 & b_2 \\ a_3 & b_3 \end{vmatrix} \mathbf{k} \right)$$

$$= a_1 \begin{vmatrix} b_2 & c_2 \\ b_3 & c_3 \end{vmatrix} - b_1 \begin{vmatrix} a_2 & c_2 \\ a_3 & c_3 \end{vmatrix} + c_1 \begin{vmatrix} a_2 & b_2 \\ a_3 & b_3 \end{vmatrix}$$

$$= \begin{vmatrix} a_1 & b_1 & c_1 \\ a_2 & b_2 & c_2 \\ a_3 & b_3 & c_3 \end{vmatrix} = \det \begin{pmatrix} \mathbf{u} \\ \mathbf{v} \\ \mathbf{w} \end{pmatrix} \qquad \boxed{6}$$

We obtain the following formulas for area and volume.

THEOREM 3 Area and Volume via Cross Products and Determinants   Let $\mathbf{u}, \mathbf{v}, \mathbf{w}$ be nonzero vectors in $\mathbf{R}^3$. Then

  (i) The parallelogram $\mathcal{P}$ spanned by $\mathbf{v}$ and $\mathbf{w}$ has area $A = \|\mathbf{v} \times \mathbf{w}\|$.
  (ii) The parallelepiped $\mathbf{P}$ spanned by $\mathbf{u}, \mathbf{v}$, and $\mathbf{w}$ has volume

$$V = |\mathbf{u} \cdot (\mathbf{v} \times \mathbf{w})| = \left| \det \begin{pmatrix} \mathbf{u} \\ \mathbf{v} \\ \mathbf{w} \end{pmatrix} \right| \qquad \boxed{7}$$

■ **EXAMPLE 6**   Let $\mathbf{v} = \langle 1, 4, 5 \rangle$ and $\mathbf{w} = \langle -2, -1, 2 \rangle$. Calculate:

(a) The area A of the parallelogram spanned by $\mathbf{v}$ and $\mathbf{w}$
(b) The volume V of the parallelepiped in Figure 10

Solution   We compute the cross product and apply Theorem 3:

$$\mathbf{v} \times \mathbf{w} = \begin{vmatrix} 4 & 5 \\ -1 & 2 \end{vmatrix} \mathbf{i} - \begin{vmatrix} 1 & 5 \\ -2 & 2 \end{vmatrix} \mathbf{j} + \begin{vmatrix} 1 & 4 \\ -2 & -1 \end{vmatrix} \mathbf{k} = \langle 13, -12, 7 \rangle$$

(a) The area of the parallelogram spanned by $\mathbf{v}$ and $\mathbf{w}$ is

$$A = \|\mathbf{v} \times \mathbf{w}\| = \sqrt{13^2 + (-12)^2 + 7^2} = \sqrt{362} \approx 19$$

(b) The vertical leg of the parallelepiped is the vector $6\mathbf{k}$, so by Eq. (7),

$$V = |(6\mathbf{k}) \cdot (\mathbf{v} \times \mathbf{w})| = |\langle 0, 0, 6 \rangle \cdot \langle 13, -12, 7 \rangle| = 6(7) = 42 \qquad ■$$

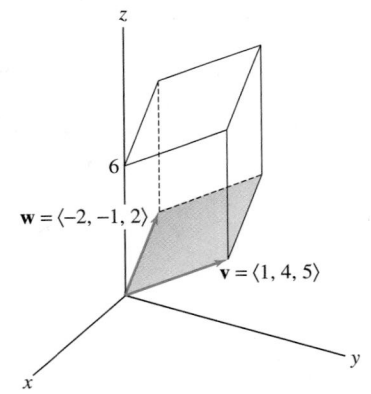

FIGURE 10

If $\mathbf{v} = \langle a, b \rangle$ and $\mathbf{w} = \langle c, d \rangle$ are vectors in the xy-plane, we may compute the area A of the parallelogram spanned by $\mathbf{v}$ and $\mathbf{w}$ as follows. We regard $\mathbf{v}$ and $\mathbf{w}$ as vectors in $\mathbf{R}^3$ with zero component in the z-direction (Figure 11). Thus, we write $\mathbf{v} = \langle a, b, 0 \rangle$ and $\mathbf{w} = \langle c, d, 0 \rangle$. The cross product $\mathbf{v} \times \mathbf{w}$ is a vector pointing in the z-direction:

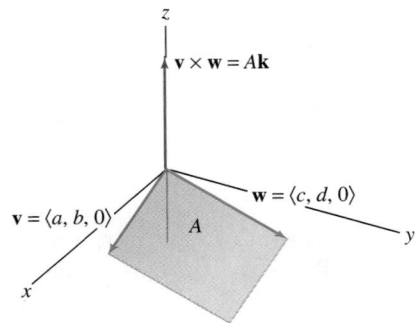

FIGURE 11 Parallelogram spanned by $\mathbf{v}$ and $\mathbf{w}$ in the xy-plane.

$$\mathbf{v} \times \mathbf{w} = \begin{vmatrix} \mathbf{i} & \mathbf{j} & \mathbf{k} \\ a & b & 0 \\ c & d & 0 \end{vmatrix} = \begin{vmatrix} b & 0 \\ d & 0 \end{vmatrix} \mathbf{i} - \begin{vmatrix} a & 0 \\ b & 0 \end{vmatrix} \mathbf{j} + \begin{vmatrix} a & b \\ c & d \end{vmatrix} \mathbf{k} = \begin{vmatrix} a & b \\ c & d \end{vmatrix} \mathbf{k}$$

By Theorem 3, the parallelogram spanned by $\mathbf{v}$ and $\mathbf{w}$ has area $A = \|\mathbf{v} \times \mathbf{w}\|$, and thus,

$$A = \left| \det \begin{pmatrix} \mathbf{v} \\ \mathbf{w} \end{pmatrix} \right| = \left| \det \begin{pmatrix} a & b \\ c & d \end{pmatrix} \right| \qquad \boxed{8}$$

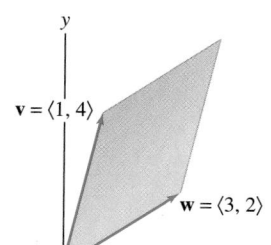

FIGURE 12

■ **EXAMPLE 7** Compute the area A of the parallelogram spanned by $\mathbf{v} = \langle 1, 4 \rangle$ and $\mathbf{w} = \langle 3, 2 \rangle$.

Solution We have $\begin{vmatrix} \mathbf{v} \\ \mathbf{w} \end{vmatrix} = \begin{vmatrix} 1 & 4 \\ 3 & 2 \end{vmatrix} = 1 \cdot 2 - 3 \cdot 4 = -10$. The area is the absolute value $A = |-10| = 10$ (Figure 12). ■

Proofs of Cross-Product Properties

We now derive the properties of the cross product listed in Theorem 1. Let

$$\mathbf{v} = \langle a_1, b_1, c_1 \rangle, \qquad \mathbf{w} = \langle a_2, b_2, c_2 \rangle$$

To show that $\mathbf{v} \times \mathbf{w}$ is orthogonal to $\mathbf{v}$, it suffices to show that $\mathbf{v} \cdot (\mathbf{v} \times \mathbf{w}) = 0$. By (6), we have

$$\mathbf{v} \cdot (\mathbf{v} \times \mathbf{w}) = \det \begin{pmatrix} \mathbf{v} \\ \mathbf{v} \\ \mathbf{w} \end{pmatrix} = a_1 \begin{vmatrix} b_1 & c_1 \\ b_2 & c_2 \end{vmatrix} - b_1 \begin{vmatrix} a_1 & c_1 \\ a_2 & c_2 \end{vmatrix} + c_1 \begin{vmatrix} a_1 & b_1 \\ a_2 & b_2 \end{vmatrix} \qquad \boxed{9}$$

Straightforward algebra (left to the reader) shows that (9) is equal to zero. Similarly, we have $\mathbf{w} \cdot (\mathbf{v} \times \mathbf{w}) = -\mathbf{w} \cdot (\mathbf{w} \times \mathbf{v}) = 0$. Thus, $\mathbf{w}$ is also orthogonal to $\mathbf{v} \times \mathbf{w}$. This proves part (i) of Theorem 1. To prove (ii), we use the following identity:

$$\|\mathbf{v} \times \mathbf{w}\|^2 = \|\mathbf{v}\|^2 \|\mathbf{w}\|^2 - (\mathbf{v} \cdot \mathbf{w})^2 \qquad \boxed{10}$$

To verify this identity, we compute $\|\mathbf{v} \times \mathbf{w}\|^2$ as the sum of the squares of the components of $\mathbf{v} \times \mathbf{w}$:

$$\|\mathbf{v} \times \mathbf{w}\|^2 = \begin{vmatrix} b_1 & c_1 \\ b_2 & c_2 \end{vmatrix}^2 + \begin{vmatrix} a_1 & c_1 \\ a_2 & c_2 \end{vmatrix}^2 + \begin{vmatrix} a_1 & b_1 \\ a_2 & b_2 \end{vmatrix}^2$$

$$= (b_1 c_2 - c_1 b_2)^2 + (a_1 c_2 - c_1 a_2)^2 + (a_1 b_2 - b_1 a_2)^2 \qquad \boxed{11}$$

On the other hand, by definition,

$$\|\mathbf{v}\|^2 \|\mathbf{w}\|^2 - (\mathbf{v} \cdot \mathbf{w})^2 = (a_1^2 + b_1^2 + c_1^2)^2 (a_2^2 + b_2^2 + c_2^2)^2 - (a_1 a_2 + b_1 b_2 + c_1 c_2)^2$$

$$\boxed{12}$$

Again, algebra (left to the reader) shows that (11) is equal to (12).

Now let θ be the angle between $\mathbf{v}$ and $\mathbf{w}$. By (10),

$$\|\mathbf{v} \times \mathbf{w}\|^2 = \|\mathbf{v}\|^2 \|\mathbf{w}\|^2 - (\mathbf{v} \cdot \mathbf{w})^2 = \|\mathbf{v}\|^2 \|\mathbf{w}\|^2 - \|\mathbf{v}\|^2 \|\mathbf{w}\|^2 \cos^2 \theta$$

$$= \|\mathbf{v}\|^2 \|\mathbf{w}\|^2 (1 - \cos^2 \theta) = \|\mathbf{v}\|^2 \|\mathbf{w}\|^2 \sin^2 \theta$$

Therefore, $\|\mathbf{v} \times \mathbf{w}\| = \|\mathbf{v}\| \|\mathbf{w}\| \sin \theta$. Note that $\sin \theta \geq 0$ since, by convention, θ lies between 0 and π. This proves (ii).

Part (iii) of Theorem 1 asserts that $\{\mathbf{v}, \mathbf{w}, \mathbf{v} \times \mathbf{w}\}$ is a right-handed system. This is a more subtle property that cannot be verified by algebra alone. We must rely on the

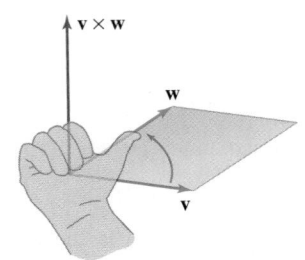

FIGURE 13 Both $\{\mathbf{v} \times \mathbf{w}, \mathbf{v}, \mathbf{w}\}$ and $\{\mathbf{v}, \mathbf{w}, \mathbf{v} \times \mathbf{w}\}$ are right-handed.

following relation between right-handedness and the sign of the determinant which can only be established using continuity:

$$\det \begin{pmatrix} \mathbf{u} \\ \mathbf{v} \\ \mathbf{w} \end{pmatrix} > 0 \quad \text{if and only if } \{\mathbf{u}, \mathbf{v}, \mathbf{w}\} \text{ is a right-handed system}$$

Granting this and using Eq. (6), we see that $\{\mathbf{v} \times \mathbf{w}, \mathbf{v}, \mathbf{w}\}$ is right-handed:

$$\det \begin{pmatrix} \mathbf{v} \times \mathbf{w} \\ \mathbf{v} \\ \mathbf{w} \end{pmatrix} = (\mathbf{v} \times \mathbf{w}) \cdot (\mathbf{v} \times \mathbf{w}) = \|\mathbf{v} \times \mathbf{w}\|^2 > 0$$

It follows that $\{\mathbf{v}, \mathbf{w}, \mathbf{v} \times \mathbf{w}\}$ is also right-handed (Figure 13).

12.4 SUMMARY

- The determinants of 2×2 and 3×3 matrices are defined by the formulas

$$\begin{vmatrix} a_{11} & a_{12} \\ a_{21} & a_{22} \end{vmatrix} = a_{11}a_{22} - a_{12}a_{21}$$

$$\begin{vmatrix} a_{11} & a_{12} & a_{13} \\ a_{21} & a_{22} & a_{23} \\ a_{31} & a_{32} & a_{33} \end{vmatrix} = a_{11} \begin{vmatrix} a_{22} & a_{23} \\ a_{32} & a_{33} \end{vmatrix} - a_{12} \begin{vmatrix} a_{21} & a_{23} \\ a_{31} & a_{33} \end{vmatrix} + a_{13} \begin{vmatrix} a_{21} & a_{22} \\ a_{31} & a_{32} \end{vmatrix}$$

- The *cross product* of $\mathbf{v} = \langle a_1, b_1, c_1 \rangle$ and $\mathbf{w} = \langle a_2, b_2, c_2 \rangle$ is the symbolic determinant

$$\mathbf{v} \times \mathbf{w} = \begin{vmatrix} \mathbf{i} & \mathbf{j} & \mathbf{k} \\ a_1 & b_1 & c_1 \\ a_2 & b_2 & c_2 \end{vmatrix} = \begin{vmatrix} b_1 & c_1 \\ b_2 & c_2 \end{vmatrix} \mathbf{i} - \begin{vmatrix} a_1 & c_1 \\ a_2 & c_2 \end{vmatrix} \mathbf{j} + \begin{vmatrix} a_1 & b_1 \\ a_2 & b_2 \end{vmatrix} \mathbf{k}$$

- Properties of the cross product:

   (i) $\mathbf{w} \times \mathbf{v} = -\mathbf{v} \times \mathbf{w}$
   (ii) $\mathbf{v} \times \mathbf{w} = 0$ if and only if $\mathbf{w} = \lambda \mathbf{v}$ for some scalar or $\mathbf{v} = \mathbf{0}$
   (iii) $(\lambda \mathbf{v}) \times \mathbf{w} = \mathbf{v} \times (\lambda \mathbf{w}) = \lambda (\mathbf{v} \times \mathbf{w})$
   (iv) $(\mathbf{u} + \mathbf{v}) \times \mathbf{w} = \mathbf{u} \times \mathbf{w} + \mathbf{v} \times \mathbf{w}$

   $\mathbf{v} \times (\mathbf{u} + \mathbf{w}) = \mathbf{v} \times \mathbf{u} + \mathbf{v} \times \mathbf{w}$

- The cross product $\mathbf{v} \times \mathbf{w}$ is the unique vector with the following three properties:

   (i) $\mathbf{v} \times \mathbf{w}$ is orthogonal to $\mathbf{v}$ and $\mathbf{w}$.
   (ii) $\mathbf{v} \times \mathbf{w}$ has length $\|\mathbf{v}\| \, \|\mathbf{w}\| \, |\sin \theta|$ (θ is the angle between $\mathbf{v}$ and $\mathbf{w}$).
   (iii) $\{\mathbf{v}, \mathbf{w}, \mathbf{v} \times \mathbf{w}\}$ is a right-handed system.

- The *vector triple product* is defined by $\mathbf{u} \cdot (\mathbf{v} \times \mathbf{w})$. We have

$$\mathbf{u} \cdot (\mathbf{v} \times \mathbf{w}) = \det \begin{pmatrix} \mathbf{u} \\ \mathbf{v} \\ \mathbf{w} \end{pmatrix}$$

The volume of the parallelepiped spanned by $\mathbf{u}$, $\mathbf{v}$, and $\mathbf{w}$ is equal to the absolute value $|\mathbf{u} \cdot (\mathbf{v} \times \mathbf{w})|$.
- $\|\mathbf{v} \times \mathbf{w}\|$ is equal to the area of the parallelogram spanned by $\mathbf{v}$ and $\mathbf{w}$.
- Cross-product identity: $\|\mathbf{v} \times \mathbf{w}\|^2 = \|\mathbf{v}\|^2 \|\mathbf{w}\|^2 - (\mathbf{v} \cdot \mathbf{w})^2$.

12.4 EXERCISES

Preliminary Questions

1. What is the (1, 3) minor of the matrix $\begin{vmatrix} 3 & 4 & 2 \\ -5 & -1 & 1 \\ 4 & 0 & 3 \end{vmatrix}$?

2. The angle between two unit vectors $\mathbf{e}$ and $\mathbf{f}$ is $\frac{\pi}{6}$. What is the length of $\mathbf{e} \times \mathbf{f}$?

3. What is $\mathbf{u} \times \mathbf{w}$, assuming that $\mathbf{w} \times \mathbf{u} = \langle 2, 2, 1 \rangle$?

4. Find the cross product without using the formula:

(a) $\langle 4, 8, 2 \rangle \times \langle 4, 8, 2 \rangle$ (b) $\langle 4, 8, 2 \rangle \times \langle 2, 4, 1 \rangle$

5. What are $\mathbf{i} \times \mathbf{j}$ and $\mathbf{i} \times \mathbf{k}$?

6. When is the cross product $\mathbf{v} \times \mathbf{w}$ equal to zero?

Exercises

In Exercises 1–4, calculate the 2 × 2-determinant.

1. $\begin{vmatrix} 1 & 2 \\ 4 & 3 \end{vmatrix}$

2. $\begin{vmatrix} \frac{2}{3} & \frac{1}{6} \\ -5 & 2 \end{vmatrix}$

3. $\begin{vmatrix} -6 & 9 \\ 1 & 1 \end{vmatrix}$

4. $\begin{vmatrix} 9 & 25 \\ 5 & 14 \end{vmatrix}$

In Exercises 5–10, calculate the 3 × 3-determinant.

5. $\begin{vmatrix} 1 & 2 & 1 \\ 4 & -3 & 0 \\ 1 & 0 & 1 \end{vmatrix}$

6. $\begin{vmatrix} 1 & 0 & 1 \\ -2 & 0 & 3 \\ 1 & 3 & -1 \end{vmatrix}$

7. $\begin{vmatrix} 1 & 0 & 1 \\ 1 & 3 & 1 \\ -2 & 0 & 3 \end{vmatrix}$

8. $\begin{vmatrix} 0 & 1 & 0 \\ 5 & 93 & 6 \\ 4 & 78 & 5 \end{vmatrix}$

9. $\begin{vmatrix} 1 & 0 & 0 \\ 0 & 1 & 0 \\ 0 & 0 & 1 \end{vmatrix}$

10. $\begin{vmatrix} 1 & 2 & 3 \\ 2 & 4 & 6 \\ -3 & -4 & 2 \end{vmatrix}$

In Exercises 11–16, calculate $\mathbf{v} \times \mathbf{w}$.

11. $\mathbf{v} = \langle 1, 2, 1 \rangle$, $\mathbf{w} = \langle 3, 1, 1 \rangle$

12. $\mathbf{v} = \langle 2, 0, 0 \rangle$, $\mathbf{w} = \langle -1, 0, 1 \rangle$

13. $\mathbf{v} = \left\langle \frac{2}{3}, 1, \frac{1}{2} \right\rangle$, $\mathbf{w} = \langle 4, -6, 3 \rangle$

14. $\mathbf{v} = \langle 0, 1, -1 \rangle$, $\mathbf{w} = \langle 1, -1, 0 \rangle$

15. $\mathbf{v} = \left\langle \frac{1}{3}, 1, \frac{1}{3} \right\rangle$, $\mathbf{w} = \langle -1, -1, 2 \rangle$

16. $\mathbf{v} = \langle 1, 1, 0 \rangle$, $\mathbf{w} = \langle 0, 1, 1 \rangle$

In Exercises 17–20, calculate the cross product.

17. $(\mathbf{i} + \mathbf{j}) \times \mathbf{k}$

18. $(\mathbf{j} - \mathbf{k}) \times (\mathbf{j} + \mathbf{k})$

19. $(\mathbf{i} + 2\mathbf{k}) \times (\mathbf{j} - \mathbf{k})$

20. $(2\mathbf{i} - 3\mathbf{j} + 4\mathbf{k}) \times (\mathbf{i} + \mathbf{j} - 7\mathbf{k})$

In Exercises 21–26, calculate the cross product assuming that $\mathbf{u} \times \mathbf{v} = \langle 1, 1, 0 \rangle$, $\mathbf{u} \times \mathbf{w} = \langle 0, 3, 1 \rangle$, and $\mathbf{v} \times \mathbf{w} = \langle 2, -1, 1 \rangle$.

21. $\mathbf{v} \times \mathbf{u}$

22. $\mathbf{w} \times (4\mathbf{w})$

23. $(3\mathbf{u} + 4\mathbf{w}) \times \mathbf{w}$

24. $(\mathbf{u} - 2\mathbf{v}) \times (\mathbf{u} + 2\mathbf{v})$

25. $\mathbf{w} \times (\mathbf{u} + \mathbf{v})$

26. $(\mathbf{v} + \mathbf{w}) \times (3\mathbf{u} + 2\mathbf{v})$

27. Let $\mathbf{v} = \langle a, b, c \rangle$. Calculate $\mathbf{v} \times \mathbf{i}$, $\mathbf{v} \times \mathbf{j}$, and $\mathbf{v} \times \mathbf{k}$.

28. Find $\mathbf{v} \times \mathbf{w}$, where $\mathbf{v}$ and $\mathbf{w}$ are vectors of length 3 in the xz-plane, oriented as in Figure 14, and the angle between $\mathbf{v}$ and $\mathbf{w}$ is $\theta = \frac{\pi}{6}$.

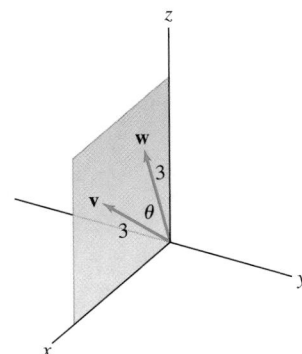

FIGURE 14

In Exercises 29–30, refer to Figure 15.

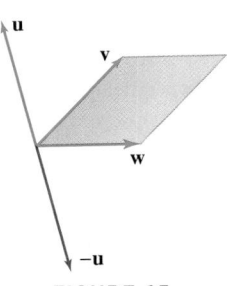

FIGURE 15

29. Which of $\mathbf{u}$ and $-\mathbf{u}$ is equal to $\mathbf{v} \times \mathbf{w}$?

30. Which of the following form a right-handed system?

(a) $\{\mathbf{v}, \mathbf{w}, \mathbf{u}\}$ (b) $\{\mathbf{w}, \mathbf{v}, \mathbf{u}\}$ (c) $\{\mathbf{v}, \mathbf{u}, \mathbf{w}\}$
(d) $\{\mathbf{u}, \mathbf{v}, \mathbf{w}\}$ (e) $\{\mathbf{w}, \mathbf{v}, -\mathbf{u}\}$ (f) $\{\mathbf{v}, -\mathbf{u}, \mathbf{w}\}$

31. Let $\mathbf{v} = \langle 3, 0, 0 \rangle$ and $\mathbf{w} = \langle 0, 1, -1 \rangle$. Determine $\mathbf{u} = \mathbf{v} \times \mathbf{w}$ using the geometric properties of the cross product rather than the formula.

32. What are the possible angles θ between two unit vectors $\mathbf{e}$ and $\mathbf{f}$ if $\|\mathbf{e} \times \mathbf{f}\| = \frac{1}{2}$?

33. Show that if $\mathbf{v}$ and $\mathbf{w}$ lie in the yz-plane, then $\mathbf{v} \times \mathbf{w}$ is a multiple of $\mathbf{i}$.

34. Find the two unit vectors orthogonal to both $\mathbf{a} = \langle 3, 1, 1 \rangle$ and $\mathbf{b} = \langle -1, 2, 1 \rangle$.

35. Let $\mathbf{e}$ and $\mathbf{e}'$ be unit vectors in $\mathbf{R}^3$ such that $\mathbf{e} \perp \mathbf{e}'$. Reason geometrically to find the cross product $\mathbf{e} \times (\mathbf{e}' \times \mathbf{e})$.

36. Calculate the force $\mathbf{F}$ on an electron (charge $q = -1.6 \times 10^{-19}$C) moving with velocity 10^5 m/s in the direction $\mathbf{i}$ in a uniform magnetic field $\mathbf{B}$, where $\mathbf{B} = 0.0004\mathbf{i} + 0.0001\mathbf{j}$ (units of teslas) (see Example 5).

37. An electron moving with velocity $\mathbf{v}$ in the plane experiences a force $\mathbf{F} = q(\mathbf{v} \times \mathbf{B})$, where q is the charge on the electron and $\mathbf{B}$ is a uniform magnetic field pointing directly out of the page. Which of the two vectors $\mathbf{F}_1$ or $\mathbf{F}_2$ in Figure 16 represents the force on the electron? Remember that q is negative.

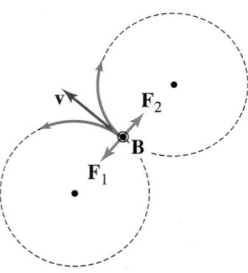

FIGURE 16 The magnetic field vector $\mathbf{B}$ points directly out of the page.

38. Calculate the scalar triple product $\mathbf{u} \cdot (\mathbf{v} \times \mathbf{w})$, where $\mathbf{u} = \langle 1, 1, 0 \rangle$, $\mathbf{v} = \langle 3, -2, 2 \rangle$, and $\mathbf{w} = \langle 4, -1, 2 \rangle$.

39. Verify identity (10) for vectors $\mathbf{v} = \langle 3, -2, 2 \rangle$ and $\mathbf{w} = \langle 4, -1, 2 \rangle$.

40. Find the volume of the parallelepiped spanned by $\mathbf{u}$, $\mathbf{v}$, and $\mathbf{w}$ in Figure 17.

41. Find the area of the parallelogram spanned by $\mathbf{v}$ and $\mathbf{w}$ in Figure 17.

42. Calculate the volume of the parallelepiped spanned by

$$\mathbf{u} = \langle 2, 2, 1 \rangle, \qquad \mathbf{v} = \langle 1, 0, 3 \rangle, \qquad \mathbf{w} = \langle 0, -4, 0 \rangle$$

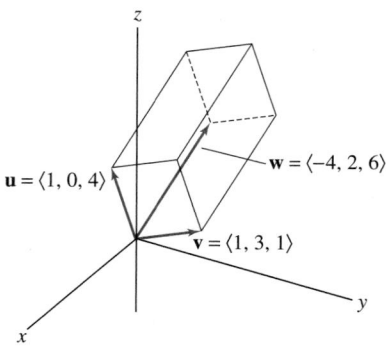

FIGURE 17

43. Sketch and compute the volume of the parallelepiped spanned by

$$\mathbf{u} = \langle 1, 0, 0 \rangle, \qquad \mathbf{v} = \langle 0, 2, 0 \rangle, \qquad \mathbf{w} = \langle 1, 1, 2 \rangle$$

44. Sketch the parallelogram spanned by $\mathbf{u} = \langle 1, 1, 1 \rangle$ and $\mathbf{v} = \langle 0, 0, 4 \rangle$, and compute its area.

45. Calculate the area of the parallelogram spanned by $\mathbf{u} = \langle 1, 0, 3 \rangle$ and $\mathbf{v} = \langle 2, 1, 1 \rangle$.

46. Find the area of the parallelogram determined by the vectors $\langle a, 0, 0 \rangle$ and $\langle 0, b, c \rangle$.

47. Sketch the triangle with vertices O, $P = (3, 3, 0)$, and $Q = (0, 3, 3)$, and compute its area using cross products.

48. Use the cross product to find the area of the triangle with vertices $P = (1, 1, 5)$, $Q = (3, 4, 3)$, and $R = (1, 5, 7)$ (Figure 18).

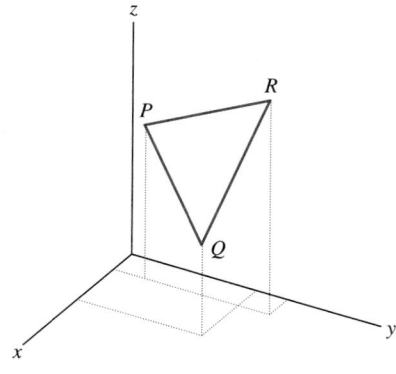

FIGURE 18

In Exercises 49–51, verify the identity using the formula for the cross product.

49. $\mathbf{v} \times \mathbf{w} = -\mathbf{w} \times \mathbf{v}$

50. $(\lambda \mathbf{v}) \times \mathbf{w} = \lambda (\mathbf{v} \times \mathbf{w})$ (λ a scalar)

51. $(\mathbf{u} + \mathbf{v}) \times \mathbf{w} = \mathbf{u} \times \mathbf{w} + \mathbf{v} \times \mathbf{w}$

52. Verify the relations (5).

53. Show that $(\mathbf{i} \times \mathbf{j}) \times \mathbf{j} \neq \mathbf{i} \times (\mathbf{j} \times \mathbf{j})$. Conclude that the Associative Law does not hold for cross products.

54. Use Eq. (9) to verify that $\mathbf{v} \cdot (\mathbf{v} \times \mathbf{w}) = 0$ for all $\mathbf{v}$ and $\mathbf{w}$.

55. Use Theorem 1(ii) to show that $\mathbf{v} \times \mathbf{w} = \mathbf{0}$ if and only if $\mathbf{w} = \lambda \mathbf{v}$ for some scalar λ or $\mathbf{v} = \mathbf{0}$.

56. The components of the cross product have a geometric interpretation. Let $\mathbf{v}_0$, $\mathbf{w}_0$ be the projections onto the xy-plane of vectors $\mathbf{v}$, $\mathbf{w}$ as in Figure 19. Show that the area of the parallelogram spanned by $\mathbf{v}_0$ and $\mathbf{w}_0$ is equal to the absolute value of the $\mathbf{k}$-component of $\mathbf{v} \times \mathbf{w}$.

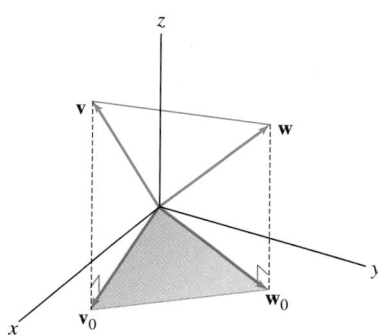

FIGURE 19

57. Formulate and prove analogs of the result in Exercise 56 for the $\mathbf{i}$ and $\mathbf{j}$ components of $\mathbf{v} \times \mathbf{w}$.

58. Determine the angle θ between two vectors $\mathbf{v}$ and $\mathbf{w}$ if $\mathbf{v} \cdot \mathbf{w} = \|\mathbf{v} \times \mathbf{w}\|$.

59. Show that three points P, Q, R are collinear (lie on a line) if and only if $\overrightarrow{PQ} \times \overrightarrow{PR} = \mathbf{0}$.

60. Use the result of Exercise 59 to determine whether the points P, Q, and R are collinear and, if not, find a vector normal to the plane containing them.

(a) $P = (2, 1, 0)$, $Q = (1, 5, 2)$, $R = (-1, 13, 6)$
(b) $P = (2, 1, 0)$, $Q = (-3, 21, 10)$, $R = (5, -2, 9)$
(c) $P = (1, 1, 0)$, $Q = (1, -2, -1)$, $R = (3, 2, -4)$

61. Find a vector $\mathbf{X}$ such that $\langle 1, 1, 1 \rangle \times \mathbf{X} = \langle 1, -1, 0 \rangle$.

62. Explain geometrically why the equation $\langle 1, 1, 1 \rangle \times \mathbf{X} = \langle 1, 0, 0 \rangle$ has no solution.

63. Suppose that vectors $\mathbf{u}$, $\mathbf{v}$, and $\mathbf{w}$ are mutually orthogonal, that is, $\mathbf{u} \perp \mathbf{v}$, $\mathbf{u} \perp \mathbf{w}$, and $\mathbf{v} \perp \mathbf{w}$. Prove that $(\mathbf{u} \times \mathbf{v}) \times \mathbf{w} = \mathbf{0}$ and $\mathbf{u} \times (\mathbf{v} \times \mathbf{w}) = \mathbf{0}$.

*Exercises 64–67 deal with torque: When a force $\mathbf{F}$ acts on an object with position vector $\mathbf{r}$, the **torque** about the origin O is the vector quantity $\tau = \mathbf{r} \times \mathbf{F}$. If several forces $\mathbf{F}_j$ act at positions $\mathbf{r}_j$, then the*

net torque is the sum $\tau = \sum \mathbf{r}_j \times \mathbf{F}_j$. By Newton's Laws, τ (units: N-m or lb-ft) is equal to the rate of change of angular momentum. Torque is a measure of how much the force causes the object to rotate.

64. Calculate the torque τ about O acting at the point P on the mechanical arm in Figure 20(A), assuming that a 25-N force acts as indicated. Ignore the weight of the arm itself.

65. Calculate the net torque about O at P, assuming that a 30-kg mass is attached at P [Figure 20(B)]. The force $\mathbf{F}_g$ due to gravity on a mass m has magnitude $9.8m$ m/s^2 in the downward direction.

(A)

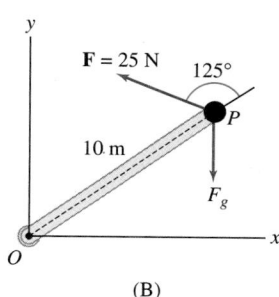

(B)

FIGURE 20

66. Let τ be the net torque about O acting on the robotic arm of Figure 21. Assume that the two segments of the arms have weight m_1 and m_2 (in pounds) and that a weight of m_3 pounds is located at the endpoint P. In calculating the torque, we may assume that the entire mass of each arm segment lies at the midpoint of the arm (its center of mass). Show that the position vectors of the masses m_1, m_2, and m_3 are

$$\mathbf{r}_1 = \frac{1}{2} L_1 (\sin \theta_1 \mathbf{i} + \cos \theta_1 \mathbf{j})$$

$$\mathbf{r}_2 = L_1 (\sin \theta_1 \mathbf{i} + \cos \theta_1 \mathbf{j}) + \frac{1}{2} L_2 (\sin \theta_2 \mathbf{i} - \cos \theta_2 \mathbf{j})$$

$$\mathbf{r}_3 = L_1 (\sin \theta_1 \mathbf{i} + \cos \theta_1 \mathbf{j}) + L_2 (\sin \theta_2 \mathbf{i} - \cos \theta_2 \mathbf{j})$$

Then show that

$$\tau = -\left(L_1 \left(\frac{1}{2} m_1 + m_2 + m_3 \right) \sin \theta_1 + L_2 \left(\frac{1}{2} m_2 + m_3 \right) \sin \theta_2 \right) \mathbf{k}$$

To simplify the computation, note that all three gravitational forces act in the $-\mathbf{j}$ direction, so the $\mathbf{j}$ components of the position vectors $\mathbf{r}_i$ do not contribute to the torque.

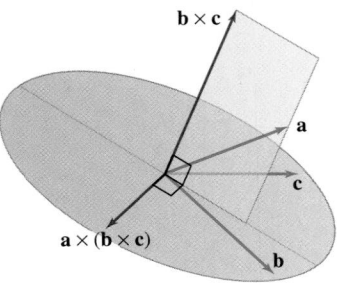

FIGURE 21

67. Continuing with Exercise 66, suppose that $L_1 = 5$ ft, $L_2 = 3$ ft, and $m_1 = 30$ lb, $m_2 = 20$ lb, $m_3 = 50$ lb. If the angles θ_1, θ_2 are equal (say, to θ), what is the maximum allowable value of θ if we assume that the robotic arm can sustain a maximum torque of 400 ft-lb?

Further Insights and Challenges

68. Show that 3×3 determinants can be computed using the **diagonal rule**: Repeat the first two columns of the matrix and form the products of the numbers along the six diagonals indicated. Then add the products for the diagonals that slant from left to right and subtract the products for the diagonals that slant from right to left.

$$\det(A) = \begin{vmatrix} a_{11} & a_{12} & a_{13} & a_{11} & a_{12} \\ a_{21} & a_{22} & a_{23} & a_{21} & a_{22} \\ a_{31} & a_{32} & a_{33} & a_{31} & a_{32} \end{vmatrix}$$

$$= a_{11}a_{22}a_{33} + a_{12}a_{23}a_{31} + a_{13}a_{21}a_{32}$$

$$- a_{13}a_{22}a_{31} - a_{11}a_{23}a_{32} - a_{12}a_{21}a_{33}$$

69. Use the diagonal rule to calculate $\begin{vmatrix} 2 & 4 & 3 \\ 0 & 1 & -7 \\ -1 & 5 & 3 \end{vmatrix}$.

70. Prove that $\mathbf{v} \times \mathbf{w} = \mathbf{v} \times \mathbf{u}$ if and only if $\mathbf{u} = \mathbf{w} + \lambda\mathbf{v}$ for some scalar λ. Assume that $\mathbf{v} \neq \mathbf{0}$.

71. Use Eq. (10) to prove the Cauchy–Schwarz inequality:

$$|\mathbf{v} \cdot \mathbf{w}| \leq \|\mathbf{v}\| \, \|\mathbf{w}\|$$

Show that equality holds if and only if $\mathbf{w}$ is a multiple of $\mathbf{v}$ or at least one of $\mathbf{v}$ and $\mathbf{w}$ is zero.

72. Show that if $\mathbf{u}$, $\mathbf{v}$, and $\mathbf{w}$ are nonzero vectors and $(\mathbf{u} \times \mathbf{v}) \times \mathbf{w} = \mathbf{0}$, then either: (i) $\mathbf{u}$ and $\mathbf{v}$ are parallel, or (ii) $\mathbf{w}$ is orthogonal to $\mathbf{u}$ and $\mathbf{v}$.

73. Suppose that $\mathbf{u}$, $\mathbf{v}$, $\mathbf{w}$ are nonzero and

$$(\mathbf{u} \times \mathbf{v}) \times \mathbf{w} = \mathbf{u} \times (\mathbf{v} \times \mathbf{w}) = \mathbf{0}$$

Show that $\mathbf{u}$, $\mathbf{v}$, and $\mathbf{w}$ are either mutually parallel or mutually perpendicular. *Hint:* Use Exercise 72.

74. Let $\mathbf{a}$, $\mathbf{b}$, $\mathbf{c}$ be nonzero vectors and set

$$\mathbf{v} = \mathbf{a} \times (\mathbf{b} \times \mathbf{c}), \qquad \mathbf{w} = (\mathbf{a} \cdot \mathbf{c})\mathbf{b} - (\mathbf{a} \cdot \mathbf{b})\mathbf{c}$$

(a) Prove that

(i) $\mathbf{v}$ lies in the plane spanned by $\mathbf{b}$ and $\mathbf{c}$.

(ii) $\mathbf{v}$ is orthogonal to $\mathbf{a}$.

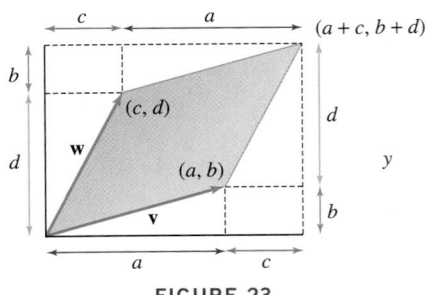

FIGURE 22

(b) Prove that $\mathbf{w}$ also satisfies (i) and (ii). Conclude that $\mathbf{v}$ and $\mathbf{w}$ are parallel.

(c) Show algebraically that $\mathbf{v} = \mathbf{w}$ (Figure 22).

75. Use Exercise 74 to prove the identity

$$(\mathbf{a} \times \mathbf{b}) \times \mathbf{c} - \mathbf{a} \times (\mathbf{b} \times \mathbf{c}) = (\mathbf{a} \cdot \mathbf{b})\mathbf{c} - (\mathbf{b} \cdot \mathbf{c})\mathbf{a}$$

76. Assume that $\mathbf{v}$ and $\mathbf{w}$ lie in the first quadrant in $\mathbf{R}^2$ as in Figure 23. Use geometry to prove that the area of the parallelogram is equal to $\det\begin{pmatrix} \mathbf{v} \\ \mathbf{w} \end{pmatrix}$.

FIGURE 23

77. Show that if $\mathbf{a}$, $\mathbf{b}$ are nonzero vectors such that $\mathbf{a} \perp \mathbf{b}$, then there exists a vector $\mathbf{x}$ such that

$$\mathbf{x} \times \mathbf{a} = \mathbf{b} \qquad \boxed{13}$$

Hint: Show that if $\mathbf{x}$ is orthogonal to $\mathbf{b}$ and is not a multiple of $\mathbf{a}$, then $\mathbf{x} \times \mathbf{a}$ is a multiple of $\mathbf{b}$.

78. Show that if **a**, **b** are nonzero vectors such that $\mathbf{a} \perp \mathbf{b}$, then the set of all solutions of Eq. (13) is a line with **a** as direction vector. *Hint:* Let $\mathbf{x}_0$ be any solution (which exists by Exercise 77) and show that every other solution is of the form $\mathbf{x}_0 + \lambda\mathbf{a}$ for some scalar λ.

79. Consider the tetrahedron spanned by vectors **a**, **b**, and **c** as in Figure 24(A). Let A, B, C be the faces containing the origin O and let D be the fourth face opposite O. For each face F, let $\mathbf{v}_F$ be the vector normal to the face, pointing outside the tetrahedron, of magnitude equal to twice the area of F. Prove the relations

$$\mathbf{v}_A + \mathbf{v}_B + \mathbf{v}_C = \mathbf{a} \times \mathbf{b} + \mathbf{b} \times \mathbf{c} + \mathbf{c} \times \mathbf{a}$$

$$\mathbf{v}_A + \mathbf{v}_B + \mathbf{v}_C + \mathbf{v}_D = 0$$

Hint: Show that $\mathbf{v}_D = (\mathbf{c} - \mathbf{b}) \times (\mathbf{b} - \mathbf{a})$.

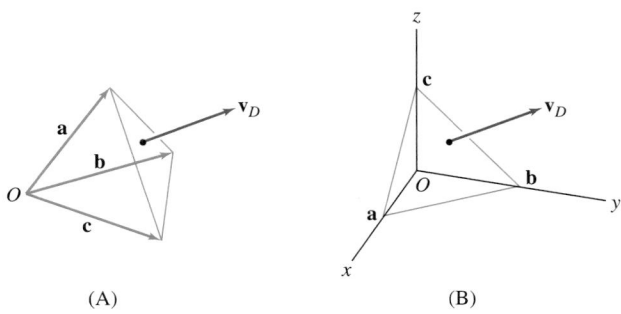

(A)   (B)

FIGURE 24 The vector $\mathbf{v}_D$ is perpendicular to the face.

80. In the notation of Exercise 79, suppose that **a**, **b**, **c** are mutually perpendicular as in Figure 24(B). Let S_F be the area of face F. Prove the following three-dimensional version of the Pythagorean Theorem:

$$S_A^2 + S_B^2 + S_C^2 = S_D^2$$

12.5 Planes in Three-Space

REMINDER the term "normal" is another word for "perpendicular."

As we know, a line in $\mathbf{R}^2$ is defined by a linear equation $ax + by = c$. In this section, we discuss planes in $\mathbf{R}^3$. Our first goal is to show that a plane is defined by a linear equation $ax + by + cz = d$ in three variables.

Consider a plane $\mathcal{P}$ passing through a point $P_0 = (x_0, y_0, z_0)$ as in Figure 1. Although infinitely many planes pass through P_0, we can identify the particular plane $\mathcal{P}$ by specifying a vector $\mathbf{n} = \langle a, b, c \rangle$ normal to $\mathcal{P}$. Only one plane through P_0 is perpendicular to $\mathbf{n}$. Furthermore, we see in Figure 2 that a point $P = (x, y, z)$ lies on $\mathcal{P}$ if and only if $\overrightarrow{P_0P}$ is perpendicular to $\mathbf{n}$. Thus, the dot product is zero:

$$\mathbf{n} \cdot \overrightarrow{P_0P} = 0 \qquad \boxed{1}$$

To write this out explicitly, let $\mathbf{r}_0 = \langle x_0, y_0, z_0 \rangle$ and $\mathbf{r} = \langle x, y, z \rangle$ be the position vectors of P_0 and P. Then

$$\overrightarrow{P_0P} = \mathbf{r} - \mathbf{r}_0 = \langle x - x_0, y - y_0, z - z_0 \rangle$$

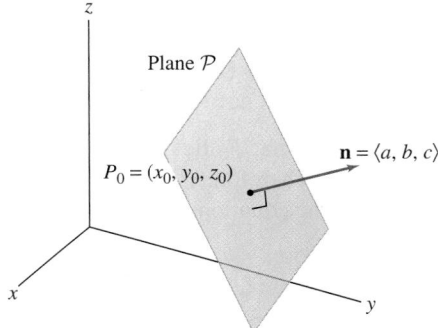

FIGURE 1 A plane through P_0 is uniquely determined by the normal vector **n**.

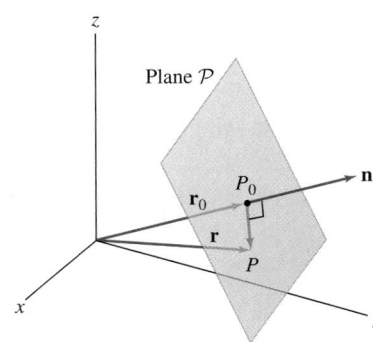

FIGURE 2 A point P lies on the plane $\mathcal{P}$ if $\overrightarrow{P_0P}$ is orthogonal to the normal vector **n**.

Equation (1) becomes

$$\mathbf{n} \cdot (\mathbf{r} - \mathbf{r}_0) = 0 \quad \text{or} \quad \langle a, b, c \rangle \cdot \langle x - x_0, y - y_0, z - z_0 \rangle = 0$$

This yields the equation

$$a(x - x_0) + b(y - y_0) + c(z - z_0) = 0$$

Alternatively, we may write $\mathbf{n} \cdot (\mathbf{r} - \mathbf{r}_0) = 0$ as

$$\mathbf{n} \cdot \mathbf{r} = \mathbf{n} \cdot \mathbf{r}_0 \quad \text{or} \quad \langle a, b, c \rangle \cdot \langle x, y, z \rangle = \langle a, b, c \rangle \cdot \langle x_0, y_0, z_0 \rangle$$

This yields the equivalent equation

$$ax + by + cz = \underbrace{ax_0 + by_0 + cz_0}_{\text{Call this } d}$$

Keep in mind that the equation of a plane is not unique since we are free to multiply the equation by any nonzero scalar. This amounts to multiplying $\mathbf{n}$ by a nonzero scalar.

THEOREM 1 Equation of a Plane in Point-Normal Form An equation of the plane through $P_0 = (x_0, y_0, z_0)$ with (nonzero) normal vector $\mathbf{n} = \langle a, b, c \rangle$ can be written in the following three ways:

Vector form:	$\mathbf{n} \cdot \langle x, y, z \rangle = d$	**2**
Scalar forms:	$a(x - x_0) + b(y - y_0) + c(z - z_0) = 0$	**3**
	$ax + by + cz = d$	**4**

where $d = \mathbf{n} \cdot \langle x_0, y_0, z_0 \rangle = ax_0 + by_0 + cz_0$.

■ **EXAMPLE 1** Find an equation of the plane through $P_0 = (3, 1, 0)$ with normal vector $\mathbf{n} = \langle 3, 2, -5 \rangle$.

Solution By Eq. (2), the vector equation is

$$\underbrace{\langle 3, 2, -5 \rangle \cdot \langle x, y, z \rangle}_{\mathbf{n} \cdot \langle x, y, z \rangle} = \underbrace{\langle 3, 2, -5 \rangle \cdot \langle 3, 1, 0 \rangle = 11}_{d}$$

This yields the scalar equation $3x + 2y - 5z = 11$. We can also write the equation in the form (3)

$$3(x - 3) + 2(y - 1) - 5z = 0 \qquad ■$$

■ **EXAMPLE 2** An Equation for the xy-Plane Find an equation of the plane $\mathcal{P}$ through $P_0 = (1, 2, 0)$ with normal vector $\mathbf{n} = \langle 0, 0, 3 \rangle$.

Solution Note that P_0 lies in the xy-plane and $\mathbf{n}$ points along the z-axis, orthogonal to the xy-plane. Therefore, $\mathcal{P}$ must be the xy-plane itself (Figure 3). We confirm this by determining the equation in vector form:

$$\mathbf{n} \cdot \langle x, y, z \rangle = d = \mathbf{n} \cdot \langle 1, 2, 0 \rangle$$

$$\langle 0, 0, 3 \rangle \cdot \langle x, y, z \rangle = \langle 0, 0, 3 \rangle \cdot \langle 1, 2, 0 \rangle$$

$$3z = 0 \quad \text{or} \quad z = 0$$

As expected, $z = 0$ is the equation of the xy-plane. ■

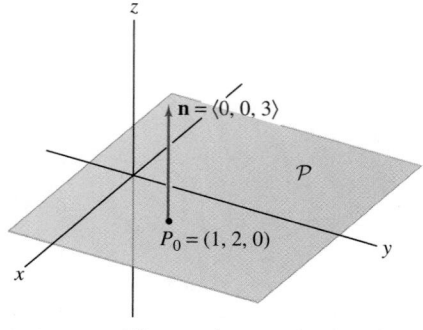

FIGURE 3 The xy-plane can be described as the plane through $P_0 = (1, 2, 0)$ with normal vector $\mathbf{n} = \langle 0, 0, 3 \rangle$.

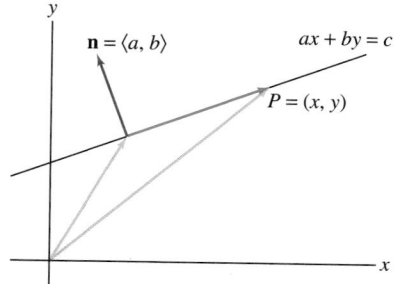

FIGURE 4 The line with normal vector **n**.

FIGURE 5

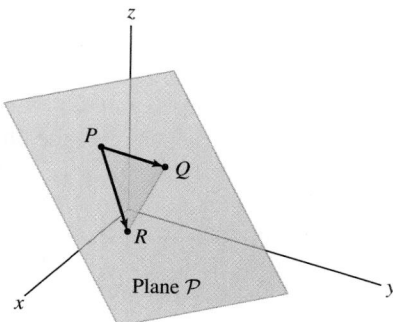

FIGURE 6 Three points P, Q, and R determine a plane (assuming they do not lie in a straight line).

In Example 4, we could just as well have used the vectors $\mathbf{a} = \overrightarrow{QP}$ and $\mathbf{b} = \overrightarrow{QR}$ (or $\overrightarrow{RP}$ and $\overrightarrow{RQ}$) to find a normal vector $\mathbf{n}$.

CONCEPTUAL INSIGHT The direction of a line in the plane is specified by the slope m, which is a number. By contrast, it takes a vector rather than a single number to specify the orientation of a plane in $\mathbf{R}^3$. Thus, the normal vector serves as a substitute for the slope. Note that a line in $\mathbf{R}^2$ may also be described via normal vectors: The line with equation $ax + by = c$ has normal vector $\mathbf{n} = \langle a, b \rangle$ (Figure 4).

Suppose we are given a linear equation

$$ax + by + cz = d$$

This equation defines a plane $\mathcal{P}$ by Theorem 1. Furthermore, the vector of coefficients $\mathbf{n} = \langle a, b, c \rangle$ is a normal vector to $\mathcal{P}$. For example, the plane $2x - 5y + z = 2$ has normal vector $\mathbf{n} = \langle 2, -5, 1 \rangle$. Now observe that two planes are *parallel* if and only if they have the same normal vector. Therefore, the planes parallel to $\mathcal{P}$ are precisely the planes with equation $ax + by + cz = d'$ for some d'. The unique plane parallel to $\mathcal{P}$ passing through the origin has equation $ax + by + cz = 0$.

■ **EXAMPLE 3** **Planes Parallel to a Given Plane** Let $\mathcal{P}$ be the plane with equation $7x - 4y + 2z = -10$. Find an equation of the plane parallel to $\mathcal{P}$ and passing through:

(a) The origin. **(b)** The point $Q = (2, -1, 3)$.

Solution As noted in the previous paragraph, the planes parallel to $\mathcal{P}$ are the planes with an equation of the form

$$7x - 4y + 2z = d' \qquad \boxed{5}$$

(a) The plane parallel to $\mathcal{P}$ passing through the origin has equation $7x - 4y + 2z = 0$.
(b) The point $Q = (2, -1, 3)$ satisfies Eq. (5) with

$$d' = \langle 7, -4, 2 \rangle \cdot \langle 2, -1, 3 \rangle = 7(2) - 4(-1) + 2(3) = 24$$

Therefore, the plane parallel to $\mathcal{P}$ passing through Q has equation $7x - 4y + 2z = 24$ (Figure 5). ■

Another way of specifying a plane is to choose three points P, Q, and R that do not lie on a line. In this case, there is a unique plane passing through P, Q, and R (Figure 6). The next example shows how to find an equation of this plane.

■ **EXAMPLE 4** **Equation of a Plane Determined by Three Points** Find an equation of the plane $\mathcal{P}$ determined by the points

$$P = (1, 0, -1), \quad Q = (2, 2, 1), \quad R = (4, 1, 2)$$

Solution

Step 1. Find a normal vector.
The vectors $\mathbf{a} = \overrightarrow{PQ}$ and $\mathbf{b} = \overrightarrow{PR}$ lie in $\mathcal{P}$ (Figure 7), so the cross product $\mathbf{n} = \mathbf{a} \times \mathbf{b}$ is normal to $\mathcal{P}$:

$$\mathbf{a} = \overrightarrow{PQ} = \langle 2, 2, 1 \rangle - \langle 1, 0, -1 \rangle = \langle 1, 2, 2 \rangle$$

$$\mathbf{b} = \overrightarrow{PR} = \langle 4, 1, 2 \rangle - \langle 1, 0, -1 \rangle = \langle 3, 1, 3 \rangle$$

$$\mathbf{n} = \mathbf{a} \times \mathbf{b} = \begin{vmatrix} \mathbf{i} & \mathbf{j} & \mathbf{k} \\ 1 & 2 & 2 \\ 3 & 1 & 3 \end{vmatrix} = 4\mathbf{i} + 3\mathbf{j} - 5\mathbf{k} = \langle 4, 3, -5 \rangle$$

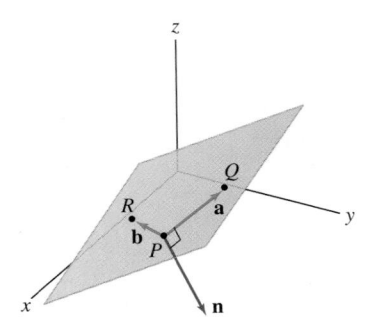

FIGURE 7

CAUTION When you find a normal vector to the plane containing points P, Q, R, be sure to compute a cross product such as $\overrightarrow{PQ} \times \overrightarrow{PR}$. A common mistake is to use a cross product such as $\overrightarrow{OP} \times \overrightarrow{OQ}$ or $\overrightarrow{OP} \times \overrightarrow{OR}$, which need not be normal to the plane.

Step 2. **Choose a point on the plane and compute d.**
Now choose any one of the three points on the plane, say, $P = (1, 0, -1)$. By Eq. (4), the vector equation of $\mathcal{P}$ is $4x + 3y - 5z = d$, where

$$d = \mathbf{n} \cdot \overrightarrow{OP} = \langle 4, 3, -5 \rangle \cdot \langle 1, 0, -1 \rangle = 9$$

Therefore, $\mathcal{P}$ has equation $4x + 3y - 5z = 9$. ∎

■ **EXAMPLE 5** Intersection of a Plane and a Line Find the intersection of the plane $3x - 9y + 2z = 7$ and the line $\mathbf{r}(t) = \langle 1, 2, 1 \rangle + t \langle -2, 0, 1 \rangle$.

Solution The line has parametric equations

$$x = 1 - 2t, \qquad y = 2, \qquad z = 1 + t$$

We want to find a value of t for which $\mathbf{r}(t)$ lies on the plane, so we substitute the parametric equations in the equation of the plane and solve for t:

$$3x - 9y + 2z = 7$$
$$3(1 - 2t) - 9(2) + 2(1 + t) = 7$$
$$-4t - 13 = 7 \qquad \text{(simplify)}$$

The last equation yields $-4t = 20$ or $t = -5$. Therefore, the point P of intersection has coordinates

$$x = 1 - 2(-5) = 11, \qquad y = 2, \qquad z = 1 + (-5) = -4$$

and thus, $P = (11, 2, -4)$. ∎

The intersection of a plane $\mathcal{P}$ with a coordinate plane or a plane parallel to a coordinate plane is called a **trace**. The trace is a line unless $\mathcal{P}$ is parallel to the coordinate plane (in which case the trace is empty or $\mathcal{P}$ itself).

■ **EXAMPLE 6** Traces of the Plane Find the traces of the plane $-2x + 3y + z = 6$ in the coordinate planes.

Solution We obtain the trace in the xy-plane by setting $z = 0$ in the equation of the plane. Thus, the trace is the line $-2x + 3y = 6$ in the xy-plane (Figure 8).
Similarly, the trace in the xz-plane is obtained by setting $y = 0$, which gives the line $-2x + z = 6$ in the xz-plane. Finally, the trace in the yz-plane is $3y + z = 6$. ∎

12.5 SUMMARY

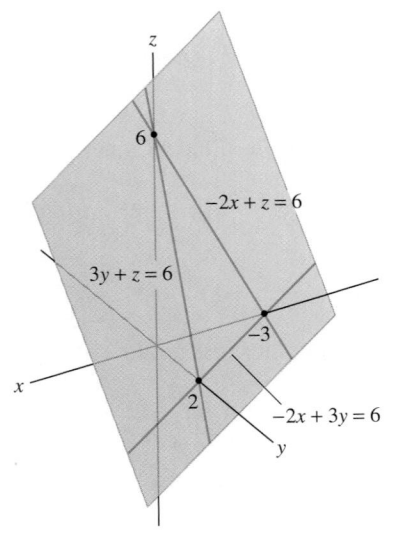

FIGURE 8 Traces of plane $-2x + 3y + z = 6$ in the coordinate planes.

• An equation of the plane through the point $P_0 = (x_0, y_0, z_0)$ with normal vector $\mathbf{n} = \langle a, b, c \rangle$ can be written in the following three ways:

Vector form: $\mathbf{n} \cdot \langle x, y, z \rangle = d$

Scalar forms: $a(x - x_0) + b(y - y_0) + c(z - z_0) = 0$
$$ax + by + cz = d$$

where $d = \mathbf{n} \cdot \langle x_0, y_0, z_0 \rangle = ax_0 + by_0 + cz_0$.

- A linear equation $ax + by + cz = d$ defines a plane $\mathcal{P}$ with normal vector $\mathbf{n} = \langle a, b, c \rangle$. The planes parallel to $\mathcal{P}$ are the planes with equation $ax + by + cz = d'$ for some d'.
- To find an equation of the plane through three points P, Q, R not on a line, first compute a normal vector as a cross product $\mathbf{n} = \overrightarrow{PQ} \times \overrightarrow{PR}$. Then, if $P = (x_0, y_0, z_0)$, an equation is $\mathbf{n} \cdot \langle x, y, z \rangle = d$, where $d = \mathbf{n} \cdot \langle x_0, y_0, z_0 \rangle$.
- The intersection of a plane $\mathcal{P}$ with a coordinate plane or a plane parallel to a coordinate plane is called a *trace*. The trace in the yz-plane is obtained by setting $x = 0$ in the equation of the plane (and similarly for the xz and xy traces).

12.5 EXERCISES

Preliminary Questions

1. What is the equation of the plane parallel to $3x + 4y - z = 5$ passing through the origin?

2. The vector $\mathbf{k}$ is normal to which of the following planes?

(a) $x = 1$ **(b)** $y = 1$ **(c)** $z = 1$

3. Which of the following planes are not parallel to the plane $x + y + z = 1$?

(a) $2x + 2y + 2z = 1$ **(b)** $x + y + z = 3$

(c) $x - y + z = 0$

4. To which coordinate plane is the plane $y = 1$ parallel?

5. Which of the following planes contains the z-axis?

(a) $z = 1$ **(b)** $x + y = 1$ **(c)** $x + y = 0$

6. Suppose that a plane $\mathcal{P}$ with normal vector $\mathbf{n}$ and a line $\mathcal{L}$ with direction vector $\mathbf{v}$ both pass through the origin and that $\mathbf{n} \cdot \mathbf{v} = 0$. Which of the following statements is correct?

(a) $\mathcal{L}$ is contained in the $\mathcal{P}$.

(b) $\mathcal{L}$ is orthogonal to $\mathcal{P}$.

Exercises

In Exercises 1–8, write an equation of the plane with normal vector $\mathbf{n}$ passing through the given point in vector form and the scalar forms (3) and (4).

1. $\mathbf{n} = \langle 1, 3, 2 \rangle$, $(4, -1, 1)$

2. $\mathbf{n} = \langle -1, 2, 1 \rangle$, $(3, 1, 9)$

3. $\mathbf{n} = \langle -1, 2, 1 \rangle$, $(4, 1, 5)$

4. $\mathbf{n} = \langle 2, -4, 1 \rangle$, $\left(\frac{1}{3}, \frac{2}{3}, 1 \right)$

5. $\mathbf{n} = \mathbf{i}$, $(3, 1, -9)$ **6.** $\mathbf{n} = \mathbf{j}$, $\left(-5, \frac{1}{2}, \frac{1}{2} \right)$

7. $\mathbf{n} = \mathbf{k}$, $(6, 7, 2)$ **8.** $\mathbf{n} = \mathbf{i} - \mathbf{k}$, $(4, 2, -8)$

9. Find an equation of any plane with normal $\mathbf{n} = \langle 2, 1, 1 \rangle$ *not* passing through the point $P = (0, 1, 1)$.

10. Find an equation of any plane passing through the origin but not passing through $P = (2, 1, 1)$.

In Exercises 11–14, find an equation of the plane passing through the three points given.

11. $P = (2, -1, 4)$, $Q = (1, 1, 1)$, $R = (3, 1, -2)$

12. $P = (5, 1, 1)$, $Q = (1, 1, 2)$, $R = (2, 1, 1)$

13. $P = (1, 0, 0)$, $Q = (0, 1, 1)$, $R = (2, 0, 1)$

14. $P = (2, 0, 0)$, $Q = (0, 4, 0)$, $R = (0, 0, 2)$

In Exercises 15–20, find a vector normal to the plane with the given equation.

15. $9x - 4y - 11z = 2$ **16.** $3x + 25y + 9z = 7$

17. $3x + 25y + 9z = 45$ **18.** $4y - 8z = -1$

19. $x - z = 0$ **20.** $x = 1$

In Exercises 21–26, find the equation of the plane with the given description.

21. Passes through O and is parallel to $4x - 9y + z = 3$

22. Passes through $(4, 1, 9)$ and is parallel to $x + y + z = 3$

23. Passes through $(4, 1, 9)$ and is parallel to $x = 3$

24. Passes through $(-2, -3, 5)$ and has normal vector $\mathbf{i} + \mathbf{k}$

25. Contains the lines $\mathbf{r}_1(t) = \langle t, 2t, 3t \rangle$ and $\mathbf{r}_2(t) = \langle 3t, t, 8t \rangle$

26. Contains $P = (-1, 0, 1)$ and $\mathbf{r}(t) = \langle t + 1, 2t, 3t - 1 \rangle$

27. Are the planes $\frac{1}{2}x + 2x - y = 5$ and $3x + 12x - 6y = 1$ parallel?

28. Let a, b, c be constants. Which two of the following equations define the plane passing through $(a, 0, 0)$, $(0, b, 0)$, $(0, 0, c)$?

(a) $ax + by + cz = 1$

(b) $bcx + acy + abz = abc$

(c) $bx + cy + az = 1$

(d) $\dfrac{x}{a} + \dfrac{y}{b} + \dfrac{z}{c} = 1$

29. Find an equation of the plane $\mathcal{P}$ in Figure 9.

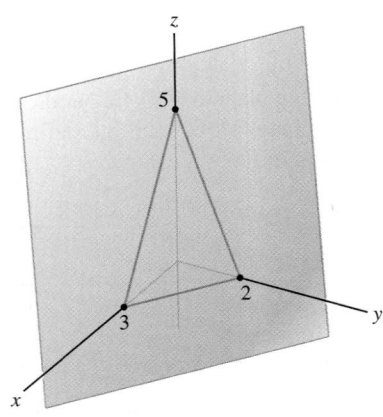

FIGURE 9

In Exercises 30–33, find the intersection of the line and plane.

30. $x + y + z = 14$, $\mathbf{r}(t) = \langle 1, 1, 0 \rangle + t \langle 0, 2, 4 \rangle$

31. $2x + y = 3$, $\mathbf{r}(t) = \langle 2, -1, -1 \rangle + t \langle 1, 2, -4 \rangle$

32. $z = 12$, $\mathbf{r}(t) = t \langle -6, 9, 36 \rangle$

33. $x - z = 6$, $\mathbf{r}(t) = \langle 1, 0, -1 \rangle + t \langle 4, 9, 2 \rangle$

34. Verify that the plane $x - y + 5z = 10$ and the line $\mathbf{r}(t) = \langle 1, 0, 1 \rangle + t \langle -2, 1, 1 \rangle$ intersect at $P = (-3, 2, 3)$.

35. The plane $3x - 4y + 2z = 8$ intersects the xy-plane in a line. What is the equation of this line?

In Exercises 36–41, find the trace of the plane in the given coordinate plane.

36. $3x - 9y + 4z = 5$, yz **37.** $3x - 9y + 4z = 5$, xz

38. $3x + 4z = -2$, xy **39.** $3x + 4z = -2$, xz

40. $-x + y = 4$, xz **41.** $-x + y = 4$, yz

42. Does the plane $x = 5$ have a trace in the yz-plane? Explain.

43. Give equations for two distinct planes whose trace in the xy-plane has equation $4x + 3y = 8$.

44. Find parametric equations for the line through $P_0 = (3, -1, 1)$ perpendicular to the plane $3x + 5y - 7z = 29$.

45. Find all planes in $\mathbf{R}^3$ whose intersection with the xy-plane is the line $\mathbf{r}(t) = t \langle 2, 1, 0 \rangle$.

46. Find all planes in $\mathbf{R}^3$ whose intersection with the xz-plane is the line with equation $3x + 2z = 5$.

47. [icon] Let $\mathcal{P}$ be the plane $\mathbf{n} \cdot \langle x, y, z \rangle = d$, where $\mathbf{n} \neq \mathbf{0}$, and let $\mathcal{P}_1$ be the parallel plane $\mathbf{n} \cdot \langle x, y, z \rangle = d_1$ (Figure 10).

(a) Show that the line through $\mathbf{n}$ intersects $\mathcal{P}$ at the terminal point of the vector $\left(\dfrac{d}{\|\mathbf{n}\|} \right) \mathbf{e_n}$.

(b) Show that the distance between $\mathcal{P}$ and $\mathcal{P}_1$ is $\dfrac{|d - d_1|}{\|\mathbf{n}\|}$.

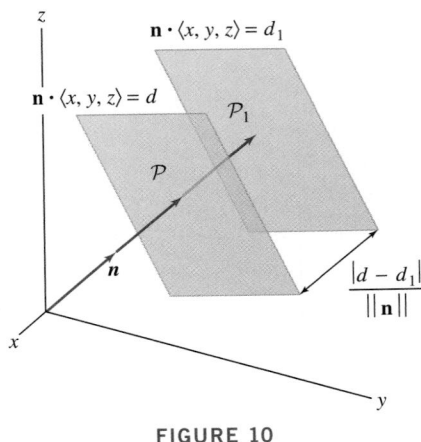

FIGURE 10

48. Use Exercise 47 to compute the distance between the parallel planes $x + 2y + 3z = 5$ and $x + 2y + 3z = 9$.

In Exercises 49–54, compute the angle between the two planes, defined as the angle θ (between 0 and π) between their normal vectors (Figure 11).

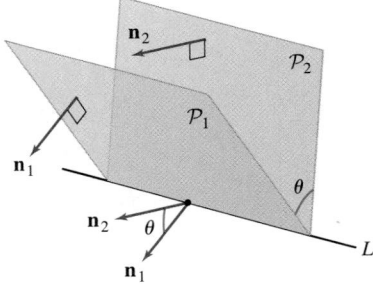

FIGURE 11 By definition, the angle between two planes is the angle between their normal vectors.

49. Planes with normals $\mathbf{n}_1 = \langle 1, 0, 1 \rangle$, $\mathbf{n}_2 = \langle -1, 1, 1 \rangle$

50. Planes with normals $\mathbf{n}_1 = \langle 1, 2, 1 \rangle$, $\mathbf{n}_2 = \langle 4, 1, 3 \rangle$

51. Planes with normals $\mathbf{n}_1 = \langle 1, 0, 0 \rangle$, $\mathbf{n}_2 = \langle 0, 1, 0 \rangle$

52. $2x + 3y + 7z = 2$ and $4x - 2y + 2z = 4$

53. $x - 3y + z = 3$ and $2x - 3z = 4$

54. The plane through $(1, 0, 0)$, $(0, 1, 0)$, and $(0, 0, 1)$ and the yz-plane

55. Find an equation of a plane making an angle of $\frac{\pi}{2}$ with the plane $3x + y - 4z = 2$.

56. Let $\mathcal{P}_1$ and $\mathcal{P}_2$ be planes with normal vectors $\mathbf{n}_1$ and $\mathbf{n}_2$. Assume that the planes are not parallel and let $\mathcal{L}$ be their intersection (a line). Show that $\mathbf{n}_1 \times \mathbf{n}_2$ is a direction vector for $\mathcal{L}$.

57. Find a plane that is perpendicular to the two planes $x + y = 3$ and $x + 2y - z = 4$.

58. Let $\mathcal{L}$ be the intersection of the planes $x + y + z = 1$ and $x + 2y + 3z = 1$. Use Exercise 56 to find a direction vector for $\mathcal{L}$. Then find a point P on $\mathcal{L}$ by *inspection* and write down the parametric equations for $\mathcal{L}$.

59. Let $\mathcal{L}$ be the intersection of the planes $x - y - z = 1$ and $2x + 3y + z = 2$. Find parametric equations for the line $\mathcal{L}$. *Hint:* To find a point $\mathcal{L}$, substitute an arbitrary value for z (say, $z = 2$) and then solve the resulting pair of equations for x and y.

60. Find parametric equations for the intersection of the planes $2x + y - 3z = 0$ and $x + y = 1$.

61. Find parametric equations for the line that is perpendicular to the plane $3x + 5y - 7z = 29$ and passes through the point $P_0 = (3, -1, 1)$.

Further Insights and Challenges

In Exercises 62–72, if $\mathcal{P}$ is a plane and Q is a point not lying on $\mathcal{P}$, then the nearest point to Q on $\mathcal{P}$ is the unique point P on $\mathcal{P}$ such that $\overline{PQ}$ is orthogonal to $\mathcal{P}$ (Figure 12).

62. Let $\mathcal{P}$ be the plane $ax + by + cz = d$ with normal $\mathbf{n} = \langle a, b, c \rangle$ and let Q be a point not on $\mathcal{P}$. Show that the point P on $\mathcal{P}$ closest to Q is determined by the equation

$$\overrightarrow{OP} = \overrightarrow{OQ} + \left(\frac{d - \overrightarrow{OQ} \cdot \mathbf{n}}{\mathbf{n} \cdot \mathbf{n}} \right) \mathbf{n} \qquad \boxed{6}$$

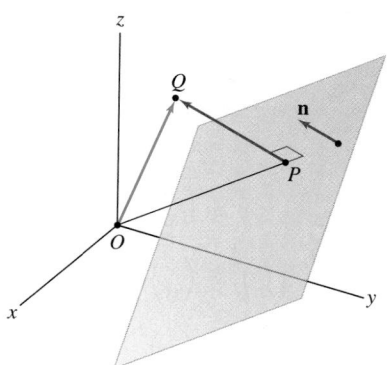

FIGURE 12

63. Use Eq. (6) to find the nearest point to $Q = (2, 1, 2)$ on the plane $x + y + z = 1$.

64. Find the nearest point to $Q = (-1, 3, -1)$ on the plane

$$x - 4z = 2$$

65. In the notation of Exercise 62, the *distance* from Q to $\mathcal{P}$ is the length $\|\overrightarrow{PQ}\|$. Show that if $Q = (x_1, y_1, z_1)$, then

$$\text{Distance from } Q \text{ to } \mathcal{P} = \frac{|ax_1 + by_1 + cz_1 - d|}{\|\mathbf{n}\|} \qquad \boxed{7}$$

66. Use Eq. (7) to find the distance from $Q = (1, 1, 1)$ to the plane $2x + y + 5z = 2$.

67. Find the distance from $Q = (1, 2, 2)$ to the plane $\mathbf{n} \cdot \langle x, y, z \rangle = 3$, where $\mathbf{n} = \left(\frac{3}{5}, \frac{4}{5}, 0 \right)$.

68. What is the distance from $P = (a, b, c)$ to the plane $x = 0$? Visualize your answer geometrically and explain without computation. Then verify that Eq. (7) yields the same answer.

69. Show that the point Q on the plane $\mathbf{n} \cdot \langle x, y, z \rangle = d$ nearest the origin lies on the ray through $\mathbf{n}$. More precisely, show that Q is the terminal point of $(d / \|\mathbf{n}\|^2)\mathbf{n}$.

70. The equation of a plane $\mathbf{n} \cdot \langle x, y, z \rangle = d$ is said to be in **normal form** if $\mathbf{n}$ is a unit vector. Show that in this case, $|d|$ is the distance from the plane to the origin. Write the equation of the plane $4x - 2y + 4z = 24$ in normal form.

71. Let $\mathbf{n} = \overrightarrow{OP}$, where $P = (x_0, y_0, z_0)$ is a point on the sphere $x^2 + y^2 + z^2 = r^2$, and let $\mathcal{P}$ be the plane with equation $\mathbf{n} \cdot \langle x, y, z \rangle = r^2$. Show that the point on $\mathcal{P}$ nearest the origin is P itself and conclude that $\mathcal{P}$ is tangent to the sphere at P (Figure 13).

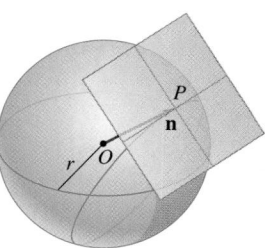

FIGURE 13 The terminal point of $\mathbf{n}$ lies on the sphere of radius r.

72. Use Exercise 71 to find the equation of the plane tangent to the unit sphere at $P = \left(\frac{1}{\sqrt{3}}, \frac{1}{\sqrt{3}}, \frac{1}{\sqrt{3}} \right)$.

12.6 A Survey of Quadric Surfaces

Quadric surfaces are the surface analogs of conic sections. Recall that a conic section is a curve in the $\mathbf{R}^2$ defined by a quadratic equation in two variables. A quadric surface is defined by a quadratic equation in *three* variables:

$$Ax^2 + By^2 + Cz^2 + Dxy + Eyz + Fzx + ax + by + cz + d = 0 \qquad \boxed{1}$$

To insure that Eq. (1) is genuinely quadratic, we assume that the degree-two coefficients A, B, C, D, E, F are not all zero.

Like conic sections, quadric surfaces are classified into a small number of types. When the coordinate axes are chosen to coincide with the axes of the quadric, the equation of the quadric has a simple form. The quadric is then said to be in **standard position**. In this short survey of quadric surfaces, we restrict our attention to quadrics in standard position.

The surface analogs of ellipses are the egg-shaped **ellipsoids** (Figure 1). In standard form, an ellipsoid has the equation

$$\left(\frac{x}{a}\right)^2 + \left(\frac{y}{b}\right)^2 + \left(\frac{z}{c}\right)^2 = 1$$

For $a = b = c$, this equation is equivalent to $x^2 + y^2 + z^2 = a^2$ and the ellipsoid is a sphere of radius a.

Surfaces are often represented graphically by a mesh of curves called **traces**, obtained by intersecting the surface with planes parallel to one of the coordinate planes. Algebraically, this corresponds to **freezing** one of the three variables (holding it constant). For example, the intersection of the horizontal plane $z = z_0$ with the surface is a horizontal trace curve (Figure 2).

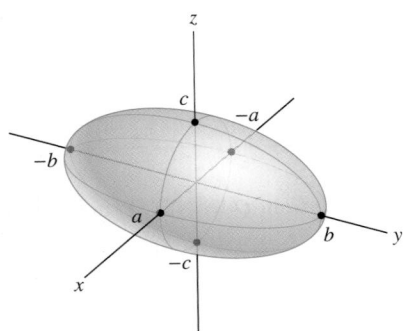

FIGURE 1 Ellipsoid with equation $\left(\frac{x}{a}\right)^2 + \left(\frac{y}{b}\right)^2 + \left(\frac{z}{c}\right)^2 = 1.$

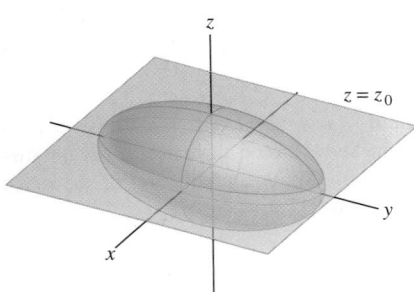

FIGURE 2 The intersection of the plane $z = z_0$ with an ellipsoid is an ellipse.

■ **EXAMPLE 1** The Traces of an Ellipsoid Describe the traces of the ellipsoid

$$\left(\frac{x}{5}\right)^2 + \left(\frac{y}{7}\right)^2 + \left(\frac{z}{9}\right)^2 = 1$$

Solution First we observe that the traces in the coordinate planes are ellipses (Figure 3A):

$$xy\text{-trace (set } z = 0\text{):} \qquad \left(\frac{x}{5}\right)^2 + \left(\frac{y}{7}\right)^2 = 1$$

$$yz\text{-trace (set } x = 0\text{):} \qquad \left(\frac{y}{7}\right)^2 + \left(\frac{z}{9}\right)^2 = 1$$

$$xz\text{-trace (set } y = 0\text{):} \qquad \left(\frac{x}{5}\right)^2 + \left(\frac{z}{9}\right)^2 = 1$$

In fact, all the traces of an ellipsoid are ellipses. For example, the horizontal trace defined by setting $z = z_0$ is the ellipse [Figure 3(B)]

$$\text{Trace at height } z_0\text{:} \qquad \left(\frac{x}{5}\right)^2 + \left(\frac{y}{7}\right)^2 + \left(\frac{z_0}{9}\right)^2 = 1 \quad \text{or} \quad \frac{x^2}{25} + \frac{y^2}{49} = \underbrace{1 - \frac{z_0^2}{81}}_{\text{A constant}}$$

For $z_0 = 9$, we obtain $\dfrac{x^2}{25} + \dfrac{y^2}{49} = 0$, and the trace reduces to a single point $(0, 0, 9)$. Similarly, for $z_0 = -9$, the trace is the point $(0, 0, -9)$. If $|z_0| > 9$, the constant $1 - z_0^2/81$ is negative and the trace has no points. The traces in the vertical planes $x = x_0$ and $y = y_0$ have a similar description [Figure 3(C)]. ■

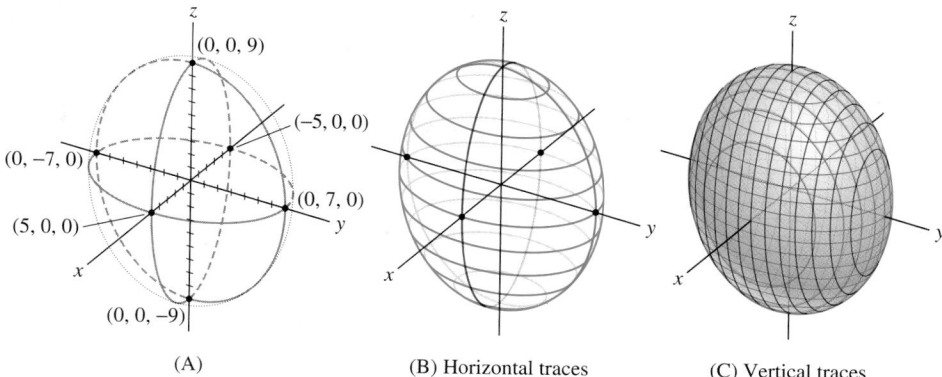

FIGURE 3 The ellipsoid $\left(\dfrac{x}{5}\right)^2 + \left(\dfrac{y}{7}\right)^2 + \left(\dfrac{z}{9}\right)^2 = 1.$

(A)   (B) Horizontal traces   (C) Vertical traces

The analogs of the hyperbolas are the **hyperboloids**, which come in two types, depending on whether the surface has one or two components. We refer to these types as hyperboloids of one or two sheets (Figure 4). Their equations in standard position are

$$\underbrace{\left(\frac{x}{a}\right)^2 + \left(\frac{y}{b}\right)^2 = \left(\frac{z}{c}\right)^2 + 1,}_{\textbf{Hyperboloid of one sheet}} \qquad \underbrace{\left(\frac{x}{a}\right)^2 + \left(\frac{y}{b}\right)^2 = \left(\frac{z}{c}\right)^2 - 1}_{\textbf{Hyperboloid of two sheets}} \qquad \boxed{2}$$

Notice that a hyperboloid of two sheets does not contain any points whose z-coordinate satisfies $-c < z < c$ because the right-hand side $\left(\dfrac{z}{c}\right)^2 - 1$ is then negative, but the left-hand side of the equation is greater than or equal to zero.

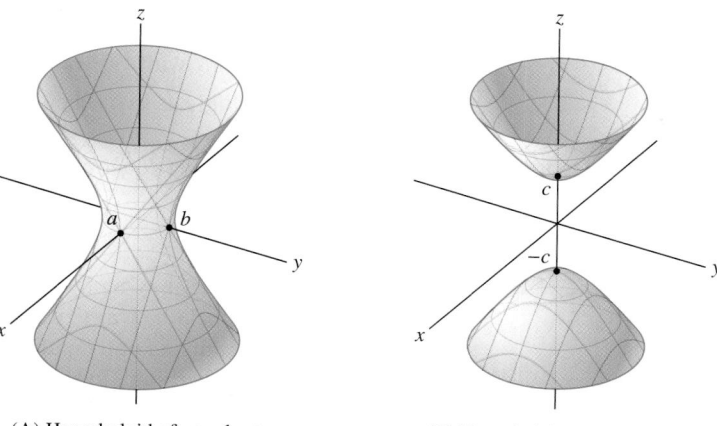

FIGURE 4 Hyperboloids of one and two sheets.

(A) Hyperboloid of one sheet   (B) Hyperboloid of two sheets

■ **EXAMPLE 2** **The Traces of a Hyperboloid of One Sheet** Determine the traces of the hyperboloid $\left(\dfrac{x}{2}\right)^2 + \left(\dfrac{y}{3}\right)^2 = \left(\dfrac{z}{4}\right)^2 + 1.$

Solution Unlike the ellipsoid, the traces of a hyperboloid are of two types. The horizontal traces are ellipses and the two families of vertical traces (parallel to the yz- and xz-planes) are hyperbolas (Figure 5):

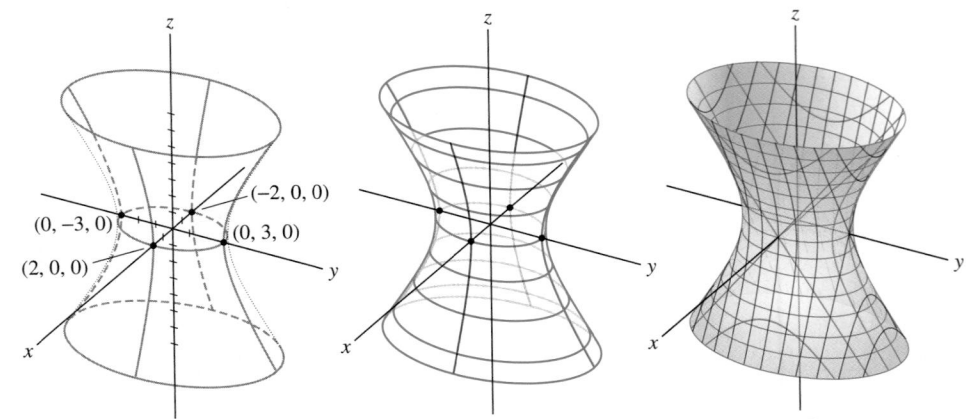

FIGURE 5 The hyperboloid
$$\left(\frac{x}{2}\right)^2 + \left(\frac{y}{3}\right)^2 = \left(\frac{z}{4}\right)^2 + 1.$$

Trace $z = z_0$ (ellipse): $\left(\frac{x}{2}\right)^2 + \left(\frac{y}{3}\right)^2 = \left(\frac{z_0}{4}\right)^2 + 1$

Trace $x = x_0$ (hyperbola): $\left(\frac{y}{3}\right)^2 - \left(\frac{z}{4}\right)^2 = 1 - \left(\frac{x_0}{2}\right)^2$

Trace $y = y_0$ (hyperbola): $\left(\frac{x}{2}\right)^2 - \left(\frac{z}{4}\right)^2 = 1 - \left(\frac{y_0}{3}\right)^2$ ∎

■ **EXAMPLE 3** Hyperboloid of Two Sheets Symmetric About the y-axis Show that $\left(\frac{x}{a}\right)^2 + \left(\frac{z}{c}\right)^2 = \left(\frac{y}{b}\right)^2 - 1$ has no points for $-b < y < b$.

Solution This equation does not have the same form as Eq. (2) because the variables y and z have been interchanged. This hyperboloid is symmetrical about the y-axis rather than the z-axis (Figure 6). The left-hand side of the equation is always ≥ 0. Thus, there are no solutions with $|y| < b$ because the right-hand side is $\left(\frac{y}{b}\right)^2 - 1 < 0$. Therefore, the hyperboloid has two sheets, corresponding to $y \geq b$ and $y \leq -b$.   ∎

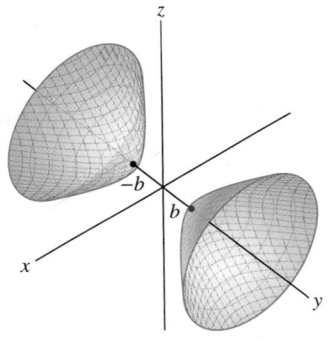

Hyperboloid of two sheets

FIGURE 6 The two-sheeted hyperboloid
$$\left(\frac{x}{a}\right)^2 + \left(\frac{z}{c}\right)^2 = \left(\frac{y}{b}\right)^2 - 1.$$

The following equation defines an **elliptic cone** (Figure 7):

$$\left(\frac{x}{a}\right)^2 + \left(\frac{y}{b}\right)^2 = \left(\frac{z}{c}\right)^2$$

An elliptic cone may be thought of as a limiting case of a hyperboloid of one sheet in which we "pinch the waist" down to a point.

The third main family of quadric surfaces are the **paraboloids**. There are two types—elliptic and hyperbolic. In standard position, their equations are

$$z = \left(\frac{x}{a}\right)^2 + \left(\frac{y}{b}\right)^2, \qquad z = \left(\frac{x}{a}\right)^2 - \left(\frac{y}{b}\right)^2$$

Elliptic paraboloid   Hyperbolic paraboloid

In both cases, the vertical trace curves are parabolas. For example, the traces defined by $x = x_0$ are the parabolas

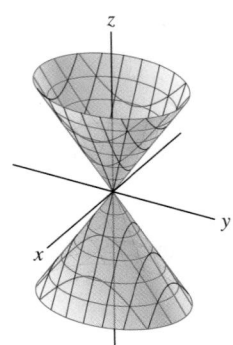

FIGURE 7 Elliptic cone
$$\left(\frac{x}{a}\right)^2 + \left(\frac{y}{b}\right)^2 = \left(\frac{z}{c}\right)^2.$$

$$z = \left(\frac{y}{b}\right)^2 + C, \qquad z = -\left(\frac{y}{b}\right)^2 + C$$

Trace $x = x_0$ of elliptic paraboloid   Trace $x = x_0$ of hyperbolic paraboloid where $C = (x_0/a)^2$. The two types of paraboloids differ in their horizontal traces. As the names suggest, the horizontal traces of an elliptic paraboloid are ellipses and the horizontal traces of a hyperbolic paraboloid are hyperbolas (Figure 8).

Paraboloids play an important role in the optimization of functions of two variables. The elliptic paraboloid in Figure 8 has a local minimum at the origin. The hyperbolic paraboloid is a "saddle shape" at the origin, which is an analog for surfaces of a point of inflection.

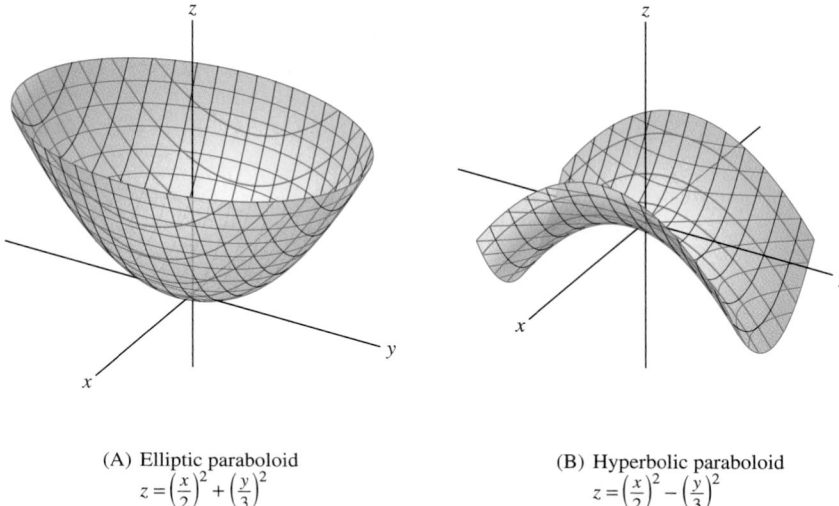

(A) Elliptic paraboloid
$$z = \left(\frac{x}{2}\right)^2 + \left(\frac{y}{3}\right)^2$$

(B) Hyperbolic paraboloid
$$z = \left(\frac{x}{2}\right)^2 - \left(\frac{y}{3}\right)^2$$

FIGURE 8

■ **EXAMPLE 4** Alternate Form of a Hyperbolic Paraboloid   Show that $z = 4xy$ is a hyperbolic paraboloid by writing the equation in terms of the variables $u = x + y$ and $v = x - y$.

Solution Note that $u + v = 2x$ and $u - v = 2y$. Therefore,

$$4xy = (u + v)(u - v) = u^2 - v^2$$

and thus the equation takes the form $z = u^2 - v^2$ in the coordinates $\{u, v, z\}$. The coordinates $\{u, v, z\}$ are obtained by rotating the coordinates $\{x, y, z\}$ by $45°$ about the z-axis (Figure 9).   ■

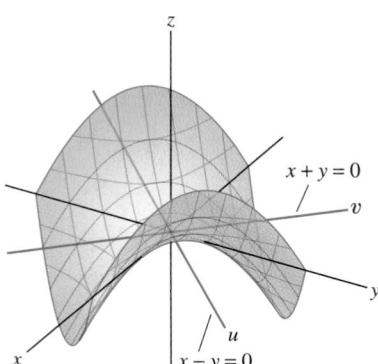

FIGURE 9 The hyperbolic paraboloid is defined by $z = 4xy$ or $z = u^2 - v^2$.

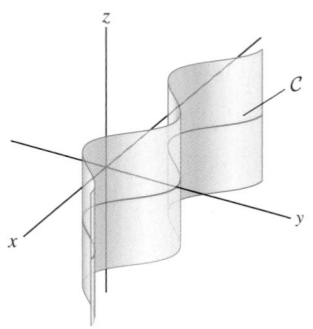

FIGURE 10 The cylinder with base C.

Further examples of quadric surfaces are the **quadratic cylinders**. We use the term *cylinder* in the following general sense: Given a curve C in the xy-plane, the cylinder with base C is the surface consisting of all vertical lines passing through C (see Figure 10). The equation $x^2 + y^2 = r^2$ defines a circular cylinder of radius r with the z-axis as central axis. Figure 11 shows a circular cylinder and three other types of quadratic cylinders.

The ellipsoids, hyperboloids, paraboloids, and quadratic cylinders are called **nondegenerate** quadric surfaces. There are also a certain number of "degenerate" quadric surfaces. For example, $x^2 + y^2 + z^2 = 0$ is a quadric that reduces to a single point $(0, 0, 0)$, and $(x + y + z)^2 = 1$ reduces to the union of the two planes $x + y + z = \pm 1$.

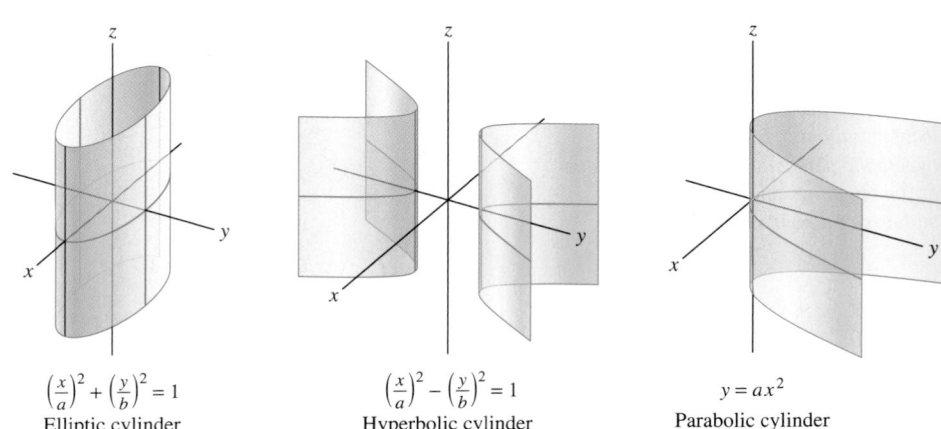

$$x^2 + y^2 = r^2$$
Circular cylinder of radius r

$$\left(\frac{x}{a}\right)^2 + \left(\frac{y}{b}\right)^2 = 1$$
Elliptic cylinder

$$\left(\frac{x}{a}\right)^2 - \left(\frac{y}{b}\right)^2 = 1$$
Hyperbolic cylinder

$$y = ax^2$$
Parabolic cylinder

FIGURE 11

12.6 SUMMARY

- A *quadric surface* is a surface defined by a quadratic equation in three variables in which the coefficients A–F are not all zero:

$$Ax^2 + By^2 + Cz^2 + Dxy + Eyz + Fzx + ax + by + cz + d = 0$$

- Quadric surfaces in standard position:

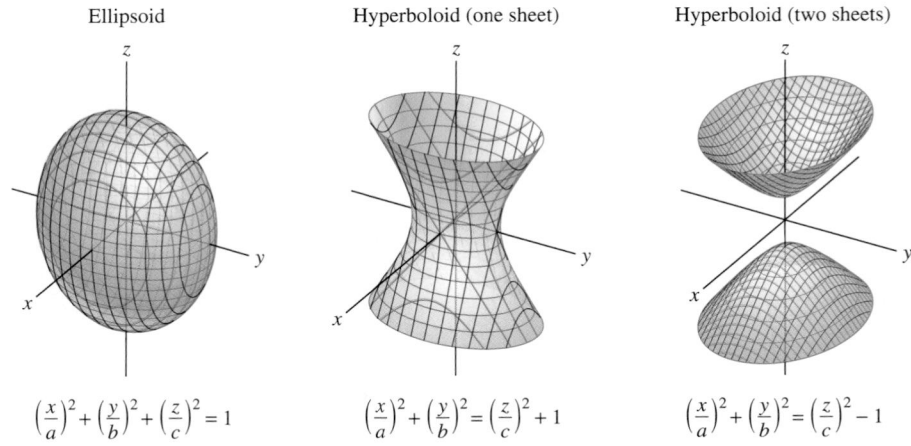

Ellipsoid

Hyperboloid (one sheet)

Hyperboloid (two sheets)

$$\left(\frac{x}{a}\right)^2 + \left(\frac{y}{b}\right)^2 + \left(\frac{z}{c}\right)^2 = 1$$

$$\left(\frac{x}{a}\right)^2 + \left(\frac{y}{b}\right)^2 = \left(\frac{z}{c}\right)^2 + 1$$

$$\left(\frac{x}{a}\right)^2 + \left(\frac{y}{b}\right)^2 = \left(\frac{z}{c}\right)^2 - 1$$

Paraboloid (elliptic)

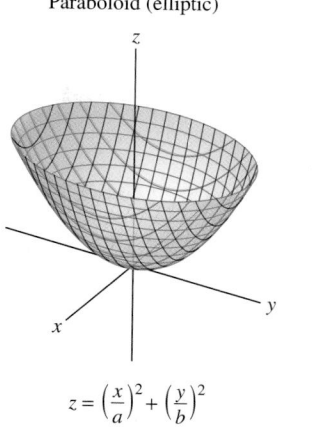

$$z = \left(\frac{x}{a}\right)^2 + \left(\frac{y}{b}\right)^2$$

Paraboloid (hyperbolic)

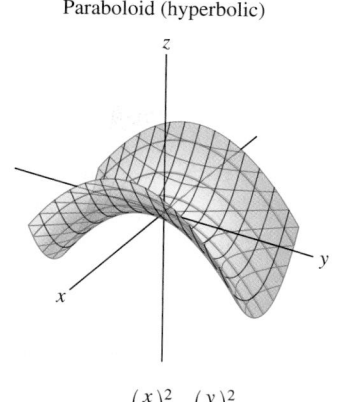

$$z = \left(\frac{x}{a}\right)^2 - \left(\frac{y}{b}\right)^2$$

Cone (elliptic)

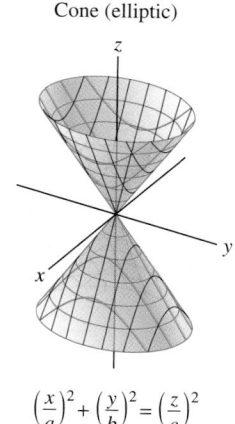

$$\left(\frac{x}{a}\right)^2 + \left(\frac{y}{b}\right)^2 = \left(\frac{z}{c}\right)^2$$

- A (vertical) cylinder is a surface consisting of all vertical lines passing through a curve (called the base) in the xy-plane. A quadratic cylinder is a cylinder whose base is a conic section. There are three types:

Elliptic cylinder

Hyperbolic cylinder

Parabolic cylinder

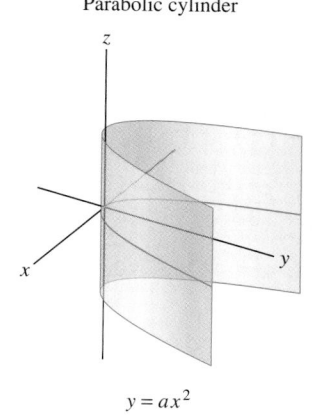

$$\left(\frac{x}{a}\right)^2 + \left(\frac{y}{b}\right)^2 = 1 \qquad\qquad \left(\frac{x}{a}\right)^2 - \left(\frac{y}{b}\right)^2 = 1 \qquad\qquad y = ax^2$$

12.6 EXERCISES

Preliminary Questions

1. True or false: All traces of an ellipsoid are ellipses.

2. True or false: All traces of a hyperboloid are hyperbolas.

3. Which quadric surfaces have both hyperbolas and parabolas as traces?

4. Is there any quadric surface whose traces are all parabolas?

5. A surface is called **bounded** if there exists $M > 0$ such that every point on the surfaces lies at a distance of at most M from the origin. Which of the quadric surfaces are bounded?

6. Give an equation for a quadric surface that consists of two separate components.

7. What is the definition of a parabolic cylinder?

Exercises

In Exercises 1–6, state whether the given equation defines an ellipsoid or hyperboloid, and if a hyperboloid, whether it is of one or two sheets.

1. $\left(\frac{x}{2}\right)^2 + \left(\frac{y}{3}\right)^2 + \left(\frac{z}{5}\right)^2 = 1$

2. $\left(\frac{x}{5}\right)^2 + \left(\frac{y}{5}\right)^2 - \left(\frac{z}{7}\right)^2 = 1$

3. $x^2 + 3y^2 + 9z^2 = 1$

4. $-\left(\dfrac{x}{2}\right)^2 - \left(\dfrac{y}{3}\right)^2 + \left(\dfrac{z}{5}\right)^2 = 1$

5. $x^2 - 3y^2 + 9z^2 = 1$

6. $x^2 - 3y^2 - 9z^2 = 1$

In Exercises 7–12, state whether the given equation defines an elliptic paraboloid, hyperbolic paraboloid, or elliptic cone.

7. $z = \left(\dfrac{x}{4}\right)^2 + \left(\dfrac{y}{3}\right)^2$

8. $z^2 = \left(\dfrac{x}{4}\right)^2 + \left(\dfrac{y}{3}\right)^2$

9. $z = \left(\dfrac{x}{9}\right)^2 - \left(\dfrac{y}{12}\right)^2$

10. $4z = 9x^2 + 5y^2$

11. $3x^2 - 7y^2 = z$

12. $3x^2 + 7y^2 = 14z^2$

In Exercises 13–20, state the type of the quadric surface and describe the trace obtained by intersecting with the given plane.

13. $x^2 + \left(\dfrac{y}{4}\right)^2 + z^2 = 1, \quad y = 0$

14. $x^2 + \left(\dfrac{y}{4}\right)^2 + z^2 = 1, \quad y = 5$

15. $x^2 + \left(\dfrac{y}{4}\right)^2 + z^2 = 1, \quad z = \dfrac{1}{4}$

16. $\left(\dfrac{x}{2}\right)^2 + \left(\dfrac{y}{5}\right)^2 - 5z^2 = 1, \quad x = 0$

17. $\left(\dfrac{x}{3}\right)^2 + \left(\dfrac{y}{5}\right)^2 - 5z^2 = 1, \quad y = 1$

18. $4x^2 - \left(\dfrac{y}{3}\right)^2 - 2z^2 = 1, \quad z = 1$

19. $y = 3x^2, \quad z = 27$

20. $y = 3x^2, \quad y = 27$

21. Match the ellipsoids in Figure 12 with the equation:
(a) $x^2 + 4y^2 + 4z^2 = 16$
(b) $4x^2 + y^2 + 4z^2 = 16$
(c) $4x^2 + 4y^2 + z^2 = 16$

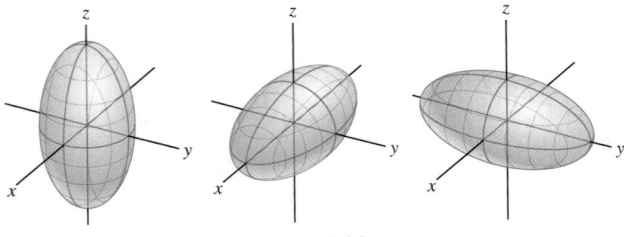

FIGURE 12

22. Describe the surface obtained if, in the equation $\pm 8x^2 \pm 3y^2 \pm z^2 = 1$, we choose (a) all plus signs, (b) one minus sign, and (c) two minus signs.

23. What is the equation of the surface obtained when the elliptic paraboloid $z = \left(\dfrac{x}{2}\right)^2 + \left(\dfrac{y}{4}\right)^2$ is rotated about the x-axis by $90°$? Refer to Figure 13.

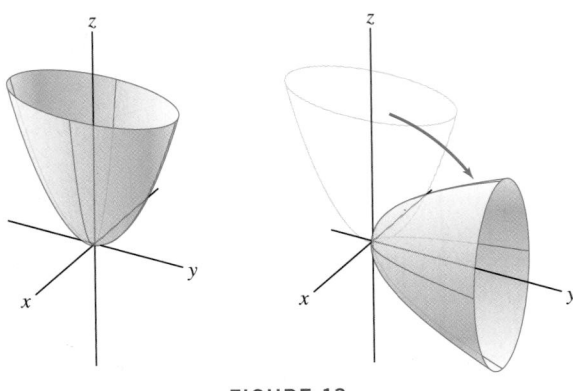

FIGURE 13

24. Describe the intersection of the horizontal plane $z = h$ and the hyperboloid $-x^2 - 4y^2 + 4z^2 = 1$. For which values of h is the intersection empty?

25. For which values of h is the intersection of the horizontal plane $x = h$ and the hyperboloid $\left(\dfrac{x}{2}\right)^2 - \left(\dfrac{y}{4}\right)^2 - \left(\dfrac{z}{9}\right)^2 = 1$ empty?

26. For which values of h is the intersection of the horizontal plane $x = h$ and the elliptic paraboloid $\left(\dfrac{y}{2}\right)^2 + \left(\dfrac{z}{4}\right)^2 = \dfrac{x}{9}$ empty?

In Exercises 27–32, sketch the given surface.

27. $x^2 + y^2 - z^2 = 1$

28. $\left(\dfrac{x}{4}\right)^2 + \left(\dfrac{y}{8}\right)^2 + \left(\dfrac{z}{12}\right)^2 = 1$

29. $z = \left(\dfrac{x}{4}\right)^2 + \left(\dfrac{y}{8}\right)^2$

30. $z = \left(\dfrac{x}{4}\right)^2 - \left(\dfrac{y}{8}\right)^2$

31. $z^2 = \left(\dfrac{x}{4}\right)^2 + \left(\dfrac{y}{8}\right)^2$

32. $z = -x^2$

33. Find the equation of the ellipsoid passing through the points marked in Figure 14(A).

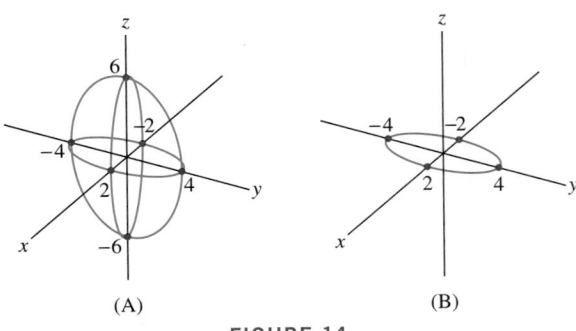

(A)   (B)

FIGURE 14

34. Find the equation of the elliptic cylinder passing through the points marked in Figure 14(B).

35. Find the equation of the hyperboloid shown in Figure 15(A).

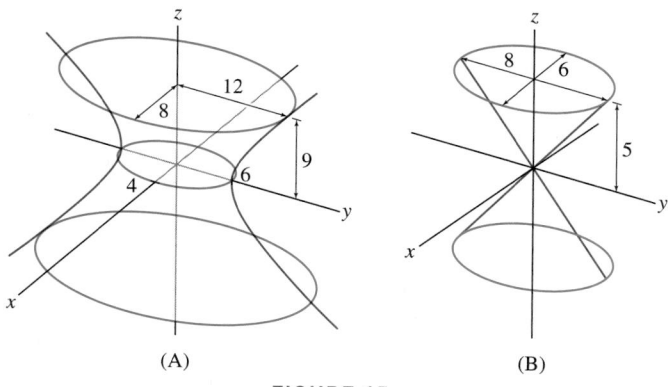

(A)   (B)

FIGURE 15

36. Find the equation of the quadric surface shown in Figure 15(B).

37. Determine the vertical traces of elliptic and parabolic cylinders in standard form.

38. What is the equation of a hyperboloid of one or two sheets in standard form if every horizontal trace is a circle?

39. Let C be an ellipse in a horizonal plane lying above the xy-plane. Which type of quadric surface is made up of all lines passing through the origin and a point on C?

40. Show that the horizontal traces of the ellipsoid

$$\left(\frac{x}{a}\right)^2 + \left(\frac{y}{b}\right)^2 + \left(\frac{z}{c}\right)^2 = 1$$

are ellipses of the same eccentricity (apart from the traces at height $h = \pm c$, which reduce to a single point). Find the eccentricity.

Further Insights and Challenges

41. Let $Q = (m, n, r)$ be a point on the ellipsoid with equation $\left(\frac{x}{a}\right)^2 + \left(\frac{y}{b}\right)^2 + \left(\frac{z}{c}\right)^2 = 1$ (Figure 16). Let $\mathcal{P}$ be the plane with equation

$$\frac{mx}{a^2} + \frac{ny}{b^2} + \frac{rz}{c^2} = 1$$

Show that Q lies on $\mathcal{P}$ and that the tangent lines to the trace curves through Q are each contained in $\mathcal{P}$. This shows that $\mathcal{P}$ is the tangent plane at Q, as defined in Section 14.4.

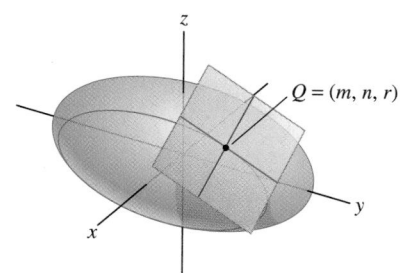

FIGURE 16 Ellipsoid $\left(\frac{x}{a}\right)^2 + \left(\frac{y}{b}\right)^2 + \left(\frac{z}{c}\right)^2 = 1$.

42. Let S be the hyperboloid $x^2 + y^2 = z^2 + 1$ and let $P = (\alpha, \beta, 0)$ be a point on S in the (x, y)-plane. Show that there are precisely two lines through P entirely contained in S (Figure 17). *Hint:* Consider the line $\mathbf{r}(t) = \langle \alpha + at, \beta + bt, t \rangle$ through P. Show that $\mathbf{r}(t)$ is contained in S if (a, b) is one of the two points on the unit circle obtained by rotating (α, β) through $\pm \frac{\pi}{2}$. This proves that a hyperboloid of one sheet is a **doubly ruled surface**, which means that it can be swept out by moving a line in space in two different ways.

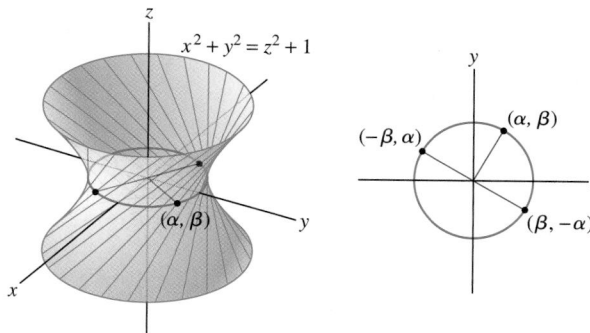

FIGURE 17

In Exercises 43–44, let C be a curve in $\mathbf{R}^3$ not passing through the origin. The cone on C is the surface consisting of all lines passing through the origin and a point on C.

43. Show that the elliptic cone $\left(\frac{z}{c}\right)^2 = \left(\frac{x}{a}\right)^2 + \left(\frac{y}{b}\right)^2$ is, in fact, a cone on the ellipse C consisting of all points (x, y, c) such that $\left(\frac{x}{a}\right)^2 + \left(\frac{y}{b}\right)^2 = 1$.

44. In this exercise, we show that if C is a parabola, then the cone on C is also an elliptic cone. Let S be the cone on the curve $C = \{(x, y, c) : y = ax^2\}$, where a, c are constants.

(a) Show that S has equation $yz = acx^2$.

(b) Show that under the change of variables $y = u + v$ and $z = u - v$, this equation becomes $acx^2 = u^2 - v^2$ or $u^2 = acx^2 + v^2$ (the equation of an elliptic cone in the variables x, v, u).

12.7 Cylindrical and Spherical Coordinates

This section introduces two generalizations of polar coordinates to $\mathbf{R}^3$: cylindrical and spherical coordinates. These coordinate systems are commonly used in problems having symmetry about an axis or rotational symmetry. For example, the magnetic field generated by a current flowing in a long straight wire is conveniently expressed in cylindrical coordinates (Figure 1). We will also see the benefits of cylindrical and spherical coordinates when we study change of variables for multiple integrals.

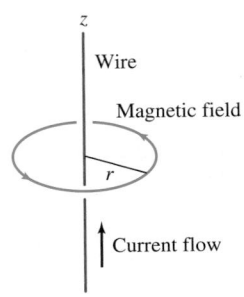

Wire

Magnetic field

r

Current flow

FIGURE 1

Cylindrical Coordinates

In cylindrical coordinates, we replace the x- and y-coordinates of a point $P = (x, y, z)$ by polar coordinates. Thus, the **cylindrical coordinates** of P are (r, θ, z), where (r, θ) are polar coordinates of (x, y). Note that (r, θ) are polar coordinates of the projection Q of P onto the xy-plane (Figure 2). We convert between rectangular and cylindrical coordinates using the rectangular-polar formulas of Section 11.3.

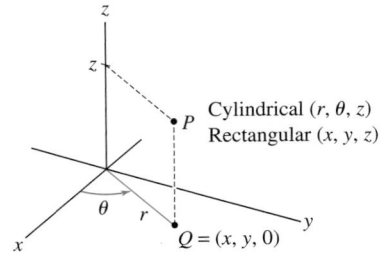

Cylindrical (r, θ, z)
Rectangular (x, y, z)

$Q = (x, y, 0)$

FIGURE 2 Cylindrical coordinates.

Cylindrical to Rectangular	Rectangular to Cylindrical
$x = r \cos \theta$	$r = \sqrt{x^2 + y^2}$
$y = r \sin \theta$	$\tan \theta = \dfrac{y}{x}$
$z = z$	$z = z$

■ EXAMPLE 1 Converting from Cylindrical to Rectangular Coordinates   Convert $(r, \theta, z) = (2, \frac{3\pi}{4}, 5)$ to rectangular coordinates.

Solution Conversion to rectangular coordinates is straightforward (Figure 3):

$$x = r \cos \theta = 2 \cos \frac{3\pi}{4} = 2\left(-\frac{\sqrt{2}}{2}\right) = -\sqrt{2}$$

$$y = r \sin \theta = 2 \sin \frac{3\pi}{4} = 2\left(\frac{\sqrt{2}}{2}\right) = \sqrt{2}$$

The z-coordinate is unchanged, so $(x, y, z) = (-\sqrt{2}, \sqrt{2}, 5)$.   ■

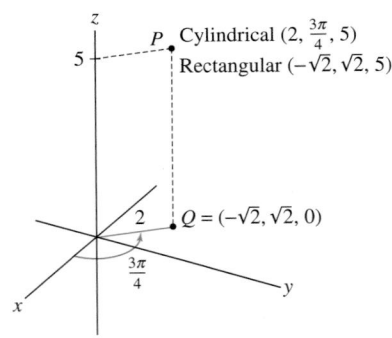

P Cylindrical $(2, \frac{3\pi}{4}, 5)$
Rectangular $(-\sqrt{2}, \sqrt{2}, 5)$

$Q = (-\sqrt{2}, \sqrt{2}, 0)$

$\frac{3\pi}{4}$

FIGURE 3

■ EXAMPLE 2 Converting from Rectangular to Cylindrical Coordinates   Convert the rectangular coordinates $(x, y, z) = (-3\sqrt{3}, -3, 5)$ to cylindrical coordinates.

Solution We have $r = \sqrt{x^2 + y^2} = \sqrt{(-3\sqrt{3})^2 + (-3)^2} = 6$ and $\tan \theta = y/x = 1/\sqrt{3}$. But note that θ is determined by the projection $Q = (-3\sqrt{3}, -3, 0)$. Although

$$\tan^{-1}\left(\frac{y}{x}\right) = \tan^{-1}\left(\frac{-3}{-3\sqrt{3}}\right) = \tan^{-1}\frac{1}{\sqrt{3}} = \frac{\pi}{6}$$

the angle θ is not equal to $\frac{\pi}{6}$ because Q lies in the third quadrant (Figure 4). The correct angle is $\theta = \frac{\pi}{6} + \pi = \frac{7\pi}{6}$ and the cylindrical coordinates are $(r, \theta, z) = (6, \frac{7\pi}{6}, 5)$.   ■

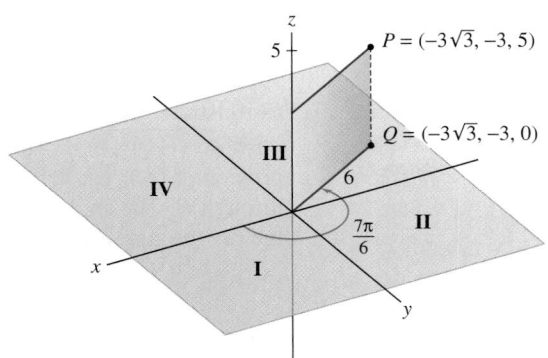

FIGURE 4 The projection Q lies in the third quadrant. Therefore,
$\theta = \tan^{-1} \frac{1}{\sqrt{3}} + \pi = \frac{7\pi}{6}$.

The **level surfaces** of a coordinate system are the surfaces obtained by setting one of the coordinates equal to a constant. In rectangular coordinates, the level surfaces are the planes $x = x_0$, $y = y_0$, and $z = z_0$. In cylindrical coordinates, the level surfaces come in three types (Figure 5). The surface $r = R$ is the cylinder of radius R consisting of all points located a distance R from the z-axis. The equation $\theta = \theta_0$ defines the half-plane of all points that project onto the ray $\theta = \theta_0$ in the (x, y)-plane. Finally, $z = c$ is the horizontal plane at height c.

Cylindrical coordinates are named after the surfaces $r = R$, which are cylinders.

Level Surfaces in Cylindrical Coordinates

$r = R$ *Cylinder of radius R with the z-axis as axis of symmetry*

$\theta = \theta_0$ *Half-plane through the z-axis making an angle θ_0 with the xz-plane*

$z = c$ *Horizontal plane at height c*

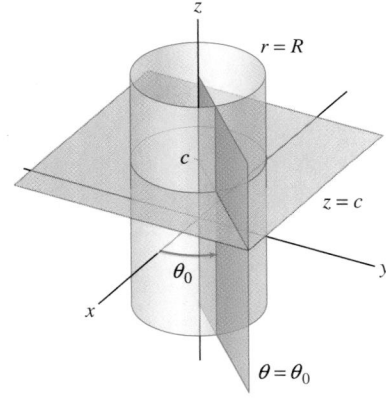

FIGURE 5 Level surfaces in cylindrical coordinates.

■ **EXAMPLE 3** Equations in Cylindrical Coordinates Find an equation of the form $z = f(r, \theta)$ for the surfaces

(a) $x^2 + y^2 + z^2 = 9$ **(b)** $x + y + z = 1$

Solution We use the formulas

$$x^2 + y^2 = r^2, \qquad x = r \cos \theta, \qquad y = r \sin \theta$$

(a) The equation $x^2 + y^2 + z^2 = 9$ becomes $r^2 + z^2 = 9$ or $z = \pm\sqrt{9 - r^2}$. This is a sphere of radius 3.

(b) The plane $x + y + z = 1$ becomes

$$z = 1 - x - y = 1 - r \cos \theta - r \sin \theta \qquad \text{or} \qquad z = 1 - r(\cos \theta + \sin \theta) \qquad ■$$

Spherical Coordinates

The symbol ϕ (usually pronounced "fee," but sometimes pronounced "fie") is the twenty-first letter of the Greek alphabet.

A point $P = (x, y, z)$ in 3-space is described in spherical coordinates by a triple (ρ, θ, ϕ) as illustrated in Figure 6. The radial coordinate ρ is the distance from P to the origin: $\rho = \sqrt{x^2 + y^2 + z^2}$. The other two coordinates are angular: θ is the polar angle of the projection $Q = (x, y, 0)$ and ϕ is the angle between the z-axis and the ray from the origin through P. The angle ϕ is called the **angle of declination** because it measures how much the ray through P declines from the vertical.

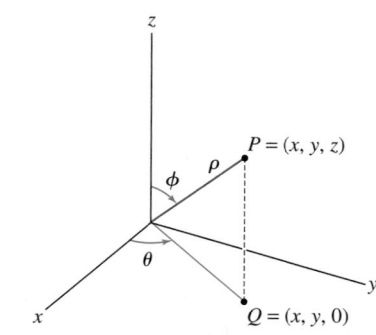

FIGURE 6 Spherical coordinates (ρ, θ, ϕ).

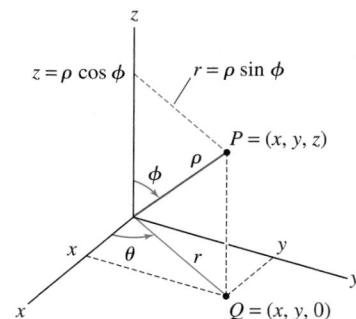

FIGURE 7 Use trigonometry to compute x, y, and z in terms of ρ, θ, and ϕ.

We use trigonometry to convert between spherical and rectangular coordinates. From Figure 7, we see that

$$\rho = \sqrt{x^2 + y^2 + z^2}, \qquad \tan\theta = \frac{y}{x}, \qquad \cos\phi = \frac{z}{\rho} \qquad \boxed{1}$$

The radial coordinate r of $Q = (x, y, 0)$ is $r = \rho \sin\phi$, and therefore,

$$x = r\cos\theta = \rho\cos\theta\sin\phi, \qquad y = r\cos\theta = \rho\sin\theta\sin\phi, \qquad z = \rho\cos\phi$$

Spherical Coordinates

ρ = distance from origin

θ = polar angle in the xy-plane

ϕ = angular declination from the vertical

In some textbooks, θ is referred to as the azimuthal angle *and ϕ as the* polar angle.

Spherical to Rectangular	Rectangular to Spherical
$x = \rho\cos\theta\sin\phi$	$\rho = \sqrt{x^2 + y^2 + z^2}$
$y = \rho\sin\theta\sin\phi$	$\tan\theta = \dfrac{y}{x}$
$z = \rho\cos\phi$	$\cos\phi = \dfrac{z}{\rho}$

■ **EXAMPLE 4** **From Spherical to Rectangular Coordinates** Find the rectangular coordinates of $P = (\rho, \theta, \phi) = (3, \frac{\pi}{3}, \frac{\pi}{4})$ and find the radial coordinate r of its projection Q onto the xy-plane.

Solution By the formulas above,

$$x = \rho\cos\theta\sin\phi = 3\cos\frac{\pi}{3}\sin\frac{\pi}{4} = 3\left(\frac{1}{2}\right)\frac{\sqrt{2}}{2} = \frac{3\sqrt{2}}{4}$$

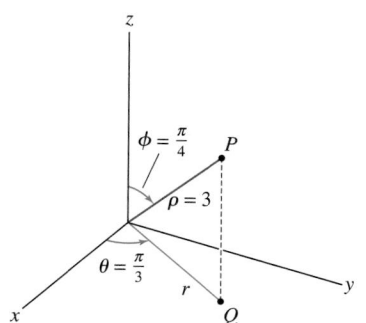

FIGURE 8 Point with spherical coordinates $(3, \frac{\pi}{3}, \frac{\pi}{4})$.

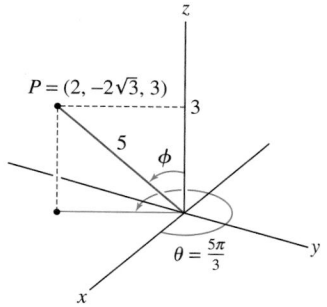

FIGURE 9 Point with rectangular coordinates $(2, -2\sqrt{3}, 3)$.

Level Surfaces in Spherical Coordinates

$\rho = R$ Sphere of radius R with its center at the origin

$\theta = \theta_0$ Half-plane through the z-axis making an angle θ_0 with the xz-plane

$\phi = \phi_0$ Right-circular cone with the following three exceptions:

$\phi = 0$ Positive z-axis

$\phi = \frac{\pi}{2}$ xy-plane

$\phi = \pi$ Negative z-axis

$$y = \rho \sin\theta \sin\phi = 3\sin\frac{\pi}{3}\sin\frac{\pi}{4} = 3\left(\frac{\sqrt{3}}{2}\right)\frac{\sqrt{2}}{2} = \frac{3\sqrt{6}}{4}$$

$$z = \rho\cos\phi = 3\cos\frac{\pi}{4} = 3\frac{\sqrt{2}}{2} = \frac{3\sqrt{2}}{2}$$

Now consider the projection $Q = (x, y, 0) = \left(\frac{3\sqrt{2}}{4}, \frac{3\sqrt{6}}{4}, 0\right)$ (Figure 8). The radial coordinate r of Q satisfies

$$r^2 = x^2 + y^2 = \left(\frac{3\sqrt{2}}{4}\right)^2 + \left(\frac{3\sqrt{6}}{4}\right)^2 = \frac{9}{2}$$

Therefore, $r = \dfrac{3}{\sqrt{2}}$. ∎

■ **EXAMPLE 5** **From Rectangular to Spherical Coordinates** Find the spherical coordinates of the point $P = (x, y, z) = (2, -2\sqrt{3}, 3)$.

Solution The radial coordinate is $\rho = \sqrt{2^2 + (-2\sqrt{3})^2 + 3^2} = \sqrt{25} = 5$. The angular coordinate θ satisfies

$$\tan\theta = \frac{y}{x} = \frac{-2\sqrt{3}}{2} = -\sqrt{3} \quad\Rightarrow\quad \theta = \frac{2\pi}{3} \text{ or } \frac{5\pi}{3}$$

Since the point $(x, y) = (2, -2\sqrt{3})$ lies in the fourth quadrant, the correct choice is $\theta = \frac{5\pi}{3}$ (Figure 9). Finally, $\cos\phi = \dfrac{z}{\rho} = \dfrac{3}{5}$ and so $\phi = \cos^{-1}\dfrac{3}{5} \approx 0.93$. Therefore, P has spherical coordinates $(5, \frac{5\pi}{3}, 0.93)$. ∎

Figure 10 shows the three types of level surfaces in spherical coordinates. Notice that if $\phi \neq \frac{\pi}{2}$ or π, then the level surface $\phi = \phi_0$ is the right-circular cone consisting of points P such that $\overline{OP}$ makes an angle ϕ_0 with the z-axis.

For fixed radius $\rho = R$, the angular coordinates (θ, ϕ) provide useful coordinates on the sphere of radius R. They are closely related to the longitude-latitude system used to identify points on the surface of the earth (Figure 11). In this system, $\theta = \theta_0$ defines a

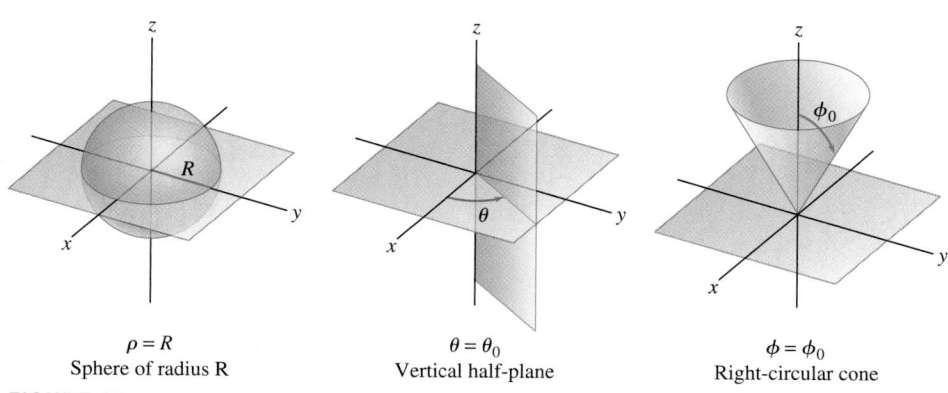

$\rho = R$
Sphere of radius R

$\theta = \theta_0$
Vertical half-plane

$\phi = \phi_0$
Right-circular cone

FIGURE 10

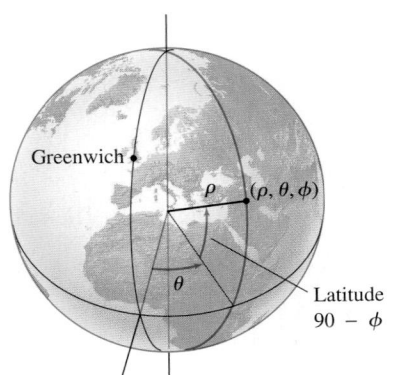

FIGURE 11 Longitude and latitude provide spherical coordinates on the surface of the earth.

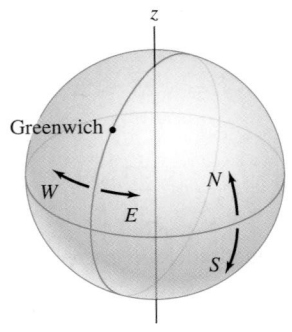

FIGURE 12 Latitude is measured from the equator and is labeled N (north) in the upper hemisphere, and S (south) in the lower hemisphere.

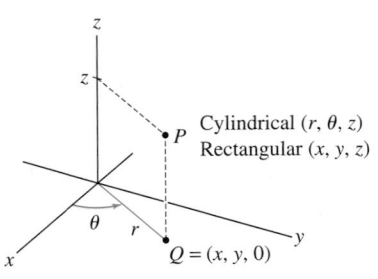

FIGURE 13 Cylindrical coordinates (r, θ, z).

longitude, that is, a half-circle stretching from the North to the South Pole (Figure 12). By convention, the axes are chosen so that $\theta = 0$ passes through Greenwich, England (this longitude is called the *prime meridian*). The longitude of a point is given as an angle between 0 and 180° and is designated E or W according to whether the point lies to the east or west of the prime meridian. The set of points on the sphere satisfying $\phi = \phi_0$ is a horizontal circle called a **latitude**. By convention, on a globe we measure latitudes from the equator and use the labels N or S to specify the Northern or Southern Hemisphere. Thus, the latitude at $\phi = \phi_0$ is $90° - \phi_0$ if $\theta \le \phi_0 \le 90°$, and it is $\phi_0 - 90°$ if $90° \le \phi_0 \le 180°$.

■ **EXAMPLE 6** Spherical Coordinates via Longitude and Latitude Find the angles (θ, ϕ) for Nairobi (1.17° S, 36.48° E) and Ottawa (45.27° N, 75.42° W).

Solution For Nairobi, $\theta = 36.48°$ since the longitude lies to the east of Greenwich. Nairobi's latitude is south of the equator, so $\phi = 90 + 1.17 = 91.17°$.

For Ottawa, we have $\theta = 360 - 75.42 = 284.58°$ since 75.42° W refers to 75.42 degrees in the negative θ direction. Since the latitude of Ottawa is north of the equator, $\phi = 90 - 45.27 = 44.73°$. ■

■ **EXAMPLE 7** Finding an Equation in Spherical Coordinates Find an equation of the form $\rho = f(\theta, \phi)$ for the surfaces:

(a) $x^2 + y^2 + z^2 = 9$ **(b)** $z = x^2 - y^2$

Solution

(a) The equation $x^2 + y^2 + z^2 = 9$ defines the sphere of radius 3 centered at the origin. Since $\rho^2 = x^2 + y^2 + z^2$, the equation in spherical coordinates is $\rho = 3$.

(b) To convert $z = x^2 - y^2$ to spherical coordinates, we substitute the formulas for x, y, and z in terms of ρ, θ, and ϕ:

$$\overbrace{\rho\cos\phi}^{z} = \overbrace{(\rho\cos\theta\sin\phi)^2}^{x^2} - \overbrace{(\rho\sin\theta\sin\phi)^2}^{y^2}$$

$$\cos\phi = \rho\sin^2\phi(\cos^2\theta - \sin^2\theta) \qquad \text{(divide by } \rho \text{ and factor)}$$

$$\cos\phi = \rho\sin^2\phi\cos 2\theta \qquad \text{(since } \cos^2\theta - \sin^2\theta = \cos 2\theta\text{)}$$

Solving for ρ, we obtain $\rho = \dfrac{\cos\phi}{\sin^2\phi\cos 2\theta}$. ■

12.7 SUMMARY

- Conversion from rectangular to cylindrical and spherical coordinates:

Cylindrical	Spherical
$r = \sqrt{x^2 + y^2}$	$\rho = \sqrt{x^2 + y^2 + z^2}$
$\tan\theta = \dfrac{y}{x}$	$\tan\theta = \dfrac{y}{x}$
$z = z$	$\cos\phi = \dfrac{z}{\rho}$

The angles are chosen so that

$$0 \le \theta < 2\pi \quad \text{(cylindrical or spherical)}, \qquad 0 \le \phi \le \pi \quad \text{(spherical)}$$

- Conversion to rectangular coordinates:

Cylindrical (r, θ, z)	Spherical (ρ, θ, ϕ)
$x = r \cos \theta$	$x = \rho \cos \theta \sin \phi$
$y = r \sin \theta$	$y = \rho \sin \theta \sin \phi$
$z = z$	$z = \rho \cos \phi$

- Level surfaces:

Cylindrical		Spherical	
$r = R$:	Cylinder of radius R	$\rho = R$:	Sphere of radius R
$\theta = \theta_0$:	Vertical half-plane	$\theta = \theta_0$:	Vertical half-plane
$z = c$:	Horizontal plane	$\phi = \phi_0$:	Right-circular cone

FIGURE 14 Spherical coordinates (ρ, θ, ϕ).

12.7 EXERCISES

Preliminary Questions

1. Describe the surfaces $r = R$ in cylindrical coordinates and $\rho = R$ in spherical coordinates.

2. Which statement about the cylindrical coordinates is correct?
(a) If $\theta = 0$, then P lies on the z-axis.
(b) If $\theta = 0$, then P lies in the xz-plane.

3. Which statement about spherical coordinates is correct?

(a) If $\phi = 0$, then P lies on the z-axis.
(b) If $\phi = 0$, then P lies in the xy-plane.

4. The level surface $\phi = \phi_0$ in spherical coordinates, usually a cone, reduces to a half-line for two values of ϕ_0. Which two values?

5. For which value of ϕ_0 is $\phi = \phi_0$ a plane? Which plane?

Exercises

In Exercises 1–4, convert from cylindrical to rectangular coordinates.

1. $(4, \pi, 4)$

2. $\left(2, \frac{\pi}{3}, -8\right)$

3. $\left(0, \frac{\pi}{5}, \frac{1}{2}\right)$

4. $\left(1, \frac{\pi}{2}, -2\right)$

In Exercises 5–10, convert from rectangular to cylindrical coordinates.

5. $(1, -1, 1)$
6. $(2, 2, 1)$
7. $(1, \sqrt{3}, 7)$

8. $\left(\frac{3}{2}, \frac{3\sqrt{3}}{2}, 9\right)$
9. $\left(\frac{5}{\sqrt{2}}, \frac{5}{\sqrt{2}}, 2\right)$
10. $(3, 3\sqrt{3}, 2)$

In Exercises 11–16, describe the set in cylindrical coordinates.

11. $x^2 + y^2 \le 1$
12. $y^2 + z^2 \le 4, \quad x = 0$

13. $x^2 + y^2 + z^2 = 4, \quad x \ge 0, \quad y \ge 0, \quad z \ge 0$

14. $x^2 + y^2 = 9, \quad x \ge 0, \quad y \ge 0, \quad x = y$

15. $y^2 + z^2 \le 9, \quad x \ge y$
16. $x^2 + y^2 + z^2 \le 1$

In Exercises 17–19, sketch the level surface.

17. $r = 4$
18. $\theta = \frac{\pi}{3}$
19. $z = -2$

In Exercises 20–23, sketch the set (described in cylindrical coordinates).

20. $1 \le r \le 3, \quad 0 \le z \le 4$
21. $z^2 + r^2 \le 4$

22. $1 \le r \le 3, \quad 0 \le \theta \le \frac{\pi}{2}, \quad 0 \le z \le 4$

23. $r \le 3, \quad \pi \le \theta \le \frac{3\pi}{2}, \quad z = 4$

In Exercises 24–29, find an equation of the form $r = f(\theta, z)$ in cylindrical coordinates for the following surfaces.

24. $z = x + y$
25. $x^2 + y^2 + z^2 = 4$

26. $\frac{x^2}{yz} = 1$
27. $x^2 - y^2 = 4$

28. $x^2 + y^2 = 4$
29. $z = 3xy$

In Exercises 30–35, convert from spherical to rectangular coordinates.

30. $\left(3, 0, \dfrac{\pi}{2}\right)$

31. $\left(2, \dfrac{\pi}{4}, \dfrac{\pi}{3}\right)$

32. $(3, \pi, 0)$

33. $\left(5, \dfrac{3\pi}{4}, \dfrac{\pi}{4}\right)$

34. $\left(6, \dfrac{\pi}{6}, \dfrac{5\pi}{6}\right)$

35. $(0.5, 3.7, 2)$

In Exercises 36–41, convert from rectangular to spherical coordinates.

36. $(\sqrt{3}, 0, 1)$

37. $\left(\dfrac{\sqrt{3}}{2}, \dfrac{3}{2}, 1\right)$

38. $(1, 1, 1)$

39. $(1, -1, 1)$

40. $\left(\dfrac{1}{2}, \dfrac{\sqrt{3}}{2}, \sqrt{3}\right)$

41. $\left(\dfrac{\sqrt{2}}{2}, \dfrac{\sqrt{2}}{2}, \sqrt{3}\right)$

In Exercises 42–47, describe the given set in spherical coordinates.

42. $x^2 + y^2 + z^2 \le 1$

43. $x^2 + y^2 + z^2 = 1, \quad z \ge 0$

44. $x^2 + y^2 + z^2 = 1, \quad x \ge 0, \quad y \ge 0, \quad z \ge 0$

45. $x^2 + y^2 + z^2 \le 1, \quad x = y, \quad x \ge 0, \quad y \ge 0$

46. $y^2 + z^2 \le 4, \quad x = 0$

47. $x^2 + y^2 = 3z^2$

In Exercises 48–50, sketch the level surface.

48. $\rho = 4$

49. $\theta = \dfrac{\pi}{3}$

50. $\phi = \dfrac{\pi}{4}$

In Exercises 51–54, sketch the set of points.

51. $\rho = 2, \quad 0 \le \phi \le \dfrac{\pi}{2}$

52. $\theta = \dfrac{\pi}{2}, \quad \phi = \dfrac{\pi}{4}, \quad \rho \ge 1$

53. $\rho \le 2, \quad 0 \le \theta \le \dfrac{\pi}{2}, \quad \dfrac{\pi}{2} \le \phi \le \pi$

54. $\rho = 1, \quad \dfrac{\pi}{3} \le \phi \le \dfrac{2\pi}{3}$

In Exercises 55–60, find an equation of the form $\rho = f(\theta, \phi)$ in spherical coordinates for the following surfaces.

55. $z = 2$

56. $z^2 = 3(x^2 + y^2)$

57. $x = z^2$

58. $z = x^2 + y^2$

59. $x^2 - y^2 = 4$

60. $xy = z$

61. [icon] Which of (a)–(c) is the equation of the cylinder of radius R in spherical coordinates? Refer to Figure 15.

(a) $R\rho = \sin\phi$　　**(b)** $\rho \sin\phi = R$　　**(c)** $\rho = R\sin\phi$

62. Let $P_1 = (1, -\sqrt{3}, 5)$ and $P_2 = (-1, \sqrt{3}, 5)$ in rectangular coordinates. In which quadrants do the projections of P_1 and P_2 onto the xy-plane lie? Find the polar angle θ of each point.

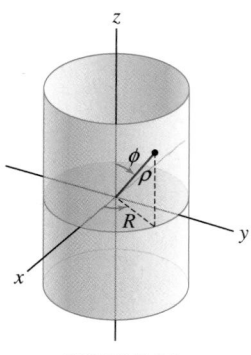

FIGURE 15

63. Consider a rectangular coordinate system with origin at the center of the earth, z-axis through the North Pole, and x-axis through the prime meridian. Find the rectangular coordinates of Sydney, Australia ($34°$ S, $151°$ E), and Bogota, Colombia ($4°$ $32'$ N, $74°$ $15'$ W). A minute is $1/60°$. Assume that the earth is a sphere of radius $R = 6{,}367$ km.

64. Find the equation in rectangular coordinates of the quadric surface consisting of the two cones $\phi = \dfrac{\pi}{4}$ and $\phi = \dfrac{3\pi}{4}$.

65. Find an equation of the form $z = f(r, \theta)$ in cylindrical coordinates for $z^2 = x^2 - y^2$.

66. Show that $\rho = 2\cos\phi$ is the equation of a sphere with its center on the z-axis. Find its radius and center.

67. [icon] Explain the following statement: If the equation of a surface in cylindrical or spherical coordinates does not involve the coordinate θ, then the surface is rotationally symmetric with respect to the z-axis.

68. *CAS* Plot the surface $\rho = 1 - \cos\phi$. Then plot the trace of S in the xz-plane and explain why S is obtained by rotating this trace.

69. Find equations $r = g(\theta, z)$ (cylindrical) and $\rho = f(\theta, \phi)$ (spherical) for the hyperboloid $x^2 + y^2 = z^2 + 1$ (Figure 16). Do there exist points on the hyperboloid with $\phi = 0$ or π? Which values of ϕ occur for points on the hyperboloid?

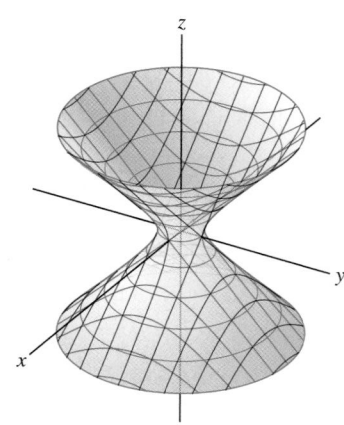

FIGURE 16 The hyperboloid $x^2 + y^2 = z^2 + 1$.

Further Insights and Challenges

*In Exercises 70–74, a **great circle** on a sphere S with center O is a circle obtained by intersecting S with a plane that passes through O (Figure 17). If P and Q are not antipodal (on opposite sides), there is a unique great circle through P and Q on S (intersect S with the plane through O, P, and Q). The geodesic distance from P to Q is defined as the length of the smaller of the two circular arcs of this great circle.*

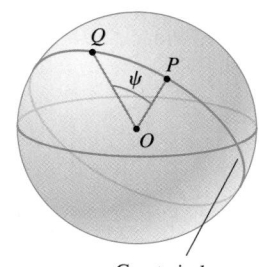

Smaller circle

Great circle through P and Q

FIGURE 17

70. Show that the geodesic distance from P to Q is equal to $R\psi$, where ψ is the *central angle* between P and Q (the angle between the vectors $\mathbf{v} = \overrightarrow{OP}$ and $\mathbf{u} = \overrightarrow{OQ}$).

71. Show that the geodesic distance from $Q = (a, b, c)$ to the North Pole $P = (0, 0, R)$ is equal to $R\cos^{-1}\left(\dfrac{c}{R}\right)$.

72. The coordinates of Los Angeles are 34° N and 118° W. Find the geodesic distance from the North Pole to Los Angeles, assuming that the earth is a sphere of radius 3,960 miles.

73. Show that the central angle ψ between points P and Q on a sphere (of any radius) with angular coordinates (θ, ϕ) and (θ', ϕ') is equal to

$$\psi = \cos^{-1}\left(\sin\phi\,\sin\phi'\,\cos(\theta - \theta') + \cos\phi\,\cos\phi'\right)$$

Hint: Compute the dot product of $\overrightarrow{OP}$ and $\overrightarrow{OQ}$. Check this formula by computing the geodesic distance between the North and South Poles.

74. Use Exercise 73 to find the geodesic distance between Los Angeles (34° N, 118° W) and Bombay (19° N, 72.8° E).

CHAPTER REVIEW EXERCISES

In Exercises 1–6, let $\mathbf{v} = \langle -2, 5 \rangle$, *and* $\mathbf{w} = \langle 3, -2 \rangle$.

1. Calculate $5\mathbf{w} - 3\mathbf{v}$ and $5\mathbf{v} - 3\mathbf{w}$.

2. Sketch $\mathbf{v}$, $\mathbf{w}$, and $2\mathbf{v} - 3\mathbf{w}$.

3. Find the unit vector in the direction of $\mathbf{v}$.

4. Find the length of $\mathbf{v} + \mathbf{w}$.

5. Express $\mathbf{i}$ as a linear combination $r\mathbf{v} + s\mathbf{w}$.

6. Find a scalar α such that $\|\mathbf{v} + \alpha\mathbf{w}\| = 6$.

7. If $P = (1, 4)$ and $Q = (-3, 5)$, what are the components of $\overrightarrow{PQ}$? What is the length of $\overrightarrow{PQ}$?

8. Let $A = (2, -1)$, $B = (1, 4)$, and $P = (2, 3)$. Find the point Q such that $\overrightarrow{PQ}$ is equivalent to $\overrightarrow{AB}$. Sketch $\overrightarrow{PQ}$ and $\overrightarrow{AB}$.

9. Find the vector with length 3 making an angle of $\frac{7\pi}{4}$ with the positive x-axis.

10. Calculate $3(\mathbf{i} - 2\mathbf{j}) - 6(\mathbf{i} + 6\mathbf{j})$.

11. Find the value of β for which $\mathbf{w} = \langle -2, \beta \rangle$ is parallel to $\mathbf{v} = \langle 4, -3 \rangle$.

12. Let $P = (1, 4, -3)$.
(a) Find the point Q such that $\overrightarrow{PQ}$ is equivalent to $\langle 3, -1, 5 \rangle$.
(b) Find a unit vector $\mathbf{e}$ equivalent to $\overrightarrow{PQ}$.

13. Let $\mathbf{w} = \langle 2, -2, 1 \rangle$ and $\mathbf{v} = \langle 4, 5, -4 \rangle$. Solve for $\mathbf{u}$ if $\mathbf{v} + 5\mathbf{u} = 3\mathbf{w} - \mathbf{u}$.

14. Let $\mathbf{v} = 3\mathbf{i} - \mathbf{j} + 4\mathbf{k}$. Find the length of $\mathbf{v}$ and the vector $2\mathbf{v} + 3(4\mathbf{i} - \mathbf{k})$.

15. Find a parametrization $\mathbf{r}_1(t)$ of the line passing through $(1, 4, 5)$ and $(-2, 3, -1)$. Then find a parametrization $\mathbf{r}_2(t)$ of the line parallel to $\mathbf{r}_1$ passing through $(1, 0, 0)$.

16. Let $\mathbf{r}_1(t) = \mathbf{v}_1 + t\mathbf{w}_1$ and $\mathbf{r}_2(t) = \mathbf{v}_2 + t\mathbf{w}_2$ be the parametrizations of lines $\mathcal{L}_1$ and $\mathcal{L}_2$. For each statement (a)–(e), either prove it or provide a counterexample showing that it is false.
(a) If $\mathcal{L}_1 = \mathcal{L}_2$, then $\mathbf{v}_1 = \mathbf{v}_2$ and $\mathbf{w}_1 = \mathbf{w}_2$.
(b) If $\mathcal{L}_1 = \mathcal{L}_2$ and $\mathbf{v}_1 = \mathbf{v}_2$, then $\mathbf{w}_1 = \mathbf{w}_2$.
(c) If $\mathcal{L}_1 = \mathcal{L}_2$ and $\mathbf{w}_1 = \mathbf{w}_2$, then $\mathbf{v}_1 = \mathbf{v}_2$.
(d) If $\mathcal{L}_1$ is parallel to $\mathcal{L}_2$, then $\mathbf{w}_1 = \mathbf{w}_2$.
(e) If $\mathcal{L}_1$ is parallel to $\mathcal{L}_2$, then $\mathbf{w}_1 = \lambda\mathbf{w}_2$ for some scalar λ.

17. Find a and b such that the lines $\mathbf{r}_1 = \langle 1, 2, 1 \rangle + t\langle 1, -1, 1 \rangle$ and $\mathbf{r}_2 = \langle 3, -1, 1 \rangle + t\langle a, b, -2 \rangle$ are parallel.

18. Find a such that the lines $\mathbf{r}_1 = \langle 1, 2, 1 \rangle + t\langle 1, -1, 1 \rangle$ and $\mathbf{r}_2 = \langle 3, -1, 1 \rangle + t\langle a, 4, -2 \rangle$ intersect.

19. Sketch the vector sum $\mathbf{v} = \mathbf{v}_1 - \mathbf{v}_2 + \mathbf{v}_3$ for the vectors in Figure 1(A).

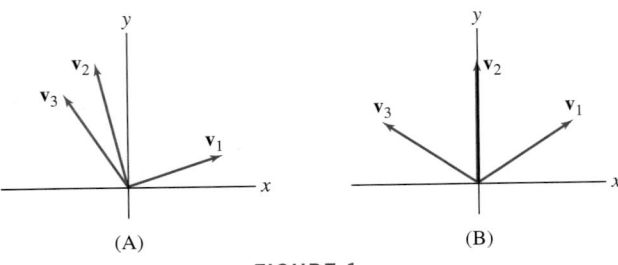

FIGURE 1

20. Sketch the sums $v_1 + v_2 + v_3$, $v_1 + 2v_2$, and $v_2 - v_3$ for the vectors in Figure 1(B).

21. Use vectors to prove that the line connecting the midpoints of two sides of a triangle is parallel to the third side.

In Exercises 22–27, let $\mathbf{v} = \langle 1, 3, -2 \rangle$ *and* $\mathbf{w} = \langle 2, -1, 4 \rangle$.

22. Compute $\mathbf{v} \cdot \mathbf{w}$.

23. Compute the angle between $\mathbf{v}$ and $\mathbf{w}$.

24. Compute $\mathbf{v} \times \mathbf{w}$.

25. Find the area of the parallelogram spanned by $\mathbf{v}$ and $\mathbf{w}$.

26. Find the volume of the parallelepiped spanned by $\mathbf{v}$, $\mathbf{w}$, and $\mathbf{u} = \langle 1, 2, 6 \rangle$.

27. Find all the vectors orthogonal to both $\mathbf{v}$ and $\mathbf{w}$.

28. Let $\mathbf{v} = \langle 1, -1, 3 \rangle$ and $\mathbf{w} = \langle 4, -2, 1 \rangle$.
(a) Find the decomposition $\mathbf{v} = \mathbf{v}_\parallel + \mathbf{v}_\perp$ with respect to $\mathbf{w}$.
(b) Find the decomposition $\mathbf{w} = \mathbf{w}_\parallel + \mathbf{w}_\perp$ with respect to $\mathbf{v}$.

29. A 50-kg wagon is pulled to the right by a force $\mathbf{F}_1$ making an angle of $30°$ with the ground. At the same time the wagon is pulled to the left by a horizontal force $\mathbf{F}_2$.
(a) Find the magnitude of $\mathbf{F}_1$ in terms of the magnitude of $\mathbf{F}_2$ if the wagon does not move.
(b) What is the maximal magnitude of $\mathbf{F}_1$ that can be applied to the wagon without lifting it?

30. Let $\mathbf{v}$, $\mathbf{w}$, and $\mathbf{u}$ be the vectors in $\mathbf{R}^3$. Which of the following is a scalar?
(a) $\mathbf{v} \times (\mathbf{u} + \mathbf{w})$
(b) $(\mathbf{u} + \mathbf{w}) \cdot (\mathbf{v} \times \mathbf{w})$
(c) $(\mathbf{u} \times \mathbf{w}) + (\mathbf{w} - \mathbf{v})$

In Exercises 31–34, let $\mathbf{v} = \langle 1, 2, 4 \rangle$, $\mathbf{u} = \langle 6, -1, 2 \rangle$, *and* $\mathbf{w} = \langle 1, 0, -3 \rangle$. *Calculate the given quantity.*

31. $\mathbf{v} \times \mathbf{w}$

32. $\mathbf{w} \times \mathbf{u}$

33. $\det \begin{pmatrix} \mathbf{u} \\ \mathbf{v} \\ \mathbf{w} \end{pmatrix}$

34. $\mathbf{v} \cdot (\mathbf{u} \times \mathbf{w})$

35. Use the cross product to find the area of the triangle whose vertices are $(1, 3, -1)$, $(2, -1, 3)$, and $(4, 1, 1)$.

36. Calculate $\|\mathbf{v} \times \mathbf{w}\|$ if $\|\mathbf{v}\| = 2$, $\mathbf{v} \cdot \mathbf{w} = 3$, and the angle between $\mathbf{v}$ and $\mathbf{w}$ is $\frac{\pi}{6}$.

37. Show that if the vectors $\mathbf{v}$, $\mathbf{w}$ are orthogonal, then $\|\mathbf{v} + \mathbf{w}\|^2 = \|\mathbf{v}\|^2 + \|\mathbf{w}\|^2$.

38. Find the angle between $\mathbf{v}$, $\mathbf{w}$ if $\|\mathbf{v} + \mathbf{w}\| = \|\mathbf{v}\| = \|\mathbf{w}\|$.

39. Show that the equation $\langle 1, 2, 3 \rangle \times \mathbf{v} = \langle -1, 2, a \rangle$ has no solution for $a \ne -1$.

40. Prove with a diagram the following: If $\mathbf{e}$ is a unit vector orthogonal to $\mathbf{v}$, then $\mathbf{e} \times (\mathbf{v} \times \mathbf{e}) = (\mathbf{e} \times \mathbf{v}) \times \mathbf{e} = \mathbf{v}$.

41. Use the identity

$$\mathbf{u} \times (\mathbf{v} \times \mathbf{w}) = (\mathbf{u} \cdot \mathbf{w})\,\mathbf{v} - (\mathbf{u} \cdot \mathbf{v})\,\mathbf{w}$$

to prove that

$$\mathbf{u} \times (\mathbf{v} \times \mathbf{w}) + \mathbf{v} \times (\mathbf{w} \times \mathbf{u}) + \mathbf{w} \times (\mathbf{u} \times \mathbf{v}) = 0$$

42. Find an equation of the plane through $(1, -3, 5)$ with normal vector $\mathbf{n} = \langle 2, 1, -4 \rangle$.

43. Write the equation of the plane $\mathcal{P}$ with vector equation

$$\langle 1, 4, -3 \rangle \cdot \langle x, y, z \rangle = 7$$

in the form

$$a\,(x - x_0) + b\,(y - y_0) + c\,(z - z_0) = 0$$

Hint: You must find a point $P = (x_0, y_0, z_0)$ on $\mathcal{P}$.

44. Find all the planes parallel to the plane passing through the points $(1, 2, 3)$, $(1, 2, 7)$, and $(1, 1, -3)$.

45. Find the plane through $P = (4, -1, 9)$ containing the line $\mathbf{r}(t) = \langle 1, 4, -3 \rangle + t\langle 2, 1, 1 \rangle$.

46. Find the intersection of the line $\mathbf{r}(t) = \langle 3t + 2, 1, -7t \rangle$ and the plane $2x - 3y + z = 5$.

47. Find the trace of the plane $3x - 2y + 5z = 4$ in the xy-plane.

48. Find the intersection of the planes $x + y + z = 1$ and $3x - 2y + z = 5$.

In Exercises 49–54, determine the type of the quadric surface.

49. $\left(\dfrac{x}{3}\right)^2 + \left(\dfrac{y}{4}\right)^2 + 2z^2 = 1$

50. $\left(\dfrac{x}{3}\right)^2 - \left(\dfrac{y}{4}\right)^2 + 2z^2 = 1$

51. $\left(\dfrac{x}{3}\right)^2 + \left(\dfrac{y}{4}\right)^2 - 2z = 0$

52. $\left(\dfrac{x}{3}\right)^2 - \left(\dfrac{y}{4}\right)^2 - 2z = 0$

53. $\left(\dfrac{x}{3}\right)^2 - \left(\dfrac{y}{4}\right)^2 - 2z^2 = 0$

54. $\left(\dfrac{x}{3}\right)^2 - \left(\dfrac{y}{4}\right)^2 - 2z^2 = 1$

55. Determine the type of the quadric surface $ax^2 + by^2 - z^2 = 1$ if:

(a) $a < 0, \quad b < 0$

(b) $a > 0, \quad b > 0$

(c) $a > 0, \quad b < 0$

56. Describe the traces of the surface

$$\left(\dfrac{x}{2}\right)^2 - y^2 + \left(\dfrac{z}{2}\right)^2 = 1$$

in the three coordinate planes.

57. Convert $(x, y, z) = (3, 4, -1)$ from rectangular to cylindrical and spherical coordinates.

58. Convert $(r, \theta, z) = \left(3, \frac{\pi}{6}, 4\right)$ from cylindrical to spherical coordinates.

59. Convert the point $(\rho, \theta, \phi) = \left(3, \frac{\pi}{6}, \frac{\pi}{3}\right)$ from spherical to cylindrical coordinates.

60. Describe the set of all points $P = (x, y, z)$ satisfying $x^2 + y^2 \le 4$ in both cylindrical and spherical coordinates.

61. Sketch the graph of the cylindrical equation $z = 2r \cos \theta$ and write the equation in rectangular coordinates.

62. Write the surface $x^2 + y^2 - z^2 = 2(x + y)$ as an equation $r = f(\theta, z)$ in cylindrical coordinates.

63. Show that the cylindrical equation

$$r^2(1 - 2\sin^2 \theta) + z^2 = 1$$

is a hyperboloid of one sheet.

64. Sketch the graph of the spherical equation $\rho = 2 \cos \theta \sin \phi$ and write the equation in rectangular coordinates.

65. Describe how the surface with spherical equation

$$\rho^2(1 + A \cos^2 \phi) = 1$$

depends on the constant A.

66. Show that the spherical equation $\cot \phi = 2 \cos \theta + \sin \theta$ defines a plane through the origin (with the origin excluded). Find a normal vector to this plane.

67. Let c be a scalar, let $\mathbf{a}$ and $\mathbf{b}$ be vectors, and let $\mathbf{X} = \langle x, y, z \rangle$. Show that the equation $(\mathbf{X} - \mathbf{a})(\mathbf{X} - \mathbf{b}) = c^2$ defines a sphere with center $\mathbf{m} = \frac{1}{2}(\mathbf{a} + \mathbf{b})$ and radius R, where $R^2 = c^2 + \|\frac{1}{2}(\mathbf{a} - \mathbf{b})\|^2$.

13 | CALCULUS OF VECTOR-VALUED FUNCTIONS

The mathematics of space flight is based on vector calculus. In this photo, Astronauts Piers Sellers and Michael Fossum work on the space shuttle's Remote Manipulator System.

In this chapter, we study vector-valued functions and their derivatives, and use them to analyze curves and motion in three-space. Although many techniques from single-variable calculus carry over to the vector setting, there is an important new aspect to the derivative. A real-valued function $f(x)$ can change in just one of two ways: it can increase or decrease. By contrast, a vector-valued function can change, not just in magnitude but also in direction, and the rate of change is not a single number but is itself a vector. To develop these new concepts, we begin with an introduction to vector-valued functions.

13.1 Vector-Valued Functions

Consider a particle moving in $\mathbf{R}^3$ and suppose that its coordinates at time t are $(x(t), y(t), z(t))$. It is convenient to represent the particle's path by the **vector-valued function**

Functions $f(x)$ (with real number values) are often called scalar-valued, *to distinguish them from vector-valued functions.*

$$\boxed{\mathbf{r}(t) = \langle x(t), y(t), z(t)\rangle = x(t)\mathbf{i} + y(t)\mathbf{j} + z(t)\mathbf{k}} \quad \boxed{1}$$

The vector $\mathbf{r}(t)$ points from the origin to the position of the particle at time t (Figure 1).

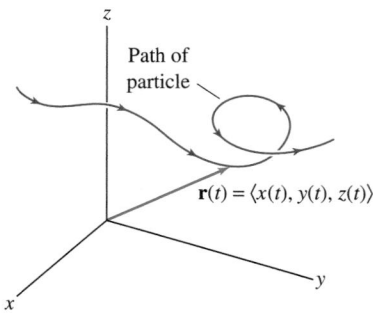

FIGURE 1

The letter t is a convenient choice for the parameter (which often represents time), but we are free to use any other variable such as s or θ. It is best to avoid writing $\mathbf{r}(x)$ or $\mathbf{r}(y)$ to prevent confusion with the x- and y-components of $\mathbf{r}$.

More generally, a vector-valued function is any function $\mathbf{r}(t)$ of the form (1) whose domain $\mathcal{D}$ is a set of real numbers and whose range is a set of position vectors. The variable t is called a **parameter**, and the functions $x(t), y(t), z(t)$ are called the **components** or **coordinate functions**. We usually take as domain the set of all values of t for which $\mathbf{r}(t)$ is defined, that is, all values of t that belong to the domains of all three coordinate functions $x(t), y(t), z(t)$. For example,

$$\mathbf{r}(t) = \langle t^2, e^t, 4 - 7t\rangle \quad \text{domain } \mathcal{D} = \mathbf{R}$$

$$\mathbf{r}(s) = \langle \sqrt{s}, e^s, s^{-1}\rangle \quad \text{domain } \mathcal{D} = \{s \in \mathbf{R} : s > 0\}$$

If the components of $\mathbf{r}(t)$ are continuous functions, then the terminal point of $\mathbf{r}(t)$ traces a path in $\mathbf{R}^3$ as t varies, and we refer to $\mathbf{r}(t)$ as a **vector parametrization** of this path. The set of all points $(x(t), y(t), z(t))$ for t in the domain of $\mathbf{r}$ is called a **space curve** $\mathcal{C}$.

The description of a path by a vector-valued function $\mathbf{r}(t) = \langle x(t), y(t) \rangle$ is equivalent to the description as a parametrized curve $c(t) = (x(t), y(t))$ that we considered in Chapter 11. The difference lies only in whether we visualize the path as traced by a "moving vector" $\mathbf{r}(t)$ or a "moving point" $c(t)$. The vector form $\mathbf{r}(t)$ is more natural in the context of vector-valued derivatives, and we will use it rather than $c(t)$ in this chapter.

We treated a special case of vector-valued functions in Chapter 12 when we studied vector parametrizations of lines. Recall that the line through $P_0 = (x_0, y_0, z_0)$ with direction vector $\mathbf{v} = \langle a, b, c \rangle$ has the vector parametrization

$$\mathbf{r}(t) = \langle x_0, y_0, z_0 \rangle + t\mathbf{v} = \langle x_0 + ta, y_0 + tb, z_0 + tc \rangle$$

In some cases, it is helpful to consider the projections onto the coordinate planes. The projection of $\mathbf{r}(t) = \langle x(t), y(t), z(t) \rangle$ onto the xy-plane is the path $\mathbf{p}(t) = \langle x(t), y(t), 0 \rangle$ (Figure 2). Similarly, the projections onto the yz- and xz-planes are the paths $\langle 0, y(t), z(t) \rangle$ and $\langle x(t), 0, z(t) \rangle$, respectively.

■ **EXAMPLE 1** Helix   The curve traced by $\mathbf{r}(t) = \langle -\sin t, \cos t, t \rangle$ is a helix. Describe this curve for $t \geq 0$ and its projections onto the coordinate planes.

Solution The projection onto the xy-plane is the path $\mathbf{p}(t) = \langle -\sin t, \cos t, 0 \rangle$, which describes a point moving counterclockwise around the unit circle with $\mathbf{p}(0) = (0, 1, 0)$. The function $\mathbf{r}(t)$ itself describes a point whose projection traces out the circle while the height $z = t$ increases linearly with time, resulting in the helix of Figure 2. The

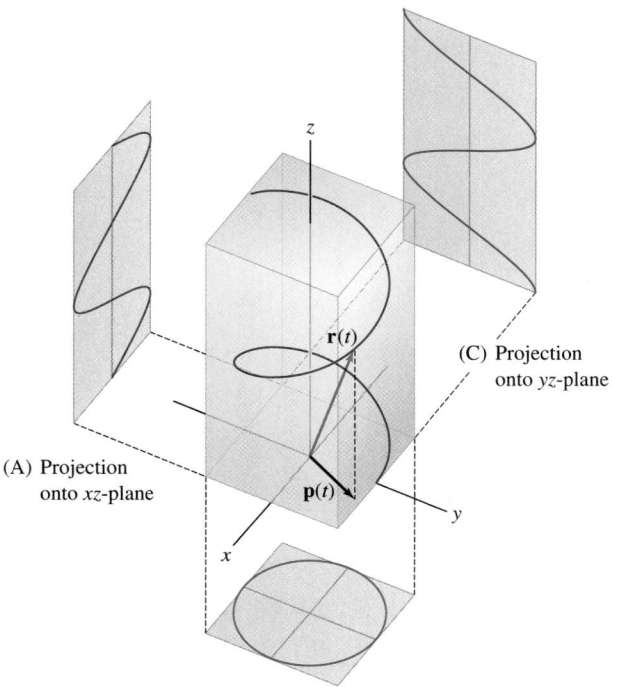

(C) Projection onto yz-plane

(A) Projection onto xz-plane

(B) Projection onto xy-plane

FIGURE 2 Projections of the helix $\mathbf{r}(t) = \langle \cos t, \sin t, t \rangle$.

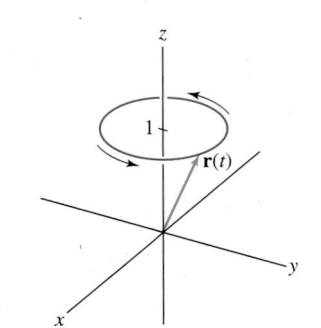

FIGURE 3 Plot of $\mathbf{r}(t) = \langle \cos t, \sin t, 1 \rangle$.

projection onto the xz-plane is the path $\langle -\sin t, 0, t \rangle$, which is a wave moving in the positive z-direction. Similarly, the projection onto the yz-plane is the wave $\langle 0, \cos t, t \rangle$. ■

It is important to distinguish between the path parametrized by $\mathbf{r}(t)$ and the underlying space curve $\mathcal{C}$. The path is a particular way of traversing the curve; it may traverse the curve several times, reverse direction, or move back and forth, etc. For example, the path $\mathbf{r}(t) = \langle \cos t, \sin t, 1 \rangle$ moves around a unit circle at height $z = 1$ infinitely many times as t varies from $-\infty$ to ∞ (Figure 3).

In general, space curves may be quite complicated and difficult to sketch by hand. Fortunately, computers produce good graphics from different perspectives (Figure 4). As an aid to visualization, it is helpful to plot a "thickened" curve as in Figures 4 and 5, but keep in mind that space curves are one-dimensional and have no thickness.

FIGURE 4 The curve
$\mathbf{r}(t) = \langle t \sin t \cos t, t \sin^2 t, t \cos t \rangle$ for
$\pi \le t \le 4\pi$, seen from three different viewpoints.

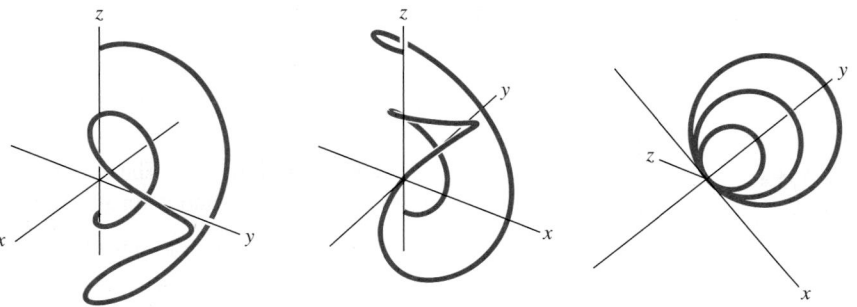

Keep in mind that every curve can be parametrized in infinitely many ways. The next example describes two very different parametrizations of the same curve.

■ **EXAMPLE 2** Parametrizing the Intersection of Surfaces Parametrize the intersection of the surfaces $x^2 - y^2 = z - 1$ and $x^2 + y^2 = 4$ (Figure 5).

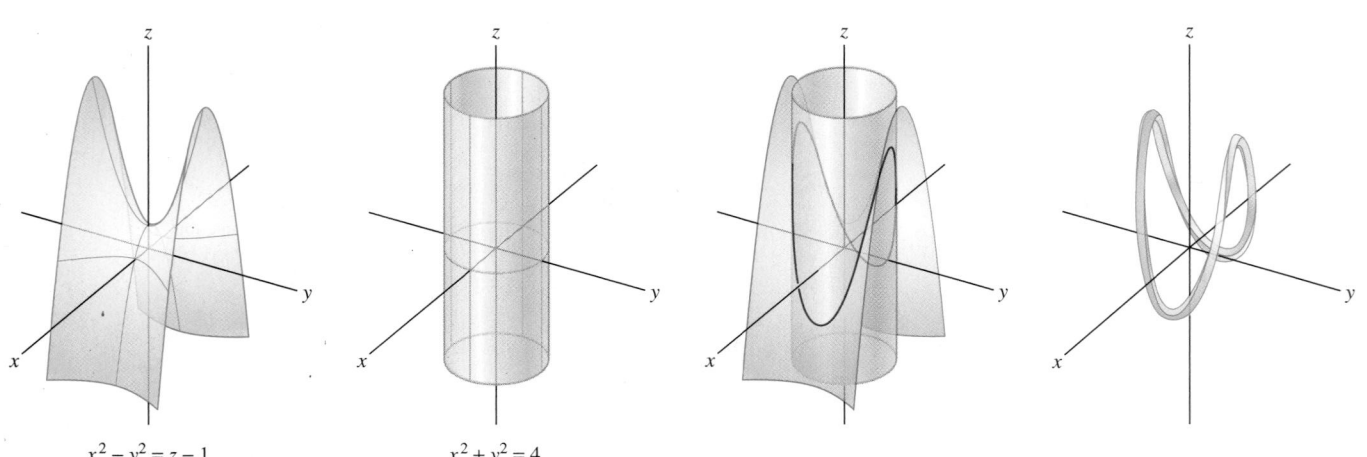

$$x^2 - y^2 = z - 1 \qquad\qquad x^2 + y^2 = 4$$

FIGURE 5 Intersection of surfaces $x^2 - y^2 = z - 1$ and $x^2 + y^2 = 4$.

Solution Our goal is to express the coordinates of a point on the curve as functions of a parameter t. We solve this problem in two ways. First, let's solve for y and z in terms of x. The two equations may be rewritten $y^2 = 4 - x^2$ and $z = x^2 - y^2 + 1$. Thus,

$$y = \pm\sqrt{4 - x^2}, \qquad z = x^2 - y^2 + 1 = x^2 - (4 - x^2) + 1 = 2x^2 - 3$$

Taking $t = x$ as parameter, we have $y = \pm\sqrt{4 - t^2}$, $z = 2t^2 - 3$. The two signs of the square root correspond to the two halves of the curve where $y > 0$ and $y < 0$ (Figure 6). Therefore, we need two vector-valued functions to parametrize the entire curve:

$$\mathbf{r}_1(t) = \left\langle t, \sqrt{4 - t^2}, 2t^2 - 3 \right\rangle, \quad \mathbf{r}_2(t) = \left\langle t, -\sqrt{4 - t^2}, 2t^2 - 3 \right\rangle, \qquad -2 \le t \le 2$$

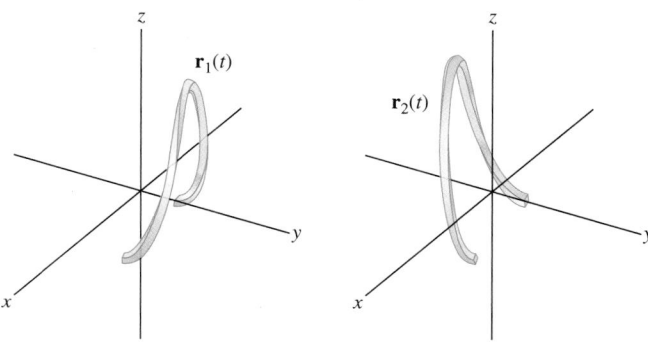

Part of curve where $y > 0$ Part of curve where $y < 0$

FIGURE 6 Two halves of the curve of intersection in Example 2.

A second and perhaps better parametrization is obtained by noting that the equation $x^2 + y^2 = 4$ has a trigonometric parametrization: $x = 2\cos t$, $y = 2\sin t$. The second equation yields

$$z = x^2 - y^2 + 1 = 4\cos^2 t - 4\sin^2 t + 1 = 4\cos 2t + 1$$

Thus, we may parametrize the entire curve by a single vector-valued function:

$$\mathbf{r}(t) = \langle 2\cos t, 2\sin t, 4\cos 2t + 1 \rangle, \qquad 0 \le t < 2\pi \qquad\blacksquare$$

■ **EXAMPLE 3** Parametrize the circle of radius 3 with center $P = (2, 6, 8)$ located in a plane:

(a) Parallel to the xy-plane   **(b)** Parallel to the xz-plane

Solution A circle of radius R in the xy-plane with center at the origin has parametrization $\langle R\cos t, R\sin t \rangle$. To place the circle in a three-dimensional coordinate system, we use the parametrization $\langle R\cos t, R\sin t, 0 \rangle$.

(a) The circle of radius 3 centered at the origin in the xy-plane has parametrization $\langle 3\cos t, 3\sin t, 0 \rangle$. To move the circle in a parallel fashion so that its center lies at $P = (2, 6, 8)$, we translate by the vector $\langle 2, 6, 8 \rangle$:

$$\mathbf{r}_1(t) = \langle 2, 6, 8 \rangle + \langle 3\cos t, 3\sin t, 0 \rangle = \langle 2 + 3\cos t, 6 + 3\sin t, 8 \rangle$$

(b) The parametrization $\langle 3\cos t, 0, 3\sin t \rangle$ gives us a circle of radius 3 centered at the origin in the xz-plane. To move the circle in a parallel fashion so that its center lies at $(2, 6, 8)$, we translate by the vector $\langle 2, 6, 8 \rangle$:

$$\mathbf{r}_2(t) = \langle 2, 6, 8 \rangle + \langle 3\cos t, 0, 3\sin t \rangle = \langle 2 + 3\cos t, 6, 8 + 3\sin t \rangle$$

These two circles are shown in Figure 7.   ■

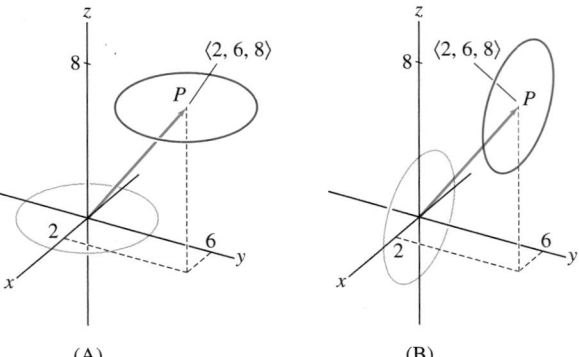

FIGURE 7 Horizontal and vertical circles of radius 3 and center $P = (2, 6, 8)$ obtained by translation.

(A)   (B)

13.1 SUMMARY

- A *vector-valued function* is a function of the form

$$\mathbf{r}(t) = \langle x(t), y(t), z(t) \rangle = x(t)\mathbf{i} + y(t)\mathbf{j} + z(t)\mathbf{k}$$

We often think of the parameter t as time and $\mathbf{r}(t)$ as a "moving vector" whose terminal point traces out a path as a function of time. We call $\mathbf{r}(t)$ a *vector parametrization* of the path.
- If the components $x(t)$, $y(t)$, and $z(t)$ are continuous, the set of points $(x(t), y(t), z(t))$ in $\mathbf{R}^3$ for t in the domain of $\mathbf{r}(t)$ is called a *space curve*.
- Every space curve can be parametrized in infinitely many ways.
- The projection of $\mathbf{r}(t)$ onto the xy-plane is the curve traced by $\langle x(t), y(t), 0 \rangle$. The projection onto the xz-plane is $\langle x(t), 0, z(t) \rangle$ and the projection onto the yz-plane is $\langle 0, y(t), z(t) \rangle$.

13.1 EXERCISES

Preliminary Questions

1. Which one of the following does *not* parametrize a line?
(a) $\mathbf{r}_1(t) = \langle 8 - t, 2t, 3t \rangle$
(b) $\mathbf{r}_2(t) = t^3\mathbf{i} - 7t^3\mathbf{j} + t^3\mathbf{k}$
(c) $\mathbf{r}_3(t) = \langle 8 - 4t^3, 2 + 5t^2, 9t^3 \rangle$

2. What is the projection of $\mathbf{r}(t) = t\mathbf{i} + t^4\mathbf{j} + e^t\mathbf{k}$ onto the xz-plane?

3. Which projection of $\langle \cos t, \cos 2t, \sin t \rangle$ is a circle?

4. What is the center of the circle with parametrization

$$\mathbf{r}(t) = (-2 + \cos t)\mathbf{i} + 2\mathbf{j} + (3 - \sin t)\mathbf{k}?$$

5. How do the paths $\mathbf{r}_1(t) = \langle \cos t, \sin t \rangle$ and $\mathbf{r}_2(t) = \langle \sin t, \cos t \rangle$ around the unit circle differ?

6. Which three of the following vector-valued functions parametrize the same space curve?
(a) $(-2 + \cos t)\mathbf{i} + 9\mathbf{j} + (3 - \sin t)\mathbf{k}$
(b) $(2 + \cos t)\mathbf{i} - 9\mathbf{j} + (-3 - \sin t)\mathbf{k}$
(c) $(-2 + \cos 3t)\mathbf{i} + 9\mathbf{j} + (3 - \sin 3t)\mathbf{k}$
(d) $(-2 - \cos t)\mathbf{i} + 9\mathbf{j} + (3 + \sin t)\mathbf{k}$
(e) $(2 + \cos t)\mathbf{i} + 9\mathbf{j} + (3 + \sin t)\mathbf{k}$

Exercises

1. What is the domain of $\mathbf{r}(t) = e^t\mathbf{i} + \frac{1}{t}\mathbf{j} + (t + 1)^{-3}\mathbf{k}$?

2. What is the domain of $\mathbf{r}(s) = e^s\mathbf{i} + \sqrt{s}\mathbf{j} + \cos s\mathbf{k}$?

3. Find a vector parametrization of the line through $P = (3, -5, 7)$ in the direction $\mathbf{v} = \langle 3, 0, 1 \rangle$.

4. Find a direction vector for the line with parametrization $\mathbf{r}(t) = (4 - t)\mathbf{i} + (2 + 5t)\mathbf{j} + \frac{1}{2}t\mathbf{k}$.

5. Match the space curves in Figure 8 with their projections onto the xy-plane in Figure 9.

6. Match the space curves in Figure 8 with the following vector-valued functions:

(a) $\mathbf{r}_1(t) = \langle \cos 2t, \cos t, \sin t \rangle$

(b) $\mathbf{r}_2(t) = \langle t, \cos 2t, \sin 2t \rangle$

(c) $\mathbf{r}_3(t) = \langle 1, t, t \rangle$

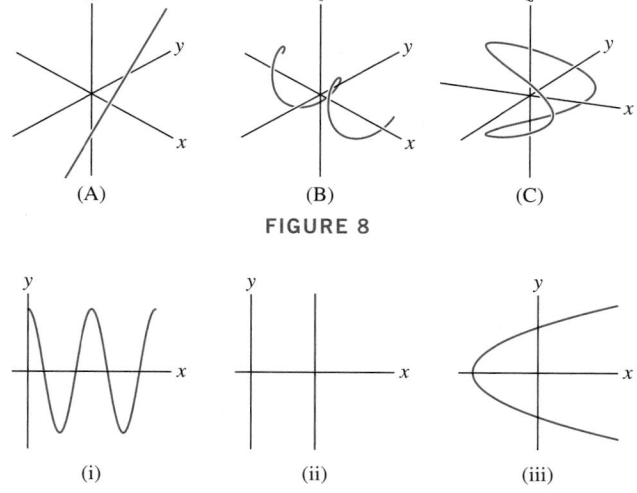

(A)   (B)   (C)

FIGURE 8

(i)   (ii)   (iii)

FIGURE 9

7. Match the vector-valued functions (a)–(f) with the space curves (i)–(vi) in Figure 10.

(a) $\mathbf{r}(t) = \langle t + 15, e^{0.08t} \cos t, e^{0.08t} \sin t \rangle$

(b) $\mathbf{r}(t) = \langle \cos t, \sin t, \sin 12t \rangle$

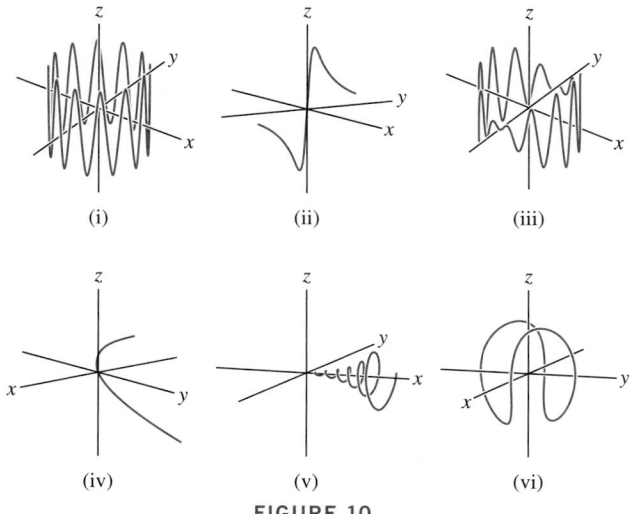

(i)   (ii)   (iii)

(iv)   (v)   (vi)

FIGURE 10

(c) $\mathbf{r}(t) = \left\langle t, t, \dfrac{25t}{1 + t^2} \right\rangle$

(d) $\mathbf{r}(t) = \langle \cos^3 t, \sin^3 t, \sin 2t \rangle$

(e) $\mathbf{r}(t) = \langle t, t^2, 2t \rangle$

(f) $\mathbf{r}(t) = \langle \cos t, \sin t, \cos t \sin 12t \rangle$

8. Which of the following curves have the same projection onto the xy-plane?

(a) $\mathbf{r}_1(t) = \langle t, t^2, e^t \rangle$

(b) $\mathbf{r}_2(t) = \langle e^t, t^2, t \rangle$

(c) $\mathbf{r}_3(t) = \langle t, t^2, \cos t \rangle$

9. Match the space curves (A)–(C) in Figure 11 with their projections (i)–(iii) onto the xy-plane.

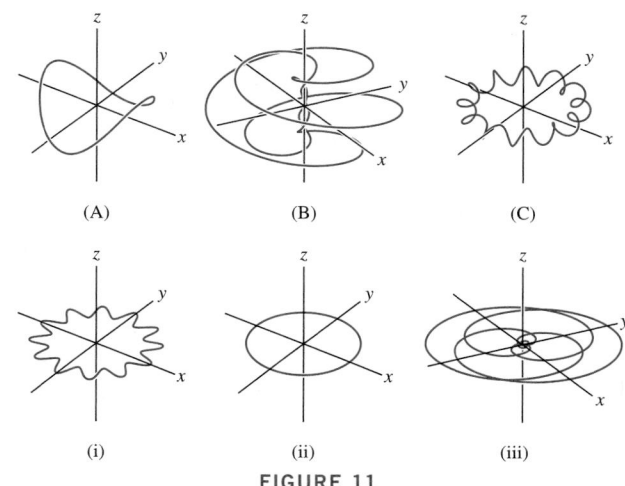

(A)   (B)   (C)

(i)   (ii)   (iii)

FIGURE 11

In Exercises 10–13, the function $\mathbf{r}(t)$ traces a circle. Determine the radius, center, and plane containing the circle.

10. $\mathbf{r}(t) = (9 \cos t)\mathbf{i} + (9 \sin t)\mathbf{j}$

11. $\mathbf{r}(t) = 7\mathbf{i} + (12 \cos t)\mathbf{j} + (12 \sin t)\mathbf{k}$

12. $\mathbf{r}(t) = \langle \sin t, 0, 4 + \cos t \rangle$

13. $\mathbf{r}(t) = \langle 6 + 3 \sin t, 9, 4 + 3 \cos t \rangle$

14. Describe the projections of the circle $\mathbf{r}(t) = \langle \sin t, 0, 4 + \cos t \rangle$ onto the coordinate planes.

15. Do either of $P = (4, 11, 20)$ or $Q = (-1, 6, 16)$ lie on the curve $\mathbf{r}(t) = \langle 1 + t, 2 + t^2, t^4 \rangle$?

16. (a) Describe the curve $\mathbf{r}(t) = \langle t \cos t, t \sin t, t \rangle$ and its projections onto the xy- and xz-planes.

(b) $\boxed{CAS}$ Plot $\mathbf{r}(t)$ with a computer algebra system if you have one.

17. Find the points where the path $\mathbf{r}(t) = \langle \sin t, \cos t, \sin t \cos 2t \rangle$ intersects the xy-plane.

18. Parametrize the intersection of the surfaces

$$y^2 - z^2 = x - 2, \qquad y^2 + z^2 = 9$$

using $t = y$ as the parameter (two vector functions are needed as in Example 2).

19. Find a parametrization of the curve in Exercise 18 using trigonometric functions.

20. Viviani's Curve $\mathcal{C}$ is the intersection of the surfaces $x^2 + y^2 = z^2$, $y = z^2$ (Figure 12).

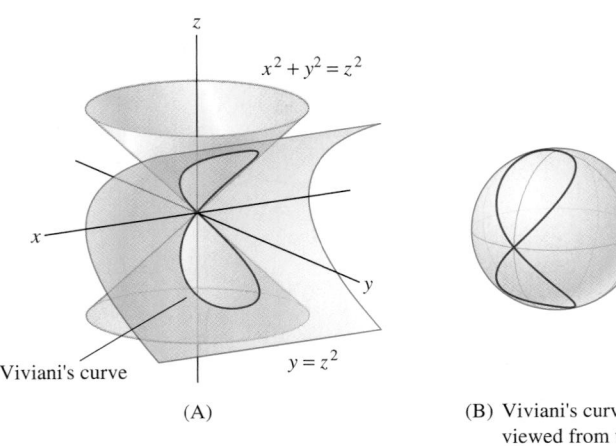

Viviani's curve

(A)

(B) Viviani's curve viewed from the negative y-axis

FIGURE 12 Viviani's curve is the intersection of the surfaces $x^2 + y^2 = z^2$ and $y = z^2$.

(a) Parametrize each of the two parts of $\mathcal{C}$ corresponding to $x \geq 0$ and $x \leq 0$ taking $t = z$ as parameter.

(b) Describe the projection of $\mathcal{C}$ onto the xy-plane.

(c) Show that $\mathcal{C}$ lies on the sphere of radius 1 with center $(0, 1, 0)$. This curve looks like a figure eight lying on a sphere [Figure 12(B)].

21. Show that any point on $x^2 + y^2 = z^2$ can be written in the form $(z \cos \theta, z \sin \theta, z)$ for some θ. Use this to find a parametrization of Viviani's curve (Exercise 20) with θ as parameter.

22. Use sine and cosine to parametrize the intersection of the cylinders $x^2 + y^2 = 1$ and $x^2 + z^2 = 1$ (use two vector-valued functions). Then describe the projections of this curve on the three coordinate planes.

23. $\mathcal{CAS}$ Use sine and cosine to parametrize the intersection of the surfaces $x^2 + y^2 = 1$ and $z = 4x^2$, and plot this curve using a CAS (Figure 13).

24. Use hyperbolic functions to parametrize the intersection of the surfaces $x^2 - y^2 = 4$, $z = xy$.

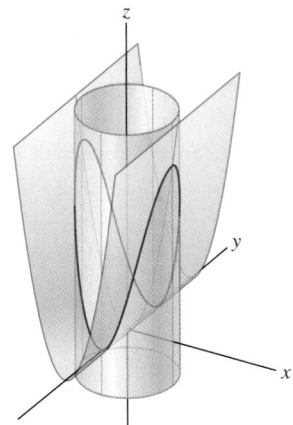

FIGURE 13 Intersection of the surfaces $x^2 + y^2 = 1$ and $z = 4x^2$.

In Exercises 25–34, find a parametrization of the curve.

25. The vertical line passing through the point $(3, 2, 0)$

26. The line passing through $(1, 0, 4)$ and $(4, 1, 2)$

27. The line through the origin whose projection on the xy-plane is a line of slope 3 and on the yz-plane is a line of slope 5 (i.e., $\Delta y / \Delta z = 5$)

28. The horizontal circle of radius 1 with center $(2, -1, 4)$

29. The circle of radius 2 with center $(1, 2, 5)$ in a plane parallel to the yz-plane

30. The ellipse $\left(\dfrac{x}{2}\right)^2 + \left(\dfrac{y}{3}\right)^2 = 1$ in the xy-plane, translated to have center $(9, -4, 0)$

31. The intersection of the plane $y = \frac{1}{2}$ with the sphere $x^2 + y^2 + z^2 = 1$

32. The intersection of the surfaces

$$z = x^2 - y^2 \qquad \text{and} \qquad z = x^2 + xy - 1$$

33. The ellipse $\left(\dfrac{x}{2}\right)^2 + \left(\dfrac{z}{3}\right)^2 = 1$ in the xz-plane, translated to have center $(3, 1, 5)$ [Figure 14(A)]

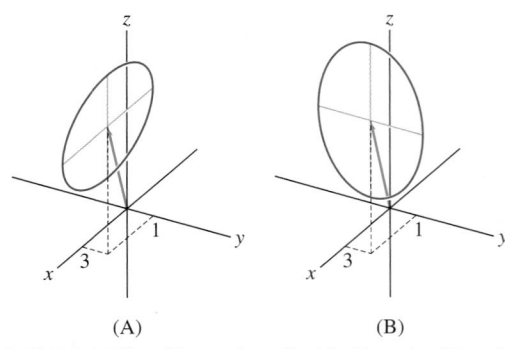

(A)

(B)

FIGURE 14 The ellipses described in Exercise 33 and 34.

34. The ellipse $\left(\dfrac{y}{2}\right)^2 + \left(\dfrac{z}{3}\right)^2 = 1$, translated to have center $(3, 1, 5)$ [Figure 14(B)]

In Exercises 35–37, assume that two paths $\mathbf{r}_1(t)$ and $\mathbf{r}_2(t)$ intersect if there is a point P lying on both curves. We say that $\mathbf{r}_1(t)$ and $\mathbf{r}_2(t)$ collide if $\mathbf{r}_1(t_0) = \mathbf{r}_2(t_0)$ at some time t_0.

35. Which of the following are true?
(a) If $\mathbf{r}_1$ and $\mathbf{r}_2$ intersect, then they collide.
(b) If $\mathbf{r}_1$ and $\mathbf{r}_2$ collide, then they intersect.
(c) Intersection depends only on the underlying curves traced by $\mathbf{r}_1$ and $\mathbf{r}_2$ but collision depends on the actual parametrizations.

36. Determine whether $\mathbf{r}_1$ and $\mathbf{r}_2$ collide or intersect:

$$\mathbf{r}_1(t) = \langle t^2 + 3, t + 1, 6t^{-1}\rangle$$
$$\mathbf{r}_2(t) = \langle 4t, 2t - 2, t^2 - 7\rangle$$

37. Determine whether $\mathbf{r}_1$ and $\mathbf{r}_2$ collide or intersect:

$$\mathbf{r}_1(t) = \langle t, t^2, t^3\rangle, \qquad \mathbf{r}_2(t) = \langle 4t + 6, 4t^2, 7 - t\rangle$$

Further Insights and Challenges

38. Sketch the curve parametrized by $\mathbf{r}(t) = \langle |t| + t, |t| - t\rangle$.

39. Find the maximum height above the xy-plane of a point on $\mathbf{r}(t) = \langle e^t, \sin t, t(4 - t)\rangle$.

40. Let $\mathcal{C}$ be the curve obtained by intersecting a cylinder of radius r and a plane. Insert two spheres of radius r into the cylinder above and below the plane, and let F_1 and F_2 be the points where the plane is tangent to the sphere [Figure 15(A)]. Let K be the vertical distance between the equators of the two spheres. Rediscover Archimedes's proof that $\mathcal{C}$ is an ellipse by showing that every point

P on $\mathcal{C}$ satisfies

$$PF_1 + PF_2 = K \qquad \boxed{2}$$

Hint: If two lines through a point P are tangent to a sphere and intersect the sphere at Q_1 and Q_2 as in Figure 15(B), then the segments $\overline{PQ_1}$ and $\overline{PQ_2}$ have equal length. Use this to show that $PF_1 = PR_1$ and $PF_2 = PR_2$.

41. *CAS* Now reprove the result of Exercise 40 using vector geometry. Assume that the cylinder has equation $x^2 + y^2 = r^2$ and the plane has equation $z = ax + by$.
(a) Show that the upper and lower spheres in Figure 15 have centers

$$C_1 = \left(0, 0, r\sqrt{a^2 + b^2 + 1}\right)$$
$$C_2 = \left(0, 0, -r\sqrt{a^2 + b^2 + 1}\right)$$

(b) Show that the points where the plane is tangent to the sphere are

$$F_1 = \frac{r}{\sqrt{a^2 + b^2 + 1}} \langle a, b, a^2 + b^2\rangle$$
$$F_2 = \frac{-r}{\sqrt{a^2 + b^2 + 1}} \langle a, b, a^2 + b^2\rangle$$

Hint: Show that $\overline{C_1 F_1}$ and $\overline{C_2 F_2}$ have length r and are orthogonal to the plane.

(c) Verify, with the aid of a computer algebra system, that Eq. (2) holds with $K = 2r\sqrt{a^2 + b^2 + 1}$. To simplify the algebra, observe that since a and b are arbitrary, it suffices to verify Eq. (2) for the point $P = (r, 0, ar)$.

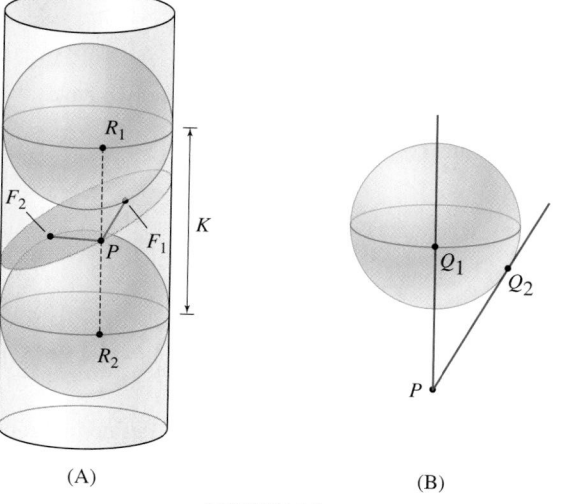

(A)   (B)

FIGURE 15

13.2 Calculus of Vector-Valued Functions

In this section, we extend differentiation and integration to vector-valued functions and we discuss the geometric interpretation of the derivative as a tangent vector. The first step is to define limits of vector-valued functions.

FIGURE 1 The vector-valued function $\mathbf{r}(t)$ approaches $\mathbf{u}$ as $t \to t_0$.

> **DEFINITION Limit of a Vector-Valued Function**   A vector-valued function $\mathbf{r}(t)$ approaches the limit $\mathbf{u}$ (a vector) as t approaches t_0 if $\lim_{t \to t_0} \|\mathbf{r}(t) - \mathbf{u}\| = 0$. In this case, we write
>
> $$\lim_{t \to t_0} \mathbf{r}(t) = \mathbf{u}$$

We may visualize the limit of a vector-valued function as a vector $\mathbf{r}(t)$ "moving" toward the limit $\mathbf{v}$ (Figure 1). According to the following theorem, vector limits may be computed componentwise.

> **THEOREM 1 Vector-Valued Limits Are Computed Componentwise**   A vector-valued function $\mathbf{r}(t) = \langle x(t), y(t), z(t) \rangle$ approaches a limit as $t \to t_0$ if and only if each component approaches a limit, and in this case,
>
> $$\lim_{t \to t_0} \mathbf{r}(t) = \left\langle \lim_{t \to t_0} x(t), \lim_{t \to t_0} y(t), \lim_{t \to t_0} z(t) \right\rangle \qquad \boxed{1}$$

The Limit Laws of scalar functions remain valid in the vector-valued case. They are verified by applying the Limit Laws to the components.

Proof   Suppose that $\lim_{t \to t_0} \mathbf{r}(t) = \mathbf{v}$, where $\mathbf{v} = \langle a, b, c \rangle$. Then the length $\|\mathbf{r}(t) - \mathbf{v}\|$ tends to zero and its square

$$\|\mathbf{r}(t) - \mathbf{v}\|^2 = (x(t) - a)^2 + (y(t) - b)^2 + (z(t) - c)^2 \qquad \boxed{2}$$

also tends to zero as $t \to t_0$. Since the terms on the right-hand side of Eq. (2) are non-negative, each term individually tends to zero. Thus,

$$\left\langle \lim_{t \to t_0} x(t), \lim_{t \to t_0} y(t), \lim_{t \to t_0} z(t) \right\rangle = \langle a, b, c \rangle = \mathbf{v} \qquad \boxed{3}$$

Conversely, if (3) holds, then $\|\mathbf{r}(t) - \mathbf{v}\|^2$ tends to zero by Eq. (2) and we have

$$\lim_{t \to t_0} \mathbf{r}(t) = \mathbf{v} \qquad \blacksquare$$

■ **EXAMPLE 1**   Calculate $\lim_{t \to 3} \mathbf{r}(t)$, where $\mathbf{r}(t) = \langle t^2, 1 - t, t^{-1} \rangle$.

Solution   By Theorem 1,

$$\lim_{t \to 3} \mathbf{r}(t) = \lim_{t \to 3} \langle t^2, 1 - t, t^{-1} \rangle = \left\langle \lim_{t \to 3} t^2, \lim_{t \to 3} (1 - t), \lim_{t \to 3} t^{-1} \right\rangle = \left\langle 9, -2, \frac{1}{3} \right\rangle \quad ■$$

Continuity and the derivative for vector-valued functions are defined in much the same way as in the scalar case. A vector-valued function $\mathbf{r}(t) = \langle x(t), y(t), z(t) \rangle$ is **continuous** at t_0 if

$$\lim_{t \to t_0} \mathbf{r}(t) = \mathbf{r}(t_0)$$

By Theorem 1, $\mathbf{r}(t)$ is continuous at t_0 if and only if the components $x(t)$, $y(t)$, $z(t)$ are continuous at t_0. The derivative of $\mathbf{r}(t)$ is the limit of the difference quotient:

$$\mathbf{r}'(t) = \frac{d}{dt}\mathbf{r}(t) = \lim_{h \to 0} \frac{\mathbf{r}(t + h) - \mathbf{r}(t)}{h} \qquad \boxed{4}$$

The terminology and notation of differential calculus carry over to the vector-valued setting. We say that $\mathbf{r}(t)$ is *differentiable* at t if the limit in Eq. (4) exists. In Leibniz notation, the derivative is written $\dfrac{d\mathbf{r}}{dt}$. Higher-order derivatives are defined by repeated differentiation:

$$\mathbf{r}''(t) = \frac{d}{dt}\,\mathbf{r}'(t), \qquad \mathbf{r}'''(t) = \frac{d}{dt}\,\mathbf{r}''(t), \ \ldots$$

The components of the difference quotient are difference quotients:

$$\lim_{h\to 0} \frac{\mathbf{r}(t+h) - \mathbf{r}(t)}{h} = \lim_{h\to 0} \left\langle \frac{x(t+h) - x(t)}{h}, \ \frac{y(t+h) - y(t)}{h}, \ \frac{z(t+h) - z(t)}{h} \right\rangle$$

It follows from Theorem 1 that $\mathbf{r}(t)$ is differentiable if and only if the components are differentiable, and in this case, $\mathbf{r}'(t)$ is equal to the vector of derivatives $\langle x'(t), y'(t), z'(t)\rangle$.

> *Theorems 1 and 2 show that vector-valued limits and derivatives may be computed "componentwise" and thus are not more difficult to compute than ordinary limits and derivatives.*

THEOREM 2 Vector-Valued Derivatives Are Computed Componentwise   A vector-valued function $\mathbf{r}(t) = \langle x(t), y(t), z(t)\rangle$ is differentiable if and only if each component is differentiable. In this case,

$$\mathbf{r}'(t) = \frac{d}{dt}\mathbf{r}(t) = \langle x'(t), y'(t), z'(t)\rangle$$

Here are some vector-valued derivatives, computed componentwise:

$$\frac{d}{dt}\langle t^2, t^3, \sin t\rangle = \langle 2t, 3t^2, \cos t\rangle, \qquad \frac{d}{dt}\langle \cos t, -1, e^{2t}\rangle = \langle -\sin t, 0, 2e^{2t}\rangle$$

■ **EXAMPLE 2**   Calculate $\mathbf{r}''(3)$, where $\mathbf{r}(t) = \langle \ln t, t, t^2\rangle$.

Solution   We perform the differentiation componentwise:

$$\mathbf{r}'(t) = \frac{d}{dt}\langle \ln t, t, t^2\rangle = \langle t^{-1}, 1, 2t\rangle$$

$$\mathbf{r}''(t) = \frac{d}{dt}\langle t^{-1}, 1, 2t\rangle = \langle -t^{-2}, 0, 2\rangle$$

Therefore, $\mathbf{r}''(3) = \langle -\frac{1}{9}, 0, 2\rangle$. ■

Since vector differentiation is performed componentwise, the differentiation rules of single-variable calculus carry over to the vector setting.

Differentiation Rules   Assume that $\mathbf{r}(t)$, $\mathbf{r}_1(t)$, and $\mathbf{r}_2(t)$ are differentiable. Then

- **Sum Rule:** $(\mathbf{r}_1(t) + \mathbf{r}_2(t))' = \mathbf{r}_1'(t) + \mathbf{r}_2'(t)$
- **Constant Multiple Rule:** For any constant c, $(c\,\mathbf{r}(t))' = c\,\mathbf{r}'(t)$.
- **Product Rule:** For any differentiable scalar-valued function $f(t)$,

$$\frac{d}{dt}\big(f(t)\mathbf{r}(t)\big) = f(t)\mathbf{r}'(t) + f'(t)\mathbf{r}(t)$$

> *We do not state a Quotient Rule because $\mathbf{r}(t)/g(t)$ equals $g(t)^{-1}\mathbf{r}(t)$ and may be differentiated using the Product Rule.*

- **Chain Rule:** For any differentiable scalar-valued function $g(t)$,

$$\frac{d}{dt}\,\mathbf{r}(g(t)) = g'(t)\mathbf{r}'(g(t))$$

Proof Each rule is proved by applying the differentiation rules to the components. For example, to prove the Product Rule (we consider vector-valued functions in the plane, to keep the notation simple), we write

$$f(t)\mathbf{r}(t) = f(t)\langle x(t), y(t)\rangle = \langle f(t)x(t), f(t)y(t)\rangle$$

Now apply the Product Rule to each component:

$$\begin{aligned}\frac{d}{dt} f(t)\mathbf{r}(t) &= \left\langle \frac{d}{dt} f(t)x(t), \frac{d}{dt} f(t)y(t)\right\rangle \\ &= \langle f'(t)x(t) + f(t)x'(t), f'(t)y(t) + f(t)y'(t)\rangle \\ &= \langle f'(t)x(t), f'(t)y(t)\rangle + \langle f(t)x'(t), f(t)y'(t)\rangle \\ &= f'(t)\langle x(t), y(t)\rangle + f(t)\langle x'(t), y'(t)\rangle = f'(t)\mathbf{r}(t) + f(t)\mathbf{r}'(t)\end{aligned}$$

The remaining proofs are left as exercises (Exercises 64–65). ∎

■ **EXAMPLE 3** Let $\mathbf{r}(t) = \langle t^2, 5t, 1\rangle$ and $f(t) = e^t$. Calculate:

(a) $\dfrac{d}{dt} f(t)\mathbf{r}(t)$

(b) $\dfrac{d}{dt}\mathbf{r}(f(t))$

Solution We have $\mathbf{r}'(t) = \langle 2t, 5, 0\rangle$ and $f'(t) = e^t$.

(a) By the Product Rule,

$$\begin{aligned}\frac{d}{dt} f(t)\mathbf{r}(t) &= f(t)\mathbf{r}'(t) + f'(t)\mathbf{r}(t) = e^t\langle 2t, 5, 0\rangle + e^t\langle t^2, 5t, 1\rangle \\ &= \langle (t^2 + 2t)e^t, (5t + 5)e^t, e^t\rangle\end{aligned}$$

(b) By the Chain Rule,

$$\frac{d}{dt}\mathbf{r}(f(t)) = f'(t)\mathbf{r}'(f(t)) = e^t\mathbf{r}'(e^t) = e^t\langle 2e^t, 5, 0\rangle = \langle 2e^{2t}, 5e^t, 0\rangle$$ ■

There are three different Product Rules for vector-valued functions. We have already stated the rule for the product of a scalar function $f(t)$ and a vector-valued function $\mathbf{r}(t)$ above. The Product Rules for the dot and cross products of two vector-valued functions are even more important in applications.

CAUTION Order is important in the Product Rule for cross products. The first term is $\mathbf{r}_1(t) \times \mathbf{r}'_2(t)$ [not $\mathbf{r}'_2(t) \times \mathbf{r}_1(t)$] and the second term is $\mathbf{r}'_1(t) \times \mathbf{r}_2(t)$. Why is order not a concern for dot products?

THEOREM 3 Product Rule for Dot and Cross Products Assume that $\mathbf{r}_1(t)$ and $\mathbf{r}_2(t)$ are differentiable. Then

Dot Products: $\quad \dfrac{d}{dt}(\mathbf{r}_1(t) \cdot \mathbf{r}_2(t)) = \mathbf{r}_1(t) \cdot \mathbf{r}'_2(t) + \mathbf{r}'_1(t) \cdot \mathbf{r}_2(t)$ **5**

Cross Products: $\quad \dfrac{d}{dt}(\mathbf{r}_1(t) \times \mathbf{r}_2(t)) = [\mathbf{r}_1(t) \times \mathbf{r}'_2(t)] + [\mathbf{r}'_1(t) \times \mathbf{r}_2(t)]$ **6**

Proof We verify (5) for vector-valued functions in the plane for simplicity of notation. If $\mathbf{r}_1(t) = \langle x_1(t), y_1(t)\rangle$ and $\mathbf{r}_2(t) = \langle x_2(t), y_2(t)\rangle$, then

$$\frac{d}{dt}\big(\mathbf{r}_1(t) \cdot \mathbf{r}_2(t)\big) = \frac{d}{dt}\big(x_1(t)x_2(t) + y_1(t)y_2(t)\big)$$

$$= x_1(t)x_2'(t) + x_1'(t)x_2(t) + y_1(t)y_2'(t) + y_1'(t)y_2(t)$$

$$= \big(x_1(t)x_2'(t) + y_1(t)y_2'(t)\big) + \big(x_1'(t)x_2(t) + y_1'(t)y_2(t)\big)$$

$$= \mathbf{r}_1(t) \cdot \mathbf{r}_2'(t) + \mathbf{r}_1'(t) \cdot \mathbf{r}_2(t)$$

The proof of (6) is left as an exercise (Exercise 66).   ∎

In the next example and throughout this chapter, *all vector-valued functions are assumed differentiable, unless otherwise stated.*

■ **EXAMPLE 4** Prove the formula $\dfrac{d}{dt}\big(\mathbf{r}(t) \times \mathbf{r}'(t)\big) = \mathbf{r}(t) \times \mathbf{r}''(t)$.

Solution By the Product Formula for cross products,

$$\frac{d}{dt}\big(\mathbf{r}(t) \times \mathbf{r}'(t)\big) = \mathbf{r}(t) \times \mathbf{r}''(t) + \underbrace{\mathbf{r}'(t) \times \mathbf{r}'(t)}_{\text{Equals }\mathbf{0}} = \mathbf{r}(t) \times \mathbf{r}''(t)$$

Here, $\mathbf{r}' \times \mathbf{r}' = \mathbf{0}$ because the cross product of a vector with itself is the zero vector.   ∎

The Derivative as a Tangent Vector

The vector $\mathbf{r}'(t_0)$ has an important geometric property: It points in the direction tangent to the path traced by $\mathbf{r}(t)$ at $t = t_0$. To understand why, consider the difference quotient where we write $\Delta\mathbf{r} = \mathbf{r}(t_0 + h) - \mathbf{r}(t_0)$ and $\Delta t = h$ with $h \neq 0$:

$$\frac{\Delta\mathbf{r}}{\Delta t} = \frac{\mathbf{r}(t_0 + h) - \mathbf{r}(t_0)}{h} \qquad \boxed{7}$$

The vector $\Delta\mathbf{r}$ points from the terminal point of $\mathbf{r}(t)$ to the terminal point of $\mathbf{r}(t + h)$ as in Figure 2(A). The difference quotient $\Delta\mathbf{r}/\Delta t$ is a scalar multiple of $\Delta\mathbf{r}$ and therefore points in the same direction, along the secant line, in the direction from $\mathbf{r}(t_0)$ to $\mathbf{r}(t_0 + h)$ [Figure 2(B)].

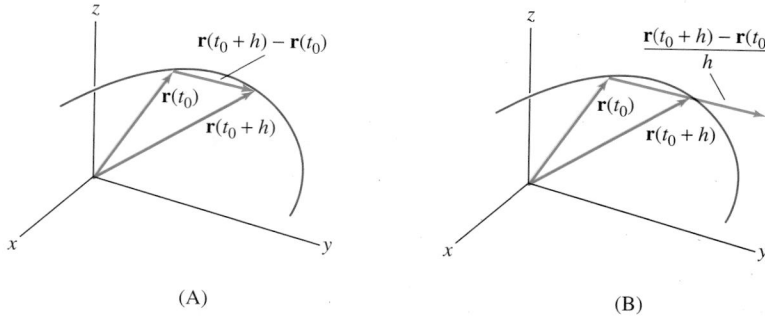

FIGURE 2 The difference quotient points in the direction of $\Delta\mathbf{r} = \mathbf{r}(t_0 + h) - \mathbf{r}(t_0)$.

(A)   (B)

As $h \to 0$, $\Delta\mathbf{r}$ itself tends to zero but the quotient $\Delta\mathbf{r}/\Delta t$ approaches a vector $\mathbf{r}'(t_0)$ which, if nonzero, points in the direction tangent to the curve (Figure 3). For this reason, $\mathbf{r}'(t_0)$ is called the **tangent vector** at $\mathbf{r}(t_0)$. We also refer to $\mathbf{r}'(t_0)$ as the **velocity vector**.

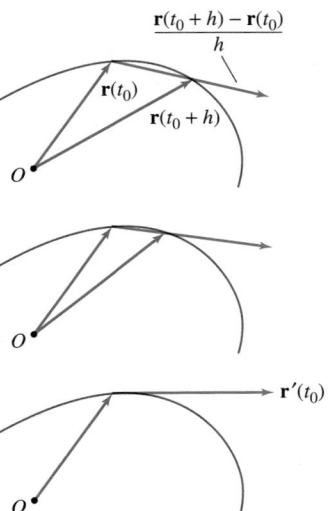

Although it has been our convention so far to regard all vectors as based at the origin, we usually view the tangent vector $\mathbf{r}'(t)$ as a vector based at the terminal point of $\mathbf{r}(t)$. This is more natural because $\mathbf{r}'(t)$ then appears as a vector tangent to the curve (Figure 3).

FIGURE 3 The difference quotient converges to a vector $\mathbf{r}'(t)$, tangent to the curve $\mathbf{r}(t)$.

If $\mathbf{r}'(t_0) \neq \mathbf{0}$, then the tangent line itself has vector parametrization:

$$\boxed{\text{Tangent line at } \mathbf{r}(t_0): \qquad \mathbf{L}(t) = \mathbf{r}(t_0) + t\mathbf{r}'(t_0)} \qquad \boxed{8}$$

Thus, $\mathbf{r}'(t_0)$ is a direction vector for the tangent line.

In Exercise 63, we verify that for a plane curve $y = f(x)$, the tangent line defined by (8) coincides with the tangent line defined using $f'(x)$ in single-variable calculus.

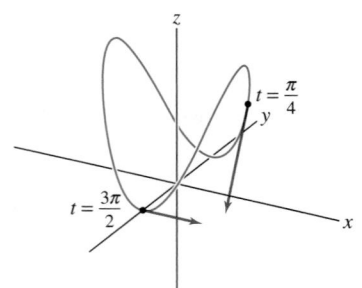

FIGURE 4 Tangent vectors to $\mathbf{r}(t) = \langle \cos t, \sin t, 4\cos^2 t \rangle$ at $t = \frac{\pi}{4}$ and $t = \frac{3\pi}{2}$.

■ **EXAMPLE 5** Plotting Tangent Vectors $\mathcal{CAS}$ Plot $\mathbf{r}(t) = \langle \cos t, \sin t, 4\cos^2 t \rangle$ together with its tangent vectors at $t = \frac{\pi}{4}, \frac{3\pi}{2}$. Find a parametrization of the tangent line at $t = \frac{\pi}{4}$.

Solution Figure 4 shows the plot of $\mathbf{r}(t)$ produced using the parametric plot function on a computer algebra system. Observe that

$$\mathbf{r}'(t) = \langle -\sin t, \cos t, -8\cos t \sin t \rangle$$

and

$$\mathbf{r}\left(\frac{\pi}{4}\right) = \left\langle \frac{\sqrt{2}}{2}, \frac{\sqrt{2}}{2}, 2 \right\rangle, \qquad \mathbf{r}'\left(\frac{\pi}{4}\right) = \left\langle -\frac{\sqrt{2}}{2}, \frac{\sqrt{2}}{2}, -4 \right\rangle$$

$$\mathbf{r}\left(\frac{3\pi}{2}\right) = \langle 0, -1, 0 \rangle, \qquad \mathbf{r}'\left(\frac{3\pi}{2}\right) = \langle 1, 0, 0 \rangle$$

We plot the vector $\mathbf{r}'(\frac{\pi}{4})$ based at $\mathbf{r}(\frac{\pi}{4})$ and the vector $\mathbf{r}'(\frac{3\pi}{2})$ based at $\mathbf{r}(\frac{3\pi}{2})$ as in Figure 4. The tangent line at $t = \frac{\pi}{4}$ has parametrization

$$\mathbf{L}(t) = \mathbf{r}\left(\frac{\pi}{4}\right) + t\mathbf{r}'\left(\frac{\pi}{4}\right) = \left\langle \frac{\sqrt{2}}{2}, \frac{\sqrt{2}}{2}, 2 \right\rangle + t\left\langle -\frac{\sqrt{2}}{2}, \frac{\sqrt{2}}{2}, -4 \right\rangle \qquad ■$$

The next example establishes an important property of vector-valued functions that will be used in Sections 13.4–13.6.

■ **EXAMPLE 6** Orthogonality of $\mathbf{r}$ and $\mathbf{r}'$ When $\mathbf{r}$ Has Constant Length Prove that $\mathbf{r}(t)$ has constant length if and only if $\mathbf{r}(t)$ is orthogonal to $\mathbf{r}'(t)$.

Solution We may assume that $\mathbf{r}(t)$ is nonzero. By the Product Rule,

$$\frac{d}{dt}\|\mathbf{r}(t)\|^2 = \frac{d}{dt}\big(\mathbf{r}(t) \cdot \mathbf{r}(t)\big) = \mathbf{r}(t) \cdot \mathbf{r}'(t) + \mathbf{r}'(t) \cdot \mathbf{r}(t) = 2\mathbf{r}(t) \cdot \mathbf{r}'(t)$$

The derivative on the left is zero if and only if $\mathbf{r}(t)$ has constant length and the dot product on the right is zero if and only if $\mathbf{r}'(t)$ is orthogonal to $\mathbf{r}(t)$ [or $\mathbf{r}'(t) = \mathbf{0}$]. The result follows. ∎

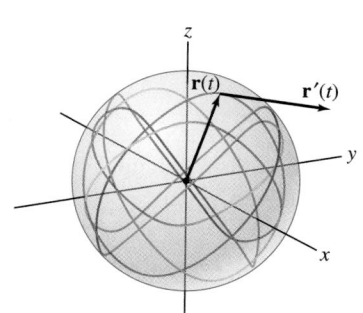

FIGURE 5

> **GRAPHICAL INSIGHT** The result of Example 6 has a geometric explanation. If $\mathbf{r}(t)$ has constant length R, then it traces a curve on the surface of a sphere of radius R with center at the origin (Figure 5). The vector $\mathbf{r}'(t)$ is then tangent to this sphere. However, any segment tangent to a sphere at a point P is orthogonal to the radial vector through P and thus $\mathbf{r}'(t)$ is orthogonal to $\mathbf{r}(t)$.

To highlight the difference between vector- and scalar-valued derivatives, consider the problem of finding points on a plane curve where the tangent line is horizontal. The tangent line to the curve $y = f(x)$ is horizontal at x_0 if $f'(x_0) = 0$. However, the tangent vector $\mathbf{r}'(t_0) = \langle x'(t_0), y'(t_0) \rangle$ is horizontal and nonzero if $y'(t_0) = 0$ but $x'(t_0) \neq 0$.

■ **EXAMPLE 7** Horizontal Tangent Vector on the Cycloid The function

$$\mathbf{r}(t) = \langle t - \sin t, 1 - \cos t \rangle$$

traces a cycloid. Find the points where:

(a) $\mathbf{r}'(t)$ is horizontal and nonzero.   **(b)** $\mathbf{r}'(t)$ is the zero vector.

Solution The y-component of the tangent vector $\mathbf{r}'(t) = \langle 1 - \cos t, \sin t \rangle$ is zero if $\sin t = 0$, that is, if $t = 0, \pi, 2\pi, \ldots$. We have

$$\mathbf{r}(0) = \langle 0, 0 \rangle, \quad \mathbf{r}'(0) = \langle 1 - \cos 0, \sin 0 \rangle = \langle 0, 0 \rangle \quad \text{(zero vector)}$$

$$\mathbf{r}(\pi) = \langle \pi, 2 \rangle, \quad \mathbf{r}'(\pi) = \langle 1 - \cos \pi, \sin \pi \rangle = \langle 2, 0 \rangle \quad \text{(horizontal)}$$

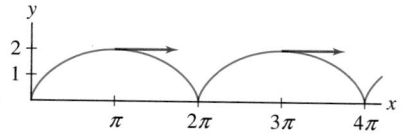

FIGURE 6 Points on the cycloid where the tangent vector is horizontal.

By periodicity, we conclude that $\mathbf{r}'(t)$ is nonzero and horizontal for $t = \pi, 3\pi, 5\pi, \ldots$ and $\mathbf{r}'(t) = \mathbf{0}$ for $t = 0, 2\pi, 4\pi, \ldots$ (Figure 6). ∎

> **CONCEPTUAL INSIGHT** The cycloid in Figure 6 shows another difference between vector and scalar derivatives. If we describe the cycloid as the graph of a function $y = f(x)$, then $f'(x)$ does not exist at the points $x = 0, 2\pi, 4\pi, \ldots$, where the curve has a sharp point called a **cusp**. However, the vector parametrization of the cycloid $\mathbf{r}(t) = \langle t - \sin t, 1 - \cos t \rangle$ is differentiable for all t and $\mathbf{r}'(t)$ is the zero vector at the cusps.

Vector-Valued Integration

The integral of a vector-valued function may be defined in terms of Riemann sums as in Chapter 5. We will define it more simply via componentwise integration (the two definitions are equivalent). In other words, we define

$$\int_a^b \mathbf{r}(t)\, dt = \left\langle \int_a^b x(t)\, dt, \int_a^b y(t)\, dt, \int_a^b z(t)\, dt \right\rangle$$

The integral exists if each of the components $x(t)$, $y(t)$, $z(t)$ is integrable. For example,

$$\int_0^\pi \langle 1, t, \sin t\rangle \, dt = \left\langle \int_0^\pi 1\,dt, \int_0^\pi t\,dt, \int_0^\pi \sin t\,dt\right\rangle = \left\langle \pi, \tfrac{1}{2}\pi^2, 2\right\rangle$$

Vector-valued integrals obey the linearity rules as scalar-valued integrals (see Exercise 68).

An **antiderivative** of $\mathbf{r}(t)$ is a vector-valued function $\mathbf{R}(t)$ such that $\mathbf{R}'(t) = \mathbf{r}(t)$. In the single-variable case, two functions $f_1(x)$ and $f_2(x)$ with the same derivative differ by a constant. Similarly, two vector-valued functions with the same derivative differ by a *constant vector* (i.e., a vector that does not depend on t). This is proved by applying the scalar result to each component of $\mathbf{r}(t)$.

> **THEOREM 4** Let $\mathbf{R}_1(t)$ and $\mathbf{R}_2(t)$ be differentiable vector-valued functions such that $\mathbf{R}_1'(t) = \mathbf{R}_2'(t)$. Then $\mathbf{R}_1(t) = \mathbf{R}_2(t) + \mathbf{c}$ for some constant vector $\mathbf{c}$.

The general antiderivative of $\mathbf{r}(t)$ is written

$$\int \mathbf{r}(t)\, dt = \mathbf{R}(t) + \mathbf{c}$$

where $\mathbf{c} = \langle c_1, c_2, c_3\rangle$ is an arbitrary constant vector. For example,

$$\int \langle 1, t, \sin t\rangle\, dt = \left\langle t, \tfrac{1}{2}t^2, -\cos t\right\rangle + \mathbf{c} = \left\langle t + c_1, \tfrac{1}{2}t^2 + c_2, -\cos t + c_3\right\rangle$$

> **Fundamental Theorem of Calculus for Vector-Valued Functions** Let $\mathbf{r}(t)$ be a continuous vector-valued function on the interval $[a, b]$, and let $\mathbf{R}(t)$ be an antiderivative of $\mathbf{r}(t)$. Then
> $$\int_a^b \mathbf{r}(t)\, dt = \mathbf{R}(b) - \mathbf{R}(a)$$

■ **EXAMPLE 8** Finding Position via Vector-Valued Differential Equations The path of a particle satisfies $\dfrac{d\mathbf{r}}{dt} = \langle 8t, 5 - 8\sin 2t\rangle$. Where is the particle located at $t = 3$ if $\mathbf{r}(0) = \langle 0, 10\rangle$?

Note that in Example 8, $d\mathbf{r}/dt$ is the particle's velocity function, and just as in the single-variable case, the integral of the velocity yields the position $\mathbf{r}(t)$.

Solution The general solution is obtained by integration:

$$\mathbf{r}(t) = \int \langle 8t, 5 - 8\sin 2t\rangle\, dt = \langle 4t^2, 5t + 4\cos 2t\rangle + \mathbf{c}$$

We have

$$\mathbf{r}(0) = \langle 4(0)^2, 5(0) + 4\cos 0\rangle + \mathbf{c} = \langle 0, 4\rangle + \mathbf{c}$$

Using the initial condition $\mathbf{r}(0) = \langle 0, 4\rangle + \mathbf{c} = \langle 0, 10\rangle$, we obtain $\mathbf{c} = \langle 0, 6\rangle$ and

$$\mathbf{r}(t) = \langle 4t^2, 5t + 4\cos 2t\rangle + \langle 0, 6\rangle = \langle 4t^2, 5t + 4\cos 2t + 6\rangle$$

The particle's position at $t = 3$ is

$$\mathbf{r}(3) = \langle 4(3)^2, 5(3) + 4\cos 6 + 6\rangle \approx \langle 36, 24.84\rangle$$

13.2 SUMMARY

- Limits, differentiation, and integration of vector-valued functions are performed componentwise.
- The differentiation rules apply in the vector-valued setting, but there are three Product Rules for vector-valued functions. If $f(t)$, $\mathbf{r}(t)$, $\mathbf{r}_1(t)$, and $\mathbf{r}_2(t)$ are differentiable, then

Scalar times vector: $\qquad \dfrac{d}{dt}\big(f(t)\mathbf{r}(t)\big) = f(t)\mathbf{r}'(t) + f'(t)\mathbf{r}(t)$

Dot product: $\qquad \dfrac{d}{dt}\big(\mathbf{r}_1(t) \cdot \mathbf{r}_2(t)\big) = \mathbf{r}_1(t) \cdot \mathbf{r}_2'(t) + \mathbf{r}_1'(t) \cdot \mathbf{r}_2(t)$

Cross product: $\qquad \dfrac{d}{dt}\big(\mathbf{r}_1(t) \times \mathbf{r}_2(t)\big) = \big[\mathbf{r}_1(t) \times \mathbf{r}_2'(t)\big] + \big[\mathbf{r}_1'(t) \times \mathbf{r}_2(t)\big]$

- The derivative $\mathbf{r}'(t_0)$ is referred to as the *tangent vector* or *velocity vector* at $\mathbf{r}(t_0)$. If $\mathbf{r}'(t_0)$ is nonzero, then it points in the direction tangent to the curve at $\mathbf{r}(t_0)$.
- If $\mathbf{r}'(t_0) \neq \mathbf{0}$, then the tangent line at t_0 has vector parametrization

$$\mathbf{L}(t) = \mathbf{r}(t_0) + t\mathbf{r}'(t_0)$$

- If $\mathbf{R}_1(t)$ and $\mathbf{R}_2(t)$ are differentiable, vector-valued functions such that $\mathbf{R}_1'(t) = \mathbf{R}_2'(t)$, then $\mathbf{R}_2(t) = \mathbf{R}_1(t) + \mathbf{c}$ for some constant vector $\mathbf{c}$.
- The Fundamental Theorem for vector-valued functions: If $\mathbf{r}(t)$ is continuous on the interval $[a, b]$ and if $\mathbf{R}(t)$ is an antiderivative of $\mathbf{r}(t)$, then

$$\int_a^b \mathbf{r}(t)\,dt = \mathbf{R}(b) - \mathbf{R}(a)$$

13.2 EXERCISES

Preliminary Questions

1. State the three forms of the Product Rule for vector-valued functions.

In Questions 2–6, indicate whether true or false and if false, provide a correct statement.

2. The derivative of a vector-valued function is defined as the limit of the difference quotient, just as in the scalar-valued case.

3. There are two Chain Rules for vector-valued functions, one for the composite of two vector-valued functions and one for the composite of a vector-valued and scalar-valued function.

4. The terms "velocity vector" and "tangent vector" for a path $\mathbf{r}(t)$ mean one and the same thing.

5. The derivative of a vector-valued function is the slope of the tangent line, just as in the scalar case.

6. The derivative of the cross product is the cross product of the derivatives.

7. State whether the following derivatives of vector-valued functions $\mathbf{r}_1(t)$ and $\mathbf{r}_2(t)$ are scalars or vectors:

(a) $\dfrac{d}{dt}\mathbf{r}_1(t)$ 　　　　　　　　**(b)** $\dfrac{d}{dt}\big(\mathbf{r}_1(t) \cdot \mathbf{r}_2(t)\big)$

(c) $\dfrac{d}{dt}\big(\mathbf{r}_1(t) \times \mathbf{r}_2(t)\big)$

Exercises

In Exercises 1–4, evaluate the limit.

1. $\displaystyle\lim_{t \to 3}\left\langle t^2, 4t, \frac{1}{t}\right\rangle$

2. $\displaystyle\lim_{t \to \pi}\sin 2t\,\mathbf{i} + \cos t\,\mathbf{j} + \tan 4t\,\mathbf{k}$

3. $\displaystyle\lim_{t \to 0}e^{2t}\mathbf{i} + \ln(t+1)\mathbf{j} + 4\mathbf{k}$

4. $\displaystyle\lim_{t \to 0}\left\langle \frac{1}{t+1}, \frac{e^t - 1}{t}, 4t\right\rangle$

5. Evaluate $\displaystyle\lim_{h \to 0}\dfrac{\mathbf{r}(t+h) - \mathbf{r}(t)}{h}$ for $\mathbf{r}(t) = \left\langle t^{-1}, \sin t, 4\right\rangle$.

6. Evaluate $\lim\limits_{t\to 0}\dfrac{\mathbf{r}(t)}{t}$ for $\mathbf{r}(t)=\langle \sin t, 1-\cos t, -2t\rangle$.

In Exercises 7–14, compute the derivative.

7. $\mathbf{r}(t)=\langle t, t^2, t^3\rangle$

8. $\mathbf{v}(t)=\langle \sin 3t, \cos 3t\rangle$

9. $\mathbf{w}(s)=\langle e^s, e^{-2s}\rangle$

10. $\mathbf{r}(\theta)=\langle \tan\theta, 4\theta-2, \sin\theta\rangle$

11. $\mathbf{r}(t)=\langle t-t^{-1}, 4t^2, 8\rangle$

12. $\mathbf{c}(t)=t^{-1}\mathbf{i}-e^{2t}\mathbf{k}$

13. $\mathbf{a}(\theta)=(\cos 2\theta)\mathbf{i}+(\sin 2\theta)\mathbf{j}+(\sin 4\theta)\mathbf{k}$

14. $\mathbf{b}(t)=\langle e^{4t-3}, \sin(t^2), (4t+3)^{-1}\rangle$

15. Calculate $\mathbf{r}'(t)$ and $\mathbf{r}''(t)$ for $\mathbf{r}(t)=\langle t, t^2, t^3\rangle$.

16. Sketch the curve $\mathbf{r}(t)=\langle 1-t^2, t\rangle$ for $-1\le t\le 1$. Compute the tangent vector at $t=1$ and add it to the sketch.

17. Sketch the curve $\mathbf{r}_1(t)=\langle t, t^2\rangle$ together with its tangent vector at $t=1$. Then do the same for $\mathbf{r}_2(t)=\langle t^3, t^6\rangle$.

18. Sketch the cycloid $\mathbf{r}(t)=\langle t-\sin t, 1-\cos t\rangle$ together with its tangent vectors at $t=\frac{\pi}{3}$ and $\frac{3\pi}{4}$.

In Exercises 19–22, use the appropriate Product Rule to evaluate the derivative, where

$$\mathbf{r}_1(t)=\langle 8t, 4, -t^3\rangle, \qquad \mathbf{r}_2(t)=\langle 0, e^t, -6\rangle$$

19. $\dfrac{d}{dt}\big(\mathbf{r}_1(t)\cdot\mathbf{r}_2(t)\big)$

20. $\dfrac{d}{dt}\big(t^4\mathbf{r}_1(t)\big)$

21. $\dfrac{d}{dt}\big(\mathbf{r}_1(t)\times\mathbf{r}_2(t)\big)$

22. $\dfrac{d}{dt}\big(\mathbf{r}_1(t)\cdot\mathbf{r}_3(t)\big)\big|_{t=5}$, assuming that $\mathbf{r}_3(5)=\langle 3,1,2\rangle$ and $\mathbf{r}_3'(5)=\langle -1,2,7\rangle$.

In Exercises 23–25, let

$$\mathbf{r}_1(t)=\langle t^2, t^3, 4t\rangle, \qquad \mathbf{r}_2(t)=\langle t^{-1}, 1+t, 2\rangle$$

23. Let $F(t)=\mathbf{r}_1(t)\cdot\mathbf{r}_2(t)$.
(a) Calculate $F'(t)$ using the Product Rule.
(b) Expand the product $\mathbf{r}_1(t)\cdot\mathbf{r}_2(t)$ and differentiate. Compare with part (a).

24. Let $G(t)=\mathbf{r}_1(t)\times\mathbf{r}_2(t)$.
(a) Calculate $G'(t)$ using the Product Rule.
(b) Expand the cross product $\mathbf{r}_1(t)\times\mathbf{r}_2(t)$ and differentiate. Compare with part (a).

25. Find the rate of change of the angle θ between $\mathbf{r}_1(t)$ and $\mathbf{r}_2(t)$ at $t=2$, assuming that t is measured in seconds.

In Exercises 26–29, evaluate $\dfrac{d}{dt}\mathbf{r}(g(t))$ using the Chain Rule.

26. $\mathbf{r}(t)=\langle t^2, 2t, 4\rangle$, $\ g(t)=e^t$

27. $\mathbf{r}(t)=\langle e^t, e^{2t}, 4\rangle$, $\ g(t)=4t+9$

28. $\mathbf{r}(t)=\langle 4\sin 2t, 6\cos 2t\rangle$, $\ g(t)=t^2$

29. $\mathbf{r}(t)=\langle 3^t, \tan^{-1}t\rangle$, $\ g(t)=\sin t$

30. Let $\mathbf{v}(s)=s^2\mathbf{i}+2s\mathbf{j}+9s^{-2}\mathbf{k}$. Evaluate $\dfrac{d}{ds}\mathbf{v}(g(s))$ at $s=4$, assuming that $g(4)=3$ and $g'(4)=-9$.

31. Let $\mathbf{r}(t)=\langle t^2, 1-t, 4t\rangle$. Calculate the derivative of $\mathbf{r}(t)\cdot\mathbf{a}(t)$ at $t=2$, assuming that $\mathbf{a}(2)=\langle 1,3,3\rangle$ and $\mathbf{a}'(2)=\langle -1,4,1\rangle$.

32. Let $\mathbf{r}(t)=\langle t^2, t^3, e^t\rangle$. Use Example 4 to calculate $\dfrac{d}{dt}(\mathbf{r}\times\mathbf{r}')$.

In Exercises 33–37, find a parametrization of the tangent line at the point indicated.

33. $\mathbf{r}(t)=\langle 1-t^2, 5t, 2t^3\rangle$, $\ t=2$

34. $\mathbf{r}(t)=\langle \cos 2t, \sin 3t, \sin 4t\rangle$, $\ t=\frac{\pi}{4}$

35. $\mathbf{r}(s)=4s^{-1}\mathbf{i}-8s^{-3}\mathbf{k}$, $\ s=2$

36. $\mathbf{r}(t)=\langle t^2, t^4\rangle$, $\ t=1$

37. $\mathbf{r}(s)=\ln s\,\mathbf{i}+s^{-1}\mathbf{j}+9s\mathbf{k}$, $\ s=1$

38. Let $\mathbf{r}(t)=\langle \sin 2t\cos t, \sin 2t\sin t, \cos 2t\rangle$. Show that $\|\mathbf{r}(t)\|$ is constant and conclude using Example 6 that $\mathbf{r}(t)$ and $\mathbf{r}'(t)$ are orthogonal. Then compute $\mathbf{r}'(t)$ and verify directly that it is orthogonal to $\mathbf{r}(t)$.

39. Show, by finding a counterexample, that in general $\|\mathbf{r}'(t)\|$ need not equal $\|\mathbf{r}(t)\|'$.

In Exercises 40–45, evaluate the integrals.

40. $\displaystyle\int_0^1 \langle 2t, 4t, -\cos 3t\rangle\, dt$

41. $\displaystyle\int_1^4 \left(t^{-1}\mathbf{i}+4\sqrt{t}\,\mathbf{j}-8t^{3/2}\mathbf{k}\right) dt$

42. $\displaystyle\int_0^1 \left(te^{-t^2}, t\ln(t^2+1)\right) dt$

43. $\displaystyle\int_{-2}^2 (u^3\mathbf{i}+u^5\mathbf{j}+u^7\mathbf{k})\, du$

44. $\displaystyle\int_0^1 \left\langle \frac{1}{1+s^2}, \frac{s}{1+s^2}\right\rangle ds$

45. $\displaystyle\int_0^t (3s\mathbf{i}+6s^2\mathbf{j}+9\mathbf{k})\, ds$

In Exercises 46–53, find the general solution $\mathbf{r}(t)$ *of the differential equation and the solution with the given initial condition.*

46. $\dfrac{d\mathbf{r}}{dt} = \langle 1 - 2t, 4t \rangle, \quad \mathbf{r}(0) = \langle 3, 1 \rangle$

47. $\mathbf{r}'(t) = \mathbf{i} - \mathbf{j}, \quad \mathbf{r}(0) = 2\mathbf{i} + 3\mathbf{k}$

48. $\mathbf{r}'(t) = t^2\mathbf{i} + 5t\mathbf{j} + \mathbf{k}, \quad \mathbf{r}(0) = \mathbf{j} + 2\mathbf{k}$

49. $\mathbf{r}'(t) = \langle \sin 3t, \sin 3t, t \rangle, \quad \mathbf{r}(0) = \langle 0, 1, 8 \rangle$

50. $\dfrac{d\mathbf{r}}{dt} = \langle e^{2t}, e^t, e^{-2t} \rangle, \quad \mathbf{r}(0) = \langle 4, -2, 3 \rangle$

51. $\mathbf{r}''(t) = 16\mathbf{k}, \quad \mathbf{r}(0) = \langle 1, 0, 0 \rangle, \quad \mathbf{r}'(0) = \langle 0, 1, 0 \rangle$

52. $\mathbf{r}''(t) = \langle 0, 0, 1 \rangle, \quad \mathbf{r}(0) = \langle 2, 1, 1 \rangle, \quad \mathbf{r}'(0) = \langle 3, 1, 1 \rangle$

53. $\mathbf{r}''(t) = \langle e^t, \sin t, \cos t \rangle, \quad \mathbf{r}(0) = \langle 1, 0, 1 \rangle, \quad \mathbf{r}'(0) = \langle 0, 2, 2 \rangle$

54. Show that $\mathbf{w}(t) = \langle \sin(3t + 4), \sin(3t - 2), \cos 3t \rangle$ satisfies the differential equation $\mathbf{w}''(t) = -9\mathbf{w}(t)$.

55. The path $\mathbf{r}(t)$ of a particle satisfies $\dfrac{d\mathbf{r}}{dt} = \langle 8, 5 - 3t, 4t^2 \rangle$. Where is the particle located at $t = 4$ if $\mathbf{r}(0) = \langle 1, 6, 0 \rangle$?

56. A fighter plane, which can only shoot bullets straight ahead, travels along the path $\mathbf{r}(t) = \langle 5 - t, 21 - t^2, 3 - t^3/27 \rangle$. Show that there is precisely one time t at which the pilot can hit a target located at the origin.

57. Find all solutions to $\mathbf{r}'(t) = \mathbf{v}$, where $\mathbf{v}$ is a constant vector in $\mathbf{R}^3$.

58. Let $\mathbf{u}$ be a constant vector in $\mathbf{R}^3$. Find the solution of $\mathbf{r}'(t) = (\sin t)\mathbf{u}$ satisfying $\mathbf{r}'(0) = 0$.

59. Find all solutions to $\mathbf{r}'(t) = 2\mathbf{r}(t)$ where $\mathbf{r}(t)$ is a vector-valued function in three-space.

60. Show that $\dfrac{d}{dt}(\mathbf{a} \times \mathbf{r}) = \mathbf{a} \times \mathbf{r}'$ for any constant vector $\mathbf{a}$.

61. Prove that $\mathbf{r}(t_0)$ and $\mathbf{r}'(t_0)$ are orthogonal at values $t = t_0$ where $\|\mathbf{r}(t)\|$ takes on a local minimum or maximum value. Explain how this result is related to Figure 7. *Hint:* In the figure, $\|\mathbf{r}(t_0)\|$ is a minimum and the path $\mathbf{r}(t)$ intersects the sphere of radius $\|\mathbf{r}(t_0)\|$ in a single point (and hence is tangent at that point).

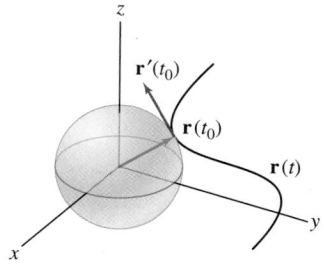

FIGURE 7

62. Newton's Second Law of Motion in vector form states that $\mathbf{F} = \dfrac{d\mathbf{p}}{dt}$ where $\mathbf{F}$ is the force acting on an object of mass m and $\mathbf{p} = m\mathbf{r}'(t)$ is the object's momentum. The analogs of force and momentum for rotational motion are the **torque** $\tau = \mathbf{r} \times \mathbf{F}$ and **angular momentum** $\mathbf{J} = \mathbf{r}(t) \times \mathbf{p}(t)$. Use the Second Law to prove that $\tau = \dfrac{d\mathbf{J}}{dt}$. This is a rotational version of the Second Law.

Further Insights and Challenges

63. In this exercise, we verify that the definition of the tangent line using vector-valued functions agrees with the usual definition in terms of the scalar derivative in the case of a plane curve. Suppose that $\mathbf{r}(t) = \langle x(t), y(t) \rangle$ traces a plane curve $\mathcal{C}$.

(a) Show that $\dfrac{dy}{dx} = \dfrac{y'(t)}{x'(t)}$ at any point such that $x'(t) \neq 0$. *Hint:* By the Chain Rule, $\dfrac{dy}{dt} = \dfrac{dy}{dx}\dfrac{dx}{dt}$.

(b) Show that if $x'(t_0) \neq 0$, then the line $\mathbf{L}(t) = \mathbf{r}(t_0) + t\mathbf{r}'(t_0)$ passes through $\mathbf{r}(t_0)$ and has slope $\dfrac{dy}{dx}\bigg|_{t=t_0}$.

64. Verify the Sum and Product Rules for derivatives of vector-valued functions.

65. Verify the Chain Rule for vector-valued functions.

66. Verify the Product Rule for cross products [Eq. (6)].

67. Prove that $\dfrac{d}{dt}(\mathbf{r} \cdot (\mathbf{r}' \times \mathbf{r}'')) = \mathbf{r} \cdot (\mathbf{r}' \times \mathbf{r}''')$

Exercises 68–71 establish additional properties of vector-valued integrals. Assume that all functions are integrable.

68. Prove the linearity properties

$$\int c\mathbf{r}(t)\, dt = c \int \mathbf{r}(t)\, dt \qquad (c \text{ any constant})$$

$$\int (\mathbf{r}_1(t) + \mathbf{r}_2(t))\, dt = \int \mathbf{r}_1(t)\, dt + \int \mathbf{r}_2(t)\, dt$$

69. Prove the Substitution Rule [where $g(t)$ is a differentiable scalar function]:

$$\int_a^b \mathbf{r}(g(t))g'(t)\, dt = \int_{g(a)}^{g(b)} \mathbf{r}(u)\, du$$

70. Formulate and verify a version of Integration by Parts for vector-valued integrals.

71. Show that if $\|\mathbf{r}(t)\| \leq K$ for $t \in [a, b]$, then

$$\left\| \int_a^b \mathbf{r}(t)\, dt \right\| \leq K(b - a)$$

13.3 Arc Length and Speed

◄··· REMINDER *The length of a path or curve is referred to as the* arc length.

In Section 11.2, we derived a formula for the arc length of a plane curve. That discussion applies with little change to paths in three-space. Recall that arc length is defined as the limit of polygonal approximations. A polygonal approximation to a path

$$\mathbf{r}(t) = \langle x(t), y(t), z(t) \rangle, \qquad a \le t \le b$$

is obtained by choosing a partition $a = t_0 < t_1 < t_2 < \cdots < t_N = b$ and joining the terminal points of the vectors $\mathbf{r}(t_j)$ by segments, as in Figure 1. Arguing as in Section 11.2, we find that if $\mathbf{r}'(t)$ exists and is continuous on $[a, b]$, then the lengths of the polygonal approximations approach a limit L as the maximum of the widths $|t_j - t_{j-1}|$ tends to zero. This limit is the arc length and is given by the formula

$$L = \int_a^b \sqrt{x'(t)^2 + y'(t)^2 + z'(t)^2}\, dt \qquad \boxed{1}$$

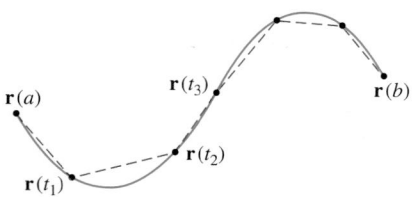

FIGURE 1 Polygonal approximation to the arc $\mathbf{r}(t)$ for $a \le t \le b$.

The quantity L in Eq. (2) is the distance traveled by a particle following the trajectory $\mathbf{r}(t)$. However, $\mathbf{r}(t)$ may traverse all or part of a curve more than once and it may reverse direction. Thus, L is equal to the length of the underlying curve only if the curve is traversed once without reversal of direction for $a \le t \le b$.

Arc Length of a Path   Assume that $\mathbf{r}(t)$ is differentiable and that $\mathbf{r}'(t)$ is continuous on $[a, b]$. Then the length L of the path $\mathbf{r}(t)$ for $a \le t \le b$ is equal to

$$L = \int_a^b \|\mathbf{r}'(t)\|\, dt = \int_a^b \sqrt{x'(t)^2 + y'(t)^2 + z'(t)^2}\, dt \qquad \boxed{2}$$

■ **EXAMPLE 1**   Find the arc length L of $\mathbf{r}(t) = \langle \cos 3t, \sin 3t, 3t \rangle$ for $0 \le t \le 2\pi$.

Solution   The derivative is $\mathbf{r}'(t) = \langle -3 \sin 3t, 3 \cos 3t, 3 \rangle$, and

$$\|\mathbf{r}'(t)\|^2 = 9 \sin^2 3t + 9 \cos^2 3t + 9 = 9(\sin^2 3t + \cos^2 3t) + 9 = 18$$

Therefore, $L = \int_0^{2\pi} \|\mathbf{r}'(t)\|\, dt = \int_0^{2\pi} \sqrt{18}\, dt = 6\sqrt{2}\pi.$ ■

As we noted in the previous section, the derivative $\mathbf{r}'(t)$ is referred to as both the *tangent vector* and the *velocity vector* of the path $\mathbf{r}(t)$. The term "tangent vector" seems justified because $\mathbf{r}'(t)$ is indeed tangent to the path when it is nonzero. To justify the term "velocity vector," assume that t is a time parameter and let us relate $\mathbf{r}'(t)$ to the speed of the path. By definition, the speed of a path, denoted $v(t)$, is the rate of change of distance traveled with respect to t. To calculate the speed, we define the arc length function:

$$s(t) = \int_a^t \|\mathbf{r}'(u)\|\, du$$

Then $s(t)$ is the distance traveled during the time interval $[a, t]$ and the derivative $v(t) = \dfrac{ds}{dt}$ is the speed. By the Fundamental Theorem of Calculus,

$$v(t) = \text{speed at time } t = \frac{ds}{dt} = \|\mathbf{r}'(t)\|$$

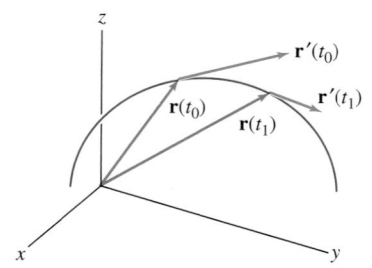

FIGURE 2 The particle is moving faster at t_0 than t_1 since the velocity vector is longer at t_0.

Thus, speed is the length of the velocity vector (Figure 2). In summary, we have the following interpretation of the velocity vector:

- $\mathbf{r}'(t)$ points in the instantaneous direction of motion (when it is nonzero).
- $v(t) = \|\mathbf{r}'(t)\|$ is the speed at time t.

■ **EXAMPLE 2** Find the speed of a particle at time $t = 2$ s if the position vector of the particle is $\mathbf{r}(t) = t^3\mathbf{i} - e^t\mathbf{j} + 4t\mathbf{k}$ (in units of feet).

Solution   The velocity vector is $\mathbf{r}'(t) = 3t^2\mathbf{i} - e^t\mathbf{j} + 4\mathbf{k}$, and at $t = 2$,

$$\mathbf{r}'(2) = 12\mathbf{i} - e^2\mathbf{j} + 4\mathbf{k}$$

The particle's speed is $\|\mathbf{r}'(2)\| = \sqrt{12^2 + (-e^2)^2 + 4^2} \approx 14.65$ ft/s. ■

Arc Length Parametrization

Keep in mind that a parametrization $\mathbf{r}(t)$ does more than just describe a curve. It also tells you how a particle traverses the curve, possibly speeding up, slowing down, or reversing direction along the way. Changing the parametrization amounts to describing a different way of traversing the same underlying curve.

We have seen that parametrizations are not unique. For example, $\mathbf{r}_1(t) = \langle t, t^2 \rangle$ and $\mathbf{r}_2(s) = \langle s^3, s^6 \rangle$ both parametrize the parabola $y = x^2$. Notice that $\mathbf{r}_2(s)$ is obtained by substituting $t = s^3$ in $\mathbf{r}_1(t)$.

In general, we obtain a new parametrization by making a substitution $t = \varphi(s)$, that is, by replacing $\mathbf{r}(t)$ with $\mathbf{r}_1(s) = \mathbf{r}(\varphi(s))$. If $\mathbf{r}(t)$ and $\varphi(s)$ are differentiable, then $\mathbf{r}_1(s)$ is also differentiable. Furthermore, if $t = \varphi(s)$ increases from a to b as s varies from c to d, then path $\mathbf{r}(t)$ for $a \leq t \leq b$ is also parametrized by $\mathbf{r}_1(s)$ for $c \leq s \leq d$.

■ **EXAMPLE 3** Parametrize the path $\mathbf{r}(t) = (t^2, \sin t, t)$ for $3 \leq t \leq 9$ in terms of the parameter s, where $t = \varphi(s) = e^s$.

Solution   Substituting $t = e^s$ in $\mathbf{r}(t)$, we obtain the parametrization

$$\mathbf{r}_1(s) = \mathbf{r}(\varphi(s)) = \langle e^{2s}, \sin e^s, e^s \rangle$$

Furthermore, $s = \ln t$, so t varies from 3 to 9 as s varies from $\ln 3$ to $\ln 9$. Therefore, the path is parametrized by $\mathbf{r}_1(s)$ for $\ln 3 \leq s \leq \ln 9$. ■

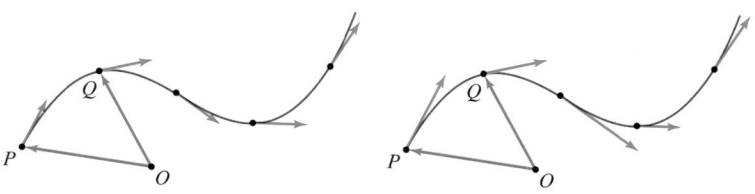

(A) An arc length parametrization
(all tangent vectors have length 1)

(B) Not an arc length parametrization
(tangent vectors' lengths vary)

FIGURE 3

A special type of parametrization, called an **arc length parametrization**, plays an important role in the study of curves. It is defined by the condition $\|\mathbf{r}'(t)\| = 1$ for all t [Figure 3(A)]. An arc length parametrization is also called the **unit speed parametrization** because the speed $\|\mathbf{r}'(t)\|$ is equal to 1 for all t. Informally speaking, *the arc length parametrization corresponds to walking along the curve at "1 m/s," and the distance*

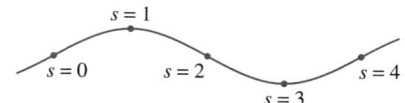

FIGURE 4 In an arc length parametrization, the particle is located s units past $\mathbf{r}(0)$ at time s.

traveled over any interval is equal to the length of the interval (Figure 4):

$$\underbrace{\int_a^b \|\mathbf{r}'(u)\| \, du}_{\text{Distance traveled during interval } [a, b]} = \int_a^b du = b - a$$

How do we find an arc length parametrization? Let $\mathbf{r}(t)$ be any parametrization and consider the arc length function:

$$s(t) = \int_a^t \|\mathbf{r}'(u)\| \, du$$

Then $s'(t) = \|\mathbf{r}'(t)\| \geq 0$. If we assume that $\mathbf{r}'(t) \neq \mathbf{0}$ for all t, then $s(t)$ is an increasing function of t. In this case, $s(t)$ has an inverse, which we denote $t = \varphi(s)$. We claim that $\mathbf{r}_1(s) = \mathbf{r}(\varphi(s))$ is an arc length parametrization. To check this, we must verify that $\|\mathbf{r}_1'(s)\| = 1$. By the formula for the derivative of an inverse,

$$\varphi'(s) = \frac{dt}{ds} = \frac{1}{s'(t)} = \frac{1}{s'(\varphi(s))} = \frac{1}{\|\mathbf{r}'(\varphi(s))\|}$$

Hence, by the Chain Rule,

$$\|\mathbf{r}_1'(s)\| = \|\mathbf{r}'(\varphi(s))\varphi'(s)\| = \|\mathbf{r}'(\varphi(s))\|\varphi'(s) = \|\mathbf{r}'(\varphi(s))\| \cdot \frac{1}{\|\mathbf{r}'(\varphi(s))\|} = 1$$

It is an important fact that arc length parametrizations exist (we will use them to define curvature in Section 13.4). However, it is often difficult or impossible to find an arc length parametrization explicitly. This requires finding the inverse of the arc length function $s(t)$ and can only be done in special cases.

The letter s is often used as the parameter in an arc length parametrization. Thus, $\mathbf{r}_1(s)$ is the position vector of the point located at a distance s along the path from $\mathbf{r}_1(0)$.

■ **EXAMPLE 4** Finding an Arc Length Parametrization Find the arc length parametrization of the line $\mathbf{r}(t) = \langle 2t, 1 - 2t, t \rangle$.

Solution

***Step 1.* Find the inverse of the arc length function.**
We have

$$\|\mathbf{r}'(t)\| = \|\langle 2, -2, 1 \rangle\| = \sqrt{2^2 + (-2)^2 + 1^2} = \sqrt{9} = 3$$

Thus, the arc length function is

$$s(t) = \int_0^t \|\mathbf{r}'(t)\| \, dt = \int_0^t 3 \, dt = 3t$$

The inverse of $s(t) = 3t$ is $t = \varphi(s) = \dfrac{s}{3}$.

***Step 2.* Reparametrize the curve.**
As explained above, an arc length parametrization is

$$\mathbf{r}_1(s) = \mathbf{r}(\varphi(s)) = \mathbf{r}\left(\frac{s}{3}\right) = \left\langle \frac{2s}{3}, 1 - \frac{2s}{3}, \frac{s}{3} \right\rangle$$

As a check, we verify that $\mathbf{r}_1(s)$ has unit speed: $\mathbf{r}_1'(s) = \left\langle \frac{2}{3}, -\frac{2}{3}, \frac{1}{3} \right\rangle$ and

$$\|\mathbf{r}_1'(s)\| = \left\| \left\langle \frac{2}{3}, -\frac{2}{3}, \frac{1}{3} \right\rangle \right\| = \sqrt{\frac{4}{9} + \frac{4}{9} + \frac{1}{9}} = 1 \qquad ■$$

> **CONCEPTUAL INSIGHT**   Although the arc length L of a curve is defined as the limit of polygonal approximations, we never compute this limit directly. Instead, we choose a parametrization $\mathbf{r}(t)$ that traverses the curve exactly once for $a \leq t \leq b$ and determine L by evaluating the arc length integral. It follows that the value of the arc length integral does not depend on the choice of parametrization (the value is L for all choices of parametrization). In Exercise 26, we use the Change of Variables formula to give another proof that the arc length integral is independent of the parametrization.

13.3 SUMMARY

- The length L of a path $\mathbf{r}(t) = \langle x(t), y(t), z(t) \rangle$ for $a \leq t \leq b$ is

$$L = \int_a^b \|\mathbf{r}'(t)\| \, dt = \int_a^b \sqrt{x'(t)^2 + y'(t)^2 + z'(t)^2} \, dt$$

- The distance traveled by a particle following the trajectory $\mathbf{r}(t)$ over the time interval $[a, t]$ is

$$s(t) = \int_a^t \|\mathbf{r}'(u)\| \, du$$

The particle's speed $v(t)$ is the rate of change of distance traveled with respect to time:

$$v(t) = \text{speed at time } t = \frac{ds}{dt} = \|\mathbf{r}'(u)\|$$

- The velocity vector $\mathbf{r}'(t)$ points in the direction of motion [provided that $\mathbf{r}'(t) \neq \mathbf{0}$] and its magnitude $\|\mathbf{r}'(t)\|$ is the particle's speed.
- We say that $\mathbf{r}(s)$ is an *arc length parametrization* or *unit speed parametrization* if $\|\mathbf{r}'(s)\| = 1$ for all s. In this case, the length of the path for $a \leq s \leq b$ is $b - a$.
- If $\mathbf{r}(t)$ is a parametrization such that $\mathbf{r}'(t) \neq \mathbf{0}$ for all t, then $\mathbf{r}_1(s) = \mathbf{r}(\varphi(s))$ is an arc length parametrization, where $t = \varphi(s)$ is the inverse of the arc length function

$$s(t) = \int_a^t \|\mathbf{r}'(u)\| \, du$$

13.3 EXERCISES

Preliminary Questions

1. At a given instant, a car on a roller coaster has velocity vector $\mathbf{r}' = \langle 25, -35, 10 \rangle$ (in miles per hour). What would the velocity vector be if the speed were doubled? What would it be if the car's direction were reversed but its speed remained unchanged?

2. Two cars travel in the same direction along the same roller coaster (at different times). Which of the following statements about their velocity vectors at a given point P on the roller coaster are true?

(a) The velocity vectors are identical.

(b) The velocity vectors point in the same direction but may have different lengths.

(c) The velocity vectors may point in opposite directions.

3. A mosquito flies along a parabola with speed $v(t) = t^2$. Let $L(t)$ be the total distance traveled at time t.

(a) How fast is $L(t)$ changing at $t = 2$?

(b) Is $L(t)$ equal to the mosquito's distance from the origin?

4. What is the length of the path traced by $\mathbf{r}(t)$ for $4 \leq t \leq 10$ if $\mathbf{r}(t)$ is an arc length parametrization?

Exercises

In Exercises 1–6, compute the length of the curve over the given interval.

1. $\mathbf{r}(t) = \langle 3t, 4t - 3, 6t + 1 \rangle$, $0 \le t \le 3$

2. $\mathbf{r}(t) = 2t\mathbf{i} - 3t\mathbf{k}$, $11 \le t \le 15$

3. $\mathbf{r}(t) = \langle 2t, \ln t, t^2 \rangle$, $1 \le t \le 4$

4. $\mathbf{r}(t) = \langle 2t^2 + 1, 2t^2 - 1, t^3 \rangle$, $0 \le t \le 4$

5. $\mathbf{r}(t) = t\mathbf{i} + 2t\mathbf{j} + (t^2 - 3)\mathbf{k}$, $0 \le t \le 2$. *Hint:*

$$\int \sqrt{t^2 + a^2}\, dt = \frac{1}{2} t\sqrt{t^2 + a^2} + \frac{1}{2} a^2 \ln\!\left(t + \sqrt{t^2 + a^2}\right)$$

6. $\mathbf{r}(t) = \langle t \cos t, t \sin t, 3t \rangle$, $0 \le t \le 2\pi$

7. Compute $s(t) = \displaystyle\int_0^t \|\mathbf{r}'(u)\|\, du$ for $\mathbf{r}(t) = \langle t^2, 2t^2, t^3 \rangle$.

In Exercises 8–11, find the speed at the given value of t.

8. $\mathbf{r}(t) = \langle 2t + 3, 4t - 3, 5 - t \rangle$, $t = 4$

9. $\mathbf{r}(t) = \langle e^{t-3}, 12, 3t^{-1} \rangle$, $t = 3$

10. $\mathbf{r}(t) = \langle \sin 3t, \cos 4t, \cos 5t \rangle$, $t = \frac{\pi}{2}$

11. $\mathbf{r}(t) = \langle \cosh t, \sinh t, t \rangle$, $t = 0$

12. What is the velocity vector of a particle traveling to the right along the hyperbola $y = x^{-1}$ with constant speed 5 cm/s when the particle's location is $(2, \frac{1}{2})$?

13. A bee with velocity vector $\mathbf{r}'(t)$ starts out at the origin at $t = 0$ and flies around for T seconds. Where is the bee located at time T if $\int_0^T \mathbf{r}'(u)\, du = \mathbf{0}$? What does the quantity $\int_0^T \|\mathbf{r}'(u)\|\, du$ represent?

14. Which of the following is an arc length parametrization of a circle of radius 4 centered at the origin?

(a) $\mathbf{r}_1(t) = \langle 4 \sin t, 4 \cos t \rangle$

(b) $\mathbf{r}_2(t) = \langle 4 \sin 4t, 4 \cos 4t \rangle$

(c) $\mathbf{r}_3(t) = \langle 4 \sin \frac{t}{4}, 4 \cos \frac{t}{4} \rangle$

15. Let $\mathbf{r}(t) = \langle 3t + 1, 4t - 5, 2t \rangle$.

(a) Calculate $s(t) = \int_0^t \|\mathbf{r}'(u)\|\, du$ as a function of t.

(b) Find the inverse $\varphi(s) = t(s)$ and show that $\mathbf{r}_1(s) = \mathbf{r}(\varphi(s))$ is an arc length parametrization.

16. Find an arc length parametrization of the circle in the plane $z = 9$ with radius 4 and center $(1, 4, 9)$.

17. Find a path that traces the circle in the plane $y = 10$ with radius 4 and center $(2, 10, -3)$ with constant speed 8.

18. Show that one arch of the cycloid $\mathbf{r}(t) = \langle t - \sin t, 1 - \cos t \rangle$ has length 8. Find the value of t in $[0, 2\pi]$ where the speed is at a maximum.

19. Find an arc length parametrization of $\mathbf{r}(t) = \langle e^t \sin t, e^t \cos t, e^t \rangle$.

20. Find an arc length parametrization of $\mathbf{r}(t) = \langle t^2, t^3 \rangle$.

21. Express the arc length L of $y = x^3$ for $0 \le x \le 8$ as an integral in two ways, using the parametrizations $\mathbf{r}_1(t) = \langle t, t^3 \rangle$ and $\mathbf{r}_2(t) = \langle t^3, t^9 \rangle$. Do not evaluate the integrals, but use substitution to show that they yield the same result.

22. Show that a helix of radius R and height h making N complete turns has the parametrization

$$\left\langle R \cos\left(\frac{2\pi N t}{h}\right), R \sin\left(\frac{2\pi N t}{h}\right), t \right\rangle, \qquad 0 \le t \le h$$

23. Consider the two springs in Figure 5. One has radius 5 cm, height 4 cm, and makes three complete turns. The other has height 3 cm, radius 4 cm, and makes five complete turns.

(a) Take a guess as to which spring uses more wire.

(b) Compute the lengths of the two springs (use Exercise 22) and compare.

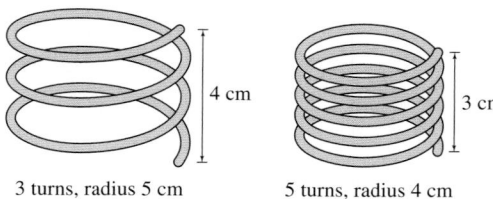

3 turns, radius 5 cm   5 turns, radius 4 cm

FIGURE 5 Which spring uses more wire?

24. Use Exercise 22 to find a general formula for the length of a helix of radius R and height h that makes N complete turns.

25. Evaluate $s(t) = \displaystyle\int_{-\infty}^t \|\mathbf{r}'(u)\|\, du$ for the **Bernoulli spiral** $\mathbf{r}(t) = \langle e^t \cos 4t, e^t \sin 4t \rangle$ (Figure 6). It is convenient to take $-\infty$ as the lower limit since $s(-\infty) = 0$. Then:

(a) Use s to obtain an arc length parametrization of $\mathbf{r}(t)$.

(b) Prove that the angle between the position vector and the tangent vector is constant.

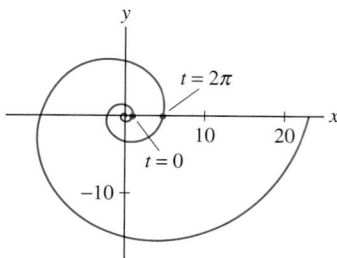

FIGURE 6 Bernoulli spiral.

Further Insights and Challenges

26. 📖 Prove that the length of a curve as computed using the arc length integral does not depend on its parametrization. More precisely, let $\mathcal{C}$ be the curve traced by $\mathbf{r}(t)$ for $a \leq t \leq b$. Let $\varphi(s)$ be a differentiable function such that $\varphi'(s) > 0$ and that $\varphi(c) = a$ and $\varphi(d) = b$. Then $\mathbf{r}_1(s) = \mathbf{r}(\varphi(s))$ parametrizes $\mathcal{C}$ for $c \leq s \leq d$. Verify that

$$\int_a^b \|\mathbf{r}'(t)\| \, dt = \int_c^d \|\mathbf{r}_1'(s)\| \, ds$$

27. Show that path $\mathbf{r}(t) = \left\langle \dfrac{1-t^2}{1+t^2}, \dfrac{2t}{1+t^2} \right\rangle$ parametrizes the unit circle with the point $(-1, 0)$ excluded for $-\infty < t < \infty$. Use this parametrization to compute the length of the unit circle as an improper integral. *Hint:* The expression for $\|\mathbf{r}'(t)\|$ simplifies.

28. The involute of a circle is the curve traced by a point at the end of a thread unwinding from a circular spool of radius R. Parametrize the involute by finding the position vector $\mathbf{r}(t)$ of the point P in Figure 7 as a function of t. Determine the arc length function and the arc length parametrization.

29. The curve $\mathbf{r}(t) = \langle t - \tanh t, \operatorname{sech} t \rangle$ is called a **tractrix**.

(a) Show that the arc length function $s(t) = \displaystyle\int_0^t \|\mathbf{r}'(u)\| \, du$ is equal to $s(t) = \ln(\cosh t)$.

(b) Show that $t = \varphi(s) = \ln(e^s + \sqrt{e^{2s} - 1})$ is an inverse of $s(t)$ and verify that

$$\mathbf{r}_1(s) = \left\langle \tanh^{-1}\left(\sqrt{1 - e^{-2s}}\right) - \sqrt{1 - e^{-2s}}, e^{-s} \right\rangle$$

is an arc length parametrization of the tractrix.

(c) Plot the tractrix if you have a computer algebra system.

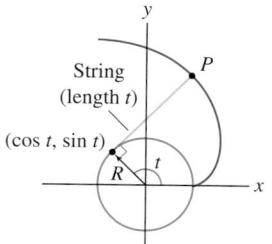

FIGURE 7 The involute of a circle.

13.4 Curvature

In this section, we apply the vector tools we have developed to study curvature which, as the name suggests, is a measure of how much a curve bends. Curvature is used to study geometric properties of curves and motion along curves. It has applications in diverse areas such as highway and roller coaster design (Figure 1), computer-aided design, optics, and even eye surgery (see Exercise 58). Biochemists have found that the curvature of DNA strands plays a role in biological processes in ways that are not yet fully understood (Figure 2).

In Chapter 4, we used the second derivative $f''(x_0)$ to measure the bending or concavity of the graph of $y = f(x)$ at $x = x_0$. It might seem natural, therefore, to take $f''(x)$ as our definition of curvature. However, there are two reasons why this will not work. First, $f''(x)$ only makes sense for a graph $y = f(x)$ in the plane, and our goal is to define curvature for curves in three-space. A second more serious problem is that $f''(x)$ does not truly capture the curvature of a curve, as we see by considering the unit circle. The circle is symmetrical, so the curvature ought to be the same at every point (Figure 3). But note that the upper semicircle is the graph of $f(x) = (1 - x^2)^{1/2}$. The second derivative $f''(x) = -(1 - x^2)^{-3/2}$ does not have the same value at every point and thus is not a good measure of curvature.

To arrive at the proper definition of curvature, let us consider a path in vector form $\mathbf{r}(t) = \langle x(t), y(t), z(t) \rangle$. The parametrization is called **regular** if $\mathbf{r}'(t) \neq \mathbf{0}$ for all t in the domain of $\mathbf{r}(t)$. We assume that $\mathbf{r}(t)$ is regular and define the **unit tangent vector** in the direction of $\mathbf{r}'(t)$, denoted $\mathbf{T}(t)$:

FIGURE 1 Curvature is a key ingredient in roller coaster design.

FIGURE 2 Biochemists study the effect of the curvature of DNA strands on biological processes.

$$\text{Unit tangent vector} = \mathbf{T}(t) = \frac{\mathbf{r}'(t)}{\|\mathbf{r}'(t)\|}$$

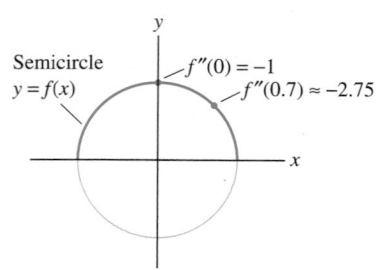

FIGURE 3 The second derivative of $f(x) = \sqrt{1 - x^2}$ does not capture the curvature of the circle, which by symmetry should be the same at all points.

Now imagine walking along a path and observing how the unit tangent vector $\mathbf{T}(t)$ changes direction (Figure 4). A change in $\mathbf{T}(t)$ indicates that the path is bending, and the more rapidly $\mathbf{T}(t)$ changes, the more the path is bent. Thus, it might seem reasonable to define curvature at $\mathbf{r}(t_0)$ as the magnitude $\|\mathbf{T}'(t)\|$ of the derivative of $\mathbf{T}(t)$ at $t = t_0$. This is almost correct, but $\|\mathbf{T}'(t)\|$ also depends on how fast you walk. To eliminate the effect of your walking speed and capture the "intrinsic" bending rate of the curve itself, we define curvature as the magnitude $\kappa(t) = \|\mathbf{T}'(t)\|$ for an *arc length parametrization* (i.e., we assume that you walk with unit speed). In an arc length parametrization, $\|\mathbf{r}'(t)\| = 1$ and $\mathbf{T}(t) = \mathbf{r}'(t)$.

DEFINITION Curvature   Let $\mathbf{r}(s)$ be an arc length parametrization and $\mathbf{T}$ the unit tangent vector. The **curvature** at $\mathbf{r}(s)$ is the quantity (denoted by a Greek letter "kappa")

$$\kappa(s) = \left\| \frac{d\mathbf{T}}{ds} \right\| \qquad \boxed{1}$$

Let us compute the curvature in the two basic cases of a line and a circle.

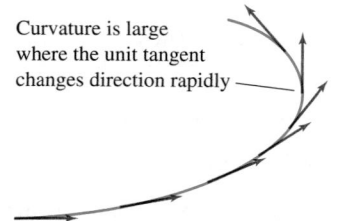

FIGURE 4 The unit tangent vector varies in direction but not length.

■ **EXAMPLE 1   A Line Has Zero Curvature**   Let $\mathcal{L}$ be the line $\mathbf{r}(s) = \langle x_0, y_0, z_0 \rangle + s\mathbf{u}$, where $\|\mathbf{u}\| = 1$. Compute the unit tangent vector and curvature at each point.

Solution   First, we note that $\mathbf{r}(s)$ is an arc length parametrization because $\mathbf{r}'(s) = \mathbf{u}$ and $\|\mathbf{r}'(s)\| = \|\mathbf{u}\| = 1$. In an arc length parametrization, $\mathbf{T}(s) = \mathbf{r}'(s)$. Therefore, $\mathbf{T}(s) = \mathbf{u}$ is constant and the curvature is zero at all points:

$$\kappa(s) = \left\| \frac{d\mathbf{T}}{ds} \right\| = \left\| \frac{d}{ds} \mathbf{u} \right\| = 0 \qquad ■$$

■ **EXAMPLE 2   Curvature of a Circle of Radius R Is $1/R$**   Compute the curvature of a circle of radius R.

Solution   Assume the circle is centered at the origin and consider the parametrization $\mathbf{r}(\theta) = \langle R\cos\theta, R\sin\theta \rangle$ (Figure 5). The tangent vector $\mathbf{r}'(\theta) = \langle -R\sin\theta, R\cos\theta \rangle$ has length R, so $\mathbf{r}(\theta)$ is not an arc length parametrization unless $R = 1$. Recall from Section 13.3 that to find an arc length parametrization, we compute the arc length function:

$$s(\theta) = \int_0^\theta \|\mathbf{r}'(t)\| \, dt = \int_0^\theta R \, dt = R\theta$$

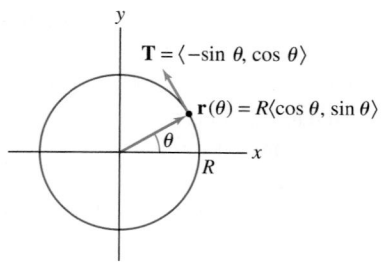

FIGURE 5 The unit tangent vector at a point on a circle of radius R.

Since a circle of radius R has curvature $1/R$, a large circle has small curvature. This makes sense if you consider an arc of fixed length 1 on a circle of radius R. As R increases, this arc gets straighter.

and form the inverse of $s = R\theta$, which is $\theta = \varphi(s) = \frac{s}{R}$. The arc length parametrization is

$$\mathbf{r}_1(s) = \mathbf{r}(\varphi(s)) = \mathbf{r}\left(\frac{s}{R}\right) = \left\langle R\cos\frac{s}{R}, R\sin\frac{s}{R} \right\rangle$$

and we have

$$\mathbf{T}(s) = \mathbf{r}_1'(s) = \left\langle -\sin\frac{s}{R}, \cos\frac{s}{R} \right\rangle$$

$$\frac{d\mathbf{T}}{ds} = -\frac{1}{R} \left\langle \cos\frac{s}{R}, \sin\frac{s}{R} \right\rangle$$

$$\kappa(s) = \left\| \frac{d\mathbf{T}}{ds} \right\| = \frac{1}{R}$$

This shows that the curvature is $1/R$ at all points on the circle. ∎

Formula (1) for curvature is often hard to use because it requires an arc length parametrization. Fortunately, there are formulas for the curvature $\kappa(t)$ in terms of any regular parametrization $\mathbf{r}(t)$. Let $s(t) = \displaystyle\int_a^t \|\mathbf{r}'(u)\| \, du$ be the arc length function (with any lower limit a) and let $t = \varphi(s)$ be the inverse of $s(t)$. As shown in Section 13.3, $\mathbf{r}(\varphi(s))$ is an arc length parametrization. Note that $\mathbf{T}(\varphi(s))$ is the unit tangent vector along $\mathbf{r}(\varphi(s))$ and thus $\kappa(t) = \left\| \dfrac{d\mathbf{T}}{ds} \right\|$. By the Chain Rule,

> **REMINDER** $\mathbf{T}(t)$ is the unit tangent vector
>
> $$\mathbf{T}(t) = \frac{\mathbf{r}'(t)}{\|\mathbf{r}'(t)\|}$$

$$\mathbf{T}'(t) = \frac{d\mathbf{T}}{ds}\frac{ds}{dt} = \frac{d\mathbf{T}}{ds} \|\mathbf{r}'(t)\|$$

$$\|\mathbf{T}'(t)\| = \left\| \frac{d\mathbf{T}}{ds} \right\| \|\mathbf{r}'(t)\| = \kappa(t) \|\mathbf{r}'(t)\|$$

This yields the formula

$$\boxed{\kappa(t) = \frac{\|\mathbf{T}'(t)\|}{\|\mathbf{r}'(t)\|}} \qquad \boxed{2}$$

The following theorem provides another useful formula for curvature directly in terms of the path $\mathbf{r}(t)$.

> *To apply Eq. (3) to plane curves, replace $\mathbf{r}(t) = \langle x(t), y(t) \rangle$ by $\mathbf{r}(t) = \langle x(t), y(t), 0 \rangle$ and compute the cross product.*

> **THEOREM 1 Formula for Curvature** If $\mathbf{r}(t)$ is a regular parametrization, then the curvature at $\mathbf{r}(t)$ is
>
> $$\kappa(t) = \frac{\|\mathbf{r}'(t) \times \mathbf{r}''(t)\|}{\|\mathbf{r}'(t)\|^3} \qquad \boxed{3}$$

Proof The vectors $\mathbf{T}(t)$ and $\mathbf{T}'(t)$ are orthogonal. Indeed, $\mathbf{T}(t)$ is a unit vector and, as we showed in Example 6 of Section 13.2, any vector-valued function of constant length is orthogonal to its derivative (see the marginal note). Set $v(t) = \|\mathbf{r}'(t)\|$. Then $\|\mathbf{T}'(t)\| = \kappa(t)v(t)$ by Eq. (2), so we have

> > **REMINDER** To prove that $\mathbf{T}(t)$ and $\mathbf{T}'(t)$ are orthogonal, differentiate the equation $\mathbf{T}(t) \cdot \mathbf{T}(t) = 1$:
> >
> > $$\frac{d}{dt}\mathbf{T}(t) \cdot \mathbf{T}(t) = 2\mathbf{T}(t) \cdot \mathbf{T}'(t) = 0$$
> >
> > Thus, $\mathbf{T}(t) \cdot \mathbf{T}'(t) = 0$. In Eq. (4), we use the formula
> >
> > $$\|\mathbf{v} \times \mathbf{w}\| = \|\mathbf{v}\| \|\mathbf{w}\| \sin\theta$$
> >
> > to conclude that $\|\mathbf{v} \times \mathbf{w}\| = \|\mathbf{v}\| \|\mathbf{w}\|$ if $\mathbf{v}$ and $\mathbf{w}$ are orthogonal.

$$\|\mathbf{T}(t) \times \mathbf{T}'(t)\| = \|\mathbf{T}(t)\| \, \|\mathbf{T}'(t)\| = \|\mathbf{T}'(t)\| = \kappa(t)\|\mathbf{r}'(t)\| = \kappa(t)v(t) \qquad \boxed{4}$$

The next step is to express this cross product in terms of $\mathbf{r}(t)$ and $\mathbf{r}'(t)$. By definition, $\mathbf{r}'(t) = v(t)\mathbf{T}(t)$, and by the Product Rule,

$$\mathbf{r}''(t) = \frac{d}{dt}\mathbf{r}'(t) = v'(t)\mathbf{T}(t) + v(t)\mathbf{T}'(t)$$

Using $\mathbf{T}(t) \times \mathbf{T}(t) = \mathbf{0}$, we find that

$$\mathbf{r}'(t) \times \mathbf{r}''(t) = v(t)\mathbf{T}(t) \times \big(v'(t)\mathbf{T}(t) + v(t)\mathbf{T}'(t)\big) = v(t)^2(\mathbf{T}(t) \times \mathbf{T}'(t))$$

To finish the proof, compute the length and apply Eq. (4):

$$\|\mathbf{r}'(t) \times \mathbf{r}''(t)\| = v(t)^2 \|\mathbf{T}(t) \times \mathbf{T}'(t)\| = \kappa(t)v(t)^3$$

We obtain $\kappa(t) = \dfrac{\|\mathbf{r}'(t) \times \mathbf{r}''(t)\|}{v(t)^3}$ as claimed. ∎

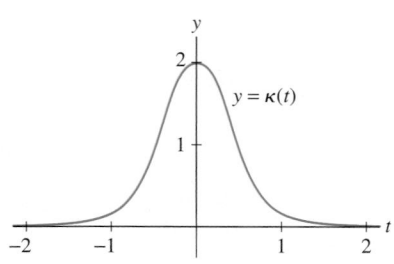

FIGURE 6 Graph of the curvature $\kappa(t)$ of the twisted cubic $\mathbf{r}(t) = \langle t, t^2, t^3 \rangle$.

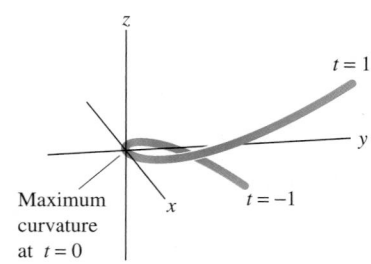

FIGURE 7 Graph of twisted cubic $\mathbf{r}(t) = \langle t, t^2, t^3 \rangle$ colored by curvature.

■ **EXAMPLE 3** Twisted Cubic Curve *CAS* Calculate the curvature $\kappa(t)$ of the twisted cubic $\mathbf{r}(t) = \langle t, t^2, t^3 \rangle$. Then plot the graph of $\kappa(t)$ and determine where the curvature is largest.

Solution We use Eq. (3):

$$\mathbf{r}'(t) = \langle 1, 2t, 3t^2 \rangle, \qquad \mathbf{r}''(t) = \langle 0, 2, 6t \rangle$$

$$\mathbf{r}'(t) \times \mathbf{r}''(t) = \begin{vmatrix} \mathbf{i} & \mathbf{j} & \mathbf{k} \\ 1 & 2t & 3t^2 \\ 0 & 2 & 6t \end{vmatrix} = 6t^2\mathbf{i} - 6t\mathbf{j} + 2\mathbf{k}$$

$$\kappa(t) = \frac{\|\mathbf{r}'(t) \times \mathbf{r}''(t)\|}{\|\mathbf{r}'(t)\|^3} = \frac{\sqrt{36t^4 + 36t^2 + 4}}{(1 + 4t^2 + 9t^4)^{3/2}}$$

The graph of $\kappa(t)$ in Figure 6 shows that the curvature is largest at $t = 0$. The curve $\mathbf{r}(t)$ is illustrated in Figure 7. The plot is colored by curvature, with large curvature represented in blue, small curvature in green.   ■

In the second paragraph of this section, we pointed out that the curvature of a graph $y = f(x)$ must involve more than just the second derivative $f''(x)$. We now show that the curvature can be expressed in terms of both $f''(x)$ and $f'(x)$.

> **THEOREM 2 Curvature of a Graph in the Plane** The curvature at the point $(x, f(x))$ on the graph of $y = f(x)$ is equal to
>
> $$\kappa(x) = \frac{|f''(x)|}{(1 + f'(x)^2)^{3/2}} \qquad \boxed{5}$$

Proof The curve $y = f(x)$ is parametrized by $\mathbf{r}(x) = \langle x, f(x) \rangle$. Therefore, $\mathbf{r}'(x) = \langle 1, f'(x) \rangle$ and $\mathbf{r}''(x) = \langle 0, f''(x) \rangle$. To apply Theorem 1, we treat $\mathbf{r}'(x)$ and $\mathbf{r}''(x)$ as vectors in $\mathbf{R}^3$ with z-component equal to zero. Then

$$\mathbf{r}'(x) \times \mathbf{r}''(x) = \begin{vmatrix} \mathbf{i} & \mathbf{j} & \mathbf{k} \\ 1 & f'(x) & 0 \\ 0 & f''(x) & 0 \end{vmatrix} = f''(x)\mathbf{k}$$

Since $\|\mathbf{r}'(t)\| = \|\langle 1, f'(x) \rangle\| = (1 + f'(x)^2)^{1/2}$, Eq. (3) yields

$$\kappa(x) = \frac{\|\mathbf{r}'(x) \times \mathbf{r}''(x)\|}{\|\mathbf{r}'(x)\|^3} = \frac{|f''(x)|}{(1 + f'(x)^2)^{3/2}} \qquad ■$$

■ **EXAMPLE 4** Compute the curvature of $f(x) = x^3 - 3x^2 + 4$ at $x = 0, 1, 2, 3$.

Solution We apply Eq. (5):

$$f'(x) = 3x^2 - 6x = 3x(x - 2), \qquad f''(x) = 6x - 6$$

$$\kappa(x) = \frac{|f''(x)|}{(1 + f'(x)^2)^{3/2}} = \frac{|6x - 6|}{(1 + 9x^2(x - 2)^2)^{3/2}}$$

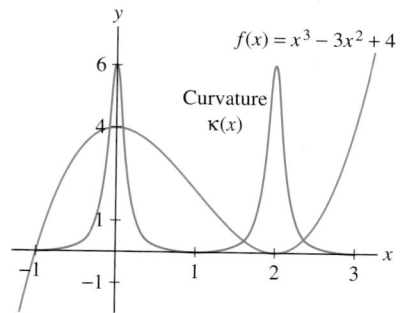

FIGURE 8 Graph of
$f(x) = x^3 - 3x^2 + 4$ and the curvature
$\kappa(x)$.

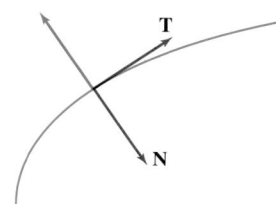

FIGURE 9 For a plane curve, the unit
normal points in the direction of bending.

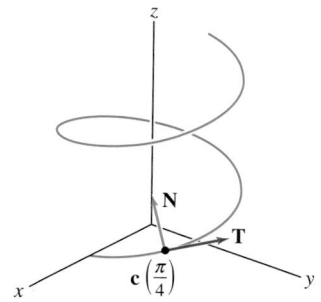

FIGURE 10 Unit tangent and normal
vectors at $t = \frac{\pi}{4}$ on the helix in Example 5.

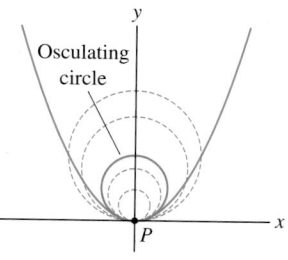

FIGURE 11 Among all circles tangent to
the curve at P, the osculating circle is the
"best fit" to the curve.

We obtain the following values:

$$\kappa(0) = \frac{6}{(1+0)^{3/2}} = 6, \quad \kappa(1) = \frac{0}{(1+9)^{3/2}} = 0$$

$$\kappa(2) = \frac{6}{(1+0)^{3/2}} = 6, \quad \kappa(3) = \frac{12}{82^{3/2}} \approx 0.016$$

Figure 8 shows that the graph bends more where the curvature is large. ∎

Unit Normal Vector

The unit tangent vector $\mathbf{T}(t)$ is orthogonal to its derivative $\mathbf{T}'(t)$, as we observed in the proof of Theorem 1. Assuming that $\mathbf{T}'(t)$ is nonzero, the unit vector in the direction $\mathbf{T}'(t)$ is called the **unit normal vector**; it is denoted $\mathbf{N}(t)$ or simply $\mathbf{N}$:

$$\boxed{\text{Unit normal vector} = \mathbf{N}(t) = \frac{\mathbf{T}'(t)}{\|\mathbf{T}'(t)\|}} \qquad \boxed{6}$$

Intuitively speaking, the unit normal $\mathbf{N}$ indicates the direction in which the curve is turning (Figure 10). This is particularly clear for a plane curve. In this case, there are two unit vectors orthogonal to $\mathbf{T}$ (Figure 9), and of these two, $\mathbf{N}$ is the vector that points to the "inside" of the curve.

■ **EXAMPLE 5** Unit Normal to a Helix Find the unit normal vector to the helix $\mathbf{r}(t) = \langle \cos t, \sin t, t \rangle$ at $t = \frac{\pi}{4}$.

Solution The tangent vector $\mathbf{r}'(t) = \langle -\sin t, \cos t, 1 \rangle$ has length $\sqrt{2}$, so

$$\mathbf{T}(t) = \frac{\mathbf{r}'(t)}{\|\mathbf{r}'(t)\|} = \frac{1}{\sqrt{2}} \langle -\sin t, \cos t, 1 \rangle$$

$$\mathbf{T}'(t) = \frac{1}{\sqrt{2}} \langle -\cos t, -\sin t, 0 \rangle$$

$$\mathbf{N}(t) = \frac{\mathbf{T}'(t)}{\|\mathbf{T}'(t)\|} = \langle -\cos t, -\sin t, 0 \rangle$$

Hence, $\mathbf{N}\left(\frac{\pi}{4}\right) = \left\langle -\frac{\sqrt{2}}{2}, -\frac{\sqrt{2}}{2}, 0 \right\rangle$ (Figure 10). ∎

We conclude this section by mentioning an important geometric interpretation of curvature in terms of the "best-fitting circle." Let P be a point on a plane curve $\mathcal{C}$ with parametrization $\mathbf{r}(t)$ and let κ_P be the curvature at P. If $\kappa_P \neq 0$, then there is a unique circle through P, called the **osculating circle** and denoted Osc_P, such that

(i) Osc_P and $\mathcal{C}$ have the same tangent line and unit normal vectors at P.
(ii) Both Osc_P and $\mathcal{C}$ have curvature κ_P at P.

Since the curvature of a circle is the reciprocal of its radius (by Example 2), the osculating circle has radius $R = 1/\kappa_P$. We refer to $R = 1/\kappa_P$ as the **radius of curvature** at P. The center of Osc_P is called the **center of curvature** at P. Among all circles tangent to $\mathcal{C}$ at P, Osc_P is the circle that "best fits" the curve (Figure 11; see also Exercise 70).

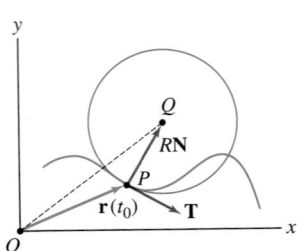

FIGURE 12 The center Q of the osculating circle at P lies at a distance $R = \kappa_P^{-1}$ from P in the normal direction.

Let Q be the center of Osc_P. We can determine Q by observing that it is located at a distance $R = 1/\kappa_P$ from P in the normal direction $\mathbf{N}$ (Figure 12). Therefore, if P is the terminal point of $\mathbf{r}(t)$ at $t = t_0$, then

$$\overrightarrow{OQ} = \mathbf{r}(t_0) + \kappa_P^{-1}\mathbf{N} \qquad \boxed{7}$$

■ **EXAMPLE 6** Parametrize the osculating circle to the parabola $y = x^2$ at $x = \frac{1}{2}$.

Solution Let $f(x) = x^2$. We use the parametrization

$$\mathbf{r}(x) = \langle x, f(x) \rangle = \left\langle x, x^2 \right\rangle$$

and proceed by the following steps.

Step 1. **Find κ and N.**
We apply Eq. (5) to $f(x) = x^2$ to compute the curvature:

$$\kappa(x) = \frac{|f''(x)|}{(1 + f'(x)^2)^{3/2}} = \frac{2}{(1 + 4x^2)^{3/2}}, \qquad \kappa\left(\frac{1}{2}\right) = \frac{2}{2^{3/2}} = \frac{1}{\sqrt{2}}$$

There is an easy way to find $\mathbf{N}$ for a plane curve that saves us the trouble of computing $\mathbf{T}'$. The tangent vector is $\mathbf{r}'(x) = \langle 1, 2x \rangle$, and we simply observe that $\langle 2x, -1 \rangle$ is orthogonal to $\mathbf{r}'(t) = \langle 1, 2x \rangle$ (since their dot product is zero). Therefore, $\mathbf{N}(x)$ is the unit vector in one of the two directions $\pm \langle 2x, -1 \rangle$. Figure 13 shows that the unit normal vector points in the positive y-direction (the direction of bending), so we conclude that

$$\mathbf{N}(x) = \frac{\langle -2x, 1 \rangle}{\|\langle -2x, 1 \rangle\|} = \frac{\langle -2x, 1 \rangle}{\sqrt{1 + 4x^2}}, \qquad \mathbf{N}\left(\frac{1}{2}\right) = \frac{1}{\sqrt{2}}\langle -1, 1 \rangle$$

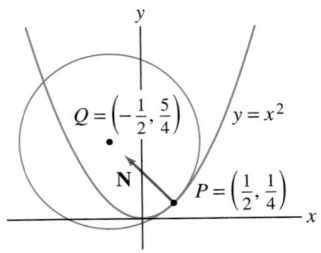

FIGURE 13 The osculating circle to $y = x^2$ at $x = 1/2$ has center Q and radius $R = \sqrt{2}$.

Step 2. **Find the center of the osculating circle.**
To find the center Q of the osculating circle at $\mathbf{r}(\frac{1}{2}) = \langle \frac{1}{2}, \frac{1}{4} \rangle$, we use Eq. (7). The position vector of Q is

$$\overrightarrow{OQ} = \mathbf{r}\left(\frac{1}{2}\right) + \kappa\left(\frac{1}{2}\right)^{-1} \mathbf{N}\left(\frac{1}{2}\right) = \left\langle \frac{1}{2}, \frac{1}{4} \right\rangle + \sqrt{2}\left(\frac{\langle -1, 1 \rangle}{\sqrt{2}}\right) = \left\langle -\frac{1}{2}, \frac{5}{4} \right\rangle$$

Step 3. **Parametrize the osculating circle.**
The osculating circle has radius $R = 1/\kappa = \sqrt{2}$ and parametrization

$$\mathbf{c}(t) = \underbrace{\left\langle -\frac{1}{2}, \frac{5}{4} \right\rangle}_{\text{Center}} + \sqrt{2}\,\langle \cos t, \sin t \rangle \qquad ■$$

If a curve C lies in a plane, then this plane is the osculating plane. For a general curve in three-space, the osculating plane varies from point to point.

To define the osculating circle for a space curve C, we must first specify the plane in which the circle lies. The **osculating plane** at a point P on C is the plane through P determined by the unit tangent $\mathbf{T}$ and the unit normal $\mathbf{N}$ at P (we assume that $\mathbf{T}' \neq 0$ so $\mathbf{N}$ is defined at P). Intuitively, the osculating plane is the plane that "most nearly" contains the curve C near P. The osculating circle is then defined as the circle of radius $R = 1/\kappa_P$ through P in the osculating plane having $\mathbf{T}$ and $\mathbf{N}$ as its unit tangent and normal vectors. Equation (7) remains valid for space curves. Figure 14 shows the osculating circle at two points on the curve $\mathbf{r}(t) = \langle \cos t, \sin t, \sin 2t \rangle$. The osculating plane at each of these points is the plane containing the circle.

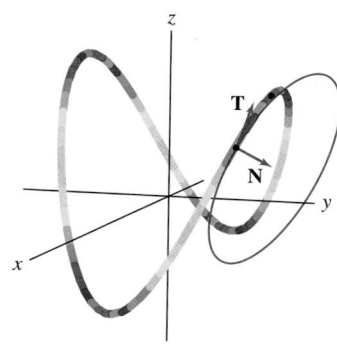

FIGURE 14 Osculating circles to $\mathbf{r}(t) = \langle \cos t, \sin t, \sin 2t \rangle$.

(A) Osculating circle at $t = \frac{\pi}{4}$; curvature is $\kappa \approx 4.12$.

(B) Osculating circle at $t = \frac{\pi}{8}$; curvature is $\kappa \approx 0.64$.

13.4 SUMMARY

- A parametrization $\mathbf{r}(t)$ is called *regular* if $\mathbf{r}'(t) \neq \mathbf{0}$ for all t. If $\mathbf{r}(t)$ is regular, we define the *unit tangent vector* $\mathbf{T}(t) = \dfrac{\mathbf{r}'(t)}{\|\mathbf{r}'(t)\|}$.

- *Curvature* is defined by $\kappa(s) = \left\| \dfrac{d\mathbf{T}}{ds} \right\|$, where $\mathbf{r}(s)$ is an arc length parametrization. In practice, we compute curvature using the following formulas, valid for arbitrary regular parametrizations:

$$\kappa(t) = \frac{\|\mathbf{T}'(t)\|}{\|\mathbf{r}'(t)\|} = \frac{\|\mathbf{r}'(t) \times \mathbf{r}''(t)\|}{\|\mathbf{r}'(t)\|^3}$$

- The curvature at a point on a graph $y = f(x)$ in the plane is

$$\kappa(x) = \frac{|f''(x)|}{(1 + f'(x)^2)^{3/2}}$$

- If $\|\mathbf{T}'(t)\| \neq 0$, we define the *unit normal vector* $\mathbf{N}(t) = \dfrac{\mathbf{T}'(t)}{\|\mathbf{T}'(t)\|}$.
- The *osculating plane* at a point P on a curve $\mathcal{C}$ is the plane through P determined by the vectors $\mathbf{T}$ and $\mathbf{N}$. It is defined only if the curvature κ_P at P is nonzero.
- The *osculating circle at* P is the circle Osc_P of radius $R = 1/\kappa_P$ in the osculating plane having $\mathbf{T}$ and $\mathbf{N}$ as its unit tangent and normal vectors. The center of Osc_P is called the *center of curvature* and R is called the *radius of curvature*.
- If $\overrightarrow{OP} = \mathbf{r}(t_0)$, the center Q of Osc_P satisfies $\overrightarrow{OQ} = \mathbf{r}(t_0) + \kappa_P^{-1}\mathbf{N}$.

13.4 EXERCISES

Preliminary Questions

1. What is the unit tangent vector of a line with direction vector $\mathbf{v} = \langle 2, 1, -2 \rangle$?

2. What is the curvature of a circle of radius 4?

3. Which has larger curvature, a circle of radius 2 or a circle of radius 4?

4. What is the curvature of $\mathbf{r}(t) = \langle 2 + 3t, 7t, 5 - t \rangle$?

5. What is the curvature at a point where $\mathbf{T}'(s) = \langle 1, 2, 3 \rangle$ in an arc length parametrization $\mathbf{r}(s)$?

6. What is the radius of curvature of a circle of radius 4?

7. What is the radius of curvature at P if $\kappa_P = 9$?

Exercises

In Exercises 1–6, calculate $\mathbf{r}'(t)$ and $\mathbf{T}(t)$, and evaluate $\mathbf{T}(1)$.

1. $\mathbf{r}(t) = \langle 12t^3, 18t^2, 9t^4 \rangle$

2. $\mathbf{r}(t) = \langle \cos \pi t, \sin \pi t, t \rangle$

3. $\mathbf{r}(t) = \langle 3 + 4t, 3 - 5t, 9t \rangle$

4. $\mathbf{r}(t) = \langle 1 + 2t, t^2, 3 - t^2 \rangle$

5. $\mathbf{r}(t) = \langle 4t^2, 9t \rangle$

6. $\mathbf{r}(t) = \langle e^t, t^2 \rangle$

In Exercises 7–12, use Eq. (3) to calculate $\kappa(t)$.

7. $\mathbf{r}(t) = \langle 1, e^t, t \rangle$

8. $\mathbf{r}(t) = \langle \cos t, t \sin t, t \rangle$

9. $\mathbf{r}(t) = \langle 4 \cos t, t, 4 \sin t \rangle$

10. $\mathbf{r}(t) = \langle 4t + 1, 4t - 3, 2t \rangle$

11. $\mathbf{r}(t) = \langle t^{-1}, 1, t \rangle$

12. $\mathbf{r}(t) = \langle \cosh t, \sinh t, t \rangle$

In Exercises 13–16, find the curvature of the plane curve at the point indicated.

13. $y = e^t, \quad t = 3$

14. $y = \cos x, \quad x = 0$

15. $y = t^4, \quad t = 2$

16. $y = t^n, \quad t = 1$

17. Find the curvature of $\mathbf{r}(t) = \langle 2 \sin t, \cos 3t, t \rangle$ at $t = \frac{\pi}{3}$ and $t = \frac{\pi}{2}$ (Figure 15).

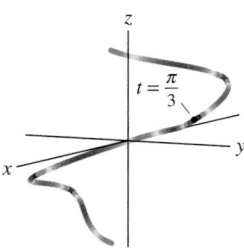

FIGURE 15 The curve $\mathbf{r}(t) = \langle 2 \sin t, \cos 3t, t \rangle$.

18. ⊂⊓⊑ Find the curvature function $\kappa(x)$ for $y = \sin x$. Use a computer algebra system to plot $\kappa(x)$ for $0 \leq x \leq 2\pi$. Prove that the curvature takes its maximum at $x = \frac{\pi}{2}$ and $\frac{3\pi}{2}$. *Hint:* To simplify the calculation, find the maximum of $\kappa(x)^2$.

19. Show that curvature at an inflection point of a plane curve $y = f(x)$ is zero.

20. Show that the tractrix $\mathbf{r}(t) = \langle t - \tanh t, \operatorname{sech} t \rangle$ has the curvature function $\kappa(t) = \operatorname{sech} t$.

21. Find the value of α such that the curvature of $y = e^{\alpha x}$ at $x = 0$ is as large as possible.

22. Find the point of maximum curvature on $y = e^x$.

23. Show that the curvature function of the parametrization $\mathbf{r}(t) = \langle a \cos t, b \sin t \rangle$ of the ellipse $\left(\frac{x}{a} \right)^2 + \left(\frac{y}{b} \right)^2 = 1$ is

$$\kappa(t) = \frac{ab}{(b^2 \cos^2 t + a^2 \sin^2 t)^{3/2}} \qquad \boxed{8}$$

24. Use a sketch to predict where the points of minimal and maximal curvature occur on an ellipse. Then use Eq. (8) to confirm or refute your prediction.

25. In the notation of Exercise 23, assume that $a \geq b$. Show that $b/a^2 \leq \kappa(t) \leq a/b^2$ for all t.

26. Use Eq. (3) to prove that for a plane curve $\mathbf{r}(t) = \langle x(t), y(t) \rangle$,

$$\kappa(t) = \frac{|x'(t)y''(t) - x''(t)y'(t)|}{(x'(t)^2 + y'(t)^2)^{3/2}} \qquad \boxed{9}$$

In Exercises 27–30, use Eq. (9) to compute the curvature at the given point.

27. $\langle t^2, t^3 \rangle, \quad t = 2$

28. $\langle \cosh s, s \rangle, \quad s = 0$

29. $\langle t \cos t, \sin t \rangle, \quad t = \pi$

30. $\langle \sin 3s, 2 \sin 4s \rangle, \quad s = \frac{\pi}{2}$

31. Let $s(t) = \int_{-\infty}^{t} \| \mathbf{r}'(u) \| \, du$ for the Bernoulli spiral $\mathbf{r}(t) = \langle e^t \cos 4t, e^t \sin 4t \rangle$ (see Exercise 25 in Section 13.3). Show that the radius of curvature is proportional to $s(t)$.

32. The **Cornu spiral** is the plane curve $\mathbf{r}(t) = \langle x(t), y(t) \rangle$, where

$$x(t) = \int_0^t \sin \frac{u^2}{2} \, du, \qquad y(t) = \int_0^t \cos \frac{u^2}{2} \, du$$

Verify that $\kappa(t) = |t|$. Since the curvature increases linearly, the Cornu spiral is used in highway design to create transitions between straight and curved road segments (Figure 16).

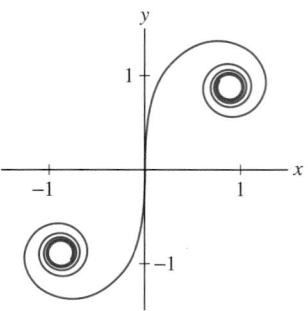

FIGURE 16 Cornu spiral.

33. ⊂⊓⊑ Plot and compute the curvature $\kappa(t)$ of the **clothoid** $\mathbf{r}(t) = \langle x(t), y(t) \rangle$, where

$$x(t) = \int_0^t \sin \frac{u^3}{3} \, du, \qquad y(t) = \int_0^t \cos \frac{u^3}{3} \, du$$

34. Find the unit normal vector $N(\theta)$ to $r(\theta) = R \langle \cos \theta, \sin \theta \rangle$, the circle of radius R. Does $N(\theta)$ point inside or outside the circle? Draw $N(\theta)$ at $\theta = \frac{\pi}{4}$ with $R = 4$.

35. Find the unit normal vector $N(t)$ to $r(t) = \langle 4, \sin 2t, \cos 2t \rangle$.

36. Sketch the graph of $r(t) = \langle t, t^3 \rangle$. Since $r'(t) = \langle 1, 3t^2 \rangle$, the unit normal $N(t)$ points in one of the two directions $\pm \langle -3t^2, 1 \rangle$. Which sign is correct at $t = 1$? Which is correct at $t = -1$?

37. Find the normal vectors to $r(t) = \langle t, \cos t \rangle$ at $t = \frac{\pi}{4}$ and $t = \frac{3\pi}{4}$.

38. Find the unit normal to the Cornu spiral (Exercise 32) at $t = \sqrt{\pi}$.

39. Find the unit normal to the clothoid (Exercise 33) at $t = \pi^{1/3}$.

40. Method for Computing N Let $v(t) = \|r'(t)\|$. Show that

$$N(t) = \frac{r''(t) - v'(t)T(t)}{\|r''(t) - v'(t)T(t)\|} \qquad \boxed{10}$$

Hint: Differentiate $r'(t) = v(t)T(t)$ and note that N is the unit vector in the direction T'.

In Exercises 41–46, use Eq. (10) to find N at the point indicated.

41. $\langle 1 + t^2, 2t, t^3 \rangle$, $\quad t = 1$

42. $\langle t, e^t, t \rangle$, $\quad t = 2$

43. $\langle t - \sin t, 1 - \cos t \rangle$, $\quad t = \pi$

44. $\langle t^2, t^3 \rangle$, $\quad t = 1$

45. $\langle t^{-1}, t, t^2 \rangle$, $\quad t = -1$

46. $\langle \cosh t, \sinh t, t \rangle$, $\quad t = 0$

47. Let $f(x) = x^2$. Show that the center of the osculating circle at (x_0, x_0^2) is given by $\left(-4x_0^3, \frac{1}{2} + 3x_0^2 \right)$.

48. Find a parametrization of the osculating circle to $y = x^2$ at $x = 1$.

49. Find a parametrization of the osculating circle to $y = \sin x$ at $x = \frac{\pi}{2}$.

50. Use Eq. (7) to find the center of curvature to $r(t) = \langle t^2, t^3 \rangle$ at $t = 1$.

In Exercises 51–55, find a parametrization of the osculating circle at the point indicated.

51. $\langle \cos t, \sin t \rangle$, $\quad t = \frac{\pi}{4}$

52. $\langle \sin t, \cos t \rangle$, $\quad t = 0$

53. $\langle t - \sin t, 1 - \cos t \rangle$, $\quad t = \pi$ (use Exercise 43)

54. $\langle 1 + t^2, 2t, t^3 \rangle$, $\quad t = 1$ (use Exercise 41)

55. $r(t) = \langle \cosh t, \sinh t, t \rangle$, $\quad t = 0$ (use Exercise 12)

56. Find the curvature and unit normal vector to the helix $r(t) = \langle \cos t, \sin t, t \rangle$ at $t = 0$. Then use Eq. (7) to find the center of curvature and a parametrization of the osculating circle at $t = 0$.

57. Figure 17 shows the graph of the half-ellipse $y = \pm\sqrt{2rx - px^2}$, where r and p are positive constants. Show that the radius of curvature at the origin is equal to r. *Hint:* One way of proceeding is to write the ellipse in the form of Exercise 23 and apply Eq. (8).

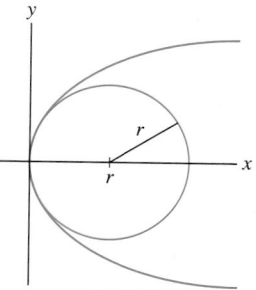

FIGURE 17 The curve $y = \sqrt{2rx - px^2}$ and the osculating circle at the origin.

58. In a recent study of laser eye surgery by Gatinel, Hoang-Xuan, and Azar, a vertical cross section of the cornea is modeled by the half-ellipse of Exercise 57. Show that the half-ellipse can be written in the form $x = f(y)$, where $f(y) = p^{-1}(r - \sqrt{r^2 - py^2})$. During surgery, tissue is removed to a depth $t(y)$ at height y for $-S \le y \le S$, where $t(y)$ is given by Munnerlyn's equation (for some $R > r$):

$$t(y) = \sqrt{R^2 - S^2} - \sqrt{R^2 - y^2} - \sqrt{r^2 - S^2} + \sqrt{r^2 - y^2}$$

After surgery, the cross section of the cornea has the shape $x = f(y) + t(y)$ (Figure 18). Show that after surgery, the radius of curvature at the point P (where $y = 0$) is R.

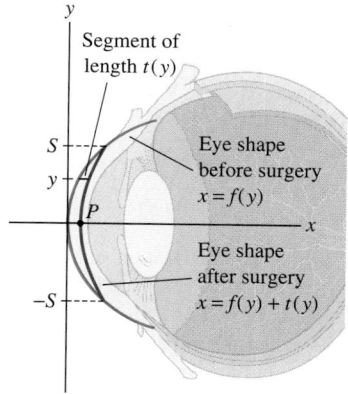

FIGURE 18 Contour of cornea before and after surgery.

59. The **angle of inclination** of a plane curve with parametrization $r(t)$ is defined as the angle $\theta(t)$ between the unit tangent vector $T(t)$ and the x-axis (Figure 19). Show that $\|T'(t)\| = |\theta'(t)|$ and conclude that if $r(s)$ is a parametrization by arc length, then

$$\kappa(s) = \left| \frac{d\theta}{ds} \right| \qquad \boxed{11}$$

Hint: Observe that $T(t) = \langle \cos \theta(t), \sin \theta(t) \rangle$.

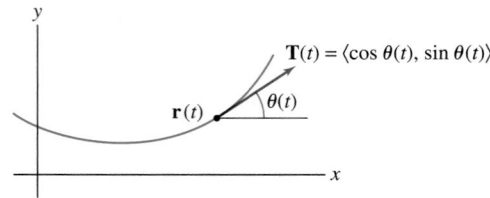

FIGURE 19 The curvature is the quantity $|d\theta/ds|$.

60. A particle moves along the path $y = x^3$ with unit speed. How fast is the tangent turning (i.e., how fast is the angle of inclination changing) when the particle passes through the point $(2, 8)$?

61. Verify Eq. (11) for a circle of radius R. Suppose that a particle traverses the circle at unit speed. Show that the change in the angle during an interval of length Δt is $\Delta\theta = \Delta t/R$ and conclude that $\theta'(s) = 1/R$.

62. Let $\theta(x)$ be the angle of inclination at a point on the graph $y = f(x)$ (see Exercise 59).

(a) Use the relation $f'(x) = \tan\theta$ to prove that $\dfrac{d\theta}{dx} = \dfrac{f''(x)}{(1 + f'(x)^2)}$.

(b) Use the arc length integral to show that $\dfrac{ds}{dx} = \sqrt{1 + f'(x)^2}$.

(c) Now give a proof of Eq. (5) using Eq. (11).

63. Use the parametrization $\mathbf{r}(\theta) = \langle f(\theta)\cos\theta, f(\theta)\sin\theta \rangle$ to show that a curve $r = f(\theta)$ in polar coordinates has curvature

$$\kappa(\theta) = \frac{|f(\theta)^2 + 2f'(\theta)^2 - 2f(\theta)f''(\theta)|}{(f(\theta)^2 + f'(\theta)^2)^{3/2}} \qquad \boxed{12}$$

Further Insights and Challenges

72. Show that the curvature of Viviani's curve, given by $\mathbf{r}(t) = \langle 1 + \cos t, \sin t, 2\sin(t/2) \rangle$, is

$$\kappa(t) = \frac{\sqrt{13 + 3\cos t}}{(3 + \cos t)^{3/2}}$$

73. Let $\mathbf{r}(s)$ be an arc length parametrization of a closed curve $\mathcal{C}$ of length L. We call $\mathcal{C}$ an **oval** if $d\theta/ds > 0$ (see Exercise 59). Observe that $-\mathbf{N}$ points to the *outside* of $\mathcal{C}$. For $k > 0$, the curve $\mathcal{C}_1$ defined by $\mathbf{r}_1(s) = \mathbf{r}(s) - k\mathbf{N}$ is called the expansion of $c(s)$ in the normal direction.

(a) Show that $\|\mathbf{r}_1'(s)\| = \|\mathbf{r}'(s)\| + k\kappa(s)$.

(b) As P moves around the oval counterclockwise, θ increases by 2π [Figure 20(A)]. Use this and a change of variables to prove that $\displaystyle\int_0^L \kappa(s)\, ds = 2\pi$.

(c) Show that $\mathcal{C}_1$ has length $L + 2\pi k$.

In Exercises 74–81, let $\mathbf{B}$ denote the binormal vector at a point on a space curve $\mathcal{C}$, defined by $\mathbf{B} = \mathbf{T} \times \mathbf{N}$.

74. Show that $\mathbf{B}$ is a unit vector.

In Exercises 64–66, use Eq. (12) to find the curvature of the curve given in polar form.

64. $f(\theta) = 2\cos\theta$ **65.** $f(\theta) = \theta$ **66.** $f(\theta) = e^\theta$

67. Use Eq. (12) to find the curvature of the general Bernoulli spiral $r = ae^{b\theta}$ in polar form (a and b are constants).

68. Show that both $\mathbf{r}'(t)$ and $\mathbf{r}''(t)$ lie in the osculating plane for a vector function $\mathbf{r}(t)$. *Hint:* Differentiate $\mathbf{r}'(t) = v(t)\mathbf{T}(t)$.

69. Show that

$$\gamma(s) = \frac{1}{\kappa}\mathbf{N} + \frac{1}{\kappa}\left((\sin\kappa s)\mathbf{T} - (\cos\kappa s)\mathbf{N}\right)$$

is an arc length parametrization of the osculating circle.

70. Two vector-valued functions $\mathbf{r}_1(s)$ and $\mathbf{r}_2(s)$ are said to *agree to order 2* at s_0 if

$$\mathbf{r}_1(s_0) = \mathbf{r}_2(s_0), \quad \mathbf{r}_1'(s_0) = \mathbf{r}_2'(s_0), \quad \mathbf{r}_1''(s_0) = \mathbf{r}_2''(s_0)$$

Let $\mathbf{r}(s)$ be an arc length parametrization of a path $\mathcal{C}$ and let P be the terminal point of $\mathbf{r}(0)$. Let $\gamma(s)$ be the arc length parametrization of the osculating circle given in Exercise 69. Show that $\mathbf{r}(s)$ and $\gamma(s)$ agree to order 2 at $s = 0$ (in fact, the osculating circle is the unique circle that approximates $\mathcal{C}$ to order 2 at P).

71. Let $\mathbf{r}(t) = \langle x(t), y(t), z(t) \rangle$ be a path with curvature $\kappa(t)$ and define the scaled path $\mathbf{r}_1(t) = \langle \lambda x(t), \lambda y(t), \lambda z(t) \rangle$, where $\lambda \neq 0$ is a constant. Prove that curvature varies inversely with the scale factor, that is, the curvature $\kappa_1(t)$ of $\mathbf{r}_1(t)$ is $\kappa_1(t) = \lambda^{-1}\kappa(t)$. This explains why the curvature of a circle of radius R is proportional to $1/R$ (in fact, it is equal to $1/R$). *Hint:* Use Eq. (3).

(A) An oval

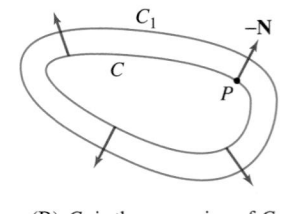

(B) $\mathcal{C}_1$ is the expansion of $\mathcal{C}$ in the normal direction.

FIGURE 20 As P moves around the oval, θ increases by 2π.

75. Follow steps (a)–(c) to prove that there is a number τ (lowercase Greek tau) called the **torsion** such that

$$\frac{d\mathbf{B}}{ds} = -\tau\mathbf{N} \qquad \boxed{13}$$

(a) Show that $\dfrac{d\mathbf{B}}{ds} = \mathbf{T} \times \dfrac{d\mathbf{N}}{ds}$ and conclude that $d\mathbf{B}/ds$ is orthogonal to $\mathbf{T}$.

(b) Differentiate $\mathbf{B} \cdot \mathbf{B} = 1$ with respect to s to show that $d\mathbf{B}/ds$ is orthogonal to $\mathbf{B}$.

(c) Conclude that $d\mathbf{B}/ds$ is a multiple of $\mathbf{N}$.

76. Show that if $\mathcal{C}$ is contained in a plane $\mathcal{P}$, then $\mathbf{B}$ is a unit vector normal to $\mathcal{P}$. Conclude that $\tau = 0$ for a plane curve.

77. Torsion means "twisting." Is this an appropriate name for τ? Explain by interpreting τ geometrically.

78. Use the identity

$$\mathbf{a} \times (\mathbf{b} \times \mathbf{c}) = (\mathbf{a} \cdot \mathbf{c})\mathbf{b} - (\mathbf{a} \cdot \mathbf{b})\mathbf{c}$$

to prove

$$\mathbf{N} \times \mathbf{B} = \mathbf{T}, \qquad \mathbf{B} \times \mathbf{T} = \mathbf{N} \qquad \boxed{14}$$

79. Follow steps (a)–(b) to prove

$$\frac{d\mathbf{N}}{ds} = -\kappa\mathbf{T} + \tau\mathbf{B} \qquad \boxed{15}$$

(a) Show that $d\mathbf{N}/ds$ is orthogonal to $\mathbf{N}$. Conclude that $d\mathbf{N}/ds$ lies in the plane spanned by $\mathbf{T}$ and $\mathbf{B}$, and hence, $d\mathbf{N}/ds = a\mathbf{T} + b\mathbf{B}$ for some scalars a, b.

(b) Use $\mathbf{N} \cdot \mathbf{T} = 0$ to show that $\mathbf{T} \cdot \dfrac{d\mathbf{N}}{ds} = -\mathbf{N} \cdot \dfrac{d\mathbf{T}}{ds}$ and compute a. Compute b similarly.

Equations (13) and (15) together with $d\mathbf{T}/dt = \kappa\mathbf{N}$ are called the **Frenet formulas** and were discovered by the French geometer Jean Frenet (1816–1900).

80. Show that $\mathbf{r}' \times \mathbf{r}''$ is a multiple of $\mathbf{B}$. Conclude that

$$\mathbf{B} = \frac{\mathbf{r}' \times \mathbf{r}''}{\|\mathbf{r}' \times \mathbf{r}''\|} \qquad \boxed{16}$$

81. The vector $\mathbf{N}$ can be computed using $\mathbf{N} = \mathbf{B} \times \mathbf{T}$ [Eq. (14)] with $\mathbf{B}$, as in Eq. (16). Use this method to find $\mathbf{N}$ in the following cases:

(a) $\mathbf{r}(t) = \langle \cos t, t, t^2 \rangle$ at $t = 0$

(b) $\mathbf{r}(t) = \langle t^2, t^{-1}, t \rangle$ at $t = 1$

13.5 Motion in Three-Space

In this section, we study the motion of a particle traveling along a path $\mathbf{r}(t)$. Recall that the velocity vector, which we now denote $\mathbf{v}(t)$, is the derivative

$$\mathbf{v}(t) = \mathbf{r}'(t) = \lim_{h \to 0} \frac{\mathbf{r}(t+h) - \mathbf{r}(t)}{h}$$

As we have seen, the velocity vector points in the direction of motion (if it is nonzero) and its magnitude $v(t) = \|\mathbf{v}(t)\|$ is the particle's speed. The **acceleration vector** is the second derivative $\mathbf{r}''(t)$, which we shall denote $\mathbf{a}(t)$. In summary,

$$\boxed{\mathbf{v}(t) = \mathbf{r}'(t), \qquad v(t) = \|\mathbf{v}(t)\|, \qquad \mathbf{a}(t) = \mathbf{r}''(t)}$$

■ **EXAMPLE 1** Find the acceleration vector of $\mathbf{r}(t) = t\mathbf{i} - t^3\mathbf{j} + \ln t\,\mathbf{k}$ at $t = 1$.

Solution

$$\mathbf{v}(t) = \mathbf{r}'(t) = \mathbf{i} - 3t^2\mathbf{j} + t^{-1}\mathbf{k}$$

$$\mathbf{a}(t) = \mathbf{r}''(t) = -6t\mathbf{j} - t^{-2}\mathbf{k}$$

Therefore, $\mathbf{a}(1) = -6\mathbf{j} - \mathbf{k}$. ■

In linear motion, acceleration is the scalar derivative of velocity. The acceleration is the rate at which the object is speeding up or slowing down, so it is zero if the object travels with constant speed. One of the surprising things about acceleration in two or three dimensions is that an object may have nonzero acceleration even when its speed is constant. This happens when $v(t) = \|\mathbf{v}(t)\|$ is constant but the *direction* of the velocity vector $\mathbf{v}(t)$ is changing. The simplest example is uniform circular motion, in which an object travels in a circular path at constant speed. In this case, $\mathbf{v}(t)$ has constant length

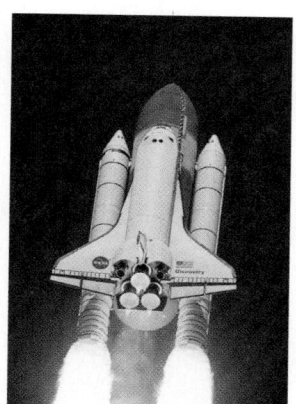

FIGURE 1 The flight of the space shuttle is analyzed using vector calculus.

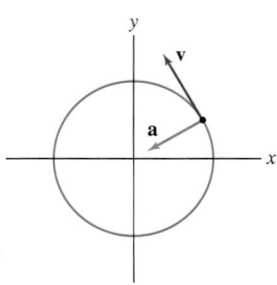

FIGURE 2 In uniform circular motion, the acceleration vector points toward the center of motion.

The constant ω is called the angular speed *because the particle's angle along the circle changes at a rate of ω radians per unit time.*

According to Newton's Law of Motion $\mathbf{F} = m\mathbf{a}$, a force is required to make an object change direction. In uniform circular motion, the force $m\mathbf{a}(t)$, which points in the normal direction by Example 2, is called the "centripetal force."

but turns continuously as the object moves around the circle. This change in direction is measured by the acceleration vector $\mathbf{a}(t)$, which points in the normal direction, toward the center of the circle (Figure 2). We will see that general motion combines the two types of acceleration: changing speed and changing direction.

■ **EXAMPLE 2** Acceleration Vector for Uniform Circular Motion Find $\mathbf{a}(t)$ and $\|\mathbf{a}(t)\|$ for motion around a circle of radius R with constant speed v.

Solution The term **uniform circular motion** refers to motion around a circle at constant speed (Figure 2). If the circle is centered at the origin and the particle begins at position $(R, 0)$ at time $t = 0$, then for some constant ω (lowercase Greek omega) we have

$$\mathbf{r}(t) = R\langle \cos \omega t, \sin \omega t\rangle, \qquad \mathbf{v}(t) = R\omega\langle -\sin \omega t, \cos \omega t\rangle$$

To determine ω, we observe that $v = \|\mathbf{v}(t)\| = R|\omega|$. Thus, $|\omega| = v/R$, and accordingly,

$$\mathbf{a}(t) = \mathbf{v}'(t) = -R\omega^2\langle \cos \omega t, \sin \omega t\rangle, \qquad \|\mathbf{a}(t)\| = R\omega^2 = R\left(\frac{v}{R}\right)^2 = \frac{v^2}{R}$$

We see that $\mathbf{a}(t)$ *has length* $\dfrac{v^2}{R}$ *and points in toward the origin* [since $\mathbf{a}(t)$ is a negative multiple of $\mathbf{r}(t)$] as in Figure 2. ■

If the acceleration $\mathbf{a}(t) = \mathbf{r}''(t)$ of an object is given, we may solve for $\mathbf{v}(t)$ and $\mathbf{r}(t)$ by integrating:

$$\mathbf{v}(t) = \int_0^t \mathbf{a}(u)\, du + \mathbf{v}_0 \qquad (\mathbf{v}_0 = \text{initial velocity vector})$$

$$\mathbf{r}(t) = \int_0^t \mathbf{v}(u)\, du + \mathbf{v}_0 t + \mathbf{r}_0 \qquad (\mathbf{r}_0 = \text{initial position})$$

■ **EXAMPLE 3** Find $\mathbf{r}(t)$ if $\mathbf{a}(t) = 2\mathbf{i} + 12t\mathbf{j}$, $\mathbf{v}(0) = 7\mathbf{i}$, and $\mathbf{r}(0) = 2\mathbf{i} + 9\mathbf{k}$.

Solution We have

$$\mathbf{v}(t) = \int_0^t \mathbf{a}(u)\, du = 2t\mathbf{i} + 6t^2\mathbf{j} + \mathbf{v}_0$$

$$\mathbf{r}(t) = \int_0^t \mathbf{v}'(u)\, du = t^2\mathbf{i} + 2t^3\mathbf{j} + \mathbf{v}_0 t + \mathbf{r}_0$$

The initial conditions give $\mathbf{v}(0) = \mathbf{v}_0 = 7\mathbf{i}$ and $\mathbf{r}(0) = \mathbf{r}_0 = 2\mathbf{i} + 9\mathbf{k}$, so

$$\mathbf{r}(t) = (t^2\mathbf{i} + 2t^3\mathbf{j}) + 7t\mathbf{i} + (2\mathbf{i} + 9\mathbf{k}) = (t^2 + 7t + 2)\mathbf{i} + 2t^3\mathbf{j} + 9\mathbf{k}$$ ■

Newton's Second Law of Motion is often stated in the scalar form $F = ma$, but a more general statement is the vector law $\mathbf{F} = m\mathbf{a}$, where $\mathbf{F}$ is the net force vector acting on the object and $\mathbf{a}$ is the acceleration vector. When the force varies from position to position, we write $\mathbf{F}(\mathbf{r}(t))$ for the force acting on a particle with position vector $\mathbf{r}(t)$ at time t. Then Newton's Second Law reads

$$\boxed{\mathbf{F}(\mathbf{r}(t)) = m\mathbf{a}(t) \qquad \text{or} \qquad \mathbf{F}(\mathbf{r}(t)) = m\mathbf{r}''(t)}$$ $\boxed{1}$

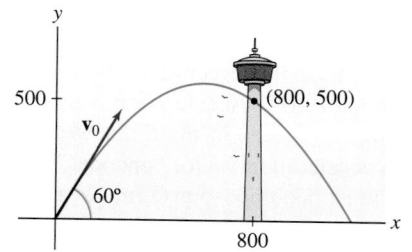

FIGURE 3 Trajectory of the bullet.

■ **EXAMPLE 4** A bullet is fired from the ground at an angle of 60° above the horizontal. What initial speed v_0 must the bullet have in order to hit a point 500 ft high on a tower located 800 ft away (ignoring air resistance)?

Solution Place the gun at the origin, and let $\mathbf{r}(t)$ be the position vector of the bullet (Figure 3).

***Step 1.* Use Newton's Law.**

Gravity exerts a downward force of magnitude mg, where m is the mass of the bullet and $g = 32$ ft/s^2. In vector form,

$$\mathbf{F} = \langle 0, -mg \rangle = m\langle 0, -g \rangle$$

In this case, Newton's Second Law $\mathbf{F} = m\mathbf{r}''(t)$ reduces to $\mathbf{r}''(t) = \langle 0, -g \rangle$. We determine $\mathbf{r}(t)$ by integrating twice:

$$\mathbf{r}'(t) = \int_0^t \mathbf{r}''(u)\, du = \int_0^t \langle 0, -32 \rangle\, du = \langle 0, -32t \rangle + \mathbf{v}_0$$

$$\mathbf{r}(t) = \int_0^t \mathbf{r}'(u)\, du = \int_0^t \left(\langle 0, -32u \rangle + \mathbf{v}_0 \right) du = \langle 0, -16t^2 \rangle + t\mathbf{v}_0 + \mathbf{r}_0$$

Here, $\mathbf{r}_0$ is the initial position and $\mathbf{v}_0$ is the initial velocity. By our choice of coordinates, $\mathbf{r}_0 = \mathbf{0}$. The initial velocity $\mathbf{v}_0$ has unknown length v_0, but we know that it points in the direction of the unit vector $\langle \cos 60°, \sin 60° \rangle$. Therefore,

$$\mathbf{v}_0 = v_0 \langle \cos 60°, \sin 60° \rangle = v_0 \left\langle \frac{1}{2}, \frac{\sqrt{3}}{2} \right\rangle$$

$$\mathbf{r}(t) = \langle 0, -16t^2 \rangle + t v_0 \left\langle \frac{1}{2}, \frac{\sqrt{3}}{2} \right\rangle$$

***Step 2.* Solve for v_0.**

The position vector of the point on the tower is $\langle 800, 500 \rangle$, so the bullet will hit a point on the tower 500 ft high if there exists a time t such that

$$\mathbf{r}(t) = \langle 0, -16t^2 \rangle + t v_0 \left\langle \frac{1}{2}, \frac{\sqrt{3}}{2} \right\rangle = \langle 800, 500 \rangle$$

Equating components, we obtain the equations

$$\frac{1}{2}t v_0 = 800, \qquad -16t^2 + \frac{\sqrt{3}}{2}t v_0 = 500$$

The first equation yields $t = \dfrac{1,600}{v_0}$. Now substitute in the second equation and solve:

$$-16 \left(\frac{1,600}{v_0} \right)^2 + \frac{\sqrt{3}}{2} \left(\frac{1,600}{v_0} \right) v_0 = 500$$

$$\left(\frac{1,600}{v_0} \right)^2 = \frac{800\sqrt{3} - 500}{16}$$

$$v_0^2 = \frac{16(1,600)^2}{800\sqrt{3} - 500}$$

$$v_0 = \frac{6,400}{\sqrt{800\sqrt{3} - 500}} \approx 215 \text{ ft/s} \qquad ■$$

Understanding the Acceleration Vector

The velocity vector $\mathbf{v}(t)$ has a clear interpretation: It points in the direction of motion and its length is the speed. What about the acceleration vector $\mathbf{a}(t)$? Acceleration is the rate of change of velocity $\mathbf{v}(t)$, but to understand $\mathbf{a}(t)$, we must keep in mind that $\mathbf{v}(t)$ can change in two ways: in length and in direction. The acceleration vector "encodes" both types of change, but to extract this information, we must decompose $\mathbf{a}(t)$ into a sum of tangential and normal components.

Recall the definition of unit tangent and unit normal vectors:

$$\mathbf{T}(t) = \frac{\mathbf{v}(t)}{\|\mathbf{v}(t)\|}, \qquad \mathbf{N}(t) = \frac{\mathbf{T}'(t)}{\|\mathbf{T}'(t)\|}$$

Thus, $\mathbf{v}(t) = v(t)\mathbf{T}(t)$, where $v(t) = \|\mathbf{v}(t)\|$, and by the Product Rule,

$$\mathbf{a}(t) = \frac{d\mathbf{v}}{dt} = \frac{d}{dt} v(t)\mathbf{T}(t) = v'(t)\mathbf{T}(t) + v(t)\mathbf{T}'(t) \qquad \boxed{2}$$

On the other hand, $\|\mathbf{T}'(t)\| = \kappa(t)v(t)$, where $\kappa(t)$ is the curvature, by Eq. (2) of Section 13.4. Therefore,

$$v(t)\mathbf{T}'(t) = v(t)\|\mathbf{T}'(t)\| \left(\frac{\mathbf{T}'(t)}{\|\mathbf{T}'(t)\|} \right) = \kappa(t)v(t)^2\mathbf{N}(t)$$

We may rewrite Eq. (2) in the form

$$\boxed{\mathbf{a}(t) = v'(t)\mathbf{T}(t) + \kappa(t)v(t)^2\,\mathbf{N}(t)} \qquad \boxed{3}$$

We call $a_{\mathbf{T}}(t) = v'(t)$ the **tangential component** and $a_{\mathbf{N}}(t) = \kappa(t)v(t)^2$ the **normal component** of acceleration (Figure 4). We often write Eq. (3) more simply as

$$\boxed{\mathbf{a} = a_{\mathbf{T}}\mathbf{T} + a_{\mathbf{N}}\mathbf{N}}$$

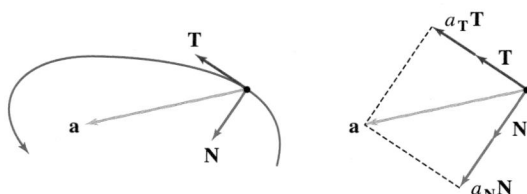

FIGURE 4 Decomposition of $\mathbf{a}$ into tangential and normal components.

For a good example of normal acceleration, imagine making a left turn in an automobile. If you drive at constant speed, your tangential acceleration is zero and you will not be pushed back against your seat. But the car seat (via friction) pushes you to the left toward the car door, causing you to accelerate in the normal direction (due to inertia, you feel as if you are being pushed to the right toward the passenger's seat). This force is proportional to κv^2, so a sharp turn (large κ) or high speed (large v) produces a strong normal force.

CONCEPTUAL INSIGHT The decomposition of Eq. (3) separates out the two ways in which velocity $\mathbf{v}(t)$ can change. The tangential component $a_{\mathbf{T}} = v'(t)$ describes the rate at which the *speed* changes, and the normal component $a_{\mathbf{N}} = \kappa(t)v(t)^2$ describes the rate at which the *direction* changes [this rate depends on both the curvature $\kappa(t)$ and the speed $v(t)$ at which you travel around the curve]. Consider the following scenarios:

- A particle travels in a straight line. Then $\kappa(t) = 0$ and $\mathbf{a}(t) = v'(t)\mathbf{T}$. Thus, $\mathbf{a}(t)$ points in the direction of motion if the particle is speeding up [$v'(t) > 0$] and opposite to the direction of motion if the particle is slowing down [$v'(t) < 0$].
- A particle travels with constant speed along a curved path. Then $v'(t) = 0$ and the acceleration $\mathbf{a}(t) = \kappa(t)v(t)^2\mathbf{N}$ is normal to the direction of motion.

General motion is a combination, involving both tangential and normal acceleration.

FIGURE 5 The Giant Ferris Wheel in Vienna, Austria, erected in 1897 to celebrate the 50th anniversary of the coronation of Emperor Franz Joseph I.

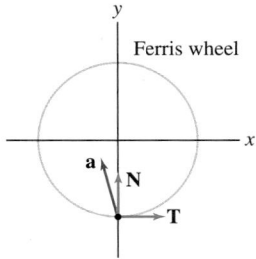

FIGURE 6

■ **EXAMPLE 5** The Giant Ferris Wheel in Vienna has radius $R = 30$ m (Figure 5). Assume that at time $t = t_0$, the wheel rotates counterclockwise with a speed of 40 m/min, and is slowing at a rate of 15 m/min^2. Find the acceleration vector **a** for a person seated in a car at the lowest point of the wheel (Figure 6).

Solution At the bottom of the wheel, $\mathbf{T} = \langle 1, 0 \rangle$ and $\mathbf{N} = \langle 0, 1 \rangle$. We are told that $a_{\mathbf{T}} = v' = -15$ at time t_0. The curvature of the wheel is $\kappa = 1/R = 1/30$, so the normal component is $a_{\mathbf{N}} = \kappa v^2 = v^2/R = (40)^2/30 \approx 53.3$. Therefore,

$$\mathbf{a} \approx -15\mathbf{T} + 53.3\mathbf{N} = \langle -15, 53.3 \rangle \text{ m/min}^2 \qquad ■$$

The following theorem provides some useful formulas for the tangential and normal components of the acceleration vector.

THEOREM 1 Tangential and Normal Components of Acceleration Let $\mathbf{a} = \mathbf{r}''(t)$ and $\mathbf{v} = \mathbf{r}'(t)$ be the acceleration and velocity vectors of a path $\mathbf{r}(t)$ at time t. Let $v(t) = \|\mathbf{v}(t)\|$ be the speed at time t. Then the coefficients $a_{\mathbf{T}} = v'(t)$ and $a_{\mathbf{N}} = \kappa(t)v(t)^2$ in the decomposition

$$\mathbf{a} = a_{\mathbf{T}}\mathbf{T} + a_{\mathbf{N}}\mathbf{N}$$

are given by the formulas

$$a_{\mathbf{T}} = \frac{\mathbf{a} \cdot \mathbf{v}}{\|\mathbf{v}\|} \qquad \boxed{4}$$

$$a_{\mathbf{N}} = \frac{\|\mathbf{a} \times \mathbf{v}\|}{\|\mathbf{v}\|} = \sqrt{\|\mathbf{a}\|^2 - |a_{\mathbf{T}}|^2} \qquad \boxed{5}$$

Proof Recall that $\mathbf{T} = \dfrac{\mathbf{v}}{\|\mathbf{v}\|}$. Since $\mathbf{T} \cdot \mathbf{T} = 1$ and $\mathbf{N} \cdot \mathbf{T} = 0$, we obtain (4) as follows:

$$\frac{\mathbf{a} \cdot \mathbf{v}}{\|\mathbf{v}\|} = \mathbf{a} \cdot \mathbf{T} = (a_{\mathbf{T}}\mathbf{T} + a_{\mathbf{N}}\mathbf{N}) \cdot \mathbf{T} = a_{\mathbf{T}}\mathbf{T} \cdot \mathbf{T} + a_{\mathbf{N}}\mathbf{N} \cdot \mathbf{T} = a_{\mathbf{T}}$$

On the other hand, $\mathbf{T} \times \mathbf{T} = 0$, so

$$\frac{\mathbf{a} \times \mathbf{v}}{\|\mathbf{v}\|} = \mathbf{a} \times \mathbf{T} = (a_{\mathbf{T}}\mathbf{T} + a_{\mathbf{N}}\mathbf{N}) \times \mathbf{T} = a_{\mathbf{T}}\mathbf{T} \times \mathbf{T} + a_{\mathbf{N}}\mathbf{N} \times \mathbf{T} = a_{\mathbf{N}}(\mathbf{N} \times \mathbf{T})$$

◄·· *REMINDER The cross product $\mathbf{v} \times \mathbf{w}$ has length*

$$\|\mathbf{v}\| \|\mathbf{w}\| \sin\theta$$

where θ is the angle between $\mathbf{v}$ and $\mathbf{w}$. If $\mathbf{v}$ and $\mathbf{w}$ are unit vectors and are orthogonal to each other ($\theta = \frac{\pi}{2}$), then $\mathbf{v} \times \mathbf{w}$ has length 1.

Note that $\mathbf{N} \times \mathbf{T}$ is a unit vector since $\mathbf{T}$ and $\mathbf{N}$ are orthogonal (see the marginal note). Thus

$$\frac{\|\mathbf{a} \times \mathbf{v}\|}{\|\mathbf{v}\|} = a_{\mathbf{N}}\|\mathbf{N} \times \mathbf{T}\| = a_{\mathbf{N}}$$

This proves the first equality in (5). Finally, since $\mathbf{T}$ and $\mathbf{N}$ are orthogonal unit vectors, $\mathbf{a}$ is the hypotenuse of a right triangle with sides of length $a_{\mathbf{T}}$ and $a_{\mathbf{N}}$ (see Figure 4), and the Pythagorean Theorem gives

$$\|\mathbf{a}\|^2 = |a_{\mathbf{T}}|^2 + |a_{\mathbf{N}}|^2$$

This yields the second equality in (5), namely $a_{\mathbf{N}} = \sqrt{\|\mathbf{a}\|^2 - |a_{\mathbf{T}}|^2}$. ■

Equation (5) is convenient if you only need to compute a_N. To compute both a_N and the vector $\mathbf{N}$, we observe that

$$a_N \mathbf{N} = \mathbf{a} - a_T \mathbf{T} = \mathbf{a} - \left(\frac{\mathbf{a} \cdot \mathbf{v}}{\|\mathbf{v}\|}\right) \frac{\mathbf{v}}{\|\mathbf{v}\|} = \mathbf{a} - \left(\frac{\mathbf{a} \cdot \mathbf{v}}{\mathbf{v} \cdot \mathbf{v}}\right) \mathbf{v} \qquad \boxed{6}$$

■ **EXAMPLE 6** Decompose the acceleration vector $\mathbf{a}$ of $\mathbf{r}(t) = \langle t^2, 2t, \ln t \rangle$ into tangential and normal components at $t = \frac{1}{2}$ (Figure 7).

Solution

Step 1. **Compute T and a_T.**
We have $\mathbf{v}(t) = \mathbf{r}'(t) = \langle 2t, 2, t^{-1} \rangle$ and $\mathbf{a}(t) = \mathbf{r}''(t) = \langle 2, 0, -t^{-2} \rangle$. Thus at $t = \frac{1}{2}$,

$$\mathbf{v} = \mathbf{r}'\left(\frac{1}{2}\right) = \left\langle 2\left(\frac{1}{2}\right), 2, \left(\frac{1}{2}\right)^{-1} \right\rangle = \langle 1, 2, 2 \rangle$$

$$\mathbf{a} = \mathbf{r}''\left(\frac{1}{2}\right) = \left\langle 2, 0, -\left(\frac{1}{2}\right)^{-2} \right\rangle = \langle 2, 0, -4 \rangle$$

$$\mathbf{T} = \frac{\mathbf{v}}{\|\mathbf{v}\|} = \frac{\langle 1, 2, 2 \rangle}{\sqrt{1^2 + 2^2 + 2^2}} = \left\langle \frac{1}{3}, \frac{2}{3}, \frac{2}{3} \right\rangle$$

$$a_T = \mathbf{a} \cdot \mathbf{T} = \langle 2, 0, -4 \rangle \cdot \left\langle \frac{1}{3}, \frac{2}{3}, \frac{2}{3} \right\rangle = \frac{-6}{3} = -2 \quad \text{[by Eq. (4)]}$$

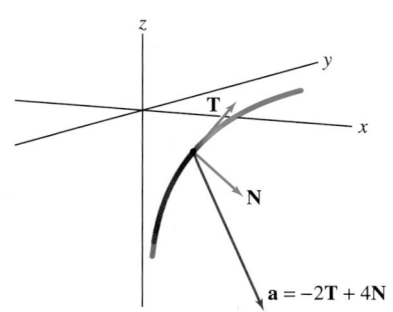

FIGURE 7 The vectors $\mathbf{T}$, $\mathbf{N}$, and $\mathbf{a}$ at $t = \frac{1}{2}$ on the curve $\mathbf{r}(t) = \langle t^2, 2t, \ln t \rangle$.

Here is a summary of the steps in Example 6:

$$a_T = \mathbf{a} \cdot \mathbf{T}$$

$$a_N \mathbf{N} = \mathbf{a} - a_T \mathbf{T}$$

$$a_N = \|a_N \mathbf{N}\| = \|\mathbf{a} - a_T \mathbf{T}\|$$

$$\mathbf{N} = \frac{\mathbf{a} - a_T \mathbf{T}}{a_N}$$

Step 2. **Compute a_N and N.**
By Eq. (6),

$$a_N \mathbf{N} = \mathbf{a} - a_T \mathbf{T} = \langle 2, 0, -4 \rangle - (-2)\left\langle \frac{1}{3}, \frac{2}{3}, \frac{2}{3} \right\rangle = \left\langle \frac{8}{3}, \frac{4}{3}, -\frac{8}{3} \right\rangle$$

Since $\mathbf{N}$ is a unit vector and $a_N \geq 0$,

$$a_N = \|a_N \mathbf{N}\| = \left\| \left\langle \frac{8}{3}, \frac{4}{3}, -\frac{8}{3} \right\rangle \right\| = \sqrt{\frac{64}{9} + \frac{16}{9} + \frac{64}{9}} = \sqrt{\frac{144}{9}} = 4$$

$$\mathbf{N} = \frac{a_N \mathbf{N}}{a_N} = \frac{\left\langle \frac{8}{3}, \frac{4}{3}, -\frac{8}{3} \right\rangle}{4} = \left\langle \frac{2}{3}, \frac{1}{3}, -\frac{2}{3} \right\rangle$$

Step 3. **Write the decomposition.**
Since $\mathbf{a} = \langle 2, 0, -4 \rangle$, $a_T = -2$, and $a_N = 4$, the decomposition of $\mathbf{a}$ into tangential and normal components is

$$\mathbf{a} = \langle 2, 0, -4 \rangle = -2\mathbf{T} + 4\mathbf{N}$$

where $\mathbf{T} = \left\langle \frac{1}{3}, \frac{2}{3}, \frac{2}{3} \right\rangle$ and $\mathbf{N} = \left\langle \frac{2}{3}, \frac{1}{3}, -\frac{2}{3} \right\rangle$. ■

■ **EXAMPLE 7** Nonuniform Circular Motion Figure 8 shows the acceleration vectors of three particles moving *counterclockwise* around a circle. In each case, state whether the particle's speed v is increasing, decreasing, or momentarily constant.

Solution In (A), the tangential component $a \cdot T$ of the acceleration vector is negative, and since $v' = a \cdot T$, the particle's speed is decreasing ($v' < 0$). In (B), the acceleration

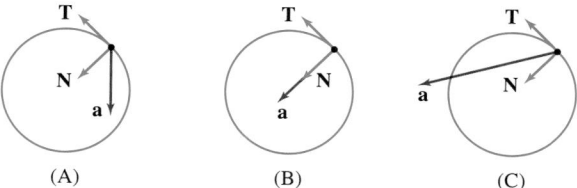

FIGURE 8 Acceleration vectors of particles moving counterclockwise (in the direction of **T**) around a circle.

(A)   (B)   (C)

vector points in the normal direction, so the particle's speed is constant at this moment, that is, $v' = 0$. In (C), the tangential component $a \cdot T$ of the acceleration vector is positive, so the particle's speed is increasing ($v' = a \cdot T > 0$). ∎

13.5 SUMMARY

- The path of an object in 3-space is described by a vector-valued function $\mathbf{r}(t)$.
- The *velocity vector* $\mathbf{v}(t) = \mathbf{r}'(t)$ points in the direction of motion and its length $v(t) = \|\mathbf{v}(t)\|$ is the object's speed.
- The *acceleration vector* is the second derivative $\mathbf{a}(t) = \mathbf{r}''(t)$.
- The acceleration vector $\mathbf{a}$ is the sum of a tangential component (reflecting change in speed) and a normal component (reflecting change in direction).

Unit tangent vector	$\mathbf{T}(t) = \dfrac{\mathbf{v}(t)}{\|\mathbf{v}(t)\|}$
Unit normal vector	$\mathbf{N}(t) = \dfrac{\mathbf{T}'(t)}{\|\mathbf{T}'(t)\|}$
Decomposition of acceleration	$\mathbf{a}(t) = a_{\mathbf{T}}(t)\mathbf{T}(t) + a_{\mathbf{N}}(t)\mathbf{N}(t)$
Tangential component	$a_{\mathbf{T}} = v'(t) = \dfrac{\mathbf{a} \cdot \mathbf{v}}{\|\mathbf{v}\|}$
Normal component	$a_{\mathbf{N}} = \kappa(t)v(t)^2 = \dfrac{\|\mathbf{a} \times \mathbf{v}\|}{\|\mathbf{v}\|}$
	$a_{\mathbf{N}}\mathbf{N} = \mathbf{a} - a_{\mathbf{T}}\mathbf{T} = \mathbf{a} - \left(\dfrac{\mathbf{a} \cdot \mathbf{v}}{\mathbf{v} \cdot \mathbf{v}}\right)\mathbf{v}$

13.5 EXERCISES

Preliminary Questions

1. If a particle travels with constant speed, must its acceleration vector be zero? Explain.

2. For a particle in uniform circular motion around a circle, which of the vectors $\mathbf{v}(t)$ or $\mathbf{a}(t)$ always points toward the center of the circle?

3. Two objects travel to the right along the parabola $y = x^2$ with nonzero speed. Which of the following must be true?

(a) Their velocity vectors point in the same direction.
(b) Their velocity vectors have the same length.
(c) Their acceleration vectors point in the same direction.

4. Use the decomposition of acceleration into tangential and normal components to explain the following statement: If the speed is constant, then the acceleration and velocity vectors are orthogonal.

5. If a particle travels along a straight line, then the acceleration and velocity vectors are (choose the correct statement):

(a) Orthogonal   **(b)** Parallel

6. What is the length of the acceleration vector of a particle traveling around a circle of radius 2 cm with constant velocity 4 cm/s?

7. Two cars are racing around a circular track. If, at a certain moment, both of their speedometers read 110 mph. then the two cars have the same (choose one):

(a) a_T (b) a_N

Exercises

1. Use the table here to calculate the difference quotients $\dfrac{\mathbf{r}(1+h) - \mathbf{r}(1)}{h}$ for $h = -0.2, -0.1, 0.1, 0.2$. Then estimate the velocity and speed at $t = 1$.

$\mathbf{r}(0.8)$	$\langle 1.557, 2.459, -1.970 \rangle$
$\mathbf{r}(0.9)$	$\langle 1.559, 2.634, -1.740 \rangle$
$\mathbf{r}(1)$	$\langle 1.540, 2.841, -1.443 \rangle$
$\mathbf{r}(1.1)$	$\langle 1.499, 3.078, -1.035 \rangle$
$\mathbf{r}(1.2)$	$\langle 1.435, 3.342, -0.428 \rangle$

2. Draw the vectors $\mathbf{r}(2+h) - \mathbf{r}(2)$ and $\dfrac{\mathbf{r}(2+h) - \mathbf{r}(2)}{h}$ for $h = 0.5$ for the path in Figure 9. Draw $\mathbf{v}(2)$ (using a rough estimate for its length).

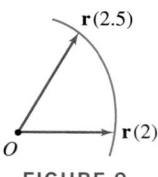

FIGURE 9

In Exercises 3–6, calculate the velocity and acceleration vectors and the speed at the time indicated.

3. $\mathbf{r}(t) = \langle t^3, 1 - t, 4t^2 \rangle$, $t = 1$

4. $\mathbf{r}(t) = e^t \mathbf{j} - \cos(2t)\mathbf{k}$, $t = 0$

5. $\mathbf{r}(\theta) = \langle \sin \theta, \cos \theta, \cos 3\theta \rangle$, $\theta = \frac{\pi}{3}$

6. $\mathbf{r}(s) = \left(\dfrac{1}{1+s^2}, \dfrac{s}{1+s^2} \right)$, $s = 2$

7. Find $\mathbf{a}(t)$ for a particle moving around a circle of radius 8 cm at a constant speed of $v = 4$ cm/s (see Example 2). Draw the path and acceleration vector at $t = \frac{\pi}{4}$.

8. Sketch the path $\mathbf{r}(t) = \langle 1 - t^2, 1 - t \rangle$ for $-2 \le t \le 2$, indicating the direction of motion. Draw the velocity and acceleration vectors at $t = 0$ and $t = 1$.

9. Sketch the path $\mathbf{r}(t) = \langle t^2, t^3 \rangle$ together with the velocity and acceleration vectors at $t = 1$.

10. [icon] The paths $\mathbf{r}(t) = \langle t^2, t^3 \rangle$ and $\mathbf{r}_1(t) = \langle t^4, t^6 \rangle$ trace the same curve and $\mathbf{r}_1(1) = \mathbf{r}(1)$. Do you expect either the velocity or acceleration vectors of these paths at $t = 1$ to point in the same direction? Compute these vectors and draw them on a single plot of the path.

In Exercises 11–14, find $\mathbf{v}(t)$ given $\mathbf{a}(t)$ and the initial velocity.

11. $\mathbf{a}(t) = \langle t, 4 \rangle$, $\mathbf{v}(0) = \langle \frac{1}{3}, -2 \rangle$

12. $\mathbf{a}(t) = \langle e^t, 0, t+1 \rangle$, $\mathbf{v}(0) = \langle 1, -3, \sqrt{2} \rangle$

13. $\mathbf{a}(t) = \mathbf{k}$, $\mathbf{v}(0) = \mathbf{i}$

14. $\mathbf{a}(t) = t^2 \mathbf{k}$, $\mathbf{v}(0) = \mathbf{i} - \mathbf{j}$

In Exercises 15–18, find $\mathbf{r}(t)$ and $\mathbf{v}(t)$ given $\mathbf{a}(t)$ and the initial velocity and position.

15. $\mathbf{a}(t) = \langle t, 4 \rangle$, $\mathbf{v}(0) = \langle 3, -2 \rangle$, $\mathbf{r}(0) = \langle 0, 0 \rangle$

16. $\mathbf{a}(t) = \langle e^t, 2t, t+1 \rangle$, $\mathbf{v}(0) = \langle 1, 0, 1 \rangle$, $\mathbf{r}(0) = \langle 2, 1, 1 \rangle$

17. $\mathbf{a}(t) = t\mathbf{k}$, $\mathbf{v}(0) = \mathbf{i}$, $\mathbf{r}(0) = \mathbf{j}$

18. $\mathbf{a}(t) = \cos t\mathbf{k}$, $\mathbf{v}(0) = \mathbf{i} - \mathbf{j}$, $\mathbf{r}(0) = \mathbf{i}$

19. A bullet is fired from the ground at an angle of $45°$. What initial speed must the bullet have in order to hit the top of a 400-ft tower located 600 ft away?

20. A bullet is fired from the ground at an angle of $60°$ with initial speed $v_0 = 30$ ft/s. How far does the bullet travel? What is the bullet's velocity vector and speed when it hits the ground?

21. A projectile fired at an angle of $60°$ lands 1,200 ft away. What was its initial speed?

22. Show that a projectile fired at an angle θ with initial speed v_0 travels a total distance $(v_0^2/g) \sin 2\theta$ before hitting the ground. Conclude that the maximum distance (for a given v_0) is attained for $\theta = 45°$.

23. One player throws a baseball to another player standing 80 ft away with initial speed 60 ft/s. Use the result of Exercise 22 to find two angles θ at which the ball can be released. Which angle gets the ball there faster?

24. Show that a bullet fired at an angle θ will hit the top of an h-ft tower located d ft away if its initial speed is

$$v_0 = \frac{4d \sec \theta}{\sqrt{d \tan \theta - h}}$$

25. At a certain moment, a moving particle has velocity $\mathbf{v} = \langle 2, 2, -1 \rangle$ and $\mathbf{a} = \langle 0, 4, 3 \rangle$. Find $\mathbf{T}$, $\mathbf{N}$, and the decomposition of $\mathbf{a}$ into tangential and normal components.

26. At a certain moment, a moving particle has velocity $\mathbf{v} = \langle 12, 20, 20 \rangle$ and acceleration $\mathbf{a} = \langle 2, 1, -3 \rangle$. Is the particle speeding up or slowing down?

27. A particle follows a path $\mathbf{r}(t)$ for $0 \le t \le T$, beginning at the origin O. The vector $\bar{\mathbf{v}} = \dfrac{1}{T} \displaystyle\int_0^T \mathbf{r}'(t)\, dt$ is called the **average velocity**

vector. Suppose that $\bar{\mathbf{v}} = \mathbf{0}$. Answer and explain the following:

(a) Where is the particle located at time T if $\bar{\mathbf{v}} = \mathbf{0}$?

(b) Is the particle's average speed necessarily equal to zero?

28. 📖 One consequence of Kepler's Laws of Planetary Motion is that a planet moves faster when it is closer to the sun. Which of Figure 10(A) or 10(B) represents a planetary orbit (with the velocity vectors as shown)?

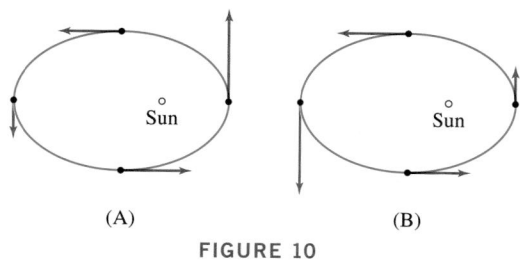

(A)　　　　　(B)

FIGURE 10

29. A space shuttle orbits the earth at an altitude 200 miles above the earth's surface, with constant speed $v = 17,000$ mph. Find the magnitude of the shuttle's acceleration (in ft/s^2), assuming that the radius of the earth is 4,000 miles (Figure 11).

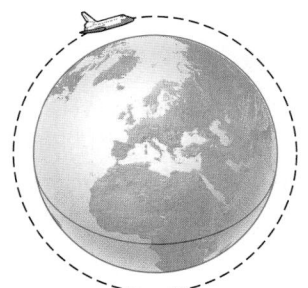

FIGURE 11 Space shuttle orbit.

In Exercises 30–33, use (4) and (5) to find $a_\mathbf{T}$ and $a_\mathbf{N}$ as a function of t or at the point indicated.

30. $\mathbf{r}(t) = \langle t^2, t^3 \rangle$

31. $\mathbf{r}(t) = \langle t, \cos t, \sin t \rangle$

32. $\mathbf{r}(t) = \langle t^{-1}, \ln t, t^2 \rangle$, $t = 1$

33. $\mathbf{r}(t) = \langle e^t, t, e^{-t} \rangle$, $t = 0$

In Exercise 34–41, use (4) and (6) to find the decomposition of $\mathbf{a}(t)$ into tangential and normal components at the point indicated, as in Example 6.

34. $\mathbf{r}(t) = \langle 4 - t, t + 1, t^2 \rangle$, $t = 1$

35. $\mathbf{r}(t) = \langle t, e^t, te^t \rangle$, $t = 0$

36. $\mathbf{r}(t) = \langle t, \frac{1}{2}t^2, \frac{1}{6}t^3 \rangle$, $t = 1$

37. $\mathbf{r}(t) = \langle t, \frac{1}{2}t^2, \frac{1}{6}t^3 \rangle$, $t = 4$

38. $\mathbf{r}(t) = \langle e^t, 1 - t \rangle$, $t = 0$

39. $\mathbf{r}(\theta) = \langle \cos \theta, \sin \theta, \theta \rangle$, $\theta = 0$

40. $\mathbf{r}(t) = \langle \frac{1}{3}t^3, 1 - 3t \rangle$, $t = -1$

41. $\mathbf{r}(t) = \langle t, \cos t, t \sin t \rangle$, $t = \frac{\pi}{2}$

42. Let $\mathbf{r}(t) = \langle t^2, 4t - 3 \rangle$. Find $\mathbf{T}(t)$ and $\mathbf{N}(t)$, and show that the decomposition of $\mathbf{a}(t)$ into tangential and normal components is

$$ \mathbf{a}(t) = \left(\frac{2t}{\sqrt{t^2 + 4}} \right) \mathbf{T} + \left(\frac{4}{\sqrt{t^2 + 4}} \right) \mathbf{N} $$

43. Find the components $a_\mathbf{T}$ and $a_\mathbf{N}$ of the acceleration vector of a particle moving along a circular path of radius $R = 100$ cm with constant velocity $v_0 = 5$ cm/s.

44. At time t_0, a moving particle has velocity vector $\mathbf{v} = 2\mathbf{i}$ and acceleration vector $\mathbf{a} = 3\mathbf{i} + 18\mathbf{k}$. Determine the curvature $\kappa(t_0)$ of the particle's path at time t_0.

45. A car proceeds along a circular path of radius $R = 1,000$ ft centered at the origin. Starting at rest, its speed increases at a rate of t ft/s^2. Find the acceleration vector $\mathbf{a}$ at time $t = 3$ s and determine its decomposition into normal and tangential components.

46. In the notation of Example 5, find the acceleration vector for a person seated in a car at (a) the highest point of the ferris wheel and (b) the two points level with the center of the wheel.

47. Suppose that $\mathbf{r} = \mathbf{r}(t)$ lies on a sphere of radius R for all t. Let $\mathbf{J} = \mathbf{r} \times \mathbf{r}'$. Show that $\mathbf{r}' = (\mathbf{J} \times \mathbf{r})/\|\mathbf{r}\|^2$. *Hint:* Observe that $\mathbf{r}$ and $\mathbf{r}'$ are perpendicular.

48. 📖 A particle moves counterclockwise around a circle. Which of the vectors in Figure 12 is not a possible acceleration vector? Explain. For the two possible acceleration vectors, state whether the particle is speeding up or slowing down.

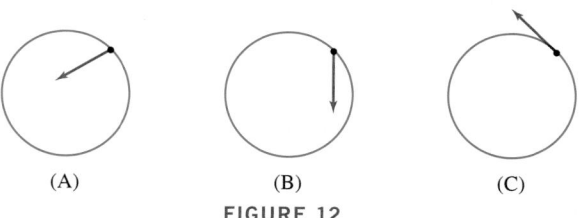

(A)　　　　　(B)　　　　　(C)

FIGURE 12

Further Insights and Challenges

49. 📐 The orbit of a planet is an ellipse with the sun at one focus. The sun's gravitational force acts along the radial line from the planet to the sun (the dashed lines in Figure 13), and by Newton's Second Law, the acceleration vector points in the same direction. Explain in words why the planet must slow down in the upper half of the orbit (as it moves away from the sun) and speed up in the lower half. Kepler's Second Law, discussed in the next section, gives a more precise version of this qualitative conclusion. *Hint:* Consider the decomposition of **a** into normal and tangential components.

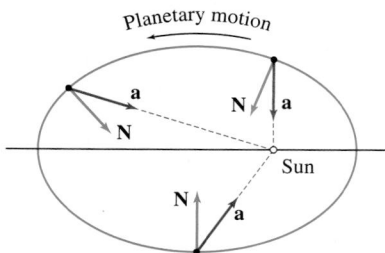

FIGURE 13 Elliptical orbit of a planet around the sun.

In Exercises 50–54, consider a car of mass m traveling along a curved but level road. To avoid skidding, the road must supply a frictional force **F** = m**a**, *where* **a** *is the car's acceleration vector. The maximum magnitude of the frictional force is* μmg, *where* μ *is the coefficient of friction and* $g = 32$ ft/s². *Let* v *be the car's speed in feet per second.*

50. Show that the car will not skid if the curvature κ of the road is such that (with $R = 1/\kappa$)

$$(v')^2 + \left(\frac{v^2}{R}\right)^2 < (\mu g)^2 \qquad \boxed{7}$$

Note that braking ($v' < 0$) and speeding up ($v' > 0$) contribute equally to skidding.

51. Suppose that the maximum radius of curvature along a curved highway is $R = 600$ ft. How fast can a car travel (at constant speed) along the highway without skidding if the coefficient of friction is $\mu = 0.5$?

52. Beginning at rest, a car drives around a circular track of radius $R = 1,000$ ft, accelerating at a rate of 1 ft/s². After how many seconds will the car begin to skid if the coefficient of friction is $\mu = 0.6$?

53. You want to reverse your direction in the shortest possible time by driving around a semicircular bend (Figure 14). If you travel at the maximum possible *constant speed* v that will not cause skidding, is it faster to hug the inside curve (radius r) or the outside curb (radius R)? *Hint:* Use Eq. (7) to show that at maximum speed, the time required to drive around the semicircle is proportional to the square root of the radius.

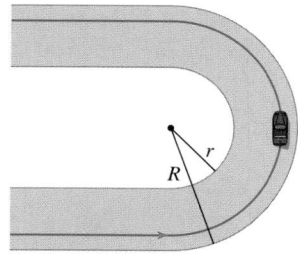

FIGURE 14 Car going around the bend.

54. What is the smallest radius R about which a car can turn without skidding at 60 mph if $\mu = 0.75$ (a typical value)?

13.6 Planetary Motion According to Kepler and Newton

No events are more emblematic of the scientific revolution of the seventeenth century than the discovery of the laws of planetary motion by the German astronomer Johannes Kepler around 1609 and Isaac Newton's publication in 1687 of a mathematical derivation of these laws from the universal laws of motion and gravitation. This dramatic success of mathematics in rendering the natural world comprehensible made a great impression on succeeding generations of scientists and encouraged them to search for mathematical laws governing other phenomena such as electricity and magnetism, aerodynamics, and atomic processes.

To formulate Kepler's Laws, we recall that an ellipse is the set of all points P such that $F_1 P + F_2 P = K$, where K is a constant and the points F_1 and F_2 are the **foci** of the ellipse. As we will state formally in a moment, the planetary orbits are ellipses with the sun at one focus. Furthermore, if we imagine a radial vector $\mathbf{r}(t)$ pointing from the sun to the planet as in Figure 1, then this radial vector sweeps out area at a constant rate or, as Kepler stated in his Second Law, it sweeps out equal areas in equal times (Figure

FIGURE 1 The planet travels along an ellipse with the sun at one focus. As it travels, the radial vector $\mathbf{r}(t)$ sweeps out area at a constant rate.

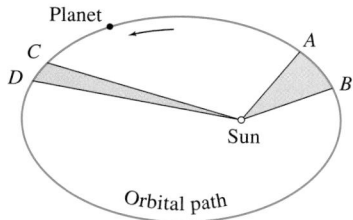

FIGURE 2 The two shaded regions have equal areas, and by Kepler's Second Law, the planet sweeps them out in equal times. To do so, the planet must travel faster along the path from A to B than from C to D.

2). The Second Law governs how speed changes along the orbit: To sweep out area at a constant rate, the planet must move faster when it is near the sun and slower when it is farther away. Kepler's Third Law concerns the **period** T of the orbit, defined as the time required to complete one full revolution.

Kepler's Three Laws

   (i) **Law of Ellipses:** The orbit of a planet is an ellipse with the sun at one focus.

  (ii) **Law of Equal Area in Equal Time:** The position vector pointing from the sun to the planet sweeps out equal areas in equal times.

 (iii) **Law of the Period of Motion:** $T^2 = \left(\dfrac{4\pi^2}{GM}\right)a^3$, where a is the length of the semimajor axis of the ellipse (see Figure 1).

In the Third Law:

- G is Newton's universal gravitational constant: 6.673×10^{-11} m^3 kg^{-1} s^{-2}.
- M is the mass of the sun, approximately 1.989×10^{30} kg.

Kepler's original version of the Third Law stated only that T^2 is proportional to a^3. It was Newton who discovered that the constant of proportionality is $4\pi^2/GM$. Newton further realized that if you can measure T and a through observation, then you can use the Third Law to solve for the mass M. This method is used by astronomers to find the masses of the planets (by measuring T and a for moons revolving around the planet) as well as the masses of binary stars and galaxies. See Exercises 4–6, 13.

Our treatment of Kepler's Laws makes several simplifying assumptions. We treat the sun and planet as point masses and ignore the gravitational attraction of the planets on each other. Furthermore, both the sun and the planet revolve around their mutual center of mass. However, the sun is much more massive than the planet, and it is justifiable to ignore the sun's motion and assume that the planet revolves around the center of the sun.

We place the sun at the origin of the coordinate system and let $\mathbf{r} = \mathbf{r}(t)$ denote the vector from the origin to a planet of mass m, as in Figure 1. Newton's Universal Law of Gravitation tells us that the sun attracts the planet with a gravitational force of magnitude $GMm/\|\mathbf{r}(t)\|^2$ in the direction of the vector $-\mathbf{r}(t)$ pointing from the planet to the sun. Therefore,

$$\mathbf{F}(\mathbf{r}(t)) = -\left(\frac{GMm}{\|\mathbf{r}(t)\|^2}\right)\underbrace{\frac{\mathbf{r}(t)}{\|\mathbf{r}(t)\|}}_{\text{Unit radial vector}} = -\frac{GMm}{\|\mathbf{r}(t)\|^3}\mathbf{r}(t)$$

Combining this with Newton's Second Law of Motion, $\mathbf{F}(\mathbf{r}(t)) = m\mathbf{r}''(t)$, we obtain

$$m\mathbf{r}''(t) = -\frac{GMm}{\|\mathbf{r}(t)\|^3}\mathbf{r}(t) \qquad \boxed{1}$$

To simplify the notation, we set $k = GM$. Then (1) yields

The constant k has the value

$$k = GM \approx 1.327 \times 10^{20} \text{ m}^3 \text{ s}^{-2}$$

$$\boxed{\mathbf{r}''(t) = -\frac{k}{\|\mathbf{r}(t)\|^3}\mathbf{r}(t)} \qquad \boxed{2}$$

Each of Kepler's Laws is a consequence of this *differential equation* for the vector-valued function $\mathbf{r}(t)$.

Proof of Kepler's Second Law

Let $A(t)$ be the area swept out by the radial vector in the time interval $[0, t]$ (Figure 3). Kepler's Second Law states that the rate of change $\dfrac{dA(t)}{dt}$ is constant. We prove this by expressing the rate of change as a cross product.

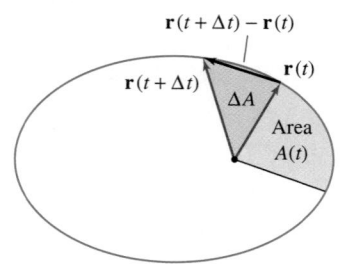

FIGURE 3 Area swept out during time interval Δt is approximately equal to the area of the triangle.

The area swept out in a small time interval $[t, t + \Delta t]$ is

$$\Delta A = A(t + \Delta t) - A(t)$$

Figure 3 shows that ΔA is approximately equal to the area of the triangle whose sides are the vectors $\mathbf{r}(t)$, $\mathbf{r}(t) + \Delta t$, and $\mathbf{r}(t + \Delta t) - \mathbf{r}(t)$. By Theorem 3 of Section 12.4, this area is equal to one-half the length of the cross product:

$$\Delta A \approx \frac{1}{2} \| \mathbf{r}(t) \times (\mathbf{r}(t + \Delta t) - \mathbf{r}(t)) \|$$

Thus, we have

$$\frac{dA}{dt} = \lim_{\Delta t \to 0} \frac{\Delta A}{\Delta t} = \frac{1}{2} \lim_{\Delta t \to 0} \left\| \mathbf{r}(t) \times \left(\frac{\mathbf{r}(t + \Delta t) - \mathbf{r}(t)}{\Delta t} \right) \right\|$$

$$= \frac{1}{2} \| \mathbf{r}(t) \times \mathbf{r}'(t) \|$$

Define

See Exercise 24 for a more rigorous derivation of Eq. (3).

$$\boxed{\mathbf{J} = \mathbf{r}(t) \times \mathbf{r}'(t)}$$

We have shown that

$$\frac{dA}{dt} = \frac{1}{2} \| \mathbf{J} \| \qquad \boxed{3}$$

To finish the proof, it is necessary to show that $\| \mathbf{J} \|$ is constant. In fact, we prove that $\mathbf{J}$ itself is constant by showing that $\mathbf{J}'(t) = 0$ for all t. By the Product Rule,

$$\frac{d\mathbf{J}}{dt} = \frac{d}{dt} \left(\mathbf{r}(t) \times \mathbf{r}'(t) \right) = \mathbf{r}(t) \times \mathbf{r}''(t) + \mathbf{r}'(t) \times \mathbf{r}'(t) \qquad \boxed{4}$$

◀··· REMINDER *Recall that* $\mathbf{a} \times \mathbf{b} = \mathbf{0}$ *if* $\mathbf{a}$ *and* $\mathbf{b}$ *are proportional.*

In physics, $m\mathbf{J}$ is called the angular momentum vector, and when $m\mathbf{J}$ is constant, we say that angular momentum is conserved. Our proof actually shows that angular momentum is conserved whenever the force on an object acts in the radial direction (whether or not its magnitude follows an inverse square law).

Equation (2) tells us that $\mathbf{r}''(t)$ and $\mathbf{r}(t)$ are proportional, so $\mathbf{r}(t) \times \mathbf{r}''(t) = \mathbf{0}$. Similarly, $\mathbf{r}'(t) \times \mathbf{r}'(t) = \mathbf{0}$, and thus we obtain the desired result

$$\boxed{\frac{d\mathbf{J}}{dt} = 0} \qquad \boxed{5}$$

Equation (5) has another important consequence. The vectors $\mathbf{r}(t)$ and $\mathbf{r}'(t)$ are orthogonal to $\mathbf{J} = \mathbf{r}(t) \times \mathbf{r}'(t)$ for all t. Since $\mathbf{J}$ is constant, we may conclude that both $\mathbf{r}(t)$ and $\mathbf{r}'(t)$ are confined to the plane orthogonal to $\mathbf{J}$ (Figure 4). In other words, the *motion of a planet around the sun takes place in a plane.*

Proof of the Law of Ellipses

In the centuries since Newton's original work on Kepler's Laws, several different proofs of the the law of ellipses have been devised. One of the most elegant methods, presented here, is based on the **Lenz vector**:

$$\mathbf{L} = \left(\frac{1}{k} \mathbf{r}' \times \mathbf{J} \right) - \frac{\mathbf{r}}{\| \mathbf{r} \|}$$

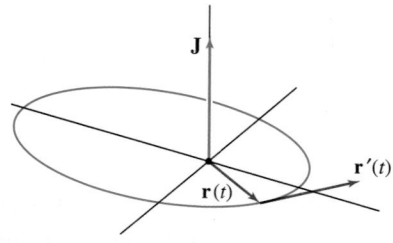

FIGURE 4 The orbit is contained in the plane orthogonal to $\mathbf{J}$.

The first step is to show that $\mathbf{L}$ is constant (thus $\mathbf{L}$, like $\mathbf{J}$, is conserved). Although it is far from obvious at this stage, we will eventually show that $\mathbf{L}$ points in the direction of the semimajor axis of the orbit.

THEOREM 1 The Lenz Vector Is Conserved

$$\frac{d\mathbf{L}}{dt} = \mathbf{0}$$

Equation (6) can be verified by direct calculation (see Exercise 74 in Section 12.4). Equation (7) may be verified using Eq. (6) in Section 12.4.

Proof We use two cross product identities in this proof and the discussion below:

$$\mathbf{u} \times (\mathbf{v} \times \mathbf{w}) = (\mathbf{u} \cdot \mathbf{w})\mathbf{v} - (\mathbf{u} \cdot \mathbf{v})\mathbf{w} \qquad \boxed{6}$$

$$\mathbf{u} \cdot (\mathbf{v} \times \mathbf{w}) = \mathbf{v} \cdot (\mathbf{w} \times \mathbf{u}) = \mathbf{w} \cdot (\mathbf{u} \times \mathbf{v}) \qquad \boxed{7}$$

By the definition of **L**,

$$\frac{d\mathbf{L}}{dt} = \overbrace{\frac{1}{k}\frac{d}{dt}(\mathbf{r}' \times \mathbf{J})}^{\text{Call this } \mathbf{A}.} - \overbrace{\frac{d}{dt}\left(\frac{\mathbf{r}}{\|\mathbf{r}\|}\right)}^{\text{Call this } \mathbf{B}.} = \mathbf{A} - \mathbf{B}$$

We compute **A** and **B** separately. First, we have

$$\mathbf{A} = \frac{d}{dt}\left(\frac{1}{k}\mathbf{r}' \times \mathbf{J}\right) = \left(\frac{1}{k}\mathbf{r}'' \times \mathbf{J}\right) + \left(\frac{1}{k}\mathbf{r}' \times \mathbf{J}'\right) = \frac{1}{k}\mathbf{r}'' \times \mathbf{J} \qquad \text{(since } \mathbf{J}' = \mathbf{0}\text{)}$$

$$= \frac{1}{k}\mathbf{r}'' \times (\mathbf{r} \times \mathbf{r}') \qquad \text{(since } \mathbf{J} = \mathbf{r} \times \mathbf{r}'\text{)}$$

$$= -\frac{1}{\|\mathbf{r}\|^3}\mathbf{r} \times (\mathbf{r} \times \mathbf{r}') \qquad \text{(since } \mathbf{r}'' = -k\mathbf{r}/\|\mathbf{r}\|^3\text{)}$$

$$= -\frac{1}{\|\mathbf{r}\|^3}\left((\mathbf{r} \cdot \mathbf{r}')\mathbf{r} - (\mathbf{r} \cdot \mathbf{r})\mathbf{r}'\right) \qquad \text{[by Eq. (6)]}$$

$$= \frac{\mathbf{r}'}{\|\mathbf{r}\|} - \frac{(\mathbf{r} \cdot \mathbf{r}')\mathbf{r}}{\|\mathbf{r}\|^3} \qquad \boxed{8}$$

Before continuing, we compute the following derivative:

$$\frac{d}{dt}\|\mathbf{r}\| = \frac{d}{dt}(\mathbf{r} \cdot \mathbf{r})^{1/2} = \frac{1}{2}(\mathbf{r} \cdot \mathbf{r})^{-1/2}\frac{d}{dt}\mathbf{r} \cdot \mathbf{r} = \frac{1}{2}(\mathbf{r} \cdot \mathbf{r})^{-1/2}(\mathbf{r} \cdot \mathbf{r}' + \mathbf{r}' \cdot \mathbf{r}) = \frac{\mathbf{r} \cdot \mathbf{r}'}{\|\mathbf{r}\|} \qquad \boxed{9}$$

Using Eq. (9) and the Quotient Rule,

$$\mathbf{B} = \frac{d}{dt}\left(\frac{\mathbf{r}}{\|\mathbf{r}\|}\right) = \frac{\|\mathbf{r}\|\mathbf{r}' - (d\|\mathbf{r}\|/dt)\mathbf{r}}{\|\mathbf{r}\|^2} = \frac{\|\mathbf{r}\|\mathbf{r}' - \|\mathbf{r}\|^{-1}(\mathbf{r} \cdot \mathbf{r}')\mathbf{r}}{\|\mathbf{r}\|^2}$$

$$= \frac{\mathbf{r}'}{\|\mathbf{r}\|} - \frac{(\mathbf{r} \cdot \mathbf{r}')\mathbf{r}}{\|\mathbf{r}\|^3} \qquad \boxed{10}$$

Comparing Eq. (8) and Eq. (10), we see that $\mathbf{A} = \mathbf{B}$ and therefore $\frac{d}{dt}\mathbf{L} = \mathbf{0}$ as claimed. ∎

To derive an equation of the planet's orbit, assume first that $\mathbf{L} \neq \mathbf{0}$. Choose coordinates so that **L** points in the direction of the positive x-axis and let (r, θ) be the polar

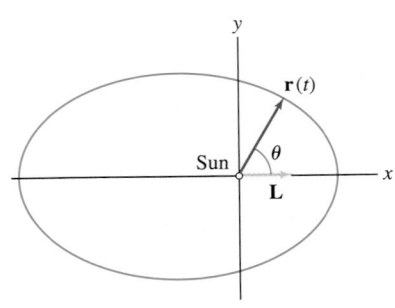

FIGURE 5 The Lenz vector **L** points from the sun to the perihelion (closest point) of the orbit. The length of **L** is the eccentricity of the orbit.

coordinates of the planet (Figure 5). Define

$$e = \|\mathbf{L}\|, \qquad p = \frac{\|\mathbf{J}\|^2}{k}, \qquad r = \|\mathbf{r}\|$$

Then the dot product $\mathbf{r} \cdot \mathbf{L}$ is equal to

$$\mathbf{r} \cdot \mathbf{L} = r\, e \cos \theta$$

On the other hand, by the definition of **L** and identity (7),

$$\mathbf{r} \cdot \mathbf{L} = \frac{1}{k}\mathbf{r} \cdot (\mathbf{r}' \times \mathbf{J}) - \frac{\mathbf{r} \cdot \mathbf{r}}{\|\mathbf{r}\|} = \frac{1}{k}\mathbf{J} \cdot (\mathbf{r} \times \mathbf{r}') - \|\mathbf{r}\| = \frac{1}{k}\mathbf{J} \cdot \mathbf{J} - \|\mathbf{r}\| = p - r$$

This equation shows that if $\mathbf{L} = \mathbf{0}$, then $r = p$ and the orbit is a circle of radius p. Assuming $\mathbf{L} \neq \mathbf{0}$, we equate the two expressions for $\mathbf{r} \cdot \mathbf{L}$ to obtain $r e \cos \theta = p - r$ or

$$r = \frac{p}{1 + e \cos \theta}$$

This is the polar equation of a conic section (ellipse, parabola, or hyperbola) with eccentricity e and focus at the origin [see Eq. (12) in Section 11.5]. If we assume that the

The Hubble Space Telescope produced this image of the Antenna galaxies, a pair of spiral galaxies that began to collide hundreds of millions of years ago.

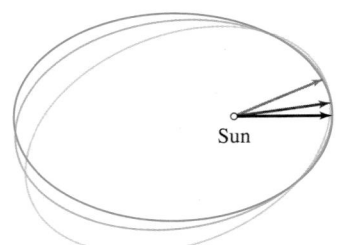

FIGURE 6 The perihelion of an orbit shifts slowly over time. For Mercury, the semimajor axis makes a full rotation approximately once every 24,000 years.

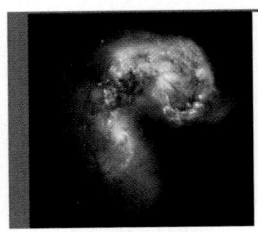

HISTORICAL PERSPECTIVE

The astronomers of the ancient world (Babylon, Egypt, and Greece) mapped out the nighttime sky with impressive accuracy, but their models of planetary motion were based on the erroneous assumption that the planets revolve around the earth. Although the Greek astronomer Aristarchus (310–230 BCE) had suggested that the earth revolves around the sun, this idea was rejected and forgotten for more than fifteen centuries, until the Polish astronomer Nicolaus Copernicus (1473–1543) introduced a revolutionary set of ideas about the solar system, including the hypothesis that the planets revolve around the sun. The ideas of Copernicus, although not entirely correct, paved the way for the next generation, most notably Tycho Brahe (1546–1601), Galileo Galilei (1564–1642), and Johannes Kepler (1571–1630).

The German astronomer Johannes Kepler was the son of a mercenary soldier who died when Johannes was 5. He was raised by his mother in his grandfather's inn. His mathematical brilliance won him a scholarship to the University of Tübingen. At the age of 29 he went

to work for Tycho Brahe, a Danish astronomer who had revolutionized the art of astronomical measurement and compiled the most complete and accurate data on the planetary orbits then available. When Brahe died in 1601, Kepler succeeded him as "Imperial Mathematician" to the Holy Roman Emperor. Kepler continued to study Brahe's data, and in 1609, he formulated the first two of his laws of planetary motion in a work entitled *Astronomia Nova* (New Astronomy).

In the centuries since Kepler's death, as the observational data improved, astronomers found that planetary orbits are not exactly elliptical. Furthermore, the perihelion (the point on the orbit closest to the sun) shifts slowly over time as shown in Figure 6. Most of these deviations can be explained by the mutual pull of the planets, but the perihelion shift of Mercury is larger than can be accounted for by Newton's Laws. On November 18, 1915, Albert Einstein made a discovery about which he later wrote to a friend, "I was beside myself with ecstasy for days." He had been working for a decade on his famous **General Theory of Relativity**, a theory that would replace Newton's law of gravitation with a new set of much more complicated equations called the Einstein Field Equations. On that 18th of November, Einstein showed that Mercury's perihelion shift was accurately explained by his new theory. At the time, this was the only substantial piece of evidence that the General Theory of Relativity was correct.

planet travels around the sun in a bounded orbit, this orbit must be an ellipse. We also see that there are "open orbits," parabolic and hyperbolic, that could describe a comet which passes by the sun and then continues into space never to return. If $\mathbf{J} = 0$, then the orbit is a straight line and the planet falls directly into the sun.

> **CONCEPTUAL INSIGHT**   You may not have noticed, but we managed to derive Kepler's Laws without ever finding a formula for the position vector $\mathbf{r}(t)$ of the planet. In fact, $\mathbf{r}(t)$ cannot be expressed in terms of elementary functions. This illustrates an important principle: We may be able to understand how the solutions to a differential equation behave even if it is hard or impossible to determine the solutions explicitly.

13.6 SUMMARY

- Kepler's three laws of planetary motion are the Law of Ellipses, the Law of Equal Area in Equal Time, and the Law of the Period $T^2 = \left(\dfrac{4\pi^2}{GM} \right) a^3$.

- Let $k = GM$. According to Newton's Laws of Motion and Gravitation, the position vector of a planet satisfies the differential equation

$$\mathbf{r}''(t) = -\frac{k}{\|\mathbf{r}\|^3}\, \mathbf{r}(t)$$

- We define the vector $\mathbf{J} = \mathbf{r}(t) \times \mathbf{r}'(t)$. The planet's position vector $\mathbf{r}(t)$ sweeps out area at the rate $\dfrac{dA}{dt} = \dfrac{1}{2}\|\mathbf{J}\|$. Using vector differentiation and the differential equation, we show that $\dfrac{d\mathbf{J}}{dt} = \mathbf{0}$, which implies Kepler's Second Law.

- The Lenz vector $\mathbf{L} = \dfrac{1}{k}\mathbf{r}' \times \mathbf{J} - \dfrac{\mathbf{r}}{\|\mathbf{r}\|}$ is also a constant of planetary motion. It points along the semimajor axis of the elliptical orbit and $\|\mathbf{L}\|$ is the orbit eccentricity.

13.6 EXERCISES

Preliminary Questions

1. Describe the relation between the vector $\mathbf{J} = \mathbf{r} \times \mathbf{r}'$ and the rate at which the radial vector sweeps out area.

2. Equation (1) shows that $\mathbf{r}''$ is proportional to $\mathbf{r}$. Explain how this fact is used to prove Kepler's Second Law.

3. How is the period T affected if the semimajor axis a is increased four-fold?

Exercises

1. Kepler's Third Law states that T^2/a^3 has the same value for each planetary orbit. Do the data in the following table support this conclusion? Estimate the length of Jupiter's period, assuming that $a = 77.8 \times 10^{10}$ m.

Planet	Mercury	Venus	Earth	Mars
$a\ (10^{10}$ m$)$	5.79	10.8	15.0	22.8
T (years)	0.241	0.615	1.00	1.88

2. A satellite has initial position $\mathbf{r} = \langle 1,000, 2,000, 0 \rangle$ and initial velocity $\mathbf{r}' = \langle 1, 2, 2 \rangle$ (units of kilometrs and seconds). Find the equation of the plane containing the satellite's orbit. *Hint:* This plane is orthogonal to $\mathbf{J}$.

3. The earth's orbit is nearly circular with radius $R = 93 \times 10^6$ miles (the eccentricity is $e = 0.017$). Find the rate at which the earth's radial vector sweeps out area in units of ft^2/s. What is the magnitude of the vector $\mathbf{J} = \mathbf{r} \times \mathbf{r}'$ for the earth (in units of squared feet per second)?

4. Finding the Mass of a Star Using Kepler's Third Law, show that if a planet revolves around a star with period T and semimajor axis a, then the mass of the star is $M = \left(\dfrac{4\pi^2}{G}\right)\left(\dfrac{a^3}{T^2}\right)$.

5. Ganymede, one of Jupiter's moons discovered by Galileo, has an orbital period of 7.154 days and a semimajor axis of 1.07×10^9 m. Use Exercise 4 to estimate the mass of Jupiter.

6. An astronomer observes a planet orbiting a star with a period of 9.5 years and a semimajor axis of 3×10^8 km. Find the mass of the star.

7. Use the fact that $\mathbf{J}$ is constant to show that a planet in a circular orbit travels at constant speed.

8. Prove that if a planetary orbit is circular of radius R, then $vT = 2\pi R$, where v is the planet's speed (constant by Exercise 7) and T is the period. Then use Kepler's Third Law to prove that $v = \sqrt{\dfrac{GM}{R}}$.

9. Show directly that the circular orbit

$$\mathbf{r}(t) = \langle R \cos \omega t, \; R \sin \omega t \rangle$$

satisfies the differential equation, Eq. (2), provided that $\omega^2 = kR^{-3}$. Then deduce Kepler's Third Law $T^2 = \left(\dfrac{4\pi^2}{k}\right) R^3$ for this orbit.

10. The orbit of a satellite orbiting above the equator of the earth is called **geosynchronous** if the period is $T = 24$ hours (in this case, the satellite stays over a fixed point on the equator). Use Kepler's Third Law to find the altitude h above the earth's surface of a geosynchronous orbit. The earth has mass $M \approx 5.974 \times 10^{24}$ kg and radius $R \approx 6{,}371$ km.

11. Use the results of Exercises 8 and 10 to find the velocity of a satellite in geosynchronous orbit.

12. Show that if a planet revolves around a star of mass M in a circular orbit of radius R with speed v, then $M = \dfrac{Rv^2}{G}$.

13. Mass of the Milky Way The sun revolves around the center of mass of the Milky Way galaxy in an orbit that is approximately circular, of radius $a \approx 2.8 \times 10^{17}$ km and velocity $v \approx 250$ km/s. Use the result of Exercise 12 to estimate the mass of the portion of the Milky Way inside the sun's orbit (place all of this mass at the center of the orbit).

14. Conservation of Energy The total mechanical energy (kinetic plus potential) of a planet of mass m orbiting a sun of mass M with position $\mathbf{r}$ and speed $v = \|\mathbf{r}'\|$ is

$$E = \frac{1}{2}mv^2 - \frac{GMm}{\|\mathbf{r}\|} \qquad \boxed{11}$$

Use (2) and (9) to show that E is conserved, that is, $\dfrac{dE}{dt} = 0$.

15. Show that the total energy (11) of a planet in a circular orbit of radius R is $E = -GMm/(2R)$. *Hint:* Use Exercise 8.

In Exercises 16–20, we consider a planetary orbit with orbital parameters p and e. The perihelion and aphelion of the orbits are the points on the orbit closest to and farthest from the sun (Figure 7). Denote the distances from the sun at the perihelion and aphelion by r_{per} and r_{ap} and the speeds of the planet at the perihelion and aphelion by v_{per} and v_{ap}.

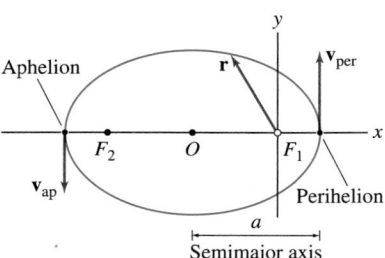

FIGURE 7 $\mathbf{r}$ and $\mathbf{r}'$ are perpendicular at the perihelion and aphelion.

16. Use the polar equation

$$r = \frac{p}{1 + e \cos \theta}$$

to show that $r_{per} = a(1 - e)$ and $r_{ap} = a(1 + e)$.

17. Compute r_{per} and r_{ap} for the orbit of Mercury, which has eccentricity $e = 0.244$ (see the table in Exercise 1 for the semimajor axis).

18. Prove the formulas

$$e = \frac{r_{ap} - r_{per}}{r_{ap} + r_{per}}, \qquad p = \frac{2r_{ap}r_{per}}{r_{ap} + r_{per}}$$

19. Prove that $v_{per}(1 - e) = v_{ap}(1 + e)$. *Hint:* $\mathbf{r} \times \mathbf{r}'$ is constant by Eq. (5). Compute this cross product at the perihelion and aphelion, noting that $\mathbf{r}$ is perpendicular to $\mathbf{r}'$ at these two points.

20. Prove that $v_{per} = \sqrt{\left(\dfrac{GM}{a}\right)\dfrac{1+e}{1-e}}$ as follows:

(a) Use Conservation of Energy (see Exercise 14) to show that

$$v_{per}^2 - v_{ap}^2 = 2GM(r_{per}^{-1} - r_{ap}^{-1})$$

(b) Show $r_{per}^{-1} - r_{ap}^{-1} = \dfrac{2e}{a(1 - e^2)}$ using Exercise 16.

(c) Show that $v_{per}^2 - v_{ap}^2 = 4\dfrac{e}{(1+e)^2}v_{per}^2$ using Exercise 19. Then solve for v_{per} using (a) and (b).

21. Show that the total mechanical energy E of a planet in an elliptical orbit with semimajor axis a is $E = -\dfrac{GMm}{2a}$. *Hint:* Use Exercise 20 to compute the total energy at the perihelion.

22. Prove that $v^2 = GM\left(\dfrac{2}{r} - \dfrac{1}{a}\right)$ at any point on an elliptical orbit with semimajor axis a, where $r = \|\mathbf{r}\|$.

23. Two space shuttles A and B orbit the earth along the solid trajectory in Figure 8. Hoping to catch up to B, the pilot of A applies a forward thrust to increase her shuttle's kinetic energy. Use Exercise 21 to show that shuttle A will move off into a larger orbit as shown in the figure. Then use Kepler's Third Law to show that A's orbital period T will increase (and she will fall farther and farther behind B)!

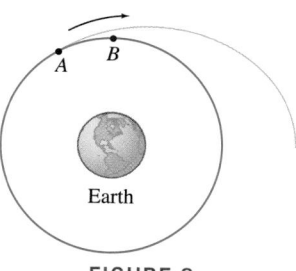

FIGURE 8

Further Insights and Challenges

24. In this exercise, we prove Eq. (3) in a rigorous fashion. Let $r(t) = \|\mathbf{r}(t)\|$ and let $\theta(t)$ be the angle between $\mathbf{r}(t)$ and the x-axis. Then $\mathbf{r} = r(t) \langle \cos \theta(t), \sin \theta(t) \rangle$.

(a) Prove that $\dfrac{dA}{dt} = \dfrac{1}{2} r(t)^2 \theta'(t)$ by applying the Fundamental Theorem of Calculus to the formula for area in polar coordinates:

$$A(\theta) = \frac{1}{2} \int_0^\theta r(u)^2 \, du$$

(b) Show that $\|\mathbf{J}\| = \|\mathbf{r} \times \mathbf{r}'\| = r(t)^2 \theta'(t)$.

(c) Conclude that $\dfrac{dA}{dt} = \dfrac{1}{2} \|\mathbf{J}\|$.

In Exercises 25–26, we prove Kepler's Third Law. Figure 9 shows an elliptical orbit with polar equation

$$r = \frac{p}{1 + e \cos \theta}$$

where $p = \|\mathbf{J}\|^2 / k$. Let a and b be the semimajor and semiminor axes, respectively. The origin is located at F_1.

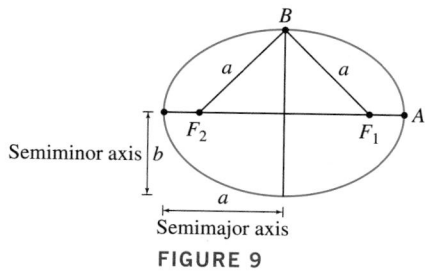

FIGURE 9

25. The goal of this exercise is to show that $b = \sqrt{pa}$.
(a) Show that $F_1 A + F_2 A = 2a$. Conclude that $F_1 B + F_2 B = 2a$ and hence $F_1 B = F_2 B = a$.
(b) Show that $F_1 A = \dfrac{p}{1 + e}$ and $F_2 A = \dfrac{p}{1 - e}$, and conclude that $a = \dfrac{p}{1 - e^2}$.
(c) Use the Pythagorean Theorem to prove that

$$b = \frac{p}{\sqrt{1 - e^2}} = \sqrt{pa}$$

26. We prove Kepler's Third Law by computing the area A of the ellipse in two ways.
(a) Use Exercise 25 to show that $A = (\pi \sqrt{p}) a^{3/2}$.
(b) Use Kepler's First Law to show that $A = \frac{1}{2} \|\mathbf{J}\| T$.
(c) Deduce that $T^2 = \dfrac{4\pi^2}{GM} a^3$.

27. Let $\mathbf{e_r} = \langle \cos \theta, \sin \theta \rangle$ and $\mathbf{e_\theta} = \langle - \sin \theta, \cos \theta \rangle$. Write the position vector of a planet as $\mathbf{r} = r \mathbf{e_r}$, where $r = \|\mathbf{r}\|$.
(a) Show that $\dfrac{d\mathbf{e_\theta}}{d\theta} = -\mathbf{e_r}$.
(b) Write Eq. (2) in the form $\dfrac{d\mathbf{v}}{dt} = -\dfrac{k}{r^2} \mathbf{e_r}$ and use the Chain Rule to show that

$$\frac{d\mathbf{v}}{d\theta} \frac{d\theta}{dt} = \frac{k}{r^2} \frac{d\mathbf{e_\theta}}{d\theta}$$

(c) Show that $\dfrac{d\mathbf{v}}{d\theta} = \dfrac{k}{\|\mathbf{J}\|} \dfrac{d\mathbf{e_\theta}}{d\theta}$. *Hint: Use Exercise 24 to show that* $\dfrac{d\theta}{dt} = \|\mathbf{J}\|/r^2$.
(d) Conclude that there is a constant vector $\mathbf{w}$ such that

$$\mathbf{v}(\theta) = \frac{k}{\|\mathbf{J}\|} \mathbf{e_\theta} + \mathbf{w}$$

This shows that as θ varies from 0 to 2π, the velocity vector $\mathbf{v}$ traces out a circle of radius $k/\|\mathbf{J}\|$ with center at the terminal point of $\mathbf{w}$ (Figure 10).

Planetary orbit

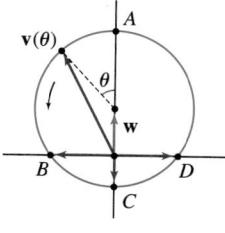

Velocity circle

FIGURE 10 The terminal point of the velocity vector traces out the circle as the planet travels along its orbit.

CHAPTER REVIEW EXERCISES

1. Determine the domains of the vector-valued functions.

(a) $\mathbf{r}_1(t) = \langle t^{-1}, (t+1)^{-1}, \sin^{-1} t \rangle$

(b) $\mathbf{r}_2(t) = \langle \sqrt{8 - t^3}, \ln t, e^{\sqrt{t}} \rangle$

2. Sketch the paths $\mathbf{r}_1(\theta) = \langle \theta, \cos \theta \rangle$ and $\mathbf{r}_2(\theta) = \langle \cos \theta, \theta \rangle$ in the xy-plane.

3. Find a vector parametrization of the intersection of the surfaces $x^2 + y^4 + 2z^3 = 6$ and $x = y^2$ in $\mathbf{R}^3$.

4. Find a vector parametrization using trigonometric functions of the intersection of the plane $x + y + z = 1$ and the elliptical cylinder $\left(\frac{y}{3}\right)^2 + \left(\frac{z}{8}\right)^2 = 1$ in $\mathbf{R}^3$.

In Exercises 5–10, calculate the derivative indicated.

5. $\mathbf{r}'(t)$, where $\mathbf{r}(t) = \langle 1 - t, t^{-2}, \ln t \rangle$

6. $\mathbf{r}'''(1)$, where $\mathbf{r}(t) = \langle t^3, 4t^2, 7t \rangle$

7. $\mathbf{r}'(0)$, where $\mathbf{r}(t) = \langle e^{2t}, e^{-4t^2}, e^{6t} \rangle$

8. $\mathbf{r}''(2)$, where $\mathbf{r}(t) = \langle e^{2t}, e^{-4t^2}, e^{6t} \rangle$

9. $\dfrac{d}{dt} e^t \langle 1, t, t^2 \rangle$

10. $\dfrac{d}{d\theta} \mathbf{r}(\cos \theta)$, where $\mathbf{r}(s) = \langle s, 2s, s^2 \rangle$

In Exercises 11–14, calculate the derivative at $t = 3$ assuming that

$$\mathbf{r}_1(3) = \langle 1, 1, 0 \rangle, \qquad \mathbf{r}_2(3) = \langle 1, 1, 0 \rangle$$
$$\mathbf{r}_1'(3) = \langle 0, 0, 1 \rangle, \qquad \mathbf{r}_2'(3) = \langle 0, 2, 4 \rangle$$

11. $\dfrac{d}{dt}(6\mathbf{r}_1(t) - 4 \cdot \mathbf{r}_2(t))$

12. $\dfrac{d}{dt}(e^t \mathbf{r}_2(t))$

13. $\dfrac{d}{dt}(\mathbf{r}_1(t) \cdot \mathbf{r}_2(t))$

14. $\dfrac{d}{dt}(\mathbf{r}_1(t) \times \mathbf{r}_2(t))$

15. Calculate $\displaystyle\int_0^3 \langle 4t + 3, t^2, -4t^3 \rangle \, dt$.

16. Calculate $\displaystyle\int_0^\pi \langle \sin \theta, \theta, \cos 2\theta \rangle \, d\theta$.

17. Find the unit tangent vector to $\mathbf{r}(t) = \langle \sin t, t, \cos t \rangle$ at $t = \pi$.

18. Find the unit tangent vector to $\mathbf{r}(t) = \langle t^2, \tan^{-1} t, t \rangle$ at $t = 1$.

19. A particle located at $(1, 1, 0)$ at time $t = 0$ follows a path whose velocity vector is $\mathbf{v}(t) = \langle 1, t, 2t^2 \rangle$. Find the particle's location at $t = 2$.

20. Find the vector-valued function $\mathbf{r}(t) = \langle x(t), y(t) \rangle$ in $\mathbf{R}^2$ satisfying $\mathbf{r}'(t) = -\mathbf{r}(t)$ with initial conditions $\mathbf{r}(0) = \langle 1, 2 \rangle$.

21. Compute the length of the path $\mathbf{r}(t) = \langle \sin 2t, \cos 2t, 3t - 1 \rangle$ for $1 \le t \le 3$.

22. $\boxed{\text{CAS}}$ Express the length of the path $\mathbf{r}(t) = \langle \ln t, t, e^t \rangle$ for $1 \le t \le 2$ as a definite integral and use a computer algebra system to find its value to two decimal places.

23. A string in the shape of a helix has a height of 20 cm and makes four full rotations over a circle of radius 5 cm. Find a parametrization $\mathbf{r}(t)$ of the string and compute its length.

24. Find the minimum speed of a particle with trajectory $\mathbf{r}(t) = \langle t, e^{t-3}, e^{4-t} \rangle$.

25. Calculate the curvature $\kappa(t)$ for $\mathbf{r}(t) = \langle t^{-1}, \ln t, t \rangle$ and find the unit tangent and normal vectors at $t = 1$.

26. A specially trained mouse runs counterclockwise in a circle of radius 2 ft on the floor of an elevator with speed 1 ft/s while the elevator ascends from ground level (along the z-axis) at a speed of 40 ft/s. Find the mouse's acceleration vector as a function of time. Assume that the circle is centered at the origin of the xy-plane and the mouse is at $(2, 0, 0)$ at $t = 0$.

In Exercises 27–30, let $\mathbf{r}(t) = \langle t, e^{-t^2} \rangle$.

27. Compute the curvature function $\kappa(t)$.

28. $\boxed{\text{GU}}$ Plot $\mathbf{r}(t)$ and $\kappa(t)$ on the same set of axes and estimate the values of t where the curvature has a maximum value.

29. Find the unit tangent and normal vectors at $t = 0$ and $t = 1$.

30. Write the acceleration vector at $t = 1$ as a sum of tangential and normal components.

31. Find the curvature $\kappa(t)$ and unit normal vector $\mathbf{N}(t)$ for $\mathbf{r}(t) = \langle \sin t, \sin t, \cos t \rangle$.

32. Find the curvature $\kappa(t)$ and unit normal vector $\mathbf{N}(t)$ for $\mathbf{r}(t) = \langle \ln t, t^{-1}, t^{-2} \rangle$.

In Exercises 33–34, write the acceleration vector $\mathbf{a}$ at the point indicated as a sum of tangential and normal components.

33. $\mathbf{r}(\theta) = \langle \cos \theta, \sin 2\theta \rangle, \quad \theta = \frac{\pi}{4}$

34. $\mathbf{r}(t) = \langle t^2, t^3 \rangle, \quad t = 2$

35. Find the osculating circle to the curve $y = e^{-x^2}$ at $x = 0$.

36. Find the osculating circle to the curve $y = \ln x$ at $x = 1$.

37. If a planet is in "orbit" around a sun whose mass is zero, Newton's Laws imply that the position vector of the plane satisfies $\mathbf{r}''(t) = \mathbf{0}$. Show that in this case, the orbit is the straight line with parametrization $\mathbf{r}(t) = \mathbf{r} + t\mathbf{v}$, where $\mathbf{r} = \mathbf{r}(0)$ and $\mathbf{v} = \mathbf{r}'(0)$ (Figure 1).

38. Continuing with Exercise 37, show that the area A swept out by the radial vector $\mathbf{r}(t)$ over the time interval $[0, t]$ is equal to $A = \frac{1}{2}\|\mathbf{r}(0) \times \mathbf{v}\|t$. Thus, Kepler's Second Law continues to hold since area is swept out at the constant rate $\frac{1}{2}\|\mathbf{r}(0) \times \mathbf{v}\|$.

FIGURE 1

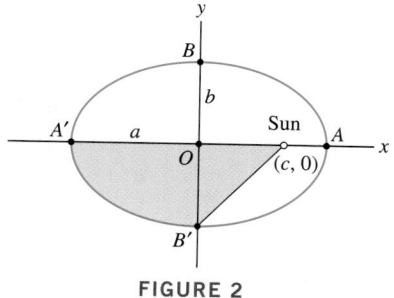

FIGURE 2

39. Suppose that the planetary orbit in Figure 2 is an ellipse with eccentricity e (by definition, $e = c/a$). Use Kepler's Second Law to show that if the period of the orbit is T, the the time it takes for a planet to travel from A' to B' is equal to

$$\left(\frac{1}{4} + \frac{e}{2\pi}\right) T$$

40. The period of Mercury is approximately 88 days and its orbit has eccentricity 0.205. How much longer does it take Mercury to travel from A' to B' than from B' to A (Figure 2)?

14 DIFFERENTIATION IN SEVERAL VARIABLES

The goal of this chapter is to extend the concepts and techniques of differential calculus to functions of several variables. As we will see, a function f that depends on two or more variables has not just one derivative but rather a set of *partial derivatives*, one for each variable. The partial derivatives make up the components of the gradient vector, which provides important geometric insight into the behavior of the function. In the last two sections of the chapter, we apply the insights gained to optimization in several variables.

Yosemite National Park. The steepness at a point in a mountain range is measured by the gradient vector, a concept defined in this chapter.

14.1 Functions of Two or More Variables

Functions of several variables are a basic tool in the study of natural phenomena. For example, scientists study variations in the density ρ of seawater (mass per unit volume). These variations are responsible for deep ocean currents that distribute heat throughout the earth and influence the global climate (Figure 1). However, ρ depends on several variables, of which the most important are salinity S and temperature T. We write $\rho = f(S, T)$ to indicate that ρ is a function of S and T. There is no simple formula for $f(S, T)$, but scientists have measured $f(S, T)$ experimentally (Figure 2). According to Table 1, the density at $S = 32$ (parts per thousand or ppt) and $T = 10°C$ is

$$\rho = f(32, 10) = 1.0246 \text{ kg/m}^3$$

FIGURE 1 Ocean "conveyer belt," the system of deep currents driven by variations in seawater density.

TABLE 1 Seawater Density ρ (kg/m^3) as a Function of Temperature and Salinity.

°C	Salinity (ppt)		
	32	32.5	33
5	1.0253	1.0257	1.0261
10	1.0246	1.0250	1.0254
15	1.0237	1.0240	1.0244
20	1.0224	1.0229	1.0232

A function of n-variables is a real-valued function $f(x_1, \ldots, x_n)$ whose domain $\mathcal{D}$ is a set of n-tuples $(x_1, \ldots, x_n)$ in $\mathbf{R}^n$. When f is defined by a formula, we usually take as domain the set of all n-tuples for which $f(x_1, \ldots, x_n)$ is defined. The range of f is the set of all values $f(x_1, \ldots, x_n)$ for $(x_1, \ldots, x_n)$ in the domain. For instance, $f(x_1, x_2, x_3) = x_1 + x_2 e^{x_3}$ has domain $\mathcal{D} = \mathbf{R}^3$ since f is defined for all (x_1, x_2, x_3). The range is $\mathbf{R}$, as we may show by observing that $f(x_1, 0, 0) = x_1$. Thus, f takes on every value. We will often focus our attention on functions of two and three variables.

FIGURE 2 A researcher lowers a Conductivity-Temperature-Depth (CDT) instrument into the ocean to measure seawater variables such as density, temperature, pressure, and salinity.

In these cases, we generally use the variables (x, y) and (x, y, z). We also write $f(P)$ to denote the value of a function at a point $P = (x, y)$ or (x, y, z).

■ **EXAMPLE 1** Sketch the domains of:

(a) $f(x, y) = \sqrt{9 - x^2 - y^2}$

(b) $g(x, y, z) = x\sqrt{y} + \ln(z - 1)$

Solution

(a) $f(x, y) = \sqrt{9 - x^2 - y^2}$ is defined when $9 - x^2 - y^2 \geq 0$, so the domain of f is $\mathcal{D} = \{(x, y) : x^2 + y^2 \leq 9\}$, a disk of radius 3 centered at the origin [Figure 3(A)].

(b) The function $g(x, y, z)$ is defined if both $\sqrt{y}$ and $\ln(z - 1)$ are defined. Thus we require that $y \geq 0$ and $z > 1$, and the domain is $\{(x, y, z) : y \geq 0, z > 1\}$ [Figure 3(B)]. ■

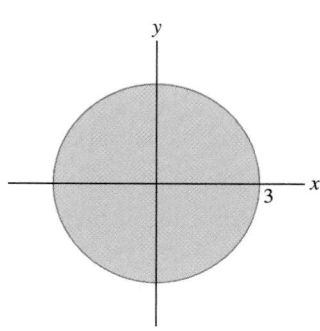

(A) The domain of $f(x, y) = \sqrt{9 - x^2 - y^2}$
is the disk of radius 3.

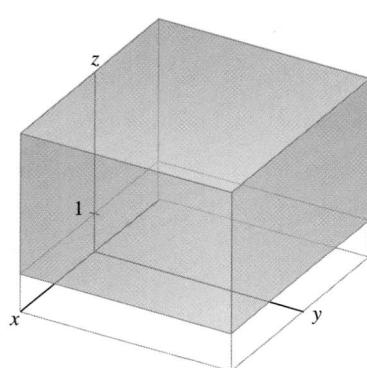

(B) Domain of $g(x, y, z) = x\sqrt{y} + \ln(z - 1)$
is the set of points with $y \geq 0$ and $z > 1$.

FIGURE 3

Graphing Functions of Two Variables

In our study of single-variable calculus, we have seen that a graph displays the important features of a function at a glance. Graphs are also useful for visualizing functions of two variables. The graph of $f(x, y)$ is the set of all points $(a, b, f(a, b))$ in $\mathbf{R}^3$ for (a, b) in the domain $\mathcal{D}$ of f. If we assume that f is continuous (as defined in the next section), then the graph is a surface whose *height* above or below the xy-plane at (a, b) is the function value $f(a, b)$ (Figure 4). Note that the graph of $f(x, y)$ is also the graph of the equation $z - f(x, y) = 0$.

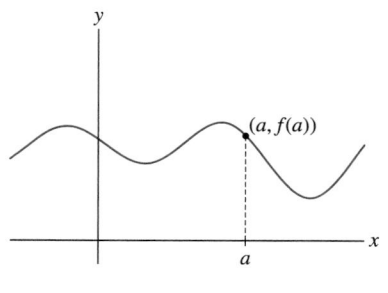

(A) Graph of $y = f(x)$

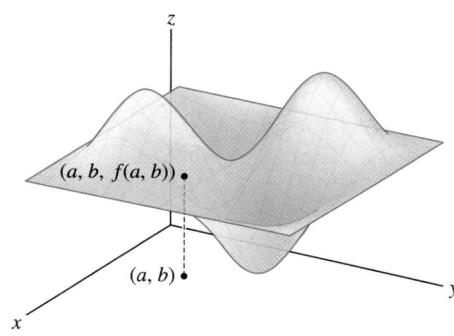

(B) Graph of $z = f(x, y)$

FIGURE 4

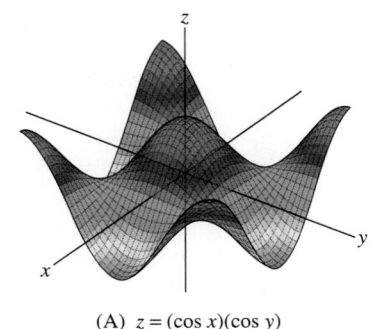

(A) $z = (\cos x)(\cos y)$

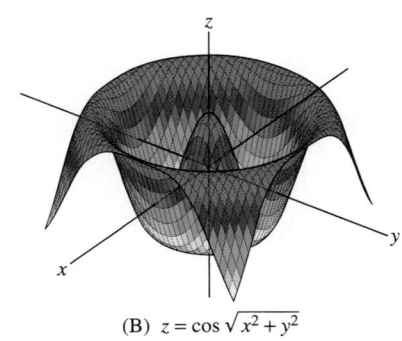

(B) $z = \cos \sqrt{x^2 + y^2}$

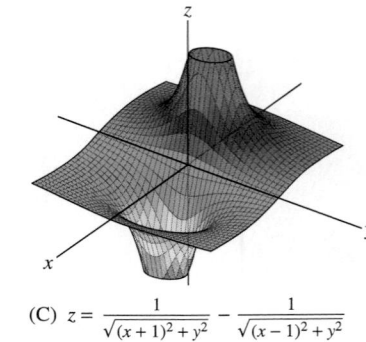

(C) $z = \dfrac{1}{\sqrt{(x+1)^2 + y^2}} - \dfrac{1}{\sqrt{(x-1)^2 + y^2}}$

FIGURE 5 Gallery of graphs.

Plotting graphs of functions of two variables by hand is often a difficult task. Fortunately, computer algebra systems eliminate much of the labor and greatly enhance our ability to explore functions graphically from different perspectives (Figure 5).

To help analyze graphs, we define the **traces**, which are the curves obtained by intersecting the graph with planes parallel to a coordinate plane (Figure 6). There are three types of traces:

- **Horizontal trace at height c:** The intersection of the graph with the horizontal plane $z = c$, consisting of points (x, y, c) such that $f(x, y) = c$. This trace projects down to the curve $f(x, y) = c$ in the xy-plane.
- **Vertical trace in the plane $x = a$:** The intersection of the graph with the vertical plane $x = a$, consisting of points $(a, y, f(a, y))$.
- **Vertical trace in the plane $y = b$:** The intersection of the graph with the vertical plane $y = b$, consisting of points $(x, b, f(x, b))$.

(A) Horizontal traces

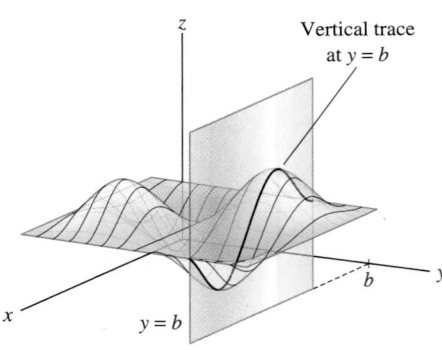

(B) Vertical traces parallel to yz-plane

(C) Vertical traces parallel to xz-plane

FIGURE 6

■ **EXAMPLE 2** Traces of a Saddle-Shaped Surface Describe the horizontal and vertical traces of $f(x, y) = x^2 - y^2$.

Solution The horizontal trace at height c lies above the hyperbola $f(x, y) = x^2 - y^2 = c$ in the xy-plane. In other words, the trace is the hyperbola raised to height c [Figure 7(A)]. The vertical traces are parabolas. The trace in the plane $x = a$ is the parabola

$z = a^2 - y^2$ moved to the plane $x = a$. The trace in the plane $y = b$ is the parabola $z = x^2 - b^2$ moved to the plane $y = b$ [Figure 7(B)]. ∎

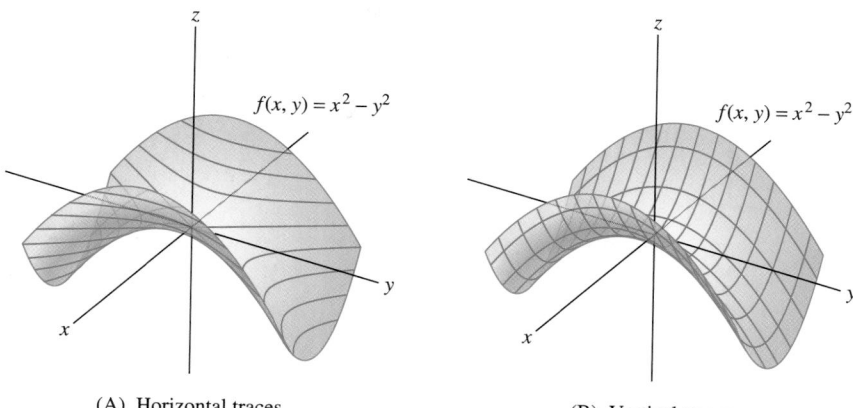

FIGURE 7 The graph of $f(x, y) = x^2 - y^2$ has a "saddle" shape.

(A) Horizontal traces

(B) Vertical traces

←·· **REMINDER** *A linear function of two variables is a function of the form* $f(x, y) = mx + ny + r$, *where* m, n, *and* r *are constants. The graph of a linear function is a plane.*

∎ **EXAMPLE 3** **The Traces of a Linear Function Are Lines** Describe the traces of the linear function $f(x, y) = 4 - 2x - y$.

Solution The graph of f is the plane with equation $z = 4 - 2x - y$. The horizontal trace at height c projects to the line $f(x, y) = c$ or $4 - 2x - y = c$. The trace itself is obtained by raising this line to height c [Figure 8(A)]. There are two families of vertical traces:

• Trace in $x = a$: The line $z = 4 - 2a - y$ in the plane $x = a$ [Figure 8(B)]
• Trace in $y = b$: The line $z = 4 - 2x - b$ in the plane $y = b$ [Figure 8(C)] ∎

(A) Horizontal traces

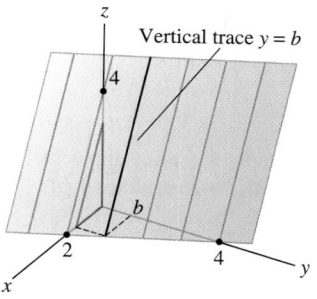

(B) Vertical traces parallel to the yz-plane

(C) Vertical traces parallel to the xz-plane

FIGURE 8 Graph of $z = 4 - 2x - y$.

Contour Maps

Another useful way of displaying the main features of a function is to draw a contour map showing level curves. The **level curves** of a function $f(x, y)$ are the curves in the xy-plane with equation $f(x, y) = c$. Thus, the level curve $f(x, y) = c$ is the projection of the horizontal trace at height c onto the xy-plane.

A **contour map** shows the level curves of $f(x, y)$ for equally spaced values of c. The spacing m is called the **contour interval**. Figure 9 is a contour map showing the level curves with contour interval $m = 10$.

■ **EXAMPLE 4** Draw a contour map for $f(x, y) = x^2 y$ with contour interval $m = 2$.

Solution The level curves are the curves with equation $x^2 y = c$ or $y = cx^{-2}$. In Figure 10, we plot $y = cx^{-2}$ for $c = 0, \pm 2, \pm 4, \pm 6, \pm 8, \pm 10$ (contour interval $m = 2$). ■

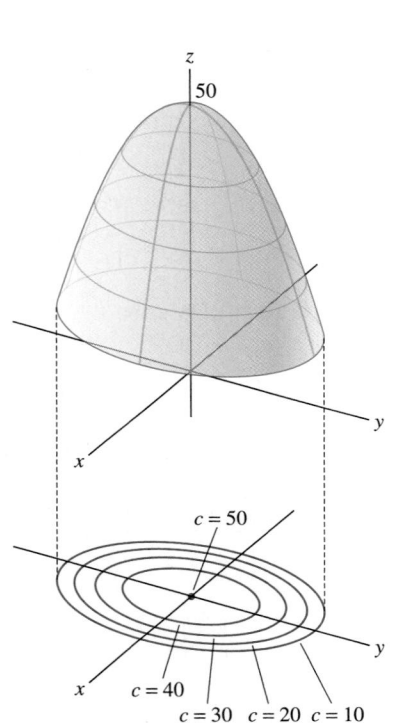

FIGURE 9 Graph of f lying above its contour map with the contour interval $m = 10$.

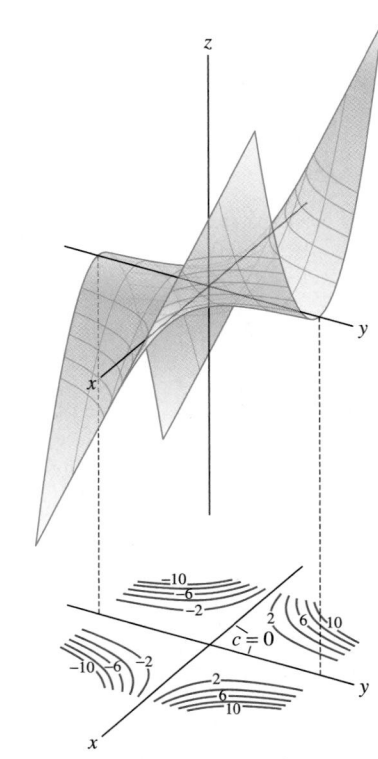

FIGURE 10 Contour map of $f(x, y) = x^2 y$.

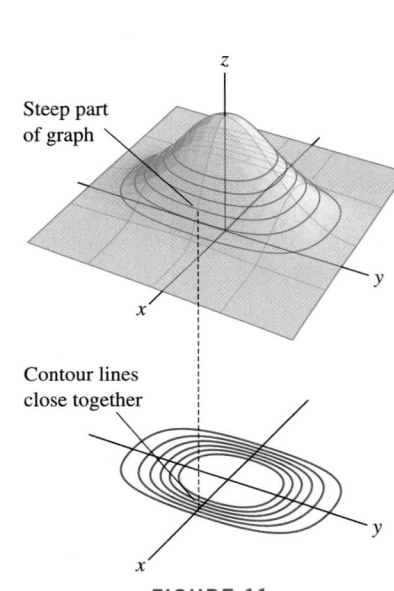

FIGURE 11

The level curve $f(x, y) = c$ is the subset of the plane where the function takes the value c. When you move from one level curve to another, the value of $f(x, y)$ changes by $\pm m$, where m is the contour interval. Therefore, *the distance between the level curves in a contour map indicates the steepness of the graph* (Figure 11). The level curves are close together where f changes quickly, and thus, where the graph is steep.

■ **EXAMPLE 5** Spacing of Level Curves ⌐A⌐5 Draw a contour map for the function $f(x, y) = x^2 + 3y^2$ with contour interval $m = 10$. Discuss the spacing of the level curves.

Solution The graph of $f(x, y)$ is a paraboloid and, for $c > 0$, the level curve is the ellipse $f(x, y) = x^2 + 3y^2 = c$ (Figure 12). The level curve is empty if $c < 0$, and for $c = 0$, the level curve reduces to the point $(0, 0)$. Since the contour interval is $m = 10$, the function changes by 10 when we go from one level curve to the next. As we move away from the origin, $f(x, y)$ increases more rapidly and the level curves get closer together.

FIGURE 12 Graph of
$f(x, y) = x^2 + 3y^2$.

FIGURE 13 Graph of
$g(x, y) = x^2 - 3y^2$.

■ **EXAMPLE 6** Contour Map of a Saddle $\mathcal{CAS}$ Draw a contour map for the function $g(x, y) = x^2 - 3y^2$.

Solution The surface $z = g(x, y)$ is a hyperbolic paraboloid and for $c \neq 0$, the level curve is the hyperbola $x^2 - 3y^2 = c$ (Figure 13). The level curve for $c = 0$ consists of the two lines $x = \pm\sqrt{3}y$ because the equation $g(x, y) = 0$ factors:

$$x^2 - 3y^2 = 0 = (x - \sqrt{3}y)(x + \sqrt{3}y) = 0$$

Notice that when you stand at the origin, $g(x, y)$ is increasing in either direction along the x-axis and decreasing in either direction along the y-axis. The resulting graph is called a *saddle* because of its saddle-like shape. ■

■ **EXAMPLE 7** Contour Map of a Linear Function Draw a contour map for the function $f(x, y) = 12 - 2x - 3y$ with contour interval $m = 4$.

Solution The level curves of a linear function $f(x, y) = mx + ny + r$ are the straight lines with equation $mx + ny + r = c$. Therefore, *the contour map of a linear function consists of equally spaced parallel lines*. In our case, the level curves are the lines $12 - 2x - 3y = c$ or $2x + 3y = 12 - c$ (Figure 14). ∎

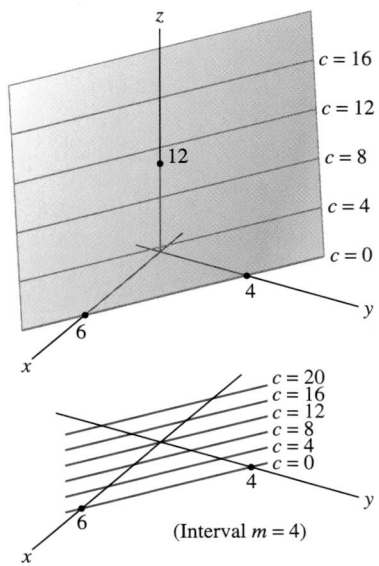

FIGURE 14 Graph and contour map of $f(x, y) = 12 - 2x - 3y$.

Two things to keep in mind when reading a contour map:

- *Your altitude does not change if you hike along a level curve.*
- *Your altitude increases or decreases by m (the contour interval) when you hike from one level curve to the next.*

To examine some further properties of contour maps, let's imagine that the surface $z = f(x, y)$ represents a mountain range. We place the xy-plane at sea level, so that $f(a, b)$ is the height (also called altitude or elevation) of the mountain above sea level at the point above (a, b) in the plane. The contour map in Figure 15 shows the elevation of the Mount Whitney range in California.

FIGURE 15 Contour map of the Mount Whitney Range in the Sierra Nevada Mountains.

Consider the "mountain" in Figure 16. To measure steepness quantitatively, we define the **average rate of change (ROC)** between points P and Q:

$$\text{Average ROC from } P \text{ to } Q = \frac{\Delta \text{ altitude}}{\Delta \text{ horizontal distance}}$$

Keep in mind that P and Q are points in the domain of f in the xy-plane and the change in altitude is the difference in heights of the points P' and Q' on the mountain (graph) that lie above P and Q.

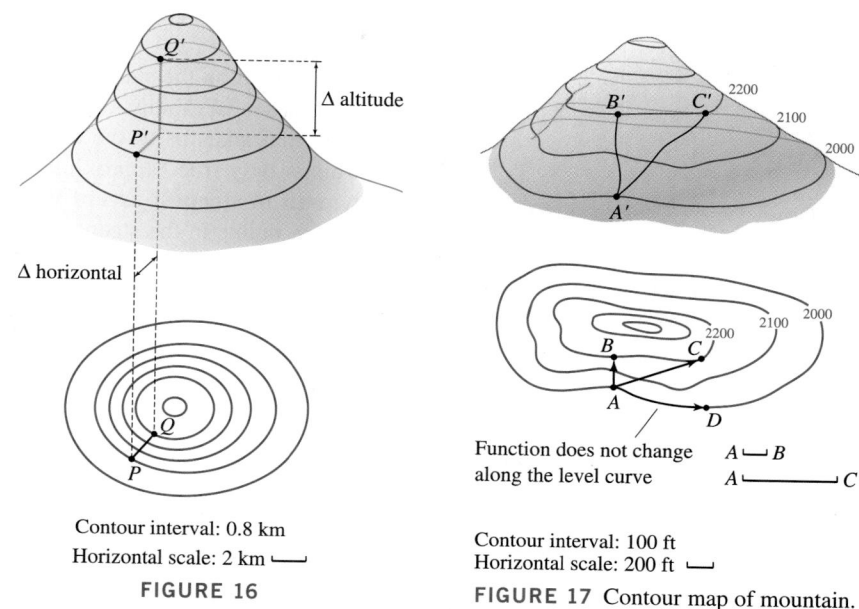

Contour interval: 0.8 km
Horizontal scale: 2 km ⌐⌐

FIGURE 16

Function does not change $A \,\rule[0.5ex]{1em}{0.4pt}\, B$
along the level curve $A \,\rule[0.5ex]{2em}{0.4pt}\, C$

Contour interval: 100 ft
Horizontal scale: 200 ft ⌐⌐

FIGURE 17 Contour map of mountain.

■ **EXAMPLE 8** Calculating Average ROC   Calculate the average ROC of $f(x, y)$ from P to Q for the function whose contour map is shown in Figure 16.

Solution The segment $\overline{PQ}$ spans three level curves and the contour interval is $m = 0.8$, so the change in altitude from P' to Q' is $3(0.8) = 2.4$ km. From the horizontal scale of the contour map, we see that the horizontal distance from P to Q is 2 km, so

$$\text{Average ROC from } P \text{ to } Q = \frac{\Delta \text{ altitude}}{\Delta \text{ horizontal distance}} = \frac{3(0.8)}{2} = 1.2 \qquad ■$$

CONCEPTUAL INSIGHT   In single-variable calculus, we measure the ROC by the derivative $f'(a)$. In the multivariable case, there is no single ROC because the change in $f(x, y)$ depends on the direction. For example, $f(x, y)$ is constant along level curves so its ROC along a level curve is zero. On the other hand, the ROC is nonzero in a direction pointing from one level curve to the next (Figure 17). We will discuss this again when we come to directional derivatives in Section 14.5.

■ **EXAMPLE 9** Average ROC Depends on Direction   Compute the average ROC from A to the points B, C, and D in Figure 17.

Solution Both segments $\overline{AB}$ and $\overline{AC}$ span two level curves, so the change in altitude is 200 ft in both cases. The horizontal scale shows that $\overline{AB}$ is two units long, corresponding to a horizontal change of 400 ft, and $\overline{AC}$ is four units long, corresponding to a horizontal change of 800 ft. On the other hand, there is no change in altitude from A to D. Therefore:

$$\text{Average ROC from } A \text{ to } B = \frac{\Delta \text{ altitude}}{\Delta \text{ horizontal distance}} \approx \frac{200}{400} = 0.5$$

$$\text{Average ROC from } A \text{ to } C = \frac{\Delta \text{ altitude}}{\Delta \text{ horizontal distance}} \approx \frac{200}{800} = 0.25$$

$$\text{Average ROC from } A \text{ to } D = \frac{\Delta \text{ altitude}}{\Delta \text{ horizontal distance}} = 0 \qquad ■$$

We obtain a path of steepest descent by following a path of steepest ascent in the opposite direction. Water flowing down the side of a mountain follows a path of steepest descent.

When we stand on the side of a mountain, there is a *steepest* direction in which the altitude increases most rapidly. On a contour map, the steepest direction is approximately the direction that takes us to the closest point on the next highest level curve [Figure 18(A)]. This is only true "approximately" because the terrain may vary between level curves. A **path of steepest ascent** is a path that begins at a point P and, everywhere along the way, points in the steepest direction. Again, we may approximate the path of steepest ascent by drawing a sequence of segments that move as directly as possible from one level curve to the next. Figure 18(B) shows two paths from P to Q. The solid path is a path of steepest ascent but the dashed path is not, because it does not move from one level curve to the next along the shortest possible segment.

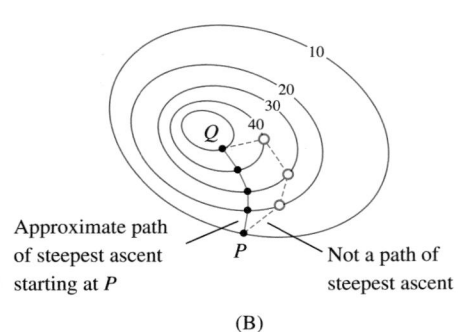

(A) Vectors pointing approximately in the direction of steepest ascent

(B)

Approximate path of steepest ascent starting at P

Not a path of steepest ascent

FIGURE 18

More Than Two Variables

It is not possible to draw the graph of a function of more than two variables. The graph of a function $f(x, y, z)$ would consist of the set of points $(x, y, z, f(x, y, z))$ in four-dimensional space $\mathbf{R}^4$. However, it is possible to draw the **level surfaces** of a function of three variables $f(x, y, z)$. These are the surfaces with equation $f(x, y, z) = c$. For example, the level surfaces of

$$f(x, y, z) = x^2 + y^2 + z^2$$

are the spheres with equation $x^2 + y^2 + z^2 = c$ (Figure 19). For functions of four or more variables, we can no longer visualize the graph or the level surfaces. We must rely on intuition developed through the study of functions of two and three variables.

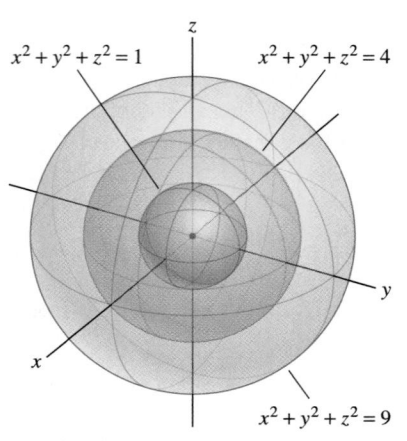

FIGURE 19 The level surfaces of $f(x, y, z) = x^2 + y^2 + z^2$ are spheres.

■ **EXAMPLE 10**   Describe the level surfaces of $g(x, y, z) = x^2 + y^2 - z^2$.

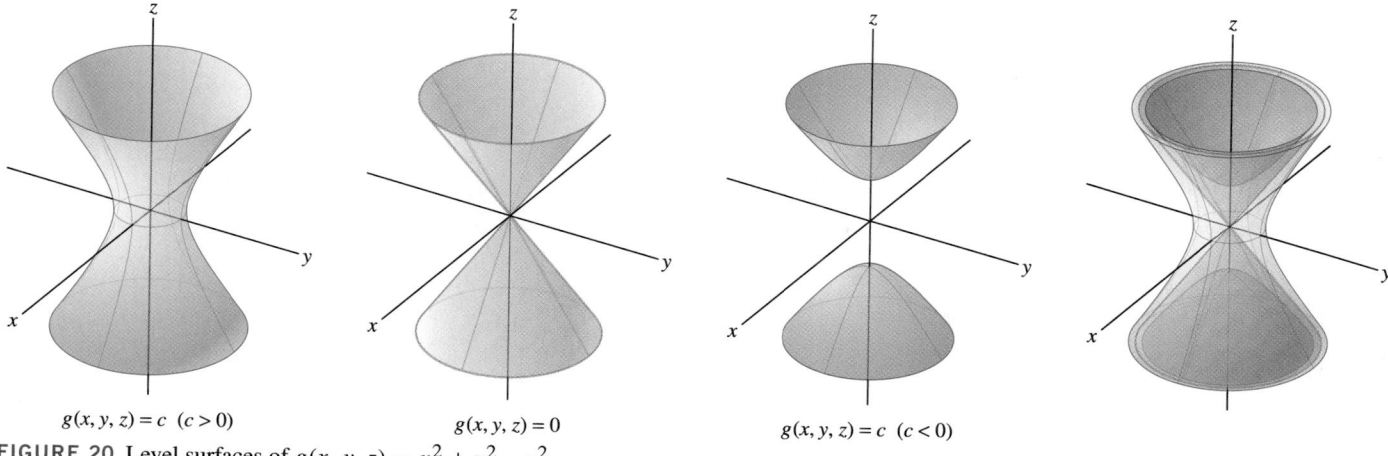

$g(x, y, z) = c$ $(c > 0)$ $g(x, y, z) = 0$ $g(x, y, z) = c$ $(c < 0)$

FIGURE 20 Level surfaces of $g(x, y, z) = x^2 + y^2 - z^2$.

Solution The level surface for $c = 0$ is the cone $x^2 + y^2 - z^2 = 0$. For $c \neq 0$, the level surfaces are the hyperboloids $x^2 + y^2 - z^2 = c$. The hyperboloid has one sheet if $c > 0$ and two sheets if $c < 0$ (Figure 20). ∎

14.1 SUMMARY

- The domain $\mathcal{D}$ of a function $f(x_1, \ldots, x_n)$ of n variables is the set of n-tuples $(a_1, \ldots, a_n)$ in $\mathbf{R}^n$ for which $f(a_1, \ldots, a_n)$ is defined. The range of f is the set of values taken by f.
- The graph of a continuous real-valued function $f(x, y)$ is the surface in $\mathbf{R}^3$ consisting of the points $(a, b, f(a, b))$ for (a, b) in the domain $\mathcal{D}$ of f.
- A *trace* is a curve obtained by intersecting the graph with a horizontal plane $z = c$ or a vertical plane $x = a$ or $y = b$.
- A *level curve* is a curve in the xy-plane defined by an equation $f(x, y) = c$. The level curve $f(x, y) = c$ is the projection onto the xy-plane of the trace curve at height $z = c$.
- A *contour map* shows the level curves $f(x, y) = c$ for equally spaced values of c. The spacing m is called the *contour interval*. The level curves are closer together where the graph is steeper.
- A direction of steepest ascent at a point P is a direction along which $f(x, y)$ increases most rapidly. The steepest direction is obtained (approximately) by drawing the vector from P to the nearest point on the next level curve.

14.1 EXERCISES

Preliminary Questions

1. What is the difference between a horizontal trace and a level curve? How are they related?

2. Describe the trace of $f(x, y) = x^2 - \sin(x^3 y)$ in the xz-plane.

3. Is it possible for two different level curves of a function to intersect? Explain.

4. Describe the contour map of $f(x, y) = x$ with contour interval 1.

5. How will the contour maps of $f(x, y) = x$ and $g(x, y) = 2x$ with contour interval 1 look different?

Exercises

In Exercises 1–4, evaluate the function at the specified points.

1. $f(x, y) = x + yx^3$, $\quad (2, 2), (-1, 4), (6, \frac{1}{2})$

2. $g(x, y) = \dfrac{y}{x^2 + y^2}$, $\quad (1, 3), (3, -2)$

3. $h(x, y, z) = xyz^{-2}$, $\quad (3, 8, 2), (3, -2, -6)$

4. $Q(y, z) = y^2 + y \sin z$, $\quad (y, z) = (2, \frac{\pi}{2}), (-2, \frac{\pi}{6})$

In Exercises 5–16, sketch the domain of the function.

5. $f(x, y) = 4x - 7y$

6. $f(x, y) = \sqrt{9 - x^2}$

7. $f(x, y) = \ln(y - 2x)$

8. $h(x, t) = \dfrac{1}{x + t}$

9. $G(x, t) = e^{1/(x+t)}$

10. $g(y, z) = \dfrac{1}{z + y^2}$

11. $f(x, y) = \sin \dfrac{y}{x}$

12. $H(r, s) = 3r^2 s^4$

13. $F(I, R) = \sqrt{IR}$

14. $f(x, y) = \cos^{-1}(x + y)$

15. $g(r, t) = \dfrac{1}{r^2 - t}$

16. $f(x, y) = \ln(4x^2 - y)$

In Exercises 17–19, describe the domain and range of the function.

17. $f(x, y, z) = xz + e^y$

18. $P(r, s) = e^{r/s}$

19. $f(x, y, z) = \sqrt{9 - x^2 - y^2 - z^2}$

20. Match the functions (a)–(f) with their graphs (A)–(F) in Figure 21.

(a) $f(x, y) = |x| + |y|$

(b) $f(x, y) = \cos(x - y)$

(c) $f(x, y) = \dfrac{-1}{1 + 9x^2 + y^2}$

(d) $f(x, y) = \cos(y^2)e^{-0.1(x^2+y^2)}$

(e) $f(x, y) = \dfrac{-1}{1 + 9x^2 + 9y^2}$

(f) $f(x, y) = \cos(x^2 + y^2)e^{-0.1(x^2+y^2)}$

In Exercises 21–30, sketch the graph and describe the vertical and horizontal traces.

21. $f(x, y) = 12 - 3x - 4y$

22. $f(x, y) = x^2 + y$

23. $f(x, y) = x^2 + 4y^2$

24. $f(x, y) = y^2$

25. $f(x, y) = \dfrac{1}{x^2 + y^2 + 1}$

26. $f(x, y) = 7$

27. $f(x, y) = x + |y|$

28. $f(x, y) = 9 - x^2 - y^2$

(A)   (B)

(C)   (D)

 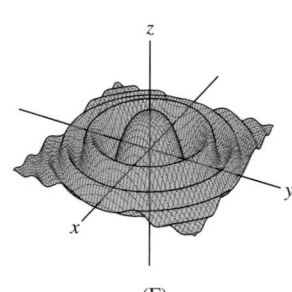

(E)   (F)

FIGURE 21

29. $f(x, y) = \sqrt{1 - x^2 - y^2}$ **30.** $f(x, y) = \sqrt{1 + x^2 + y^2}$

31. Sketch the contour maps of $f(x, y) = x + y$ with contour intervals $m = 1$ and 2.

32. Sketch the contour map of $f(x, y) = x^2 + y^2$ with level curves $c = 0, 4, 8, 12, 16$.

33. The function $f(x, t) = t^{-1/2}e^{-x^2/t}$, whose graph is shown in Figure 22, models the temperature along a metal bar after an intense burst of heat is applied at its center point.

(a) Sketch the graphs of the vertical traces at times $t = 0.5, 1, 1.5, 2$. What do these traces tell us about the way heat diffuses through the bar?

(b) Sketch the vertical trace at $x = 0$. Describe how temperature varies in time at the center point.

(c) Sketch the vertical trace $x = c$ for $c = \pm 0.2, \pm 0.4$. Describe how temperature varies in time at points near the center.

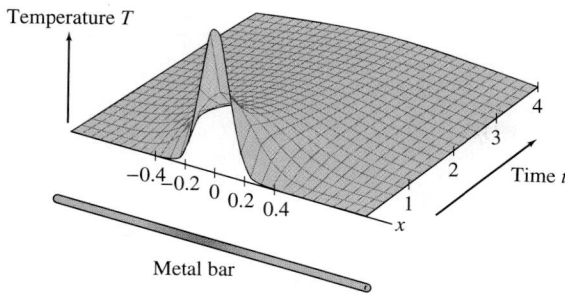

Temperature T

Metal bar

FIGURE 22 Graph of $f(x, t) = t^{-1/2}e^{-x^2/t}$ beginning shortly after $t = 0$.

In Exercises 34–43, draw a contour map of $f(x, y)$ with an appropriate contour interval, showing at least six level curves.

34. $f(x, y) = \dfrac{y}{x}$

35. $f(x, y) = e^{y/x}$

36. $f(x, y) = \dfrac{y}{x^2}$

37. $f(x, y) = xy$

38. $f(x, y) = x - y$

39. $f(x, y) = x + 2y - 1$

40. $f(x, y) = x^2 - y$

41. $f(x, y) = x^2$

42. $f(x, y) = 3x^2 - y^2$

43. $f(x, y) = \dfrac{10}{1 + x^2 + y^2}$

44. Match the contour maps (A) and (B) in Figure 23 with the two functions $f(x, y) = x - 2y$ and $g(x, y) = 2x - y$.

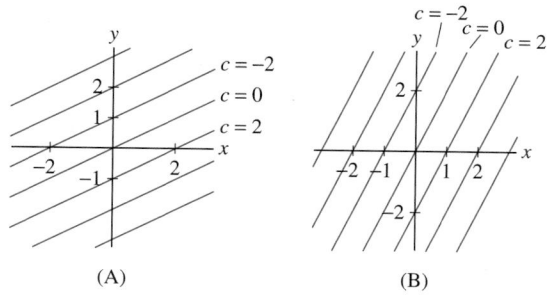

FIGURE 23

45. Which linear function has the contour map shown in Figure 24 (with level curve $c = 0$ as indicated), assuming that the contour interval is $m = 6$? What if $m = 3$?

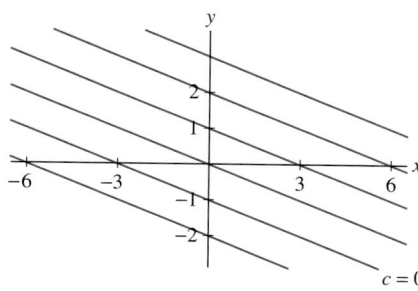

FIGURE 24

In Exercises 46–49, $f(S, T)$ denotes the density of seawater at salinity level S (parts per thousand) and temperature T (degrees Celsius). Refer to the contour map of $f(S, T)$ in Figure 25.

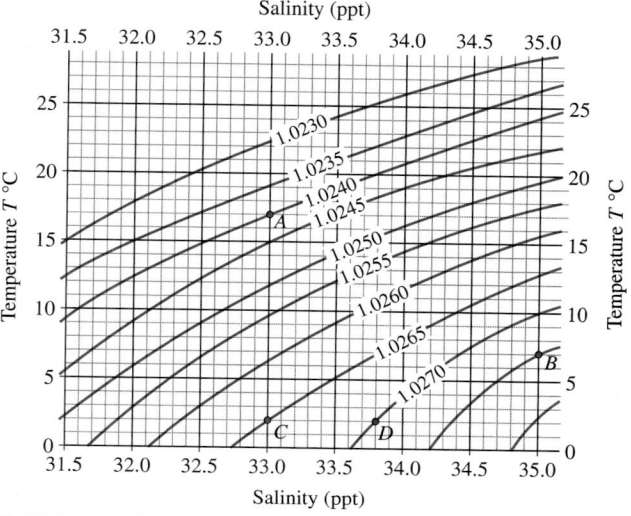

FIGURE 25 Contour map of seawater density $f(S, T)$ (kg/m^3).

46. Calculate the average ROC of density with respect to temperature from C to A.

47. Calculate the average ROC of density with respect to salinity from C to D.

48. At a fixed level of salinity, is seawater density an increasing or decreasing function of temperature?

49. Does water density appear to be more sensitive to a change in temperature at point A or point B?

In Exercises 50–53, refer to Figure 26.

50. Find the change in elevation from A and B.

51. Estimate the average ROC from A and B and from A to C.

52. Estimate the average ROC from P to points i, ii, iii, and iv.

53. Sketch the paths of steepest ascent beginning at D and E.

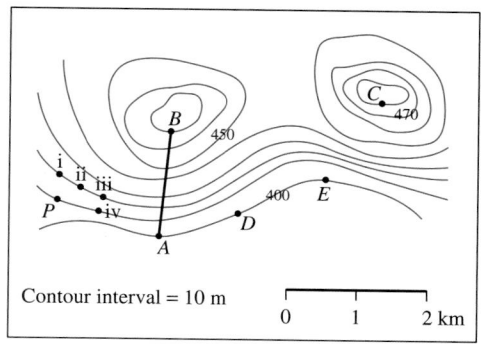

Contour interval = 10 m

0   1   2 km

FIGURE 26 Contour map of mountain.

54. Refer to Figure 27 to answer the following questions.

(a) At which of (A)–(E) is temperature increasing in the easterly direction?

(b) At which of (A)–(E) is temperature decreasing most rapidly in the northern direction?

(c) In which direction at (B) is temperature increasing most rapidly?

SurfaceTemperature (F)   **12Z Sun Sep 18 2005**

WW2010   (http://ww2010.atmos.uiuc.edu/)
Atmospheric Sciences, University of Illinois at Urbana-Champaign

FIGURE 27

Further Insights and Challenges

55. Let $f(x, y) = \dfrac{x}{\sqrt{x^2 + y^2}}$ for $(x, y) \neq 0$. Write f as a function $f(r, \theta)$ in polar coordinates and use this to find the level curves of f.

56. ⌐A5 Use a computer algebra system to draw the graph of $f(x, y) = \dfrac{xy}{x^2 + y^2 + 1}$ and its contour map.

(a) Show that the level curve $f(x, y) = c$ has equation

$$\left(\frac{1}{4c^2} - 1\right) y^2 - \left(x - \frac{1}{2c}y\right)^2 = 1$$

(b) Show that the level curve is empty if $|c| > \frac{1}{2}$ and is a hyperbola if $|c| < \frac{1}{2}$. What is the level curve for $c = \frac{1}{2}$?

14.2 Limits and Continuity in Several Variables

In this section, we extend the basic concepts of limit and continuity to the multivariable setting, focusing on functions of two variables (the cases of three or more variables are similar).

To formulate the limit definition, we define the **open disk** $D(P, r)$ and the **punctured disk** $D^*(P, r)$ of radius r and center $P = (a, b)$ (Figure 1):

$$D(P, r) = \{(x, y) \in \mathbf{R}^2 : (x - a)^2 + (y - b)^2 < r^2\}$$
$$D^*(P, r) = \{(x, y) \in \mathbf{R}^2 : 0 < (x - a)^2 + (y - b)^2 < r^2\}$$

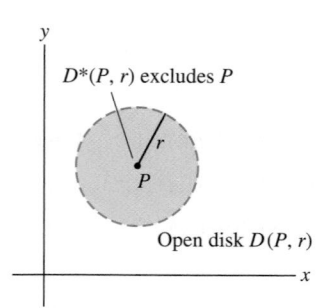

$D^*(P, r)$ excludes P

Open disk $D(P, r)$

FIGURE 1 The open disk $D(P, r)$ does not include the boundary circle.

Observe that $D^*(P, r)$ consists of all points in $D(P, r)$ other than P itself.

Recall that in one variable, $\lim\limits_{x \to a} f(x) = L$ if $|f(x) - L|$ is as small as desired for x sufficiently close (but not equal) to a. In two variables, we say that a function $f(x, y)$ with domain $\mathcal{D}$ approaches the limit L as (x, y) approaches $P = (a, b)$ if $|f(x, y) - L|$ is as small as desired for all (x, y) in $\mathcal{D}$ lying in a sufficiently small punctured disk centered at P [Figure 2(C)]. We write

$$\lim_{(x,y) \to P} f(x, y) = \lim_{(x,y) \to (a,b)} f(x, y) = L$$

Here is the formal ϵ-δ definition.

> **DEFINITION Limit**   Let $P = (a, b)$ be a point in the domain $\mathcal{D}$ of $f(x, y)$. Then
> $$\lim_{(x,y)\to(a,b)} f(x, y) = L$$
> if, for any $\epsilon > 0$, there exists $\delta > 0$ such that
> $$|f(x, y) - L| < \epsilon \quad \text{for all} \quad (x, y) \in \mathcal{D} \text{ lying in } D^*(P, \delta)$$

Observe that in a one-variable limit, $f(x)$ approaches L as x approaches a from the left or right as in Figure 2(A). In a multivariable limit, $f(x, y)$ must approach L no matter how (x, y) approaches (a, b) [Figure 2(B)].

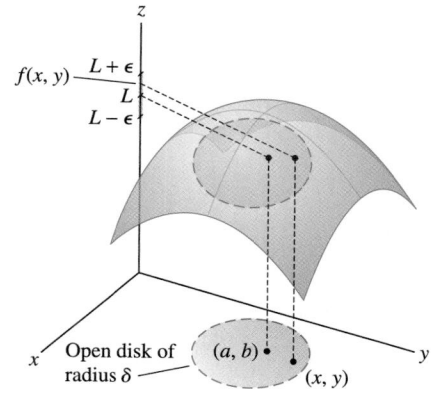

(A) In one variable, we can approach $x = a$ from only two possible directions.

(B) In two variables, (x, y) can approach $P = (a, b)$ along any direction or path.

(C) $|f(x, y) - L| < \epsilon$ for all (x, y) inside the disk

FIGURE 2

■ **EXAMPLE 1**   Show that for all (a, b), **(a)** $\displaystyle\lim_{(x,y)\to(a,b)} x = a$ and **(b)** $\displaystyle\lim_{(x,y)\to(a,b)} y = b$.

Solution   To verify (a), let $f(x, y) = x$ and $L = a$. We must show that for any $\epsilon > 0$, we can find $\delta > 0$ such that

$$|f(x, y) - L| = |x - a| < \epsilon \quad \text{for all} \quad (x, y) \in D^*(P, \delta) \qquad \boxed{1}$$

But if $(x, y) \in D^*(P, \delta)$, then

$$(x - a)^2 + (y - b)^2 < \delta^2 \quad \Rightarrow \quad (x - a)^2 < \delta^2 \quad \Rightarrow \quad |x - a| < \delta$$

Therefore, (1) holds with $\delta = \epsilon$:

$$|x - a| < \epsilon \quad \text{for all} \quad (x, y) \in D^*(P, \epsilon)$$

This proves (a). The limit (b) is similar.   ■

The following theorem lists the basic laws for working with limits. The proofs are similar to the proofs in the single-variable Limit Laws and are omitted.

THEOREM 1 **Limit Laws**   Assume that $\lim\limits_{(x,y)\to(a,b)} f(x, y)$ and $\lim\limits_{(x,y)\to(a,b)} g(x, y)$ exist. Then:

(i) **Sum Law:**

$$\lim_{(x,y)\to(a,b)} (f(x, y) + g(x, y)) = \lim_{(x,y)\to(a,b)} f(x, y) + \lim_{(x,y)\to(a,b)} g(x, y)$$

(ii) **Constant Multiple Law:** For any number k,

$$\lim_{(x,y)\to(a,b)} kf(x, y) = k \lim_{(x,y)\to(a,b)} f(x, y)$$

(iii) **Product Law:**

$$\lim_{(x,y)\to(a,b)} f(x, y)\, g(x, y) = \left(\lim_{(x,y)\to(a,b)} f(x, y) \right)\left(\lim_{(x,y)\to(a,b)} g(x, y) \right)$$

(iv) **Quotient Law:** If $\lim\limits_{(x,y)\to(a,b)} g(x, y) \neq 0$, then

$$\lim_{(x,y)\to(a,b)} \frac{f(x, y)}{g(x, y)} = \frac{\lim\limits_{(x,y)\to(a,b)} f(x, y)}{\lim\limits_{(x,y)\to(a,b)} g(x, y)}$$

Continuity for functions of two variables is defined by imitating the single-variable case: f is continuous at (a, b) if $f(x, y)$ approaches the function value $f(a, b)$ as $(x, y) \to (a, b)$.

DEFINITION **Continuity**   A function $f(x, y)$ is **continuous** at a point $P = (a, b)$ in its domain if

$$\lim_{(x,y)\to(a,b)} f(x, y) = f(a, b)$$

We say that f is continuous on a set $\mathcal{D}$ if f is continuous at each point of $\mathcal{D}$. We say simply that f is continuous if f is continuous on its domain.

It follows from the Limit Laws that arbitrary sums, multiples, and products of continuous functions are continuous. For instance, $f(x, y) = x$ and $g(x, y) = y$ are continuous by Example 1, so we may conclude that the power functions $f(x, y) = x^m y^n$ are continuous for all whole numbers m, n. We may also conclude that polynomials and rational functions [quotients $P(x, y)/Q(x, y)$, where P and Q are polynomials] are continuous at all points (a, b) where $Q(a, b) \neq 0$. As in one variable, we may evaluate limits of continuous functions using substitution.

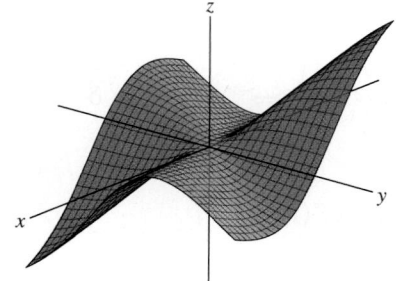

FIGURE 3 Graph of
$$f(x, y) = \frac{2x^2 y - 3xy^2}{x^2 + y^2 + 1}.$$

■ **EXAMPLE 2** Evaluating Limits by Substitution   Show that $f(x, y) = \dfrac{2x^2 y - 3xy^2}{x^2 + y^2 + 1}$ is continuous. Evaluate $\lim\limits_{(x,y)\to(1,2)} f(x, y)$.

Solution The function $f(x, y)$ is continuous everywhere because it is a rational function whose denominator $Q(x, y) = x^2 + y^2 + 1$ is never zero (Figure 3). We evaluate the limit using substitution:

$$\lim_{(x,y)\to(1,2)} \frac{2x^2y - 3xy^2}{x^2 + y^2 + 1} = \frac{2(1^2)(2) - 3(1)(2^2)}{1^2 + 2^2 + 1} = -\frac{4}{3} \qquad \blacksquare$$

Consider the special case that $f(x, y)$ is a product $f(x, y) = h(x)g(y)$, where $h(x)$ and $g(y)$ are continuous. By the Product Law, the limit is equal to a product:

$$\lim_{(x,y)\to(a,b)} f(x, y) = \lim_{(x,y)\to(a,b)} h(x)g(y) = \left(\lim_{x\to a} h(x)\right)\left(\lim_{y\to b} g(y)\right)$$

■ **EXAMPLE 3** **Product of Functions of One Variable** Evaluate $\lim\limits_{(x,y)\to(-1,1)} e^x \tan^{-1} y$.

Solution We evaluate the limit as a product of limits:

$$\lim_{(x,y)\to(-1,1)} e^x \tan^{-1} y = \left(\lim_{x\to -1} e^x\right)\left(\lim_{y\to 1} \tan^{-1} y\right) = e^{-1}(\tan^{-1} 1) = \frac{e^{-1}\pi}{4} \qquad \blacksquare$$

Composition is another important way to build functions. If $f(x, y)$ is a function of two variables and $G(u)$ a function of one variable, then the composite $G \circ f$ is the function $G(f(x, y))$. According to the next theorem, a composite of continuous functions is again continuous (the proof is omitted).

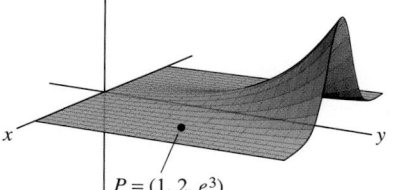

P = (1, 2, e³)

FIGURE 4 Graph of $H(x, y) = e^{-x^2+2y}$.

> **THEOREM 2** **A Composite of Continuous Functions Is Continuous** If $f(x, y)$ is continuous at (a, b) and $G(u)$ is continuous at $c = f(a, b)$, then the composite function $G(f(x, y))$ is continuous at (a, b).

■ **EXAMPLE 4** Write $H(x, y) = e^{-x^2+2y}$ as a composite function and evaluate

$$\lim_{(x,y)\to(1,2)} H(x, y)$$

Solution We have $H(x, y) = G \circ f$, where $G(u) = e^u$ and $f(x, y) = -x^2 + 2y$ (Figure 4). Since G and f (and hence H) are continuous, we may evaluate the limit using substitution:

$$\lim_{(x,y)\to(1,2)} H(x, y) = \lim_{(x,y)\to(1,2)} e^{-x^2+2y} = e^{-(1)^2+2(2)} = e^3 \qquad \blacksquare$$

We have emphasized that if $\lim\limits_{(x,y)\to(a,b)} f(x, y) = L$, then $f(x, y)$ must tend to L as (x, y) approaches (a, b) along any path. In the next example, we prove that a limit *does not exist* by showing that $f(x, y)$ approaches *different limits* along lines through the origin (Figure 5).

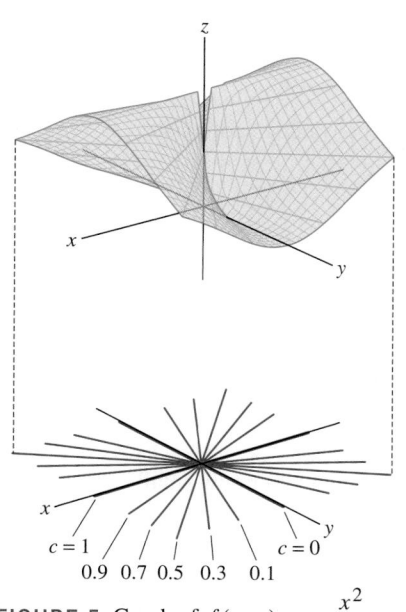

c = 1   c = 0
0.9   0.7   0.5   0.3   0.1

FIGURE 5 Graph of $f(x, y) = \dfrac{x^2}{x^2 + y^2}$.

■ **EXAMPLE 5** **Showing a Limit Does Not Exist** Examine $\lim\limits_{(x,y)\to(0,0)} \dfrac{x^2}{x^2 + y^2}$ numerically. Prove that the limit does not exist.

Solution First, let's examine a table of values of $f(x, y) = \dfrac{x^2}{x^2 + y^2}$. Table 1 suggests that $f(x, y)$ takes on all values between 0 and 1, no matter how close (x, y) gets to $(0, 0)$.

TABLE 1   Values of $f(x,y) = \dfrac{x^2}{x^2+y^2}$

y \ x	−0.5	−0.4	−0.3	−0.2	−0.1	0	0.1	0.2	0.3	0.4	0.5
0.5	**0.5**	0.39	0.265	0.138	0.038	**0**	0.038	0.138	0.265	0.39	**0.5**
0.4	0.61	**0.5**	0.36	0.2	0.059	**0**	0.059	0.2	0.36	**0.5**	0.61
0.3	0.735	0.64	**0.5**	0.308	0.1	**0**	0.1	0.308	**0.5**	0.64	0.735
0.2	0.862	0.8	0.692	**0.5**	0.2	**0**	0.2	**0.5**	0.692	0.8	0.862
0.1	0.962	0.941	0.9	0.8	**0.5**	**0**	**0.5**	0.8	0.9	0.941	0.962
0	1	1	1	1	1		1	1	1	1	1
−0.1	0.962	0.941	0.9	0.8	**0.5**	**0**	**0.5**	0.8	0.9	0.941	0.962
−0.2	0.862	0.8	0.692	**0.5**	0.2	**0**	0.2	**0.5**	0.692	0.8	0.862
−0.3	0.735	0.640	**0.5**	0.308	0.1	**0**	0.1	0.308	**0.5**	0.640	0.735
−0.4	0.610	**0.5**	0.360	0.2	0.059	**0**	0.059	0.2	0.36	**0.5**	0.61
−0.5	**0.5**	0.39	0.265	0.138	0.038	**0**	0.038	0.138	0.265	0.390	**0.5**

For example,

$$f(0.1,0) = 1, \qquad f(0.1,0.1) = 0.5, \qquad f(0,0.1) = 0$$

Thus, $f(x,y)$ does not seem to approach any fixed value L as $(x,y) \to (0,0)$.

Now let's prove that the limit does not exist by showing that $f(x,y)$ approaches different limits along the x- and y-axes (Figure 5). Since $f(x,0) = x^2/x^2 = 1$ for $x \neq 0$ and $f(0,y) = 0/y^2 = 0$ for $y \neq 0$,

$$\text{Limit along } x\text{-axis} = \lim_{x\to 0} f(x,0) = \lim_{x\to 0} 1 = 1$$

$$\text{Limit along } y\text{-axis} = \lim_{y\to 0} f(0,y) = \lim_{y\to 0} 0 = 0$$

These two limits are different and hence $\displaystyle\lim_{(x,y)\to(0,0)} f(x,y)$ does not exist. ■

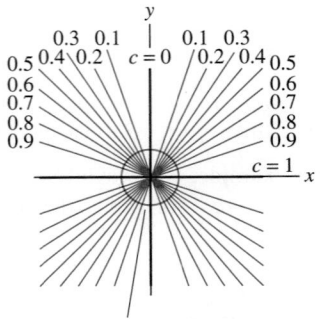

$f(x, y)$ takes on all values between 0 and 1 in every small disk around the origin

FIGURE 6 Contour map of $f(x, y) = \dfrac{x^2}{x^2 + y^2}$.

GRAPHICAL INSIGHT   To better understand the behavior of $f(x,y) = \dfrac{x^2}{x^2 + y^2}$ near $(0,0)$, observe that f is constant on each line $y = mx$:

$$f(x, mx) = \frac{x^2}{x^2 + (mx)^2} = \frac{1}{m^2 + 1} \quad \text{(for } x \neq 0\text{)}$$

In other words, the lines through the origin $y = mx$ are the level curves of f (Figure 6). As m varies, $(m^2 + 1)^{-1}$ varies between 0 and 1, and thus, $f(x,y)$ takes on all values between 0 and 1 in every disk around the origin $(0,0)$, no matter how small. This shows in greater detail that there is no number L such that $|f(x,y) - L|$ is arbitrarily small for all (x,y) close to $(0,0)$.

As we know from single-variable calculus, there is no general method for computing limits. The next example shows that polar coordinates are useful in some cases.

■ **EXAMPLE 6** Two Methods for Verifying a Limit Prove that the following function is continuous:

$$f(x, y) = \begin{cases} \dfrac{xy^2}{x^2 + y^2} & \text{for } (x, y) \neq (0, 0) \\ 0 & \text{for } (x, y) = (0, 0) \end{cases}$$

Solution The denominator of f is nonzero for $(x, y) \neq (0, 0)$, so f is continuous on the domain $\{(x, y) \neq (0, 0)\}$ (Figure 7). To prove that $f(x, y)$ is also continuous at $(0, 0)$, we must verify that

$$\lim_{(x,y)\to(0,0)} \frac{xy^2}{x^2 + y^2} = f(0, 0) = 0$$

First Method We use the inequality

$$|xy^2| = |x||y|^2 \leq |x|x^2 + |x||y|^2 = |x|(x^2 + y^2)$$

The inequality holds because $|x|x^2 \geq 0$. For $(x, y) \neq (0, 0)$, division by $(x^2 + y^2)$ yields

$$0 \leq \left| \frac{xy^2}{x^2 + y^2} \right| \leq |x|$$

Now use the Squeeze Theorem (which is valid for limits in several variables):

$$\lim_{(x,y)\to(0,0)} 0 \leq \lim_{(x,y)\to(0,0)} \left| \frac{xy^2}{x^2 + y^2} \right| \leq \lim_{(x,y)\to(0,0)} |x| = 0$$

We conclude that $f(x, y)$ approaches $f(0, 0) = 0$ as $(x, y) \to (0, 0)$.

Second Method Use polar coordinates: $x = r\cos\theta$, $y = r\sin\theta$. Then $x^2 + y^2 = r^2$ and for $r \neq 0$,

$$0 \leq \left| \frac{xy^2}{x^2 + y^2} \right| = \left| \frac{(r\cos\theta)(r^2\sin^2\theta)}{r^2} \right| = r|\cos\theta\sin^2\theta| \leq r$$

As (x, y) approaches $(0, 0)$, the variable r also approaches 0, so again, the desired conclusion follows from the Squeeze Theorem:

$$0 \leq \lim_{(x,y)\to(0,0)} \left| \frac{xy^2}{x^2 + y^2} \right| \leq \lim_{r\to 0} r = 0 \qquad ■$$

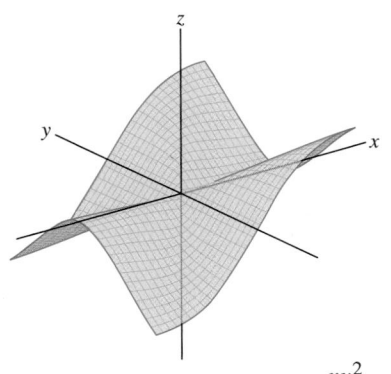

FIGURE 7 Graph of $f(x, y) = \dfrac{xy^2}{x^2 + y^2}$.

14.2 SUMMARY

- The *open disk* of radius r centered at $P = (a, b)$ is defined by

$$D(P, r) = \{Q = (x, y) \in \mathbf{R}^2 : (x - a)^2 + (y - b)^2 < r^2\}$$

The *punctured disk* $D^*(P, r)$ is $D(P, r)$ with the point P removed.
- Suppose $P = (a, b)$ belongs to the domain $\mathcal{D}$ of $f(x, y)$. Then

$$\lim_{(x,y)\to(a,b)} f(x, y) = L$$

if, for any $\epsilon > 0$, there exists $\delta > 0$ such that

$$|f(x, y) - L| < \epsilon \quad \text{for all} \quad (x, y) \in \mathcal{D} \text{ lying in } D^*(P, \delta)$$

• A function $f(x, y)$ is *continuous* at $P = (a, b)$ if

$$\lim_{(x,y)\to(a,b)} f(x, y) = f(a, b)$$

14.2 EXERCISES

Preliminary Questions

1. What is the difference between $D(P, r)$ and $D^*(P, r)$?

2. Suppose that $f(x, y)$ is continuous at $(2, 3)$ and that $f(2, y) = y^3$ for $y \neq 3$. What is the value $f(2, 3)$?

3. Suppose that $Q(x, y)$ is a function such that $\dfrac{1}{Q(x, y)}$ is continuous for all (x, y). Which of the following statements are true?

(a) $Q(x, y)$ is continuous at all (x, y).

(b) $Q(x, y)$ is continuous for $(x, y) \neq (0, 0)$.

(c) $Q(x, y) \neq 0$ for all (x, y).

4. Suppose that $f(x, 0) = 3$ for all $x \neq 0$ and $f(0, y) = 5$ for all $y \neq 0$. What can you conclude about $\lim\limits_{(x,y)\to(0,0)} f(x, y)$?

Exercises

In Exercises 1–10, use continuity to evaluate the limit.

1. $\lim\limits_{(x,y)\to(1,2)} (x^2 + y)$

2. $\lim\limits_{(x,y)\to(\frac{4}{9},\frac{2}{9})} \dfrac{x}{y}$

3. $\lim\limits_{(x,y)\to(3,-1)} (3x^2 y - 2xy^3)$

4. $\lim\limits_{(x,y)\to(-2,4)} \dfrac{x^2 - 3y^2}{4x + y}$

5. $\lim\limits_{(x,y)\to(1,2)} \dfrac{x^2 + y}{x + y^2}$

6. $\lim\limits_{(x,y)\to(\frac{\pi}{4},0)} \tan x \cos y$

7. $\lim\limits_{(x,y)\to(4,4)} \arctan \dfrac{y}{x}$

8. $\lim\limits_{(x,y)\to(0,0)} e^{x^2 - y^2}$

9. $\lim\limits_{(x,y)\to(1,1)} \dfrac{e^{x^2} - e^{-y^2}}{x + y}$

10. $\lim\limits_{(x,y)\to(1,0)} \ln(x - y)$

In Exercises 11–18, evaluate the limit or determine that it does not exist.

11. $\lim\limits_{(x,y)\to(0,1)} \dfrac{x}{y}$

12. $\lim\limits_{(x,y)\to(1,0)} \dfrac{x}{y}$

13. $\lim\limits_{(x,y)\to(1,1)} e^{xy} \ln(1 + xy)$

14. $\lim\limits_{(x,y)\to(0,0)} \dfrac{x^2 + y^2}{1 + y^2}$

15. $\lim\limits_{(x,y)\to(-1,-2)} x^2 |y|^3$

16. $\lim\limits_{(x,y)\to(-1,-2)} \dfrac{xy^2}{|x|}$

17. $\lim\limits_{(x,y)\to(4,2)} \dfrac{y - 2}{\sqrt{x^2 - 4}}$

18. $\lim\limits_{(x,y)\to(\pi,0)} \dfrac{\sin x}{\sin y}$

In Exercises 19–22, assume that

$$\lim_{(x,y)\to(2,5)} f(x, y) = 3, \qquad \lim_{(x,y)\to(2,5)} g(x, y) = 7$$

19. $\lim\limits_{(x,y)\to(2,5)} \big(f(x, y) + 4g(x, y)\big)$

20. $\lim\limits_{(x,y)\to(2,5)} f(x, y)g(x, y)^2$

21. $\lim\limits_{(x,y)\to(2,5)} e^{f(x,y)}$

22. $\lim\limits_{(x,y)\to(2,5)} \ln\big(g(x, y) - 2f(x, y)\big)$

In Exercises 23–32, evaluate the limit or determine that the limit does not exist. You may evaluate the limit of a product function as a product of limits as in Example 3.

23. $\lim\limits_{(x,y)\to(0,0)} \dfrac{(\sin x)(\sin y)}{xy}$

24. $\lim\limits_{(x,y)\to(2,1)} e^{x^2 - y^2}$

25. $\lim\limits_{(z,w)\to(-1,2)} (z^2 w - 9z)$

26. $\lim\limits_{(x,y)\to(0,0)} \sin x \cos \dfrac{1}{y}$

27. $\lim\limits_{(h,k)\to(2,0)} h^4 \dfrac{(2 + k)^2 - 4}{k}$

28. $\lim\limits_{(h,k)\to(0,0)} \dfrac{h(k^2 + 4)}{k}$

29. $\lim\limits_{(x,y)\to(0,0)} e^{1/x} \tan^{-1} \dfrac{1}{y}$

30. $\lim\limits_{(x,k)\to(1,0)} x^2 \left(\dfrac{e^k - 1}{k}\right)$

31. $\lim\limits_{(x,y)\to(0,0)} x \ln y$

32. $\lim\limits_{(x,y)\to(0,0)} \left(\dfrac{1}{2xy} - \dfrac{1}{xy(x + 2)}\right)$

33. Let $f(x, y) = \dfrac{x^3 + y^3}{x^2 + y^2}$.

(a) Show that

$$|x^3| \le |x|(x^2 + y^2), \quad |y^3| \le |y|(x^2 + y^2)$$

(b) Show that $|f(x, y)| \le |x| + |y|$.
(c) Use (b) and the formal definition of the limit to prove that $\lim\limits_{(x,y)\to(0,0)} f(x, y) = 0$.
(d) Verify the conclusion of (c) again using polar coordinates as in Example 6.

34. Show that $\lim\limits_{(x,y)\to(0,0)} \dfrac{y^2}{x^2 + y^2}$ does not exist.

35. Show that $\lim\limits_{(x,y)\to(0,0)} \dfrac{|x|}{|x| + |y|}$ does not exist. *Hint:* Consider the limits along the lines $y = mx$.

36. Show that $\lim\limits_{(x,y)\to(0,0)} \dfrac{x}{\sqrt{x^2 + y^2}}$ does not exist. *Hint:* Use polar coordinates.

37. Let $a, b \ge 0$. Show that $\lim\limits_{(x,y)\to(0,0)} \dfrac{x^a y^b}{x^2 + y^2} = 0$ if $a + b > 2$ and the limit does not exist if $a + b \le 2$.

38. Evaluate $\lim\limits_{(x,y)\to(0,0)} \dfrac{x^2 + y^2}{\sqrt{x^2 + y^2 + 1} - 1}$.

39. Figure 8 shows the contour maps of two functions with contour interval $m = 2$. Explain why the limit $\lim\limits_{(x,y)\to P} f(x, y)$ does not exist. Does $\lim\limits_{(x,y)\to P} g(x, y)$ appear to exist in (B)? If so, what is its limit?

(A) Contour map of $f(x, y)$ (B) Contour map of $g(x, y)$

FIGURE 8

Further Insights and Challenges

40. Evaluate $\lim\limits_{(x,y)\to(0,2)} (1 + x)^{y/x}$.

41. *CAS* The function $f(x, y) = \dfrac{\sin(xy)}{xy}$ is defined for $xy \ne 0$.
(a) Is it possible to extend the domain of $f(x, y)$ to all of $\mathbf{R}^2$ so that the result is a continuous function?
(b) Use a computer algebra system to plot $f(x, y)$. Does the result support your conclusion in (a)?

42. *CAS* Repeat Exercise 41 for the functions $f(x, y) = \dfrac{xy}{x + y}$ and $g(x, y) = \dfrac{xy}{x^2 + y^2}$.

43. The function $f(x, y) = \dfrac{x^2 y}{x^4 + y^2}$ provides an interesting example where the limit as $(x, y) \to (0, 0)$ does not exist, even though the limit along every line $y = mx$ exists and is zero (Figure 9).
(a) Show that the limit along any line $y = mx$ exists and is equal to 0.
(b) Calculate $f(x, y)$ at the points $(10^{-1}, 10^{-2})$, $(10^{-5}, 10^{-10})$, $(10^{-20}, 10^{-40})$. Do not use a calculator.
(c) Show that $\lim\limits_{(x,y)\to(0,0)} f(x, y)$ does not exist. *Hint:* Compute the limit along the parabola $y = x^2$.

44. Is the following function continuous?

$$f(x, y) = \begin{cases} x^2 + y^2 & \text{if } x^2 + y^2 < 1 \\ 1 & \text{if } x^2 + y^2 \ge 1 \end{cases}$$

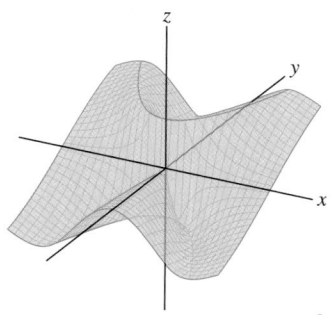

FIGURE 9 Graph of $f(x, y) = \dfrac{x^2 y}{x^4 + y^2}$.

14.3 Partial Derivatives

In calculus of one variable, the derivative $f'(a)$ is the rate of change of $f(x)$ at $x = a$. By contrast, a function f of two or more variables does not have a unique rate of change because each variable may affect f in different ways. For example, the current I in a circuit is a function of both voltage V and resistance R as given by Ohm's Law:

$$I(V, R) = \frac{V}{R}$$

The current I is *increasing* as function of V but *decreasing* as a function of R.

The **partial derivatives** are the rates of change with respect to each variable separately. A function $f(x, y)$ of two variables has two partial derivatives, denoted f_x and f_y, defined by the following limits (if they exist):

$$f_x(a, b) = \lim_{h \to 0} \frac{f(a + h, b) - f(a, b)}{h}, \qquad f_y(a, b) = \lim_{k \to 0} \frac{f(a, b + k) - f(a, b)}{k}$$

Thus, f_x is the derivative at $f(x, b)$ as a function of x alone and f_y is the derivative at $f(a, y)$ as a function of y alone. The Leibniz notation for partial derivatives is

The partial derivative symbol ∂ is a rounded "d." The symbols $\dfrac{\partial f}{\partial x}$ and $\dfrac{\partial f}{\partial y}$ are read as follows: "dee-eff dee-ex" and "dee-eff dee-why."

$$\frac{\partial f}{\partial x} = f_x, \quad \frac{\partial f}{\partial y} = f_y \quad \text{and} \quad \frac{\partial f}{\partial x}\bigg|_{(a,b)} = f_x(a, b), \quad \frac{\partial f}{\partial y}\bigg|_{(a,b)} = f_y(a, b)$$

If $z = f(x, y)$, then we also write $\dfrac{\partial z}{\partial x}$ and $\dfrac{\partial z}{\partial y}$.

Partial derivatives are computed just like ordinary derivatives in one variable with this difference: To compute f_x, treat y as a constant, and to compute f_y, treat x as a constant.

■ **EXAMPLE 1** Compute the partial derivatives of $f(x, y) = x^2 y^5$.

Solution

$$\frac{\partial f}{\partial x} = \underbrace{\frac{\partial}{\partial x}(x^2 y^5) = y^5 \frac{\partial}{\partial x}(x^2)}_{\text{Treat } y^5 \text{ as a constant}} = y^5(2x) = 2xy^5$$

$$\frac{\partial f}{\partial y} = \underbrace{\frac{\partial}{\partial y}(x^2 y^5) = x^2 \frac{\partial}{\partial x}(y^5)}_{\text{Treat } x^2 \text{ as a constant}} = x^2(5y^4) = 5x^2 y^4$$

■

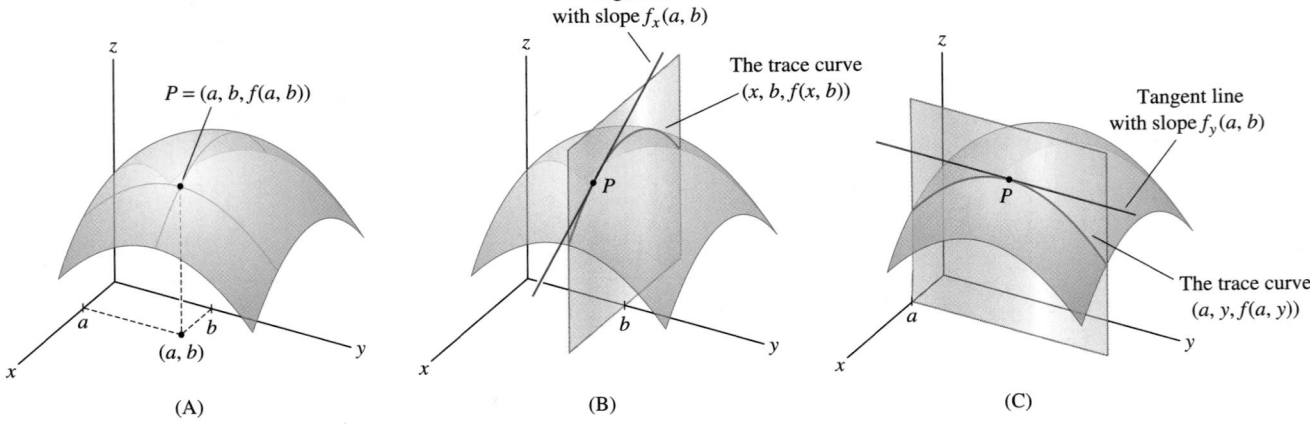

FIGURE 1 Graph of $z = f(x, y)$ and its vertical trace curves.

GRAPHICAL INSIGHT The partial derivatives at $P = (a, b)$ are the slopes of the tangent lines to the vertical trace curves passing through the point $(a, b, f(a, b))$ on the graph of $z = f(x, y)$ [Figure 1(A)]. Thus, the trace curve in the vertical plane $y = b$ is the set of points $(x, b, f(x, b))$, and $f_x(a, b)$ is the slope of the tangent line to this curve at $x = a$ [Figure 1(B)]. Similarly, $f_y(a, b)$ is the slope of the tangent line at $y = b$ to the trace curve obtained by intersecting with the vertical plane $x = a$ [Figure 1(C)].

All of the differentiation rules from calculus of one variable are valid for partial derivatives.

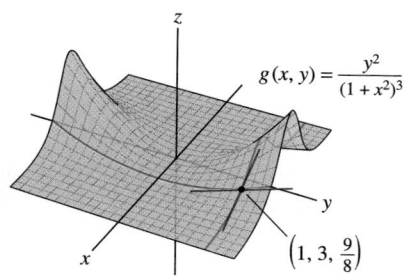

$$g(x, y) = \frac{y^2}{(1 + x^2)^3}$$

$$\left(1, 3, \frac{9}{8}\right)$$

FIGURE 2 The slopes of the tangent lines to the trace curves are $g_x(1, 3)$ and $g_y(1, 3)$.

■ **EXAMPLE 2** Calculate $g_x(1, 3)$ and $g_y(1, 3)$, where $g(x, y) = \dfrac{y^2}{(1 + x^2)^3}$.

Solution To calculate $\dfrac{\partial g}{\partial x}$, treat y (and therefore y^2) as a constant:

$$g_x(x, y) = \frac{\partial}{\partial x} y^2 (1 + x^2)^{-3} = y^2 \frac{\partial}{\partial x} (1 + x^2)^{-3} = \frac{-6xy^2}{(1 + x^2)^4}$$

$$g_x(1, 3) = \frac{-6(1)3^2}{(1 + 1^2)^4} = -\frac{27}{8}$$

To differentiate with respect to y, treat x (and therefore $1 + x^2$) as a constant:

CAUTION Do not *use the Quotient Rule to compute the partial derivative in Eq. (1). The denominator does not depend on y, so we treat it as a constant when differentiating with respect to y.*

$$g_y(x, y) = \frac{\partial}{\partial y} \frac{y^2}{(1 + x^2)^3} = \frac{1}{(1 + x^2)^3} \frac{\partial}{\partial y} y^2 = \frac{2y}{(1 + x^2)^3} \qquad \boxed{1}$$

$$g_y(1, 3) = \frac{2(3)}{(1 + 1^2)^3} = \frac{3}{4}$$

See Figure 2 for the geometric interpretation of these partial derivatives. ■

We use the Chain Rule in the usual way to compute the partial derivative of a composite function $f(x, y) = F(g(x, y))$, where $F(u)$ is a function of one variable and $u = g(x, y)$:

$$\frac{\partial f}{\partial x} = \frac{dF}{du} \frac{\partial u}{\partial x}, \qquad \frac{\partial f}{\partial y} = \frac{dF}{du} \frac{\partial u}{\partial y}$$

■ **EXAMPLE 3** **The Chain Rule in Partial Derivatives** Compute $\dfrac{\partial}{\partial x} \sin(x^2 y^3)$.

Solution We have $\sin(x^2 y^3) = F(u)$, where $F(u) = \sin u$ and $u = x^2 y^3$, so

$$\underbrace{\frac{\partial}{\partial x} \sin(x^2 y^3) = \frac{dF}{du} \frac{\partial u}{\partial x} = \cos(x^2 y^3) \frac{\partial}{\partial x} x^2 y^3}_{\text{Chain Rule}} = 2xy^3 \cos(x^2 y^3) \qquad ■$$

Partial derivatives are defined for functions of any number of variables. We compute the partial derivative with respect to any one of the variables by holding the remaining variables constant.

■ **EXAMPLE 4** More Than Two Variables Calculate $f_z(0, 1, 0, 1)$, where

$$f(x, y, z, w) = \frac{e^{xz+yw}}{z + w}$$

In Example 4, note that the calculation

$$\frac{\partial}{\partial z} e^{xz+yw} = x\, e^{xz+yw}$$

follows from the Chain Rule, just like

$$\frac{d}{dz} e^{az+b} = a\, e^{az+b}$$

Solution Use the Quotient Rule, treating x, y, and w as constants:

$$f_z(x, y, z, w) = \frac{\partial}{\partial z}\left(\frac{e^{xz+yw}}{z + w}\right) = \frac{(z + w)\frac{\partial}{\partial z}e^{xz+yw} - e^{xz+yw}\frac{\partial}{\partial z}(z + w)}{(z + w)^2}$$

$$= \frac{(z + w)x\, e^{xz+yw} - e^{xz+yw}}{(z + w)^2}$$

$$f_z(0, 1, 0, 1) = \frac{0 - e^1}{(0 + 1)^2} = -e$$ ■

Since $f_x(a, b)$ is the derivative of the function $f(x, b)$, we can use the partial derivative to estimate the change in $f(x, b)$ when x changes from a to $a + \Delta x$ (but $y = b$ is fixed) as in the single-variable case. Similarly, we can estimate the change in $f(a, y)$:

$$f(a + \Delta x, b) - f(a, b) \approx f_x(a, b)\Delta x$$

$$f(a, b + \Delta y) - f(a, b) \approx f_y(a, b)\Delta y$$

This approximation applies in any number of variables.

Solder ball of radius R

Chip

L

Circuit board

FIGURE 3 A BGA package. Temperature variations strain the BGA and may cause it to fail because the chip and board expand at different rates.

■ **EXAMPLE 5** Testing Microchips for Reliability A **ball grid array** (BGA) consists of a microchip joined to a circuit board by small solder balls of radius R mm separated by a distance L mm (Figure 3). Manufacturers test the reliability of BGA's by subjecting them to repeated cycles in which the temperature is varied from $0°$ to $100°$ over a 40-min period. According to one model, the average number N of cycles before the chip fails is

$$N = \left(\frac{2,200R}{Ld}\right)^{1.9}$$

where d is the difference between the coefficients of expansion of the chip and the board. Estimate the change in N when $d = 10$, $R = 0.12$, and L is increased from 0.4 to 0.42.

Solution Since we are interested in the effect on N of a change in L, we calculate the partial derivative with respect to L at the point $(L, R, d) = (0.4, 0.12, 10)$:

$$\frac{\partial N}{\partial L} = \frac{\partial}{\partial L}\left(\frac{2,200R}{Ld}\right)^{1.9} = \left(\frac{2,200R}{d}\right)^{1.9}\frac{\partial}{\partial L}L^{-1.9} = -1.9\left(\frac{2,200R}{d}\right)^{1.9}L^{-2.9}$$

$$\frac{\partial N}{\partial L}\bigg|_{(L,R,d)=(0.4,0.12,10)} = -1.9\left(\frac{2,200(0.12)}{10}\right)^{1.9}(0.4)^{-2.9} \approx -13{,}609$$

The average number of cycles before the chip fails decreases, changing by the amount

$$\Delta N \approx \frac{\partial N}{\partial L}\Delta L = -13{,}609(0.02) \approx -272 \text{ cycles}$$ ■

In the next example, we estimate a partial derivative numerically. Since f_x and f_y are limits of difference quotients, we have the following approximations when h and k

are "small" (just how small depends on the function and the accuracy required):

$$f_x(a, b) \approx \frac{\Delta f}{\Delta x} = \frac{f(a + h, b) - f(a, b)}{h}$$

$$f_y(a, b) \approx \frac{\Delta f}{\Delta y} = \frac{f(a, b + k) - f(a, b)}{k}$$

A similar approximation is valid in any number of variables.

■ **EXAMPLE 6** Estimating Partial Derivatives Using Contour Maps   The density of seawater ρ (kg/m^3) depends on salinity S (parts per thousand or ppt) and the temperature T (°C). Use the contour map in Figure 4 to estimate $\dfrac{\partial \rho}{\partial T}$ and $\dfrac{\partial \rho}{\partial S}$ at A.

Solution   Point A has coordinates $(S, T) = (33, 15)$ and lies on the level curve $\rho = 1.0245$. We estimate $\dfrac{\partial \rho}{\partial T}$ at A in two steps.

Step 1.   **Move vertically from A to the next level curve.**
Since T varies in the vertical direction, we move vertically from point A to the point B on the next higher level curve, where $\rho = 1.0240$. The point B has coordinates $(S, T) = (33, 17)$. Note that in moving from A to B, we have kept S constant since both points have salinity $S = 33$.

Step 2.   **Compute the changes and the difference quotient.**

$$\Delta \rho = \text{change in } \rho \text{ from } A \text{ to } B = 1.0240 - 1.0245 = -0.0005$$

$$\Delta T = 17 - 15 = 2$$

$$\left.\frac{\partial \rho}{\partial T}\right|_A \approx \frac{\Delta \rho}{\Delta T} \approx \frac{-0.0005}{2} = -0.00025 \text{ kg-m}^{-3}/°C$$

To estimate $\dfrac{\partial \rho}{\partial S}$, we move horizontally to point C on the next level curve where $\rho = 1.0250$, and compute the changes as before. Since C has coordinates $(S, T) \approx (33.7, 15)$,

$$\Delta \rho = \text{change in } \rho \text{ from } A \text{ to } C = 1.0250 - 1.0245 = 0.0005$$

$$\Delta S \approx 33.7 - 33 = 0.7$$

$$\left.\frac{\partial \rho}{\partial S}\right|_A \approx \frac{\Delta \rho}{\Delta S} \approx \frac{0.0005}{0.7} \approx 0.0007 \text{ kg-m}^{-3}/\text{ppt} \qquad \blacksquare$$

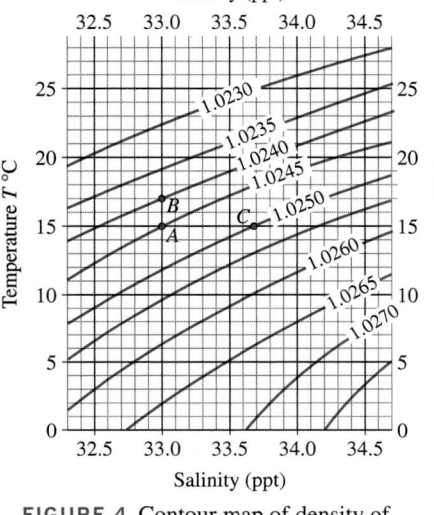

FIGURE 4 Contour map of density of seawater as a function of temperature and salinity.

Higher-Order Partial Derivatives

The higher-order partial derivatives of a function $f(x, y)$ are the derivatives of the partial derivatives f_x and f_y. For instance, the *second-order* partial derivatives are the derivatives of f_x and f_y with respect to x or y. We write f_{xx} for the x-derivative of f_x and f_{yy} for the y-derivative of f_y:

$$f_{xx} = \frac{\partial}{\partial x}\left(\frac{\partial f}{\partial x}\right), \qquad f_{yy} = \frac{\partial}{\partial y}\left(\frac{\partial f}{\partial y}\right)$$

We also have the "mixed partials," which are the derivatives

$$f_{xy} = \frac{\partial}{\partial y}\left(\frac{\partial f}{\partial x}\right), \qquad f_{yx} = \frac{\partial}{\partial x}\left(\frac{\partial f}{\partial y}\right)$$

The process can be continued. For example, f_{xyx} is the x-derivative of f_{xy} and f_{xyy} is the y-derivative of f_{xy} (perform the differentiation in the order of the subscripts from left to right). Leibniz notation for the higher-order partial derivatives is

$$f_{xx} = \frac{\partial^2 f}{\partial x^2}, \qquad f_{xy} = \frac{\partial^2 f}{\partial y \partial x}, \qquad f_{yx} = \frac{\partial^2 f}{\partial x \partial y}, \qquad f_{yy} = \frac{\partial^2 f}{\partial y^2}$$

Higher partial derivatives are defined for functions of three or more variables in a similar manner.

■ **EXAMPLE 7** Calculate the second-order partials and f_{xyy} of $f(x, y) = x^3 + y^2 e^x$.

Solution

Remember how the subscripts are used in partial derivatives. The notation f_{xyy} means "first differentiate with respect to x and then differentiate twice with respect to y."

$$f_x(x, y) = \frac{\partial}{\partial x}(x^3 + y^2 e^x) = 3x^2 + y^2 e^x \qquad f_y(x, y) = \frac{\partial}{\partial y}(x^3 + y^2 e^x) = 2ye^x$$

$$f_{xx}(x, y) = \frac{\partial}{\partial x} f_x = \frac{\partial}{\partial x}(3x^2 + y^2 e^x) \qquad f_{yy}(x, y) = \frac{\partial}{\partial y} f_y = \frac{\partial}{\partial y} 2ye^x$$

$$= 6x + y^2 e^x \qquad\qquad\qquad = 2e^x$$

$$f_{xy}(x, y) = \frac{\partial f_x}{\partial y} = \frac{\partial}{\partial y}(3x^2 + y^2 e^x) \qquad f_{yx}(x, y) = \frac{\partial f_y}{\partial x} = \frac{\partial}{\partial x} 2ye^x$$

$$= 2ye^x \qquad\qquad\qquad\qquad = 2ye^x \qquad \boxed{2}$$

Now we have $f_{xyy} = \frac{\partial}{\partial y} f_{xy} = \frac{\partial}{\partial y} 2ye^x = 2e^x$. ■

Observe in (2) that $f_{xy} = f_{yx}$ for $f(x, y) = x^3 + y^2 e^x$. It is a pleasant circumstance that the equality $f_{xy} = f_{yx}$ holds for all $f(x, y)$, provided that the mixed partials are continuous. See Appendix D for a proof of the following theorem.

The hypothesis of Clairaut's Theorem, that f_{xy} and f_{yx} are continuous, is almost always satisfied in practice, but see Exercise 81 for an example where the mixed partials are not equal.

THEOREM 1 Clairaut's Theorem: Equality of Mixed Partials   If f_{xy} and f_{yx} are both continuous functions on a disk D, then $f_{xy}(a, b) = f_{yx}(a, b)$ for all $(a, b) \in D$. In other words,

$$\boxed{\frac{\partial^2 f}{\partial x \, \partial y} = \frac{\partial^2 f}{\partial y \, \partial x}}$$

Although Clairaut's Theorem is stated for f_{xy} and f_{yx}, it implies more generally that partial differentiation may be carried out any order, provided that the derivatives in question are continuous (see Exercise 73). For example, we may compute f_{xyxy} by differentiating f twice with respect to x and twice with respect to y, in any order. Thus,

$$f_{xyxy} = f_{xxyy} = f_{yyxx} = f_{yxyx} = f_{xyyx} = f_{yxxy}$$

■ **EXAMPLE 8** Check that $\dfrac{\partial^3 W}{\partial V \partial T^2} = \dfrac{\partial^3 W}{\partial T^2 \partial V}$ for $W = PV^{4/3}T^{-1}$.

Solution We compute both derivatives and observe that they are equal:

$$\frac{\partial W}{\partial T} = -PV^{4/3}T^{-2} \qquad\qquad \frac{\partial W}{\partial V} = \frac{4}{3}PV^{1/3}T^{-1}$$

$$\frac{\partial^2 W}{\partial T^2} = \frac{\partial}{\partial T}\frac{\partial W}{\partial T} = 2PV^{4/3}T^{-3} \qquad \frac{\partial^2 W}{\partial T\partial V} = \frac{\partial}{\partial T}\frac{\partial W}{\partial V} = -\frac{4}{3}PV^{1/3}T^{-2}$$

$$\frac{\partial^3 W}{\partial V\partial T^2} = \frac{\partial}{\partial V}\frac{\partial^2 W}{\partial T^2} = \frac{8}{3}PV^{1/3}T^{-3} \qquad \frac{\partial^3 W}{\partial T^2\partial V} = \frac{\partial}{\partial T}\frac{\partial^2 W}{\partial T\partial V} = \frac{8}{3}PV^{1/3}T^{-3}$$

■ **EXAMPLE 9** Choosing the Order Wisely Calculate the derivative g_{zzwx}, where $g(x, y, z, w) = x^3w^2z^2 + \sin\left(\dfrac{xy}{z^2}\right)$.

Solution Let's take advantage of the fact that the derivatives may be calculated in any order to differentiate with respect to w first. This causes the second term, which does not depend on w, to disappear:

$$g_w = \frac{\partial}{\partial w}\left(x^3w^2z^2 + \sin\left(\frac{xy}{z^2}\right)\right) = 2x^3wz^2$$

Next, differentiate twice with respect to z and once with respect to x:

$$g_{wz} = \frac{\partial}{\partial z}2x^3wz^2 = 4x^3wz$$

$$g_{wzz} = \frac{\partial}{\partial z}4x^3wz = 4x^3w$$

$$g_{wzzx} = \frac{\partial}{\partial x}4x^3w = 12x^2w$$

We conclude that $g_{zzwx} = g_{wzzx} = 12x^2w$. ■

A **partial differential equation** (PDE) is a differential equation involving functions of several variables and their partial derivatives. The heat equation in the next example is a PDE that models the change in temperature as heat spreads through an object (Figure 6).

■ **EXAMPLE 10** The Heat Equation Show that $f(x, t) = \dfrac{1}{\sqrt{t}}e^{-(x^2/4t)}$ satisfies the heat equation

$$\frac{\partial f}{\partial t} = \frac{\partial^2 f}{\partial x^2}$$

Solution We compute $\dfrac{\partial^2 f}{\partial x^2}$ and $\dfrac{\partial f}{\partial t}$ using the Product Rule:

$$\frac{\partial f}{\partial x} = \frac{\partial}{\partial x}t^{-1/2}e^{-(x^2/4t)} = -\frac{1}{2}xt^{-3/2}e^{-(x^2/4t)}$$

$$\frac{\partial^2 f}{\partial x^2} = \frac{\partial}{\partial x}\left(-\frac{1}{2}xt^{-3/2}e^{-(x^2/4t)}\right) = -\frac{1}{2}t^{-3/2}e^{-(x^2/4t)} + \frac{1}{4}x^2t^{-5/2}e^{-(x^2/4t)}$$

FIGURE 5 Alexis Clairaut (1713–1765) was a brilliant French mathematician who presented his first paper to the Paris Academy of Sciences at the age of 13. In 1752, Clairaut won a prize for an essay on lunar motion that Euler praised (surely an exaggeration) as "the most important and profound discovery that has ever been made in mathematics."

Maxwell's equations of electromagnetism and Schrödinger's equation in quantum mechanics are famous examples of PDE's in physics. Partial differential equations are an active field of research with applications in such diverse fields as weather prediction and financial risk analysis.

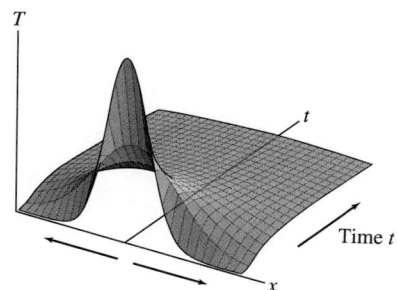

Spatial distribution of heat

FIGURE 6 The plot of $f(x, t) = \dfrac{1}{\sqrt{t}}e^{-(x^2/4t)}$ illustrates the diffusion of heat over time.

As required, $\dfrac{\partial^2 f}{\partial x^2}$ is equal to

$$\frac{\partial f}{\partial t} = \frac{\partial}{\partial t}\left(t^{-1/2}e^{-(x^2/4t)}\right) = -\frac{1}{2}t^{-3/2}e^{-(x^2/4t)} + \frac{1}{4}x^2 t^{-5/2}e^{-(x^2/4t)} \qquad \blacksquare$$

14.3 SUMMARY

- The partial derivatives of $f(x, y)$ are defined as the limits

$$f_x(a, b) = \left.\frac{\partial f}{\partial x}\right|_{(a,b)} = \lim_{h\to 0}\frac{f(a+h, b) - f(a, b)}{h}$$

$$f_y(a, b) = \left.\frac{\partial f}{\partial y}\right|_{(a,b)} = \lim_{k\to 0}\frac{f(a, b+k) - f(a, b)}{k}$$

- Compute f_x by holding y constant and compute f_y by holding x constant.
- The partial derivative $f_x(a, b)$ is the slope of the tangent line to the trace of the graph of $z = f(x, y)$ in the vertical plane $y = b$ at $x = a$. Similarly, $f_y(a, b)$ is the slope of the tangent line to the trace of the graph in the vertical plane $x = a$ at $y = b$.
- The second-order partial derivatives are

$$\frac{\partial^2}{\partial x^2}f = f_{xx}, \qquad \frac{\partial}{\partial y\,\partial x}f = f_{xy}, \qquad \frac{\partial}{\partial x\,\partial y}f = f_{yx}, \qquad \frac{\partial}{\partial y^2}f = f_{yy}$$

- Clairaut's Theorem states that mixed partials are equal, that is, $f_{xy} = f_{yx}$ (provided that f_{xy} and f_{yx} are continuous). More generally, higher partial derivatives may be computed in any order. For example, $f_{xyyz} = f_{yxzy}$ if f is a function of x, y, z whose fourth-order partial derivatives are continuous.

14.3 EXERCISES

Preliminary Questions

1. Patricia derived the following *incorrect* formula by misapplying the Product Rule:

$$\frac{\partial}{\partial x}(x^2 y^2) = x^2(2y) + y^2(2x)$$

What was her mistake and what is the correct calculation?

2. Explain why it is not necessary to use the Quotient Rule to compute $\dfrac{\partial}{\partial x}\left(\dfrac{x+y}{y+1}\right)$. Should the Quotient Rule be used to compute $\dfrac{\partial}{\partial y}\left(\dfrac{x+y}{y+1}\right)$?

3. Which of the following partial derivatives should be evaluated without using the Quotient Rule?

(a) $\dfrac{\partial}{\partial x}\dfrac{xy}{y^2+1}$ **(b)** $\dfrac{\partial}{\partial y}\dfrac{xy}{y^2+1}$ **(c)** $\dfrac{\partial}{\partial x}\dfrac{y^2}{y^2+1}$

4. What is f_x, where $f(x, y, z) = (\sin yz)e^{z^3 - z^{-1}}\sqrt{y}$?

5. Which of the following partial derivatives are equal to f_{xxy}?

(a) f_{xyx} **(b)** f_{yyx} **(c)** f_{xyy} **(d)** f_{yxx}

Exercises

1. Use the limit definition of the partial derivative to verify the formulas

$$\frac{\partial}{\partial x}xy^2 = y^2, \qquad \frac{\partial}{\partial y}xy^2 = 2xy$$

2. Use the Product Rule to compute $\dfrac{\partial}{\partial y}(x^2 + y)(x + y^4)$.

3. Use the Quotient Rule to compute $\dfrac{\partial}{\partial y}\dfrac{y}{x+y}$.

4. Use the Chain Rule to compute $\dfrac{\partial}{\partial u}\ln(u^2 + uv)$.

5. Calculate $f_z(2, 3, 1)$, where $f(x, y, z) = xyz$.

6. ✏ Explain the relation between the two formulas (c is a constant)

$$\frac{d}{dx}\sin(cx) = c\cos(cx), \qquad \frac{\partial}{\partial x}\sin(xy) = y\cos(xy)$$

7. The plane $y = 1$ intersects the surface $z = x^4 + 6xy - y^4$ in a certain curve (Figure 7). Find the slope of the tangent line to this curve at the point $P = (1, 1, 6)$.

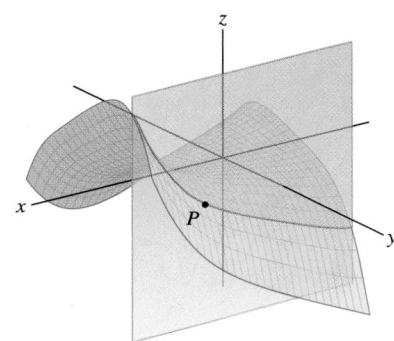

FIGURE 7 Graph of $f(x, y) = x^4 + 6xy - y^4$.

8. Determine if the partial derivatives $\dfrac{\partial f}{\partial x}$ and $\dfrac{\partial f}{\partial y}$ are positive or negative at the point P in Figure 7.

In Exercises 9–11, refer to Figure 8.

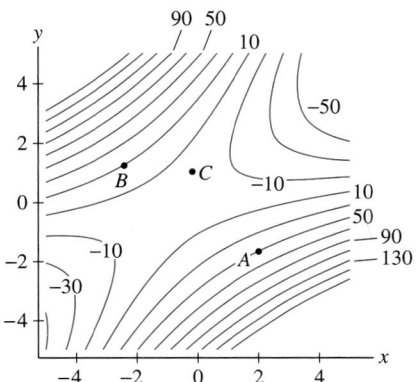

FIGURE 8 Contour map of $f(x, y)$.

9. Estimate f_x and f_y at point A.

10. Starting at point B, in which direction does f increase most rapidly?

11. At which of A, B, or C is f_y smallest?

In Exercises 12–39, compute the partial derivatives.

12. $z = x^2 + y^2$

13. $z = x^4 y^3$

14. $z = x^4 y + xy^{-2}$

15. $V = \pi r^2 h$

16. $z = \dfrac{x}{y}$

17. $z = \dfrac{x}{x - y}$

18. $z = \sqrt{9 - x^2 - y^2}$

19. $z = \dfrac{x}{\sqrt{x^2 + y^2}}$

20. $z = (\sin x)(\sin y)$

21. $z = \sin(u^2 v)$

22. $z = \tan\dfrac{x}{y}$

23. $S = \tan^{-1}(wz)$

24. $z = \ln(x + y)$

25. $z = \ln(x^2 + y^2)$

26. $W = e^{r+s}$

27. $z = e^{xy}$

28. $R = e^{-v^2/k}$

29. $z = e^{-x^2 - y^2}$

30. $P = e^{\sqrt{y^2 + z^2}}$

31. $z = y^x$

32. $z = \cosh(3x + 2y)$

33. $z = \sinh(x^2 y)$

34. $w = xy^2 z^3$

35. $w = \dfrac{x}{y + z}$

36. $Q = re^\theta$

37. $w = \dfrac{x}{(x^2 + y^2 + z^2)^{3/2}}$

38. $A = \sin(4\theta - 9t)$

39. $U = \dfrac{e^{-rt}}{r}$

In Exercises 40–44, compute the given partial derivatives.

40. $f(x, y) = 3x^2 y + 4x^3 y^2 - 7xy^5$, $\quad f_x(1, 4)$

41. $f(x, y) = \sin(x^2 - y)$, $\quad f_y(0, \pi)$

42. $g(u, v) = u\ln(u + v)$, $\quad g_u(1, 2)$

43. $h(x, z) = e^{xz - x^2 z^3}$, $\quad h_z(2, 1)$

44. $h(x, z) = e^{xz - x^2 z^3}$, $\quad h_z(1, 0)$

45. Calculate $\dfrac{\partial W}{\partial E}$ and $\dfrac{\partial W}{\partial T}$, where $W = e^{-E/kT}$.

46. According to the ideal gas law, $PV = nRT$, where P, V, and T are the pressure, volume, and temperature (R and n are constants). View P as a function of V and T, and compute $\dfrac{\partial P}{\partial T}$ and $\dfrac{\partial P}{\partial V}$.

47. The volume of a right-circular cone of radius r and height h is $V = \frac{\pi}{3}r^2 h$. Calculate $\dfrac{\partial V}{\partial r}$ and $\dfrac{\partial V}{\partial h}$.

48. A right-circular cone has $r = h = 12$ cm. What leads to a greater increase in V, a 1-cm increase in r or 1-cm increase in h? Argue using partial derivatives.

49. 📖 Use the contour map of $f(x, y)$ in Figure 9 to explain why the following are true:

(a) f_y is larger at P than at Q and f_x is smaller (more negative) at P than at Q.

(b) $f_y(x, y)$ is decreasing as a function of y, that is, for any x, $f_y(x, b_1) > f_y(x, b_2)$ if $b_1 < b_2$.

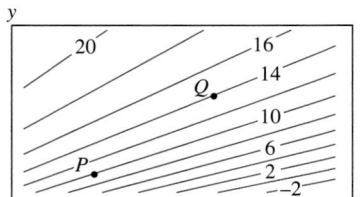

FIGURE 9 Contour interval 2.

50. Over most of the earth, a magnetic compass does not point to true (geographic) north; instead, it points at some angle east or west of true north. The angle D between magnetic north and true north is called the **magnetic declination**. Use Figure 10 to determine which of the following statements is true. Note that because of the way longitude is measured, the x-axis increases from right to left.

(a) $\left.\dfrac{\partial D}{\partial y}\right|_A > \left.\dfrac{\partial D}{\partial y}\right|_B$ **(b)** $\left.\dfrac{\partial D}{\partial x}\right|_C > 0$ **(c)** $\left.\dfrac{\partial D}{\partial y}\right|_C > 0$

(d) D is $28°$ larger at A than at C.

Magnetic Declination for the U.S. 2004

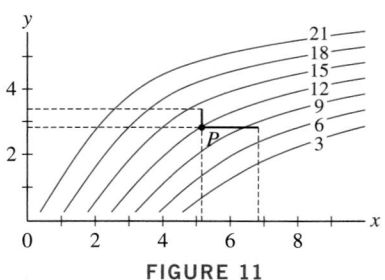

FIGURE 10 Contour interval $1°$.

51. Seawater Density 📖 Refer to Example 6 and Figure 4.

(a) Estimate $\dfrac{\partial \rho}{\partial T}$ and $\dfrac{\partial \rho}{\partial S}$ at points B and C.

(b) Which has a greater effect on density ρ at B, a $1°$ increase in temperature or a 1-ppt increase in salinity?

(c) True or false: The density of warm seawater is more sensitive to a change in temperature than the density of cold seawater? Explain.

52. Estimate the partial derivatives at P of the function whose contour map is shown in Figure 11.

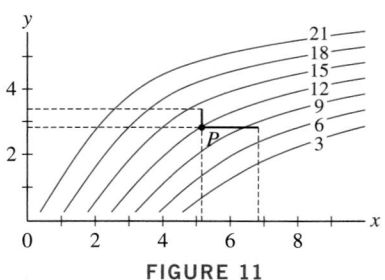

FIGURE 11

In Exercises 53–58, compute the derivative indicated.

53. $f(x, y) = 3x^2 y - 6xy^4$, $\dfrac{\partial^2 f}{\partial x^2}$ and $\dfrac{\partial^2 f}{\partial y^2}$

54. $f(x, y) = x \ln(y^2)$, $f_{yy}(2, 3)$

55. $g(x, y) = \dfrac{xy}{x - y}$, $\dfrac{\partial^2 g}{\partial x \, \partial y}$

56. $g(x, y) = xye^{-y}$, $g_{yy}(1, 0)$

57. $h(u, v) = \dfrac{u}{u + v}$, $h_{uv}(u, v)$

58. $h(x, y) = \ln(x^3 + y^3)$, $h_{xy}(x, y)$

In Exercises 59–60, use Table 1 to answer the following questions.

TABLE 1 **Seawater Density ρ as a Function of Temperature T and Salinity S**

T \ S	30	31	32	33	34	35	36
12	22.75	23.51	24.27	25.07	25.82	26.6	27.36
10	23.07	23.85	24.62	25.42	26.17	26.99	27.73
8	23.36	24.15	24.93	25.73	26.5	27.28	29.09
6	23.62	24.44	25.22	26	26.77	27.55	28.35
4	23.85	24.62	25.42	26.23	27	27.8	28.61
2	24	24.78	25.61	26.38	27.18	28.01	28.78
0	24.11	24.92	25.72	26.5	27.34	28.12	28.91

59. Estimate $\dfrac{\partial \rho}{\partial T}$ and $\dfrac{\partial \rho}{\partial S}$ at $(S, T) = (34, 2)$ and $(35, 10)$.

60. 📖 For fixed salinity $S = 33$, is ρ concave up or concave down as a function of T? *Hint:* Determine if the quotients $\Delta\rho / \Delta T$ are increasing or decreasing. What can you conclude about the sign of $\dfrac{\partial^2 \rho}{\partial T^2}$?

61. Compute f_{xyz} for

$$f(x, y, z) = \sin(yx) + \tan\left(\frac{z + z^{-1}}{x - x^{-1}}\right)$$

Hint: Use a well-chosen order of differentiation on each term.

62. Let

$$f(x, y, u, v) = \frac{x^2 + e^y v}{3y^2 + \ln(2 + u^2)}$$

What is the fastest way to show that $f_{uvxyvu}(x, y, u, v) = 0$ for all (x, y, u, v)?

In Exercises 63–70, compute the derivative indicated.

63. $f(u, v) = \cos(u + v^2)$, f_{uuv}

64. $g(x, y, z) = x^4 y^5 z^6$, g_{xxyz}

65. $F(r, s, t) = r(s^2 + t^2)$, F_{rst}

66. $g(x, y, z) = x^4 y^5 z^3$, g_{xxyz}

67. $u(x, t) = t^{-1/2} e^{-(x^2/4t)}$, u_{xx}

68. $R(u, v, w) = \dfrac{u}{v + w}$, R_{uvw}

69. $g(x, y, z) = \sqrt{x^2 + y^2 + z^2}$, g_{xyz}

70. $u(x, t) = \text{sech}^2(x - t)$, u_{xxx}

71. Find (by guessing) a function such that $\dfrac{\partial f}{\partial x} = 2xy$ and $\dfrac{\partial f}{\partial y} = x^2$.

72. 📖 Prove that there does not exist any function $f(x, y)$ such that $\dfrac{\partial f}{\partial x} = xy$ and $\dfrac{\partial f}{\partial y} = x^2$. *Hint:* Show that f cannot satisfy Clairaut's Theorem.

73. Assume that f_{xy} and f_{yx} are continuous and that f_{yxx} exists. Show that f_{xyx} also exists and that $f_{yxx} = f_{xyx}$.

74. Show that $u(x, t) = \sin(nx) e^{-n^2 t}$ satisfies the heat equation (Example 10) for any constant n:

$$\frac{\partial f}{\partial t} = \frac{\partial^2 f}{\partial x^2} \qquad \boxed{3}$$

75. Find all values of A and B such that $f(x, t) = e^{Ax + Bt}$ satisfies Eq. (3).

*In Exercises 76–79, the **Laplace operator** Δ is defined by $\Delta f = f_{xx} + f_{yy}$. A function $u(x, y)$ satisfying the Laplace equation $\Delta u = 0$ is called **harmonic**.*

76. Show that the following functions are harmonic:

(a) $u(x, y) = x$ (b) $u(x, y) = e^x \cos y$

(c) $u(x, y) = \tan^{-1}\dfrac{y}{x}$ (d) $u(x, y) = \ln(x^2 + y^2)$

77. Find all harmonic polynomials $u(x, y)$ of degree three, that is, $u(x, y) = ax^3 + bx^2 y + cxy^2 + dy^3$.

78. Show that if $u(x, y)$ is harmonic, then the partial derivatives $\dfrac{\partial u}{\partial x}$ and $\dfrac{\partial u}{\partial y}$ are harmonic.

79. Find all constants a, b such that $u(x, y) = \cos(ax)e^{by}$ is harmonic.

80. Show that $u(x, t) = \text{sech}^2(x - t)$ satisfies the **Korteweg–deVries equation** (which arises in the study of water waves)

$$4u_t + u_{xxx} + 12uu_x = 0$$

Further Insights and Challenges

81. *CAS* **Assumptions Matter** This exercise shows that the hypotheses of Clairaut's Theorem are needed. Let $f(x, y) = xy\dfrac{x^2 - y^2}{x^2 + y^2}$ for $(x, y) \neq (0, 0)$ and $f(0, 0) = 0$.

(a) Use a computer algebra system to verify the following formulas for $(x, y) \neq (0, 0)$:

$$f_x(x, y) = \frac{y(x^4 + 4x^2 y^2 - y^4)}{(x^2 + y^2)^2}$$

$$f_y(x, y) = \frac{x(x^4 - 4x^2 y^2 - y^4)}{(x^2 + y^2)^2}$$

(b) Use the limit definition of the partial derivative to show that $f_x(0, 0) = f_y(0, 0) = 0$ and that $f_{yx}(0, 0)$ and $f_{xy}(0, 0)$ both exist but are not equal.

(c) Use a computer algebra system to show that for $(x, y) \neq (0, 0)$:

$$f_{xy}(x, y) = f_{yx}(x, y) = \frac{x^6 + 9x^4 y^2 - 9x^2 y^4 - y^6}{(x^2 + y^2)^3}$$

Show that f_{xy} is not continuous at $(0, 0)$. *Hint:* Show that $\lim_{h \to 0} f_{xy}(h, 0) \neq \lim_{h \to 0} f_{xy}(0, h)$.

(d) Explain why the result of (b) does not contradict Clairaut's Theorem.

14.4 Differentiability, Linear Approximation, and Tangent Planes

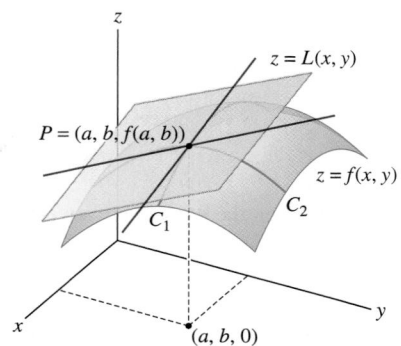

FIGURE 1 The tangent plane to a surface.

In this section, we generalize two key concepts from single-variable calculus: the linear approximation and the tangent line (which becomes the *tangent plane* for functions of two variables; see Figure 1). Recall that in one variable:

- The linearization of $f(x)$ at $x = a$ is the linear function

$$L(x) = f(a) + f'(a)(x - a)$$

- The tangent line to $y = f(x)$ at $x = a$ is the line $y = L(x)$.

The **linearization** of a function $f(x, y)$ of two variables at (a, b), assuming that $f_x(a, b)$ and $f_y(a, b)$ exist, is the linear function

$$\boxed{L(x, y) = f(a, b) + f_x(a, b)(x - a) + f_y(a, b)(y - b)}$$

To justify this definition, we observe first that the point $P = (a, b, f(a, b))$ lies on the plane $z = L(x, y)$ since $L(a, b) = f(a, b)$. Furthermore, $z = L(x, y)$ is the unique plane containing the tangent lines to the two vertical trace curves through P (Figure 2). For instance, the trace in $y = b$ is the curve $z = f(x, b)$ whose tangent line at P is

$$\underbrace{z = L(x, b) = f(a, b) + f_x(a, b)(x - a)}_{\text{Equation of tangent line to vertical trace } z \,=\, f(x, b)}$$

FIGURE 2 The tangent lines to the trace curves C_1 and C_2 lie in the tangent plane (and span it).

■ **EXAMPLE 1**   Find the linearization of $f(x, y) = 5x + 4y^2$ at $(a, b) = (2, 1)$.

Solution   First evaluate $f(x, y)$ and its partial derivatives at $(a, b) = (2, 1)$:

$$f(x, y) = 5x + 4y^2, \qquad f_x(x, y) = 5, \qquad f_y(x, y) = 8y$$
$$f(2, 1) = 14, \qquad\qquad f_x(2, 1) = 5, \qquad f_y(2, 1) = 8$$
$$L(x, y) = \underbrace{14 + 5(x - 2) + 8(y - 1)}_{f(a,b)+f_x(a,b)(x-a)+f_y(a,b)(y-b)} = -4 + 5x + 8y$$

Observe that the graph $z = L(x, y)$ is the tangent plane at $P = (2, 1, 14)$ (see Figure 3). ■

Our next goal is to define differentiability. It might seem appropriate to say that $f(x, y)$ is differentiable at (a, b) if the partial derivatives $f_x(a, b)$ and $f_y(a, b)$ exist. It turns out, however, that this is not sufficient to guarantee that $f(x, y)$ has the smoothness we expect of a differentiable function. What is needed is the additional requirement of local linearity.

Recall that in one variable, local linearity at a point $x = a$ means that the linear approximation $L(x)$ approximates the function $f(x)$ at a to *first-order*. In other words, the error $E(x) = f(x) - L(x)$ tends to zero faster than the distance $|x - a|$ as $x \to a$. We express this in a precise fashion by saying that there exists a function $\epsilon(x)$ such that

$$f(x) = L(x) + \epsilon(x)|x - a|, \qquad \text{where } \lim_{x \to a} \epsilon(x) = 0$$

The multivariable case is similar. A function $f(x, y)$ is locally linear at (a, b) if $|f(x, y) - L(x, y)|$ tends to zero faster than the distance $\sqrt{(x - a)^2 + (y - b)^2}$ from (x, y) to (a, b).

> **DEFINITION Local Linearity** Assume that $f(x, y)$ is defined in a disk D containing (a, b) and that $f_x(a, b)$ and $f_y(a, b)$ exist. We say that $f(x, y)$ is **locally linear** at (a, b) if $L(x, y)$ approximates $f(x, y)$ at (a, b) to **first order**, that is, if
>
> $$f(x, y) = L(x, y) + \underbrace{\epsilon(x, y)\sqrt{(x - a)^2 + (y - b)^2}}_{\text{The error } E(x, y)} \quad \text{for } (x, y) \in D \qquad \boxed{1}$$
>
> where $\epsilon(x, y)$ is a function such that $\displaystyle\lim_{(x,y)\to(a,b)} \epsilon(x, y) = 0$.

■ **EXAMPLE 2** Verifying a First-Order Approximation Show that $f(x, y) = 5x + 4y^2$ is locally linear at $P = (2, 1)$.

Solution In Example 1, we showed that the linearization of $f(x, y)$ at $P = (2, 1)$ is $L(x, y) = -4 + 5x + 8y$. To verify local linearity, we compute the error

$$E(x, y) = f(x, y) - L(x, y) = (5x + 4y^2) - (-4 + 5x + 8y)$$
$$= 4y^2 - 8y + 4 = 4(y - 1)^2$$

For $(x, y) \neq (2, 1)$, this may be written

$$E(x, y) = 4(y - 1)^2 = \frac{4(y - 1)^2}{\sqrt{(x - 2)^2 + (y - 1)^2}} \sqrt{(x - 2)^2 + (y - 1)^2}$$

So if we set

$$\epsilon(x, y) = \frac{4(y - 1)^2}{\sqrt{(x - 2)^2 + (y - 1)^2}}$$

then

$$f(x, y) = L(x, y) + \epsilon(x, y)\sqrt{(x - 2)^2 + (y - 1)^2}$$

To check that $L(x, y)$ yields a first-order approximation, we must verify that

$$\lim_{(x,y)\to(2,1)} \epsilon(x, y) = 0$$

Observe that

$$\epsilon(x, y) = \frac{4(y - 1)^2}{\sqrt{(x - 2)^2 + (y - 1)^2}} \leq \frac{4(y - 1)^2}{\sqrt{(y - 1)^2}} = 4|y - 1| \qquad \boxed{2}$$

Thus, $0 \leq \epsilon(x, y) \leq 4|y - 1|$, and since $\displaystyle\lim_{(x,y)\to(2,1)} 4|y - 1| = 0$, the Squeeze Theorem implies that $\displaystyle\lim_{(x,y)\to(2,1)} \epsilon(x, y) = 0$ as desired. ■

Does the linearization $L(x, y)$ always approximate $f(x, y)$ to first order? The answer is *"yes" for most functions that arise in practice.* However, the mere existence of $f_x(a, b)$ and $f_y(a, b)$ does not guarantee that $f(x, y)$ is locally linear at (a, b). For this reason, we define differentiability by requiring that $f(x, y)$ be locally linear in addition to having partial derivatives at (a, b).

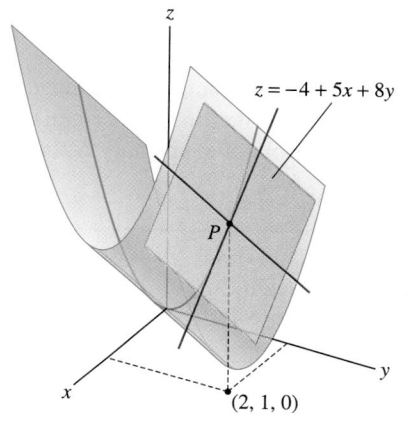

FIGURE 3 Graph of $f(x, y) = 5x + 4y^2$ and the tangent plane at $P = (2, 1, 14)$.

The inequality (2) is valid because
$$\sqrt{(x - 2)^2 + (y - 1)^2} \geq \sqrt{(y - 1)^2}$$

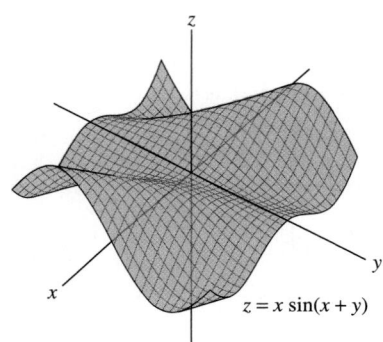

FIGURE 4 The function
$f(x, y) = x \sin(x + y)$ is differentiable.

The definition of differentiability extends to functions of n-variables and Theorem 1 holds in this setting: If all of the partial derivatives of $f(x_1, \ldots, x_n)$ exist and are continuous on an open domain $\mathcal{D}$, then $f(x_1, \ldots, x_n)$ is differentiable on $\mathcal{D}$.

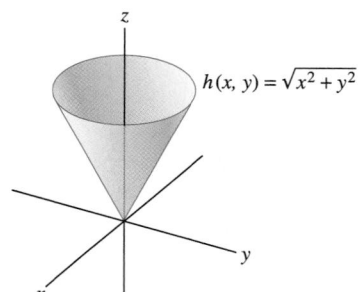

FIGURE 5 The function
$h(x, y) = \sqrt{x^2 + y^2}$ is differentiable except at the origin.

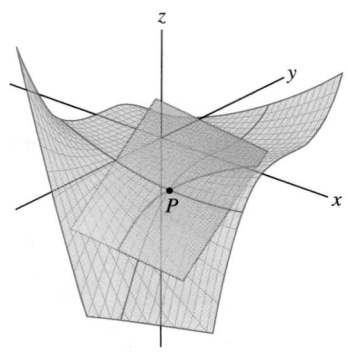

FIGURE 6 Tangent plane to the surface
$f(x, y) = xy^3 + x^2$ at $(2, -2)$.

DEFINITION Differentiability and the Tangent Plane Assume that $f(x, y)$ is defined in a disk D containing (a, b). We say that $f(x, y)$ is **differentiable** at (a, b) if:

- $f_x(a, b)$ and $f_y(a, b)$ exist, and
- $f(x, y)$ is locally linear at (a, b).

In this case, the **tangent plane** to the surface $z = f(x, y)$ at $(a, b, f(a, b))$ is the plane with equation $z = L(x, y)$. Explicitly,

$$z = f(a, b) + f_x(a, b)(x - a) + f_y(a, b)(y - b) \qquad \boxed{3}$$

If $f(x, y)$ is differentiable at all points in a domain $\mathcal{D}$, we say that $f(x, y)$ is differentiable on $\mathcal{D}$.

It is cumbersome to check differentiability using the above definition, but fortunately, this is usually not necessary. The following theorem provides a criterion for differentiability that is easy to apply. It assures us that most functions built out of elementary functions are differentiable on their domains. See Appendix D for a proof.

THEOREM 1 Criterion for Differentiability If $f_x(x, y)$ and $f_y(x, y)$ exist and are continuous on an open disk D, then $f(x, y)$ is differentiable on D.

■ **EXAMPLE 3** Checking Differentiability Show that $f(x, y) = x \sin(x + y)$ is differentiable.

Solution The function $f(x, y)$ (Figure 4) is differentiable by Theorem 1 because its partial derivatives are continuous:

$$f_x(x, y) = x \cos(x + y) + \sin(x + y), \qquad f_y(x, y) = x \cos(x + y) \qquad ■$$

■ **EXAMPLE 4** Checking Differentiability Where is $h(x, y) = \sqrt{x^2 + y^2}$ differentiable?

Solution The function $h(x, y) = \sqrt{x^2 + y^2}$ is differentiable for $(x, y) \neq (0, 0)$ because the partial derivatives exist and are continuous except at $(0, 0)$ (Figure 5):

$$h_x(x, y) = \frac{x}{\sqrt{x^2 + y^2}}, \qquad h_y(x, y) = \frac{y}{\sqrt{x^2 + y^2}}$$

However, $h(x, y)$ is not differentiable at $(0, 0)$ because $h_x(0, 0)$ and $h_y(0, 0)$ do not exist.

■

■ **EXAMPLE 5** Find the tangent plane to the graph of $f(x, y) = xy^3 + x^2$ at $(2, -2)$.

Solution We have

$$f_x(x, y) = y^3 + 2x, \qquad f_x(2, -2) = -4$$
$$f_y(x, y) = 3xy^2, \qquad f_y(2, -2) = 24$$

Since $f(2, -2) = -12$, the tangent plane has equation

$$z = -12 - 4(x - 2) + 24(y + 2)$$

This can be rewritten as $z = 44 - 4x + 24y$ (Figure 6).

■

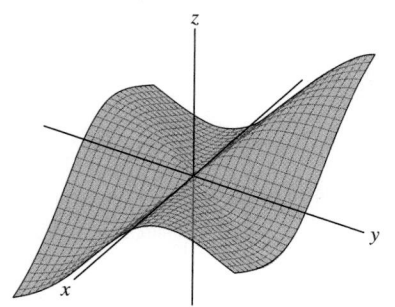

FIGURE 7 The graph of
$g(x, y) = \dfrac{2xy(x + y)}{x^2 + y^2}$. The partial
derivatives at $(0, 0)$ exist, but the graph is
not locally linear at $(0, 0)$.

*BMI is one factor used by health care
professionals to assess the risk of certain
diseases such as diabetes or high blood
pressure. The range $18.5 \leq I \leq 24.9$ is
considered normal for adults over 20 years
of age.*

Assumptions Matter Differentiability is a stricter requirement in several variables than in one variable. In one variable, if $f'(a)$ exists, then $f(x)$ is automatically locally linear at $x = a$. As noted above, however, in two variables, the existence of the partial derivatives $f_x(a, b)$ and $f_y(a, b)$ does not guarantee that $f(x, y)$ is locally linear at (a, b). Figure 7 shows the graph of a continuous function $g(x, y)$ that is not locally linear at $P = (0, 0)$ (and hence not differentiable), even though both $g_x(0, 0)$ and $g_y(0, 0)$ exist (see Exercise 42).

There are two useful ways of rewriting the linearization as a **linear approximation**. First, set $x = a + h$ and $y = b + k$. Then $h = x - a$, $k = y - b$, and we have

$$ f(a + h, b + k) \approx f(a, b) + f_x(a, b)h + f_y(a, b)k $$ **4**

The linear approximation can also be written in terms of the *change in* f:

$$ \Delta f = f(x, y) - f(a, b), \qquad \Delta x = x - a, \qquad \Delta y = y - b $$

$$ \Delta f \approx f_x(a, b)\Delta x + f_y(a, b)\Delta y $$ **5**

■ **EXAMPLE 6** Estimating Change: Body Mass Index   A person's Body Mass Index is $I = \dfrac{W}{H^2}$, where W is the body weight (in kilograms) and H is the body height (in meters). The Body Mass Index (BMI) for a child of mass $W = 17$ kg and height $H = 1.05$ m is $I = 17/1.05^2 \approx 15.4$. Estimate ΔI if (W, H) changes to $(18.5, 1.07)$.

Solution We compute the partial derivatives at $(W, H) = (17, 1.05)$:

$$ \frac{\partial I}{\partial W} = \frac{\partial}{\partial W}\left(\frac{W}{H^2}\right) = \frac{1}{H^2}, \qquad \frac{\partial I}{\partial H} = \frac{\partial}{\partial H}\left(\frac{W}{H^2}\right) = -\frac{2W}{H^3} $$

$$ \frac{\partial I}{\partial W}\bigg|_{(17, 1.05)} = \frac{1}{1.05^2} \approx 0.91, \qquad \frac{\partial I}{\partial H}\bigg|_{(17, 1.05)} = -\frac{2(17)}{1.05^3} \approx -29.37 $$

The linear approximation in delta form is

$$ \Delta I \approx 0.91 \Delta W - 29.37 \Delta H $$

If (W, H) changes from $(17, 1.05)$ to $(18.5, 1.07)$, then $\Delta W = 1.5$ and $\Delta H = 0.02$. Thus,

$$ \Delta I \approx 0.91(1.5) - 29.37(0.02) \approx 0.8 $$ ■

The linear approximation applies to functions of more than two variables. In three variables, the linear approximation at (a, b, c) is

$$ f(a + h, b + k, c + \ell) \approx f(a, b, c) + f_x(a, b, c)h + f_y(a, b, c)k + f_z(a, b, c)\ell $$

Furthermore, the linear approximation is often expressed in terms of differentials:

$$ df = f_x(x, y, z)\,dx + f_y(x, y, z)\,dy + f_z(x, y, z)\,dz = \frac{\partial f}{\partial x}dx + \frac{\partial f}{\partial y}dy + \frac{\partial f}{\partial x}dz $$

In this text, we use Δ-notation rather than differentials:

$$ \Delta f \approx f_x(x, y, z)\Delta x + f_y(x, y, z)\Delta y + f_z(x, y, z)\Delta z $$

■ **EXAMPLE 7** Estimate $(3.99)^3(1.01)^4(1.98)^{-1}$ using the linear approximation and use a calculator to find the percentage error.

Solution Think of $(3.99)^3(1.01)^4(1.98)^{-1}$ as a value of $f(x, y, z) = x^3 y^4 z^{-1}$. Then

$$f(3.99, 1.01, 1.98) = (3.99)^3(1.01)^4(1.98)^{-1}$$

and it makes sense to use the linear approximation at $(4, 1, 2)$. We have

$$f(x, y, z) = x^3 y^4 z^{-1}, \qquad f(4, 1, 2) = (4^3)(1^4)(2^{-1}) = 32$$
$$f_x(x, y, z) = 3x^2 y^4 z^{-1}, \qquad f_x(4, 1, 2) = 24$$
$$f_y(x, y, z) = 4x^3 y^3 z^{-1}, \qquad f_y(4, 1, 2) = 128$$
$$f_z(x, y, z) = -x^3 y^4 z^{-2}, \qquad f_z(4, 1, 2) = -16$$

The linear approximation yields

$$f(4 + h, 1 + k, 2 + \ell) \approx 32 + 24h + 128k - 16\ell \qquad \boxed{6}$$

We apply (6) with $h = -0.01$, $k = 0.01$, and $\ell = -0.02$:

$$(3.99)^3(1.01)^4(1.98)^{-1} \approx 32 + 24(-0.01) + 128(0.01) - 16(-0.02) = 33.36$$

Using a calculator, we find that $(3.99)^3(1.01)^4(1.98)^{-1} \approx 33.384$, so the error is less than 0.025 and the percentage error is at most

$$\text{Percentage error} \approx \frac{(0.025)}{33.384} \times 100 \approx 0.075\% \qquad ■$$

14.4 SUMMARY

- The *linearization* of $f(x, y)$ at (a, b) is the linear function

$$L(x, y) = f(a, b) + f_x(a, b)(x - a) + f_y(a, b)(y - b)$$

- The linearization in three variables is

$$f(x, y, z) \approx f(a, b, c) + f_x(a, b, c)(x - a) + f_y(a, b, c)(y - b) + f_z(a, b, c)(z - c)$$

- A function $f(x, y)$ is *locally linear* at (a, b) if $f_x(a, b)$ and $f_y(a, b)$ exist and $L(x, y)$ approximates $f(x, y)$ to first order at (a, b). This means that the error $E(x, y) = f(x, y) - L(x, y)$ has the form

$$E(x, y) = \epsilon(x, y)\sqrt{(x - a)^2 + (y - b)^2}, \quad \text{where} \quad \lim_{(x,y)\to(a,b)} \epsilon(x, y) = 0$$

- Equivalent forms of the linear approximation:

$$f(x, y) \approx f(a, b) + f_x(a, b)(x - a) + f_y(a, b)(y - b)$$
$$f(a + h, b + k) \approx f(a, b) + f_x(a, b)h + f_y(a, b)k$$

Or setting $\Delta f = f(x, y) - f(a, b)$, $\Delta x = x - a$, $\Delta y = y - b$,

$$\boxed{\Delta f \approx f_x(a, b)\, \Delta x + f_y(a, b)\, \Delta y}$$

- By definition, $f(x, y)$ is differentiable at (a, b) if $f_x(a, b)$ and $f_y(a, b)$ exist and $f(x, y)$ is locally linear at (a, b).
- In practice, the following result is used to check differentiability: *If $f_x(x, y)$ and $f_y(x, y)$ exist and are continuous in a disk D containing (a, b), then $f(x, y)$ is differentiable at (a, b).*
- An equation of the tangent plane to $z = f(x, y)$ at (a, b) is

$$z = f(a, b) + f_x(a, b)(x - a) + f_y(a, b)(y - b)$$

14.4 EXERCISES

Preliminary Questions

1. How is the linearization of $f(x, y)$ at (a, b) defined?

2. Define local linearity for functions of two variables.

In Questions 3–5, assume that

$$f(2, 3) = 8, \qquad f_x(2, 3) = 5, \qquad f_y(2, 3) = 7$$

3. Which of (a)–(b) is the linearization of f at $(2, 3)$?

   (a) $L(x, y) = 8 + 5x + 7y$

   (b) $L(x, y) = 8 + 5(x - 2) + 7(y - 3)$

4. Estimate $f(2, 3.1)$.

5. Estimate Δf at $(2, 3)$ if $\Delta x = -0.3$ and $\Delta y = 0.2$.

6. Which theorem allows us to conclude that $f(x, y) = x^3 y^8$ is differentiable?

Exercises

1. Let $f(x, y) = x^2 y^3$.

(a) Find the linearization of f at $(a, b) = (2, 1)$.

(b) Use the linear approximation to estimate $f(2.01, 1.02)$ and $f(1.97, 1.01)$, and compare your estimates with the values obtained using a calculator.

2. Write the linear approximation to $f(x, y) = x(1 + y)^{-1}$ at $(a, b) = (8, 1)$ in the form

$$f(a + h, b + k) \approx f(a, b) + f_x(a, b)h + f_y(a, b)k$$

Use it to estimate $7.98/2.02$ and compare the estimate with the value obtained using a calculator.

3. Let $f(x, y) = x^3 y^{-4}$. Use Eq. (5) to estimate the change $\Delta f = f(2.03, 0.9) - f(2, 1)$.

4. Assume that $f(3, 2) = 2$, $f_x(3, 2) = -1$, and $f_y(3, 2) = 3$. Use the linear approximation to estimate $f(3.1, 1.8)$.

5. Use the linear approximation of $f(x, y) = e^{x^2+y}$ at $(0, 0)$ to estimate $f(0.01, -0.02)$. Compare with the value obtained using a calculator.

6. Let $f(x, y) = \dfrac{x^2}{y^2 + 1}$. Use the linear approximation at an appropriate point (a, b) to estimate $f(4.01, 0.98)$.

7. Use Eq. (3) to find an equation of the tangent plane to the graph of $f(x, y) = 2x^2 - 4xy^2$ at the point $(-1, 2)$.

8. Let $f(x, y) = e^{x/y}$ and $P = (2, 1, e^2)$.

(a) Find an equation of the tangent plane to the graph of $f(x, y)$ at P.

(b) *C A S* Plot $f(x, y)$ and the tangent plane at P on the same screen to illustrate local linearity.

9. Find the linear approximation to $f(x, y, z) = \dfrac{xy}{z}$ at the point $(2, 1, 2)$.

10. Assume that $f(1, 0, 0) = -3$, $f_x(1, 0, 0) = -2$, $f_y(1, 0, 0) = 4$, and $f_z(1, 0, 0) = 2$. Use the linear approximation to estimate $f(1.02, 0.01, -0.03)$.

In Exercises 11–18, find an equation of the tangent plane at the given point.

11. $f(x, y) = x^2 y + xy^3$, $(2, 1)$

12. $f(x, y) = \dfrac{x}{\sqrt{y}}$, $(4, 4)$

13. $f(x, y) = x^2 + y^{-2}$, $(4, 1)$

14. $G(u, w) = \sin(uw)$, $\left(\frac{\pi}{6}, 1\right)$

15. $F(r, s) = r^2 s^{-1/2} + s^{-3}$, $(2, 1)$

16. $H(s, t) = se^{st}$, $(0, 0)$

17. $f(x, y) = e^x \ln y$, $(0, 1)$

18. $f(x, y) = \ln(x^2 + y^2)$, $(1, 1)$

19. Find the points on the graph of $z = 3x^2 - 4y^2$ at which the vector $\mathbf{n} = \langle 3, 2, 2 \rangle$ is normal to the tangent plane.

20. Find the points on the graph of $z = x^2 e^y$ at which the tangent plane is parallel to $5x - 2y + \frac{1}{2}z = 0$.

21. **CAS** Use a computer algebra system to plot the graph of $f(x, y) = x^2 + 3xy + y$ together with the tangent plane at $(x, y) = (1, 1)$ on the same screen.

22. Suppose that the plane tangent to the surface $z = f(x, y)$ at $(-2, 3, 4)$ has equation $z + 4x + 2y = 2$. Estimate $f(-2.1, 3.1)$.

23. The following values are given:

$$f(1, 2) = 10, \qquad f(1.1, 2.01) = 10.3, \qquad f(1.04, 2.1) = 9.7$$

Find an approximation to the equation of the tangent plane to the graph of f at $(1, 2, 10)$.

24. Use the linear approximation to $f(x, y) = \sqrt{x/y}$ at $(9, 4)$ to estimate $\sqrt{9.1/3.9}$.

In Exercises 25–30, use the linear approximation to estimate the value. Compare with the value given by a calculator.

25. $(2.01)^3(1.02)^2$

26. $4.1/7.9$

27. $\sqrt{3.01^2 + 3.99^2}$

28. $\dfrac{0.98^2}{2.01^3 + 1}$

29. $\sqrt{(1.9)(2.02)(4.05)}$

30. $\dfrac{8.01}{\sqrt{(1.99)(2.01)}}$

31. Estimate $f(2.1, 3.8)$ given that $f(2, 4) = 5$, $f_x(2, 4) = 0.3$, and $f_y(2, 4) = -0.2$.

In Exercises 32–34, let $I = W/H^2$ denote the BMI described in Example 6.

32. A boy has weight $W = 34$ kg and height $H = 1.3$ m. Use the linear approximation to estimate the change in I if (W, H) changes to $(36, 1.32)$.

33. Suppose that $(W, H) = (34, 1.3)$ and W increases to 35. Use the linear approximation to estimate the increase in H required to keep I constant.

34. **(a)** Show that $\Delta I \approx 0$ if $\Delta H/\Delta W \approx H/2W$.
(b) Suppose that $(W, H) = (25, 1.1)$. What increase in H will leave I (approximately) constant if W is increased by 1 kg?

35. The volume of a cylinder of radius r and height h is $V = \pi r^2 h$.
(a) Use the linear approximation to show that

$$\frac{\Delta V}{V} \approx \frac{2\Delta r}{r} + \frac{\Delta h}{h}$$

(b) Calculate the percentage increase in V if r and h are each increased by 2%.
(c) The volume of a certain cylinder V is determined by measuring r and h. Which will lead to a greater error in V: a 1% error in r or a 1% error in h?

36. Use the linear approximation to show that if $I = x^a y^b$, then

$$\frac{\Delta I}{I} \approx a\frac{\Delta x}{x} + b\frac{\Delta y}{y}$$

37. The monthly payment for a home loan is given by a function $f(P, r, N)$, where P is the principal (the initial size of the loan), r the interest rate, and N the length of the loan in months. Interest rates are expressed as a decimal: A 6% interest rate is denoted by $r = 0.06$. If $P = \$100{,}000$, $r = 0.06$, and $N = 240$ (a 20-year loan), then the monthly payment is $f(100{,}000, 0.06, 240) = 716.43$. Furthermore, with these values, we have

$$\frac{\partial f}{\partial P} = 0.0071, \qquad \frac{\partial f}{\partial r} = 5{,}769, \qquad \frac{\partial f}{\partial N} = -1.5467$$

Estimate:
(a) The change in monthly payment per \$1,000 increase in loan principal.
(b) The change in monthly payment if the interest rate increases to $r = 6.5\%$ and $r = 7\%$.
(c) The change in monthly payment if the length of the loan increases to 24 years.

38. Automobile traffic passes a point P on a road of width w ft at an average rate of R vehicles per second. The arrival of automobiles is irregular, however, and traffic engineers have found that the formula $T = te^{Rt}$ gives a good model for the average waiting time T until there is a gap in traffic of at least t seconds. A pedestrian walking at a speed of 3.5 ft/s (5.1 mph) requires $t = w/3.5$ s to cross the road. Therefore, the average time the pedestrian will have to wait before crossing is $f(w, R) = (w/3.5)e^{wR/3.5}$ s.
(a) What is the pedestrian's average waiting time if $w = 25$ ft and $R = 0.2$ vehicles per second?
(b) Use the linear approximation to estimate the increase in waiting time if w is increased to 27 ft.
(c) Estimate the waiting time if the width is increased to 27 ft and R decreases to 0.18.
(d) What is the rate of increase in waiting time per 1-ft increase in width when $w = 30$ ft and $R = 0.3$ vehicles per second?

39. The volume V of a cylinder is computed using the values 3.5 m for diameter and 6.2 m for height. Use the linear approximation to estimate the maximum error in V if each of these values has a possible error of at most 5%.

Further Insights and Challenges

40. Show that if $f(x, y)$ is differentiable at (a, b), then the function of one variable $f(x, b)$ is differentiable at $x = a$. Use this to prove that $f(x, y) = \sqrt{x^2 + y^2}$ is *not* differentiable at $(0, 0)$.

41. Assume that $f(0) = 0$. By the discussion in this section, $f(x)$ is differentiable at $x = 0$ if there is a constant M such that

$$f(h) = Mh + h\epsilon(h) \qquad \boxed{7}$$

where $\lim_{h \to 0} \epsilon(h) = 0$ [in this case, $M = f'(0)$].

(a) Use this definition to verify that $f(x) = 1/(x + 1)$ is differentiable with $M = -1$ and $E(h) = h/(h + 1)$.

(b) Use this definition to verify that $f(x) = x^{3/2}$ is differentiable with $M = 0$.

(c) Show that $f(x) = x^{1/2}$ is not differentiable at $x = 0$ by showing that if M is any constant, and if we write $\sqrt{h} = Mh + h\epsilon(h)$, then the function $\epsilon(h)$ does not approach zero as $h \to 0$.

42. Assumptions Matter Recall that if $f(x, y)$ is differentiable at (a, b), then the partial derivatives $f_x(a, b)$ and $f_y(a, b)$ exist. This exercise shows that the mere existence of $f_x(a, b)$ and $f_y(a, b)$ does not imply $f(x, y)$ is differentiable. Define $g(x, y) = \dfrac{2xy(x + y)}{x^2 + y^2}$ for $(x, y) \neq 0$ and $g(0, 0) = 0$.

(a) Prove that $|g(x, y)| \leq |x + y|$. *Hint:* Use the inequality

$$(x \pm y)^2 \geq 0$$

to show that $\left| \dfrac{2xy}{x^2 + y^2} \right| \leq 1$.

(b) Use (a) to show that $g(x, y)$ is continuous at $(0, 0)$.

(c) Use the limit definitions to show that $g_x(0, 0)$ and $g_y(0, 0)$ exist and both are equal to zero.

(d) Show that if a function $f(x, y)$ is locally linear at $(0, 0)$ and $f(0, 0) = 0$, then $f(h, h)/h$ approaches 0 as $h \to 0$. Show that $g(x, y)$ does not have this property and hence is not differentiable at $(0, 0)$.

(e) Explain why $g_x(x, y)$ and $g_y(x, y)$ cannot both be continuous at $(0, 0)$. In fact, since $g(x, y)$ is symmetric in x and y, both partial derivatives are discontinuous at $(0, 0)$.

(f) Show directly that $\lim_{(x,y) \to (0,0)} g_x(x, y)$ does not exist by verifying that $g_x(x, 0) = 0$ for $x \neq 0$ and $g_x(0, y) = 2$ for $y \neq 0$.

14.5 The Gradient and Directional Derivatives

In previous sections, we have emphasized that the rate of change of a function f of several variables depends on a choice of direction. Since directions are indicated by vectors, it is natural to use vectors to describe the derivative of f in a specified direction.

To do this, we first introduce the **gradient** ∇f_P, which is the vector whose components are the partial derivatives of f at P.

The gradient of a function of n variables is the vector
$$\nabla f = \left\langle \frac{\partial f}{\partial x_1}, \frac{\partial f}{\partial x_2}, \dots, \frac{\partial f}{\partial x_n} \right\rangle$$

DEFINITION The Gradient The gradient of a function $f(x, y)$ at a point $P = (a, b)$ is the vector
$$\nabla f_P = \langle f_x(a, b), f_y(a, b) \rangle$$
In three variables, if $P = (a, b, c)$,
$$\nabla f_P = \langle f_x(a, b, c), f_y(a, b, c), f_z(a, b, c) \rangle$$

The symbol ∇, called "del," is an upside-down Greek delta. The use of ∇ for the gradient was popularized by the Scottish physicist P. G. Tait (1831–1901) who called the symbol "nabla." Nabla means harp in Hebrew and refers to ∇'s resemblance to an ancient ten-string harp. The great physicist James Clerk Maxwell was reluctant to adopt this term and referred to the gradient simply as the "slope." He wrote jokingly to his friend Tait in 1871: "Still harping on that nabla?"

We also write $\nabla f_{(a,b)}$ or $\nabla f(a, b)$ for the gradient. Sometimes, we omit reference to the point P and write
$$\nabla f = \left\langle \frac{\partial f}{\partial x}, \frac{\partial f}{\partial y} \right\rangle \qquad \text{or} \qquad \nabla f = \left\langle \frac{\partial f}{\partial x}, \frac{\partial f}{\partial y}, \frac{\partial f}{\partial z} \right\rangle$$

The gradient ∇f "assigns" a vector ∇f_P to each point in the domain of f.

■ **EXAMPLE 1 Drawing Gradient Vectors** Find the gradient of $f(x, y) = x^2 + y^2$ at $P = (1, 1)$ and draw several gradient vectors.

Solution The partial derivatives are $f_x(x, y) = 2x$ and $f_y(x, y) = 2y$, so
$$\nabla f = \langle 2x, 2y \rangle$$

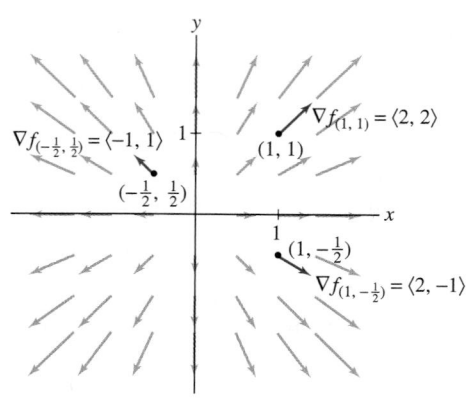

FIGURE 1 Gradient vectors of $f(x, y) = x^2 + y^2$ at several points (vectors not drawn to scale).

At $P = (1, 1)$, we have $\nabla f_P = \nabla f(1, 1) = \langle 2, 2 \rangle$. Figure 1 shows gradient vectors of $f(x, y)$ at several points. ∎

■ **EXAMPLE 2** Gradient in Three Variables Calculate the vector $\nabla f_{(3, -2, 4)}$, where $f(x, y, z) = ze^{2x+3y}$.

Solution The partial derivatives are

$$\frac{\partial f}{\partial x} = 2ze^{2x+3y}, \qquad \frac{\partial f}{\partial y} = 3ze^{2x+3y}, \qquad \frac{\partial f}{\partial z} = e^{2x+3y}$$

Therefore, $\nabla f = \langle 2ze^{2x+3y}, 3ze^{2x+3y}, e^{2x+3y} \rangle$, and (see Figure 2)

$$\nabla f_{(3, -2, 4)} = \langle 2 \cdot 4e^0, 3 \cdot 4e^0, e^0 \rangle = \langle 8, 12, 1 \rangle$$ ∎

The following theorem lists some useful properties of the gradient. The proofs are left as exercises (see Exercises 61–63).

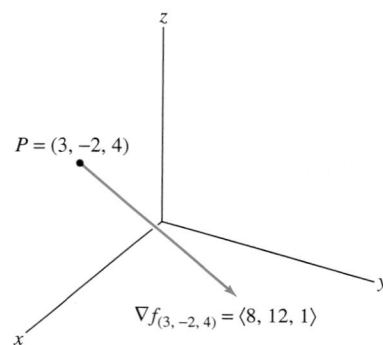

FIGURE 2 The gradient of $f(x, y, z) = ze^{2x+3y}$ at $P = (3, -2, 4)$.

THEOREM 1 Properties of the Gradient If $f(x, y, z)$ and $g(x, y, z)$ are differentiable and c is a constant, then:

(i) $\nabla(f + g) = \nabla f + \nabla g$
(ii) $\nabla(cf) = c\nabla f$
(iii) Product Rule for Gradients: $\nabla(fg) = f\nabla g + g\nabla f$
(iv) Chain Rule for Gradients: If $F(t)$ is a differentiable function of one variable, then

$$\nabla(F(f(x, y, z))) = F'(f(x, y, z))\nabla f \qquad \boxed{1}$$

■ **EXAMPLE 3** Using the Chain Rule for Gradients Find the gradient of $g(x, y, z) = (x^2 + y^2 + z^2)^8$.

Solution Apply Eq. (1) with $F(t) = t^8$ and $f(x, y, z) = x^2 + y^2 + z^2$. Then $g(x, y, z) = F(f(x, y, z))$ and

$$\nabla g = \nabla\big((x^2 + y^2 + z^2)^8\big) = 8(x^2 + y^2 + z^2)^7 \nabla(x^2 + y^2 + z^2)$$

$$= 8(x^2 + y^2 + z^2)^7 \langle 2x, 2y, 2z \rangle = 16(x^2 + y^2 + z^2)^7 \langle x, y, z \rangle$$ ∎

The Chain Rule for Paths

As a first application of the gradient, we discuss the Chain Rule for Paths. Recall that in Chapter 13, we represented a path (or parametrized curve) in $\mathbf{R}^3$ by a vector-valued function $\mathbf{r}(t) = \langle x(t), y(t), z(t) \rangle$. In this chapter, we change notation and represent such a path instead by a function $\mathbf{c}(t) = (x(t), y(t), z(t))$ (Figure 3). We think of $\mathbf{c}(t)$ as a moving point rather than a moving vector. As in Chapter 13, the derivative $\mathbf{c}'(t)$ is the tangent or "velocity" vector that points in the direction of motion tangent to the path:

$$\mathbf{c}'(t) = \langle x'(t), y'(t), z'(t) \rangle$$

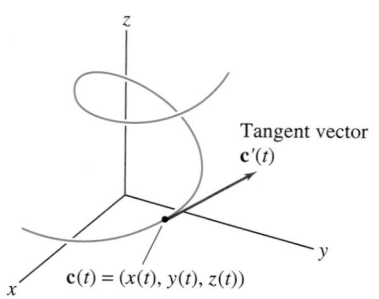

FIGURE 3 Tangent vector $\mathbf{c}'(t)$ to a path $\mathbf{c}(t) = (x(t), y(t), z(t))$.

We use similar notation for paths in $\mathbf{R}^2$.

The Chain Rule for Paths is different from the more elementary Chain Rule for Gradients stated in Eq. (1) above. Suppose that $f(x, y)$ is a function of two variables and let

$\mathbf{c}(t) = (x(t), y(t))$ be a path in $\mathbf{R}^2$. The composite is a function of t:

$$f(\mathbf{c}(t)) = f(x(t), y(t))$$

What is the idea behind this type of composite function? Imagine riding a bike along the path $\mathbf{c}(t) = (x(t), y(t))$ and let $T(x, y)$ be the outside temperature at location (x, y) (Figure 4). Then $T(\mathbf{c}(t))$ is the temperature at your location $\mathbf{c}(t)$ at time t. The derivative $\dfrac{d}{dt} T(\mathbf{c}(t))$ is the rate at which the outside temperature changes as you ride along the path. The Chain Rule states that you will observe the temperature changing at a rate equal to the dot product of the temperature gradient ∇T and your velocity vector $\mathbf{c}'(t)$.

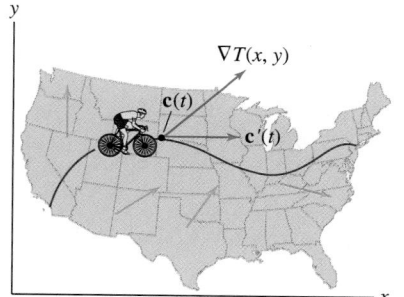

FIGURE 4 By the Chain Rule, a biker feels the temperature T changing at a rate equal to the dot product of the temperature gradient ∇T and the velocity vector $\mathbf{c}'(t)$.

THEOREM 2 Chain Rule for Paths If f is a differentiable function and $\mathbf{c}(t)$ is a differentiable path, then

$$\frac{d}{dt} f(\mathbf{c}(t)) = \nabla f_{\mathbf{c}(t)} \cdot \mathbf{c}'(t)$$

Explicitly, in the case of two variables, if $\mathbf{c}(t) = (x(t), y(t))$ is a differentiable path in $\mathbf{R}^2$, then

$$\frac{d}{dt} f(\mathbf{c}(t)) = \left\langle \frac{\partial f}{\partial x}, \frac{\partial f}{\partial y} \right\rangle \cdot \langle x'(t), y'(t) \rangle = \frac{\partial f}{\partial x} \frac{dx}{dt} + \frac{\partial f}{\partial y} \frac{dy}{dt}$$

Proof We derive this result using the linear approximation. By definition,

$$\frac{d}{dt} f(\mathbf{c}(t)) = \lim_{h \to 0} \frac{f(x(t + h), y(t + h)) - f(x(t), y(t))}{h}$$

To calculate this derivative, set

$$\Delta x = x(t + h) - x(t), \qquad \Delta y = y(t + h) - y(t)$$

Since $f(x, y)$ is differentiable, we may apply the linear approximation (Section 14.4):

$$f(x(t + h), y(t + h)) - f(x(t), y(t))$$

$$= f_x(x(t), y(t))\Delta x + f_y(x(t), y(t))\Delta y + \epsilon(\Delta x, \Delta y)\sqrt{\Delta x^2 + \Delta y^2}$$

where $\epsilon(\Delta x, \Delta y) \to 0$ as Δx and Δy tend to zero. Now set $h = \Delta t$ and for $\Delta t \neq 0$, divide by Δt:

$$\frac{f(x(t + h), y(t + h)) - f(x(t), y(t))}{h}$$

$$= f_x(x(t), y(t))\frac{\Delta x}{\Delta t} + f_y(x(t), y(t))\frac{\Delta y}{\Delta t} + \epsilon(\Delta x, \Delta y)\sqrt{\left(\frac{\Delta x}{\Delta t}\right)^2 + \left(\frac{\Delta y}{\Delta t}\right)^2}$$

As $h = \Delta t$ tends to zero, the left-hand side approaches $\dfrac{d}{dt} f(\mathbf{c}(t))$ and the difference quotients $\Delta x / \Delta t$ and $\Delta y / \Delta t$ on the right approach dx/dt and dy/dt, respectively. Since $\epsilon(\Delta x, \Delta y)$ approaches zero, we obtain the desired result:

$$\frac{d}{dt} f(\mathbf{c}(t)) = \lim_{h \to 0} \frac{f(x(t+h), y(t+h)) - f(x(t), y(t))}{h}$$

$$= f_x(x(t), y(t)) \lim_{\Delta t \to 0} \frac{\Delta x}{\Delta t} + f_y(x(t), y(t)) \lim_{\Delta t \to 0} \frac{\Delta y}{\Delta t}$$

$$= f_x(x(t), y(t)) \frac{dx}{dt} + f_y(x(t), y(t)) \frac{dy}{dt} = \nabla f_{\mathbf{c}(t)} \cdot \mathbf{c}'(t) \qquad \blacksquare$$

■ **EXAMPLE 4** A bug carries a tiny thermometer along the path

$$\mathbf{c}(t) = (\cos(t - 2), \sin 2t)$$

(t in seconds) as in Figure 5. How fast is the temperature changing at $t = 2$ if the temperature at location (x, y) is $T(x, y) = x^2 - xy + 60\,°F$?

Solution The bug's location at $t = 2$ is $\mathbf{c}(2) = (\cos 0, \sin 4) \approx (1, -0.76)$. We compute the gradient ∇T and evaluate it at $\mathbf{c}(2)$:

$$\nabla T = \langle 2x - y, -x \rangle, \qquad \nabla T_{\mathbf{c}(2)} = \langle 2 - (-0.76), -1 \rangle = \langle 2.76, -1 \rangle$$

Next, we compute the tangent vector

$$\mathbf{c}'(t) = \langle -\sin(t - 2), 2\cos 2t \rangle, \qquad \mathbf{c}'(2) = \langle 0, 2\cos 4 \rangle \approx \langle 0, -1.31 \rangle$$

According to the Chain Rule, at time $t = 2$ s, the temperature changes at the rate

$$\left. \frac{d}{dt} T(\mathbf{c}(t)) \right|_{t=2} = \nabla T_{\mathbf{c}(2)} \cdot \mathbf{c}'(2) = \langle 2.76, -1 \rangle \cdot \langle 0, -1.31 \rangle = 1.31\,°F/s \qquad \blacksquare$$

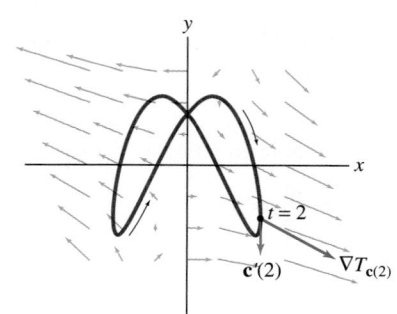

FIGURE 5 The path
$\mathbf{c}(t) = (\cos(t - 2), \sin 2t)$.

The Chain Rule holds for functions of n variables. If $f(x_1, \ldots, x_n)$ is differentiable and if $\mathbf{c}(t) = (x_1(t), \ldots, x_n(t))$ is a differentiable path, then

$$\frac{d}{dt} f(\mathbf{c}(t)) = \nabla f \cdot \mathbf{c}'(t) = \frac{\partial f}{\partial x_1} \frac{dx_1}{dt} + \frac{\partial f}{\partial x_2} \frac{dx_2}{dt} + \cdots + \frac{\partial f}{\partial x_n} \frac{dx_n}{dt}$$

■ **EXAMPLE 5** Calculate $\left. \dfrac{d}{dt} f(\mathbf{c}(t)) \right|_{t=\pi/2}$, where

$$f(x, y, z) = xy + z^2 \qquad \text{and} \qquad \mathbf{c}(t) = (\cos t, \sin t, t)$$

Solution First, note that $\mathbf{c}(\frac{\pi}{2}) = (\cos \frac{\pi}{2}, \sin \frac{\pi}{2}, \frac{\pi}{2}) = (0, 1, \frac{\pi}{2})$. Next, evaluate ∇f and $\mathbf{c}'(t)$ at $t = \frac{\pi}{2}$:

$$\nabla f = \left(\frac{\partial f}{\partial x}, \frac{\partial f}{\partial y}, \frac{\partial f}{\partial z} \right) = \langle y, x, 2z \rangle, \qquad \nabla f_{\mathbf{c}(\pi/2)} = \nabla f \left(0, 1, \frac{\pi}{2} \right) = \langle 1, 0, \pi \rangle$$

$$\mathbf{c}'(t) = \langle -\sin t, \cos t, 1 \rangle, \qquad \mathbf{c}' \left(\frac{\pi}{2} \right) = \left\langle -\sin \frac{\pi}{2}, \cos \frac{\pi}{2}, 1 \right\rangle = \langle -1, 0, 1 \rangle$$

By the Chain Rule,

$$\left. \frac{d}{dt} f(\mathbf{c}(t)) \right|_{t=\pi/2} = \nabla f_{\mathbf{c}(\pi/2)} \cdot \mathbf{c}' \left(\frac{\pi}{2} \right) = \langle 1, 0, \pi \rangle \cdot \langle -1, 0, 1 \rangle = \pi - 1 \qquad \blacksquare$$

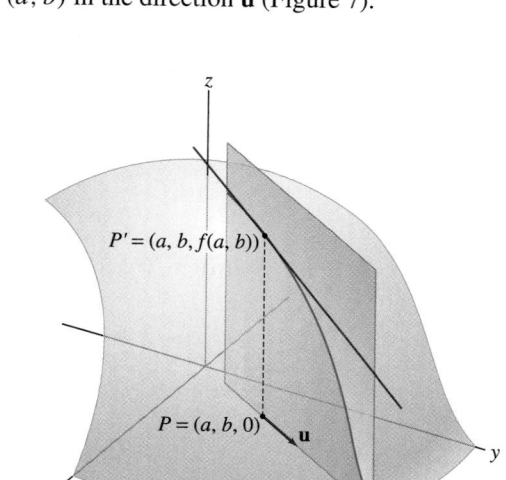

FIGURE 6 $D_{\mathbf{v}} f(a, b)$ is the rate of change of $f(\mathbf{c}(t))$ at $t = 0$.

Directional Derivatives

As mentioned at the beginning of this section, we are interested in studying the derivative of a function f in the direction of an arbitrary vector $\mathbf{v}$. To define this derivative, let $P = (a, b)$ and $\mathbf{v} = \langle h, k \rangle$. The path

$$\mathbf{c}(t) = (a + th, b + tk)$$

parametrizes the line through P in the direction of $\mathbf{v}$ with $\mathbf{c}(0) = P$ (Figure 6). The derivative of f with respect to $\mathbf{v}$, denoted $D_{\mathbf{v}} f$, is the derivative of f along this path at $t = 0$:

$$D_{\mathbf{v}} f(P) = D_{\mathbf{v}} f(a, b) = \frac{d}{dt} f(\mathbf{c}(t))\Big|_{t=0} = \lim_{t \to 0} \frac{f(\mathbf{c}(t)) - f(\mathbf{c}(0))}{t}$$

$$= \lim_{t \to 0} \frac{f(a + th, b + tk) - f(a, b)}{t}$$

The derivative $D_{\mathbf{v}} f$ is defined for any vector $\mathbf{v}$, but when $\mathbf{u}$ is a unit vector, $D_{\mathbf{u}} f$ is called the *directional derivative* in the direction of $\mathbf{u}$. In this case, $D_{\mathbf{u}} f$ is the rate of change of f per *unit change* in the direction of $\mathbf{u}$. We can also interpret $D_{\mathbf{u}} f(a, b)$ as the slope of the tangent line to the trace curve through $P' = (a, b, f(a, b))$ in the vertical plane through $P = (a, b)$ in the direction $\mathbf{u}$ (Figure 7).

FIGURE 7 For a unit vector $\mathbf{u}$, $D_{\mathbf{u}} f(a, b)$ is the slope of the tangent line to the trace curve through $P' = (a, b, f(a, b))$ in the vertical plane through $P = (a, b)$ in the direction $\mathbf{u}$.

DEFINITION Directional Derivative The derivative of $f(x, y)$ at $P = (a, b)$ with respect to a nonzero vector $\mathbf{v} = \langle h, k \rangle$ is the limit (assuming it exists):

$$D_{\mathbf{v}} f(P) = D_{\mathbf{v}} f(a, b) = \lim_{t \to 0} \frac{f(a + th, b + tk) - f(a, b)}{t}$$

If $\mathbf{u}$ is a unit vector, $D_{\mathbf{u}} f$ is called the **directional derivative** in the direction $\mathbf{u}$.

By definition, $D_{\mathbf{v}} f(a, b)$ is the derivative at $t = 0$ of the composite function $f(\mathbf{c}(t))$ where $\mathbf{c}(t) = (a + th, b + tk)$. Therefore, if $f(x, y)$ is differentiable, we may compute $D_{\mathbf{v}} f(a, b)$ using the Chain Rule for Paths. We have $\mathbf{c}'(t) = \langle h, k \rangle = \mathbf{v}$, so

$$D_{\mathbf{v}} f(a, b) = \frac{d}{dt} f(\mathbf{c}(t)) \bigg|_{t=0} = \nabla f_{(a,b)} \cdot \mathbf{c}'(0) = \nabla f_{(a,b)} \cdot \mathbf{v}$$

We also see that for any scalar λ, $D_{\lambda \mathbf{v}} f(P) = \nabla f_P \cdot (\lambda \mathbf{v}) = \lambda \nabla f_P \cdot \mathbf{v}$. Therefore,

$$D_{\lambda \mathbf{v}} f(P) = \lambda D_{\mathbf{v}} f(P)$$

In particular, if $\mathbf{v} \neq \mathbf{0}$, then $\mathbf{u} = \dfrac{\mathbf{v}}{\|\mathbf{v}\|}$ is the unit vector in the direction $\mathbf{v}$ and the directional derivative in the direction of $\mathbf{v}$ is

$$D_{\mathbf{u}} f(P) = \frac{1}{\|\mathbf{v}\|} D_{\mathbf{v}} f(P) \qquad \boxed{2}$$

THEOREM 3 Evaluating the Derivative with Respect to v   If $f(x, y)$ is differentiable, then $D_{\mathbf{v}} f(a, b)$ exists for any vector $\mathbf{v}$ and

$$D_{\mathbf{v}} f(a, b) = \nabla f_{(a,b)} \cdot \mathbf{v}$$

Similarly, in three variables, $D_{\mathbf{v}} f(a, b, c) = \nabla f_{(a,b,c)} \cdot \mathbf{v}$. These formulas can be written $D_{\mathbf{v}} f(P) = \nabla f_P \cdot \mathbf{v}$.

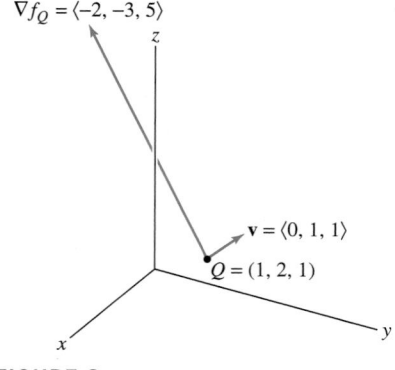

$\nabla f_Q = \langle -2, -3, 5 \rangle$

$\mathbf{v} = \langle 0, 1, 1 \rangle$

$Q = (1, 2, 1)$

FIGURE 8

■ **EXAMPLE 6** Find the rate of change of pressure (in millibars per meter) at the point $Q = (1, 2, 1)$ in the direction of $\mathbf{v} = \langle 0, 1, 1 \rangle$, where the pressure (in millibars) is given by

$$f(x, y, z) = 1{,}000 + yz^2 + x^2 z - xy^2 \qquad (x, y, z \text{ in meters})$$

Solution We compute the gradient at $Q = (1, 2, 1)$ and apply Theorem 3:

$$\nabla f = \left\langle \frac{\partial f}{\partial x}, \frac{\partial f}{\partial y}, \frac{\partial f}{\partial z} \right\rangle = \left\langle 2xz - y^2, z^2 - 2xy, 2yz + x^2 \right\rangle$$

$$\nabla f_Q = \nabla f_{(1,2,1)} = \langle -2, -3, 5 \rangle$$

$$D_{\mathbf{v}} f(Q) = \nabla f_Q \cdot \mathbf{v} = \langle -2, -3, 5 \rangle \cdot \langle 0, 1, 1 \rangle = -3 + 5 = 2$$

The rate of change per unit meter is the directional derivative with respect to the unit vector $\mathbf{u} = \dfrac{\mathbf{v}}{\|\mathbf{v}\|}$ (Figure 8). Since $\|\mathbf{v}\| = \sqrt{2}$, Eq. (2) yields

$$D_{\mathbf{u}} f(Q) = \frac{1}{\|\mathbf{v}\|} D_{\mathbf{v}} f(Q) = \frac{2}{\sqrt{2}} \approx 1.4 \text{ mb/m} \qquad ■$$

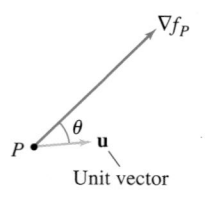

∇f_P

P θ $\mathbf{u}$

Unit vector

FIGURE 9 $\nabla f_P \cdot \mathbf{u} = \|\nabla f_P\| \cos \theta$.

▸▸▸ **REMINDER** *For any vectors* $\mathbf{u}$ *and* $\mathbf{v}$*, the dot product is*

$$\mathbf{v} \cdot \mathbf{u} = \|\mathbf{v}\| \|\mathbf{u}\| \cos \theta$$

where θ *is the angle between* $\mathbf{v}$ *and* $\mathbf{u}$.

We now discuss two important geometric properties of the gradient. First, suppose that $\nabla f_P \neq \mathbf{0}$ and let $\mathbf{u}$ be a unit vector (Figure 9). By the properties of the dot product,

$$D_{\mathbf{u}} f(P) = \nabla f_P \cdot \mathbf{u} = \|\nabla f_P\| \cos \theta \qquad \boxed{3}$$

where θ is the angle between ∇f_P and $\mathbf{u}$. Note that $\cos \theta$ takes its maximum value when $\theta = 0$. Therefore, $D_{\mathbf{u}} f(P)$ has the largest possible value when $\theta = 0$, that is, when $\mathbf{u}$ points in the direction of ∇f_P. It follows that the *gradient vector points in the direction of the maximum rate of increase*. Similarly, f decreases most rapidly in the opposite direction, $-\nabla f_P$, since $\cos\theta = -1$ for $\theta = \pi$. In the earlier scenario of a biker measuring outside temperature, the rate of change of temperature depends on the angle between the temperature gradient ∇T and the biker's direction of motion. Observe that the ROC of temperature is zero in a direction orthogonal to the gradient since $\cos\theta = 0$ for $\theta = \pi/2$ (Figure 10).

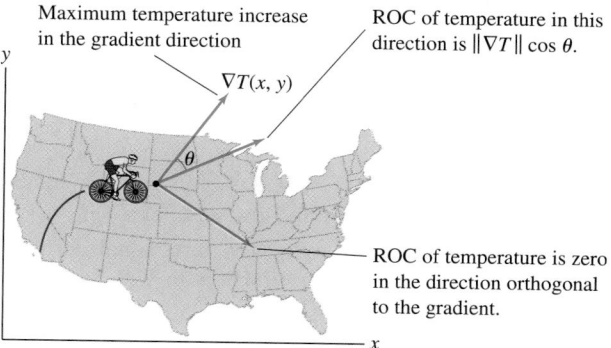

Maximum temperature increase in the gradient direction

$\nabla T(x, y)$

ROC of temperature in this direction is $\|\nabla T\|\cos\theta$.

ROC of temperature is zero in the direction orthogonal to the gradient.

FIGURE 10 The biker should travel in the direction of the gradient to increase temperature most rapidly.

Next, suppose that $f(P) = k$, so that P lies on the level curve $f(x, y) = k$, and let us parametrize this level curve by a path $\mathbf{c}(t)$ such that $\mathbf{c}(0) = P$ and $\mathbf{c}'(0) \neq \mathbf{0}$ (this is possible whenever $\nabla f_P \neq \mathbf{0}$). Then $f(\mathbf{c}(t)) = k$ for all t, and by the Chain Rule,

$$\nabla f_P \cdot \mathbf{c}'(0) = \frac{d}{dt} f(\mathbf{c}(t))\Big|_{t=0} = \frac{d}{dt} k = 0$$

This shows that the vector ∇f_P is perpendicular to $\mathbf{c}'(0)$. However, $\mathbf{c}'(0)$ is tangent to the level curve, so we may conclude that ∇f_P is perpendicular to the level curve through P. For functions of three variables, ∇f_P is normal to the level surface $f(x, y, z) = k$ through P.

THEOREM 4 Interpretation of the Gradient Vector   Assume that $\nabla f_P \neq \mathbf{0}$. Let $\mathbf{u}$ be a unit vector making an angle θ with ∇f_P. Then

$$\boxed{D_{\mathbf{u}} f(P) = \|\nabla f_P\| \cos\theta} \qquad \boxed{4}$$

- ∇f_P points in the direction of maximum rate of increase of f and $-\nabla f_P$ points in the direction of maximum rate of decrease.
- ∇f_P is normal to the level curve (or surface) of f at P.

GRAPHICAL INSIGHT   On a contour map, the gradient direction at P is perpendicular to the level curve through P. This is the direction along which we cross the largest number of contour lines per unit distance. In Figure 11, ∇f_P spans three contour lines, while vectors of the same length pointing in other directions cross fewer lines. Thus, f increases more slowly in these other directions.

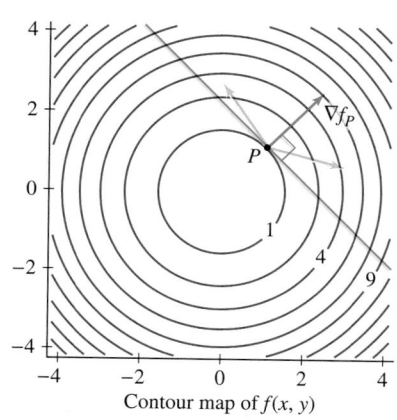

Contour map of $f(x, y)$

FIGURE 11 The gradient ∇f_P points in the direction of maximal increase and is perpendicular to the level curve of f at P.

■ **EXAMPLE 7**   Figure 12 shows the contour map of $f(x, y) = 25 + (x + y^2)e^{-0.3y^2}$ describing the temperature (in degrees Celsius) of a rectangular metal plate (x, y in inches). Find the directional derivative of temperature at $P = (-1, -1)$ in the direction of:

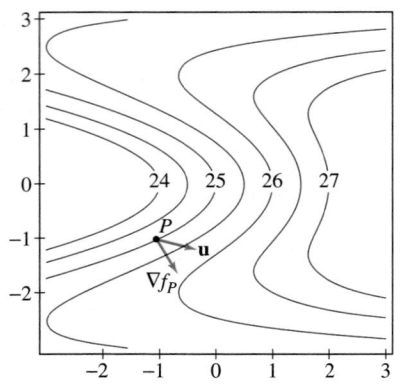

FIGURE 12

By Eq. (4), the directional derivative in the direction of the gradient itself ($\theta = 0$) is equal to $\|\nabla f_P\|$.

(a) The gradient.

(b) The unit vector $\mathbf{u}$ making an angle of $\theta = 45°$ with the gradient.

Solution To apply Eq. (4), we compute the length of the gradient at $P = (-1, -1)$:

$$f_x(x, y) = e^{-0.3y^2}, \qquad f_y(x, y) = y(2 - 0.6x - 0.6y^2)e^{-0.3y^2}$$

$$f_x(P) = e^{-0.3} \approx 0.74, \qquad f_y(P) = -2e^{-0.3} \approx -1.48$$

Hence, $\nabla f_P \approx \langle 0.74, -1.48 \rangle$ and

$$\|\nabla f_P\| \approx \sqrt{0.74^2 + (-1.48)^2} \approx 1.65°C/in.$$

(a) Applying Eq. (4) with $\theta = 0$, we see that the directional derivative in the direction of the gradient is $\|\nabla f_P\| \approx 1.65°C/in.$

(b) The directional derivative in the direction of a unit vector $\mathbf{u}$ making an angle of $\theta = 45°$ with ∇f_P is

$$D_{\mathbf{u}} f(P) = \|\nabla f_P\| \cos 45° \approx 1.65 \left(\frac{\sqrt{2}}{2} \right) \approx 1.17°C/in. \qquad \blacksquare$$

Another application of the gradient is to finding an equation for the tangent plane at a point $P = (a, b, c)$ on a surface defined implicitly by an equation $F(x, y, z) = k$, where k is a constant. Assume that $\nabla F_P \neq \mathbf{0}$. According to Theorem 4, ∇F_P is a normal vector to the level surface $F(x, y, z) = k$. Therefore, the tangent plane has equation

$$\nabla F_P \cdot \langle x - a, y - b, z - c \rangle = 0$$

or

$$\boxed{F_x(a, b, c)(x - a) + F_y(a, b, c)(y - b) + F_z(a, b, c)(z - c) = 0}$$

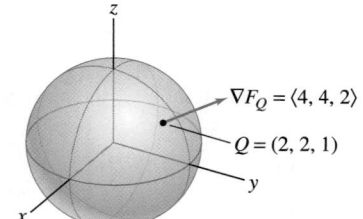

FIGURE 13 The sphere $F(x, y, z) = x^2 + y^2 + z^2 - 9 = 0$.

■ **EXAMPLE 8** **Finding a Normal Vector** Let P be a point on the sphere with equation $x^2 + y^2 + z^2 = 9$. Show that the radial vector $\overrightarrow{OP}$ is normal to the sphere. Find a vector normal at $Q = (2, 2, 1)$.

Solution Let $F(x, y, z) = x^2 + y^2 + z^2$ and $P = (x, y, z)$. Then

$$\nabla F_P = \langle 2x, 2y, 2z \rangle = 2 \langle x, y, z \rangle$$

Since ∇F_P is normal to the sphere at P and is a multiple of the radial vector $\overrightarrow{OP} = \langle x, y, z \rangle$, we conclude that $\overrightarrow{OP}$ is normal to the sphere. A normal vector at $Q = (2, 2, 1)$ is $\nabla F_Q = \langle 4, 4, 2 \rangle$ (Figure 13). ■

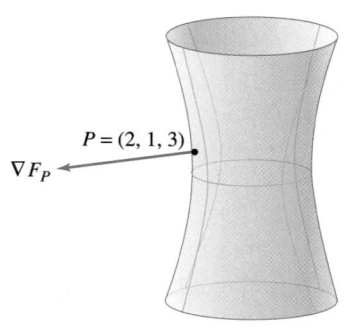

FIGURE 14 The gradient vector ∇F_P is normal to the surface at P.

■ **EXAMPLE 9** **Finding the Tangent Plane** Find an equation of the tangent plane to the surface $4x^2 + 9y^2 - z^2 = 16$ at $P = (2, 1, 3)$.

Solution Let $F(x, y, z) = 4x^2 + 9y^2 - z^2$. Then

$$\nabla F = \langle 8x, 18y, -2z \rangle, \qquad \nabla F_P = \nabla F_{(2,1,3)} = \langle 16, 18, -6 \rangle$$

The vector ∇F_P is normal to the surface $F(x, y, z) = 16$ (Figure 14), so the tangent plane has equation

$$16(x - 2) + 18(y - 1) - 6(z - 3) = 0 \qquad \text{or} \qquad 16x + 8y - 6z = 32 \qquad \blacksquare$$

The symbol ψ (pronounced "p-sigh" or "p-see") is the lowercase Greek letter psi.

CONCEPTUAL INSIGHT   The directional derivative has an interpretation as $\tan \psi$, where ψ is the **angle of inclination**. Think of the graph of $z = f(x, y)$ as a mountain lying over the xy-plane (Figure 15). Let $P = (a, b)$ be a point in the domain of $f(x, y)$ and $P' = (a, b, f(a, b))$ the point on the mountain above P. Given a unit vector $\mathbf{u} = \langle u_1, u_2 \rangle$, you can move up the mountain "in the direction $\mathbf{u}$," that is, such that your horizontal displacement occurs in the direction $\mathbf{u}$. This amounts to moving along the trace curve obtained by intersecting the graph with the vertical plane through P' in the direction $\mathbf{u}$. The directional derivative $D_{\mathbf{u}} f$ is the slope of the tangent line to this trace curve at P' and the vector $\mathbf{w} = \langle u_1, u_2, D_{\mathbf{u}} f \rangle$ is tangent to the trace curve. The angle of inclination ψ, which measures the steepness of the terrain in the direction $\mathbf{u}$, is defined by

$$\tan \psi = \text{slope of tangent in direction } \mathbf{u} = D_{\mathbf{u}} f \qquad \boxed{5}$$

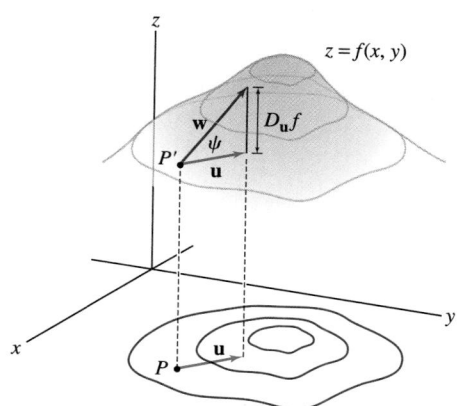

FIGURE 15 The vector $\mathbf{w}$ is tangent to $z = f(x, y)$ at P'.

■ **EXAMPLE 10** Angle of Inclination   You are standing on the side of a mountain [a surface $z = f(x, y)$] at a point $P' = (a, b, f(a, b))$, where $\nabla f_{(a,b)} = \langle 0.4, 0.02 \rangle$. Find the angle of inclination:

(a) In the direction of the gradient.

(b) In a direction making an angle of $\theta = 60°$ with the gradient.

Solution

(a) Let $\mathbf{u}$ be the unit vector in the direction of $\nabla f_{(a,b)}$. Then

$$D_{\mathbf{u}} f(a, b) = \|\nabla f_{P'}\| \cos 0 = \sqrt{(0.4)^2 + (0.02)^2} \approx 0.4$$

The angle of inclination ψ satisfies $\tan \psi = D_{\mathbf{u}} f(a, b) \approx 0.4$. Therefore,

$$\psi \approx \tan^{-1} 0.4 \approx 21.8°$$

(b) Now let $\mathbf{u}$ be a unit vector making an angle of $\theta = 60°$ with $\nabla f_{(a,b)}$. Then

$$D_{\mathbf{u}} f(a, b) = \|\nabla f_{(a,b)}\| \cos 60° \approx (0.4)(0.5) \approx 0.2$$

The angle of inclination ψ in the direction of $\mathbf{u}$ satisfies $\tan \psi = 0.2$, so $\psi = \tan^{-1} 0.2 \approx 11.3°$.   ■

14.5 SUMMARY

- The *gradient* of a function f is the vector of partial derivatives:

$$\nabla f = \left\langle \frac{\partial f}{\partial x}, \frac{\partial f}{\partial y} \right\rangle \quad \text{or} \quad \nabla f = \left\langle \frac{\partial f}{\partial x}, \frac{\partial f}{\partial y}, \frac{\partial f}{\partial z} \right\rangle$$

- The *Chain Rule for Paths:* If $\mathbf{c}(t)$ is a path, then the derivative of the composite function $f(\mathbf{c}(t))$ is $\dfrac{d}{dt} f(\mathbf{c}(t)) = \nabla f_{\mathbf{c}(t)} \cdot \mathbf{c}'(t)$.
- For any vector $\mathbf{v} = \langle v_1, v_2 \rangle$, we define the *derivative of f with respect to* $\mathbf{v}$:

$$D_{\mathbf{v}} f(a, b) = \lim_{t \to 0} \frac{f(a + tv_1, b + tv_2) - f(a, b)}{t}$$

If $\mathbf{u}$ is a *unit vector*, we call $D_{\mathbf{u}} f$ the *directional derivative* of f in the direction $\mathbf{u}$. These definitions extend to three or more variables.
- For any vector $\mathbf{v}$, $D_{\mathbf{v}} f(a, b) = \nabla f_{(a,b)} \cdot \mathbf{v}$.
- The gradient has two basic geometric properties: If $\nabla f_P \neq \mathbf{0}$, then ∇f_P points in the direction of maximum rate of increase and ∇f_P is orthogonal to the level curve (or surface) through P.
- We can write the equation of the tangent plane to the level surface $F(x, y, z) = k$ at a point $P = (a, b, c)$ in the following ways:

$$\nabla F_P \cdot \langle x - a, y - b, z - c \rangle = 0$$

$$F_x(a, b, c)(x - a) + F_y(a, b, c)(y - b) + F_z(a, b, c)(z - c) = 0$$

14.5 EXERCISES

Preliminary Questions

1. Which of the following is a possible value of the gradient ∇f of a function $f(x, y)$ of two variables?

(a) 5   **(b)** $\langle 3, 4 \rangle$ **(c)** $\langle 3, 4, 5 \rangle$

2. True or false: A differentiable function increases at the rate $\| \nabla f_P \|$ in the direction of ∇f_P?

3. Describe the two main geometric properties of the gradient ∇f.

4. Express the partial derivative $\dfrac{\partial f}{\partial x}$ as a directional derivative $D_{\mathbf{u}} f$ for some unit vector $\mathbf{u}$.

5. You are standing at point where the temperature gradient vector is pointing in the northeast (NE) direction. In which direction(s) should you walk to avoid a change in temperature?

(a) NE   **(b)** NW   **(c)** SE   **(d)** SW

6. What is the rate of change of $f(x, y)$ at $(0, 0)$ in the direction making an angle of $45°$ with the x-axis if $\nabla f(0, 0) = \langle 2, 4 \rangle$?

Exercises

1. Let $f(x, y) = xy^2$ and $\mathbf{c}(t) = (\frac{1}{2}t^2, t^3)$.

(a) Calculate $\nabla f \cdot \mathbf{c}'(t)$.

(b) Use the Chain Rule for Paths to evaluate $\dfrac{d}{dt} f(\mathbf{c}(t))$ at $t = 1$ and $t = -1$.

2. Let $f(x, y) = e^{xy}$ and $\mathbf{c}(t) = (t^3, 1 + t)$.

(a) Calculate ∇f and $\mathbf{c}'(t)$.

(b) Use the Chain Rule for Paths to calculate $\dfrac{d}{dt} f(\mathbf{c}(t))$ as $\nabla f \cdot \mathbf{c}'(t)$.

(c) Write out the composite $f(\mathbf{c}(t))$ as a function of t and differentiate. Check that the result agrees with (b).

3. Let $f(x, y) = x^2 + y^2$ and $\mathbf{c}(t) = (\cos t, \sin t)$.

(a) Find $\dfrac{d}{dt} f(\mathbf{c}(t))$ without making any calculations. Explain.

(b) Verify your answer to (a) using the Chain Rule.

4. Figure 16 shows the level curves of a function $f(x, y)$ and a path $c(t)$, traversed in the direction indicated. State whether the derivative $\dfrac{d}{dt} f(c(t))$ is positive, negative, or zero at points A–D.

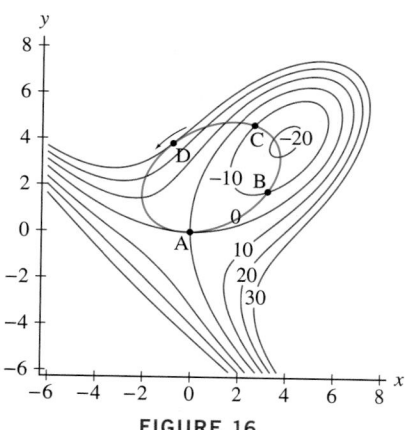

FIGURE 16

In Exercises 5–8, calculate the gradient.

5. $f(x, y) = \cos(x^2 + y)$

6. $g(x, y) = \dfrac{x}{x^2 + y^2}$

7. $h(x, y, z) = xyz^{-3}$

8. $r(x, y, z, w) = xze^{yw}$

In Exercises 9–20, use the Chain Rule to calculate $\dfrac{d}{dt} f(c(t))$.

9. $f(x, y) = 3x - 7y$, $\quad c(t) = (\cos t, \sin t)$, $\quad t = 0$

10. $f(x, y) = 3x - 7y$, $\quad c(t) = (t^2, t^3)$, $\quad t = 2$

11. $f(x, y) = x^2 - 3xy$, $\quad c(t) = (\cos t, \sin t)$, $\quad t = 0$

12. $f(x, y) = x^2 - 3xy$, $\quad c(t) = (\cos t, \sin t)$, $\quad t = \frac{\pi}{2}$

13. $f(x, y) = \sin(xy)$, $\quad c(t) = (e^{2t}, e^{3t})$, $\quad t = 0$

14. $f(x, y) = \cos(y - x)$, $\quad c(t) = (e^t, e^{2t})$, $\quad t = \ln 3$

15. $f(x, y) = x - xy$, $\quad c(t) = (t^2, t^2 - 4t)$, $\quad t = 4$

16. $f(x, y) = xe^y$, $\quad c(t) = (t^2, t^2 - 4t)$, $\quad t = 0$

17. $f(x, y) = \ln x + \ln y$, $\quad c(t) = (\cos t, t^2)$, $\quad t = \frac{\pi}{4}$

18. $g(x, y, z) = xye^z$, $\quad c(t) = (t^2, t^3, t - 1)$, $\quad t = 1$

19. $g(x, y, z) = xyz^{-1}$, $\quad c(t) = (e^t, t, t^2)$, $\quad t = 1$

20. $g(x, y, z, w) = x + 2y + 3z + 5w$, $\quad c(t) = (t^2, t^3, t, t-2)$, $t = 1$

In Exercises 21–30, calculate the directional derivative in the direction of $\mathbf{v}$ at the given point. Remember to normalize the direction vector or use Eq. (2).

21. $f(x, y) = x^2 + y^3$, $\quad \mathbf{v} = \langle 4, 3 \rangle$, $\quad P = (1, 2)$

22. $f(x, y) = x^2 y^3$, $\quad \mathbf{v} = \mathbf{i} + \mathbf{j}$, $\quad P = (-2, 1)$

23. $f(x, y) = x^2 y^3$, $\quad \mathbf{v} = \mathbf{i} + \mathbf{j}$, $\quad P = (\frac{1}{6}, 3)$

24. $f(x, y) = \sin(x - y)$, $\quad \mathbf{v} = \langle 1, 1 \rangle$, $\quad P = (\frac{\pi}{2}, \frac{\pi}{6})$

25. $f(x, y) = \tan^{-1}(xy)$, $\quad \mathbf{v} = \langle 1, 1 \rangle$, $\quad P = (3, 4)$

26. $f(x, y) = e^{xy - y^2}$, $\quad \mathbf{v} = \langle 12, -5 \rangle$, $\quad P = (2, 2)$

27. $f(x, y) = \ln(x^2 + y^2)$, $\quad \mathbf{v} = 3\mathbf{i} - 2\mathbf{j}$, $\quad P = (1, 0)$

28. $g(x, y, z) = z^2 - xy^2$, $\quad \mathbf{v} = \langle -1, 2, 2 \rangle$, $\quad P = (2, 1, 3)$

29. $g(x, y, z) = xe^{-yz}$, $\quad \mathbf{v} = \langle 1, 1, 1 \rangle$, $\quad P = (1, 2, 0)$

30. $g(x, y, z) = x \ln(y + z)$, $\quad \mathbf{v} = 2\mathbf{i} - \mathbf{j} + \mathbf{k}$, $\quad P = (2, e, e)$

31. Find the directional derivative of $f(x, y) = x^2 + 4y^2$ at $P = (3, 2)$ in the direction pointing to the origin.

32. Find the directional derivative of $f(x, y, z) = xy + z^3$ at $P = (3, -2, -1)$ in the direction pointing to the origin.

33. A bug located at $(3, 9, 4)$ begins walking in a straight line toward $(5, 7, 3)$. At what rate is the bug's temperature changing if the temperature is $T(x, y, z) = xe^{y-z}$? Units are in meters and degrees Celsius.

34. Suppose that $\nabla f_P = \langle 2, -4, 4 \rangle$. Is f increasing or decreasing at P in the direction $\mathbf{v} = \langle 2, 1, 3 \rangle$?

35. Let $f(x, y) = xe^{x^2 - y}$ and $P = (1, 1)$.
(a) Calculate $\|\nabla f_P\|$.
(b) Find the rate of change of f in the direction ∇f_P.
(c) Find the rate of change of f in the direction of a vector making an angle of $45°$ with ∇f_P.

36. Let $f(x, y, z) = \sin(xy + z)$ and $P = (0, -1, \pi)$. Calculate $D_{\mathbf{u}} f(P)$ where $\mathbf{u}$ is a unit vector making an angle $\theta = 30°$ with ∇f_P.

37. Let $T(x, y)$ be the temperature at location (x, y). Assume that $\nabla T = \langle y - 4, x + 2y \rangle$. Let $c(t) = (t^2, t)$ be a path in the plane. Find the values of t such that

$$\frac{d}{dt} T(c(t)) = 0$$

38. Find a vector normal to the surface $x^2 + y^2 - z^2 = 6$ at $P = (3, 1, 2)$.

39. Find a vector normal to the surface $3z^3 + x^2 y - y^2 x = 1$ at $P = (1, -1, 1)$.

40. Find the two points on the ellipsoid $\dfrac{x^2}{4} + \dfrac{y^2}{9} + z^2 = 1$ where the tangent plane is normal to $\mathbf{v} = \langle 1, 1, -2 \rangle$.

In Exercises 41–44, find an equation of the tangent plane to the surface at the given point.

41. $x^2 + 3y^2 + 4z^2 = 20$, $\quad P = (2, 2, 1)$

42. $xz + 2x^2 y + y^2 z^3 = 11$, $\quad P = (2, 1, 1)$

43. $x^2 + z^2 e^{y-x} = 13$, $P = \left(2, 3, \dfrac{3}{\sqrt{e}}\right)$

44. $\ln[1 + 4x^2 + 9y^4] - 0.1z^2 = 0$, $P = (3, 1, 6.1876)$

45. Verify what is clear from Figure 17: Every tangent plane to the cone $x^2 + y^2 - z^2 = 0$ passes through the origin.

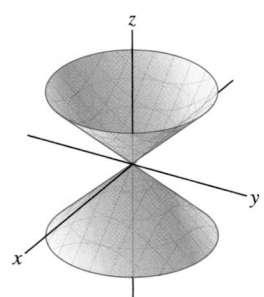

FIGURE 17 Graph of $x^2 + y^2 - z^2 = 0$.

46. *CAS*   Use a computer algebra system to produce a contour plot of $f(x, y) = x^2 - 3xy + y - y^2$ together with its gradient vector field on the domain $[-4, 4] \times [-4, 4]$.

47. Find a function $f(x, y, z)$ such that ∇f is the constant vector $\langle 1, 3, 1 \rangle$.

48. Find a function $f(x, y, z)$ such that $\nabla f = \langle 2x, 1, 2 \rangle$.

49. Find a function $f(x, y, z)$ such that $\nabla f = \langle x, y^2, z^3 \rangle$.

50. Find a function $f(x, y, z)$ such that $\nabla f = \langle z, 2y, x \rangle$.

51. Find a function $f(x, y)$ such that $\nabla f = \langle y, x \rangle$.

52. Show that there does not exist a function $f(x, y)$ such that $\nabla f = \langle y^2, x \rangle$. *Hint:* Use Clairaut's Theorem $f_{xy} = f_{yx}$.

53. Let $\Delta f = f(a + h, b + k) - f(a, b)$ be the change in f at $P = (a, b)$. Set $\Delta \mathbf{v} = \langle h, k \rangle$. Show that the linear approximation can be written

$$\Delta f \approx \nabla f_P \cdot \Delta \mathbf{v} \qquad \boxed{6}$$

54. Use Eq. (6) to estimate

$$\Delta f = f(3.53, 8.98) - f(3.5, 9)$$

assuming that $\nabla f_{(3.5,9)} = \langle 2, -1 \rangle$.

55. Find the unit vector $\mathbf{n}$ normal to the surface $z^2 - 2x^4 - y^4 = 16$ at $P = (2, 2, 8)$ that points in the direction of the xy-plane.

56. Suppose, in the previous exercise, that a particle located at the point $P = (2, 2, 8)$ travels toward the xy-plane in the direction normal to the surface.

(a) Through which point Q on the xy-plane will the particle pass?

(b) Suppose the axes are calibrated in centimeters. Determine the path $\mathbf{c}(t)$ of the particle if it travels at a constant speed of 8 cm/s. How long will it take the particle to reach Q?

57. Let $f(x, y) = \tan^{-1} \dfrac{x}{y}$ and $\mathbf{u} = \left\langle \dfrac{\sqrt{2}}{2}, \dfrac{\sqrt{2}}{2} \right\rangle$.

(a) Calculate the gradient of f.

(b) Calculate $D_{\mathbf{u}} f(1, 1)$ and $D_{\mathbf{u}} f(\sqrt{3}, 1)$.

(c) Show that the lines $y = mx$ for $m \neq 0$ are level curves for f.

(d) Verify that ∇f_P is orthogonal to the level curve through P for $P = (x, y) \neq (0, 0)$.

58.   Suppose that the intersection of two surfaces $F(x, y, z) = 0$ and $G(x, y, z) = 0$ is a curve C and let P be a point on C. Explain why the vector $\mathbf{v} = \nabla F_P \times \nabla G_P$ is a direction vector for the tangent line to C at P.

59. Let C be the curve of intersection of the two surfaces $x^2 + y^2 + z^2 = 3$ and $(x - 2)^2 + (y - 2)^2 + z^2 = 3$. Use the result of Exercise 58 to find parametric equations of the tangent line to C at $P = (1, 1, 1)$.

60. Let C be the curve of intersection of the two surfaces $x^3 + 2xy + yz = 7$ and $3x^2 - yz = 1$. Find the parametric equations of the tangent line to C at $P = (1, 2, 1)$.

61. Verify the linearity relations for gradients:

(a) $\nabla(f + g) = \nabla f + \nabla g$

(b) $\nabla(cf) = c\nabla f$

62. Prove the Chain Rule for gradients (Theorem 1).

63. Prove the Product Rule for gradients (Theorem 1).

Further Insights and Challenges

64. Let $\mathbf{u}$ be a unit vector. Show that the directional derivative $D_{\mathbf{u}} f$ is equal to the component of ∇f along $\mathbf{u}$.

65. Let $f(x, y) = (xy)^{1/3}$.

(a) Use the limit definition to show that $f_x(0, 0) = f_y(0, 0) = 0$.

(b) Use the limit definition to show that the directional derivative $D_{\mathbf{u}} f(0, 0)$ does not exist for any unit vector $\mathbf{u}$ other than $\mathbf{i}$ and $\mathbf{j}$.

(c) Is f differentiable at $(0, 0)$?

66. Use the definition of differentiability to show that if $f(x, y)$ is differentiable at $(0, 0)$ and

$$f(0, 0) = f_x(0, 0) = f_y(0, 0) = 0$$

then

$$\lim_{(x,y) \to (0,0)} \frac{f(x, y)}{\sqrt{x^2 + y^2}} = 0 \qquad \boxed{7}$$

67. This exercise shows that there exists a function which is not differentiable at $(0, 0)$ even though all directional derivatives at $(0, 0)$ exist. Define $f(x, y) = \dfrac{x^2 y}{x^2 + y^2}$ for $(x, y) \neq 0$ and $f(0, 0) = 0$.

(a) Use the limit definition to show that $D_{\mathbf{v}} f(0, 0)$ exists for all vectors $\mathbf{v}$. Show that $f_x(0, 0) = f_y(0, 0) = 0$.

(b) Prove that f is *not* differentiable at $(0, 0)$ by showing that Eq. (7) does not hold.

68. Prove that if $f(x, y)$ is differentiable and $\nabla f_{(x, y)} = \mathbf{0}$ for all (x, y), then f is constant.

69. Prove the following Quotient Rule where f, g are differentiable:

$$\nabla \left(\frac{f}{g} \right) = \frac{g \nabla f - f \nabla g}{g^2}$$

In Exercises 70–72, a path $\mathbf{c}(t) = (x(t), y(t))$ *follows the gradient of a function* $f(x, y)$ *if the tangent vector* $\mathbf{c}'(t)$ *points in the direction of* ∇f *for all t. In other words,* $\mathbf{c}'(t) = k(t) \nabla f_{\mathbf{c}(t)}$ *for some positive function* $k(t)$. *Note that in this case,* $\mathbf{c}(t)$ *crosses each level curve of* $f(x, y)$ *at a right angle.*

70. Show that if the path $\mathbf{c}(t) = (x(t), y(t))$ follows the gradient of $f(x, y)$, then $\dfrac{y'(t)}{x'(t)} = \dfrac{f_y}{f_x}$.

71. Find a path of the form $\mathbf{c}(t) = (t, g(t))$ passing through $(1, 2)$ that follows the gradient of $f(x, y) = 2x^2 + 8y^2$ (Figure 18). *Hint:* Use Separation of Variables.

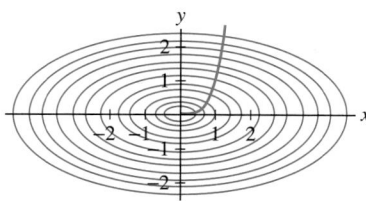

FIGURE 18 The path $\mathbf{c}(t)$ is orthogonal to the level curves of $f(x, y) = 2x^2 + 8y^2$.

72. *CAS* Find the curve $y = g(x)$ passing through $(0, 1)$ that crosses each level curve of $f(x, y) = y \sin x$ at a right angle. If you have a computer algebra system, graph $y = g(x)$ together with the level curves of f.

14.6 The Chain Rule

In the previous section, we proved the Chain Rule for Paths and used it to derive the main properties of the gradient. There is a more general version of the Chain Rule that applies to general composite functions. Suppose, for example, that x, y, z are differentiable functions of s and t, say, $x = x(s, t)$, $y = y(s, t)$, and $z = z(s, t)$. Then the composite

$$f(x(s, t), y(s, t), z(s, t)) \qquad \boxed{1}$$

is a function of s and t.

■ **EXAMPLE 1** Find the composite function where $f(x, y, z) = xy + z$ and $x = s^2$, $y = st$, $z = t^2$.

Solution The composite function is

$$f(x(s, t), y(s, t), z(s, t)) = xy + z = (s^2)(st) + t^2 = s^3 t + t^2 \qquad ■$$

For the composite function (1) above, the general Chain Rule has the form

$$\frac{\partial f}{\partial s} = \frac{\partial f}{\partial x} \frac{\partial x}{\partial s} + \frac{\partial f}{\partial y} \frac{\partial y}{\partial s} + \frac{\partial f}{\partial z} \frac{\partial z}{\partial s} \qquad \boxed{2}$$

$$\frac{\partial f}{\partial t} = \frac{\partial f}{\partial x} \frac{\partial x}{\partial t} + \frac{\partial f}{\partial y} \frac{\partial y}{\partial t} + \frac{\partial f}{\partial z} \frac{\partial z}{\partial t} \qquad \boxed{3}$$

We prove these formulas by showing that they follow from the Chain Rule for Paths. Fix a point (s_0, t_0). The partial derivative $\partial f / \partial s$ at (s_0, t_0) is the derivative of the function $f(x(s, t_0), y(s, t_0), z(s, t_0))$ with respect to s. To evaluate it, we define a path

$\mathbf{c}(s) = (x(s, t_0), y(s, t_0), z(s, t_0))$. The tangent vector is

$$\mathbf{c}'(s) = \left\langle \frac{\partial x}{\partial s}(s, t_0), \frac{\partial y}{\partial s}(s, t_0), \frac{\partial z}{\partial s}(s, t_0) \right\rangle$$

By the Chain Rule for Paths,

$$\frac{\partial f}{\partial s}\bigg|_{(s_0,t_0)} = \frac{d}{ds} f(\mathbf{c}(s))\bigg|_{s=s_0} = \nabla f \cdot \mathbf{c}'(s_0) = \frac{\partial f}{\partial x}\frac{\partial x}{\partial s} + \frac{\partial f}{\partial y}\frac{\partial y}{\partial s} + \frac{\partial f}{\partial z}\frac{\partial z}{\partial s}$$

where the partial derivatives on the right are evaluated at (s_0, t_0). This proves (2). Equation (3) is proved similarly. Furthermore, this argument generalizes to functions $f(x_1, \ldots, x_n)$ of n variables that depend on other variables $t_1, \ldots, t_m$. We call $t_1, \ldots, t_m$ the **independent variables**.

THEOREM 1 General Version of Chain Rule Let $f(x_1, \ldots, x_n)$ be a differentiable function of n variables. Suppose that each of the variables $x_1, \ldots, x_n$ is a differentiable function of m independent variables $t_1, \ldots, t_m$. Then for $k = 1, \ldots, m$,

$$\frac{\partial f}{\partial t_k} = \frac{\partial f}{\partial x_1}\frac{\partial x_1}{\partial t_k} + \frac{\partial f}{\partial x_2}\frac{\partial x_2}{\partial t_k} + \cdots + \frac{\partial f}{\partial x_n}\frac{\partial x_n}{\partial t_k} \qquad \boxed{4}$$

As an aid to remembering the general Chain Rule, we will refer to the derivatives $\frac{\partial f}{\partial x_1}, \ldots, \frac{\partial f}{\partial x_n}$ as the **primary derivatives**. If $t = t_k$ is any one of the independent variables, then $\frac{\partial f}{\partial t_k}$ is equal to a sum of n terms, one for each primary derivative:

$$j\text{th term:} \quad \frac{\partial f}{\partial x_j}\frac{\partial x_j}{\partial t_k} \quad \text{for } j = 1, 2, \ldots, n$$

The term "primary derivative" is not standard. We use it in this section to clarify the structure of the Chain Rule.

Note that the primary derivatives are the components of the gradient, so we can express Eq. (4) as a dot product:

$$\frac{\partial f}{\partial t_k} = \underbrace{\left\langle \frac{\partial f}{\partial x_1}, \frac{\partial f}{\partial x_2}, \ldots, \frac{\partial f}{\partial x_n} \right\rangle}_{\nabla f} \cdot \left\langle \frac{\partial x_1}{\partial t_k}, \frac{\partial x_2}{\partial t_k}, \ldots, \frac{\partial x_n}{\partial t_k} \right\rangle$$

■ **EXAMPLE 2** Using the Chain Rule Let $f(x, y, z) = xy + z$. Calculate $\frac{\partial f}{\partial s}$, where $x = s^2, y = st, z = t^2$.

Solution

Step 1. **Compute the primary derivatives.**

$$\frac{\partial f}{\partial x} = y, \qquad \frac{\partial f}{\partial y} = x \qquad \frac{\partial f}{\partial z} = 1$$

Step 2. **Apply the Chain Rule:**

$$\frac{\partial f}{\partial s} = \frac{\partial f}{\partial x}\frac{\partial x}{\partial s} + \frac{\partial f}{\partial y}\frac{\partial y}{\partial s} + \frac{\partial f}{\partial z}\frac{\partial z}{\partial s} = y\frac{\partial}{\partial s}s^2 + x\frac{\partial}{\partial s}st + \frac{\partial}{\partial s}t^2$$

$$= (y)(2s) + (x)(t) + 0 = 2sy + xt$$

Step 3. **Express the answer in terms of the independent variables.**

If desired, we may use $x = s^2$ and $y = st$ to express the derivative in terms of s and t:

$$\frac{\partial f}{\partial s} = 2ys + xt = 2(st)s + (s^2)t = 3s^2t$$

As a check of this result, recall that in Example 1, we computed the composite function

$$f(x(s, t), y(s, t), z(s, t)) = f(s^2, st, t^2) = s^3t + t^2$$

From this we see directly that $\dfrac{\partial f}{\partial s} = 3s^2t$, confirming our result obtained using the Chain Rule.   ■

■ **EXAMPLE 3** Evaluating the Derivative at a Point Let $f(x, y) = e^{xy}$. Evaluate $\dfrac{\partial f}{\partial t}$ at $(s, t, u) = (-1, 3, 5)$, where

$$x = st, \qquad y = u - t$$

Solution First, we calculate the primary derivatives:

$$\frac{\partial f}{\partial x} = ye^{xy}, \qquad \frac{\partial f}{\partial y} = xe^{xy}$$

Since $\dfrac{\partial x}{\partial t} = s$ and $\dfrac{\partial y}{\partial t} = -1$, using the Chain Rule, we have

$$\frac{\partial f}{\partial t} = \frac{\partial f}{\partial x}\frac{\partial x}{\partial t} + \frac{\partial f}{\partial y}\frac{\partial y}{\partial t} = ye^{xy}(s) + xe^{xy}(-1) = (ys - x)e^{xy}$$

It is not necessary to express $\dfrac{\partial f}{\partial t}$ in terms of s, t, u. For $(s, t, u) = (-1, 3, 5)$, we have $x = st = -3$ and $y = u - t = 2$. Therefore,

$$\left.\frac{\partial f}{\partial t}\right|_{(-1,3,5)} = (ys - x)e^{xy}\Big|_{(-1,3,5)} = (2(-1) - (-3))e^{(-3)(2)} = e^{-6}$$ ■

■ **EXAMPLE 4** Polar Coordinates Let $f(x, y)$ be a function of two variables and let (r, θ) be polar coordinates.

(a) Express $\dfrac{\partial f}{\partial \theta}$ in terms of $\dfrac{\partial f}{\partial x}$ and $\dfrac{\partial f}{\partial y}$.

(b) Evaluate $\dfrac{\partial f}{\partial \theta}$ at $(x, y) = (1, 1)$ if $f(x, y) = yx^{-1}$.

Solution

(a) Since $x = r\cos\theta$ and $y = r\sin\theta$,

$$\frac{\partial x}{\partial \theta} = -r\sin\theta, \qquad \frac{\partial y}{\partial \theta} = r\cos\theta$$

By the Chain Rule,

$$\frac{\partial f}{\partial \theta} = \frac{\partial f}{\partial x}\frac{\partial x}{\partial \theta} + \frac{\partial f}{\partial y}\frac{\partial y}{\partial \theta} = -r\sin\theta\frac{\partial f}{\partial x} + r\cos\theta\frac{\partial f}{\partial y} \qquad \boxed{5}$$

(b) Applying Eq. (5) to $f(x, y) = yx^{-1}$, we obtain

$$\frac{\partial f}{\partial \theta} = -r\sin\theta\frac{\partial f}{\partial x} + r\cos\theta\frac{\partial f}{\partial y} = (-r\sin\theta)(-yx^{-2}) + (r\cos\theta)(x^{-1})$$

The point $(x, y) = (1, 1)$ has polar coordinates $r = \sqrt{2}$ and $\theta = \tan^{-1} 1 = \frac{\pi}{4}$, so

$$\frac{\partial f}{\partial \theta}\bigg|_{(1,1)} = (-r \sin \theta)(-yx^{-2}) + (r \cos \theta)(x^{-1})$$

$$= \left(-\sqrt{2} \sin \frac{\pi}{4}\right)(-1) + \left(\sqrt{2} \cos \frac{\pi}{4}\right)(1)$$

$$= \sqrt{2}\left(\frac{\sqrt{2}}{2}\right) + \sqrt{2}\left(\frac{\sqrt{2}}{2}\right) = 2 \qquad \blacksquare$$

Implicit Differentiation

In calculus of one variable, implicit differentiation is used to compute $\dfrac{dy}{dx}$ in cases where y is defined implicitly as a function of x through an equation $f(x, y) = 0$. This extends to functions of several variables. Consider a surface defined by an equation

$$F(x, y, z) = 0$$

where $F(x, y, z)$ is differentiable. We may differentiate both sides with respect to x using the Chain Rule:

$$\frac{\partial F}{\partial x}\frac{\partial x}{\partial x} + \frac{\partial F}{\partial y}\frac{\partial y}{\partial x} + \frac{\partial F}{\partial z}\frac{\partial z}{\partial x} = 0$$

Since $\partial x/\partial x = 1$ and $\partial y/\partial x = 0$, this equation reduces to

$$\frac{\partial F}{\partial x} + \frac{\partial F}{\partial z}\frac{\partial z}{\partial x} = F_x + F_z \frac{\partial z}{\partial x} = 0$$

If $F_z \neq 0$, we may solve for $\dfrac{\partial z}{\partial x}$ $\left(\dfrac{\partial z}{\partial y} \text{ is computed similarly}\right)$:

$$\boxed{\frac{\partial z}{\partial x} = -\frac{F_x}{F_z}, \qquad \frac{\partial z}{\partial y} = -\frac{F_y}{F_z}} \qquad \boxed{6}$$

■ **EXAMPLE 5** Calculate $\partial z/\partial x$ and $\partial z/\partial y$ at $P = (1, 1, 1)$, where (Figure 1)

$$F(x, y, z) = x^2 + y^2 - 2z^2 + 12x - 8z - 4 = 0$$

What is the graphical interpretation of these partial derivatives?

Solution We have

$$F_x = 2x + 12, \qquad F_y = 2y, \qquad F_z = -4z - 8$$

and hence,

$$\frac{\partial z}{\partial x} = -\frac{F_x}{F_z} = \frac{2x + 12}{4z + 8}, \qquad \frac{\partial z}{\partial y} = -\frac{F_y}{F_z} = \frac{2y}{4z + 8}$$

$$\frac{\partial z}{\partial x}\bigg|_{(1,1,1)} = \frac{14}{12} = \frac{7}{6}, \qquad \frac{\partial z}{\partial y}\bigg|_{(1,1,1)} = \frac{2}{12} = \frac{1}{6}$$

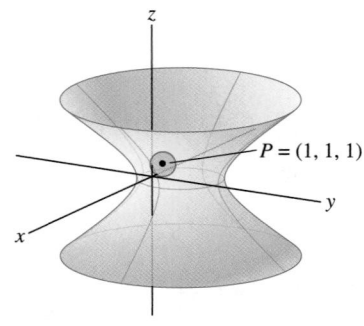

FIGURE 1 The surface $x^2 + y^2 - 2z^2 + 12x - 8z - 4 = 0$. In a small patch around P, the surface can be represented as the graph of a function.

Figure 1 shows the surface $F(x, y, z) = 0$. The surface as a whole is not the graph of a function (it fails the Vertical Line Test). However, a small patch near P may be represented as a graph of some function $z = f(x, y)$, and the partial derivatives $\dfrac{\partial z}{\partial x}$ and $\dfrac{\partial z}{\partial y}$ are equal to the partial derivatives f_x and f_y. Implicit differentiation has allowed us to compute these partial derivatives without explicitly finding $f(x, y)$. ∎

Assumptions Matter Implicit differentiation is based on the assumption that the equation $F(x, y, z) = 0$ defines z as a function $z = f(x, y)$, at least near a given point $P = (a, b, c)$. According to the Implicit Function Theorem of advanced calculus, such a function $f(x, y)$ exists if F has continuous partial derivatives and if $F_z(P) \neq 0$ [this is also the condition under which the formulas (6) are valid]. Note that $F_z(P) \neq 0$ implies that the tangent plane at P is nonvertical. To see what can go wrong when $F_z(P) = 0$, consider the cylinder (Figure 2)

$$F(x, y, z) = x^2 + y^2 - 1 = 0$$

In this case, $F_z = 0$. The z-coordinate on the cylinder does not depend on x or y, so it is impossible to represent the cylinder as a graph $z = f(x, y)$.

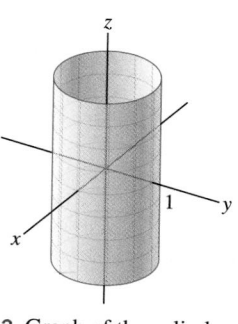

FIGURE 2 Graph of the cylinder $x^2 + y^2 - 1 = 0$.

14.6 SUMMARY

- If $f(x, y, z)$ is a function of variables x, y, z, and x, y, z depend on two other variables, say, s and t, then

$$f(x, y, z) = f(x(s, t), y(s, t), z(s, t))$$

is a composite function of s and t. We refer to s and t as the *independent variables*.
- The *Chain Rule* expresses the partial derivatives with respect to the independent variables s and t in terms of the *primary derivatives*:

$$\frac{\partial f}{\partial x}, \qquad \frac{\partial f}{\partial y}, \qquad \frac{\partial f}{\partial z}$$

Namely,

$$\frac{\partial f}{\partial s} = \frac{\partial f}{\partial x}\frac{\partial x}{\partial s} + \frac{\partial f}{\partial y}\frac{\partial y}{\partial s} + \frac{\partial f}{\partial z}\frac{\partial z}{\partial s}, \qquad \frac{\partial f}{\partial t} = \frac{\partial f}{\partial x}\frac{\partial x}{\partial t} + \frac{\partial f}{\partial y}\frac{\partial y}{\partial t} + \frac{\partial f}{\partial z}\frac{\partial z}{\partial t}$$

- The general case is similar. Let $f(x_1, \ldots, x_n)$ be a function of n variables. If $x_1, \ldots, x_n$ depend on the independent variables $t_1, \ldots, t_m$, then

$$\frac{\partial f}{\partial t_k} = \frac{\partial f}{\partial x_1}\frac{\partial x_1}{\partial t_k} + \frac{\partial f}{\partial x_2}\frac{\partial x_2}{\partial t_k} + \cdots + \frac{\partial f}{\partial x_n}\frac{\partial x_n}{\partial t_k}$$

This relation can be expressed as a dot product:

$$\frac{\partial f}{\partial t_k} = \underbrace{\left\langle \frac{\partial f}{\partial x_1}, \frac{\partial f}{\partial x_2}, \ldots, \frac{\partial f}{\partial x_n} \right\rangle}_{\nabla f} \cdot \left\langle \frac{\partial x_1}{\partial t_k}, \frac{\partial x_2}{\partial t_k}, \ldots, \frac{\partial x_n}{\partial t_k} \right\rangle$$

- Implicit differentiation is used to find the partial derivatives $\dfrac{\partial z}{\partial x}$ and $\dfrac{\partial z}{\partial y}$ when z is defined implicitly by an equation $F(x, y, z) = 0$:

$$\frac{\partial z}{\partial x} = -\frac{F_x}{F_z}, \qquad \frac{\partial z}{\partial y} = -\frac{F_y}{F_z}$$

14.6 EXERCISES

Preliminary Questions

1. Consider a function $f(x, y)$ where $x = uv$ and $y = u/v$.

(a) What are the primary derivatives of f?

(b) What are the independent variables?

In Questions 2–4, suppose that $f(u, v) = ue^v$, where $u = rs$ and $v = r + s$.

2. The composite function $f(u, v)$ is equal to:

(a) rse^{r+s} **(b)** re^s **(c)** rse^{rs}

3. What is the value of $f(u, v)$ at $(r, s) = (1, 1)$?

4. According to the Chain Rule, $\dfrac{\partial f}{\partial r}$ is equal to (choose correct answer):

(a) $\dfrac{\partial f}{\partial x} \dfrac{\partial x}{\partial r} + \dfrac{\partial f}{\partial x} \dfrac{\partial x}{\partial s}$

(b) $\dfrac{\partial f}{\partial x} \dfrac{\partial x}{\partial r} + \dfrac{\partial f}{\partial y} \dfrac{\partial y}{\partial r}$

(c) $\dfrac{\partial f}{\partial r} \dfrac{\partial r}{\partial x} + \dfrac{\partial f}{\partial s} \dfrac{\partial s}{\partial x}$

5. Suppose that x, y, z are functions of the independent variables u, v, w. Given a function $f(x, y, z)$, which of the following terms appear in the Chain Rule expression for $\dfrac{\partial f}{\partial w}$?

(a) $\dfrac{\partial f}{\partial v} \dfrac{\partial x}{\partial v}$

(b) $\dfrac{\partial f}{\partial w} \dfrac{\partial w}{\partial x}$

(c) $\dfrac{\partial f}{\partial z} \dfrac{\partial z}{\partial w}$

(d) $\dfrac{\partial f}{\partial v} \dfrac{\partial v}{\partial w}$

6. With notation as in the previous question, does $\dfrac{\partial x}{\partial v}$ appear in the Chain Rule expression for $\dfrac{\partial f}{\partial u}$?

Exercises

1. Let $f(x, y, z) = x^2 y^3 + z^4$ and $x = s^2$, $y = st^2$, and $z = s^2 t$.

(a) Calculate the primary derivatives $\dfrac{\partial f}{\partial x}, \dfrac{\partial f}{\partial y}, \dfrac{\partial f}{\partial z}$.

(b) Calculate $\dfrac{\partial x}{\partial s}, \dfrac{\partial y}{\partial s}, \dfrac{\partial z}{\partial s}$.

(c) Compute $\dfrac{\partial f}{\partial s}$ using the Chain Rule:

$$\frac{\partial f}{\partial s} = \frac{\partial f}{\partial x} \frac{\partial x}{\partial s} + \frac{\partial f}{\partial y} \frac{\partial y}{\partial s} + \frac{\partial f}{\partial z} \frac{\partial z}{\partial s}$$

Express the answer in terms of the independent variables s, t.

2. Let $f(x, y) = x \cos(y)$ and $x = u^2 + v^2$ and $y = u - v$.

(a) Calculate the primary derivatives $\dfrac{\partial f}{\partial x}, \dfrac{\partial f}{\partial y}$.

(b) Use the Chain Rule to calculate $\dfrac{\partial f}{\partial v}$. Leave the answer in terms of both the dependent and independent variables.

(c) Determine (x, y) for $(u, v) = (2, 1)$ and evaluate $\dfrac{\partial f}{\partial v}$ at $(u, v) = (2, 1)$.

In Exercises 3–10, use the Chain Rule to calculate the partial derivatives. Express the answer in terms of the independent variables.

3. $\dfrac{\partial f}{\partial s}, \dfrac{\partial f}{\partial r}$: $f(x, y, z) = xy + z^2$, $x = s^2$, $y = 2rs$, $z = r^2$

4. $\dfrac{\partial f}{\partial r}, \dfrac{\partial f}{\partial t}$: $f(x, y, z) = xy + z^2$, $x = r + s - 2t$, $y = 3rt$, $z = s^2$

5. $\dfrac{\partial g}{\partial u}, \dfrac{\partial g}{\partial v}$: $g(x, y) = \cos(x^2 - y^2)$, $x = 2u - 3v$, $y = -5u + 8v$

6. $\dfrac{\partial h}{\partial t_2}$: $h(x, y) = \dfrac{x}{y}$, $x = t_1 t_2$, $y = t_1^2 t_2$

7. $\dfrac{\partial F}{\partial x}, \dfrac{\partial F}{\partial y}$: $F(u, v) = e^{u+v}$, $u = x^2$, $v = xy$

8. $\dfrac{\partial f}{\partial u}$: $f(x, y) = x^2 + y^2$, $x = e^{u+v}$, $y = u + v$

9. $\dfrac{\partial f}{\partial x}, \dfrac{\partial f}{\partial y}$: $f(r, \theta) = r \sin^2 \theta$, $x = r \cos \theta$, $y = r \sin \theta$

10. $\dfrac{\partial f}{\partial \theta}, \dfrac{\partial f}{\partial r}$: $f(x, y, z) = xy - z^2$, $x = r \cos \theta$, $y = \cos^2 \theta$, $z = r$

In Exercises 11–16, use the Chain Rule to evaluate the partial derivative at the point specified.

11. $\dfrac{\partial f}{\partial u}$ and $\dfrac{\partial f}{\partial v}$ at $(u, v) = (-1, -1)$, where $f(x, y, z) = x^3 + yz^2$, $x = u^2 + v$, $y = u + v^2$, $z = uv$.

12. $\dfrac{\partial f}{\partial s}$ at $(r, s) = (1, 0)$, where $f(x, y) = \ln(xy)$, $x = 3r + 2s$, $y = 5r + 3s$.

13. $\dfrac{\partial g}{\partial \theta}$ at $(r, \theta) = (2\sqrt{2}, \frac{\pi}{4})$, where $g(x, y) = \dfrac{1}{x + y^2}$, $x = r\sin\theta$, $y = r\cos\theta$.

14. $\dfrac{\partial g}{\partial s}$ at $s = 4$, where $g(x, y) = x^2 - y^2$, $x = s^2 + 1$, $y = 1 - 2s$.

15. $\dfrac{\partial g}{\partial u}$ at $(u, v) = (0, 1)$, where $g(x, y) = x^2 - y^2$, $x = e^u \cos v$, $y = e^u \sin v$.

16. $\dfrac{\partial h}{\partial q}$ at $(q, r) = (3, 2)$, where $h(u, v) = ue^v$, $u = q^3$, $v = qr^2$.

17. The temperature at a point (x, y) is $T(x, y) = 20 + 0.1(x^2 - xy)$ (degrees Celsius). A particle moves clockwise along the unit circle at unit speed (1 cm/s), starting at $(1, 0)$ at time $t = 0$. How fast is the particle's temperature changing at time $t = \pi$?

18. Let $u = u(x, y)$ and let (r, θ) be polar coordinates. Express u_x and u_y in terms of u_r and u_θ. Then show that

$$\|\nabla u\|^2 = u_r^2 + \frac{1}{r^2}u_\theta^2 \qquad \boxed{7}$$

19. Let $u(r, \theta) = r^2 \cos^2 \theta$. Use Eq. (7) to compute $\|\nabla u\|^2$. Then compute $\|\nabla u\|^2$ directly by observing that $u(x, y) = x^2$ and compare.

20. Let $x = s + t$ and $y = s - t$. Show that for any differentiable function $f(x, y)$,

$$\left(\frac{\partial f}{\partial x}\right)^2 - \left(\frac{\partial f}{\partial y}\right)^2 = \frac{\partial f}{\partial s}\frac{\partial f}{\partial t}$$

21. Express the derivatives $\dfrac{\partial f}{\partial \rho}, \dfrac{\partial f}{\partial \theta}, \dfrac{\partial f}{\partial \phi}$ in terms of $\dfrac{\partial f}{\partial x}, \dfrac{\partial f}{\partial y}, \dfrac{\partial f}{\partial z}$, where (ρ, θ, ϕ) are spherical coordinates.

22. Suppose that z is defined implicitly as a function of x and y by the equation $F(x, y, z) = xz^2 + y^2z + xy - 1 = 0$.
(a) Calculate F_x, F_y, F_z.
(b) Use Eq. (6) to calculate $\dfrac{\partial z}{\partial x}$ and $\dfrac{\partial z}{\partial y}$.

23. Calculate $\dfrac{\partial z}{\partial x}$ and $\dfrac{\partial z}{\partial y}$ at the points $(3, 2, 1)$ and $(3, 2, -1)$, where z is defined implicitly by the equation $z^4 + z^2x^2 - y - 8 = 0$.

In Exercises 24–29, calculate the derivative using implicit differentiation.

24. $\dfrac{\partial z}{\partial x}$, $\quad x^2y + y^2z + xz^2 = 10$

25. $\dfrac{\partial w}{\partial z}$, $\quad x^2w + w^3 + wz^2 + 3yz = 0$

26. $\dfrac{\partial z}{\partial y}$, $\quad e^{xy} + \sin(xz) + y = 0$

27. $\dfrac{\partial r}{\partial t}$ and $\dfrac{\partial t}{\partial r}$, $\quad r^2 = te^{s/r}$

28. $\dfrac{\partial w}{\partial y}$, $\quad \dfrac{1}{w^2 + x^2} + \dfrac{1}{w^2 + y^2} = 1$ at $(x, y, w) = (1, 1, 1)$

29. $\dfrac{\partial U}{\partial T}$ and $\dfrac{\partial T}{\partial U}$, $\quad (TU - V)^2 \ln(W - UV) = 1$ at $(T, U, V, W) = (1, 1, 2, 4)$

30. The pressure P, volume V, and temperature T of a van der Waals gas with n molecules (n constant) are related by the equation

$$\left(P + \frac{an^2}{V^2}\right)(V - nb) = nRT$$

where a, b, and R are constant. Calculate $\dfrac{\partial P}{\partial T}$ and $\dfrac{\partial V}{\partial P}$.

31. Let $f(x, y, z) = F(r)$, where $r = \sqrt{x^2 + y^2 + z^2}$. Show that

$$\nabla f = F'(r)e_{\mathbf{r}} \qquad \boxed{8}$$

where $e_{\mathbf{r}} = \dfrac{\mathbf{r}}{\|\mathbf{r}\|}$ and $\mathbf{r} = \langle x, y, z \rangle$.

32. Let $f(x, y, z) = e^{-x^2 - y^2 - z^2} = e^{-r^2}$, with r as in Exercise 31. Compute ∇f directly and using Eq. (8).

33. Use Eq. (8) to compute $\nabla\left(\dfrac{1}{r}\right)$ and $\nabla(\ln r)$.

34. Show that if $f(x)$ is differentiable and $c \neq 0$ is a constant, then $u(x, t) = f(x - ct)$ satisfies the so-called **advection equation**

$$\frac{\partial u}{\partial t} + c\frac{\partial u}{\partial x} = 0$$

35. Jessica and Matthew are running toward the point P along the straight paths that make a fixed angle of θ (Figure 3). Suppose that Matthew runs with velocity v_a m/s and Jessica with velocity v_b m/s.

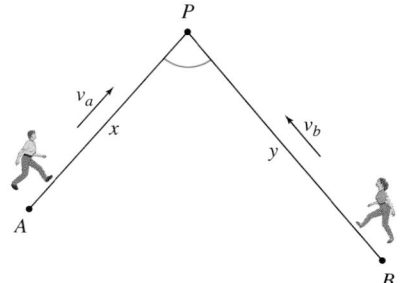

FIGURE 3

Let $f(x, y)$ be the distance from Matthew to Jessica when Matthew is x meters from P and Jessica is y meters from P.

(a) Show that $f(x, y) = \sqrt{x^2 + y^2 - 2xy \cos \theta}$.

(b) Assume that $\theta = \pi/3$. Use the Chain Rule to determine the rate at which the distance between Matthew and Jessica is changing when $x = 30$, $y = 20$, $v_a = 4$ m/s, and $v_b = 3$ m/s.

Further Insights and Challenges

36. The Law of Cosines states that $c^2 = a^2 + b^2 - 2ab \cos \theta$, where a, b, c are the sides of a triangle and θ is the angle opposite the side of length c.

(a) Use implicit differentiation to compute the derivatives $\dfrac{\partial \theta}{\partial a}$, $\dfrac{\partial \theta}{\partial b}$, and $\dfrac{\partial \theta}{\partial c}$.

(b) Suppose that $a = 10$, $b = 16$, $c = 22$. Estimate the change in θ if a and b are increased by 1 and c is increased by 2.

*In Exercises 37–40, a function $f(x, y, z)$ is called **homogeneous of degree n** if $f(\lambda x, \lambda y, \lambda z) = \lambda^n f(x, y, z)$ for all $\lambda \in \mathbf{R}$.*

37. Show that the following functions are homogeneous and determine their degree.

(a) $f(x, y, z) = x^2 y + xyz$

(b) $f(x, y, z) = 3x + 2y - 8z$

(c) $f(x, y, z) = \ln \left(\dfrac{xy}{z^2} \right)$

(d) $f(x, y, z) = z^4$

38. Prove that if $f(x, y, z)$ is homogeneous of degree n, then $f_x(x, y, z)$ is homogeneous of degree $n - 1$. *Hint:* Either use the limit definition or apply the Chain Rule to $f(\lambda x, \lambda y, \lambda z)$.

39. Prove that if $f(x, y, z)$ is homogeneous of degree n, then

$$x \frac{\partial f}{\partial x} + y \frac{\partial f}{\partial y} + z \frac{\partial f}{\partial z} = nf \qquad \boxed{9}$$

Hint: Let $F(t) = f(tx, ty, tz)$ and calculate $F'(1)$ using the Chain Rule.

40. Verify Eq. (9) for the functions in Exercise 37.

41. Suppose that $x = g(t, s)$, $y = h(t, s)$. Show that f_{tt} is equal to

$$f_{xx} \left(\frac{\partial x}{\partial t} \right)^2 + 2 f_{xy} \left(\frac{\partial x}{\partial t} \right) \left(\frac{\partial y}{\partial t} \right) + f_{yy} \left(\frac{\partial y}{\partial t} \right)^2$$

$$+ f_x \frac{\partial^2 x}{\partial t^2} + f_y \frac{\partial^2 y}{\partial t^2} \qquad \boxed{10}$$

42. Let $r = \sqrt{x_1^2 + \cdots + x_n^2}$ and let $g(r)$ be a function of r. Prove the formulas

$$\frac{\partial g}{\partial x_i} = \frac{x_i}{r} g_r, \qquad \frac{\partial^2 g}{\partial x_i^2} = \frac{x_i^2}{r^2} g_{rr} + \frac{r^2 - x_i^2}{r^3} g_r$$

43. Prove that if $g(r)$ is a function of r as in Exercise 42, then

$$\frac{\partial^2 g}{\partial x_1^2} + \cdots + \frac{\partial^2 g}{\partial x_n^2} = g_{rr} + \frac{n - 1}{r} g_r$$

*In Exercises 44–48, the **Laplace operator** is defined by $\Delta f = f_{xx} + f_{yy}$. A function $f(x, y)$ satisfying the Laplace equation $\Delta f = 0$ is called **harmonic**. A function $f(x, y)$ is called **radial** if $f(x, y) = g(r)$, where $r = \sqrt{x^2 + y^2}$.*

44. Use Eq. (10) to prove that in polar coordinates (r, θ),

$$\Delta f = f_{rr} + \frac{1}{r^2} f_{\theta\theta} + \frac{1}{r} f_r \qquad \boxed{11}$$

45. Use Eq. (11) to show that $f(x, y) = \ln r$ is harmonic.

46. Verify that $f(x, y) = x$ and $f(x, y) = y$ are harmonic using both the rectangular and polar expressions for Δf.

47. Verify that $f(x, y) = \tan^{-1} \dfrac{y}{x}$ is harmonic using both the rectangular and polar expressions for Δf.

48. Use the Product Rule to show that

$$f_{rr} + \frac{1}{r} f_r = r^{-1} \frac{\partial}{\partial r} \left(r \frac{\partial f}{\partial r} \right)$$

Use this formula to show that if f is a radial harmonic function, then $r f_r = C$ for some constant C. Conclude that $f(x, y) = C \ln r + b$ for some constant b.

49. Figure 4 shows the graph of the equation

$$F(x, y, z) = x^2 + y^2 - z^2 - 12x - 8z - 4 = 0$$

(a) Use the quadratic formula to solve for z as a function of x and y. This gives two formulas, depending on the choice of a sign.

(b) Which formula defines the portion of the surface satisfying $z \geq -4$? Which formula defines the portion satisfying $z \leq -4$?

(c) Calculate $\dfrac{\partial z}{\partial x}$ using the formula $z = f(x, y)$ (for both choices of sign) and again via implicit differentiation. Verify that the two answers agree.

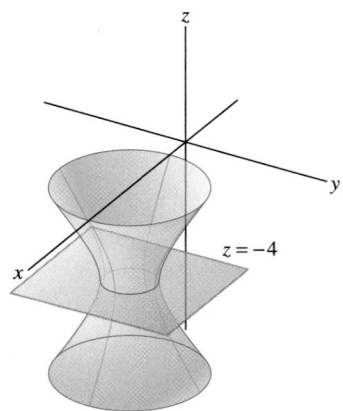

FIGURE 4 Graph of $x^2 + y^2 - z^2 - 12x - 8z - 4 = 0$.

50. When x, y, and z are related by an equation $F(x, y, z) = 0$, we sometimes write $\left(\dfrac{\partial z}{\partial x}\right)_y$ in place of $\dfrac{\partial z}{\partial x}$ to indicate that in the differentiation, z is treated as a function of x with y held constant (and similarly for the other variables).

(a) Use Eq. (6) to prove the **cyclic relation**

$$\left(\frac{\partial z}{\partial x}\right)_y \left(\frac{\partial x}{\partial y}\right)_z \left(\frac{\partial y}{\partial z}\right)_x = -1 \qquad \boxed{12}$$

(b) Verify Eq. (12) for $F(x, y, z) = x + y + z = 0$.

(c) Verify the cyclic relation for the variables P, V, T in the ideal gas law $PV - nRT = 0$ (n and R are constants).

14.7 Optimization in Several Variables

Recall that optimization is the process of finding the extreme values of a function. This amounts to finding the highest and lowest points on the graph over a given domain. As we saw in the one-variable case, it is important to distinguish between *local* and *global* extreme values. A local extreme value is a value $f(a, b)$ that is a maximum or minimum in some small open disk around (a, b) (Figure 1).

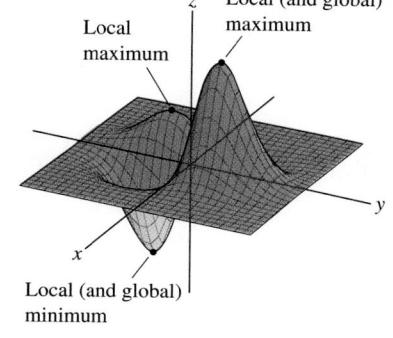

FIGURE 1 Graph of $f(x, y) = (x^2 + 40xy + 20y + 5y^2)e^{-0.4(2x^2+y^2)}$.

> **DEFINITION Local Extreme Values** A function $f(x, y)$ has a **local extremum** at $P = (a, b)$ if there exists an open disk $D(P, r)$ such that:
>
> - **Local maximum:** $f(x, y) \le f(a, b)$ for all $(x, y) \in D(P, r)$.
> - **Local minimum:** $f(x, y) \ge f(a, b)$ for all $(x, y) \in D(P, r)$.

←·· REMINDER *The term "extremum" (the plural is "extrema") means a minimum or maximum value.*

In one variable, Fermat's Theorem states that if $f(a)$ is a local extreme value, then a is a critical point [either $f'(a) = 0$ or $f'(a)$ does not exist]. Thus, the tangent line at a local extremum is horizontal (if the tangent line exists). We can expect a similar result for functions of two variables, but in this case, it is the *tangent plane* that must be horizontal at a local extremum (Figure 2). Recall that the tangent plane to the surface $z = f(x, y)$ at $P = (a, b)$ has equation

$$z = f(a, b) + f_x(a, b)(x - a) + f_y(a, b)(y - b)$$

We see that the tangent plane is horizontal if $f_x(a, b) = f_y(a, b) = 0$, that is, if the equation reduces to $z = f(a, b)$.

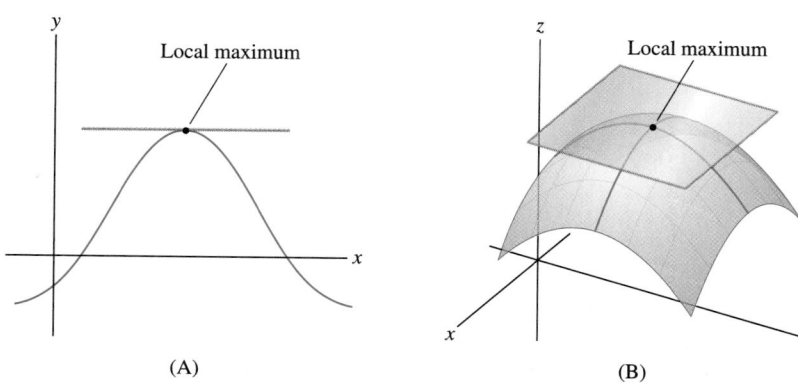

FIGURE 2 The tangent line or plane is horizontal at a local extremum.

(A)

(B)

> **DEFINITION Critical Point** A point $P = (a, b)$ in the domain of $f(x, y)$ is called a **critical point** if:
>
> - $f_x(a, b) = 0$ and $f_y(a, b) = 0$, or
> - At least one of the partial derivatives $f_x(a, b)$, $f_y(a, b)$ does not exist.

According to the next theorem, local extrema occur at critical points, as in the single-variable case.

THEOREM 1 Fermat's Theorem in Several Variables   If $f(x, y)$ has a local minimum or maximum at $P = (a, b)$, then $P = (a, b)$ is a critical point of $f(x, y)$.

Proof   Assume that $f(x, y)$ has a local minimum at $P = (a, b)$ (a local maximum is similar). Then $f(x, y) \geq f(a, b)$ for (x, y) in a small disk $D(r, P)$. In particular, $f(x, b) \geq f(a, b)$ if $|x - a| < r$. This shows that the function $g(x) = f(x, b)$ has a local minimum at $x = a$, and therefore, either $g'(a) = 0$ or $g'(a)$ does not exist. Since $g'(a) = f_x(a, b)$, we conclude that either $f_x(a, b) = 0$ or $f_x(a, b)$ does not exist. Similarly, $f_y(a, b) = 0$ or $f_y(a, b)$ does not exist. Therefore, (a, b) is a critical point.   ∎

More generally, we say that $(a_1, \ldots, a_n)$ is a critical point of $f(x_1, \ldots, x_n)$ if all the partial derivatives are zero:

$$f_{x_j}(a_1, \ldots, a_n) = 0$$

or if at least one of the partial derivatives $f_{x_j}(a_1, \ldots, a_n)$ does not exist. Theorem 1 holds in any number of variables: Local extrema occur at critical points.

❚ **EXAMPLE 1   Finding Critical Points**   Show that $f(x, y) = 11x^2 - 2xy + 2y^2 + 3y$ has one critical point. Use Figure 3 to determine whether it corresponds to local minima or maxima.

Solution   We set the partial derivatives equal to zero and solve the resulting pair of equations:

$$f_x(x, y) = 22x - 2y = 0$$
$$f_y(x, y) = -2x + 4y + 3 = 0$$

By the first equation, $y = 11x$. Substituting $y = 11x$ in the second equation gives

$$-2x + 4y + 3 = -2x + 4(11x) + 3 = 42x + 3 = 0$$

Thus $x = -\frac{1}{14}$ and $y = -\frac{11}{14}$. There is just one critical point, $P = \left(-\frac{1}{14}, -\frac{11}{14}\right)$. Figure 3 shows that $f(x, y)$ has a local minimum at P.   ∎

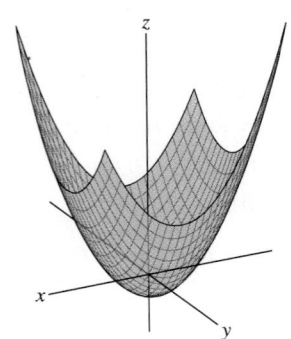

FIGURE 3 Graph of $f(x, y) = 11x^2 - 2xy + 2y^2 + 3y$.

As we saw in the previous example, finding the critical points involves solving the simultaneous equations $f_x(x, y) = 0$ and $f_x(x, y) = 0$. It is not always possible to find the solutions exactly, but we can use a computer to find numerical approximations.

❚ **EXAMPLE 2 ⌂𝘈𝘚   Numerical Example**   Use a computer algebra system to find numerical approximations to the critical points of

$$f(x, y) = \frac{x - y}{2x^2 + 8y^2 + 3}$$

Refer to Figure 4 to determine whether they correspond to local minima or maxima.

Solution   We use a CAS to compute the partial derivatives and solve

$$f_x(x, y) = \frac{-2x^2 + 8y^2 + 4xy + 3}{(2x^2 + 8y^2 + 3)^2} = 0$$

$$f_x(x, y) = \frac{-2x^2 + 8y^2 - 16xy - 3}{(2x^2 + 8y^2 + 3)^2} = 0$$

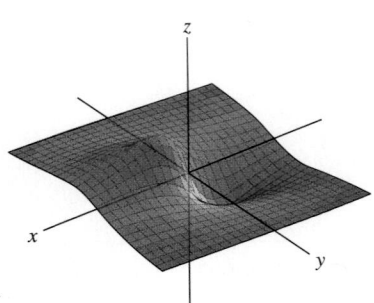

FIGURE 4 Plot of $f(x, y) = \dfrac{x - y}{2x^2 + 8y^2 + 3}$.

Figure 4 suggests that $f(x, y)$ has two critical points near $(1, 0)$ and $(-1, 0)$. We use a CAS to search for solutions near these points. It suffices to set numerators of f_x and f_y equal to zero, and we do so to simplify the problem. The Mathematica command

```
[FindRoot[{-2x^2+8y^2+4xy+3 == 0, -2x^2+8y^2-16xy-3 == 0},
  {{x,1},{y,0}}]
```

returns the answer

```
{x -> 1.095, y -> -0.274}
```

Thus, $(1.095, -0.274)$ is an approximate critical point at which f has a local maximum, as we see in Figure 4. A search for the second critical point near $(-1, 0)$ yields $(-1.095, 0.274)$, which approximates the critical point where $f(x, y)$ takes on a local minimum. ∎

A basic problem in optimization is determining whether a critical point yields a local minimum or maximum or neither. In one variable, $f(x)$ may have a point of inflection rather than a local extremum at a critical point. A similar phenomenon occurs in several variables. Figure 5(C) shows a saddle point. The origin is a critical point, but it yields neither a local minimum nor a local maximum. If you stand at the saddle point and begin walking, some directions take you uphill and other directions take you downhill.

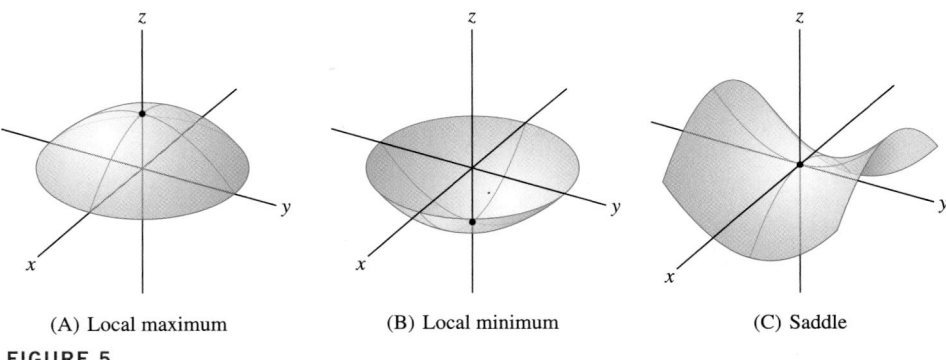

(A) Local maximum   (B) Local minimum   (C) Saddle

FIGURE 5

■ **EXAMPLE 3** Find the critical points of $f(x, y) = x^3 + y^3 - 12xy$ and use Figure 6 to determine whether they correspond to extrema.

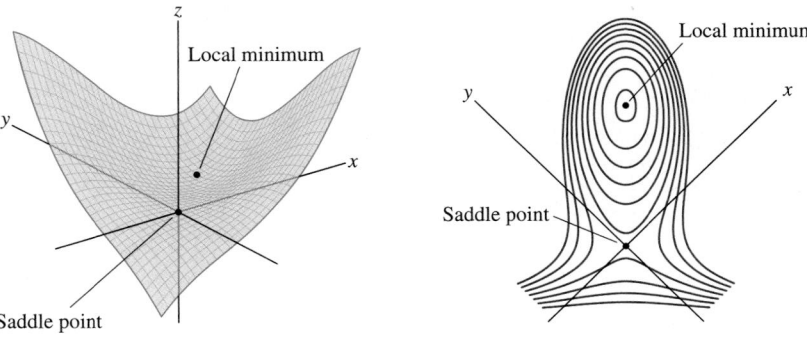

FIGURE 6 The function $f(x, y) = x^3 + y^3 - 12xy$ has two critical points.

Solution Again, we set the partial derivatives equal to zero and solve:

$$f_x(x, y) = 3x^2 - 12y = 0 \quad \Rightarrow \quad y = \frac{1}{4}x^2$$

$$f_y(x, y) = 3y^2 - 12x = 0$$

Substituting $y = \frac{1}{4}x^2$ in the second equation yields

$$3y^2 - 12x = 3\left(\frac{1}{4}x^2\right)^2 - 12x = \frac{3}{16}x(x^3 - 64) = 0 \quad \Rightarrow \quad x = 0, 4$$

Since $y = \frac{1}{4}x^2$, the critical points are $(0, 0)$ and $(4, 4)$. Figure 6 shows that $f(x, y)$ has a local minimum at $(4, 4)$ and saddle point at $(0, 0)$. ▪

GRAPHICAL INSIGHT   We can read off the type of a critical point from the contour map. Typically, if $f(P)$ is a local minimum or maximum, then the nearby level curves are closed curves encircling P as in Figure 7(A). The contour plot shows that $f(x, y)$ increases in all directions emanating from P. In Figure 7(B), $g(x, y)$ has a saddle point at Q. The level curves of $g(x, y)$ through Q consist of two intersecting lines that divide the neighborhood near Q into four regions. The contour map shows that $g(x, y)$ is decreasing in the x-direction and increasing in the y-direction. There also exist more general types of saddle points. The graph of $h(x, y)$ in Figure 7(C) is called a "monkey saddle" (because a monkey can sit on this saddle with room for his tail in the back).

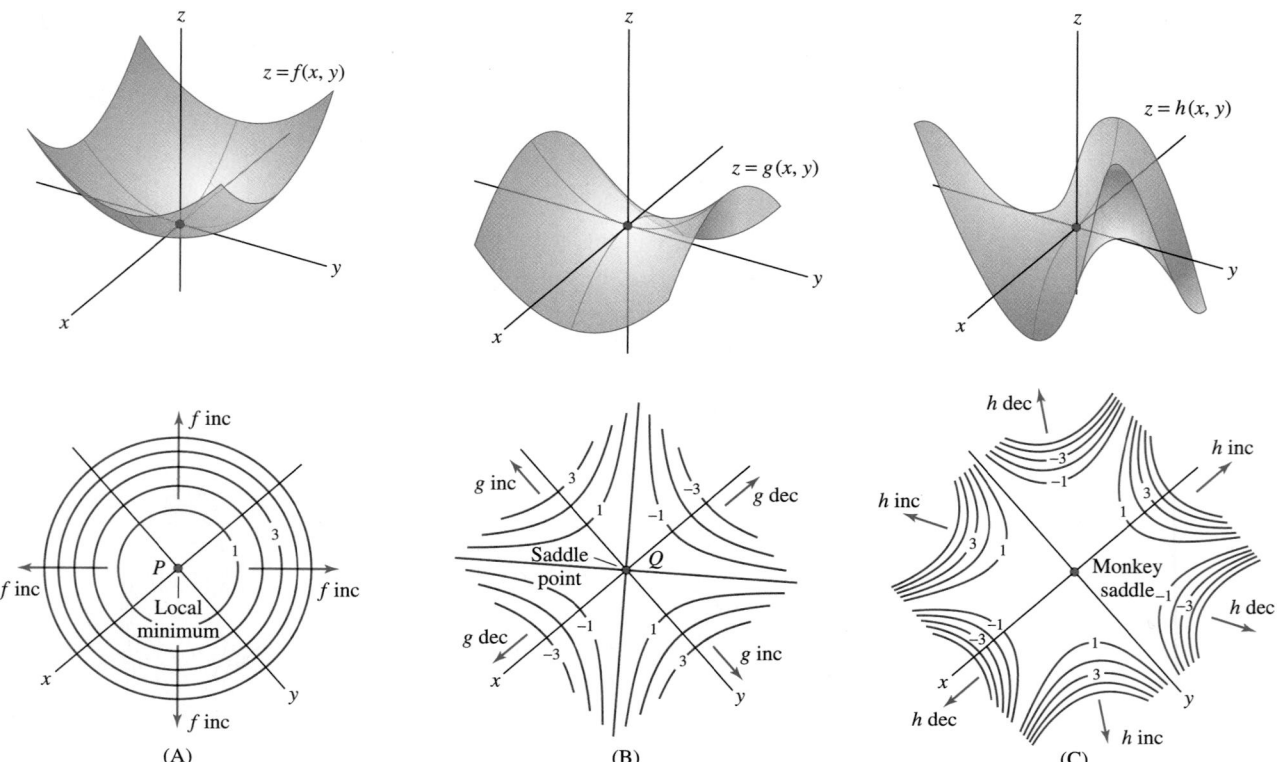

FIGURE 7 Level curves near a local minimum or saddle point.

Recall that if c is a critical point of $f(x)$ and $f''(c) \neq 0$, then the sign of $f''(c)$ determines whether $f(c)$ is a local minimum or maximum. There is also a Second Derivative Test for determining the type of a critical point (a, b) of a function $f(x, y)$ in two variables. It relies on the sign of the **discriminant** $D = D(a, b)$, defined as follows:

$$D = D(a, b) = f_{xx}(a, b)f_{yy}(a, b) - f_{xy}^2(a, b)$$

THEOREM 2 Second Derivative Test   Let $P = (a, b)$ be a critical point of $f(x, y)$. Assume that the second derivatives f_{xx}, f_{yy}, f_{xy} are continuous near P. Then:

(i) If $D > 0$, then $f(a, b)$ is a local minimum or local maximum.
(ii) If $D < 0$, then (a, b) is a saddle point.
(iii) If $D = 0$, the test is inconclusive.

Furthermore,

In applying (iv) and (v), note that if $D > 0$, then $f_{xx}(a, b)$ and $f_{yy}(a, b)$ must have the same sign.

(iv) If $D > 0$ and $f_{xx}(a, b) > 0$, then $f(a, b)$ is a local minimum.
(v) If $D > 0$ and $f_{xx}(a, b) < 0$, then $f(a, b)$ is a local maximum.

A proof of this theorem is discussed at the end of this section.

■ **EXAMPLE 4** Applying the Second Derivative Test   Find the critical points of the function $f(x, y) = (x^2 + y^2)e^{-x}$ and analyze them using the Second Derivative Test.

Solution

Step 1. Find the critical points.
Set the partial derivatives equal to zero and solve:

$$f_x(x, y) = -(x^2 + y^2)e^{-x} + 2xe^{-x} = (2x - x^2 - y^2)e^{-x} = 0$$
$$f_y(x, y) = 2ye^{-x} = 0 \quad \Rightarrow \quad y = 0$$

Substituting $y = 0$ in the first equation then gives

$$(2x - x^2 - y^2)e^{-x} = (2x - x^2)e^{-x} = 0 \quad \Rightarrow \quad x = 0, 2$$

The critical points are $(0, 0)$ and $(2, 0)$ (Figure 8).

Step 2. Compute the second-order partials and the discriminant.

$$f_{xx}(x, y) = \frac{\partial}{\partial x}(2x - x^2 - y^2)e^{-x} = (2 - 4x + x^2 + y^2)e^{-x}$$
$$f_{yy}(x, y) = \frac{\partial}{\partial y}2ye^{-x} = 2e^{-x}$$
$$f_{xy}(x, y) = f_{yx}(x, y) = \frac{\partial}{\partial x}2ye^{-x} = -2ye^{-x}$$
$$D(x, y) = f_{xx}f_{yy} - f_{xy}^2 = 2(2 - 4x + x^2 + y^2)e^{-2x} - 4y^2e^{-2x}$$
$$= (4 - 8x + 2x^2 - 2y^2)e^{-2x}$$

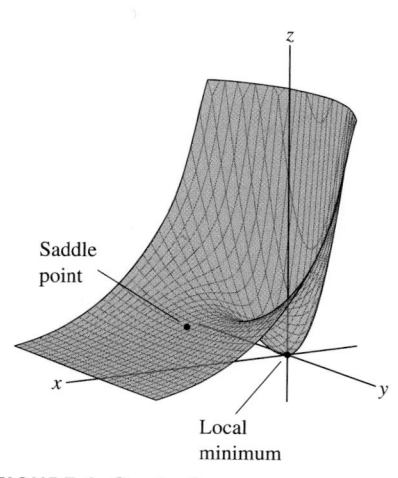

Saddle point

Local minimum

FIGURE 8 Graph of $f(x, y) = (x^2 + y^2)e^{-x}$.

Step 3. **Apply the Second Derivative Test.**

Critical Point	f_{xx}	f_{yy}	f_{xy}	Discriminant D	Type
$(0, 0)$	2	2	0	$2(2) - 0^2 = 4$	Local minimum since $D > 0$ and $f_{xx} > 0$
$(2, 0)$	$-2e^{-2}$	$2e^{-2}$	0	$-2e^{-2}(2e^{-2}) - 0^2 = -4e^{-2}$	Saddle since $D < 0$

■

Global Extrema

In many optimization problems, we are interested in finding the minimum or maximum value of a function f on a given domain $\mathcal{D}$. These values are called the **global** or **absolute extreme values** of f on $\mathcal{D}$. The existence of global extrema depends on the domain $\mathcal{D}$. For example, $f(x, y) = x + y$ has a maximum value $f(1, 1) = 2$ on the unit square $\mathcal{D}_1$ in Figure 9, but it has no maximum value on the entire plane $\mathbf{R}^2$.

To state conditions that guarantee the existence of global extrema, a few definitions are needed. First, we say that a domain $\mathcal{D}$ is **bounded** if it is contained in a disk of radius M centered at the origin for some $M > 0$. In other words, no point of $\mathcal{D}$ is more than a distance M from the origin [Figures 10(A) and (C)]. Next, a point P is called an:

- **Interior point** of $\mathcal{D}$ if $\mathcal{D}$ contains some open disk $D(P, r)$ centered at P.
- **Boundary point** of $\mathcal{D}$ if every disk centered at P contains points in $\mathcal{D}$ and points not in $\mathcal{D}$.

The **interior** of $\mathcal{D}$ is the set of all interior points and the **boundary** of $\mathcal{D}$ is the set of all boundary points. In Figure 10(C), the boundary is the curve surrounding the domain and the interior consists of all points in the domain not lying on the boundary curve.

A domain $\mathcal{D}$ is called **closed** if $\mathcal{D}$ contains all its boundary points; $\mathcal{D}$ is called **open** if every point of $\mathcal{D}$ is an interior point. The domain in Figure 10(A) is closed because the domain includes its boundary curve. In Figure 10(C), some of the boundary points are excluded and the domain is neither open nor closed.

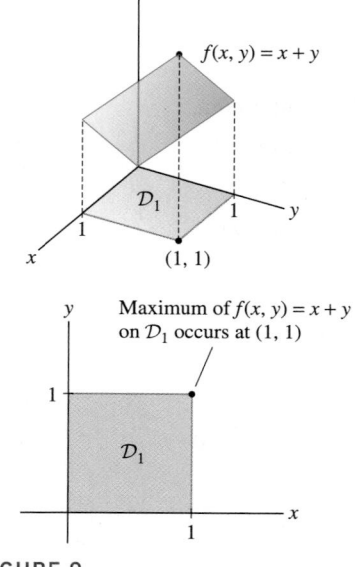

$f(x, y) = x + y$

$\mathcal{D}_1$

$(1, 1)$

Maximum of $f(x, y) = x + y$ on $\mathcal{D}_1$ occurs at $(1, 1)$

$\mathcal{D}_1$

FIGURE 9

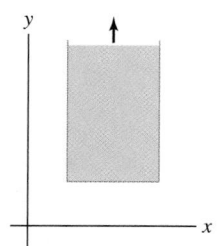

(A) This domain is bounded (all points within bounded distance from the origin) and closed (contains all boundary points).

Interior point

Boundary point

(B) An unbounded domain (contains points arbitrarily far from the origin).

(C) A non-closed domain (contains some but not all boundary points).

Boundary point not in $\mathcal{D}$

FIGURE 10 Domains in $\mathbf{R}^2$.

In Section 4.2, we stated two basic results on optimization. First, every continuous function $f(x)$ on a *closed, bounded interval* $[a, b]$ takes on both a minimum and a maximum value on $[a, b]$. Second, these extreme values occur either at critical points in the interior (a, b) or at the endpoints. Analogous results hold in several variables.

> **THEOREM 3 Existence of Global Extrema** Assume that $f(x, y)$ is defined and continuous on a closed, bounded domain $\mathcal{D}$ in $\mathbf{R}^2$. Then:
>
> **(i)** $f(x, y)$ takes on both a minimum and a maximum value on $\mathcal{D}$.
> **(ii)** The extreme values occur either at critical points in the interior of $\mathcal{D}$ or at points on the boundary of $\mathcal{D}$.

■ **EXAMPLE 5** Find the maximum of $f(x, y) = 2x + y - 3xy$ on the unit square $\mathcal{D} = \{(x, y) : 0 \le x, y \le 1\}$ (Figure 11).

Solution By Theorem 3, the maximum occurs either at a critical point or on the boundary of the square.

***Step 1.* Examine the critical points.**
Set the partial derivatives equal to zero and solve:

$$f_x(x, y) = 2 - 3y = 0 \Rightarrow y = \frac{2}{3}, \qquad f_y(x, y) = 1 - 3x = 0 \Rightarrow x = \frac{1}{3}$$

There is a unique critical point $P = (\frac{1}{3}, \frac{2}{3})$ and

$$f(P) = f\left(\frac{1}{3}, \frac{2}{3}\right) = 2\left(\frac{1}{3}\right) + \left(\frac{2}{3}\right) - 3\left(\frac{1}{3}\right)\left(\frac{2}{3}\right) = \frac{2}{3}$$

***Step 2.* Find the maximum on the boundary.**
We must find the maximum value on the boundary of the square. We do this by checking each of the four edges separately. The bottom edge is described by $y = 0$, $0 \le x \le 1$, and $f(x, 0) = 2x$. Hence, the maximum value of f on this edge occurs at $x = 1$, where $f(1, 0) = 2$. Proceeding in a similar fashion with the other edges, we obtain the following:

Edge	Restriction of $f(x, y)$ to Edge	Maximum of $f(x, y)$ on Edge
Lower: $y = 0, 0 \le x \le 1$	$f(x, 0) = 2x$	$f(1, 0) = 2$
Upper: $y = 1, 0 \le x \le 1$	$f(x, 1) = 1 - x$	$f(0, 1) = 1$
Left: $x = 0, 0 \le y \le 1$	$f(0, y) = y$	$f(0, 1) = 1$
Right: $x = 1, 0 \le y \le 1$	$f(1, y) = 2 - 2y$	$f(1, 0) = 2$

The maximum of f on the boundary is $f(1, 0) = 2$. This is larger than the value $f(P) = \frac{2}{3}$ at the critical point, so the maximum of f on the unit square is 2. ■

■ **EXAMPLE 6** **Box of Maximum Volume** Find the volume of the largest box that can be inscribed in the tetrahedron bounded by the coordinate planes and the plane

$$\frac{1}{3}x + y + z = 1$$

Solution

***Step 1.* Find a function to be maximized.**
Let $P = (x, y, z)$ be the corner of the box lying on the front face of the tetrahedron (Figure 12). Then the box has sides of lengths x, y, z and volume $V = xyz$. We use

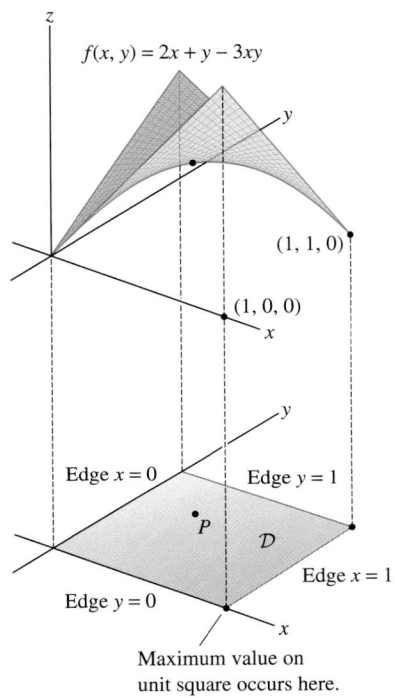

$f(x, y) = 2x + y - 3xy$

Edge $x = 0$ Edge $y = 1$

P $\mathcal{D}$

Edge $y = 0$ Edge $x = 1$

Maximum value on unit square occurs here.

FIGURE 11

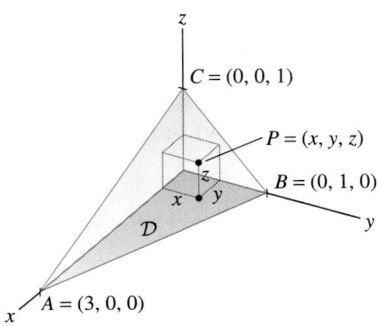

FIGURE 12 The shaded triangle is the domain of $V(x, y)$.

the equation $\frac{1}{3}x + y + z = 1$, or $z = 1 - \frac{1}{3}x - y$, to express V in terms of x and y:

$$V(x, y) = xyz = xy\left(1 - \frac{1}{3}x - y\right) = xy - \frac{1}{3}x^2 y - xy^2$$

Step 2. Determine the domain.

Let $\mathcal{D}$ be the shaded triangle $\triangle OAB$ in the xy-plane in Figure 12. The corner point $P = (x, y, z)$ projects to the point (x, y) in $\mathcal{D}$, so our problem is to find the maximum of $V(x, y)$ on $\mathcal{D}$. Since $\mathcal{D}$ is a closed, bounded region, this maximum occurs at a critical point inside $\mathcal{D}$ or on the boundary of $\mathcal{D}$.

Step 3. Examine the critical points.

First, set the partial derivatives equal to zero and solve:

$$\frac{\partial V}{\partial x} = y - \frac{2}{3}xy - y^2 = y\left(1 - \frac{2}{3}x - y\right) = 0$$

$$\frac{\partial V}{\partial y} = x - \frac{1}{3}x^2 - 2xy = x\left(1 - \frac{1}{3}x - 2y\right) = 0$$

If $x = 0$ or $y = 0$, then (x, y) lies on the boundary of $\mathcal{D}$, so assume that x and y are both nonzero. Then we may divide by y and x in the above equations to obtain

$$1 - \frac{2}{3}x - y = 0 \quad \Rightarrow \quad y = 1 - \frac{2}{3}x$$

$$1 - \frac{1}{3}x - 2y = 0$$

Substituting $y = 1 - \frac{2}{3}x$ in the second equation yields

$$1 - \frac{1}{3}x - 2\left(1 - \frac{2}{3}x\right) = 0 \quad \Rightarrow \quad x - 1 = 0 \quad \Rightarrow \quad x = 1$$

Therefore, $(1, \frac{1}{3})$ is a critical point, and

$$V\left(1, \frac{1}{3}\right) = (1)\frac{1}{3} - \frac{1}{3}(1)^2\frac{1}{3} - (1)\left(\frac{1}{3}\right)^2 = \frac{1}{9}$$

Step 4. Check the boundary.

We have $V(x, y) = 0$ for all points (x, y) on the boundary of $\mathcal{D}$ (because the three edges of the boundary are defined by $x = 0$, $y = 0$, or $1 - \frac{1}{3}x - y = 0$). Clearly, then, the maximum occurs at the critical point and the maximum volume is $\frac{1}{9}$. ∎

Proof of the Second Derivative Test We discuss the main ideas in the proof of the Second Derivative Test, leaving the full details to the Internet supplement. The proof is based on an algebraic fact about quadratic forms (Theorem 4 below). A **quadratic form** is a function in two variables of the type

$$Q(h, k) = ah^2 + 2bhk + ck^2 \quad (a, b, c \text{ constants, not all zero})$$

A quadratic form with a nonzero discriminant D is often called "nondegenerate."

The discriminant of Q is the quantity $D = ac - b^2$. For example,

If $Q(h, k) = h^2 - k^2$,   then $D = (1)(-1) - 0 = -1$

If $Q(h, k) = h^2 + 2hk + 5k^2$,   then $D = (1)(5) - 1 = 4$

Some quadratic forms, such as $Q(h, k) = h^2 - k^2$, take on both positive and negative values. Other quadratic forms, such as $Q(h, k) = h^2 + k^2$, satisfy $Q(h, k) > 0$ for $(h, k) \neq 0$. According to the next theorem, the sign of the discriminant determines which of these two possibilities occurs (Figure 13).

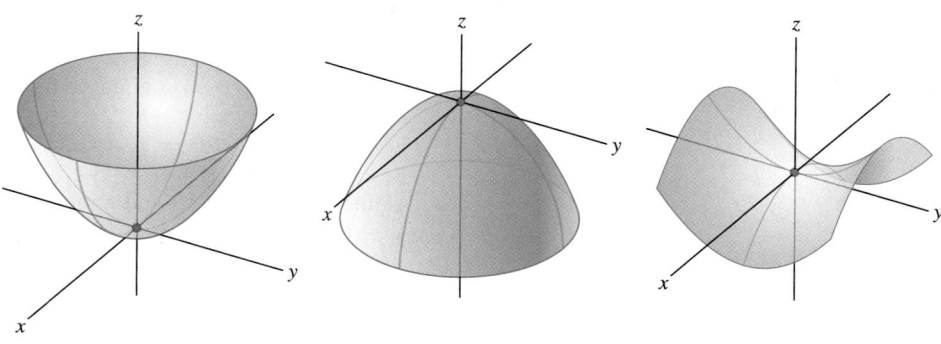

FIGURE 13 Graphs of $Q(x, y)$ with $D > 0$ (local minimum) and $D < 0$ (local maximum).

(A) $D > 0, a > 0$
local minimum

(B) $D > 0, a < 0$
local maximum

(C) $D < 0$, neither local minimum nor local maximum

THEOREM 4 Let $Q(h, k) = ah^2 + 2bhk + ck^2$ be a quadratic form with discriminant $D = ac - b^2$. Then:

(i) If $D > 0$ and $a > 0$, then $Q(h, k) > 0$ for $(h, k) \neq (0, 0)$.
(ii) If $D > 0$ and $a < 0$, then $Q(h, k) < 0$ for $(h, k) \neq (0, 0)$.
(iii) If $D < 0$, then $Q(h, k)$ takes on both positive and negative values.

Proof Assume first that $a \neq 0$. Then we may rewrite $Q(h, k)$ by "completing the square":

$$Q(h, k) = ah^2 + 2bhk + ck^2 = a\left(h + \frac{b}{a}k\right)^2 + \left(c - \frac{b^2}{a}\right)k^2$$

$$= a\left(h + \frac{b}{a}k\right)^2 + \frac{D}{a}k^2 \qquad \boxed{1}$$

If $D > 0$ and $a > 0$, then $D/a > 0$ and both terms in (1) are positive. Hence, $Q(h, k) \geq 0$. Furthermore, if $Q(h, k) = 0$, then $k = 0$ and $h + \frac{b}{a}k = 0$, in which case $h = 0$. Thus, $Q(h, k) > 0$ if $(h, k) \neq 0$. This proves (i) and (ii) follows similarly. To prove (iii), note that if $a \neq 0$ and $D < 0$, then coefficients of the squared terms in Eq. (1) have opposite signs and $Q(h, k)$ takes on both positive and negative values. If $a = 0$ and $D < 0$, then $Q(h, k) = 2bhk + ck^2$ with $b \neq 0$, and $Q(h, k)$ again takes on both positive and negative values. ∎

The Second Derivative Test is based on the second-order approximation of a function $f(x, y)$ of two variables (see the Internet supplement):

$$f(a + h, b + k) \approx f(a, b) + f_x(a, b)h + f_y(a, b)k$$

$$+ \frac{1}{2}\left(f_{xx}(a, b)h^2 + 2f_{xy}(a, b)hk + f_{yy}(a, b)k^2\right) + E_2(h, k) \qquad \boxed{2}$$

where $E_2(h, k)$ is the error. If (a, b) is a critical point, then the linear term is zero since $f_x(a, b) = f_y(a, b) = 0$. The behavior of f near (a, b) is then determined by the quadratic term $Q(h, k)$. That is, for $|h|$ and $|k|$ small we have

$$f(a + h, b + k)) \approx f(a, b) + \frac{1}{2} \overbrace{\left(f_{xx}(a, b)h^2 + 2f_{xy}(a, b)hk + f_{yy}(a, b)k^2 \right)}^{Q(h, k)}$$

The discriminant of $Q(h, k)$ is equal to the discriminant of f at (a, b):

$$D = f_{xx}(a, b)f_{yy}(a, b) - f_{xy}(a, b)^2$$

Now assume $(h, k) \neq (0, 0)$ and apply Theorem 4. If $D > 0$ and $f_{xx}(a, b) > 0$, then $Q(h, k) > 0$ and we see that $f(a + h, b + k) > f(a, b)$ when $|h|$ and $|k|$ are small. Thus, $f(a, b)$ is a local minimum. Similarly, $f(a, b)$ is a local maximum if $D > 0$ and $f_{xx}(a, b) < 0$. If $D < 0$, then $Q(h, k)$ takes on both positive and negative values, in which case (a, b) is a saddle point. To complete this argument, we would have to prove the second-order approximation in the multivariable setting and verify that the error is sufficiently small to ignore.

14.7 SUMMARY

- We say that $P = (a, b)$ is a *critical point* of $f(x, y)$ if either

$$f_x(a, b) = f_y(a, b) = 0$$

or at least one of the two partial derivatives $f_x(a, b)$, $f_y(a, b)$ does not exist. In n-variables, $P = (a_1, \dots, a_n)$ is a critical point of $f(x_1, \dots, x_n)$ if $f_{x_j}(a_1, \dots, a_n) = 0$ for $j = 1, \dots, n$, or if $f_{x_j}(a_1, \dots, a_n)$ does not exist for some j.
- If $f(P)$ is a local minimum or maximum value of f, then P is a critical point.
- The *discriminant* of $f(x, y)$ at $P = (a, b)$ is the quantity

$$D(a, b) = f_{xx}(a, b)f_{yy}(a, b) - f_{xy}(a, b)^2$$

- *Second Derivative Test:* If (a, b) is a critical point of $f(x, y)$ and the second derivatives f_{xx}, f_{yy}, f_{xy} are continuous near P, then

$$D(a, b) > 0 \quad \Rightarrow \quad f(a, b) \text{ is a local minimum or maximum}$$

$$D(a, b) < 0 \quad \Rightarrow \quad (a, b) \text{ is a saddle point}$$

$$D(a, b) = 0 \quad \Rightarrow \quad \text{test inconclusive}$$

Furthermore, if $D(a, b) > 0$, then $f(a, b)$ is a local minimum if $f_{xx}(a, b) > 0$, and it is a local maximum if $f_{xx}(a, b) < 0$.
- Let $\mathcal{D}$ be a domain. A point P in $\mathcal{D}$ is an *interior* point if $\mathcal{D}$ contains some open disk $D(P, r)$ centered at P. A point P is a *boundary point* of $\mathcal{D}$ if every open disk $D(P, r)$ contains points in $\mathcal{D}$ and points not in $\mathcal{D}$. The *interior* of $\mathcal{D}$ is the set of all interior points and the *boundary* is the set of all boundary points. A domain is *closed* if it contains all its boundary points and *open* if it is equal to its interior.
- *Existence of Global Extrema:* A continuous function f takes on both a minimum and a maximum value on any closed, bounded domain $\mathcal{D}$.

• *Location of Extreme Values:* The extreme values of f on a closed, bounded domain $\mathcal{D}$ occur either at critical points in the interior of $\mathcal{D}$ or at points on the boundary of $\mathcal{D}$. To determine the extreme values, first find the critical points in the interior of $\mathcal{D}$. Then compare the values of f at the critical points with the minimum and maximum values of f on the boundary.

14.7 EXERCISES

Preliminary Questions

1. The functions $f(x, y) = x^2 + y^2$ and $g(x, y) = x^2 - y^2$ both have a critical point at $(0, 0)$. How is the behavior of the two functions at the critical point different?

2. Identify the points indicated in the contour maps as local minima, maxima, saddle points, or neither (Figure 14).

  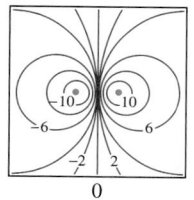

0   0   0

FIGURE 14

3. Let $f(x, y)$ be a continuous function on a domain $\mathcal{D}$ in $\mathbf{R}^2$. Determine which of the following statements are true:

(a) If $\mathcal{D}$ is closed and bounded, then f takes on a maximum value on $\mathcal{D}$.

(b) If $\mathcal{D}$ is neither closed nor bounded, then f does not take on a maximum value of $\mathcal{D}$.

(c) $f(x, y)$ need not have a maximum value on $\mathcal{D} = \{(x, y) : 0 \le x, y \le 1\}$.

(d) A continuous function takes on neither a minimum nor a maximum value on the open quadrant $\{(x, y) : x > 0, y > 0\}$.

Exercises

1. Let $P = (a, b)$ be a critical point of $f(x, y) = x^2 + y^4 - 4xy$.

(a) First use $f_x(, x, y) = 0$ to show that $a = 2b$. Then use $f_y(x, y) = 0$ to show that $P = (0, 0)$, $(2\sqrt{2}, \sqrt{2})$, or $(-2\sqrt{2}, -\sqrt{2})$.

(b) Referring to Figure 15, determine the local minima and saddle points of $f(x, y)$ and find the absolute minimum value of $f(x, y)$.

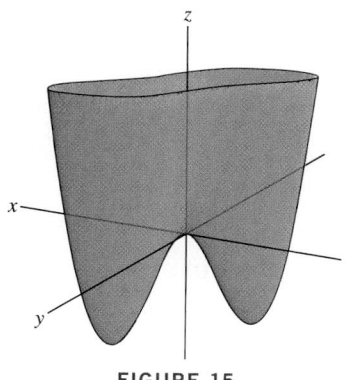

FIGURE 15

2. Find the critical points of the functions

$$f(x, y) = x^2 + 2y^2 - 4y + 6x, \ g(x, y) = x^2 - 12xy + y$$

Use the Second Derivative Test to determine the local minimum, maximum, and saddle points. Match $f(x, y)$ and $g(x, y)$ with their graphs in Figure 16.

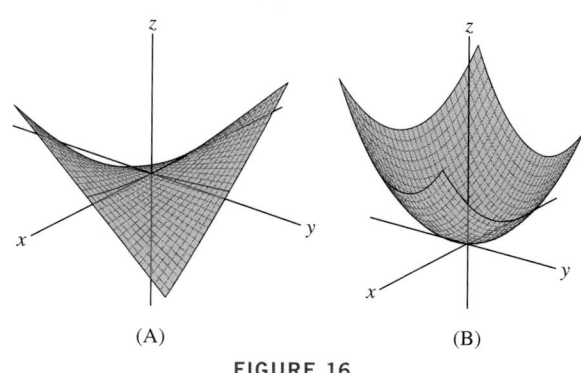

(A)   (B)

FIGURE 16

3. Find the critical points of

$$f(x, y) = 8y^4 + x^2 + xy - 3y^2 - y^3$$

Use the contour map in Figure 17 to determine their nature (minimum, maximum, saddle point).

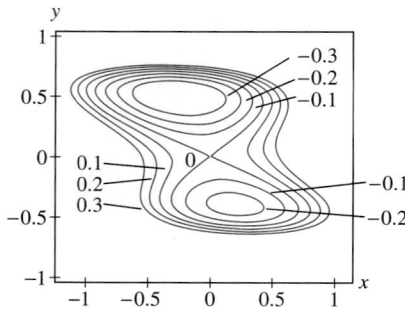

FIGURE 17 Contour map of $f(x, y) = 8y^4 + x^2 + xy - 3y^2 - y^3$.

4. Let $f(x, y) = y^2x - yx^2 + xy$.

(a) Show that the critical points (x, y) satisfy the equations

$$y(y - 2x + 1) = 0, \qquad x(2y - x + 1) = 0$$

(b) Show that there are three critical points with either $x = 0$ or $y = 0$ (or both) and one critical point with $x, y \neq 0$.

(c) Use the second derivative to determine the nature of the critical points.

In Exercises 5–20, find the critical points of the function. Then use the Second Derivative Test to determine whether they are local minima or maxima (or state that the test fails).

5. $f(x, y) = x^2 + y^2 - xy + x$

6. $f(x, y) = x^3 - xy + y^3$

7. $f(x, y) = x^3y + 12x^2 - 8y$

8. $f(x, y) = x^3 + 2xy - 2y^2 - 10x$

9. $f(x, y) = 4x - 3x^3 - 2xy^2$

10. $f(x, y) = x^3 + y^4 - 6x - 2y^2$

11. $f(x, y) = x^4 + y^4 - 4xy$

12. $f(x, y) = e^{x^2 - y^2 + 4y}$

13. $f(x, y) = xye^{-x^2 - y^2}$

14. $f(x, y) = x \ln(x + y)$

15. $f(x, y) = e^x - xe^y$

16. $f(x, y) = (x + 3y)e^{y - x^2}$

17. $f(x, y) = \ln x + 2 \ln y - x - 4y$

18. $f(x, y) = (x^2 + y^2)e^{-x^2 - y^2}$

19. $f(x, y) = x - y^2 - \ln(x + y)$

20. $f(x, y) = (x - y)e^{x^2 - y^2}$

21. Show that $f(x, y) = \sqrt{x^2 + y^2}$ has one critical point P and that f is nondifferentiable at P. Show that $f(P)$ is an absolute minimum value.

22. Let $f(x, y) = (x^2 + y^2 - 1)^2$.

(a) Find the critical points of $f(x, y)$.

(b) $\boxed{CAS}$ Use a computer algebra system to graph $f(x, y)$ and identify the critical points on the graph.

(c) Find all points where f takes on a minimum value. Does the Second Derivative Test apply?

23. $\boxed{CAS}$ Use a computer algebra system to find numerical approximations to the critical points of

$$f(x, y) = (1 - x + x^2)e^{y^2} + (1 - y + y^2)e^{x^2}$$

Use Figure 18 to determine whether they correspond to local minima or maxima.

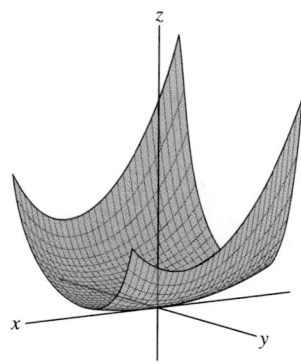

FIGURE 18 Plot of the function
$f(x, y) = (1 - x + x^2)e^{y^2} + (1 - y + y^2)e^{x^2}$.

24. Use the contour map in Figure 19 to determine whether the critical points A, B, C, D are local minima, maxima, or saddle points.

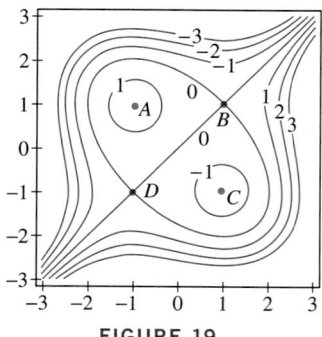

FIGURE 19

25. Which of the following domains are closed and which are bounded?

(a) $\{(x, y) \in \mathbf{R}^2 : x^2 + y^2 \leq 1\}$

(b) $\{(x, y) \in \mathbf{R}^2 : x^2 + y^2 < 1\}$

(c) $\{(x, y) \in \mathbf{R}^2 : x \geq 0\}$

(d) $\{(x, y) \in \mathbf{R}^2 : x > 0, y > 0\}$

(e) $\{(x, y) \in \mathbf{R}^2 : 1 \le x \le 4, 5 \le y \le 10\}$

(f) $\{(x, y) \in \mathbf{R}^2 : x > 0, x^2 + y^2 \le 10\}$

In Exercises 26–29, determine the global extreme values of the function on the given set without using calculus.

26. $f(x, y) = x + y$, $\quad 0 \le x, y \le 1$

27. $f(x, y) = 2x - y$, $\quad 0 \le x \le 1, 0 \le y \le 3$

28. $f(x, y) = (x^2 + y^2 + 1)^{-1}$, $\quad 0 \le x \le 3, 0 \le y \le 5$

29. $f(x, y) = e^{-x^2-y^2}$, $\quad x^2 + y^2 \le 1$

30. Assumptions Matter Show that $f(x, y) = x + y$ has no global minimum or maximum on the domain $0 < x, y < 1$. Does this contradict Theorem 3?

31. Let $\mathcal{D} = \{(x, y) : x > 0, y > 0\}$. Show that $\mathcal{D}$ is not closed. Find a continuous function that does not have a global minimum value on $\mathcal{D}$.

32. The goal is to find the extreme values of $f(x, y) = x^2 - 2xy + 2y$ on the square $0 \le x, y \le 2$ (Figure 20).

(a) Evaluate f at the critical points that lie in the square.

(b) On the bottom edge of the square, $y = 0$ and $f(x, 0) = x^2$. Find the extreme values of f on the bottom edge.

(c) Find the extreme values of f on the remaining edges of the square.

(d) Find the largest and smallest values of f among the values computed in (a), (b), and (c).

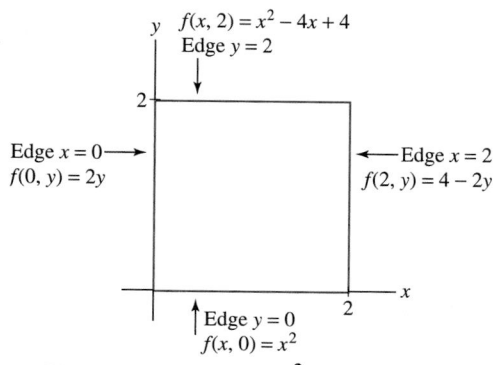

FIGURE 20 The function $f(x, y) = x^2 - 2xy + 2y$ on the four boundary segments of the square $0 \le x \le 2$.

33. Let $f(x, y) = (x + y) \ln(x^2 + y^2)$, defined for $(x, y) \ne (0, 0)$.

(a) Show that if (x, y) is a critical point of $f(x, y)$, then either $x = y$ or $x = -y$. Determine the critical points (four in all).

(b) CAS Plot a graph of $f(x, y)$ and use it to determine whether these critical points are maxima, minima, or neither.

34. Let $f(x, y) = x^4 - 4xy + 2y^2$.

(a) Find the critical points of $f(x, y)$.

(b) CAS Use a plot to determine whether the critical points are maxima, minima, or saddle points. Then confirm this result using the Second Derivative Test.

In Exercises 35–41, determine the global extreme values of the function on the given domain.

35. $f(x, y) = x^3 - 2y$, $\quad 0 \le x, y \le 1$

36. $f(x, y) = 4x + 3y$, $\quad 0 \le x, y \le 1$

37. $f(x, y) = x^2 + 2y^2$, $\quad 0 \le x, y \le 1$

38. $f(x, y) = (4y^2 - x^2)e^{-x^2-y^2}$, $\quad x^2 + y^2 \le 2$

39. $f(x, y) = x^3 + x^2y + 2y^2$, $\quad x, y \ge 0, x + y \le 1$

40. $f(x, y) = x^3 + y^3 - 3xy$, $\quad 0 \le x, y \le 1$

41. CAS $f(x, y) = x^3 y^5$, the set bounded by $x = 0$, $y = 0$, and $y = 1 - \sqrt{x}$. *Hint:* Use a computer algebra system to find the minimum along the boundary curve $y = 1 - \sqrt{x}$, which is parametrized by $(t, 1 - \sqrt{t})$ for $0 \le t \le 1$.

42. Show that the rectangular box (including a top and bottom) with fixed volume V with the smallest possible surface area is a cube (Figure 21).

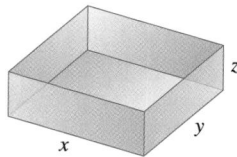

FIGURE 21 Rectangular box with sides x, y, z.

43. Consider a rectangular box B with a bottom and sides but no top such that B has minimal surface area among all boxes with fixed volume V.

(a) Do you think B is a cube as in the solution to Exercise 42? If not, how would you make its shape differ from a cube?

(b) Find the dimensions of B and compare with your response to (a).

Further Insights and Challenges

44. Given n data points $(x_1, y_1), \ldots, (x_n, y_n)$, we may seek a linear function $f(x) = mx + b$ that best fits the data. The **linear least-squares fit** is the linear function $f(x) = mx + b$ that minimizes the

sum of the squares (Figure 22)

$$E(m, b) = \sum_{j=1}^{n} (y_j - f(x_j))^2$$

Show that E is minimized for m and b satisfying

$$m \sum_{j=1}^{n} x_j + bn = \sum_{j=1}^{n} y_j$$

$$m \sum_{j=1}^{n} x_j^2 + b \sum_{j=1}^{n} x_j = \sum_{j=1}^{n} x_j y_j$$

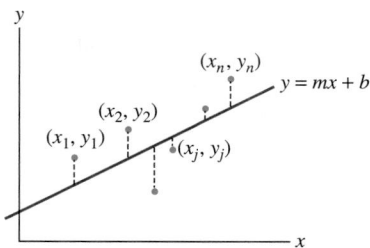

FIGURE 22 The linear least-squares fit minimizes the sum of the squares of the vertical distances from the data points to the line.

45. The power (in microwatts) of a laser is measured as a function of current (in milliamps). Find the linear least-squares fit (Exercise 44) for the data points.

Current (mA)	1.0	1.1	1.2	1.3	1.4	1.5
Laser power (μW)	0.52	0.56	0.82	0.78	1.23	1.50

46. Prove that if $\alpha, \beta \geq 1$ are numbers such that $\alpha^{-1} + \beta^{-1} = 1$, then the following inequality holds for all $x, y \geq 0$:

$$\frac{1}{\alpha} x^\alpha + \frac{1}{\beta} x^\beta \geq xy$$

For example,

$$\frac{1}{3} x^3 + \frac{2}{3} y^{3/2} \geq xy, \qquad \frac{2}{5} x^{5/2} + \frac{3}{5} y^{5/3} \geq xy, \quad \text{etc.}$$

Hint: Find the critical points of $f(x, y) = \alpha^{-1} x^\alpha + \beta^{-1} x^\beta - xy$.

47. The following problem was proposed by Fermat as a challenge to the Italian scientist Evangelista Torricelli (1608–1647), a student of Galileo and inventor of the barometer. Given three points $A = (a_1, a_2)$, $B = (b_1, b_2)$, and $C = (c_1, c_2)$ in the plane, find the point $P = (x, y)$ that minimizes the sum of the distances

$$f(x, y) = AP + BP + CP$$

(a) Write out $f(x, y)$ as a function of x and y, and show that $f(x, y)$ is differentiable except at the points A, B, C.

(b) Define the unit vectors

$$\mathbf{e} = \frac{\overrightarrow{AP}}{\|\overrightarrow{AP}\|}, \qquad \mathbf{f} = \frac{\overrightarrow{BP}}{\|\overrightarrow{BP}\|}, \qquad \mathbf{g} = \frac{\overrightarrow{CP}}{\|\overrightarrow{CP}\|}$$

Show that the condition $\nabla f = 0$ is equivalent to

$$\mathbf{e} + \mathbf{f} + \mathbf{g} = 0 \qquad \boxed{3}$$

Prove that Eq. (3) holds if and only if the mutual angles between the unit vectors are all 120°.

(c) Define the Fermat point to be the point P such that angles between the segments $\overrightarrow{AP}$, $\overrightarrow{BP}$, $\overrightarrow{CP}$ are all 120°. Conclude that the **Fermat point** solves the minimization problem (Figure 23).

(d) Show that the Fermat point does not exist if one of the angles in $\triangle ABC$ is $> 120°$. Where does the minimum occur in this case? *Hint:* The minimum must occur at a point where $f(x, y)$ is not differentiable.

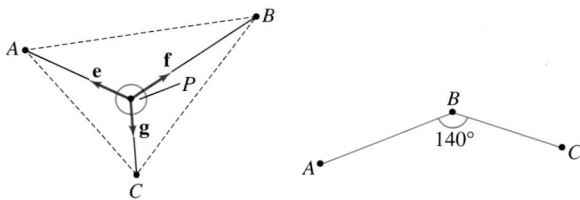

(A) P is the Fermat point (the angles between $\mathbf{e}$, $\mathbf{f}$, and $\mathbf{g}$ are all 120°).

(B) Fermat point does not exist.

FIGURE 23

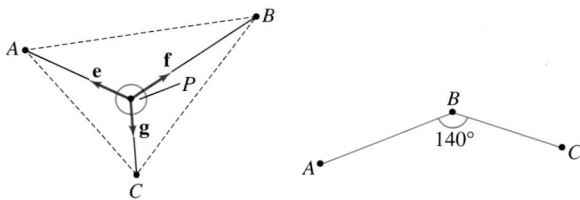

FIGURE 1 Optimization with a constraint: find the minimum of $f(x, y) = \sqrt{x^2 + y^2}$ on the line $x + y = 4$.

14.8 Lagrange Multipliers: Optimizing with a Constraint

Some optimization problems involve finding the minimum or maximum of a function $f(x, y)$ subject to a constraint $g(x, y) = 0$ (Figure 1). An example is the problem of minimizing

$$f(x, y) = \sqrt{x^2 + y^2} \quad \text{subject to} \quad g(x, y) = x + y - 4 = 0$$

Since $f(x, y)$ is the distance from (x, y) to the origin, this problem asks us to find the point $P = (x, y)$ on the line $x + y = 4$ closest to the origin. The method of **Lagrange**

multipliers is a general procedure for finding extreme values of $f(x, y)$ on a constraint curve $g(x, y) = 0$. The main idea is as follows.

GRAPHICAL INSIGHT   Suppose we are standing at point Q in Figure 2(A), and we wish to move along the constraint curve in a direction that increases the value of f. The gradient vector ∇f_Q points in the direction of maximal increase of f (the northeast direction). Although moving in the gradient direction would take us off the constraint curve, we can still increase f by moving to the right along the constraint curve.

We keep moving to the right until we arrive at the point P, where ∇f_P is orthogonal to the constraint curve [Figure 2(B)]. Once at P, we cannot increase f further by moving to the right or left along the constraint curve, and thus, $f(P)$ is a local maximum (subject to the constraint). Now observe that ∇f_P and ∇g_P are both normal to the constraint curve, and hence they must point in the same or opposite directions. In other words, $\nabla f_P = \lambda \nabla g_P$ for some scalar λ (called a **Lagrange multiplier**). We state this conclusion formally in Theorem 1 below. Note also that level curve of f through P and the constraint curve are tangent to each other at P.

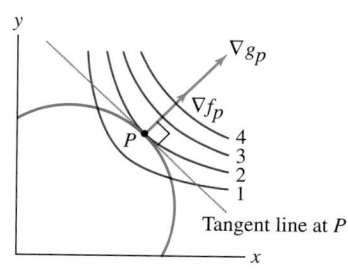

(A) The gradient vector ∇f_Q shows that f increases as we move to the right along the constraint curve.

(B) The local maximum of f on the constraint curve $g(x, y) = 0$ occurs at a point P where ∇f_P and ∇g_P point in the same direction.

FIGURE 2

THEOREM 1 Lagrange Multipliers   Assume that $f(x, y)$ and $g(x, y)$ are differentiable functions. If $f(x, y)$ has a local minimum or maximum on the constraint curve $g(x, y) = 0$ at $P = (a, b)$, and if $\nabla g_P \neq \mathbf{0}$, then there is a scalar λ such that

$$\nabla f_P = \lambda \nabla g_P \qquad \boxed{1}$$

In Theorem 1, the assumption $\nabla g_P \neq \mathbf{0}$ guarantees (by the Implicit Function Theorem of advanced calculus) that we can parametrize the curve $g(x, y) = 0$ near P by a path $\mathbf{c}$ such that $\mathbf{c}(0) = P$ and $\mathbf{c}'(0) \neq \mathbf{0}$.

Proof   Let $\mathbf{c}(t)$ be a parametrization of the constraint curve $g(x, y) = 0$ near P such that $\mathbf{c}(0) = P$ and $\mathbf{c}'(0) \neq \mathbf{0}$. Then $f(\mathbf{c}(0)) = f(P)$, and $f(\mathbf{c}(t))$ has a local minimum or maximum at $t = 0$. Thus, $t = 0$ is a critical point of $f(\mathbf{c}(t))$ and

$$\underbrace{\frac{d}{dt} f(\mathbf{c}(t)) \Big|_{t=0} = \nabla f_P \cdot \mathbf{c}'(0) = 0}_{\text{Chain Rule}}$$

This shows that ∇f_P is orthogonal to the tangent vector $\mathbf{c}'(0)$ to the curve $g(x, y) = 0$. The gradient ∇g_P is also orthogonal to $\mathbf{c}'(0)$ since ∇g_P is orthogonal to the level curve $g(x, y) = 0$ at P. We conclude that ∇f_P and ∇g_P are proportional as claimed. ∎

We refer to Eq. (1) as the **Lagrange condition**. When we write this condition in terms of components, we obtain the **Lagrange equations**:

$$f_x(a, b) = \lambda g_x(a, b)$$
$$f_y(a, b) = \lambda g_y(a, b)$$

The point $P = (a, b)$ is called a critical point for the optimization problem with constraint and $f(a, b)$ is called a critical value.

■ **EXAMPLE 1** Find the extreme values of the function $f(x, y) = x + 2y$ on the elliptic cylinder $3x^2 + 4y^2 = 3$.

Solution

Step 1. Write out the Lagrange equations.
The constraint curve is $g(x, y) = 0$, where $g(x, y) = 3x^2 + 4y^2 - 3$. The gradients are $\nabla f = \langle 1, 2 \rangle$ and $\nabla g = \langle 6x, 8y \rangle$, so the Lagrange condition gives us

$$\langle 1, 2 \rangle = \lambda \langle 6x, 8y \rangle \quad \Rightarrow \quad 1 = \lambda(6x), \qquad 2 = \lambda(8y) \qquad \boxed{2}$$

Step 2. Solve for λ in terms of x and y.
We use (2) to obtain two equations for λ:

$$\lambda = \frac{1}{6x}, \qquad \lambda = \frac{2}{8y} \qquad \boxed{3}$$

To justify dividing by x and y, we observe that x and y must be nonzero since $x = 0$ or $y = 0$ would violate (2).

Step 3. Solve for x and y using the constraint.
The two expressions for λ must be equal, so we obtain $\dfrac{1}{6x} = \dfrac{2}{8y}$ or $y = \frac{3}{2}x$. Now substitute $y = \frac{3}{2}x$ in the constraint equation:

$$g(x, y) = 3x^2 + 4y^2 - 3 = 3x^2 + 4\left(\frac{3}{2}x\right)^2 - 3 = 12x^2 - 3 = 0$$

Thus, $x = \frac{1}{2}$ or $-\frac{1}{2}$ and $y = \frac{3}{2}x = \frac{3}{4}$ or $-\frac{3}{4}$.

Step 4. Calculate the critical values.
Now evaluate $f(x, y)$ at the critical points $P = (\frac{1}{2}, \frac{3}{4})$ and $Q = (-\frac{1}{2}, -\frac{3}{4})$:

$$f(P) = \frac{1}{2} + 2\left(\frac{3}{4}\right) = 2, \qquad f(Q) = -\frac{1}{2} + 2\left(-\frac{3}{4}\right) = -2$$

We conclude that the maximum of $f(x, y)$ on the ellipse is 2 and the minimum is -2 (Figure 3). ■

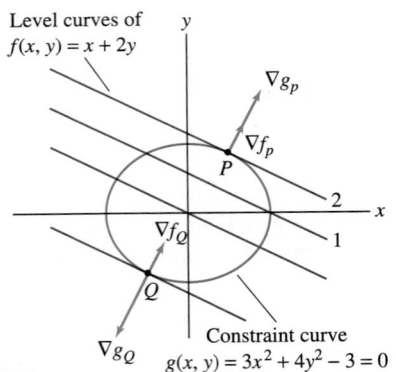

FIGURE 3 The minimum and maximum occur where the level curve of $f(x, y) = x + 2y$ is tangent to the constraint curve $3x^2 + 4y^2 = 3$.

GRAPHICAL INSIGHT In an ordinary optimization problem without a constraint, the maximum value of $f(x, y)$ corresponds to the highest point on the surface $z = f(x, y)$. In the case of optimization subject to a constraint $g(x, y) = 0$, consider the cylinder $\mathcal{S}$ whose base is the constraint curve, that is, the cylinder consisting of all points (x, y, z) such that $g(x, y) = 0$ (Figure 3). The maximum subject to the constraint corresponds to the highest point on the intersection of the surface $z = f(x, y)$ and the cylinder $\mathcal{S}$.

FIGURE 4 Economist Paul Douglas and mathematician Charles Cobb arrived at the production functions $P(x, y) = Cx^a y^b$ in 1927 by fitting data gathered on the relationship between labor, capital, and the output in an industrial economy. The photo shows Douglas, who was a professor at the University of Chicago and also served as U.S. Senator from Illinois from 1949–1967.

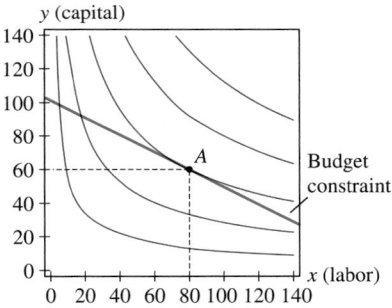

FIGURE 5 Contour plot of the Cobb–Douglas production function $P(x, y) = 50x^{0.4}y^{0.6}$.

■ **EXAMPLE 2** Cobb–Douglas Production Function By investing x units of labor and y units of capital, a low-end watch manufacturer can produce $P(x, y) = 50x^{0.4}y^{0.6}$ watches. Find the maximum number of watches that can be produced on a budget of $20,000 if labor costs $100 per unit and capital costs $200 per unit.

Solution The total cost of x units of labor and y units of capital is $100x + 200y$, so the budget constraint is $100x + 200y = 20,000$. Thus, our task is to maximize the function $P(x, y) = 50x^{0.4}y^{0.6}$ subject to the constraint (Figure 5)

$$g(x, y) = 100x + 200y - 20,000 = 0 \qquad \boxed{4}$$

Step 1. Write out the Lagrange equations.
The Lagrange equations are

$$P_x(x, y) = \lambda g_x(x, y): \quad 20x^{-0.6}y^{0.6} = 100\lambda$$

$$P_y(x, y) = \lambda g_y(x, y): \quad 30x^{0.4}y^{-0.4} = 200\lambda$$

Step 2. Solve for λ in terms of x and y.
These equations yield two expressions for λ that must be equal:

$$\lambda = \frac{1}{5}x^{-0.6}y^{0.6} = \frac{3}{20}x^{0.4}y^{-0.4} \qquad \boxed{5}$$

Step 3. Solve for x and y using the constraint.
Multiply Eq. (5) by $5x^{0.6}y^{0.4}$ to obtain $y = \frac{3}{4}x$, and substitute in Eq. (4):

$$100x + 200y = 100x + 200\left(\frac{3}{4}x\right) = 20,000 \quad \Rightarrow \quad 250x = 20,000$$

We obtain $x = 20,000/250 = 80$ and $y = \frac{3}{4}x = 60$. Therefore, the critical point is $A = (80, 60)$.

Step 4. Calculate the Critical Values.
Since $P(x, y)$ is increasing as a function of x and y, ∇P points to the northeast and it is clear that $P(x, y)$ takes on a maximum value at A (Figure 5). The maximum is $P(80, 60) = 50(80)^{0.6}(60)^{0.4} = 3,365.87$, or roughly 3,365 watches, with a cost per watch of $20,000/3,365$ or $5.94. ■

The method of Lagrange multipliers is valid in any number of variables. The next example illustrates the method for three variables.

■ **EXAMPLE 3** Lagrange Multipliers in Three Variables Find the point on the plane $\frac{x}{2} + \frac{y}{4} + \frac{z}{4} = 1$ closest to the origin in $\mathbf{R}^3$.

Solution Our task is to minimize the distance $d = \sqrt{x^2 + y^2 + z^2}$ subject to the constraint $\frac{x}{2} + \frac{y}{4} + \frac{z}{4} = 1$. When the distance d is at a minimum, the *square* d^2 is also at a minimum (since the function d^2 is increasing). Thus, it suffices to find the minimum of the square of the distance $f(x, y, z) = x^2 + y^2 + z^2$.

Step 1. Write out the Lagrange equations.

The Lagrange condition is

$$\underbrace{\langle 2x, 2y, 2z \rangle}_{\nabla f} = \lambda \underbrace{\left\langle \frac{1}{2}, \frac{1}{4}, \frac{1}{4} \right\rangle}_{\nabla g}$$

Step 2. Solve for λ in terms of x, y, and z.

We obtain

$$\lambda = 4x = 8y = 8z \quad \Rightarrow \quad z = y = \frac{x}{2}$$

Step 3. Solve for x, y, and z using the constraint.

Substitute in the constraint equation:

$$\frac{x}{2} + \frac{y}{4} + \frac{z}{4} = \frac{2z}{2} + \frac{z}{4} + \frac{z}{4} = \frac{3z}{2} = 1 \quad \Rightarrow \quad z = \frac{2}{3}$$

We obtain $x = 2z = \frac{4}{3}$ and $y = z = \frac{2}{3}$. It is clear geometrically that the critical point corresponds to a minimum of f. Hence, the point on the plane closest to the origin is $P = (\frac{4}{3}, \frac{2}{3}, \frac{2}{3})$ (Figure 6). ∎

FIGURE 6 Point P closest to the origin on the plane.

The method of Lagrange multipliers applies when there is more than one constraint equation, but we must add another multiplier for each constraint. For example, if the problem is to minimize $f(x, y, z)$ subject to constraints $g(x, y, z) = 0$ and $h(x, y, z) = 0$, the Lagrange condition is

$$\nabla f = \lambda \nabla g + \mu \nabla h$$

■ **EXAMPLE 4** Lagrange Multipliers with Multiple Constraints Find the minimum of $f(x, y, z) = x^2 y^2 + z^2$ subject to $x + y = 2$ and $y + z = 4$.

Solution The constraint equations are

$$g(x, y) = x + y - 2 = 0, \qquad h(x, y) = y + z - 4 = 0$$

Step 1. Write out the Lagrange equations.

We have $\nabla f = \langle 2xy^2, 2x^2 y, 2z \rangle$, so the Lagrange condition is

$$\underbrace{\langle 2xy^2, 2x^2 y, 2z \rangle}_{\nabla f} = \lambda \underbrace{\langle 1, 1, 0 \rangle}_{\nabla g} + \mu \underbrace{\langle 0, 1, 1 \rangle}_{\nabla h} = \langle \lambda, \lambda + \mu, \mu \rangle$$

Step 2. Solve for λ and μ.

We obtain $\lambda = 2xy^2$, $\lambda + \mu = 2x^2 y$, and $\mu = 2z$. Using the first and third equation in the middle equation, we obtain

$$2xy^2 + 2z = 2x^2 y \quad \Rightarrow \quad z = x^2 y - xy^2 \qquad \boxed{6}$$

Step 3. Solve for x, y, and z using the constraints.

First, use the constraint equations to express y and z in terms of x:

$$y = 2 - x \quad \text{and} \quad z = 4 - y = 4 - (2 - x) = x + 2$$

Then substitute in Eq. (6) to obtain an equation for x:

$$x + 2 = x^2(2 - x) - x(2 - x)^2 \quad \Rightarrow \quad 2x^3 - 6x^2 + 5x + 2 = 0$$

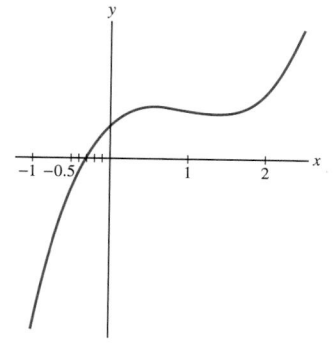

FIGURE 7 Plot of
$F(x) = 2x^3 - 6x^2 + 5x + 2.$

The plot of the function $F(x) = 2x^3 - 6x^2 + 5x + 2$ in Figure 7 shows that $F(x) = 0$ has a unique real root $x \approx -0.3$. We then have

$$y = 2 - x \approx 2 - (-0.3) = 2.3, \qquad z = x + 2 \approx -0.3 + 2 = 1.7$$

The unique critical point is $P \approx (-0.3, 2.3, 1.7)$.

Step 4. Calculate the critical values.

The value of f at the unique critical point is approximately $f(-0.3, 2.3, 1.7) \approx 3.4$. To show that this is a minimum value, observe that $y = 2 - x$ and $z = x + 2$ and $f(x, y, z) = x^2(2 - x)^2 + (x + 2)^2$. Thus $f(x, y, z)$ tends to ∞ as x grows large and the critical point must yield a minimum. In fact, we could have found this minimum by optimizing $g(x) = x^2(2 - x)^2 + (x + 2)^2$, but our purpose was to illustrate the method of Lagrange multipliers. ■

Assumptions Matter Constrained optimization problems do not always have solutions. For example, $f(x, y) = x$ has no maximum or minimum subject to the constraint $g(x, y) = x - y = 0$ because $P = (a, a)$ satisfies the constraint and $f(a, a) = a$ can be arbitrarily large or small. However, a minimum and maximum are guaranteed to exist if the constraint curve $g(x, y) = 0$ is *bounded*. For example, extreme values exist if the constraint curve is an ellipse. This follows from Theorem 3 in Section 14.7, which states that a continuous function on a closed, bounded domain takes on extreme values.

14.8 SUMMARY

- According to the method of *Lagrange multipliers*, the extreme values of $f(x, y)$ subject to a constraint $g(x, y) = 0$ occur at points P (called critical points) satisfying the Lagrange condition $\nabla f_P = \lambda \nabla g_P$. This condition is equivalent to the *Lagrange equations*

$$f_x(x, y) = \lambda g_x(x, y), \qquad f_y(x, y) = \lambda g_y(x, y)$$

- Theorem 1 does not guarantee that a minimum or maximum value exists, but if it does exist, it must occur at a critical point.
- If the constraint curve $g(x, y) = 0$ is bounded [e.g., if $g(x, y) = 0$ is a circle or ellipse], then minimum and maximum values exist and occur at a critical point.
- The Lagrange condition for the extreme values of a function of three variables $f(x, y, z)$ subject to two constraints $g(x, y, z) = 0$ and $h(x, y, z) = 0$ is

$$\nabla f = \lambda \nabla g + \mu \nabla h$$

14.8 EXERCISES

Preliminary Questions

1. Suppose that the maximum of $f(x, y)$ subject to the constraint $g(x, y) = 0$ occurs at a point $P = (a, b)$ such that $\nabla f_P \neq 0$. Which of the following are true?

(a) ∇f_P is tangent to $g(x, y) = 0$ at P.

(b) ∇f_P is orthogonal to $g(x, y) = 0$ at P.

2. Figure 8 shows a constraint $g(x, y) = 0$ and the level curves of a function f. In each case, determine whether f has a local minimum, local maximum, or neither at the labeled point.

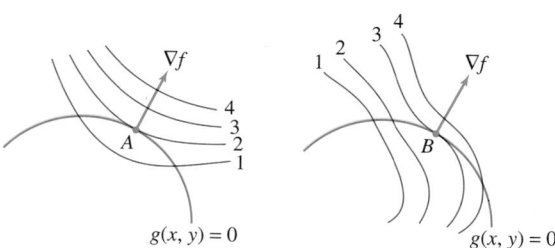

FIGURE 8

3. On the contour map in Figure 9:

(a) Identify the points where $\nabla f = \lambda \nabla g$ for some scalar λ.

(b) Identify the minimum and maximum values of $f(x, y)$ subject to $g(x, y) = 0$.

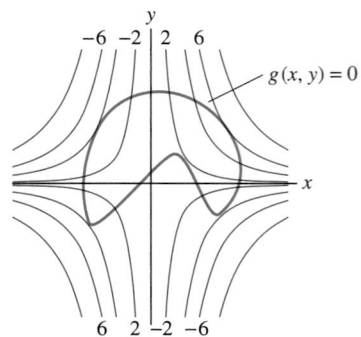

FIGURE 9 Contour map of $f(x, y)$; contour interval 2.

Exercises

1. Use Lagrange multipliers to find the extreme values of the function $f(x, y) = 2x + 4y$ subject to the constraint $g(x, y) = x^2 + y^2 - 5 = 0$.

(a) Show that the Lagrange equation $\nabla f = \lambda \nabla g$ gives $\lambda x = 1$ and $\lambda y = 2$.

(b) Show that these equations imply $\lambda \neq 0$ and $y = 2x$.

(c) Use the constraint equation to determine the possible critical points (x, y).

(d) Evaluate $f(x, y)$ at the critical points and determine the minimum and maximum values.

2. Find the extreme values of $f(x, y) = x^2 + 2y^2$ subject to the constraint $g(x, y) = 4x - 6y = 25$.

(a) Show that the Lagrange equations yield $2x = 4\lambda$, $4y = -6\lambda$.

(b) Show that if $x = 0$ or $y = 0$, then $\lambda = 0$ and the Lagrange equations give $x = y = 0$. Since $(0, 0)$ does not satisfy the constraint, you may assume that x and y are nonzero.

(c) Use the Lagrange equations to show that $y = -\frac{3}{4}x$.

(d) Substitute in the constraint equation to show that there is a unique critical point P.

(e) Does P correspond to a minimum or maximum value of f? Refer to Figure 10 to justify your answer. *Hint:* Do the values of $f(x, y)$ increase or decrease as (x, y) moves away from P along the line $g(x, y) = 0$?

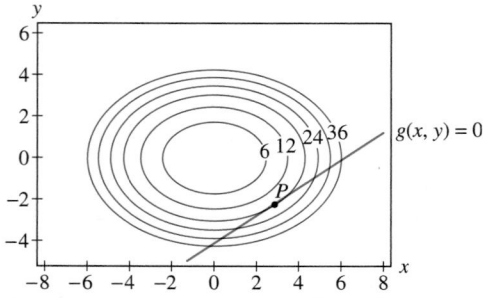

FIGURE 10 Level curves of $f(x, y) = x^2 + 2y^2$ and constraint $g(x, y) = 4x - 6y - 25 = 0$.

3. Apply the method of Lagrange multipliers to the function $f(x, y) = (x^2 + 1)y$ subject to the constraint $x^2 + y^2 = 5$. *Hint:* First show that $y \neq 0$; then treat the cases $x = 0$ and $x \neq 0$ separately.

In Exercises 4–13, find the minimum and maximum values of the function subject to the given constraint.

4. $f(x, y) = 2x + 3y, \quad x^2 + y^2 = 4$

5. $f(x, y) = x^2 + y^2, \quad 2x + 3y = 6$

6. $f(x, y) = 4x^2 + 9y^2, \quad xy = 4$

7. $f(x, y) = xy, \quad 4x^2 + 9y^2 = 32$

8. $f(x, y) = x^2 y + x + y, \quad xy = 4$

9. $f(x, y) = x^2 + y^2, \quad x^4 + y^4 = 1$

10. $f(x, y) = x^2 y^4, \quad x^2 + 2y^2 = 6$

11. $f(x, y, z) = 3x + 2y + 4z, \quad x^2 + 2y^2 + 6z^2 = 1$

12. $f(x, y, z) = x^2 - y - z, \quad x^2 - y^2 + z = 0$

13. $f(x, y, z) = xy + 3xz + 2yz, \quad 5x + 9y + z = 10$

14. [icon] Let $f(x, y) = x^3 + xy + y^3, g(x, y) = x^3 - xy + y^3$.

(a) Show that there is a unique point $P = (a, b)$ on $g(x, y) = 1$ where $\nabla f_P = \lambda \nabla g_P$ for some scalar λ.

(b) Refer to Figure 11 to determine whether $f(P)$ is a local minimum or maximum of f subject to the constraint.

(c) Does Figure 11 suggest that $f(P)$ is a global extremum subject to the constraint?

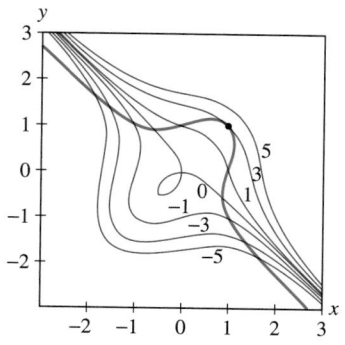

FIGURE 11 Contour map of $f(x, y) = x^3 + xy + y^3$ and graph of $g(x, y) = x^3 - xy + y^3 = 1$.

15. Use Lagrange multipliers to find the point (a, b) on the graph of $y = e^x$, where the value ab is as small as possible.

16. Find the rectangular box of maximum volume if the sum of the lengths of the edges is 300 cm.

17. The surface area of a right-circular cone of radius r and height h is $S = \pi r \sqrt{r^2 + h^2}$, and its volume is $V = \frac{1}{3}\pi r^2 h$.
(a) Determine the ratio h/r for the cone with given surface area S and maximal volume V.
(b) What is the ratio h/r for a cone with given volume V and minimal surface area S?
(c) Does a cone with given volume V and maximal surface area exist?

18. In Example 1, we found the maximum of $f(x, y) = x + 2y$ on the ellipse $3x^2 + 4y^2 = 3$. Solve this problem again without using Lagrange multipliers. First, show that the ellipse is parametrized by $x = \cos t$, $y = \frac{\sqrt{3}}{2}\sin t$. Then find the maximum value of $f\left(\cos t, \frac{\sqrt{3}}{2}\sin t\right)$ using single-variable calculus. Which method do you find easier?

19. Use Lagrange multipliers to find the maximum area of a rectangle inscribed in the ellipse (Figure 12):

$$\frac{x^2}{a^2} + \frac{y^2}{b^2} = 1$$

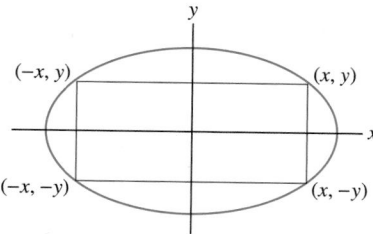

FIGURE 12 Rectangle inscribed in the ellipse $\frac{x^2}{a^2} + \frac{y^2}{b^2} = 1$.

20. Show that the point (x_0, y_0) closest to the origin on the line $ax + by = c$ has coordinates

$$x_0 = \frac{ac}{a^2 + b^2}, \qquad y_0 = \frac{bc}{a^2 + b^2}$$

21. Find the maximum value of $f(x, y) = x^a y^b$ for $x, y \geq 0$ on the unit circle, where $a, b > 0$ are constants.

22. Find the maximum value of $f(x, y) = x^a y^b$ for $x, y \geq 0$ on the line $x + y = 1$, where $a, b > 0$ are constants.

23. Find the maximum value of $f(x, y, z) = x^a y^b z^c$ for $x, y, z \geq 0$ on the unit sphere, where $a, b, c > 0$ are constants.

24. Show that the Lagrange equations for $f(x, y, z) = x^2 y + zy^2$ subject to the constraint $g(x, y) = x + yz = 4$ have no solution. What can you conclude about the minimum and maximum values of f subject to $g = 0$?

25. Let $f(x, y, z) = y + z - x^2$.
(a) Find the solutions to the Lagrange equations for f subject to the constraint $g(x, y, z) = x^2 - y^2 + z^3 = 0$. *Hint:* Show that at a critical point, λ, y, z must be nonzero and $x = 0$.
(b) Show that f has no minimum or maximum subject to the constraint. *Hint:* Consider the values of f at the points $(0, y^3, y^2)$, which satisfy the constraint.
(c) Does (b) contradict Theorem 1?

26. Use Lagrange multipliers to find the point $P = (x_0, y_0, z_0)$ on the plane $ax + by + cz = d$ closest to the origin. Then calculate the distance from P to O.

27. ✎ Let Q be the point on an ellipse closest to a given point P outside the ellipse. It was known to the Greek mathematician Apollonius (third century BCE) that $\overline{PQ}$ is perpendicular to the tangent at Q (Figure 13). Explain in words why this conclusion is a consequence of the method of Lagrange multipliers. *Hint:* The circles centered at P are level curves of the function to be minimized.

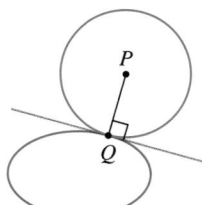

FIGURE 13

28. Antonio has $5.00 to spend on a lunch consisting of hamburgers ($1.50 each) and French fries ($1.00 per order). Antonio's satisfaction from eating x_1 hamburgers and x_2 orders of French fries is measured by a function $U(x_1, x_2) = \sqrt{x_1 x_2}$. How much of each type of food should he purchase to maximize his satisfaction (assume that fractional amounts of each food can be purchased)?

29. Find the maximum value of $f(x, y, z) = xy + xz + yz - 4xyz$ subject to the constraints $x + y + z = 1$ and $x, y, z \geq 0$.

30. A plane with equation $\dfrac{x}{a} + \dfrac{y}{b} + \dfrac{z}{c} = 1$ $(a, b, c > 0)$ together with the positive coordinate planes forms a tetrahedron of volume $V = \frac{1}{6}abc$ (Figure 14). Find the plane that minimizes V if the plane is constrained to pass through the point $P = (1, 1, 1)$.

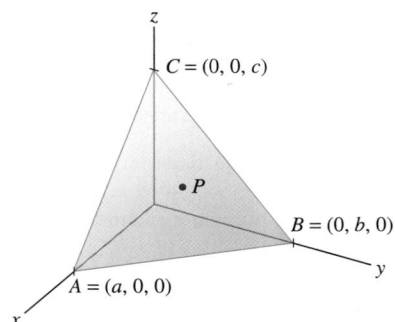

FIGURE 14

31. With the same set-up as in the previous problem, find the plane that minimizes V if the plane is constrained to pass through a point $P = (\alpha, \beta, \gamma)$ with $\alpha, \beta, \gamma > 0$.

32. In a contest, a runner starting at A must touch a point P along a river and then run to B in the shortest time possible (Figure 15). The runner should choose the point P minimizing the total length of the path.

(a) Define a function $f(x, y) = AP + PB$, where $P = (x, y)$. Rephrase the runner's problem as a constrained optimization problem, assuming that the river is given by an equation $g(x, y) = 0$.

(b) Explain why the level curves of $f(x, y)$ are ellipses.

(c) Use Lagrange multipliers to justify the following statement: The ellipse through the point P minimizing the length of the path is tangent to the river.

(d) Identify the point on the river in Figure 15 for which the length is minimal.

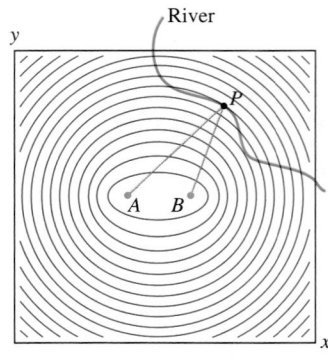

FIGURE 15

33. Let L be the minimum length of a ladder that can reach over a fence of height h to a wall located a distance b behind the wall.

(a) Use Lagrange multipliers to show that $L = (h^{2/3} + b^{2/3})^{3/2}$ (Figure 16). *Hint:* Show that the problem amounts to minimizing $f(x, y) = (x + b)^2 + (y + h)^2$ subject to $y/b = h/x$ or $xy = bh$.

(b) Show that the value of L is also equal to the radius of the circle with center $(-b, -h)$ that is tangent to the graph of $xy = bh$.

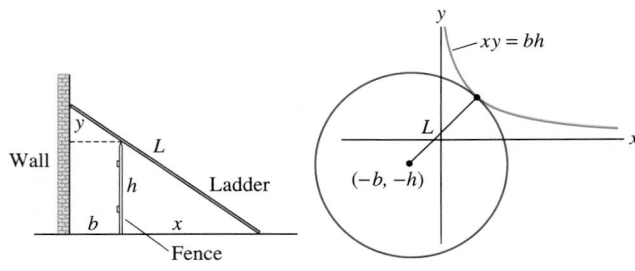

FIGURE 16

34. Find the minimum and maximum of $f(x, y, z) = y + 2z$ subject to two constraints, $2x + z = 4$ and $x^2 + y^2 = 1$.

35. Find the minimum value of $f(x, y, z) = x^2 + y^2 + z^2$ subject to two constraints, $x + 2y + z = 3$ and $x - y = 4$.

Further Insights and Challenges

36. Suppose that both $f(x, y)$ and the constraint function $g(x, y)$ are linear. Use contour maps to explain why $f(x, y)$ does not have a maximum subject to $g(x, y) = 0$ unless $g = af + b$ for some constants a, b.

37. Assumptions Matter *CAS* Consider the problem of minimizing $f(x, y) = x$ subject to $g(x, y) = (x - 1)^3 - y^2 = 0$.

(a) Show, without using calculus, that the minimum occurs at $P = (1, 0)$.

(b) Show that the Lagrange condition $\nabla f_P = \lambda \nabla g_P$ is not satisfied for any value of λ.

(c) Does this contradict Theorem 1?

38. Marginal Utility Goods 1 and 2 are available at prices (in dollars) of p_1 per unit of good 1 and p_2 per unit of good 2. A utility function $U(x_1, x_2)$ is a function representing the **utility** or benefit of consuming x_j units of good j. The **marginal utility** of the jth good is $\partial U / \partial x_j$, the rate of increase in utility per unit increase in the jth good. Prove the following law of economics: Given a budget of L dollars, utility is maximized at the consumption level (a, b) where the ratio of marginal utility is equal to the ratio of prices:

$$\frac{\text{Marginal utility of good 1}}{\text{Marginal utility of good 2}} = \frac{U_{x_1}(a, b)}{U_{x_2}(a, b)} = \frac{p_1}{p_2}$$

39. Consider the utility function $U(x_1, x_2) = x_1 x_2$ with budget constraint $p_1 x_1 + p_2 x_2 = c$.

(a) Show that the maximum of $U(x_1, x_2)$ subject to the budget constraint is equal to $c^2/(4 p_1 p_2)$.

(b) Calculate the value of the Lagrange multiplier λ occurring in (a).

(c) Prove the following interpretation: λ is the rate of increase in utility per unit increase of total budget c.

40. This exercise shows that the multiplier λ may be interpreted as a rate of change in general. Assume that the maximum value of $f(x, y)$ subject to $g(x, y) = c$ occurs at a point P. Then P depends on the value of c, so we may write $P = (x(c), y(c))$ and we have $g(x(c), y(c)) = c$.

(a) Show that

$$\nabla g(x(c), y(c)) \cdot \langle x'(c), y'(c) \rangle = 1$$

Hint: Use the Chain Rule to differentiate the equation $g(x(c), y(c)) = c$ with respect to c.

(b) Show that

$$\frac{d}{dc} f(x(c), y(c)) = \lambda$$

Hint: Use the Chain Rule and the Lagrange Condition $\nabla f_P = \lambda \nabla g_P$.

(c) Conclude that λ is the rate of increase in f per unit increase in the "budget level" c.

41. Let $B > 0$. Show that the maximum of $f(x_1, \dots, x_n) = x_1 x_2 \cdots x_n$ subject to the constraints $x_1 + \cdots + x_n = B$ and $x_j \geq 0$ for $j = 1, \dots, n$ occurs for $x_1 = \cdots = x_n = B/n$. Use this to conclude that

$$(a_1 a_2 \cdots a_n)^{1/n} \leq \frac{a_1 + \cdots + a_n}{n}$$

for all positive numbers $a_1, \dots, a_n$.

42. Let $B > 0$. Show that the maximum of $f(x_1, \dots, x_n) = x_1 + \cdots + x_n$ subject to $x_1^2 + \cdots + x_n^2 = B^2$ is $\sqrt{n}B$. Conclude that

$$|a_1| + \cdots + |a_n| \leq \sqrt{n}(a_1^2 + \cdots + a_n^2)^{1/2}$$

for all numbers $a_1, \dots, a_n$.

43. Given constants E, E_1, E_2, E_3, consider the maximum of

$$S(x_1, x_2, x_3) = x_1 \ln x_1 + x_2 \ln x_2 + x_3 \ln x_3$$

subject to two constraints:

$$x_1 + x_2 + x_3 = N, \quad E_1 x_1 + E_2 x_2 + E_3 x_3 = E$$

Show that there is a constant μ such that $x_i = A^{-1} e^{\mu E_i}$ for $i = 1, 2, 3$, where $A = N^{-1}(e^{\mu E_1} + e^{\mu E_2} + e^{\mu E_3})$.

44. Boltzmann Distribution   Generalize Exercise 43 to n variables: Show that there is a constant μ such that the maximum of

$$S = x_1 \ln x_1 + \cdots + x_n \ln x_n$$

subject to the constraints

$$x_1 + \cdots + x_n = N, \quad E_1 x_1 + \cdots + E_n x_n = E$$

occurs for $x_i = A^{-1} e^{\mu E_i}$, where

$$A = N^{-1}(e^{\mu E_1} + \cdots + e^{\mu E_n})$$

This result is used in physics to determine the distribution of velocities of gas molecules at temperature T; x_i is the number of molecules with kinetic energy E_i; $\mu = -(kT)^{-1}$, where k is Boltzmann's constant. The quantity S is called the **entropy**.

CHAPTER REVIEW EXERCISES

1. Given $f(x, y) = \dfrac{\sqrt{x^2 - y^2}}{x + 3}$,

(a) Sketch the domain of f.

(b) Calculate $f(3, 1)$ and $f(-5, -3)$.

(c) Find a point satisfying $f(x, y) = 1$.

2. Find the domain and range of:

(a) $f(x, y, z) = \sqrt{x - y} + \sqrt{y - z}$

(b) $f(x, y) = \ln(4x^2 - y)$

3. Sketch the graph $f(x, y) = x^2 - y + 1$ and describe its vertical and horizontal traces.

4. CAS   Use a graphing utility to draw the graph of the function $\cos(x^2 + y^2)e^{1 - xy}$ in the domains $[-1, 1] \times [-1, 1]$, $[-2, 2] \times [-2, 2]$, and $[-3, 3] \times [-3, 3]$, and explain its behavior.

5. Match the functions (a)–(d) with their graphs in Figure 1.

(a) $f(x, y) = x^2 + y$

(b) $f(x, y) = x^2 + 4y^2$

(c) $f(x, y) = \sin(4xy)e^{-x^2 - y^2}$

(d) $f(x, y) = \sin(4x)e^{-x^2 - y^2}$

(a)   (b)

FIGURE 1

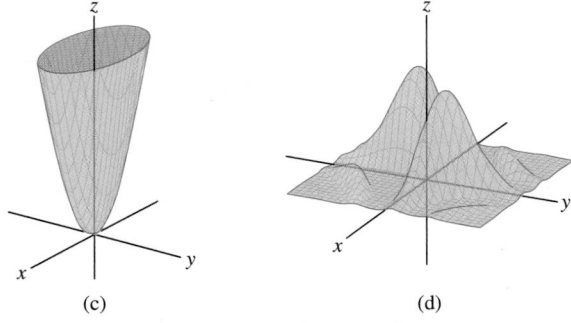

(c)   (d)

FIGURE 1 (*continued*)

6. Referring to the contour map in Figure 2, estimate:

(a) The average rate of change of elevation from A to C and from A to D.

(b) The directional derivative at A in the direction of **v**.

Contour interval = 50 meters

0   1   2 km

FIGURE 2

7. Describe the level curves of:

(a) $f(x, y) = e^{4x-y}$ **(b)** $f(x, y) = \ln(4x - y)$

(c) $f(x, y) = 3x^2 - 4y^2$ **(d)** $f(x, y) = x + y^2$

8. Match each function (a)–(c) with its contour graph (i)–(iii) in Figure 3:

(a) $f(x, y) = xy$

(b) $f(x, y) = e^{xy}$

(c) $f(x, y) = \sin(xy)$

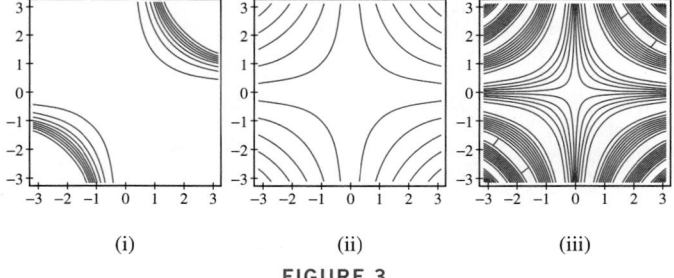

(i)   (ii)   (iii)

FIGURE 3

In Exercises 9–14, evaluate the limit or state that it does not exist.

9. $\displaystyle\lim_{(x,y)\to(1,-3)} (xy + y^2)$

10. $\displaystyle\lim_{(x,y)\to(1,-3)} \ln(3x + y)$

11. $\displaystyle\lim_{(x,y)\to(0,0)} \frac{xy + xy^2}{x^2 + y^2}$

12. $\displaystyle\lim_{(x,y)\to(0,0)} \frac{x^3y^2 + x^2y^3}{x^4 + y^4}$

13. $\displaystyle\lim_{(x,y)\to(1,-3)} (2x + y)e^{-x+y}$

14. $\displaystyle\lim_{(x,y)\to(0,2)} \frac{(e^x - 1)(e^y - 1)}{x}$

15. Let

$$f(x, y) = \begin{cases} \dfrac{(xy)^p}{x^4 + y^4} & (x, y) \ne (0, 0) \\ 0 & (x, y) = (0, 0) \end{cases}$$

Use polar coordinates to show that $f(x, y)$ is continuous at all (x, y) if $p > 2$, but discontinuous at $(0, 0)$ if $p \le 2$.

16. Calculate $f_x(1, 3)$ and $f_y(1, 3)$ for $f(x, y) = \sqrt{7x + y^2}$.

In Exercises 17–20, compute f_x and f_y.

17. $f(x, y) = 2x + y^2$ **18.** $f(x, y) = 4xy^3$

19. $f(x, y) = \sin(xy)e^{-x-y}$ **20.** $f(x, y) = \ln(x^2 + xy^2)$

21. Calculate f_{xxyz} for $f(x, y, z) = y\sin(x + z)$.

22. Fix $c > 0$. Show that for any constants α, β, the function $u(t, x) = \sin(\alpha ct + \beta)\sin(\alpha x)$ satisfies the wave equation

$$\frac{\partial^2 u}{\partial t^2} = c^2 \frac{\partial^2 u}{\partial x^2}$$

23. Find an equation of the tangent to the graph of $f(x, y) = xy^2 - xy + 3x^3y$ at $P = (1, 3)$.

24. Suppose that $f(4, 4) = 3$ and $f_x(4, 4) = f_y(4, 4) = -1$. Use the linear approximation to estimate $f(4.1, 4)$ and $f(3.88, 4.03)$.

25. Estimate $\sqrt{7.1^2 + 4.9^2 + 70.1}$ using the linear approximation. Compare with a calculator value.

26. Suppose that the plane $z = 2x - y - 3$ is tangent to the graph of $z = f(x, y)$ at $P = (2, 4)$.

(a) Determine $f(2, 4)$, $f_x(2, 4)$, and $f_y(2, 4)$.

(b) Approximate $f(2.2, 3.9)$.

27. Jason earns $S(h, c) = 20h\left(1 + \dfrac{c}{100}\right)^{1.5}$ dollars per month at a used car lot, where h is the number of hours worked and c is the number of cars sold. He has already worked 160 hours and sold 69 cars. Right now Jason wants to go home but wonders how much more he might earn if he stays another 10 minutes with a customer who is considering buying a car. Use the linear approximation to estimate how much extra money Jason will earn if he sells his 70th car during these 10 minutes.

In Exercises 28–31, compute $\dfrac{d}{dt} f(\mathbf{c}(t))$ at the given value of t.

28. $f(x, y) = x + e^y$, $\mathbf{c}(t) = (3t - 1, t^2)$ at $t = 2$

29. $f(x, y, z) = xz - y^2$, $\mathbf{c}(t) = (t, t^3, 1 - t)$

30. $f(x, y) = xe^{3y} - ye^{3x}$, $\mathbf{c}(t) = (e^t, \ln t)$ at $t = 1$

31. $f(x, y) = \tan^{-1} \dfrac{y}{x}$, $\mathbf{c}(t) = (\cos t, \sin t)$, $t = \dfrac{\pi}{3}$

In Exercises 32–35, compute the directional derivative at P in the direction of $\mathbf{v}$.

32. $f(x, y) = x^3 y^4$, $P = (3, -1)$, $\mathbf{v} = 2\mathbf{i} + \mathbf{j}$

33. $f(x, y, z) = zx - xy^2$, $P = (1, 1, 1)$, $\mathbf{v} = \langle 2, -1, 2 \rangle$

34. $f(x, y) = e^{x^2 + y^2}$, $P = \left(\dfrac{\sqrt{2}}{2}, \dfrac{\sqrt{2}}{2} \right)$, $\mathbf{v} = \langle 3, -4 \rangle$

35. $f(x, y, z) = \sin(xy + z)$, $P = (0, 0, 0)$, $\mathbf{v} = \mathbf{j} + \mathbf{k}$

36. Find the unit vector $\mathbf{e}$ at $P = (0, 0, 1)$ pointing in the direction along which $f(x, y, z) = xz + e^{-x^2 + y}$ increases most rapidly.

37. Find an equation of the tangent plane at $P = (0, 3, -1)$ to the surface with equation

$$ze^x + e^{z+1} = xy + y - 3$$

38. Let $n \neq 0$ be an integer and r an arbitrary constant. Show that the tangent plane to the surface $x^n + y^n + z^n = r$ at $P = (a, b, c)$ has equation

$$a^{n-1}x + b^{n-1}y + c^{n-1}z = r$$

39. Let $f(x, y) = (x - y)e^x$. Use the Chain Rule to calculate $\dfrac{\partial f}{\partial u}$ and $\dfrac{\partial f}{\partial v}$, where $x = u - v$ and $y = u + v$.

40. Let $f(x, y, z) = x^2 y + y^2 z$. Use the Chain Rule to calculate $\dfrac{\partial f}{\partial s}$ and $\dfrac{\partial f}{\partial t}$, where

$$x = s + t, \quad y = st, \quad z = 2s - t$$

41. Express the partial derivatives $\dfrac{\partial f}{\partial r}$ and $\dfrac{\partial f}{\partial \theta}$ of a function $f(x, y, z)$ in terms of $\dfrac{\partial f}{\partial x}$, $\dfrac{\partial f}{\partial y}$, and $\dfrac{\partial f}{\partial z}$, where (r, θ, z) are cylindrical coordinates.

42. Let $f(x, y, z) = (x^2 + y^2)e^{-xz}$ and $P = (1, 0, 2)$. Use the result of Exercise 41 to calculate $\dfrac{\partial f}{\partial r}$ and $\dfrac{\partial f}{\partial \theta}$ at the point P.

43. Let $g(u, v) = f(u^3 - v^3, v^3 - u^3)$. Prove that

$$v^2 \dfrac{\partial g}{\partial u} - u^2 \dfrac{\partial g}{\partial v} = 0$$

44. Let $f(x, y) = g(u)$, where $u = x^2 + y^2$ and $g(u)$ is a differentiable function in one variable. Prove that

$$\left(\dfrac{\partial f}{\partial x} \right)^2 + \left(\dfrac{\partial f}{\partial y} \right)^2 = 4u \left(\dfrac{\partial g}{\partial u} \right)^2$$

45. Calculate $\dfrac{\partial z}{\partial x}$, where $xe^z + ze^y = x + y$.

46. Let $f(x, y) = x^4 - 2x^2 + y^2 - 6y$.
(a) Find the critical points of f and use the Second Derivative Test to determine whether they are local minima or maxima.
(b) Find the minimum value of f without calculus by completing the square.

In Exercises 47–50, find the critical points of the function and analyze them using the Second Derivative Test.

47. $f(x, y) = x^2 + 2y^2 - 4xy + 6x$

48. $f(x, y) = x^3 + 2y^3 - xy$

49. $f(x, y) = e^{x+y} - xe^{2y}$

50. $f(x, y) = \sin(x + y) - \dfrac{1}{2}(x + y^2)$

51. Prove that $f(x, y) = (x + 2y)e^{xy}$ has no critical points.

52. Find the global extrema of $f(x, y) = x^3 - xy - y^2 + y$ on the square $[0, 1] \times [0, 1]$.

53. Find the global extrema of $f(x, y) = 2xy - x - y$ on the domain $\{y \leq 4, y \geq x^2\}$.

54. Find the maximum of $f(x, y, z) = xyz$ subject to the constraint $g(x, y, z) = 2x + y + 4z = 1$.

55. Use Lagrange multipliers to find the minimum and maximum value of $f(x, y) = 3x - 2y$ on the circle $x^2 + y^2 = 4$.

56. Find the minimum value of $f(x, y) = xy$ subject to the constraint $5x - y = 4$ in two ways: using Lagrange multipliers and setting $y = 5x - 4$ in $f(x, y)$.

57. Find the minimum and maximum values of $f(x, y) = x^2 y$ on the ellipse $4x^2 + 9y^2 = 36$.

58. Find the point in the first quadrant on the curve $y = x + x^{-1}$ closest to the origin.

59. Find the extreme values of $f(x, y, z) = x + 2y + 3z$ subject to the two constraints $x + y + z = 1$ and $x^2 + y^2 + z^2 = 1$.

60. Use Lagrange multipliers to find the dimensions of a cylindrical can of fixed volume V with minimal surface area (including the top and bottom of the can).

61. Find the dimensions of the box of maximum volume with its sides parallel to the coordinate planes that can be inscribed in the ellipsoid (Figure 4)

$$\left(\frac{x}{a}\right)^2 + \left(\frac{y}{b}\right)^2 + \left(\frac{z}{c}\right)^2 = 1$$

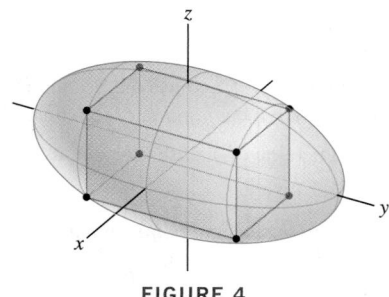

FIGURE 4

62. Given n nonzero numbers $\sigma_1, \dots, \sigma_n$, show that the minimum value of

$$f(x_1, \dots, x_n) = x_1^2 \sigma_1^2 + \cdots + x_n^2 \sigma_n^2$$

subject to $x_1 + \cdots + x_n = 1$ is c, where $c = \left(\displaystyle\sum_{j=1}^{n} \sigma_j^{-2}\right)^{-1}$.

63. A bead hangs on a string of length ℓ whose ends are fixed by thumbtacks located at points $(0, 0)$ and (a, b) on a bulletin board (Figure 5). The bead rests in the position that minimizes its height y. Use Lagrange multipliers to show that the two sides of the string make equal angles with the horizontal, that is, show that $\theta_1 = \theta_2$.

As an aside, note that the locus of the bead when pulled taut is an ellipse with foci at O and P, and the tangent line at the lowest point is horizontal. Therefore, this exercise provides another proof of the reflective property of the ellipse (a light ray emanating from one focus and bouncing off the ellipse is reflected to the other focus).

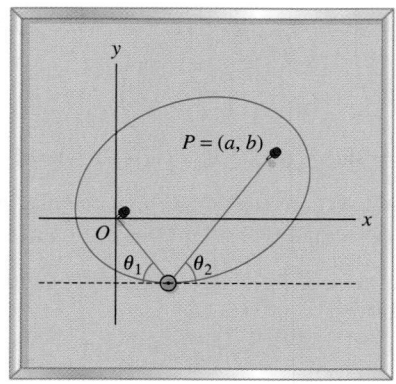

FIGURE 5

15 | MULTIPLE INTEGRATION

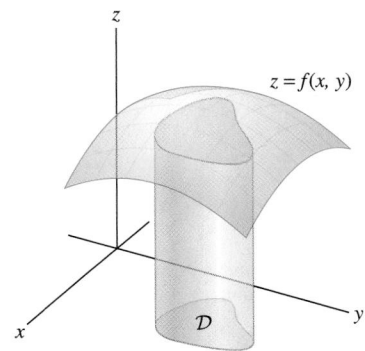

The integral of a function of several variables is called a **multiple integral**. In this chapter, we define multiple integrals and develop techniques for computing them. We also discuss some of their applications. Like the one-variable definite integral, the multiple integral is used to compute many important and diverse quantities, such as volumes, surface areas, centers of mass, probabilities, and average values.

Ranger Uranium Mine in Kakadu National Park, Australia. The volume of ore removed is one type of quantity that is expressed by a multiple integral.

15.1 Integration in Several Variables

In the first two sections of this chapter, we study integration of functions $f(x, y)$ of two variables. An integral of $f(x, y)$ is called a **double integral**.

There are many similarities between double integrals and the single integrals with which you are familiar:

- Double and single integrals are both defined as limits of Riemann sums.
- A double integral represents signed volume, just as a single integral represents signed area.
- We evaluate double integrals using the Fundamental Theorem of Calculus (FTC), but we have to apply it twice (see the discussion of iterated integrals below).
- Double integrals can be approximated using numerical methods similar to the Endpoint and Midpoint Rules for single integrals.

An important feature in the multivariable case is that the domain of integration plays a more prominent role. The domain of integration of a single integral $\int_a^b f(x)\,dx$ is an interval $[a, b]$, but in two variables, the domain is a region $\mathcal{D}$ in the plane with a more general boundary curve (Figure 1).

In this section, we restrict our attention to the simplest case in which $\mathcal{D}$ is a rectangle, leaving more general domains for discussion in Section 15.2. The notation

$$\mathcal{R} = [a, b] \times [c, d]$$

denotes the rectangle (Figure 2) consisting of all points (x, y) such that

$$\mathcal{R}: \quad a \le x \le b, \qquad c \le y \le d$$

As mentioned above, the double integral is a limit of Riemann sums. To form a Riemann sum, we choose positive integers N and M and partitions of the intervals $[a, b]$ and $[c, d]$:

$$a = x_0 < x_1 < \cdots < x_N = b, \qquad c = y_0 < y_1 < \cdots < y_M = d$$

Let $\Delta x_i = x_i - x_{i-1}$ and $\Delta y_j = y_j - y_{j-1}$ be the widths of the subintervals defined by these partitions. We then partition $\mathcal{R}$ into NM subrectangles $\mathcal{R}_{ij}$, whose area we denote

In the context of multiple integration, we often refer to the definite integral of a function $f(x)$ of one variable as a "single integral."

FIGURE 1 Region between the graph of $f(x, y)$ and the xy-plane over the domain $\mathcal{D}$.

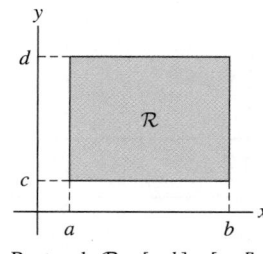

FIGURE 2

by ΔA_{ij} [Figure 3(B)]:

$$\mathcal{R}_{ij} = [x_{i-1}, x_i] \times [y_{j-1}, y_j], \qquad \text{Area}(\mathcal{R}_{ij}) = \Delta A_{ij} = \Delta x_i \, \Delta y_j$$

The partition is called **regular** if the intervals $[a, b]$ and $[c, d]$ are both divided into subintervals of equal length. In other words, the partition is regular if $\Delta x_i = \Delta x$ and $\Delta y_j = \Delta y$ where

$$\Delta x = \frac{b - a}{N}, \qquad \Delta y = \frac{d - c}{M}$$

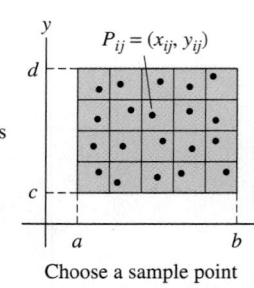

Rectangle $\mathcal{R} = [a, b] \times [c, d]$

Create $N \times M$ grid of subrectangles

Choose a sample point P_{ij} in each rectangle

FIGURE 3

Next, we choose a **sample point** $P_{ij} = (x_{ij}, y_{ij})$ in each subrectangle $\mathcal{R}_{ij}$ [Figure 3(C)] and construct a box of height $f(P_{ij})$ above $\mathcal{R}_{ij}$ as in Figure 4(B). This box has signed volume

$$f(P_{ij}) \, \Delta A_{ij} = f(P_{ij}) \, \Delta x_i \, \Delta y_j = \underbrace{\text{height} \times \text{area}}_{\text{Signed volume of box}}$$

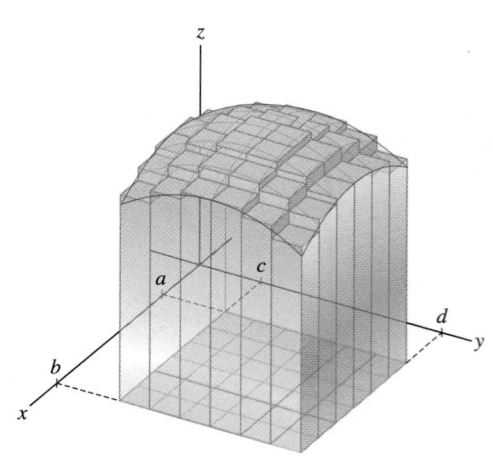

(A) In one variable, a Riemann sum approximates the area under the curve by a sum of areas of rectangles.

(B) The volume of the box is $f(P_{ij})\Delta A$ where $\Delta A = \Delta x \Delta y$.

(C) The Riemann sum $S_{N, M}$ is the sum of the volumes of the boxes.

FIGURE 4

Here, by definition, the signed volume of a region is positive if it lies above the xy-plane and negative if it lies below the xy-plane. If $f(P_{ij}) < 0$, the box extends below the xy-

plane and the signed volume is negative. The Riemann sum $S_{N,M}$ is the sum of the signed volumes of the boxes [Figure 4(C)]:

$$S_{N,M} = \sum_{i=1}^{N} \sum_{j=1}^{M} f(P_{ij}) \, \Delta A_{ij} = \sum_{i=1}^{N} \sum_{j=1}^{M} f(P_{ij}) \, \Delta x_i \, \Delta x_j$$

This double summation runs over all i and j in the ranges $1 \le i \le N$ and $1 \le j \le M$, a total of NM terms.

The Riemann sums $S_{N,M}$ yield approximations to the (signed) volume of the region between the graph of $z = f(x, y)$ and the xy-plane. They are a natural generalization of the rectangular approximations to area provided by Riemann sums in the single variable case [Figure 4(A)].

Let $\mathcal{P} = \{\{x_i\}, \{y_j\}\}$ be the set of endpoints of the partition and let $\|\mathcal{P}\|$ be the maximum of the widths $\Delta x_i, \Delta y_j$. As $\|\mathcal{P}\|$ tends to zero, the boxes become thinner and they approximate the signed volume more closely (Figure 5). Therefore, we may expect $S_{N,M}$ to converge to the signed volume under the graph as $\|\mathcal{P}\|$ approaches zero. Note that N and M both approach ∞ as $\|\mathcal{P}\| \to 0$. The precise definition of the limit is as follows:

> *The Riemann sum $S_{N,M}$ approaches a limit L as $\|\mathcal{P}\| \to 0$ if, for all $\epsilon > 0$, there exists $\delta > 0$ such that $|L - S_{N,M}| < \epsilon$ for all partitions satisfying $\|\mathcal{P}\| < \delta$ and all choices of sample points.*

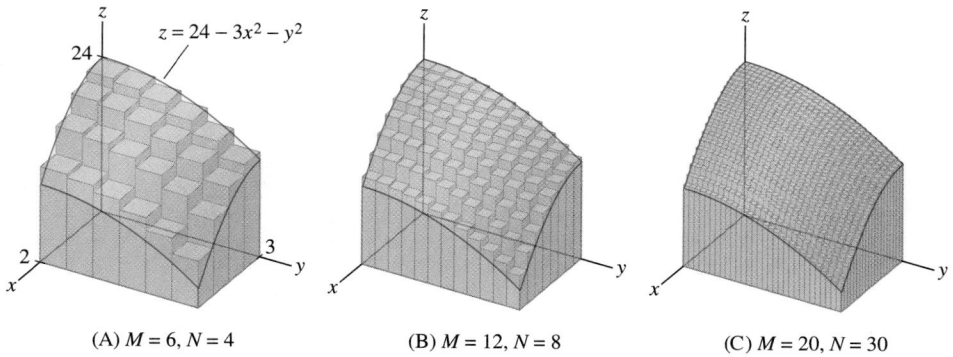

(A) $M = 6, N = 4$ (B) $M = 12, N = 8$ (C) $M = 20, N = 30$

In the margin:

Keep in mind that a Riemann sum depends on the choice of partition and sample points. It would be more proper to write

$$S_{N,M}(\{P_{ij}\}, \{x_i\}, \{y_j\})$$

but we write $S_{N,M}$ to simplify the notation.

FIGURE 5 Midpoint approximations to the volume under $z = 12 + 3x^2 + 5y^2$.

In this case, we write

$$\lim_{\|\mathcal{P}\|\to 0} S_{N,M} = \lim_{\|\mathcal{P}\|\to 0} \sum_{i=1}^{N} \sum_{j=1}^{M} f(P_{ij}) \, \Delta A_{ij} = L$$

This limit L, if it exists, is called the double integral of $f(x, y)$ over $\mathcal{R}$ (Figure 6), and is denoted $\displaystyle\iint_{\mathcal{R}} f(x, y) \, dA$.

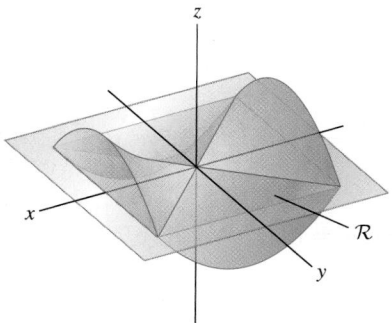

FIGURE 6 The double integral $\displaystyle\iint_{\mathcal{R}} f(x, y) \, dA$ is the signed volume between the surface $z = f(x, y)$ and the xy-plane over the rectangle $\mathcal{R}$.

DEFINITION **Double Integral over a Rectangle** The double integral of $f(x, y)$ over a rectangle $\mathcal{R}$ is defined as the limit

$$\iint_{\mathcal{R}} f(x, y) \, dA = \lim_{\|\mathcal{P}\|\to 0} \sum_{i=1}^{N} \sum_{j=1}^{M} f(P_{ij}) \Delta A_{ij}$$

If this limit exists, we say that $f(x, y)$ is **integrable** over $\mathcal{R}$.

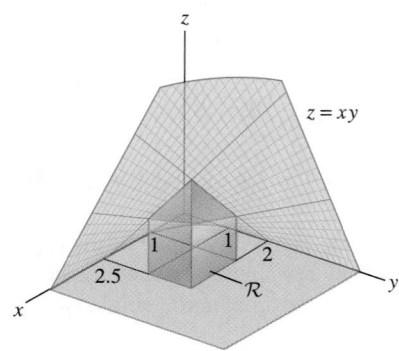

FIGURE 7 Graph of $z = xy$.

■ **EXAMPLE 1**   Estimating a Double Integral   Calculate $S_{3,2}$ for the integral $\iint_{\mathcal{R}} xy\,dA$, where $\mathcal{R} = [1, 2.5] \times [1, 2]$ (Figure 7). Use the regular partition and the following two choices of sample points:

(a) Lower-left vertex   **(b)** Midpoint of rectangle

Solution   Figures 8 and 9 show a 3×2 grid of subrectangles on the rectangle $\mathcal{R}$. Since we use the regular partition, each subrectangle has sides of length

$$\Delta x = \frac{2.5 - 1}{3} = \frac{1}{2}, \qquad \Delta y = \frac{2 - 1}{2} = \frac{1}{2}$$

and area $\Delta A = \Delta x\,\Delta y = \frac{1}{4}$. The corresponding Riemann sum is

$$S_{3,2} = \sum_{i=1}^{3} \sum_{j=1}^{2} f(P_{ij})\,\Delta A = \frac{1}{4} \sum_{i=1}^{3} \sum_{j=1}^{2} f(P_{ij})$$

(a)   If we use the lower-left vertices shown in Figure 8, the Riemann sum is

$$S_{3,2} = \frac{1}{4}\big(f(1, 1) + f(1, 1.5) + f(1.5, 1) + f(1.5, 1.5) + f(2, 1) + f(2, 1.5)\big)$$

$$= \frac{1}{4}\big(1 + 1.5 + 1.5 + 2.25 + 2 + 3\big) = \frac{1}{4}(11.25) = 2.8125$$

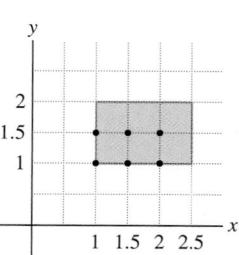

FIGURE 8   The sample points are the lower-left vertices of each square.

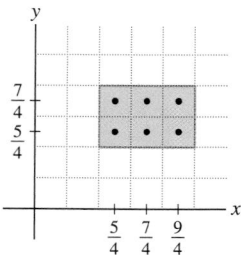

FIGURE 9   The sample points are the centers of each square.

(b)   Using the midpoints of the rectangles shown in Figure 9, we obtain

$$S_{3,2} = \frac{1}{4}\left(f\!\left(\frac{5}{4}, \frac{5}{4}\right) + f\!\left(\frac{5}{4}, \frac{7}{4}\right) + f\!\left(\frac{7}{4}, \frac{5}{4}\right) + f\!\left(\frac{7}{4}, \frac{7}{4}\right) + f\!\left(\frac{9}{4}, \frac{5}{4}\right) + f\!\left(\frac{9}{4}, \frac{7}{4}\right)\right)$$

$$= \frac{1}{4}\left(\frac{25}{16} + \frac{35}{16} + \frac{35}{16} + \frac{49}{16} + \frac{45}{16} + \frac{63}{16}\right) = \frac{1}{4}\left(\frac{252}{16}\right) = 3.9375 \quad ■$$

■ **EXAMPLE 2**   Evaluate $\iint_{\mathcal{R}} (8 - 2y)\,dA$, where $\mathcal{R} = [0, 3] \times [0, 4]$.

Solution   The graph of $z = 8 - 2y$ is the plane shown in Figure 10. The double integral represents the volume V of a solid of length $\ell = 3$ whose vertical cross section is a triangle of height 8 and base 4, and area $A = \frac{1}{2}(8)4 = 16$. The volume of the solid is $V = \ell A = 3(16) = 48$. Therefore,

$$\iint_{\mathcal{R}} (8 - 2y)\,dA = 48 \quad ■$$

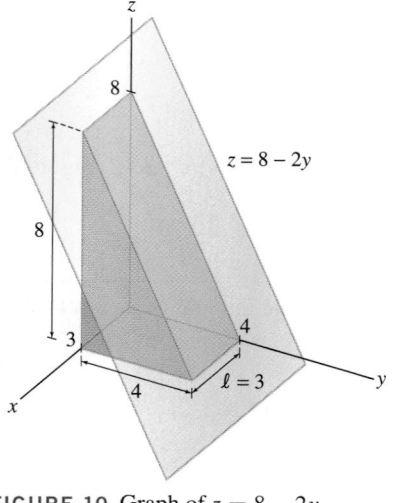

FIGURE 10   Graph of $z = 8 - 2y$.

The following theorem states that continuous functions are integrable. The proof is similar to the single-variable case and is omitted.

THEOREM 1 Continuous Functions Are Integrable   If $f(x, y)$ is continuous on a rectangle $\mathcal{R}$, then $f(x, y)$ is integrable over $\mathcal{R}$.

As in the single-variable case, we often make use of the linearity properties of the double integral. These follow from the definition of the double integral as a limit of Riemann sums.

THEOREM 2 Linearity of the Double Integral   Assume that $f(x, y)$ and $g(x, y)$ are integrable over a rectangle $\mathcal{R}$. Then:

(i) $\displaystyle\iint_{\mathcal{R}} \big(f(x, y) + g(x, y)\big)\, dA = \iint_{\mathcal{R}} f(x, y)\, dA + \iint_{\mathcal{R}} g(x, y)\, dA$

(ii) For any constant C, $\displaystyle\iint_{\mathcal{R}} C f(x, y)\, dA = C \iint_{\mathcal{R}} f(x, y)\, dA$

If $f(x, y) = C$ is a constant function, then

$$\iint_{\mathcal{R}} C\, dA = C \cdot \mathrm{Area}(\mathcal{R})$$

In this case, the double integral is the signed volume of the box of base $\mathcal{R}$ and height $|C|$ (Figure 11).

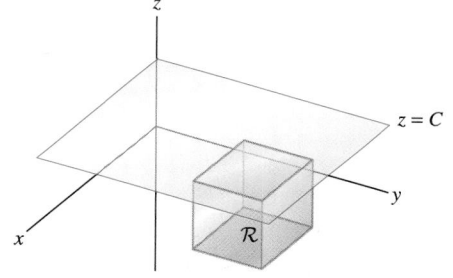

FIGURE 11 The double integral of $f(x, y) = C$ over a rectangle $\mathcal{R}$ is $C \cdot \mathrm{Area}(\mathcal{R})$.

■ **EXAMPLE 3 Arguing by Symmetry** Use symmetry to justify $\displaystyle\iint_{\mathcal{R}} xy^2\, dA = 0$, where $\mathcal{R} = [-1, 1] \times [-1, 1]$.

Solution The double integral is the signed volume of the region between the graph of $f(x, y) = xy^2$ and the xy-plane. Our function satisfies

$$f(-x, y) = -xy^2 = -f(x, y)$$

so the region below the xy-plane (where $-1 \le x \le 0$) cancels with the region above the xy-plane (where $0 \le x \le 1$) (Figure 12). ■

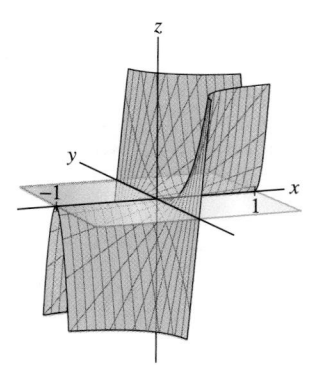

FIGURE 12

Iterated Integrals

Our main technique for evaluating double integrals is based on the FTC, as in the single-variable case. To use the FTC, we must first express the double integral as an iterated integral. An **iterated integral** is an expression of the form

$$\int_a^b \left(\int_c^d f(x, y)\, dy \right) dx$$

It is evaluated in two steps. First, hold x constant and evaluate the inner integral with respect to y:

$$S(x) = \int_c^d f(x, y)\, dy$$

Then integrate the resulting function $S(x)$ with respect to x.

Notice that the inner integral $\displaystyle\int_c^d$ specifies the limits of integration for the y-variable, and the outer integral $\displaystyle\int_a^b$ specifies the limits for the x-variable.

EXAMPLE 4 Evaluate $\displaystyle\int_2^4 \left(\int_1^9 ye^x \, dy \right) dx$.

Solution First evaluate the inner integral, treating x as a constant:

$$S(x) = \int_1^9 ye^x \, dy = e^x \int_1^9 y \, dy = e^x \left(\frac{1}{2} y^2 \right) \Big|_{y=1}^9 = e^x \left(\frac{81 - 1}{2} \right) = 40e^x$$

Then integrate $S(x)$ with respect to x:

$$\int_2^4 \left(\int_1^9 ye^x \, dy \right) dx = \int_2^4 40e^x \, dx = 40e^x \Big|_2^4 = 40(e^4 - e^2) \qquad \blacksquare$$

If an iterated integral is written with dx preceding dy, then we integrate first with respect to x:

$$\int_c^d \int_a^b f(x, y) \, dx \, dy = \int_{y=c}^d \left(\int_{x=a}^b f(x, y) \, dx \right) dy$$

Here, for clarity, we have included the variables in the limits of integration of the iterated integral on the right.

We often omit the parentheses in the notation for an iterated integral:

$$\int_a^b \int_c^d f(x, y) \, dy \, dx$$

The order of the variables in $dy \, dx$ tells us to integrate first with respect to y between the limits $y = c$ and $y = d$.

EXAMPLE 5 Evaluate $\displaystyle\int_0^{\pi/2} \int_0^{\pi/2} \sin(2x + y) \, dx \, dy$.

Solution Compute the inner integral, treating y as a constant:

$$\int_0^{\pi/2} \sin(2x + y) \, dx = -\frac{1}{2} \cos(2x + y) \Big|_{x=0}^{\pi/2} = -\frac{1}{2} \cos(\pi + y) + \frac{1}{2} \cos y$$

$$\int_0^{\pi/2} \int_0^{\pi/2} \sin(2x + y) \, dx \, dy = -\frac{1}{2} \int_0^{\pi/2} \left(\cos(\pi + y) - \cos y \right) dy$$

$$= -\frac{1}{2} \left(\sin(\pi + y) - \sin(y) \right) \Big|_{y=0}^{\pi/2}$$

$$= -\frac{1}{2} \left(\sin \frac{3\pi}{2} - \sin \frac{\pi}{2} \right) + \frac{1}{2} (\sin \pi - \sin 0) = 1 \qquad \blacksquare$$

EXAMPLE 6 Reversing the Order of Integration   Verify that

$$\int_0^1 \int_2^4 x^2 y^3 \, dy \, dx = \int_2^4 \int_0^1 x^2 y^3 \, dx \, dy$$

Solution Both integrals have the value 20:

$$\int_0^1 \int_2^4 x^2 y^3 \, dy \, dx = \int_0^1 \left(\frac{1}{4} x^2 y^4 \Big|_{y=2}^4 \right) dx = \int_0^1 \left(64x^2 - 4x^2 \right) dx$$

$$= \int_0^1 60x^2 \, dx = 20x^3 \Big|_0^1 = 20$$

$$\int_2^4 \int_0^1 x^2 y^3 \, dx \, dy = \int_2^4 \left(\frac{1}{3} x^3 y^3 \Big|_{x=0}^1 \right) dy = \frac{1}{3} \int_2^4 y^3 \, dy$$

$$= \frac{1}{12} y^4 \Big|_2^4 = \frac{1}{12} (256 - 16) = 20 \qquad \blacksquare$$

The previous example illustrates a general fact: The value of an iterated integral does not depend on the order in which the integration is performed. This is part of Fubini's Theorem. Even more important, Fubini's Theorem states that a double integral over a rectangle can be evaluated as an iterated integral.

THEOREM 3 Fubini's Theorem   The double integral of a continuous function $f(x, y)$ over a rectangle $\mathcal{R} = [a, b] \times [c, d]$ is equal to the iterated integral (in either order):

$$\iint_{\mathcal{R}} f(x, y) \, dA = \int_a^b \int_c^d f(x, y) \, dy \, dx = \int_c^d \int_a^b f(x, y) \, dx \, dy$$

Proof   We sketch the proof. Consider a Riemann sum $S_{N,M}$ for a regular partition. Then $S_{N,M}$ is the sum of values $f(P_{ij}) \, \Delta x \, \Delta y$, where

$$\Delta x = \frac{b - a}{N} \qquad \text{and} \qquad \Delta y = \frac{d - c}{M}$$

Fubini's Theorem ultimately stems from the elementary fact that we may sum these values in any order. Think of the values $f(P_{ij})$ as listed in an $N \times M$ array. For example, the values for $S_{3,3}$ are listed in the array in the margin. If we first sum the columns and then add up the column sums, we obtain

$$S_{N,M} = \underbrace{\sum_{i=1}^N \sum_{j=1}^M f(P_{ij}) \, \Delta x \, \Delta y}_{\text{Sum in any order}} = \underbrace{\sum_{i=1}^N \left(\sum_{j=1}^M f(P_{ij}) \, \Delta y \right) \Delta x}_{\substack{\text{First sum the columns,} \\ \text{then add up the column sums}}}$$

Let us choose sample points of the form $P_{ij} = (x_i, y_j)$, where $\{x_i\}$ are sample points for the regular partition on $[a, b]$ and $\{y_j\}$ are sample points for the regular partition of $[c, d]$. We observe that Δx and Δy approach zero as M and N tend to infinity. Therefore, the limit of the sums $S_{N,M}$ as $\|\mathcal{P}\| \to 0$ may be expressed as a limit as $N, M \to \infty$:

$$\iint_{\mathcal{R}} f(x, y) \, dA = \lim_{N,M \to \infty} \sum_{i=1}^N \left(\sum_{j=1}^M f(x_i, y_j) \Delta y \right) \Delta x$$

The inner sum on the right is a Riemann sum that approaches the single integral $\int_c^d f(x_i, y) \, dy$ as M approaches ∞:

$$\lim_{M \to \infty} S_{N,M} = \lim_{M \to \infty} \sum_{j=1}^M f(x_i, y_j) = \int_c^d f(x_i, y) \, dy$$

To complete the proof, we take two facts for granted. First is the fact that the function $S(x) = \int_c^d f(x, y) \, dy$ is continuous for $a \le x \le b$. Second, the limit $\lim_{N,M \to \infty} S_{N,M}$ may be computed by first taking the limit with respect to M and then with respect to N.

(Margin notes, left column:)

CAUTION When you reverse the order of integration in an iterated integral, remember to interchange the limits of integration (the inner limits become the outer limits).

j \ i	1	2	3
3	$f(P_{13})$	$f(P_{23})$	$f(P_{33})$
2	$f(P_{12})$	$f(P_{22})$	$f(P_{32})$
1	$f(P_{11})$	$f(P_{21})$	$f(P_{31})$

Then

$$\iint_{\mathcal{R}} f(x, y)\, dA = \lim_{N, M \to \infty} S_{N,M} = \lim_{N \to \infty} \left(\lim_{M \to \infty} S_{N,M} \right)$$

$$= \lim_{N \to \infty} \sum_{i=1}^{N} \left(\int_{c}^{d} f(x_i, y)\, dy \right) \Delta x \qquad \boxed{1}$$

$$= \int_{a}^{b} \left(\int_{c}^{d} f(x, y)\, dy \right) dx \qquad \boxed{2}$$

Notice that (2) follows from (1) because the sum in (1) is a Riemann sum for $S(x)$ over the interval $[a, b]$. This proves Fubini's Theorem for the order $dy\, dx$. A similar argument applies to the order $dx\, dy$. ∎

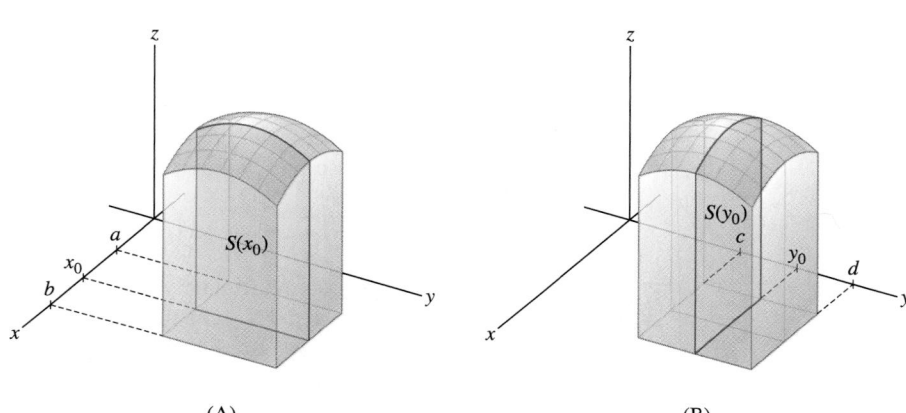

(A)   (B)

FIGURE 13

GRAPHICAL INSIGHT   The double integral $\displaystyle\iint_{\mathcal{R}} f(x, y)\, dA$ is the signed volume V of the solid region S between the xy-plane and the graph of $f(x, y)$. When we write a double integral as an iterated integral in the order $dy\, dx$, then for each fixed value $x = x_0$, the inner integral is the area of the cross section of S in the vertical plane $x = x_0$ perpendicular to the x-axis (Figure 13):

$$S(x_0) = \int_{c}^{d} f(x_0, y)\, dy = \begin{array}{l} \text{(area of cross section in vertical plane} \\ x = x_0 \text{ perpendicular to the } x\text{-axis)} \end{array}$$

Fubini's Theorem thus asserts that the volume V of S may be calculated as the integral of cross-sectional area $S(x)$:

$$V = \int_{a}^{b} \int_{c}^{d} f(x, y)\, dy\, dx = \int_{a}^{b} S(x)\, dx = \text{integral of cross-sectional area}$$

Similarly, when we write V as an iterated integral in the order $dx\, dy$, we are calculating V as the integral of cross sections perpendicular to the y-axis.

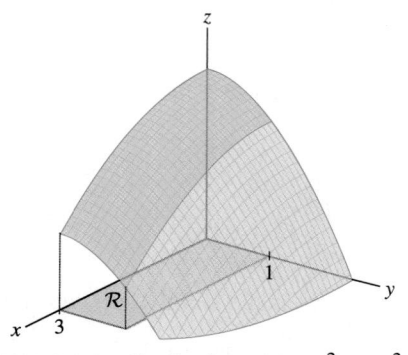

FIGURE 14 Graph of $z = 16 - x^2 - 3y^2$ over $\mathcal{R} = [0, 3] \times [0, 1]$.

■ **EXAMPLE 7**   Find the volume V between the surface $z = 16 - x^2 - 3y^2$ and the rectangle $\mathcal{R} = [0, 3] \times [0, 1]$ (Figure 14).

Solution The volume V is equal to a double integral, which we write as an iterated integral:

$$V = \iint_{\mathcal{R}} (16 - x^2 - 3y^2)\, dA = \int_{x=0}^{3} \int_{y=0}^{1} (16 - x^2 - 3y^2)\, dy\, dx$$

We evaluate the inner integral and compute V:

$$\int_{y=0}^{1} (16 - x^2 - 3y^2)\, dy = (16y - x^2 y - y^3)\Big|_{y=0}^{1} = 15 - x^2$$

$$V = \iint_{\mathcal{R}} (16 - x^2 - 3y^2)\, dA = \int_{x=0}^{3} (15 - x^2)\, dx = \left(15x - \frac{1}{3}x^3\right)\Big|_{0}^{3} = 36 \quad \blacksquare$$

■ **EXAMPLE 8** Calculate $\iint_{\mathcal{R}} \dfrac{dA}{(x+y)^2}$, where $\mathcal{R} = [1, 2] \times [0, 1]$ (Figure 15).

Solution

$$\iint_{\mathcal{R}} \frac{dA}{(x+y)^2} = \int_{x=1}^{2} \left(\int_{y=0}^{1} \frac{dy}{(x+y)^2}\right) dx = \int_{1}^{2} \left(-\frac{1}{x+y}\Big|_{y=0}^{1}\right) dx$$

$$= \int_{1}^{2} \left(-\frac{1}{x+1} + \frac{1}{x}\right) dx = \Big(\ln x - \ln(x+1)\Big)\Big|_{1}^{2}$$

$$= (\ln 2 - \ln 3) - (\ln 1 - \ln 2) = \ln \frac{4}{3} \quad \blacksquare$$

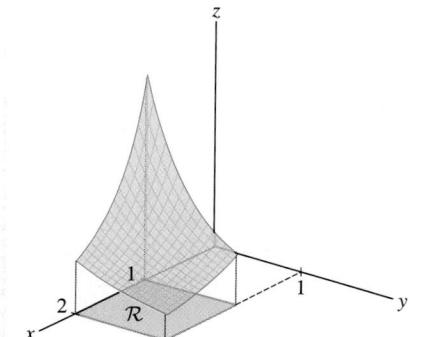

FIGURE 15 Graph of $z = (x+y)^{-2}$ over $\mathcal{R} = [1, 2] \times [0, 1]$.

When the function is a product $f(x, y) = g(x)h(y)$, the double integral over a rectangle is simply the product of the single integrals. This follows by writing the double integral as an iterated integral:

$$\iint_{\mathcal{R}} g(x)h(y)\, dA = \int_{a}^{b} \left(\int_{c}^{d} g(x)h(y)\, dy\right) dx = \int_{a}^{b} g(x) \left(\int_{c}^{d} h(y)\, dy\right) dx$$

$$= \left(\int_{a}^{b} g(x)\, dx\right)\left(\int_{c}^{d} h(y)\, dy\right) \qquad \boxed{3}$$

■ **EXAMPLE 9** Iterated Integral of a Product Function Calculate

Solution The integrand $f(x, y) = e^x \cos y$ is a product, so we obtain

$$\int_{0}^{2} \int_{0}^{\pi/2} e^x \cos y\, dy\, dx = \left(\int_{0}^{2} e^x\, dx\right)\left(\int_{0}^{\pi/2} \cos y\, dy\right) = \left(e^x\Big|_{0}^{2}\right)\left(\sin y\Big|_{0}^{\pi/2}\right)$$

$$= (e^2 - 1)(1) = e^2 - 1 \quad \blacksquare$$

15.1 SUMMARY

- A *Riemann sum* for $f(x, y)$ on a rectangle $\mathcal{R} = [a, b] \times [c, d]$ is a sum of the form

$$S_{N,M} = \sum_{i=1}^{N} \sum_{j=1}^{M} f(P_{ij}) \, \Delta x_i \, \Delta y_j$$

where P_{ij} is a sample point in the subrectangle $\mathcal{R}_{ij}$ (see Figure 3).
- The double integral of $f(x, y)$ over $\mathcal{R}$ is defined as the limit (if it exists):

$$\iint_{\mathcal{R}} f(x, y) \, dA = \lim_{M,N \to \infty} \sum_{i=1}^{N} \sum_{j=1}^{M} f(P_{ij}) \, \Delta x_i \, \Delta y_j$$

We say that $f(x, y)$ is *integrable* over $\mathcal{R}$ if this limit exists. Every continuous function on a rectangle $\mathcal{R}$ is integrable.
- The double integral is equal to the *signed volume* of the region between the graph of $z = f(x, y)$ and the rectangle $\mathcal{R}$. The signed volume of a region is positive if it lies above the xy-plane and negative if it lies below the xy-plane.
- If $f(x, y) = C$ is a constant function, then

$$\iint_{\mathcal{R}} C \, dA = C \cdot \text{Area}(\mathcal{R})$$

- By Fubini's Theorem, the double integral of a continuous function $f(x, y)$ over a rectangle $\mathcal{R} = [a, b] \times [c, d]$ may be evaluated as an iterated integral (in either order):

$$\iint_{\mathcal{R}} f(x, y) \, dA = \int_a^b \int_c^d f(x, y) \, dy \, dx = \int_c^d \int_a^b f(x, y) \, dx \, dy$$

15.1 EXERCISES

Preliminary Questions

1. In the Riemann sum $S_{8,4}$ for a double integral over $\mathcal{R} = [1, 5] \times [2, 10]$, what is the area of each subrectangle and how many subrectangles are there?

2. Estimate the double integral of a continuous function f over the small rectangle $\mathcal{R} = [0.9, 1.1] \times [1.9, 2.1]$ if $f(1, 2) = 4$.

3. What is the integral of the constant function $f(x, y) = 5$ over the rectangle $[-2, 3] \times [2, 4]$?

4. What is the interpretation of $\iint_{\mathcal{R}} f(x, y) \, dA$ if $f(x, y)$ takes on both positive and negative values on $\mathcal{R}$?

5. Which of (a) or (b) is equal to $\int_1^2 \int_4^5 f(x, y) \, dy \, dx$?

(a) $\int_1^2 \int_4^5 f(x, y) \, dx \, dy$ **(b)** $\int_4^5 \int_1^2 f(x, y) \, dx \, dy$

6. For which of the following functions is the double integral over the rectangle in Figure 16 equal to zero? Explain your reasoning.

(a) $f(x, y) = x^2 y$ **(b)** $f(x, y) = xy^2$

(c) $f(x, y) = \sin x$ **(d)** $f(x, y) = e^x$

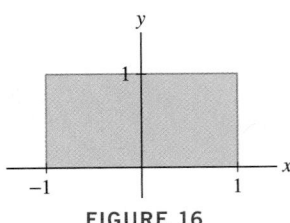

FIGURE 16

Exercises

1. Compute the Riemann sum $S_{4,3}$ to estimate the double integral of $f(x, y) = xy$ over $\mathcal{R} = [1, 3] \times [1, 2.5]$. Use the regular partition and upper-right vertices of the subrectangles as sample points.

2. Compute the Riemann sum with $N = M = 2$ to estimate the integral of $\sqrt{x + y}$ over $\mathcal{R} = [0, 1] \times [0, 1]$. Use the regular partition and midpoints of the subrectangles as sample points.

In Exercises 3–6, compute the Riemann sums for the double integral
$$\iint_{\mathcal{R}} f(x, y)\, dA, \quad \text{where } \mathcal{R} = [1, 4] \times [1, 3], \text{ for the grid and two}$$
choices of sample points shown in Figure 17.

(A)

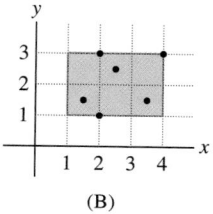

(B)

FIGURE 17

3. $f(x, y) = 2x + y$

4. $f(x, y) = 7$

5. $f(x, y) = 4x$

6. $f(x, y) = x - 2y$

7. Let $\mathcal{R} = [0, 1] \times [0, 1]$. Estimate $\displaystyle\iint_{\mathcal{R}} (x + y)\, dA$ by computing two different Riemann sums, each with at least six rectangles.

8. Evaluate $\displaystyle\iint_{\mathcal{R}} 4\, dA$, where $\mathcal{R} = [2, 5] \times [4, 7]$.

9. Evaluate $\displaystyle\iint_{\mathcal{R}} (-5)\, dA$, where $\mathcal{R} = [2, 5] \times [4, 7]$.

10. Use the following table to compute a Riemann sum $S_{3,3}$ for $f(x, y)$ on the square $\mathcal{R} = [0, 1.5] \times [0.5, 2]$. Use the regular partition and sample points of your choosing.

Values of f(x, y)					
2	2.6	2.17	1.86	1.62	1.44
1.5	2.2	1.83	1.57	1.37	1.22
1	1.8	1.5	1.29	1.12	1
0.5	1.4	1.17	1	0.87	0.78
0	1	0.83	0.71	0.62	0.56
y / x	0	0.5	1	1.5	2

11. The following table gives the approximate height at 1-ft intervals of a 5 × 4-ft mound of gravel. Estimate the volume of the mound by computing the average of the four Riemann sums $S_{5,4}$ with lower left, lower right, upper left, and upper right vertices of the subrectangles as sample points.

4	0.4	0.4	0.6	0.8	0.6	0.6
3	0.8	1.6	1.8	2.5	2.1	0.8
2	0.5	1.5	3.2	3.5	2.1	0.6
1	0.4	0.8	1.3	1.5	1.4	0.6
0	0.3	0.3	0.5	0.8	0.5	0.4
y / x	0	1	2	3	4	5

12. Figure 18 shows a grid with values of $f(x, y)$ at sample points in each square. Estimate the double integral of $f(x, y)$ over the rectangles:

(a) $[0.25, 1] \times [0.5, 1]$

(b) $[-0.5, 0.5] \times [0, 1]$

(c) $[-1, -0.5] \times [-1, -0.5]$

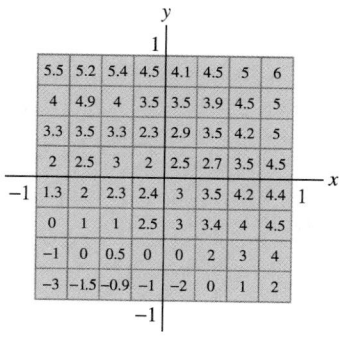

FIGURE 18

13. $\boxed{CAS}$ Let $S_{N,N}$ be the Riemann sum for $\displaystyle\int_0^1 \int_0^1 e^{x^3 - y^3}\, dy\, dx$ using the regular partition and the lower left-hand vertex of each subrectangle as sample points. Use a computer algebra system to calculate $S_{N,N}$ for $N = 25, 50, 100$.

14. $\boxed{CAS}$ Let $S_{N,M}$ be the Riemann sum for
$$\int_0^4 \int_0^2 \ln(1 + x^2 + y^2)\, dy\, dx$$
using the regular partition and the upper right-hand vertex of each subrectangle as sample points. Use a computer algebra system to calculate $S_{2N,N}$ for $N = 25, 50, 100$.

In Exercises 15–16, use symmetry to evaluate the double integral.

15. $\displaystyle\iint_{\mathcal{R}} \sin x\, dA, \quad \mathcal{R} = [0, 2\pi] \times [0, 2\pi]$

16. $\displaystyle\iint_{\mathcal{R}} (2 + x^2 y)\, dA, \quad \mathcal{R} = [0, 1] \times [-1, 1]$

In Exercises 17–32, evaluate the iterated integral.

17. $\displaystyle\int_1^3 \int_0^2 x^3 y\, dy\, dx$

18. $\displaystyle\int_0^2 \int_1^3 x^3 y\, dx\, dy$

19. $\displaystyle\int_0^2 \int_1^3 x^3 y \, dy \, dx$

20. $\displaystyle\int_{-1}^1 \int_0^\pi x^2 \sin y \, dy \, dx$

21. $\displaystyle\int_2^6 \int_1^4 x^2 \, dx \, dy$

22. $\displaystyle\int_2^6 \int_1^4 y^2 \, dx \, dy$

23. $\displaystyle\int_0^1 \int_0^2 (x + 4y^3) \, dx \, dy$

24. $\displaystyle\int_0^2 \int_0^2 (x^2 - y^2) \, dy \, dx$

25. $\displaystyle\int_0^1 \int_2^3 \sqrt{x + 4y} \, dx \, dy$

26. $\displaystyle\int_0^{\pi/4} \int_0^{\pi/2} \cos(2x + y) \, dy \, dx$

27. $\displaystyle\int_1^2 \int_0^4 \frac{dy \, dx}{x + y}$

28. $\displaystyle\int_0^1 \int_0^2 e^{2x+3y} \, dy \, dx$

29. $\displaystyle\int_0^1 \int_2^3 \frac{1}{(x + 4y)^3} \, dx \, dy$

30. $\displaystyle\int_0^1 \int_0^1 \frac{y}{x^2 + 1} \, dy \, dx$

31. $\displaystyle\int_1^2 \int_1^2 \ln(xy) \, dy \, dx$

32. $\displaystyle\int_1^4 \int_0^2 y^{-2} e^{2x} \, dx \, dy$

33. Let $f(x, y) = mxy^2$, where m is a constant. Find a value of m such that $\displaystyle\iint_{\mathcal{R}} f(x, y) \, dx \, dy = 1$, where $\mathcal{R} = [0, 1] \times [0, 2]$.

In Exercises 34–37, evaluate the double integral of the function over the rectangle.

34. $\displaystyle\iint_{\mathcal{R}} (2x + 6y) \, dA, \quad \mathcal{R} = [0, 3] \times [0, 2]$

35. $\displaystyle\iint_{\mathcal{R}} x^2 y \, dA, \quad \mathcal{R} = [-1, 1] \times [0, 2]$

36. $\displaystyle\iint_{\mathcal{R}} \cos x \sin 2y \, dA, \quad \mathcal{R} = [0, \frac{\pi}{2}] \times [0, \frac{\pi}{2}]$

37. $\displaystyle\iint_{\mathcal{R}} \frac{y}{x + 1} \, dA, \quad \mathcal{R} = [0, 2] \times [0, 4]$

In Exercises 38–41, use Eq. (3) to evaluate the integral.

38. $\displaystyle\iint_{\mathcal{R}} e^x \sin y \, dA, \quad \mathcal{R} = [0, 2] \times [0, \frac{\pi}{4}]$

39. $\displaystyle\iint_{\mathcal{R}} e^{2x+3y} \, dA, \quad \mathcal{R} = [0, 1] \times [0, 2]$

40. $\displaystyle\int_0^1 \int_0^2 x^4 y^3 \, dy \, dx$

41. $\displaystyle\int_1^2 \int_0^2 \frac{x}{y} \, dx \, dy$

42. Evaluate $I = \displaystyle\int_1^3 \int_0^1 ye^{xy} \, dy \, dx$. You will need Integration by Parts and the formula

$$\int e^x (x^{-1} - x^{-2}) \, dx = x^{-1} e^x + C$$

Then evaluate I again using Fubini's Theorem to change the order of integration (i.e., integrate first with respect to y). Which method is easier?

43. **CAS** Use the inequality $0 \le \sin x \le x$ for $x \ge 0$ to show that

$$\int_0^1 \int_0^1 \sin(xy) \, dx \, dy \le \frac{1}{4}$$

Then use a computer algebra system to evaluate the double integral to three decimal places.

44. Evaluate $\displaystyle\int_0^1 \int_0^1 \frac{y}{1 + xy} \, dy \, dx$. *Hint:* Change the order of integration.

45. Calculate a Riemann sum $S_{3,3}$ on the square $\mathcal{R} = [0, 3] \times [0, 3]$ for the function $f(x, y)$ whose contour plot is shown in Figure 19. Choose sample points and use the plot to find the values of $f(x, y)$ at these points.

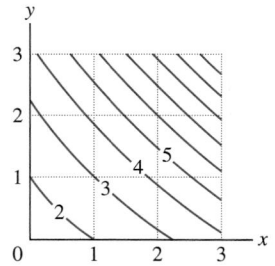

FIGURE 19 Contour plot of $f(x, y)$.

46. ✏️ Using Fubini's Theorem, argue that the solid in Figure 20 has volume AL.

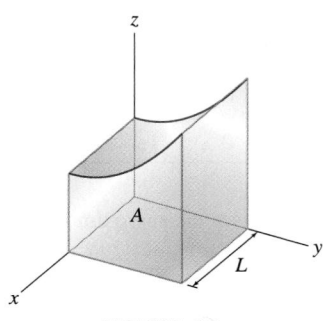

FIGURE 20

Further Insights and Challenges

47. Prove the following extension of the FTC to two variables: If $\dfrac{\partial^2 F}{\partial x\, \partial y} = f(x, y)$, then

$$\iint_{\mathcal{R}} f(x, y)\, dA = F(b, d) - F(a, d) - F(b, c) + F(a, c)$$

where $\mathcal{R} = [a, b] \times [c, d]$.

48. Let $F(x, y) = x^{-1}e^{xy}$. Show that $\dfrac{\partial^2 F}{\partial x\, \partial y} = ye^{xy}$ and use the result of Exercise 47 to evaluate $\displaystyle\iint_{\mathcal{R}} ye^{xy}\, dA$ for the rectangle $\mathcal{R} = [1, 3] \times [0, 1]$.

49. Find a function $F(x, y)$ satisfying $\dfrac{\partial^2 F}{\partial x\, \partial y} = 6x^2 y$ and use the result of Exercise 47 to evaluate $\displaystyle\iint_{\mathcal{R}} 6x^2 y\, dA$ for the rectangle $\mathcal{R} = [0, 1] \times [0, 4]$.

50. In this exercise, we use double integration to evaluate the following improper integral for $a > 0$ a positive constant:

$$I(a) = \int_0^{\infty} \frac{e^{-x} - e^{-ax}}{x}\, dx$$

(a) Use L'Hôpital's Rule to show that $f(x) = \dfrac{e^{-x} - e^{-ax}}{x}$, though not defined at $x = 0$, can be made continuous by assigning the value $f(0) = a - 1$. Then use the Comparison Theorem to show that $I(a)$ converges. *Hint:* $f(x) \le e^{-x} + e^{-ax}$ for $x > 1$.

(b) Show that $I(a) = \displaystyle\int_0^{\infty} \int_1^{a} e^{-xy}\, dy\, dx$.

(c) Prove, by interchanging the order of integration, that

$$I(a) = \ln a - \lim_{T \to \infty} \int_1^{a} \frac{e^{-Ty}}{y}\, dy \qquad \boxed{4}$$

(d) Use the Comparison Theorem to show that the limit in Eq. (4) is zero. *Hint:* If $a \ge 1$, use $e^{-Ty} \le e^{-T}$ for $y \ge a$, and if $a < 1$, use $e^{-Ty} \le e^{-aT}$ for $a \le y \le 1$. Conclude that $I(a) = \ln a$ (Figure 21).

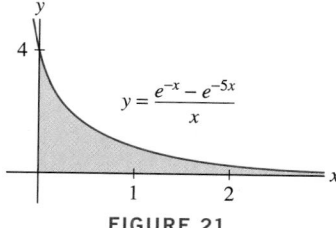

$$y = \frac{e^{-x} - e^{-5x}}{x}$$

FIGURE 21

15.2 Double Integrals over More General Regions

In the previous section, we restricted our attention to rectangular domains. We now treat double integrals over more general domains $\mathcal{D}$ whose boundaries are simple closed curves (a curve is *simple* if it does not intersect itself). We assume that $\mathcal{D}$ is closed, that is, $\mathcal{D}$ contains its boundary, and that the boundary is either smooth as in Figure 1(A) or has at most finitely many points of nondifferentiability as in Figure 1(B). A boundary curve of this type is called **piecewise smooth**.

FIGURE 1

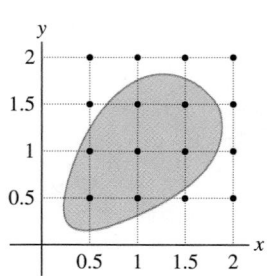

FIGURE 2 The function $\tilde{f}$ is zero outside of $\mathcal{D}$.

We shall define the double integral of a function $f(x, y)$ on a general domain $\mathcal{D}$ in terms of integration over rectangles. Choose a rectangle $\mathcal{R}$ containing $\mathcal{D}$ and define a new function $\tilde{f}(x, y)$ that agrees with $f(x, y)$ on $\mathcal{D}$ and is zero off of $\mathcal{D}$ (Figure 2):

$$\tilde{f}(x, y) = \begin{cases} f(x, y) & \text{if } (x, y) \in \mathcal{D} \\ 0 & \text{if } (x, y) \notin \mathcal{D} \end{cases}$$

We then define the double integral over $\mathcal{D}$ as the integral of $\tilde{f}$ over $\mathcal{R}$:

$$\boxed{\iint_{\mathcal{D}} f(x, y)\, dA = \iint_{\mathcal{R}} \tilde{f}(x, y)\, dA} \qquad \boxed{1}$$

We say that f is integrable over $\mathcal{D}$ if the integral of $\tilde{f}$ over $\mathcal{R}$ exists. The value of the integral does not depend on the choice of $\mathcal{R}$ because $\tilde{f}$ is zero outside of $\mathcal{D}$.

This definition seems reasonable because the integral of $\tilde{f}$ only "picks up" the values of f on $\mathcal{D}$. There is a possible difficulty, however. The function $\tilde{f}$ is usually discontinuous along the boundary of $\mathcal{D}$ since its values suddenly jump to zero when we cross the boundary. The following theorem, which we state without proof, tells us that these possible discontinuities do not affect the integrability of $\tilde{f}$ if f is continuous.

THEOREM 1   Let $\mathcal{D}$ be a domain whose boundary is a simple, closed, piecewise smooth curve. If $f(x, y)$ is continuous on $\mathcal{D}$, then $\displaystyle\iint_{\mathcal{D}} f(x, y)\, dA$ exists.

We may approximate the double integral of $f(x, y)$ over a nonrectangular domain $\mathcal{D}$ by Riemann sums. Choose a rectangle $\mathcal{R}$ containing $\mathcal{D}$ and subdivide $\mathcal{R}$ into MN subrectangles $\mathcal{R}_{ij}$ of size $\Delta A = \Delta x \Delta y$ using a regular partition, as in Section 15.1. Then choose a sample point P_{ij} in each $\mathcal{R}_{ij}$. Since $\tilde{f}(P_{ij}) = 0$ unless P_{ij} lies in $\mathcal{D}$, the Riemann sum reduces to a sum over those sample points that lie in $\mathcal{D}$:

$$S_{N,M} = \sum_{i=1}^{N}\sum_{j=1}^{M} \tilde{f}(P_{ij})\, \Delta x\, \Delta y = \underbrace{\sum f(P_{ij})\, \Delta x\, \Delta y}_{\substack{\text{Sum only over points} \\ P_{ij} \text{ that lie in } \mathcal{D}.}} \approx \iint_{\mathcal{D}} f(x, y)\, dA \qquad \boxed{2}$$

■ **EXAMPLE 1** Let $\mathcal{D}$ be the shaded region in Figure 3 and let $f(x, y) = x + y$. Approximate $\displaystyle\iint_{\mathcal{D}} f(x, y)\, dA$ by computing $S_{4,4}$ for $\displaystyle\iint_{\mathcal{R}} \tilde{f}(x, y)\, dA$ using the regular partition of $\mathcal{R} = [0, 2] \times [0, 2]$ and the upper right-hand corners of the squares as sample points.

Solution The subrectangles in Figure 3 have sides of length $\Delta x = \Delta y = \frac{1}{2}$ and area $\Delta A = \frac{1}{4}$. Of the 16 sample points in Figure 3, only seven lie in $\mathcal{D}$, so

$$S_{4,4} = \sum_{i=1}^{4}\sum_{j=1}^{4} \tilde{f}(P_{ij})\, \Delta x\, \Delta y = \frac{1}{4}\big(f(0.5, 0.5) + f(1, 0.5) + f(0.5, 1) + f(1, 1)$$

$$+\, f(1.5, 1) + f(1, 1.5) + f(1.5, 1.5)\big)$$

$$= \frac{1}{4}\big(1 + 1.5 + 1.5 + 2 + 2.5 + 2.5 + 3\big) = \frac{7}{2} \qquad ■$$

FIGURE 3

The linearity properties of the double integral carry over to general domains: If $f(x, y)$ and $g(x, y)$ are integrable and C is a constant, then

$$\iint_{\mathcal{D}} (f(x, y) + g(x, y)) \, dA = \iint_{\mathcal{D}} f(x, y) \, dA + \iint_{\mathcal{D}} g(x, y) \, dA$$

$$\iint_{\mathcal{D}} Cf(x, y) \, dA = C \iint_{\mathcal{D}} f(x, y) \, dA$$

We define the area of a domain $\mathcal{D}$ to be the double integral of the constant function $f(x, y) = 1$:

$$\boxed{\text{Area}(\mathcal{D}) = \iint_{\mathcal{D}} 1 \, dA} \qquad \boxed{3}$$

More generally, for any constant C,

$$\iint_{\mathcal{D}} C \, dA = C \, \text{Area}(\mathcal{D}) \qquad \boxed{4}$$

CONCEPTUAL INSIGHT According to Eq. (3), the area of a domain $\mathcal{D}$ is equal to the limit of Riemann sums for the function $f(x, y) = 1$. By Eq. (2), such a Riemann sum is obtained by covering $\mathcal{D}$ with a grid of rectangles (Figure 4) and summing up the areas of those rectangles that intersect $\mathcal{D}$ (the rectangles on the boundary are included only if the sample point lies in $\mathcal{D}$). In Exercise 35, we show that Eq. (3) yields the same value for the area as a single integral in the case of a region between two graphs.

Regions Between Curves

When the domain $\mathcal{D}$ is the region between two curves, we can evaluate the double integral as an iterated integral. A region of this type is called **simple**. More precisely, we say that $\mathcal{D}$ is **vertically simple** if $\mathcal{D}$ is the region between the graphs of continuous functions $y = \alpha(x)$ and $y = \beta(x)$ (Figure 5):

$$\mathcal{D} = \{(x, y) : a \le x \le b, \quad \alpha(x) \le y \le \beta(x)\}$$

Similarly, $\mathcal{D}$ is **horizontally simple** if

$$\mathcal{D} = \{(x, y) : c \le y \le d, \quad \alpha(y) \le x \le \beta(y)\}$$

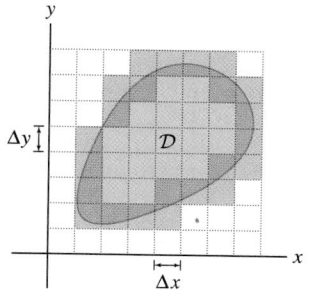

FIGURE 4 The area of $\mathcal{D}$ is approximated by the sum of the areas of rectangles that intersect $\mathcal{D}$.

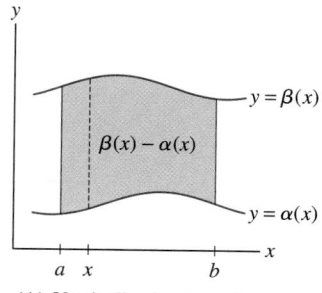

(A) Vertically simple region
$a \le x \le b, \quad \alpha(x) \le y \le \beta(x)$

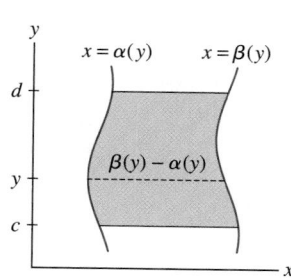

(B) Horizontally simple region
$c \le y \le d, \quad \alpha(y) \le x \le \beta(y)$

FIGURE 5

THEOREM 2 If $\mathcal{D}$ is a vertically simple region with description

$$a \leq x \leq b, \qquad \alpha(x) \leq y \leq \beta(x)$$

where $\alpha(x)$ and $\beta(x)$ are continuous, then

$$\iint_{\mathcal{D}} f(x, y)\, dA = \int_a^b \int_{\alpha(x)}^{\beta(x)} f(x, y)\, dy\, dx$$

If $\mathcal{D}$ is a horizontally simple region with description

$$c \leq y \leq d, \qquad \alpha(y) \leq x \leq \beta(y)$$

then

$$\iint_{\mathcal{D}} f(x, y)\, dA = \int_c^d \int_{\alpha(y)}^{\beta(y)} f(x, y)\, dx\, dy$$

Proof We sketch the proof. Assume that $\mathcal{D}$ is vertically simple (the horizontally simple case is similar). Let $\mathcal{R} = [a, b] \times [c, d]$ be a rectangle containing $\mathcal{D}$. Then

$$\iint_{\mathcal{D}} f(x, y)\, dA = \iint_{\mathcal{R}} \tilde{f}(x, y)\, dA = \int_a^b \int_c^d \tilde{f}(x, y)\, dy\, dx \qquad \boxed{5}$$

Although $\tilde{f}$ need not be continuous, the use of Fubini's Theorem in Eq. (5) can be justified. In particular, the integral $\int_c^d \tilde{f}(x, y)\, dy$ exists and is a continuous function of x.

By definition, $\tilde{f}(x, y)$ is zero outside $\mathcal{D}$, so for fixed x, $\tilde{f}(x, y)$ is zero unless y satisfies $\alpha(x) \leq y \leq \beta(x)$. Therefore,

$$\int_c^d \tilde{f}(x, y)\, dy = \int_{\alpha(x)}^{\beta(x)} f(x, y)\, dy$$

Substituting in Eq. (5), we obtain the desired equality:

$$\iint_{\mathcal{D}} f(x, y)\, dA = \int_a^b \int_{\alpha(x)}^{\beta(x)} f(x, y)\, dy\, dx \qquad ∎$$

Integration over a simple region is similar to integration over a rectangle with one difference: The limits of the inner integral may be functions instead of constants.

■ **EXAMPLE 2** Evaluate $\displaystyle\iint_{\mathcal{D}} x^3\, dA$, where $\mathcal{D}$ is the region in Figure 6.

Solution We may describe $\mathcal{D}$ as a vertically simple region:

$$\underbrace{1 \leq x \leq 3,}_{\substack{\text{Limits of outer} \\ \text{integral}}} \qquad \underbrace{2 - x^2 \leq y \leq 6 - x^2}_{\substack{\text{Limits of inner} \\ \text{integral}}}$$

Step 1. **Set up the iterated integral.**

$$\iint_{\mathcal{D}} x^3\, dA = \int_1^3 \int_{y=2-x^2}^{6-x^2} x^3\, dy\, dx \qquad \boxed{6}$$

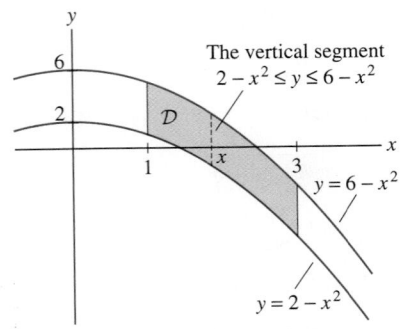

FIGURE 6 The region $1 \leq x \leq 3$, $2 - x^2 \leq y \leq 6 - x^2$. The inner integral in Eq. (6) is the integral of $f(x, y) = x^3$ over the vertical segment between the curves at x.

Step 2. **Evaluate the inner integral.**

As usual, we evaluate the inner integral by treating x as a constant, but now the upper and lower limits depend on x:

$$\int_{y=2-x^2}^{6-x^2} x^3\, dy = x^3 y \Big|_{y=2-x^2}^{6-x^2} = x^3(6-x^2) - x^3(2-x^2) = 4x^3$$

Step 3. **Complete the computation.**

$$\iint_{\mathcal{D}} x^3\, dx\, dy = \int_1^3 4x^3\, dx = x^4 \Big|_1^3 = 3^4 - 1^4 = 80 \qquad \blacksquare$$

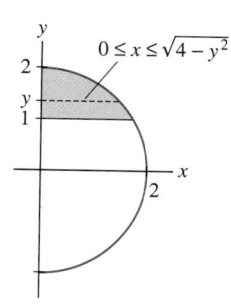

$$0 \le x \le \sqrt{4 - y^2}$$

FIGURE 7

■ **EXAMPLE 3** **Integration over Part of a Circle** Evaluate $\iint_{\mathcal{D}} \dfrac{x}{y}\, dA$, where $\mathcal{D}$ is the shaded part of the semicircle of radius 2 in Figure 7.

Solution The domain $\mathcal{D}$ may be described as a horizontally simple region:

$$1 \le y \le 2, \qquad 0 \le x \le \sqrt{4 - y^2}$$

These inequalities give us the limits of integration:

$$\iint_{\mathcal{D}} \frac{x}{y}\, dA = \int_{y=1}^{2} \int_{x=0}^{\sqrt{4-y^2}} \frac{x}{y}\, dx\, dy$$

Evaluate the inner integral, treating y as a constant:

$$\int_{x=0}^{\sqrt{4-y^2}} \frac{x}{y}\, dx = \frac{x^2}{2y} \Big|_{x=0}^{\sqrt{4-y^2}} = \frac{4-y^2}{2y} = \frac{2}{y} - \frac{y}{2}$$

The double integral equals

$$\iint_{\mathcal{D}} \frac{x}{y}\, dA = \int_{y=1}^{2} \left(\frac{2}{y} - \frac{y}{2} \right) dy = \left(2\ln y - \frac{y^2}{4} \right) \Big|_{y=1}^{2}$$

$$= (2\ln 2 - 1) - \left(0 - \frac{1}{4} \right) = 2\ln 2 - \frac{3}{4} \qquad \blacksquare$$

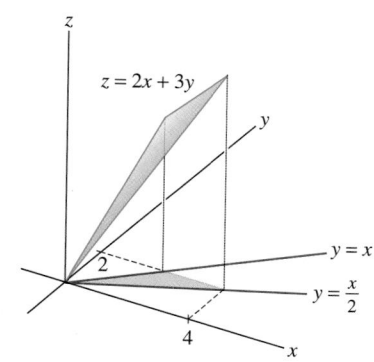

FIGURE 8

■ **EXAMPLE 4** **Calculating a Volume** Find the volume V of the region between the plane $z = 2x + 3y$ and the xy-plane lying over the triangle $\mathcal{D}$ in Figure 8.

Solution The triangular base $\mathcal{D}$ is the horizontally simple region

$$\mathcal{D}: \quad 0 \le y \le 2, \quad y \le x \le 2y$$

The volume V is the integral of $z = 2x + 3y$ over $\mathcal{D}$:

$$V = \iint_{\mathcal{D}} z\, dA = \int_0^2 \int_{x=y}^{2y} (2x + 3y)\, dx\, dy = \int_0^2 (x^2 + 3yx) \Big|_{x=y}^{2y} dy$$

$$= \int_0^2 \left((4y^2 + 6y^2) - (y^2 + 3y^2) \right) dy = \int_0^2 6y^2\, dy = 2y^3 \Big|_0^2 = 16 \qquad \blacksquare$$

When a domain is both vertically and horizontally simple, we may express the double integral as an iterated integral in two ways. In some cases, one iterated integral is easier to evaluate than the other.

■ **EXAMPLE 5** Choosing the Best Iterated Integral Evaluate $\iint_{\mathcal{D}} e^{y^2}\, dA$ for $\mathcal{D}$ as in Figure 9.

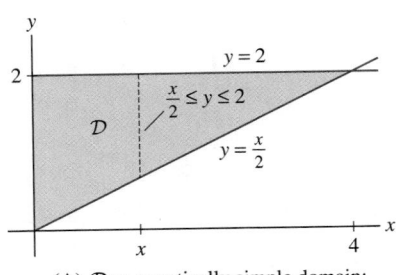

(A) $\mathcal{D}$ as a vertically simple domain:
$0 \le x \le 4,\ \ x/2 \le y \le 2$

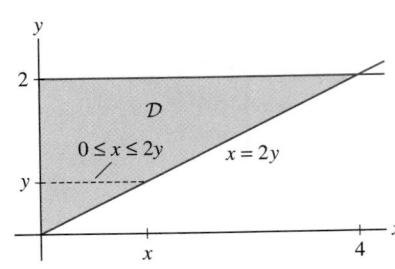

(B) $\mathcal{D}$ as a horizontally simple domain:
$0 \le y \le 2,\ \ 0 \le x \le 2y$

FIGURE 9 The region $\mathcal{D}$ is horizontally and vertically simple.

Solution First, let's try describing $\mathcal{D}$ as a vertically simple domain. Referring to Figure 9(A), we obtain

$$\mathcal{D}:\ \ 0 \le x \le 4,\ \ \frac{1}{2}x \le y \le 2 \quad \Rightarrow \quad \iint_{\mathcal{D}} e^{y^2}\, dA = \int_{x=0}^{4} \int_{y=x/2}^{2} e^{y^2}\, dy\, dx$$

The inner integral cannot be evaluated because we have no explicit antiderivative for e^{y^2}. Therefore, we try describing $\mathcal{D}$ as horizontally simple [Figure 9(B)]:

$$\mathcal{D}:\ \ 0 \le y \le 2, \qquad 0 \le x \le 2y$$

This leads to an iterated integral that can be evaluated:

$$\int_{0}^{2} \int_{x=0}^{2y} e^{y^2}\, dx\, dy = \int_{0}^{2} \left(xe^{y^2}\Big|_{x=0}^{2y} \right) dy = \int_{0}^{2} 2ye^{y^2}\, dy$$

$$= e^{y^2}\Big|_{0}^{2} = e^4 - 1 \qquad ■$$

■ **EXAMPLE 6** Changing the Order of Integration Sketch the domain $\mathcal{D}$ of integration corresponding to

$$\int_{1}^{9} \int_{\sqrt{y}}^{3} xe^y\, dx\, dy$$

Then change the order of integration and evaluate.

Solution The limits of integration give us inequalities that describe the domain $\mathcal{D}$:

$$1 \le y \le 9, \qquad \sqrt{y} \le x \le 3$$

From the sketch of $\mathcal{D}$ in Figure 10, we see that $\mathcal{D}$ can also be expressed as a vertically simple region:

$$1 \le x \le 3, \qquad 1 \le y \le x^2$$

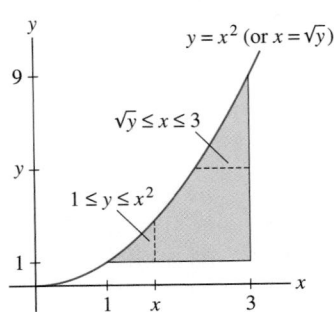

FIGURE 10 Describing $\mathcal{D}$ as a horizontally or vertically simple region.

so we may rewrite our integral and evaluate:

$$\int_1^9 \int_{\sqrt{y}}^3 xe^y \, dx \, dy = \int_1^3 \int_1^{x^2} xe^y \, dy \, dx = \int_1^3 \left(\int_{y=1}^{x^2} xe^y \, dy \right) dx$$

$$= \int_1^3 \left(xe^y \Big|_{y=1}^{x^2} \right) dx = \int_1^3 (xe^{x^2} - ex) \, dx = \frac{1}{2}(e^{x^2} - ex^2) \Big|_1^3$$

$$= \frac{1}{2}(e^9 - 9e) - 0 = \frac{1}{2}(e^9 - 9e)$$ ∎

◄·· **REMINDER** Equation (7) is similar to the definition of an average value in one variable:

$$\overline{f} = \frac{1}{b-a} \int_a^b f(x) \, dx = \frac{\int_a^b f(x) \, dx}{\int_a^b 1 \, dx}$$

The **average value** of a function $f(x, y)$ on a domain $\mathcal{D}$ is defined by

$$\boxed{\overline{f} = \frac{1}{\text{Area}(\mathcal{D})} \iint_{\mathcal{D}} f(x, y) \, dA = \frac{\iint_{\mathcal{D}} f(x, y) \, dA}{\iint_{\mathcal{D}} 1 \, dA}}$$ [7]

Note that the integral of $f(x, y)$ over $\mathcal{D}$ has the same value as the integral of the constant function $\overline{f}(x, y) = \overline{f}$ (Figure 11).

FIGURE 11

FIGURE 12 Pagoda with ceiling $H(x, y) = 32 - x^2 - y^2$.

■ **EXAMPLE 7** Computing Average Value An architect needs to know the average height $\overline{H}$ of the ceiling of a pagoda whose base $\mathcal{D}$ is the square $[-4, 4] \times [-4, 4]$ and roof is the graph of $H(x, y) = 32 - x^2 - y^2$ where distances are in feet (Figure 12). Find this average height.

Solution First, we compute the integral of $H(x, y)$ over $\mathcal{D}$:

$$\iint_{\mathcal{D}} (32 - x^2 - y^2) \, dA = \int_{-4}^4 \int_{-4}^4 (32 - x^2 - y^2) \, dy \, dx$$

$$= \int_{-4}^4 \left(\left(32y - x^2 y - \frac{1}{3}y^3 \right) \Big|_{-4}^4 \right) dx = \int_{-4}^4 \left(\frac{640}{3} - 8x^2 \right) dx$$

$$= \left(\frac{640}{3}x - \frac{8}{3}x^3 \right) \Big|_{-4}^4 = \frac{4,096}{3}$$

The area of $\mathcal{D}$ is $8 \times 8 = 64$, so the average height of the pagoda's ceiling is

$$\overline{H} = \frac{1}{\text{Area}(\mathcal{D})} \iint_{\mathcal{D}} H(x, y) \, dA = \frac{1}{64} \left(\frac{4{,}096}{3} \right) = \frac{64}{3} \approx 21.3 \text{ ft} \qquad \blacksquare$$

In the previous example, the average $\overline{H}$ lies between the minimum height $H = 0$ and maximum height $H = 32$. According to the first part of the next theorem, it is true in general that the average value lies between the minimum and maximum values of the function on the domain. The second part tells us that integration "preserves order" for functions f and g such that $f(x, y) \le g(x, y)$ on $\mathcal{D}$.

THEOREM 3   Let $f(x, y)$ and $g(x, y)$ be integrable functions on $\mathcal{D}$. Assume that $m \le f(x, y) \le M$ for all $(x, y) \in \mathcal{D}$, then

$$m \, \text{Area}(\mathcal{D}) \le \iint_{\mathcal{D}} f(x, y) \, dA \le M \, \text{Area}(\mathcal{D}) \qquad \boxed{8}$$

If $f(x, y) \le g(x, y)$ for all $(x, y) \in \mathcal{D}$, then

$$\iint_{\mathcal{D}} f(x, y) \, dA \le \iint_{\mathcal{D}} g(x, y) \, dA \qquad \boxed{9}$$

Proof   We prove (9) first. If $f(x, y) \le g(x, y)$, then every Riemann sum for $f(x, y)$ is less than or equal to the corresponding Riemann sum for g:

$$\sum f(P_{ij}) \, \Delta x_i \, \Delta y_j \le \sum g(P_{ij}) \, \Delta x_i \, \Delta y_j$$

We obtain (9) by taking the limit. Now suppose that $f(x, y) \le M$ and apply (9) with $g(x, y) = M$:

$$\iint_{\mathcal{D}} f(x, y) \, dA \le \iint_{\mathcal{D}} M \, dA = M \, \text{Area}(\mathcal{D})$$

This proves half of Eq. (8). The other half follows similarly. $\blacksquare$

If $\mathcal{D}$ is a nonconnected domain consisting of two or more subdomains bounded by simple closed curves [as in Figure 13(B)], then the integral $\iint_{\mathcal{D}} f(x, y) \, dA$ is defined as the sum of the integrals over the subdomains.

According to the Mean Value Theorem for Integrals (Section 6.2), if $f(x)$ is a continuous function with average value $\overline{f}$ on $[a, b]$, then there exists $c \in [a, b]$ such that $f(c) = \overline{f}$. A similar statement holds for functions of two or more variables on a domain $\mathcal{D}$, provided that $\mathcal{D}$ is closed, bounded, and connected. By definition, $\mathcal{D}$ is **connected** if any two points can be joined by a curve lying entirely in $\mathcal{D}$ [Figure 13(A)]. The domain in Figure 13(B) is not connected. See Exercise 72 for a proof of the MVT for Double Integrals.

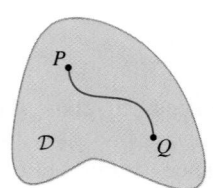

(A) Connected domain: Any two points can be joined by a curve lying entirely in $\mathcal{D}$.

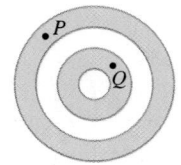

(B) Nonconnected domain.

FIGURE 13

THEOREM 4 Mean Value Theorem for Double Integrals   If $f(x, y)$ is continuous and $\mathcal{D}$ is closed, bounded, and connected, then there exists a point $P \in \mathcal{D}$ such that

$$\iint_{\mathcal{D}} f(x, y) \, dA = f(P) \, \text{Area}(\mathcal{D}) \qquad \boxed{10}$$

Equivalently, $f(P) = \overline{f}$, where $\overline{f}$ is the average value of f on $\mathcal{D}$.

Decomposing the Domain into Smaller Domains

Double integrals are additive with respect to the domain: If $\mathcal{D}$ is the union of domains $\mathcal{D}_1, \mathcal{D}_2, \ldots, \mathcal{D}_N$ that do not overlap except possibly on boundary curves (Figure 14), then

$$\iint_{\mathcal{D}} f(x, y)\, dA = \iint_{\mathcal{D}_1} f(x, y)\, dA + \cdots + \iint_{\mathcal{D}_N} f(x, y)\, dA$$

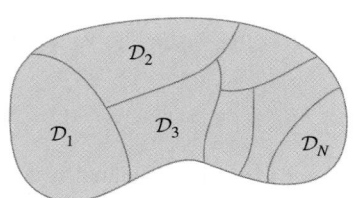

FIGURE 14 The region $\mathcal{D}$ is a union of smaller domains.

Additivity may be used to evaluate double integrals over domains $\mathcal{D}$ that are not simple but can be decomposed into finitely many simple domains.

We close this section with a simple remark that will be useful later on. If $f(x, y)$ is a continuous function on a *small* domain $\mathcal{D}$, then

In general, the approximation (11) is useful only if $\mathcal{D}$ is small in both width and length, that is, if $\mathcal{D}$ is contained in a circle of small radius. If $\mathcal{D}$ has small area but is very long and thin, then f may be far from constant on $\mathcal{D}$.

$$\iint_{\mathcal{D}} f(x, y)\, dx\, dy \approx \underbrace{f(P)\, \text{Area}(\mathcal{D})}_{\text{Function value} \times \text{area}} \qquad \boxed{11}$$

where P is any sample point in $\mathcal{D}$. In fact, by the MVT for Integrals, it is possible to choose P so that (11) holds with equality. If $\mathcal{D}$ is small, then f is nearly constant on $\mathcal{D}$ and (11) holds as a good approximation for all $P \in \mathcal{D}$.

If the domain $\mathcal{D}$ is not small, we may partition it into N smaller subdomains $\mathcal{D}_1, \ldots, \mathcal{D}_N$ and choose sample points P_j in $\mathcal{D}_j$. By additivity,

$$\iint_{\mathcal{D}} f(x, y)\, dx\, dy = \sum_{j=1}^{N} \iint_{\mathcal{D}_j} f(x, y)\, dx\, dy = \sum_{j=1}^{N} f(P_j)\, \text{Area}(\mathcal{D}_j)$$

and thus we have the approximation

$$\iint_{\mathcal{D}} f(x, y)\, dx\, dy \approx \sum_{j=1}^{N} f(P_j)\, \text{Area}(\mathcal{D}_j) \qquad \boxed{12}$$

Observe that Eq. (12) is a generalization of the Riemann sum approximation. In the case of a Riemann sum, the $\mathcal{D}_j$ are rectangles of area $\Delta A_{ij} = \Delta x_i\, \Delta y_j$.

■ **EXAMPLE 8** Estimate $\displaystyle\iint_{\mathcal{D}} f(x, y)\, dA$ for the domain $\mathcal{D}$ in Figure 15, using the areas and function values given there and the accompanying table.

Solution

$$\iint_{\mathcal{D}} f(x, y)\, dA \approx \sum_{j=1}^{4} f(P_j)\, \text{Area}(\mathcal{D}_j)$$

$$= (1.8)(1) + (2.2)(1) + (2.1)(0.9) + (2.4)(1.2) \approx 8.8 \qquad ■$$

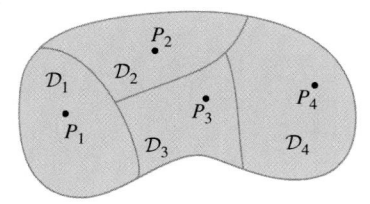

FIGURE 15

j	1	2	3	4
Area($\mathcal{D}_j$)	1	1	0.9	1.2
$f(P_j)$	1.8	2.2	2.1	2.4

15.2 SUMMARY

- Let $\mathcal{D}$ be a domain whose boundary is a simple closed curve that is either smooth or has a finite number of corners. To define the integral of $f(x, y)$ over $\mathcal{D}$, choose a rectangle $\mathcal{R}$ containing $\mathcal{D}$ and set

$$\iint_{\mathcal{D}} f(x, y)\, dA = \iint_{\mathcal{R}} \tilde{f}(x, y)\, dA$$

Here, $\tilde{f}(x, y) = f(x, y)$ if $(x, y) \in \mathcal{D}$ and $\tilde{f}(x, y) = 0$ otherwise. The value of the integral does not depend on the choice of $\mathcal{R}$.

- For any constant C, $\iint_{\mathcal{D}} C\, dA = C \cdot \text{Area}(\mathcal{D})$.

- If $\mathcal{D}$ is vertically or horizontally simple, $\iint_{\mathcal{D}} f(x, y)\, dA$ may be evaluated as an iterated integral:

Vertically simple domain $a \le x \le b,\quad \alpha(x) \le y \le \beta(x)$	$\displaystyle\int_a^b \int_{\alpha(x)}^{\beta(x)} f(x, y)\, dy\, dx$
Horizontally simple domain $c \le y \le d,\quad \alpha(y) \le x \le \beta(y)$	$\displaystyle\int_c^d \int_{\alpha(y)}^{\beta(y)} f(x, y)\, dx\, dy$

- The *average value* of f on $\mathcal{D}$ is $\overline{f} = \dfrac{1}{\text{Area}(\mathcal{D})} \iint_{\mathcal{D}} f(x, y)\, dA$.

- If $f(x, y) \le g(x, y)$ on $\mathcal{D}$, then $\iint_{\mathcal{D}} f(x, y)\, dA \le \iint_{\mathcal{D}} g(x, y)\, dA$.

- *Additivity with respect to the domain:* If $\mathcal{D}$ is a union of nonoverlapping (except possibly on their boundaries) domains $\mathcal{D}_1, \ldots, \mathcal{D}_N$, then

$$\iint_{\mathcal{D}} f(x, y)\, dA = \sum_{j=1}^{N} \iint_{\mathcal{D}_j} f(x, y)\, dA$$

- If the domains $\mathcal{D}_1, \ldots, \mathcal{D}_N$ are small and P_j is a sample point in $\mathcal{D}_j$, then

$$\iint_{\mathcal{D}} f(x, y)\, dA \approx \sum_{j=1}^{N} f(P_j)\text{Area}(\mathcal{D}_j)$$

15.2 EXERCISES

Preliminary Questions

1. Which of the following expressions do not make sense?

(a) $\displaystyle\int_0^1 \int_1^y f(x, y)\, dy\, dx$

(b) $\displaystyle\int_0^1 \int_1^x f(x, y)\, dy\, dx$

(c) $\displaystyle\int_1^y \int_0^1 f(x, y)\, dy\, dx$

(d) $\displaystyle\int_0^1 \int_x^1 f(x, y)\, dy\, dx$

2. Draw a domain in the plane that is neither vertically nor horizontally simple.

3. Which of the four regions in Figure 16 is the domain of integration

for $\displaystyle\int_{-\sqrt{2}/2}^{0} \int_{-x}^{\sqrt{1-x^2}} f(x, y)\, dy\, dx$?

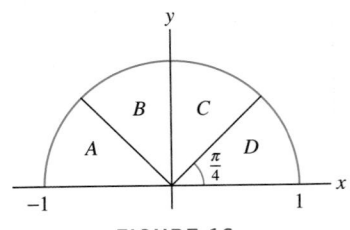

FIGURE 16

4. Let $\mathcal{D}$ be the unit circle. If the maximum value of $f(x, y)$ on $\mathcal{D}$ is 4, then the largest possible value of $\iint_{\mathcal{D}} f(x, y)\, dx\, dy$ is (choose the correct answer):

(a) 4   **(b)** 4π **(c)** $\dfrac{4}{\pi}$

Exercises

1. Calculate the Riemann sum for $f(x, y) = x - y$ and domain $\mathcal{D}$ in Figure 17 with two choices of sample points, ● and ○. Which do you think is a better approximation to the integral of f over $\mathcal{D}$? Why?

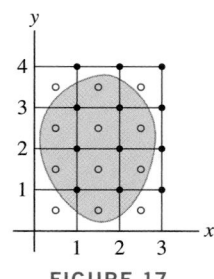

FIGURE 17

2. Approximate values of $f(x, y)$ at sample points on a grid are given in Figure 18. Estimate $\iint_{\mathcal{D}} f(x, y)\, dx\, dy$ by computing the Riemann sum with the given sample points.

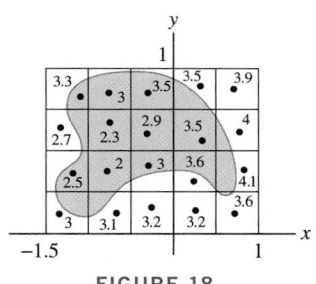

FIGURE 18

3. Let $\mathcal{D}$ be the domain defined by $0 \le x \le 1$, $x^2 \le y \le 4 - x^2$. Sketch $\mathcal{D}$ and express $\iint_{\mathcal{D}} y\, dA$ as an iterated integral and evaluate the result.

4. Express the domain $\mathcal{D}$ in Figure 19 as both a vertically and horizontally simple region and evaluate the integral of $f(x, y) = xy$ over $\mathcal{D}$ as an iterated integral in two ways.

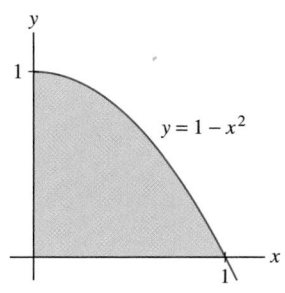

FIGURE 19

In Exercises 5–7, compute the double integral of $f(x, y) = x^2 y$ over the given shaded domain in Figure 20.

FIGURE 20

5. (A)   **6.** (B)   **7.** (C)

8. Sketch the domain $\mathcal{D}$ defined by $x + y \le 12$, $x \ge 4$, $y \ge 4$ and compute $\iint_{\mathcal{D}} e^{x+y}\, dA$.

9. Integrate $f(x, y) = x$ over the region bounded by $y = x^2$ and $y = x + 2$.

10. Calculate the double integral of $f(x, y) = y^2$ over the rhombus $\mathcal{R}$ in Figure 21.

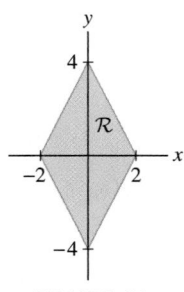

FIGURE 21

11. Sketch the region $\mathcal{D}$ between $y = x^2$ and $y = x(1 - x)$. Express $\mathcal{D}$ as an elementary region and calculate the integral of $f(x, y) = 2y$ over $\mathcal{D}$.

12. Find the double integral of $f(x, y) = x^3 y$ over the region between the curves $y = x^2$ and $y = x(1 - x)$.

13. Calculate the integral of $f(x, y) = x$ over the region $\mathcal{D}$ bounded above by $y = x(2 - x)$ and below by $x = y(2 - y)$. *Hint:* Apply the quadratic formula to the lower boundary curve to solve for y as a function of x.

14. Evaluate $\iint_{\mathcal{D}} (x + y)\, dx\, dy$, where $\mathcal{D} = \{(x, y) : x^2 + y^2 \le 4, y \ge 0\}$.

15. Integrate $f(x, y) = (x + y + 1)^{-2}$ over the triangle with vertices $(0, 0)$, $(4, 0)$, and $(0, 8)$.

16. Integrate $f(x, y) = x + y$ over:

(a) The region between $y = x$ and $y = 4x - x^2$.

(b) The region with boundary $y = x$, $y = 4x - x^2$, and $y = 0$. *Hint:* Write the answer as a sum of two integrals.

In Exercises 17–30, compute the integral of the function over the given region. Sketch the region.

17. $f(x, y) = x^3$, $0 \le x \le 2$, $x^2 \le y \le 4$

18. $f(x, y) = xy$, $0 \le x \le 2$, $0 \le y \le x^2$

19. $f(x, y) = x^2y$, $1 \le x \le 3$, $x \le y \le 2x + 1$

20. $f(x, y) = x + y$, $0 \le x \le 3$, $0 \le y \le \sqrt{9 - x^2}$

21. $f(x, y) = 1$, $0 \le x \le 1$, $1 \le y \le e^x$

22. $f(x, y) = \sin x$, $0 \le x \le \frac{\pi}{2}$, $0 \le y \le \cos x$

23. $f(x, y) = x$, $0 \le x \le 1$, $1 \le y \le e^{x^2}$

24. $f(x, y) = \sin(x + y)$, $0 \le x \le \frac{\pi}{2}$, $0 \le y \le 2x$

25. $f(x, y) = \cos(2x + y)$, $\frac{1}{2} \le x \le \frac{\pi}{2}$, $1 \le y \le 2x$

26. $f(x, y) = \frac{y}{x}$, $0 \le y \le 1$, $1 \le x \le e^y$

27. $f(x, y) = 2xy$, $0 \le y \le 1$, $y^2 \le x \le y$

28. $f(x, y) = e^{x+y}$, $2 \le y \le 4$, $y + 1 \le x \le 12 - y$

29. $f(x, y) = (x + y)^{-1}$, $1 \le y \le e$, $0 \le x \le y$

30. $f(x, y) = \dfrac{\ln y}{x^2}$, $1 \le y \le 2$, $y \le x \le 2y$

In Exercises 31–34, sketch the domain of integration and express the iterated integral in the opposite order.

31. $\displaystyle\int_0^4 \int_x^4 f(x, y)\, dy\, dx$

32. $\displaystyle\int_4^9 \int_{\sqrt{y}}^3 f(x, y)\, dx\, dy$

33. $\displaystyle\int_4^9 \int_2^{\sqrt{y}} f(x, y)\, dx\, dy$

34. $\displaystyle\int_0^1 \int_{e^x}^e f(x, y)\, dy\, dx$

35. According to Eq. (3), the area of a domain $\mathcal{D}$ is equal to $\displaystyle\iint_{\mathcal{D}} 1\, dA$. Prove that if $\mathcal{D}$ is the region between two curves $y = \alpha(x)$ and $y = \beta(x)$ for $a \le x \le b$, then

$$\iint_{\mathcal{D}} 1\, dA = \int_a^b (\beta(x) - \alpha(x))\, dx$$

The integral on the right is the area of $\mathcal{D}$ as defined in Chapter 6.

36. Sketch the domain $\mathcal{D}$ corresponding to

$$\int_0^4 \int_{\sqrt{y}}^2 \sqrt{x^2 + y}\, dx\, dy$$

Then change the order of integration and evaluate.

In Exercises 37–44, for each double integral, sketch the region. Then change the order of integration and evaluate. Explain the simplification achieved by interchanging the order.

37. $\displaystyle\int_0^1 \int_y^1 \frac{\sin x}{x}\, dx\, dy$

38. $\displaystyle\int_0^4 \int_{\sqrt{y}}^2 \sqrt{x^3 + 1}\, dx\, dy$

39. $\displaystyle\int_0^1 \int_{y=x}^1 xe^{y^3}\, dy\, dx$

40. $\displaystyle\int_0^1 \int_{y=x^{2/3}}^1 xe^{y^4}\, dy\, dx$

41. $\displaystyle\int_0^1 \int_0^{\pi/2} x\cos(xy)\, dx\, dy$

42. $\displaystyle\int_0^2 \int_0^3 y^3 \cos(xy^2)\, dy\, dx$

43. $\displaystyle\int_0^9 \int_0^{\sqrt{y}} \frac{x^3\, dx\, dy}{(3x^2 + y)^{1/2}}$

44. $\displaystyle\int_{-1}^1 \int_{-\sqrt{1-x^2}}^{\sqrt{1-x^2}} \sqrt{1 - y^2}\, dy\, dx$

45. Sketch the region $\mathcal{D}$ bounded by the curves $y = e^x$, $y = e^{\sqrt{x}}$, $y = 1$, and $y = 2$, and compute $\displaystyle\iint_{\mathcal{D}} (\ln y)^{-1}\, dA$ by writing $\mathcal{D}$ as a horizontally simple region.

46. Sketch $\mathcal{D}$ where $0 \le x, y \le 2$ and x or y is greater than 1, and compute $\displaystyle\iint_{\mathcal{D}} e^{x+y}\, dA$.

In Exercises 47–50, calculate the double integral of $f(x, y)$ over the triangle indicated in Figure 22.

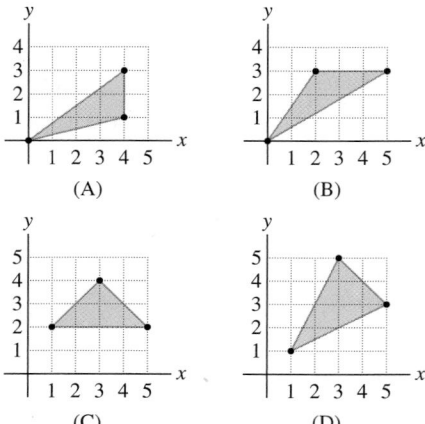

(A)

(B)

(C)

(D)

FIGURE 22

47. $f(x, y) = ye^x$, (A)

48. $f(x, y) = 1 - 2x$, (B)

49. $f(x, y) = \dfrac{x}{y^2}$, (C)

50. $f(x, y) = \dfrac{x}{y^2}$, (D)

51. Calculate the double integral $f(x, y) = \dfrac{\sin y}{y}$ over the region $\mathcal{D}$ in Figure 23.

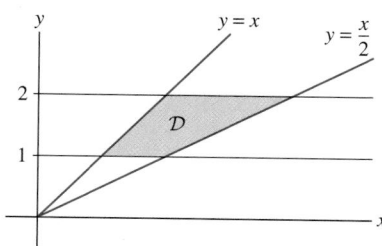

FIGURE 23

52. Evaluate $\displaystyle\iint_{\mathcal{D}} x \, dA$ for $\mathcal{D}$ in Figure 24.

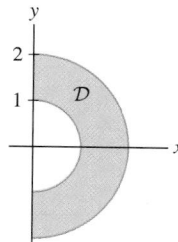

FIGURE 24

53. Find the volume of the region in Figure 25 bounded by the paraboloids $z = x^2 + y^2$ and $z = 8 - x^2 - y^2$.

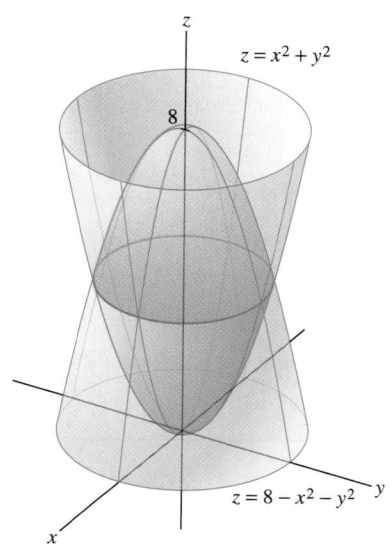

FIGURE 25

54. Find the volume of the region bounded by $z = 8 - y^2$, $y = 8 - x^2$, $x = 0$, $y = 0$, and $z = 0$.

55. Find the volume of the region enclosed by $z = 1 - y^2$ and $z = y^2 - 1$ for $0 \le x \le 2$.

56. Find the total mass of the square disk $0 \le x, y \le 1$ if the mass density is $\rho(x, y) = x^2 + y^2$.

57. A plate in the shape of the region bounded by $y = x^{-1}$ and $y = 0$ for $1 \le x \le 4$ has mass density $\rho(x, y) = y/x$. Calculate the total mass of the plate.

58. Calculate the average value of the function $f(x, y) = e^{x+y}$ on the square $[0, 1] \times [0, 1]$.

59. Calculate the average height above the x-axis of a point in the region $0 \le x \le 1, 0 \le y \le x^2$.

60. Find the average height of the "ceiling" in Figure 26 defined by $z = y^2 \sin x$ for $0 \le x \le \pi, 0 \le y \le 1$.

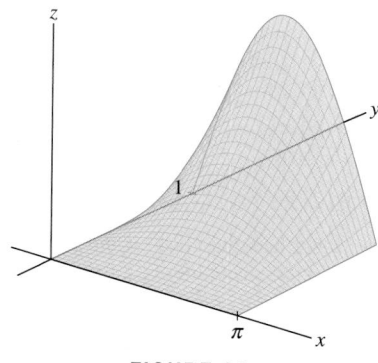

FIGURE 26

61. Calculate the average value of the x-coordinate of a point on the semicircle $x^2 + y^2 \le R^2$, $x \ge 0$. What is the average value of the y-coordinate?

62. What is the average value of a constant function $f(x, y) = C$ over a domain $\mathcal{D}$?

63. What is the average value of the linear function $f(x, y) = mx + ny + p$ on the ellipse $\left(\dfrac{x}{a}\right)^2 + \left(\dfrac{y}{b}\right)^2 \le 1$? Argue by symmetry rather than calculation.

64. Find the average square distance from the origin to a point in the domain $\mathcal{D}$ in Figure 27.

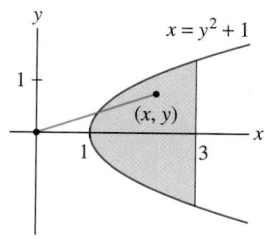

FIGURE 27

65. Show that the average value of $f(x, y) = y$ (the y-coordinate of the centroid) of the sector in Figure 28 is equal to
$$\bar{y} = \left(\frac{2R}{3}\right)\left(\frac{\sin \theta}{\theta}\right).$$

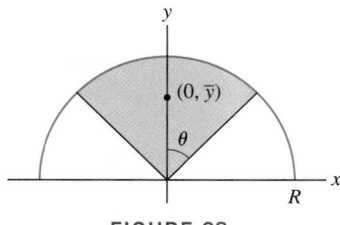

FIGURE 28

66. Prove the inequality $\displaystyle\iint_{\mathcal{D}} \frac{dA}{4 + x^2 + y^2} \le \pi$, where $\mathcal{D}$ is the disk $x^2 + y^2 \le 4$.

67. Find a point P in $\mathcal{D} = [0, 1] \times [0, 4]$ such that $f(P) = \overline{f}$, where $\overline{f}$ is the average of $f(x, y) = xy^2$ on $\mathcal{D}$ (the existence of such a point is guaranteed by the MVT for Double Integrals).

68. Verify the MVT for Double Integrals for $f(x, y) = e^{x-y}$ on the triangle bounded by $y = 0$, $x = 1$, and $y = x$.

In Exercises 69–70, use Eq. (12) to estimate the double integral.

69. The following table lists the areas of the subdomains $\mathcal{D}_j$ of the domain $\mathcal{D}$ in Figure 29 and the values of a function $f(x, y)$ at sample points $P_j \in \mathcal{D}_j$. Estimate $\displaystyle\iint_{\mathcal{D}} f(x, y)\, dA$.

j	1	2	3	4	5	6
Area($\mathcal{D}_j$)	1.2	1.1	1.4	0.6	1.2	0.8
$f(P_j)$	9	9.1	9.3	9.1	8.9	8.8

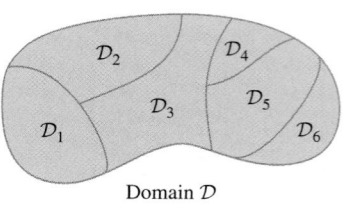

Domain $\mathcal{D}$

FIGURE 29

70. The domain $\mathcal{D}$ between the circles of radii 5 and 5.2 in the first quadrant in Figure 30 is divided into six subdomains of angular width $\Delta\theta = \frac{\pi}{12}$, and the values of a function $f(x, y)$ at sample points are given. Compute the area of the subdomains and estimate $\displaystyle\iint_{\mathcal{D}} f(x, y)\, dA$.

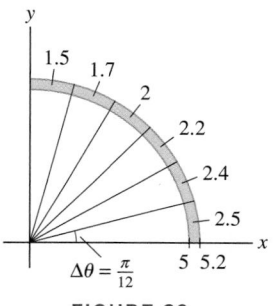

FIGURE 30

Further Insights and Challenges

71. Suppose that $f(x, y)$ is a continuous function on a closed connected domain $\mathcal{D}$. The Intermediate Value Theorem (IVT) states that if $f(P) = a$ and $f(Q) = b$, where $P, Q \in \mathcal{D}$, then $f(x, y)$ takes on every value between a and b at some point in $\mathcal{D}$.
(a) Show, by constructing a counterexample, that the IVT is false if $\mathcal{D}$ is not connected.
(b) Prove the IVT in two variables as follows: Let $\mathbf{c}(t)$ be a path such that $\mathbf{c}(0) = P$ and $\mathbf{c}(1) = Q$ (such a path exists because $\mathcal{D}$ is connected). Apply the IVT in one variable to the composite function $f(\mathbf{c}(t))$.

72. Use the IVT and Theorem 3 to prove the Mean Value Theorem for Double Integrals.

73. Let $f(y)$ be a function of y alone and set $G(t) = \displaystyle\int_0^t \int_0^x f(y)\, dy\, dx$.
(a) Use the FTC to prove that $G''(t) = f(t)$.
(b) Show, by changing the order in the double integral, that
$$G(t) = \int_0^t (t - y) f(y)\, dy.$$
This shows that the "second antiderivative" of $f(y)$ can be expressed as a single integral.

15.3 Triple Integrals

Triple integrals of functions $f(x, y, z)$ of three variables are a fairly straightforward generalization of double integrals. Instead of a rectangle in the plane, we consider a box

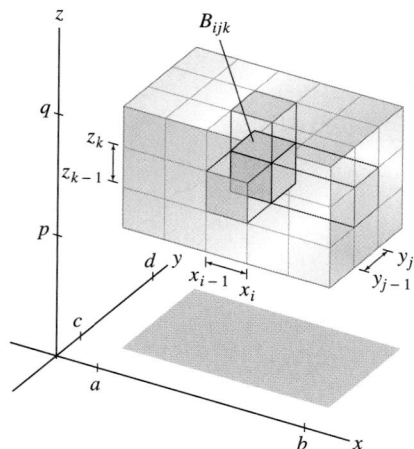

FIGURE 1 The box
$\mathcal{B} = [a, b] \times [c, d] \times [p, q]$ decomposed into subboxes.

(Figure 1)

$$\mathcal{B} = [a, b] \times [c, d] \times [p, q]$$

consisting of all points (x, y, z) in $\mathbf{R}^3$ such that

$$a \le x \le b, \qquad c \le y \le d, \qquad p \le z \le q$$

To form a Riemann sum, we choose positive integers N, M, L and partitions of the three intervals

$$a = x_0 < x_1 < \cdots < x_N = b$$

$$c = y_0 < y_1 < \cdots < y_M = d$$

$$p = z_0 < z_1 < \cdots < z_L = q$$

Let $\Delta x_i = x_i - x_{i-1}$, $\Delta y_j = y_j - y_{j-1}$, $\Delta z_k = z_k - z_{k-1}$ be the widths of the subintervals defined by these partitions. We then partition $\mathcal{B}$ into NML smaller boxes $\mathcal{B}_{ijk}$, whose volume we denote by ΔV_{ijk} (Figure 1):

$$\mathcal{B}_{ijk} = [x_{i-1}, x_i] \times [y_{j-1}, y_j] \times [z_{k-1}, z_k],$$

$$\text{Volume}(\mathcal{B}_{ijk}) = \Delta V_{ijk} = \Delta x_i \, \Delta y_j \, \Delta z_k$$

This partition is called **regular** if $[a, b]$, $[c, d]$, and $[p, q]$ are each divided into subintervals of equal lengths Δx, Δy, and Δz, where

$$\Delta x = \frac{b - a}{N}, \qquad \Delta y = \frac{d - c}{M}, \qquad \Delta z = \frac{q - p}{L}$$

For any choice of sample points P_{ijk} in $\mathcal{B}_{ijk}$, we define the Riemann sum:

$$S_{N,M,L} = \sum_{i=1}^{N} \sum_{j=1}^{M} \sum_{k=1}^{L} f(P_{ijk}) \, \Delta V_{ijk}$$

As in the previous section, let $\mathcal{P} = \{\{x_i\}, \{y_j\}, \{z_k\}\}$ be the set of endpoints and let $\|\mathcal{P}\|$ be the maximum of the widths Δx_i, Δy_j, Δz_k. If the sums $S_{N,M,L}$ approach a limit as $\|\mathcal{P}\| \to 0$ for arbitrary choices of sample points, we say that f is **integrable** over $\mathcal{B}$. The limit value is denoted

$$\iiint_{\mathcal{B}} f(x, y, z) \, dV = \lim_{\|\mathcal{P}\| \to 0} S_{N,M,L}$$

Triple integrals obey the same linearity properties as double and single integrals. The following theorem assures us that continuous functions are integrable over a box and that the triple integral can be evaluated as an iterated integral.

THEOREM 1 Fubini's Theorem for Triple Integrals If $f(x, y, z)$ is continuous on $\mathcal{B} = [a, b] \times [c, d] \times [p, q]$, then the triple integral exists and is equal to the iterated integral:

$$\iiint_{\mathcal{B}} f(x, y, z) \, dV = \int_{x=a}^{b} \int_{y=c}^{d} \int_{z=p}^{q} f(x, y, z) \, dz \, dy \, dx$$

Furthermore, the iterated integral may be evaluated in any order.

The notation dA, used in the previous section, suggests area and occurs in double integrals over domains in the plane. Similarly, dV suggests volume and occurs in triple integrals over regions in $\mathbf{R}^3$.

As noted in the theorem, we are free to evaluate the iterated integral in any order. For instance,

$$\int_{x=a}^{b}\int_{y=c}^{d}\int_{z=p}^{q} f(x, y, z)\,dz\,dy\,dx = \int_{z=p}^{q}\int_{y=c}^{d}\int_{x=a}^{b} f(x, y, z)\,dx\,dy\,dz$$

There are in total six different orders given by

$$dz\,dy\,dx \quad dy\,dz\,dx \quad dz\,dx\,dy \quad dx\,dz\,dy \quad dy\,dx\,dz \quad dx\,dy\,dz$$

■ **EXAMPLE 1** Integration over a Box Calculate the integral $\iiint_{\mathcal{B}} x^2 e^{y+3z}\,dV$, where $\mathcal{B} = [1, 4] \times [0, 3] \times [2, 6]$.

Solution We write this triple integral as an iterated integral:

$$\iiint_{\mathcal{B}} x^2 e^{y+3z}\,dV = \int_{1}^{4}\int_{0}^{3}\int_{2}^{6} x^2 e^{y+3z}\,dz\,dy\,dx$$

Step 1. **Evaluate the inner integral with respect to z, holding x and y constant.**

$$\int_{z=2}^{6} x^2 e^{y+3z}\,dz = \frac{1}{3}x^2 e^{y+3z}\Big|_{2}^{6} = \frac{1}{3}x^2 e^{y+18} - \frac{1}{3}x^2 e^{y+6} = \frac{1}{3}(e^{18} - e^6)x^2 e^y$$

Step 2. **Evaluate the middle integral with respect to y, holding x constant.**

$$\int_{y=0}^{3} \frac{1}{3}(e^{18} - e^6)x^2 e^y\,dy = \frac{1}{3}(e^{18} - e^6)x^2 \int_{y=0}^{3} e^y\,dy = \frac{1}{3}(e^{18} - e^6)(e^3 - 1)x^2$$

Step 3. **Evaluate the outer integral respect to x.**

$$\iiint_{\mathcal{B}} (x^2 e^{y+3z})\,dV = \frac{1}{3}(e^{18} - e^6)(e^3 - 1)\int_{x=1}^{4} x^2\,dx = 7(e^{18} - e^6)(e^3 - 1) \quad ■$$

Next, we consider triple integrals over solid regions $\mathcal{W}$ that are *simple* in the following sense: $\mathcal{W}$ consists of all points $P = (x, y, z)$ that lie between the graphs of continuous functions $\psi(x, y)$ and $\phi(x, y)$ over a domain $\mathcal{D}$ in the xy-plane (Figure 2). Assume that $\psi(x, y) \le \phi(x, y)$. Then

$$\mathcal{W} = \{(x, y, z) : (x, y) \in \mathcal{D} \quad \text{and} \quad \psi(x, y) \le z \le \phi(x, y)\} \qquad \boxed{1}$$

The domain $\mathcal{D}$ is the **projection** of $\mathcal{W}$ onto the xy-plane. As in the case of double integrals, we define the triple integral of $f(x, y, z)$ over $\mathcal{W}$ by

$$\iiint_{\mathcal{W}} f(x, y, z)\,dV = \iiint_{\mathcal{B}} \tilde{f}(x, y, z)\,dV$$

where $\mathcal{B}$ is a box containing $\mathcal{W}$ and $\tilde{f}$ is the function that is equal to f on $\mathcal{W}$ and equal to zero outside of $\mathcal{W}$. In practice, instead of using this formal definition, we evaluate triple integrals as iterated integrals. This is justified by the following theorem, whose proof is similar to that of Theorem 2 in Section 15.2.

*Sidebar (left margin):*

When the integrand is a product $f(x, y, z) = a(x)b(y)c(z)$ and the domain of integration is a box, then the triple integral factors as a product of three integrals. For example, the function in Example 1

$$f(x, y, z) = x^2 e^{y+3z} = x^2 e^y e^{3z}$$

is a product, and

$$\iiint_{\mathcal{B}} x^2 e^y e^{3z}\,dV$$

can be evaluated more simply as the product

$$\left(\int_{1}^{4} x^2\,dx\right)\left(\int_{0}^{3} e^y\,dy\right)\left(\int_{2}^{6} e^{3z}\,dz\right)$$

$$= (21)(e^3 - 1)\left(\frac{e^{18} - e^6}{3}\right)$$

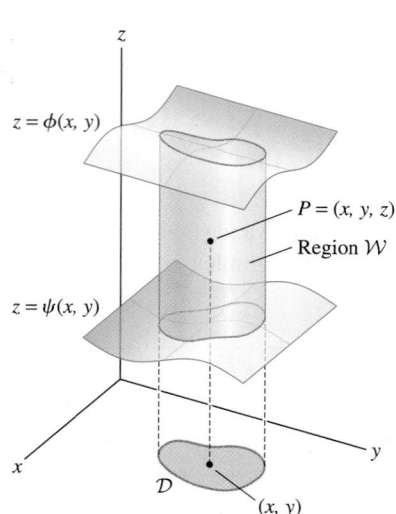

FIGURE 2 The point $P = (x, y, z)$ in the simple region $\mathcal{W}$ if $(x, y) \in \mathcal{D}$ and $\psi(x, y) \le z \le \phi(x, y)$.

THEOREM 2 Let $\mathcal{D}$ be a region in the xy-plane. Assume that $\psi(x, y)$ and $\phi(x, y)$ are continuous with $\psi(x, y) \leq \phi(x, y)$ for $(x, y) \in \mathcal{D}$. Then the triple integral of a continuous function $f(x, y, z)$ over the domain

$$\mathcal{W} = \{(x, y, z) : (x, y) \in \mathcal{D} \quad \text{and} \quad \psi(x, y) \leq z \leq \phi(x, y)\} \qquad \boxed{2}$$

exists and is equal to the iterated integral:

$$\iiint_{\mathcal{W}} f(x, y, z) \, dV = \iint_{\mathcal{D}} \left(\int_{z=\psi(x,y)}^{\phi(x,y)} f(x, y, z) \, dz \right) dA$$

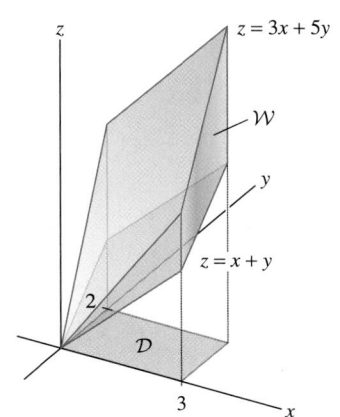

FIGURE 3 Region $\mathcal{W}$ between the planes $z = x + y$ and $z = 3x + 5y$ lying over $\mathcal{D} = [0, 3] \times [0, 2]$.

■ **EXAMPLE 2** Integrating over a Solid Region with a Rectangular Base Evaluate $\iiint_{\mathcal{W}} z \, dV$, where $\mathcal{W}$ is the region between the planes $z = x + y$ and $z = 3x + 5y$ lying over the rectangle $\mathcal{D} = [0, 3] \times [0, 2]$ (Figure 3).

Solution We apply Theorem 2 with $\psi(x, y) = x + y$ and $\phi(x, y) = 3x + 5y$ to write the triple integral as an iterated integral and evaluate:

$$\iiint_{\mathcal{W}} z \, dV = \iint_{\mathcal{D}} \left(\int_{z=x+y}^{3x+5y} z \, dz \right) dA = \int_{x=0}^{3} \int_{y=0}^{2} \int_{z=x+y}^{3x+5y} z \, dz \, dy \, dx$$

Note that the integral over $\mathcal{D}$ has been converted to an iterated integral.

Step 1. **Evaluate the inner integral with respect to z, holding x and y constant.**

$$\int_{z=x+y}^{3x+5y} z \, dz = \frac{1}{2} z^2 \Big|_{z=x+y}^{3x+5y} = \frac{1}{2}(3x + 5y)^2 - \frac{1}{2}(x + y)^2 = 4x^2 + 14xy + 12y^2$$

$$\boxed{3}$$

Step 2. **Evaluate the middle integral with respect to y, holding x constant.**

$$\int_{y=0}^{2} (4x^2 + 14xy + 12y^2) \, dy = (4x^2 y + 7xy^2 + 4y^3) \Big|_{y=0}^{2} = 8x^2 + 28x + 32$$

Step 3. **Evaluate the outer integral with respect to x.**

$$\iiint_{\mathcal{W}} z \, dV = \int_{x=0}^{3} (8x^2 + 28x + 32) \, dx = \left(\frac{8}{3} x^3 + 14x^2 + 32x \right) \Big|_{0}^{3}$$

$$= 72 + 126 + 96 = 294$$ ■

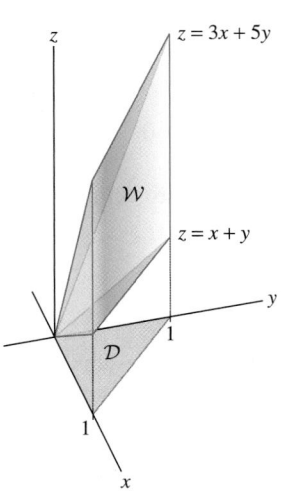

FIGURE 4 Region $\mathcal{W}$ between the planes $z = x + y$ and $z = 3x + 5y$ lying over the triangle $\mathcal{D}$.

■ **EXAMPLE 3** Integrating over a Solid Region with a Triangular Base Evaluate $\iiint_{\mathcal{W}} z \, dV$, where $\mathcal{W}$ is the region in Figure 4:

$$\mathcal{W} = \{(x, y, z) : x + y \leq z \leq 3x + 5y, x \geq 0, y \geq 0, x + y \leq 1\}$$

Solution $\mathcal{W}$ is the region between the two planes $z = x + y$ and $z = 3x + 5y$ lying over the triangle $\mathcal{D}$ in the xy-plane defined by

$$0 \leq x \leq 1, \qquad 0 \leq y \leq 1 - x$$

Thus, the triple integral is equal to the iterated integral:

$$\iiint_{\mathcal{W}} z \, dV = \iint_{\mathcal{D}} \left(\int_{z=x+y}^{3x+5y} z \, dz \right) dA = \underbrace{\int_{x=0}^{1} \int_{y=0}^{1-x} \int_{z=x+y}^{3x+5y} z \, dz \, dy \, dx}_{\text{Integral over triangle}}$$

The inner integral is the same as in the previous example [see Eq. (3)]:

$$\int_{z=x+y}^{3x+5y} z \, dz = \frac{1}{2} z^2 \Big|_{x+y}^{3x+5y} = 4x^2 + 14xy + 12y^2$$

Next, we integrate with respect to y:

$$\int_{y=0}^{1-x} (4x^2 + 14xy + 12y^2) \, dy = (4x^2 y + 7xy^2 + 4y^3) \Big|_{y=0}^{1-x}$$
$$= 4x^2(1-x) + 7x(1-x)^2 + 4(1-x)^3$$
$$= 4 - 5x + 2x^2 - x^3$$

and finally,

$$\iiint_{\mathcal{W}} z \, dV = \int_{x=0}^{1} (4 - 5x + 2x^2 - x^3) \, dx = 4 - \frac{5}{2} + \frac{2}{3} - \frac{1}{4} = \frac{23}{12} \qquad \blacksquare$$

In the examples so far, we have evaluated triple integrals by projecting the region $\mathcal{W}$ onto a domain $\mathcal{D}$ in the xy-plane. We can integrate equally well by projecting onto domains in the xz- or yz-planes. For example, if $\mathcal{W}$ is the region between the graphs of $x = \psi(y, z)$ and $x = \phi(y, z)$, where $\psi(y, z) \le \phi(y, z)$, lying over a domain $\mathcal{D}$ in the yz-plane (Figure 5), then

$$\iiint_{\mathcal{W}} f(x, y, z) \, dV = \iint_{\mathcal{D}} \left(\int_{x=\psi(y,z)}^{\phi(y,z)} f(x, y, z) \, dx \right) dA$$

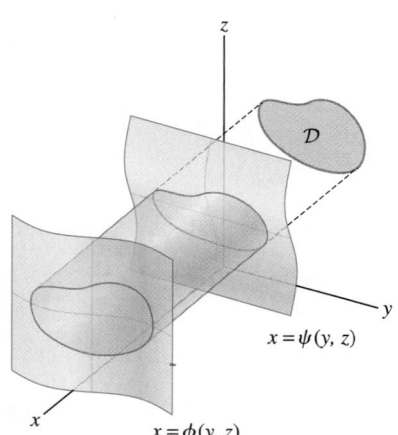

$x = \psi(y, z)$

$x = \phi(y, z)$

FIGURE 5 $\mathcal{D}$ is the projection of $\mathcal{W}$ onto the yz-plane.

■ **EXAMPLE 4** Writing a Triple Integral in Three Ways   Let $\mathcal{W}$ be the region bounded by (Figure 6)

$$z = 4 - y^2, \qquad y = 2x, \qquad z = 0, \qquad x = 0$$

Express $\displaystyle\iiint_{\mathcal{W}} xyz \, dV$ as an iterated integral in three ways, by projecting onto each of the three coordinate planes (but do not evaluate).

Solution   We consider each coordinate plane separately.

Step 1. The xy-plane.
The upper face $z = 4 - y^2$ intersects the first quadrant of the xy-plane ($z = 0$) in the line $y = 2$ [Figure 6(A)]. Therefore, the projection of $\mathcal{W}$ onto the xy-plane is a triangle $\mathcal{D}$ defined by $0 \le x \le 1$, $2x \le y \le 2$, and

$$\mathcal{W}: \quad 0 \le x \le 1, \qquad 2x \le y \le 2, \qquad 0 \le z \le 4 - y^2$$

$$\iiint_{\mathcal{W}} xyz \, dV = \int_{x=0}^{1} \int_{y=2x}^{2} \int_{z=0}^{4-y^2} xyz \, dz \, dy \, dx \qquad \boxed{4}$$

You can check that all three ways of writing the triple integral in Example 4 as an iterated integral yield the same answer:

$$\iiint xyz \, dV = \frac{2}{3}$$

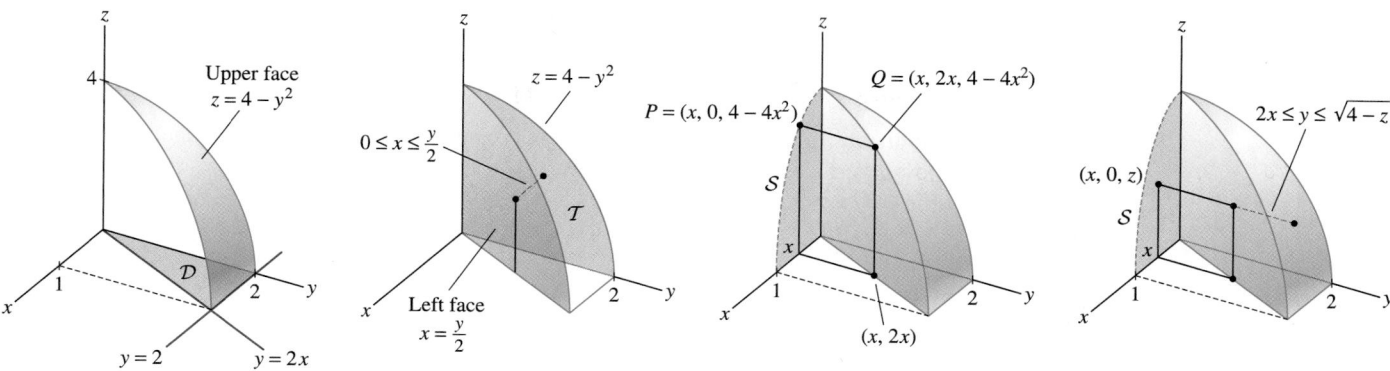

(A) Projection to xy-plane   (B) Projection to yz-plane   (C) Projection to xz-plane

FIGURE 6 Projections of $\mathcal{W}$ onto the coordinate planes.

Step 2. The yz-plane.

The projection of $\mathcal{W}$ onto the yz-plane is the domain $\mathcal{T}$ [Figure 6(B)]:

$$\mathcal{T}: \quad 0 \le y \le 2, \qquad 0 \le z \le 4 - y^2$$

The region $\mathcal{W}$ consists of all points lying between $\mathcal{T}$ and the "left face" $y = 2x$. In other words, the x-coordinate must satisfy $0 \le x \le \frac{1}{2}y$. Thus,

$$\mathcal{W}: \quad 0 \le y \le 2, \qquad 0 \le z \le 4 - y^2, \qquad 0 \le x \le \frac{1}{2}y$$

$$\iiint_{\mathcal{W}} xyz \, dV = \int_{y=0}^{2} \int_{z=0}^{4-y^2} \int_{x=0}^{y/2} xyz \, dx \, dz \, dy$$

Step 3. The xz-plane.

This is the most difficult case. First, we must describe the projection S of $\mathcal{W}$ onto the xz-plane. To do this, we need the equation of the dashed curve in the xz-plane in Figure 6(C). We claim that a point P on this dashed curve has coordinates $P = (x, 0, 4 - 4x^2)$. To check this, observe that P is the projection of a point $Q = (x, y, z)$ on the boundary of the left face. Since Q lies on both the plane $y = 2x$ and the surface $z = 4 - y^2$, $Q = (x, 2x, 4 - 4x^2)$ and $P = (x, 0, 4 - 4x^2)$, as claimed. We see that the projection of $\mathcal{W}$ onto the xz-plane is the domain

$$S: \quad 0 \le x \le 1, \qquad 0 \le z \le 4 - 4x^2$$

We have limits for x and z variables, so the triple integral can be written

$$\iiint_{\mathcal{W}} xyz \, dV = \int_{x=0}^{1} \int_{z=0}^{4-4x^2} \int_{y=??}^{??} xyz \, dy \, dz \, dx$$

What are the limits for y? The equation of the upper face $z = 4 - y^2$ may be written $y = \sqrt{4 - z}$. Thus, $\mathcal{W}$ is bounded by the left face $y = 2x$ and the upper face $y = \sqrt{4 - z}$, and the y-coordinate of a point in $\mathcal{W}$ satisfies

$$2x \le y \le \sqrt{4 - z}$$

Now we can write the triple integral as the following iterated integral:

$$\iiint_{\mathcal{W}} xyz \, dV = \int_{x=0}^{1} \int_{z=0}^{4-4x^2} \int_{y=2x}^{\sqrt{4-z}} xyz \, dy \, dz \, dx \qquad \blacksquare$$

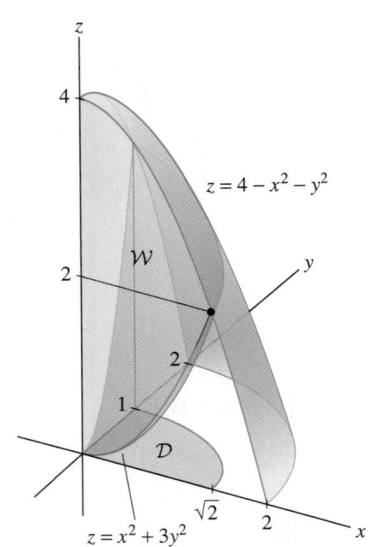

FIGURE 7 Region
$x^2 + 3y^2 \leq z \leq 4 - x^2 - y^2$ lying over
the ellipse $x^2 + 2y^2 = 2$ in the octant
$x, y, z \geq 0$.

■ **EXAMPLE 5** Integration over the Region Between Intersecting Surfaces Integrate $f(x, y, z) = x$ over the region $\mathcal{W}$ bounded above by $z = 4 - x^2 - y^2$ and below by $z = x^2 + 3y^2$ in the octant $x, y, z \geq 0$.

Solution First, we must determine the projection of $\mathcal{W}$ onto the xy-plane. The upper and lower surfaces intersect in the curve

$$x^2 + 3y^2 = 4 - x^2 - y^2 \qquad \text{or} \qquad x^2 + 2y^2 = 2$$

Therefore, as we see in Figure 7, $\mathcal{W}$ projects onto the domain $\mathcal{D}$ in the first quadrant bounded by the ellipse $x^2 + 2y^2 = 2$, and we have

$$\mathcal{W} = \{(x, y) \in \mathcal{D}, \ x^2 + 3y^2 \leq z \leq 4 - x^2 - y^2, x \geq 0, y \geq 0\}$$

$$\iiint_{\mathcal{W}} x \, dV = \iint_{\mathcal{D}} \int_{z=x^2+3y^2}^{4-x^2-y^2} x \, dz \, dA$$

Next, we express the quarter ellipse $\mathcal{D}$ in the first quadrant as a simple domain:

$$0 \leq y \leq 1, \qquad 0 \leq x \leq \sqrt{2 - 2y^2}$$

and write the triple integral as an iterated integral:

$$\iiint_{\mathcal{W}} x \, dV = \int_0^1 \int_{x=0}^{\sqrt{2-2y^2}} \int_{z=x^2+3y^2}^{4-x^2-y^2} x \, dz \, dx \, dy$$

Here are the results of evaluating the integrals in order:

Inner integral: $\displaystyle \int_{z=x^2+y^2}^{4-x^2-y^2} x \, dz = xz \Big|_{z=x^2+3y^2}^{4-x^2-y^2} = 4x - 2x^3 - 4y^2 x$

Middle integral:

$$\int_{x=0}^{\sqrt{2-2y^2}} (4x - 2x^3 - 4y^2 x) \, dx = \left(2x^2 - \frac{1}{2}x^4 - 2x^2 y^2\right)\Big|_{x=0}^{\sqrt{2-2y^2}}$$
$$= 2 - 4y^2 + 2y^4$$

Triple integral: $\displaystyle \iiint_{\mathcal{W}} x \, dV = \int_0^1 (2 - 4y^2 + 2y^4) \, dy = 2 - \frac{4}{3} + \frac{2}{5} = \frac{16}{15}$ ■

We conclude this section by mentioning some applications of triple integrals. First, we note that the triple integral of the function $f(x, y, z) = 1$ is the volume

$$\boxed{\text{Volume}(\mathcal{W}) = \iiint_{\mathcal{W}} 1 \, dV}$$

More generally, if we imagine $\mathcal{W}$ as a solid object with variable but continuous mass density $\rho(x, y, z)$ (in units of mass per unit volume), then the total mass of $\mathcal{W}$ is

$$\text{Total mass of } \mathcal{W} = \iiint_{\mathcal{W}} \rho(x, y, z)\, dV$$

We may justify this formula when $\mathcal{W}$ is a box by noting that the triple integral is a limit of Riemann sums:

$$S_{N,M,L} = \sum_{i=1}^{N} \sum_{j=1}^{M} \sum_{k=1}^{L} \underbrace{\rho(P_{ijk})\, \Delta V}_{\text{Approximate mass of } \mathcal{B}_{ijk}}$$

Here, $\mathcal{W}$ is divided into small boxes $\mathcal{B}_{ijk}$ with sides of lengths Δx, Δy, Δz and volume $\Delta V = \Delta x\, \Delta y\, \Delta z$, and P_{ijk} is a sample point in $\mathcal{B}_{ijk}$. If $\rho(x, y, z)$ is continuous, then the density is nearly constant on each small box, so the mass of $\mathcal{B}_{ijk}$ is approximately $\rho(P_{ijk})\, \Delta V$ (Figure 8). In the limit, we obtain the exact mass of $\mathcal{W}$. This argument applies in a straightforward way to more general regions.

The **average value** of $f(x, y, z)$ over a region $\mathcal{W}$ with volume V is defined by

$$\overline{f} = \frac{1}{V} \iiint_{\mathcal{W}} f(x, y, z)\, dV = \frac{\iiint_{\mathcal{W}} f(x, y, z)\, dV}{\iiint_{\mathcal{W}} 1\, dV}$$

The **centroid** of a region $\mathcal{W}$ is the point $P = (\overline{x}, \overline{y}, \overline{z})$ whose coordinates are the averages of x, y, and z over $\mathcal{W}$. More generally, if $\mathcal{W}$ represents a solid with mass density $\rho(x, y, z)$, then the **center of mass** of $\mathcal{W}$ is the point $P_{\text{CM}} = (x_{\text{CM}}, y_{\text{CM}}, z_{\text{CM}})$, where x_{CM} is defined as follows (and y_{CM}, z_{CM} are defined similarly):

$$x_{\text{CM}} = \frac{\iiint_{\mathcal{W}} x\rho(x, y, z)\, dV}{\iiint_{\mathcal{W}} \rho(x, y, z)\, dV}$$

■ **EXAMPLE 6** Center of Mass Find the z-coordinate of the center of mass of the first octant $\mathcal{W}$ of the unit sphere, assuming a mass density of $\rho(x, y, z) = y$ (Figure 9).

Solution The numerator of z_{CM} is

$$\iiint_{\mathcal{W}} z\, \rho(x, y, z)\, dV = \int_{x=0}^{1} \int_{y=0}^{\sqrt{1-x^2}} \int_{z=0}^{\sqrt{1-x^2-y^2}} zy\, dz\, dy\, dx$$

$$= \int_{x=0}^{1} \int_{y=0}^{\sqrt{1-x^2}} \left(\frac{1}{2} yz^2 \Big|_{z=0}^{\sqrt{1-x^2-y^2}} \right) dy\, dx$$

$$= \int_{x=0}^{1} \int_{y=0}^{\sqrt{1-x^2}} \frac{1}{2} y \left(1 - x^2 - y^2 \right) dy\, dx$$

$$= \int_{x=0}^{1} \left(\frac{1}{4} y^2 - \frac{1}{4} x^2 y^2 - \frac{1}{8} y^4 \right) \Big|_{y=0}^{\sqrt{1-x^2}} dx$$

$$= \int_{x=0}^{1} \left(\frac{1}{8} x^4 - \frac{1}{4} x^2 + \frac{1}{8} \right) dx = \frac{1}{15}$$

The total mass of the $\mathcal{W}$ is equal to the integral of the mass density $\rho(x, y, z)$ over $\mathcal{W}$. In the last step of the following computation, we use the formula stated in the margin:

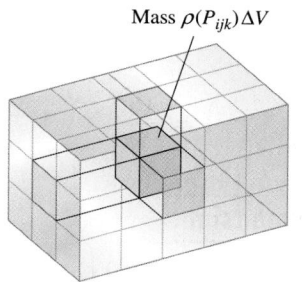

Mass $\rho(P_{ijk})\Delta V$

FIGURE 8 The mass of a small cube is approximately $\rho(P)\, \Delta V$.

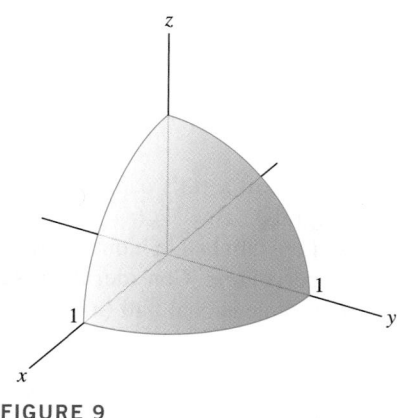

FIGURE 9

The following formula may be derived using trigonometric substitution $x = \sin\theta$:

$$\int (1 - x^2)^{3/2}\, dx$$

$$= \frac{1}{8}\left(x(5 - x^2)\sqrt{1 - x^2} + 3\sin^{-1}x\right) + C$$

$$m = \iiint_{\mathcal{W}} \rho(x, y, z)\, dV = \int_{x=0}^{1}\int_{y=0}^{\sqrt{1-x^2}}\int_{z=0}^{\sqrt{1-x^2-y^2}} y\, dz\, dy\, dx$$

$$= \int_{x=0}^{1}\int_{y=0}^{\sqrt{1-x^2}} y\sqrt{1 - x^2 - y^2}\, dy\, dx$$

$$= \int_{x=0}^{1}\left(-\frac{1}{3}(1 - x^2 - y^2)^{3/2}\right)\Bigg|_{y=0}^{\sqrt{1-x^2}} dx$$

$$= \frac{1}{3}\int_{x=0}^{1}(1 - x^2)^{3/2}\, dx = \frac{\pi}{16}$$

We conclude that

$$z_{\text{CM}} = \frac{1}{m}\iiint_{\mathcal{W}} z\, \rho(x, y, z)\, dV = \frac{16}{\pi}\left(\frac{1}{15}\right) = \frac{16}{15\pi}$$

Excursion: Volume of the Sphere in Higher Dimensions

Archimedes (287–212 BCE) discovered the beautiful formula $V = \frac{4}{3}\pi r^3$ for the volume of a sphere nearly 2,000 years before calculus was invented. Archimedes gave a brilliant geometric argument showing that the volume of a sphere is equal to two-thirds the volume of the circumscribed cylinder. He valued this achievement so highly that he requested that a sphere with circumscribed cylinder be engraved on his tomb. We can use integration to generalize Archimedes's formula to n dimensions. The ball of radius r in $\mathbf{R}^n$, denoted $B_n(r)$, is the set of points $(x_1, \ldots, x_n)$ in $\mathbf{R}^n$ such that

$$x_1^2 + x_2^2 + \cdots + x_n^2 \le r^2$$

$B_1(r)$ Interval of radius r

$B_2(r)$ Disk of radius r

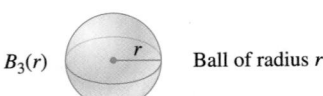
$B_3(r)$ Ball of radius r

FIGURE 10 Balls of radius r in dimensions $n = 1, 2, 3$.

The balls $B_n(r)$ in dimensions 1, 2, and 3 are the interval, disk, and ball shown in Figure 10. In dimensions $n \ge 4$, the ball $B_n(r)$ is hard, if not impossible, to visualize.

Let $V_n(r)$ be the volume (also called the **hypervolume**) of $B_n(r)$. We know that

$$V_1(r) = 2r, \qquad V_2(r) = \pi r^2, \qquad V_3(r) = \frac{4}{3}\pi r^3$$

It is reasonable to assume that $V_n(r)$ is proportional to r^n for all n (this can be proved using the Change of Variables Formula for multiple integrals). Thus $V_n(r) = A_n r^n$, where A_n is a constant. In fact, $A_n = V_n(1)$ is the volume of the unit ball in n dimensions.

To compute A_n, we shall integrate cross-sectional hypervolume. Consider the case $n = 3$, where the horizontal slice at height $z = c$ is a two-dimensional ball (a disk) of radius $\sqrt{r^2 - c^2}$ (Figure 11). The volume $V_3(r)$ is equal to the integral of these horizontal slices:

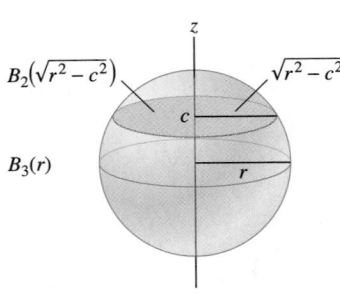
$B_2(\sqrt{r^2 - c^2})$
$\sqrt{r^2 - c^2}$
c
$B_3(r)$
r

FIGURE 11 The volume $V_3(r)$ is the integral of cross-sectional area $V_2(\sqrt{r^2 - c^2})$.

$$V_3(r) = \int_{z=-r}^{r} V_2\left(\sqrt{r^2 - z^2}\right) dz = \int_{z=-r}^{r} \pi(r^2 - z^2)\, dz = \frac{4}{3}\pi r^3$$

The same approach works in higher dimensions. The slice of $B_n(r)$ at height $x_n = c$ has equation

$$x_1^2 + x_2^2 + \cdots + x_{n-1}^2 + c^2 = r^2$$

This slice is the ball $B_{n-1}\left(\sqrt{r^2 - c^2}\right)$ of radius $\sqrt{r^2 - c^2}$, and $V_n(r)$ is obtained by integrating the hypervolume of these slices:

$$V_n(r) = \int_{x_n=-r}^{r} V_{n-1}\left(\sqrt{r^2 - x_n^2}\right) dx_n = A_{n-1} \int_{x_n=-r}^{r} \left(\sqrt{r^2 - x_n^2}\right)^{n-1} dx_n$$

Using the substitution $x_n = r \sin \theta$ and $dx_n = r \cos \theta \, d\theta$, we have

$$V_n(r) = A_{n-1} r^n \int_{-\pi/2}^{\pi/2} \cos^n \theta \, d\theta = A_{n-1} C_n r^n$$

where $C_n = \int_{\theta=-\pi/2}^{\pi/2} \cos^n \theta \, d\theta$. Setting $r = 1$, we obtain [since $V_n(1) = A_n$]

$$\boxed{A_n = A_{n-1} C_n} \qquad \boxed{5}$$

In Exercise 41, you are asked to use Integration by Parts to verify the relation

$$C_n = \left(\frac{n-1}{n}\right) C_{n-2} \qquad \boxed{6}$$

It is easy to check directly that $C_0 = \pi$ and $C_1 = 2$. By Eq. (6), $C_2 = \frac{1}{2}C_0 = \frac{\pi}{2}$, $C_3 = \frac{2}{3}(2) = \frac{4}{3}$, and so on. Here are the first few values of C_n:

n	0	1	2	3	4	5	6	7
C_n	π	2	$\frac{\pi}{2}$	$\frac{4}{3}$	$\frac{3\pi}{8}$	$\frac{16}{15}$	$\frac{5\pi}{16}$	$\frac{32}{35}$

We also have $A_1 = 2$ and $A_2 = \pi$, so we can use the values of C_n together with Eq. (5) to obtain the values of A_n in Table 1. Using induction and formulas (5) and (6), one can verify the following general formulas:

$$A_{2m} = \frac{\pi^m}{m!}, \qquad A_{2m+1} = \frac{2^{m+1}\pi^m}{1 \cdot 3 \cdot 5 \cdots (2m+1)}$$

This sequence of numbers A_n has a curious property. Recall that A_n is the volume of the unit ball in n dimensions. From Table 1, it appears that the hypervolumes increase up to dimension 5 and then begin to decrease. In Exercise 42, you are asked to verify that the five-dimensional ball has the largest volume. Furthermore, the volumes A_n tend to 0 as $n \to \infty$.

TABLE 1

n	A_n
1	2
2	$\pi \approx 3.14$
3	$\frac{4}{3}\pi \approx 4.19$
4	$\frac{\pi^2}{2} \approx 4.93$
5	$\frac{8\pi^2}{15} \approx 5.26$
6	$\frac{\pi^3}{6} \approx 5.17$
7	$\frac{16\pi^3}{105} \approx 4.72$

15.3 SUMMARY

- The triple integral over a box $\mathcal{B} = [a,b] \times [c,d] \times [p,q]$ is equal to the iterated integral

$$\iiint_{\mathcal{B}} f(x,y,z)\, dV = \int_{x=a}^{b} \int_{y=c}^{d} \int_{z=p}^{q} f(x,y,z)\, dz\, dy\, dx$$

The iterated integral may be written in any one of six possible orders, for example,

$$\int_{z=p}^{q} \int_{y=c}^{d} \int_{x=a}^{b} f(x,y,z)\, dx\, dy\, dz$$

- A *simple region* W in $\mathbf{R}^3$ is a region consisting of the points (x, y, z) between two surfaces $z = \psi(x, y)$ and $z = \phi(x, y)$, where $\psi(x, y) \le \phi(x, y)$, lying over a domain $\mathcal{D}$ in the xy-plane. In other words, W is defined by

$$(x, y) \in \mathcal{D}, \qquad \psi(x, y) \le z \le \phi(x, y)$$

The triple integral over W is equal to an iterated integral:

$$\iiint_W f(x, y, z)\, dV = \iint_{\mathcal{D}} \left(\int_{z=\psi(x,y)}^{\phi(x,y)} f(x, y, z)\, dz \right) dA$$

- The *average value* of $f(x, y, z)$ on a region W of volume V is the quantity

$$\overline{f} = \frac{1}{V} \iiint_W f(x, y, z)\, dV, \qquad V = \iiint_W 1\, dV$$

- If a solid W has continuous mass density $\rho(x, y, z)$, then the *center of mass* of W is $P_{\text{CM}} = (x_{\text{CM}}, y_{\text{CM}}, z_{\text{CM}})$, where x_{CM} is ($y_{\text{CM}}, z_{\text{CM}}$ are defined similarly):

$$x_{\text{CM}} = \frac{\iiint_W x\rho(x, y, z)\, dV}{\iiint_W \rho(x, y, z)\, dV}$$

The *centroid* of W is the center of mass with $\rho = 1$.

15.3 EXERCISES

Preliminary Questions

1. Which of (a)–(c) is not equal to $\displaystyle\int_0^1 \int_3^4 \int_6^7 f(x, y, z)\, dz\, dy\, dx$?

(a) $\displaystyle\int_6^7 \int_0^1 \int_3^4 f(x, y, z)\, dy\, dx\, dz$

(b) $\displaystyle\int_3^4 \int_0^1 \int_6^7 f(x, y, z)\, dz\, dx\, dy$

(c) $\displaystyle\int_0^1 \int_3^4 \int_6^7 f(x, y, z)\, dx\, dz\, dy$

2. Which of the following does not represent a meaningful triple integral?

(a) $\displaystyle\int_0^1 \int_0^x \int_{x+y}^{2x+y} e^{x+y+z}\, dz\, dy\, dx$

(b) $\displaystyle\int_0^1 \int_0^z \int_{x+y}^{2x+y} e^{x+y+z}\, dz\, dy\, dx$

3. Describe the projection of the region of integration W onto the xy-plane:

(a) $\displaystyle\int_0^1 \int_0^x \int_0^{x^2+y^2} f(x, y, z)\, dz\, dy\, dx$

(b) $\displaystyle\int_0^1 \int_0^{\sqrt{1-x^2}} \int_2^4 f(x, y, z)\, dz\, dy\, dx$

Exercises

In Exercises 1–10, evaluate $\displaystyle\iiint_{\mathcal{B}} f(x, y, z)\, dV$ *for the specified function f and box $\mathcal{B}$.*

1. $f(x, y, z) = z^4;\quad 2 \le x \le 8, 0 \le y \le 5, 0 \le z \le 1$

2. $f(x, y, z) = xyz;\quad 0 \le x, y, z \le 1$

3. $f(x, y, z) = xz^2;\quad [0, 2] \times [1, 6] \times [3, 4]$

4. $f(x, y, z) = (x - y)(y - z);\quad [0, 1] \times [0, 3] \times [0, 3]$

5. $f(x, y, z) = xe^{y-2z};\quad 0 \le x \le 2, 0 \le y, z \le 1$

6. $f(x, y, z) = \dfrac{x}{(y + z)^2};\quad [0, 2] \times [2, 4] \times [-1, 1]$

7. $f(x, y, z) = \dfrac{z}{x};\quad 1 \le x \le 3, 0 \le y \le 2, 0 \le z \le 4$

8. $f(x, y, z) = xy^2 e^{xyz};\quad 0 \le x, z \le 1, 0 \le y \le 2$

9. $f(x, y, z) = (x + z)^3;\quad [0, a] \times [0, b] \times [0, c]$

10. $f(x, y, z) = (2x + 3y - z)^4;\quad [0, a] \times [0, b] \times [0, c]$

In Exercises 11–14, evaluate $\displaystyle\iiint_W f(x, y, z)\, dV$ *for the function f and region W specified.*

11. $f(x, y, z) = x + y;\quad W: y \le z \le x, 0 \le y \le x, 0 \le x \le 1$

12. $f(x, y, z) = e^{x+y+z};\quad W: 0 \le z \le 1, 0 \le y \le x, 0 \le x \le 1$

13. $f(x, y, z) = xyz$; $\mathcal{W}: 0 \le z \le 1, 0 \le y \le \sqrt{1 - x^2}, 0 \le x \le 1$

14. $f(x, y, z) = e^{2x - z}$; $\mathcal{W}: x + y + z \le 1, x, y, z \ge 0$

15. Calculate the integral of $f(x, y, z) = z$ over the region $\mathcal{W}$ below the upper hemisphere of the sphere $x^2 + y^2 + z^2 = 9$ lying over the unit square $0 \le x, y \le 1$ in Figure 12.

16. Integrate $f(x, y, z) = z$ over the region $\mathcal{W}$ below the upper hemisphere of radius 3 as in Figure 12, but lying over the triangle in the xy-plane bounded by the lines $x = 1$, $y = 0$, and $x = y$.

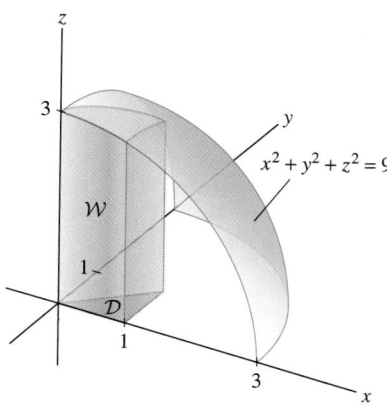

FIGURE 12

17. Calculate the integral of $f(x, y, z) = e^{x+y+z}$ over the tetrahedron $\mathcal{W}$ in Figure 13.

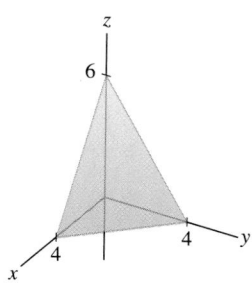

FIGURE 13

18. Describe the domain of integration and evaluate:

$$\int_0^3 \int_0^{\sqrt{9-x^2}} \int_0^{\sqrt{9-x^2-y^2}} xy \, dz \, dy \, dx$$

19. Integrate $f(x, y, z) = xz$ over the region in the first octant $(x, y, z \ge 0)$ above the parabolic cylinder $z = y^2$ and below the paraboloid $z = 8 - 2x^2 - y^2$.

20. Compute the integral of $f(x, y, z) = (y - x)z$ over the region within the cylinder $x^2 + y^2 = 16$ between the planes $z = 0$ and $z = y$.

21. Find the triple integral of the function z over the ramp in Figure 14. Here, z is the height above the ground.

FIGURE 14

22. Let W be the region in the first octant $(x \ge 0, y \ge 0, z \ge 0)$ of $\mathbf{R}^3$ bounded by $x + y + z = 1$ and $x^2 + y^2 = z - 2$. Express the triple integral (for an arbitrary function $f(x, y, z)$)

$$\iiint_W f(x, y, z) \, dV$$

as an iterated integral.

23. Find the volume of the solid in $\mathbf{R}^3$ bounded by $y = x^2$, $x = y^2$, $z = x + y + 5$, and $z = 0$.

24. Find the volume of the solid in the octant $x, y, z \ge 0$ bounded by $x + y + z = 1$ and $x + y + 2z = 1$.

25. Calculate $\iiint_{\mathcal{W}} y \, dV$, where $\mathcal{W}$ is the region above $z = x^2 + y^2$ and below $z = 5$, and bounded by $y = 0$ and $y = 1$.

26. Evaluate $\iiint_{\mathcal{W}} y \, dV$, where $\mathcal{W}$ is the domain bounded by the elliptic cylinder $\dfrac{x^2}{4} + \dfrac{y^2}{9} = 1$, the sphere $x^2 + y^2 + z^2 = 16$ in the first octant $x \ge 0, y \ge 0, z \ge 0$ (Figure 15).

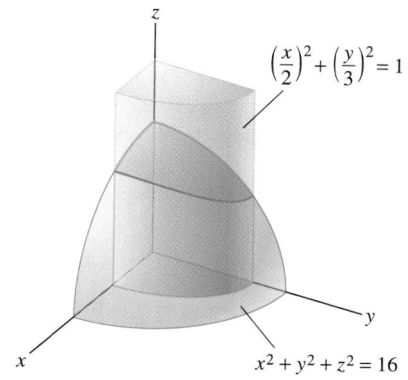

FIGURE 15

27. The solution to Example 4 expresses a triple integral in the three orders $dz \, dy \, dx$, $dx \, dz \, dy$, and $dy \, dz \, dx$. Write the integral in the three other orders: $dz \, dx \, dy$, $dx \, dy \, dz$, and $dy \, dx \, dz$.

28. Let $\mathcal{W}$ be the region bounded by $y + z = 2$, $2x = y$, $x = 0$, and $z = 0$ (Figure 16). Express and evaluate the triple integral of xe^z by projecting $\mathcal{W}$ onto the:

(a) xy-plane   **(b)** yz-plane   **(c)** xz-plane

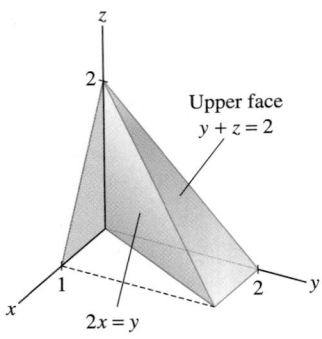

FIGURE 16

29. Express the triple integral $\iiint_{W} f(x, y, z)\,dV$ as an iterated integral in the two orders $dz\,dy\,dx$ and $dx\,dy\,dz$, where (Figure 17)

$$W = \left\{(x, y, z): \sqrt{x^2 + y^2} \le z \le 1\right\}$$

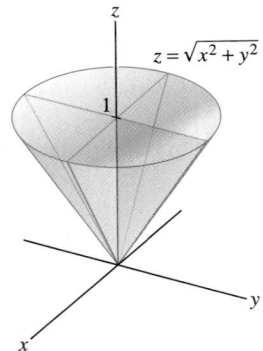

FIGURE 17

30. Let W be the region bounded by the cylinders $z = 1 - y^2$ and $y = x^2$, and the planes $z = 0$ and $y = 1$. Calculate the volume of W as a triple integral in the three orders $dz\,dy\,dx$, $dx\,dz\,dy$, and $dy\,dz\,dx$.

31. Find the average of $f(x, y, z) = xy\sin(\pi z)$ over the cube $0 \le x, y, z \le 1$.

32. Let W be the region bounded by the planes $y = z$, $2y + z = 3$, and $z = 0$ for $0 \le x \le 4$. Express the triple integral $\iiint_{W} f(x, y, z)\,dV$ as an iterated integral in the order $dx\,dz\,dy$ (project W onto the yz-plane). Then find the centroid of W.

33. Find the average of $f(x, y, z) = x^2 + y^2 + z^2$ over the region bounded by the planes $2y + z = 1$, $x = 0$, $x = 1$, $z = 0$, and $y = 0$.

34. Find the total mass of the part of the solid cylinder $x^2 + y^2 \le 4$ such that $x^2 \le z \le 9 - x^2$, assuming that the mass density is $\rho(x, y, z) = |y|$.

35. Calculate the centroid of the tetrahedron in Figure 18.

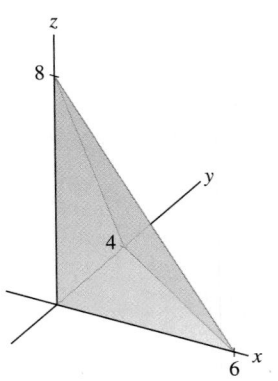

FIGURE 18

36. Find the centroid of the region described in Exercise 30.

37. Find the center mass of the solid bounded by planes $x + y + z = 1$, $x = 0$, $y = 0$, and $z = 0$, assuming a mass density of $\rho(x, y, z) = \sqrt{z}$.

38. Find the center mass of the cylinder $x^2 + y^2 = 1$ for $0 \le z \le 1$, assuming a mass density of $\rho(x, y, z) = x^2 + y^2$.

In Exercises 39–40, let $I = \int_0^1 \int_0^1 \int_0^1 f(x, y, z)\,dV$ and let $S_{N,N,N}$ be the Riemann sum approximation

$$S_{N,N,N} = \frac{1}{N^3} \sum_{i=1}^{N} \sum_{j=1}^{N} \sum_{k=1}^{N} f\left(\frac{i}{N}, \frac{j}{N}, \frac{k}{N}\right)$$

39. *CAS* Calculate $S_{N,N,N}$ for $f(x, y, z) = e^{x^2 - y - z}$ for $N = 10$, 20, 30. Then evaluate I and find an N such that $S_{N,N,N}$ approximates I to two decimal places.

40. *CAS* Calculate $S_{N,N,N}$ for $f(x, y, z) = \sin(xyz)$ for $N = 10$, 20, 30. Then use a computer algebra system to calculate I numerically and estimate the error $|I - S_{N,N,N}|$.

Further Insights and Challenges

41. Use Integration by Parts to verify Eq. (6).

42. Compute the volume A_n of the unit ball in $\mathbf{R}^n$ for $n = 8, 9, 10$.

Show that $C_n \le 1$ for $n \ge 6$ and use this to prove that of all unit balls, the five-dimensional ball has the largest volume. Can you explain why A_n tends to 0 as $n \to \infty$?

15.4 Integration in Polar, Cylindrical, and Spherical Coordinates

We have seen many examples in single-variable calculus where a well-chosen change of variables transforms a complicated integral into a simpler one. Change of variables is also useful in multivariable calculus, but with a slight change of emphasis. In the single-variable case, the goal is to find a change of variables that simplifies the integrand, whereas in the multivariable case, we are usually interested in simplifying the domain of integration. In this section, we treat the change of variables for polar, cylindrical, and spherical coordinates. These are the three most important cases. They are especially useful when the domain has radial or spherical symmetry (for example, a disk, ball, or cylinder). The general Change of Variables Formula is discussed in the next section.

Double Integrals in Polar Coordinates

Polar coordinates are convenient when the domain of integration is an angular sector or a **polar rectangle** (Figure 2):

$$\text{Polar rectangle } \mathcal{R}: \qquad \theta_1 \leq \theta \leq \theta_2, \qquad r_1 \leq r \leq r_2 \qquad \boxed{1}$$

We assume throughout that $r_1 \geq 0$ and that all radial coordinates are nonnegative. Recall that rectangular and polar coordinates are related by

$$x = r \cos \theta, \qquad y = r \sin \theta$$

Thus, we write a function $f(x, y)$ in polar coordinates as $f(r \cos \theta, r \sin \theta)$. We will prove the following Change of Variables Formula for the polar rectangle $\mathcal{R}$:

$$\boxed{\iint_{\mathcal{R}} f(x, y) \, dA = \int_{\theta_1}^{\theta_2} \int_{r_1}^{r_2} f(r \cos \theta, r \sin \theta) \, r \, dr \, d\theta} \qquad \boxed{2}$$

Notice the extra factor r in the integrand on the right.

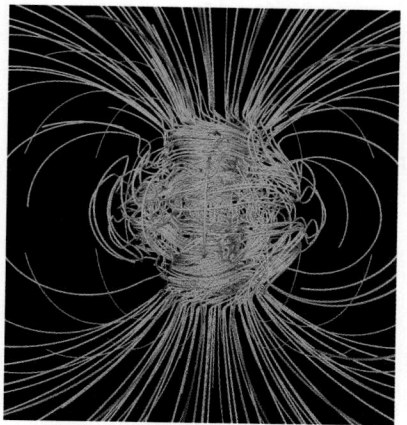

FIGURE 1 Spherical coordinates are used in mathematical models of the earth's magnetic field. This computer simulation, based on the Glatzmaier–Roberts model, shows the magnetic lines of force, representing inward and outward directed field lines in blue and yellow, respectively.

Equation (2) expresses the integral of $f(x, y)$ over a sector as the integral of a new function $f(r \cos \theta, r \sin \theta)$ over the rectangle $[\theta_1, \theta_2] \times [r_1, r_2]$. In this sense, the change of variables "simplifies" the domain of integration.

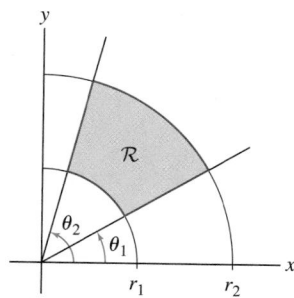

FIGURE 2 Polar rectangle.

FIGURE 3

To verify Eq. (2), we first compute the area ΔA of the small polar rectangle as in Figure 3 (recall that the area of a sector is $\frac{1}{2} r^2 \Delta \theta$):

$$\Delta A = \frac{1}{2}(r + \Delta r)^2 \Delta \theta - \frac{1}{2} r^2 \Delta \theta = r(\Delta r \, \Delta \theta) + \frac{1}{2}(\Delta r)^2 \Delta \theta \approx r(\Delta r \, \Delta \theta)$$

Next, we decompose $\mathcal{R}$ into a grid of $N \times M$ small polar subrectangles $\mathcal{R}_{ij}$, each of angular width $\Delta \theta = (\theta_2 - \theta_1)/N$ and radial width $\Delta r = (r_2 - r_1)/M$. If $\mathcal{R}_{ij}$ is small and $f(x, y)$ is continuous, then

$$\iint_{\mathcal{R}_{ij}} f(x, y)\, dx\, dy \approx f(P_{ij})\, \text{Area}(\mathcal{R}_{ij}) \approx f(P_{ij})\, r_{ij}\, \Delta r\, \Delta\theta \qquad \boxed{3}$$

◄·· REMINDER *According to (11) in Section 15.2, if f is continuous and D is a small domain,*

$$\iint_{\mathcal{D}} f(x, y)\, dA \approx f(P)\, \text{Area}(\mathcal{D})$$

where P is any sample point in D.

where $P_{ij} = (r_{ij}, \theta_{ij})$ is a sample point in $\mathcal{R}_{ij}$ (Figure 4). [In fact, we may choose P_{ij} so that (3) is an equality by the Mean Value Theorem for Double Integrals.] The integral over $\mathcal{R}$ is the sum of integrals (3); thus,

$$\iint_{\mathcal{R}} f(x, y)\, dx\, dy = \sum_{i=1}^{N} \sum_{j=1}^{M} \iint_{\mathcal{R}_{ij}} f(x, y)\, dx\, dy$$

$$\approx \sum_{i=1}^{N} \sum_{j=1}^{M} f(P_{ij})\, \text{Area}(\mathcal{R}_{ij})$$

$$\approx \sum_{i=1}^{N} \sum_{j=1}^{M} f(r_{ij} \cos\theta_{ij},\, r_{ij} \sin\theta_{ij})\, r_{ij}\, \Delta r\, \Delta\theta \qquad \boxed{4}$$

Observe that (4) is a Riemann sum for the integral of $r f(r \sin\theta, r \cos\theta)$ over the region $r_1 \leq r \leq r_2$, $\theta_1 \leq \theta \leq \theta_2$. As $N, M \to \infty$, this Riemann sum approaches the integral and we obtain Eq. (2).

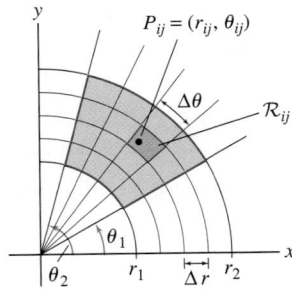

FIGURE 4 Decomposition of a polar rectangle into subrectangles.

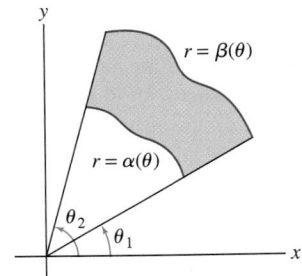

FIGURE 5 General polar region.

A similar derivation is valid for more general polar regions that may be described as the region between two polar curves $r = \alpha(\theta)$ and $r = \beta(\theta)$ (Figure 5).

Equation (5) is summarized in the symbolic expression for the "area element" dA in polar coordinates:

$$dA = r\, dr\, d\theta$$

THEOREM 1 Double Integral in Polar Coordinates   Let $\mathcal{D}$ be a domain of the form $\theta_1 \leq \theta \leq \theta_2$, $\alpha(\theta) \leq r \leq \beta(\theta)$, and assume that $f(x, y)$ is continuous on $\mathcal{D}$. Then

$$\iint_{\mathcal{D}} f(x, y)\, dA = \int_{\theta_1}^{\theta_2} \int_{r=\alpha(\theta)}^{\beta(\theta)} f(r \cos\theta, r \sin\theta)\, r\, dr\, d\theta \qquad \boxed{5}$$

■ **EXAMPLE 1** Use polar coordinates to compute $\displaystyle\iint_{\mathcal{D}} (x + y)\, dA$, where $\mathcal{D}$ is the quarter annulus $4 \leq x^2 + y^2 \leq 16$, $x, y \geq 0$.

Solution

Step 1. Describe $\mathcal{D}$ and f in polar coordinates.
The quarter annulus $\mathcal{D}$ is defined by the inequalities (Figure 6)

$$0 \le \theta \le \frac{\pi}{2}, \qquad 2 \le r \le 4$$

$$f(x, y) = x + y = r \cos \theta + r \sin \theta = r(\cos \theta + \sin \theta)$$

Step 2. Change variables in the integral and evaluate.
In converting to polar coordinates, we replace dA by $r\, dr\, d\theta$:

$$\iint_{\mathcal{D}} (x + y)\, dA = \int_0^{\pi/2} \int_2^4 r(\cos \theta + \sin \theta)\, r\, dr\, d\theta$$

The inner integral is

$$\int_2^4 (\cos \theta + \sin \theta)\, r^2\, dr = (\cos \theta + \sin \theta)\left(\frac{4^3}{3} - \frac{2^3}{3}\right) = \frac{56}{3}(\cos \theta + \sin \theta)$$

and

$$\iint_{\mathcal{D}} (x + y)\, dA = \frac{56}{3} \int_0^{\pi/2} (\cos \theta + \sin \theta)\, d\theta = \frac{56}{3}(\sin \theta - \cos \theta)\Big|_0^{\pi/2} = \frac{112}{3} \quad \blacksquare$$

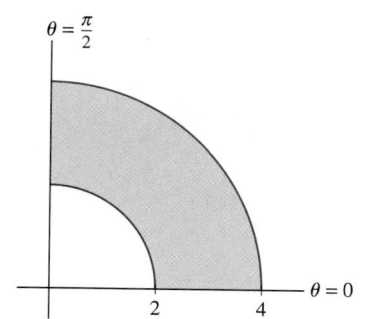

FIGURE 6 Quarter annulus $0 \le \theta \le \frac{\pi}{2}$, $2 \le r \le 4$.

■ **EXAMPLE 2** Calculate $\displaystyle\iint_{\mathcal{D}} (x^2 + y^2)^{-2}\, dA$ for the shaded domain $\mathcal{D}$ in Figure 7.

Solution

Step 1. Describe $\mathcal{D}$ and f in polar coordinates.
Recall from Section 11.3 (Example 7) that:

- The vertical line $x = 1$ has polar equation $r \cos \theta = 1$ or $r = \sec \theta$.
- The circle of radius 1 and center $(1, 0)$ has polar equation $r = 2 \cos \theta$.

The quarter circle lies in the angular sector $0 \le \theta \le \frac{\pi}{4}$ [note that the point $P = (1, 1)$ has polar angle $\frac{\pi}{4}$]. A ray of angle θ intersects $\mathcal{D}$ for r between $\sec \theta$ and $2 \cos \theta$. Therefore,

$$\mathcal{D}: \quad 0 \le \theta \le \frac{\pi}{4}, \qquad \sec \theta \le r \le 2 \cos \theta$$

On the other hand, $f(x, y) = (x^2 + y^2)^{-2} = (r^2)^{-2} = r^{-4}$.

Step 2. Change variables in the integral and evaluate.
By the Change of Variables Formula,

$$\iint_{\mathcal{D}} (x^2 + y^2)^{-2}\, dA = \int_0^{\pi/4} \int_{\sec \theta}^{2 \cos \theta} r^{-4}\, r\, dr\, d\theta = \int_0^{\pi/4} \int_{\sec \theta}^{2 \cos \theta} r^{-3}\, dr\, d\theta$$

The inner integral is

$$\int_{\sec \theta}^{2 \cos \theta} r^{-3}\, dr = -\frac{1}{2} r^{-2}\Big|_{\sec \theta}^{2 \cos \theta} = -\frac{1}{8} \sec^2 \theta + \frac{1}{2} \cos^2 \theta$$

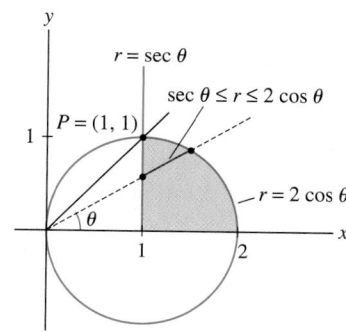

FIGURE 7

←·· *REMINDER*

$$\cos^2 \theta = \frac{1}{2}(1 + \cos 2\theta)$$

$$\int \cos^2 \theta\, d\theta = \frac{1}{2}\left(\theta + \frac{1}{2}\sin 2\theta\right) + C$$

$$\int \sec^2 \theta\, d\theta = \tan \theta + C$$

Therefore,

$$\iint_{\mathcal{D}} (x^2 + y^2)^{-2}\, dA = \int_0^{\pi/4} \left(\frac{1}{2} \cos^2 \theta - \frac{1}{8} \sec^2 \theta \right) d\theta$$

$$= \left(\frac{1}{4}(\theta + \frac{1}{2} \sin 2\theta) - \frac{1}{8} \tan \theta \right) \bigg|_0^{\pi/4}$$

$$= \frac{1}{4} \left(\frac{\pi}{4} + \frac{1}{2} \sin \frac{\pi}{2} \right) - \frac{1}{8} \tan \frac{\pi}{4} = \frac{\pi}{16} \qquad \blacksquare$$

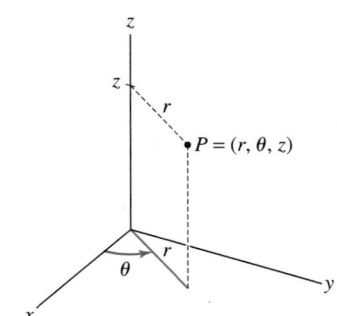

FIGURE 8 Cylindrical coordinates.

Triple Integrals in Cylindrical Coordinates

Cylindrical coordinates, introduced in Section 12.7, are useful when the domain has **axial symmetry**, that is, symmetry with respect to an axis. In cylindrical coordinates (r, θ, z), the axis of symmetry is the z-axis. Recall the relations (Figure 8)

$$x = r \cos \theta, \quad y = r \sin \theta, \quad z = z$$

To set up a triple integral in cylindrical coordinates, we assume that the domain of integration $\mathcal{W}$ can be described as the region between two surfaces (Figure 9)

$$z_1(r, \theta) \leq z \leq z_2(r, \theta)$$

lying over a domain $\mathcal{D}$ in the xy-plane with polar description

$$\mathcal{D}: \quad \theta_1 \leq \theta \leq \theta_2, \quad \alpha(\theta) \leq r \leq \beta(\theta)$$

By Fubini's Theorem,

$$\iiint_{\mathcal{W}} f(x, y, z)\, dV = \iint_{\mathcal{D}} \left(\int_{z=z_1(r,\theta)}^{z_2(r,\theta)} f(x, y, z)\, dz \right) dA$$

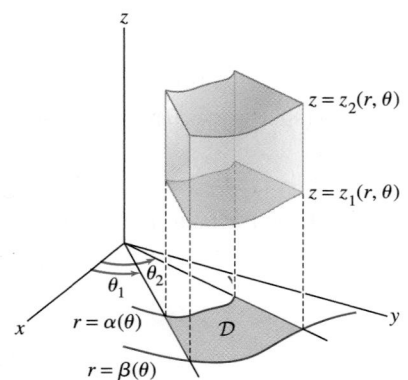

FIGURE 9 Region described in cylindrical coordinates.

Expressing the integral over $\mathcal{D}$ in polar coordinates, we obtain the following Change of Variables Formula.

We summarize Eq. (6) in a formula for the volume element (Figure 10):

$$dV = r\, dz\, dr\, d\theta$$

THEOREM 2 Triple Integrals in Cylindrical Coordinates   For a region $\mathcal{W}$ of the form

$$\theta_1 \leq \theta \leq \theta_2, \quad \alpha(\theta) \leq r \leq \beta(\theta), \quad z_1(r, \theta) \leq z \leq z_2(r, \theta)$$

the triple integral $\displaystyle\iiint_{\mathcal{W}} f(x, y, z)\, dV$ is equal to:

$$\int_{\theta_1}^{\theta_2} \int_{r=\alpha(\theta)}^{\beta(\theta)} \int_{z=z_1(r,\theta)}^{z_2(r,\theta)} f(r \cos \theta, r \sin \theta, z)\, r\, dz\, dr\, d\theta \qquad \boxed{6}$$

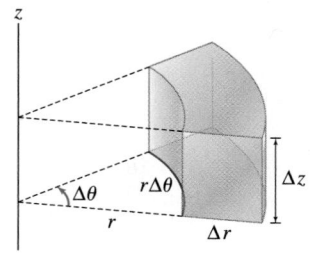

FIGURE 10

■ **EXAMPLE 3** Integrate $f(x, y, z) = z\sqrt{x^2 + y^2}$ over the cylinder $x^2 + y^2 \leq 4$ for $1 \leq z \leq 5$ (Figure 11).

Solution The domain of integration $\mathcal{W}$ lies above the circle of radius 2, so in cylindrical coordinates,

$$\mathcal{W}: \quad 0 \leq \theta \leq 2\pi, \quad 0 \leq r \leq 2, \quad 1 \leq z \leq 5$$

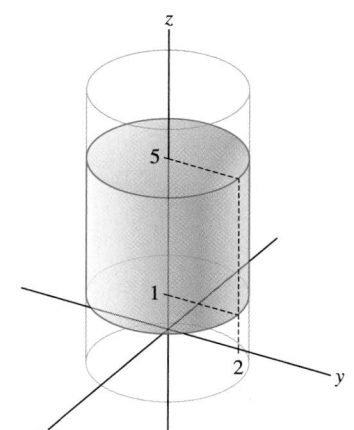

FIGURE 11 The cylinder $x^2 + y^2 \le 4$.

We write the function in cylindrical coordinates:

$$f(x, y, z) = z\sqrt{x^2 + y^2} = zr$$

and apply Eq. (6), noting that the resulting triple integral is equal to a product of single integrals:

$$\iiint_{\mathcal{W}} z\sqrt{x^2 + y^2} \, dV = \int_0^{2\pi} \int_{r=0}^2 \int_{z=1}^5 (zr)r \, dz \, dr \, d\theta$$

$$= \left(\int_0^{2\pi} d\theta \right) \left(\int_{r=0}^2 r^2 \, dr \right) \left(\int_{z=1}^5 z \, dz \right)$$

$$= (2\pi) \left(\frac{2^3}{3} \right) \left(\frac{5^2 - 1^2}{2} \right) = 64\pi \qquad \blacksquare$$

■ **EXAMPLE 4** Compute the total mass of the region $\mathcal{W}$ within the cylinder $x^2 + y^2 \le 4$ where $0 \le z \le y$, assuming a mass density of $\rho(x, y, z) = z$.

Solution

Step 1. **Express $\mathcal{W}$ in cylindrical coordinates.**
The condition $0 \le z \le y$ tells us that $y \ge 0$, so $\mathcal{W}$ projects onto the semicircle $\mathcal{D}$ in the xy-plane of radius 2 where $y \ge 0$, as in Figure 12. In polar coordinates,

$$\mathcal{D}: \quad 0 \le \theta \le \pi, \qquad 0 \le r \le 2$$

The upper boundary of $\mathcal{W}$ is the surface $z = y = r\sin\theta$ and the lower boundary is $z = 0$, so

$$\mathcal{W}: \quad 0 \le \theta \le \pi, \qquad 0 \le r \le 2, \qquad 0 \le z \le r\sin\theta$$

Step 2. **Set up the integral in cylindrical coordinates and evaluate.**
The total mass is the integral of mass density:

$$\iiint_{\mathcal{W}} \rho(x, y, z) \, dV = \int_0^\pi \int_{r=0}^2 \int_{z=0}^{r\sin\theta} zr \, dz \, dr \, d\theta$$

$$= \int_0^\pi \int_0^2 \frac{1}{2}(r\sin\theta)^2 r \, dr \, d\theta = \frac{1}{2} \int_0^\pi \int_0^2 r^3 \sin^2\theta \, dr \, d\theta$$

$$= \frac{1}{2}\left(\frac{2^4}{4}\right) \int_0^\pi \sin^2\theta \, d\theta = \left(\theta - \frac{1}{2}\sin 2\theta\right)\Big|_0^\pi = \pi \qquad \blacksquare$$

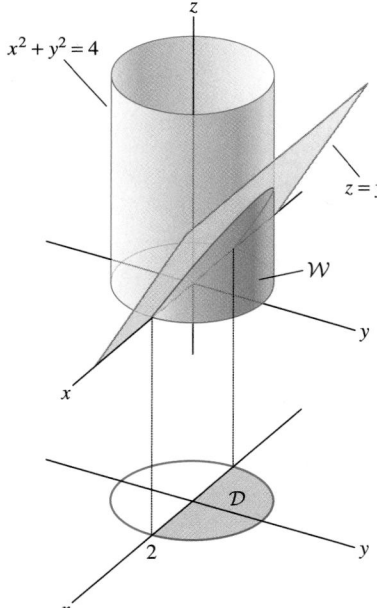

FIGURE 12

◀··· *REMINDER*

$$\int \sin^2\theta \, d\theta = \frac{1}{2}\left(\theta - \frac{1}{2}\sin 2\theta\right) + C$$

Triple Integrals in Spherical Coordinates

We noted above that the Change of Variables Formula in cylindrical coordinates is summarized by the symbolic equation $dV = r \, dr \, d\theta \, dz$. The analog in spherical coordinates (Figure 13) is the formula

$$\boxed{dV = \rho^2 \sin\phi \, d\rho \, d\phi \, d\theta}$$

FIGURE 13 Spherical coordinates.

FIGURE 15 Decomposition of a spherical wedge into subwedges.

where we recall (Section 12.7)

$$x = \rho \cos \theta \sin \phi, \qquad y = \rho \sin \theta \sin \phi, \qquad z = \rho \cos \phi$$

The key step in deriving this formula is to compute the volume of a small **spherical wedge** $\mathcal{W}$, defined by the inequalities

$$\mathcal{W}: \quad \theta_1 \leq \theta \leq \theta_2, \quad \phi_1 \leq \phi \leq \phi_2, \quad \rho_1 \leq \rho \leq \rho_2 \qquad \boxed{7}$$

Referring to Figure 14, we see that when the increments

$$\Delta\theta = \theta_2 - \theta_1, \quad \Delta\phi = \phi_2 - \phi_1, \quad \Delta\rho = \rho_2 - \rho_1$$

are small, the spherical wedge is nearly a box with sides $\Delta\rho$, $\rho_1 \Delta\phi$, and $\rho_1 \sin\phi_1 \Delta\theta$ and volume

$$\text{Volume}(\mathcal{W}) \approx \rho_1^2 \sin\phi_1 \, \Delta\rho \, \Delta\phi \, \Delta\theta \qquad \boxed{8}$$

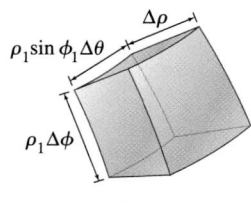

For small increments, the wedge is nearly a rectangular box with dimensions $\rho_1 \sin\phi_1 \Delta\theta \times \rho_1 \Delta\phi \times \Delta\rho$.

FIGURE 14 Spherical wedge.

Now we may proceed as in our discussion of polar coordinates. Decompose $\mathcal{W}$ into N^3 spherical subwedges $\mathcal{W}_i$ (Figure 15) with increments

$$\Delta\theta = \frac{\theta_2 - \theta_1}{N}, \qquad \Delta\phi = \frac{\phi_2 - \phi_1}{N}, \qquad \Delta\rho = \frac{\rho_2 - \rho_1}{N}$$

If N is large, then $\mathcal{W}_i$ is small and if $f(x, y, z)$ is continuous, we have the approximation

$$\iiint_{\mathcal{W}_i} f(x, y, z)\, dx\, dy\, dz \approx f(P_i)\text{Volume}(\mathcal{W}_i) \approx f(P_i)\rho_i^2 \sin\phi_i \, \Delta\rho_i \, \Delta\theta_i \, \Delta\phi_i$$

where $P_i = (\rho_i, \theta_i, \phi_i)$ is any sample point in $\mathcal{W}_i$. Thus,

$$\iiint_{\mathcal{W}} f(x, y, z)\, dx\, dy\, dz \approx \sum_i f(P_i)\rho_i^2 \sin\phi_i \, \Delta\rho_i \, \Delta\theta_i \, \Delta\phi_i \qquad \boxed{9}$$

The sum on the right is a Riemann sum for the function

$$f(\rho \cos\theta \sin\phi, \rho \sin\theta \sin\phi, \rho \cos\phi)\, \rho^2 \sin\phi$$

on the domain $\mathcal{W}$. The Change of Variables Formula follows by passing to the limit as $N \to \infty$. This argument applies more generally to regions in which the radial variable ρ is defined by an inequality $\rho_1(\theta, \phi) \leq \rho \leq \rho_2(\theta, \phi)$.

> **THEOREM 3 Triple Integrals in Spherical Coordinates**   For a region $\mathcal{W}$ defined by
> $$\theta_1 \le \theta \le \theta_2, \qquad \phi_1 \le \phi \le \phi_2, \qquad \rho_1(\theta, \phi) \le \rho \le \rho_2(\theta, \phi)$$
> the triple integral $\iiint_{\mathcal{W}} f(x, y, z) \, dV$ is equal to
>
> $$\int_{\theta_1}^{\theta_2} \int_{\phi=\phi_1}^{\phi_2} \int_{\rho=\rho_1(\theta,\phi)}^{\rho_2(\theta,\phi)} f(\rho \cos\theta \sin\phi, \rho \sin\theta \sin\phi, \rho \cos\phi) \, \rho^2 \sin\phi \, d\rho \, d\phi \, d\theta$$
>
> $\boxed{10}$

■ **EXAMPLE 5** Find the mass of a sphere S of radius 4 and center at the origin with mass density $f(x, y, z) = x^2 + y^2$ (Figure 16).

Solution We compute total mass by integrating mass density in spherical coordinates. First, we write $f(x, y, z)$ in spherical coordinates:

$$f(x, y, z) = x^2 + y^2 = (\rho \cos\theta \sin\phi)^2 + (\rho \sin\theta \sin\phi)^2$$
$$= \rho^2 \sin^2\phi(\cos^2\theta + \sin^2\theta) = \rho^2 \sin^2\phi$$

Since we are integrating over the entire sphere S of radius 4, r varies from 0 to 4, θ from 0 to 2π, and ϕ from 0 to π. In the following computation, we carry out the integration first with respect to θ:

$$\iiint_S (x^2 + y^2) \, dV = \int_0^{2\pi} \int_{\phi=0}^{\pi} \int_{\rho=0}^{4} (\rho^2 \sin^2\phi) \, \rho^2 \sin\phi \, d\rho \, d\phi \, d\theta$$

$$= 2\pi \int_{\phi=0}^{\pi} \int_{\rho=0}^{4} \rho^4 \sin^3\phi \, d\rho \, d\phi = 2\pi \int_0^{\pi} \left(\frac{\rho^5}{5} \bigg|_0^4 \right) \sin^3\phi \, d\phi$$

$$= \frac{2{,}048\pi}{5} \int_0^{\pi} \sin^3\phi \, d\phi$$

$$= \frac{2{,}048\pi}{5} \left(\frac{1}{3} \cos^3\phi - \cos\phi \right) \bigg|_0^{\pi} = \frac{8{,}192\pi}{15} \qquad ■$$

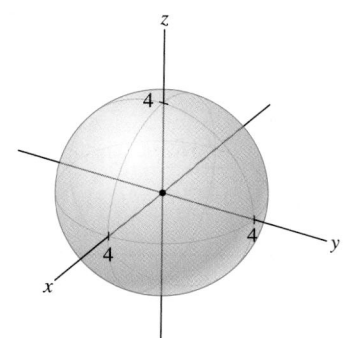

FIGURE 16 Sphere of radius 4.

◄·· **REMINDER**

$$\int \sin^3\phi \, d\phi = \frac{1}{3} \cos^3\phi - \cos\phi + C$$

[write $\sin^3\phi = \sin\phi(1 - \cos^2\phi)$].

■ **EXAMPLE 6** Find the centroid of the ice cream cone $\mathcal{W}$ consisting of the region below the hemisphere $x^2 + y^2 + z^2 = R^2$, $z \ge 0$, and above the cone $x^2 + y^2 = z^2$.

Solution First, we write equation $x^2 + y^2 = z^2$ in spherical coordinates:

$$(\rho \cos\theta \sin\phi)^2 + (\rho \sin\theta \sin\phi)^2 = (\rho \cos\phi)^2$$
$$\rho^2 \sin^2\phi(\cos^2\theta + \sin^2\theta) = \rho^2 \cos^2\phi$$
$$\sin^2\phi = \cos^2\phi$$

Thus, $\sin\phi = \pm\cos\phi$ or $\phi = \frac{\pi}{4}, \frac{3\pi}{4}$, and the equation of the upper branch of the cone is $\phi = \frac{\pi}{4}$ (Figure 17). The ice cream cone has the description

$$\mathcal{W}: \quad 0 \le \theta \le 2\pi, \qquad 0 \le \phi \le \frac{\pi}{4}, \qquad 0 \le \rho \le R$$

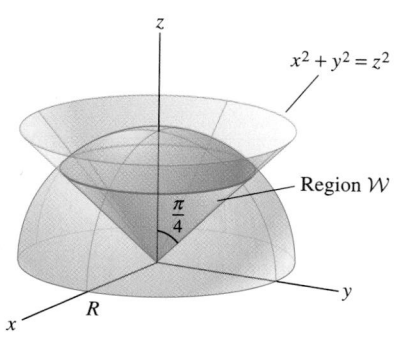

FIGURE 17 Ice cream cone defined by $0 \le \rho \le R$, $0 \le \phi \le \pi/4$.

To find the centroid, we first compute the volume V by integrating $f(x, y, z) = 1$ over W. In the following computation, we carry out the integration in the order θ, ρ, ϕ:

$$V = \iiint_W 1\, dV = \int_0^{2\pi} \int_{\phi=0}^{\pi/4} \int_{\rho=0}^R \rho^2 \sin\phi\, d\rho\, d\phi\, d\theta$$

$$= 2\pi \int_{\phi=0}^{\pi/4} \int_{\rho=0}^R \rho^2 \sin\phi\, d\rho\, d\phi = \frac{2\pi R^3}{3} \int_0^{\pi/4} \sin\phi\, d\phi = \frac{\pi R^3(2 - \sqrt{2})}{3}$$

The region W is symmetric with respect to the z-axis, so its centroid lies on the z-axis and we need only find the z-coordinate z_{CM}. We have

$$\iiint_W z\, dV = \int_0^{2\pi} \int_{\phi=0}^{\pi/4} \int_{\rho=0}^R (\rho \cos\phi)\rho^2 \sin\phi\, d\rho\, d\phi\, d\theta$$

$$= 2\pi \int_{\phi=0}^{\pi/4} \int_{\rho=0}^R \rho^3 \cos\phi \sin\phi\, d\rho\, d\phi = \frac{\pi R^4}{2} \int_0^{\pi/4} \sin\phi \cos\phi\, d\phi = \frac{\pi R^4}{8}$$

$$z_{CM} = \frac{\iiint_W z\, dV}{\iiint_W 1\, dV} = \left(\frac{3}{\pi R^3(2 - \sqrt{2})}\right)\left(\frac{\pi R^4}{8}\right) = \frac{3R}{8(2 - \sqrt{2})} \quad \blacksquare$$

15.4 SUMMARY

- Double integral in *polar coordinates:*

$$\iint_D f(x, y)\, dA = \int_{\theta_1}^{\theta_2} \int_{r=\alpha(\theta)}^{\beta(\theta)} f(r\cos\theta, r\sin\theta)\, r\, dr\, d\theta$$

- Triple integral $\iiint_R f(x, y, z)\, dV$

  - *Cylindrical coordinates:*

$$\int_{\theta_1}^{\theta_2} \int_{r=\alpha(\theta)}^{\beta(\theta)} \int_{z=z_1(r,\theta)}^{z_2(r,\theta)} f(r\cos\theta, r\sin\theta, z)\, r\, dz\, dr\, d\theta$$

  - *Spherical coordinates:*

$$\int_{\theta_1}^{\theta_2} \int_{\phi=\phi_1}^{\phi_2} \int_{\rho=\rho_1(\theta,\phi)}^{\rho_2(\theta,\phi)} f(\rho\cos\theta\sin\phi, \rho\sin\theta\sin\phi, \rho\cos\phi)\, \rho^2 \sin\phi\, d\rho\, d\phi\, d\theta$$

15.4 EXERCISES

Preliminary Questions

1. Which of the following represent the integral of $f(x, y) = x^2 + y^2$ over the unit circle?

(a) $\int_0^1 \int_0^{2\pi} r^2\, dr\, d\theta$

(b) $\int_0^{2\pi} \int_0^1 r^2\, dr\, d\theta$

(c) $\int_0^1 \int_0^{2\pi} r^3\, dr\, d\theta$

(d) $\int_0^{2\pi} \int_0^1 r^3\, dr\, d\theta$

2. What are the limits of integration in $\iiint f(r, \theta, z)r\, dr\, d\theta\, dz$ if the integration extends over the following region?

(a) $x^2 + y^2 \le 4$, $-1 \le z \le 2$

(b) lower hemisphere of the sphere of radius 2, center at origin

3. What are the limits of integration in

$$\iiint f(\rho, \phi, \theta)\rho^2 \sin\phi \, d\rho \, d\phi \, d\theta$$

if the integration extends over the following spherical regions centered at the origin?

(a) Sphere of radius 4

(b) Region between the spheres of radii 4 and 5

(c) Lower hemisphere of the sphere of radius 2

4. An ordinary rectangle of sides Δx and Δy has area $\Delta x \, \Delta y$, no matter where it is located in the plane. However, the area of a polar rectangle of sides Δr and $\Delta\theta$ depends on its distance from the origin. How is this difference reflected in the Change of Variables Formula for polar coordinates?

Exercises

In Exercises 1–6, sketch the $\mathcal{D}$ indicated and integrate $f(x, y)$ over $\mathcal{D}$ using polar coordinates.

1. $f(x, y) = \sqrt{x^2 + y^2}, \quad x^2 + y^2 \le 2$

2. $f(x, y) = x^2 + y^2, \quad 1 \le x^2 + y^2 \le 4$

3. $f(x, y) = xy; \qquad x \ge 0, \quad y \ge 0, \quad x^2 + y^2 \le 4$

4. $f(x, y) = y(x^2 + y^2)^3; \qquad y \ge 0, \quad x^2 + y^2 \le 1$

5. $f(x, y) = y(x^2 + y^2)^{-1}; \qquad y \ge \frac{1}{2}, \quad x^2 + y^2 \le 1$

6. $f(x, y) = e^{x^2 + y^2}, \quad x^2 + y^2 \le R$

7. Find the volume of the wedge-shaped region (Figure 18) contained in the cylinder $x^2 + y^2 = 9$ and bounded above by the plane $z = x$ and below by the xy-plane.

FIGURE 19

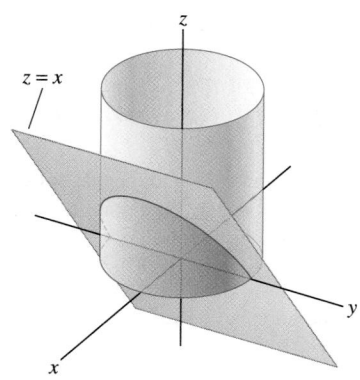

FIGURE 18

8. Let $\mathcal{W}$ be the region above the sphere $x^2 + y^2 + z^2 = 6$ and below the paraboloid $z = 4 - x^2 - y^2$.

(a) Show that the projection of $\mathcal{W}$ on the xy-plane is the disk $x^2 + y^2 \le 2$ (Figure 19).

(b) Compute the volume of $\mathcal{W}$ using polar coordinates.

9. Use polar coordinates to compute the volume of the region defined by $4 - x^2 - y^2 \le z \le 10 - 4x^2 - 4y^2$.

10. Compute the total mass of the plate in Figure 20, assuming a mass density of $f(x, y) = x^{-1}y$ g/cm^2.

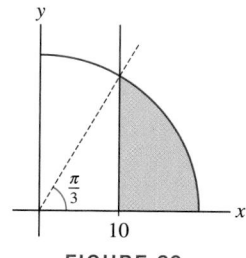

FIGURE 20

11. Use polar coordinates to find the centroid of a quarter circle of radius R.

12. Find the center of mass of the semicircle $x^2 + y^2 \le R^2$, $y \ge 0$, if the mass density is $\rho(x, y) = y$.

In Exercises 13–20, sketch the region of integration and evaluate by changing to polar coordinates.

13. $\displaystyle\int_{-2}^{2} \int_{0}^{\sqrt{4-x^2}} (x^2 + y^2) \, dy \, dx$

14. $\displaystyle\int_{0}^{3} \int_{0}^{\sqrt{9-y^2}} \sqrt{x^2 + y^2} \, dx \, dy$

15. $\displaystyle\int_{0}^{1/2} \int_{\sqrt{3x}}^{\sqrt{1-x^2}} x \, dy \, dx$

16. $\displaystyle\int_{-1}^{2}\int_{0}^{\sqrt{4-x^2}}(x^2+y^2)\,dy\,dx$

17. $\displaystyle\int_{0}^{4}\int_{0}^{\sqrt{16-x^2}}\tan^{-1}\frac{y}{x}\,dy\,dx$

18. $\displaystyle\int_{0}^{2}\int_{x}^{\sqrt{3}x}x\,dy\,dx$

19. $\displaystyle\int_{1}^{2}\int_{0}^{\sqrt{2x-x^2}}\frac{1}{\sqrt{x^2+y^2}}\,dy\,dx$

20. $\displaystyle\int_{0}^{2}\int_{x}^{\sqrt{3}x}y\,dy\,dx$

In Exercises 21–26, calculate the integral over the given region by changing to polar coordinates.

21. $f(x,y)=(x^2+y^2)^{-2}; \quad x^2+y^2\le 2, \quad x\ge 1$

22. $f(x,y)=x, \quad 2\le x^2+y^2\le 4$

23. $f(x,y)=|xy|, \quad x^2+y^2\le 1$

24. $f(x,y)=(x^2+y^2)^{-3/2}; \quad x^2+y^2\le 1, \quad x+y\ge 1$

25. $f(x,y)=x-y; \quad x^2+y^2\le 1, \quad x+y\ge 1$

26. $f(x,y)=y; \quad x^2+y^2\le 1, \quad (x-1)^2+y^2\le 1$

27. Evaluate $\displaystyle\iint_{\mathcal{D}}\sqrt{x^2+y^2}\,dA$, where $\mathcal{D}$ is the domain in Figure 21. *Hint:* Find the equation of the inner circle in polar coordinates and treat the right and left parts of the region separately.

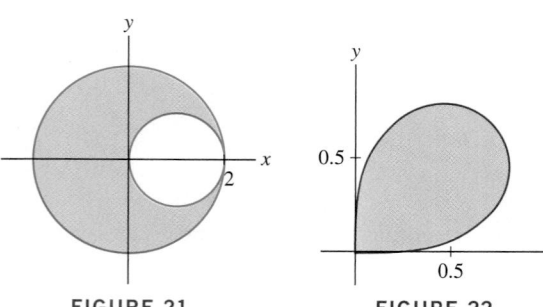

FIGURE 21   FIGURE 22

28. Evaluate $\displaystyle\iint_{\mathcal{D}}x^2\,dA$, where $\mathcal{D}$ is the region in the first quadrant enclosed by the lemniscate $r^2=\sin 2\theta$ (Figure 22).

29. Let $\mathcal{W}$ be the region between the paraboloids $z=x^2+y^2$ and $z=8-x^2-y^2$.
(a) Describe $\mathcal{W}$ in cylindrical coordinates.
(b) Use cylindrical coordinates to compute the volume of $\mathcal{W}$.

30. Use cylindrical coordinates to calculate the integral of $f(x,y,z)=z$ over the region above the disk $x^2+y^2=1$ in the xy-plane and below the surface $z=4+x^2+y^2$.

In Exercises 31–36, use cylindrical coordinates to calculate $\displaystyle\iiint_{\mathcal{W}}f(x,y,z)\,dV$ for the given function and region.

31. $f(x,y,z)=x^2+y^2; \quad x^2+y^2\le 9, \quad 0\le z\le 5$

32. $f(x,y,z)=xz; \quad x^2+y^2\le 1, \quad x\ge 0, \quad 0\le z\le 2$

33. $f(x,y,z)=y; \quad x^2+y^2\le 1, \quad x\ge 0, \quad y\ge 0, \quad 0\le z\le 2$

34. $f(x,y,z)=z\sqrt{x^2+y^2}, \quad x^2+y^2\le z\le 8-(x^2+y^2)$

35. $f(x,y,z)=z, \quad x^2+y^2\le z\le 9$

36. $f(x,y,z)=z, \quad 0\le z\le x^2+y^2\le 9$

37. Find the height of the centroid (average value of the y-coordinate) for the region $\mathcal{W}$ in Figure 19, lying above the unit sphere $x^2+y^2+z^2=6$ and below the paraboloid $z=4-x^2-y^2$.

In Exercises 38–41, express the triple integral in cylindrical coordinates.

38. $\displaystyle\int_{-1}^{1}\int_{y=-\sqrt{1-x^2}}^{\sqrt{1-x^2}}\int_{z=0}^{4}f(x,y,z)\,dz\,dy\,dx$

39. $\displaystyle\int_{0}^{1}\int_{y=-\sqrt{1-x^2}}^{\sqrt{1-x^2}}\int_{z=0}^{4}f(x,y,z)\,dz\,dy\,dx$

40. $\displaystyle\int_{-1}^{1}\int_{y=0}^{\sqrt{1-x^2}}\int_{z=0}^{x^2+y^2}f(x,y,z)\,dz\,dy\,dx$

41. $\displaystyle\int_{0}^{2}\int_{y=0}^{\sqrt{2x-x^2}}\int_{z=0}^{\sqrt{x^2+y^2}}f(x,y,z)\,dz\,dy\,dx$

42. Use cylindrical coordinates to find the mass of a cylinder of radius 4 and height 10 if the mass density at a point is equal to the square of the distance from the cylinder's central axis.

43. Find the equation of the right-circular cone in Figure 23 in cylindrical coordinates and compute its volume.

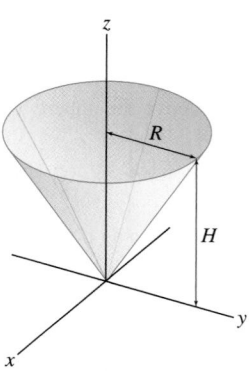

FIGURE 23

44. Let $\mathcal{W}$ be the portion of the half-cylinder $x^2 + y^2 \le 4, x \ge 0$ such that $0 \le z \le 3y$. Use cylindrical coordinates to compute the total mass of $\mathcal{W}$ if the mass density is $\rho(x, y, z) = z^2$.

45. Use cylindrical coordinates to integrate $f(x, y, z) = z$ over the intersection of the sphere $x^2 + y^2 + z^2 = 4$ and the cylinder $x^2 + y^2 = 1$.

46. Use spherical coordinates to evaluate the triple integral of $f(x, y, z) = z$ over the region

$$0 \le \theta \le \frac{\pi}{3}, \qquad 0 \le \phi \le \frac{\pi}{2}, \qquad 1 \le \rho \le 2 \qquad \boxed{11}$$

47. Evaluate the triple integral of $f(x, y, z) = x^2$ for region (11).

48. Calculate the volume of the sphere $x^2 + y^2 + z^2 = a^2$, using both spherical and cylindrical coordinates.

49. Find the volume of the region lying above the cone $\phi = \phi_0$ and below the sphere $\rho = R$.

50. Let $\mathcal{W}$ be the region within the cylinder $x^2 + y^2 = 2$ between $z = 0$ and the cone $z = \sqrt{x^2 + y^2}$. Calculate the integral of $f(x, y, z) = x^2 + y^2$ over $\mathcal{W}$, using both spherical and cylindrical coordinates.

In Exercises 51–56, use spherical coordinates to calculate the triple integral of $f(x, y, z)$ over the given region.

51. $f(x, y, z) = y;$ $x^2 + y^2 + z^2 \le 1,$ $x, y, z \le 0$

52. $f(x, y, z) = \sqrt{x^2 + y^2 + z^2},$ $x^2 + y^2 + z^2 \le 2z$

53. $f(x, y, z) = \rho^{-3},$ $2 \le x^2 + y^2 + z^2 \le 4$

54. $f(x, y, z) = x^2 + y^2,$ $\rho \le 1$

55. $f(x, y, z) = 1;$ $x^2 + y^2 + z^2 \le 4z,$ $z \ge \sqrt{x^2 + y^2}$

56. $f(x, y, z) = \rho;$ $x^2 + y^2 + z^2 \le 4,$ $z \le 1$ $x \ge 0$

57. Find the centroid of the region $\mathcal{W}$ bounded by the cone $\phi = \phi_0$ and the sphere $\rho = R$ as a function of ϕ_0 and R.

58. Find the total mass of the portion of the sphere $x^2 + y^2 + z^2 \le 12$ with $x, y, z \ge 0$ if the mass density is $f(x, y, z) = x(x^2 + y^2 + z^2)^{-1/2}$.

59. Evaluate the triple integral of $f(x, y, z) = z(x^2 + y^2 + z^2)^{-3/2}$ over the part of the ball $x^2 + y^2 + z^2 \le 16$ defined by $z \ge 2$.

60. Calculate the volume of the cone in Figure 23 using spherical coordinates.

61. Find the center of mass of a cylinder of radius 2 and height 4 and mass density e^{-z}, where z is the height above the base.

62. Find the center of mass of a cylinder defined by $0 \le r \le 2$ and $2 \le z \le 4$ with mass density $\rho(r, \theta, z) = r$.

In Exercises 63–65, compute the centroid of the shapes in Figure 24.

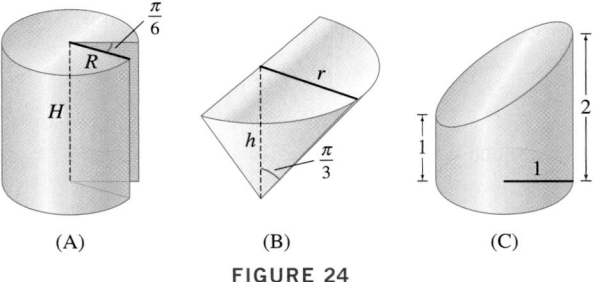

(A)   (B)   (C)

FIGURE 24

63. (A)   **64.** (B)   **65.** (C)

66. According to Coulomb's Law, the force between two electric charges of magnitude q_1 and q_2 separated by a distance r is Cq_1q_2/r^2 acting along the line through the two charges (C is a constant). Suppose that a circular disk of radius R has a uniformly distributed charge of density k Coulombs per square centimeter (Figure 25). Let F be the net force on a charged particle P of magnitude Q coulombs located at a distance d cm vertically above the center of the disk (by symmetry, F acts in the vertical direction).

(a) Let $\mathcal{R}$ be a small polar rectangle of size $\Delta r \times \Delta \theta$ located at distance r. Show that $\mathcal{R}$ exerts a force on P whose vertical component is

$$\left(\frac{kCQd}{(r^2 + d^2)^{3/2}} \right) r \, \Delta r \, \Delta \theta$$

(b) Explain why F is equal to the following double integral and evaluate:

$$F = kCQd \int_0^{2\pi} \int_0^R \frac{r \, dr \, d\theta}{(r^2 + d^2)^{3/2}}$$

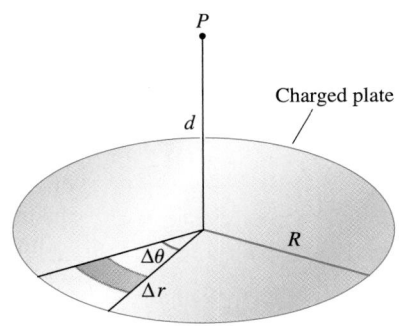

FIGURE 25

Further Insights and Challenges

67. Area Under the Bell-Shaped Curve Let $I = \int_{-\infty}^{\infty} e^{-x^2}\,dx$ be the area under the bell-shaped curve (Figure 26). By Fubini's Theorem, $I^2 = J$, where J is the improper double integral:

$$J = \int_{-\infty}^{\infty}\int_{-\infty}^{\infty} e^{-x^2-y^2}\,dx\,dy$$

Write J in polar coordinates and evaluate, showing that $J = \pi$ and $I = \sqrt{\pi}$.

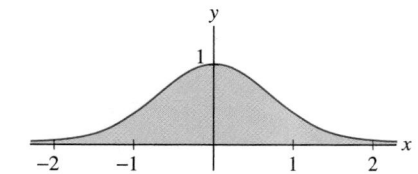

FIGURE 26 The bell-shaped curve $y = e^{-x^2}$.

68. Let D be a plate in the shape of an ellipse with the polar equation

$$r^2 = \left(\frac{1}{6}\sin^2\theta + \frac{1}{9}\cos^2\theta\right)^{-1}$$

with the disk $x^2 + y^2 \leq 1$ removed (Figure 27). Assume that D is charged with a charge density $\rho(r, \theta) = 3r^{-4}$ Coulombs per square centimeter. Calculate the total charge on D.

69. Calculate the integral of $\cos z$ over the ball $x^2 + y^2 + z^2 \leq 1$. *Hint:* Use cylindrical coordinates and integrate in the order $d\theta\,dz\,dr$.

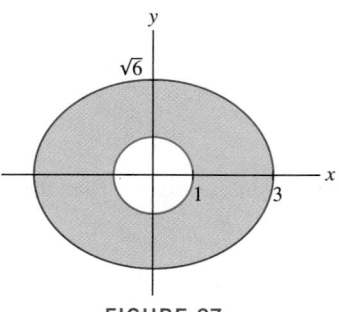

FIGURE 27

70. An Improper Multiple Integral Show that a triple integral of $(x^2 + y^2 + z^2 + 1)^{-2}$ over all of $\mathbf{R}^3$ is equal to π^2. This is an improper integral, so integrate first over $\rho \leq R$ and let $R \to \infty$. *Hint:* Use trigonometric substitution to evaluate the integral with respect to ρ.

71. Prove the formula

$$\iint_{\mathcal{D}} \ln r\,dA = -\frac{\pi}{2}$$

where $r = \sqrt{x^2 + y^2}$ and $\mathcal{D}$ is the unit disk $x^2 + y^2 \leq 1$. This is an improper integral since $\ln r$ is not defined at $(0, 0)$, so integrate first over the annulus $a \leq r \leq 1$ and let $a \to 0$.

72. Recall that the improper integral $\int_0^1 x^{-a}\,dx$ converges if $a < 1$.

For which values of a does $\iint_{\mathcal{D}} r^{-a}\,dA$ converge, where $r = \sqrt{x^2 + y^2}$ and $\mathcal{D}$ is the unit disk $x^2 + y^2 \leq 1$?

15.5 Change of Variables

In the previous section, we discussed the Change of Variables Formulas for the three most common non-Cartesian coordinate systems: polar, cylindrical, and spherical. These formulas are instances of the general Change of Variables Formula for multiple integrals. In this section, we shall use the term **map**, which is another name for a function. It is commonly used when the domain of the function is a domain in $\mathbf{R}^n$ or some set other than a subset of real numbers. A map is also called a **mapping**.

Maps from $\mathbf{R}^2$ to $\mathbf{R}^2$

A map or mapping is a function $\Phi : X \to Y$ from a set X (the domain) to a set Y. For $P \in X$, the element $\Phi(P)$ in Y is called the **image** of P. If Z is a subset of X, the image of Z is the set $\Phi(Z)$ consisting of all images $\Phi(z)$ for $z \in Z$. The set $\mathcal{R} = \Phi(X)$ is called the image or **range** of Φ.

We will study maps $\Phi : \mathcal{D} \to \mathbf{R}^2$ defined on a domain $\mathcal{D}$ in $\mathbf{R}^2$. To prevent confusion, we use different variables for the domain and range. For instance, we may view Φ as a map from the uv-plane to the xy-plane (Figure 1). Explicitly, the map Φ has the form $\Phi(u, v) = (\phi(u, v), \psi(u, v))$, where ϕ and ψ are the **component functions**. We also

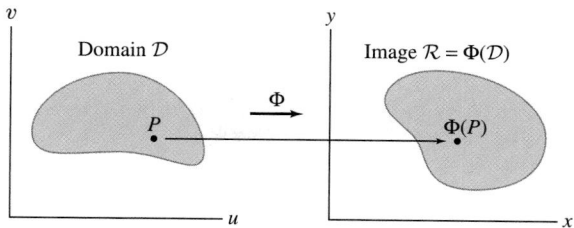

FIGURE 1 Φ maps $\mathcal{D}$ to $\mathcal{R}$.

write

$$x = \phi(u, v), \qquad y = \psi(u, v)$$

One map with which we are already familiar is the map defining polar coordinates. For this map, we use variables r, θ instead of u, v.

■ **EXAMPLE 1** **Polar Coordinates Map** Describe the image of a rectangle $[r_1, r_2] \times [\theta_1, \theta_2]$ under the polar coordinates map $\Phi : \mathbf{R}^2 \to \mathbf{R}^2$ defined by

$$\Phi(r, \theta) = (r \sin \theta, r \cos \theta)$$

Solution First, we determine the images of horizontal and vertical lines. Referring to Figure 2, we see that:

- The vertical line $r = r_0$ is mapped to the set of points with radial coordinate r_0, that is, the circle of radius r_0.
- The horizontal line $\theta = \theta_0$ is mapped to the line through the origin of angle θ_0.

It follows that the image of the rectangle $[r_1, r_2] \times [\theta_1, \theta_2]$ is the polar rectangle in the xy-plane defined by $r_1 \leq r \leq r_2$, $\theta_1 \leq \theta \leq \theta_2$. ■

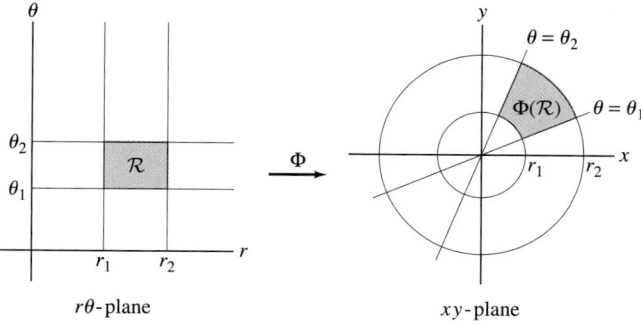

FIGURE 2 The polar coordinate map $\Phi(r, \theta) = (r \cos \theta, r \sin \theta)$.

General mappings may be quite complicated, so it is useful to study the simplest type of maps, the linear maps. A map $\Phi(u, v)$ is called **linear** if it has the form

$$\Phi(u, v) = (Au + Cv, Bu + Dv) \quad (A, B, C, D \text{ constants})$$

It is convenient to regard a linear map Φ as a map from vectors in the uv-plane to vectors in the xy-plane. Then Φ satisfies the following linearity properties (see Exercise 46):

$$\Phi(u_1 + u_2, v_1 + v_2) = \Phi(u_1, v_1) + \Phi(u_2, v_2) \qquad \boxed{1}$$

$$\Phi(cu, cv) = c\Phi(u, v) \quad (c \text{ any constant}) \qquad \boxed{2}$$

A key property of linear maps is that they map lines to lines. More precisely, *the segment joining any two points P and Q is mapped to the segment joining $\Phi(P)$ and $\Phi(Q)$* (Figure 3).

To check this, recall that in vector notation, the segment $\overline{PQ}$ has the parametrization

$$(1-t)\overrightarrow{OP} + t\overrightarrow{OQ} \quad \text{for} \quad 0 \le t \le 1$$

Using properties (1) and (2), we see that the image of $\overline{PQ}$ is the segment from $\Phi(P)$ to $\Phi(Q)$:

$$\Phi\left((1-t)\overrightarrow{OP} + t\overrightarrow{OQ}\right) = \Phi\left((1-t)\overrightarrow{OP}\right) + \Phi\left(t\overrightarrow{OQ}\right) = (1-t)\Phi(\overrightarrow{OP}) + t\Phi(\overrightarrow{OQ})$$

This gives us a clear picture of the linear mapping $\Phi(u, v) = (Au + Cv, Bu + Dv)$. The images of the basis vectors $\mathbf{i} = \langle 1, 0 \rangle$ and $\mathbf{j} = \langle 0, 1 \rangle$ are

$$\mathbf{r} = \Phi(\langle 1, 0 \rangle) = \langle A, B \rangle, \qquad \mathbf{s} = \Phi(\langle 0, 1 \rangle) = \langle C, D \rangle$$

Thus, Φ maps the grid of horizontal and vertical lines in the uv-plane generated by $\mathbf{i}$ and $\mathbf{j}$ to the grid generated by the vectors $\mathbf{r}$ and $\mathbf{s}$ in the xy-plane (Figure 3).

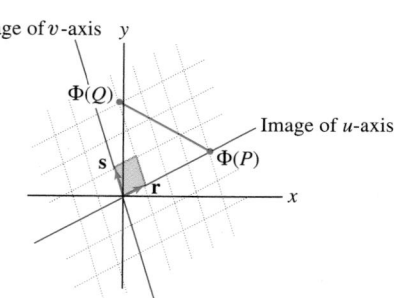

FIGURE 3 A linear mapping Φ maps the grid of horizontal and vertical lines in the uv-plane generated by $\mathbf{i}$ and $\mathbf{j}$ to the grid generated by $\mathbf{r}$ and $\mathbf{s}$ in the xy-plane.

■ **EXAMPLE 2** Image of a Triangle Under a Linear Map Find the image of the triangle $\mathcal{T}$ with vertices $(1, 2)$, $(2, 1)$, $(3, 4)$ under the mapping $\Phi(u, v) = (2u - v, u + v)$.

Solution The image of any polygon $\mathcal{P}$ under a linear map Φ is the polygon whose vertices are the images of the vertices of $\mathcal{P}$. Thus, to find the image of $\mathcal{T}$, we compute the images of the vertices under $\Phi(u, v) = (2u - v, u + v)$:

$$\Phi(1, 2) = (0, 3), \qquad \Phi(2, 1) = (3, 3), \qquad \Phi(3, 4) = (2, 7)$$

These are the vertices of $\Phi(\mathcal{T})$ as shown in Figure 4. ■

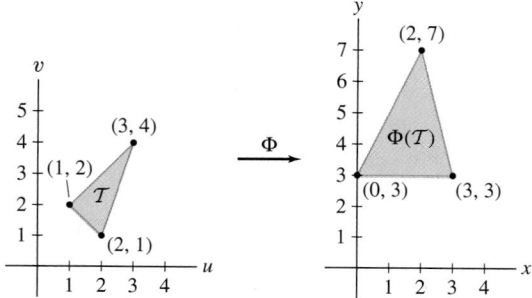

FIGURE 4 The map $\Phi(u, v) = (2u - v, u + v)$.

To understand a nonlinear map, it is usually helpful to determine the images of horizontal and vertical lines as we did for the polar coordinate mapping.

■ **EXAMPLE 3** Let $\Phi(u, v) = (uv^{-1}, uv)$ for $u, v > 0$. Determine the images of:

(a) The lines $u = c$ and $v = c$ **(b)** $[1, 2] \times [1, 2]$

Find the inverse map Φ^{-1}.

Solution

(a) In this map, we have $x = uv^{-1}$ and $y = uv$. Thus $xy = u^2$ and $y/x = v^2$. It follows that Φ maps the vertical line $u = c$ to the hyperbola $xy = c^2$. Similarly, Φ maps the horizontal line $v = c$ to the line $y/x = c^2$, or $y = c^2 x$.

(b) The image of $[1, 2] \times [1, 2]$ is the *curvilinear* rectangle bounded by the four curves which are the images of the lines $u = 1$, $u = 2$, and $v = 1$, $v = 2$. By (a), the lines $u = 1$ and $u = 2$ are mapped by Φ to the hyperbolas $xy = 1$ and $xy = 4$. Similarly, the lines $v = 1$ and $v = 2$ are mapped to the lines through the origin $y = x$ and $y = 4x$. Therefore, Φ maps $[1, 2] \times [1, 2]$ to the shaded curvilinear rectangle on the right in Figure 5. This image is defined by the inequalities

$$1 \le xy \le 4, \qquad 1 \le \frac{y}{x} \le 4$$

To find Φ^{-1}, we solve for u and v in terms of x and y. We noted that $xy = u^2$ and $y/x = v^2$. Therefore, $u = \sqrt{xy}$ and $v = \sqrt{y/x}$ and the inverse map is $\Phi^{-1}(x, y) = \left(\sqrt{xy}, \sqrt{y/x} \right)$.

> The term "curvilinear rectangle" refers to a region bounded on four sides by curves as in Figure 5.

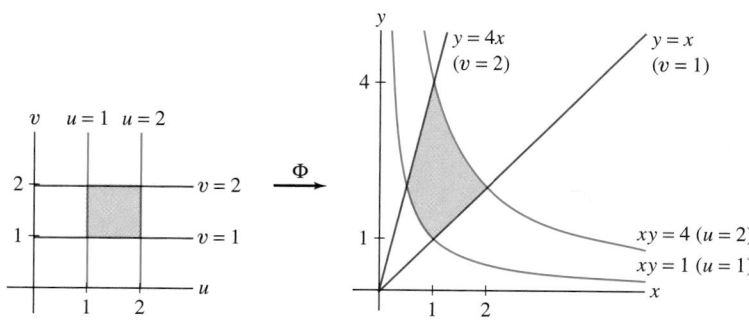

FIGURE 5 The mapping $\Phi(u, v) = (uv^{-1}, uv)$.

How Area Changes Under a Mapping: The Jacobian Determinant

The **Jacobian determinant** (or simply "Jacobian") of a map

$$\Phi(u, v) = (\phi(u, v), \psi(u, v))$$

is the determinant

> ◄·· **REMINDER** The definition of a 2×2 determinant is
> $$\begin{vmatrix} a & b \\ c & d \end{vmatrix} = ad - bc \qquad \boxed{3}$$

$$\mathrm{Jac}(\Phi) = \begin{vmatrix} \dfrac{\partial \phi}{\partial u} & \dfrac{\partial \phi}{\partial v} \\[2mm] \dfrac{\partial \psi}{\partial u} & \dfrac{\partial \psi}{\partial v} \end{vmatrix}$$

Since x and y are functions of u and v:

$$x = \phi(u, v), \qquad y = \psi(u, v)$$

we may also write the Jacobian determinant $\mathrm{Jac}(\Phi)$ in the notation

$$\frac{\partial(x, y)}{\partial(u, v)} = \begin{vmatrix} \dfrac{\partial x}{\partial u} & \dfrac{\partial x}{\partial v} \\[2mm] \dfrac{\partial y}{\partial u} & \dfrac{\partial y}{\partial v} \end{vmatrix} = \frac{\partial x}{\partial u}\frac{\partial y}{\partial v} - \frac{\partial x}{\partial v}\frac{\partial y}{\partial u}$$

Keep in mind that $\mathrm{Jac}(\Phi)$ is a function of u and v.

■ **EXAMPLE 4** Evaluate the Jacobian of $\Phi(u, v) = (u^3 + v, uv)$ at $(u, v) = (2, 1)$.

Solution We have $x = u^3 + v$ and $y = uv$, so

$$\mathrm{Jac}(\Phi) = \frac{\partial(x, y)}{\partial(u, v)} = \begin{vmatrix} \dfrac{\partial x}{\partial u} & \dfrac{\partial x}{\partial v} \\[2mm] \dfrac{\partial y}{\partial u} & \dfrac{\partial y}{\partial v} \end{vmatrix} = \begin{vmatrix} 3u^2 & 1 \\ v & u \end{vmatrix} = 3u^3 - v$$

The value of the Jacobian at $(2, 1)$ is

$$\mathrm{Jac}(\Phi)(2, 1) = 3(2)^3 - 1 = 23$$ ■

The Jacobian of a linear map $\Phi(u, v) = (Au + Cv, Bu + Dv)$ has two special properties:

• The Jacobian is *constant* with value $\mathrm{Jac}(\Phi) = \begin{vmatrix} A & C \\ B & D \end{vmatrix} = AD - BC.$
• Under Φ, the area of a region is multiplied by the factor $|\mathrm{Jac}(\Phi)|$.

The first property follows by direct calculation: Since $x = Au + Cv$ and $y = Bu + Dv$, the partial derivatives in Jacobians are the constants A, B, C, D.

The claim in the second property is that if $\mathcal{D}$ is a domain in the uv-plane, then the area of its image $\Phi(\mathcal{D})$ is (Figure 6):

$$\boxed{\mathrm{Area}(\Phi(\mathcal{D})) = |\mathrm{Jac}(\Phi)|\,\mathrm{Area}(\mathcal{D})} \qquad \boxed{4}$$

This is certainly true for the unit rectangle $\mathcal{D} = [1, 0] \times [0, 1]$ because $\Phi(\mathcal{D})$ is the parallelogram spanned by $\langle A, B \rangle$ and $\langle C, D \rangle$ (Figure 6) and this parallelogram has area $|\mathrm{Jac}(\Phi)| = |AD - BC|$ by Theorem 3 in Section 12.4. Similarly, we can check that Eq. (4) holds for arbitrary parallelograms (see Exercise 47). However, Eq. (4) also holds for general domains because a general domain may be approximated as closely as desired by a union of rectangles in a fine grid of lines parallel to the u- and v-axes.

We cannot expect Eq. (4) to hold for a nonlinear map $\Phi(u, v) = (\phi(u, v), \psi(u, v))$. In fact, it would not make sense as stated because $\mathrm{Jac}(\Phi)(P)$ may vary from point to

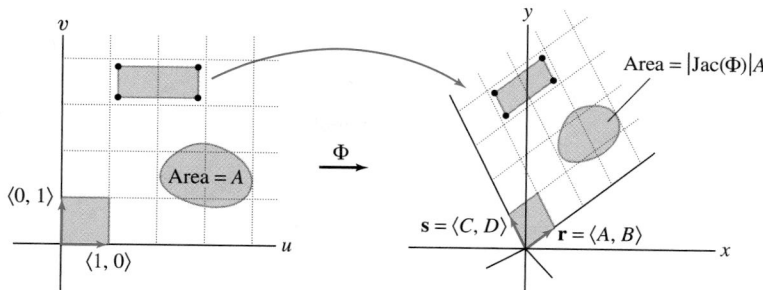

FIGURE 6 A linear map Φ expands (or shrinks) area by the factor $|\text{Jac}(\Phi)|$.

point. However, it is *approximately true* if the domain $\mathcal{D}$ is small and P is a sample point in $\mathcal{D}$:

$$\text{Area}(\Phi(\mathcal{D})) \approx |\text{Jac}(\Phi)(P)|\text{Area}(\mathcal{D})$$

$$\boxed{5}$$

This result may be stated more precisely as the limit relation:

$$|\text{Jac}(\Phi)(P)| = \lim_{|\mathcal{D}| \to 0} \frac{\text{Area}(\Phi(\mathcal{D}))}{\text{Area}(\mathcal{D})}$$

$$\boxed{6}$$

Here, we write $|\mathcal{D}| \to 0$ to indicate the limit as the diameter of $\mathcal{D}$ (the maximum distance between two points in $\mathcal{D}$) tends to zero.

CONCEPTUAL INSIGHT   Although a rigorous proof of Eq. (6) is somewhat technical, we can understand the underlying idea as an application of linear approximation. Consider a rectangle $\mathcal{R}$ with vertex at $P = (u, v)$ and sides of lengths Δu and Δv, assumed to be small as in Figure 7. The image $\Phi(\mathcal{R})$ is not a parallelogram, but it is approximated well by the parallelogram spanned by the vectors $\mathbf{A}$ and $\mathbf{B}$ in the figure:

$$\mathbf{A} = \Phi(u + \Delta u, v) - \Phi(u, v)$$

$$\mathbf{B} = \Phi(u, v + \Delta v) - \Phi(u, v)$$

The linear approximation applied to the components of Φ yields

$$\mathbf{A} = \langle \phi(u + \Delta u, v), \psi(u + \Delta u, v) \rangle - \langle \phi(u, v), \psi(u, v) \rangle$$

$$= \langle \phi(u + \Delta u, v) - \phi(u, v), \psi(u + \Delta u, v) - \psi(u, v) \rangle$$

$$\approx \langle \phi_u(u, v)\Delta u, \psi_u(u, v)\Delta u \rangle$$

$$\boxed{7}$$

$$= \langle \phi_u(u, v), \psi_u(u, v) \rangle \Delta u$$

◄·· **REMINDER** *Equation (7) uses the linear approximation for $\phi(u, v)$ and $\psi(u, v)$:*

$$\phi(u + \Delta u, v) \approx \phi_u(u, v)\Delta u$$

$$\psi(u + \Delta u, v) \approx \psi_u(u, v)\Delta u$$

FIGURE 7 The image of a small rectangle under a nonlinear map can be approximated by a parallelogram whose sides are determined by the linear approximation.

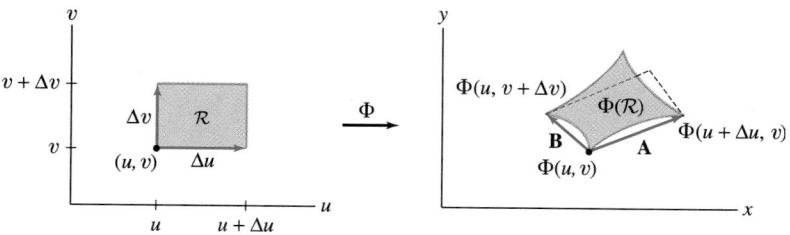

Similarly,

$$\mathbf{B} = \Phi(u, v + \Delta v) - \Phi(u, v) \approx \langle \phi_v(u, v), \psi_v(u, v) \rangle \, \Delta v$$

Since the area of $\mathcal{R}$ is $\Delta u \, \Delta v$, we obtain the desired approximation:

$$\text{Area}(\Phi(\mathcal{R})) \approx \left| \det \begin{pmatrix} \mathbf{A} \\ \mathbf{B} \end{pmatrix} \right| = \left| \det \begin{pmatrix} \phi_u(u, v)\Delta u & \psi_u(u, v)\Delta u \\ \phi_v(u, v)\Delta v & \psi_v(u, v)\Delta v \end{pmatrix} \right|$$

$$= |\phi_u(u, v)\psi_v(u, v) - \psi_u(u, v)\phi_v(u, v)| \, \Delta u \, \Delta v$$

$$= |\text{Jac}(\Phi)(P)| \text{Area}(\mathcal{R})$$

This justifies (5) in the case of a rectangle. For more general regions $\mathcal{D}$, we justify (5) by approximating $\mathcal{D}$ by rectangles.

The Change of Variables Formula

Recall the formula for integration in polar coordinates:

$$\iint_{\mathcal{D}} f(x, y) \, dx \, dy = \int_{\theta_1}^{\theta_2} \int_{r_1}^{r_2} f(r \cos \theta, r \sin \theta) \, r \, dr \, d\theta \qquad \boxed{8}$$

where $\mathcal{D}$ is the image of a rectangle $\mathcal{R} = [\theta_1, \theta_2] \times [r_1, r_2]$ in the $r\theta$-plane (Figure 8).

Polar coordinates map

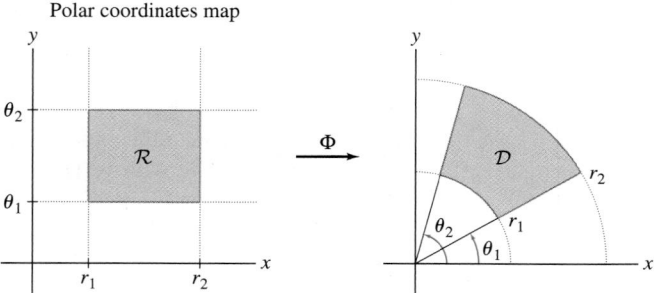

FIGURE 8

The general Change of Variables Formula has a similar form. We consider a one-to-one map $\Phi : \mathcal{D}_0 \to \mathcal{D}$ from a domain in the uv-plane to a domain in the xy-plane (Figure 9). The Change of Variables Formula expresses an integral over $\mathcal{D}$ as an integral over $\mathcal{D}_0$. The Jacobian appears instead of the factor r on the right-hand side of Eq. (8):

◄·· **REMINDER** Φ is called "one-to-one" if $\Phi(P) = \Phi(Q)$ only for $P = Q$.

$$\iint_{\mathcal{D}} f(x, y) \, dx \, dy = \iint_{\mathcal{D}_0} f(x(u, v), y(u, v)) \left| \frac{\partial(x, y)}{\partial(u, v)} \right| \, du \, dv$$

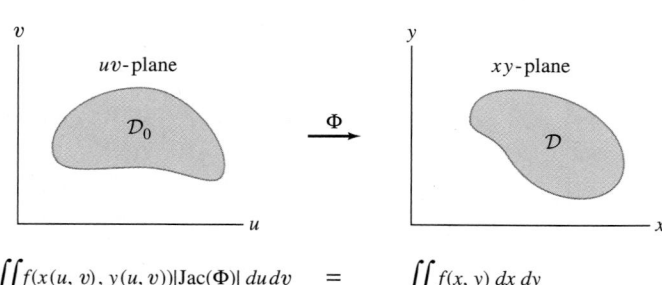

$$\iint_{\mathcal{D}_0} f(x(u, v), y(u, v)) |\text{Jac}(\Phi)| \, du \, dv \quad = \quad \iint_{\mathcal{D}} f(x, y) \, dx \, dy$$

FIGURE 9 The Change of Variables Formula expresses a double integral over $\mathcal{D}$ as a double integral over $\mathcal{D}_0$.

To justify this formula (without attempting to give a fully rigorous proof), suppose first that the domain $\mathcal{D}$ is small and let $P \in \mathcal{D}$ be a sample point. There is a unique point $P_0 \in \mathcal{D}_0$ such that $\Phi(P_0) = P$ since Φ is one-to-one. Assume $f(x, y)$ is continuous and consider the following sequence of approximations:

REMINDER If $\mathcal{D}$ is a domain of small diameter, $P \in \mathcal{D}$ is a sample point, and $f(x, y)$ is continuous, then (see Section 15.2)

$$\iint_{\mathcal{D}} f(x, y)\, dA \approx f(P)\text{Area}(\mathcal{D})$$

We use this in (9) and (11).

$$\iint_{\mathcal{D}} f(x, y)\, dx\, dy \approx f(P)\text{Area}(\mathcal{D}) \tag{9}$$

$$\approx f(\Phi(P_0))\, |\text{Jac}(\Phi)(P_0)|\, \text{Area}(\mathcal{D}_0) \tag{10}$$

$$\approx \iint_{\mathcal{D}_0} f(\Phi(u, v))\, |\text{Jac}(\Phi)(u, v)|\, du\, dv \tag{11}$$

If $\mathcal{D}$ is not small, divide it into small subdomains $D_j = \Phi(\mathcal{D}_{0j})$ and choose sample points $P_j = \Phi(P_{0j})$ where $P_{0j} \in \mathcal{D}_{0j}$ (Figure 10). Then apply the approximation to each subdomain and take the sum:

$$\iint_{\mathcal{D}} f(x, y)\, dx\, dy = \sum_j \iint_{\mathcal{D}_j} f(x, y)\, dx\, dy$$

$$\approx \sum_j \iint_{\mathcal{D}_{0j}} f(\Phi(u, v)))\, |\text{Jac}(\Phi)(u, v)|\, du\, dv$$

$$= \iint_{\mathcal{D}_0} f(\Phi(u, v))\, |\text{Jac}(\Phi)(u, v)|\, du\, dv$$

By carefully estimating the error in each step of this argument, we can show that the error tends to zero as the maximum of the diameters of the subdomains $\mathcal{D}_j$ tends to zero. This yields the Change of Variables Formula. To state it precisely, we assume that Φ is a $\mathbf{C^1}$ **map**, by which we mean that the component functions ϕ and ψ have continuous partial derivatives. We also assume that the domains are either vertically or horizontally simple (see Section 15.2). Although we assumed earlier that Φ is one-to-one, in fact it suffices to assume that Φ is one-to-one on the interior of $\mathcal{D}_0$.

Domain $\mathcal{D}_0$ — Rectangle $\mathcal{D}_{0j}$, Sample point P_{0j}

Domain $\mathcal{D} = \Phi(\mathcal{D}_0)$ — Curvilinear rectangle $\mathcal{D}_j = \Phi(\mathcal{D}_{0j})$, Sample point $P_j = \Phi(P_{0j})$

FIGURE 10 Φ maps a rectangular grid on $\mathcal{D}_0$ to a curvilinear grid on $\mathcal{D}$.

Equation (12) is summarized by the symbolic equality

$$dx\, dy = \left| \frac{\partial(x, y)}{\partial(u, v)} \right|\, du\, dv$$

THEOREM 1 General Change of Variables Let $\Phi : \mathcal{D}_0 \to \mathcal{D}$ be a C^1 mapping of vertically or horizontally simple regions and assume that Φ is one-to-one on the interior of $\mathcal{D}_0$. Let $f(x, y)$ be a continuous function on $\mathcal{D}$. Then

$$\iint_{\mathcal{D}} f(x, y)\, dx\, dy = \iint_{\mathcal{D}_0} f(x(u, v), y(u, v)) \left| \frac{\partial(x, y)}{\partial(u, v)} \right|\, du\, dv \tag{12}$$

■ **EXAMPLE 5** Polar Coordinates Revisited   Use the general Change of Variables Formula to derive the formula for integration in polar coordinates.

Solution   The Jacobian of the polar coordinate map $\Phi(r, \theta) = (r\cos\theta, r\sin\theta)$ is

$$\operatorname{Jac}(\Phi) = \begin{vmatrix} \dfrac{\partial x}{\partial r} & \dfrac{\partial x}{\partial \theta} \\[1.2em] \dfrac{\partial y}{\partial r} & \dfrac{\partial y}{\partial \theta} \end{vmatrix} = \begin{vmatrix} \cos\theta & -r\sin\theta \\ \sin\theta & r\cos\theta \end{vmatrix} = r(\cos^2\theta + \sin^2\theta) = r$$

If $\mathcal{D}$ is the polar rectangle defined by $r_0 \le r \le r_1$, $\theta_0 \le \theta \le \theta_1$, then Eq. (12) yields

$$\iint_{\mathcal{D}} f(x, y)\, dA = \int_{\theta_0}^{\theta_1} \int_{r_0}^{r_1} f(r\cos\theta, r\sin\theta) r\, dr\, d\theta \qquad \boxed{13}$$

■

Assumptions Matter   In Theorem 1, we assume that Φ is one-to-one on the interior but not necessarily on the boundary of the domain. Thus, it is valid to apply Eq. (12) to the polar coordinates map Φ on the rectangle $\mathcal{D}_0 = [0, 1] \times [0, 2\pi]$. In this case, Φ is one-to-one on the interior but not on the boundary of $\mathcal{D}_0$ since $\Phi(0, \theta) = (0, 0)$ for all θ and $\Phi(r, 0) = \Phi(r, 2\pi)$ for all r. On the other hand, Eq. (13) cannot be applied to Φ on the rectangle $[0, 1] \times [0, 4\pi]$ because it is not one-to-one on the interior.

■ **EXAMPLE 6**   Use the Change of Variables Formula to calculate $J = \displaystyle\iint_{\mathcal{P}} e^{4x-y}\, dA$, where $\mathcal{P}$ is the parallelogram spanned by the vectors $\langle 4, 1 \rangle$, $\langle 3, 3 \rangle$ in Figure 11.

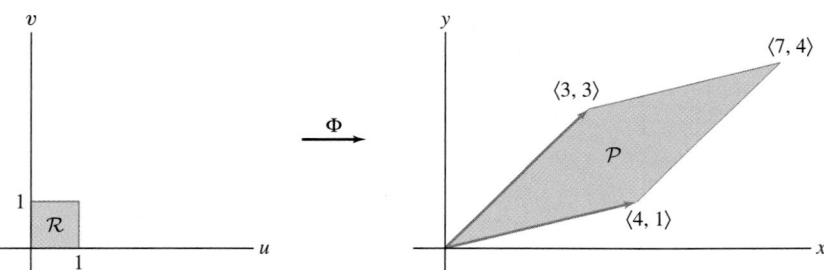

FIGURE 11 The map
$\Phi(u, v) = (4u + 3v, u + 3v)$.

In general (Figure 3), the map

$$\Phi(u, v) = (Au + Cv, Bu + Dv)$$

satisfies

$$\Phi(1, 0) = (A, B), \quad \Phi(0, 1) = (C, D)$$

Solution   It would not be too hard to evaluate this double integral directly, but our purpose here is to illustrate the Change of Variables Formula.

Step 1. **Define a map.**
We look for a linear map Φ that sends the unit square $\mathcal{R} = [0, 1] \times [0, 1]$ to $\mathcal{P}$. Observe that (Figure 11)

$$\Phi(u, v) = (4u + 3v, u + 3v)$$

satisfies $\Phi(1, 0) = (4, 1)$ and $\Phi(0, 1) = (3, 3)$. Thus, Φ maps the vectors $\langle 1, 0 \rangle$ and $\langle 0, 1 \rangle$ to $\langle 4, 1 \rangle$ and $\langle 3, 3 \rangle$. By linearity, Φ maps the unit square $\mathcal{R}$ to $\mathcal{P}$.

Step 2. **Express $f(x, y)$ in terms of the new variables.**
Since $x = 4u + 3v$ and $y = u + 3v$, we have

$$e^{4x-y} = e^{4(4u+3v)-(u+3v)} = e^{15u+9v}$$

Step 3. **Compute the Jacobian.**

$$\text{Jac}(\Phi) = \begin{vmatrix} \dfrac{\partial x}{\partial u} & \dfrac{\partial x}{\partial v} \\ \dfrac{\partial y}{\partial u} & \dfrac{\partial y}{\partial v} \end{vmatrix} = \begin{vmatrix} 4 & 3 \\ 1 & 3 \end{vmatrix} = 9$$

Step 4. **Apply the Change of Variables Formula.**

The Change of Variables Formula tells us that $dA = 9\, du\, dv$:

$$\iint_{\mathcal{P}} e^{4x-y}\, dA = \iint_{\mathcal{R}} e^{15u+9v}\, |\text{Jac}(\Phi)|\, du\, dv = 9 \int_0^1 \int_0^1 e^{15u+9v}\, du\, dv$$

$$= 9\left(\int_0^1 e^{15u}\, du\right)\left(\int_0^1 e^{9v}\, dv\right) = \frac{1}{15}(e^{15}-1)(e^9-1) \quad\blacksquare$$

■ **EXAMPLE 7** Use the Change of Variables $x = uv^{-1}$, $y = uv$ to compute

$$\iint_{\mathcal{D}} (x^2 + y^2)\, dA$$

where $\mathcal{D}$ is the domain $1 \le xy \le 4$, $1 \le y/x \le 4$.

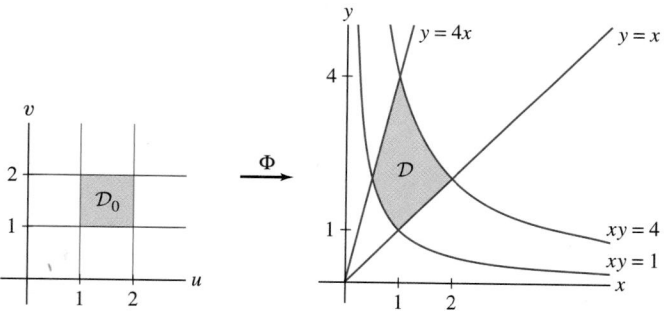

FIGURE 12 The map $\Phi(u, v) = (uv^{-1}, uv)$.

Solution We studied the map $\Phi(u, v) = (uv^{-1}, uv)$ in Example 3 (Figure 12), where we showed that $u = \sqrt{xy}$ and $v = \sqrt{y/x}$. The inequalities $1 \le xy \le 4$ and $1 \le y/x \le 4$ are thus equivalent to $1 \le u \le 2$ and $1 \le v \le 2$. Therefore, Φ maps the rectangle $\mathcal{D}_0 = [1, 2] \times [1, 2]$ to the domain $\mathcal{D}$. The Jacobian of Φ is

$$\frac{\partial(x, y)}{\partial(u, v)} = \begin{vmatrix} \dfrac{\partial x}{\partial u} & \dfrac{\partial x}{\partial v} \\ \dfrac{\partial y}{\partial u} & \dfrac{\partial y}{\partial v} \end{vmatrix} = \begin{vmatrix} v^{-1} & -uv^{-2} \\ v & u \end{vmatrix} = \frac{2u}{v}$$

To apply the Change of Variables, we write $f(x, y)$ in terms of u and v:

$$f(x, y) = x^2 + y^2 = \left(\frac{u}{v}\right)^2 + (uv)^2 = u^2(v^{-2} + v^2)$$

By the Change of Variables Formula,

$$\iint_{\mathcal{D}} (x^2 + y^2)\, dA = \iint_{\mathcal{D}_0} u^2(v^{-2} + v^2)\left|\frac{2u}{v}\right|\, du\, dv$$

$$= 2 \int_{v=1}^2 \int_{u=1}^2 u^3(v^{-3} + v)\, du\, dv$$

$$= 2 \left(\int_{u=1}^{2} u^3 \, du \right) \left(\int_{v=1}^{2} (v^{-3} + v) \, dv \right)$$

$$= 2 \left(\frac{2^4 - 1^4}{4} \right) \left(\left(\frac{1}{2} v^2 - \frac{1}{2} v^{-2} \right) \Big|_{1}^{2} \right) = \frac{225}{16} \qquad \blacksquare$$

The Jacobian of the inverse map Φ^{-1} is the reciprocal of $\mathrm{Jac}(\Phi)$. This can be verified using the Chain Rule (see Exercises 48–50). Thus, we have $\mathrm{Jac}(\Phi^{-1}) = \mathrm{Jac}(\Phi)^{-1}$, which may be written

$$\frac{\partial(x, y)}{\partial(u, v)} = \left(\frac{\partial(u, v)}{\partial(x, y)} \right)^{-1} \qquad \boxed{14}$$

■ **EXAMPLE 8** Using the Inverse Map Use Change of Variables to integrate the function $f(x, y) = x^2 + y^2$ over the domain

$$\mathcal{D}: \quad -3 \le x^2 - y^2 \le 3, \qquad 1 \le xy \le 4$$

Solution This is different from Example 7 because it is not clear which map to use to change variables. What we can do is define a map Φ in the wrong direction (Figure 13), from $\mathcal{D}$ to the rectangle $\mathcal{R} = [-3, 3] \times [1, 4]$ in the uv-plane:

$$\Phi : \mathcal{D} \to \mathcal{R}$$

$$(x, y) \to (x^2 - y^2, xy)$$

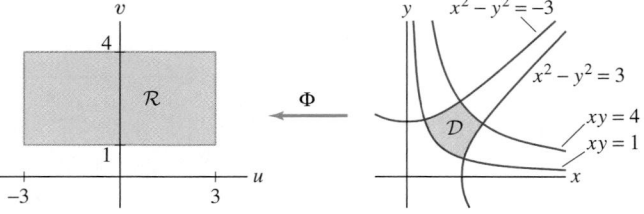

FIGURE 13 The map Φ. Note that Φ goes in the "wrong" direction.

Since we want to convert an integral over $\mathcal{D}$ into an integral over the rectangle $\mathcal{R}$, we must apply the Change of Variables Formula to the inverse mapping:

$$\Phi^{-1} : \mathcal{R} \to \mathcal{D}$$

To compute the Jacobian of Φ^{-1}, we first compute the Jacobian of Φ. Since $u = x^2 - y^2$ and $v = xy$, we have

$$\mathrm{Jac}(\Phi) = \frac{\partial(u, v)}{\partial(x, y)} = \begin{vmatrix} \dfrac{\partial u}{\partial x} & \dfrac{\partial u}{\partial y} \\ \dfrac{\partial v}{\partial x} & \dfrac{\partial v}{\partial y} \end{vmatrix} = \begin{vmatrix} 2x & -2y \\ y & x \end{vmatrix} = 2(x^2 + y^2)$$

Then, by Eq. (14),

$$\mathrm{Jac}(\Phi^{-1}) = \frac{\partial(x, y)}{\partial(u, v)} = \left(\frac{\partial(u, v)}{\partial(x, y)} \right)^{-1} = \frac{1}{2(x^2 + y^2)}$$

Normally, the next step would be to express $f(x, y)$ in terms of u and v. To do so, we would have to express x and y in terms of u and v (this amounts to finding Φ^{-1}). Fortunately, this is not necessary in our case because the Jacobian cancels with $f(x, y)$:

$$\iint_{\mathcal{D}} (x^2 + y^2)\, dx\, dy = \iint_{\mathcal{R}} f(x(u, v), y(u, v)) \left| \text{Jac}(\Phi^{-1}) \right| du\, dv$$

$$= \iint_{\mathcal{R}} (x^2 + y^2) \frac{1}{2(x^2 + y^2)}\, du\, dv$$

$$= \frac{1}{2} \iint_{\mathcal{R}} du\, dv = \frac{1}{2} \int_{-3}^{3} \int_{1}^{4} dv\, du = \frac{1}{2}(6)(3) = 9 \quad \blacksquare$$

Change of Variables in Three Variables

The Change of Variables Formula has the same form in three (or more) variables as in two variables. Let

$$\Phi : \mathcal{W}_0 \to \mathcal{W}$$

be a mapping from a three-dimensional region $\mathcal{W}_0$ in (u, v, w)-space to a region $\mathcal{W}$ in (x, y, z)-space, say,

$$x = \phi_1(u, v, w), \qquad y = \phi_2(u, v, w), \qquad z = \phi_3(u, v, w)$$

←·· REMINDER 3×3-determinants are defined in Eq. (2) of Section 12.4.

The Jacobian $\text{Jac}(\Phi)$ is the 3×3-determinant:

$$\text{Jac}(\Phi) = \frac{\partial(x, y, z)}{\partial(u, v, w)} = \begin{vmatrix} \dfrac{\partial x}{\partial u} & \dfrac{\partial x}{\partial v} & \dfrac{\partial x}{\partial w} \\[2mm] \dfrac{\partial y}{\partial u} & \dfrac{\partial y}{\partial v} & \dfrac{\partial y}{\partial w} \\[2mm] \dfrac{\partial z}{\partial u} & \dfrac{\partial z}{\partial v} & \dfrac{\partial z}{\partial w} \end{vmatrix} \qquad \boxed{15}$$

The Change of Variables Formula amounts to the relation

$$dx\, dy\, dz = \left| \frac{\partial(x, y, z)}{\partial(u, v, w)} \right| du\, dv\, dw$$

Assume that Φ is C^1 and one-to-one on the interior of $\mathcal{W}_0$, and that f is continuous. Then

$$\iiint_{\mathcal{W}} f(x, y, z)\, dx\, dy\, dz$$

$$= \iiint_{\mathcal{W}_0} f(x(u, v, w), y(u, v, w), z(u, v, w)) \left| \frac{\partial(x, y, z)}{\partial(u, v, w)} \right| du\, dv\, dw \qquad \boxed{16}$$

In Exercises 42 and 43, you are asked to use the general Change of Variables Formula to derive the formulas for integration in cylindrical and spherical coordinates developed in Section 15.4.

15.5 SUMMARY

• Let $\Phi(u, v) = (\phi(u, v), \psi(u, v))$ be a mapping. We also write $x = \phi(u, v)$ and $y = \psi(u, v)$. The Jacobian of Φ is the determinant

$$\mathrm{Jac}(\Phi) = \frac{\partial(x, y)}{\partial(u, v)} = \begin{vmatrix} \dfrac{\partial x}{\partial u} & \dfrac{\partial x}{\partial v} \\ \dfrac{\partial y}{\partial u} & \dfrac{\partial y}{\partial u} \end{vmatrix}$$

- If $\Phi : \mathcal{D}_0 \to \mathcal{D}$ is C^1 and one-to-one on the interior of $\mathcal{D}_0$, and if f is continuous, then

$$\iint_{\mathcal{D}} f(x, y)\, dA = \iint_{\mathcal{D}_0} f(x(u, v), y(u, v)) \left| \frac{\partial(x, y)}{\partial(u, v)} \right| du\, dv$$

- $\dfrac{\partial(u, v)}{\partial(x, y)} = \left(\dfrac{\partial(x, y)}{\partial(u, v)} \right)^{-1}$

15.5 EXERCISES

Preliminary Questions

1. Which of the following maps is linear?
(a) (uv, v) **(b)** $(u + v, u)$ **(c)** $(3, e^u)$

2. Suppose that Φ is a linear map such that $\Phi(2, 0) = (4, 0)$ and $\Phi(0, 3) = (-3, 9)$. Find the images of:
(a) $\Phi(1, 0)$ **(b)** $\Phi(1, 1)$ **(c)** $\Phi(2, 1)$

3. What is the area of $\Phi(\mathcal{R})$ if $\mathcal{R}$ is a rectangle of area 9 and Φ is a mapping whose Jacobian has constant value 4?

4. Estimate the area of $\Phi(\mathcal{R})$, where $\mathcal{R} = [1, 1.2] \times [3, 3.1]$ and Φ is a mapping such that $\mathrm{Jac}(\Phi)(1, 3) = 3$.

Exercises

1. Determine the image under $\Phi(u, v) = (2u, u + v)$ of the following sets:
(a) The u- and v-axes
(b) The rectangle $\mathcal{R} = [0, 5] \times [0, 7]$
(c) The line segment joining $(1, 2)$ and $(5, 3)$
(d) The triangle with vertices $(0, 1)$, $(1, 0)$, and $(1, 1)$

2. Describe [in the form $y = f(x)$] the images of the lines $u = c$ and $v = c$ under the mapping $\Phi(u, v) = (u/v, u^2 - v^2)$.

3. Let $\Phi(u, v) = (u^2, v)$. Is Φ one-to-one? If not, determine a domain on which Φ is one-to-one. Find the image under Φ of:
(a) The u- and v-axes
(b) The rectangle $\mathcal{R} = [-1, 1] \times [-1, 1]$
(c) The line segment joining $(0, 0)$ and $(1, 1)$
(d) The triangle with vertices $(0, 0)$, $(0, 1)$, and $(1, 1)$

4. Let $\Phi(u, v) = (e^u, e^{u+v})$.
(a) Is Φ one-to-one? What is the image of Φ?
(b) Describe the images of the vertical lines $u = c$ and horizontal lines $v = c$.

In Exercises 5–12, let $\Phi(u, v) = (2u + v, 5u + 3v)$ be a map from the uv-plane to the xy-plane.

5. Show that the image of the horizontal line $v = c$ is the line $y = \frac{5}{2}x + \frac{1}{2}c$. What is the image (in slope-intercept form) of the vertical line $u = c$?

6. Describe the image of the line through the points $(u, v) = (1, 1)$ and $(u, v) = (1, -1)$ under Φ in slope-intercept form.

7. Describe the image of the line $v = 4u$ under Φ in slope-intercept form.

8. Show that Φ maps the line $v = mu$ to the line of slope $(5 + 3m)/(2 + m)$ through the origin in the xy-plane.

9. Show that the inverse of Φ is

$$\Phi^{-1}(x, y) = (3x - y, -5x + 2y)$$

Hint: Show that $\Phi(\Phi^{-1}(x, y)) = (x, y)$ and $\Phi^{-1}(\Phi(u, v)) = (u, v)$.

10. Use the inverse in Exercise 9 to find:
(a) A point in the uv-plane mapping to $(2, 1)$.
(b) A segment in the uv-plane mapping to the segment joining $(-2, 1)$ and $(3, 4)$.

11. Calculate $\mathrm{Jac}(\Phi) = \dfrac{\partial(x, y)}{\partial(u, v)}$.

12. Calculate $\mathrm{Jac}(\Phi^{-1}) = \dfrac{\partial(u, v)}{\partial(x, y)}$.

In Exercises 13–18, compute the Jacobian (at the point, if indicated).

13. $\Phi(u, v) = (3u + 4v, u - 2v)$

14. $\Phi(r, s) = (rs, r + s)$

15. $\Phi(u, v) = (ue^v, ve^{3u})$, $(u, v) = (1, 2)$

16. $\Phi(r, t) = (r \sin t, r - \cos t)$, $(r, t) = (1, \pi)$

17. $\Phi(r, \theta) = (r \cos \theta, r \sin \theta)$

18. $\Phi(u, v) = (ue^v, e^u)$

19. Find a linear mapping Φ that maps $[0, 1] \times [0, 1]$ to the parallelogram in the xy-plane spanned by the vectors $\langle 2, 3 \rangle$ and $\langle 4, 1 \rangle$.

20. Find a linear mapping Φ that maps $[0, 1] \times [0, 1]$ to the parallelogram in the xy-plane spanned by the vectors $\langle -2, 5 \rangle$ and $\langle 1, 7 \rangle$.

21. Let $\mathcal{D}$ be the parallelogram in Figure 14. Apply the Change of Variables Formula to the $\Phi(u, v) = (5u + 3v, u + 4v)$ to evaluate $\iint_{\mathcal{D}} xy \, dA$ as an integral over $\mathcal{D}_0 = [0, 1] \times [0, 1]$.

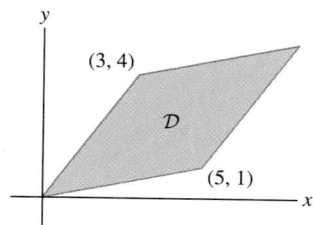

FIGURE 14

22. Let $\Phi(u, v) = (u - uv, uv)$.

(a) Determine the images of the horizontal and vertical lines in the uv-plane.

(b) Compute the Jacobian of Φ.

(c) Observe that by the formula for the area of a triangle, the region $\mathcal{D}$ in Figure 15 has area $\frac{1}{2}(b^2 - a^2)$. Compute this area again using Change of Variables Formula applied to Φ.

(d) Calculate $\iint_{\mathcal{D}} xy \, dA$.

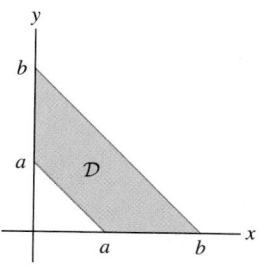

FIGURE 15

23. Let $\Phi(u, v) = (3u + v, u - 2v)$. Use the Jacobian to determine the area of $\Phi(\mathcal{R})$ for:

(a) $\mathcal{R} = [0, 3] \times [0, 5]$

(b) $\mathcal{R} = [2, 5] \times [1, 7]$

24. Find a linear map T that maps $[0, 1] \times [0, 1]$ to the parallelogram $\mathcal{P}$ in the xy-plane with vertices $(0, 0), (2, 2), (1, 4), (3, 6)$. Then calculate the double integral of e^{2x-y} over $\mathcal{P}$ via change of variables.

25. With Φ as in Example 3, use the Change of Variables Formula to compute the area of the image of $[1, 4] \times [1, 4]$.

In Exercises 26–28, let $\mathcal{R}_0 = [0, 1] \times [0, 1]$ be the unit square. The translate of a map $\Phi_0(u, v) = (\phi(u, v), \psi(u, v))$ is a map

$$\Phi(u, v) = (a + \phi(u, v), b + \psi(u, v))$$

where a, b are constants. Observe that the map Φ_0 in Figure 16 maps $\mathcal{R}_0$ to the parallelogram $\mathcal{P}_0$ and the translate

$$\Phi_1(u, v) = (2 + 4u + 2v, 1 + u + 3v)$$

maps $\mathcal{R}_0$ to $\mathcal{P}_1$.

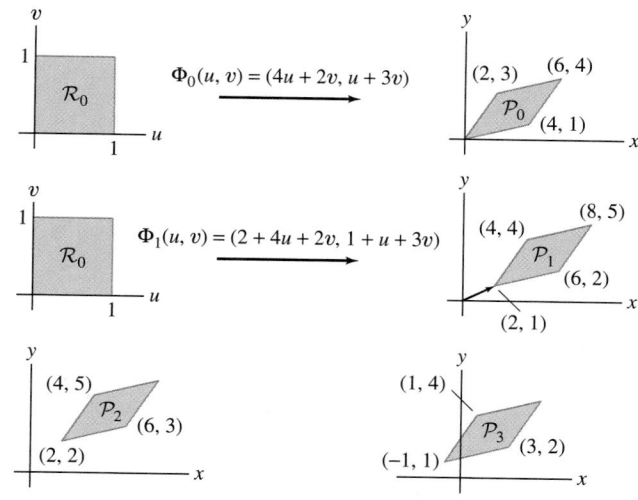

FIGURE 16

26. Find translates Φ_2 and Φ_3 of the mapping Φ_0 in Figure 16 that map the unit square $\mathcal{R}_0$ to the parallelograms $\mathcal{P}_2$ and $\mathcal{P}_3$.

27. Sketch the parallelogram $\mathcal{P}$ with vertices $(1, 1), (2, 4), (3, 6), (4, 9)$ and find the translate of a linear mapping that maps $\mathcal{R}_0$ to $\mathcal{P}$.

28. Find the translate of a linear mapping that maps $\mathcal{R}_0$ to the parallelogram spanned by the vectors $\langle 3, 9 \rangle$ and $\langle -4, 6 \rangle$ based at $(4, 2)$.

29. Let $\mathcal{D} = \Phi(\mathcal{R})$, where $\Phi(u, v) = (u^2, u + v)$ and $\mathcal{R} = [1, 2] \times [0, 6]$. Calculate $\iint_{\mathcal{D}} y \, dA$. *Note:* It is not necessary to describe $\mathcal{D}$.

30. Let $\mathcal{D}$ be the image of $\mathcal{R} = [1, 4] \times [1, 4]$ under the map $\Phi(u, v) = (u^2/v, v^2/u)$.

(a) Compute Jac(Φ).

(b) Sketch $\mathcal{D}$.

(c) Use the Change of Variables Formula to compute Area($\mathcal{D}$) and $\iint_{\mathcal{D}} f(x, y) \, dA$, where $f(x, y) = x^2 + y^2$.

31. Compute $\iint_{\mathcal{D}} (x + 3y)\,dx\,dy$, where $\mathcal{D}$ is the shaded region in Figure 17. *Hint:* Use the map $\Phi(u, v) = (u - 2v, v)$.

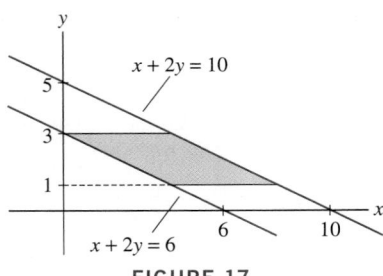

FIGURE 17

32. Use the map $\Phi(u, v) = \left(\dfrac{u}{v+1}, \dfrac{uv}{v+1} \right)$ to compute

$$\iint_{\mathcal{D}} (x + y)\,dx\,dy$$

where $\mathcal{D}$ is the shaded region in Figure 18.

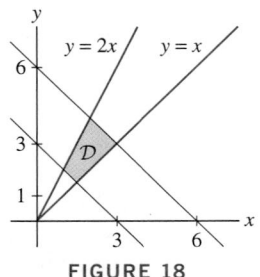

FIGURE 18

33. Show that $T(u, v) = (u^2 - v^2, 2uv)$ maps the triangle $\mathcal{D}_0 = \{(u, v) : 0 \le v \le u \le 1\}$ to the domain $\mathcal{D}$ bounded by $x = 0$, $y = 0$, and $y^2 = 4 - 4x$. Use T to evaluate

$$\iint_{\mathcal{D}} \sqrt{x^2 + y^2}\,dx\,dy$$

34. Find a mapping Φ that maps the disk $u^2 + v^2 \le 1$ onto the interior of the ellipse $\left(\dfrac{x}{a} \right)^2 + \left(\dfrac{y}{b} \right)^2 \le 1$. Then use the Change of Variables Formula to prove that the area of the ellipse is πab.

35. Calculate $\iint_{\mathcal{D}} e^{9x^2 + 4y^2}\,dA$, where $\mathcal{D}$ is the interior of the ellipse $\left(\dfrac{x}{2} \right)^2 + \left(\dfrac{y}{3} \right)^2 \le 1$.

36. Compute the area of the region enclosed by the ellipse $x^2 + 2xy + 2y^2 - 4y = 8$ as an integral in the variables $u = x + y$, $v = y$.

37. Sketch the domain $\mathcal{D}$ bounded by $y = x^2$, $y = \frac{1}{2}x^2$, and $y = x$. Use a change of variables with the map $x = uv$, $y = u^2$ to calculate

$$\iint_{\mathcal{D}} y^{-1}\,dx\,dy$$

This is an improper integral since $f(x, y) = y^{-1}$ is undefined at $(0, 0)$, but it becomes proper after changing variables.

38. Find an appropriate change of variables to evaluate

$$\iint_{\mathcal{R}} (x + y)^2 e^{x^2 - y^2}\,dx\,dy$$

where $\mathcal{R}$ is the square with vertices $(1, 0)$, $(0, 1)$, $(-1, 0)$, $(0, -1)$.

39. Let Φ be the inverse of the map $T(x, y) = (xy, x^2 y)$ from the xy-plane to the uv-plane. Let $\mathcal{D}$ be the domain in Figure 19. Show, by applying the Change of Variables Formula to the inverse $\Phi = T^{-1}$, that

$$\iint_{\mathcal{D}} e^{xy}\,dA = \int_{10}^{20} \int_{20}^{40} e^u v^{-1}\,dv\,du$$

and evaluate this result. *Hint:* Use Eq. (14) to compute Jac(Φ). It is not necessary to determine Φ explicitly.

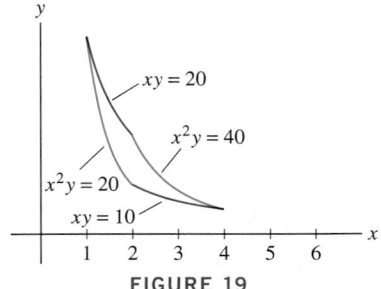

FIGURE 19

40. Sketch the domain

$$\mathcal{D} = \{(x, y) : 1 \le x + y \le 4, \ -4 \le y - 2x \le 1\}$$

(a) Let T be the map $u = x + y$, $v = y - 2x$ from the xy-plane to the uv-plane and let Φ be its inverse. Use Eq. (14) to compute Jac(Φ).
(b) Compute $\iint_{\mathcal{D}} e^{x+y}\,dA$ using Change of Variables with the map Φ. *Hint:* It is not necessary to solve for Φ explicitly.

41. Let $I = \iint_{\mathcal{D}} (x^2 - y^2)\,dx\,dy$, where

$$\mathcal{D} = \{(x, y) : 2 \le xy \le 4, 0 \le x - y \le 3, x \ge 0, y \ge 0\}$$

(a) Show that the mapping $u = xy$, $v = x - y$ maps $\mathcal{D}$ to the rectangle $\mathcal{R} = [2, 4] \times [0, 3]$.
(b) Compute $\partial(x, y)/\partial(u, v)$ by first computing $\partial(u, v)/\partial(x, y)$.
(c) Use the Change of Variables Formula to show that I is equal to the integral of $f(u, v) = v$ over $\mathcal{R}$ and evaluate.

42. Derive formula (6) in Section 15.4 for integration in cylindrical coordinates from the general Change of Variables Formula.

43. Derive formula (10) in Section 15.4 for integration in spherical coordinates from the general Change of Variables Formula.

44. Use the Change of Variables Formula in three variables to prove that the volume of the ellipsoid $\left(\dfrac{x}{a} \right)^2 + \left(\dfrac{y}{b} \right)^2 + \left(\dfrac{z}{c} \right)^2 = 1$ is equal to $abc \times$ the volume of the unit sphere.

Further Insights and Challenges

45. Use the map

$$x = \frac{\sin u}{\cos v}, \qquad y = \frac{\sin v}{\cos u}$$

to evaluate the integral

$$\int_0^1 \int_0^1 \frac{dx\,dy}{1 - x^2 y^2}$$

This is an improper integral since the integrand is infinite if $x = \pm 1$, $y = \pm 1$, but the Change of Variables shows that the result is finite.

46. Verify properties (1) and (2) for linear functions and show that any map satisfying these two properties is linear.

47. 🖉 Let Φ be a linear map. Prove Eq. (4) in the following steps.

(a) For any set $\mathcal{D}$ in the uv-plane and any vector $\mathbf{u}$, let $\mathcal{D} + \mathbf{u}$ be the set obtained by translating all points in $\mathcal{D}$ by $\mathbf{u}$. By linearity, Φ

maps $\mathcal{D} + \mathbf{u}$ to the translate $\Phi(\mathcal{D}) + \Phi(\mathbf{u})$ [Figure 20(C)]. Therefore, if Eq. (4) holds for $\mathcal{D}$, it also holds for $\mathcal{D} + \mathbf{u}$.

(b) In the text, we verified Eq. (4) for the unit rectangle. Use linearity to show that Eq. (4) also holds for all rectangles with vertex at the origin and sides parallel to the axes. Then argue that it also holds for each triangular half of such a rectangle, as in Figure 20(A).

(c) Figure 20(B) shows that the area of a parallelogram is a difference of the areas of rectangles and triangles covered by steps (a) and (b). Use this to prove Eq. (4) for arbitrary parallelograms.

48. The product of 2×2 matrices A and B is the matrix AB defined by

$$\underbrace{\begin{pmatrix} a & b \\ c & d \end{pmatrix}}_{A} \underbrace{\begin{pmatrix} a' & b' \\ c' & d' \end{pmatrix}}_{B} = \underbrace{\begin{pmatrix} aa' + bc' & ab' + bd' \\ ca' + dc' & cb' + dd' \end{pmatrix}}_{AB}$$

The (i, j)-entry of A is the **dot product** of the ith row of A and jth column of B. Prove that $\det(AB) = \det(A)\det(B)$.

49. Let $\Phi_1 : \mathcal{D}_1 \to \mathcal{D}_2$ and $\Phi_2 : \mathcal{D}_2 \to \mathcal{D}_3$ be C^1 maps, and let $\Phi_2 \circ \Phi_1 : \mathcal{D}_1 \to \mathcal{D}_3$ be the composite map. Use the Multivariable Chain Rule and Exercise 48 to show that

$$\text{Jac}(\Phi_2 \circ \Phi_1) = \text{Jac}(\Phi_2)\text{Jac}(\Phi_1)$$

50. Use Exercise 49 to prove that $\text{Jac}(\Phi^{-1}) = \text{Jac}(\Phi)^{-1}$. *Hint:* Verify that $\text{Jac}(I) = 1$, where I is the identity map $I(u, v) = (u, v)$.

51. Let $(\overline{x}, \overline{y})$ be the centroid of a domain $\mathcal{D}$. For $\lambda > 0$, let $\lambda \mathcal{D}$ be the **dilate** of $\mathcal{D}$, defined by

$$\lambda \mathcal{D} = \{(\lambda x, \lambda y) : (x, y) \in \mathcal{D}\}$$

Use the Change of Variables Formula to prove that the centroid of $\lambda \mathcal{D}$ is $(\lambda \overline{x}, \lambda \overline{y})$.

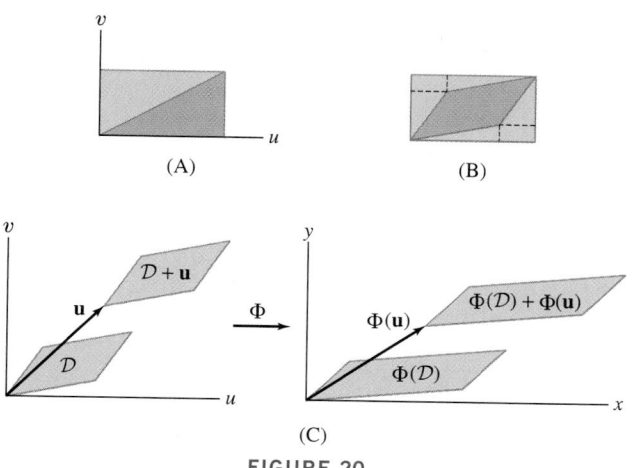

(A)   (B)

(C)

FIGURE 20

CHAPTER REVIEW EXERCISES

1. Calculate the Riemann sum $S_{3,4}$ for $\int_1^2 \int_2^3 x^2 y \, dx \, dy$ using two choices of sample points:

(a) Lower-left vertex

(b) Midpoint of rectangle

Then calculate the exact value of the double integral.

2. Let $S_{N,N}$ be the Riemann sum for $\int_0^1 \int_0^1 \cos(xy) \, dx \, dy$ using midpoints as sample points.

(a) Calculate $S_{4,4}$.

(b) *CAS* Use a computer algebra system to calculate $S_{N,N}$ for $N = 10, 50, 100$.

3. Let $\mathcal{D}$ be the shaded domain in Figure 1.

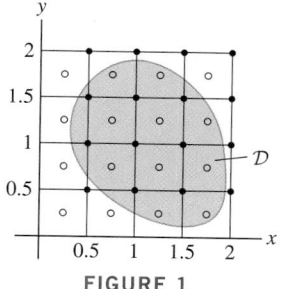

FIGURE 1

Estimate $\iint_{\mathcal{D}} xy \, dA$ by the Riemann sum whose sample points are the midpoints of the squares in the grid.

4. Explain the following:

(a) $\int_{-1}^{1} \int_{-1}^{1} \sin(xy) \, dx \, dy = 0$

(b) $\int_{-1}^{1} \int_{-1}^{1} \cos(xy) \, dx \, dy > 0$

In Exercises 5–8, evaluate the iterated integral.

5. $\int_{0}^{2} \int_{3}^{5} y(x - y) \, dx \, dy$

6. $\int_{1/2}^{0} \int_{0}^{\pi/6} e^{2y} \sin 3x \, dx \, dy$

7. $\int_{0}^{\pi/3} \int_{0}^{\pi/6} \sin(x + y) \, dx \, dy$

8. $\int_{1}^{2} \int_{1}^{2} \frac{y \, dx \, dy}{x + y^2}$

In Exercises 9–14, sketch the domain $\mathcal{D}$ and calculate $\iint_{\mathcal{D}} f(x, y) \, dA$.

9. $\mathcal{D} = \{0 \le x \le 4, \ 0 \le y \le x\}, \quad f(x, y) = \cos y$

10. $\mathcal{D} = \{0 \le x \le 2, \ 0 \le y \le 2x - x^2\}, \quad f(x, y) = \sqrt{xy}$

11. $\mathcal{D} = \{0 \le x \le 1, \ 1 - x \le y \le 2 - x\} \quad f(x, y) = e^{x+2y}$

12. $\mathcal{D} = \{1 \le x \le 2, \ 0 \le y \le \frac{1}{x}\}, \quad f(x, y) = \cos(xy)$

13. $\mathcal{D} = \{0 \le y \le 1, \ 0.5y^2 \le x \le y^2\} \quad f(x, y) = ye^{1+x}$

14. $\mathcal{D} = \{1 \le y \le e, \ y \le x \le 2y\}, \quad f(x, y) = \ln(x + y)$

15. Express $\int_{-3}^{3} \int_{0}^{9-x^2} f(x, y) \, dy \, dx$ as an iterated integral in the order $dx \, dy$.

16. Let $\mathcal{D}$ be the domain between $y = x$ and $y = \sqrt{x}$. Calculate $\iint_{\mathcal{D}} xy \, dA$ as an iterated integral in the order $dx \, dy$ and $dy \, dx$.

17. Verify directly that

$$\int_{2}^{3} \int_{0}^{2} \frac{dy \, dx}{1 + x - y} = \int_{0}^{2} \int_{2}^{3} \frac{dx \, dy}{1 + x - y}$$

18. Prove the formula

$$\int_{0}^{1} \int_{0}^{y} f(x) \, dx \, dy = \int_{0}^{1} (1 - x) f(x) \, dx$$

Then use it to calculate $\int_{0}^{1} \int_{0}^{y} \frac{\sin x}{1 - x} \, dx \, dy$.

19. Rewrite $\int_{0}^{1} \int_{-\sqrt{1-y^2}}^{\sqrt{1-y^2}} \frac{y \, dx \, dy}{(1 + x^2 + y^2)^2}$ by interchanging the order of integration and evaluate.

20. Evaluate $\iint_{\mathcal{D}} x \, dA$, where $\mathcal{D}$ is the shaded domain in Figure 2.

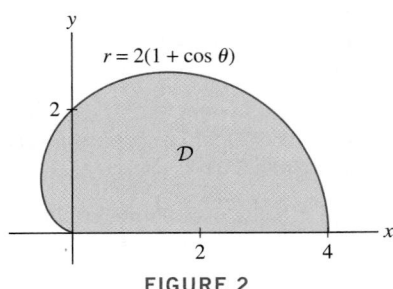

$r = 2(1 + \cos \theta)$

FIGURE 2

21. Find the center of mass of the sector of central angle 2θ (symmetric with respect to the y-axis) in Figure 3, assuming that the mass density is $\rho(x, y) = x^2$.

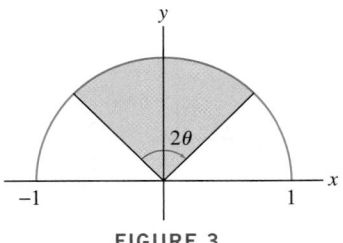

2θ

FIGURE 3

22. Find the volume of the region between the graph of the function $f(x, y) = 1 - (x^2 + y^2)$ and the xy-plane.

23. ⎡AS Express the average value of $f(x, y) = e^{xy}$ over the ellipse $\frac{x^2}{2} + y^2 = 1$ as an iterated integral and evaluate numerically using a computer algebra system.

24. Evaluate $\int_{0}^{3} \int_{1}^{4} \int_{2}^{5} (x^3 + y^2 + z) \, dx \, dy \, dz$.

25. Calculate $\iiint_{\mathcal{B}} (xy + z) \, dV$, where

$$\mathcal{B} = \{0 \le x \le 2, \ 0 \le y \le 1, \ 1 \le z \le 3\}$$

as an iterated integral in two different ways.

26. Calculate $\iiint_{\mathcal{W}} xyz\,dV$, where

$$\mathcal{W} = \{0 \le x \le 1,\ x \le y \le 1,\ x \le z \le x+y\}$$

27. Evaluate $I = \int_{-1}^{1}\int_{0}^{\sqrt{1-x^2}}\int_{0}^{1}(x+y+z)\,dz\,dy\,dx$. Then rewrite I as an iterated integral in the order $dy\,dz\,dx$ and evaluate a second time.

28. Find the average value of $f(x,y,z) = xy^2z^3$ on the box $[0,1] \times [0,2] \times [0,3]$.

29. Find the center of mass of the first octant of the ball $x^2 + y^2 + z^2 = 1$, assuming a mass density of $\rho(x,y,z) = x$.

30. Find the centroid of the tetrahedron with vertices $(0,0,0)$, $(2,0,0)$, $(0,4,0)$, and $(0,0,6)$.

31. Use polar coordinates to calculate $\iint_{\mathcal{D}} \sqrt{x^2+y^2}\,dA$, where $\mathcal{D}$ is the region in the first quadrant bounded by the spiral $r = \theta$, the circle $r = 1$, and the x-axis.

32. Calculate $\iint_{\mathcal{D}} \sin(x^2+y^2)\,dA$, where

$$\mathcal{D} = \left\{ \frac{\pi}{2} \le x^2 + y^2 \le \pi \right\}$$

33. Use cylindrical coordinates to find the total mass of the solid bounded by $z = 8 - x^2 - y^2$ and $z = x^2 + y^2$, assuming a mass density of $f(x,y,z) = (x^2+y^2)^{1/2}$.

34. Describe a region whose volume is equal to:

(a) $\int_{0}^{2\pi}\int_{0}^{\pi/2}\int_{4}^{9}\rho^2\sin\phi\,d\rho\,d\phi\,d\theta$

(b) $\int_{-2}^{1}\int_{\pi/3}^{\pi/4}\int_{0}^{2} r\,dr\,d\theta\,dz$

(c) $\int_{0}^{2\pi}\int_{0}^{3}\int_{-\sqrt{9-r^2}}^{0} r\,dz\,dr\,d\theta$

35. Find the volume of the solid contained in the cylinder $x^2 + y^2 = 1$ below the curve $z = (x+y)^2$ and above the curve $z = -(x-y)^2$.

36. Use polar coordinates to evaluate $\iint_{\mathcal{D}} x\,dA$, where $\mathcal{D}$ is the shaded region between the two circles of radius 1 in Figure 4.

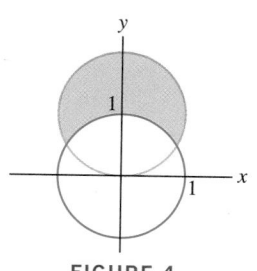

FIGURE 4

37. Express in cylindrical coordinates and evaluate:

$$\int_{0}^{1}\int_{0}^{\sqrt{1-x^2}}\int_{0}^{\sqrt{x^2+y^2}} z\,dz\,dy\,dx$$

38. Use spherical coordinates to calculate the triple integral of $f(x,y,z) = x^2 + y^2 + z^2$ over the region

$$1 \le x^2 + y^2 + z^2 \le 4$$

39. Convert to spherical coordinates and evaluate:

$$\int_{-2}^{2}\int_{-\sqrt{4-x^2}}^{\sqrt{4-x^2}}\int_{0}^{\sqrt{4-x^2-y^2}} e^{-(x^2+y^2+z^2)^{3/2}}\,dz\,dy\,dx$$

40. Using cylindrical coordinates, prove that the centroid of a right-circular cone of height h and radius R is located at height $\frac{h}{4}$ on the central axis.

41. Compute the Jacobian of the map

$$\Phi(r,s) = \left(e^r\cosh(s),\ e^r\sinh(s)\right)$$

42. Find a linear mapping $\Phi(u,v)$ that maps the unit square to the parallelogram in the xy-plane spanned by the vectors $\langle 3,-1\rangle$ and $\langle 1,4\rangle$. Then, use the Jacobian to find the area of $\Phi(\mathcal{R})$, where $\mathcal{R} = [0,4] \times [0,3]$.

43. Use the map

$$\Phi(u,v) = \left(\frac{u+v}{2}, \frac{u-v}{2}\right)$$

to compute $\iint_{\mathcal{R}} \big((x-y)\sin(x+y)\big)^2\,dx\,dy$, where $\mathcal{R}$ is the square with vertices $(\pi,0)$, $(2\pi,\pi)$, $(\pi,2\pi)$, and $(0,\pi)$.

44. Let $\mathcal{D}$ be the shaded region in Figure 5 and let Φ be the map

$$u = -y + x^2, \qquad v = y - x^3$$

(a) Show that Φ maps $\mathcal{D}$ to a rectangle $\mathcal{R}$ in the uv-plane.
(b) Apply Eq. (5) in Section 15.5 with $P = (1,7)$ to estimate Area$(\mathcal{D})$.

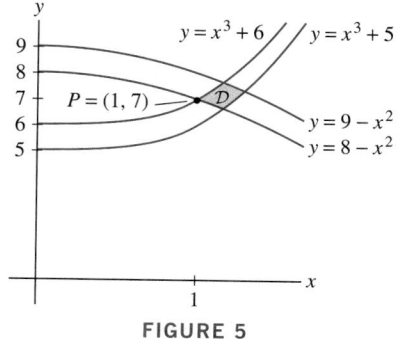

FIGURE 5

45. Calculate the integral of $f(x, y) = e^{3x-2y}$ over the parallelogram in Figure 6.

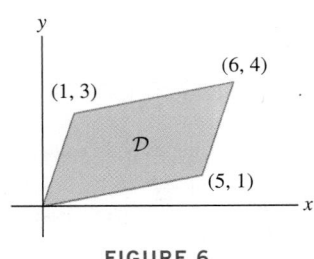

(1, 3)

(6, 4)

$\mathcal{D}$

(5, 1)

y

x

FIGURE 6

46. Sketch the region $\mathcal{D}$ bounded by the curves $y = \dfrac{2}{x}$, $y = \dfrac{1}{2x}$, $y = 2x$, $y = \dfrac{x}{2}$ in the first quadrant. Let Φ be the map $u = xy$, $v = y/x$ from the xy-plane to the uv-plane.

(a) Find the image of $\mathcal{D}$ under Φ.

(b) Show that $\left| \text{Jac}(\Phi^{-1}) \right| = \dfrac{1}{2|v|}$.

(c) Apply the Change of Variables Formula to prove the formula

$$\iint_{\mathcal{D}} f\left(\frac{y}{x}\right) dA = \frac{3}{4} \int_{1/2}^{2} \frac{f(v)\, dv}{v}$$

(d) Apply (c) to evaluate $\displaystyle\iint_{\mathcal{D}} \frac{ye^{y/x}}{x}\, dA$.

16 | LINE AND SURFACE INTEGRALS

In the previous chapter, we generalized integration from one variable to several variables. Our goal in this chapter is to generalize integration still further to include integration over curves or paths and surfaces. We will define integration not just of functions but also of vector fields. Integrals of vector fields are particularly important in applications involving the "field theories" of physics, such as the theory of electromagnetism, heat transfer, fluid dynamics, and aerodynamics. To lay the groundwork, this chapter begins with a discussion of vector fields.

16.1 Vector Fields

There are situations in which it is natural to consider a vector attached to each point in a given region. Such a collection of vectors is called a **vector field**. For example, Figure 2 shows the velocity vector field of wind off the coast of California at Los Angeles. The vector at each point is the velocity vector of the wind at that point.

We represent a vector field in three-space by a vector whose components are functions:

$$\mathbf{F}(x, y, z) = \langle F_1(x, y, z), F_2(x, y, z), F_3(x, y, z) \rangle$$

We also write

$$\mathbf{F} = F_1\mathbf{i} + F_2\mathbf{j} + F_3\mathbf{k}$$

and we denote the value of $\mathbf{F}$ at a point P by $\mathbf{F}(P)$. A vector field in the plane has the form $\mathbf{F}(x, y) = \langle F_1(x, y), F_2(x, y) \rangle$ or $F_1\mathbf{i} + F_2\mathbf{j}$. *In this chapter, we assume that the components F_j are smooth functions (i.e., have partial derivatives of all orders) on their domains.*

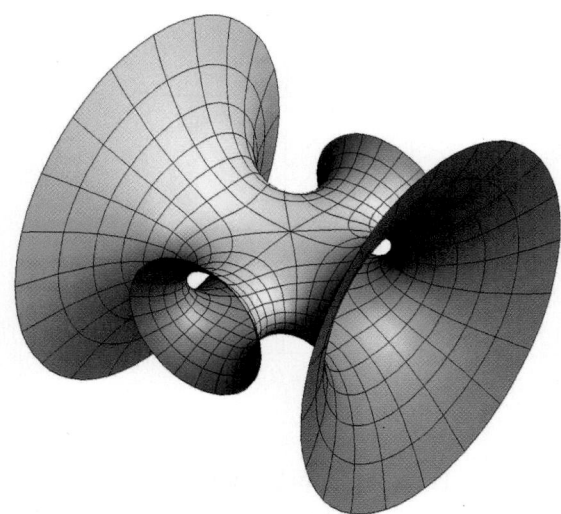

This image of the Symmetric Four-Noid surface was produced by R. Palais using the mathematical visualization program 3D-XplorMath.

"Fluid dynamics" is the study of the equations that must be satisfied by the components of velocity fields of fluids. Many interesting problems remain unsolved, especially the problem of describing turbulent flow (Figure 1). Researchers use high-speed computers to simulate the complex dynamics of turbulence.

FIGURE 1 Turbulent flow of water around the hull of a submarine.

FIGURE 2 Vector field representing horizontal wind velocity off the coast of California at Los Angeles.

■ **EXAMPLE 1** Sketch the vector fields $\mathbf{F} = \langle -y, x \rangle$ and $\mathbf{G} = \mathbf{i} + x\mathbf{j}$.

Solution These vector fields are sketched in Figure 3. Observe that $\mathbf{F} = \langle -y, x \rangle$ is perpendicular to the position vector $\langle x, y \rangle$ and has the same length. Thus, we may describe $\mathbf{F}$ as follows: The vectors along a circle of radius r centered at the origin are tangent to the circle and have length r.

The vector field $\mathbf{G} = \mathbf{i} + x\mathbf{j}$ depends only on x, so it assigns the same vector to points with the same x-coordinate. In other words, the vectors do not change along vertical lines. As $|x|$ increases, the vector $\langle 1, x \rangle$ grows longer and its slope is positive or negative, depending on the sign of x. ■

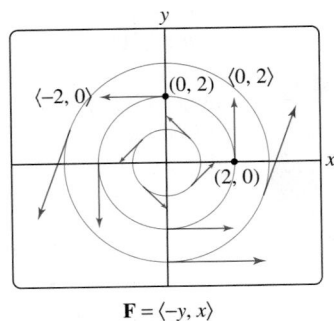
$$\mathbf{F} = \langle -y, x \rangle$$

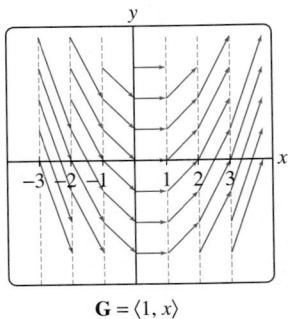
$$\mathbf{G} = \langle 1, x \rangle$$

FIGURE 3

Although it is generally not practical to sketch vector fields in three dimensions by hand, a computer algebra system can be used to produce useful visual representations (Figure 4). Note that the vector field in Figure 4(B) is a **constant vector field** whose value at every point is the vector $\langle 1, -1, 3 \rangle$.

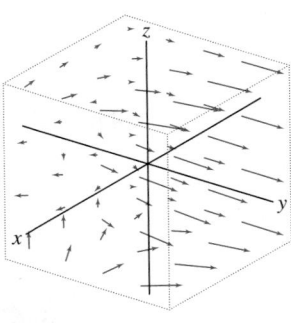
(A) $\mathbf{F} = \langle x \sin z, y^2, x/(z^2 + 1) \rangle$

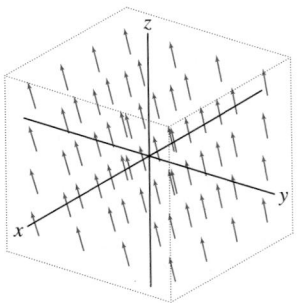
(B) Constant vector field $\mathbf{F} = \langle 1, -1, 3 \rangle$

FIGURE 4

A **unit vector field** is a vector field $\mathbf{F}$ such that $\|\mathbf{F}(P)\| = 1$ for all points P. Important examples are the unit radial vector fields in three and two dimensions [Figures 5(A) and 5(B)]:

$$\mathbf{e}_r = \left\langle \frac{x}{r}, \frac{y}{r}, \frac{z}{r} \right\rangle \qquad \text{or} \qquad \mathbf{e}_r = \left\langle \frac{x}{r}, \frac{y}{r} \right\rangle \qquad \boxed{1}$$

Here $r = (x^2 + y^2 + z^2)^{1/2}$ if $n = 3$ and $r = (x^2 + y^2)^{1/2}$ if $n = 2$. At each point P, $\mathbf{e}_r(P)$ is a unit vector pointing in the direction of the radial vector $\overrightarrow{OP}$.

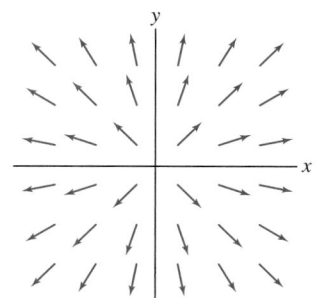

(A) Unit radial vector field in 3-space
$\mathbf{e}_r = \langle x/r,\, y/r,\, z/r \rangle$

(B) Unit radial vector field in the plane
$\mathbf{e}_r = \langle x/r,\, y/r \rangle$

FIGURE 5

Gradient Vector Fields

One way of producing a vector field is to take the gradient of a differentiable function φ:

$$\mathbf{F} = \nabla\varphi = \left\langle \frac{\partial\varphi}{\partial x}, \frac{\partial\varphi}{\partial y}, \frac{\partial\varphi}{\partial z} \right\rangle$$

A vector field of this type is called a **gradient vector field** and the function φ is called a **potential function** for $\mathbf{F}$. The same definitions apply in two variables or, more generally, in n variables. We will see that gradient vector fields have important properties when we study line integrals in Section 16.3.

■ **EXAMPLE 2** Show that $\varphi(x, y, z) = xy + yz^2$ is a potential function for the vector field $\mathbf{F} = \langle y, x + z^2, 2yz \rangle$.

Solution We compute the gradient of φ:

$$\frac{\partial\varphi}{\partial x} = y, \quad \frac{\partial\varphi}{\partial y} = x + z^2, \quad \frac{\partial\varphi}{\partial z} = 2yz \quad \Rightarrow \quad \nabla\varphi = \langle y, x + z^2, 2yz \rangle$$

Thus, $\nabla\varphi = \mathbf{F}$ as claimed. ■

Is every vector field equal to the gradient of some function? The answer is no, as we will show in Example 3; the reason is that the components of a gradient vector field have the special property that the "cross partials are equal." Suppose that $\mathbf{F} = \nabla\varphi$, where φ has continuous second partial derivatives:

$$\mathbf{F} = \langle F_1, F_2, F_3 \rangle = \left\langle \frac{\partial\varphi}{\partial x}, \frac{\partial\varphi}{\partial y}, \frac{\partial\varphi}{\partial z} \right\rangle$$

Now differentiate the x-component with respect to y and the y-component with respect to x:

$$\frac{\partial F_1}{\partial y} = \frac{\partial}{\partial y}\left(\frac{\partial\varphi}{\partial x}\right) = \frac{\partial^2\varphi}{\partial y\,\partial x}, \qquad \frac{\partial F_2}{\partial x} = \frac{\partial}{\partial x}\left(\frac{\partial\varphi}{\partial y}\right) = \frac{\partial^2\varphi}{\partial x\,\partial y}$$

Recall that by Clairaut's Theorem in Section 14.3, the mixed partial derivatives are equal: $\dfrac{\partial^2\varphi}{\partial y\,\partial x} = \dfrac{\partial^2\varphi}{\partial x\,\partial y}$. It follows that for a gradient vector field,

The English physicist Paul Dirac (1902–1984), who received the Nobel Prize in 1933, introduced a generalization of vectors called "spinors" to unify the special theory of relativity with quantum mechanics. This led to the discovery of the positron, an elementary particle used in the PET-scan imaging to detect cancer and study metabolic processes in the body.

$$\frac{\partial F_1}{\partial y} = \frac{\partial F_2}{\partial x}$$

Similarly, the other cross partials of **F** are equal:

$$\frac{\partial F_1}{\partial z} = \frac{\partial F_3}{\partial x}, \qquad \frac{\partial F_2}{\partial z} = \frac{\partial F_3}{\partial y}$$

A vector **F** that does not satisfy these equations cannot be a gradient vector field. This proves the following theorem.

THEOREM 1 Cross Partials of a Gradient Vector Field Are Equal   Let

$$\mathbf{F} = \langle F_1, F_2, F_3 \rangle$$

be a gradient vector field whose components have continuous partial derivatives. Then the cross partials are equal:

$$\frac{\partial F_1}{\partial y} = \frac{\partial F_2}{\partial x}, \quad \frac{\partial F_2}{\partial z} = \frac{\partial F_3}{\partial y}, \quad \frac{\partial F_3}{\partial x} = \frac{\partial F_1}{\partial z}$$

Similarly, if the vector field in the plane $\mathbf{F} = \langle F_1, F_2 \rangle$ is a gradient vector field, then $\dfrac{\partial F_1}{\partial y} = \dfrac{\partial F_2}{\partial x}$.

■ **EXAMPLE 3** Show that $\mathbf{F} = \langle e^x, y^2, xz \rangle$ is not a gradient vector field.

Solution The cross partials for x and z are not equal:

$$\frac{\partial F_1}{\partial z} = \frac{\partial}{\partial z} e^x = 0 \quad \text{is not equal to} \quad \frac{\partial F_3}{\partial x} = \frac{\partial}{\partial x} xz = z$$

Thus, **F** is not a gradient vector field, even though the other cross partials agree:

$$\frac{\partial F_1}{\partial y} = \frac{\partial F_2}{\partial x} = 0 \quad \text{and} \quad \frac{\partial F_2}{\partial z} = \frac{\partial F_3}{\partial y} = 0 \qquad ■$$

Although not every vector field has a potential function, the next theorem shows that if a potential function does exist, then it is unique up to a constant.

THEOREM 2 Uniqueness of Potential Functions   If **F** is a gradient vector field on a ball $\mathcal{D}$ in $\mathbf{R}^3$ (or a disk $\mathcal{D}$ in $\mathbf{R}^2$), then any two potential functions differ by a constant.

Proof If both φ_1 and φ_2 are potential functions of **F**, then the gradient of the difference $\varphi = \varphi_1 - \varphi_2$ is zero:

$$\nabla\varphi = \nabla\varphi_1 - \nabla\varphi_2 = \mathbf{F} - \mathbf{F} = \mathbf{0}$$

However, a function whose gradient is zero on a ball is a constant function (this generalizes the fact from single-variable calculus that a function on an interval with zero derivative is a constant function—see Exercise 33). Therefore, $\varphi(x, y, z) = C$ and $\varphi_1 = \varphi_2 + C$ for some constant C. ■

◀··· *REMINDER In* $\mathbf{R}^3$,

$$\mathbf{e}_r = \left\langle \frac{x}{r}, \frac{y}{r}, \frac{z}{r} \right\rangle$$

where $r = (x^2 + y^2 + z^2)^{1/2}$. *In* $\mathbf{R}^2$,

$$\mathbf{e}_r = \left\langle \frac{x}{r}, \frac{y}{r} \right\rangle$$

where $r = (x^2 + y^2)^{1/2}$.

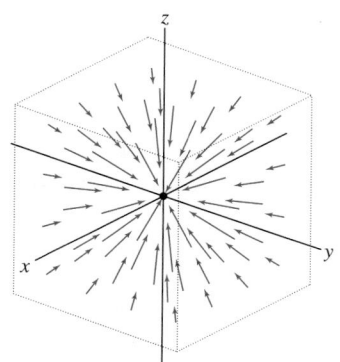

FIGURE 6 The vector field $\dfrac{-\mathbf{e}_r}{r^2}$, which represents the gravitational force field due to a point mass (up to a scalar multiple).

In the next example, we show that the inverse-square radial field is a gradient vector field. This field is of great importance because it describes the gravitational force due to a point mass or the electrostatic force due to a point charge. If the mass or charge is located at the origin, then there is a constant k such that the force field is

$$\mathbf{F} = \frac{k}{r^2}\mathbf{e}_r = k \left\langle \frac{x}{r^3}, \frac{y}{r^3}, \frac{z}{r^3} \right\rangle$$

This vector field is not defined at the origin $r = 0$.

■ **EXAMPLE 4** Potentials for the Unit and Inverse Square Radial Vector Fields Show that $\mathbf{e}_r = \nabla r$ and $\dfrac{\mathbf{e}_r}{r^2} = \nabla \left(\dfrac{-1}{r} \right)$.

Solution We treat the case of radial vector fields in $\mathbf{R}^3$ (Figure 6). The computation is similar for $\mathbf{R}^2$. Since $r = (x^2 + y^2 + z^2)^{1/2}$, we have

$$\frac{\partial r}{\partial x} = \frac{\partial}{\partial x}(x^2 + y^2 + z^2)^{1/2} = \frac{1}{2}(x^2 + y^2 + z^2)^{-1/2}(2x) = \frac{x}{r}$$

Similarly, $\dfrac{\partial r}{\partial y} = \dfrac{y}{r}$ and $\dfrac{\partial r}{\partial z} = \dfrac{z}{r}$. Therefore,

$$\nabla r = \left\langle \frac{x}{r}, \frac{y}{r}, \frac{z}{r} \right\rangle = \mathbf{e}_r$$

Using the Chain Rule for gradients (Theorem 1 in Section 14.5), we obtain

$$\nabla(-r^{-1}) = r^{-2}\nabla r = r^{-2}\mathbf{e}_r \qquad ■$$

16.1 SUMMARY

- A *vector field* assigns a vector to each point in a domain in $\mathbf{R}^3$ or $\mathbf{R}^2$. A vector field in $\mathbf{R}^3$ is represented by a triple of functions: $\mathbf{F} = \langle F_1, F_2, F_3 \rangle$ and in $\mathbf{R}^2$ by a pair of functions $\mathbf{F} = \langle F_1, F_2 \rangle$. We assume that the components F_j are smooth functions on their domains.
- A *gradient vector field* is a vector field of the form $\mathbf{F} = \nabla \varphi$, where $\varphi(x, y, z)$ is a differentiable function.
- If $\mathbf{F} = \nabla \varphi$, then φ is called a *potential function* for $\mathbf{F}$. Any two potential functions for a gradient vector field differ by a constant.
- If $\mathbf{F} = \langle F_1, F_2, F_3 \rangle = \nabla \varphi$ is a gradient vector field, then the cross partials are equal:

$$\frac{\partial F_1}{\partial y} = \frac{\partial F_2}{\partial x}, \qquad \frac{\partial F_1}{\partial z} = \frac{\partial F_3}{\partial x}, \qquad \frac{\partial F_2}{\partial z} = \frac{\partial F_3}{\partial y}$$

If the components of $\mathbf{F}$ do not satisfy these equations, then $\mathbf{F}$ is not a gradient vector field.
- The unit and inverse square radial vector fields are gradient vector fields

$$\mathbf{e}_r = \left\langle \frac{x}{r}, \frac{y}{r}, \frac{z}{r} \right\rangle = \nabla r, \qquad \frac{\mathbf{e}_r}{r^2} = \left\langle \frac{x}{r^3}, \frac{y}{r^3}, \frac{z}{r^3} \right\rangle = \nabla(-r^{-1})$$

where $r = (x^2 + y^2 + z^2)^{1/2}$. In $\mathbf{R}^2$, $\mathbf{e}_r = \left\langle \dfrac{x}{r}, \dfrac{y}{r} \right\rangle$, where $r = (x^2 + y^2)^{1/2}$.

16.1 EXERCISES

Preliminary Questions

1. Which of the following is a unit vector field in the plane?

(a) $\mathbf{F} = \langle y, x \rangle$

(b) $\mathbf{F} = \left\langle \dfrac{y}{\sqrt{x^2 + y^2}}, \dfrac{x}{\sqrt{x^2 + y^2}} \right\rangle$

(c) $\mathbf{F} = \left\langle \dfrac{y}{x^2 + y^2}, \dfrac{x}{x^2 + y^2} \right\rangle$

2. Sketch an example of a nonconstant vector field in the plane in which each vector is parallel to $\langle 1, 1 \rangle$.

3. Show that the vector field $\mathbf{F} = \langle -z, 0, x \rangle$ is orthogonal to the position vector $\overrightarrow{OP}$ at each point P. Give an example of another vector field with this property.

4. Give an example of a potential function for $\langle yz, xz, xy \rangle$ other than $\varphi(x, y, z) = xyz$.

Exercises

1. Compute and sketch the vector assigned to the points $P = (1, 2)$ and $Q = (-1, -1)$ by the vector field $\mathbf{F} = \langle x^2, x \rangle$.

2. Compute and sketch the vector assigned to the points $P = (1, 2)$ and $Q = (-1, -1)$ by the vector field $\mathbf{F} = \langle -y, x \rangle$.

3. Compute and sketch the vector assigned to the points $P = (0, 1, 1)$ and $Q = (2, 1, 0)$ by the vector field $\mathbf{F} = \langle xy, z^2, x \rangle$.

4. Compute the vector assigned to the points $P = (1, 1, 0)$ and $Q = (2, 1, 2)$ by the vector fields $\mathbf{e}_r$, $\dfrac{\mathbf{e}_r}{r}$, and $\dfrac{\mathbf{e}_r}{r^2}$.

In Exercises 5–13, sketch the following planar vector fields by drawing the vectors attached to points with integer coordinates in the rectangle $-3 \le x, y \le 3$. Instead of drawing the vectors with their true lengths, scale them if necessary to avoid overlap.

5. $\mathbf{F} = \langle 1, 0 \rangle$

6. $\mathbf{F} = x\mathbf{i}$

7. $\mathbf{F} = \langle 0, x \rangle$

8. $\mathbf{F} = y\mathbf{j}$

9. $\mathbf{F} = \langle 1, 1 \rangle$

10. $\mathbf{F} = x^2\mathbf{i} + y\mathbf{j}$

11. $\mathbf{F} = \left\langle \dfrac{x}{x^2 + y^2}, \dfrac{y}{x^2 + y^2} \right\rangle$

12. $\mathbf{F} = \left\langle \dfrac{-y}{x^2 + y^2}, \dfrac{x}{x^2 + y^2} \right\rangle$

13. $\mathbf{F} = \left\langle \dfrac{y}{x^2 + y^2}, \dfrac{-x}{x^2 + y^2} \right\rangle$

In Exercises 14–17, match the planar vector field with the corresponding plot in Figure 7.

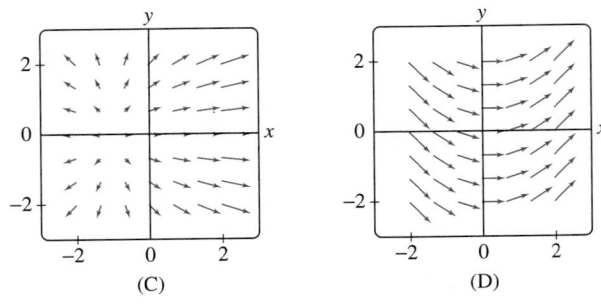

FIGURE 7

14. $\mathbf{F} = \langle 2, x \rangle$

15. $\mathbf{F} = \langle 2x + 2, y \rangle$

16. $\mathbf{F} = \langle y, \cos x \rangle$

17. $\mathbf{F} = \langle x + y, x - y \rangle$

In Exercises 18–21, match the three-dimensional vector field with the corresponding plot in Figure 8.

FIGURE 8

(A)

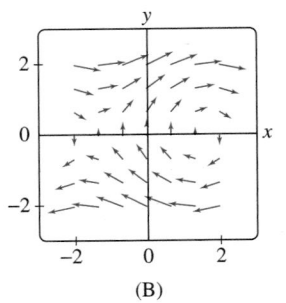

(B)

18. $\mathbf{F} = \langle x, y, z \rangle$

19. $\mathbf{F} = \langle x, 0, z \rangle$

20. $\mathbf{F} = \left\langle \dfrac{x}{\sqrt{x^2 + y^2 + z^2}}, \dfrac{y}{\sqrt{x^2 + y^2 + z^2}}, \dfrac{z}{\sqrt{x^2 + y^2 + z^2}} \right\rangle$

21. $\mathbf{F} = \langle 1, 1, 1 \rangle$

*In Exercises 22–25, find a potential function for the vector field **F** by inspection.*

22. $\mathbf{F} = \langle x, y \rangle$

23. $\mathbf{F} = \langle ye^{xy}, xe^{xy} \rangle$

24. $\mathbf{F} = \langle yz^2, xz^2, 2xyz \rangle$

25. $\mathbf{F} = \langle 2xze^{x^2}, 0, e^{x^2} \rangle$

26. Find potential functions for the vector fields $\mathbf{F} = \dfrac{\mathbf{e}_r}{r^3}$ and $\mathbf{G} = \dfrac{\mathbf{e}_r}{r^4}$ in $\mathbf{R}^3$.

27. Which of (A) or (B) in Figure 9 is the contour plot of a potential function for the vector field **F**? Recall that the gradient vectors are perpendicular to the level curves.

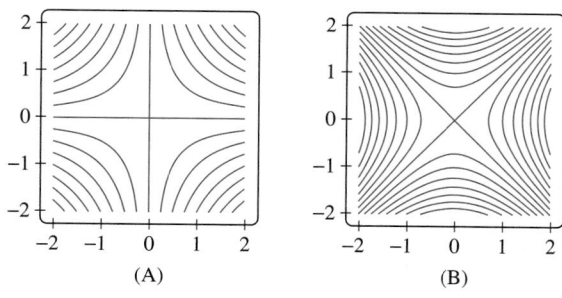

(A)   (B)

FIGURE 9

28. Which of (A) or (B) in Figure 10 is the contour plot of a potential function of the vector field **F**?

29. Let $\varphi = \ln r$, where $r = \sqrt{x^2 + y^2}$. Express $\nabla \varphi$ in terms of the unit radial vector $\mathbf{e}_r$ in $\mathbf{R}^2$.

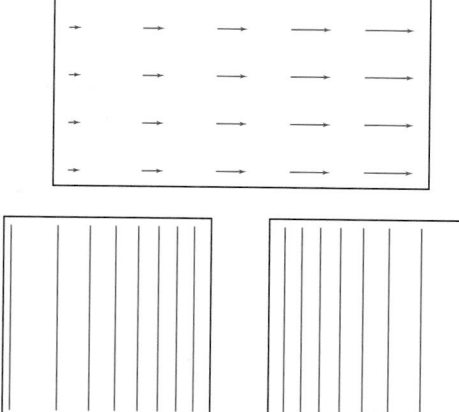

(A)   (B)

FIGURE 10

30. Match the force fields (a)–(c) with (A)–(C) in Figure 11. Note that the vectors (A)–(C) indicate direction but are not drawn to indicate magnitude.

(a) One positive and one negative charge

(b) Two positive charges

(c) Two positive charges and one negative charge

(A)

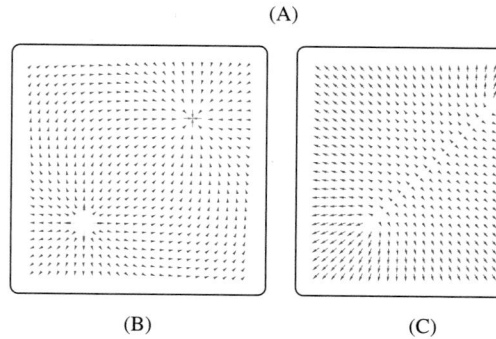

(B)   (C)

FIGURE 11

Further Insights and Challenges

31. Show that any vector field of the form $\mathbf{F} = \langle f(x), g(y), h(z) \rangle$ has a potential function. Assume that f, g, and h are continuous.

32. In this exercise, we show that the vector field **F** in Figure 12 is not a gradient vector field. Suppose that a potential function φ did exist.

(a) Argue that the level curves of φ would have to be vertical lines.

(b) Argue that the level curves would have to grow farther apart as y increases.

(c) Explain why the conclusions of (a) and (b) are incompatible.

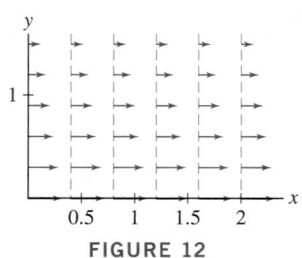

FIGURE 12

33. Show that if $\nabla\varphi(x, y) = \mathbf{0}$ for all (x, y) in a disk $\mathcal{D}$ in $\mathbf{R}^2$, then φ is constant on $\mathcal{D}$. *Hint:* Given points $P = (a, b)$ and $Q = (c, d)$

in $\mathcal{D}$, let $R = (c, b)$ (Figure 13). Use single-variable calculus to show that φ is constant along the segments $\overline{PR}$ and $\overline{RQ}$ and conclude that $\varphi(P) = \varphi(R)$.

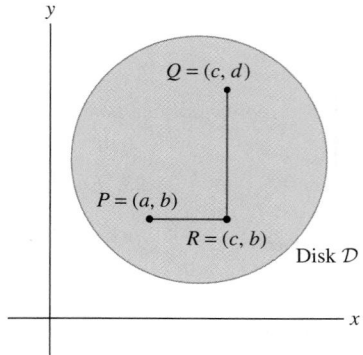

FIGURE 13

16.2 Line Integrals

In this section, we introduce integration over curves in $\mathbf{R}^3$. Integrals over curves are traditionally called **line integrals**, although it would be more appropriate to call them "curve integrals." As mentioned in the introduction to this chapter, we define line integrals both of functions and vector fields. The line integral of a function $f(x, y, z)$ over a curve $\mathcal{C}$ is called a **scalar line integral** and is denoted $\int_{\mathcal{C}} f(x, y, z)\, ds$.

All integrals, line integrals included, are defined as limits of suitable Riemann sums. To define the scalar line integral, we form a Riemann sum by dividing $\mathcal{C}$ into N consecutive arcs $\mathcal{C}_1, \ldots, \mathcal{C}_N$ (Figure 1). Denote the length of the arc $\mathcal{C}_i$ by Δs_i. Within each arc $\mathcal{C}_i$, choose a sample point $P_i \in \mathcal{C}_i$ and consider the Riemann sum:

$$\sum_{i=1}^{N} f(P_i)\, \text{length}(\mathcal{C}_i) = \sum_{i=1}^{N} f(P_i)\, \Delta s_i$$

The line integral of f over $\mathcal{C}$ is the limit (if it exists) of these Riemann sums as the maximum of the lengths Δs_i approaches zero:

$$\boxed{\int_{\mathcal{C}} f(x, y, z)\, ds = \lim_{\{\Delta s_i\} \to 0} \sum_{i=1}^{N} f(P_i)\, \Delta s_i} \qquad \boxed{1}$$

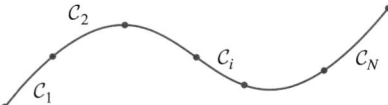

FIGURE 1 The curve $\mathcal{C}$ divided into N small arcs.

Partition of $\mathcal{C}$ into N small arcs

Choice of sample points P_i in each arc

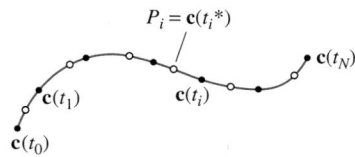

Here we write $\{\Delta s_i\} \to 0$ to indicate that the limit is taken over all Riemann sums as the maximum of the lengths Δs_i tends to zero.

The scalar line integral is a generalization of the arc length integral discussed in Section 13.3. If we take the constant function $f(x, y, z) = 1$, then all Riemann sums have the same value:

$$\sum_{i=1}^{N} 1\,\Delta s_i = \sum_{i=1}^{N} \text{length}(\mathcal{C}_i) = \text{length}(\mathcal{C})$$

and thus $\int_{\mathcal{C}} 1\,ds = \text{length}(\mathcal{C})$.

The definition of the scalar line integral as a limit in Eq. (1) is rarely if ever used in computations. As we now show, line integrals may be computed using parametrizations of the curve.

In this chapter, we use the notation $\mathbf{c}(t) = (x(t), y(t), z(t))$ to denote a path in $\mathbf{R}^3$. This notation was used in Section 14.5. We think of $\mathbf{c}(t)$ as a point moving in space as a function of time t. The derivative $\mathbf{c}'(t)$ is the tangent vector (also called the velocity vector)

$$\mathbf{c}'(t) = \langle x'(t), y'(t), z'(t) \rangle$$

Recall that if $\mathbf{c}'(t) \neq \mathbf{0}$, then $\mathbf{c}'(t)$ is tangent to the path, points in the direction of motion, and its length $\|\mathbf{c}'(t)\|$ is the speed at time t.

Now assume that $\mathcal{C}$ has a parametrization $\mathbf{c}(t)$ for $a \leq t \leq b$ with continuous derivative $\mathbf{c}'(t)$. Choose a partition of the interval $[a, b]$:

$$a = t_0 < t_1 < \cdots < t_{N-1} < t_N = b$$

and let $\mathcal{C}_i$ be the portion of $\mathcal{C}$ parametrized by $\mathbf{c}(t)$ for $t_{i-1} \leq t \leq t_i$ (Figure 2). According to the arc length formula,

$$\text{Length}(\mathcal{C}_i) = \Delta s_i = \int_{t_{i-1}}^{t_i} \|\mathbf{c}'(t)\|\,dt$$

By the Mean Value Theorem for Integrals, there exists an intermediate point t_i^* in $[t_{i-1}, t_i]$ such that

$$\Delta s_i = \int_{t_{i-1}}^{t_i} \|\mathbf{c}'(t)\|\,dt = \|\mathbf{c}'(t_i^*)\|\,\Delta t_i \qquad \Delta t_i = t_i - t_{i-1}$$

Taking $P_i = \mathbf{c}(t_i^*)$ as our intermediate point in $\mathcal{C}_i$, we see from Eq. (1) that

$$\int_{\mathcal{C}} f(x, y, z)\,ds = \lim_{\{\Delta t_i\} \to 0} \sum_{i=1}^{N} f(\mathbf{c}(t_i^*))\|\mathbf{c}'(t_i^*)\|\,\Delta t_i \qquad \boxed{2}$$

The limit is taken over all partitions as the maximum of the lengths Δt_i tend to zero. Since the sums on the right are Riemann sums for the integral

$$\int_a^b f(\mathbf{c}(t))\|\mathbf{c}'(t)\|\,dt$$

we obtain Eq. (3) in the following theorem. Furthermore, our proof does not depend on the choice of parametrization. It follows that you can use any parametrization to compute a scalar line integral.

FIGURE 2 Partition of parametrized curve $\mathbf{c}(t)$.

The arc length formula from Section 13.3 states that the length of $\mathbf{c}(t)$ for $a \leq t \leq b$ is given by the integral

$$\int_a^b \|\mathbf{c}'(t)\|\,dt$$

THEOREM 1 Computing a Scalar Line Integral   Let $\mathbf{c}(t)$ be a parametrization of a curve C for $a \leq t \leq b$. Assume that $f(x, y, z)$ and $\mathbf{c}'(t)$ are continuous. Then

$$\int_C f(x, y, z)\, ds = \int_a^b f(\mathbf{c}(t)) \|\mathbf{c}'(t)\|\, dt \qquad \boxed{3}$$

The value of the integral on the right does not depend on the choice of parametrization. For $f(x, y, z) = 1$, we obtain the length of C:

$$\text{Length of } C = \int_C \|\mathbf{c}'(t)\|\, dt \qquad \boxed{4}$$

We also note that if $\mathbf{c}(t) = (x(t), y(t), z(t))$, then

$$\|\mathbf{c}'(t)\| = \sqrt{x'(t)^2 + y'(t)^2 + z'(t)^2}$$

and the scalar line integral may be written explicitly as

$$\int_C f(x, y, z)\, ds = \int_a^b f(\mathbf{c}(t)) \sqrt{x'(t)^2 + y'(t)^2 + z'(t)^2}\, dt$$

The symbol ds is intended to suggest arc length s and is often referred to as the **line element** or **arc length differential**. When we evaluate the line integral in terms of a parametrization, we replace ds by $\|\mathbf{c}'(t)\|\, dt$.

■ **EXAMPLE 1** **Integrating along a Helix**   Calculate $\displaystyle\int_C (x + y + z)\, ds$, where C is the helix $\mathbf{c}(t) = (\cos t, \sin t, t)$ for $0 \leq t \leq \pi$ (Figure 3).

Solution

Step 1. Compute $ds = \|\mathbf{c}'(t)\|\, dt$.
We have $\mathbf{c}'(t) = \langle -\sin t, \cos t, 1 \rangle$, so

$$\|\mathbf{c}'(t)\| = \sqrt{(-\sin t)^2 + \cos^2 t + 1} = \sqrt{2}$$

$$ds = \|\mathbf{c}'(t)\|\, dt = \sqrt{2}\, dt$$

Step 2. Write out $f(\mathbf{c}(t))$ and evaluate the line integral.
Let $f(x, y, z) = x + y + z$. Then

$$f(\mathbf{c}(t)) = f(\cos t, \sin t, t) = \cos t + \sin t + t$$

By Eq. (3),

$$\int_C f(x, y, z)\, ds = \int_0^\pi f(\mathbf{c}(t)) \|\mathbf{c}'(t)\|\, dt = \int_0^\pi (\cos t + \sin t + t)\sqrt{2}\, dt$$

$$= \sqrt{2}\left(\sin t - \cos t + \frac{1}{2}t^2\right)\Bigg|_0^\pi$$

$$= \sqrt{2}\left(0 + 1 + \frac{1}{2}\pi^2\right) - \sqrt{2}(0 - 1 + 0) = 2\sqrt{2} + \frac{\sqrt{2}}{2}\pi^2 \quad ■$$

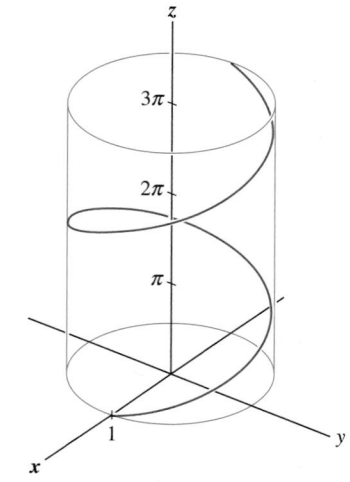

FIGURE 3 The helix $\mathbf{c}(t) = (\cos t, \sin t, t)$.

The scalar line integral can be used to express the **total mass** of a curve (think of the curve as a wire) in terms of its **mass density** $\rho(x, y, z)$, given in units of mass per unit length:

$$\text{Total mass of } C = \int_C \rho(x, y, z)\, ds \qquad \boxed{5}$$

To justify this interpretation, divide C into N arcs C_i of length Δs_i as above. If Δs_i is small, then the mass density along C_i is nearly constant and the mass of C_i is approximately equal to $\rho(P_i)\,\Delta s_i$, where P_i is any sample point on C_i (Figure 4). Thus, we have the following approximation for the total mass:

$$\text{Total mass of } C = \sum_{i=1}^{N} \text{mass of } C_i \approx \sum_{i=1}^{N} \rho(P_i)\, \Delta s_i$$

In the limit as the maximum of the lengths Δs_i tends to zero, the sums on the right approach the line integral of $\rho(x, y, z)$ and we obtain Eq. (5).

■ **EXAMPLE 2** Scalar Line Integrals as Total Mass   A wire in the shape of the parabola $y = x^2$ for $1 \le x \le 4$ has mass density $\rho(x, y) = y/x$ in units of grams per centimeter. Find the total mass of the wire.

Solution   The arc of the parabola is parametrized by $\mathbf{c}(t) = (t, t^2)$ for $1 \le t \le 4$.

Step 1. **Compute** $ds = \|\mathbf{c}'(t)\|\, dt$.

$$\mathbf{c}'(t) = \langle 1, 2t \rangle$$

$$ds = \|\mathbf{c}'(t)\|\, dt = \sqrt{1 + 4t^2}\, dt$$

Step 2. **Write out** $\rho(\mathbf{c}(t))$ **and evaluate the line integral.**

$$\rho(\mathbf{c}(t)) = \rho(t, t^2) = \frac{t^2}{t} = t$$

$$\int_C \rho(x, y)\, ds = \int_1^4 \rho(\mathbf{c}(t)) \|\mathbf{c}'(t)\|\, dt = \int_1^4 t\sqrt{1 + 4t^2}\, dt$$

Now use the substitution $u = 1 + 4t^2$, $du = 8t\, dt$. Then the new limits of integration are $u(1) = 5$ and $u(4) = 65$, and

$$\int_1^4 t\sqrt{1 + 4t^2} = \frac{1}{8} \int_5^{65} \sqrt{u}\, du = \frac{1}{12} u^{3/2} \Big|_5^{65} = \frac{1}{12}(65^{3/2} - 5^{3/2}) \approx 42.74$$

The total mass of the wire is approximately 42.7 g.   ■

The Vector Line Integral

We now discuss the line integral of a vector field $\mathbf{F}$ along a curve C, denoted $\int_C \mathbf{F} \cdot d\mathbf{s}$. We make the standing assumption that $\mathbf{F} = \langle F_1, F_2, F_3 \rangle$ is smooth, that is, the components F_1, F_2, F_3 are smooth and, unless otherwise stated, that the curve C is smooth.

Intuitively speaking, the vector line integral is the integral of the **tangential component** of the vector field along the curve. When the context is clear, we refer to a vector line integral more simply as a line integral. However, there is an important difference between vector and scalar line integrals: To define a vector line integral, we must specify

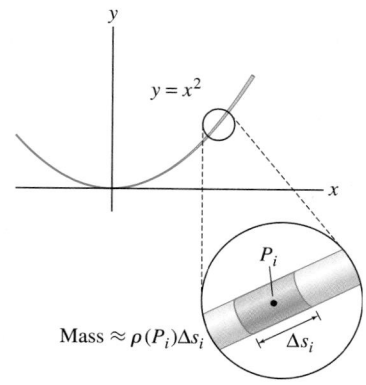

$y = x^2$

Mass $\approx \rho(P_i)\Delta s_i$

P_i

Δs_i

FIGURE 4

We defined line integrals for functions of three variables. The line integral of a function $f(x, y)$ of two variables over a curve in $\mathbf{R}^2$ is defined and evaluated in a similar fashion (but the z-component does not appear in the formulas).

a *direction* along the path or curve. A curve C can be traversed in one of two directions, and we say that C is **oriented** if one of these two directions is specified. We refer to this direction as the **forward direction** along the curve (Figure 5).

Oriented path from P to Q A closed oriented path

FIGURE 5 An oriented curve is a curve with a specified direction.

Given an oriented curve C in the plane or in 3-space, let $\mathbf{T} = \mathbf{T}(P)$ denote the unit tangent vector pointing in the forward direction along the curve. If $\mathbf{F}$ is a vector field and P a point on C, the dot product $\mathbf{F}(P) \cdot \mathbf{T}(P)$ represents the tangential component of $\mathbf{F}$ along C at P, that is, the length of the projection of $\mathbf{F}(P)$ along $\mathbf{T}(P)$ (Figure 6). Recall that, since $\|\mathbf{T}(P)\| = 1$, we have

The unit tangent vector $\mathbf{T}$ varies from point to point along the curve. When it is necessary to stress this dependence, we write $\mathbf{T}(P)$ or $\mathbf{T}(\mathbf{c}(t))$.

$$\mathbf{F}(P) \cdot \mathbf{T}(P) = \|\mathbf{F}(P)\| \cos \theta$$

where θ is the angle between $\mathbf{F}(P)$ and $\mathbf{T}(P)$. As P varies along the curve, $\mathbf{F} \cdot \mathbf{T}$ defines a scalar function on C and the scalar line integral of this function is the vector line integral of $\mathbf{F}$.

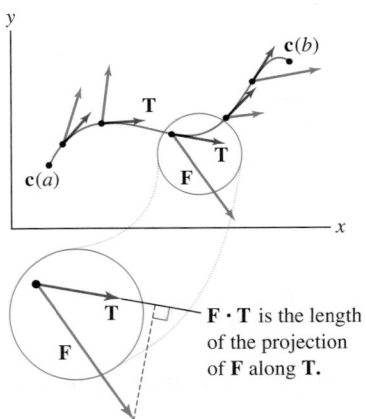

FIGURE 6 The line integral is the integral of the tangential component of $\mathbf{F}$ along C.

> **DEFINITION Vector Line Integral** Let C be an oriented curve and let $\mathbf{T}$ denote the unit tangent vector pointing in the forward direction along C. The line integral of a vector field $\mathbf{F}$ along C is the integral of the tangential component of $\mathbf{F}$:
>
> $$\int_C \mathbf{F} \cdot d\mathbf{s} = \int_C (\mathbf{F} \cdot \mathbf{T}) \, ds \qquad \boxed{6}$$

As in the scalar case, we evaluate vector line integrals by using parametrizations. However, in the vector case, we must choose a parametrization $\mathbf{c}(t)$ that traces C in the forward direction. We also assume that $\mathbf{c}(t)$ is regular, that is, $\mathbf{c}'(t) \neq 0$ for $a \leq t \leq b$. In this case, $\mathbf{c}'(t)$ is a nonzero tangent vector pointing in the forward direction and the unit tangent vector is

$$\mathbf{T} = \text{unit tangent vector} = \frac{\mathbf{c}'(t)}{\|\mathbf{c}'(t)\|}$$

The tangential component of $\mathbf{F}$ at $\mathbf{c}(t)$ is $\mathbf{F}(\mathbf{c}(t)) \cdot \mathbf{T}(\mathbf{c}(t))$, and thus,

$$\int_C \mathbf{F} \cdot d\mathbf{s} = \int_C (\mathbf{F} \cdot \mathbf{T}) \, ds = \int_a^b \mathbf{F}(\mathbf{c}(t)) \cdot \mathbf{T}(\mathbf{c}(t)) \, \|\mathbf{c}'(t)\| \, dt$$

$$= \int_a^b \mathbf{F}(\mathbf{c}(t)) \cdot \left(\frac{\mathbf{c}'(t)}{\|\mathbf{c}'(t)\|} \right) \|\mathbf{c}'(t)\| \, dt$$

$$= \int_a^b \mathbf{F}(\mathbf{c}(t)) \cdot \mathbf{c}'(t) \, dt$$

THEOREM 2 Computing a Vector Line Integral Let $\mathbf{c}(t)$ be a regular parametrization of an oriented curve $\mathcal{C}$ for $a \leq t \leq b$. The line integral of a vector field $\mathbf{F}$ over a curve $\mathcal{C}$ is equal to

$$\int_{\mathcal{C}} \mathbf{F} \cdot d\mathbf{s} = \int_{a}^{b} \mathbf{F}(\mathbf{c}(t)) \cdot \mathbf{c}'(t) \, dt \qquad \boxed{7}$$

It is useful to think of $d\mathbf{s}$ as a "vector line element" or "vector differential" whose expression is

$$d\mathbf{s} = \mathbf{c}'(t) \, dt$$

when we choose a parametrization. The vector line integral is the integral of the dot product of $\mathbf{F}$ and the vector differential $d\mathbf{s}$.

■ **EXAMPLE 3** Let $\mathbf{F} = \langle z, y^2, x \rangle$. Evaluate $\displaystyle\int_{\mathcal{C}} \mathbf{F} \cdot d\mathbf{s}$, where $\mathcal{C}$ is parametrized by $\mathbf{c}(t) = (t + 1, e^t, t^2)$ for $0 \leq t \leq 2$.

Solution There are two steps in evaluating a line integral.

Step 1. Calculate the integrand $\mathbf{F}(\mathbf{c}(t)) \cdot \mathbf{c}'(t)$.

$$\mathbf{c}(t) = (t + 1, e^t, t^2)$$

$$\mathbf{F}(\mathbf{c}(t)) = \langle z, y^2, x \rangle = \langle t^2, e^{2t}, t + 1 \rangle$$

$$\mathbf{c}'(t) = \langle 1, e^t, 2t \rangle$$

The integrand is the dot product:

$$\mathbf{F}(\mathbf{c}(t)) \cdot \mathbf{c}'(t) = \langle t^2, e^{2t}, t + 1 \rangle \cdot \langle 1, e^t, 2t \rangle = e^{3t} + 3t^2 + 2t$$

Step 2. Evaluate the line integral.

$$\int_{\mathcal{C}} \mathbf{F} \cdot d\mathbf{s} = \int_{0}^{2} \mathbf{F}(\mathbf{c}(t)) \cdot \mathbf{c}'(t) \, dt$$

$$= \int_{0}^{2} (e^{3t} + 3t^2 + 2t) \, dt = \left(\frac{1}{3} e^{3t} + t^3 + t^2 \right) \Big|_{0}^{2}$$

$$= \left(\frac{1}{3} e^6 + 8 + 4 \right) - \frac{1}{3} = \frac{1}{3} \left(e^6 + 35 \right) \qquad ■$$

GRAPHICAL INSIGHT To reinforce the concept of the line integral as the integral of the tangential component, consider the line integral of the vector field $\mathbf{F} = \langle 2y, -3 \rangle$ around the ellipse in Figure 7. Observe in part (A) that the dot product of $\mathbf{F} \cdot \mathbf{T}$ at each point along the top part of the ellipse is negative because $\mathbf{F} \cdot \mathbf{T} = \|\mathbf{F}\| \cos\theta$ and the angle θ between $\mathbf{F}$ and $\mathbf{T}$ is obtuse. Therefore, the line integral in part (A) is negative. Similarly, in Figure 7(B), $\mathbf{F} \cdot \mathbf{T}$ is positive and the line integral is positive. The line integral around the entire ellipse in Figure 7(C) appears to be negative because the negative tangential components from the upper part of the curve appear to dominate the positive contribution of the tangential components from the lower part. We verify this in the next example.

Before proceeding, we introduce another standard notation for the line integral $\displaystyle\int_{\mathcal{C}} \mathbf{F} \cdot d\mathbf{s}$ of a vector field $\mathbf{F} = \langle F_1, F_2, F_3 \rangle$:

$$\int_{\mathcal{C}} F_1 \, dx + F_2 \, dy + F_3 \, dz$$

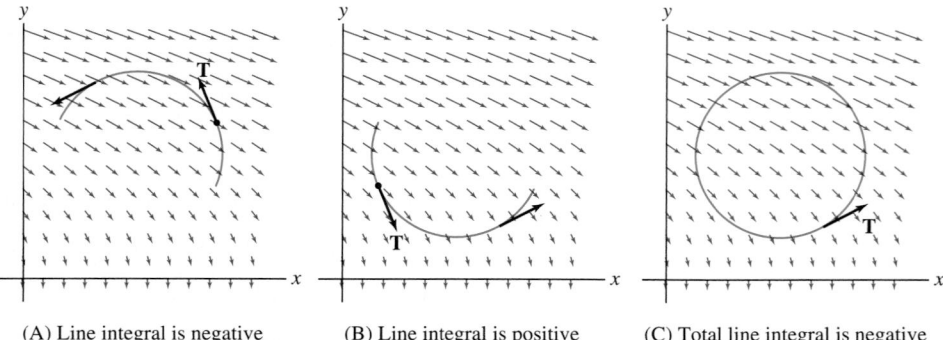

The dot product $\mathbf{T} \cdot \mathbf{F}$ is negative because the angle between the vectors is obtuse.

Here, the dot product $\mathbf{T} \cdot \mathbf{F}$ is positive because the angle between the vectors is acute.

FIGURE 7 The vector field $\mathbf{F} = \langle 2y, -3 \rangle$.

(A) Line integral is negative   (B) Line integral is positive   (C) Total line integral is negative

To evaluate the vector line integral in this notation, we choose a parametrization $\mathbf{c}(t) = (x(t), y(t), z(t))$ and replace dx by $\dfrac{dx}{dt}\,dt$, dy by $\dfrac{dy}{dt}\,dt$, and dz by $\dfrac{dz}{dt}\,dt$:

$$\underbrace{\int_a^b \left(F_1(\mathbf{c}(t)) \frac{dx}{dt} + F_2(\mathbf{c}(t)) \frac{dy}{dt} + F_3(\mathbf{c}(t)) \frac{dz}{dt} \right) dt}_{\mathbf{F}(\mathbf{c}(t)) \cdot \mathbf{c}'(t)}$$

■ **EXAMPLE 4** Calculate $\displaystyle\int_C 2y\,dx - 3\,dy$, where C is the ellipse in Figure 7 parametrized counterclockwise by $\mathbf{c}(t) = (4 + 3\cos\theta, 3 + 2\sin\theta)$ for $0 \le \theta < 2\pi$.

In Example 4, keep in mind that

$$\int_C 2y\,dx - 3\,dy$$

is another notation for the line integral of the vector field $\mathbf{F} = \langle 2y, -3 \rangle$ over C.

Solution We have $x(\theta) = 4 + 3\cos\theta$ and $y(\theta) = 3 + 2\sin\theta$, and

$$\frac{dx}{d\theta} = -3\sin\theta, \qquad \frac{dy}{d\theta} = 2\cos\theta$$

The integrand of the line integral is

$$2y\,dx - 3\,dy = \left(2y\frac{dx}{d\theta} - 3\frac{dy}{d\theta} \right) d\theta$$

$$= \big(2(3 + 2\sin\theta)(-3\sin\theta) - 3(2\cos\theta)\big)\,d\theta$$

$$= -\big(6\cos\theta + 18\sin\theta + 12\sin^2\theta\big)\,d\theta$$

We evaluate the line integral, noting that the integrals of $\cos\theta$ and $\sin\theta$ over $[0, 2\pi]$ are zero:

$$\int_C 2y\,dx - 3\,dy = -\int_0^{2\pi} \big(6\cos\theta + 18\sin\theta + 12\sin^2\theta\big)\,d\theta$$

$$= -12\int_0^{2\pi} \sin^2\theta\,d\theta = -12\int_0^{2\pi} \left(\frac{1 - \cos 2\theta}{2} \right) d\theta$$

$$= (-12)(\pi) = -12\pi \qquad ■$$

We now state some basic properties of vector line integrals. First, we note that on every curve there are two possible orientations corresponding to the two directions along the curve (Figure 8). We write $-C$ to denote the curve C with the opposite orientation. If we reverse the orientation of the curve, the unit tangent vector changes sign from $\mathbf{T}$ to $-\mathbf{T}$, and it follows that the vector line integral changes sign:

$$\int_{-C} \mathbf{F} \cdot d\mathbf{s} = -\int_{C} \mathbf{F} \cdot d\mathbf{s}$$

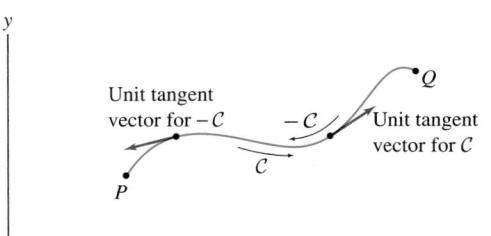

FIGURE 8 The path from P to Q has two possible orientations.

Next, suppose that C is the disjoint union of n curves $C_1, \dots, C_n$. We indicate this using the notation

$$C = C_1 + \cdots + C_n$$

In this case, the line integral over C is equal to the sum of line integrals:

$$\int_{C} \mathbf{F} \cdot d\mathbf{s} = \int_{C_1} \mathbf{F} \cdot d\mathbf{s} + \cdots + \int_{C_n} \mathbf{F} \cdot d\mathbf{s}$$

We use this formula to define the line integral when C is not smooth, but can be written as a union of smooth curves $C_1, \dots, C_n$. In this case, C is called **piecewise smooth**. The next theorem summarizes the main properties of line integrals.

THEOREM 3 Properties of Line Integrals   Let C be a smooth oriented curve and let $\mathbf{F}$ and $\mathbf{G}$ be vector fields.

(i) **Linearity:** $\displaystyle\int_{C} (\mathbf{F} + \mathbf{G}) \cdot d\mathbf{s} = \int_{C} \mathbf{F} \cdot d\mathbf{s} + \int_{C} \mathbf{G} \cdot d\mathbf{s}$

$\displaystyle\int_{C} k\mathbf{F} \cdot d\mathbf{s} = k \int_{C} \mathbf{F} \cdot d\mathbf{s}$ (k a constant)

(ii) **Reversing orientation:** $\displaystyle\int_{-C} \mathbf{F} \cdot d\mathbf{s} = -\int_{C} \mathbf{F} \cdot d\mathbf{s}$

(iii) **Additivity:** If C is a union of n smooth curves $C_1 + \cdots + C_n$, then

$$\int_{C} \mathbf{F} \cdot d\mathbf{s} = \int_{C_1} \mathbf{F} \cdot d\mathbf{s} + \cdots + \int_{C_n} \mathbf{F} \cdot d\mathbf{s}$$

■ **EXAMPLE 5** Compute $\displaystyle\int_{C} \mathbf{F} \cdot d\mathbf{s}$, where $\mathbf{F} = \langle e^z, e^y, x + y \rangle$ and C is the triangle joining $(1, 0, 0)$, $(0, 1, 0)$, and $(0, 0, 1)$ oriented in the counterclockwise direction when viewed from above.

Solution With vertices labeled as in Figure 9, we evaluate the line integral as the sum of line integrals over the edges of the triangle:

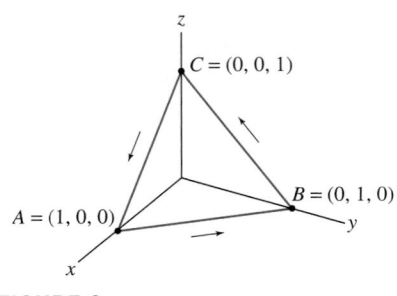

FIGURE 9

$$\int_C \mathbf{F} \cdot d\mathbf{s} = \int_{\overline{AB}} \mathbf{F} \cdot d\mathbf{s} + \int_{\overline{BC}} \mathbf{F} \cdot d\mathbf{s} + \int_{\overline{CA}} \mathbf{F} \cdot d\mathbf{s}$$

The segment $\overline{AB}$ is parametrized by the path $\mathbf{c}(t) = (1 - t, t, 0)$ for $0 \le t \le 1$. We have $\mathbf{c}'(t) = \langle -1, 1, 0 \rangle$ and

$$\mathbf{F}(\mathbf{c}(t)) \cdot \mathbf{c}'(t) = \mathbf{F}(1 - t, t, 0) \cdot \langle -1, 1, 0 \rangle$$
$$= \langle e^0, e^t, 1 \rangle \cdot \langle -1, 1, 0 \rangle = -1 + e^t$$

Thus,

$$\int_{\overline{AB}} \mathbf{F} \cdot d\mathbf{s} = \int_0^1 (e^t - 1)\, dt = (e^t - t)\Big|_0^1 = (e - 1) - 1 = \boxed{e - 2}$$

Similarly, $\overline{BC}$ is parametrized by $\mathbf{c}(t) = (0, 1 - t, t)$ for $0 \le t \le 1$ and

$$\mathbf{F}(\mathbf{c}(t)) \cdot \mathbf{c}'(t) = \langle e^t, e^{1-t}, 1 - t \rangle \cdot \langle 0, -1, 1 \rangle = -e^{1-t} + 1 - t$$

$$\int_{\overline{BC}} \mathbf{F} \cdot d\mathbf{s} = \int_0^1 (-e^{1-t} + 1 - t)\, dt = \left(e^{1-t} + t - \frac{1}{2}t^2 \right)\Big|_0^1 = \boxed{\frac{3}{2} - e}$$

Finally, $\overline{CA}$ is parametrized by $\mathbf{c}(t) = (t, 0, 1 - t)$ for $0 \le t \le 1$ and

$$\mathbf{F}(\mathbf{c}(t)) \cdot \mathbf{c}'(t) = \langle e^{1-t}, 1, t \rangle \cdot \langle 1, 0, -1 \rangle = e^{1-t} - t$$

$$\int_{\overline{CA}} \mathbf{F} \cdot d\mathbf{s} = \int_0^1 (e^{1-t} - t)\, dt = \left(-e^{1-t} - \frac{1}{2}t^2 \right)\Big|_0^1 = \boxed{-\frac{3}{2} + e}$$

The total line integral is equal to

$$\int_C \mathbf{F} \cdot d\mathbf{s} = (e - 2) + \left(\frac{3}{2} - e \right) + \left(-\frac{3}{2} + e \right) = e - 2 \qquad \blacksquare$$

Physical Interpretation of the Line Integral

Recall that in physics, the term "work" refers to the energy expended when a force is applied to an object to move it along a path. If we move the object along the straight line from P to Q by applying a constant force $\mathbf{F}$ [Figure 10(A)], then the work W is equal to the product:

$$W = (\text{tangential component of force}) \text{ times } (\text{distance from } P \text{ to } Q)$$

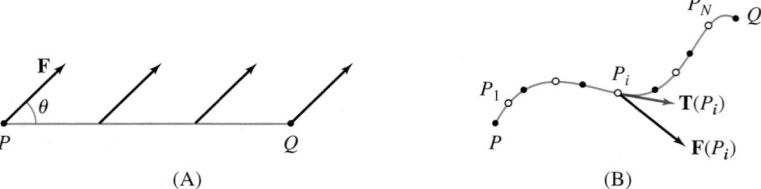

(A)   (B)

FIGURE 10

The distance from P to Q is the length $\|\mathbf{v}\|$ of the vector $\mathbf{v} = \overrightarrow{PQ}$, and the tangential component of the force is $\mathbf{F} \cdot (\mathbf{v}/\|\mathbf{v}\|)$. We see that the work is simply the dot product

$$W = \left(\mathbf{F} \cdot \frac{\mathbf{v}}{\|\mathbf{v}\|} \right) \|\mathbf{v}\| = \mathbf{F} \cdot \mathbf{v}$$

When we move the object along a curved path C and when the applied force $\mathbf{F}$ is no longer constant, it makes sense to define the work performed by $\mathbf{F}$ as the line integral of the tangential component of $\mathbf{F}$ along the path [Figure 10(B)]:

$$\boxed{\text{Work } W = \int_C \mathbf{F} \cdot d\mathbf{s}}$$

In other words, work is equal to the integral of the tangential component of force along the direction of motion. To justify this definition, we divide C into N consecutive arcs $C_1, \ldots, C_N$, where C_i has length Δs_i. The work W_i required to move the object along C_i is approximately equal to $(\mathbf{F}(P_i) \cdot \mathbf{T}(P_i)) \Delta s_i$, where P_i is a sample point in C_i. This gives us the approximation to the total work:

$$W = \sum_{i=1}^{N} W_i \approx \sum_{i=1}^{N} (\mathbf{F}(P_i) \cdot \mathbf{T}(P_i)) \Delta s_i$$

The right-hand side approaches the line integral $\displaystyle\int_C \mathbf{F} \cdot d\mathbf{s}$ as the maximum of the lengths Δs_i tends to zero.

We often wish to calculate the work required to move an object along a path in the presence of a force field $\mathbf{F}$ (such as an electrical or gravitational field). In this case, $\mathbf{F}$ acts on the object and we must work *against* the force field to move the object. The work required to move the object is the negative of the line integral:

$$\text{Work performed against } \mathbf{F} = -\int_C \mathbf{F} \cdot d\mathbf{s}$$

■ **EXAMPLE 6** Calculating Work   The force field $\mathbf{F} = \langle x^2, -z, -yz^{-1} \rangle$ acts on a particle. Calculate the work performed moving the particle from $P = (0, 0, 0)$ to $Q = (1, 1, 1)$ along the path $\mathbf{c}(t) = (t^2, t^3, t)$ for $0 \le t \le 1$. Assume that $\mathbf{F}$ is given in newtons and the unit of distance is the meter.

Solution   In this case, $\mathbf{F}(\mathbf{c}(t)) = \langle t^4, -t, -t^2 \rangle$ and $\mathbf{c}'(t) = \langle 2t, 3t^2, 1 \rangle$, so

$$\mathbf{F}(\mathbf{c}(t)) \cdot \mathbf{c}'(t) = \langle t^4, -t, -t^2 \rangle \cdot \langle 2t, 3t^2, 1 \rangle = 2t^5 - 3t^3 - t^2$$

Since we must work against the force field, the work performed in joules is

$$W = -\int_C \mathbf{F} \cdot d\mathbf{s} = -\int_0^1 (2t^5 - 3t^3 - t^2) \, dt = -\left(\frac{1}{3} - \frac{3}{4} - \frac{1}{3} \right) = \frac{3}{4} \text{ J} \qquad ■$$

◄·· **REMINDER** *Work has units of energy. The SI unit of force is the newton and the unit of energy is the joule, defined as 1 newton-meter. The British unit is the foot-pound.*

16.2 SUMMARY

• We write $\mathbf{c}(t) = (x(t), y(t), z(t))$ for the parametrization of a curve in $\mathbf{R}^3$. The derivative $\mathbf{c}'(t) = \langle x'(t), y'(t), z'(t) \rangle$ is the tangent or velocity vector, which points in the direction of motion.

• The *line integral of a scalar function* $f(x, y, z)$ over a curve C is evaluated using the formula

$$\int_C f(x, y, z)\, ds = \int_a^b f(\mathbf{c}(t))\, \|\mathbf{c}'(t)\|\, dt$$

where $\mathbf{c}(t)$ is a parametrization of C for $a \le t \le b$. We have the symbolic equality $ds = \|\mathbf{c}'(t)\|\, dt$.

- An *oriented curve* C is a curve in which one of the two possible directions along C (called the *forward direction*) is chosen.

- The *line integral of a vector field* $\mathbf{F}$ along an oriented curve C is defined as the integral of the tangential component $\mathbf{F} \cdot \mathbf{T}$, where $\mathbf{T}$ is the unit tangent vector pointing in the forward direction of C:

$$\int_C \mathbf{F} \cdot d\mathbf{s} = \int_C (\mathbf{F} \cdot \mathbf{T})\, ds = \int_a^b \mathbf{F}(\mathbf{c}(t)) \cdot \mathbf{c}'(t)\, dt$$

Here $\mathbf{c}(t)$ is a regular parametrization ($\mathbf{c}'(t) \ne \mathbf{0}$) that traces the oriented curve C in the forward direction. We have the symbolic equality of vector differentials, $d\mathbf{s} = \mathbf{c}'(t)\, dt$.

- Let $-C$ be the curve C with the opposite orientation. Then

$$\int_C \mathbf{F} \cdot d\mathbf{s} = -\int_{-C} \mathbf{F} \cdot d\mathbf{s}$$

- When a variable force $\mathbf{F}$ is exerted on an object to move it along a curve C, the work performed by $\mathbf{F}$ is equal to $\int_C \mathbf{F} \cdot d\mathbf{s}$.

16.2 EXERCISES

Preliminary Questions

1. What is the line integral of the constant function $f(x, y, z) = 10$ over a curve C of length 5?

2. Which of the following have a zero line integral over the vertical segment from $(0, 0)$ to $(0, 1)$?

(a) $f(x, y) = x$ **(b)** $f(x, y) = y$

(c) $\mathbf{F} = \langle x, 0 \rangle$ **(d)** $\mathbf{F} = \langle y, 0 \rangle$

(e) $\mathbf{F} = \langle 0, x \rangle$ **(f)** $\mathbf{F} = \langle 0, y \rangle$

3. State whether true or false. If false, give the correct statement.

(a) The scalar line integral does not depend on how you parametrize the curve.

(b) If you reverse the orientation of the curve, neither the vector nor the scalar line integral changes sign.

4. Let C be a curve of length 5. What is the value of $\int_C \mathbf{F} \cdot d\mathbf{s}$ if

(a) $\mathbf{F}(P)$ is normal to C at all points P on C?

(b) $\mathbf{F}(P) = \mathbf{T}(P)$ at all points P on C, where $\mathbf{T}(P)$ is the unit tangent vector pointing in the forward direction along the curve?

Exercises

1. Let $f(x, y, z) = x + yz$ and let C be the line segment from $P = (0, 0, 0)$ to $(6, 2, 2)$.

(a) Calculate $f(\mathbf{c}(t))$ and $ds = \|\mathbf{c}'(t)\|\, dt$ for the parametrization $\mathbf{c}(t) = (6t, 2t, 2t)$ for $0 \le t \le 1$.

(b) Evaluate $\int_C f(x, y, z)\, ds$.

2. Repeat Exercise 1 with the parametrization $\mathbf{c}(t) = (3t, t, t)$ for $0 \le t \le 2$.

3. Let $\mathbf{F} = \langle y^2, x^2 \rangle$ and let C be the $y = x^{-1}$ for $1 \le x \le 2$, oriented from left to right.

(a) Calculate $\mathbf{F}(\mathbf{c}(t))$ and $d\mathbf{s} = \mathbf{c}'(t)\, dt$ for the parametrization $\mathbf{c}(t) = (t, t^{-1})$.

(b) Calculate the dot product $\mathbf{F}(\mathbf{c}(t)) \cdot \mathbf{c}'(t)\, dt$ and evaluate $\int_C \mathbf{F} \cdot d\mathbf{s}$.

4. Let $\mathbf{F} = \langle z^2, x, y \rangle$ and let C be the path $\mathbf{c}(t) = (3 + 5t^2, 3 - t^2, t)$ for $0 \le t \le 2$.

(a) Calculate $\mathbf{F}(\mathbf{c}(t))$ and $d\mathbf{s} = \mathbf{c}'(t)\, dt$.

(b) Calculate the dot product $\mathbf{F}(\mathbf{c}(t)) \cdot \mathbf{c}'(t)\, dt$ and evaluate $\int_C \mathbf{F} \cdot d\mathbf{s}$.

In Exercises 5–8, calculate the integral of the given scalar function or vector field over the curve $\mathbf{c}(t) = (\cos t, \sin t, t)$ for $0 \le t \le \pi$.

5. $f(x, y, z) = x^2 + y^2 + z^2$ **6.** $f(x, y, z) = xy + z$

7. $\mathbf{F} = \langle x, y, z \rangle$ **8.** $\mathbf{F} = \langle xy, 2, z^3 \rangle$

9. Calculate the total mass of a circular piece of wire of radius 4 cm centered at the origin whose mass density is $\rho(x, y) = x^2$ g/cm.

10. Calculate the total mass of a metal tube in the helical shape $\mathbf{c}(t) = (\cos t, \sin t, t^2)$ (distance in centimeters) for $0 \le t \le 2\pi$ if the mass density is $\rho(x, y, z) = \sqrt{z}$ g/cm.

11. The values of a function $f(x, y, z)$ and vector field $\mathbf{F}(x, y, z)$ are given at six sample points along the path ABC in Figure 11. Estimate the line integrals of f and $\mathbf{F}$ along ABC.

Point	$f(x, y, z)$	$\mathbf{F}(x, y, z)$
$(1, \frac{1}{6}, 0)$	3	$\langle 1, 0, 2 \rangle$
$(1, \frac{1}{2}, 0)$	3.3	$\langle 1, 1, 3 \rangle$
$(1, \frac{5}{6}, 0)$	3.6	$\langle 2, 1, 5 \rangle$
$(1, 1, \frac{1}{6})$	4.2	$\langle 3, 2, 4 \rangle$
$(1, 1, \frac{1}{2})$	4.5	$\langle 3, 3, 3 \rangle$
$(1, 1, \frac{5}{6})$	4.2	$\langle 5, 3, 3 \rangle$

FIGURE 11

12. Estimate the line integrals of $f(x, y)$ and $\mathbf{F}(x, y)$ along the quarter circle (oriented counterclockwise) in Figure 12 using the values at the three sample points along each path.

Point	$f(x, y)$	$\mathbf{F}(x, y)$
A	1	$\langle 1, 2 \rangle$
B	-2	$\langle 1, 3 \rangle$
C	4	$\langle -2, 4 \rangle$

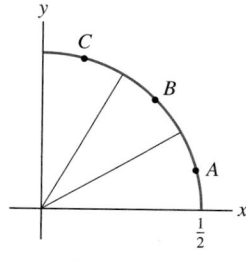

FIGURE 12

13. Figure 13 shows three vector fields. In each case, determine whether the line integral around the circle (oriented counterclockwise) is positive, negative, or zero.

(A)

(B)

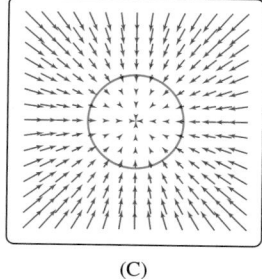

(C)

FIGURE 13

14. Determine whether the line integrals around the curves (oriented counterclockwise) in Figures 14(A) and (B) are positive or negative. What is the value of the line integral in Figure 14(C)? Explain.

(A)

(B)

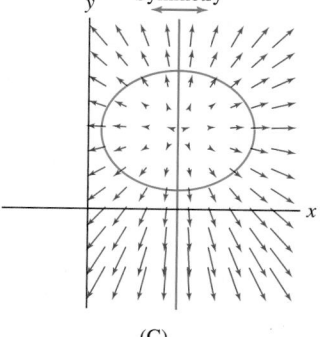

(C)

FIGURE 14

In Exercises 15–22, compute the line integral of the scalar function over the curve.

15. $f(x, y, z) = z^2$, $\mathbf{c}(t) = (2t, 3t, 4t)$ for $0 \le t \le 2$

16. $f(x, y, z) = 3x - 2y + z$, $\mathbf{c}(t) = (2 + t, 2 - t, 2t)$ for $-2 \le t \le 1$

17. $f(x, y) = \sqrt{1 + 9xy}$, $y = x^3$ for $0 \le x \le 1$

18. $f(x, y) = \dfrac{y^3}{x^7}$, $y = \frac{1}{4}x^4$ for $1 \le x \le 2$

19. $f(x, y, z) = xe^{z^2}$,   piecewise linear path from $(0, 0, 1)$ to $(0, 2, 0)$ to $(1, 1, 1)$.

20. $f(x, y, z) = x^2 z$, $\mathbf{c}(t) = (e^t, \sqrt{2}t, e^{-t})$ for $0 \le t \le 1$

21. $f(x, y, z) = 2x^2 + 8z$, $\mathbf{c}(t) = (e^t, t^2, t)$, $0 \le t \le 1$

22. $f(x, y, z) = 3zy - 1 + 12xz$, $\mathbf{c}(t) = (t, \frac{t^2}{\sqrt{2}}, \frac{t^3}{3})$, $0 \le t \le 2$

In Exercises 23–35, compute the line integral of the vector field over the oriented curve.

23. $\mathbf{F} = \langle x^2, xy \rangle$,   line segment from $(0, 0)$ to $(2, 2)$

24. $\mathbf{F} = \langle 4, y \rangle$,   quarter circle $x^2 + y^2 = 1$ with $x, y \le 0$ oriented counterclockwise

25. $\mathbf{F} = \langle x^2, xy \rangle$,   circle $x^2 + y^2 = 9$ oriented clockwise

26. $\mathbf{F} = \langle e^{y-x}, e^{2x} \rangle$,   piecewise linear path from $(1, 1)$ to $(2, 2)$ to $(0, 2)$

27. $\mathbf{F} = \langle xy, x + y \rangle$, $\mathbf{c}(t) = (1 + t^{-1}, t^2)$ for $1 \le t \le 2$

28. $\mathbf{F} = \langle y + z, x, y \rangle$, $\mathbf{c}(t) = (t + t^2, \frac{1}{3}t^3, 2 + t)$ for $0 \le t \le 2$

29. $\mathbf{F} = \langle 3zy^{-1}, 4x, -y \rangle$, $\mathbf{c}(t) = (e^t, e^t, t)$ for $-1 \le t \le 1$

30. $\mathbf{F} = \langle x - y, y - z, z \rangle$,   line segment from $(0, 0, 0)$ to $(1, 4, 4)$

31. $\mathbf{F} = \left\langle \dfrac{-y}{(x^2 + y^2)^2}, \dfrac{x}{(x^2 + y^2)^2} \right\rangle$,   circle of radius R with center at the origin oriented counterclockwise

32. $\mathbf{F} = \left\langle \dfrac{-y}{x^2 + y^2}, \dfrac{x}{x^2 + y^2} \right\rangle$,   the square with vertices $(1, 1)$, $(1, -1)$, $(-1, -1)$, and $(-1, 1)$ in the counterclockwise direction

33. $\mathbf{F} = \left\langle z^2, x, y \right\rangle$, $\mathbf{c}(t) = (\cos t, \tan t, t)$ for $0 \le t \le \frac{\pi}{4}$

34. $\mathbf{F} = \left\langle \dfrac{1}{y^3 + 1}, \dfrac{1}{z + 1}, 1 \right\rangle$, $\mathbf{c}(t) = (t^3, 2t, t^2)$ for $0 \le t \le 1$

35. $\mathbf{F} = \left\langle z^3, yz, x \right\rangle$,   circle of radius 2 in the yz-plane with center at the origin oriented clockwise when viewed from the positive x-axis

36. **CAS**   Let $f(x, y, z) = x^{-1}yz$ and let $\mathcal{C}$ be the curve parametrized by $\mathbf{c}(t) = (\ln t, t, t^2)$ for $2 \le t \le 4$. Use a computer algebra system to calculate $\displaystyle\int_{\mathcal{C}} f(x, y, z)\, ds$ to four decimal places.

37. **CAS**   Use a CAS to calculate $\displaystyle\int_{\mathcal{C}} \langle e^{x-y}, e^{x+y} \rangle \cdot d\mathbf{s}$ to four decimal places, where $\mathcal{C}$ is the curve $y = \sin x$ for $0 \le x \le \pi$, oriented from left to right.

In Exercises 38–39, calculate the line integral of $\mathbf{F} = \langle e^z, e^{x-y}, e^y \rangle$ over the given path.

38. The path from P to Q in Figure 15

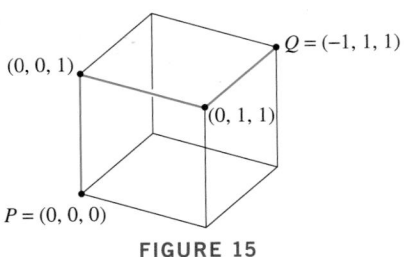

FIGURE 15

39. The path ABC in Figure 16

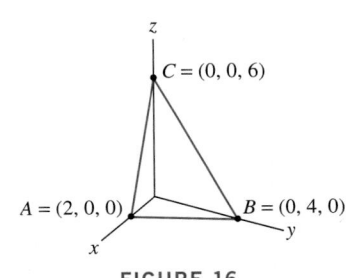

FIGURE 16

40. Let $\mathcal{C}$ be the path from P to Q in Figure 17 that traces $\mathcal{C}_1, \mathcal{C}_2$, and $\mathcal{C}_3$ in the orientation indicated. Suppose that

$$\int_{\mathcal{C}} \mathbf{F} \cdot d\mathbf{s} = 5, \qquad \int_{\mathcal{C}_1} \mathbf{F} \cdot d\mathbf{s} = 8, \qquad \int_{\mathcal{C}_3} \mathbf{F} \cdot d\mathbf{s} = 8$$

Determine:

(a) $\displaystyle\int_{-\mathcal{C}_3} \mathbf{F} \cdot d\mathbf{s}$ **(b)** $\displaystyle\int_{\mathcal{C}_2} \mathbf{F} \cdot d\mathbf{s}$ **(c)** $\displaystyle\int_{-\mathcal{C}_1 - \mathcal{C}_3} \mathbf{F} \cdot d\mathbf{s}$

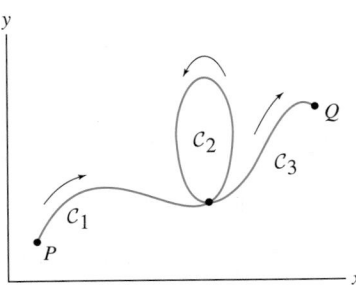

FIGURE 17

(d) What is the value of the line integral of **F** over the path that traverses the loop C_2 four times in the clockwise direction?

In Exercises 41–44, let **F** *be the* **vortex vector field** *(so-called because it swirls around the origin as shown in Figure 18)*

$$\mathbf{F} = \left\langle \frac{-y}{x^2 + y^2}, \frac{x}{x^2 + y^2} \right\rangle$$

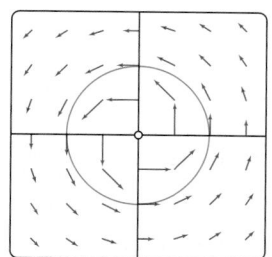

FIGURE 18 Vector field $\mathbf{F} = \left\langle \dfrac{-y}{x^2 + y^2}, \dfrac{x}{x^2 + y^2} \right\rangle$.

41. Let $I = \displaystyle\int_C \mathbf{F} \cdot d\mathbf{s}$, where C is the circle of radius 2 centered at the origin oriented counterclockwise (Figure 18).

(a) Do you expect I to be positive, negative, or zero?

(b) Evaluate I.

(c) Verify that I changes sign when C is oriented in the clockwise direction.

42. Compute $\displaystyle\int_{C_R} \mathbf{F} \cdot d\mathbf{s}$, where C_R is the circle of radius R centered at the origin oriented counterclockwise. Show that the result is independent of R.

43. Calculate $\displaystyle\int_{\mathcal{A}} \mathbf{F} \cdot d\mathbf{s}$, where $\mathcal{A}$ is the arc of angle θ_0 on the circle of radius R centered at the origin oriented counterclockwise. *Note:* $\mathcal{A}$ begins at $(R, 0)$ and ends at $(R \cos \theta_0, R \sin \theta_0)$.

44. Let $a > 0, b < c$. Show that the integral of **F** along the segment from $P = (a, b)$ to $Q = (a, c)$ is equal to the angle $\angle POQ$ (O is the origin).

45. Calculate the line integral of the constant vector field $\mathbf{F} = \langle 2, -1, 4 \rangle$ along the segment $\overrightarrow{PQ}$, where:

(a) $P = (0, 0, 0)$, $Q = (1, 0, 0)$

(b) $P = (0, 0, 0)$, $Q = (4, 3, 5)$

(c) $P = (3, 2, 3)$, $Q = (4, 8, 12)$

46. Show that if **F** is a constant vector field and C is any oriented path from P to Q, then

$$\int_C \mathbf{F} \cdot d\mathbf{s} = \mathbf{F} \cdot \overrightarrow{PQ}$$

47. Figure 19 shows the vector field $\mathbf{F}(x, y) = \langle x, x \rangle$.

(a) Are $\displaystyle\int_{AB} \mathbf{F} \cdot d\mathbf{s}$ and $\displaystyle\int_{DC} \mathbf{F} \cdot d\mathbf{s}$ equal? If not, which is larger?

(b) Which is smaller: the line integral of **F** over the path ADC or ABC?

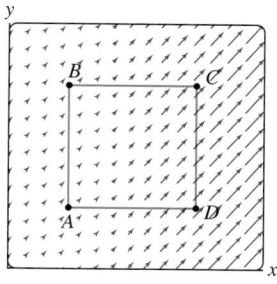

FIGURE 19

48. Calculate the work done by a field $\mathbf{F} = \langle x + y, x - y \rangle$ when an object moves from $(0, 0)$ to $(1, 1)$ along each of the paths $y = x^2$ and $x = y^2$.

49. Calculate the work done by the force field $\mathbf{F} = \langle x, y, z \rangle$ along the path $(\cos t, \sin t, t)$ for $0 \le t \le 3\pi$.

50. Let $I = \displaystyle\int_C f(x, y, z)\, ds$. Assume that $f(x, y, z) \ge m$ for some number m and all points (x, y, z) on C. Which of the following conclusions is correct? Explain.

(a) $I \ge m$

(b) $I \ge mL$, where L is the length of C

Further Insights and Challenges

51. As observed in the text, the value of a scalar line integral does not depend on the choice of parametrization. Prove this directly. Namely, suppose that $\mathbf{c}_1(t)$ and $\mathbf{c}(t)$ are two parametrizations of C and that $\mathbf{c}_1(t) = \mathbf{c}(\varphi(t))$, where $\varphi(t)$ is an increasing function. Use the Change of Variables Formula to verify that

$$\int_c^d f(\mathbf{c}_1(t)) \|\mathbf{c}_1'(t)\|\, dt = \int_a^b f(\mathbf{c}(t)) \|\mathbf{c}'(t)\|\, dt$$

where $a = \varphi(c)$ and $b = \varphi(d)$.

52. Suppose we wish to compute the *average value* of a continuous function $f(x, y, z)$ along a curve C of length L. Given a large number N, divide C into N consecutive arcs $C_1, \dots, C_N$, each of length L/N, and let P_i be a sample point in C_i (Figure 20). Then the points P_i are somewhat evenly spaced along the curve and the sum $\dfrac{1}{N} \displaystyle\sum_{i=1}^{N} f(P_i)$ may be considered an approximation to the average value of f along C. We define the average to be the limit (if it exists) as $N \to \infty$:

$$\text{Av}(f) = \lim_{N \to \infty} \frac{1}{N} \sum_{i=1}^{N} f(P_i)$$

Prove that

$$\mathrm{Av}(f) = \frac{1}{L}\int_C f(x, y, z)\, ds \qquad \boxed{8}$$

Hint: Show that $\dfrac{L}{N}\displaystyle\sum_{i=1}^{N} f(P_i)$ is a Riemann sum approximation to the line integral of f along C.

FIGURE 20

53. Use Eq. (8) to calculate the average value of $f(x, y) = x - y$ along the segment from $P = (2, 1)$ to $Q = (5, 5)$.

54. Use Eq. (8) to calculate the average value of xy along the curve $y = x^2$ for $0 \le x \le 1$.

55. The temperature (in degrees centigrade) at a point P on a circular wire of radius 2 cm centered at the origin is equal to the square of the distance from P to $P_0 = (2, 0)$. Compute the average temperature along the wire.

56. Let $\mathbf{F} = \langle x, 0 \rangle$. Prove that if C is any path from (a, b) to (c, d), then

$$\int_C \mathbf{F} \cdot d\mathbf{s} = \frac{1}{2}(c^2 - a^2)$$

57. Let $\mathbf{F} = \langle y, x \rangle$. Prove that if C is any path from (a, b) to (c, d), then

$$\int_C \mathbf{F} \cdot d\mathbf{s} = cd - ab$$

16.3 Conservative Vector Fields

Some vector fields possess a special property called **path independence**. By this we mean that when the vector field is integrated along a path from P to Q, the result depends only on the endpoints P and Q, not on the path followed. A vector field with this property is called **conservative**.

If $\mathbf{c}(t)$ is a path from P to Q, we write $\displaystyle\int_{\mathbf{c}} \mathbf{F} \cdot d\mathbf{s}$ for the line integral $\displaystyle\int_{C} \mathbf{F} \cdot d\mathbf{s}$ along the curve C parametrized by $\mathbf{c}(t)$. We use these two notations for the line integral interchangeably. Then a vector field $\mathbf{F}$ defined on a domain $\mathcal{D}$ is conservative if, for all P and Q in $\mathcal{D}$ and for any two paths $\mathbf{c}_1(t)$ and $\mathbf{c}_2(t)$ in $\mathcal{D}$ from P to Q (Figure 1), we have

$$\int_{\mathbf{c}_1} \mathbf{F} \cdot d\mathbf{s} = \int_{\mathbf{c}_2} \mathbf{F} \cdot d\mathbf{s}$$

FIGURE 1 If $\mathbf{F}$ is conservative, then the line integrals over $\mathbf{c}_1$ and $\mathbf{c}_2$ are equal.

When C is a closed curve, we sometimes use the symbol $\oint$ and write the line integral as $\displaystyle\oint_C \mathbf{F} \cdot d\mathbf{s}$. This line integral is called the **circulation** of $\mathbf{F}$ around C.

Recall that $\mathbf{F}$ is a gradient vector field if $\mathbf{F} = \nabla \varphi$ for some function φ (called a *potential function*). The next theorem shows that the line integral of $\mathbf{F}$ along any path from P to Q is equal to $\varphi(Q) - \varphi(P)$. In particular, the line integral is independent of the path and hence, *all gradient vector fields are conservative.*

THEOREM 1 Fundamental Theorem for Gradient Vector Fields If $\mathbf{F} = \nabla \varphi$ on a domain $\mathcal{D}$, then for every oriented curve C in $\mathcal{D}$ with initial point P and terminal point Q,

$$\int_C \mathbf{F} \cdot d\mathbf{s} = \varphi(Q) - \varphi(P) \qquad \boxed{1}$$

If C is closed (i.e., if $P = Q$), then $\displaystyle\oint_C \mathbf{F} \cdot d\mathbf{s} = 0$.

Proof Let $c(t)$ be a parametrization of C for $a \leq t \leq b$. Then $c(a) = P$ and $c(b) = Q$. According to the Chain Rule for Paths (Theorem 2 in Section 14.5),

$$\frac{d}{dt}\varphi(c(t)) = \nabla\varphi(c(t)) \cdot c'(t) = F(c(t)) \cdot c'(t)$$

Therefore,

$$\int_C F \cdot ds = \int_a^b F(c(t)) \cdot c'(t)\, dt = \int_a^b \frac{d}{dt}\varphi(c(t))\, dt$$

Equation (1) follows from the Fundamental Theorem of Calculus:

$$\int_a^b \frac{d}{dt}\varphi(c(t))\, dt = \varphi(c(t))\Big|_a^b = \varphi(c(b)) - \varphi(c(a)) = \varphi(Q) - \varphi(P)$$

If C is a closed curve, then the initial and terminal points coincide, that is, $P = Q$. In this case, the line integral is equal to $\varphi(Q) - \varphi(P) = 0$. ∎

■ **EXAMPLE 1** Let $F = \langle 2xy + z, x^2, x \rangle$. Verify that $\varphi(x, y, z) = x^2 y + xz$ is a potential function for F and evaluate the line integrals of F over the following paths:

(a) $c(t) = (e^t, e^{2t}, t^2)$ for $0 \leq t \leq 2$

(b) Any path from $(1, -1, 2)$ to $(2, 2, 2)$

Solution The partial derivatives of φ are the components of F:

$$\frac{\partial\varphi}{\partial x} = 2xy + z, \qquad \frac{\partial\varphi}{\partial y} = x^2, \qquad \frac{\partial\varphi}{\partial z} = x$$

Therefore, $\nabla\varphi = \langle 2xy + z, x^2, x \rangle = F$.

(a) The initial and terminal points of the path $c(t) = (e^t, e^{2t}, t^2)$ for $0 \leq t \leq 2$ are

$$P = c(0) = (1, 1, 0), \qquad Q = c(2) = (e^2, e^4, 4)$$

The values of φ at the endpoints are

$$\varphi(Q) = \varphi(e^2, e^4, 4) = e^8 + 4e^2, \qquad \varphi(P) = \varphi(1, 1, 0) = 1$$

By the Fundamental Theorem for Gradient Vector Fields,

$$\int_c F \cdot ds = \varphi(Q) - \varphi(P) = (e^8 + 4e^2) - 1 = e^8 + 4e^2 - 1$$

(b) The line integral of F along any path c from $(1, -1, 2)$ to $(2, 2, 2)$ is

$$\int_c F \cdot ds = \varphi(Q) - \varphi(P) = \varphi(2, 2, 2) - \varphi(1, -1, 2) = 12 - 1 = 11 \qquad ■$$

■ **EXAMPLE 2** Let $F = \langle 2x + y, x \rangle$. Evaluate $\displaystyle\int_c F \cdot ds$, where c is a path from $(1, 1)$ to $(3, -1)$, by finding a potential function for F.

Solution A general method for finding potential functions is discussed later in this section. For $F = \langle 2x + y, x \rangle$, we see by inspection that $\varphi(x, y) = x^2 + xy$ is a potential function. Indeed,

$$\frac{\partial\varphi}{\partial x} = \frac{\partial}{\partial x}(x^2 + xy) = 2x + y, \qquad \frac{\partial\varphi}{\partial y} = \frac{\partial}{\partial y}(x^2 + xy) = x$$

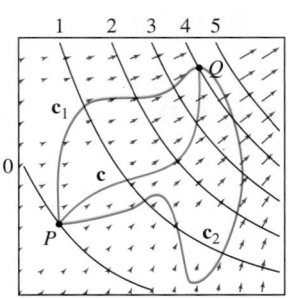

FIGURE 2 The quantity $\varphi(Q) - \varphi(P)$ is the "net" number of level curves of φ that you have to cross to get from P to Q.

Therefore, for any path $\mathbf{c}$ from $(1, 1)$ to $(3, -1)$,

$$\int_{\mathbf{c}} \mathbf{F} \cdot d\mathbf{s} = \int_{\mathbf{c}} \nabla\varphi \cdot d\mathbf{s} = \varphi(3, -1) - \varphi(1, 1) = (3^2 - 3) - (1^2 + 1) = 4 \qquad \blacksquare$$

CONCEPTUAL INSIGHT We have seen that if φ is a potential function for a gradient vector field $\mathbf{F} = \nabla\varphi$ and $\mathbf{c}(t)$ is a path from P to Q, then

$$\int_{\mathbf{c}} \nabla\varphi \cdot d\mathbf{s} = \varphi(Q) - \varphi(P) \qquad \boxed{2}$$

We shall interpret this formula in terms of the contour map of φ. If we assume that the contour interval is 1, as in Figure 2, then the potential φ changes by 1 when we move from one level curve to the next. Therefore, the net change in φ along the path $\mathbf{c}(t)$ is equal to the number of level curves crossed by $\mathbf{c}(t)$. More precisely,

$$\varphi(Q) - \varphi(P) = \text{net number of level curves crossed by } \mathbf{c} \qquad \boxed{3}$$

For example, in Figure 2, $\varphi(P) = 0$ and $\varphi(Q) = 4$, so every path from P to Q begins at the level curve $c = 0$ and ends at the level curve $c = 4$. Along the way, the path makes four crossings from one level curve to the next higher level curve. This is true even if the curve backtracks, provided that we count the "net" number of crossings. In the figure, $\mathbf{c}$ and $\mathbf{c}_1$ travel consistently in the direction of increasing potential whereas $\mathbf{c}_2$ backtracks.

Now we can explain Eq. (2). For any path $\mathbf{c}(t)$ from P to Q for $a \le t \le b$, the composite function $\varphi(\mathbf{c}(t))$ tells us which level curve of φ we are on at time t. Its derivative is the rate at which we cross level curves (assuming the contour interval is 1):

$$\frac{d}{dt}\varphi(\mathbf{c}(t)) = \nabla\varphi(\mathbf{c}(t)) \cdot \mathbf{c}'(t) = \begin{array}{l} \text{number of level curves} \\ \text{crossed per unit time} \end{array}$$

The line integral is nothing more than the integral of this rate, which is equal to the net number of crossings:

$$\int_{\mathbf{c}} \nabla\varphi \cdot d\mathbf{s} = \int_{a}^{b} \nabla\varphi(\mathbf{c}(t)) \cdot \mathbf{c}'(t)\, dt = \int_{a}^{b} \frac{d}{dt}\varphi(\mathbf{c}(t))\, dt = \begin{array}{l} \text{net number of} \\ \text{level curves crossed} \end{array} \qquad \boxed{4}$$

The net number of crossings depends only on the endpoints of the path, not on the path itself, and hence the line integral is path-independent. However, this interpretation of the line integral is valid only for a gradient vector field—otherwise, we cannot speak of the level curves of a potential function.

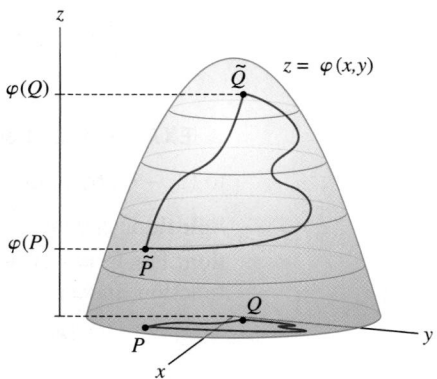

FIGURE 3 The line integral is independent of path.

In terms of the graph of a potential function $z = \varphi(x, y)$, the line integral is simply the change in height from the point $\tilde{P}$ above P to the point $\tilde{Q}$ above Q (Figure 3).

In the Conceptual Insight, we discussed gradient vector fields in the plane. A standard example of a conservative field in $\mathbf{R}^3$ is the force field $\mathbf{F} = -mg\mathbf{k}$ due to gravity acting on an object of mass m near the surface of the earth. This force field is constant, and $\mathbf{F} = -\nabla\varphi$ where $\varphi(x, y, z) = mgz$. When you hike up a mountain along a path $\mathbf{c}(t)$ beginning at altitude z_1 and ending at altitude z_2, your work against gravity is independent of the particular path and is equal to

$$-\int_{\mathbf{c}} \mathbf{F} \cdot d\mathbf{s} = \varphi(z_2) - \varphi(z_1) = mg(z_2 - z_1)$$

■ **EXAMPLE 3** Integral around a Closed Path   Let $\varphi(x, y, z) = xy \sin(yz)$. Evaluate $\int_{C} \nabla\varphi \cdot d\mathbf{s}$, where C is the closed curve in Figure 4.

Solution   By Theorem 1, the integral of a gradient vector around any closed path is zero. In other words, $\int_{C} \nabla\varphi \cdot d\mathbf{s} = 0$.   ■

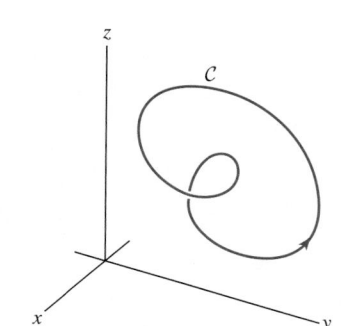

FIGURE 4 The line integral of a gradient vector field around a closed curve is zero.

In the next theorem, we show that the condition of path independence can be stated in terms of closed paths.

THEOREM 2 Equivalent Conditions for Path Independence   Let $\mathbf{F}$ be a vector field defined on a domain $\mathcal{D}$. The following two conditions are equivalent:

(a) For any two paths C_1 and C_2 in $\mathcal{D}$ with the same initial and terminal points,

$$\int_{C_1} \mathbf{F} \cdot d\mathbf{s} = \int_{C_2} \mathbf{F} \cdot d\mathbf{s}$$

(b) For all closed paths C in $\mathcal{D}$, $\oint_{C} \mathbf{F} \cdot d\mathbf{s} = 0$.

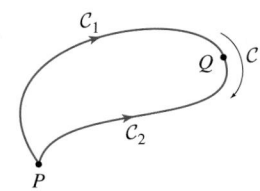

FIGURE 5 The closed loop C is represented as $C_1 - C_2$.

Proof   Given any closed path C, we may choose distinct points P and Q along C and let C_1 and C_2 be the two paths from P to Q as in Figure 5. Then

$$C = C_1 - C_2$$

since $-C_2$ is the path from Q back to P along C_2 in the opposite direction. Thus,

$$\oint_{C} \mathbf{F} \cdot d\mathbf{s} = \int_{C_1} \mathbf{F} \cdot d\mathbf{s} - \int_{C_2} \mathbf{F} \cdot d\mathbf{s} \qquad \boxed{5}$$

If we assume (a), then the right-hand term in Eq. (5) is zero. Thus the left-hand term is zero and (b) holds. On the other hand, Eq. (5) is valid for any two paths C_1 and C_2 from P to Q. If we assume (b), then the left-hand side of Eq. (5) is zero. Thus, the right-hand term is zero and (a) holds as well.   ■

The Fundamental Theorem (Theorem 1) tells us that gradient vector fields are conservative. We now prove the converse statement: *Every* conservative vector field is a gradient

Connected region

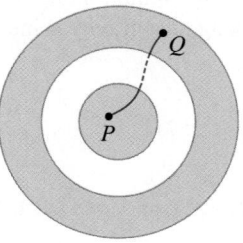

Nonconnected region

FIGURE 6 $\mathcal{D}$ is connected if any two points in $\mathcal{D}$ can be joined by a path entirely contained in $\mathcal{D}$.

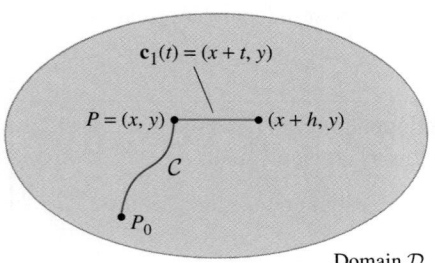

Domain $\mathcal{D}$

FIGURE 7

vector field. Recall that a domain $\mathcal{D}$ is said to be connected if any two points in $\mathcal{D}$ can be joined by a path in $\mathcal{D}$ (Figure 6) and that $\mathcal{D}$ is open if for all $P \in \mathcal{D}$, there exists an open disk centered at P contained in $\mathcal{D}$.

THEOREM 3   Every conservative vector field $\mathbf{F}$ on an open connected domain is a gradient vector field, that is, $\mathbf{F} = \nabla \varphi$ for some potential function φ.

Proof   We treat the case of a planar vector field $\mathbf{F} = \langle F_1, F_2 \rangle$. The proof for vector fields in $\mathbf{R}^3$ is similar. Choose a point P_0 in $\mathcal{D}$ and for any point $P = (x, y) \in \mathcal{D}$, define

$$\varphi(P) = \varphi(x, y) = \int_{\mathcal{C}} \mathbf{F} \cdot d\mathbf{s}$$

where $\mathcal{C}$ is any path beginning at P_0 and ending at P (Figure 7). For a general vector field, this definition would not be meaningful because the line integral on the right would depend on the choice of $\mathcal{C}$. However, since $\mathbf{F}$ is assumed conservative, the integral is path-independent and φ is defined unambiguously. We must show that $\dfrac{\partial \varphi}{\partial x} = F_1$ and $\dfrac{\partial \varphi}{\partial y} = F_2$. We prove the first equation since the second may be verified in a similar manner.

To compute $\dfrac{\partial \varphi}{\partial x}$, let $\mathcal{C}_1$ be the horizontal segment from (x, y) to $(x + h, y)$. For $|h|$ small enough, $\mathcal{C}_1$ lies inside $\mathcal{D}$ and the sum $\mathcal{C} + \mathcal{C}_1$ is a curve beginning at P_0 and ending at $(x + h, y)$ (Figure 7). Therefore,

$$\varphi(x + h, y) - \varphi(x, y) = \int_{\mathcal{C} + \mathcal{C}_1} \mathbf{F} \cdot d\mathbf{s} - \int_{\mathcal{C}} \mathbf{F} \cdot d\mathbf{s}$$

$$\left(\int_{\mathcal{C}} \mathbf{F} \cdot d\mathbf{s} + \int_{\mathcal{C}_1} \mathbf{F} \cdot d\mathbf{s} \right) - \int_{\mathcal{C}} \mathbf{F} \cdot d\mathbf{s} = \int_{\mathcal{C}_1} \mathbf{F} \cdot d\mathbf{s}$$

We may parametrize $\mathcal{C}_1$ by $\mathbf{c}_1(t) = (x + t, y)$ for $0 \leq t \leq h$. Then $\mathbf{c}'(t) = \langle 1, 0 \rangle$ and

$$\mathbf{F}(\mathbf{c}(t)) \cdot \mathbf{c}'(t) = \langle F_1(x + t, y), F_2(x + t, y) \rangle \cdot \langle 1, 0 \rangle = F_1(x + t, y)$$

$$\varphi(x + h, y) - \varphi(x, y) = \int_{\mathbf{c}_1} \mathbf{F} \cdot d\mathbf{s} = \int_0^h F_1(x + t, y)\, dt$$

Therefore (making the substitution $u = x + t$ in the last step),

$$\frac{\varphi(x + h, y) - \varphi(x, y)}{h} = \frac{1}{h} \int_0^h F_1(x + t, y)\, dt = \frac{1}{h} \int_x^{x+h} F_1(u, y)\, du$$

Now take the limit as $h \to 0$:

$$\frac{\partial \varphi}{\partial x} = \lim_{h \to 0} \frac{\varphi(x + h, y) - \varphi(x, y)}{h} = \underbrace{\lim_{h \to 0} \frac{1}{h} \int_x^{x+h} F_1(u, y)\, du}_{\text{Equal to } F_1(x, y) \text{ by the FTC}}$$

The limit on the right is equal to the derivative of the integral of $F_1(x, y)$ with respect to x (for fixed y) and is therefore equal to $F_1(x, y)$ itself by the Fundamental Theorem of Calculus. This completes the proof. ∎

Finding a Potential Function

Having shown that a vector field **F** is conservative if and only if $\mathbf{F} = \nabla\varphi$ for some potential function φ, we now address the following two problems:

- How to determine whether a vector field is conservative
- Given a conservative vector field, how to find a potential function

Suppose that $\mathbf{F} = \langle F_1, F_2, F_3 \rangle$. We already know, by Theorem 1 of Section 16.1, that if **F** has a potential function, then the cross partials of the components are equal (Figure 8):

$$\frac{\partial F_1}{\partial y} = \frac{\partial F_2}{\partial x}, \qquad \frac{\partial F_2}{\partial z} = \frac{\partial F_3}{\partial y}, \qquad \frac{\partial F_3}{\partial x} = \frac{\partial F_1}{\partial z} \qquad \boxed{6}$$

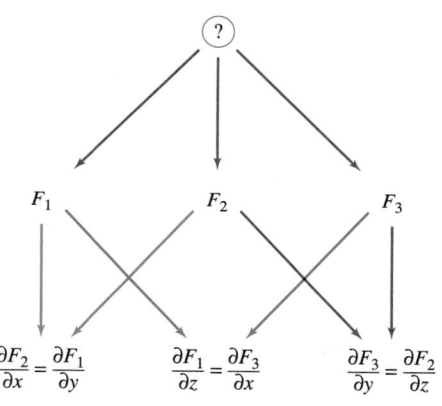

FIGURE 8

Now suppose that **F** satisfies the condition of equal cross partials. Does it follow that **F** has a potential function? The answer is yes, provided that the domain $\mathcal{D}$ has a property called simple connectedness. Roughly speaking, a domain $\mathcal{D}$ in the plane is **simply connected** if it does not have any "holes" (Figure 9). More precisely, in a simply connected domain $\mathcal{D}$, every loop in $\mathcal{D}$ can be shrunk or "contracted" down to a point while staying within $\mathcal{D}$. Observe that the disk with a point removed in Figure 10 is not simply connected. The loop shown in the figure cannot be shrunk down to a point without passing through the point that was removed. Standard examples of simply-connected domains in

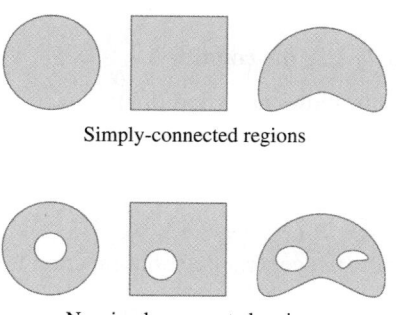

Simply-connected regions

Nonsimply-connected regions

FIGURE 9 Simple connectedness means "no holes."

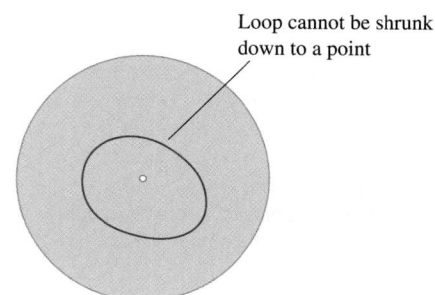

Loop cannot be shrunk down to a point

FIGURE 10 A disk with one point removed is not simply connected.

$\mathbf{R}^2$ are disks, rectangles, and the entire xy-plane. In three-space, the interiors of balls and boxes are simply connected, as is the entire space $\mathbf{R}^3$.

THEOREM 4 Existence of a Potential Function Let $\mathbf{F} = \langle F_1, F_2, F_3 \rangle$ be a vector field on a simply-connected domain $\mathcal{D}$. If the cross partials of $\mathbf{F}$ are equal,

$$\frac{\partial F_1}{\partial y} = \frac{\partial F_2}{\partial x}, \qquad \frac{\partial F_2}{\partial z} = \frac{\partial F_3}{\partial y}, \qquad \frac{\partial F_3}{\partial x} = \frac{\partial F_1}{\partial z} \qquad \boxed{7}$$

then $\mathbf{F} = \nabla\varphi$ for some function φ.

Rather than prove Theorem 4, we indicate a procedure for finding a potential function in the next two examples. This procedure always works "locally"; that is, it allows us to find a potential function near a given point in the domain $\mathcal{D}$. The more subtle issue of proving that it always works on the entire domain relies on the fact that $\mathcal{D}$ is simply connected. A rigorous proof uses the theorems of Green and Stokes from Chapter 17.

■ **EXAMPLE 4** Finding a Potential Function Verify that the cross partials are equal for the vector field

$$\mathbf{F} = \left\langle 2xy + y^3 + 2, \ x^2 + 3xy^2 + 2y \right\rangle$$

and find a potential function in the domain $\mathcal{D} = \mathbf{R}^2$.

Solution Let $\mathbf{F} = \langle F_1, F_2 \rangle$. We first check that the cross partials are equal:

$$\frac{\partial F_1}{\partial y} = \frac{\partial}{\partial y}(2xy + y^3 + 2) \quad = 2x + 3y^2$$

$$\frac{\partial F_2}{\partial x} = \frac{\partial}{\partial x}(x^2 + 3xy^2 + 2y) = 2x + 3y^2$$

By Theorem 4, there exists a potential function φ such that $\dfrac{\partial\varphi}{\partial x} = F_1$ and $\dfrac{\partial\varphi}{\partial y} = F_2$.

Step 1. Use the condition $\dfrac{\partial\varphi}{\partial x} = F_1$.

The condition $\dfrac{\partial\varphi}{\partial x} = F_1$ tells us that φ is an antiderivative of $F_1(x, y)$ when y is fixed:

$$\varphi(x, y) = \int F_1(x, y)\,dx = \underbrace{\int (2xy + y^3 + 2)\,dx = x^2 y + xy^3 + 2x}_{\text{Integrate with respect to } x} + g(y)$$

Note that the antiderivative is only determined up to a constant of integration. This "constant" may depend on y (but not on x), so we write it as $g(y)$. Thus,

$$\varphi(x, y) = x^2 y + xy^3 + 2x + g(y)$$

Step 2. Use the condition $\dfrac{\partial\varphi}{\partial y} = F_2$.

In summary, to find a potential function for $\mathbf{F} = \langle F_1, F_2 \rangle$, integrate $F_1(x, y)$ with respect to x and "adjust" the result by adding on a function $g(y)$ of y alone. Alternatively, integrate $F_2(x, y)$ with respect to y and adjust the result by adding on a function $f(x)$ of x alone.

This allows us to find $g(y)$:

$$F_2(x, y) = \frac{\partial\varphi}{\partial y} = \frac{\partial}{\partial y}\left(x^2 y + xy^3 + 2x + g(y)\right) = x^2 + 3xy^2 + g'(y)$$

Since $F_2(x, y) = x^2 + 3xy^2 + 2y$, we obtain $g'(y) = 2y$ or $g(y) = y^2 + C$. Therefore, the general potential function for $\mathbf{F}$ is

$$\varphi(x, y) = x^2 y + xy^3 + 2x + y^2 + C \qquad ■$$

The same method works for vector fields in three-space.

■ **EXAMPLE 5** Finding a Potential Function Verify that the following vector field has a potential function φ and find one:

$$\mathbf{F} = \langle 2xyz^{-1}, z + x^2z^{-1}, y - x^2yz^{-2} \rangle$$

Solution First, we check that the cross partials are equal by verifying that their differences are zero:

$$\frac{\partial F_1}{\partial y} - \frac{\partial F_2}{\partial x} = \frac{\partial}{\partial y}2xyz^{-1} - \frac{\partial}{\partial x}(z + x^2z^{-1}) = 2xz^{-1} - 2xz^{-1} = 0$$

$$\frac{\partial F_2}{\partial z} - \frac{\partial F_3}{\partial y} = \frac{\partial}{\partial z}(z + x^2z^{-1}) - \frac{\partial}{\partial y}(y - x^2yz^{-2}) = (1 - x^2z^{-2}) - (1 - x^2z^{-2}) = 0$$

$$\frac{\partial F_3}{\partial x} - \frac{\partial F_1}{\partial z} = \frac{\partial}{\partial x}(y - x^2yz^{-2}) - \frac{\partial}{\partial z}2xyz^{-1} = -2xyz^{-2} - (-2xyz^{-2}) = 0$$

By Theorem 4, there exists a φ such that $\dfrac{\partial\varphi}{\partial x} = F_1$, $\dfrac{\partial\varphi}{\partial y} = F_2$, and $\dfrac{\partial\varphi}{\partial z} = F_3$.

Step 1. **Use the condition** $\dfrac{\partial\varphi}{\partial x} = F_1$.

The function φ is an antiderivative of F_1 when y and z are fixed:

$$\varphi(x, y, z) = \underbrace{\int F_1(x, y, z)\, dx}_{\text{Integrate with respect to } x} = \int 2xyz^{-1}\, dx = x^2yz^{-1} + \underbrace{g(y, z)}_{\substack{\text{Constant of} \\ \text{integration}}} \qquad \boxed{8}$$

Therefore, $\varphi(x, y, z) = x^2yz^{-1} + g(y, z)$, where $g(y, z)$ does not depend on x.

Step 2. **Use the condition** $\dfrac{\partial\varphi}{\partial y} = F_2$.

This allows us to determine $g(y, z)$ up to a function depending on z alone:

$$F_2(x, y, z) = \frac{\partial\varphi}{\partial y} = \frac{\partial}{\partial y}(x^2yz^{-1} + g(y, z)) = x^2z^{-1} + \frac{\partial g}{\partial y} \qquad \boxed{9}$$

Since $F_2(x, y, z) = z + x^2z^{-1}$, we conclude that $\dfrac{\partial g}{\partial y} = z$ and hence

$$g(y, z) = \int z\, dy = yz + h(z)$$

Step 3. **Use the condition** $\dfrac{\partial\varphi}{\partial z} = F_3$.

At this point, by Eqs. (8) and (9), we know that

$$\varphi(x, y, z) = x^2yz^{-1} + g(y, z) = x^2yz^{-1} + yz + h(z)$$

We use this last condition to identify $h(z)$:

$$\frac{\partial\varphi}{\partial z} = \frac{\partial}{\partial z}(x^2yz^{-1} + zy + h(z)) = -x^2yz^{-2} + y + \frac{dh}{dz}$$

Since $F_3(x, y, z) = -x^2yz^{-2} + y$, it follows that $\dfrac{dh}{dz} = 0$. Therefore, $h(z)$ is a constant C and the general potential function is

$$\varphi(x, y, z) = x^2yz^{-1} + zy + C \qquad\qquad ■$$

Assumptions Matter We cannot expect the above method for finding a potential function to work if **F** does not satisfy the cross-partials condition. After all, in this case, no potential function exists. For example, $\mathbf{F} = \langle y, 0 \rangle$ does not satisfy the cross-partials condition. If we attempt to find a potential function, we first use the condition $\partial \varphi / \partial x = y$ to obtain $\varphi(x, y) = xy + g(y)$. But the second condition yields an equation with no solution:

$$\frac{\partial \varphi}{\partial y} = x + \frac{dg}{dy} = 0 \quad \Rightarrow \quad \frac{dg}{dy} = -x \quad \text{(no solution)}$$

Indeed, $g(y)$ depends only on y, so its derivative cannot depend on x.

CONCEPTUAL INSIGHT In physics, two important conservative force fields **F** are the gravitational force and electrostatic force. It is customary to change sign and write $\mathbf{F} = -\nabla \varphi$ and to define the **potential energy** (PE) of an object at location (x, y, z) to be $\varphi(x, y, z)$. The work required to move an object from point P to point Q along any path **c** is equal to the change in potential energy:

$$\text{Work against } \mathbf{F} = -\int_{\mathbf{c}} \mathbf{F} \cdot d\mathbf{s} = \underbrace{\varphi(Q) - \varphi(P)}_{\text{Change in potential energy}}$$

The name "conservative field" stems from the Law of Conservation of Energy. Suppose that an object subject to a conservative force **F** travels along a path $\mathbf{c}(t)$ with velocity $\mathbf{v} = \mathbf{c}'(t)$. By definition, the object's **kinetic energy** (KE) is $\frac{1}{2}m\mathbf{v} \cdot \mathbf{v}$ and its total energy at time t is the sum

$$E = KE + PE = \frac{1}{2}m\mathbf{v} \cdot \mathbf{v} + \varphi(\mathbf{c}(t))$$

The Law of Conservation of Energy states that E is constant in time. To prove it, we use Newton's Law $\mathbf{F} = m\mathbf{a}$ (where $\mathbf{a} = \mathbf{v}'(t)$):

$$\frac{dE}{dt} = \frac{d}{dt}\left(\frac{1}{2}m\mathbf{v} \cdot \mathbf{v} + \varphi(\mathbf{c}(t))\right)$$

$$= m\mathbf{v} \cdot \mathbf{a} + \nabla \varphi(\mathbf{c}(t)) \cdot \mathbf{c}'(t) \qquad \text{(by the Chain Rule)}$$

$$= m\mathbf{v} \cdot \mathbf{a} + \nabla \varphi(\mathbf{c}(t)) \cdot \mathbf{v}$$

$$= m\mathbf{v} \cdot \mathbf{a} - \mathbf{F} \cdot \mathbf{v} \qquad \text{(since } \mathbf{F} = -\nabla \varphi\text{)}$$

$$= \mathbf{v} \cdot (m\mathbf{a} - \mathbf{F}) = 0 \qquad \text{(since } \mathbf{F} = m\mathbf{a}\text{)}$$

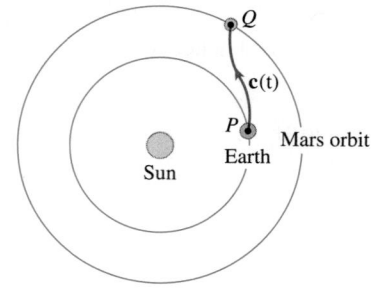

FIGURE 11

The gravitational force exerted by the sun (located at the origin) on a mass the size of the earth is $\mathbf{F} = -k\dfrac{\mathbf{e}_r}{r^2}$, where

$$\mathbf{e}_r = \frac{1}{r}\langle x, y, z \rangle \qquad \left(r = \sqrt{x^2 + y^2 + z^2}\right)$$

and $k \approx 7.9 \times 10^{44}$ N-m^2.

■ **EXAMPLE 6** Work against Gravity Compute the work W (in joules) required to move the earth, along any path, from its current orbit to the orbit of Mars. The average distance from the sun to the earth is 1.5×10^{11} m and to Mars is 2.28×10^{11} m.

Solution The gravitational force of the sun on the earth is described by the inverse square field $\mathbf{F} = -k\dfrac{\mathbf{e}_r}{r^2}$ (see the marginal note). The computation in Example 4 of Section 16.1 showed that $\mathbf{F} = -\nabla \varphi$ where $\varphi = -kr^{-1}$. If $\mathbf{c}(t)$ is a path from a point P on the orbit of the earth to a point Q on the orbit of Mars (Figure 11), the work W performed *against* gravity in moving the earth along **c** is the line integral

$$W = -\int_{\mathbf{c}} \mathbf{F} \cdot d\mathbf{s} = \int_{\mathbf{c}} \nabla \varphi \cdot d\mathbf{s} = \varphi(Q) - \varphi(P) = -kr^{-1}\Big|_P^Q$$

$$= -kr^{-1}\Big|_{1.5\times10^{11}}^{2.28\times10^{11}} = -(7.9\times10^{44})\left(\frac{1}{2.28\times10^{11}} - \frac{1}{1.5\times10^{11}}\right) \approx 1.8\times10^{33} \text{ J} \blacksquare$$

Assumptions Matter A Close Examination of the Vortex Vector Field Theorem 4 states that a vector field **F** is conservative if the cross partials of its components are equal. However, Theorem 4 assumes that the domain $\mathcal{D}$ is simply connected. To see why this assumption is necessary, consider the vortex vector field **F** defined by (Figure 12)

$$\mathbf{F} = \left\langle \frac{-y}{x^2 + y^2}, \frac{x}{x^2 + y^2} \right\rangle$$

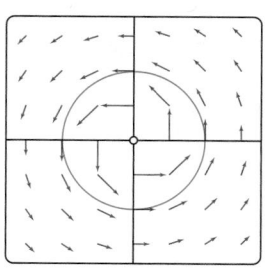

FIGURE 12 The vortex vector field.

This vector field has three properties that seem contradictory.

(i) The cross partials are equal:

$$\frac{\partial}{\partial x}\left(\frac{x}{x^2 + y^2}\right) = \frac{(x^2 + y^2) - x(\partial/\partial x)(x^2 + y^2)}{(x^2 + y^2)^2} = \frac{y^2 - x^2}{(x^2 + y^2)^2}$$

$$\frac{\partial}{\partial y}\left(\frac{-y}{(x^2 + y^2)}\right) = \frac{-(x^2 + y^2) + y(\partial/\partial y)(x^2 + y^2)}{(x^2 + y^2)^2} = \frac{y^2 - x^2}{(x^2 + y^2)^2}$$

(ii) But **F** is not conservative because the line integral around the unit circle $\mathcal{C}$ is not zero. Indeed, using the parametrization $\mathbf{c}(\theta) = (\cos\theta, \sin\theta)$, we have

$$F(\mathbf{c}(\theta)) \cdot \mathbf{c}'(\theta) = \langle -\sin\theta, \cos\theta \rangle \cdot \langle -\sin\theta, \cos\theta \rangle = \sin^2\theta + \cos^2\theta = 1$$

$$\int_{\mathcal{C}} \mathbf{F} \cdot d\mathbf{s} = \int_0^{2\pi} \mathbf{F}(\mathbf{c}(\theta)) \cdot \mathbf{c}'(\theta)\, d\theta = \int_0^{2\pi} d\theta = 2\pi \neq 0 \qquad \boxed{10}$$

(iii) Yet **F** seems to have a potential function, namely $\varphi(x, y) = \tan^{-1}\frac{y}{x}$ (defined for $x \neq 0$). Indeed, the formula

$$\frac{d}{dt}\tan^{-1} t = \frac{1}{1 + t^2}$$

and the Chain Rule give us:

$$\frac{\partial\varphi}{\partial x} = \frac{\partial}{\partial x}\tan^{-1}\frac{y}{x} = \frac{1}{1 + (y/x)^2}\left(-\frac{y}{x^2}\right) = \frac{-y}{x^2 + y^2}$$

$$\frac{\partial\varphi}{\partial y} = \frac{\partial}{\partial y}\tan^{-1}\frac{y}{x} = \frac{1}{1 + (y/x)^2}\left(\frac{1}{x}\right) = \frac{x}{x^2 + y^2}$$

In other words, $\nabla\varphi = \mathbf{F}$.

These contradictions are resolved when we take the domain $\mathcal{D}$ of **F** into account (Figure 13). The vector field **F** is not defined at the origin, so $\mathcal{D} = \{(x, y) \neq (0, 0)\}$. Since this domain is not simply connected, Theorem 4 does not apply and the cross-partial condition does not guarantee that **F** is conservative.

Now observe that the potential function may be interpreted as the angular coordinate of polar coordinates:

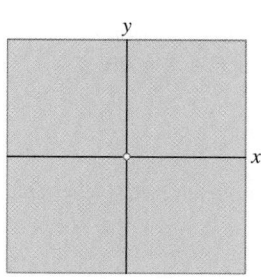

FIGURE 13 The domain $\mathcal{D}$ of the vortex vector **F** is the plane with the origin removed. This domain is not simply connected.

$$\varphi(x, y) = \theta = \tan^{-1}\frac{y}{x}$$

Therefore, the line integral of $\mathbf{F}$ along a curve C from P to Q as in Figure 14(A) is equal to the change in angle along the path:

$$\int_C \mathbf{F} \cdot d\mathbf{s} = \int_C \nabla\theta \cdot d\mathbf{s} = \theta_2 - \theta_1 = \text{the change in angle along } C$$

But recall that the angle θ is only defined up to integer multiples of 2π. If we move along a path that goes all the way around the origin, the angle increases by 2π. This explains why the integral of $\mathbf{F}$ around the unit circle is 2π rather than zero. The key point is that $\varphi = \theta$ may be defined locally, but it cannot be defined as a continuous potential function on the entire domain $\mathcal{D}$ of $\mathbf{F}$. In general, if a closed curve C winds around the origin n times (where n is negative if the curve winds in the clockwise direction), then [see Figures 14(C) and (D)]:

$$\oint \mathbf{F} \cdot d\mathbf{s} = 2\pi n$$

The number n is called the *winding number* of the curve.

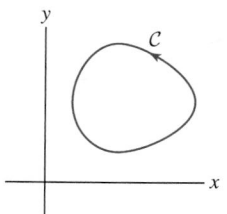

(A) $\int_C \mathbf{F} \cdot d\mathbf{s} = \theta_2 - \theta_1$

(B) $\int_C \mathbf{F} \cdot d\mathbf{s} = \theta_2 - \theta_1 + 2\pi$

(C) C goes around the origin twice, so $\int_C \mathbf{F} \cdot d\mathbf{s} = 4\pi$.

(D) C does not go around the origin, so $\int_C \mathbf{F} \cdot d\mathbf{s} = 0$.

FIGURE 14 The line integral of the vortex vector field $\mathbf{F} = \nabla\theta$ is equal to the change in θ along the curve.

16.3 SUMMARY

- A vector field $\mathbf{F}$ on a domain $\mathcal{D}$ is called *conservative* or *path-independent* if

$$\int_{c_1} \mathbf{F} \cdot d\mathbf{s} = \int_{c_2} \mathbf{F} \cdot d\mathbf{s}$$

for any two paths c_1 and c_2 in $\mathcal{D}$ with the same initial and terminal points.
- A vector field $\mathbf{F}$ on a domain $\mathcal{D}$ is called a *gradient vector field* if there exists a function φ such that $\nabla\varphi = \mathbf{F}$ on $\mathcal{D}$. The function φ is called a *potential function* of $\mathbf{F}$.
- The *Fundamental Theorem for Gradient Vector Fields*: If $\mathbf{F} = \nabla\varphi$, then

$$\int_c \mathbf{F} \cdot d\mathbf{s} = \varphi(Q) - \varphi(P)$$

where c is a path from P to Q in the domain of $\mathbf{F}$. If c is a *closed path* ($P = Q$), then

$$\oint_c \mathbf{F} \cdot d\mathbf{s} = 0$$

In particular, gradient vector fields are conservative.
- Also, the converse is true: For an open connected domain, a conservative vector field is a gradient vector field.

- All conservative vector fields satisfy the cross partials condition

$$\frac{\partial F_1}{\partial y} = \frac{\partial F_2}{\partial x}, \qquad \frac{\partial F_2}{\partial z} = \frac{\partial F_3}{\partial y}, \qquad \frac{\partial F_3}{\partial x} = \frac{\partial F_1}{\partial z} \qquad \boxed{11}$$

- Equality of the cross partials guarantees that $\mathbf{F}$ is conservative if the domain $\mathcal{D}$ is simply connected; that is, if any loop in $\mathcal{D}$ can be shrunk to a point.

16.3 EXERCISES

Preliminary Questions

1. The following statement is false. *If $\mathbf{F}$ is a gradient vector field, then the line integral of $\mathbf{F}$ along every curve is zero.* Which single word must be added to make it true?

2. Which of the following statements are true for all vector fields, which are true only for conservative vector fields?
(a) The line integral along a path from P to Q does not depend on which path is chosen.
(b) The line integral over an oriented curve $\mathcal{C}$ does not depend on how the $\mathcal{C}$ is parametrized.
(c) The line integral around a closed curve is zero.
(d) The line integral changes sign if the orientation is reversed.
(e) The line integral is equal to the difference of a potential function at the two endpoints.
(f) The line integral is equal to the integral of the tangential component along the curve.
(g) The cross partials of the components are equal.

3. Let $\mathbf{F}$ be a vector field on an open, connected domain $\mathcal{D}$. Which of the following statements are always true, and which are true under additional hypotheses on $\mathcal{D}$?

(a) If $\mathbf{F}$ has a potential function, then $\mathbf{F}$ is conservative.
(b) If $\mathbf{F}$ is conservative, then the cross partials of $\mathbf{F}$ are equal.
(c) If the cross partials of $\mathbf{F}$ are equal, then $\mathbf{F}$ is conservative.

4. Let $\mathcal{C}$, $\mathcal{D}$, and $\mathcal{E}$ be the oriented curves in Figure 15 and let $\mathbf{F} = \nabla\varphi$ be a gradient vector field such that $\int_{\mathcal{C}} \mathbf{F} \cdot d\mathbf{s} = 4$. What are the values of the following integrals?

(a) $\displaystyle\int_{\mathcal{D}} \mathbf{F} \cdot d\mathbf{s}$ \qquad **(b)** $\displaystyle\int_{\mathcal{E}} \mathbf{F} \cdot d\mathbf{s}$

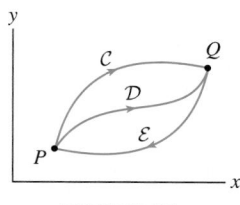

FIGURE 15

Exercises

1. Let $\varphi(x, y, z) = xy\sin(yz)$. Evaluate $\int_{\mathbf{c}} \nabla\varphi \cdot d\mathbf{s}$, where $\mathbf{c}$ is any path from $(0, 0, 0)$ to $(1, 1, \pi)$.

2. Let $\mathbf{F}$ be the gradient of $\varphi(x, y, z) = xy + z^2$. Compute the line integral of $\mathbf{F}$ along:
(a) The line segment from $(1, 1, 1)$ to $(1, 2, 2)$
(b) A circle in the xz-plane
(c) The upper half of a circle of radius 4 with its center $(0, 1, 1)$ in the yz-plane, oriented clockwise

In Exercises 3–8, verify that $\mathbf{F} = \nabla\varphi$ and evaluate the line integral of $\mathbf{F}$ over the given path.

3. $\mathbf{F} = \langle 3, 6y \rangle$, $\varphi(x, y, z) = 3x + 3y^2$; $\mathbf{c}(t) = (t, 2t^{-1})$ for $1 \le t \le 4$

4. $\mathbf{F} = \langle \sin y^2, 2xy\cos y^2 \rangle$, $\varphi(x, y) = x\sin y^2$; line segment from $(2, 0)$ to $(3, \sqrt{\pi})$

5. $\mathbf{F} = \langle xy^2, x^2y \rangle$, $\varphi(x, y) = \frac{1}{2}x^2y^2$; upper half of the unit circle centered at the origin oriented counterclockwise

6. $\mathbf{F} = \langle y + 2xz, x, x^2 \rangle$, $\varphi(x, y, z) = xy + x^2z$; any path from $(1, 2, 1)$ to $(3, 1, 0)$

7. $\mathbf{F} = \langle ye^z, xe^z, xye^z \rangle$, $\varphi(x, y, z) = xye^z$; $\mathbf{c}(t) = (t^2, t^3, t - 1)$ for $1 \le t \le 2$

8. $\mathbf{F} = \left\langle \dfrac{x}{x - y}, \dfrac{z}{y - z}, \ln(x - y) \right\rangle$, $\varphi(x, y, z) = z\ln(x - y)$; ellipse $2x^2 + 3(y - 4)^2 = 12$ in the clockwise direction

9. Find a potential function for $\mathbf{F} = \langle 2xy + 5, x^2 - 4z, -4y \rangle$ and evaluate

$$\int_{\mathbf{c}} \mathbf{F} \cdot d\mathbf{s}$$

where $\mathbf{c}(t) = (t^2, \sin(\pi t), e^{t^2 - 2t})$ for $0 \le t \le 2$.

10. Verify the cross-partials condition and find a potential function for

$$\mathbf{F} = \langle 2xyz^{-1} + yz, x^2z^{-1} + xz, -x^2yz^{-2} + xy \rangle$$

on the domain $\{z \ne 0\}$.

11. Find the line integral of $\mathbf{F} = \langle 2xyz, x^2z, x^2y \rangle$ over any path from $(0, 0, 0)$ to $(3, 2, 1)$.

In Exercises 12–17, determine whether the vector field is conservative and, if so, find a potential function.

12. $\mathbf{F} = \langle z, 1, x \rangle$

13. $\mathbf{F} = \langle 0, x, y \rangle$

14. $\mathbf{F} = \langle y^2, 2xy + e^z, ye^z \rangle$

15. $\mathbf{F} = \langle y, x, z^3 \rangle$

16. $\mathbf{F} = \langle \cos(xz), \sin(yz), xy \sin z \rangle$

17. $\mathbf{F} = \langle \cos z, 2y, -x \sin z \rangle$

18. Calculate the work expended when a particle is moved from O to Q along segments $\overline{OP}$ and $\overline{PQ}$ in Figure 16 in the presence of the force field $\mathbf{F} = \langle x^2, y^2 \rangle$. How much work is expended moving in a complete circuit around the square?

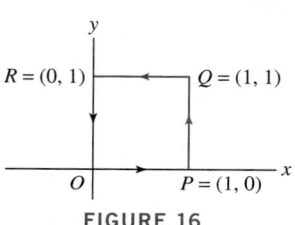

FIGURE 16

19. Let $\mathbf{F} = \langle \dfrac{1}{x}, \dfrac{-1}{y} \rangle$. Calculate the work against F required to move an object from $(1, 1)$ to $(3, 4)$ along any path in the first quadrant.

20. Let $\mathbf{F}$ be the vector field in Figure 17. Let $\overline{PQ}$, $\overline{QR}$, and $\overline{PR}$ be the segments in the figure with the orientations indicated.

(a) Is $\displaystyle\int_{\overline{PR}} \mathbf{F} \cdot d\mathbf{s}$ positive, negative, or zero?

(b) What is the value of $\displaystyle\int_{\overline{QR}} \mathbf{F} \cdot d\mathbf{s}$?

21. 📝   The vector field $\mathbf{F}$ in Figure 17 is horizontal and appears to depend on only the x-coordinate. Suppose that $\mathbf{F} = \langle g(x), 0 \rangle$. Prove that

$$\int_{\overline{PR}} \mathbf{F} \cdot d\mathbf{s} = \int_{\overline{PQ}} \mathbf{F} \cdot d\mathbf{s}$$

by showing that both integrals are equal to $\displaystyle\int_a^b g(x)\, dx$.

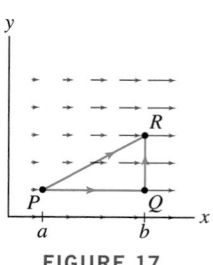

FIGURE 17

22. Let $\mathbf{F} = \langle x^{-1}z^2, y^{-1}z^2, 2z \log(xy) \rangle$.

(a) Verify that $\mathbf{F} = \nabla\varphi$, where $\varphi(x, y, z) = z^2 \log(xy)$.

(b) Evaluate $\displaystyle\int_{\mathbf{c}} \mathbf{F} \cdot d\mathbf{s}$, where $\mathbf{c}(t) = \langle e^t, e^{2t}, t^2 \rangle$ for $1 \leq t \leq 3$.

(c) Evaluate $\displaystyle\int_{\mathbf{c}} \mathbf{F} \cdot d\mathbf{s}$ for any path $\mathbf{c}$ from $P = (\frac{1}{2}, 4, 2)$ to $Q = (2, 3, 3)$ contained in the region $x > 0$, $y > 0$, $z > 0$.

(d) Why was it necessary to specify that the path lie in the region where x, y, and z are positive?

23. How much energy (in joules) does it take to carry a 2-kg object from sea level along any path to the top of a hill that is 1,000 m high? Assume that the force of gravity $\mathbf{F}$ is constant $-mg$ in the vertical direction, where $g = 9.8$ m/s^2. *Hint:* Find a potential function for $\mathbf{F}$.

24. Let $\mathbf{F} = \left\langle \dfrac{-y}{x^2 + y^2}, \dfrac{x}{x^2 + y^2} \right\rangle$ be the vortex vector field. Determine $\displaystyle\int_{\mathbf{c}} \mathbf{F} \cdot d\mathbf{s}$ for each of the paths in Figure 18.

(A)   (B)   (C)

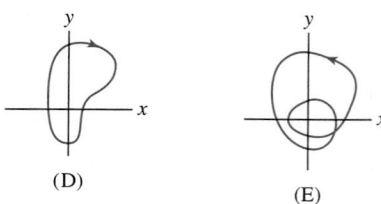

(D)   (E)

FIGURE 18

Further Insights and Challenges

25. The vector field $\mathbf{F} = \left\langle \dfrac{x}{x^2 + y^2}, \dfrac{y}{x^2 + y^2} \right\rangle$ is defined on the domain $\mathcal{D} = \{(x, y) \neq (0, 0)\}$.

(a) Show that $\mathbf{F}$ satisfies the cross-partials condition on $\mathcal{D}$.

(b) Show that $\varphi(x, y) = \frac{1}{2} \ln(x^2 + y^2)$ is a potential function for $\mathbf{F}$.

(c) Is $\mathcal{D}$ simply connected?

(d) Do these results contradict Theorem 4?

16.4 Parametrized Surfaces and Surface Integrals

We now develop the concept of a surface integral, focusing in this section on the scalar case. The first step is to introduce parametrized surfaces which play a role analogous to parametrized curves in line integrals.

A **parametrized surface** S is a surface whose points are described in the form

$$\Phi(u, v) = (x(u, v), y(u, v), z(u, v))$$

where the variables u, v (called parameters) vary in a region $\mathcal{D}$ called the **parameter domain**. Note that two parameters u and v are needed to parametrize a surface because the surface is two-dimensional, whereas just one parameter is needed for a curve. We shall always assume that Φ is **continuously differentiable**, by which we mean that the functions $x(u, v)$, $y(u, v)$, and $z(u, v)$ have continuous partial derivatives.

Figure 1 shows a plot of the surface S with the parametrization

$$\Phi = (u + v, u^3 - v, v^3 - u)$$

This surface consists of all points (x, y, z) in $\mathbf{R}^3$ such that

$$x = u + v, \qquad y = u^3 - v, \qquad z = v^3 - u$$

for (u, v) in $\mathcal{D} = \mathbf{R}^2$.

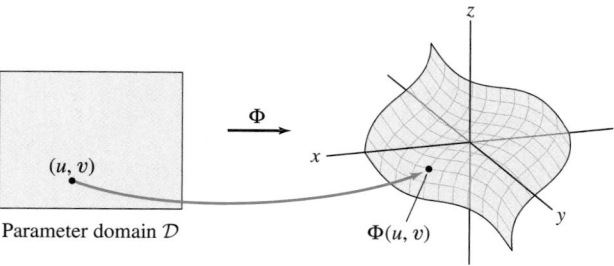

FIGURE 1 The parametric surface $\Phi(u, v) = (u + v, u^3 - v, v^3 - u)$.

Parameter domain $\mathcal{D}$

$\Phi(u, v)$

■ **EXAMPLE 1** Parametrization of a Cone   Find a parametrization of the portion S of the cone with equation $x^2 + y^2 = z^2$ lying above or below the disk $x^2 + y^2 \le 4$. Specify the domain $\mathcal{D}$ of the parametrization.

Solution This surface $x^2 + y^2 = z^2$ is a cone whose horizonal cross section at height $z = u$ is the circle $x^2 + y^2 = u^2$ of radius u (Figure 2). So the coordinates of a point on the cone at height u are of the form $(u \cos v, u \sin v, u)$ for some angle v. Thus, the cone has the parametrization

$$\Phi(u, v) = (u \cos v, u \sin v, u)$$

Since we are interested in the portion of the cone where $x^2 + y^2 = u^2 \le 4$, the height variable u satisfies $-2 \le u \le 2$. The angular variable v varies in the interval $[0, 2\pi)$, and therefore, the parameter domain is $\mathcal{D} = [-2, 2] \times [0, 2\pi)$.   ■

Three standard parametrizations arise often in computations. First, the **cylinder of radius R** with equation $x^2 + y^2 = R^2$ is conveniently parametrized in cylindrical coordinates (Figure 3). The cylinder consists of all points with cylindrical coordinates (R, θ, z), so it makes sense to parametrize the cylinder using θ and z as parameters (with fixed R):

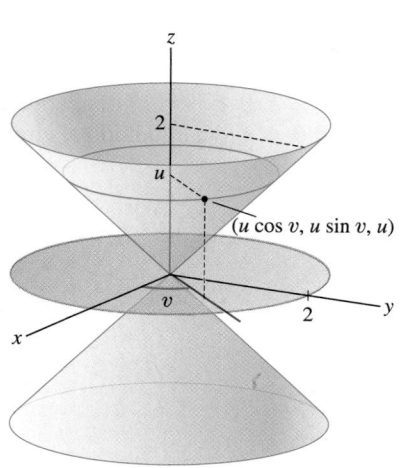

FIGURE 2 The cone $x^2 + y^2 = z^2$.

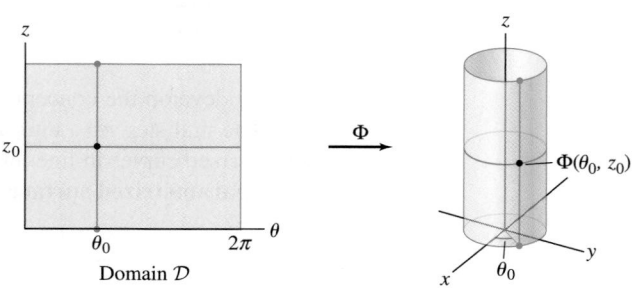

FIGURE 3 The parametrization of a cylinder by cylindrical coordinates amounts to wrapping the rectangle around the cylinder.

Domain $\mathcal{D}$

If necessary, review cylindrical and spherical coordinates in Section 12.7. They are used often in surface calculations.

$$\Phi(\theta, z) = (R\cos\theta,\ R\sin\theta,\ z),\qquad 0 \le \theta < 2\pi,\quad -\infty < z < \infty$$

The sphere of radius R with center at the origin may be parametrized using spherical coordinates (ρ, θ, ϕ) with $\rho = R$ (Figure 4):

$$\Phi(\theta, \phi) = (R\cos\theta\sin\phi,\ R\sin\theta\sin\phi,\ R\cos\phi),\qquad 0 \le \theta < 2\pi,\qquad 0 \le \phi \le \pi$$

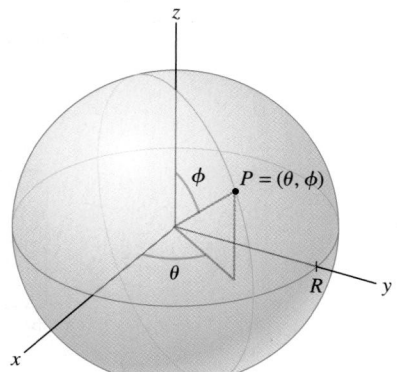

FIGURE 4 Spherical coordinates on a sphere of radius R.

The North and South Poles correspond to $\phi = 0$ and $\phi = \pi$ with any value of θ (the map Φ fails to be one-to-one at the poles):

$$\text{North Pole: } \Phi(\theta, 0) = (0, 0, R),\qquad \text{South Pole: } \Phi(\theta, \pi) = (0, 0, -R)$$

As shown in Figure 5, Φ maps each horizontal segment $\phi = c$ $(0 < c < \pi)$ to a latitude (a circle parallel to the equator) and each vertical segment $\theta = c$ to a longitudinal circle passing through the North and South Poles.

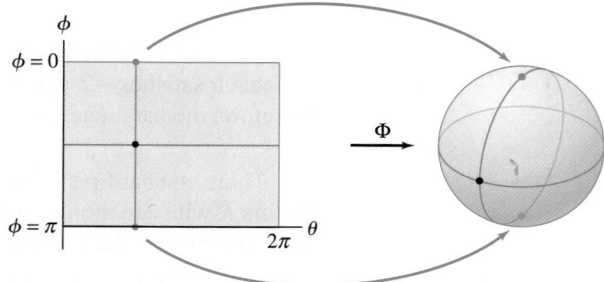

FIGURE 5 The parametrization of the sphere by spherical coordinates amounts to wrapping the rectangle around the sphere. The top and bottom edges of the rectangle are collapsed to the North and South Poles.

Finally, there is a simple way to parametrize the **graph of a function** $z = f(x, y)$ (Figure 6). We set

$$\Phi(x, y) = (x, y, f(x, y))$$

In this case, the parameters are x and y.

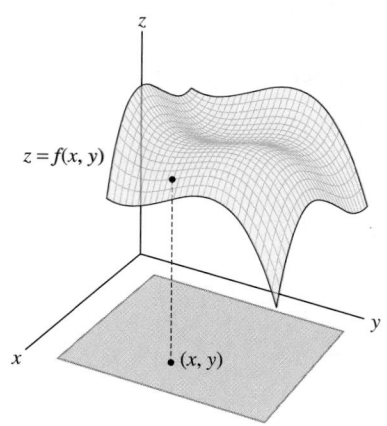

FIGURE 6 Parametrization of the graph $z = f(x, y)$.

Grid Curves, Normal Vectors, and the Tangent Plane

Suppose that a surface $\mathcal{S}$ has a parametrization

$$\Phi(u, v) = (x(u, v), y(u, v), z(u, v))$$

that is one-to-one on a domain $\mathcal{D}$. Then each point P on $\mathcal{S}$ corresponds to a unique pair (u_0, v_0) such that $P = \Phi(u_0, v_0)$ and it is useful to think of (u_0, v_0) as the "coordinates" of P determined by the parametrization. Coordinates of this type are sometimes called **curvilinear coordinates**.

In the uv-plane, we have a grid formed by the families of lines parallel to the coordinates axes. The parametrization maps these grid lines to a system of **grid curves** on the surface (Figure 7). The horizontal line $v = v_0$ maps to the curve $\Phi(u, v_0)$ (the grid curve in the u-direction) and the vertical line $u = u_0$ maps to the curve $\Phi(u_0, v)$ (the grid curve in the v-direction).

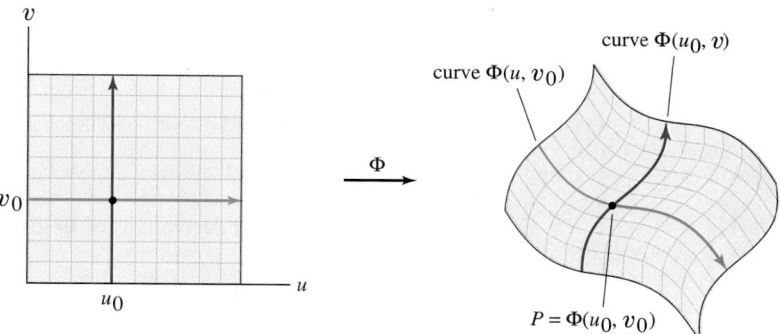

FIGURE 7 Grid curves.

Through each point $P = \Phi(u_0, v_0)$ on $\mathcal{S}$ there pass two grid curves, namely, the curve $\Phi(u, v_0)$ in the u-direction and the curve $\Phi(u_0, v)$ in the v-direction. The tangent vectors to these grid curves are

$$\mathbf{T}_u(P) = \frac{\partial \Phi}{\partial u}(u_0, v_0) = \left\langle \frac{\partial x}{\partial u}(u_0, v_0), \frac{\partial y}{\partial u}(u_0, v_0), \frac{\partial z}{\partial u}(u_0, v_0) \right\rangle$$

$$\mathbf{T}_v(P) = \frac{\partial \Phi}{\partial v}(u_0, v_0) = \left\langle \frac{\partial x}{\partial v}(u_0, v_0), \frac{\partial y}{\partial v}(u_0, v_0), \frac{\partial z}{\partial v}(u_0, v_0) \right\rangle$$

Observe that these vectors are tangent to the surface $\mathcal{S}$ at P because they are tangent to grid curves that lie on $\mathcal{S}$ (Figure 8). Set

$$\mathbf{n}(P) = \mathbf{n}(u_0, v_0) = \mathbf{T}_u(P) \times \mathbf{T}_v(P)$$

If this cross product is nonzero, then it is normal (perpendicular) to $\mathcal{S}$. In this case, $\mathbf{T}_u$ and $\mathbf{T}_v$ span the tangent plane at P and $\mathbf{n}(u, v)$ is a **normal vector** to the tangent plane. We often write $\mathbf{n}$ instead of $\mathbf{n}(u, v)$ or $\mathbf{n}(P)$, but it is understood that the vector $\mathbf{n}$ varies from point to point on the surface. Similarly, we often denote the tangent vectors by $\mathbf{T}_u$ and $\mathbf{T}_v$. The parametrization Φ is called **regular** at P if $\mathbf{n}(u_0, v_0) \neq 0$. Note that $\mathbf{T}_u$, $\mathbf{T}_v$, and $\mathbf{n}$ need not be unit vectors (thus the notation here differs from that in Sections 13.4, 13.5, and 16.2, where $\mathbf{T}$ and $\mathbf{n}$ denote unit vectors).

At each point on a surface, the normal vector points in one of two opposite directions. If we change the parametrization, the length of $\mathbf{n}$ may change and its direction may be reversed.

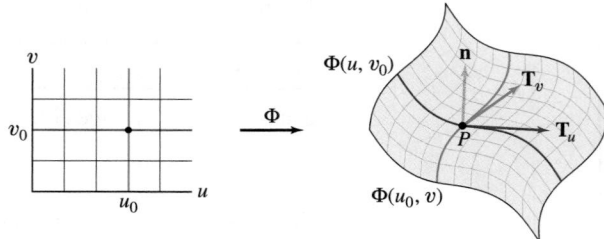

FIGURE 8 The vectors $\mathbf{T}_u$ and $\mathbf{T}_v$ are tangent to the grid curves through $P = \Phi(u_0, v_0)$.

■ **EXAMPLE 2** Consider the parametrization $\Phi(\theta, z) = (2 \cos \theta, 2 \sin \theta, z)$ of the cylinder $x^2 + y^2 = 4$:

(a) Describe the grid curves.

(b) Compute $\mathbf{T}_\theta$, $\mathbf{T}_z$, and $\mathbf{n}(\theta, z)$.

(c) Find an equation of the tangent plane at $P = \Phi(\frac{\pi}{4}, 5)$.

Solution

(a) The grid curves on the cylinder through $P = (\theta_0, z_0)$ are (Figure 9)

$$\theta\text{-grid curve:}\quad (2 \cos \theta, 2 \sin \theta, z_0) = \text{circle of radius 2 at height } z_0$$

$$z\text{-grid curve:}\quad (2 \cos \theta_0, 2 \sin \theta_0, z) = \text{vertical line through } P$$

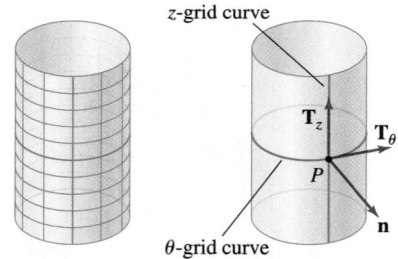

FIGURE 9 Grid curves on the cylinder.

(b) The partial derivatives of Φ give us the tangent vectors at P:

$$\mathbf{T}_\theta = \frac{\partial \Phi}{\partial \theta} = \frac{\partial}{\partial \theta}(2 \cos \theta, 2 \sin \theta, z) = \langle -2 \sin \theta, 2 \cos \theta, 0 \rangle$$

$$\mathbf{T}_z = \frac{\partial \Phi}{\partial z} = \frac{\partial}{\partial z}(2 \cos \theta, 2 \sin \theta, z) = \langle 0, 0, 1 \rangle$$

As shown in Figure 9, $\mathbf{T}_\theta$ is tangent to the θ-grid curve (a circle through P) and $\mathbf{T}_z$ points in the direction of the z-grid curve (the vertical line through P). The normal vector is

$$\mathbf{n}(\theta, z) = \mathbf{T}_\theta \times \mathbf{T}_z = \begin{vmatrix} \mathbf{i} & \mathbf{j} & \mathbf{k} \\ -2 \sin \theta & 2 \cos \theta & 0 \\ 0 & 0 & 1 \end{vmatrix} = 2 \cos \theta \mathbf{i} + 2 \sin \theta \mathbf{j}$$

Observe that $\mathbf{n}$ is horizontal and points directly out of the cylinder.

(c) For $u = \frac{\pi}{4}$, $v = 5$,

$$P = \Phi\left(\frac{\pi}{4}, 5\right) = \langle \sqrt{2}, \sqrt{2}, 5 \rangle, \qquad \mathbf{n} = \mathbf{n}\left(\frac{\pi}{4}, 5\right) = \langle \sqrt{2}, \sqrt{2}, 0 \rangle$$

◀┄ *REMINDER An equation of the plane through $P = (x_0, y_0, z_0)$ with normal vector $\mathbf{n}$ is*

$$\langle x - x_0, y - y_0, z - z_0 \rangle \cdot \mathbf{n} = 0$$

The tangent plane through P has normal vector $\mathbf{n}$ and thus has equation

$$\langle x - \sqrt{2}, y - \sqrt{2}, z - 5 \rangle \cdot \langle \sqrt{2}, \sqrt{2}, 0 \rangle = 0$$

This can be written

$$\sqrt{2}(x - \sqrt{2}) + \sqrt{2}(y - \sqrt{2}) = 0 \qquad \text{or} \qquad x + y = 2\sqrt{2}$$

As we might have expected, the tangent plane is vertical (since z does not appear in the equation). ∎

■ EXAMPLE 3 *CAS* **Helicoid Surface** Describe the surface $\mathcal{S}$ with parametrization

$$\Phi(u, v) = (u \cos v, u \sin v, v), \qquad 0 \le u \le 1, \quad 0 \le v < 2\pi$$

(a) Use a computer algebra system to plot $\mathcal{S}$.
(b) Compute $\mathbf{n}(u, v)$ at $u = \frac{1}{2}, v = \frac{\pi}{2}$.

Solution For each fixed value $u = a$, the curve $\Phi(a, v) = (a \cos v, a \sin v, v)$ is a helix of radius a. Therefore, as u varies from 0 to 1, $\Phi(u, v)$ describes a family of helices of radius u. The resulting surface is a "helical ramp."

(a) A typical command for plotting a parametric surface with a computer algebra system is

```
ParametricPlot3D[{u*Cos[v],u*Sin[v],v},{u,0,1},{v,0,2Pi}]
```

We obtain the plot in Figure 10.

(b) The tangent and normal vectors are

$$\mathbf{T}_u = \frac{\partial \Phi}{\partial u} = \langle \cos v, \sin v, 0 \rangle$$

$$\mathbf{T}_v = \frac{\partial \Phi}{\partial v} = \langle -u \sin v, u \cos v, 1 \rangle$$

$$\mathbf{n}(u, v) = \mathbf{T}_u \times \mathbf{T}_v = \begin{vmatrix} \mathbf{i} & \mathbf{j} & \mathbf{k} \\ \cos v & \sin v & 0 \\ -u \sin v & u \cos v & 1 \end{vmatrix}$$

$$= (\sin v)\mathbf{i} - (\cos v)\mathbf{j} + (u \cos^2 v + u \sin^2 v)\mathbf{k}$$

$$= (\sin v)\mathbf{i} - (\cos v)\mathbf{j} + u\mathbf{k}$$

At $u = \frac{1}{2}, v = \frac{\pi}{2}$, we have $\mathbf{n} = \mathbf{i} + \frac{1}{2}\mathbf{k}$. ∎

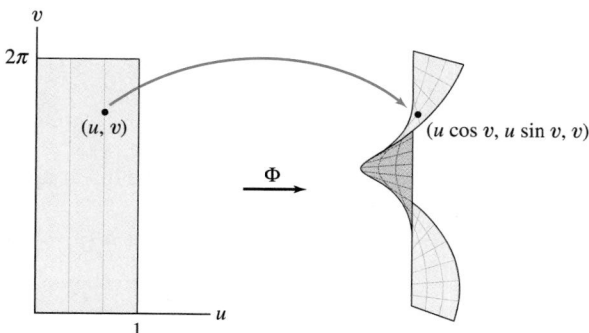

FIGURE 10 Helicoid.

Surface Area

Let $\Phi(u, v)$ be a parametrization of a surface $\mathcal{S}$ with domain $\mathcal{D}$. Since we want Φ to cover the surface $\mathcal{S}$ just once, we assume that Φ is one-to-one, except possibly on the boundary of $\mathcal{D}$. We also assume that Φ is regular, except possibly on the boundary of $\mathcal{D}$. Recall that Φ is regular at (u, v) if the normal vector $\mathbf{n} = \mathbf{T}_u \times \mathbf{T}_v$ is nonzero.

The length $\|\mathbf{n}\|$ of the normal vector has an important interpretation in terms of area. To discuss this interpretation, we assume, for simplicity, that the domain $\mathcal{D}$ is a rectangle but our argument applies to more general domains. Let us divide $\mathcal{D}$ into a grid of small rectangles $\mathcal{R}_{ij}$ of size $\Delta u \times \Delta v$, as in Figure 11. Each rectangle $\mathcal{R}_{ij}$ is mapped by Φ to a "curved" parallelogram $\mathcal{S}_{ij}$ on the surface $\mathcal{S}$. Referring to the figure, suppose that $\mathcal{R}_{ij}$ is the rectangle with vertices $P'Q'R'S'$, where $P' = (u_{ij}, v_{ij})$. Let $\mathcal{S}_{ij}$ be the image $\Phi(\mathcal{R}_{ij})$ of $\mathcal{R}_{ij}$. Let P, Q, R, and S be the images of P', Q', R', and S'. Thus, $P = \Phi(P') = \Phi(u_{ij}, v_{ij})$.

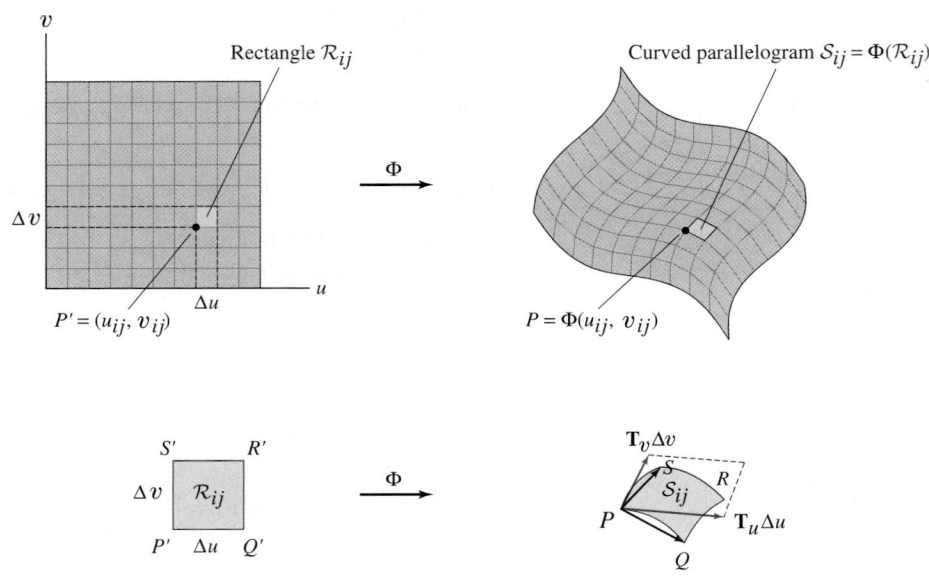

FIGURE 11

Let's examine how the area of $\mathcal{S}_{ij}$ is related to the area of $\mathcal{R}_{ij}$. First, we observe that if Δu and Δv are small, then the "curved" parallelogram $\mathcal{S}_{ij}$ has approximately the same area as the "genuine" parallelogram with sides $\overrightarrow{PQ}$ and $\overrightarrow{PS}$. Since the area of the parallelogram spanned by two vectors is the length of their cross product, we obtain the approximation

$$\text{Area}(\mathcal{S}_{ij}) \approx \text{area of parallelogram spanned by } \overrightarrow{PQ} \text{ and } \overrightarrow{PS}$$

$$= \|\overrightarrow{PQ} \times \overrightarrow{PS}\|$$

Next, we use the linear approximation to estimate the vectors $\overrightarrow{PQ}$ and $\overrightarrow{PS}$:

$$\overrightarrow{PQ} = \Phi(u_{ij} + \Delta u, v_{ij}) - \Phi(u_{ij}, v_{ij}) \approx \frac{\partial \Phi}{\partial u}(u_{ij}, v_{ij})\Delta u = \mathbf{T}_u \Delta u$$

$$\overrightarrow{PS} = \Phi(u_{ij}, v_{ij} + \Delta v) - \Phi(u_{ij}, v_{ij}) \approx \frac{\partial \Phi}{\partial v}(u_{ij}, v_{ij})\Delta v = \mathbf{T}_v \Delta v$$

We obtain

$$\text{Area}(\mathcal{S}_{ij}) \approx \|\overrightarrow{PQ} \times \overrightarrow{PS}\| \approx \|\mathbf{T}_u \Delta u \times \mathbf{T}_v \Delta v\| = \|\mathbf{T}_u \times \mathbf{T}_v\| \, \Delta u \, \Delta v$$

Observing that $\mathbf{n}(u_{ij}, v_{ij}) = \mathbf{T}_u \times \mathbf{T}_v$ and $\text{Area}(\mathcal{R}_{ij}) = \Delta u \Delta v$, we may write our result as follows:

$$\boxed{\text{Area}(\mathcal{S}_{ij}) \approx \|\mathbf{n}(u_{ij}, v_{ij})\| \text{Area}(\mathcal{R}_{ij})} \qquad \boxed{1}$$

The approximation (1) is valid for any small region $\mathcal{R}$ in the uv-plane:

$$\text{Area}(\mathcal{S}) \approx \|\mathbf{n}(u_0, v_0))\| \text{Area}(\mathcal{R})$$

where $\mathcal{S} = \Phi(\mathcal{R})$ and (u_0, v_0) is any sample point in $\mathcal{R}$. Here, "small" means contained in a small disk. We do not allow $\mathcal{R}$ to be very thin and wide.

Thus, $\|\mathbf{n}\|$ *is a distortion factor that tells us how the area of a small rectangle $\mathcal{R}_{ij}$ is altered under the map* Φ.

The entire surface $\mathcal{S}$ is the union of the small patches $\mathcal{S}_{ij}$, so we may approximate the total area of $\mathcal{S}$ by the sum:

$$\text{Area}(\mathcal{S}) = \sum_{ij} \text{Area}(\mathcal{S}_{ij}) \approx \sum_{ij} \|\mathbf{n}(u_{ij}, v_{ij})\| \Delta u \Delta v \qquad \boxed{2}$$

The sum on the right is a Riemann sum for the double integral of $\|\mathbf{n}(u, v)\|$ over the parameter domain $\mathcal{D}$. As Δu and Δv tend to zero, these Riemann sums converge to the following double integral, which we take as the definition of surface area:

$$\text{Area}(\mathcal{S}) = \iint_{\mathcal{D}} \|\mathbf{n}(u, v)\| \, du \, dv$$

A modification of this formula leads to the notion of the **surface integral** of a function $f(x, y, z)$ over $\mathcal{S}$. We consider sums of the form

$$\sum_{i,j} f(P_{ij}) \text{Area}(\mathcal{S}_{ij}) \qquad \boxed{3}$$

where $P_{ij} = \Phi(u_{ij}, v_{ij})$ is a sample point in $\mathcal{S}_{ij}$. If we think of $f(P)$ as a continuous mass density on the surface, then

$$\text{Mass of } \mathcal{S}_{ij} \approx \text{density} \times \text{area} \approx f(P_{ij}) \text{Area}(\mathcal{S}_{ij})$$

and (3) is an approximation to the total mass of the surface. In any case, the limit of the sums as Δu and Δv tend to zero (if it exists) is called the surface integral of f over $\mathcal{S}$; it is denoted

$$\iint_{\mathcal{S}} f(x, y, z) \, dS = \lim_{\Delta u, \Delta v \to 0} \sum_{i,j} f(P_{ij}) \text{Area}(\mathcal{S}_{ij})$$

Applications of surface integrals: If $\mathcal{S}$ is a surface with mass density $\rho(x, y, z)$, then

$$\text{Mass of } \mathcal{S} = \iint_{\mathcal{S}} \rho(x, y, z) \, dS$$

Similarly, if an electric charge is distributed over $\mathcal{S}$ with charge density $\rho(x, y, z)$, then this integral yields the total charge on $\mathcal{S}$.

To evaluate the surface integral, observe that the approximation (1) yields

$$\sum_{ij} f(P_{ij}) \text{Area}(\mathcal{S}_{ij}) \approx \sum_{i,j} f(\Phi(u_{ij}, v_{ij})) \|\mathbf{n}(u_{ij}, v_{ij})\| \, \Delta u \, \Delta v$$

The sum on the right is a Riemann sum for the double integral of

$$f(\Phi(u, v)) \|\mathbf{n}(u, v)\|$$

over the parameter domain $\mathcal{D}$. Under our assumption that Φ is continuously differentiable, we can show the Riemann sums on the right approach the same limit as the sums on the left (the proof is omitted), yielding the formula in the next theorem.

It is interesting to note that Eq. (4) includes the Change of Variables Formula for double integrals (Theorem 1 in Section 15.5) as a special case. If the surface S is a domain in the xy-plane [in other words, $z(u, v) = 0$], then the integral over S reduces to the double integral of the function $f(x, y, 0)$. We may view $\Phi(u, v)$ as a mapping from the uv-plane to the xy-plane, and we find that $\|\mathbf{n}(u, v)\|$ is the Jacobian of this mapping.

THEOREM 1 Surface Integrals and Surface Area   Let $\Phi(u, v)$ be a parametrization of a surface S with parameter domain $\mathcal{D}$. Assume that Φ is continuously differentiable, one-to-one, and regular (except possibly at the boundary of $\mathcal{D}$). Then

$$\iint_S f(x, y, z)\, dS = \iint_{\mathcal{D}} f(\Phi(u, v))\|\mathbf{n}(u, v)\|\, du\, dv \qquad \boxed{4}$$

For $f(x, y, z) = 1$, we obtain the surface area of S:

$$\text{Area}(S) = \iint_{\mathcal{D}} \|\mathbf{n}(u, v)\|\, du\, dv$$

Equation (4) is summarized by the symbolic expression for the "surface element":

$$dS = \|\mathbf{n}(u, v)\|\, du\, dv$$

EXAMPLE 4   Calculate Area(S) and $\iint_S x^2 z\, dS$, where S is the portion of the cone $x^2 + y^2 = z^2$ lying above the disk $x^2 + y^2 \le 4$ (Figure 12).

Solution   A parametrization of the cone was found in Example 1. Using the variables θ and t, this parametrization is

$$\Phi(\theta, t) = (t \cos \theta, t \sin \theta, t), \qquad 0 \le t \le 2, \quad 0 \le \theta < 2\pi$$

Step 1. **Compute the tangent and normal vectors.**

$$\mathbf{T}_\theta = \frac{\partial \Phi}{\partial \theta} = \langle -t \sin \theta, t \cos \theta, 0 \rangle, \qquad \mathbf{T}_t = \frac{\partial \Phi}{\partial t} = \langle \cos \theta, \sin \theta, 1 \rangle$$

$$\mathbf{n} = \mathbf{T}_\theta \times \mathbf{T}_t = \begin{vmatrix} \mathbf{i} & \mathbf{j} & \mathbf{k} \\ -t \sin \theta & t \cos \theta & 0 \\ \cos \theta & \sin \theta & 1 \end{vmatrix}$$

$$= t \cos \theta\, \mathbf{i} + t \sin \theta\, \mathbf{j} - t\, \mathbf{k}$$

The normal vector has length

$$\|\mathbf{n}\| = \sqrt{t^2 \cos^2 \theta + t^2 \sin^2 \theta + (-t)^2} = \sqrt{2t^2} = \sqrt{2}\,|t|$$

Thus, $dS = \sqrt{2}\,|t|\, d\theta\, dt$. In our case, we are integrating over a region where $t \ge 0$, so we may drop the absolute value.

Step 2. **Calculate the surface area.**

$$\text{Area}(S) = \iint_{\mathcal{D}} \|\mathbf{n}\|\, du\, dv = \int_0^2 \int_0^{2\pi} \sqrt{2}\, t\, d\theta\, dt = \sqrt{2}\pi t^2 \Big|_0^2 = 4\sqrt{2}\pi$$

Step 3. **Calculate the surface integral.**
We express $f(x, y, z) = x^2 z$ in terms of the parameters t and θ:

$$f(\Phi(\theta, t)) = f(t \cos \theta, t \sin \theta, t) = (t \cos \theta)^2 t = t^3 \cos^2 \theta$$

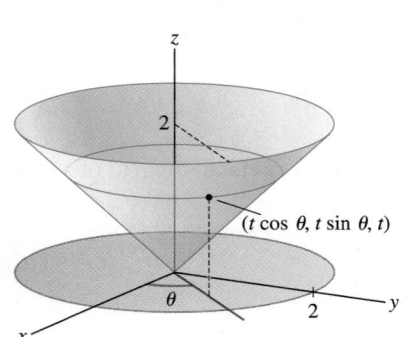

FIGURE 12 Portion S of the cone $x^2 + y^2 = z^2$ lying over the disk $x^2 + y^2 \le 4$.

← ·· REMINDER

$$\int_0^{2\pi} \cos^2 \theta \, d\theta = \int_0^{2\pi} \frac{1 + \cos 2\theta}{2} \, d\theta = \pi$$

Then

$$\iint_S f(x, y, z) \, dS = \int_{t=0}^2 \int_{\theta=0}^{2\pi} f(\Phi(\theta, t)) \, \|\mathbf{n}(\theta, t)\| \, d\theta \, dt$$

$$= \int_{t=0}^2 \int_{\theta=0}^{2\pi} (t^3 \cos^2 \theta)(\sqrt{2}\,t) \, d\theta \, dt$$

$$= \sqrt{2} \left(\int_0^2 t^4 \, dt \right) \left(\int_0^{2\pi} \cos^2 \theta \, d\theta \right)$$

$$= \sqrt{2} \left(\frac{32}{5} \right) (\pi) = \frac{32\sqrt{2}\pi}{5} \qquad \blacksquare$$

For use in the next example and for future reference, we compute the normal vector in the parametrization of the sphere of radius R centered at the origin by spherical coordinates, with outward-pointing normal (Figure 13):

$$\boxed{\Phi(\phi, \theta) = (R \cos \theta \sin \phi, \, R \sin \theta \sin \phi, \, R \cos \phi)}$$

We have

$$\mathbf{T}_\phi = \langle R \cos \theta \cos \phi, \, R \sin \theta \cos \phi, \, -R \sin \phi \rangle$$

$$\mathbf{T}_\theta = \langle -R \sin \theta \sin \phi, \, R \cos \theta \sin \phi, \, 0 \rangle$$

$$\mathbf{n} = \mathbf{T}_\phi \times \mathbf{T}_\theta = \begin{vmatrix} \mathbf{i} & \mathbf{j} & \mathbf{k} \\ R \cos \theta \cos \phi & R \sin \theta \cos \phi & -R \sin \phi \\ -R \sin \theta \sin \phi & R \cos \theta \sin \phi & 0 \end{vmatrix}$$

$$= R^2 \cos \theta \sin^2 \phi \, \mathbf{i} + R^2 \sin \theta \sin^2 \phi \, \mathbf{j} + R^2 \cos \phi \sin \phi \, \mathbf{k}$$

$$= R^2 \sin \phi \, \langle \cos \theta \sin \phi, \, \sin \theta \sin \phi, \, \cos \phi \rangle \qquad \boxed{5}$$

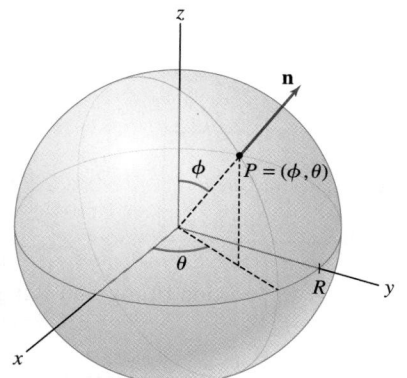

FIGURE 13

Note that the vector $\langle \cos \theta \sin \phi, \, \sin \theta \sin \phi, \, \cos \phi \rangle$ lies on the unit sphere and is, in fact, the radial unit vector $\mathbf{e}_r$ that points in the direction from the origin to $\Phi(\phi, \theta)$. Therefore, for the parametrization by spherical coordinates with outward-pointing normal, we have

$$\boxed{\mathbf{n} = \mathbf{T}_\phi \times \mathbf{T}_\theta = (R^2 \sin \phi) \, \mathbf{e}_r, \qquad \text{and} \qquad \|\mathbf{n}\| = R^2 \sin \phi} \qquad \boxed{6}$$

The normal vector $\mathbf{T}_\theta \times \mathbf{T}_\phi$ is an inward-pointing normal.

■ **EXAMPLE 5** Total Charge on a Surface Find the total charge (in coulombs) on a sphere S of radius 5 cm whose charge density in spherical coordinates is $\rho(\phi, \theta) = 0.003 \cos^2 \phi$ C/cm^2.

Solution Integrals over the sphere are usually easier to evaluate in spherical coordinates, so we parametrize S by

$$\Phi(\phi, \theta) = (5 \cos \theta \sin \phi, \, 5 \sin \theta \sin \phi, \, 5 \cos \phi)$$

By Eq. (6), $\|\mathbf{n}\| = 5^2 \sin \phi$ and

$$\text{Total charge} = \iint_S \rho(\phi, \theta)\, dS = \int_{\theta=0}^{2\pi} \int_{\phi=0}^{\pi} \rho(\phi, \theta)\|\mathbf{n}\|\, d\phi\, d\theta$$

$$= \int_{\theta=0}^{2\pi} \int_{\phi=0}^{\pi} (0.003\cos^2\phi)(25\sin\phi)\, d\phi\, d\theta$$

$$= (0.075)(2\pi) \int_{\phi=0}^{\pi} \cos^2\phi \sin\phi\, d\phi$$

$$= 0.15\pi \left(-\frac{\cos^3\phi}{3} \right) \Bigg|_0^{\pi} = 0.15\pi \left(\frac{2}{3} \right) \approx 0.3 \text{ C} \qquad \blacksquare$$

When a graph $z = g(x, y)$ is parametrized by $\Phi(x, y) = (x, y, g(x, y))$, the tangent vectors are

$$\mathbf{T}_x = (1, 0, g_x), \qquad \mathbf{T}_y = (0, 1, g_y)$$

and the normal vector is

$$\mathbf{n} = \mathbf{T}_x \times \mathbf{T}_y = \begin{vmatrix} \mathbf{i} & \mathbf{j} & \mathbf{k} \\ 1 & 0 & g_x \\ 0 & 1 & g_y \end{vmatrix} = -g_x\mathbf{i} - g_y\mathbf{j} + \mathbf{k} \qquad \boxed{7}$$

Thus, $\|\mathbf{n}\| = \sqrt{1 + g_x^2 + g_y^2}$, and the surface integral over the portion of a graph lying over a domain $\mathcal{D}$ in the xy-plane is

$$\text{Surface integral over a graph} = \iint_{\mathcal{D}} f(x, y, g(x, y))\sqrt{1 + g_x^2 + g_y^2}\, dx\, dy \qquad \boxed{8}$$

■ **EXAMPLE 6** Calculate the surface integral $\iint_S (z - x)\, dS$, where S is the portion of the graph of $z = x + y^2$ where $0 \le x \le y \le 1$ (Figure 14).

Solution Let $z = g(x, y) = x + y^2$. Then $g_x = 1$ and $g_y = 2y$, and

$$dS = \sqrt{1 + g_x^2 + g_y^2}\, dx\, dy = \sqrt{1 + 1 + 4y^2}\, dx\, dy = \sqrt{2 + 4y^2}\, dx\, dy$$

We must express $f(x, y, z) = z - x$ in terms of the parameters x and y:

$$f(x, y, z) = f(x, y, x + y^2) = z - x = (x + y^2) - x = y^2$$

Then we may use Eq. (8) with the domain $0 \le x \le y, 0 \le y \le 1$:

$$\iint_S f(x, y, z)\, dS = \int_{y=0}^{1} \int_{x=0}^{y} y^2 \sqrt{2 + 4y^2}\, dx\, dy$$

$$= \int_{y=0}^{1} \left(y^2 \sqrt{2 + 4y^2} \right) x \Bigg|_{x=0}^{y} dy = \int_0^1 y^3 \sqrt{2 + 4y^2}\, dy$$

Let $u = 2 + 4y^2$, $du = 8y\, dy$. Then $y^2 = \frac{1}{4}(u - 2)$, and

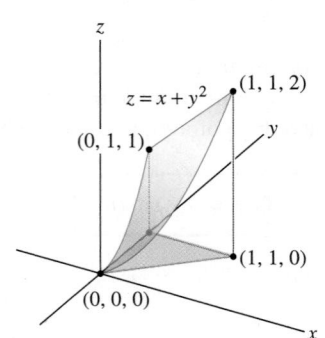

FIGURE 14 The surface $z = x + y^2$ over $0 \le x \le y \le 1$.

$$\int_0^1 y^3 \sqrt{2 + 4y^2}\, dy = \frac{1}{8} \int_2^6 \frac{1}{4}(u-2)\sqrt{u}\, du = \frac{1}{32} \int_2^6 (u^{3/2} - 2u^{1/2})\, du$$

$$= \frac{1}{32}\left(\frac{2}{5}u^{5/2} - \frac{4}{3}u^{3/2}\right)\Big|_2^6 = \frac{1}{30}(6\sqrt{6} + \sqrt{2}) \approx 0.54 \qquad \blacksquare$$

Excursion

The French mathematician Pierre Simon, Marquis de Laplace (1749–1827) showed that the gravitational potential satisfies the Laplace equation $\Delta \varphi = 0$, where Δ is the Laplace operator

$$\Delta \varphi = \frac{\partial^2 \varphi}{\partial x^2} + \frac{\partial^2 \varphi}{\partial y^2} + \frac{\partial^2 \varphi}{\partial z^2}$$

This equation plays an important role in more advanced branches of math and physics.

It is a basic fact of phyics that the gravitational force field $\mathbf{F}$ corresponding to any arrangement of masses may be expressed as a gradient $\mathbf{F} = -\nabla \varphi$ (recall that the minus sign is a convention of physicists, chosen so that positive work is required to achieve a gain in potential energy). We call φ the **gravitational potential**. If a mass m is located at a point Q, then the force on a unit mass at a point P is $\mathbf{F} = -\dfrac{Gm}{r^2}\mathbf{e}_r$, where $\mathbf{e}_r$ is the unit vector pointing from Q to P and $r = |P - Q|$ is the distance from P to Q. We saw in Example 4 of Section 16.1 that $\mathbf{F} = -\nabla \varphi$, where $\varphi(P) = -\dfrac{Gm}{r} = -\dfrac{Gm}{|P - Q|}$. If, instead of a single mass, we have N point masses $m_1, \ldots, m_N$ located at $Q_1, \ldots, Q_N$, then the gravitational potential is the sum

$$\varphi(P) = -G \sum_{i=1}^N \frac{m_i}{|P - Q_i|} \qquad \boxed{9}$$

It is often more realistic to think of mass as distributed continuously in a region according to some density function $\rho(x, y, z)$. If the mass is distributed over a thin surface $\mathcal{S}$, we obtain the potential function by replacing the sum in Eq. (9) with a surface integral:

$$\varphi(P) = -G \iint_{\mathcal{S}} \frac{\rho(x, y, z)\, dS}{|P - Q|} = -G \iint_{\mathcal{S}} \frac{\rho(x, y, z)\, dS}{\sqrt{(x - a)^2 + (y - b)^2 + (z - c)^2}} \qquad \boxed{10}$$

where $P = (a, b, c)$. This surface integral is hard or impossible to evaluate explicitly unless the surface and mass distribution are sufficiently symmetrical. A basic case is that of a hollow sphere of uniform mass density (Figure 15).

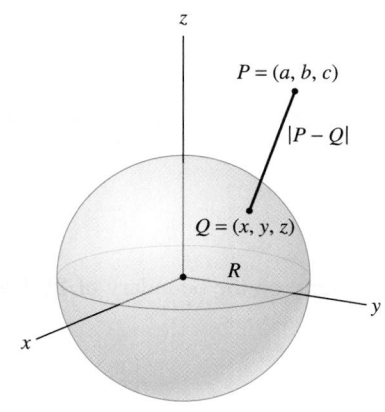

FIGURE 15

THEOREM 2 Gravitational Potential of a Uniform Hollow Sphere The gravitation potential φ due to a hollow sphere of radius R with uniform mass distribution of total mass m at a point P located at a distance r from the center is

$$\varphi(P) = \begin{cases} \dfrac{-Gm}{r} & \text{if } r > R \quad (P \text{ outside the sphere}) \\[2mm] \dfrac{-Gm}{R} & \text{if } r < R \quad (P \text{ inside the sphere}) \end{cases} \qquad \boxed{11}$$

We leave this calculation as an exercise (Exercise 50). We will prove it again with much less calculation using Gauss's Law in Section 17.3.

In his magnum opus, *Principia Mathematica*, Isaac Newton proved that a sphere of uniform mass density (whether hollow or solid) attracts a particle outside the sphere as if the entire mass were concentrated at the center. In other words, a uniform sphere behaves like a point mass as far as gravity is concerned. Furthermore, if the sphere is hollow, then the sphere exerts no gravitational force on a particle inside it. This result follows immediately from Eq. (11). Indeed, outside the sphere, φ has the same formula as the potential due to a point mass. Inside the sphere, the potential is *constant* with a value of $-Gm/R$. But constant potential means zero force because the force is the gradient of the potential.

This discussion applies equally well to the electrostatic force between charged objects because this force also obeys an inverse square law (Coulomb's Law). Thus, for example, a uniformly charged sphere behaves like a point charge (when seen from outside the sphere).

16.4 SUMMARY

- A *parametrized surface* is a surface S whose points are described in the form $\Phi(u, v) = (x(u, v), y(u, v), z(u, v))$, where the *parameters* u and v vary in a domain $\mathcal{D}$ in the uv-plane. We assume that Φ is continuously differentiable.
- At each point on S, we define the tangent and normal vectors

$$\mathbf{T}_u = \frac{\partial \Phi}{\partial u} = \left\langle \frac{\partial x}{\partial u}, \frac{\partial y}{\partial u}, \frac{\partial z}{\partial u} \right\rangle, \qquad \mathbf{T}_v = \frac{\partial \Phi}{\partial v} = \left\langle \frac{\partial x}{\partial v}, \frac{\partial y}{\partial v}, \frac{\partial z}{\partial v} \right\rangle$$

$$\mathbf{n} = \mathbf{n}(u, v) = \mathbf{T}_u \times \mathbf{T}_v$$

The parametrization is called *regular* at (u, v) if $\mathbf{n}(u, v) \neq 0$.
- The quantity $\|\mathbf{n}\|$ may be thought of as a "distortion factor." If $\mathcal{D}$ is a small region in the uv-plane and $S = \Phi(\mathcal{D})$, then

$$\text{Area}(S) \approx \|\mathbf{n}(u_0, v_0)\| \text{Area}(\mathcal{D})$$

where (u_0, v_0) is any sample point in $\mathcal{D}$.
- Assume that Φ is one-to-one and regular, except possibly on the boundary of $\mathcal{D}$. The surface area and surface integral are given by the formulas

$$\text{Area}(S) = \iint_{\mathcal{D}} \|\mathbf{n}(u, v)\| \, du \, dv$$

$$\iint_S f(x, y, z) \, dS = \iint_{\mathcal{D}} f(\Phi(u, v)) \, \|\mathbf{n}(u, v)\| \, du \, dv$$

- Some standard parametrizations:

  - Cylinder of radius R (with the z-axis as central axis) with outward-pointing normal:

$$\Phi(\theta, z) = (R \cos \theta, R \sin \theta, z)$$

$$\mathbf{n} = \mathbf{T}_\theta \times \mathbf{T}_z = R \langle \cos \theta, \sin \theta, 0 \rangle$$

$$dS = \|\mathbf{n}\| \, d\theta \, dz = R \, d\theta \, dz$$

  - Sphere of radius R centered at the origin with outward-pointing normal:

$$\Phi(\phi, \theta) = (R \cos \theta \sin \phi, R \sin \theta \sin \phi, R \cos \phi)$$

$$\mathbf{n} = \mathbf{T}_\phi \times \mathbf{T}_\theta = R^2 \sin \phi \langle \cos \theta \sin \phi, \sin \theta \sin \phi, \cos \phi \rangle = (R^2 \sin \phi) \, \mathbf{e}_r$$

$$dS = \|\mathbf{n}\| \, d\phi \, d\theta = R^2 \sin \phi \, d\phi \, d\theta$$

– Graph of $z = g(x, y)$:

$$\Phi(x, y) = (x, y, g(x, y))$$

$$\mathbf{n} = \mathbf{T}_x \times \mathbf{T}_y = \langle -g_x, -g_y, 1 \rangle$$

$$dS = \|\mathbf{n}\| \, dx \, dy = \sqrt{1 + g_x^2 + g_y^2} \, dx \, dy$$

16.4 EXERCISES

Preliminary Questions

1. What is the surface integral of the function $f(x, y, z) = 10$ over a surface of total area 5?

2. What interpretation can we give to the length $\|\mathbf{n}\|$ of the normal vector for a parametrization $\Phi(u, v)$?

3. A parametrization maps a rectangle of size 0.01×0.02 in the uv-plane onto a small patch $\mathcal{S}$ of a surface. Estimate Area($\mathcal{S}$) if $\mathbf{T}_u \times \mathbf{T}_v = \langle 1, 2, 2 \rangle$ at a sample point in the rectangle.

4. A small surface $\mathcal{S}$ is divided into three small pieces, each of area 0.2. Estimate $\iint_{\mathcal{S}} f(x, y, z) \, dS$ if $f(x, y, z)$ takes the values 0.9, 1, and 1.1 at sample points in these three pieces.

5. A surface $\mathcal{S}$ has a parametrization whose domain is the square $0 \le u, v \le 2$ such that $\|\mathbf{n}(u, v)\| = 5$ for all (u, v). What is Area($\mathcal{S}$)?

6. What is the outward-pointing unit normal to the sphere of radius 3 centered at the origin at $P = (2, 2, 1)$?

Exercises

1. Match the parametrization with the surface in Figure 16.

(a) $(u, \cos v, \sin v)$ **(b)** $(u, u + v, v)$

(c) (u, v^3, v)

(d) $(\cos u \sin v, 3 \cos u \sin v, \cos v)$

(e) $(u, u(2 + \cos v), u(2 + \sin v))$

(i)   (ii)   (iii)

(iv)   (v)

FIGURE 16

2. Show that $\Phi(r, \theta) = (r \cos \theta, r \sin \theta, 1 - r^2)$ parametrizes the paraboloid $z = 1 - x^2 - y^2$. Describe the grid curves of this parametrization.

3. Show that $\Phi(u, v) = (2u + 1, u - v, 3u + v)$ parametrizes the plane $2x - y - z = 2$. Then:

(a) Calculate $\mathbf{T}_u$, $\mathbf{T}_v$, and $\mathbf{n}(u, v)$.

(b) Find the area of $\mathcal{S} = \Phi(\mathcal{D})$, where $\mathcal{D} = \{(u, v) : 0 \le u \le 2, 0 \le v \le 1\}$.

(c) Express $f(x, y, z) = yz$ in terms of u and v and evaluate $\iint_{\mathcal{S}} f(x, y, z) \, dS$.

4. Let $\mathcal{S} = \Phi(\mathcal{D})$, where $\mathcal{D} = \{(u, v) : u^2 + v^2 \le 1, u \ge 0, v \ge 0\}$ and Φ is as defined in Exercise 3.

(a) Calculate the surface area of $\mathcal{S}$.

(b) Evaluate $\iint_{\mathcal{S}} (x - y) \, dS$. *Hint:* Use polar coordinates.

5. Let $\Phi(x, y) = (x, y, xy)$.

(a) Calculate $\mathbf{T}_x$, $\mathbf{T}_y$, and $\mathbf{n}(x, y)$.

(b) Let S be the part of the surface with parameter domain $\mathcal{D} = \{(x, y) : x^2 + y^2 \le 1, x \ge 0, y \ge 0\}$. Verify the following formula and evaluate using polar coordinates:

$$\iint_{S} 1 \, dS = \iint_{\mathcal{D}} \sqrt{1 + x^2 + y^2} \, dx \, dy$$

(c) Verify the following formula and evaluate:

$$\iint_{S} z \, dS = \int_{0}^{\pi/2} \int_{0}^{1} (\sin \theta \cos \theta) r^3 \sqrt{1 + r^2} \, dr \, d\theta$$

6. A surface $\mathcal{S}$ has a parametrization $\Phi(u, v)$ whose domain $\mathcal{D}$ is the square in Figure 17. Suppose that Φ has the following normal vectors:

$$\mathbf{n}(A) = \langle 2, 10 \rangle, \quad \mathbf{n}(B) = \langle 1, 3, 0 \rangle$$

$$\mathbf{n}(C) = \langle 3, 0, 1 \rangle, \quad \mathbf{n}(D) = \langle 2, 0, 1 \rangle$$

Estimate $\int_S f(x, y, z)\,dS$, where f is a function such that $f(\Phi(u, v)) = u + v$.

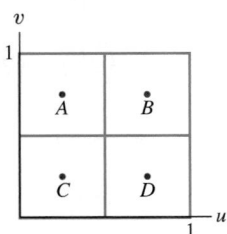

v

FIGURE 17

In Exercises 7–10, calculate $\mathbf{T}_u$, $\mathbf{T}_v$, and $\mathbf{n}(u, v)$ for the parametrized surface at the given point. Then find the equation of the tangent plane to the surface at that point.

7. $\Phi(u, v) = (2u + v, u - 4v, 3u)$; $u = 1,\quad v = 4$

8. $\Phi(u, v) = (u^2 - v^2, u + v, u - v)$; $u = 2,\quad v = 3$

9. $\Phi(\theta, \phi) = (\cos\theta\sin\phi, \sin\theta\sin\phi, \cos\phi)$; $\theta = \frac{\pi}{2}$, $\phi = \frac{\pi}{4}$

10. $\Phi(r, \theta) = (r\cos\theta, r\sin\theta, 1 - r^2)$; $r = \frac{1}{2}$, $\theta = \frac{\pi}{4}$

11. Use the normal vector computed in Exercise 8 to estimate the area of the small patch of the surface $\Phi(u, v) = (u^2 - v^2, u + v, u - v)$ defined by

$$2 \le u \le 2.1,\qquad 3 \le v \le 3.2$$

12. Sketch the small patch of the sphere whose spherical coordinates satisfy

$$\frac{\pi}{2} - 0.15 \le \theta \le \frac{\pi}{2} + 0.15,\qquad \frac{\pi}{4} - 0.1 \le \phi \le \frac{\pi}{4} + 0.1$$

Use the normal vector computed in Exercise 9 to estimate its area.

13. A surface S has a parametrization $\Phi(u, v)$ with domain $0 \le u \le 2, 0 \le v \le 4$ such that the following partial derivatives are constant:

$$\frac{\partial\Phi}{\partial u} = \langle 2, 0, 1\rangle,\qquad \frac{\partial\Phi}{\partial v} = \langle 4, 0, 3\rangle$$

What is the surface area of S?

14. Show that the ellipsoid

$$\left(\frac{x}{a}\right)^2 + \left(\frac{y}{b}\right)^2 + \left(\frac{z}{c}\right)^2 = 1$$

is parametrized by

$$\Phi(\phi, \theta) = (a\cos\theta\sin\phi, b\sin\theta\sin\phi, c\cos\phi)$$

Express the surface area of the ellipsoid as an integral but do not attempt to evaluate it.

15. $\boxed{CAS}$ Let S be the surface with parametrization

$$\Phi(u, v) = \big((3 + \sin v)\cos u, (3 + \sin v)\sin u, v\big)$$

for $0 \le u, v \le 2\pi$. Using a computer algebra system:
(a) Plot S from several different viewpoints. Is S best described as a "vase that holds water" or a "bottomless vase"?
(b) Calculate the normal vector $\mathbf{n}(u, v)$.
(c) Calculate the surface area of S to four decimal places.

16. $\boxed{CAS}$ Let S be the surface $z = \ln(5 - x^2 - y^2)$ for $0 \le x, y \le 1$. Using a computer algebra system:
(a) Calculate the surface area of S to four decimal places.
(b) Calculate $\iint_S x^2 y^3\,dS$ to four decimal places.

17. Use spherical coordinates to compute the surface area of a sphere of radius R.

18. Compute the integral of z over the upper hemisphere of a sphere of radius R centered at the origin.

19. Compute the integral of x^2 over the octant of the unit sphere centered at the origin, where $x, y, z \ge 0$.

20. Show that the hemisphere $x^2 + y^2 + z^2 = R^2$, $z \ge 0$ is parametrized by

$$\Phi(r, \theta) = (r\cos\theta, r\sin\theta, \sqrt{R^2 - r^2})$$

for $0 \le \theta \le 2\pi, 0 \le r \le R$. Compute the surface area of the hemisphere using this parametrization.

In Exercises 21–32, calculate $\iint_S f(x, y, z)\,dS$ for the given surface and function.

21. $\Phi(u, v) = (u\cos v, u\sin v, u)$, $0 \le u, v \le 1$; $f(x, y, z) = z(x^2 + y^2)$

22. $\Phi(r, \theta) = (r\cos\theta, r\sin\theta, \theta)$, $0 \le r \le 1,\ 0 \le \theta \le 2\pi$; $f(x, y, z) = \sqrt{x^2 + y^2}$

23. $x^2 + y^2 = 4$, $0 \le z \le 4$; $f(x, y, z) = e^{-z}$

24. $z = 4 - x^2 - y^2$, $z \ge 0$; $f(x, y, z) = z$

25. $z = 4 - x^2 - y^2$, $z \ge 0$; $f(x, y, z) = z(x^2 + y^2)$

26. $y = 9 - z^2$, $0 \le x \le z \le 3$; $f(x, y, z) = 1$

27. $y = 9 - z^2$, $0 \le x, z \le 3$; $f(x, y, z) = z$

28. $\Phi(u, v) = (u, v^3, u + v)$, $0 \le u, v \le 1$; $f(x, y, z) = y$

29. Part of the plane $x + y + z = 1$, where $x, y, z \ge 0$; $f(x, y, z) = z$

30. Region in the plane $x + y + z = 0$ contained in the cylinder $x^2 + y^2 = 1$; $f(x, y, z) = xz$

31. Part of the surface $x = z^3$, where $0 \le x, y \le 1$; $f(x, y, z) = x$

32. Part of the unit sphere centered at the origin, where $x \ge 0$ and $|y| \le x$; $f(x, y, z) = x$

33. Let S be the sphere of radius R centered at the origin. Explain the following equalities using symmetry:

(a) $\iint_S x \, dS = \iint_S y \, dS = \iint_S z \, dS = 0$

(b) $\iint_S x^2 \, dS = \iint_S y^2 \, dS = \iint_S z^2 \, dS$

Then show, by adding the three integrals in part (b), that $\iint_S x^2 \, dS = \frac{4}{3}\pi R^4$.

34. Calculate $\iint_S (xy + e^z) \, dS$, where S is the triangle in Figure 18 with vertices $(0, 0, 3)$, $(1, 0, 2)$, and $(0, 4, 1)$. *Hint:* Find the equation of the plane containing the triangle.

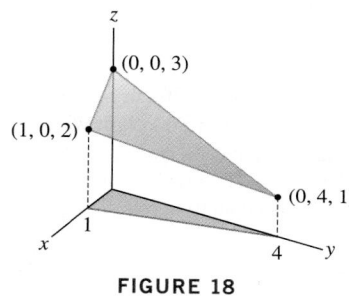

FIGURE 18

35. Find the area of the portion of the plane $2x + 3y + 4z = 28$ lying above the rectangle $1 \le x \le 3, 2 \le y \le 5$ in the xy-plane.

36. What is the area of the portion of the plane $2x + 3y + 4z = 28$ lying above the domain $\mathcal{D}$ in the xy-plane in Figure 19 if Area$(\mathcal{D}) = 5$?

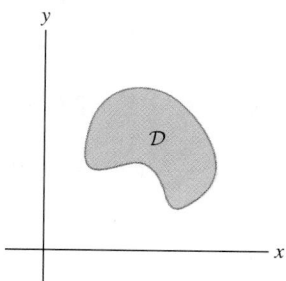

FIGURE 19

37. Compute the integral of $f(x, y, z) = z^2 (x^2 + y^2 + z^2)^{-1}$ over the cap of the sphere $x^2 + y^2 + z^2 = 4$ defined by $z \ge 1$.

38. Calculate $\iint_S f(x, y, z) = z(x^2 + y^2) \, dS$, where S is the hemisphere $x^2 + y^2 + z^2 = R^2, z \ge 0$.

39. Let S be the portion of the sphere $x^2 + y^2 + z^2 = 9$, where $1 \le x^2 + y^2 \le 4$ and $z \ge 0$ (Figure 20). Find a parametrization of S in polar coordinates and use it to compute:

(a) The area of S

(b) $\iint_S z^{-1} \, dS$

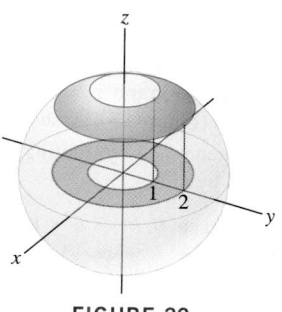

FIGURE 20

40. Find the surface area of the part of the cone $x^2 + y^2 = z^2$ between the planes $z = 2$ and $z = 5$.

41. Find the surface area of the portion S of the cone $z^2 = x^2 + y^2$, where $z \ge 0$, contained within the cylinder $y^2 + z^2 \le 1$.

42. Find the integral of e^{2x+yz} over the following faces of the cube of side 2 in Figure 21.

(a) The top face   **(b)** The face $PQRS$

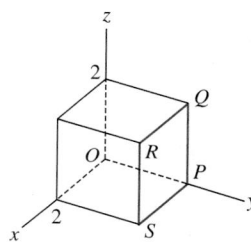

FIGURE 21

43. Prove a famous result of Archimedes: The surface area of the portion of the sphere of radius r between two horizontal planes $z = a$ and $z = b$ is equal to the surface area of the corresponding portion of the circumscribed cylinder (Figure 22).

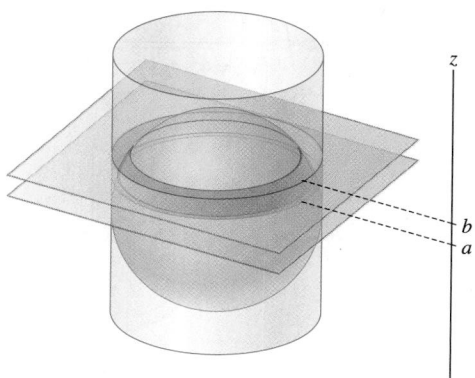

FIGURE 22

Further Insights and Challenges

44. Surfaces of Revolution Let S be the surface formed by revolving the region underneath the graph $z = g(y)$ in the yz-plane for $c \leq y \leq d$ about the z-axis (Figure 23). Assume that $c \geq 0$.

(a) Show that the circle generated by rotating a point $(0, a, b)$ about the z-axis is parametrized by

$$(a \cos \theta, a \sin \theta, b), \quad 0 \leq \theta \leq 2\pi$$

(b) Show that S is parametrized by

$$\Phi(y, \theta) = (y \cos \theta, y \sin \theta, g(y)) \qquad \boxed{12}$$

for $c \leq y \leq d, 0 \leq \theta \leq 2\pi$.

(c) Show that

$$\text{Area}(S) = 2\pi \int_c^d y \sqrt{1 + g'(y)^2} \, dy \qquad \boxed{13}$$

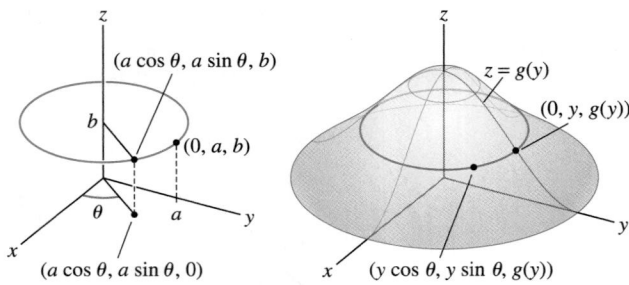

FIGURE 23

45. Use Eq. (13) to compute the surface area of $z = 4 - y^2$ for $0 \leq y \leq 2$ rotated about the z-axis.

46. Describe the upper half of the cone $x^2 + y^2 = z^2$ for $0 \leq z \leq d$ as a surface of revolution (Figure 2) and use Eq. (13) to compute its surface area.

47. Area of a Torus Let T be the torus obtained by rotating the circle in the yz-plane of radius a centered at $(0, b, 0)$ about the z-axis (Figure 24). We assume that $b > a > 0$.

(a) Use Eq. (13) to show that

$$\text{Area}(T) = 4\pi \int_{b-a}^{b+a} \frac{ay}{\sqrt{a^2 - (b - y)^2}} \, dy$$

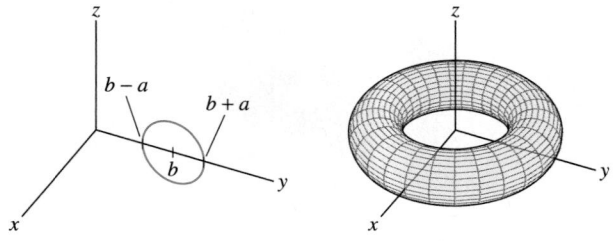

FIGURE 24 The torus obtained by rotating a circle of radius a.

(b) Show that $\text{Area}(T) = 4\pi^2 ab$. *Hint:* Rewrite the integral using substitution.

48. Pappus's Theorem Also called **Guldin's Rule**, Pappus's Theorem states that the area of a surface of revolution S is equal to the length L of the generating curve times the distance traversed by the center of mass. Use Eq. (13) to prove Pappus's Theorem. If C is the graph $z = g(y)$ for $c \leq y \leq d$, then the center of mass is defined as the point $(\bar{y}, \bar{z})$ with

$$\bar{y} = \frac{1}{L} \int_C y \, ds, \qquad \bar{z} = \frac{1}{L} \int_C z \, ds$$

49. Compute the surface area of the torus in Exercise 47 using Pappus's Theorem.

50. ✏ **Potential due to a Uniform Sphere** Let S be a hollow sphere of radius R with center at the origin with a uniform mass distribution of total mass m [since S has surface area $4\pi R^2$, the mass density is $\rho = m/(4\pi R^2)$]. The gravitational potential $\varphi(P)$ due to S at a point $P = (a, b, c)$ is equal to

$$-G \iint_S \frac{\rho \, dS}{\sqrt{(x - a)^2 + (y - b)^2 + (z - c)^2}}$$

(a) Use symmetry to conclude that the potential depends only on the distance r from P to the center of the sphere. Therefore, it suffices to compute $\varphi(P)$ for a point $P = (0, 0, r)$ on the z-axis (with $r \neq R$).

(b) Use spherical coordinates to show that $\varphi(0, 0, r)$ is equal to

$$\frac{-Gm}{4\pi} \int_0^\pi \int_0^{2\pi} \frac{\sin \phi \, d\theta \, d\phi}{\sqrt{R^2 + r^2 - 2Rr \cos \phi}}$$

(c) Use the substitution $u = R^2 + r^2 - 2Rr \cos \phi$ to show that

$$\varphi(0, 0, r) = \frac{-mG}{2Rr} \left(|R + r| - |R - r| \right)$$

(d) Verify formula (11) for φ.

51. Calculate the gravitational potential φ for a hemisphere of radius R with uniform mass distribution.

52. The surface of a cylinder of radius R and length L has a uniform mass distribution ρ (the top and bottom of the cylinder are excluded). Use Eq. (10) to find the gravitational potential at a point P located along the axis of the cylinder.

53. Let S be the part of the graph $z = g(x, y)$ lying over a domain $\mathcal{D}$ in the xy-plane. Let $\phi = \phi(x, y)$ be the angle between the normal to S and the vertical. Prove the formula

$$\text{Area}(S) = \iint_\mathcal{D} \frac{dA}{|\cos \phi|}$$

16.5 Surface Integrals of Vector Fields

In this section, we study integrals of vector fields over surfaces. These integrals represent flux, which arises often in physics and engineering as the rate at which a quantity flows across a surface.

As a preliminary step, it is necessary to introduce the concept of an **oriented surface.** An *orientation* of a surface S is a continuously varying choice of unit normal vector $\mathbf{e_n}(P)$ at each point P on the surface (Figure 1). Since there are two normal directions at each point on S, an orientation serves to specify one of the two "sides" of the surface. For example, a sphere has two orientations, depending on whether the unit normal vector points to the outside or to the inside of the sphere. If $\mathbf{e_n}(P)$ is the unit normal vector determined by an orientation, then $-\mathbf{e_n}(P)$ is the unit normal vector determined by the *opposite orientation.*

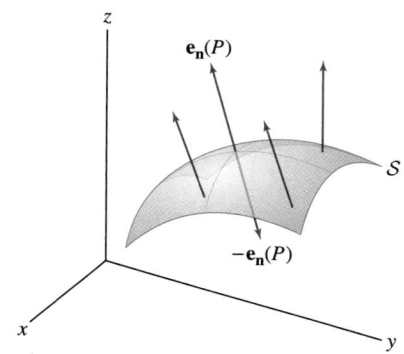

FIGURE 1 Two possible unit normal vectors. An orientation is a continuously varying choice of normal direction at each point.

The **normal component** of a vector field $\mathbf{F}$ at a point P on an oriented surface is the dot product $\mathbf{F}(P) \cdot \mathbf{e_n}(P) = \|\mathbf{F}(P)\| \cos \theta$, where θ is the angle between $\mathbf{F}(P)$ and $\mathbf{e_n}(P)$ (Figure 2). We usually write $\mathbf{e_n}$ instead of $\mathbf{e_n}(P)$, but it is understood that $\mathbf{e_n}$ varies from point to point on the surface. The surface integral of $\mathbf{F}$ is defined as the integral of this normal component:

$$\text{Vector surface integral:} \quad \iint_S \mathbf{F} \cdot d\mathbf{S} = \iint_S (\mathbf{F} \cdot \mathbf{e_n}) \, dS$$

The surface integral of $\mathbf{F}$ is also called the **flux** of $\mathbf{F}$ across S.

As in the scalar case, we use a parametrization $\Phi(u, v)$ of S to evaluate the surface integral. Assume that Φ is regular, so that $\mathbf{n} \neq 0$, and define a unit normal vector:

$$\mathbf{e_n} = \mathbf{e_n}(u, v) = \frac{\mathbf{n}}{\|\mathbf{n}\|}$$

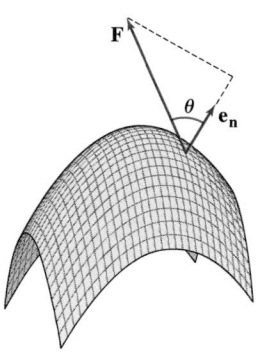

FIGURE 2 The normal component of a vector to a surface.

Compare Eq. (1) with the formula for a scalar surface integral:

$$\iint_S f(x, y, z) \, dS$$

$$= \iint f(\Phi(u, v)) \|\mathbf{n}(u, v)\| \, du \, dv$$

Orientation plays no role in a scalar surface integral because the integral involves the length $\|\mathbf{n}\|$ but not the direction of $\mathbf{n}$.

We assume that Φ parametrizes the oriented surface S, that is, we assume that $\mathbf{e_n}$ points in the normal direction specified by the orientation of S. In this case,

$$\iint_S \mathbf{F} \cdot d\mathbf{S} = \iint_D (\mathbf{F} \cdot \mathbf{e_n}) \|\mathbf{n}(u, v)\| \, du \, dv$$

$$= \iint_D \mathbf{F}(\Phi(u, v)) \cdot \left(\frac{\mathbf{n}(u, v)}{\|\mathbf{n}(u, v)\|} \right) \|\mathbf{n}(u, v)\| \, du \, dv$$

$$= \iint_D \mathbf{F}(\Phi(u, v)) \cdot \mathbf{n}(u, v) \, du \, dv \qquad \boxed{1}$$

This formula remains valid even if $\mathbf{n}(u, v)$ is zero at points on the boundary of the parameter domain. If we reverse the orientation of S in a vector surface integral, $\mathbf{n}(u, v)$ is replaced by $-\mathbf{n}(u, v)$ and the integral changes sign.

> **THEOREM 1 Vector Surface Integral** Let $\Phi(u, v)$ be a parametrization of an oriented surface S with parameter domain $\mathcal{D}$. Assume that Φ is one-to-one and regular, except possibly at points on the boundary of $\mathcal{D}$. Let $\mathbf{n}(u, v)$ be the normal vector defined by Φ. Then
>
> $$\iint_S \mathbf{F} \cdot d\mathbf{S} = \iint_{\mathcal{D}} \mathbf{F}(\Phi(u, v)) \cdot \mathbf{n}(u, v)\, du\, dv \qquad \boxed{2}$$
>
> If the orientation of S is reversed, the surface integral changes sign.

We may interpret the symbol $d\mathbf{S}$ as a "vector surface element" that is expressed in terms of a parametrization by

$$d\mathbf{S} = \mathbf{n}(u, v)\, du\, dv$$

■ **EXAMPLE 1** Calculate $\iint_S \mathbf{F} \cdot d\mathbf{S}$, where $\mathbf{F} = \langle 0, 0, x \rangle$ and S is the surface with parametrization

$$\Phi(u, v) = (u^2, v, u^3 - v^2), \qquad 0 \le u, v \le 1$$

oriented with an upward-pointing normal.

Solution

Step 1. **Compute the tangent and normal vectors.**

$$\mathbf{T}_u = \langle 2u, 0, 3u^2 \rangle, \qquad \mathbf{T}_v = \langle 0, 1, -2v \rangle$$

$$\mathbf{n}(u, v) = \mathbf{T}_u \times \mathbf{T}_v = \begin{vmatrix} \mathbf{i} & \mathbf{j} & \mathbf{k} \\ 2u & 0 & 3u^2 \\ 0 & 1 & -2v \end{vmatrix}$$

$$= -3u^2\mathbf{i} + 4uv\mathbf{j} + 2u\mathbf{k} = \langle -3u^2, 4uv, 2u \rangle$$

The z-component of $\mathbf{n}$ is positive on the domain $0 \le u \le 1$, so $\mathbf{n}$ is the upward-pointing normal (Figure 3).

Step 2. **Evaluate the dot product $\mathbf{F} \cdot \mathbf{n}$.**
Write $\mathbf{F}$ in terms of the parameters $x = u^2$, $y = v$, and $z = u^3 - v^2$:

$$\mathbf{F}(\Phi(u, v)) = \langle 0, 0, x \rangle = \langle 0, 0, u^2 \rangle$$

and compute the dot product:

$$\mathbf{F}(\Phi(u, v)) \cdot \mathbf{n}(u, v) = \langle 0, 0, u^2 \rangle \cdot \langle -3u^2, 4uv, 2u \rangle = 2u^3$$

Step 3. **Evaluate the surface integral.**
The surface integral is equal to a double integral over the rectangle $0 \le u, v \le 1$:

$$\iint_S \mathbf{F} \cdot d\mathbf{S} = \int_{u=0}^{1} \int_{v=0}^{1} \mathbf{F}(\Phi(u, v)) \cdot \mathbf{n}(u, v)\, dv\, du$$

$$= \int_{u=0}^{1} \int_{v=0}^{1} 2u^3\, dv\, du = \int_{u=0}^{1} 2u^3\, du = \frac{1}{2} \qquad ■$$

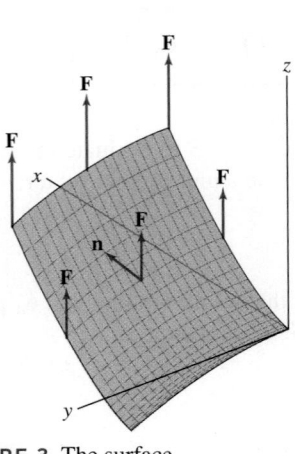

FIGURE 3 The surface $\Phi(u, v) = (u^2, v, u^3 - v^2)$ with an upward-pointing normal. The vector field $\mathbf{F} = \langle 0, 0, x \rangle$ points in the vertical direction.

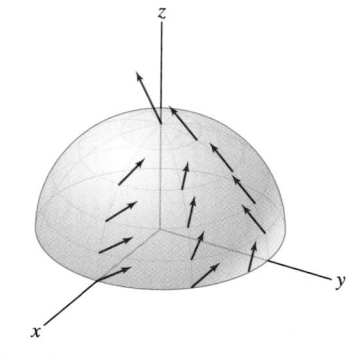

FIGURE 4

■ **EXAMPLE 2** Integral over a Hemisphere   Calculate the flux of $\mathbf{F} = \langle z, x, 1 \rangle$ across the upper hemisphere $\mathcal{S}$ of the sphere $x^2 + y^2 + z^2 = 1$, oriented with outward-pointing normal vectors (Figure 4).

Solution   The angle of declination ϕ varies between 0 and $\frac{\pi}{2}$ in the upper hemisphere, so we may parametrize $\mathcal{S}$ by

$$\Phi(\phi, \theta) = (\cos\theta \sin\phi, \sin\theta \sin\phi, \cos\phi), \qquad 0 \le \phi \le \frac{\pi}{2}, \quad 0 \le \theta < 2\pi$$

Step 1. Compute the normal vector.

According to Eq. (5) in Section 16.4, the outward-pointing normal vector is

$$\mathbf{n} = \mathbf{T}_\phi \times \mathbf{T}_\theta = \sin\phi \langle \cos\theta \sin\phi, \sin\theta \sin\phi, \cos\phi \rangle$$

Step 2. Evaluate the dot product $\mathbf{F} \cdot \mathbf{n}$.

Express the vector field in terms of the parameters and evaluate the dot product:

$$\mathbf{F}(\Phi(\phi, \theta)) = \langle z, x, 1 \rangle = \langle \cos\phi, \cos\theta \sin\phi, 1 \rangle$$

$$\mathbf{F}(\Phi(\phi, \theta)) \cdot \mathbf{n}(\phi, \theta) = \langle \cos\phi, \cos\theta \sin\phi, 1 \rangle \cdot \langle \cos\theta \sin^2\phi, \sin\theta \sin^2\phi, \cos\phi \sin\phi \rangle$$

$$= \cos\theta \sin^2\phi \cos\phi + \cos\theta \sin\theta \sin^3\phi + \cos\phi \sin\phi$$

Step 3. Evaluate the surface integral.

$$\iint_{\mathcal{S}} \mathbf{F} \cdot d\mathbf{S} = \int_{\phi=0}^{\pi/2} \int_{\theta=0}^{2\pi} \mathbf{F}(\Phi(\phi, \theta)) \cdot \mathbf{n}(\phi, \theta)\, d\theta\, d\phi$$

$$= \int_{\phi=0}^{\pi/2} \int_{\theta=0}^{2\pi} \underbrace{(\cos\theta \sin^2\phi \cos\phi + \cos\theta \sin\theta \sin^3\phi}_{\text{Integral over } \theta \text{ is zero}} + \cos\phi \sin\phi)\, d\theta\, d\phi$$

The integrals of $\cos\theta$ and $\cos\theta \sin\theta$ over $0 \le \theta < 2\pi$ are both zero, so we are left with

$$\int_{\phi=0}^{\pi/2} \int_{\theta=0}^{2\pi} \cos\phi \sin\phi\, d\theta\, d\phi = 2\pi \int_{\phi=0}^{\pi/2} \cos\phi \sin\phi\, d\phi = 2\pi \frac{\sin^2\phi}{2} \Big|_0^{\pi/2} = \pi \quad ■$$

■ **EXAMPLE 3**   Surface Integral over a Graph   Calculate the flux of $\mathbf{F} = 2\mathbf{i} - y\mathbf{j} + x\mathbf{k}$ through the surface $y = 1 + x^2 + z^2$ for $1 \le y \le 5$, oriented with normal pointing in the negative y-direction.

Solution   This surface is the graph of the function $y = 1 + x^2 + z^2$, where x and z are the independent variables.

Step 1. Find a parametrization.

Since y is given explicitly in terms of x and z, we may use x and z as parameters. The surface is parametrized by

$$\Phi(x, z) = (x, 1 + x^2 + z^2, z)$$

What is the parameter domain? Since $y = 1 + x^2 + z^2$, the condition $1 \le y \le 5$ is equivalent to $1 \le 1 + x^2 + z^2 \le 5$ or $0 \le x^2 + z^2 \le 4$. In other words, the parameter domain is $\mathcal{D} = \{(x, z) : x^2 + z^2 \le 4\}$.

Step 2. **Compute the tangent and normal vectors.**

$$\mathbf{T}_x = \langle 1, 2x, 0 \rangle, \qquad \mathbf{T}_z = \langle 0, 2z, 1 \rangle$$

$$\mathbf{n} = \mathbf{T}_x \times \mathbf{T}_z = \begin{vmatrix} \mathbf{i} & \mathbf{j} & \mathbf{k} \\ 1 & 2x & 0 \\ 0 & 2z & 1 \end{vmatrix} = 2x\mathbf{i} - \mathbf{j} + 2z\mathbf{k}$$

Observe that $\mathbf{n}$ points in the negative y-direction.

Step 3. **Evaluate the dot product $\mathbf{F} \cdot \mathbf{n}$.**

Express the vector field in terms of the parametrization and compute the dot product:

$$\mathbf{F}(\Phi(x, z)) = \langle 2, -y, x \rangle = \langle 2, -(1 + x^2 + z^2), x \rangle$$

$$\mathbf{F}(\Phi(x, z)) \cdot \mathbf{n} = \langle 2, -(1 + x^2 + z^2), x \rangle \cdot \langle 2x, -1, 2z \rangle$$

$$= 4x + (1 + x^2 + z^2) + 2xz$$

Step 4. **Evaluate the surface integral.**

$$\iint_{\mathcal{S}} \mathbf{F} \cdot d\mathbf{S} = \iint_{\mathcal{D}} (x^2 + z^2 + 2xz + 4x + 1) \, dx \, dz$$

Since $\mathcal{D}$ is the circle of radius 2 in the xz-plane, it is convenient to use polar coordinates

$$x = r \cos \theta, \qquad z = r \sin \theta$$

Then $x^2 + z^2 = r^2$ and the integrand becomes

$$x^2 + z^2 + 2xz + 4x + 1 = r^2 + 2r^2 \sin \theta \cos \theta + 4r \cos \theta + 1$$

◄⋯ **REMINDER** *The Change of Variables Formula for polar coordinates reads*

$$\iint f(x, y) \, dx \, dy$$

$$= \iint f(r \sin \theta, r \cos \theta) \, r \, dr \, d\theta$$

We apply it in this example with variables x and z.

By the Change of Variables Formula for polar coordinates, $dx \, dz$ becomes $r \, dr \, d\theta$ and

$$\iint_{\mathcal{S}} \mathbf{F} \cdot d\mathbf{S} = \int_{\theta=0}^{2\pi} \int_{r=0}^{2} (r^2 + 2r^2 \sin \theta \cos \theta + 4r \cos \theta + 1) \, r \, dr \, d\theta$$

$$= 2\pi \int_0^2 (r^2 + 1) r \, dr = 2\pi \left(\frac{2^4}{4} + \frac{2^2}{2} \right) = 12\pi \qquad \blacksquare$$

The word "flux" is derived from the Latin word fluere *meaning "to flow."*

Fluid Flux

The term *flux* comes from physics, where it is often used to denote the rate of transfer of fluid, particles, or energy across a given surface. Imagine dipping a net into a stream of flowing water (Figure 5). The flux is the rate at which water flows through the net per unit time.

To compute this flux or flow rate, let $\mathbf{v}$ be the velocity vector field (Figure 5). At each point P, $\mathbf{v}(P)$ gives the speed and direction of the water particle located at the point P. If we think of the net as a surface $\mathcal{S}$, then the rate of flow through $\mathcal{S}$ (in volume per time) is equal to the surface integral of $\mathbf{v}$ over $\mathcal{S}$.

To explain why, suppose first that $\mathbf{v}$ is constant with value $\mathbf{v}_0$ and that our "net" $\mathcal{S}$ is a rectangle of area A (Figures 6 and 7). Then each water particle flows in the same direction with the same speed $\|\mathbf{v}_0\|$ (say, in meters per second). If $\mathbf{v}_0$ is perpendicular to $\mathcal{S}$, then a water particle passes through $\mathcal{S}$ within a 1-s time interval if its distance to $\mathcal{S}$ is at most $\|\mathbf{v}_0\|$ meters. In other words, the water flowing through $\mathcal{S}$ in a 1-s time interval is

FIGURE 5 Velocity field of a fluid flow.

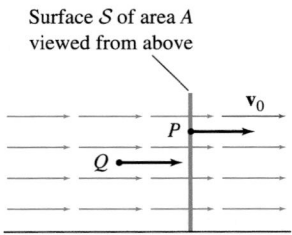

FIGURE 6 Water flowing at constant velocity $\mathbf{v}_0$ perpendicular to a rectangular surface S.

a box of volume $\|\mathbf{v}_0\| A$, and thus,

$$\text{Flow rate} = \text{velocity} \times \text{area} = \|\mathbf{v}_0\| A$$

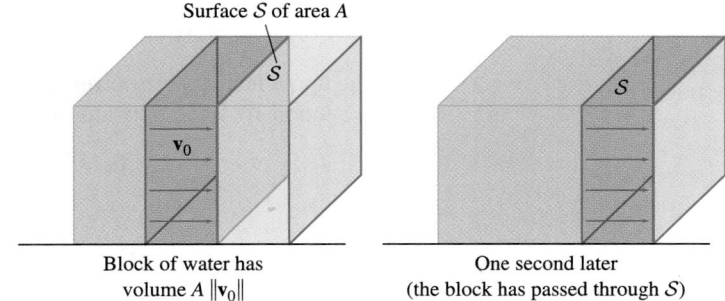

Surface S of area A

Block of water has volume $A\|\mathbf{v}_0\|$

One second later (the block has passed through S)

FIGURE 7

If the fluid has constant velocity $\mathbf{v}_0$ but flows at an angle θ relative to S, then the block of water flowing past S in a 1-s interval is no longer a rectangular box but a parallelepiped of volume $A\|\mathbf{v}_0\| \cos \theta$ (Figure 8). If $\mathbf{n}$ is a vector normal to S of length equal to the area A, then we may write the flow rate as a dot product:

$$\text{Flow rate} = A\|\mathbf{v}_0\| \cos \theta = \mathbf{v}_0 \cdot \mathbf{n}$$

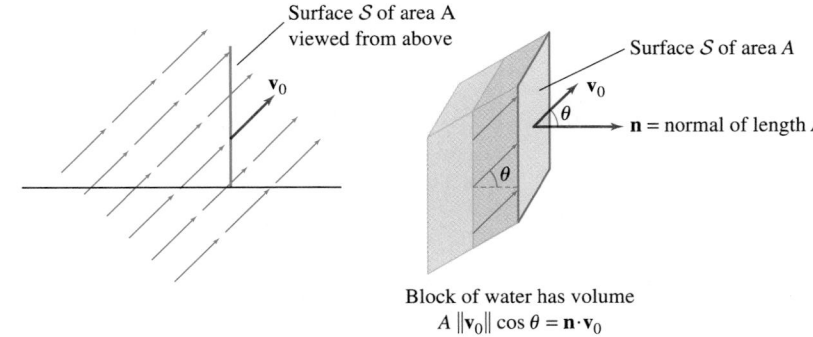

Surface S of area A viewed from above

Surface S of area A

$\mathbf{n} = $ normal of length A

Block of water has volume $A\|\mathbf{v}_0\| \cos \theta = \mathbf{n} \cdot \mathbf{v}_0$

FIGURE 8 Water flowing at constant velocity $\mathbf{v}_0$, making an angle θ with a rectangular surface.

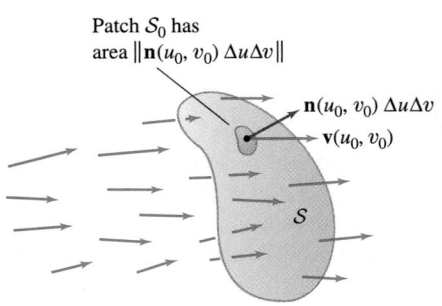

Patch S_0 has area $\|\mathbf{n}(u_0, v_0) \Delta u \Delta v\|$

$\mathbf{n}(u_0, v_0) \Delta u \Delta v$

$\mathbf{v}(u_0, v_0)$

FIGURE 9 The fluid flow across a small patch S_0 is approximately equal to $\mathbf{v}(u_0, v_0) \cdot \mathbf{n}(u_0, v_0) \Delta u \, \Delta v$.

In the general case, the velocity field $\mathbf{v} = \mathbf{v}(u, v)$ need not be constant and the surface S may be curved. However, suppose that S has a parametrization $\Phi(u, v)$ that maps a small rectangle of size $\Delta u \times \Delta v$ to a small patch S_0 of S. For any sample point $\Phi(u_0, v_0)$ in S_0, the vector $\mathbf{n}(u_0, v_0) \Delta u \, \Delta v$ is a normal vector whose length is approximately equal to the area of S_0 [Eq. (1) in Section 16.4]. This patch is nearly rectangular, and thus, the flow rate through S_0 is approximately $\mathbf{v}(u_0, v_0) \cdot \mathbf{n}(u_0, v_0) \Delta u \, \Delta v$. We find that the total flux is equal to the integral of $\mathbf{v}(u, v) \cdot \mathbf{n}(u, v)$, which is the surface integral of $\mathbf{v}$ over S.

Flow Rate of a Fluid Through a Surface   The flow rate (in volume per unit time) of a fluid across a surface S is equal to the surface integral of the velocity vector field $\mathbf{v}$:

$$\text{Flow rate across } S \text{ (volume per unit time)} = \iint_S \mathbf{v} \cdot d\mathbf{S} \qquad \boxed{3}$$

■ **EXAMPLE 4** Let $\mathbf{v} = \langle x^2 + y^2, 0, z^2 \rangle$ be the velocity field (in centimeters per second) of a fluid in $\mathbf{R}^3$. Compute the flow rate through the upper hemisphere $\mathcal{S}$ of the unit sphere centered at the origin.

Solution We use spherical coordinates:

$$x = \cos\theta \sin\phi, \qquad y = \sin\theta \sin\phi, \qquad z = \cos\phi$$

The upper hemisphere corresponds to the ranges $0 \le \phi \le \frac{\pi}{2}$ and $0 \le \theta \le 2\pi$. By Eq. (5) in Section 16.4, the outward-pointing normal is

$$\mathbf{n} = \mathbf{T}_\phi \times \mathbf{T}_\theta = \sin\phi \langle \cos\theta \sin\phi, \sin\theta \sin\phi, \cos\phi \rangle$$

We have

$$x^2 + y^2 = (\cos\theta \sin\phi)^2 + (\sin\theta \sin\phi)^2 = \sin^2\phi$$

Therefore,

$$\mathbf{v} = \langle x^2 + y^2, 0, z^2 \rangle = \langle \sin^2\phi, 0, \cos^2\phi \rangle$$

$$\mathbf{v} \cdot \mathbf{n} = \sin\phi \langle \sin^2\phi, 0, \cos^2\phi \rangle \cdot \langle \cos\theta \sin\phi, \sin\theta \sin\phi, \cos\phi \rangle$$

$$= \sin^4\phi \cos\theta + \sin\phi \cos^3\phi$$

The flux is equal to

$$\iint_{\mathcal{S}} \mathbf{v} \cdot d\mathbf{S} = \int_{\phi=0}^{\pi/2} \int_{\theta=0}^{2\pi} (\sin^4\phi \cos\theta + \sin\phi \cos^3\phi)\, d\theta\, d\phi$$

The integral of the first term, $\sin^4\phi \cos\theta$, with respect to θ is zero, so we are left with

$$\int_{\phi=0}^{\pi/2} \int_{\theta=0}^{2\pi} \sin\phi \cos^3\phi\, d\theta\, d\phi = 2\pi \int_{\phi=0}^{\pi/2} \cos^3\phi \sin\phi\, d\phi$$

$$= 2\pi \left(-\frac{\cos^4\phi}{4} \right) \Big|_{\phi=0}^{\pi/2} = \frac{\pi}{2}\ \mathrm{cm}^3/\mathrm{s} \qquad ■$$

Electric and Magnetic Fields

The laws of electricity and magnetism are expressed in terms of two vector fields, the electric field $\mathbf{E}$ and the magnetic field $\mathbf{B}$, whose properties are summarized by four equations known as Maxwell's Equations. One of these equations expresses **Faraday's Law of Induction**, which can be formulated either as a partial differential equation or in the following "integral form":

$$\int_{\mathcal{C}} \mathbf{E} \cdot d\mathbf{s} = -\frac{d}{dt} \iint_{\mathcal{S}} \mathbf{B} \cdot d\mathbf{S} \qquad \boxed{4}$$

In this equation, $\mathcal{S}$ is a surface with boundary curve $\mathcal{C}$. We orient $\mathcal{S}$ using the unit normal vectors whose direction is determined by the orientation of $\mathcal{C}$ via the right-hand rule (Figure 10). The line integral of $\mathbf{E}$ around the boundary curve $\mathcal{C}$ is equal to the voltage drop around the curve (the work performed by $\mathbf{E}$ moving a positive unit charge around $\mathcal{C}$).

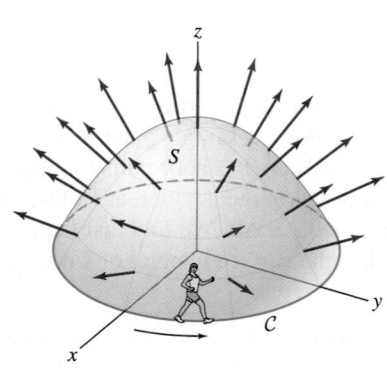

FIGURE 10 If she walks along $\mathcal{C}$ in the positive direction, with the surface to her left, then her head points in the outward direction.

■ **EXAMPLE 5** A varying current of magnitude $i = 28\cos\left(\dfrac{30t}{\pi}\right)$ amperes (t in seconds) flows through a straight wire, producing a magnetic field $\mathbf{B}$ whose magnitude at

*The electric field **E** is conservative when the charges are stationary (or more generally, when the magnetic field is constant). When the magnetic field **B** varies in time, the integral on the right in Eq. (4) is nonzero for some surface, and hence the circulation of **E** around the closed curve C is nonzero. This shows that **E** is not conservative when the magnetic field varies.*

a distance r m from the wire is $\|\mathbf{B}\| = \dfrac{\mu_0 i}{2\pi r}$ teslas (where $\mu_0 = 4\pi \cdot 10^{-7}$ T-m/A). Furthermore, at any point P, $\mathbf{B}$ points in the direction tangent to the circle through P perpendicular to the wire as in Figure 11.

(a) Calculate the flux $\Phi(t)$ at time t of $\mathbf{B}$ through the rectangle $\mathcal{R}$ shown in Figure 11, oriented with the normal vector pointing out of the page. Assume that $L = 1.2$ m, $H = 0.7$ m, and $d = 0.1$ m.

(b) Use Faraday's Law to determine the voltage drop (in volts) around the rectangular loop $\mathcal{C}$ as a function of t. Assume that $L = 1.2$ m.

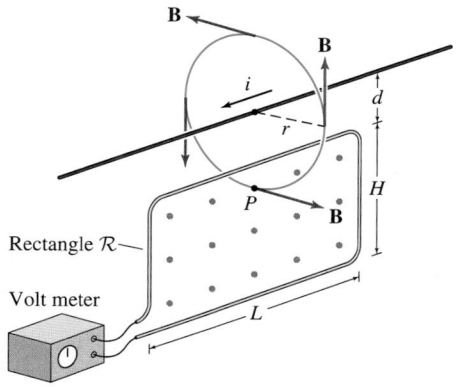

FIGURE 11

Solution Referring to Figure 11, we choose coordinates (x, y) so that the rectangle $\mathcal{R}$ is the region

$$0 \le x \le L, \qquad d \le y \le H + d$$

The distance from a point $P = (x, y)$ in $\mathcal{R}$ to the wire is y. The vector field $\mathbf{B}$ points directly out of the page, so if we orient $\mathcal{R}$ with the unit normal $\mathbf{n}$ pointing out of the page, then $\mathbf{B}$ and $\mathbf{n}$ point in the same direction and

$$\mathbf{B} \cdot \mathbf{n} = \frac{\mu_0 i}{2\pi y}$$

(a) The flux $\Phi(t)$ of $\mathbf{B}$ through $\mathcal{R}$ at time t is then

$$\Phi(t) = \iint_{\mathcal{R}} \mathbf{B} \cdot d\mathbf{S} = \int_{x=0}^{L} \int_{y=d}^{H+d} \frac{\mu_0 i}{2\pi y}\, dy\, dx = \frac{\mu_0 L i}{2\pi} \int_{y=d}^{H+d} \frac{dy}{y}$$

$$= \frac{\mu_0 L}{2\pi} \left(\ln \frac{H+d}{d} \right) i$$

$$= \frac{4\pi \cdot 10^{-7}(1.2)}{2\pi} \left(\ln \frac{0.8}{0.1} \right) 28 \cos \left(\frac{30t}{\pi} \right)$$

$$\approx 1.4 \times 10^{-5} \cos \left(\frac{30t}{\pi} \right) \text{ T-m}^2$$

(b) By Faraday's Law [Eq. (4)], the voltage drop around the rectangular loop $\mathcal{C}$, oriented in the counterclockwise direction, is

$$\int_{\mathcal{C}} \mathbf{E} \cdot d\mathbf{s} = -\frac{d\Phi}{dt} \approx -(1.4 \times 10^{-5}) \left(\frac{30}{\pi} \right) \sin \left(\frac{30t}{\pi} \right) \approx -0.00013 \sin \left(\frac{30t}{\pi} \right) \text{ V} \quad \blacksquare$$

CONCEPTUAL INSIGHT We have seen that a vector surface integral depends on the choice of orientation of the surface. However, some surfaces, such as the Möbius strip (discovered in 1858), cannot be oriented because they are one-sided. You can construct a Möbius strip M with a rectangular strip of paper: Join the two ends of the strip together with a 180° twist. Unlike an ordinary two-sided strip, the Möbius strip M has only one side and it is impossible to specify an outward direction in a consistent manner (Figure 12). If you choose a unit normal vector at a point P and carry that unit vector around M, when you return to P, the vector will point in the opposite direction. Therefore, we cannot integrate a vector field over a Möbius strip. On the other hand, it is possible to integrate a scalar function. For example, the integral of mass density would equal the total mass of the Möbius strip.

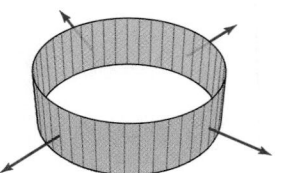

Möbius strip   Ordinary (untwisted) band

FIGURE 12 It is not possible to choose a continuously varying unit normal vector on a Möbius strip.

16.5 SUMMARY

- A surface S is *oriented* if a continuously varying unit normal vector $\mathbf{e_n}(P)$ is specified at each point on S. This distinguishes an "outward" direction on the surface.
- The integral of a vector field $\mathbf{F}$ over an oriented surface S is defined as the integral of the normal component $\mathbf{F} \cdot \mathbf{e_n}$ over S.
- Vector surface integrals are computed using the formula

$$\iint_S \mathbf{F} \cdot d\mathbf{S} = \iint_D \mathbf{F}(\Phi(u, v)) \cdot \mathbf{n}(u, v)\, du\, dv$$

Here, $\Phi(u, v)$ is a parametrization of S such that $\mathbf{n}(u, v) = \mathbf{T}_u \times \mathbf{T}_v$ points in the direction of the unit normal vector specified by the orientation.

- The surface integral of a vector field $\mathbf{F}$ over S is also called the *flux* of $\mathbf{F}$ through Φ. If $\mathbf{F}$ is the velocity field of a fluid, then the flux $\iint_S \mathbf{F} \cdot d\mathbf{S}$ is the rate at which fluid flows through S per unit time.

16.5 EXERCISES

Preliminary Questions

1. Let $\mathbf{F}$ be a vector field and $\Phi(u, v)$ a parametrization of a surface S, and set $\mathbf{n} = \mathbf{T}_u \times \mathbf{T}_v$. Which of the following is the normal component of $\mathbf{F}$?

(a) $\mathbf{F} \cdot \mathbf{n}$ **(b)** $\mathbf{F} \cdot \mathbf{e_n}$

2. The vector surface integral $\iint_S \mathbf{F} \cdot d\mathbf{S}$ is equal to the scalar surface integral of the function (choose the correct answer):

(a) $\|\mathbf{F}\|$

(b) $\mathbf{F} \cdot \mathbf{n}$, where $\mathbf{n}$ is a normal vector

(c) $\mathbf{F} \cdot \mathbf{e_n}$, where $\mathbf{e_n}$ is the unit normal vector

3. $\iint_{\mathcal{S}} \mathbf{F} \cdot d\mathbf{S}$ is zero if (choose the correct answer):

(a) $\mathbf{F}$ is tangent to $\mathcal{S}$ at every point.

(b) $\mathbf{F}$ is perpendicular to $\mathcal{S}$ at every point.

4. If $\mathbf{F}(P) = \mathbf{e_n}(P)$ at each point on $\mathcal{S}$, then $\iint_{\mathcal{S}} \mathbf{F} \cdot d\mathbf{S}$ is equal to (choose the correct answer):

(a) Zero   **(b)** Area($\mathcal{S}$)   **(c)** Neither

5. Let $\mathcal{S}$ be the disk $x^2 + y^2 \le 1$ in the xy-plane oriented with normal in the positive z-direction. Determine $\iint_{\mathcal{S}} \mathbf{F} \cdot d\mathbf{S}$ for each of the following vector constant fields:

(a) $\mathbf{F} = \langle 1, 0, 0 \rangle$

(b) $\mathbf{F} = \langle 0, 0, 1 \rangle$

(c) $\mathbf{F} = \langle 1, 1, 1 \rangle$

6. Estimate $\iint_{\mathcal{S}} \mathbf{F} \cdot d\mathbf{S}$, where $\mathcal{S}$ is a tiny oriented surface of area 0.05 and the value of $\mathbf{F}$ at a sample point in $\mathcal{S}$ is a vector of length 2 making an angle $\frac{\pi}{4}$ with the normal to the surface.

7. A small surface $\mathcal{S}$ is divided into three pieces of area 0.2. Estimate $\iint_{\mathcal{S}} \mathbf{F} \cdot d\mathbf{S}$ if $\mathbf{F}$ is a unit vector field making angles of 85, 90, and 95° with the normal at sample points in these three pieces.

Exercises

1. Let $\mathbf{F} = \langle y, z, x \rangle$ and let $\mathcal{S}$ be the oriented surface parametrized by $\Phi(u, v) = (u^2 - v, u + v, v^2)$ for $0 \le u \le 2$, $-1 \le v \le 1$. Calculate:

(a) $\mathbf{n}$ and $\mathbf{F} \cdot \mathbf{n}$ as functions of u and v

(b) The normal component of $\mathbf{F}$ to the surface at $P = (3, 3, 1) = \Phi(2, 1)$

(c) $\iint_{\mathcal{S}} \mathbf{F} \cdot d\mathbf{S}$

2. Compute the surface integral of the vector field $\mathbf{F} = \langle x, y, x + y \rangle$ over the portion $\mathcal{S}$ of the paraboloid $z = x^2 + y^2$ lying over the disk $x^2 + y^2 \le 1$.

3. Let $\mathcal{S}$ be the square in the xy-plane shown in Figure 13, oriented with the normal pointing in the positive z-direction. Estimate

$$\iint_{\mathcal{S}} \mathbf{F} \cdot d\mathbf{S}$$

where $\mathbf{F}$ is a vector field whose values at the labeled points are

$$\mathbf{F}(A) = \langle 2, 6, 4 \rangle, \qquad \mathbf{F}(B) = \langle 1, 1, 7 \rangle$$
$$\mathbf{F}(C) = \langle 3, 3, -3 \rangle, \qquad \mathbf{F}(D) = \langle 0, 1, 8 \rangle$$

4. Suppose that $\mathcal{S}$ is a surface in $\mathbf{R}^3$ with a parametrization Φ whose domain $\mathcal{D}$ is the square in Figure 13. The values of a function f, a vector field $\mathbf{F}$, and the normal vector $\mathbf{n} = \mathbf{T}_u \times \mathbf{T}_v$ at $\Phi(P)$ are given for the four sample points in $\mathcal{D}$ in the following table. Estimate the surface integrals of f and $\mathbf{F}$ over $\mathcal{S}$.

Point P in $\mathcal{D}$	f	$\mathbf{F}$	$\mathbf{n}$
A	3	$\langle 2, 6, 4 \rangle$	$\langle 1, 1, 1 \rangle$
B	1	$\langle 1, 1, 7 \rangle$	$\langle 1, 1, 0 \rangle$
C	2	$\langle 3, 3, -3 \rangle$	$\langle 1, 0, -1 \rangle$
D	5	$\langle 0, 1, 8 \rangle$	$\langle 2, 1, 0 \rangle$

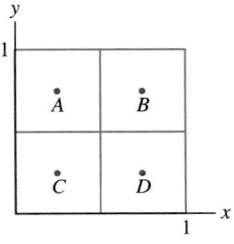

FIGURE 13

In Exercises 5–17, compute the surface integral over the given oriented surface.

5. $\mathbf{F} = \langle y, z, x \rangle$,   plane $3x - 4y + z = 1$, $0 \le x, y \le 1$,   upward-pointing normal

6. $\mathbf{F} = \langle e^z, z, x \rangle$, $\Phi(r, s) = (rs, r + s, r)$, $0 \le r \le 1, 0 \le s \le 1$,   oriented by $\mathbf{T}_r \times \mathbf{T}_s$

7. $\mathbf{F} = \langle 0, 3, x^2 \rangle$,   hemisphere $x^2 + y^2 + z^2 = 9$, $z \ge 0$,   outward-pointing normal

8. $\mathbf{F} = \langle x, y, z \rangle$,   part of sphere $x^2 + y^2 + z^2 = 1$, where $\frac{1}{2} \le z \le \frac{\sqrt{3}}{2}$,   inward-pointing normal

9. $\mathbf{F} = \langle e^z, z, x \rangle$, $z = 9 - x^2 - y^2, z \ge 0$,   upward-pointing normal

10. $\mathbf{F} = \langle \sin y, \sin z, yz \rangle$,   rectangle $0 \le y \le 2, 0 \le z \le 3$ in the (y, z)-plane,   normal pointing in negative x-direction

11. $\mathbf{F} = y^2 \mathbf{i} + 2\mathbf{j} - x\mathbf{k}$,   portion of the plane $x + y + z = 1$ in the octant $x, y, z \ge 0$,   upward-pointing normal

12. $\mathbf{F} = \langle x, y, e^z \rangle$,   cylinder $x^2 + y^2 = 4, 1 \le z \le 5$,   outward-pointing normal

13. $\mathbf{F} = \langle xz, yz, z^{-1} \rangle$,   disk of radius 3 at height 4 parallel to the xy-plane,   upward-pointing normal

14. $\mathbf{F} = \langle xy, y, 0 \rangle$,   cone $z^2 = x^2 + y^2, x^2 + y^2 \le 4, z \ge 0$,   downward-pointing normal

15. $\mathbf{F} = \langle 0, 0, e^{y+z} \rangle$, boundary of unit cube $0 \le x, y, z \le 1$, outward-pointing normal

16. $\mathbf{F} = \langle 0, 0, z^2 \rangle$, $\Phi(u, v) = (u \cos v, u \sin v, v)$, $0 \le u \le 1$, $0 \le v \le 2\pi$, upward-pointing normal

17. $\mathbf{F} = \langle y, z, 0 \rangle$, $\Phi(u, v) = (u^3 - v, u + v, v^2)$, $0 \le u \le 2$, $0 \le v \le 3$, downward-pointing normal

18. Let S be the oriented half-cylinder in Figure 14. In (a)–(f), determine whether $\iint_S \mathbf{F} \cdot d\mathbf{S}$ is positive, negative, or zero. Explain your reasoning.

(a) $\mathbf{F} = \mathbf{i}$ (b) $\mathbf{F} = \mathbf{j}$

(c) $\mathbf{F} = \mathbf{k}$ (d) $\mathbf{F} = y\mathbf{i}$

(e) $\mathbf{F} = -y\mathbf{j}$ (f) $\mathbf{F} = x\mathbf{j}$

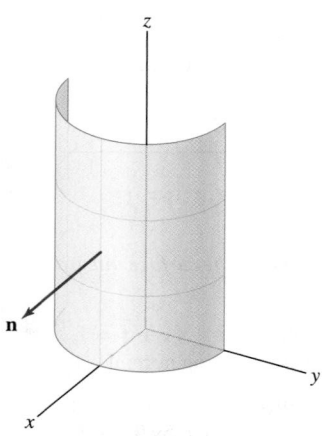

FIGURE 14

19. Let $\mathbf{e}_r$ be the unit radial vector and $r = \sqrt{x^2 + y^2 + z^2}$. Calculate the integral of $\mathbf{F} = e^{-r}\mathbf{e}_r$ over:

(a) The upper-hemisphere of $x^2 + y^2 + z^2 = 9$, outward-pointing normal

(b) The octant $x, y, z \ge 0$ of the unit sphere centered at the origin

20. Show that the flux of $\mathbf{F} = \dfrac{\mathbf{e}_r}{r^2}$ through a sphere centered at the origin does not depend on the radius of the sphere.

21. The electric field due to a point charge located at the origin is $\mathbf{E} = k\dfrac{\mathbf{e}_r}{r^2}$, where k is a constant. Calculate the flux of $\mathbf{E}$ through the disk D of radius 2 parallel to the xy-plane with center $(0, 0, 3)$.

22. Let S be the ellipsoid $\left(\dfrac{x}{4}\right)^2 + \left(\dfrac{y}{3}\right)^2 + \left(\dfrac{z}{2}\right)^2 = 1$. Calculate the flux of $\mathbf{F} = \langle z, 1, 0 \rangle$ over the portion of S where $x, y, z \le 0$ with upward-pointing normal. *Hint:* Parametrize S using a modified form of spherical coordinates (θ, ϕ).

23. Let $\mathbf{v} = \langle x, 0, z \rangle$ be the velocity field (in ft/s) of a fluid in $\mathbf{R}^3$. Calculate the flow rate (in ft^3/s) through the upper hemisphere of the sphere $x^2 + y^2 + z^2 = 1$ ($z \ge 0$).

24. Calculate the flow rate through the upper hemisphere of the sphere $x^2 + y^2 + z^2 = R^2$ ($z \ge 0$) for $\mathbf{v}$ as in Exercise 23.

25. Calculate the flow rate of a fluid with velocity field $\mathbf{v} = \langle x, y, x^2 y \rangle$ (in ft/s) through the portion of the ellipse $\left(\dfrac{x}{2}\right)^2 + \left(\dfrac{y}{3}\right)^2 = 1$ in the xy-plane, where $x, y \ge 0$, oriented with the normal in the positive z-direction.

In Exercises 26–27, let $\mathcal{T}$ be the triangular region with vertices $(1, 0, 0)$, $(0, 1, 0)$, and $(0, 0, 1)$ oriented with upward-pointing normal vector (Figure 15). Assume distances are in meters.

26. A fluid flows with constant velocity field $\mathbf{v} = 2\mathbf{k}$ (m/s). Calculate:

(a) The flow rate through $\mathcal{T}$

(b) The flow rate through the projection of $\mathcal{T}$ onto the xy-plane [the triangle with vertices $(0, 0, 0)$, $(1, 0, 0)$, and $(0, 1, 0)$]

27. Calculate the flow rate through $\mathcal{T}$ if $\mathbf{v} = -\mathbf{j}$ m/s.

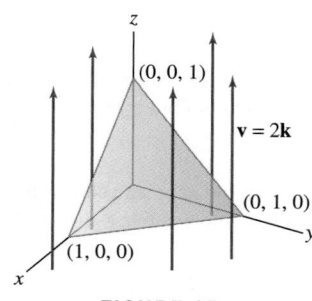

FIGURE 15

28. Prove that if S is the part of a graph $z = g(x, y)$ lying over a domain $\mathcal{D}$ in the xy-plane, then

$$\iint_S \mathbf{F} \cdot d\mathbf{S} = \iint_{\mathcal{D}} \left(-F_1 \frac{\partial g}{\partial x} - F_2 \frac{\partial g}{\partial y} + F_3 \right) dx\, dy$$

In Exercises 29–30, a varying current $i(t)$ flows through a long straight wire in the xy-plane as in Example 5. The current produces a magnetic field $\mathbf{B}$ whose magnitude at a distance r from the wire is $\|\mathbf{B}\| = \dfrac{\mu_0 i}{2\pi r}$ T, where $\mu_0 = 4\pi \cdot 10^{-7}$ T-m/A. Furthermore, $\mathbf{B}$ points into the page at points P in the xy-plane.

29. Assume that $i(t) = t(12 - t)$ A (t in seconds). Calculate the flux $\Phi(t)$, at time t, of $\mathbf{B}$ through a rectangle of dimensions $L \times H = 3 \times 2$ m, whose top and bottom edges are parallel to the wire and whose bottom edge is located $d = 0.5$ m above the wire (similar to Figure 11). Then use Faraday's Law to determine the voltage drop around the rectangular loop (the boundary of the rectangle) at time t.

30. Assume that $i = 10e^{-0.1t}$ A (t in seconds). Calculate the flux $\Phi(t)$, at time t, of **B** through the isosceles triangle of base 12 cm and height 6 cm, whose bottom edge is 3 cm from the wire, as in Figure 16. Assume the triangle is oriented with normal vector pointing out of the page. Use Faraday's Law to determine the voltage drop around the triangular loop (the boundary of the triangle) at time t.

FIGURE 16

Further Insights and Challenges

31. A point mass m is located at the origin. Let Q be the flux of the gravitational field $\mathbf{F} = -Gm\dfrac{\mathbf{e}_r}{r^2}$ through the cylinder $x^2 + y^2 = R^2$ for $a \leq z \leq b$, including the top and bottom (Figure 17). Show that $Q = -4\pi Gm$ if $a < 0 < b$ (m lies inside the cylinder) and $Q = 0$ if $0 < a < b$ (m lies outside the cylinder).

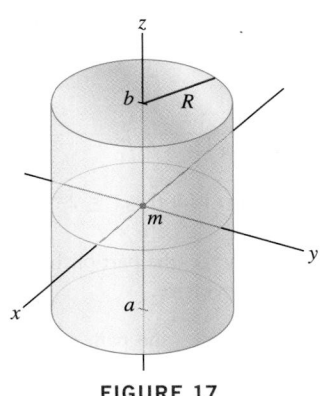

FIGURE 17

In Exercises 32–33, let S be the surface with parametrization

$$\Phi(u, v) = \left(\left(1 + v\cos\frac{u}{2}\right)\cos u, \left(1 + v\cos\frac{u}{2}\right)\sin u, v\sin\frac{u}{2}\right)$$

for $0 \leq u \leq 2\pi$, $-\frac{1}{2} \leq v \leq \frac{1}{2}$.

32. **CAS** Use a computer algebra system.
(a) Plot S and confirm visually that S is a Möbius strip.
(b) The intersection of S with the xy-plane is the unit circle $\Phi(u, 0) = (\cos u, \sin u, 0)$. Verify that the normal vector along this circle is

$$\mathbf{n}(u, 0) = \left(\cos u \sin\frac{u}{2}, \sin u \sin\frac{u}{2}, -\cos\frac{u}{2}\right)$$

(c) As u varies from 0 to 2π, the point $\Phi(u, 0)$ moves once around the unit circle, beginning and ending at $\Phi(0, 0) = \Phi(2\pi, 0) = (1, 0, 0)$. Verify that $\mathbf{n}(u, 0)$ is a unit vector which varies continuously but that $\mathbf{n}(2\pi, 0) = -\mathbf{n}(0, 0)$. This shows that S is not orientable, that is, it is not possible to choose a nonzero normal vector at each point on S in a continuously varying manner (if it were possible, the unit normal vector returns to itself rather than its negative when carried around the circle).

33. **CAS** It is not possible to integrate a vector field over S because S is not orientable. However, it is possible to integrate functions over S. Using a computer algebra system:
(a) Verify that

$$\|\mathbf{n}(u, v)\|^2 = 1 + \frac{3}{4}v^2 + 2v\cos\frac{u}{2} + \frac{1}{2}v^2\cos u$$

(b) Compute the surface area of S to four decimal places.
(c) Compute $\displaystyle\iint_S (x^2 + y^2 + z^2)\, dS$ to four decimal places.

CHAPTER REVIEW EXERCISES

1. Compute the vector assigned to the point $P = (-3, 5)$ by the vector field:

(a) $\mathbf{F} = \langle xy, y - x \rangle$

(b) $\mathbf{F} = \langle 4, 8 \rangle$

(c) $\mathbf{F} = \langle 3^{x+y}, \log_2(x + y) \rangle$

2. Find a vector field $\mathbf{F}$ in the plane such that $\|\mathbf{F}(x, y)\| = 1$ and $\mathbf{F}(x, y)$ is orthogonal to $\mathbf{G}(x, y) = \langle x, y \rangle$ for all x, y.

In Exercises 3–6, sketch the vector field.

3. $\mathbf{F}(x, y) = \langle y, 1 \rangle$

4. $\mathbf{F}(x, y) = \langle 4, 1 \rangle$

5. $\nabla\varphi$, where $\varphi(x, y) = x^2 - y$

6. $\mathbf{F}(x, y) = \left\langle \dfrac{4y}{\sqrt{x^2 + 16y^2}}, \dfrac{-x}{\sqrt{x^2 + 16y^2}} \right\rangle$

Hint: Show that $\mathbf{F}$ is a unit vector field tangent to the family of ellipses $x^2 + 4y^2 = c^2$.

In Exercises 7–14, determine whether or not the vector field is conservative and, if so, find a potential function.

7. $\mathbf{F}(x, y, z) = \langle \sin x, e^y, z \rangle$

8. $\mathbf{F}(x, y, z) = \langle 2, 4, e^z \rangle$

9. $\mathbf{F}(x, y, z) = \langle xyz, \frac{1}{2}x^2 z, 2z^2 y \rangle$

10. $\mathbf{F}(x, y, z) = \langle xy^2 z, x^2 yz, \frac{1}{2}x^2 y^2 \rangle$

11. $\mathbf{F}(x, y, z) = \left(\dfrac{y}{1 + x^2}, \tan^{-1} x, 2z \right)$

12. $\mathbf{F}(x, y) = \langle x^2 y, y^2 x \rangle$

13. $\mathbf{F}(x, y, z) = \langle xe^{2x}, ye^{2z}, ze^{2y} \rangle$

14. $\mathbf{F}(x, y) = \langle y^4 x^3, x^4 y^3 \rangle$

15. Calculate $\displaystyle\int_{\mathbf{c}} \nabla\varphi \cdot d\mathbf{s}$, where $\varphi(x, y, z) = x^4 y^3 z^2$ and $\mathbf{c}(t) = (t^2, 1 + t, t^{-1})$ for $1 \le t \le 3$.

16. Find a gradient vector field of the form $\mathbf{F} = \langle g(y), h(x) \rangle$ such that $\mathbf{F}(0, 0) = \langle 1, 1 \rangle$, where $g(y)$ and $h(x)$ are differentiable functions. Determine all such vector fields.

In Exercises 17–20, compute the line integral $\displaystyle\int_{\mathcal{C}} f(x, y)\, ds$ for the given function and path or curve.

17. $f(x, y) = xy$, the path $\mathbf{c}(t) = (t, 2t - 1)$ for $0 \le t \le 1$

18. $f(x, y) = x - y$, the unit semicircle $x^2 + y^2 = 1$, $y \ge 0$

19. $f(x, y, z) = e^x - \dfrac{y}{2\sqrt{2z}}$, the path $\mathbf{c}(t) = \left(\ln t, \sqrt{2}t, \frac{1}{2}t^2 \right)$ for $1 \le t \le 2$

20. $f(x, y, z) = x + 2y + z$, the helix $\mathbf{c}(t) = (\cos t, \sin t, t)$ for $-1 \le t \le 3$

21. Find the total mass of an L-shaped rod consisting of the segments $(2t, 2)$ and $(2, 2 - 2t)$ for $0 \le t \le 1$ (length in centimeters) with mass density $\rho(x, y) = x^2 y$ g/cm.

22. Calculate $\mathbf{F} = \nabla\varphi$, where $\varphi(x, y, z) = xye^z$, and compute $\displaystyle\int_{\mathcal{C}} \mathbf{F} \cdot d\mathbf{s}$, where

(a) $\mathcal{C}$ is any curve from $(1, 1, 0)$ to $(3, e, -1)$.

(b) $\mathcal{C}$ is the boundary of the square $0 \le x, y \le 1$ oriented counterclockwise.

23. Calculate $\displaystyle\int_{\mathcal{C}_1} y^3\, dx + x^2 y\, dy$, where $\mathcal{C}_1$ is the oriented curve in Figure 1(A).

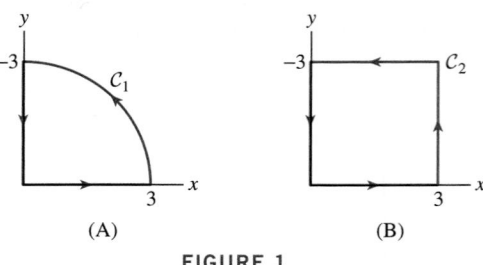

(A)   (B)

FIGURE 1

24. Let $\mathbf{F}(x, y) = \langle 9y - y^3, e^{\sqrt{y}}(x^2 - 3x) \rangle$ and let $\mathcal{C}_2$ be the oriented curve in Figure 1(B).

(a) Show that $\mathbf{F}$ is not conservative.

(b) Show that $\displaystyle\int_{\mathcal{C}_2} \mathbf{F} \cdot d\mathbf{s} = 0$ without explicitly computing the integral. *Hint:* Consider the direction of $\mathbf{F}$ along the edges of the square.

In Exercises 25–28, compute the line integral $\displaystyle\int_{\mathbf{c}} \mathbf{F} \cdot d\mathbf{s}$. for the given vector field and path.

25. $\mathbf{F}(x, y) = \left\langle \dfrac{2y}{x^2 + 4y^2}, \dfrac{x}{x^2 + 4y^2} \right\rangle$, the path $\mathbf{c}(t) = \left(\cos t, \frac{1}{2}\sin t \right)$ for $0 \le t \le 2\pi$

26. $\mathbf{F}(x, y) = \langle 2xy, x^2 + y^2 \rangle$, the part of the unit circle in the first quadrant oriented counterclockwise.

27. $\mathbf{F}(x, y) = \langle x^2 y, y^2 z, z^2 x \rangle$, the path $\mathbf{c}(t) = (e^{-t}, e^{-2t}, e^{-3t})$ for $0 \le t < \infty$

28. $\mathbf{F} = \nabla\varphi$, where $\varphi(x, y, z) = 4x^2 \ln(1 + y^4 + z^2)$, the path $\mathbf{c}(t) = (t^3, \ln(1 + t^2), e^t)$ for $0 \le t \le 1$

29. Consider the line integrals $\displaystyle\int_{\mathbf{c}} \mathbf{F} \cdot d\mathbf{s}$ for the vector fields $\mathbf{F}$ and paths $\mathbf{c}$ in Figure 2. Which two of the line integrals appear to have a value of zero? Which of the other two is negative?

(A)   (B)

FIGURE 2

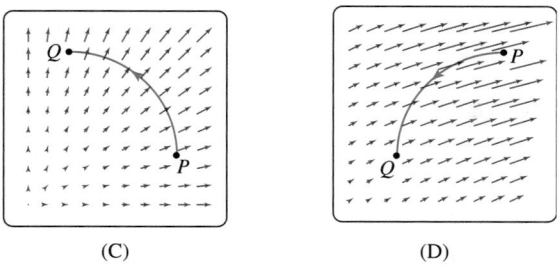

(C)   (D)

FIGURE 2 (*continued*)

30. Calculate the work required to move an object from $P = (1, 1, 1)$ to $Q = (3, -4, -2)$ against the force field $\mathbf{F}(x, y, z) = -12r^{-4}\langle x, y, z\rangle$ (distance in meters, force in Newtons), where $r = \sqrt{x^2 + y^2 + z^2}$. *Hint:* Find a potential function for $\mathbf{F}$.

31. Find constants a, b, c such that

$$\Phi(u, v) = (u + av, bu + v, 2u - c)$$

parametrizes the plane $3x - 4y + z = 5$. Calculate $\mathbf{T}_u$, $\mathbf{T}_v$, and $\mathbf{n}(u, v)$.

32. Calculate the integral of $f(x, y, z) = e^{z-y}$ over the portion of the plane $6x + 4y + z = 24$, where $x, y, z \geq 0$.

33. Let $\mathcal{S}$ be the surface parametrized by

$$\Phi(u, v) = \left(2u \sin \frac{v}{2}, 2u \cos \frac{v}{2}, 3v\right)$$

for $0 \leq u \leq 1$ and $0 \leq v \leq 2\pi$.
(a) Calculate the tangent vectors $\mathbf{T}_u$ and $\mathbf{T}_v$, and normal vector $\mathbf{n}(u, v)$ at $P = \Phi(1, \frac{\pi}{3})$.
(b) Find the equation of the tangent plane at P.
(c) Compute the surface area of $\mathcal{S}$.

34. ⌘⌘⌘ Plot the surface with parametrization

$$\Phi(u, v) = (u + 4v, 2u - v, 5uv)$$

for $-1 \leq v \leq 1$, $-1 \leq u \leq 1$. Express the surface area as a double integral and use a computer algebra system to compute the area numerically.

35. ⌘⌘⌘ Express the surface area of the surface $z = 10 - x^2 - y^2$, $-1 \leq x \leq 1$, $-3 \leq y \leq 3$ as a double integral. Evaluate the integral numerically using a CAS.

36. Evaluate $\iint_{\mathcal{S}} x^2 y \, dS$, where $\mathcal{S}$ is the surface $z = \sqrt{3}x + y^2$, $-1 \leq x \leq 1$, $0 \leq y \leq 1$.

37. Calculate $\iint_{\mathcal{S}} \left(x^2 + y^2\right) e^{-z} \, dS$, where $\mathcal{S}$ is the cylinder with equation $x^2 + y^2 = 9$ for $0 \leq z \leq 10$.

38. Let $\mathcal{S}$ be the upper hemisphere $x^2 + y^2 + z^2 = 1$, $z \geq 0$. For each of the functions (a)–(d), determine whether $\iint_{\mathcal{S}} f \, dS$ is positive, zero, or negative (without evaluating the integral). Explain your reasoning.

(a) $f(x, y, z) = y^3$
(b) $f(x, y, z) = z^3$
(c) $f(x, y, z) = xyz$
(d) $f(x, y, z) = z^2 - 2$

39. Let $\mathcal{S}$ be a small patch of surface with a parametrization $\Phi(u, v)$ for $0 \leq u, v \leq 0.1$ such that the normal vector $\mathbf{n}(u, v)$ for $(u, v) = (0, 0)$ is $\mathbf{n} = \langle 2, -2, 4\rangle$. Use Eq. (1) in Section 16.4 to estimate the surface area of $\mathcal{S}$.

40. Let $\mathcal{S}$ be the upper hemisphere of the sphere $x^2 + y^2 + z^2 = 9$ and let $\Phi(r, \theta) = (r \cos \theta, r \sin \theta, \sqrt{9 - r^2})$ be its parametrization by cylindrical coordinates (Figure 3).
(a) Calculate the normal vector $\mathbf{n} = \mathbf{T}_r \times \mathbf{T}_\theta$ at the point $\Phi(2, \frac{\pi}{3})$.
(b) Use Eq. (1) in Section 16.4 to estimate the surface area of $\Phi(\mathcal{R})$, where $\mathcal{R}$ is the small domain defined by

$$2 \leq r \leq 2.1, \qquad \frac{\pi}{3} \leq \theta \leq \frac{\pi}{3} + 0.05$$

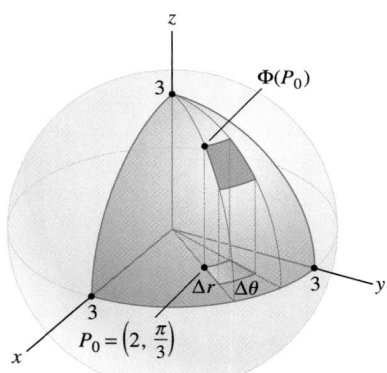

FIGURE 3

In Exercises 41–46, compute $\iint_{\mathcal{S}} \mathbf{F} \cdot d\mathbf{S}$ *for the given oriented surface or parametrized surface.*

41. $\mathbf{F}(x, y, z) = \langle y, x^2 y, e^{xz}\rangle$, $\quad x^2 + y^2 = 9, -3 \leq z \leq 3$, outward-pointing normal

42. $\mathbf{F}(x, y, z) = \langle -y, z, -x\rangle$, $\quad \Phi(u, v) = (u + 3v, v - 2u, 2v + 5)$, $0 \leq u, v \leq 1$, upward-pointing normal

43. $\mathbf{F}(x, y, z) = \langle x^2, y^2, x^2 + y^2\rangle$, $\quad x^2 + y^2 + z^2 = 4, z \geq 0$, outward-pointing normal

44. $\mathbf{F}(x, y, z) = \langle z, 0, z^2\rangle$, $\quad \Phi(u, v) = (v \cosh u, v \sinh u, v)$, $0 \leq u, v \leq 1$, upward-pointing normal

45. $\mathbf{F}(x, y, z) = \langle 0, 0, xze^{xy}\rangle$, $\quad z = xy, 0 \leq x, y \leq 1$, upward-pointing normal

46. $\mathbf{F}(x, y, z) = \langle 0, 0, z\rangle$, $\quad 3x^2 + 2y^2 + z^2 = 1, z \geq 0$, upward-pointing normal

47. Calculate the total charge on the cylinder

$$x^2 + y^2 = R^2, \qquad 0 \le z \le H$$

if the charge density in cylindrical coordinates is $\rho(\theta, z) = Kz^2 \cos^2 \theta$, where K is a constant.

48. Find the flow rate of a fluid with velocity field $\mathbf{v} = \langle 2x, y, xy \rangle$ m/s across the part of the cylinder $x^2 + y^2 = 9$ where $x, y \ge 0$, and $0 \le z \le 4$ (distance in meters).

49. With $\mathbf{v}$ as in Exercise 48, calculate the flow rate across the part of the elliptic cylinder $\dfrac{x^2}{4} + y^2 = 1$ where $x, y \ge 0$, and $0 \le z \le 4$.

50. Calculate the flux of the vector field $\mathbf{E}(x, y, z) = \langle x, 0, z \rangle$ through the part of the ellipsoid

$$4x^2 + 9y^2 + z^2 = 36$$

where $z \ge 3$. *Hint:* Use the parametrization

$$\Phi(r, \theta) = \left(3r \cos \theta, 2r \sin \theta, 6\sqrt{1 - r^2}\right)$$

17 FUNDAMENTAL THEOREMS OF VECTOR ANALYSIS

In this final chapter, we study the three main theorems of vector analysis: Green's Theorem, Stokes' Theorem, and the Divergence Theorem. This is a fitting conclusion to the text because each of these theorems is a vector generalization of the Fundamental Theorem of Calculus. This chapter is thus the culmination of our efforts to extend the concepts and methods of single-variable calculus to the multivariable setting. However, far from being a terminal point, vector analysis is the gateway to the field theories of mathematics, physics, and engineering. This includes, first and foremost, the theory of electricity and magnetism as expressed by the famous Maxwell equations. It also includes fluid dynamics, aerodynamics, analysis of continuous matter, and at a more advanced level, fundamental physical theories such as general relativity and the theory of elementary particles.

A kayaker on the Kananaskis River in Alberta, Canada, contends with the velocity vector field of a turbulent vortex.

17.1 Green's Theorem

Green's Theorem provides a new perspective on line integrals. We shall use the term *circulation* to denote a line integral around a closed curve. In Section 16.3, we proved that the circulation of a gradient vector field $\nabla \varphi$ around a closed curve is equal to zero. Green's Theorem expresses the circulation of a general vector field **F** as a double integral, which need not be zero if **F** is not a gradient vector field.

Before stating Green's Theorem, we recall a notation for line integrals introduced in Section 16.2. The line integral of $\mathbf{F} = \langle P, Q \rangle$ along an oriented curve $\mathcal{C}$ is denoted by

$$\underbrace{\int_{\mathcal{C}} P \, dx + Q \, dy}_{\text{Another notation for } \int_{\mathcal{C}} \mathbf{F} \cdot d\mathbf{s}}$$

"Circulation" is another term for the line integral of a vector field over a closed curve.

The symbolic differential $P \, dx + Q \, dy$ is interpreted in terms of a parametrization as follows. If $\mathbf{c}(t) = (x(t), y(t))$ parametrizes $\mathcal{C}$ (in the direction of its orientation) for $a \leq t \leq b$, we use the symbolic relations

$$dx = \frac{dx}{dt} \, dt \qquad \text{and} \qquad dy = \frac{dy}{dt} \, dt$$

$$P \, dx + Q \, dy = P(x(t), y(t)) \frac{dx}{dt} \, dt + Q(x(t), y(t)) \frac{dy}{dt} \, dt$$

to express the line integral as an ordinary integral with respect to t:

$$\int_{\mathcal{C}} P \, dx + Q \, dy = \int_a^b \left(P(x(t), y(t)) \frac{dx}{dt} + Q(x(t), y(t)) \frac{dy}{dt} \right) dt$$

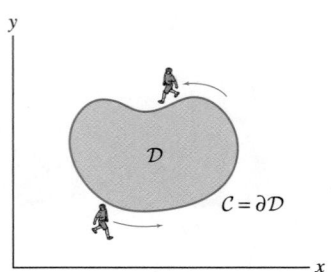

FIGURE 1 The boundary of the domain $\mathcal{D}$ is a simple closed curve $\mathcal{C}$, oriented in the counterclockwise direction.

Recall that if $\mathbf{F} = \nabla\varphi$, *then*

$$\frac{\partial Q}{\partial x} - \frac{\partial P}{\partial y} = 0$$

In this case, Green's Theorem merely confirms what we already know, namely that the circulation of a gradient vector field around any closed curve is zero.

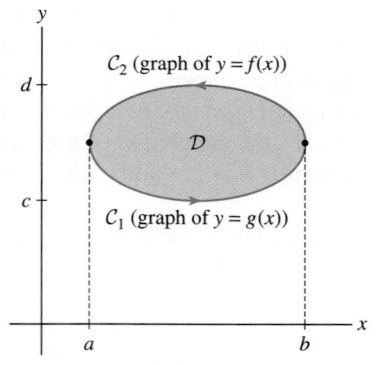

FIGURE 2 The boundary curve $\partial\mathcal{D}$ is the union of the graph of $y = g(x)$ and the graph of $y = f(x)$ oriented counterclockwise.

The integrand on the right is $\mathbf{F} \cdot \mathbf{c}'(t)$, so this coincides with our previous definition of the line integral $\int_{\mathcal{C}} \mathbf{F} \cdot d\mathbf{s}$.

To state Green's Theorem, recall that a **simple closed curve** is a closed curve $\mathcal{C}$ that does not intersect itself. In this case, $\mathcal{C}$ encloses a domain $\mathcal{D}$ (Figure 1). We assume that $\mathcal{C}$ has a parametrization $\mathbf{c}(t)$ such that $\mathbf{c}'(t)$ exists and is continuous with the exception of at most finitely many values of t where the components of $\mathbf{c}'(t)$ may have jump discontinuities (and $\mathcal{C}$ may have a corner). There are two ways to orient $\mathcal{C}$, namely clockwise or counterclockwise. We refer to the counterclockwise orientation as the **boundary orientation**. When you traverse $\mathcal{C}$ in the counterclockwise direction, the domain $\mathcal{D}$ enclosed by $\mathcal{C}$ lies to your left. We write $\partial\mathcal{D}$ to denote the boundary $\mathcal{C}$ with its counterclockwise orientation.

THEOREM 1 Green's Theorem Let $\mathcal{D}$ be a domain whose boundary $\partial\mathcal{D}$ is a simple closed curve, oriented counterclockwise. If $P(x, y)$ and $Q(x, y)$ are differentiable and have continuous first partial derivatives, then

$$\oint_{\partial\mathcal{D}} P\,dx + Q\,dy = \iint_{\mathcal{D}} \left(\frac{\partial Q}{\partial x} - \frac{\partial P}{\partial y} \right) dA \qquad \boxed{1}$$

Proof A complete proof of Green's Theorem for a general domain is beyond the scope of this text. To illustrate the main idea, we prove it for a domain $\mathcal{D}$ as in Figure 2, whose boundary can be described both as the union of two graphs $y = f(x)$ and $y = g(x)$ with $f(x) \geq g(x)$ and also as the union of two graphs $x = f_1(y)$ and $x = g_1(y)$, with $f_1(y) \geq g_1(y)$.

Equation (1) in Green's Theorem breaks up into two separate equalities, one involving P and one involving Q. Indeed, by additivity,

$$\oint_{\partial\mathcal{D}} P\,dx + Q\,dy = \oint_{\partial\mathcal{D}} \langle P(x, y), 0 \rangle \cdot d\mathbf{s} + \oint_{\partial\mathcal{D}} \langle 0, Q(x, y) \rangle \cdot d\mathbf{s}$$

Therefore, Green's Theorem follows if we prove

$$\oint_{\partial\mathcal{D}} \langle P(x, y), 0 \rangle \cdot d\mathbf{s} = -\iint_{\mathcal{D}} \frac{\partial P}{\partial y}\,dA \qquad \boxed{2}$$

$$\oint_{\partial\mathcal{D}} \langle 0, Q(x, y) \rangle \cdot d\mathbf{s} = \iint_{\mathcal{D}} \frac{\partial Q}{\partial x}\,dA \qquad \boxed{3}$$

To prove Eq. (2), we write the line integral as a sum:

$$\oint_{\partial\mathcal{D}} \langle P(x, y), 0 \rangle \cdot d\mathbf{s} = \oint_{\mathcal{C}_1} \langle P(x, y), 0 \rangle \cdot d\mathbf{s} + \oint_{\mathcal{C}_2} \langle P(x, y), 0 \rangle \cdot d\mathbf{s}$$

where $\mathcal{C}_1$ is the graph of $y = g(x)$ and $\mathcal{C}_2$ is the graph of $y = f(x)$, oriented as in Figure 2. We parameterize the graphs from left to right using x as parameter:

$$\text{Graph of } y = g(x): \qquad \mathbf{c}_1(x) = (x, g(x)), \qquad a \leq x \leq b$$

$$\text{Graph of } y = f(x): \qquad \mathbf{c}_2(x) = (x, f(x)), \qquad a \leq x \leq b$$

Since $\mathcal{C}_2$ is oriented from right to left, the line integral over $\partial\mathcal{D}$ is the difference of the integral over $\mathbf{c}_1$ and $\mathbf{c}_2$:

$$\oint_{\partial \mathcal{D}} \langle P(x, y), 0 \rangle \cdot d\mathbf{s} = \int_{\mathbf{c}_1} \langle P(x, y), 0 \rangle \cdot d\mathbf{s} - \int_{\mathbf{c}_2} \langle P(x, y), 0 \rangle \cdot d\mathbf{s}$$

$$= \int_{x=a}^{b} \langle P(x, g(x)), 0 \rangle \cdot \mathbf{c}_1'(x) \, dx$$

$$- \int_{x=a}^{b} \langle P(x, f(x)), 0 \rangle \cdot \mathbf{c}_2'(x) \, dx$$

Now we have

$$\langle P(x, g(x)), 0 \rangle \cdot \mathbf{c}_1'(x) = \langle P(x, g(x)), 0 \rangle \cdot \langle 1, g'(x) \rangle = P(x, g(x))$$

$$\langle P(x, f(x)), 0 \rangle \cdot \mathbf{c}_2'(x) = \langle P(x, f(x)), 0 \rangle \cdot \langle 1, f'(x) \rangle = P(x, f(x))$$

and thus

$$\oint_{\partial \mathcal{D}} \langle P(x, y), 0 \rangle \cdot d\mathbf{s} = \int_{x=a}^{b} P(x, g(x)) \, dx - \int_{x=a}^{b} P(x, f(x)) \, dx \qquad \boxed{4}$$

The key step is to apply the Fundamental Theorem of Calculus to $\dfrac{\partial P}{\partial y}(x, y)$ as a function of y with x held constant:

$$P(x, f(x)) - P(x, g(x)) = \int_{y=g(x)}^{f(x)} \frac{\partial P}{\partial y}(x, y) \, dy$$

We obtain Eq. (2) by substituting the integral on the right in Eq. (4):

$$\oint_{\partial \mathcal{D}} \langle P(x, y), 0 \rangle \cdot d\mathbf{s} = -\int_{x=a}^{b} \int_{y=g(x)}^{f(x)} \frac{\partial P}{\partial y}(x, y) \, dy \, dx = -\iint_{\mathcal{D}} \frac{\partial P}{\partial y} \, dA$$

Equation (3) is proved in a similar fashion, by expressing $\partial \mathcal{D}$ as the union of the graphs of $x = g_1(y)$ and $x = g_2(y)$. ∎

■ **EXAMPLE 1** Verifying Green's Theorem Verify Green's Theorem for the line integral

$$\oint_{\mathcal{C}} xy^2 \, dx + x \, dy$$

where $\mathcal{C}$ is the unit circle, oriented counterclockwise (Figure 3).

Solution

Step 1. **Evaluate the line integral.**

We use the standard parametrization of the unit circle:

$$x = \cos \theta, \qquad\qquad y = \sin \theta$$
$$dx = -\sin \theta \, d\theta, \qquad dy = \cos \theta \, d\theta$$

The integrand in the line integral becomes

$$xy^2 \, dx + x \, dy = \cos \theta \sin^2 \theta (-\sin \theta \, d\theta) + \cos \theta (\cos \theta \, d\theta)$$

$$= (-\cos \theta \sin^3 \theta + \cos^2 \theta) \, d\theta$$

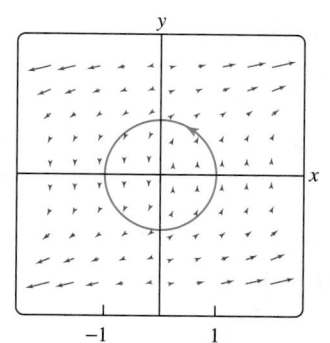

FIGURE 3 The vector field $\mathbf{F} = \langle xy^2, x \rangle$.

◄⋅⋅ **REMINDER** To integrate $\cos^2 \theta$, use the identity $\cos^2 \theta = \frac{1}{2}(1 + \cos 2\theta)$.

and

$$\oint_{C} xy^2 \, dx + x \, dy = \int_0^{2\pi} \left(-\cos \theta \sin^3 \theta + \cos^2 \theta \right) d\theta$$

$$= -\left. \frac{\sin^4 \theta}{4} \right|_0^{2\pi} + \frac{1}{2} \left(\theta + \frac{1}{2} \sin 2\theta \right) \Big|_0^{2\pi}$$

$$= 0 + \frac{1}{2}(2\pi + 0) = \boxed{\pi}$$

Step 2. Evaluate the double integral.

In this example, $P = xy^2$ and $Q = x$, so

$$\frac{\partial Q}{\partial x} - \frac{\partial P}{\partial y} = \frac{\partial}{\partial x} x - \frac{\partial}{\partial y} xy^2 = 1 - 2xy$$

The double integral in Green's Theorem is taken over the domain bounded by C, which is the disk $\mathcal{D}$ defined by $x^2 + y^2 \leq 1$:

$$\iint_{\mathcal{D}} \left(\frac{\partial Q}{\partial x} - \frac{\partial P}{\partial y} \right) dA = \iint_{\mathcal{D}} (1 - 2xy) \, dA$$

By symmetry, the integral of $2xy$ over $\mathcal{D}$ is zero because the contributions for positive and negative x cancel each other (you can also check this directly). Therefore,

$$\iint_{\mathcal{D}} \left(\frac{\partial Q}{\partial x} - \frac{\partial P}{\partial y} \right) dA = \iint_{\mathcal{D}} 1 \, dA = \text{Area}(\mathcal{D}) = \boxed{\pi}$$

Step 3. Compare.

Both the line integral and double integral have the value π, so Green's Theorem is verified in this case.   ∎

It is convenient to refer to the quantity appearing in the double integral in Green's Theorem as the **curl** or **scalar curl**. For $\mathbf{F} = \langle P, Q \rangle$, we write

$$\boxed{\operatorname{curl}_z(\mathbf{F}) = \frac{\partial Q}{\partial x} - \frac{\partial P}{\partial y}}$$

Notice the attached subscript z in curl_z. In the next section, we will define a vector quantity $\operatorname{curl}(\mathbf{F})$ for vector fields in $\mathbf{R}^3$. Our scalar quantity $\operatorname{curl}_z(\mathbf{F})$ is the z-component of the vector curl. With this notation, we may write Green's Theorem as

$$\boxed{\oint_{\partial \mathcal{D}} \mathbf{F} \cdot d\mathbf{s} = \iint_{\mathcal{D}} \operatorname{curl}_z(\mathbf{F}) \, dA}$$

■ **EXAMPLE 2** Computing a Line Integral Using Green's Theorem   Use Green's Theorem to compute $\oint_{C} \sin x \, dx + x^2 y^3 \, dy$, where C is the triangular path in Figure 4.

Solution We apply Green's Theorem to $\mathbf{F} = \langle P, Q \rangle = \langle \sin x, x^2 y^3 \rangle$:

$$\operatorname{curl}_z(\mathbf{F}) = \frac{\partial Q}{\partial x} - \frac{\partial P}{\partial y} = \frac{\partial}{\partial x} x^2 y^3 - \frac{\partial}{\partial y} \sin x = 2xy^3$$

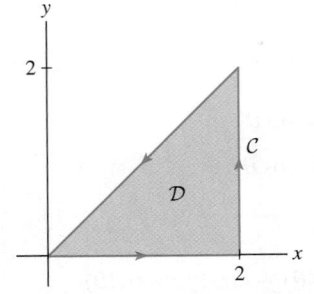

FIGURE 4 The region $\mathcal{D}$ is described by $0 \leq x \leq 2, 0 \leq y \leq x$.

The domain $\mathcal{D}$ enclosed by the triangle is described by $0 \le x \le 2, 0 \le y \le x$, and thus,

$$\oint_C \sin x \, dx + x^2 y^3 \, dy = \iint_{\mathcal{D}} \text{curl}_z(\mathbf{F}) \, dA$$

$$= \iint_{\mathcal{D}} 2xy^3 \, dA = \int_0^2 \int_{y=0}^x 2xy^3 \, dy \, dx$$

$$= \int_0^2 \left(\frac{1}{2} xy^4 \Big|_0^x \right) dx = \frac{1}{2} \int_0^2 x^5 \, dx = \frac{1}{12} x^6 \Big|_0^2 = \frac{16}{3} \quad \blacksquare$$

Green's Theorem applied to the vector field $\mathbf{F} = \langle -y, x \rangle$ leads to an interesting formula for the area of the region $\mathcal{D}$ enclosed by a simple closed curve C (Figure 5). In this case, $P = -y$ and $Q = x$, and

$$\text{curl}_z(\mathbf{F}) = \frac{\partial Q}{\partial x} - \frac{\partial P}{\partial y} = \frac{\partial}{\partial x} x - \frac{\partial}{\partial y}(-y) = 2$$

$$\oint_C -y \, dx + x \, dy = \iint_{\mathcal{D}} 2 \, dx \, dy = 2 \, \text{Area}(\mathcal{D})$$

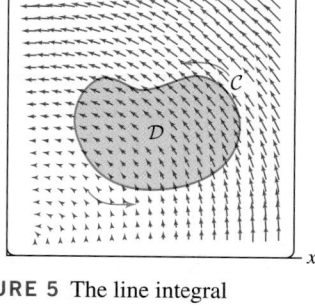

FIGURE 5 The line integral $\oint_C x \, dy - y \, dx$ is equal to twice the area enclosed by C.

We obtain the formula

$$\boxed{\text{Area enclosed by } C = \frac{1}{2} \oint_C x \, dy - y \, dx} \qquad \boxed{5}$$

A practical application of this relation is the **planimeter**, a device that computes the area of an irregular shape when you trace the boundary with a pointer at the end of a movable arm (Figure 6). The mathematical principle behind the planimeter is Eq. (5).

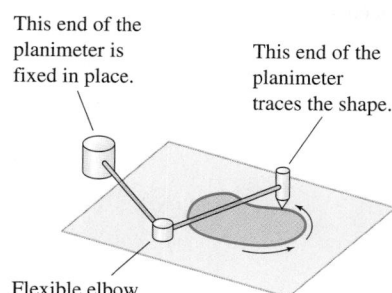

This end of the planimeter is fixed in place.

This end of the planimeter traces the shape.

Flexible elbow

FIGURE 6 Equation (5) is the basis of the planimeter, a mechanical device used to measure the areas of irregular shapes.

■ **EXAMPLE 3** Computing Area via Green's Theorem Compute the area of the ellipse $\left(\frac{x}{a}\right)^2 + \left(\frac{y}{b}\right)^2 = 1$ as a line integral.

Solution We parametrize the boundary of the ellipse by

$$x = a \cos \theta, \qquad y = b \sin \theta, \qquad 0 \le \theta < 2\pi$$

and compute the area using Eq. (5):

$$x \, dy - y \, dx = (a \cos \theta)(b \cos \theta \, d\theta) - (b \sin \theta)(-a \sin \theta \, d\theta)$$

$$= ab(\cos^2 \theta + \sin^2 \theta) \, d\theta = ab \, d\theta$$

$$\text{Enclosed area} = \frac{1}{2}\oint_C x\,dy - y\,dx = \frac{1}{2}\int_0^{2\pi} ab\,d\theta = \pi ab$$

This is the well-known formula for the area of an ellipse. ■

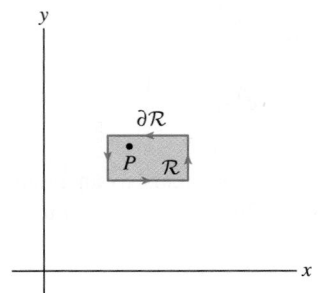

FIGURE 7 The circulation of **F** around $\partial\mathcal{R}$ is approximately $\text{curl}_z(\mathbf{F})(P) \cdot \text{Area}(\mathcal{R})$.

CONCEPTUAL INSIGHT   What does the quantity $\text{curl}_z(\mathbf{F})$ represent? To answer this question, let us apply Green's Theorem to a small region $\mathcal{R}$ whose boundary $\mathcal{C} = \partial\mathcal{R}$ is a simple closed curve. Since $\mathcal{R}$ is small and **F** is continuous, $\text{curl}_z(\mathbf{F})$ is nearly constant on $\mathcal{R}$, and to a first approximation, we may replace the function $\text{curl}_z(\mathbf{F})$ by the constant value $\text{curl}_z(\mathbf{F})(P)$, where P is any sample point in $\mathcal{R}$ (Figure 7). Green's Theorem yields the following approximation for the circulation of **F** around $\mathcal{C}$:

$$\oint_C \mathbf{F}\cdot d\mathbf{s} = \iint_{\mathcal{R}} \text{curl}_z(\mathbf{F})\,dA \approx \text{curl}_z(\mathbf{F})(P)\cdot\text{Area}(\mathcal{R})$$

In other words, *the circulation around a small region $\mathcal{R}$ is, to a first-order approximation, equal to the curl times the area.* We may think of $\text{curl}_z(\mathbf{F})(P)$ as the **circulation per unit area**.

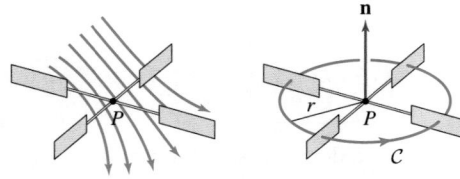

FIGURE 8 The curl at P is approximately equal to one-half the angular velocity (in radians per unit time) of a small paddle wheel placed at P.

GRAPHICAL INSIGHT   More geometrically, we may think of the curl as a measure of a vector field's "tendency to rotate." To make this idea precise, let's think of **F** as the velocity field of a fluid. The particles of fluid flow along curves called **integral curves** (or **stream lines**) that are tangent to the vector field at each point. We can measure the curl at a point P by placing a small paddle wheel in the stream at P and observing how fast it rotates (Figure 8).

To justify this statement, suppose that the paddle wheel has radius r and let $\mathcal{C}_r$ be a circle of radius r centered at P. Each paddle on the wheel is constrained to move along the circle $\mathcal{C}_r$ and is pushed to move with a velocity equal to the tangential component of **F**. We may assume that the paddle wheel itself rotates with a velocity v_a equal to the average of the tangential components along the circle. The average is equal to the circulation divided by the length $2\pi r$ of the circle:

$$v_a = \text{velocity of paddle wheel} = \frac{1}{2\pi r}\oint_{\mathcal{C}_r}\mathbf{F}\cdot d\mathbf{s}$$

According to the Conceptual Insight, if r is small, then the circulation is approximately equal to $\text{curl}_z(\mathbf{F})(P)$ times the area πr^2 of the circle. Thus we obtain

$$v_a \approx \frac{1}{2\pi r}(\pi r^2)\text{curl}_z(\mathbf{F})(P) = \left(\frac{1}{2}r\right)\text{curl}_z(\mathbf{F})(P)$$

Angular Velocity   On a circle of radius r meters, an arc of ℓ meters has radian measure ℓ/r. Therefore, an object moving along the circle with a speed of v meters per second travels v/r radians per second. In other words, the object has angular velocity v/r.

Now if an object moves in a circle with speed v_a, then its angular velocity (in radians per unit time) is $v_a/r \approx \frac{1}{2}\text{curl}_z(\mathbf{F})(P)$. Thus, *the angular velocity of the paddle wheel is approximately one-half the curl.*

Figures 9(A)–(C) show vector fields with nonzero constant curl. The vector field in (A) has positive curl and describes a fluid rotating counterclockwise around the origin. The vector field in (B) has negative curl and describes a fluid that spirals into the origin in the clockwise direction. However, a nonzero curl does not mean that the fluid is necessarily rotating—only that a small paddle wheel would rotate if placed in the fluid. The fluid in Figure 9(C) does not rotate even though its curl is nonzero (this is an example of **shear flow**, also known as a Couette flow). Compare with the vector fields in Figures 9(D) and (E), which have zero curl.

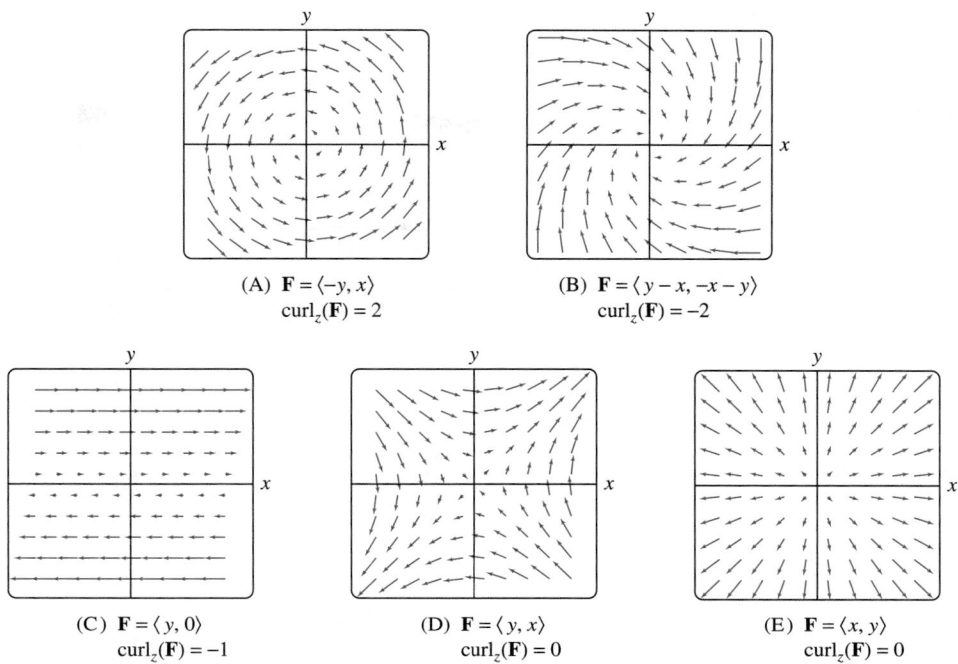

(A) $\mathbf{F} = \langle -y, x \rangle$
$\operatorname{curl}_z(\mathbf{F}) = 2$

(B) $\mathbf{F} = \langle y - x, -x - y \rangle$
$\operatorname{curl}_z(\mathbf{F}) = -2$

(C) $\mathbf{F} = \langle y, 0 \rangle$
$\operatorname{curl}_z(\mathbf{F}) = -1$

(D) $\mathbf{F} = \langle y, x \rangle$
$\operatorname{curl}_z(\mathbf{F}) = 0$

(E) $\mathbf{F} = \langle x, y \rangle$
$\operatorname{curl}_z(\mathbf{F}) = 0$

FIGURE 9

Circulation around a closed curve has an important additivity property: If we decompose a domain $\mathcal{D}$ into two (or more) domains $\mathcal{D}_1$ and $\mathcal{D}_2$ that intersect only on part of their boundaries as in Figure 10, then

$$\oint_{\partial \mathcal{D}} \mathbf{F} \cdot d\mathbf{s} = \oint_{\partial \mathcal{D}_1} \mathbf{F} \cdot d\mathbf{s} + \oint_{\partial \mathcal{D}_2} \mathbf{F} \cdot d\mathbf{s} \qquad \boxed{6}$$

To verify this equation, note first that

$$\oint_{\partial \mathcal{D}} \mathbf{F} \cdot d\mathbf{s} = \int_{\mathcal{C}_{\text{top}}} \mathbf{F} \cdot d\mathbf{s} + \int_{\mathcal{C}_{\text{bot}}} \mathbf{F} \cdot d\mathbf{s}$$

with $\mathcal{C}_{\text{top}}$ and $\mathcal{C}_{\text{bot}}$ as in Figure 10. Then observe that the dashed segment $\mathcal{C}_{\text{middle}}$ occurs in both $\partial \mathcal{D}_1$ and $\partial \mathcal{D}_2$ but with opposite orientations. With $\mathcal{C}_{\text{middle}}$ oriented right to left,

$$\oint_{\partial \mathcal{D}_1} \mathbf{F} \cdot d\mathbf{s} = \int_{\mathcal{C}_{\text{top}}} \mathbf{F} \cdot d\mathbf{s} - \int_{\mathcal{C}_{\text{middle}}} \mathbf{F} \cdot d\mathbf{s}$$

$$\oint_{\partial \mathcal{D}_2} \mathbf{F} \cdot d\mathbf{s} = \int_{\mathcal{C}_{\text{bot}}} \mathbf{F} \cdot d\mathbf{s} + \int_{\mathcal{C}_{\text{middle}}} \mathbf{F} \cdot d\mathbf{s}$$

$\partial \mathcal{D}$

$\mathcal{D}$

$\mathcal{C}_{\text{top}}$

$\mathcal{D}_1$

$=$

$\mathcal{C}_{\text{middle}}$

$+$

$\mathcal{C}_{\text{middle}}$

$\mathcal{D}_2$

$\mathcal{C}_{\text{bottom}}$

FIGURE 10

We obtain Eq. (6) by adding these two equations:

$$\oint_{\partial\mathcal{D}_1} \mathbf{F} \cdot d\mathbf{s} + \oint_{\partial\mathcal{D}_2} \mathbf{F} \cdot d\mathbf{s} = \int_{C_{\text{top}}} \mathbf{F} \cdot d\mathbf{s} + \int_{C_{\text{bot}}} \mathbf{F} \cdot d\mathbf{s} = \oint_{\partial\mathcal{D}} \mathbf{F} \cdot \mathbf{s}$$

More General Form of Green's Theorem

Green's Theorem remains valid when $\mathcal{D}$ is a region whose boundary consists of more than one simple closed curve as in Figure 11, provided that we orient the boundary in the correct way. *Each boundary curve must be oriented so that the region lies to your left as you traverse the curve in the direction specified by the orientation.* We write $\partial\mathcal{D}$ for the boundary of $\mathcal{D}$ with this "boundary orientation." For the domains in Figure 11,

$$\partial\mathcal{D}_1 = C_1 + C_2, \qquad \partial\mathcal{D}_2 = C_1 + C_2 - C_3$$

The curve C_3 occurs with a minus sign because it is oriented counterclockwise, but the boundary orientation requires a clockwise orientation.

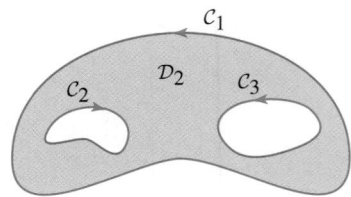

(A) Oriented boundary of $\mathcal{D}_1$ is $C_1 + C_2$ (B) Oriented boundary of $\mathcal{D}_2$ is $C_1 + C_2 - C_3$

FIGURE 11

With this definition of $\partial\mathcal{D}$, Green's Theorem takes the form

$$\boxed{\oint_{\partial\mathcal{D}} \mathbf{F} \cdot d\mathbf{s} = \iint_{\mathcal{D}} \text{curl}_z(\mathbf{F}) \, dA} \qquad \boxed{7}$$

This general form of Green's Theorem may be proved using additivity. Consider the region $\mathcal{D}$ in Figure 12. We may decompose $\mathcal{D}$ into domains $\mathcal{D}_1$ and $\mathcal{D}_2$, each of which is bounded by a simple closed curve. Then,

$$\partial\mathcal{D} = \partial\mathcal{D}_1 + \partial\mathcal{D}_2$$

because the edges common to $\partial\mathcal{D}_1$ and $\partial\mathcal{D}_2$ occur with opposite orientation and therefore cancel. Our previous version of Green's Theorem applies to both $\mathcal{D}_1$ and $\mathcal{D}_2$ and thus

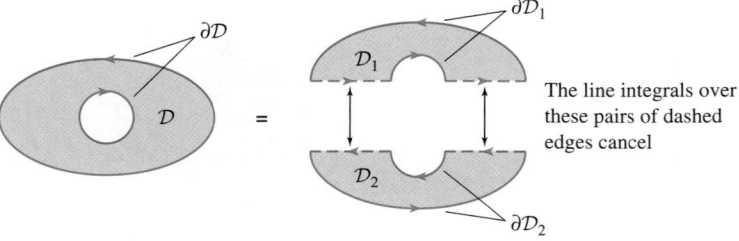

The line integrals over these pairs of dashed edges cancel

FIGURE 12 The boundary of $\partial\mathcal{D}$ is the sum $\partial\mathcal{D}_1 + \partial\mathcal{D}_2$ because the straight edges cancel.

$$\oint_{\partial \mathcal{D}} \mathbf{F} \cdot d\mathbf{s} = \int_{\partial \mathcal{D}_1} \mathbf{F} \cdot d\mathbf{s} + \int_{\partial \mathcal{D}_2} \mathbf{F} \cdot d\mathbf{s}$$

$$= \iint_{\mathcal{D}_1} \mathrm{curl}_z(\mathbf{F})\, dA + \iint_{\mathcal{D}_2} \mathrm{curl}_z(\mathbf{F})\, dA = \iint_{\mathcal{D}} \mathrm{curl}_z(\mathbf{F})\, dA$$

■ **EXAMPLE 4** Using the General Form of Green's Theorem The domain $\mathcal{D}$ in Figure 13 has area 8. Calculate $\oint_{\mathcal{C}_1} \mathbf{F} \cdot d\mathbf{s}$, where $\mathbf{F} = \langle x - y, x + y^3 \rangle$ and $\mathcal{C}_1$ is the outer boundary curve of $\mathcal{D}$ oriented counterclockwise.

Solution We cannot compute the line integral over $\mathcal{C}_1$ directly because the curve $\mathcal{C}_1$ is not specified. However $\partial \mathcal{D} = \mathcal{C}_1 - \mathcal{C}_2$, so Green's Theorem yields

$$\oint_{\mathcal{C}_1} \mathbf{F} \cdot d\mathbf{s} - \oint_{\mathcal{C}_2} \mathbf{F} \cdot d\mathbf{s} = \iint_{\mathcal{D}} \mathrm{curl}_z(\mathbf{F})\, dA$$

or

$$\oint_{\mathcal{C}_1} \mathbf{F} \cdot d\mathbf{s} = \oint_{\mathcal{C}_2} \mathbf{F} \cdot d\mathbf{s} + \iint_{\mathcal{D}} \mathrm{curl}_z(\mathbf{F})\, dA \qquad \boxed{8}$$

We compute each of the integrals on the right. We have

$$\mathrm{curl}_z(\mathbf{F}) = \frac{\partial}{\partial x}(x + y^3) - \frac{\partial}{\partial y}(x - y) = 1 - (-1) = 2$$

and therefore,

$$\iint_{\mathcal{D}} \mathrm{curl}_z(\mathbf{F})\, dA = 2 \iint_{\mathcal{D}} dA = 2\,\mathrm{Area}(\mathcal{D}) = 2(8) = 16$$

To compute the line integral over $\mathcal{C}_2$, we parametrize the circle by $\mathbf{c}(t) = (\cos\theta, \sin\theta)$. Then

$$\mathbf{F} \cdot \mathbf{c}'(t) = \langle \cos\theta - \sin\theta, \cos\theta + \sin^3\theta \rangle \cdot \langle -\sin\theta, \cos\theta \rangle$$

$$= -\sin\theta\cos\theta + \sin^2\theta + \cos^2\theta + \sin^3\theta\cos\theta$$

$$= 1 - \sin\theta\cos\theta + \sin^3\theta\cos\theta$$

The integrals of $\sin\theta\cos\theta$ and $\sin^3\theta\cos\theta$ over $[0, 2\pi]$ are both zero, so

$$\oint_{\mathcal{C}_2} \mathbf{F} \cdot d\mathbf{s} = \int_0^{2\pi}(1 - \sin\theta\cos\theta + \sin^3\theta\cos\theta)\, d\theta = \int_0^{2\pi} d\theta = 2\pi$$

Thus, Eq. (8) yields $\oint_{\mathcal{C}_1} \mathbf{F} \cdot d\mathbf{s} = 16 + 2\pi$. ■

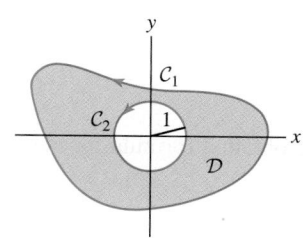

FIGURE 13 $\mathcal{D}$ has area 8 and $\mathcal{C}_2$ is a circle of radius 1.

17.1 SUMMARY

• We have two notations for the line integral of a vector field $\mathbf{F} = \langle P, Q \rangle$ over a curve $\mathcal{C}$:

$$\int_{\mathcal{C}} \mathbf{F} \cdot d\mathbf{s} \qquad \text{and} \qquad \int_{\mathcal{C}} P\, dx + Q\, dy$$

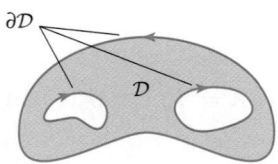

FIGURE 14 The boundary orientation is chosen so that the region lies to your left as you walk along the curve.

- We write $\partial\mathcal{D}$ for the boundary of a domain $\mathcal{D}$ with its boundary orientation (Figure 14).
- Green's Theorem is the equality

$$\oint_{\partial\mathcal{D}} P\,dx + Q\,dy = \iint_{\mathcal{D}} \left(\frac{\partial Q}{\partial x} - \frac{\partial P}{\partial y}\right) dA$$

- The quantity appearing in the double integral of Green's Theorem is called the *curl* or *scalar curl*:

$$\operatorname{curl}_z(\mathbf{F}) = \frac{\partial Q}{\partial x} - \frac{\partial P}{\partial y}$$

- Taking $P = -y$ and $Q = x$ in Green's Theorem, we obtain a formula for the area of the region $\mathcal{D}$ enclosed by $\mathcal{C}$:

$$\operatorname{Area}(\mathcal{D}) = \frac{1}{2}\oint_{\mathcal{C}} x\,dy - y\,dx$$

17.1 EXERCISES

Preliminary Questions

1. Which vector field $\mathbf{F}$ is being integrated in the line integral $\oint x^2\,dy - e^y\,dx$?

2. Draw a domain in the shape of an ellipse and indicate with an arrow the boundary orientation of the boundary curve. Do the same for the annulus (the region between two concentric circles).

3. The circulation of a gradient vector field around a closed curve is zero. Is this fact consistent with Green's Theorem? Explain.

4. Which of the following vector fields possess the following property: For every simple closed curve $\mathcal{C}$, $\int_{\mathcal{C}} \mathbf{F} \cdot d\mathbf{s}$ is equal to the area enclosed by $\mathcal{C}$?
(a) $\mathbf{F} = \langle -y, 0\rangle$
(b) $\mathbf{F} = \langle x, y\rangle$
(c) $\mathbf{F} = \langle \sin(x^2), x + e^{y^2}\rangle$

Exercises

1. Verify Green's Theorem for the line integral $\oint_{\mathcal{C}} xy\,dx + y\,dy$, where $\mathcal{C}$ is the unit circle, oriented counterclockwise.

2. Let $I = \oint_{\mathcal{C}} \mathbf{F} \cdot d\mathbf{s}$, where $\mathbf{F} = \langle y + \sin x^2, x^2 + e^{y^2}\rangle$ and $\mathcal{C}$ is the circle of radius 4 centered at the origin.
(a) Which is easier: evaluating I directly or using Green's Theorem?
(b) Evaluate I using the easier method.

In Exercises 3–11, use Green's Theorem to evaluate the line integral. Orient the curve counterclockwise unless otherwise indicated.

3. $\oint_{\mathcal{C}} y^2\,dx + x^2\,dy$, where $\mathcal{C}$ is the boundary of the unit square $0 \le x \le 1, 0 \le y \le 1$

4. $\oint_{\mathcal{C}} e^{2x+y}\,dx + e^{-y}\,dy$, where $\mathcal{C}$ is the triangle with vertices $(0, 0)$, $(1, 0)$, and $(1, 1)$

5. $\oint_{\mathcal{C}} x^2 y\,dx$, where $\mathcal{C}$ is the unit circle centered at the origin

6. $\oint_{\mathcal{C}} \mathbf{F} \cdot d\mathbf{s}$, where $\mathbf{F} = \langle x + y, x^2 - y\rangle$ and $\mathcal{C}$ is the boundary of the region enclosed by $y = x^2$ and $y = \sqrt{x}$ for $0 \le x \le 1$.

7. $\oint_{\mathcal{C}} \mathbf{F} \cdot d\mathbf{s}$, where $\mathbf{F} = \langle x^2, x^2\rangle$ and $\mathcal{C}$ consists of the arcs $y = x^2$ and $y = x$ for $0 \le x \le 1$

8. $\oint_{\mathcal{C}} (\ln x + y)\,dx - x^2\,dy$, where $\mathcal{C}$ is the rectangle with vertices $(1, 1)$, $(3, 1)$, $(1, 4)$, and $(3, 4)$

9. The line integral of $\mathbf{F} = \langle x^3, 4x\rangle$ around the boundary of the parallelogram in Figure 15 (note the orientation)

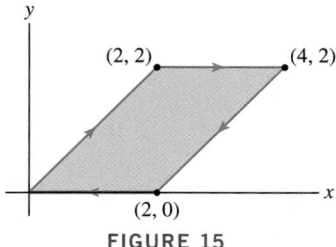

FIGURE 15

10. The line integral of $\mathbf{F} = \langle e^{x+y}, e^{x-y}\rangle$ around the boundary of the parallelogram in Figure 15

11. $\int_C xy\,dx + (x^2 + x)\,dy$, where C is the path in Figure 16

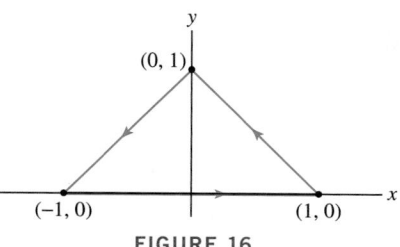

(0, 1)

(−1, 0)   (1, 0)

FIGURE 16

12. Let C_R be the circle of radius R centered at the origin. Use Green's Theorem to determine $\oint_{C_2} \mathbf{F} \cdot d\mathbf{s}$, where $\mathbf{F}$ is a vector field such that $\oint_{C_1} \mathbf{F} \cdot d\mathbf{s} = 9$ and $\text{curl}_z(\mathbf{F}) = x^2 + y^2$ for $1 \le x^2 + y^2 \le 4$.

In Exercises 13–16, use Eq. (5) to calculate the area of the given region.

13. The circle of radius 3 centered at the origin

14. The triangle with vertices $(0, 0)$, $(1, 0)$, and $(1, 1)$

15. The region between the x-axis and the cycloid parametrized by $\mathbf{c}(t) = (t - \sin t, 1 - \cos t)$ for $0 \le t \le 2\pi$ (Figure 17)

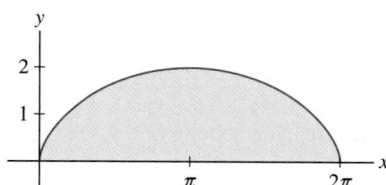

FIGURE 17 Cycloid.

16. The region between the graph of $y = x^2$ and the x-axis for $0 \le x \le 2$

17. Let $x^3 + y^3 = 3xy$ be the **folium of Descartes** (Figure 18).

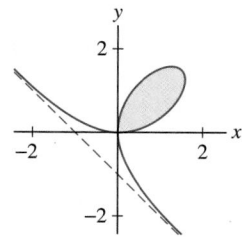

FIGURE 18 Folium of Descartes.

(a) Show that the folium has a parametrization in terms of $t = y/x$ given by

$$x = \frac{3t}{1+t^3}, \qquad y = \frac{3t^2}{1+t^3} \qquad (-\infty < t < \infty) \quad (t \ne -1)$$

(b) Show that

$$x\,dy - y\,dx = \frac{9t^2}{(1+t^3)^2}\,dt$$

Hint: By the Quotient Rule,

$$x^2 d\left(\frac{y}{x}\right) = x\,dy - y\,dx$$

(c) Find the area of the loop of the folium.

18. Follow the procedure of Exercise 17 to find the area of the loop of the lemniscate curve with equation $(x^2 + y^2)^2 = xy$ (Figure 19).

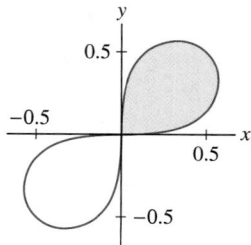

FIGURE 19 Lemniscate.

19. Show that if C is a simple closed curve, then

$$\oint_C -y\,dx = \oint_C x\,dy$$

and both integrals are equal to the area enclosed by C.

20. For the vector fields (A)–(D) in Figure 20, state whether the curl at the origin appears to be positive, negative, or zero.

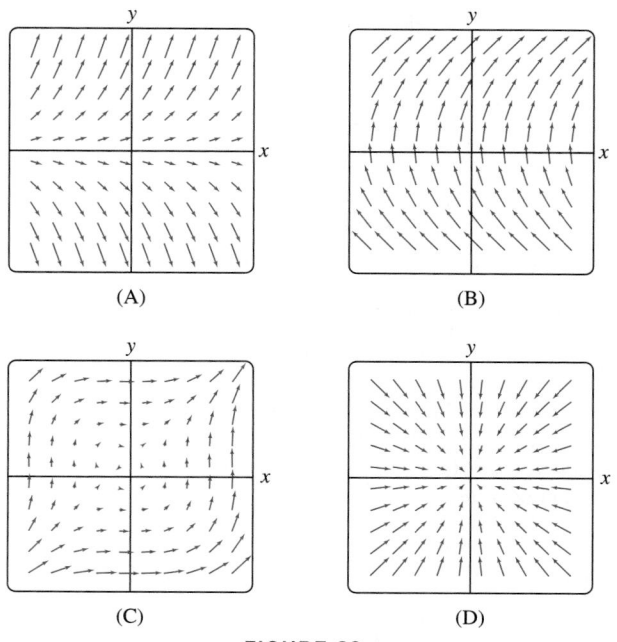

(A)

(B)

(C)

(D)

FIGURE 20

21. Let $\mathbf{F} = \langle 2xe^y, x + x^2 e^y \rangle$ and let C be the quarter-circle path from A to B in Figure 21. Evaluate $I = \oint_C \mathbf{F} \cdot d\mathbf{s}$ as follows:

(a) Find a function $\varphi(x, y)$ such that $\mathbf{F} = \mathbf{G} + \nabla\varphi$, where $\mathbf{G} = \langle 0, x \rangle$.

(b) Show that the line integrals of $\mathbf{G}$ along the segments $\overline{OA}$ and $\overline{OB}$ are zero.

(c) Use Green's Theorem to show that

$$I = \varphi(B) - \varphi(A) + 4\pi$$

and evaluate I.

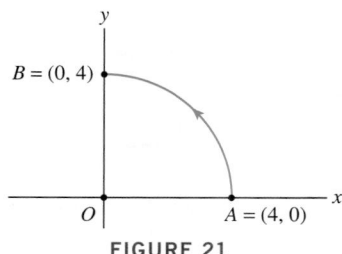

FIGURE 21

22. Compute the line integral of $\mathbf{F} = \langle x^3, 4x \rangle$ along the path from A to B in Figure 22. *Hint:* To save work, use Green's Theorem to relate this line integral to the line integral along the vertical path from B to A.

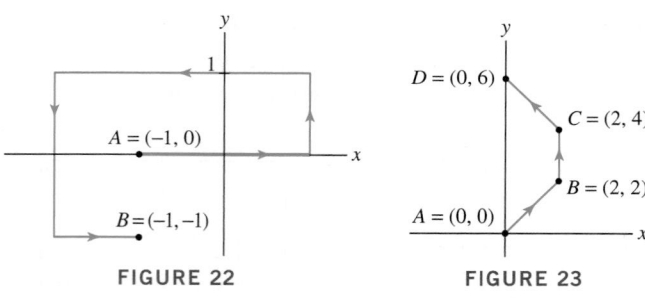

FIGURE 22   **FIGURE 23**

23. Evaluate $I = \int_C (\sin x + y)\,dx + (3x + y)\,dy$ for the nonclosed path $ABCD$ in Figure 23. *Hint:* Use the method of Exercise 22.

24. Estimate the circulation of a vector field $\mathbf{F}$ around a circle of radius $R = 0.1$, assuming that $\mathrm{curl}_z(\mathbf{F})$ takes the value 4 at the center of the circle.

25. Let $\mathbf{F}$ be the velocity field. Estimate the circulation of $\mathbf{F}$ around a circle of radius $R = 0.05$ with center P, assuming that $\mathrm{curl}_z(\mathbf{F})(P) = -3$. In which direction would a small paddle placed at P rotate? How fast would it rotate (in radians per second) if $\mathbf{F}$ is expressed in meters per second?

26. Referring to Figure 24, suppose that $\oint_{C_2} \mathbf{F} \cdot d\mathbf{s} = 12$. Use Green's Theorem to determine $\int_{C_1} \mathbf{F} \cdot d\mathbf{s}$, assuming that $\mathrm{curl}_z(\mathbf{F}) = -3$ in $\mathcal{D}$.

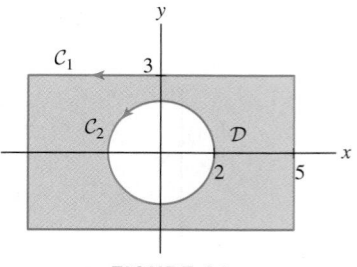

FIGURE 24

27. Referring to Figure 25, suppose that

$$\oint_{C_2} \mathbf{F} \cdot d\mathbf{s} = 3\pi, \qquad \oint_{C_3} \mathbf{F} \cdot d\mathbf{s} = 4\pi$$

Use Green's Theorem to determine the circulation of $\mathbf{F}$ around C_1, assuming that $\mathrm{curl}_z(\mathbf{F}) = 9$ on the shaded region.

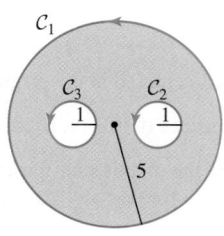

FIGURE 25

28. Area of a Polygon Green's Theorem leads to a convenient formula for the area of a polygon.

(a) Let C be the line segment joining (x_1, y_1) to (x_2, y_2). Show that

$$\frac{1}{2}\oint_C -y\,dx + x\,dy = \frac{1}{2}(x_1 y_2 - x_2 y_1)$$

(b) Prove that the area of the polygon with vertices $(x_1, y_1), (x_2, y_2), \ldots, (x_n, y_n)$ is equal [where we set $(x_{n+1}, y_{n+1}) = (x_1, y_1)$] to

$$\frac{1}{2}\sum_{i=1}^{n}(x_i y_{i+1} - x_{i+1}y_i)$$

29. Use the result of Exercise 28 to compute the areas of the polygons in Figure 26. Check your result for the area of the triangle in (A) using geometry.

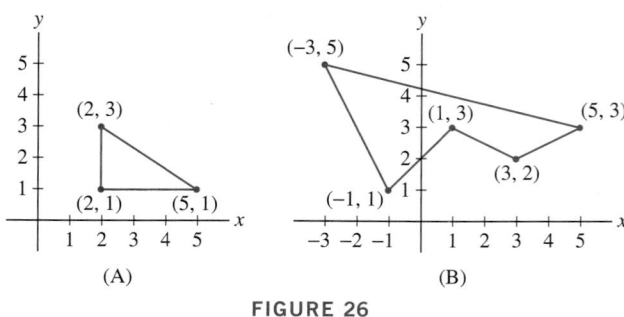

(A)   **(B)**

FIGURE 26

Further Insights and Challenges

In Exercises 30–31, let **F** be the vortex vector field [defined for $(x, y) \neq (0, 0)$]:

$$\mathbf{F} = \left\langle \frac{-y}{x^2 + y^2}, \frac{x}{x^2 + y^2} \right\rangle$$

30. (a) Show that $\text{curl}_z (\mathbf{F})(P) = 0$ for all $P \neq (0, 0)$.

(b) Show that the circulation of **F** around the circle $\mathcal{C}_R$ of radius R centered at the origin is equal to 2π for all R.

31. Prove that if $\mathcal{C}$ is a simple closed curve whose interior contains the origin, then $\oint_{\mathcal{C}} \mathbf{F} \cdot d\mathbf{s} = 2\pi$ (Figure 27). *Hint:* Apply Green's Theorem to the domain between $\mathcal{C}$ and $\mathcal{C}_R$ where R is so small that $\mathcal{C}_R$ is contained in $\mathcal{C}$.

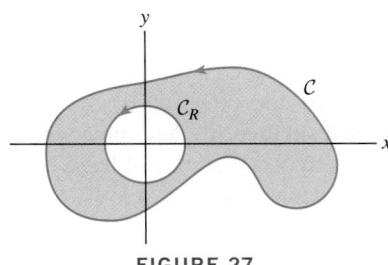

FIGURE 27

In Exercises 32–34, the *conjugate* of a vector field $\mathbf{F} = \langle P, Q \rangle$ is the vector field $\mathbf{F}^* = \langle -Q, P \rangle$.

32. Explain the following statement: $\mathbf{F}^*$ is the vector field obtained by rotating the vectors of **F** counterclockwise through an angle of $\frac{\pi}{2}$.

33. The normal component of **F** at a point P on a simple closed $\mathcal{C}$ is the quantity $\mathbf{F}(P) \cdot \mathbf{n}(P)$, where $\mathbf{n}(P)$ is the outward-pointing unit normal vector. The **flux** of **F** across curve $\mathcal{C}$ is defined as the line integral of the normal component around $\mathcal{C}$ (Figure 28). Show that the flux across $\mathcal{C}$ is equal to $\oint_{\mathcal{C}} \mathbf{F}^* \cdot d\mathbf{s}$.

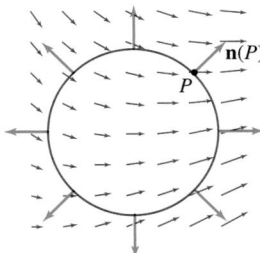

FIGURE 28 The flux of **F** is the integral of the normal component $\mathbf{F} \cdot \mathbf{n}$ around the curve.

34. Define $\text{div}(\mathbf{F}) = \dfrac{\partial P}{\partial x} + \dfrac{\partial Q}{\partial y}$. Use Green's Theorem to prove that for any simple closed curve $\mathcal{C}$,

$$\text{Flux across } \mathcal{C} = \iint_{\mathcal{D}} \text{div}(\mathbf{F}) \, dA$$

where $\mathcal{D}$ is the region enclosed by $\mathcal{C}$. This is a two-dimensional version of the **Divergence Theorem** discussed in Section 17.3.

In Exercises 35–38, the **Laplace operator** Δ is defined by

$$\Delta \varphi = \frac{\partial^2 \varphi}{\partial x^2} + \frac{\partial^2 \varphi}{\partial y^2} \qquad \boxed{9}$$

35. Let $\mathbf{F} = \nabla \varphi$. Show that $\text{curl}_z (\mathbf{F}^*) = \Delta \varphi$, where $\mathbf{F}^*$ is the conjugate vector field (defined in Exercises 32–34).

36. Let **n** denote the outward-pointing unit normal vector to a simple closed $\mathcal{C}$. The **normal derivative** of a function φ, denoted $\dfrac{\partial \varphi}{\partial \mathbf{n}}$, is the directional derivative $D_{\mathbf{n}}(\varphi) = \nabla \varphi \cdot \mathbf{n}$. Prove the formula

$$\oint_{\mathcal{C}} \frac{\partial \varphi}{\partial \mathbf{n}} \, ds = \iint_{\mathcal{D}} \Delta \varphi \, dA$$

where $\mathcal{D}$ is the domain enclosed by $\mathcal{C}$. *Hint:* Let $\mathbf{F} = \nabla \varphi$. Show that $\dfrac{\partial \varphi}{\partial \mathbf{n}} = \mathbf{F}^* \cdot \mathbf{T}$, where **T** is the unit tangent vector pointing in the counterclockwise direction along $\mathcal{C}$, and apply Green's Theorem.

37. Let $P = (a, b)$ and let $\mathcal{C}(r)$ be the circle of radius r centered at P. The average value of a continuous function φ on $\mathcal{C}(r)$ is defined as the integral

$$I_\varphi(r) = \frac{1}{2\pi} \int_0^{2\pi} \varphi(a + r\cos\theta, b + r\sin\theta) \, d\theta$$

(a) Show that

$$\frac{\partial \varphi}{\partial \mathbf{n}}(a + r\cos\theta, b + r\sin\theta)$$

$$= \frac{\partial \varphi}{\partial r}(a + r\cos\theta, b + r\sin\theta)$$

(b) Use differentiation under the integral sign to prove that

$$\frac{d}{dr} I_\varphi(r) = \frac{1}{2\pi r} \int_{\mathcal{C}(r)} \frac{\partial \varphi}{\partial \mathbf{n}} \, ds$$

(c) Use Exercise 36 to conclude that

$$\frac{d}{dr} I_\varphi(r) = \frac{1}{2\pi r} \iint_{\mathcal{D}(r)} \Delta \varphi \, dA$$

where $\mathcal{D}(r)$ is the interior of $\mathcal{C}(r)$.

38. Prove that $m(r) \leq I_\varphi(r) \leq M(r)$, where $m(r)$ and $M(r)$ are the minimum and maximum values of φ on $\mathcal{C}(r)$. Then use the continuity of φ to prove that $\lim_{r \to 0} I_\varphi(r) = \varphi(P)$.

In Exercises 39–40, let $\mathcal{D}$ be the region bounded by a simple closed curve $\mathcal{C}$. A function $\varphi(x, y)$ on $\mathcal{D}$ (whose second-order partial derivatives exist and are continuous) is called **harmonic** if $\Delta \varphi = 0$, where $\Delta \varphi$ is the Laplace operator defined in Eq. (9).

39. Use the results of Exercises 37 and 38 to prove the **mean-value property** of harmonic functions: If φ is harmonic, then $I_\varphi(r) = \varphi(P)$ for all r.

40. Show that $f(x, y) = x^2 - y^2$ is harmonic. Verify the mean-value property for $f(x, y)$ directly [expand $f(a + r\cos\theta, b + r\sin\theta)$ as a function of θ and compute $I_\varphi(r)$]. Show that $x^2 + y^2$ is not harmonic and does not satisfy the mean-value property.

17.2 Stokes' Theorem

Stokes' Theorem is a generalization of Green's Theorem to three dimensions. Whereas Green's Theorem relates a line integral to a double integral in the plane, Stokes' Theorem relates a line integral in $\mathbf{R}^3$ to a surface integral.

A few preliminaries are needed before we can state Stokes' Theorem. Recall that when applying Green's Theorem to a domain $\mathcal{D}$, we must orient the boundary curve $\partial \mathcal{D}$ so that $\mathcal{D}$ lies to the left when you traverse the boundary in the positive direction (i.e., counterclockwise if $\mathcal{D}$ is bounded by a simple closed curve). In Stokes' Theorem, we must take into account both the orientation of the surface $\mathcal{S}$ and the orientation of its boundary $\partial \mathcal{S}$. First, let us clarify what is meant by the boundary of a surface.

Figure 1 illustrates several possibilities: The boundary may be a simple closed curve as in (A), or it may consist of several closed curves as in (B). If the surface has *no boundary* as in (C), we say that $\mathcal{S}$ is a **closed surface**. To determine whether $\mathcal{S}$ is closed or not, ask yourself if it can hold air. If so, then $\mathcal{S}$ is closed. If it has a boundary, then air can escape through the hole cut off by the boundary curve.

FIGURE 1 Surfaces and their boundaries.

(A) Boundary consists of a single closed curve

(B) Boundary consists of three closed curves

(C) Closed surface (the boundary is empty)

Recall from Section 16.5 that a surface $\mathcal{S}$ is *oriented* if a unit normal vector at each point has been specified in a continuously varying manner. An oriented surface has two sides and a choice of normal vectors specifies one of these two sides. Once we fix an orientation on $\mathcal{S}$, we may induce a **boundary orientation** of $\partial \mathcal{S}$ as follows: If you imagine yourself as the normal vector walking along the boundary curve, then the surface lies to your left. Figure 2 shows two orientations of a surface whose boundary consists of two curves, $\mathcal{C}_1$ and $\mathcal{C}_2$. Note how the orientation of the boundary depends on the orientation of the surface.

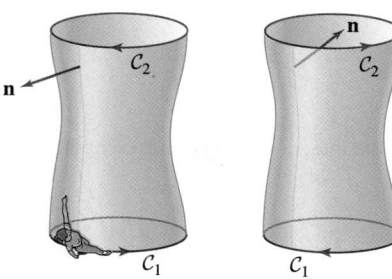

FIGURE 2 The orientation of the boundary ∂S for each of the two possible orientations of the surface $\mathcal{S}$.

The last ingredient in Stokes' Theorem is the **curl** of a vector field $\mathbf{F} = \langle F_1, F_2, F_3 \rangle$, defined by the symbolic determinant

$$\text{curl}(\mathbf{F}) = \begin{vmatrix} \mathbf{i} & \mathbf{j} & \mathbf{k} \\ \dfrac{\partial}{\partial x} & \dfrac{\partial}{\partial y} & \dfrac{\partial}{\partial z} \\ F_1 & F_2 & F_3 \end{vmatrix}$$

$$= \left(\frac{\partial F_3}{\partial y} - \frac{\partial F_2}{\partial z} \right) \mathbf{i} - \left(\frac{\partial F_3}{\partial x} - \frac{\partial F_1}{\partial z} \right) \mathbf{j} + \left(\frac{\partial F_2}{\partial x} - \frac{\partial F_1}{\partial y} \right) \mathbf{k}$$

In terms of components,

$$\text{curl}(\mathbf{F}) = \left\langle \frac{\partial F_3}{\partial y} - \frac{\partial F_2}{\partial z}, \; \frac{\partial F_1}{\partial z} - \frac{\partial F_3}{\partial x}, \; \frac{\partial F_2}{\partial x} - \frac{\partial F_1}{\partial y} \right\rangle$$

Keep in mind that $\text{curl}(\mathbf{F})$ is again a vector field. The scalar curl defined in the previous section is equal to the z-component of the vector curl. It is convenient to define the del "operator" (also called "nabla"):

$$\nabla = \left\langle \frac{\partial}{\partial x}, \frac{\partial}{\partial y}, \frac{\partial}{\partial z} \right\rangle$$

We may think of $\text{curl}(\mathbf{F})$ as the symbolic cross product of ∇ and $\mathbf{F}$:

$$\text{curl}(\mathbf{F}) = \nabla \times \mathbf{F}$$

It is straightforward to check that taking the curl is a **linear operation** in the following sense:

$$\text{curl}(\mathbf{F} + \mathbf{G}) = \text{curl}(\mathbf{F}) + \text{curl}(\mathbf{G})$$

$$\text{curl}(a\mathbf{F}) = a \, \text{curl}(\mathbf{F}) \quad (a \text{ any constant})$$

■ **EXAMPLE 1** Calculating the Curl   Calculate the curl of $\mathbf{F} = \langle xy, e^x, y + z \rangle$.

Solution We compute the curl as a symbolic determinant:

$$\text{curl}(\mathbf{F}) = \begin{vmatrix} \mathbf{i} & \mathbf{j} & \mathbf{k} \\ \dfrac{\partial}{\partial x} & \dfrac{\partial}{\partial y} & \dfrac{\partial}{\partial z} \\ xy & e^x & y + z \end{vmatrix}$$

$$= \left(\frac{\partial}{\partial y}(y + z) - \frac{\partial}{\partial z} e^x \right) \mathbf{i} - \left(\frac{\partial}{\partial x}(y + z) - \frac{\partial}{\partial z} xy \right) \mathbf{j} + \left(\frac{\partial}{\partial x} e^x - \frac{\partial}{\partial y} xy \right) \mathbf{k}$$

$$= \mathbf{i} + (e^x - x)\mathbf{k} \qquad\qquad ■$$

We now state Stokes' Theorem for an oriented surface $\mathcal{S}$ in $\mathbf{R}^3$ bounded by a closed curve or union of closed curves. We must require that $\mathcal{S}$ be smooth or piecewise smooth, in a suitable sense, but it is rather technical to make this requirement precise. It suffices to assume that $\mathcal{S}$ has a one-to-one, regular, continuously differentiable parametrization

$\Phi : \mathcal{D} \to \mathcal{S}$, where $\mathcal{D}$ is a domain in the plane bounded by a smooth, simple closed curve. More generally, Stokes' Theorem is valid for smooth surfaces such as a sphere or torus or piecewise smooth surfaces such as the surface of a cube or tetrahedron.

THEOREM 1 Stokes' Theorem   Assume that $\mathcal{S}$ is an oriented, piecewise smooth surface bounded by a closed curve or union of closed curves as described above. Let $\partial \mathcal{S}$ be the boundary of $\mathcal{S}$ with its boundary orientation. Let $\mathbf{F}$ be a vector field whose components have continuous partial derivatives. Then

$$\oint_{\partial \mathcal{S}} \mathbf{F} \cdot d\mathbf{s} = \iint_{\mathcal{S}} \operatorname{curl}(\mathbf{F}) \cdot d\mathbf{S} \qquad \boxed{1}$$

If $\mathcal{S}$ is closed (i.e., $\partial \mathcal{S}$ is empty), then the surface integral on the right is zero.

Proof   Although a complete proof of Stokes' Theorem is beyond the scope of this text, we shall prove it here in the special case whereby $\mathcal{S}$ is the graph of a function $z = f(x, y)$ lying over a domain $\mathcal{D}$ in the xy-plane. Orient $\mathcal{S}$ with upward-pointing normal.

Let $\mathcal{C} = \partial \mathcal{S}$ be the boundary curve of $\mathcal{S}$. We may write each side of Eq. (1) as a sum over components:

$$\oint_{\mathcal{C}} (F_1 \mathbf{i} + F_2 \mathbf{j} + F_3 \mathbf{k}) \cdot d\mathbf{s} = \oint_{\mathcal{C}} F_1 \mathbf{i} \cdot d\mathbf{s} + \oint_{\mathcal{C}} F_2 \mathbf{j} \cdot d\mathbf{s} + \oint_{\mathcal{C}} F_3 \mathbf{k} \cdot d\mathbf{s}$$

$$\iint_{\mathcal{S}} \operatorname{curl}(F_1 \mathbf{i} + F_2 \mathbf{j} + F_3 \mathbf{k}) \cdot d\mathbf{S} = \iint_{\mathcal{S}} \operatorname{curl}(F_1 \mathbf{i}) \cdot d\mathbf{S} + \iint_{\mathcal{S}} \operatorname{curl}(F_2 \mathbf{j}) \cdot d\mathbf{S}$$

$$+ \iint_{\mathcal{S}} \operatorname{curl}(F_3 \mathbf{k}) \cdot d\mathbf{S}$$

It suffices to show that the terms corresponding to the $\mathbf{i}$-, $\mathbf{j}$-, and $\mathbf{k}$-components are separately equal. We carry this out for the $\mathbf{i}$-component. The calculations for the $\mathbf{j}$-components are similar and we leave the equality for the $\mathbf{k}$-components as an exercise (Exercise 31). Thus, we shall prove that

$$\oint_{\mathcal{C}} F_1(x, y, z)\mathbf{i} \cdot d\mathbf{s} = \iint_{\mathcal{S}} \operatorname{curl}(F_1(x, y, z)\mathbf{i}) \cdot d\mathbf{S} \qquad \boxed{2}$$

Let $\mathcal{C}_0$ be the boundary of the domain $\mathcal{D}$ in the xy-plane, oriented counterclockwise (Figure 3). Observe that the boundary $\mathcal{C}$ of $\mathcal{S}$ projects onto $\mathcal{C}_0$. Let $\mathbf{c}_0(t) = (x(t), y(t))$ (for $a \le t \le b$) be a counterclockwise parametrization of $\mathcal{C}_0$. Then $\mathcal{C}$ with its boundary orientation is parametrized by the path $\mathbf{c}(t) = (x(t), y(t), f(x(t), y(t)))$. Thus, we have

$$\mathbf{c}(t) = (x(t), y(t), f(x(t), y(t)))$$

$$\mathbf{c}'(t) = \left\langle x'(t), y'(t), \frac{d}{dt} f(x(t), y(t)) \right\rangle$$

and so

$$F_1 \mathbf{i} \cdot \mathbf{c}'(t) = \langle F_1(x(t), y(t), f(x(t), y(t))), 0, 0 \rangle \cdot \left\langle x'(t), y'(t), \frac{d}{dt} f(x(t), y(t)) \right\rangle$$

$$= F_1(x(t), y(t), f(x(t), y(t))) x'(t)$$

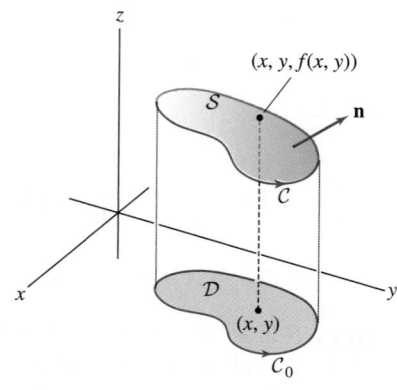

FIGURE 3

Therefore, the left-hand side of Eq. (2) is equal to

$$\oint_{\mathcal{C}} F_1\mathbf{i} \cdot d\mathbf{s} = \int_a^b F_1\big(x(t), y(t), f(x(t), y(t))\big) x'(t)\, dt$$

Now we observe that the integral on the right is equal to the line integral of the planar vector field $F_1(x, y, f(x, y))\mathbf{i}$ over $\mathcal{C}_0$, so we may apply Green's Theorem. By the Chain Rule,

$$\frac{\partial}{\partial y} F_1(x, y, f(x, y)) = F_{1y}(x, y, f(x, y)) + F_{1z}(x, y, f(x, y)) f_y(x, y)$$

Green's Theorem yields

$$\oint_{\mathcal{C}} F_1\mathbf{i} \cdot d\mathbf{s} = -\iint_{\mathcal{D}} \Big(F_{1y}(x, y, f(x, y)) + F_{1z}(x, y, f(x, y)) f_y(x, y) \Big) dA \qquad \boxed{3}$$

To finish the proof, we compute the surface integral of $\mathrm{curl}(F_1\mathbf{i})$ using the parametrization $\Phi(x, y) = (x, y, f(x, y))$ of $\mathcal{S}$. We obtain

$$\mathbf{n} = \big\langle -f_x(x, y), -f_y(x, y), 1 \big\rangle$$

$$\mathrm{curl}(F_1\mathbf{i}) \cdot \mathbf{n} = \big\langle 0, F_{1z}, -F_{1y} \big\rangle \cdot \big\langle -f_x(x, y), -f_y(x, y), 1 \big\rangle$$

$$= -F_{1z}(x, y, f(x, y)) f_y(x, y) - F_{1y}(x, y, f(x, y))$$

$$\iint_{\mathcal{S}} \mathrm{curl}(F_1\mathbf{i}) \cdot d\mathbf{S} = -\iint_{\mathcal{D}} \Big(F_{1z}(x, y, z) f_y(x, y) + F_{1y}(x, y, f(x, y)) \Big) dA \qquad \boxed{4}$$

The right-hand sides of (3) and (4) are equal. Therefore, Eq. (2) holds as desired.

■ **EXAMPLE 2** Verifying Stokes' Theorem Verify Stokes' Theorem for

$$\mathbf{F} = \langle -y, 2x, x + z \rangle$$

and the upper hemisphere:

$$\mathcal{S} = \{(x, y, z) : x^2 + y^2 + z^2 = 1, z \geq 0\}$$

oriented by outward-pointing normal vectors (Figure 4).

Solution We compute both sides of Eq. (1) in Stokes' Theorem.

Step 1. Compute the line integral around the boundary curve.
The boundary of $\mathcal{S}$ is the unit circle oriented in the counterclockwise direction with parametrization $\mathbf{c}(t) = (\cos t, \sin t, 0)$. Thus,

$$\mathbf{c}'(t) = \langle -\sin t, \cos t, 0 \rangle$$

$$\mathbf{F}(\mathbf{c}(t)) = \langle -\sin t, 2\cos t, \cos t \rangle$$

$$\mathbf{F}(\mathbf{c}(t)) \cdot \mathbf{c}'(t) = \langle -\sin t, 2\cos t, \cos t \rangle \cdot \langle -\sin t, \cos t, 0 \rangle$$

$$= \sin^2 t + 2\cos^2 t = 1 + \cos^2 t$$

$$\oint_{\partial\mathcal{S}} \mathbf{F} \cdot d\mathbf{s} = \int_0^{2\pi} (1 + \cos^2 t)\, dt = 2\pi + \pi = \boxed{3\pi} \qquad \boxed{5}$$

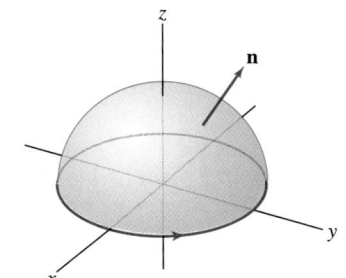

FIGURE 4 Upper hemisphere with oriented boundary.

◀··· **REMINDER** In Eq. (5), we use

$$\int_0^{2\pi} \cos^2 t\, dt = \int_0^{2\pi} \frac{1 + \cos 2t}{2}\, dt = \pi$$

Step 2. Compute the curl.

$$\text{curl}(\mathbf{F}) = \begin{vmatrix} \mathbf{i} & \mathbf{j} & \mathbf{k} \\ \dfrac{\partial}{\partial x} & \dfrac{\partial}{\partial y} & \dfrac{\partial}{\partial z} \\ -y & 2x & x+z \end{vmatrix}$$

$$= \left(\frac{\partial}{\partial y}(x+z) - \frac{\partial}{\partial z}2x \right)\mathbf{i} - \left(\frac{\partial}{\partial x}(x+z) - \frac{\partial}{\partial z}(-y) \right)\mathbf{j}$$

$$+ \left(\frac{\partial}{\partial x}2x - \frac{\partial}{\partial y}(-y) \right)\mathbf{k}$$

$$= \langle 0, -1, 3 \rangle$$

Step 3. Compute the flux of the curl through the surface.

We parametrize the hemisphere using spherical coordinates:

$$\Phi(\phi, \theta) = (\cos\theta \sin\phi, \sin\theta \sin\phi, \cos\phi)$$

As we have seen in previous examples [Eq. (5) in Section 16.4], the outward-pointing normal vector at the point $\Phi(\phi, \theta)$ is

$$\mathbf{n} = \sin\phi \, \langle \cos\theta \sin\phi, \sin\theta \sin\phi, \cos\phi \rangle$$

We have

$$\text{curl}(\mathbf{F}) \cdot \mathbf{n} = \sin\phi \, \langle 0, -1, 3 \rangle \cdot \langle \cos\theta \sin\phi, \sin\theta \sin\phi, \cos\phi \rangle$$

$$= -\sin\theta \sin^2\phi + 3\cos\phi \sin\phi$$

The upper hemisphere S corresponds to $0 \le \phi \le \frac{\pi}{2}$, so the flux of $\text{curl}(\mathbf{F})$ through S is equal to

$$\iint_S \text{curl}(\mathbf{F}) \cdot d\mathbf{S} = \int_{\phi=0}^{\pi/2} \int_{\theta=0}^{2\pi} (-\sin\theta \sin^2\phi + 3\cos\phi \sin\phi) \, d\theta \, d\phi$$

$$= 0 + 2\pi \int_{\phi=0}^{\pi/2} 3\cos\phi \sin\phi \, d\phi = 2\pi \left(\frac{3}{2}\sin^2\phi \right) \Big|_{\phi=0}^{\pi/2}$$

$$= \boxed{3\pi}$$

Step 4. Compare the answers.

Both the line integral and flux of the curl are equal to 3π. Thus Stokes' Theorem is verified in this case. ∎

In computing the curl of a vector field $\mathbf{F} = \langle F_1, F_2, F_3 \rangle$, it is helpful to observe that the partial derivatives $\dfrac{\partial F_1}{\partial y}$ and $\dfrac{\partial F_1}{\partial z}$ appear in $\text{curl}(\mathbf{F})$ but $\dfrac{\partial F_1}{\partial x}$ does not. Thus, if $F_1 = F_1(x)$ is a function of x alone, then $\dfrac{\partial F_1}{\partial y} = \dfrac{\partial F_1}{\partial z} = 0$ and F_1 does not contribute to the curl. The same holds for the other components. In other words, if F_1, F_2, and F_3 depend only on their corresponding variables x, y, and z, then

$$\text{curl}(\langle F_1(x), F_2(y), F_3(z) \rangle) = 0$$

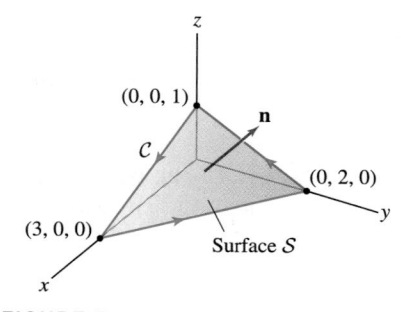

(0, 0, 1)

n

$\mathcal{C}$

(0, 2, 0)

(3, 0, 0)

y

Surface $\mathcal{S}$

x

z

FIGURE 5

■ **EXAMPLE 3** Using Stokes' Theorem to Compute a Line Integral   Use Stokes' Theorem to compute $I = \oint_{\mathcal{C}} \mathbf{F} \cdot d\mathbf{s}$, where

$$\mathbf{F} = \left\langle \sin x^2, e^{y^2} + x^2, z^4 + 2x^2 \right\rangle$$

and $\mathcal{C}$ is the boundary of the triangle in Figure 5 with the indicated orientation.

Solution   We apply Stokes' Theorem to the triangular surface $\mathcal{S}$ in Figure 5. The plane containing $\mathcal{S}$ has equation

$$\frac{x}{3} + \frac{y}{2} + z = 1$$

This plane is the graph of the function $z = f(x, y) = 1 - \frac{x}{3} - \frac{y}{2}$ and $\mathcal{S}$ is the part of the graph lying over a triangle $\mathcal{D}$ in the xy-plane. It is not hard to describe $\mathcal{D}$ explicitly, but we do not need to do this to solve our problem. Observe that the normal vector to $\mathcal{S}$ in the parametrization $\Phi(x, y) = (x, y, f(x, y))$ is

$$\mathbf{n} = \langle -f_x, -f_y, 1 \rangle = \left\langle \frac{1}{3}, \frac{1}{2}, 1 \right\rangle$$

To simplify the computation of curl($\mathbf{F}$), we use the remark in the preceding paragraph according to which the terms $\sin x^2$, e^{y^2}, and z^4 may be dropped:

$$\text{curl}\left(\left\langle \sin x^2, e^{y^2} + x^2, z^4 + 2x^2 \right\rangle\right)$$

$$= \text{curl}\left(\overbrace{\left\langle \sin x^2, e^{y^2}, z^4 \right\rangle}^{\text{Automatically zero}}\right) + \text{curl}\left(\left\langle 0, x^2, 2x^2 \right\rangle\right) = \text{curl}\left(\left\langle 0, x^2, 2x^2 \right\rangle\right)$$

It is then straightforward to compute

$$\text{curl}\left(\left\langle 0, x^2, 2x^2 \right\rangle\right) = \left\langle 0, -\frac{\partial}{\partial x} 2x^2, \frac{\partial}{\partial x} x^2 \right\rangle = \langle 0, -4x, 2x \rangle$$

Now we observe that curl($\mathbf{F}$) is orthogonal to $\mathbf{n}$:

$$\text{curl}(\mathbf{F}) \cdot \mathbf{n} = \langle 0, -4x, 2x \rangle \cdot \left\langle \frac{1}{3}, \frac{1}{2}, 1 \right\rangle = 0$$

Therefore, by Stokes' Theorem,

$$I = \oint_{\mathcal{C}} \mathbf{F} \cdot d\mathbf{s} = \iint_{\mathcal{S}} \text{curl}(\mathbf{F}) \cdot d\mathbf{S} = \iint_{\mathcal{D}} \overbrace{\text{curl}(\mathbf{F}) \cdot \mathbf{n}}^{\text{Equal to 0}} \, dA = 0 \qquad \blacksquare$$

CONCEPTUAL INSIGHT   There is an interesting and important analogy between Stokes' Theorem and the Fundamental Theorem for Gradient Vector Fields stated in Section 16.3.

• The Fundamental Theorem tells us that if $\mathbf{F} = \nabla \varphi$, then

$$\int_{\mathcal{C}_1} \mathbf{F} \cdot d\mathbf{s} = \int_{\mathcal{C}_2} \mathbf{F} \cdot d\mathbf{s} = \varphi(Q) - \varphi(P)$$

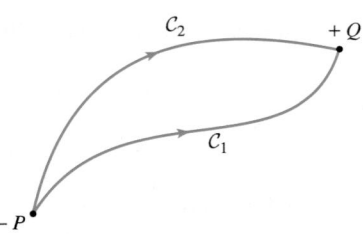

FIGURE 6 Two paths with the same boundary $Q - P$.

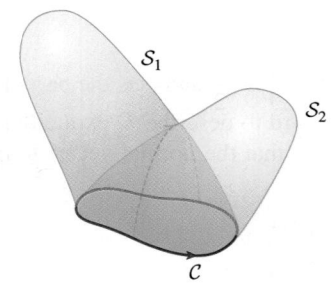

FIGURE 7 Surfaces S_1 and S_2 have the same oriented boundary.

for any two paths C_1 and C_2 from P to Q (Figure 6). As we have noted, this shows that the line integral of a gradient vector field is independent of the path and depends only on the *oriented boundary* of the path (think of the oriented boundary as the difference of the endpoints $Q - P$).

- For surface integrals, instead of asking whether $\mathbf{F} = \nabla \varphi$ for some potential function φ, we ask if $\mathbf{F} = \text{curl}(\mathbf{A})$ for some vector field $\mathbf{A}$. Such a vector field, when it exists, is often called a **vector potential** for $\mathbf{F}$. If $\mathbf{F}$ has a vector potential $\mathbf{A}$, then according to Stokes' Theorem applied to $\mathbf{A}$,

$$\iint_S \mathbf{F} \cdot d\mathbf{S} = \oint_{\partial S} \mathbf{A} \cdot d\mathbf{s}$$

The right-hand side depends only on the oriented boundary ∂S, not on S itself. In other words, if S_1 and S_2 are any two surfaces with the same oriented boundary curve C (Figure 7), then

$$\iint_{S_1} \mathbf{F} \cdot d\mathbf{S} = \iint_{S_2} \mathbf{F} \cdot d\mathbf{S} \qquad \text{if} \quad \mathbf{F} = \text{curl}(\mathbf{A})$$

since both surface integrals are equal to $\oint_C \mathbf{A} \cdot d\mathbf{s}$. This is the analog of path independence of gradient vector fields. In particular, if S is closed, then the boundary curve ∂S is empty. The circulation of $\mathbf{A}$ around the empty curve is zero, so

$$\iint_S \mathbf{F} \cdot d\mathbf{S} = 0 \qquad \text{if} \quad \mathbf{F} = \text{curl}(\mathbf{A}) \text{ and } S \text{ is closed}$$

A vector potential of a vector field $\mathbf{F}$ is a kind of "antiderivative," and, like all antiderivatives, it is not unique. If $\mathbf{F} = \text{curl}(\mathbf{A})$, then $\mathbf{F} = \text{curl}(\mathbf{A} + \mathbf{B})$ for any vector field $\mathbf{B}$ such that $\text{curl}(\mathbf{B}) = \mathbf{0}$.

THEOREM 2 Surface Independence for Curl Vector Fields   If $\mathbf{F} = \text{curl}(\mathbf{A})$, then the flux of $\mathbf{F}$ through a surface S depends only on the oriented boundary ∂S and not on the surface itself. We have

$$\iint_S \mathbf{F} \cdot d\mathbf{S} = \oint_{\partial S} \mathbf{A} \cdot d\mathbf{s}$$

In particular, if S is *closed* (i.e., ∂S is empty), then

$$\iint_S \mathbf{F} \cdot d\mathbf{S} = 0$$

■ **EXAMPLE 4** Let $\mathbf{F} = \text{curl}(\mathbf{A})$, where $\mathbf{A} = \langle y + z, \sin(xy), e^{xyz} \rangle$. Determine the flux of $\mathbf{F}$ through the surfaces S_1 and S_2 whose common boundary C is the unit circle in the xz-plane as in Figure 8.

Solution When you traverse the circle C in the direction of the arrow, the surface S_1 lies to the left. Therefore, C is oriented as the boundary of S_1, and by Stokes' Theorem,

$$\oint_C \mathbf{A} \cdot d\mathbf{s} = \iint_{S_1} \mathbf{F} \cdot d\mathbf{S}$$

The parametrization $\mathbf{c}(t) = (\cos t, 0, \sin t)$ traces C in the direction of the arrow because it begins at $\mathbf{c}(0) = (1, 0, 0)$ and moves in the direction of $\mathbf{c}(\frac{\pi}{2}) = (0, 0, 1)$. We have

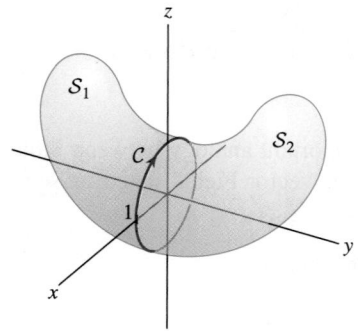

FIGURE 8

$$\mathbf{A}(\mathbf{c}(t)) = \langle 0 + \sin t, \sin(0), e^0 \rangle = \langle \sin t, 0, 1 \rangle$$

$$\mathbf{A}(\mathbf{c}(t)) \cdot \mathbf{c}'(t) = \langle \sin t, 0, 1 \rangle \cdot \langle -\sin t, 0, \cos t \rangle = -\sin^2 t + \cos t$$

$$\oint_C \mathbf{A} \cdot d\mathbf{s} = \int_0^{2\pi} (-\sin^2 t + \cos t)\, dt = -\pi$$

We conclude that $\iint_{S_1} \mathbf{F} \cdot d\mathbf{S} = -\pi$. On the other hand, S_2 lies on the right as you traverse C and therefore $-C$ is oriented as the boundary of S_2. We obtain

$$\iint_{S_2} \mathbf{F} \cdot d\mathbf{S} = \oint_{-C} \mathbf{A} \cdot d\mathbf{s} = -\oint_C \mathbf{A} \cdot d\mathbf{s} = \pi \qquad\blacksquare$$

CONCEPTUAL INSIGHT In Section 17.1, we interpreted the scalar curl as a measure of the "tendency to rotate." We saw that if $\mathbf{F}$ is the velocity field of a fluid, then the curl of $\mathbf{F}$ at a point P is measured by placing a small paddle wheel at P and determining how fast it rotates. Can we extend this interpretation to three dimensions? Notice an important difference between two and three dimensions. In the plane, there is only one way to place the paddle wheel in the fluid. In three dimensions, we can orient the paddle wheel at P in any direction and the speed of rotation depends on how the wheel is oriented [Figure 9(A)]. Let us show that the vector curl($\mathbf{F}$) simultaneously encodes the tendency to rotate in all possible directions.

Let C be the boundary circle of the paddle wheel and $\mathcal{R}$ the enclosed region [Figure 9(B)]. If C is sufficiently small and curl($\mathbf{F}$) is continuous, then curl($\mathbf{F}$) is nearly constant on $\mathcal{R}$ with value $\mathbf{v}_P = \text{curl}(\mathbf{F})(P)$. If $\mathbf{e_n}$ is a unit normal vector to $\mathcal{R}$ at P, then Stokes' Theorem yields the approximation

$$\oint_C \mathbf{F} \cdot d\mathbf{s} = \iint_{\mathcal{R}} \text{curl}(\mathbf{F}) \cdot d\mathbf{S} \approx \iint_{\mathcal{R}} \mathbf{v}_P \cdot d\mathbf{S} = (\mathbf{v}_P \cdot \mathbf{e_n})\,\text{Area}(\mathcal{R})$$

This approximation holds for any small region $\mathcal{R}$ with boundary C. Our result may be expressed in terms of the angle θ between $\mathbf{v}_P$ and $\mathbf{e_n}$:

$$\boxed{\int_C \mathbf{F} \cdot d\mathbf{s} \approx (\mathbf{v}_P \cdot \mathbf{e_n})\,\text{Area}(\mathcal{R}) = \|\mathbf{v}_P\|\,\text{Area}(\mathcal{R})\cos\theta} \qquad \boxed{6}$$

This is a remarkable result. If we fix the size of the circle C but change its orientation in space, then the circulation around C varies, but (to a first-order approximation) it varies in a simple way, namely as the cosine of the angle θ. In particular, the maximum circulation occurs when $\mathbf{e_n}$ and $\mathbf{v}_P$ point in the same direction, while the circulation is zero (to a first-order approximation) when $\mathbf{e_n}$ is perpendicular to $\mathbf{v}_P$.

(A)   (B)

FIGURE 9

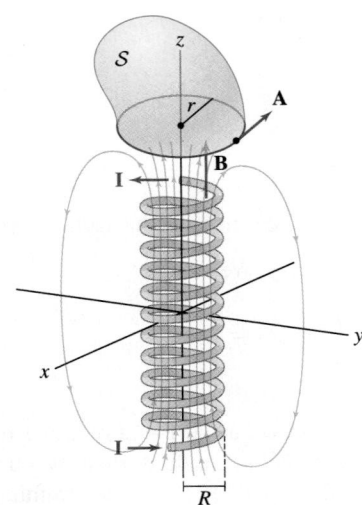

FIGURE 10 The magnetic field of a long solenoid is nearly uniform inside and weak outside. In practice, we treat the solenoid as "infinitely long" if it is very long in comparison with its radius.

■ **EXAMPLE 5** Vector Potential for a Solenoid A magnetic field **B** is created when a current flows through a solenoid (a tightly wound spiral of wire; see Figure 10). If we assume that the solenoid is infinitely long, with radius R and the z-axis as central axis, then **B** is zero outside the solenoid and has value $B\mathbf{k}$ inside the solenoid, where B is a constant that depends on the current strength and spacing of the turns of wire.

(a) Show that a vector potential for **B** is

$$\mathbf{A} = \begin{cases} \dfrac{1}{2}R^2 B\left\langle -\dfrac{y}{r^2}, \dfrac{x}{r^2}, 0\right\rangle & \text{if } r > R \\[2ex] \dfrac{1}{2}B\left\langle -y, x, 0\right\rangle & \text{if } r < R \end{cases}$$

where $r = (x^2 + y^2)^{1/2}$ is the distance to the z-axis.

(b) Calculate the flux of **B** through the surface $\mathcal{S}$ in Figure 10 whose boundary is a circle of radius $r > R$.

Solution

(a) We must verify that $\mathbf{B} = \text{curl}(\mathbf{A})$. To simplify the calculation, note that for a vector field whose z-component is zero, the curl reduces to

$$\text{curl}(\langle f, g, 0\rangle) = \langle -g_z, f_z, g_x - f_y\rangle$$

In our case, for $r > R$, we have

$$\text{curl}(\mathbf{A}) = \frac{1}{2}R^2 B\left\langle 0, 0, \frac{\partial}{\partial x}xr^{-2} - \frac{\partial}{\partial y}(-yr^{-2})\right\rangle$$

However, the z-component is also zero:

$$\frac{\partial}{\partial x}xr^{-2} + \frac{\partial}{\partial y}yr^{-2} = \frac{\partial}{\partial x}\left(\frac{x}{x^2 + y^2}\right) + \frac{\partial}{\partial y}\left(\frac{y}{x^2 + y^2}\right)$$

$$= \frac{(x^2 + y^2) - x(2x)}{(x^2 + y^2)^2} + \frac{(x^2 + y^2) - y(2y)}{(x^2 + y^2)^2}$$

$$= 0$$

The vector potential **A** is not differentiable on the cylinder $r = R$, that is, on the solenoid itself (Figure 11). The magnetic field $\mathbf{B} = \text{curl}(\mathbf{A})$ has a jump discontinuity where $r = R$. We take for granted the fact that Stokes' Theorem remains valid in this setting.

Thus, $\text{curl}(\mathbf{A}) = \mathbf{0}$ when $r > R$ as required. For $r < R$, we have

$$\text{curl}(\mathbf{A}) = \frac{1}{2}B\left\langle 0, 0, \frac{\partial}{\partial x}x - \frac{\partial}{\partial y}(-y)\right\rangle = \langle 0, 0, B\rangle = B\mathbf{k}$$

(b) By Stokes' Theorem, the flux of **B** through $\mathcal{S}$ is equal to the circulation of **A** through the boundary $\partial \mathcal{S}$, which is a horizontal circle of radius r oriented counterclockwise. We parametrize the circle by $\mathbf{c}(t) = (r\cos t, r\sin t, 0)$. Then

$$\mathbf{c}'(t) = \langle -r\sin t, r\cos t, 0\rangle$$

$$\mathbf{A}(\mathbf{c}(t)) = \frac{1}{2}R^2 B r^{-1}\langle -\sin t, \cos t, 0\rangle$$

$$\mathbf{A}(\mathbf{c}(t)) \cdot \mathbf{c}'(t) = \frac{1}{2}R^2 B\left((-\sin t)^2 + \cos^2 t\right) = \frac{1}{2}R^2 B$$

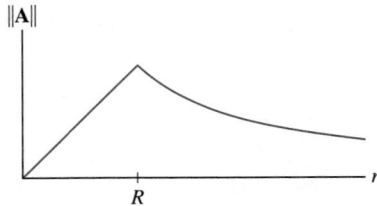

FIGURE 11 The length $\|\mathbf{A}\|$ of the vector potential as a function of distance r to the z-axis.

Therefore, the flux of **B** through $\mathcal{S}$ is equal to

$$\iint_{\mathcal{S}} \mathbf{B} \cdot d\mathbf{S} = \oint_{\partial\mathcal{S}} \mathbf{A} \cdot d\mathbf{s} = \int_0^{2\pi} \mathbf{A}(\mathbf{c}(t)) \cdot \mathbf{c}'(t)\, dt$$

$$= \frac{1}{2} R^2 B \int_0^{2\pi} dt = \pi R^2 B \qquad \blacksquare$$

CONCEPTUAL INSIGHT There is an interesting difference between scalar and vector potentials. If $\mathbf{F} = \nabla \varphi$, then the scalar potential φ is constant in regions where the field **F** is zero (since a function with zero gradient is constant). This is not true for vector potentials. As we saw in Example 4, the magnetic field **B** produced by a solenoid is zero everywhere outside the solenoid, but the vector potential **A** is not constant outside the solenoid. In fact, **A** is proportional to $\left\langle -\dfrac{y}{r^2}, \dfrac{x}{r^2}, 0 \right\rangle$. This is related to an intriguing phenomenon in physics called the *Aharonov-Bohm (AB) effect*, first proposed on theoretical grounds in the 1940s.

According to electromagnetic theory, a magnetic field **B** exerts a force on a moving electron, causing a deflection in the electron's path. We do not expect any deflection when an electron moves past a solenoid because **B** is zero outside the solenoid (in practice, the field is not actually zero but it is very small—we ignore this difficulty). However, according to quantum mechanics, electrons have both particle and wave properties. In a double slit experiment, a stream of electrons passing through two small slits creates a wavelike interference pattern on a detection screen (Figure 12). The AB effect predicts that if we place a small solenoid between the slits as in the figure (the solenloid is so small that the electrons never pass through it), then the interference pattern will shift slightly. It is as if the elections are "aware" of the magnetic field inside the solenoid even though they never encounter the field directly.

The existence of the AB effect was hotly debated until it was confirmed definitively in 1985, in experiments carried out by a team of Japanese physicists led by Akira Tonomura. It appeared to contradict "classical" electromagnetic theory, where the trajectory of an electron is determined by **B** alone. There is no such contradiction in quantum mechanics, because the behavior of the electrons is governed not by **B** but by a "wave function" that is derived from the nonconstant vector potential **A**.

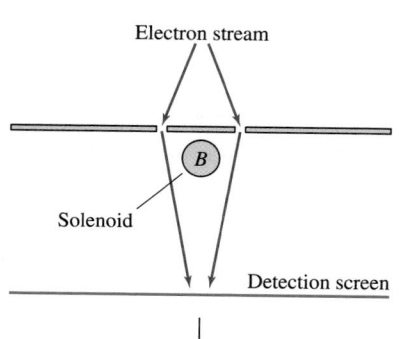

Electron stream

B

Solenoid

Detection screen

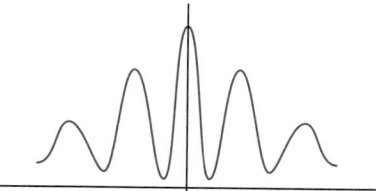

FIGURE 12 A stream of electrons passing through a double slit produces an interference pattern on the detection screen. The pattern shifts slightly when an electric current flows through the solenoid.

17.2 SUMMARY

- The *boundary* of a surface $\mathcal{S}$ is denoted $\partial\mathcal{S}$. We say that $\mathcal{S}$ is *closed* if $\partial\mathcal{S}$ is empty.
- Suppose that $\mathcal{S}$ is oriented (a continuously varying unit normal is specified at each point of $\mathcal{S}$). We define the *boundary orientation* of $\partial\mathcal{S}$ by the following condition: If you walk along the boundary in the positive direction with your head pointing in the normal direction, then the surface is on your left.
- Stokes' Theorem relates the circulation around the boundary to the surface integral of the curl:

$$\boxed{\oint_{\partial\mathcal{S}} \mathbf{F} \cdot d\mathbf{s} = \iint_{\mathcal{S}} \operatorname{curl}(\mathbf{F}) \cdot d\mathbf{S}}$$

- If $\mathcal{R}$ is a small region with boundary $\mathcal{C}$ and P is a sample point in $\mathcal{R}$ with normal vector $\mathbf{e_n}$ at P, then

$$\int_{\mathcal{C}} \mathbf{F} \cdot d\mathbf{s} \approx (\text{curl}(\mathbf{F})(P) \cdot \mathbf{e_n}) \, \text{Area}(\mathcal{R})$$

17.2 EXERCISES

Preliminary Questions

1. Indicate with an arrow the boundary orientation of the boundary curves of the surfaces in Figure 13, oriented by the outward-pointing normal vectors.

(A)   (B)

FIGURE 13

2. Let $\mathbf{F} = \text{curl}(\mathbf{A})$. Which of the following are related by Stokes' Theorem?

(a) The circulation of $\mathbf{A}$ and flux of $\mathbf{F}$.

(b) The circulation of $\mathbf{F}$ and flux of $\mathbf{A}$.

3. What is the definition of a vector potential?

4. Which of the following statements is correct?

(a) The flux of $\text{curl}(\mathbf{F})$ through every oriented surface is zero.

(b) The flux of $\text{curl}(\mathbf{F})$ through every closed, oriented surface is zero.

5. Which condition on $\mathbf{F}$ guarantees that the flux through $\mathcal{S}_1$ is equal to the flux through $\mathcal{S}_2$ for any two oriented surfaces $\mathcal{S}_1$ and $\mathcal{S}_2$ with the same oriented boundary?

Exercises

In Exercises 1–4, calculate curl(**F**).

1. $\mathbf{F} = \langle z - y^2, x + z^3, y + x^2 \rangle$

2. $\mathbf{F} = \left\langle \dfrac{y}{x}, \dfrac{y}{z}, \dfrac{z}{x} \right\rangle$

3. $\mathbf{F} = \langle e^y, \sin x, \cos x \rangle$

4. $\mathbf{F} = \left\langle \dfrac{x}{x^2 + y^2}, \dfrac{y}{x^2 + y^2}, 0 \right\rangle$

In Exercises 5–8, verify Stokes' Theorem for the given vector field and surface, oriented with an upward-pointing normal.

5. $\mathbf{F} = \langle 2xy, x, y + z \rangle$, the surface $z = 1 - x^2 - y^2$ for $x^2 + y^2 \leq 1$

6. $\mathbf{F} = \langle yz, 0, x \rangle$, the portion of the plane $\dfrac{x}{2} + \dfrac{y}{3} + z = 1$ where $x, y, z \geq 0$

7. $\mathbf{F} = \langle e^{y-z}, 0, 0 \rangle$, the square with vertices $(1, 0, 1)$, $(1, 1, 1)$, $(0, 1, 1)$, and $(0, 0, 1)$

8. $\mathbf{F} = \langle -y, 2x, x + z \rangle$, the upper hemisphere $x^2 + y^2 + z^2 = 1$, $z \geq 0$

In Exercises 9–10, use Stokes' Theorem to compute the flux of curl(**F**) *through the given surface.*

9. $\mathbf{F} = \langle z, y, x \rangle$, the hemisphere $x^2 + y^2 + z^2 = 1, x \geq 0$

10. $\mathbf{F} = \left\langle x^2 + y^2, x + z^2, 0 \right\rangle$, the part of the cone $z^2 = x^2 + y^2$ such that $2 \leq z \leq 4$

11. Let $\mathcal{S}$ be the surface of the cylinder (not including the top and bottom) of radius 2 for $1 \leq z \leq 6$, oriented with outward-pointing normal (Figure 14).

(a) Indicate with an arrow the orientation of $\partial \mathcal{S}$ (the top and bottom circles).

(b) Verify Stokes' Theorem for $\mathcal{S}$ and $\mathbf{F} = \langle yz^2, 0, 0 \rangle$.

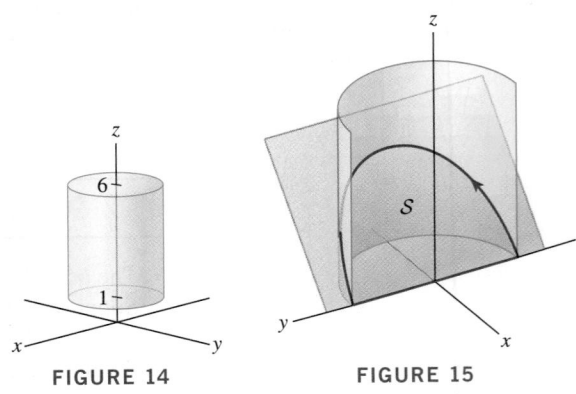

FIGURE 14   FIGURE 15

12. Let $\mathcal{S}$ be the portion of the plane $z = x$ contained in the half-cylinder of radius R depicted in Figure 15. Use Stokes' Theorem to calculate the circulation of $\mathbf{F} = \langle z, x, y + 2z \rangle$ around the boundary of $\mathcal{S}$ (a half-ellipse) in the counterclockwise direction when viewed from above. *Hint:* Show that curl(**F**) is orthogonal to the normal vector to the plane.

13. Let I be the flux of $\mathbf{F} = \langle e^y, 2xe^{x^2}, z^2 \rangle$ through the upper hemisphere S of the unit sphere.

(a) Let $\mathbf{G} = \langle e^y, 2xe^{x^2}, 0 \rangle$. Find a vector field $\mathbf{A}$ such that $\text{curl}(\mathbf{A}) = \mathbf{G}$.

(b) Use Stokes' Theorem to show that the flux of $\mathbf{G}$ through S is zero. *Hint:* Calculate the circulation of $\mathbf{A}$ around ∂S.

(c) Calculate I. *Hint:* Use (b) to show that I is equal to the flux of $\langle 0, 0, z^2 \rangle$ through S.

14. Let $\mathbf{F} = \langle 0, -z, 1 \rangle$ and let S be the spherical cap $x^2 + y^2 + z^2 \le 1$, where $z \ge \frac{1}{2}$. Evaluate $\iint_S \mathbf{F} \cdot d\mathbf{S}$ directly as a surface integral. Then verify that $\mathbf{F} = \text{curl}(\mathbf{A})$, where $\mathbf{A} = (0, x, xz)$ and evaluate the surface integral again using Stokes' Theorem.

15. Let $\mathbf{A}$ be the vector potential and $\mathbf{B}$ the magnetic field of the infinite solenoid of radius R in Example 5. Use Stokes' Theorem to compute:

(a) The flux of $\mathbf{B}$ through a circle in the xy-plane of radius $r < R$

(b) The circulation of $\mathbf{A}$ around the boundary C of a surface lying outside the solenoid

16. The magnetic field $\mathbf{B}$ due to a small current loop (which we place at the origin) is called a magnetic dipole (Figure 16). Let $\rho = (x^2 + y^2 + z^2)^{1/2}$. For ρ large, $\mathbf{B} = \text{curl}(\mathbf{A})$, where

$$\mathbf{A} = \left\langle -\frac{y}{\rho^3}, \frac{x}{\rho^3}, 0 \right\rangle$$

(a) Let C be a horizontal circle of radius R located far from the origin with center on z-axis. Show that $\mathbf{A}$ is tangent to C.

(b) Use Stokes' Theorem to calculate the flux of $\mathbf{B}$ through C.

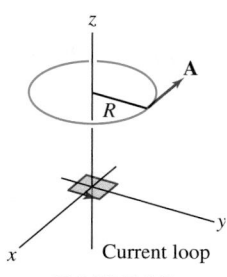

FIGURE 16

17. A uniform magnetic field $\mathbf{B}$ has constant strength b teslas in the z-direction [i.e., $\mathbf{B} = \langle 0, 0, b \rangle$].

(a) Verify that $\mathbf{A} = \frac{1}{2}\mathbf{B} \times \mathbf{r}$ is a vector potential for $\mathbf{B}$, where $\mathbf{r} = \langle x, y, 0 \rangle$.

(b) Calculate the flux of $\mathbf{B}$ through the rectangle with vertices A, B, C, and D in Figure 17.

18. Let $\mathbf{F} = \langle -x^2 y, x, 0 \rangle$. Referring to Figure 17, let C be the closed path $ABCD$. Use Stokes' Theorem to evaluate $\int_C \mathbf{F} \cdot d\mathbf{s}$ in two ways. First, regard C as the boundary of the rectangle with vertices A, B, C,

and D. Then treat C as the boundary of the wedge-shaped box with open top.

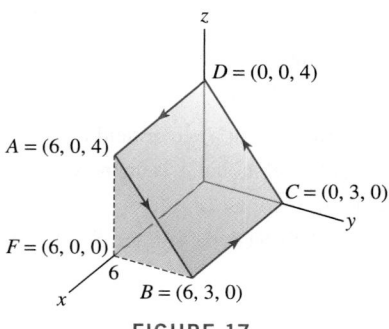

FIGURE 17

19. Let $\mathbf{F} = \langle -z^2, 2zx, 4y - x^2 \rangle$ and let C be a simple closed curve in the plane $x + y + z = 4$ that encloses a region of area 16 (Figure 18). Calculate $\oint_C \mathbf{F} \cdot d\mathbf{s}$, where C is oriented in the counterclockwise direction (when viewed from above the plane).

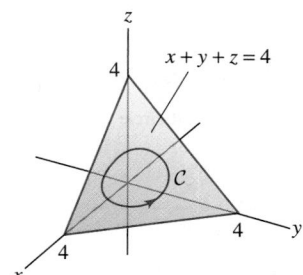

FIGURE 18

20. Let $\mathbf{F} = \langle y^2, x^2, z^2 \rangle$. Show that

$$\int_{C_1} \mathbf{F} \cdot d\mathbf{s} = \int_{C_2} \mathbf{F} \cdot d\mathbf{s}$$

for any two closed curves lying on a cylinder whose central axis is the z-axis (Figure 19).

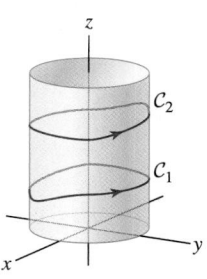

FIGURE 19

21. Let C be the triangular boundary of the portion of the plane $\frac{x}{a} + \frac{y}{b} + \frac{z}{c} = 1$ lying in the octant $x, y, z \ge 0$. Use Stokes'

Theorem to find positive constants a, b, c such that the line integral of $\mathbf{F} = \langle y^2, 2z + x, 2y^2 \rangle$ around C is zero. *Hint:* Choose constants so that $\mathrm{curl}(\mathbf{F})$ is orthogonal to the normal vector.

22. The curl of a vector field $\mathbf{F}$ at the origin is $\mathbf{v}_0 = \langle 3, 1, 4 \rangle$. Estimate the circulation around the small parallelogram spanned by the vectors $\mathbf{A} = \langle 0, \frac{1}{2}, \frac{1}{2} \rangle$ and $\mathbf{B} = \langle 0, 0, \frac{1}{3} \rangle$.

23. You know two things about a vector field $\mathbf{F}$:

(a) $\mathbf{F}$ has a vector potential $\mathbf{A}$ (but $\mathbf{A}$ is unknown).

(b) $\mathbf{F}(x, y, 0) = \langle 0, 0, 1 \rangle$ for all (x, y).

Determine the flux of $\mathbf{F}$ through the surface S in Figure 20.

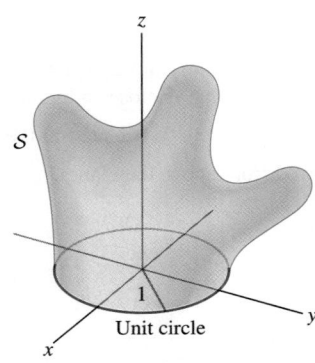

FIGURE 20

24. Find the flux of $\mathbf{F}$ through the surface S in Figure 20, assuming that $\mathbf{F}$ has a vector potential and $\mathbf{F}(x, y, 0) = \langle \cos x, 0, 0 \rangle$.

25. Use Eq. (8) to prove that if $\mathbf{a}$ is a constant vector, then $\mathrm{curl}(\varphi \mathbf{a}) = \nabla \varphi \times \mathbf{a}$.

26. A vector field $\mathbf{F}$ is called **radial** if it is of the form $\mathbf{F} = \varphi(\rho) \langle x, y, z \rangle$ for some function $\varphi(\rho)$, where $\rho = \sqrt{x^2 + y^2 + z^2}$. Show that the curl of a radial vector field is zero. *Hint:* It is enough to show that one component of the curl is zero, since the calculation for the other two components is similar by symmetry.

27. Verify the identity

$$\boxed{\mathrm{curl}(\nabla(\varphi)) = \mathbf{0}} \qquad \boxed{7}$$

28. Prove the Product Rule

$$\mathrm{curl}(\varphi \mathbf{F}) = \varphi\,\mathrm{curl}(\mathbf{F}) + \nabla \varphi \times \mathbf{F} \qquad \boxed{8}$$

29. Assume that the second partial derivatives of φ and ψ exist and are continuous. Use (7) and (8) to prove that

$$\oint_{\partial S} \varphi \nabla(\psi) \cdot d\mathbf{s} = \int_S \nabla(\varphi) \times \nabla(\psi) \cdot d\mathbf{s}$$

where S is a smooth surface with boundary ∂S.

30. ✎ Explain carefully why Green's Theorem is a special case of Stokes' Theorem.

Further Insights and Challenges

31. Complete the proof of Theorem 1 by proving the equality

$$\oint_C F_3(x, y, z)\mathbf{k} \cdot d\mathbf{s} = \int_S \mathrm{curl}(F_3(x, y, z)\mathbf{k}) \cdot d\mathbf{S}$$

where S is the graph of a function $z = f(x, y)$ over a domain $\mathcal{D}$ in the xy-plane whose boundary is a simple closed curve.

32. Let $\mathbf{F}$ be a continuously differentiable vector field in $\mathbf{R}^3$, Q a point, and S a plane containing Q with unit normal vector $\mathbf{e}$. Let C_r be a circle of radius r centered at Q in S and let S_r be the disk enclosed by C_r. Assume S_r is oriented with unit normal vector $\mathbf{e}$.

(a) Let $m(r)$ and $M(r)$ be the minimum and maximum values of $\mathrm{curl}(\mathbf{F}(P)) \cdot \mathbf{e}$ for $P \in S_r$. Prove that

$$m(r) \le \frac{1}{\pi r^2} \iint_{S_r} \mathrm{curl}(\mathbf{F}) \cdot d\mathbf{S} \le M(r)$$

(b) Prove that

$$\mathrm{curl}(\mathbf{F}(Q)) \cdot \mathbf{e} = \lim_{r \to 0} \frac{1}{\pi r^2} \int_{C_r} \mathbf{F} \cdot d\mathbf{s}$$

This proves that $\mathrm{curl}(\mathbf{F}(Q)) \cdot \mathbf{e}$ is the circulation per unit area in the plane S.

17.3 Divergence Theorem

Before stating the third and last of the fundamental theorems of vector analysis, we take a moment to put matters in perspective. First of all, we observe that each main theorem is a relation of the type:

$$\text{Integral of a derivative on a domain} = \text{Integral over the } \textit{oriented } \textit{boundary} \text{ of the domain}$$

In single-variable calculus, the Fundamental Theorem of Calculus (FTC) relates the integral of $f'(x)$ over an interval to the "integral" of $f(x)$ over the boundary:

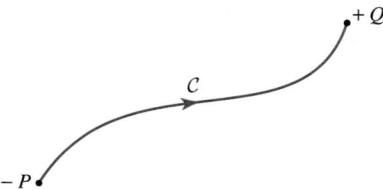

FIGURE 1 The oriented boundary of $\mathcal{C}$ is $Q - P$.

$$\underbrace{\int_a^b f'(x)\,dx}_{\text{Integral of derivative over } [a,b]} = \underbrace{f(b) - f(a)}_{\text{"Integral" over the boundary of } [a,b]}$$

We orient the boundary of the interval $[a, b]$ by assigning a plus sign to b and a minus sign to a.

The Fundamental Theorem for Line Integrals is a direct generalization, where we integrate over a path from P to Q (Figure 1). Instead of the ordinary derivative, we have the gradient:

$$\underbrace{\int_{\mathcal{C}} \nabla\varphi \cdot d\mathbf{s}}_{\text{Integral of derivative over a curve}} = \underbrace{\varphi(Q) - \varphi(P)}_{\substack{\text{Integral of } \varphi \text{ over} \\ \text{the boundary } Q - P}}$$

Stokes' Theorem is a two-dimensional version of the FTC that relates the integral over a surface to an integral over its boundary (Figure 2). In this case, the appropriate derivative is the curl:

FIGURE 2 The oriented boundary of $\mathcal{S}$ is $\partial\mathcal{S}$.

$$\underbrace{\iint_{\mathcal{S}} \text{curl}(\mathbf{F}) \cdot d\mathbf{S}}_{\text{Integral of derivative over surface}} = \underbrace{\int_{\partial\mathcal{S}} \mathbf{F} \cdot d\mathbf{s}}_{\text{Integral over boundary}}$$

The Divergence Theorem follows the same pattern, where the domain is a region $\mathcal{W}$ in $\mathbf{R}^3$ whose boundary is a surface $\partial\mathcal{W}$. Figure 3 shows two examples: a ball whose boundary is a sphere and a cube whose boundary is the surface of the cube.

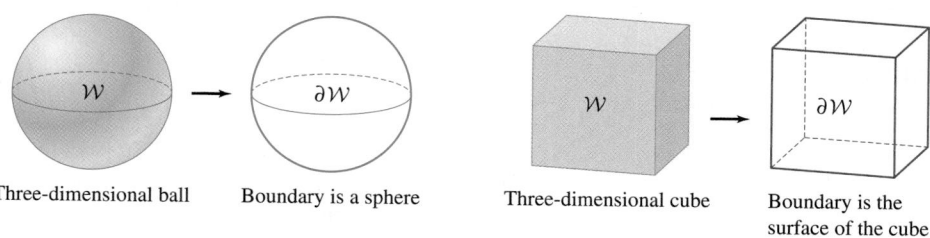

Three-dimensional ball   Boundary is a sphere   Three-dimensional cube   Boundary is the surface of the cube

FIGURE 3

The derivative appearing in the Divergence Theorem is the **divergence** of a vector field $\mathbf{F} = \langle F_1, F_2, F_3 \rangle$, defined by

More advanced treatments of vector calculus use the theory of "differential forms" to formulate a general version of Stokes' Theorem that is valid in all dimensions and includes each of our main theorems (Green, Stokes, Divergence) as a special case.

$$\boxed{\text{div}(\mathbf{F}) = \frac{\partial F_1}{\partial x} + \frac{\partial F_2}{\partial y} + \frac{\partial F_3}{\partial z}}$$ 1

The divergence is often denoted $\nabla \cdot \mathbf{F}$, which we view as a symbolic dot product:

$$\nabla \cdot \mathbf{F} = \left\langle \frac{\partial}{\partial x}, \frac{\partial}{\partial y}, \frac{\partial}{\partial z} \right\rangle \cdot \langle F_1, F_2, F_3 \rangle = \frac{\partial F_1}{\partial x} + \frac{\partial F_2}{\partial y} + \frac{\partial F_3}{\partial z}$$

Keep in mind that, unlike the gradient and curl, the divergence of a vector field is a scalar function. Furthermore, taking the divergence is a linear operation:

$$\text{div}(\mathbf{F} + \mathbf{G}) = \text{div}(\mathbf{F}) + \text{div}(\mathbf{G}), \qquad \text{div}(c\mathbf{F}) = c\,\text{div}(\mathbf{F}) \quad \text{(for any constant } c\text{)}$$

■ **EXAMPLE 1** Evaluate the divergence of $\mathbf{F} = \langle e^{xy}, xy, z^4 \rangle$ at $P = (1, 0, 2)$.

Solution The divergence is the function

$$\text{div}(\mathbf{F}) = \frac{\partial}{\partial x} e^{xy} + \frac{\partial}{\partial y} xy + \frac{\partial}{\partial z} z^4 = ye^{xy} + x + 4z^3$$

and $\text{div}(\mathbf{F})(P) = 0 \cdot e^0 + 1 + 4 \cdot 2^3 = 33$.   ■

The Divergence Theorem is also referred to as "Gauss's Theorem" or the "Gauss-Ostrogradsky Theorem."

> **THEOREM 1 Divergence Theorem** Let $\mathcal{W}$ be a region in $\mathbf{R}^3$ whose boundary $\partial\mathcal{W}$ is a piecewise smooth surface, oriented so that the normal vectors to $\partial\mathcal{W}$ point outside of $\mathcal{W}$. Let $\mathbf{F}$ be a vector field whose domain contains $\mathcal{W}$ and whose components have continuous partial derivatives. Then
>
> $$\iint_{\partial\mathcal{W}} \mathbf{F} \cdot d\mathbf{S} = \iiint_{\mathcal{W}} \text{div}(\mathbf{F}) \, dV \qquad \boxed{2}$$

Proof We prove the Divergence Theorem for the special case in which $\mathcal{W}$ is a box $[a, b] \times [c, d] \times [e, f]$ as in Figure 4. The proof can be modified to treat more general regions such as the interiors of spheres and cylinders.

Both sides of the equality in the Divergence Theorem are additive in the following sense. If $\mathbf{F} = \langle F_1, F_2, F_3 \rangle$, then

$$\iint_{\partial\mathcal{W}} (F_1\mathbf{i} + F_2\mathbf{j} + F_3\mathbf{k}) \cdot d\mathbf{S} = \iint_{\partial\mathcal{W}} F_1\mathbf{i} \cdot d\mathbf{S} + \iint_{\partial\mathcal{W}} F_2\mathbf{j} \cdot d\mathbf{S} + \iint_{\partial\mathcal{W}} F_3\mathbf{k} \cdot d\mathbf{S}$$

$$\iiint_{\mathcal{W}} \text{div}(F_1\mathbf{i} + F_2\mathbf{j} + F_3\mathbf{k}) \, dV = \iiint_{\mathcal{W}} \text{div}(F_1\mathbf{i}) \, dV + \iiint_{\mathcal{W}} \text{div}(F_2\mathbf{j}) \, dV$$

$$+ \iiint_{\mathcal{W}} \text{div}(F_3\mathbf{k}) \, dV$$

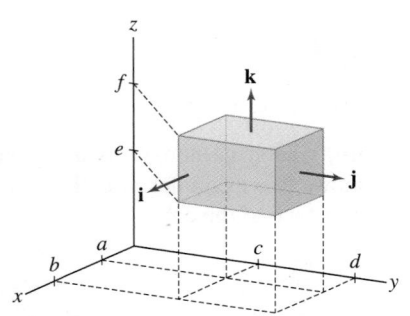

FIGURE 4

As in the proof of Green's and Stokes' Theorems, we may prove the Divergence Theorem by showing that the terms corresponding to the $\mathbf{i}$-, $\mathbf{j}$-, and $\mathbf{k}$-components are separately equal. We carry out the argument for the $\mathbf{i}$-component (the other two components are similar).

Assume that $\mathbf{F} = F_1\mathbf{i}$ and let $\mathcal{S} = \partial\mathcal{W}$ be the boundary of the box. Then $\mathcal{S}$ consists of six faces. The integral of $\mathbf{F}$ over $\mathcal{S}$ is the sum of the integrals over the six faces. However, $\mathbf{F} = F_1\mathbf{i}$ is orthogonal to the normal vectors to the top and bottom as well as the two side faces since

$$\mathbf{F} \cdot \mathbf{j} = \mathbf{F} \cdot \mathbf{k} = 0$$

Therefore, the integrals over the top, bottom, and two side faces are zero. Nonzero contributions come only from the front and back faces, which we denote $\mathcal{S}_f$ and $\mathcal{S}_b$ (Figure 5):

$$\iint_{\mathcal{S}} \mathbf{F} \cdot d\mathbf{S} = \iint_{\mathcal{S}_f} \mathbf{F} \cdot d\mathbf{S} + \iint_{\mathcal{S}_b} \mathbf{F} \cdot d\mathbf{S}$$

To evaluate these integrals, we parametrize $\mathcal{S}_f$ and $\mathcal{S}_b$ by

$$\Phi_f(y, z) = (b, y, z), \qquad c \le y \le d, \ e \le z \le f$$

$$\Phi_b(y, z) = (a, y, z), \qquad c \le y \le d, \ e \le z \le f$$

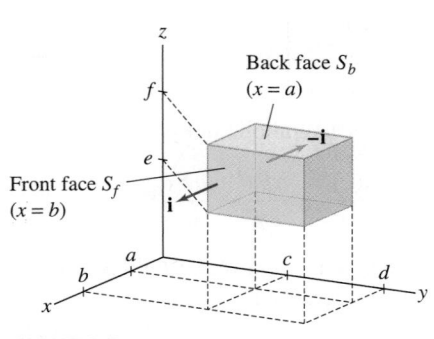

FIGURE 5

The normal vectors for these parametrizations are

$$\frac{\partial \Phi_f}{\partial y} \times \frac{\partial \Phi_f}{\partial z} = \mathbf{j} \times \mathbf{k} = \mathbf{i}$$

$$\frac{\partial \Phi_b}{\partial y} \times \frac{\partial \Phi_b}{\partial z} = \mathbf{j} \times \mathbf{k} = \mathbf{i}$$

However, the outward-pointing normal for the face $\mathcal{S}_b$ is the vector $-\mathbf{i}$ and hence a minus sign occurs when we compute the surface integral of $\mathbf{F}$ over $\mathcal{S}_b$ using the parametrization Φ_b:

$$\iint_{\mathcal{S}_f} \mathbf{F} \cdot d\mathbf{S} + \iint_{\mathcal{S}_b} \mathbf{F} \cdot d\mathbf{S} = \int_e^f \int_c^d F_1(b, y, z) \, dy \, dz - \int_e^f \int_c^d F_1(a, y, z) \, dy \, dz$$

$$= \int_e^f \int_c^d \left(F_1(b, y, z) - F_1(a, y, z) \right) dy \, dz$$

By the FTC in one variable,

$$F_1(b, y, z) - F_1(a, y, z) = \int_a^b \frac{\partial F_1}{\partial x}(x, y, z) \, dx$$

Since $\operatorname{div}(\mathbf{F}) = \operatorname{div}(F_1 \mathbf{i}) = \dfrac{\partial F_1}{\partial x}$, we obtain the desired result:

$$\iint_{\mathcal{S}} \mathbf{F} \cdot d\mathbf{S} = \int_e^f \int_c^d \int_a^b \frac{\partial F_1}{\partial x}(x, y, z) \, dx \, dy \, dz = \iiint_{\mathcal{W}} \operatorname{div}(\mathbf{F}) \, dV \qquad \blacksquare$$

■ **EXAMPLE 2** Verifying the Divergence Theorem Verify the Divergence Theorem for the vector field $\mathbf{F} = \langle y, yz, z^2 \rangle$ and the portion $\mathcal{S}$ of the cylinder $x^2 + y^2 = 4$ between $z = 0$ and $z = 5$ (Figure 6).

Solution We calculate both sides of Eq. (2).

Step 1. **Integrate over the side of the cylinder.**
We parametrize the side of the cylinder by $\Phi(\theta, z) = (2 \cos \theta, 2 \sin \theta, z)$ for $0 \le \theta < 2\pi$ and $0 \le z \le 5$. Then

$$\mathbf{n} = \mathbf{T}_\theta \times \mathbf{T}_z = \langle -2 \sin \theta, 2 \cos \theta, 0 \rangle \times \langle 0, 0, 1 \rangle = \langle 2 \cos \theta, 2 \sin \theta, 0 \rangle$$

$$\mathbf{F}(\Phi(\theta, z)) \cdot \mathbf{n} = \langle 2 \sin \theta, 2z \sin \theta, z^2 \rangle \cdot \langle 2 \cos \theta, 2 \sin \theta, 0 \rangle$$

$$= 4 \cos \theta \sin \theta + 4z \sin^2 \theta$$

$$\iint_{\text{side}} \mathbf{F} \cdot d\mathbf{S} = \int_0^5 \int_0^{2\pi} (4 \cos \theta \sin \theta + 4z \sin^2 \theta) \, d\theta \, dz$$

$$= 0 + 4\pi \int_0^5 z \, dz = 4\pi \left(\frac{25}{2} \right) = 50\pi \qquad \boxed{3}$$

Step 2. **Integrate over the top and bottom of the cylinder.**
The top of the cylinder is parametrized by $\Phi(x, y) = (x, y, 5)$ with parameter domain $\mathcal{D} = \{(x, y) : x^2 + y^2 \le 4\}$. We obtain

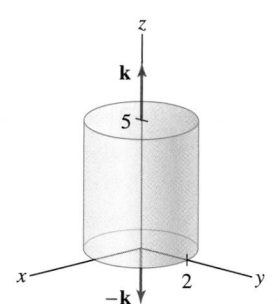

FIGURE 6 Cylinder of radius 2 and height 5.

◄·· *REMINDER* In Eq. (3), we use

$$\int_0^{2\pi} \cos \theta \sin \theta \, d\theta = 0$$

$$\int_0^{2\pi} \sin^2 \theta \, d\theta = \pi$$

$$\mathbf{F}(\Phi(x, y)) = \langle y, 5y, 5^2 \rangle$$

$$\mathbf{n} = \mathbf{T}_x \times \mathbf{T}_y = \langle 1, 0, 0 \rangle \times \langle 0, 1, 0 \rangle = \langle 0, 0, 1 \rangle$$

$$\mathbf{F}(\Phi(x, y)) \cdot \mathbf{n} = \langle y, 5y, 5^2 \rangle \cdot \langle 0, 0, 1 \rangle = 25$$

Since $\mathcal{D}$ has area 4π,

$$\iint_{\text{top}} \mathbf{F} \cdot d\mathbf{S} = \iint_{\mathcal{D}} 25 \, dA = 25 \, \text{Area}(\mathcal{D}) = 25(4\pi) = 100\pi$$

The integral over the bottom of the cylinder is zero. Indeed, along the bottom where $z = 0$, the vector field $\mathbf{F}(x, y, 0) = \langle y, 5y, 0 \rangle$ is orthogonal to the normal vector $-\mathbf{k}$.

Step 3. **Find the total flux.**

$$\iint_{\mathcal{S}} \mathbf{F} \cdot d\mathbf{S} = \text{sides} + \text{top} + \text{bottom} = 50\pi + 100\pi + 0 = \boxed{150\pi}$$

Step 4. **Compare with integral of divergence.**

$$\text{div}(\mathbf{F}) = \text{div}(\langle y, yz, z^2 \rangle) = \frac{\partial}{\partial x} y + \frac{\partial}{\partial y}(yz) + \frac{\partial}{\partial z} z^2 = 0 + z + 2z = 3z$$

The region $\mathcal{W}$ consists of points (x, y, z), where $(x, y) \in \mathcal{D}$ and $0 \le z \le 5$:

$$\iiint_{\mathcal{W}} \text{div}(\mathbf{F}) \, dV = \iint_{\mathcal{D}} \int_{z=0}^{5} 3z \, dV = \iint_{\mathcal{D}} \frac{75}{2} \, dV$$

$$= \left(\frac{75}{2}\right)(\text{Area}(\mathcal{D})) = \left(\frac{75}{2}\right)(4\pi) = \boxed{150\pi}$$

The flux is equal to the integral of divergence, thus verifying the Divergence Theorem. ∎

■ **EXAMPLE 3** **Using the Divergence Theorem** Use the Divergence Theorem to evaluate $\iint_{\mathcal{S}} \langle x^2, z^4, e^z \rangle \cdot d\mathbf{S}$, where $\mathcal{S}$ is the boundary of the box $\mathcal{W} = [0, 2] \times [0, 3] \times [0, 1]$ (Figure 7).

Solution

$$\text{div}(\langle x^2, z^4, e^z \rangle) = \frac{\partial}{\partial x} x^2 + \frac{\partial}{\partial y} z^4 + \frac{\partial}{\partial z} e^z = 2x + e^z$$

By the Divergence Theorem,

$$\iint_{\mathcal{S}} \langle x^2, z^4, e^z \rangle \cdot d\mathbf{S} = \iiint_{\mathcal{W}} (2x + e^z) \, dV = \int_0^2 \int_0^3 \int_0^1 (2x + e^z) \, dz \, dy \, dx$$

$$= 3\int_0^2 2x \, dx + 6 \int_0^1 e^z \, dz = 12 + 6(e - 1) = 6e + 6 \quad ■$$

FIGURE 7

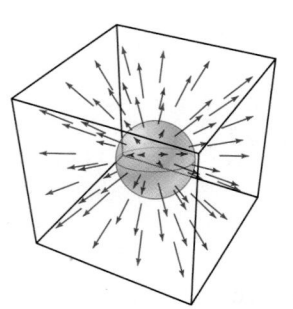

FIGURE 8 A vector field with nonzero divergence.

GRAPHICAL INSIGHT **The Interpretation of Divergence** As in our discussion of curl, let's assume that $\mathbf{F}$ is the velocity field of a fluid. Then the flux of $\mathbf{F}$ through a surface $\mathcal{S}$ is the volume of fluid passing through $\mathcal{S}$ per unit time (Figure 8). Suppose that $\mathcal{S}$ encloses a small region $\mathcal{W}$ containing a point P (e.g., a ball of small radius). Then $\text{div}(\mathbf{F})$

is nearly constant on $\mathcal{W}$ with value div$(\mathbf{F})(P)$, and the Divergence Theorem yields the approximation

$$\text{Flux across } \mathcal{S} = \iiint_{\mathcal{W}} \text{div}(\mathbf{F}) \, dV \approx \text{div}(\mathbf{F})(P) \, \text{Vol}(\mathcal{W})$$

In other words, the *flux is approximately equal to the divergence times the enclosed volume*. We may think of div$(\mathbf{F})(P)$ as the "flux per unit volume" at P. In particular,

- If div$(\mathbf{F})(P) > 0$, there is a net outflow across any small closed surface enclosing P. Thus, fluid is "produced" at P.
- If div$(\mathbf{F})(P) < 0$, there is a net inflow across any small closed surface enclosing P. Thus, fluid is "consumed" at P.
- If div$(\mathbf{F})(P) = 0$, then to a first-order approximation, there is no net flow, in or out, across any small closed surface enclosing P.

These cases are more easily visualized in two dimensions, where we define div$((P, Q)) = \dfrac{\partial P}{\partial x} + \dfrac{\partial Q}{\partial y}$. The vector field in Figure 9(A) has positive divergence and there is a positive net flow of fluid across every circle per unit time. Similarly, Figure 9(B) shows a vector field with negative divergence. By contrast, the vector field in Figure 9(C) has zero divergence. The fluid flowing into every circle is balanced by the fluid flowing out. In general, a vector field $\mathbf{F}$ such that div$(\mathbf{F}) = 0$ is called *incompressible*. If $\mathbf{F}$ is an incompressible vector field defined on $\mathbf{R}^3$, then the flux of $\mathbf{F}$ through every closed surface is zero.

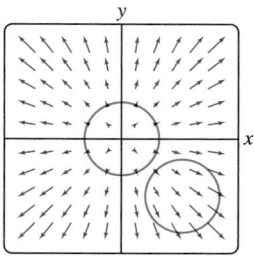

(A) The field $\mathbf{F} = \langle x, y \rangle$ with div$(\mathbf{F}) = 2$. There is a net outflow through every circle.

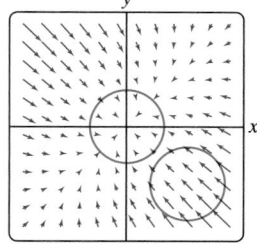

(B) The field $\mathbf{F} = \langle y - 2x, x - 2y \rangle$ with div$(\mathbf{F}) = -4$. There is a net inflow into every circle.

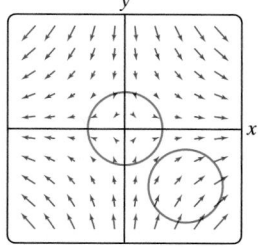

(C) The field $\mathbf{F} = \langle x, -y \rangle$ with div$(\mathbf{F}) = 0$. The flux through every circle is zero.

FIGURE 9

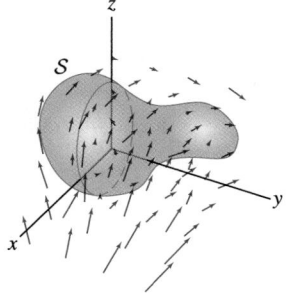

FIGURE 10

■ **EXAMPLE 4** A Vector Field with Zero Divergence Compute the flux of

$$\mathbf{F} = \left\langle z^2 + xy^2, \cos(x + z), e^{-y} - zy^2 \right\rangle$$

through the boundary of the surface $\mathcal{S}$ in Figure 10.

Solution The divergence of $\mathbf{F}$ is zero:

$$\text{div}(\mathbf{F}) = \frac{\partial}{\partial x}(z^2 + xy^2) + \frac{\partial}{\partial y}\cos(x + z) + \frac{\partial}{\partial z}(e^{-y} - zy^2) = y^2 - y^2 = 0$$

By the Divergence Theorem, the flux of **F** through *every* closed surface $\mathcal{S}$ enclosing a region $\mathcal{W}$ is zero:

$$\iint_{\mathcal{S}} \mathbf{F} \cdot d\mathbf{S} = \iiint_{\mathcal{W}} \operatorname{div}(\mathbf{F}) \, dV = \iiint_{\mathcal{W}} 0 \, dV = 0$$

In particular, the flux through $\mathcal{S}$ in Figure 10 is zero. ∎

CONCEPTUAL INSIGHT   The Divergence Theorem and Stokes' Theorem give us two ways of showing that a vector field **F** has zero flux through closed surfaces:

- If $\operatorname{div}(\mathbf{F}) = 0$, then $\iint_{\mathcal{S}} \mathbf{F} \cdot d\mathbf{S} = 0$ for every closed surface $\mathcal{S}$ that is the boundary of a region $\mathcal{W}$ contained in the domain of **F**.
- If $\mathbf{F} = \operatorname{curl}(\mathbf{A})$ for some vector potential **A**, then $\iint_{\mathcal{S}} \mathbf{F} \cdot d\mathbf{S} = 0$ for every closed surface contained in the domain of **F** (by Stokes' Theorem; see Theorem 2 in Section 17.2).

This raises the following question: Are the two conditions $\operatorname{div}(\mathbf{F}) = 0$ and $\mathbf{F} = \operatorname{curl}(\mathbf{A})$ related? Half of this question is easy to answer. In Exercise 22, you are asked to verify the identity

$$\boxed{\operatorname{div}(\operatorname{curl}(\mathbf{A})) = 0}$$

So if $\mathbf{F} = \operatorname{curl}(\mathbf{A})$, then $\operatorname{div}(\mathbf{F})$ is automatically zero. The other half is not as straightforward. If $\operatorname{div}(\mathbf{F}) = 0$, we cannot be sure that **F** has a vector potential unless the domain of **F** is a region with "no holes" such as a ball or box or all of $\mathbf{R}^n$ (a precise definition of "no holes" would lead us into the area of mathematics called topology). If a domain has no holes, then every vector field with zero divergence has a vector potential (see Exercise 34). By contrast, if the domain has one or more holes (such as $\mathbf{R}^3$ with a sphere or even just a point removed), then there exist vector fields with zero divergence but nonzero flux through some closed surface. This is illustrated in the next example. Such a vector field cannot have a vector potential because $\mathbf{F} = \operatorname{curl}(\mathbf{A})$ implies that the flux through *every* closed surface is zero.

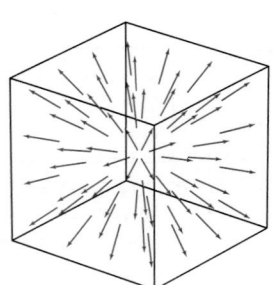

FIGURE 11 Unit radial vector field $\mathbf{e}_r$.

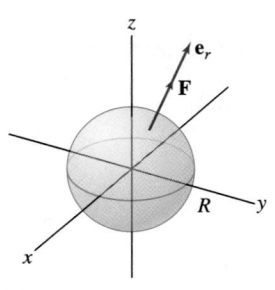

FIGURE 12

Let $\rho = \sqrt{x^2 + y^2 + z^2}$ and let $\mathbf{e}_r$ denote the unit radial vector field (Figure 11):

$$\boxed{\mathbf{e}_r = \frac{\langle x, y, z \rangle}{\rho} = \frac{\langle x, y, z \rangle}{\sqrt{x^2 + y^2 + z^2}} \qquad (\text{for } \rho \neq 0)}$$ **4**

■ **EXAMPLE 5**   **The Inverse-Square Radial Vector Field**   Let $\mathbf{F} = \dfrac{\mathbf{e}_r}{\rho^2}$ (defined for $\rho \neq 0$). Verify that:

(a) $\operatorname{div}(\mathbf{F}) = 0$

(b) $\iint_{\mathcal{S}_R} \mathbf{F} \cdot d\mathbf{S} = 4\pi$, where $\mathcal{S}_R$ is the sphere of radius R centered at the origin (Figure 12).

(c) $\iint_S \mathbf{F} \cdot d\mathbf{S} = 0$ for any closed surface S that does not contain the origin.

Solution   Write the field as $\mathbf{F} = \langle F_1, F_2, F_3 \rangle = \langle x\rho^{-3}, y\rho^{-3}, z\rho^{-3} \rangle$.

(a) To compute the divergence, we first note that

$$\frac{\partial \rho}{\partial x} = \frac{\partial}{\partial x}(x^2 + y^2 + z^2)^{1/2} = \frac{1}{2}(x^2 + y^2 + z^2)^{-1/2}(2x) = \frac{x}{\rho}$$

Therefore,

$$\frac{\partial F_1}{\partial x} = \frac{\partial}{\partial x}x\rho^{-3} = \rho^{-3} - 3x\rho^{-4}\frac{\partial \rho}{\partial x} = \rho^{-3} - (3x\rho^{-4})\frac{x}{\rho} = \frac{\rho^2 - 3x^2}{\rho^5}$$

The derivatives $\dfrac{\partial F_2}{\partial y}$ and $\dfrac{\partial F_3}{\partial z}$ are similar, so we have

$$\text{div}(\mathbf{F}) = \frac{\rho^2 - 3x^2}{\rho^5} + \frac{\rho^2 - 3y^2}{\rho^5} + \frac{\rho^2 - 3z^2}{\rho^5} = \frac{3\rho^2 - 3(x^2 + y^2 + z^2)}{\rho^5} = 0$$

(b) We compute the flux through S_R using the parametrization by spherical coordinates $\Phi(\phi, \theta) = (R \cos \theta \sin \phi, R \sin \theta \sin \phi, R \cos \phi)$. As we saw in Section 16.4, Eq. (5), the outward-pointing normal is

$$\mathbf{n} = \mathbf{T}_\phi \times \mathbf{T}_\theta = (R^2 \sin \phi)\mathbf{e}_r$$

Furthermore, on the sphere of radius R, $\mathbf{F}(\Phi(\phi, \theta)) = R^{-2}\mathbf{e}_r$ and therefore,

$$\mathbf{F} \cdot \mathbf{n} = (R^{-2}\mathbf{e}_r) \cdot (R^2 \sin \phi \mathbf{e}_r) = \sin \phi(\mathbf{e}_r \cdot \mathbf{e}_r) = \sin \phi$$

Since the flux of $\mathbf{F} = \rho^{-2}\mathbf{e}_r$ through a sphere centered at the origin is nonzero, $\mathbf{F}$ cannot have a vector potential even though $\text{div}(\mathbf{F}) = 0$ on its domain.

Now we can verify that the flux is 4π:

$$\iint_{S_R} \mathbf{F} \cdot d\mathbf{S} = \int_0^{2\pi} \int_0^\pi \mathbf{F} \cdot \mathbf{n} \, d\phi \, d\theta = 2\pi \int_0^\pi \sin \phi \, d\phi = 4\pi$$

(c) If S is a closed surface not containing the origin, then $\mathbf{F}$ is defined at every point in the region $\mathcal{W}$ enclosed by S and $\text{div}(\mathbf{F}) = 0$ on $\mathcal{W}$. The Divergence Theorem yields

$$\iint_S \mathbf{F} \cdot d\mathbf{S} = \iiint_{\mathcal{W}} \text{div}(\mathbf{F}) \, dV = 0 \qquad \blacksquare$$

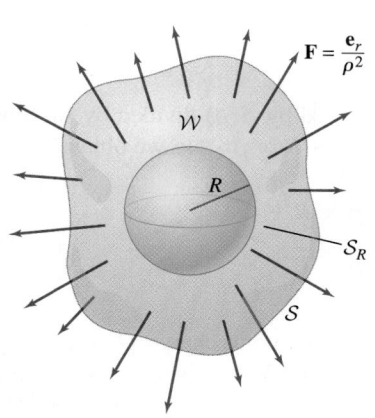

In the previous example, we showed that the flux of $\mathbf{F} = \dfrac{\mathbf{e}_r}{\rho^2}$ through S_R is equal to 4π. We can prove that this result remains true for any surface S containing the origin as follows. Choose $R > 0$ small enough so that S_R is contained inside S and let $\mathcal{W}$ be the region between S_R and S. Then the oriented boundary of $\mathcal{W}$ is the difference $S - S_R$, that is, the boundary consists of S with an outward-pointing normal and S_R with an inward-pointing normal (Figure 13). The Divergence Theorem is valid for the region $\mathcal{W}$, and since $\text{div}(\mathbf{F}) = 0$ on $\mathcal{W}$, we have

$$\iint_S \mathbf{F} \cdot d\mathbf{S} - \iint_{S_R} \mathbf{F} \cdot d\mathbf{S} = \iiint_{\mathcal{W}} \text{div}(\mathbf{F}) \, dV = 0$$

FIGURE 13

It follows that the flux of $\mathbf{F}$ through S is equal to the flux through S_R.

This result plays a fundamental role in the theory of electrostatics. The electric field **E** due to a point charge of magnitude q coulombs placed at the origin in a vacuum is

$$\mathbf{E} = \left(\frac{q}{4\pi\epsilon_0}\right)\frac{\mathbf{e}_r}{\rho^2}$$

where $\epsilon_0 = 8.85 \times 10^{-12}$ C^2/N-m^2 is the permittivity constant. Since **E** and $\mathbf{e}_r/\rho^2$ differ by the constant $q/(4\pi\epsilon_0)$, we have the following general result for any closed surface $\mathcal{S}$:

$$\text{Flux of } \mathbf{E} \text{ through } \mathcal{S} = \begin{cases} 0 & \text{if } q \text{ is outside } \mathcal{S} \\ \dfrac{q}{\epsilon_0} & \text{if } q \text{ is inside } \mathcal{S} \end{cases}$$

Now, instead of placing just one point charge at the origin, we may distribute a finite number N of point charges q_i at different points in space. The resulting electric field **E** is the sum of the fields $\mathbf{E}_i$ due to the individual charges and we have

$$\iint_{\mathcal{S}} \mathbf{E} \cdot d\mathbf{S} = \iint_{\mathcal{S}} \mathbf{E}_1 \cdot d\mathbf{S} + \cdots + \iint_{\mathcal{S}} \mathbf{E}_N \cdot d\mathbf{S}$$

Each integral on the right is either 0 or $\dfrac{q_i}{\epsilon_0}$, according to whether or not $\mathcal{S}$ contains q_i, so we conclude that

$$\iint_{\mathcal{S}} \mathbf{E} \cdot d\mathbf{S} = \frac{\text{total charge enclosed by } \mathcal{S}}{\epsilon_0} \qquad \boxed{5}$$

This relation is called **Gauss's Law**. A limiting argument may be used to show that Eq. (5) remains valid for the electric field due to a *continuous* distribution of charge.

Let's examine Gauss's Law for a uniformly charged hollow sphere of radius R centered at the origin (Figure 14). Let Q be the total quantity of charge on the sphere. Let P be a point not lying on the sphere. By symmetry, the electric field **E** must be directed radially away from the origin and its magnitude depends only on the distance ρ from P to the origin. Thus, $\mathbf{E} = E(\rho)\mathbf{e}_r$ for some function $E(\rho)$. The flux of **E** through the sphere $\mathcal{S}_\rho$ is

$$\iint_{\mathcal{S}_\rho} \mathbf{E} \cdot d\mathbf{S} = E(\rho)\underbrace{\iint_{\mathcal{S}_\rho} \mathbf{e}_r \cdot d\mathbf{S}}_{\text{Surface area of sphere}} = 4\pi\rho^2 E(\rho)$$

By Gauss's Law, this flux is equal to the charge enclosed by $\mathcal{S}(\rho)$ divided by ϵ_0. Since the enclosed charge is either Q or zero, according as $\rho > R$ or $\rho < R$, we obtain

$$E(\rho) = \begin{cases} 0 & \text{if } \rho < R \\ \dfrac{Q}{4\pi\epsilon_0\rho^2} & \text{if } \rho > R \end{cases} \qquad \boxed{6}$$

In other words, **E** coincides with the electric field due to a point charge Q at the origin outside the sphere, and inside the sphere the field **E** is zero. We obtained this same result for the gravitational field by a laborious calculation in Exercise 50 of Section 16.4. Here, we have derived it from Gauss's Law and a simple appeal to symmetry.

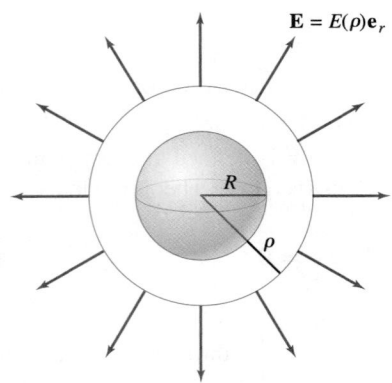

$\mathbf{E} = E(\rho)\mathbf{e}_r$

FIGURE 14 The electric field due to a uniformly charged sphere.

HISTORICAL PERSPECTIVE

James Clerk Maxwell
(1831–1879)

The theorems of vector analysis that we have studied in this chapter were developed in the nineteenth century, in large part, to express the laws of electricity and magnetism. Electromagnetism was studied intensively in the period 1750–1890, culminating in the famous Maxwell Equations, which provide a unified understanding in terms of two vector fields: the electric field **E** and the magnetic field **B**. In a region of empty space (where there are no charged particles), Maxwell's Equations are

$$\text{div}(\mathbf{E}) = 0, \qquad \text{div}(\mathbf{B}) = 0$$
$$\text{curl}(\mathbf{E}) = -\frac{\partial \mathbf{B}}{\partial t}, \qquad \text{curl}(\mathbf{B}) = \mu_0 \epsilon_0 \frac{\partial \mathbf{E}}{\partial t}$$

where μ_0 and ϵ_0 are experimentally determined constants. In MKS units,

$$\mu_0 = 4\pi \times 10^{-7} \text{ henries/m}$$

and

$$\epsilon_0 \approx 8.85 \times 10^{-12} \text{ farads/m}$$

On the basis of these four equations, Maxwell was led to make two predictions of fundamental importance: (1) That electromagnetic waves exist (this was confirmed by H. Hertz in 1887), and (2) that light is an electromagnetic wave.

How do Maxwell's equations suggest that electromagnetic waves exist? And why did Maxwell conclude that light is an electromagnetic wave? It was known to mathematicians in the eighteenth century that waves traveling with velocity c may be described by functions $\varphi(x, y, z, t)$ that satisfy the *wave equation*

$$\boxed{\Delta \varphi = \frac{1}{c^2} \frac{\partial^2 \varphi}{\partial t^2}} \qquad \boxed{7}$$

where Δ is the Laplace operator (or "Laplacian") $\Delta \varphi = \frac{\partial^2 \varphi}{\partial x^2} + \frac{\partial^2 \varphi}{\partial y^2} + \frac{\partial^2 \varphi}{\partial z^2}$.

We will show that the components of **E** satisfy this wave equation. Take the curl of both sides of Maxwell's third equation:

$$\text{curl}(\text{curl}(\mathbf{E})) = \text{curl}\left(-\frac{\partial \mathbf{B}}{\partial t}\right) = -\frac{\partial}{\partial t}\text{curl}(\mathbf{B})$$

Then apply Maxwell's fourth equation to obtain

$$\text{curl}(\text{curl}(\mathbf{E})) = -\frac{\partial}{\partial t}\left(\mu_0 \epsilon_0 \frac{\partial \mathbf{E}}{\partial t}\right)$$
$$= -\mu_0 \epsilon_0 \frac{\partial^2 \mathbf{E}}{\partial t^2} \qquad \boxed{8}$$

Finally, let us define the Laplacian of a vector field

$$\mathbf{F} = \langle F_1, F_2, F_3 \rangle$$

by applying Δ to each component, $\Delta\mathbf{F} = \langle \Delta F_1, \Delta F_2, \Delta F_3 \rangle$. Then the following identity holds (see Exercise 26):

$$\text{curl}(\text{curl}(\mathbf{F})) = \nabla(\text{div}(\mathbf{F})) - \Delta\mathbf{F}$$

Applying this identity to **E**, we obtain $\text{curl}(\text{curl}(\mathbf{E})) = -\Delta\mathbf{E}$ since $\text{div}(\mathbf{E}) = 0$ by Maxwell's first equation. Thus, Eq. (8) yields

$$\boxed{\Delta\mathbf{E} = \mu_0 \epsilon_0 \frac{\partial^2 \mathbf{E}}{\partial t^2}}$$

In other words, each component of the electric field satisfies the wave equation (7), with $c = (\mu_0\epsilon_0)^{-1/2}$. This tells us that the **E**-field (and similarly the **B**-field) can propagate through space like a wave, giving rise to electromagnetic radiation (Figure 15). Maxwell noticed that the velocity of an electromagnetic wave is

$$c = (\mu_0\epsilon_0)^{-1/2} \approx 3 \times 10^8 \text{ m/s}$$

This value of c turned out be suspiciously close to the velocity of light (first measured by Olaf Römer in 1676). As Maxwell wrote in 1862, "We can scarcely avoid the conclusion that light consists in the transverse undulations of the same medium which is the cause of electric and magnetic phenomena." Needless to say, our modern world relies at every turn on the unseen electromagnetic radiation whose existence was first predicted by Maxwell on theoretical grounds.

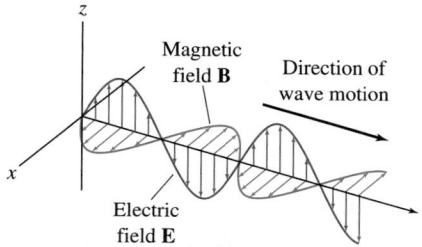

FIGURE 15 The **E** and **B** fields of an electromagnetic wave along an axis of motion.

17.3 SUMMARY

- The *divergence* of a vector field $\mathbf{F} = \langle F_1, F_2, F_3 \rangle$ is defined by

$$\text{div}(\mathbf{F}) = \frac{\partial F_1}{\partial x} + \frac{\partial F_2}{\partial y} + \frac{\partial F_3}{\partial z}$$

- The Divergence Theorem applies to a region $\mathcal{W}$ in $\mathbf{R}^3$ whose boundary $\partial\mathcal{W}$ is a surface, oriented by normal vectors pointing outside $\mathcal{W}$. Assume that $\mathbf{F}$ is defined and has continuous partial derivatives on $\mathcal{W}$. Then

$$\iint_{\partial\mathcal{W}} \mathbf{F} \cdot d\mathbf{S} = \iiint_{\mathcal{W}} \text{div}(\mathbf{F})\, dV$$

- The Divergence Theorem has the following corollary: If $\text{div}(\mathbf{F}) = 0$ and $\mathcal{W}$ is a region contained in the domain of $\mathbf{F}$, then the flux of $\mathbf{F}$ through the boundary $\partial\mathcal{W}$ is zero.
- The vector field $\mathbf{F} = \dfrac{\mathbf{e}_r}{\rho^2}$ in Example 5, defined for $\rho \neq 0$, has zero divergence. However, the flux of $\mathbf{F}$ through every closed surface S containing the origin is equal to 4π. This does not contradict the Divergence Theorem because $\mathbf{F}$ is not defined at the origin.

17.3 EXERCISES

Preliminary Questions

1. What is the flux of $\mathbf{F} = \langle 1, 0, 0 \rangle$ through a closed surface?

2. Justify the following statement: The flux of $\mathbf{F} = \langle x^3, y^3, z^3 \rangle$ through every closed surface is positive.

3. Which of the following expressions are meaningful (where $\mathbf{F}$ is a vector field and φ is a function)? Of those that are meaningful, which are automatically zero?

(a) $\text{div}(\nabla\varphi)$ **(b)** $\text{curl}(\nabla\varphi)$ **(c)** $\nabla\text{curl}(\varphi)$
(d) $\text{div}(\text{curl}(\mathbf{F}))$ **(e)** $\text{curl}(\text{div}(\mathbf{F}))$ **(f)** $\nabla(\text{div}(\mathbf{F}))$

4. Which of the following statements is correct (where $\mathbf{F}$ is a continuously differentiable vector field defined everywhere)?
(a) The flux of $\text{curl}(\mathbf{F})$ through all surfaces is zero.
(b) If $\mathbf{F} = \nabla\varphi$, then the flux of $\mathbf{F}$ through all surfaces is zero.
(c) The flux of $\text{curl}(\mathbf{F})$ through all closed surfaces is zero.

5. How does the Divergence Theorem imply that the flux of $\mathbf{F} = \langle x^2, y - e^z, y - 2zx \rangle$ through a closed surface is equal to the enclosed volume?

Exercises

In Exercises 1–4, compute the divergence of the vector field.

1. $\mathbf{F} = \langle xy, yz, y^2 - x^3 \rangle$

2. $x\mathbf{i} + y\mathbf{j} + z\mathbf{k}$

3. $\mathbf{F} = \langle x - 2zx^2, z - xy, z^2x^2 \rangle$

4. $\sin(x+z)\mathbf{i} - ye^{xz}\mathbf{k}$

In Exercises 5–8, verify the Divergence Theorem for the vector field and region.

5. $\mathbf{F} = \langle z, x, y \rangle$ and the box $[0,4] \times [0,2] \times [0,3]$

6. $\mathbf{F} = \langle y, x, z \rangle$ and the region $x^2 + y^2 + z^2 \leq 4$

7. $\mathbf{F} = \langle 2x, 3z, 3y \rangle$ and the region $x^2 + y^2 \leq 1, 0 \leq z \leq 2$

8. $\mathbf{F} = \langle x, 0, 0 \rangle$ and the region $x^2 + y^2 \leq z \leq 4$

In Exercises 9–16, use the Divergence Theorem to evaluate the surface integral $\iint_S \mathbf{F} \cdot d\mathbf{S}$.

9. $\mathbf{F} = \langle x, y, z \rangle$, S is the sphere $x^2 + y^2 + z^2 = 1$.

10. $\mathbf{F} = \langle y, z, x \rangle$, S is the sphere $x^2 + y^2 + z^2 = 1$.

11. $\mathbf{F} = \langle x^3, 0, z^3 \rangle$, S is the sphere $x^2 + y^2 + z^2 = 4$.

12. $\mathbf{F} = \langle x, -y, z \rangle$, S is the boundary of the unit cube $0 \leq x, y, z \leq 1$.

13. $\mathbf{F} = \langle x, y^2, z + y \rangle$, S is the boundary of the region contained in the cylinder $x^2 + y^2 = 4$ between the planes $z = x$ and $z = 8$.

14. $\mathbf{F} = \langle x^2 - z^2, e^{z^2} - \cos x, y^3 \rangle$, S is the boundary of the region bounded by $x + 2y + 4z = 12$ and the coordinate planes in the first octant.

15. $\mathbf{F} = \langle x + y, z, z - x \rangle$, S is the boundary of the region between the paraboloid $z = 9 - x^2 - y^2$ and the xy-plane.

16. $\mathbf{F} = \langle e^{z^2}, \sin(x^2 z), \sqrt{x^2 + 9y^2} \rangle$, S is the region $x^2 + y^2 \leq z \leq 8 - x^2 - y^2$.

17. Let $\mathcal{W}$ be the region in Figure 16 bounded by the cylinder $x^2 + y^2 = 9$, the plane $z = x + 1$, and the xy-plane. Use the Divergence Theorem to compute the flux of $\mathbf{F} = \langle z, x, y + 2z \rangle$ through the boundary of $\mathcal{W}$.

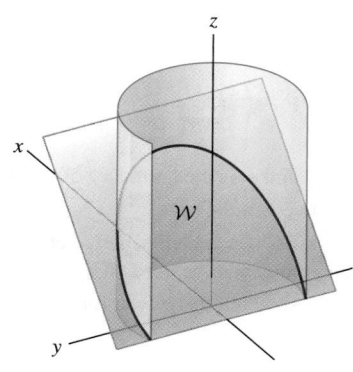

FIGURE 16

18. Find a constant c for which the velocity field

$$\mathbf{F} = (cx - y)\mathbf{i} + (y - z)\mathbf{j} + (3x + 4cz)\mathbf{k}$$

of a fluid is incompressible [meaning that $\operatorname{div}(\mathbf{F}) = 0$].

19. Volume as a Surface Integral Let $\mathbf{F} = \langle x, y, z \rangle$. Prove that if $\mathcal{W}$ is a region $\mathbf{R}^3$ with a smooth boundary S, then

$$\operatorname{Volume}(\mathcal{W}) = \frac{1}{3} \iint_S \mathbf{F} \cdot d\mathbf{S} \qquad \boxed{9}$$

20. Use Eq. (9) to calculate the volume of the unit ball as a surface integral over the unit sphere.

21. Show that $\Phi(a \cos \theta \sin \phi, b \sin \theta \sin \phi, c \cos \phi)$ is a parametrization of the ellipsoid

$$\left(\frac{x}{a}\right)^2 + \left(\frac{y}{b}\right)^2 + \left(\frac{z}{c}\right)^2 = 1$$

Then use Eq. (9) to calculate the volume of the ellipsoid as a surface integral over its boundary.

22. Prove the identity

$$\boxed{\operatorname{div}(\operatorname{curl}(\mathbf{F})) = 0}$$

23. Find and prove a Product Rule expressing $\operatorname{div}(f\mathbf{F})$ in terms of $\operatorname{div}(\mathbf{F})$ and ∇f.

24. Prove the identity

$$\operatorname{div}(\mathbf{F} \times \mathbf{G}) = \operatorname{curl}(\mathbf{F}) \cdot \mathbf{G} - \mathbf{F} \cdot \operatorname{curl}(\mathbf{G})$$

Then prove that the cross product of two irrotational vector fields is incompressible [$\mathbf{F}$ is called **irrotational** if $\operatorname{curl}(\mathbf{F}) = 0$ and **incompressible** if $\operatorname{div}(\mathbf{F}) = 0$].

25. Prove that $\operatorname{div}(\nabla f \times \nabla g) = 0$.

In Exercises 26–28, let Δ denote the Laplace operator defined by

$$\Delta \varphi = \frac{\partial^2 \varphi}{\partial x^2} + \frac{\partial^2 \varphi}{\partial y^2} + \frac{\partial^2 \varphi}{\partial z^2}$$

26. Prove the identity

$$\operatorname{curl}(\operatorname{curl}(\mathbf{F})) = \nabla(\operatorname{div}(\mathbf{F})) - \Delta \mathbf{F}$$

where $\Delta \mathbf{F}$ denotes $\langle \Delta F_1, \Delta F_2, \Delta F_3 \rangle$.

27. A function φ satisfying $\Delta \varphi = 0$ is called **harmonic**.
(a) Show that $\Delta \varphi = \operatorname{div}(\nabla \varphi)$ for any function φ.
(b) Show that φ is harmonic if and only if $\operatorname{div}(\nabla \varphi) = 0$.
(c) Show that if $\mathbf{F}$ is the gradient of a harmonic function, then $\operatorname{curl}(F) = 0$ and $\operatorname{div}(F) = 0$.
(d) Show $\mathbf{F} = \left(xz, -yz, \frac{1}{2}(x^2 - y^2) \right)$ is the gradient of a harmonic function. What is the flux of $\mathbf{F}$ through a closed surface?

28. Let $\mathbf{F} = \rho^n \mathbf{e}_r$, where n is any number, $\rho = (x^2 + y^2 + z^2)^{1/2}$, and $\mathbf{e}_r = \rho^{-1} \langle x, y, z \rangle$ is the unit radial vector.
(a) Calculate $\operatorname{div}(\mathbf{F})$.
(b) Calculate the flux of $\mathbf{F}$ through the surface of a sphere of radius R centered at the origin. For which values of n is this flux independent of R?
(c) Prove that $\nabla(\rho^n) = n\rho^{n-1}\mathbf{e}_r$.
(d) Use (c) to show that $\mathbf{F}$ is a gradient vector field for $n \neq -1$. Then show that $\mathbf{F} = \rho^{-1}\mathbf{e}_r$ is also a gradient vector field by computing the gradient of $\ln \rho$.
(e) What is the value of $\int_C \mathbf{F} \cdot d\mathbf{s}$, where C is a closed curve?
(f) Find the values of n for which the function $\varphi = \rho^n$ is harmonic.

29. The electric field due to a unit electric dipole oriented in the $\mathbf{k}$ direction is $\mathbf{E} = \nabla\left(\dfrac{z}{\rho^3} \right)$, where $\rho = (x^2 + y^2 + z^2)^{1/2}$ (Figure 17).
Let $\mathbf{e}_r = \rho^{-1} \langle x, y, z \rangle$.
(a) Show that $\mathbf{E} = \rho^{-3}\mathbf{k} - 3z\rho^{-4}\mathbf{e}_r$.
(b) Calculate the flux of $\mathbf{E}$ through a sphere centered at the origin.
(c) Calculate $\operatorname{div}(\mathbf{E})$.
(d) ✎ Can we use the Divergence Theorem to compute the flux of $\mathbf{E}$ through a sphere centered at the origin?

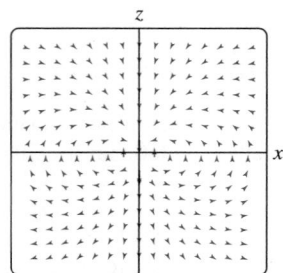

FIGURE 17 The dipole vector field restricted to the xz-plane.

Charged rod

FIGURE 18

30. Let $\mathbf{E}$ be the electric field due to a long, uniformly charged rod of radius R with charge density δ per unit length (Figure 18). By symmetry, we may assume that $\mathbf{E}$ is everywhere perpendicular to the rod and its magnitude $E(r)$ depends only on the distance r to the rod (strictly speaking, this would hold only if the rod were infinite, but it is nearly true if the rod is long enough). Show that $E(r) = \dfrac{\delta}{2\pi\epsilon_0 r}$ for $r > R$.

Hint: Apply Gauss's Law to a cylinder of radius R and of unit length with its axis along the rod.

31. Let $I = \displaystyle\iint_S \mathbf{F} \cdot d\mathbf{S}$, where

$$\mathbf{F} = \left\langle \frac{2yz}{\rho^2}, -\frac{xz}{\rho^2}, -\frac{xy}{\rho^2} \right\rangle$$

$(\rho = \sqrt{x^2 + y^2 + z^2})$ and S is the boundary of a region $\mathcal{W}$.

(a) Check that $\mathbf{F}$ is divergence-free.

(b) Show that $I = 0$ if S is a sphere centered at the origin. Note that the Divergence Theorem cannot be used. Why not?

(c) Give an argument showing that $I = 0$ for all S.

Further Insights and Challenges

32. Let S be the boundary surface of a region $\mathcal{W}$ in $\mathbf{R}^3$ and let $D_{\mathbf{e}_n}\varphi$ denote the directional derivative of φ, where $\mathbf{e}_n$ is the outward unit normal vector. Let Δ be the Laplace operator defined earlier.

(a) Use the Divergence Theorem to prove

$$\iint_S D_{\mathbf{e}_n}\varphi \, dS = \iiint_{\mathcal{W}} \Delta\varphi \, dV$$

(b) Show that if φ is a harmonic function (defined in Exercise 27), then

$$\iint_S D_{\mathbf{e}_n}\varphi \, dS = 0$$

33. Assume that φ is harmonic. Show that $\operatorname{div}(\varphi\nabla\varphi) = \|\nabla\varphi\|^2$ and conclude that

$$\iint_S \varphi D_{\mathbf{e}_n}\varphi \, dS = \iiint_{\mathcal{W}} \|\nabla\varphi\|^2 \, dV$$

34. Let $\mathbf{F} = \langle P, Q, R\rangle$ be a vector field defined on $\mathbf{R}^3$ such that $\operatorname{div}(\mathbf{F}) = 0$. Use the following steps to show that $\mathbf{F}$ has a vector potential.

(a) Let $\mathbf{A} = \langle f, 0, g\rangle$. Show that

$$\operatorname{curl}(\mathbf{A}) = \left\langle \frac{\partial g}{\partial y}, \frac{\partial f}{\partial z} - \frac{\partial g}{\partial x}, -\frac{\partial f}{\partial y} \right\rangle$$

(b) Fix any value y_0 and show that if we define

$$f(x, y, z) = -\int_{y_0}^{y} R(x, t, z)\, dt + \alpha(x, z)$$

$$g(x, y, z) = \int_{y_0}^{y} P(x, t, z)\, dt + \beta(x, z)$$

where α and β are any functions of x and z, then $\partial g/\partial y = P$ and $-\partial f/\partial y = R$.

(c) It remains for us to show that α and β can be chosen so $Q = \partial f/\partial z - \partial g/\partial x$. Verify that the following choice works (for any choice of z_0):

$$\alpha(x, z) = \int_{z_0}^{z} Q(x, y_0, t)\, dt, \qquad \beta(x, z) = 0$$

Hint: You will need to use the relation $\operatorname{div}(\mathbf{F}) = 0$.

35. Show that $\mathbf{F} = \langle 2y - 1, 3z^2, 2xy\rangle$ has a vector potential and find one.

36. Show that $\mathbf{F} = \langle 2ye^z - xy, y, yz - z\rangle$ has a vector potential and find one.

37. A vector field with a vector potential has zero flux through every closed surface in its domain. In the text, we observed that although the inverse-square radial vector field $\mathbf{F} = \dfrac{\mathbf{e}_r}{\rho^2}$ satisfies $\operatorname{div}(\mathbf{F}) = 0$, $\mathbf{F}$ cannot have a vector potential on its domain $\{(x, y, z) \neq (0, 0, 0)\}$ because the flux of $\mathbf{F}$ through a sphere containing the origin is nonzero.

(a) Show that the method of Exercise 34 produces a vector potential $\mathbf{A}$ such that $\mathbf{F} = \operatorname{curl}(\mathbf{A})$ on the restricted domain $\mathcal{D}$ consisting of $\mathbf{R}^3$ with the y-axis removed.

(b) Show that $\mathbf{F}$ also has a vector potential on the domains obtained by removing either the x-axis or the z-axis from $\mathbf{R}^3$.

(c) Does the existence of a vector potential on these restricted domains contradict the fact that the flux of $\mathbf{F}$ through a sphere containing the origin is nonzero?

CHAPTER REVIEW EXERCISES

1. Let $\mathbf{F}(x, y) = \langle x + y^2, x^2 - y \rangle$ and let C be the unit circle, oriented counterclockwise. Evaluate $\oint_C \mathbf{F} \cdot d\mathbf{s}$ directly as a line integral and using Green's Theorem.

2. Let $\partial\mathcal{R}$ be the boundary of the rectangle in Figure 1 and let $\partial\mathcal{R}_1$ and $\partial\mathcal{R}_2$ be the boundaries of the two triangles, all oriented counterclockwise.

(a) Determine $\oint_{\partial\mathcal{R}_1} \mathbf{F} \cdot d\mathbf{s}$ if $\oint_{\partial\mathcal{R}} \mathbf{F} \cdot d\mathbf{s} = 4$ and $\oint_{\partial\mathcal{R}_2} \mathbf{F} \cdot d\mathbf{s} = -2$.

(b) What is the value of $\oint_{\partial\mathcal{R}} \mathbf{F}\, d\mathbf{s}$ if $\partial\mathcal{R}$ is oriented clockwise?

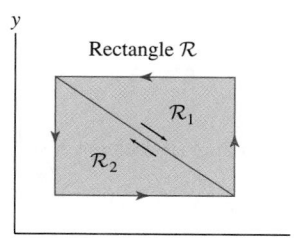

FIGURE 1

In Exercises 3–6, use Green's Theorem to evaluate the line integral around the given closed curve.

3. $\oint_C xy^3\, dx + x^3 y\, dy$, where C is the rectangle $-1 \leq x \leq 2$, $-2 \leq y \leq 3$, oriented counterclockwise.

4. $\oint_C (3x + 5y - \cos y)\, dx + x \sin y\, dy$, where C is any closed curve enclosing a region with area 4, oriented counterclockwise.

5. $\oint_C y^2\, dx - x^2\, dy$, where C consists of the arcs $y = x^2$ and $y = \sqrt{x}$, $0 \leq x \leq 1$, oriented clockwise.

6. $\oint_C ye^x\, dx + xe^y\, dy$, where C is the triangle with vertices $(-1, 0)$, $(0, 4)$, and $(0, 1)$, oriented counterclockwise.

7. Let $\mathbf{c}(t) = \left(t^2(1 - t), t(t - 1)^2 \right)$.

(a) ⌐GU⌐ Plot the path $\mathbf{c}(t)$ for $0 \leq t \leq 1$.

(b) Calculate the area of the region enclosed by $\mathbf{c}(t)$ for $0 \leq t \leq 1$.

8. In (a)–(d), state whether the equation is an identity (valid for all $\mathbf{F}$ or φ). If it is not, provide an example in which the equation does not hold.

(a) $\operatorname{curl}(\nabla\varphi) = 0$ **(b)** $\operatorname{div}(\nabla\varphi) = 0$

(c) $\operatorname{div}(\operatorname{curl}(\mathbf{F})) = 0$ **(d)** $\nabla(\operatorname{div}(\mathbf{F})) = 0$

In Exercises 9–12, calculate the curl and divergence of the vector field.

9. $\mathbf{F} = y\mathbf{i} - z\mathbf{k}$

10. $\mathbf{F} = \langle e^{x+y}, e^{y+z}, xyz \rangle$

11. $\mathbf{F} = \nabla(e^{-x^2 - y^2 - z^2})$

12. $\mathbf{e}_\rho = \rho^{-1} \langle x, y, z \rangle\ (\rho = \sqrt{x^2 + y^2 + z^2})$

13. Recall that if F_1, F_2, and F_3 are differentiable functions of one variable, then

$$\operatorname{curl}\left(\langle F_1(x), F_2(y), F_3(z) \rangle \right) = 0$$

Use this to calculate the curl of

$$\mathbf{F} = \langle x^2 + y^2,\ \ln y + z^2,\ z^3 \sin(z^2)e^{z^3} \rangle$$

14. Give an example of a nonzero vector field $\mathbf{F}$ such that $\operatorname{curl}(\mathbf{F}) = 0$ and $\operatorname{div}(\mathbf{F}) = 0$.

15. Verify the identities of Exercises 22 and 24 in Section 17.3 for the vector fields $\mathbf{F} = \langle xz, ye^x, yz \rangle$ and $\mathbf{G} = \langle z^2, xy^3, x^2 y \rangle$.

16. Suppose that $\mathcal{S}_1$ and $\mathcal{S}_2$ are surfaces with the same oriented boundary curve $\mathcal{C}$. Which of the following conditions guarantees that the flux of $\mathbf{F}$ through $\mathcal{S}_1$ is equal to the flux of $\mathbf{F}$ through $\mathcal{S}_2$?

(a) $\mathbf{F} = \nabla\varphi$ for some function φ

(b) $\mathbf{F} = \operatorname{curl}(\mathbf{G})$ for some vector field $\mathbf{G}$

17. Prove that if $\mathbf{F}$ is a gradient vector field, then the flux of $\operatorname{curl}(\mathbf{F})$ through a smooth surface $\mathcal{S}$ (whether closed or not) is equal to zero.

18. Verify Stokes' Theorem for $\mathbf{F} = \langle y, z - x, 0 \rangle$ and the surface $z = 4 - x^2 - y^2$, $z \geq 0$, oriented by outward-pointing normals.

In Exercises 19–20, let $\mathbf{F} = \langle z^2, x + z, y^2 \rangle$ and let $\mathcal{S}$ be the upper half of the ellipsoid $\dfrac{x^2}{4} + y^2 + z^2 = 1$, oriented by outward-pointing normals.

19. Use Stokes' Theorem to compute $\displaystyle\iint_{\mathcal{S}} \operatorname{curl}(\mathbf{F}) \cdot d\mathbf{S}$.

20. Compute $\displaystyle\iint_{\mathcal{S}} \mathbf{F} \cdot d\mathbf{S}$. *Hint:* Find a vector potential $\mathbf{A}$, that is, a vector field $\mathbf{A}$ such that $\mathbf{F} = \operatorname{curl}(\mathbf{A})$, and use Stokes' Theorem.

21. Use Stokes' Theorem to evaluate $\oint_C \langle y, z, x \rangle \cdot d\mathbf{s}$, where C is the curve in Figure 2.

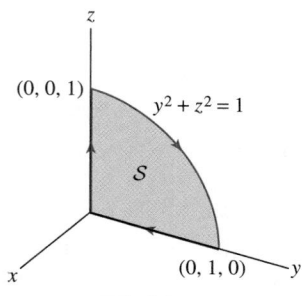

FIGURE 2

22. Verify the Divergence Theorem for $\mathbf{F} = \langle 0, 0, z \rangle$ and the region $x^2 + y^2 + z^2 = 1$.

In Exercises 23–26, use the Divergence Theorem to calculate $\iint_S \mathbf{F} \cdot d\mathbf{S}$ *for the given vector field and surface.*

23. $\mathbf{F} = \langle xy, yz, x^2z + z^2 \rangle$, S is the boundary of the box $[0, 1] \times [2, 4] \times [1, 5]$.

24. $\mathbf{F} = \langle xy, yz, x^2z + z^2 \rangle$, S is the boundary of the unit sphere.

25. $\mathbf{F} = \langle xyz + xy, \frac{1}{2}y^2(1 - z) + e^x, e^{x^2 + y^2} \rangle$, S is the boundary of the solid bounded by the cylinder $x^2 + y^2 = 16$ and the planes $z = 0$ and $z = y - 4$.

26. $\mathbf{F} = \langle \sin(yz), \sqrt{x^2 + z^4}, x \cos(x - y) \rangle$, S is any smooth closed surface that is the boundary of a region in $\mathbf{R}^3$.

27. Find the volume of a region $\mathcal{W}$ if

$$\iint_{\partial \mathcal{W}} \left\langle x + xy + z, x + 3y - \frac{1}{2}y^2, 4z \right\rangle \cdot d\mathbf{S} = 16$$

28. Show that the circulation of $\mathbf{F} = \langle x^2, y^2, z(x^2 + y^2) \rangle$ around any curve $\mathcal{C}$ on the surface of the cone $z^2 = x^2 + y^2$ is equal to zero (Figure 3).

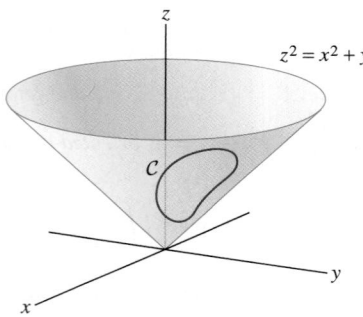

$z^2 = x^2 + y^2$

FIGURE 3

In Exercises 29–32, let $\mathbf{F}$ be a vector field whose curl and divergence at the origin are

$$\operatorname{curl}(\mathbf{F})(0, 0, 0) = \langle 2, -1, 4 \rangle, \qquad \operatorname{div}(\mathbf{F})(0, 0, 0) = -2$$

29. Estimate $\oint_{\mathcal{C}} \mathbf{F} \cdot d\mathbf{s}$, where $\mathcal{C}$ is the circle of radius 0.03 in the xy-plane centered at the origin.

30. Estimate $\oint_{\mathcal{C}} \mathbf{F} \cdot d\mathbf{s}$, where $\mathcal{C}$ is the boundary of the square of side 0.03 in the yz-plane centered at the origin. Does the estimate depend on how the square is oriented within the yz-plane? Might the actual circulation depend on how it is oriented?

31. Suppose that $\mathbf{F}$ is the velocity field of a fluid and imagine placing a small paddle wheel at the origin. Find the equation of the plane in which the paddle wheel should be placed to make it rotate as quickly as possible.

32. Estimate the flux of $\mathbf{F}$ through the box of side 0.5 in Figure 4. Does the result depend on how the box is oriented relative to the coordinate axes?

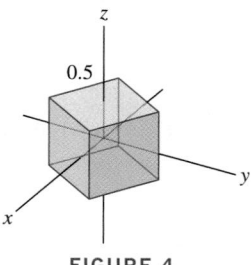

FIGURE 4

33. The velocity field of a fluid (in meters per second) is

$$\mathbf{F} = \langle x^2 + y^2, 0, z^2 \rangle$$

Let $\mathcal{W}$ be the region between the hemisphere

$$S = \{(x, y, z) : x^2 + y^2 + z^2 = 1, \quad x, y, z \geq 0\}$$

and the disk

$$\mathcal{D} = \{(x, y, 0) : x^2 + y^2 \leq 1\}$$

in the xy-plane.

(a) Show that no fluid flows across $\mathcal{D}$.

(b) Use (a) to show that the rate of fluid flow across S is equal to $\iiint_{\mathcal{W}} \operatorname{div}(\mathbf{F}) \, dV$. Compute this triple integral using spherical coordinates.

34. The velocity field of a fluid (in meters per second) is

$$\mathbf{F} = (3y - 4)\mathbf{i} + e^{-y(z+1)}\mathbf{j} + (x^2 + y^2)\mathbf{k}$$

(a) Estimate the flow rate (in cubic meters per second) through a small surface S around the origin if S encloses a region of volume 0.01 m^3.

(b) Estimate the circulation of $\mathbf{F}$ about a circle in the xy-plane of radius $r = 0.1$ m centered at the origin (oriented counterclockwise when viewed from above).

(c) Estimate the circulation of $\mathbf{F}$ about a circle in the yz-plane of radius $r = 0.1$ m centered at the origin (oriented counterclockwise when viewed from the positive x-axis).

35. Let $\varphi(x, y) = x + \dfrac{x}{x^2 + y^2}$. The vector field $\mathbf{F} = \nabla \varphi$ (Figure 5) provides a model in the plane of the velocity field of an incompressible, irrotational fluid flowing past a cylindrical obstacle (in this case, the obstacle is the unit circle $x^2 + y^2 = 1$).

(a) Verify that $\mathbf{F}$ is irrotational [by definition, $\mathbf{F}$ is irrotational if $\mathrm{curl}(\mathbf{F}) = \mathbf{0}$].

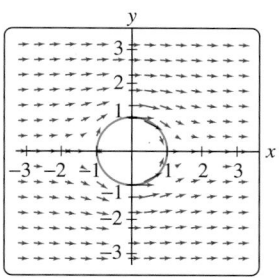

FIGURE 5 The vector field $\nabla\varphi$ for $\varphi(x, y) = x + \dfrac{x}{x^2 + y^2}$.

(b) Verify that $\mathbf{F}$ is tangent to the unit circle at each point along the unit circle except $(1, 0)$ and $(-1, 0)$ (where $\mathbf{F} = \mathbf{0}$).

(c) What is the circulation of $\mathbf{F}$ around the unit circle?

(d) Calculate the line integral of $\mathbf{F}$ along the upper and lower halves of the unit circle separately.

36. Figure 6 shows the vector field $\mathbf{F} = \nabla\varphi$, where $\varphi(x, y) = \ln(x^2 + (y - 1)^2) + \ln(x^2 + (y + 1)^2)$, which is the velocity field for the flow of a fluid with sources of equal strength at $(0, \pm 1)$ (note that φ is undefined at these two points). Show that $\mathbf{F}$ is both irrotational and incompressible, that is, $\mathrm{curl}_z(\mathbf{F}) = 0$ and $\mathrm{div}(\mathbf{F}) = 0$ [in computing $\mathrm{div}(\mathbf{F})$, treat $\mathbf{F}$ as a vector field in $\mathbf{R}^3$ with a zero z-component]. Is it necessary to compute $\mathrm{curl}_z(\mathbf{F})$ to conclude that it is zero?

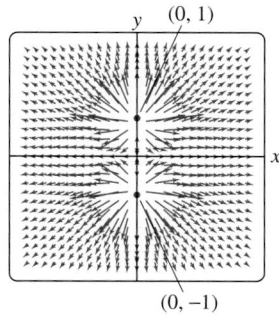

FIGURE 6

37. In Section 17.1, we showed that if $\mathcal{C}$ is a simple closed curve, oriented counterclockwise, then the line integral is

$$\text{Area enclosed by } \mathcal{C} = \frac{1}{2} \oint_{\mathcal{C}} x \, dy - y \, dx \qquad \boxed{1}$$

Suppose that $\mathcal{C}$ is a path from P to Q that is not closed but has the property that every line through the origin intersects $\mathcal{C}$ in at most one point, as in Figure 7. Let $\mathcal{R}$ be the region enclosed by $\mathcal{C}$ and the two radial segments joining P and Q to the origin. Show that the line integral in Eq. (1) is equal to the area of $\mathcal{R}$. *Hint:* Show that the line integral of $\mathbf{F} = \langle -y, x \rangle$ along the two radial segments is zero and apply Green's Theorem.

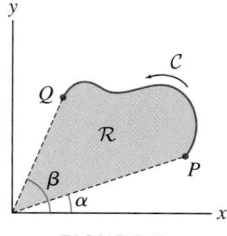

FIGURE 7

38. Suppose that the curve $\mathcal{C}$ in Figure 7 has the polar equation $r = f(\theta)$.

(a) Show that $\mathbf{c}(\theta) = (f(\theta)\cos\theta, f(\theta)\sin\theta)$ is a counterclockwise parametrization of $\mathcal{C}$.

(b) In Section 11.4, we showed that the area of the region $\mathcal{R}$ is given by the formula

$$\text{Area of } \mathcal{R} = \frac{1}{2}\int_{\alpha}^{\beta} f(\theta)^2 \, d\theta$$

Use the result of Exercise 37 to give a new proof of this formula. *Hint:* Evaluate the line integral in (1) using $\mathbf{c}(\theta)$.

39. Prove the following generalization of Eq. (1). Let $\mathcal{C}$ be a simple closed curve in the plane (Figure 8)

$$\mathcal{S}: \quad ax + by + cz + d = 0$$

Then the area of the region R enclosed by $\mathcal{C}$ is equal to

$$\frac{1}{2\|\mathbf{n}\|} \oint_{\mathcal{C}} (bz - cy)\, dx + (cx - az)\, dy + (ay - bx)\, dz$$

where $\mathbf{n} = \langle a, b, c \rangle$ is the normal to $\mathcal{S}$ and $\mathcal{C}$ is oriented as the boundary of $\mathcal{R}$ (relative to the normal vector $\mathbf{n}$). *Hint:* Apply Stokes' Theorem to $\mathbf{F} = \langle bz - cy, cx - az, ay - bx \rangle$.

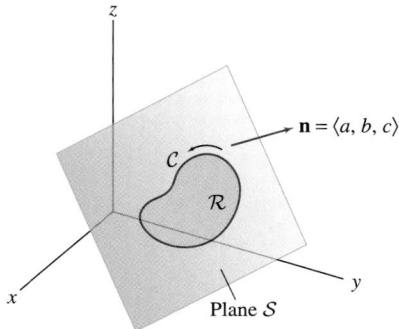

FIGURE 8

40. Use the result of Exercise 39 to calculate the area of the triangle with vertices $(1, 0, 0)$, $(0, 1, 0)$, and $(0, 0, 1)$ as a line integral. Verify your result using geometry.

A | THE LANGUAGE OF MATHEMATICS

One of the challenges in learning calculus is growing accustomed to its precise language and terminology, especially in the statements of theorems. In this section, we analyze a few details of logic that are helpful, and indeed essential, in understanding and applying theorems properly.

Many theorems in mathematics involve an **implication**. If A and B are statements, then the implication $A \implies B$ is the assertion that A implies B:

$$A \implies B: \qquad \textit{If A is true, then B is true.}$$

Statement A is called the **hypothesis** (or premise) and statement B the **conclusion** of the implication. Here is an example: *If m and n are even integers, then m + n is an even integer*. This statement may be divided into a hypothesis and conclusion:

$$\underbrace{m \text{ and } n \text{ are even integers}}_{A} \implies \underbrace{m + n \text{ is an even integer}}_{B}$$

In everyday speech, implications are often used in a less precise way. An example is: *If you work hard, then you will succeed*. Furthermore, some statements that do not initially have the form $A \implies B$ may be restated as implications. For example, the statement "Cats are mammals" can be rephrased:

$$\text{Let } X \text{ be an animal.} \quad \underbrace{X \text{ is a cat}}_{A} \implies \underbrace{X \text{ is a mammal}}_{B}$$

When we say that an implication $A \implies B$ is true, we do not claim that A or B is necessarily true. Rather, we are making the conditional statement that *if* A happens to be true, *then* B is also true. In the above, if X does not happen to be a cat, the implication tells us nothing.

The **negation** of a statement A is the assertion that A is false and is denoted $\neg A$.

Statement A	Negation $\neg A$
X lives in California.	X does not live in California.
$\triangle ABC$ is a right triangle.	$\triangle ABC$ is not a right triangle.

The negation of the negation is the original statement: $\neg(\neg A) = A$. To say that X does *not not live in California* is the same as saying that X *lives in California*.

■ **EXAMPLE 1** State the negation of:

(a) The door is open and the dog is barking.

(b) The door is open or the dog is barking (or both).

Solution

(a) The first statement is true if two conditions are satisfied (door open and dog barking), and it is false if at least one of these conditions is not satisfied. So the negation is

Either the door is not open *OR* the dog is not barking *(or both)*.

(b) The second statement is true if at least one of the conditions (door open or dog barking) is satisfied, and it is false if neither condition is satisfied. So the negation is

The door is not open AND the dog is not barking.   ∎

Contrapositive and Converse

Keep in mind that when we form the contrapositive, we reverse the order of A and B. The contrapositive of A ⟹ B is NOT ¬A ⟹ ¬B.

Two important operations are the formation of the contrapositive and converse of a statement. The **contrapositive** of $A \Longrightarrow B$ is the statement "If B is false, then A is false":

$$\boxed{\text{The contrapositive of } \quad A \Longrightarrow B \quad \text{is} \quad \neg B \Longrightarrow \neg A.}$$

Here are some examples:

Statement	Contrapositive
If X is a cat, then X is a mammal.	If X is not a mammal, then X is not a cat.
If you work hard, then you will succeed.	If you did not succeed, then you did not work hard.
If m and n are both even, then $m + n$ is even.	If $m + n$ is not even, then m and n are not both even.

A key observation is this:

The contrapositive and the original implication are equivalent.

The fact that A ⟹ B is equivalent to its contrapositive is a general rule of logic that does not depend on what A and B happen to mean. This rule belongs to the subject of "formal logic," which deals with logical relations between statements without concern for the actual content of these statements.

In other words, if an implication is true, then its contrapositive is automatically true and vice versa. In essence, an implication and its contrapositive are two ways of saying the same thing. For example, the contrapositive "If X is not a mammal, then X is not a cat" is a roundabout way of saying that cats are mammals.

The **converse** of $A \Longrightarrow B$ is the *reverse* implication $B \Longrightarrow A$:

Implication: $A \Longrightarrow B$	Converse $B \Longrightarrow A$
If A is true, then B is true.	If B is true, then A is true.

The converse plays a very different role than the contrapositive because *the converse is NOT equivalent to the original implication*. The converse may be true or false, even if the original implication is true. Here are some examples:

True Statement	Converse	Converse True or False?
If X is a cat, then X is a mammal.	If X is a mammal, then X is a cat.	False
If m is even, then m^2 is even.	If m^2 is even, then m is even.	True

■ **EXAMPLE 2** **An Example Where the Converse Is False** Show that the converse of "If m and n are even, then $m + n$ is even" is false.

A counterexample is an example that satisfies the hypothesis but not the conclusion of a statement. If a single counterexample exists, then the statement is false. However, we cannot prove that a statement is true merely by giving an example.

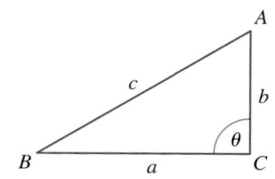

FIGURE 1

Solution The converse is "If $m + n$ is even, then m and n are even." To show that the converse is false, we display a counterexample. Take $m = 1$ and $n = 3$ (or any other pair of odd numbers). The sum is even (since $1 + 3 = 4$) but neither 1 nor 3 is even. Therefore, the converse is false. ∎

■ **EXAMPLE 3** An Example Where the Converse Is True State the contrapositive and converse of the Pythagorean Theorem. Are either or both of these true?

Solution Consider a triangle with sides a, b, and c, and let θ be the angle opposite the side of length c as in Figure 1. The Pythagorean Theorem states that if $\theta = 90°$, then $a^2 + b^2 = c^2$. Here are the contrapositive and converse:

Pythagorean Theorem	$\theta = 90° \implies a^2 + b^2 = c^2$	True
Contrapositive	$a^2 + b^2 \neq c^2 \implies \theta \neq 90°$	Automatically true
Converse	$a^2 + b^2 = c^2 \implies \theta = 90°$	True (but not automatic)

The contrapositive is automatically true because it is just another way of stating the original theorem. The converse is not automatically true since there could conceivably exist a nonright triangle that satisfies $a^2 + b^2 = c^2$. However, the converse of the Pythagorean Theorem is, in fact, true. This follows from the Law of Cosines (see Exercise 38). ∎

When both a statement $A \implies B$ and its converse $B \implies A$ are true, we write $A \iff B$. In this case, A and B are **equivalent**. We often express this with the phrase

$$A \iff B \qquad A \text{ is true } \textit{if and only if } B \text{ is true.}$$

For example,

$$a^2 + b^2 = c^2 \qquad \text{if and only if} \qquad \theta = 90°$$

$$\text{It is morning} \qquad \text{if and only if} \qquad \text{the sun is rising.}$$

We mention the following variations of terminology involving implications that you may come across:

Statement	Is Another Way of Saying
A is true <u>if</u> B is true.	$B \implies A$
A is true <u>only if</u> B is true.	$A \implies B$ (A cannot be true unless B is also true.)
For A to be true, <u>it is necessary</u> that B be true.	$A \implies B$ (A cannot be true unless B is also true.)
For A to be true, <u>it is sufficient</u> that B be true.	$B \implies A$
For A to be true, it is <u>necessary and sufficient</u> that B be true.	$B \iff A$

Analyzing a Theorem

To see how these rules of logic arise in the study of calculus, consider the following result from Section 4.2.

> **THEOREM 1 Existence of a Maximum on a Closed Interval**   If $f(x)$ is a continuous function on a closed (bounded) interval $I = [a, b]$, then $f(x)$ takes on a maximum value on I (Figure 2).

To analyze this theorem, let's write out the hypotheses and conclusion separately:

Hypotheses A: $f(x)$ is continuous and I is closed.

Conclusion B: $f(x)$ takes on a maximum value on I.

A first question to ask is: "Are the hypotheses necessary?" Is the conclusion still true if we drop one or both assumptions? To show that both hypotheses are necessary, we provide the counterexamples:

- **The continuity of $f(x)$ is a necessary hypothesis.** Figure 3(A) shows the graph of a function on a closed interval $[a, b]$ that is not continuous. This function has no maximum value on $[a, b]$, which shows that the conclusion may fail if the continuity hypothesis is not satisfied.
- **The hypothesis that I is closed is necessary.** Figure 3(B) shows the graph of a continuous function on an *open interval* (a, b). This function has no maximum value, which shows that the conclusion may fail if the interval is not closed.

We see that both hypotheses in Theorem 1 are necessary. In stating this, we do not claim that the conclusion *always* fails when one or both of the hypotheses are not satisfied. We claim only that the conclusion *may* fail when the hypotheses are not satisfied. Next, let's analyze the contrapositive and converse:

- **Contrapositive $\neg B \Rightarrow \neg A$ (automatically true):** If $f(x)$ does not have a maximum value on I, then either $f(x)$ is not continuous or I is not closed (or both).
- **Converse $B \Rightarrow A$ (in this case, false):** If $f(x)$ has a maximum value on I, then $f(x)$ is continuous and I is closed.

Figure 2 (left margin)

y

Maximum value

a b x

FIGURE 2 A continuous function on a closed interval $I = [a, b]$ has a maximum value.

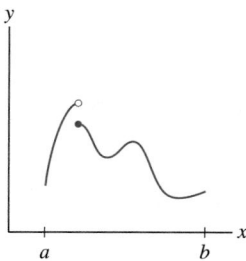

(A) The interval is closed but the function is not continuous. The function has no maximum value.

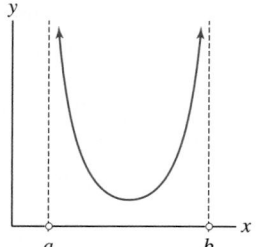

(B) The function is continuous but the interval is open. The function has no maximum value.

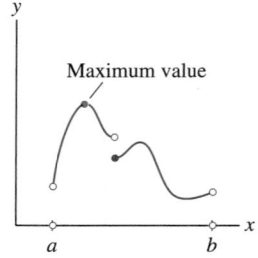

(C) This function is not continuous and the interval is not closed, but the function does have a maximum value.

FIGURE 3

The technique of proof by contradiction is also known by its Latin name reductio ad absurdum or "reduction to the absurd." The ancient Greek mathematicians used proof by contradiction as early as the fifth century BC, and Euclid (325–265 BC) employed it in his classic treatise on geometry entitled The Elements. A famous example is the proof that $\sqrt{2}$ is irrational in Example 4. The philosopher Plato (427–347 BC) wrote: "He is unworthy of the name of man who is ignorant of the fact that the diagonal of a square is incommensurable with its side."

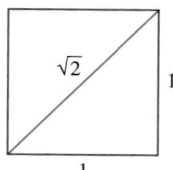

FIGURE 4 The diagonal of the unit square has length $\sqrt{2}$.

One of the most famous problems in mathematics is known as "Fermat's Last Theorem." It states that the equation

$$x^n + y^n = z^n$$

has no solutions in positive integers if $n \geq 3$. In a marginal note written around 1630, Fermat claimed to have a proof, and over the centuries, that assertion was verified for many values of the exponent n. However, only in 1994 did the British-American mathematician Andrew Wiles, working at Princeton University, find a complete proof.

As we know, the contrapositive is merely a way of restating the theorem, so it is automatically true. The converse is not automatically true, and in fact, in this case it is false. The function in Figure 3(C) provides a counterexample to the converse: $f(x)$ has a maximum value on $I = (a, b)$, but $f(x)$ is not continuous and I is not closed.

Mathematicians have devised various general strategies and methods for proving theorems. The method of proof by induction is discussed in Appendix C. Another important method is **proof by contradiction**, also called **indirect proof**. Suppose our goal is to prove that $A \implies B$. In a proof by contradiction, we start by assuming that A is true but B is false, and we then show that this leads to a contradiction. Therefore, B must be true (to avoid the contradiction).

■ **EXAMPLE 4** Proof by Contradiction   The number $\sqrt{2}$ is irrational (Figure 4).

Solution   Assume that the theorem is false, namely that $\sqrt{2} = p/q$, where p and q are whole numbers. We may assume that p/q is in lowest terms and, therefore, at most one of p and q is even.

The relation $\sqrt{2} = p/q$ implies that $2 = p^2/q^2$ or $p^2 = 2q^2$. This shows that p must be even. But if p is even, say, $p = 2m$, then $4m^2 = 2q^2$ or $q^2 = 2m^2$. This shows that q is also even. This contradicts our assumption that p/q is in lowest terms. We conclude that our original assumption, that $\sqrt{2} = p/q$, must be false. Therefore, $\sqrt{2}$ is irrational. ■

CONCEPTUAL INSIGHT   The hallmark of mathematics is precision and rigor. A theorem is established, not through observation or experimentation, but by a proof that consists of a chain of reasoning with no gaps.

This approach to mathematics comes down to us from the ancient Greek mathematicians, especially Euclid, and it remains the standard in contemporary research. In recent decades, the computer has become a powerful tool for mathematical experimentation and data analysis. Researchers may use experimental data to discover potential new mathematical facts, but the title "theorem" is not bestowed until someone writes down a proof.

This insistence on theorems and proofs distinguishes mathematics from the other sciences. In the natural sciences, facts are established through experiment and are subject to change or modification as more knowledge is acquired. In mathematics, theories are also developed and expanded, but previous results are not invalidated. The Pythagorean Theorem was discovered in antiquity and is a cornerstone of plane geometry. In the nineteenth century, mathematicians began to study more general types of geometry (of the type that eventually led to Einstein's four-dimensional space-time geometry in the Theory of Relativity). The Pythagorean Theorem does not hold in these more general geometries, but its status in plane geometry is unchanged.

A. SUMMARY

- The implication $A \implies B$ is the assertion "If A is true, then B is true."
- The *contrapositive* of $A \implies B$ is the implication $\neg B \implies \neg A$, which says "If B is false, then A is false." An implication and its contrapositive are equivalent (one is true if and only if the other is true).
- The *converse* of $A \implies B$ is $B \implies A$. An implication and its converse are not necessarily equivalent. One may be true and the other false.
- A and B are *equivalent* if $A \implies B$ and $B \implies A$ are both true.

• In a proof by contradiction (in which the goal is to prove statement A), we start by assuming that A is false and show that this assumption leads to a contradiction.

A. EXERCISES

Preliminary Questions

1. Which is the contrapositive of $A \implies B$?

(a) $B \implies A$ **(b)** $\neg B \implies A$

(c) $\neg B \implies \neg A$ **(d)** $\neg A \implies \neg B$

2. Which of the choices in Question 1 is the converse of $A \implies B$?

3. Suppose that $A \implies B$ is true. Which is then automatically true, the converse or the contrapositive?

4. Restate as an implication: "A triangle is a polygon."

Exercises

1. Which is the negation of the statement "The car and the shirt are both blue"?

(a) Neither the car nor the shirt is blue.

(b) The car is not blue and/or the shirt is not blue.

2. Which is the contrapositive of the implication "If the car has gas, then it will run"?

(a) If the car has no gas, then it will not run.

(b) If the car will not run, then it has no gas.

In Exercises 3–8, state the negation.

3. The time is 4 o'clock.

4. $\triangle ABC$ is an isosceles triangle.

5. m and n are odd integers.

6. Either m is odd or n is odd.

7. x is a real number and y is an integer.

8. $f(x)$ is a linear function.

In Exercises 9–14, state the contrapositive and converse.

9. If m and n are odd integers, then mn is odd.

10. If today is Tuesday, then we are in Belgium.

11. If today is Tuesday, then we are not in Belgium.

12. If $x > 4$, then $x^2 > 16$.

13. If m^2 is divisible by 3, then m is divisible by 3.

14. If $x^2 = 2$, then x is irrational.

In Exercise 15–18, give a counterexample to show that the converse of the statement is false.

15. If m is odd, then $2m + 1$ is also odd.

16. If $\triangle ABC$ is equilateral, then it is an isosceles triangle.

17. If m is divisible by 9 and 4, then m is divisible by 12.

18. If m is odd, then $m^3 - m$ is divisible by 3.

In Exercise 19–22, determine whether the converse of the statement is false.

19. If $x > 4$ and $y > 4$, then $x + y > 8$.

20. If $x > 4$, then $x^2 > 16$.

21. If $|x| > 4$, then $x^2 > 16$.

22. If m and n are even, then mn is even.

In Exercises 23–24, state the contrapositive and converse (it is not necessary to know what these statements mean).

23. If $f(x)$ and $g(x)$ are differentiable, then $f(x)g(x)$ is differentiable.

24. If the force field is radial and decreases as the inverse square of the distance, then all closed orbits are ellipses.

*In Exercises 25–28, the **inverse** of $A \implies B$ is the implication $\neg A \implies \neg B$.*

25. Which of the following is the inverse of the implication "If she jumped in the lake, then she got wet"?

(a) If she did not get wet, then she did not jump in the lake.

(b) If she did not jump in the lake, then she did not get wet.

Is the inverse true?

26. State the inverses of these implications:

(a) If X is a mouse, then X is a rodent.

(b) If you sleep late, you will miss class.

(c) If a star revolves around the sun, then it's a planet.

27. Explain why the inverse is equivalent to the converse.

28. State the inverse of the Pythagorean Theorem. Is it true?

29. Theorem 1 in Section 2.4 states the following: "If $f(x)$ and $g(x)$ are continuous functions, then $f(x) + g(x)$ is continuous." Does it follow logically that if $f(x)$ and $g(x)$ are not continuous, then $f(x) + g(x)$ is not continuous?

30. Write out a proof by contradiction for this fact: There is no smallest positive rational number. Base your proof on the fact that if $r > 0$, then $0 < r/2 < r$.

31. Use proof by contradiction to prove that if $x + y > 2$, then $x > 1$ or $y > 1$ (or both).

In Exercises 32–35, use proof by contradiction to show that the number is irrational.

32. $\sqrt{\dfrac{1}{2}}$ **33.** $\sqrt{3}$ **34.** $\sqrt[3]{2}$ **35.** $\sqrt[4]{11}$

36. An isosceles triangle is a triangle with two equal sides. The following theorem holds: If $\triangle$ is a triangle with two equal angles, then $\triangle$ is an isosceles triangle.

(a) What is the hypothesis?

(b) Show by providing a counterexample that the hypothesis is necessary.

(c) What is the contrapositive?

(d) What is the converse? Is it true?

37. Consider the following theorem: Let $f(x)$ be a quadratic polynomial with a positive leading coefficient. Then $f(x)$ has a minimum value.

(a) What are the hypotheses?

(b) What is the contrapositive?

(c) What is the converse? Is it true?

Further Insights and Challenges

38. Let a, b, and c be the sides of a triangle and let θ be the angle opposite c. Use the Law of Cosines (Theorem 1 in Section 1.4) to prove the converse of the Pythagorean Theorem.

39. Carry out the details of the following proof by contradiction, due to R. Palais, that $\sqrt{2}$ is irrational. If $\sqrt{2}$ is rational, then $n\sqrt{2}$ is a whole number for some whole number n. Let n be the smallest such whole number and let $m = n\sqrt{2} - n$.

(a) Prove that $m < n$.

(b) Prove that $m\sqrt{2}$ is a whole number.

Explain why (a) and (b) imply that $\sqrt{2}$ is irrational.

40. Generalize the argument of Exercise 39 to prove that $\sqrt{A}$ is irrational if A is a whole number but not a perfect square. *Hint:* Choose n

as before and let $m = n\sqrt{A} - n[\sqrt{A}]$, where $[x]$ is the greatest integer function.

41. Generalize further and show that for any whole number r, the rth root $\sqrt[r]{A}$ is irrational unless A is an rth power. *Hint:* Let $x = \sqrt[r]{A}$. Show that if x is rational, then we may choose a smallest whole number n such that nx^j is a whole number for $j = 1, \ldots, r - 1$. Then consider $m = nx - n[x]$ as before.

42. [image] Given a finite list of prime numbers $p_1, \ldots, p_N$, let $M = p_1 \cdot p_2 \cdots p_N + 1$. Show that M is not divisible by any of the primes $p_1, \ldots, p_N$. Use this and the fact that every number has a prime factorization to prove that there exist infinitely many prime numbers. This argument was advanced by Euclid in *The Elements*.

B | PROPERTIES OF REAL NUMBERS

FIGURE 1 The real number line.

In this appendix, we discuss the basic properties of real numbers. First, let us recall that a real number is a number that may be represented by a finite or infinite decimal expansion. The set of all real numbers is denoted **R** and is often visualized as the "number line" (Figure 1). The two algebraic operations of addition and multiplication are defined on **R** and satisfy the Commutative, Associative, and Distributive Laws (Table 1).

TABLE 1	Algebraic Laws
Commutative Laws:	$a + b = b + a, \quad ab = ba$
Associative Laws:	$(a + b) + c = a + (b + c), \quad (ab)c = a(bc)$
Distributive Law:	$a(b + c) = ab + ac$

Every real number x has an additive inverse $-x$ such that $x + (-x) = 0$ and every nonzero real number x has a multiplicative inverse x^{-1} such that $x(x^{-1}) = 1$. We do not regard subtraction and division as separate algebraic operations because they are defined in terms of inverses. By definition, the difference $x - y$ is equal to $x + (-y)$ and the quotient x/y is equal to $x(y^{-1})$ for $y \neq 0$.

In addition to the algebraic operations, there is an **order relation** on **R**: For any two real numbers a and b, precisely one of the following is true:

$$\text{Either} \quad a = b, \quad \text{or} \quad a < b, \quad \text{or} \quad a > b$$

To distinguish between the conditions $a \leq b$ and $a < b$, we often refer to $a < b$ as a **strict inequality**. Similar conventions hold for $>$ and $\geq$. The rules given in Table 2 allow us to manipulate inequalities.

TABLE 2	Order Properties	
If $a < b$ and $b < c$,	then $a < c$.	
If $a < b$ and $c < d$,	then $a + c < b + d$.	
If $a < b$ and $c > 0$,	then $ac < bc$.	
If $a < b$ and $c < 0$,	then $ac > bc$.	

The last order property says that an inequality reverses direction when multiplied by a negative number c. For example,

$$-2 < 5 \quad \text{but} \quad (-3)(-2) > (-3)5$$

The algebraic and order properties of real numbers are certainly familiar. We now discuss the less familiar **Least Upper Bound (LUB) Property** of the real numbers. This property is one way of expressing the so-called **completeness** of the real numbers. There are other ways of formulating completeness (such as the so-called nested interval property discussed in any book on analysis) that are equivalent to the LUB Property and serve the same purpose. Completeness is used in calculus to construct rigorous proofs of basic theorems about continuous functions such as the Intermediate Value Theorem (IVT) or the existence of extreme values on a closed interval. The underlying idea is that the real number line "has no holes." We elaborate on this idea below. First, we introduce the necessary definitions.

Suppose that S is a nonempty set of real numbers. A number M is called an **upper bound** for S if

$$x \leq M \qquad \text{for all } x \in S$$

If S has an upper bound, we say that S is **bounded above**. A **least upper bound** L is an upper bound for S such that every other upper bound M satisfies $M \geq L$. For example (Figure 2),

FIGURE 2 $M = 3$ is an upper bound for the set $S = (-2, 1)$. The LUB is $L = 1$.

- $M = 3$ is an upper bound for the open interval $S = (-2, 1)$.
- $L = 1$ is the LUB for $S = (-2, 1)$.

We now state the LUB Property of the real numbers.

THEOREM 1 Existence of a Least Upper Bound Let S be a nonempty set of real numbers that is bounded above. Then S has an LUB.

In a similar fashion, we say that a number B is a **lower bound** for S if $x \geq B$ for all $x \in S$. We say that S is **bounded below** if S has a lower bound. A **greatest lower bound** (GLB) is a lower bound M such that every other lower bound B satisfies $B \leq M$. The set of real numbers also has the GLB Property: If S is a nonempty set of real numbers that is bounded below, then S has a GLB. This may be deduced immediately from Theorem 1. For any nonempty set of real numbers S, let $-S$ be the set of numbers of the form $-x$ for $x \in S$. Then $-S$ has an upper bound if S has a lower bound. Consequently, $-S$ has an LUB L by Theorem 1, and $-L$ is a GLB for S.

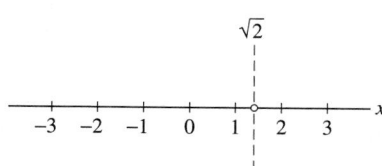

FIGURE 3 The rational numbers have a "hole" at the location $\sqrt{2}$.

CONCEPTUAL INSIGHT Theorem 1 may appear quite reasonable, but perhaps it is not clear why it is useful. We suggested above that the LUB Property expresses the idea that **R** is "complete" or "has no holes." To illustrate this idea, let's compare **R** to the set of rational numbers, denoted **Q**. Intuitively, **Q** is not complete because the irrational numbers are missing. For example, **Q** has a "hole" where the irrational number $\sqrt{2}$ should be located (Figure 3). This hole divides **Q** into two disconnected halves (the half to the left and the half to the right of $\sqrt{2}$). Because of this, **Q** is not a truly "continuous" object. In the following example, we will see that the existence of $\sqrt{2}$ as a real number is directly related to the LUB Property. Another consequence of completeness is the IVT, whose proof below relies on the LUB Property.

■ **EXAMPLE 1** Show that 2 has a square root by applying the LUB Property to the set

$$S = \{x : x^2 < 2\}$$

Solution First, we note that S is bounded with the upper bound $M = 2$. Indeed, if $x > 2$, then x satisfies $x^2 > 4$, and hence, x does not belong to S. By the LUB Property, S has a least upper bound. Call it L. We claim that $L = \sqrt{2}$ or, equivalently, that $L^2 = 2$ (Figure 4). We prove this by showing that $L^2 \geq 2$ and $L^2 \leq 2$.

If $L^2 < 2$, let $b = L + h$, where $h > 0$. Then

$$b^2 = L^2 + 2Lh + h^2 = L^2 + h(2L + h) \qquad \boxed{1}$$

FIGURE 4 The set $S = \{x : x^2 < 2\}$ is bounded with the LUB $L = \sqrt{2}$.

We can make the quantity $h(2L + h)$ as small as desired by choosing $h > 0$ small enough. In particular, we may choose a positive h so that $h(2L + h) < 2 - L^2$. For this choice, $b^2 < L^2 - (2 - L)^2 = 2$ by Eq. (1). Therefore, $b \in S$. But $b > L$ since $h > 0$, and thus L is not an upper bound for S, in contradiction to our hypothesis on L. We conclude that $L^2 \geq 2$.

If $L^2 > 2$, let $b = L - h$, where $h > 0$. Then

$$b^2 = L^2 - 2Lh + h^2 = L^2 - h(2L + h)$$

Now choose h positive but small enough so that $h(2L + h) < L^2 - 2$. Then $b < L$ and $b^2 > L^2 - (L^2 - 2) = 2$. But in this case, b is a smaller lower bound for S. Indeed, if $x \geq b$, then $x^2 \geq b^2 > 2$, and x does not belong to S. This contradicts our hypothesis that L is the LUB. We conclude that $L^2 \leq 2$, and since we have already shown that $L^2 \geq 2$, we have $L^2 = 2$ as claimed. ∎

We now prove three important theorems, the third of which is used in the proof of the LUB Property below.

THEOREM 2 Bolzano–Weierstrass Theorem   Let S be a bounded, infinite set of real numbers. Then there exists a sequence of distinct elements $\{a_n\}$ in S such that the limit $L = \lim\limits_{n \to \infty} a_n$ exists.

Proof For simplicity of notation, we assume that S is contained in the unit interval $[0, 1]$ (a similar proof works in general). If $k_1, k_2, \dots, k_n$ is a sequence of n digits (i.e., each k_j is a whole number and $0 \leq k_j \leq 9$), let

$$S(k_1, k_2, \dots, k_n)$$

be the set of $x \in S$ whose decimal expansion begins $0.k_1 k_2 \dots k_n$. The set S is the union of the subsets $S(0), S(1), \dots, S(9)$, and since S is infinite, at least one of these subsets must be infinite. Therefore, we may choose k_1 so that $S(k_1)$ is infinite. In a similar fashion, at least one of the set $S(k_1, 0), S(k_2, 1), \dots, S(k_1, 9)$ must be infinite, so we may choose k_2 so that $S(k_1, k_2)$ is infinite. Continuing in this way, we obtain an infinite sequence $\{k_n\}$ such that $S(k_1, k_2, \dots, k_n)$ is infinite for all n. We may choose a sequence of elements $a_n \in S(k_1, k_2, \dots, k_n)$ with the property that a_n differs from $a_1, \dots, a_{n-1}$ for all n. Let L be the infinite decimal $0.k_1 k_2 k_3 \dots$. Then $\lim\limits_{n \to \infty} a_n = L$ since $|L - a_n| < 10^{-n}$ for all n. ∎

We use the Bolzano–Weierstrass Theorem to prove two important results about sequences $\{a_n\}$. Recall that an upper bound for $\{a_n\}$ is a number M such that $a_j \leq M$ for all j. If an upper bound exists, $\{a_n\}$ is said to be bounded from above. Lower bounds are defined similarly and $\{a_n\}$ is said to be bounded from below if a lower bound exists. A sequence is bounded if it is bounded from above and below. A **subsequence** of $\{a_n\}$ is a sequence of elements $a_{n_1}, a_{n_2}, a_{n_3}, \dots$, where $n_1 < n_2 < n_3 < \cdots$.

Now consider a bounded sequence $\{a_n\}$. If infinitely many of the a_n are distinct, the Bolzano–Weierstrass Theorem implies that there exists a subsequence $\{a_{n_1}, a_{n_2}, \dots\}$ such that $\lim\limits_{n \to \infty} a_{n_k}$ exists. Otherwise, infinitely many of the a_n must coincide and these terms form a convergent subsequence. This proves the next result.

THEOREM 3   Every bounded sequence has a convergent subsequence.

THEOREM 4 Bounded Monotonic Sequences Converge

- If $\{a_n\}$ is increasing and $a_n \leq M$ for all n, then $\{a_n\}$ converges and $\lim_{n \to \infty} a_n \leq M$.
- If $\{a_n\}$ is decreasing and $a_n \geq M$ for all n, then $\{a_n\}$ converges and $\lim_{n \to \infty} a_n \geq M$.

Proof Suppose that $\{a_n\}$ is increasing and bounded above by M. Then $\{a_n\}$ is automatically bounded below by $m = a_1$ since $a_1 \leq a_2 \leq a_3 \cdots$. Hence, $\{a_n\}$ is bounded, and by Theorem 3, we may choose a convergent subsequence $a_{n_1}, a_{n_2}, \ldots$. Let

$$L = \lim_{k \to \infty} a_{n_k}$$

Observe that $a_n \leq L$ for all n. For if not, then $a_n > L$ for some n and then $a_{n_k} \geq a_n > L$ for all k such that $n_k \geq n$. But this contradicts that $a_{n_k} \to L$. Now, by definition, for any $\epsilon > 0$, there exists $N_\epsilon > 0$ such that

$$|a_{n_k} - L| < \epsilon \qquad \text{if } n_k > N_\epsilon$$

Choose m such that $n_m > N_\epsilon$. If $n \geq n_m$, then $a_{n_m} \leq a_n \leq L$, and therefore,

$$|a_n - L| \leq |a_{n_m} - L| < \epsilon \qquad \text{for all } n \geq n_m$$

This proves that $\lim_{n \to \infty} a_n = L$ as desired. It remains to prove that $L \leq M$. If $L > M$, let $\epsilon = (L - M)/2$ and choose N so that

$$|a_n - L| < \epsilon \qquad \text{if } k > N$$

Then $a_n > L - \epsilon = M + \epsilon$. This contradicts our assumption that M is an upper bound for $\{a_n\}$. Therefore, $L \leq M$ as claimed. ∎

Proof of Theorem 1 We now use Theorem 4 to prove the LUB Property (Theorem 1). If x is a real number, we write $x(d)$ for the real number obtained by truncating the decimal expansion after the dth digit to the right of the decimal point. We call $x(d)$ the truncation of x of length d. Thus,

$$\text{If } x = 1.41569, \text{ then } x(3) = 1.415.$$

We say that x is a *decimal of length d* if $x = x(d)$. Any two distinct decimals of length d differ by at least 10^{-d}. It follows that for any two real numbers $A < B$, there are at most finitely many decimals of length d between A and B.

Now let S be a nonempty set of real numbers with an upper bound M. We shall prove that S has an LUB. Let $S(d)$ be the set of truncations of length d:

$$S(d) = \{x(d) : x \in S\}$$

We claim that $S(d)$ has a maximum element. To verify this, choose any $a \in S$. If $x \in S$ and $x(d) > a(d)$, then

$$a(d) \leq x(d) \leq M$$

Thus, by the remark of the previous paragraph, there are at most finitely many values of $x(d)$ in $S(d)$ larger than $a(d)$. The largest of these is the maximum element in $S(d)$.

For $d = 1, 2, \ldots$, choose an element x_d such that $x_d(d)$ is the maximum element in $S(d)$. By construction, $\{x_d(d)\}$ is an increasing sequence (since the largest dth truncation cannot get smaller as d increases). Furthermore, $x_d(d) \leq M$ for all d. We now apply

Theorem 4 to conclude that $\{x_d(d)\}$ converges to a limit L. We claim that L is the LUB of S. Observe first that L is an upper bound for S. Indeed, if $x \in S$, then $x(d) \leq L$ for all d and thus $x \leq L$. To show that L is the LUB, suppose that M is an upper bound such that $M < L$. Then $x_d \leq M$ for all d and hence $x_d(d) \leq M$ for all d. But then

$$L = \lim_{d \to \infty} x_d(d) \leq M$$

This is a contradiction since $M < L$. Therefore L is the LUB of S. ■

As mentioned above, the LUB Property is used in calculus to establish certain basic theorems about continuous functions. As an example, we prove the IVT. Another example is the theorem on the existence of extrema on a closed interval (see Appendix D).

THEOREM 5 Intermediate Value Theorem   If $f(x)$ is continuous on a closed interval $[a, b]$ and $f(a) \neq f(b)$, then for every value M between $f(a)$ and $f(b)$, there exists at least one value $c \in (a, b)$ such that $f(c) = M$.

Proof We may replace $f(x)$ by $f(x) - M$ to reduce to the case $M = 0$. Similarly, replacing $f(x)$ by $-f(x)$ if necessary, we may assume that $f(a) < 0$ and $f(b) > 0$. Now let

$$S = \{x \in [a, b] : f(x) < 0\}$$

Then $a \in S$ since $f(a) < 0$ and thus S is nonempty. Clearly, b is an upper bound for S. Therefore, by the LUB Property, S has an LUB L. We claim that $f(L) = 0$. If not, set $r = f(L)$. Assume first that $r > 0$.

Since $f(x)$ is continuous, there exists a number $\delta > 0$ such that

$$|f(x) - f(L)| = |f(x) - r| < \frac{1}{2}r \qquad \text{if} \qquad |x - L| < \delta$$

Equivalently,

$$\frac{1}{2}r < f(x) < \frac{3}{2}r \qquad \text{if} \qquad |x - L| < \delta$$

The number $\frac{1}{2}r$ is positive so we conclude that

$$f(x) > 0 \qquad \text{if} \qquad L - \delta < x < L + \delta$$

By definition of L, $f(x) \geq 0$ for all $x \in [a, b]$ such that $x > L$, and thus $f(x) \geq 0$ for all $x \in [a, b]$ such that $x > L - \delta$. Thus, $L - \delta$ is an upper bound for S. This is a contradiction since L is the LUB of S and it follows that $r = f(L)$ cannot satisfy $r > 0$. Similarly, r cannot satisfy $r < 0$. We conclude that $f(L) = 0$ as claimed. ■

C | INDUCTION AND THE BINOMIAL THEOREM

The Principle of Induction is a method of proof that is widely used to prove that a given statement $P(n)$ is valid for all natural numbers $n = 1, 2, 3, \ldots$. Here are two statements of this kind:

- $P(n)$: The sum of the first n odd numbers is equal to n^2.
- $P(n)$: $\dfrac{d}{dx}x^n = nx^{n-1}$.

The first statement claims that for all natural numbers n,

$$\underbrace{1 + 3 + \cdots + (2n - 1)}_{\text{Sum of first } n \text{ odd numbers}} = n^2 \qquad \boxed{1}$$

We can check directly that $P(n)$ is true for the first few values of n:

$$P(1) \text{ is the equality:} \qquad 1 = 1^2 \quad \text{(true)}$$
$$P(2) \text{ is the equality:} \qquad 1 + 3 = 2^2 \quad \text{(true)}$$
$$P(3) \text{ is the equality:} \qquad 1 + 3 + 5 = 3^2 \quad \text{(true)}$$

The Principle of Induction may be used to establish $P(n)$ for all n.

The Principle of Induction applies if $P(n)$ is an assertion defined for $n \geq n_0$, where n_0 is a fixed integer. Assume that

(i) *Initial step:* $P(n_0)$ *is true.*

(ii) *Induction step:* *If $P(n)$ is true for $n = k$, then $P(n)$ is also true for $n = k + 1$.*

Then $P(n)$ is true for all $n \geq n_0$.

THEOREM 1 Principle of Induction Let $P(n)$ be an assertion that depends on a natural number n. Assume that:

(i) **Initial step:** $P(1)$ is true.

(ii) **Induction step:** If $P(n)$ is true for $n = k$, then $P(n)$ is also true for $n = k + 1$.

Then $P(n)$ is true for all natural numbers $n = 1, 2, 3, \ldots$.

■ **EXAMPLE 1** Prove that $1 + 3 + \cdots + (2n - 1) = n^2$ for all natural numbers n.

Solution As above, we let $P(n)$ denote the equality

$$P(n): \qquad 1 + 3 + \cdots + (2n - 1) = n^2$$

Step 1. **Initial step: Show that $P(1)$ is true.**
We checked this above. $P(1)$ is the equality $1 = 1^2$.

Step 2. **Induction step: Show that if $P(n)$ is true for $n = k$, then $P(n)$ is also true for $n = k + 1$.**
Assume that $P(k)$ is true. Then

$$1 + 3 + \cdots + (2k - 1) = k^2$$

Add $2k + 1$ to both sides:

$$\left[1 + 3 + \cdots + (2k - 1)\right] + (2k + 1) = k^2 + 2k + 1 = (k + 1)^2$$

$$1 + 3 + \cdots + (2k + 1) = (k + 1)^2$$

This is precisely the statement $P(k + 1)$. Thus, $P(k + 1)$ is true whenever $P(k)$ is true. By the Principle of Induction, $P(k)$ is true for all k. ∎

The intuition behind the Principle of Induction is the following. If $P(n)$ were not true for all n, then there would exist a smallest natural number k such that $P(k)$ is false. Furthermore, $k > 1$ since $P(1)$ is true. Thus $P(k - 1)$ is true [otherwise, $P(k)$ would not be the smallest "counterexample"]. On the other hand, if $P(k - 1)$ is true, then $P(k)$ is also true by the induction step. This is a contradiction. So $P(k)$ must be true for all k.

■ **EXAMPLE 2** Use Induction and the Product Rule to prove that for all whole numbers n,

$$\frac{d}{dx}x^n = nx^{n-1}$$

Solution Let $P(n)$ be the formula $\dfrac{d}{dx}x^n = nx^{n-1}$.

Step 1. **Initial step: Show that $P(1)$ is true.**
We use the limit definition to verify $P(1)$:

$$\frac{d}{dx}x = \lim_{h \to 0}\frac{(x + h) - x}{h} = \lim_{h \to 0}\frac{h}{h} = \lim_{h \to 0} 1 = 1$$

Step 2. **Induction step: Show that if $P(n)$ is true for $n = k$, then $P(n)$ is also true for $n = k + 1$.**
To carry out the induction step, assume that $\dfrac{d}{dx}x^k = kx^{k-1}$, where $k \geq 1$. Then, by the Product Rule,

$$\frac{d}{dx}x^{k+1} = \frac{d}{dx}(x \cdot x^k) = x\frac{d}{dx}x^k + x^k\frac{d}{dx}x = x(kx^{k-1}) + x^k$$

$$= kx^k + x^k = (k + 1)x^k$$

This shows that $P(k + 1)$ is true.

By the Principle of Induction, $P(n)$ is true for all $n \geq 1$. ■

As another application of induction, we prove the Binomial Theorem, which describes the expansion of the binomial $(a + b)^n$. The first few expansions are familiar:

$$(a + b)^1 = a + b$$

$$(a + b)^2 = a^2 + 2ab + b^2$$

$$(a + b)^3 = a^3 + 3a^2b + 3ab^2 + b^3$$

In general, we have an expansion

$$(a + b)^n = a^n + \binom{n}{1}a^{n-1}b + \binom{n}{2}a^{n-2}b^2 + \binom{n}{3}a^{n-3}b^3 + \cdots + \binom{n}{n-1}ab^{n-1} + b^n$$

$$\boxed{2}$$

In Pascal's Triangle, the nth row displays the coefficients in the expansion of $(a + b)^n$:

n													
0							1						
1						1		1					
2					1		2		1				
3				1		3		3		1			
4			1		4		6		4		1		
5		1		5		10		10		5		1	
6	1		6		15		20		15		6		1

The triangle is constructed as follows: Each entry is the sum of the two entries above it in the previous line. For example, the entry 15 in line $n = 6$ is the sum $10 + 5$ of the entries above it in line $n = 5$. The recursion relation guarantees that the entries in the triangle are the binomial coefficients.

where the coefficient of $x^{n-k}x^k$, denoted $\binom{n}{k}$, is called the **binomial coefficient**. Note that the first term in (2) corresponds to $k = 0$ and the last term to $k = n$; thus, $\binom{n}{0} = \binom{n}{n} = 1$. In summation notation,

$$(a + b)^n = \sum_{k=0}^{n} \binom{n}{k} a^k b^{n-k}$$

Pascal's Triangle (described in the marginal note) can be used to compute binomial coefficients if n and k are not too large. The Binomial Theorem provides the following general formula:

$$\binom{n}{k} = \frac{n!}{k!\,(n-k)!} = \frac{n(n-1)(n-2)\cdots(n-k+1)}{k(k-1)(k-2)\cdots 2 \cdot 1}$$

| 3 |

Before proving this formula, we prove a recursion relation for binomial coefficients. Note, however, that (3) is certainly correct for $k = 0$ and $k = n$ (recall that by convention, $0! = 1$):

$$\binom{n}{0} = \frac{n!}{(n-0)!\,0!} = \frac{n!}{n!} = 1, \qquad \binom{n}{n} = \frac{n!}{(n-n)!\,n!} = \frac{n!}{n!} = 1$$

THEOREM 2   Recursion Relation for Binomial Coefficients

$$\binom{n}{k} = \binom{n-1}{k} + \binom{n-1}{k-1} \qquad \text{for } 1 \le k \le n-1$$

Proof   We write $(a + b)^n$ as $(a + b)(a + b)^{n-1}$ and expand in terms of binomial coefficients:

$$(a + b)^n = (a + b)(a + b)^{n-1}$$

$$\sum_{k=0}^{n} \binom{n}{k} a^{n-k} b^k = (a + b) \sum_{k=0}^{n-1} \binom{n-1}{k} a^{n-1-k} b^k$$

$$= a \sum_{k=0}^{n-1} \binom{n-1}{k} a^{n-1-k} b^k + b \sum_{k=0}^{n-1} \binom{n-1}{k} a^{n-1-k} b^k$$

$$= \sum_{k=0}^{n-1} \binom{n-1}{k} a^{n-k} b^k + \sum_{k=0}^{n-1} \binom{n-1}{k} a^{n-(k+1)} b^{k+1}$$

Replacing k by $k - 1$ in the second sum, we obtain

$$\sum_{k=0}^{n} \binom{n}{k} a^{n-k} b^k = \sum_{k=0}^{n-1} \binom{n-1}{k} a^{n-k} b^k + \sum_{k=1}^{n} \binom{n-1}{k-1} a^{n-k} b^k$$

On the right-hand side, the first term in the first sum is a^n and the last term in the second sum is b^n. Thus, we have

$$\sum_{k=0}^{n} \binom{n}{k} a^{n-k} b^k = a^n + \left(\sum_{k=1}^{n-1} \left(\binom{n-1}{k} + \binom{n-1}{k-1} \right) a^{n-k} b^k \right) + b^n$$

The recursion relation follows because the coefficients of $a^{n-k}b^k$ on the two sides of the equation must be equal. ∎

We now use induction to prove Eq. (3). Let $P(n)$ be the claim:

$$\binom{n}{k} = \frac{n!}{k!\,(n-k)!} \qquad \text{for } 0 \le k \le n$$

We have $\binom{1}{0} = \binom{1}{1} = 1$ since $(a+b)^1 = a + b$, so $P(1)$ is true. Furthermore,

$\binom{n}{n} = \binom{n}{0} = 1$ as observed above, since a^n and b^n have coefficient 1 in the expansion of $(a+b)^n$. For the inductive step, assume that $P(n)$ is true. By the recursion relation, for $1 \le k \le n$, we have

$$\binom{n+1}{k} = \binom{n}{k} + \binom{n}{k-1} = \frac{n!}{k!\,(n-k)!} + \frac{n!}{(k-1)!\,(n-k+1)!}$$

$$= n!\left(\frac{n+1-k}{k!\,(n+1-k)!} + \frac{k}{k!\,(n+1-k)!}\right) = n!\left(\frac{n+1}{k!\,(n+1-k)!}\right)$$

$$= \frac{(n+1)!}{k!\,(n+1-k)!}$$

Thus, $P(n+1)$ is also true and the Binomial Theorem follows by induction. ∎

■ **EXAMPLE 3** Use the Binomial Theorem to expand $(x+y)^5$ and $(x+2)^3$.

Solution The fifth row in Pascal's Triangle yields

$$(x+y)^5 = x^5 + 5x^4 y + 10x^3 y^2 + 10x^2 y^3 + 5xy^4 + y^5$$

The third row in Pascal's Triangle yields

$$(x+2)^3 = x^3 + 3x^2(2) + 3x(2)^2 + 2^3 = x^3 + 6x^2 + 12x + 8$$ ■

C. EXERCISES

In Exercises 1–4, use the Principle of Induction to prove the formula for all natural numbers n.

1. $1 + 2 + 3 + \cdots + n = \dfrac{n(n+1)}{2}$

2. $1^3 + 2^3 + 3^3 + \cdots + n^3 = \dfrac{n^2(n+1)^2}{4}$

3. $\dfrac{1}{1\cdot 2} + \dfrac{1}{2\cdot 3} + \cdots + \dfrac{1}{n(n+1)} = \dfrac{n}{n+1}$

4. $1 + x + x^2 + \cdots + x^n = \dfrac{1 - x^{n+1}}{1 - x}$ for any $x \ne 1$

5. Let $P(n)$ be the statement $2^n > n$.
(a) Show that $P(1)$ is true.
(b) Observe that if $2^n > n$, then $2^n + 2^n > 2n$. Use this to show that if $P(n)$ is true for $n = k$, then $P(n)$ is true for $n = k+1$. Conclude that $P(n)$ is true for all n.

6. Use induction to prove that $n! > 2^n$ for $n \ge 4$.

Let $\{F_n\}$ be the Fibonacci sequence, defined by the recursion formula

$$F_n = F_{n-1} + F_{n-2}, \qquad F_1 = F_2 = 1$$

The first few terms are 1, 1, 2, 3, 5, 8, 13, …. In Exercises 7–10, use induction to prove the identity.

7. $F_1 + F_2 + \cdots + F_n = F_{n+2} - 1$

8. $F_1^2 + F_2^2 + \cdots + F_n^2 = F_{n+1} F_n$

9. $F_n = \dfrac{R_+^n - R_-^n}{\sqrt{5}}$, where $R_\pm = \dfrac{1 \pm \sqrt{5}}{2}$

10. $F_{n+1} F_{n-1} = F_n^2 + (-1)^n$. *Hint:* For the induction step, show that

$$F_{n+2} F_n = F_{n+1} F_n + F_n^2$$

$$F_{n+1}^2 = F_{n+1} F_n + F_{n+1} F_{n-1}$$

11. Use induction to prove that $f(n) = 8^n - 1$ is divisible by 7 for all natural numbers n. *Hint:* For the induction step, show that

$$8^{k+1} - 1 = 7 \cdot 8^k + (8^k - 1)$$

12. Use induction to prove that $n^3 - n$ is divisible by 3 for all natural numbers n.

13. Use induction to prove that $5^{2n} - 4^n$ is divisible by 7 for all natural numbers n.

14. Use Pascal's Triangle to write out the expansions of $(a + b)^6$ and $(a - b)^4$.

15. Expand $(x + x^{-1})^4$.

16. What is the coefficient of x^9 in $(x^3 + x)^5$?

17. Let $S(n) = \sum_{k=0}^{n} \binom{n}{k}$.

(a) Use Pascal's Triangle to compute $S(n)$ for $n = 1, 2, 3, 4$.

(b) Prove that $S(n) = 2^n$ for all $n \geq 1$. *Hint:* Expand $(a + b)^n$ and evaluate at $a = b = 1$.

18. Let $T(n) = \sum_{k=0}^{n} (-1)^k \binom{n}{k}$.

(a) Use Pascal's Triangle to compute $T(n)$ for $n = 1, 2, 3, 4$.

(b) Prove that $T(n) = 0$ for all $n \geq 1$. *Hint:* Expand $(a + b)^n$ and evaluate at $a = 1, b = -1$.

ADDITIONAL PROOFS

In this appendix, we provide proofs of several theorems that were stated or used in the text.

I *Section 2.3*

THEOREM 1 **Basic Limit Laws** Assume that $\lim\limits_{x \to c} f(x)$ and $\lim\limits_{x \to c} g(x)$ exist. Then:

(i) $\lim\limits_{x \to c} \left(f(x) + g(x) \right) = \lim\limits_{x \to c} f(x) + \lim\limits_{x \to c} g(x)$

(ii) For any number k, $\lim\limits_{x \to c} kf(x) = k \lim\limits_{x \to c} f(x)$.

(iii) $\lim\limits_{x \to c} f(x)g(x) = \left(\lim\limits_{x \to c} f(x) \right) \left(\lim\limits_{x \to c} g(x) \right)$

(iv) If $\lim\limits_{x \to c} g(x) \neq 0$, then

$$\lim_{x \to c} \frac{f(x)}{g(x)} = \frac{\lim\limits_{x \to c} f(x)}{\lim\limits_{x \to c} g(x)}$$

Proof Let $L = \lim\limits_{x \to c} f(x)$ and $M = \lim\limits_{x \to c} g(x)$. The Sum Law (i) was proved in Section 2.6. Observe that (ii) is a special case of (iii), where $g(x) = k$ is a constant function. Thus, it will suffice to prove the Product Law (iii). We write

$$f(x)g(x) - LM = f(x)(g(x) - M) + M(f(x) - L)$$

and apply the Triangle Inequality to obtain

$$|f(x)g(x) - LM| \leq |f(x)(g(x) - M)| + |M(f(x) - L)| \qquad \boxed{1}$$

By the limit definition, we may choose $\delta > 0$ so that

$$|f(x) - L| < 1 \qquad \text{if } 0 < |x - c| < \delta$$

If follows that $|f(x)| < |L| + 1$ for $0 < |x - c| < \delta$. Now choose any number $\epsilon > 0$. Applying the limit definition again, we see that by choosing a smaller δ if necessary, we may also ensure that if $0 < |x - c| < \delta$, then

$$|f(x) - L| \leq \frac{\epsilon}{2(|M| + 1)} \quad \text{and} \quad |g(x) - M| \leq \frac{\epsilon}{2(|L| + 1)}$$

Using (1), we see that if $0 < |x - c| < \delta$, then

$$|f(x)g(x) - LM| \leq |f(x)|\,|g(x) - M| + |M|\,|f(x) - L|$$

$$\leq (|L| + 1)\frac{\epsilon}{2(|L| + 1)} + |M|\frac{\epsilon}{2(|M| + 1)}$$

$$\leq \frac{\epsilon}{2} + \frac{\epsilon}{2} = \epsilon$$

Since ϵ is arbitrary, this proves that $\lim_{x \to c} f(x)g(x) = LM$. To prove the Quotient Law (iv), it suffices to verify that if $M \neq 0$, then

$$\lim_{x \to c} \frac{1}{g(x)} = \frac{1}{M} \qquad \boxed{2}$$

For if (2) holds, then we may apply the Product Law to $f(x)$ and $g(x)^{-1}$ to obtain the Quotient Law:

$$\lim_{x \to c} \frac{f(x)}{g(x)} = \lim_{x \to c} f(x) \frac{1}{g(x)} = \left(\lim_{x \to c} f(x)\right)\left(\lim_{x \to c} \frac{1}{g(x)}\right)$$

$$= L\left(\frac{1}{M}\right) = \frac{L}{M}$$

We now verify (2). Since $g(x)$ approaches M and $M \neq 0$, we may choose $\delta > 0$ so that $|g(x)| \geq |M|/2$ if $0 < |x - c| < \delta$. Now choose any number $\epsilon > 0$. By choosing a smaller δ if necessary, we may also ensure that

$$|M - g(x)| < \epsilon |M| \left(\frac{|M|}{2}\right) \qquad \text{for } 0 < |x - c| < \delta$$

Then

$$\left|\frac{1}{g(x)} - \frac{1}{M}\right| = \left|\frac{M - g(x)}{Mg(x)}\right| \leq \left|\frac{M - g(x)}{M(M/2)}\right| \leq \frac{\epsilon |M|(|M|/2)}{|M|(|M|/2)} = \epsilon$$

Since ϵ is arbitrary, the limit (2) is proved. ∎

The following result was used in the text.

THEOREM 2 Limits Preserve Inequalities Let (a, b) be an open interval and let $c \in (a, b)$. Suppose that $f(x)$ and $g(x)$ are defined on (a, b), except possibly at c. Assume that

$$f(x) \leq g(x) \qquad \text{for } x \in (a, b), \quad x \neq c$$

and that the limits $\lim_{x \to c} f(x)$ and $\lim_{x \to c} g(x)$ exist. Then

$$\lim_{x \to c} f(x) \leq \lim_{x \to c} g(x)$$

Proof Let $L = \lim_{x \to c} f(x)$ and $M = \lim_{x \to c} g(x)$. To show that $L \leq M$, we use proof by contradiction. If $L > M$, let $\epsilon = \frac{1}{2}(L - M)$. By the formal definition of limits, we may choose $\delta > 0$ so that the following two conditions are satisfied:

$$|M - g(x)| < \epsilon \qquad \text{if } |x - c| < \delta$$

$$|L - f(x)| < \epsilon \qquad \text{if } |x - c| < \delta$$

But then

$$f(x) > L - \epsilon = M + \epsilon > g(x)$$

This is a contradiction since $f(x) \leq g(x)$. We conclude that $L \leq M$. ∎

THEOREM 3 Limit of a Composite Function   Assume that the following limits exist:

$$L = \lim_{x \to c} g(x) \qquad \text{and} \qquad M = \lim_{x \to L} f(x)$$

Then $\lim_{x \to c} f(g(x)) = M$.

Proof Let $\epsilon > 0$ be given. By the limit definition, there exists $\delta_1 > 0$ such that

$$|f(x) - M| < \epsilon \qquad \text{if } 0 < |x - L| < \delta_1 \qquad \boxed{3}$$

Similarly, there exists $\delta > 0$ such that

$$|g(x) - L| < \delta_1 \qquad \text{if } 0 < |x - c| < \delta \qquad \boxed{4}$$

We replace x by $g(x)$ in (3) and apply (4) to obtain

$$|f(g(x)) - M| < \epsilon \qquad \text{if } 0 < |x - c| < \delta$$

Since ϵ is arbitrary, this proves that $\lim_{x \to c} f(g(x)) = M$.   ∎

| Section 2.4

THEOREM 4 Continuity of Composite Functions   Let $F(x) = f(g(x))$ be a composite function. If g is continuous at $x = c$ and f is continuous at $x = g(c)$, then $F(x)$ is continuous at $x = c$.

Proof By definition of continuity,

$$\lim_{x \to c} g(x) = g(c) \qquad \text{and} \qquad \lim_{x \to g(c)} f(x) = f(g(c))$$

Therefore, we may apply Theorem 3 to obtain

$$\lim_{x \to c} f(g(x)) = f(g(c))$$

This proves that $f(g(x))$ is continuous at $x = c$.   ∎

| Section 2.6

THEOREM 5 Squeeze Theorem   Assume that for $x \neq c$ (in some open interval containing c),

$$l(x) \leq f(x) \leq u(x) \qquad \text{and} \qquad \lim_{x \to c} l(x) = \lim_{x \to c} u(x) = L$$

Then $\lim_{x \to c} f(x)$ exists and

$$\lim_{x \to c} f(x) = L$$

Proof Let $\epsilon > 0$ be given. We may choose $\delta > 0$ such that

$$|l(x) - L| < \epsilon \quad \text{and} \quad |u(x) - L| < \epsilon \qquad \text{if } 0 < |x - c| < \delta$$

In principle, a different δ may be required to obtain the two inequalities for $l(x)$ and $u(x)$, but we may choose the smaller of the two deltas. Thus, if $0 < |x - c| < \delta$, we have

$$L - \epsilon < l(x) < L + \epsilon$$

and

$$L - \epsilon < u(x) < L + \epsilon$$

Since $f(x)$ lies between $l(x)$ and $u(x)$, it follows that

$$L - \epsilon < l(x) \le f(x) \le u(x) < L + \epsilon$$

and therefore $|f(x) - L| < \epsilon$ if $0 < |x - c| < \delta$. Since ϵ is arbitrary, this proves that $\lim_{x \to c} f(x) = L$ as desired.   ■

I *Section 3.9*

> **THEOREM 6 Derivative of the Inverse**   Assume that $f(x)$ is differentiable and one-to-one with inverse $g(x)$. If b belongs to the domain of $g(x)$ and $f'(g(b)) \ne 0$, then $g'(b)$ exists and
>
> $$g'(b) = \frac{1}{f'(g(b))}$$

Proof The function $f(x)$ is one-to-one and continuous (since it is differentiable). It follows that $f(x)$ is monotonic increasing or decreasing. For if not, then $f(x)$ would have a local minimum or maximum at some point $x = x_0$. But then $f(x)$ would not be one-to-one in a small interval around x_0 by the IVT.

Suppose that $f(x)$ is increasing (the decreasing case is similar). We shall prove that $g(x)$ is continuous at $x = b$. Suppose that $f(a) = b$. Fix a small number $\epsilon > 0$. Since $f(x)$ is an increasing function, it maps the open interval $(a - \epsilon, a + \epsilon)$ to the open interval $(f(a - \epsilon), f(a + \epsilon))$ containing $f(a) = b$. We may choose a number $\delta > 0$ so that $(b - \delta, b + \delta)$ is contained in $(f(a - \epsilon), f(a + \epsilon))$. Then $g(x)$ maps $(b - \delta, b + \delta)$ back into $(a - \epsilon, a + \epsilon)$. It follows that

$$|g(y) - g(b)| < \epsilon \qquad \text{if } 0 < |y - b| < \delta$$

This proves that g is continuous at $x = b$.

To complete the proof, we must show that the following limit exists and is equal to $1/f'(g(b))$:

$$g'(a) = \lim_{y \to b} \frac{g(y) - g(b)}{y - b}$$

Let $a = g(b)$. By the inverse relationship, if $y = f(x)$, then $g(y) = x$, and since $g(y)$ is continuous, x approaches a as y approaches b. Thus, since $f(x)$ is differentiable and $f'(a) \ne 0$,

$$\lim_{y \to b} \frac{g(y) - g(b)}{y - b} = \lim_{x \to a} \frac{x - a}{f(x) - f(a)} = \frac{1}{f'(a)} = \frac{1}{f'(g(b))} \qquad ■$$

I *Section 4.2*

> **THEOREM 7 Existence of Extrema on a Closed Interval**   If $f(x)$ is a continuous function on a closed (bounded) interval $I = [a, b]$, then $f(x)$ takes on a minimum and a maximum value on I.

Proof We prove that $f(x)$ takes on a maximum value in two steps (the case of a minimum is similar).

Step 1. **Prove that $f(x)$ is bounded from above.**

We use proof by contradiction. If $f(x)$ is not bounded from above, then there exist points $a_n \in [a, b]$ such that $f(a_n) \geq n$ for $n = 1, 2, \ldots$. By Theorem 3 in Appendix B, we may choose a subsequence of elements $a_{n_1}, a_{n_2}, \ldots$ that converges to a limit in $[a, b]$, say, $\lim_{k \to \infty} a_{n_k} = L$. Since $f(x)$ is continuous, there exists $\delta > 0$ such that

$$|f(x) - f(L)| < 1 \quad \text{if} \quad x \in [a, b] \quad \text{and} \quad |x - L| < \delta$$

Therefore,

$$f(x) < f(L) + 1 \quad \text{if} \quad x \in [a, b] \quad \text{and} \quad x \in (L - \delta, L + \delta) \qquad \boxed{5}$$

For k sufficiently large, a_{n_k} lies in $(L - \delta, L + \delta)$ because $\lim_{k \to \infty} a_{n_k} = L$. By (5), $f(a_{n_k})$ is bounded by $f(L) + 1$. However, $f(a_{n_k}) = n_k$ tends to infinity as $k \to \infty$. This is a contradiction. Hence, our assumption that $f(x)$ is not bounded from above is false.

Step 2. **Prove that $f(x)$ takes on a maximum value.**

The range of $f(x)$ on $I = [a, b]$ is the set

$$S = \{f(x) : x \in [a, b]\}$$

By the previous step, S is bounded from above and therefore has a least upper bound M by the LUB Property. To complete the proof, we show that $f(c) = M$ for some $c \in [a, b]$.

By definition, $M - 1/n$ is not an upper bound for $n \geq 1$, and therefore, we may choose a point b_n in $[a, b]$ such that

$$M - \frac{1}{n} \leq f(b_n) \leq M$$

Again by Theorem 3 in Appendix B, there exists a subsequence of elements $\{b_{n_1}, b_{n_2}, \ldots\}$ in $\{b_1, b_2, \ldots\}$ that converges to a limit, say,

$$\lim_{k \to \infty} b_{n_k} = c$$

To show that $f(c) = M$, let $\epsilon > 0$. Since $f(x)$ is continuous, we may choose k so large that the following two conditions are satisfied: $|f(c) - f(b_{n_k})| < \epsilon/2$ and $n_k > 2/\epsilon$. Then

$$|f(c) - M| \leq |f(c) - f(b_{n_k})| + |f(b_{n_k}) - M| \leq \frac{\epsilon}{2} + \frac{1}{n_k} \leq \frac{\epsilon}{2} + \frac{\epsilon}{2} = \epsilon$$

Thus, $|f(c) - M|$ is smaller than ϵ for all positive numbers ϵ. But this is not possible unless $|f(c) - M| = 0$. Thus $f(c) = M$ as desired. ∎

❚ *Section 5.2*

THEOREM 8 Continuous Functions Are Integrable   If $f(x)$ is continuous on $[a, b]$, then $f(x)$ is integrable over $[a, b]$.

Proof   We shall make the simplifying assumption that $f(x)$ is differentiable and that its derivative $f'(x)$ is bounded. In other words, we assume that $|f'(x)| \leq K$ for some constant K. This assumption is used to show that $f(x)$ cannot vary too much in a small interval. More precisely, let us prove that if $[a_0, b_0]$ is any closed interval contained in

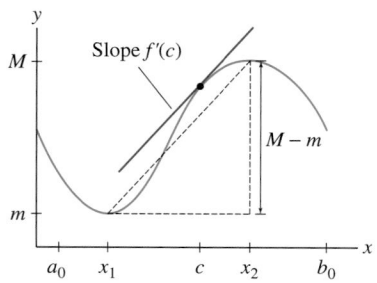

FIGURE 1 Since $M - m = f'(c)(x_2 - x_1)$, we conclude that $M - m \leq K(b_0 - a_0)$.

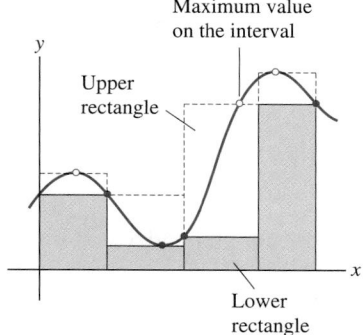

FIGURE 2 Lower and upper rectangles for a partition of length $N = 4$.

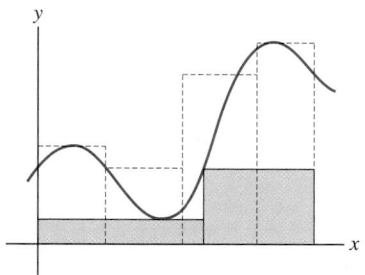

FIGURE 3 The lower rectangles always lie below the upper rectangles, even when the partitions are different.

$[a, b]$ and if m and M are the minimum and maximum values of $f(x)$ on $[a_0, b_0]$, then

$$|M - m| \leq K|b_0 - a_0| \qquad \boxed{6}$$

Figure 1 illustrates the idea behind this inequality. Suppose that $f(x_1) = m$ and $f(x_2) = M$, where x_1 and x_2 lie in $[a_0, b_0]$. If $x_1 \neq x_2$, then by the Mean Value Theorem (MVT), there is a point c between x_1 and x_2 such that

$$\frac{M - m}{x_2 - x_1} = \frac{f(x_2) - f(x_1)}{x_2 - x_1} = f'(c)$$

Since x_1, x_2 lie in $[a_0, b_0]$, we have $|x_2 - x_1| \leq |b_0 - a_0|$, and thus,

$$|M - m| = |f'(c)|\,|x_2 - x_1| \leq K|b_0 - a_0|$$

This proves (6).

We divide the rest of the proof into two steps. Consider a partition P:

$$P: \qquad x_0 = a < x_1 < \quad \cdots \quad < x_{N-1} < x_N = b$$

Let m_i be the minimum value of $f(x)$ on $[x_{i-1}, x_i]$ and M_i the maximum on $[x_{i-1}, x_i]$. We define the *lower* and *upper* Riemann sums

$$L(f, P) = \sum_{i=1}^{N} m_i \Delta x_i, \qquad U(f, P) = \sum_{i=1}^{N} M_i \Delta x_i,$$

These are the particular Riemann sums in which the intermediate point in $[x_{i-1}, x_i]$ is the point where $f(x)$ takes on its minimum or maximum on $[x_{i-1}, x_i]$. Figure 2 illustrates the case $N = 4$.

Step 1. **Prove that the lower and upper sums approach a limit.**
We observe that

$$L(f, P_1) \leq U(f, P_2) \quad \text{for any two partitions } P_1 \text{ and } P_2 \qquad \boxed{7}$$

Indeed, if a subinterval I_1 of P_1 overlaps with a subinterval I_2 of P_2, then the minimum of f on I_1 is less than or equal to the maximum of f on I_2 (Figure 3). In particular, the lower sums are bounded above by $U(f, P)$ for all partitions P. Let L be the least upper bound of the lower sums. Then for all partitions P,

$$L(f, P) \leq L \leq U(f, P) \qquad \boxed{8}$$

According to Eq. (6), $|M_i - m_i| \leq K \Delta x_i$ for all i. Since $\|P\|$ is the largest of the widths Δx_i, we see that $|M_i - m_i| \leq K\|P\|$ and

$$|U(f, P) - L(f, P)| \leq \sum_{i=1}^{N} |M_i - m_i| \Delta x_i$$

$$\leq K\|P\| \sum_{i=1}^{N} \Delta x_i = K\|P\|\,|b - a| \qquad \boxed{9}$$

Let $c = K\,|b - a|$. Using (8), we obtain

$$|L - U(f, P)| \leq |U(f, P) - L(f, P)| \leq c\|P\|$$

and we conclude that $\lim\limits_{||P|| \to 0} |L - U(f, P)| = 0$. Similarly,

$$|L - L(f, P)| \leq c||P||$$

and

$$\lim\limits_{||P|| \to 0} |L - L(f, P)| = 0$$

Thus, we have

$$\lim\limits_{||P|| \to 0} U(f, P) = \lim\limits_{||P|| \to 0} L(f, P) = L$$

Step 2. **Prove that** $\displaystyle\int_a^b f(x)\,dx$ **exists and has value** ***L.***

Recall that for any choice C of intermediate points $c_i \in [x_{i-1}, x_i]$, we define the Riemann sum by

$$R(f, P, C) = \sum_{i=1}^{N} f(x_i)\Delta x_i$$

We have

$$L(f, P) \leq R(f, P, C) \leq U(f, P)$$

Indeed, since $c_i \in [x_{i-1}, x_i]$, we have $m_i \leq f(c_i) \leq M_i$, for all i and

$$\sum_{i=1}^{N} m_i \, \Delta x_i \leq \sum_{i=1}^{N} f(c_i)\,\Delta x_i \leq \sum_{i=1}^{N} M_i \, \Delta x_i$$

It follows that

$$|L - R(f, P, C)| \leq |U(f, P) - L(f, P)| \leq c||P||$$

This shows that $R(f, P, C)$ converges to L as $||P|| \to 0$. ∎

| Section 10.1

THEOREM 9 If $f(x)$ is continuous and $\{a_n\}$ is a sequence such that the limit $\lim\limits_{n \to \infty} a_n = L$ exists, then

$$\lim\limits_{n \to \infty} f(a_n) = f(L)$$

Proof Choose any $\epsilon > 0$. Since $f(x)$ is continuous, there exists $\delta > 0$ such that

$$|f(x) - f(L)| < \epsilon \qquad \text{if } 0 < |x - L| < \delta$$

Since $\lim\limits_{n \to \infty} a_n = L$, there exists $N > 0$ such that $|a_n - L| < \delta$ for $n > N$. Thus,

$$|f(a_n) - f(L)| < \epsilon \qquad \text{for } n > N$$

It follows that $\lim\limits_{n \to \infty} f(a_n) = f(L)$. ∎

| Section 14.3

THEOREM 10 Clairaut's Theorem   If f_{xy} and f_{yx} are both continuous functions on a disk D, then $f_{xy}(a, b) = f_{yx}(a, b)$ for all $(a, b) \in D$.

Proof We prove that both $f_{xy}(a, b)$ and $f_{yx}(a, b)$ are equal to the limit

$$L = \lim_{h \to 0} \frac{f(a + h, b + h) - f(a + h, b) - f(a, b + h) + f(a, b)}{h^2}$$

Let $F(x) = f(x, b + h) - f(x, b)$. The numerator in the limit is equal to

$$F(a + h) - F(a)$$

and $F'(x) = f_x(x, b + h) - f_x(x, b)$. By the MVT, there exists a_1 between a and $a + h$ such that

$$F(a + h) - F(a) = h F'(a_1) = h(f_x(a_1, b + h) - f_x(a_1, b))$$

By the MVT applied to f_x, there exists b_1 between b and $b + h$ such that

$$f_x(a_1, b + h) - f_x(a_1, b) = h f_{xy}(a_1, b_1)$$

Thus,

$$F(a + h) - F(a) = h^2 f_{xy}(a_1, b_1)$$

and

$$L = \lim_{h \to 0} \frac{h^2 f_{xy}(a_1, b_1)}{h^2} = \lim_{h \to 0} f_{xy}(a_1, b_1) = f_{xy}(a, b)$$

The last equality follows from the continuity of f_{xy} since (a_1, b_1) approaches (a, b) as $h \to 0$. To prove that $L = f_{yx}(a, b)$, repeat the argument using the function $F(y) = f(a + h, y) - f(a, y)$, with the roles of x and y reversed. ∎

I Section 14.4

THEOREM 11 **Criterion for Differentiability** If $f_x(x, y)$ and $f_y(x, y)$ exist and are continuous on an open disk D, then $f(x, y)$ is differentiable on D.

Proof Let $(a, b) \in D$ and set

$$L(x, y) = f(a, b) + f_x(a, b)(x - a) + f_y(a, b)(y - b)$$

To prove that $f(x, y)$ is differentiable, we must show that

$$f(x, y) = L(x, y) + \epsilon(x, y)\sqrt{(x - a)^2 + (y - b)^2}$$

where $\epsilon(x, y)$ is a function such that $\lim\limits_{(x,y) \to (a,b)} \epsilon(x, y) = 0$. It is convenient to switch to the variables h and k, where $x = a + h$ and $y = b + k$. Set

$$\Delta f = f(a + h, b + k) - f(a, b)$$

Then

$$L(x, y) = f(a, b) + f_x(a, b)h + f_y(a, b)k$$

and we may define the function

$$E(h, k) = f(x, y) - L(x, y) = \Delta f - (f_x(a, b)h + f_y(a, b)k)$$

Thus, $\epsilon(x, y) = E(h, k)/\sqrt{h^2 + k^2}$ and it is necessary to show that

$$\lim_{(h,k) \to (0,0)} \frac{E(h, k)}{\sqrt{h^2 + k^2}} = 0$$

To do this, we write Δf as a sum of two terms:

$$\Delta f = (f(a+h, b+k) - f(a, b+k)) + (f(a, b+k) - f(a, b))$$

and apply the MVT to each term separately. We find that there exist a_1 between a and $a+h$ and b_1 between b and $b+k$ such that

$$f(a+h, b+k) - f(a, b+k) = hf_x(a_1, b+k)$$

$$f(a, b+k) - f(a, b) = kf_y(a, b_1)$$

Therefore,

$$E(h, k) = h(f_x(a_1, b+k) - f_x(a, b)) + k(f_y(a, b_1) - f_y(a, b))$$

and

$$\left| \frac{E(h, k)}{\sqrt{h^2 + k^2}} \right| = \left| \frac{h(f_x(a_1, b+k) - f_x(a, b)) + k(f_y(a, b_1) - f_y(a, b))}{\sqrt{h^2 + k^2}} \right|$$

$$\leq \left| \frac{h(f_x(a_1, b+k) - f_x(a, b))}{\sqrt{h^2 + k^2}} \right| + \left| \frac{k(f_y(a, b_1) - f_y(a, b))}{\sqrt{h^2 + k^2}} \right|$$

$$\leq \left| \frac{h(f_x(a_1, b+k) - f_x(a, b))}{h} \right| + \left| \frac{k(f_y(a, b_1) - f_y(a, b))}{k} \right|$$

$$= |f_x(a_1, b+k) - f_x(a, b)| + |f_y(a, b_1) - f_y(a, b)|$$

In the second line, we use the Triangle Inequality, and we may pass to the third line because $|h|$ and $|k|$ are both smaller than $\sqrt{h^2 + k^2}$. Both terms in the last line tend to zero as $(h, k) \to (0, 0)$ because f_x and f_y are assumed to be continuous. This completes the proof that $f(x, y)$ is differentiable. ∎

ANSWERS TO ODD-NUMBERED EXERCISES

Chapter 1

Section 1.1 Preliminary

1. $a = -3, b = 1$

2. The numbers $a \geq 0$ satisfy $|a| = a$ and $|-a| = a$. The numbers $a \leq 0$ satisfy $|a| = -a$.

3. $a = -3, b = 1$ **4.** $(9, -4)$

5. **(a)** First quadrant   **(b)** Second quadrant   **(c)** Fourth quadrant   **(d)** Third quadrant

6. The radius is 3.   **7.** **(b)**

8. Symmetry with respect to the origin.

Section 1.1 Exercises

1. $r = 9.8696$ **3.** $|x| \leq 2$ **5.** $|x - 2| < 2$ **7.** $|x - 3| \leq 2$

9. $-8 < x < 8$ **11.** $-3 < x < 2$ **13.** $(-4, 4)$ **15.** $(2, 6)$

17. $\left[-\frac{7}{4}, \frac{9}{4}\right]$ **19.** $(-\infty, 2) \cup (6, \infty)$

21. $(-\infty, -\sqrt{3}) \cup (\sqrt{3}, \infty)$

23. **(a)** (i)   **(b)** (iii)   **(c)** (v)   **(d)** (vi)   **(e)** (ii)   **(f)** (iv)

27. $|a + b - 13| = |(a - 5) + (b - 8)| \leq |a - 5| + |b - 8|$
$$< \frac{1}{2} + \frac{1}{2} = 1$$

29. **(a)** 9 **(b)** $|x^2 - 16| = |x - 4| \cdot |x + 4| \leq 1 \cdot 9 = 9$

31. $r_1 = \frac{3}{11}, r_2 = \frac{4}{15}$

33. Let $a = 1$ and $b = .\overline{9}$. The decimal expansions of a and b do not agree, but $1 - .\overline{9} = 0 < 10^{-k}$ for any k.

35. $(x - 2)^2 + (y - 4)^2 = 3^2 = 9$

37. Located a distance 5 from the origin:

$(5, 0)$	$(-5, 0)$	$(0, 5)$	$(0, -5)$
$(3, 4)$	$(-3, 4)$	$(3, -4)$	$(-3, -4)$
$(4, 3)$	$(-4, 3)$	$(4, -3)$	$(-4, -3)$

Located a distance 5 from the point $(2, 3)$:

$(7, 3)$	$(-3, 3)$	$(2, 8)$	$(2, -2)$
$(5, 7)$	$(-1, 7)$	$(5, -1)$	$(-1, -1)$
$(6, 6)$	$(-2, 6)$	$(6, 0)$	$(-2, 0)$

39. Example: $f : \{a, b, c\} \to \{1, 2\}$ where $f(a) = 1, f(b) = 1, f(c) = 2$. There is no function whose domain has two elements and range has three elements.

41. D : All reals; $R : \{y : y \geq 0\}$

43. $D : \{t : t \leq 2\}; R : \{y : y \geq 0\}$

45. $D : \{s : s \neq 0\}; R : \{y : y \neq 0\}$

47. D : All reals; $R : \{y : y \geq \sqrt{2}\}$ **49.** $(-1, \infty)$ **51.** $(0, \infty)$

53. Zeros: ± 2; Inc: $x > 0$; Dec: $x < 0$; Sym: $f(-x) = f(x)$ (even function); y-axis symmetry.

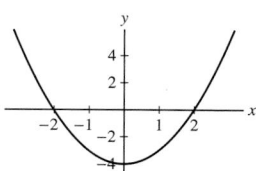

55. Zeros: $0, \pm 2$; Sym: $f(-x) = -f(x)$ (odd function); origin symmetry.

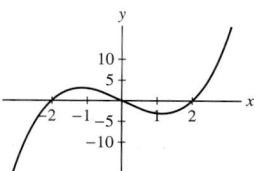

57. Zero: $x = \sqrt[3]{2}$; this is an x-axis reflection of x^3 moved up 2 units.

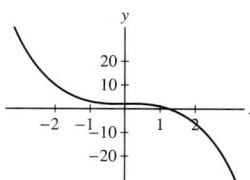

59. B   **61.** $D : [0, 4]; R : [0, 4]$

63.

$f(2x)$

$f(x/2)$

$2f(x)$

65.

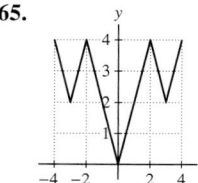

67. (a) $D : [4, 8]$, $R : [5, 9]$ **(b)** $D : [1, 5]$, $R : [2, 6]$
(c) $D : \left[\frac{4}{3}, \frac{8}{3}\right]$, $R : [2, 6]$ **(d)** $D : [4, 8]$, $R : [6, 18]$
69. (a) $h(x) = \sin(2x - 10)$ **(b)** $h(x) = \sin(2x - 5)$
71.

$f(2x)$

$f(x/2)$

73.

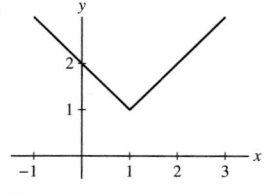

$D : \mathcal{R}$, $R : \{y | y \geq 1\}$
$f(x) = |x - 1| + 1$

75. Even:

$$(f + g)(-x) = f(-x) + g(-x) \overset{\text{even}}{=} f(x) + g(x)$$
$$= (f + g)(x)$$

Odd:

$$(f + g)(-x) = f(-x) + g(-x) \overset{\text{odd}}{=} -f(x) + -g(x)$$
$$= -(f + g)(x)$$

77. A circle of radius 1 with its center at the origin.
81. (a) An example:

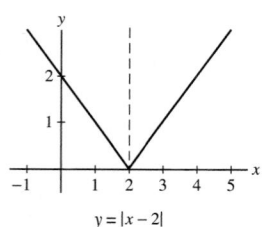

$y = |x - 2|$

(b) Let $g(x) = f(x + a)$. Then
$g(-x) = f(-x + a) = f(a + x) = g(x)$. Thus, $g(x)$ is even.

Section 1.2 Preliminary

1. -4 **2.** No
3. Parallel to the y-axis when $b = 0$ and parallel to the x-axis when $a = 0$.
4. $\Delta y = 9$ **5.** -4 **6.** $(x - 0)^2 + 1$

Section 1.2 Exercises

1. $m = 3$, $y = 12$, $x = -4$ **3.** $m = -\frac{4}{9}$, $y = \frac{1}{3}$, $x = \frac{3}{4}$
5. $m = 3$ **7.** $m = -\frac{3}{4}$ **9.** $y = 3x + 8$ **11.** $y = 3x - 12$
13. $y = -2$ **15.** $y = 3x - 2$ **17.** $y = \frac{5}{3}x - \frac{1}{3}$ **19.** $y = 4$
21. $y - 3 = -2(x - 3)$ **23.** $3x + 4y = 12$

25. (a) $c = -\frac{1}{4}$ **(b)** $c = -2$ **(c)** Such a constant does not exist.
(d) $c = 0$
27. (a) $L = 40.0248$ cm **(b)** $L = 64.9597$ in
(c) $L = 65(1 + \alpha(T - 100))$
29. $b = 4$
31. No, because the slopes between consecutive data points are not equal.
33. (a) $x = 1$ or $-\frac{1}{4}$ **(b)** $x = 1 \pm \sqrt{2}$
35. Minimum value is 0. **37.** Minimum value is -7.
39. Maximum value is $\frac{137}{16}$. **41.** Maximum value is $\frac{1}{3}$.
43.

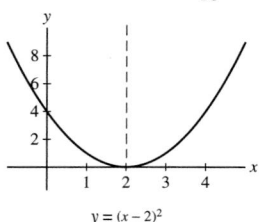

$y = (x - 2)^2$

45. A double root occurs when $c = \pm 2$. There are no real roots when $-2 < c < 2$.
47. For all $x > 0$, $0 \leq (x^{1/2} - x^{-1/2})^2 = x - 2 + \frac{1}{x}$.
51. $4 + 2\sqrt{2}$ and $4 - 2\sqrt{2}$
55. For x^2, $\frac{\Delta y}{\Delta x} = \frac{x_2^2 - x_1^2}{x_2 - x_1} = x_2 + x_1$.
59. $(x - \alpha)(x - \beta) = x^2 - \alpha x - \beta x + \alpha \beta$
$$= x^2 + (-\alpha - \beta)x + \alpha \beta$$

Section 1.3 Preliminary

1. An example: $\frac{3x^2 - 2}{7x^3 + x - 1}$
2. $|x|$ is not a polynomial. $|x^2 + 1|$ is a polynomial.
3. The domain of $f(g(x))$ is the empty set.
4. Decreasing **5.** An example: $f(x) = e^x - \sin x$

Section 1.3 Exercises

1. $x \geq 0$ **3.** All real numbers **5.** $t \neq -2$ **7.** $u \neq \pm 2$
9. $x \neq 0, 1$ **11.** $y > 0$ **13.** Polynomial **15.** Algebraic
17. Transcendental **19.** Rational **21.** Transcendental
23. Rational **25.** Yes
27. $f(g(x)) = \sqrt{x + 1}$; $D: x \geq -1$; $g(f(x)) = \sqrt{x} + 1$; $D: x \geq 0$
29. $f(g(x)) = 2^{x^2}$; $D: \mathcal{R}$; $g(f(x)) = (2^x)^2 = 2^{2x}$; $D: \mathcal{R}$
31. $f(g(x)) = \cos(x^3 + x^2)$; $D: \mathcal{R}$
$g(f(\theta)) = \cos^3 \theta + \cos^2 \theta$; $D: \mathcal{R}$
33. $f(g(t)) = \frac{1}{\sqrt{-t^2}}$; D: Not valid for any t

$g(f(t)) = -\left(\frac{1}{\sqrt{t}}\right)^2 = -\frac{1}{t}$; $D: t > 0$

35. $P(t + 10) = 30 \cdot 2^{0.1(t+10)} = 30 \cdot 2^{0.1t+1}$
$$= 2(30 \cdot 2^{0.1t}) = 2P(t)$$

$g\left(t + \frac{1}{k}\right) = a2^{k(t+1/k)} = a2^{kt+1}$
$$= 2a2^{kt} = 2g(t)$$

37. $f(x) = x^2$: $\delta f(x) = f(x+1) - f(x)$
$$= (x+1)^2 - x^2 = 2x + 1$$
$f(x) = x$: $\delta f(x) = x + 1 - x = 1$
$f(x) = x^3$: $\delta f(x) = (x+1)^3 - x^3 = 3x^2 + 3x + 1$

39. $\delta(f + g) = (f(x+1) + g(x+1)) - (f(x) - g(x))$
$$= (f(x+1) - f(x)) + (g(x+1) - g(x))$$
$$= \delta f(x) + \delta g(x)$$
$$\delta(cf) = cf(x+1) - cf(x) = c(f(x+1) - f(x))$$
$$= c\delta f(x).$$

Section 1.4 Preliminary

1. Two rotations that differ by a whole number of full revolutions will have the same ending radius.

2. $\frac{9\pi}{4}$ and $\frac{41\pi}{4}$ **3.** $-\frac{5\pi}{3}$ **4. (a)**

5. Let O denote the center of the unit circle, and let P be a point on the unit circle such that the radius $\overline{OP}$ makes an angle θ with the positive x-axis. Then, $\sin\theta$ is the y-coordinate of the point P.

Section 1.4 Exercises

1. $5\pi/4$

3. (a) $\frac{180°}{\pi} \approx 57.1°$ **(b)** $60°$ **(c)** $\frac{75°}{\pi} \approx 23.87°$ **(d)** $-135°$

5. $s = r\theta = 3.6$; $s = r\phi = 8$

7.

θ	$(\cos\theta, \sin\theta)$	θ	$(\cos\theta, \sin\theta)$
$\frac{\pi}{2}$	$(0, 1)$	$\frac{5\pi}{4}$	$\left(\frac{-\sqrt{2}}{2}, \frac{-\sqrt{2}}{2}\right)$
$\frac{2\pi}{3}$	$\left(\frac{-1}{2}, \frac{\sqrt{3}}{2}\right)$	$\frac{4\pi}{3}$	$\left(\frac{-1}{2}, \frac{-\sqrt{3}}{2}\right)$
$\frac{3\pi}{4}$	$\left(\frac{-\sqrt{2}}{2}, \frac{\sqrt{2}}{2}\right)$	$\frac{3\pi}{2}$	$(0, -1)$
$\frac{5\pi}{6}$	$\left(\frac{-\sqrt{3}}{2}, \frac{1}{2}\right)$	$\frac{5\pi}{3}$	$\left(\frac{1}{2}, \frac{-\sqrt{3}}{2}\right)$
π	$(-1, 0)$	$\frac{7\pi}{4}$	$\left(\frac{\sqrt{2}}{2}, \frac{-\sqrt{2}}{2}\right)$
$\frac{7\pi}{6}$	$\left(\frac{-\sqrt{3}}{2}, \frac{-1}{2}\right)$	$\frac{11\pi}{6}$	$\left(\frac{\sqrt{3}}{2}, \frac{-1}{2}\right)$

9. $\theta = \frac{\pi}{3}, \frac{5\pi}{3}$ **11.** $\theta = \frac{3\pi}{4}, \frac{7\pi}{4}$ **13.** $x = \frac{\pi}{3}, \frac{2\pi}{3}$

15.

θ	$\frac{\pi}{6}$	$\frac{\pi}{4}$	$\frac{\pi}{3}$	$\frac{\pi}{2}$	$\frac{2\pi}{3}$	$\frac{3\pi}{4}$	$\frac{5\pi}{6}$
$\tan\theta$	$\frac{1}{\sqrt{3}}$	1	$\sqrt{3}$	und	$-\sqrt{3}$	-1	$-\frac{1}{\sqrt{3}}$
$\sec\theta$	$\frac{2}{\sqrt{3}}$	$\sqrt{2}$	2	und	-2	$-\sqrt{2}$	$-\frac{2}{\sqrt{3}}$

17. $\cos\theta = \frac{1}{\sec\theta} = \frac{1}{\sqrt{1+\tan^2\theta}} = \frac{1}{\sqrt{1+c^2}}$

19. $\sin\theta = \frac{12}{13}$ and $\tan\theta = \frac{12}{5}$

21. $\sin\theta = \frac{2}{\sqrt{53}}$ and $\sec\theta = \frac{\sqrt{53}}{7}$

23. $\cos 2\theta = 23/25$

25. $\cos\theta = -\frac{\sqrt{21}}{5}$ and $\tan\theta = -\frac{2\sqrt{21}}{21}$

27. Figure 23(A):
Point in first quadrant: $\sin\theta = 0.918$, $\cos\theta = 0.3965$, and $\tan\theta = \frac{0.918}{0.3965} = 2.3153$.
Point in second quadrant: $\sin\theta = 0.3965$, $\cos\theta = -0.918$, and $\tan\theta = \frac{0.3965}{-0.918} = -0.4319$.
Point in third quadrant: $\sin\theta = -0.918$, $\cos\theta = -0.3965$, and $\tan\theta = \frac{-0.918}{-0.3965} = 2.3153$.
Point in fourth quadrant: $\sin\theta = -0.3965$, $\cos\theta = 0.918$, and $\tan\theta = \frac{-0.3965}{0.918} = -0.4319$.

Figure 23(B):
Point in first quadrant: $\sin\theta = 0.918$, $\cos\theta = 0.3965$, and $\tan\theta = \frac{0.918}{0.3965} = 2.3153$.
Point in second quadrant: $\sin\theta = 0.918$, $\cos\theta = -0.3965$, and $\tan\theta = \frac{0.918}{0.3965} = -2.3153$.
Point in third quadrant: $\sin\theta = -0.918$, $\cos\theta = -0.3965$, and $\tan\theta = \frac{-0.918}{-0.3965} = 2.3153$.
Point in fourth quadrant: $\sin\theta = -0.918$, $\cos\theta = 0.3965$, and $\tan\theta = \frac{-0.918}{0.3965} = -2.3153$.

29. $\cos\psi = 0.3$, $\sin\psi = \sqrt{0.91}$, $\cot\psi = \frac{0.3}{\sqrt{0.91}}$, and $\csc\psi = \frac{1}{\sqrt{0.91}}$

31.   **33.**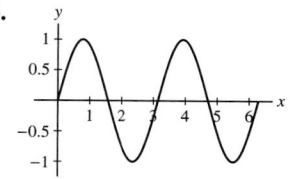

35. If $|c| > 1$, the horizontal line and the graph never intersect. If $|c| = 1$, then they intersect once. If $|c| < 1$, they intersect twice.

37. $\theta = 0, \frac{2\pi}{5}, \frac{4\pi}{5}, \pi, \frac{6\pi}{5}, \frac{8\pi}{5}$ **39.** $\theta = \frac{\pi}{6}, \frac{\pi}{2}, \frac{5\pi}{6}, \frac{7\pi}{6}, \frac{3\pi}{2}, \frac{11\pi}{6}$

41. Starting from the double angle formula for cosine, $\cos^2\theta = \frac{1}{2}(1 + \cos 2\theta)$, solve for $\cos 2\theta$.

43. Substitute $x = \theta/2$ into the double angle formula for sine, $\sin^2 x = \frac{1}{2}(1 - \cos 2x)$, and then take the square root of both sides.

45. $\cos(\theta + \pi) = \cos\theta\cos\pi - \sin\theta\sin\pi = \cos\theta(-1) = -\cos\theta$

47. $\tan(\pi - \theta) = \frac{\sin(\pi - \theta)}{\cos(\pi - \theta)} = \frac{-\sin(-\theta)}{-\cos(-\theta)} = \frac{\sin\theta}{-\cos\theta} = -\tan\theta$

49. $\frac{\sin 2x}{1 + \cos 2x} = \frac{2\sin x\cos x}{1 + 2\cos^2 x - 1} = \frac{2\sin x\cos x}{2\cos^2 x} = \frac{\sin x}{\cos x} = \tan x$

53. 16.928

Section 1.5 Preliminary

1. (a) Yes **(b)** Yes **(c)** No **(d)** No **(e)** No **(f)** Yes

2. No

3. The function is not one-to-one because many teenagers have the same last name.

4. Yes; $f^{-1}(6:27) =$ Hamilton Township.

5. The graph of the inverse is the reflection of the graph of $y = f(x)$ through the line $y = x$.

6. (a) $-\pi/6$ **(b)** Undefined **(c)** Undefined **(d)** $\pi/6$

7. Any angle $\theta < 0$ or $\theta > \pi$ will work. No, this does not contradict the definition of inverse function.

Section 1.5 Exercises

1. $f^{-1}(x) = \frac{1}{7}(x+4)$ **3.** $[-\pi/2, \pi/2]$

5. $f(g(x)) = \left((x-3)^{1/3}\right)^3 + 3 = x - 3 + 3 = x$
  $g(f(x)) = (x^3 + 3 - 3)^{1/3} = (x^3)^{1/3} = x$

7. $v^{-1}(R) = \frac{2GM}{R^2}$

9. One-to-one for all real numbers; $f^{-1}(x) = 4 - x$

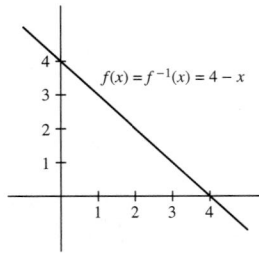

11. One-to-one on its entire domain, $\{x : x \neq \frac{3}{7}\}$; $f^{-1}(x) = \frac{1}{7x} + \frac{3}{7}$

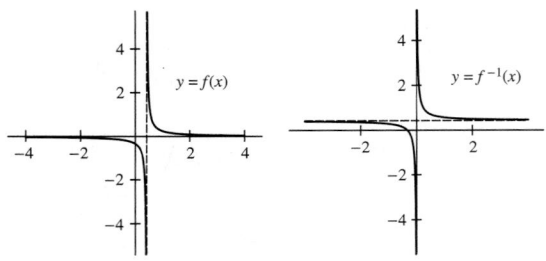

13. If we restrict the domain to $\{x : x \geq 0\}$, then $f^{-1}(x) = \frac{\sqrt{1-x^2}}{x}$;
If we restrict the domain to $\{x : x \leq 0\}$, then $f^{-1}(x) = -\frac{\sqrt{1-x^2}}{x}$

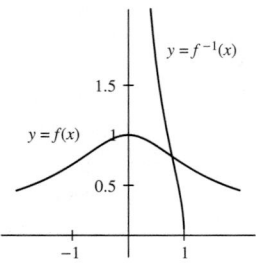

15. One-to-one on its entire domain, $\{x : x \geq -9^{1/3}\}$; $f^{-1}(x) = (x^2 - 9)^{1/3}$

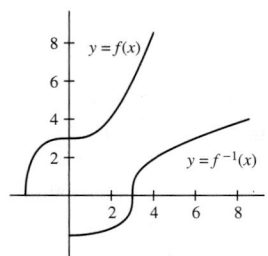

17. Figures (B) and (C) would not change when reflected around the line $y = x$. Therefore, these two satisfy $f^{-1} = f$.

19. **(a)** $f'(x) = 7x^6 + 1 > 0$ for all x. Thus, $f(x)$ is a strictly increasing function and by Example 3, it is one-to-one. Because f is one-to-one, by Theorem 1, f^{-1} exists.
(b) The domain of $f^{-1}(x)$ is the range of $f(x)$: $(-\infty, \infty)$.
(c) Note that $f(1) = 1^7 + 1 + 1 = 3$; therefore, $f^{-1}(3) = 1$.

21. If the domain of f is restricted to $x \leq 1$, then $f^{-1}(x) = 1 - \sqrt{x-1}$; if the domain of f is restricted to $x \geq 1$, then $f^{-1}(x) = 1 + \sqrt{x-1}$.

23. 0 **25.** $\frac{\pi}{4}$ **27.** $\frac{\pi}{3}$ **29.** $\frac{\pi}{3}$ **31.** $\frac{\pi}{2}$ **33.** $-\frac{\pi}{4}$ **35.** π

37. Undefined **39.** $\frac{\sqrt{1-x^2}}{x}$ **41.** $\frac{1}{\sqrt{x^2-1}}$ **43.** $\frac{\sqrt{5}}{3}$ **45.** $\frac{4}{3}$

47. $\sqrt{3}$ **49.** $\frac{1}{20}$

Section 1.6 Preliminary

1. **(a)** Correct **(b)** Correct **(c)** Incorrect **(d)** Correct
2. 2 **3.** $0 < x < 1$ **4.** Not defined
5. The phrase "The logarithm converts multiplication into addition" is a verbal translation of the property $\log(ab) = \log a + \log b$
6. Domain: $x > 0$; range: all real numbers
7. $\cosh x$ and $\operatorname{sech} x$ **8.** $\sinh x$ and $\tanh x$
9. Parity, identities, and derivative formulas

Section 1.6 Exercises

1. **(a)** 1 **(b)** 29 **(c)** 1 **(d)** 81 **(e)** 16 **(f)** 0
3. $x = 1$ **5.** $x = -1/2$ **7.** $x = -1/3$ **9.** $k = 9$
11. 3 **13.** $\frac{5}{3}$ **15.** $\frac{1}{3}$ **17.** $\frac{5}{6}$ **19.** 1 **21.** 7
23. **(a)** $\ln 1{,}600$ **(b)** $\ln 9x^{7/2}$
25. $t = \frac{1}{5}\ln\left(\frac{100}{7}\right)$ **27.** $x = -1, x = 3$ **29.** $x = e$

31.

	-3	0	5
$\sinh x$	-10.0179	0	74.203
$\cosh x$	10.0677	1	74.210

33. $\sinh x$ is increasing for all x; $\cosh x$ is decreasing for $x < 0$ and increasing for $x > 0$.
35. $t = \frac{\ln 2}{0.06} \approx 11.55$ yrs
37. **(a)** $E = 10^{4.8+1.5M}$ **(c)** $\frac{dE}{dM} = 1.5\ln 10 \cdot 10^{4.8+1.5M}$
39. $\cosh x = \frac{\sqrt{41}}{5}$, $\tanh x = \frac{4\sqrt{41}}{41}$
41. $\sinh(2x) = \sinh x \cosh x + \cosh x \sinh x = 2 \sinh x \cosh x$
  $\cosh(2x) = \cosh x \cosh x + \sinh x \sinh x = \cosh^2 x + \sinh^2 x$
43. $\log_a b \cdot \log_b a = \frac{\ln b}{\ln a} \cdot \frac{\ln a}{\ln b} = 1$

Section 1.7 Preliminary

1. No

2. (a) The screen will show nothing.

(b) The screen will show the portion of the parabola between the points (0, 3) and (1, 4).

3. No

4. Experiment with the viewing window to zoom in on the lowest point on the graph of the function. The y-coordinate of the lowest point on the graph is the minimum value of the function.

Section 1.7 Exercises

1.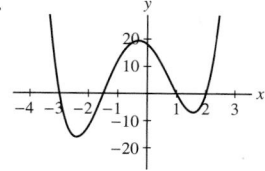
$x = -3, x = -1.5,$
$x = 1,$ and $x = 2$

3. Two positive solutions   **5.** There are no solutions.

7. The display will show nothing. An appropriate viewing rectangle: $[-50, 150]$ by $[1,000, 2,000]$

9.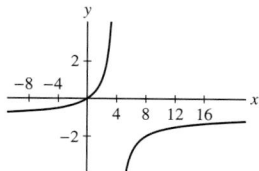
Asymptotes are at: $x = 4, y = -1.$

11.

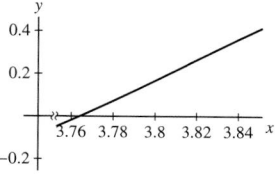

13. The following table and graphs suggest that as n gets large, $n^{1/n}$ approaches 1.

n	$n^{1/n}$
10	1.258925412
10^2	1.047128548
10^3	1.006931669
10^4	1.000921458
10^5	1.000115136
10^6	1.000013816

  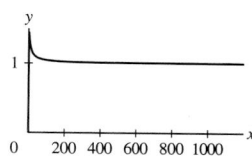

15. The following table and graphs suggest that as n gets large, $f(n)$ tends toward ∞.

n	$\left(1 + \frac{1}{n}\right)^{n^2}$
10	13780.61234
10^2	$1.635828711 \times 10^{43}$
10^3	$1.195306603 \times 10^{434}$
10^4	$5.341783312 \times 10^{4342}$
10^5	$1.702333054 \times 10^{43429}$
10^6	$1.839738749 \times 10^{434294}$

  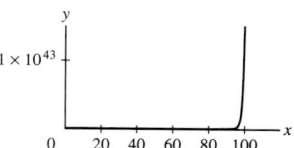

17. The following table and graphs suggest that as x gets large, $f(x)$ approaches 1.

x	$\left(x \tan \frac{1}{x}\right)^x$
10	1.033975759
10^2	1.003338973
10^3	1.000333389
10^4	1.000033334
10^5	1.000003333
10^6	1.000000333

19.

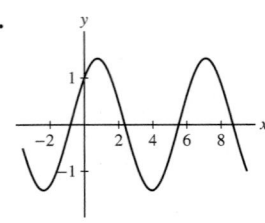

(A, B) = (1, 1)   (A, B) = (1, 2)

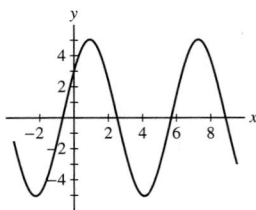

(A, B) = (3, 4)

21. $x \in (-2, 0) \cup (3, \infty)$

23. $f_3(x) = \frac{x^2+6x+1}{4(x+1)}$, $f_4(x) = \frac{x^4+28x^3+70x^2+28x+1}{8(1+x)(1+6x+x^2)}$, and $f_5(x) =$
$\frac{1+120x+1,820x^2+8,008x^3+12,870x^4+8,008x^5+1,820x^6+120x^7+x^8}{16(1+x)(1+6x+x^2)(1+28x+70x^2+28x^3+x^4)}$.
It seems as if the f_n are asymptotic to $\sqrt{x}$.

Chapter 1 Review

1. $\{x : |x - 7| < 3\}$ **3.** $\{x : 2 \le |x - 1| \le 6\} = [-5, -1] \cup [3, 7]$
5. $(x, 0)$ with $x \ge 0$. $(0, y)$ with $y < 0$
7.

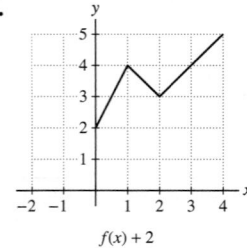

f(x) + 2   f(x + 2)

9.

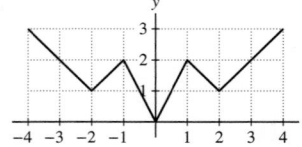

11. $D : \{x : x \ge -1\}$, $R : \{y : y \ge 0\}$
13. $D : \{x : x \ne 3\}$, $R : \{y : y \ne 0\}$
15. (a) Decreasing **(b)** Neither **(c)** Neither **(d)** Increasing
17. $2x - 3y = -14$ **19.** $6x - y = 53$ **21.** $x + y = 5$ **23.** Yes

25. Roots: $x = -2$, $x = 0$, and $x = 2$. Function is decreasing when $x < -1.4$ or $0 < x < 1.4$.

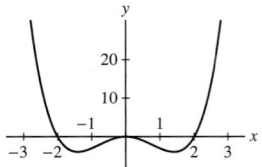

27. $f(x) = 10x^2 + 2x + 5$. Its minimum value is $\frac{49}{10}$.
29.   **31.**

33.

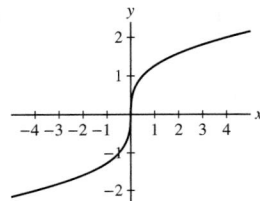

35. Let $g(x) = f\left(\frac{1}{3}x\right)$. Then

$$g(x - 3b) = f\left(\frac{1}{3}(x - 3b)\right) = f\left(\frac{1}{3}x - b\right).$$

The graph of $y = |\frac{1}{3}x - 4|$:

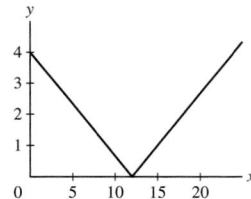

37. $f(t) = t^4$, $g(t) = 12t + 9$ **39.** 4π
41. (a) $a = b = \pi/2$ **(b)** $a = \pi$
43. $x = \pi/2$, $x = 7\pi/6$, $x = 3\pi/2$, and $x = 11\pi/6$
45. There are no solutions.
47. (a) No match **(b)** No match **(c) (i):** $(2^a)^b = 2^{ab}$ **(d) (iii):**
$2^{a-b}3^{b-a} = 2^{a-b}\left(\frac{1}{3}\right)^{a-b} = \left(\frac{2}{3}\right)^{a-b}$
49. $f^{-1}(x) = \sqrt[3]{x + 8}$; domain and range are all real numbers
51. If we restrict the domain to $\{t : t \ge 3\}$, then $h^{-1}(t) = 3 + \sqrt{t}$; if we restrict the domain to $\{t : t \le 3\}$, then $h^{-1}(t) = 3 - \sqrt{t}$
53. (a) (iii) (b) (iv) (c) (ii) (d) (i)

Chapter 2

Section 2.1 Preliminary

1. The ratio of distance traveled to time elapsed.

2. On the graph of position as a function of time.

3. No, it is defined as the limit of average velocity as time elapsed shrinks to zero.

4. The slope of the line tangent to the graph of position as a function of time at $t = t_0$.

5. The slope of the secant line over the interval $[x_0, x_1]$ approaches the slope of the tangent line at $x = x_0$.

6. The graph of atmospheric temperature as a function of altitude. Possible units for this rate of change are °F/ft or °C/m.

Section 2.1 Exercises

1. (a) $\Delta s = 36$ ft   (b) $\frac{\Delta s}{\Delta t} = 72$ ft/s

(c)

Time interval	[2, 2.01]	[2, 2.005]	[2, 2.001]	[2, 2.00001]
Average velocity	64.16	64.08	64.016	64.00016

The instantaneous velocity is 64 ft/s.

3. The instantaneous rate of change is approximately 0.57735 m/(s · K).

5. $\frac{\Delta h}{\Delta t} = 0.3$

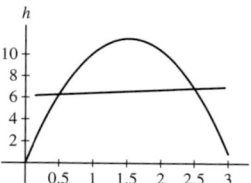

7. (a) Dollars/year

(b)

Time interval	[0, 0.5]	[0, 1]
Average rate of change	7.8461	8

(c) The rate of change at $t = 0.5$ is approximately \$8/yr.

9. 16   **11.** −0.062   **13.** 1   **15.** 0.3333

17. The ROC is a constant −.00356 degrees per foot since T is a linear function of h with slope −0.00356.

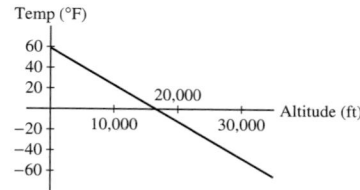

19. (a) ROC = 3,000 cells/hour

(b) $m = \frac{4,000}{3}$ cells/hour, which represents the instantaneous ROC at $t = 1$ hour.

21. (a) Seconds per meter; measures the amount by which the period changes when the length is changed

(b) The slope of the line B represents the average rate of change in T from $L = 1$ m to $L = 3$ m. The slope of the line A represents the instantaneous rate of change of T at $L = 3$ m.

(c) 0.4330 s/m

23. (a) D   (b) B   (c) A   (d) C

25. (a) The slope of line A is the average rate of change over the interval [4, 6], whereas the slope of the line B is the instantaneous rate of change at $t = 6$. Thus, the slope of the line $A \approx (0.28 - 0.19)/2 = 0.045$, whereas the slope of the line $B \approx (0.28 - 0.15)/6 = 0.0217$.

(b) $t = 3$ (c) $t = 4$

27.

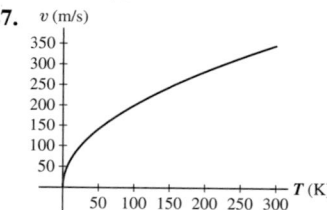

As the graph progresses to the right, the graph bends progressively downward, meaning that the ROC of v with respect to T is lower at high temperatures.

29.

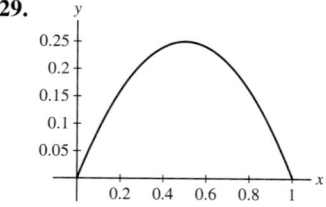

(a) 0   (b) 0   (c) $0 < x < 0.5$

31. (a) $h(1) = 48$ ft;
$h(t) - h(1) = -16t^2 + 64t - 48 = -16(t - 1)(t - 3)$
(b) $\frac{h(t)-h(1)}{t-1} = \frac{-16(t-1)(t-3)}{t-1} = -16(t - 3)$ (c) 32 ft/s

33. $\frac{f(x)-f(1)}{x-1} = \frac{x^3-1}{x-1} = \frac{(x-1)(x^2+x+1)}{x-1} = x^2 + x + 1$

At $x = 1$, the instantaneous ROC is $1^2 + 1 + 1 = 3$.

Section 2.2 Preliminary

1. $\lim\limits_{x \to \pi} 1 = 1$ **2.** $\lim\limits_{t \to \pi} t = \pi$ **3.** Yes, $\lim\limits_{x \to 1} \frac{x^2-1}{x-1}$

4. $\lim\limits_{x\to 10} 20 = 20$ **5.** $\lim\limits_{x\to 1-} f(x) = \infty$ and $\lim\limits_{x\to 1+} f(x) = 3$

6. No, to determine whether $\lim\limits_{x\to 5} f(x)$ exists, we must examine values of $f(x)$ on both sides of $x = 5$.

7. Yes   **8.** The information in (a), (b), or (c) would be sufficient.

Section 2.2 Exercises

1.

x	0.998	0.999	0.9995	0.99999
$f(x)$	1.498501	1.499250	1.499625	1.499993

x	1.00001	1.0005	1.001	1.002
$f(x)$	1.500008	1.500375	1.500750	1.501500

$\lim\limits_{x\to 1} f(x) = \frac{3}{2}$

3.

y	1.998	1.999	1.9999	2.0001	2.001	2.02
$f(y)$	0.59984	0.59992	0.599992	0.600008	0.60008	0.601594

$\lim\limits_{y\to 2} f(y) = \frac{3}{5}$

5.

x	-0.5	-0.1	-0.05	-0.01
$f(x)$	0.426123	0.483742	0.491770	0.498337

x	0.01	0.05	0.1	0.5
$f(x)$	0.501671	0.508439	0.517092	0.594885

$\lim\limits_{x\to 1} f(x) = \frac{1}{2}$

7. $\lim\limits_{x\to 0.5} f(x) = 1.5$ **9.** $\lim\limits_{x\to 21} f(x) = 21$ **21.** $\lim\limits_{x\to 1} f(x) = \frac{1}{2}$

23. $\lim\limits_{x\to 2} f(x) = \frac{5}{3}$ **25.** $\lim\limits_{x\to 0} f(x) = 2$

27. The limit does not exist.   **29.** The limit does not exist.

31. $\lim\limits_{h\to 0} f(h) = 0.693$. (The exact answer is ln 2.)

33. $\lim\limits_{x\to 1+} f(x) = 1.414$. (The exact answer is $\sqrt{2}$.)

35. The limit does not exist.

37. $\lim\limits_{x\to 2-} f(x) = 2$, $\lim\limits_{x\to 2+} f(x) = 1$

39. (a) The one-sided limits exist for all real values of c.
(b) $\lim\limits_{x\to c} [x]$ exists when c is a noninteger.

41. $\lim\limits_{x\to 0-} f(x) = -1$, $\lim\limits_{x\to 0+} f(x) = 1$

43. $\lim\limits_{x\to 0-} f(x) = \infty$, $\lim\limits_{x\to 0+} f(x) = \frac{1}{6}$

45. $\lim\limits_{x\to 2-} f(x) = \infty$, $\lim\limits_{x\to 2+} f(x) = \infty$, $\lim\limits_{x\to 4-} f(x) = -\infty$, and $\lim\limits_{x\to 4+} f(x) = 10$

47.

49.

51.

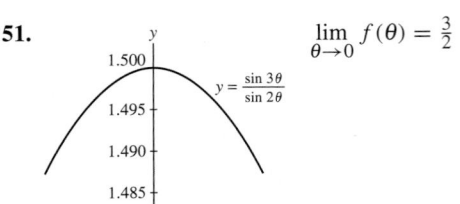

$\lim\limits_{\theta\to 0} f(\theta) = \frac{3}{2}$

53.

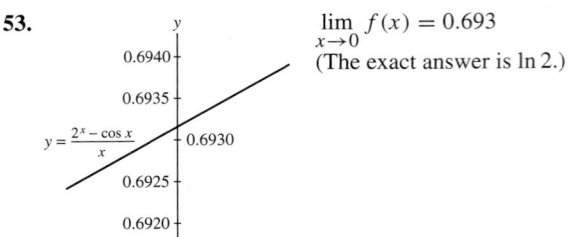

$\lim\limits_{x\to 0} f(x) = 0.693$
(The exact answer is ln 2.)

55.

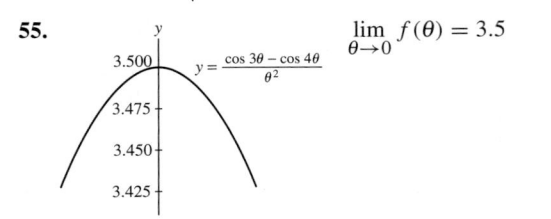

$\lim\limits_{\theta\to 0} f(\theta) = 3.5$

57. If $R = 2$,

θ	-0.01	-0.005	0.005	0.01
$I(\theta)$	0.998667 I_m	0.9999667 I_m	0.9999667 I_m	0.9998667 I_m

If $R = 3$, the table becomes:

θ	-0.01	-0.005	0.005	0.01
$I(\theta)$	0.999700 I_m	0.999925 I_m	0.999925 I_m	0.999700 I_m

59.

x	-0.1	-0.01	-0.001	0.001	0.01	0.1
$\frac{5^x-1}{x}$	1.486601	1.596556	1.608144	1.610734	1.622459	1.746189

Note ln 5 ≈ 1.6094.

x	-0.1	-0.01	-0.001	0.001	0.01	0.1
$\frac{3^x-1}{x}$	1.040415	1.092600	1.098009	1.099216	1.104669	1.161232

Note ln 3 ≈ 1.0986.

61. For $k = 1$, the limit is 0; for $k = 2$, the limit is 1; for odd $k > 2$, the limit does not exist; for even $k > 2$, the limit is ∞.

63. (a)

x	-0.3	-0.2	-0.1	0.1	0.2	0.3
$f(x)$	-0.980506	-0.998049	-0.999998	0.999998	0.998049	0.980506

(b) As $x \to 0-$, $f(x) \to -1$, whereas as $x \to 0+$, $f(x) \to 1$.

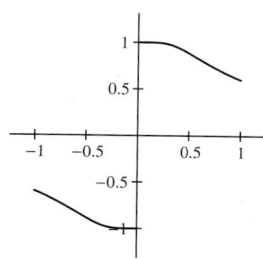

Section 2.3 Preliminary

1. Suppose $\lim\limits_{x \to c} f(x)$ and $\lim\limits_{x \to c} g(x)$ both exist. The Sum Law states that $\lim\limits_{x \to c}(f(x) + g(x)) = \lim\limits_{x \to c} f(x) + \lim\limits_{x \to c} g(x)$. Provided $\lim\limits_{x \to c} g(x) \neq 0$, the Quotient Law states that $\lim\limits_{x \to c} \frac{f(x)}{g(x)} = \frac{\lim\limits_{x \to c} f(x)}{\lim\limits_{x \to c} g(x)}$.

2. (b)   **3.** (a) and (d)   **4.** (a) and (c)

Section 2.3 Exercises

1. $\lim\limits_{x \to 9} x = 9$ **3.** $\lim\limits_{x \to 9} 14 = 14$ **5.** $\lim\limits_{x \to -3}(3x + 4) = -5$

7. $\lim\limits_{y \to -3}(y + 14) = 11$ **9.** $\lim\limits_{t \to 4}(3t - 14) = -2$

11. $\lim\limits_{x \to \frac{1}{2}}(4x + 1)(2x - 1) = 0$ **13.** $\lim\limits_{x \to 2} x(x + 1)(x + 2) = 24$

15. $\lim\limits_{t \to 9} \frac{t}{t+1} = \frac{9}{10}$ **17.** $\lim\limits_{x \to 3} \frac{1-x}{1+x} = -\frac{1}{2}$ **19.** $\lim\limits_{t \to 2} t^{-1} = \frac{1}{2}$

21. $\lim\limits_{x \to 3}(x^2 + 9x^{-3}) = \frac{28}{3}$

23. $\lim\limits_{x \to c}\left(\frac{1}{f(x)}\right) = \frac{\left(\lim\limits_{x \to c} 1\right)}{\left(\lim\limits_{x \to c} f(x)\right)} = \frac{1}{\lim\limits_{x \to c} f(x)}$

25. $\lim\limits_{x \to -4} f(x)g(x) = 3$ **27.** $\lim\limits_{x \to -4} \frac{g(x)}{x^2} = \frac{1}{16}$

29. No, because $\lim\limits_{x \to 0} x = 0$.

31. $f(x) = 1/x$, $g(x) = -1/x$

35. $\lim\limits_{x \to c} f(x) = \lim\limits_{x \to c} \frac{f(x)g(x)}{g(x)} = \frac{\lim\limits_{x \to c} f(x)g(x)}{\lim\limits_{x \to c} g(x)} = \frac{L}{M}$

37. $\lim\limits_{t \to 3} h(t) = \lim\limits_{t \to 3} t \cdot \frac{h(t)}{t} = \left(\lim\limits_{t \to 3} t\right)\left(\lim\limits_{t \to 3} \frac{h(t)}{t}\right) = 3 \cdot 5 = 15$

41. $\lim\limits_{h \to 0} \frac{f(ah)}{h} = \lim\limits_{h \to 0}\left(a \cdot \frac{f(ah)}{ah}\right) = a \lim\limits_{h \to 0} \frac{f(ah)}{ah} = aL$, using the result from the previous exercise.

43. **(b)** is the correct law; $\lim\limits_{x \to 2} \sin(g(x)) = \lim\limits_{x \to \pi/6} \sin x = \frac{1}{2}$

Section 2.4 Preliminary

1. We can conclude that $\lim\limits_{x \to 2} x^3 = 8$ because x^3 is continuous at $x = 2$.

2. $f(3) = \frac{1}{2}$ **3.** f cannot be continuous at $x = 0$.   **4.** No

5. **(a)** False: "$f(x)$ is continuous at $x = a$ if the left- and right-hand limits of $f(x)$ as $x \to a$ exist and equal $f(a)$."

(b) True

(c) False: "If the left- and right-hand limits of $f(x)$ as $x \to a$ are equal but not equal to $f(a)$, then f has a removable discontinuity at $x = a$."

(d) True

(e) False: "If $f(x)$ and $g(x)$ are continuous at $x = a$ and $g(a) \neq 0$, then $f(x)/g(x)$ is continuous at $x = a$."

Section 2.4 Exercises

1. Left-continuous at $x = 1$; neither left-continuous nor right-continuous at $x = 3$; left-continuous at $x = 5$.

3. $c = 3$; $f(c) = f(3) = 4.5$ makes f continuous at $x = 3$.

5. **(a)** $x = 0$; $\lim\limits_{x \to 0-} f(x) = \infty$; $\lim\limits_{x \to 0+} f(x) = 2$
$x = 2$; $\lim\limits_{x \to 2-} f(x) = 6$; $\lim\limits_{x \to 2+} f(x) = 6$

(b) The discontinuity at $x = 2$ is removable; $f(2) = 6$.

7. x and $\sin x$ are continuous, so $x + \sin x$ is continuous by Continuity Law (i).

9. x and $\sin x$ are continuous, so $3x$ and $4 \sin x$ are continuous by Continuity Law (iii). Thus $3x + 4 \sin x$ is continuous by Continuity Law (i).

11. x is continuous, so x^2 is continuous by Continuity Law (ii). Constant functions are continuous, so $x^2 + 1$ is continuous. Finally, $\frac{1}{x^2+1}$ is continuous by Continuity Law (iv) because $x^2 + 1$ is never 0.

13. The functions 3^x, 1 and 4^x are each continuous. Therefore, $1 + 4^x$ is continuous by Continuity Law (i). Because $1 + 4^x$ is never zero, it follows that $\frac{3^x}{1+4^x}$ is continuous by Continuity Law (iv).

15. e^x and $\cos 3x$ are continuous, so $e^x \cos 3x$ is continuous by Continuity Law (ii).

17. Infinite discontinuity at $x = 0$.

19. Infinite discontinuity at $x = 1$.

21. Jump discontinuities (right-continuous and not left-continuous) at $x = n$ for every integer n.

23. Infinite discontinuities at $t = -1$ and $t = 1$.

25. Right-continuous at $x = 0$ (not defined for $x < 0$).

27. Infinite discontinuities at $z = -2$ and $z = 3$.

29. Jump discontinuity at $x = 2$.

31. Infinite discontinuities at $x = \pm\sqrt{n\pi}$ where n is a positive integer.

33. Continuous everywhere

35. Infinite discontinuity at $x = 0$

37. Continuous on $|x| \leq 3$. Not defined elsewhere.

39. Continuous on $x \geq 0$. Not defined elsewhere.

41. Continuous for all real numbers.

43. Continuous on $x \neq 0$. Not defined at $x = 0$.

45. Continuous on $x \neq \pm(2n - 1)\pi/2$ where n is a positive integer. Not defined elsewhere.

47. Continuous for all real numbers.

49. Continuous on $x \neq \pm 1$. Not defined elsewhere.

51. $f(0) = \lim\limits_{x \to 0-} f(x) = -4$; $f(0) = \lim\limits_{x \to 0+} f(x) = 2$

53.

55.

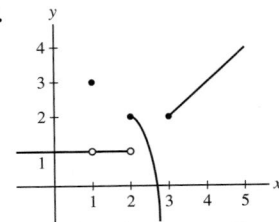

59. $\lim\limits_{x\to 5} x^2 = 25$ **61.** $\lim\limits_{x\to -1} (2x^3 - 4) = -6$

63. $\lim\limits_{x\to 0} \frac{x+9}{x-9} = -1$ **65.** $\lim\limits_{x\to \pi} \sin\left(\frac{x}{2} - \pi\right) = -1$

67. $\lim\limits_{x\to \frac{\pi}{4}} \tan(3x) = -1$ **69.** $\lim\limits_{x\to 4} x^{-5/2} = \frac{1}{32}$

71. $\lim\limits_{x\to -1} (1 - 8x^3)^{3/2} = 27$ **73.** $\lim\limits_{x\to 3} 10^{x^2 - 2x} = 1{,}000$

75. $\lim\limits_{x\to 1} e^{x^2 - x} = 1$ **77.** $\lim\limits_{x\to 4^-} \sin^{-1}\left(\frac{x}{4}\right) = \frac{\pi}{2}$

79.

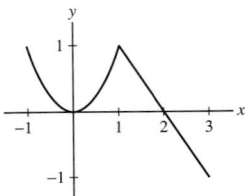

The function f is continuous everywhere.

81.

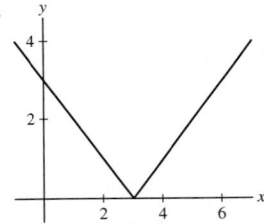

The function f is continuous everywhere.

83. $c = \frac{5}{3}$ **85.** $a - b = \frac{8 - 2\sqrt{2}}{\pi}$

87.

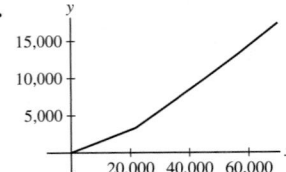

89. Answers may vary. $f(x) = C$ and $g(x)$ is defined for all x.

Section 2.5 Preliminary

1. $\frac{x^2 - 1}{\sqrt{x+3} - 2}$

2. (a) Let $f(x) = \frac{x^2 - 1}{x - 1}$. At $x = 1$, f is indeterminate of the form $\frac{0}{0}$ but

$$\lim_{x\to 1^-} \frac{x^2 - 1}{x - 1} = \lim_{x\to 1^-} (x + 1) = 2 = \lim_{x\to 1^+} (x + 1) = \lim_{x\to 1^+} \frac{x^2 - 1}{x - 1}.$$

(b) Again, let $f(x) = \frac{x^2 - 1}{x - 1}$. Then

$$\lim_{x\to 1} f(x) = \lim_{x\to 1} \frac{x^2 - 1}{x - 1} = \lim_{x\to 1} (x + 1) = 2$$

but $f(1)$ is indeterminate of the form $\frac{0}{0}$.

(c) Let $f(x) = \frac{1}{x}$. Then f is undefined at $x = 0$ but does not have an indeterminate form at $x = 0$.

3. The "simplify and plug-in" strategy is based on simplifying a function which is indeterminate to a continuous function. Once the simplification has been made, the limit of the remaining continuous function is obtained by evaluation.

Section 2.5 Exercises

1. $\lim\limits_{x\to 5} \frac{x^2 - 25}{x - 5} = \lim\limits_{x\to 5} (x + 5) = 10$ **3.** $\lim\limits_{t\to 7} \frac{2t - 14}{5t - 35} = \lim\limits_{t\to 7} \frac{2}{5} = \frac{2}{5}$

5. $\lim\limits_{x\to 8} \frac{x^2 - 64}{x - 8} = 16$ **7.** $\lim\limits_{x\to 2} \frac{x^2 - 3x + 2}{x - 2} = 1$ **9.** $\lim\limits_{x\to 2} \frac{x - 2}{x^3 - 4x} = \frac{1}{8}$

11. $\lim\limits_{h\to 0} \frac{(1+h)^3 - 1}{h} = 3$ **13.** $\lim\limits_{x\to 2} \frac{3x^2 - 4x - 4}{2x^2 - 8} = 1$

15. $\lim\limits_{y\to 2} \frac{(y-2)^3}{y^3 - 5y + 2} = 0$ **17.** $\lim\limits_{h\to 0} \frac{\frac{1}{3+h} - \frac{1}{3}}{h} = -\frac{1}{9}$

19. The limit does not exist.

21. $\lim\limits_{x\to 2} \frac{x - 2}{\sqrt{x} - \sqrt{4 - x}} = \sqrt{2}$ **23.** $\lim\limits_{x\to 2} \frac{\sqrt{x^2 - 1} - \sqrt{x + 1}}{x - 3} = 0$

25. $\lim\limits_{x\to 4} \left(\frac{1}{\sqrt{x} - 2} - \frac{4}{x - 4}\right) = \frac{1}{4}$ **27.** $\lim\limits_{x\to 0} \frac{\cot x}{\csc x} = 1$

29. $\lim\limits_{x\to \frac{\pi}{4}} \frac{\sin x - \cos x}{\tan x - 1} = \frac{\sqrt{2}}{2}$ **31.** $\lim\limits_{x\to 0} \frac{e^x - e^{2x}}{1 - e^x} = 1$

33. $\lim\limits_{x\to \frac{\pi}{3}} \frac{2\cos^2 x + 3\cos x - 2}{2\cos x - 1} = \frac{5}{2}$

35. $\lim\limits_{x\to 2} f(x) \approx 1.41 \approx \sqrt{2}$

37. $\lim\limits_{x\to 1} \frac{x^3 - 1}{x - 1} = 3$ **39.** $\lim\limits_{x\to 1} \frac{x^2 - 3x + 2}{x^3 - 1} = -\frac{1}{3}$

41. $\lim\limits_{x\to 1} \frac{x^4 - 1}{x^3 - 1} = \frac{4}{3}$ **43.** $\lim\limits_{x\to 0} (2a + x) = 2a$

45. $\lim\limits_{t\to -1} (4t - 2at + 3a) = -4 + 5a$ **47.** $\lim\limits_{x\to 0} \frac{2(x+h)^2 - 2x^2}{h} = 2h$

49. $\lim\limits_{x\to a} \frac{\sqrt{x} - \sqrt{a}}{x - a} = \frac{1}{2\sqrt{a}}$ **51.** $\lim\limits_{x\to 0} \frac{(x+a)^3 - a^3}{x} = 3a^2$

53. $c = -1, c = 6$ **55.** +

Section 2.6 Preliminary

1. $\lim\limits_{x\to 0} f(x) = 0$; because $\lim\limits_{x\to 1/2} -x^4 = -\frac{1}{16} \neq \frac{1}{4} = \lim\limits_{x\to 1/2} x^2$,
we do not have enough information to determine $\lim\limits_{x\to 1/2} f(x)$.

2. Assume that for $x \neq c$ (in some open interval containing c),
$l(x) \leq f(x) \leq u(x)$ and that $\lim\limits_{x\to c} l(x) = \lim\limits_{x\to c} u(x) = L$. Then
$\lim\limits_{x\to c} f(x)$ exists and $\lim\limits_{x\to c} f(x) = L$.

3. Yes **4.** (a)

Section 2.6 Exercises

1. *Squeezed* at $x = 3$ and *trapped* at $x = 2$.

3. The Squeeze Theorem guarantees that $\lim\limits_{x\to 7} f(x) = 6$ provided
$f(x)$ is squeezed in an open interval containing $x = 7$.

5. $\lim\limits_{x\to 0} x \cos\frac{1}{x} = 0$ **7.** $\lim\limits_{x\to 0+} \sqrt{x}e^{\cos(\pi/x)} = 0$

9. $\lim\limits_{x\to 0} \frac{\sin x \cos x}{x} = 1$ **11.** $\lim\limits_{t\to 0} \frac{\sin^2 t}{t} = 0$ **13.** $\lim\limits_{x\to 0} \frac{x^2}{\sin^2 x} = 1$

15. $\lim\limits_{t\to \frac{\pi}{4}} \frac{\sin t}{t} = \frac{2\sqrt{2}}{\pi}$

17. $\lim\limits_{x\to 0} \frac{\sin 10x}{x} = \lim\limits_{\theta\to 0} \frac{\sin\theta}{(\theta/10)} = \lim\limits_{\theta\to 0} \cdot 10\frac{\sin\theta}{\theta}$
$= \lim\limits_{\theta\to 0} 10 \cdot \frac{\sin\theta}{\theta} = 10$

19. $\lim\limits_{h\to 0} \frac{\sin 6h}{h} = 6$ **21.** $\lim\limits_{h\to 0} \frac{\sin 6h}{6h} = 1$ **23.** $\lim\limits_{x\to 0} \frac{\sin 7x}{3x} = \frac{7}{3}$

25. $\lim\limits_{x\to 0} \frac{\tan 4x}{9x} = \frac{4}{9}$ **27.** $\lim\limits_{t\to 0} \frac{\tan 4t}{t \sec t} = 4$ **29.** $\lim\limits_{z\to 0} \frac{\sin(z/3)}{\sin z} = \frac{1}{3}$

31. $\lim\limits_{x\to 0} \frac{\tan 4x}{\tan 9x} = \frac{4}{9}$ **33.** $\lim\limits_{x\to 0} \frac{\sin 5x \sin 2x}{\sin 3x \sin 5x} = \frac{2}{3}$

35. $\lim\limits_{h\to 0} \frac{1-\cos 2h}{h} = 0$ **37.** $\lim\limits_{t\to 0} \frac{1-\cos t}{\sin t} = 0$

39. $\lim\limits_{h\to \frac{\pi}{2}} \frac{1-\cos 3h}{h} = \frac{2}{\pi}$

41. (a) $\lim\limits_{x\to 0+} \frac{\sin x}{|x|} = 1$ (b) $\lim\limits_{x\to 0-} \frac{\sin x}{|x|} = -1$

43. Since $\left|\cos\left(\frac{1}{x}\right)\right| \leq 1$, it follows that $\left|\tan x \cos\left(\frac{1}{x}\right)\right| \leq |\tan x|$,
which is equivalent to $-|\tan x| \leq \tan x \cos\left(\frac{1}{x}\right) \leq |\tan x|$;
$\lim\limits_{x\to 0} \tan x \cos\left(\frac{1}{x}\right) = 0$

45.

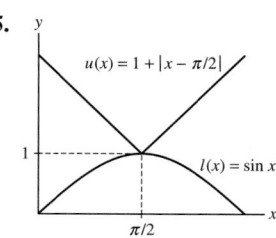

$u(x) = 1 + |x - \pi/2|$

$l(x) = \sin x$

Any function $f(x)$ satisfying $l(x) \leq f(x) \leq u(x)$ for all x near
$\pi/2$ will satisfy $\lim\limits_{x\to \pi/2} f(x) = 1$.

47.

h	-0.1	-0.01	0.01	0.1
$\frac{1-\cos h}{h^2}$	0.499583	0.499996	0.499996	0.499583

The limit is $\frac{1}{2}$.

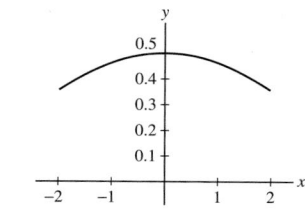

$$\lim\limits_{h\to 0} \frac{1-\cos h}{h^2} = \lim\limits_{h\to 0} \frac{1-\cos^2 h}{h^2(1+\cos h)}$$
$$= \lim\limits_{h\to 0} \left(\frac{\sin h}{h}\right)^2 \frac{1}{1+\cos h} = \frac{1}{2}$$

49. $\lim\limits_{h\to 0} \frac{\cos 3h-1}{\cos 2h-1} = \frac{9}{4}$

53. (a) The area of each triangular piece of the n-gon is $\frac{1}{2}\sin\frac{2\pi}{n}$, so
the area of the n-gon is $\frac{1}{2}n\sin\frac{2\pi}{n}$.
(b) As n increases, the difference between the n-gon and the unit
circle shrinks to zero; hence, as n increases, $A(n)$ approaches the area
of the unit circle.
(c) $\lim\limits_{n\to\infty} A(n) = \pi$

Section 2.7 Preliminary

1. $f(x) = x^2$ is continuous on $[0, 1]$ with $f(0) = 0$ and $f(1) = 1$.
Because $f(0) < 0.5 < f(1)$, the Intermediate Value Theorem
guarantees there is a $c \in [0, 1]$ such that $f(c) = 0.5$.

2. We must assume that temperature is a continuous function of
time.

3. If f is continuous on $[a, b]$, then the horizontal line $y = k$ for
every k between $f(a)$ and $f(b)$ intersects the graph of $y = f(x)$ at
least once.

4.

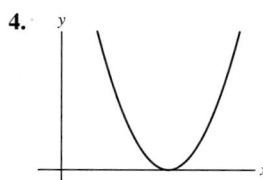

Section 2.7 Exercises

1. f is continuous everywhere with $f(1) = 2$ and $f(2) = 10$.

3. g is continuous for all $t \in [0, \frac{\pi}{4}]$ with $g(0) = 0$ and
$g(\frac{\pi}{4}) = \frac{\pi^2}{16} > \frac{1}{2}$.

5. $f(x) = x - \cos x$ is continuous everywhere with $f(0) = -1$ and
$f(1) = 1 - \cos 1 \approx 0.46$.

7. $f(x) = \sqrt{x} + \sqrt{x+1} - 2$ is continuous on $\left[\frac{1}{4}, 1\right]$ with
$f(\frac{1}{4}) = \sqrt{\frac{1}{4}} + \sqrt{\frac{5}{4}} - 2 \approx -0.38$ and $f(1) = \sqrt{2} - 1 \approx 0.41$.

11. $f(x) = x^k - \cos x$ is continuous on $[0, \frac{\pi}{2}]$ with $f(0) = -1$ and
$f(\frac{\pi}{2}) = (\frac{\pi}{2})^k > 0$.

15. $f(x) = e^x + \ln x$ is continuous on $[0.1, 1]$ with $f(0.1) < 0$ and $f(1) = e > 0$.

17. Note that $f(x)$ is continuous for all x.
(a) $f(1) = 1$, $f(1.5) = 2^{1.5} - (1.5)^3 < 3 - 3.375 < 0$. Hence, $f(x) = 0$ for some x between 1 and 1.5.
(b) $f(1.25) \approx 0.4253 > 0$ and $f(1.5) < 0$. Hence, $f(x) = 0$ for some x between 1.25 and 1.5.
(c) $f(1.375) \approx -0.0059$. Hence, $f(x) = 0$ for some x between 1.25 and 1.375.

19. $[0, 0.25]$

21.

23.
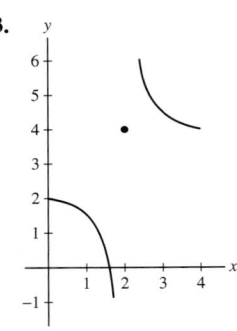

25. If $[a, b]$ contains $[-1, 1]$, $f(a) > 0$ and $f(b) > 0$. If we were using the corollary, we would guess that $f(x)$ has no roots in $[a, b]$.

Section 2.8 Preliminary

1. (c)   **2.** (b) and (d)

Section 2.8 Exercises

1. (a) $|f(x) - 35| = |8x + 3 - 35| = |8x - 32| = |8(x - 4)|$
$= 8|x - 4|$.
(b) Given $\epsilon > 0$, let $\delta = \epsilon/8$ and suppose $|x - 4| < \delta$. By part **(a)**, $|f(x) - 35| = 8|x - 4| < 8\delta$. Substituting $\delta = \epsilon/8$, we see $|f(x) - 35| < 8\epsilon/8 = \epsilon$.

3. (a) If $0 < |x - 2| < \delta = 0.01$, then $|x| < 3$ and
$|x^2 - 4| = |x - 2||x + 2| \le |x - 2|(|x| + 2) < 5|x - 2| < 0.05$
(b) If $0 < |x - 2| < \delta = 0.0002$, then $|x| < 2.0002$ and
$|x^2 - 4| \le |x - 2|(|x| + 2) < 4.0002|x - 2| < 0.00080004 < 0.0009$
(c) $\delta = 10^{-5}$

5. $\delta = 6 \times 10^{-4}$

7. $\delta = 0.005$

9. For $\epsilon = 0.5$, take $\delta = 0.18$; for $\epsilon = 0.2$, take $\delta = 0.075$; for $\epsilon = 0.1$, take $\delta = 0.035$

11. (b) $\delta = 0.0002$

13. $c = 3$, $L = 10.1$, $\epsilon = 0.3$, $\delta = 0.1$

15. Given $\epsilon > 0$, let $\delta = \epsilon/|a|$. Then, whenever $|x - c| < \delta$,
$$|f(x) - (ac + b)| = |a||x - c| < |a| \cdot \epsilon/|a| = \epsilon.$$

Chapter 2 Review

1. Average velocity: $\frac{\sqrt{26} - \sqrt{5}}{3} \approx 0.954$ m/s; instantaneous velocity: 0.894 m/s.

3. The tangent line drawn in the figure appears to pass through the points $(15, 140)$ and $(10.5, 40)$; therefore, ROC $\approx \frac{200}{9}$.

5. $\lim_{x \to 0} \frac{1 - \cos^3(x)}{x^2} \approx 1.50$ **7.** $\lim_{x \to 2} \frac{x^x - 4}{x^2 - 4} \approx 1.69$

9. $\lim_{x \to 4} \left(3 + x^{\frac{1}{2}}\right) = 5$ **11.** $\lim_{x \to 2} \frac{4}{x^3} = -\frac{1}{2}$ **13.** $\lim_{x \to -1} \frac{x^3 - x}{x - 1} = 2$

15. $\lim_{t \to 9} \frac{\sqrt{t} - 3}{t - 9} = \frac{1}{6}$ **17.** $\lim_{x \to 3} \frac{\sqrt{x+1} - 2}{x - 3} = \frac{1}{4}$

19. $\lim_{h \to 0} \frac{2(a+h)^2 - 2a^2}{h} = 4a$ **21.** The two-sided limit does not exist.

23. $\lim_{a \to b} \frac{a^2 - 3ab + 2b^2}{a - b} = -b$ **25.** $\lim_{x \to 1} \frac{x^3 - ax^2 + ax - 1}{x - 1} = 3 - a$

27. The limit does not exist.   **29.** $\lim_{x \to 4.3} \frac{1}{x - [x]} = \frac{10}{3}$

31. $\lim_{x \to 0+} \frac{[x]}{x} = 0$ **33.** The limit does not exist.

35. $\lim_{\theta \to 0} \frac{\sin 5\theta}{\theta} = 5$ **37.** The limit does not exist.

39. The limit does not exist.   **41.** $\lim_{x \to 0} \frac{\sin 7x}{\sin 3x} = \frac{7}{3}$

43. $\lim_{\theta \to 0} \frac{\tan \theta - \sin \theta}{\sin^3 \theta} = \frac{1}{2}$

45.

47.

49. Because e^x and $\sin x$ are continuous for all real numbers, their composition, $e^{\sin x}$ is continuous for all real numbers. Moreover, x is continuous for all real numbers, so the product $xe^{\sin x}$ is continuous for all real numbers. Thus, $f(x) = xe^{\sin x}$ is continuous for all real numbers.

51. (a) $\lim\limits_{x\to3}(f(x) - 2g(x)) = -2$ (b) $\lim\limits_{x\to3} x^2 f(x) = 54$

(c) $\lim\limits_{x\to3}\frac{f(x)}{g(x)+x} = \frac{6}{7}$ (d) $\lim\limits_{x\to3}(2g(x)^3 - g(x)^2) = 112$

53. Let $f(x) = \frac{1}{(x-a)^3}$ and $g(x) = \frac{1}{(x-a)^5}$. Then, neither A nor B exists but $L = 0$.

57. Let $r_1 = \frac{20}{1+\sqrt{1-20a}}$ and $r_2 = \frac{20}{1-\sqrt{1-20a}}$.

59. $\delta = 0.55$

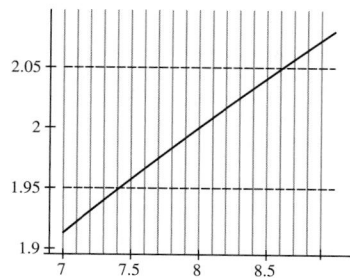

61. Given $\epsilon > 0$, let $\delta = \min\{1, \epsilon/6\}$. Because $\delta \le 1$, $|x - 3| < \delta$ guarantees $|x + 2| < 6$. Then, whenever $|x - 3| < \delta$,

$$|f(x) - 6| = |x - 3|\,|x + 2| < 6|x - 3| < 6\delta \le \epsilon.$$

63. $f(x)$ is continuous on $[0, 2]$ with $f(0) = -\frac{2}{3} < 0$ and $f(2) \approx 4.21 > 0$

65. 2.65

Chapter 3

Section 3.1 Preliminary

1. A and D **2.** $\frac{f(x)-f(a)}{x-a}$ or $\frac{f(a+h)-f(a)}{h}$ **3.** $x = 7$

4. (a) The difference in height between the points $(0.9, \sin 0.9)$ and $(1.3, \sin 1.3)$.

(b) The slope of the secant line between the points $(0.9, \sin 0.9)$ and $(1.3, \sin 1.3)$ on the graph.

(c) The slope of the tangent line to the graph at $x = 1.3$.

5. $a = 3$ and $h = 2$ **6.** $f(x) = \tan x$ at $x = \frac{\pi}{4}$

Section 3.1 Exercises

1. $f(2 + h) = 3(2 + h)^2 = 3(4 + 4h + h^2) = 12 + 12h + 3h^2$;
$\frac{f(2+h)-f(2)}{h} = \frac{12+12h+3h^2-12}{h} = \frac{12h+3h^2}{h} = 12 + 3h$;
$f'(2) = 12$

3. $f'(0) = 9$ **5.** $f'(-1) = -2$

7. $f(0) \approx 6$ and $f(2.5) \approx 2$, so the slope is -1.6.

9. For $h = 0.5$, the estimate is 2 [greater than $f'(2)$]; for $h = -0.5$, the estimate is 1 [smaller than $f'(2)$].

11. $f'(1) = f'(2) = 0$; $f'(4) = \frac{1}{2}$; $f'(7) = 0$

13. $f'(5.5)$ is larger than $f'(6.5)$. **15.** $f'(x) = 3$ for all x

17. $f'(t) = -1$ for all t

19. $f(-2 + h) = \frac{1}{-2+h}$. The difference quotient is $-\frac{1}{3}$.

21. $f'(9) = \frac{1}{6}$ **23.** $f'(2) = 14$; $y = 14x - 12$

25. $f'(2) = 12$; $y = 12x - 16$ **27.** $f'(0) = 1$; $y = x$

29. $f'(3) = -\frac{1}{9}$; $y = \left(\frac{1}{3}\right) - \left(\frac{1}{9}\right)(x - 3)$

31. $f'(-7) = 9$; $y = 9x - 4$ **33.** $f'(-2) = -1$; $y = -x - 1$

35. $f'(-1) = \frac{1}{2}$; $y = \frac{x}{2} + \frac{3}{2}$ **37.** $f'(1) = -3$; $y = 4 - 3t$

39. $f'(9) = -\frac{1}{54}$; $y = \left(-\frac{x}{54}\right) + \left(\frac{1}{2}\right)$ **41.** $y = 5 + 2(x - 3)$

43. The tangent at any point is $y = 2x + 8$.

45. $f(1) = \frac{1}{1+1^2} = \frac{1}{2}$; $L(1) = \frac{1}{2} + m(1 - 1) = \frac{1}{2}$;

$f'(1) = -\frac{1}{2}$

47.

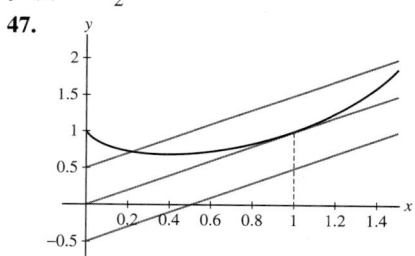

(a) The graph of $y = x$ appears to be tangent to the graph of $f(x)$ at $x = 1$. We therefore estimate that $f'(1) = 1$.

(b) With $x_0 = 1$, we generate the table

h	0.01	0.001	0.0001
$(f(x_0 + h) - f(x_0))/h$	1.010050	1.001001	1.000100

49. (a) $P'(350)$ is larger than $P'(300)$.

(b) $P'(303) \approx 0.00265$ atm/K;
$P'(313) \approx 0.004145$ atm/K;
$P'(323) \approx 0.006295$ atm/K;
$P'(333) \approx 0.00931$ atm/K;
$P'(343) \approx 0.013435$ atm/K.

51. $V'(120) \approx 9.75$ **53.** $f(x) = x^3$ and $a = 5$

55. $f(x) = \sin x$ and $a = \frac{\pi}{6}$ **57.** $f(x) = 5^x$ and $a = 2$

59. $f'\left(\frac{\pi}{2}\right) = 0$

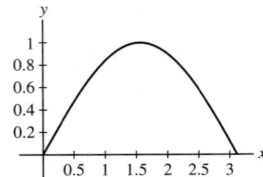

h	-0.01	-0.001	-0.0001	0.0001	0.001	0.01
$\frac{\sin\left(\frac{\pi}{2}+h\right)-1}{h}$	0.005	0.0005	0.00005	-0.00005	-0.0005	-0.005

61. (a) $f'(0) \approx -0.68$

 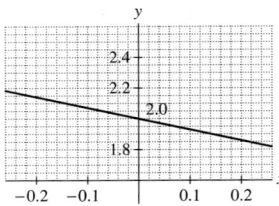

(b) $y = 2 - 0.68x$

63. $f'\left(\frac{\pi}{2}\right) \approx -1$

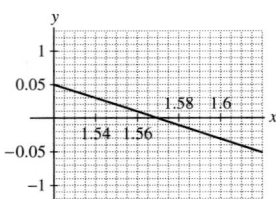

65. $f'\left(\frac{\pi}{4}\right) \approx 0.7071$

67. (b) $f'(4) \approx 20.0000$

(c) $y = 20x - 48$

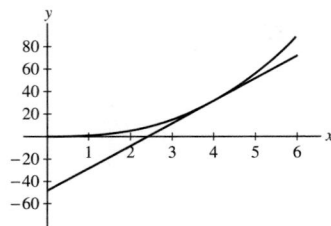

69. $W'(4) \approx 0.9$ kg/yr; horizontal tangent: $t = 10$ and $t = 11.6$; negative slope: $10 < t < 11.6$

71. $v'(10) \approx 7.9125$; $i(10) = 0.495485$

77. Figure (A) satisfies the inequality.

Section 3.2 Preliminary

1. The slope is 8.

2. $(f - g)'(1) = -2$; $(3f + 2g)'(1) = 19$; can't evaluate $(fg)'(1)$

3. (a) Yes **(b)** No **(c)** Yes **(d)** No **(e)** No **(f)** Yes

4. $x^n - a^n = (x - a)(x^{n-1} + x^{n-2}a + x^{n-3}a^2 + \cdots + xa^{n-2} + a^{n-1})$

5. Yes **6.** Vertical tangent

7. The line tangent to $f(x) = e^x$ at $x = 0$ has slope equal to 1.

8. 1; 7

Section 3.2 Exercises

1. $f'(x) = 4$ **3.** $f'(x) = -4x$ **5.** $f'(x) = -\frac{1}{x^2}$

7. $f'(x) = \frac{1}{2\sqrt{x}}$ **9.** $\frac{d}{dx}x^4\big|_{x=-2} = -32$ **11.** $\frac{d}{dt}t^{2/3}\big|_{t=8} = \frac{1}{3}$

13. $\frac{d}{dx}x^{0.35} = 0.35x^{-0.65}$ **15.** $\frac{d}{dt}t^{\sqrt{17}} = \sqrt{17}t^{\sqrt{17}-1}$

17. $f'(1) = 5$; $y = 5(x - 1) + 1$

19. $f'(9) = \frac{17}{2}$; $y = \frac{17}{2}x + \frac{9}{2}$

21. (a) $\frac{d}{dx}9e^x = 9e^x$ **(b)** $\frac{d}{dt}(3t - 4e^t) = 3 - 4e^t$

(c) $\frac{d}{dt}e^{t+2} = e^{t+2}$

23. $\frac{d}{dx}(x^3 + x^2 - 12) = 3x^2 + 2x$

25. $\frac{d}{dx}(2x^3 - 10x^{-1}) = 6x^2 + 10x^{-2}$

27. $\frac{d}{dz}(7z^{-3} + z^2 + 5) = 2z - 21z^{-4}$

29. $f'(s) = \frac{1}{4}s^{-3/4} + \frac{1}{3}s^{-2/3}$

31. $\frac{d}{dx}((x + 1)^3) = 3x^2 + 6x + 3$

33. $\frac{d}{dz}((3z - 1)(2z + 1)) = 12z + 1$

35. $\frac{d}{dx}e^2 = 0$ **37.** $\frac{d}{dt}5e^{t-3} = 5e^{t-3}$

39. $f'(2) = -\frac{3}{8}$ **41.** $\frac{dT}{dC}\big|_{C=8} = 1$

43. $\frac{ds}{dz}\big|_{z=2} = -60$ **45.** $\frac{dp}{dh}\big|_{h=4} = 7e^2$

47. Graph (A) matches the graph in (III).
Graph (B) matches the graph in (I).
Graph (C) matches the graph in (II).
Graph (D) matches the graph in (III).
The graph in (D) is just a vertical translation of the graph in (A) which means the two functions differ by a constant. The derivative of a constant is zero, so the two functions end up with the same derivative.

49.

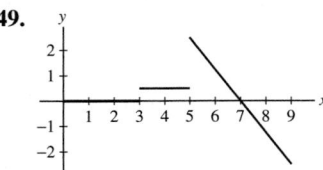

51. (a) $\frac{d}{dR}R = 1$ **(b)** $\frac{d}{dR}r = 0$ **(c)** $\frac{d}{dR}r^2R^3 = 3r^2R^2$
53. $x = \frac{1}{2}$ **55.** $x = 0$ or $x = \frac{3}{4}$ **57.** $a = 2$ and $b = -3$
59. The two parallel tangent lines with slope $m = 2$ are shown with the graph of $f(x)$ here:

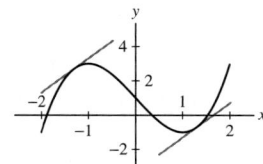

61. $f'(x) = -\frac{2}{x^3}$ **63.** $m_4 \approx 1.386$; $y'(0) \approx 1.386$; $y'(2) \approx 22.176$
65. $\frac{d}{dT}v_{avg}\big|_{T=300°K} = 0.74234$
67. $\frac{dK}{dm}\big|_{m=68} = 0.00315966$; $\frac{d}{dm}\left(\frac{K}{m}\right)\big|_{m=68} = -8.19981 \times 10^{-6}$
69. The area is 2 regardless.
71. Graph (A) matches (III). Graph (B) matches (I). Graph (C) matches (II).
73.

75. $g(x)$ is the derivative of $f(x)$.

77. $x = 1$ **79.** $x = 0$ **81.** $x = -1$; $x = 1$
83. f is not differentiable at $x = 0$. The tangent line does not exist.

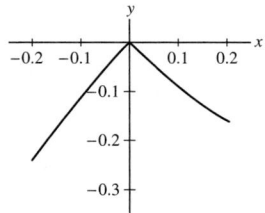

85. f is not differentiable at $x = 3$. The tangent line appears to be vertical.

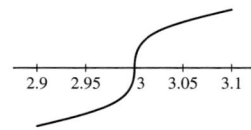

87. f is not differentiable at $x = 0$. The tangent line does not exist at this point.

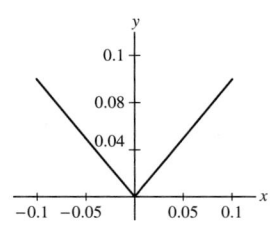

89. This statement is false. The function $|x|$ is not differentiable at $x = 0$, but it is continuous.
91. tangent line: $y = 2ax - a^2$; x-intercept: $\frac{a}{2}$

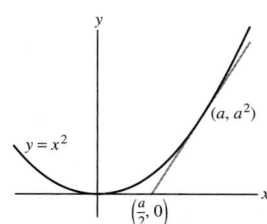

93.

The tangent line to f at $x = a$ is

$$y = \left(4 - a^{2/3}\right)^{3/2} - \frac{\sqrt{4-a^{2/3}}(x-a)}{a^{1/3}}$$

The y-intercept of this line is the point $P = \left(0, 4\sqrt{4 - a^{2/3}}\right)$, its x-intercept is the point $Q = \left(4a^{1/3}, 0\right)$, and the distance between P and Q is 8.
95. Radius $\leq 1/2$ **97.** $\frac{d}{dx}(x^k) = kx^{k-1}$ for negative integers k.
101. (a) The equation has a unique solution when $\lambda < 0$ and when $\lambda = e$.
(b) The equation has at least one solution when $\lambda < 0$ and when $\lambda \geq e$.

Section 3.3 Preliminary

1. (a) False: The notation fg denotes the function whose value at x is $f(x)g(x)$.
(b) True **(c)** False: The derivative of the product fg is $f'(x)g(x) + f(x)g'(x)$.
2. (a) False: $\frac{d}{dx}(fg)\big|_{x=4} = f(4)g'(4) + g(4)f'(4)$.
(b) False: $\frac{d}{dx}(f/g)\big|_{x=4} = [g(4)f'(4) - f(4)g'(4)]/g(4)^2$.
(c) True
3. $\frac{d}{dx}(f/g)\big|_{x=1} = -1$ **4.** $g(1) = 5$

Section 3.3 Exercises

1. $f'(x) = 3x^2 + 1$ 3. $f'(x) = e^x(x^3 + 3x^2 - 1)$

5. $\left.\frac{dy}{dt}\right|_{t=3} = 82$ 7. $f'(x) = \frac{-2}{(x-2)^2}$ 9. $\left.\frac{dg}{dt}\right|_{t=-2} = \frac{8}{9}$

11. $g'(x) = -\frac{e^x}{(1+e^x)^2}$ 13. $f'(t) = 6t^2 + 2t - 4$

15. $g'(x) = 2x + 2 - 6x^{-3}$

17. $f'(x) = 6x^5 + 5x^4 + 4x^3 - 8x - 4$ 19. $\left.\frac{dy}{dx}\right|_{x=2} = -\frac{1}{36}$

21. $f'(x) = 1$ 23. $\left.\frac{dy}{dx}\right|_{x=2} = -80$ 25. $\left.\frac{dz}{dx}\right|_{x=1} = -\frac{3}{4}$

27. $h'(t) = -\frac{10t^{11}+8t^9+3t^4+t^2}{(t^{11}+t^9+t^4+t^2)^2}$ 29. $f'(t) = 0$

31. $f'(x) = 3x^2 - 6x - 13$ 33. $f'(x) = \frac{e^x(x-e^x)}{(e^x+1)^2(x+1)^2}$

35. $g'(z) = 2z - 1$ 37. $\frac{d}{dt}\left(\frac{xt-4}{t^2-x}\right) = \frac{-xt^2+8t-x^2}{(t^2-x)^2}$

39. $f'(x) > 0$ for $-1 < x < 1$, $f'(x) < 0$ for $|x| > 1$. $f'(x)$ is plotted on the left, $f(x)$ on the right.

 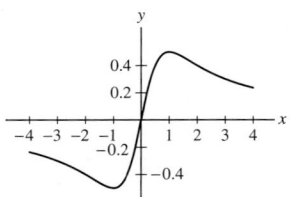

41. $\frac{dP}{dr} = -\frac{2V^2R}{(R+r)^3}$

43. (a) $\left.\frac{dI}{dR}\right|_{R=6} = -\frac{2}{3}$ (b) $\left.\frac{dV}{dR}\right|_{R=6} = 4$

45. At $x = -1$, the tangent line is $y = \frac{1}{2}x + 1$; at $x = 1$, the tangent line is $y = 1 - \frac{1}{2}x$.

47. $(f^2)' = (ff)' = ff' + ff' = 2ff'$

49. $(fg)'(4) = -17$; $\left(\frac{f}{g}\right)'(4) = -\frac{13}{25}$

51. $G'(4) = -58$ 53. $F'(0) = -7$

55. $(fgh)' = f(gh' + hg') + ghf' = fgh' + fg'h + f'gh$

59. Let $g(x) = 1/f(x)$. Then by the quotient rule,

$$g'(x) = \frac{f(x)\cdot 0 - f'(x)}{f^2(x)} = -\frac{f'(x)}{f^2(x)}.$$

63. (a) $c = -1$ is a multiple root
(b) $c = -1$ is a root, but not a multiple root

Section 3.4 Preliminary

1. (a) atmospheres/meter   (b) moles/(liter · hour)
2. $f(5) = 13$ 3. 90 mph   4. $f(26) \approx 43.75$
5. (a) Rate of change of the population of Freedonia in the year 1933.
(b) If $P'(1933) = 0.2$, then $P(1934) \approx 5.2$ million. If $P'(1933) = 0$, then $P(1934) \approx 5$ million.

Section 3.4 Exercises

1. When $s = 3$, the area changes at a rate of 6 square units per unit increase. When $s = 5$, the area changes at a rate of 10 square units per unit increase.
3. If $x = 1$, the ROC is -1. If $x = 10$, the ROC is $-\frac{1}{100}$.
5. $dV/dr = 3\pi r^2$ 7. $dV/dr = 4\pi r^2$
9. (a) Average velocity over the interval $[0.5, 1]$ is 50 km/h.
(b) Average velocity is greater over $[2, 3]$.   (c) $t = 2.5$ h
11. The tangent line sketched in the figure below appears to pass through the points $(10, 3)$ and $(30, 4)$. Thus, the ROC of voltage at $t = 20$ seconds is approximately

$$\frac{4-3}{30-10} = 0.05 \text{ V/s}.$$

As we move to the right of the graph, the tangent lines to it grow shallower, indicating that the voltage changes more slowly as time goes on.

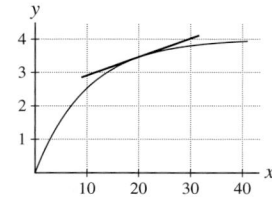

13. (a) 35.25 ft   (b) -7 ft/s   (c) ≈ 2.36 s
15. -15 degrees Fahrenheit per minute
17. $F'(6.77 \times 10^6) = -1.93 \times 10^{-4}$ N/m
19. $P'(3) = -0.1312$ W/Ω; $P'(5) = -0.0609$ W/Ω
21. (a) $\frac{dV}{dv} = -1$ (b) $\frac{dV}{dt} = -4$
23. (a) 1.0618 million; 10,600 people per year   (b) Year 2010
27. (a)

t	Jan	Feb	Mar	Apr	May	Jun	Jul	Aug	Sep	Oct	Nov
$P'(t)$	146	155	175	191	210	227	252	242	233	219	185

(b) Here is a plot of these estimates ($1 = $ Jan, $2 = $ Feb, etc.)

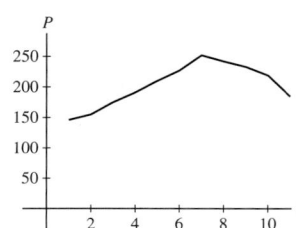

(c) "U.S. Growth Rate Declines After Midsummer Peak"
29. Denote $f(w) = \sqrt{hw}/60$ so that: $f'(w) = \frac{\sqrt{5}}{20\sqrt{w}}$;
$f'(70) = \frac{\sqrt{14}}{280} \approx 0.0133631 \frac{m^2}{kg}$; $f'(80) = \frac{1}{80}\frac{m^2}{kg}$
31. Falling at 76.68 m/s
33. $v_0 = 19.6$ m/s. Peak height is 19.6 m.

35. Either the 16th or 17th floor.

39. Approximation: $\sqrt{2} - \sqrt{1} \approx \frac{1}{2}$

Value up to six decimal places: $\sqrt{2} - \sqrt{1} = 0.414214$

Approximation: $\sqrt{101} - \sqrt{100} \approx 0.05$

Value up to six decimal places: $\sqrt{101} - \sqrt{100} = 0.0498756$

41. $F(65) = 198.25$ ft

Approximation: $F(66) - F(65) \approx F'(65) = 5$ ft

Actual value: $F(66) - F(65) = 5.03$ ft

43. $C(2000) = \$796$

Approximation: $C(2001) - C(2000) \approx C'(2000) = \$0.244$

Actual value: $C(2001) - C(2000) = \$0.244$

45. $D(40) = 22.5$ barrels per year. An increase in oil prices of a dollar leads to a decrease in demand of 0.5625 barrels a year, and a decrease of a dollar leads to an increase in demand of 0.5625 barrels a year.

49. (a) 7.09306 km/yr **(b)** 0.75992 kg

51. (a) Growing when $P = 250$; Shrinking when $P = 350$

(b) dP/dt **(c)** Graph (A)

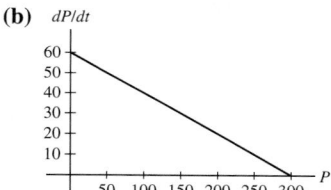

53. At point A, average cost is greater than marginal cost.
At point B, average cost is greater than marginal cost.
At point C, average cost and marginal cost are nearly the same.
At point D, average cost is less than marginal cost.

Section 3.5 Preliminary

1. The second derivative of GNP is negative.
The first derivative is positive.

2. The first derivative of stock prices is postive.
The second derivative is negative.

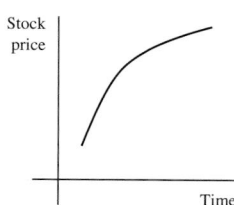

3. True **4.** Polynomials of the form $ax + b$. **5.** e^x

Section 3.5 Exercises

1. $y'' = 28$ and $y''' = 0$ **3.** $y'' = 12x^2 - 50$ and $y''' = 24x$

5. $y'' = 8\pi r$ and $y''' = 8\pi$

7. $y'' = -\frac{16}{5}t^{-6/5} + \frac{4}{3}t^{-4/3}$ and $y''' = \frac{96}{25}t^{-11/5} - \frac{16}{9}t^{-7/3}$

9. $y'' = -2z^{-3}$ and $y''' = 6z^{-4}$

11. $y'' = 20x^3 + 12x^2 + 2$ and $y''' = 60x^2 + 24x$

13. $y'' = e^x$ and $y''' = e^x$

15. $f^{(4)}(1) = 24$ **17.** $\left.\frac{d^2 y}{dt^2}\right|_{t=1} = 54$ **19.** $h'''(9) = \frac{1}{648}$

21. $\left.\frac{d^4 x}{dt^4}\right|_{t=16} = \frac{3,465}{134,217,728}$ **23.** $g''(1) = -\frac{1}{4}$ **25.** $h''(1) = \frac{1}{8}$

27. $f''(0) = -\frac{1}{2}$

29. $y^{(0)}(0) = d$, $y^{(1)}(0) = c$, $y^{(2)}(0) = 2b$, $y^{(3)}(0) = 6a$, $y^{(4)}(0) = 24$, and $y^{(5)}(0) = 0$

31. $\frac{d^6}{dx^6}x^{-1} = 720x^{-7}$ **33.** $f^{(n)}(x) = (-1)^n n!(x + 1)^{-(n+1)}$

35. $f^{(n)}(x) = (-1)^n \frac{(2n-1)\times(2n-3)\times\cdots\times 1}{2^n}x^{-(2n+1)/2}$

37. (a) $a(5) = -90$ ft/min^2

(b) As the acceleration of the helicopter is negative, the velocity of the helicopter must be decreasing. Because the velocity is positive, the helicopter is slowing down.

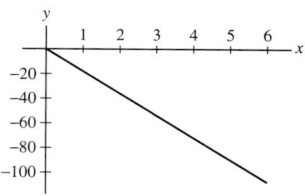

39. (A) f'' **(B)** f' **(C)** f

41. From time 15 to time 20 and from time 32 on.

43. $f(x) = x^2 - 2x$

45. (a) Does not apply. **(b)** Can apply. **(c)** Can apply.

47. $\frac{dS}{dQ} = -2882Q^{-2} - 0.052$; $\frac{d^2 S}{dQ^2} = 5764Q^{-3}$

49. (a) Set $\frac{d^2 s}{dt^2} = 0$ and solve for $v = \frac{ds}{dt}$.

(b) If $D = 0.003$ ft, then $v_{term} \approx 13.8564$ ft/s; if $D = 0.0003$ ft, then $v_{term} \approx 4.38178$ ft/s.

(c) Larger velocities correspond to lower acceleration.

51. $f'(x) = -\frac{3}{(x-1)^2} = (-1)^1 \frac{3\cdot 1}{(x-1)^{1+1}}$;

$f''(x) = \frac{6}{(x-1)^3} = (-1)^2 \frac{3\cdot 2\cdot 1}{(x-1)^{2+1}}$;

$f'''(x) = -\frac{18}{(x-1)^4} = (-1)^3 \frac{3\cdot 3!}{(x-1)^{3+1}}$; and

$f^{(4)}(x) = \frac{72}{(x-1)^5} = (-1)^4 \frac{3\cdot 4!}{(x-1)^{4+1}}$.

We conjecture the general formula for $f^{(k)}(x)$ is
$f^{(k)}(x) = (-1)^k \frac{3\cdot k!}{(x-1)^{k+1}}$.

53. $p^{(99)}(x) = 99!$

55. $(fg)''' = f'''g + 3f''g' + 3f'g'' + fg'''$; We conjecture the general formula for $(fg)^{(n)}$ is $(fg)^{(n)} = \sum_{k=0}^{n} \binom{n}{k} f^{(n-k)} g^{(k)}$.

57. $f'(x) = (x^2 + 2x)e^x$;

$f''(x) = (x^2 + 4x + 2)e^x$;

$f'''(x) = (x^2 + 6x + 6)e^x$; and

$f^{(4)}(x) = (x^2 + 8x + 12)e^x$;

We conjecture the general formula for $f^{(n)}(x)$ is
$f^{(n)}(x) = (x^2 + 2nx + n(n - 1))e^x$.

Section 3.6 Preliminary

1. (a) $\frac{d}{dx}(\sin x + \cos x) = -\sin x + \cos x$

(b) $\frac{d}{dx}\tan x = \sec^2 x$ (c) $\frac{d}{dx}\sec x = \sec x \tan x$

(d) $\frac{d}{dx}\cot x = -\csc^2 x$

2. (a) and (c) using the Product Rule.

3. $\frac{d}{dx}(\sin^2 x + \cos^2 x) = 0$

4. The difference quotient for the function $\sin x$ involves the expression $\sin(x + h)$. The addition formula for the sine function is used to expand this expression as $\sin(x + h) = \sin x \cos h + \sin h \cos x$.

Section 3.6 Exercises

1. $y - \frac{\sqrt{2}}{2} = \frac{\sqrt{2}}{2}\left(x - \frac{\pi}{4}\right)$ 3. $y - 1 = 2\left(x - \frac{\pi}{4}\right)$

5. $f'(x) = -\sin^2 x + \cos^2 x$ 7. $f'(x) = 2\sin x \cos x$

9. $f'(x) = x^3 \cos x + 3x^2 \sin x$

11. $f'(\theta) = (\tan^2 \theta + \sec^2 \theta)\sec \theta$

13. $h'(\theta) = -e^{-\theta}\cos \theta(\cos \theta + 2\sin \theta)$

15. $f'(x) = (x - x^2)(-\csc^2 x) + \cot x(1 - 2x)$

17. $f'(x) = \frac{x \sec x \tan x - 2 \sec x}{x^3}$

19. $g'(t) = \cos t - 2\sec t \tan t$

21. $f'(x) = \frac{2+\sin x - x \cos x}{(2+\sin x)^2}$ 23. $f'(x) = \frac{2\cos x}{(1-\sin x)^2}$

25. $g'(x) = \frac{(x \tan x - 1)\sec x}{x^2}$ 27. $f''(x) = -3\sin x - 4\cos x$

29. $g''(\theta) = 2\cos \theta - \theta \sin \theta$ 31. $y = x$

33. $y = \left(1 - \frac{3\sqrt{3}}{2}\right)x + \sqrt{3} + \frac{3}{2} + \frac{\sqrt{3}}{2}\pi - \frac{\pi}{3}$

35. $y = (2 - \sqrt{2})x + \sqrt{2} - 1 + \frac{\pi}{4}(\sqrt{2} - 2)$

37. $y = x + 1$

43. $f'(x) = -\sin x$, $f''(x) = -\cos x$, $f'''(x) = \sin x$, $f^{(4)}(x) = \cos x$, and $f^{(5)}(x) = -\sin x$; $f^{(8)}(x) = \cos x$; $f^{(37)}(x) = -\sin x$.

45. $f''(x) = 2\sec^2 x \tan x$
$f'''(x) = 2\sec^4 x + 4\sec^2 x \tan^2 x$

47. (a)
(b) $c_0 = 4.493409$

(c)

49. $v\left(\frac{\pi}{3}\right) = 20$ cm/s; $a\left(\frac{\pi}{3}\right) = -20\sqrt{3}$ cm/s^2

55. (a) $f'(x) = x \cos x + (\sin x) \cdot 1 = g(x) + \sin x$ and
$g'(x) = (x)(-\sin x) + (\cos x) \cdot 1 = -f(x) + \cos x$

(b) $f''(x) = g'(x) + \cos x = -f(x) + 2\cos x$ and
$g''(x) = -f'(x) - \sin x = -g(x) - 2\sin x$

Section 3.7 Preliminary

1. (a) Outer function is $\sqrt{x}$, inner function is $4x + 9x^2$.

(b) Outer function is $\tan x$, inner function is $x^2 + 1$.

(c) Outer function is x^5, inner function is $\sec x$.

(d) Outer function is x^4, inner function is $1 + e^x$.

2. The functions $\frac{x}{x+1}$, $\sqrt{x} \cdot \sec x$, and xe^x.

3. (b) $5f'(5x)$

4. (a) Once (b) Twice (c) Three times

(d) Twice

5. We do not have enough information—value of $f'(1)$ is missing.

Section 3.7 Exercises

1.

$f(g(x))$	$f'(u)$	$f'(g(x))$	$g'(x)$	$(f \circ g)'$
$(x^4 + 1)^{3/2}$	$\frac{3}{2}u^{1/2}$	$\frac{3}{2}(x^4 + 1)^{1/2}$	$4x^3$	$6x^3(x^4 + 1)^{1/2}$

3.

$f(g(x))$	$f'(u)$	$f'(g(x))$	$g'(x)$	$(f \circ g)'$
$\tan(x^4)$	$\sec^2 u$	$\sec^2(x^4)$	$4x^3$	$4x^3 \sec^2(x^4)$

5. $\frac{d}{dx}(x + \sin x)^4 = 4(x + \sin x)^3(1 + \cos x)$

7. (a) $\frac{d}{dx}\cos(9 - x^2) = 2x \sin(9 - x^2)$

(b) $\frac{d}{dx}\cos(x^{-1}) = \frac{\sin(x^{-1})}{x^2}$

(c) $\frac{d}{dx}\cos(\tan x) = -\sec^2 x \sin(\tan x)$

9. $\frac{d}{dx}(x^2 + 9)^4 = 4(x^2 + 9)^3(2x)$ 11. $\frac{d}{dx}\sqrt{11x + 4} = \frac{11}{2\sqrt{11x+4}}$

13. $\frac{d}{dx}e^{10-x^2} = -2xe^{10-x^2}$ 15. $\frac{d}{dx}\sin(2x + 1) = 2\cos(2x + 1)$

17. $\frac{d}{dx}e^{x+x^{-1}} = \left(1 - x^{-2}\right)e^{x+x^{-1}}$

19. $\frac{d}{dx}f(g(x)) = -\sin(x^2 + 1)(2x)$; $\frac{d}{dx}g(f(x)) = -2\sin x \cos x$

21. $y' = 2x \cos(x^2)$ 23. $y' = -8t \csc^2(4t^2 + 9)$

25. $y' = -\frac{5(2t+3)}{2(t^2+3t+1)^{7/2}}$ 27. $y' = \frac{8(1+x)^3}{(1-x)^5}$

29. $y' = -12e^{4\theta}\cos^2(e^{4\theta})\sin(e^{4\theta})$ 31. $y' = \frac{x \sin(x^2)}{(1+\cos(x^2))^{3/2}}$

33. $y' = -\frac{1}{x^2}e^{1/x}$ 35. $\frac{dy}{dx} = 5\sec^2(5x)$

37. $y' = 3x \sin(1 - 3x) + \cos(1 - 3x)$ 39. $\frac{dy}{dt} = 2(4t + 9)^{-1/2}$

41. $y' = 4(\sin x - 3x^2)(x^3 + \cos x)^{-5}$ 43. $y' = \frac{\sqrt{2}\cos 2x}{2\sqrt{\sin 2x}}$

45. $\frac{dy}{dz} = (z + 1)^3(2z - 1)^2(14z + 2)$

47. $y' = \frac{1}{2x^2\sqrt{x+1}}(3x^3 + 2x^2 - x - 2)$ 49. $y' = \frac{x\cos(x^2) - 3\sin 6x}{\sqrt{\cos 6x + \sin(x^2)}}$

51. $y' = 3(x^2 \sec^2(x^3) + \sec^2 x \tan^2 x)$ 53. $\frac{dy}{dz} = \frac{-1}{\sqrt{z+1}(z-1)^{3/2}}$

55. $\frac{dy}{dx} = \frac{\sin(-1) - \sin(1+x)}{(1+\cos x)^2}$ 57. $\frac{dy}{dx} = -35x^4 \cot^6(x^5)\csc^2(x^5)$

59. $\frac{dy}{dx} = 30x(1 + (x^2 + 2)^5)^2(x^2 + 2)^4$

61. $\frac{dy}{dx} = -4e^{-x} - 14e^{-2x}$

63. $\frac{dy}{dx} = 24(2e^{3x} + 3e^{-2x})^3(e^{3x} - e^{-2x})$

65. $\frac{dy}{dt} = e^{-2t}(2t - 1)\sin(te^{-2t})$

67. $\frac{dy}{dx} = 4(x+1)(x^2 + 2x + 3)e^{(x^2+2x+3)^2}$

69. $\frac{dy}{dx} = \dfrac{1}{8\sqrt{x}\sqrt{1+\sqrt{x}}\sqrt{1+\sqrt{1+\sqrt{x}}}}$

71. $y' = \frac{k}{2\sqrt{kx+b}}$ **73.** $\frac{df}{dx} = 12$

75. (a) $\frac{d}{d\theta}\sin\theta\big|_{\theta=60°} = \frac{\pi}{360}$

(b) $\frac{d}{d\theta}(\theta + \tan\theta)\big|_{\theta=45°} = 1 + \frac{\pi}{90}$

77. $\frac{d}{dx}h(\sin x)\big|_{x=\frac{\pi}{6}} = 5\sqrt{3}$ **79.** $\frac{d}{dx}f(g(x))\big|_{x=6} = 12$

81. $\frac{d}{dx}g(\sqrt{x})\big|_{x=16} = \frac{1}{16}$

83. $\frac{d^2}{dx^2}\sin(x^2) = 2\cos(x^2) - 4x^2\sin(x^2)$

85. $\frac{d^3}{dx^3}(3x+9)^{11} = 26{,}730(3x+9)^8$

87. (a) $f'(x) = -2x\csc^2(x^2)$
$f''(x) = 2\csc^2(x^2)(4x^2\cot(x^2) - 1)$
$f'''(x) = -8x\csc^2(x^2)(6x^2\cot^2(x^2) - 3\cot(x^2) + 2x^2)$

(b) $f'(x) = \dfrac{3x^2}{2\sqrt{x^3+1}}$

$f''(x) = \dfrac{3x(x^3+4)}{4(x^3+1)^{3/2}}$

$f'''(x) = -\dfrac{3(x^6 + 20x^3 - 8)}{8(x^3+1)^{5/2}}$

89. $\frac{d}{dx}\sin(g(x))\big|_{x=2} = -11\sqrt{2}$ **91.** $\frac{d}{dt}(Ri^2)\big|_{t=2} = 0$

93. $\frac{dF}{dt}\big|_{t=10} = -1{,}600/(41^3) \approx -0.0232$ m/s

Section 3.8 Preliminary

1. The Chain Rule

2. (c) is incorrect; $\frac{d}{dx}\sin(y^2) = 2y\cos(y^2)\frac{dy}{dx}$.

3. He made two mistakes.
He did not use the product rule on the second term:
$\frac{d}{dx}(2xy) = 2x\frac{dy}{dx} + 2y$.
He did not use the general power rule on the third term:
$\frac{d}{dx}y^3 = 3y^2\frac{dy}{dx}$.

4. (b)

Section 3.8 Exercises

1. $y' = \frac{-2x}{6y^2}$. At $(2,1)$, $y' = -\frac{2}{3}$

3. $\frac{d}{dx}(x^2y^3) = 3x^2y^2y' + 2xy^3$

5. $\frac{d}{dx}(x^2+y^2)^{3/2} = 3(x + yy')\sqrt{x^2+y^2}$

7. $\frac{d}{dx}(z+z^2) = z' + 2z(z')$ **9.** $y' = -\frac{2x}{9y^2}$

11. $y' = \frac{1-2xy-2y^2}{x^2+4xy-1}$ **13.** $y' = \frac{3-2xy}{x^2+4y^3}$ **15.** $z' = -x^3/z^3$

17. $y' = \sqrt{xy^5}$ **19.** $s' = -\dfrac{s^2(x^2+2\sqrt{x+s})}{x^2(s^2+2\sqrt{x+s})}$ **21.** $y' = \frac{y}{x-y^2}$

23. $y' = \frac{1-\cos(x+y)}{\cos(x+y)+\sin y}$ **25.** $y' = \frac{e^y-2y}{2x+3y^2-xe^y}$

27. $y' = \frac{1}{4}$ **29.** $y = 4 - x$ **31.** $y = 2 - x$ **33.** $y = \frac{8}{5} - \frac{3}{5}x$

35. (b) Setting $y' = 0$ we have $0 = 3x^2 - 3$, so $x = 1$ or $x = -1$.
(c) If we return to the equation $y^2 = x^3 - 3x + 1$ and substitute $x = 1$, we obtain the equation $y^2 = -1$, which has no real solutions.
(d) The tangent line is horizontal at the points $(-1, \sqrt{3})$ and $(-1, -\sqrt{3})$.

37. $y' = \frac{2x-3y}{3x-2y} = 0$ when $y = \frac{2}{3}x$. Substituting $y = \frac{2}{3}x$ into the equation of the implicit curve gives $-\frac{5}{9}x^2 = 1$, which has *no* real solutions.

39. $\frac{dx}{dy} = \frac{y(2y^2-1)}{x}$. The points where the tangent line is vertical are:
$(1,0), (-1,0), \left(\frac{\sqrt{3}}{2}, \frac{\sqrt{2}}{2}\right), \left(-\frac{\sqrt{3}}{2}, \frac{\sqrt{2}}{2}\right), \left(\frac{\sqrt{3}}{2}, -\frac{\sqrt{2}}{2}\right),$
$\left(-\frac{\sqrt{3}}{2}, -\frac{\sqrt{2}}{2}\right).$

41. $\frac{dy}{dt} = -\frac{x^2+y^2}{2xy}\frac{dx}{dt}$ **43.** $\frac{dy}{dt} = -\frac{x+y}{x+2y^3}\frac{dx}{dt}$ **45.** $y = \frac{4}{5}x + \frac{4}{5}$

47. (a)

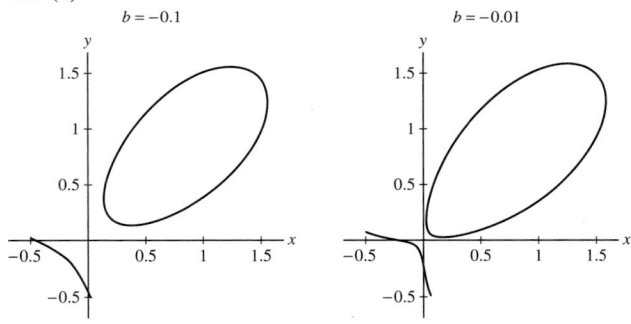

$b = -0.1$ $b = -0.01$

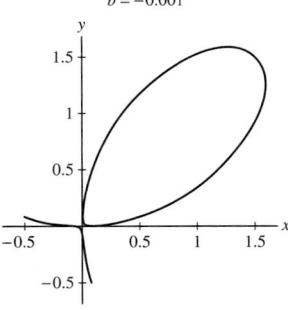

$b = -0.001$

$b = 0.1$

$b = 0.01$

$b = 0.001$

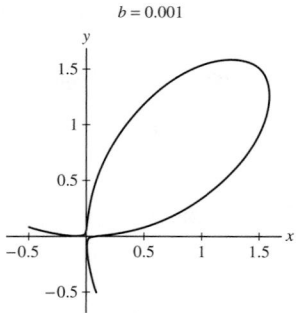

(b) At the point where $y = 0$, $y' = \sqrt[3]{b}$. Therefore, $y' \to \infty$ when $b \to \infty$.

$b = 01$

$b = 10$

$b = 100$

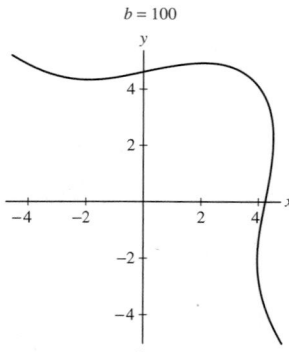

49. (b) $x = \frac{1}{2}, 1 \pm \sqrt{2}$

51. At $(1, 2)$, $y = \frac{1}{3}(x - 1) + 2$.
At $(1, -2)$, $y = -\frac{1}{3}(x - 1) - 2$.
At $\left(1, \frac{1}{2}\right)$, $y = \frac{11}{12}(x - 1) + \frac{1}{2}$.

At $\left(1, -\frac{1}{2}\right)$, $y = -\frac{11}{12}(x - 1) - \frac{1}{2}$.

53.

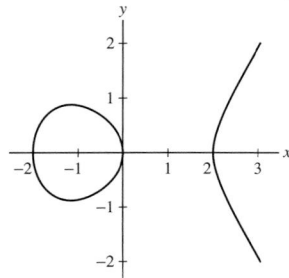

$\frac{dx}{dy} = \frac{2y}{3x^2 - 4} = 0$ when $y = 0$, so the tangent line to this curve is vertical at the points where the curve intersects the x-axis.

55. $y' = -\frac{x}{y}$; thus

$$y'' = -\frac{y \cdot 1 - xy'}{y^2} = -\frac{y - x\left(-\frac{x}{y}\right)}{y^2} = -\frac{y^2 + x^2}{y^3}$$
$$= -\frac{1}{y^3} = -y^{-3}$$

57. At the point $(\frac{2}{3}, \frac{4}{3})$, $y'' = -\frac{162}{125}$

59. $\left(-\frac{\sqrt{6}}{2}, -\frac{\sqrt{2}}{2}\right), \left(-\frac{\sqrt{6}}{2}, \frac{\sqrt{2}}{2}\right), \left(\frac{\sqrt{6}}{2}, -\frac{\sqrt{2}}{2}\right)$, and $\left(\frac{\sqrt{6}}{2}, \frac{\sqrt{2}}{2}\right)$

Section 3.9 Preliminary

1. 2   **2.** $\frac{1}{3}$ **3.** $g(x) = \tan^{-1} x$
4. Angles whose sine and cosine are x are complementary.

Section 3.9 Exercises

1. $g(x) = \sqrt{x^2 - 9}$, $g'(x) = \frac{x}{\sqrt{x^2 - 9}}$ **3.** $g'(x) = \frac{1}{7}$
5. $g'(x) = -\frac{1}{5}x^{-6/5}$ **7.** $g'(x) = (1 - x)^{-2}$ **9.** $g(7) = 1$,
$g'(7) = \frac{1}{5}$ **11.** $g(1) = 0$, $g'(1) = 1$
13. $g(4) = 2$, $g'(4) = \frac{4}{5}$ **15.** $g(\frac{1}{4}) = 3$, $g'(\frac{1}{4}) = -16$
17. $g'(x) = \frac{1}{n}x^{1/n - 1}$; agrees with the Power Rule
19. $y'(\frac{1}{2}) = \frac{4}{5}$ **21.** $y'(\frac{1}{5}) = -\frac{20}{3}$ **23.** $y' = \frac{3}{x^2 + 9}$
25. $y' = \frac{1}{|t + 1|\sqrt{t^2 + 2t}}$ **27.** $y' = -\frac{\pi}{2\sqrt{1 - x^2}(\sin^{-1} x)^2}$
29. $y' = \frac{1}{\sqrt{1 - x^2}}$ **31.** $y' = \frac{1}{t^2 + 1}$ **33.** $y' = -\frac{e^{\cos^{-1} x}}{\sqrt{1 - x^2}}$
35. $y' = \frac{1}{\sqrt{1 - t^2} \sin^{-1} t}$ **37.** $y' = x^{\sin^{-1} x}\left(\frac{\sin^{-1} x}{x} + \frac{\ln x}{\sqrt{1 - x^2}}\right)$

Section 3.10 Preliminary

1. $\ln 4$ **2.** $b = e^2$ **3.** $\frac{1}{10}$ **4.** $b = e^3$
5. $y^{(100)}(x) = \cosh x$; $y^{(101)}(x) = \sinh x$

Section 3.10 Exercises

1. $\ln x + 1$ **3.** $\frac{2}{x} \ln x$ **5.** $\frac{3x^2 + 3}{x^3 + 3x + 1}$ **7.** $\frac{\cos t}{\sin t + 1}$ **9.** $\frac{1 - \ln x}{x^2}$
11. $\frac{1}{x \ln x}$ **13.** $\frac{3}{x \ln x}$ **15.** $\frac{1}{\sin x \cos x}$ **17.** $5^x \ln 5$
19. $f'(x) = \frac{1}{x \ln 2}$ **21.** $f'(3) = \frac{1}{3 \ln 5}$

23. $y = 64 \ln 4(x - 3) + 64$

25. $y = 7 \ln 3 \cdot 3^{14}(t - 2) + 3^{14}$ 27. $y = 5^8$

29. $y = 1 + \ln 4 - t$ 31. $y = \frac{9}{5}(x - 2) + 4 \ln 20$

33. $y = \frac{1}{2 \ln 5}(x - 2) + \frac{\ln 2}{\ln 5}$ 35. $y' = x^{2x}(2 + 2 \ln x)$

37. $y' = x^{e^x}\left(\frac{e^x}{x} + e^x \ln x\right)$ 39. $y' = x^{2^x}\left(\frac{2^x}{x} + 2^x \ln 2 \cdot \ln x\right)$

41. $y' = 2x + 6$ 43. $y' = \frac{(x+1)^3}{(3x-1)^2} + \frac{3x(x+1)^2}{(3x-1)^2} - \frac{6x(x+1)^3}{(3x-1)^3}$

45. $y' = 4x^2(2x + 1)\sqrt{x - 9}\left(\frac{2}{2x+1} + \frac{2}{x} + \frac{1}{2(x-9)}\right)$

47. $y' = (x^2 + 1)(x^2 + 2)(x^2 + 3)^2\left(\frac{2x}{x^2+1} + \frac{2x}{x^2+2} + \frac{4x}{x^2+3}\right)$

49. $(\ln x)' = \frac{1}{x}$ and $(\ln 2x)' = \frac{2}{2x} = \frac{1}{x}$; note $\ln 2x = \ln 2 + \ln x$

51. $y' = 2x \cosh(x^2)$ 53. $y' = 2t \operatorname{sech}^2(t^2 + 1)$

55. $y' = \sinh x + \tanh x \operatorname{sech} x$ 57. $y' = \tanh x$

59. $y' = \frac{\cosh(\ln x)}{x}$ 61. $y' = e^x \operatorname{sech}^2 e^x$

63. $y' = \sinh x \cosh(\cosh x)$ 65. $y' = -2x \operatorname{csch}^2 x^2$

67. $y' = -\operatorname{csch} x \coth x$ 69. $y' = x^{\sinh x}\left(\cosh x \ln x + \frac{\sinh x}{x}\right)$

71. $y' = \frac{2x}{\sqrt{x^4+1}}$ 73. $y' = \frac{e^{\cosh^{-1} x}}{\sqrt{x^2-1}}$ 75. $y' = \frac{1}{(1-x^2)\tanh^{-1} x}$

79. (a) $E = 10^{4.8+1.5M}$ (c) $\frac{dE}{dM} = (1.5 \ln 10) 10^{4.8+1.5M}$

Section 3.11 Preliminary

1. The restatement is "Determine $\frac{dV}{dt}$ if $\frac{ds}{dt} = 0.5$ cm/s."

2. $\frac{dV}{dt} = 4\pi r^2 \frac{dr}{dt}$.

3. The restatement is "Determine $\frac{dh}{dt}$ if $\frac{dV}{dt} = 2$ cm^3/min."

4. The restatement is "Determine $\frac{dV}{dt}$ if $\frac{dh}{dt} = 1$ cm/min."

Section 3.11 Exercises

1. 0.039 ft/min

3. (a) $100\pi \approx 314.16$ m^2/min   (b) $24\pi \approx 75.40$ m^2/min

5. $3,584\pi$ in^3/min   7. 896π in^2/min   9. $18\sqrt{10}$ mph

11. 1.68 m^3/min

13. (a) Issac traveled $\frac{27}{5}$ miles. Sonya traveled $\frac{88}{15}$ miles.

(b) $\frac{5,003}{\sqrt{14,305}} \approx 41.83$ mph

15. (a) 499.86 mph   (b) 0 mph

17. 0.61 mi/min   19. $\frac{-11}{\sqrt{15}}$ ft/s

21. $x = 4\sqrt{15}$ ft; $\frac{dx}{dt} = \frac{4}{\sqrt{15}} \approx 1.03$ ft/s

23. $\frac{dh}{dt} = -\frac{x}{h}\frac{dx}{dt}$; as $\frac{dx}{dt}$ is constant and positive and x approaches the length of the ladder as it falls, $-\frac{x}{h}\frac{dx}{dt}$ gets arbitrarily large as $h \to 0$.

25. $1,047.20$ cm^3/s   27. $4,800\pi \approx 15,079.64$ mph

29. 495.25 mph   31. $\frac{dy}{dt} = -10.5$ ft/s   33. $\frac{dP}{dt} = -1.92$ kPa/min

35. $l'(t) = \frac{9x(2x^2+3)}{\sqrt{x^4+3x^2+1}}$ ft/s (where x is a function of t).

37. $-\frac{1}{1,440\pi} \approx -2.21 \times 10^{-4}$ cm/min

39. (a) $\frac{dx}{dt} = \frac{h}{x}\frac{dh}{dt}$ where the rate at which Henry pulls the rope is $-dh/dt$.

(b) $-\sqrt{13}/2 \approx -1.80$ ft/s

41. 300 mph   43. -8.94 ft/s   45. ≈ -4.68 m/s

Chapter 3 Review

1. ROC of $f(x)$ over $[0, 2]$ is 3.

3. The estimation is $\frac{8}{3}$ and it is larger than $f'(0.7)$.

5. $f'(1) = 1$; Tangent line: $y = x - 1$

7. $f'(4) = -\frac{1}{16}$; Tangent line: $y = \frac{1}{2} - \frac{1}{16}x$

9. $\frac{dy}{dx} = -2x$ 11. $\frac{dy}{dx} = \frac{1}{(2-x)^2}$ 13. $f'(1)$ where $f(x) = \sqrt{x}$

15. $f'(\pi)$ where $f(t) = \sin(t)\cos(t)$ 17. $f'(4) = 3$; $f(4) = -2$

19. (C)

21. (a) 8.05 cm/yr   (b) Larger over the first half.

(c) $h'(3) \approx 7.8$ cm/yr; $h'(8) \approx 6.0$ cm/yr

23. $A'(t)$ is the rate of change in automobile production in the United States.
$A'(1971) \approx 0.25$ million automobiles/year.
$A'(1974)$ would be negative.

25. (b) $(\ln 2)2^x$

27. $g'(8) = \frac{1}{12}$ 29. $y' = -6x^{-5/2}$ 31. $y' = 8x + 2x^{-3}$

33. $y' = -\frac{19}{(4t-9)^2}$ 35. $y' = 6(6t - 60t^{-4})(3t^2 + 20t^{-3})^5$

37. $y' = (7x + 16)(x + 1)^2(x + 4)^3$ 39. $y' = -\frac{3}{x^2}\left(1 + \frac{1}{x}\right)^2$

41. $y' = \frac{5-3x}{2(1-x)^2(2-x)^{3/2}}$ 43. $y' = 12\sin(2 - 3x)$

45. $y' = \frac{x}{\sqrt{x^2+1}}\cos(\sqrt{x^2 + 1})$ 47. $y' = -\frac{4}{\theta^2}\cos\left(\frac{4}{\theta}\right)$

49. $y' = -9z \csc(9z + 1)\cot(9z + 1) + \csc(9z + 1)$

51. $y' = -\sin x \sec^2(\cos x)$

53. $y' = -\sin(\cos(\cos \theta))\sin(\cos \theta)\sin \theta$

55. $f'(x) = \frac{8x}{4x^2+1}$ 57. $f'(x) = \frac{1+e^x}{x+e^x}$ 59. $G'(s) = \frac{2}{s}$

61. $g'(t) = (2t - 1)e^{1/t}$ 63. $f'(\theta) = \frac{\cos(\ln \theta)}{\theta}$

65. $f'(x) = 2\sin x \cos x e^{\sin^2 x}$ 67. $h'(y) = \frac{2e^y}{(1-e^y)^2}$

69. $G'(s) = \frac{1}{2\sqrt{s}(1+s)}$ 71. $f'(x) = \frac{e^{\sec^{-1} x}}{|x|\sqrt{x^2-1}}$

73. $h'(y) = 4y \operatorname{sech}^2(4y) + \tanh(4y)$

75. $g'(t) = \sqrt{\frac{t^2-1}{t^2+1}} + \frac{t\sinh^{-1} t}{\sqrt{t^2-1}}$ 79. $S'(2) = -27$

81. $R'(2) = -\frac{57}{16}$ 83. $F'(2) = -18$ 85. $(-1, -1)$ and $(3, 7)$

87. $b = -76, 32$

89. (a) $a = \frac{1}{6}$ (b)

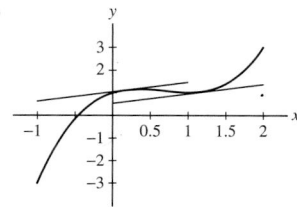

91. $y'' = 72x - 10$ 93. $y'' = -(2x + 3)^{-3/2}$

95. $y'' = 8x^2 \sec^2(x^2)\tan(x^2) + 2\sec^2(x^2)$

97. For the plot on the left, the red, green and blue curves, respectively, are the graphs of f, f' and f''. For the plot on the right, the green, red and blue curves, respectively, are the graphs of f, f' and f''.

99. **(b)** $E(8) = -3$ **101.** $\dfrac{dy}{dx} = \dfrac{x^2}{y^2}$ **103.** $\dfrac{dy}{dx} = \dfrac{y^2+4x}{1-2xy}$

105. $\dfrac{dy}{dx} = \dfrac{y-2x}{y-x}$ **107.** $\dfrac{dy}{dx} = \dfrac{y-\sec^2(x+y)}{\sec^2(x+y)-x}$

109. $(1,1), \left(-\sqrt[3]{3}, \sqrt[3]{9}\right)$ **111.** $\alpha = 0, \alpha > 1$

113. $\frac{1}{15}$ m/min **115.** 0.75 ft/s

117. **(a)** -0.72 rad/s **(b)** -1.50 cm/s

119. $y' = \dfrac{(x+1)(x+2)^2}{(x+3)(x+4)}\left(\dfrac{1}{x+1} + \dfrac{2}{x+2} - \dfrac{1}{x+3} - \dfrac{1}{x+4}\right)$

121. $y' = \dfrac{e^x\sin^{-1}x}{\ln x}\left(1 + \dfrac{1}{\sqrt{1-x^2}\sin^{-1}x} - \dfrac{1}{x\ln x}\right)$

123. $y' = x^{\sqrt{x}}\left(x^{\ln x}\right)\left(\dfrac{\ln x}{2\sqrt{x}} + \dfrac{1}{\sqrt{x}} + \dfrac{2\ln x}{x}\right)$

Chapter 4

Section 4.1 Preliminary

1. $g(1.2) - g(1) \approx -0.8$ **2.** $f(2.1) \approx 1.3$ **3.** 55 ft
5. $f'(2) = 2; f(2) = 8$

Section 4.1 Exercises

1. $\Delta f \approx 0.12$ **3.** $\Delta f \approx -0.00222$

5. $\Delta f \approx 0.04e^6 \approx 16.137$ **7.** $\Delta f \approx \dfrac{1}{10e^2} \approx .013534$

9. $\Delta f \approx 0.166667$; error: 0.004389; percentage error: 2.70%

11. $\Delta f \approx 0.02$; error: 1.3×10^{-6}; percentage error: 0.0065%

13. $\Delta f \approx -0.021213$; error: 0.000315; percentage error: 1.46%

15. $\Delta f \approx 0.133333$; error: 0.007513; percentage error: 5.33%

17. $\Delta f \approx 0.015625$. The error is ≈ 0.00018.

19. $\Delta f \approx 0.001$. The error is $\approx 1.53 \times 10^{-5}$.

21. $\Delta f \approx 0.023$. The error is $\approx 2.03 \times 10^{-6}$.

23. $\Delta f \approx -0.03$. The error is $\approx 4.59 \times 10^{-4}$.

25. $\Delta f \approx 0.0074074$ **27.** $\Delta f \approx 0.008727$

29. **(a)** $\Delta P \approx -0.434906$ kilopascals
(b) $P(20.5) - P(20) = -0.418274$ kilopascals; percentage error in linear approximation: 3.98%

31. $\Delta A \approx 0.5$ ft^2

33. **(a)** $\Delta h \approx 0.78125$ ft **(b)** $\Delta h \approx 0.9375$ ft

35. $\Delta F \approx 4.88$ ft; $\Delta F \approx 7.04$ ft

37. Weight loss ≈ 0.066 pounds; pilot weighs 129.5 pounds at an altitude of approximately 7.6 miles.

39. **(a)** $P = 4.8$ atm, $\Delta P \approx 0.48$ atm **(b)** $|\Delta V| < 0.520833$ L

41. $L(x) = x$ **43.** $L(x) = -\frac{1}{2}x + 1$ **45.** $L(x) = 1$

47. $L(x) = \frac{2\sqrt{3}}{3}\left(x - \frac{1}{2}\right) + \frac{\pi}{6}$ **49.** $L(x) = e(x - 1)$

51. $\sqrt{16.2} \approx 4.025$

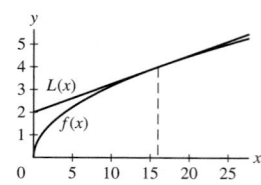

53. $\sqrt{17} \approx 4.125$. The percentage error is $\approx 0.046\%$.

55. $17^{1/4} \approx 2.03125$. The percentage error is $\approx 0.035\%$.

57. $27.001^{\frac{1}{3}} \approx 3.0000370370370$. The percentage error is 0.000000015%.

59. $\ln(1.07) \approx 0.07$. The percentage error is $\approx 3.46\%$.

61. $\cos^{-1}(0.52) \approx 1.024104$. The percentage error is $\approx 0.015\%$.

65. For $v = 25$ ft/s, $\Delta x \approx -0.118$ ft; for $v = 30$ ft/s, $\Delta x \approx -0.170$ ft. The shot is more sensitive to angle at larger velocities.

67.

h	$E = \left|\sqrt{9+h} - 3 - \frac{1}{6}h\right|$	$0.006h^2$
10^{-1}	4.604×10^{-5}	6.00×10^{-5}
10^{-2}	4.627×10^{-7}	6.00×10^{-7}
10^{-3}	4.629×10^{-9}	6.00×10^{-9}
10^{-4}	4.627×10^{-11}	6.00×10^{-11}

69. $(1.02)^{0.7} \approx 1.014$; $(1.02)^{-0.3} \approx 0.994$

73. **(a)** $dy/dx = -1$; $L(x) = 4 - x$
(b)

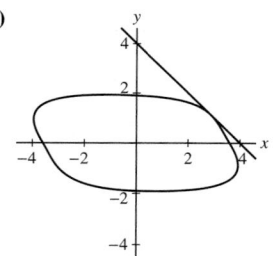

(c) $y \approx 0.9$; too large
(d) $y = 0.890604$; percentage error in approximation from part (c): 1.06%

Section 4.2 Preliminary

1. (c) **2.** (b) **4.** True **6.** (b)

Section 4.2 Exercises

1. **(a)** Three **(b)** 6 **(c)** 5 at $x = 5$
(d) One example is $[4, 8]$. **(e)** One example is $[0, 2]$.

3. $x = 1$ **5.** $x = -3$ and $x = 6$ **7.** $x = \pm 1$ **9.** $x = 0$

11. $x = 1/e$ **13.** $x = \pm\dfrac{\sqrt{3}}{2}$

15. **(a)** $c = 2$; $f(c) = -3$ **(b)** $f(0) = f(4) = 1$
(c) Maximum 1; Minimum -3 **(d)** Maximum 1; Minimum -2

17. $\tan^{-1}\left(-\frac{1}{4}\right) \approx -0.244979$

19. Critical points at $t = 0$ and $t = \pm 1$; on $[0, 1]$, the maximum value is $h(1) = 0$ and the minimum is $h(0) = -1$; on $[0, 2]$, the maximum value is $h(2) = 3^{1/3} \approx 1.44225$ and the minimum is $h(0) = -1$

21. Critical points at approximately $x = 0.2$, $x = 1.4$ and $x = 4.8$; No minimum value as $\lim_{x \to 0+} f(x) = -\infty$. The maximum value is $f(4.8) = 6.5$.

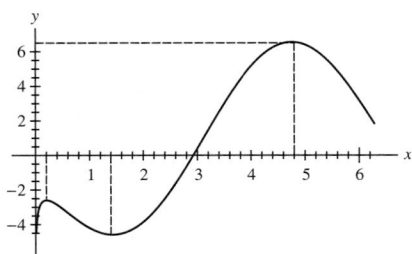

23. The minimum is $f(1) = 0$. The maximum is $f(3) = 8$.

25. The minimum is $f(2) = -9$. The maximum is $f(-2) = 15$.

27. The minimum is $f(-1) = -3$. The maximum is $f(\frac{3}{8}) = 4.5625$.

29. The minimum is $f(1) = -4$. The maximum is $f(-1) = 6$.

31. The minimum is $f(1) = -3$. The maximum is $f(-1) = 13$.

33. The minimum is $f(1) = -3$. The maximum is $f(4) = 78$.

35. The minimum is $f(-1) = -4$.
The maximum value is $f(5) = 3{,}050$.

37. The minimum is $f(6) = 18.5$. The maximum is $f(5) = 26$.

39. The minimum is $f(1) = -1$. The maximum is $f(0) = f(3) = 0$.

41. The minimum is $f(0) = 2\sqrt{6}$. The maximum is $f(2) = 4\sqrt{2}$.

43. The minimum is $f\left(\frac{\sqrt{3}}{2}\right) \approx -0.589980$.
The maximum is $f(4) \approx 0.472136$.

45. The minimum is $f(0) = f\left(\frac{\pi}{2}\right) = 0$. The maximum is $f\left(\frac{\pi}{4}\right) = \frac{1}{2}$.

47. The minimum is $f(0) = -1$.
The maximum is $f\left(\frac{\pi}{4}\right) = \sqrt{2}\left(\frac{\pi}{4} - 1\right)$.

49. The minimum is $g\left(\frac{\pi}{3}\right) = \frac{\pi}{3} - \sqrt{3}$.
The maximum is $g\left(\frac{5}{3}\pi\right) = \frac{5}{3}\pi + \sqrt{3}$.

51. The minimum is $f\left(\frac{\pi}{4}\right) = 1 - \frac{\pi}{2}$. The maximum is $f(0) = 0$.

53. The minimum is $f(1) = 0$.
The maximum is $f(e) = 1/e \approx 0.367879$.

55. The minimum is $f(5) = 5\tan^{-1} 5 - 5 \approx 1.867004$.
The maximum is $f(2) = 5\tan^{-1} 2 - 2 \approx 3.535744$.

57. **(d)** The critical points in the range $[0, 2\pi]$ are $\frac{\pi}{6}, \frac{\pi}{2}, \frac{5\pi}{6}, \frac{7\pi}{6}, \frac{3\pi}{2}$, and $\frac{11\pi}{6}$. On this interval, the maximum value is $f\left(\frac{\pi}{6}\right) = f\left(\frac{7\pi}{6}\right) = \frac{3\sqrt{3}}{2}$ and the minimum value is $f\left(\frac{5\pi}{6}\right) = f\left(\frac{11\pi}{6}\right) = -\frac{3\sqrt{3}}{2}$.

(e)

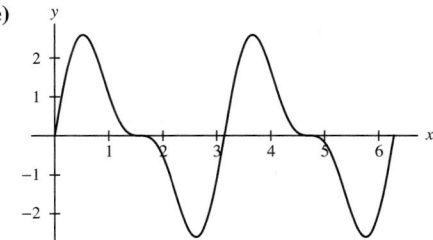

59. critical point: $x = 2$; minimum is $f(2) = 0$; maximum is $f(0) = 2$

61. critical point $x = 2$; minimum is $f(2) = 0$; maximum is $f(0) = 12$

63. $f(-2) = -6 + 8 = 2$ and $f(1) = 3 - 1 = 2$; therefore, there exists a $c \in (-2, 1)$ at which $f'(c) = 0$; take $c = -1$

65. $f\left(\frac{\pi}{4}\right) = f\left(\frac{3\pi}{4}\right) = \frac{\sqrt{2}}{2}$; take $c = \frac{\pi}{2}$

67. $f\left(\frac{\pi}{4}\right) = f\left(\frac{3\pi}{4}\right) = 0$; take $c = \frac{\pi}{2}$

73. **(a)** critical points: $v = 0$ and $v = 3v_r/2$
(b) Take $v_r = 10$. Then the positive critical point is at $v = 15$.

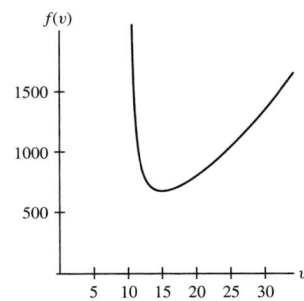

75. critical point: $x = 1$; minimum is $f(-5) = \frac{1}{7}$; maximum is $f(1) = 1$

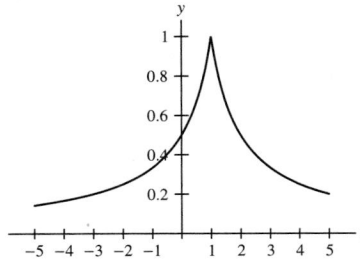

77. critical points: $x = \pm 1$ and $x = \pm 2$; minimum is $f(2) = -\frac{2}{3}$; maximum is $f(-2) = \frac{2}{3}$

79. **81.**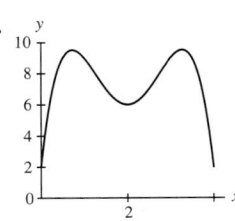

87. $b \geq \frac{1}{4}a^2$

Section 4.3 Preliminary

1. $m = 3$ **2.** (c)
3. Yes

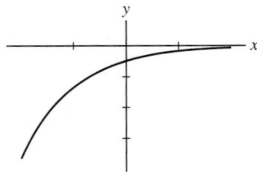

4. (a) $f(c)$ is a local maximum.   **(b)** No

Section 4.3 Exercises

1. $c = 2$ **3.** $c = 2$ **5.** $c = 2\sqrt{7} - 1$ **7.** $c = \frac{4}{e}$
9. $c = 0$ **11.** $c \approx 0.62$
13. $f(x)$ increasing over $(0, 2)$ and $(4, 6)$, and decreasing over $(2, 4)$.
15.   **17.**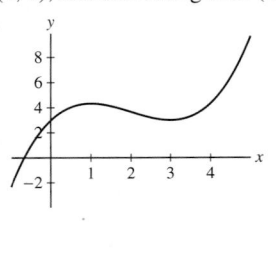

19. Maximum   **21.** Minimum
23. $f(2)$ is a local maximum. $f(4)$ is a local minimum.
25. The critical point is $c = \frac{7}{2}$.

x	$\left(-\infty, \frac{7}{2}\right)$	$\frac{7}{2}$	$\left(\frac{7}{2}, \infty\right)$
f'	+	0	−
f	↗	M	↘

27. The critical points are $c = 0, 4$.

x	$(-\infty, 0)$	0	$(0, 4)$	4	$(4, \infty)$
f'	+	0	−	0	+
f	↗	M	↘	m	↗

29. The critical points are $c = -2, -1, 1$.

x	$(-\infty, -2)$	-2	$(-2, -1)$	-1	$(-1, 1)$	1	$(1, \infty)$
f'	−	0	+	0	−	0	+
f	↘	m	↗	M	↘	m	↗

31. The critical points are $c = -2, -1$.

x	$(-\infty, -2)$	-2	$(-2, -1)$	-1	$(-1, \infty)$
f'	+	0	−	0	+
f	↗	M	↘	m	↗

33. The critical points are $c = 0, -\frac{3}{4}$.

x	$\left(-\infty, -\frac{3}{4}\right)$	$-\frac{3}{4}$	$\left(-\frac{3}{4}, 0\right)$	0	$(0, \infty)$
f'	−	0	+	0	+
f	↘	m	↗	¬	↗

35. The critical point is $c = 0$.

x	$(-\infty, 0)$	0	$(0, \infty)$
f'	+	0	−
f	↗	M	↘

37. The critical point is $c = 1$.

x	$(0, 1)$	1	$(1, \infty)$
f'	−	0	+
f	↘	m	↗

39. The critical point is $x = \frac{16}{25}$.

x	$\left(0, \frac{16}{25}\right)$	$\frac{16}{25}$	$\left(\frac{16}{25}, \infty\right)$
f'	−	0	+
f	↘	m	↗

41. The critical points are $\theta = \frac{\pi}{4}$, $\theta = \frac{3\pi}{4}$, $\theta = \frac{5\pi}{4}$, and $\theta = \frac{7\pi}{4}$.

x	$\left(0, \frac{\pi}{4}\right)$	$\frac{\pi}{4}$	$\left(\frac{\pi}{4}, \frac{3\pi}{4}\right)$	$\frac{3\pi}{4}$	$\left(\frac{3\pi}{4}, \frac{5\pi}{4}\right)$	$\frac{5\pi}{4}$	$\left(\frac{5\pi}{4}, \frac{7\pi}{4}\right)$	$\frac{7\pi}{4}$	$\left(\frac{7\pi}{4}, 2\pi\right)$
f'	$+$	0	$-$	0	$+$	0	$-$	0	$+$
f	↗	M	↘	m	↗	M	↘	m	↗

43. The critical point is $x = \frac{\pi}{2}$.

x	$\left[0, \frac{\pi}{2}\right)$	$\frac{\pi}{2}$	$\left(\frac{\pi}{2}, 2\pi\right]$
f'	$+$	0	$+$
f	↗	¬	↗

45. The critical point is $x = 0$.

x	$(-\infty, 0)$	0	$(0, \infty)$
f'	$-$	0	$+$
f	↘	m	↗

47. The critical point is $x = -\frac{\pi}{4}$.

x	$\left[-\frac{\pi}{2}, -\frac{\pi}{4}\right)$	$-\frac{\pi}{4}$	$\left(-\frac{\pi}{4}, \frac{\pi}{2}\right]$
f'	$+$	0	$-$
f	↗	M	↘

49. The critical points are $x = \pm 1$.

x	$(-\infty, -1)$	-1	$(-1, 1)$	1	$(1, \infty)$
f'	$-$	0	$+$	0	$-$
f	↘	m	↗	M	↘

51. The critical point is $x = 1$.

x	$(0, 1)$	1	$(1, \infty)$
f'	$-$	0	$+$
f	↘	m	↗

55. $a \geq 0$

57. **(a)** MVT **(b)** IVT

63. The MVT, applied to the interval $[2, 4]$, guarantees there exists a $c \in (2, 4)$ such that $f'(c) = \frac{f(4) - f(2)}{4 - 2}$ or $f(4) - f(2) = 2f'(c)$. Because $f'(x) \geq 5$, it follows that $f(4) - f(2) \geq 10$, or $f(4) \geq f(2) + 10 = 8$.

65. Let $f(x) = x^3 + ax^2 + bx + c$. Then $f'(x) = 3x^2 + 2ax + b = 3\left(x + \frac{a}{3}\right)^2 - \frac{a^2}{3} + b > 0$ for all x if $b - \frac{a^2}{3} > 0$. Therefore, if $b > a^2/3$, then $f(x)$ is increasing on $(-\infty, \infty)$.

67. Let $f(x) = x$ and $g(x) = \sin x$. Then $f(0) = g(0) = 0$ and $f'(x) = 1 \geq \cos x = g'(x)$ for $x \geq 0$. Apply the result of the previous exercise to conclude that $x \geq \sin x$ or $\sin x \leq x$ for $x \geq 0$.

71. (b) No

Section 4.4 Preliminary

1. (a) **2.** (b) **3.** False **4.** True **5.** False **6.** True
7. True **8.** True **9.** (c) and (d)

Section 4.4 Exercises

1. **(a)** C **(b)** A **(c)** B **(d)** D

3. Increasing function (left); Decreasing function (right)

 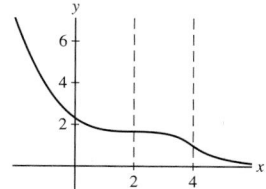

5. $b, e; b < x < e$

7. Concave up everywhere; no points of inflection.

9. $x = \frac{1}{2}(2n + 1)\pi$; concave up $\left(-\frac{\pi}{2} + 2n\pi, \frac{\pi}{2} + 2n\pi\right)$; concave down $\left(\frac{\pi}{2} + 2n\pi, \frac{3\pi}{2} + 2n\pi\right)$

11. Concave down for $0 < x < 9$; concave up for $x > 9$; inflection point at $x = 9$

13. Concave down for $|x| < 1$; concave up for $|x| > 1$; inflection point at $x = \pm 1$

15. Concave down for $x < 2$; concave up for $x > 2$; $x = 2$

17. Concave down for $-2 - \sqrt{5} < x < -2 + \sqrt{5}$; concave up for $x < -2 - \sqrt{5}$ and for $x > -2 + \sqrt{5}$; $x = -2 \pm \sqrt{5}$

19. f has no points of inflection.

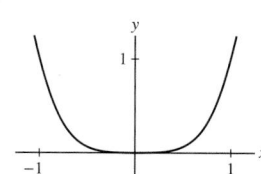

21. The growth rate at the point of inflection is 5.5 cm/day. First derivative (left); Second derivative (right)

 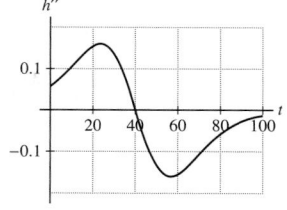

23. $f(3)$ is a local maximum, and $f(5)$ is a local minimum.

25. $f(0)$ is a local minimum. Second derivative test is inconclusive for $x = 1$.

27. $f\left(\sqrt{\frac{3}{5}}\right)$ is a local minimum. $f\left(-\sqrt{\frac{3}{5}}\right)$ is a local maximum.

29. $f(x)$ has local minima when x is an even multiple of π. $f(x)$ has local maxima when x is an odd multiple of π.

31. $f\left(\frac{1}{9}\right)$ is a local minimum. Second derivative test cannot be applied at $x = 0$.

33. $f\left(\frac{\sqrt{2}}{2}\right)$ is a local maximum, and $f\left(-\frac{\sqrt{2}}{2}\right)$ is a local minimum.

35. $f(e^{-1/3})$ is a local minimum.

37.

x	$\left(-\infty, \frac{1}{3}\right)$	$\frac{1}{3}$	$\left(\frac{1}{3}, 1\right)$	1	$(1, \infty)$
f'	$+$	0	$-$	0	$+$
f	↗	M	↘	m	↗

x	$\left(-\infty, \frac{2}{3}\right)$	$\frac{2}{3}$	$\left(\frac{2}{3}, \infty\right)$
f''	$-$	0	$+$
f	⌢	I	⌣

39.

t	$(-\infty, 0)$	0	$\left(0, \frac{2}{3}\right)$	$\frac{2}{3}$	$\left(\frac{2}{3}, \infty\right)$
f'	$-$	0	$+$	0	$-$
f	↘	m	↗	M	↘

t	$\left(-\infty, \frac{1}{3}\right)$	$\frac{1}{3}$	$\left(\frac{1}{3}, \infty\right)$
f''	$+$	0	$-$
f	⌣	I	⌢

41.

x	0	$(0, 0.396850)$	0.396850	$(0.396850, \infty)$
f'	U	$-$	0	$+$
f	M	↘	m	↗

Concave up everywhere; no points of inflection.

43.

t	$(-\infty, 0)$	0	$(0, \infty)$
f'	$+$	0	$-$
f	↗	M	↘

t	$\left(-\infty, -\sqrt{\frac{1}{3}}\right)$	$-\sqrt{\frac{1}{3}}$	$\left(-\sqrt{\frac{1}{3}}, \sqrt{\frac{1}{3}}\right)$	$\sqrt{\frac{1}{3}}$	$\left(\sqrt{\frac{1}{3}}, \infty\right)$
f''	$+$	0	$-$	0	$+$
f	⌣	I	⌢	I	⌣

45.

θ	$[0, \pi)$	π	$(\pi, 2\pi]$
f'	$+$	0	$+$
f	↗	¬	↗

θ	0	$(0, \pi)$	π	$(\pi, 2\pi)$	2π
f''	0	$-$	0	$+$	0
f	¬	⌢	I	⌣	¬

47.

x	$(0, 2\pi)$
f'	$+$
f	↗

x	$(0, \pi)$	π	$(\pi, 2\pi)$
f''	$+$	0	$-$
f	⌣	I	⌢

49.

x	$\left(-\frac{\pi}{2}, -\frac{\pi}{4}\right)$	$-\frac{\pi}{4}$	$\left(-\frac{\pi}{4}, \frac{3\pi}{4}\right)$	$\frac{3\pi}{4}$	$\left(\frac{3\pi}{4}, \frac{3\pi}{2}\right)$
f'	$+$	0	$-$	0	$+$
f	↗	M	↘	m	↗

x	$\left(-\frac{\pi}{2}, 0\right)$	0	$(0, \pi)$	π	$\left(\pi, \frac{3\pi}{2}\right)$
f''	$-$	0	$+$	0	$-$
f	⌢	I	⌣	I	⌢

51. (a) Near the beginning, R is concave up. Near the end, R is concave down.

(b) "Epidemic subsiding: number of new cases declining."

53. The point of inflection should occur when the water level is equal to the radius of the sphere.

55. $\frac{d}{dx}\tan^{-1}x = \frac{1}{1+x^2} > 0$ for all x, so $\tan^{-1}x$ is always increasing. $\frac{d^2}{dx^2}\tan^{-1}x = \frac{-2x}{(1+x^2)^2}$, so $\tan^{-1}x$ is concave down for $x > 0$ and is concave up for $x < 0$.

57.

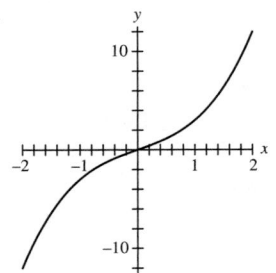

59. (a) False. If $f(x)$ is increasing, then $f^{-1}(x)$ is increasing.

(b) True.

(c) False. If $f(x)$ is concave up, the $f^{-1}(x)$ is concave down.

(d) True.

Section 4.5 Preliminary

1.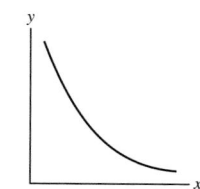

2. (c)

3. (a) $\lim\limits_{x \to \infty} x^3 = \infty$ **(b)** $\lim\limits_{x \to -\infty} x^3 = -\infty$

(c) $\lim\limits_{x \to -\infty} x^4 = \infty$

4. Negative **5.** Negative

6. The function f does not have a point of inflection at $x = 4$ because $x = 4$ is not in the domain of f.

Section 4.5 Exercises

1. A, $f' < 0$ and $f'' > 0$. B, $f' > 0$ and $f'' > 0$. C, $f' > 0$ and $f'' < 0$. D, $f' < 0$ and $f'' < 0$. E, $f' < 0$ and $f'' > 0$. F, $f' > 0$ and $f'' > 0$. G, $f' > 0$ and $f'' < 0$.

3.   **5.**

7.   **9.**

11.   **13.**

15.   **17.**

19.   **21.**

23.   **25.**

27.   **29.**

31.

33.

35.

37.

39.

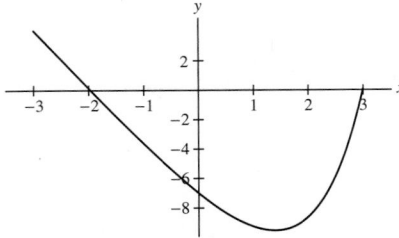

The graph has a local minimum at $2\ln 2$.

41.

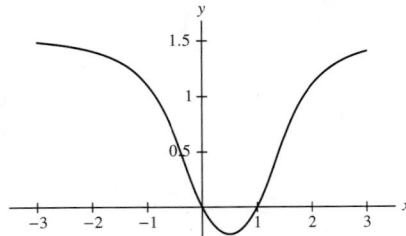

The graph has a local minimum at $\frac{1}{2}$ and inflection points at approximately -0.327199 and 1.327199.

43.

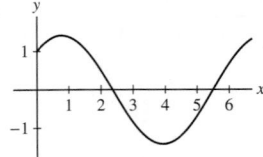

The graph has a local maximum at $\frac{\pi}{4}$, a local minimum at $\frac{5\pi}{4}$, and inflection points at $\frac{3\pi}{4}$ and $\frac{7\pi}{4}$.

45.

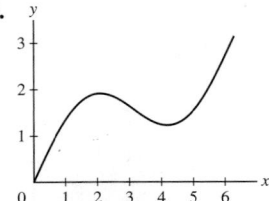

The graph has a local maximum at $\frac{2\pi}{3}$, a local minimum at $\frac{4\pi}{3}$, and an

inflection point at π.

47.

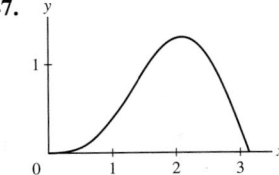

The graph has a local maximum at $x = \frac{2\pi}{3}$ and a point of inflection at $x = \cos^{-1}\frac{1}{4}$.

49. (a)

(b)

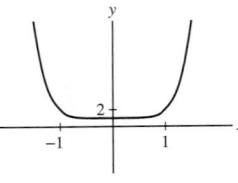

51. $\lim\limits_{x\to\infty} \frac{x}{x+9} = 1$ **53.** $\lim\limits_{x\to\infty} \frac{3x^2+20x}{2x^4+3x^3-29} = 0$

55. $\lim\limits_{x\to\infty} \frac{7x-9}{4x+3} = \frac{7}{4}$ **57.** $\lim\limits_{x\to-\infty} \frac{7x^2-9}{4x+3} = -\infty$

59. $\lim\limits_{x\to-\infty} \frac{x^2-1}{x+4} = -\infty$ **61.** $\lim\limits_{x\to\infty} \frac{\sqrt{x^2+1}}{x+1} = 1$

63. $\lim\limits_{x\to\infty} \frac{x+1}{\sqrt[3]{x^2+1}} = \infty$ **65.** $\lim\limits_{x\to-\infty} \frac{x}{(x^6+1)^{1/3}} = 0$

67. $\lim\limits_{x\to0+} \frac{\ln x}{x^2} = -\infty$ **69.** $\lim\limits_{x\to\infty} \frac{e^{-x^2}}{x^2-5} = 0$ **71.** (A)

73. (a) D **(b)** A **(c)** B **(d)** C

75.

77.

79.

81.

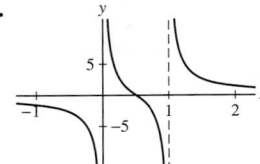

The graph has a point of inflection at $x = \frac{1}{2}$.

83.

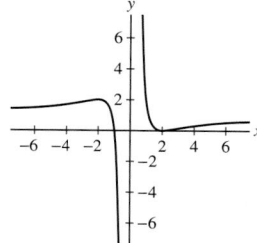

The graph has a local maximum at $x = -2$, a local minimum at $x = 2$ and points of inflection at $x = \pm 2\sqrt{2}$.

85.

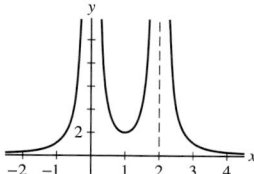

The graph has a local minimum at $x = 1$.

87.

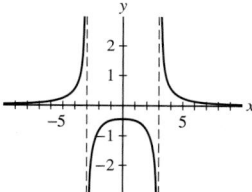

The graph has a local maximum at $x = 0$.

89.

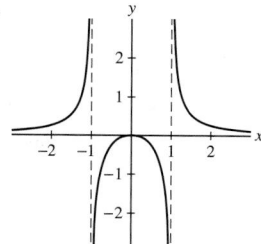

The graph has a local maximum at $x = 0$.

91.

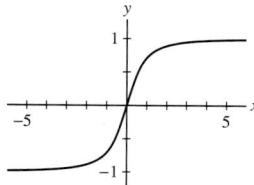

The graph has a point of inflection at $x = 0$.

93. (a) $y = x - 2$ **(b)** $y = x - 1$

95. $f(x)$ approaches the slant asymptote from above in both directions.

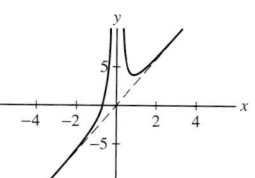

Section 4.6 Preliminary

1. $b + h + \sqrt{b^2 + h^2} = 10$ **2.** r, h **3.** No

4. If the function tends to infinity at the endpoints of the interval, then the function must take on a minimum value at a critical point.

Section 4.6 Exercises

1. (a) $y = 25 - x$ **(b)** $25x - x^2$ **(c)** Closed interval

(d) $x = y = \frac{25}{2}$ in

3. $x = 1$ **5.** $x = y = 4$

7. (a) $x = 10$ and $y = 10$ **(b)** No

9. width $= \frac{1,200}{4+\pi}$ and length $= \frac{600}{4+\pi}$

11. length of brick wall $= 10\sqrt{5}$ ft, length of adjacent side $= 20\sqrt{5}$ ft

13. $P \approx (0.59, 0.35)$

15. square base of side length 3.42 ft, height of 1.71 ft

17. $a = 4 + \frac{4\sqrt{3}}{3}, b = 4 - \frac{4\sqrt{3}}{3}$ **19.** $r = \frac{1}{3^{1/4}\sqrt{\pi}}, h = \frac{\sqrt{2}}{3^{1/4}\sqrt{\pi}}$

21. height $= \frac{2R}{\sqrt{3}}$, radius $= \sqrt{\frac{2}{3}}R$

23. $\theta = \cos^{-1}\left(\frac{1-\sqrt{3}}{2}\right)$ **25.** $A = \frac{18,000,000}{n+1}$

27. $A = 8 - 4\sqrt{3}$ **29.** $A = \frac{1}{2}$

31. (b) Net cash intake is $C(r) = 200r - \frac{1}{10}r^2 - 19,000$; maximized when $r = \$1,000$.

33. For $n = 2$, $x = \frac{1}{2}(x_1 + x_2)$; for $n = 3$, $x = \frac{1}{3}(x_1 + x_2 + x_3)$; for general n, $x = \frac{1}{n}\sum_{k=1}^{n} x_k$.

35. (a) $v = \frac{175\sqrt{6}}{3+1.5\sqrt{6}} \approx 64.2$ mph

(b) The cost as a function of speed is shown below for $L = 100$. The optimal speed is clearly around 64 mph.

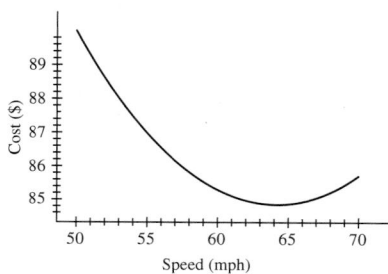

(c) We expect v to increase if P goes down to $2 per gallon.

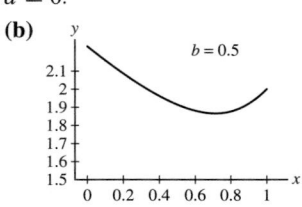

Cost ($) vs Speed (mph) graph with curves P = $3 and P = $2.

37. $r = \frac{1}{(4\pi)^{1/3}}, h = 4^{2/3}\pi^{-1/3}$

39. $h = 3$; dimensions: $9 \times 18 \times 3$ **41.** $A = B = 12$ in

45. $N = 9$ **47.** $A = \frac{3\sqrt{3}}{4}r^2$

51. (b) The underwater observer sees what appears to be a silvery mirror due to total internal reaction.

55. (a) If $b < \sqrt{3}a$, then $d = a - b/\sqrt{3} > 0$. If $b \geq \sqrt{3}a$, then $d = 0$.

(b)

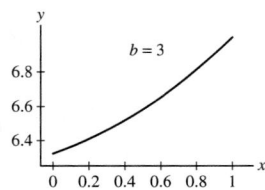

For $b = 3$ and $b = \sqrt{3}$, the minimum is 0. For $b = 0.5$, the minimum is 0.711.

59. $4\sqrt{1 + 2^{2/3}}(2^{2/3} + 1)$ **61.** $\theta = \pi/6$

67. (a) $d \approx 630.13$ ft   **(b)** That a different model is required.

69. (e) When $r = 1.5$ cm, the minimum is $S(0) = 8.4954$ cm^2; when $r = 2$ cm, the minimum is $S(x_2) = 12.852$ cm^2.

Section 4.7 Preliminary

1. $\lim_{x\to 4} \frac{3x-12}{x^2-16}$

2. $\lim_{x\to 0} \frac{x^2-2x}{3x-2}$ is not indeterminate, so L'Hôpital's Rule does not apply.

Section 4.7 Exercises

1. 5   **3.** $\frac{1}{15}$ **5.** $\frac{1}{114}$ **7.** 2   **9.** 0   **11.** $-\frac{1}{4}$ **13.** 0   **15.** 0

17. $\frac{4}{3}$ **19.** 1   **21.** 0   **23.** $-\frac{3}{5}$ **25.** $-\frac{7}{3}$ **27.** $\frac{3}{4}$ **29.** $\frac{1}{2}$

31. -1 **33.** 2   **35.** e **37.** $\frac{1}{2}$ **39.** 0   **41.** $\frac{1}{15}$ **43.** 1

45. 0   **47.** $\frac{9}{7}$ **49.** 1   **51.** $\frac{1}{\pi}$

53. $\lim_{x\to\pi/2} \frac{\cos mx}{\cos nx} = \begin{cases} (-1)^{(m-n)/2}, & m, n \text{ even} \\ \text{does not exist}, & m \text{ even}, n \text{ odd} \\ 0 & m \text{ odd}, n \text{ even} \\ (-1)^{(m-n)/2}\frac{m}{n}, & m, n \text{ odd} \end{cases}$

55. Since $\sin(1/x)$ oscillates as $x \to 0+$, L'Hôpital's Rule cannot be applied. Both numerical and graphical investigations suggest that the limit does not exist due to the oscillation.

x	1	0.1	0.01	0.001	0.0001	0.00001
$x^{\sin(1/x)}$	1	3.4996	10.2975	0.003316	16.6900	0.6626

57. (a) $\lim_{x\to 0+} f(x) = 0$; $\lim_{x\to\infty} f(x) = 1$

(b) $f(x)$ is increasing for $0 < x < e$, is decreasing for $x > e$, and achieves a maximum at $x = e$. The maximum value is $f(e) = e^{1/e} \approx 1.444668$

(c) Because $(e, e^{1/e})$ is the only maximum, no solution exists for $c > e^{1/e}$ and only one solution exists for $c = e^{1/e}$. Moreover, because $f(x)$ increases from 0 to $e^{1/e}$ as x goes from 0 to e and then decreases from $e^{1/e}$ to 1 as x goes from e to $+\infty$, it follows that there are two solutions for $1 < c < e^{1/e}$, but only one solution for $0 < c \leq 1$.

(d) Observe that if we sketch the horizontal line $y = c$, this line will intersect the graph of $y = f(x)$ only once for $0 < c \leq 1$ and $c = e^{1/e}$ and will intersect the graph of $y = f(x)$ twice for $1 < c < e^{1/e}$. There are no points of intersection for $c > e^{1/e}$.

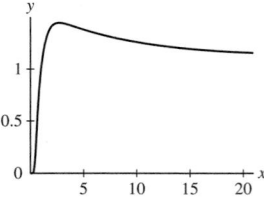

63. $\lim_{x\to\infty} x^n e^{-x} = \lim_{x\to\infty} \frac{x^n}{e^x} = \lim_{x\to\infty} \frac{nx^{n-1}}{e^x}$

$= \lim_{x\to\infty} \frac{n(n-1)x^{n-2}}{e^x}$

$= \cdots = \lim_{x\to\infty} \frac{n!}{e^x} = 0.$

65. Because we are interested in the limit as $t \to +\infty$, we will restrict attention to $t > 1$. Then, for all k,

$$0 \leq t^k e^{-t^2} \leq t^k e^{-t}$$

As $\lim\limits_{t\to\infty} t^k e^{-t} = 0$, it follows from the Squeeze Theorem that

$$\lim_{t\to\infty} t^k e^{-t^2} = 0$$

71. (a) $\lim\limits_{\omega\to\lambda} y(t) = C\frac{\lambda t \cos(\lambda t) - \sin(\lambda t)}{-2\lambda}$

(c) The graphs below were produced with $C = 1$ and $\lambda = 1$. Moving from left to right and from top to bottom, $\omega = 0.5, 0.8, 0.9, 0.99, 0.999, 1$.

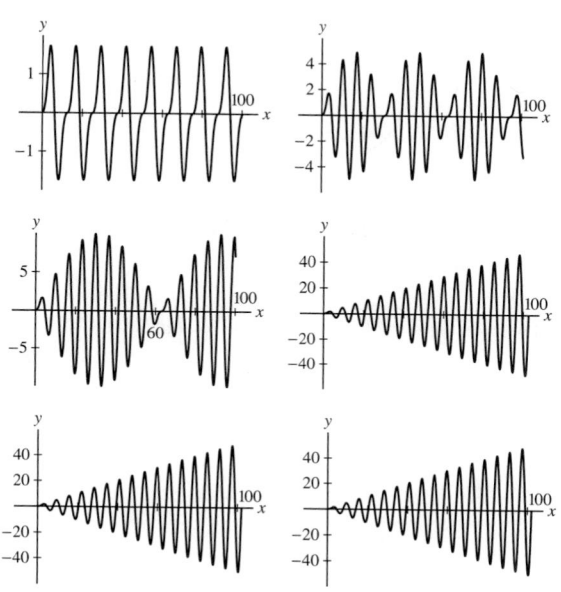

75. (a) $e^{-1/6}$ **(b)** $\frac{1}{3}$

Section 4.8 Preliminary

1. One **2.** Sequence will remain x_0. **3.** x_1 will not be defined.
4. Yes

Section 4.8 Exercises

1.

n	1	2	3
x_n	1.5	1.416666667	1.414215686

3.

n	1	2	3
x_n	1.717708333	1.710010702	1.709975947

5. $x_0 = -1.4$

n	1	2	3
x_n	-1.330964467	-1.328272820	-1.328268856

7. The roots are approximately 0.259 and 2.543.

9.

n	1	2	3
x_n	3.16667	3.162280702	3.16227766

3.16227766

11.

n	1	2	3	4
x_n	1.75	1.710884354	1.709976429	1.709975947

1.709975947

13. 2.093064358

15. $x_0 = 1$

n	1	2	3
x_n	1.25	1.189379699	1.184171279

17. $\theta_0 = 0.8$; The smallest positive solution is 0.787.
19. $x_1 \approx 3.0$, $x_2 \approx 2.2$ **21.** $x = 4.49341$

 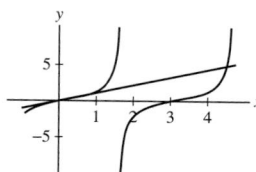

23. (a) $P \approx \$156.69$ **(b)** $b \approx 1.02121$; $r \approx 0.25452$
25. (c) $\theta = 1.76696$. Yes.
27. The sequence of iterates is $-2, 2, -2, 2, \dots$, which diverges by oscillation.
29. -3.58291867 for $x_0 = 1.85$ and $x_0 = 1.7$; 0.251322863 for $x_0 = 1.55$.
31. $\theta = 1.2757$; $h = L\frac{1-\cos\theta}{2\sin\theta} \approx 1.11181$
33. (a) $a \approx 46.95$ **(b)** $s \approx 29.23$
35. (a) $a \approx 28.46$
(b) For $\Delta L = 1$, $\Delta s \approx 0.61$; for $\Delta L = 5$, $\Delta s \approx 3.04$
(c) $s(161) - s(160) \approx 0.6068$; $s(165) - s(160) \approx 3.0155$

Section 4.9 Preliminary

1. $\ln(-x)$ **2.** Any constant function **3.** No difference
4. No **5.** $\sin x + C$; at $x = 0$, the value of the antiderivative is C.
6. (a) False **(b)** True **(c)** False
7. $y = x^2$ is *not* a solution.

Section 4.9 Exercises

1. $F(x) = 6x^2 + C$ **3.** $F(x) = \frac{1}{3}x^3 + \frac{3}{2}x^2 + 2x + C$
5. $F(x) = -\frac{8}{3}x^{-3} + C$ **7.** $F(x) = 5e^x + \frac{1}{3}x^3 + C$
9. (b) **11. (a)**
13. $F(x) = \frac{1}{2}x^2 + x + C$ **15.** $F(t) = \frac{1}{6}t^6 + \frac{3}{2}t^2 + 2t + C$
17. $F(t) = -\frac{5}{4}t^{-4/5} + C$ **19.** $F(x) = 2x + C$
21. $F(t) = \frac{5}{2}t^2 - 9t + C$ **23.** $F(x) = -\frac{1}{x} + C$
25. $F(x) = -\frac{1}{x+3} + C$ **27.** $F(z) = -\frac{3}{4}z^{-4} + C$
29. $F(x) = \frac{2}{5}x^{5/2} - \frac{2}{3}x^{3/2} + C$ **31.** $F(t) = \frac{2}{3}t^{3/2} - 14t^{1/2} + C$
33. $F(x) = -4\cos x - 3\sin x + C$ **35.** $F(t) = \frac{1}{6}\sin(6t + 4) + C$
37. $F(t) = -\frac{1}{4}\sin(3 - 4t) + C$ **39.** $F(x) = \sin x - e^x + C$

41. $F(x) = 5e^{5x} + C$ **43.** (A)

45. (a) $F(x) = \frac{1}{3}\tan(3x) + C$ (b) $F(x) = \sec(x + 3) + C$

47. $y = \frac{1}{2}\sin 2x + 3$ **49.** $y = \frac{1}{2}x^2 + 5$

51. $y = 5t - \frac{2}{3}t^3 - \frac{7}{3}$ **53.** $y = 2t^2 + 9t + 1$

55. $y = 1 - \cos x$ **57.** $y = 3 + \frac{1}{5}\sin 5x$

59. $y = e^x + 3$ **61.** $y = \frac{1}{5}e^{5x} - \frac{16}{5}$

63. $f(x) = \frac{1}{6}x^3 + x$ **65.** $f(x) = \frac{1}{20}x^5 - \frac{1}{3}x^3 + \frac{1}{2}x^2 - \frac{1}{4}x + \frac{121}{30}$

67. $f(\theta) = -\cos\theta + 6$

71. $\frac{ds}{dt} = \frac{1}{2}t^2 - t$, $s(0) = 0$, $s(t) = \frac{1}{6}t^3 - \frac{1}{2}t^2$

73. (a) $s(t) = \frac{25}{2}t^2 - \frac{1}{3}t^3 + 5$ (b) $s(t) = \frac{25}{2}t^2 - \frac{1}{3}t^3 - \frac{127}{3}$

75. $t = 6$ s, $s = 252$ ft **77.** $v(500) \approx 764$ m/s

79. $c_1 = c_2 = 1$

83. (a) $\frac{d}{dx}\left(\frac{1}{2}F(2x)\right) = \frac{1}{2}F'(2x)\cdot 2 = F'(2x) = f(2x)$

(b) $\frac{1}{k}F(kx) + C$

Chapter 4 Review

1. $8.1^{1/3} - 2 \approx 0.00833333$. The error is 3.445×10^{-5}.

3. $625^{1/4} - 624^{1/4} \approx 0.002$. The error is 1.201×10^{-6}.

5. $\frac{1}{1.02} \approx L(1.02) = 0.98$. The error is 3.922×10^{-4}.

7. $L(x) = 5 + \frac{1}{10}(x - 25)$ **9.** $L(r) = 36\pi(r - 2)$

11. $L(x) = \frac{1}{\sqrt{e}}(2 - x)$ **13.** $\Delta s \approx 0.632$

15. 60 MP3 players will be sold per week when the price is $80; 104 MP3 players will be sold per week when the price is $69.

17. $\sqrt{26} \approx 5.1$. The error is 9.80×10^{-4}.

23. $f\left(\frac{2}{3}\right)$ is a local maximum. $f(2)$ is a local minimum.

25. $f(-2)$ is neither a local maximum nor a local minimum, $f\left(-\frac{4}{5}\right)$ is a local maximum, and $f(0)$ is a local minimum.

27. The maximum value is 21. The minimum value is -11.

29. The minimum value is -1. The maximum value is $\frac{5}{4}$.

31. The minimum value is -1. The maximum value is 3.

33. The minimum value is $12 - 12\ln 12 \approx -17.818880$. The maximum value is $40 - 12\ln 40 \approx -4.266553$.

35. Critical points $x = 1$, $x = 3$. The minimum value is 2. The maximum value is 17.

37. $x = \frac{4}{3}$ **39.** $x = \pm\frac{2}{\sqrt{3}}$ **41.** $x = 1$ and $x = 4$

43. (a) ii (b) i (c) iii

45. $L = \infty$ **47.** $L = \infty$ **49.** $L = \frac{1}{4}$ **51.** $L = \infty$

53. $L = \frac{1}{2}$ **55.** $L = -\infty$ **57.** $L = -\infty$ **59.** $L = 0$

61. **63.**

65.

67.

69.

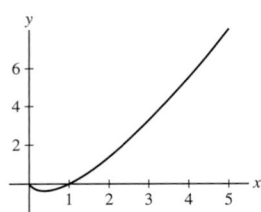

71. $V = \frac{4}{81}\pi R^2 H$

77. $\sqrt[3]{25} = 2.9240$ **79.** $F(x) = x^4 - \frac{2}{3}x^3 + C$

81. $F(\theta) = -\cos(\theta - 8) + C$ **83.** $F(t) = -2t^{-2} + 4t^{-3} + C$

85. $F(x) = \tan x + C$

87. $F(y) = \frac{1}{5}(y + 2)^5$

89. $F(x) = e^x - \frac{1}{2}x^2 + C$ **91.** $F(x) = 4\ln|x| + C$

93. $y(x) = x^4 + 3$ **95.** $y(x) = 2x^{1/2} - 1$ **97.** $y(x) = 4 - e^{-x}$

99. $f(t) = \frac{1}{2}t^2 - \frac{1}{3}t^3 - t + 2$ **101.** $f(0) = \frac{2}{e}$ is a local minimum

103. Local minimum at $x = e^{-1}$; no points of inflection;
$\lim_{x\to 0+} x\ln x = 0$; $\lim_{x\to\infty} x\ln x = \infty$

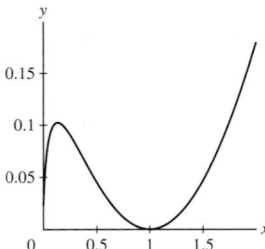

105. Local minimum at $x = 1$; local maximum at
$x = 10^{-2/\ln 10} = e^{-2}$; point of inflection at $x = 10^{-1/\ln 10} = \frac{1}{e}$;
$\lim_{x\to 0+} x(\log x)^2 = 0$; $\lim_{x\to\infty} x(\log x)^2 = \infty$

107. As $x \to \infty$, both $2x - \sin x$ and $3x + \cos 2x$ tend toward infinity, so L'Hôpital's Rule applies to $\lim_{x\to\infty} \frac{2x - \sin x}{3x + \cos 2x}$; however, the resulting limit, $\lim_{x\to\infty} \frac{2 - \cos x}{3 - 2\sin 2x}$, does not exist due to the oscillation of $\sin x$ and $\cos x$. $\lim_{x\to\infty} \frac{2x - \sin x}{3x + \cos 2x} = \frac{2}{3}$

109. 4 **111.** 0 **113.** 3 **115.** $\ln 2$ **117.** $\frac{1}{6}$ **119.** 2

Chapter 5

Section 5.1 Preliminary

1. The right-endpoints are $\frac{5}{2}$, 3, $\frac{7}{2}$, 4, $\frac{9}{2}$, 5. The left-endpoints are 2, $\frac{5}{2}$, 3, $\frac{7}{2}$, 4, $\frac{9}{2}$.
2. L_2 **3.** (b)
5. (a) 1, $\frac{15}{4}$ (b) $\frac{5}{4}$, $\frac{3}{2}$
6. (a) False   (b) True   (c) True

Section 5.1 Exercises

1. $10\frac{1}{2}$ mi   **3.** 11.4 in.   **5.** $R_6 = 54.5$ ft, $L_6 = 44.5$ ft, $M_3 = 51$ ft
7. (a) $L_6 = 16.5$, $R_6 = 19.5$ (b) exact area: $A = 18$;
$L_6 - A = -1.5$; $R_6 - A = 1.5$
9. $L_6 = 2.925$, $R_6 = 2.4875$, $M_6 = 2.75$
11.

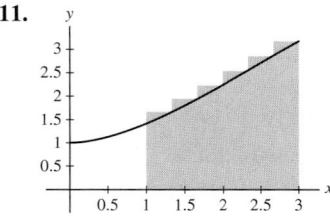

13. $R_8 = 5.75$ **15.** $M_4 = 0.328125$ **17.** $R_6 = 47.5$
19. $L_5 \approx 0.745635$ **21.** $L_4 \approx 0.361372$
23. $M_6 \approx 0.386871$ **25.** $0.518 \le A \le 0.768$
29. $L_{100} = 0.6614629$, $R_{100} = 0.6714629$; $L_{150} = 0.6632220$, $R_{150} = 0.6698887$
31. (a) 15   (b) 18   (c) 99   (d) -1 (e) $\frac{11}{6}$ (f) 40
33. 15,050   **35.** $\sum_{k=2}^{5} (k^2 + k)$ **37.** $\sum_{k=1}^{n} \sqrt{k + k^3}$
39. $\sum_{k=1}^{n} e^{\pi/k}$ **41.** 440   **43.** 379,507,500   **45.** $\frac{46,168}{3}$
47. 26,475   **49.** 3   **51.** 11   **53.** $\frac{1}{4}$ **55.** $-\frac{79}{4}$ **57.** 20   **59.** 0.5
61. $R_N = \frac{1}{4} + \frac{1}{2N} + \frac{1}{4N^2}$, $\lim_{N \to \infty} R_N = \frac{1}{4}$
63. $R_N = 1 - \frac{1}{4} - \frac{1}{2N} - \frac{1}{4N^2}$, $\lim_{N \to \infty} R_N = \frac{3}{4}$
65. $R_N = 128 + \frac{136}{N} + \frac{32}{N^2}$, $\lim_{N \to \infty} R_N = 128$
67. $R_N = \frac{56}{3} + \frac{12}{N} + \frac{4}{3N^2}$, $\lim_{N \to \infty} R_N = \frac{56}{3}$
69.
$$R_N = a^2(b - a) + a(b - a)^2 + \frac{a(b - a)^2}{N} + \frac{(b - a)^3}{3}$$
$$+ \frac{(b - a)^3}{2N} + \frac{(b - a)^3}{6N^2},$$
$$\lim_{N \to \infty} R_N = \frac{1}{3}b^3 - \frac{1}{3}a^3$$
73. The area between the graph of $f(x) = x^4$ and the x-axis over the interval $[2, 5]$.
75. The area between the graph of $f(x) = \sin x$ and the x-axis over the interval $\left[\frac{\pi}{3}, \frac{5\pi}{6}\right]$.

77. $\lim_{N \to \infty} \frac{\pi}{N} \sum_{k=1}^{N} \sin\left(\frac{k\pi}{N}\right)$

79. $\lim_{N \to \infty} \frac{1}{2N} \sum_{j=1}^{N} \tan\left(\frac{1}{2} + \frac{1}{2N}\left(j - \frac{1}{2}\right)\right)$

81. $\lim_{N \to \infty} \frac{7\pi}{8N} \sum_{j=0}^{N-1} \cos\left(\frac{\pi}{8} + j\frac{7\pi}{8N}\right)$ **85.** R_{501}, L_{500}, M_{10}

87. Left/right-endpoint approximation, $n = 2$ No.

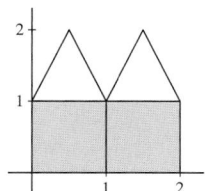

93. $N > 30,000$

Section 5.2 Preliminary

1. $b - a$
2. (a) False   (b) True   (c) True
4. No   **5.** 2

Section 5.2 Exercises

1. Area $= 0$ **3.** Area $= 7.5$

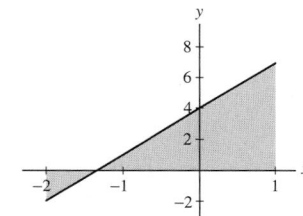

5. Area $= 0$ **7.** Area $= \frac{25\pi}{4}$

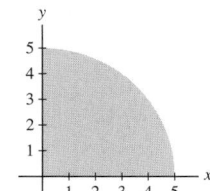

9. Area $= 4$ **11.** Area $= 6$

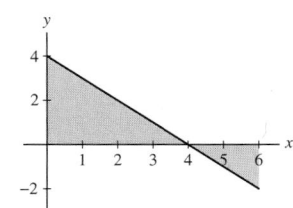

13. (a) $-\frac{\pi}{2}$ (b) $\frac{3\pi}{2}$ (c) $\frac{3}{4}\pi$ (d) $\frac{9}{4}\pi$
15. $a = 4$, $b = 1$, $c = 4$

17.

19.

21.

23. Positive   **25.** Negative

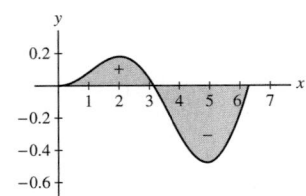

27. $R(f, P, C) = 1.59$ **29.** $R(f, P, C) = 0.24$

   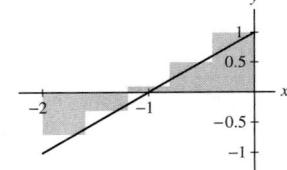

31. $\frac{64}{3}$ **33.** $\frac{51}{2}$ **35.** $-\frac{2}{3}$ **37.** $\frac{1}{3}a^3 - \frac{1}{2}a^2 + \frac{5}{6}$ **39.** $e^4 - 1$

43. $\frac{20}{3}$ **45.** -2 **47.** -24 **49.** $-\frac{9}{4}$ **51.** 53 **53.** -45

55. 8 **57.** -7 **59.** $\int_0^7 f(x)\,dx$ **61.** $\int_5^9 f(x)\,dx$ **63.** $\frac{2}{3}$

65. -3

69. Here is a graphical example of this phenomenon.

   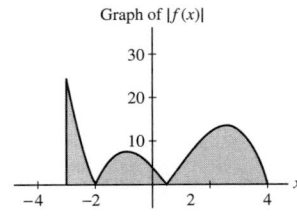

Graph of $f(x)$ Graph of $|f(x)|$

71. 9   **73.** 0.5

75. On the interval $[0, 1]$, $x^5 \le x^4$, so $\int_0^1 x^5\,dx \le \int_0^1 x^4\,dx$. On the other hand, $x^4 \le x^5$ for $x \in [1, 2]$, so $\int_1^2 x^4\,dx \le \int_1^2 x^5\,dx$.

81. The assertion is false. A counterexample: $a = 0, b = 1$, $f(x) = x, g(x) = 2$.

Section 5.3 Preliminary

1. 4

2. Signed area between the graph of $y = f(x)$ and the x-axis over the interval $[1, 4]$.

3. $\int_0^7 f(x)\,dx = 6$, $\int_2^7 f(x)\,dx = 2$

4. (a) False   (b) False   (c) False

5. 0

Section 5.3 Exercises

1. Area $= \frac{1}{3}$ **3.** Area $= 1$

   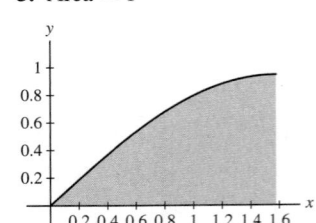

5. $\frac{27}{2}$ **7.** $\frac{35}{3}$ **9.** $e^5 - e^3$ **11.** $2e^{-2} - 8$ **13.** $\frac{34}{3}$ **15.** $\frac{133}{3}$

17. 0 **19.** $\frac{162\sqrt{3}}{5} - \frac{82}{45}$ **21.** $\frac{3}{4}$ **23.** 60 **25.** 4 **27.** $\frac{1}{3}\ln 50$

29. $-\frac{3}{8}$ **31.** $60\sqrt{3} - \frac{8}{3}$ **33.** 2 **35.** $\sqrt{2}$ **37.** 1 **39.** $2 - \frac{2}{3}\sqrt{3}$

41. $\int_{-2}^0 -x\,dx + \int_0^1 x\,dx = \frac{5}{2}$ **43.** $\int_{-2}^0(-x^3)\,dx + \int_0^3 x^3\,dx = \frac{97}{4}$

45. $\int_0^{\pi/2}\cos x\,dx + \int_{\pi/2}^\pi(-\cos x)\,dx = 2$

47. $\frac{1}{4}(b^4 - 1)$ **49.** $\frac{1}{6}(b^6 - 1)$ **51.** $\ln 5$ **57.** $\frac{707}{12}$

Section 5.4 Preliminary

1. $A(-2) = 0$

2. (a) No   (b) Yes

3. (c)   **4.** False   **5.** (b), (e), (f) are correct

6. $\frac{d}{dx}\left(\int_1^3 t^3\,dt\right)\Big|_{x=2} = 0$

Section 5.4 Exercises

1. $A(x) = \int_{-2}^x(2t + 4)\,dt = x^2 + 4x + 4$

3. (a) $G(1) = \int_1^1(t^2 - 2)\,dt = 0$ (b) $G'(1) = -1$ and $G'(2) = 2$
(c) $G(x) = \frac{1}{3}x^3 - 2x + \frac{5}{3}$

5. $G(1) = 0, G'(0) = 0, G'(\frac{\pi}{4}) = 1$ **7.** $F(x) = \frac{1}{4}x^4 - 4$

9. $F(x) = \frac{1}{2}x^4 - \frac{1}{2}$ **11.** $F(x) = e^5 - e^x$

13. $F(x) = \tan x + 1$ **15.** $F(x) = \int_3^x \sqrt{t^4 + 1}\,dt$

17. $F(x) = \int_0^x \sec t\,dt$ **19.** $\frac{d}{dx}\int_0^x(t^3 - t)\,dt = x^3 - x$

21. $\frac{d}{dt}\int_{100}^t \cos 5x\,dx = \cos 5t$

23.

25.

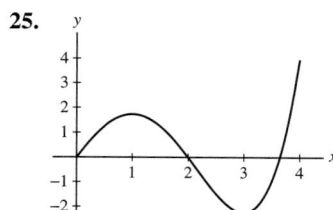

27. $G'(x) = 3x^2 \tan x^3$

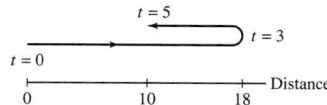

13. 132 liters **15.** 205 ft/s **17.** 9,200 cars

19. The total cost of reducing the amount of CO_2 by 3 tons.

21. (a) Net migration during 1988–1991. **(b)** Net outflow of 66,500 people. **(c)** 1989

23. (b) $\frac{1,300 - 100\sqrt{2}}{3} \approx 386.19$ J

25. 3,517 families

29. $\frac{d}{dx} \int_0^{x^2} \sin^2 t\, dt = 2x \sin^2(x^2)$

31. $\frac{d}{ds} \int_{-6}^{\cos s} (u^4 - 3u)\, du = -\sin s (\cos^4 s - 3\cos s)$

33. $\frac{d}{dx} \int_{x^3}^0 \sin^2 t\, dt = -3x^2 \sin^2 x^3$

35. $\frac{d}{dx} \int_{\sqrt{x}}^{x^2} \tan t\, dt = 2x \tan x^2 - \frac{\tan(\sqrt{x})}{2\sqrt{x}}$

37. Minimum: $A(1.5) = -1.25$; maximum: $A(4.5) = 1.25$

39. (a) No **(b)** R **(c)** S **(d)** True

43. (a) Increasing on $(0, 4)$ and $(8, 12)$. Decreasing on $(4, 8)$ and $(12, \infty)$.

(b) Local minimum at $x = 8$. Local maximum at $x = 4$ and at $x = 12$.

(c) $x = 2$, $x = 6$, and $x = 10$

(d) Concave up $(0, 2)$ and $(6, 10)$. Concave down $(2, 6)$ and $(10, \infty)$.

45.

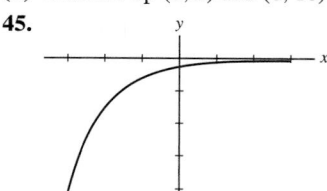

47. $x = \pi^{2/3}$, $F(x)$ changes from concave down to concave up.

49. $g(x) = 2x + 1$, $c = 2$, or $c = -3$

53. $a = -3$ and $b = 3$

Section 5.5 Preliminary

1. 350 mi **2.** The total drop in temperature

3. (a) and **(c)** represented as derivatives. **(b)** and **(d)** represented as integrals.

4. Difference in distance

Section 5.5 Exercises

1. 15,062.5 gal **3.** 682 insects **5.** 186 bicycles **7.** 12 ft

9. Displacement $= 10$ ft, distance $= 26$ ft

11. Displacement $= 0$ m, distance $= 1$ m

Section 5.6 Preliminary

1. (a), (b)

2. (a) $c = \frac{1}{2}$, $g(u) = u^4$, and $u(x) = x^2 + 9$

(b) $c = \frac{1}{3}$, $g(u) = \sin u$, and $u(x) = x^3$

(c) $c = -1$, $g(u) = u^2$, and $u(x) = \cos x$

3. (c)

Section 5.6 Exercises

1. $du = -2x\, dx$ **3.** $du = 3x^2\, dx$ **5.** $du = -2x \sin(x^2)\, dx$

7. $\int u^3\, du = \frac{1}{4}(x - 7)^4 + C$ **9.** $\int u^{-2}\, du = -\frac{1}{x+1} + C$

11. $\frac{1}{2} \int \sin u\, du = -\frac{1}{2} \cos(2x - 4) + C$

13. $\frac{1}{2} \int \frac{1}{u^3}\, du = \frac{-1}{4(x^2+2x)^2} + C$

15. $\frac{1}{4} \int u^{1/2}\, du = \frac{1}{6}(4x - 1)^{3/2} + C$

17. $\frac{1}{64} \int \left(u^{5/2} + 2u^{3/2} + u^{1/2}\right) du =$
$\frac{1}{224}(4x - 1)^{7/2} + \frac{1}{160}(4x - 1)^{5/2} + \frac{1}{96}(4x - 1)^{3/2} + C$

19. $\int u^2\, du = \frac{1}{3} \sin^3 x + C$

21. $-\frac{1}{2} \int \frac{du}{u} = -\frac{1}{2} \ln |\cos 2x| + C$

23. $-\frac{1}{2} \int e^u\, du = -\frac{1}{2} e^{-x^2} + C$ **25.** $\int \frac{du}{(u+1)^2} = -\frac{1}{e^t+1} + C$

27. $\int u^{-2}\, du = -\frac{1}{\ln x} + C$ **29.** Take $u = x^4$ **31.** Take $u = x^{3/2}$

33. $\int (4x + 3)^4\, dx = \frac{1}{20}(4x + 3)^5 + C$

35. $\int \frac{1}{\sqrt{x-7}}\, dx = 2\sqrt{x - 7} + C$

37. $\int x\sqrt{x^2 - 4}\, dx = \frac{1}{3}(x^2 - 4)^{3/2} + C$

39. $\int \frac{dx}{(x+9)^2}\, dx = -\frac{1}{x+9} + C$

41. $\int \frac{2x^2+x}{(4x^3+3x^2)^2}\, dx = -\frac{1}{6}(4x^3 + 3x^2)^{-1} + C$

43. $\int \frac{5x^4+2x}{(x^5+x^2)^3}\, dx = -\frac{1}{2} \frac{1}{(x^5+x^2)^2} + C$

45. $\int (3x + 9)^{10}\, dx = \frac{1}{33}(3x + 9)^{11} + C$

47. $\int x(x + 1)^{1/4}\, dx = \frac{4}{9}(x + 1)^{9/4} - \frac{4}{5}(x + 1)^{5/4} + C$

49. $\int x^3(x^2 - 1)^{3/2}\, dx = \frac{1}{7}(x^2 - 1)^{7/2} + \frac{1}{5}(x^2 - 1)^{5/2} + C$

51. $\int \sin^5 x \cos x\, dx = \frac{1}{6} \sin^6 x + C$

53. $\int \tan 3x\, dx = -\frac{1}{3} \ln |\cos 3x| + C$

55. $\int \sec^2(4x + 9)\, dx = \frac{1}{4} \tan(4x + 9) + C$

57. $\int \frac{\cos 2x}{(1+\sin 2x)^2}\, dx = -\frac{1}{2}(1 + \sin 2x)^{-1} + C$

59. $\int \cos x (3 \sin x - 1)\, dx = \frac{1}{6}(3 \sin x - 1)^2 + C$

61. $\int \sec^2 x (4 \tan^3 x - 3 \tan^2 x)\, dx = \tan^4 x - \tan^3 x + C$

63. $\int (x + 1)e^{x^2+2x}\, dx = \frac{1}{2}e^{x^2+2x} + C$

65. $\int \frac{e^x\, dx}{(e^x+1)^4} = -\frac{1}{3}(e^x + 1)^{-3} + C$

67. $\int \frac{(\ln x)^4\, dx}{x} = \frac{1}{5}(\ln x)^5 + C$

69. $\int \frac{dx}{x \ln x} = \ln|\ln x| + C$

71. $\int \sqrt{x^3 + 1}\, x^5\, dx = \frac{2}{15}(x^3 + 1)^{5/2} - \frac{2}{9}(x^3 + 1)^{3/2} + C$

77. (d) **79.** $\frac{38}{3}$ **81.** $\frac{42}{5}$ **83.** $\frac{72}{289}$ **85.** 3 **87.** $\frac{1}{2}\ln(\sec 1)$

89. $-\frac{1}{3}$ **91.** $\frac{1}{3}$ **93.** $\frac{20}{3}\sqrt{5} - \frac{32}{5}\sqrt{3}$

95. $\int \frac{f'(x)}{f(x)^2}\, dx = -\frac{1}{f(x)} + C$ **97.** $\frac{1}{n+1}$ **99.** $\frac{\pi}{4}$

Section 5.7 Preliminary

1. (a) $\frac{2^x}{\ln 2} + C$ (b) $\ln|x| + C$ (c) $\sin^{-1} x + C$

2. (a) $b = 3$ (b) $b = e^3$

3. $b = \sqrt{3}$ **4.** (b): Use the substitution $u = 3x$ **5.** $x = 4u$

Section 5.7 Exercises

1. $\ln 2$ **3.** 1 **5.** -1 **7.** $\frac{\pi}{6}$ **9.** $\frac{\pi}{6}$ **13.** $\frac{\pi}{8}$

15. $\sin^{-1}\left(\frac{t}{4}\right) + C$ **17.** $\frac{1}{2}\sin^{-1}\left(\frac{2t}{5}\right) + C$ **19.** $\frac{1}{2}\sin^{-1} 2x + C$

21. $-\sqrt{1 - x^2} + \sin^{-1} x + C$ **23.** $\tan^{-1} e^x + C$

25. $\frac{1}{2}(\tan^{-1} x)^2 + C$ **27.** $\frac{2}{\ln 3}$ **29.** $\frac{1}{\ln 2}$ **31.** $-\frac{1}{\ln 9}\cos(9^x) + C$

33. $e^x + 2x + C$ **35.** $-\frac{7^{-x}}{\ln 7} + C$ **37.** $\frac{1}{4}e^{4x} + x + C$

39. $-\frac{1}{9}e^{-9t} + C$ **41.** $\frac{1}{4}\sin^{-1}(4x) + C$ **43.** $\frac{2}{3}(e^t + 1)^{3/2} + C$

45. $7x - \frac{1}{10}e^{10x} + C$ **47.** $\sec^{-1}(5x) + C$ **49.** $-\frac{1}{8}e^{-4x^2} + C$

51. $2\sqrt{e^x + 1} + C$ **53.** $\frac{1}{2}\ln|2x + 4| + C$ **55.** $\frac{1}{3}\ln|x^3 + 2| + C$

57. $-\frac{1}{4}\ln|\cos(4x + 1)| + C$ **59.** $\frac{1}{2}\ln(2 \sin x + 3) + C$

61. $\frac{1}{8}(4 \ln x + 5)^2 + C$ **63.** $\ln|\ln x| + C$ **65.** $\frac{1}{2}\left(\ln(\ln x)\right)^2 + C$

67. $\frac{3^x}{\ln 3} + C$ **69.** $\frac{3^{\sin x}}{\ln 3} + C$

75. Assume $f(x) = e^x$ is a polynomial function of degree n. Then $f^{(n+1)}(x) = 0$. But we know that any derivative of e^x is e^x and $e^x \ne 0$. Hence, e^x cannot be a polynomial function.

81.

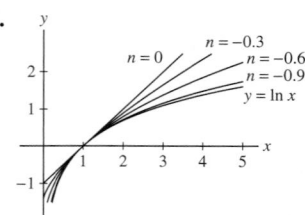

Section 5.8 Preliminary

1. The quantity with $k = 3.4$ doubles more rapidly. **2.** No

3. They take the same amount of time. **4.** $\frac{dS}{dn} = \frac{1}{2}S$

5. Too ancient **6.** 12% compounded quarterly

7. Continuously: 1.09417; quarterly: 1.09308

8. (b) The amount you would have to invest today in order to receive N dollars at time T.

9. Decrease

10. If the interest rate goes up, the present value of $1,000 one year from today decreases. Therefore, Xavier will be sad if the interest rate has just increased from 6 to 7%.

Section 5.8 Exercises

1. (a) $P(0) = 2,000$ (b) $t = \frac{1}{1.3}\ln 5 \approx 1.24$ hr

3. $N'(t) = \frac{\ln 2}{3}N(t)$; after 10 minutes, there are 10 molecules.

5. $\frac{\ln 2}{0.13} \approx 5.33$ yr **7.** $y(t) = Ce^{-5t}$; $y(t) = 3.4e^{-5t}$

9. $y(t) = 4e^{3(t-2)}$

11. (a) $\frac{\ln 2}{0.06} \approx 11.55$ yr (b) $\frac{\ln 3}{0.06} \approx 18.31$ yr (c) 23.10 yr

13. Data I **17.** $k = \frac{\ln(3/10)}{-17} \approx 0.071$ yr^{-1}

19. $10^{-12}e^{-0.000121(32,400)} \approx 1.98 \times 10^{-14}$;
$10^{-12}e^{-0.000121(29,700)} \approx 2.75 \times 10^{-14}$

21. (a) $k = -\frac{1}{10}\ln\left(\frac{2.13}{14.7}\right) \approx 0.193$ mi^{-1}

(b) $P(15) \approx 0.813$ lb/in^2

23. $t = \frac{1}{0.1155}\ln 1.5 \approx 3.5$ hr

25. (a) Yes; $k \approx \frac{\ln 18,666.67}{29} \approx 0.339$. Note: A better estimate can be found by calculating k for each time period and then averaging the k values.

(b)

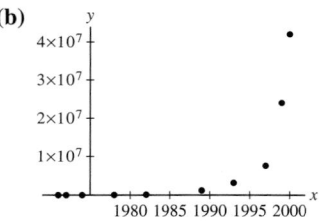

(c) $N(t) = 2,250e^{0.339t}$

(d) The doubling time is $\ln 2/0.339 \approx 2.04$ yr.

(e) $2,250e^{0.339(39)} \approx 1,241,623,327$

(f) No, you can't make a microchip smaller than an atom.

29. (a) Starting from $t_0 = 1$, P doubles when $t = \sqrt{2}$. Starting from $t_0 = 2$, P doubles when $t = 2\sqrt{2}$. Finally, starting from $t_0 = 3$, P doubles when $t = 3\sqrt{2}$.

(b) Starting from $t = t_0$, P doubles when $t = t_0\sqrt{2}$.

31. (a) $4,870.38 (b) $4,902.71 (c) $4,919.21

33. (a) 1.05 (b) 1.0508 (c) 1.0513

35. The account doubles when $P(t) = 2P_0 = P_0e^{rt}$ so $2 = e^{rt}$ and $t = \frac{\ln 2}{r}$.

37. $26,629.59

39. $r = 0.08$: receive $1,000 now; $r = 0.03$: receive $1,300 four years from now

41. Yes, the investment is worthwhile.

43. (a) $22,252,915.21 (b) $2,747,084.79

45. $39,346.93   **47.** $3,296.80   **51.** $R = \$1,200$
53. $71,460.53   **55.** Potassium-49
57. (d) Let $g(c) = c \ln c - c$. Then,

c	0.01	0.001	0.0001	0.00001
$g(c)$	-0.056052	-0.007908	-0.001021	-0.000125

Thus, as $c \to 0+$, it appears that $g(c) \to 0$.
(e) $M = 1/k$ **(f)** $M = 5.52$ days

Chapter 5 Review

1. $L_4 = \frac{23}{4}, M_4 = 7$ **3.** $[0, 2]$ for R_4, $[2, 3]$ for L_4
5. $R_6 = \frac{463}{8}$ $M_6 = \frac{815}{16}$

$L_6 = \frac{355}{8}$

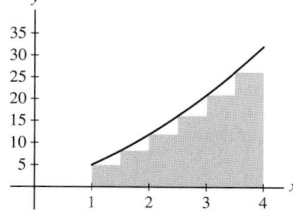

7. $L_N = \frac{32}{3} - \frac{12}{N} + \frac{4}{3N^2} + \frac{8(N-1)}{N}$, $\int_0^2 f(x)\,dx = \frac{32}{3}$
9. $R_6 \approx 0.742574$, $M_6 \approx 0.785977$, $L_6 \approx 0.825907$
11. $R_5, R_5 = L_5 = 90$
13. $\int (6x^3 - 9x^2 + 4x)\,dx = \frac{3}{2}x^4 - 3x^3 + 2x^2 + C$
15. $\int (2x^3 - 1)^2\,dx = \frac{4}{7}x^7 - x^4 + x + C$
17. $\int \frac{x^4+1}{x^2}\,dx = \frac{1}{3}x^3 - x^{-1} + C$ **19.** 30
21. $\int \csc^2\theta\,d\theta = -\cot\theta + C$
23. $\int \sec^2(9t - 4)\,dt = \frac{1}{9}\tan(9t - 4) + C$
25. $\int (9t - 4)^{11}\,dt = \frac{1}{108}(9t - 4)^{12} + C$
27. $\int \sin^2(3\theta)\cos(3\theta)\,d\theta = \frac{1}{9}\sin^3(3\theta) + C$
29. $\int \frac{2x^3+3x}{(3x^4+9x^2)^5}\,dx = -\frac{1}{24}(3x^4 + 9x^2)^{-4} + C$
31. $\int \sin\theta\sqrt{4 - \cos\theta}\,d\theta = \frac{2}{3}(4 - \cos\theta)^{3/2} + C$
33. $\int y\sqrt{2y + 3}\,dy = \frac{1}{10}(2y + 3)^{5/2} - \frac{1}{2}(2y + 3)^{3/2} + C$
35. $\int_{-2}^{6} f(x)\,dx$ **37.** $x = \pm 1$

39. Daily consumption is 9.312 million gallons. 18:00–24:00 consumption is 1.68 million gallons.
41. $208,245   **45.** $\int_0^2 \sin x\,dx = 1 - \cos 2$
47. $\int_{\pi/2}^{3\pi/2} \sin x\,dx = 0$ **49.** $\int_0^1 x^k\,dx = \frac{1}{k+1}$
51. $\int_0^1 f(x)\,dx = \frac{13}{2}$ **53.** $\int x\cos x\,dx = x\sin x + \cos x + C$
55.

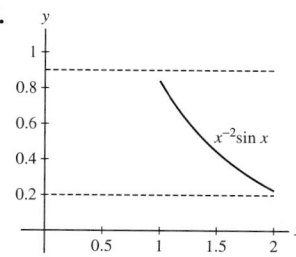

57. $A'(x) = \sin(x^3)$ **59.** $\frac{d}{dy}\int_{-2}^{y} 3^x\,dx = 3^y$ **61.** $G'(2) = 36$
65. $\frac{1}{3}(\ln x)^3 + C$ **67.** $-\sin^{-1}(e^x) + C$ **69.** $\tan^{-1}(\ln t) + C$
71. $-\frac{1}{2}e^{9-2x} + C$ **73.** $\frac{1}{2}\cos(e^{-2x}) + C$ **75.** $\frac{1}{2}$
77. $\sin^{-1}\left(\frac{2}{3}\right) - \sin^{-1}\left(\frac{1}{3}\right)$ **79.** $\ln 2$ **81.** $\frac{1}{2}\sin 2$ **83.** $\frac{1}{2}\ln 2$
85. $\frac{1}{2}\sin^{-1} x^2 + C$ **87.** $\frac{1}{2}(\sin^{-1})^2 + C$ **89.** $-\frac{1}{4}\cos^4 x + C$
91. $\frac{\sqrt{2}}{2}\tan^{-1}(4\sqrt{2})$
95. Because $t > \ln t$ for $t > 2$,

$$F(x) = \int_2^x \frac{dt}{\ln t} > \int_2^x \frac{dt}{t} > \ln x$$

Thus, $F(x) \to \infty$ as $x \to \infty$. Moreover,

$$\lim_{x\to\infty} G(x) = \lim_{x\to\infty} \frac{1}{1/x} = \lim_{x\to\infty} x = \infty$$

Thus, $\lim_{x\to\infty} \frac{F(x)}{G(x)}$ is of the form ∞/∞ and L'Hôpital's Rule applies. Finally,

$$L = \lim_{x\to\infty} \frac{F(x)}{G(x)} = \lim_{x\to\infty} \frac{\ln x}{\ln x - 1} = 1$$

97. approximately 6,066 years   **99.** $12,809.44
101. $\frac{dk}{dT} \approx 12.27$ hr^{-1}-K^{-1}; approximate change in k when T is raised from 500 to 510: 122.7 hr^{-1}.
103. If $r = 5\%$, this is a good investment. The largest interest rate that would make the investment worthwhile is $r \approx 10.13\%$.

Chapter 6

Section 6.1 Preliminary

1. Area of the region bounded between the graphs of $y = f(x)$ and $y = g(x)$, bounded on the left by the vertical line $x = a$ and on the right by the vertical line $x = b$.
2. Yes   **3.** $\int_0^3 (f(x) - g(x))\,dx + \int_3^5 (g(x) - f(x))\,dx$
4. Negative

Section 6.1 Exercises

1. 102

3. (a) $(-2, -2)$ and $(1, 1)$ **(b)** 4.5

5. $\sqrt{2} - 1$ **7.** $2\sqrt{2}$ **9.** $\frac{343}{3}$ **11.** $\frac{1}{2}e^2 - e + \frac{1}{2}$

13. $\frac{2\pi}{3}$

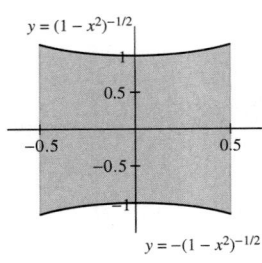

15. $\frac{160}{3}$ **17.** $\frac{5}{24}$ **19.** $\frac{81}{2}$ **21.** $2 - \frac{\pi}{2}$ **23.** $\frac{1,331}{6}$ **25.** $\frac{32}{3}$

27. $\frac{64}{3}$

29. $\frac{64}{3}$ **31.** $2 - \frac{\pi}{2}$

33. 2 **35.** $\frac{128\sqrt{2}}{15}$

37. $\frac{44}{3}$ **39.** 12

41. $\frac{32}{3}$ **43.** $\frac{16}{3}$

45. $2 - \sqrt{2}$

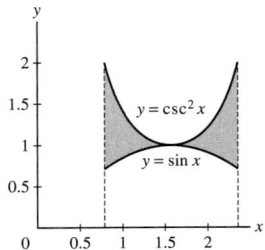

47. The region is $\frac{1}{4}$ of the unit circle; $\frac{\pi}{4}$

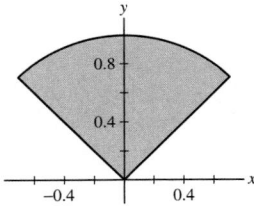

49. $\int_0^3 (f(x) - g(x))\, dx + \int_3^5 f(x)\, dx - \int_5^9 f(x)\, dx$

51. $\int_{-1}^1 \left(\sqrt{2 - x^2} - (1 - \sqrt{1 - x^2}) \right) dx$

53. $x = 0.8278585215$. The area is 0.8244398727.

55. $x = 0.6662394325$. The area is 0.09393667698.

57. $m = 1 - \left(\frac{1}{2}\right)^{1/3} \approx 0.206299$

Section 6.2 Preliminary

1. 3 **2.** 15 **3.** Circles, circles

4. $2\pi \int_0^R r\rho(r)\, dr$, where $\rho(r)$ is the radial population density function.

5. Flow rate is the volume of fluid that passes through a cross-sectional area at a given point per unit time.

6. The fluid velocity depended only on the radial distance from the center of the tube.

Section 6.2 Exercises

1. (a) $\frac{4}{25}(20 - y)^2$ **(b)** $V = \frac{1,280}{3}$

3. $V = \frac{\pi r^2 h}{3}$ **5.** $V = \pi\left(Rh^2 - \frac{h^3}{3}\right)$ **7.** $V = \frac{1}{6}abc$ **9.** $V = \frac{8}{3}$

11. $V = 36$ **13.** $V = 18$ **15.** $V = \frac{\pi}{3}$ **17.** $V = \frac{s^3\sqrt{2}}{12}$

21. (a) $w = 2\sqrt{r^2 - y^2}$ (b) $4(r^2 - y^2)$ (c) $V = \frac{16}{3}r^3$
23. $V = 160\pi$ **25.** $M = 2$ kg **27.** $P \approx 4{,}423.59$ thousand
29. $P = 463$ **31.** $P \approx 61$ deer **33.** $Q = 128\pi$ cm^3/s
35. $Q = \frac{8\pi}{3}$ cm^3/s **37.** $\frac{1}{4}$ **39.** $\frac{2}{\pi}$ **41.** 0.1 **43.** 3 **45.** $\frac{\pi}{4}$
47. Mean Value Theorem for Integrals; $c = \frac{A}{\sqrt[3]{4}}$
49. Over $[0, 1]$, $f(x)$; over $[1, 2]$, $g(x)$.
51. Many solutions exist. One could be:

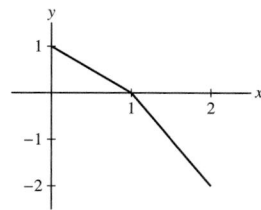

53. Over $[2, 6]$, the average temperature is 72.387324; over $[0, 24]$, the average temperature is 70.
55. $\frac{\pi}{3}$
57. Average acceleration $= \frac{1}{4}$ m/s^2; average velocity $= \frac{7}{60}$ m/s
59. $c = \frac{1{,}444}{225}$ **61.** $v_0/2$

Section 6.3 Preliminary

1. (a), (c) **2.** True
3. False. The cross sections will be washers. **4.** (b)

Section 6.3 Exercises

1. (a)

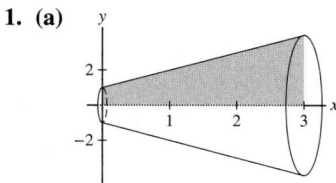

(b) Disk with radius $x + 1$ (c) $V = 21\pi$
3. (a)

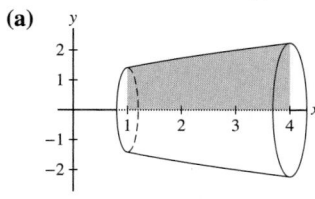

(b) Disk with radius $\sqrt{x + 1}$ (c) $V = \frac{21\pi}{2}$
5. $V = \frac{81\pi}{10}$ **7.** $V = \frac{24{,}573\pi}{13}$ **9.** $V = \pi$ **11.** $\frac{\pi}{2}(e^2 - 1)$
13. (a)

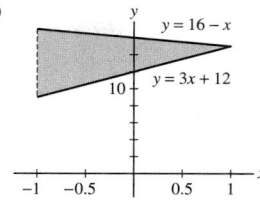

(b) A washer with outer radius $R = 10 - x^2$ and inner radius $r = x^2 + 2$.
(c) $V = 256\pi$
15. (a)

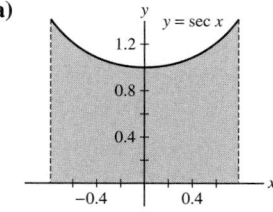

(b) A washer with outer radius $R = 16 - x$ and inner radius $r = 3x + 12$.
(c) $V = \frac{656\pi}{3}$
17. (a)

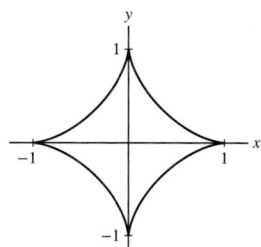

(b) A circular disk with radius $R = \sec x$. (c) $V = 2\pi$
19. $V = \frac{15\pi}{2}$ **21.** $V = \frac{3\pi}{10}$ **23.** $V = \frac{704\pi}{15}$ **25.** $V = \frac{128\pi}{5}$
27. $V = 40\pi$ **29.** $V = \frac{376\pi}{15}$ **31.** $V = \frac{824\pi}{15}$ **33.** $V = \frac{32\pi}{3}$
35. $V = \frac{1{,}872\pi}{5}$ **37.** $V = \frac{4\pi}{3}$ **39.** $V = \frac{1{,}472\pi}{15}$ **41.** $V = \frac{9\pi}{8}$
43. $V = \frac{32\pi}{35}$ **45.** $V = 7\pi(1 - \ln 2)$ **47.** $V = \frac{32\pi}{3}$
49. $V = \frac{32\pi}{105}$

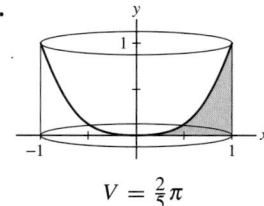

51. $V = 4\pi\sqrt{3}$ **53.** $V = \frac{4}{3}\pi a^2 b$ **55.** $V = \frac{\pi}{3}(1 - c^2)^{3/2}$

Section 6.4 Preliminary

1. (a) Radius h and height r. (b) Radius r and height h.
2. (a) With respect to x. (b) With respect to y.

Section 6.4 Exercises

1.

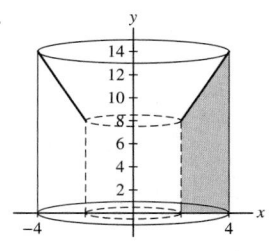

$V = \frac{2}{5}\pi$

3.

$V = 136\pi$

5.

$$V = 8\pi$$

7.

$$V = 2\pi$$

9.

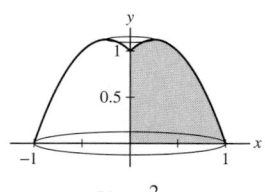

$$V = \tfrac{2}{3}\pi$$

11. $V = 16\pi$

13. $V = \tfrac{3}{10}\pi$

15. The point of intersection is $x = 1.376769504$; $V = 1.321975576$.

17.

$$V = \tfrac{3\pi}{5}$$

19.

$$V = \tfrac{280\pi}{81}$$

21.

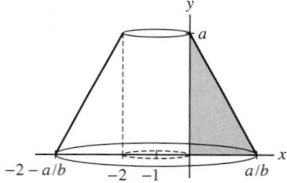

$$V = \frac{a^2(a+3b)}{3b^2}\pi$$

23. $V = \tfrac{\pi}{3}$

25. $V = \tfrac{38\pi}{5}$ **27.** $V = \tfrac{7\pi}{192}$

29. (a) $V = \tfrac{576\pi}{7}$ **(b)** $V = \tfrac{96\pi}{5}$

31. (a) The segment $\overline{AB}$ gives a disk with radius $R = h(y)$; the segment $\overline{CB}$ gives a shell with radius x and height $f(x)$.
(b) Shell method: $V = 2\pi \int_0^2 x f(x)\, dx$. Disk method: $V = \pi \int_0^{1.3} (h(y))^2\, dy$.

33. $V = 8\pi$ **35.** $V = \tfrac{40\pi}{3}$ **37.** $V = \tfrac{1{,}024\pi}{15}$ **39.** $V = 16\pi$

41. $V = \tfrac{32\pi}{3}$ **43.** $V = \tfrac{776\pi}{15}$ **45.** $V = \tfrac{4}{3}\pi r^3$ **47.** $V = 2\pi^2 a r^2$

49. (a) $V = \tfrac{563\pi}{30}$ **(b)** $V = \tfrac{1{,}748\pi}{35}$

Section 6.5 Preliminary

1. Because the required force is not constant through the stretching process.

2. The force involved in lifting a tank is the weight of the tank, which is constant.

3. $\tfrac{1}{2}kx^2$

Section 6.5 Exercises

1. $W = 627.2$ J **3.** $W = 1.08$ J **5.** $W = 1.5$ J **7.** $W = 11.25$ J

9. 6 in. **11.** $W = 1.6 \times 10^6$ ft-lb **13.** $W = \tfrac{128{,}000\pi}{3}$ ft-lb

15. $W = 1.399 \times 10^{12}$ ft-lb **17.** $W \approx 1.18 \times 10^8$ J

19. $W = 9{,}800\ell \pi r^3$ J **21.** $W = 2.94 \times 10^6$ J

23. $W = \tfrac{1{,}684{,}375\pi}{2}$ J $\approx 2.65 \times 10^6$ J

25. (a) $F(y) = 9{,}800\pi(1{,}000 - y^3) - \tfrac{1{,}225\pi}{2}(10{,}000 - y^4)$ J
(b)

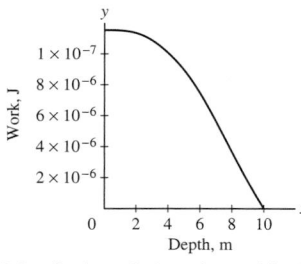

(c) Marginal work function **(d)** 6.91 m

27. $W = 176.4$ J **29.** $W = 600$ ft-lb **31.** $W = 34{,}500$ ft-lb

33. $W \approx 1.99 \times 10^{10}$ J

35. (a) $k = 2.517382 \times 10^6$ lb-in^2 **(b)** $F = k(9\pi)^{-0.4}x^{-1.4}$
(c) $W = 74{,}677.8$ ft-lb

39. $v_0 = \sqrt{2GM_e\left(\tfrac{1}{r_e} - \tfrac{1}{r+r_e}\right)}$ m/s **41.** $v_{esc} = \sqrt{\tfrac{2GM_e}{r_e}}$ m/s

Chapter 6 Review

1. $3\sqrt{2} - 1$ **3.** $\tfrac{9}{4}$ **5.** $e - \tfrac{3}{2}$ **7.** $\tfrac{1}{4}\pi - \tfrac{5}{12}$ **9.** 0.5

11. $x = 0.7145563847$, Area $= 0.08235024596$

13. $\tfrac{33}{2}$ lb **15.** $\tfrac{9}{4}$ **17.** 2 **19.** $\tfrac{1}{2}\sinh 1$ **21.** 27 **23.** $V = \tfrac{2\pi}{15}$ m^5

25. (a) $\int_0^1 \left(\sqrt{1 - (x - 1)^2} - \left(1 - \sqrt{1 - x^2}\right)\right) dx$

(b) $\pi \int_0^1 \left((1 - (x - 1)^2) - \left(1 - \sqrt{1 - x^2}\right)^2\right) dx$

27. $V = \tfrac{2{,}048\pi}{3}$ **29.** $V = 8\pi$ **31.** $V = \tfrac{2\pi}{35}(12{,}032\sqrt{2} - 7{,}083)$

33. $V = \tfrac{56\pi}{15}$ **35.** $V = 2\pi\left(16\sqrt{3} - \tfrac{40}{3}\right)$

37. x-axis: $V = c\pi$; y-axis: $V = \tfrac{2\pi}{3}\left((1 + c^2)^{3/2} - c^3 - 1\right)$

39. $V = \tfrac{\pi}{6}$ **41.** $W = 74{,}900\pi$ ft-lb **43.** $W = 83{,}040$ ft-lb

Chapter 7

Section 7.1 Preliminary

1. $T_1 = 6$; $T_2 = 7$

2. T_N overestimates the integral of $y = g(x)$; M_N overestimates the integral of $y = f(x)$.

3. 0; the Trapezoidal Rule integrates a linear function exactly.

4. $\tfrac{9}{32}$

5. The sum of the areas of the midpoint rectangles or the sum of the areas of the tangential trapezoids.

Section 7.1 Exercises

1. $T_4 = 2.75$; $M_4 = 2.625$ 3. $T_6 = 64.6875$; $M_6 = 63.28125$

5. $T_6 \approx 1.40536$; $M_6 \approx 1.37693$

7. $T_6 \approx 0.883387$; $M_6 \approx 0.880369$

9. $T_5 \approx 1.12096$; $M_5 \approx 1.11716$

11. $T_8 \approx 2.96581$; $M_8 \approx 2.95302$ 13. $S_4 \approx 5.25221$

15. $S_4 \approx 1.08055$ 17. $S_6 \approx 0.746830$ 19. $S_8 \approx 2.54499$

21. $S_{10} \approx 0.881377$ 23. $V \approx 2.46740$ 25. $V \approx 1.87691$

27. $\int_0^{\pi/2} \frac{\sin x}{x}\, dx \approx 1.37076$

29. $M_6 \approx 1.0028615$

(a) Because $f(x) = \cos x$ is concave down on $[0, \pi/2]$, M_6 is too large.

(b) $\text{Error}(M_6) \le \frac{K_2(b-a)^3}{24N^2} = \frac{\pi^3}{6,912} \approx 0.00448586$

(c) $\int_0^{\pi/2} \cos x\, dx = 1$; $\text{Error}(M_6) = |M_6 - 1| \approx 0.00286$

31. T_{20} overestimates the integral; $\text{Error}(T_{20}) \le 1.04167 \times 10^{-4}$

33. M_{20} overestimates the integral; $\text{Error}(M_{20}) \le 5.04659 \times 10^{-5}$

35. $N \ge 109.37$; $T_{110} \approx 0.4999990559$;
$\left| \int_0^{\pi/6} \cos x\, dx - T_{110} \right| \approx 9.441 \times 10^{-7}$.

37. $N \ge 1,500$; $T_{1,500} \approx 0.9502132468$;
$\left| \int_0^3 e^{-x}\, dx - T_{1,500} \right| \approx 3.15 \times 10^{-7}$.

39. (a) $S_6 \approx 0.432361$

(b) $\text{Error}(S_6) \le \frac{K_4(b-a)^5}{180N^4} = \frac{1}{14,580} \approx 6.85871 \times 10^{-5}$

(c) $\int_0^1 e^{-2x}\, dx = \frac{1-e^{-2}}{2} \approx 0.432332$; $\text{Error}(S_6) \approx 2.93 \times 10^{-5}$

41. $S_8 \approx 4.046655$; $\text{Error}(S_8) \le \frac{1}{120} \approx 0.0083333$; $N = 78$

43. With $K_4 = 15$, $\text{Error}(S_{40}) \le \frac{5}{49,152} \approx 1.0173 \times 10^{-4}$.

45. $N = 396$ 47. $N = 186$

49. $\int_0^1 \frac{dx}{1+x^2} = \tan^{-1} x \big|_0^1 = \tan^{-1} 1 - \tan^{-1} 0 = \frac{\pi}{4}$

(a) Let $f(x) = (1 + x^2)^{-1}$. Then $|f^{(4)}(x)| \le 24$ on $[0, 1]$.

(b) $N = 20$; $S_{20} \approx 0.785398163242$ and
$\left| 0.785398163242 - \frac{\pi}{4} \right| \approx 1.55 \times 10^{-10}$.

51.

Approximation	Error	Error Ratio
2.052344	0.052344	
2.012909	0.012909	0.246619
2.0032164	0.0032164	0.249160
2.00080342	0.00080342	0.249789
2.00020081	0.00020081	0.249944

53. Note $\text{Error}(S_{2N}) = \frac{K_4(b-a)^5}{180(2N)^4} = \frac{1}{16} \cdot \frac{K_4(b-a)^5}{180N^4} = \frac{1}{16} \text{Error}(S_N)$.

Approximation	Error	Error Ratio
2.004560	4.560×10^{-3}	
2.0002692	2.692×10^{-4}	0.059035
2.00001659	1.659×10^{-5}	0.061627
2.000001033	1.033×10^{-6}	0.062266
2.00000006453	6.453×10^{-8}	0.062469

55. $T_{\text{ave}} = 21.2°\text{C}$

61. Let $f(x) = ax^3 + bx^2 + cx + d$, with $a \ne 0$, be any cubic polynomial. Then, $f^{(4)}(x) = 0$, so we can take $K_4 = 0$. This yields $\text{Error}(S_N) \le \frac{0}{180N^4} = 0$. In other words, S_N is exact for all cubic polynomials for all N.

Section 7.2 Preliminary

1. Integration by parts is derived from the Product Rule.

2. (a) $\int x \cos(x^2)\, dx$; use the substitution $u = x^2$.

(b) $\int x \cos x\, dx$; use Integration by Parts.

(c) $\int x^2 e^x\, dx$; use Integration by Parts.

(d) $\int x e^{x^2}\, dx$; use the substitution $u = x^2$.

3. Transforming $v' = x$ into $v = \frac{1}{2}x^2$ increases the power of x and makes the new integral harder than the original.

Section 7.2 Exercises

1. $-x \cos x + \sin x + C$ 3. $e^x(2x + 9) + C$

5. $\frac{x^4}{16}(4 \ln x - 1) + C$ 7. $-e^{-x}(3x + 2) + C$

9. $e^x(x^2 - 2x + 2) + C$ 11. $\frac{1}{2}x \sin 2x + \frac{1}{4}\cos 2x + C$

13. $-\frac{1}{2}e^{-x}(\sin x + \cos x) + C$ 15. $\frac{x^2}{4}(2 \ln x - 1) + C$

17. $-\frac{1}{8x^8}\left(\ln x + \frac{1}{8}\right) + C$ 19. $-x \sin(2 - x) + \cos(2 - x) + C$

21. $\frac{2^x}{\ln 2}\left(x - \frac{1}{\ln 2}\right) + C$ 23. $x\left((\ln x)^2 - 2 \ln x + 2\right) + C$

25. $x \sin^{-1} x + \sqrt{1 - x^2} + C$ 27. $\frac{5^x}{\ln 5}\left(x - \frac{1}{\ln 5}\right) + C$

29. $\frac{1}{2}x \sinh(2x) - \frac{1}{4}\cosh(2x) + C$ 31. $x \sinh^{-1} x - \sqrt{1 + x^2} + C$

33. $2e^{\sqrt{x}}\left(\sqrt{x} - 1\right) + C$ 35. $\frac{1}{4}x \sin 4x + \frac{1}{16}\cos 4x + C$

37. $\frac{2}{3}(x + 1)^{3/2} - 2(x + 1)^{1/2} + C$ 39. $\sin x \ln(\sin x) - \sin x + C$

41. $2 \sin \sqrt{x} - 2\sqrt{x} \cos \sqrt{x} + C$ 43. $\frac{1}{4}(\ln x)^2\left(2 \ln(\ln x) - 1\right) + C$

45. $\frac{1}{81}(17e^{18} + 1)$ 47. $128/15$ 49. $\frac{16}{3} \ln 4 - \frac{28}{9}$

51. $e^x(x^4 - 4x^3 + 12x^2 - 24x + 24) + C$

53. $\int x^n e^{-x}\, dx = -x^n e^{-x} + n \int x^{n-1} e^{-x}\, dx$

55. Integration by Parts with $u = \ln x$ and $v' = \sqrt{x}$.

57. Substitution with $u = 4 - x^2$, $du = -2x\, dx$ and $x^2 = 4 - u$. Follow this with algebraic manipulation.

59. Substitution with $u = x^2 + 3x + 6$ and $du = (2x + 3)\, dx$.

61. Integration by Parts with $u = x$ and $v' = \sin(3x + 4)$.

63. $x(\sin^{-1} x)^2 + 2\sqrt{1 - x^2} \sin^{-1} x - 2x + C$

65. $\frac{x^4}{4} \sin(x^4) + \frac{1}{4}\cos(x^4) + C$ 67. $V = 2\pi(e^2 + 1)$

69. $PV = \frac{5,000r + 100 - e^{-10r}(6,000r + 100)}{r^2}$

71. $\int (\ln x)^2\, dx = x(\ln x)^2 - 2x \ln x + 2x + C$

$\int (\ln x)^3\, dx = x(\ln x)^3 - 3x(\ln x)^2 + 6x \ln x - 6x + C$

77. (d) $I(1, 1) = \frac{1}{6}$; $I(3, 2) = \frac{1}{60}$

79. (c) $I_3 = \frac{1}{2}x^2 \sin(x^2) + \frac{1}{2}\cos(x^2) + C$

Section 7.3 Preliminary

1. Rewrite $\sin^5 x = \sin x \sin^4 x = \sin x (1 - \cos^2 x)^2$ and then substitute $u = \cos x$.

2. Use a reduction formula.

3. No, a reduction formula is not needed because the sine function is raised to an odd power.

4. Write $\cos^2 x = 1 - \sin^2 x$ and then apply the reduction formula for powers of the sine function.

5. The first integral can be evaluated using a simple substitution; the second requires use of reduction formulas.

Section 7.3 Exercises

1. $\sin x - \frac{1}{3}\sin^3 x + C$ **3.** $-\frac{1}{3}\cos^3\theta + \frac{1}{5}\cos^5\theta + C$

5. $-\frac{1}{4}\cos^4 t + \frac{1}{6}\cos^6 t + C$ **7.** $A = 2$

9. $\frac{1}{4}\cos^3 y \sin y + \frac{3}{8}\cos y \sin y + \frac{3}{8}y + C$

11. $\frac{1}{6}\sin^5 x \cos x - \frac{1}{24}\sin^3 x \cos x - \frac{1}{16}\sin x \cos x + \frac{1}{16}x + C$

13. $\frac{1}{3}\tan^3 x + C$ **15.** $\frac{1}{5}\tan^5 x + C$ **17.** $\tan t - t + C$

19. $\frac{1}{2}\tan x \sec x + \frac{1}{2}\ln|\sec x + \tan x| + C$ **21.** $-\frac{1}{6}\cos^6 x + C$

23. $\frac{1}{12}\cos^3(3x)\sin(3x) + \frac{1}{8}\cos(3x)\sin(3x) + \frac{3}{8}x + C$

25. $\frac{1}{5\pi}\sin^5\pi\theta - \frac{1}{7\pi}\sin^7\pi\theta + C$

27. $-\frac{1}{12}\sin^3(3x)\cos(3x) - \frac{1}{8}\cos(3x)\sin(3x) + \frac{3}{8}x + C$

29. $\frac{1}{7}\ln|\sec 7t + \tan 7t| + C$ **31.** $\frac{1}{2}\tan^2 x + C$

33. $\frac{1}{8}\sec^8 x - \frac{1}{3}\sec^6 x + \frac{1}{4}\sec^4 x + C$

35. $\frac{1}{4}\tan x \sec^3 x - \frac{5}{8}\tan x \sec x + \frac{3}{8}\ln|\sec x + \tan x| + C$

37. $\frac{1}{6}\tan x \sec^5 x - \frac{7}{24}\tan x \sec^3 x + \frac{1}{16}\tan x \sec x$
$+ \frac{1}{16}\ln|\sec x + \tan x| + C$

39. $\frac{1}{4}\sin^2 2x + C$ **41.** $\frac{1}{4}\cos 2x - \frac{1}{12}\cos 6x + C$

43. $\frac{1}{2}\tan^2(\ln t) - \ln|\sec(\ln t)| + C$ **45.** π **47.** $\frac{5}{24}$

49. $\ln(\sqrt{2}+1) \approx 0.88137$ **51.** $\frac{1}{2}\ln 2 - \frac{1}{4} \approx 0.096574$ **53.** $-\frac{6}{7}$

57. $\csc x - \frac{1}{3}\csc^3 x + C$ **59.** $-\frac{1}{3}\cot^3 x + C$ **61.** $V = \frac{\pi^2}{2}$

63. $\frac{1}{8}x - \frac{1}{16}\sin 2x \cos 2x + C$

65. $\frac{1}{16}x - \frac{1}{48}\sin 2x - \frac{1}{32}\sin 2x \cos 2x + \frac{1}{48}\cos^2 2x \sin 2x + C$

69. $\int \csc x \, dx = \int \frac{\csc x(\csc x - \cot x)}{\csc x - \cot x}\,dx = \ln|\csc x - \cot x| + C$

73. $\pi/2$

Section 7.4 Preliminary

1. Substitute $u = 9 - x^2$; then, $du = -2x\,dx$.

2. (a) $x = 3\sin\theta$ (b) $x = 4\sec\theta$ (c) $x = 4\tan\theta$
(d) $x = \sqrt{5}\sec\theta$

3. Triangle (A)   **4.** $\frac{x}{\sqrt{9-x^2}}$ **5.** $\frac{3}{x}$

6. $\sin 2\theta = 2\sin\theta\cos\theta = 2x\sqrt{1-x^2}$

Section 7.4 Exercises

1. (b) $\sin^{-1}\left(\frac{x}{3}\right) + C$ **3.** (c) $\ln\left|x + \sqrt{x^2+9}\right| + C$

5. $2\sin^{-1}\left(\frac{x}{2}\right) + \frac{1}{2}x\sqrt{4-x^2} + C$ **7.** $\frac{1}{3}\sec^{-1}\left(\frac{x}{3}\right) + C$

9. $-\frac{x}{4\sqrt{x^2-4}} + C$

11. The substitution $u = x^2 - 4$ is not effective. Using the substitution $x = 2\sec\theta$,

$$\int \frac{x^2}{\sqrt{x^2-4}}\,dx = \frac{1}{2}x\sqrt{x^2-4} + 2\ln\left|x + \sqrt{x^2-4}\right| + C.$$

13. $\frac{9}{2}\sin^{-1}\left(\frac{x}{3}\right) - \frac{1}{2}x\sqrt{9-x^2} + C$

15. $x\sqrt{x^2+3} + 3\ln\left|\sqrt{x^2+3}+x\right| + C$

17. $\frac{t}{4\sqrt{4-t^2}} + C$ **19.** $-\frac{\sqrt{5-y^2}}{5y} + C$

21. $\frac{1}{16}\sec^{-1}\left(\frac{z}{2}\right) + \frac{\sqrt{z^2-4}}{8z^2} + C$

23. $\frac{1}{2}x\sqrt{x^2+1} - \frac{1}{2}\ln\left|\sqrt{x^2+1}+x\right| + C$

25. $\frac{1}{54}\tan^{-1}\left(\frac{x}{3}\right) + \frac{x}{18(x^2+9)} + C$

27. $-\frac{x}{8(x^2-4)} - \frac{1}{16}\ln\left|\frac{x-2}{\sqrt{x^2-4}}\right| + C$

29. $-3(9-x^2)^{3/2} + \frac{1}{5}(9-x^2)^{5/2} + C$

33. (b) $I = \ln\left|\sqrt{(x-2)^2+4}+x-2\right| + C$

35. $\ln\left|\sqrt{x^2+4x+13}+x+2\right| + C$

37. $\ln\left|2x+1+2\sqrt{x^2+x}\right| + C$

39. $\frac{1}{2}(x-2)\sqrt{x^2-4x+3} - \frac{1}{2}\ln\left|x-2+\sqrt{x^2-4x+3}\right| + C$

41. $x\sec^{-1}x - \ln\left|x + \sqrt{x^2-1}\right| + C$

43. First, complete the square. Next, substitute $u = x+3, du = dx$. Finally, use the trigonometric substitution $u = \sqrt{21}\sin\theta$.

45. Integration by Parts with $u = x$ and $v' = \sec^2 x$.

47. First, substitute $u = e^{2x}, du = 2e^{2x}\,dx$. Then, use the integration formula for the inverse tangent.

49. Use the integration formula for the cosecant function.

51. First, substitute $u = x+2, du = dx$. Then, use the integration formula for power functions.

53. (a) Use the substitution $u = 1 - x^2$. Then $du = -2x\,dx$, $x^2 = 1 - u$ and $\int x^3\sqrt{1-x^2}\,dx = -\frac{1}{2}\int(1-u)u^{1/2}\,du$.

(b) Let $x = \sin\theta$. Then $dx = \cos\theta\,d\theta$, $1 - x^2 = \cos^2\theta$ and $\int x^2\sqrt{1-x^2}\,dx = \int \sin^2\theta\cos^2\theta\,d\theta$.

(c) Let $x = \sin\theta$. Then $dx = \cos\theta\,d\theta$, $1 - x^2 = \cos^2\theta$ and $\int \frac{x^4}{\sqrt{1-x^2}}\,dx = \int \sin^4\theta\,d\theta$.

(d) Use the substitution $u = 1 - x^2$. Then $du = -2x\,dx$, and $\int \frac{x}{\sqrt{1-x^2}} = -\frac{1}{2}\int u^{-1/2}\,du$.

55. $V = \pi^2/8$ **57.** $V = 8\pi\sqrt{3} - 6\pi\ln(2+\sqrt{3}) \approx 18.707$

59. $x = 2.385$ in

Section 7.5 Preliminary

1. (a) $x = \sinh t$ (b) $x = 3\sinh t$ (c) $3x = \sinh t$

2. $\sinh x$ and $\tanh x$ 3. $\frac{1}{2}\ln\left|\frac{1+x}{1-x}\right|$

Section 7.5 Exercises

1. $\frac{1}{3}\sinh 3x + C$ 3. $\frac{1}{2}\cosh(x^2 + 1) + C$

5. $-\frac{1}{2}\tanh(1 - 2x) + C$ 7. $\frac{1}{2}\tanh^2 x + C$ 9. $\ln\cosh x + C$

11. $\ln|\sinh x| + C$ 13. $\frac{1}{16}\sinh(8x - 18) - \frac{1}{2}x + C$

15. $\frac{1}{32}\sinh(4x) - \frac{x}{8} + C$ 17. $\cosh^{-1} 4 - \cosh^{-1} 2$

19. $\sinh^{-1}\left(\frac{x}{4}\right) + C$ 21. $\frac{1}{2}x\sqrt{x^2 - 1} - \frac{1}{2}\cosh^{-1} x + C$

23. $2\tanh^{-1}\left(\frac{1}{2}\right)$ 25. $\sinh^{-1} 1$

27. $\frac{1}{4}\ln 3 - \frac{1}{4}\ln(4 + \sqrt{17})$ 29. $\cosh^{-1} x - \frac{\sqrt{x^2-1}}{x} + C$

33. Using trigonometric substitution:
$\frac{1}{2}x\sqrt{x^2 + 16} + 8\ln|x + \sqrt{x^2 + 16}| + C$

35. $\frac{1}{2}\ln\left|\frac{1+x}{1-x}\right| + C$

37. tangent line: $y = 1.048 - 0.976 * (x - 0.8)$

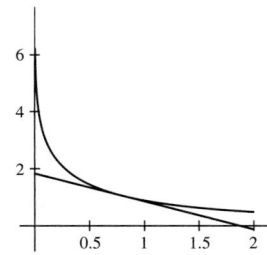

39. $\frac{1}{4}\cosh^3 x \sinh x + \frac{3}{8}\cosh x \sinh x + \frac{3}{8}x + C$

41. $x\sinh^{-1} x - \sqrt{x^2 + 1} + C$

43. $\frac{1}{2}x^2\tanh^{-1} x - \frac{1}{2}\tanh^{-1} x + \frac{x}{2} + C$

45. $2\tan^{-1}(\tanh(x/2)) + C$

Section 7.6 Preliminary

1. No, $f(x)$ cannot be a rational function because the integral of a rational function cannot contain a term with a non-integer exponent.

2. (a) No   (b) Yes   (c) Yes   (d) No

3. (a) Square is already completed; irreducible
(b) Square is already completed; factors as $(x - \sqrt{5})(x + \sqrt{5})$
(c) $x^2 + 4x + 6 = (x + 2)^2 + 2$; irreducible
(d) $x^2 + 4x + 2 = (x + 2)^2 - 2$; factors as $(x + 2 - \sqrt{2})(x + 2 + \sqrt{2})$

4. (a) is a sum of logarithmic terms $A_i \ln(x - a_i)$ for some constants A_i.

Section 7.6 Exercises

1. (a) $\frac{x^2+4x+12}{(x+2)(x^2+4)} = \frac{1}{x+2} + \frac{4}{x^2+4}$

(b) $\frac{2x^2+8x+24}{(x+2)^2(x^2+4)} = \frac{1}{x+2} + \frac{2}{(x+2)^2} + \frac{-x+2}{x^2+4}$

(c) $\frac{x^2-4x+8}{(x-1)^2(x-2)^2} = \frac{-8}{x-2} + \frac{4}{(x-2)^2} + \frac{8}{x-1} + \frac{5}{(x-1)^2}$

(d) $\frac{x^4-4x+8}{(x+2)(x^2+4)} = x - 2 + \frac{4}{x+2} - \frac{4x-4}{x^2+4}$

3. $B = -2$

5. $\frac{x}{3x-9} = \frac{1}{3} + \frac{1}{x-3}$; so $\int \frac{x}{3x-9}\,dx = \frac{1}{3}x + \ln|x - 3| + C$

7. $\frac{x^3+x+1}{x-2} = x^2 + 2x + 5 + \frac{11}{x-2}$; so
$\int \frac{x^3+x+1}{x-2}\,dx = \frac{1}{3}x^3 + x^2 + 5x + 11\ln|x - 2| + C$

9. $-\frac{1}{2}\ln|x - 2| + \frac{1}{2}\ln|x - 4| + C$

11. $\ln|x| - \ln|2x + 1| + C$

13. $-3\ln|x - 2| + 5\ln|x - 3| + C$

15. $2\ln|x + 3| - \ln|x + 5| - \frac{2}{3}\ln|3x - 2| + C$

17. $3\ln|x - 1| - 2\ln|x + 1| - \frac{5}{x+1} + C$

19. $2\ln|x - 1| - \frac{1}{x-1} - 2\ln|x - 2| - \frac{1}{x-2} + C$

21. $3\ln|x| - 3\ln|x + 4| + \frac{12}{x+4} + C$

23. $-\frac{1}{9}\ln|x - 4| - \frac{1}{3(x-4)} + \frac{1}{9}\ln|x - 1| + C$

25. $\frac{11}{3}\ln|x| - \frac{2}{x} - \frac{9}{2}\ln|x - 1| + \frac{5}{6}\ln|x - 3| + C$

27. $\frac{3}{2}x^2 + 12x + 46\ln|x - 4| + C$

29. $\ln|x| - \frac{1}{2}\ln(x^2 + 1) + C$ 31. $x - \sqrt{3}\tan^{-1}\left(\frac{x}{\sqrt{3}}\right) + C$

33. $\frac{1}{2}\ln|x + 1| + \frac{1}{4}\ln(x^2 + 1) - \frac{1}{2}\tan^{-1} x + C$

35. $\frac{1}{27}\ln|3x + 7| + \frac{14}{27}(3x + 7)^{-1} - \frac{49}{54}(3x + 7)^{-2} + C$

37. $-\frac{1}{25x} - \frac{1}{125}\tan^{-1}\left(\frac{x}{5}\right) + C$

39. $\frac{1}{10}\ln|x + 1| - \frac{1}{20}\ln(x^2 + 9) + \frac{7}{135}\tan^{-1}\left(\frac{x}{3}\right) + \frac{x+9}{18(x^2+9)} + C$

41. $3\ln|x - 3| - \frac{3}{2}\ln(x^2 + 1) - 4\tan^{-1} x + \frac{5x+15}{x^2+1} + C$

43. $\ln|x + 1| - \frac{1}{2}\ln(x^2 - 2x + 6) + \frac{2}{\sqrt{5}}\tan^{-1}\left(\frac{x-1}{\sqrt{5}}\right) + C$

45. $\frac{3\sqrt{2}}{4}\tan^{-1}\left(\frac{x+1}{\sqrt{2}}\right) + \frac{x+3}{2(x^2+2x+3)} + C$

47. $\frac{1}{2}\ln|x - 1| - \frac{1}{2}\ln|x + 1| + C = \ln\left|\frac{x-1}{\sqrt{x^2-1}}\right| + C$

49. $2\sqrt{x} + \ln|\sqrt{x} - 1| - \ln|\sqrt{x} + 1| + C$

51. $-\frac{\sqrt{4-x^2}}{4x} + C$ 53. $\ln|x| - \ln|x - 1| - \frac{1}{x-1} + C$

55. $\frac{1}{54}\tan^{-1}\left(\frac{x}{3}\right) + \frac{x}{18(x^2+9)} + C$

57. $\frac{1}{5}\sec^5 x - \frac{2}{3}\sec^3 x + \sec x + C$

59. $\ln|x + \sqrt{x^2 - 1}| - \frac{x}{\sqrt{x^2-1}} + C$

61. $-\ln|x| + \frac{1}{2}\ln|x - 1| + \frac{1}{2}\ln|x + 1| + C$

63. $\frac{2}{3}\tan^{-1}(x^{3/2}) + C$

65. $-\frac{1}{3}\ln|x^3| - \frac{1}{3x^3} + \frac{1}{3}\ln|x^3 + 1| + C$

67. $\frac{4}{5}\ln\left|2\tan\left(\frac{\theta}{2}\right) + 1\right| - \frac{4}{5}\ln\left|\tan\left(\frac{\theta}{2}\right) - 2\right| + C$

71. (b) $\frac{3x-2}{x^2-4x-12} = \frac{2}{x-6} + \frac{1}{x+2}$

Section 7.7 Preliminary

1. (a) Converges (b) Diverges (c) Diverges (d) Converges
2. Yes, this is an improper integral because $\cot x$ is undefined at $x = 0$.
3. One choice is $b = 2$. **4.** $\int_0^\infty \frac{dx}{x+e^x} < \int_0^\infty e^{-x}\,dx$
5. $\frac{e^{-x}}{x} < \frac{1}{x}$ for $1 \le x < \infty$, but $\int_1^\infty \frac{dx}{x}$ diverges.

Section 7.7 Exercises

1. (a) Improper: The function $x^{-1/3}$ is infinite at $x = 0$.
(b) Improper: Infinite limit of integration.
(c) Improper: Infinite limit of integration.
(d) Proper: The function e^{-x} is continuous on the finite interval $[0, 1]$.
(e) Improper: The function $\sec x$ is infinite at $x = \pi/2$.
(f) Improper: Infinite limit of integration.
(g) Proper: The function $\sin x$ is continuous on the finite interval $[0, 1]$.
(h) Proper: The function $(3 - x^2)^{-1/2}$ is continuous on the finite interval $[0, 1]$.
(i) Improper: Infinite limit of integration.
(j) Improper: The function $\ln x$ is infinite at $x = 0$.
5. Integral does not converge. **7.** $10{,}000e^{0.0004}$
9. Integral does not converge. **11.** 4 **13.** $\frac{1}{8}$ **15.** 2
17. 1.25 **19.** $\frac{1}{3}e^{-12}$ **21.** $\frac{1}{3}$ **23.** $\frac{1}{375}$ **25.** $2\sqrt{2}$
27. Integral does not converge **29.** $\frac{\pi}{4}$ **31.** $\frac{1}{2}$ **33.** $\frac{\pi}{2}$
35. Integral does not converge. **37.** Integral does not converge.
39. Integral does not converge. **41.** Integral does not converge.
43. $-\frac{1}{4}$ **45.** Integral does not converge.
49. Integral does not converge. **51.** π **53.** 0 **55.** $a < 0$
59. The integral $\int_1^\infty x^{-3}\,dx$ converges because $3 > 1$. Since $x^3 + 4 \ge x^3$, it follows that $\frac{1}{x^3+4} \le \frac{1}{x^3}$. Therefore, by the comparison test $\int_1^\infty \frac{dx}{x^3+4}$ converges.
63. Let $f(x) = \frac{1-\sin x}{x^2}$. Because $f(x) \le \frac{2}{x^2}$ and $\int_1^\infty 2x^{-2}\,dx = 2$, it follows by the comparison test that $\int_1^\infty f(x)\,dx$ converges.
65. Converges **67.** Diverges **69.** Converges **71.** Diverges
73. Converges **75.** \$3,571.43 **77.** \$2,000,000
79. (a) $V = \pi$
83. $A = \frac{1}{k}$; For Radon-222, $A = \frac{1}{k} = \frac{3.825}{\ln 2} \approx 5.518$ days
85. $J_4 = \frac{24}{\alpha^5}$
87. Let $\alpha = h/kT$. Then $E = \frac{8\pi h}{c^3}\int_0^\infty \frac{v^3}{e^{\alpha v}-1}\,dv$. Since $\alpha > 0$ and $8\pi h/c^3$ is a constant, it follows from the previous exercise that E is finite.
89. $C = k; \mu = \frac{1}{k}$ **93.** $\frac{s}{s^2+\alpha^2}$

Chapter 7 Review

1. $T_2 = -\frac{9}{8}$; $M_3 = -2$; $T_6 = -\frac{13}{8}$; $S_6 = -\frac{21}{12}$ **3.** 2.19 in
5. $T_3 \approx 25.976514$ **7.** $T_6 \approx 0.358016$ **9.** $S_8 \approx 0.608711$
11. With $K_2 = 1.5$, Error$(T_9) \le 9$.
13. With $K_2 = 30$, Error$(T_{24}) \le 0.016822$. **15.** $N = 4$
17. $\frac{1}{9}\sin^9\theta - \frac{1}{11}\sin^{11}\theta + C$
19. $\frac{1}{6}\tan\theta\sec^5\theta - \frac{7}{24}\tan\theta\sec^3\theta + \frac{1}{16}\tan\theta\sec\theta + \frac{1}{16}\ln|\sec\theta + \tan\theta| + C$
21. $-\frac{1}{\sqrt{x^2-1}} - \sec^{-1}x + C$ **23.** $\frac{1}{2}\ln(x^2+1) + C$
25. $\frac{1}{6}\ln|x-1| - \frac{1}{6}\ln|x+5| + C$
27. $-\frac{1}{2}x\sqrt{9-x^2} + \frac{9}{2}\sin^{-1}\left(\frac{x}{3}\right) + C$ **29.** $\frac{1}{5}\tan^5\theta + C$
31. $\frac{1}{4}\ln|t+1| - \frac{1}{4}\ln|t-1| - \frac{1}{4(t-1)} - \frac{1}{4(t+1)} + C$
33. $\frac{1}{2}\sin^4\theta + C$
35. $(x+1)(\ln(x+1))^2 - 2(x+1)\ln(x+1) + 2x + C$
37. $\frac{3}{8}x + \frac{1}{36}\sin(18x-4) + \frac{1}{288}\sin(36x-8) + C$
39. $2\ln|\sec x| + C$ **41.** $2\tan x + 2\sec x - x + C$
43. $-2\cot\left(\frac{\theta}{2}\right) - \theta + C$
45. $\frac{1}{49}\ln|t+4| - \frac{1}{49}\ln|t-3| - \frac{1}{7(t-3)} + C$
47. $\frac{1}{2}\sec^{-1}\left(\frac{x}{2}\right) + C$ **49.** $\frac{2}{\sqrt{a}}\tan^{-1}\left(\frac{\sqrt{x}}{\sqrt{a}}\right) + C$ (assuming $a > 0$)
51. $\ln|x+2| + \frac{5}{x+2} - \frac{3}{(x+2)^2} + C$
53. $-\frac{2}{x-2} + \frac{1}{2}\ln(x^2+4) - \ln|x-2| + C$
55. $\frac{1}{3}\tan^{-1}\left(\frac{x+4}{3}\right) + C$ **57.** $\ln|x+2| + \frac{5}{x+2} - \frac{3}{(x+2)^2} + C$
59. $\frac{(x^2-2)\sqrt{x^2+4}}{24x^3} + C$ **61.** $-\frac{1}{9}(4+3x)e^{4-3x} + C$
63. $-\frac{\tan^{-1}x}{x} - \frac{1}{2}\ln(1+x^2) + \ln|x| + C$
65. $\frac{1}{3}x^3(\ln x)^2 - \frac{2}{9}x^3\ln x + \frac{2}{27}x^3 + C$ **67.** $\frac{1}{2}(\tan^{-1}t)^2 + C$
69. $\frac{1}{2}\sin x\sinh x - \frac{1}{2}\cos x\cosh x + C$ **71.** $\frac{1}{5}\ln|\cosh 5x| + C$
73. $t + \frac{1}{4}\coth(1-4t) + C$ **75.** $\frac{\pi}{3}$ **77.** $\tan^{-1}(\tanh t) + C$
81. $a = 0$ **83.** Integral does not converge.
85. $\frac{5}{189}3^{1/5}$ **87.** $\frac{1}{4}e^{36}$ **89.** $\ln\left(\frac{6}{5}\right)$ **91.** $\frac{3}{2}3^{2/3}$
93. Converges **95.** Diverges **97.** Converges
99. (a) Infinite (b) Finite: π (c) Infinite
101. $\bar{E} = kT$ **103.** $\frac{2}{(s-\alpha)^3}$

Chapter 8

Section 8.1 Preliminary

1. $\int_0^\pi \sqrt{1+\sin^2 x}\,dx$
2. The graph of $y = f(x) + C$ is a vertical translation of the graph of $y = f(x)$; hence, the two graphs should have the same arc length. We can explicitly establish this as follows:

Length of $y = f(x) + C = \int_a^b \sqrt{1 + \left[\dfrac{d}{dx}(f(x) + C)\right]^2}\, dx$

$= \int_a^b \sqrt{1 + [f'(x)]^2}\, dx$

$=$ length of $y = f(x)$.

Section 8.1 Exercises

1. $L = \int_2^6 \sqrt{1 + 16x^6}\, dx$ **3.** $\frac{13}{12}$ **5.** $3\sqrt{10}$

7. $\frac{1}{27}(22\sqrt{22} - 13\sqrt{13})$ **9.** $e^2 + \frac{\ln 2}{2} + \frac{1}{4}$

11. $\int_1^2 \sqrt{1 + x^6}\, dx \approx 3.957736$

13. $\int_1^2 \sqrt{1 + \frac{1}{x^4}}\, dx \approx 1.132123$ **15.** 6   **17.** $2\sqrt{3}$

19. Let h denote the length of the hypotenuse. Then, by the Pythagorean Theorem, $h^2 = (b - a)^2 + m^2(b - a)^2$
$= (b - a)^2(1 + m^2)$, or $h = (b - a)\sqrt{1 + m^2}$ since $b > a$.
Moreover, $(f'(x))^2 = m^2$, so

$$L = \int_a^b \sqrt{1 + m^2}\, dx = (b - a)\sqrt{1 + m^2}.$$

25. $\sqrt{1 + e^{2a}} + \frac{1}{2}\ln\frac{\sqrt{1+e^{2a}}-1}{\sqrt{1+e^{2a}}+1} - \sqrt{2} + \frac{1}{2}\ln\frac{1+\sqrt{2}}{\sqrt{2}-1}$

27. $\ln(1 + \sqrt{2})$ **31.** 1.552248   **33.** $60\pi\sqrt{17}$ **35.** $\frac{384\pi}{5}$

37. $\pi\left(e\sqrt{1 + e^2} + \ln\left(e + \sqrt{1 + e^2}\right) - \sqrt{2} - \ln(1 + \sqrt{2})\right)$

39. $2\pi\left(\sqrt{2} + \ln(1 + \sqrt{2})\right)$

43. $\frac{\pi}{32}\left(1{,}032\sqrt{65} - \ln(8 + \sqrt{65})\right)$

45. $2\pi \int_0^{\pi/4} \tan x\sqrt{1 + \sec^4 x}\, dx \approx 3.839077$ **47.** $n = 13$

51. $2\pi b^2 + \frac{2\pi b a^2}{\sqrt{b^2 - a^2}}\ln\left|\frac{\sqrt{b^2 - a^2}}{a} + \frac{b}{a}\right|$

Section 8.2 Preliminary

1. Pressure is defined as force per unit area.

2. The factor of proportionality is the weight density of the fluid, $w = \rho g$.

3. Fluid force acts in the direction perpendicular to the side of the submerged object.

4. Pressure depends only on depth and does not change horizontally at a given depth.

5. When a plate is submerged vertically, the pressure is not constant along the plate, so the fluid force is not equal to the pressure times the area.

Section 8.2 Exercises

1. (a) Top: $F = 1{,}125$ lb; bottom: $F = 4{,}500$ lb

(b) $F \approx \sum_{j=1}^{N} 3wy_j\, \Delta y$ **(c)** $F = \int_2^8 3wy\, dy$ **(d)** $F = 5{,}625$ lb

3. (a) The width of the triangle varies linearly from 0 at a depth of $y = 3$ ft to 1 at a depth of $y = 5$ ft. Thus, $w = \frac{1}{2}(y - 3)$.

(b) The area of the strip at depth y is $\frac{1}{2}(y - 3)\Delta y$, and the pressure at depth y is wy, where $w = 62.5$ lb/ft^3. Thus, the fluid force acting on the strip at depth y is approximately equal to $\frac{1}{2}wy(y - 3)\Delta y$.

(c) $F \approx \sum_{j=1}^{N} \frac{1}{2}wy_j(y_j - 3)\,\Delta y \rightarrow \int_3^5 \frac{1}{2}wy(y - 3)\, dy$

(d) $F = \frac{1{,}625}{6}$ lb

5. (b) $F = \frac{19{,}600}{3}r^3$ N

7. $F = \frac{19{,}600}{3}r^3 + 4{,}900\pi m r^2$ N   **9.** $F = 441{,}000$ N

11. $F = \frac{64}{5}w$ **13.** $F = \left(1 - \frac{\pi}{8}\right)w$ **15.** $F = w$

17. $F = 595{,}350$ N   **19.** $F = \frac{200w\sqrt{3}}{3}$

21. $F = \frac{45}{2}hd\sqrt{(b - a)^2 + 4h^2}$ lb

23. Front and back: $F = \frac{62.5\sqrt{3}}{9}y^3$; slanted sides: $F = \frac{62.5\sqrt{3}}{3}\ell y^2$ where ℓ denotes the length of the trough.

Section 8.3 Preliminary

1. $M_x = M_y = 0$ **2.** $M_x = 21$ **3.** $M_x = 5$; $M_y = 10$

4. Because a rectangle is symmetric with respect to both the vertical line and the horizontal line through the center of the rectangle, the Symmetry Principle guarantees that the centroid of the rectangle must lie along both these lines. The only point in common to both lines of symmetry is the center of the rectangle, so the centroid of the rectangle must be the center of the rectangle.

Section 8.3 Exercises

1. (a) $\left(\frac{9}{4}, 1\right)$ **(b)** $\left(\frac{46}{17}, \frac{14}{17}\right)$

5. A sketch of the lamina is shown here.

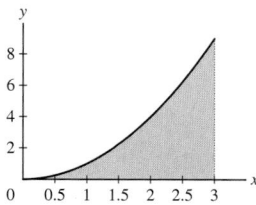

(a) $M_x = \frac{243}{10}$; $M_y = \frac{243}{4}$

(b) Area $= 27$; center of mass: $\left(\frac{9}{4}, \frac{9}{10}\right)$

7. $M_x = \frac{64\rho}{7}$; $M_y = \frac{32\rho}{5}$; center of mass : $\left(\frac{8}{5}, \frac{16}{7}\right)$

9. (a) $M_x = 24$ **(b)** $M = 12$, so $y_{cm} = 2$; center of mass: $(0, 2)$

11. $\left(\frac{633}{95}, \frac{195}{152}\right)$ **13.** $\left(\frac{9}{8}, \frac{18}{5}\right)$ **15.** $\left(\frac{1-5e^{-4}}{1-e^{-4}}, \frac{1-e^{-8}}{4(1-e^{-4})}\right)$

17. $\left(\frac{\pi}{2}, \frac{\pi}{8}\right)$

19. A sketch of the region is shown here.

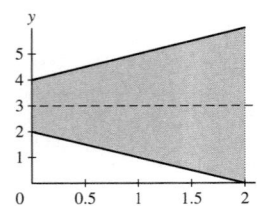

The region is clearly symmetric about the line $y = 3$, so we expect the centroid of the region to lie along this line. We find $M_x = 24$, $M_y = \frac{28}{3}$, centroid: $\left(\frac{7}{6}, 3\right)$.

21. $\left(\frac{9}{20}, \frac{9}{20}\right)$ 23. $\left(\frac{1}{2(e-2)}, \frac{e^2-3}{4(e-2)}\right)$ 25. $\left(\frac{\pi\sqrt{2}-4}{4(\sqrt{2}-1)}, \frac{1}{4(\sqrt{2}-1)}\right)$

27. A sketch of the region is shown here. Centroid: $\left(0, \frac{4}{7}\right)$

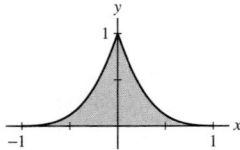

29. $\left(0, \frac{4b}{3\pi}\right)$ 31. $\left(\frac{4}{3\pi}, \frac{4}{3\pi}\right)$

33. $\left(0, \dfrac{\frac{2}{3}(r^2-h^2)^{3/2}}{r^2 \sin^{-1}\sqrt{1-h^2/r^2}-h\sqrt{r^2-h^2}}\right)$; with $r = 1$ and $h = \frac{1}{2}$:

$\left(0, \frac{3\sqrt{3}}{4\pi-3\sqrt{3}}\right) \approx (0, 0.71)$

35. $\left(0, \frac{16+3\pi\sqrt{3}}{\pi+2\sqrt{3}}\right)$

37. (c) $M_y^{\text{big}} = 0$, $M_y^{\text{small}} = -\pi r^3$, $M_y^S = \pi r^3$ (d) $\left(\frac{r}{3}, 0\right)$

41. Let A be the cross-sectional area of the rod. Place the rod with the m_1 at the origin and the rod lying along the positive x-axis. Then, the center of mass is located at

$$x = \frac{dm_2 + \frac{1}{2}\rho A d^2}{m_1 + m_2 + \rho A d}$$

Section 8.4 Preliminary

1. $T_3(x) = 9 + 8(x-3) + 2(x-3)^2 + 2(x-3)^3$

2. The polynomial graphed on the right is a Maclaurin polynomial.

3. A Maclaurin polynomial gives the value of $f(0)$ exactly.

4. The correct statement is (b): $\left|T_3(2) - f(2)\right| \le \frac{2^3}{6}$.

Section 8.4 Exercises

1. $T_2(x) = x$; $T_3(x) = x - \frac{x^3}{6}$

3. $T_2(x) = 1 - x + x^2$; $T_3(x) = 1 - x + x^2 - x^3$

5. $T_2(x) = x$; $T_3(x) = x + \frac{x^3}{3}$

7. $T_2(x) = 1 - x^2$; $T_3(x) = 1 - x^2$

9. $T_2(x) = 1 + x + \frac{x^2}{2}$; $T_3(x) = 1 + x + \frac{x^2}{2} + \frac{x^3}{6}$

11. $T_2(x) = 2 - 3x + \frac{5x^2}{2}$; $T_3(x) = 2 - 3x + \frac{5x^2}{2} - \frac{3x^3}{2}$

13. $T_2(x) = x - \frac{x^2}{2}$; $T_3(x) = x - \frac{x^2}{2} + \frac{x^3}{3}$

15. $T_2(x) = 1 + x + \frac{x^2}{2}$; $\left|T_2(-0.5) - f(-0.5)\right| = 0.018531$

17. $T_2(x) = 1 - \frac{3(x-1)}{2} + \frac{15(x-1)^2}{8}$; $\left|T_2(0.3) - f(0.3)\right| = 3.11706$

19. Let $f(x) = e^x$. Then, for all n,

$$f^{(n)}(x) = e^x \quad \text{and} \quad f^{(n)}(0) = 1.$$

It follows that

$$T_n(x) = 1 + \frac{x}{1!} + \frac{x^2}{2!} + \cdots + \frac{x^n}{n!}.$$

21. $T_n(x) = 1 + x + x^2 + x^3 + \cdots + x^n$

23. $T_n(x) = e + e(x-1) + \frac{e(x-1)^2}{2!} + \cdots + \frac{e(x-1)^n}{n!}$

25. $T_N(x) = 4\sqrt{2} + 5\sqrt{2}(x-2) + \frac{15(x-2)^2}{4\sqrt{2}} + \frac{5(x-2)^3}{16\sqrt{2}} + \cdots$

$$+ \frac{15}{8}(-1)^{n-3}\frac{1}{2^{n-3}}\left(\frac{(2n-7)!}{2^{n-3}(n-4)!n!}\right)2^{-(2n-7)/2}(x-2)^n$$

27. The odd polynomials are concave down on the left and lie beneath the graph of $y = e^x$; the even polynomials are concave up on the left and lie above the graph of $y = e^x$.

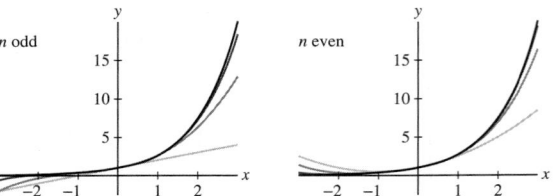

29. Error bound $= \frac{|0.3|^6}{6!} \approx 1.0125 \times 10^{-6}$; $\left|\cos 0.3 - T_5(0.3)\right| = 1.010874 \times 10^{-6}$

31. $n = 3$

33. $T_3(x) = x - \frac{x^3}{3}$; $T_3\left(\frac{1}{2}\right) = \frac{11}{24}$. With $K = 5$,

$$\left|T_3\left(\tfrac{1}{2}\right) - \tan^{-1}\tfrac{1}{2}\right| \le \frac{5\left(\frac{1}{2}\right)^4}{4!} = \frac{5}{384}.$$

37. $n = 6$ 39. $n = 9$

41. $1.5^{1.6} \approx T_2(0.5) = \frac{307}{288} \approx 1.065872$

43. $T_4(x) = x - \frac{2}{3}x^3$

45. $T_n(x) = 1 - x^2 + x^4 - x^6 + \cdots + (-x^2)^n$

47. At $a = 0$,

$$T_1(x) = -4 - x$$
$$T_2(x) = -4 - x + 2x^2$$
$$T_3(x) = -4 - x + 2x^2 + 3x^3 = f(x)$$
$$T_4(x) = T_3(x)$$
$$T_5(x) = T_3(x)$$

At $a = 1$

$$T_1(x) = 12(x - 1)$$

$$T_2(x) = 12(x - 1) + 11(x - 1)^2$$

$$T_3(x) = 12(x - 1) + 11(x - 1)^2 + 3(x - 1)^3$$

$$= -4 - x + 2x^2 + 3x^3 = f(x)$$

$$T_4(x) = T_3(x)$$

$$T_5(x) = T_3(x)$$

49. (b) By substitution

$$L = \lim_{x \to 0} \frac{x^2 + g_1(x)}{x^2/2 + g_2(x)} = \lim_{x \to 0} \frac{1 + g_1(x)/x^2}{1/2 + g_2(x)/x^2} = 2.$$

(c) Using L'Hôpital's Rule

$$L = \lim_{x \to 0} \frac{e^x + e^{-x} - 2}{1 - \cos x} = \lim_{x \to 0} \frac{e^x - e^{-x}}{\sin x} = \lim_{x \to 0} \frac{e^x + e^{-x}}{\cos x} = 2.$$

51. (c) $R_1 \approx \sqrt{\lambda L} = 0.04873$ cm and
$R_{100} \approx \sqrt{100 \lambda L} = 0.4873$ cm

57. $\int_0^{1/2} \sin(x^2)\,dx \approx \int_0^{1/2} x^2\,dx = \frac{1}{24}$; on $[0, 1/2]$,

$$\left| \sin(x^2) - T_4(x) \right| \le \frac{(1/2)^6}{6},$$

so the error in the approximation is bounded by

$$\frac{(1/2)^7}{6} = 1.116 \times 10^{-3}$$

Chapter 8 Review

1. $\frac{2{,}918{,}074}{10{,}125}$ **3.** $4\sqrt{17}$ **7.** $24\pi\sqrt{2}$

9. $\frac{67\pi}{36}$ **11.** $8\pi^2 + 12\pi$ **13.** 1,125 lb

15. $M_x = \frac{256}{3}$; $M_y = \frac{320}{3}$; center of mass: $\left(2, \frac{8}{5}\right)$

17. $\left(0, \frac{2}{\pi}\right)$ **19.** $\left(-\frac{44}{80 + 15\pi}, 0\right)$

21. $T_3(x) = 3(x + 2)^3 - 5(x + 2)$

23. $T_3(x) = 2 + \frac{1}{4}(x - 2) - \frac{1}{32}(x - 2)^2 + \frac{5}{768}(x - 2)^3$

25. $T_3(x) = -\frac{1}{2}x^2$

27. $\sqrt{e} \approx 1.6486979$; error $\approx 2.335 \times 10^{-5}$

29. $\left| f(17) - T_4(17) \right| \le \frac{105/32}{5!}4^{-9} \approx 1.043 \times 10^{-7}$

31. $\left| f(1.2) - T_4(1.2) \right| \le \frac{6}{5!}(0.2)^5 = 1.6 \times 10^{-5}$

Chapter 9

Section 9.1 Preliminary

1. (a) First order **(b)** First order **(c)** Order 3

2. Yes **3.** Example: $y' = y^2$ **4.** Example: $y' = y^2$

5. Example: $y' + xy = x^2$

Section 9.1 Exercises

1. (a) First order **(b)** Not first order **(c)** First order
(d) First order **(e)** Not first order **(f)** First order

3. Let $y = 4x^2$. Then $y' - 8x = 8x - 8x = 0$.

5. Let $y = \sqrt{12 - 4x^2}$. Then

$$yy' + 4x = \sqrt{12 - 4x^2}\,\frac{-4x}{\sqrt{12 - 4x^2}} + 4x$$

$$= -4x + 4x = 0$$

11. (b) $e^{-y} = -\sin x + C$ **(d)** $C = 1$

13. $y = \dfrac{-1}{\left(\frac{1}{2}\right)x^2 + C}$, where C is an arbitrary constant.

15. $y = Ae^{9x}$, where A is an arbitrary constant.

17. $y = Ae^{-3x} - \frac{2}{3}$, where A is an arbitrary constant.

19. $y = -\ln\left|C + \frac{1}{2}t^2\right|$, where C is an arbitrary constant.

21. $y = \dfrac{1}{(1/3)x^3 - x + C}$, where C is an arbitrary constant.

23. $x = \tan(\tan^{-1} t + C)$ OR $x = \frac{t + C}{1 - Ct}$ using the sum identities for tangents, where C is an arbitrary constant.

25. $y = \sin^{-1}\left(\frac{1}{2}x^2 + C\right)$, where C is an arbitrary constant.

27. $y = C \sec t$, where C is an arbitrary constant.

29. $y = 12e^{-2x}$ **31.** $y = -\sqrt{\ln(x^2 + e)}$ **33.** $y = e^{(1/2)x^2 - x} + 2$

35. $y = (e)e^{-e^{-t}} = e^{1 - e^{-t}}$ **37.** $y = \frac{et}{e^{1/t}} - 1$ **39.** $y = \dfrac{1}{1 - \sin^{-1} x}$

41. $y = \dfrac{1}{\cos x - (2/3)}$ **43.** $a = -4, 2$

45. (a) $t = (45/2)(\sqrt{10} - \sqrt{5}) \approx 20.840$ s
(b) $t = (45/2)\sqrt{10} \approx 71.15$ s

47. $y(t) = 20 - \left(\frac{t}{320} + 10^{3/2}\right)^{2/3}$ and

$t_e = 320(20^{3/2} - 10^{3/2}) \approx 18502.4$ s or 308.4 min

49. (a) $q = CV - Ae^{-t/(RC)}$
(b) As $t \to \infty$, $e^{-t/(RC)} \to 0$, so $q(t) \to CV$.
(c) $q(t) = CV - CVe^{-t/(RC)}$ and
$q(\tau) = CV(1 - e^{-RC/RC}) \approx CV(0.6321)$

51. $V = (kt/3 + C)^3$, V increases roughly with the cube of time.

53. $g(x) = Ae^{2x}$; $g(x) = \frac{A}{1-x}$ **57.** $y = \pm\sqrt{x^2 + C}$ and $y = A/x$

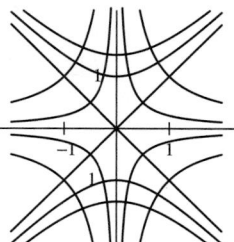

59. (a) $v(t) = -\alpha\left(\dfrac{1 - e^{-(2k\alpha/m)t}}{1 + e^{-(2k\alpha/m)t}}\right)$

61. (a) $(y')^2 + y^2 \ge 0$ and equals zero if and only if $y' = 0$ and $y = 0$
(b) $(y')^2 + y^2 + 1 \ge 1 > 0$ for all y' and y, so $(y')^2 + y^2 + 1 = 0$ has no solution

63. (c) $C = \frac{7\pi}{120B} R^{5/2}$

Section 9.2 Preliminary

1. $y(t) = b + ce^{-kt}$, where c is an arbitrary constant.

2. $y(t) = 5 - ce^{4t}$ for any positive constant c.

3. No   **4.** True   **5.** Material A

Section 9.2 Exercises

1. General solution: $y(t) = 10 + ce^{2t}$; solution satisfying $y(0) = 25$: $y(t) = 10 + 15e^{2t}$; solution satisfying $y(0) = 5$: $y(t) = 10 - 5e^{2t}$

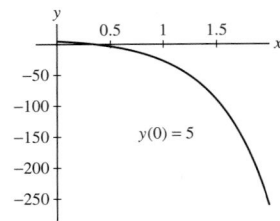

5. (a) $k = \frac{\ln 2}{40} \approx 0.017 \text{ s}^{-1}$ (b) $F'(t) = -0.017(F(t) - 60)$
(c) $F(t) = 60 + 56.2e^{-0.017t}$ (d) $116.2°\text{F}$

7. $50\frac{3^{4/5} - 2}{3^{4/5} - 1} \approx 14.49°\text{C}$

9. (a) $5.4°\text{C/min}$ (b) $0.54°\text{C}$ (c) $\frac{1}{0.09}\ln\left(\frac{70}{45}\right) \approx 4.91 \text{ min}$

11. $-58.8 \text{ m/s} \approx -131.5 \text{ mph}$ **13.** -25.0 ft/s

15. (a) $t = \frac{1}{0.05}\ln 4 \approx 28 \text{ yr}$ (b) $P_0 = \$20{,}000$

17. $\$10{,}000$ **19.** 10%

21. (b) $t = \frac{1}{0.09}\ln\left(\frac{13{,}333.33}{3{,}333.33}\right) \approx 15.4 \text{ yr}$ (c) No

23. (a) $b = \frac{V}{R}$ (b) $I(t) = \frac{V}{R}(1 - e^{-kt})$

Section 9.3 Preliminary

1. 7   **2.** $y = \pm\sqrt{t+1}$ **3.** False   **4.** Yes   **5.** $n = 20$

Section 9.3 Exercises

1.   **3.**

5.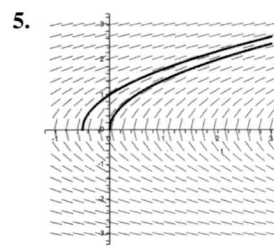

7. With $y(0) > 0$, as $t \to \infty$, y increases at an increasing rate, so that $\lim_{t\to\infty} y = \infty$. On the other hand, if $y(0) < 0$, then y decreases at an ever faster rate as $t \to \infty$, so $\lim_{t\to\infty} y = -\infty$.

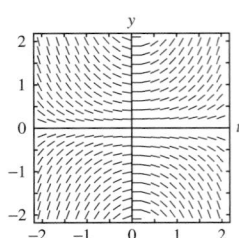

9. For $y' = t^2$, y' only depends on t. The isoclines of any slope c will be the two vertical lines $t = \pm\sqrt{c}$. This indicates that the slope field will be the one given in Figure 12(A). The solutions are sketched in the following figure.

For $y' = y^2$, y' depends only on y. The isoclines of any slope c will be the two *horizontal* lines $y = \pm\sqrt{c}$. The solutions are sketched in the following figure.

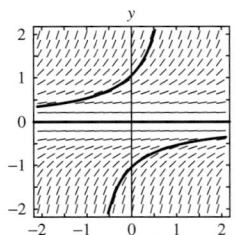

11. This differential equation cannot be solved by separation of variables. See the solutions sketched in the following figure.

13. (a) Since the step width is $h = 0.1$, $y(1.0) \approx y_{10}$. We compute:

k	t_k	y_k
0	0	0
1	0.1	0
2	0.2	0.01
3	0.3	0.029801
4	0.4	0.0589202
5	0.5	0.0966314
6	0.6	0.142026
7	0.7	0.194082
8	0.8	0.251733
9	0.9	0.313929
10	1.0	0.379681

(b) $y = \ln\left|\frac{1}{2}t^2 + 1\right|$

(c) $\left|y(0.1) - y_1\right| = |0.00498754 - 0| = 0.00498754$

$\left|y(0.5) - y_5\right| = |0.117783 - 0.0966314| = 0.0211516$

$\left|y(1.0) - y_{10}\right| = |0.405465 - 0.379681| = 0.257844$

15. $y(1) \approx 0$ **17.** $y(1.5) \approx 2.478982$ **19.** $y(0.5) \approx 0.68098$

21. Euler's method: $y(1.5) \approx 4.177248$; Euler's midpoint method:
$y(1.5) \approx 4.471304$; $y(t) = e^t$; Error from Euler's method: 0.304441;
Error from Euler's midpoint method: 0.010385

23. $y(0.5) \approx 0$ **25.** $y(1) \approx 0.714081$

Section 9.4 Preliminary

1. (a) No **(b)** Yes **(c)** No **(d)** Yes

2. False **3.** True

4. (a) $\lim_{t \to \infty} y(t) = 3$ **(b)** $\lim_{t \to \infty} y(t) = 3$ **(c)** $\lim_{t \to \infty} y(t) = -\infty$

Section 9.4 Exercises

1. $y = \dfrac{5}{1 - e^{-3t}/C}$ and $y = \dfrac{5}{1 + (3/2)e^{-3t}}$ **3.** $\lim_{t \to \infty} y(t) = 2$

5. (a) $P(t) = \dfrac{2,000}{1 + 3e^{-0.6t}}$ **(b)** $t = \frac{1}{0.6} \ln 3 \approx 1.83$ yrs

7. $k = \ln \frac{81}{31} \approx 0.96$ yrs^{-1}; $t = \dfrac{\ln 9}{2 \ln 9 - \ln 31} \approx 2.29$ yrs

9. After $t = \dfrac{4 \ln(207/2)}{\ln(23/2)} \approx 7.6$ hours, or at 3:36 PM

11. (a) $y_1(t) = \dfrac{10}{10 - 9e^{-t}}$ and $y_2(t) = \dfrac{1}{1 - 2e^{-t}}$

(b) $t = \ln \frac{9}{8}$ **(c)** $t = \ln 2$

13. (a) $u = \sqrt{M} \dfrac{Ce^{(k/\sqrt{M})t} - 1}{Ce^{(k/\sqrt{M})t} + 1}$

17. (d) $t = \frac{1}{k} \ln \dfrac{A - y_0}{y_0}$

Section 9.5 Preliminary

1. (a) Yes **(b)** No **(c)** Yes **(d)** No

2. The answer is (b).

Section 9.5 Exercises

1. (c) $y = \frac{x^4}{5} + \frac{C}{x}$ **(d)** $y = \frac{x^4}{5} - \frac{1}{5x}$

5. $y = \frac{1}{2}x + \frac{C}{x}$ **7.** $y = -\frac{1}{4}x^{-1} + Cx^{1/3}$

9. $y = \frac{1}{5}x^2 + \frac{1}{3} + Cx^{-3}$ **11.** $y = -x \ln|x| + Cx$

13. $y = \frac{1}{2}e^x + Ce^{-x}$ **15.** $y = x \cos x + C \cos x$

17. $y = x^x + Cx^x e^{-x}$ **19.** $y = \frac{1}{5}e^{2x} - \frac{6}{5}e^{-3x}$

21. $y = \dfrac{\ln|x|}{x+1} - \dfrac{1}{x(x+1)} + \dfrac{5}{x+1}$ **23.** $y = -\cos x + \sin x$

25. $y = \tanh x + 3 \operatorname{sech} x$

27. For $m \neq -n$: $y = \dfrac{1}{m+n}e^{mx} + Ce^{-nx}$; for $m = -n$:
$y = (x + C)e^{-nx}$

29. $y'(t) = 8 - \frac{y}{5}$, $y(0) = 10$; $y(t) = 40 - 30e^{-t/5}$; 0.4 lbs per gal

31. (a) $\dfrac{dV}{dt} = \dfrac{20}{1+t} - 5$ and $V(t) = 20 \ln(1 + t) - 5t + 100$

(b) The maximum value is $V(3) = 20 \ln 4 - 15 + 100 \approx 112.726$.

(c) Tank is empty after $t \approx 34.25$ min.

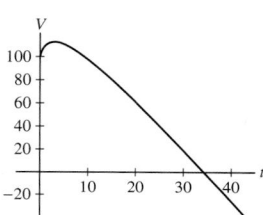

33. $I(t) = \dfrac{2 - 2e^{-20t}}{20} = \dfrac{1 - e^{-20t}}{10}$

35. (a) $I(t) = \dfrac{V}{R} - \dfrac{V}{R}e^{-(R/L)t}$

(c) Approximately 0.0184 s

37. (b) $y_2(t) = -4e^{-t/10} + 4e^{-t/20}$

$= 4e^{-t/20}(1 - e^{-t/20})$

(c) See accompanying figure.

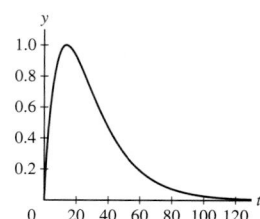

(d) $\frac{1}{200}$

Chapter 9 Review

1. (a) No, first order **(b)** Yes, first order

(c) No, order 3 **(d)** Yes, order 2

3. $y = \left(\frac{4}{3}t^3 + C\right)^{1/4}$, where C is an arbitrary constant.

5. $y = Cx - 1$, where C is an arbitrary constant.

7. $y = \frac{1}{2}\left(x + \frac{1}{2}\sin 2x\right) + \frac{\pi}{4}$ **9.** $y = \dfrac{2}{2 - x^2}$

11.

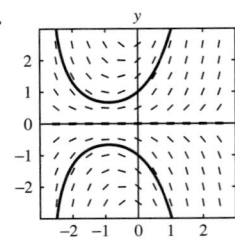

13. $y(t) = \tan t$

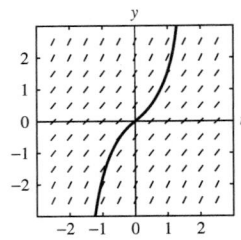

15. $y(0.1) \approx 1.1$; $y(0.2) \approx 1.209890$; $y(0.3) \approx 1.329919$

17. $y = x^2 + 2x$ **19.** $y = \frac{1}{2} + e^{-x} - \frac{11}{2}e^{-2x}$

21. $y = \cos x \sin x + C \cos x$, where C is an arbitrary constant

23. $\frac{1}{2}y^2 - y = \frac{1}{2}t^2 + 7$, or since we know that $y(1) < 0$,

$y = 1 - \sqrt{15 + t^2}$ **25.** $w = \tan(k \ln x + \frac{\pi}{4})$

27. $y = -\cos x + \frac{\sin x}{x} + \frac{C}{x}$, where C is an arbitrary constant.

29. Solution satisfying $y(0) = 3$: $y(t) = 4 - e^{-2t}$; solution

satisfying $y(0) = 4$: $y(t) = 4$

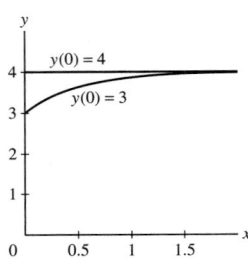

31. (a) 12

(b) ∞, if $y(0) > 12$; 12, if $y(0) = 12$; $-\infty$, if $y(0) < 12$

(c) -3

33. $40,000 - 35,000e^{0.1} \approx \$1,319.02$ **35.** $\$40,000$

39. $y(t) = \frac{4}{1-(1/2)e^{-1.6t}}$ **41.** $k = -\frac{\ln 2}{10} \approx -0.0693$

43. (a) $\frac{dP}{dt} = 0.12P - r$, $P(0) = 150$ **(b)** $r = 18$

45. (a) $\frac{dy}{dt} = 1.5 - \frac{4y}{20-t}$, $y(0) = 0$

(b) $y(t) = 10 - \frac{1}{2}t - \frac{(20-t)^4}{16,000}$

(c) Amount of toxin is maximal at $t = 20 - 10\sqrt[3]{2} \approx 7.4$ minutes.

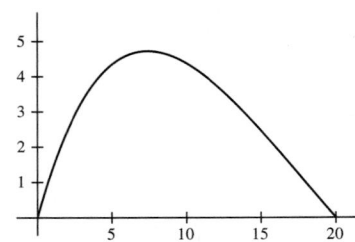

Chapter 10

Section 10.1 Preliminary

1. $a_4 = 12$ **2.** (c)   **3.** $\lim_{n\to\infty} a_n = \sqrt{2}$ **4.** (b)

5. (a) False. Counterexample: $a_n = \cos \pi n$.   **(b)** True

(c) False. Counterexample: $a_n = (-1)^n$.

Section 10.1 Exercises

1. (a) (iv)   **(b)** (i)   **(c)** (iii)   **(d)** (ii)

3. $c_1 = 2, c_2 = 2, c_3 = \frac{4}{3}, c_4 = \frac{2}{3}$

5. $a_1 = 3, a_2 = 10, a_3 = 101, a_4 = 10,202$

7. $c_1 = 1, c_2 = \frac{3}{2}, c_3 = \frac{11}{6}, c_4 = \frac{25}{12}$

9. $b_1 = 2, b_2 = 5, b_3 = 9, b_4 = 19$

11. (a) $a_n = \frac{(-1)^{n+1}}{n^3}$ **(b)** $a_n = \frac{n+1}{5+n}$

13. $\lim_{n\to\infty} a_n = 4$ **15.** $\lim_{n\to\infty} \left(5 - \frac{9}{n^2}\right) = 5$

17. $\lim_{n\to\infty} (-2^{-n}) = 0$ **19.** $\lim_{n\to\infty} \left(\frac{1}{3}\right)^n = 0$

21. The sequence diverges.   **23.** $\lim_{n\to\infty} \frac{n}{\sqrt{n^3+1}} = 0$

25. The sequence diverges.

27. (a) $M = 999$ **(b)** $M = 99,999$

31. $\lim_{n\to\infty} 2^{1/n} = 1$ **33.** $\lim_{n\to\infty} n^{1/n} = 1$ **35.** $\lim_{n\to\infty} \left(1 + \frac{1}{n}\right)^n = e$

37. $\lim_{n\to\infty} a_n = 0$ **39.** $\lim_{n\to\infty} n \sin \frac{1}{n} = 1$ **41.** (b)

43. $\lim_{n\to\infty} \frac{3n^2+n+2}{2n^2-3} = \frac{3}{2}$ **45.** $\lim_{n\to\infty} 3 + \left(-\frac{1}{2}\right)^n = 3$

47. $\lim_{n\to\infty} \tan^{-1}(1 - \frac{2}{n}) = \frac{\pi}{4}$ **49.** $\lim_{n\to\infty} \ln \left(\frac{2n+1}{3n+4}\right) = \ln \frac{2}{3}$

51. $\lim_{n\to\infty} \frac{e^n+3^n}{5^n} = 0$ **53.** $\lim_{n\to\infty} \frac{e^n}{2^{n^2}} = 0$

55. $\lim_{n\to\infty} \frac{n^3+2e^{-n}}{3n^3+4e^{-n}} = \frac{1}{3}$ **57.** The sequence diverges.

59. $\lim_{n\to\infty} \frac{(-4,000)^n}{n!} = 0$ **61.** $\lim_{n\to\infty} \cos \frac{\pi}{n} = 1$ **63.** $\lim_{n\to\infty} \sqrt[n]{n} = 1$

65. Any number greater than or equal to 3 is an upper bound.

71. Example: $a_n = -n^2, b_n = n^2 + \frac{1}{n}$

75. $\lim_{n\to\infty} a_n = \frac{1+\sqrt{21}}{2}$ **77.** $\lim_{n\to\infty} c_n = 1$

81. (a) $c_1 = \frac{3}{2}; c_2 = \frac{13}{12}; c_3 = \frac{19}{20}; c_4 = \frac{743}{840}$ **(c)** $\lim_{n\to\infty} c_n = \ln 2$

Section 10.2 Preliminary

1. The sum of an infinite series is defined as the limit of the sequence of partial sums.

2. $S = \frac{1}{2}$ **3.** The result is negative hence the formula is invalid.

4. No   **5.** No.   **6.** $N = 13$ **7.** No   **8.** Example: $\sum_{n=1}^{\infty} \frac{1}{n^{9/10}}$.

Section 10.2 Exercises

1. (a) $a_n = \frac{1}{3^n}$ (b) $a_n = \left(\frac{5}{2}\right)^{n-1}$

(c) $a_n = (-1)^{n+1} \frac{n^n}{n!}$ (d) $a_n = \frac{1 + \frac{(-1)^{n+1}+1}{2}}{n^2+1}$

3. $S_2 = \frac{5}{4}$, $S_4 = \frac{205}{144}$, $S_6 = \frac{5{,}369}{3{,}600}$ **5.** $S_2 = \frac{2}{3}$, $S_4 = \frac{4}{5}$, $S_6 = \frac{6}{7}$

7. $S_5 = 0.035672$, $S_{10} = 0.035352$, $S_{15} = 0.035413$. Yes.

9. $S_3 = \frac{3}{10}$, $S_4 = \frac{1}{3}$, $S_5 = \frac{5}{14}$; $\sum\limits_{n=1}^{\infty} \left(\frac{1}{n+1} - \frac{1}{n+2}\right) = \frac{1}{2}$

11. $\sum\limits_{n=3}^{\infty} \frac{1}{n(n-1)} = \sum\limits_{n=3}^{\infty} \left(\frac{1}{n-1} - \frac{1}{n}\right) = \frac{1}{2}$

13. $\lim\limits_{n\to\infty} (-1)^n n^2 \neq 0$

17. (a) **19.** $S = \frac{1}{18}$ **21.** $S = \frac{27}{968}$ **23.** $S = \frac{7}{5}$ **25.** $S = \frac{e}{e^2-1}$

27. The series diverges. **29.** $S = \frac{8}{5}$ **31.** $S = \frac{512}{49}$ **33.** (b), (c)

35. (a) $\sum\limits_{n=1}^{10} a_n = \frac{249}{50}$, $\sum\limits_{n=4}^{16} a_n = \frac{494}{2{,}304}$ (b) $a_3 = \frac{5}{18}$

(c) $a_n = \frac{2(2n-1)}{n^2(n-1)^2}$ (d) $\sum\limits_{n=1}^{\infty} a_n = 5$

39. 50 ft **41.** $S = \frac{1}{2}$ **45.** $2 + \sqrt{2}$

Section 10.3 Preliminary

1. (b) **2.** f must be positive, continuous and decreasing

3. Integral Test **4.** Comparison Test

5. No; $\sum \frac{1}{n}$ diverges, but since $\frac{e^{-n}}{n} < \frac{1}{n}$, this tells us nothing about $\sum \frac{e^{-n}}{n}$.

Section 10.3 Exercises

1. $\int_1^\infty \frac{dx}{x^4}$ converges, so the series converges.

3. $\int_1^\infty x^{-1/3}\, dx = \infty$. The series diverges.

5. $\int_{25}^\infty \frac{x^2}{(x^3+9)^{5/2}}\, dx$ converges, so the series converges.

7. $\int_1^\infty \frac{dx}{x^2+1}$ converges, so the series converges.

9. $\int_1^\infty x e^{-x^2}\, dx$ converges, so the series converges.

11. $\int_1^\infty \frac{1}{2^{\ln x}}\, dx = \infty$. The series diverges.

13. $\int_1^\infty \frac{\ln x}{x^2}\, dx = \frac{1+\ln 2}{2}$. The series converges.

15. $\frac{1}{n^3+8n} \leq \frac{1}{n^3}$. The series converges.

19. $\frac{1}{n 2^n} \leq \frac{1}{2^n}$. The series converges.

21. $\frac{k^{1/3}}{k^2+k} \leq \frac{k^{1/3}}{k^2} = \frac{1}{k^{5/3}}$. The series converges.

23. $\frac{1}{\sqrt{n^3+1}} \leq \frac{1}{\sqrt{n^3}} = \frac{1}{n^{3/2}}$. The series converges.

25. $0 \leq \frac{\sin^2 k}{k^2} \leq \frac{1}{k^2}$. The series converges.

27. $\frac{4}{m!+4^m} \leq \frac{4}{4^m} = 4 \cdot \left(\frac{1}{4}\right)^m$. The series converges.

29. $\frac{1}{2^{k^2}} \leq \frac{1}{2^k} = \left(\frac{1}{2}\right)^k$. The series converges.

31. $0 \leq \frac{\ln n}{n^3+3\ln n} \leq \frac{\ln n}{n^3}$. The series converges.

33. The general term of the series does not approach zero and the series diverges.

35. The series converges. **37.** The series diverges.

39. The series converges. **41.** The series converges.

47. The series converges for $a > 1$ and diverges for $a \leq 1$.

49. The series converges. **51.** The series diverges.

53. The series converges. **55.** The series converges.

57. The series diverges. **59.** The series converges.

61. The series converges. **63.** The series converges.

65. The series converges. **67.** The series converges.

69. The series converges. **71.** The series converges.

73. The series converges. **81.** $\sum\limits_{n=1}^{\infty} \frac{1}{n^5} \approx 1.0369$

87. (c) $\sum\limits_{n=1}^{1{,}000} \frac{1}{n^2} = 1.643934568$; $1 + \sum\limits_{n=1}^{100} \frac{1}{n^2(n+1)} = 1.644884890$;

the latter sum is a better approximation to $\pi^2/6$

Section 10.4 Preliminary

1. (c) **2.** (b) **3.** No **4.** Example: $\sum \frac{(-1)^n}{\sqrt[3]{n}}$.

Section 10.4 Exercises

3. Converges conditionally **5.** Converges absolutely

7. Converges conditionally **9.** Converges absolutely

11. Converges absolutely

13. (a)

n	S_n
1	1.000000
2	0.875000
3	0.912037
4	0.896412
5	0.904412
6	0.899782
7	0.902698
8	0.900745
9	0.902116
10	0.901116

15. $S_5 = 0.947539$ **17.** $S_{44} = 0.0656746$

19. Converges (by Comparison Test)

21. Converges (by Leibniz Test)

23. Converges (by Linearity of Series)

25. Converges (by Leibniz Test)

27. Converges conditionally

Section 10.5 Preliminary

1. $\rho = \lim\limits_{n\to\infty} \left|\frac{a_{n+1}}{a_n}\right|$

2. The Ratio Test is conclusive for $\sum\limits_{n=1}^{\infty} \frac{1}{2^n}$ and inconclusive for $\sum\limits_{n=1}^{\infty} \frac{1}{n}$.

3. No

Section 10.5 Exercises

1. Converges (by the Ratio Test)

3. Converges absolutely (by the Ratio Test)

5. The Ratio Test is inconclusive. **7.** Diverges (by the Ratio Test)

9. Converges (by the Ratio Test)

11. Converges (by the Ratio Test)

13. Diverges (by the Ratio Test) **15.** The Ratio Test is inconclusive.

17. Converges (by the Ratio Test) **19.** $\lim\limits_{n\to\infty}\left|\frac{a_{n+1}}{a_n}\right|=\frac{1}{3}<1$

21. $\lim\limits_{n\to\infty}\left|\frac{a_{n+1}}{a_n}\right|=2|x|$ **23.** $\lim\limits_{n\to\infty}\left|\frac{a_{n+1}}{a_n}\right|=|r|$

27. Converges absolutely (by the Ratio Test)

29. The Ratio Test is inconclusive and the series may diverge/converge, depending on a_n.

31. Converges absolutely (by the Ratio Test)

33. The Ratio Test is inconclusive.

35. Converges (by the Root Test) **37.** Converges (by the Root Test)

39. Converges (by the Root Test)

41. Converges (by the Ratio Test or by Linearity)

43. Converges (by the Ratio Test)

45. Converges (by the Limit Comparison Test)

47. Converges (by the Limit Comparison Test)

49. Converges (by the Limit Comparison Test)

51. Diverges (by the Divergence Test)

Section 10.6 Preliminary

1. Yes. The series converges for both $x=4$ and $x=-3$.

2. Converges at $x=8$; cannot determine convergence for $x=0$, $x=2$, $x=12$.

3. $R=4$ **4.** $\sum\limits_{n=0}^{\infty}(n+1)^2x^n$; $R=1$

Section 10.6 Exercises

1. $R=2$ **3.** $R=3$ for all series.

7. Converges in $|x|<1$ and diverges elsewhere.

9. Converges in $-\frac{1}{2}\le x<\frac{1}{2}$ and diverges elsewhere.

11. Converges in $-1\le x<1$ and diverges elsewhere.

13. Converges for all x.

15. Converges in $-5<x<-3$ and diverges elsewhere.

17. Converges in $-2<x<2$ and diverges elsewhere.

19. Converges in $3\le x\le5$ and diverges elsewhere.

21. Converges for $x=-5$ and diverges for $x\ne-5$.

23. Converges for $2-\frac{1}{e}<x<2+\frac{1}{e}$ and diverges elsewhere.

25. Converges in $-1\le x<1$ and diverges elsewhere.

27. Expansion: $\sum\limits_{n=0}^{\infty}3^nx^n$; the expansion is valid for $|x|<\frac{1}{3}$.

29. Expansion: $\sum\limits_{n=0}^{\infty}\frac{x^n}{3^{n+1}}$; the expansion is valid for $|x|<3$.

31. Expansion: $\sum\limits_{n=0}^{\infty}(-1)^nx^{9n}$; the expansion is valid for $|x|<1$.

33. Expansion: $\sum\limits_{n=0}^{\infty}(-1)^n3^nx^{7n}$; the expansion is valid for $|x|<\frac{1}{3^{1/7}}$.

37. $\frac{1}{4-x}=\sum\limits_{n=0}^{\infty}(-1)^{n+1}(x-5)^n$ on the interval $(4,6)$.

39. Example: $\sum\limits_{n=1}^{\infty}\frac{(x-4)^n}{n2^n}$ **41.** N is at least 7; $S_7=0.405804$.

45. N is at least 7; $S_7=0.368056$.

47. $1-\frac{1}{2}x^2-\sum\limits_{n=2}^{\infty}\frac{1\cdot3\cdot5\cdots(2n-3)}{(2n)!}x^{2n}$; $R=\infty$

53. Converges for $-\frac{1}{3}<x<\frac{1}{3}$

Section 10.7 Preliminary

1. $f(0)=3$ and $f'''(0)=30$ **2.** $f(-2)=0$; $f^{(4)}(-2)=48$

3. Substituting x^2 for the MacLaurin series for $\sin(x)$.

4. $f(x)=4+\sum\limits_{n=1}^{\infty}\frac{(x-3)^{n+1}}{n(n+1)}$ **5.** (a) and (c)

Section 10.7 Exercises

1. $f(x)=2+3x+2x^2+2x^3+\cdots$

3. $\frac{1}{1-2x}=\sum\limits_{n=0}^{\infty}2^nx^n$ for $|x|<\frac{1}{2}$

5. $\cos(3x)=\sum\limits_{n=0}^{\infty}\frac{(-1)^n9^n}{(2n)!}x^{2n}$ for all x

7. $\sin(x^2)=\sum\limits_{n=0}^{\infty}\frac{(-1)^n}{(2n+1)!}x^{4n+2}$ for all x

9. $\ln(1-x^2)=-\sum\limits_{n=1}^{\infty}\frac{x^{2n}}{n}$ for $|x|<1$

11. $\tan^{-1}(x^2)=\sum\limits_{n=0}^{\infty}\frac{(-1)^nx^{4n+2}}{2n+1}$ for $|x|\le1$

13. $e^{x-2}=\sum\limits_{n=0}^{\infty}\frac{1}{n!e^2}x^n$ for all x

15. $\ln(1-5x)=-\sum\limits_{n=1}^{\infty}\frac{5^n}{n}x^n$ for $|x|<\frac{1}{5}$ and $x=-\frac{1}{5}$

17. $\sinh(x)=\sum\limits_{n=0}^{\infty}\frac{1}{(2n+1)!}x^{2n+1}$ for all x

19. $\frac{1-\cos(x^2)}{x}=\sum\limits_{n=1}^{\infty}\frac{(-1)^{n+1}}{(2n)!}x^{4n-1}$ for $x\ne0$

21. $e^x\sin(x)=x+x^2+\frac{x^3}{3}-\frac{x^5}{30}+\cdots$

23. $e^x\ln(1-x)=-x-\frac{3x^2}{2}-\frac{4x^3}{3}-x^4+\cdots$

25. $(1+x)^{-3/2}=1-\frac{3x}{2}+\frac{15x^2}{8}-\frac{35x^3}{16}+\frac{315x^4}{128}-\cdots$

27. $e^{e^x}=e+ex+ex^2+\frac{5e}{6}x^3+\cdots$

29. $\sin(x)=\sum\limits_{n=0}^{\infty}\frac{(-1)^n}{(2n)!}\left(x-\frac{\pi}{2}\right)^{2n}$ **31.** $\frac{1}{x}=\sum\limits_{n=0}^{\infty}(-1)^n(x-1)^n$

33. $\frac{1}{1-x} = \sum\limits_{n=0}^{\infty} \frac{(-1)^{n+1}(x-5)^n}{4^{n+1}}$

35. $x^4 + 3x - 1$

$$= 21 + 35(x-2) + 24(x-2)^2 + 8(x-2)^3 + (x-2)^4$$

37. $e^{3x} = \sum\limits_{n=0}^{\infty} \frac{3^n e^{-3}}{n!}(x+1)^n$

39. $\frac{1}{1-x^2} = \sum\limits_{n=0}^{\infty} \frac{(-1)^{n+1}(2^{n+1}-1)}{2^{2n+3}}(x-3)^n$, for $|x-3| < 2$

41. $f(x) = 1 + \sum\limits_{n=0}^{\infty} \frac{1 \cdot 3 \cdot 5 \cdots (2n-1)}{2^n n!}(3x)^{2n}$ for $|x| < \frac{1}{3}$

43. $\sin^{-1}\frac{1}{2} \approx 0.52358519539$

47. (a) 5 **(b)** $\sum\limits_{n=0}^{4} \frac{(-1)^n}{(2n+1)n!} = 0.747487$

49. $\int_0^1 \cos(x^2)\,dx = \sum\limits_{n=0}^{\infty} \frac{(-1)^n}{(4n+1)(2n)!}$;

$\sum\limits_{n=0}^{3} \frac{(-1)^n}{(4n+1)(2n)!} = 0.904522792$

51. $\int_0^1 e^{-x^3}\,dx = \sum\limits_{n=0}^{\infty} \frac{(-1)^n}{(3n+1)n!}$; $\sum\limits_{n=0}^{24} \frac{(-1)^n}{(3n+1)n!} = 0.807511$

53. $\int_0^x \frac{1-\cos t}{t}\,dt = \sum\limits_{n=1}^{\infty} \frac{(-1)^n}{2n(2n)!}x^{2n}$

55. $\int_0^x \ln(1+t^2)\,dt = \sum\limits_{n=1}^{\infty} \frac{(-1)^{n-1}}{n(2n+1)}x^{2n+1}$

57. $f(x) = \frac{1}{1+2x}$ **59.** $x^2 - \frac{2}{3}x^6 + \frac{2}{15}x^{10} - \frac{4}{315}x^{14}$

61. $1 + x - \frac{1}{2}x^2 - \frac{1}{8}x^4$ **63.** $f(x) = e^{x^3}$

65. $f(x) = 1 - 5x + \sin(5x)$ **67.** $I(t) = \frac{V}{R}\sum\limits_{n=1}^{\infty} \frac{(-1)^{n+1}}{n!}\left(\frac{Rt}{L}\right)^n$

69. $f(x) = \sum\limits_{n=0}^{\infty} \frac{(-1)^n x^n}{(2n)!}$ and $f^{(5)}(0) = -\frac{1}{30,240}$

71. $f^{(8)}(0) = -1,575$

73. The expansion is not valid since the series only converges for $|x| < 1$.

77. 4; $\sum\limits_{n=1}^{4} \frac{(-1)^{n-1}(0.2)^n}{n} = 0.182267$

79. $\frac{1}{(1-x)(1-2x)} = \sum\limits_{n=0}^{\infty} (-1)^n \left(1 - \left(\frac{2}{3}\right)^{n+1}\right)(x-2)^n$, for $|x-2| < 1$

83. If a positive whole, $R = \infty$; if negative whole, $R = 1$.

Chapter 10 Review

1. (a) $a_1^2 = 4, a_2^2 = \frac{1}{4}, a_3^2 = 0$

(b) $b_1 = \frac{1}{24}, b_2 = \frac{1}{60}, b_3 = \frac{1}{240}$

(c) $a_1 b_1 = -\frac{1}{12}, a_2 b_2 = -\frac{1}{120}, a_3 b_3 = 0$

(d) $2a_2 - 3a_1 = 5, 2a_3 - 3a_2 = 1.5, 2a_4 - 3a_3 = \frac{1}{12}$

3. $\lim\limits_{n \to \infty}(5a_n - 2a_n^2) = 2$ **5.** $\lim\limits_{n \to \infty} e^{a_n} = e^2$

7. The limit does not exist. **9.** $\lim\limits_{n \to \infty} a_n = 0$

11. $\lim\limits_{n \to \infty} a_n = 1$ **13.** The sequence diverges.

15. $\lim\limits_{n \to \infty} b_n = \frac{\pi}{4}$ **17.** $\lim\limits_{n \to \infty} b_n = \frac{1}{2}$ **19.** $\lim\limits_{n \to \infty} a_n = 0$

21. $\lim\limits_{n \to \infty} c_n = e^3$

25. (a) $\lim\limits_{n \to \infty} a_n = \infty$ **(b)** $\lim\limits_{n \to \infty} \frac{a_{n+1}}{a_n} = 3$

27. $S_4 = -\frac{11}{60} = -0.183333$; $S_7 = \frac{41}{630} = 0.0650794$

29. $\sum\limits_{n=2}^{\infty} \left(\frac{2}{3}\right)^n = \frac{4}{3}$ **31.** $\sum\limits_{n=0}^{\infty} \frac{2^{n+1}}{3^n} = 6$

33. Example: $a_n = \left(\frac{1}{2}\right)^n + 1, b_n = -1$

35. The sequence diverges.

37. $\int_1^{\infty} \frac{x^3}{e^{x^4}}\,dx = \frac{e^{-1}}{4}$. The sequence converges.

39. $\frac{1}{n^2} > \frac{1}{(n+1)^2}$. The sequence converges.

41. $\frac{n^2+1}{n^{3.5}-2} < \frac{4}{n^{1.5}}$. The sequence converges.

43. Since $\frac{\ln n}{1.5^n} < \frac{n}{1.5^n}$ for all $n > 1$. The sequence converges.

45. $\frac{1}{3^n - 2^n} < \frac{1}{2^n}$ for all $n \geq 2$. The sequence converges.

49. $0.3971162690 \leq S \leq 0.3971172688$. The maximum size of the error 10^{-6}.

51. Converges absolutely. **53.** The series diverges.

55. 22; $K \approx \sum\limits_{k=0}^{21} \frac{(-1)^k}{(2k+1)^2} = 0.91571$

57. (a) Converges **(b)** Converges **(c)** Diverges **(d)** Converges

59. The Ratio Test is inconclusive. **61.** The series converges.

63. The Ratio Test is inconclusive. **65.** The series converges.

67. The series converges. **69.** The series converges.

71. The series converges. **73.** The series converges.

75. The series converges by Leibniz Test.

77. The series converges by Leibniz Test.

79. The series converges. **81.** The series diverges.

83. The series converges. **85.** The series converges for $x \in \mathbf{R}$.

87. The series converges for $x \in [2, 4]$.

89. The series converges for $x = 0$.

91. $\frac{2}{4-3x} = \frac{1}{2}\sum\limits_{n=0}^{\infty} \left(\frac{3}{4}\right)^n x^n$. The series converges for $|x| < \frac{4}{3}$.

93. (c)

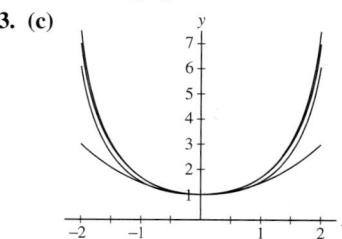

95. $\lim\limits_{x \to 0} \frac{1 - \frac{x^2}{2} - \cos x}{x^4} = -\frac{1}{24}$ **97.** $f(x) = e^{-2}\sum\limits_{n=0}^{\infty} \frac{2^n(x+1)^n}{n!}$

99. $f(x) = \sum_{n=1}^{\infty} \frac{(-1)^{n+1}(x-2)^n}{n2^n}$

101. $f(x) = \sum_{n=0}^{\infty} \frac{(-1)^n x^{n+1}}{2n+1}$ **103.** $f(x) = \sum_{n=0}^{\infty} \frac{(x+1)^n}{n!e^2}$

105. $f^{(4)}(0) = -\frac{1}{2}$ **107.** $F(x) = \sum_{n=1}^{\infty} \frac{x^n}{n \cdot n!}$

Chapter 11

Section 11.1 Preliminary

1. A circle of radius 3 centered at the origin.

2. The center is at $(4, 5)$.   **3.** Maximum height: 4

4. Yes; no   **5.** (a) ↔ (iii), (b) ↔ (ii), (c) ↔ (i)

Section 11.1 Exercises

1. $(t = 0)$ $(1, 9)$; $(t = 2)$ $(9, -3)$; $(t = 4)$ $(65, -39)$

5. (a)   (b)

(c)   (d)

    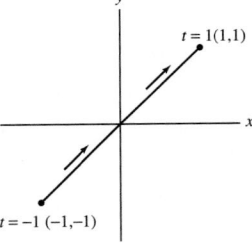

7. $y = 4x - 12$ **9.** $y = \tan^{-1}(x^3 + e^x)$

11. $y = \frac{6}{x^2}$ (where $x > 0$)   **13.** $y = 2 - e^x$

15.   **17.**

    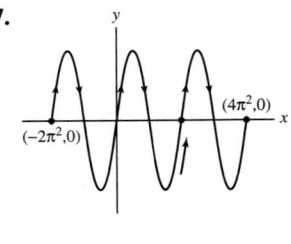

19. (a) ↔ (iv), (b) ↔ (ii), (c) ↔ (iii), (d) ↔ (i)

21. $\pi \le t \le 2\pi$ **23.** $c(t) = (t, 9 - 4t)$ **25.** $c(t) = \left(\frac{5+t^2}{4}, t\right)$

27. $c(t) = (-9 + 7\cos t, 4 + 7\sin t)$ **29.** $c(t) = \left(2 + t, 5 - \frac{1}{3}t\right)$

31. $c(t) = (4\cos t, 9\sin t)$ **33.** $c(t) = (t + 2, t^2 + 3)$

35. $y = 1 - x$ $(0 \le x \le 1)$ **37.** $c(t) = (t + 2, 3t + 2)$

39. $\frac{dy}{dx} = \coth t$; $x = a \cosh t$, $y = b \sinh t$ **41.** (I) $y(t)$; (II) $x(t)$

43.

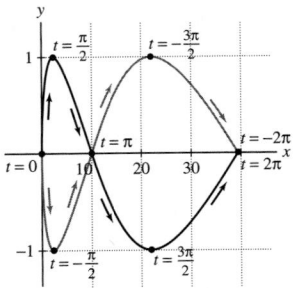

45. $\frac{dy}{dx}\big|_{t=-4} = -\frac{1}{6}$ **47.** $\frac{dy}{dx}\big|_{s=-1} = -\frac{3}{4}$

49. $y = -\frac{9}{2}x + \frac{11}{2}$; $\frac{dy}{dx} = -\frac{9}{2}$

51. $y = x^2 + x^{-1}$; $\frac{dy}{dx} = 2x - \frac{1}{x^2}$

53.

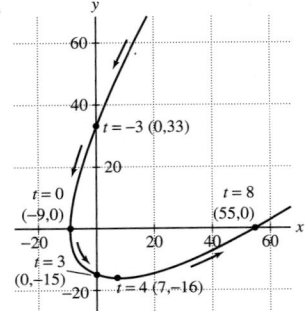

The graph is in: quadrant (i) for $t < -3$ or $t > 8$, quadrant (ii) for $-3 < t < 0$, quadrant (iii) for $0 < t < 3$, quadrant (iv) for $3 < t < 8$.

55. $(55, 0)$ **57.** $\left(\frac{x}{4}\right)^2 + \left(\frac{y}{7}\right)^2 = 1$; $m = -\frac{7}{4}$

59. $c(t) = (3 - 9t + 24t^2 - 16t^3, 2 + 6t^2 - 4t^3)$, $0 \le t \le 1$

63.

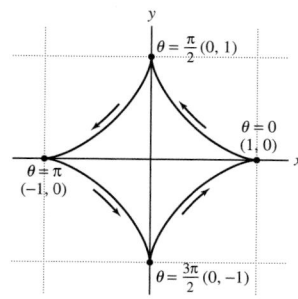

$y = -\sqrt{3}x + \frac{\sqrt{3}}{2}$

65. $((2k-1)\pi, 2)$, $k = 0, \pm 1, \pm 2, \ldots$

75. The coordinates of P, $(R \cos \theta, r \sin \theta)$ describe an ellipse for $0 \le \theta \le 2\pi$.

79. $\left. \frac{d^2 y}{dx^2} \right|_{t=2} = -\frac{21}{512}$ **81.** $\left. \frac{d^2 y}{dx^2} \right|_{t=-3} = 0$ **83.** Concave up: $t > 0$

87. Formula for the area under the graph of a positive function.

Section 11.2 Preliminary

1. $S = \int_a^b \sqrt{x'(t)^2 + y'(t)^2}\, dt$ **2.** The speed at time t

3. Displacement: 5; no   **4.** $L = 180$ cm

Section 11.2 Exercises

1. $S = 3\pi$ **3.** $S = 10$ **5.** $S = 16\sqrt{13}$

7. $S = \frac{1}{2}(65^{3/2} - 5^{3/2}) \approx 256.43$ **9.** $S = 3\pi$ **11.** $S = 2$

15. $S = \pi\sqrt{1 + 4\pi^2} + \frac{1}{2}\ln(2\pi + \sqrt{1 + 4\pi^2}) \approx 21.256$

17. $\left. \frac{ds}{dt} \right|_{t=\frac{\pi}{4}} = 5\sqrt{9 + 55 \sin^2\left(5 \cdot \frac{\pi}{4}\right)} \approx 30.21$ m/s

19. $\left. \frac{ds}{dt} \right|_{t=1} = \sqrt{10} \approx 3.16$ m/s   **21.** $\left(\frac{ds}{dt}\right)_{\min} \approx \sqrt{4.89} \approx 2.21$

23. $\frac{ds}{dt} = 8$

25.

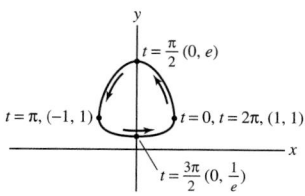

$M_{10} = 6.903734$, $M_{20} = 6.915035$, $M_{30} = 6.914949$, $M_{50} = 6.914951$

27.

$M_{10} = 25.528309$,

$M_{20} = 25.526999$, $M_{30} = 25.526999$, $M_{50} = 25.526999$

31. $S = \frac{6\pi}{5}$ **33.** $S = \frac{64\pi}{3}$

37. $R_m = 6,378 \left(\frac{39,499.2}{84.5}\right)^{2/3} \approx 384{,}154$ km

Section 11.3 Preliminary

1. (b)   **2.** Positive: $(r, \theta) = \left(1, \frac{\pi}{2}\right)$; Negative: $(r, \theta) = \left(-1, \frac{3\pi}{2}\right)$

3. Exactly one

4. (a) Equation of the circle of radius 2 centered at the origin.

(b) Equation of the circle of radius $\sqrt{2}$ centered at the origin.

(c) Equation of the vertical line through the point $(2, 0)$.

5. (a)

Section 11.3 Exercises

1. (a) $\left(3\sqrt{2}, \frac{3\pi}{4}\right)$ (b) $(3, \pi)$ (c) $(\sqrt{5}, \pi + 0.46) \approx (\sqrt{5}, 3.6)$

(d) $\left(\sqrt{2}, \frac{5\pi}{4}\right)$ (e) $\left(\sqrt{2}, \frac{\pi}{4}\right)$ (f) $\left(4, \frac{\pi}{6}\right)$ (g) $\left(4, \frac{11\pi}{6}\right)$

3. (a) $(1, 0)$ (b) $\left(\sqrt{12}, \frac{\pi}{6}\right)$ (c) $\left(\sqrt{8}, \frac{3\pi}{4}\right)$ (d) $\left(2, \frac{2\pi}{3}\right)$

5. (a) $\left(\frac{3\sqrt{3}}{2}, \frac{3}{2}\right)$ (b) $\left(-\frac{6}{\sqrt{2}}, \frac{6}{\sqrt{2}}\right)$ (c) $(0, -5)$

7. (b), (c)   **9.** $\theta \approx 0.46$ **11.** $r = 2 \csc \theta$

13. $x^2 + \left(y - \frac{1}{2}\right)^2 = \left(\frac{1}{2}\right)^2$ **15.** $y = 2$ **17.** $\frac{\left(x - \frac{1}{3}\right)^2}{\frac{4}{9}} + \frac{y^2}{\frac{1}{3}} = 1$

19. $r = 5 \sec \theta$ **21.** $r^2 = 2 \csc 2\theta$

23. A, $\theta = 0$; B, $\theta = \frac{\pi}{4}$; C, $\theta = \frac{\pi}{2}, \frac{3\pi}{2}$; D, $\theta = \frac{7\pi}{4}$

25. (a) ↔ (iv), (b) ↔ (iii), (c) ↔ (i), (d) ↔ (ii)

27. A, $\left(\frac{\pi}{2}, 1\right)$; B, $\left(\frac{3\pi}{4}, 0\right)$; C, $(0, 1)$; D, $\left(\frac{\pi}{4}, \sqrt{2}\right)$

29.

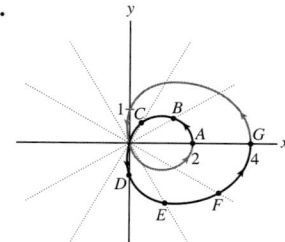

31. (a) A, $\theta = 0$, $r = 0$; B, $\theta = \frac{\pi}{4}$, $r = \sin\frac{2\pi}{4} = 1$; C, $\theta = \frac{\pi}{2}$, $r = 0$; D, $\theta = \frac{3\pi}{4}$, $r = \sin\frac{2 \cdot 3\pi}{4} = -1$; E, $\theta = \pi$, $r = 0$; F, $\theta = \frac{5\pi}{4}$, $r = 1$; G, $\theta = \frac{3\pi}{2}$, $r = 0$; H, $\theta = \frac{7\pi}{4}$, $r = -1$; I, $\theta = 2\pi$, $r = 0$

(b) $0 \le \theta \le \frac{\pi}{2}$ is in the first quadrant. $\frac{\pi}{2} \le \theta \le \pi$ is in the fourth quadrant. $\pi \le \theta \le \frac{3\pi}{2}$ is in the third quadrant. $\frac{3\pi}{2} \le \theta \le 2\pi$ is in the second quadrant.

33.

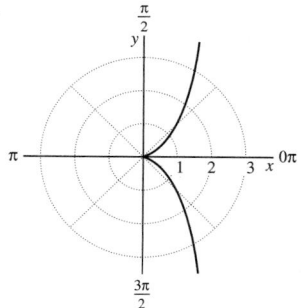

35. $\left(x - \frac{a}{2}\right)^2 + \left(y - \frac{b}{2}\right)^2 = \frac{a^2+b^2}{4}$, $r = \frac{\sqrt{a^2+b^2}}{2}$, centered at the point $\left(\frac{a}{2}, \frac{b}{2}\right)$.

37. $r^2 = \sec 2\theta$ **39.** $\left(x^2 + y^2\right)^2 = x^3 - 3y^2x$

41. $r = 2\sec\left(\theta - \frac{\pi}{9}\right)$ **43.** $r = 2\sqrt{10}\sec(\theta - 4.39)$

45. $r = \frac{9}{4\cos\theta - \sin\theta}$ **53.** $y = \frac{x}{\sqrt{3}}$

Section 11.4 Preliminary

1. False   **2.** (b)   **3.** False   **4.** (c)

Section 11.4 Exercises

1. $A = \frac{1}{2}\int_{\pi/2}^{\pi} r^2\, d\theta = \frac{25\pi}{4}$

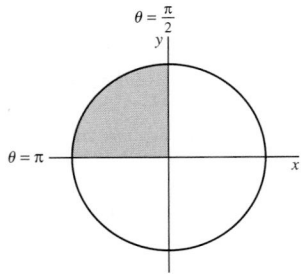

3. $A = \frac{1}{2}\int_0^\pi r^2\, d\theta = 4\pi$ **5.** $A = \frac{3\pi}{2}$

7. $A = \frac{\pi}{8} \approx 0.39$ **9.** $A = 1$ **11.** $A = \frac{3\pi a^2}{2}$

13. $A = \frac{\sqrt{15}}{2} + 7\cos^{-1}\left(\frac{1}{4}\right) \approx 11.163$

15. $A = \pi - \frac{3\sqrt{3}}{2} \approx 0.54$ **17.** $A = \frac{\pi}{8} - \frac{1}{4} \approx 0.14$

19. $A = 4\pi$ **21.** $A = \frac{9\pi}{2} - 4\sqrt{2}$ **23.** $A = \frac{\pi^3}{48} - \frac{\pi}{8} \approx 0.25$

25. $L = \frac{A}{2}\sqrt{A^2 + 1} + \frac{1}{2}\ln\left|A + \sqrt{A^2 + 1}\right|$

27. $L = \tan A$

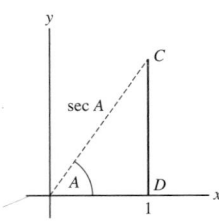

29. $L = 8$ **31.** $L = \sqrt{2}(e^{2\pi} - 1) \approx 755.9$

33. $L = \int_0^\pi \sqrt{1 + a^2} e^{a\theta}\, d\theta$

35. $L = \int_0^{2\pi} \sqrt{5 - 4\cos\theta}(2 - \cos\theta)^{-2}\, d\theta$

Section 11.5 Preliminary

1. (a) Hyperbola   (b) Parabola   (c) Ellipse
(d) Not a conic section
2. Hyperbola   **3.** The points $(0, c)$ and $(0, -c)$
5. $\pm\frac{b}{a}$ are the slopes of the two asymptotes of the hyperbola.

Section 11.5 Exercises

1. $F_1 = (-\sqrt{65}, 0)$, $F_2 = (\sqrt{65}, 0)$; The vertices are $(9, 0)$, $(-9, 0)$, $(0, 4)$, $(0, -4)$.
3. $F_1 = (-\sqrt{5}, 0)$, $F_2 = (\sqrt{5}, 0)$; The vertices are $(3, 0)$, $(-3, 0)$, $(0, 2)$, $(0, -2)$.
5. $F_1 = (\sqrt{97}, 0)$, $F_2 = (-\sqrt{97}, 0)$; The vertices are $(4, 0)$ and $(-4, 0)$.
7. $F_1 = (\sqrt{65} + 3, -1)$, $F_2 = (-\sqrt{65} + 3, -1)$; The vertices are $(10, -1)$ and $(-4, -1)$.
9. $\frac{x^2}{5^2} + \frac{y^2}{7^2} = 1$ **11.** $\frac{(x-12)^2}{5^2} + \frac{(y-6)^2}{7^2} = 1$
13. $\frac{x^2}{9} + \frac{y^2}{16} = 1$ **15.** $\left(\frac{x}{4}\right)^2 + \left(\frac{y}{2\sqrt{13}}\right)^2 = 1$ **17.** $\frac{x^2}{9} - \frac{y^2}{16} = 1$
19. $\left(\frac{x}{4}\right)^2 - \left(\frac{y}{12}\right)^2 = 1$ **21.** $\left(\frac{y+(1/2)}{9/2}\right)^2 - \left(\frac{x}{6}\right)^2 = 1$
23. $x = \frac{y^2}{8}$ **25.** $y = \frac{1}{20}x^2$ **27.** $y = \frac{x^2}{16}$ **29.** $x = \frac{y^2}{8}$
31. Vertices $(\pm 4, 0)$, $(0, \pm 2)$. Foci $\left(-\sqrt{12}, 0\right)$, $\left(\sqrt{12}, 0\right)$. Focal axis is the x-axis. Conjugate axis is the y-axis. Centered at the origin.
33. Vertices $(\pm 2, 0)$, $(0, \pm 4)$. Foci $\left(0, \pm\sqrt{12}\right)$. Focal axis is the y-axis. Conjugate axis is the x-axis. Centered at the origin.
35. Vertices $(5, 5)$, $(-7, 5)$. Foci $\left(\sqrt{84} - 1, 5\right)$, $\left(-\sqrt{84} - 1, 5\right)$. Focal axis: $y = 5$. Conjugate axis: $x = -1$. Asymptotes $y = -1.15x + 3.85$, $y = 1.15x + 6.15$. Centered at $(-1, 5)$.
37. Vertex $(4, 0)$. Focus $\left(4, \frac{1}{16}\right)$. The axis is the vertical line $x = 4$.
39. Foci $\left(-\frac{\sqrt{21}}{2} + 1, \frac{1}{5}\right)$, $\left(\frac{\sqrt{21}}{2} + 1, \frac{1}{5}\right)$. Focal axis: $y = 0.2$. Conjugate axis: $x = 1$. Centered at $(1, 0.2)$.
41. $D = -87$; ellipse   **43.** $D = 88$; hyperbola
47. Focus $(0, c)$. Directrix $y = -c$.
49. $\overline{A'F_0} = a + c = 3.6 + 0.9 = 4.5$ billion mi
51. $r = \frac{3}{2 + \cos\theta}$ **53.** $r = \frac{4}{1 + \cos\theta}$
55. Hyperbola, $e = 4$; directrix $x = 2$
57. Ellipse, $e = \frac{3}{4}$; directrix $x = \frac{8}{3}$
61. $\left(\frac{x + \frac{4}{3}}{\frac{8}{3}}\right)^2 + \left(\frac{y}{\frac{4}{\sqrt{3}}}\right)^2 = 1$

Chapter 11 Review

1. a, c
3. $c(t) = (1 + 2\cos t, 1 + 2\sin t)$. The intersection points with the y-axis are $(0, 1 \pm \sqrt{3})$. The intersection points with the x-axis are $(1 \pm \sqrt{3}, 0)$.
5. $c(\theta) = (\cos(\theta + \pi), \sin(\theta + \pi))$ **7.** $c(t) = (1 + 2t, 3 + 4t)$
9. $y = -\frac{x}{4} + \frac{37}{4}$ **11.** $y = \frac{8}{(3-x)^3} + \frac{3-x}{2}$
13. $(\pi^2 k^2, 0)$, where $k \in \mathbb{Z}$

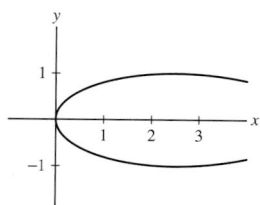

15. $\left.\dfrac{dy}{dx}\right|_{\theta=\frac{\pi}{4}} - \dfrac{1}{4\sqrt{2}}$ **17.** $\left.\dfrac{dy}{dx}\right|_{P} = 5$

19. Vertical: $t = \pi + 2\pi k$ where $k \in \mathbb{Z}$; horizontal: $t = \pm\frac{\pi}{3} + 2\pi k$ where $k \in \mathbb{Z}$

21. $\left.\dfrac{ds}{dt}\right|_{t=\frac{\pi}{4}} = \sqrt{20.5} \approx 4.53$ **23.** $L = 5(e - 1)$

25. $t_0 = \frac{3}{4}\pi$; $\left.\dfrac{ds}{dt}\right|_{t=\frac{3}{4}\pi} = e^{-3\pi/4}\sqrt{2}$

27. $\left(\sqrt{10}, 5.034\right)$; $\left(\sqrt{10}, -0.321\right)$ **29.** $r^2 = \dfrac{12}{2+\sin 2\theta}$

31. $7x - y = 4$ **33.** $A = \frac{9}{4}\left(\frac{\pi}{3} + \frac{\sqrt{3}}{2}\right)$ **35.** $A = \frac{\pi}{2}$

37. $A = \frac{5\pi}{8} - 1$ **39.** $L_{\text{inner}} = 7.5087$; $L_{\text{outer}} = 36.121$

41. Foci $\left(\pm\sqrt{6}, 0\right)$. Vertices $(\pm 2, 0)$. Hyperbola centered at $(0, 0)$.

43. Hyperbola shifted 3 units on the y-axis. Foci $\left(\pm\sqrt{\frac{3}{2}}, 3\right)$. Vertices $\left(\pm\frac{1}{\sqrt{2}}, 3\right)$.

45. $\left(\frac{x}{8}\right)^2 - \left(\frac{y}{6}\right)^2 = 1$ **47.** $\left(\frac{x}{64}\right)^2 + \left(\frac{y}{8\sqrt{63}}\right)^2 = 1$

51. $e = \frac{38}{778} \approx 0.0488$

Chapter 12

Section 12.1 Preliminary

1. (a) True   (b) False   (c) True   (d) True
2. $\|-3\mathbf{a}\| = 15$ **3.** The components are not changed.   **4.** $\langle 0, 0 \rangle$
5. (a) True   (b) False

Section 12.1 Exercises

1. $\mathbf{v}_1 = \langle 2, 0 \rangle$, $\|\mathbf{v}_1\| = 2$ $\mathbf{v}_2 = \langle 2, 0 \rangle$, $\|\mathbf{v}_2\| = 2$

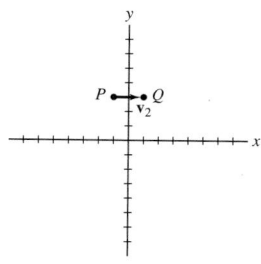

$\mathbf{v}_3 = \langle 3, 1 \rangle$, $\|\mathbf{v}_3\| = \sqrt{10}$ $\mathbf{v}_4 = \langle 2, 2 \rangle$, $\|\mathbf{v}_4\| = 2\sqrt{2}$

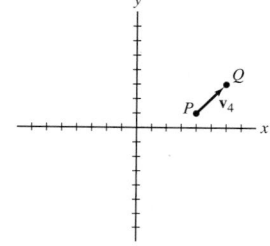

3. $(3, 5)$ **5.** $\overrightarrow{PQ} = \langle -1, 5 \rangle$

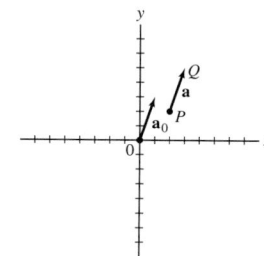

7. $\overrightarrow{PQ} = \langle -2, -9 \rangle$ **9.** $\langle 5, 5 \rangle$ **11.** $\langle 30, 10 \rangle$ **13.** $\left(\frac{5}{2}, 5\right)$

15. (B)

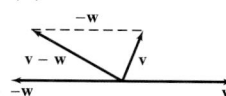

17. $2\mathbf{v} = \langle 4, 6 \rangle$ $-\mathbf{w} = \langle -4, -1 \rangle$

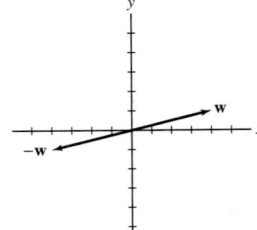

$2\mathbf{v} - \mathbf{w} = \langle 0, 5 \rangle$ $\mathbf{v} + \mathbf{w} = \langle 6, 4 \rangle$

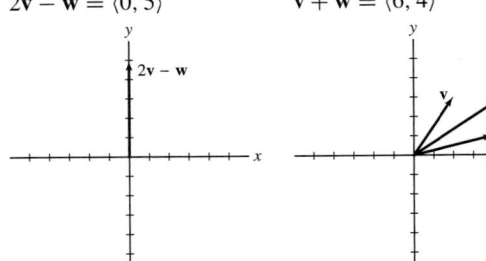

19. $3\mathbf{v} + \mathbf{w} = \langle -2, 10 \rangle$, $2\mathbf{v} - 2\mathbf{w} = \langle 4, -4 \rangle$

21.   **23.** (b), (c)

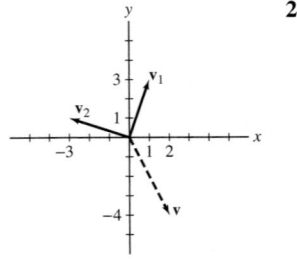

25. $\overrightarrow{AB} = \langle 2, 6 \rangle$ **27.** $\overrightarrow{AB} = \langle 3, -2 \rangle$
$\overrightarrow{PQ} = \langle 2, 6 \rangle$ $\overrightarrow{PQ} = \langle 3, -2 \rangle$
The vectors are equivalent.   The vectors are equivalent.

29. $\overrightarrow{AB} = \langle 2, 3 \rangle$
$\overrightarrow{PQ} = \langle 6, 9 \rangle$
The vectors are parallel.

31. $\overrightarrow{AB} = \langle -8, 1 \rangle$
$\overrightarrow{PQ} = \langle 8, -1 \rangle$
The vectors are parallel and point in opposite directions.

33. $\|\overrightarrow{OR}\| = \sqrt{53}$ **35.** $P = (0, 0)$ **37.** $\mathbf{e_v} = \frac{1}{5}\langle 3, 4 \rangle$

39. $\mathbf{e_u} = \frac{1}{\sqrt{2}}\langle -1, -1 \rangle$ **41.** $-\mathbf{e_v} = \left\langle \frac{1}{\sqrt{5}}, -\frac{2}{\sqrt{5}} \right\rangle$

43. $\left\langle \sqrt{3}, 1 \right\rangle$ **45.** $\lambda = \pm\frac{1}{\sqrt{13}}$ **47.** $a = 0; b = 4$

49. (a) $\langle 4, 3 \rangle$; $\|4\mathbf{i} + 3\mathbf{j}\| = 5$ (b) $\langle 2, -3 \rangle$; $\|2\mathbf{i} - 3\mathbf{j}\| = \sqrt{13}$
(c) $\langle 1, 1 \rangle$; $\|\mathbf{i} + \mathbf{j}\| = \sqrt{2}$ (d) $\langle 1, -3 \rangle$; $\|\mathbf{i} - 3\mathbf{j}\| = \sqrt{10}$

51. $8\mathbf{i} + 46\mathbf{j}$ **53.** $24\mathbf{i} - 4\mathbf{j}$

55. $s = -1, r = 2$ **57.** $\mathbf{u} = \frac{11}{18}\langle 1, 4 \rangle + \frac{5}{18}\langle 5, 2 \rangle$

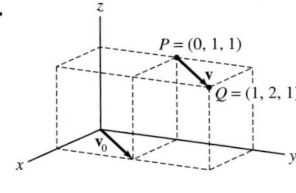

59. $f_1 = 27.3$ lb, $f_2 = 24.5$ lb
61. $\mathbf{r} = \langle L_1 \sin\theta_1 + L_2 \sin\theta_2, L_1 \cos\theta_1 - L_2 \cos\theta_2 \rangle$
63. $\left(\frac{x}{8}\right)^2 + \left(\frac{y}{2}\right)^2 = 1$

Section 12.2 Preliminary

1. $Q = (4, 3, 2)$
2. $\mathbf{v} = \langle 3, 2, 1 \rangle$ regardless of the base point.
3. (a)   **4.** (c)   **5.** Infinitely many direction vectors.   **6.** True

Section 12.2 Exercises

1.   **3.**

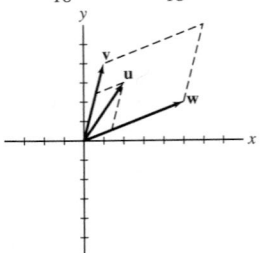

$\|\mathbf{v}\| = \sqrt{14}$
5. $\overrightarrow{PQ} = \langle 1, 1, -1 \rangle$ **7.** $\overrightarrow{PQ} = \left\langle -\frac{9}{2}, -\frac{3}{2}, 1 \right\rangle$
9. $\|\overrightarrow{OR}\| = \sqrt{26} \approx 5.1$ **11.** $P = (-2, 6, 0)$
13. (a) Same direction   (b) Not parallel
(c) Opposite directions   (d) Not parallel
15. Not equivalent   **17.** Not equivalent
19. $\langle -8, -18, -2 \rangle$ **21.** $\langle 4, 0, -5 \rangle$ **23.** $\mathbf{e_w} = \left\langle \frac{4}{\sqrt{21}}, \frac{-2}{\sqrt{21}}, \frac{-1}{\sqrt{21}} \right\rangle$
25. $-\mathbf{e_v} = \left\langle \frac{2}{3}, -\frac{2}{3}, -\frac{1}{3} \right\rangle$ **27.** $\mathbf{r}(t) = \langle 1 + 2t, 2 + t, -8 + 3t \rangle$
29. $\mathbf{r}(t) = \langle 4 + 7t, 0, 8 + 4t \rangle$ **31.** $\mathbf{r}(t) = \langle 1 + 2t, 1 - 6t, 1 + t \rangle$
33. $\mathbf{r}(t) = \langle 4t, t, t \rangle$ **35.** $\mathbf{r}(t) = \langle 0, 0, t \rangle$ **37.** $\mathbf{r}(t) = \langle t, 0, 2 \rangle$
39. $\mathbf{r}(t) = \langle 2 + 2t, 3 + t, 1 - 2t \rangle$ **41.** $\mathbf{r}\left(\frac{1}{2}\right) = (3, 4, 3)$
43. $R = (6, 13, 15)$
47. $\mathbf{r}_1(t) = \langle 5, 5, 2 \rangle + t\langle 0, -2, 1 \rangle$; $\mathbf{r}_2(t) = \langle 5, 5, 2 \rangle + t\langle 0, -20, 10 \rangle$
51. The point of intersection is $(3, 4, 7)$.   **53.** $\mathbf{v} = \left\langle 0, \frac{1}{2}, -\frac{1}{2} \right\rangle$
57. $\mathbf{r}(t) = \langle 5, -3, 10 \rangle + t\langle 9, 7, 1 \rangle$

Section 12.3 Preliminary

1. Scalar   **2.** The angle between **a** and **b** is obtuse.
3. Distributive Law   **4.** $\text{proj}_v(\mathbf{v}) = \mathbf{v}$ **5.** No difference
6. (a) $\text{proj}_{2\mathbf{v}}(\mathbf{u}) = \mathbf{v}$ **(b)** $\text{proj}_{\mathbf{v}}(2\mathbf{u}) = 2\mathbf{v}$
7. (b) $\mathbf{u}_\parallel$
8. (c) $\mathbf{e_u} \cdot \mathbf{e_v}$

Section 12.3 Exercises

1. 15   **3.** 41   **5.** 5   **7.** 0   **9.** 1   **11.** 0   **13.** Obtuse
15. Acute   **17.** Obtuse   **19.** $\theta = 90°$ **21.** $\cos\theta = \frac{1}{\sqrt{10}}$
23. $\theta = \cos^{-1} 0.997 \approx 4.4°$ **25.** $\theta \approx 70.53°$
27. $\theta = \cos^{-1} 0.724 \approx 43.57°$
29. (a) $b = -\frac{1}{2}$ **(b)** $b = 0$ or $b = \frac{1}{2}$
31. $\mathbf{v}_1 = \langle 0, 1, 0\rangle$, $\mathbf{v}_2 = \langle 3, 2, 2\rangle$ **33.** 8   **35.** 2   **37.** $\|\mathbf{v}\|^2$
39. $\|\mathbf{v}\|^2 - \|\mathbf{w}\|^2$ **41.** $\text{proj}_{\mathbf{v}}(\mathbf{u}) = \left\langle -\frac{4}{5}, 0, -\frac{2}{5}\right\rangle$
43. $\text{proj}_{\mathbf{v}}(\mathbf{u}) = \langle 1, 1, 0\rangle$ **45.** $\text{proj}_{\mathbf{v}}(\mathbf{u}) = -4\mathbf{k}$
47. $\text{proj}_{\mathbf{v}}(\mathbf{u}) = a\mathbf{i}$ **49.** $\mathbf{a} = \langle -2, 0, 0\rangle + \langle 0, 1, 1\rangle$
51. $\mathbf{a} = \left\langle \frac{1}{2}, \frac{1}{2}\right\rangle + \left\langle \frac{1}{2}, -\frac{1}{2}\right\rangle$ **53.** $\mathbf{a}_\parallel = \mathbf{0}$, $\mathbf{a}_\perp = \mathbf{a} = \langle 4, 2\rangle$
57. $\|\mathbf{v} + \mathbf{w}\| = \sqrt{2 + \sqrt{3}} \approx 1.93$
59. (a) $\mathbf{v} \cdot \mathbf{w} = -3$ **(b)** $\|2\mathbf{v} + \mathbf{w}\| = \sqrt{13} \approx 3.61$
(c) $\|2\mathbf{v} - 3\mathbf{w}\| = \sqrt{133} \approx 11.53$
61. $\theta_1 \approx 47.47°$, $\theta_2 \approx 71.58°$, $\theta_3 \approx 60.95°$ **63.** $\beta = 60°$
65. $\text{proj}_{\overrightarrow{AB}}(\overrightarrow{AD}) = \left\langle \frac{1}{2}, 0, -\frac{1}{2}\right\rangle$
67. Counterexample: $\mathbf{a} = \langle 1, 0, 1\rangle$, $\mathbf{v} = \langle 3, 1, 1\rangle$, $\mathbf{w} = \langle 4, 1, 0\rangle$
69. The force required is 68.07 N.
71. $\mathbf{F}_1 = -62.1\mathbf{i} + 35.9\mathbf{j}$, $\mathbf{F}_2 = 62.1\mathbf{i} + 62.1\mathbf{j}$, in Newtons
85. $\cos\alpha = \frac{3}{7}$, $\cos\beta = \frac{6}{7}$, $\cos\gamma = -\frac{2}{7}$

Section 12.4 Preliminary

1. $\begin{vmatrix} -5 & -1 \\ 4 & 0 \end{vmatrix}$ **2.** $\|\mathbf{e} \times \mathbf{f}\| = \frac{1}{2}$ **3.** $\mathbf{u} \times \mathbf{w} = \langle -2, -2, -1\rangle$
4. (a) 0   **(b)** 0
5. $\mathbf{i} \times \mathbf{j} = \mathbf{k}$ and $\mathbf{i} \times \mathbf{k} = -\mathbf{j}$
6. **v** or **w** (or both) is the zero vector, or if **v** and **w** are parallel vectors.

Section 12.4 Exercises

1. -5 **3.** -15 **5.** -8 **7.** 15   **9.** 1   **11.** $\mathbf{v} \times \mathbf{w} = \mathbf{i} + 2\mathbf{j} - 5\mathbf{k}$
13. $\mathbf{v} \times \mathbf{w} = 6\mathbf{i} - 8\mathbf{k}$ **15.** $\mathbf{v} \times \mathbf{w} = \frac{7}{3}\mathbf{i} - \mathbf{j} + \frac{2}{3}\mathbf{k}$ **17.** $-\mathbf{j} + \mathbf{i}$
19. $-2\mathbf{i} + \mathbf{j} + \mathbf{k}$ **21.** $\langle -1, -1, 0\rangle$ **23.** $\langle 0, 9, 3\rangle$
25. $\langle -2, -2, -2\rangle$
27. $\mathbf{v} \times \mathbf{i} = \langle 0, c, -b\rangle$; $\mathbf{v} \times \mathbf{j} = \langle -c, 0, a\rangle$; $\mathbf{v} \times \mathbf{k} = \langle b, -a, 0\rangle$
29. $\mathbf{v} \times \mathbf{w}$ equals $-\mathbf{u}$ **31.** $\mathbf{u} = \mathbf{v} \times \mathbf{w} = \langle 0, 3, 3\rangle$ **35.** $\mathbf{e}'$
37. $\mathbf{F}_1$ **41.** The area is $2\sqrt{138}$.

43.

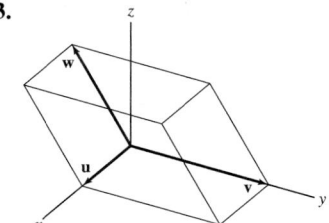

The volume is 4.
45. $A = \sqrt{35} \approx 5.92$.   **47.** $S = \frac{9\sqrt{3}}{2} \approx 7.8$.
61. $\mathbf{X} = \langle 1, 1, 2\rangle$ **65.** $\tau = (204.79 - 2{,}940\cos\theta)\mathbf{k}$
67. $\theta = 41.39°$ **69.** 107

Section 12.5 Preliminary

1. $3x + 4y - z = 0$ **2. (c)** $z = 1$ **3.** Plane (c)
4. xz-plane   **5. (c)** $x + y = 0$ **6. (a)** is correct

Section 12.5 Exercises

1. $\langle 1, 3, 2\rangle \cdot \langle x, y, z\rangle = 3$
$x + 3y + 2z = 3$
$(x - 4) + 3(y + 1) + 2(z - 1) = 0$
3. $\langle -1, 2, 1\rangle \cdot \langle x, y, z\rangle = 3$
$-x + 2y + z = 3$
$-(x - 4) + 2(y - 1) + (z - 5) = 0$
5. $\langle 1, 0, 0\rangle \cdot \langle x, y, z\rangle = 3$
$x = 3$
$(x - 3) + 0 \cdot (y - 1) + 0 \cdot (z + 9) = 0$
7. $\langle 0, 0, 1\rangle \cdot \langle x, y, z\rangle = 2$
$z = 2$
$0(x - 6) + 0(y - 7) + (z - 2) = 0$
9. $2x + y + z = 1$ **11.** $6x + 9y + 4z = 19$ **13.** $x + 2y - z = 1$
15. $\mathbf{n} = \langle 9, -4, -11\rangle$ **17.** $\mathbf{n} = \langle 3, 25, 9\rangle$ **19.** $\mathbf{n} = \langle 1, 0, -1\rangle$
21. $4x - 9y + z = 0$ **23.** $x = 4$ **25.** $13x + y - 5z = 0$
27. The planes are parallel.   **29.** $10x + 15y + 6z = 30$
31. $P = (2, -1, -1)$ **33.** $P = (9, 18, 3)$ **35.** $3x - 4y = 8$
37. $3x + 4z = 5$ **39.** $3x + 4z = -2$ **41.** $y = 4$
43. $4x + 3y + z = 8$, $4x + 3y - 5z = 8$
45. $ax - 2ay + cz = 0$, $a \neq 0$ **49.** $\theta = \frac{\pi}{2}$ **51.** $\theta = \frac{\pi}{2}$
53. $\theta = 94.80°$ **55.** $x + y + z = 1$
57. $x - y - z = f$ **59.** $x = \frac{9}{5} + 2t$, $y = -\frac{6}{5} - 3t$, $z = 2 + 5t$
61. $x = 3 + 3t$, $y = -1 + 5t$, $z = 1 - 7t$
63. $P = \left(\frac{2}{3}, -\frac{1}{3}, \frac{2}{3}\right)$ **67.** $\ell = \frac{4}{5}$

Section 12.6 Preliminary

1. True, mostly, except at $x = \pm a$, $y = \pm b$, or $z = \pm c$.
2. False   **3.** Hyperbolic paraboloid   **4.** No   **5.** Ellipsoid
6. $-\left(\frac{x}{2}\right)^2 - \left(\frac{y}{3}\right)^2 + \left(\frac{z}{4}\right)^2 = 1$
7. All vertical lines passing through a parabola $\mathcal{C}$ in the xy-plane.

Section 12.6 Exercises

1. Ellipsoid **3.** Ellipsoid **5.** Hyperboloid of one sheet

7. Elliptic paraboloid **9.** Hyperbolic paraboloid

11. Hyperbolic paraboloid **13.** Ellipsoid, a circle on the xz-plane.

15. Ellipsoid, the trace is an ellipse.

17. Hyperboloid of one sheet, the trace is a hyperbola.

19. Parabolic cylinder, the trace is the parabola $y = 3x^2$.

21. a ↔ Figure b; b ↔ Figure c; c ↔ Figure a

23. $y = \left(\frac{x}{2}\right)^2 + \left(\frac{z}{4}\right)^2$ **25.** $|h| < 2$

27. **31.**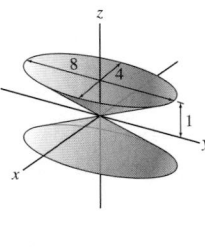

33. $\left(\frac{x}{2}\right)^2 + \left(\frac{y}{4}\right)^2 + \left(\frac{z}{6}\right)^2 = 1$ **35.** $\left(\frac{x}{4}\right)^2 + \left(\frac{y}{6}\right)^2 - \left(\frac{z}{3\sqrt{3}}\right)^2 = 1$

39. The upper part of an elliptic cone

Section 12.7 Preliminary

1. Cylinder of radius R whose axis is the z-axis.
Sphere of radius R centered at the origin.

2. (b) **3.** (a) **4.** $\phi = 0, \pi$ **5.** $\phi = \frac{\pi}{2}$, the xy-plane

Section 12.7 Exercises

1. $(-4, 0, 4)$ **3.** $(x, y, z) = \left(0, 0, \frac{1}{2}\right)$

5. $(r, \theta, z) = \left(\sqrt{2}, \frac{7\pi}{4}, 1\right)$ **7.** $(r, \theta, z) = \left(2, \frac{\pi}{3}, 7\right)$

9. $(r, \theta, z) = \left(5, \frac{\pi}{4}, 2\right)$ **11.** $r^2 \le 1$

13. $z = \sqrt{4 - r^2}, 0 \le \theta \le \frac{\pi}{2}$

15. $r^2 \sin^2 \theta + z^2 \le 9, 0 \le \theta \le \frac{\pi}{4}$ and $\frac{5\pi}{4} \le \theta \le 2\pi$

17. **19.**

21. **23.**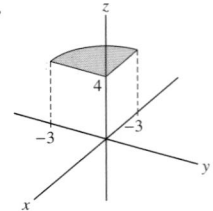

25. $r = \sqrt{4 - z^2}$ **27.** $r = \frac{2}{\sqrt{\cos 2\theta}}$ **29.** $r = \sqrt{\frac{2z}{3 \sin 2\theta}}$

31. $(x, y, z) = \left(\frac{\sqrt{6}}{2}, \frac{\sqrt{6}}{2}, 1\right)$ **33.** $(x, y, z) = (-2.5, 2.5, 2.5\sqrt{2})$

35. $(x, y, z) = (-0.386, -0.241, -0.208)$

37. $(\rho, \theta, \phi) = \left(2, \frac{\pi}{3}, \frac{\pi}{3}\right)$ **39.** $(\rho, \theta, \phi) = \left(\sqrt{3}, \frac{7\pi}{4}, 0.955\right)$

41. $(\rho, \theta, \phi) = \left(2, \frac{\pi}{4}, \frac{\pi}{6}\right)$ **43.** $\rho = 1, 0 \le \phi \le \frac{\pi}{2}$

45. $\left\{(\rho, \theta, \phi) : 0 \le \rho \le 1, \ \theta = \frac{\pi}{4}\right\}$ **47.** $\phi = \frac{\pi}{3}$ or $\phi = \frac{2\pi}{3}$

49. **51.**

53. 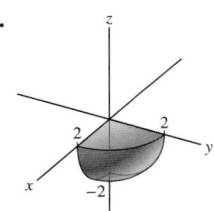 **55.** $\rho = \frac{2}{\cos \phi}$

57. $\rho = \frac{\cos \theta \tan \phi}{\cos \phi}$ **59.** $\rho = \frac{2}{\sin \phi \sqrt{\cos 2\theta}}$ **61.** (b)

63. Sydney: $(-4,616.7, 2,559, -3,560.4)$
Bogota: $(1,722.9, -6,108.8, 502.9)$

65. $z = \pm r \sqrt{\cos 2\theta}$

Chapter 12 Review

1. $\langle 21, -25 \rangle$ and $\langle -19, 31 \rangle$ **3.** $\left\langle \frac{-2}{\sqrt{29}}, \frac{5}{\sqrt{29}} \right\rangle$ **5.** $\mathbf{i} = \frac{2}{11}\mathbf{v} + \frac{5}{11}\mathbf{w}$

7. $\overrightarrow{PQ} = \langle -4, 1 \rangle; \|\overrightarrow{PQ}\| = \sqrt{17}$ **9.** $\left\langle \frac{3}{\sqrt{2}}, -\frac{3}{\sqrt{2}} \right\rangle$

11. $\beta = \frac{3}{2}$ **13.** $\mathbf{u} = \left\langle \frac{1}{3}, \frac{-11}{6}, \frac{7}{6} \right\rangle$

15. $\mathbf{r}_1(t) = \langle 1 + 3t, 4 + t, 5 + 6t \rangle; \mathbf{r}_2(t) = \langle 1 + 3t, t, 6t \rangle$

17. $a = -2, b = 2$ **19.**

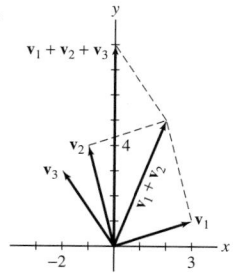

23. $\theta = 2.123$ rad **25.** $A = \sqrt{213} \approx 14.59$

27. $\mathbf{u} = -\frac{c}{7}\langle 10, -8, -7 \rangle$ **29.** $\|\mathbf{F}_1\| = \frac{2\|\mathbf{F}_2\|}{\sqrt{3}}$; $\|\mathbf{F}_1\| = 980$ N

31. $\mathbf{v} \times \mathbf{w} = \langle -6, 7, -2 \rangle$ **33.** $\det \begin{pmatrix} \mathbf{u} \\ \mathbf{v} \\ \mathbf{w} \end{pmatrix} = -47$

35. $S = 5\sqrt{2}$ **43.** $(x - 0) + 4(y - 1) - 3(z + 1) = 0$

45. $17x - 21y - 13z = -28$ **47.** $3x - 2y = 4$

49. Ellipsoid **51.** Elliptic paraboloid **53.** Elliptic cone

55. (a) Empty set (b) Hyperboloid of one sheet

(c) Hyperboloid of two sheets

57. $(\rho, \theta, \phi) = \left(\sqrt{26}, \tan^{-1} \frac{4}{3}, \cos^{-1}\left(\frac{-1}{\sqrt{26}}\right)\right)$,

$(r, \theta, z) = \left(5, \tan^{-1}\left(\frac{4}{3}\right), -1\right)$

59. $(r, \theta, z) = \left(\frac{3\sqrt{3}}{2}, \frac{\pi}{6}, \frac{3}{2}\right)$ **61.** $z = 2x$

65. $A < -1$: Hyperboloid of one sheet.

$A = -1$: Cylinder with the z-axis as its central axis.

$A > 1$: Ellipsoid.

Chapter 13

Section 13.1 Preliminary

1. (c) **2.** The curve $z = e^x$

3. The projection onto the xz-plane **4.** The point $(-2, 2, 3)$

5. As t increases from 0 to 2π, a point on $\sin t \mathbf{i} + \cos t \mathbf{j}$ moves clockwise and a point on $\cos t \mathbf{i} + \sin t \mathbf{j}$ moves counterclockwise.

6. (a), (c), and (d)

Section 13.1 Exercises

1. $D = \{t \in \mathbf{R}, t \neq 0, t \neq -1\}$

3. $\mathbf{r}(t) = (3 + 3t)\mathbf{i} - 5\mathbf{j} + (7 + t)\mathbf{k}$

5. A ↔ ii, B ↔ i, C ↔ iii

7. $(a) = (v), (b) = (i), (c) = (ii), (d) = (vi), (e) = (iv), (f) = (iii)$

9. C ↔ i, A ↔ ii, B ↔ iii

11. Plane $x = 7$, circle centered at $(7, 0, 0)$ with radius $\sqrt{144} = 12$.

13. Plane $y = 9$, circle centered at $(6, 9, 4)$ with $r = 3$.

15. Q lies on the curve.

17. $(0, 1, 0), (0, -1, 0), \left(\frac{1}{\sqrt{2}}, \frac{1}{\sqrt{2}}, 0\right), \left(\frac{1}{\sqrt{2}}, -\frac{1}{\sqrt{2}}, 0\right),$

$\left(-\frac{1}{\sqrt{2}}, -\frac{1}{\sqrt{2}}, 0\right), \left(-\frac{1}{\sqrt{2}}, \frac{1}{\sqrt{2}}, 0\right)$

19. $\mathbf{r}(t) = \langle 2 + 9\cos 2t, 3\cos t, 3\sin t \rangle$

23. $\mathbf{r}(t) = \left\langle \cos t, \sin t, 4\cos^2 t \right\rangle$

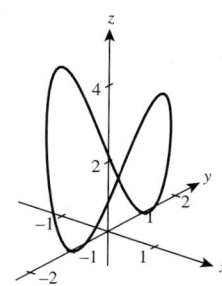

25. $\mathbf{r}(t) = \langle 3, 2, t \rangle, -\infty < t < \infty$ **27.** $\mathbf{r}(t) = \langle t, 3t, 15t \rangle$

29. $\mathbf{r}(t) = \langle 1, 2 + 2\cos t, 5 + 2\sin t \rangle$

31. $\mathbf{r}(t) = \left\langle \frac{\sqrt{3}}{2}\cos t, \frac{1}{2}, \frac{\sqrt{3}}{2}\sin t \right\rangle$

33. $\mathbf{r}(t) = \langle 3 + 2\cos t, 1, 5 + 3\sin t \rangle$

35. (a) False (b) True (c) True

37. The paths intersect. **39.** $z_{\max} = 4$

Section 13.2 Preliminary

1. $\dfrac{d}{dt} f(t)\mathbf{r}(t) = f(t)\mathbf{r}'(t) + f'(t)\mathbf{r}(t)$

$\dfrac{d}{dt}\mathbf{r}_1(t) \cdot \mathbf{r}_2(t) = \mathbf{r}_1(t) \cdot \mathbf{r}_2'(t) + \mathbf{r}_1'(t) \cdot \mathbf{r}_2(t)$

$\dfrac{d}{dt}\mathbf{r}_1(t) \times \mathbf{r}_2(t) = \mathbf{r}_1(t) \times \mathbf{r}_2'(t) + \mathbf{r}_1'(t) \times \mathbf{r}_2(t)$

2. True **3.** False **4.** True **5.** False **6.** False

7. (a) Vector (b) Scalar (c) Vector

Section 13.2 Exercises

1. $\displaystyle\lim_{t \to 3}\left\langle t^2, 4t, \frac{1}{t} \right\rangle = \left\langle 9, 12, \frac{1}{3} \right\rangle$

3. $\displaystyle\lim_{t \to 0}\left(e^{2t}\mathbf{i} + \ln(t + 1)\mathbf{j} + 4\mathbf{k}\right) = \mathbf{i} + 4\mathbf{k}$

5. $\displaystyle\lim_{h \to 0}\frac{\mathbf{r}(t+h) - \mathbf{r}(t)}{h} = \left\langle -\frac{1}{t^2}, \cos t, 0 \right\rangle$ **7.** $\dfrac{d\mathbf{r}}{dt} = \left\langle 1, 2t, 3t^2 \right\rangle$

9. $\mathbf{w}'(s) = \left\langle e^s, -2e^{-2s} \right\rangle$ **11.** $\mathbf{r}'(t) = \left\langle 1 + t^{-2}, 8t, 0 \right\rangle$

13. $\mathbf{a}'(\theta) = (-2\sin 2\theta)\mathbf{i} + (2\cos 2\theta)\mathbf{j} + (4\cos 4\theta)\mathbf{k}$

15. $\mathbf{r}'(t) = \left\langle 1, 2t, 3t^2 \right\rangle$

$\mathbf{r}''(t) = \langle 0, 2, 6t \rangle$

17.

19. $\dfrac{d}{dt}\left(\mathbf{r}_1(t) \cdot \mathbf{r}_2(t)\right) = 4e^t + 18t^2$

21. $\dfrac{d}{dt}\left(\mathbf{r}_1(t) \cdot \mathbf{r}_2(t)\right) = \left\langle t^2 e^t (t + 3), 48, 8e^t (t + 1) \right\rangle$

23. $F'(t) = 4t^3 + 3t^2 + 9$ **27.** $\frac{d}{dt}\mathbf{r}(g(t)) = \langle 4e^{4t+9}, 8e^{8t+18}, 0 \rangle$

29. $\frac{d}{dt}\mathbf{r}(g(t)) = \langle 3^{\sin t}\cos t \ln 3, \frac{\cos t}{1+\sin^2 t} \rangle$

31. $\frac{d}{dt}\mathbf{r}(t) \cdot \mathbf{a}(t)\big|_{t=2} = 13$

33. $\ell(t) = \langle -3 - 4t, 10 + 5t, 16 + 24t \rangle$

35. $\ell(t) = (2-s)\mathbf{i} + \left(\frac{3}{2}s - 1\right)\mathbf{k}$

37. $\ell(s) = s\mathbf{i} + (1-s)\mathbf{j} + (9+9s)\mathbf{k}$

41. $(\ln 4)\mathbf{i} + \frac{56}{3}\mathbf{j} - \frac{496}{5}\mathbf{k}$ **43.** $0\mathbf{i} + 0\mathbf{j} + 0\mathbf{k}$

45. $\left(\frac{3}{2}t^2\right)\mathbf{i} + (2t^3)\mathbf{j} + (9t)\mathbf{k}$

47. $\mathbf{r}(t) = t\mathbf{i} - t\mathbf{j} + \mathbf{c}$, with initial conditions
$\mathbf{r}(t) = (t+2)\mathbf{i} - t\mathbf{j} + 3\mathbf{k}$.

49. $\mathbf{r}(t) = \left\langle -\frac{1}{3}\cos 3t, -\frac{1}{3}\cos 3t, \frac{1}{2}t^2 \right\rangle + \mathbf{c}$, with initial conditions
$\mathbf{r}(t) = \left\langle \frac{1}{3}(1 - \cos 3t), \frac{1}{3}(4 - \cos 3t), 8 + \frac{1}{2}t^2 \right\rangle$.

51. $\mathbf{r}(t) = (8t^2)\mathbf{k} + \mathbf{c}_1 t + \mathbf{c}_2$, with initial conditions
$\mathbf{r}(t) = \mathbf{i} + t\mathbf{j} + (8t^2)\mathbf{k}$.

53. $\mathbf{r}(t) = \langle e^t, -\sin t, -\cos t \rangle + \mathbf{c}_1 t + \mathbf{c}_2$, with initial conditions
$\mathbf{r}(t) = \langle e^t - t, -\sin t + 3t, -\cos t + 2t + 2 \rangle$.

55. $\mathbf{r}(4) = \left\langle 33, 2, \frac{256}{3} \right\rangle$ **57.** $\mathbf{r}(t) = \mathbf{c} + t\mathbf{v}$ **59.** $\mathbf{r}(t) = e^{2t}\mathbf{c}$

Section 13.3 Preliminary

1. $2\mathbf{r}' = \langle 50, -70, 20 \rangle$, $-\mathbf{r}' = \langle -25, 35, -10 \rangle$

2. (b) True

3. (a) $L'(2) = 4$

(b) $L(t)$ is the distance along the path traveled, which is usually different from the distance from the origin.

4. 6

Section 13.3 Exercises

1. $L = 3\sqrt{61}$ **3.** $L = 15 + \ln 4$ **5.** $L \approx 6.26$

7. $s(t) = \frac{1}{27}\left((20 + 9t^2)^{3/2} - 20^{3/2}\right)$ **9.** $v(3) \approx 1.05$

11. $v(0) = \sqrt{2}$

13. At the origin. The integral $\int_0^T \|\mathbf{r}'(u)\| du$ is the length of the path traveled in the time interval $0 \le t \le T$.

15. (a) $s(t) = \sqrt{29}\,t$ (b) $t = \varphi(s) = \frac{s}{\sqrt{29}}$

17. $\mathbf{r}_1(s) = \langle 2 + 4\cos(2s), 10, -3 + 4\sin(2s) \rangle$

19. $\mathbf{r}(\varphi(s)) =$
$\left(1 + \frac{s}{\sqrt{3}}\right)\left\langle \sin\left(\ln\left(1 + \frac{s}{\sqrt{3}}\right)\right), \cos\left(\ln\left(1 + \frac{s}{\sqrt{3}}\right)\right), 1 \right\rangle$

21. $L = \int_0^8 \sqrt{1 + 9t^4}\,dt$, $L = \int_0^2 \sqrt{1 + 9t^{12}}\,3t^2\,dt$

23. (b) $L_1 \approx 94.3$, $L_2 \approx 125.7$

25. (a) $\mathbf{r}_1(s) = \frac{s}{\sqrt{17}}\left\langle \cos\left(4\ln \frac{s}{\sqrt{17}}\right), \sin\left(4\ln \frac{s}{\sqrt{17}}\right) \right\rangle$

27. $L = 2\pi$

29.

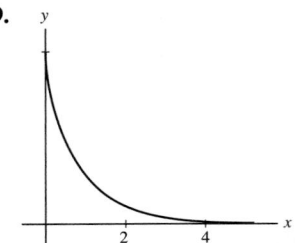

Section 13.4 Preliminary

1. $\mathbf{T}(t) = \frac{\mathbf{v}}{\|\mathbf{v}\|} = \left\langle \frac{2}{3}, \frac{1}{3}, -\frac{2}{3} \right\rangle$ **2.** $\frac{1}{4}$ **3.** A circle of radius 2

4. Zero curvature **5.** $\kappa = \sqrt{14}$ **6.** 4 **7.** $R = \frac{1}{9}$

Section 13.4 Exercises

1. $\mathbf{r}'(t) = \langle 36t^2, 36t, 36t^3 \rangle$, $\mathbf{T}(t) = \frac{t}{|t|\sqrt{1+t^2+t^4}}\langle t, 1, t^2 \rangle$,
$\mathbf{T}(1) = \left\langle \frac{1}{\sqrt{3}}, \frac{1}{\sqrt{3}}, \frac{1}{\sqrt{3}} \right\rangle$

3. $\mathbf{r}'(t) = \langle 4, -4, 9 \rangle$, $\mathbf{T}(t) = \left\langle \frac{4}{\sqrt{122}}, -\frac{5}{\sqrt{122}}, \frac{9}{\sqrt{122}} \right\rangle$, $\mathbf{T}(1) = \mathbf{T}(t)$

5. $\mathbf{r}'(t) = \langle 8t, 9 \rangle$, $\mathbf{T}(t) = \frac{1}{\sqrt{64t^2+81}}\langle 8t, 9 \rangle$, $\mathbf{T}(1) = \left\langle \frac{8}{\sqrt{145}}, \frac{9}{\sqrt{145}} \right\rangle$

7. $\kappa = \frac{e^t}{(1+e^{2t})^{3/2}}$ **9.** $\kappa = \frac{4}{17}$ **11.** $\kappa(t) = \frac{2|t|^3}{(1+|t|^4)^{3/2}}$

13. $\kappa(3) \approx 0.0025$ **15.** $\kappa(2) \approx 0.0015$

17. $\kappa\left(\frac{\pi}{3}\right) \approx 4.54$, $\kappa\left(\frac{\pi}{2}\right) = 0.2$ **21.** $\alpha = \pm\sqrt{2}$

27. $\kappa(2) \approx 0.012$ **29.** $\kappa(\pi) \approx 1.11$

33. $\kappa(t) = t^2$

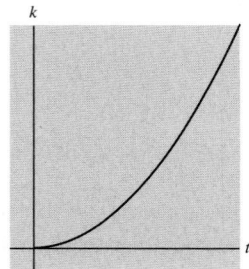

35. $\mathbf{N}(t) = \langle 0, -\sin 2t, -\cos 2t \rangle$

37. $\mathbf{T}'\left(\frac{\pi}{4}\right) = \left\langle -\frac{\sqrt{2}}{3\sqrt{3}}, -\frac{2}{3\sqrt{3}} \right\rangle$, $\mathbf{T}'\left(\frac{3\pi}{4}\right) = \left\langle \frac{\sqrt{2}}{3\sqrt{3}}, \frac{2}{3\sqrt{3}} \right\rangle$

39. $\mathbf{N}\left(\pi^{1/3}\right) = \left\langle \frac{1}{2}, -\frac{\sqrt{3}}{2} \right\rangle$ **41.** $\mathbf{N}(1) = \frac{1}{7\sqrt{17}}\langle -5, -22, 18 \rangle$

43. $\mathbf{N}(\pi) = \frac{\langle 0, -1 \rangle}{1} = \langle 0, -1 \rangle$ **45.** $\mathbf{N}(-1) = \frac{1}{\sqrt{66}}\langle -7, 1, 4 \rangle$

49. $\mathbf{c}(t) = \left\langle \frac{\pi}{2}, 0 \right\rangle + \langle \cos t, \sin t \rangle$ **51.** $\langle \cos t, \sin t \rangle$

53. $\mathbf{c}(t) = \langle \pi, -2 \rangle + 4\langle \cos t, \sin t \rangle$

55. $\mathbf{c}(t) = \langle 3, 0, 0 \rangle + 2\langle 1, 0, 0 \rangle \cos t + \frac{2}{\sqrt{2}}\langle 0, 1, 1 \rangle \sin t$

65. $\kappa(\theta) = \frac{\theta^2+2}{(\theta^2+1)^{3/2}}$ **67.** $\kappa(\theta) = \frac{1}{a\sqrt{1+b^2}}e^{-b\theta}$

81. (a) $\mathbf{N}(0) = \frac{1}{\sqrt{5}}\langle -1, 0, 2 \rangle$ (b) $\mathbf{N}(1) = \frac{1}{\sqrt{66}}\langle 4, 7, -1 \rangle$

Section 13.5 Preliminary

1. No, since the particle may change its direction.   2. $\mathbf{a}(t)$

3. (a) Their velocity vectors point in the same direction.

5. (b) Parallel

6. $\|\mathbf{a}(t)\| = 8$ cm/s^2

7. (b) $a_{\mathbf{N}}$

Section 13.5 Exercises

1. $h = -0.2$: $\dfrac{\mathbf{r}(1 - 0.2) - \mathbf{r}(1)}{-0.2} = \langle -0.085, 1.91, 2.635 \rangle$

$h = -0.1$: $\dfrac{\mathbf{r}(1 - 0.1) - \mathbf{r}(1)}{-0.1} = \langle -0.19, 2.07, 2.97 \rangle$

$h = 0.1$: $\dfrac{\mathbf{r}(1 + 0.1) - \mathbf{r}(1)}{0.1} = \langle -0.41, 2.37, 4.08 \rangle$

$h = 0.2$: $\dfrac{\mathbf{r}(1 + 0.2) - \mathbf{r}(1)}{0.2} = \langle -0.525, 2.505, 5.075 \rangle$

$\mathbf{v}(1) \approx \langle -0.3, 2.2, 3.5 \rangle$, $v(1) \cong 4.1$

3. $\mathbf{v}(1) = \langle 3, -1, 8 \rangle$, $\mathbf{a}(1) = \langle 6, 0, 8 \rangle$, $v(1) = \sqrt{74}$

5. $\mathbf{v}\left(\frac{\pi}{3}\right) = \left\langle \frac{1}{2}, -\frac{\sqrt{3}}{2}, 0 \right\rangle$, $\mathbf{a}\left(\frac{\pi}{3}\right) = \left\langle -\frac{\sqrt{3}}{2}, -\frac{1}{2}, 9 \right\rangle$, $v\left(\frac{\pi}{3}\right) = 1$

7. $\mathbf{a}(t) = -2\left\langle \cos \frac{t}{2}, \sin \frac{t}{2} \right\rangle$

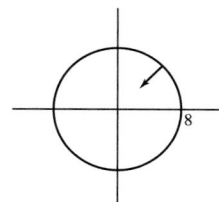

$\mathbf{r}(t) = 8\left\langle \cos \frac{t}{2}, \sin \frac{t}{2} \right\rangle$

9.

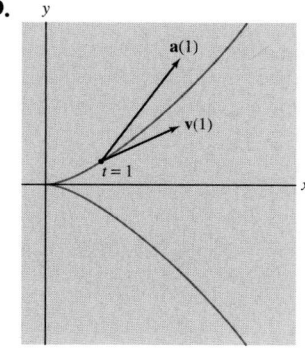

$r(t) = (t^2, t^3)$

11. $\mathbf{v}(t) = \left\langle \frac{3t^2+2}{6}, 4t - 2 \right\rangle$ 13. $\mathbf{v}(t) = \mathbf{i} + t\mathbf{k}$

15. $\mathbf{v}(t) = \left\langle \frac{t^2}{2} + 3, 4t - 2 \right\rangle$, $\mathbf{r}(t) = \left\langle \frac{t^3}{6} + 3t, 2t^2 - 2t \right\rangle$

17. $\mathbf{v}(t) = \mathbf{i} + \frac{t^2}{2}\mathbf{k}$, $\mathbf{r}(t) = t\mathbf{i} + \mathbf{j} + \frac{t^3}{6}\mathbf{k}$ 19. $v_0 = 240$ ft/s

21. $v_0 \approx 210$ ft/s

23. $\theta \approx 22.66°$ or $\theta \approx 67.34°$

$\theta = 22.66°$ will get the ball faster to the other player.

25. $\mathbf{T} = \frac{1}{3}\langle 2, 2, -1 \rangle$, $\mathbf{N} = \frac{1}{15\sqrt{2}}\langle -5, 13, 16 \rangle$, $\mathbf{a} = \frac{5}{3}\mathbf{T} + \frac{10\sqrt{2}}{3}\mathbf{N}$

27. (a) At its original position   (b) No

29. $\|\mathbf{a}\| = 27.6$ ft/s^2 31. $a_{\mathbf{T}} = 0$, $a_{\mathbf{N}} = 1$

33. $a_{\mathbf{T}}(t = 0) = 0$, $a_{\mathbf{N}}(t = 0) = \sqrt{2}$

35. $\mathbf{a}(0) = \sqrt{3}\mathbf{T} + \sqrt{2}\mathbf{N}$, with $\mathbf{T} = \frac{1}{\sqrt{3}}\langle 1, 1, 1 \rangle$ and

$\mathbf{N} = \frac{1}{\sqrt{2}}\langle -1, 0, 1 \rangle$

37. $a_{\mathbf{T}}(4) = 4$, $a_{\mathbf{N}}(4) = 1$, so $\mathbf{a} = 4\mathbf{T} + \mathbf{N}$, with $\mathbf{T} = \left\langle \frac{1}{9}, \frac{4}{9}, \frac{8}{9} \right\rangle$ and

$\mathbf{N} = \left\langle -\frac{4}{9}, -\frac{7}{9}, \frac{4}{9} \right\rangle$

39. $a_{\mathbf{T}}(t) = 0$, $a_{\mathbf{N}}(0) = 1$, so $\mathbf{a} = \mathbf{N}$, with $\mathbf{N} = \langle -1, 0, 0 \rangle$

41. $\mathbf{a}\left(\frac{\pi}{2}\right) = -\frac{\pi}{2\sqrt{3}}\mathbf{T} + \frac{\pi}{\sqrt{6}}\mathbf{N}$, with $\mathbf{T} = \frac{1}{\sqrt{3}}\langle 1, -1, 1 \rangle$ and

$\mathbf{N} = \frac{1}{\sqrt{6}}\langle 1, -1, -2 \rangle$

43. $a_{\mathbf{T}} = 0$, $a_{\mathbf{N}} = 0.25$ cm/s^2

45. $\mathbf{a}(3) = 3\mathbf{T} + 0.02025\mathbf{N}$, where $\mathbf{T} \approx \langle -0.0045, 1.0000 \rangle$,

$\mathbf{N} \approx \langle -1.0000, -0.0045 \rangle$

51. The maximum speed (in case of constant speed) is about 98 ft/s.

53. It is faster to hug the inside curve of radius r, $(r < R)$.

Section 13.6 Preliminary

1. $\frac{dA}{dt} = \frac{1}{2}\|\mathbf{J}\|$ 3. The period is increased eightfold.

Section 13.6 Exercises

3. $|\mathbf{J}| = 4.78 \times 10^{16}$ ft^2/s   5. $M = 1.897 \times 10^{27}$ kg

11. $v = 11{,}060$ km/h

13. $M = 2.6225 \times 10^{41}$ kg (132 billions of sun)

17. $r_{\text{per}} = 4.377 \cdot 10^7$ km, $r_{\text{ap}} = 7.203 \cdot 10^7$ km

Chapter 13 Review

1. (a) $-1 < t < 0$ or $0 < t \le 1$ (b) $0 < t \le 2$

3. $\mathbf{r}(t) = \left\langle t^2, t, \sqrt[3]{3 - t^4} \right\rangle$ 5. $\mathbf{r}'(t) = \left\langle -1, -2t^{-3}, \frac{1}{t} \right\rangle$

7. $\mathbf{r}'(0) = \langle 2, 0, 6 \rangle$.   9. $\frac{d}{dt} e^t \left\langle 1, t, t^2 \right\rangle = e^t \left\langle 1, 1 + t, 2t + t^2 \right\rangle$

11. $\frac{d}{dt}\left(6\mathbf{r}_1(t) - 4\mathbf{r}_2(t)\right)\Big|_{t=3} = \langle 0, -8, -10 \rangle$

13. $\frac{d}{dt}\left(\mathbf{r}_1(t) \cdot \mathbf{r}_2(t)\right)\Big|_{t=3} = 2$

15. $\int_0^3 \left\langle 4t + 3, t^2, -4t^3 \right\rangle dt = \langle 27, 9, -81 \rangle$

17. $\mathbf{T}(\pi) = \left\langle \frac{-1}{\sqrt{2}}, \frac{1}{\sqrt{2}}, 0 \right\rangle$ 19. $\left(3, 3, 5\frac{1}{3}\right)$ 21. $L = 2\sqrt{13}$

23. $\mathbf{r}(t) = \left\langle \frac{5}{2}\cos\frac{2\pi t}{5}, \frac{5}{2}\sin\frac{2\pi t}{5}, t \right\rangle$, $0 \le t \le 20$, $L \approx 127.2$

25. $\mathbf{T}(1) = \left\langle -\frac{1}{\sqrt{3}}, \frac{1}{\sqrt{3}}, \frac{1}{\sqrt{3}} \right\rangle$, $\mathbf{N}(1) = \frac{3}{\sqrt{18}}\langle 1, 0, 1 \rangle$,

$\kappa(t) = \dfrac{t^2\sqrt{t^4 + 4t^2 + 1}}{(t^4 + t^2 + 1)^{3/2}}$

27. $\kappa(t) = \dfrac{|4t^2 - 2|e^{-t^2}}{\left(1 + 4t^2 e^{-2t^2}\right)^{3/2}}$

29. $\mathbf{T}(0) = \langle 1, 0 \rangle$, $\mathbf{T}(1) = \left\langle \frac{e}{\sqrt{e^2+4}}, \frac{-2}{\sqrt{e^2+4}} \right\rangle$, $\mathbf{N}(0) = \langle 0, -1 \rangle$,

$\mathbf{N}(1) = \left\langle \frac{2}{\sqrt{e^2+4}}, \frac{e}{\sqrt{e^2+4}} \right\rangle$

31. $\kappa(t) = \frac{\sqrt{2}}{(1+\cos^2 t)^{3/2}}$, $\mathbf{N}(t) = \frac{-\langle \sin t, \sin t, 2\cos t \rangle}{\sqrt{2} \cdot \sqrt{1+\cos^2 t}}$

33. $\mathbf{a} = \frac{1}{\sqrt{2}} \mathbf{T} + 4\mathbf{N}$, where $\mathbf{T} = \langle -1, 0 \rangle$ and $\mathbf{N} = \langle 0, -1 \rangle$.

35. $\mathbf{c}(t) = \left\langle 0, \frac{1}{2} \right\rangle + \frac{1}{2} \langle \cos t, \sin t \rangle$, $0 \le t \le 2\pi$

Chapter 14

Section 14.1 Preliminary

1. The two curves have the same shape but are located in parallel planes.

2. A parabola $z = x^2$ in the xz-plane.

3. Two different level curves of a function do not intersect.

4. The vertical lines $x = c$ with distance of 1 unit between adjacent lines.

5. In the contour map of $g(x, y) = 2x$, the distance between two adjacent vertical lines is $\frac{1}{2}$.

Section 14.1 Exercises

1. $f(2, 2) = 18$, $f(-1, 4) = -5$, $f\left(6, \frac{1}{2}\right) = 114$

3. $h(3, 8, 2) = 6$; $h(3, -2, -6) = -\frac{1}{6}$

5. The domain is the entire xy-plane.

7.

9.

$x \ne -t$

11.

$x \ne 0$

13.

$IR \ge 0$

15.
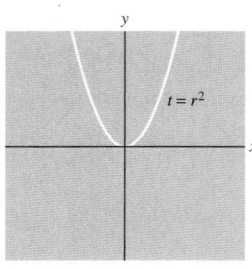

$t \ne r^2$

17. Domain: entire (x, y, z)-space. Range: entire real line.

19. $\left\{(x, y, z) : x^2 + y^2 + z^2 \le 9\right\}$, $\{w \in \mathbf{R} : 0 \le w \le 3\}$.

21.
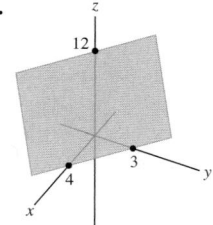

Horizontal trace: $3x + 4y = 12 - c$ in the plane $z = c$.
Vertical trace: $z = (12 - 3a) - 4y$ and $z = -3x + (12 - 4a)$ in the planes $x = a$ and $y = a$, respectively.

23.
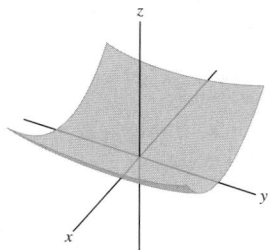

The horizontal traces are ellipses for $c > 0$.
The vertical trace in the plane $x = a$ is the parabola $z = a^2 + 4y^2$.
The vertical trace in the plane $y = a$ is the parabola $z = x^2 + 4a^2$.

25.
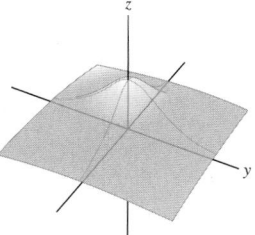

Horizontal trace in the plane $z = c$: $x^2 + y^2 = \frac{1}{c} - 1$.
Vertical trace in the plane $x = a$: $z = \frac{1}{(1+a^2)+y^2}$.
Vertical trace in the plane $y = a$: $z = \frac{1}{x^2+a^2+1}$

27.

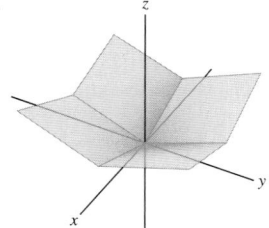

Horizontal trace: $|x| + |y| = c$.
Vertical traces: $z = |a| + |y|$ or $z = |x| + |a|$, in the planes $x = a$ and $y = a$, respectively.

29.

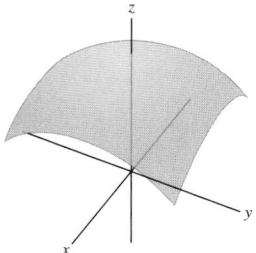

Horizontal trace: $x^2 + y^2 = 1 - c^2$.
Vertical traces: $z = \sqrt{1 - a^2 - y^2}$ in the plane $x = a$ and $z = \sqrt{1 - a^2 - x^2}$ in the plane $y = a$.

31. $m = 1$: $m = 2$:

33. (a)

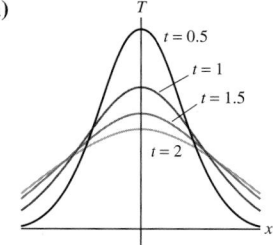

For each t, the temperature decreases as we move away from the center point. Also, as t increases, the temperature at each point in the bar (except at the middle) increases and then decreases.

(b)

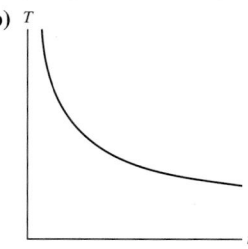

The temperature is decreasing with time at the center point, according to $\dfrac{1}{\sqrt{t}}$.

(c)

For small values of t, the temperature increases fast and then slowly decreases as t increases.

35. **37.**

39. **41.**

43.

45. $m = 6 : f(x, y) = 2x + 6y + 6$
    $m = 3 : f(x, y) = x + 3y + 3$

47. Average ROC from C to $D = 0.000625$ kg/m$^3 \cdot$ ppt

49. At point A

51. Average ROC from A to $B \approx 0.047$ m/km
Average ROC from A to $C \approx 0.023$ m/km

53.

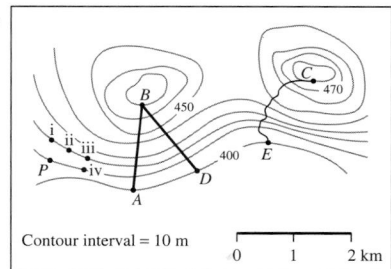

55. $f(r, \theta) = \cos \theta$

Section 14.2 Preliminary

1. $D^*(p, r)$ consists of all points in $D(p, r)$ other than p itself.
2. $f(2, 3) = 27$ **3.** All three statements are true.
4. $\lim\limits_{(x,y)\to(0,0)} f(x, y)$ does not exist.

Section 14.2 Exercises

1. $\lim\limits_{(x,y)\to(1,2)} (x^2 + y) = 3$

3. $\lim\limits_{(x,y)\to(3,-1)} (3x^2y - 2xy^3) = -21$

5. $\lim\limits_{(x,y)\to(1,2)} \frac{x^2+y}{x+y^2} = \frac{3}{5}$ **7.** $\lim\limits_{(x,y)\to(4,4)} \arctan\frac{y}{x} = \frac{\pi}{4}$

9. $\lim\limits_{(x,y)\to(1,1)} \frac{e^{x^2}-e^{-y^2}}{x+y} = \frac{1}{2}(e - e^{-1})$

11. $\lim\limits_{(x,y)\to(0,1)} \frac{x}{y} = 0$ **13.** $\lim\limits_{(x,y)\to(1,1)} e^{xy}\ln(1 + xy) = e\ln 2$

15. $\lim\limits_{(x,y)\to(-1,-2)} x^2|y|^3 = 8$ **17.** $\lim\limits_{(x,y)\to(4,2)} \frac{y-2}{\sqrt{x^2-4}} = 0$

19. $\lim\limits_{(x,y)\to(2,5)} (f(x, y) + 4g(x, y)) = 31$

21. $\lim\limits_{(x,y)\to(2,5)} e^{f(x,y)} = e^3$ **23.** $\lim\limits_{(x,y)\to(0,0)} \frac{(\sin x)(\sin y)}{xy} = 1$

25. $\lim\limits_{(z,w)\to(-1,2)} (z^2w - 9z) = 11$

27. $\lim\limits_{(h,k)\to(2,0)} h^4\frac{(2+k)^2-4}{k} = 64$

29. The limit does not exist. **31.** The limit does not exist.
39. In (B), $\lim\limits_{(x,y)\to P} g(x, y)$ appears to exist. If so, it must be 6, which is the level curve of P.
41. (a) Yes
(b)

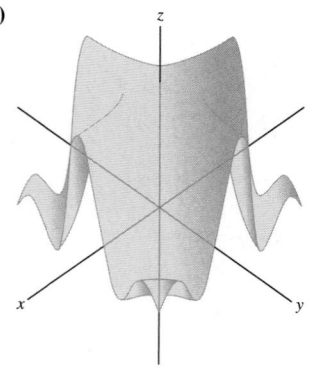

Near the axes, the values of $f(x, y)$ are approaching 1.
43. $f(10^{-1}, 10^{-2}) = \frac{1}{2}$, $f(10^{-5}, 10^{-10}) = \frac{1}{2}$, $f(10^{-20}, 10^{-40}) = \frac{1}{2}$

Section 14.3 Preliminary

1. $\frac{\partial}{\partial x}(x^2y^2) = 2xy^2$
2. In this case, the Constant Multiple Rule can be used. In the second part, since y appears in both the numerator and the denominator, then the Quotient Rule is preferred.
3. (a), (c) **4.** $f_x = 0$ **5.** (a), (d)

Section 14.3 Exercises

3. $\frac{\partial}{\partial y}\frac{y}{x+y} = \frac{x}{(x+y)^2}$ **5.** $f_z(2, 3, 1) = 6$ **7.** $m = 10$

9. $f_x(A) \approx 16$, $f_y(A) \approx -33.3$ **11.** f_y is smallest at C.

13. $\frac{\partial}{\partial x}(x^4y^3) = 4x^3y^3$, $\frac{\partial}{\partial y}(x^4y^3) = 3x^4y^2$

15. $\frac{\partial V}{\partial r} = 2\pi rh$, $\frac{\partial V}{\partial h} = \pi r^2$

17. $\frac{\partial}{\partial x}\left(\frac{x}{x-y}\right) = -\frac{y}{(x-y)^2}$, $\frac{\partial}{\partial y}\left(\frac{x}{x-y}\right) = \frac{x}{(x-y)^2}$

19. $\frac{\partial z}{\partial x} = \frac{y^2}{(x^2+y^2)^{3/2}}$, $\frac{\partial z}{\partial y} = \frac{-xy}{(x^2+y^2)^{3/2}}$

21. $\frac{\partial}{\partial u}\sin(u^2v) = 2uv\cos(u^2v)$, $\frac{\partial}{\partial v}\sin(u^2v) = u^2\cos(u^2v)$

23. $\frac{dS}{dw} = \frac{z}{1+w^2z^2}$, $\frac{dS}{dz} = \frac{w}{1+w^2z^2}$

25. $\frac{\partial z}{\partial x} = \frac{2x}{x^2+y^2}$, $\frac{\partial z}{\partial y} = \frac{2y}{x^2+y^2}$

27. $\frac{\partial}{\partial x}e^{xy} = ye^{xy}$, $\frac{\partial}{\partial y}e^{xy} = xe^{xy}$

29. $\frac{\partial z}{\partial x} = -2xe^{-x^2-y^2}$, $\frac{\partial z}{\partial y} = -2ye^{-x^2-y^2}$

31. $\frac{\partial z}{\partial y} = xy^{x-1}$, $\frac{\partial z}{\partial x} = y^x\ln y$

33. $\frac{\partial}{\partial x}\sinh(x^2y) = 2xy\cosh(x^2y)$, $\frac{\partial}{\partial y}\sinh(x^2y) = x^2\cosh(x^2y)$

35. $\frac{\partial w}{\partial x} = \frac{1}{y+z}$, $\frac{\partial w}{\partial y} = \frac{-x}{(y+z)^2}$, $\frac{\partial w}{\partial z} = \frac{-x}{(y+z)^2}$

37. $\frac{\partial w}{\partial x} = \frac{y^2+z^2-2x^2}{(x^2+y^2+z^2)^{5/2}}$, $\frac{\partial w}{\partial y} = -\frac{3xy}{(x^2+y^2+z^2)^{5/2}}$, $\frac{\partial w}{\partial z} = -\frac{3xz}{(x^2+y^2+z^2)^{5/2}}$

39. $\frac{\partial U}{\partial t} = -e^{-rt}$, $\frac{\partial U}{\partial r} = \frac{-e^{-rt}(rt+1)}{r^2}$

41. $f_y(0, \pi) = 1$ **43.** $h_z(2, 1) = -10e^{-2}$

45. $\frac{\partial W}{\partial E} = -\frac{1}{kT}e^{-\frac{E}{kT}}$, $\frac{\partial W}{\partial T} = \frac{E}{kT^2}e^{-\frac{E}{kT}}$

47. $\frac{\partial V}{\partial r} = \frac{2\pi hr}{3}$, $\frac{\partial V}{\partial h} = \frac{\pi}{3}r^2$

51. (a) $\frac{\partial\rho}{\partial T}\big|_B = -2\cdot 10^{-4}$ kg/m^3 °C, $\frac{\partial\rho}{\partial S}\big|_C = 0.001$ kg/m^3 ppt
(b) At the point C, the effect caused by the change in salinity is greater than the effect caused by the change in temperature.
(c) True

53. $\frac{\partial^2 f}{\partial x^2} = 6y$; $\frac{\partial^2 f}{\partial y^2} = -72xy^2$ **55.** $\frac{\partial^2 g}{\partial x\partial y} = -\frac{2xy}{(x-y)^3}$

57. $h_{uv}(u, v) = \frac{u-v}{(u+v)^3}$

59. $\frac{\partial\rho}{\partial T}(34, 2) = -0.09$, $\frac{\partial\rho}{\partial T}(35, 10) = -0.195$, $\frac{\partial\rho}{\partial S}(34, 2) = 0.83$, $\frac{\partial\rho}{\partial S}(35, 10) = 0.74$

61. $f_{xyz}(x, y, z) = 0$ **63.** $f_{uuv} = 2v\sin(u + v^2)$ **65.** $F_{rst} = 0$

67. $u_{xx} = -\frac{1}{2}t^{-3/2}e^{-x^2/4t}\left(1 - \frac{x^2}{2t}\right)$

69. $g_{xyz} = \frac{3xyz}{(x^2+y^2+z^2)^{5/2}}$ **71.** $f(x, y) = x^2y$ **75.** $B = A^2$

77. $u(x, y) = ax^3 - 3dx^2y - 3axy^2 + dy^3$

Section 14.4 Preliminary

1. $L(x, y) = f(a, b) + f_x(a, b)(x - a) + f_y(a, b)(y - b)$

2. $f(x, y) - L(x, y) = \epsilon(x, y)\sqrt{(x - a)^2 + (y - b)^2}$, for all (x, y) in an open disk containing (a, b)

3. (b) $L(x, y) = 8 + 5(x - 2) + 7(y - 3)$

4. $f(2, 3.1) \approx 8.7$ **5.** $\Delta f \approx -0.1$

6. Criterion for Differentiability

Section 14.4 Exercises

1. (a) $f(x, y) = -16 + 4x + 12y$
(b) $f(2.01, 1.02) \approx 4.28$, $f(1.97, 1.01) \approx 4$

3. $\Delta f \approx 3.56$ **5.** $f(0.01, -0.02) \approx 0.98$

7. $z = -34 - 20x + 16y$ **9.** $\frac{xy}{z} \approx \frac{1}{2}x + y - \frac{1}{2}z$

11. $z = 5x + 10y - 14$ **13.** $z = 8x - 2y - 13$

15. $z = 4r - 5s + 2$ **17.** $z = y - 1$ **19.** $\left(-\frac{1}{4}, \frac{1}{8}, \frac{1}{8}\right)$

23. $z = 3x - 3y + 13$ **25.** Estimate: 8.44

27. Estimate: 4.998   **29.** Estimate: 3.945   **31.** 5.07   **33.** 0.0191

35. (b) 6%   **(c)** 1% error in r

37. (a) \$7.1   **(b)** \$28.845, \$57.69   **(c)** $-\$74.2416$

Section 14.5 Preliminary

1. (b)   **2.** False

3. The gradient at P points in the direction of maximum rate of increase of the function, and is perpendicular to the level curve (or surface) at P.

4. $\frac{\partial f}{\partial x}(a, b) = D_i f(a, b)$ **5.**
(b) NW   **(c)** SE

6. $3\sqrt{2}$

Section 14.5 Exercises

1. (a) $\nabla f = \langle y^2, 2xy \rangle$, $\mathbf{c}'(t) = \langle t, 3t^2 \rangle$, $\nabla f \cdot \mathbf{c}'(t) = y^2 t + 6xyt^2$
(b) $\nabla f \cdot \mathbf{c}'(t) = 4t^7$
(c) $\frac{d}{dt}\left(f(\mathbf{c}(t))\right)\Big|_{t=1} = 4$; $\frac{d}{dt}\left(f(\mathbf{c}(t))\right)\Big|_{t=-1} = -4$

3. (a) $\frac{d}{dt} f(\mathbf{c}(t)) = 0$

5. $\nabla f = -\sin(x^2 + y)\langle 2x, 1 \rangle$

7. $\nabla h = \langle yz^{-3}, xz^{-3}, -3xyz^{-4} \rangle$

9. $\frac{d}{dt} f(\mathbf{c}(t))\Big|_{t=0} = -7$ **11.** $\frac{d}{dt} f(\mathbf{c}(t))\Big|_{t=0} = -3$

13. $\frac{d}{dt} f(\mathbf{c}(t))\Big|_{t=0} = 5\cos(1)$ **15.** $\frac{d}{dt} f(\mathbf{c}(t))\Big|_{t=4} = -56$

17. $\frac{d}{dt} f(\mathbf{c}(t))\Big|_{t=\frac{\pi}{4}} \approx 1.546$ **19.** $\frac{d}{dt} g(\mathbf{c}(t))\Big|_{t=1} = 0$

21. $D_{\mathbf{u}} f(1, 2) = 8.8$

23. $D_{\mathbf{u}} f\left(\frac{1}{6}, 3\right) = \frac{39}{4\sqrt{2}}$ **25.** $D_{\mathbf{u}} f(3, 4) = \frac{7\sqrt{2}}{290}$

27. $D_{\mathbf{u}} f(1, 0) = \frac{6}{\sqrt{13}}$ **29.** $D_{\mathbf{u}} f(1, 2, 0) = -\frac{1}{\sqrt{3}}$

31. $D_{\mathbf{u}} f(3, 2) = \frac{-50}{\sqrt{13}}$ **33.** $D_{\mathbf{u}} f(P) = -\frac{e^5}{3} \approx -49.47$

35. (a) $\|\nabla f_P\| = \sqrt{10}$ **(b)** $\|\nabla f_P\| = \sqrt{10}$
(c) $D_{\mathbf{e_v}} f(P) = \sqrt{5} \approx 2.236$

37. $t_1 = 0, t_2 = 2$ **39.** $\nabla f_P = \langle -3, 3, 9 \rangle$ **41.** $x + 3y + 2z = 10$

43. $-5x + 9y + 6\sqrt{e}z = 35$ **47.** $f(x, y, z) = x + 3y + z$

49. $f(x, y, z) = \frac{x^2}{2} + \frac{y^3}{3} + \frac{z^4}{4}$ **55.** $\frac{1}{\sqrt{21}} \langle 4, 2, -1 \rangle$

59. $x = 1 + 8t, y = 1 - 8t, z = 1$

65. (c) No

71. $\mathbf{c}(t) = (t, 2t^4)$

Section 14.6 Preliminary

1. (a) $\frac{\partial f}{\partial x}$ and $\frac{\partial f}{\partial y}$ **(b)** u and v

2. (a)   **3.** $f(u, v)|_{(r,s)} = e^2$ **4. (b)**   **5. (c)**   **6.** No

Section 14.6 Exercises

1. (a) $\frac{\partial f}{\partial x} = 2xy^3$, $\frac{\partial f}{\partial y} = 3x^2y^2$, $\frac{\partial f}{\partial z} = 4z^3$
(b) $\frac{\partial x}{\partial s} = 2s$, $\frac{\partial y}{\partial s} = t^2$, $\frac{\partial z}{\partial s} = 2st$ **(c)** $\frac{\partial f}{\partial s} = 7s^6t^6 + 8s^7t^4$

3. $\frac{\partial f}{\partial s} = 6rs^2$, $\frac{\partial f}{\partial r} = 2s^3 + 4r^3$

5. $\frac{\partial g}{\partial u} = (42u - 68v)\sin(68uv - 21u^2 - 55v^2)$
$\frac{\partial g}{\partial v} = (110v - 68u)\sin(68uv - 21u^2 - 55v^2)$

7. $\frac{\partial F}{\partial x} = (2x + y)e^{x^2 + xy}$, $\frac{\partial F}{\partial y} = xe^{x^2 + xy}$

9. $\frac{\partial f}{\partial x} = \frac{-xy^2}{(x^2 + y^2)^{3/2}}$ **11.** $\frac{\partial f}{\partial u}\Big|_{(u,v)=(-1,-1)} = 1$
$\frac{\partial f}{\partial y} = \frac{y(y^2 + 2x^2)}{(x^2 + y^2)^{3/2}}$ $\frac{\partial f}{\partial v}\Big|_{(u,v)=(-1,-1)} = -2$

13. $\frac{\partial g}{\partial \theta}\Big|_{(r,\theta)=\left(2\sqrt{2}, \frac{\pi}{4}\right)} = \frac{1}{6}$ **15.** $\frac{\partial g}{\partial u}\Big|_{(u,v)=(0,1)} = 2\cos 2$

17. -0.1 degrees per second   **19.** $\|\nabla u\|^2 = 4r^2\cos^2\theta$

21. $\frac{\partial f}{\partial \rho} = (\sin\phi\cos\theta)\frac{\partial f}{\partial x} + (\sin\phi\sin\theta)\frac{\partial f}{\partial y} + (\cos\phi)\frac{\partial f}{\partial z}$
$\frac{\partial f}{\partial \phi} = (\rho\cos\phi\cos\theta)\frac{\partial f}{\partial x} + (\rho\cos\phi\sin\theta)\frac{\partial f}{\partial y} - (\rho\sin\phi)\frac{\partial f}{\partial z}$
$\frac{\partial f}{\partial \theta} = (-\rho\sin\phi\sin\theta)\frac{\partial f}{\partial x} + (\rho\sin\phi\cos\theta)\frac{\partial f}{\partial y}$

23. $\frac{\partial z}{\partial x}\Big|_{(3,2,1)} = -\frac{3}{11}$, $\frac{\partial z}{\partial y}\Big|_{(3,2,1)} = \frac{1}{22}$;
$\frac{\partial z}{\partial x}\Big|_{(3,2,-1)} = \frac{3}{11}$, $\frac{\partial z}{\partial y}\Big|_{(3,2,-1)} = -\frac{1}{22}$

25. $\frac{\partial w}{\partial z} = -\frac{2wz+3y}{x^2+3w^2+z^2}$ **27.** $\frac{\partial r}{\partial t} = \frac{r^2 e^{s/r}}{2r^3 + ste^{s/r}}$, $\frac{\partial t}{\partial r} = 2re^{-s/r} + \frac{st}{r^2}$

29. $\frac{\partial U}{\partial T}\Big|_{(1,1,2,4)} = -\frac{2\ln 2}{1+2\ln 2}$, $\frac{\partial T}{\partial U}\Big|_{(1,1,2,4)} = -\frac{1+2\ln 2}{2\ln 2}$

33. $\nabla\left(\frac{1}{r}\right) = -\frac{1}{r^3}\mathbf{r}$, $\nabla(\ln r) = \frac{\mathbf{r}}{r^2}$ **35.** $\frac{-18+17\cos\theta}{\sqrt{13-12\cos\theta}}$

Section 14.7 Preliminary

1. f has a local (and global) min at $(0, 0)$; g has a saddle point at $(0, 0)$.

Section 14.7 Preliminary

1. f has a local (and global) min at $(0, 0)$; g has a saddle point at $(0, 0)$.

2.

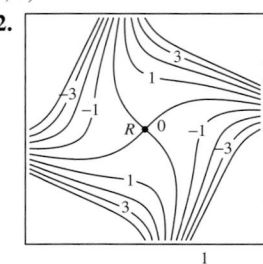

R is a saddle point.

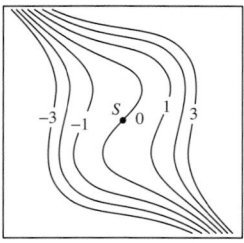

Point S is neither a local extremum nor a saddle point.

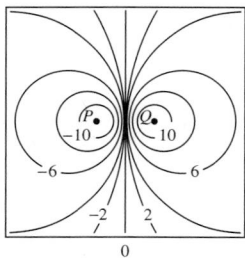

Local minimum at P, local maximum at Q.

3. (a)

Section 14.7 Exercises

1. (b) $P_1 = (0, 0)$ saddle point, $P_2 = (2\sqrt{2}, \sqrt{2})$,
$P_3 = (-2\sqrt{2}, -\sqrt{2})$ local minima. Absolute minimum value of f is -4.

3. $(0, 0)$, $\left(\frac{13}{64}, -\frac{13}{32}\right)$, $\left(-\frac{1}{4}, \frac{1}{2}\right)$ **5.** $\left(-\frac{2}{3}, -\frac{1}{3}\right)$ local minimum

7. $(2, -4)$ saddle point

9. $\left(0, \pm\sqrt{2}\right)$ saddle points, $\left(\frac{2}{3}, 0\right)$ local maximum, $\left(-\frac{2}{3}, 0\right)$ local minimum.

11. $(0, 0)$ saddle point, $(1, 1)$ and $(-1, -1)$ local minima.

13. $(0, 0)$ saddle point, $\left(\frac{1}{\sqrt{2}}, \frac{1}{\sqrt{2}}\right)$ local maximum, $\left(\frac{1}{\sqrt{2}}, -\frac{1}{\sqrt{2}}\right)$ local minimum, $\left(-\frac{1}{\sqrt{2}}, \frac{1}{\sqrt{2}}\right)$ local minimum, $\left(-\frac{1}{\sqrt{2}}, -\frac{1}{\sqrt{2}}\right)$ local maximum.

15. $(0, 0)$ saddle point **17.** $\left(1, \frac{1}{2}\right)$ local maximum

19. $\left(\frac{3}{2}, -\frac{1}{2}\right)$ saddle point **23.** $x = y = 0.27788$ local minimum

25. (a) Bounded and closed (b) Bounded

(c) Closed (d) Neither bounded nor closed

(e) Bounded and closed (f) Bounded

27. Global maximum 2, global minimum -3

29. Maximum value $f(0, 0) = 1$, minimum value e^{-1}

31. $f(x, y) = -\frac{1}{x+y}$

35. Global minimum $f(0, 1) = -2$, global maximum $f(1, 0) = 1$

37. Global maximum 3, global minimum 0

39. Global minimum $f(0, 0) = 0$, global maximum $f(0, 1) = 2$

41. Minimum value 0, maximum value $f\left(\frac{36}{121}, \frac{5}{11}\right) = 0.0005$

43. (a) In the box B with minimal surface area, z is smaller than $\sqrt[3]{V}$, which is the side of a cube with volume V.

(b) Width: $x = (2V)^{1/3}$; length: $y = (2V)^{1/3}$, height: $z = \left(\frac{V}{4}\right)^{1/3}$

45. $f(x) = 1.9629x - 1.5519$

47. (a) $f_x(x, y) =$
$$\frac{x-a_1}{\sqrt{(x-a_1)^2+(y-a_2)^2}} + \frac{x-b_1}{\sqrt{(x-b_1)^2+(y-b_2)^2}} + \frac{x-c_1}{\sqrt{(x-c_1)^2+(y-c_2)^2}}$$

$f_y(x, y) =$
$$\frac{y-a_2}{\sqrt{(x-a_1)^2+(y-a_2)^2}} + \frac{y-b_2}{\sqrt{(x-b_1)^2+(y-b_2)^2}} + \frac{y-c_2}{\sqrt{(x-c_1)^2+(y-c_2)^2}}$$

(b) $\mathbf{e} = \frac{\langle x-a_1, y-a_2 \rangle}{\sqrt{(x-a_1)^2+(y-a_2)^2}}$, $\mathbf{f} = \frac{\langle x-b_1, y-b_2 \rangle}{\sqrt{(x-b_1)^2+(y-b_2)^2}}$,
$\mathbf{g} = \frac{\langle x-c_1, y-c_2 \rangle}{\sqrt{(x-c_1)^2+(y-c_2)^2}}$

Section 14.8 Preliminary

1. (b)

2. f has a local maximum 2, under the constraint, at A. $f(B)$ is neither local minimum nor local maximum of f.

3. (a)

(b) Global minimum -4, global maximum 6

Section 14.8 Exercises

1. (c) $(-1, -2)$ and $(1, 2)$ (d) Maximum: 10, minimum: -10

3. Maximum: $4\sqrt{2}$, minimum: $-4\sqrt{2}$

5. Minimum: $\frac{468}{169}$, no maximum value

7. Maximum: $\frac{8}{3}$, minimum: $-\frac{8}{3}$ **9.** Maximum: $\sqrt{2}$, minimum: 1

11. Minimum: $-\sqrt{\frac{41}{3}}$, maximum: $\sqrt{\frac{41}{3}}$

13. No maximum and minimum values **15.** $(-1, e^{-1})$

17. (a) $h = \sqrt{\frac{2}{\sqrt{3}\pi}} \approx 0.6$, $r = \sqrt{\frac{1}{\sqrt{3}\pi}} \approx 0.43$ (b) $\frac{h}{r} = \sqrt{2}$

(c) There is no cone of volume 1 and maximal surface area.

19. $2ab$ **21.** $\sqrt{\dfrac{a^a b^b}{(a+b)^{a+b}}}$ **23.** $\sqrt{\dfrac{a^a b^b c^c}{(a+b+c)^{a+b+c}}}$

25. (a) $\left(0, -\dfrac{8}{27}, \dfrac{4}{9}\right)$ **(c)** No

29. $\dfrac{3}{16}$ **31.** $\dfrac{x}{\alpha} + \dfrac{y}{\beta} + \dfrac{z}{\gamma} = 3$ **35.** Minimum: $\dfrac{138}{11}$

39. (b) $\lambda = \dfrac{c}{2p_1 p_2}$

Chapter 14 Review

1. (a)

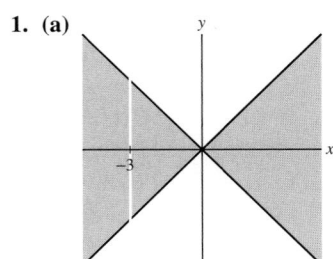

(b) $f(3,1) = \dfrac{\sqrt{2}}{3}$, $f(-5,-3) = -2$ **(c)** $\left(-\dfrac{5}{3}, 1\right)$

3.

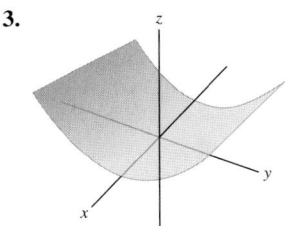

Vertical and horizontal traces: the line $z = (c^2 + 1) - y$ in the plane $x = c$, the parabola $z = x^2 - c + 1$ in the plane $y = c$.

5. (a) (b) **(b)** (c) **(c)** (d) **(d)** (a)

7. (a) Parallel lines $4x - y = \ln c$, $c > 0$, in the xy-plane.
(b) Parallel lines $4x - y = e^c$, in the xy-plane.
(c) Hyperbolas $3x^2 - 4y^2 = c$ in the xy-plane.
(d) Parabolas $x = c - y^2$ in the xy-plane.

9. $\displaystyle\lim_{(x,y)\to(1,-3)} (xy + y^2) = 6$ **11.** The limit does not exist.

13. $\displaystyle\lim_{(x,y)\to(1,-3)} (2x + y)e^{-x+y} = -e^{-4}$ **17.** $f_x = 2$, $f_y = 2y$

19. $f_x = e^{-x-y}\left(y\cos(xy) - \sin(xy)\right)$
$\quad f_y = e^{-x-y}\left(x\cos(yx) - \sin(yx)\right)$

21. $f_{xxyz} = -\cos(x + z)$ **23.** $z = 33x + 8y - 42$

25. $\sqrt{7.1^2 + 4.9^2 + 70.1} \approx 12.020833$

27. Approximately \$69.72 **29.** $\dfrac{d}{dt} f(\mathbf{c}(t)) = -6t^5 - 2t + 1$

31. $\dfrac{d}{dt} f(\mathbf{c}(t))\Big|_{t=\frac{\pi}{3}} = 1$ **33.** $D_{\mathbf{u}} f(1,1,1) = \dfrac{4}{3}$

35. $D_{\mathbf{v}} f(P) = \dfrac{1}{\sqrt{2}}$ **37.** $-4x - y + 2z = -5$

39. $\dfrac{\partial f}{\partial u} = -2v e^{u-v}$, $\dfrac{\partial f}{\partial v} = 2e^{u-v}(v - 1)$

41. $\dfrac{\partial f}{\partial r} = \dfrac{\partial f}{\partial x}\cos\theta + \dfrac{\partial f}{\partial y}\sin\theta$, $\quad \dfrac{\partial f}{\partial \theta} = -\dfrac{\partial f}{\partial x} r\sin\theta + \dfrac{\partial f}{\partial y} r\cos\theta$

45. $\dfrac{\partial z}{\partial x} = -\dfrac{e^z - 1}{xe^z + e^y}$ **47.** $(3, 3)$ saddle point

49. $\left(\dfrac{1}{2}, \dfrac{1}{2}\right)$ saddle point

53. $f(2,4) = 10$ global maximum, $f(-2,4) = -18$ global minimum

55. Maximum: $\dfrac{26}{\sqrt{13}}$, minimum: $-\dfrac{26}{\sqrt{13}}$

57. Maximum: $\dfrac{12}{\sqrt{3}}$, minimum: $-\dfrac{12}{\sqrt{3}}$

61. $x = \dfrac{a}{\sqrt{3}}$, $y = \dfrac{b}{\sqrt{3}}$, $z = \dfrac{c}{\sqrt{3}}$

Chapter 15

Section 15.1 Preliminary

1. $\Delta A = 1$, the number of subrectangles is 32.
2. $\iint_{\mathcal{R}} f\, dA \approx S_{1,1} = 0.16$ **3.** $\iint_{\mathcal{R}} 5\, dA = 50$
4. The signed volume between the graph $z = f(x, y)$ and the xy-plane. The region below the xy-plane is treated as negative volume.
5. (b) **6.** (c), (d)

Section 15.1 Exercises

1. $S_{4,3} = 13.5$ **3. (A)** $S_{32} = 42$ **(B)** $S_{32} = 43.5$
5. (A) $S_{32} = 60$ **(B)** $S_{32} = 62$
7. (A) $S_{32} \approx 1.069$ **(B)** $S_{23} \approx 1.097$
9. $\iint_{\mathcal{R}} (-5)\, dA = -45$ **11.** $S_{54} = 24.2$
13. $1.0731, 1.0783, 1.0809$ **15.** $\iint_{\mathcal{R}} \sin x\, dA = 0$
17. $\int_1^3 \int_0^2 x^3 y\, dy\, dx = 40$ **19.** $\int_0^2 \int_1^3 x^3 y\, dy\, dx = 16$
21. $\int_2^6 \int_1^4 x^2\, dx\, dy = 84$ **23.** $\int_0^1 \int_0^2 (x + 4y^3)\, dx\, dy = 4$
25. $\int_0^1 \int_2^3 \sqrt{x + 4y}\, dx\, dy = \dfrac{1}{15}(7^{5/2} - 6^{5/2} - 3^{5/2} + 2^{5/2}) \approx 2.102$
27. $\int_1^2 \int_0^4 \dfrac{dy\, dx}{x+y} = 6\ln 6 - 2\ln 2 - 5\ln 5 \approx 1.31$
29. $\int_0^1 \int_2^3 \dfrac{1}{(x+4y)^3}\, dx\, dy = \dfrac{1}{56}$
31. $\int_1^2 \int_1^2 \ln(xy)\, dy\, dx = 4\ln 2 - 2 \approx 0.773$ **33.** $m = \dfrac{3}{4}$
35. $\iint_{\mathcal{R}} x^2 y\, dA = \dfrac{4}{3}$ **37.** $\iint_{\mathcal{R}} \dfrac{y}{x+1}\, dA = 8\ln 3 \approx 8.79$
39. $\iint_{\mathcal{R}} e^{2x+3y}\, dA = \dfrac{1}{6}(e^2 - 1)(e^6 - 1) \approx 428.52$
41. $\int_1^2 \int_0^2 \dfrac{x}{y}\, dx\, dy = 2\ln 2 \approx 1.39$
43. $\int_0^1 \int_0^1 \sin(xy)\, dx\, dy \approx 0.240$ **45.** $S_{33} = 44$
49. $F(x, y) = y^2 x^3$, $\iint_{\mathcal{R}} 6x^2 y\, dA = 16$

Section 15.2 Preliminary

1. (a)
2.

3.

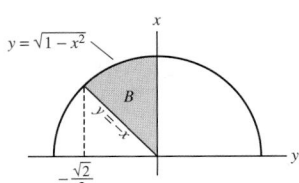

4. (b)

Section 15.2 Exercises

1. (a) Sample points •: $S_{34} = -3$

(b) Sample points ○: $S_{34} = -4$

3.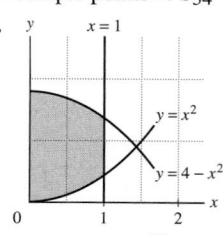

$\iint_D y \, dA = \frac{20}{3} \approx 6.67$

5. $\iint_D x^2 y \, dA = 38.4$ 7. $\iint_D x^2 y \, dA = \frac{608}{15} \approx 40.53$

9. $\iint_D x \, dA = 2\frac{1}{4}$

11.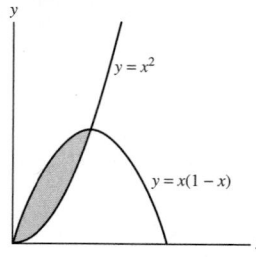

The region D is defined by the following inequalities: $0 \le x \le \frac{1}{2}$, $x^2 \le y \le x(1-x)$, $\iint_D 2y \, dA = \frac{1}{96} \approx 0.010$

13. $\iint_D x \, dA = \frac{11}{60}$ 15. $\iint_D f(x, y) \, dA = \ln \frac{25}{9} \approx 1.02$

17.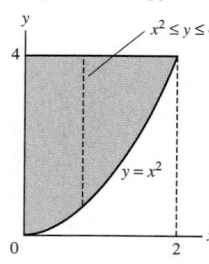

$\iint_D x^3 \, dA = \frac{16}{3} \approx 5.33$

19.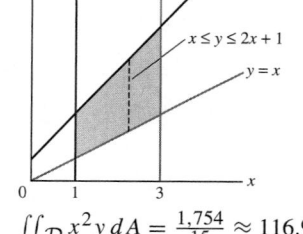

$\iint_D x^2 y \, dA = \frac{1,754}{15} \approx 116.93$

21.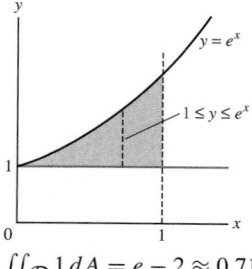

$\iint_D 1 \, dA = e - 2 \approx 0.718$

23.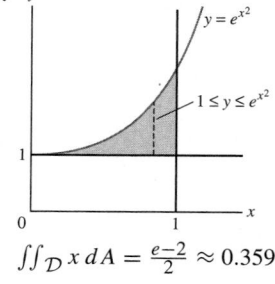

$\iint_D x \, dA = \frac{e-2}{2} \approx 0.359$

25.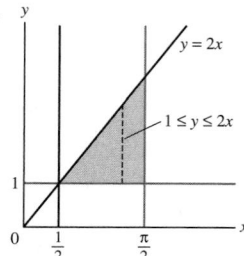

$\iint_D \cos(2x + y) \, dA = -0.416$

27.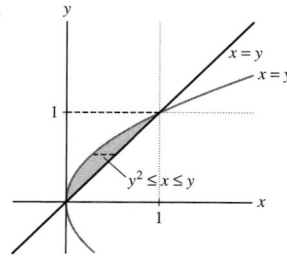

$\iint_D 2xy \, dA = \frac{1}{12}$

29.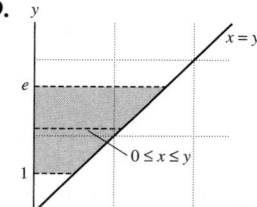

$\iint_D f(x, y) \, dA = (e - 1) \ln 2 \approx 1.19$

31.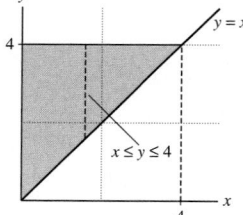

$\int_0^4 \int_x^4 f(x, y) \, dy \, dx = \int_0^4 \int_0^y f(x, y) \, dx \, dy$

33.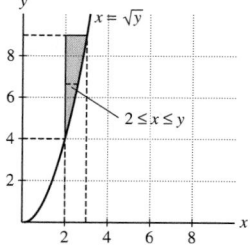

$\int_4^9 \int_2^{\sqrt{y}} f(x, y) \, dx \, dy = \int_2^3 \int_{x^2}^9 f(x, y) \, dy \, dx$

37.

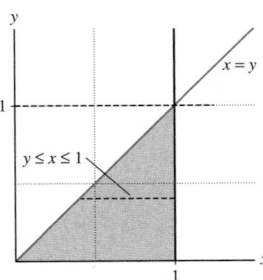

$$\int_0^1 \int_0^x \frac{\sin x}{x}\, dy\, dx = 1 - \cos 1 \approx 0.46$$

39.

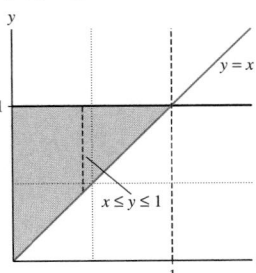

$$\int_0^1 \int_0^y x e^{y^3}\, dx\, dy = \frac{e-1}{6} \approx 0.286$$

41.

$$\int_0^{\pi/2} \int_0^1 x \cos(xy)\, dy\, dx = 1$$

43.

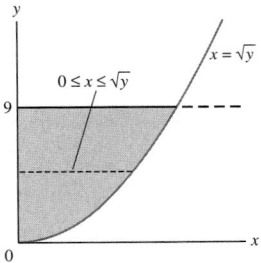

$$\int_0^3 \int_{x^2}^9 \frac{x^3\, dy\, dx}{(3x^2+y)^{1/2}} = 14.4$$

45.

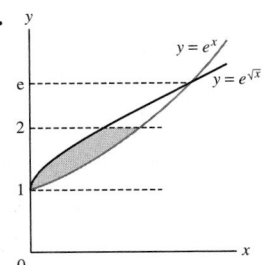

$$\iint_{\mathcal{D}} (\ln y)^{-1}\, dA = 2 - 2\ln 2 \approx 0.614$$

47. $\int_0^4 \int_{x/4}^{3x/4} y e^x\, dy\, dx = \frac{5e^4}{2} - \frac{1}{2} \approx 136$

49. $\int_2^4 \int_{y-1}^{7-y} \frac{x}{y^2}\, dx\, dy = 6 - 6\ln 2 = 1.84$

51. $\iint_{\mathcal{D}} \frac{\sin y}{y}\, dA = \cos 1 - \cos 2 \approx 0.956$ **53.** $V = 16\pi$

55. $V = \frac{16}{3} \approx 5.33$ **57.** $M = \frac{15}{64}$ **59.** $\overline{H} = \frac{3}{10}$

61. $\overline{x} = \frac{4R}{3\pi}, \overline{y} = 0$ **63.** $\overline{f} = p$ **67.** $P = \left(\frac{2}{3}, 2\right)$

69. $\iint_{\mathcal{D}} f(x, y)\, dA \approx 57.01$

Section 15.3 Preliminary

1. (c)   **2.** (b)

3. (a) $\mathcal{D} = \{(x, y) : 0 \le x \le 1, 0 \le y \le x\}$

(b) $\mathcal{D} = \left\{(x, y) : 0 \le x \le 1, 0 \le y \le \sqrt{1 - x^2}\right\}$

Section 15.3 Exercises

1. $\iiint_{\mathcal{B}} z^4\, dV = 6$ **3.** $\iint_{\mathcal{B}} xz^2\, dV = 123\frac{1}{3}$

5. $\iiint_{\mathcal{B}} x e^{y-2z}\, dV = (e - 1)(1 - e^{-2})$

7. $\iiint_{\mathcal{B}} \frac{z}{x}\, dV = 16\ln 3$

9. $\iiint_{\mathcal{B}} (x + z)^3\, dV = \frac{b}{20}\left[(a + c)^5 - a^5 - c^5\right]$

11. $\iiint_{\mathcal{W}} (x + y)\, dV = \frac{1}{6}$ **13.** $\iiint_{\mathcal{W}} xyz\, dV = \frac{1}{16}$

15. $\iiint_{\mathcal{W}} z\, dV = 4\frac{1}{6}$

17. $\iiint_{\mathcal{V}} e^{x+y+z}\, dV = -15e^4 + 4e^6 - 1 \approx 793.74$

19. $\iiint_{\mathcal{W}} xz\, dV = 24\frac{8}{21}$ **21.** $\iiint_{\mathcal{W}} z\, dV = 2$

23. Volume$(\mathcal{W}) = 1\frac{29}{30}$

25. $\iiint_{\mathcal{W}} y\, dV = \frac{4}{15}(5^{5/2} - 4^{5/2}) \approx 6.37$

27. $\iiint_{\mathcal{W}} xyz^2\, dV = \int_0^2 \int_0^{y/2} \int_0^{4-y^2} xyz^2\, dz\, dx\, dy$

$$\iiint_{\mathcal{W}} xyz^2\, dV = \int_0^4 \int_0^{\sqrt{4-z}} \int_0^{y/2} xyz^2\, dx\, dy\, dz$$

$$\iiint_{\mathcal{W}} xyz^2\, dV = \int_0^4 \int_0^{\sqrt{1-\frac{z}{4}}} \int_{2x}^{\sqrt{4-z}} xyz^2\, dy\, dx\, dz$$

29. $\iiint_{\mathcal{W}} f(x, y, z)\, dV$

$$= \int_{-1}^1 \int_{-\sqrt{1-x^2}}^{\sqrt{1-x^2}} \int_{\sqrt{x^2+y^2}}^1 f(x, y, z)\, dz\, dy\, dx$$

$$\iiint_{\mathcal{W}} f(x, y, z)\, dV$$

$$= \int_0^1 \int_{-z}^z \int_{-\sqrt{z^2-y^2}}^{\sqrt{z^2-y^2}} f(x, y, z)\, dx\, dy\, dz$$

31. $\overline{f} = \frac{1}{2\pi}$ **33.** $\overline{f} = \frac{13}{24}$ **35.** $P = (1.5, 1, 2)$

37. $(x_{\text{CM}}, y_{\text{CM}}, z_{\text{CM}}) = \left(\frac{2}{9}, \frac{2}{9}, \frac{1}{3}\right)$

39. $S_{N,N,N} \approx 0.561, 0.572,$ and 0.576 for $N = 10, 20, 30,$ respectively. $I \approx 0.584,$ for $N = 100$ $S_{N,N,N} \approx 0.582.$

Section 15.4 Preliminary

1. (d)

2. (a) $\int_{-1}^2 \int_0^{2\pi} \int_0^2 f(r, \theta, z) r\, dr\, d\theta\, dz$

(b) $\int_{-2}^0 \int_0^{2\pi} \int_0^{\sqrt{4-z^2}} f(r, \theta, z) r\, dr\, d\theta\, dz$

3. (a) $\int_0^{2\pi} \int_0^{\pi} \int_0^4 f(P)\rho^2 \sin\phi\, d\rho\, d\phi\, d\theta$

(b) $\int_0^{2\pi} \int_0^{\pi} \int_4^5 f(P)\rho^2 \sin\phi\, d\rho\, d\phi\, d\theta$

(c) $\int_0^{2\pi} \int_{\pi/2}^{\pi} \int_0^2 f(P)\rho^2 \sin\phi\, d\rho\, d\phi\, d\theta$

4. $\Delta A \approx r(\Delta r \Delta\theta)$

Section 15.4 Exercises

1.

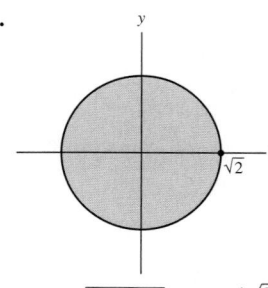

$$\iint_{\mathcal{D}} \sqrt{x^2 + y^2}\, dA = \frac{4\sqrt{2}\pi}{3}$$

3.

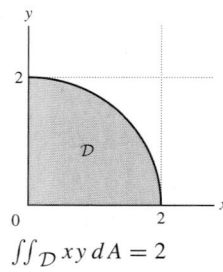

$$\iint_{\mathcal{D}} xy\, dA = 2$$

5.

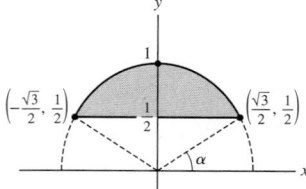

$$\iint_{\mathcal{D}} y(x^2 + y^2)^{-1}\, dA = \sqrt{3} - \frac{\pi}{3} \approx 0.685$$

7. $V = 18$ **9.** $V = 6\pi$ **11.** $P = \left(\frac{4R}{3\pi}, \frac{4R}{3\pi}\right)$

13.

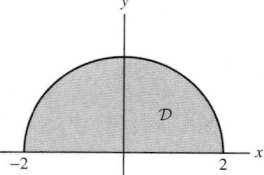

$$\int_{-2}^{2} \int_{0}^{\sqrt{4-x^2}} (x^2 + y^2)\, dy\, dx = 4\pi$$

15.

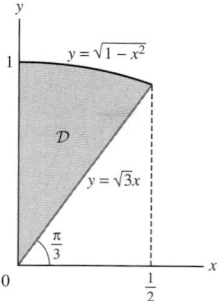

$$\int_{0}^{1/2} \int_{\sqrt{3}x}^{\sqrt{1-x^2}} x\, dy\, dx = \frac{1}{3}\left(1 - \frac{\sqrt{3}}{2}\right) \approx 0.045$$

17. π^2

19.

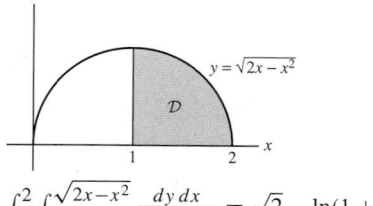

$$\int_{1}^{2} \int_{0}^{\sqrt{2x-x^2}} \frac{dy\, dx}{\sqrt{x^2+y^2}} = \sqrt{2} - \ln(1 + \sqrt{2}) \approx 0.533$$

21. $\iint_{\mathcal{D}} (x^2 + y^2)^{-2}\, dA = \frac{1}{4}$ **23.** $\iint_{\mathcal{D}} |xy|\, dA = \frac{1}{2}$

25. $\iint_{\mathcal{D}} (x - y)\, dA = 0$

27. $\iint_{\mathcal{D}} \sqrt{x^2 + y^2}\, dA = \frac{48\pi - 32}{9} \approx 13.2$

29. (a) $\mathcal{W}: 0 \le \theta \le 2\pi, 0 \le r \le 2, r^2 \le z \le 8 - r^2$
(b) Volume$(\mathcal{W}) = 16\pi$

31. $\iiint_{\mathcal{W}} (x^2 + y^2)\, dV = \frac{5 \cdot 3^4 \pi}{2} \approx 636.17$ **33.** $\iiint_{\mathcal{W}} y\, dV = \frac{2}{3}$

35. $\iiint_{\mathcal{W}} z\, dV = 243\pi$ **37.** $P = (0, 0, 2.81)$

39. $\int_{-\pi/2}^{\pi/2} \int_{0}^{1} \int_{0}^{4} f(r\cos\theta, r\sin\theta, z)r\, dz\, dr\, d\theta$

41. $\int_{0}^{\pi/2} \int_{0}^{2\cos\theta} \int_{0}^{r} f(r\cos\theta, r\sin\theta, z)r\, dz\, dr\, d\theta$

43. $z = \frac{H}{R}r, V = \frac{\pi R^2 H}{3}$ **45.** $\iiint_{\mathcal{W}} z\, dV = 0$

47. $\iiint_{\mathcal{W}} x^2\, dV = \frac{62}{15}\left(\frac{\pi}{6} + \frac{\sqrt{3}}{8}\right) \approx 3.06$

49. $V = \frac{2\pi R^3 (1 - \cos\phi_0)}{3}$ **51.** $\iiint_{\mathcal{W}} y\, dV = -\frac{\pi}{16}$

53. $\iiint_{\mathcal{W}} f(x, y, z)\, dV = 2\pi \ln 2$ **55.** $\iiint_{\mathcal{W}} 1\, dV = 8\pi$

57. $P = \left(0, 0, \frac{3R(1 - \cos 2\phi_0)}{16(1 - \cos\phi_0)}\right)$ **59.** $\iiint_{\mathcal{W}} f(x, y, z)\, dV = \pi$

61. $(x_{CM}, y_{CM}, z_{CM}) = (0, 0, 0.91)$

63. $(x_{CM}, y_{CM}, z_{CM}) = \left(\frac{2R}{\pi}, \frac{4R}{\pi}\left(1 - \frac{\sqrt{3}}{2}\right), \frac{H}{2}\right)$

65. $(x_{CM}, y_{CM}, z_{CM}) = \left(0, \frac{1}{12}, \frac{37}{48}\right)$

69. $4\pi(\sin 1 - \cos 1)$

Section 15.5 Preliminary

1. (b)
2. (a) $\Phi(1, 0) = (2, 0)$ (b) $\Phi(1, 1) = (1, 3)$
(c) $\Phi(2, 1) = (3, 3)$
3. Area$(\Phi(\mathcal{R})) = 36$ **4.** Area$(\Phi(\mathcal{R})) = 0.06$

Section 15.5 Exercises

1. (a) The image of the u-axis is the line $y = \frac{1}{2}x$; the image of the v-axis is the line $x = 0$.
(b) The parallelogram with vertices $(0, 0)$, $(10, 5)$, $(10, 12)$, $(0, 7)$.
(c) A segment in the xy-plane joining the points $(2, 3)$ and $(10, 8)$.
(d) A triangle in the xy-plane whose vertices are at the points $(0, 1)$, $(2, 1)$, and $(2, 2)$.
3. Φ is one-to-one on the domain $\{(u, v)\, u \ge 0\}$ and on the domain $\{(u, v)\, u \le 0\}$.
(a) The positive x-axis and the y-axis, respectively.
(b) The rectangle $[0, 1] \times [-1, 1]$
(c) The curve $y = \sqrt{x}$ for $0 \le x \le 1$
(d)

y
$(0, 1)$ ────── $(1, 1)$
        $y = \sqrt{x}$
$(0, 0)$ ───────────── x

5. $y = \frac{5x}{2} + \frac{c}{2}$ **7.** $y = \frac{17}{6}x$ **11.** Jac$(\Phi) = 1$

13. $\text{Jac}(\Phi) = -10$ **15.** $\text{Jac}(\Phi) = (1 - 3uv)e^{v+3u}$

17. $\text{Jac}(\Phi) = r$ **19.** $\Phi(u, v) = (4u + 2v, u + 3v)$

21. $\iint_D xy\, dA = \frac{2{,}329}{12} \approx 194.08$

23. (a) $\text{Area}(\Phi(\mathcal{R})) = 105$ (b) $\text{Area}(\Phi(\mathcal{R})) = 126$

25. $\text{Jac}(T) = \frac{2u}{v}$, $\iint_{\Phi(T)} 1\, dA = 15\ln 4$

27.

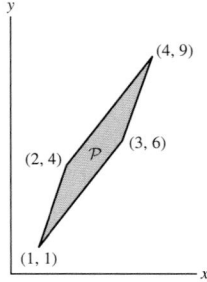

$\Phi(u, v) = (1 + 2u + v, 1 + 5u + 3v)$

29. $\iint_D y\, dA = 82$ **31.** $\iint_D (x + 3y)\, dx\, dy = 80$

33. $\iint_D \sqrt{x^2 + y^2}\, dx\, dy = \frac{56}{45}$

35. $\iint_D e^{9x^2+4y^2}\, dA = \frac{\pi(e^{36}-1)}{6}$

37.

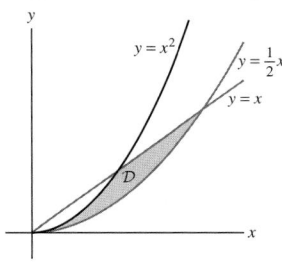

$\iint_D y^{-1}\, dx\, dy = 1$

39. $\iint_D e^{xy}\, dA = (e^{20} - e^{10})\ln 2$

41. (b) $\frac{\partial(x,y)}{\partial(u,v)} = -\frac{1}{x+y}$ (c) $I = 9$

43. $\iiint_{\mathcal{W}} f(x, y, z)\, dv$
$$= \int_{\theta_1}^{\theta_2} \int_{\phi_1}^{\phi_2} \int_{\rho_1(\theta,\phi)}^{\rho_2(\theta,\phi)} f(\rho\cos\theta\sin\phi, \rho\sin\theta\sin\phi, \rho\cos\phi)$$
$$\cdot \rho^2\sin\phi\, d\rho\, d\phi\, d\theta$$

45. $\int_0^1 \int_0^1 \frac{dx\,dy}{1-x^2y^2} = \frac{\pi^2}{8}$

Chapter 15 Review

1. (a) $S_{3,4} \approx 7.588$ (b) $S_{3,4} \approx 9.486$

(c) $\int_1^2 \int_2^3 x^2 y\, dx\, dy = 9.5$

3. $S_{44} = 2.9375$ **5.** $\int_0^2 \int_3^5 y(x - y)\, dx\, dy = \frac{32}{3}$

7. $\int_0^{\pi/3} \int_0^{\pi/6} \sin(x + y)\, dx\, dy = \frac{\sqrt{3}-1}{2}$

9.

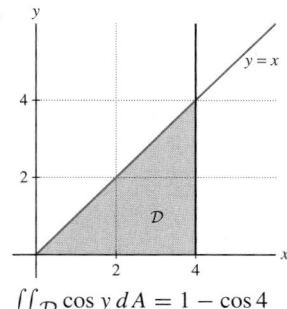

$\iint_D \cos y\, dA = 1 - \cos 4$

11.

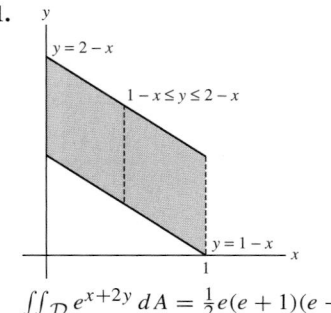

$\iint_D e^{x+2y}\, dA = \frac{1}{2}e(e + 1)(e - 1)^2$

13.

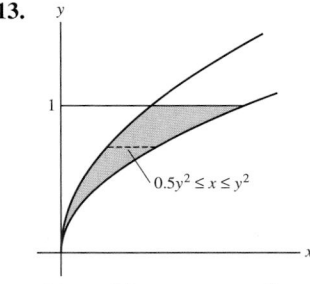

$\iint_D ye^{1+x}\, dA = 0.5(e^2 - 2e^{1.5} + e)$

15. $\int_{-2}^2 \int_0^{9-x^2} f(x, y)\, dy\, dx =$
$\int_5^9 \int_{-\sqrt{9-y}}^{\sqrt{9-y}} f(x, y)\, dx\, dy + \int_0^5 \int_{-2}^2 f(x, y)\, dx\, dy$

19. $\iint_D \frac{y}{(1+x^2+y^2)}\, dx\, dy = \frac{\pi}{4} - \frac{1}{2}$

21. $x_{\text{CM}} = 0$, $y_{\text{CM}} = \frac{2}{3}\frac{\sin\theta}{\theta}$

23. $\overline{f} = \frac{1}{\pi\sqrt{2}} \int_{-\sqrt{2}}^{\sqrt{2}} \int_{-\sqrt{1-x^2/2}}^{\sqrt{1-x^2/2}} e^{xy}\, dy\, dx \approx 1.0421$

25. $\iiint_B (xy + z)\, dV = 10$ **27.** $I = \frac{\pi}{4} + \frac{2}{3}$

29. $P = \left(\frac{8}{15}, \frac{16}{15\pi}, \frac{16}{15\pi}\right)$ **31.** $\iint_D \sqrt{x^2 + y^2}\, dA = \frac{1}{4}$

33. $M = \frac{256}{15}\pi \approx 53.62$ **35.** $V = \pi$

37. $\int_0^{\pi/2} \int_0^1 \int_0^r zr\, dz\, dr\, d\theta = \frac{\pi}{16}$

39. $2\pi \int_0^2 e^{-\rho^3}\rho^2\, d\rho = \frac{2\pi(-1+e^8)}{3e^8}$ **41.** $\text{Jac}(\Phi) = e^{2r}$

43. $\iint_{\mathcal{R}} (x - y)^2 \sin^2(x + y)\, dx\, dy = \frac{\pi^4}{3}$

45. $\iint_D e^{3x-2y}\, dx\, dy = \frac{14}{39}(1 - e^{-3})(e^{13} - 1)$

Chapter 16

Section 16.1 Preliminary

1. (b) **2.** The vector $\langle x, x \rangle$ **3.** $\mathbf{F} = \langle 0, -z, y \rangle$
4. $\phi_1(x, y, z) = xyz + 1$

Section 16.1 Exercises

1. $\mathbf{F}(1, 2) = \langle 1, 1 \rangle, \mathbf{F}(-1, -1) = \langle 1, -1 \rangle$

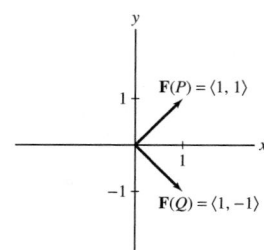

3. $\mathbf{F}(P) = \langle 0, 1, 0 \rangle, \mathbf{F}(Q) = \langle 2, 0, 2 \rangle$

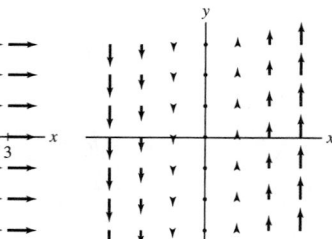

5. $\mathbf{F} = \langle 1, 0 \rangle$ **7.** $\mathbf{F}(x, y) = \langle 0, x \rangle$

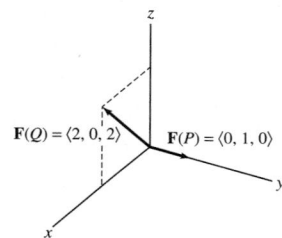

9. $\mathbf{F} = \langle 1, 1 \rangle$

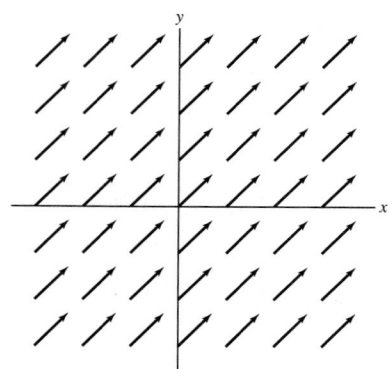

11. $\mathbf{F} = \left\langle \dfrac{x}{x^2+y^2}, \dfrac{y}{x^2+y^2} \right\rangle$

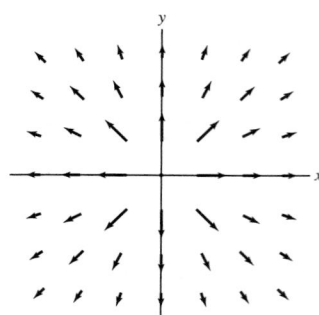

13. $\mathbf{F} = \left\langle \dfrac{y}{x^2+y^2}, \dfrac{-x}{x^2+y^2} \right\rangle$

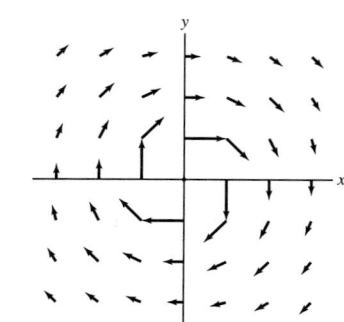

15. Plot (C) **17.** Plot (A) **19.** Plot (A) **21.** Plot (C)
23. $\varphi(x, y) = e^{xy}$ **25.** $\varphi(x, y, z) = ze^{x^2}$ **27.** Plot (B)
29. $\nabla \varphi = \dfrac{\mathbf{e}_r}{r}$

Section 16.2 Preliminary

1. $\int_C 10 \, ds = 10 \int_C 1 \, ds = 50$ **2.** (a), (c), (d), (e)
3. (a) True
(b) False. Reversing the orientation of the curve changes the sign of the vector line integral.
4. (a) $\int_C \mathbf{F} \cdot d\mathbf{s} = 0$ **(b)** $\int_C \mathbf{F} \cdot d\mathbf{s} = 5$

Section 16.2 Exercises

1. (a) $f(\mathbf{c}(t)) = 6t + 4t^2, ds = 2\sqrt{11} \, dt$
(b) $\int_C f(x, y, z) \, ds = \dfrac{26\sqrt{11}}{3}$
3. (a) $\mathbf{F}(\mathbf{c}(t)) = \langle t^{-2}, t^2 \rangle, d\mathbf{s} = \langle 1, -t^{-2} \rangle \, dt$
(b) $\int_C \mathbf{F} \cdot d\mathbf{s} = -\dfrac{1}{2}$
5. $\int_C (x^2 + y^2 + z^2) \, ds = \sqrt{2} \left(\pi + \dfrac{\pi^3}{3} \right)$
7. $\int_C \mathbf{F} \, ds = \dfrac{\pi^2}{2}$ **9.** $M = 64\pi$ g
11. $\int_{ABC} f(x, y, z) \, ds \approx 7.6, \quad \int_{ABC} \mathbf{F} \cdot d\mathbf{s} \approx 4$
13. (A) Zero (B) Negative (C) Zero

15. $\int_C f(x, y, z)\,ds = \frac{128\sqrt{29}}{3} \approx 229.8$

17. $\int_C f(x, y)\,ds = 2.8$

21. $\int_C f(x, y, z)\,ds = \frac{2}{3}\left((e^2 + 5)^{3/2} - 2^{3/2}\right)$

23. $\int_C \mathbf{F} \cdot d\mathbf{s} = \frac{16}{3}$ **25.** $\int_C \mathbf{F} \cdot d\mathbf{s} = 0$ **27.** $\int_C \mathbf{F} \cdot d\mathbf{s} = 11.5 - \ln 2$

29. $\int_C \mathbf{F} \cdot d\mathbf{s} = 2(e^2 - e^{-2}) - (e - e^{-1}) \approx 12.157$

31. $\int \mathbf{F} \cdot d\mathbf{s} = \frac{2\pi}{R^2}$

33. $\int_C \mathbf{F} \cdot d\mathbf{s} = \frac{\pi(\pi-8)}{16\sqrt{2}} + \ln(2 + \sqrt{2}) + 2 - \sqrt{2} \approx 1.139$

35. $\int_C \mathbf{F} \cdot d\mathbf{s} = 0$ **37.** $\int_C \langle e^{x-y}, e^{x+y} \rangle \cdot d\mathbf{s} \approx -4.5088$

39. $\int_C \mathbf{F} \cdot d\mathbf{s} = \frac{1}{3}e^6 + \frac{3}{2}e^4 + \frac{2}{3}e^2 - \frac{65}{6} + \frac{1}{3}e^{-4}$

41. (a) Positive   (b) $\int_C \mathbf{F} \cdot d\mathbf{s} = 2\pi$

43. $\int_A \mathbf{F} \cdot d\mathbf{s} = 0_0$

45. (a) $\int_C \mathbf{F} \cdot d\mathbf{s} = 2$ (b) $\int_C \mathbf{F} \cdot d\mathbf{s} = 25$ (c) $\int_C \mathbf{F} \cdot d\mathbf{s} = 32$

47. (a) $\int_{AB} \mathbf{F} \cdot d\mathbf{s} < \int_{DC} \mathbf{F} \cdot d\mathbf{s}$ (b) $\int_{ABC} \mathbf{F} \cdot d\mathbf{s} < \int_{ADC} \mathbf{F} \cdot d\mathbf{s}$

49. $W = \frac{9\pi^2}{2}$ **55.** $\mathrm{Av}(T) = 8°C$

Section 16.3 Preliminary

1. Closed

2. (a) Conservative vector fields   (b) All vector fields
(c) Conservative vector fields   (d) All vector fields
(e) Conservative vector fields   (f) All vector fields
(g) Conservative vector fields

3. (a) Always true
(b) True under additional hypotheses on $\mathcal{D}$
(c) True under additional hypotheses on $\mathcal{D}$

4. (a) $\int_D \mathbf{F} \cdot d\mathbf{s} = 4$ (b) $\int_E \mathbf{F} \cdot d\mathbf{s} = -4$

Section 16.3 Exercises

1. $\int_c \nabla\varphi \cdot d\mathbf{s} = 0$ **3.** $\int_c \mathbf{F} \cdot d\mathbf{s} = -\frac{9}{4}$ **5.** $\int_c \mathbf{F} \cdot d\mathbf{s} = 0$
7. $\int_c \mathbf{F} \cdot d\mathbf{s} = 32e - 1$

9. $\varphi(x, y, z) = x^2 y + 5x - 4zy$, $\int_c \mathbf{F} \cdot d\mathbf{s} = 20$

11. $\int_c \mathbf{F} \cdot d\mathbf{s} = 18$ **13.** $\mathbf{F}$ is not conservative.

15. $\varphi(x, y, z) = yx + \frac{z^4}{4}$ **17.** $\varphi(x, y, z) = x \cos z + y^2$

19. $W = \ln 4 - \ln 3$ **23.** $W = 19,600$ joules

25. (c) $\mathcal{D}$ is not simply connected.   (d) No

Section 16.4 Preliminary

1. $\iint_S f(x, y, z)\,dS = 50$

2. A distortion factor that indicates how much the area of R_{ij} is altered under the map ϕ.

3. Area$(S) \approx 0.0006$ **4.** $\iint_S f(x, y, z)\,dS \approx 0.6$

5. Area$(S) = 20$ **6.** $\left\langle \frac{2}{3}, \frac{2}{3}, \frac{1}{3} \right\rangle$

Section 16.4 Exercises

1. (a) v   (b) iii   (c) i   (d) iv   (e) ii

3. (a) $\mathbf{T}_u = \langle 2, 1, 3 \rangle$, $\mathbf{T}_v = \langle 0, -1, 1 \rangle$, $\mathbf{n}(u, v) = \langle 4, -2, -2 \rangle$

(b) Area$(S) = 4\sqrt{6}$ (c) $\iint_S f(x, y, z)\,dS = \frac{32\sqrt{6}}{3}$

5. (a) $\mathbf{T}_x = \langle 1, 0, y \rangle$, $\mathbf{T}_y = \langle 0, 1, x \rangle$, $\mathbf{n}(x, y) = \langle -y, -x, 1 \rangle$

(b) $\iint_S 1\,dS = \frac{(2\sqrt{2}-1)\pi}{6}$ (c) $\iint_S z\,dS = \frac{\sqrt{2}+1}{15}$

7. $\mathbf{T}_u = \langle 2, 1, 3 \rangle$, $\mathbf{T}_v = \langle 1, -4, 0 \rangle$, $\mathbf{n}(u, v) = 3\langle 4, 1, -3 \rangle$,
$4x + y - 3z = 0$

9. $\mathbf{T}_\theta = \langle -\sin\theta \sin\phi, \cos\theta \sin\phi, 0 \rangle$
$\mathbf{T}_\phi = \langle \cos\theta \cos\phi, \sin\theta \cos\phi, -\sin\phi \rangle$
$\mathbf{n}(\theta, \phi) = -(\cos\theta \sin^2\phi)\mathbf{i} - (\sin\theta \sin^2\phi)\mathbf{j} - (\sin\phi \cos\phi)\mathbf{k}$,
$y + z = \sqrt{2}$

11. Area$(S) \approx 0.2078$ **13.** Area$(S) = 16$

15. (a)

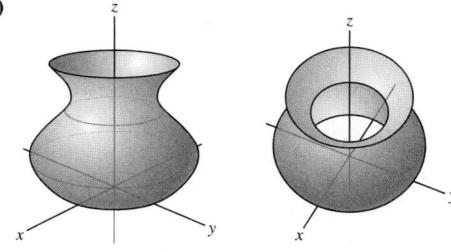

"Bottomless vase"

(b) $\mathbf{n}(u, v) =$
$\left((3 + \sin v)\cos u\right)\mathbf{i} + \left((3 + \sin v)\sin u\right)\mathbf{j} - \left((3 + \sin v)\cos v\right)\mathbf{k}$

(c) Area$(S) \approx 144.0181$

17. Area$(S) = 4\pi R^2$ **19.** $\iint_S x^2\,dS = \frac{\pi}{6}$

21. $\iint_S f(x, y, z)\,dS = \frac{\sqrt{2}}{5}$

23. $\iint_S f(x, y, z)\,dS = 4\pi(1 - e^{-4})$

25. $\iint_S f(x, y, z)\,dS = \frac{(3,179\sqrt{17}+29)\pi}{420} \approx 98.26$

27. $\iint_S f(x, y, z)\,dS = \frac{37\sqrt{37}-1}{4} \approx 56.02$

29. $\iint_S f(x, y, z)\,dS = \frac{\sqrt{3}}{6}$

31. $\iint_S f(x, y, z)\,dS = \frac{1}{54}(10\sqrt{10} - 1) \approx 0.567$

35. Area$(S) = \frac{3\sqrt{29}}{2}$ **37.** $\iint_S f(x, y, z)\,dS = \frac{7\pi}{3}$

39. (a) Area$(S) = 6\pi(\sqrt{8} - \sqrt{5}) \approx 11.166$
(b) $\iint_S z^{-1}\,dS = 3\pi \ln 1.6 \approx 4.43$

41. Area$(S) = \frac{\pi}{2}$ **45.** Area$(S) = \frac{\pi}{6}(17\sqrt{17} - 1) \approx 36.18$

49. $4\pi^2 ab$ **51.** $\varphi(r) = -\frac{Gm}{2Rr}\left(\sqrt{R^2 + r^2} - |R - r|\right)$

Section 16.5 Preliminary

1. (b)   **2.** (c)   **3.** (a)   **4.** (b)

5. (a) $\iint_S \mathbf{F} \cdot d\mathbf{S} = 0$ (b) $\iint_S \mathbf{F} \cdot d\mathbf{S} = \pi$ (c) $\iint_S \mathbf{F} \cdot d\mathbf{S} = \pi$

6. $\iint_S \mathbf{F} \cdot d\mathbf{S} \approx 0.05\sqrt{2} \approx 0.0707$ **7.** $\iint_S \mathbf{F} \cdot d\mathbf{S} = 0$

Section 16.5 Exercises

1. (a) $\mathbf{n} = \langle 2v, -4uv, 2u+1 \rangle$
$\mathbf{F} \cdot \mathbf{n} = 2u^3 - 4uv^3 + 2v^2 + u^2 - v$
(b) $\mathbf{F}(P) \cdot \mathbf{e_n}(P) = \frac{13}{\sqrt{93}}$ **(c)** $\iint_S \mathbf{F} \cdot d\mathbf{S} = 24$

3. $\iint_S \mathbf{F} \cdot d\mathbf{S} = 4$ **5.** $\iint_S \mathbf{F} \cdot d\mathbf{S} = -4$ **7.** $\iint_S \mathbf{F} \cdot d\mathbf{S} = \frac{9\pi}{4}$

9. $\iint_S \mathbf{F} \cdot d\mathbf{S} = 0$ **11.** $\iint_S \mathbf{F} \cdot d\mathbf{S} = -\frac{11}{12}$

13. $\iint_S \mathbf{F} \cdot d\mathbf{S} = \frac{9\pi}{4}$ **15.** $\iint_S \mathbf{F} \cdot d\mathbf{S} = (e-1)^2$

17. $\iint_S \mathbf{F} \cdot d\mathbf{S} = 270$

19. (a) $\iint_S \mathbf{F} \cdot d\mathbf{S} = 18\pi e^{-3}$ **(b)** $\iint_S \mathbf{F} \cdot d\mathbf{S} \approx 1.46456$

21. $\iint_D \mathbf{E} \cdot d\mathbf{S} = (2 - \frac{6}{\sqrt{13}})\pi k$ **25.** $\iint_S \mathbf{v} \cdot d\mathbf{S} = 4.8 \text{ ft}^3/\text{s}$

27. $\iint_S \mathbf{v} \cdot d\mathbf{S} = -\frac{1}{2} \text{ m}^3/\text{s}$

29. $\Phi(t) = (-9.65 \times 10^{-7}) \cdot t(12-t) \text{ T/m}^2$
$\int_C \mathbf{E} \cdot d\mathbf{S} = 1.93 \times 10^{-6} \cdot (6-t) \text{ V}$

Chapter 16 Review

1. (a) $\mathbf{F} = \langle -15, 8 \rangle$ **(b)** $\mathbf{F} = \langle 4, 8 \rangle$ **(c)** $\mathbf{F} = \langle 9, 1 \rangle$

3. $\mathbf{F} = \langle y, 1 \rangle$ **5.** $\mathbf{F}(x, y) = \langle 2x, -1 \rangle$

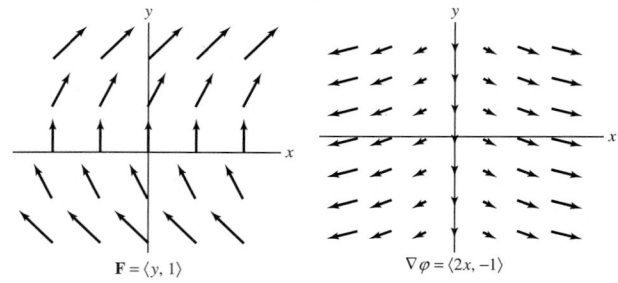

$\mathbf{F} = \langle y, 1 \rangle$ $\nabla\varphi = \langle 2x, -1 \rangle$

7. $\varphi(x, y, z) = -\cos x + e^y + \frac{z^2}{2}$ **9.** $\mathbf{F}$ is not conservative.

11. $\varphi(x, y, z) = y \tan^{-1}(x) + z^2$ **13.** $\mathbf{F}$ is not conservative.

15. $\int_c \nabla\varphi \cdot d\mathbf{s} = 46,648$ **17.** $\int_C f(x, y) \, ds = \frac{\sqrt{5}}{6}$

19. $\int_C f(x, y) \, ds = \frac{11}{6}$ **21.** $M = 13\frac{1}{3}$

23. $\int_{C_1} \mathbf{F} \cdot d\mathbf{s} = -\frac{243\pi}{16} + 20.25 \approx -27.463$

25. $\int_C \mathbf{F} \cdot d\mathbf{s} = -\frac{\pi}{2}$ **27.** $\int_C \mathbf{F} \cdot d\mathbf{s} = -\frac{13}{18}$

29. (B) and (C) Zero, (D) Negative

31. $a = \frac{4}{3}, b = \frac{5}{4}, c = -5, \mathbf{T}_u = \langle 1, \frac{5}{4}, 2 \rangle; \mathbf{T}_v = \langle \frac{4}{3}, 1, 0 \rangle,$
$\mathbf{n} = \langle -2, \frac{8}{3}, -\frac{2}{3} \rangle$

33. (a) $\mathbf{T}_u \left(1, \frac{\pi}{3}\right) = \langle 1, \sqrt{3}, 0 \rangle, \mathbf{T}_v \left(1, \frac{\pi}{3}\right) = \langle \frac{\sqrt{3}}{2}, -\frac{1}{2}, 3 \rangle,$
$\mathbf{n}\left(1, \frac{\pi}{3}\right) = \langle 3\sqrt{3}, -3, -2 \rangle$

(b) $3\sqrt{3}x - 3y - 2z + 2\pi = 0$ **(c)** Area$(S) \approx 38.4$

35. Area$(S) = \int_{-3}^{3} \int_{-1}^{1} \sqrt{1 + 4x^2 + 4y^2} \, dx \, dy \approx 41.8525$

37. $\iint_S (x^2 + y^2)e^{-z} dS = 54\pi(-e^{-10} + 1) \approx 54\pi$

39. Area$(S) = 0.02\sqrt{6} \approx 0.049$ **41.** $\iint_S \mathbf{F} \cdot d\mathbf{S} = \frac{243\pi}{2} \approx 381.7$

43. $\iint_S \mathbf{F} \cdot d\mathbf{S} = 8\pi$ **45.** $\iint_S \mathbf{F} \cdot d\mathbf{S} = 3 - e$

47. $\iint_S \rho \cdot d\mathbf{S} = \frac{\pi}{3} K H^3 R$ **49.** $\iint_S \mathbf{v} \cdot d\mathbf{S} = 6\pi$

Chapter 17

Section 17.1 Preliminary

1. $\mathbf{F} = \langle -e^y, x^2 \rangle$

2.

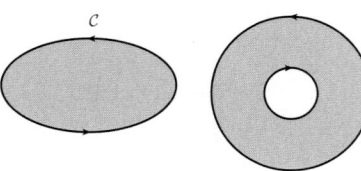

3. Yes **4.** (a), (c)

Section 17.1 Exercises

3. $\int_C y^2 \, dx + x^2 \, dy = 0$ **5.** $\int_C x^2 y \, dx = -\frac{\pi}{4}$ **7.** $\int_C \mathbf{F} \cdot d\mathbf{s} = \frac{1}{6}$

9. $\int_C x^3 \, dx + 4x \, dy = -4$ **11.** $\int_C xy \, dx + (x^2 + x) \, dy = 1$

13. $A = 9\pi$ **15.** $A = 3\pi$

17. (c) $A = \frac{3}{2}$

21. (a) $\varphi(x, y) = x^2 e^y + C$ **(c)** $I = 4\pi - 16$

23. $I = 34$

25. $\int_C \mathbf{F} \cdot d\mathbf{s} \approx -0.024$, CW direction, $\omega = 1.5$ radians per second

27. $\int_{C_1} \mathbf{F} \cdot d\mathbf{s} = 214\pi$ **29.** $A = 12$

Section 17.2 Preliminary

1.

 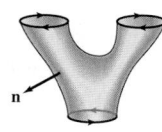

(A)   (B)

2. (a)

3. A vector field $\mathbf{A}$ such that $\mathbf{F} = \text{curl}(\mathbf{A})$ is a vector potential for $\mathbf{F}$.

4. (b) **5.** $\iint_{S_1} \mathbf{F} \cdot d\mathbf{S} = \iint_{S_2} \mathbf{F} \cdot d\mathbf{S}$ for $\mathbf{F} = \text{curl}(\mathbf{A})$

Section 17.2 Exercises

1. $\text{curl}(\mathbf{F}) = \langle 1 - 3z^2, 1 - 2x, 1 + 2y \rangle$

3. $\text{curl}(\mathbf{F}) = \langle 0, \sin x, \cos x - e^y \rangle$

5. $\int_C \mathbf{F} \cdot d\mathbf{s} = \iint_S \text{curl}(\mathbf{F}) \cdot d\mathbf{S} = \pi$

7. $e^{-1} - 1$ **9.** $\iint_S \text{curl}(\mathbf{F}) \cdot d\mathbf{S} = 0$

11. (a)   **(b)** 140π

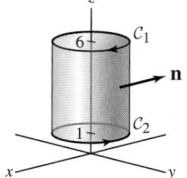

13. (a) $\mathbf{A} = \langle 0, 0, e^y - e^{x^2} \rangle$ **(c)** $\iint_S \mathbf{F} \cdot d\mathbf{S} = \frac{\pi}{2}$

15. (a) $\iint_S \mathbf{B} \cdot d\mathbf{S} = r^2 B \pi$ **(b)** $\int_{\partial S} \mathbf{A} \cdot d\mathbf{s} = 0$

17. (b) $\iint_S \mathbf{B} \cdot d\mathbf{S} = 18b$

19. $\int_C \mathbf{F} \cdot d\mathbf{s} = 32$ **21.** $a = 2c$, b any positive number.

23. $\iint_S \mathbf{F} \cdot d\mathbf{S} = 2\pi$

Section 17.3 Preliminary

1. $\iint_S \mathbf{F} \cdot d\mathbf{S} = 0$

2. $\iint_S \mathbf{F} \cdot d\mathbf{S} = \iiint_W (3x^2 + 3y^2 + 3z^2)\, dV$ positive for all $(x, y, z) \neq (0, 0, 0)$.

3. a, b, d, f meaningful; b, d automatically zero vector and zero scalar, respectively.

4. (c)

5. $\text{div}(\mathbf{F}) = 2x + 1 - 2x = 1$ and flux $= \int \text{div}(\mathbf{F})\, dV =$ volume

Section 17.3 Exercises

1. $\text{div}(\mathbf{F}) = y + z$ **3.** $\text{div}(\mathbf{F}) = 1 - 4zx - x + 2zx^2$ **5.** 0

7. 4π **9.** $\iint_S \mathbf{F} \cdot d\mathbf{S} = 4\pi$ **11.** $\iint_S \mathbf{F} \cdot d\mathbf{S} = \frac{256\pi}{5}$

13. $\iint_S \mathbf{F} \cdot d\mathbf{S} = 64\pi$ **15.** $\iint_S \mathbf{F} \cdot d\mathbf{S} = 81\pi$

17. $\iint_S \mathbf{F} \cdot d\mathbf{S} = 70.22$ **21.** Volume$(\mathcal{W}) = \frac{4\pi abc}{3}$

23. $\text{div}(f\mathbf{F}) = f\,\text{div}(\mathbf{F}) + \mathbf{F} \cdot \nabla f$

29. (b) $\iint_S \mathbf{E} \cdot d\mathbf{S} = 0$ **(c)** $\text{div}(\mathbf{E}) = 0$ **(d)** No

35. $\mathbf{A} = \left\langle z^3 - xy^2, 0, y^2 - y \right\rangle$

Chapter 17 Review

1. $\int_C \mathbf{F} \cdot d\mathbf{s} = 0$ **3.** $\int_C xy^3\, dx + x^3 y\, dy = -30$

5. $\int_C y^2\, dx - x^2\, dy = \frac{3}{5}$

7. (a)

(b) $A = \frac{1}{60}$

9. $\text{curl}(\mathbf{F}) = -\mathbf{k}$, $\text{div}(\mathbf{F}) = -1$

11. $\text{curl}(\mathbf{F}) = \mathbf{0}$, $\text{div}(\mathbf{F}) = 2e^{-x^2-y^2-z^2}\left(2(x^2 + y^2 + z^2) - 3\right)$

13. $\text{curl}(\mathbf{F}) = \langle -2z, 0, -2y \rangle$ **19.** $\iint_S \text{curl}(\mathbf{F}) \cdot d\mathbf{S} = 2\pi$

21. $\int_C \langle y, z, x \rangle\, d\mathbf{s} = \frac{\pi}{4}$ **23.** $\iint_S \left\langle xy, yz, x^2 z + z^2 \right\rangle \cdot d\mathbf{S} = 142$

25. $\iint_S \mathbf{F} \cdot d\mathbf{S} = -128\pi$ **27.** Volume$(\mathcal{W}) = 2$

29. $\int_C \mathbf{F} \cdot d\mathbf{s} \approx 0.0113$ **31.** $2x - y + 4z = 0$

33. (b) $\iint_S \mathbf{F} \cdot d\mathbf{S} = \iiint_W \text{div}(\mathbf{F})\, dV = \frac{\pi}{2}$

35. (c) $\int_C \mathbf{F} \cdot d\mathbf{s} = 0$ **(d)** $\int_{C_1} \mathbf{F} \cdot d\mathbf{s} = -4$, $\int_{C_2} \mathbf{F} \cdot d\mathbf{s} = 4$

REFERENCES

The online source MacTutor History of Mathematics Archive (www-history.mcs. st-and.ac.uk) has been a valuable source of historical information.

Section 1.1

1. (EX 73) Adapted from *Calculus Problems for a New Century*, Robert Fraga, ed., Mathematical Association of America, Washington, DC, 1993, p. 9.

Section 1.2

1. (EX 25) Adapted from *Calculus Problems for a New Century*, Robert Fraga, ed., Mathematical Association of America, Washington, DC, 1993, p. 9.

Section 1.7

2. (EXMP 4) Adapted from B. Waits and F. Demana, "The Calculator and Computer Pre-Calculus Project," in *The Impact of Calculators on Mathematics Instruction*, University of Houston, 1994.

2. (EX 12) Adapted from B. Waits and F. Demana, "The Calculator and Computer Pre-Calculus Project," in *The Impact of Calculators on Mathematics Instruction*, University of Houston, 1994.

Section 2.1

3. (CO2) From Z. Toroczkai, G. Karolyi, Á. Péntek, T. Tè, C. Grebogi, and J. A. Yorke, "Wada Dye Boundaries in Open Hydrodynamical Flows," *Physica A* 239:235 (1997).

Section 2.2

4. (EX 61) Adapted from *Calculus Problems for a New Century*, Robert Fraga, ed., Mathematical Association of America, Washington, DC, 1993, Note 28.

Section 2.3

4. (EX 38) Adapted from *Calculus Problems for a New Century*, Robert Fraga, ed., Mathematical Association of America, Washington, DC, 1993, Note 28.

Chapter 2 Review

4. (EX 54) Adapted from *Calculus Problems for a New Century*, Robert Fraga, ed., Mathematical Association of America, Washington, DC, 1993, Note 28.

Section 3.1

5. (EX 42) Problem suggested by Dennis DeTurck, University of Pennsylvania.

5. (EX 75) Problem suggested by Dennis DeTurck, University of Pennsylvania.

Section 3.2

6. (EX 94) Problem suggested by Chris Bishop, SUNY Stony Brook.

6. (EX 95) Problem suggested by Chris Bishop, SUNY Stony Brook.

Section 3.4

7. (PQ 2) From *Calculus Problems for a New Century*, Robert Fraga, ed., Mathematical Association of America, Washington, DC, 1993, p. 25.

8. (EX 49) Karl J. Niklas and Brian J. Enquist, "Invariant Scaling Relationships for Interspecific Plant Biomass Production Rates and Body Size," *Proc. Natl. Acad. Sci.* 98, no. 5:2922–2927 (February 27, 2001).

Section 3.5

9. (EX 46) Adapted from a contribution by Jo Hoffacker, University of Georgia.

10. (EX 47–48) Adapted from a contribution by Thomas M. Smith, University of Illinois at Chicago, and Cindy S. Smith, Plainfield High School.

11. (EX 49) Adapted from Walter Meyer, *Falling Raindrops*, in *Applications of Calculus*, P. Straffin, ed., Mathematical Association of America, Washington, DC, 1993.

6. (EX 52, 56) Problem suggested by Chris Bishop, SUNY Stony Brook.

Section 3.6

12. (EX 56) Adapted from J. M. Gelfand and M. Saul, *Trigonometry*, Birkhäuser, Boston, 2001.

Section 3.11

13. (EX 34) Adapted from *Calculus Problems for a New Century*, Robert Fraga, ed., Mathematical Association of America, Washington, DC, 1993.

14. (EX 38) Problem suggested by Kay Dundas.

13. (EX 43, 45) Adapted from *Calculus Problems for a New Century*, Robert Fraga, ed., Mathematical Association of America, Washington, DC, 1993.

Chapter 3 Review

6. (EX 89–90, 111, 114) Problem suggested by Chris Bishop, SUNY Stony Brook.

Section 4.2

15. (MN, p. 225) From Pierre Fermat, *On Maxima and Minima and on Tangents*, translated by D. J. Struik (ed.), *A Source Book in Mathematics, 1200–1800*, Princeton University Press, Princeton, N.J., 1986.

Section 4.5

16. (EX 49, 98) Adapted from *Calculus Problems for a New Century*, Robert Fraga, ed., Mathematical Association of America, Washington, DC, 1993, p. 63.

Section 4.6

13. (EX 29–30) Adapted from *Calculus Problems for a New Century*, Robert Fraga, ed., Mathematical Association of America, Washington, DC, 1993.

17. (EX 32) Problem suggested by John Haverhals, Bradley University. *Source:* Illinois Agrinews.

13. (EX 35–36) Adapted from *Calculus Problems for a New Century*, Robert Fraga, ed., Mathematical Association of America, Washington, DC, 1993.

18. (EX 61) From Michael Helfgott, *Thomas Simpson and Maxima and Minima, Convergence Magazine*, published online by the Mathematical Association of America.

19. (EX 65–67) Adapted from B. Noble, *Applications of Undergraduate Mathematics in Engineering*, Macmillan, New York, 1967.

20. (EX 69) Adapted from Roger Johnson, "A Problem in Maxima and Minima," *American Mathematical Monthly*, 35:187–188 (1928).

Section 4.7

31. (EX 64) Adapted from Robert J. Bumcrot, "Some Subtleties in L'Hôpital's Rule," in *A Century of Calculus*, Part II, Mathematical Association of America, Washington, DC, 1992.

Section 4.8

21. (EX 19) Adapted from *Calculus Problems for a New Century*, Robert Fraga, ed., Mathematical Association of America, Washington, DC, 1993, p. 52.

22. (EX 30–31) Adapted from E. Packel and S. Wagon, *Animating Calculus*, Springer-Verlag, New York, 1997, p. 79.

Chapter 4 Review

13. (EX 76) Adapted from *Calculus Problems for a New Century*, Robert Fraga, ed., Mathematical Association of America, Washington, DC, 1993.

Section 5.1

23. (EX 3) Problem suggested by John Polhill, Bloomsburg University.

Section 5.2

5. (EX 79) Problem suggested by Dennis DeTurck, University of Pennsylvania.

6. (EX 84) Problem suggested by Chris Bishop, SUNY Stony Brook.

Section 5.4

24. (PQ 6) Adapted from *Calculus Problems for a New Century*, Robert Fraga, ed., Mathematical Association of America, Washington, DC, 1993, p. 103.

25. (EX 48–49) Adapted from *Calculus Problems for a New Century*, Robert Fraga, ed., Mathematical Association of America, Washington, DC, 1993, p. 102.

5. (EX 53) Problem suggested by Dennis DeTurck, University of Pennsylvania.

Section 5.5

26. (EX 24–25) M. Newman and G. Eble, "Decline in Extinction Rates and Scale Invariance in the Fossil Record," *Paleobiology* 25:434–439 (1999).

27. (EX 27) From H. Flanders, R. Korfhage, and J. Price, *Calculus*, Academic Press, New York, 1970.

Section 5.6

28. (EX 73) Adapted from *Calculus Problems for a New Century*, Robert Fraga, ed., Mathematical Association of America, Washington, DC, 1993, p. 121.

Section 6.1

29. (EX 49) Adapted from Tom Farmer and Fred Gass, "Miami University: An Alternative Calculus," in *Priming the Calculus Pump*, Thomas Tucker, ed., Mathematical Association of America, Washington, DC, 1990, Note 17.

13. (EX 59) Adapted from *Calculus Problems for a New Century*, Robert Fraga, ed., Mathematical Association of America, Washington, DC, 1993.

Section 6.3

30. (EX 52, 56) Adapted from G. Alexanderson and L. Klosinski, "Some Surprising Volumes of Revolution," *Two-Year College Mathematics Journal* 6, 3:13–15 (1975).

Section 7.1

33. See R. Courant and F. John, *Introduction to Calculus and Analysis*, Vol. 1, Springer-Verlag, New York, 1989.

Section 7.2

34. (EX 55–62) Problems suggested by Brian Bradie, Christopher Newport University.

13. (EX 66) Adapted from *Calculus Problems for a New Century*, Robert Fraga, ed., Mathematical Association of America, Washington, DC, 1993.

35. (EX 74) Adapted from J. L. Borman, "A Remark on Integration by Parts," *American Mathematical Monthly* 51:32–33 (1944).

Section 7.4

34. (EX 43–52) Problems suggested by Brian Bradie, Christopher Newport University.

36. (EX 59) Adapted from *Calculus Problems for a New Century*, Robert Fraga, ed., Mathematical Association of America, Washington, DC, 1993, p. 118.

Section 7.7

6. (EX 79) Problem suggested by Chris Bishop, SUNY Stony Brook.

Section 8.1

37. (EX 48) Adapted from G. Klambauer, *Aspects of Calculus*, Springer-Verlag, New York, 1986, Ch. 6.

Section 9.1

38. (EX 51) Adapted from E. Batschelet, *Introduction to Mathematics for Life Scientists*, Springer-Verlag, New York, 1979.

13. (EX 53) Adapted from *Calculus Problems for a New Century*, Robert Fraga, ed., Mathematical Association of America, Washington, DC, 1993.

39. (EX 54–55) Adapted from M. Tenenbaum and H. Pollard, *Ordinary Differential Equations*, Dover, New York, 1985.

Section 10.1

40. (EX 77) Adapted from G. Klambauer, *Aspects of Calculus*, Springer-Verlag, New York, 1986, p. 393.

Section 10.2

41. (EX 35) Adapted from *Calculus Problems for a New Century*, Robert Fraga, ed., Mathematical Association of America, Washington, DC, 1993, p. 137.

42. (EX 36) Adapted from *Calculus Problems for a New Century*, Robert Fraga, ed., Mathematical Association of America, Washington, DC, 1993, p. 138.

43. (EX 49) Adapted from George Andrews, "The Geometric Series in Calculus," *American Mathematical Monthly* 105, 1:36–40 (1998).

44. (EX 51) Adapted from Larry E. Knop, "Cantor's Disappearing Table," *The College Mathematics Journal* 16, 5:398–399 (1985).

Section 10.3

45. (EX 83) Adapted from *Calculus Problems for a New Century*, Robert Fraga, ed., Mathematical Association of America, Washington, DC, 1993, p. 141.

Section 10.4

46. (EX 27) Adapted from *Calculus Problems for a New Century*, Robert Fraga, ed., Mathematical Association of America, Washington, DC, 1993, p. 145.

Section 11.2

47. (EX 36) Adapted from Richard Courant and Fritz John, *Differential and Integral Calculus*, Wiley-Interscience, New York, 1965.

Section 11.3

13. (EX 56) Adapted from *Calculus Problems for a New Century*, Robert Fraga, ed., Mathematical Association of America, Washington, DC, 1993.

Section 12.4

48. (EX 66) Adapted from Ethan Berkove and Rich Marchand, "The Long Arm of Calculus," *The College Mathematics Journal* 29, 5:376–386 (November 1998).

Section 13.3

13. (EX 23) Adapted from *Calculus Problems for a New Century*, Robert Fraga, ed., Mathematical Association of America, Washington, DC, 1993.

Section 13.4

49. (EX 58) Damien Gatinel, Thanh Hoang-Xuan, and Dimitri T. Azar, "Determination of Corneal Asphericity After Myopia Surgery with the Excimer Laser: A Mathematical Model," *Investigative Opthalmology and Visual Science* 42: 1736–1742 (2001).

Section 13.5

50. (EX 50, 53) Adapted from notes to the course "Dynamics and Vibrations" at Brown University (see http://www.engin.brown.edu/courses/en4/).

Section 14.8

51. (EX 33) Adapted from C. Henry Edwards, "Ladders, Moats, and Lagrange Multipliers," *Mathematica Journal* 4, Issue 1 (Winter 1994).

Section 15.3

52. (FIG. 11 COMPUTATION) The computation is based on Jeffrey Nunemacher, "The Largest Unit Ball in Any Euclidean Space," in *A Century of Calculus*, Part II, Mathematical Association of America, Washington, DC, 1992.

Section 15.4

53. (EX 68) Adapted from A. E. Richmond, *Calculus for Electronics*, McGraw-Hill, New York, 1958.

Section 15.5

33. (CONCEPTUAL INSIGHT) See R. Courant and F. John, *Introduction to Calculus and Analysis*, Springer-Verlag, New York, 1989, p. 534.

Section 16.2

54. Figure 8 was inspired by Tevian Dray and Corinne A. Manogue, "The Murder Mystery Method for Determining Whether a Vector Field Is Conservative," *The College Mathematics Journal*, May 2003.

Section 16.3

13. (EX 20) Adapted from *Calculus Problems for a New Century*, Robert Fraga, ed., Mathematical Association of America, Washington, DC, 1993.

Appendix D

33. [Proof of Theorem 7] A proof without this simplifying assumption can be found in R. Courant and F. John, *Introduction to Calculus and Analysis*, Vol. 1, Springer-Verlag, New York, 1989.

PHOTO CREDITS

INDEX

ELEMENTARY FUNCTIONS

Power Functions $f(x) = x^a$

$f(x) = x^n$, n a positive integer

n even

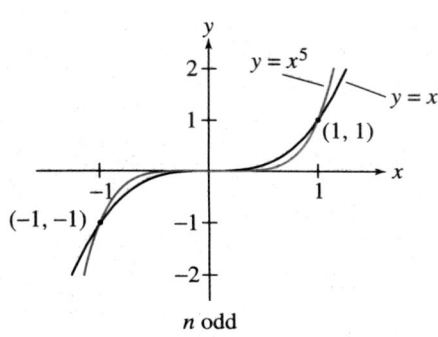

n odd

Asymptotic behavior of a polynomial function of even degree and positive leading coefficient

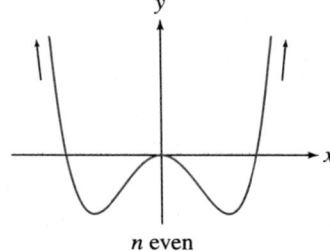

n even

Asymptotic behavior of a polynomial function of odd degree and positive leading coefficient

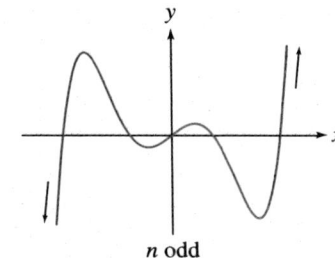

n odd

$f(x) = x^{-n} = \dfrac{1}{x^n}$

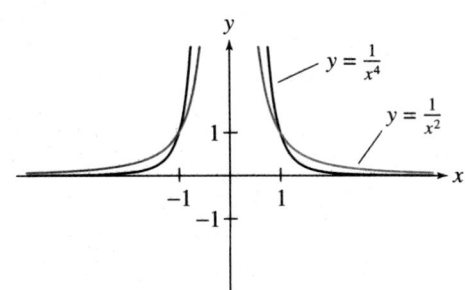

Inverse Trigonometric Functions

$\arcsin x = \sin^{-1} x = \theta$

$\Leftrightarrow \quad \sin\theta = x, \quad -\dfrac{\pi}{2} \le \theta \le \dfrac{\pi}{2}$

$\arccos x = \cos^{-1} x = \theta$

$\Leftrightarrow \quad \cos\theta = x, \quad 0 \le \theta \le \pi$

$\arctan x = \tan^{-1} x = \theta$

$\Leftrightarrow \quad \tan\theta = x, \quad -\dfrac{\pi}{2} < \theta < \dfrac{\pi}{2}$

Exponential and Logarithmic Functions

$$\boxed{\log_a x = y \quad \Leftrightarrow \quad a^y = x}$$

$$\log_a(a^x) = x \qquad a^{\log_a x} = x$$

$$\log_a 1 = 0 \qquad \log_a a = 1$$

$$\boxed{\ln x = y \quad \Leftrightarrow \quad e^y = x}$$

$$\ln(e^x) = x \qquad e^{\ln x} = x$$

$$\ln 1 = 0 \qquad \ln e = 1$$

$$\log_a(xy) = \log_a x + \log_a y$$

$$\log_a\left(\frac{x}{y}\right) = \log_a x - \log_a y$$

$$\log_a(x^r) = r \log_a x$$

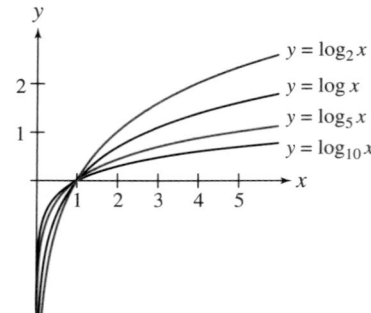

$$\lim_{x\to\infty} a^x = \infty, \quad a > 1$$

$$\lim_{x\to\infty} a^x = 0, \quad 0 < a < 1$$

$$\lim_{x\to-\infty} a^x = 0, \quad a > 1$$

$$\lim_{x\to-\infty} a^x = \infty, \quad 0 < a < 1$$

$$\lim_{x\to 0^+} \log_a x = -\infty$$

$$\lim_{x\to\infty} \log_a x = \infty$$

Hyperbolic Functions

$$\sinh x = \frac{e^x - e^{-x}}{2} \qquad \operatorname{csch} x = \frac{1}{\sinh x}$$

$$\cosh x = \frac{e^x + e^{-x}}{2} \qquad \operatorname{sech} x = \frac{1}{\cosh x}$$

$$\tanh x = \frac{\sinh x}{\cosh x} \qquad \coth x = \frac{\cosh x}{\sinh x}$$

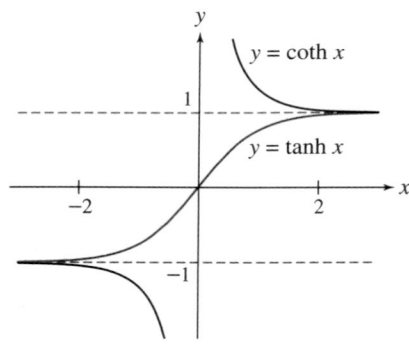

$$\sinh(x + y) = \sinh x \cosh y + \cosh x \sinh y$$

$$\cosh(x + y) = \cosh x \cosh y + \sinh x \sinh y$$

$$\sinh 2x = 2 \sinh x \cosh x$$

$$\cosh 2x = \cosh^2 x + \sinh^2 x$$

Inverse Hyperbolic Functions

$$y = \sinh^{-1} x \quad \Leftrightarrow \quad \sinh y = x$$

$$y = \cosh^{-1} x \quad \Leftrightarrow \quad \cosh y = x \text{ and } y \geq 0$$

$$y = \tanh^{-1} x \quad \Leftrightarrow \quad \tanh y = x$$

$$\sinh^{-1} x = \ln\left(x + \sqrt{x^2 + 1}\right)$$

$$\cosh^{-1} x = \ln\left(x + \sqrt{x^2 - 1}\right) \quad x > 1$$

$$\tanh^{-1} x = \frac{1}{2} \ln\left(\frac{1+x}{1-x}\right) \quad -1 < x < 1$$

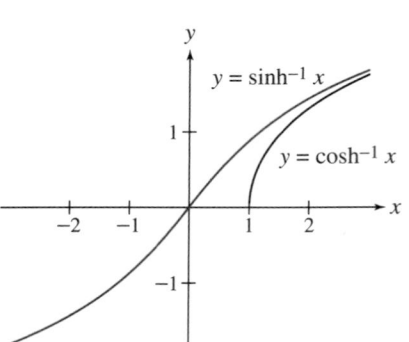

DIFFERENTIATION

Differentiation Rules

1. $\dfrac{d}{dx}(c) = 0$

2. $\dfrac{d}{dx}x = 1$

3. $\dfrac{d}{dx}(x^n) = nx^{n-1}$ (Power Rule)

4. $\dfrac{d}{dx}[cf(x)] = cf'(x)$

5. $\dfrac{d}{dx}[f(x) + g(x)] = f'(x) + g'(x)$

6. $\dfrac{d}{dx}[f(x)g(x)] = f(x)g'(x) + g(x)f'(x)$ (Product Rule)

7. $\dfrac{d}{dx}\left[\dfrac{f(x)}{g(x)}\right] = \dfrac{g(x)f'(x) - f(x)g'(x)}{[g(x)]^2}$ (Quotient Rule)

8. $\dfrac{d}{dx}f(g(x)) = f'(g(x))g'(x)$ (Chain Rule)

9. $\dfrac{d}{dx}f(x)^n = nf(x)^{n-1}f'(x)$ (General Power Rule)

10. $\dfrac{d}{dx}f(kx + b) = kf'(kx + b)$

11. $g'(x) = \dfrac{1}{f'(g(x))}$ where $g(x)$ is the inverse $f^{-1}(x)$

12. $\dfrac{d}{dx}\ln f(x) = \dfrac{f'(x)}{f(x)}$

Trigonometric Functions

13. $\dfrac{d}{dx}\sin x = \cos x$

14. $\dfrac{d}{dx}\cos x = -\sin x$

15. $\dfrac{d}{dx}\tan x = \sec^2 x$

16. $\dfrac{d}{dx}\csc x = -\csc x \cot x$

17. $\dfrac{d}{dx}\sec x = \sec x \tan x$

18. $\dfrac{d}{dx}\cot x = -\csc^2 x$

Inverse Trigonometric Functions

19. $\dfrac{d}{dx}(\sin^{-1} x) = \dfrac{1}{\sqrt{1 - x^2}}$

20. $\dfrac{d}{dx}(\cos^{-1} x) = -\dfrac{1}{\sqrt{1 - x^2}}$

21. $\dfrac{d}{dx}(\tan^{-1} x) = \dfrac{1}{1 + x^2}$

22. $\dfrac{d}{dx}(\csc^{-1} x) = -\dfrac{1}{x\sqrt{x^2 - 1}}$

23. $\dfrac{d}{dx}(\sec^{-1} x) = \dfrac{1}{x\sqrt{x^2 - 1}}$

24. $\dfrac{d}{dx}(\cot^{-1} x) = -\dfrac{1}{1 + x^2}$

Exponential and Logarithmic Functions

25. $\dfrac{d}{dx}(e^x) = e^x$

26. $\dfrac{d}{dx}(a^x) = (\ln a)a^x$

27. $\dfrac{d}{dx}\ln |x| = \dfrac{1}{x}$

28. $\dfrac{d}{dx}(\log_a x) = \dfrac{1}{(\ln a)x}$

Hyperbolic Functions

29. $\dfrac{d}{dx}(\sinh x) = \cosh x$

30. $\dfrac{d}{dx}(\cosh x) = \sinh x$

31. $\dfrac{d}{dx}(\tanh x) = \text{sech}^2 x$

32. $\dfrac{d}{dx}(\text{csch}\, x) = -\text{csch}\, x \coth x$

33. $\dfrac{d}{dx}(\text{sech}\, x) = -\text{sech}\, x \tanh x$

34. $\dfrac{d}{dx}(\coth x) = -\text{csch}^2 x$

Inverse Hyperbolic Functions

35. $\dfrac{d}{dx}(\sinh^{-1} x) = \dfrac{1}{\sqrt{1 + x^2}}$

36. $\dfrac{d}{dx}(\cosh^{-1} x) = \dfrac{1}{\sqrt{x^2 - 1}}$

37. $\dfrac{d}{dx}(\tanh^{-1} x) = \dfrac{1}{1 - x^2}$

38. $\dfrac{d}{dx}(\text{csch}^{-1} x) = -\dfrac{1}{|x|\sqrt{x^2 + 1}}$

39. $\dfrac{d}{dx}(\text{sech}^{-1} x) = -\dfrac{1}{x\sqrt{1 - x^2}}$

40. $\dfrac{d}{dx}(\coth^{-1} x) = \dfrac{1}{1 - x^2}$

INTEGRATION

Substitution

If an integrand has the form $f(u(x))u'(x)$, then rewrite the entire integral in terms of u and its differential $du = u'(x)\,dx$:

$$\int f(u(x))u'(x)\,dx = \int f(u)\,du$$

Integration by Parts Formula

$$\int u(x)v'(x)\,dx = u(x)v(x) - \int u'(x)v(x)\,dx$$

TABLE OF INTEGRALS

Basic Forms

1. $\displaystyle\int u^n\,du = \frac{u^{n+1}}{n+1} + C, \quad n \neq -1$

2. $\displaystyle\int \frac{du}{u} = \ln|u| + C$

3. $\displaystyle\int e^u\,du = e^u + C$

4. $\displaystyle\int a^u\,du = \frac{a^u}{\ln a} + C$

5. $\displaystyle\int \sin u\,du = -\cos u + C$

6. $\displaystyle\int \cos u\,du = \sin u + C$

7. $\displaystyle\int \sec^2 u\,du = \tan u + C$

8. $\displaystyle\int \csc^2 u\,du = -\cot u + C$

9. $\displaystyle\int \sec u \tan u\,du = \sec u + C$

10. $\displaystyle\int \csc u \cot u\,du = -\csc u + C$

11. $\displaystyle\int \tan u\,du = \ln|\sec u| + C$

12. $\displaystyle\int \cot u\,du = \ln|\sin u| + C$

13. $\displaystyle\int \sec u\,du = \ln|\sec u + \tan u| + C$

14. $\displaystyle\int \csc u\,du = \ln|\csc u - \cot u| + C$

15. $\displaystyle\int \frac{du}{\sqrt{a^2 - u^2}} = \sin^{-1}\frac{u}{a} + C$

16. $\displaystyle\int \frac{du}{a^2 + u^2} = \frac{1}{a}\tan^{-1}\frac{u}{a} + C$

Exponential and Logarithmic Forms

17. $\displaystyle\int ue^{au}\,du = \frac{1}{a^2}(au - 1)e^{au} + C$

18. $\displaystyle\int u^n e^{au}\,du = \frac{1}{a}u^n e^{au} - \frac{n}{a}\int u^{n-1}e^{au}\,du$

19. $\displaystyle\int e^{au}\sin bu\,du = \frac{e^{au}}{a^2 + b^2}(a\sin bu - b\cos bu) + C$

20. $\displaystyle\int e^{au}\cos bu\,du = \frac{e^{au}}{a^2 + b^2}(a\cos bu + b\sin bu) + C$

21. $\displaystyle\int \ln u\,du = u\ln u - u + C$

22. $\displaystyle\int u^n \ln u\,du = \frac{u^{n+1}}{(n+1)^2}[(n+1)\ln u - 1] + C$

23. $\displaystyle\int \frac{1}{u\ln u}\,du = \ln|\ln u| + C$

Hyperbolic Forms

24. $\displaystyle\int \sinh u\,du = \cosh u + C$

25. $\displaystyle\int \cosh u\,du = \sinh u + C$

26. $\displaystyle\int \tanh u\,du = \ln\cosh u + C$

27. $\displaystyle\int \coth u\,du = \ln|\sinh u| + C$

28. $\displaystyle\int \operatorname{sech} u\,du = \tan^{-1}|\sinh u| + C$

29. $\displaystyle\int \operatorname{csch} u\,du = \ln\left|\tanh\frac{1}{2}u\right| + C$

30. $\displaystyle\int \operatorname{sech}^2 u\,du = \tanh u + C$

31. $\displaystyle\int \operatorname{csch}^2 u\,du = -\coth u + C$

32. $\displaystyle\int \operatorname{sech} u \tanh u\,du = -\operatorname{sech} u + C$

33. $\displaystyle\int \operatorname{csch} u \coth u\,du = -\operatorname{csch} u + C$

Trigonometric Forms

34. $\displaystyle\int \sin^2 u\,du = \frac{1}{2}u - \frac{1}{4}\sin 2u + C$

35. $\displaystyle\int \cos^2 u\,du = \frac{1}{2}u + \frac{1}{4}\sin 2u + C$

36. $\displaystyle\int \tan^2 u\,du = \tan u - u + C$

37. $\displaystyle\int \cot^2 u\,du = -\cot u - u + C$

38. $\displaystyle\int \sin^3 u\,du = -\frac{1}{3}(2 + \sin^2 u)\cos u + C$

39. $\displaystyle\int \cos^3 u\,du = \frac{1}{3}(2 + \cos^2 u)\sin u + C$

40. $\displaystyle\int \tan^3 u\,du = \frac{1}{2}\tan^2 u + \ln|\cos u| + C$

41. $\displaystyle\int \cot^3 u\,du = -\frac{1}{2}\cot^2 u - \ln|\sin u| + C$

42. $\displaystyle\int \sec^3 u\,du = \frac{1}{2}\sec u \tan u + \frac{1}{2}\ln|\sec u + \tan u| + C$

43. $\int \csc^3 u \, du = -\frac{1}{2} \csc u \cot u + \frac{1}{2} \ln |\csc u - \cot u| + C$

44. $\int \sin^n u \, du = -\frac{1}{n} \sin^{n-1} u \cos u + \frac{n-1}{n} \int \sin^{n-2} u \, du$

45. $\int \cos^n u \, du = \frac{1}{n} \cos^{n-1} u \sin u + \frac{n-1}{n} \int \cos^{n-2} u \, du$

46. $\int \tan^n u \, du = \frac{1}{n-1} \tan^{n-1} u - \int \tan^{n-2} u \, du$

47. $\int \cot^n u \, du = \frac{-1}{n-1} \cot^{n-1} u - \int \cot^{n-2} u \, du$

48. $\int \sec^n u \, du = \frac{1}{n-1} \tan u \sec^{n-2} u + \frac{n-2}{n-1} \int \sec^{n-2} u \, du$

49. $\int \csc^n u \, du = \frac{-1}{n-1} \cot u \csc^{n-2} u + \frac{n-2}{n-1} \int \csc^{n-2} u \, du$

50. $\int \sin au \sin bu \, du = \frac{\sin(a-b)u}{2(a-b)} - \frac{\sin(a+b)u}{2(a+b)} + C$

51. $\int \cos au \cos bu \, du = \frac{\sin(a-b)u}{2(a-b)} + \frac{\sin(a+b)u}{2(a+b)} + C$

52. $\int \sin au \cos bu \, du = -\frac{\cos(a-b)u}{2(a-b)} - \frac{\cos(a+b)u}{2(a+b)} + C$

53. $\int u \sin u \, du = \sin u - u \cos u + C$

54. $\int u \cos u \, du = \cos u + u \sin u + C$

55. $\int u^n \sin u \, du = -u^n \cos u + n \int u^{n-1} \cos u \, du$

56. $\int u^n \cos u \, du = u^n \sin u - n \int u^{n-1} \sin u \, du$

57. $\int \sin^n u \cos^m u \, du$

$\quad = -\frac{\sin^{n-1} u \cos^{m+1} u}{n+m} + \frac{n-1}{n+m} \int \sin^{n-2} u \cos^m u \, du$

$\quad = \frac{\sin^{n+1} u \cos^{m-1} u}{n+m} + \frac{m-1}{n+m} \int \sin^n u \cos^{m-2} u \, du$

Inverse Trigonometric Forms

58. $\int \sin^{-1} u \, du = u \sin^{-1} u + \sqrt{1-u^2} + C$

59. $\int \cos^{-1} u \, du = u \cos^{-1} u - \sqrt{1-u^2} + C$

60. $\int \tan^{-1} u \, du = u \tan^{-1} u - \frac{1}{2} \ln(1+u^2) + C$

61. $\int u \sin^{-1} u \, du = \frac{2u^2-1}{4} \sin^{-1} u + \frac{u\sqrt{1-u^2}}{4} + C$

62. $\int u \cos^{-1} u \, du = \frac{2u^2-1}{4} \cos^{-1} u - \frac{u\sqrt{1-u^2}}{4} + C$

63. $\int u \tan^{-1} u \, du = \frac{u^2+1}{2} \tan^{-1} u - \frac{u}{2} + C$

64. $\int u^n \sin^{-1} u \, du = \frac{1}{n+1} \left[u^{n+1} \sin^{-1} u - \int \frac{u^{n+1} \, du}{\sqrt{1-u^2}} \right], \quad n \neq -1$

65. $\int u^n \cos^{-1} u \, du = \frac{1}{n+1} \left[u^{n+1} \cos^{-1} u + \int \frac{u^{n+1} \, du}{\sqrt{1-u^2}} \right], \quad n \neq -1$

66. $\int u^n \tan^{-1} u \, du = \frac{1}{n+1} \left[u^{n+1} \tan^{-1} u - \int \frac{u^{n+1} \, du}{1+u^2} \right], \quad n \neq -1$

Forms Involving $\sqrt{a^2 - u^2},\ a > 0$

67. $\int \sqrt{a^2-u^2} \, du = \frac{u}{2} \sqrt{a^2-u^2} + \frac{a^2}{2} \sin^{-1} \frac{u}{a} + C$

68. $\int u^2 \sqrt{a^2-u^2} \, du = \frac{u}{8}(2u^2-a^2)\sqrt{a^2-u^2} + \frac{a^4}{8} \sin^{-1} \frac{u}{a} + C$

69. $\int \frac{\sqrt{a^2-u^2}}{u} \, du = \sqrt{a^2-u^2} - a \ln \left| \frac{a+\sqrt{a^2-u^2}}{u} \right| + C$

70. $\int \frac{\sqrt{a^2-u^2}}{u^2} \, du = -\frac{1}{u} \sqrt{a^2-u^2} - \sin^{-1} \frac{u}{a} + C$

71. $\int \frac{u^2 \, du}{\sqrt{a^2-u^2}} = -\frac{u}{2} \sqrt{a^2-u^2} + \frac{a^2}{2} \sin^{-1} \frac{u}{a} + C$

72. $\int \frac{du}{u\sqrt{a^2-u^2}} = -\frac{1}{a} \ln \left| \frac{a+\sqrt{a^2-u^2}}{u} \right| + C$

73. $\int \frac{du}{u^2\sqrt{a^2-u^2}} = -\frac{1}{a^2 u} \sqrt{a^2-u^2} + C$

74. $\int (a^2-u^2)^{3/2} \, du = -\frac{u}{8}(2u^2-5a^2)\sqrt{a^2-u^2} + \frac{3a^4}{8} \sin^{-1} \frac{u}{a} + C$

75. $\int \frac{du}{(a^2-u^2)^{3/2}} = \frac{u}{a^2\sqrt{a^2-u^2}} + C$

Forms Involving $\sqrt{u^2 - a^2},\ a > 0$

76. $\int \sqrt{u^2-a^2} \, du = \frac{u}{2} \sqrt{u^2-a^2} - \frac{a^2}{2} \ln |u + \sqrt{u^2-a^2}| + C$

77. $\int u^2 \sqrt{u^2-a^2} \, du$

$\quad = \frac{u}{8}(2u^2-a^2)\sqrt{u^2-a^2} - \frac{a^4}{8} \ln |u + \sqrt{u^2-a^2}| + C$

78. $\int \frac{\sqrt{u^2-a^2}}{u} \, du = \sqrt{u^2-a^2} - a \cos^{-1} \frac{a}{|u|} + C$

79. $\int \frac{\sqrt{u^2-a^2}}{u} \, du = -\frac{\sqrt{u^2-a^2}}{u} + \ln |u + \sqrt{u^2-a^2}| + C$

80. $\int \frac{du}{\sqrt{u^2-a^2}} = \ln |u + \sqrt{u^2-a^2}| + C$

81. $\int \frac{u^2 \, du}{\sqrt{u^2-a^2}} = \frac{u}{2} \sqrt{u^2-a^2} + \frac{a^2}{2} \ln |u + \sqrt{u^2-a^2}| + C$

82. $\int \frac{du}{u^2\sqrt{u^2-a^2}} = \frac{\sqrt{u^2-a^2}}{a^2 u} + C$

83. $\int \frac{du}{(u^2-a^2)^{3/2}} = -\frac{u}{a^2\sqrt{u^2-a^2}} + C$

Forms Involving $\sqrt{a^2 + u^2},\ a > 0$

84. $\int \sqrt{a^2+u^2} \, du = \frac{u}{2} \sqrt{a^2+u^2} + \frac{a^2}{2} \ln(u + \sqrt{a^2+u^2}) + C$

85. $\int u^2 \sqrt{a^2+u^2} \, du$

$\quad = \frac{u}{8}(a^2+2u^2)\sqrt{a^2+u^2} - \frac{a^4}{8} \ln(u + \sqrt{a^2+u^2}) + C$

86. $\int \frac{\sqrt{a^2+u^2}}{u} \, du = \sqrt{a^2+u^2} - a \ln \left| \frac{a+\sqrt{a^2+u^2}}{u} \right| + C$

87. $\int \frac{\sqrt{a^2+u^2}}{u^2} \, du = -\frac{\sqrt{a^2+u^2}}{u} + \ln(u + \sqrt{a^2+u^2}) + C$

88. $\displaystyle\int \frac{du}{\sqrt{a^2 + u^2}} = \ln\left(u + \sqrt{a^2 + u^2}\right) + C$

89. $\displaystyle\int \frac{u^2\, du}{\sqrt{a^2 + u^2}} = \frac{u}{2}\sqrt{a^2 + u^2} - \frac{a^2}{2}\ln\left(u + \sqrt{a^2 + u^2}\right) + C$

90. $\displaystyle\int \frac{du}{u\sqrt{a^2 + u^2}} = -\frac{1}{a}\ln\left|\frac{\sqrt{a^2 + u^2} + a}{u}\right| + C$

91. $\displaystyle\int \frac{du}{u^2\sqrt{a^2 + u^2}} = -\frac{\sqrt{a^2 + u^2}}{a^2 u} + C$

92. $\displaystyle\int \frac{du}{(a^2 + u^2)^{3/2}} = \frac{u}{a^2\sqrt{a^2 + u^2}} + C$

Forms Involving $a + bu$

93. $\displaystyle\int \frac{u\, du}{a + bu} = \frac{1}{b^2}\left(a + bu - a\ln|a + bu|\right) + C$

94. $\displaystyle\int \frac{u^2\, du}{a + bu} = \frac{1}{2b^3}\left[(a + bu)^2 - 4a(a + bu) + 2a^2\ln|a + bu|\right] + C$

95. $\displaystyle\int \frac{du}{u(a + bu)} = \frac{1}{a}\ln\left|\frac{u}{a + bu}\right| + C$

96. $\displaystyle\int \frac{du}{u^2(a + bu)} = -\frac{1}{au} + \frac{b}{a^2}\ln\left|\frac{a + bu}{u}\right| + C$

97. $\displaystyle\int \frac{u\, du}{(a + bu)^2} = \frac{a}{b^2(a + bu)} + \frac{1}{b^2}\ln|a + bu| + C$

98. $\displaystyle\int \frac{du}{u(a + bu)^2} = \frac{1}{a(a + bu)} - \frac{1}{a^2}\ln\left|\frac{a + bu}{u}\right| + C$

99. $\displaystyle\int \frac{u^2\, du}{(a + bu)^2} = \frac{1}{b^3}\left(a + bu - \frac{a^2}{a + bu} - 2a\ln|a + bu|\right) + C$

100. $\displaystyle\int u\sqrt{a + bu}\, du = \frac{2}{15b^2}(3bu - 2a)(a + bu)^{3/2} + C$

101. $\displaystyle\int u^n\sqrt{a + bu}\, du$
$= \frac{2}{b(2n + 3)}\left[u^n(a + bu)^{3/2} - na\int u^{n-1}\sqrt{a + bu}\,\right]$

102. $\displaystyle\int \frac{u\, du}{\sqrt{a + bu}} = \frac{2}{3b^2}(bu - 2a)\sqrt{a + bu} + C$

103. $\displaystyle\int \frac{u^n\, du}{\sqrt{a + bu}} = \frac{2u^n\sqrt{a + bu}}{b(2n + 1)} - \frac{2na}{b(2n + 1)}\int \frac{u^{n-1}\, du}{\sqrt{a + bu}}$

104. $\displaystyle\int \frac{du}{u\sqrt{a + bu}} = \frac{1}{\sqrt{a}}\ln\left|\frac{\sqrt{a + bu} - \sqrt{a}}{\sqrt{a + bu} + \sqrt{a}}\right| + C, \quad \text{if } a > 0$

$= \frac{2}{\sqrt{-a}}\tan^{-1}\sqrt{\frac{a + bu}{-a}} + C, \quad \text{if } a < 0$

105. $\displaystyle\int \frac{du}{u^n\sqrt{a + bu}} = -\frac{\sqrt{a + bu}}{a(n - 1)u^{n-1}} - \frac{b(2n - 3)}{2a(n - 1)}\int \frac{du}{u^{n-1}\sqrt{a + bu}}$

106. $\displaystyle\int \frac{\sqrt{a + bu}}{u}\, du = 2\sqrt{a + bu} + a\int \frac{du}{u\sqrt{a + bu}}$

107. $\displaystyle\int \frac{\sqrt{a + bu}}{u^2}\, du = -\frac{\sqrt{a + bu}}{u} + \frac{b}{2}\int \frac{du}{u\sqrt{a + bu}}$

Forms Involving $\sqrt{2au - u^2},\ a > 0$

108. $\displaystyle\int \sqrt{2au - u^2}\, du = \frac{u - a}{2}\sqrt{2au - u^2} + \frac{a^2}{2}\cos^{-1}\left(\frac{a - u}{a}\right) + C$

109. $\displaystyle\int u\sqrt{2au - u^2}\, du$
$= \frac{2u^2 - au - 3a^2}{6}\sqrt{2au - u^2} + \frac{a^3}{2}\cos^{-1}\left(\frac{a - u}{a}\right) + C$

110. $\displaystyle\int \frac{du}{\sqrt{2au - u^2}} = \cos^{-1}\left(\frac{a - u}{a}\right) + C$

111. $\displaystyle\int \frac{du}{u\sqrt{2au - u^2}} = -\frac{\sqrt{2au - u^2}}{au} + C$

ESSENTIAL THEOREMS

Intermediate Value Theorem

If $f(x)$ is continuous on a closed interval $[a, b]$ and $f(a) \neq f(b)$, then for every value M between $f(a)$ and $f(b)$, there exists at least one value $c \in (a, b)$ such that $f(c) = M$.

Mean Value Theorem

If $f(x)$ is continuous on a closed interval $[a, b]$ and differentiable on (a, b), then there exists at least one value $c \in (a, b)$ such that

$$f'(c) = \frac{f(b) - f(a)}{b - a}$$

Extreme Values on a Closed Interval

If $f(x)$ is continuous on a closed interval $[a, b]$, then $f(x)$ attains both a minimum and a maximum value on $[a, b]$. Furthermore, if $c \in [a, b]$ and $f(c)$ is an extreme value (min or max), then c is either a critical point or one of the endpoints a or b.

The Fundamental Theorem of Calculus, Part I

Assume that $f(x)$ is continuous on $[a, b]$ and let $F(x)$ be an antiderivative of $f(x)$ on $[a, b]$. Then

$$\int_a^b f(x)\, dx = F(b) - F(a)$$

Fundamental Theorem of Calculus, Part II

Assume that $f(x)$ is a continuous function on $[a, b]$. Then the area function $A(x) = \displaystyle\int_a^x f(t)\, dt$ is an antiderivative of $f(x)$, that is,

$$A'(x) = f(x) \quad \text{or equivalently} \quad \frac{d}{dx}\int_a^x f(t)\, dt = f(x)$$

Furthermore, $A(x)$ satisfies the initial condition $A(a) = 0$.